28th International DAAAM Symposium on Intelligent Manufacturing and Automation 2017

Zadar, Croatia
8-11 November 2017

Volume 1 of 2

Editors:

B. Katalinic

ISBN: 978-1-5108-5327-0

Printed from e-media with permission by:

Curran Associates, Inc.
57 Morehouse Lane
Red Hook, NY 12571

Printed by Curran Associates, Inc. (2018)

DAAAM International Vienna
Bahnallee 10
A-2120 Wolkersdorf
Austria, EU

Phone: +436646424398

secretary@daaam.info

Additional copies of this publication are available from:

Curran Associates, Inc.
57 Morehouse Lane
Red Hook, NY 12571 USA
Phone: 845-758-0400
Fax: 845-758-2633
Email: curran@proceedings.com
Web: www.proceedings.com

ISSN 1726-9679

DANUBE ADRIA ASSOCIATION FOR AUTOMATION & MANUFACTURING

DAAAM INTERNATIONAL VIENNA

www.daaam.info

PROCEEDINGS

OF THE 28TH INTERNATIONAL DAAAM SYMPOSIUM
"INTELLIGENT MANUFACTURING & AUTOMATION
08-11TH NOVEMBER 2017
ZADAR, CROATIA

ORGANIZED BY:
DAAAM INTERNATIONAL VIENNA

IN COOPERATION WITH:
ÖIAV 1848, VIENNA UNIVERSITY OF TECHNOLOGY, INTERNATIONAL ACADEMY OF ENGINEERING
AND UNIVERSITY OF APPLIED SCIENCES – TECHNIKUM WIEN

UNDER THE AUSPICES OF THE DANUBE RECTORS' CONFERENCE &
RECTORS' AND PRESIDENTS' HONOR COMMITTEE OF DAAAM INTERNATIONAL

EDITOR: B.[RANKO] KATALINIC

IMPRESSUM 2017

Proceedings of the 28th International DAAAM Symposium "Intelligent Manufacturing & Automation"
08-11th November 2017, Zadar, Croatia, Volume 28, No.1,
Abbreviated key title: ***Ann. DAAAM 2017***

ISSN 1726-9679

Ed. B. Katalinic,

Published by DAAAM International, Vienna, Austria, EU, 2017
Layout & Design: B[ranko] Katalinic/ D[amir] Haskovic
Technical Editors: D[amir] Haskovic / I[lija] Zec / R[oberta] Cvetkovska / I[lya] Kukushkin

Publisher:

DAAAM International Vienna
TU Wien, Getreidemarkt 9, 1060 Wien, Austria
E-mails: president@daaam.com / branko.katalinic@tuwien.ac.at
DAAAM homepage: http://www.daaam.info

CONTENTS

AUTHOR INDEX

A

B

ORGANIZING COMMITTEE FOR 2017

COVIC, DRAGAN – PRESIDENT OF ORGANIZING COMMITTEE
KATALINIC, BRANKO – PRESIDENT OF DAAAM INTERNATIONAL
TOMIC, ZORAN – RECTOR OF THE UNIVERSITY OF MOSTAR
GUSEV, BORIS – PRESIDENT OF INTERNATIONAL ACADEMY OF ENGINEERING
UGLESIC, ANTE – COORDINATOR OF ORGANIZING COMMITTEE

BUCHMEISTER, BORUT
COLAK, IVO
COSIC, ILIJA
CVETKOVSKA, ROBERTA
EGOROV, SERGEY
FILARETOV, VLADIMIR
HASKOVIC, DAMIR (SECRETARY)
HOELLER, LIANE
KUKUSHKIN, ILYA
LEKO, TOMISLAV
MAJSTOROVIC, VLADO
NAUMBURGER, JOERG
PANFILOV, PETER
PARK, HONG-SEOK
PRYANICHNIKOV, VALENTIN
STAZHKOV, SERGEY
STOJKIC, INJA
STOPPER, MARKUS
TEKIC, ZELJKO
UDILJAK, TOMA
ZAVRAZHINA, ALINA
ZEC, ANTE
ZEC, ILIJA
ZEC, LUKA

RECTORS' AND PRESIDENTS' HONOR COMMITTEE

ADAMCZAK, STANISLAW	RECTOR, TECHNICAL UNIVERSITY OF KIELCE
AAVIKSOO, JAAK	RECTOR, TALLINN UNIVERSITY OF TECHNOLOGY
ABRUDAN, IOAN VASILE	RECTOR, TRANSILVANIA UNIVERSITY OF BRASOV
ALEXANDROV, ANATOLIY	RECTOR, BAUMAN MOSCOW STATE TECHNICAL UNIVERSITY
ANDJELINOVIC, SIMUN	RECTOR, UNIVERSITY OF SPLIT
ANISIMOV, NIKITA YU.	PRESIDENT, FAR EASTERN FEDERAL UNIVERSITY
ANTRANIKIAN, GARABED	PRESIDENT, HAMBURG UNIVERSITY OF TECHNOLOGY
BAAIJENS, FRANK	RECTOR, EINDHOVEN UNIVERSITY OF TECHNOLOGY
BALLUERKA LASA, NEKANE	RECTOR, UNIVERSITY OF THE BASQUE COUNTRY
BEIBST, GABRIELE	RECTOR, JENA UNIVERSITY OF APPLIED SCIENCE
BORAS, DAMIR	RECTOR, UNIVERSITY OF ZAGREB
BRUNGS, ATTILA	VICE-CHANCELLOR AND PRESIDENT, UNIVERSITY OF TECHNOLOGY, SYDNEY
BUCHVALD VIND, THOMAS	DIRECTOR, UNIVERSITY OF SOUTHERN DENMARK
CHEON, YEON	PRESIDENT, UNIVERSITY OF ULSAN
CISNEROS PÉREZ, GUILLERMO	RECTOR, UNIVERSIDAD POLITECNICA DE MADRID
COSTOIU, MIHNEA	RECTOR, UNIVERSITY POLITEHNICA OF BUCHAREST
DANIELS, MITCHELL	PRESIDENT, PURDUE UNIVERSITY
DEMA, MARJAN	RECTOR, UNIVERSITY OF PRISHTINA
DI SCIASCIO, EUGENIO	RECTOR, POLYTECHNIC OF BARI
FALKOV, VALERIY	RECTOR, TYUMEN STATE UNIVERSITY
FRUNZAVERDE, DOINA	RECTOR, "EFTIMIE MURGU" UNIVERSITY OF RESITA
GALIJEV, ILJHAM	PRESIDENT, OMSK STATE TRANSPORT UNIVERSITY
GARCÍA MARÍN, JUAN FRANCISCO	RECTOR, UNIVERSITY OF LEON
GELENCSER, ANDRAS	RECTOR, UNIVERSITY OF PANNONIA
GILLI, MARCO	RECTOR, POLITECNICO DI TORINO
GOLIAS, IOANNIS	RECTOR, NATIONAL TECHNICAL UNIVERSITY OF ATHENS

INTERNATIONAL SCIENTIFIC COMMITTEE

INTERNATIONAL PROGRAM COMMITTEE

Abram Zver, Marta
Acko, Bojan
Afzulpurkar, Nitin
Agic, Darko
Ahuja, Harish
Albers, Albert
Alekseyev, Wladimir A.
Amberg, Michael
Badida, Miroslav
Bajic, Drazen
Balic, Joze
Banyai, Agota
Banyai, Tamas
Baritz, Mihaela Ioana
Barreiro, Joaquin
Begic, Derzija
Bijelonja, Izet
Bleicher, Friedrich
Bley, Helmut
Botezatu, Cornelia Paulina
Botezatu, Mihai
Brajlih, Tomaz
Brezak, Danko
Brezocnik, Miran
Brnic, Josip
Buchmeister, Borut
Bucuresteanu, Anca
Butala, Peter
Campbell, R. Ian
Car, Zlatan
Cekic, Ahmet
Celar, Stipe
Chan, Chee Seng
Chaus, Alexander
Chramcov, Bronislav
Cohodar, Maida
Coleman, J. J.
Cosic, Ilija Petar
Cosic, Pejo
Cosic, Predrag
Cosma, Smaranda
Cotet, Costel Emil
Cubonova, Nadezda
Curtu, Ioan
Cus, Franci
Damic, Vjekoslav
Denkena, Berend
Devahastin, Sakamon
Dolga, Lia
Dolga, Valer
Dolinay, Jan
Dolinay, Viliam
Dostal, Petr
Dostalek, Petr
Dragcevic, Zvonko
Dragicevic, Marija
Drstvensek, Igor
Menz, Peter
Michal, Bliznak
Micieta, Branislav
Miclosi, Viorel
Duflou, Joost
Duhovnik, Joze
Egorova, Tatyana
Ejub Dzaferovic,
Filaretov, Vladimir
Filetin, Tomislav
First Rogale, Snjezana
Fujii, Norio
Gheorghiu, Horia
Gojo, Miroslav
Gomez, Antonio
Gordin, Doina Margareta
Hlava, Jaroslav
Hofreiter, Milan
Hrubina, Kamil
Hruska, Frantisek
Hruskova, Erika
Ibishi, Ismet M.
Janacova, Dagmar
Janssens, Thierry
Jarvis, Christine W.
Javorova, Angela
Jerbic, Bojan
Juraga, Ivan
Jurum-Kipke, Jasna
Karaulova, Tatyana
Katalinic, Branko
Kawalec, Andrzej
Khanna, Ashutosh
Khomchenko, Vasily
Kirk, Chris. S.
Kiss, Imre
Kittl, Burkhard
Klicek, Bozidar
Klimenko, S. A.
Kljajin, Milan
Knego, Nikola
Knezevic, Petar
Kohlert, Helmut
Kolakovic, Marko
Kolibal, Zdenek
Kolomaznik, Karel
Kondo, I.
Korbel, Jiri
Kosmol, Jan
Kostal, Peter
Koukhta, Sergey
Kovac, Karol
Kovarikova, Ingrid
Kralikova, Ruzena
Kramar, Vadym A.
Krasniqi, Z.
Kremljak, Zvonko
Krucinska, Izabella
Kubalcik, M.
Kubinger, Wilfried
Kulenovic, Malik
Pechacek, Frantisek
Pelaez Lourido, Gustavo
Pencea, Ion
Peric, Nedjeljko
Kundrak, Janos
Kunica, Zoran
Kuzman, Karl
Lache, Simona
Lampinen, Juoni
Leasu, Florin Gabriel
Lee, Kyo Il
Lee, Moonyong
Legutko, Stanislaw
Leinonen, T.
Lesconi Frumusanu,
Li, Guihua
Lica, Dumitru
Lipa, Zdenko
Liu, Huaming
Ljoljic, Berislav
Lo Valvo, Ernesto
Loghin, Carmen Maria
Lopez De Lacalle, Luis Norberto
Low, Kin Huat
Luca Motoc, Dana I.
Ludovico, Antonio D.
Lujic, Roberto
Lukic, Ivana
Lupea, Iulian
Lupu, Ciprian
Lupulescu, Nouras Barbu
Machedon Pisu, Teodor
Macku, Lubomir
Macurova, Anna
Magheti, Ioan
Majetic, Dubravko
Malisa, Viktorio
Malopolski, Waldemar
Man, Mariana
Mandic, Lidija
Mandru, Dan
Manescu, Tiberiu Stefan
Marascu Klein, Vladimir
Marcos, Mariano
Marcu, Marius
Marek, Kubalcik
Marian Tolnay
Marin, Dumitru
Marin, Mihnea
Marinescu, Niculae Ion
Marjanovic, Dorian
Marn, Ferdinand
Maronek, Milan
Marsland, David
Marta, Constantin
Martinez, Susana
Mastacan, Lucian
Matusova, Miriam
Matusu, Radek
Mehta, Narinder Kumar
Meier, Horst
Sisek, Boris
Situm, Zeljko
Skoczynski, Waclaw
Skolud, Bozena

MOHORA, CRISTINA
MOK, H.
MORARU, SORIN AUREL
MORIWAKI, T.
MORTOIU, DOINA
MOTOMANCEA, ADRIAN
MRSA, ZORAN
MRVAC, NIKOLA
MUNTEAN, MIHAELA
MUNTEAN, SEBASTIAN
MUNTEANU, ANCA
MUSCA, GAVRIL
MUSTAFIZUL, KARIM
MYSHLYAVTSEV, ALEXANDR
NAGI, MIHAI
NAGLA, K. S.
NAKAMOTO, NOBUYUKI
NEUMANN, PETR
NICA COTET, GABRIELA BEATRICE
NIKOLIC, GOJKO
NKUMBWA, RODGER
NOH, SANG DO
NOVAC, OVIDIU
NOVAKOVIC, BRANKO M.
NOVOSELOVIC, DANIEL
NURUL AMIN A.K.M.
OANCEA, GHEORGHE I.
OBAD, MILENKO
OKAMOTO, J.
OLARU, ADRIAN, DUMITRU
OLARU, ONISIFOR
OLEKSIK, VALENTIN
OLTEANU, CIPRIAN
OMEROVIC, MIDHAT R.
OPRAN, CONSTANTIN
OPREAN, AUREL
OPRUTA, DAN
OSMANI, HYSNI
OSMERA, PAVEL
OTTOSSON, S.
PACAIOVA, HANA
PALCIC, IZTOK
PALOS, RAMONA
PAMINTAS, EUGEN
PANFILOV, PETER
PARASCHIV, NICOLAE
PARK, HONG-SEOK
PARPALA, LIDIA FLORENTINA
PASCUAL, R.COMPAMYS
PASIC, MUGDIM
PASTRAMA, STEFAN DAN
PATACHIA, SILVIA FLORICA CRISTINA
PAULIKOVA, ALENA
PAULOVA, IVETA
PAUNESCU, CARMEN MONICA
PAVIC, IVAN
PERJU, DAN
PERNIU, DANA
PERUTKA, KAREL
PETERS, JACQUES
PETKOVIC, DARKO
PETRE, CRISTIAN
PETRE, MARIAN
PETRESCU, VALENTIN
PINCA-BRETOTEAN, CAMELIA
PINTILIE, GHEORGHE
PIRTEA, GABRIEL MARILEN
PIVONKA, PETR
PJERO, ELENICA
PLESA, ANGELA
PODURAEV, JURY
POKORNY, PAVEL
POLJAK, KRUNOSLAV
POMAZAN, VALENTINA
POPA, CICERONE LAURENTIU
POPPEOVA, VIERA
PRODAN, DAN
PROKOP, ROMAN
PROSTREDNIK, DANIEL
PRYANICHNIKOV, VALENTIN
PYTKO, STANISLAW
QUINN, ROGER D.
RADIC, JURE
RAMAJ, VEHBI XH.
RAOS, PERO
RAVAS, RUDOLF
RAY, T. S.
RIZESCU, SABIN VASILE
ROGALE, DUBRAVKO
ROHDE, ULRICH L.
ROMERO, FERNANDO
ROS, OLIMPIA
ROSADO, V.O.G.
RUDAS, IMRE J.
RUGESCU, DRAGOS RADU DAN
RUJA, IOAN
RUSU ANGHEL, NICOLAE
RUSU, LIGIA
RUSU-CASANDRA, AURELIA
RUTEMI, GANI BRAHIM
RUZIC, DRAGO
SALIHU, AVDI HAJDIN
SAMEK, DAVID
SCHREIBER, PETER
SELVAKUMAR, ARUL LAWRENCE
SEONG, POONG HYUN
SERCER, MLADEN
SHIBASAKA, TOSHIROH
SHIH, MING CHANG
SIEMIATKOWSKI, M.
SIHN, WILFRIED
SINAY, JURAJ
SLUGA, ALOJZ
SMIRNOV, ALEXANDER
STEFANIC, NEDELJKO
STOPPER, MARKUS
STRGAR KURECIC, MAJA
SUBRAMANIAN, ARUNKUMAR
SUGAR, PETER
SYROVA, LIVIA
SYSALA, TOMAS
TADAMURA, KATSUMI
TAKALA, JOSU
TAMASILA, MATEI
TAN, KOK KIONG
THIERHEIMER, WALTER WILHELM
TORIMS, TOMS
TURKALJ, GORAN
ULRICH, GUENTHER
VAARIO, JARI
VASEK, VLADIMIR
VELISEK, KAROL
VEZA, IVICA
VISEKRUNA, VOJO
VRANJES, BOZO
VUJASINOVIC, EDITA
WANG, JYHWEN
WEIGANG, WU
WEIGL, K. HEINZ
WESSELY, EMIL
WIENER, PETER
WINKLER, T.
WOZNIAK, ADAM
XING, KE
XUE, D.
YAMAZAKI, K.
YORDANOV, RUMEN STOYANOV
YOSHIDA, SUSUMU
YU, Y.-C.
ZAJAC, JERZY
ZAJAC, JOZEF
ZAPCIU, MIRON
ZAVOIANU, ION
ZEQIRI, HAKIF
ZGODAVOVA, KRISTINA
ZHIRABOK, ALEXEY
ZICHIL, VALENTIN
ZIRRA, DANIELA
ZUEV, ALEXANDER
ZUPERL, UROS

REVIEWING COMMITTEE

28TH DAAAM INTERNATIONAL SYMPOSIUM ON INTELLIGENT MANUFACTURING AND AUTOMATION

Preface

The 28th DAAAM International Symposium on Intelligent Manufacturing and Automation took place in Zadar, Croatia between the 08th and 11th November 2017, during the DAAAM International Week. The Symposium was organized by DAAAM International Vienna in cooperation with ÖIAV 1848, Vienna University of Technology, International Academy of Engineering and University of Applied Sciences – Technikum Wien and Under the Auspices of the Danube Rectors' Conference & Rectors' and Presidents' Honor Committee of DAAAM International for 2017. The Symposium took place in Zadar, Croatia. This year symposium aimed at continuing the success of the previous year's focusing on the five-fold traditional objectives of the symposium: the presentation of the most recent high-quality results, support of development of young scientists and researchers, organization of international (summer) doctoral school, inauguration of new members of Central European Branch of International Academy of Engineering and the provision of the necessary setting for stimulating discussions, brainstorming and networking among European and international researchers coming both from the academia government agencies and industry.

DAAAM International Symposium has been established as the leading Central European conference devoted to the cutting-edge advancements in the field of modern production and automation. Symposium provides an international forum for engineers, scientists and industrial managers to present and discuss the current status and impact of advanced manufacturing and automation. The international series of Symposium was initiated by the first meeting in Vienna (1990) and continued in Strbske Pleso (1991), Budapest (1992), Brno (1993), Maribor (1994), Cracow (1995), Vienna (1996), Dubrovnik (1997), Cluj - Napoca (1998), Vienna (1999), Opatija (2000), Jena (2001), Vienna (2002), Sarajevo (2003), Vienna (2004), Opatija (2005), Vienna (2006), Zadar (2007), Trnava (2008), Vienna (2009), Zadar (2010), Vienna (2011), Zadar (2012), Zadar (2013), Vienna (2014), Zadar (2015), Mostar (2016), Zadar (2017).

Regular organization of the annual Symposium is one of the central activities of the DAAAM International association. The association was founded on 5. November 1990 in Vienna on the meeting organised by Professor B. Katalinic in order to celebrate the 175 Years of Vienna University of Technology. The main idea of association was to establish network for new kind of international scientific and academic cooperation in the fields of intelligent manufacturing and automation. Our primer focus is on: all aspects of production of technical products made out of parts and corresponding automation and knowledge, trans-disciplinary application of methods and all kind of education.

DAAAM International activities, during the last twenty-eight years, have been supported by scientists and experts from over fifty countries all over the world. In the course of the last decade, DAAAM International has focused activities on the regional and global long-term co-operation between (technical) universities. The result of this initiative is the founding of the University Network of DAAAM International Vienna and DAAAM International Network for Advanced Technologies which was founded on 2002-10-25 in Vienna.

The main activities of DAAAM are:

- Organisation of annual DAAAM International Symposium.
- To maintain and improve the University Network of DAAAM International Vienna as a strategic alliance and permanent international open platform for long-term co-operation and partnership in all fields of technology, research and education.
- Organisation of long-term projects, initially known as Inter-University Doctoral Studies.
- Organisation of long-term projects, initially known as the Virtual University.
- Publishing of international journals, books and proceedings.
- Promotion of the exchange of people, scientific, and technical information.
- Promotion of international co-operation among scientists and engineers.
- Promotion of R & D, education and industrial application.
- To co-operate with other national and international organisations in matters of common interest.
- Organisation of International Doctoral School as platform for the support of next generation of engineers at the highest level.

This proceeding contains 172 refereed papers from 407 authors presented at the 28th DAAAM 2017. In the DAAAM proceedings since first symposium in 1990 we published papers from 8870 authors.

All the papers included in these proceedings are reviewed by the experts in double blind peer review way, and finally revised by the authors for the improvement according to referee's reports and advices.

The papers discuss many aspects of modern manufacturing and automation such as: Algorithms, Artificial Intelligence, CAX, Computer Integration, Control, Cutting Tools, Design, FEM, Invited Lectures, Knowledge, Management, Manufacturing System, Mechatronics, Methodology, Methods, Modelling, Optimization, Robotics, Simulation, Technical Solutions, Technology and Trends.

Appreciation is forwarded to reviewers who contributed to improve the quality of papers reviewing the papers, providing constructive critical comments, improvements and corrections to the authors. We also recognize the support of the reviewers, who were in charge of acceptance/rejection process. We wish to thank the Scientific, Program and Organizing Committees, Session Chairs, student helpers and Administrative assistants, for contributing to the success of this Conference. They did an excellent job.

Finally, we wish to thank all the authors who submitted papers, making this conference possible and the authors of accepted papers for making appropriate modifications and corrections with care and patience and preparing the final manuscript. It is the quality of their presentations and their passion to communicate with the others participants that really makes this conference series a success.

Special thanks to Festo company for support of our activities and support of young researchers and scientists in the form of Festo Prize and Festo scholarships for the participants of the 6th DAAAM International Doctoral School, which was held parallel to the symposium and interacted positively each other.

We would like to thank all members of the International Academy of Engineering – Central European Branch, Mayor of City of Zadar – Dr Branko Dukic as well as Rector of the University of Mostar Professor Dr. Sc. Zoran Tomic and his team, management and all those nice colleagues for their excellent support in the organization of 28th DAAAM International Symposium and the 6th DAAAM International Doctoral School.

Vienna, 2017-12-04
Editor
Chair of the 28th DAAAM 2017

Branko Katalinic

Professor Dr.sc. Dr.mult.h.c.Prof.h.c. Branko Katalinic
President of DAAAM International
TU Wien, IFLT-IMS, Getreidemarkt 9
1060 Wien, Austria, Europe
URL: http://www.daaam.info
E-mail: president@daaam.com
branko.katalinic@tuwien.ac.at

DOI: 10.2507/28th.daaam.proceedings.001

DESIGN OF A CONSENSUS BASED FLOCKING CONTROL OF MULTIPLE AUTONOMOUS UNDERWATER VEHICLES USING SLIDING MODE APPROACH

Bidyadhar Subudhi[1], Madan Mohan Rayguru[1],
Vladimir Filaretov[2,3], Aleksandr Zuev[2,3]

[1] Dept. of Electrical Engineering, National Institute of Technology, Rourkela-769008, India
[2] Far Esatern Federal University, Vladivostok, Russia
[3] Institute of Automation and Control Processes FEB RAS, Vladivostok, Russia

This Publication has to be referred as: Subudhi, B[idyadhar]; Rayguru, M[adan] M[ohan]; Filaretov, V[ladimir] & Zuev, A[lexander] (2017). Design of a Consensus Based Flocking Control of Multiple Autonomous Underwater Vehicles Using Sliding Mode Approach, Proceedings of the 28th DAAAM International Symposium, pp.0004-0013, B. Katalinic (Ed.), Published by DAAAM International, ISBN 978-3-902734-11-2, ISSN 1726-9679, Vienna, Austria
DOI: 10.2507/28th.daaam.proceedings.001

Abstract

This paper proposes a flocking control algorithm for a team of autonomous underwater vehicles (AUVs) deployed for group motion in a network where only local communication exists among the participating AUVs. The control scheme is proposed with two levels where the inner level utilizes agreement or consensus protocol to derive reference positions for the follower AUVs whereas the outer loop employs a sliding mode controller (SMC) to reduce the tracking error to zero. SMC being a suitable candidate for handling is chosen because of its inherent robustness to any uncertainties and disturbances, it is intended to design a SMC to address the problem of handling uncertainties in the hydrodynamic parameters of the AUV dynamics and has been used to steer each AUV in the flock to their desired states. By exploiting the consensus protocol and SMC, it can be ensured that once an AUV comes into the neighbour zone (maximum range for information exchange for a member AUV outside which no communication is possible) cannot be separated from the group. Therefore, this controller structure is suitable for flocking of multiple AUVs where communication constraints play a vital role in efficient performance of control algorithms. The effectiveness of the proposed controller scheme is demonstrated through MATLAB simulation.

Keywords: AUV; Co-operative control; Flocking; Consensus; SMC

1. Introduction

Attention on cooperative control of robotic systems has grown enormously in the control community because of its increasing use in intelligence, surveillance, mine countermeasures, antisubmarine warfare, oceanography, payload delivery, information operations and time critical strikes [1-3]. Formation control is an approach to cooperative control strategy where a spatial structure of the group of agents/robots/AUVs in motion remains constant throughout the mission whereas flocking control is an approach where structure of the group in motion is dynamic or time varying. Flocking of multi agent systems with communication constraints is an emerging area of research now-a-days.

Formation behavior has been studied in robotics usually with different type structures namely behavior-based structure, leader–follower structure etc. [8, 17]. This work is concerned with leader follower structure only. Leader follower flocking is very similar to leader follower formation technique except that in the former the followers do not have to follow the leader with rigid position constraints [5]. However, the leader follower formation keeping leads to a poor disturbance rejection property due to absence of feedback from followers to leader [18]. By adding the algebraic graph theory and consensus protocol to the conservative control techniques we can confirm connectedness of flocking network and can model different communication topology with time varying structure easily [6, 7, 9-12]. Single and double integrator consensus has been effectively exploited in flocking of mobile robots and unmanned aerial vehicles (UAV) [12]. The flocking algorithms used in [8, 9, 12] composed of a feedback structure utilizing consensus protocol and derive control strategies using this feedback information to keep the participating members of the flock cohesive. These controller structures [24] are not effective in case of flocking of AUVs as these algorithms are not robust enough to uncertainties and nonlinearities present in the dynamics of the AUVs. One needs robust low-level controllers with high level planners using above cited algorithms so that the total structure become robust.

In recent years, leader following flocking control of multiple agents and UAVs has been developed using graph theory based control techniques [12]. Virtual leader based formation control methodology has been used to derive formation control for multiple AUVs [21]. In this approach approximation techniques are used to derive final control law which is not robust enough to parameter uncertainties. In potential energy based formation control techniques [22] also the controller is not robust to uncertainties. Some adaptive controllers [23] for formation keeping can be found in literature but these controllers are not derived using local information exchange between neighbor AUVs. However, a limited amount of literature is available on flocking of AUVs with limited communication facility which makes the information flow more complex and the required flocking controller should be able to sort out this problem while preserving connectedness. Therefore, flocking control of AUVs is a challenging task due to its non-linear dynamics and underwater communication constraint. A group of AUVs has been considered in which the leader only has the knowledge of the desired path. The problem is to make the group cohesive as the followers have the potential to diverge away from the group and to make the group follow a desired path.

The motivation of this paper lies in developing co-operative control algorithm of a group of AUVs where there is severe communication constraint compared to surface vessels. It is difficult to control a group of AUVs with low communication bandwidth because centralizing co-operative control techniques require information about the whole group at each instant and so it require a larger bandwidth which is not readily available in underwater environment. Therefore, the need of distributed control approach arises [4].

In this work a new flocking algorithm has been developed for a group of AUVs in network where only neighbor to neighbor (local) communication is available and the desired path is only known to the group leader. The new proposed flocking controller structure consists of two loops namely an inner loop one that calculates the reference position for the followers using single and double integrator consensus respectively. In the outer loop, sliding mode control is used to control each AUV to its desired track. Sliding mode control has been chosen for its robustness to any nonlinearity and uncertainties present in the AUV dynamics [13-16]. SMC is employed for tracking a reference path and distributed consensus is applied for determining the reference position for followers which is used for design of error vector and consequently sliding surface.

2. Problem Statement

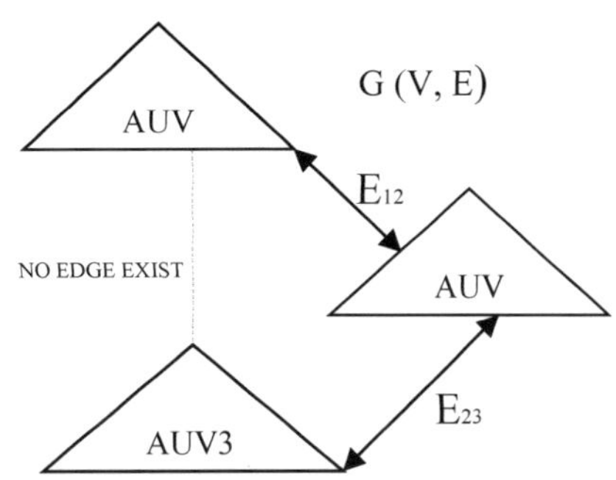

Fig. 1. Group of AUVs in a dynamic graph with each AUV represented as a node

Let G (V, E) denotes a dynamic graph undirected graph represents a group of AUVs (consisting three AUVs) in a flock (see Fig. 1). Each AUV is represented as a node V of this graph G and an edge E_{ij} exists between the ith node and j^{th}

node at any time instant if a network connection exists between the two AUVs present at those nodes. One of the AUVs in the group knows the desired path is termed as leader and other AUVs are termed as followers. The follower AUVs do not know which AUV of them is the leader and they do not know the desired path also. Let each AUV in the graph has a common dynamics of motion [1]. The objective is to develop a decentralized flocking control algorithm for the above group of AUVs where each AUV is controlled separately through a controller such that each of these AUVs remain in the flock and move in a predefined path with common forward velocity. In flocking problem for a group of AUVs, one important task is to achive convergence of the heading angle of each AUV in the group so that they will follow the same path keeping forward velocity constant.

3. System Description

The motion of the body-fixed frame of reference is described relative to an inertial or earth-fixed reference frame (Fig. 2).

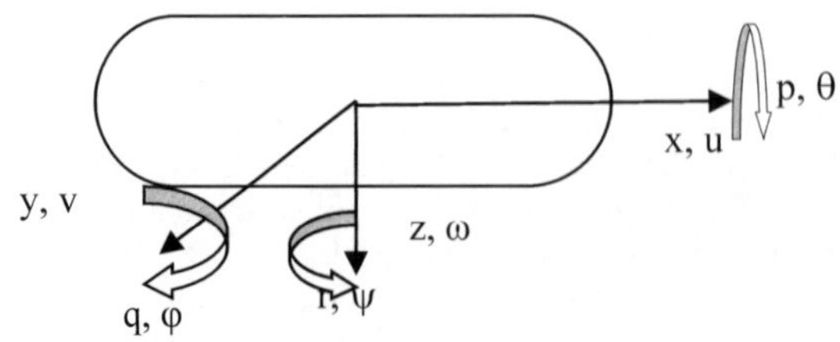

Fig. 2. Degrees of freedom of AUV

The notations different variables used are defined as follows:

- x, u are inertial x coordinate position and body fixed surge velocity;
- y, v are inertial y coordinate position and body fixed sway velocity;
- z, ω are inertial z coordinate position and body fixed heave velocity;
- p, θ are angular orientation (pitch) and rate of change of orientation along x axis;
- q, φ are angular orientation (roll) and rate of change of orientation along y axis;
- r, ψ are angular orientation (yaw) and rate of change of orientation along z axis.

The position and orientation are described in earth fixed reference frame and velocities (both linear angular) are described in body fixed reference frame. The equations of motion of an AUV in x-y plane can be written as (neglecting pitch and roll):

$$\underbrace{\begin{bmatrix} m-X_{\dot{u}} & 0 & 0 & 0 & 0 & 0 \\ 0 & m-Y_{\dot{v}} & 0 & 0 & 0 & 0 \\ 0 & 0 & I_{zz}-N_{\dot{r}} & 0 & 0 & 0 \\ 0 & 0 & 0 & 1 & 0 & 0 \\ 0 & 0 & 0 & 0 & 1 & 0 \\ 0 & 0 & 0 & 0 & 0 & 1 \end{bmatrix}}_{M} \underbrace{\begin{bmatrix} \dot{u} \\ \dot{v} \\ \dot{r} \\ \dot{\psi} \\ \dot{x} \\ \dot{y} \end{bmatrix}}_{\dot{X}} = \underbrace{\begin{bmatrix} mvr+X_{uu}u|u|+X_{vr}vr+X_{rr}rr \\ -mur+Y_{vv}v|v|+Y_{rr}r|r|+Y_{ur}ur+Y_{uv}uv \\ N_{vv}v|v|+N_{rr}r|r|+N_{ur}ur+N_{uv}uv \\ r \\ u\cos\psi-v\sin\psi \\ u\sin\psi+v\cos\psi \end{bmatrix}}_{F} + \underbrace{\begin{bmatrix} 1 & 0 \\ 0 & Y_{uudr}u^2 \\ 0 & N_{uudr}u^2 \\ 0 & 0 \\ 0 & 0 \\ 0 & 0 \end{bmatrix}}_{G} \underbrace{\begin{bmatrix} X_{prop} \\ \delta_r \end{bmatrix}}_{\tau} \tag{1}$$

$$Y = \underbrace{\begin{bmatrix} 0 & 0 & 0 & 0 & 1 & 0 \\ 0 & 0 & 0 & 0 & 0 & 1 \end{bmatrix}}_{H} [x],\ X = [u, v, r, \psi, x, y]^T \tag{2}$$

where Y is the otput matrix, X_{prop} is the thruster force, δ is the rudder angle, m is the mass of the vehicle, I_{zz} is moment of inertia along z axis and other parameters are known as hydrodynamic parameters are given in appendix. AUVs in flock are free to move without rigid position constraints so they can move sidewise (x-y plane) and in flock they always maintain a minimum distance among them. Therefore even if the flocking algorithm does not consider z-plane motion, the flock remains intact.

4. Control Law Development

The controller structure is presented in Fig. 3; it is divided in two levels. The group of AUVs which are in the flock can communicate locally so they can transmit their position and velocity (x_i, y_i, u_i, v_i) to the local neighbors. Each AUV can determine the desired position and velocity to remain in the flock using this transmitted information from neighbors using consensus algorithm. Finally using the desired states a feedback control law for each AUV is developed through sliding control law.

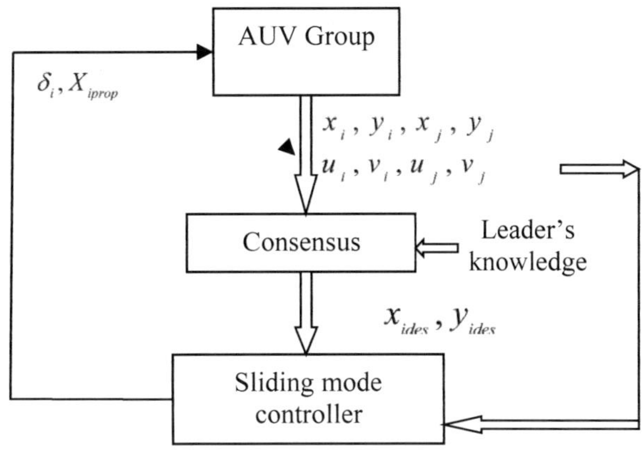

Fig. 3. Controller structure for flocking of AUVs

4.1. Sliding Surface Design

Dynamics of AUV in (1) can be written as

$$M\dot{X} = F(X) + G(X, \tau) \tag{3}$$

The sliding surface is chosen as

$$\sigma = \tilde{Y} + \gamma\dot{\tilde{Y}}, \text{ where } \gamma = \begin{bmatrix} \gamma_1 & 0 \\ 0 & \gamma_2 \end{bmatrix} \tag{4}$$

For leader AUV ($\tilde{Y} = HX - Y_d$) and follower AUVs ($\tilde{Y} = HX - Y_{df}$) Y_d is the desired track chosen. The leader leader AUV should know the desired track but the desired track Y_{df} for followers is derived from consensus algorithm. γ is chosen as diagonal matrix whose elements are positive constants and it weights the velocity error term compared to position error. It has a direct effect on convergence of state dynamics into sliding surface.

4.2. Graph Theory and Laplacian

A network of group of agents can be modeled using algebraic graph theory. Graph can be used to represent interconnections between nodes. A vertex of the graph corresponds to a node, and the edges of the graph capture the dependence of interconnections.

Formally, a graph $G = (V, E)$ consists of a set of vertices, $V = [1, 2 \ldots N]$ and set of edges E={(i, j)∈ V×V} containing unordered pairs of distinct vertices. The graph has no self-loop, i.e., (i, j) ∈E implies i≠j. A graph is connected if, for any vertices (i, j) ∈V, there exists a path of edges in E from to i to j. The set of neighbors of vertex i is defined as N_i = {j∈V :(i, j) ∈E}.

The degree of vertex is defined as $d_i = N_i$, and the maximum degree is $d_{\max} = \max(d_i)$. Let D be the degree matrix [D=diag(d_i)]. The adjacency matrix A is the integer matrix with rows and columns indexed by the vertices, e.g., i-j entry of A is equal to the number of edges from i to j. The Laplacian matrix L of a graph G is defined as L = D-A

For a connected graph, the Laplacian matrix L is symmetric and positive semi definite. Its minimum eigenvalue is zero, and the corresponding eigenvector is $[1, 1 \ldots 1]^T$.

4.3. Command Angle and Reference Speed Generation

Consensus or agreement protocol can easily model a network of multiple agents without really going into the detail about information flow between the agents. The network is modeled as a undirected graph (weighted) or as a directed graph. Participating agents in the flock can be viewed as nodes where as their information flow with their neighbors are represented with directed or undirected edges having specific weights. Control strategy applied for flocking control uses the information states in consensus to design feedback structure for formation tracking, maneuvering etc. This algorithm adapts the first derivative of the agreement information state on each node based on the value of the agreement variable of its neighbors. In this work we use the consensus algorithm to derive the desired heading for followers and apply a sliding mode control strategy to generate control input (in this case rudder angle). This idea is helpful when communication bandwidth is very small and only local neighbor to neighbor communication is possible (Information constrained network).

Desired position and rate of its change for the followers are decided by single integrator consensus and double integrator consensus algorithms. Consensus algorithm abstracts the communication topology in to a simple information sharing protocol in a static or dynamic network. The convergence of information exchange occurs if there is a minimal spanning tree always present in the dynamic graph [17]. As the problem defined in this paper is to converge the position of each AUV in the group to the desired position which is given only to the leader. If the dynamics of AUV is neglected then the output of each AUV should change according to eq. (5), (6) to reach consensus among the group [14, 19].

$$\dot{Y}_{id} = -\sum_{j\in N}(Y_{id} - Y_{jd}) - (\alpha_i - \alpha_j)) \tag{5}$$

$$\ddot{Y}_{id} = -\sum_{j\in N}(Y_{id} - Y_{jd}) - \sum_{j\in N}\beta_{ij}(\dot{Y}_{id} - \dot{Y}_{jd}) \tag{6}$$

The equation described above are extensions of single and double integrator consensus algorithms which are thoroughly discussed in various literature [19-21]. These equations give a desired rate of change of output position for the follower AUVs and a sliding controller will be used to force the actual position of each AUV to their desired one. Let's Y_{id} is the desired position for follower i and is chosen from eq. (5) and (6), N is the neighbor set of ith AUV in a dynamic graph consisting of all participating AUVs. α_i-α_j is the required difference in position between i^{th} and j^{th} AUV for avoiding collision and $\beta_{ij}$is a gain which is arbitrary chosen as 1 here [19], [20]. By the consensus algorithm given by eq. (5) difference between information state of i_{th} and j_{th} AUV will converge to zero [19].

$$\sum_{t\to\infty} Y_{id} - Y_{jd} = \alpha_i - \alpha_j,\ \sum_{t\to\infty} \dot{Y}_{id} - \dot{Y}_{jd} = 0 \tag{7}$$

Theorem 1: The consensus eq. (5) will converge asymptotically if and only if a minimal spanning tree always exist in the information exchange topology [19]

Proof: Eq. (5) can be written in the form of [19]

$$\dot{Y}_{id} = -\sum_{j\in N}(Y^*_{id} - Y^*_{jd}),\ Y^*_{id} = Y_{id} - \alpha_i,\ Y^*_{jd} = Y_{jd} - \alpha_j \tag{8}$$

$$\dot{Y}_d = -LY_d$$

$$Y_d = \left[Y^*_{id},\ Y^*_{jd},\ \ldots\right]^T$$

We have taken the network graph of AUVs to be connected, for this case the solution to eq. (8) is

$$Y(t) = e^{Lt}Y(0) \tag{9}$$

Time is denoted as 't' and The spectrum of Laplacian matrix 'L' of a undirected connected sub graph is [20]

$$0 = \lambda_1 < \lambda_2 \leq \ldots\ldots\ldots \leq \lambda_N \tag{10}$$

The lowest eigen value λ_1 which is zero corresponds to a eigen vector 1, the vector of all ones. L is symmetric and L1= 0 for an arbitrary undirected *G*. Let the matrix L has mutually orthogonal and normalized eigen vectors Q = [Q_1 Q_2 ...Q_N] corresponding to its ordered eigen values. Furthermore, set

$$\Lambda = \mathbf{Diag}(\lambda_1, \lambda_2, \ldots\ldots\lambda_N) \tag{11}$$

$$Y(t)=e^{-\lambda_1 t}(Q_1^T Y_0)Q_1 + e^{-\lambda_2 t}(Q_2^T Y_0)Q_2 + \cdots e^{-\lambda_N t}(Q_N^T Y_0)Q_N \tag{12}$$

For a connected graph λ_1 is zero and $\lambda_i \geq 0$ for $i \geq 2$ therefore as [20]

$$t \to \infty, Y(t) \to Q_1^T Y_0 Q_1 = \frac{1^T Y_0}{N} 1 \tag{13}$$

and λ_2 the smallest positive value of graph Laplacian dictates the slowest mode of convergence. As a positive value of λ_2 suffice the asymptotic convergence and connectivity it can be concluded that a minimum order structure needed for convergence of the equation is a graph containing a spanning tree. Similarly eq. (6) can be written as

$$\ddot{Y} = -LY - \beta L\dot{Y} \tag{14}$$

L=Laplacian matrix of dynamic graph (flocking network) and $\beta = diag(\beta_{ij})$.
Convergence of consensus algorithm requires [12] a spanning tree to be present in the information topology and the weighting factors (β) should be greater than some constant which is given by

$$\beta_{ij} > \max_{i=2\cdots N} \sqrt{\frac{2}{|\lambda_i|\cos(\frac{\pi}{2} - \tan^{-1}\frac{\mathrm{Re}(\lambda_i)}{\mathrm{Im}(\lambda_i)})}} \tag{15}$$

$Y_i - Y_j \to 0, \dot{Y}_i - \dot{Y}_j \to 0, \forall i \neq j$, as $t \to \infty$ for any $Y_i(0), \dot{Y}_i(0)$

4.4. Control Law Development

A sliding control law should be chosen such that the rate of change of state vector of the AUV system in the flock will always point to the sliding surface. So that once in the vicinity of the sliding surface it will always remain inside this surface. A Lyaponov stability criterion is used to generate a control law to stabilize the plant (AUV) in case of uncertainties. Let V is a Lyaponov candidate function defined as

$$V = \frac{1}{2}\sigma^2 \tag{16}$$

Then asymptotic stability of error vector between actual states and reference states of the AUV is guaranteed if [14]

$$\dot{V} \leq -\mu^2 \mathrm{sign}(\sigma) \tag{17}$$

where μ is a positive quantity which should be chosen properly by trade of between magnitude of control forces generated and rate of convergence of state dynamics of each AUV into the sliding plane. A higher value would generate a higher magnitude of control action which is always undesired. Using eq. (4) the dynamics of leader AUV in the sliding plane is [14]

$$\dot{\sigma} = \dot{\tilde{Y}} - \ddot{\tilde{Y}} = H\dot{X} - \dot{Y}_d) + \gamma(H\ddot{X} - \ddot{Y}_d) \tag{18}$$

For follower AUVs the sliding plane is derived from eq. (5) and (6)

$$\dot{\sigma} = \dot{\tilde{Y}} - \ddot{\tilde{Y}} = (H\dot{X} - \dot{Y}_{id}) + \gamma(H\ddot{X} - \ddot{Y}_{id}) \tag{19}$$

To reduce chattering a boundary layer B with a continuous hyperbolic tangent function is used in place of hard limiting signum function

$$\begin{aligned}\dot{\sigma} &= (H(M^{-1}F(X) + M^{-1}G\tau) - \dot{Y}_d) + \gamma(H(M^{-1}\dot{F}(X) + M^{-1}G\dot{\tau} + M^{-1}\dot{G}\tau) - \ddot{Y}_d) \\ &= -\mu^2 \tanh(\frac{\sigma}{B})\end{aligned} \tag{20}$$

$$HM^{-1}G = HM^{-1}\dot{G} = \begin{bmatrix} 0 & 0 \\ 0 & 0 \end{bmatrix}^{-1} \tag{21}$$

$$\dot{F}(X) = \frac{\partial F}{\partial X}\frac{\partial X}{\partial t}$$
$$\dot{F}(X) = \frac{\partial F}{\partial X}\dot{X} \tag{22}$$

$$\frac{\partial F}{\partial X} = F_d = \begin{bmatrix} F_{d1} & F_{d2} \\ F_{d3} & F_{d4} \end{bmatrix} \tag{23}$$

$$F_{d1} = \begin{bmatrix} 2X_{uu}u & mr + X_{vr}r & mv + X_{vr}v + 2X_{rr}r \\ -mr + Y_{ur}r + Y_{uv}v & 2Y_{vv}v + Y_{uv}u & -mu + 2Y_{rr}r + Y_{ur}u \\ N_{uv}v & 2N_{vv}v + N_{uv}u & 2N_{rr}r + N_{ur}u \end{bmatrix} \tag{24}$$

$$F_{d2} = \text{zeros}(3 \times 3)$$

$$F_{d3} = \begin{bmatrix} 0 & 0 & 1 \\ \cos\psi & -\sin\psi & 0 \\ \sin\psi & \cos\psi & 0 \end{bmatrix}$$
$$F_{d4} = \begin{bmatrix} 0 & 0 & 0 \\ 0 & -u\sin\psi - v\sin\psi & 0 \\ 0 & u\cos\psi - v\sin\psi & 0 \end{bmatrix} \tag{25}$$

From eq. (20), (21) and (23)

$$\tau = (\lambda HM^{-1}\dot{F}(X)M^{-1}G)^{-1}\left[(\dot{Y}_d - HM^{-1}F(X)) + \gamma(\ddot{Y}_d - HM^{-1}\dot{F}(X)MF) - \mu^2 \tanh(\frac{\sigma}{B})\right] \tag{26}$$

The sliding mode controller developed above can be made robust enough to deal with uncertainties present in the model.

5. Results and Discussion

Effectiveness of the proposed SMC flocking control of multiple AUVs is presented in this section. A flocking situation is simulated considering three similar REMUS AUVs. The parameters used are same as in Prestro et al [1]. The kinematic and dynamic equations of the AUVs given in eq. (1) and (2) are used to apply the proposed control eq. (26).

The simulations were performed by using MATLAB and SIMULINK. In this paper we have chosen a local communication radius of a distance 3 m. That means whenever two AUVs come within this range only local communication is possible and outside this range no connection exist. An edge exists between two nodes of the graph iff distance between them is less than 3 meter. Other factors are chosen as follows:

$$\lambda = \begin{bmatrix} 2 & 0 \\ 0 & 2 \end{bmatrix} \qquad \phi = \begin{bmatrix} 5 & 0 \\ 0 & 5 \end{bmatrix}$$
$$\mu = \begin{bmatrix} 40 & 0 \\ 0 & 40 \end{bmatrix} \qquad Y_d = \begin{bmatrix} 20\cos(0.1t) \\ 20\sin(0.1t) \end{bmatrix}$$

where λ is chosen as diagonal matrix and it weights the velocity error term compared to position error. This parameter decides the position or its rate dominant the sliding surface and it has a direct effect on time constant of particular output variable. The smoothening constant 'ϕ' should be chosen as a tradeoff between robustness and high frequency switching which can lead to excitation of unmodeled dynamics.

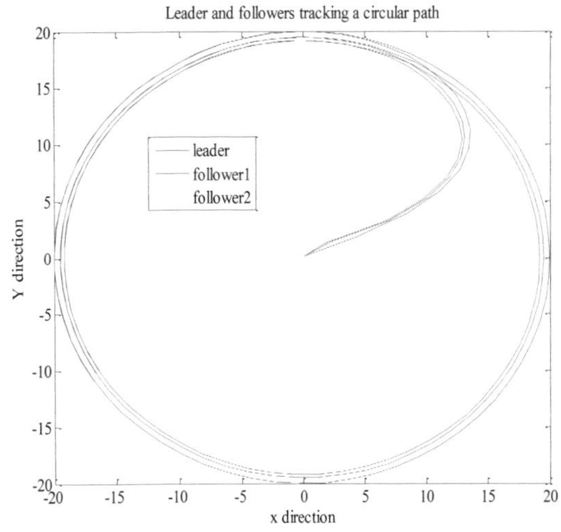

Fig. 4. Tracking of a circular path

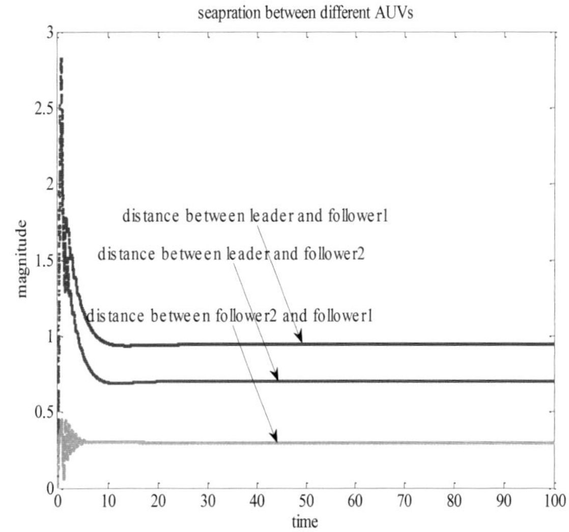

Fig. 5. Distance between the AUVs during flocking motion

A circular path of radius 20 m is specified to the leader. The proposed flocking controller is able to keep the group together with tracking the desired track. A separation of 1 m is used in consensus protocol which combined with sliding mode controller was successful in keeping the AUVs separate at desired distance.

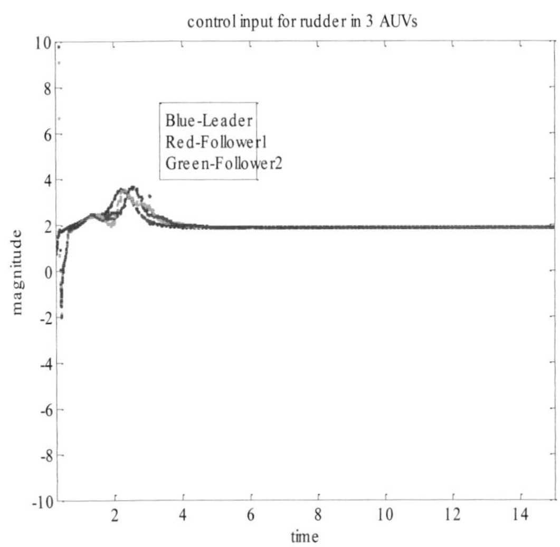

Fig. 6. Rudder angle input for three AUVs

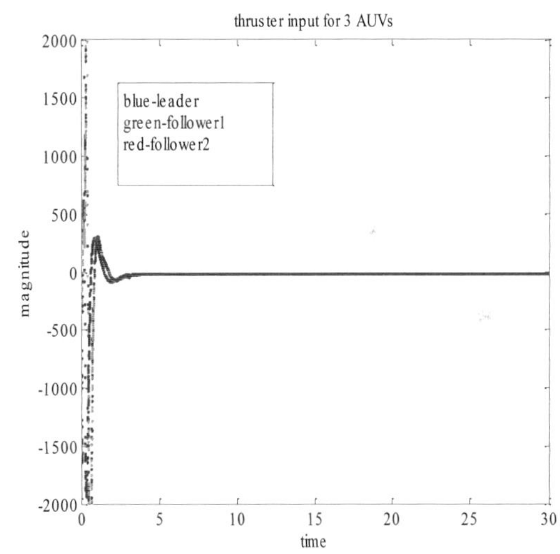

Fig. 7. Thruster input for three AUVs

Initially the thruster forces and rudder inputs required as shown in Fig.6 and Fig. 7 are high but practically realizable and after the AUVs hit the sliding surface the control inputs required are pretty much small. In above simulation eq (7) and (26) are utilized to follow a circular track of radius 20m.

From the simulation result, it is verified that a constant rudder angle is generated from the proposed controller to follow a circular track and meanwhile the thruster input is of a constant magnitude so that the AUVs track the desired track in approximately constant speed.

Fig.8, Fig.9 and Fig. 10 envisage that the proposed controller provide constant angular velocity (0.1), constant surge and sway velocities for tracking a circular path even with uncertainty in the model.

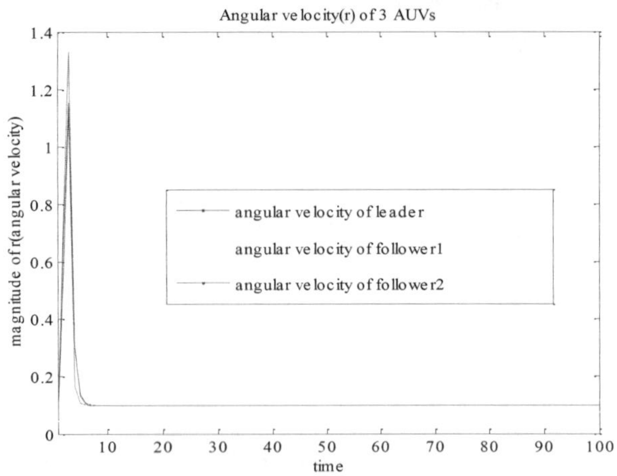

Fig. 8. Angular velocity while flocking for 3 AUVs

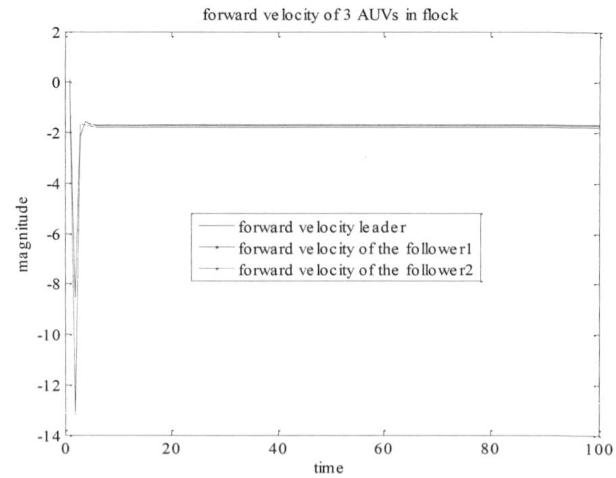

Fig. 9. Surge velocity while flocking for 3 AUVs

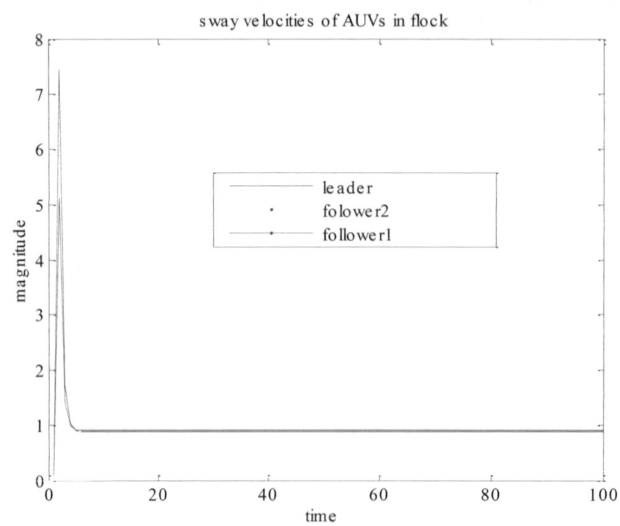

Fig. 10. Sway velocity while flocking for 3 AUVs

6. Conclusions

We proposed a sliding mode approach to design a flocking control algorithms for achieving an effcient cooperative motion of three AUVs. We have demnostaed that by employing consensus algorithm in the feedback structure provides a simple but effective tool for maintaining a flock of AUVs with local communication facilities. The proposed new SMC based robust controller provides excellent leader follower flocking performance. This control schem is not only simple but also robust enough to handle a large amount of uncertainties present in the AUV dynamics. Consensus algorithm (single and double integrator) is exploited for generation of the reference command for the follower AUVs in an information constrained network (local communication and low bandwidth) whereas sliding mode control handles robustness and nonlinearity present in the AUV dynamics.

Future work will focus in designing a switching network and deal with communication loss in the flocking control.

7. Acknowledgments

This research was supported by joint Russian-Indian project (RFBR 17-57-45055) and project MK-8536.2016.8.

8. References

[1] Timothy, P. (2001). Verification of a six-degree of freedom simulation model of REMUS autonomous underwater vehicle, M.S. Dissertation, Department of Ocean Engineering, Massachusetts Institute of Technology and Dept. of Mechanical Engineering, Woods Hole Oceanographic Institution.

[2] Filaretov, V.; Konoplin, A., Konoplin, N. & Gorbachev, G. (2016). Control system for underwater vehicle with multilink manipulator for automatic manipulation operations, Proceedings of the 27th DAAAM International Symposium, 2016, Vienna, Austria, ISBN 978-3-902734-08-2, ISSN 1726-9679, Katalinic, B. (Ed.), pp.0714-0720, DOI: 10.2507/27th.daaam.proceedings.103

[3] Filaretov, V. & Pryanichnikov, V. (2015) Autonomous mobile university robots amur: technology and applications to extreme robotics. Procedia Engineering, Vol. 100, (2015), pp. 269-277, ISSN 1877-7058.

[4] Winfield, A. (2000). Distributed sensing and data collection via broken ad hoc wireless connected networks of mobile robots, In: Distributed Autonomous Robotic Systems 4, L.E. Parker, G. Bekey and J. Barhen, (Ed.), 273–282, Springer-Verlag, ISBN 978-4-431-67919-6, New York.

[5] Desai, J. P., Kumar, V. & Ostrowski, J. P. (2001). Modeling and control of formations of nonholonomic mobile robots. IEEE Transactions on Robotics and Automation, Vol. 17, No. 7, (2001), pp. 905–908, ISSN 1042-296X.

[6] Fax, J. A. & Murray, R. M. (2004). Information flow and cooperative control of vehicle formations. IEEE Transactions on Automatic Control, Vol. 49, No. 9, (2004), pp.1465–1476, ISSN 0018-9286.

[7] Das, B., Subudhi, B. & Pati, B. B. (2016). Adaptive sliding mode formation control of multiple underwater robots. Archives of Control Sciences, Vol. 24(LX), No. 4, (2014), pp. 515-543, ISSN 2300-2611.

[8] Gu, D. & Wang Z. (2009). Leader–follower flocking: algorithms and experiments. IEEE Transactions on Control Systems Technology, Vol. 17, No. 5, (2009), pp. 1211 – 1219, ISSN 1063-6536.

[9] Ji, M. & Egerstedt, M. (2007). Distributed coordination control of multi agent systems while preserving connectedness. IEEE Transactions on Robotics, Vol. 23, No. 4, (2007), pp. 693 – 703, ISSN 1552-3098.

[10] Tanner, H. G., Jadbabaie, A. & Pappas, G. J. (2007). Flocking in fixed and switching networks. IEEE Transactions on Automatic Control, Vol. 52, No. 5, (2007), pp.863–868, ISSN 0018-9286.

[11] Das, B., Subudhi, B. & Pati, B. B. (2016). Co-operative control coordination of a team of underwater vehicles with communication constraints. Transactions of the Institute of Measurement and Control, Vo. 38, No. 4, (2016), pp. 463-481, ISSN 1477-0369.

[12] Ren, W., & Atkins, E.M. (2007). Distributed multi-vehicle coordinated control via local information exchange. International Journal of Robust and Nonlinear Control, Vol. 17, (2007), pp. 1002–1033, ISSN 1099-1239.

[13] Ren, W. (2007). Distributed attitude alignment in spacecraft formation flying. International journal of adaptive control and signal processing, Vol. 21, No. 2–3, (2007), pp.95–113, ISSN 1099-1115.

[14] Fahimi, F. (2007). Sliding-mode formation control for under actuated surface vessels. IEEE Transactions on Robotics, Vol.23, No.3, (2007), pp.617–622, ISSN 1552-3098.

[15] Healey, A J. & Lienard, D. (1993). Multivariable sliding mode control for autonomous diving and steering of unmanned underwater vehicles. IEEE Journal of Oceanic Engineering, Vol. 18, No. 3, (1993), pp. 1-13, ISSN 0364-9059.

[16] Utkin,V. I. (1977). Variable structure systems with sliding modes. IEEE Transactions on Automatic Control, Vol. 22 No. 2, (1977), pp.212-222, ISSN 0018-9286.

[17] Balch, T. & Arkin, C. (1998). Behaviour based formation control of multi robot teams. IEEE Trans. on Robotics and Automation, Vol.14, No. 2, (1998), pp. 926 – 939, ISSN 1042-296X.

[18] Das, B., Subudhi, B. & Pati, B. B. (2016). Co-operative formation control of autonomous underwater vehicles: an overview. International Journal of Automation and computing, Vo. 13, No. 3, (2016), pp. 199-225, ISSN 1751-8520.

[19] Ren, W. (2008). On consensus algorithms for double-integrator dynamics. IEEE Transactions On Automatic Control, Vol. 53, No. 6, (2008), pp. 1503 – 1509, ISSN 0018-9286.

[20] Zhu, J., Tian, Y. & Kuang, J. (2009). On the general consensus protocol of multi-agent system with double integrator dynamics. Linear Algebra and its Applications, Vo. 431, No. 5–7, (2009), pp. 701-715, ISSN 0024-3795.

[21] Cui, R., Ge, S. S. & How, B. V. E. (2010). Leader–follower formation control of under actuated autonomous underwater Vehicles. Ocean Engineering, Vol. 37, No. 17-18, (2010), pp. 1491-1502, ISSN 0029-8018.

[22] Das, B., Subudhi, B. & Pati, B. B. (2016). Co-operative control of a team of autonomous underwater vehicles in an obstacle-rich environment. Journal of Marine Engineering & Technology, Vol. 15, No. 3, (2016), pp. 135-151, ISSN 2046-4177.

[23] Das, B., Subudhi, B. & Pati, B. B. (2016). Employing nonlinear observer for formation control of AUVs under communication constraints. International Journal of Intelligent Unmanned Systems, Vol. 3, No. 2/3, (2015), pp. 122-155, ISSN 2049-6427.

[24] Subudhi, B., Mukherjee, K. & Ghosh, S. (2013). A static output feedback control design for path following of autonomous underwater vehicle in vertical plane. Ocean Engineering, Vol. 63, No. 1, (2013), pp. 72–76, ISSN 0029-8018.

DOI: 10.2507/28th.daaam.proceedings.002

DEVELOPMENT OF FAULT ACCOMMODATION SYSTEM FOR ACTUATORS OF UNDERWATER MANIPULATORS

Vladimir Filaretov[1, 2], Alexander Zuev[1, 2], Alexey Zhirabok[2, 3], Alexander Konoplin[2, 3], Alexander Protcenko[2, 3], Subudhi Bidyadhar[4]

[1] Institute of Automation and Control Processes FEB RAS, Vladivostok, Russia
[2] Far Eastern Federal University, Vladivostok, Russia
[3] Institute of Marine Technology Problems, Vladivostok, Russia
[4] National Institute of Technology, Rourkela-769008, India

This Publication has to be referred as: Filaretov, V[ladimir]; Zuev, A[lexander]; Zhirabok, A[lexey]; Konoplin, A[lexander]; Protcenko, A[leksandr] & Subudhi, B[idyadhar] (2017). Development of Fault Accommodation System for Actuators of Underwater Manipulators, Proceedings of the 28th DAAAM International Symposium, pp.0014-0022, B. Katalinic (Ed.), Published by DAAAM International, ISBN 978-3-902734-11-2, ISSN 1726-9679, Vienna, Austria
DOI: 10.2507/28th.daaam.proceedings.002

Abstract

The paper presents a fault accommodation system for electric servo actuators of underwater manipulators. These actuators are described by nonlinear equations with variable parameters. It is assumed, that the following typical faults are possible in electric servo actuators: fault caused by changing of viscous friction moment and error in the sensor, which measures the position of output shaft of gear. The proposed system is based on using of diagnostic observers with residual signal feedback. It is proposed the methodology of synthesis of such observers for detection and identification of occurred faults. The accommodation to faults is ensured due to the generation of additional control signals based on identified value of faults. This provides the reliability of underwater working of manipulators at faults. The efficiency of the proposed system is confirmed by mathematical modelling.

Keywords: underwater manipulator; diagnosis; fault; accommodation; actuators

1. Introduction

Now autonomous underwater vehicles (AUV) are the most effective research instrument of the World Ocean and performance of various underwater and technical operations [1, 2]. At this, fitting multilink underwater manipulators (UM) on AUVs is a very perspective tendency of development [3-6]. The manipulator-equipped AUVs can be effectively used for fulfillment of research operations in hydrogeology, hydrobiology, geophysics, hydrochemistry, in oil and gaze extraction. In these cases the following operations are carried out: selective sampling of soil probes and geological rocks; collecting bio-organisms; set-up and servicing of different tools; definition of composition and density of the soil using special probes and drills; taking precipitation samples by means of hermetically sealed soil tubes;

measurements with thermistor sensors in different layers of sedimentary soil etc. All it considerably expands the field of effective use of AUVs and allows to exclude the human at a research of sea depths.

An important task at autonomous operation of the UMs is to increase in their reliability and safety. Any fault in the actuators of UMs leading significant reduction of accuracy and therefore quality of performance. Moreover, at some faulty cases the expensive vehicle can be lost. The task of increase of reliability can be solved by using modern fault diagnosis and accommodation systems [7, 8]. Fault accommodation is based on forming additional control signals for automatic stabilization of main parameters of the actuators when faults occur. For forming of such control signals all possible faults should be timely detected and their values should be estimated.

As a rule, the actuators of UMs are described by essentially nonlinear equations. Therefore, it is necessary to use nonlinear methods of fault diagnosis and accommodation. However, most of papers dealing with this problem consider nonlinear systems with differential nonlinearities [9, 10] therefore they cannot be used in our case. Today, a several approaches for fault diagnosis and accommodation for systems with no differentiable nonlinearities are known. Among them there are methods using the algebra of functions [11], the logic-dynamic approach (LDA) [12], and others [13, 14]. Whereas the algebra of functions demands complex analytical calculations using symbolic mathematical packages, the LDA allows using the methods of linear algebra for nonlinear systems and provides relatively simple procedure for synthesis of diagnostic observers. There are several fault accommodation systems based on LDA observers [15, 16]. These systems provide exact faults identification and accommodation for only actuators of manipulators with fully measured state vector. This condition requires installation of additional sensors for actuators of UMs, which in turn reduces reliability and is not always possible constructively.

When UMs move in water a significant generalized moments act in all its degrees of freedom. These moments are caused by inertial and gravitational forces and forces determined by interaction of working UMs and viscous environment [17]. At this case, the actuators of UMs will be described by nonlinear equations with variable parameters. It should be considering at synthesis of diagnostic observers and accommodation control law.

Thus, the problem of development of highly efficient and easily implementing fault accommodation systems for actuators of UMs is still topical.

This paper considers a solution to the problem of development of the effective accommodation system for faults caused by errors in the sensor, which measure the position of output shaft of gear and by changing the value of viscous friction coefficient in electric servo actuators of the UMs described by mathematical model including differential equations with nonlinearity and variable parameters and not fully known state vector.

2. Description of model of the actuators of UMs

Let's consider the AUV with UM fixed in point O (see Fig. 1). In general, this UM can have n degrees of freedom. The UM is able to move in forward hemisphere in front of the AUV. Each degree of freedom is actuated by corresponding actuator.

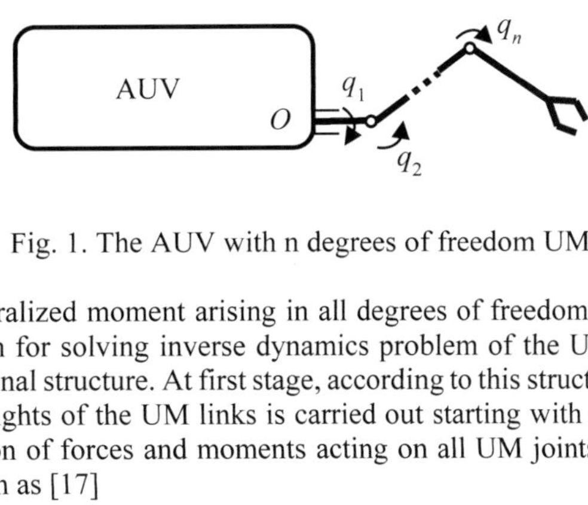

Fig. 1. The AUV with n degrees of freedom UM

For definition of external generalized moment arising in all degrees of freedom of UM at the motion in water, it is possible to use recurrent algorithm for solving inverse dynamics problem of the UM [17]. This algorithm is based on well-known two-staged computational structure. At first stage, according to this structure the calculation of angular speeds and accelerations of centers of weights of the UM links is carried out starting with first and finishing with last the UM link. At second stage, the definition of forces and moments acting on all UM joints is carried out in the reverse order. This algorithm can be written down as [17]

$$
\begin{aligned}
&\omega_i = A_i^{i-1}\omega_{i-1} + e_i\dot{q}_i\,\bar{\sigma}_i,\ \ \omega_0 = \omega_0^*,\ i = \overline{1,n};\\
&\dot{\omega}_i = A_i^{i-1}\dot{\omega}_{i-1} + [(A_i^{i-1}\omega_{i-1})\times e_i\dot{q}_i + e_i\ddot{q}_i]\bar{\sigma}_i,\ \ \dot{\omega}_0 = \dot{\omega}_0^*,\ i = \overline{1,n};\\
&\ddot{P}_i = A_i^{i-1}(\ddot{P}_{i-1} + \delta_{i-1}p_{i-1}^*) + (2\dot{q}_i\omega_i \times e_i + \ddot{q}_i e_i)\sigma_i,\ \ddot{P}_0 = P_0^*;\\
&\ddot{r}_{mi} = \ddot{P}_i + \delta_i r_i^*,\ i = \overline{1,n};\\
&\mathrm{v}_i = A_i^{i-1}(\mathrm{v}_{i-1} + \omega_{i-1}\times p_{i-1}^*),\ \mathrm{v}_1 = \mathrm{v}_0,\ i = \overline{2,n};\ \ \mathrm{v}_{Ai} = \mathrm{v}_i + \omega_i \times r_i^*;\\
&\psi_i = \arccos\frac{\mathrm{v}_{Ai}p_i^*}{|\mathrm{v}_{Ai}||p_i^*|},\ \alpha_i^* = \arccos\frac{\mathrm{v}_i p_i^*}{|\mathrm{v}_i||p_i^*|},\ \beta_i^* = \arccos\frac{\omega_i p_i^*}{|\omega_i||p_i^*|},\ i = \overline{1,n};
\end{aligned}
\tag{1}
$$

$$r_{pi} = r_i^* + K_{Ai}\mathrm{v}_{Ai},\ i=\overline{1,n};$$
$$\mathrm{v}_j^* = \mathrm{v}_i + \omega_i \times h_j^*,\ \omega_{Li} = |\omega_i| \mathrm{e}_{Li}\cos\beta_i^*,\ i=\overline{1,n},\ j=\overline{1,N};$$
$$\mathrm{v}_{Li}^* = |\mathrm{v}_i| \mathrm{e}_{Li}\cos\alpha_i^*,\ \mathrm{v}_{pj}^* = \mathrm{v}_j^* - \mathrm{v}_{Li}^*, i=\overline{2,n},\ j=\overline{1,N};$$
$$\text{if } \frac{\rho \mathrm{v}_{Li}^* l_i}{\eta} \le 10^3 \text{ then } F_{RLi} = k_{Li}\eta \mathrm{v}_{Li}^*,\ i=\overline{n,1};$$
$$\text{if } \frac{\rho \mathrm{v}_{Li}^* l_i}{\eta} > 10^3 \text{ then } F_{RLi} = \frac{1}{2}\rho k_{Li} s_i \mathrm{v}_{Li}^{*\,2},\ i=\overline{n,1};$$
$$\text{if } \frac{2\rho \mathrm{v}_{pi}^* r_i}{\eta} \le 10^3 \text{ then } F_{Rpj} = k_i^* \eta \mathrm{v}_{pj}^*,\ i=\overline{n,1},\ j=\overline{1,N};$$
$$\text{if } \frac{2\rho \mathrm{v}_{pi}^* r_i}{\eta} > 10^3 \text{ then } F_{Rpj} = \frac{1}{2}\rho k_i^* d_j \mathrm{v}_{pj}^{*\,2},\ i=\overline{n,1},\ j=\overline{1,N};$$
$$F_{Rpi} = \sum_{j=1}^{N} F_{Rpj},\ M_{Rpj} = h_j^* \times F_{Rpj},\ M_{Rpi} = \sum_{j=1}^{N} M_{Rpj},\ M_{Li} = k_{Li}^* \eta r_i \omega_{\mathrm{L}i},\ i=\overline{n,1},\ j=\overline{1,N};$$
$$F_i = A_i^{i+1}F_{i+1} + (m_i + \Pi_{mi})\ddot{r}_{mi} + F_{RLi} + F_{Rpi},\ F_{n+1}=0,\ i=\overline{n,1};$$
$$M_i = A_i^{i+1}M_{i+1} + p_i^* \times (A_i^{i+1}F_{i+1}) + r_i^* \times (m_i\ddot{r}_{mi}) + r_{pi}\times(\Pi_{mi}\ddot{r}_{mi}) +$$
$$+(\tau_i + T_i)\dot{\omega}_i + \omega_i \times ((\tau_i + T_i)\omega_i) + M_{Rpi} + M_{Li},\ M_{n+1}=0,\ i=\overline{n,1};$$
$$\hat{M}_i = e_i^{\mathrm{T}}(F_i\sigma_i + M_i\overline{\sigma}_i),$$

where $\delta_i = \begin{bmatrix} -(\omega_{i(2)}^2 + \omega_{i(3)}^2) & \omega_{i(1)}\cdot\omega_{i(2)} - \dot{\omega}_{i(3)} & \omega_{i(1)}\cdot\omega_{i(3)} + \dot{\omega}_{i(2)} \\ \omega_{i(1)}\cdot\omega_{i(2)} + \dot{\omega}_{i(3)} & -(\omega_{i(1)}^2 + \omega_{i(3)}^2) & \omega_{i(2)}\cdot\omega_{i(3)} - \dot{\omega}_{i(1)} \\ \omega_{i(1)}\cdot\omega_{i(3)} - \dot{\omega}_{i(2)} & \omega_{i(2)}\cdot\omega_{i(3)} + \dot{\omega}_{i(1)} & -(\omega_{i(1)}^2 + \omega_{i(2)}^2) \end{bmatrix}$ (in matrix δ_i lower index indicates number of element in respective vectors); A_i^{i-1} is matrix of vectors translation from (*i*-1)- th to *i*- th coordinate system; v_0 is linear velocity of the UM base; $e_i = (0\ 0\ 1)^T$ is vector directed along axis of joint *i*; $\omega_0^* \in R^3$ is angular velocity of AUV rotation; q_i is angle of joint i rotation; $\omega_i \in R^3$ is angular velocity of link *i*; $p_i^* \in R^3$ is concerned with longitudinal axis of link i vector defining position of joint (*i*+1) relative to joint *i*; $e_{Li} \in R^3$ is unitary vector directed along longitudinal axis of link *i*; $h_j^* \in R^3$ is vector defining centre of mass of segment j of length δh_i^* of link *i* concerning joint *i*; $\alpha_i^* = \arccos\frac{\mathrm{v}_i \cdot p_i^*}{|\mathrm{v}_i||p_i^*|}$ is an angle between v_i and p_i^*; l_i and r_i are length and radius of link *i*, respectively; ρ and η are density and viscosity of water, respectively; v and *d* are linear velocity and characteristic size of body, respectively); $\mathrm{v}_{Ai} \in R^3$ is linear velocity of center of size of UM link *i*; $\sigma_i = 1$ if joint *i* is translational; $\sigma_i = 0$ if joint *i* is rotational ($\overline{\sigma}_i = 1 - \sigma_i$); m_i is mass of link *i*; $\dot{\omega}_i \in R^3$ is angular acceleration of link *i*; $\omega_{Li} \in R^3$ is component of angular velocity ω_i parallel to longitudinal axis of link *i*; Π_{mi} are added masses of fluid attached to link *i*; $\ddot{r}_{mi} \in R^3$ is linear acceleration of center of mass of link *i*; $r_{pi} \in R^3$ is vector defining position of center of added mass of fluid Π_{mi} relative to joint *i*; $r_i^* \in R^3$ is vector defining position of center of mass of link relative to joint *i*; k_{Li}^* is experimentally determined viscous friction coefficient; K_{Ai} is parameter depending from $|\mathrm{v}_{Ai}|$ and angle ψ_i; $\tau_i \in R^{3\times3}$ is inertia tensor of link *i* concerning its centre of mass; $T_i \in R^{3\times3}$ is inertia tensor of fluid masses attached to UM link *i*; $F_i, M_i \in R^3$ are force and torque acting on joint *i*, respectively; $P_0^* = -g + \ddot{P}_{UV}$ if centres of mass of UM links do not coincide with their centres of size; $P_0^* = g\left(\frac{w_i}{m_i} - 1\right) + \ddot{P}_{UV}$ if centres of mass of all UM links coincide with their centres of size; $\ddot{P}_{UV} \in R^3$ is linear acceleration of UM base; *g* is intensity of gravity; w_i is mass of fluid forced out by link; $\hat{M}_i$ is generalized moment that arises in *i*-th degree of freedom of the UM at the motion in water. Further for simplicity we will not write the index *i*.

Each degree of the UM has an electric servo actuator that includes DC motor, reducing gear and sensors measuring the angular position and angular speed of output shaft of reducing gear.

It is assumed that the following typical faults are possible in actuators of the UM:

- the fault $d_1(t)$ caused by error $\tilde{\alpha}_r$ of sensor of angular position;
- the fault $d_2(t)$ caused by change of value of viscous friction coefficient $\tilde{k}_v$;

Considering these faults and (1), model of each actuator of the UM can be presented by following differential equations:

$$\dot{x}_1 = \frac{1}{i_r} x_2 + d_1(t),$$
$$\dot{x}_2 = -\frac{k_v}{J} x_2 + \frac{k_m}{J} x_3 - \frac{M_{cf} + \hat{M}^*}{J} + d_2(t), \qquad (2)$$
$$\dot{x}_3 = -\frac{k_\omega}{L} x_2 - \frac{R}{L} x_3 + \frac{k_u}{L} u,$$

where $x_1 = \alpha_r$ is the angle of output shaft of reducing gear; $x_2 = \dot{\alpha}$ is the speed of DC motor shaft; R, L and $x_3 = I$, are the resistance, inductance and current of electric motor rotor circuits, accordingly; k_ω is the coefficient of counter-EMF; k_m is the coefficient of motor moment; k_v is the coefficient of viscous friction; J is the rotational inertia of rotating parts of reduction drive and DC motor; i_r is the coefficient of the reducing gear; u is the control signal; k_u is the amplifier gain; $\hat{M}^* = \frac{\hat{M}}{i_r}$; $M_{cf} = M_{cf0} sign(\dot{a})$ is moment of Coulomb friction; M_{cf0} is coefficient of Coulomb friction.

The variables d_1 and d_2 in (1) represent the effects of faults on the system and can be presented in next form $d_1(t) = \dot{\tilde{\alpha}}_r$, $d_2(t) = -\frac{\tilde{k}_v \dot{\alpha}}{J}$.

It should be noted, that (1) has nonlinearities and significantly changing parameters because of changes in parameter $\hat{M}^*$.

Equation (2) can be represented by the differential equation in a matrix form as follows:

$$\dot{x}(t) = Fx(t) + Gu(t) + B(x) + Ed,$$
$$y(t) = Nx(t), \qquad (3)$$

where $B(x) \in R^3$ is a vector corresponding to the system nonlinearities, $x \in R^3$, $y \in R^2$, $d \in R^2$, $F \in R^{3\times3}$, $G \in R^3$, $E \in R^{3\times2}$, $N \in R^{2\times3}$,

$$F = \begin{bmatrix} 0 & 1/i_r & 0 \\ 0 & -k_v/J & k_m/J \\ 0 & -k_\omega/L & -R/L \end{bmatrix}, \quad x = \begin{bmatrix} x_1 \\ x_2 \\ x_3 \end{bmatrix}, \quad E = \begin{bmatrix} 1 & 0 \\ 0 & 1 \\ 0 & 0 \end{bmatrix}, \quad N = \begin{bmatrix} 1 & 0 & 0 \\ 0 & 1 & 0 \end{bmatrix}, \quad B(x) = \begin{bmatrix} 0 \\ -(M_{cf} + \hat{M}^*)/J \\ 0 \end{bmatrix}, \quad G = \begin{bmatrix} 0 \\ 0 \\ k_u/L \end{bmatrix},$$
$$y = \begin{bmatrix} \alpha_r \\ \dot{\alpha} \end{bmatrix}, \quad d = \begin{bmatrix} d_1 \\ d_2 \end{bmatrix}.$$

If faults occur during the work of the UM, it should be promptly detected and identified. Then the fault accommodation can be used. The traditional methods of diagnosis that focuses on systems described by linear models cannot provide the solution to the problem of synthesis of accommodation systems for electric servo actuators of the UM described by model with the presence of the nonlinear component $B(x)$.

In this paper, another synthesis method of fault accommodation systems is proposed. This method includes two basic steps: 1) using LDA to construct the diagnostic observers [12] with residual feedback [18] which can detect occurrence of specific faults and identification of their values; 2) forming of control signal which can provide fault tolerance to actuators of UM. Below these stages are considered in more detail.

3. Synthesis of fault accommodation system

The LDA includes following steps [12]. 1. Replacing the initial nonlinear system (3) by linear system. 2. Solving the accommodation problem for the linear system with some additional restrictions and obtaining the bank of linear observers. 3. Transforming the linear observers into the nonlinear ones. The feedback signals by residual are introduced in observers for fault identification [18]. The nonlinear diagnostic observers with residual feedbacks detecting the faults have following general form

$$\dot{x}_*(t) = F_*x_*(t) + G_*u(t) + B_*(x_*) + Vy(t) + w(r(t)),$$
$$y_*(t) = N_*x_*(t), \tag{4}$$

where $x_*(t) \in R^k$ is the state vector of the observer, $y_*(t)$ is the output signal, $F_* \in R^{k\times k}$, $G_* \in R^k$, $B_* \in R^k$, $V \in R^{k\times 2}$ and $N_* \in R^k$ are constant matrices, k is dimension of the observer; w is feedback vector.

It is assumed that the matrices F_* and N_* can be represented in a canonical form [12]: $F_* = \begin{bmatrix} 0 & 1 & 0 & \dots & 0 \\ 0 & 0 & 1 & \dots & 0 \\ . & . & . & & . \\ 0 & 0 & 0 & \dots & 0 \end{bmatrix}$, $N_* = [1 \quad 0 \quad 0 \quad \dots \quad 0]$.

The observer (4) generates the residual $r(t) = Cy(t) - y_*(t)$ for certain row matrix $C \in R^2$ which should be determined. If there are no faults and $d = 0$, then $r(t) = 0$, if a fault occurs, $r(t) \neq 0$. It is well-known [12] that for the observer design, the matrix Φ such that $\Phi x(t) = x_*(t)$ in the unfaulty case after the response to unlike conditions have died out plays the main role. In the absence of faults, the following well-known set of equations holds [12]:

$$N_*\Phi = CN,$$
$$\Phi F = F_*\Phi + VN,$$
$$G_* = \Phi G,$$
$$B_*(x_*) = \Phi B(x). \tag{5}$$

For each fault d, the diagnostic observer should be constructed. Matrices and vectors Φ, F_*, $B_*(x_*)$, G_*, V and C describing the observer can be found from (5). As a result, the observers O_1 and O_2 for the faults $d_1(t)$ and $d_2(t)$ were obtained:

$$O_1: \Phi = \begin{bmatrix} 1 & 0 & 0 \\ -1 & 0 & 0 \end{bmatrix}, \; G_* = \begin{bmatrix} 0 \\ 0 \end{bmatrix}, \; V = \begin{bmatrix} 1 & 1/i_r \\ 0 & -1/i_r \end{bmatrix}, \; C = [1 \quad 0], \; B_* = \begin{bmatrix} 0 \\ 0 \end{bmatrix};$$

$$O_2: \Phi = \begin{bmatrix} 0 & 1 & 0 \\ 0 & R/L & k_m/J \end{bmatrix}, \; G_* = \begin{bmatrix} 0 \\ \frac{k_u k_m}{LJ} \end{bmatrix}, \; V = \begin{bmatrix} 0 & (-k_v/J)-(R/L) \\ 0 & -k_\omega k_m/LJ - k_v R/LJ \end{bmatrix}, \; C = [0 \quad 1], \tag{6}$$

$$B_* = \begin{bmatrix} -\hat{M}^*/J \\ -\hat{M}^* R/JL \end{bmatrix}.$$

For solving of identification problem for observers (6) which have dimension $k = 2$ it is proposed to used approach developed in paper [18]. This approach uses residual signal feedbacks for elimination of mismatch vector between diagnosed object and observer, i.e. $e(t) = \Phi x(t) - x_*(t) = 0$, even in the presence of faults. The procedure for the synthesis of such feedbacks is shown in detail in the paper [18].

After applying this known procedure [18] for observers O_1 and O_2 it is possible to receive the following feedbacks for fault identification:

$$w = [w_1 \quad w_2]^T,$$
$$w_1 = \frac{T_1}{T_0} r(t),$$
$$w_2 = \frac{T_2}{T_0} r(t) + \int \frac{1}{T_0} r(t) dt,$$

where T_0, T_1 and T_2 are feedback coefficients. These coefficients are selected in such a way as to guarantee stability to observers. Finally, it is possible to identify the faults value for the electric servo actuators of the UM [18]:

$$\tilde{a}_r = -\frac{1}{f_1 T_0}\iint r dt, \quad \tilde{k}_v = -\frac{J}{f_2 T_0 \dot{\alpha}}\int r dt, \tag{7}$$

where $f_1 = -1$ and $f_2 = R/L$. Thus, the problem of identification of value of faults has been solved.

The next step of synthesis of the fault accommodation system is forming a control law for the compensation of the detected fault. Actuators of the UM are described by nonlinear differential equations with variable parameters. The control law for such objects can be formed using an approach of synthesis of corrective devices, which provides stabilization of the parameters of the control object to the nominal values [19].

At first the desired dynamic properties of actuators of the UM should be defined. Dynamic properties for actuators with DC motors can be defined by the equation $LJ\dddot{\alpha} + RJ\ddot{\alpha} + k_\omega k_m \dot{\alpha} = k_u k_m \tilde{u}$, which is the differential equation with constant coefficients, where $\tilde{u}$ is a control signal of adaptive regulator.

Next, variable $\ddot{\alpha}$, which is the highest derivative of the object output signal, should be expressed in this equation. Desired control law can be obtained by the substitution of this variable into the initial model of actuator. Thus, control law obtained by this method is a function of the output of the system and its lower derivatives. Control law of this form ensures the required performance and dynamic properties of the actuators of the UM. For stabilizing the properties of actuators in case of faults, estimates of these faults obtained by observers (6) with feedbacks (7) should be used in the control law. Thus, the accommodation to faults in actuators of the UM can be achieved by the following control law:

$$u = P(\varepsilon)[a_r^d - a_r + \tilde{a}_r] + \frac{L(k_v + \tilde{k}_v)}{k_u k_m}\ddot{\alpha} + \frac{R(k_v + \tilde{k}_v)}{k_u k_m}\dot{\alpha} + \frac{R}{k_u k_m}[\hat{M}^* + M_{cf}] + \frac{L\dot{\hat{M}}^*}{k_u k_m}, \tag{8}$$

where $P(\varepsilon)$ is a function of typical regulator, and a_r^d is a desired position of the output shaft of actuator reduction gear.

Block diagram of the actuator with the synthesized fault accommodation system is shown in Fig. 2.

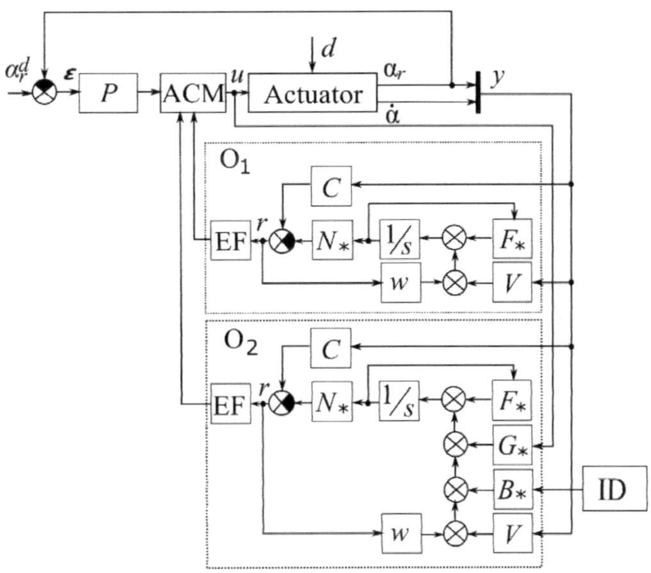

ACM – accommodation block (8), EF – block of estimation of faults value (7), ID – block of solving inverse dynamics problem of UM (1)

Fig. 2. Block diagram of the actuator with the synthesized fault accommodation system

The synthesis of proposed control law (8), which provides accommodation to faults for all actuators of the UM, is quite simple. Such control law can be realized on the base of common microcontrollers. In such way, it is possible to provide accurate stabilization of the dynamic properties of the actuators of underwater multilink manipulators described by complicated differential equations with variable parameters and nonlinearity, by using additional signals of residuals in the law of self-tuning, even in the presence of a fault.

4. Simulation of performance of fault accommodation system

Analysis of effectiveness of developed fault accommodation system was implemented using model of UM with three degrees of freedom presented on Fig. 3. The accommodation system was synthesized for electric servo actuator of second degree of freedom of this UM. The results for other degree of freedom of the UM are similar.

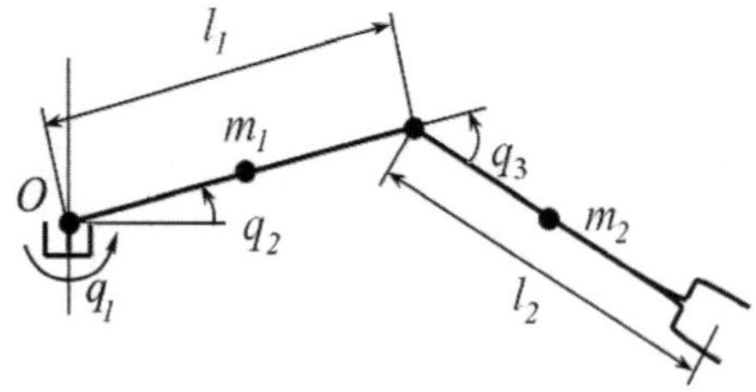

Fig. 3. The kinematic scheme of UM with 3 degrees of freedom

Parameters of UM have the following values: $l_1 = l_2 = 0.5$ m are lengths of UM links; $m_1 = m_2 = 3.9$ kg are masses of links. The links centers of size coincide with their centers of weights. Links have neutral buoyancy and cylindrical form with base radius $r_1 = r_2 = 0.05$ m. Tensors of inertia of these links concerning their centers of weights are diagonal and $\tau_{1_{11}} = \tau_{2_{11}} = 5 \cdot 10^{-3}$, $\tau_{1_{22}} = \tau_{1_{33}} = \tau_{2_{22}} = \tau_{2_{33}} = 8.4 \cdot 10^{-2}$ are elements of inertia tensors of the first τ_1 and second τ_2 links, respectively. Besides $\Pi_1 = \Pi_2 = 1.075$ kg; $|r_1^*| = |r_2^*| = 0.25$ m; $K_{A2} = \sin \psi_2$; $\rho = 10^3$ kg/m³; $k_{L1} = k_{L2} = 0.012$; $k_1^* = k_2^* = 0.89$; $T_{1_{11}} = T_{2_{11}} = 1.4 \cdot 10^{-3}$; $T_{1_{22}} = T_{1_{33}} = T_{2_{22}} = T_{2_{33}} = 2.6 \cdot 10^{-2}$ are diagonal elements of inertia tensors of fluid masses attached to appropriate UM links.

The electric drive servo actuator is described by following set of parameters: $R = 0.5\ \Omega$, $L = 0.0005$ H, $k_\omega = 0.04$ V·s, $k_m = 0.04$ N·m/A, $J =$ 10-3 kg·m², $i_r = 100$, $k_v = 0.005$ N·m·s/rad, $M_{cf0} = 0.02$ N·m, $k_u = 800$.

The input signal $q_2 = \alpha_r^d = \sin(1.5t)$ was applied to the electric servo actuator. The parameters of feedbacks for observers were chosen as follows: $T_0 = 4.63 \cdot 10^{-6}$, $T_1 = 8.33 \cdot 10^{-4}$, $T_2 = 0.05$.

Faults were presented by changing the following values:

- adding the error of position sensor signal $\tilde{a}_r = 0.1$ rad at the time $t_1 = 3$ s;
- increasing viscous friction coefficient by 50% $\tilde{k}_v = 0.0025$ N·m·s/rad at the time $t_2 = 5$ s.

Values of introduced faults (curve 1) and estimated value of fault (curve 2) are shown on the Fig. 4 and Fig.5. It follows from these figures that the constructed system provides accurate estimation of occurred faults. Fig. 6 shows the value of dynamic error of position of actuator in absence (curve 1) and presence (curve 2) of faults. Fig. 7 shows the value of dynamic error of position of actuator with accommodation system (curve 1) and without accommodation system (curve 2).

The results of simulation confirm the efficiency of the proposed fault accommodation system for actuators of UMs. Realization of the synthesized accommodation systems does not present any difficulties. These systems can be realized with the help of serial microprocessors.

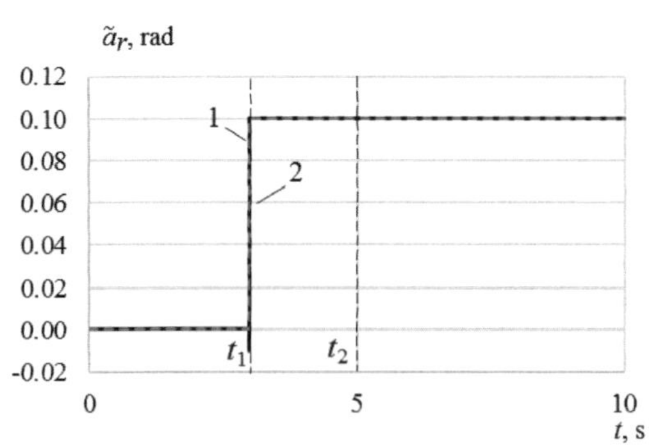

Fig. 4. Estimation of fault $\tilde{a}_r$ value by proposed fault accommodation system

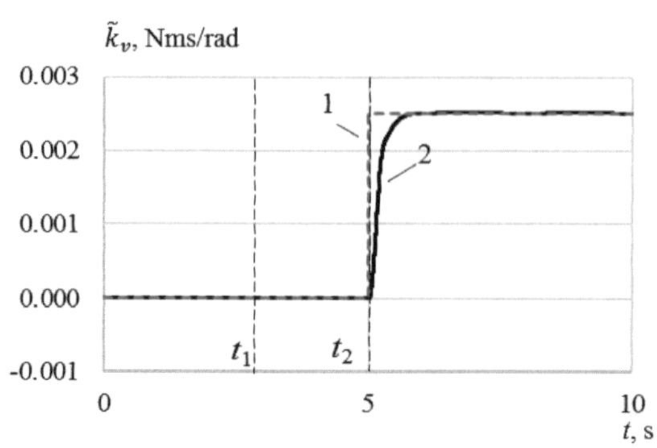

Fig. 5. Estimation of fault $\tilde{k}_v$ value by proposed fault accommodation system

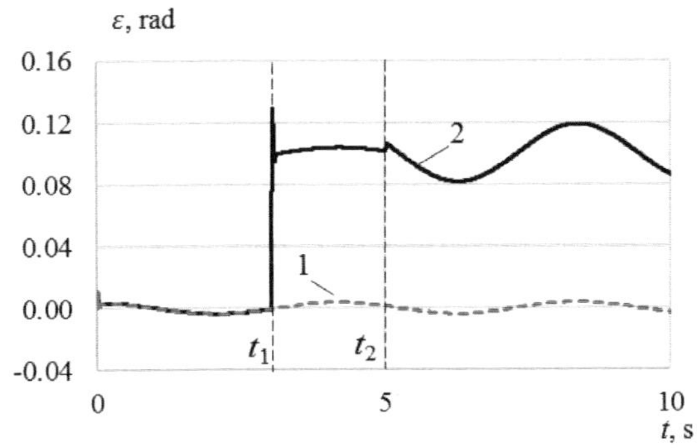

Fig. 6. Dynamic error of position of actuator in absence and presence of faults

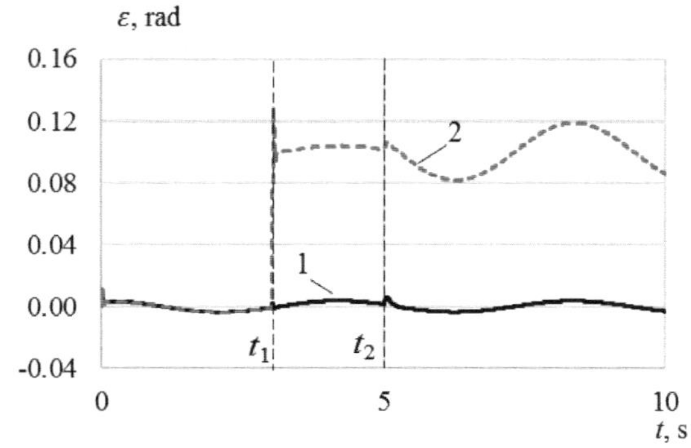

Fig. 7. Dynamic error of position of actuator with fault accommodation system and without fault accommodation system

5. Conclusion

An effective fault accommodation system for electric servo actuators of UMs is suggested. Two typical faults in actuators of UM were considered: fault caused by changing of viscous friction moment and error in the sensor, which measures the position of output shaft of gear. For diagnosis and identification of these faults were synthesized observers with residual signal feedbacks. It was proposed to realize the accommodation to faults by generation of additional control signals based on identified value of faults. As a result, it is possible provide the reliability of underwater working of manipulators in the event of faults.

The mathematical modeling confirms the efficiency of proposed fault accommodation system for actuators of UMs. The future work will relate to solving of problem of exact fault identification at the acting of unknown disturbance.

6. Acknowledgment

This research was supported by grant of President of Russia (MK-8536.2016.8) and grant of RFBR (16-29-04195).

7. References

[1] Wynn, R., Huvenne, V., Le Bas, T., Murton, B. and et al. (2014). Autonomous Underwater Vehicles (AUVs): Their past, present and future contributions to the advancement of marine geoscience. Marine Geology, Vol.352, (2014), pp. 451-468, ISSN 0025-3227.

[2] Bogue R. (2015). Underwater robots: a review of technologies and applications. Industrial Robot: An International Journal, Vol. 42, No. 3, (2015), pp.186-191, ISSN 0143-991X.

[3] Kim, T.W., Marani, G. & Yuh, J. (2016). Underwater vehicle manipulators, In: Handbook of Ocean Engineering, Dhanak M.R., Xiros N.I. (Ed.), pp. 407-422, Springer, ISBN 978-3-319-16648-3.

[4] Yuh, J., Marani, G. & Blidberg, D. (2011). Applications of marine robotic vehicles. Intelligent Service Robotics, Vol. 4, No. 4, (2011), pp. 221-231, ISSN 1861-2776.

[5] Filaretov, V. & Konoplin, A. (2016). Development of control systems for implementation of manipulative operations in hovering mode of underwater vehicle, Proceedings of OCEANS 2016, 2016, Shanghai, China, ISBN 978-146739724-7, pp. 1-5, DOI: 10.1109/OCEANSAP.2016.7485692.

[6] Filaretov, V.; Konoplin, A; Konoplin, N. & Gorbachev, G. (2016). Control system for underwater vehicle with multilink manipulator for automatic manipulation operations, Proceedings of the 27th DAAAM International Symposium, pp.0714-0720, B. Katalinic (Ed.), Published by DAAAM International, ISBN 978-3-902734-08-2, ISSN 1726-9679, Vienna, Austria, DOI: 10.2507/27th.daaam.proceedings.103.

[7] Blanke, M.; Kinnaert, M.; Lunze, J. & Staroswiecki, M. (2003). Diagnosis and Fault Tolerant Control, Springer-Verlag, ISBN 978-3-540-35652-3, Berlin, Heidelberg.

[8] Frank, P. (1990). Fault diagnosis in dynamic systems using analytical and knowledge-based re-dundancy – A survey and some new results, Automatica, vol. 26, pp. 459-474.

[9] Schreier, G.; Ragot, J.; Patton, R. & Frank, P. (1997). Observer design for class of nonlinear systems, Proceedings of the IFAC Symposium SAFEPROCESS'97, UK, Hull, pp. 498-503.

[10] De Persis, C. & Isidori, A. (2001). A geometric approach to nonlinear fault detection and isolation, IEEE Trans. Automat. Contr., Vol. AC-46, No. 6, pp. 853-865.

[11] Shumsky, A.Ye. & Zhirabok, A.N. (2006). Nonlinear diagnostic filter design: Algebraic and geo-metric points of view, Int. J. Applied Mathematics and Computer Science, Vol. 16, No. 1, pp. 115-127.
[12] Zhirabok, A. & Usoltsev, S. (2000). Linear diagnosis methods for nonlinear systems, Automation and Remote Control, Vol. 7, pp. 149-159.
[13] Isermann, R. (2006). Fault-Diagnosis Systems: An Introduction from Fault Detection to Fault Tolerance, Springer-Verlag, Berlin.
[14] Staroswiecki, M.; Yang, H. & Jiang, B. (2006). Progressive accommodation of aircraft actuator faults, Proceedings of IFAC Symposium Safeprocess, Beijing, China, pp. 877–882.
[15] Filaretov, V.F.; Zhirabok, A.N.; Zuev, A.V. & Procenko, A.A. (2013). Synthesis Method of Fault Tolerant Control System for Manipulators, Advanced Materials Research, Vol. 717, pp 551-556.
[16] Filaretov, V.F.; Zirabok, A.N.; Zuev, A.V. & Protsenko, A.A. (2012). The Development of the Faults Accommodation System for Actuators of Multilink Manipulator, Proceedings of the 23th DAAAM International Symposium, pp.575-578, B. Katalinic (Ed.), Published by DAAAM International, ISSN 1726-9687, ISBN 978-3-901509-86-5, Zadar, Croatia.
[17] Filaretov, V.F. & Konoplin, A.Yu. (2015). System of Automatic Stabilization of Underwater Vehicle in Hang Mode with Working Multilink Manipulator, International Conference on Computer, Control, Informatics and Its Applications, Bandung, Indonesia, pp. 132-137.
[18] Filaretov, V.; Zhirabok, A.; Zuev, A.; Protcenko, A. (2016). Fault identification in nonlinear dynamic systems, Proceedings of the 5th International Conference on Systems and Control, Marrakesh, Morocco, pp. 273-277.
[19] Filaretov, V.F. (2007). Synthesis of Adaptive Control Systems for Electric Servo Actuators of Manipulators, Proceedings of the 28th DAAAM International Symposium, pp. 277- 279, B. Katalinic (Ed.), Published by DAAAM International, ISSN 1726-9679, ISBN 3-901509-58-5, Zadar, Croatia.

DOI: 10.2507/28th.daaam.proceedings.003

THE PRINCIPLE OF FULL FUNCTIONALITY – THE BASIS FOR RAPID RECONFIGURATION IN HETEROGENEOUS MODULAR MOBILE ROBOTS

Victor Andreev, Valerii Kim & Pavel Pletenev

This Publication has to be referred as: Andreev, V[iktor]; Kim, V[alerii] & Pletenev, P[avel] (2017). The Principle of Full Functionality – the Basis for Rapid Reconfiguration in Heterogeneous Modular Mobile Robots, Proceedings of the 28th DAAAM International Symposium, pp.0023-0028, B. Katalinic (Ed.), Published by DAAAM International, ISBN 978-3-902734-11-2, ISSN 1726-9679, Vienna, Austria
DOI: 10.2507/28th.daaam.proceedings.003

Abstract

This paper proposes to design a heterogeneous modular mobile robot control system based on the principle of full functionality of the modules. The full functionality of a mechatronic device is the ability to perform its goal function using only its own facilities for executing instructions from an external control system. The external control system is a module-supervisor of modular robot in relation to the other modules of the robot. This module forms the control goal and verifies only the result of its achievement by executive module, but doesn't control the task execution process by the module. The full functionality of robot modules is achieved by designing each module as a full-featured electronic or mechatronic device with its own the information-measuring and control system that containing all necessary components: sensors, computing devices and actuators. The use of a network principle of the robot's information-measuring and control system construction and the Ethernet standards provides the capability of direct inter-module information interaction and the ability to rapidly form and alter the structure of the robot directly on the scene of the works and during the works themselves.

Keywords: mobile robot; modular robot; heterogeneous robot; control system; reconfiguration

1. Introduction

The maximum uncertainty in terms of future work in case of emergency situations, accidents, catastrophes, and active countermeasures makes it no possible to determine in advance the composition and configuration of necessary robotics [1]. Therefore, mobile robots (MR) that are focused on the use in extreme situations, must possess the ability to rapidly form and alter its structure directly on the works location and during the works themselves. In other words, the robots with variable structure must be used to perform such works, i.e. a ***reconfigurable*** modular mobile robots. At now the modular mobile robotics is at an early stage of its development. One of the main difficulties of creating the reconfigurable systems is to develop a plug and play control systems. In such robots, an automatic reorganization of the overall control system structure should be realized in accordance with the rapidly changing composition and configuration of the robotic device. Such a requirement leads to a significant increase in the system complexity of the coordinated control of all modules of a heterogeneous robot.

The purpose of the study is to suggest a new structure of the software and hardware part of the modular mobile robot control system, which would be simple enough and possess the ability of automatic reorganization. The use of the ***principle*** of the ***full functionality*** for the modules design makes it possible to simplify such control system.

The heterogeneous modular mobile robot and its modules are mechatronic devices. ***The full functionality of a mechatronic device is the ability to perform its goal function using only its own facilities for executing instructions from an external control system.*** The main control system of a modular robot is a module-supervisor in relation to the other modules of the heterogeneous modular mobile robot. The full functionality of robot modules is achieved by designing each module as a **full-featured electronic or mechatronic device** with its own the information-measuring and control system (IMCS) that containing all necessary components: sensors, computing devices and actuators.

This functional completeness **is the main difference** of our solution from the other heterogeneous robots where module-supervisor controls the operation of all modules at the executive level. In our solution the module-supervisor forms the control goal and verifies only the result of its achievement by executive module, but doesn't control the task execution process by the module. As a consequence of this approach, a ***distributed control*** is realized. Such distributed control allows parallelizing the computational process of realisation of the robot goal function by dividing the process into the functional subtasks and their distributing among the microprocessors of modules. Thus the module-supervisor computer performs only the following functions: splitting the process into functional sub-tasks, distribution of tasks between the modules, control of their execution, the processes synchronization and the solution of behavioural tasks. The implementation of the functional subtask is performed by the corresponding full-featured electronic or mechatronic robot module. As a result, the module-supervisor computational power requirements are reduced.

Task 1. To implement distributed control it's necessary to develop a language for inter-module communication.

Another problem of creating reconfigurability is to develop means of mechanical, electrical and information interaction between the modules, i.e. the corresponding interfaces. On the one hand, each interface should be simple, and on the another – universal. This will ensure the ease of the robot assembly from the modules developed by different manufacturers in the same way as it is done in computer technology. For example, printers, scanners and other external devices from different manufacturers are easily connected to a computer equipped with an appropriate operating system (OS) and drivers. We believe that the modular robot should be assembled similarly, i.e. on a plug and play method.

Task 2. It is necessary to define the communication environment (electrical interface) and to develop software (software interface) that will provide an assembly of the robot from the modules of different manufacturers in the "plug and play" mode. At this stage of the project the mechanical interface we are not yet considering.

2. Modern solutions in the modular mobile robotics

In the course of works of accident consequences elimination at the Chernobyl nuclear power plant (Soviet Union, 26 April 1986), a new direction in modular robotics has been appeared – ***reconfigurable modular robots***.

Currently, modular mobile robots are divided into two types: *homogeneous* and *heterogeneous*. In ***homogeneous*** robots most of the modules are *identical in structure and functionality*. Essentially this is a technical analog of colonial organisms. The main objective of such systems is to provide autonomous reconfiguration and adaptation to any environmental conditions. Homogeneous robots are modular robots such as M-TRAN [2] (2002 – 2008), ATRON [3] (2006), Transmote [4] (2012).

Homogeneous modular robots have several unique properties that distinguish them among the diversity of mobile robots. Those properties are as follows:

- adaptation to different and sometimes extreme environmental conditions thanks to automatic reconfiguration;
- high failure resistance of individual modules during execution of work;
- scalability of modular systems.

To the disadvantages of homogeneous robots can be attributed:

- each module should have all equipment for movement and navigation;
- complexity of the coordinated control of a modular system;
- high requirements to the design and control system of the each module;
- system scalability is limited by the capabilities of software and hardware implementation of the modules.

In addition, when structure and functionality of all modules are the same, one module is usually different from the rest – this module is needed for a common control and communication with external supervisor.

A distinctive feature of ***heterogeneous*** robots is *the differentiation of modules according to their functionality*. Each module is responsible for its own limited set of tasks, so the modules in heterogeneous systems unequal. The heterogeneous robots can be exemplified by the modular robots such as Thor [5] (2013), SMART [6] (2012), and modular robot for space exploration [7] (2006).

Despite the lower level of autonomy of heterogeneous systems compared to homogeneous systems, many practical problems are solved more efficiently by systems with modules of different functionality.

Heterogeneous modular robots have the following advantages:
- relatively low requirements to the control system of modules and their design;
- scalability of the robot is practically unlimited;
- the ability to perform a wide range of tasks;
- the ability to adapt to environmental conditions.

To the disadvantages of heterogeneous robots can be attributed:
- the complexity of automatic reconfiguration implementation;
- low failure tolerance of individual modules;
- the complexity of the coordinated control of all modules.

It should be noted that in the control system of considered homogeneous and heterogeneous mobile robots it is observed the presence of elements of distributed control.

From the position of creating mobile robots with variable structure, i.e. reconfigurable modular robots, focused on the use in case of extreme situations, we believe that to date heterogeneous robots have the advantage compared to homogeneous robots. Also we believe that the automatic reconfiguration of robot in general, while solving the aforesaid tasks is not actual, since it can be done manually, but automatic rebuilding of all structures of the robot IMCS to the new composition and configuration needs to be solved necessarily. Low failure resistance of individual modules is usually solved by duplicating the systems and replacement of faulty modules on the same or modified. But such replacement requires the capability to quickly adapt the robot IMCS to its new structure. This leads to increase of the complexity of the coordinated control of all modules. Therefore, the desired solution of the robot IMCS should allow the ***rapid reconfiguration of the robot***.

3. The principle of the full functionality of modules

Each module should be a *full-featured electronic or mechatronic device* with its own the information-measuring and control system. This means that each module of the robot needs to fulfill its mission in *any convenient to it way* – it follows that the module design must meet the functional purpose of the module, and the module IMCS must fully provide information and executive functionality [8].

Consider the principle of full functionality by the example of the transport module (Fig.1). Let the robot consists of the following modules: module-supervisor, transport module (TM) and sensor modules. Module-supervisor receives from an external supervisor (e.g., human operator) common task. The software of module-supervisor breaks down the task into subtasks and distributes them to the appropriate modules. Module-supervisor sends to TM the control goal – a vector of desired robot position $[X_G , Y_G , \varphi_G]^T$, where X_G , Y_G – the robot characteristic point coordinates with respect to the fixed world coordinate frame, φ_G – the robot orientation angle. The transport module should **independently** achieve this target location, using the methods of control laid down in its control system, and information from its sensory systems.

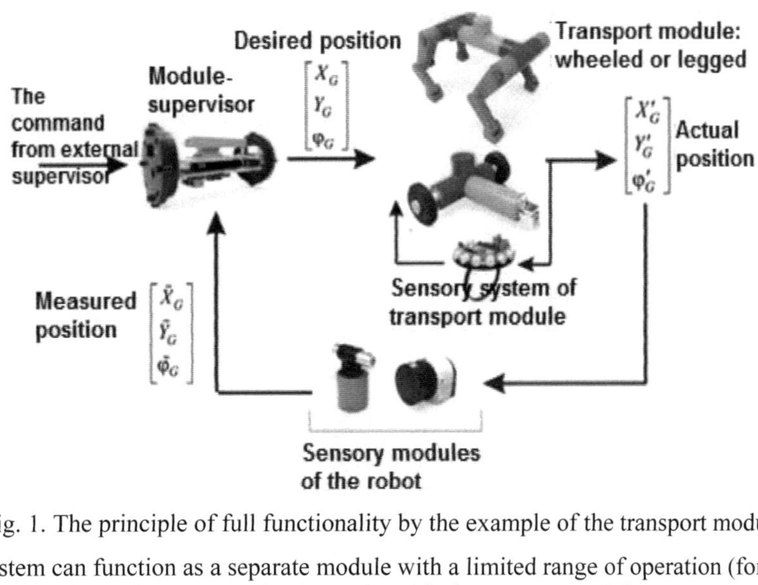

Fig. 1. The principle of full functionality by the example of the transport module

The TM sensor system can function as a separate module with a limited range of operation (for example, in the range of 5 ÷ 100 cm) and have information communication only with the computer of the transport module, i.e. the sensor system is a part of the transport module IMCS [9]. Due to the natural position errors of trajectory following by the transport module, its actual position, and therefore the position of whole robot, specified by the vector $[X'_G, Y'_G, \varphi'_G]^T$, will differ from the desired position. Here, the accuracy of achievement of the control goal is tested by the module-supervisor with the help of sensor modules of the robot. A sensor modules of the robot that use more sophisticated sensors (television, ultrasonic, infrared, etc.), provide affixment of the robot to the fixed world coordinate frame, for example, from landmarks in the environment. Comparing the vector of measured position $[\tilde{X}_G, \tilde{Y}_G, \tilde{\varphi}_G]^T$ with desired, the module-supervisor if necessary, corrects the actual position of the robot, setting a new goal.

In this scheme, the transport module may be implemented as the wheeled, legged robot, or any other mobile platform, but it must accomplish the control goal given by a module-supervisor, regardless of its mechanical construction, the structure of the IMCS and the control algorithm of the module.

The *distributed control* based on the principle of full functionality is implemented. Such distributed control allows parallelizing the computational process of the robot target function executing by dividing the process into the functional subtasks and their distributing among the microprocessors of modules. Then the computer of module-supervisor performs only the following functions: splitting the process into functional sub-tasks, distribution of tasks between the modules, control of their execution, processes synchronization and the solution of behavioral tasks. In addition, the module-supervisor carries out information interaction with the external supervisor, for example, the operator of a mobile robot.

To implement such distributed control it's necessary to develop *a language of inter-module communication* and software that supports dynamic reconfiguration of the robot IMCS.

4. The inter-module communication

Modern modular robots (homogeneous and heterogeneous) for inter-module communication, as a rule, use different serial bus: I2C – modules of RTC; CAN – in robots M-TRAN, SMART; IrDA (essentially RS-232) – ATRON; RS-485 – in robots Thor, Odin. Serial bus is used on the robot TRANSMOTE – it uses ZigBee wireless interface. In such systems the role of inter-module interaction is to ensure the most rapid and effective transfer of commands to the executive or tactical level from an external computer to the modules, and transmission of sensor data. In this case, the information flow in the process of inter-module interaction is small, because on each module there are no sensors or many sensors cannot be installed. As a result, it is necessary basically to transmit only the executive level commands.

Our solution involves the use of a duplex sensor system (sensor modules near and far radius of action) with a large variety of sensors, including multi-camera vision system. In this case, the system inter-module communication must have a high throughput. Listed interfaces do not have sufficient bandwidth. It is proposed to organize inter-module communication via the interface, developed on the basis of *Ethernet*. It is a broadband, reliable and affordable interface. Its modification EtherCAT (for real-time) does not allow the creation of horizontal linkages – all inter-module communication must be through the "master" bus, which eliminates the possibility of direct inter-module interaction. Since the use of the principle of full functionality of modules eliminates the need to transfer commands of the executive level, as noted previously, it is allowed more soft timing requirements of message delivery and the use of interfaces with the "soft real time". In our case, all inter-module communication is restricted to small dimensions of the robot, so we choose the topology of the network of type "star", which corresponds to a typical Ethernet network topology.

Currently, for control of dissimilar mechatronic devices the software framework ROS (Robotics Operating System) is widely used. However, a direct application of ROS is "heavy" – you need at least a computer running a full Linux OS, you must additionally set a large set of various software, which requires a high level of input use (for example, dialects of XML are using for the configuration). In the modules of the robot the microcontrollers or, as a maximum, single board computers are typically used; ROS is not focused to work directly on the microcontroller [10]. Therefore, we must create a software framework that would be compatible with ROS, but it would be easy to implement on embedded systems.

The choice of the Ethernet standard allows realizing the modular robot information-measuring and control system in the form of local area network, the nodes of which are IMCS of individual modules [11]. This solution allows transferring such network properties as scalability and reconfiguration to the network structure of the robot's IMCS. Experience in the development and long-term exploitation of the network training Internet laboratory convinces authors in the correctness of the choice made [12]. The method of geographically distributed control designed for the laboratory forms the basis of the method of inter-module information interaction.

The approach proposed in [12] is based on the library Zmq_robot [13], which was based on ZMQ library. Library Zmq_robot allows you to create software as for modules, and the control programs for a supervisor. However, this library has the same disadvantage that ROS – the complicated software that only runs on a full OS and not working on embedded systems must be installed on the device. To solve this problem the method for specification of inter-module interaction that takes into account the modules full functionality property was developed. For the organization of inter-module interaction we propose to use the ZMQ library as the basis. But in an addition to the ZMQ the UDP protocol broadcast messages, which allows implementing a specification for embedded systems are used. Unlike ROS, in this software each module has a single interface that is common for all modules and provides general information about itself – the module name, its description, functional interfaces, etc. The proposed approach of the inter-module information interaction is described in more detail in [14].

5. Conclusion

A new structure of the hardware and software part of the control system of the heterogeneous modular mobile robots is proposed. This structure is fairly simple to implement and has the ability of automatic reorganization. The basis of this structure is the principle of the full functionality of the modules – each module is designed as a full-featured electronic or mechatronic device with its own information-measuring and control system. Automatic reconfiguration of the robot control system is achieved by using:

a) a network structure of the information-measuring and control system based on the Ethernet technology, in which each module is a node of the local area network;
b) the developed specification for the method of an inter-module communication, based on the principles of the Robotic Operating System, but implemented in the embedded computing systems;
c) the developed language for an inter-module communication.

This approach provides:

d) parallelizing of the computational process due to its distribution between the calculation devices of the modules, which simplifies the overall control task and reduces the requirements for the modules calculation devices;
e) the ability to quickly assemble a robot from modules of different functional purpose and different manufacturers by using the appropriate drivers.

Currently, these studies are at the most basic level. The use of the principle of full functionality for the design of the IMCS of a heterogeneous modular mobile robot was tested on a laboratory working model consisting of the intelligent control module (module-supervisor), the wheeled transport module, the short-range sensors module and the power supply module (the batteries). The focus at this stage was paid to ensuring the requirements of the full functionality of the transport module and the power supply module. The functions of all modules are implemented in the minimal version. For the transport module, a novel motion planning method for a differential drive mobile robot using two-dimensional vector fields is proposed. It is assumed that the sensor module consists only of odometers, distance sensors and a microcontroller that processes signals from sensors. The robot information-measuring and control system is implemented as a local area network Ethernet. The software for inter-module information interaction is built on the basis of the ZMQ library using the UDP network protocol with broadcast messages.

Experiments have shown the promise of this area of research. In the near future we are going to increase the functional composition of the robot modules, increase the number of remote sensors and their diversity in the short-range sensor module, and also complicate the functionality of the module-supervisor.

We believe that the proposed principle of modular design of robotic systems can be successfully used in the Bionic Assembly System [15, 16].

6. Acknowledgments

The authors are expressing their sincere gratitude to S. Kuvshinov, Director of International Institute of the New Educational Technologies of Russian State University for the Humanities, for assistance in construction of the experimental equipment.

This work was supported by the Russian Foundation for Basic Research, grants no. 16-07-00811a & 16-07-01264a.

7. References

[1] Lopota, A.V. & Yurevich, E.I. (2013). Stages and development prospects of robotic systems design modular principle. St. Petersburg State Polytechnical University Journal. Computer Science. Telecommunication and Control Systems. No.1(164), 2013, pp. 98–103, ISSN 1994-2354.

[2] Murata, S.; Yoshida, E.; Kamimura, A.; Kurokawa, H.; Tomita, K. & Kokaji, S. (2002). M-TRAN: selfreconfigurable modular robotic system. IEEE/ASME Transactions on Mechatronics, n.7(4), 2002, pp.432–441.

[3] Østergaard, E.H.; Kassow, K.; Beck, R. & Lund, H.H. (2006). Design of the ATRON lattice-based self-reconfigurable robot. Autonomous Robots, No.21(2), 2006, pp.165–183.

[4] Guifang Qiao, Guangming Song, Jun Zhang, Hongtao Sun, Weiguo Wang & Aiguo Song (2012). Design of Transmote: a Modular Self-Reconfigurable Robot with Versatile Transformation Capabilities. Proceedings of the 2012 IEEE International Conference on Robotics and Biomimetics, pp.1331-1336.

[5] Lyder, A.H.; Stoy, K.; Mendoza-Garcia, R.-F.; Larsen, J.C. & Hermansen, P. (2013). On sub-modularization and morphological heterogeneity in modular robotics. Intelligent Autonomous Systems, No.12, V.193 of Advances in Intelligent Systems and Computing, Springer Berlin Heidelberg, 2013, pp. 649-661.

[6] Baca, J. ; Ferre, M. & Aracil, R. (2012). A heterogeneous modular robotic design for fast response to a diversity of tasks. Robotics and Autonomous Systems, vol. 60, no.4, 2012, pp. 522–531.

[7] Hancher, M.D. & Hornby, G.S. (2006). A modular robotic system with applications to space exploration. 2nd IEEE International Conference on Space Mission Challenges for Information Technology (SMC-IT'06). - Pasadena, CA: Publisher «IEEE», 2006, pp.132-140.

[8] Andreev, V. & Kim, V. (2017)., The Application of Artificial Vector Fields for Motion Control of a Heterogeneous Modular Mobile Robot, Proceedings of the 28th DAAAM International Symposium, pp.xxxx-xxxx, B. Katalinic

(Ed.), Published by DAAAM International, ISBN 978-3-902734-xx-x, ISSN 1726-9679, Vienna, Austria, OI: 10.2507/28th.daaam.proceedings.xxx, in press.

[9] Andreev, V. & Kim, V. (2016). Control system and design of the motion module of a heterogeneous modular mobile robot, 27th International DAAAM Symposium on Intelligent Manufacturing and Automation 2016, Proceedings of a meeting held 26-29 October 2016, Mostar, Bosnia and Herzegovina, B. Katalinic (Ed.), ISBN 978-1-5108-3300-5. Curran Associates, Inc., NY 12571 (Jan 2017), pp.0586-0594.

[10] Andreev, V.; Pletenev, P. & Poduraev, Y. (2015). The method of network control of robotic systems of different models and manufacturers, Proceedings of the 26th DAAAM International Symposium., B. Katalinic (Ed.), Published by DAAAM International, ISBN: 978-1-5108-1839-2, ISSN 1726-9679, Vienna, Austria. 2015. Curran Associates, Inc., New York, Feb. 2016, pp.0727-0730. DOI: 10.2507/26th.daaam.proceedings.101.

[11] Andreev, V. & Poduraev, Y. (2016). Network–based Design of Heterogeneous Modular Mobile Robotic Systems, 27th International DAAAM Symposium on Intelligent Manufacturing and Automation 2016, Proceedings of a meeting held 26-29 October 2016, Mostar, Bosnia and Herzegovina, B. Katalinic (Ed.), ISBN 978-1-5108-3300-5. Curran Associates, Inc., NY 12571 (Jan 2017), pp.0004-0009.

[12] Andreev, V.; Pryanichnikov, V. & Kirsanov, K. (2015). Geographically distributed "multi-operator" Control for Mechatronic Devices via the Internet, Proceedings of the 26th DAAAM International Symposium., B. Katalinic (Ed.), Published by DAAAM International, ISBN: 978-1-5108-1839-2, ISSN 1726-9679, Vienna, Austria. 2015. Curran Associates, Inc., New York, Feb. 2016, pp. 0004-0011. DOI: 10.2507/26th.daaam.proceedings.001.

[13] Kirsanov, K. (2014). Software architecture of control system for heterogeneous group of mobile robots, 25th DAAAM Int. Symp. on Intelligent Manufacturing and Automation, 2014. Procedia Engineering, Elsevier BV. ISSN 1877-7058, 2015, V.100, pp.216–221. DOI: 10.1016/j.proeng.2015.01.339.

[14] Andreev, V. & Pletenev, P. (2017). Organizing Intermodular Communication for Heterogeneous Modular Mobile Robot, Proceedings of the 28th DAAAM International Symposium, pp.xxxx-xxxx, B. Katalinic (Ed.), Published by DAAAM International, ISBN 978-3-902734-xx-x, ISSN 1726-9679, Vienna, Austria, OI: 10.2507/28th.daaam.proceedings.xxx, in press.

[15] Katalinic, B.; Kukushkin, I.; Pryanichnikov, V. & Haskovic, D. (2014). Cloud Communication Concept for Bionic Assembly System. Procedia Engineering, 1877-7058, 69 (2014), pp.1562–1568 doi:10.1016/j.proeng. 2014.03.156.

[16] Katalinic, B.; Kukushkin, I. & Haskovic D. (2014). Bionic Assembly System Cloud: Functions, Information Flow and Behavior. In 9th International Conference of DAAAM Baltic, Industrial Engineering, ISBN 978-9949-23-620-6, ISSN 2346-6138, pp.103–108, ed. T.Otto, Tallinn, Estonia, 2014.

DOI: 10.2507/28th.daaam.proceedings.004

IMPLEMENTATION OF CORRELATION ANALYSIS TO WORK SAMPLING RESULTS

Borut Buchmeister & Palcic Iztok

This Publication has to be referred as: Buchmeister, B[orut] & Palcic, I[ztok] (2017). Implementation of Correlation Analysis to Work Sampling Results, Proceedings of the 28th DAAAM International Symposium, pp.0029-0034, B. Katalinic (Ed.), Published by DAAAM International, ISBN 978-3-902734-11-2, ISSN 1726-9679, Vienna, Austria
DOI: 10.2507/28th.daaam.proceedings.004

Abstract

Work sampling is the process that estimates the proportion of time of the workday spent by workers (or machines) on different activities (work or non-work), based on sufficient random observations of activities over a period of time. When jobs do not have short cycle times or high repetition rates the use of such statistical technique is required, because work sampling data can provide information that can be used to establish standards. The work sampling procedure is well standardized by the recommended steps, but some contributions are possible in the evaluation of the observations.
In the paper we present our contribution in the improvement of the decision process based on the work sampling data. We introduce the correlation comparison of the measured hourly proportions of all pairs of activities to check, if there are some interrelationships. The results enable easier decision making (conclusions) about the influence of the selected activities to trigger the others. This is bringing us to better production system's efficiency and productivity.

Keywords: Work sampling; Activities; Observations; Correlation analysis; Decision making

1. Introduction

In many industries the presence of labour with high productivity at each stage plays a significant role in achieving success [1]. There are many jobs where time study, standard data, and predetermined time standards just are not useful, productive ways to set standards. When jobs do not have short cycle times or high repetition rates, such as those in maintenance, material handling, clerical and the like, these methods are not effective [2]. Jobs with long cycle times and low frequency of repetition require the use of work sampling [3, 4, 5].

Work sampling (introduced by L. H. C. Tippett in 1934 in the textile industry) is the process of making sufficient random observations of an operator's activities to determine the relative amount of time the operator spends on the various activities associated with the job. Work sampling method is used for collection of empirical data. Although it is not the express purpose of work sampling to determine how long specific tasks should take, work sampling data, when coupled with historical production data for key volume indicators and performance levelling, can provide information that can be used to establish standards [6, 7]. The major goal of work sampling, however, is to determine the proportion of the workday spent on certain types of work. To set the amount of time required to perform a task is not a goal of work sampling [8, 9, 10]. Work sampling can also give us an insight to activities structure and related physical exposures which may cause heavy injuries [11, 12]. So it is useful for indirect assessment of workplace exposures and injury risk factors. With work sampling approach we can also record workers' satisfaction levels (when performing some tasks) to study how different tasks influence workers' satisfaction with their work [13].

Rather than observing the worker constantly for many days, the data collector may observe the workers at 15 to 30 instances during each day for two weeks or more. These observations are simply snapshots of what the workers are doing at that moment. If enough observations are made over a sufficiently long and representative period of time, the fraction of time workers spend on various activities can be estimated statistically. Work sampling estimates include some degree of error. The same number of observations taken at different times during the week will produce different estimates, and all estimates will usually differ from the actual (unknown) values.

The main advantages of work sampling are [14]:

- Observations are spread out over a period of time, making results less susceptible to short-term fluctuations.
- There is little or no disruption of work.
- Workers are less resentful.
- Studies are less costly and less time-consuming, and the skill requirements of the analyst are much less than at other methods.
- The study can be interrupted without affecting the results.
- Many different studies can be conducted simultaneously.
- No timing device is required.

And some disadvantages of work sampling [14]:

- There is much less detail on the elements of a job.
- Workers may alter their work patterns when they spot the observer, thereby invalidating the results.
- It is not well suited for short, repetitive tasks.
- Much time may be required to move from one workplace to another and back to satisfy the randomness requirement.

2. Methodology

The method to conduct a work sampling study is as follows [3, 15]:

- Establish the purpose of the study. Is the intent to set a standard to drive staffing levels or is it to identify the proportion of time spent on each major category of work?
- Identify the subjects. Who is doing the work within the organization? Careful problem definition can prevent mistakes such as studying the wrong worker or wrong activity.
- Identify the measures of output. This is essential if the objective of the study is to develop standards.
- Establish a time period for the study. The period of the study should be long enough to be representative of normal operations.
- Define the activities. This includes a definition of the major tasks performed. An example of some work definitions is included in a case study in section 4.
- Determine the number of observations needed. After the work elements are defined, the number of observations for the desired accuracy and confidence level must be determined.
- Schedule the observations. If an analyst will record the data, use a random number table or random number generator. Only randomly distributed observation times will reflect typical performance.
- Inform the personnel involved. As in any productivity measurement study this part of the procedure is important. Workers and their supervisors might think that they personally are being measured rather than the work they are doing.
- Record the raw data (manually written on the sheet of paper or by notepad, by portable computer, by special recording device etc.).
- Summarize the data. Determine the proportion of time spent on each activity.

Work sampling studies are easy and inexpensive to perform. One observer can monitor and sample the behaviour of several workers during the same period. To carry out data collection, each observer selects a number of workers performing the same operation. A work sampling study usually requires a substantial period of time to complete. The minimum number of observations (N_{obs}) necessary to achieve the desired level of accuracy and confidence can be computed by:

$$N_{obs} = (z^2 \cdot \dot{p} \cdot (1 - \dot{p})) / e^2 \quad (1)$$

where $\dot{p}$ is the proportion of observations during which the target activity is performed, z-value is a factor for the confidence level (typically $z = 1.96$ for 95 % confidence level), e is the maximum absolute error desired (between 0.01 and 0.05).

As mentioned before the most important issue is what fraction of time is spent doing each type of activity, or what fraction of time is spent on productive work versus idle time or unproductive work [16]. Reports include: activities' proportion (for whole study), activities' proportion during a work day (for hours 1 to 8), activities' proportion related to days of a week (Monday to Friday), and activities' proportion related to work shifts (1 to 3).

3. Inclusion of the correlation analysis

Within the results we have also the proportions of all observed activities during workday hours (1 to 8). We can presume that sometimes specific activities trigger the occurrence of another activity (or more activities). If such activities are unproductive, then we are able to prevent the appearance of selected activities (mainly in the field of idle time) with the goal to reduce losses, caused by such interrelationships.

So we compare the hourly proportions (8 data) of all pairs of activities using the equation for the correlation coefficient:

$$r = \frac{8 \cdot \Sigma xy - (\Sigma x) \cdot (\Sigma y)}{\sqrt{(8 \cdot \Sigma x^2 - (\Sigma x)^2) \cdot (8 \cdot \Sigma y^2 - (\Sigma y)^2)}} \tag{2}$$

where r is the correlation coefficient, x is the proportion of the first compared activity and y the same for the second activity (both for 8 hours). The result is between -1 and 1, where: 1 indicates a strong positive relationship, -1 indicates a strong negative relationship, and a result of zero indicates no relationship at all. We suggest that all results lower than -0.6 and higher than 0.6 should be considered. This additional calculation in results' analysis is the main contribution of our paper.

4. Case study

In a textile company we have made the necessary observations in the frame of work sampling project. For the purpose of this paper the assembly workplaces' data (1600 observations) for 17 activities were used for the analysis. The activities and their proportions during the 8-hours workday (with a 30 minutes break in the 4th hour) are collected in Table 1.

Activity	**Proportion of time** in the hour of a workday [%]							
	1	2	3	4	5	6	7	8
1. Complete all parts	14.00	15.00	17.00	18.00	12.50	18.00	12.00	13.50
2. Assembly the parts	58.50	56.00	53.50	41.50	42.50	53.50	49.00	45.50
3. Depositing the parts	14.00	17.50	14.50	5.50	15.50	17.50	14.00	13.00
4. Annotation of work results	1.50	0.50	0.00	0.50	1.00	0.50	1.00	5.00
5. Cleaning the machine	0.00	0.00	0.00	0.00	0.00	0.00	0.00	1.50
6. Keeping workplace in order	0.00	0.00	0.00	0.00	0.00	0.00	0.00	8.50
7. Conversation about tasks	0.00	0.00	0.50	1.50	0.00	0.00	0.00	2.50
8. Idle time (waiting for duties)	2.50	2.00	1.00	1.00	1.00	0.00	0.50	0.50
9. Waiting time – bad quality	0.00	0.00	1.00	0.00	0.00	0.00	0.00	0.00
10. Waiting time – missing device	0.00	0.00	0.00	0.00	0.00	0.00	0.00	0.00
11. Machine failure	0.00	0.00	0.00	0.00	0.00	0.00	0.50	0.00
12. Repair of defects	0.00	0.00	0.50	0.00	0.00	0.50	0.00	0.00
13. Business meeting	0.00	0.00	0.00	1.50	2.50	0.00	0.00	0.00
14. Work on other workplace	0.00	0.00	0.50	1.50	0.00	0.00	0.00	0.00
15. Absence from company	4.00	4.00	6.00	6.00	6.00	5.00	4.50	6.00
16. Private conversation	2.00	0.50	0.50	0.50	0.00	1.00	3.50	2.00
17. Private absence from workplace	3.50	4.50	5.00	22.50	19.00	4.00	15.00	2.00
Total: **Productive work**	86.50	88.50	85.00	65.00	70.50	89.00	75.00	72.00
Total: **Losses**	13.50	11.50	15.00	35.00	29.50	11.00	25.00	28.00

Table 1. Time usage within a workday (hourly report)

The results collected in Table 1 are partly presented in graphical form (see Fig. 1) showing proportions of time in the hours of a workday for individual activities.

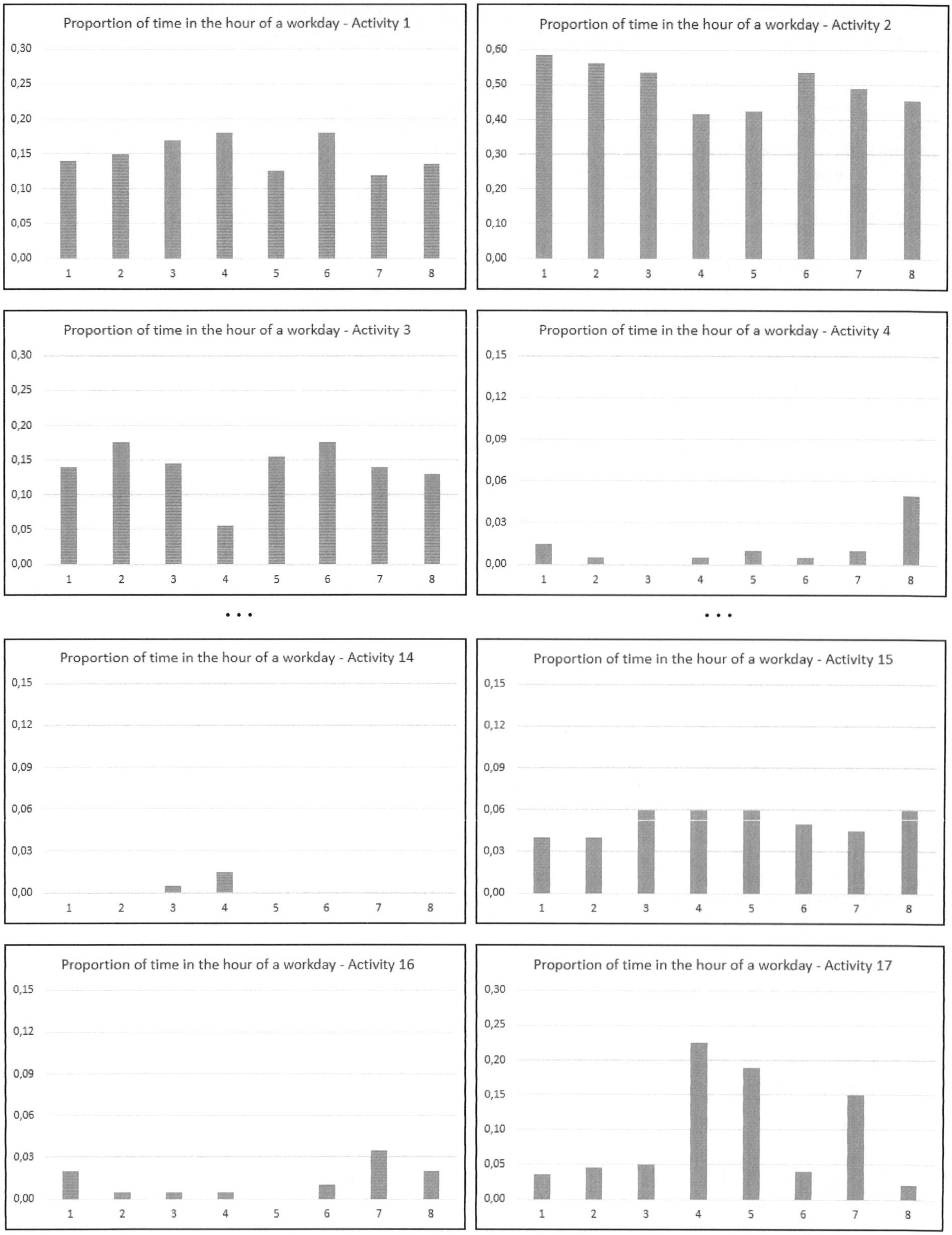

Fig. 1. Proportions of time in the hours of a workday for the selected activities 1, 2, 3, 4, 14, 15, 16 and 17

Similar course of productive work time proportions at specific activities may show us possible interrelationships between these activities.

The results of the pairwise correlation calculation using (2) are summarized in the Table 2. All values are multiplied by 10 and rounded (so, values between -9 and +9 appear).

Act.	1	2	3	4	5	6	7	8	9	10	11	12	13	14	15	16	17
1	/	1	-2	-4	-2	-2	1	-1	3	0	-5	6	-1	6	2	-5	0
2		/	5	-2	-2	-2	-5	4	2	0	0	3	-7	-4	-7	1	-7
3			/	0	-1	-1	-5	0	0	0	0	3	-3	-8	-4	0	-5
4				/	9	9	7	-1	-3	0	0	-3	-1	-2	2	3	-3
5					/	9	8	-2	-1	0	-1	-2	-2	-1	3	2	-3
6						/	8	-2	-1	0	-1	-2	-2	-1	3	2	-3
7							/	-2	0	0	-2	-2	0	3	6	0	0
8								/	0	0	-2	-4	0	0	-5	-1	-1
9									/	0	-1	6	-2	1	3	-2	-2
10										/	0	0	0	0	0	0	0
11											/	-2	-2	-1	-3	7	2
12												/	-3	0	2	-2	-3
13													/	3	5	-5	7
14														/	4	-3	5
15															/	-4	3
16																/	-1

Table 2. Coefficients of correlation between the activities (for the presented case)

4.1. Discussion

When checking, if there are some interrelationships between the activities, we have to discuss these combinations, based on the results in Table 2 (for values lower than -6 or higher than 6):

- Assembly the parts : Absence from company – of course: more productive work, less absence and loses (and vice versa).
- Assembly the parts : Private absence from workplace – the same comment as above.
- Depositing the parts : Work on other workplace – the same comment as above.
- Annotation of work results : Cleaning the machine – normal sequence after the job's completion.
- Annotation of work results : Keeping workplace in order – after the job's completion some maintenance is normally needed.
- Annotation of work results : Conversation about tasks – it is evident that before starting new task some instructions from foreman are usually needed; maybe some improvements in working documentation are required (useful comment no. 1).
- Cleaning the machine : Keeping workplace in order – this activities are connected; no changes.
- Cleaning the machine : Conversation about tasks – maybe the information about the machine maintenance is insufficient or working documentation is incomplete (useful comment no. 2).
- Keeping workplace in order : Conversation about tasks – the same comment as above.
- Machine failure : Private conversation – during the machine downtime the worker spends time in private chatting; time to repair should be as short as possible; maybe we should hire more maintenance staff or improve knowledge of the existent service technicians to be faster (useful comment no. 3).
- Business meeting : Private absence from workplace – obviously the workers take the opportunity of absence; when they are outside the workshop hall it is possible to return later than adequately; maybe we should change the room for business meetings, selecting closer places or some corner in the workshop directly (useful comment no. 4).

It is evident that with the correlation test we are getting an additional tool for work sampling results evaluation. In our simple case we have 4 comments, which are giving the basis for further investigation. Even in simpler cases some hints for critical reflection are present. Sometimes they are correct (activities' interrelationship exists) and require some changes in the production process, but it may also happen that they will be useless.

5. Conclusion

Work sampling is the statistical technique for determining the proportion of time spent by workers in various defined categories of activity. Workers are observed at random times throughout the work period, recording what they are doing each time. Work sampling permits quick analysis, recognition, and enhancement of job responsibilities, tasks, performance competencies, and organizational work flows. Its primary advantage is in studying nonrepetitive activities, but it can also be used to develop time standards for repetitive jobs. In such cases a lot of observations (1000 and more) are required to achieve acceptable accuracy.

In the paper we introduce the pairwise correlation comparison of the measured hourly proportions of all activities to check, if there are some interrelationships. The results of the work sampling method are given in tables (hard to read, pure statistics), now some additional calculations are made to enable the discussion between the members of the work design team about the necessary process improvements. With the correlation analysis we are getting the first signal that some activities trigger the occurrence of another (maybe non-value-adding) activity. This is the essential contribution in the decision making process.

The implementation of correlation analysis is of immense benefit to managers in enhancing companies work performance and productivity. This paper contributes both theoretically (new idea) and empirically (test case) to the finding on labour productivity. Future research should include more sophisticated statistical methods.

6. Acknowledgement

The authors acknowledge the financial support from the Slovenian Research Agency (research core funding No. P2-0190). Thanks also to the researchers in the textile company giving us the work sampling data.

7. References

[1] Hajikazemi, S.; Andersen, B. & Langlo, J. A. (2017). Analyzing electrical installation labor productivity through work sampling, International Journal of Productivity and Performance Management, Vol. 66, No. 4, pp. 539-553, ISSN 1741-0401, DOI:10.1108/IJPPM-06-2016-0122

[2] Saric, T.; Simunovic, G.; Simunovic, K. & Svalina, I. (2016). Estimation of machining time for CNC manufacturing using neural computing, International Journal of Simulation Modelling, Vol. 15, No. 4, pp. 663-675, ISSN 1726-4529, DOI:10.2507/IJSIMM15(4)7.359

[3] Geng, H. (Ed.) (2004). Manufacturing Engineering Handbook, McGraw-Hill, ISBN 978-0071398251, New York

[4] Antinaho, T.; Kivinen, T.; Hurunen, H. & Partanen, P. (2015). Nurses' working time use – how value adding it is?, Journal of Nursing Management, Vol. 23, No. 8, pp. 1094-1105, ISSN 0966-0429, DOI:10.1111/jonm.12258

[5] Stanisavljev, S.; Cockalo, D.; Klarin, M.; Brkic, V. S. & Dordevic, D. (2015). Stohastic model to determine the elements of the production cycle time: case of Serbian textile industry, Fibres & Textiles in Eastern Europe, Vol. 23, No. 5, pp. 23-29, ISSN 1230-3666, DOI:10.5604/12303666.1161752

[6] Tahriri, F.; Mousavi, M.; Yap, H. J.; Siti Zawiah, M. D. & Taha, Z. (2015). Optimizing the robot arm movement time using virtual reality robotic teaching system, International Journal of Simulation Modelling, Vol. 14, No. 1, pp. 28-38, ISSN 1726-4529, DOI:10.2507/IJSIMM14(1)3.273

[7] Haskovic, D.; Katalinic, B.; Zec, I.; Kukushkin, I. & Zavrazhina, A. (2016). Structure and working modes of the intelligent adviser module, Proceedings of the 27th DAAAM International Symposium, pp.0866-0875, B. Katalinic (Ed.), DAAAM International Vienna, ISBN 978-3-902734-08-2, ISSN 1726-9679, Vienna

[8] Car, M. & Adzam, A. (2009). Influence of the technological process structure elements model on approaches of time standards determining, Proceedings of the 13th International Research/Expert Conference TMT 2009, pp. 197-200, Hammamet, Tunisia

[9] Unal, C. (2013). A new line balancing algorithm for manufacturing cell transformation in apparel industry, Industria Textila, Vol. 64, No. 3, pp. 155-162, ISSN 1222-5347

[10] Roche, M. A.; Friedman, S.; Duffield, C.; Twigg, D. E. & Cook, R. (2017). A comparison of nursing tasks undertaken by regulated nurses and nursing support workers: a work sampling study, Journal of Advanced Nursing, Vol. 73, No. 6, pp. 1421-1432, ISSN 0309-2402, DOI:10.1111/jan.13224

[11] Trask, C.; Teschke, K.; Morrison, J.; Village, J.; Johnson, P. & Koehoorn, M. (2010). Using observation and self-report to predict mean, 90th percentile, and cumulative low back muscle activity in heavy industry workers, The Annals of Occupational Hygiene, Vol. 54, No. 5, pp. 595-606, ISSN 0003-4878, DOI:10.1093/annhyg/meq011

[12] Pinosova, M.; Andrejiova, M.; Kralikova, R.; Hricova, B.; Lumnitzer, E. & Wessely, E. (2015). Assessment of hearing impairment risk from the point of view of long-term exposure to noise in working environment, Chapter 30 in DAAAM International Scientific Book 2015, pp. 345-358, B. Katalinic (Ed.), DAAAM International Vienna, ISBN 978-3-902734-05-1, ISSN 1726-9687, Vienna

[13] Robinson, M. A. (2012). How design engineers spend their time: job content and task satisfaction, Design Studies, Vol. 33, No. 4, pp. 391-425, ISSN 0142-694X, DOI:10.1016/j.destud.2012.03.002

[14] Stevenson, W. J. (1993). Production / Operations Management, 4th edition, Irwin, ISBN 0-256-10511-1, Homewood

[15] Robinson, M. A. (2010). Work sampling: methodological advances and new applications, Human Factors and Ergonomics in Manufacturing & Service Industries, Vol. 20, No. 1, pp. 42-60, ISSN 1090-8471, DOI:10.1002/hfm.20186

[16] Gouett, M. C.; Haas, C. T.; Goodrum, P. M. & Caldas, C. H. (2011). Activity analysis for direct-work rate improvement in construction, Journal of Construction Engineering and Management – ASCE, Vol. 137, No. 12, pp. 1117-1124, ISSN 0733-9364, DOI:10.1061/(ASCE)CO.1943-7862.0000375

DOI: 10.2507/28th.daaam.proceedings.005

ANALYIS OF WEBSITES IN EVENT MANAGEMENT

Juraj Puljić & Božidar Jaković & Ivan Strugar

This Publication has to be referred as: Puljic, J[uraj]; Jakovic, B[ozidar] & Strugar, I[van] (2017). Analysis of Websites in Event Management, Proceedings of the 28th DAAAM International Symposium, pp.0035-0043, B. Katalinic (Ed.), Published by DAAAM International, ISBN 978-3-902734-11-2, ISSN 1726-9679, Vienna, Austria
DOI: 10.2507/28th.daaam.proceedings.005

Abstract

The event management industry has become very close to technology recently, and there are no signs of that relationship slowing down. Electronic and other online services have become a necessity, and are no longer considered luxury, because we are living life too fast. The services that event management organizations offer today are mostly transacted online. This paper will show a detailed analysis that provides answers to various user experience related questions in the interaction with Croatian and global event management related websites and pages. The research also displays the top Croatian and global organizations that meet the criteria presented in the paper. The study is based on an analysis of a large number of global websites – both national and international – associated with event organization, and it answers some questions users might have during the use of the website or mobile application. The contents of 200 websites and applications related to event management have been analyzed for the purposes of showing the trends and possibilities in the digital event organization industry, as well as the availability and functionality of the services the organizations provide. This was done by analyzing over 100 attributes for each website.

Keywords: Internet; Digital Business; Event Management; Event Organization

1. Introduction

Today, companies across all industries are required to go digital with their businesses in some part, if not completely, and so is the event management industry. The purpose of this thesis is to analyze different Web services associated with event management and show the possibilities of a digitized event organization and management. This thesis additionally shows the development of a mobile application whereby every user can get an insight into every aspect of the event they are attending, giving them the chance to fully immerse themselves in the event instead of wasting time on secondary activities, e.g. standing in lines while waiting for a drink. The possibilities include: checking for and booking of the closest service required through the use of sensors (e.g. locking the closest toilet), looking at the situation of the event through drone cameras (especially useful if there are multiple stages during a concert, to see what one is missing), etc.

2. Development and application of electronic business

The notion of business as we knew it for thousands of years started to change around 1995. Namely, the Internet had been around for some time, but no one really understood its possible applications in real life, except for connecting military and educational centers within countries, and later around the world [1]. However, at that time, due to an exponential rise in the use of personal computers, e-business had started to prosper: it had come to be a modern business model that applies IT and Internet technology to transact and perform key business activities [2]. The term e-business was coined by IBM as an attempt to brand their Internet business services, but the term had become so wildly popular and ubiquitous that they could no longer differentiate by using it, since a whole new industry had been born with that name.

Even though the terms digital business and e-commerce are often used interchangeably, since both of them are understood as using the electronic technology to do business, this use is incorrect [3]. That is to say, digital business is the broader concept of the two, as it covers all business processes that take action online, while e-commerce is limited to buying and selling through the use of electronic technology [4]. An example of e-commerce would be a Web page or Webshop that lets its users buy online, where a company on the one side sells a product or service, and the buyer buys it [5]. On the other hand, e-business would also entail e-mail marketing to existing or potential new buyers; online system that keeps track of the warehouse conditions and sends signals electronically; CMS (content management systems) that maintain and manage the flow of information between editors, publishers, programmers and managers; online program that automatizes the induction of new employees into the register and supplies information, etc. [6][7] Therefore, e-business is a broad term that includes any business process that is done through electronic technology, while e-commerce is but a part of e-business that deals with electronic transactions, i.e. buying and selling online [8]. As technology and business develop, we slowly but surely cross from e-business toward d-business, i.e. digital business. That would entail the creation of new revenue and value through the yet non-existent combinations of people, businesses and things [9]. For example, if Nike somehow received the information about the number of miles their user had run in one day or through a period of time, and sent that same user a new pair of sneakers when the time comes for a new pair (fitting them personally) based on that information, that would be d-business [10][11]. Even though the advent of electronic business has enabled many economic advancements, there are some negative aspects as well [12].

One of the negative sides of electronic business is the ease of business consolidation and the disappearance of SMEs and brick-and-mortar businesses, which brings about income disparity [13]. As an adequate example, Hilton employs more than 150,000 people and had a market value only two times higher than Airbnb as of 2014, and Airbnb only employs 700 people.

There is also the question of Internet security [14]. Since the access to electronic businesses has been made easy thanks to the Internet, there are more people who can get inside and that carries more security risk [15]. All stakeholders of a business, from buyers and suppliers to employees, use and have access to the electronic system, and there is a lot of data circulating that should be saved safely [16]. Therefore, hackers pose a considerable threat for e-business since they can get unauthorized access to sensitive data and endanger the business and all of its stakeholders. For that reason, companies that use electronic business should put in effort and resources to disable and prevent such hacking attacks to enable secure data transmission. This is where firewall usually comes in handy [17].

The most frequent ways of protection are antivirus programs, and they are very necessary today. They protect data and enable secure data transmission [18]. To protect a business from unauthorized third-party access, firewall is frequently used. It logs login attempts and notifies the administrators, so it limits the access on private networks, as well as public networks that businesses use [19].

Encryption is the part of cryptography that deals with coding, i.e. turning messages into unreadable code. For those messages to be read or usable, decryption is needed. Thanks to encryption, it is easier to make sure that the data sent will be used fort he intended purposes, as the addressee receives a key which decrypts the message. Digital certificates are a way to determine the identity of the document owner [20]. In that way, it is ensured that the employees will not open a document (a potential threat) that doesn't belong to the company, as hackers could be given a way in without anyone knowing. There are many other ways to protect an e-business, but these were the most usual ones [21].

Electronic business is one of the latest examples that show that business is not a static, but a very dynamic process where constant innovation and investment in research is a necessity [22]. Technology is making adaptation very hard, as it changes faster and faster every day and requires an even faster environment change. For that reason, e-business is making life easier and faster, i.e. making business simpler and faster by eliminating unnecessary bureaucracy, leaving us with more free time. Competition adds to that, since it makes businesses better and more efficient by trying to outdo them, which is a healthy phenomenon. The more competition there is, the fewer resources buyers have to give up to get what they want, and businesses have to try harder for the buyers to choose them instead of someone else.

One of the main technologies that came to be in the digital revolution is cloud computing. That is remote data storage, whereby data can be accessed through computer networks, storage media and software applications, which makes it so that users don't have to carry their own hard disks with them wherever they go, but they can access their data anywhere in the world as long as they are connected to the Internet. That means less investment in servers and their maintenance for companies that only used them as a support for their key activities. They can now pay for the cloud server service and focus on their business [23].

Web 2.0 is a development of the World Wide Web, which was static in the beginning and could be meddled with only by the technically sound users. With Web 2.0, creation and collaboration were channeled to average users, where they lead the way. The focus is on social media and communication, as opposed to solely reading the existent information. So, the content is created by the end users, and examples include Facebook, where anyone can post a status update that another person can like, tag, add hashtags, etc. Web 2.0. simply is a collaborative medium where businesses can get closer to their buyers [24].

The Internet of Things is another modern technology that is associated with electronic business, and it is special because it connects machines to machines, so that they can communicate without the need for humans [25]. For now, this is limited to smart homes: for example, a fridge connected to a phone can find out what frequent groceries are missing and add them to a smartphone's to-do list, which enables the connection between the seller and buyer, where the seller can send the buyer the groceries at that moment, without the buyer even being aware of all the processes that had happened. The possibilities are endless: in the future, people won't need to program machines, as they will be able to program and fix and update themselves. On the other hand, this brings about the scary thought of machine superiority over humans. But the focus should remain on the positive, such as smart cities that would eliminate and reduce waste and increase efficiency in every aspect of our lives.
Another big technology that is applicable in electronic business is mobile. Surprisingly, the continent that makes mostuse of this technology is Africa, where m-banking has developed due to a safer, cheaper and easier money transfer [26][27]. Namely, mobile technology enables micro-loans, which are especially useful in the rural parts of the Earth and enable SMEs. All transactions go through digitally so there is no waiting, as everything is happening in real time [28]. Mobile or smart phone could be also used as a remote real system for monitoring and control [29].

3. Methodology and Sample Description

As the basis for conducting this research, use was made of the list of event management websites. The list was made by authors using Web mining techniques. Most of the data were collected from the websites of the event management websites covered by this research. All the data were collected in the period from 01 August 2017 (the beginning of the research) to 15 September 2017. There are used descriptive statistics methods, as well as inductive conclusions. A new model for analysing event management websites was presented by authors (Figure 1). Similar model Spremić et al. presented in their paper, but it was applied on hotel websites [30]. All event management websites should be viewed through five main forms: (1) Website content, (2) Social Networks and Web 2.0, (3) Public Relations and User Support, (4) Multimedia and Mobile technologies and (5) User Experience and Technical Specification. The central part of the research analyses the websites of event management websites with particular regard to that model.

Fig. 1. Model for analysing Event Management Websites

The research was done so that every out of 200 websites which were found on the Google Search Engine (and suited the content which was collected on other similar websites) while searching for the terms such as "event organizations in Croatia", etc. were allocated 103 attributes or criteria, and were graded with the Boolean expression, i.e. 0 or 1, if they didn't or did have a certain attribute.

In the end, the result was the list of the best and the worst graded websites, along with their respective points. Altogether, there were 17 organisations' websites analyzed from the Republic of Croatia, 150 organizations from the United States of America, 16 websites from the United Kingdom, 5 websites from Australia, 5 websites from Canada, 3 websites from the Republic of Ireland, 2 organisation websites from the South African Republic, 1 organisation website from Austria and 1 organisation website from India.

4. Event Management Websites – Research Results

This research shows the suitability and the availability of services connected to event organisation through the Internet around the world. The observed criteria/attributes are the following: Visible logo, Search Engine, About the organisation, Contact, Slogan, Telephone number, Toll free phone, Fax, E-mail, Headquarters, existence of a physical office, Locations, Locations with a built in map, Account log in (sign up/log in), Pop-up window, Facebook link on the website, Twitter link on the website, Google+ link on the website, LinkedIn link on the website, YouTube link on the website, Instagram link on the website, Vimea link on the website, Pinterest link on the website, Identification Number, Personal Identification Number, Giro-account, Working hours, sitemap, Useful links, Praises, Suggestions and complaints (feedback), Calendar of events, Notices about upcoming events, Notices about finished events, Terms of use, Privacy,

Facebook, Number of Facebook friends (in thousands), Twitter, Number of Twitter followers (in thousands), Google+, Number of Google+ followers (in hundreds), LinkedIn, Number of LinkedIn followers (in hundreds), Instagram, Number of Instagram followers (in thousands), RSS, Flickr, Youtube, Number of YouTube views (in thousands), Chat, Blog, Promptness of the blog, English, German, French, Spanish, Italian, organisation's domain, .hr, .com, .com.hr, .co.uk, .com.au, .ca, and others, Links to your website, to other Web pages, Media communication, Newsletter, Corporate Social Responsibility, Customer Support, Information department, Through chat service, Through the telephone, Through e-mail, Support 24/7, Mission, Vision, Goals, Photos, Video content, Animations, Existence of the organisation's logo, Functionality on mobile devices, Special adaptation of websites for mobile devices, iOS, Android, Windows phone, Performance grade is above 80%, Loading time of the homepage is under 3 seconds, Size of the homepage is under 5 MB, Bounce rate is under 50%, Global rank is under 500.000, Rank in the country of origin is under 5.000, Authority of the homepage is over 50, Authority of the domain is over 50, HTML, ASP, PHP, Flash, Java, Secure HTTP (HTTPS) and other.

User experience included an analysis using the tools like http://tools.pingdom.com, https://www.alexa.com/siteinfo and http://moz.com. The goal of the analysis was to process and show my own research in detail and give recommendations based on the conducted research (Figures 2-7).

Figure 2 shows the disambiguation of the websites' content, and we can see that pop-up windows pop up on 46 (25%) of websites, as opposed to 0 in Croatia. The pop-up windows give an opportunity to the organizations that use them to collect information, such as e-mail addresses, from their users.

The privacy policy is present on 152 websites (83%), and terms & conditions are present on 163 websites (89%) – numbers that clearly show that both are taken more seriously in the rest of the world than in Croatia. A field with events that are coming up, as well as past events, was present on 43 websites (23%), and an events calendar was shown on 31 websites (17%). The ability to give feedback, complain and commend was present on 41 websites (22%), which shows that it is another factor taken more seriously outside of Croatia. Further, 21 websites (11%) saw the importace of links when ranking in search engines, and the sitemap was used in 11 cases (6%).

The working time was shown on 17 websites (9%), the giro-account information on 13 websites (7%), PIN on 22 websites (12%), and the ID number on 45 websites (45). According to this information, it is obvious that foreign websites leave less space for the bureaucratic information, and focus more on the Web site user. The reason for that could be the foreign legislation, as in some instances it is still necessary for Croatian companies to show the bureaucratic information online, which is in reality useless to the users.

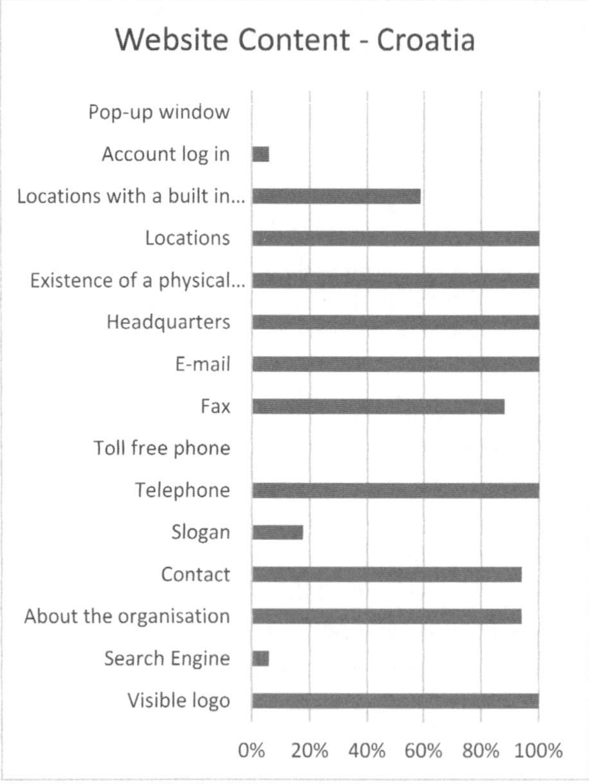

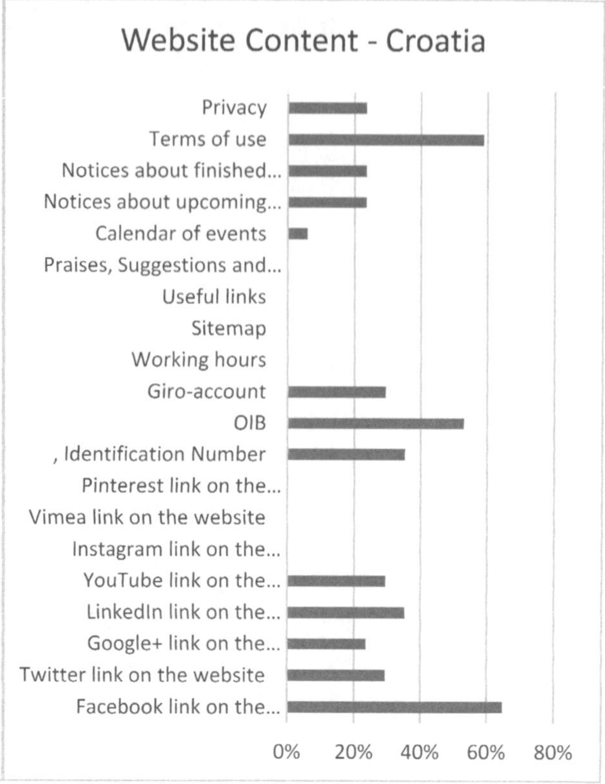

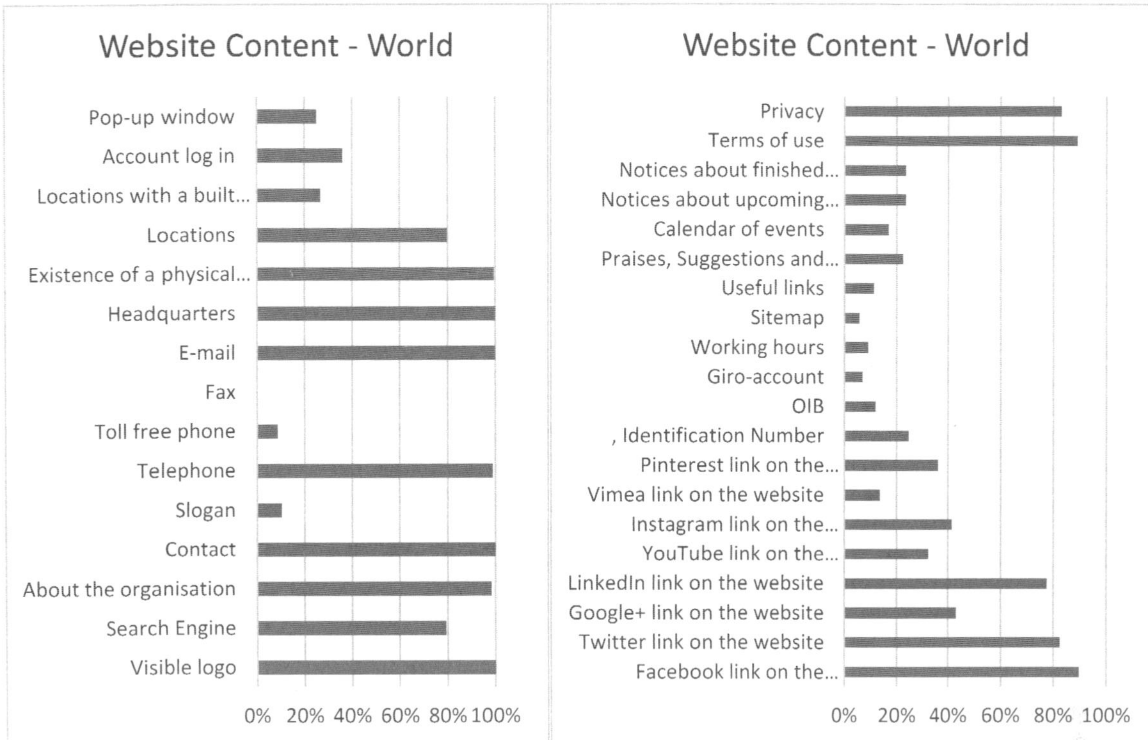

Fig. 2. Website Content

As for social media links, it can be seen that the foreign companies, as opposed to the Croatian ones, have found value in using the newer applications, such as Pinterest and Vimeo. 66 websites (36%) had a Pinterest link on their websites, and 25 (14%) had a Vimeo link. Instagram is another popular service, as 76 websites (42%) had an Instagram link on their websites, which is almost half of all the analyzed foreign websites. YouTube was linked by 59 websites (32%), Google+ by 79 (43%), and the most linked social media were Facebook with 164 websites (90%) and LinkedIn with 142 (78%).

Figure 3 shows the organizational presence on social media and Web 2.0. As per these data, the most organizations had a Twitter profile – 169 (92%), followed by Facebook – 164 (90%) and LinkedIn – 148 (81%). As in Croatia, many organizations have Google+ profiles – 88 (48%). Every organization that had an Instagram account linked to it – 76 (42%). 66 organizations had a YouTube account (36%), 1 had Flickr (1% – which is 100% more than the Croatian organizations). 28 websites (15%) had the chat option, and 183 websites (100%, all of them) had a blog. However, only 142 of those (78%) had an up to date blog, which means at least one post in August 2017.

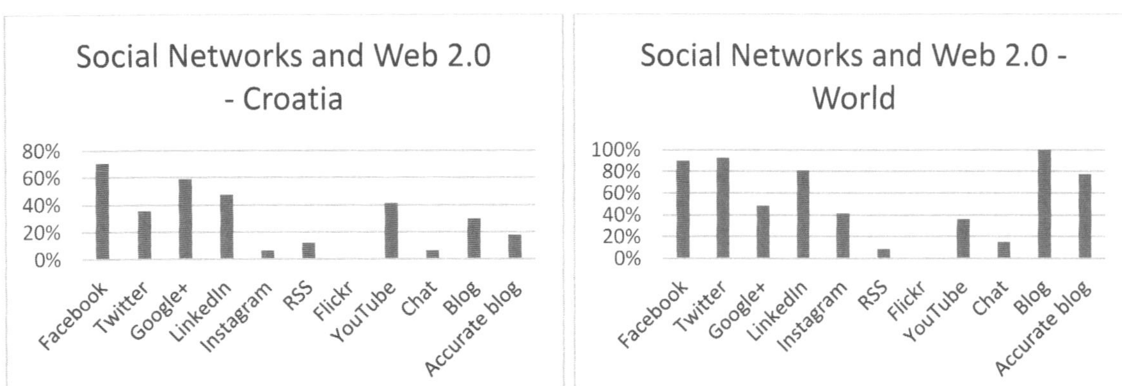

Fig. 3. Social Networks and Web 2.0

As for the international relations – according to figure 4, 25 websites (14%) had a special space for media releases, and 151 (83%) had a newsletter sign-up field. 8 websites showed information regarding their corporate social responsibility (4%).

24/7 customer support information was not present on either foreign Web site (0%), which shows that it is undersood by everyone today that communication is constant. Difficulties reporting was possible on 133 websites (73%), and 16 websites (9%) had a special information department online. Customer support was mostly available through e-mail (183, 100%), phone (181, 99%), and chat (29, 16%).

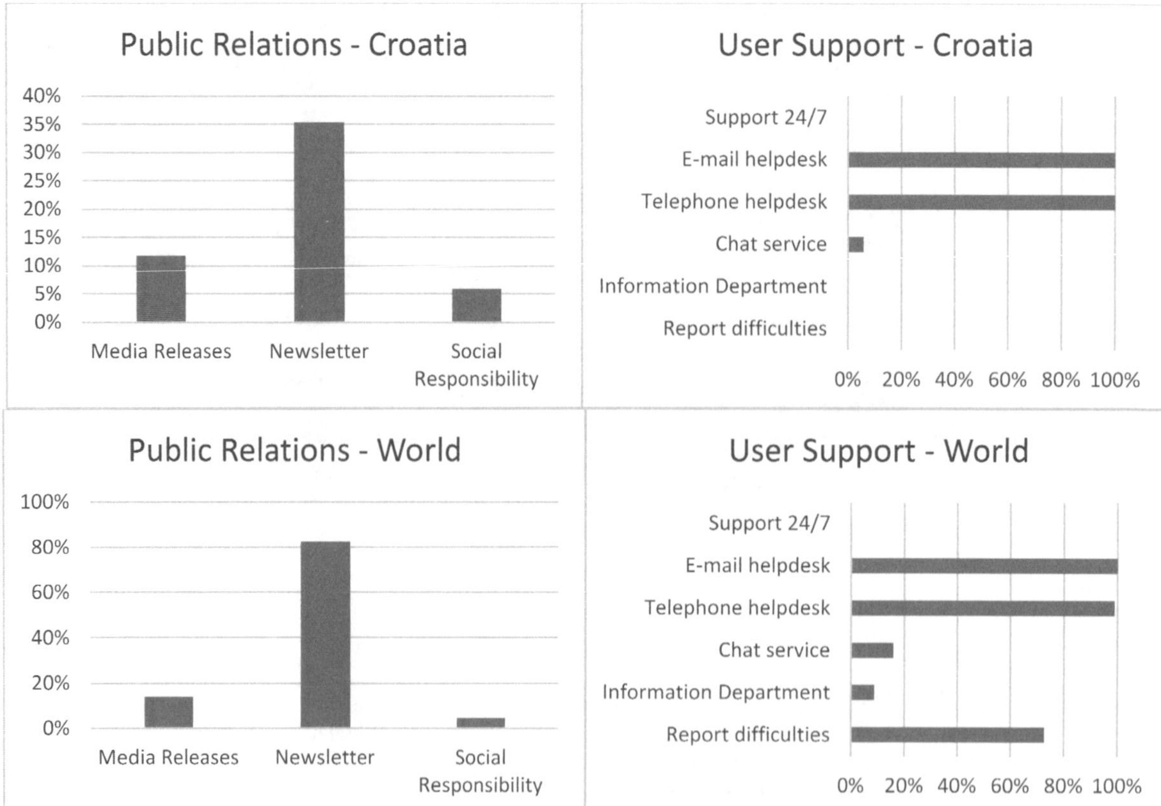

Fig. 4. Public Relations and User Support

All of the foreign websites that were analyzed (183, 100%), like in Croatia, as seen in figure 5, use photos on their websites, 157 (86%) use animations, and 35 (19%) have video content. All of the websites had a logo.

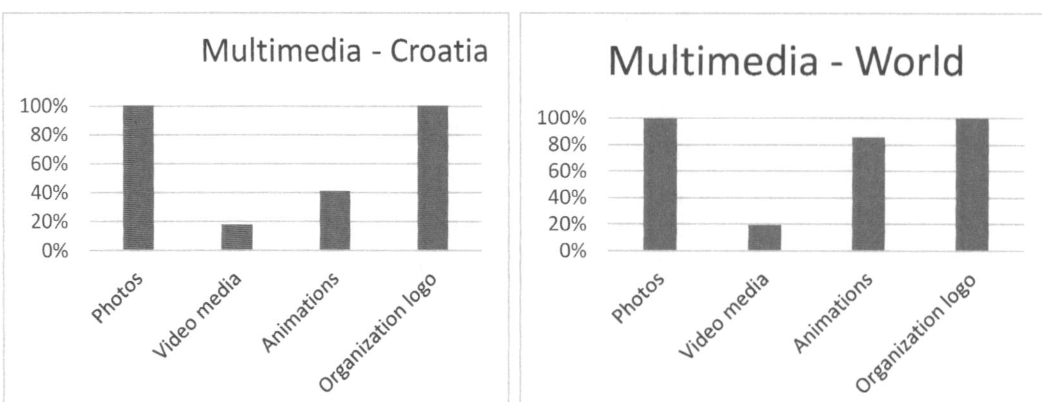

Fig. 5. Multimedia

46 websites (25%) had a domain authority over 50, and 39 (21%) had a page authority over 50. 18 websites (10%) had an originating country rank under 5.000, and 41 (22%) had a global rank under 500.000. 71 websites (39%) had a bounce rate under 50%. 73% of websites had a landing page size under 5 MB, and 143 (78%) of the landing pages loaded in under 3 seconds. 103 websites (56%) had a performance rate of over 80% (figure 6).

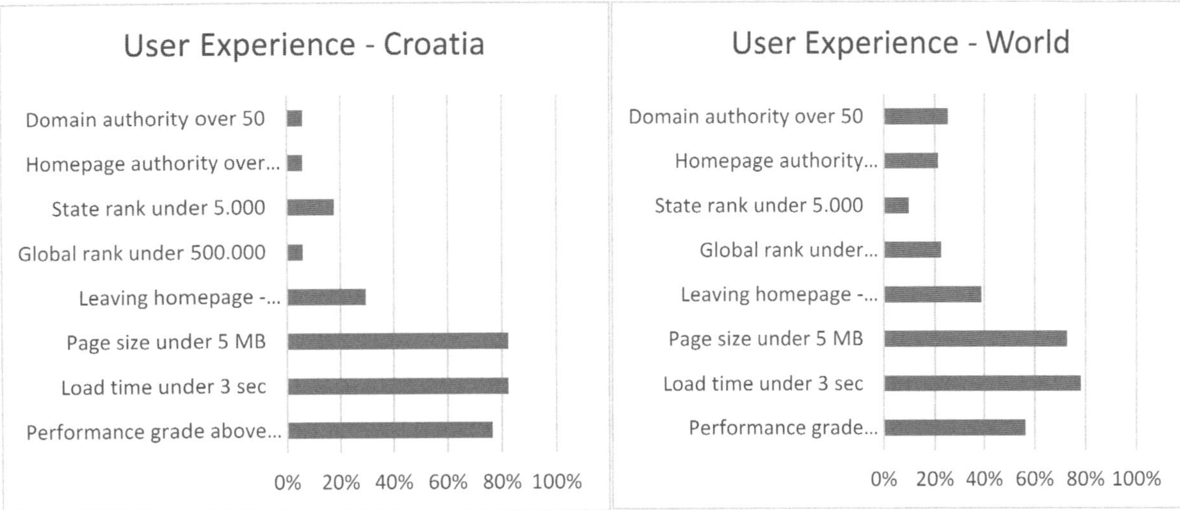

Fig. 6. User Experience

All of the foregin websites (183, 100%) used HTML, as well as Java (like in Croatia). 3 used ASP (2%), 70 used PHP (38%) and 102 used Flash (56%). 50 websites (27%) are available in more than 1 language, and 8 websites (4%) use secure HTTP (https).

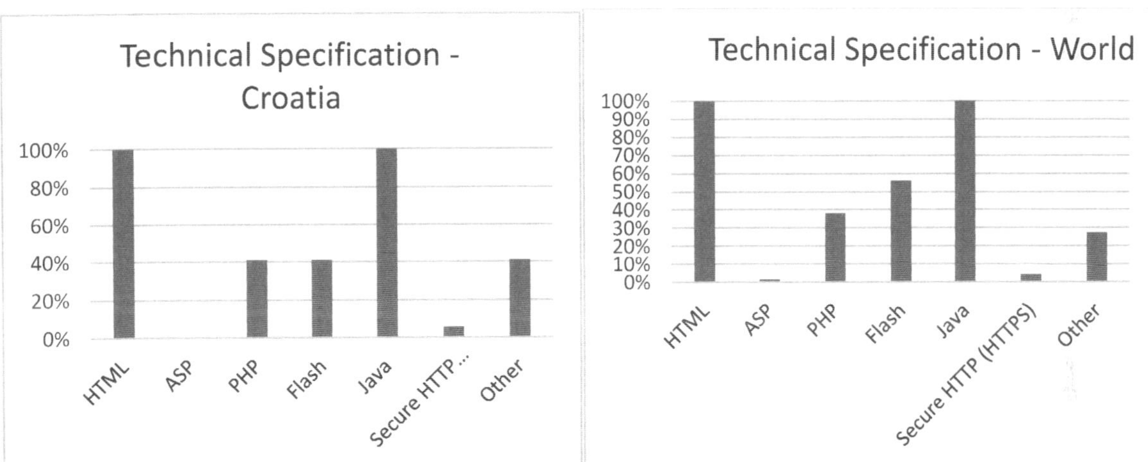

Fig. 7. Technical Specification

5. Conclusion

Activities of the organisation and event managements are changing at an accelerated rate, but because of that speed change, they can only be hard to understand and notice, especially in a situation when we need to plan and execute a lot of events and activites connected with them during the upcoming year.

New data suggest that users of events, i.e. those who attend them and consume them, increasingly interact online, so in that way they reserve or check information using the Internet and e-business.

No matter if it's through video content, web pages or event places, through social media or in other ways, event organisers who are capable of creating a high quality online presence will be the most successful in their intents.

Technology isn't just something which we will sometimes use when we want anymore, yet it is omnipresent and we use it 100% of the time nowadays, in other words, we could say that it is using us. It is necessary to use smartphones nowadays, tablets or computers / laptops in the process of event organisations, just like in many other activities. It is Internet and innovative digital technology which enabled replacing meeting in real life, but not fully. We need to incorporate technology to the real world and contact so that we could extract and get the maximum out of event organisations and real and virtual world.

In this work the research and results based on analysis of services and informations collected on web pages and organisation websites connected to the event organisations are presented. Analysis included 200 websites of the mentioned organisations, of which 17 are Croatian and 183 foreign websites, i.e. websites which operate outside Republic of Croatia. 183 of foreign websites include organisation websites which operate in United Kingdom, Republic of Ireland, Canada, Australia, India, Austria and South African Republic, since the presence on the Internet is of very high importance nowadays, it is important to present yourself well on your website, i.e. web pages. Users of services of mentioned organisations connected with event organisations use the Internet more and more in the process of using the services or searching the services, so there's nothing left to do other than to innovate and move from physical advertising and content presentations at any time, without the need for humans.

The listed results show how there's a lot of space for improvement for Croatian organisations which operate in the event organisation industry, since the first Croatian organisaton on the list of this research is on the 40th spot out of 200 in total, and the last i.e. the 200th is a Croatian organisation.

The one thing we need to work the most on is satisfying the user experience, and that can be done so that we decrease the loading time of the web page or website of important content on websites, and remove the unnecessary content. Also, presence on social media is very important nowadays, domain where Croatian organisations really falls short, in regard to foreign organisations. Generally, all the organisations fear the changes on social media, and that can be seen through the fact that Facebook is still the most popular social media for presentation from the aspect of event organisers, while the youth are focusing more and more on new social media like Instagram.

Furthermore, adaptation of websites is clean on mobile devices, but we are honestly dissapointed with low number of independent mobile applications which are especially in todays world very widespread.

Mobile devices and tablets are slowly taking the lead over Personal Computers and laptops because of their mobility, and every foreign start-up has atleast one application on mobile markets nowadays. That points towards the global trend which has to be tracked if the organisations want to stay relevant in the future.

Multimedia content is key in event organisations because people today look at experiences from previous events and based on them they judge the potential level of fun and profitability of events in the future.

Photography really is a massively used multimedia content because it is so widespread and easy to use, and it doesn't affect the size of a webpage and the loading time. Although video content is also very important, it affects the size of websites a lot so most of organisation outsource the video content on Internet tools like YouTube. That way they solve the loading problem, while the content is easily searchable on the whole internet.

Limitation of this paper is in the research sample due to fact that only webistes from selected countries were selected. Future analyses should be done from the perspective of customers. It should investigate the relationship between marketing activities and main strategies and activities of event management organizations.

6. References

[1] Tapscott, D. (1997). The Digital Economy: Promise and Peril In The Age of Networked Intelligence, McGraw-Hill, pp. 15.

[2] Spremić, M. (2017). Digitalna transformacija poslovanja. Sveučilište u Zagrebu / Ekonomski fakultet, pp. 38.

[3] Brynjolfsson, E. (2002). Understanding the Digital Economy: Data, Tools, and Research, MIT Press, pp. 9.

[4] Tapscott, D. (1997). Growing Up Digital: The Rise of the Net Generation, McGraw-Hill, pp. 34.

[5] Kehal, H. S.; Singh, V. P. (2005). Digital Economy: Impacts, Influences, and Challenges, Idea Group Pub., pp. 54.

[6] Waldfogel, J.; Peitz, M. (2012). The Oxford Handbook of the Digital Economy, Oxford University Press, pp. 118.

[7] Panian, Ž.; Strugar, I. „Informatizacija poslovanja". Ekonomski fakultet Zagreb, Zagreb, 2013., pp. 27.

[8] Khurana, A. (2017). E-business vs. E-commerce: Understanding the Difference. [online]. The Balance. Available on https://www.thebalance.com/ebusiness-vs-ecommerce-1141573 Accessed on: 2017-08-27

[9] Bertels, A. „The Difference Between E-Business and e-Commerce" [online], Computer World. Available on https://www.computerworld.com/article/2588708/e-commerce/e-commerce-the-difference-between-e-business-and-e-commerce.html Accessed on: 2017-08-29

[10] McCall, T. (2015). Is Your Organization an e-Business or d-Business?. [online]. Gartner. Available on http://www.gartner.com/smarterwithgartner/is-your-organization-an-e-business-or-d-business/ Accessed on: 2017-08-29

[11] Greenstein, S; Tucker, C.; Goldfarb, A. (2015) Economic Analysis of the Digital Economy, National Bureau of Economic Research, pp. 26.

[12] Brynjolfsson, E.; McAfee, A. (2011). Race Against the Machine, MIT Press, pp. 119.

[13] Harris, M. (2015). Book review: 'The Internet Is Not the Answer' by Andrew Keen. Washington Post, pp. 79.

[14] Schultz, J.; Perzanowski, A. (2016). The End of Ownership: Personal Property in the Digital Economy, MIT Press, pp. 14.

[15] Huws, U. (2014). Labor in the Global Digital Economy: The Cybertariat Comes of Age, NYU Press, pp. 26.

[16] Kermek, D. (1997). Databases on the Internet. Journal of Information and Organizational Sciences [online], 21(2). Available on http://hrcak.srce.hr/index.php?show=clanak&id_clanak_jezik=117455 Accessed on: 2017-08-11

[17] University of Pittsburgh. (2003). E-Business Resource Group Security Guidelines. [online]. Available on http://www.cfo.pitt.edu/ebusiness/arEBSecurityGuide.pdf, Accessed on: 2017-08-28
[18] Brynjolfsson, E. (2014). The Second Machine Age: Work, Progress, and Prosperity in a Time of Brilliant Technologies, MIT Press, pp. 46.
[19] Westfall, J. (2010). Privacy: Electronic Information and the Individual. Santa Clara University, Markkula Center for Applied Ethics, pp. 187.
[20] Ibrahimpašić, B., Liđan, E. (2011). Digitalni potpis. Osječki matematički list [online], 10(2). Available on http://hrcak.srce.hr/index.php?show=clanak&id_clanak_jezik=100091 Accessed on: 2017-07-23
[21] Karlsson, C. (2010). The Emerging Digital Economy: Entrepreneurship, Clusters, and Policy, Springer Berlin Heidelberg, pp. 26.
[22] Coltman, T., Devinney, T., M., Latukefu, A., Midgley, D.,F. (2001). E – business: Revolution, Evolution or Hype?. [online]. JSTOR. Dostupno na: http://www.jstor.org/stable/41166111 Accessed on: 2017-08-02
[23] Jamsa, K. (2013). Cloud Computing. Jones & Bartlett Learning, Burlington, MA, pp. 34.
[24] O'Reilly, T. (2005). What Is Web 2.0. [online]. O'Reilly Network. Available on http://www.oreilly.com/pub/a/web2/archive/what-is-web-20.html Accessed on: 2017-08-28
[25] Panian, Ž. (2000). Bogatsvo Interneta. Zagreb: Strijelac
[26] Spremić, M. (2017). Digitalna transformacija poslovanja. Zagreb: Ekonomski fakultet u Zagrebu, pp. 38.
[27] Mandić, M. (2009). Privacy and security in E-commerce. Tržište [online], 21(2). Available on http://hrcak.srce.hr/index.php?show=clanak&id_clanak_jezik=81070 Accessed on: 2017-07-18
[28] Rowley, M. (2017). How mobile technology is transforming lives in rural India. [online]. TechCrunch. Available on https://techcrunch.com/2017/02/19/how-mobile-technology-is-transforming-lives-in-rural-india/ Accessed on: 2017-08-28
[29] Matysek, M., Adamek, M., Karafiat, T. (2016). The Mobile Monitoring and Control of Real Systems, Proceedings of the 26th DAAAM International Symposium, pp.0104-0108, B. Katalinic (Ed.), Published by DAAAM International, ISBN 978-3-902734-07-5, ISSN 1726-9679, Vienna, Austria
[30] Spremic, M., Kokolek, N., Jakovic, B., Simurina, J. (2015). Marketing and Commercial Activities Offered on Websites of European Union Banks. International journal of social, behavioral, educational, economics and management engineering. 9, 6; pp. 1757-1763.

DOI: 10.2507/28th.daaam.proceedings.006

DISSEMINATION OF ENGINEERING EDUCATION AT SCHOOLS AND ITS ADJUSTMENT TO THE NEEDS OF ENTERPRISES

Eduard Shevtshenko, Tatyana Karaulova, Maris Igavens, Gunars Strods, Ilona Tandzegolskiene, Vidmantas Tutlys, Tavahodi Seyed & Vladimir Kuts

This Publication has to be referred as: Shevtshenko, E[duard]; Karaulova, T[atyana]; Igavens, M[aris]; Strods, G[unar]; Tandzegolskiene, I[lona], Tutlys V[idmantas], Tavahodi S[eyed] & Kuts, V[ladimir] (2017). Dissemination of Engineering Education at Schools and its Adjustment to Needs of Enterprises, Proceedings of the 28th DAAAM International Symposium, pp.0044-0053, B. Katalinic (Ed.), Published by DAAAM International, ISBN 978-3-902734-11-2, ISSN 1726-9679, Vienna, Austria
DOI: 10.2507/28th.daaam.proceedings.006

Abstract

In recent years, there has been a constant deficit of students in technical specialities in higher educational institutions. This problem is especially significant for industrial engineering area. To solve this issue, the authors initiated the joint project with participation of three universities in the Baltic region. This project aims to develop a new approach for preparing schoolchildren to choose the right profession through collective efforts of university, schools and enterprises. Authors address the critical issue in this paper, the gap between the needs of companies in the region and university education. Authors introduce this problem based on the questionnaire answers provided by the metal and machinery industries of the Baltic countries. Results of the current analysis may be useful to other nations and readers can apply them to other sectors.

Keywords: CDIO; IMATEII; teaching strategy; design, technologies and economics competencies; machinery industry; questionnaire

1. Introduction

In the new market relation, a central task for the schools is to provide children with a powerful platform of knowledge. This knowledge can help children to find worthy places in the society and grantee their successfulness in the future. There is evidence that teaching method plays a crucial role in developing children's interest toward the choosing of future profession. As a result, the elements of novelty in performing learning tasks are among the most critical factors for this development.

Recent developments in the higher education led to growing trend in youth's orientations, toward top statuses in the society. The presence of a diploma of higher education is one of the leading motives for admission to universities. Therefore, a significant proportion of young people tempts to select prestigious profession in their opinion in the future such as a programmer, a lawyer, financier, economist, and entrepreneur. However, labour market, in addition to these high statuses jobs, urgently requests for qualified engineers and technicians, and not always with higher education. As a result, often qualification of young specialists is inadequate to fulfil the real needs of the labour market. Investigations through manufacturing and industrial sectors have shown that engineers, technologists and other technical jobs are among the most popular professions [1].

Despite recent developments in higher education, the significant number of children still have problems with choosing their future career. In fact, their interest in professions is mainly unformed and blurred, and merely small parts of them think about future job and employment after graduation. The misconception about engineering profession has led to the lack of positive motivation toward these fields and their unpopularity between young people.

After Baltic countries have joined the EU, students can improve their professional level in other European countries. Often people leave their own country because they have not complete information about the opportunities to obtain the appropriate speciality in own country. Therefore, students usually break studies and leave their country seeking for the suitable position abroad, because they have no information about of all aspects selected profession. For schoolchildren to orient themselves in the professional sphere and realise themselves in the interests of the country, it is necessary to help them solve one of the most difficult and important questions, „who I want to be?". The correct choice of high school students of their future profession will make the education process full and productive. A significant role in professional orientation belongs to higher education institutions.

In recent years there has been a constant shortage of students in higher education institutions for engineering specialities. For example, in Tallinn University of Technology (TUT) for several years the number of offered places in groups is higher than the number of students admitted (see Fig. 1). However, the number of student embraces the students from foreign country mostly Asian and African countries. This year TUT statistic shows the number of 53 for Industrial Engineering specialisation students, but only 11 of them were from Estonia. However, the role of demographic decline and openness of borders in the EU is undeniable, and this shortage emanates from low awareness of school children about the benefits of obtaining an education in own country.

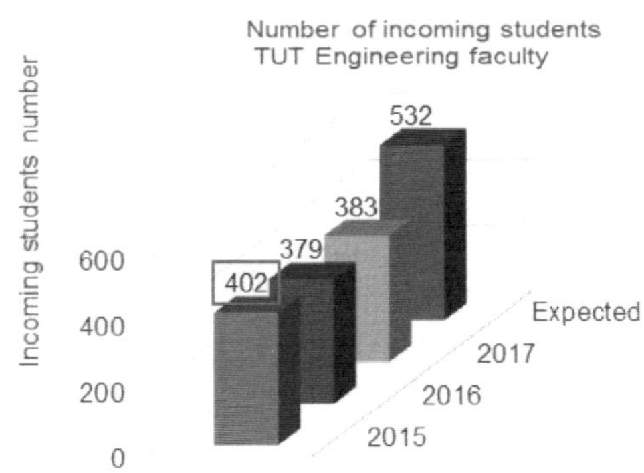

Fig. 1. Number of students coming to Engineering Faculty according to TUT statistic data

The other serious problem is the gap between the needs of the regions and education response in the area of the Career Guidance at secondary schools and introduction of youth with professions demanded in selected industries. Universities mostly focus on developing and assessing the knowledge competencies of future engineers, while employers needs skills that allow university graduates solving actual engineering problems immediately upon graduation.

For solving these problems the universities of the Baltic region: Rezekne Academy of Technologies, Vitautas Magnus University and Tallinn University of Technology starts the project "Innovative methods and technologies of education for implementing interdisciplinary (IMATEII) in career guidance. The purpose of this investigation is optimisation of opportunities in labour market by increasing the number of professionals in the strategic business of Baltic Sea countries.

The researchers set out the following objectives for the current project:

- develop and implement innovative practices in career education at secondary school,
- promote secondary school teachers the career guidance competence based on interdisciplinary among design, technology and economics,
- facilitate motivated and targeted career choice of learners in selected fields of industries.

2. CDIO modern approach to engineering education

The CDIO (Conceive - Design - Implement - Operate) initiative is an international collaboration between engineering schools for engineering education reconstruction to fulfil the industry requirements. Massachusetts Institute of Technology (MIT) has founded the basis of CDIO at in the late 1990s, since then, it was significantly enlarged and include many schools across North America, Europe, Asia, Africa and Australia and Russia. The Department of Mechanical and Materials Engineering (MME) is an official collaborator of the CDIO Initiative since 2003 [2]. CDIO initiative became the most exhaustive and holistic system of standards among others. Many universities have become the participants of CDIO world initiative applying and implementing similar approaches. In addition, many universities have disseminated these methods in schools and students now are involved in scientific and project activities under the supervision of both university and school teachers [3]. CDIO initiative sheds new light on "how universities can continually provide high-quality education in technical fields, while simultaneously imparting a sense of engineering professions in students." Primary concern of CDIO is narrowing the existing gap between engineering education and practice [4], [5].

The CDIO process includes [6], [7]:

"Conceive" stage is a creative process of conceiving new structures for technical devises and systems. At this stage should be considered: customer needs, technology, enterprise strategy, regulations, and conceptual, professional and business plans. Many universities using TRIZ in their educational practice for the rising of students' ability to generate ideas, which helps to find an exciting solution, create new product or system [8], [9], [10]. In addition, TRIZ is widely used by many leading transnational corporations for inventive solutions development.

At *„Design"* stage, teachers should consider plans, drawings, and algorithms that identify the necessary items of this process. Usually, teachers employ CAD, CAM, and CAE programs at this stage. In addition to calculations, "Design" stage can also include the experimental part supported by the systems of multidimensional simulation [11]. To simplify the understanding of Design stage for particular domain it is also possible to create the virtual 3D rooms for professional activities [12].

"Implement" stage is an applying the design in product, process, or system which embraces manufacturing, coding, testing and validation. *"Operate"* stage is an applying the design in product, process, or system which embraces manufacturing, coding, testing and validation. "Operate" stage includes the implementation of product or process that presenting the intended value. This phase consists of the maintaining, evolving and retiring the system. The university can successfully implement the last two stages in industrial premises in collaboration with partner enterprises.

Applying CDIO principles in engineering education significantly change the conventional approach to educational programs development and implementation and, as a result, enhance the competence of the graduates and increase their competitiveness in the international engineering labour market.

Selected authors have introduced ideas and approaches on emerging skills development [13]. Taajamaa et al. [14] have suggested adding "design thinking" methods [15] to the front and end of CDIO process, thereby exposing and providing students with the tools for dealing with complex, multidisciplinary problems. This method is significantly helpful when problem identification itself is quite challenging. Campbell and Beck [16] proposed a CDIO standard for internationalisation and mobility. Enelund et al. [17] suggested a concept for a computational mathematics curriculum, for engineering students.

Teachers have conducted two first projects in Scandinavia and the Baltic Sea region (*Turku University of Applied Sciences and Helsinki Metropolia University of Applied Sciences* from Finland, *Royal Institute of Technology* from Sweden and *Technical University* of Denmark, the *Tartu University* from Estonia and *Vilnius University of Applied Sciences* from Lithuania). By offering new innovative models and tools, these projects promote "*the quality assurance work and continuous curriculum development in the field of engineering science in Baltic partner universities"* [18].

The authors of the current project with participants from three Baltic state countries: Latvia, Lithuania and Estonia are also using CDIO principles with some particular differences. In addition to CDIO guidelines, it also includes economic aspect. Most exciting part is the assessment of such factors as external senses and working conditions during the enterprise visits (visual impression, noise, smell, staffs' attitude etc.), what gives the full expression ("colour") about the selected profession.

3. Project implementation

Adequate training of qualified specialists for the metal and machinery industry is possible only in close cooperation between the schools, universities and enterprises. University educational programs should correspond to the needs of the industrial companies developed in particular region. Therefore, there is the need for the constant monitoring of the competencies of the participants in the educational and the production systems.

The task for project first activity was to collect information on employers needs corresponding to their staff competencies in design, technology and economics areas and elaborating recommendations for career counsellor in-service training program development. The project team have used the questionnaire for data collection.

Researchers have divided the analysis tasks between universities by several industrial domains: forestry and wood, metal and machinery, agriculture and food sectors of industry. The project work was carried out in all areas according to the general scheme presented in Fig. 2. The current paper introduces the domain of metal and machinery industry, which is one of the most important growth areas in Estonia. There are more than 2000 metalworking and machine building companies, employing more than 35000 employees, but the average percentage of dropouts in Estonian higher schools was 20% during the last three years. The sector needs more engineers whose skills and knowledge are suitable for starting work after graduating, for developing the business that creates bigger added value [19].

3.1. Dissemination of the engineering education in schools

The primary objective of this project is to attract and train students in schools for admission to the Higher Educational Institutions with technical specialisation. Within the framework of this project, the main emphasis is on familiarizing the students with the realities of the chosen profession through visits to enterprises and master classes in school and university laboratories. It is essential to show the students how companies make the selected product. Fig. 3 shows the primary collaboration function between the project group members from university and schools.

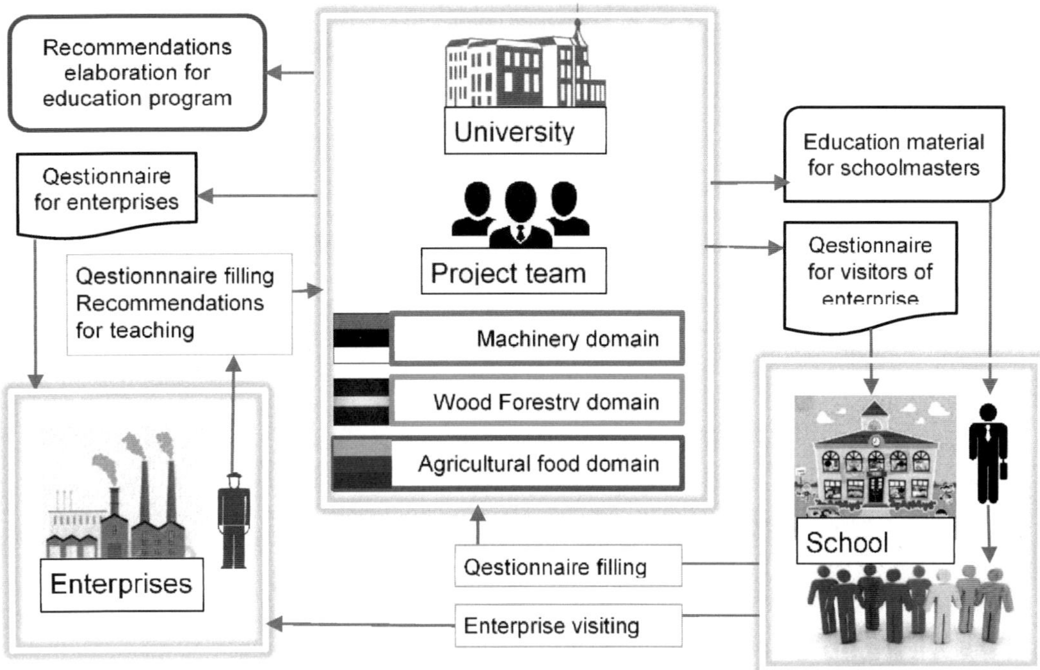

Fig. 2. General scheme of project implementation

The first step is to establish the contacts with schools and enterprises of a particular area. Project leaders must consider a possible list of products to show lifecycle from design to end-product may (products library) to schoolchildren (products library). The students select the most exciting product for studying by themselves. The project team can choose different company products to show, but highlight those products, which production process teachers will show youngsters while visiting university partners companies.

Each product from „Product library" must have the description of it technological process (Technologies library). It must be done on the reasonable level for schoolchildren, as short description, film, illustration. University lecturers and researchers prepare slides and training materials to give over to schoolteachers the technological knowledge and skills. Schoolmaster explain to the students based on practical lessons how to design and manufacture the selected product, supported by the university lecturers.

In addition, project leaders must organise visits to the companies, which produce the previously studied products, so that the pupils have a complete view of a real manufacturing process and production environment. It is essential because often working conditions and sensorial feelings are playing the critical role in choosing a future profession. Our project team elaborated the questionnaire for self-checking of the sensual impressions for company visitors.

Sense	**Attitude**				**Comments**
	Like	Mostly like	Mostly don't like	Don't like	
Smell					
Touch					
Visual					
Temperature					
Wet					
Dust					
Safety					
Staff attitude					

Table 1. Questionnaire for enterprise visitors

Work in classroom with schoolchildren must base on teaching and learning strategies [20]. In these recommendations given samples how to implement critical thinking teaching and learning strategies in career guidance and to allow students from secondary school to introduce with local industries through Design, Technologies and Economics competencies development.

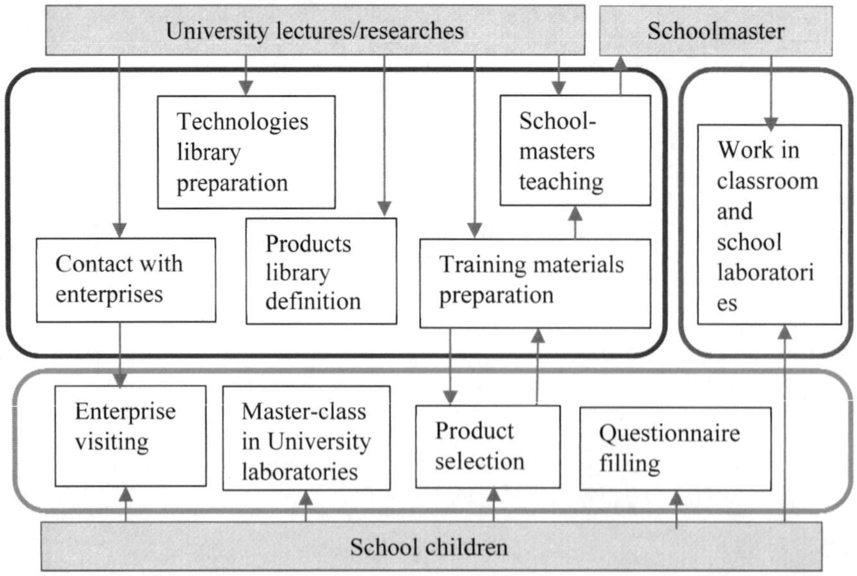

Fig. 3. The primary functions of university and school in frame of current project

3.2. Analysis of Enterprises needs for machinery industry in the Baltic region.

The target of the project is to synchronise the needs of engineering enterprises with the curricula of universities in the Baltic Sea area. For this purpose, authors have surveyed firms from three Baltic state countries on three core competencies identified by the teaching and learning strategies. The researchers have prepared the questionnaire for enterprises skills in design, technology and economics based on Latvian professional standard [21]. In this project, each country has selected the area of primary competence, Estonian researchers have selected the machinery domain; Latvian project team have chosen the wood industry, and Lithuania have the most significant experience in the food industry. In this article, authors are focusing only on machinery domain. The questionnaire is universal, and readers can apply it to other sectors.

Nr.	Competences related to DESIGN
1	Creative and independent development of existing and new products
2	Exploring the needs of the ultimate users and forecasting all the experience-based design aspects, technology, material and fashion trends.
3	Creating long-term product: innovative, functional, produce in a rational and cheaper way, convenient and easy to use, with aesthetically high-quality forms.
4	Formulating basic design tasks and to planning artistic design and construction works.
5	Analysis and definition of logistical, functional, aesthetical, and ergonomic set of qualities.
6	Navigating and managing used materials, to analyse their improvement and development trends, and to be able to use them at work.
7	Setting products' manufacturing restrictions and choosing a set of possible solutions
8	Developing products, including sketches of collection or series conceptual solution, maps and digital three-dimensional visualisations and organisation and carrying out model
9	Developing of projects by using CAD and CAM technologies.
10	Preparation of project materials for presentation
11	Developing equipment monitoring and visualization schemes.
12	Designing in compliance with the universal design principles, applying innovative approaches and technological developments.
13	Carrying out project presentation and coordination with the contractor and responsible departments in various product development stages.
14	Preparing and conducting presentations, managing shows and exhibitions

Table 2. Competencies related to design

The purpose of the questionnaire is to identify the most significant aspects of preparing the specialists for machinery industry. To collect information on requirements that employers request to their staff in design, technology and economics and elaborate the recommendations for career counsellor in-service training program development.

The authors have collected the 30 answers from machinery domain companies of three countries: 11 respondents from 8 enterprises in Latvia; 4 respondents from 2 firms in Lithuania; 15 respondents from 7 businesses in Estonia. Each respondent was asked to assess the relevance of various work processes related to the company's activities and their future development. The questionnaire includes 66 items: 14 questions related to design (see Table 2), 30 related to technology (see Table 3) and 22 connected to the economy (see Table 4). Authors have asked to assess each skill described in the questionnaire on a 10-point scale, where 10 "very, very necessary" and 0 "not necessary". Respondents have provided information on their company name, position, age and level of education. The authors are introducing the results of the questionnaire in tables (Table 2, Table 3 Table 4, Table 5, Table 6) and related diagrams (Fig. 4, Fig. 5, Fig. 6).

			Number of activity for DESIGN competence													
	Enterprise	**Position**	**1**	**2**	**3**	**4**	**5**	**6**	**7**	**8**	**9**	**10**	**11**	**12**	**13**	**14**
EE	Enterprise 1	Top manager	7	5	7	7	10	7	7	7	8	7	9	7	8	7
EE	Enterprise 2	Head of dep.	10	10	10	10	8	10	8	8	10	7	7	7	8	7
EE	Enterprise 2	Specialist	8	8	8	8	7	10	10	10	9	7	8	9	8	7
EE	Enterprise 3	Head of dep.	8	8	10	10	7	10	7	7	8	6	8	8	8	5
EE	Enterprise 3	Top manager	10	6	10	10	10	10	10	10	6	6	6	10	10	8
EE	Enterprise 3	Specialist	9	10	8	6	5	8	3	9	10	9	9	10	8	8
EE	Enterprise 4	Specialist	8	9	10	10	9	8	9	8	8	8	9	10	9	8
EE	Enterprise 5	Head of dep.	8	10	8	5	5	8	7	7	8	8	8	6	9	7
EE	Enterprise 5	Specialist	10	5	10	10	6	10	10	6	10	10	6	10	6	6
EE	Enterprise 5	Top manager	4	5	6	2	3	6	5	5	5	6	9	6	6	6
EE	Enterprise 6	Head of Dep.	4	6	3	3	2	5	3	3	8	8	8	5	4	4
EE	Enterprise 6	Specialist	8	7	9	6	10	5	5	10	8	4	6	9	10	10
EE	Enterprise 7	Top manager	6	7	7	7	6	7	8	7	10	8	9	8	8	8
EE	Enterprise 7	Head of Dep.	6	8	8	6	8	10	9	6	8	6	8	7	6	5
EE	Enterprise 7	Specialist	10	4	3	6	6	10	10	10	10	7	8	10	8	5
Average in Estonia			7,73	7,2	7,8	7,07	6,8	8,27	7,4	7,53	8,4	7,13	7,87	8,13	7,73	6,73
ET	Enterprise 8	Specialist	9	10	8	6	5	8	3	9	10	9	9	10	8	8
ET	Enterprise 8	Head of dep.	10	10	10	10	8	10	8	8	10	7	7	7	8	7
ET	Enterprise 9	Top manager	7	5	7	7	10	7	7	7	8	7	9	7	8	7
ET	Enterprise 9	Specialist	8	8	8	8	7	10	10	10	9	7	8	9	8	7
Average in Lithuania			8,75	8,25	7,8	8,3	8,8	8	7,8	8,3	8,3	7,5	8,3	8,8	8,5	7,8
LV	Enterprise 10	Specialist	8	7	6	3	5	8	5	9	10	8	7	7	7	5
LV	Enterprise 10	Top manager	1	1	1	1	1	1	1	1	5	1	1	1	1	1
LV	Enterprise 11	Top manager	4	3	5		4	6	6	4	3	3	5	2	4	4
LV	Enterprise 12	Specialist	7	7	7	7	7	8	8	8	8	8	8	8	8	8
LV	Enterprise 12	Top manager	7	7	7	7	7	7	8	8	8	8	7	8	8	8
LV	Enterprise 12	Head of dep	7	7	8	8	9	8	9	9	9	9	9	8	8	9
LV	Enterprise 13	Head of dep	10	10	7	8	10	10	10	7	10	8	8	7	9	8
LV	Enterprise 13	Top manager	10	8	7	7	8	9	7	8	9	8	7	9	8	9
LV	Enterprise 13	Specialist	8	9	9	3	9	9	7	3	3	3	1	1	1	1
LV	Enterprise 14	Top manager	10	8	7	5	5	9	8	5	10	3	3	3	5	5
LV	Enterprise 14	Specialist	9	9	9	3	9	9	7	3	3	3	1	1	1	1
Average in Latvia			7	6,7	6,4	5,44	6,5	7,5	6,9	6,2	7,5	5,9	5,6	5,4	5,9	5,8

Table 3. Questionnaire results in competence related to design

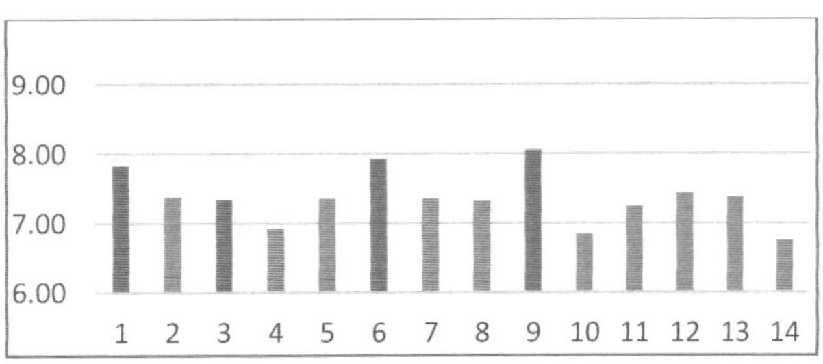

Fig. 4. Results of design activities estimation (by red are marked activities with more high grade)

Recommendations for design competence:

- Navigating and managing used materials, to analyse their improvement and development trends.
- Developing of projects by using computer-assisted design and computer-aided manufacturing technologies.
- Creative and independent development of existing and new products.

- Creating long-term product: innovative, functional, produce in a rational and cheaper way.
- Setting products' manufacturing restrictions and choosing a set of possible solutions (possible material usage and compatibility as wells as possible constructive solution options).
- Analysis and definition of logistical, functional, aesthetical, and ergonomic set of qualities.

The authors have not included into the questioner the following design skills, which companies have marked as important: custom-made products design, standardisation, design for manufacturing and Assembly (DFMA), knowhow of reliability tools (DFEA, FMEA), quality management.

Nr.	Competencies related to TECHNOLOGY
15	Developing a map of detail inspection, choosing the necessary instruments and calculating their amount, aligning their conditions of use and storage to ensure the necessary quality.
16	Diagnosing the service and repairs of assisting devices and ensuring their production, mechanisms, and tools.
17	Managing electrical and system production and installation; determining control samples and technological processes to ensure the safety and efficiency of operation of electrical systems, engines and equipment.
18	Developing technological planning of production
19	Evaluating the level of production automation
20	Performing linking of automated electrical equipment to technologies.
21	Managing the essential requirements for equipment (mechanical strength and stability (resilience), environmental friendliness, safety of usage, application compliance, access to structures, rational use of natural resources).
22	Managing the production technology of Products and Service
23	Contributing to introduction of new technologies
24	Developing, optimisation, and improvement of current product technologies and systems for generation, distribution, transportation, utilization, and disposal.
25	Conducting calculations and drawings independently by using necessary software.
26	Conducting scientifically valuable research and implementing it into work.
27	Planning, organization and control of work methods and technologies.
28	Choosing most rational technique and technology.
29	Managing technological processes
30	Developing proposals which are directed to more efficient and useful recourse utilization
31	Determining technological installations necessary for work implementation and planning their placement.
32	Using latest information technologies
33	Collecting and summarising necessary professional information and following topicalities of the field
34	Conducting a professional work assignment analysis,
35	Applying the principles of environmental friendly production
36	Making strategic and operative decisions by summarizing and analysing production information.
37	Adjustment of production regimes by applying knowledge of production process. Organising technological processes in complying with environmental laws and regulations.
38	Organising technological processes in complying with environmental laws and regulations.
39	Forming a safe workplace in complying with fire safety regulations in the production facility.
40	Using rationally and recycling of sub-products acquired in the process.
41	Elaborating normative documents (technological instructions, quality manuals, technical regulations, etc.) according to company specifics.
42	Development of hygiene requirements for the production process.
43	Using product labelling (by applying knowledge of the EAN).
44	Analysing causes of the production incompatibility and taking preventive actions to address them.

Table 4. Competencies related to technology

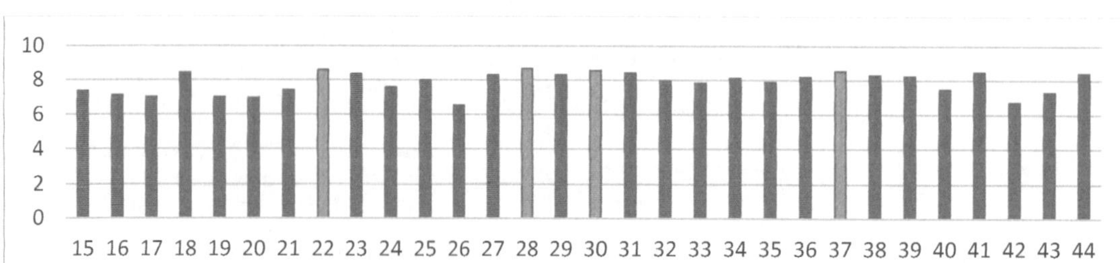

Fig. 5. Results of technology activities estimation (by green are marked activities with more high grade)

Recommendations for technology competence:
- Choosing the most rational technique and technology.
- Contributing to the introduction of new technologies.
- Developing the technological planning of production
- Managing the production technology of Products and Services
- Forming a safe workplace, which meets the fire safety regulations in the production facility.
- Adjustment of the production regimes by applying knowledge of production process.

The authors have not included into the questionnaire the following technical competencies, which companies have marked as important: ability to perform the overview of scientific research and implement it in today's business; Poka-Yoke and 8D improvements; ability to learn from mistakes, analysis of the dimensional system.

Nr.	Competencies related to ECONOMY
45	Evaluating the calculations of the cost of services to be performed, the necessary investments and workforce consumption.
46	Applying different management models.
47	Navigating the European Union sector-specific policies and its implementation methods
48	Preparation of costs for designing or manufacturing equipment and determining the time of expenditure reimbursement.
49	Development of the most cost-effective technical solutions.
50	Choosing and aligning materials with a client and planning economic material consumption
51	Defining of strategic objectives.
52	Running and planning operating strategy sector.
53	Planning and organisation of complex resource use.
54	Dealing with sector market
55	Drawing up a business plan and estimates
56	Handling problem situations adequately and timely adopting necessary decisions
57	Defining tasks in achieving set targets in order to produce products with high added value
58	Using the necessary technical and normative documentation for the work, carrying out applied and professional documents.
59	Analysing sector benefits and assessing the competitiveness of the holding
60	Calculation and evaluation of economic performance of the holding
61	Assessment of risk factors of the particular holding and assessment of the measures to reduce the risk
62	Navigating and using management accounting capabilities
63	Development and management of projects
64	Organising and controlling economic activities of the enterprise.
65	Developing and implementing quality management and self-checking system
66	Controlling the material and financial cash flow of the enterprise.

Table 5. Competencies related to economy

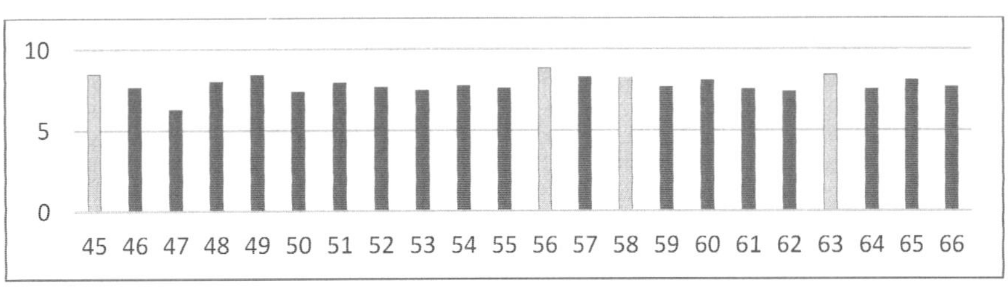

Fig. 6. Results of economy activities estimation (by yellow are marked activities with more high grade)

Recommendations for economy competence:
- Handling problem situations adequately and timely adopting necessary decisions.
- Evaluating the calculations of the cost of services to be performed, the necessary investments and workforce consumption.
- Development and management of projects.
- Using the necessary technical and normative documentation for the work, carrying out applied and professional documents
- Defining of strategic objectives.
- Development of the most cost-effective technology solutions.

The economy competencies that are important, but authors have not included in the questionnaire are: defining of most cost-effective solutions; analyse the production processes by cost-intensive processes; understand the connections between different functions and departments; ability to see the whole Value Chain from Sales till Support operations; risk analysis; control plans and perform the process mapping.

Authors have introduced other critical skills in machinery sector in Table 6. From the overall results given in Table 6, it can be seen that enterprises are mostly consistent with suggested results.

		Avg	Nr	Competencies
DESIGN	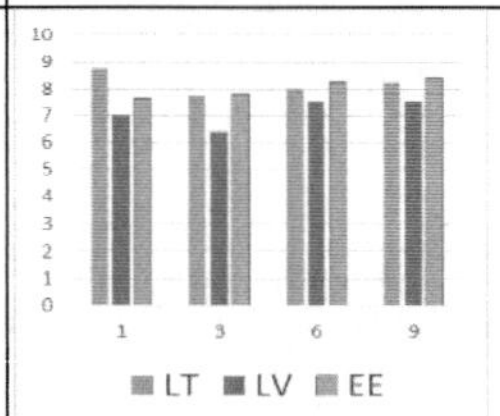	8,03	6	Navigating and managing used materials, to analyse their improvement and development trends, and to be able to use them at work.
		8,03	1	Creative and independent development of existing and new products.
		7,6	3	Creating long-term product: innovative, functional, produce in a rational and cheaper way, convenient and easy to use, with aesthetically high-quality forms.
		7,57	9	Developing of projects by using CAD and CAM technologies
TECHNOLOGY	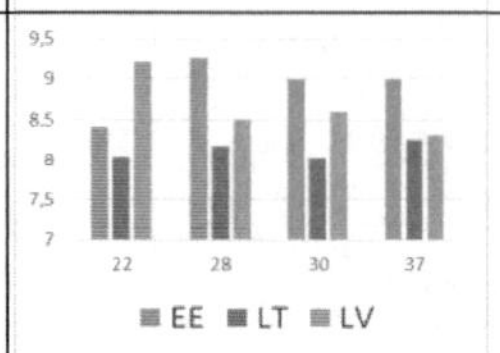	8,64	28	Choosing most rational technique and technology
		8,54	30	Developing proposals which are directed to more efficient and useful recourse utilization
		8,54	22	Managing the production technology of Products and Services.
		8,50	37	Adjustment of production regimes by applying knowledge of production process
ECONOMY	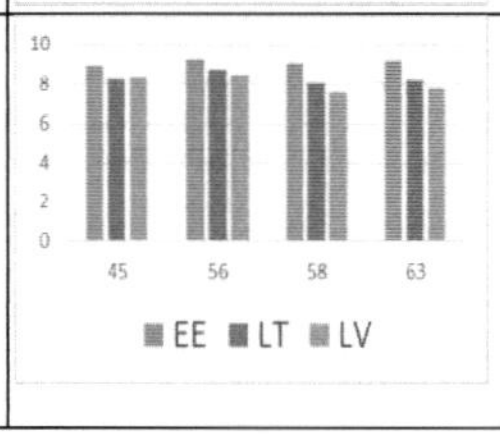	8,80	56	Handling problem situations adequately and timely adopting necessary decisions
		8,48	45	Evaluating the calculations of the cost of services to be performed, the necessary investments and workforce consumption
		8,38	63	Development and management of projects
		8,20	58	Using the necessary technical and normative documentation for the work, carrying out applied and professional documents

Table 6. Four high ranked competencies in design, technology and economics in machinery sector

4. Conclusions

Since this project is not yet finished and the work in schools is still conducted by the project participants, authors are not ready to present the final results. The authors will receive the result by analysing the enrolment statistics of schools graduates into engineering faculty of Tallinn Technical University during the two year period. At the current stage of the project, authors see that the project will create the new and upgrade the existing technical knowledge and educational materials for qualified professional training teachers from schools and universities. The possibility to communicate with the enterprises will give a realistic impression of a selected profession to schoolchildren.

The analysis of the enterprise's needs revealed the most important aspects in the companies for studying, and it is essential to prepare the qualified specialists to solve them. The universities can use the analysis of project results during the development of educational programs.

The results of the project will be:

- Particularization of the gap situation between the needs of the regions and education response in the area of the Career Guidance at secondary schools and introduce youth with professions which are demanded in selected industries;
- The Career Counsellors and secondary school teacher in-service training course "Innovative methods and technologies of education for implementing interdisciplinary in career guidance (IMATEII)";
- Results of IMATEII piloting in secondary school;
- The delivery of new and upgraded knowledge together with interdisciplinary methods in secondary school classes;
- Promoted career counsellors and class teachers competencies in career education and guidance at secondary schools;
- The results of the implementation of new interdisciplinary methods disseminated;
- Best practice samples offered for career education and guidance practitioners.

5. Acknowledgments

This research supported by Erasmus + project VERT17065 - Innovative methods for implementing interdisciplinary in career counselling (IMATEII) and F15027 - Smart manufacturing and materials technologies competence centre.

6. References

[1] Gorina E. & Shikina E. (2013) Features of the occupational choice: results of the trend research, УДК 316.334.2+331.5, Available from https://cyberleninka.ru/article/n/osobennosti-vybora-buduschey-professii-rezultaty-trendovogo-issledovaniya Accessed: 2017-10-01

[2] Berggren, K.; Brodeur, D.; Crawley, E.; Ingemarsson, I.; Litant, W.; Malmqvist, J. & Östlund, S. (2003) CDIO: An international initiative for reforming engineering education, World Transactions on Engineering and Technology Education Vol.2, No.1, 2003, pp 49 – 52.

[3] Sidorkina, O. & Pogrebnaya, T. (2014) CDIO within the system of continuous education "From school to Higher Education Institution (HEI)": Stage "Conceive" at school. Journal of the Association for Engineering Education of Russia. Vol 16, 2014, pp.42 – 47.

[4] Woollacott, L. (2007) The goals of engineering education: a rationale for a universal document based on the CDIO syllabus and the taxonomy of engineering competencies, Proceedings of the 3rd International CDIO Conference, MIT, Cambridge, Massachusetts, June 11-14, 2007. Available from http://rocketship.cdio.org/files/document/file/T2A2Woollacott.pdf, Accessed: 2017-10-01.

[5] Crawley E. (2002)Creating the CDIO syllabus, a universal template for engineering education, 32nd ASEE/IEEE Frontiers in Education Conference, Nov 6-9, 2002, Boston.

[6] Leong, H. (2007) Designing a CDIO programme: The CDIO syllabus and standards. Singapore Polytechnic. Available from http://www.kanazawa-it.ac.jp/cdio/english/file/slide10_leong.pdf. Accessed: 2017-10-01

[7] Podlesny, S. & Kozlov, A. (2014) CDIO: objectives and means of achievement, Journal of the Association for Engineering Education of Russia. Vol 16, 2014 pp.7 – 11.

[8] Jiang Jin-gang, Zhang Yong-de, Shao Jun-peng, Sui Xiu-lin, Zhang Jian-yi (2014) Education and cultivation research of engineering undergraduate's innovative ability based on TRIZ-CDIO theory, The 9th International Conference on Computer Science & Education (ICCSE 2014) August 22-24, 2014. Vancouver, Canada.

[9] Xun-tao Liu, Cun-you Zhao, Peng Xu (2011) TRTZ theory and students innovative ability cultivation, Higher Education Forum, No. 3, pp. 29-31, 2011.

[10] Jiang, F.; Sun, H. & Wang, Y. (2010) Application of TRIZ theory in mechanical principle experiment teaching and management, Experiment Science & Technology, No. 8, pp. 140-143, 2010.

[11] Chuchalin, A. (2014) Modernization of engineering education based on international CDIO standards, Engineering Education, Vol 16, 2014, pp 14 – 27. Available from http://aeer.ru/filesen/io/m16/IO-2014_16_eng.pdf. Accessed: 2017-10-01.

[12] Andreev, V., Pryanichnikov, V., Poduraev, Y., and Kuvshinov, S. (2014). Education on the basis of virtual learning robotics laboratory and group-controlled robots, in 24th DAAAM Int. Symp. on Intelligent Manufacturing and Automation, 2013, Procedia Engineering, 2014, vol. 69, pp. 35–40

[13] Malmqvist, J.; Edström, K. & Hugo, R. (2017) A proposal for introducing optional cdio standards Proceedings of the 13th International CDIO Conference, University of Calgary, Calgary, Canada, June 18-22, 2017

[14] Taajamaa, V., Eskandari, M., Karanian, B., Airola, A., Pahikkala, T. & Salakoski, T. (2016). O-CDIO: Emphasizing Design Thinking in the CDIO Engineering Cycle, International Journal of Engineering Education, 32(3B [14] Rowe, P. (1991). Design thinking. Cambridge, MA: MIT Press.

[15] Chong Siew Ping, Patrick Chow & Christopher Teoh (2011) The use of design thinking in C-D-I-O projects, Proceedings of the 7th International CDIO Conference, Technical University of Denmark, Copenhagen, June 2011. Available from: http://www.wwwwww.cdio.org/files/document/file/70_paper.pdf. Accessed: 2017-10-01.

[16] Campbell, D. & Beck, H. (2010). Toward internationalized engineering curriculum and student mobility. Proceedings of the 6th International CDIO Conference, Montréal, Canada.

[17] Enelund, M., Larsson, S., & Malmqvist, J. (2011). Integration of computational mathematics in the mechanical engineering curriculum. Proceedings of the 7th International CDIO Conference, Lyngby, Denmark.

[18] Kontio, J. (2016) Enhancing quality together with CDIO community, Proceedings of the 12th International CDIO Conference, Turku University of Applied Sciences, Finland, June 12-16, 2016. Available from http://www.cdio.org/files/document/cdio2016/158/158_Paper_PDF.pdf. Accessed: 2017-10-01.

[19] Summary of proposals of the OSKA expert panel on the manufacturing of metal products, machinery and equipment (2016). Available from: http://oska.kutsekoda.ee/wp-content/uploads/2016/05/MMT_ENG.pdf. Accessed: 2017-10-01

[20] Alan Crawford, A.; Mathews, S.; Makinster, J. & Saul, W. (2005) Teaching and learning strategies for the thinking classroom. New York: The International Debate Education Association. Available from: https://pendidikanmatematikausn.files.wordpress.com/2015/11/teaching-and-learning-strategies.pdf. Accessed: 2017-10-01.

[21] Professional standards approved by the Cabinet of ministers of Republic of Latvia on May 18th, 2010 by the rules Nr.461

DOI: 10.2507/28th.daaam.proceedings.007

ASSESSMENT OF SHIELDING EFFECTIVENESS AND PERCOLATION THRESHOLD OF COMPOSITE MATERIALS

Skočík Petr & Korbel Jiří

This Publication has to be referred as: Skocik, P[etr] & Korbel, J[iri] (2017). Assessment of Shielding Effectiveness and Percolation Threshold of Composite Materials, Proceedings of the 28th DAAAM International Symposium, pp.0054-0062, B. Katalinic (Ed.), Published by DAAAM International, ISBN 978-3-902734-11-2, ISSN 1726-9679, Vienna, Austria
DOI: 10.2507/28th.daaam.proceedings.007

Abstract

This paper deals with the protection of avionics of small airplanes in an electromagnetic environment. Various composite materials are used for this purpose. The mechanical properties of these materials are very well described. On the other hand, the electrical properties are described insufficiently. A fabric with long copper-plated polyamide fibers was selected as a suitable shielding material. The aim is to verify the shielding effectiveness of the fabric by measuring the transverse electromagnetic wave on the coaxial line. Further, determine the flat fiber concentration required to achieve a certain shielding efficiency of given composite sample and its percolation threshold which is an important parameter determining the conductivity of the composite material. It is therefore possible to determine the critical content of the filler in which the composite material starts to act as a conductor.

Keywords: Shielding effectiveness; Percolation threshold; Composite materials; Coaxial line

1. Introduction

The development of society brings new demands in all areas of human research. At present, considerable progress has been made in the field of information technology, which is related to developments in the field of microelectronics, computing and control technology. Unfortunately, such development also has its negatives. One of the notable problems is the reliability of these devices. This is influenced not only by the quality of production itself but also by external influences on the equipment. At present, this problem escalates as the environment is overwhelmed by a variety of interfering signals. These signals may have an undesirable effect on any device resulting in technological failures or accidents. Typical representatives of such devices are, for example, devices used in air traffic control or the aerospace industry itself. It is therefore a prerequisite to prevent possible failure of a device during its design and construction, ie to design the equipment in a way that it can operate properly even in the environment with any source of electromagnetic interference. The development of interference resistant devices is noticeable especially since the nineties. Of course, this has economic consequences as well. For example, designing the aircraft without consideration of the electromagnetic interference can lead to failure in the safety test during the final certification process, including the measurement of avionics resistance to electromagnetic interference. Which implies the return of the prototype back to the initial phase of development with all associated costs.

A significant part of today's airplanes is made of composite materials. The aim of these materials is to achieve a lighter and more durable design with longer life and reliability. These materials are very well described in terms of mechanical properties, but unfortunately only at least in terms of electrical properties. This work is focused on verifying the shielding efficiency of a suitable composite material and to determine its percolation threshold which is an important parameter determining the conductivity of the composite material. It defines the content of the filler in which the composite starts to act as a conductor.

2. Electromagnetic compatibility

Electromagnetic Compatibility (EMC) is defined as the ability of a device or system to perform properly in an environment where other sources of electromagnetic signals (natural or artificial) operate. Additionaly, these devices or systems must not radiate signals that are interfering with other equipment [5]. EMC has two main areas, as can be seen in Figure 1.

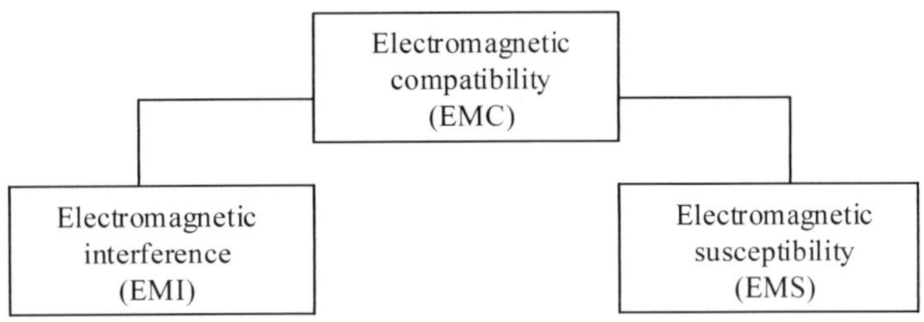

Fig. 1. Main areas of the EMC

Electromagnetic interference (EMI) is the process by which the signal generated by a source of interference is transmitted through electromagnetic coupling to interfering systems. It deals with identification of interference sources, describtion and measurement of interfering signals and identification of parasitic transmission paths. Compatibility of the whole system is achieved by technical measures, especially on the side of the sources of interference and their transmission paths. The EMI mainly concerns the causes of interference and its removal [1].

Electromagnetic susceptibility (EMS) is the ability of a device or system to operate without failures in an environment where electromagnetic interference occurs. EMS is therefore focused on technical measures that increase the electromagnetic immunity to interference disturbances at the receiver. The EMS thus concerns with elimination of the interference consequences without removing its causes [1].

3. Electromagnetic shielding

Electromagnetic shielding is one of the means of electromagnetic compatibility to reduce the emission-side of interference sources and to increase electromagnetic immunity on the receiver side. Shielding from a technical point of view is a structural means (shielding cover) for weakening the field of disturbing signals. It is used to protect electrical and electronic components, devices or systems that can be both sources and receivers of electromagnetic interference. It is one of the highly efficient ways of electromagnetic protection against power interference [3], [5].

The shielding itself is defined as the ratio of the intensity of the electromagnetic field in front of and behind the shielding material. A prerequisite is that the shielding material is an infinite plane located between the source and the receiver of the electromagnetic radiation. The region around the source of electromagnetic radiation can be divided into three areas [3]:

- distant field - field properties depend on the environment in which the wave propagates,
- near field - field properties depend primarily on the properties of sources of electromagnetic interference and the surroundings,
- transition area - located between the previous two areas.

The theoretical boundary between the distant and near field is assumed at $r = \lambda/2\pi$ (approximately λ/π from wavelength λ). In practice, when talking about a distant field, the distance $r > 5\lambda/2\pi$ is assumed. If the shielding material is in the near field, the distance of the wave source from the shielding material is much smaller than the wavelength λ, more precisely $r << \lambda/2\pi$ [3].

3.1. Electromagnetic shielding efficiency

The effect of electromagnetic shielding as a linear system can be characterized by the Shielding Effectivity (SE), which is defined by the ratio of the electric field intensity E_t or the magnetic field intensity H_t at a certain point of the shielded space to the intensity E_i (H_i) of the field impacting on the shielding material or bulkhead [5]. The logarithmic SE rate is determined by [1] as:

$$SE = 20 \cdot \log\left|\frac{E_t}{E_i}\right| = 20 \cdot \log\left|\frac{H_t}{H_i}\right| \quad [dB] \tag{1}$$

If the incident wave is homogeneous and the environment on both sides of the shielding bulkhead is the same, both definitions SE (1) are the same, since the *E* and *H* fields are mutually bound by the same impedance of the environment. Otherwise, the equation (1) give different results [5].

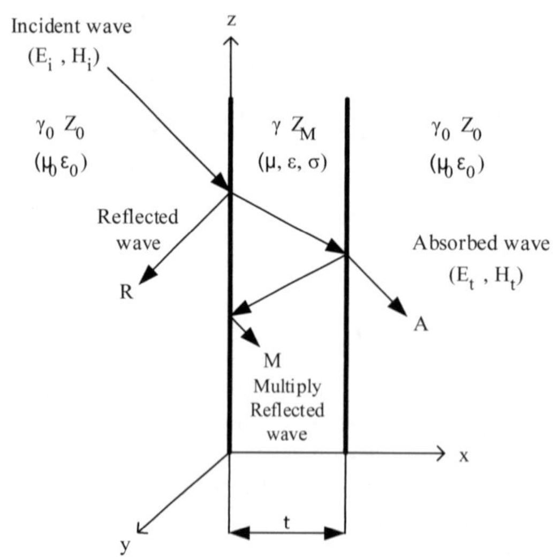

Fig. 2. The physics of the shielding effectiveness [3], [5]

Calculating the SE in Figure 2 needs to determine the components of the electromagnetic wave penetrating the shielded area at a known magnitude of the incident wave E_i, H_i, known parameters of the shielding bulkhead (ε, μ, σ, t) and known parameters of the surrounding environment (ε_0, μ_0). The solution is based on Maxwell's equations and after mathematical modifications it can be obtained as [1], [5]:

$$SE = 20 \cdot \log\left|\frac{(Z_0 + Z_M)^2}{4Z_0 Z_M} \cdot e^{\gamma t} \cdot \left[1 - \left(\frac{Z_0 - Z_M}{Z_0 + Z_M}\right)^2 \cdot e^{-2\gamma t}\right]\right| \quad [dB] \tag{2}$$

Z_0 is the impedance of the free environment before and behind the shielding bulkhead and Z_M id the impedance of the conductive environment of the bulkhead. For these parameters applies [5]

$$Z_0 = \sqrt{\frac{\mu_0}{\varepsilon_0}} \quad [\Omega] \quad and \quad Z_M = \sqrt{\frac{j\omega\mu}{\sigma}} \quad [\Omega] \qquad (3), (4)$$

The parameter γ expresses the propagation constant of planar electromagnetic wave in the conductive environment of the shielding bulkhead [5]

$$\gamma = \sqrt{j\omega\mu\sigma} = (1 + j) \cdot \sqrt{\frac{\omega\mu\sigma}{2}} = a + j\beta \tag{5}$$

The relation (2) for the SE in the distant electromagnetic field can be rewritten to the form which better describes the physical principle of the shielding efficiency presented in Fig. 2 [1], [5]:

$$SE = R + A + M \qquad [dB] \tag{6}$$

where R represents attenuation caused by reflection, A describes the attenuation caused by absorbtion of part of the energy of the electromagnetic wave when passing through the shielding bulkhead, and M represents the multiple reflections that occur within the bulkhead. Reflection and absorbtion are most important for determining overall shielding efficiency, as can be seen in Fig. 2.

4. Verification of the shielding efficiency of the selected composite material

Composite materials are heterogeneous materials composed of two or more phases that differ significantly from each other by their mechanical, physical and chemical properties. The composite materials are produced by mechanical mixing of the individual components. This differs, for example, from alloys which are also heterogeneous [6]. Composite materials are characterized by synergism, which means that the properties of the composite material are higher than would correspond to mere proportional summation of the properties of the individual components. The existence of synergism is very important as it leads to the acquisition of new features. A typical example of synergistic behavior is a composite material composed of a ceramic matrix reinforced by ceramic fibers. Although both the matrix and the fibers are individually very fragile, the resulting composite material is characterized by a certain degree of toughness. The spreading fracture of the composite material is inhibited at the matrix and fiber interface. There is a deflection of the crack propagation direction and the intense mutual friction between the matrix and the pulling fibers. The quality of the interface between the matrix and the reinforcement therefore has a major effect on the properties of the resulting composite material [6], [11].

Composite material was selected from the three available samples of metallized fabric shown in Figure 3. The selection was made on the assumption that the sample would be used and would serve as a suitable shielding material [12]. This is a copper-plated polyamide fiber fabric labeled PBN-II-1.3-Cu-Sn (see Figure 3) by the Hungarian firm LORIX [8]. The metallized layer is very thin, usually around 0.1 μm.

Fig. 3. Available samples of composite material

Measurements of the sample were carried out on a co-axial line supporting Transverse electromagnetic mode (TEM). The inlet and outlet sections of the line connect an annular flange through the adapter. The adapter consists of two symmetrical parts. The body of the adapter, including the flange, was made of brass and the parameters were designed with respect to the 50 Ω input and output impedance of the line. The coaxial line description is given in [3] and [7]. Configuration of the assembly is shown in Figure 4.

The measuring assembly consists of the coaxial line, the SMR20 signal generator (in Figure 4, bottom right) and the spectrophotometer FSP40, both from Rohde & Schwarz. The measured sample was clamped into a stainless steel collar specially designed and manufactured to minimize unintentional signal flow from the generator to the analyzer. Nevertheless, it is necessary to consider the integrity of the measured sample of the fabric and hence the occurring gaps between the fibers. The whole measurement process has been automated using the laptop and Agilent VEE Pro 7.5 software. It is a graphical development environment with an open structure allowing to work with hardware and software from almost any manufacturer.

Measurement of shielding efficiency of the sample was performed over a frequency range of 1 to 10 GHz. The resulting graphical dependence is shown in Figure 5.

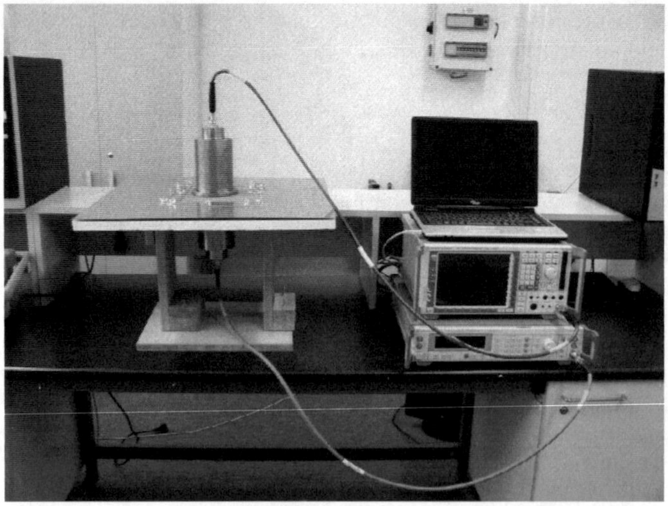

Fig. 4. Configuration of the measuring device

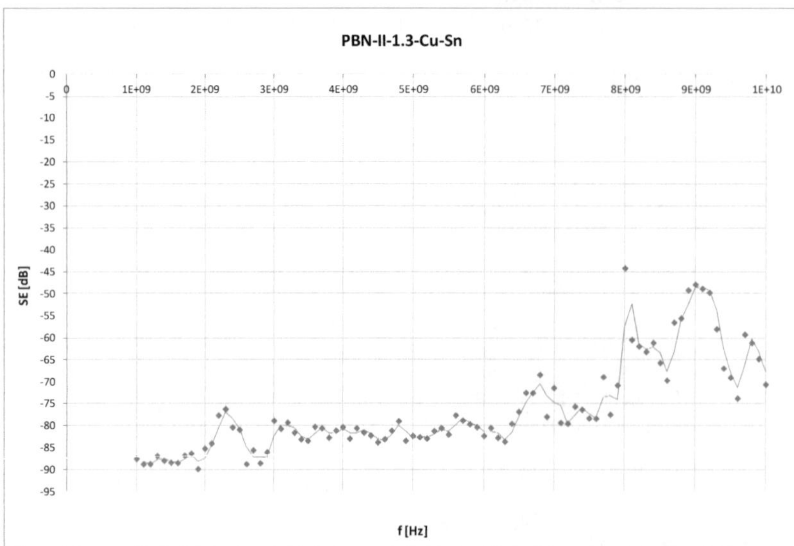

Fig. 5. Shielding efficiency of PBN-II-1.3-Cu-Sn

The comparison of the sample SE values supplied by the manufacturer and obtained by the measurement is performed only for 1 GHz frequency because the measurement was performed for higher frequencies than 1 GHz. It can be seen from Figure 5 that the measured SE at a given frequency is about -87 dB. From the data supplied by the manufacturer, the SE value for 1 GHz can be determined to about -75 dB [8]. The difference between the values is obvious. By verification, the SE of the given sample was found to be higher than the manufacturer states, which may not be a false result. On the contrary, it may be an indication of the real SE of the given sample. It can be assumed that the manufacturer gives the minimum guaranteed SE of the sample or the possibility that the sample may have a higher metallization thickness.

5. Determination of the percolation threshold of the composite material

The percolation threshold is an important parameter determining the conductivity of the composite material. It determines the critical content of the filler in which the composite begins to act as a conductor. Corresponding information on the conductivity of the composite material gives a curve, which is often referred to as percolation curve. It is the dependence of the electrical conductivity on the filler concentration and is usually of a sowing shape. At the inflection point of the curve, a critical concentration of the composite filler can be determined. In this point a conductive path is formed, percolation threshold occurs. The relationships between the conductivity and the concentration of the randomly dispersed filler particles are discussed in percolation theory [10], [13].

In the percolation threshold, the number of contacts between the filler particles rises. Further increase in the particle concentration results in a step increase of conductivity up to several orders, especially for low frequencies. If the concentration of particles is less than critical, the conductivity of the composite material changes slowly because the distance between the particles is large [13].

To theoretically determine the percolation threshold of the composite sample, it will be based on the simulation of the fabric SE, depending on the surface concentration of the fiber in its structure. The calculation will then determine the performance parameters of PWR_T transmission and PWR_R reflection. After plotting these parameters into graphical dependence on the fiber concentration in the fabric, the percolation threshold will be estimated. This problem is dealt with in publication of Youngs [4]. A jump change from 0 to 1 for PWR_T and from 1 to 0 for PWR_R will be sought.

5.1. Structure of the selected composite

For the simulation, it was necessary to determine the diameter of the metalized fiber used in the sample and to get an idea of the sample structure itself before the determination of the fiber concentration S_V of composite material required to achieve a certain shielding efficiency. For this purpose, a compact stereoscopic microscope Olympus SZX7 with expanded configuration needed for the analysis of the monitored object was used. A complete set of devices is shown in Figure 6.

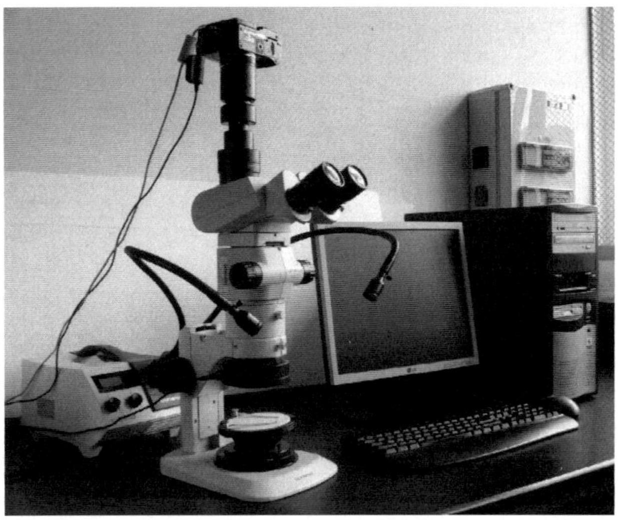

Fig. 6. Configuration of the microscope Olympus SZX7

The image in Figure 7 was used to determine the diameter of the metallised fiber of the fabric. As can be seen in the picture, these are long threads that are limited by sample size. Fibers are stored in several layers of irregular structure and thus their area concentration in the sample differs. From the measured values of the fiber diameters t, their mean value and the standard deviation 27,7 ± 0,7 μm were determined. The rounded value of fiber diameter (28 μm) was used for the following simulations.

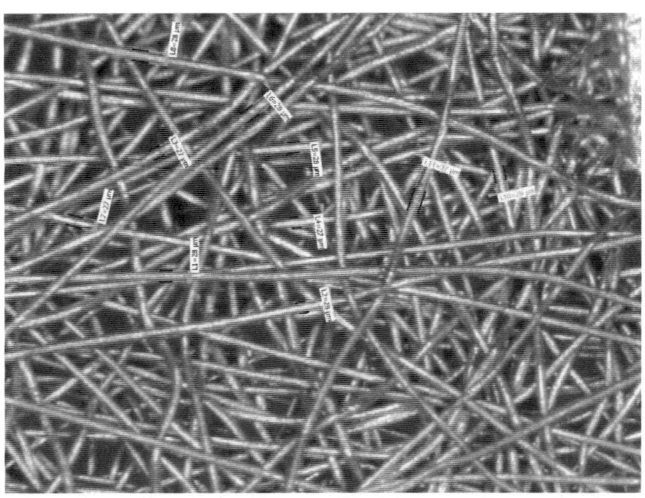

Fig. 7. Image of the fabric structure

5.2. Determination of S_V to achieve the required SE of the composite material

Determination of the fiber concentration S_V needed to achieve the specific shielding efficiency was done by simulation in the CST Microwave Studio. To simplify the simulations, only the wiregrid structure of the fabric model [2] was considered. It is not a composite structure with several layers of fibers but with only one layer of crossed fibers with a diameter $t = 28$ µm. Furthermore, the square-shaped fiber was considered and the results of the simulations were analyzed only for the 1 GHz frequency and the case of the perpendicular electromagnetic wave.

The S_V parameter was chosen within the range shown in table 1. It is assumed that the cell size parameter $a = b$.

$$S_V = \frac{2ta - t^2}{a^2} \quad [-] \tag{7}$$

$$a = \frac{t}{S_V} \cdot \left(1 + \sqrt{1 - S_V}\right) \quad [m] \tag{8}$$

After another parameterization of the sample and verification of all settings, an analysis was performed, from which the SE values for the desired frequency could be read. The SE values obtained from the simulations are shown in table 1. The graphical dependence of SE on the S_V fiber concentration in the model structure of the fabric is shown in Fig. 8.

S_V [-]	*a* [mm]	*a/t* [-]	*SE* [dB]	PWR_T [-]	PWR_R [-]
0,8	0,051	1,809	-118	1,585E-12	9,999999999984E-01
0,7	0,062	2,211	-102	6,310E-11	9,999999999369E-01
0,6	0,076	2,721	-91	7,943E-10	9,999999992057E-01
0,5	0,096	3,414	-81	7,943E-09	9,999999920570E-01
0,4	0,124	4,436	-73	5,012E-08	9,999999498800E-01
0,3	0,171	6,122	-65	3,162E-07	9,999996838000E-01
0,2	0,265	9,472	-58	1,585E-06	9,999984150000E-01
0,1	0,546	19,487	-45	3,162E-05	9,999683800000E-01
0,05	1,106	39,494	-36	2,512E-04	9,997488113568E-01
0,02	2,786	99,497	-25	3,162E-03	9,968380000000E-01
0,01	5,586	199,499	-18	1,585E-02	9,841500000000E-01
0,005	11,186	399,499	-16	2,512E-02	9,748800000000E-01
0,002	27,986	999,500	-10	1,000E-01	9,000000000000E-01
0,001	55,986	1999,500	-5	3,311E-01	6,689000000000E-01
0,0005	111,986	3999,500	-1	7,943E-01	2,057000000000E-01
0,0004	139,986	4999,500	-0,5	8,872E-01	1,128000000000E-01
0,0003	186,653	6666,167	-0,3	9,397E-01	0,603000000000E-01

Table 1. Values of SE from CST simulations depending on the S_V

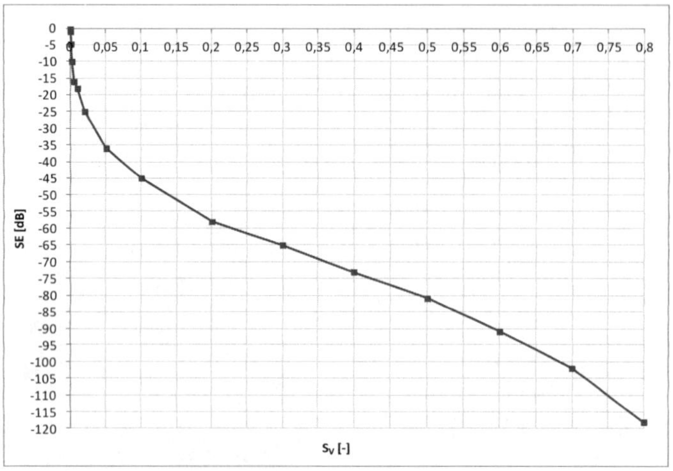

Fig. 8. Graphical representation of SE and S_V dependencies from table 1

The simulations were performed only up to $S_V = 0.8$, which means 80 % of the fiber content in the simulated fabric structure. Further simulations would not be relevant as the measurement approaches the noise level, which ranges from -113 to -130 dB according to the measurement method.

5.3. Calculation of transmission and reflection performance parameters

The electromagnetic wave falls on the surface of the fabric sample of a given thickness t. Part of the energy is reflected (PWR_R) from the surface, part of the energy is absorbed (PWR_D) and the remaining energy passes through (PWR_T) [9]. In general [1]:

$$1 = PWR_R + PWR_D + PWR_T \quad [-] \tag{9}$$

The assumption for the above problem is the loss-free case where $PWR_D = 0$. Then PWR_T and PWR_R are [1]:

$$PWR_T = |T|^2 \; [-] \quad and \quad PWR_R = |R|^2 \; [-] \tag{10}$$

where

$$|T| = 10^{\frac{-SE}{20}} \; [-] \quad and \quad |R|^2 = 1 - |T|^2 \; [-] \tag{11}$$

The resulting PWR_T and PWR_R values obtained from the SE simulations are shown in tab. 1.

5.4. Determining the percolation threshold of the composite model

A jumping change from 0 to 1 for PWR_T and 1 to 0 for PWR_R is sought. The S_V value at which these large jump changes occur can be considered as the assumed percolation threshold. The significance of this value is mainly in the field of electrically conductive materials for antistatic purposes, protection materials for electronics and electrotechnics.

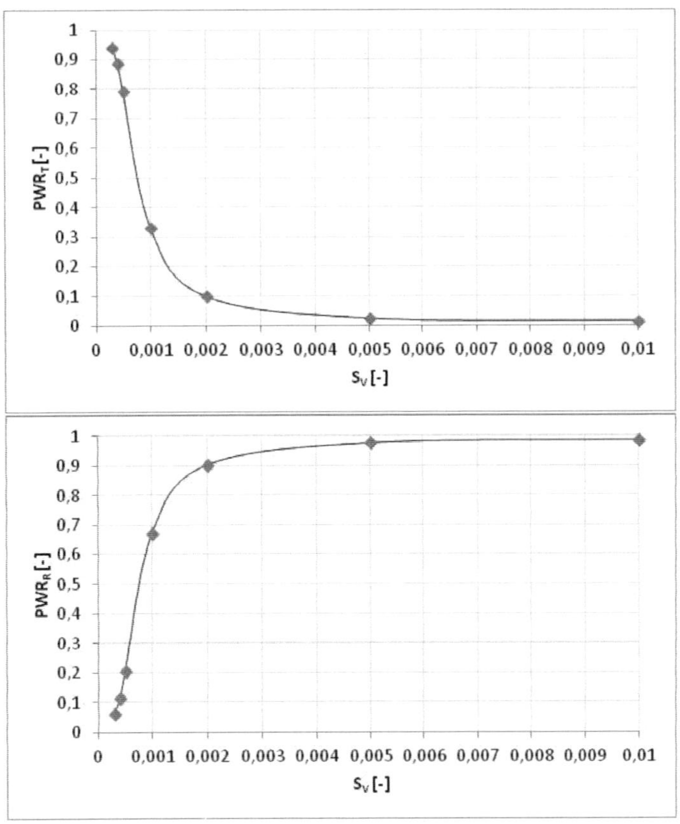

Fig. 9. Dependency of PWR_T and PWR_R on S_V

6. Conclusion

A metallized fabric from LORIX was chosen as a suitable composite material. This is a fabric with long copper-plated polyamide fibers designated by PBN-II-1.3-Cu-Sn. In order to verify its shielding efficiency, measurements were made on coaxial lines for a frequency range of 1 to 10 GHz. Comparison of the shielding effects of the measured sample supplied by the manufacturer and obtained by the measurements was performed only for the 1 GHz frequency due to the known manufacturer's data for this sample. The shielding efficiency was measured at a given 1 GHz frequency of -87 dB. The manufacturer supplied a lower value of -75 dB. It does not need to be a false result. It may be an indication of the real shielding efficiency of the sample. It can be assumed that the manufacturer supplied the minimum guaranteed shielding efficiency. Furthermore, for the theoretical determination of the percolation threshold of the composite sample, the simulation of the shielding effect of the fabric was based on the surface concentration of the fiber in its structure. Simulations were performed in a CST Microwave Studio to 80 % fiber concentration in a simulated fabric structure. At this value, the fabric achieves nearly -120 dB, so further simulations leading in this direction would not be significant in the case of real measurements, given the dynamic range of the device used. Transmission and reflection parameters were determined by calculation. Determining the percolation threshold was made from the plot of the transmission and/or reflection parameters depending on the surface fiber concentration in the fabric. A jumping change from 0 to 1 for PWR_T and 1 to 0 for PWR_R was sought. The percolation threshold was determined to 0.03 % of the fiber content in the simulated model fabric sample.

The suitability of the method used to determine the percolation threshold will be discussed in future work which will focus on determining the percolation threshold from the AC conductivity dependence to the concentration of the metallic fibers of the composite material. The selected frequency range will be determined in correlation of usage of a composite material.

7. References

[1] Paul, C. R. (2006). Introduction to Electromagnetic Compatibility. New Jersey: John Wiley & Sons, Inc., 2006. ISBN 0-471-75500-1.

[2] Volski, V.; Aerts, W.; Vasylchenko, O. & Vandenbosch, G. (2006). Analysis of Composite Textiles Filled with Arbitrarily Oriented Conducting Fibers using a Periodic Model for Crossed Strips. Proceedings of International Conference on Mathematical Methods in Electromagnetic Theory, Ukraine, Kharkiv, 2006, pp. 58-63. ISBN 1-4244-0490-8.

[3] Wieckowski, T. W. & Janukiewicz, J. M. (2006). Methods for Evaluating the Shielding Effectiveness of Textiles. Fibres & Textiles in Eastern, 2006, Vol. 14, No. 5, pp. 18-22. ISSN 1230-3666.

[4] Youngs, I. J. (2007). The Shielding Effectiveness of Composite Media – a Parametric Analysis Supporting an Engineering Perspective. IEEE Transactions on Dielectrics and Electrical Insulation, 2007, Vol. 14, No. 5, pp. 1145-1153. ISSN 1070-9878.

[5] Svačina, J. (2002). Electromagnetic Compatibility: Lectures. Czech Republic, Brno: Brno University of Technology, 2002. ISBN 80-214-1573-8.

[6] Michna, Š. & al. (2007). Aluminium Materials and Technologies from A to Z. Slovakia, Prešov: Adin, s.r.o., 2007. ISBN 978-80-89244-18-8.

[7] Dřínovský, J.; Kejík, Z. (2009). Electromagnetic Shielding Efficiency Measurement of Composite Materials. Measurement Science Review, 2009, Vol. 9, No. 4, pp. 109-112.

[8] Lorix Ltd. (2011). Metallised textile. Available from: http://www.lorix.shp.hu/hpc/web.php?a=lorix&o=950XGqatYS: 2011-12-19.

[9] Vojtěch, L. a Hájek, J. (2010). Measurement of Electromagnetic Shielding Efficiency of Planar Textiles in Frequency Range 100 kHz - 1,5 GHz. Access Server, 2010. ISSN 1214-9675.

[10] Vilčáková, J. (2007). Electrical and Magnetic Properties of Polymer Composites: Habilitation thesis summary. Czech Republic, Zlín: Tomas Bata University in Zlín, 2007. ISBN 978-80-7318-519-0.

[11] Michna, Š. (2008). Composite Materials. Available from: http://www.stefanmichna.com: 2011-11-18.

[12] Skočík, P.; Goňa, S. & Křesálek, V. (2011). Fullwave Modelling of Shielding Effectiveness of Metalized Textiles, Annals of DAAAM for 2011 & Proceedings of the 22nd International DAAAM Symposium, 23-26th November 2011, Vienna, Austria, Volume 22, No. 1, ISSN 1726-9679, ISBN 978-3-901509-83-4, Katalinic, B. (Ed.), pp. 0101-0102, Published by DAAAM International Vienna, Vienna.

[13] Římská, Z.; Křesálek, V. & Špaček, J. (2002). AC Conductivity of Carbon Fiber - Polymer Matrix. Polymer Composites, 2002, Vol. 23, No. 1, pp. 95-103. ISSN 0272-8397.

DOI: 10.2507/28th.daaam.proceedings.008

MATHEMATICAL MODEL OF WASHING OF FILTRATION CAKE

Dagmar Janacova, Pavel Mokrejs, Vladislav Sviatskii,
Vladimir Vasek, Ondrej Liska & Karel Kolomaznik

This Publication has to be referred as: Janacova, D[agmar]; Mokrejs, P[avel]; Sviatskii, V[ladislav]; Vasek, V[ladimir]; Liska, O[ndrej] & Kolomaznik, K[arel] (2017). Mathematical Model of Washing of Filtration Cake, Proceedings of the 28th DAAAM International Symposium, pp.0063-0069, B. Katalinic (Ed.), Published by DAAAM International, ISBN 978-3-902734-11-2, ISSN 1726-9679, Vienna, Austria
DOI: 10.2507/28th.daaam.proceedings.008

Abstract

The issue of this article is one stage washing of reaction blend after enzyme hydrolysis under which we are able to recycle almost all contained chromium in solid leather waste. During hydrolysis we gain protein hydrolysate as a filtrate that is appropriate substance for use. On the basis of equation for streaming of liquids through layer of substance we determined optimal pressure differential for filtering and washing of reaction mix and we applied mathematical description valid for filtering and one stage washing and on the basis of this models we able to optimize washing process that is add to filtering and which aim is to reduce content of protein hydrolysate in filter cake.

Keywords: filtration process; one stage washing; optimization; filtration cake; protein hydrolysate

1. Introduction

In leather industry are created more than five thousand tone chrome tanning waste per year and we can suppose that its amount is going to grow up. The waste is removed to waste dump very often. There chromium poses a threat to ground water because it can escape there. Chromium can be oxidized to hexavalent chromium, its compounds are carcinogenic. It is more than evident that ecological treatment of chrome tanning parings is urgent. Many methods of gaining chromium from chrome tanning parings were conceived.

As the most convenient way is to deal with the waste by enzymatic hydrolysis [1] and during this we are able to recycle almost all used chromium. During hydrolysis we obtain protein hydrolysate as a filtrate. To reduce of the protein hydrolysate content we included washing of the filter cake. Enzyme hydrolysis is environmentally friendly. During this process we fully uses chromium cake and high-quality protein hydrolysate for wide using.

2. Washing of filter cake

The product of filter process is filter cake which contains residual amount of filtrate and other substances that are adverse. Washing is the process during it we want to remove a maximum of this substance. During optimization of washing process is important to know whether the washing component belongs to solid phase or not. We must establish sorption isotherm. Common sorption isotherm is depicted in the picture 1.

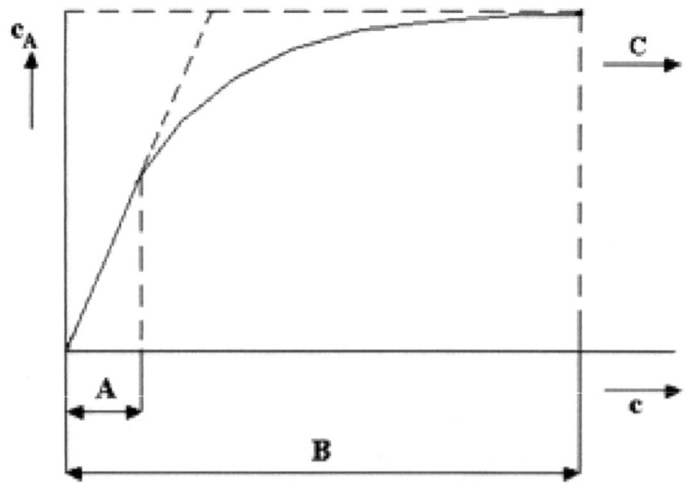

Fig. 1. Langmuir´s sorption isotherm

Essential for other process of rationalizing washing process is finding in which part of sorption isotherm we can find the condition of washing substance. According to the picture 1 is the condition "C" or "B". Around "C" is washing substance free. Around "B" is washing substance connected to solid phase. In this area we can introduce zone "A" in which is sorption addiction linear [2].

We divide the washing process in two basic categories:

- Washing of untied constituent: washing liquid is usually water.
- Washing of tied constituent: washing liquid is substance in which the tied constituent melts (e.g. washing MgO diluted by solution of sulphuric acid.)

According to technical process we distinguish bath washing. Success rate when is from the cake removed washout substance is the dependence of washing process level y on soaking number Na [8]:

$$y = \frac{c_0 V_0}{c_p V} = C_0 \cdot Na \tag{1}$$

where: y *is* ratio of removed part of the substance to its whole amount that was in the cake before washing, The value is in interval 0-1, Na is *r*atio of washing liquid volume to solid phase volume, this characterizes usage of washing liquid, c_o is concentration of the scrubbed components in the bath, c_p is the initial concentration of the scrubbed components in the solid phase (filtration cake), C_o is dimensionless concentration of the scrubbed components in the bath.

3. Mathematic model of filter cake washing

During this way of washing is filter cake removed from filter apparatus and mixed with such amount of washing liquid to get lower then critical concentration of washing substance. This solution is then filtrated.
Special case of bath washing is decanting washing [5, 7]. During bath washing is the whole amount of washing liquid is used/consumed during one process/operation. Decanting washing means that we use washing liquid in several operations. Bath washing is made several time with lower capacity of washing liquid so long when the figure of concentration of washing constituent is lower than critical figure.

This method is used in case when we wash from cake constituents tied and untied because every type of constituents is washed by different washing liquid [1, 9]. Mathematical model is based on equation which describes one-dimensional concentration field of washing constituent in solid phase.

$$D\frac{d^2c}{dx^2} = \frac{dc}{dt}(x,t) \qquad t > 0, \quad 0 \le x \le b \tag{2}$$

and from conditions for this equation: Concentration field in solid phase is symmetrical:

$$\frac{dc}{dx}(0,t)=0 \tag{3}$$

Bath is perfectly mixed:

$$c(b,t)=\varepsilon.c_0(t) \tag{4}$$

The speed of substance of washing constituent on the surface of solid phase is equal to accumulation of this substance in the bath:

$$-DS\frac{dc}{dx}(b,t)=V_0\frac{dc_0}{dt}(t) \tag{5}$$

At the beginning of washing is constant difference of concentration in filter cake:

$$c(x,0)=c_p \tag{6}$$

In washing liquid is zero concentration of washing constituent:

$$c_0(0)=0 \tag{7}$$

For general expression we introduce dimensionless quantities:

$$C=\frac{c}{c_p},\quad C_0=\frac{c_0}{c_p},\quad F_0=\frac{Dt}{b^2},\quad X=\frac{x}{b},\quad Na=\frac{V_0}{V} \qquad 8\ (a,b,c,d,e)$$

where: V is je capacity of solid phase.

The result of equation (2) under the conditions (3-7) we accomplished by Laplace transformation [2, 8] and then we get: Dimensionless concentration field in solid phase:

$$C(X,F_0)=\frac{\varepsilon}{Na+\varepsilon}-2Na\sum_{n=1}^{\infty}\frac{\cos(Xq_n).e^{-F_0q_n^2}}{\cos(q_n)[Na+\varepsilon]-q_nNa.\sin(q_n)} \tag{9}$$

Time course of concentration of washing constituent in filtrate

$$C_0(F_0)=\frac{1}{Na+\varepsilon}-\frac{2Na}{\varepsilon}\sum_{n=1}^{\infty}\frac{e^{-F_0q_n^{\ 2}}}{\varepsilon+Na+\dfrac{Na^2q_n^{\ 2}}{\varepsilon}} \tag{10}$$

Degree of washing y we determined wih using of equation (1) [3, 7].

4. Filter and wash apparatus

Filter apparatus where I do all my measurements is formed for heated filter funnel and filter receptacle. Vacuum pump is connected to filter receptacle that enables creation of under pressure behind filter divider.

Converters for capacity sensor of water-level, resistance thermometer Pt100 and pressure sensor are together with regulator connected to temperature sensor situated in converter and regulatory cabinet.

Connection of regulatory and converter cabinet to computer enables well arranged monitoring of filter process and archiving of measured data [3, 4].

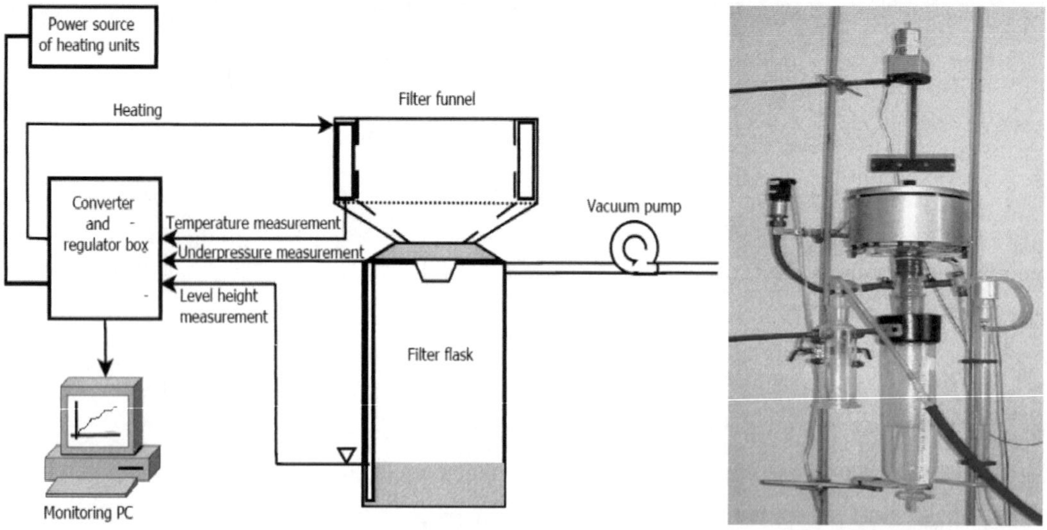

Fig. 2. Filter apparatus [3]

5. Experimental part of washing

Filtration cake, which was created from filtration of chromium sludge, was washed. This washing was composed of two phases:

1. Washing of the cake by water in the purpose of removing residual proteins.

2. The cake that was washed was poured by distilled water and was carried out the modification of it pH by sulphuric acid. This sludge was put to filter funnel and filtered during pouring of distilled water. The aim was to remove residual magnesium oxide from filter cake. From particular filtered cakes and filtrates we removed samples for later analysis which was done in work [1, 5].

Experiment No.1:

The cake was situated in heated filter funnel where was constantly poured distilled water. During stirring the cake was firstly washed by water, then the pH was changed to 6 by sulphuric acid and then again washed by water. Attributes of entering cake (FK1), cake after washing by water (FK2) and final cake after washing with changed pH (FK3) are depicted in Table 1.

Sample		Weight m	Dry part	Ash	Content of C	Content of N	Content of Cr	Content of Mg
FK 1	[% / dry]	-	18,31	28,37	29,74	10,04	5,11	5,05
	[g]							
FK 2	[% / dry]	-	12,37	41,63	21,2		9,87	8,84
	[g]	840	103,91	43,26	22,03		10,25	9,18
FK3	[% / dry	-	17,1	34,77	24,47	8,14	13,2	3,63
	[g]	450	76,95	26,75	18,83	6,26	10,15	2,79

Table 1. Attributes of filter cake No.1

Entering cake FK1 was not weighted therefore it is not able exactly balanced washing of particular parts. (its weight was approximately same as during following attempt: around 1000 - 1150 g)

200ml samples of filtration were taken. After taking first 10 samples during washing by water, the filter cake was removed to beaker, it was weighted, poured by distilled water, and then the change of pH to 6 by sulphuric acid was carried out (dilute 1:10). Sludge was again removed to filter funnel and distilled water was poured. 10 samples of 200ml filtration were taken.

Filter funnel was heated to the temperature 55°C during washing by water, the temperature of the cake was not measured. When water started to flow in the cake, some amount of water flow in to sampling receptacle therefore there is lower concentration of washed substances in the first sampling filtrate. We stirred the cake because otherwise the cake started to crack and started to separate from walls of filter funnel. During the second part of washing, when diluted sulhpuric acid was added to the cake, the filter funnel was not heated.

Equivalent amount of sulphuric acid was not added to the cake so therefore an amount of magnesium was still in the cake. We found out appropriate conditions during this experiment (sampling, regulation of flow and temperature). Experiences were used during following experiment.

5.1 Experiment No.2:

During second experiment samples of filtrate cumulated with the amount of 500ml and we took 6 in each period of experiment. The process was the same as during previous case. The temperature of cake in filter funnel was 40°C the same as during washing by water and washing after changing pH. Characteristic of filter cake is described in Table 2

Sample		**Weight m**	**Dry part**	**Ash**	**Content of C**	**Content of N**	**Content of Cr**	**Content of Mg**
FK 1	[% / dry]	-	19,4	26,52	29,64	9,58	5,45	6,02
	[g]	970	188,18	49,91	55,77	18,03	10,25	11,32
FK 2	[% / dry]	-	15,69	28,37	27,95	-	7,84	7,24
	[g]	810	127,09	36,05	37,72	-	9,96	10,3
FK3	[% / dry	-	17,86	19,65	32,94	13,99	8,86	1,38
	[g]	585	104,48	20,53	36,87	14,62	9,26	2,39

Table 2. Attributes of filter cake No.2

5.2 Evaluation of measured data

For evaluation of process for washing the filter cake we used data from experiment No.2 because during the first try the entering cake was not weighted. Therefore we could not precisely evaluate it. The filter cake was weighted only at the beginning and in the end of the process therefore figures for particular parts of the process are not measured. We used average figures. Inaccuracy is not great.

The ratio of filtration volume to solid phase volume we got the figure. The ratio of washing substances to its original amount we defined a degree of washing y. In each chart we have graph dependence *y* on *Na* (dependence washing quality on output of washing liquid).

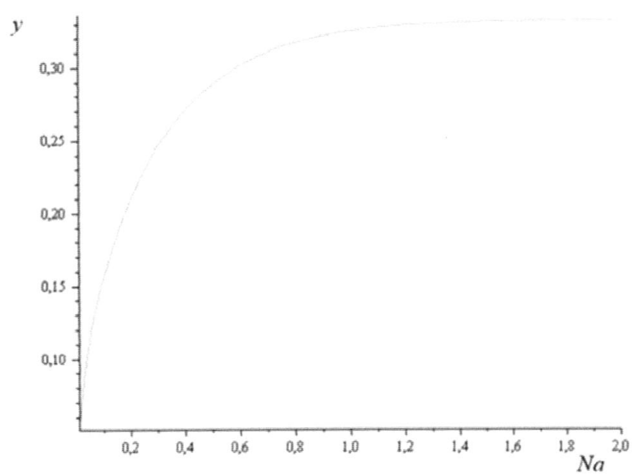

Fig. 3. Dependence y on *Na* during washing of protein by water.

6. Optimization of filtration cake washing

For washing required amount of undesired substance from filter cake, it is necessary particular amount of washing liquid and energy. In this case it was energy for heating the filter funnel. To have the lowest cost for washing the filter cake, we tried to find minimal volume of washing liquid, that will be sufficient for washing required amount of given substance. The amount of washing substance is characterized by washing level *y* [3, 6]. When we want to find out minimum volume we chose appropriate soaking numbers during which we get required degree of washing and then we get optimal volume. The volume which is equal to equation minimum is optimum volume and total costs are optimal costs needed for washing required amount of undesirable substance [10].

To wash certain amount of undesirable substance from filter cake, it is necessary a certain amount of washing liquid and also a certain amount of energy. In our case it was the energy for heating the filter funnel.

To have the cost of the washing of the filter cake as low as possible, we try to find a minimum volume of washing liquid, which will be enough for washing the desired amount of the substance. The amount of eluted substances is characterized by the degree of washing y.

When we search for the minimum volume we choose the appropriate soaking numbers Na, when will be achieved the desired degree of washing, and subsequently we determine the optimal volume of the cost function. The volume corresponding to the minimum function is the searched optimal volume and the total costs corresponding to this minimum are the optimal cost, necessary to wash the desired amount of undesirable substance.

7. Conclusion

We carried out filtration and we washed several reaction mixes formed after the enzyme hydrolysis of chrome tanning parings and hydrolyzed bowels under defined conditions. With the help of existing mathematical models, we have expressed the specific resistances of filter cakes, which came into existence by filtration of the aforementioned suspensions. From the dependences of these resistances on developing under pressure, we set the compressible and incompressible cake and on the basis of these findings we have set the appropriate parameters of filtration.

On the basis of analyses of samples taken during the washing of the filter cake, we then determined the dependence of washout quality on the consumption of washout liquids for eluting the bound and unbound components. For eluting protein from the filter cake, we calculated optimal figure of washing liquid volume during which are minimal costs for washing required amount of undesirable component. Enzyme hydrolysis is environmentally friendly, during this process we fully uses chromium cake and high-quality protein hydrolysate.

In the next phase of our research, we will also focus on filter cake washing after enzymatic hydrolysis of other biowaste - for example, waste feathers to increase the yield of valuable protein.

8. List of symbols

Symbol	Meaning	Unit
b	Half thickness of washed solid raw hide	[m]
c	Concentration of unbound component in solid phase	[$kg.m^{-3}$]
C	Non-dimensional concentration field of washed component in solid phase	[1]
c_0	Concentration of component in washing bath	[$kg.m^{-3}$]
C_0	Non-dimensional concentration of the component in bath	[1]
c_p	Initial concentration of component in solid phase	[$kg.m^{-3}$]
$c(t)$	Concentration of salt that by diffusion goes into the soaking liquid	[$kg.m^{-3}$]
c_∞	Concentration of salt that by diffusion goes to soaking liquid in infinite time	[$kg.m^{-3}$]
D	Effective value of diffusion coefficient	[$m^2\ s^{-1}$]
ε	Porosity of solid phase	[1]
F_0	Non-dimensional time	[1]
Na	Soak number, non-dimensional consumption of liquid	[1]
q_n	Roots of transcendental equation	[1]
S	Surface of raw hide	[m^2]
t	Time	[s]
V	Volume of the solid phase	[m^{-3}]
V_0	Volume of the washing bath	[m^{-3}]
x	Area coordinates	[m]
X	Non-dimensional distance	[1]
y	Washing process stage	[1]

9. Acknowledgements

This work was supported by the Ministry of Education, Youth and Sports of the Czech Republic within the National Sustainability Programme Project No. LO1303 (MSMT-7778/2014) and also by the European Regional Development Fund under the project CEBIA-Tech No. CZ.1.05/2.1.00/03.0089

10. References

[1] Kolomazník, K. (1996). Recovery and recycling of chromium and protein from liquid and solid wastes from leather industry, ERB 3512 PL 940974 (COP 974).

[2] Janacova, D., Kolomaznik, K. & Vasek, V. (2005). Optimization of liquids consumption at washing. 6th International Carpathian Control Conference. Miskolc-Lilafuered, Hungary, ISBN 963-661-644-2.

[3] Vasek, V., Kolomaznik, K. & Janacova, D. (2006). Optimization and Automatic Control of Chromium Recycling Technology, WSEAS Transactions on Systems, 5 (11), ISSN 2651-2654.

[4] Sunaoka, Y., Kitazato, K., & Tsuda, S. (1992). U.S. Patent No. 5,151,191. Washington, DC: U.S. Patent and Trademark Office.

[5] Tsubaki, Y., Kitajima, K., Hidetoshi, I., Hayashi, S., Ueda, A., Yagami, K., & Yamada, S. (1991). U.S. Patent No. 5,050,259. Washington, DC: U.S. Patent and Trademark Office.

[6] Vasek, L. & Dolinay V. (2010). International Journal of Mathematical Models and Methods in Applied Sciences, Vol. 4 (4), p. 240-248.

[7] Pitel, J., & Mižák, J. (2013). Computational intelligence and low cost sensors in biomass combustion process. In Computational Intelligence in Control and Automation (CICA), 2013 IEEE Symposium on (pp. 181-184). IEEE.

[8] Božek, P. (2013). In Advances in Intelligent Systems and Computing: Proceedings of the International Conference on Systems Science, ICSS 2013; Wroclaw; Poland; p. 229-240. ISSN 2194-5357.

[9] Janacova, D., Kolomaznik, K., Mokrejs, P., Vašek, V. & Líška, O. (2015). Optimization of one-stage extraction process. In *Recent Advances in Systems*. New Jersey, Piscataway: IEEE, p. 88-91. ISSN 1790-5117. ISBN 978-1-61804-321-4.

[10] Sambaer, W., Zatloukal, M., & Kimmer, D. (2011). 3D modeling of filtration process via polyurethane nanofiber based nonwoven filters prepared by electrospinning process. Chemical Engineering Science, 66(4), 613-623.

DOI: 10.2507/28th.daaam.proceedings.009

NEW COMPUTER GAME IN MATLAB FOR EDUCATIONAL PURPOSES

Karel Perutka & Lenka Sarmanova

This Publication has to be referred as: Perutka, K[arel] & Sarmanova, L[enka] (2017). New Computer Game in MATLAB for Educational Purposes, Proceedings of the 28th DAAAM International Symposium, pp.0070-0078, B. Katalinic (Ed.), Published by DAAAM International, ISBN 978-3-902734-11-2, ISSN 1726-9679, Vienna, Austria
DOI: 10.2507/28th.daaam.proceedings.009

Abstract

This paper deals with the development environment for 2D computer game in the MATLAB. The theoretical part briefly describes the MATLAB environment and several games developed in this program, which are available on the MATLAB Central. The rest of the paper, the main part, is oriented practically, it contains the description of the game design. The game description is made in two versions, in a programmer and user perspective. The main part includes chapters about the game specification, game realization and the game from the user´s point of view. The game is de facto original set of mazes and going through the maze is done after the right answer of question. The database of questions was created in purpose to practice the theoretical knowledge of MATLAB environment.

Keywords: MATLAB; GUI; Programming; Game

1. Introduction

Education in the sphere of scientific disciplines is a foundation in training of experts for high-tech industries. There are many tools and programs that have been developer and they can be used for educational purposes. It was followed by the usage of e-learning. After that, the educational tools were created. Teaching and practice using specific computer games became common during very last years.

Common education using lectures was presented for example by Goodwin. The lecture is an educational tool that allows a teacher to cover topic of a course that is to follow. It is a means of imparting knowledge by telling an audience (which can be of considerable size) a mixture of facts, opinions and interpretations [14]. Educational videos became the effective tool to improve training and education, for example in the medical area [17].

We can also mention some selected practical tool created in purpose to make the educational process much easier, such as graphical- based educational simulation tool called Gbest-WSN for simulating routing protocols of the static and mobile, homogeneous and heterogeneous WSNs [18]. Or, there was presented the tool that provides an easy and intuitive remote interaction with a team of mobile robots through Internet. It was designed with the aim that students of robotics carry out training with mobile robots from their own home, independently of the operating system they use and with a minimum installation in their computer [9]. RobotiCad is another example of such tool. It is a user-friendly Matlab/Simulink toolbox for the modeling and simulation of robotic manipulators [10]. There were also created the newly developed educational tools to encourage high school students' interest in STEM (Science, Technology, Engineering, Mathematics) disciplines. These tools include laboratory courses developed in the fields of physics, information technologies and mathematics. These activities were aimed at increasing the quality of STEM disciplines learning which resulted in higher quality of training of future engineers [11]. Next example of created educational tools is from the area of the artificial intelligence (AI). Artificial neural networks are some kind of data processing systems, which try to simulate features of the human brain and its learning process. So, they are widely used by researchers to solve different problems in optimization, classification, pattern recognition, associative memory and control. There was created the educational tool, which can be used to work on different kinds of neural network models and learn fundamentals of the artificial neural network [15]. Next mentioned example is from the area of uncertainty measurements. Uncertainty of measurements and parameters is frequently neglected by practitioners in the design of systems even in safety critical applications. Although there exists numerous tools to support uncertainty calculation, reasons for limited usage in early design phases may be low awareness of the existence of the tools and insufficient training in the practical application. Therefore, the teaching concept for uncertainty calculation in measurement science education was created. It is directly based on the utilization of software tools [16].

The games are the important aspect of the modern education. There was also suggested that online multiplayer educational games should be approached as a complex learning system, based on the principles of activity theory, where the Subjects would interact with other Subjects, Objects and Tools of the game, under specified Rules and create Communities through division of labor, leading to the expected learning outcome. Thus, there was suggest taking into account some important issues concerning the Subjects that the activity theory refers to, such as gender differences in playing games, academic performance, self-esteem and computer self-efficacy [12].

Augmented reality plays more important role in the educational process. The implementation of Mobile Pedestrian Navigation and Augmented Reality in mobile learning contexts shows new forms of interaction when students are taught by means of learning activities in formal settings. There was presented the educational, quantitative, and qualitative evaluation of an Augmented Reality and Mobile Pedestrian Navigation application. The software was designed for mobile learning in an educational context, to evaluate its effectiveness when applied as a teaching tool, in comparison to similar tools such as e-learning [13].

The name MATLAB originated from two English words, matrix and laboratory, which means "matrix and lab". MATLAB is the program environment that allows us to count a variety of mathematical operations, it can handle with vectors and matrices, plot charts, and create user interface applications. MATLAB is an advanced technical computing language and environment, which includes many tools to develop algorithms and perform data analysis and complex numeric computations. MATLAB has been chosen due to the high usage in education and research as well as in industry. There were formulated the following questions: Are existing tools suitable and do they include all topics we teach at our university? Is it necessary to create new teaching aid or game?

Based on the previous wide research in the area of modern teaching methods and tools, the necessity of new educational tool to practice MATLAB knowledge in attractive form appeared. The purpose of the study was to create supporting document for the practical realization. After that, the new computer game practicing the knowledge of MATLAB was created.

The paper is organized as it is followed. Firstly, the brief study of created MATLAB games was created. This is followed by the new game description. The advantages of created game are evaluated in the conclusion.

2. Brief study of created games in MATLAB

2.1. Computer Games in MATLAB – MATLAB Central

MATLAB Central is the main source of created projects that MATLAB users share. The games from the MATLAB Central servers are freely downloadable with permission of the authors. There are various categories of games such as logical, casual, shooting and 3D games.

2048

This game had become very popular in recent years and is the expansion of 1024 with the same rules. User @bmtran transferred it to MATLAB and published it on MATLAB Central.

The game is controlled using the arrow keys on the keyboard. The goal of the game is to combine the same numbers together and reach the final 2048 value. Pressing the arrow on the keypad specifies the direction that all the tiles come out in the 4x4 field until they are stopped by another tile or edge.

Tiles of the same value are moved together and added together. At each turn or move in a field, a new tile with a minimum value of 2 appears, so if the player does not choose a good strategy, the whole field can be filled with tiles with different values that cannot be combined and the player cannot move in such a situation. The game is downloaded directly as an application, so its code cannot be seen. [1]

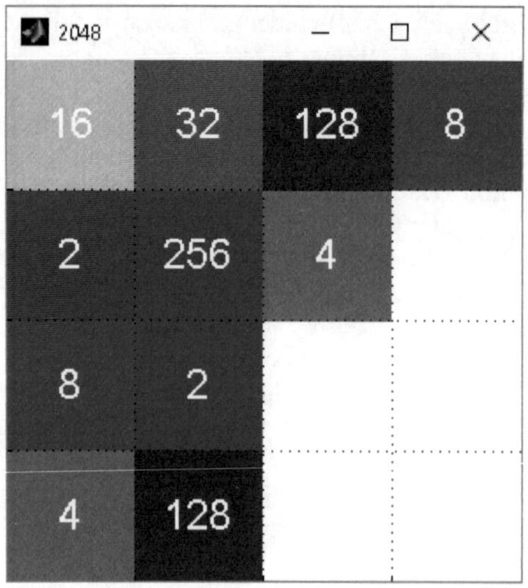

Fig. 1. 2048 game

Sudoku

The author of this famous MATLAB game is Stepen Sahrun. The goal of Sudoku is to fill in the empty spaces in the 9x9 field using numbers 1-9, so that in each row, column, and square the numbers 1-9 are not repeated. There is a choice of three difficulty options - light, medium or hard. There is also the option to reset the play - run it from the beginning, or run the solver to solve the game for us. Once the game has been completed, we have the option to run a game review to evaluate whether the field is filled correctly and, if not, to highlight the error boxes. Enter the numbers in each field using the numeric keypad. We use the mouse to select the game difficulty or reset, check, or running the solver.

The program contains two M-files, one for the game and the other for the sudoku solver. Together, both files have about 850 rows of code. [2]

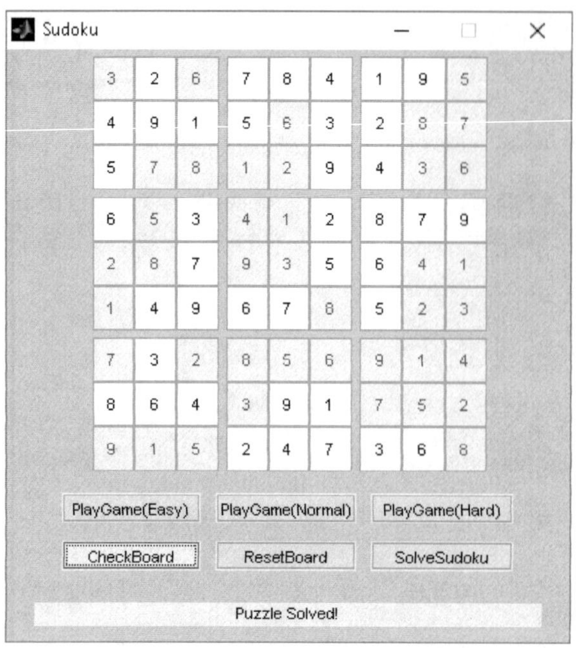

Fig. 2. Sudoku game

Matris

The Matris game is the similar game to the well-known Tetris game created by the MATLAB program, written by Mingjing Zhang. Matris is controlled using the mouse and keys. To start the game, click the Start button. The shapes fall randomly from the database, and the player's task is to fold the falling shapes side by side to fill the line and then disappear.

With shapes, you can move left and right with the left and right arrows, use the up arrow to rotate the shape, and use the down arrow to speed it down to the ground. In the top right corner, a shape is displayed, which starts in the field as the next one, which allows the player to tact forward. In the right part of the window we also see information about the results of the game, namely our score and the number of disappeared rows.

With the increasing score of the game, the rate of crash of individual shapes increases, making the game more difficult. If a player fails to arrange shapes and some of the shapes touch the upper axis, the game ends. The game can also be paused at any time. The program contains one M-file with approximately 850 lines of code. The game has very nice graphics. [3]

Fig. 3. Matris game

2.2. *Computer Games in MATLAB – Games created by students*

Ludo

The author of this classic board game is David Fiala, who created in 2013/2014 it as his thesis [6] on Mathematics 2D Computer Game.

The game is for 2-4 players, with players being either real people or a computer. The aim of the game is not to go through the player four pieces until the target field is determined by the same color as the player's pieces before the other players. Individual players alternate in a dice roll, the throw result indicating the number of shift boxes. If there is not yet one of our gaming figures in the field, but it's in the house, we have the option of 3x throwing a dice until we hit number 6, so we can put a piece from the house into the field.

If we have more than one piece in the field, we can select the number of pieces that will be moved by the number of squares given by the throw. If our figurine appears on a field where another player's figure is already in place, the opponent's figure must be returned to the house.

You can stop the game at any time, save it and run it again. We can also look into the rules during the game. The author has worked on creating the graphics of the game, so the game is graphically very good. In MATLAB CommandWindow, the game commentaries appear throughout the game, and on the right of the game window, we have an overview of the current state of the game, specifically which player is on the move. The program contains about 50 M-files, most of which are scripts by name, clearly defining functionality.

Fig. 4. Ludo game – first screen (left) and main dialog (right)

Risk

In 2005/2006, Michal Heczko, as a part of his work [5], he created the Tutorial and Testing Program for the PPAR, so called as Risk, and as the application aimed for MATLAB practicing.

The game is for 1-3 players. As the title suggests, the rules of the game are the same as the Risk television entertainment program. Specifically, in this game, the themes are focused on describing the MATLAB environment, operators, string operations, complex numbers, arrays, and matrices. The game is controlled by mouse and keyboard. The player on the move selects the topic and value of the question he / she wants to answer. If the answer is correct, it is added to the score. The game has a time limit for both the player's response and the lap length, both of which are represented by the progress bar. After the timeout, a player with more points wins.

The game itself contains about 40 M-files and 5 files with FIG. Since the entire app is a learning program, this game is de facto nice application for users who can verify their MATLAB knowledge in a funny way.

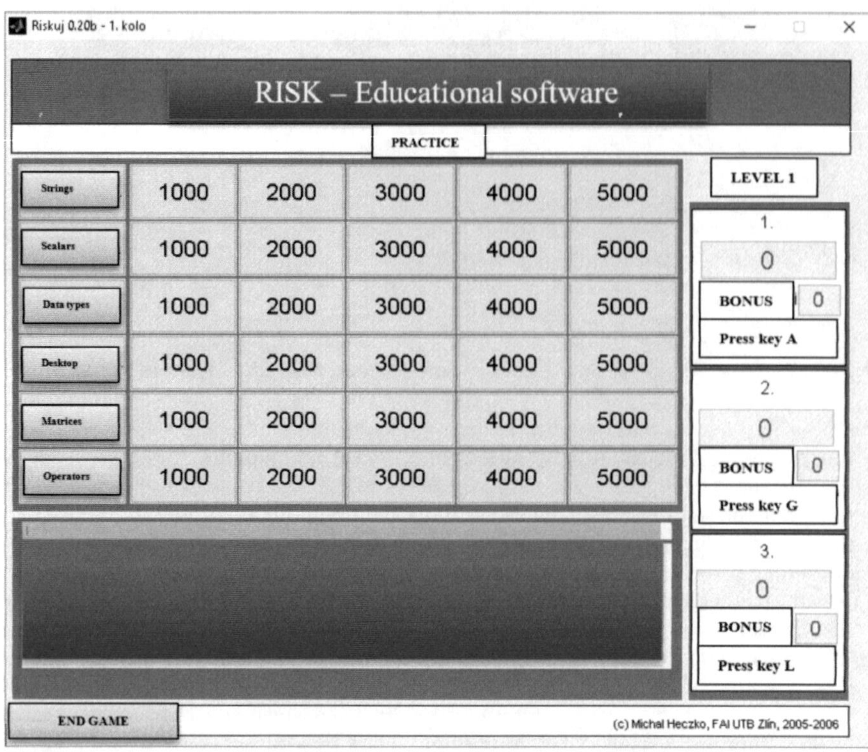

Fig. 5. Risk game

Other nice research papers were published [7], [8]. The results presented in these papers can be extended to educational purposes, too.

3. New Game Specification

3.1. Game description and parameters

The new created game for educational purposes is named as Labyrinth of MATLAB. The game is for single player only [4]. The game contains a total of 10 rooms that the player has to pass within a given time limit. In each room, there is a labyrinth in which there are obstacles in the form of so-called gates. The player passes through the gateway if he correctly answers the question from the MATLAB usage.

The player's position in the labyrinth is indicated by a green square that you can move with the arrow keys. We answer the questions by entering a letter on the keyboard, such as a, b, c or d.

The Labyrinth of MATLAB can be played by anyone who is able to answer the questions correctly, it is necessary to have the basic experience with MATLAB. The game is based on the multiple activities at the same time when the player has to look for the right way to go through the labyrinth, keep track of time, collect as many points as possible and also show his/her knowledge of MATLAB.

You can run the game in the Command Window of MATLAB using the LabyrinthOfMatlab.m file, i.e. calling the LabyrinthOfMatlab command.

3.2 Game rules

The goal of the game is to go through ten rooms. In each room, there is a labyrinth that a player must pass through and reach the destination, or another room, within a two minute time limit.

In the labyrinth, obstacles in the form of so-called gates are placed during the journey to the goal, which the player removes only with the correct answers to questions about MATLAB. If a player responds badly or not at all, then the primary (yellow) gate will become a warning (orange) gate. In this situation, the player has only one attempt to get the right answer and the last option to remove the obstacle. If a player replies badly or not at all, a blinded wall becomes a barrier and the player blocks one of the possible paths to the target.

If a player answers the question correctly for the first time, he will not only remove the gambling but get 20 points for the correct answer, with the correct answer until the second try, the player gets 10 points. If a player walks through a room earlier than a given time limit, he earns 1 point for each remaining second.

The database has 200 questions. Each room has 20 questions, and labyrinths are designed to give the player more of his ability to reach the end, and he does not have to answer all the questions the room contains, at least at the first, simpler levels. In more sophisticated rooms, a player may block the wrong way to the finish and lose everything. The player has up to two attempts to destroy the gate with the correct answer. Questions are randomly generated at each game and their responses are always mixed, ensuring the uniqueness of each game.

The points earned after passing the room are counted and after the successful completion of the game, the player is entered into the table of the best players if the resulting score is among the top ten.

The game can be stopped, paused, or stored in memory and run at any time.

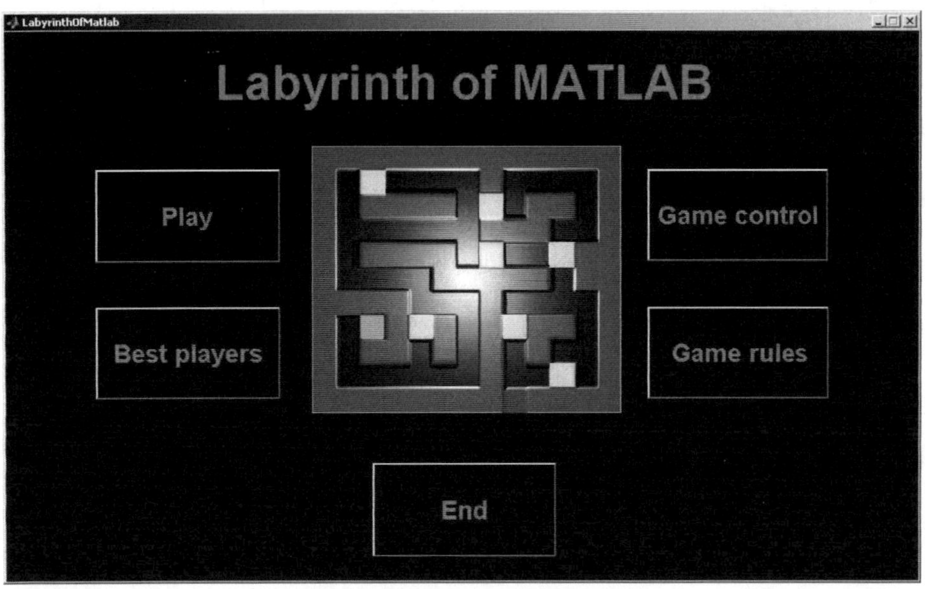

Fig. 6. Main screen of new game

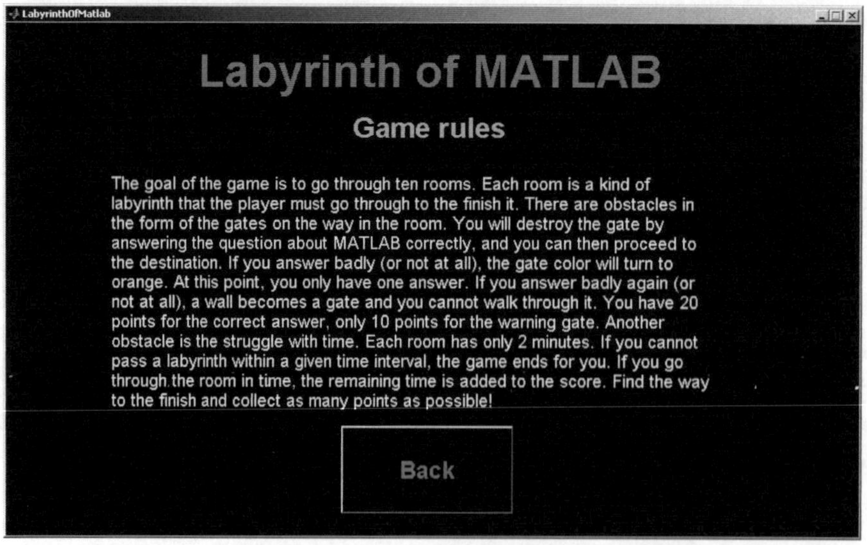

Fig. 7. Brief description of the rules directly in the game

4. Game Realization

4.1. Game Description of game creation

At first, we done the research about the MATLAB Central games and student papers, which are already described in the theoretical part of the thesis.

We then proposed the structure of the game. The game was developed in MALTAB version R2013a. We created two GUI files with which two main M-files are connected, and we added another eight M-files to make the program more transparent, so we did not repeat the source code.

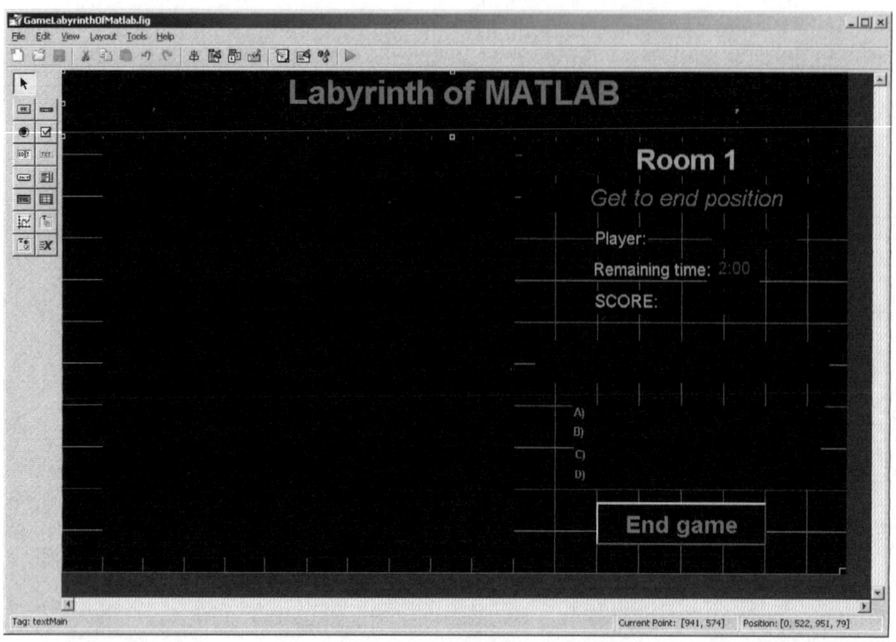

Fig. 8. Game GUI created using GUIDE

The main M-files LabyrinthOfMatlab.m and GameLabyrinthOfMatlab.m are linked so that when one ends, the second one starts and vice versa (except for the direct termination of the entire program in LabyrinthOfMatlab.m)

The structure of the main M-files is described by the following flow charts.

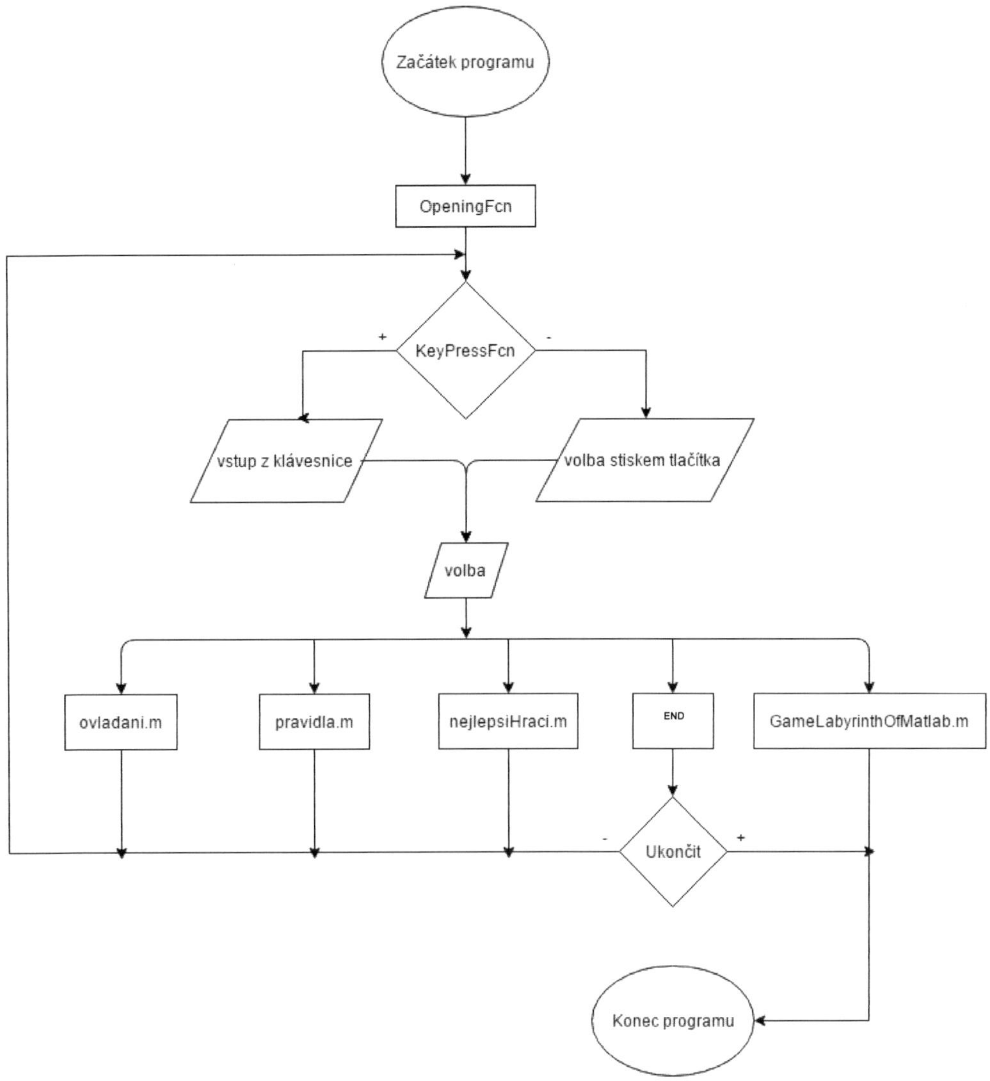

Fig. 9. Flowchart of LabyrinthOfMatlab.m file, the main file of the game

4.2. Blocked path

The game is programmed so that the user searches for the path to the labyrinth goal, so that he can evaluate whether to continue or not in the game when he / she blocks all the way to the target.

This problem was also a frequent reminder, although the game was designed in this way. However, this problem could easily be solved. The solution would be a function that keeps important gate positions for each room for each room while checking whether or not these gates are blocked. The following figure shows the solution.

In the following picture, you can see the first room in which an important gate is highlighted. The new function would therefore keep this gateway in line with its position and continuously monitor whether he pass or not. If a blocking has been registered, the game would alert the player that he no longer has the opportunity to reach the target. In other more difficult rooms, it would no longer matter just to one gate, as in the first room, but to more, which would not be a problem to control, but the algorithm for the other rooms would have been more complicated.

5. Evaluation of created solution

The created game became the part of new modern educational aids in according to the modern strategy of teaching that implements the lectures and practices in the attractive form. Games are leading applications in this area. The students confirm that they are more attract to studies. The tests that verified the MATLAB knowledge of students confirmed the better knowledge of students who played the game in comparison to those who did not use this game.

The issue of the paper was to solve the problem – to find something to increase the knowledge of MATLAB programming at the faculty in the given curricula. It is important to understand the game functionality and to have the knowledge of MATLAB core to successfully finish the created game. The contents of the game is limited to MATLAB core only.

6. Limitations

Although the paper has reached its aims, there were some limitations. First, because the game is new, it was verified only by small group of students who was attending the classes. Therefore, it is necessary to use it for some years to generalize the result and to find what to improve and enlarge. Second, the content of the game is specific. It was created for the given curriculum, so it is necessary to modify or enlarge it in purpose to use this game by users from other faculty.

7. Conclusion

The paper presented newly created computer game in MATLAB that was created for practicing the knowledge of MATLAB programming in the attractive way. There was no specific tool practicing the lectures from the area of MATLAB programming at my faculty. It was the main problem. This problem was solved by new project, and lead to the creation of the new single player game in MATLAB. Before the problem solution, the research of existing games in MATLAB was done and this was followed by new game creation. This game was used during teaching lessons before final test and students like it. Future work will be focused on the comments of the testers and to enlarge the number of the rooms in purpose to practice more topics from MATLAB, not only the MATLAB core.

8. References

[1] 2048. MathWorks [online]. @bmtran, c1994-2017 [cit. 2017-09-29]. Available from: https://www.mathworks.com/matlabcentral/fileexchange/46028-2048

[2] Sudoku. MathWorks [online]. Stepen Sahrun, c1994-2017 [cit. 2017-09-29]. Available from: https://www.mathworks.com/matlabcentral/fileexchange/34988-sudoku

[3] MATRIS v0.10.10.09sb for Windows. MathWorks [online]. Mingjing Zhang, c1994-2017 [cit. 2017-09-29]. Available from: https://www.mathworks.com/matlabcentral/fileexchange/28985-matris-v0-10-10-09sb-for-windows

[4] Sarmanova, L. (2016). The Development of a Computer Game Environment in MATLAB, thesis, UTB in Zlin, Zlin.

[5] Heczko, M. (2006). Tutorial and testing software for the subject "Computer aid of automatic control", thesis, UTB in Zlin, Zlin.

[6] Fiala, D. (2014). A 2D Computer Game in the MATLAB Environment, thesis, UTB in Zlin, Zlin.

[7] Haskovic, D., Katalinic, B. & Kukushkin I. (2015). Role of the Adviser Module in the Hybrid Assembly Subordinating Control Structure, Procedia Engineering, Volume 100, ISSN 1877-7058, pp. 1706-1713.

[8] Gastermann, B., Stopper, M., Kossik, A. & Katalinic, B. (2015). Secure Implementation of an On-premises Cloud Storage Service for Small and Medium-sized Enterprises, Procedia Engineering, Volume 100, ISSN 1877-7058, pp. 574-583.

[9] Payá, L., Reinoso, O., Sánchez, A., Gil, A. & Fernandéz, L. (2010). An educational tool for mobile robots remote interaction, IFAC Proceedings Volume, Volume 42, Issue 24, ISSN 1474-6670, pp. 180-185.

[10] Falconi, R. & Melchiorri, C. (2008). RobotiCad: an Educational Tool for Robotics, IFAC Proceedings Volume, Volume 41, Issue 2, ISSN 1474-6670, pp. 9111-9116.

[11] Mayorova, V., Grishko, D. & Leonov, V. (2017). New educational tools to encourage high-school students‘ activity in stem, Advances in Space Research, In press, ISSN 0273-1177.

[12] Paraskeva, F., Mysirlaki, S. & Papagianni, A. (2010). Multiplayer online games as educational tools: Facing new challenges in learning, Computers & Education, Volume 54, Issue 2, ISSN 0360-1315, pp. 498-505.

[13] Joo-Nagata, J., Martinez Abad, M., García-Bermejo Giner, J. & García-Peñalvo, F.J,. (2017). Augmented reality and pedestrian navigation through its implementation in m-learning and e-learning: Evaluation of an educational program in Chile, Computers & Education, Volume 111, ISSN 0360-1315, pp. 1-17.

[14] Goodwin, A.P.L. (2003). The lecture as an educational tool—a personal view, Current Anaesthesia & Critical Care, Volume 14, Issue 4, ISSN 0953-7112, pp. 170-172.

[15] Deperlioglu, O. & Kose, U. (2011). An educational tool for artificial neural networks, Computers & Electrical Engineering, Volume 37, Issue 3, ISSN 0045-7906, pp. 392-402.

[16] Zangl, H. & Hoermaier, K. (2017). Educational aspects of uncertainty calculation with software tools, Measurement, Volume 101, ISSN 0263-2241, pp. 257-264.

[17] Phillips, J., Brook, A., Tseng, I., Sharpe, R.E., Fein-Zachary, V., Slanetz, P.J. & Mehta, T.S. (2016). Educational Videos: An Effective Tool to Improve Training in Interventional Breast Procedures, Journal of the American College of Radiology, Volume 13, Issue 6, ISSN 1546-1440, pp. 719-724.

[18] Sabor, N., Sasaki, S., Abo-Zahhad, M. & Ahmed, S.M. (2016). A Graphical-based educational simulation tool for Wireless Sensor Networks, Simulation Modelling Practice and Theory, Volume 69, ISSN 1569-190X, pp. 55-79.

DOI: 10.2507/28th.daaam.proceedings.010

NEW SOFTWARE SUPPORTING TEACHING OF SIMULINK FOR FULL-TIME CURRICULUM

Karel Perutka & Aneta Hribnakova

This Publication has to be referred as: Perutka, K[arel] & Hribnakova, A[neta] (2017). New Software Supporting Teaching of SIMULINK for Full-time Curriculum, Proceedings of the 28th DAAAM International Symposium, pp.0079-0086, B. Katalinic (Ed.), Published by DAAAM International, ISBN 978-3-902734-11-2, ISSN 1726-9679, Vienna, Austria
DOI: 10.2507/28th.daaam.proceedings.010

Abstract

This paper deals with the environment of the program MATLAB – SIMULINK. The paper describes the new software created in purpose to support the teaching of the subject called SIMULINK. It contains the description of the program, the specification of the work with the program from the very beginning and the explanations of the several basic blocks, which are commonly used when studying this subject. Theoretical part mainly analyses already existing teaching supports and it tries to explain the program to the user, who deals with SIMULINK for the first time. Practical part mainly focuses on the description of the application, which is created as the support for teaching the subject MATLAB and SIMULINK, specifically the part of subject dealing with Simulink. The software was used during teaching and positively evaluated by students.

Keywords: MATLAB; SIMULINK; Educational Tool

1. Introduction

The contemporary conception of education is in a large measure based on accentuation of the particular individual's needs in relation to his role in society. Each individual can fully develop his capabilities only in society able to learn and to response to changes in a flexible manner. Education as a demonstration of socio-cultural and economic dimension is in all aspects and fields interconnected with the utilization of modern technologies [9].

The last research in modern education incorporates the usage of cartoons as the specific form of presentation that attracts student more than classical books or lectures.

Are the cartoons another educational tool or aid and why? The childhood of a person is shaped according to the conditions of his/her community. However, the childhood in our technology-based era is highly overwhelmed by the ubiquitous communication devices. As a pioneering type, television achieves in grabbing children's attention by using its multi-colored and animated world. What is more, cartoons provide the audience a great load of new ideas, allowing them to enrich their dream world as well as to improve their vocabulary and learn new games. These developments are then turned to permanent behaviors [10].

The usage of specific tools, mostly games, is also important for teaching people with some learning difficulties.

ICT educational tools can play an essential role in the cognitive development and inclusion of disabled people. The increasing adoption of ICT has created the conditions for significant improvements in the ability to acquire, produce and exchange information. This is of particular importance because better education and fluency in ICT meets the objectives of e-inclusion. Starting from the mathematics learning difficulties of motor-impaired people, it was argued that in this field specific educational tools are propaedeutic to a more informal learning approach based on serious games. For disabled people ICT educational tools can be in many cases essential to overcome motor, perceptual or cognitive difficulties that prevent or at least make difficult their inclusion in the society [11].

There exist the concept of the inverted classroom and it was implemented for example to educate MATLAB in chemical process control. Integrating mathematical simulation software such as MATLAB in chemical engineering curricula has been attracting widespread interest during recent decades, especially for courses involve process design. Research has been conducted in comparing different software packages including POLYMATH, Mathcad, MATLAB, Mathematica, Maple and Excel for educational use in chemical engineering, and results indicated that MATLAB is especially excellent for demonstrating matrix operations and multi-variable iterative processes. In addition, MATLAB-based simulation has been verified for promoting deeper understanding of process modeling techniques [16].

MATLAB is widely used software at technical universities to create and practice technical solution of several technical tasks [8].

Identification of system is one of such tasks. For example, the identification of continuous-time (CT) models is a problem of considerable importance that has applications in virtually all disciplines of science. It is therefore important to introduce data-based continuous-time model identification to engineering students in the user friendly form. For that purpose, Continuous-Time System Identification (CONTSID) toolbox and its graphical user interface were created in MATLAB [13]. MATLAB also offers its web server and JAVA implementation. Experiences with MATLAB web server and JAVA implantation were also published [14].

Implementation at overhead power lines using MATLAB is another example. The electrical power transmission is realised in most of the cases by overhead power lines long of hundreds of kilometres and operating at different high voltage levels. The load of these overhead power lines is widely varying depending both on the evolution of consumers (the consumers load curve) and on the operating conditions of the power system elements. Therefore, the 400 kV overhead power line was modelled using MATLAB and SIMULINK and the operating conditions were analyzed. The presented model was used in teaching activity during the power grids laboratory lessons for power engineering students of the university curriculum [17].

Next example of educational tool in the area of the electrical engineering is the following one. It is the educational software package called TSCOM (thyristor switched reactive power compensators). It contains simulation models of thyristor switched capacitor and thyristor switched reactor-based static VAR compensator which are two of the shunt flexible AC transmission systems devices. The design was realized using MATLAB and SIMULINK. The proposed package helped to design the practical prototypes for students and researchers [12].

MATLAB application in the area of global navigation satellite systems education is another example. It is used by students in aerospace engineering and was created using Graphical User Interface Design Environment. It highlights the space segment of these navigation systems and concepts such as visibility, ground-tracks and coverage. Using this graphic tool, students can visualize the real position of all satellites in the GPS constellation at any time since its start of operation, analyze the number of satellites that can be visualized from any point in Earth, or study the quality of the navigational solution which depends on the relative geometry of the user and satellites. The tool also allows to perform these studies for a flight between any two points on Earth, or for the user-defined region [18].

Educational games help to increase the motivation of the students and to enhance their learning performance. For instance, educational games illustrate the link between theoretical methods presented in class and practical applications. There were created games in MATLAB in purpose to help teaching, such as the one created at the University of Stuttgart to practice the knowledge of advanced automatic control, the airplane behavior [12].

It is important to have the software tool that practices the knowledge of this software. For that purpose and directly according to the contents of the labs from SIMULINK, the new software supporting teaching of SIMULINK was created and it is described in this paper. The paper is organized in the following way. After introduction part, the short description of MATLAB and SIMULINK software is presented. It is followed by the description of the newly created application with depiction of main dialog and selected practical example. Finally, the new application is evaluated.

2. Brief intro about MATLAB and SIMULINK

2.1. MATLAB in general

The name MATLAB originated from two English words, matrix and laboratory, which means "matrix and lab". MATLAB is the program environment that allows us to count a variety of mathematical operations, it can handle with vectors and matrices, plot charts, and create user interface applications [1], [2].

The program itself contains predefined features that user calls from the shortcut menu. MATLAB is mainly used for scientific and technical purposes or in colleges for teaching purposes.

2.2. MATLAB environment

When the MATLAB is started, the basic window will be displayed and divided into several main sections. The sections are:

Menu - the main menu bar at the top of the window which contains three tabs - Home for basic file operations, variables, help browser, and others, Chart Style Plots and Apps where you can download various additional applications to ease the work with MATLAB [4].

Current Folder – displays list of all files in the current folder, MATLAB is the default folder that is displayed. The files in this window are triggered by double-clicking with the mouse, the file opens in the given editor according to its extension.

Command Window - A window into which the user enters commands, runs functions and scripts, and results are displayed [7].

Workspace - this window is empty when MATLAB is started, and then all the variables and their values that were used during the work with MATLAB are displayed. With more variables, there is better orientation using Workspace window than just using the command window.

Command History - A list of commands that were used during last MATLAB calls including the date and time, the history of the commands can be easily reused by double-clicking on the command or set of commands by the mouse. The Command History can be saved in the new source file [1, 4, 8].

2.3. SIMULINK in general

The word SIMULINK comes from two English words - simulation and link. It is the main extension of the MATLAB build-in environment, and it was designed to model and simulate dynamic systems, so it is necessary to have MATLAB installed to work with SIMULINK. While commands are entered in the Command Window in MATLAB, SIMULINK combines blocks that have different functions. User can create his/her own blocks or edit existing blocks. To work in SIMULINK, it is appropriate for the user to have minimal basic knowledge of MATLAB [2, 3]. Other good research papers were published [3], [5], which might be transformed for educational purposes.

2.4. Running SIMULINK

SIMULINK can be started in several ways. For example, by typing the text "simulink" in the MATLAB Command Window and pressing the ENTER key. Another option to open SIMULINK is via the Simulink Library button, which is located in the main ribbon section and SIMULINK starts as was described before. After SIMULINK calling new window appears, where all the standard SIMULINK libraries are located, as well as those that the user had additionally installed or created by himself. Libraries have blocks sorted alphabetically, according to their area of use.

2.5. SIMULINK Environment

The SIMULINK window is divided into two main sections. In the left part of the window, there are all available libraries and in the right part of the window, there are the blocks that the selected library contains. Blocks can be searched either by the category they are divided into, or if you know at least part of the name, go to the search box at the top of the window in the Enter search term edit box. This window contains two important buttons, namely the creation of a new model and the opening of existing model, located at the top of the window on the left.

2.6. Creating new model

At the beginning, you need to open new window to create a model. This can be done in several ways. The first option is to click on File -> New -> Simulink Model in the MATLAB environment, the second one is by clicking the SIMULINK icon with the "+" sign located on the right side in the library window, and the last method is, as in many Microsoft Windows programs, shortcut Ctrl + N. After each of these operations, the blank window opens, where you can create the new model. The model is simply created by dragging the blocks from the Simulink Library Browser to the desktop of the window that is currently opened. If you need to use a block multiple times, you can simply copy it by right-clicking and dragging it, holding Ctrl and dragging, or copying using the Ctrl + C and Ctrl + V shortcut keys. Within one model, you cannot find blocks with the same name, so the same blocks are created with names like "Block, Block1, Block2, etc." You can simply rename the block by clicking on its name and overwriting it. The interconnection of blocks is done by dragging the mouse from the output of one block to the input of another block, or vice versa. If you need to connect a block to an existing connection, simply drag the input / output of this block to the junction and automatically create a new node.

Double-clicking on a block opens a dialog for parameters setting. Each block has specific settings. The model starts after clicking the Run button at the top of the window. The simulation time can also be set in the Simulation stop time edit box.

2.7. Creating subsystems

As an example for creating a subsystem, there is a model for calculating the roots of the quadratic equation found also in the newly created application. First, it is necessary to create the entire model and then drag the mouse or hold down the Shift key and click on the blocks to mark the ones to create the subsystem.

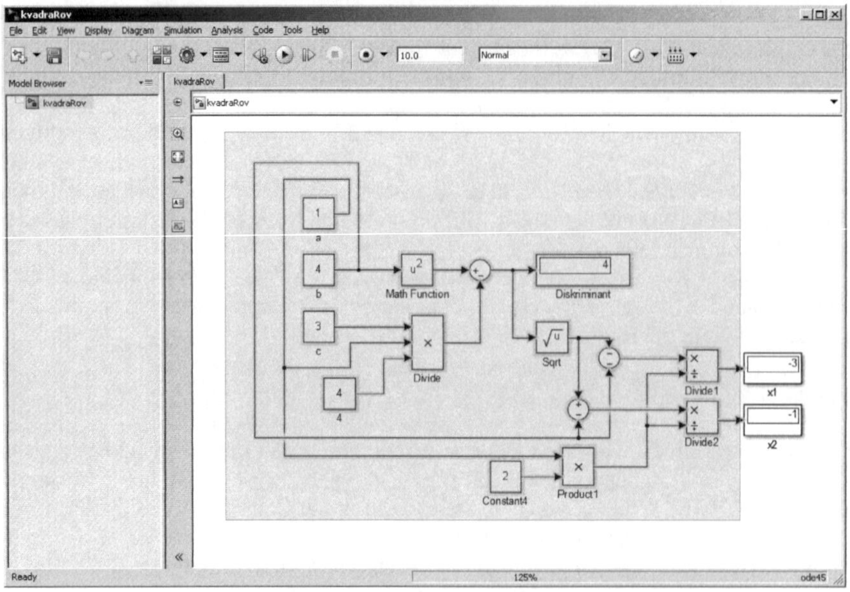

Fig. 1. Marking the blocks in SIMULINK model

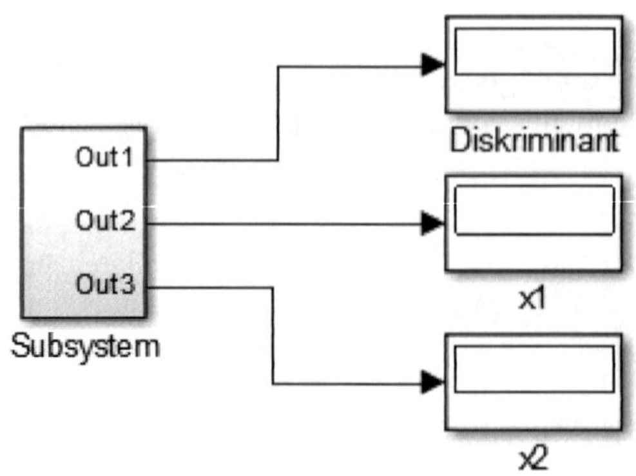

Fig. 2. The newly created subsystem from the marked blocks

3. Newly created application

3.1. Creating new GUI for application

The main dialog of the application contains five objects. There are two Static Text objects, one Axes object and two Push Buttons objects [6].

Object Static Text displays static text, which is in the form of the title and subtitle of the application, after double-clicking the mouse, the Property Inspector window will appear to set almost all parameters.

For static text, it's usually enough to set the size, font and color of the text, and last but not least, write the text itself that the object will display.

The Axes object in the created dialog is used to display the image, specifically the faculty logo. It does not generate any code into source file, so the command displaying the picture, must be written manually.

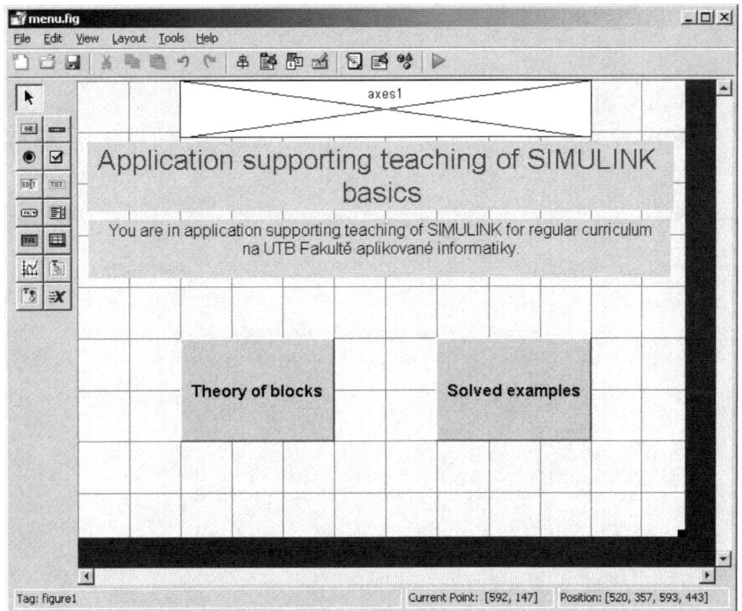

Fig. 3. The main dialog layout of the newly created application

3.2. Blocks – applications

The bloky.fig file contains two Static Text objects, the title of the application and the title of the current slide, and five Push Button objects, one of which refers back to the main menu, and four refer to libraries. Button functions are the same as in the previous main menu. The portion of the code is automatically generated and the rest is manually completed.

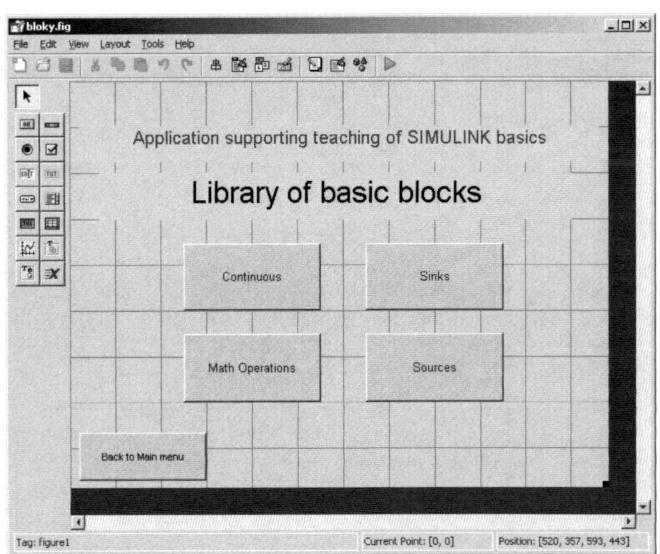

Fig. 4. The scheme of blocks describing and practicing most commonly used blocks

3.3. Examples – applications

The priklady.fig file contains Static Text and Push Button objects. One button moves back to the main menu and the other displays the given example. Clicking one of the examples closes the current window and opens the window with the selected example. As the first example of the window with examples, the input is in Static Text object and there are also two Push Buttons, one returns to the previous dialog and the other refers to the next figure. The following objects are the dialog: the Axes object where you can see an example of solved task and two push-buttons, one returns to the main menu and closes the current window, and the other opens an example in SIMULINK, so the .slx file, but the current window remains open.

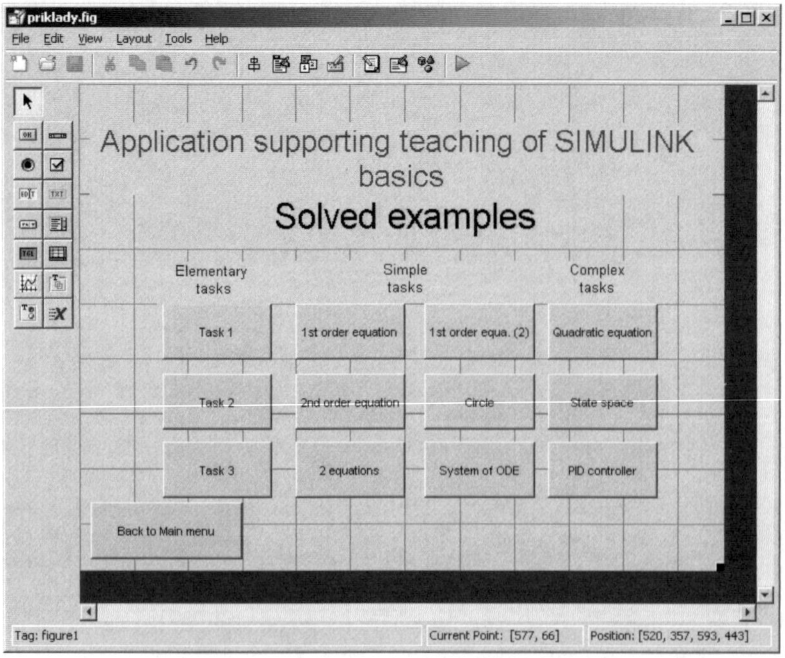

Fig. 5. GUI layout of examples dialog

When a user selects one of the buttons in the "Basic tasks" section, a simple entry dialog is displayed to check that he knows where to find the blocks, how to set them and then link them. For example, there is a simple mathematical example √ (6-12 + 22), and when it is called, a picture is displayed, in which is described how to create the scheme. This image also contains the "Solution in SIMULINK" button, which opens the completed schema in the SIMULINK environment.

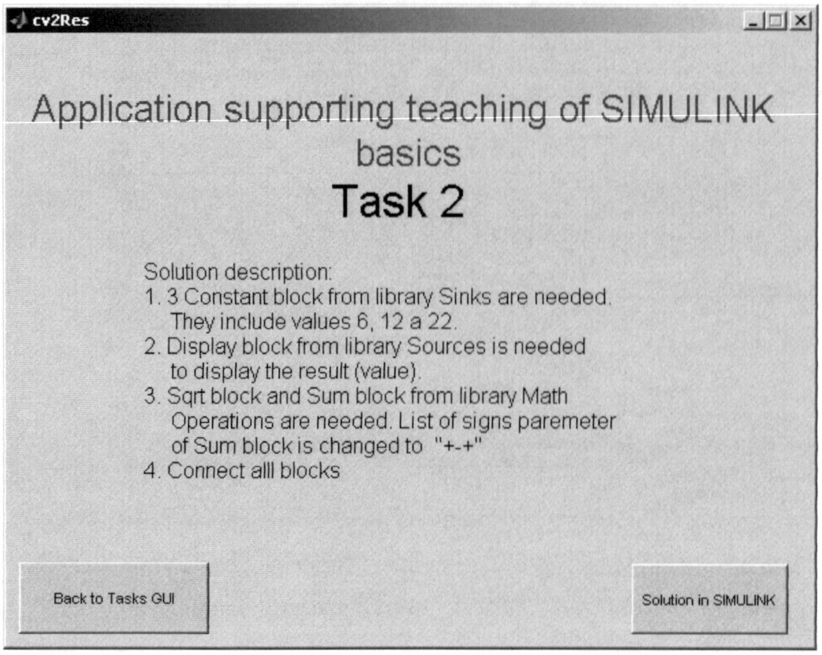

Fig. 6. GUI describing how to solve selected task

Clicking on the "Solution in SIMULINK" button opens this schema in SIMULINK, where it can be edited and run. The second example in this section is focused on the first-order differential equation, it is otherwise specified and contains some other blocks, such as the square root block.

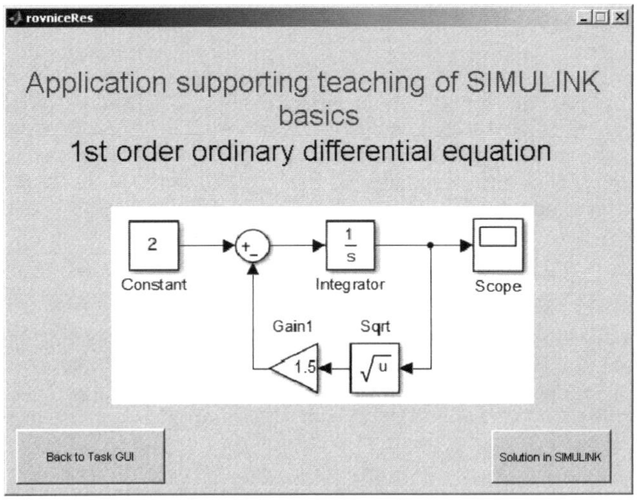

Fig. 7. Scheme used to solve the differential equation.

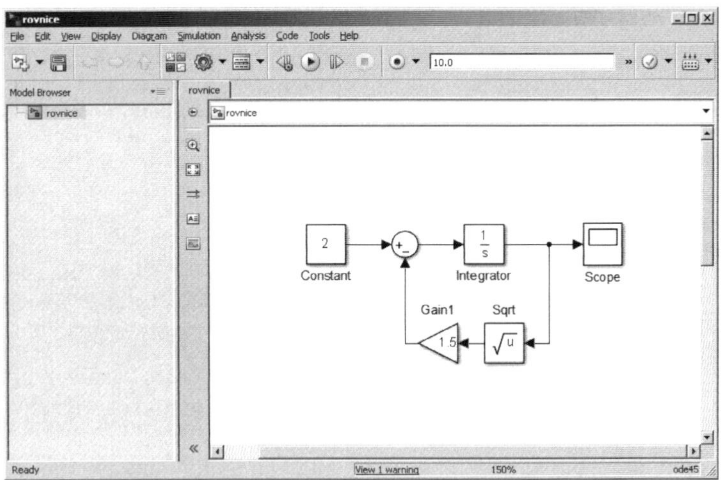

Fig. 8. Model used to solve the differential equation.

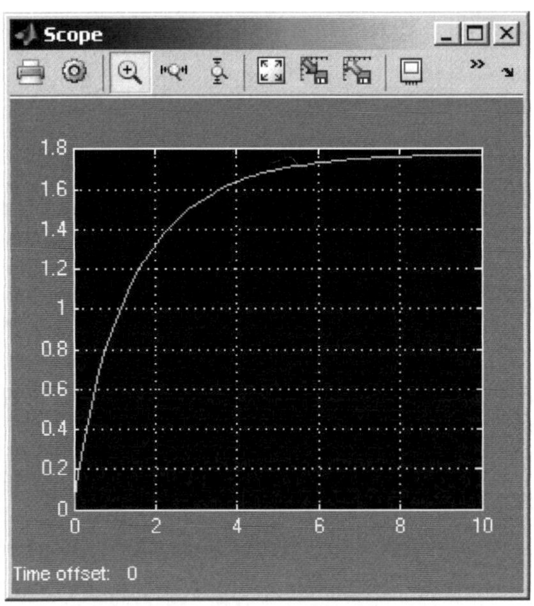

Fig. 9. Solution of differential equation.

4. Evaluation of created solution

The created educational tool became the part of new modern educational aids in according to the modern strategy of teaching that implements the lectures and practices in the attractive form. The students confirm that they are more attract to studies. The tests that verified the SIMULINK knowledge of students confirmed the better knowledge of students who worked with this new application in comparison to those who did not use it. The issue of the paper was to solve the problem – to find something to increase the knowledge of SIMULINK usage at the faculty in the given curricula. It is important to understand the application functionality and to have the knowledge of SIMULINK. The contents of the teaching aid is limited to SIMULINK applications used during classes only.

5. Limitations

Although the paper has reached its aims, there were some limitations. First, because the teaching tool is new, it was verified only by small group of students who was attending the classes at the laboratories. Therefore, it is necessary to use it for some years to generalize the result and to find what to improve and enlarge. Second, the content of the created programme was prepared for the given curriculum, so it is necessary to modify or enlarge it in purpose to use the new programme by users from other university.

6. Conclusion

The paper presented newly created application in SIMULINK that was created for practicing the knowledge of SIMULINK usage in the attractive way. There was no specific tool practicing the lectures from the area of SIMULINK at my faculty. It was the main problem. This problem was solved by new project, and lead to the creation of the new tool in SIMULINK. Before the problem solution, the necessary introduction to SIMULINK was done and this was followed by the creation of new application practicing the student knowledge of Simulink and it was positively evaluated by students. Students used it before final test from SIMULINK. It covers the whole introductive course to SIMULINK and examples how to solve set of algebraic equations, differential equation, the quadratic equation, drawing circle in SIMULINK etc. Future work will be oriented on more complex enlargement of the newly created application.

7. References

[1] Perutka, K.(2005). MATLAB – Basics for students of automation and IT, UTB ve Zline, ISBN 80-7318-355-2, Zlin.
[2] Dušek, F. (2001) Simulink – Introduction for users, University of Pardubice, ISBN 80-7194-273-1, Pardubice.
[3] Haskovic, D., Katalinic, B. & Kukushkin I. (2015). Role of the Adviser Module in the Hybrid Assembly Subordinating Control Structure, Procedia Engineering, Volume 100, ISSN 1877-7058, pp. 1706-1713.
[4] Zaplatílek, K. & Doňar ,B. (2005). MATLAB for beginners. BEN – technical literature. ISBN 80-7300-175-6, Brno.
[5] Gastermann, B., Stopper, M., Kossik, A. & Katalinic, B. (2015). Secure Implementation of an On-premises Cloud Storage Service for Small and Medium-sized Enterprises, Procedia Engineering, Volume 100, ISSN 1877-7058, pp. 574-583.
[6] Hribnakova, A. (2017). Application design to support Simulink teaching for full-time curriculum. Thesis. Tomas Bata University in Zlin, Zlin
[7] Kozák, Š. Kajan, S. (1999). MATLAB-SIMULINK I. Bratislava: Slovak University of Technology in Bratislava, ISBN 80-277-1213-2, Bratislava.
[8] Kupka, L. (2006). MATLAB Simulink Introduction for users . SOŠ a SOU Lanškroun, ISBN 978-80-239-8871-0, Lanškroun.
[9] Dostál, J., Serafín, Č., Havelka, M. & Minarčík, J. (2012). Assessment of Quality of Material Educational Tools for Technical Education, Procedia - Social and Behavioral Sciences, Volume 69, ISSN 1877-0428, pp. 709-718.
[10] Özer, D. & Avcı, İ.,B.. (2015). Cartoons as Educational Tools and the Presentation of Cultural Differences Via Cartoons, Procedia - Social and Behavioral Sciences, Volume 191, ISSN 1877-0428, pp. 418-423.
[11] Curatelli, F. & Martinengo, C. (2012). Design Criteria for Educational Tools to Overcome Mathematics Learning Difficulties, Procedia Computer Science, Volume 15, ISSN 1877-0509, pp. 99-102.
[12] Münz, U., Böhm, Ch., Eck, J., Reble, M., Schumm, P. & Allgöver, F. (2010). A Matlab-Based Game for Advanced Automatic Control Education, IFAC Proceedings Volumes, Volume 42, Issue 24, ISSN 1474-6670, pp. 140-145.
[13] Garnier, H. (2011). Teaching Data-based Continuous-time Model Identification with the CONTSID Toolbox for Matlab, IFAC Proceedings Volumes, Volume 44, Issue 1, ISSN 1474-6670, pp. 6373-6378.
[14] Hölttä, V. & Hyötyniemi, H. (2003). Experiences with Matlab Web Server and Java, IFAC Proceedings Volumes, Volume 36, Issue 10, ISSN 1474-6670, pp. 47-52.
[15] Gelen, A. & Yalcinoz, T. (2010). An educational software package for Thyristor Switched Reactive Power Compensators using Matlab/Simulink, Simulation Modelling Practice and Theory, Volume 18, Issue 3, ISSN 1569-190X, pp. 366-377.
[16] Li, X. & Huang, Z.(J.) (2017). An inverted classroom approach to educate MATLAB in chemical process control, Education for Chemical Engineers, Volume 19, ISSN 1749-7728, pp. 1-12.
[17] Baloi, A., Pana, A. & Molnar-Matei, F. (2015). Advantages of Using MatLab Simulink in Laboratory Lessons on Operating Conditions of Overhead Power Lines, Procedia - Social and Behavioral Sciences, Volume 191, ISSN 1877-0428, pp. 179-184.
[18] Lanagran-Soler, F., Vazquez, R. & Arahal, M.R. (2015). A Matlab Educational GUI for Analysis of GNSS Coverage and Precision, IFAC-PapersOnLine, Volume 48, Issue 29, ISSN 2405-8963, pp. 93-98.

DOI: 10.2507/28th.daaam.proceedings.011

FORMATION OF EQUIVALENT MODELS OF STRUCTURED MULTIRATE SYSTEMS IN THE FORM OF SIGNAL GRAPH

Vadim Kramar

This Publication has to be referred as: Kramar, V[adim] (2017). Formation of Equivalent Models of Structured Multirate Systems in the Form of Signal Graph, Proceedings of the 28th DAAAM International Symposium, pp.0087-0092, B. Katalinic (Ed.), Published by DAAAM International, ISBN 978-3-902734-11-2, ISSN 1726-9679, Vienna, Austria
DOI: 10.2507/28th.daaam.proceedings.011

Abstract

The usage of microprocessors in the measuring and processing channels along with continuously working arrangement is typical for modern control systems. Similar control systems carry out measurements and processing of some signals sharing time. Mathematical models of such systems are represented in a multivariate multistep discrete control systems. Investigations of similar systems lead to the need of their model development in a complex range, with transfer functions as main modelling elements. Signal graph provides evident representation of system variables and their interaction.
This article gives a way to the creation of a signal graph one-time uninterruptedly discrete system for which the Mason equation can be applied. Thus, the problem of process analysis in given multirate systems becomes reduced to the problem of rational functions.

Keywords: Multidimensional systems; digital-analogue systems; signal graph; Mason equation; equivalent model

1. Introduction

The usage of microprocessors in the measuring and processing channels along with continuously working arrangement is typical for modern automatic control systems [1-4]. As a rule, similar control systems carry out measurements and processing of some signals sharing time. Mathematical models of such systems are represented in a multivariate multistep uninterruptedly discrete automatic control systems. Investigations of similar systems lead to the need of their model development in a complex range, with transfer functions as main modelling elements. To obtain control systems models in the form of transfer functions, methods using signal graph have found a wide applicationx[5,6]. Signal graph provides evident representation of system variables and their interaction. It is well known that to determine transfer functions of linear permanent systems the Mason equation can be applied. In [7] approaches of Mason equation [8] application in one-time uninterruptedly discrete systems are observed. If there are some various times, the Coffi and Williams matrix approach should be applied. This article gives a way to the creation of a signal graph one-time uninterruptedly discrete system for which the Manson equation can be applied.

While constructing signal graph of multirate uninterrupted discrete system, let us consider the symbol system of Sedler and Becky [9, 10]. White knot in the graph is used to define uninterrupted variable system. Black knot is used to define discrete variable and quantum operations, variable significance represented by any black knot being discrete form of sum transformation of all variables, being in the knot, according to a certain rate. Because in uninterrupted discrete, as a rule, it is impossible to outline inlet variable in presentation for, thus it is worthwhile to coordinate inlet influence. It is carried out by introduction of branches with transfer function equal to inlet variables so that entire transfer function belonging to the single output, transfer function and output become equal.

2. Formation of equivalent mathematic model of structured multirate systems

Linear uninterrupted discrete system with some quantifiers with discrete time $T_1, \dots, T_N$ can be described by the system of linear algebraic equations in certain fields

$$A(s)x(s) = \sum_{i=1}^{N} B_i x^{Ti}(s) + R \tag{1}$$

where: $x(s) - n$ is vector of variable systems represented according to Laplace; $A(s)$ - $n\times n$ uninterrupted transfer function matrix; $x^{Ti}(s)$ - discrete transformation in times T_i, $i = 1, \dots, N$ vector $x(s)$; $R - n$ – standardized input vector which, according to system linearity, can be considered as vector $R'=[1\ 0\ 0\ \dots\ 0]'$ where ' – here and further means transportation; B_i - $n\times n$ – matrixes characterizing quantifier presence with various discrete times in system, they consist of 0 and 1 and represent quantification.

To simplify (1) it is worthwhile to write first of all equations for uninterrupted, and then for quantum variables by turns for each discrete time T_i, $i = 1, \dots, N$ [1]. $A(s)$ is of block type

$$A(s) = \begin{bmatrix} W(s) & \Phi_1(s) & \cdots & \Phi_N(s) \\ 0 & I_1 & \cdots & 0 \\ \vdots & \vdots & \ddots & \vdots \\ 0 & 0 & \cdots & I_N \end{bmatrix} \tag{2}$$

where: N – quantity of different measure point; $W(s)$ - $l\times l$ – transfer function matrix; l – uninterrupted variables number; $\Phi_i(s)$ - $l\times m_i$ – transfer function matrix; m_i – quantifier number with discrete time T_i, $i = 1, \dots, N$; I_i – identity matrixes $m_i \times m_i$.

Let us define

$$y_i^{T_i}(s) = B_i x^{T_i}(s), \quad i = 1, \dots, N. \tag{3}$$

Vector $y_i^{T_i}(s)$ consists of the m_i vector component x^{T_i}, complying with quantifier outputs with discrete time T_i.

From (1), taking into account (3) (considering the presence of $A^{-1}(s)$), we obtain

$$x(s) = A^{-1}(s)\sum_{j=1}^{N} y_j^{T_j}(s) + A^{-1}(s)R \tag{4}$$

Elements a_{ij}^{-1} of matrix A^{-1} are transfer functions from j knot to i knot of initial system with all open quantifiers. Actually, considering quantifiers and vectors $y_i^{T_i}(s)$ to be independent input effects which can be defined as zero, from (4) we obtain

$$x_i(s) = \sum_{j=1}^{N} a_{ij}^{-1} r_j \tag{5}$$

where: r_j – vector elements of normalized input R.

On the basis of (4), uninterrupted discrete system 'complex graph' can be created, which gives us system transfer functions by applying the Manson formula. A complex graph is uninterrupted discrete system initial graph and discrete graph combination, made up $y_i^{T_i}(s)$. In (4), variables $y_i^{T_i}(s)$ are input signals of keys which we will define as input variables. Having formed N discrete graph for $y_i^{T_i}(s)$, we may delete quantifiers from the initial graph and connect input knots of keys with equivalent knots of discrete graphs by means of branches with identity intensification coefficients. In this way, we obtain a complex system graph to which the Manson formula is applicable. Consider uninterrupted discrete system with N quantizers, so that discreteness es are as follows

$$\frac{T_i}{T_{i+1}} = \frac{b_i}{q_i} \qquad i = 1,\ldots,N-1 \tag{6}$$

The system of linear algebraic equations with the time rate for this system is shown in (4). Multiplying (4) to the correspondent B_i, we obtain a ratio system

$$y_i(s) = B_i A^{-1}(s)\sum_{j=1}^{N} y_j^{T_j} + B_i A^{-1}(s)R \quad i = 1,\ldots,N \tag{7}$$

Let T – discrete time – be equal to the least common multiple time of all N times T_i.

We can carry out discrete conversion of vector $x(s)$ to least common multiple – time T

$$x^T(s) = (A^{-1}(s)\sum_{j=1}^{N} y_j^{T_j}(s))^T + (A^{-1}(s)R)^T \tag{8}$$

Assuming the meaning of $\frac{T}{T_i} = n_i$, and using discrete transformer property $[g(s)z^T(s)]^{nT} = \sum_{i=o}^{n-1}(g(s)e^{-iTs})^{nT}(z(s)e^{iTs})^{nT}$, we can write

$$\begin{aligned} x^T(s) = &\sum_{i_1=0}^{n_1-1}(A^{-1}(s)e^{-i_1T_1s})^T(y_1(s)e^{i_1T_1s})^T + \ldots + \\ &+ \sum_{i_N=0}^{n_N-1}(A^{-1}(s)e^{-i_NT_Ns})^T(y_N(s)e^{i_NT_Ns})^T + \\ &+ (A^{-1}(s)R)^T \end{aligned} \tag{9}$$

Let us identify

$$\begin{aligned} &Y_{1(i_1)} = y_1(s)e^{i_1T_1s} \ldots Y_{N(i_N)} = y_N(s)e^{i_NT_Ns}; \\ &i_j = 0,\ldots,(n_j-1); \quad j = 1,\ldots,N. \end{aligned}$$

We obtain equation of vectors $Y_{1(i)}^T,\ldots,Y_{N(i)}^T$, then multiply in (8) each equation to $e^{i_jT_is}$ $(i_j = 0,\ldots,(n_j-1),\ j = 1,\ldots,N)$.

$$\begin{cases} Y_{1(i_1)} = (B_1A^{-1}(s)e^{i_1T_1s}\sum_{j=1}^{N} y_j^{T_j}(s)) + (B_1A^{-1}(s)\mathrm{R}e^{i_1T_1s}) \\ i_1 = 0,\ldots,(n_1-1) \\ \vdots \\ Y_{N(i_N)} = (B_NA^{-1}(s)e^{i_NT_Ns}\sum_{j=1}^{N} y_j^{T_j}(s)) + (B_NA^{-1}(s)\mathrm{R}e^{i_NT_Ns}) \\ i_N = 0,\ldots,(n_N-1) \end{cases} \tag{10}$$

Let us carry out discrete transformation (10) to the least common multiple – time T

$$\begin{cases} Y^T_{1(i_1)} = \sum_{k_1=0}^{n_1-1} (\overline{W}_{1(i_1)} e^{-k_1 T_1 s})^T Y^T_{1(k_1)} + \\ \quad + \sum_{k_2=0}^{n_2-1} (\overline{W}_{1(i_1)} e^{-k_2 T_2 s})^T Y^T_{2(k_2)} + \ldots + \\ \quad + \sum_{k_N=0}^{n_N-1} (\overline{W}_{1(i_1)} e^{-k_N T_N s})^T Y^T_{N(k_N)} + \Phi^T_{(1)} \\ i_1 = 0, \ldots, (n_1 - 1) \\ \vdots \\ Y^T_{N(i_N)} = \sum_{k_1=0}^{n_1-1} (\overline{W}_{N(i_N)} e^{-k_1 T_1 s})^T Y^T_{1(k_1)} + \\ \quad + \sum_{k_2=0}^{n_2-1} (\overline{W}_{N(i_N)} e^{-k_2 T_2 s})^T Y^T_{2(k_2)} + \ldots + \\ \quad + \sum_{k_N=0}^{n_N-1} (\overline{W}_{N(i_N)} e^{-k_N T_N s})^T Y^T_{N(k_N)} + \Phi^T_{(N)} \\ i_N = 0, \ldots, (n_N - 1) \end{cases} \tag{11}$$

where:

$$\begin{aligned} \overline{W}_{j(i_j)} &= B_j A^{-1}(s) e^{i_j T_j s} \quad j = 1, \ldots, N, \\ \Phi_j &= B_j A^{-1}(s) R e^{i_j T_j s} \quad j = 1, \ldots, N. \end{aligned} \tag{12}$$

Calculating together (12) and (10) and introducing Y_i we may obtain input-output relation for all variable systems. This system contains

$$n + \sum_{j=1}^{N} m_j n_j$$

equation, where: n – vector dimension $x(s)$; m_j – number of keys with times T_j; n_j – number equal to the relation of the least common multiple time to time T_j. Ratio (10), after introducing Y_i, can be accepted as discrete system description, inputs of which along with outputs of initial uninterruptedly discrete system are variables $Y^T_{j(i_j)}$ $j = 1, \ldots, N;\ i_j = 0, \ldots, (n_j - 1)$. As their representation is

$$y_j^{T_j} = \sum_{i_j=0}^{n_j-1} e^{-i_j T_j s} [y_j e^{i_j T_j s}]^T = \sum_{i_j=0}^{n_j-1} e^{-i_j T_j s} Y^T_{j(i_j)} \tag{13}$$

it can be considered to be a result of input signals parallelization of keys with time T_j in the initial system on n_j branches, each of which has transfer coefficient

$$e^{-i_j T_j s} \quad i_j = 0, \ldots, (n_j - 1).$$

Then, the initial system in (9) can be specified by the discrete graph in which keys are open and their outputs are input signals into the system along with initial input signals. At the same time, each of these new inputs from quantifiers T_j ($j=1,\ldots,N$) can be represented as a set of inputs $Y^T_{j(i_j)}$ $j = 1, \ldots, N;\ i_j = 0, \ldots, (n_j - 1)$ and, by connecting certain knots by identity connections, we can formulate the input-output ratio according to the final discrete graph.

Discrete graphs for $Y^T_{j(i_j)}$ are built on the basis of (12). There should be $k = \sum_{j=1}^{N} n_j$ discrete graphs.

3. Algorithm for uninterruptedly discrete system signal graph

The construction of algorithm for uninterruptedly discrete system signal graph on the basis of equations (10) – (12) will be as follows:

1. On the basis of structure scheme an uninterruptedly discrete system initial graph is formed. All quantifiers in it, T_j $j = 1,...,N$, are considered to be regulated by diminution open, at the same time output signals of keys are considered to be input ones into the system where $\sum_{j=1}^{N} m_j$ is the quantity of black knots and m_j – key number with time T_j.

2. Form n_j $j = 1,...,N$ of the discrete graph, corresponding to time T_j on the following procedures:

a1) $j = 1;$

a) in the initial graph we compose only knots corresponding to quantifiers input signals with time T_j and connected with them input knots (initial and from quantifiers). As a result we obtain the intermediate graph;

a2) $i_j = 0;$

b) in the intermediate graph we replace all transfer functions of links W with $We^{i_j T_j s}$;

c) we make input signals parallel from keys T_k, $k = 1,...,N$, $k \neq j$ to n_k branches, substituting each transfer function of the branch W for $We^{-l_k T_k s}$ $l_k = 0,\ldots,(n_k - 1)$, $k = 1,\ldots N$, $k \neq j$.

d) we substitute knots for black, and transfer functions for their discrete transformation in time T. We enter input knots in the Table through $Y^T_{k(i_k)}$, then state conformity between knots x^T и $Y^T_{k(i_k)}$ by means of identity connections, where it is necessary. We obtain a discrete graph corresponding to time T_j.

e) points b) ÷ d) are now executed n_j times, supposing $i_j = i_j + 1$; this results in discrete graph n_j, corresponding to time T_j, $i_j = 0,...,(n_j - 1);$

f) j – we increase by one and repeat points a) ÷ e) for discrete graph construction for all N quantifiers, i.e. $j = 1,...,N$.

3. Now we turn back to the initial graph and make all input signals parallel from keys $y_j^{T_j}$, $j = 1,\ldots,N$, into n_j corresponding branches $j = 1,...,N$.

We transfer each function from n_j branches W and substitute for $We^{-k_j T_j s}$ $k_j = 0,\ldots,(n_j - 1)$, $j = 1,\ldots N$, respectively.

We now enter the input knots in the Table through variables $Y^T_{j(i_j)}$, $i_j = 0,\ldots,(n_j - 1)$.

We connect to the marked inputs all of discrete graphs, got in item 2, by single connections. We set, where necessary, other single connections, proper identical knots. We get the component graph.

4. We substitute all the knots with black, and transfer functions on their discrete transformations on time the least common multiple – on time of T. We get the final discrete graph of the system.

5. By applying the Mason rule we determined the necessity of input-output correlation for the system variables of x^T for the obtained graph of the system.

Now we can formulate the algorithm of receipt of signal graph of multivariate of the continuously-discrete system for the case of multiple times.

1. On the basis of the flow diagram of the system, an initial graph containing white and black knots is formed. Black knots have indexes of Ti, $I = 1,.,N$, proper to the value of time of discreteness. Times are considered well-organized on a decrease. In the initial graph keys are considered broken; here the output signals of keys are considered entrance knots in the system and black knots correspond to them.
2. On the initial graph of the system, through the application of the Mason algorithm, the discrete graph of 1st level is formed in conformance with the following:

All the knots, being by an entrance for the keys with the smallest time of *TN*, get out in the initial graph. They are considered output knots. All the entrance knots, related to the indicated output, get out then. Entrance knots can be the entrance signals of the initial system, and also outputs of keys with large times. In the Mason algorithm, connections between these knots are determined and the intermediate count of 1st level is formed.

Further, all white knots of the intermediate graph are replaced with black knots, with the index of variables of *TN*, which corresponds to discrete transformation of variables in time of *TN*, and the transmission functions of connections are replaced by their discrete transformations in time of *TN*. Black knots that are proper discrete variables for large times remain unchanged. The discrete count of the system of 1st level is formed in the same manner. We set, where necessary, single connections for proper identical knots.

3. The component graph of 1st level is formed. It turns out to be a combination of the initial graph of the system and discrete graph of 1st level. Thus the proper black knots of the graph, being weakened signals of keys *Ti*, are united by single connections, $i = 1,.,N$. A component graph of 1st level is the basis for the construction of discrete graph of 2nd level.

4. Algorithms for the discrete graph of 2nd level, et cetera till *N*-th number of levels 2, are formed as per item a, with the change that in place of keys with time of *TN*, keys are utilized accordingly with time of *TN-1* et cetera, till time of T_1. As a result of the expounded procedure, *N* number of discrete counts of the system will be formed.

5. The final count of the system is further formed, which turns out to be a combination of the initial and *N* of discrete counts of the system. By single connections, the accordance of knots of discrete counts is set with the entrance knots of the initial count.

6. In the algorithm of Mason, the input-output correlations are determined for the system variables of multirate continuously-discrete system with multiple times.

4. Conclusion

The offered method provides a formalised procedure of construction of input-output correlations of multirate continuously-discrete systems. The equivalent presentation of the mathematical model of multivariative multi-stage system makes it possible to provide the analysis for the one- time systems, and it simplifies greatly the whole process of research and allows developing the elements of computerized analysis systems and systems of designing multivariative multi-stage discrete-continuous control systems.

Observed multirate systems applications in the form of equivalent one-time models give common reasons for process analysis in given one-time systems. While observing resultant correlation, it is not difficult to see that in all cases impulse images of equivalent models outputs have the form of rational functions of given variables. Turning to Z–images we will have an equation defining the output images as of z. They correspond to:

$$Y(kT) = \frac{1}{2\pi j}\oint f(z)z^{k-1}dz \tag{14}$$

The closed contour of integration in equation (14) covers all poles f(z) [11]. Thus, the problem of process analysis in given multirate systems becomes reduced to the problem of rational functions of pole distribution analysis f(z) concerning a single circle.

5. References

[1] Araki M., Yamamoto K. (1986). Multivariable multirate sampled-data systems: statespace description, transfer characteristics, and Nyquist criterion, IEEE Trans. On Control, vol.31.

[2] Sauro Longhi (1994). Structural properties of multirate sampled-data systems, IEEE Transaction on Automatic Control, vol. 39, No. 3.

[3] Berg M.C., Amit N., Powell J.D. (1988). Multirate digital control system design, IEEE Trans. Auto. Contr., vol.AC-33, No.12.

[4] Svetlichny, P., Kramar, V. (2009). Building of festo servo-motor imitation model using linear ARX-model, 2009 Annals of DAAAM and Proceedings of the International DAAAM Symposium

[5] Sekala A., Swider J. (2005). Hybrid graphs in modeling and analysis of discrete – continuous mechanical systems Proc. 13th International Scientific Conference on Achievements in Mechanical and Materials Engineering.

[6] Sedlar M. Bekey G. (1967). Signal Flow Graphs of Sampled–Data Systems: A New formulation IEEE Trans. on Automatic Control, AC-12, 2.

[7] Coffey T.C., Williams I.J. (1966). Stability Analysis of Multiloop, Multirate, Sampled Systems AIAA Journal, 4.

[8] Mason S.J. (1956). Feedback Theory – Further Properties of Signal Flow Graphs Proc. IRE, 41.

[9] Kuo B. (1986).Teorija i proektirovanie cifrovykh sistem upravlenija.-M.:Mashinostroenie.

[10] Kuo B.C. (1961). Composite Flow Graph Technique for the Solution of Multiloop, Multisampler, Sampled Systems IRE Trans. on Automatic Control, 1961.-Vol. AC-6, , pp.343-344

[11] Sain, M. K., Schrader C. B. (1990). The role of zeros in the performance of multi-input, multi-outputfeedback systems IEEE Transactions on Education №33.

DOI: 10.2507/28th.daaam.proceedings.012

MOBILE ROBOT FOR EXPLORATION AND MANIPULATION INSIDE A NUCLEAR POWER PLANT

Niklas Ritzberger, Simon Emenshuber, Gerald Schubert,
Florian Voglsinger, Wilfried Kubinger & Florian Fuchslocher

This Publication has to be referred as: Ritzberger, N[iklas]; Emenshuber, S[imon]; Schubert, G[erald]; Voglsinger, F[lorian]; Kubinger, W[ilfried] & Fuchslocher, F[lorian] (2017). Mobile Robot for Exploration and Manipulation Inside a Nuclear Power Plant, Proceedings of the 28th DAAAM International Symposium, pp.0093-100, B. Katalinic (Ed.), Published by DAAAM International, ISBN 978-3-902734-11-2, ISSN 1726-9679, Vienna, Austria
DOI: 10.2507/28th.daaam.proceedings.012

Abstract

It is not easy to handle a nuclear accident in a nuclear power plant and keep the damage as little as possible. To alleviate the accident of Chernobyl, personnel were harmed. In case of Fukushima people were also sent into dangerous areas in the power plant. The first European Robotics Hackathon (EnRicH) in June 2017 was a robotics competition inside of a nuclear power plant in Austria, which was never put into operation. Creating a nuclear heat map within a two-dimensional map, a three-dimensional map and the manipulation of nuclear sources were the main tasks of the competition. A mobile robot called taurob tracker from the taurob GmbH was equipped with a SICK TIM551 laser scanner, a Kinect for XBoxOne, a nuclear sensor, cameras and a gripper. The 2D-mapping was accurate if enough features were recognized within an area of maximum 10 meters around the robot. With the Kinect a coloured point cloud was created. Successfully gripping a single nuclear source took about one minute. With the described robotic system, the given tasks of EnRicH 2017 have been successfully completed.

Keywords: EnRicH; mobile robot; mapping; ROS; teleoperation

1. Introduction

All over the world nuclear power plants are used to provide energy for industry, cities and peoples private consumption. Producing such type of energy can be very dangerous for the area around the power plant. It is not only that contaminated water could escape out of the power plant. The one of the worst cases is a meltdown of a reactor, like the Chernobyl or Fukushima disaster. The emergency workers had been harmed and died from the acute radiation syndrome. The area around the power plant is also still contaminated. To avoid humans getting harmed during the emergency work, robots will do that until it is save enough to enter the area. For this a mobile robotic system is needed. It must be a teleoperated system so the operator can stay in a save area. Additionally it should create a map, do measurements and is able to interact with his environment, for example with a gripper.

1.1. Motivation

To evaluate the state of art for such robotic systems, European Robotics in cooperation with Fraunhofer FKIE, Energy-Supply-Lower-Austria (EVN) and the Austrian Armaments and Defence Technology Agency (ADTA), initiate the first European Robotics Hackathon (EnRicH) in June 2017. It is a type of competition which took place inside of the nuclear power plant in Austria which was never put into operation. There have been three tasks to fulfil. The first was to create a 3D-map of the scenario. Create a Radiation Map where sources have to be detected and marked inside the map. The Last task was to detect and pick up radiation sources and put them into a bin [3].

1.2. Problem description

A Chain-Driven robot with a four-axis arm with 4 degrees of freedom (DOF) has to be equipped with the adequate sensors and actors. So a gripper has to be developed and sensors have to be chosen to create the maps. To teleoperate the robot also a visual system has to be mounted. An autonomous driving mode shall be implemented which navigates the system autonomous through the map.

1.3. State of the art

To run such a robotic systems the popular Robot Operating System (ROS) is used. It provides a bunch of driver for sensors and actors but also functions for teleoperation, mapping, navigation and a lot more. ROS can also run across multiple machines, which is needed if the robot has a build in embedded pc and the operation is done with the operator pc. For example the map will be calculated on the embedded computer because it provides more performance and then it will be shown on the operator's pc so an orientation is possible.

Taurob provides several ROS packages. These packages enable the control of the robot and the arm. The body of the robot can be controlled ether via the "teleop_twist_keyboard" or the "teleop_twist_joy" package. In both cases the key bindings can be set within a ".yaml" file. The key difference is as the names suggest that the "teleop_twist_keyboard" package enables the control via a keyboard and the "teleop_twist_joy" package via a controller. The used controller can be set within the ".yaml" file. The arm can be controlled via the standard "joint_state_publisher" package. This reads the parameter of the robot and generates a graphic user interface (GUI). The axis can be controlled via sliders. Furthermore, two buttons are generated. The first button "Centering" centres all axis. The second button "Randomize" drives each axis to a random position. To note is, that all axis are moved at the same time. The reason for this behaviour is, that the values are first written into a variable and this variable is afterwards posted into the ROS topic.

The manipulator, which is mounted on the housing o the Taurob Tracker, is a robot arm with 4 degrees of freedom. The total reach is 140 cm and the maximum payload 5 kg. The robot arm can be used to manipulate objects or open doors. In order to do this, different end effectors are necessary, which do not exist yet. The end effector is the device that achieves an effect and is located at the furthest point of a kinematic chain. In this work, a gripper is developed as an end effector, which is attached to the end of the robotic arm by the Taurob and serves as an interface between the robot and the object to be manipulated.

To find the objects that have to be manipulated and to fulfil the tasks of the challenge, a method of orientation for the robot or its operator in the unknown environment has to be found. This is accomplished by using a method of machine vision to create two and three dimensional maps. An overview of the machine vision technologies is given in [12]. Multiple of these technologies are also already described in use cases similar to the ones at the EnRicH 2017 [2]. Approaches the task with stereo vision and inertial measurements to determine the position of the robot and its surroundings. Another considered method was the use of multiple laser scanners, similar to [10]. In this work the use of a time of flight system, namely the Kinect for X-Box One was used to test if such a system can work on a very small budget.

Navigation of a Robot means the autonomous movement through an area without moving orders from the operator, only the target has to be defined. Navigation is possible through programmed behaviour or path planning. The combination of these two are the state of art. To realize the navigation a classification of the environment and the robotic system has to be done. It is necessary to know the operational area of the robot. Is it an unknown or known area, is it static or dynamic and is it structured or not. A static area means that no moving objects, like humans or vehicles are expected. Structured area for example are buildings or mostly differed by indoor and outdoor operation. The robot itself is defined as a non-holonomic robot, with differential drive. For a navigation a model of the environment is needed. Either there is already a model of the area or the robot has to create his own. The robot has to localize itself inside the model trough camera systems, laser scanner, triangulation or a combination of these systems. With the model and the localization the robot can calculate a path and follow it with the programmed behaviour.

2. System Concept

The used robot is a so called tracker from taurob GmbH. It is a Chain-Driven platform with a mounted 4-axis arm. It is designed to use it in difficult terrain and there is also an ATEX certificated variant to protect the robot against explosions. It is shown in figure 1.

Fig. 1. Tracker robot without additional equipment

Inside of the white body are:
- electric motors
- WLAN Station
- one camera located at the glass in the middle
- another camera on the other side of the body
- LED lights at the decentred glasses
- the battery
- controller

A special feature of this robot is that the Chains can lift or lower partly to drive over stairs and obstacles. The arm has four degrees of freedom, turning around the z-axis and x-axis and movement along the y-axis and the z-axis. At the end of the arm there is a dummy sensor head [8].

2.1. Additional Hardware

To pass all the tasks sensors are needed and a gripper for manipulation. To get an overview over the environment of the robot two cameras got attached to the dummy sensor head at the arm. These help to drive through terrain with less space. A LED headlight was also installed to handle low light situations. For the 3D-mapping task the Microsoft Kinect was also attached to the head. The radioactive module SSM1+ which was provided was fixed on the arm and with a cable the sensor was fixed beneath the headlight. The At least the gripper got attached in front of the head. In figure 2 the head is shown but normally the Kinect is mounted on the side without the gripper.

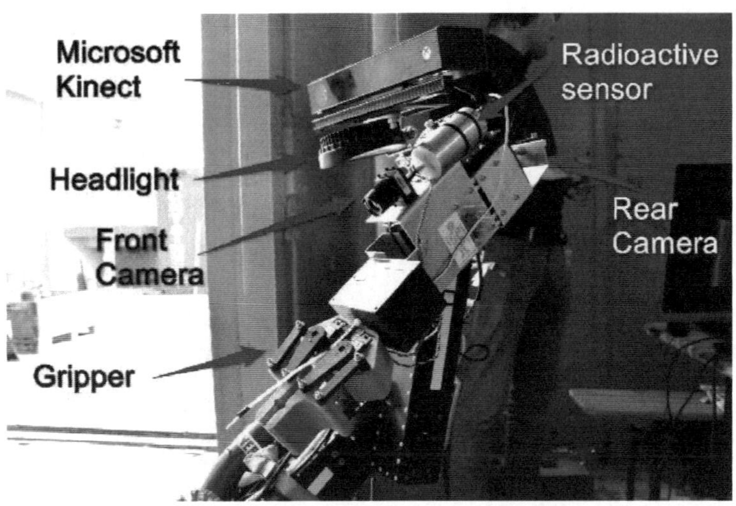

Fig. 2. Additional hardware at the end of the arm

On the Body an embedded pc is mounted. It controls the gripper, provides the data of the cameras and builds the 3D-map. To get a 2D-map a LIDAR from SICK is mounted in front of the robot above the chains. With the range of ten meters and angular resolution of one degree an accurate map can be built.

2.2. Robot Operating System

The Open Source framework 'Robot Operating System' (ROS) is the base to control the robot. Developed at the University of Stanford with the main targets:

- Peer-to-peer
- Tools-based
- Multi-lingual
- Thin
- Free and Open-Source

So ROS is a multi-host platform where a functions can work with all components. Functions can be written in different programming languages, the most common and best documented ones are Python and C++. Thin means to keep the function as simple and reusable as possible. Instead of running and building ROS functions in a large environment, a lot of small tools are doing this to keep the process as simple as possible [13].

2.3. EnRicH 2017

The EnRicH, which took place in the nuclear power plant (NPP) in Zwentendorf, Austria, is a competition which set the focus on detecting radiation sources, mapping and manipulation inside of a power plant. The testing area was the reactor hall on the last floor of the NPP. Around the fuel assembly changing basin and the fuel element cooling basin it was able act with the robot. The floor can be seen in the virtual tour on the Webpage of the NPP. The robot was brought up with a crane. Upstairs a Wireless Network was built up to stay in connection with the robot.

To create detailed 3D and 2D maps it was necessary to drive a complete round. The radiation sources should be measured and marked on a map. At one side of the basins radioactive sources and dummy sources where scattered and a marked bin stands beside. The challenge here was to differ between the real sources and the dummy ones. The next step was to pick up the real sources and throw them into the bin.

2.4. Strategy

Two valued runs were able to pass so the plan was to do the mapping and drive a whole round in the first run. The focus for the second run was on the manipulation task. Each run lasts 20 minutes. For the mapping challenge the Kinect was mounted without the gripper. One trial run was available which should be used to check if the time suffices to drive one time around the basins and create a sufficient map. The manipulation task has to be trained in the Pit lane with dummy sources.

3. Methods

With the chosen Hardware and ROS as base the needed parts got developed and implemented. The teleoperation, the gripper and the 3D-mapping was needed to complete all tasks. Additionally, an autonomous driving mode was implemented to drive back to start without orders from the operator.

3.1. Teleoperation

The teleoperation of the tracker can be split in three parts. These are the:

- Control of the body
- Control of the arm
- Control of the gripper

The control of the body two standard ROS-packages have been checked and compared. The "teleop_twist_keyboard" package enabled the control of the body through the keyboard. The speed of the robot can be changed via the push of a button, but the operator cannot vary the speed continuously [4]. The "teleop_twist_joy" package enables the control via X-Box controller. The controller can regulate the speed via an analogue stick [5]. Furthermore, a mode to halve the maximum speed of the robot was implemented by pressing the left bumper to get a more precise control.

The arm is controlled via a GUI. It allows the operator to individually position each axis. Furthermore, buttons with predetermined potions have been implemented. To achieve this the "joint_state_publisher" package has been used. It was altered to suit the robot and the given tasks.

The tree servomotors of the gripper were controlled via an Arduino micro. The Arduino was via the "Arduino" package integrated into the ROS-System. This enabled it to receive the servo positions via the UART bus. These positions were determined via a GUI. The distance between the jaws is displayed in millimetres in the GUI. To achieve this the servo angles are interpolated.

3.2. Gripper

In the development of a gripper for mobile and service robotics, in contrast to automation technology, neither the properties of the object nor its position and orientation in space are known. After consultation with the organizers, a photo of the object that has to be manipulated was published. This is a metal rod with the approximate dimensions of D = 12 mm x L = 110 mm and a mass of 0.1 kg. The rod is hold with a force-fit in the z-direction. The necessary gripping force can be calculated with equation 1.

$$F_G = m \cdot (a_z + g) \cdot \frac{\sin(\propto)}{2 \cdot \mu} \cdot S \tag{1}$$

With an acceleration of 1 m/s², a coefficient of friction of 0.5 (equivalent to rubber on steel), an angle α of 45° and a safety of 2, the gripping force is 1.53 N. Now that the gripping force has been calculated, the necessary motor torque can be calculated. The equation 2 calculates the motor torque.

$$M_M = F_G \cdot L \tag{2}$$

In this case, it is important to ensure that the length L of a parallel jaw gripper with circular movement changes with opening and closing. Therefore, the maximum length that occurs when the gripper is closed must be used. With a length L of 0.09 m, the resulting motor torque is 0.14 Nm. Now three Hitec HS-645MG servomotors have been selected, which provide a torque of 0.96 Nm.

The gripper was developed in the 3D CAD (Computer aided design) program SolidWorks 2016. During the construction of the individual parts, special attention was paid to the fact that the parts will be later produced with a Fused Deposition Modelling (FDM) 3D printer. The gripper essentially consists of two assemblies. The gripper itself and the rotary unit that allows to rotate the gripper by 90°.

3.3. 3D map

For the creation of the three dimensional map, a "Kinect for X-Box One" was used. This version of the Kinect uses multiple cameras, a time-of-flight version for the depth information and a colour camera for colour information. This two separate information get then combined via an internal chip to create an RGB-D image. RGB-D stands for Red-Green-Blue-Depth and is basically a coloured point cloud. The algorithm that produces this is assisted by the third built in camera, which senses infra-red light and helps separating background and foreground. To make use of this information that is provided over the USB3.0 port of the Kinect, the open-source driver for ROS "libfreenect2" is used [15]. This library converts all the data of the Kinect to ROS standards and therefore further usage is enabled. The information provided by this can be directly used by several ROS programs, like the visualization RVIZ which is used for the 2D-map and navigation. To generate 3D-maps the ROS package "rtabmap" was used. Like all the used ROS packages it is open-source. The program uses SIFT, Scale Invariant Feature Transform [9], and SURF, Speeded-Up Robust Features [1], for Feature detection in the RGB pictures. With this features it can determine which pixels are the same between multiple images and merge the respective point clouds. These merged clouds form the three dimensional map. The map can then be saved and viewed or loaded back into the program in a future mapping run and be continued or refined.

3.4. 2D map and Autonomous Navigation

The autonomous navigation was realized with a laser based Simultaneous Localization and Mapping (SLAM) and the algorithm called A Star Best first Search (A*) to navigate. SLAM is using the data of the laser scanner to create a map and locate the robot inside it. At Start-up of the system the robot creates a coordinate system with the centre point depending on his actual position. The algorithm matches the old and the new data from the laser scanner and a map is created when the robot is moving. The ROS-package 'Hector-SLAM' from university Darmstadt in Germany was implemented [6]. To create the heat map QGIS with the add-on 'heatmap' was used.

The algorithm A* calculates a path from goal to start. For that the map gets transformed to an occupancy grid. The fields of the grid can be occupied, free or unknown. If a field of the grid is partly covered by an object, it also counts as occupied [14]. The algorithm calculates always the fields around the actual field starting with the goal point. Each field get three values, the costs 'g(n)' from goal to the actual field, the estimated costs 'h(n)' from the actual field to the start and the calculated costs 'f(n)' for the complete way as it is shown in equation 3.

$$f(n) = g(n) + h(n) \tag{3}$$

The algorithm is used by the ROS standard package for navigation [7]. The package was created during a project where a robot should navigate through an office. It got implemented in the ROS System and calibrated for the robot [11].

4. Results

All tasks had been passed successfully. During the first run the Kinect malfunctioned, so only the 2D-Map with the radiation mapping was created which can be seen in figure 4. The second run was passed with mounted gripper and Kinect as it is shown in figure 1. Ten radioactive sources lied on the ground, five real and five dummy sources. Seven sources were lifted and five dummy sources were thrown into the bin by the operator. The way from start to the sources was mapped by the Kinect a picture of the 3D map is in figure 3.

4.1. Teleoperation

The two robot controls have been tested and their functions documented. The control via the keyboard allows the operator to change the maximum speed, but the robot can only drive at that speed. In case of the controller the maximum speed of the robot had to be set in the configuration files and the operator can drive at any speed between standing still and the maximum speed. To ensure the quality of the generated maps the maximum speed is capped at 40% of the maximum speed of the robot. This means the second, more precise mode runs at 20% speed.

Through tests the used GUI has been adapted to suit the robot and the challenges better. To prevent collision with the robot peripherals the end stops of the fist and the second axis had to be adjusted. The new end stop is for both axis is 0,2rad. Buttons with predetermined positions have been implemented. These are a button to centre each axis individually, a button for the camera position while driven and two buttons to position the gripper close to the floor. Furthermore, the arms reaction in case the connection between the operator and the robot is lost have been tested. The tests showed that the arm will always drive to the last received position and no further movement will be made until the connection is re-established. Than the arm will drive to the current position on the GUI. The end stops of the gripper have been determined. They were set in a way that the jaws meet exactly in the middle of the gripper. When no object is grabbed, the jaws have a gape of about 0,5millimeter between each other in the end position.

4.2. Gripper

To be able to test the parallel jaw gripper, steel pins were manufactured which are similar to the object EnRicH objects. The manipulating task of the EnRicH was trained at the military test ground Tritolwerk. The steel pins were placed vertically in a hole on a table. Now the gripper was used to pick up the steel pin and place it in a box. In order to find the exact position of the gripper in which the object can be gripped, a marked cable tie was attached to the gripper. The manipulation of the steel pin worked relatively well. But sometimes there were vibrations in the whole arm of the Taurob Tracker, which lead to shaking movements of the servomotors which are built into the gripper.

4.3. 3D map

The map could be created and downloaded from the robot via Wi-Fi or a USB-stick. Static environments, like in the testing scenario at EnRicH could be mapped without to many problems as long as the operator of the robot drives carefully. The mapping is volatile if the robot makes some jerky movements and the operator has to prevent these as good as possible. Additionally, areas with lots of open space should be avoided, because the depth sensors of the Kinect only have a limited range of between three and five meters. Another weak point of the program used to generate is, that it assumes that there are no dynamic surroundings.

Fig. 3. 3D map from start to the hazard area

4.4. 2D map and Autonomous Navigation

The 2D map was created with the SICK-Laser scanner. Because of the low range and the large distances a lot of problems came up especially when passing areas where only fence posts had been in range. In figure 4 the heat map is shown which was created during the first run. Two times the operator had driven too fast and too less reference points for the mapping tool had been in range so the map got a bend.

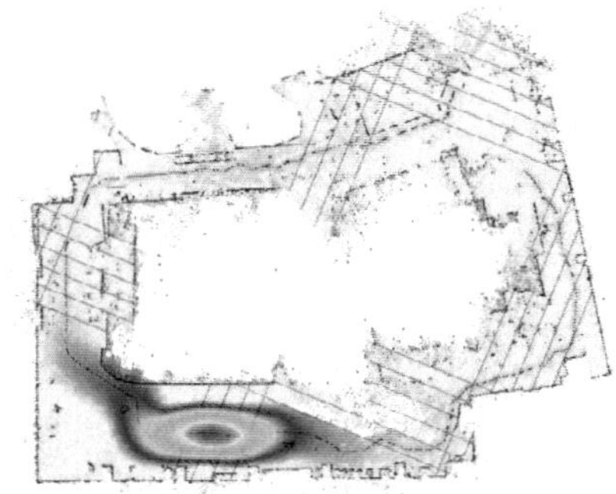

Fig. 4. 2D map of the reactor hall

The navigation wasn't used during the competition, because there have been too much complex objects like the fence posts and the danger to crash into a fence near the basins. Before the competition took place the navigation was tested with rooms and corridors. It worked after a lot of tests and calibrations.

5. Discussion

The setup of the robot was quite good. At the manipulation task the same amount of sources had been lifted compared to the winning team with automated manipulation functions. The predefined arm configurations made the whole process of grabbing a source very efficient. The Kinect worked well at the second run and the coloured point cloud is detailed and also small details can be identified.

5.1. Teleoperation

Through tests the handling via the X-Box controller proved to be more precise. Furthermore, the movement is smoother. This allowed for easier handling of objects and the map generation proofed to be more reliable. The GUI used to control the arm, proved to be reliable. The predetermined positions allowed the user to quickly move to often used arm configurations. To compensate a small offset between the two jaws, the endpoints of the jaws were adjusted. To ensure the jaws do not apply unnecessary force against each other while driving, the gap has been implemented.

The implementation of an inverse kinematic for the arm would be obvious. At the moment not all buttons on the controller are used. The implementation via the controller would assist the operator in controlling the robot.

5.2. Gripper

The stability of the 3D printed parallel jaw gripper was surprising. The vibrations occurred because the 0-axis of the robot arm temporarily readjusts its position. Due to the long lever, this leads to a swinging movement of the gripper. The servo motors tried to compensate these vibrations which resulted in shaking movements of the gripper jaws. To solve this problem, the mounting plate was reworked for the EnRicH. It was shown that the shaking movements do not occur when the gripper holds the steel pin under force fit.

5.3. 3D map

To improve the quality and reliability of the existing mapping solution, there are some parameters that could be tweaked. In the program there are options like the frame rate at which the map tries to refresh and how many common features are needed for two areas to count as identical. This tweaks come at the price of computation power. A more precise and reliable map generation therefore needs either a slower driving speed of the robot, which would have not been possible, given the time restriction of the challenge, or a more powerful PC mounted to the robot. More computing power comes with higher energy use and therefore shortens the time the robot can be operated between having to load its batteries. If the robot has to map a dynamic area, a new algorithm would be needed that can distinguish between moving and non-moving objects.

Another way of optimizing the results could be to combine this method with other kinds of localization, like the also implemented 2D SLAM, to get a better estimation of the robots position and therefore an easier recognition of images that could show the same features. Another solution would be to use 3D laser scanners, which are more precise and can work on greater distances but the map would lose the coloured point clouds.

5.4. *2D map and Autonomous Navigation*

The creation of the 2D map with the laser based SLAM is very efficient. The range of the laser scanner was a problem at the EnRicH. With a range above 20 meters the results would be much better, because in such an area more features can be recognized and the algorithm would be able to expand the map efficiently. Additionally the localization would be better which leads to better navigation results.

6. Conclusion

The first European Robotics Hackathon EnRicH 2017 was a competition where mobile robots had to solve tasks inside of a nuclear power plant with real radioactive sources. These tasks were to manipulation of objects, to create a 3D map of the scenario and to map the radioactive sources. The taurob tracker was equipped with an SICK TIM551 laser scanner, a Kinect for X-Box One, a nuclear sensor, cameras and a gripper. By the implementation of ROS the robot was controlled and all needed functions were available. The drive control was realized through an X-Box controller, the control of the arm and the gripper was done by a GUI on the operator computer. ROS packages had been used to create the 2D and 3D maps. All three tasks were successfully completed. The 2D map got a bend because of a driving mistake of the operator, but the map was still useable and the radioactive sources were correctly located. The Kinect had a failure at the first run. At the second run the Kinect mapped the way from the starting point to the hazard area where the manipulation task took place. Thanks to the Buttons in the arm control GUI the operator was fast enough to lift seven sources and throw five of them into the bin. With a laser scanner with more range a more precise map could be created easily because the operator would not have to be as carful while driving. The implementation of an inverse kinematic would make the arm control more intuitive because the operator can directly control the arm in three dimensions. This would mean the results would be much better. Also an Embedded PC with more power mounted on the robot would improve the quality of the 3D map.

7. Acknowledgments

This project has been funded by MA23 - City of Vienna within the Project Call 16-02 "Photonics: Foundations and industrial applications".

8. References

[1] Bay, H., et al, (2008). Surf: Speeded up robust features. In: Computer Vision and Image Understanding (CVIU). pp. 346 – 359.
[2] Dias, J., Lobo, J., & Queiroz, C. (2003). World feature detection and mapping using stereovision and inertial sensors. Robotics and Autonomous Systems, 44, 69-81. DOI: 10.1016/S0921-8890(03)00011-3
[3] http://enrich.european-robotics.eu/, (2016). European Robotics, The First European Robotics Hackathon, Accessed on: 30.08.2017
[4] http://wiki.ros.org/teleop_twist_keyboard, (2016). Jay, G., teleop_twist_keyboard, Accessed on: 2017-06-02
[5] http://wiki.ros.org/teleop_twist_joy, (2016). Purvis, M., teleop_twist_joy, Accessed on: 2017-06-02
[6] http://wiki.ros.org/hector_mapping, (2012). Kohlbrecher S., hector_mapping, Accessed on: 30.08.2017
[7] http://wiki.ros.org/navigation, (2017). Marder-Eppstein, E., navigation, Accessed on: 01.09.2017
[8] http://taurob.com/produkte/ugv-taurob-tracker, (2016). Taurob GmbH, UGV – taurob tracker, Accessed on: 30.08.2017
[9] Lowe, D. G., (2004). Distinctive image features from scale-invariant keypoints. In: International Journal of Computer Vision. pp. 91 – 110.
[10] Lepej, P., Lakota, M. & Rakun, J., (2016). Simultaneous Localization, Mapping and Scene Reconstruction, Chapter 07 in DAAAM International Scientific Book 2016, pp.069-076, B. Katalinic (Ed.), Published by DAAAM International, ISBN 978-3-902734-09-9, ISSN 1726-9687, Vienna, Austria DOI: 10.2507/daaam.scibook.2016.07
[11] Marder-Eppstein E., et al,(2010). The office marathon: Robust navigation in an indoor office environment, International Conference on Robotics and Automation, in press
[12] Pérez, L. et al., (2016). Robot Guidance Using Machine Vision Techniques in Industrial Environments: A Comparative Review. Sensors, 16(3), p.335. Available at: http://dx.doi.org/10.3390/s16030335.
[13] Quigley, M., et al, (2009). Ros: an open-source robot operating system, ICRA Workshop on Open Source Software, in press
[14] Thrun, S., Burgard, W. & Fox, D., (2006). Probabilistic Robotics, The MIT Press, ISBN 978-0-262-20162-9, Cambridge, US
[15] Xiang, L., et al, (2016). libfreenect2: Release 0.2. Available from: https://doi.org/10.5281/zenodo.50641 Accessed 2017-09-07

DOI: 10.2507/28th.daaam.proceedings.013

A COMPARISON OF TWO PARAMETER SAME SLOPE SEASONALITY AND HOLT-WINTERS EXPONENTIAL SMOOTHING MODELS

Hadis Bajric, Edin Kadric, Mugdim Pasic & Izet Bijelonja

This Publication has to be referred as: Bajric, H[adis]; Kadric, E[din]; Pasic, M[ugdim]; Pasic, M[ugdim] & Bijelonja, I[zet] (2017). A Comparison of Two Parameter Same Slope Seasonality and Holt-Winters Exponential Smoothing Models, Proceedings of the 28th DAAAM International Symposium, pp.0101-0108, B. Katalinic (Ed.), Published by DAAAM International, ISBN 978-3-902734-11-2, ISSN 1726-9679, Vienna, Austria
DOI: 10.2507/28th.daaam.proceedings.013

Abstract

In this paper two parameter Same Slope Seasonality forecasting model was developed. Analytical expressions for determining optimal parameter values of the developed two parameter Same Slope Seasonality forecasting model were developed too. Furthermore, performance of Two Parameter Same Slope Seasonality model was compared with performance of Holt-Winters exponential model as well as with performance of Same Slope Seasonality model. All tests were performed on M2-Competition time series. To determine the parameters of Holt-Winters exponential model, nonlinear mathematical programming was used. Performed tests proved that Two Parameter Same Slope Seasonality model is more successful than Holt-Winters exponential model and Same Slope Seasonality model. Expressions developed for determination parameters of two parameter Same Slope Seasonality forecasting model give parameter values identical to the values obtained by using Solver in MS Excel, set to solve nonlinear programming model, in a way that the problem is set up to search the parameter values which minimizes the mean squared error of the model.

Keywords: Holt-Winters model; Exponential smoothing; M2-Competition; Two Parameter Same Slope Seasonality forecasting model

1. Introduction

According to study of Green and Armstrong [1], regardless of the forecasting method increasing complexity leads to decrease in reliability. Authors further state that increase in complexity increases forecasting error for about 25 %. They suggest forecasters to use forecasting models they understand.
The reasons why Holt-Winters model (HW model) was taken for comparing to developed Two Parameter Same Slope Seasonality model (2pSSS model) and to Same Slope forecasting model (SS model) are:

- In time series analysis exponential smoothing methods are popular because they are straightforward and the whole forecasting procedure can happen automatically [2]. Yet it is often used in practice where it shows good performance.
- The HW model, also referred to as double exponential smoothing, is an extension of exponential smoothing designed for trended and seasonal time series. HW model is a widely used tool for forecasting business data that contain seasonality, changing trends and seasonal correlation [3].

- HW model is very robust and reliable, which in many studies shows close or better performances even than very complex ARIMA models [4, 5, 6, 7].
- HW model which is one of the most popular forecasting methods is utilized in cases where data show seasonality and/or trend. Probably the most important problem with HW model (apart from initialization) is setting the smoothing constants [8].
- Many companies use the HW model to produce short-term demand forecasts when their sales data contain a trend and a seasonal pattern. Fifty years old this year, the method is popular because it is simple, has low data-storage requirements, and is easily automated. It also has the advantage of being able to adapt to changes in trends and seasonal patterns in sales when they occur [9].

SS model and SSS model are distinguished by their simplicity and these models show better performances then some more complex exponential smoothing models. SS model, developed in [10], is one parameter model. SS model is of the same complexity as simple exponential smoothing model. In [10] testing and comparisons of SS model to fifteen exponential smoothing models are performed. Research results show that SS model is more efficient than simple exponential smoothing model. In [11] definition and test results of SSS model are presented. SSS model is one parameter model, which uses same logic as SS model, successfully models seasonality and with selection of adequate parameter value it successfully models trend component in available data. Research results show that developed model generates forecasts of the same reliability level as more complex exponential smoothing models.

In [12] analytical expression for estimate of optimal parameter value of SSS model are derived and presented. With this analytical expression SSS model has become even simpler for use. In the same research authors have compared performances of SSS model to HW model. Tests are performed on time series from M2-Competition [13] and SSS model has shown more reliable forecasts even SSS model is simpler than HW model. SSS model is one parameter model, while HW model is three parameter model. SSS model is defined by one expression and one parameter for level, trend and seasonality, while HW model uses three expressions and three parameters for every time series component. SSS model uses analytical expression for optimal parameter value estimation, while HW model requires use of nonlinear mathematical programming for estimation of optimal parameter values.

In this paper Two Parameter Same Slope Seasonality forecasting model (2pSSS model) is developed. This model should give better forecasts than SSS model because of its higher flexibility. In addition, analytical expressions for determination of both parameters are developed. Evaluation of developed model is performed on time series from M2-Competition [13]. Performances of developed model are compared to performances of SSS model and HW model.

2. Two Parameter Same Slope Seasonality model

SS model is based on idea that time series will have the same gradient (slope) in the next time period like in previous. Model is defined by (1), whose settings are presented in [10]:

$$\hat{X}_t(1) = X_t - \beta(X_{t-1} - X_t) \tag{1}$$

Where is:
X_t - Observed value of the time series in time t,
$\hat{X}_t(1)$ - Predicted time series value one step forward,
β - Coefficient (parameter) of forecasting model.

SSS model is based on the same idea that change in time series between t and $t+m$ will be the same to the change time series had in time period between $t-p$ and $t-p+m$. Graphical presentation of logic of SSS model is given on Fig. 1.

SSS model is also single parameter forecasting model defined by (2), whose settings are presented in [11]:

$$\hat{X}_t(m) = X_t - \beta\left(X_{t-p} - X_{t-p+m}\right) \tag{2}$$

Where is:
X_t - Observed value of the time series in time t,
$\hat{X}_t(m)$ - Predicted time series value m step(s) forward, forecasted at time t ($m \le p$),
m - Forecast horizon,
p - Number of seasons,
β - Coefficient (parameter) of forecasting models,

With respect to parameter β following cases can be considered:
- $\beta < 1$ - forecast smaller change rate in the period t to $t+m$ than in the prior period, e.g. forecasts smaller change rate in the period t to $t+m$ than in the period $t-p$ to $t-p+m$.

- $\beta = 1$ - forecast the same change rate in the period t to $t+m$ as in the period $t-p$ to $t-p+m$.
- $\beta > 1$ - forecast bigger change rate in the period t to $t+m$ than in the period $t-p$ to $t-p+m$.

Considering (1) and (2) it can be concluded that in the case $p = 1$, SSS model becomes SS model.

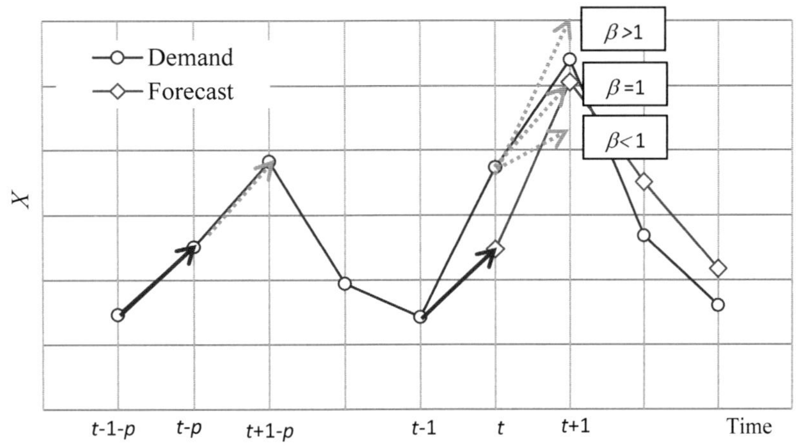

Fig. 1. Graphical presentation of logic of SSS model

Because of its mathematical simplicity SSS model allows finding analytical expression for determining optimal β parameter. Analytical expression for calculating β parameter is developed by [12]:

$$\beta = \frac{\sum_{t=p+1}^{n+m-1}(X_t - X_{t+m}) \cdot (X_{t-p} - X_{t-p+m})}{\sum_{t=p+1}^{n+m-1}(X_{t-p} - X_{t-p+m})^2} \tag{3}$$

Where n is number of data in time-series.

The value of β parameter calculated by (3) corresponds to the value determined using nonlinear mathematical programming, where standard deviation or MSE of model is minimized.

SSS model defined by expression (2) can be even more flexible if another coefficient is introduced. Adding another coefficient to SSS model, we define new model named Two Parameter SSS model by (4):

$$\hat{X}_t(m) = \alpha X_t - \beta(X_{t-p} - X_{t-p+m}) \tag{4}$$

Where is:
X_t - Observed value of the time series in time t,
$\hat{X}_t(m)$ - Predicted time series value m step(s) forward, forecasted at time t $(m \leq p)$,
m - Forecast horizon,
p - Number of seasons,
α - Coefficient (parameter) of forecasting models,
β - Coefficient (parameter) of forecasting models.

With respect to parameter α following cases can be considered:

- $\alpha < 1$ - The last available data has impact on the next forecast in the lower rate than its value is.
- $\alpha = 1$ - The last available data has impact on the next forecast in the same rate than its value is. In this case 2pSSS model become SSS model.
- $\alpha > 1$ - The last available data has impact on the next forecast in the bigger rate than its value is.

Considering (1), (2) and (3) it can be concluded that in the case $\alpha = 1$, 2pSSS model becomes SSS model. In the case $\alpha = 1$and $\beta = 1$ 2pSSS model becomes SS model.

Graphical presentation of the logic of SSS model is given on Fig. 2

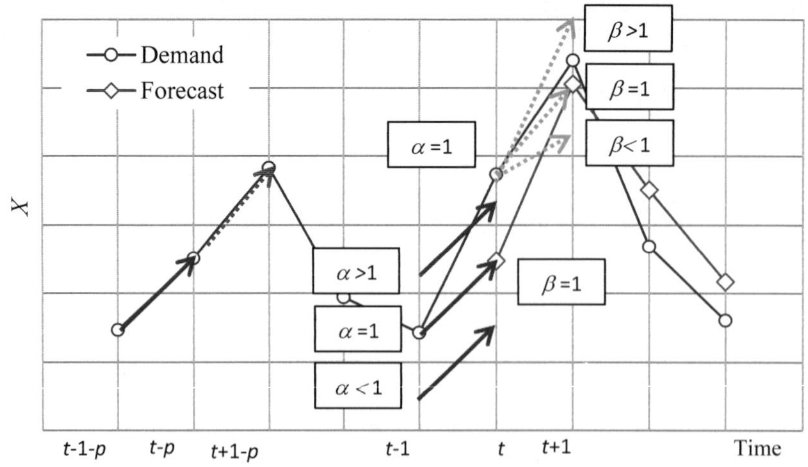

Fig. 2. Graphical presentation of logic of 2pSSS model

3. Derivation of analytical expressions for determination of optimal parameter values of 2pSSS model

Because of its mathematical simplicity, 2pSSS model allows derivation of expression for determining optimal α and β parameters. It is necessary to formulate function of squared deviations between forecasted and true values, dependent on α and β parameters, given by:

$$f(\alpha,\beta) = \sum_{t=p+1}^{n+m-1} \left[X_{t+m} - \hat{X}_t(m)\right]^2 \tag{5}$$

Where n is number of data in time-series.

Substituting (4) into (5) we get expression for sum of squared errors as the function of α and β parameters:

$$f(\alpha,\beta) = \sum_{t=p+1}^{n+m-1} \left[X_{t+m} - X_t - \beta\left(X_{t-p} - X_{t-p+m}\right)\right]^2 \tag{6}$$

In order to determine values for α and β parameters, equation (6) is differentiated with respect to each coefficient. The first partial derivative of the function of squared errors $f(\alpha,\beta)$ with respect to α is given by:

$$\frac{\partial f(\alpha,\beta)}{\partial\alpha} = \sum_{t=p+1}^{n+m-1} 2\cdot\left[X_{t+m} - \alpha X_t + \beta\left(X_{t-p} - X_{t-p+m}\right)\right]\cdot(-X_t) \tag{7}$$

The first partial derivative of the function of squared errors $f(\alpha,\beta)$ with respect to β is given by:

$$\frac{\partial f(\alpha,\beta)}{\partial\beta} = \sum_{t=p+1}^{n+m-1} 2\cdot\left[X_{t+m} - \alpha X_t + \beta\left(X_{t-p} - X_{t-p+m}\right)\right]\cdot\left(X_{t-p} - X_{t-p+m}\right) \tag{8}$$

Setting derivatives (7) and (8) equal to zero we get set of two linear equation with two unknowns (α and β):

$$\sum_{t=p+1}^{n+m-1} \left[X_{t+m}\cdot X_t - \alpha X_t^2 + \beta\left(X_{t-p} - X_{t-p+m}\right)\cdot X_t\right] = 0 \tag{9}$$

$$\sum_{t=p+1}^{n+m-1} 2\cdot\left[X_{t+m} - \alpha X_t + \beta\left(X_{t-p} - X_{t-p+m}\right)\right]\cdot\left(X_{t-p} - X_{t-p+m}\right) = 0 \tag{10}$$

Solution of this two linear equation system provides optimal α and β parameter values in explicit analytical form, defined by (11) and (12):

$$\alpha = \frac{\sum_{t=p+1}^{n+m-1} X_{t+m} \cdot X_t \cdot \sum_{t=p+1}^{n+m-1} (X_{t-p} - X_{t-p+m})^2}{\sum_{t=p+1}^{n+m-1} X_t^2 \sum_{t=p+1}^{n+m-1} (X_{t-p} - X_{t-p+m})^2 - \left[\sum_{t=p+1}^{n+m-1} X_t \cdot (X_{t-p} - X_{t-p+m})\right]^2} - \frac{\sum_{t=p+1}^{n+m-1} X_{t+m} \cdot (X_{t-p} - X_{t-p+m}) \cdot \sum_{t=p+1}^{n+m-1} (X_{t-p} - X_{t-p+m}) \cdot X_t}{\sum_{t=p+1}^{n+m-1} X_t^2 \sum_{t=p+1}^{n+m-1} (X_{t-p} - X_{t-p+m})^2 - \left[\sum_{t=p+1}^{n+m-1} X_t \cdot (X_{t-p} - X_{t-p+m})\right]^2} \tag{11}$$

$$\beta = \frac{\sum_{t=p+1}^{n+m-1} (\alpha X_t - X_{t+m}) \cdot (X_{t-p} - X_{t-p+m})}{\sum_{t=p+1}^{n+m-1} (X_{t-p} - X_{t-p+m})^2} \tag{12}$$

The value of α and β parameters calculated by (11) and (12) respectively corresponds to the value determined using nonlinear mathematical programming, where mean squared error or MSE of model is minimized.

4. Research methodology

Research methodology in this paper is same to the research methodology used in [12]. But here performances of 2pSSS model are compared with the performances of SSS model and with performances of three parameter HW model with additive trend and additive seasonal component, which is listed in standard list of exponential smoothing models [14]:

$$\left.\begin{aligned} S_t &= \alpha(X_t - I_{t-p}) + (1-\alpha)(S_{t-1} + T_{t-1}) \\ T_t &= \gamma(S_t - S_{t-1}) + (1-\gamma)T_{t-1} \\ I_t &= \delta(X_t - S_t) + (1-\delta)I_{t-p} \\ \hat{X}_t(m) &= S_t + mT_t + I_{t-p+m} \end{aligned}\right\} \tag{13}$$

Where is:
X_t - Observed value of the time series in time t,
$\hat{X}_t(m)$ - Predicted time series value m step(s) forward, forecasted at time t ($m \le p$),
m - Forecast horizon,
p - Number of seasons,
S_t - Smoothed level of the series, computed after X_t is observed,
T_t - Smoothed additive trend at the end of period t,
I_t - Smoothed seasonal index at the end of period t,
α - Smoothing parameter for the level of the series,
γ - Smoothing parameter for the trend,
δ - Smoothing parameter for seasonal indices.

Tests are performed on times series “M2CAll” from M2-Competition available in [15]. Last year data are used to estimate MAPE of the considered models, and all other data are used to estimate optimal values of forecasting models parameters.

Two Parameter SSS model parameters are determined using (11) and (12) for the case $m = 1$. SSS model parameter is determined using (3) also for the case $m = 1$. To determine HW model parameters, nonlinear mathematical programming is used. Nonlinear mathematical programming models are solved using Solver in Microsoft Excel application. The model standard deviation is used as the objective function of nonlinear mathematical programming model, also for the case $m = 1$.

Forecasts are generated for time horizon of one year. After forecasts are generated, true time series values are used to estimate reliability measures. To compare model performances MAPE is used as a measure of model reliability. Forecasts are generated using only quantitative data and without any available qualitative information.

Tests are performed on 28 of 29 time series from M2-Competition. Time series INTERSAL haven’t been used in tests, because its value in the last year is 0, which prevents calculating MAPE value. Among 28 analyzed, 22 were monthly and 6 quarterly time series.

5. Results

The developed expressions for determination α and β parameters in 2pSSS model, for each of the tested time series, give identical values for both α and β parameters to the values obtained by using Solver in MS Excel. The value of α and β parameters calculated by (11) and (12) respectively corresponds to the value determined using nonlinear mathematical programming, where mean squared error or MSE of model is minimized.

MAPE values for both models and all analyzed time series are given in Table 1. It can be seen from Table 1. that 2pSSS model has better performances for 15 time series, while HW model has for 13. Also it can be seen that 2pSSS model has better performances for 16 time series, while HW model has for 12.

Average MAPE values for 2pSSS, SSS and HW are 25,56%, 24,25%, 24,03% respectively. If three time series (PANTER, PHARMA11 and TRADUSAL), for which both models have high uncertainty levels, are omitted from analysis, then average MAPE values for 2pSSS, SSS and HW are 11,64%, 11,80% and 12,41%, respectively.

Series	HW model	SSS model	2pSSS model
PANTER	284,54%	259,65%	253,75%
CHEETAH	18,39%	17,43%	14,66%
LION	29,45%	37,15%	36,31%
BIGCAT	13,50%	15,34%	15,01%
ANIMAL	14,47%	16,94%	16,14%
CARINDS	13,03%	12,85%	12,53%
REALGNP	0,45%	7,06%	2,11%
IPDGNP	0,17%	0,79%	0,29%
IPDIMP	2,55%	0,97%	1,46%
FIXINV	4,60%	4,43%	4,73%
COMPIND	1,44%	0,35%	0,35%
BUSINV	10,24%	7,48%	13,22%
OEMUSAL	18,00%	10,32%	10,86%
TRADUSAL	36,32%	58,16%	44,33%
FLMARKT	22,54%	8,26%	8,97%
RESDVSAL	10,87%	7,79%	7,33%
INDSHIPF	10,49%	9,30%	9,27%
PHARMA11	41,73%	102,92%	90,02%
PHARMA22	30,04%	21,50%	23,21%
PHARMA33	28,48%	23,59%	24,10%
METOGEN	24,25%	23,64%	22,18%
CARDERS	23,23%	22,40%	21,09%
NEWRX	3,66%	2,97%	2,98%
REFILLS	4,32%	2,73%	2,73%
UNCPAPEUR	7,08%	18,37%	17,92%
CPAPEUR	8,59%	10,21%	10,20%
UNCPAPUSA	5,34%	6,99%	7,16%
CPAPUSA	5,18%	6,17%	6,13%

Table 1. Comparative display of MAPE values for all analyzed time-series

It can be concluded, based on test results shown in Table 1 that 2pSSS forecasting model shows better performances than more complex HW forecasting model on respectable number of time series. The same can be concluded for 2pSSS with regard to SSS model. Important advantage of 2pSSS in relation to HW model is existence of analytical expressions for direct calculation of optimal α and β parameter values. For estimation of optimal parameter values of HW model it is necessary to use nonlinear mathematical programming and appropriate solver applications. Importance of developed model is even bigger if it is known that HW model is very robust and reliable, which in many studies shows close or better performances even than very complex ARIMA models [4, 5, 6, 7].

Tests show that 2pSSS model is less sensitive on current observation in time series which is main disadvantage of SSS model. So, 2pSSS model obtains better results than SSS model, in almost all time series with outliers. For example, see MAPE values for time series PANTER, LION, TRADUSAL, PHARMA 11, in Table 1.

Even accuracies of 2pSSS model and SSS model forecasts significantly depend on current observation in time series, and it does not mean that models are not usable in such situations. We consider that forecasting of such time series requires inclusion of some qualitative information. For comparison, HW model which usually has more stabile forecasts, when forecasting such time series can result with quite poor forecasts. One of examples of such time series is PANTER, shown on Fig. 3. For PANTER time series, MAPE values for SSS model, 2pSSS model and HW model are 259,65%, 253,75%, 284,54%, respectively. From the MAPE values it can be seen that HW performed with highest value, eg. with poorest forecasts.

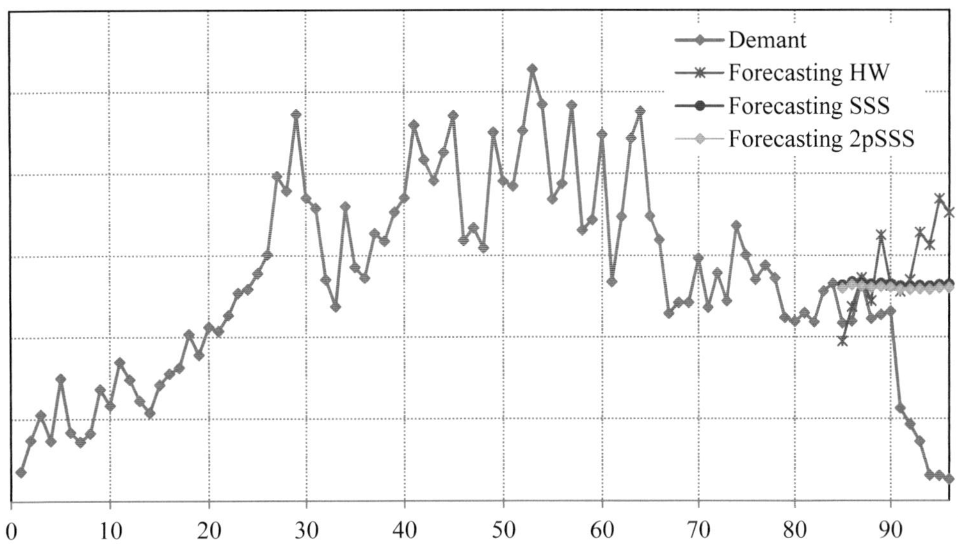

Fig. 3. PANTER Time series

6. Conclusion

The new 2pSSS model, which was developed in this paper, is a very simple and easy to use model. The research results have shown that new 2pSSS model has great potential which worth and which need to be examined in different situations. Two Parameter SSS model shows better results than the famous HW model in the tests performed on 28 M2-Competition time series. Two Parameter SSS model has less MAPE value on 15, while HW model has on 13 time series. Also 2pSSS model shows better results than SSS model. Two Parameter SSS model has less MAPE value on 16, while SSS model has on 12 time series.

In this paper, the analytical expressions for determining optimal parameter values of the developed Two Parameter Same Slope Seasonality forecasting model were developed too. The idea for developing these analytical expressions is simple and it has same logic as a least squares method. Determine the value of parameters which give the best fit forecast. These analytical expressions, for each of the tested time series, give α and β values identical to the values obtained by using Solver in MS Excel, set to solve nonlinear programming model, in a way that the problem is set up to search the α and β parameters which minimizes the mean squared error of the model.

Comparative advantage of 2pSSS model to HW model especially is important when considering costs of parameter determination. Parameters of 2pSSS model are simply determined using analytical expressions, while parameters determination in case of HW model requires solving nonlinear mathematical programming model. Besides this obvious advantage, 2pSSS model is easier to understand and implement than HW model, and also it requires less computational operations and efforts in forecast generations. Forecasts generated by 2pSSS model are of the same or better reliability levels as forecasts generated by more complex HW model.

Two Parameter SSS model performs well in forecasting time series with dominant seasonal, as well as trend component. Unreliable forecasts by 2pSSS model can be expected in case of non-standard observations in the time series, because 2pSSS model forecasts are very dependent on current observation for which forecasting value is generated. However, tests have shown that 2pSSS model is less sensitive on current observation in time series which is main disadvantage of SSS model. Finally, developed 2pSSS model performs better than SSS model. Advantage of 2pSSS model is especially emphasized in time series with outliers and in time series with short term interventions (quantum change of time series level).

It can be concluded that developed 2pSSS model, as well as SSS model, could be more effective if, for different seasons, different β parameter values are used. Use of this approach would not make models more complex, but it requires estimation of number of different β parameter values equivalent to number of different seasons. Even with additional calculations, 2pSSS model and SSS model would be simpler than HW model. Besides, the future research should be focused on testing developed model on other cases in different fields.

7. References

[1] Green, C. K. & Armstrong, J. S. (2015). Simple versus complex forecasting: The evidence. Journal of Business Research, Volume 68, Issue 8, pp. 1678–1685

[2] Crevits, R. & Croux, C. (2016). Forecasting with Robust Exponential Smoothing with Damped Trend and Seasonal Components, Working Paper, Available from: https://www.researchgate.net/profile/Ruben_Crevits/publication/310638788_Forecasting_with_Robust_Exponential_Smoothing_with_Damped_Trend_and_Seasonal_Components/links/58347e5808ae102f073957f1.pdf Accessed: 2017-07-05, DOI: 10.13140/RG.2.2.11791.18080

[3] Gelper, S.; Fried, R. & Croux, C. (2010). Robust Forecasting with Exponential and Holt-Winters Smoothing. Journal of Forecasting, 29, 285–300.

[4] Makridakis, S. & Hibon, M. (2000). The M3-Competition: results, conclusions and implications. International Journal of Forecasting, 16 (2000) 451-476.

[5] Puthran, D.; Shivaprasad, H. C.; Keerthesh Kumar, K. S. & Manjunath, M. (2014). Comparing SARIMA and Holt-Winters' forecasting accuracy with respect to Indian motorcycle industry, Transactions on Engineering and Sciences 2 (5) (2014)

[6] Tanvir, K. (2011). Identifying an Appropriate Forecasting Model for Forecasting Total Import of Bangladesh, International Journal of Trade, Economics and Finance, 2 (3) (2011)

[7] Walters, Q. C. (2008). Investigating the use of Holt-Winters time series model for forecasting population at the state and sub-state levels. Demographics and Workforce Section, Weldon Cooper Center for Public Service, University of Virginia, 2008.

[8] Williams, T. M. (1987). Adaptive Holt-Winters Forecasting, Journal of the Operational Research Society 38 (6) pp. 553–560.

[9] Goodwin, P., 2010. The Holt-Winters approach to exponential smoothing: 50 years old and going strong. Foresight: The International Journal of Applied Forecasting 19, 30–33.

[10] Pašić, M.; Bijelonja, I.; Šunje, A. & Bajrić, H. (2007). Same Slope Forecasting Method, Proceedings of the 18th International DAAAM Symposium, B. Katalinic, (Ed.), (2007) 547-548.

[11] Bajrić, H.; Bijelonja, I. & Pašić, M. (2009). A Comparison of Same Slope Seasonality and Exponential Smoothing Forecasting Models, Proceedings of the 20th International DAAAM Symposium, B. Katalinic, (Ed.), (2009) 1343-1344.

[12] Bajric, H.; Kadric, E. & Pasic, M. (2016). A Comparison of Same Slope Seasonality and Holt-Winters Smoothing Forecasting Models, Proceedings of the 26th DAAAM International Symposium, pp.0062-0068, B. Katalinic (Ed.), Published by DAAAM International, ISBN 978-3-902734-07-5, ISSN 1726-9679, Vienna, Austria, DOI:10.2507/26th.daaam.proceedings.009

[13] Makridakis, S.; Chatfield, C.; Hibon, M.; Lawrence, M.; Mills, T.; Ord, K. & Simmons, L. F. (1993). The M2-Competition: A real-time judgmentally based forecasting study, International Journal of Forecasting, 9 (1993) 5-22.

[14] E. S. Gardner, Exponential smoothing: The state of the art – Part II, International Journal of Forecasting, 22 (4) (2006) 637-666.

[15] IIF, International Institute of Forecasters, M2-Competition, Available from: http://forecasters.org/resources/time-series-data/m2-competition/ Accessed: 2015-05-30

DOI: 10.2507/28th.daaam.proceedings.014

PROPOSAL FOR THE IMPLEMENTATION OF THE EUROPEAN RAIL TRAFFIC MANAGEMENT SYSTEM (ERTMS) TO THE RAILWAY NETWORK IN THE REPUBLIC OF CROATIA

Tomislav Josip Mlinaric, Bozica Rados & Marko Vajdic

This Publication has to be referred as: Mlinaric, T[omislav] J[osip]; Rados, B[ozica] & Vajdic, M[arko] (2017). Proposal for the Implementation of the European Rail Traffic Management System (ERTMS) to the Railway Network in the Republic of Croatia, Proceedings of the 28th DAAAM International Symposium, pp.0109-0117, B. Katalinic (Ed.), Published by DAAAM International, ISBN 978-3-902734-11-2, ISSN 1726-9679, Vienna, Austria
DOI: 10.2507/28th.daaam.proceedings.014

Abstract

The European Rail Traffic Management System (ERTMS) is a project promoted by the European Union (EU) with aim to increase security and interoperability and reduce the cost of maintaining railway systems in Europe. The aim of ERTMS is to replace the traditional signal-security systems of individual countries and to enable the creation of a single integrated European rail system which would result in increased competitiveness of European railways at the interstate level. For Croatia as an EU member through the European Corridors TEN-T, implementation of ERTMS is necessary in order to better integrate the Croatian railway network into a single European network. This work will present a deep analysis of the current status of the railway network as well as strategies for further implementation and integration of the ERTMS system into the railway network in Croatia.

Keywords: ERTMS; GSM-R; ETSC; interoperability; safety

1. Introduction

European Rail Traffic Management System (ERTMS) is a project induced by the EU with an objective to increase security and interoperability, and to reduce maintenance costs of railway systems in Europe, thus harmonizing the technical railway standards between the Member States. The reduction of railway traffic in international traffic is caused by the popularity increase of road and air traffic, whose advantages with border crossings have made the railways uncompetitive. This policy's objective is to bridge the gaps between traffic networks of Member States, to remove the bottlenecks that still hamper the smooth functioning of the internal market and to overcome technical obstacles.

The main goal of ERTMS is to replace the traditional signal-security systems of individual states and to enable the creation of a unified integrated European railway system that would result in the increase of European railway competitiveness on international level. [1]

After analysing of the current situation and future capacity needs and demands expected from the railway in Croatia, it can be concluded that the implementation of the ERTMS is one of the necessary measures with the aim to eliminate potential bottlenecks on the network, increases current capacity of the railway network in whole, improve the interoperability of railway transport, increase of competitiveness of the railway network and significantly raise the level of quality and safety. Moreover, in order to integrate the national railway network into unified European network and satisfy the defined technical conditions for the base and overall TEN-T corridors (the Mediterranean Corridor and the Rhine-Danube Corridor), the authors of this article proposed the phasing (working steps) and optimal time schedule for the implementation of the ERTMS system on the Croatian railway network.

2. ERTMS-European Rail Traffic Management System

European Rail Traffic Management System (ERTMS) is composed out of two systems: European Train Control System (ETCS) and Global System for Mobile Communications - Railways (GSM-R).

European Union has launched, with Directive 96/48/EZ from 23 July 1996 about the interoperability of Trans–European railway system of fast trains and Directive 2001/16/EZ from 19. March 2001 about the interoperability of Trans – European conventional railway system, a procedure that would enable daily use of different railway systems of Member States, thus enabling safer and undisturbed passage from one national railway network to another. Within the scope of these Directives, several technical solutions were devised (TSIa). Those solutions primarily refer to basic elements, such as train operation and their safety, telematics applications in freight transport, professional qualifications of employees in international traffic, freight wagons and sound emissions.

In order to reduce technical obstacles of train interoperability the representatives of the railway sector and the Commission have signed, in 2005, 2008 and 2012, Memorandums about introduction and development of the European system of railway signalization ERTMS/ETCS. They were aiming to harmonize twenty different systems of signalization, which currently exist in Europe, and to implement an automatic and uniform speed check, which is based on technological achievements in the field of telecommunications. In 2009, the European Commission adopted a European plan for the gradual introduction of the ERTMS system into the major European corridors within a decade.

During 2013 the European Commission and the ministers of traffic of Member States of the EU have made a final decision about the guidelines of EU's new traffic policy. The Regulation was adopted, (EU) No. 1316/2013 of the European parliament and the Council from 11th December 2013, about the establishment of an Instrument for Connecting Europe, the amendment of the Regulation(EU) No. 913/2010 and the repeal of the Regulations (EC) No. 680/2007 and (EC) No. 67/2010. This Regulation has connected the existing traffic network of European roads into a Trans-European traffic network [2], [3].

The aim of creating a unified traffic network is to remove bottlenecks on European traffic routes, improve the infrastructure and connect different types of transport into a multimodal traffic across the EU.

Considering that the Directive 2008/57/EZ has been significantly changed several times, due to clarity and understanding of further changes, the Directive on 11th May 2016 has been modified into Directive (EU) 2016/797. This Directive determines the conditions that must be fulfilled in order to achieve interoperability in the EU's railway system in accordance with the Directive (EU) 2016/798 in order to define the optimal level of technical harmonization, enable facilitation, improvement and development of railway transport in the EU and with other non-Member States, thus contributing to the completion of a unified European railway space and a gradual creation of internal market. These conditions are referring to planning, construction, putting in use, modernization, renewal, work and maintenance of parts of that system, professional qualifications, health and safety conditions which are applied to employees and benefit their work and maintenance. This Directive for each subsystem sets out the provisions on the interoperability constituents, interfaces and procedures and the conditions for the overall compatibility of the railway system of the Union which are necessary to achieve interoperability.

ERTMS is the European standard for ATP that enables the interoperable system of railways in Europe.

ERTMS and ATP are security systems that enable the conformity of trains with speed limitations and signalization.

The purpose of the system is to help in the area of signalization and in the system of train management, which represents one of the biggest obstacles in development of railway traffic. Unification of multiple signalization systems will improve competitiveness and enable better interoperability of freight and passenger traffic. Furthermore, it will also encourage European market of railway equipment, reduce costs and improve overall quality and security of railway traffic.

ETCS is an automated system of train protection that is based on the signalization in driver's cabin and constant data transmission between the tracks and the train. [4].

It secures safe train traffic in all conditions by supplying the driver directly with data on the screen in driver's cabin and by constantly monitoring driver's actions.

The result of work by the European Commission and other telecommunications entities involved was the definition of GSM for railway (GSM-R) as digital radio platforms for railway's internal voice and data transmission, and enabling the function of provider for ETCS as a common CCS.

GSM-R and ETCS are integrated in a system of communication and signalization ERTMS.

Current mobile network GSM-R is a result of years of collaboration between several Europe's railway operators, railway industry and standardization bodies. Common goal is to achieve interoperability throughout Europe via one communication platform.

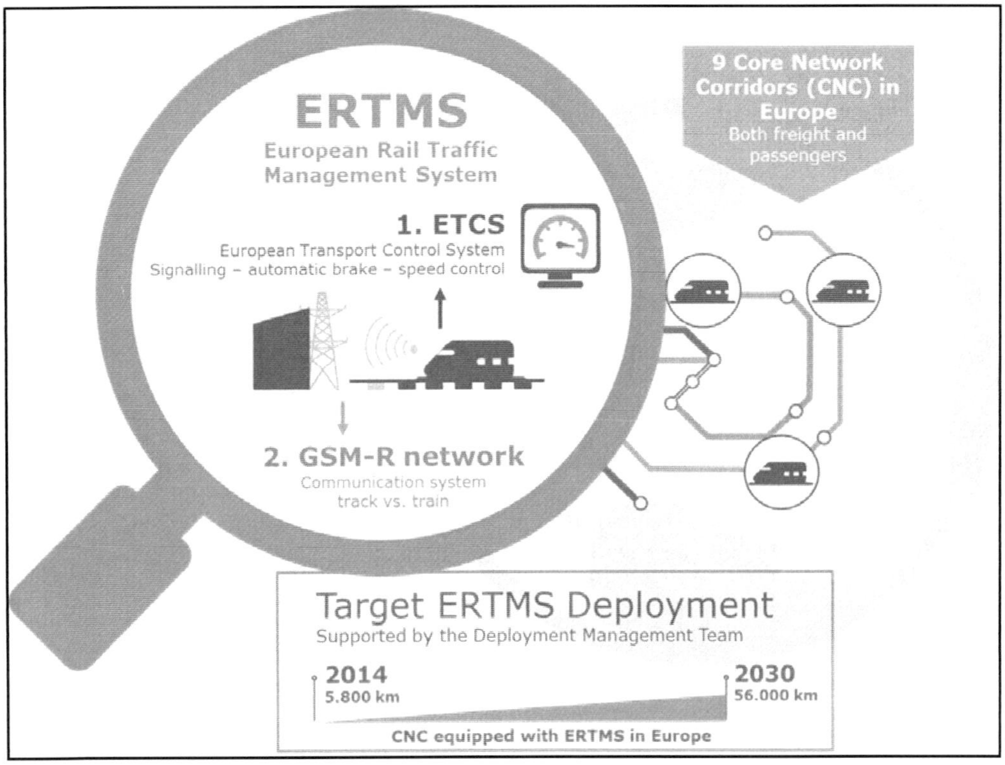

Fig. 1. ERTMS - European Rail Traffic Management System
Source: European Commission – Mobility and Transport – 2016

The current GSM-R standard encompasses the key functionalities and experiences of analogue radio systems that were used throughout Europe.

GSM-R platform enables a safe platform for voice and data transmission between the staff and the railway operator. The network supports advanced features such as dual voice calls, conference calls, location-based connections, priority calls, and emergency call retrieval. These functions significantly improve communication, collaboration and security aspects for railway operative staff.

As a part of ERTMS standard, GSM-R signal transmits information directly to the unit in the train, enabling higher train speeds and traffic density with high level of security.

The communication's platform for railway networks uses GSM technologies and specific applications for railway traffic management. GSM-R is developed under the sponsorship of the EU in order to successfully resolve the challenges of railway development, such as:

- interoperability with national and international communication systems of railway networks
- improved efficiency, security and reliability, with it passenger satisfaction
- differentiation of value added services in order to increase customer loyalty and research of other sources of revenue

GSM-R has already fulfilled the expectations in the first and second phase of implementation in Europe, and now operators in numerous countries around the world are implementing it.

GSM-railway is an international standard for wireless railway communication and applications. System ERTMS is used for the communication between the train and railway's control centres. It is based, inter alia, on the GSM technology and specifications of EIRENE and MORANE, which ensure accessibility even when the vehicle is moving up to 500 km/h without loss of communication.

3. ERTMS-corridors

"Corridors of base network" are introduced as a platform for a unified basic network[5]. They encompass public and private funds concentrating the support of the EU from CEF (Connecting Europe Facility), especially regarding the following:

- removal of bottlenecks
- construction of missing cross-border connection
- promotion of modal integration and interoperability

Their goals also include:

- integration of railway corridors for freight traffic (as a permanent modal measure, these corridors will be integrated into the multimodal TEN-T network)
- promotion of clean fuels
- other innovative traffic solutions
- improvement of telematic applications for efficient use of infrastructure
- integration of urban areas into a TEN-T network
- improvement of security

Nine corridors of the core network, which are set out in the Amendment of the Regulation about the establishment of the Instrument for Connecting Europe that includes the list of projects predetermined for potential financing from the EU during a period from 2104 to 2020, are determined based on their additional value for the development of TEN-T and on their maturity status.

TEN-T program:

The European Commission has created the TEN-T program in order to support the construction and the renewal of traffic infrastructure throughout the European Union.

TEN-T program is dedicating financial support in the realisation of important projects of traffic infrastructure – in accordance with the general scope of European competitiveness, job creation and cohesion.

Advantages of railway line equipped with ERTMS:

- The advantages that are expected from ERTMS can be summarized as follows: [6]
- Cross-border interoperability;
- Improving security in national and international railway traffic;
- Improvement of the management of international passenger and freight railway traffic;
- A shorter time interval of trains on railway lines with dense traffic, driving in APB mode, enabling the usage of maximum railway line capacity;
- The possibility of gradual implementation of a new technology;
- Enabling a Pan-European market competition among ERTMS/ETCS component manufacturers. Strengthening the position of the European railway industry in the world market;
- Enabling the preconditions for future harmonization in other areas of rail traffic management;
- The possibility to optimize traffic capacity.

4. The position of the Republic of Croatia in trans-European traffic network of Europe (TEN-T)

Connecting the core network of traffic infrastructure with trans-European networks and corridors is one of the main goals and an important prerequisite for an even development of all EU Member States. Thus, the EU continually invests efforts, in sync with the increasing number of Member States, in enablement of the construction of necessary roads and in integration of national road networks into a Trans-European road network TEN-T. Formation of this European road network will remove bottlenecks and connect distant regions into a common system of roads. TEN-T, in accordance with the proposition of Regulation about EU guidelines for the development of Trans-European road network from 19 October 2011 and with CEF, should be developed based on dual-layer approach, consisting from overall and base network.

The overall network represents the general layer of TEN-T and includes all the existing and planned infrastructure that complies with the prerequisites of the Guidelines, and needs to be finished as late as 31 December 2050. According to earlier interpretations of the European Commission, for Croatia it should be the network agreed in the frame P21 of the Trans-European network.

The basic network includes only the parts of the overall network that are strategically most important ones, and should be finished as late as 31 December 2030. The Decision of the European Commission from 18 October 2013 defines nine corridors of the Basic EU road network. They act as a basis for the connection of 94 main European ports and 38 key airports with railway and road in the capitals of European countries (Port of Rijeka and Zagreb's airport are among them), for the development of 15 thousand kilometres of railway infrastructure capacitated for achieving satisfactory speeds for freight and passenger trains and for 35 thousand border crossings [7], [8].

The Mediterranean corridor connects the ports of the Iberian Peninsula that are Algeciras, Cartagena, Valencia, Tarragona and Barcelona via southern France, with the connection to Marseille and Lyon towards northern Italy and Slovenia, and via a branch through Croatia and Hungary to the Ukrainian border. It covers railways, roads, airports, ports and internal waterways of the river Po. Key projects are the railway in Spain according to UIC norm, the railway tunnel Lyon-Torino and the railway across the karst area of Trieste/Kopar - Ljubljana.

The Rhine-Danube corridor connects Strasbourg and Mannheim via two parallel axes in southern Germany, one alongside the Main river and Danube river and other via Stuttgart and München, with a branch towards Prague and Žilina towards the border between Slovakia and Ukraine, through Austria, Slovakia and Hungary to Romanian ports of Constanta and Galati. It covers railways, roads, airports, ports, railway-road terminals and systems of internal waterways of river Main, channel Main-Danube, the entire stream of Danube downstream of Kelheim and of river Sava. With key projects the bottlenecks alongside the internal waterways and sections of railway lines Stuttgart-Ulm and München – Freilassing are removed. [5].

5. ERTMS implementation objectives in the Republic of Croatia

Objectives of implementing ERTMS:

It is very important that the objectives by the implementation of the ERTMS in every country of the European Union are the same.

Croatia is an official member of the European Union since the 1st. of July 2013. Because of this is important that the objectives will be implemented.

Through the abolition of national Systems (i. e. signaling, train control and other properties) will be guaranteed the continuous train traffic in Europe and Croatia.

The following table is an investment plan by the European Union. This includes measure for the implementation:

- The enhancement in signalling (especially ERTMS)
- Renewal of tracks
- The renewal of signallong system
- The renewal of tunnel, bridge etc.
- The electrification
- The creation of siding, passing tracks, extra tracks
- The creation of a new structure (line, bridge, tunnel,)
- Adjustment of gauge
- The track enhancement
- The level crossings
- The noise reduction
- And other projects

The objectives of implementing the ERTMS in Croatia were viewed through three key goals [3]:

- Establishment of interoperability
- Safety of railway system
- Capacity increase

ERTMS implementation in Croatia will result in numerous benefits: it will contribute to the harmonization of security systems of railway lines in Europe (LEU, signalling marks, balise, RBC and Euroloop) and of trains (EVC, interface driver-machine, balise, odometer).

The national APB system will no longer be needed. The trains that are coming from Croatia will be able to cross the border faster since it will no longer be required to change the system of APB whilst crossing the border. Furthermore, drivers will no longer have to be changed at the border crossing. ETCS enables higher density of trains and increased competitiveness of the railway system. Interoperability in our case means that trains have the possibility of driving on every section of the railway line within the EU and that they can use railway systems in the EU.

The benefits for the railway sector in the Republic of Croatia will be demonstrated through the following parameters:

- greater competitiveness for the rail sector
- unified signalization standard
- facilitation of cross-border traffic

Benefits for HŽI, the infrastructure manager in the Republic of Croatia, will be manifested through the following parameters:

- increased security
- limitation removal
- higher traffic speeds
- increased capacity on the lines
- high reliability

The benefit for railway operators in Republic of Croatia will be demonstrated by the fact that trains will be able to move across the European network equipped with ERTMS with only one signalling system, unlike the existing state in which several ATP systems have to be installed. This will reduce the costs of training and the training of train drivers and a waste of time at the borders because of the formalities and technical limitations.

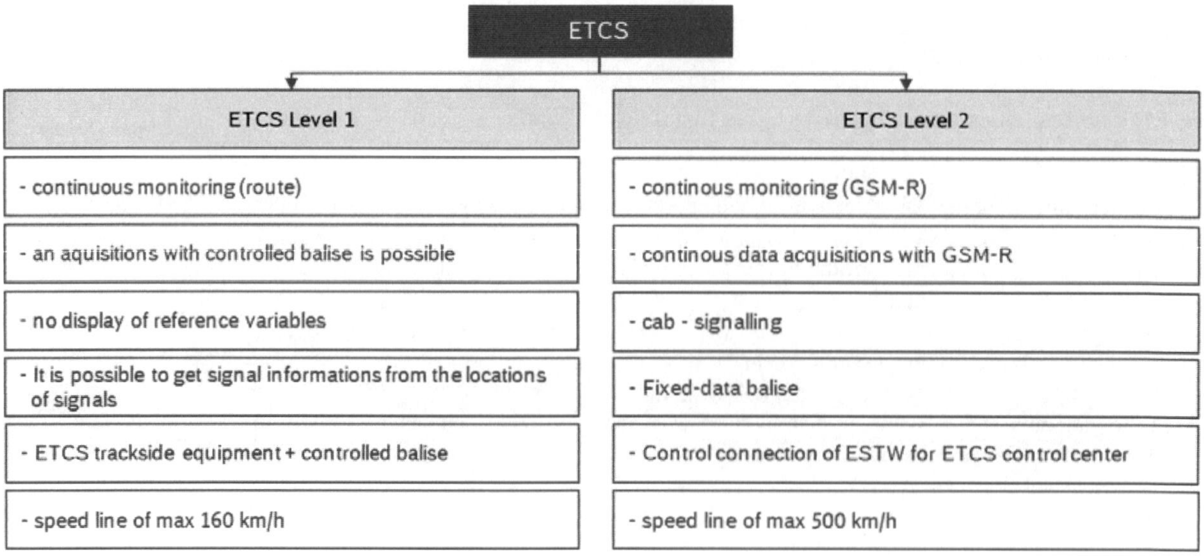

Fig. 2. Compare ETCS Level 1 and ETCS Level 2
Source: Study of implementing ERTMS in Croatia, 2016

5.1. Gradualness (phase) of ERTMS introduction in the network of the Republic of Croatia

Due to the aforementioned, [9], [10] in order to implement ERTMS in the Republic of Croatia the following items have to be considered:

1. Given the state of the signal-safety and traffic-control systems, as well as the currently open projects that are in the development stage of project documentation, it is possible to immediately access the implementation of ERTMS / ETCS level 1 on railway lines of the corridor. However, this solution does not allow, due to deterioration of SSD, a further upgrade to ERTMS / ETCS level 2.

2. The inexistence of the GSM-R network is complicating any further development or upgrade. Thus, it is recommended that the prerequisites be achieved before anything else, which are provided below in text for PHASE 1 and PHASE 2, after which it would be possible to attempt the completion of ERTMS network in Croatia. Relating to the GSM-R network, it is recommended to immediately approach the preparation / design of the network for voice and data system. Below is a description of the ERTMS implementation – choosing the right option of the proposed level, which is, as a solution, ready for only voice and/or voice + data upgrade as a solution.

ERTMS implementation in phase 1

ETCS Level 1

- Goal is to provide main international corridors RH 1 and RH 2 with continuous ETCS L1
- Design, procurement and construction of ETCS L1 in network gaps
- GSM-R (voice only)
- Parallel roll-out of GSM-R for main international corridors RH 1 and RH 2
- Setup of GSM-R core system with scalable upgrade option for nationwide roll-out
- Start and progress of roll-out depending on available funding

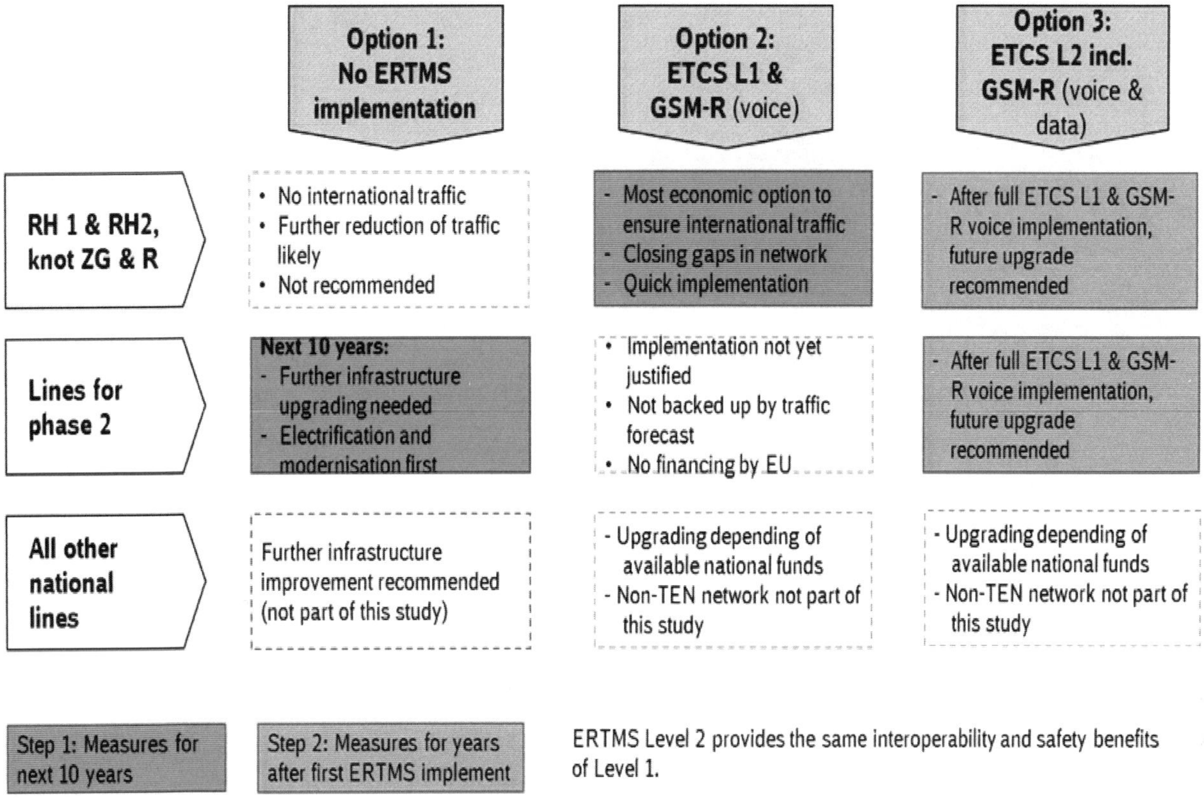

Fig. 3. ERTMS implementation objectives in the Republic of Croatia
Source: Study of implementing ERTMS in Croatia, 2016

5.2. ERTMS in Croatia – implementation plan for first 10 years until 2025: closing gaps in network to ensure international traffic

GSM-R (voice only)

- Parallel roll-out of GSM-R for main international corridors RH 1 and RH 2
- Setup of GSM-R core system with scalable upgrade option for nationwide roll-out
- Start and progress of roll-out depending on available funding

ERTMS implementation in phase 2

Establishing interoperability

- *Enable cross-border traffic without need to install many national ATP systems*
- *Save time when cargo or passenger trains cross border*
- *Increase competitiveness for railway operators and infrastructure managers*
- *reduce development time and investment costs for cross-border transport vehicles*
- *simplify vehicle registration, reducing of vehicles for international traffic*

Increasing capacity

- *Increasing line capacity with ERTMS Level 2:*
- *higher operational speeds*
- *reduced headways*
- *• increase the line speed*
- *• Using "moving block" technology*
- *• Still in conceptual phase*
- *• No capacity increase with ERTMS Level 1, but same safety and interoperability benefits*

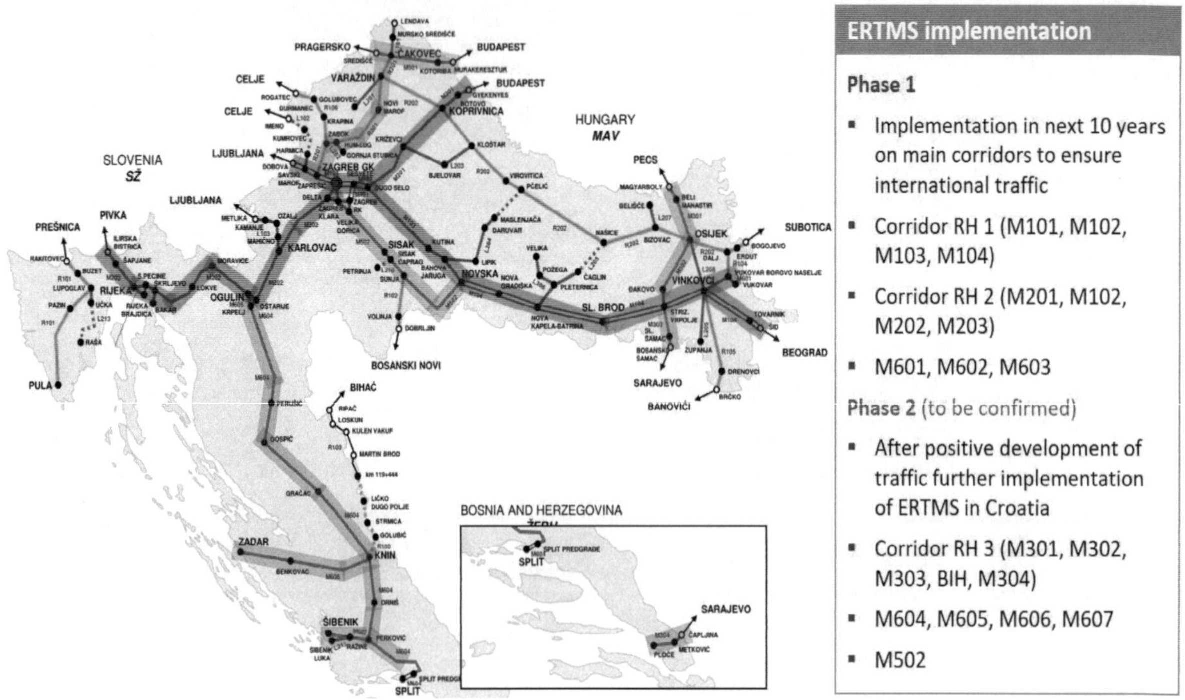

Fig. 4. ERTMS in Croatia – proposal on final implementation after 30 years
Source: Study of implementing ERTMS in Croatia, 2016

Enhancing safety and quality

- One signalling and safety standard in the European Union
- replace older ATP systems with a more secure option
- One radio communication platform for all railway functions, applications and services
- Decrease risk of railway traffic accidents through continuous communication
- High security in operational communication

5.3. Benefits for the railway sector: ERTMS introduction in the network of the Republic of Croatia

ERTMS can make the railway sector more competitive. One of the aims of ERTMS is to facilitate cross-border traffic by standardizing the signalling systems under just one standard, and to also reduce the technical and operational complexity of trains and wagons/carriages. The traditional drawbacks of international rail traffic l (e.g.: different gauge, different voltage, different signalling systems at each country…) makes traffic between European states more difficult.

ERTMS is intended to facilitate cross-border traffic and interoperability, at least in signalling. Regarding the European supplier industry, ERTMS is a European open standard that has promoted the European industry worldwide due to the existence of ERTMS deployment both in Europe and on other continents.

Safety Increase: ERTMS has a higher safety level compared with most of the existing national ATP systems. This is mainly achieved through the continuous supervision of the speed of trains Higher speeds: ERTMS allows operation speeds of up to 500km/h. This means that the signaling system installed along the side of the rail track will not hinder the maximum operation speed for ERTMS equipped vehicles. The first benefit is for the Infrastructure Manager given the elimination of this restriction. Higher capacity on lines: Depending on the previously installed system, ERTMS can reduce the minimum headway between trains (i.e.: the minimum distance or time between commercial service vehicles). This allows the Infrastructure Managers to manage a greater number of trains at the same time, and thus the railway system is able to transport of passengers or goods. Potential maintenance cost reduction: Since ERTMS (especially pure Level 2 installation) requires fewer signals at the side of the track, there is a possibility of reducing maintenance costs.

High reliability rates: this will involve achieving better punctuality for both freight and passenger services, since failures (and consequent delays) due to the signalling system can be even less likely thanks to the high safety standards of ERTMS components.

6. Conclusion

ERTMS implementation in Croatia will result in numerous advantages: it will contribute to the harmonization of safety systems of railway lines in Europe (LEU, signalling marks, balise, RBC and Euroloop) and of trains (EVC, interface driver-machine, balise, odometer) and it will increase the overall availability of the railway traffic infrastructure.

After the analysis of status of HŽI network and after the processing of collected data the conclusions, upon which the following recommendations were made, are the following:

Network analysis shows that currently there is sufficient capacity for planned traffic load, but it also shows that the status of railway line's infrastructure is, on individual sections, insufficient. This significantly influences the capacity itself (reducing speed of traffic and creating bottlenecks on corridor railway lines).

Given the current state of the SS device, and the inability to repair it, the guidelines and suggestions of the author to implement the ERTMS imply the development in two phases, and according to the previously described dynamics. Concerning ETCS level 2, it is recommended for both development phases, with the note that in phase 1, where the basic corridor network is closed, it is necessary to prepare i.e. to integrate already existing installed ETCS level 1 systems in order to pass onto ETCS levels 2.

The problem of integration and transfer to interoperability are inadequate platforms, landscaped grounds and unsatisfactory access for passengers, which are a potential threat in case of unoccupied facilities. An insight into the existing state of the building and the surface for receiving and shipping passengers on railway lines reveals the diversity in the level of equipment and arrangement of these elements. Work on existing platforms and landscaped areas that are reconstructed and/or upgraded, wherever possible, is being pursued to meet the requirements, to the fullest possible extent, of the special regulations regulating railway infrastructure subsystems as well as the relevant provisions of regulations which regulate spatial planning and construction. The investment plans provide for a solution to these problems. It is therefore recommended that the prerequisites for ETSC Level 2 and GSM-R be fulfilled: construction work to be done in terms of achieving interoperability requirements (reconstruction of the railway station, etc.), in order to install electronic devices.

Finally, the authors give their recommendation, while the final decision on the development strategy will be on the side of the Investor and the relevant Ministry, as well as decision on funding sources.

7. References

[1] Smith, P., Majumdar, A., Ochieng, W. Y. (2012). An overview of lessons learnt from ERTMS implementation in European railways. Journal of Rail Transport Planning & Management, Vol. 2, Issue 4, (December 2012) pp. 79-87, ISSN: 2210-9706

[2] Haramina, H., Brabec, D, Grgic, D. (2012). Influence of train control system on railway track capacity, In: DAAAM international scientific book 2012, Katalinić, B. (Ed.), pp. 419-426, DAAAM International Publishing, DAAAM International Vienna, ISBN 978-3-901509-86-5, Vienna

[3] Study of implementing ERTMS in Croatia, 2016, DB E&C, HZ Infrastructura

[4] https://ec.europa.eu/transport/modes/rail/ertms_enht, 2016, Mobility and transport, European Commission, Accessed on: 01-10-17

[5] https://ec.europa.eu/transport/themes/infrastructure/mediterranean_en , 2016, Mobility and transport, European Commission, Accessed on: 01-10-17

[6] https://ec.europa.eu/transport/themes/infrastructure/rhine-danube_en , 2016, Mobility and transport, European Commission, Accessed on: 01-10-17

[7] Haramina. H.; Brabec, D.; Štefančić, I. (2011). Influence of Train Control System Characteristic on Railway Infrastructure Capacity, Proceedings of 22nd DAAAM World Symposium "Intelligent Manufacturing & Automation: Power of Knowledge and Creativity"Katalinic, B.(Ed)., pp. 0357-0358, ISBN 978-3901509-83-4,23-26th November 2011, Vienna, Austria

[8] Zahar, N. A., Khan, S. A., Araki, K. (2012). Towards the Safety Properties of Moving Block Railway Interlocking System. International journal of innovative computing, Information and Control, Vol. 8, No. 8, (August 2012) pp. 5703-5713, ISSN 13494198

[9] UIC (2008). ETCS Implementation Handbook, ERTMS Platform, Infrastructure Department, ISBN 2-7461-1499-2, Paris UIC, International railways statistics, 2012

[10] Dincel, E., Kurtulan, S. (2012). Interlocking and Automatic Operating System Design with Automaton Method, Proceedings of the 13th IFSC Symposium on Control in Transportation Systems, The International Federation of Automatic Control, Stoilov, T. (Ed.), pp.191-196, ISBN 978-392823-13-7, Sofia, Bulgaria, September 2012

DOI: 10.2507/28th.daaam.proceedings.015

MEASURING OF FRICTION FACTOR OF CUTTING ENVIRONMENT

Marian Králik, Michal Bachratý, Marcela Pokusová & Numan M. Durakbasa

This Publication has to be referred as: Kralik, M[arian]; Bachraty, M[ichal]; Pokusova, M[arcela] & Durakbasa, N[uman] (2017). Measuring of Friction Factor of Cutting Environment, Proceedings of the 28th DAAAM International Symposium, pp.0118-0123, B. Katalinic (Ed.), Published by DAAAM International, ISBN 978-3-902734-11-2, ISSN 1726-9679, Vienna, Austria
DOI: 10.2507/28th.daaam.proceedings.015

Abstract

At metals machining the cutting environment has a significant impact on the quality of the workpiece surface, a tool life and energy consumption. The paper deals with the results of research focusing on monitoring the friction properties of the most widely used cutting fluids. Cutting fluids appear to be an important means of increasing the productivity of technological processes, and require a complex knowledge of problems in the field of tribology, the theory of machining and chemical-physical processes. For this purpose the design of a tribometer and the methodology for evaluating the measured values is described. The measurement results contained in the table can be assumed to functionality and reliability of the proposed tribological device.
The experimental part of the research forms the basis for further research into the use of cutting fluids in the machining of components. It will analyse the impact of cutting fluids on the quality of the surfaces of work-pieces and the energy consumption during machining.

Keywords: machining; cutting environment; cutting fluid; sliding friction; friction force; surface roughness; tool life

1. Introduction

The importance of cutting environment for metals machining continues to grow and has a significant impact on improving the quality of the surface of the workpiece, tool life and reduce energy consumption [1 - 3]. Cutting fluids appear to be an important means of increasing the productivity of technological processes, and require a complex knowledge of problems in the field of tribology, the theory of machining and chemical-physical processes [4 - 6]. Lubricating effect of cutting environment in recent years appears to be the dominant effect, which significantly influences the economics of the cutting process [7]. As a criterion for evaluation of these properties is chosen factor of sliding friction, whose value can be determined experimentally on the equipment for simulation of the process for the two contact surfaces. The values of friction factor depend mainly on the particular pair of contacting surfaces, the material properties and the surface roughness of the bodies between the frictions occurs [8, 9].

2. Description the experimental equipment for monitoring of sliding friction factor

Relatively many authors deal with the design and construction of tribometers to monitor the parameters of the cutting area [10 - 13]. Most authors characterize the requirements for their construction as follows: obtaining the characteristic traces of wear scar on the test sample, which is most important with regard to the analysis of friction properties. The test sample is usually always pressed onto the surface of the rotating ring mechanism that its kinematics, shape and the flow of cutting fluid is closest to a given cutting operation of machining [14-17]. Lubricating effect of cutting medium can be accurately entered using the resulting wear scar traces of the sample. Simple design, easy and quick replacement of cutting fluids and testing samples, minimizing the volume of tested cutting fluids, easy to adjust the contact force, the minimum eccentricity of the cylindrical surface of the ring etc. are other important requirements that should fulfill design and requirements for maintenance [16, 17]. Test equipment, which was developed at the workplace of authors, is based on measurements of lubricating properties of cutting medium and the relationship between the load vector and velocity vector of motion. Tribotester was designed so that the axes of rotation of the two cylindrical surfaces of the friction are skew. The projection of these axes is 90 degrees. The projection of these axes is 90 degrees (Fig. 1).

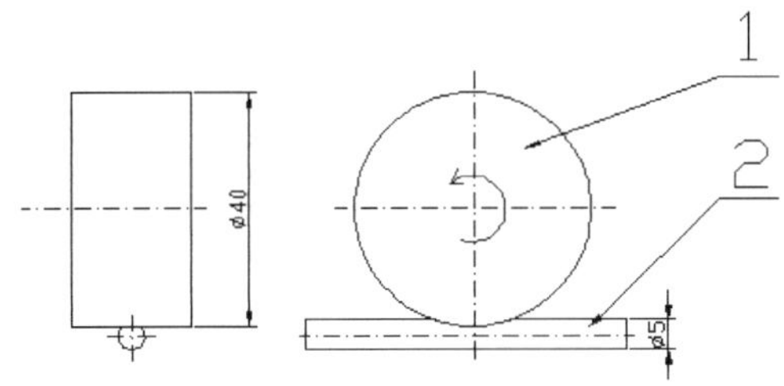

Fig. 1. Schematic representation of tribometer (1 - dynamometer 2 - test sample)

The ring makes a rotating movement around its own axis and lying in a skew position is pressed against the sample by means of a lever mechanism, fixed to the ring by gravity (Fig. 2). As an evaluation criterion was chosen off pressure contact between the frictions surfaces on which the continuous lubricating film, that prevents their direct contact adversely wear of test sample.

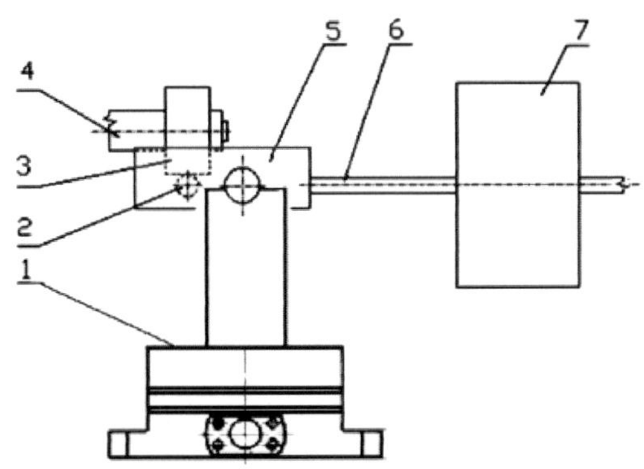

Fig. 2. Schematic representation of tribometer
(1 - dynamometer 2 - test sample, 3 - rotating ring, 4 – arbor work, 5 - container rocker, 6 - rocker arm, 7 – moving weight).

Fig. 3. Look at the fixing of tribometer in the spindle and longitudinal support of the lathe SV 18RD

The ring of 40 mm diameter using a collet set on said driving thorns. This is clamped via cone M50 in the spindle lathe SV 18RD and insured against moving in the axial direction with a screw. The test sample is firmly clamped in a rocker and pushed against the rotating ring pinch. Tribometer cradle is rotatable mounted on the holder fixed on the dynamometer, which is bolted to the longitudinal traction support of the lathe (Fig. 3). An important element of tribometer is the elimination eccentricity, which, together with the inertia of the rocker arm may draw additional dynamic load of the system and thereby influence the value of the frictional power. The elimination of this negative phenomenon can be achieved by high surface quality with the final grinding and polishing the ring, which is clamped in the spindle lathe directly on a lathe.

2.1 Test conditions of experiments

The principal factors that influence the progress and results of the measurements on the laboratory tribotester are: pinch ring test sample, wear conditions, cutting environment, measuring devices, equipment and instruments. Pressure ring was made from the steel STN 19 865 with respect to your specific requirements and function. The steel was hardened and tempered to 64 HRC with the following parameters:

Inner grip diameter:	$d = 20\ mm$
Outside diameter:	$D = 40\ mm$
Width:	$w = 22\ mm$

The roughness of the pressure ring was checked in transverse profile curve and the criterion of completion of surface preparation was the value of the arithmetical mean deviation $Ra \leq 0.07$ *microns*.

Test of the sample - a cylindrical steel rod diameter $d = 5\ mm$, was made from the material STN 12 050.1 and cut into in length of $l = 100\ mm$.

2.2 Methodology of experimental measurements

In the first phase of the experiment, we have created a track wear - the intersection of two cylindrical surfaces based on kinematic pairs - hardened abrasive ring and the sample, where was used as the loading force of $F_n = 300N$. The path length of abrasion was determined by measurement of expansion of the abrasive surface samples, depending on the path length of the ring in the sample. The size of abrasive surface was determined microscopically on the SmartScope MVP 200 (Fig. 4).

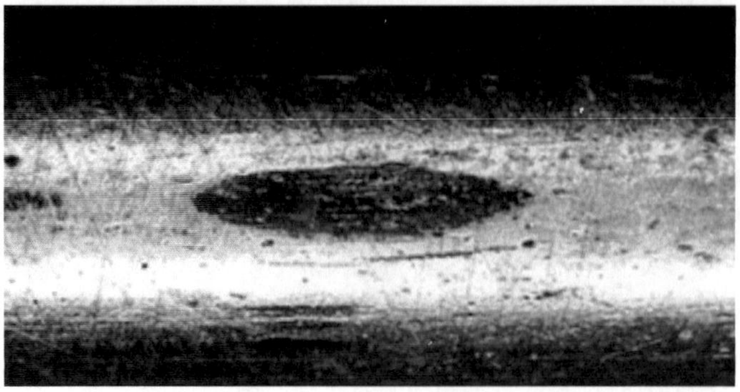

Fig. 4. Abrasive area on the sample surface

Measuring conditions:
- cutting environment: CF 1-26
- revolution per minute of the spindle $n = 200$ r / min
- burdensome (normal) force of: $F_n = 300$ N
- material of the ring: STN 19 856
- material of the sample: STN 12 050.1

In the second stage of the experiment was reduced the frictional force at such a value as not to up edge formation. Normal force was reduced from the value $F_n = 300$N to $F_n = 100$N. Reducing forces was simultaneously reduced contact pressure, thus producing the secure lubrication film. Lubrication cutting medium from various manufacturers has been examined on this film (RP 1-26).

Size component friction force was captured by the three - component dynamometer KISTLER the type 9257. The bases of the dynamometer are sensors in the bridge connection placed in each sensing axis, which measures changes the electrical potential in relation to the energy burdening of the piezo crystal.

Normal forces are adjusted by means of adjustable weights, and were used calibrated weights 100N and 300N for the calibration.

By means of measured values of the friction force F_t and the known normal force F_n = 100N is then possible to determine the factor of friction for each cutting environment. Degree of lubricity cutting medium was the size the factor of friction. It can be expressed by the formula:

$$f = \frac{F_t}{F_n} \tag{1}$$

F_t – the shear frictional force on (N)
F_n – the normal force between bodies (eg. body weight) (N)

2.3 Statistical analysis and evaluation of the obtained results

The result of measurement of the frictional force F_t is an estimate of the actual value of the measured quantity (this corresponds on the dynamometer for component of force F_y in the kinematic system of the machine tool); what for *i - th* measurement can be expressed by equation:

$$\overline{F_{t_i}} = \frac{1}{n}\sum_{i=1}^{n} F_{t_i} \tag{2}$$

Where n is the number of the repeated measurements.

The evaluation of the standard uncertainty with the method A is based on the statistical analysis of measured data. We assume that the measured values are obtained by the independent direct measurements under the same measurement conditions. Standard measurement uncertainty is evaluated by an estimation method A, which is expressed relations (2) and it is equal to the standard deviation the arithmetic average.

$$u_A(F_{t_i}) = \sqrt{\frac{1}{n(n-1)}\sum_{i=1}^{n}\left(F_{t_i} - \overline{F_{t_i}}\right)^2} \tag{3}$$

Measurement uncertainties are evaluated by the method B. They are linked on well-known and identifiable quantified sources, which are mainly linked to the measuring means. In measuring the friction force F_t has been used the measuring device with constant measurement uncertainty at various subscale of the measuring instrument. The standard uncertainty of measurement, as assessed by using of the method B, is determined from the relationship:

$$u_B(z_i) = \frac{z_{i\max}}{k} \tag{4}$$

Where k is the value which is a matter for the approximation of the normal distribution.

Consequently, it is possible to express the combined the measurement uncertainty with relations:

$$u_c(F_{t_i}) = \sqrt{u_A^2(F_{t_i}) + u_B^2(z_i)} \tag{5}$$

The standard measurement uncertainties overlap the actual measured value with only a small probability (around 66%). A larger number of expected values are covered by the expanded uncertainty, which is obtained by using the factor k.

$$U = k.u_c(F_{t_i}) \tag{6}$$

Examples of full record the results of measuring the friction force F_t is as follows

$$F_{t_i} = \left(\bar{F}_{t_i} \pm U\right) \quad [N] \tag{7}$$

Statistical evaluation of the results of measurement of F_t and the results of the friction factor are given in Table 1, and the graphical presentation of dependence between the estimated values of the frictional force F_t and the used cutting fluids can be seen in Fig. 5.

Num.	Max Y [N]	Min Y [N]	Average Y [N]	UA (F_t) [N]	UB (F_t) [N]	UC (F_t) [N]	f
1	-6,012202	-54,824664	-29,048982	6,310236	1,154700538	6,415014553	0,29049
2	-11,342898	-76,356503	-33,76166	7,003226	1,154700538	7,097781889	0,337617
3	-4,235303	-64,440825	-28,539748	7,6109	1,154700538	7,697995333	0,285397
4	-13,119797	-71,130328	-37,598341	9,407398	1,154700538	9,477999286	0,375983
5	-17,823354	-65,695107	-41,412462	6,541256	1,154700538	6,642391391	0,414125
6	-14,583126	-81,896248	-45,305565	8,989203	1,154700538	9,063062612	0,453056
7	-3,921732	-72,07104	-41,632422	12,597702	1,154700538	12,65051102	0,416324
8	-26,916896	-83,882194	-49,809314	8,3088	1,154700538	8,3886525	0,498093
9	1,408964	-54,615617	-26,853241	8,315234	1,154700538	8,395025301	0,268532
10	-1,204122	-42,804464	-21,963729	5,607136	1,154700538	5,724797591	0,219637
11	1,095394	-49,493967	-22,689293	6,877333	1,154700538	6,973596097	0,226893
12	-11,133852	-55,242758	-34,973398	5,813571	1,154700538	5,927135995	0,349734
13	-11,342898	-48,239685	-29,382542	5,025589	1,154700538	5,156537417	0,293825
14	-20,43644	-60,677979	-40,795055	6,004212	1,154700538	6,11423708	0,407951
15	-7,266483	-44,999457	-26,740433	5,236117	1,154700538	5,361926386	0,267404
16	-4,130779	-40,818518	-23,866435	4,914895	1,154700538	5,048715301	0,238664
17	7,575849	-73,325321	-43,846772	11,865426	1,154700538	11,92147925	0,438468
18	-10,715758	-55,242758	-33,760549	5,892287	1,154700538	6,004363365	0,337605
19	-8,416241	-54,824664	-32,024988	7,371016	1,154700538	7,460912156	0,32025
20	-0,681505	-40,713994	-23,164837	4,816616	1,154700538	4,953092269	0,231648
21	-7,520765	-66,158436	-31,237965	8,222961	1,154700538	8,303639018	0,31238
22	-3,921732	-61,409644	-27,695846	6,617638	1,154700538	6,717623541	0,276958
23	-14,583126	-51,68896	-32,737425	5,879984	1,154700538	5,992290478	0,327374
24	-20,750011	-74,684127	-47,903293	7,646252	1,154700538	7,732949177	0,479033
25	-8,520765	-67,158436	-31,879182	8,222961	1,154700538	8,303639018	0,318792
26	-24,094762	-66,217724	-46,450293	6,285635	1,154700538	6,390816903	0,464503

Table 1. Statistical evaluation of the results of measurement of F_t and the results of the friction factor

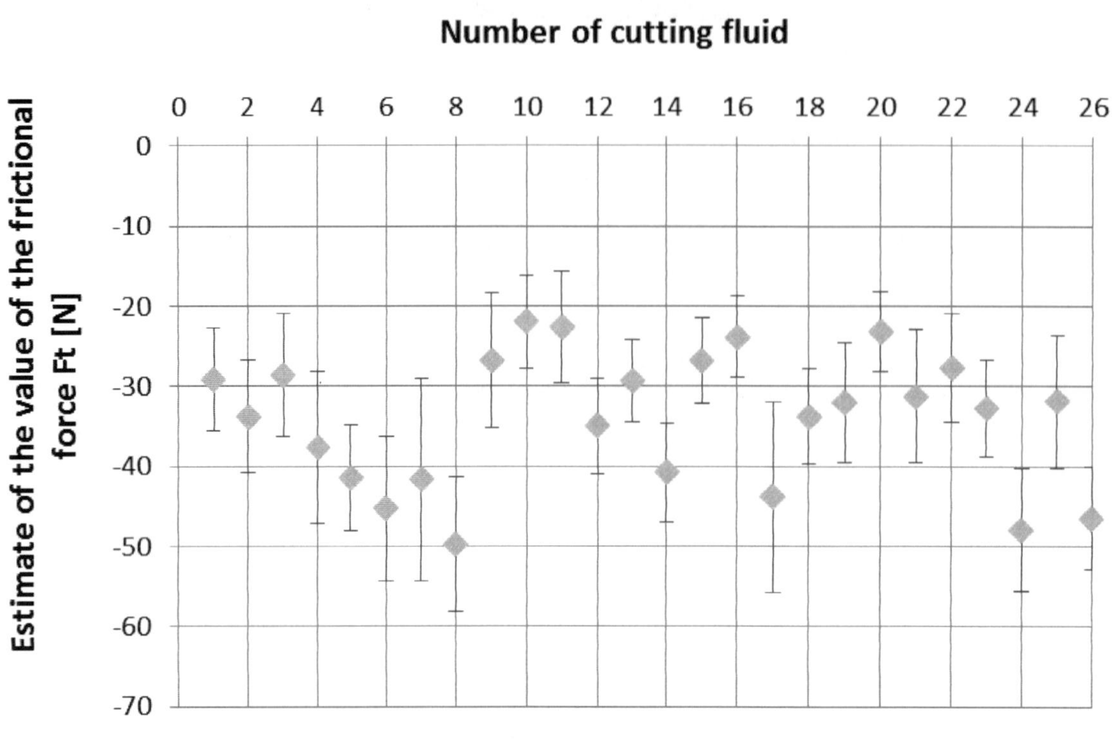

Fig. 5. Dependence of estimated value of the frictional force F_t and the use of cutting fluid

3. Conclusion

The achieved results can be considered criterial, therefore, an indicator of lubricating properties of the cutting fluids. Thus, they are an indicator of lubricating properties of the cutting fluids. It is therefore appropriate to take the further measurements of the available cutting fluids under the same measuring conditions. The results can create a database of indicators of their suitability for the machining process.

Tests of cutting fluids will be performed in the machining process itself – turning of the external rotating surfaces.

The methodology of influence of the cutting fluids on the electric energy consumption will be determined for selected machining parameters. The cutting environment also influences the quality of the surface layer, the residual stress in the surface layers, the stability of the machining process, etc. These areas will be the subject of further research, the aim of which will be selecting the optimal cutting fluids so as to achieve the selected criteria relating to the quality of the work-pieces and the energy consumption during machining.

4. Acknowledgments

The performed research was the content of the project the Ministry of Education, Science, Research and Sport of the Slovak Republic KEGA 1/1056/12 "The introduction of progressive educational methods for manufacturing systems to car production" that also provided the background for our paper and the bilateral project APVV SK-AT-2015-0023 "Improvement of production quality in selected application".

5. References

[1] Vasilko, K. (1997). Znižovanie energetickej a ekologickej náročnosti procesu obrábania. Zborník z medzinárodnej konferencie. Perspektívy a horizonty SR. Trenčianske Teplice. 22. - 23. október 1997. pp. 175 - 177.
[2] Kocman, K. (2011). Technologické procesy obrábění CERM, 2011, Brno, ISBN 978-80-7204-722-2
[3] Prokop, J. & Kocman, K. (2005). Technologie obrábění, CERM, 2005, Brno, ISBN 80-214-3068-0.
[4] Neslušan, M. & Čilliková, M. Teória obrábania (2007). Žilina : Žilinská univerzita, 2007, ISBN 978-80-8070-790-3.
[5] Camposeco-Negrete, C. (2013) Optimization of cutting parameters for minimizing energy consumption in turning of AISI 6061 T6 using Taguchi methodology and ANOVA. Journal of Cleaner Production 53, (2013.), p.195-203, ISSN: 0959-6526.
[6] Ghani, J.A. & Choudhury, I.A. & Hassan, H.H. (2004) Application of Taguchi method in the optimization of end milling parameters. Journal of Materials Processing Technology 145, (2004.), p. 84-92, ISSN: 0924-0136.
[7] Fratila & Domnita & Caizar & Cristian. (2011). Application of Taguchi method to selection of optimal lubrication and cutting conditions in face milling of AlMg3. Journal of Cleaner Production 19, (2011.), p. 640-645, ISSN: 0959-6526.
[8] Kara, S. & Li, W. (2011) Unit process energy consumption models for material removal processes. CIRP Annals e Manufacturing Technology 60, (2011.), p. 37-40, ISSN: 0007-8506.
[9] Muminovic, A. & Repcic, N. & Colic, M. (2014) Thermo Elasto Hydrodynamic Lubrication Model of Mixed Friction. 24th DAAAM International Symposium on Intelligent Manufacturing and Automation. In Procedia Engineering 69, (2014.), p. 49 – 56, ISSN: 1877-7058
[10] Behrendt, T., Zein, A., Min, S. (2012). Development of an energy consumption monitoring procedure for machine tools. CIRP Annals - Manufacturing Technology 61, (2012.), p.43–46, ISSN: 0007-8506.
[11] Kuhrke, B. & Schrems, S. & Eisele, C. & Abele, E. (2010). Methodology to Assess the Energy Consumption of Cutting Machine Tools. Conference Proceedings LCE – 2010, 17th CIRP International Conference on Life Cycle Engineering (Hefei, China), (2010.) p. 76–82, ISBN: 978-75-6500-186-4.
[12] Verl, A. & Abele, E. & Heisel, U. & Dietmair, A. & Eberspächer, P. & Rahäuser, R. & Schrems, S. & Braun, S. (2011). Modular Modeling of Energy Consumption for Monitoring and Control. In: Glocalized Solutions for Sustainability in Manufacturing, (Hesselbach, J., et al.) p. 341-346, Springer-Verlag, Berlin, Heidelberg, Germany, Print ISBN 978-3-642-19691-1.
[13] He, Yan & Liu, Bo & Zhang, Xiaodong & Gao, Huai & Liu, Xuehui. (2012) A modeling method of task-oriented energy consumption for machining manufacturing system. Journal of Cleaner Production 23 (2012.), 167-174, ISSN: 0959-6526.
[14] Li, W. & Kara, S. (2011). An Empirical Model for Predicting Energy Consumption of Manufacturing Processes: A Case of Turning Process, In: Proceedings of the Institution of Mechanical Engineers, Part B: Journal of Engineering Manufacture 255, (2011.), p. 1636-1646, ISSN: 09544054.
[15] Quintana, G. & Ciurana, J. & Ribatallada, J. (2011). Modeling power consumption in ball-end milling operations. Materials and Manufacturing Processes 26, (2011.), p. 746–756, Online ISSN: 1532-2475.
[16] Eberspächer, P. & Verl, A. (2013). Realizing energy reduction of machine tools through a control integrated consumption graph-based optimization method. Forty Sixth CIRP Conference on Manufacturing Systems 2013, In. Procedia CIRP 7 (2013.), p. 640 – 645, ISSN: 2212-8271.
[17] Bi, Z.M. & Wang, L. (2012). Optimization of machining processes from the perspective of energy consumption: A case study. In. Journal of Manufacturing Systems 31, (2012.), p. 420– 428, ISSN: 0278-6125.

DOI: 10.2507/28th.daaam.proceedings.016

EKOLABELING AS DEVELOPMENT FACILITATOR OF THE ENVIRONMENTALLY ORIENTED PRODUCT POLICY

Ruzena Kralikova, Miroslav Rusko & Emil Wessely

This Publication has to be referred as: Kralikova, R[uzena]; Rusko, M[iroslav] & Wessely, E[mil] (2017). Ekolabeling as Development Facilitator of the Environmentally Oriented Product Policy, Proceedings of the 28th DAAAM International Symposium, pp.0124-0130, B. Katalinic (Ed.), Published by DAAAM International, ISBN 978-3-902734-11-2, ISSN 1726-9679, Vienna, Austria
DOI: 10.2507/28th.daaam.proceedings.016

Abstract

Products available on the market should be healthiness and should not threaten human life and the overall safety of the consumer and does not pollute the environment. This means that security specifications and standards have to exist and to be adopted by each country for any product. Consumers must be protected from any damage that could be caused by the product. Fundamental right of every consumer should be the availability of necessary information to protect their own health and safety. Systems of environmental evaluation and labelling are implemented, which claim that the product (in concordance with the recent knowledge) process parameters minimizing or eliminating its negative environmental impacts. Product labelling is an option to reduce the risk and provide some form of assurance that products are safe. This paper discuses the present state and perspectives of ecolabeling lined from research hypothesis based on the results of a study from surveys oriented to interest of customers for buying environmental suitable products.

Keywords: labelling; environment; quality; safety; product; policy

1. Introduction

Quality and safety of products determines the success of all manufacturers, too. Therefore is important to ensure that the produced products have to the high quality and do not pose a danger either to people or to the environment. The product design becomes a key issue focused on development of such parameters, which will minimize or eliminate negative impacts to the environment even before its placement to the market [2].

The decisive factors are quality, compatibility, transparency and trust ability of the programmes, increase of the environmental awareness of the producers, consumers as well as the public, information availability, and mutual acceptance of the programmes in the international frame. Systems of environmental evaluation and labelling are implemented, which claim that the product process parameters minimizing or eliminating its negative environmental impacts.

2. Product policy

Product policy results from the main elements, giving its conceptual development, political elaboration and implementation, i.e. aims and strategy, tools, subjects and products [13]. All those main elements influence selection of the parameters, which are supposed to minimize or eliminate negative environmental effects of a product. Environmental product policy aims and strategy are focused to the sustainable development principles at present, as they were defined by the report „Our Common Future" (Brundtland G.H., 1991) and confirmed by the Rio summit in 1992. Certain change of the model concepts appears at present as described by the following Figure 1.

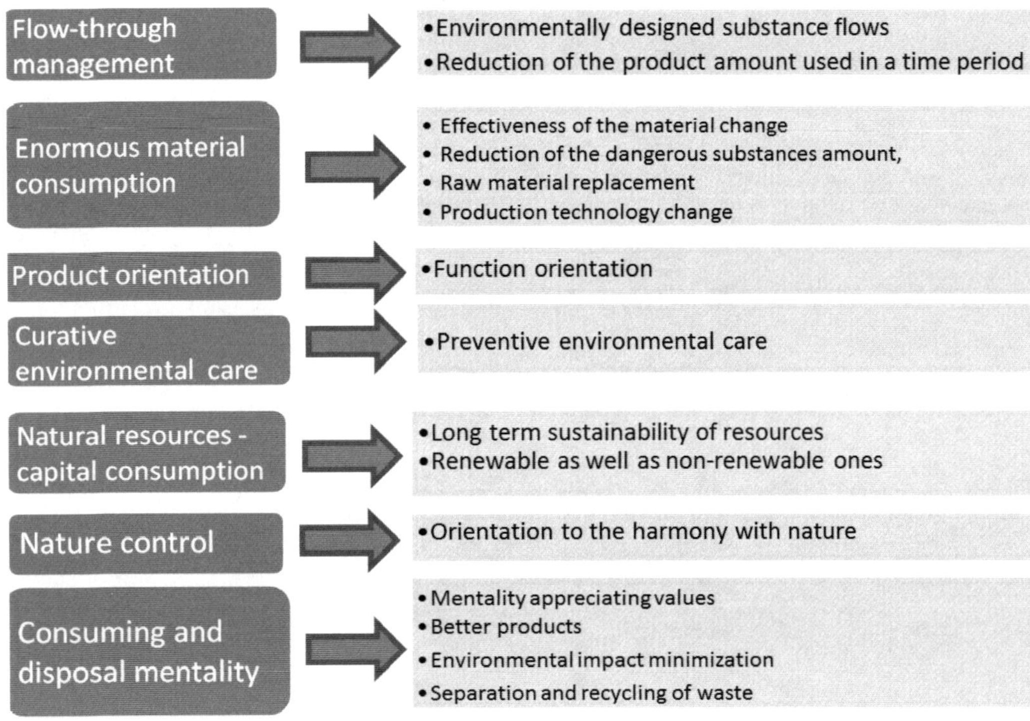

Fig. 1. Change of the model concepts in the environmental policy Equations

2.1. Environmental policy

Broadening trends of the indirect tools and emphasis to the voluntary information tools appear in the environmental product policy. These tools are extensively based on the voluntary activities of the producers, as well as increased environmental awareness of the public. Systems of environmental evaluation and labelling are implemented, which claim that the product (in concordance with the recent knowledge) process parameters minimizing or eliminating its negative environmental impacts. Implementation of such systems results not only from initiatives and requirements of the state administration, but also from the engagement of the producers and consumers [4].

The questions of different Eco labelling programmes development are dealt by several important international organizations, e.g. International Chamber of Commerce – ICC, World Trade Organization – WTO, United Nations Environmental Programme - UNEP, who assess and investigate the implemented environmental labelling programmes and who also evaluate their efficiency and effectiveness. Ecolabelling is now a very useful tool to encourage environmental practices for governments and to identify and establish markets i.e. domestic and international for industries to promote their environmentally preferable products. Many countries have adopted some kind of eco mark, while others are considering program development.

2.2. Integrated product policy

In recent times, the European Commission (EC) has been placing an emphasis on promoting a so-called Integrated Product Policy (IPP). IPP as defined by the EC aims to support the realisation of environmental product innovations in order to achieve a broad reduction of all environmental impacts along a product's life cycle (EU, 2001). The background is that all industrial goods cause environmental degradation in some way, whether from their manufacturing, use or disposal. IPP might be better termed "environmental product policy". It is an attempt by the European Commission to create conditions in which environment-friendly products, or those with a reduced impact on the environment, will gain widespread acceptance among the European Union's member states and 380 million consumers. Aspects of IPP:

- Involves many different actors such as designers, industry, marketing people, retailers and costumers
- Attempts to simulate each part of these individual phases to improve their environmental performance
- Is based on a variety of tools – both voluntary and mandatory
- Include e.g. economic instruments, environmental labelling and product design guidelines.

Integrated product policy is a key strategy for EU environmental policy to minimise environmental impacts by looking of all phases of a product life cycle and taking action it is most effective. The Cradle to cradle philosophy is an important part of the integrated approach that help to [9]:

- Increase reuse and recycling
- Increase the use of renewable energy and reduce CO_2 emissions
- Increase sources, material and water recycling
- Develop and produce innovative and greener products.

Initiatives concerning IPP are largely due to the fact that traditional additive environmental protection has increasingly reached technical and economic limits. Moreover, increasing importance of the phases of consumption, disposal, extraction of raw materials and the using waste of production process, Figure 2.

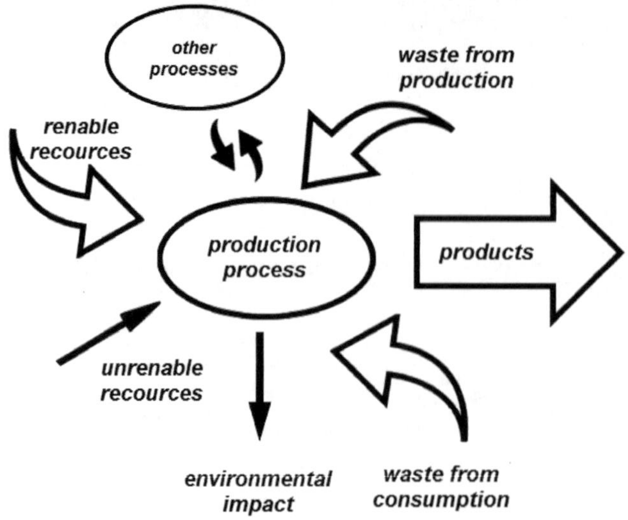

Fig. 2. Environmental suitable production process

Due to the heterogeneity of products, IPP cannot just consist of one general policy instrument, of course. Instead, there is a whole variety of measures that can be used to achieve the IPP objective, e.g., standards or so-called soft environmental policy instruments such as activities regarding voluntary agreements or environmental labelling. Concerning the latter, some companies already apply some of these voluntary environmental organisational measures. Indeed, despite the Green Paper on IPP published by the EC, which outlines proposals for the promotion of an IPP, it is still not well understood which factors and environmental policy instruments influence the environmental performance of products in general (EU, 2001 and EU, 2003), and thus environmental product innovations in particular [7].

3. Marks and labelling

The label may by one of the instruments of sustainable development, environmental and safety policy. In the field of environment and safety, there is an effort to make marks universal, e.i. , to have the same meaning in different social cultures.Marks notice about the level of quality (should be stable) and the reliability of the product (product, respectively service), and thus help to consumers make decisions what to buy more comfortable. Among other things, it also contributes to increase loyalty of customers. Marks make the cooperation and communication between dealers, distributors and producers more effective[10]. The symbol has the form of a geometric element. It represents a medium of receiving the relevant information to mediate the knowledge. Symbols indicate the importance assigned to them in opinion system (labelling scheme) to the consumer [1].

Products which are intended entirely to be used and get on the consumer market should meet requirements of Directive 2001/95/EC. Every product which is placed on the market must be labelled. [12]

European Conformity marking (CE) is a certification mark that indicates conformity with health, safety, and environmental protection standards for products sold within the European Economic Area (EEA). The CE marking is the manufacturer's declaration that the product meets the requirements of the applicable EC directives. The letters "CE" are the abbreviation of French phrase "Conformité Européene" which literally means "European Conformity". The term initially used was "EC Mark" and it was officially replaced by "CE Marking" in the Directive 93/68/EEC in 1993.The CE mark certifies compliance of the product with the EU legislation and thus enables free movement of product within the European market. By placing the CE marking on a product producer declares his responsibility that the product complies with all the legal requirements for CE marking, which means that the product can be sold by throughout the European Economic Area (EEA, 28), EU Member States and countries of the European Free Trade Association (EFTA), namely Iceland, Norway and Liechtenstein. The European Commission thus refers to the CE Marking as a "Passport" which allows products to be freely circulated within the EU single market.

The EC declaration of conformity is the written statement and the a single declaration drawn up by the manufacturer to demonstrate the fulfilment of the EU requirements relating to a product bearing the CE marking he has manufactured. The declaration shall be in respect of all Community acts applicable to the product containing all information required for the identification of Community harmonisation legislation to which the declaration relates. It concerns the European Union (EU) Ecolabel which is a voluntary environmental labelling scheme. By means of transparent ecological criteria, it enables consumers to make conscious choices without compromising on the quality of the products. Other very important labelling of products placed on the market are:

- CE "European Conformity" marking
- Slovak Conformity mark CSK (in Slovakia)
- Quality brand SK.

Slovak Conformity mark (CSK) particular product certifies that the properties of a product meets the technical requirements for the product and in the case of construction products and technical requirements have been complied with and the conformity assessment procedures. Quality brand SK is owned by the Ministry of Agriculture and Rural Development, and was put to use in 2004, when program to support agricultural products and foodstuffs was established [11].

3.1. Ecolabelling

A 2010 Commission decision (Decision 2010/709/EU) was established the European Union Ecolabelling Board EUEB. Its members are appointed by the Commission. It is composed of representatives from the EU countries and the European Economic Area, as well as from certain European organisations representing, for example, consumers, business and environmental concerns. The Commission consults the EUEB when developing or revising the award criteria and requirements of the label. The Commission has adopted a series of decisions establishing ecological criteria for the awarding of the EU Ecolabel to different types of products; for several of these, the validity period expired by the end of December 2016. The Ecolabel covers a wide range of product groups such as cleaning products, appliances, paper products, textile and home and garden products, lubricants and services such as tourist accommodations [8]. To qualify for the Ecolabel, products have to comply with a tough set of criteria[15]. These environmental criteria, set by a panel of experts from a number of stakeholders, including consumer organisations and industry, take the whole product life cycle into account – from the extraction of the raw materials, to production, packaging and transport, right through to use and disposal/recycling. This life cycle approach guarantees that the products' main environmental impacts are reduced in comparison to similar products on the market [14].

The EU Ecolabel is a voluntary scheme, established in 1992 to encourage businesses to market products and services that are kinder to the environment. Countries or regions have their own eco-labels, which help customers to identify environmentally friendly products, Fig.3.

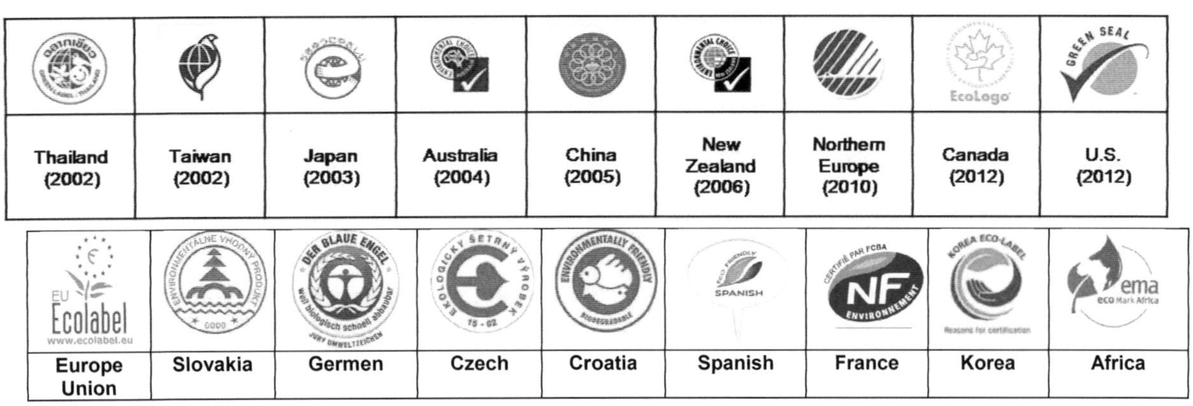

Fig. 3. Examples of eco-labels in the world

4. Knowledge and Information about the Ecological Behaviours, Research hypothesis

Based on the theoretical perspective, research hypothesis are drawn in order to compare the results of a study from surveys oriented to interest of customers for buying environmental suitable products. Several surveys have been carried out to determine the level of consumer awareness of environmental labels in the SR [5], [11]. Respondents answered the question "In which area would you prefer environmentally friendly products?" From a questionnaire surveys conducted in Slovakia in 2012, which dealt with the perception of environmental labelling of products by Slovak consumers suggests that most respondents would prefer environmentally-friendly products in vehicle, paint and varnish groups, cleaners, batteries and services (Figure 4).

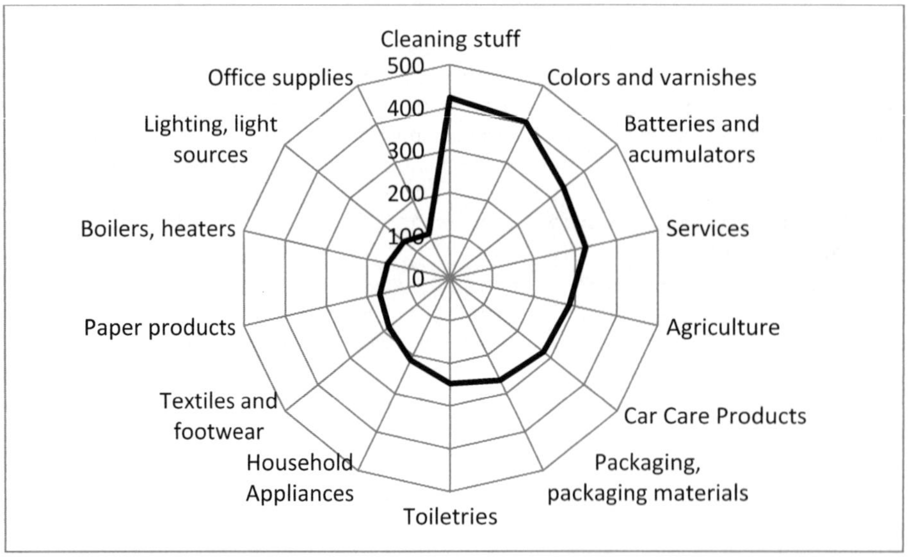

Fig. 4. Preferences of environmentally friendly products according to use field

As another example in Europe we introduce the survey in Germany (2009, Fig. 5) ilustrating answers of costumers on question: "Are you willing to pay more evironmentaly bening products". Note form questonary survey shows, that does not mean, that green products have to more costly [5].

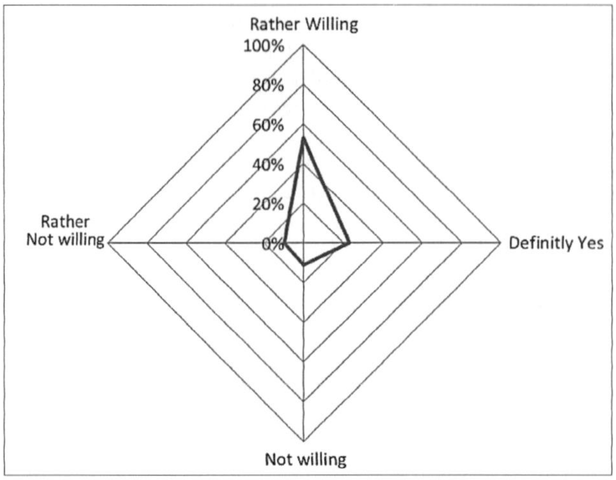

Fig. 5. Results of a study in Germany

Next, Nielsen's 2014 Global Survey shows that 55% of global online consumers across 60 countries are willing to pay more for products and services from companies that are committed to positive social and environmental impact [6]: Globally, more than half (52%) of respondents said they have purchased at least one product or service in the past six months from a socially responsible company. Four in 10 respondents in North America and Europe said that they have made a sustainable purchase in the past six months. 52 % of global respondents said that their purchase decisions partly depended on packaging, i.e. they check labelling to ensure that the brand is committed to positive social and environmental impact. This impacted decisions to a lesser degree in North America (32%) and Europe (36%).

Consumers in Asia-Pacific response that they had spent more on products and services from socially responsible companies. In Peru 62% of respondents said they would be willing to spend more, just 42 percent said they had actually done so. In Hong Kong 52% respondents said they would be willing to spend more, just 32% said they had actually done so. In Slovakia, while 50% of respondents said they would be willing to spend more, just 22 % said they had actually done so. Similar spreads existed in Bulgaria (53% willing, but 31% had).

Nielsen found that 43% of global respondents claim to have actually spent more on products and services from companies that have implemented programs to give back to society (7 % fewer than those who say they would simply be willing to pay). According age of respondents were results:

- Millennials (age 21-34) constituted 51% of global respondents who would pay more for sustainable products and check packaging for sustainable labelling.
- Regionally, in the developing world (Asia, Middle East, Africa), there were wide gaps with respondents (age 21-34) in favour of sustainability actions on average three times more agreeable to sustainability actions than respondents (age 35- 49) and 12 times more agreeable than (age 50-64) respondents.

From the studies based on questionnaire surveys has been report that higher price outweighed ethical considerations in case of purchase of green products. Lack of economic resources of consumers was found to magnify the effect of price and act as a barrier to purchase of green products and low price sensitivity of consumers was found to positively affect green purchase behaviour. Conversely, high price sensitivity negatively affected green purchase behaviour of consumers. It is thus clear that high price negatively influences green purchase intention and behaviour.

Previous studies revealed that, the consumers 'purchasing decisions are significantly affected by the consideration of the product impacts on the environment, and based on this empirical evidence should be highlighting the significant role of eco-labels in increasing the perceived behavioural control in consumers, too [3].

5. Conclusion

The main objective of environmental labelling is to generate demand for products with the lowest environmental impact. The aim of the research was to point to the ever-increasing environmental awareness of consumers who often have the problem of buying what they really want. In today's competitive environment, almost every organization seeks to promote its product as "environmentally-friendly", so governments, the industry and environmental NGOs have developed eco-labels to help consumers choose the "right" product.

Recent emergence of green trends has created a brand-new generation of environmentally concerned consumers and attracted attention of many companies worldwide. Green consumers have brought new market niche for business. Eco-labelling is known as one of the most popular tools of green marketing worldwide. Customers should trust the company offering eco-labelled products, because such company should be green in its everyday activities, i.e. to reduce the waste, reduce, and recycle. Therefore, the role of consumers in such a scenario becomes crucial in guiding the production paths [1], since these can directly influence the environmental impact of a product in the phases of use and end of life (way of functioning, separation of waste, collection, recycling), and exert pressures on producers to take environmental criteria into account starting from the design and production processes of the environmentally friendly products.

Fundamental right of every consumer should be availability of necessary information to protect their own health. Information should be obtained from the marks referred to the product which it comes into contact. Marks are an integral part of everyone's life. We can see various types of marks at home, in stores and everywhere around us. Symbols indicate the importance assigned to them in opinion system (labelling scheme) to the consumer.

Eco-design, LCA, Eco labelling and green marketing as a voluntary regulative tools of the environmental policy are significant instrument for the increase of abilities of the organization competitive. Their using is important for the organizations with established EMS by ISO 14001, thus EMAS, which contracted to the continual release of negative impacts of its activities on the environment. Abroad, during the period of its existence, Eco labelling has entered not only the minds of customers, but also became an inevitable part of advertisements of producers and importers introducing appropriate brands of environment-friendly products into the market.

Products available on the market should be healthiness and should not threaten human life and the overall safety of the consumer and does not pollute the environment. This means that security specifications and standards have to exist and to be adopted by each country for any product. Consumers must be protected from any damage that could be caused by the product. Based on the knowledge the consumer creates an own personal value of product quality which helps during the process of making decision about buying the product. All external and internal features and characteristics of the product affect the consumer behavior .Important role in consumer behaviour is a safety and environmental labelling and brand. Provides a means of receiving the relevant information for the purpose of mediation of knowledge. Basic right of every consumer's need for information to enable it to protect their own health. Information obtained from the brands mentioned on the product with which it comes into contact.

Supervision and inspection the market is more and more difficult. It has to be clear that products situated in the market and stores are not absolutely risk free. Not sufficient or incorrect information can be present on the product which can endanger the customer´s health.

Many countries have adopted some kind of eco mark and the results achieved show that well organized programmes of environmental evaluation and labelling may serve as an effective tool not only for the environmental protection policy, but also for the trade policy.

Despite the fact that in SR there was no campaign about the Eco labelling programme it is sensible to label the products because it strengthens the feeling of the public about the environmental protection. By means of environment-friendly product labelling the Slovak customers get guaranteed information about the products respecting the environment. The problem of successful implementation of environmental label in Slovakia is especially the provision and availability of information about the products. Future research will focus on monitoring the sale and purchase of environmental friendly products in Slovakia and these comparisons with the developed countries of Europe.

6. Acknowledgments

This paper is a result of research in the frame of the project KEGA 048TUKE-4/2015 "Transfer of knowledge from scientific-research activities into multimedia education process" provided by the Ministry of Education, Youth and Sports of the Slovak Republic.

We are grateful to the partner MtF of the Slovak Technical University of Trnava as member of the project VEGA 1/0990/15 "The readiness of industrial enterprises to implement the requirements of standards for quality management systems ISO 9001: 2015 and environmental management systems ISO 14001: 2014".

7. References

[1] Daddi,T., Iraldo,F., & Testa F.(2015) Environmental Certification for Organisations and Products: Management Approaches and Operational Tools, Business & Economics.232 p.

[2] Draxlerova, M., Rusko, M. & Kuracina, M. (2015): Labelling of textile products from the perspective of quality, maintenance, safety and environment. In Research papers Faculty of Materials Science and Technology Slovak University of Technology in Trnava. Vol. 23, No. 36 (2015), s. 57-64. ISSN 1336-1589.

[3] Ha H, Janda S. 2012. Predicting consumer intentions to purchase energy-efficient products. Journal of Consumer Marketing 29: 461–469.

[4] Hurt, U., Otto, T., Kaare,K., Koppel, O. (2014). New Approachto Knowledge Transfer Environment Development. In 24th DAAAM International Symposium on Intelligent Manufacturing and Automation, Procedia Engineering 69. pp. 273 – 281.

[5] http://ec.europa.eu (2012) Flash Eurobarometer 367: Attitudes of Europeans Towards Building the Single Market For Green Products, Report, 2013, Accessed on: 2016-09-05.

[6] htpp://www.nielsen.com(2015) What People Watch, Listen To and Buy. Nielsen Accessed on: 2016-08-10.

[7] Krechovska. M.; Tausl Prochazkova. P. (2014). Sustainability and its Integration Into Corporate Governance Focusing on Corporate Performance Management and Reporting. In DAAAM International Symposium on Intelligent Manufacturing and Automation, Procedia Engineering 69. pp. 1144 – 1151.

[8] Pedersen E.R, & Neergaard P.(2006.) Caveat emptor–let the buyer beware! environmental labelling and the imitations „Green“ consumerism. Business Strategy and the Environment Vol.15, pp. 15–29.

[9] Polankova, M., Manlig, F. & Kralikova, R. (2015). Environmental reporting in the enterprise and related issues. In. MM Science Journal, 10/2015, pp. 691-695.

[10] Pujari, D. (2006) Eco-innovation and New Product Development: Understanding the Influences on Market Performance. Technovation. Vol. 26. Issue 1. pp. 76–85. ISSN: 0166-4972.

[11] Rusko, M., Chovancova, J. (2008) Consumers` Awareness about Eco-Labels Type I. In Environmental Management for Education and Edification. Vol. 5, No. 2 s.23-34. ISSN 1336-5762.

[12] Rusko, M.; Kralikova, R.; Mikulova, M. & Ilko, J. (2016). Labeling of Products from the Context of Environment, Quality and Safety, Chapter 37 in DAAAM International Scientific Book 2016, pp.419-434, B. Katalinic (Ed.), ISBN 978-3-902734-09-9. ISSN 1726-9687, Vienna, Austria.

[13] Rehfeld, K.M., Rennings, K. & Ziegler, A.(2007). Integrated Product Policy and Environmental Product Innovations: An empiricalAnalysis. Ecological Economics, Volume 61, Issue 1, pp. 91–100.

[14] Testa, F., Iraldo, F., Vaccari, A. & Ferrari, E. (2015), Why Eco-labels can be Effective Marketing Tools: Evidence from a Study on Italian Consumers. Business Strategy and the Environment. Vol. 24, pp. 252–265. doi: 10.1002/bse.1821.

[15] ISO Central Secretariat. (2012) Environmental labels and declarations. How ISO standards help. Genève. 25 p. ISBN 978-92-67-10586-4.

DOI: 10.2507/28th.daaam.proceedings.017

POSSIBILITIES OF USING KINECT SENSOR IN ERGONOMIC

Frantisek Koblasa, Peter Kancian, Jaroslav Props,
Ružena Kralikova & Emil Wessely

This Publication has to be referred as: Koblasa, F[rantisek]; Kancian, P[eter]; Props, J[aroslav]; Kralikova, R[uzena] & Wessely, E[mil] (2017). Possibilites of Using Kinect Sensor in Ergonomy, Proceedings of the 28th DAAAM International Symposium, pp.0131-0135, B. Katalinic (Ed.), Published by DAAAM International, ISBN 978-3-902734-11-2, ISSN 1726-9679, Vienna, Austria
DOI: 10.2507/28th.daaam.proceedings.017

Abstract

Nowadays integral part of the innovation processes is emphasis on considering human resources. Ergonomic requirements are one of the most important requirements given on the agile and sustainable manufacturing systems. This article briefly debates necessity of ergonomic studies and reacts on some problems connected with ergonomic software. The main emphases are given to demanding design of simulation model and points out possible simplifications and ease its creation. Article focuses also on the possible improvement of capturing the position of the wrist and hand fingers. This field is due great share of handmade operations during manipulation or assembly very important and use of Kinect seams as highly promising.

Keywords: Agile manufacturing systems; ergonomics; simulation; Jack; Kinect

1. Introduction

Nowadays, the innovation, the agility and the sustainability are main requirements on the manufacturing processes. Thought, the technical i.e. the product innovation has come "hand in hand" with the organizational i.e. the process innovations, see [2]. Significant role is played also by human resources [16]. The scientific literature offers a lot of particular examples both technical e.g. [1], [3], [18], [21] and also organizational innovations e.g., [7], [13], [17]. The complex point of view which includes all aspects is rather unique, see e.g. [15], [19].

The integral part of innovation processes, as it has been indicated, is also taking in the account human resources mainly during manufacturing, where there are discussed ergonomics aspects, see [6], [14]. According to Rasa [16] there are, in Germany, health problems (diseases of tendons, muscles and bones) caused by inefficient ergonomics "… responsible for nearly one quarter of all workers incapacity…".

Not considering ergonomic requirements could lead to deterioration of the working conditions and thus the health complications (e.g. if a worker is overloaded by constant monotonous activities with short cycle time of the manufacturing operations). It is the paradox, that in many cases it is caused by effort to make the most economic manufacturing as possible [20]. However, not taking in account ergonomic could led, besides the health problems, also to decrease of manufactured quality (e.g. by increased worker fatigue caused by excessive physical decreases his attention). So instead of increasing productivity, wasting occurs and also expenses rises. Not mentioning increasing workforce fluctuation. The ergonomic requirements could be covered by e.g. ergonomic check lists or nowadays more and more important computer systems so called Digital Factory (DF). DF are connecting each design phase of the manufacturing process in to the complex systems with the goal to reduce and make more effective planning phases, preparation and launching of the production.

Systems of DF becomes the nowadays phenomenon same as CAD/CAM (Computer aided design/manufacturing) or ERP (Enterprise resource planning) during 80s. Systems of computer aided production design are mostly used in the automobile industry, where whole complex modular systems (e.g. Delmia or Technomatix), subsystems (e.g. Process designer/simulate, Plant simulation, Quest) or standalones (e.g. Jack, Witness) are used. While other computer simulation systems of manufacturing and logistics processes (e.g. Plant simulation, Witness, Simio, etc.) had already found its way to the users, complex ergonomic systems are still hardly ever used due its high price and demanding design of simulation model.

Despite the fact that ergonomic simulation systems are not cheap and making simulation model is difficult they offer, comparing with other ways of ergonomic studies, more detailed and accurate analyses. The great advantage is also visualization and possible use of before designed model to the new model. Those systems are mostly used in automobile industry.

2. Computer aided Ergonomics

One of the most famous computer systems aiding ergonomic can be considered software solution from the DF family, particularly Delmia V5 Human and Tecnomatix Jack (i.e. Tecnomatix Process Simulate Human).

It is necessary, for the further purpose of this article, to briefly describe work procedure of making model in before mentioned ergonomic systems. Tecnomatix Jack allows to design 3D real model of analysed workshop, technological operation or whole manufacturing system. Base on the particular operations of analysed worker are assigned each positions and movement sequences. It is possible to use standardized positions from database to ease model design. However, those positions are usually inefficient to describe particular workers activity, so it is necessary to edit them. Editing is done by adjustment of related joints with the respect to the physiological possibilities of given worker. The example of such created scene (model) is on Fig.1.

Fig. 1. Example of designed scene – Jack software

Thanks to such created model it is possible to simulate workload of each body parts i.e. it is possible to create number of classical ergonomic analyses as NIOSH, OWAS and RULA.

It is possible, beside the use of predefined position database, to use systems of virtual reality to ease positions and movements (motions) modelling. Nowadays, there are several systems, whose allow capturing workers body positions.

There are, in the industrial practice, used mainly specialized suits (so called Motion Capture suites), which are equipped with sensors nets and evaluation elements. Those suites are usually aided by haptic gloves. The disadvantage of suite is its high price and constrains connected with certain body type (given by size of suite). The advantage is than obtaining accurate dates of each body positions in the real time. Another option is to use optic systems to capture body positions by specialized cameras and marks. Its disadvantage, once again, is high purchase price and also its use in the manufacturing conditions is rather limited due surrounding areas (attached machines, supplies, corridors, workers etc.). Undisputable advantage of these systems is the high accuracy.

An interesting and price acceptable possibility to speed up design of ergonomic simulation model offers sensor Kinect. However, it has same constraints as other optic systems given by high requirements on the sufficient area without obstacles.

3. The example of kinect aided model design

The use of sensor Kinect is relatively widespread. Despite the fact that mentioned device was developed by Microsoft especially for entertaining business, version for operating system MS Windows allows to spread its use in to the other areas. Today's Kinect use is versatile. Some of the examples are in e.g. [10], [11], [12]. Its noticeable use is also in the field of ergonomics studies, see e.g. [4], [5], [8], [9].

The following sections are briefly describing some experiences gathered mainly throughout student grant project, which was focused on the mentioned body posture modelling and its motion and also on the further possibilities of its extension to the hand fingers and wrist movement.

3.1. The use of Kinect with Tecnomatix Jack

Software Jack, as it has been briefly mentioned the 2nd chapter is used for ergonomic analysis and simulations. The main purpose of this tool is than ergonomic evaluation on manufacturing workplaces. The building stone of those analysis is than so called scene where is necessary to include one or more people, whose interact whit this environment. The scene design in the software Jack is not heavy demanding. It is usual, that lay out is imported from some kind of CAD program, but it is necessary to set working human postures in given conditions (tools, machines, supplies).

Human model contains 135 degrees of freedom, so e.g. setting up arm in to the required position (e.g. grabbing the object) can be time consuming. The use of Kinect can be than appropriate alternative to reduce time necessary to design simulation model. It is necessary to briefly describe connection of mentioned sensor with software Jack, to discuss its use further problem.

System Jack beginning version 8.0 offers possibility of communication with sensor Kinect. It is "only" necessary to upload correct version of the library Kinect SDK for proper functionality. Simply said connecting Kinect with Jack allows capturing Human body pose by processing captured image and connect it with modelled character.

The building stone is to distinguish human body and its shape in captured image. Those shapes are than substituted by several mutually connected points, representing basic joints (ankles, knees, hips, shoulders, elbows, wrists, etc.). Thanks to that is made skeleton (wire model) of captured person (see Fig. 2)

Fig. 2. Sample of body outline recognition and wire model

Wire model is than connected to before defined body in the simulation model. Procedure of designing motions with Kinect is following. Design of simulation scene includes two persons.

The first one is generating particular positions and motions, the second one controls computer and captures each position.

One of the most critical steps is to place sensor Kinect in to the correct position. Human can move not continuously, body positions can be unrealistic or can "flow though" the scene, if sensor is not settled up correctly. The control of system Jack is very fast and simple, beginning with opening the model, which supports communication with Kinect, selecting human body (model), which will be used to assign each body postures. Each time body posture is made, it is saved in to the model. Defined Body postures are than used during ergonomic simulation (see Fig. 3).

The disadvantage of described approach is that it is not capturing fingertips and the last captured point of upper limbs is wrist. It is necessary to additionally set fingers in the desired position manually, if it is required by simulation of assembly operations. Despite possibility to use basic hand postures from the database it is very time consuming and difficult. That is why following capture is focusing on possible further research in this field, i.e. possibilities which allow use of Kinect in before mentioned field.

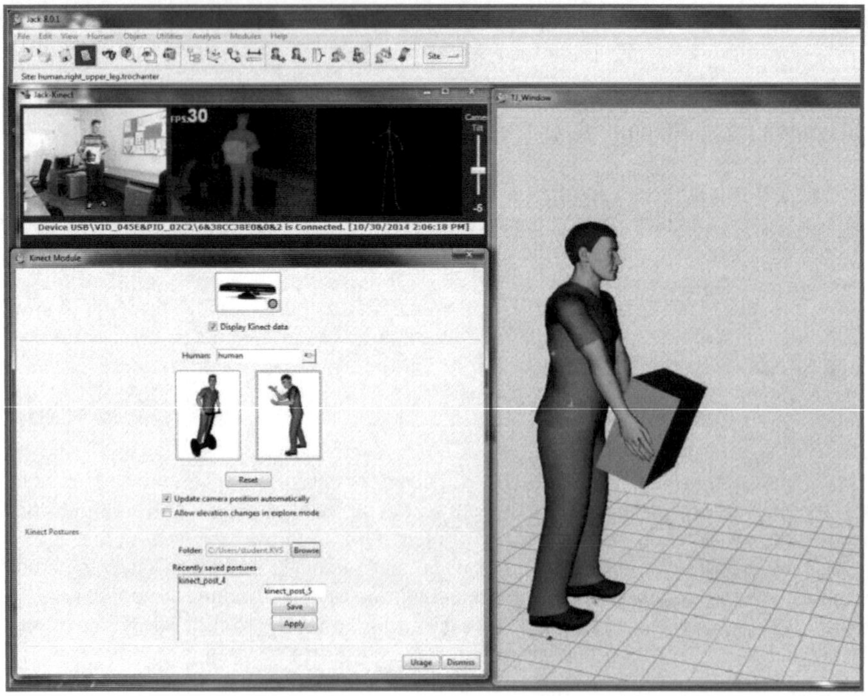

Fig. 3. Sample of the Kinect captured posture in the Jack

3.2. Expansion of the Kinect use to capture the position of the wrist and hand fingers

There was indicated the principle of recognition and character modelling with support for Kinect in chapter 3.1. This approach is possible to use also for capturing position of wrist and hand fingers, i.e.:

- recognize the contours of the wrist and each finger
- finding their main anatomical features
- designing wire model
- implementation in to the respective software.

The next section will briefly mention some specific aspects of the possible expansion of the use Kinect for sensing the position of the wrist and fingers. It can be developed custom algorithms or consider the use of some already on the market offered a solution to detect contours.

Another important step is to find a most important joints, wrists and fingers. It is necessary to correctly determine the ratio coefficients lengths of the phalanges, because the position of the joints is given by anatomical prerequisites of human body. Attention has to be paid also on solving the problem, whether it is a left or right wrist.

After finding significant points on wrist, we can proceed similarly to capture the whole body. Wrists and fingers wireframe is made by linking important points.

Generated wireframe is forwarded to Jack software, which it should process like wireframe of whole body. Basically, it is the assignment of found and replaced fingers to already predefined model of the wrist.

Nowadays, initiation phase of described project is under way. Its goal is to find basic possibilities of designing important elements of wireframe skeleton.

4. Conclusion

Modelling positions the wrist and fingers is important because of a large number of manual operations in the real production and is found in almost every ergonomic simulation model. Therefore, the use Kinect for capturing the position of individual fingers and wrists appears very promising. Linking the scanned image with respective software will accelerate the process of modelling ergonomic situations in real life conditions. This will accelerate and even make cheaper creating ergonomic study of the entire manufacturing operation.

The project presented in this paper suggests that the Kinect software can be a useful motion capture tool for ergonomic evaluation. From the ergonomic point of view, correctly using a Kinect in actual workplace conditions would enable ergonomists to analyse motions instead of isolated poses. It provides supplementary temporal information that could also be used as real-time feedbacks. The future research will oriented to methods of develop a multiple-Kinect system for tracking human movements. An important research topic will to include situation based timely information into the process.

5. Acknowledgments

This research was supported by project of the Technical University of Liberec No 21130 “Research and development in the field of 3D technology, manufacturing systems and automation" provided by the Ministry of Education, Youth and Sports of the Czech Republic in the year 2017 and by project of the Technical University of Košice VEGA No 1/0537/15 “Research of influence of chosen parameters of working environment on working power and productivity ” provided by the Ministry of Education, Youth and Sports of the Slovak Republic.

We are thankful to our colleague assoc. prof. dr. Frantisek Manlig, PhD, who provided expertise that greatly assisted the research and that improved the manuscript significantly.

References

[1] Ackermann, M., et al. (2014) Properties of Models Produced by Direct Selective Laser Melting Technology. Applied Mechanics and Materials. Novel Trends in Production Devices and Systems II. Vol. 693, pp 231-236. ISSN 1660-9336.

[2] Dvorak, R. (2015) Global world of innovation in the enterprises point of view. In : The innovation in the Lean manufacturing. Journal supplement of MM Industry spectrum, No. 4, pp II-III. ISSN 1212-2572 (in Czech).

[3] Havlik, R. et al (2014) Application of 3D Scanner for Transportation a CNC Machine Tools. Applied Mechanics and Materials. Novel Trends in Production Devices and Systems II. Vol. 693, pp. 213-218. ISSN 1660-9336.

[4] Horejsi et al. (2013)Using Kinect technology equipment for ergonomics. MM Science journal, No. 3, pp 388-391. ISSN 1803-1269.

[5] Hu M.-Ch. et al. (2015) Real-Time Human Movement Retrieval and Assessment With Kinect Sensor. IEEE Transactions On Cybernetics, April 2015, Vol. 45, No. 4, pp 742-753. ISSN 2168-2267

[6] Kancian, P. & Props, J. (2014) The use of ergonomic software. In: 8th annual international conference Manufacturing systems today and tomorrow 2014. Peer-reviewed conference proceedings [CD]. TU v Liberci, ISBN 978-80-7494-150-4. (in Czech).

[7] Kralikova, R. et al. (2014) Thermal environment evaluation according to indices in industrial workplaces, Procedia Engineering, Vol. 69, pp. 158-167.

[8] Kurkin, O. et al.(2013) The use of virtual reality and entertainment technology for manufacturing. AI magazine, No. 1, pp 58-61. ISSN 1337 - 7612 (in Czech).

[9] Nguyen T. D. et al.(2014) Ergo Assist: An assistance system to maintain ergonomic guidelines at workplaces. In : 19th IEEE International Conference on Emerging Technologies and Factory Automation, ETFA 2014, Barcelona, September 2014. ISBN: 978-147994846-8

[10] Novak-Marcincin, J. et al. (2013) Augmented Reality Aided Manufacturing. Procedia Computer Science, Vol. 25, pp 23–31. ISSN 1877-0509.

[11] Novak-Marcincin, J. et al.(2014) New possibility of visualization in mechanical engineering. Applied Mechanics and Materials, Volume 442, pp 209-215. ISSN 1660-9336.

[12] Pan, S. et al. (2015)A Kinect-Based Real-Time Compressive Tracking Prototype System for Amphibious Spherical Robots. Sensors, No. 15, pp 8232-8252. ISSN 1424-8220

[13] Pelantova,V.(2013) Knowledge Waste in Organisations. Periodica Academia, Vol. VIII.(1), pp 57–62.

[14] Props, J. & Kancian, P. (2014) Sensor Kinect and ergonomic simulation. In: 8th annual international conference Manufacturing systems today and tomorrow 2014. Peer-reviewed conference proceedings [CD]. TU v Liberci, 2014. ISBN 978-80-7494-150-4. (in Czech)

[15] Pagac, M. (2015) Kinect motion sensor helps in the nuclear industry. Konstruktér, [online 24.04.2015]. Available from <www.caxmix.cz/2015/04/13/herni-konzole-kinect-pomaha-v-jadernem-prumyslu/> (in Czech)

[16] Rasa, J. (2015) Ergonomic in the time of Industry 4.0. MM Průmyslové spektrum, No. 6, pp 80. (Based on Christoph Krieger a Holger Müller, MM Das Industriemagazin, 8/2015). ISSN 1212-2572. (in Czech).

[17] Slaichova, E. & Marsikova, K. (2013) The Effect of Implementing a Maintenance Information System on the Efficiency of Production Facilities. Journal of Competitiveness, Vol. 5, Issue 3, pp 60–75. ISSN 1804-1728.

[18] Safka, J. & Keller, P. (2010) Practical Experiences With Sandvik Coroturn HP Tool. University Review 2010, Vol. 4, No. 2. pp 33–37. ISSN 1337-6047

[19] Vach, T. (2009) Effect of product design solution in logistics. In : 4. annual internat. conference Manufacturing systems today and tomorrow. Conference proceedings [CD]. TU – Liberec. ISBN 978-80-7372-541-9

[20] Varga, G. &, Kundrak, J. (2017) Effects of technological parameters on surface characteristics in face milling Solid State Phenomena, 261 SSP, pp. 285-292.

[21] Votrubec, R. &Vavrousek, M.(2014) Control system of a rotary pneumatic motor. In: Proceedings of the 16th International Conference on Mechatronics - Mechatronika. IEEE, 2014, s. 588-593. ISBN 978-80-214-4816-2.

DOI: 10.2507/28th.daaam.proceedings.018

ECONOMY OF BIOGAS PLANTS

Zvonko Kremljak

This Publication has to be referred as: Kremljak, Z[vonko] (2017). Economy of Biogas Plants, Proceedings of the 28th DAAAM International Symposium, pp.0136-0143, B. Katalinic (Ed.), Published by DAAAM International, ISBN 978-3-902734-11-2, ISSN 1726-9679, Vienna, Austria
DOI: 10.2507/28th.daaam.proceedings.018

Abstract

At a time when worldwide energy demand grows, the care for greener environment grows as well. Due to the extremely high energy potential, along the safety and cleanness of its operations, the biogas plant as an energy solution is deemed a great investment in the future. Frequently in that regard we face the question concerning the scope of the investment and the time needed for its return.
This paper gives an example of an investment in a biogas plant and the exploitation of renewable sources, which is simultaneously an example of circular economy. The biogas plant has a power of 1 MW, which along the existing concession prices and the resulting economy justifies the validity of the decision to invest while ensuring a cleaner environment. Invested funds shall be returned within a maximum of 6 years, reflecting the extremely dynamic business process.

Keywords: Biogas plant; Costs; Forecast; Profit; Electrical energy

1. Introduction

The current energy policy requires a constant technology development towards more sustainable energy transformation systems. Cogeneration and the use of renewables are two important solutions available to achieve this goal [1]. To analyse such systems from the point of view of resource efficiency, the pure energy analysis is not sufficient since the quality of particular energy carriers is not evaluated. Biogas projects demand high investments. Biogas plants are still experiencing harsh economic conditions [2]. Financing is therefore one of the key elements in order to ensure project viability. The financing scheme of a biogas plant project differs from country to country, but in general, low interest long-term loans are used. Ordinary mortgage loans are not frequently used. The index-regulated annuity loans are low-interest loans, which secure the investor against inflation through a re-evaluation of the unpaid debts according to the inflation rate. The pay-back period is more than 20 years. This type of loans proved to be the most suitable for financing the construction of biogas plants, meeting the demands for long maturity, low interest and low initial instalments. The disadvantages of such loans are that they are raised by ordinary sales of bonds at the stock exchange market price, implying a depreciation risk that may induce some uncertainty in the planning phase [3, 4]. In countries like Denmark, biogas projects are e.g. financed by means of index-regulated annuity loans, guaranteed by the municipalities. Most of the past biogas projects received also supplementary government subsidies, representing up to 30 % of the investment costs of the project.

A single farmer, a consortium of farmers or a municipality are usually the entrepreneurs likely to implement successful biogas projects. The success of the project depends on some factors that can be controlled and influenced by strategic decisions concerning investment and operational costs. Very few research studies exist on the positive effects of biogas production on the socio-economics of rural areas [5]. Considerable changes occur regarding the employment level. Additionally, in the majority of areas there is a significant potential for further development of biogas production, which would likewise be reflected in the effects of added value and employment opportunities. Farmers have heterogeneous investment thresholds. Their investment decisions are mainly driven by capital costs and the subjective perception of the risk resulting from the investment. Other decision parameters like sustainability and non-monetary objectives play only minor roles [6]. Choosing the best technology in respect of level of investment and operational costs is very difficult. If tendering a biogas plant, it is important receive offer on operational cost like:

- Operational cost of combined heat and power plant (CHP) incl. all services and spare parts (amount/kWh),
- Maintenance costs of biogas plant in total (% of investment/year),
- Own electrical energy demand, including demand of CHP (kWh/year),
- Average working hours/day of staff (maintenance and feeding the system).

The success of the project is also influenced by some factors that cannot be controlled such as:

- Interest terms,
- Grid access and feed in tariffs,
- World market prices for feedstock (e.g. energy crops),
- Competition for feedstock from other sectors.

Industrial waste collectors face problems securing long term availability of the feedstock. This could be a problem because the waste recycling market is highly competitive and contracts with waste producers are rarely for periods longer than five years.

Quite often, before a bank offers to finance the biogas plant project, the economical long term success of the project must be proven by a study/ calculation of profitability [7]. The calculation is normally done within the preliminary planning by an experienced planning/consulting company, but in many cases, especially in the case of single based biogas projects, this work can be done by the project developer, with two consequently advantages: the project developers/partners are forced to have a very close view to the different aspects of the project and, in case of cancelling the project, no external costs have occurred.

2. Economic forecast of a biogas plant project

In a case of a biogas plant treating municipal waste, it is recommended to mandate an experienced consulting company. Waste treatment plants are much more complex regarding handling of feedstock, biological stability of the system and the whole plant design, compared to a farm based plant (Fig. 1).

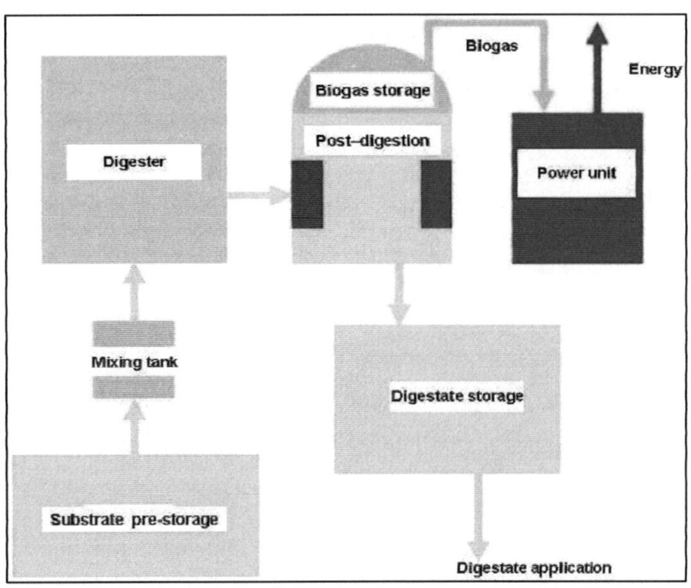

Fig. 1. Main components and general process flow of biogas production

For case specific calculation of the economic forecast, a calculation model was elaborated, allowing the preliminary estimation of costs, plant size, dimensioning, technical outline etc. finding an independent and reliable planning partner is mandatory. The framework conditions of organic biogas generation and its monetary implications on production economics must be carefully analysed [8] and [9], with risks evaluation [10]. Renewable energy production is not economically viable by its own, without considering the wastewater treatment function and the associated incomes [11].

3. Project case

For the purpose of construction of the biogas plant, land in area of 1.4 ha is being purchased. The Ministry of economy has issued a preliminary electricity permission needed in order to construct the plant with bio power of 1 MWh based on prior economic and technological analyses.

Collection of waste from farms represents an ecological problem because raw manure cannot be immediately (due to its high concentration and / or vegetation period) used on arable land as fertilizer. The waste is therefore collected or deposited along the fields in inadequate (usually permeable) reservoirs / disposal sites / lagoons from where they penetrate into the surface and ground waters, polluting them. Stench and even more important pollution from glass gases (NH_3, CO_2, CH_4, H_2S), that destroy the ozone layer, are expanding.

3.1. Project goals

Three main goals need to be achieved:

- Tackling environmental problems and at the same time exploiting "waste" traits as a renewable source of energy,
- Exploitation of processed and fermented residue as a high caloric organic fertilizer,
- Development of energy crops (silages) with the aim of improving the energy efficiency with partial exploitation of fertilizer produced on site.

The raw material for the biogas plant will be purchased in the environmental field on approximately 10,000 hectares of arable land, which provides a minimum of 10,000 tons of biomass with the price of 17 €/tonne. The production itself is agreed for a reason that it should not endanger food production and should take the arable land. Deficiencies of the contracts for inbounds are identified as price hikes of the raw materials required for the upcoming biogas plant and the price hikes of the agricultural products, corn in particular. In absence of price hikes regarding the biogas production, long-term contracts for period of 10-12 years are suggested to be signed. In terms of the biomass price the following formula is applied:

$$C_{Goie} = C_{Goie\text{-}1} \times ICM_{Goie\text{-}1} \tag{1}$$

- C_{Goie} is the purchase price for the current year,
- $C_{Goie\text{-}1}$ is the purchase price for the previous year,
- $ICM_{Goie\text{-}1}$ is an annual index of growth of the retail prices in terms of the previous year,
- *Goie* is an annual index, the minimum value in the year.

Plants connected to the distribution network with installed electrical power up to 1 MW, which use renewable sources of energy for its production, sell such energy under the price of 0,174 €/kWh, with no value added tax (VAT) calculated.

The raw material is delivered to the cleaver and the blender by pressure pipeline and/or tanks, where it is mixed with silage, other biodegradable substrates (waste material) and hot water (heated with the surplus of the heat already generated), after what it is dosed in an equalization pool, mixed with a submerging device. The liquid is being pumped from the pool into a hydrolyser, where for 2-3 days a hydrolysis is commencing on temperature of approx. 55 ^{0}C. The heating is carried out by heat exchanger of the water heated with the surplus of the produced heat (hot water). The hydrolysate is pumped onwards into a digester – 1, where for 15-20 days a thermophilic conversion of hydrolysed organic substances into biogas is commencing on temperature of approx. 55 ^{0}C [12]. The heating is carried out by heat exchanger of the water heated with the surplus of the produced heat (hot water). Ferment thickens by passing through the separator and thus condensed, pumps through the reservoir, where the conversion into biogas continues, accumulating in the gas tank / in the dome above the reservoir. In that gas tank, the biogas is dosed from the dome of the digester – 1. Reservoir – ferment will be soon built for the phase II as well, thus the initial period of retention of the ferment shall take approx. 30 days, whilst in the second phase it shall decrease [13].

The obtained biogas is purified by aerobic biological conversion of sulphide into elementary sulphur at the top of the dome and by removing the moisture by passing through the capacitors and the filters. Later on the obtained biogas is used for drive of CHP units which are used in the production of electrical and thermal energy.

Processed and fully stabilized fermenter (without odour) is transported through the same reservoirs, which are used for delivery of the substrate to the smaller reservoirs (container, metal or terrestrial), mounted on arable areas as high calorific organic fertilizer with corresponding tractor hitch. Optionally it can be one of the larger reservoirs (which holds fermenter for approx. 90 days, during the vegetation period), from which the fertilizer can be deprived if necessary.

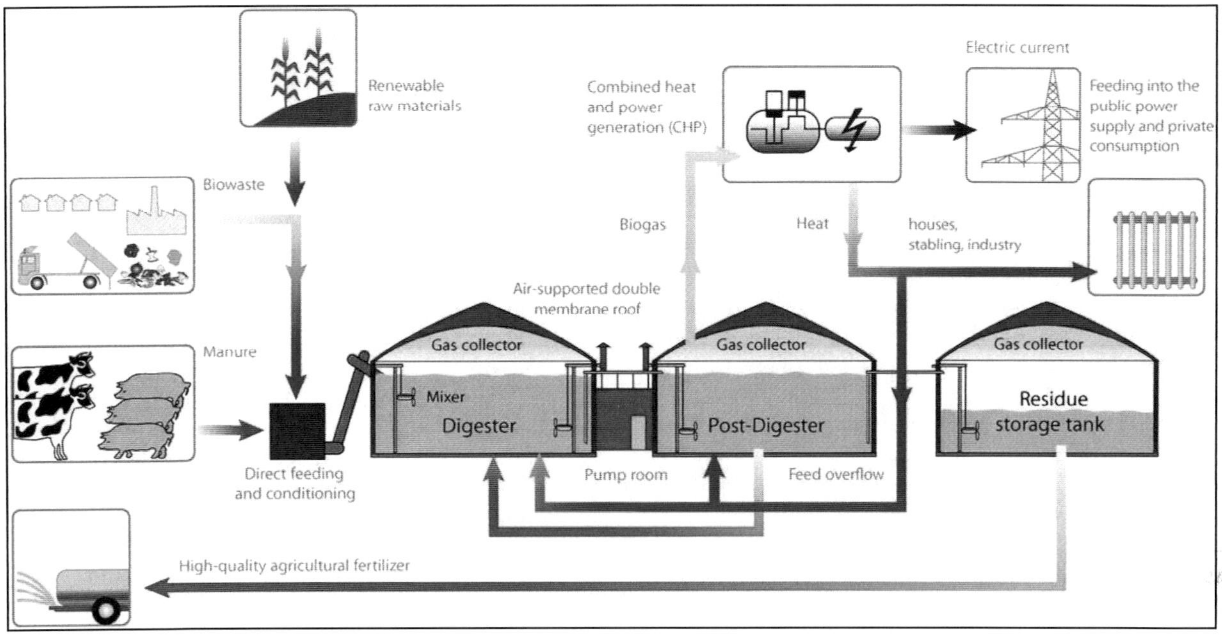

Fig. 2. Technological process of the biogas plant

The biogas plant will be installed at a location with good infrastructure in terms of logistical point of view. At 500 m distance from the bioplant there is a transformer-station with an option to connect to the network.

Possible impacts of the construction of biogas plant on the surrounding area can affect the soil, underground and groundwater, the air and the "production" waste. Negative impact can occur on soil and subsoil in case of outlet of the liquid manure. In that case the liquid manure will penetrate into the underground, in particular into the underground water or it will reach the recipient on the surface, where it can destroy the living world due to lack of oxygen. In order to prevent such negative impact, the manure is deposited into closed reservoirs.

The impact on air is unfavourable odour-stench from the biogas plant because of the silage, the livestock faeces, aggregate pits for wastewater from the silages and the reservoirs for the fermented manure. It is considered that the stench reaches an area of 500 m, whilst in terms of unfavourable weather conditions it can reach an area of up to 1 km. Closest living areas are 1 km away, which means that it shall not affect the area of residential neighbourhoods.

In order to monitor the state of the environment, regular controls and analyses on the underground water shall be performed, whilst once in two years' time analyses on the soil, which are fertilized by the fermented liquid manure from the plant shall be commenced as well. The construction of the biogas plant creates jobs, new values and raises the awareness of the usage of renewable sources.

4. Economic and financial analysis

The input data and the results of the analysis for the observed 10 years are summarised in tables 1 to 8. We have developed the cost model considering all fixed values of parameters (some of them are fixed by the government regulation or market situation) and some cost components giving us some optimisation possibilities.

Description	Equipment value	Amortization period	Annual depreciation
Construction works	707.616 €	20	35.381 €
Equipment	2.153.164 €	10	215.316 €
Cogeneration unit	739.220 €	7	105.603 €
Total	**3.600.000 €**		356.300 €

Table 1. Total investment costs

Description	Price (€)
Construction works	707.616
Electro-mechanical equipment and works	2.672.384
Engineering	220.000
Silage purchase in first year of business	220.000
Total	**3.820.000**

Table 2. Total infrastructure costs

No.	Service / Equipment	Capacity	Value
1.	Design / Construction		135.000 €
2.	Commissioning (Momentum)		65.000 €
3.	Design of plans for location permit		20.000 €
4.	Static budget		V
5.	Apparatus for receiving solids		149.838 €
6.	Reception reservoir of liquid manure		16.977 €
7.	Digester charging room		105.596 €
8.	Digester 1	3.115 m^3	406.403 €
9.	Digester 2	3.115 m^3	406.403 €
10.	Post reservoir with associated gas-tank		230.792 €
11.	CHP	Approx. 999 kW_{el}	739.220 €
12.	Safety torch		
13.	Desulfurization of biogas		
14.	Scale (quantity: 1)	30 t	
15.	Measuring, monitoring and control equipment		284.625 €
16.	Dehydration of sludge		53.590 €
17.	Laguna for post-fermented mass		100.000 €
18.	Condensing pool		
19.	Design of pipelines		117.300 €
20.	Insurance, transport		61.640 €
	Plant expenses		**2.892.384 €**
	Infrastructure expenses		**707.616 €**
1.	Silage silo		274.700 €
2.	Earthworks and asphalt bases (quantity: 2000)	40 €	80.000 €
3.	Other earthworks		54.000 €
4.	Foundations		138.966 €
5.	Substation		125.000 €
6.	Starting cost		34.950 €
	Total investment		**3.600.000 €**

Table 3. Detailed specification of equipment, material and labour

Structure \Year	1	2	3	4	5	6	7	8	9	10
Revenue										
Thermal energy	0	31.906	32.864	33.850	34.865	35.911	36.988	38.098	39.241	40.418
Electricity	730.037	1.417.548	1.460.074	1.502.601	1.545.127	1.587.654	1.630.180	1.672.707	1.715.233	1.757.760
Value of the organic fertilizers										
Total revenue	730.037	1.449.455	1.492.938	1.536.450	1.579.992	1.623.565	1.667.169	1.710.805	1.754.474	1.798.178
Costs										
Loan repayment										
Interest										
Depreciation	178.150	356.300	356.300	356.300	356.300	356.300	356.300	356.300	356.300	356.300
Maintenance of the plant		16.802	16.802	33.605	33.605	33.605	33.605	33.605	50.407	50.407
Maintenance of the CHP	47.875	119.689	123.280	126.870	130.461	134.052	137.642	141.233	144.824	148.414
Insurance	10.000	10.000	10.300	10.609	10.927	11.255	11.593	11.941	12.299	12.668
Fuel	5.833	10.000	10.300	10.609	10.927	11.255	11.593	11.941	12.299	12.668

Fermented mass supply costs		38.245	39.010	39.790	40.586	41.398	42.226	43.070	43.932	44.810
Raw materials costs (Silage)	110.000	220.000	226.600	233.398	240.400	247.612	255.040	262.692	270.572	278.689
Monitoring of the biological process		2.340	2.410	2.483	2.557	2.634	2.713	2.794	2.878	2.964
Workforce	12.775	21.900	22.557	23.234	23.931	24.649	25.388	26.150	26.934	27.742
Spare parts	9.946	19.893	20.091	20.292	20.495	20.700	20.907	21.116	21.327	21.541
Total costs	374.579	815.169	827.650	857.190	870.190	883.460	897.007	910.841	941.771	956.203
EBITDA	355.458	634.285	665.288	679.261	709.803	740.105	770.162	799.964	812.703	841.974

Table 4. Profit and loss statement (part 1)

Structure \ Year	**1**	**2**	**3**	**4**	**5**	**6**	**7**	**8**	**9**	**10**
I. Revenue	730.037	1.449.455	1.492.938	1.536.450	1.579.992	1.623.565	1.667.169	1.710.805	1.754.474	1.798.178
1. Total revenue	730.037	1.449.455	1.492.938	1.536.450	1.579.992	1.623.565	1.667.169	1.710.805	1.754.474	1.798.178
2. Rest of project	0	0	0	0	0	0	0	0	0	0
2.1. Fixed assets										
II. Expenditures	319.979	787.306	798.688	822.059	833.899	845.909	858.234	870.813	895.414	908.522
3.Fixed assets investments										
4. Operating costs	200.596	458.869	471.350	500.890	513.889	527.159	540.707	554.541	585.471	599.903
4.1.Tangible costs	177.821	426.969	438.493	467.047	479.031	491.255	503.726	516.450	546.239	559.494
4.2. Service costs	10.000	10.000	10.300	10.609	10.927	11.255	11.593	11.941	12.299	12.668
4.3. Intangible costs										
4.4. Gross salaries	12.775	21.900	22.557	23.234	23.931	24.649	25.388	26.150	26.934	27.742
5. Profit tax	79.589	218.958	218.225	214.112	213.340	212.499	211.685	210.848	206.628	205.746
6. Reserves	39.794	109.479	109.113	107.056	106.670	106.250	105.842	105.424	103.314	102.873
III. Net revenue	410.058	662.148	694.250	714.392	746.093	777.656	808.935	839.992	859.060	889.656

Table 5. Profit and loss statement (part 2)

Structure \ Year	**1**	**2**	**3**	**4**	**5**	**6**	**7**	**8**	**9**	**10**
I. Revenue	4.330.037	1.449.455	1.492.938	1.536.450	1.579.992	1.623.565	1.667.169	1.710.805	1.754.474	1.798.178
1. Total revenue	730.037	1.449.455	1.492.938	1.536.450	1.579.992	1.623.565	1.667.169	1.710.805	1.754.474	1.798.178
2. Funding sources	3.600.000	0	0	0	0	0	0	0	0	0
2.1. Own resources	720.000									
2.1. Loans	2.880.000									
3. Rest of project	0	0	0	0	0	0	0	0	0	0
3.1. Fixed assets										
II. Expenditures	200.596	883.634	896.115	925.655	938.655	951.925	965.472	979.306	1.010.237	1.024.669
4. Investments in fixed assets										
5. Operational costs	200.596	458.869	471.350	500.890	513.889	527.159	540.707	554.541	585.471	599.903
5.1. Tangible costs	177.821	426.969	438.493	467.047	479.031	491.255	503.726	516.450	546.239	559.494
5.2. Services costs	10.000	10.000	10.300	10.609	10.927	11.255	11.593	11.941	12.299	12.668
5.3. Intangible costs	0	0	0	0	0	0	0	0	0	0
5.4. Gross salaries	12.775	21.900	22.557	23.234	23.931	24.649	25.388	26.150	26.934	27.742
6. Profit tax 20 %	0	0	0	0	0	0	0	0	0	0
7. Reserves	0	0	0	0	0	0	0	0	0	0
8. Annuities	0	424.765	424.765	424.765	424.765	424.765	424.765	424.765	424.765	424.765
III. Net revenue	4.129.441	565.820	596.823	610.795	641.338	671.640	701.696	731.499	744.237	773.509
IV. Cumulative net revenue	4.129.441	4.695.261	5.292.084	5.902.880	6.544.217	7.215.858	7.917.554	8.649.053	9.393.290	10.166.799

Table 6. Economic flow of project

Month	Monthly production	Own consumption	Energy for sale	Rest
January	662.400	198.720	463.680	0
February	662.400	198.720	463.680	0
March	662.400	198.720	463.680	0
April	662.400	158.976	302.054	201.370
May	662.400	0	264.960	397.440
June	662.400	0	0	662.400
July	662.400	0	0	662.400
August	662.400	0	0	662.400
September	662.400	0	198.720	463.680
October	662.400	59.616	361.670	241.114
November	662.400	99.360	563.040	0
December	662.400	198.720	463.680	0
Total per annum	7.948.800	1.112.832	3.545.164	3.290.804

Table 7. Production balance of thermal energy (in KWh)

Consumption of thermal energy for own needs is 55 %.

Month	Monthly production	Capacity (kW)	Daily production (h)
January	674.505	999	21,78
February	609.230	999	21,78
March	674.505	999	21,78
April	652.747	999	21,78
May	674.505	999	21,78
June	652.747	999	21,78
July	674.505	999	21,78
August	674.505	999	21,78
September	652.747	999	21,78
October	674.505	999	21,78
November	652.747	999	21,78
December	674.505	999	21,78
Average	661.813	999	21,78
Total per annum	7.941.750	-	7949,7

Table 8. Production of electricity (in KWh)

Consumption of electricity for own needs of the biogas plant is 6 % or 476.505 kWh. During operation, we have a computer control loop, which indicates that we are running within planned costs, within the achievement of the planned objectives and within other technical parameters. All technical parameters and costs, eventual failures are controlled by dedicated software. Any deviations are displayed using visual warning signal, overall control can also be remotely controlled and synchronized with the network. This ensures a stable economic operation of the plant.

5. Conclusion

Production and consequently sales of energy derived from biogas production devices offers economic opportunity for treatment and recycling of many agricultural residues and by-products, a variety of bio-waste, organic waste water from industry and sewage in a sustainable and environmentally friendly manner.

Having done the pre-calculations using the calculation tool as recommended in the result is a model of the economy of the project. As stated before, the operational costs and the investment costs can be influenced by strategic decisions. For example by choosing the best adapted technology. So, if labour is cheap in your country, than it might be cheaper engaging more people than spending money for automation of a plant.

The revenue side of a project is difficult to influence. The feed in tariffs are set by the government. In case of waste treatment plants, the tipping fees are market prices. There are other possibilities of improve the revenue side:

- Using/selling the produced heat,
- Selling digestant as a fertilizer.

If the project obtains an internal rate of return (*IRR*) lower than 9 %, you should reconsider all the project premises, and improve some of them. If the *IRR* rate is higher than 9 %, the premises are good and it is worth continuing the project and moving to the next planning phase. It is important to compare the assumptions with the material reality. This helps to get a realistic idea of the biogas plant itself, the needed space, the true mass current and the real building costs. The calculation model is useful for providing the rough information which is necessary to kick start the actual planning phase. Also the earnings before interest, taxes, depreciation and amortization (*EBITDA*) are very prosperous. For the next steps of project, finding an independent and reliable planning partner is mandatory.

According to the economic indicators, return on the whole investment is anticipated in 5-6 years. The lifetime of the biogas plant is 20 years. The plant itself delivers 5 jobs, whilst it indirectly ensures 20 jobs including production of biomass on the fields. The utilization of the fermented mass as organic fertilizer improves crops contribution for 15-25 % and reduces the application of artificial fertilizers, which as a result improves the total quality of the agricultural production and decreases the artificial fertilizers expenses. In the biogas plant, the waste is used as bio-renewable source of energy.

The main paper contribution is the development of new structure of cost calculation applying a wide list of influencing parameters which consider the actual state and conditions. Another achievement is the modified system of evaluation of the eligibility of investment. In future research some model expansions are planned, including the generalization of implemented (combined) energy sources.

6. References

[1] Stanek, W.; Gazda, W. & Kostowski, W. (2015). Thermo-ecological assessment of CCHP (combined cold-heat-and-power) plant supported with renewable energy. Energy, Vol. 92, Special issue – Part 3, 279-289, ISSN 0360-5442

[2] Ahlberg-Eliasson, K.; Nadeau, E.; Leven, L. & Schnurer, A. (2017). Production efficiency of Swedish farm-scale biogas plants. Biomass & Bioenergy, Vol. 97, 27-37, ISSN 0961-9534

[3] Glavan, I.; Prelec, Z. & Pavkovic, B. (2015). Modelling, simulation and optimization of small-scale CCHP energy systems. International Journal of Simulation Modelling, Vol. 14, No. 4, 683- 696, ISSN 1726-4529

[4] Kang, J. Y.; Kang, D. W.; Kim, T. S. & Hur, K. B. (2014). Comparative economic analysis of gas turbine-based power generation and combined heat and power systems using biogas fuel. Energy, Vol. 67, 309-318, ISSN 0360-5442

[5] Guenther-Luebbers, W.; Bergmann, H. & Theuvsen, L. (2016). Potential analysis of the biogas production - as measured by effects of added value and employment. Journal of Cleaner Production, Vol. 129, 556-564, ISSN 0959-6526

[6] Reise, C.; Musshoff, O.; Granoszewski, K. & Spiller, A. (2012). Which factors influence the expansion of bioenergy? An empirical study of the investment behaviours of German farmers. Ecological Economics, Vol. 73, 133-141, ISSN 0921-8009

[7] Lai, X. D.; Wu, G.-D.; Shi, J. G.; Wang, H. M. & Kong, Q. S. (2015). Project value-adding optimization of project-based supply chain under dynamic reputation incentives. International Journal of Simulation Modelling, Vol. 14, No. 1, 121-133, ISSN 1726-4529

[8] Blumenstein, B.; Siegmeier, T. & Moeller, D. (2016). Economics of anaerobic digestion in organic agriculture: Between system constraints and policy regulations. Biomass & Bioenergy, Vol. 86, 105-119, ISSN 0961-9534

[9] Krajnc, M.; Dolsak, B. (2013). Computer and experimental simulation of biomass production using drum chipper. International Journal of Simulation Modelling, Vol. 12, No. 1, 39-49, ISSN 1726-4529

[10] Kremljak, Z. (2016). Risk analysis of specific project problems. Proceedings of the 27th International DAAAM Symposium "Intelligent Manufacturing & Automation", October 2016, Mostar, Bosnia & Herzegovina, ISSN 2304-1382, ISBN 978-3-902734-3-6, Katalinic, B. (Ed.), 8 pages, DAAAM International Vienna

[11] Fersi, S.; Chtourou, N.; Jury, C. & Poncelet, F. (2015). Economic analysis of renewable heat and electricity production by sewage sludge digestion – a case study. International Journal of Energy Research, Vol. 39, No. 2, 234-243, ISSN 1099-114X

[12] Voegel, V.; Bertron, A. & Erable, B. (2016). Mechanisms of cementitious material deterioration in biogas digester. Science of The Total Environment, Vol. 571, 892-901, ISSN 0048-9697

[13] Siswantara, A. I.; Daryus, A.; Darmawan, S.; Gunadi, G. G. & Camalia, R. (2016). CFD analysis of slurry flow in an anaerobic digester. International Journal of Technology, Vol. 7, No. 2, 197-203, ISSN 2086-9614

DOI: 10.2507/28th.daaam.proceedings.019

DYNAMIC MODELLING OF MOBILE ROBOTS BASED ON ITS 3D CAD MODEL

Vjekoslav Damic, Maida Cohodar & Munir Muratovic

This Publication has to be referred as: Damic, V[jekoslav]; Cohodar, M[aida] & Muratovic, M[unir] (2017). Dynamic Modelling of Mobile Robots Based on its 3D CAD Model, Proceedings of the 28th DAAAM International Symposium, pp.0144-0149, B. Katalinic (Ed.), Published by DAAAM International, ISBN 978-3-902734-11-2, ISSN 1726-9679, Vienna, Austria
DOI: 10.2507/28th.daaam.proceedings.019

Abstract

Usage of the mobile robots is extended in all areas of human life and work. The investigation of their behaviour becomes popular among the researchers around the world. Dynamic modelling of the mobile robot can be simplified by using their 3D CAD model. In this paper, the dynamic model of mobile robot is developed by MATLAB/Simulink using previously created 3D CAD model in SOLIDWORKS. As example, the mobile robot Robotino by FESTO is chosen. Assembly of mobile robot in form of xml file is imported from SOLIDWORKS to MATLAB. Such approach provides automatically generation of dynamics of robot's mechanical subsystem. Then, the model can be modified and upgraded, for instance to organize it on several levels, to actuate the wheels by motors, etc. Validation of the dynamic model is performed by simulation.

Keywords: mobile robot; dynamic modelling; visualization; 3D CAD model

1. Introduction

Investigations of the robots attract attention of many researches around the world in different areas. Modelling and simulation are recognized as powerful tool for solving many problems in robotics. Analysis of industrial robots using bond graphs by visualization is presented in [2], where BondSim is used for dynamic analysis of robot and BondSim3D Visual application is used for its visualization. Simulation of hobby robot by MATLAB is subject of analysis in [4]. Also, the huge research area is modelling of mobile robotics behaviours. The kinematics and dynamic analysis of mobile robot was done in [5]. Investigation of behaviour of group of mobile robots is subject of research in [8,9]. A presentation of working scenarios and mobile robot working algorithms in Bionic Assembly System and their realization in AnyLogic software is given in [10].

The paper proposes procedure for the dynamic analysis of mobile robot by MATLAB/Simulink based on previously created 3D CAD model [11]. Proposed approach is explained on example of mobile robot Robotino from FESTO [6]. Robotino is subject of research of [1], where it is used for arrangement of remote education. Modelling of Robotino by bond graphs is presented in [3].

In this research, the dynamic model of Robotino is automatically developed by importing its 3D CAD model from SOLIDWORKS to MATLAB following the idea from [7]. Dimensions of robot parts are roughly measured and estimated.

The obtained robot's dynamic model is reorganized systematically into more hierarchical levels and its validation was conducted by simulation. It can be upgraded and used to test different control algorithms, trajectory planning, collision avoidance, etc.

2. 3D CAD model of Robotino

Mobile robot Robotino (Fig.1a) is equipped by three omnidirectional wheels. They are powered by dc motors, which are denoted by manufacturer FESTO as motor 1 (M1), motor 2 (M2) and motor 3 (M3). Wheels and corresponding motors are denoted inFig.1 by red, green and blue colour, respectively. Each wheel is composed of the roller supporter and six rollers, three rollers per each side, shifted by 120^0 with respect to each other. Rollers can rotate about their axes.

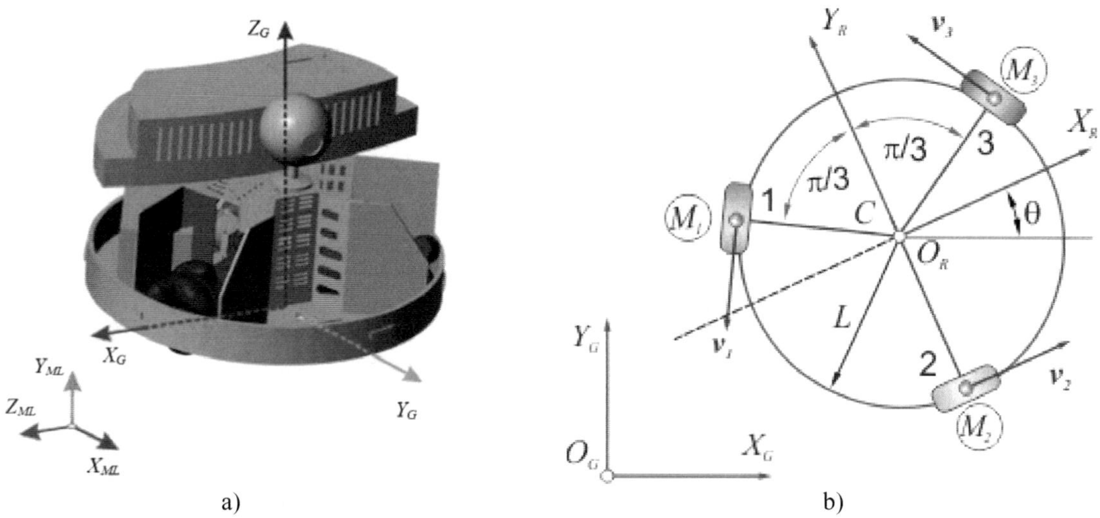

Fig. 1. Robotino with assigned coordinate frames: a) the global frame; b) the local frame

As previously emphasized, the first step to develop the mobile robot dynamic model is creating of their 3D CAD model. Using SOLIDWORKS 3D CAD models of the robot's parts are generated (Fig.2). For simplicity unnecessary details are omitted, without loss of robot functionality. 3D CAD model of robot is taken from [11].

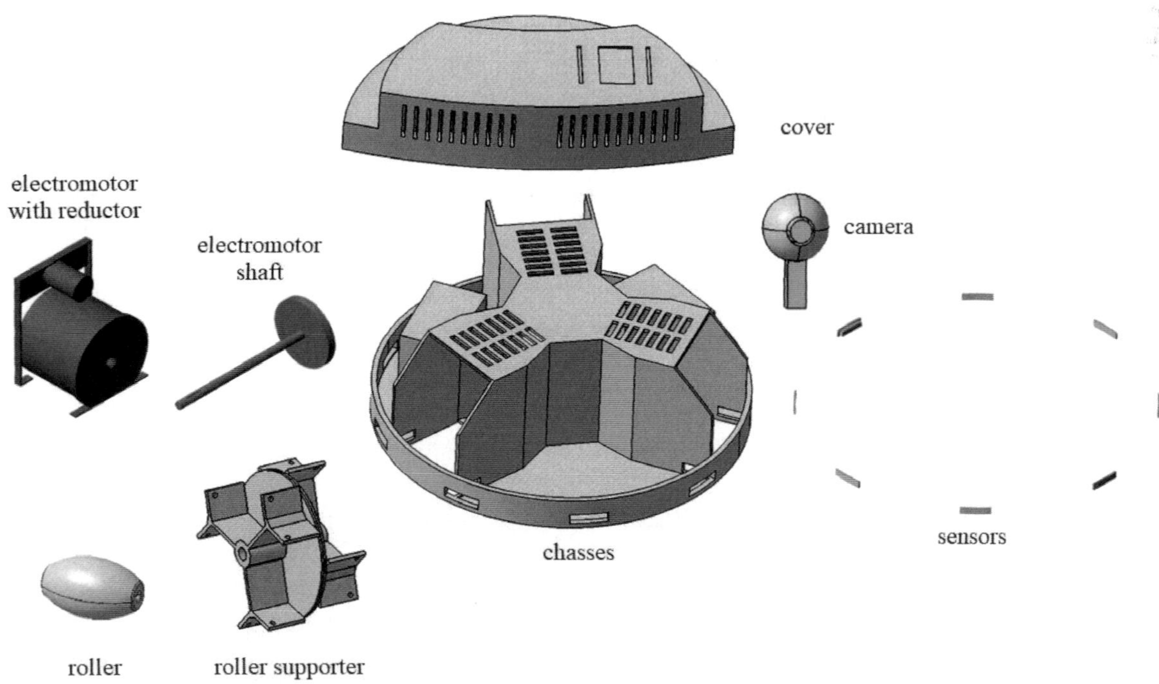

Fig. 2. Robotino parts

Robot's parts are combined in the assembly by exactly defined mating between them using the SOLIDWORKS instruction. Finally, the assembly shown in Fig.1a has been imported using SimMechanic link (ver. 4.7 of Matlab 2015b and Windows x64) in the form of xml file, and their parts as stl files.

3. Dynamic model

By importing the robot assembly into MATLAB in form of xlm file, that is obtained using SOLIDWORKS, the dynamic model of the Robotino has been automatically developed. It is realized as single level structure. Taking into account that robot consists of many parts the model is pretty unclear and complicated. Hence, it can be organized systematically into more levels by putting group of parts into subsystems, as depicted in Figs. 3 and 4.

There is several blocks only in the first level: the block which represents the dynamic model of Robotino, planar joint, two sensors to measure robot positons and blocks to define world coordinate frame, appropriate solver and the mechanical simulation environment for the robot (for instance gravity, constraint solver type, tolerances, linearization, etc.). It is important to emphasize that the coordinate system in which the robotic assembly is made, corresponds to the world coordinate system in MATLAB, $O_{ML}X_{ML}Y_{ML}Z_{ML}$.

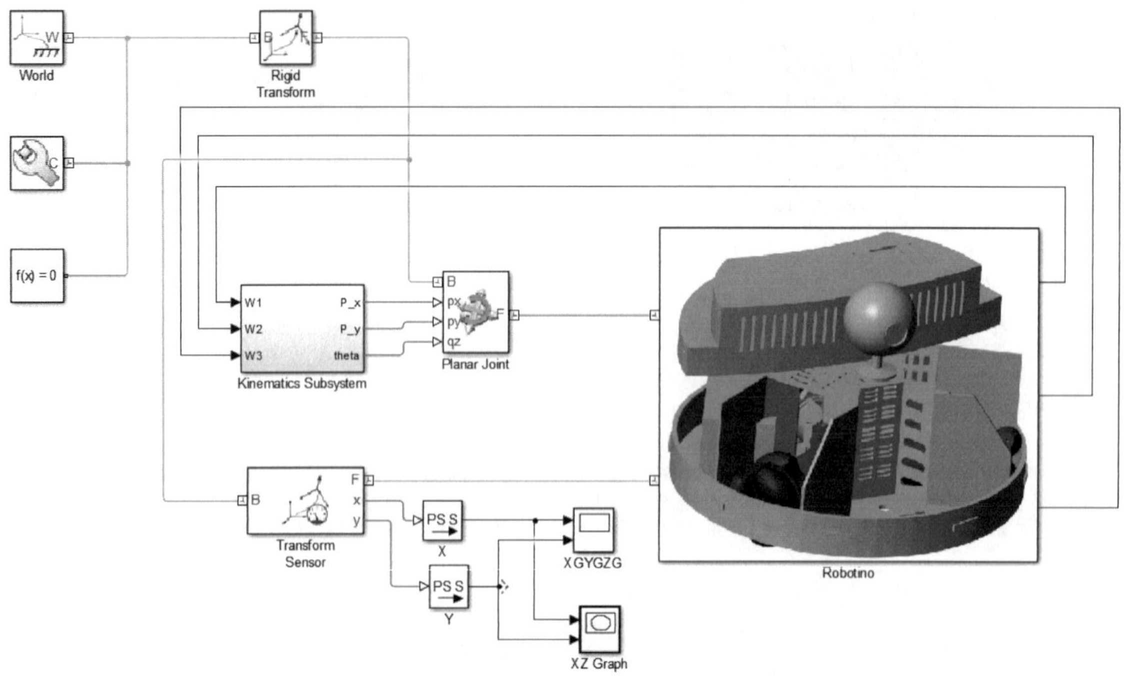

Fig. 3. The system level dynamic model of Robotino,

Due to fact that Robotino manufacturer in the robot manuals and other documentations uses the global coordinate frame $O_GX_GY_GZ_G$ (as shown in Fig.1a) all quantities should be transform to $O_{ML}X_{ML}Y_{ML}Z_{ML}$ using block 'Rigid transform', as depicted on the top of Fig.3. This transformation is given by:

$$\mathbf{R}_{ML}^{G} = \begin{bmatrix} 0 & 0 & 1 \\ 1 & 0 & 0 \\ 0 & 1 & 0 \end{bmatrix}. \tag{1}$$

Following idea from [7], the robot is connecting to the ground with the planar joint providing two translator motions p_x and p_y along X_G and Y_G-axis and rotation θ about Z_G-axis. Inputs for the planar joint are given by:

$$\begin{aligned} p_x &= \int v_x\,dt = -\int v_1 \cdot \cos\left(\theta + \frac{\pi}{3}\right) dt + \int v_2 \cdot \cos(\theta)\, dt - \int v_3 \cdot \cos\left(\theta - \frac{\pi}{3}\right) dt, \\ p_y &= \int v_y\,dt = -\int v_1 \cdot \sin\left(\theta + \frac{\pi}{3}\right) dt + \int v_2 \cdot \sin(\theta)\, dt - \int v_3 \cdot \sin\left(\theta - \frac{\pi}{3}\right) dt, \\ \theta &= \int \frac{(\omega_1 + \omega_2 + \omega_3) \cdot R}{3L} dt \end{aligned} \tag{2}$$

where v_i and ω_i are linear and angular velocities of wheels (i=1,2,3), R is wheel radius, L is radius of mobile platform.

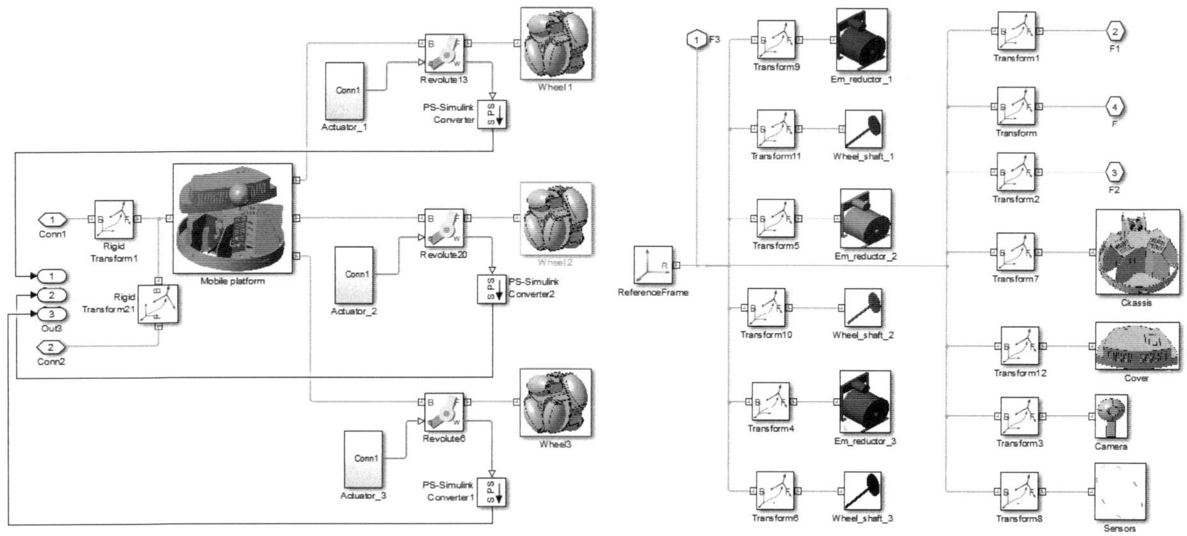

a) b)

Fig. 4. The structure: a) 'Robotino' block; b) 'Mobile platform' block

Structure of Robotino on the next level of decomposition is shown in Fig.4a. The Mobile platform, three omnidirectional wheels, and their actuators, are the basic components presented on this level. Structure of 'Mobile platform' is depicted in Fig.4b. In order to taking acount mass of electromotors and reductors, they are presented as rigid bodies. Their funcionalities can be represented by Actuator_i (i=1,2,3) blocks.

Strucuture of wheels are presented in Fig. 5a. Ahead each component there is block 'Rigid transform', which provides the appropriate coordinate transformations. Each component is realized as a rigid body (Fig.5b), whose mass and geometric properties are automatically calculated from SOLIDWORKS 3D CAD models.

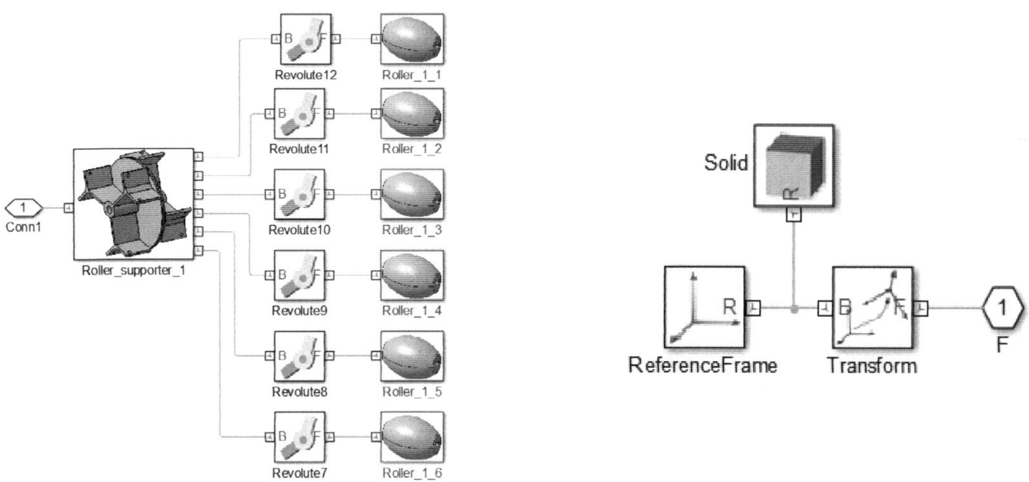

a) b)

Fig. 5. The structure of: a) 'Wheel' block; b) The rigid body block

4. Simulation results

To verify developed dynamic model of Robotino a simple experiment is performed. Wheels are powered in such a way that their angular accelerations change in the time as shown in Fig.6. Robot moves along Y_G axis during the first 4 s, because wheels 1 and 3 are actuated by the same angular velocities, but of the opposite sign. During this motion, the wheel 2 is not powered. In the next 5.1 s all three wheels rotate with the same angular velocity assuring that the robot just rotates about Z_G axis, as shown in Fig.7. Finally, during 9.1 s to 15 s (the end of the simulation time) the robot moves along Y_G axis, similarly to motion in the first phase (Fig.8).

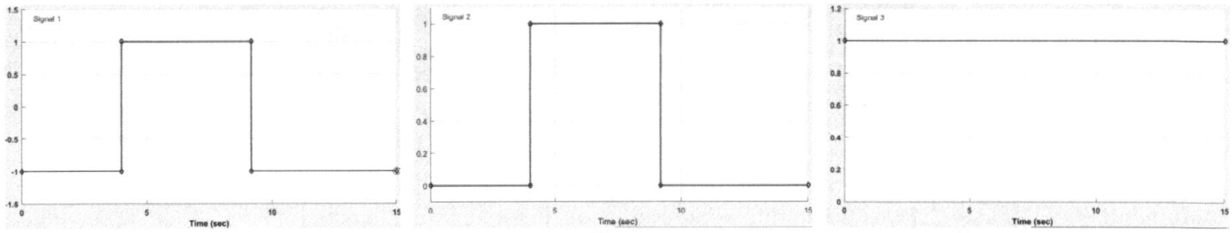

a) b) c)

Fig. 6. The angular acceleration of wheel: a) 1; b) 2; c) 3;

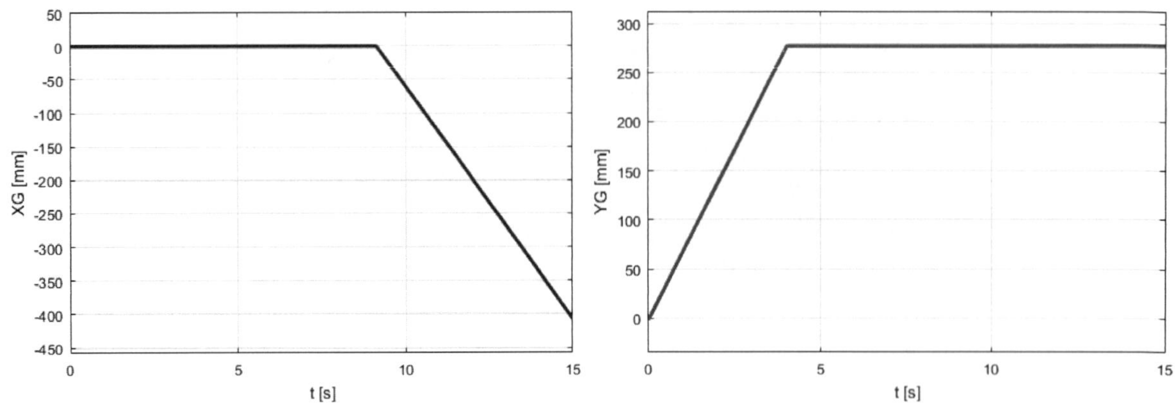

Fig. 7. X and Y coordinates of Robotino in $O_GX_GY_GZ_G$

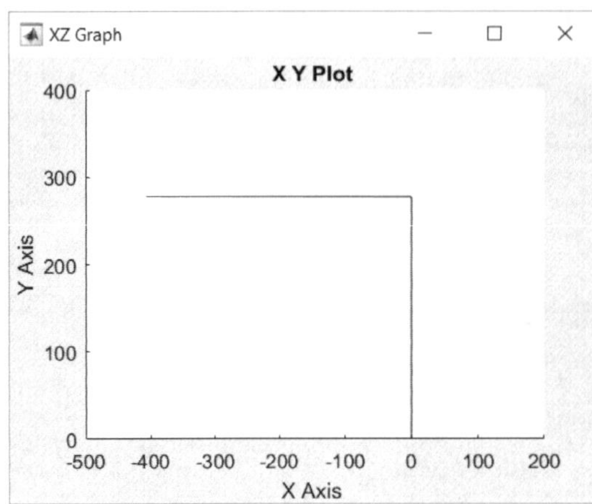

Fig. 8. The Robotino motion in plane in $O_GX_GY_G$

5. Conclusion

The procedure for development of dynamic model of mobile robot based on its 3D CAD model is explained in the paper on example of mobile robot FESTO Robotino. Firstly, the 3D CAD model is developed using SOLIDWORKS and exported in form of xml file. The file is imported to MATLAB's SimMechanics module (MATLAB/Simulink ver. 2015b). The mass and geometric parameters including the all necessary transformations are automatically calculated from the 3D CAD model (assuming the material properties of the each robot part are defined previously). This approach provides a comfortable and simple way to develop the dynamic model of the mechanical part of the mobile robot.

The obtained dynamic model is realized as a single level (flat) structure. It could be changed by grouping some of the parts into the subsystems and thus providing systematically the model representation on more hierarchical levels. Model can be also upgraded by adding appropriate control structure, the wheel actuators, etc. assuring an excellent research platform for investigation of the robot behaviour. Comparison of different control algorithms, and path planning, work in unstructured environmental would be direction of future investigation.

6. References

[1] Breido, I. V.; Markvardt, R. V. & Zarnitcin, A.(2012) Modernization of communication interface between mobile robot Robotino and Matlab software, Annals of DAAAM for 2012 & Proceedings of the 23rd International DAAAM Symposium, ISBN 978-3-901509-91-9, ISSN 2304-1382, pp 1107 - 1110, Editor B[ranko] Katalinic, Published by DAAAM International, Vienna, Austria

[2] Damic, V. & Cohodar, M. (2017). Multibody System Modeling, Simulation, and 3D Visualization, Chapter 17, pp.627-671, in book © Springer International Publishing Switzerland 2017, W. Borutzky (ed.), Bond Graphs for Modelling, Control and Fault Diagnosis of Engineering Systems, DOI 10.1007/978-3-319-47434-2_17

[3] Damic, V.; Cohodar, M. & Omerspahic, A. (2013). Dynamic Analysis of an Omni-directional Mobile Robot, Proceedings of 17th Onternational Research/Expert Conference TMT 2013, 10-11 September 2013, Istanbul, ISSN, ISBN, Katalinic, B. (Ed.), pp. 14-17, Publisher, City, DOI

[4] Damic, V.; Cohodar, M.; & Tvrtkovic, M. (2016). Inverse Dynamic Analysis of Hobby Robot uArm by Matlab/Simulink, Proceedings of the 27th DAAAM International Symposium, pp.0095-0101, B. Katalinic (Ed.), Published by DAAAM International, ISBN 978-3-902734-08-2, ISSN 1726-9679, Vienna, Austria DOI: 10.2507/27th.daaam.proceedings.014

[5] Du, Y. (2016). Kinematics and Dynamic Modeling and Simulation Analysis of Three-wheeled Mobile Robot, International Conference on Mechanics Design, Manufacturing and Automation (MDM 2016) ISBN: 978-1-60595-354-0

[6] http://www.festo-didactic.com/ov3/media/customers/1100/544305_robotino_deen2.pdf, Accessed on: 2017-09-01

[7] https://www.mathworks.com/videos/mobile-robot-simulation-for-collision-avoidance-with-simulink-90193.html, Accessed on: 2017-09-01

[8] Kirill, K. (2015). Software Architecture of Control System for Heterogeneous Group of Mobile Robos. Procedia Engineering, Vol.100, pp (278-282), doi: 10.1016/j.proeng.2015.01.368

[9] Klancar, G., Zupancic, B. & Karba.R. (2007). Modelling and simulation of a group of mobile robots. Simulation Modelling Practice and Theory, 15:647–658.

[10] Kukushkin, I. K.; Katalinic, B.; Cesarec, P.; Zdyb, D. & Kettler, R.. (2012) Modeling of mobile robot behavior for line-less Bionic Assembly System, Annals of DAAAM for 2012 & Proceedings of the 23rd International DAAAM Symposium, ISBN 978-3-901509-91-9, ISSN 2304-1382, pp 0865 - 0870, Editor B[ranko] Katalinic, Published by DAAAM International, Vienna, Austria

[11] Muratovic, M. (2017). Modeling and simulation of mobile robot by Simulink, Faculty of Mechanical Engineering, University of Sarajevo, Master thesis

DOI: 10.2507/28th.daaam.proceedings.020

FAULT ISOLATION IN TECHNICAL SYSTEMS BASED ON NON-PARAMETRIC METHOD

Alexey Zhirabok, Vladimir Filaretov & Alexey Shumsky

This Publication has to be referred as: Zhirabok, A[lexey]; Filaretov, V[ladimir] & Shumsky, A[lexey] (2017). Fault Isolation in Technical Systems Based on Non-parametric Method, Proceedings of the 28th DAAAM International Symposium, pp.0150-0159, B. Katalinic (Ed.), Published by DAAAM International, ISBN 978-3-902734-11-2, ISSN 1726-9679, Vienna, Austria
DOI: 10.2507/28th.daaam.proceedings.020

Abstract

The paper is devoted to the fault isolation problem in technical systems described by nonlinear dynamic models. So-called non-parametric method is used to solve the problem. The feature of this method is that parameters of the system under consideration may be unknown. To apply this method, the system is transformed into the input-out form. The methods to obtain such a transformation and to isolate faults are suggested. They are based on so-called logic-dynamic approach. The feature of this approach is that the system transformation and fault isolation problems for nonlinear systems are solved by methods of linear algebra. The adaptive threshold is used to avoid a false alarm. The theoretical results are illustrated by practical example of diagnosis of general electric servoactuator.

Keywords: Nonlinear models; Fault isolation; Non-parametric method; Matrix of syndromes

1. Introduction

The problem of fault diagnosis in technical systems is very important since if a fault occurs and is not timely detected, this may results in great damages and losses. This problem was extensively investigated for the past 30 years; see, e.g., the papers [7, 8, 12, 15, 17, 19, 20], the books [2, 3, 11, 13, 14, 16, 21]. Many problems have been studied and solved: different methods of residual generation and relationships among them, robustness and adaptive threshold test, fuzzy logic, sliding mode; many types of systems have been considered: linear, bilinear, linearly-analytical, descriptor, hybrid. Many practical examples have been considered; see, for example, the books devoted to industrial and mechatronic systems [3, 14].

There exists a promising method of fault diagnosis in technical systems known as non-parametric, or model-free, or data-driven method (see, for example, the paper [1, 4, 6, 9, 18, 24] and the book [5]). The feature of this method is that some parameters of the system under consideration may be unknown. Besides, the method can be used for diagnosis in systems which parameters are known with limited degree of accuracy. In this paper, this method is used to solve the fault diagnosis problem in technical systems containing non-differentiable nonlinearities.

To isolate the faults, a bank of residual generators is used. Every generator from this bank is sensitive to some faults and is invariant with respect to the rest of then. There are many methods to obtain the invariance with respect to faults. They require to obtain several variants of solution and then to analyze them in order to choose the appropriate variant. In this paper, the method allowing reducing the number of variants is suggested. It is based on so-called logic-dynamic approach. The feature of this approach is that the problems for nonlinear systems are solved by methods of linear algebra [22, 23] but it does not guarantee the minimal dimension of the system designed.

Parameters of the system under diagnosis may deviate from their nominal values in some tolerance. In this case the adaptive threshold should be used to avoid a false alarm. The new method to generate such a threshold is suggested. Consider control system described by the nonlinear equations

$$x(t+t) = Fx(t) + Gu(t) + C\begin{pmatrix} \varphi_1(A_1x(t),u(t)) \\ \vdots \\ \varphi_p(A_px(t),u(t)) \end{pmatrix} + \sum_{i=1}^{q} D_i d_i(t), \qquad y(t) = Hx(t)\,, \tag{1}$$

where $x \in R^n$, $u \in R^m$, $y \in R^l$ are vectors of state, control, and output, respectively; F, G, H, and $D_1,\ldots,D_q$ are constant matrices, $d_1(t),\ldots,d_q(t)$ are scalar functions describing the faults: if there are no faults, $d_1(t) = \ldots = d_q(t) = 0$, if the i-th fault occurs, $d_i(t)$ becomes an unknown function of time; C is $n \times p$ constant matrix, A_1, …, A_p are row matrices; φ_1, …, φ_p are arbitrary nonlinear functions. Notice that the linear terms in (1) are separated from the nonlinear ones, this can be done by methods described in [22, 23].

It is assumed that the matrices F, G, C, and A_1, …, A_p may contain the parameters $\gamma_1,\ldots,\gamma_q$, therewith if the parameter γ_i is equal to its nominal value γ_{0i}, then $d_i = 0$, otherwise $d_i \neq 0$, $i = 1,\ldots,q$.

2. Problem solution

2.1. Preliminaries

To implement the non-parametric methods, we need obtaining the input-output representation of system (1) or some its subsystem. Such a representation can be obtained by transformation of the state and output vectors of the initial system:

$$x_*(t) = \Phi x(t), \qquad y_*(t) = Ry(t)\,,$$

where Φ and R are some matrices. Such a transformation yields the system described by the equations

$$\begin{aligned} x_{*1}(t+1) &= x_{*2}(t) + J_1 y(t) + G_{*1}u(t) + \psi_1(x_{*2}, x_{*3},\ldots, y, u)\,, \\ &\cdots \\ x_{*k}(t+1) &= J_k y(t) + G_{*k}u(t) + \psi_k(y,u)\,, \\ y_*(t) &= x_{*1}(t)\,, \end{aligned} \tag{2}$$

where x_{*i} is the i-th component of the state vector x_*, J_i and G_i are the i-th rows of some matrices J and G, $\psi_1,\ldots,\psi_k$ are nonlinear functions.

Clearly, system (2) can be transformed into the input-output form via several temporal shifts and substitutions as follows:

$$y_*(t+1) = x_{*2}(t) + J_1 y(t) + G_{*1}u(t) + \psi_1(x_{*2}(t), x_{*3}(t),\ldots, y(t), u(t))\,,$$

$$y_*(t+2) = J_1 y(t+1) + G_{*1}u(t+1) + \psi_1(x_{*3}(t) + J_2 y(t) + G_{*2}u(t) + \psi_2(x_{*3}(t), x_{*4}(t),\ldots, y(t), u(t)),$$
$$x_{*3}(t+1),\ldots, y(t+1), u(t+1)) + x_{*3}(t) + J_2 y(t) + G_{*2}u(t) + \psi_2(x_{*3}(t), x_{*4}(t),\ldots, y(t), u(t)),$$

and so forth. The last expression has the form

$$y_*(t+k) = J_1 y(t+k-1) + G_{*1}u(t+k-1) + \ldots + J_k y(t) + G_{*k}u(t) + \Psi(y(t+k-1), u(t+k-1),\ldots, y(t), u(t))$$

with some function Ψ. This expression can be rewritten in the form

$$y_*(t+k) = (J_1 \quad G_{*1} \quad \cdots \quad J_k \quad G_{*k} \quad 1)\begin{pmatrix} y(t+k-1) \\ u(t+k-1) \\ \cdots \\ y(t) \\ u(t) \\ \Psi(t) \end{pmatrix} \tag{3}$$

Write down the expression for the variable y_* for T instants of time:

$$Y_T(t) = (y_*(t+k+T-1) \quad y_*(t+k+T-2) \quad \cdots \quad y_*(t+k)) = (J_1 \quad G_{*1} \quad \cdots \quad J_k \quad G_{*k} \quad 1)V_T(t) \tag{4}$$

where

$$V_T(t) = \begin{pmatrix} y(t+k+T-2) & y(t+k+T-3) & & y(t+k-1) \\ u(t+k+T-2) & u(t+k+T-3) & & u(t+k-1) \\ \cdots & \cdots & \cdots & \cdots \\ y(t+T-1) & y(t+T-2) & & y(t) \\ u(t+T-1) & u(t+T-2) & & u(t) \\ \Psi(t+T-1) & \Psi(t+T-2) & & \Psi(t) \end{pmatrix}.$$

It is assumed that the unknown parameters are in the row $(J_1 \quad G_{*1} \quad \cdots \quad J_k \quad G_{*k} \quad 1)$, the row $Y_T(t)$ and the matrix $V_T(t)$ depend on the measured inputs and outputs only. This allows to perform a diagnostic process using the row $Y_T(t)$ and the matrix $V_T(t)$, i.e. without information about values of some parameters.

2.2. The design of linear part: general solution

To transform system (1) into input-output representation (3) based on (2), the logic-dynamic approach will be used. According to this approach one has at first to transform the linear part of (1) described by the matrices F and G and then to make a transformation of the nonlinear term. Generally, the linear part of system (2) is described by the equations

$$x_*(t+1) = F_*x_*(t) + G_*u(t) + Jy(t), \qquad y_*(t) = H_*x_*(t) \tag{5}$$

with the matrices F_*, G_*, J, and H. Since $x_* = \Phi x$ and $y_*(t) = Ry(t)$, it is known [7, 10, 22] that these matrices satisfy the conditions

$$RH = H_*\Phi\,, \qquad \Phi F = F_*\Phi + JH\,, \qquad G_* = \Phi G\,. \tag{6}$$

To construct the system in the form (2), the matrices F_* and H_* are found in the canonical form

$$F_* = \begin{pmatrix} 0 & 1 & 0 & \cdots & 0 \\ 0 & 0 & 1 & \cdots & 0 \\ \cdots & \cdots & \cdots & \ddots & \cdots \\ 0 & 0 & 0 & \cdots & 0 \end{pmatrix}, \qquad H_* = (1 \quad 0 \quad 0 \quad \cdots \quad 0)\,. \tag{7}$$

Using matrices in (7), one can obtain from (6) equations for rows of the matrices Φ and J:

$$\Phi_1 = RH\,, \quad \Phi_i F = \Phi_{i+1} + J_i H\,, \quad i = 1,\ldots,k-1\,, \quad \Phi_k F = J_k H\,. \tag{8}$$

As shown in [22], equations (8) can be transformed into the single equation $RHF^k = J_1HF^{k-1} + J_2HF^{k-2} + \ldots + J_kH$. Rewrite this equation in the form

$$(R \quad -J_1 \quad -J_2 \quad \ldots \quad -J_k)V^{(k)} = 0 \tag{9}$$

where

$$V^{(k)} = \begin{pmatrix} HF^k \\ HF^{k-1} \\ \vdots \\ H \end{pmatrix}.$$

2.3. The design of linear part: fault isolation

Consider some fault described by some matrix D. The transformed system is invariant with respect to this fault if $\Phi D = 0$. To take into account this condition, write down, based on (8), the expressions for three first rows of the matrix Φ :

$$\Phi_1 = RH, \quad \Phi_2 = \Phi_1 F - J_1 H = RHF - J_1 H,$$
$$\Phi_3 = \Phi_2 F - J_2 H = (RHF - J_1 H)F - J_2 H = RHF^2 - J_1 HF - J_2 H. \tag{10}$$

The expressions for the rest of rows can be obtained by analogy. It can be shown that the condition $\Phi D = 0$ can be written in the form

$$(R \quad -J_1 \quad -J_2 \quad \ldots \quad -J_k)B^{(k)} = 0 \tag{11}$$

where

$$B^{(k)} = \begin{pmatrix} HD & HFD & HF^2 D & HF^{k-1}D \\ 0 & HD & HFD & HF^{k-2}D \\ 0 & 0 & HD & HF^{k-3}D \\ \cdots & \cdots & \cdots & \\ 0 & 0 & 0 & 0 \end{pmatrix} \tag{12}$$

Since the row $(R \quad -J_1 \quad -J_2 \quad \ldots \quad -J_k)$ has to satisfy the condition (9), one obtains from (9) and (11) the equation

$$(R \quad -J_1 \quad -J_2 \quad \ldots \quad -J_k)(V^{(k)} \quad B^{(k)}) = 0 \tag{13}$$

Equation (13) has a nontrivial solution if

$$rank(V^{(k)} \quad B^{(k)}) < l(k+1) \tag{14}$$

To isolate the faults, a bank of systems (5) should be used therewith each system is invariant with respect to some faults and sensitive to the rest of them. To construct the system invariant with respect to the *i*-th fault, the condition $\Phi D_i = 0$ has to satisfy. To construct such a system, replace in (12) the matrix D by D_i, find from (14) the minimal dimension k and find the row $(R \quad -J_1 \quad -J_2 \quad \ldots \quad -J_k)$ satisfying (13). Then calculate the rows of the matrix Φ based on (10) and set $G_* := \Phi G$.

2.4. The design of nonlinear part

To construct the nonlinear part of system (2), note that the general form of the nonlinear term is as follows:

$$C_* \begin{pmatrix} \varphi_1(A_{*1} z, u) \\ \vdots \\ \varphi_p(A_{*p} z, u) \end{pmatrix} = C_* \phi(z,u),$$

where $z = \begin{pmatrix} x_* \\ y \end{pmatrix}$. It can be shown [22] that $C_* = \Phi C$ and

$$A = A_* \begin{pmatrix} \Phi \\ H \end{pmatrix} \tag{15}$$

where $A = \begin{pmatrix} A_1 \\ \cdots \\ A_p \end{pmatrix}$, $A_* = \begin{pmatrix} A_{*1} \\ \cdots \\ A_{*p} \end{pmatrix}$.

It can be proved that if $(R^{(1)} \quad -J_1^{(1)} \quad -J_2^{(1)} \quad \ldots \quad -J_k^{(1)}), \ldots, (R^{(N)} \quad -J_1^{(N)} \quad -J_2^{(N)} \quad \ldots \quad -J_k^{(N)})$ are solutions of (13) and $\Phi^{(1)}, \ldots, \Phi^{(N)}$ are the appropriate matrices, then the arbitrary linear combination of these solutions is a solution of (13) as well. This means that if $v = (v_1 \quad v_2 \quad \ldots \quad v_N)$ is the arbitrary vector of weigh coefficients, then

$$(R \quad -J_1 \quad -J_2 \quad \ldots \quad -J_k) = v_1(R^{(1)} \quad -J_1^{(1)} \quad -J_2^{(1)} \quad \ldots \quad -J_k^{(1)}) + \ldots + v_N(R^{(N)} \quad -J_1^{(N)} \quad -J_2^{(N)} \quad \ldots \quad -J_k^{(N)})$$

and $\Phi = v_1\Phi^{(1)} + \ldots + v_N\Phi^{(N)}$ is the matrix corresponding to $(R \quad -J_1 \quad -J_2 \quad \ldots \quad -J_k)$. Rewrite (15) as

$$A = A_*^1 \begin{pmatrix} \Phi_1 \\ \vdots \\ \Phi_k \end{pmatrix} + A_*^2 H \tag{16}$$

where $A_* = (A_*^1 \quad A_*^2)$. Introducing the notations

$$\Phi_1^\Sigma = \begin{pmatrix} \Phi_1^{(1)} \\ \vdots \\ \Phi_1^{(N)} \end{pmatrix}, \ldots, \Phi_k^\Sigma = \begin{pmatrix} \Phi_k^{(1)} \\ \vdots \\ \Phi_k^{(N)} \end{pmatrix}, \quad \Phi^\Sigma = \begin{pmatrix} \Phi_1^\Sigma \\ \vdots \\ \Phi_k^\Sigma \end{pmatrix},$$

one may rewrite (16) in the form

$$A = A_*^1 \begin{pmatrix} v\Phi_1^\Sigma \\ \vdots \\ v\Phi_k^\Sigma \end{pmatrix} + A_*^2 H \tag{17}$$

This equation has a solution if

$$rank\begin{pmatrix} \Phi^\Sigma \\ H \end{pmatrix} = rank\begin{pmatrix} \Phi^\Sigma \\ H \\ A \end{pmatrix} \tag{18}$$

Assuming that (18) holds, consider initially the simple case when A is a row matrix. Here, (17) can be represented in the form $A = (a_1 v \quad \ldots \quad a_k v)\Phi^\Sigma + A_*^2 H$, where $A_*^1 = (a_1 \quad \ldots \quad a_k)$, or in the form

$$A = A_v \Phi^\Sigma + A_*^2 H \tag{19}$$

considering A_v as an unknown matrix. Solve this algebraic equation and find the matrices A_v and A_*^2. If A_v can be represented in the form $(a_1 v \quad \ldots \quad a_k v)$ for some coefficients $a_1, \ldots, a_k$ and the vector $v = (v_1 \quad v_2 \quad \ldots \quad v_N)$, then stop, the results are the matrices $A_*^1 = (a_1 \quad \ldots \quad a_k)$ and A_*^2 and the vector v of weight coefficients. Then one finds the matrix R and the rows of the matrices J and Φ from the relations

$$R = \sum_{i=1}^N v_i R^{(i)}, \quad J_j = \sum_{i=1}^N v_i J_j^{(i)}, \quad \Phi_j = \sum_{i=1}^N v_i \Phi_j^{(i)}, \quad j = 1, 2, \ldots, k, \tag{20}$$

and the matrices $G_* = \Phi G$ and $C_* = \Phi C$. As a result, the nonlinear term has been constructed.

If the matrix A has several rows, equation (19) is solved for all its rows with some coefficients $a_1, \ldots, a_k$ specific for each row and the vector v which is the same for all rows. If the condition (18) does not hold or the matrix A_v cannot be represented in the required form, the dimension k should be increased and the described procedure is repeated.

To check whether or not the obtained nonlinear term has the form shown in (2), consider at first the nonlinear part of the equation for x_{*1} formed by the first row C_{*1} of the matrix C_*. The product $C_{*1}\phi(z,u)$ contains some rows of the form A_{*j}. If the first elements of these rows are equal to zero, then the product $C_{*1}\phi(z,u)$ can be written as $\psi_1(x_{*2}, x_{*3},...,u)$ that corresponds to (2). Then consider the product $C_{*2}\phi(z,u)$ and all rows in the form A_{*j} contained in it. If the first and second elements of these rows are equal to zero, then the product $C_{*2}\phi(z,u)$ can be written as $\psi_2(x_{*3}, x_{*4},...,u)$. The rest of the products are analyzed by analogy.

In the case when all such checks are positive, one can obtain the input-output representation of the considered system and use the non-parametric method. Otherwise, generally, the input-output representation of the considered system does not exist and one has to use other methods for diagnosis. However, in some cases, in particular, when the product $C_*\phi(z,u)$ contains the only nonlinearity, the possibility exists to obtain such a representation, see Section 3.

Next, the system is transformed into the input-output form (3) and (4) by performing some temporal shifts and substitutions. To generate a residual, the matrix $V_T(t)$ is used, see Section 2.

Note that the system invariant with respect to the i-th fault may be invariant with respect to some other faults. If equations in (2) depend on the parameters γ_i, γ_j, and γ_v, then the corresponding system is sensitive to the i-th, j-th, and v-th faults and invariant with respect to others. To reflect these peculiarities and make a decision about faults, the matrix of syndromes S is used constructed as follows. If the parameter γ_i is in the j-th system, then $S[i,j]=1$, otherwise $S[i,j]=0$. If the v-th and w-th columns of this matrix are the same, then the v-th and w-th faults are indistinguishable in the framework of the considered approach.

3. Decision making

3.1. Residual generation

In [4, 5], the value of T is chosen such that

$$rank(V_T(t)) = rank(V_{T-1}(t)) .$$

It follows from this equality that the last column of the matrix $V_T(t)$ is linearly dependent on the rest of columns. That is, the vector $v(T)$ exists such that $V_T(t)v(T)=0$ meaning that the matrix $V_T(t)$ has nonempty kernel. In this case, one obtains from (4) the equality $Y_T(t)v(T)=0$. So, the residual can be generated as

$$r_T(t) = Y_T(t)v(T), \qquad v(T) \in \ker(V_T(t)) . \tag{21}$$

Clearly, the rule (21) is robust in the sense that it is independent of the values of some parameters of system (1). Note that the residual should be generated on-line.

To reduce the calculation complexity, one may take T such that $T > k(m+l)$; here the equality $rank(V_T(t)) = rank(V_{T-1}(t))$ is valid. In some cases, the value of T may be reduced (see example).

3.2. Adaptive threshold

It follows from (21) that in the unfaulty case when all parameters have their nominal values, the residual $r_T(t)=0$. In some cases, the parameters may vary their values slightly in some tolerances, and such variations must not be considered as faults. An adaptive threshold should be used to take into account this fact.

To design such a threshold, assume that the i-th parameter γ_i may be presented by random variable with normal distribution having zero expectation and variance σ_i^2, $i=1,2,...,q$. Assume also that the matrices J and G_* in (4) depend on the parameters γ_1, ..., γ_i, i.e. $J_j = J_j(\gamma_1,...,\gamma_q)$, $G_{*j} = G_{*j}(\gamma_1,...,\gamma_q)$, $j=1,2,...,k$. For uniformity set $\Gamma = (\Gamma_1, \Gamma_2, ..., \Gamma_{2k-1}, \Gamma_{2k}, 1) := (J_1, G_{*1}, ..., J_k, G_{*k}, 1)$, where $\Gamma_{2j-1} = J_j$, $\Gamma_{2j} = G_{*j}$, $j=1,2,...,k$.

Linearization of the last expressions yields

$$\Delta\Gamma_j = \sum_{i=1}^{q} \frac{\partial \Gamma_j}{\partial \gamma_i} \Delta\gamma_i , \quad j=1,2,...,2k ,$$

where $\Delta\gamma_i = \gamma_i - \gamma_{0i}$, γ_{0i} is the nominal value of the i-th parameter. Then

$$D[\Delta\Gamma_j] = \sum_{i=1}^{q} \frac{\partial \Gamma_j}{\partial \gamma_i} \sigma_i^2, \quad j = 1,2,\dots,2k.$$

Since the variable $\Delta\Gamma_j$ has normal distribution with zero expectation, its tolerance $[\Gamma_{lj}, \Gamma_{uj}]$ can be determined as follows:

$$\Gamma_{lj} := \Gamma_{0j} - 3\left(\sum_{i=1}^{q} \frac{\partial \Gamma_j}{\partial \gamma_i} \sigma_i^2\right)^{1/2}, \quad \Gamma_{uj} := \Gamma_{0j} + 3\left(\sum_{i=1}^{q} \frac{\partial \Gamma_j}{\partial \gamma_i} \sigma_i^2\right)^{1/2}, \quad j = 1,2,\dots,2k,$$

where $\Gamma_{0j} = \Gamma_j(\gamma_{01},\dots,\gamma_{0q})$. Set $\Gamma_l := (\Gamma_{l,1}, \Gamma_{l,2},\dots,\Gamma_{l,2k-1}, \Gamma_{l,2k}, 1)$, $\Gamma_u := (\Gamma_{u,1}, \Gamma_{u,2},\dots,\Gamma_{u,2k-1}, \Gamma_{u,2k}, 1)$.

To make a decision about faults, construct the matrices $V_T(t)$ and $Y_T(t)$, find the vector $v(T)$ from $\ker(V_T(t))$, and calculate the residual $r_T(t) = Y_T(t)v(T)$. If $r_T(t) = 0$, one concludes that there is no fault. Otherwise calculate two values of the output y_*: $y_{l*}(t+k+T-1) = \Gamma_l V_T^{(1)}(t)$ and $y_{u*}(t+k+T-1) = \Gamma_u V_T^{(1)}(t)$, where $V_T^{(1)}(t)$ is the first column of the matrix $V_T(t)$, and check whether or not the inclusion

$$y_*(t+k+T-1) \in [y_{l*}(t+k+T-1), y_{u*}(t+k+T-1)]$$

is true. If it is true, one concludes that there is no fault, otherwise the fault occurred.

4. Example

Consider the model of the general electric servoactuator:

$$\dot{x}_1 = \frac{1}{i_p} x_2,$$

$$\dot{x}_2 = \frac{K_M}{J_H} x_3 + \frac{M_d}{J_H} sign(x_2),$$

$$\dot{x}_3 = -\frac{K_\omega}{L} x_2 - \frac{R}{L} x_3 + \frac{K_U}{L} u.$$

Here x_1 is the output rotation angle; x_2 is the velocity at the motor output shaft; x_3 is the current through the servoactuator windings; i_p is the reducing ratio of the reducer; J_H is the moment of inertia of the electric servoactuator and of the rotating parts of the reducer; K_ω and K_M are the respective coefficients of the counter EMF and of the torque; M_d describes the Coulomb friction at the motor; R and L are the active and inductive resistances of the electric servoactuator windings, respectively; K_U is the amplification.

Sampled-data model of the servoactuator can be obtained in the form

$$\begin{aligned} x_1^+ &= k_1 x_2 + x_1, \\ x_2^+ &= k_2 x_3 + k_3 sign(x_2) + x_2, \\ x_3^+ &= k_4 x_2 + k_5 x_3 + k_6 u; \end{aligned} \tag{22}$$

for simplicity, we use the following notations: $x^+ = x(t+1)$, $x = x(t)$, and $u = u(t)$. We assume that $y_1 = x_1$, $y_2 = x_3$. It is assumed also that faults are deviation of the coefficients $k_1 \div k_6$ from their nominal values.

Matrices, describing system (22) and faults, are as follows:

$$F = \begin{pmatrix} 1 & k_1 & 0 \\ 0 & 1 & k_2 \\ 0 & k_4 & k_5 \end{pmatrix}, \quad G = \begin{pmatrix} 0 \\ 0 \\ k_6 \end{pmatrix}, \quad H = \begin{pmatrix} 1 & 0 & 0 \\ 0 & 0 & 1 \end{pmatrix}, \quad C = \begin{pmatrix} 0 \\ k_3 \\ 0 \end{pmatrix}, \quad A = (0 \quad 1 \quad 0), \quad \varphi(x) = sign(Ax),$$

$$D_1 = \begin{pmatrix} 1 \\ 0 \\ 0 \end{pmatrix}, \quad D_2 = \begin{pmatrix} 0 \\ 1 \\ 0 \end{pmatrix}, \quad D_3 = \begin{pmatrix} 0 \\ 0 \\ 1 \end{pmatrix}.$$

For simplicity, consider the second and the third faults; make the necessary calculations for the second fault:

$$HD_2 = \begin{pmatrix} 0 \\ 0 \end{pmatrix}, \quad (V^{(1)}, B^{(1)}) = \begin{pmatrix} 1 & k_1 & 0 & 0 \\ 0 & k_4 & k_5 & 0 \\ 1 & 0 & 0 & 0 \\ 0 & 0 & 1 & 0 \end{pmatrix}.$$

It can be shown that (13) has the solution $(R \;\; -J) = (k_4 \;\; -k_1 \;\; -k_4 \;\; k_1k_5)$. Then $\Phi = (k_4 \;\; 0 \;\; -k_1)$, $G_* = -k_1k_6$, $C_* = 0$. The observer is linear and is described by the equations

$$\begin{aligned} x_*^+ &= k_4y_1 - k_1k_5y_2 - k_1k_6u, \\ y_* &= x_{*1} = k_4y_1 - k_1y_2, \end{aligned} \tag{23}$$

where $x_* = k_4x_1 - k_1x_2$. The temporal shift yields the following result: $y_*(t+1) = k_4y_1(t) - k_1k_5y_2(t) - k_1k_6u(t)$. The expressions for $y_*(t+2)$, $y_*(t+3)$, and $y_*(t+4)$ can be obtained by analogy.

Write down the relations for $y_*(t+1) \div y_*(t+4)$ in the form (4):

$$(y_*(t+4) \;\; y_*(t+3) \;\; y_*(t+2) \;\; y_*(t+1)) = (k_4 \;\; -k_1k_5 \;\; -k_1k_6)\begin{pmatrix} y_1(t+3) & y_1(t+2) & y_1(t+1) & y_1(t) \\ y_2(t+3) & y_2(t+2) & y_2(t+1) & y_2(t) \\ u(t+3) & u(t+2) & u(t+1) & u(t) \end{pmatrix}.$$

Thus, $T = 4$, $Y_4^{(1)} = (y_*(t+4) \;\; y_*(t+3) \;\; y_*(t+2) \;\; y_*(t+1))$,

$$V_4^{(1)} = \begin{pmatrix} y_1(t+3) & y_1(t+2) & y_1(t+1) & y_1(t) \\ y_2(t+3) & y_2(t+2) & y_2(t+1) & y_2(t) \\ u(t+3) & u(t+2) & u(t+1) & u(t) \end{pmatrix}, \quad r^{(1)}(4) = Y_4^{(1)}v(4), \quad v(4) \in \ker(V_4^{(1)}).$$

Calculations for the third fault yield the following results:

$$HD_1 = \begin{pmatrix} 0 \\ 1 \end{pmatrix}, \quad HFD_1 = \begin{pmatrix} 0 \\ k_4 \end{pmatrix}, \quad HF^2D_1 = \begin{pmatrix} k_1k_2 \\ k_2k_3 + k_4^2 \end{pmatrix},$$

$$(V^{(2)}, B^{(2)}) = \begin{pmatrix} 1 & 2k_1 & k_1k_2 & 0 & 0 & k_1k_2 \\ 0 & k_3(1+k_4) & k_2k_3 + k_4^2 & 1 & k_4 & k_2k_3 + k_4^2 \\ 1 & k_1 & 0 & 0 & 0 & 0 \\ 0 & k_3 & k_4 & 0 & 1 & k_4 \\ 1 & 0 & 0 & 0 & 0 & 0 \\ 0 & 0 & 1 & 0 & 0 & 1 \end{pmatrix}.$$

Equation (13) has the solution $(R \;\; -J_1 \;\; -J_2) = (1 \;\; 0 \;\; -2 \;\; 0 \;\; 1 \;\; -k_1k_2)$; $\Phi_1 = (1 \;\; 0 \;\; 0)$, $\Phi_2 = (-1 \;\; k_1 \;\; 0)$, $C_* = C\Phi = \begin{pmatrix} 0 \\ k_1k_3 \end{pmatrix}$, and $G_* = 0$. Since we have a single solution, then $N = 1$; one can check that the condition (18) holds, and $A = (0 \;\; 1 \;\; 0) = (1/k_1)(\Phi_1 + \Phi_2)$. Therefore, the nonlinear term is of the form

$\phi(z,u) = sign((1/k_1)(x_{*1} + x_{*2})) = sign((1/k_1)(y_1 + x_{*2}))$. The observer is described by the equations

$$\begin{aligned} x_{*1}^+ &= x_{*2} + 2y_1 \\ x_{*2}^+ &= -y_1 + k_1k_2y_2 + k_1k_3 sign((1/k_1)(y_1 + x_{*2})) \\ y_* &= x_{*1} = y_1 \end{aligned} \tag{24}$$

where $x_{*1} = x_1$, $x_{*2} = k_1 x_2 - x_1$. The temporal shifts yield the following results:

$$y_1(t+1) = x_{*2}(t) + 2y_1(t),$$
$$y_1(t+2) = -y_1(t) + k_1k_2y_2(t) + 2y_1(t+1) + k_1k_3 sign((1/k_1)(y_1(t) + x_{*2}(t))),$$
$$y_1(t+3) = -y_1(t+1) + k_1k_2y_2(t+1) + 2y_1(t+2) + k_1k_3 sign((1/k_1)(y_1(t+1) + x_{*2}(t+1)) =$$
$$= -y_1(t+1) + k_1k_2y_2(t+1) + 2y_1(t+2) + k_1k_3 sign((1/k_1)(y_1(t+1) - y_1(t) + k_1k_2y_2(t) + k_1k_3 sign((1/k_1)(y_1(t) + x_{*2}(t)))).$$

It follows from the expression for $y_1(t+2)$ that

$$-y_1(t) + k_1k_2y_2(t) + k_1k_3 sign((1/k_1)(y_1(t) + x_{*2}(t))) = y_1(t+2) - 2y_1(t+1).$$

Substituting the right hand side of this expression in the equation for $y_1(t+3)$, one obtains

$$y_1(t+3) = -y_1(t+1) + k_1k_2y_2(t+1) + 2y_1(t+2) + k_1k_3 sign((1/k_1)(y_1(t+1) + y_1(t+2) - 2y_1(t+1))).$$

The expressions for $y_1(t+4)$, $y_1(t+5)$, $y_1(t+6)$, and $y_1(t+7)$ are evident.

Assume that the coefficient k_1 is known and denote $y_0(t) = sign((1/k_1)(y_1(t+1) + y_1(t+2) - 2y_1(t+1)))$. Write down the relations for $y_1(t+3) \div y_1(t+7)$ in the form (4) with $y_* = y_1$:

$$(y_1(t+7) \quad y_1(t+6) \quad y_1(t+5) \quad y_1(t+4) \quad y_1(t+3)) =$$
$$= (2 \quad -1 \quad k_1k_2 \quad k_1k_3)\begin{pmatrix} y_1(t+6) & y_1(t+5) & y_1(t+4) & y_1(t+3) & y_1(t+2) \\ y_2(t+5) & y_2(t+4) & y_2(t+3) & y_2(t+2) & y_2(t+1) \\ y_1(t+5) & y_1(t+4) & y_1(t+3) & y_1(t+2) & y_1(t+1) \\ y_0(t+4) & y_0(t+3) & y_0(t+2) & y_0(t+1) & y_0(t) \end{pmatrix}.$$

Thus, $T = 5$, $Y_5^{(2)} = (y_1(t+7) \quad y_1(t+6) \quad y_1(t+5) \quad y_1(t+4) \quad y_1(t+3))$,

$$V_5^{(2)} = \begin{pmatrix} y_1(t+6) & y_1(t+5) & y_1(t+4) & y_1(t+3) & y_1(t+2) \\ y_2(t+5) & y_2(t+4) & y_2(t+3) & y_2(t+2) & y_2(t+1) \\ y_1(t+5) & y_1(t+4) & y_1(t+3) & y_1(t+2) & y_1(t+1) \\ y_0(t+4) & y_0(t+3) & y_0(t+2) & y_0(t+1) & y_0(t) \end{pmatrix},$$

and $r^{(2)}(5) = Y_5^{(2)} v(5)$, $v(5) \in \ker(V_5^{(2)})$.

Since the model (23) does not depend on the parameters k_2 and k_3 and (24) does not depend on k_4, k_5, and k_6, the matrix of syndromes takes the form

$$S = \begin{pmatrix} 1 & 0 & 1 \\ 1 & 1 & 0 \end{pmatrix},$$

where rows correspond to the residuals, columns to the faults.

5. Conclusion

The paper is devoted to the problem of fault isolation in technical systems described by nonlinear models. The non-parametric method and the logic-dynamic approach to solve this problem have been used. Based on the logic-dynamic approach, the input-output representation of the systems has been obtained. The new method to generate the adaptive threshold to avoid a false alarm has been suggested. The obtained results are illustrated by practical example of diagnosis of general electric servoactuator containing non-differentiable nonlinearity.

The future plan of researches is: (1) development of methods for residual generation based on the geometric point of view; (2) development of procedure to design the adaptive threshold to take into account the disturbances in the output vector.

6. Acknowledgements

This publication is supported by Russian Scientific Foundation (project No 16-19-00046).

7. References

[1] Bittencourt, A.; Saarinen, K. & Sander-Tavallaey, S. (2012). A Data-driven method for monitoring systems that operate repetitively – applications to wear monitoring in an industrial robot joint, Proc. of 8th IFAC Symp. Safeprocess'2012, pp. 198-203, Mexico.

[2] Blanke, M.; Kinnaert, M.; Lunze, J. & Staroswiecki, M. (2006). Diagnosis and Fault-Tolerant Control, Berlin Heidelberg, Springer-Verlag.

[3] Caccavale, F. & Villiani, L. (Eds) (2002). Fault Diagnosis and Tolerance for Mechatronic Systems, Recent Advances, Berlin Heidelberg, Springer-Verlag.

[4] Ding, S.; Wang, Y.; Yin, S.; Zhang, P.; Yang, Y. & Ding, E. (2012). Date-driven design of fault-tolerant control systems, Proc. of 8th IFAC Symp. Safeprocess'2012, pp. 1323-1328, Mexico.

[5] Ding, S. (2014). Data-driven Design of Fault Diagnosis and Fault-tolerant Control Systems, London, Springer-Verlag.

[6] Filaretov, V.; Zhirabok, A. & Tkachev, D. (2012). Non-parametric method for fault diagnosis in electrical circuits, Proc. of 23d Int. DAAAM Symposium, pp. 5-8, Zadar.

[7] Frank, P. (1990). Fault diagnosis in dynamic systems using analytical and knowledge-based redundancy – A survey and some new results, Automatica, Vol. 26, 1990, pp. 459-474.

[8] Gertler, J. (1993). Residual generation in model-based fault diagnosis, Control Theory and Advanced Technology, Vol. 9, 1993, 1993, pp. 259-285.

[9] Hakem, A.; Pekpe, K. & Cocquempot, V. (2011). Sensor fault diagnosis for bilinear systems using data-based residuals, Proc. of 50th IEEE Conference on Decision and Control and European Control Conference (CDC-ECC), pp. 1576-1582, Orlando.

[10] Kwakernaak, H. & Sivan, R. (1972). Linear optimal control systems, New York, A Division of John & Sons, Inc.

[11] Noura, H.; Ponsart, D. & Chamseddine, A. (2009). Fault-tolerant Control Systems. Design and Practical Applications, London, Springer-Verlag.

[12] Patton, R. (1994). Robust model-based fault diagnosis: the state of the art, Proc. of 2nd IFAC Symp. Safeprocess'1994, pp. 1-24, Espoo.

[13] Patton, R.; Frank, P. & Clark, R. (2000). Issues of fault diagnosis for dynamic systems, London, Spring Verlag.

[14] Russell, E.; Chiang, L. & Chiang, L. (2001). Fault Detection and Diagnosis in Industrial Systems, Berlin Heidelberg, Springer-Verlag.

[15] Samy, I.; Postlethwaite, I. & Gu, D. (2011). Survey and application of sensor fault detection and isolation schemes,Control Engineering Practice, Vol. 19, 2011, pp. 658-674.

[16] Simani, S.; Fantuzzi, C. & Patton, R. (2002). Model-based Fault Diagnosis in Dynamic Systems Using Identification, Berlin Heidelberg, Springer-Verlag.

[17] Shumsky, A. & Zhirabok, A. (2006). Nonlinear diagnostic filter design: algebraic and geometric points of view, Int. Journal of Applied Mathematics and Computer Science, Vol. 16, 2006, pp. 115-127.

[18] Shumsky, A. (2009). Functional diagnosis of nonlinear time-delay dynamic systems, Automation and Remote Control, Vol. 70, 2009, pp. 172-184.

[19] Tan, C. & Edwards, C. (2009). Robust fault reconstruction using multiple sliding mode observers in cascade: development and design, Proc. of 2009 American Control Conference, pp. 3411-3416, St. Louis.

[20] Witczak, M.; Korbicz, J. & Jrozefowicz, R. (2013). Design of unknown input observers for non-linear stochastic systems and their application to robust fault diagnosis, Control and Cybernetics, Vol. 42, 2013, pp. 1-30.

[21] Yang, H.; Jiang, B. & Cocquempot, V. (2010). Fault Tolerant Control Design for Hybrid Systems, In Lecture Notes in Control and Information Sciences, Berlin Heidelberg, Springer-Verlag.

[22] Zhirabok, A. & Usoltsev, S. (2000). Linear diagnosis methods for nonlinear systems, Automation and Remote Control, Vol. 61, 2000, pp. 201-1209.

[23] Zhirabok, A.; Kucher, D. & Filaretov, V. (2010). Achieving robustness at diagnosis of nonlinear systems, Automation and Remote Control, Vol. 71, 2010, pp. 142-155.

[24] Zhirabok, A. & Pavlov, S. (2015). Data-driven method of fault detection in technical systems, Procedia Engineering, Vol. 100, 2015, pp. 242-248.

DOI: 10.2507/28th.daaam.proceedings.021

HUMAN-OPERATOR BEHAVIOUR MONITORING IN IT-SYSTEMS

Alexey Shumsky, Alexey Zhirabok & Nadezhda Kalinina

This Publication has to be referred as: Shumsky, A[lexey]; Zhirabok, A[lexey] & Kalinina, N[adezhda] (2017). Human-Operator Behaviour Monitoring in IT-Systems, Proceedings of the 28th DAAAM International Symposium, pp.0160-0167, B. Katalinic (Ed.), Published by DAAAM International, ISBN 978-3-902734-11-2, ISSN 1726-9679, Vienna, Austria
DOI: 10.2507/28th.daaam.proceedings.021

Abstract

This work deals with IT-systems that include human-operator. It is well known that just human-operator incorrect actions in many cases resulted in fatal accidents. Due to this, the problem arises to timely detect and isolate the human-operator behaviour errors. The objective of present paper is the generalization and extension of former results proposed by authors within the framework of the above problem solution. To solve the problem, the algebra of partitions is involved. The model in the form of nondeterministic finite automation is used to describe human-operator behaviour. The method for monitor design based on the nondeterministic finite automata determinization is proposed. The procedure for the determinization that guarantees the minimal loss of information is developed. The facilities of the monitor to detect and isolate human-operator errors are investigated. Illustration is given for a case of the change management process in IT-systems.

Keywords: IT-systems; Algebra of partitions; Nondeterministic finite automata; Determinization.

1. Introduction

Strict demands imposed on the modern systems of critical purposes require timely detection of such system behaviour deviation from the prescribed order that may cause the no desirable accidents. This work deals with IT-systems that include human-operator. It is well known that just human-operator incorrect actions in many cases resulted in fatal accidents. That is why this work focuses solely on the diagnosis of human-operator behaviour errors. Related problems were considered in several papers dealing with human-operator monitoring in human-machine systems (HMS). For example, [1] deals with the task to timely detect and prevent consequences of unexpected situations whose solution is based on direct measurements of human emotional and psychological states. But existing sensors technological limitations may hinder the performance of such approaches at practice. Due to this, when human-operator errors are not immediately related to human life, health or significant financial loss, so-called behavioural approach looks reasonable. So, when direct measurements are impossible or difficult, an intuitive solution is to observe variables that are directly linked to the unobservable information we need to assess. In order to do so, qualitative or quantitative models of human-machine systems are required to estimate the human state, the performance or the behaviour [2].

Within the behavioural approach the solution to the problems of unexpected situation detection and abnormal operation diagnosis in HMS was given for the train-driving context in [3-5]. The feature of this solution is the use of the human-operator behaviour model (HBM) in the form of nondeterministic finite automaton (NFA). Then, solution is reduced to this model determinization involving mathematical techniques of the pair algebra of partitions [6].

The objective of present paper is the generalization and extension of the results proposed in [4, 5] to IT-systems. At the beginning, the modelling approach proposed in above papers is generalized for a case of IT-systems. It is the first contribution of the paper. The second contribution is the condition to check the facilities of the new monitor to detect and isolate human-operator errors.

The rest of the paper is organised as follows. Firstly, in Section 2, designing procedure for HBM in the form of NFA with application to IT-systems is considered. Then, in Section 3, an approach to monitor design in the form of a bank of deterministic finite automata (FA) is developed. Section 4 is devoted to human-operator errors detection and isolation facilities analysis. All the theoretical results from Sections 2-4 are illustrated by examples related to a case of the change management process in IT-systems. Section 5 concludes the paper.

2. The HBM design for IT-systems

2.1. Nondeterministic modelling approach

An approach to design the HBM in the form of NFA [4, 5] involves the procedure including the following steps:

1) design the HBM in the form of deterministic FA which describes the "desired" behaviour according to existing operator's guidelines;

2) extend this model up to nondeterministic one, taking into account the actions of operator under possible troubles that involves the link: action – observation –reaction;

3) augment the obtained model taking into account the possible human-operator errors resulting in deviations from the prescribed operator actions.

Notice that application of the two first steps of this procedure results in HBM in the form of NFA:

$$M = (I, Q, \delta) \tag{1}$$

where I and Q are the finite sets of the NFA inputs and states respectively, δ is the function of transitions. In contrast to the conventional deterministic FA, the model (1) is nondeterministic, since for some state $q \in Q$ and some input $i \in I$ several transitions may exist and what transition should take place in real situation becomes clear only when this transition has occurred. This fact can be explained as follows: in general case one can admit existence of different results of observations and human-operator reactions which does not conflict with operator's guidelines in the main.

The function of transitions for NFA is written as follows

$$\delta(q,\ i) \in \{q^+ \mid q^+ \in Q^+_{q,i} \subseteq Q\} \tag{2}$$

where $Q^+_{q,i}$ is the NFA subset of states that are admissible after transition from the state $q \in Q$ initiated by the input $i \in I$. It is assumed that the function of transitions is specified by appropriate table or directional graph.

Human-operator errors e_j, $1 \le j \le N$, where N is a number of errors, may result in additional transitions which are not initially assumed in (2). The model with the error e_j is written in the form

$$M_j = (I, Q, \delta_j) \tag{3}$$

with the function of transitions δ_j which contains additional (with respect to the function δ) transitions corresponding to the error e_j . The model (3) is also nondeterministic.

2.2. Application to IT-systems

Consider the change management process in IT-systems. This process is the most important for IT-systems: it is responsible for living cycle management for all changes and facilitates of the useful change realization with minimal interruption of the IT-services [7]. The individuals of the process are:

1) the initiator is the member of IT department who realizes the initial processing, assignment and checking the process fulfilling;

2) the executor is an engineer who fulfills the changes in the elements of configuration or coordinates the work of the contractor;

3) the consulting committee which fulfills the scheduled estimating and planning the changes;

4) the process manager is the member of the IT department who monitors the change management process and forms

the proposals for its improving.

The prescribed order of the process assumes fulfilling the following steps:

1) the initiator formulates the task and sends it for confirmation to the consulting committee;
2) the consulting committee makes a decision to confirm this task and to send the task to the process manager or to return the task for its modification to the initiator;
3) the process manager makes appropriate mark and sends the task to the executor;
4) the executor performs the task and informs the process manager when the task is completed;
5) the process manager checks the result and completes the task or sends the information about necessary task modification to the initiator.
6) the task is completed.

Consider HBM design in detail taking into account prescribed order of the process given above. This design follows the steps of HBM designing approach described above.

Step 1: Construct the model in the form of the deterministic FA. Denote the FA states q_i, $1 \le i \le 6$, according to the steps of above order. The FA inputs generated by the process manager are introduced as follows: i_1 is generated when the task formulation is accomplished, i_2 is generated when the task is confirmed by the consulting committee, i_3 is generated when the consulting committee returns the task for its modification to the initiator, i_4 is generated when the task is sent to the executor, i_5 is generated when the task is completed by the executor, i_6 is generated when the task is completed by the process manager, and i_7 is generated when the information about necessary task modification is sent to the initiator. Appropriate model description is given in Table 1.

	i_1	i_2	i_3	i_4	i_5	i_6	i_7
q_1	q_2	–	–	–	–	–	–
q_2	–	q_3	q_1	–	–	–	–
q_3	–	–	–	q_4	–	–	–
q_4	–	–	–	–	q_5	–	–
q_5	–	–	–	–	–	q_6	q_1
q_6	–	–	–	–	–	–	–

Table 1. The deterministic model

Step 2: Extend the above deterministic model up to nondeterministic. To do this, consider possible situations which violate the prescribed order. These situations may occur because of insufficiently accurate order description, approbation absence of the order for all possible situations or private reasons of the process participants.

Due to this, additional transitions arise that correspond to the following situations:

1) the initiator does not send the task for confirmation to the consulting committee but sends it to the process manager;
2) the executer does not send the task to the process manager but completes the process;
3) the process manager returns the confirmed task for additional confirmation to the consulting committee;
4) the process manager returns the confirmed task to the initiator;
5) the executor returns the confirmed task to the initiator;
6) the executor returns the confirmed task to the consulting committee.

Additional transitions, corresponding to the situations 3-6, are the reactions of the process participants aimed at making the task clearer. So, from the deterministic model (Table 1) one obtains the basic model (Table 2).

	i_1	i_2	i_3	i_4	i_5	i_6	i_7
q_1	q_2	–	–	–	–	–	–
q_2	–	q_3	q_1	–	–	–	–
q_3	–	–	–	q_1, q_2, q_4	–	–	–
q_4	–	–	–	–	q_1, q_3, q_5	–	–
q_5	–	–	–	–	–	q_6	q_1
q_6	–	–	–	–	–	–	–

Table 2. The basic nondeterministic model

Step 3: Additional transitions, corresponding to the situations 1 and 2, may be considered as the initiator and the executer errors respectively. So, the augment of the basic model by appropriate additional transitions results in the models M_1 and M_2 related to the above errors. Descriptions for the errors containing models are given in tables 3 and 4.

	i_1	i_2	i_3	i_4	i_5	i_6	i_7
q_1	q_2, q_3	–	–	–	–	–	–
q_2	–	q_3	q_1	–	–	–	–
q_3	–	–	–	q_1, q_2, q_4	–	–	–
q_4	–	–	–	–	q_1, q_2, q_5	–	–
q_5	–	–	–	–	–	q_6	q_1
q_6	–	–	–	–	–	–	–

Table 3. The model M_1

	i_1	i_2	i_3	i_4	i_5	i_6	i_7
q_1	q_2	–	–	–	–	–	–
q_2	–	q_3	q_1	–	–	–	–
q_3	–	–	–	q_1, q_2, q_4	–	–	–
q_4	–	–	–	–	q_1, q_2, q_5, q_6	–	–
q_5	–	–	–	–	–	q_6	q_1
q_6	–	–	–	–	–	–	–

Table 4. The model M_2

3. Monitor design

3.1. Description of the monitor scheme

The monitor scheme applied below is the special case of one proposed in [8, 9]. The last scheme is an original extension of the scheme usually used for fault isolation in engineering dynamic systems (a bank of unknown input observers) [10]. The structure of the scheme in use is given in Fig. 1.

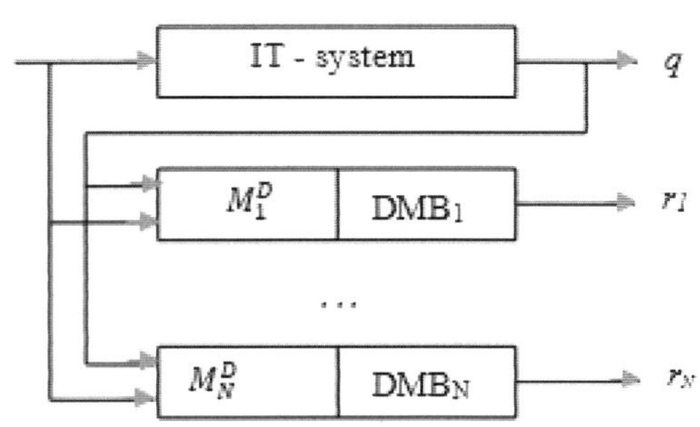

Fig. 1. The structural representation of the monitor

The deterministic FA M_j^D, $1 \le j \le N$, from this bank is constructed by determinization of the automation M_j such that the behavior of this automation is invariant (insensitive) with respect to the human-operator error e_j and sensitive (as soon as it is possible) to the rest errors. For the model M_j^D the following description is used

$$M_j^D = (I \times Q, Q^{(j)}, \delta^{(j)}) \tag{4}$$

with the transition function

$$q^{(j)+} = \delta^{(j)}(q,i), \quad q^{(j)+} \in Q^{(j)}, \quad q \in Q, \quad i \in I \tag{5}$$

and $Q^{(j)}$ is the model M_j^D states set

Consider the link between the models M_j^D and M_j. Assume existence of the map $\varphi^{(j)}: Q \to Q^{(j)}$ such that under absence of human-operator errors e_k, $k \neq j$, the following equality holds

$$q^{(j)} = \varphi^{(j)}(q). \quad (6)$$

Taking into account (2), (3), (5), and (6), one can write defining relation for the transition function of the model M_j^D:

$$\delta^{(j)} = \varphi^{(j)}(\delta_j). \quad (7)$$

Decision making block DMB_j checks equality (6). By construction, the model M_j^D is invariant to human-operator error e_j. As a result, DMB_j. forms the residual $r_j = 0$ under presence of this error. Also, if M_j^D is invariant with respect to some error e_i and this error has occurred, DMB_j forms the residual $r_j = 0$ again. To detect the rest human-operator errors via the residual r_j, DMB_j should form $r_j = 1$ for all these errors that assumes additional demand imposed on the map $\varphi^{(j)}$ (see Proposal given below). The correspondence between human-operator errors and the residuals is specified by the human-operator error syndrome matrix (ESM). An example of such matrix for $N = 3$ is given in Table 5.

	e_1	e_2	e_3
r_1	0	0	1
r_2	0	0	1
r_3	1	1	0

Table 5. Example of the ESM

The cell of ESM related to the pair r_j and e_k contains "1" if the error e_k violates equality (6), otherwise it contains "0". The ESM is a tool for decision making via residual vector $r = col(r_1, r_2, ..., r_N)$. Errors are isolated if every two columns of this matrix are different. It follows from above example that errors e_1 and e_2 are indistinguishable. As a result, the monitor scheme can be simplified by excluding one of the model (M_1^D or M_2^D) and appropriate decision making block.

To find the maps $\varphi^{(j)}$, $1 \le j \le N$, and to construct the ESM, pair algebra of partitions proposed by Hartmanis and Stearns [6] is essentially used.

3.2. Mathematical background

The main elements of algebra in use are partitions of the set Q. Remind that partition of the set Q is the set of its subsets (blocks of partition) $\{B_{\pi 1}, B_{\pi 2}, ..., B_{\pi n}\}$ such that

$$B_{\pi i} \subseteq Q, \quad B_{\pi i} \cap B_{\pi j} = \varnothing, \; i \neq j, \quad \bigcup_{i=1}^{p} B_{\pi i} = Q.$$

Let π and σ be partitions of Q. One says that partition π is less or equal to σ and denote $\pi \le \sigma$ if for every $B_{\pi i}$ there exists $B_{\sigma k}$ such that $B_{\pi i} \subseteq B_{\sigma k}$, where $B_{\pi i}$ and $B_{\sigma k}$ are appropriate blocks of the partitions π and σ, respectively. One can say that if we know the partition π, then we have information about states of the FA with an accuracy up to partition π; if $\pi \le \sigma$, then one says that the partition π has the same or more information than σ.

There are two special partitions denoted by **0** and **1**: each block of the partition **0** contains the only element of the set Q; the partition **1** has single block containing all the elements of Q. For arbitrary partition π of Q the evident inequality holds: $0 \le \pi \le 1$.

It is known [6] that a set of all partitions of Q with relation of partial order is a lattice. Therefore, for each pair of partitions (π, σ) of Q one can find two partitions *inf* (π, σ) and *sup*(π, σ). It is common practice to denote these partitions by $\pi \times \sigma$ and $\pi + \sigma$.

To compute the result of operations × and +, one may use the following simple rules [6]: each block of the partition $\pi\times\sigma$ is intersection of some blocks of the partitions π and σ; each block of the partition $\pi+\sigma$ is union of all intersected blocks of the partitions π and σ.

Let π and σ be partitions of Q. Binary relation Δ is formed from all pairs (π,σ) of partitions of Q satisfying the following condition:

$$q \equiv q'(\pi) \Rightarrow \delta(q,i) \equiv \delta(q',i)(\sigma), \quad \forall i \in I$$

In this case one says that partitions π and σ form a pair. Given a partition π, several partitions σ may exist such that $(\pi,\sigma)\in\Delta$, in particular, $(\pi,\mathbf{1})\in\Delta$ is valid for all π. Define the operator **m** as follows: **m**(π) is a partition satisfying conditions

$$(\pi, m(\pi)) \in \Delta,\ (\pi,\sigma)\in\Delta \Rightarrow m(\pi)\leq\sigma.$$

Thus, the partition **m**(π) is the smallest one which the partition π forms a pair with. Because $(\pi,\mathbf{1})\in\Delta$ for each π, the partition $\mathbf{m}(\pi)$ exists for every π, at least, $\mathbf{m}(\pi)=\mathbf{1}$. The following formula of the operator **m** computation for NFA is the extension of known (for FA) formula [6] and has been proposed in [5]:

$$\mathbf{m}(\pi)=\sum_j \sigma_j \tag{8}$$

where σ_j is the minimal partition such that

$$q \equiv q'(\pi) \Rightarrow \delta(q,i) \equiv \delta(q',i)(\sigma_j) \tag{9}$$

for all $\delta(q,i)\in Q^+_{q,i},\ \delta(q',i)\in Q^+_{q',i}$.

3.3 The model determinization

Solution of this problem for some model M_j, $1\leq j\leq N$, assumes finding the map $\varphi^{(j)}$, followed by the transition function $\delta^{(j)}$ determination from defining relation (7).

Notice, in all cases the determinization of nondeterministic model results in loss of information about the next system state as soon it is unknown what possible transition of the nondeterministic system will realize at practice. To guarantee the maximal accuracy of monitoring (human-operator errors detectability and isolability), information loss should be reduced to minimal one.

Let $\pi_{\varphi^{(j)}}$ be the partition specified by the map $\varphi^{(j)}$ according to the rule

$$q \equiv q'(\pi_{\varphi^{(j)}}) \Leftrightarrow \varphi^{(j)}(q)=\varphi^{(j)}(q') \tag{10}$$

Zero partition of the set Q contains the whole information about nondeterministic model M_j state. Thus, according to the definition of operator **m**, the rule of the next state computation for the model M_j with the maximal accuracy is

$$\pi_{\varphi^{(j)}}=\mathbf{m}(0) \tag{11}$$

where operator **m** is applied to the model M_j. So, relation (11) gives the rule for the model determinization with minimal loss of information.

3.4 *Monitor design for IT-system*

Consider the model M_1^D design on the base of the model M_1 (Table 3). The first step is computation of the partition **m(0)**. From (9) one has (rows of the Table 3 are considered; the states from the same cell of the table are combined within the same block of partition): $\sigma_1=\{(q_1);(q_2,q_3);(q_4);(q_5);(q_6)\}$; $\sigma_2=\sigma_5=\sigma_6=\{(q_1),(q_2),(q_3),(q_4),(q_5),(q_6)\}$; $\sigma_3=\{(q_1,q_2,q_4);(q_3);(q_5);(q_6)\}$; $\sigma_4=\{(q_1,q_2,q_5);(q_3);(q_4);(q_6)\}$. After this, applying formula (8) that assumes combining the intersected blocks of above partitions, one obtains $\pi_{\varphi^{(j)}}=\mathbf{m}(\mathbf{0})=$

$\{(q_1, q_2, q_3, q_4, q_5); (q_6)\}$. The second step involves finding transition function for the model M_1^D (Table 3). Solution of this task consists in:

1) put the states of the model M_1^D into correspondence to appropriate blocks of the partition $\pi_{\varphi(1)}$: $q_{1,1} \in Q^{(1)}$ corresponds to the block $(q_1, q_2, q_3, q_4, q_5)$ and $q_{1,2} \in Q^{(1)}$ corresponds to the block (q_6);

2) rewrite the matrix of transitions (Table 3) according to the new definitions (see Table 6).

	i_1	i_2	i_3	i_4	i_5	i_6	i_7
q_1	$q_{1,1}$	–	–	–	–	–	–
q_2	–	$q_{1,1}$	$q_{1,1}$	–	–	–	–
q_3	–	–	–	$q_{1,1}$	–	–	–
q_4	–	–	–	–	$q_{1,1}$	–	–
q_5	–	–	–	–	–	$q_{1,2}$	$q_{1,1}$
q_6	–	–	–	–	–	–	–

Table 6. The model M_1^D

The partition $\pi_{\varphi(2)} = \{(q_1, q_2, q_4, q_5, q_6); (q_3)\}$ and the table of transitions for M_2^D (Table 7) are given without additional illustration

	i_1	i_2	i_3	i_4	i_5	i_6	i_7
q_1	$q_{2,1}$	–	–	–	–	–	–
q_2	–	$q_{2,2}$	$q_{2,1}$	–	–	–	–
q_3	–	–	–	$q_{2,1}$	–	–	–
q_4	–	–	–	–	$q_{2,1}$	–	–
q_5	–	–	–	–	–	$q_{2,1}$	$q_{2,1}$
q_6	–	–	–	–	–	–	–

Table 7. The model M_2^D

4. Error syndrome matrix construction

Construction of the ESM is based on finding conditions that guarantee violation of (6) in a case of human-operator error. Assume that human-operator error e_k instead of transition into state q results in transition into state q'. Put these states into the same block of partition π_{e_k} whose rest of blocks contain only single state from the list that does not contain q and q'.

Proposal. Human-operator error e_k results in violation of condition (6) if

$$\pi_{e_k} \times \pi_{\varphi(j)} = 0 . \tag{12}$$

Proof. As soon by construction the states q and q' are contained in the same block of partition π_{e_k}, from (12) it follows that these states should be contained in different blocks of partition $\pi_{\varphi(j)}$. Then, from (10) one obtains

$$\varphi^{(j)}(q) \neq \varphi^{(j)}(q') .$$

Using this result, illustrate ESM construction for above IT-system. Denote human-operator errors in this system as e_1 and e_2 (initiator and executor errors respectively, see Subsection 2.2). So, the error e_1 results in transition from the state q_1 to the state q_3 under input i_1 while under absence of this error transition to the state q_2 should take place. Respectively, the error e_2 results in transition from the state q_4 to the state q_6 under input i_5 while under absence of this error transition to the state q_5 should take place. As a result, one obtains partitions $\pi_{e_1} = \{(q_1); (q_2, q_3); (q_4); (q_5); (q_6)\}$ and $\pi_{e_2} = \{(q_1); (q_2); (q_3); (q_4); (q_5, q_6)\}$. Using partitions $\pi_{\varphi(1)} = \{(q_1, q_2, q_3, q_4, q_5); (q_6)\}$ and $\pi_{\varphi(2)} =$

$\{(q_1,q_2,q_3,q_4,q_6);(q_3)\}$ obtained above, it is easily to show that $\pi_{e_1} \times \pi_{\varphi(2)} = 0$ and $\pi_{e_2} \times \pi_{\varphi(1)} = 0$. The first equality means that the first error appearance will result in violation of (6) and generation $r_2 = 1$ while the second equality means that the second error also will result in violation of (6) and generation $r_1 = 1$. Thus, ESM takes the following form:

	e_1	e_2
r_1	0	1
r_2	1	0

Table 8. ESM for IT-system

As soon the columns of this matrix are different, the isolation of human-operator errors is possible.

5. Conclusion

The problem of human-operator errors monitoring has been considered with application to IT-systems. Solution of the problem assumes the use of behavioral human-operator model in the form of NFA and involves the determinization of this model. Following new results have been obtained:

1) the method of nondeterministic behavioral human-operator model design for IT-systems;

2) the method of nondeterministic finite automata determinization that guarantees the minimal loss of information and, in contrast to known method [3, 4], allows to decrease the computation expanses;

3) the method of human-operator errors analysis. Application of all above methods has been illustrated for the change management process monitoring in IT-systems.

It is assumed to improve the obtaining results by application of the mathematical tools of the algebra of covers instead of the algebra of partitions.

6. Acknowledgments

This work was supported by Russian Scientific Foundation (project No 16-19-00046).

7. References

[1] Wilson, J.R. & Norris, B.J. (2005). Rail human factors: Past, present and future, Applied ergonomics, Vol. 36, 2005, pp.649-660.

[2] Mceod, R.W.; Walker, G.H. & Moray, N. (2005). Analyzing and modelling train driver performance. Applied ergonomics, Vol. 36, 2005, pp.671–680.

[3] Ouedraogo K.A.; Enjalbert, S. & Vanderhaegen, F. (2011). A state of the art n feedforward-feedback learning control systems for human errors prediction, Preprints of the 18th IFAC World Congress. Milano, Italy, 2011.

[4] Berdjag, D.; Vanderhagen, F.; Shumsky, A. & Zhirabok, A. (2014). Unexpected situation diagnosis: a model-based approach for human-machine systems, Proceedings of 19th IFAC Congress, Cape Town, South Africa, 2014, pp. 3545-3550.

[5] Berdjag, D.; Vanderhaegen, F.; Shumsky, A. & Zhirabok, A. (2015). Abnormal operation diagnosis in human-machine systems, Proceedings of 10th Asian Control Conference, Kota-Kinabalu, Malayzia, 2015.

[6] Hartmanis, J. & Stearns, R. (1966). The algebraic structure theory of sequential machines. N. Y.: Prentice-Hall Inc. 1966, 211 p.

[7] IT infrastructure library version 3, office of Government commerce (UK), Available from: https://www.axelos.com/best-practice-solutions/itil, Accessed: 2016-01- 26.

[8] Patton, R. (1994). Robust model-based fault diagnosis: the state of the art, Proceedings of IFAC symposium Safeprocess, Espoo, Finland, 1994, pp.1-24.

[9] Zhirabok A.; Filaretov V. & Pavlov S. (2015) Fault Isolation in Technical Systems by Data-Driven Method, CD ROM Proceedings of 26th International DAAAM Symposium, Zadar, Croatia, 2015.

[10] Zhirabok, A. & Shumsky, A. (2012). Fault accommodation in discrete-event robots, Advances in Robotics (Edited by: C. Ciufudean), iConcept Press, Hong Kong, 2012, pp.323-339.

DOI: 10.2507/28th.daaam.proceedings.022

FUNCTIONAL SAFETY CONCEPT FOR A HANDLING ROBOT BUILT ON OPTICAL SYSTEMS

Thomas Radinger, Kemajl Stuja, Walter Wölfel & Erich Markl

This Publication has to be referred as: Radinger, T[homas]; Stuja, K[emajl]; Wolfel, W[alter] & Markl, E[rich] (2017). Functional Safety Concept for a Handling Robot Built on Optical Systems, Proceedings of the 28th DAAAM International Symposium, pp.0168-0172, B. Katalinic (Ed.), Published by DAAAM International, ISBN 978-3-902734-11-2, ISSN 1726-9679, Vienna, Austria
DOI: 10.2507/28th.daaam.proceedings.022

Abstract

This work introduce a safety concept for a handling robot, built on optical system for the three-dimensional monitoring. The core system is the Pilz SafetyEye, which is intended to monitor the working area of an industrial robot in the University of Applied Sciences Technikum Wien's own digital factory. This system was extended with low-level sensor "Kinect 360" for the human detection. By entering safety area of human, the industrial robot is decelerated\ deactivated via the safety control used. In this case, a signal is sent to the robot controller, which forces the manipulator to slow down or to stop its operation immediately. Further outputs of the system are an acoustic signal, which is intended to warn the threatened person. Kinect can generate and process not only colour images, but also depth images. The advantage of this system is the integrated environment, which provides a libraries and functions for recognizing of human as a virtual image. However, this concept has been worked according to currently valid standards, which takes all safety-related aspects of the industrial robot. This system was practically implemented and on-line tested. Different data, like reaction rate and different orientation of human has been collected and presented. Further work is the integration safety and security within entire handling system.

Keywords: Functional Safety; handling Robot; SafetyEye; Kinect

1. Introduction

A typically safety concept variant to protect humans from the movements of industrial robots in the past was to place the robots in guarded robotic cells. Safety guarded robotic cells avoids not only mechanically interaction between human and robots, in the same time between the mobile and stationary robots too. For that reason, this kind of safety systems is incompatible with the future industrial concept such as industry 4.0. A main task of the new safety concept is to allow the sharing of the work-areas between mobile and stationary robots, and to avoid the mechanical contact- collisions between a humans and non-collaborative robots. Differentiation between human and mobile robots can be achieved, only with adding of very expensive third-party hardware and software components. This would be immense disadvantage for the small and medium sized companies. In order to offer a low-cost solution for mentioned companies, at the University of Applied Sciences Technikum Wien was written to works by [1] and [2] for these subject.

The knowledge of this two works would be used in this research to build the new cost-effective safety optical system concept for one pick & place robotic cell, which among other enables the possibility for recognizing of humans entering of the threatened area of the stationary robots. This system implements the Kinect low-level sensor from Microsoft, capable to recognize humans before they can enter a mentioned areas.

2. Problem statement and requirements

For the digital factory of the UAS Technikum Wien should be built and implement a safety work system based on optical system for three-dimensional workspace monitoring capable for recognizing (among others) of humans. The safety system must to take into account all relevant safety-related standards, as well as the state of the art (German Institute for Standardization, 2010). Among that, the system status should be detailed by changing of signal state, as well as to activate acoustic devices and warning lights. By default, when persons or objects do not enter the warning and protection area, a green lamp should be light up. When entering the warning area, should be switched from green to orange light. Entering the protection area by human, the orange light should be switched to red light, and the speed of the robot should be reduced to 0 mm/s. On the other hand, if the mobile robot would come into the safety area the robot should operate with programmed speed. A flashing orange should indicates a failure of the safety system. Therefore request of this work/project was, to find a suitable inexpensive real-time monitoring system on the local market capable for differentiating of humans from other moving components in the digital factory on Technikum Vienna.

3. Solution and methods

The problem was solved by combining two optical monitoring systems: the core part "SafetyEyes" provided by Pilz GmbH and add-ins part "Kinect" provided by Microsoft. The core part is responsible for protected area and second part monitors the warning area.

3.1. Components of safety system:

Using "SafetyEye" library is possible to create several virtual warning and protection areas. The hardware of this system is shown in the figure 1 and contains [3]:

- the sensor unit consisting of three cameras
- the processing unit for processing of the image data
- the safety controller for interacting with the robot and the periphery. The safety controller is relayed to the input module of the robot controller.

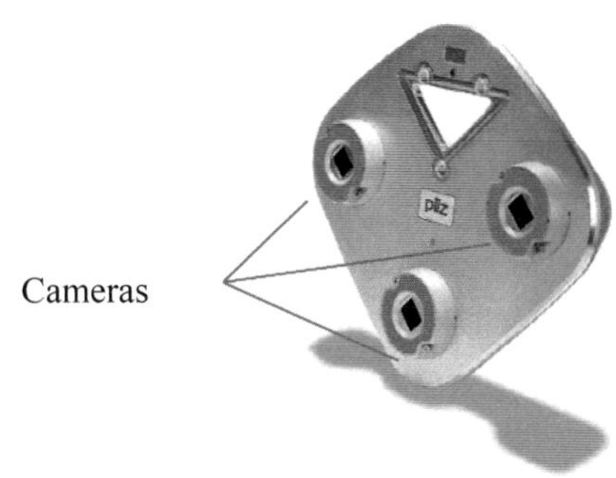

Fig. 1. "SafetyEye" – Pilz GmbH. [4]

A second part of the safety system is a Kinect sensor shown in the figure 2. The system contains [5]:

- An RGB camera for capturing a colour image that stores data in a 1280x960 resolution
- An infrared (IR) emitter, which emits infrared light beams
- An IR depth sensor. Which reads the IR beams reflected from object back to the sensor and converts into depth information measuring the distance between an object and the sensor
- A multi-array microphone for capturing sound, as well as to find the location of the sound source and the direction of the audio wave
- A 3-axis accelerometer designed for a 2G range and used to determine the current orientation of the Kinect. G is the acceleration due to gravity
- A 3-axis accelerometer is the most important sensor at Kinect. Accuracy of the accelerometer is 1 degree. The accuracy of the Kinect is slightly temperature sensitive [6], with up to 3 degree of drift. If required the user can compensate this drift by comparing the accelerometer vertical (the y-axis in the accelerometer's coordinate system) and floor plane depth data.

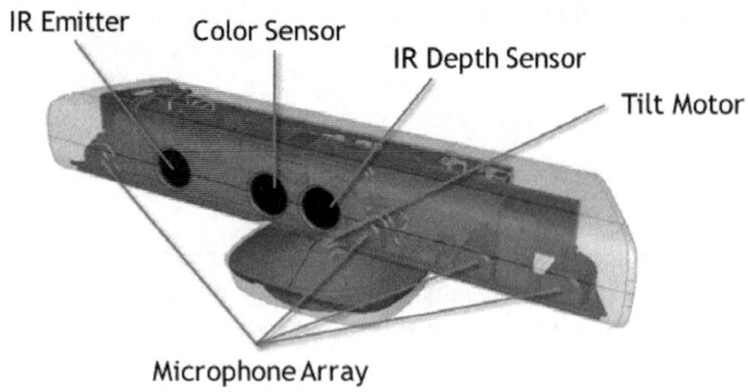

Fig. 2. Microsoft Kinect [5]

3.2. Calculation of minimum distance for mounting of „SafetyEyes"

The calculation of the minimum distance S is dependent of various statically predetermined as well as on dynamic parameters. These dynamic parameters depending on factors such as mounting and on the configuration of the system. The minimum distance S was calculated using the formula (1) by following parameters (as [1] and [3]):

$$S = K \cdot (t1+t2) + C + Zg \quad (1)$$

Where:

- The minimum distance S is determined by the approach speed K=1600 [mm/s] (Deutsches Institut für Normung e. V., 2010)
- the response time of the SafetyEyes t1=0,365 [s]
- the response and stopping time of the robot as well as the robot control t2=0,4 [s]
- the addition for the mounting height of the sensor unit for used case C=850 [mm]
- the addition for system-specific measurement tolerances Zg=235 [mm].

Using the given formula and parameters, the minimum distance S can be defined like in (2)

$$S = 1600 \cdot (0{,}365 + 0{,}4) + 850 + 235 = 2309 \text{ [mm]} \quad (2)$$

3.3. Minimum distance for mounting of „Kinect"

Kinect (in default mode) can recognize humans [5] standing between 0.8 meters and 4.0 meters away. A practical range (shown in Fig 3) is between 1.2 to 3.5 meters. High accuracy of recognizing can be achieved only in this range. Technically is possible to extend a range of safety area by using of a second camera. Though, if more than one Kinect sensor is used to light up the safety area, a reduction in the accuracy and precision tracking will be noticed due to interferences of light sources.

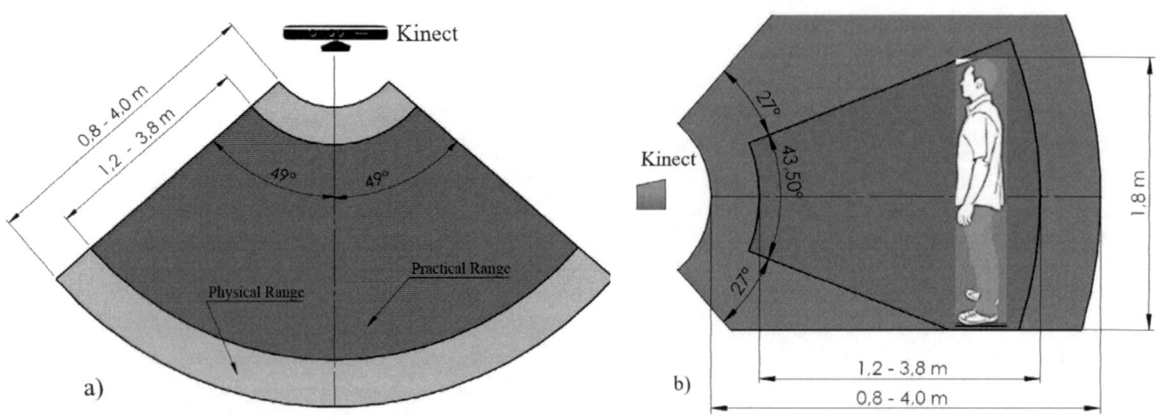

Fig. 3. Kinect recognizing range: horizontal view (a) and vertical view (b) like in [6]

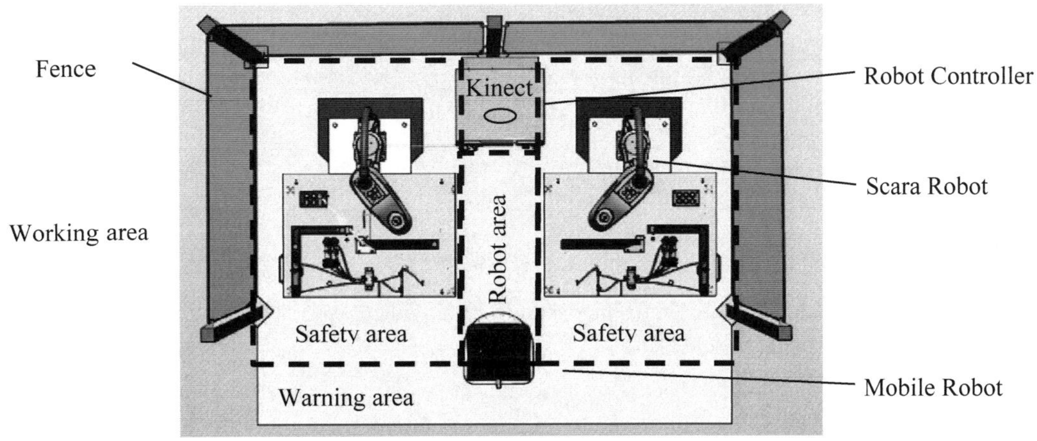

Fig. 4. Layout of robot Cell and monitoring areas

4. Results

The final implementation of the safety system is presented in Layout (Figure 4) with components used in [7] and [8]. This layout shows pick & place robot cell. In normal/working mode (see the flow chart Figure 5), when persons or objects are moving outside of warning-, safety- and robot area, a green lamp will light up. When entering the warning area by human the industrial robot decelerate to programming speed max 250 mm/s. An acoustic/light signal is activated in order to warn the threatened person. When the Robot/Safety is violated by human, the SafetyEye controller sends a signal to the robot controller, which forces the manipulator to stop its operation. The system is in Emergency mode. On the other hand, if the mobile robot enters the warning area an Arduino board relayed from Kinect sensor sends a signal to SafetyEye controller in order to deactivate Robot area an dot enables the working mode. The Safety area still remains active.

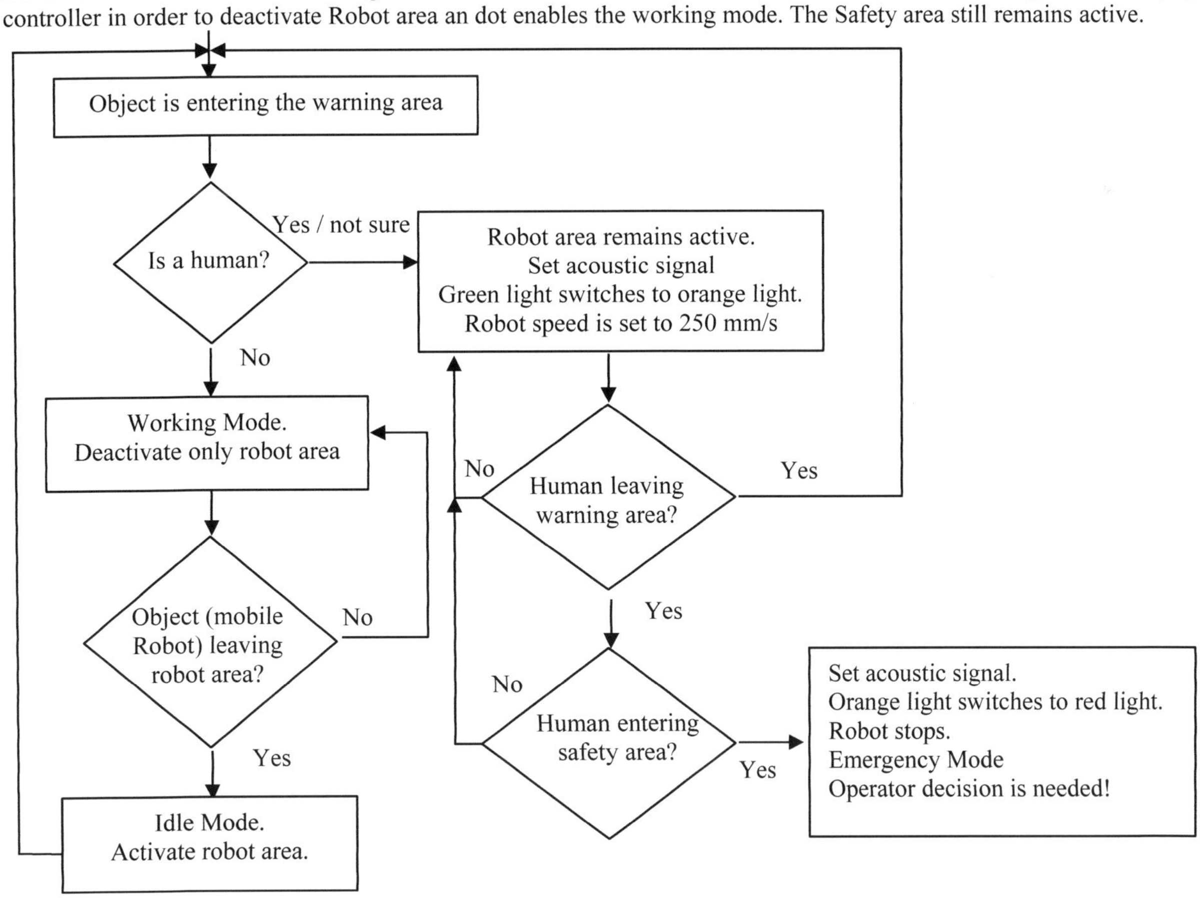

Fig. 5. Decision logic of safety system.

Furthermore, it was observed a moving persons in different orientation toward camera in the warning area. Four samples, each with 84 tests, were carried out with the following result:

Human orientation moving toward Kinect sensor:	Recognition percentage [%]
Frontal	100,00
Backward	39,29
Lateral left (Eyes along negative x-axis of Kinect)	28,57
Lateral right (Eyes along positive x-axis of Kinect)	32,14

Table 1. Recognition percentage of moving human toward Kinect sensor.

As can be seen in Table 1, the lateral movement of humans toward to Kinect is very difficult for sensing, although the probability for this kinds of moving is very low.

5. Conclusion

For the digital factory of the University of Applied Sciences Technikum- Wien in Vienna should be built and implement a safety system for the robot handling cell. The safety system must to take into account all relevant safety-related standards, as well as the state of the art (German Institute for Standardization, 2010). Among others, this system must be capable to monitor work areas in real time in order to prevent human casualties. Another very important demand for this system, is to be inexpensive. Summary, the main task of this research was, to find a suitable inexpensive real-time monitoring system on the market capable for differentiating of humans from other moving components for the robot handling system in the digital factory.

The problem was solved by combining two optical monitoring systems: the core part “SafetyEyes” provided by Pilz GmbH and add-ins part “Kinect” provided by Microsoft. The core part is responsible for recognizing of any kind objects entering the safety and robot area. The second part is responsible for recognizing of humans entering the warning area.

The designed safety system was implemented, on the digital Factory of University of Applied Sciences Technikum Wien and was tested/ analysed for functionality, failures and difficulties. Different data, like reaction rate and different orientation of human entering to the safety area has been collected and presented above. Deficiency of this system is sensitivity against lighting conditions for both cameras and lateral movement of the humans toward Kinect camera. While the first part can be resolved, more by providing of constant light conditions and less by reducing reflective surfaces of the objects. The second deficit can be resolved by mounting the second Kinect camera crosswise to the first one.

The further works of this project is to increase the level of accuracy of safety system for the poor light conditions, as well as implementing within entire manufacturing system. In order to achieve a high level for the recognizing of the human actions and behaviours, a design and implementation of advanced algorithm based on artificially intelligence for the new Kinect Version 2 sensor would be of great interest.

6. References

[1] Angel, N. (2017). Roboterüberwachung mittels Kinect Sensorsystem, University of Applied Sciences Technikum Wien, Department Robotics and Mechatronics, Vienna.

[2] Walter, R. (2017). Arbeitsraumüberwachung eines sechs-achs-Knickarmroboters mittels Pilz SafetyEye, University of Applied Sciences Technikum Wien, Department Robotics and Mechatronics, Vienna.

[3] Pilz GmbH & Co. KG, (2011). SafetyEye: Bedienungsanleitung/Dokumentation. [Datenblatt] Ostfildern, Deutschland: Pilz GmbH & Co. KG :<https://www.pilz.com/download/restricted/PSENse_Operat_Man_21743-DE-11.pdf> Accessed on 21.03.2017.

[4] Pilz GmbH & Co. KG (2017). Pilz – Sichere Automation. [Online]: <https://www.pilz.com/de-AT> Accessed on 21.03.2017.

[5] Kataumo, B. (2014). Kinect for Windows Sensor Components and Specifications. [Online]: https://msdn.microsoft.com/en-us/library/jj131033.aspx [Accessed on 01.09.2017], Microsoft, 2014.

[6] https://msdn.microsoft.com/en-us/library/jj663790.aspx. Microsoft (2014). Kinect for Windows Sensor Components and Specifications, Accelerometer. [Accessed on 01.09.2017].

[7] Stuja K., Bruqi, M., Markl E. & Aburaia, M. (2016). Lightweight 4 -Axis Scara Robot for Education and Research, Proceedings of the 27th DAAAM International Symposium,, B. Katalinic (Ed.), Published by DAAAM International, ISBN 978-3-902734-08-2, ISSN 1726-9679, Vienna, Austria.

[8] Aburaia, B.; Markl, E. & Stuja, K. (2014). New concept for design and control of 4 axis robot using the additive manufacturing technology, 25th DAAAM International Symposium on Intelligent Manufacturing and Automation, Elsevier, pp. 1364 – 136, ISSN 1877-7058, Vienna.

[9] Meisner, J. (2013). Collaboration, expertise produce enhanced sensing in Xbox One https://blogs.technet.microsoft.com/microsoft_blog/2013/10/02/collaboration-expertise-produce-enhanced-sensing-in-xbox-one. Microsoft (2013). [Accessed on 22.10.2017].

DOI: 10.2507/28th.daaam.proceedings.023

SIGNAGE AND ERGONOMICS IN PACKAGING SYSTEMS

Nicoleta Luminiţa Caruţaşu, Victor Adîr, Nicoleta-Elisabeta Pascu,
George Căruţaşu Aurel Florin Arion

This Publication has to be referred as: Carutasu, N[icoleta] L[uminita]; Adir, V[ictor]; Pascu, N[icoleta-Elisabeta]; Carutasu, G[eorge] & Florin Arion, A[urel] (2017). Signage and Ergonomics in Packaging Systems, Proceedings of the 28th DAAAM International Symposium, pp.0173-0179, B. Katalinic (Ed.), Published by DAAAM International, ISBN 978-3-902734-11-2, ISSN 1726-9679, Vienna, Austria
DOI: 10.2507/28th.daaam.proceedings.023

Abstract

This paper talks about the connection of three important elements of a manufacture process: the human factor, the equipment and the ergonomic strategy. This connection is impossible to be developed without a good signage system which may assure a healthy and secure work place for a user. We have tried to answer to a few questions concerning a manufacture process: How does it work a signage system? What is the place of ergonomics? What`s about the connection between them? And, of course, to apply in our research study the principles which lead these main characteristics into a working process (necessity, universality, usability, normativity, simplicity, perceptibility).We have analysed the main characteristics of the human factor and the influence of a lot of elements for the process in a user(human factor)-machine context. In this paper we have explained on a real system the importance of a signage system into an ergonomic strategy. Our study have proposed a few changes for this real system and some recommendations which will assure, in our opinion, a better quality of working, a safety work place and an efficiency of the whole activity of the company. In addition we have developed pictograms to be inserted for every workstation of the manufacture flow.

Keywords: Signage; ergonomics; packaging system; principles; pictograms

1. Introduction

There are questions and answers about signage system and ergonomics. What means ergonomics? What is a signage system? What is the connection between them? What about the relationship with other disciplines? A lot of papers and books talk about ergonomics and signage tied to the human factor. Ergonomics is a discipline, we may say, which is connected to engineering, design, technology, sociology, psychology and human resources. This kind of science has to exist into a system and for the system to improve quality of life, efficiency of the manufacture process, a healthy and secure work place etc. A manufacture process is tied of many problems including work places, goods, design and human factor. It is well known that the performance of human factor and the usability of products depend, generally, of ergonomics. The signage system comes to help the process to be efficient into an ergonomic strategy. In [1], [2] and [3] there are many papers which discuss about ergonomic and signage system for many cases. Also, in [4] is a lesson about human-centred design and ergonomics and in [5] there are developed examples using signage systems in an ergonomic strategy for airport, museums and exhibitions, schools and libraries, public spaces, culture, congresses and trade fairs (e.g.

Düsseldorf International Airport, Germany; Fachhochschule Osnabrück, Germany; New 42nd Street Studios/Duke Theater, New York , Schiphol Airport Amsterdam, Netherlands etc.). We affirm that for a safety and healthy work place is necessary to develop an adequate signage system and a clever ergonomic strategy. An efficient signage system has to improve a good and qualified communication using symbols/pictograms, a colour code and information panels. All the graphic representations have to convey a clear message. And a clever ergonomic strategy may include this system to create an efficient connection human factor-manufacture process.

2. About signage system and ergonomic strategy

2.1 The Signage System

As it is known such a system has to inform, to guide, to allow, to forbid, to warn in different context of the human factor-machine interaction. The designer of such a system has to answer to a few questions, as follows [6]:

- What are the symbols/pictograms for each working position and the quantity of each of them?
- What is the production flow and what kind of machines ?
- Where are the places for these symbols/pictograms concerning the human factor-machine interaction?
- What are the dimensions of the specific sign?
- What is the colour code for the signage system?
- What information panels and quantity are necesarry?
- What kind of information have to be visualized on these panels?

These are the main questions which are waiting answers from the designer. As it is known a signage system is created only if is necessary for an activity. In signage there are two kind of signs: for a general purpose and for specific one. That means general symbols/pictograms and special drawings for specific work places. What is very important is that every graphic representation has to be clear, legible, efficient, working, useful, with an attractive design and, of course, in the right place along the production flow. In this way is realized a safety, secured and healthy work place, for persons and goods. The color code is another important element in a signage system. The colors (red, yellow, blue and green) and shapes where are drawing pictograms are elements which make the difference [7]:

- Red color & circle shape is translated as interdiction;
- Red color & triangle shape is translated as compulsory;
- Yellow color & triangle shape is translated as warning, concern, caution;
- Green color & rectangle shape is translated as emergency exit, rescue, fire exit;
- Blue color & square shape is translated as indicative(give an information about something);
- Blue color & circle shape is translated as indicative;
- Blue color & rectangle shape is translated as indicative.

2.2 The Ergonomic Strategy

In the speciality literature is written like this: ”good ergonomics = good economics”. And it is true. Because a good ergonomics means: lower injures rates, safety and comfort for the workers, easier jobs, product quality and a reduced fatigue for workers. A bad ergonomics means disorder and risk factors for workers and goods. That means problems in productivity, efficiency, quality and safety. Ergonomics is a science which has the principle of adapting the object to the user. It a science centred on the user to assure a safety and comfortable work place. Ergonomics realize the connection between three main elements of the production system (the human user, the machine/equipment and the environment). It is impossible to create a good production chain without to pay attention to these factors. „The origins of ergonomics were in the study of workers activities in factories. At the begining the worker it was recepted as machine-like. As the 20 th century progressed, the idea of worker as robot has been largely discredited and ergonomics has centred on health, safety and enhancement of individuals, particularly by focusing on allowing for human differences” [4]. One of the important designers all over the world who was involved in the ergonomic design was Niels Diffrient, an American industrial designer. The Forbes magazine has described this designer as a”granddaddy of the ergonomic revolution”[4].

The ergonomic strategy for an work place is connected to:

- The Human Factor, as the User. What is the activity ruled by him;
- The Machine(equipment),as an Useful Big Tool. What and how works;
- The Environment(the work place and surrounding). The impact of the environment for workers and goods.

The goal of ergonomics is to prevent problems concerning Musculoskeletal Disorders for the workers, fatigue and discomfort for them and to create healthy and safety jobs. And, of course, to develop a competitive organization solving these by using an ergonomic strategy. That means to conceive a Culture of Ergonomics in the company [5] , [8], [9] and [10].

3. Human factor - machine interaction

3.1 The human factor, as user

A human been has a lot of abilities to solve problems at his work. For this he must have knowledge about the production flow and the machines that work. During his job may intervene disfunctions tied by the anathomy and physiology of him, fatigue, discomfort, physics conditions, social communication, psychology etc. He will try to pass over but, many times, the effort is big enough. The solution is good ergonomics to make a satisfied work place, safety, healthy and comfortable. This good ergonomics means to create such a work place knowing the skills, the abilities and the psychology of the workers, and the limits of him. Namely, the right worker in the right place. Another important aspect, connected to the worker, is the work analysis. This analysis has to identify the possible problems which may appear in the production process, the adequate technology, the instrumentation devices, the measurement theory, a signage system etc.

In figure 1 is presented the human factor surrounded by the main characteristics, as part of the human factor-machine system.

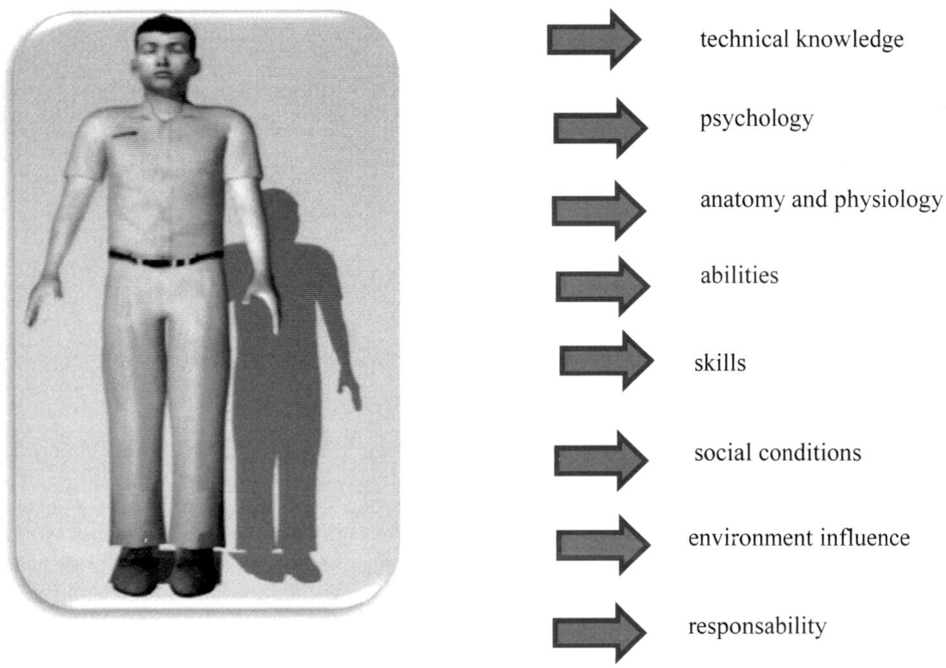

Fig. 1. The main characteristics

3.2 The machine

The Machine is a general name for any industrial equipment to product goods. Analysing this part of the system it is possible to have a technical map, concerning tools, devices, technology, systems of charging, labeling, and packaging, if it is stipulated on the production flow. It is an important part because the ergonomic designer has to involve the human factor in a friendly work place. And to design it paying attention to the effects of the environment. Figure 2 is relevant in this way.

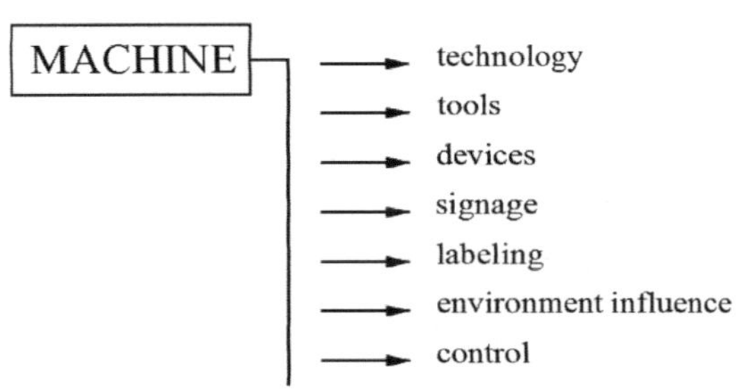

Fig. 2. System

Figure 3 represents the importance of ergonomics about the human factor - machine interaction. The ergonomics is over the components of the system because its goal is to harmonize the man and the machine in the production process.

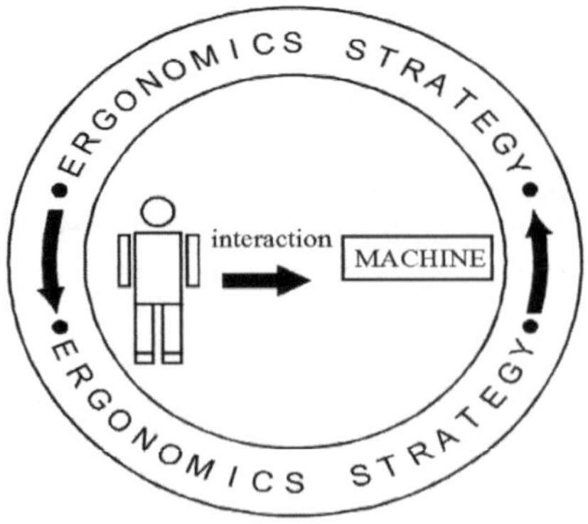

Fig. 3. The ergonomics strategy all over the human factor - machine interaction

4. Case study - a working and packaging system for walnut oil using glass bottles

This system has 8 work places, as follows (figure 5):

- 1 - Weighing and Sorting equipment for walnut;
- 2 - Press machine;
- 3 - Decantation;
- 4 - Filtration;
- 5 - Filling;
- 6 - Caping operation - closing bottles operation;
- 7 - Labeling;
- 8 - Packaging.

The bottles are washed in other room and they are brought manually to be introduced in the flow. The technological flow is not automatized, there are workers to do operations and to assure that the process goes on. It is seen that every work place is not „tied" in an integrated and automatized flow, every machine/equipment is separated from others. Only the filling and labeling operations are automatized. Even the caping operation (closing bottles) needs a worker to push a button to assure the operation, bottle by bottle. Finally, there are workers to put bottles in boxes for packaging.

The figure 4 [11], presents the production area and in figure 5 we have represented the scheme of the manufacturing flow.

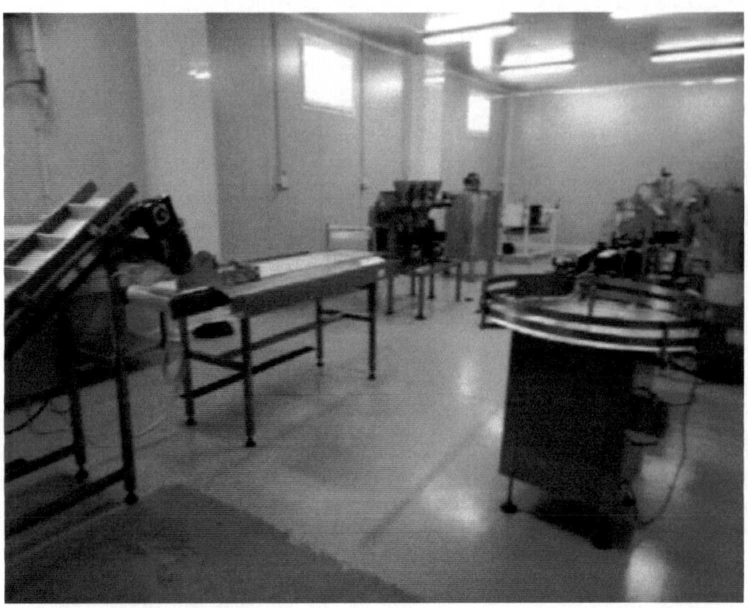

Fig. 4. The working area [11]

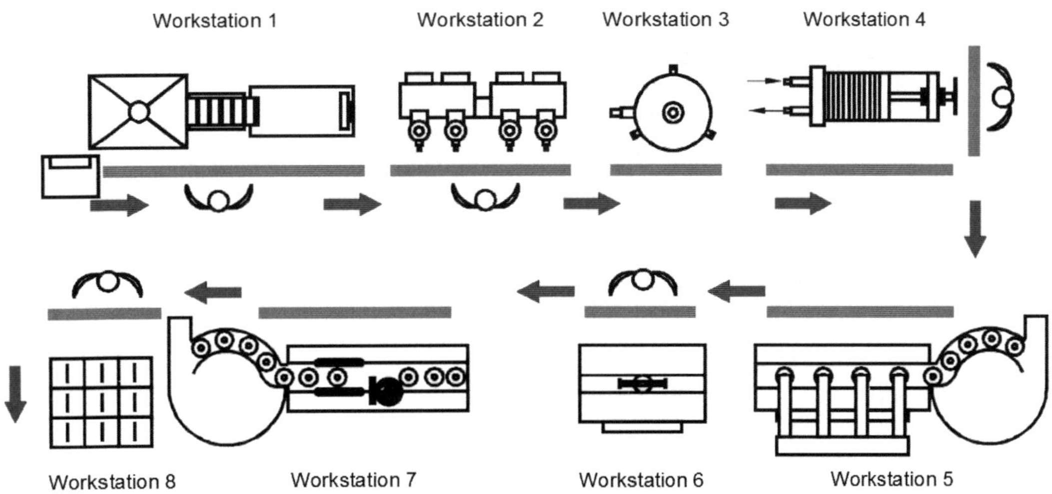

Fig. 5. The manufacturing flow

The critical examination of this process has revealed the necessity of a more implication of the signage-ergonomics concept because pictograms and information/warning panels means more work security and health for employers to avoid the accidents (table 1 and 2).

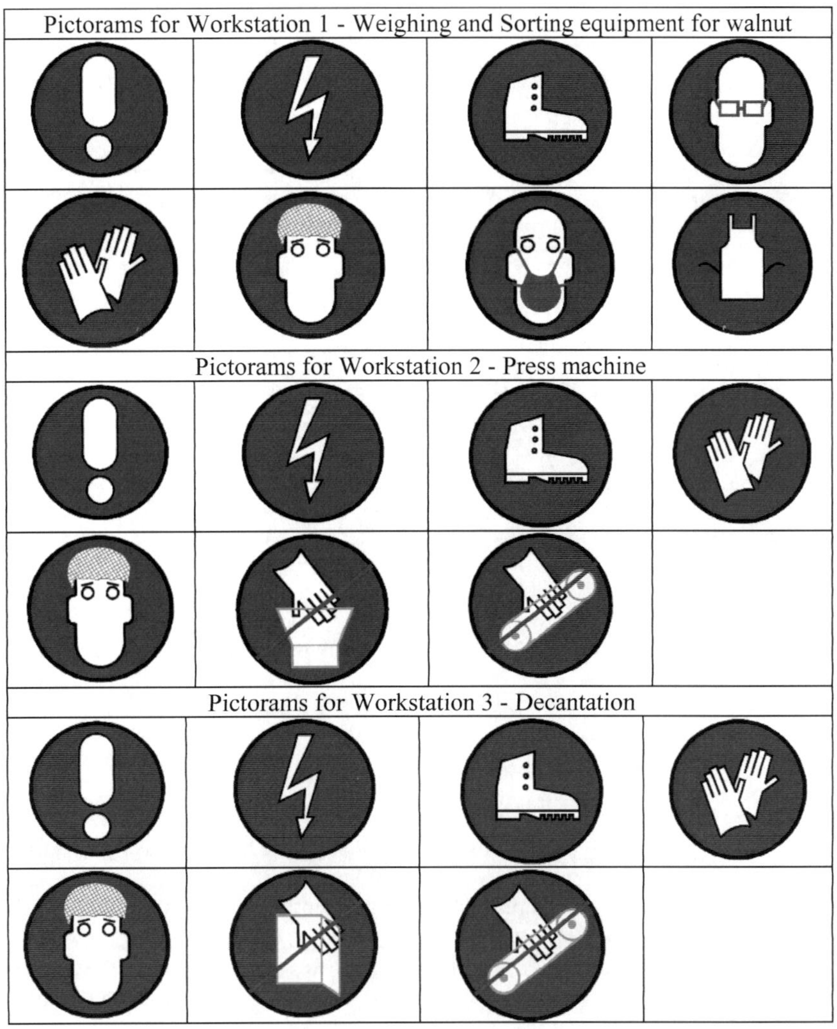

Table 1. Signage

Pictorams for Workstation 4 - Filtration			

Pictorams for Workstation 5 - Filling			

Pictorams for Workstation 6 - Caping			

Pictorams for Workstation 7 - Labeling			

Pictorams for Workstation 8 - Packaging			

Table 2. Signage

As a result of our research we have made a few recommendations to assure more information and more security, as follows:

- At the entrance to the working area is necessary to be a working flow scheme to familiarize with the workers for each work place. The adecquate pictograms and information panels for each place have to be visual and clear to be understood;
- Also, the existing pictograms in the working area have to be greater and placed in a visual location (for example: do not drop water on a fire is so small and it is hanging on the wall);
- The entrepreneur has to identify the possibility to realized a „tied flow" between filling,caping and labeling working places to develop a continuous and automatized process. Of course that means to increase the efficiency and to specialize one or two workers to serve the process;
- Another recomandation is about the possibility to assure adequate chairs for some work places which allow to sit down;

- To avoid eye fatigue is important to see if it is possibile to change the eye glasses with glass window in some work places;
- The pictograms and information panels developed for this process have to be exposed in right and visual places to avoid injuries or troubles.

These are the main conclusions from the research study concerning the importance and the consequences connected to the signage-ergonomics concept.

5. Conclusion

This research study has become visible following the visit in a SME (small manufacture entreprise) and the discussions about the process flow with the managing staff. We have realised all the process and we have seen the oportunity to improve new elements of signage into an ergonomic strategy, because we have understood the necessity to use better the human factor. It was a challenge for us. First we have established the necessary elements for a adequate signage system. For this purpose we have designed graphic representations for each workstation. We have proposed panels of information at the entrance to familiarize the workers with the signage system and with the process flow. Then we have proposed a „tied together process"of filling, caping and labeling workstations to be efficent and to reduce the working time. In this SME these three operations were prepared individually, not as an automatic flow. Also, there are workstations which need chairs for workers. It was another recommendation of us concerning the necessity to have them along the process flow, to avoid the fatigue and discomfort for workers and to create a healthy work place. Finally, we have discuss with the entrepreneur and we have analised together the recommendations and the future steps to be implemented in the factory. This research study has created for us the oportunity to develop a kind of guide for these SME, according to their working field. In a future research, of course.

6. References

[1] Soares, M.M. & Rebelo, F. (2017) Ergonomics in Design. Methods & Techniques. Edited by Marcelo M. Soares and Francisco Rebelo, Published by CRC Press, Taylor Francis Group, Boca Raton, Fl., USA

[2] Advances in Ergonomics in Design, Usability & Special Populations Part I, (2014), Edited by Marcelo Soares and Francisco Rebelo, Published by AHFE (Advances in Human Factors and Ergonomics) Conference, USA

[3] Advances in Cognitive Ergonomics, (2010), Edited by David B. Kaber and Guy Boy, Published by CRC Press, Taylor & Francis Group, Boca Raton, Fl., USA

[4] Gareth, W. (2015). DESIGN, an essential introduction, Goodman Fiell, an imprint of the Carlton Publishing Group, ISBN 978 1 78313 012 2, London

[5] Andreas Uebele, A. (2007). Signage Systems - Information Graphics. A professional Sourcebook, Thames & Hudson Ltd., ISBN 978-0-500-28848-1, London

[6] Smitshuijzen, E. (2007). Signage Design Manual, Lars Muller Publishers, ISBN 978-3-03778-096-1, Baden, Switzerland

[7] Abdulah, R. & Roger Hubner, R. (2006). Pictograms, Icons & Signs. A guide to information graphics, Thames & Hudson Ltd., ISBN 978-0-500-28635-7, London

[8] Adîr, V.; Pascu, N.E. & Adîr, G. (2016). Design de limbaj grafic-vizual. Manualul conceptiei grafice, Editura Printech, ISBN 978-606-23-0693-9, Bucuresti

[9] Berger, M.C. (2009). Wayfinding. Designing and Implementing Graphic Navigational Systems, RotoVision SA, , ISBN 978-2-88893-057-0, Switzerland

[10] Dreyfuss, H. (1984) Symbol Sourcebook. An Authoritative Guide to International Graphic Symbols, John Wiley&Sons Inc., ISBN 0-471-28872-1, New York;

[11] http://www.magicnuc.ro/, (2017)

DOI: 10.2507/28th.daaam.proceedings.024

MODELING LEAD TIME DEMAND IN CONTINUOUS REVIEW INVENTORY SYSTEMS

Edin Kadric*, Hadis Bajric & Mugdim Pasic

This Publication has to be referred as: Kadric, E[din]; Bajric, H[adis] & Pasic, M[ugdim] (2017). Modeling Lead Time Demand in Continuous Review Inventory Systems, Proceedings of the 28th DAAAM International Symposium, pp.0180-0186, B. Katalinic (Ed.), Published by DAAAM International, ISBN 978-3-902734-11-2, ISSN 1726-9679, Vienna, Austria
DOI: 10.2507/28th.daaam.proceedings.024

Abstract

Inventory systems that use continuous review policy are under risk during lead time, when stock-out can occur. Therefore, system must have enough on-hand units to prevent such situations. Generally, in inventory control literature it can be found that lead time demand follows normal distribution and all other conclusions are derived from this assumption. However, in real life this does not have to be true, so it is of crucial importance to get better estimates of stochastic demand parameters over lead time. The objective of this research is to estimate the optimal (s, Q) continuous review inventory policy parameters that reduce risk of stock-out during lead time and to enhance robustness of such estimated parameters. This is done using approach we propose for demand modeling. Performances and adequacy of the proposed approach for lead time demand modeling, with various demand patterns, and its application in (s, Q) continuous inventory models are obtained by simulation and show very good results.

Keywords: lead time; stochastic demand; continuous review inventory control; grouped demand; overlapping periods

1. Introduction

In many situations the assumption of known, deterministic demand and lead time is not necessarily realistic. The demand for each time period is not known in advance. The lead time of items can be uncertain for several reasons such as machine breakdowns, quality or transportation issues. In [1] it is shown that studying the behaviour of the demand and the lead time are essential in order to achieve a useful representation of the system to take proper decisions. Significant research is developed on optimal inventory policies by assuming that demands are independent and follow certain distributions [2]. In practice it is common to assume that the lead time demand follows a particular distributional family [3]. A simplistic approach to modelling uncertainty in the continuous review system is to assume that lead time demand is normally distributed.

Use of the normal distribution or another standard probability density function for lead time demand, or a deterministic lead time assumption, can permit an analytical solution for inventory control parameters, and such results have revealed significant insights into the nature of optimal policies in inventory systems modelled under a variety of conditions [4]. Analytical evaluation in [3] has revealed that the lead time demand should not be approximated by the Poisson distribution unless the variance-to-mean ratio is exactly 1, and that the gamma and the negative binomial distributions yield fairly good results especially when there is much more variability in data. So in order to avoid stock-out and to achieve higher target or service level it is necessary to decrease the uncertainty of the demand fluctuation with a more accurate forecasts or by increase of the safety stocks to cover the fluctuating demands [5]. Increasing safety stocks results in higher average inventory levels and higher holding costs, but at the same time it decreases risk of stock-out and penalty costs. Ratio of penalty cost to holding cost may range from 100 to 400, depending on nature and purpose of item [6]. According to findings in [7], traditional approaches in estimation of lead time demand variance, essential for safety stock calculations, can lead to safety stocks that are up to 30% too low and service levels that are up to 10% below the target. In continuous review, system is under risk during lead time, and to reduce risk and avoid stock-out it is very important to accurately predict lead time demand and estimate its parameters.

Estimating demand parameters, such as expected value and variance, for periods longer than time unit, e.g. lead time, can be achieved by multiplying expected daily demand by length of period and by variance. Problems arise when demand data are available for short time horizons. In this paper we are presenting new approach for demand modelling which increases number of demand data sufficient for reliable estimation of lead time demand parameters. First, we consider all demand data available in given time horizon, instead of just lead time demand data. This way it is obvious that more demand data is available, but it is also justified because replenishment process in continuous review models, is stochastic in nature. In traditional approach, demand parameters are estimated by considering only demand over lead time. Lead time length can be constant, but start of replenishment process is not known in advance, hence using only demand over lead time can lead to very poor estimates. Secondly, all available demand data, in given time horizon, are grouped into periods of desired length that mutually overlaps, yielding the same number of demand data. Newly generated demand data represent demand per period and contain all possible realizations of demand per period, while in traditional models demand data are given per day.

Proposed approach to lead time demand modelling is tested on fast and slow moving items using (s, Q) continuous review inventory model. Continuous review model used in analysis differs from traditional ones in terms that define demand parameters, such as average lead time demand and probability density function of lead time demand. As demand parameters obtained using our approach are more reliable and robust than those obtained using traditional approach, policy parameters of used continuous review model are also more accurate and robust. Results have shown that (s, Q) continuous review model provides optimal continuous review inventory policy parameters, which reduce risk of stock-out during lead time and enhance robustness of model parameters.

2. Modelling of the lead time demand

Lead time is defined as the time that elapses from the time order is placed and the time order is received. In continuous review inventory systems order is placed when the inventory level reaches reorder point s. During the lead time, system is under risk of stock-out due to stochastic nature of demand. To avoid stock-out and penalty costs, system must have sufficient on-hand inventories to satisfy demand over lead time. Two most important factors in estimating reorder point are lead time demand and its length. As lead time demand is more volatile and lead time length is longer then reorder point is higher, and vice versa. Expressions for calculation of lead time demand, whether lead time length is deterministic or variable, are shown below.

Case: variable demand and constant lead time
If average demand is $\bar{x}$ and its variance is v, and lead time length is constant L, then expected lead time demand $\bar{x}_L$ and its variance v_L over lead time, can be calculated using following expressions:

$$\bar{x}_L = \bar{x} \cdot L \tag{1}$$

$$v_L = v \cdot L \tag{2}$$

Case: variable demand and variable lead time
If lead time length is variable with expected length λ and variance ω, and if expected demand is $\bar{x}$ with variance v, then expected lead time demand $\bar{x}_L$ and its variance v_L over lead time, can be calculated using following expressions:

$$\bar{x}_L = \bar{x} \cdot \tau \tag{3}$$

$$v_L = \lambda \cdot v^2 + \bar{x}^2 \cdot \omega \tag{4}$$

Above expressions are appropriate and valid in situations when demand follows normal distribution. In situations when demand distribution does not follow normal distribution, expressions presented above, generally are not appropriate for estimation of demand parameters for periods longer then unit time. More details and discussions about estimation of demand parameters for time periods longer then time unit can be found in [8]. In order to obtain more robust lead time demand parameters, we present our approach to lead time demand modeling below.

If historical demand data are available, we divide original time series into periods of equal lengths L. Then we sum demand in every period and obtain demand that represent first 'period' time series. In the next step, we create second demand time series, in such way that first demand from original time series is set to the end of the original time series. New time series is then again divided into periods of length L, demand is summed, and second 'period' times series is obtained. This procedure is repeated $L-1$ times. Presented approach of grouping demand data by overlapping periods of 7 days length, is graphically presented on Fig. 1.

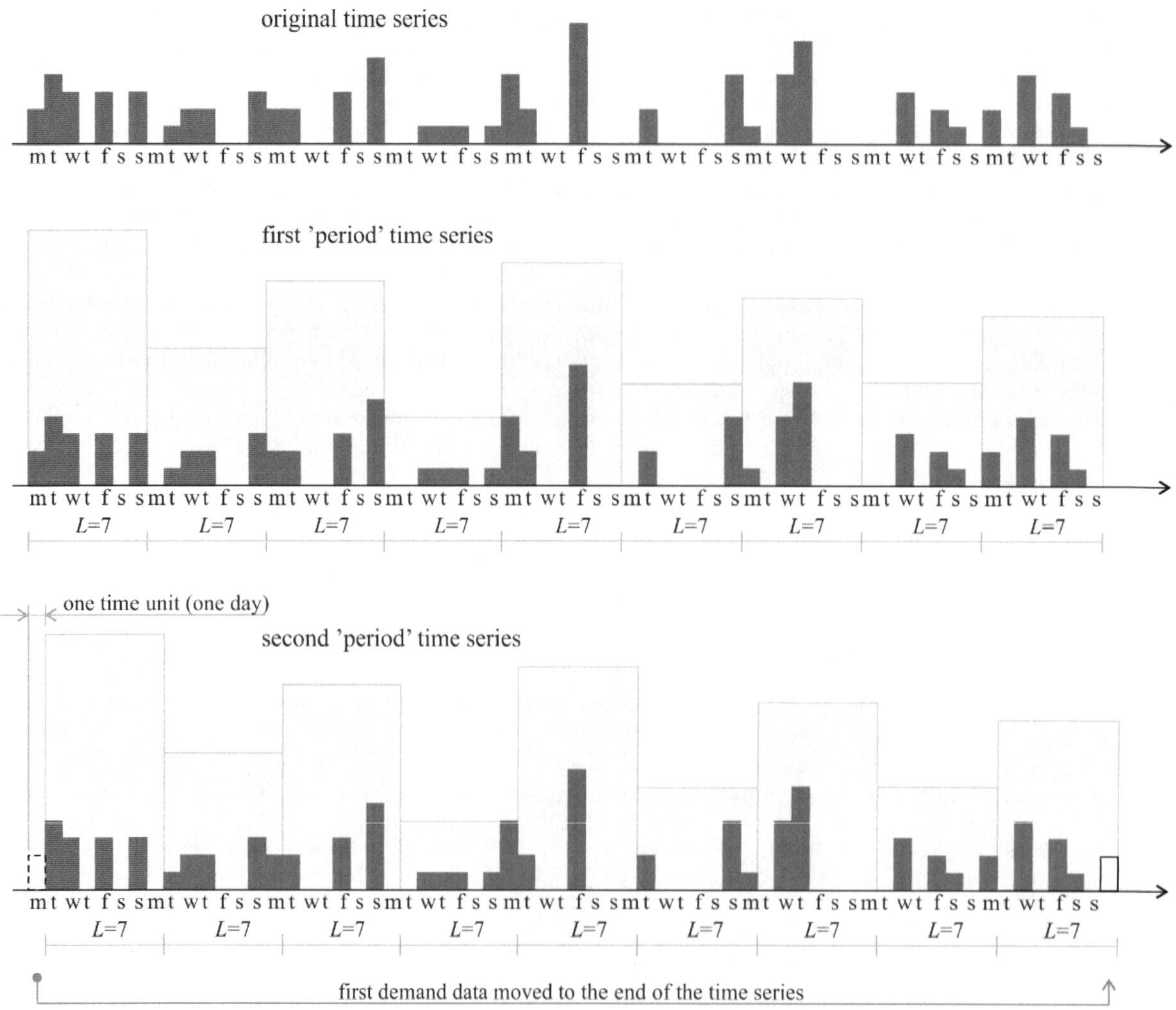

Fig. 1. Graphical presentation of demand modelling with periods that overlaps.

When all 'period' time series are constructed then it is possible to proceed with lead time demand analysis in two directions: first one uses all 'period' time series to estimate demand parameters, while the second one uses just one 'period' time series with lowest variance.

If all 'period' time series are used in analysis then we can expect that expected demand will be close to that estimated using (1), while demand variance will be higher than that estimated using (2). As lead time length is longer then lead time demand variance will be higher. Higher lead time demand variance increases reorder point level s, holding and total costs, but at the same time risk of stock-out is lesser and less affected by sudden changes in demand and supplier response and will have lower penalty costs. We propose this approach to be used in uncertain environments, where lead time and lead time demand is volatile and suppliers unreliable.

Second approach considers choosing one, among all, 'period' time series with lowest lead time demand variance. Opposite to above explained approach, this approach will result with lower reorder point level s, holding and total costs, but this approach is prone to risk. Sudden changes in demand can lead to significant variance increase, and then to stock-out occurrence. Stock-out will decrease service level and increase penalty and total costs. We propose this approach to be used in systems that operate in certain and stable environments.

3. Formulation of (s, Q) continuous inventory model

We consider single item, single location, and continuous review inventory control problem with allowed backorders. We suppose that demand in successive time periods is positive, random, independent and identically distributed variable. When system has positive inventory level, all customer demands are satisfied immediately from inventories on stock. When system is out of stock, we suppose that customers are willing to wait for next order delivery to satisfy their demand, so all unsatisfied demand is backordered. When next order arrives, backorders are satisfied first and then regular demands.

When system has positive inventory levels, then for every item unit on-hand per time unit, holding cost is charged. As exact number of item units on-hand is not known in advance, holding cost is charged on expected number of item units on-hand. For every item unit, system is charged h_d currency units per time unit. When system is out-of-stock than for every backordered unit, system is charged π currency units per item unit. We suppose that lead-time L, the time between placing and receiving an order is constant. There is no order overlapping and in any time period system has the most one outstanding order.

Proposed mathematical model of expected total cost function is defined by two parameters: reorder point s and fixed order quantity Q. The expected total cost $C(s, Q)$ for proposed continuous review inventory model, is given by expression:

$$C(s,Q) = A \cdot \frac{D_H}{Q} + h_H \cdot \left[\frac{Q}{2} + \int_0^s (s - x_L) \cdot f(x_L)\, dx_L\right] + \frac{D_H}{Q} \cdot \pi \cdot \int_s^{+\infty} (x_L - s) \cdot f(x_L)\, dx_L \tag{5}$$

Where: A = fixed ordering cost, D_H = total demand over considered time horizon, h_H = holding cost per item unit over considered time horizon, x_L = expected demand over lead-time, $f(.)$ = probability density function of demand over lead-time.

Deriving expected total cost function $C(s, Q)$, given by (5), over parameters s and Q, and equaling derivatives to zero, gives optimal values of control parameters s and Q.

$$\frac{\partial C(s,Q)}{\partial Q} = -\frac{A \cdot D_H}{Q^2} + \frac{h_H}{2} - \frac{D_H \cdot \pi}{Q^2} \cdot N_L = 0 \tag{6}$$

$$\frac{\partial C(s,Q)}{\partial s} = h_H \cdot \int_0^s f(x_L) dx_L - \frac{D_H}{Q} \cdot \pi \cdot \int_s^{+\infty} f(x_L) dx_L = 0 \tag{7}$$

From (2) follows:

$$-\frac{A \cdot D_H}{Q^2} + \frac{h_H}{2} - \frac{D_H \cdot \pi}{Q^2} \cdot N_L = 0$$

$$-A \cdot D_H + \frac{h_H \cdot Q^2}{2} - D_H \cdot \pi \cdot N_L = 0$$

$$Q = \sqrt{\frac{2 \cdot D_H \cdot [A + \pi \cdot N_L]}{h_H}} \tag{8}$$

From (3) follows:

$$h_H \cdot F_L(s) - \frac{D_H}{Q} \cdot \pi[1 - F_L(s)] = 0$$

$$F_L(s)\left[h_H + \frac{D_H}{Q} \cdot \pi\right] - \frac{D_H}{Q} \cdot \pi = 0$$

$$F_L(s) = \frac{D_H \cdot \pi}{D_H \cdot \pi + h_H \cdot Q} \tag{9}$$

Where: N_L = expected number of backorders, $F_L(.)$ = cumulative probability density function of demand over lead-time.

Using known inverse cumulative distribution function of demand over lead-time given by (9), or numerical methods, it is possible to find optimal value of reorder point s, which minimizes total cost function.

Considering expression (8) for optimal fixed order quantity Q, it can be seen that it is very similar to well known EOQ quantity. Only difference is that, in particular circumstances, fixed ordering cost A is replaced with $A + \pi \cdot N_L$ cost. Economical meaning of this difference is that ordering cycle implies not only the fixed ordering cost but also the risk of stock-out, before supplier delivers ordered units. Because of the risk of stock-out companies try to increase ordering quantity Q in order to avoid frequent orderings. Every time order is placed there is the risk that demand over lead-time will be larger than reorder point s, which leads to inventory stock-out and customer dissatisfaction. Generally, continuous inventory control models with allowed backorders will have larger ordering quantity Q then basic EOQ model. This increase in ordering quantity will be larger if penalty cost π is larger and reorder point s is smaller, because probability of stock-out will increase.

Using expressions (8) and (9) to find optimal values of s and Q is not simple because each variable is dependent on each other. Since optimal values of s and Q can not be estimated directly by analytical methods, it is possible to use iterative procedure. The essence of the iterative procedure can be roughly explained in the following way: initially an approximate real value is chosen for the order quantity, which is Q from the EOQ model. Then using this value for the optimal order quantity, which can be represented by Q_0, determine the initial estimate for the reorder point s_0. Using the initial estimate for s_0, it is possible to find better estimate for the fixed order quantity Q_1, and so on. The iterative process stops when $Q_i \approx Q_{i-1}$ or when $C_i(s,Q) - C_{i-1}(s,Q) \leq \varepsilon$, where ε = cost treshold, and i = iteration number.

4. Verification and testing of proposed approach to lead time demand modelling

To verify and test proposed approach to lead time demand modelling and its performance in continuous review inventory model it is necessary to compare values of model performance indicators to those of simulation. Values of simulation performance indicators are considered exact ones. Simulation is performed using proposed continuous review inventory model and its input and control parameters on real demand data. Performance indicators used in this experimental setting are: P_1 and P_2 service levels, average inventory level, expected value of backordered units, total and component costs.

Verification and testing are conducted on real life examples. Items used in analysis significantly differ in parameters such as: purchase and selling price, demand frequency, value and variance. Selected items have relatively low and high values of these parameters. Item 1 is from grocery group and is widely available in the most grocery stores. Data for Item 1, used in analysis, are as follows: time horizon length is 355 days, total demand in time horizon is 16424 units, daily demand is 46.20, standard deviation is 29.87, purchase price c = 1.64 €/unit, fixed ordering cost A = 10 €/order, backorder cost π = 0.70 €/unit backordered, holding cost h_d = 0.001348 €/unit/day. Frequency, demand value and variance for Item 1 are high, so it can be classified as fast moving item. Item 2 is household air-conditioning device and is available in the most home appliances stores. Data for Item 2, used in analysis, are as follows: time horizon length is 247 days, total demand in time horizon is 18 units, daily demand is 0.07, standard deviation is 0.25, purchase price c = 376.72 €/unit, fixed ordering cost A = 50 €/order, backorder cost π = 100 €/unit backordered, holding cost h_d = 0.31 €/unit/day. Frequency, demand value and variance for Item 2 are low, so it can be classified as slow moving item.

To estimate accuracy and reliability of proposed inventory model two measures have been used: error or deviation of model performance indicators to those simulated, and relative error used to estimate relative deviations of performance indicators. Error Δ_j is difference between values of j-th model and simulation performance indicator and can be calculated using following expression:

$$\Delta_j = Model_j - Simulation_j \tag{10}$$

where: $Model_j$ – value of j-th model performance indicator, $Simulation_j$– value of j-th simulation performance indicator. If error Δ_j is negative, than it means that model underestimates j-th performance indicator value, meaning real value is greater, and if error Δ_j is positive, than it means that model overestimates j-th performance indicator value, hence real value is smaller. Error shows how much expected value differs from exact one, but it does not show degree of sensitivity. To estimate sensitivity of differences we have used relative error. Relative error $\Delta_j\%$ of j-th performance indicator, can be calculated using following expression:

$$\Delta_j\% = \frac{\Delta_j}{Simulation_j} \cdot 100\% \tag{11}$$

Error and relative error of total cost, as one of the performance indicators, are especially important, because control parameters of proposed inventory model are obtained using this indicator. Errors and relative errors of other performance indicators are also important in analysis of trade-offs between average inventory levels and backorders, i.e. holding and backorder costs, and can be used to further improve inventory control model.

Optimal values of maximum reorder point s and fixed order quantity Q, as well as values of performance indicators with their relative errors are shown in tables below. Relative errors of performance indicators for which simulation value is equal to 0 could not be calculated and are represented with hyphen (-).

Lead time L	2	5	7
Reorder point s	179	366	516
Fixed order quantity Q	840	847	831
Performance indicators	**Relative error, %**		
P_1 service level	0,0000	0,0000	0,0000
P_2 service level	-0,0487	-0,0731	-0,0122
Average inventory level	0,1224	-0,2401	-0,9518
Backorders	-	-	-
Ordering cost	0,0000	0,0000	0,0000
Holding cost	0,1196	-0,2379	-0,9521
Backorder cost	-	-	-
Total cost	1,3313	1,6545	-0,2833

Table 1. Comparative view of optimal control values and values of relative errors of performance indicators of (s, Q) model for Item 1 and lead times lengths of 2, 5 and 7 days.

It can be seen from Table 1. that for Item 1 classified as fast moving item, model performance indicators match very well simulation ones, with relative errors less than 2%.

Lead time L	2	5	7
Reorder point s	0	1	1
Fixed order quantity Q	5	5	5
Performance indicators	**Relative error, %**		
P_1 service level	0,0000	0,0000	0,0000
P_2 service level	0,0000	0,0000	0,0000
Average inventory level	-20,6349	-10,8635	-9,0379
Backorders	0,0000	0,0000	0,0000
Ordering cost	0,0000	0,0000	0,0000
Holding cost	-20,6145	-10,9649	-9,1388
Backorder cost	0,0000	0,0000	0,0000
Total cost	-9,2005	-5,2528	-4,2734

Table 2. Comparative view of optimal control values and values of relative errors of performance indicators of (s, Q) model for Item 2 and lead times lengths of 2, 5 and 7 days.

Analysis of Item 2 shows lack of model to more accurately estimate on-hand units, hence relative errors of average inventory level and holding cost are some high. Relative errors of total cost do not exceed 10%, but it can be seen that as lead time length becomes longer, relative errors become smaller. When lead time is 2 days model actually becomes EOQ model.

5. Conclusion

In this paper new approach to lead time demand modeling and its application in continuous review inventory model is presented. Proposed approach to lead time demand modelling enables significant improvements of demand parameters estimation with higher reliability and robustness. Generally, advantages of proposed approach to ones typically used in inventory control literature and practice, can be summarized as follows:

- Large enough sample of grouped demand data, used for estimation of reliable and robust lead time demand parameters, even in situations when limited lead time demand data are available.
- Lead time demand data obtained using proposed approach are less sensitive to sudden changes in demand, such as unexpected increase or decrease of demand during some periods.
- Lead time demand data contain all the possible realizations of demand per period that can occur in the considered time horizon.

Relative errors of model performance indicators are acceptable, even the items used in the analysis significantly differ in demand and cost properties. Optimization results are especially good for fast moving item (Item 1), with relative errors less than 2%, for all lead time lengths used in analysis. Optimization results for slow moving item (Item 2) are poorer, with relative errors less than 10% for lead time length of 2 days, but for longer lead times as 5 and 7 days, relative errors are less than 5%.

Future research should focus on better estimation of expected number of on-hand units and to them related holding costs, in order to improve overall model, especially in case of slow moving items.

6. References

[1] Roldán, R.F., Basagoiti, R. & Coelho, L.C. (2016). A survey on the inventory-routing problem with stochastic lead times and demands, J. Appl. Log., in press

[2] Hnaien, F.& Afsar, H.M. (2017). Robust single-item lot-sizing problems with discrete-scenario lead time. International Journal of Production Economics 185 (2017) 223–229, ISSN: 0925-5273

[3] Rossetti, M.D. & Unlu, Y. (2011). Evaluating the robustness of lead time demand models. Int. J. Production Economics, 134 (2011), pp. 159-176, ISSN: 0925-5273

[4] Cobb, B.R., Johnson, A.W., Rumí, R. & Salmerón, A. (2015). Accurate Lead Time Demand Modeling and Optimal Inventory Policies in Continuous Review Systems. International Journal of Production Economics 163 (2015), pp. 124-136, ISSN: 0925-5273

[5] Korponai, J.; Banyai, A. & Illes, B. (2015). The Effect of the Demand-Changes on the Inventories, Proceedings of the 26th DAAAM International Symposium, pp.1068-1075, B. Katalinic (Ed.), Published by DAAAM International, ISBN 978-3-902734-07-5, ISSN 1726-9679, Vienna, Austria DOI: 10.2507/26th.daaam.proceedings.150

[6] Huh, W.T., Janakiraman, G., Muckstadt, J.A. & Rusmevichientong, P. (2009). Asymptotic Optimality of Order-Up-To Policies in Lost Sales Inventory Systems. Management Science, Vol. 55, No. 3, pp. 404-420, ISSN: 0025-1909

[7] Prak, D., Teunter, R. & Syntetos, A. (2016). On the calculation of safety stocks when demand is forecasted. European Journal of Operational Research, Vol. 256, Issue 2, pp. 454-461, ISSN: 0377-2217

[8] Kadric, E., Bajric, H. & Pasic, M. (2015). Demand Modeling with Overlapping Time Periods. Procedia Engineering 100 (2015) 305 – 313, doi: 10.1016/j.proeng.2015.01.372

DOI: 10.2507/28th.daaam.proceedings.025

ROMANIAN ROADMAP FOR E-CALL TECHNOLOGY ADOPTION AND FUTURE DEVELOPMENTS OF EMERGENCY SYSTEMS

George Carutasu, Cristina Coculescu & Mihai Alexandru Botezatu

This Publication has to be referred as: Carutasu, G[eorge]; Coculescu, C[ristina] & Botezatu, M[ihai] (2017). Romanian Roadmap for E-Call Technology Adoption and Future Developments of Emergency Systems, Proceedings of the 28th DAAAM International Symposium, pp.0187-0193, B. Katalinic (Ed.), Published by DAAAM International, ISBN 978-3-902734-11-2, ISSN 1726-9679, Vienna, Austria
DOI: 10.2507/28th.daaam.proceedings.025

Abstract

Reducing road fatalities by Intelligent Transport System (ITS) implementation was an objective of EU Commission through e-Safety plan, initiated in 2004. E-call technology aims to reduce the intervention time for road accidents by sending a Minimum Set of Data (MSD), containing relevant information regarding the position, car identification, time, to nearest Public Safety Answering Point (PSAP) using mobile network. The service chain planned that after MSD is received by 112, a voice channel to be established between 112 and the car, which sent the MSD, in case of ordinary emergency calls, the call initiator being the person in emergency situations. The approaching deadline of eCall implementation, on March 2018, as free European harmonized emergency service, rise a series of technical and operational issues presented in this paper. The Romanian roadmap of e-Call technology adoption is presented, together with national and EU funded project that participated to technical and operational setting for eCall service. Moreover, the adopted Romanian pilot site solution is explained in the context of emergency system organization. Current research, regarding eCall extension to long distance coaches, heavy goods vehicle (HGV), dangerous goods and motorcycles, as result of iHeERO CEF funded project and previous research is also debated

Keywords: Intelligent Transport Systems; emergency systems; eCall; iCar

1. Introduction

The statics offered by EU [1] show a significant reduction in the last ten years, for EU28. According to this source, the total number of fatalities decreased from 45 900 persons in 2005 to 26 100 persons in 2015. From this, 8% of fatalities resulted from accidents on motorways, 37% in urban areas and 55% on rural roads. Furthermore, fatalities for 2015 might by categorized on transport mode: 46% of fatalities resulted from car accidents, pedestrians 21%, moto 14%, bicycle 8%, moped 3% and other 7%. Even those numbers seem to indicate a progress on EU28 level, with 51.5 fatalities per million of habitants in 2015, it can be underlined that, in case of several countries, like Bulgaria (98 fatalities/millions of habitants) and Romania (95 fatalities/millions of habitants) remains almost four times higher than Sweden (27 fatalities/millions of habitants). If into account it is considered accumulated value for fatalities and hospitalized persons, it is emphasized that Austria reach 929 persons/ millions of habitants, the mean value for EU28 being 536.5 persons/ millions of habitants, lowest values recorded for Ireland with 134 fatalities + hospitalized / millions of habitants. Furthermore, a sample regarding accident data and related structure is presented in [2].

Besides EU countries positions in various statistics, it is recorded an advance of overall road safety because of EU and national policies. Those positive outputs are the results of emergency services improvement and eSafety measures implementation in transportation. A closer analysis show that the fatalities rates and severity of injuries depend on road structure, quality of road, safety measures implemented into road infrastructure, cars and emergency services response time and quality. The single European emergency call number, introduced in 1991, by Council Decision 91/396/EEC remains a strong step forward, unifying all emergency agencies. 112 service reduced the response time and proper synchronize all dispatched emergency force to accident site. At that time, accident location was possible by direct interview of caller. European Road Safety Action Programme 2003-2010 introduced a series of measures that aimed to halved the fatalities resulted from road accidents. So, the action program combined measures regarding technical conditions of vehicles, transports of dangerous goods, Intelligent Transport Systems (ITS) and the eSafety initiative. The ITS Directive [3] introduced a wide range of information technologies that could be used in to increase the road safety. Related to terms definitions, ITS is referring to a suite of applications aiming to offer new services to multiple transport modes, traffic management and user to be informed regarding road safety. The Directive had four areas: optimal use of road, continuity of traffic, ITS road safety and linking the vehicle with the transport infrastructure. In addition, it is decided that the access to ITS services should be interoperable, harmonized and continuous across EU. Furthermore, it is established that the geolocation needed for various ITS applications should be able to use EGNOS and Galileo systems. From this paper perspective, eCall technology is referred as ITS service by ITS directive. Regarding traffic management, general principles of material flow are applied [4], with a detailed example in [5].

2. e-Call technology

As a single post-crash ITS service, eCall aims to reduce the intervention time of emergency teams to crash site, by accurate localization and time stamping of crash, detailing vehicle involved and severity of crash. The eCall systems (fig.1) consists into an In-Vehicle System (IVS) embedded into the car. When the accident occur, the IVS, triggered by various car mounted sensors [6] or manual, reads the current position, using a GPS module, and it is combined with predetermined Vehicle Identification Number (VIN), generating a Minimum Set of Data (MSD) message. This message is carried by Mobile Network Operators (MNO) and send it to nearest Public Safety Answering Point (PSAP). After MSD is successfully received by PSAP, it is resolved by a MSD Decoder and VIN code is extracted. Using VIN code, the EUropean CAR and driving license Information System (EUCARIS) is interrogated, resulted information being displayed directly into case file window opened by PSAP operator. In the second phase, the PSAP initiates a voice call to the vehicle, to check the nature of emergency or to obtain additional information regarding the accident. The seizing and nature of emergency team dispatched to crash site might be influenced by information obtained through voice call or the number of sensors triggered. Also, by vehicle type, decoded from VIN, it is displayed the maximum number of passengers, or vehicle type (figure 1).

In comparison to ordinary emergency call, some differences should be underlined:

- Ordinary 112 call is not supporting data exchange, and MSD should be treated as emergency call, using discriminator,
- 112 call is using MNO location or location provided by PTNO (Public Telephone Network Operators) operator for each customer, in case of eCall the location is provided by IVS GSM module,
- eCall initiates backward a voice call from PSAP to the car, which suppose that also this call should be considered as emergency call and prioritized using a discriminator. In case of 112 call, the emergency was signalized when 112 was introduced into Receiver field. For eCall, discrimination must be set into Receiver and Sender field,
- eCall supply the VIN code, additional information regarding the car and owner could be obtained by resolving EUCARIS database,
- MSD contains the time stamp of the crash and provides the moving direction (very important in case of motorways.

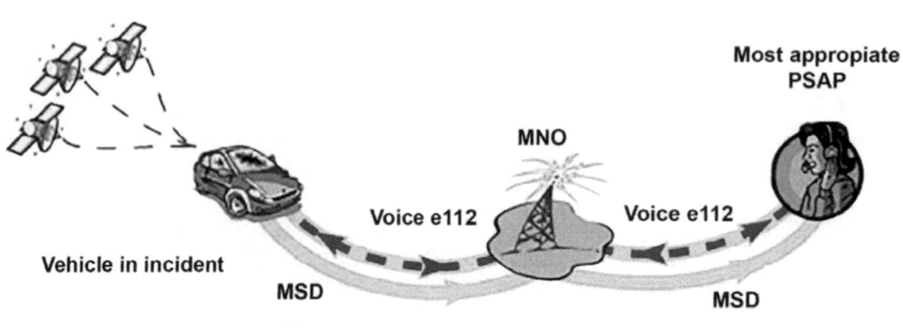

Fig. 1. eCall service chain – official representation

Localization provided by PTNO was addressed by [7], which also stipulates that 112 service should be available from any network and any device, including special feature regarding person with disabilities access, using text messages or other types. Highly important is that the Directive enforce the interoperability and harmonization of 112 service across EU member states. In case of localization provided by MNO, a close look is given by [8].

The eCall system (e112) was initially assessed by EU [9] in 2010. In this report, was estimated that eCall introduction will reduce the response time by 50% in rural areas and 40% in urban areas. However, the fatalities reduction is estimated between 2 and 10% and severity of injuries between 2 and 15%. Beyond humanitarian reasons, where no life can be measured and replaced, the cost of health treatment and person unavailability is taken into consideration, together traffic congestion savings, for the period 2005-2033, having a positive impact, balanced with PSAP upgrade costs and IVS costs.

As implementation deadline for eCall service, estimated first in 2010 by [10], was prolongated, having another proposal for 2015 [9] and finally to 2018, 31st of March [11] to be available for all new produced M1 and N1 vehicles (passenger cars and light duty vehicles).

3. Romanian roadmap for e-call technology adoption

The important events and actions carried out by Romania to implement eCall service started in 2007, when ITS Romania signed eCall MOU (Memorandum of Understanding for Realization of Interoperable In-Vehicle eCall). Later, in 2010, rallied to the action, Romanian Ministry of Communications and Information Society, National Company of Motorways and National Roads of Romania and Romanian-American University.

In Romania, the 112 service is managed by The Single National Emergency Call System – SNECS 112 implemented and administrated by STS (Special Telecommunication Service) established from 2006. Regarding administrative structure, Romania has 41 counties plus Bucharest. However, the STS organized a PSAP responsible for each county.

The research on eCall field started with the project "Study regarding eCall technology", having as partners ITS Romania, Romanian-American University and ElSol, in the period 2008-2010, funded by Romanian Ministry of Communications and Information Society. The project aimed to identify the stakeholders for eCall implementation and structural updates to existing infrastructure [12]. Furthermore, the eCall operation by private operators was taken into account and presented in [13]. The main findings of research were:

- PSAP infrastructure should be updated, to receive MSD and to be integrated into existing file case software application – stakeholder STS,
- MNOs must implement discriminator flag for signalizing backward call, from PSAP to vehicle that triggered eCall – Stakeholder local MNOs and Romanian Authority in Communication,
- TMC (Traffic Management Centre) should implement a data interface, to receive and post on existing software, the GNSS coordinates of vehicle crash, provided by MSD, initially received by PSAP and transferred to TMC – stakeholder National Company of Motorways and National Roads of Romania,
- VIN code resolving, by interrogating EUCARIS database in real time, using data embedded into MSD, and details presentations to PSAP operator- stakeholder STS and Ministry of Interior, Vehicle License and Vehicle Regime Direction,
- eCall service chain will include only public organization, Romania had no private PSAP and insurance company had no interest in eCall technology, accordingly to questionnaire filled in 2009.

The key findings of "Study regarding eCall technology" project become later, in 2010, objectives of Romanian Pilot Site of the project ICT-PSP / FP7 / No. 270906/2010 / Harmonised eCall European Pilot (HeERO), financed by EC. The project consortium covered various stakeholder from nine countries: Croatia, Czech Republic, Finland, Germany, Greece, Italy, The Netherlands, Romania and Sweden. The Romanian consortium was composed by ITS Romania, STS, National Company of Motorways and National Roads of Romania, UTI System, Romanian-American University and ElSol.

Additional objectives were added [14]:

- Interoperability and cross border, eCall service should be available to any vehicle that travels across EU and to transfer the emergency call to nearest PSAP, in case cross border situation,
- Harmonization, mapping and levelling the 112/e112 service into member states.

The technical solution adopted by Romanian Pilot Site (presented in figure 2), indicates that the MSD triggered by IVS, is carried by PLMN and routed to Telekom network. From this network, the call is diverted to closest EONES modules, that links Telekom to 112 Network, detailed in [15].

Using identification of eCall flag, the call is divertet to eCall modem. Further, the added B number, specific to county where eCall was triggered is replaced with specific code for Bucharest 021 112. This action is needed because Romania upgraded only two PSAPs with eCall function, primary Bucharest, and Brasov as operational backup. So, eCall enters into Multi Service Network (MSS) managed by STS and received into Bucharest PSAP. Here, MD110 PBX is used to commute to a VoIP server (Communication eXchange Environment -CXE).

The eCall is signalized to operator with a special ring, and displayed to operator console with data contained in MSD. To resolve VIN code, a query is sent to EUCARIS database. When the MSD reception is done and confirmed, the voice channel between 112 and IVS is opened. In the same time, resulted information from VIN interrogation (owner, car color, car type, maximum number of passengers, fuel type) is displayed into case file opened for the accident.

After voice call is established, the operator might interrogate the passengers, regarding further details regarding the emergency team proper seize the emergency teams. If no answer is given, the procedure details a standard rescue team dispatched to crash site. The developed infrastructure is highlighted with yellow. The test operated by STS in spring 2012, indicated a successfully eCall operation chain [16], Romanian PSAP readiness being set starting from November 2013. HeERO2 project involved six countries: Belgium, Bulgaria, Denmark, Luxembourg, Spain and Turkey, aligning countries PSAP to eCall requirements. To close the service chain, with two Romanian vehicle manufacturers, Renault and Ford, should homologate the vehicles produced from class M1 and N1, applicable from 31st of March 2018 [17]. Furthermore, the local MNO, Vodafone reported eCall readiness [18] for twelve countries including Romania, starting from 31st of March 2016. However, it is still an issue regarding how new SIM dedicated to eCall will be released and what cost model will be applied.

The impact assessment for eCall in Romania, may start by new car evolution in period 2018-2020. On 31st December 2016, in Romania where registered 7 010 608 vehicles, from which 5 470 578 in M1 category and 781 196 in N, representing cumulated 89.17% from total vehicles. Furthermore, 174 407 M1 under 2 years and 41 729 N under 2 years were registered in 2016. So, is expected that in next two years, to have 200 000 units equipped with eCall function (2.8% from total registered vehicles). From total 28 994 road accidents (in 2015) is expected that at least 50% that will trigger an eCall, reported to 2.8%, new vehicles it foreseen around 400 eCalls in each year for first two years. Also, in [9] is estimated an yearly rate of 11% of increasing the number of vehicles with eCall. Same source reveals that for Romania, is estimated a 6% diminution of road fatalities. The full implementation threshold of eCall is 2040.

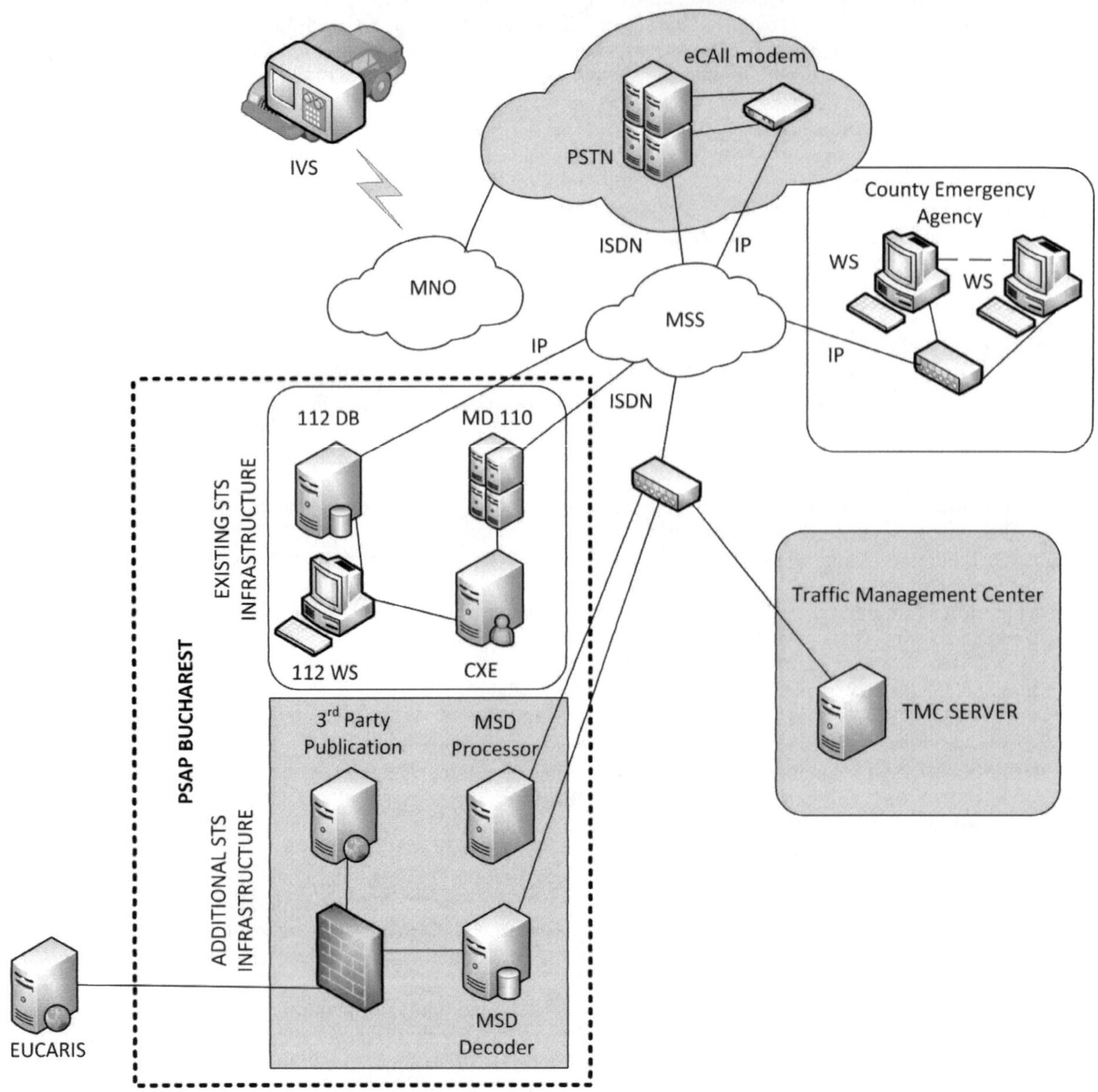

Fig. 2. eCall updates for Romanian Pilot Site

4. iHeERO and future emergency services development

The iHeERO (i from Infrastructure) project started in 2015, with a three-year duration and over 30 mil. EUR as budget, financed by EU through CEF program, including members from eleven countries (Bulgaria, Cyprus, Czech Republic, Finland, Germany, Greece, Ireland, Italy, Portugal, Romania and Slovenia) continued the previous HeERO1 and HeERO2 project to align the PSAP to eCall requirements.

New objectives were added regarding eCall implementation to new categories of vehicles. In case of previous HeERO project only M1 and N1 categories were considered. Even those categories represent almost 90% of total registered vehicles, some notorious accidents, having great impact, by number of fatalities, good losses or environmental impact, the need of eCall extension to other types of vehicle was approached. IHeERO deals with, so call, special eCall, considering changes needed into the eCall service chain, presented in figure 3.

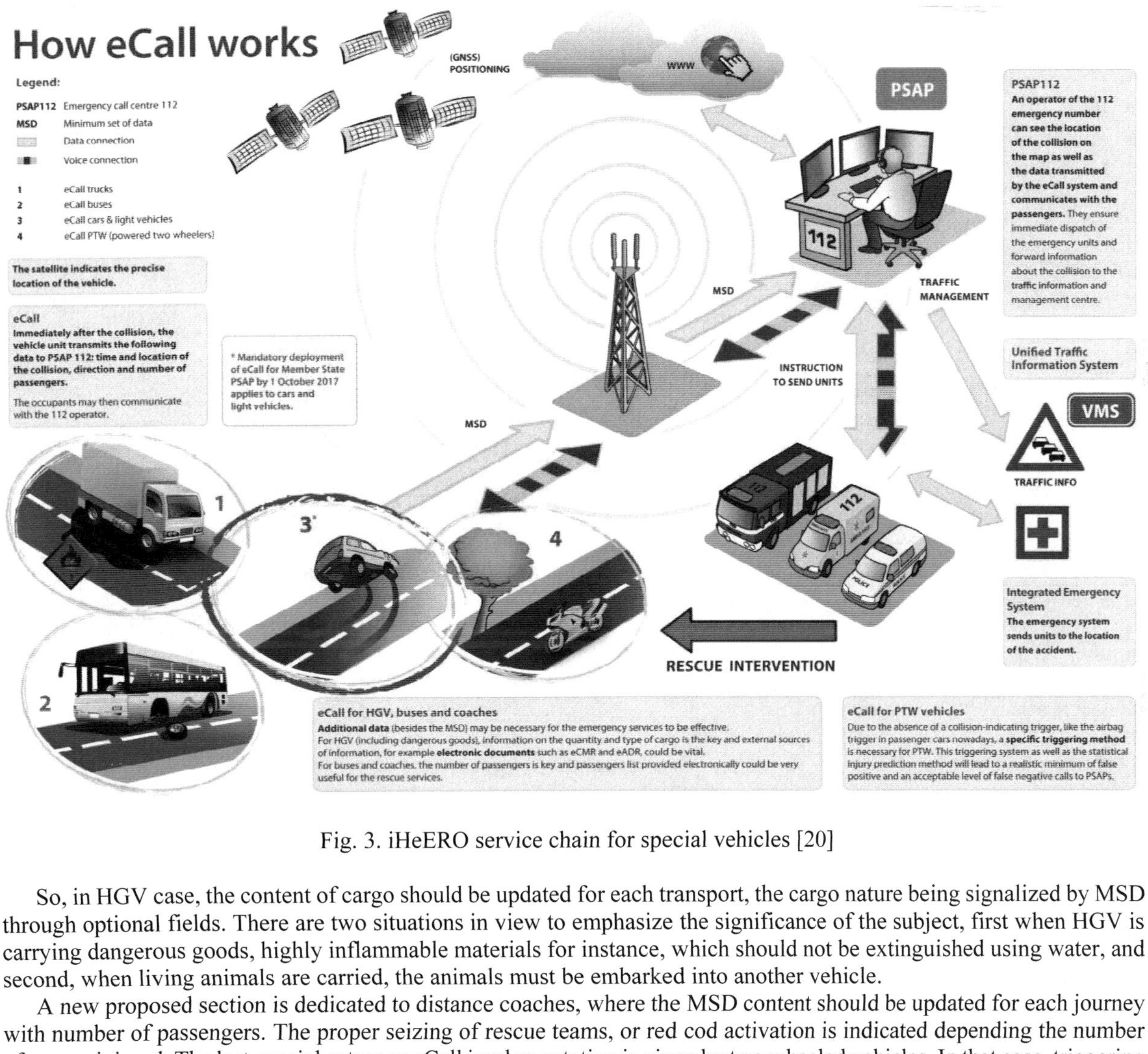

Fig. 3. iHeERO service chain for special vehicles [20]

So, in HGV case, the content of cargo should be updated for each transport, the cargo nature being signalized by MSD through optional fields. There are two situations in view to emphasize the significance of the subject, first when HGV is carrying dangerous goods, highly inflammable materials for instance, which should not be extinguished using water, and second, when living animals are carried, the animals must be embarked into another vehicle.

A new proposed section is dedicated to distance coaches, where the MSD content should be updated for each journey with number of passengers. The proper seizing of rescue teams, or red cod activation is indicated depending the number of person injured. The last special category eCall implementation is given by two wheeled vehicles. In that case, triggering eCall is impossible with standard sensors (airbag inflation), requiring a two-part architecture (helmet and body part). Such system is detailed in [19]. Also, the PSAP conformity is another objective of iHeERO. The current advance of iHeERO might be discovered in [20].

The future emergency development should consider realities of today world. Moreover, using social media environment, smartphones, video calls in ordinary conversation, must indicate a change of vision for emergency services. Besides, with a very long period from the planning to implementation risks that eCall to be obsolete before even started. So, some directions of future work could be underlined:

- Extending eCall technology for pedestrians, using a smartphone application, same information regarding incident might be transmitted to PSAP, replacing VIN code with personal number and car sensor replaced with life monitoring sensors,

- Retrofit development for old cars, including certification, the market potential for owners that wish voluntary to implement eCall,
- Further development of MNO infrastructure will allow to replace the voice call communication from PSAP to IVS with a video call and the MSD to contain multimedia message with relevant information about the crash.

5. Conclusion

Reducing road fatalities, remains an objective for EU, even the numbers halved in the last decade. The measure that should be taken by EU Commission and by the member states must envisage also the increasing of vehicle safety, like safe belt in the past, continuing with airbag displacement for rear passengers etc. Furthermore, the road quality and safety, expressed into motorways length, in case of Romania depreciate the effort of emergency services to save lives. Moreover, even Romania is a best practice for eCall implementation, the overall quality of the roads will limit the success of rescue teams. The eCall implementation should be enforced with other ITS measures, critical in avoiding and signaling dangerous situations.

After a very long journey, started with eSafety Action Plan in 2000, the eCall will become available as universal service in 2018, with an expected maturity in 2040. Moreover, with an implementation deadline prorogated several times, initially in 2010 and 2015 second, eCall passed very difficult technical and organizational issues, to align all PSAPs, MNOs, TMCs and vehicle manufactures to comply eCall standards. The HeERO projects, developed in the period 2010-2018, having a total budget of almost 50 million EUR (with 50% EU contribution), is a landmark of ITS implementation. During this period, 21 EU and associated members states participated to eCall specific activities, regarding PSAP and TMC infrastructure updates, functional and operational tests, cross border situation and cooperation with Russian similar service ERA-GLONASS. The interest of the automotive industry, MNOs and other stakeholders on HeERO projects was very high, iHeERO project alone having 69 partners (part of them being associated partners with no budget) that manifest their will to participate to project activities, iHeERO being by far the largest consortium that run a project financed by EU Commission.

Romanian expertise accumulated in HeERO1 and iHeERO, launched new opportunities for emergency services. The actual situation of terrorist attacks across Europe led to necessity of further improvement of information collection from incident site. On course projects, like NEXES, propose new methods for carrying the emergency message to PSAP, introducing the possibility of wireless communication to be accepted by PSAP and to enrich the content using multimedia files. Even the reported accident coordinates and time are accurate, the accident evaluation must be corrected using other field sources, like CCTV or satellite images, to a proper seize of rescue team.

Regarding eCall further development, the first look envisages old vehicles, being a huge market opportunity. With an estimated price of 150 EUR for IVS and moreover, the third-party services for updating and replacing IVS for special purposes (introduced by iHeERO) is another market opportunity.

The authors participated to mentioned projects and contributed to eCall Romanian pilot site development and iHeERO activities. After a decade of effort in eCall field, it is hopped that launching in 2018, eCall service to save the promised lives and to reduce cost regarding hospitalization, unavailability and traffic congestions. An impact assessment is scheduled after two years of implementation.

6. Acknowledgments

This work was supported by a grant of the European Commission, through CONNECTING EUROPE FACILITY (CEF) - TRANSPORT SECTOR program, AGREEMENT No INEA/CEF/TRAN/A2014/1031743 "I_HeERO".

7. References

[1] http://ec.europa.eu/eurostat EUROSTAT (2017) Statistics – accidents data - European Commission. Accessed on: 2017-06-23

[2] Zovak, G.; Saric, Z.; & Cop, A. (2011). Conceptual Model for Structuring Traffic Accident Data, Annals of DAAAM for 2011 & Proceedings of the 22nd International DAAAM Symposium, 23-26th November 2011, Vienna, ISBN 978-3-901509-83-4, ISSN 1726-9679, Katalinic, B. (Ed.), pp. 0523-0524, Publisher DAAAM International, .

[3] European Parliament (2010) Directive 2010/40/EU of the European Parliament and of The Council of 7 July 2010. Official Journal of European Union

[4] Cotet, C. E.; & Popescu, D. (2014). Material flow management in industrial engineering, In: Encyclopedia of Information Science and Technology, Third Edit., Khosrow-Pour, M. (Ed.), pp. 3786 – 3794, Information Science Reference, ISBN 978-1-4666-5888-2, United States of America

[5] Rados, B; & Rados, A. (2016). Implementation of Multi-Criteria Decision Making in Planning Traffic Systems, Proceedings of the 26th DAAAM International Symposium, pp.0069-0074, B. Katalinic (Ed.), Published by DAAAM International, ISBN 978-3-902734-07-5, ISSN 1726-9679, Vienna, Austria DOI:10.2507/26th.daaam.proceedings.010.

[6] Seidl, M., Carroll, J., & Cuerden, R. (2015). eCall–Defining Accident Conditions for Mandatory Triggering of Automatic Emergency Calls. 24th International Technical Conference on the Enhanced Safety of Vehicles (ESV),

Paper Number 15-0137, 8-11th June 2015, Gothenburg, Sweden. Available from: https://www-esv.nhtsa.dot.gov/proceedings/24/files/24ESV-000137.PDF, Accessed: 2017-06-11

[7] European Parliament (2002). Directive 2002/22/EC of the European Parliament and of the Council of 7 March 2002 on universal service and users' rights relating to electronic communications networks and services (Universal Service Directive). Available from: http://eur-lex.europa.eu/legal-content/EN/ALL/?uri=celex:32002L0022. Accessed: 2017-07-04

[8] Adamek, M; Neumann, P; Pospisilik, M & Matysek, M. (2016). Using of GSM Locator in Security Applications, Proceedings of the 26th DAAAM International Symposium, pp.0075-0081, B. Katalinic (Ed.), Published by DAAAM International, ISBN 978-3-902734-07-5, ISSN 1726-9679, Vienna, Austria DOI:10.2507/26th.daaam.proceedings.011

[9] European Comission (2011). eCall Impact Assessment | Digital Single Market. Available from: https://ec.europa.eu/digital-single-market/en/news/ecall-impact-assessment. Accessed: 2017-03-25

[10] eCall Driving Group (2006). Recommendations of the DG eCall for the introduction of the pan-European eCall. Available from: https://www.trafi.fi/filebank/a/1436524683/a9892dd401033f54359ff9bb16ccb93d/18185-Position_papers_DG_eCall_v2.pdf. Accessed: 2017-07-05

[11] European Parliament (2015) Regulation (EU) 2015/758 of the European Parliament and of the Council of 29 April 2015 concerning type-approval requirements for the deployment of the eCall in-vehicle system based on the 112 service and amending Directive 2007/46/EC. Available from: http://eur-lex.europa.eu/legal-content/EN/TXT/?uri=CELEX:32015R0758. Accessed: 2017-07-05

[12] Botezatu, C.; Botezatu, C. P.; Carutasu, G.; & Barca, C. (2009). eCall safety transportation management systems – features and capabilities. Annals of Oradea University. Fascicle Management and Technological Engineering, vol. 8, no. 1, pp. 118–123, ISSN 1583-0691

[13] Carutasu, G.; Botezatu, C.; Botezatu, C. P.; & Pop, P. D. (2010). Business Models For Extending of 112 Emergency Call Center Capabilities With E-Call Function Insertion. Annals of Oradea University. Economic Science, vol. 1, no. 2, pp. 1177–1182, ISSN 1582-5450

[14] http://www.heero-pilot.eu/view/en/heero/objectives.html, (2012) HeERO Consortium, HeERO - About HeERO Objectives. Accessed: 2017-07-05

[15] Dumitrescu, D.; Grososiu, A.; Dumitrescu, S.; Rosu, V.; Ristea, G. & Carutasu, G. (2012). Solution for eCall Implementation at National Level within the HeERO Project–Romania Case Study, 19th ITS World Congress EU00407, 22-26 October 2012, ITS America, Vienna, Available from: https://trid.trb.org/view.aspx?id=1264000

[16] Dumitrescu, D.; Oncioiu, G.; Ropot, R.; Mohora, A.; Ristea, G.; & Carutasu, G. (2013). eCall Operational Tests in the Live System Implemented in Romania, 9th ITS European Congress, TP 0092, 4-7 June 2013, Dublin, Ireland,

[17] European Parliament (2015). Regulation (EU) 2015/ 758 of The European Parliament concerning type-approval requirements for the deployment of the eCall in-vehicle system based on the 112 service.

[18] MacDougall, R. (2017). Recommendations on residual issues relevant to eCall, Vodafone Group Public Policy, ECC Public Workshop on numbering for eCall, 31st January 2017, Available from: https://www.cept.org/Documents/wg-nan/34819/session-3-presentation-3_robert-macdougall, Accessed: 2017-07-02

[19] Harnischmacher, F.; Cosyns, C.; Grugl, K.-M.; Moerbe, M.; Portouli, E.; & Savaresi, S. M. (2016). State of the art assessment Powered Two-Wheeler (P2W) eCall, 11th ITS European Congress, 6-9th June 2016, Glasgow, Scotland, pp. 6–9, Available from: http://iheero.eu/wp-content/uploads/sites/3/2016/06/ITS16_State_of_the_art_assessment_P2W_eCall_FINAL.pdf. Accessed: 2017-07-2

[20] http://iheero.eu/, (2016). iHeERO - I_HeERO Official website. Accessed: 2017-07-06

DOI: 10.2507/28th.daaam.proceedings.026

ROBUST STABILIZATION OF A HEATING-COOLING SYSTEM BY USING TWO FEEDBACK CONTROLLERS: A NUMERICAL-GRAPHICAL ANALYSIS

Radek Matušů, Libor Pekař & Jiří Korbel

This Publication has to be referred as: Matusu, R[adek]; Pekar, L[ibor] & Korbel, J[iri] (2017). Robust Stabilization of a Heating-Cooling System by Using Two Feedback Controllers: a Numerical-Graphical Analysis, Proceedings of the 28th DAAAM International Symposium, pp.0194-0198, B. Katalinic (Ed.), Published by DAAAM International, ISBN 978-3-902734-11-2, ISSN 1726-9679, Vienna, Austria
DOI: 10.2507/28th.daaam.proceedings.026

Abstract

This contribution is intended to present a numerical-graphical method for analyzing the robust stability of a control loop which contains two feedback controllers and a heating-cooling system with a heat exchanger. The controlled plant is described by the anisochronic model with internal delays and the parameters of this model are supposed to vary within given intervals. The applied technique for robust stability analysis is based on the numerical calculations of the value sets in combination with the graphical test via the zero exclusion condition.

Keywords: Robust Stability; Parametric Uncertainty; Anisochronic Model; Heating-Cooling System; Heat Exchanger; Two Feedback Controllers

1. Introduction

Robust stability is the key property of the control systems under uncertainty and its analysis has attracted the researchers for several decades [1], [2], [3], [4]. Generally, the more complicated type and structure of uncertainty is presented in the model, the more sophisticated and complex tools have to be used for testing the robust stability. However, the value set concept combined with the zero exclusion condition [1] represents extremely universal technique which can be applied to various ranges of parametric uncertainty structures [5], [6].

In this contribution, the robust stability of a control loop with two feedback controllers and a heating-cooling system with a heat exchanger [7], modelled as the anisochronic model with internal delays and parametric uncertainty, is analyzed with the assistance of the numerical calculations of the value sets and the graphical evaluation of the zero exclusion condition. The paper presents the modified version of the example that has been already published in the previous work [6].

2. Process Description

Assume a cooling-heating system with a through-flow air-water heat exchanger that evinces long internal delays. In [7], [8], the plant was modelled by means of the anisochronic principle – see e.g. [9]. This approach allowed including all significant delays and latencies caused by the heat transferring from a source through a piping system by using the heat transferring media (water) into a heat-consuming part.

The process was described in [7], [8] by a set of ordinary differential equations and algebraic equations. Subsequently, the same works [7], [8] derived also a linearized model formulated by the transfer function:

$$G(s)=\frac{\left(b_0+b_{0D}e^{-\Theta_0 s}\right)e^{-\Theta_b s}}{s^3+a_2s^2+a_1s+a_0+a_{0D}e^{-\Theta_a s}} \tag{1}$$

which expresses the relation between the input power to the heater and the outlet stream temperature from the heat exchanger.

The nominal parameters identified in the specific operating point are [7]:

$$\begin{aligned}
b_0 &= -2.052\times10^{-7}\\
b_{0D} &= 2.334\times10^{-6}\\
\Theta_0 &= 1.5\\
\Theta_{0b} &= 141\\
a_2 &= 0.1767\\
a_1 &= 0.08989\\
a_0 &= 1.413\times10^{-4}\\
a_{0D} &= -7.625\times10^{-5}\\
\Theta_{0a} &= 151
\end{aligned} \tag{2}$$

The experiments in [7] showed that the plant is affected by the parameter perturbations and other non-modelled internal and external influences.

The robust stability test from [6] supposed the possible variations of the size ±10% of all the nominal parameter values (2). In this contribution, the robust stability is analyzed for the variations of the size ±15%, i.e.:

$$\begin{aligned}
b_0 &\in\left[-2.3598\times10^{-7},-1.7442\times10^{-7}\right]\\
b_{0D} &\in\left[1.9839\times10^{-6},2.6841\times10^{-6}\right]\\
\Theta_0 &\in[1.275,1.725]\\
\Theta_{0b} &\in[119.85,162.15]\\
a_2 &\in[0.150195,0.203205]\\
a_1 &\in[0.0764065,0.1033735]\\
a_0 &\in\left[1.20105\times10^{-4},1.62495\times10^{-4}\right]\\
a_{0D} &\in\left[-8.76875\times10^{-5},-6.48125\times10^{-5}\right]\\
\Theta_{0a} &\in[128.35,173.65]
\end{aligned} \tag{3}$$

3. Control Loop

The control system with two feedback controllers was chosen as a convenient configuration. The scheme of this control loop is shown in Fig. 1. Two feedback controllers are represented by the blocks C_R and C_Q and a controlled plant by the block G.

The signals are denoted as follows: w – reference signal, e – tracking (control) error, u_0 – difference of controllers' outputs, u – control signal, y – controlled signal (output), v – load disturbance.

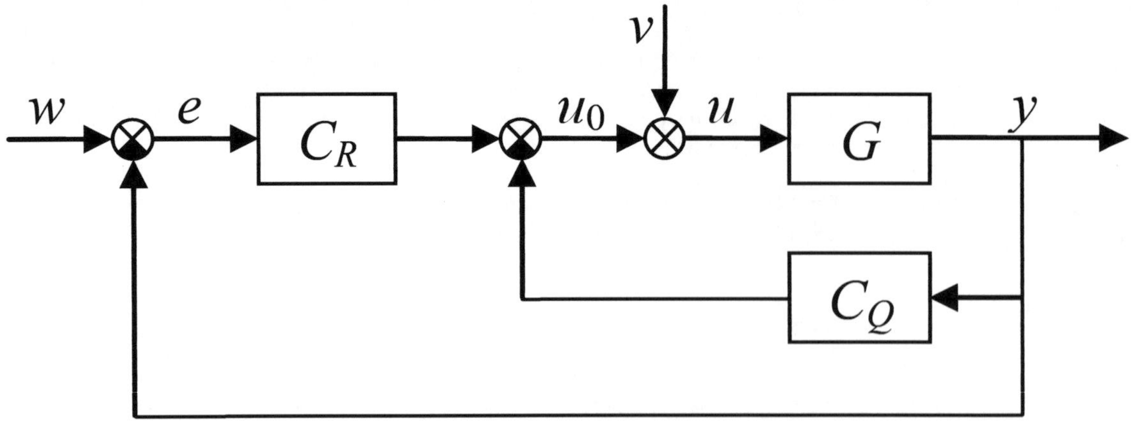

Fig. 1. Control loop with two feedback controllers

In [7], [8], the controllers:

$$C_R(s) = \frac{47.2676s^2 + 0.416887s + 6.16545\times 10^{-4}}{s^2}$$
$$C_Q(s) = 24.6114 \qquad (4)$$

were designed via the algebraic method for the system (1) with nominal parameters (2).

4. Robust Stability Analysis

Remind that the aim is to analyze the robust stability of the control loop that contains feedback controllers (4) and the heating-cooling system (1) with the uncertain parameters (3). The family of the closed-loop characteristic retarded quasi-polynomials is:

$$p_{CL}(s,b_0,b_{0D},\Theta_0,\Theta_b,a_2,a_1,a_0,a_{0D},\Theta_a) = s^3 + a_2s^2 + a_1s + a_0 + a_{0D}e^{-\Theta_a s} + \cdots$$
$$+b_0e^{-\Theta_b s}\left(47.2676s^2 + 0.416887s + 24.612016545\right) + b_{0D}e^{-(\Theta_0+\Theta_b)s}\left(47.2676s^2 + 0.416887s + 24.612016545\right) \qquad (5)$$

where all 9 uncertain parameters can vary within the intervals from (3).

The robust stability of this family with quite complicated uncertainty structure will be investigated by means of the value set concept and the zero exclusion condition [1]. Its application to various types of thermal systems with uncertain parameters can be found in [6].

The value sets are obtained numerically through sampling (gridding) the uncertain parameters and subsequent direct calculation of related partial points of the value sets for a suitable pre-selected set of non-negative frequencies. Then, the robust stability is tested graphically – the family is robustly stable if and only if the zero point is excluded from the value sets and all required preconditions are fulfilled, especially the existence of at least one stable member of the family. For details see e.g. [1], [5], [6].

The value sets of the family (5) with the uncertain parameters (3) are plotted in Fig. 2. The value sets are obtained for the frequencies from 0 to 0.5 with the step 0.01. For the calculations, 7 out of 9 uncertain parameters are sampled by 3 equidistant values and the remaining 2 parameters are sampled by 15 values which mean that the value set consists of $3^7 \cdot 15^2 = 492075$ points at each frequency.

A closer look near the complex plane origin is depicted in Fig. 3. The family (5) contains a stable member and the zero point is excluded from the value sets which means (with necessary and sufficient condition) that the family is robustly stable, i.e. two feedback controllers (4) stabilize the anisochronic model (1) of a heating-cooling plant for all possible assumed variations of the uncertain parameters (3).

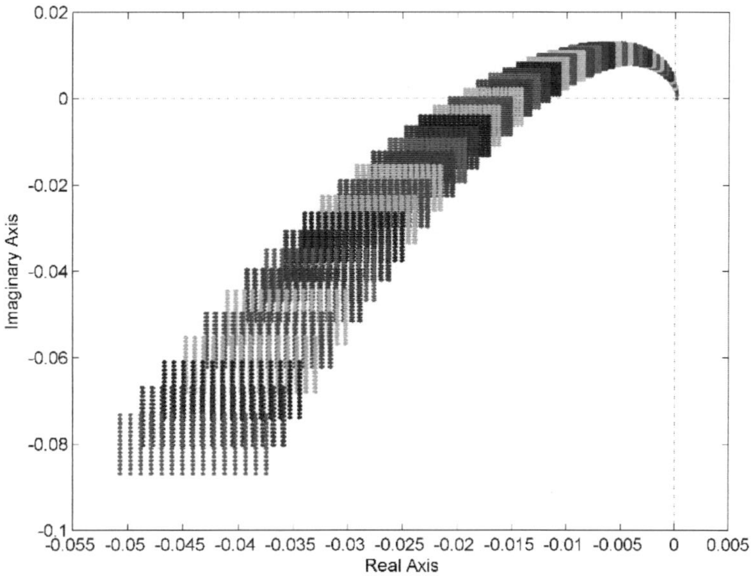

Fig. 2. The value sets of the family (5) with parameters (3) – the full view

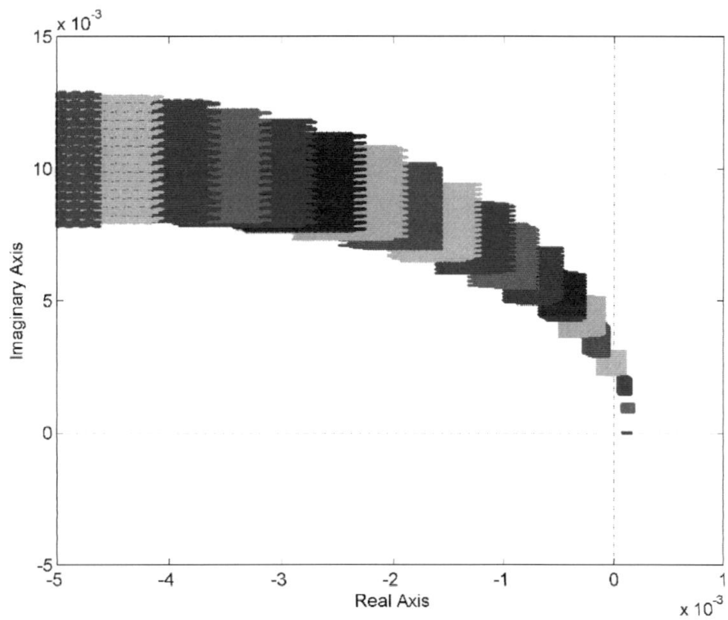

Fig. 3. The value sets of the family (5) with parameters (3) – a zoomed view

5. Conclusion

This brief contribution was focused on the numerical-graphical technique for testing the robust stability of the control system with two feedback controllers and the specific thermal plant. The controlled process is represented by a heating-cooling system with a heat exchanger that is described by the anisochronic model with internal delays and parametric uncertainty. The robust stability is analyzed by means of the numerical calculations of the value sets in combination with the graphical evaluation via the zero exclusion condition.

6. Acknowledgments

This work was supported by the Ministry of Education, Youth and Sports of the Czech Republic within the National Sustainability Programme project No. LO1303 (MSMT-7778/2014) and also by the European Regional Development Fund under the project CEBIA-Tech No. CZ.1.05/2.1.00/03.0089.

7. References

[1] Barmish, B. R. (1994). New Tools for Robustness of Linear Systems, Macmillan, New York, USA

[2] Bhattacharyya, S. P.; Chapellat, H. & Keel, L. H. (1995). Robust Control: The Parametric Approach, Prentice Hall, Englewood Cliffs, New Jersey, USA

[3] Bhattacharyya, S. P. (2017). Robust control under parametric uncertainty: An overview and recent results, Annual Reviews in Control, In Press – Corrected Proof, https://doi.org/10.1016/j.arcontrol.2017.05.001

[4] Stevanović, S.; Kasać, J. & Stepanić, J. (2012). Robust Tracking Control of a Quadrotor Helicopter without Velocity Measurement, Annals of DAAAM for 2012 & Proceedings of the 23rd International DAAAM Symposium, Vienna, Austria, ISBN 978-3-901509-91-9, ISSN 2304-1382, pp. 595-600

[5] Matušů, R. (2016). Investigation of Robust Stability for Discrete-Time Polynomials with Nonlinear Uncertainty Structure, Proceedings of 27th DAAAM International Symposium, Vienna, Austria, ISBN 978-3-902734-08-2, ISSN 1726-9679, pp. 142-146

[6] Matušů, R. & Pekař, L. (2017). Robust Stability of Thermal Control Systems with Uncertain Parameters: The Graphical Analysis Examples. Applied Thermal Engineering, Vol. 125, pp. 1157-1163

[7] Pekař, L. & Prokop, R. (2017). Algebraic robust control of a closed circuit heating-cooling system with a heat exchanger and internal loop delays. Applied Thermal Engineering, Vol. 113, pp. 1464-1474

[8] Pekař, L. (2013). Control of Time Delay Systems – An Algebraic Approach, English Doctoral Thesis, Faculty of Applied Informatics, Tomas Bata University in Zlín, Czech Republic

[9] Zítek, P. & Hlava, J. (2001). Anisochronic internal model control of time-delay systems. Control Engineering Practice, Vol. 9, No. 5, pp. 501-516

DOI: 10.2507/28th.daaam.proceedings.027

STUDY INTO MOVEMENT OF WIDE SPAN TRACTORS (VEHICLES) USED IN CONTROLLED TRAFFIC FARMING

Volodymyr Bulgakov, Valerii Adamchuk, Volodymyr Kuvachov,
Margus Arak & Jüri Olt

This Publication has to be referred as: Bulgakov, V[olodymyr]; Adamchuk, V[alerii]; Kuvachov, V[olodymyr]; Arak, M[argus] & Olt, J[ueri] (2017). Study Into Movement of Wide Span Tractors (Vehicles) Used in Controlled Traffic Farming, Proceedings of the 28th DAAAM International Symposium, pp.0199-0208, B. Katalinic (Ed.), Published by DAAAM International, ISBN 978-3-902734-11-2, ISSN 1726-9679, Vienna, Austria
DOI: 10.2507/28th.daaam.proceedings.027

Abstract

Wide span tractors/vehicles have good prospects as the mechanisation means for controlled traffic farming. The efficiency of the practical application of wide span tractors (vehicles) depends on the availability of the substantiated scientific basis or the theoretical basis for the research into their key properties, including among other factors the smoothness of movement. The aim of the research is to improve the smoothness of movement of wide span tractors (vehicles) by substantiating their configuration, design-and-process and other parameters as well as the properties of the longitudinal profile of the agricultural background irregularities in the ruts of the laid-down process track. The smoothness of movement of the wide span tractor (vehicle) as a dynamic system travelling in the ruts of the permanent process track substantially depends on the structure of the latter's longitudinal profile, the stiffness of tyres on the supporting wheels, while the dynamics of the vertical oscillations of the process section depends on its operating weight.

Keywords: wide span tractor (vehicle); smoothness; movement; dynamic system

1. Introduction

One of the ways to improve the cultural and technological level of farming is to arrange the strictly controlled (routed) traffic of all mechanical means only on specially preformed paths – permanent process tracks [1-3] and [16-19]. Their implementation facilitates the resolution of the contradiction that exists in the "running gear – soil" system. The physical essence of it is in the fact that process and transport machines need for their travelling dry and firm ground (engineering zone of field), while the efficient growth of plants takes place on broken and humid soil (agronomical zone of field) [1, 4], [8-10] and [20].

It has been known that the reduction of the area of transport and process lanes and extension of the working width of agricultural units improves the land use indices in controlled traffic farming [5]. A substantial rise of efficiency in this context is achieved through the application of wide span tractors or wide span vehicles [6, 7] and [11, 12] designed specifically for controlled traffic farming (Fig. 1).

Fig. 1. Various designs of wide span tractors (vehicles) for controlled traffic farming: Dowler wide span tractor (a) and vehicle (b); ETC CTBE wide span tractor (c); ASA-Lift WS 9600 WS wide span tractor (d); wide span vehicles Ukraine (e and f).

We assume that the subject matter of the theory of wide span tractor (vehicle) is the methods of determining its properties, including in particular the smoothness of movement.

Obviously, the dynamics of the movement of a wide span tractor (vehicle) in the vertical plane is defined by the perturbing input actions, which include the irregularities of the longitudinal profile of the permanent process track and the non-uniformity of the tractive resistance of the soil. Certainly, the internal structure characteristics of the longitudinal profile of the laid-down process track has an effect on the smoothness of movement of the wide span tractor (vehicle) and the fluctuation of its tractive force with all the negative implications.

As is known, the quality of some dynamic system's response to input variables depends on its characteristic properties. As regards the wide span tractor (vehicle), they include its configuration as well as its design and other parameters. Therefore, the correct selection of the latter with respect to the required smoothness of movement will ensure the optimal transformation by the machine of the perturbing actions that have an effect on it.

All the recognised studies on the discussed issue, for example [13-15] and [21] were aimed at the investigation of the dynamics of motion of the tractor-implement units built on the basis of conventional tractors, while the problem under consideration remained outside their scope of research. The accumulated scientific and practical experience of using conventional tractor-implement units in controlled traffic farming has facilitated the substantiation of certain requirements to the parameters of the permanent process track. Nevertheless, the said requirements do not take into account the atypical design layout of the wide span tractor (vehicle), its specific way of unitization and functioning conditions, therefore, they need adjustment. At the same time, the already established methods of selecting the structural layouts, parameters and modes of operation for tractor-implement units cannot be employed in the investigation of the dynamics of motion of wide span tractors in the vertical plane. Thereby, with regard to the efficient utilization of wide span tractors (vehicles) unresolved issues arise, which are concerned with research into the conditions stipulating their design and other parameters.

The objective of the research is to improve the smoothness of movement of wide span tractor (vehicle) by substantiating their configuration, design-and-process and other parameters as well as the properties of the longitudinal profile of the agricultural background irregularities in the ruts of the laid-down process track.

2. Materials and Methods

The basis of the research is made up of the fundamental principles of the theoretical mechanics, the statistical dynamics, the theory of the automatic control of linear dynamic systems responding to statistically random perturbing input actions, the theory of tractor. The assessment of the effect that the design and the parameters of a wide span tractor (vehicle) have on the dynamics of its motion in the vertical plane is based on the analysis of the amplitude and phase frequency response characteristics, programme design and calculations with the use of the PC.

It is expedient to consider the functional model of the wide span tractor/vehicle as a dynamic system in the form of its reaction to the input perturbation, the characteristic of the response to which unequivocally determines the smoothness of movement. In this case the perturbations are represented by the shocks and impacts caused by the profile irregularities of the permanent process track and the variation of the tractive resistance created by the agricultural tools on the process section of the wide span tractor (vehicle). The behaviour of a particular dynamic system as regards its response to the input variables depends on its specific features. As applied to the wide span tractor, such specific features include its configuration, design, process, kinematic and other parameters. Hence, the adequate selection of the latter with respect to the desired stability of motion will ensure for the wide span tractor the optimal transformation of the perturbing actions it is under the effect of.

It has been known that the transformation properties of a dynamic system can be expressed in terms of transfer functions and frequency-response characteristics. Those characteristics are exactly what provides the most complete and physical notion of the reactions of the agricultural unit to various perturbations as well as the transient and stable processes of its operation.

For the theoretical analysis of the transfer functions and frequency-response characteristics, it is necessary to generate the system of corresponding differential equations relating the original variables to the input perturbations, i.e. the mathematical model of the process. At this stage of investigation, it is appropriate to consider it in the form of a linear equation system. Such idealisation will be rather efficient in this case, because the dynamics of motion of the wide span tractor has not yet been sufficiently well studied. While the knowledge of it will provide the capability to interpret physically the obtained results and accumulate the experience of designing.

In the process of solving the problems of optimisation of parameters of the linear stationary dynamic system, amplitude (AFR) and phase (PFR) frequency response characteristics will be used in the capacity of the operators. For this purpose, after generating the corresponding differential equations and formulating transfer functions on their basis, it is necessary to calculate and analyse the AFR and PFR for both the assumed input perturbations. The following provisions and assumptions will be adopted / made in order to simplify the generation of the differential equations: angular oscillations of the process part of the wide span tractor (vehicles) are ignored; variation of the tractive resistance of the agricultural implements has no effect on the velocity of the translational movement of the wide span tractor (vehicle), therefore, it is assumed to be constant; the permanent process track profile irregularities are described by a random ergodic stationary function of the covered distance; there is no inclination of the wide span tractor (vehicle) in the transverse-longitudinal plane; resistance forces in the carrying wheel tyres are assumed to be proportional to the velocity of oscillations, while the characteristics of elastic members – linear.

The actual wide span tractor (vehicle) can be represented by an equivalent dynamic model (Fig. 2) with three degrees of freedom: vertical displacement z_T of the centre of mass (S_T), angular oscillations of the body φ_T and vertical displacement z_3 of the centre of mass of the process section (S_p). The displacement z_T and angular oscillations φ_T are related to the vertical displacements of the front and rear parts of the wide span tractor's (vehicles) body (z_1 and z_2) (Fig. 2). Therefore, the vertical displacements z_1, z_2 and z_3 of the front and rear parts of the wide span tractor (vehicle) body and the body of the process section respectively will be considered as the generalised coordinates.

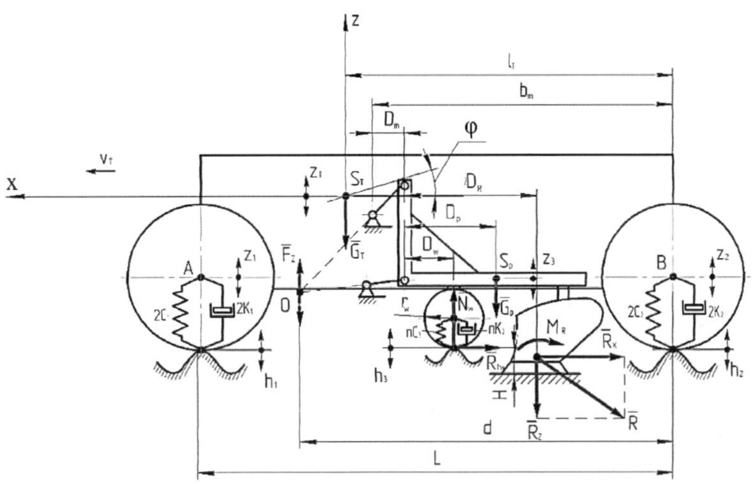

Fig. 2. Equivalent dynamic model of the wide span tractor (vehicle)

The mathematical model of the plane parallel motion of a wide span tractor (vehicle) in the vertical-longitudinal plane developed on the basis of the Lagrange equations of the second kind will appear as follows:

$$\left.\begin{aligned}
&\grave{A}_{11}\cdot\ddot{z}_1+\grave{A}_{12}\cdot\dot{z}_1+\grave{A}_{13}\cdot z_1+\grave{A}_{14}\cdot\ddot{z}_2=f_{11}\cdot\dot{h}_1+f_{12}\cdot h_1+f_{13}\cdot R_{\tilde{o}}+f_{14}\cdot R_z+f_{15}\cdot M_R+f_{16},\\
&\grave{A}_{21}\cdot\ddot{z}_2+\grave{A}_{22}\cdot\dot{z}_2+\grave{A}_{23}\cdot z_2+\grave{A}_{24}\cdot\ddot{z}_1=f_{21}\cdot\dot{h}_2+f_{22}\cdot h_2+f_{23}\cdot R_{\tilde{o}}+f_{24}\cdot R_z+f_{25}\cdot M_R+f_{26},\\
&\grave{A}_{31}\cdot\ddot{z}_3+\grave{A}_{32}\cdot\dot{z}_3+\grave{A}_{33}\cdot z_3=f_{31}\cdot\dot{h}_3+f_{32}\cdot h_3+f_{33}\cdot R_{\tilde{o}}+f_{34}\cdot R_z+f_{35}\cdot M_R+f_{36},
\end{aligned}\right\}\qquad(1)$$

where

$$\grave{A}_{11}=\left(\grave{I}_T\cdot l_T^2+J_T\right)/L^2;\qquad \grave{A}_{21}=\left(\grave{I}_T\cdot(L-l_T)^2+J_T\right)/L^2;$$

$$\grave{A}_{12}=2\hat{E}_1;\quad \grave{A}_{13}=2\tilde{N}_1;\qquad \grave{A}_{22}=2\hat{E}_2;\quad \grave{A}_{23}=2\tilde{N}_2;$$

$$\grave{A}_{14}=2\left(\hat{I}_T\cdot l_T\cdot(L-l_T)-J_T\right)/L^2;\qquad \grave{A}_{24}=2\left(\hat{I}_T\cdot l_T\cdot(L-l_T)-J_T\right)/L^2;$$

$$\grave{A}_{31}=\grave{I}_{\tilde{n}};\quad \grave{A}_{32}=n\hat{E}_3;\quad \grave{A}_{33}=n\tilde{N}_3;\qquad f_{21}=2\hat{E}_2;\quad f_{22}=2\tilde{N}_2;$$

$$f_{11}=2\hat{E}_1;\quad f_{12}=2\tilde{N}_1;\qquad f_{31}=n\hat{E}_3;\quad f_{32}=n\tilde{N}_3;$$

$$f_{13}=\frac{0{,}5\cdot H\cdot d}{L(d-b_m+D_m+D_w)};\qquad f_{23}=\frac{0{,}5\cdot H\cdot(1-d)}{L(d-b_m+D_m+D_w)};$$

$$f_{14}=\frac{d\cdot(D_R-D_w)}{L(d-b_m+D_m+D_w)};\qquad f_{24}=-\frac{(1-d)\cdot(D_R-D_w)}{L(d-b_m+D_m+D_w)};$$

$$f_{15}=\frac{d}{L(d-b_m+D_m+D_w)};\qquad f_{25}=\frac{(1-d)}{L(d-b_m+D_m+D_w)};$$

$$f_{16}=\frac{d\cdot G_p(D_p-D_w+f_w\cdot r_w)}{L(d-b_m+D_m+D_w)};\qquad f_{26}=\frac{(1-d)\cdot G_p\cdot(D_p-D_w+f_w\cdot r_w)}{L(d-b_m+D_m+D_w)};$$

$$f_{33}=\frac{0{,}5\cdot H}{(d-b_m+D_m+D_w)};\qquad f_{35}=-\frac{1}{(d-b_m+D_m+D_w)};$$

$$f_{34}=\frac{(D_R-D_w)}{(d-b_m+D_m+D_w)};\qquad f_{36}=\frac{G_p(D_p-D_w+f_w\cdot r_w)}{(d-b_m+D_m+D_w)}.$$

In the system of equations (1) the terms M_T, J_T designate the mass (kg) of the wide span tractor (vehicle) and its moment of inertia (kN·m·s^2) relative to the axis that passes through the point S_T (Fig. 2); K_1, K_2, K_3, C_1, C_2, C_3 – reduced deformation resistance factors of the dissipative (kN·s·m^{-1}) and elastic (kN·m^{-1}) members of the springing system of the wide span tractor (vehicle) and its process section;

n – number of carrying wheels under the process section of the wide span tractor (vehicle); M_p, G_p – mass (kg) and weight force (kN) of the agricultural implements;

L, l_T – wheel base and the longitudinal coordinate of the centre of mass of the wide span tractor (vehicle) (m);

H – depth of soil cultivation (m); f_w – coefficient of rolling resistance of a carrying wheel of the process section;

r_w – rolling radius of a carrying wheel; d, b_m, D_m, D_w, D_R, D_p – design parameters, the nature of which is explained by Fig. 2.

After the Laplace transformation, the system of equations (1) will assume the following form in the operator notation:

$$\left.\begin{aligned}
&K_{11}\cdot z_1(s)+K_{12}\cdot z_2(s)+K_{13}\cdot z_3(s)=F_{11}\cdot h_1(s)+F_{12}\cdot h_2(s)+\\
&+F_{13}\cdot h_3(s)+F_{14}\cdot R_X(s)+F_{15}\cdot R_Z(s)+F_{16}\cdot M_R(s)+F_{17},\\
&K_{21}\cdot z_1(s)+K_{22}\cdot z_2(s)+K_{23}\cdot z_3(s)=F_{21}\cdot h_1(s)+F_{22}\cdot h_2(s)+\\
&+F_{23}\cdot h_3(s)+F_{24}\cdot R_X(s)+F_{25}\cdot R_Z(s)+F_{26}\cdot M_R(s)+F_{27},\\
&K_{31}\cdot z_1(s)+K_{32}\cdot z_2(s)+K_{33}\cdot z_3(s)=F_{31}\cdot h_1(s)+F_{32}\cdot h_2(s)+\\
&+F_{33}\cdot h_3(s)+F_{34}\cdot R_X(s)+F_{35}\cdot R_Z(s)+F_{36}\cdot M_R(s)+F_{37},
\end{aligned}\right\}\qquad(2)$$

where:

$$\hat{E}_{11} = \grave{A}_{11} \cdot s^2 + A_{12} \cdot s + A_{13};$$
$$\hat{E}_{12} = \grave{A}_{14} \cdot s^2; \hat{E}_{21} = \grave{A}_{24} \cdot s^2;$$
$$\hat{E}_{22} = \grave{A}_{21} \cdot s^2 + A_{22} \cdot s + A_{23};$$
$$\hat{E}_{13} = \hat{E}_{23} = \hat{E}_{31} = \hat{E}_{32} = 0;$$
$$\hat{E}_{33} = \grave{A}_{31} \cdot s^2 + A_{32} \cdot s + A_{33};$$

$s = d \cdot (dt)^{-1}$ – operator of differentiation;

$$F_{11} = f_{11} \cdot s + f_{12};$$
$$F_{12} = F_{13} = 0;$$
$$F_{14} = f_{13}; F_{15} = f_{14};$$
$$F_{16} = f_{15}; F_{17} = f_{16};$$
$$F_{21} = F_{23} = 0;$$
$$F_{22} = f_{21} \cdot s + f_{22};$$
$$F_{24} = f_{23};$$
$$F_{25} = f_{24}; F_{26} = f_{25};$$
$$F_{31} = F_{32} = 0;$$
$$F_{33} = f_{31} \cdot s + f_{32};$$
$$F_{34} = f_{33}; F_{35} = f_{34};$$
$$F_{36} = f_{35}; F_{37} = f_{36}.$$

The amplitude variation in the longitudinal profile of the track under the tractor's front (h_1) and rear (h_2) wheels in the area of the permanent process track and the carrying wheels of the process section (h_3) as well as the variation of the agricultural implement's tractive resistance (R_x and R_z) and the resultant moment of resistance (M_R) (Fig. 2) are the main perturbations causing the vertical displacements of the wide span tractor/vehicle in the vertical-longitudinal plane.

The mentioned perturbations are the input parameters in the system of equations (2). The output parameters of the system include the amplitudes of displacement of the body of the wide span tractor/vehicle above the front (z_1) and rear (z_2) wheels and the oscillations of its process section body (z_3).

Considering the response of the wide span tractor/vehicle to the perturbing actions, the transfer functions chosen for the analysis will be expressed as follows:

$W_1(s) = D_{11}/D_0$ – transfer function of the longitudinal profile of the track under the rear wheels of the wide span tractor (vehicle) relative to the oscillations of the rear part of its body;

$W_2(s) = D_{12}/D_0$ – transfer function of the longitudinal profile of the track under the front wheels of the wide span tractor (vehicle) relative to the oscillations of the front part of its body;

$W_3(s) = D_{13}/D_0$ – transfer function of the longitudinal profile of the track under the carrying wheels of the wide span tractor (vehicle) process section relative to the oscillations of its body;

$W_4(s) = D_{14}/D_0$ – transfer function of the horizontal resistance reaction of the agricultural implement relative to the oscillations of the rear part of the body of the wide span tractor (vehicle);

$W_5(s) = D_{15}/D_0$ – transfer function of the vertical resistance reaction of the agricultural implement relative to the oscillations of the rear part of the body of the wide span tractor (vehicle),

where D_0, D_{11},…D_{15} – determinants defined as follows:

$$D_0 = a_6 \cdot s^6 + a_5 \cdot s^5 + a_3 \cdot s^3 + a_2 \cdot s^2 + a_1 \cdot s^1 + a_0;$$
$$D_{11} = (\grave{A}_{11} \cdot s^2 + \grave{A}_{12} \cdot s^1 + \grave{A}_{13}) \cdot (f_{21} \cdot s^1 + f_{22}) \cdot (\grave{A}_{11} \cdot s^2 + \grave{A}_{12} \cdot s^1 + \grave{A}_{13});$$
$$D_{12} = (\grave{A}_{31} \cdot s^2 + \grave{A}_{32} \cdot s^1 + \grave{A}_{33}) \cdot (f_{11} \cdot s^1 + f_{12}) \cdot (\grave{A}_{21} \cdot s^2 + \grave{A}_{22} \cdot s^1 + \grave{A}_{23});$$
$$D_{13} = (f_{31} \cdot s^1 + f_{32}) \cdot (\grave{A}_{14} \cdot s^2 \cdot \grave{A}_{24} \cdot s^2 - (\grave{A}_{11} \cdot s^2 + \grave{A}_{12} \cdot s^1 + \grave{A}_{13}) \cdot (\grave{A}_{21} \cdot s^2 + \grave{A}_{22} \cdot s^1 + \grave{A}_{23});$$
$$D_{14} = (\grave{A}_{31} \cdot s^2 + \grave{A}_{32} \cdot s^1 + \grave{A}_{33}) \cdot ((\grave{A}_{11} \cdot s^2 + \grave{A}_{12} \cdot s^1 + \grave{A}_{13}) \cdot f_{23} - A_{24} \cdot s^2 \cdot f_{13});$$
$$D_{15} = (\grave{A}_{31} \cdot s^2 + \grave{A}_{32} \cdot s^1 + \grave{A}_{33}) \cdot ((\grave{A}_{11} \cdot s^2 + \grave{A}_{12} \cdot s^1 + \grave{A}_{13}) \cdot f_{24} - A_{24} \cdot s^2 \cdot f_{14});$$

herein:

$$a_6 = (A_{11} \cdot A_{21} - A_{14} \cdot A_{24}) \cdot A_{31};$$
$$\grave{a}_5 = (\grave{A}_{12} \cdot \grave{A}_{21} + A_{11} \cdot A_{22}) \cdot \grave{A}_{31} + (\grave{A}_{11} \cdot \grave{A}_{21} - A_{14} \cdot A_{24}) \cdot \grave{A}_{32};$$
$$\grave{a}_4 = (\grave{A}_{13} \cdot \grave{A}_{21} + A_{12} \cdot A_{22} + \grave{A}_{11} \cdot \grave{A}_{23}) \cdot \grave{A}_{31} + (\grave{A}_{12} \cdot \grave{A}_{21} + A_{11} \cdot A_{22}) \cdot \grave{A}_{32} + (\grave{A}_{11} \cdot \grave{A}_{21} - A_{14} \cdot A_{24}) \cdot \grave{A}_{33};$$
$$\grave{a}_3 = (\grave{A}_{13} \cdot \grave{A}_{22} + A_{12} \cdot A_{23}) \cdot \grave{A}_{31} + (\grave{A}_{13} \cdot \grave{A}_{21} + A_{12} \cdot A_{22} + \grave{A}_{11} \cdot \grave{A}_{23}) \cdot \grave{A}_{32} + (\grave{A}_{12} \cdot \grave{A}_{21} + A_{11} \cdot A_{22}) \cdot \grave{A}_{33};$$
$$\grave{a}_2 = \grave{A}_{13} \cdot \grave{A}_{23} \cdot \grave{A}_{31} + (\grave{A}_{13} \cdot \grave{A}_{22} + A_{12} \cdot A_{23}) \cdot \grave{A}_{32} + (\grave{A}_{13} \cdot \grave{A}_{21} + A_{12} \cdot A_{22} + \grave{A}_{11} \cdot \grave{A}_{23}) \cdot \grave{A}_{33};$$
$$\grave{a}_1 = \grave{A}_{13} \cdot \grave{A}_{23} \cdot \grave{A}_{32} + (\grave{A}_{13} \cdot \grave{A}_{22} + A_{12} \cdot A_{23}) \cdot \grave{A}_{33};$$
$$\grave{a}_0 = \grave{A}_{13} \cdot \grave{A}_{23} \cdot \grave{A}_{33};$$

After substituting the above determinants into the transfer functions W_1-W_5, the general form of the latter will appear as follows:

$$W_i(s) = \frac{b_5 \cdot s^5 + b_4 \cdot s^4 + b_3 \cdot s^3 + b_2 \cdot s^2 + b_1 \cdot s^1 + b_0}{\grave{a}_6 \cdot s^6 + \grave{a}_5 \cdot s^5 + \grave{a}_4 \cdot s^4 + \grave{a}_3 \cdot s^3 + \grave{a}_2 \cdot s^2 + \grave{a}_1 \cdot s^1 + a_0},$$

where b_0…b_6 – coefficients, the nature of which is stipulated by the functional relationship between the design and process parameters of the investigated wide span tractor (vehicle) and the perturbing action characteristics.

It can be concluded, after the analysis of the mathematical model (1), that the dynamics of the vertical oscillations of the wide span tractor/vehicle depends on a number of its design parameters. The parameters of the carrying wheel pneumatic tyre, in particular its stiffness coefficient, are the most important of them. In the process of mathematical modelling of the vertical oscillations of the wide span tractor (vehicle) of our design, which was taken as the physical subject of research, three practicable options were given consideration with regard to the standard size of the tyres on its wheels: 1) 11.2R20; 2) 11.2R32; 3) 9.5R42.

The stiffness of the tyre on the i-th carrying wheel of the wide span tractor (vehicles) can be found with the use of the established formula of Heideckel:

$$C_i = \pi \cdot \rho_w \cdot \sqrt{D \cdot B}\,; \qquad (3)$$

where D – static tyre diameter (m); B – tyre section width (m); ρ_w – tyre air pressure (Pa).

3. Results and Discussion

As a result of the mathematical modelling, it has been established that the difference between the design AFRs and PFRs of the above-mentioned tyre options is virtually negligible (Graphs 1-3, Fig. 3), despite the certain difference in the values of their coefficients of stiffness C_2. At the same time, the characteristics of the response of the wide span tractor /vehicle to the variation of irregularities in the longitudinal profile of the track in the area of the permanent process track to a considerable extent depend on the perturbing action frequency. Thus, at frequencies of up to 7.0 s^{-1} the effect of this parameter is virtually unnoticeable. At $\omega > 7.0$ s^{-1}, the rise of the tyre's stiffness causes the lowering of the AFR together with the shifting of the resonance peaks to the side of higher frequencies (see Fig. 3 a). For the tyre size options under consideration, the latter are concentrated within a range of $\omega = 10\text{-}11$ s^{-1}. And that is exactly the most undesirable range of frequencies for the variation of the permanent process track's longitudinal profile irregularities, because in this range the dynamic system (i.e. the wide span tractor (vehicle) amplifies the input perturbation at a gain rate of over 6. The dynamics of the vertical oscillations of the wide span tractor (vehicle) cannot be improved even by raising the stiffness of its tyres to 450 kN·m^{-1} (Graph 5, Fig. 3 a). On the other hand, its possible reduction down to 150 kN·m^{-1} lowers the AFR and also displaces its resonance peaks to the side of low frequencies (Graph 6, Fig. 3 a). At the same time, the dispersion of oscillation of the permanent process track irregularities within a frequency range of $\omega = 13\text{-}15$ s^{-1} is the most desirable for all the three wheel tyre options under consideration, because it brings the characteristics into proximity with the ideal ones.

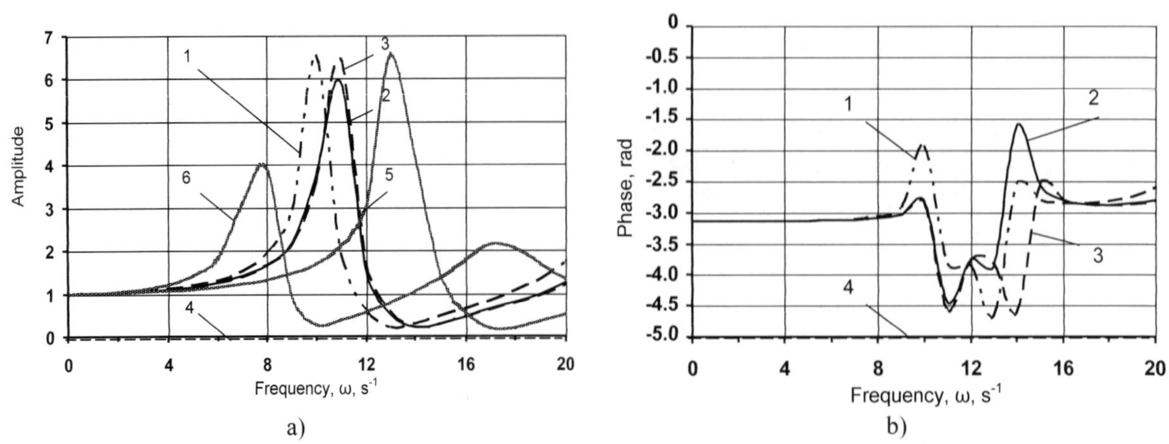

Fig. 3. AFR (a) and PFR (b) characteristics of the response of the body above the rear wheels of the wide span tractor (vehicle) to the track's longitudinal profile variation, for tyres with different stiffness factors: 1) 11.2R20 ($C_2 = 254$ kN·m^{-1}); 2) 11,2R32 ($C_2 = 292$ kN·m^{-1}); 3) 9.5R42 ($C_2 = 296$ kN·m^{-1}); 4¤ ideal characteristics; 5) $C_2 = 450$ kN·m^{-1}; 6) $C_2 = 150$ kN·m^{-1}

As regards the desirable response of the dynamic system to the considered perturbation, the significant increase of the coefficient of stiffness of the wide span tractor's tyres will be effective only in the event that the dispersion of oscillations of the permanent process track irregularities remains within a frequency range of $\omega = 0...8$ s^{-1} and $\omega = 16...20$ s^{-1}. When the perturbation frequency is below 8 s^{-1}, the vertical oscillations of the wide span tractor (vehicle) very little depend on the magnitude of the coefficient of stiffness of its tyres. Also, it is desirable in case of the considered three options of its wheel tyres that the dispersion of the permanent process track's longitudinal profile irregularities remains within a frequency range of $\omega = 13\text{-}15$ s^{-1}, where the dynamic system under consideration reacts almost not at all to the input perturbation. In practice this condition can be met by applying the corresponding technology of shaping the ruts of the permanent process track or by changing the stiffness of the pneumatic tyres through the selection of such an air pressure in them, which will ensure the minimum response of the dynamic system to the input perturbation (Fig. 4).

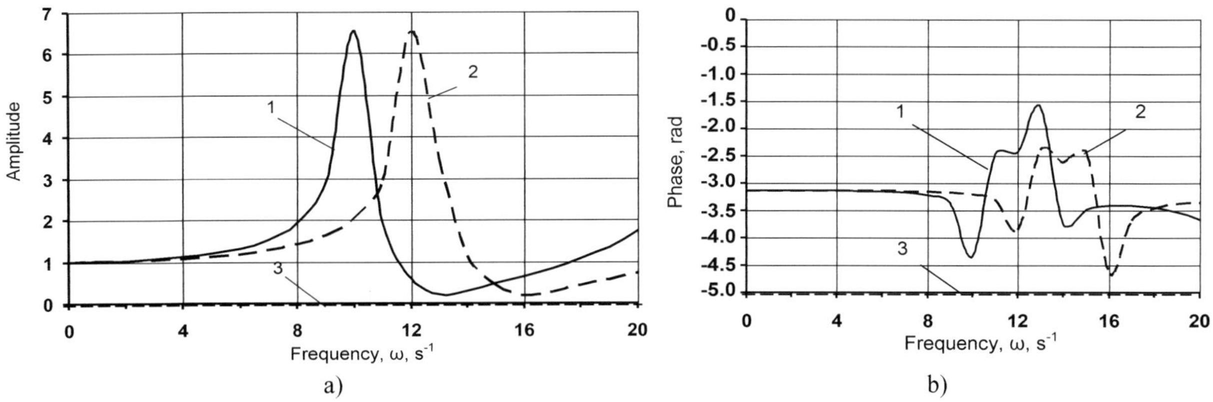

Fig. 4. AFR (a) and PFR (b) characteristics of the response of the wide span tractor (vehicle) to the track's longitudinal profile variation, for 11.2R20 tyres at different air pressures: 1) p_w = 160 kPa; 2) p_w = 240 kPa; 3) ideal characteristics

It follows from the analysis of the experimental data on the shaping of the longitudinal profiles of the ruts of the permanent process track created on uncultivated ground by the wheeled and tracked running gear of tractors that, although the longitudinal profiles of the tracks may have somewhat similar non-uniformity, their inner patterns are different. The results of the analysis of the normalized correlation functions (Fig. 5) show that the longitudinal profile of the track rut created by the wheeled tractor has the lowest-frequency characteristic. Its correlation distance is about 0.57 m, which is 8 times greater, when compared to the longitudinal profile of the track rut created by the caterpillar tractor. Even when compared to the original uncultivated agricultural background, the running gear of the caterpillar tractor generates a higher-frequency longitudinal profile of the rut with a pronounced periodic component. The source of generation of the latter can be found in the track pitch (17 cm), which is virtually equal to the wavelength of the longitudinal track profile irregularities, that is 17.4 cm. The progression of a wheeled tractor on the ruts laid down by a caterpillar tractor generates the oscillations of its body with a higher-frequency characteristic as compared to its travelling on its own track. In the former case the correlation time did not exceed 0.16 s, in the latter one it rose to 1.5 s.

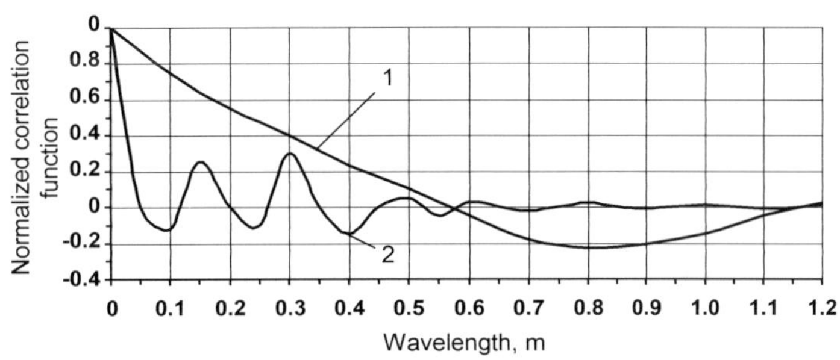

Fig. 5. Normalized correlation functions of longitudinal profiles of permanent process track ruts created on uncultivated agricultural background by the running gear of wheeled (1) and tracked (2) tractors

Basing on the above-stated analysis, it is possible to conclude that the reduction of the stiffness of tyres of the wide span tractor (vehicle) in case of a lower-frequency range of the permanent process track rut's longitudinal profile irregularities will be generally effective. For example, when the tyre stiffness of the wide span tractor (vehicle) under consideration is reduced to 150 kN m^{-1} (Graph 6, Fig. 3 a), the desirable frequency range of the process track's longitudinal profile irregularities will shift to lower values of ω = 9-13 s^{-1}.

The dynamics of the vertical oscillations of the wide span tractor (vehicle) process section is defined, according to (1), first of all by its carrying wheel tyre stiffness and its operating mass M_p. The mathematical modelling has shown that the increase of the operating mass M_p from 300 kg to 500 kg causes the undesirable rise of the AFR characteristic of the response by the process section to the variation of the track's profile together with the shift of the resonance peaks to the lower frequency side (Graphs 2 and 3, Fig. 6 a). If that is complemented with the significant reduction of the carrying wheel tyre stiffness factor to 25 kN·m^{-1}, i.e. by factor of 4 (Graph 5, Fig. 6 a), then the AFR resonance peaks will shift even more to the side of lower frequencies. Meanwhile, the rate of amplification of the input perturbation by the dynamic system at a resonance frequency will be reduced only by factor of 2. Only at frequencies over 13 s^{-1}, the AFR becomes less than a unity and the characteristics move closer to the ideal ones. By contrast, the increase of the stiffness of the carrying wheel tyres to 200 kN·m^{-1} (Graph 1, Fig. 6a) results in the situation, where within a low-frequency range (up to ω = 8..10 s^{-1}) the process section responds to the input perturbation at a virtually insignificant gain rate; only at a frequency of ω = 18 s^{-1} the gain rate reaches a level of 2.

When the variation of the permanent process track's longitudinal profile irregularities stays within a frequency range of ω = 13-15 s^{-1}, which is the most desirable for the motion of the wide span tractor (vehicle), as it was mentioned earlier, the change of the process section carrying wheel tyre stiffness results in the following. At a tyre stiffness factor of C_3 = 200 kN·m^{-1} (Graph 1, Fig. 6 a), the AFR has an input perturbation amplification rate of not more than 1.5. The further increase of the tyre stiffness up to the use of a bare band brings the AFR into proximity to 1, i.e. the dynamic system copies the track's longitudinal profile irregularities. The reduction of the tyre stiffness from 200 kN·m^{-1} to 100 kN·m^{-1} (Graph 2, Fig. 6 a) within the mentioned frequency range raises the AFR to 3, distancing it from the ideal one. The further reduction of the tyre stiffness from 100 kN·m^{-1} to 50 kN·m^{-1} (Graph 4, Fig. 6 a) makes the dynamics of motion of the process section in the vertical plane within the mentioned frequency range even worse. Only after bringing the tyre stiffness factor down to 25 kN·m^{-1}, the characteristics in a frequency range of ω = 13-15 s^{-1} approximate the ideal ones (Graph 5, Fig. 6 a).

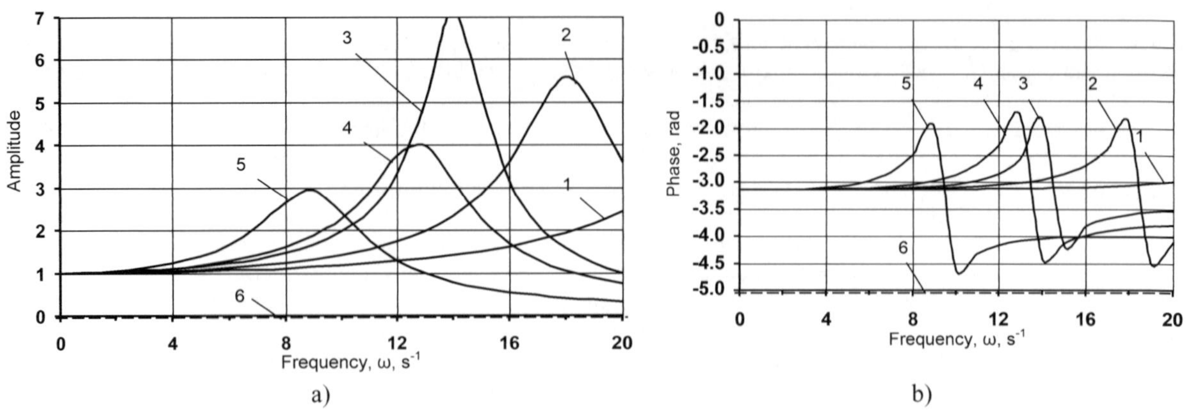

Fig. 6. AFR (a) and PFR (b) characteristics of the response of the wide span tractor (vehicle) process section to the track's longitudinal profile variation, at different values of mass (M_p) and carrying wheel tyre stiffness factor (C_3): 1) M_p = 300 kg; C_3 = 200 kN·m^{-1}; 2) M_p = 300 kg; C_3 = 100 kN·m^{-1}; 3) M_p = 500 kg; C_3 = 100 kN·m^{-1}; 4) M_p = 300 kg; C_3 = 50 kN·m^{-1}; 5) M_p = 300 kg; C_3 = 25 kN·m^{-1}; 6) ideal characteristics

Proceeding from the above-stated analysis, it can be concluded that increasing the operating mass of the process section is not advisable. When the main dispersion spectrum of the permanent process track irregularities is concentrated in a higher-frequency range of ω = 13-15 s^{-1}, the coefficient of stiffness of its tyres must not exceed 25 kN·m^{-1}. Alternatively, the use of process section carrying wheels with a tyre stiffness factor of about 200 kN·m^{-1} and higher is desirable with respect to the section's dynamics of motion in the vertical plane.

The characteristic of the response to the variation of the tractive resistance of the agricultural implement on the process section of the wide span tractor (vehicle) (Fig. 7) to a considerable extent depends on the position of the hitch mechanism relative to the axis of the wide span tractor (vehicle) rear wheels (design parameter b_m, Fig. 2). It follows from the analysis of the obtained AFR that the worst characteristics are observed, when the hitch mechanism of the wide span tractor (vehicle) is situated in the space between its wheels (Graph 3, Fig. 7 a). In that case, the perturbation gain rate at a resonance frequency of ω = 11 s^{-1} reaches almost 20 for the tractive resistance component R_x, and 30 for the vertical component R_z. The most acceptable, with respect to the dynamics of the vertical oscillations of the wide span tractor (vehicle), will be the positioning of the hitch mechanism of the wide span tractor (vehicle) behind the rear wheels of the wide span tractor (vehicle), since that arrangement brings the actual AFRs into proximity to the ideal ones (Graph 1, Fig. 7 a). In that case, considering an oscillation frequency range of 0...14.0 s^{-1}, which is characteristic of the majority of agricultural implements, each 1 kN is associated with 5 to 35 mm of the amplitude of the vertical oscillations of the wide span tractor (vehicle) (Graph 1, Fig. 7 a).

As regards the phase shift of the dynamic system's response to the perturbations in the tractive resistance of the agricultural implement, the PFR characteristic is virtually independent of the design parameter $b_н$ (Fig. 7 b). Moreover, the observed tendency of the actual PFRs to the ideal ones at a resonance frequency of ω = 11 s^{-1} is desirable.

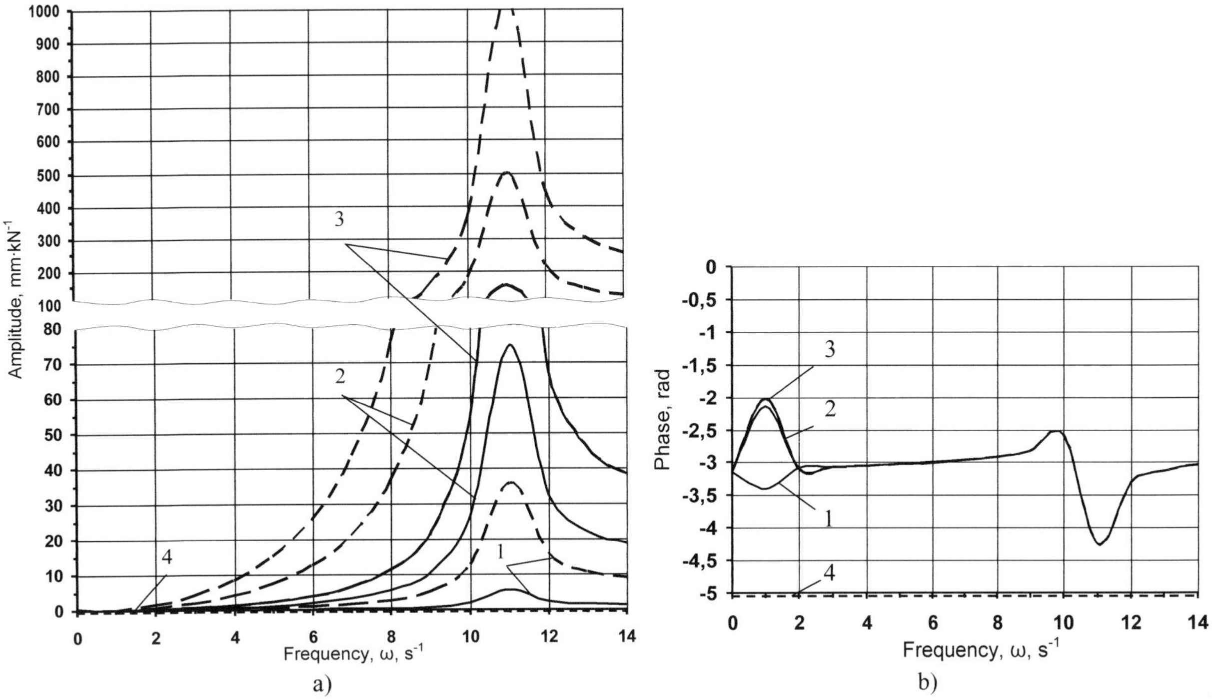

Fig. 7. FR (a) and PFR (b) characteristics of response of wide span tractor (vehicle) to variation of horizontal R_x (——) and vertical R_z (-----) components of the tractive resistance of agricultural implement, for different positions of hitch mechanism (design parameter b_m): 1) b_m = 1 m; 2) b_m = 0 m; 3) b_m = -1 m; 4) ideal characteristics

4. Conclusions

The smoothness of movement of the wide span tractor (vehicle) as a dynamic system travelling on the ruts of the permanent process track depends to a considerable extent on the characteristics of the track's longitudinal profile irregularities. The required inner pattern of the longitudinal profile of the laid-down process track can be obtained in practice with the use of an adequate technology of its shaping. The quality of the dynamic system's response to the input perturbations, which comprise the permanent process track's longitudinal profile irregularities and the variation of the soil's tractive resistance, depends on the configuration as well as the design and other parameters of the wide span tractor (vehicle). A considerable effect on the smoothness of movement of the latter has the stiffness of its carrying wheel tyres, the value of which can be controlled, within certain limits, by changing the air pressure in the tyres. The improved dynamics of motion of the process section of the wide span tractor (vehicle) is observed, when the stiffness of the tyres on its carrying wheels is increased and its operating mass is reduced.

5. References

[1] Bindi, I.; Blackwell, P., Riethmuller, G.; Davies, S.; Whitlock, A. & Neale, T. (2013). This Controlled Traffic Farming Technical Manual. It updates the Tramline Farming Systems: Technical Manual Bulletin 4607. Published by the Department of Agriculture and food, Western Australia. 78 p.

[2] Blackwell, P.; McKenzie, D; Webb, B.; Lemon, J.; Barber, P.; Fretwell, G.; Bignell, G. & Moffat, N. (2004, a). Compaction of heavy soils by cropping traffic and estimated benefits of tramline farming. Agribusiness Crop Updates, Perth.

[3] Blackwell, P.; Webb, B.; Fretwell, G.; Moffat, N. & Chappel, L. (2004, b). Tramlines for less fuel use, pollution and greener farming! Agribusiness Crop Updates, Perth.

[4] Bochtis, D.D.; Sørensen, C.G.; Busatob, P.; Hameed, I.A.; Rodias, E.; Green, O. & Papadakis, G. (2010). Tramline establishment in controlled traffic farming based on operational machinery cost. Biosystems Engineering, Vol. 107, No. 3, pp. 221–231, ISSN 1537-5110, DOI: 10.1016/j.biosystemseng.2010.08.004.

[5] Bulgakov, V.; Adamchuck, V., Kuvachov, V., Ivanovs, S. (2017). Investigation of technological properties of wide sapn tractors for controlled trafic farming. Proceedings of 16th Int. Sci. Conf. Engineering for Rural Development, 24.-26.05.2017, ISSN 1691-5976, pp. 303-308, Jelgava, DOI: 10.22616/ERDev2017.16.N060.

[6] Chamen, W.C.T. (2000). A New Methodology For Weed Control And Cereal Crop Production Based On Wide Span Vehicles And Precision Guidance: Biotrac / 4th EWRS Workshop on Physical Weed Control, , 20-22 March 2000. pp. 51-55, Elspeet, The Netherlands.

[7] Chamen T. (2015). Controlled traffic farming – from world wide research to adoption in Europe and its future prospects. Acta Technologica Agriculturae Nitra 3, pp. 64-73.

[8] Evans, T.A.; Dawes, T.Z.; Ward, P.R. & Lo, N. (2011). Ants and termites increase crop yield in a dry climate. Nature Communications, 2:262 doi: 10.1038/ncomms1257.

[9] Evans, F.; Hashim, A. & Diggle, A. (2009). Risk of glysophate resistance in wide-row lupin cropping systems. 2009 Agribusiness Crop Updates, Perth.

[10] Onal I. (2012). Controlled Traffic farming and Wide Span Tractors. Journal of Agricultural Machinery Science Vol. 8, No 4, pp. 353-364.

[11] Pedersen, H.H. (2011). Harvest Capacity Model for a Wide Span Onion Bunker Harvester. Automation and System Technology in Plant Production, CIGR section V & NJF section VII conference, pp. 27-36.

[12] Pedersen, H.H.; Sørensen, C.G.; Oudshoorn, F.W. & McPhee, J.E. (2013). User requirements for a Wide Span Tractor for Controlled Traffic Farming. International Commission of Agricultural and Biological Engineers, Section V. CIOSTA XXXV Conference “From Effective to Intelligent Agriculture and Forestry”, 3-5 July 2013. pp. 134-136, Billund, Denmark.

[13] Popp, K., Schiehler, W. (2010). Ground Vehicle Dynamics. Springer. ISBN 978-3-540-68553-1.

[14] Rajamani, R. (2012). Vehicle dynamics and control. Springer. ISBN 978-1-4614-1433-9.

[15] Tullberg J.N. (2000), Wheel traffic effects on tillage draft. Journal of Agricultural Engineering Research, Vol. 75, No. 4, pp. 375-382, ISSN 0021-8634, DOI: 10.1006/jaer.1999.0516.

[16] Tullberg, J.N.; Yule, D.F. & McGarry D. (2007). Controlled traffic farming -from research to adoption in Australia. Soil and Tillage Research, Vol. 97, No. 2, pp. 272–281, ISSN 0167-1987, DOI: 10.1016/j.still.2007.09.007.

[17] Tullberg, J.N. (2013). CTF impacts: Environment = Economic. ISTRO Controlled Traffic Conference 2013, Toowoomba.

[18] Tullberg, J.N.; Ziebarth, P.J. & Yuxia, Li. (2001). Tillage and traffic effects on runoff. Australian Journal of Soil Research, Vol. 39, pp. 249-257.

[19] Tullberg, J.N.; McHugh, A.; Ghareel Khabbaz, B.; Scheer, C. & Grace, P. (2011). Controlled Traffic/permanent bed farming reduces GHG emissions. Proceedings of 5th World Congress of Conservation Agriculture incorporating 3rd Farming Systems Design Conference, Brisbane.

[20] Webb, B., Blackwell, P., Riethmuller, G. & Lemon, J. (2004). Tramline farming systems: technical manual, Bulletin 4607, Department of Agriculture, Western Australia.

[21] Wong, J.Y. (2008). Theory of ground vehicles. John Wiley & Sons, ISBN 978-0-470-17038-0.

DOI: 10.2507/28th.daaam.proceedings.028

DETERMINING THE LOAD ON SUPPORT ROLLERS THE PIPE CONVEYOR BELT

Viacheslav Maksarov, Evgeny Zlotnikov & Jüri Olt

This Publication has to be referred as: Maksarov, V[iacheslav]; Zlotnikov, E[vgeny] & Olt, J[ueri] (2017). Determining the Load on Support Rollers the Pipe Conveyor Belt, Proceedings of the 28th DAAAM International Symposium, pp.0209-0215, B. Katalinic (Ed.), Published by DAAAM International, ISBN 978-3-902734-11-2, ISSN 1726-9679, Vienna, Austria

DOI: 10.2507/28th.daaam.proceedings.028

Abstract

The pipe conveyor belt design principles are considered in these pages. A method of calculation is suggested for the load on the base support rollers resulting from the bending of the initially plane belt into a cylindrical shell for the circular cross-section upon the elasticity theory with a non-uniform modulus, considering elongation and contraction variations in the elasticity modulus of the belt inserts which are positioned above and below the neutral surface at bending.

Keywords: pipe conveyor belt; cylindrical shell; non-uniform modulus belt; support rollers; neutral surface; elasticity modulus

1. Introduction

The application of pipe conveyor belts is a perspective trend in the development of the means of transportation in mining and agricultural enterprises. The work in this sphere started as early as in the 1970s. By that time most countries around the world introduced substantially stricter measures for environmental protection which facilitated the intensive development of closed-type technologies in the transportation of goods [1], [2], [3] and [4].

Such conveyors are used for the transportation of coal, crushed ore, gypsum, cement, pebble, sand, ash, chemical substances, paste-form, fodder and other materials [5].

A block diagram of the pipe conveyor belt is represented in Fig 1, and has the following particularities: the initially-flat belt after passing through the end-face cylinder, 1, is gradually folded and accepts a tubular form with the help of the specially-positioned crank-pin support rollers at Section **A**. In the material feeding area the belt already has a corrugated shape which provides for a larger transversal cross-section and the increasing of the linear weight of the transported granular load. Loading at this section of the path is carried out in the same way as in cases which involve a regular corrugated conveyor belt via the silo device, 2.

Along the main path and up to the unloading area, 5, the belt has a tubular shape with overlapped edges in the upper area which provide for the hermetical transportation of the material, preventing it from spilling out as well as isolating the material and the environment from each other. The circular cross-section of the conveyor belt is formed at the expense of the base support rollers, which are installed with a preset spacing, and having the rollers, 7, executed in the hexagonal form.

The folded into a tube belt which contains the load can vary the trajectory of motion along the path based on the radius curves in horizontal and vertical planes, depending upon the available terrain of the surface. The tilt angle for the pipe conveyor normally amounts to 30°; however, in cases of necessity it can be increased up to the vertical position of 90° due to the creation of the support for the transported material at the loading end and at the initial tubular section.

At the end of the path, in Section **B**, the folded tube is unfolded with the help of the specially positioned support rollers, and it then accepts first the corrugated form and then the flat form running over the driving cylinder, 6. At 5, unloading of the transported material takes place just as it does with normal belt conveyors.

On the lower branch, 4, the conveyor belt is folded into a tube in a similar manner, and the reverse branch is also used for the transportation of the load in some cases.

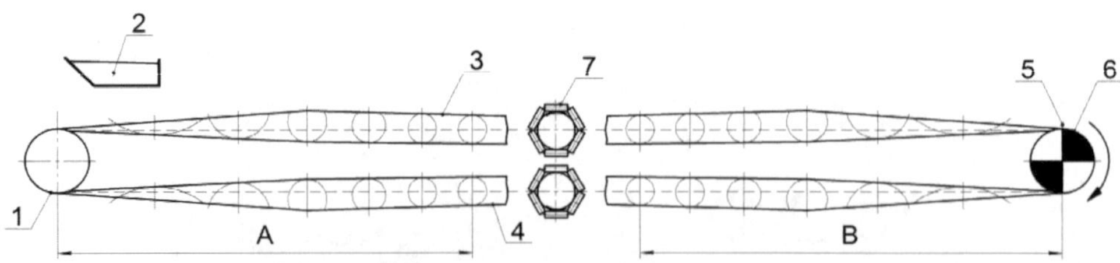

Fig. 1. Block diagram showing the pipe conveyor belt: 1 – end-face cylinder; 2 – material loading silo; 3 – upper branch; 4 – lower branch; 5 – material unloading area; 6 – driving cylinder; 7 – base support rollers; A & B – belt folding and unfolding sections.

The advantages of pipe conveyor belts include the following:

- the option of hermetically-sealed transportation for environmentally-unfriendly loads (loads which may be dusty, polluting, or corroding);
- load protection from natural effects (wind or precipitation);
- the option of establishing curvilinear spatial paths;
- the option of transporting granular loads under high angles of tilt;
- the option of transporting sticky materials due to the fact that there is no contact between the contaminated surface of the belt and the supporting elements and other elements on the conveyor;
- the pipe conveyor belt can be used for the reverse feed of the load.

The tubular shape of the belt ensures less sagging between the support rollers which decreases resistance in the motion of the belt with the load, allows the belt's speed to be increased, and decreases the risk of the transported load being crushed. The main drawbacks and difficulties which may occur in operation of the pipe conveyors can relate to the following:

- the chance that the belt may unfold, or that it may be over-tensioned, or that a corrugated surface may be caused at curvilinear sections of the path [6];
- restrictions related to filling the transversal tubular cross-section with a load for not more than 75%; therefore, at the same capacity as a traditional belt conveyor, the pipe conveyor will require a belt width, which is larger by 50%;
- the existence of the problem related to rotation in the pipe conveyor belt with respect to the longitudinal axis, especially at curvilinear sections means that the installation of a special means of control over the motion of the belt is required;
- increased capital expenditure due to increasing the quantity of idle rollers, the large dimensions of the cylinders, and the high costs related to the conveyor belt;
- the high load on the rollers crimping the tubular belt increases resistance to the belt motion along with higher consumption rates of electricity [7], [8] and [9].

2. Materials and Methods

Normal conveyor belts can be used for pipe conveyors; however, especially developed belt designs demonstrate better performance parameters and will enjoy a longer service life.

The capability of being able to accept the tubular shape is attained with the help of two basic methods:

- the belt is initially manufactured in a tubular shape and is subsequent cut in the longitudinal direction;
- the belt is manufactured with a bending rigidity varying by width.

The first method failed to become widely applied thanks to the complexity involved in manufacturing the tubular belt to a sufficiently large diameter [10].

With the second method, the bending rigidity varying by width in the rubber and fabric belts (Fig. 2) is attained thanks to a stepwise variation of the number of inserts (the number of inserts is lower near the edges, so that the belt possesses a high degree of flexibility) in rubber and steel cord conveyor belts (Fig. 2) – due to a variation in the diameter of cords and the spacing between them (due to high tension levels the cord diameters are larger near the edges, and the spacing between them is increased to provide for a higher level of flexibility). Such specialised belts for use in the pipe conveyors are manufactured by various companies such as, for example, ContiTech AG [11].

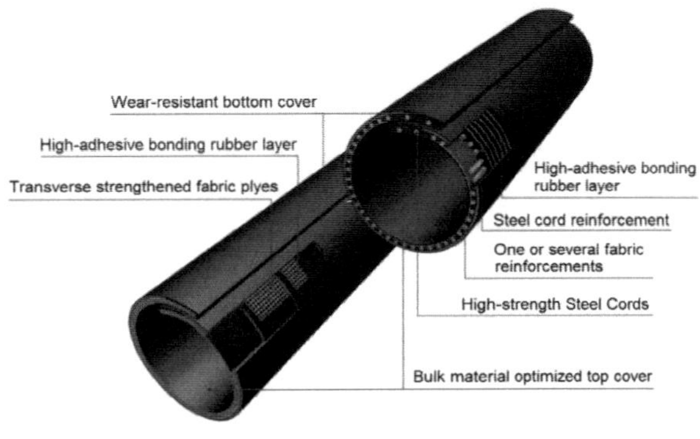

Fig. 2. Two types of design for pipe conveyor belts – rubber and fabric (on the left) and rubber and steel cord (on the right).

In the upper part of the tubular surface the conveyor belt edges are overlapped where they form up the superimposition zone, the dimension of which depends on the belt type, bulk density, and coarseness of the particles in the load being transported, as well as at the stage at which the support rollers are being installed.

When folded into a tube, the initially flat elastic conveyor belt tends to unfold, thereby creating an additional load upon the support rollers. This load is taken into consideration when solving the contact problem with the purpose of determining the force of resistance to the motion due to the support rollers being pushed into the top cover of the tubular shape belt [8], [12] and [13].

As a result of the effect of the load resulting from folding upon a large number of the support rollers directing the tubular belt along the lengthy path, additional resistance forces occur which have to be taken into consideration when calculating the conveyor's load capacity [3] and [14].

The objective of this paper is to determine the load on the support rollers, at the section which is folded into a cylindrical shell from an initially flat conveyor belt, taking into consideration any variation in belt elasticity modulus during the process of elongation and contraction.

At the direct and reverse branches of the pipe conveyor, the belt is symmetrically loaded with regards to the longitudinal plane, and it can be considered as being an orthotropic, initially-tense cylindrical shell.

The elasticity modulus of the belt inserts is varied at elongation and contraction, depending on the value and direction of the applied load [15] and [16].

3. Results and Discussion

Belt tension increases as the load moves from the conveyor's end-face to the driving cylinder, resulting in an increase in the longitudinal elasticity modulus E_1 along the path.

The transversal rigidity of the belt is determined by the transversal elasticity modulus E_2 which, where the belt folds into a tube, accepts different values for those inserts which are positioned at a higher or lower level than the neutral surface of the belt.

In cases in which the orthotropic conveyor belt is bent at the support rollers into a cylindrical shell with an arbitrary curvature radius $R = R(\varphi)$, the elasticity modulus E_2 in the direction of the axis OY will be higher at the elongation point than it is at the contraction point $E_2^+ > E_2^-$. Then, when the belt is folded into a tube, the neutral surface and its axis will be shifted by the value e, as shown in Fig 3.

Deformations during the elongation and contraction of the conveyor belt will be equal to:

$$\varepsilon^+ = \frac{(R+z)\varphi - R\varphi}{R\varphi} = \frac{z}{R}; \quad \varepsilon^- = \frac{R\varphi - (R-z)\varphi}{R\varphi} = \frac{z}{R}, \tag{1}$$

where z is the distance along the axis OZ from the considered arbitrary point of the belt to the neutral surface of the radius R.

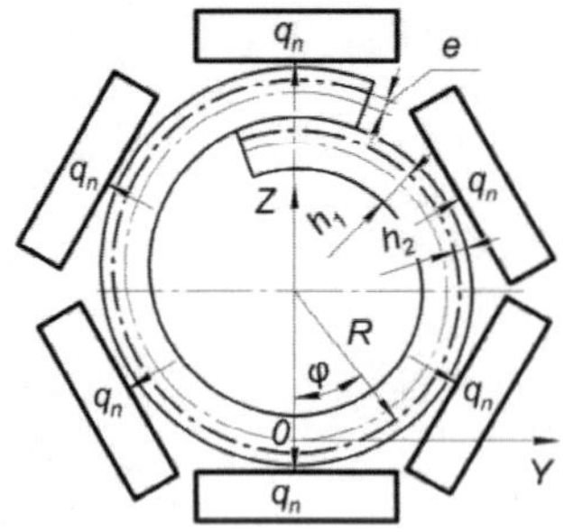

Fig. 3. Transversal bending when folding the non-uniform modulus belt into a tube.

We determine the value of the shift, e, for the neutral surface from the central position at the point at which the belt is folded. Neglecting the kinetic energy in the process of the belt's deformation, we equalise the power of elastic deformation upon overcoming the resistance of inner forces which have been spent on the elongation and contraction procedures. Using (1) and Hooke's law for the flat tense state of the orthotropic belt, we obtain:

$$\int_{R}^{R+h_2} \frac{E_2^+}{2\left(1-\nu_1^+\nu_2^+\right)}\left(\frac{z}{R}\right)^2 dz = \int_{R-h_1}^{R} \frac{E_2^-}{2\left(1-\nu_1^-\nu_2^-\right)}\left(\frac{z}{R}\right)^2 dz, \tag{2}$$

where $\nu_1^{+(-)}$, $\nu_2^{+(-)}$ are the Poisson coefficients at elongation and contraction, correspondingly. The belt thickness is equal to $h = h_1 + h_2$, where $h_1 = 0.5h + e$, $h_2 = 0.5h - e$ are the ranges by thickness from the neutral surface to the external surface, and the internal parts of the deformed belt.
Using the iteration method we integrate (2) and find:

$$e_{i+1} = \left(\begin{array}{l} \dfrac{hE_2^+}{2\left(1-\nu_1^+\nu_2^+\right)}\left[3R\left(R+\dfrac{h}{2}\right)+\left(\dfrac{h}{2}\right)^2\right] - \dfrac{hE_2^-}{2\left(1-\nu_1^-\nu_2^-\right)}\left[3R\left(R-\dfrac{h}{2}\right)+\left(\dfrac{h}{2}\right)^2\right]+ \\ +\dfrac{e_i^2E_2^+}{2\left(1-\nu_1^+\nu_2^+\right)}\left(3R+\dfrac{3h}{2}-e_i\right)-\dfrac{e_i^2E_2^-}{2\left(1-\nu_1^-\nu_2^-\right)}\left(\dfrac{3h}{2}-3R+e_i\right) \end{array} \right)\cdot\frac{1}{E}, \tag{3}$$

$$(i = 0, 1, \ldots),$$

where $$E = \frac{3E_2^+}{1-\nu_1^+\nu_2^+}\left(R+\frac{h}{2}\right)^2 + \frac{3E_2^-}{1-\nu_1^-\nu_2^-}\left(R-\frac{h}{2}\right)^2.$$

for the first approximation in (3) only the linear members can be considered with respect to *e*.

At a strong bend in its deformed state, the conveyor belt accepts the form of the cylindrical shell with a variable curvature radius, which can conveniently be described in the space it occupied by the curvilinear coordinates x, $R\varphi$. The belt equilibrium equations for the coordinates in the deformed system will be same as for the cylindrical shell equilibrium equations (Rudolphi & Reicks, 2006) in the non-deformed state:

$$\begin{array}{l} \dfrac{\partial T_1}{\partial x}+\dfrac{1}{R}\dfrac{\partial S}{\partial\varphi}=-q_1; \quad \dfrac{\partial T_2}{R\partial\varphi}+\dfrac{\partial S}{\partial x}+\dfrac{1}{R}\left(\dfrac{1}{R}\dfrac{\partial M_2}{\partial\varphi}+\dfrac{\partial H}{\partial x}\right)+\dfrac{1}{R}\dfrac{\partial H}{\partial x}=-q_2; \\ \dfrac{T_2}{R}-\dfrac{\partial}{\partial x}\left(\dfrac{\partial M_1}{\partial x}+\dfrac{1}{R}\dfrac{\partial H}{\partial\varphi}\right)-\dfrac{1}{R}\dfrac{\partial}{\partial\varphi}\left(\dfrac{1}{R}\dfrac{\partial M_2}{\partial\varphi}+\dfrac{\partial H}{\partial x}\right)=q_n, \end{array} \tag{4}$$

where T_1, T_2, M_1, M_2 are the forces and the momenta applied to the neutral surface; $S = T_{12} - M_{21}R_2^{-1} = T_{21} - M_{12}R_1^{-1}$ and $H = 0.5\left(M_{12} + M_{21}\right)$ are the symmetrical forces and the momenta suggested in an existing paper [7]; q_1, q_2, q_n is the external load, under the influence of which the deformed belt is in a state of equilibrium.

Dislocation of the arbitrary point of the belt expressed via dislocation of the neutral surface points $\tilde{u}, \tilde{v}, \tilde{w}$, correspondent to it, are equal to:

$$u = \tilde{u}(x,y) + z\vartheta(x,y); \; v = \tilde{v}(x,y) + z\psi(x,y); \; w = \tilde{w}(x,y) + z\chi(x,y), \tag{5}$$

where the parameters ϑ, ψ, χ are the functions of dislocations in the neutral surface of the belt:

$$\begin{aligned} &\vartheta = -\frac{\partial \tilde{w}}{\partial x}\left(1+\frac{\partial \tilde{v}}{\partial y}\right)+\frac{\partial \tilde{v}}{\partial x}\frac{\partial \tilde{w}}{\partial y}; \; \psi = -\frac{\partial \tilde{w}}{\partial y}\left(1+\frac{\partial \tilde{u}}{\partial x}\right)+\frac{\partial \tilde{u}}{\partial y}\frac{\partial \tilde{w}}{\partial x}; \\ &\chi = \frac{\partial \tilde{u}}{\partial x}+\frac{\partial \tilde{v}}{\partial y}+\frac{\partial \tilde{u}}{\partial x}\frac{\partial \tilde{v}}{\partial y}-\frac{\partial \tilde{u}}{\partial y}\frac{\partial \tilde{v}}{\partial x}. \end{aligned} \tag{6}$$

The coordinates of the belt's neutral surface points before and after the deformation process are expressed via the deformed state coordinates:

Before the deformation these are:

$x = x; \; y = R\varphi; \; z = 0,$

and after they are:

$\xi = x + C; \; \eta = R\sin\varphi; \; \zeta = R(1-\cos\varphi),$

where C is constant.

Hence dislocations of the belt neutral surface points are equal to:

$$\tilde{u} = \xi - x = C; \; \tilde{v} = \eta - y = R(\sin\varphi - \varphi); \; \tilde{w} = \zeta - z = R(1-\cos\varphi). \tag{7}$$

The derivatives from dislocations (7) along the coordinate y of the non-deformed state are:

$$\begin{aligned} &\frac{\partial \tilde{v}}{\partial y} = \left[R'(\sin\varphi - \varphi) + R(\cos\varphi - 1)\right]\cdot\left(R + R'\varphi\right)^{-1}; \\ &\frac{\partial \tilde{w}}{\partial y} = \left[R'(1-\cos\varphi) + R\sin\varphi\right]\cdot\left(R + R'\varphi\right)^{-1}; \\ &\frac{\partial^2 \tilde{v}}{\partial y^2} = \left[\left(R''R - R^2 - \varphi R'R - 2R'^2\right)\sin\varphi + \varphi\left(2R'^2 - R''R\right)\cos\varphi\right]\cdot\left(R + R'\varphi\right)^{-3}; \\ &\frac{\partial^2 \tilde{w}}{\partial y^2} = \begin{bmatrix} R''R - 2R'^2 - \left(R''R - R^2 - \varphi R'R - 2R'^2\right)\cos\varphi + \\ +\varphi\left(2R'^2 - R''R\right)\sin\varphi \end{bmatrix}\cdot\left(R + R'\varphi\right)^{-3}. \end{aligned} \tag{8}$$

The deformation components [9] for the belt neutral surface are equal to:

$$\begin{aligned} &\tilde{\varepsilon}_{xx} = \tilde{\varepsilon}_{xy} = æ_{xx} = æ_{xy} = \nu_{xx} = \nu_{xy} = 0; \\ &\tilde{\varepsilon}_{yy} = \frac{\partial \tilde{v}}{\partial y} + \frac{1}{2}\left[\left(\frac{\partial \tilde{v}}{\partial y}\right)^2 + \left(\frac{\partial \tilde{w}}{\partial y}\right)^2\right]; \\ &æ_{yy} = -\frac{\partial^2 \tilde{w}}{\partial y^2}\left(1+\frac{\partial \tilde{v}}{\partial y}\right) + \frac{\partial \tilde{w}}{\partial y}\frac{\partial^2 \tilde{v}}{\partial y^2}; \\ &\nu_{yy} = \frac{1}{2}\left[\left(\frac{\partial^2 \tilde{v}}{\partial y^2}\right)^2 + \left(\frac{\partial^2 \tilde{w}}{\partial y^2}\right)^2\right]. \end{aligned} \tag{9}$$

We assume that the dependence between the tensions and deformations for the non-uniform modulus orthotropic belt corresponds to the correlations for the generalised Hooke's law [4]:

$$\sigma_{xx}^{+(-)} = \frac{E_1^{+(-)}}{1-\nu_1^{+(-)}\nu_2^{+(-)}}\left(\varepsilon_{xx} + \nu_2^{+(-)}\varepsilon_{yy}\right);$$
$$\sigma_{yy}^{+(-)} = \frac{E_2^{+(-)}}{1-\nu_1^{+(-)}\nu_2^{+(-)}}\left(\varepsilon_{yy} + \nu_1^{+(-)}\varepsilon_{xx}\right); \tag{10}$$
$$\sigma_{xy}^{+(-)} = G^{+(-)}\varepsilon_{xy},$$

where $G^{+(-)}$ is the shift modulus at elongation and contraction.

Using the generalised expressions for the deformation components:

$$\varepsilon_{xx} = \tilde{\varepsilon}_{xx} + z\text{æ}_{xx} + z^2\nu_{xx};$$
$$\varepsilon_{yy} = \tilde{\varepsilon}_{yy} + z\text{æ}_{yy} + z^2\nu_{yy}; \tag{11}$$
$$\varepsilon_{xy} = \tilde{\varepsilon}_{xy} + z\text{æ}_{xy} + z^2\nu_{xy}$$

and (10), we determine the forces and momenta applied to the neutral surface:

$$T_1 = \int_0^{h_1}\sigma_{xx}^-(1+z/R)dz + \int_{-h_2}^{0}\sigma_{xx}^+(1+z/R)dz;\ T_2 = \int_0^{h_1}\sigma_{yy}^- dz + \int_{-h_2}^{0}\sigma_{yy}^+ dz;$$
$$T_{12} = \int_0^{h_1}\sigma_{xy}^-(1+z/R)dz + \int_{-h_2}^{0}\sigma_{xy}^+(1+z/R)dz;\ T_{21} = \int_0^{h_1}\sigma_{xy}^- dz + \int_{-h_2}^{0}\sigma_{xy}^+ dz;$$
$$M_1 = \int_0^{h_1}\sigma_{xx}^-(1+z/R)zdz + \int_{-h_2}^{0}\sigma_{xx}^+(1+z/R)zdz;\ M_2 = \int_0^{h_1}\sigma_{yy}^- zdz + \int_{-h_2}^{0}\sigma_{yy}^+ zdz; \tag{12}$$
$$M_{12} = \int_0^{h_1}\sigma_{xy}^-(1+z/R)zdz + \int_{-h_2}^{0}\sigma_{xy}^+(1+z/R)zdz;\ M_{21} = \int_0^{h_1}\sigma_{xy}^- zdz + \int_{-h_2}^{0}\sigma_{xy}^+ zdz.$$

When integrating (12) while taking into consideration (9) and (11), we obtain:

$$T_1 = \frac{\nu_2^- E_1^- h_1}{1-\nu_1^-\nu_2^-}\left[\tilde{\varepsilon}_{yy} + \left(\text{æ}_{yy} + \frac{\tilde{\varepsilon}_{yy}}{R}\right)\frac{h_1}{2} + \left(\nu_{yy} + \frac{\text{æ}_{yy}}{R}\right)\frac{h_1^2}{3} + \frac{\nu_{yy}}{R}\frac{h_1^3}{4}\right] + \frac{\nu_2^+ E_1^+ h_2}{1-\nu_1^+\nu_2^+}\left[\tilde{\varepsilon}_{yy} - \left(\text{æ}_{yy} + \frac{\tilde{\varepsilon}_{yy}}{R}\right)\frac{h_2}{2} + \left(\nu_{yy} + \frac{\text{æ}_{yy}}{R}\right)\frac{h_2^2}{3} - \frac{\nu_{yy}}{R}\frac{h_2^3}{4}\right];$$
$$T_2 = \frac{E_2^- h_1}{1-\nu_1^-\nu_2^-}\left(\tilde{\varepsilon}_{yy} + \text{æ}_{yy}\frac{h_1}{2} + \nu_{yy}\frac{h_1^2}{3}\right) + \frac{E_2^+ h_2}{1-\nu_1^+\nu_2^+}\left(\tilde{\varepsilon}_{yy} - \text{æ}_{yy}\frac{h_2}{2} + \nu_{yy}\frac{h_2^2}{3}\right);$$
$$M_1 = \frac{\nu_2^- E_1^- h_1^2}{1-\nu_1^-\nu_2^-}\left[\frac{1}{2}\tilde{\varepsilon}_{yy} + \left(\text{æ}_{yy} + \frac{\tilde{\varepsilon}_{yy}}{R}\right)\frac{h_1}{3} + \left(\nu_{yy} + \frac{\text{æ}_{yy}}{R}\right)\frac{h_1^2}{4} + \frac{\nu_{yy}}{R}\frac{h_1^3}{5}\right] + \frac{\nu_2^+ E_1^+ h_2^2}{1-\nu_1^+\nu_2^+}\left[-\frac{1}{2}\tilde{\varepsilon}_{yy} + \left(\text{æ}_{yy} + \frac{\tilde{\varepsilon}_{yy}}{R}\right)\frac{h_2}{3} + \left(\nu_{yy} + \frac{\text{æ}_{yy}}{R}\right)\frac{h_2^2}{4} + \frac{\nu_{yy}}{R}\frac{h_2^3}{5}\right]; \tag{13}$$
$$M_2 = \frac{E_2^- h_1^2}{1-\nu_1^-\nu_2^-}\left(\frac{1}{2}\tilde{\varepsilon}_{yy} + \text{æ}_{yy}\frac{h_1}{3} + \nu_{yy}\frac{h_1^2}{4}\right) + \frac{E_2^+ h_2^2}{1-\nu_1^+\nu_2^+}\left(\frac{1}{2}\tilde{\varepsilon}_{yy} - \text{æ}_{yy}\frac{h_2}{3} + \nu_{yy}\frac{h_2^2}{4}\right);$$
$$T_{12} = T_{21} = S = 0;\ M_{12} = M_{21} = H = 0.$$

At the point at which the belt is bent into a cylindrical shell with a radius of R = const, the neutral surface deformation components when taking into consideration (8) and (9) are equal to:

$$\tilde{\varepsilon}_{xx} = \tilde{\varepsilon}_{yy} = \tilde{\varepsilon}_{xy} = æ_{xx} = æ_{xy} = \nu_{xx} = \nu_{xy} = 0;$$
$$æ_{yy} = -R^{-1}; \qquad (14)$$
$$\nu_{yy} = \left(2R^2\right)^{-1}.$$

When substituting (14) into (13) we determine the forces and the momenta T_1, T_2, M_1, M_2, S, H, and then we can determine from (4) the load, based on which the deformed belt is in equilibrium in the form of a circular cylindrical shell:

$$q_n = \frac{E_2^-}{1-\nu_1^-\nu_2^-}\cdot\left(-1+\frac{h_1}{3R}\right)\cdot\frac{h_1^2}{2R^2} + \frac{E_2^+}{1-\nu_1^+\nu_2^+}\cdot\left(1+\frac{h_2}{3R}\right)\cdot\frac{h_2^2}{2R^2}. \qquad (15)$$

4. Conclusions

The design of the pipe conveyor belts has been duly considered, and the advantages and application particularities of it have been determined in production conditions. When considering the mechanical properties of the conveyor belt and its tense and deformed state in the operation section, it is suggested that the method for the calculation of the load upon the base support rollers resulted from bending the initially flat belt into a cylindrical shell with a circular cross-section according to the non-uniform modulus elasticity theory. The expressions obtained allow the components to be determined for the additional forces of resistance in relation to belt motion over the support rollers for the direct and reverse branches, which are subsequently considered during the calculation of the conveyance capacity of the pipe conveyor.

5. References

[1] Ansari, A.H. (2005). Pipe conveyors advantages. Chemical Engineering World 40(10), 47–49.

[2] Fedorko, G. (2010) Variable pipe conveyor. Annals of DAAAM and Proceedings of the International DAAAM Symposium 2010, 20-23 October 2010, Zadar; Croatia, B. Katalinic (Ed.), pp. 1149-1150, Published by DAAAM International, ISSN: 17269679; ISBN: 978-390150973-5.

[3] Loeffier, F. J. (2000). Pipe/tube conveyors – a modern method of bulk materials transport. bulk solids handling, Vol. 20, Issue 4, pp. 431–435, ISSN: 01739980.

[4] Staples, P. (2002). The History of Pipe Conveyors. Bulk Solids Handling 22(3), 210–213.

[5] Will, F. & Staribacher, J. (2011). Pipe conveyors transport bulk material efficiently over long distances. World of Mining – Surface and Underground, 63(3), 146–155.

[6] Del Coz Diaz, J.J., Garcia Nieto, P.J., Vilan Vilan, J.A., Martin Rodriguez, A., Prado Tamargo, J.R. & Lozano Martinez-Luengas, A. (2007). Non-linear analysis and warping of tubular pipe conveyors by the finite element method. Mathematical and Computer Modelling 46, 95–108.

[7] Fedorko, G., Ivančo, V., Molnár, V. & Husáková, N. (2012). Simulation of interaction of a pipe conveyor belt with moulding rolls. Procedia Engineering, MMaMS 2012, 6-8 November 2012, Vol. 48, pp. 129–134, ISSN: 1877-7058, DOI: 10.1016/j.proeng.2012.09.495.

[8] Wang, S., Guo, Y.-C., Li, D.-Y., Hu, K. 2016 Research of lateral force of pipe conveyor belt´s vertical transport section. Archives of Transport, Vol. 37, Issue 1, 2016, pp. 67-75, DOI: 10.5604/08669546.1203204.

[9] Zhang, Y. Steven, R. (2012). Pipe conveyor and belt: Belt construction, low rolling resistance and dynamic analysis. SME Annual Meeting and Exhibit, Feb.19–22, Seattle (WA), USA, pp. 616-619, ISBN: 978-162276089-3.

[10] Guo, Y.-C., Wang, S., Hu, K. & Li, D.-Y. (2016). Optimizing the Pipe Diameter of the Pipe Belt Conveyor Based on Discrete Element Method. 3D Research 7 (1), Article number 5, 1–9. DOI: 10.1007/s13319-016-0085-8.

[11] Mining Industry ContiTech (2015). http://www.contitech.de/pages/produkte/transportbaender/cbgmining/cbgmining_en.html

[12] Klimenda, F., Skocilasova, B. 2015. Rollers vibration of pipe conveyor. Manufacturing Technology, Vol. 15, Issue 6, pp. 991-995, ISSN: 1213-2489.

[13] Molnár, V., Fedorko, G., Stehlíková, B., Tomašková, M.,Hulínová, Z. 2014. Analysis of asymmetrical effect of tension forces in conveyor belt on the idler roll contact forces in the idler housing. Measurement: Journal of the International Measurement Confederation, Vol. 52, Issue 1, pp. 22-32, ISSN: 0263-2241, DOI: 10.1016/j.measurement.2014.02.035.

[14] Nuttall, A.J.G., Lodewijks, G. & Klein Breteler, A.G. (2006). Modelling rolling contact phenomena in a pouch belt conveyor system. *Wear* 260(9–10), 1081–1089.

[15] Rudolphi, T.J. & Reicks, A.V. (2006). Viscoelastic indentation and resistance to motion of conveyor belts using a generalized maxwell model of the backing material. *Rubber Chemistry and Technology* 79, 307–319.

[16] Michalik, P., Molnár, V. Fedorko, G. & Weiszer, M. (2013). An experimental test rig: For measuring the strength of pipe conveyor belts. Bulk Solids Handling, Vol. 33, Issue 5, pp. 52–55, ISSN: 0173-9980.

DOI: 10.2507/28th.daaam.proceedings.029

INFLUENCE OF ONLINE IDENTIFICATION METHODS ON THE NONLINEAR PROCESS CONTROL

Lubomír Macků & David Novosad

This Publication has to be referred as: Macku, L[ubomir] & Novosad, D[avid] (2017). Influence of Online Identification Methods on the Nonlinear Process Control, Proceedings of the 28th DAAAM International Symposium, pp.0216-0223, B. Katalinic (Ed.), Published by DAAAM International, ISBN 978-3-902734-11-2, ISSN 1726-9679, Vienna, Austria
DOI: 10.2507/28th.daaam.proceedings.029

Abstract

Control of systems with strong nonlinearities is impossible or difficult by common linear systems methods. Sometimes the system linearization around the working point can be a solution, but it is not always applicable. Other, better possibility, is using online system identification during the whole process and based on it the controller parameters changing. In this case, the control is closely dependent on the process identification quality. In this paper, the quality of the control process using different identification methods and their settings is observed. A nonlinear process model based on the exothermic semi-batch reactor is used and the whole process is simulated using data of real process. Results are discussed.

Keywords: nonlinear system; online identification; exponential forgetting; semi-batch reactor; process control

1. Introduction

Linear systems control is usually done by means of controllers which parameters are set to fixed values. For the PID controller, for example, individual components P, I and D are constant. It is possible here because the linear system can be described by a linear function, and therefore also the linear controller satisfies the entire range of possible input variables.

For nonlinear systems such settings are not possible. Due to the nonlinearities, for example, the controller settings at the beginning of the process may differ significantly from the settings around the permanent operating point. Sometimes it may be possible to divide the control process into smaller individual parts with different local working points and use them as the sub-controller settings. However, in the event of a controlled system deviation from the expected behaviour, a complete failure of control may occur. Sometimes robust or adaptive controllers can be used to overcome the possible problems. These controllers are able to work even with slight system parameters changing, as shown for example on a continuous stirred tank reactor by Zavacka in [1] and by Vojtesek in [2] or on a drying process by Plsek and Matusu [3]. Also the application of model based control can be very powerful for such nonlinear systems as a solid-state fermentation which was introduced by Bück [4]. Again, if the model or its parameters do not match the actual process, there is a high risk of failure.

Another, more resilient solution is the continuous adjustment of the controller based on the current behaviour of the controlled system. It concerns the controlled system parameters real time identification, so-called online identification. Thanks to online identification, it is possible to adjust the controller settings to the current state of the system.
Any system deviations from the expected behaviour then have no effect on the control itself. The above mentioned procedure is also utilized in this article. Namely, a control process running in an exothermic semi batch chemical reactor is described here. This process is also strongly nonlinear.

The temperature profile in batch and semi-batch exothermic reactors usually follows three stages [5]: (i) heating of the reaction mixture up to the desired reaction temperature, (ii) maintenance of the system at this temperature and (iii) cooling in order to minimize the by-products formation. Any controller used to control such system must be able to take into account these different stages.

In the literature some papers have been published which discuss the control of a batch or semi-batch reactor. A global linearization control strategy with online state and parameter estimation was applied for a polymerization reactor in [6]. However, the authors concluded that the implementation of the proposed method is still difficult due to the missing support of required mathematical functions. The other approach was used in the study [7], where the authors applied a dual-mode (DM) control improved by iterative learning technique. Simulations showed that the proposed method can enhance the conventional DM control with modest effort. For rapid and suitable reference-trajectory tracking a self-adaptive predictive functional control algorithm by Škrjanc was recommended [8]. This approach was successful in a reactor with switching between cold and hot water in the inlet. Neural network was applied to similar system [9] to accommodate the online identification of a nonlinear system. The authors found this strategy effective in identification and control of a class of time-varying-delayed nonlinear dynamic systems.

This paper presents results of experiments obtained by simulations of the semi batch reactor control. For the control the Ziegler-Nichols controller with filtration of D-component (ZNFD) using online identification was used. The paper is organised as follows: In section 2, the semi-batch reactor and the ZNFD controller with online identification methods are described, section 3 presents simulation results and section 4 concludes the current work and suggests new areas for investigation.

2. Methods section

2.1. Semi-batch reactor model

A mathematical model of the semi batch chemical reactor used for simulations is defined here. Its scheme is shown in Fig 1.

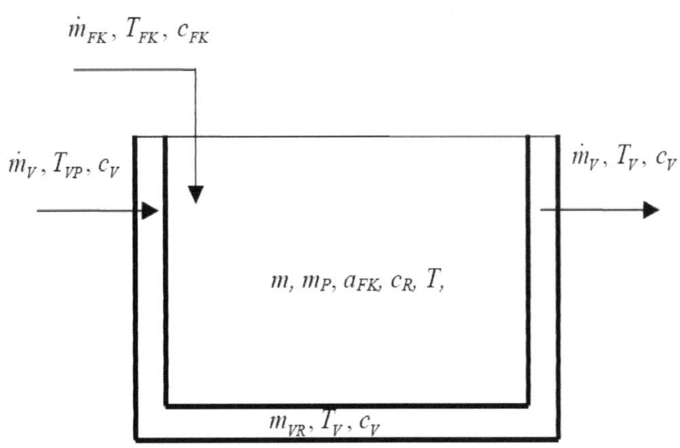

Fig. 1. Chemical reactor scheme

The differential equations 1-4 corresponding to the chemical reactor mathematical model are:

$$\dot{m}_{FK} = \frac{d}{dt} m(t) \tag{1}$$

$$\dot{m}_{FK} = k\, m(t)\, a_{FK}(t) + \frac{d}{dt}\left[m(t) a_{FK}(t)\right] \tag{2}$$

$$\dot{m}_{FK} c_{FK} T_{FK} + \Delta H_r k\, m(t)\, a_{FK}(t) = K\, S\left[T(t) - T_V(t)\right] + \frac{d}{dt}\left[m(t)\, c_R\, T(t)\right] \tag{3}$$

$$\dot{m}_V c_V T_{VP} + K\,S\left[T(t) - T_V(t)\right] = \dot{m}_V c_V T_V(t) + m_{VR} c_V\, T_V'(t) \tag{4}$$

The reactor model comprises the total mass balance (1), reaction component mass balance (2), the enthalpy balance (3) and coolant heat balance (4). Further variables and the parameters of the reactor model are listed in Table 1. In (2) k [s^{-1}] is the reaction rate constant expressed by the Arrhenius equation (5):

$$k = Ae^{-\frac{E}{RT(t)}} \tag{5}$$

(1)	$\dot{m}_{FK}$ [kg.s^{-1}]	Mass flow of the entering reaction component
	$m(t)$ [kg.s^{-1}]	Accumulation of the in-reactor content
(2)	$a_{FK}(t)$ [-]	Mass concentration of the reaction component
	$m(t)$ [kg]	Weight of all reaction components in the system
	k [s^{-1}]	The reaction rate constant
(3)	c_{FK} [J.kg^{-1}.K^{-1}]	reaction component specific heat capacity
	c_R [J.kg^{-1}.K^{-1}]	Reactor content specific heat capacity
	T_{FK} [K]	reaction component temperature
	ΔH_r [J.kg^{-1}]	Reaction heat
	K [J.m^{-2}.K^{-1}.s^{-1}]	Conduction coefficient
	S [m^2]	Heat transfer surface
	$T(t)$ [K]	Temperature of reaction components in the reactor
	$T_v(t)$ [K]	Temperature of coolant in the reactor double wall
(4)	$\dot{m}_v$ [kg.s^{-1}]	Coolant mass flow
	c_v [J.kg^{-1}.K^{-1}]	Coolant specific heat capacity
	T_{vp} [K]	Input coolant temperature
	m_{vR} [kg]	Coolant mass weight in the reactor double wall
(5)	A [s^{-1}]	Pre-exponential factor
	E [J.mol^{-1}]	Activation energy
	R [J.mol^{-1}.K^{-1}]	Gas constant

Table 1. Reactor model variables and parameters

2.2. Ziegler-Nichols controller with filtration of D-component using Tustin approximation (ZNFD)

As was already mentioned, to reach desired control properties in minimum time, ZNFD was used to calculate the temperature trajectory. Control law (6):

$$u_k = q_0 e_k + q_1 e_{k-1} + q_2 e_{k-2} - p_1 u_{k-1} - p_2 u_{k-2} \tag{6}$$

where e_k is control error ($e_k = w_k - y_k$) and controller parameters are calculated using following equations:

$$q_o = K_P \frac{1 + 2(c_f + d_d) + \frac{c_i}{2}(1 + 2c_f)}{1 + 2c_f} \tag{7}$$

$$q_1 = K_P \frac{\frac{c_i}{2} - 4(c_f + c_d)}{1 + 2c_f} \tag{8}$$

$$q_2 = K_P \frac{c_f(2 - c_i) + 2c_d + \frac{c_i}{2} - 1}{1 + 2c_f} \tag{9}$$

$$p_I = \frac{-4c_f}{1 + 2c_f} \tag{10}$$

$$p_2 = \frac{2c_f - 1}{1 + 2c_f} \tag{11}$$

$$c_f = \frac{T_f}{T_0} \tag{12}$$

$$c_i = \frac{T_0}{T_i} \tag{13}$$

$$c_d = \frac{T_D}{T_0} \tag{14}$$

$$K_P = 0.6K_{Pu} \tag{15}$$

$$T_I = 0.5T_u \tag{16}$$

$$T_D = 0.125T_u \tag{17}$$

$$T_f = \frac{T_D}{\alpha} \tag{18}$$

where K_P, T_D and T_I are controllers parameters. In (18), α is the filtration coefficient used by filter of process output signal.

2.3. Online identification methods

Two methods of identification were used for our simulations purposes, recursive least square method (LSM) and LSM with exponential forgetting. These methods can be used for the discrete on-line identification of processes that are described by the following transfer function *G(z)*:

$$G(z) = \frac{B(z^{-1})}{A(z^{-1})} = \frac{b_1 z^{-1} + b_2 z^{-2} + \ldots + b_m z^{-m}}{1 + a_1 z^{-1} + a_2 z^{-2} + \ldots + a_n z^{-n}} z^{-d} \tag{19}$$

The estimated output of the process in the step $k(\hat{y}_k)$ is computed on the basis of the previous process inputs u and outputs y according to the (20):

$$\hat{y}_k = -\hat{a}_1 y_{k-1} - \ldots - \hat{a}_n y_{k-n} + \hat{b}_1 u_{k-d-1} + \ldots + \hat{b}_m u_{k-d-m} \tag{20}$$

where $\hat{a}_1,\ldots\hat{a}_n,\hat{b}_1,\ldots,\hat{b}_m$ are the current estimations of the process parameters. This equation can be also written in vector form, which is more suitable for further work:

$$\begin{aligned} \hat{y}_k &= \Theta_{k-1}^T \cdot \Phi_k \\ \Theta_{k-1} &= \left[\hat{a}_1,\ldots\hat{a}_n,\hat{b}_1,\ldots\hat{b}_m\right]^T \\ \Phi_k &= \left[-y_{k-1},\ldots -y_{k-n}, u_{k-d-1},\ldots u_{k-d-m}\right]^T \end{aligned} \tag{21}$$

The vector Θ_{k-1} contains the process parameter estimations computed in the previous step and the vector Φ_k includes output and input values for computation of current output y_k.

2.4. Recursive least square method

Least square methods are based on minimisation of the sum of prediction errors squares:

$$J_k = \sum_{i=1}^{k} (y_i - \Theta_k^T \Phi_i)^2 \tag{22}$$

where y_i is the process output in i-th step and the product $\Theta_k^T \Phi_i$ represents the predicted process output. Solving this equation leads to the recursive version of least square method where vector of parameters estimations is updated in each step according to formula:

$$\Theta_k = \Theta_{k-1} + \frac{C_{k-1} \cdot \Phi_k}{1 + \Phi_k^T \cdot C_{k-1} \cdot \Phi_k} \cdot (y_k - \Phi_{k-1}^T \Phi_k) \tag{23}$$

The covariance matrix C is then updated in each step as defined by the equation:

$$C_k = C_{k-1} + \frac{C_{k-1} \cdot \Phi_k \cdot \Phi_k^T \cdot C_{k-1}}{1 + \Phi_k^T \cdot C_{k-1} \cdot \Phi_k} \tag{24}$$

Initial value of matrix C determines the influence of initial parameters estimations on the process identification.

2.5. Least square method with exponential forgetting

The main disadvantage of this pure recursive least square method is an absence of signal weighting. Each input and output affect the result by the same weight, but actual process parameters can change in time. Thus newer inputs and outputs should affect the output more than older values. This problem can be solved by exponential forgetting method, which uses forgetting coefficient φ and decreases the weights of the data in the previous steps. Weights φ_{k-1} are assigned to values u_i and y_i. Parameter estimations are computed according to the following relations:

$$\Theta_k = \Theta_{k-1} + \frac{C_{k-1} \cdot \Phi_k}{\varphi + \Phi_k^T \cdot C_{k-1} \cdot \Phi_k} \cdot (y_k - \Phi_{k-1}^T \Phi_k) \tag{25}$$

$$C_k = \frac{1}{\varphi}\left(C_{k-1} \frac{C_{k-1} \cdot \Phi_k \cdot \Phi_k^T \cdot C_{k-1}}{\varphi + \Phi_k^T \cdot C_{k-1} \cdot \Phi_k} \right) \tag{26}$$

3. Results section

3.1. Parameter α adjusting using recursive LSM identification

First, α filtration coefficient of ZNFD controller was adjusted. In this aspect it must be said that Chalupa [10] introduced the recommended values of parameter α in the interval ($3<\alpha<20$); however, these values were not suitable in our case. It was necessary to use a far smaller value. In Fig. 2, several temperature profiles with different α filtration coefficients are plotted. As can be seen, the performance of the ZNFD controller is the best for parameter $\alpha = 0.003$. In cases of higher α, the set point is overshot.

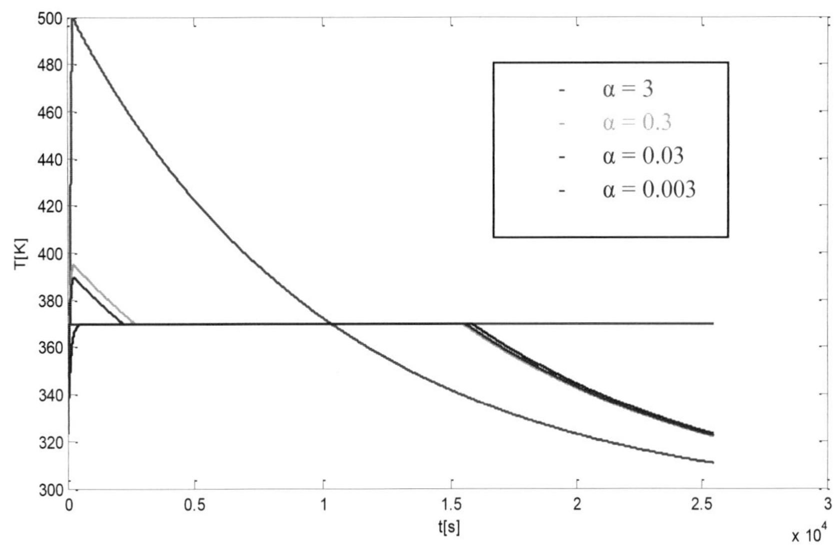

Fig. 2. Comparison of the temperature profiles for different α coefficients

3.2. Identification using LSM with exponential forgetting

In the previous section, the controller contains an online identification method - Least squares method (LSM). Here the identification method was extended by exponential forgetting. In order to assess a differences, the parameter φ influencing weights of the data from the previous steps, was changed. As can be seen in Fig. 3, the setpoint is reached very quickly without overshoot when $\varphi = 1$. It can be noticed that exponential forgetting is not applied when $\varphi = 1$, see (23) - (26).

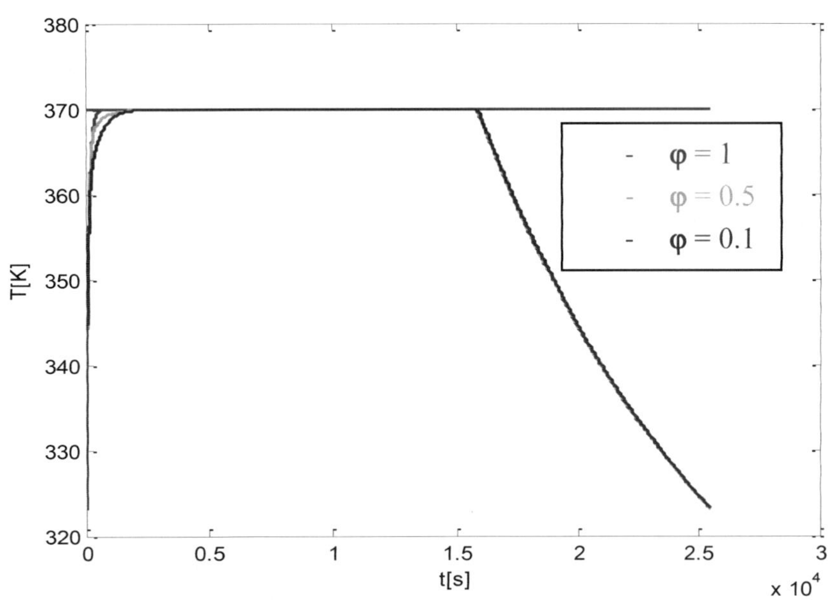

Fig. 3. Comparison of the temperature profiles for different φ coefficients

3.3. *Feeding profiles comparing*

Figure 4 shows that the feeding profile with online identification without exponential forgetting (φ = 1) is without oscillations. It can be observed that decreasing weights of the data in the previous steps did not improve the temperature and the feeding flow rate profiles in our semi-batch reactor simulations.

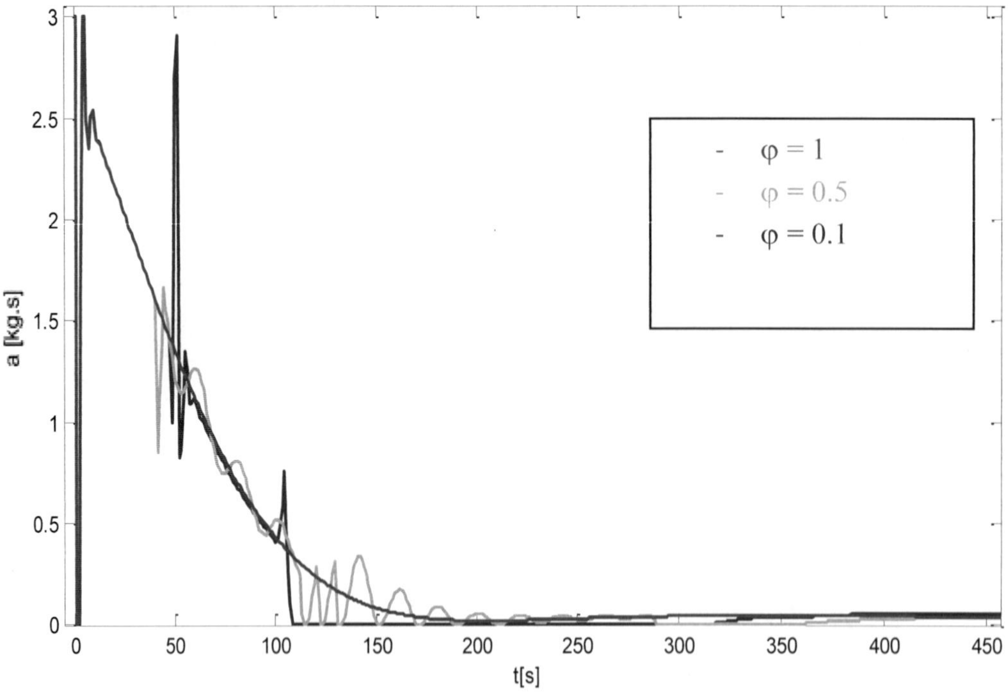

Fig. 4. Feeding profiles comparison for different φ coefficients

4. Conclusion

Usually some advanced control methods (such a model predictive control) are used for complex systems. These methods can be inevitable for example in case of systems with strong nonlinearities. But, as was shown in this article, sometimes even simple control method can deal with nonlinear systems. Here, the Ziegler-Nichols controller with filtration of D-component using online identification was able to control the semi batch exothermic process. It was visible that this control was closely dependent on the quality of the system identification. Two methods of identification were tested, recursive least square method and least square method with exponential forgetting. However, there was not any visible improvement using the second method, i.e. using the on-line identification with exponential forgetting. It may be caused by the lost ("forgotten") data from the beginning of the process. All previous steps input and output values were needed for good control performance in our semi-batch reactor system.

Future work should focus on the other identification methods verification as well as on the application of these methods to systems with different behaviour, for example flow reactors and the like. Also some general dependences and rules for application these methods to different systems could be observed (if there any).

5. References

[1] Zavacka, J. ; Bakosova, M.; Matejickova, K. (2012). Robust PID Controller Design for Unstable Processes with Parametric Uncertainty, Procedia Engineering, Volume 42, p.p. 1572-1578, ISSN 1877-7058, Available from: https://doi.org/10.1016/j.proeng.2012.07.550

[2] Vojtesek, J. (2010). Adaptive Control of Conductivity Inside CSTR, Annals of DAAAM for 2010 & Proceedings of the 21st International DAAAM Symposium, 20-23rd October 2010, Zadar, Croatia, ISSN 1726-9679, ISBN 978-3-901509-73-5, Katalinic, B. (Ed.), pp. 0207-0208, Published by DAAAM International Vienna, Vienna

[3] Plsek, S.; Matusu, R. (2015). Application of Adaptive Dead-beat Controller in Drying Process, Procedia Engineering, Volume 100, p.p. 756-764, ISSN 1877-7058, Available from: https://doi.org/10.1016/j.proeng.2015.01.429

[4] Bück, A.; Casciatori, F.P.; Thoméo, J.C.; Tsotsas, E. (2015). Model-based Control of Enzyme Yield in Solid-state Fermentation, Procedia Engineering, Volume 102, p.p. 362-371, ISSN 1877-7058, Available from: https://doi.org/10.1016/j.proeng.2015.01.163

[5] Bouhenchir, H.; Cabassud, M.; Le Lann, M.V. (2006). Predictive functional control for the temperature control of a chemical batch reactor, Computers and Chemical Engineering, Issue 30, p.p. 1141-1154

[6] Beyer, M.A.; Grote, W.; Reinig, G. (2008). Adaptive exact linearization control of batch polymerization reactors using a Sigma-Point Kalman Filter, Journal of Process Control, Issue 18, p.p. 663-675, ISSN 0959-1524

[7] Cho, W.; Edgar, T.F.; Lee, J. (2008). Iterative learning dual-mode control of exothermic batch reactors, Control Engineering Practice, 16, p.p. 1244-1249, ISSN 0967-0661

[8] Skrnajc, I. (2007). Self-adaptive supervisory predictive functional control of a hybrid semi-batch reactor with constraints, University of Ljubljana, Slovenia 2007

[9] Wu, X.; Zhang, J.; Zhu, Q. (2010). A generalized procedure in designing recurrent neural network identification and control of time-varying-delayed nonlinear dynamic systems, Neurocomputing, Issue 73, p.p. 1376-1383. ISSN 0925-2312

[10] Bobál, V., Chalupa, P., & Dostál, P. (2008). Library for design and simulation verification of self-tuning controllers, Proceedings of the 22nd European Conference on Modelling and Simulation (pp. 3-6)

DOI: 10.2507/28th.daaam.proceedings.030

DELAY SYSTEMS WITH MEROMORPHIC FUNCTIONS DESIGN

Jiří Korbel, Libor Pekař & Roman Prokop

This Publication has to be referred as: Korbel, J[iri]; Pekar, L[ibor] & Prokop, R[oman] (2017). Delay Systems with Meromorphic Functions Design, Proceedings of the 28th DAAAM International Symposium, pp.0224-0228, B. Katalinic (Ed.), Published by DAAAM International, ISBN 978-3-902734-11-2, ISSN 1726-9679, Vienna, Austria
DOI: 10.2507/28th.daaam.proceedings.030

Abstract

The contribution is focused on analysis and synthesis of delay systems in a ring of RQ-meromorphic functions. Controller design is based on solution of Diophantine equations in this ring that leads to Smith predictor like structure. Tuning of the controllers is based on pole-placement method as a desired multiple root of the characteristic closed loop equation. A stable second order transfer function is used as an example to validate the methodology. Furthermore, a biased relay experiment is combined with the proposed control design to bring an autotuning method. Matlab-Simulink examples are given to illustrate the developed approach.

Keywords: Delay systems; Diophantine equation; Meromorphic functions; Relay experiment

1. Introduction

Most of the real processes in the industry are time delay systems. Study and analysis of those systems became an attractive and active topic for researchers for many decades [1], [2]. Various approaches and methods were developed for compensating the time delay because it negatively affects system dynamics. Behavioral approach [3], ring of pseudo-polynomials [4] or fractional representation [5], [6], [7] can be mentioned as representatives of those methods.

Automatic tuning of controllers exists in many modifications and principles. During the automatic tuning procedure, the unknown system parameters are identified and a suitable controller is designed. The first part can be achieved by performing a biased relay experiment that estimates the controlled system parameters. For the following controller design, a various methods exist.

This paper uses a combination of biased relay experiment and algebraic controller design in a ring of RQ-meromorphic functions [8]. This approach enables designing the controller for time delay systems without the need of approximations of controlled system time delay term.

2. Delay systems

Algebraic tools are frequently used in control theory. These tools include, among others, also rings. Control theory uses several types of rings, e.g. the ring of proper and stable rational functions R_{PS}, the ring of polynomials R_P or the ring of proper and stable meromorphic functions R_{MS}.

Zítek and Kučera [8] developed the R_{MS} ring for delay systems. An element of R_{MS} is a ratio of two retarded quasipolynomials $y(s) / x(s)$. A retarded quasipolynomial $x(s)$ of degree n means

$$x(s) = s^n + \sum_{i=0}^{n-1}\sum_{j=1}^{h} x_{ij} s^i \exp(-\vartheta_{ij} s),\ \vartheta_{ij} \geq 0 \tag{1}$$

Quasipolynomial (1) is stable when it owns no finite zero s_0. The publication [8] deals with stability tests. When the degree of the numerator is less or equal to the degree of the denominator the ratio $y(s) / x(s)$ is proper.

A linear time-invariant delay system can be written as a ratio of two R_{MS} elements. A second order system with time delay can be expressed as

$$G(s) = \frac{\dfrac{b_0 e^{-\tau s}}{(s+m_0 e^{-\theta s})^2}}{\dfrac{(Ts+1)^2}{(s+m_0 e^{-\theta s})^2}} = \frac{\dfrac{b(s)}{m(s)}}{\dfrac{a(s)}{m(s)}},\quad m_0 > 0 \tag{2}$$

3. Controller synthesis

The controlled system is estimated by second order model (2) as a ratio of two elements $B(s)$ and $A(s)$ in R_{MS} ring, where each element's denominator $\Theta = 0$. The system parameters K, T and τ are estimated by biased relay experiment. Parameter $m_0 > 0$ is a free scalar parameter. The element $B(s)$ contains time delay term $e^{-\tau}$. The control loop is a simple feedback system with a controller $G_R(s) = Q(s) / P(s)$, which can be seen in 0

The control synthesis must achieve stabilization of the feedback loop, asymptotic tracking of the reference value and attenuation of the load disturbance.

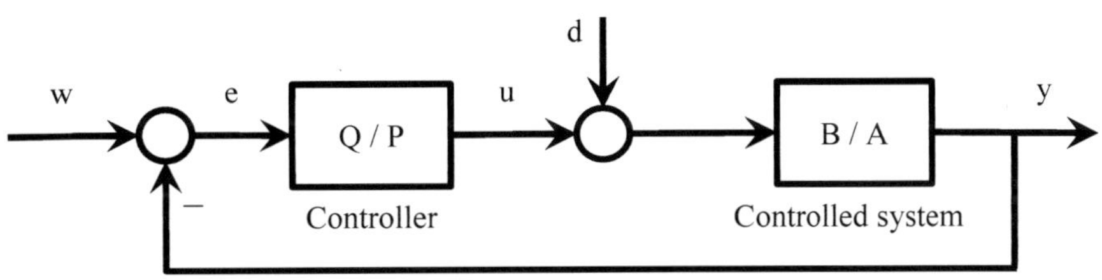

Fig. 1. Feedback control loop

Stabilization of the system can be formulated by the Diophantine equation

$$A(s)P_0(s) + B(s)Q_0(s) = 1 \tag{3}$$

where $P_0(s)$ and $Q_0(s)$ are particular solutions. All stabilizing controllers are derived from

$$\frac{Q(s)}{P(s)} = \frac{Q_0(s) + A(s)Z(s)}{P_0(s) - B(s)Z(s)},\quad P_0(s) - B(s)Z(s) \neq 0 \tag{4}$$

where $Z(s)$ is an arbitrary element of R_{MS}. The right choice of this element can ensure additional control conditions [5], [6], [8]. The reference and load disturbance can be described by equations

$$W(s) = H_W(s) / F_W(s),\ D(s) = H_D(s) / F_D(s) \tag{5}$$

Conditions for asymptotic tracking and disturbance attenuation result from expression for $E(s)$ which is

$$E(s) = \frac{A(s)P(s)}{A(s)P(s) + B(s)Q(s)} W(s) - \frac{B(s)P(s)}{A(s)P(s) + B(s)Q(s)} D(s) \tag{6}$$

It is required that $E(s)$ belongs to R_{MS}. It is necessary that both $F_W(s)$ and $F_D(s)$ divide $P(s)$. Details about divisibility in R_{MS} and R_{PS} can be found in [5], [6], [8]. Both signals $w(t)$ and $d(t)$ are step functions. Thus, for the case of R_{PS} ring, it is equivalent to reach the absolute term of $P(s)$ equal to zero. The last condition is not possible to reach in R_{MS}, due to

B(s) and/or $P_0(s)$ containing delay term *exp(-τs)*. For this case, the following expression for the absolute term in *P(s)* is required

$$\lambda\left(1-e^{-\tau s}\right) \tag{7}$$

where λ is a selected real parameter, usually $\lambda = m_0^n$, where *n* is the order of the controlled system. Equation (4) says that the controller $G_R(s)$ has integral behavior for s → 0. This condition is assured by a proper choice of *Z(s)* in (4). If *w(t)* or *d(t)* are other functions, divisibility conditions can be more complex.

For the second order system, the control design is given as follows. The equation (3) takes the form

$$(Ts+1)^2 P_0(s) + Ke^{-\tau s} Q_0(s) = (s+m_0)^2 \tag{8}$$

By the choice $Q_0(s) = 1$, the solution is obtained as

$$P_0(s) = \frac{(s+m_0)^2 - Ke^{-\tau s}}{(Ts+1)^2} \tag{9}$$

and the general solution is given by

$$\frac{Q(s)}{P(s)} = \frac{1 + \dfrac{(Ts+1)^2}{(s+m_0)^2} Z(s)}{\dfrac{(s+m_0)^2 - Ke^{-\tau s}}{(Ts+1)^2} - \dfrac{Ke^{-\tau s}}{(s+m_0)^2} Z(s)} \tag{10}$$

To obtain *P*(*s*) in a simple form, let be

$$Z(s) = \frac{\kappa (s+m_0)^2}{(Ts+1)^2} \tag{11}$$

where κ is a real free parameter. It is clear that $Z(s) \in R_{MS}$. The choice $\kappa = (m_0^2/K) - 1$ gives *P(s)* a simple finite-dimensional form with the denominator of the degree two and

$$P(s) = \frac{s^2 + 2m_0 s + m_0^2\left(1-e^{-\tau s}\right)}{(Ts+1)^2}; \quad Q(s) = \frac{m_0^2}{K} \tag{12}$$

Thus, the final form of the controller is

$$G_R(s) = \frac{m_0^2}{K} \frac{(Ts+1)^2}{s^2 + 2m_0 s + m_0^2\left(1-e^{-\tau s}\right)} \tag{13}$$

where $m_0 > 0$ is a real positive tuning parameter. It is obvious that the controller is in the quasipolynomial form. The construction of this controller is more complex than usual PI or PID controllers.

4. Biased relay feedback experiment

The method was proposed by Åström and Hägglund. It is based on a symmetrical relay feedback test when a relay of magnitude h_r is inserted in the feedback loop. The period of the limit cycle is the ultimate period T_u and a limit cycle of amplitude a_r is generated by the process output. However, there are another relays used in identification experiments. The biased (asymmetrical) relay experiment used for identification can give the final model transfer function according to [9] in the form

$$G(s) = \frac{Ke^{-\tau s}}{(Ts+1)^2} \tag{14}$$

Almost every stable industrial process can be approximated by model (14). The gain K of the process can be estimated by the expression [9]

$$K = \frac{\int_0^{iT_u} y(t)dt}{\int_0^{iT_u} u(t)dt}; \quad i = 1,2,3,... \tag{15}$$

and time constant T is given by

$$T = \frac{T_u}{2\pi} \cdot \sqrt{\frac{4 \cdot K \cdot h_r}{\pi^2 \cdot a_r^2} - 1} \tag{16}$$

Methodology in [10] also enables to estimate a delay term τ in (14) by the relation

$$\tau = \frac{T_u}{2\pi}\left[\pi - 2arctg\frac{2\pi T}{T_u} - arctg\frac{\varepsilon}{\sqrt{a_r^2 - \varepsilon^2}}\right] \tag{17}$$

where ε is a relay hysteresis. For the subsequent controller design estimated parameters K, T and τ are used.

5. Verification of the methodology

Consider a stable system with time delay described by the transfer function

$$G(s) = \frac{10}{(20s+1)(5s+1)(3s+1)} \cdot e^{-10s} \tag{18}$$

Transfer functions of the type (18) represent a wide class of frequent stable industrial processes. The biased relay experiment was used to approximate the system and relations (15) - (17) give the approximation as

$$G^*(s) = \frac{10}{(12.03s+1)^2} \cdot e^{-10.59s} \tag{19}$$

Asymmetric relay with hysteresis was used during the experiment and following parameters were set: $h_r = 0.225$ (0.25 when off, -0.2 when on), $\varepsilon = 0.05$. The comparison of step responses of systems (18) and (19) is shown in Figure 2.

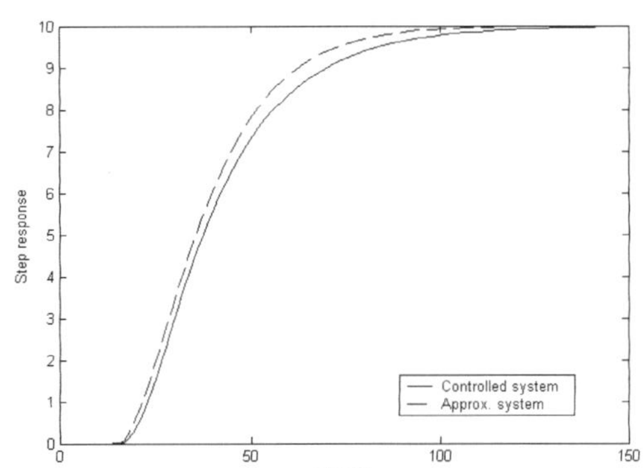

Fig. 2. Step responses of systems (18) and (19).

According to the algebraic control design in the R_{MS} ring, the following controller was obtained through (8) – (13)

$$G_R(s) = \frac{3.89s^2 + 0.65s + 0.37}{100s^2 + 10s + 0.27\left(1 - e^{-10.59s}\right)} \tag{20}$$

For the comparison, another controller was also designed by the algebraic method in the R_{PS} ring. Pade approximation of delay term $e^{-\tau s}$ was used. This approach gives a more general class of PID controllers. A suitable choice for m_0 was found experimentally by simulations.

Control responses are compared in Figure 3. The responses are similar, nevertheless, taking the controller (20) provides the shortest time to observe the reference value and gives the smallest "undershoot" after disturbance injection.

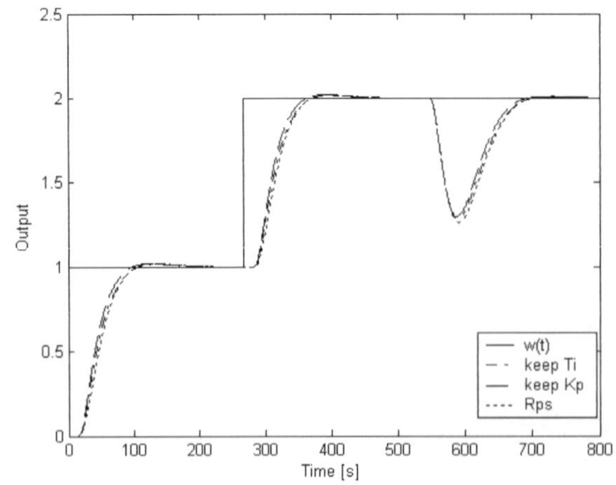

Fig. 3. Control responses for algebraic design in R_{MS} and in R_{PS} with Pade approximation.

6. Conclusion

The contribution gives a methodology for a set of delayed systems. The control synthesis was performed through a solution of a Diophantine equation in the ring of proper and stable RQ-meromorphic functions. A second order methodology generates a class of generalized Smith-like controllers with the realistic PID structure. The approach also offers a scalar tuning parameter $m_0 > 0$ which can be adjusted by a suitable principle. With a combination of biased-relay feedback identification, it brings a new autotuning method.

The developed methodology was implemented in the Matlab-Simulink environment and the approach was illustrated by an example for a higher order dynamic. It proves that the RQ-meromorphic controller is able to control the delayed systems without the need of approximation of the controlled system time delay term, like Pade approximation in R_{PS} synthesis.

7. References

[1] Kamen, E. W. (1975). On the algebraic theory of systems defined by convolution operations. Mathematical Systems Theory, Vol. 9, No. 1, pp. 57-74.

[2] Morse, A. S. (1976). Ring models for delay-differential systems. Automatica, Vol. 12, No. 5, pp. 529-531.

[3] Gluesing-Lueerssen, H. (1997). A behavioral approach to delay-differential systems. SIAM Journal on Control and Optimization, Vol. 35, No. 2, pp. 480-499.

[4] Brethé, D. & Loiseau, J. J. (1998). An effective algorithm for finite spectrum assignment of single-input systems with delays. Mathematics and Computers in Simulation, Vol. 45, No. 4-5, pp. 339-348.

[5] Vidyasagar, M. (2011). Control system synthesis: a factorization approach, Calif.: Morgan & Claypool, ISBN 9781608456611, San Rafael.

[6] Kučera, V. (1993). Diophantine equations in control - A survey. Automatica, Vol. 29, pp. 1361-75.

[7] Dlapa, M. (2010). Robust Controllers for Anisochronic Delayed Systems, Annals of DAAAM for 2010 & Proceedings of the 21st International DAAAM Symposium, 20-23rd October 2010, Zadar, Croatia, ISSN 1726-9679, ISBN 978-3-901509-73-5, Katalinic, B. (Ed.), pp. 0075-0076, Published by DAAAM International Vienna, Vienna.

[8] Zítek, P. & Kučera, V. (2003). Algebraic design of anisochronic controllers for time delay systems. International Journal of Control, Vol. 76, No. 16, pp. 905-921.

[9] Vítečková, M. & Víteček, A. (2010). Plant identification by relay methods, Proceedings of Engineering the future, Sciyo, Rijeka, Dudas, L. (Ed.), pp. 242-256.

[10] Prokop, R.; Korbel, J. & Prokopová, Z. (2010). Relay feedback autotuning – A polynomial approach, Proceedings of 24th European Conf. on modelling and Simulation, Kuala Lumpur, ISBN 978-0-9564944-1-2, pp. 290-295.

DOI: 10.2507/28th.daaam.proceedings.031

DESIGN AND CONSTRUCTION OF A MODULAR COST-EFFECTIVE CNC MACHINE

Arbnor Pajaziti, Tafil Asllani & Afrim Gjelaj

This Publication has to be referred as: Pajaziti, A[rbnor]; Asllani, T[afil] & Gjelaj, A[frim] (2017). Design And Construction of a Modular Cost-Effective CNC Machine, Proceedings of the 28th DAAAM International Symposium, pp.0229-0236, B. Katalinic (Ed.), Published by DAAAM International, ISBN 978-3-902734-11-2, ISSN 1726-9679, Vienna, Austria
DOI: 10.2507/28th.daaam.proceedings.031

Abstract

Main goal of the paper is design and construction of a prototype CNC machine which has the function of drawing, cutting, milling and drilling on a working table. We tried to describe in details all the steps we have undertaken from the beginning to the end of the work, with regards to design of mechanical parts, electrical parts and software developments. Our strength relied on the design and construction of functional prototype quickly, cost-effectively and with high precision.
To meet the above-mentioned criteria, in our study we have chosen several tasks to compile and execute the programs from the computer with the help of Lazy CAM Mach3 software, and then transfer of data on CNC machine, which performs the sequences by moving the z axis and draw or cut on x, y axes. The required critical components such as motors and linear guides have been carefully compared and selected. Personal computer has been used to design the parts and also is used to compile programs by either manual typing of G-Code.
The machine described in the paper will help to solve high precision machine tool system problem and serve those who demand to fabricate the CNC machines with low costs.

Keywords: CNC Machine design; low cost automation; CNC software; CNC graphical user interface; G-code

1. Introduction

CNC machines nowadays are widely used in manufacturing industry and are able to be controlled by programming language to carry out a wider variety of tasks with greater accuracy. The main issue during the design and construction of a CNC machine is how to eliminate unpredicted vibrations in mobile supporters x, y and z axis.

Computer Numeric Control (CNC) - brought a new opportunity to control the machine through the introduction of an integrated computer to the process control. To control the functions of the machine such as the movement of the tip of the pen, the movement of the cutting instrument, the movement of the work piece, changes in the cutting tool, etc., CNC takes instructions from the program in the form of program-sequences and translates from the control of machine to output analogue signals through postprocessors.

For modelling the movement characteristics in 1995, D. Spath [1], and in 1999, H. Weule [2] has studied the kinematics and trybo-system of circular spheres screw and has provided its specifics and priorities on growth speed of the moving table. Y. Zhou et al. [3], in 2011, proposed a detailed modelling of the support mechanism of, taking into account hardness and masses of all the mechanical elements of screw system with circulation spheres, to be used for analysis of vibrations by using the finite element method. D. Prevost et al., in 2011 [4], from simulation of contour errors, have proposed modelling of the movement of the food based on five axial milling centre for high-speed work, and have considered the factors that influence mostly in the dynamic behaviour of axes, inertia, friction and open contour compensation control [4].

To control the movement of feed rate, all elements of the feed rate system should be taken into consideration. D. Prevost [4] has proposed the allocation of process control in three levels: digital level - real-time interpretation program sequences of CL-file data to generate commands of axes.

At this level, the numerical control unit calculates position and speed kinematics limits based on each axle. The second level, the support mechanism - includes files and elements that perform axial commands, after the conversion of analogue to digital data. The third level - the kinematic level, consisting of mechanical structure of CNC machine, such as the kinematic chain, creates relative movement between the instrument and the piece.

Some authors have proposed modelling at each level (Altintas, 2005) [5], while others have preferred a specific work piece-modelling. The first model of the trajectory of movement is proposed by Lavernhe, 2008 [6]. Altintas et al. in 2000 [7], have proposed an adaptive sliding controller type (SMC-Siding Mode Controller) for the control of the movement of support mechanism and the contour movement during milling. The controller is implemented in the screw system with circulation areas and has shown more sophisticated and convenient method to control speed, better than the classic cascade controllers.

An integrated control, kinematics and dynamics of the circulation spheres screw servo-system have been given by Min-Seok Kim [8] in 2006. He has designed in details the requirements of the stability of servo-mechanism, geometric errors, stresses of the structure etc.

Implementation of artificial intelligence techniques to control support mechanism movement has given good results and is mainly observed in combination with the PID regulator. Lacerda et al. [9] in 1999 in order to minimize the contour errors has proposed a PID + Fuzzy Hybrid Controller (FHC) that manages to reduce contour errors at the 70:1 ratio.

Jin-Sung Kim et al. in 2008 [10], did the adjustment of the PID coefficients with genetic algorithm. In 2009 Isayad Naimuddin et al. [11], they have applied the Fuzzy Logic Controller (FLC) to the closed circuit inductive motor. In March 2011 F. Cus et al. [12], have published an article which represents an intelligent controller to control the milling force by combination of open and closed contour. The force is corrected from the external perturbations through closed loop, while through open loop with neural network the inverse dynamics of the production process is simplified. Meanwhile, various tests have been carried out with variants of implementation techniques. Gayathri J. et al. [13], 2011 have demonstrated the effect of different forms of membership functions to control alternating motor current, etc.

Therefore, in order to preserve the principle of simplicity of modelling the system and escape from the use of mathematics, the set of important elements that cause the destruction of stability have to be considered:

- Control of the inertia growth of the masses movement;
- Discontinuities that are raised during the goal of setting points;
- Non-linearity arising from random errors;
- Changes of the resisting forces of the cutting process.

Determining the shape of the system modelling for movement of drawing table will be explained in more details in Section III. Based on the model, the virtual simulation and the real CNC machine with the help of SOLIDWORKS has been designed and constructed.

2. Research goals

2.1 Precision increase

Recently there has been a particular interest in terms of increasing the accuracy of the metal-processing machines and their application in various fields. This has led to increased interest in processing centres of CNC machines that offer high speed machining, optimal efficiency, high precision and whose aim is to reduce the cost. High machining speed means the systems with fast support mechanism and with high cutting speed.

The z-axis speed ranges from 600 m/min and over 18,000 m/min for ultra fast. High processing speed and great accuracy is required to increase the productivity of CNC machines, to shorten working time and preparatory steps, to fabricate complicated work pieces, etc.

2.2 Questions and hypothesis

The main question would be, whether the goal would be reached only by following the correct programming on CNC machine, or there is a need to pay particular attention on implementation stages of design and production of the prototype by applying the FLC of the closed loop of inductive motor.

The hypothesis would be performance and additional information about how the machines are designed and constructed that would be used as expected results.

- *What is the purpose of creating such a device?*

The study is focused on the design, construction and testing of CNC machine which today has been implemented mostly in industry of producing the electrical items.

- *What are the expected results?*

Drawings are designed to present various contours through the Mach3 program and are ordered to be executed at the CNC machine prototype. This prototype of CNC machine is intended to achieve maximum accuracy during the execution of the program.

Upon reaching the completion of CNC machine, logic programming is a function of time and steps which is enabled from the computer via Mach3 software tool.

The Lazy CAM software is used for G-code, which enabled to convert the trajectory of the work pieces from analogue into digital way. Machine-testing is enabled by certain tasks that are correctly dimensioned, then create digital program based on the dimensions of the item, in which case the complicated mechanical details have been taken.

Thus, to achieve the programming phase of task, the machine should possess in advance all the mechanical and electrical elements ready to execute the program. In addition, how the steps have been met on creating and programming the machine movements have been described.

3. Methods and research techniques

After completion of the design, the production of mechanical components of the CNC machine has been done. In Figure 1 is presented the 3D SOLIDWORKS assembling model, where all designed elements are produces afterwards. After completion of elements for x, y and z axes, the 3D model has been obtained, as is presented in Figure 2. The model has shown the manner and format how to produce the prototype machine.

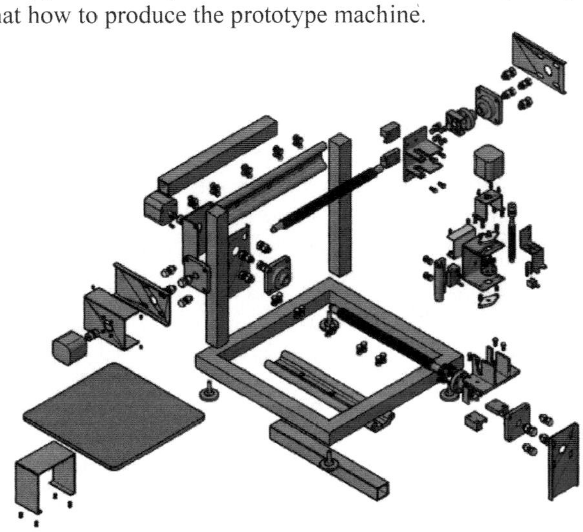

Fig. 1. The CNC machine modelling with x, y and z axes.

4. Fabrication of mechanical parts

Here is described the implementation of mechanical parts, based on the graphic design of the CNC machine. In addition, there are shown the pictures of the elements prepared and completed by drawings, defined in Section III of the graphical model.

X axis is completed by connecting all the needed elements and is tested properly. The axis is formed by:

1. The holder of the metal profile 40x40x2 (mm);
2. Slider and aluminium profiles;
3. Axes with trapeze fillets;
4. Bearing Lodge;
5. Link elements.

Also the y axis is realized similar to the x axis, but the difference between the x and y axes are that x axis is linked to z axis and the y axis and is connected to the working table.

It is implemented identically as the shape of the z axis. The latter axis is with the help of a fixed holder of x axis. The element to the z axis is formed to the fixed screw lever. Z axis in a positive/negative direction moves with the rotation of x axis by the step motor, while the positive/negative displacement of z axis moves by step motor. These movements enable the displacement of the reference point of a pencil tip.

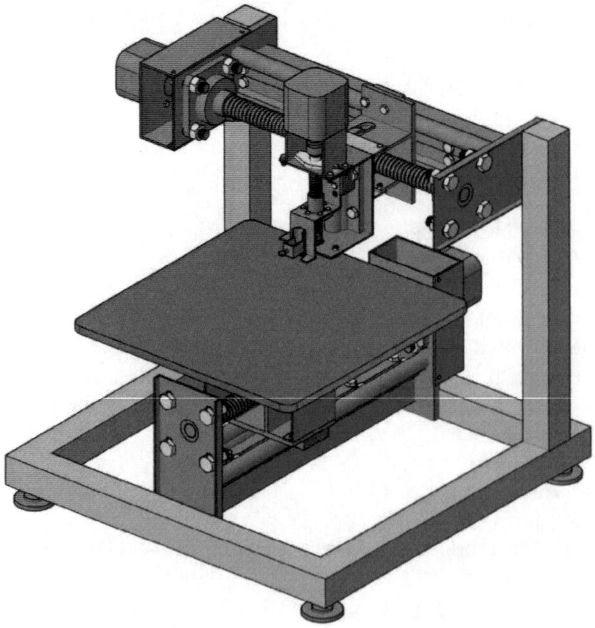

Fig. 2. CNC machine – front side.

In Figure 3 the CNC machine moves in three directions by axes of x, y and z coordinates. Its task is to draw on the working table, so that the movement of the x axis moves the z axis, and enables the drawing direction in x axis, while the movement of the y axis is determined to move the table and contact the tip of the pencil.

If it is requested to form a circle on the working table, then z axis moves positively with the numerical value assigned by the program, which values change until contacting the tip of the pen with the working table. Then, the values provided by the program by coordinating the movement of x and y axes. So forth; begin the movement of the x axis which carries the z axis, and the movement of the table from the y axis. For the circle formation, the function of two movements of x and y at the same time are needed, but with difference time of the motor step where the circle drawing is enabled.

Calculation and configuration of movements x, z and y axis is given with expression as above:

$$step\ per\ rev = (M * S)/P \quad (1)$$

Where is: M – motor steps, S – micro stepping and P – pitch for lead unscrew nuts.

Fig. 3. The X, Y and Z axes of CNC machine.

5. Electrical devices, installation and programming

In Figure 4 the cabling between motors and TB6560-4V3 controller board is provided. This enabled the transfer of the signals from the computer to the step motors. This signal is determined by the needs of the actions that are required; based on the drawing with pencil has been implemented. Thus, the controller board enables conversion of electrical signals into movements of the CNC machine.

Fig. 4. Plugging the step motors on the TB6560-4V3 control board.

With controlled movements the intention was to draw mechanical drawings with precise dimensions. The Mach3 software tool has been used for putting into operation the CNC machine and its programming. The numerical values of the commands, G-codes etc., by usage of Mach3 software tool, have been programmed. G-codes that enable the conversion of lines into numerical values for the coordinates of the x, y and z axes have been introduced. The CAM software make possibilities to optimize the tool path trajectory and choice the best start point of the CNC Machine tools [15].

A pencil that draws on desktop and performs simulation of a camera head cutter or a cutting tool for metal cutting metal has been taken as an example. The research has been focused on accuracy, speed, quality of work of a prototype machine that is managed to be accomplished.

6. Obtained results

The main task of this research work has been the realization of mechanisms for the design and production of mechanical parts of a CNC machine with the electrical items + software. This machine performs controlled movements with the help of the computer and Mach3software tool.

The outcome has been achieved by controlling the movement of the z axis. The important task was the coordination and functioning of the mechanical parts, including axles, bearings and of all the elements necessary for connecting the metal skeleton.

The result of the research was the selection of electrical control board which converts data from the computer commands to step motors which are linked to the x, y and axis of axes.

In Figure 5 is shown a drawing of the work piece-size.

The Mach3 program then is compiled for the mentioned drawing by using G-codes, Figure 6.

The same code, then in CNC machine has been executed, as is shown in Figure 7.

During the study of this prototype machine there are displayed many problems that we overcame and solved in the best engineering methods [14]. From this study one can conclude that the essence of automation mechanisms was to reduce vibration. Also, programming has particular importance; therefore it is done by using the software tools Mach3. Mach3 and CAM software enable programming and simulation of more complex work pieces. Programming enables conversion of numeric codes, which numerical value compel the electromotive to work with steps in function of time.

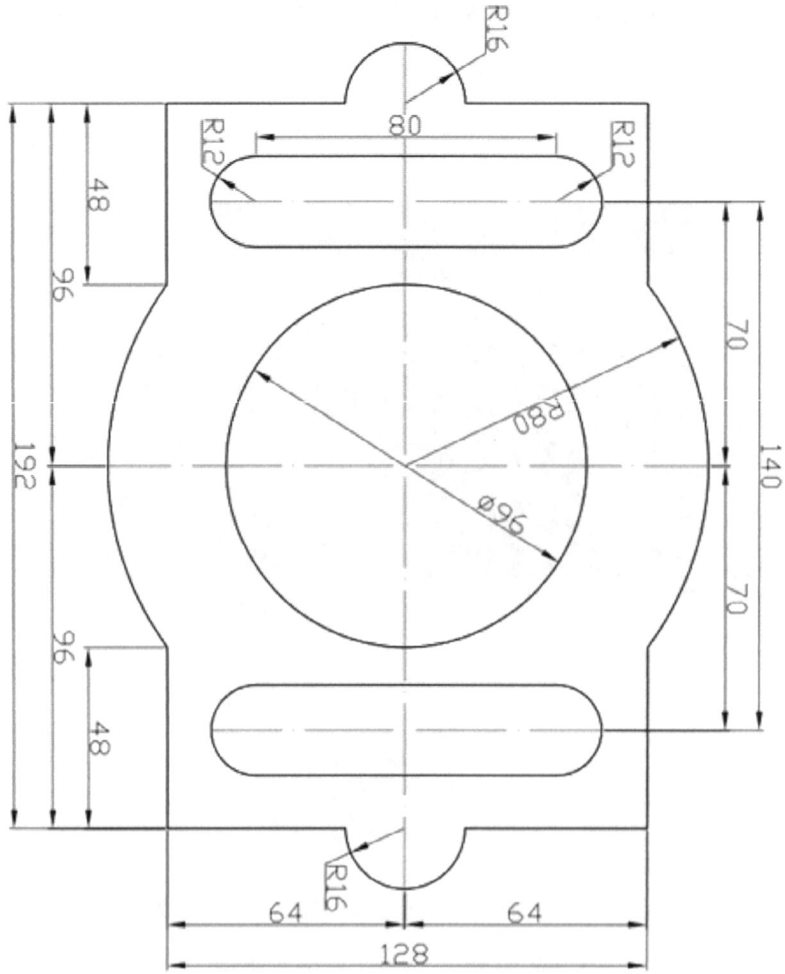

Fig. 5. Work piece drawing.

```
G-CODE CAD/CAM
N5 (FILE NAME = DETALI MEKANIK00 ON TUESDAY, MARCH 22, 2016)
N10 (DEFAULT MILL POST)
N15  G91.1
N20 G0  Z10
N25 M3
N30  X54.0000  Y40.0001
N40 G1  Z0.000  F6000
N45  Y120.0000
N50 G3  X30.0000  Y119.9999  I-12.0000  J-0.0075
N55 G1  X30.0000  Y40.0001
N60 G0  Z10
N65  X54.0000  Y39.9849
..............................................
..............................................
..............................................
N190 G0  Z0.0000
N195  X170.0000  Y120.0000
N205 G1  Z10
N210  Y40.0001
N220  X160.0000  Y80.0000
N225  Z0.000
N230 G1  Z0.0000
N235 G2  I-48.0000  J0.0000
N240 G0  Z10
N245  X194.0000  Y120.0151
N255 G1  Z0.0000
N260 G3  X170.0000  Y120.0455  I-12.0000  J0.0077
N265 G0  Z10
N270  X194.0000  Y40.0001
N280 G1  Z0.000
N285  Y120.0000
N290 G0  Z10
N295 M5
N300 M30
```

Fig. 6. Listing of the Lazy CAM Mach G-Code.

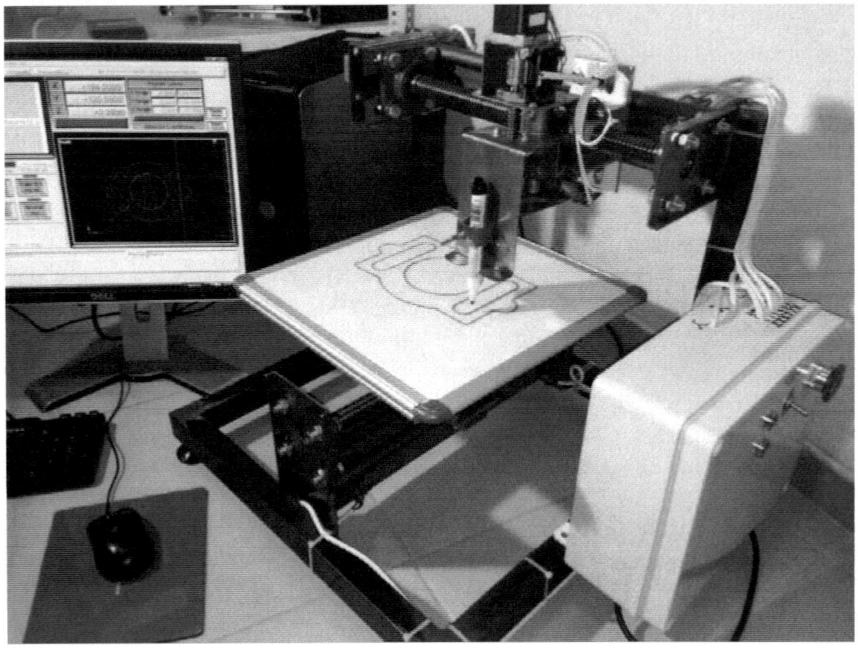

Fig. 7. Realization of the task in CNC machine.

7. Conclusions

Relying on the development of mechanical and electrical technology in recent years, now we have increased research interest of automation of the industrial machines. Nowadays it is focused on the manufacturing of a prototype machine which is capable of providing the expected results.

In particular herein is designed and constructed the machine tool with computer numerical control. Manufacturing of the prototype machine has to deal with the mechanical, electrical and software, such as at:

- Mechanical parts: the holders, axles, bearings and other supporting machine elements and connectors.
- Electrical components: Electro motors, controller board, cables and buttons (emergency, limiting and activator).
- Software: In our case we have applied the software Mach3 to machine control.

The study is based on the production of CNC machine that can move in three directions x, y and z axes. Z axis is used for the main task, as there was located a pen for drawing. One can replace the pen with any other device which makes metal cutting or drilling, since the machine enables the controlled movement along z axis, where we can set any cutting tool.

The goal of the study has been achieved, i.e., the production and programming of CNC mechanical device is accomplished. Regarding the accuracy of the drawing the trajectory, it can be achieved with the usage of certain commands in Mach3 where software automatically calibrates the ratio between motor and spindle moving step.

Future plans would be implementation of tool selection method and camera for metal processing, instead of using the pen for drawing, and calculation of all the forces caused during the metal processing. Also, the usage of Matlab software should be considered for design, optimization of machining parameters, and cutting forces.

8. References

[1] Spath, D., Rosum, J., A. Haberkern, A., (1995). Kinematics Friction a I Characteristics and Wear Reduction by PVD Coating on Ball Screw Drives, Annals of the CIRP – Manufacturing Technology, Vol. 44, Issue 1, pp. 349-352, doi.org/10.1016/S0007-8506(07)62340-3

[2] Weule H., and Frank, T., (1999). Advantages and Characteristics of a Dynamic Feeds Axis with Ball Screw Drive and Driven Nut, Annals of the CIRP Vol. 48, Issue 1, pp. 303-306., doi.org/10.1016/S0007-8506(07)63189-8

[3] Zhou, Y., Peng, F.Y., and Cao, X.H., (2011). Parameter sensitivity analysis of axial vibration for lead-screw feed drives with time-varying framework, Mechanics, Vol. 17, No. 5, pp. 523-528., ISSN 1392-1207

[4] Prévost, D., and Lavernhe, S., Dumur, D., (2011). Feed drive modelling for the simulation of tool path tracking in multi-axis High Speed Machining, International Journal of Mechatronics and Manufacturing System, Vol. 4, pp. 266-284, 10.1504/IJMMS.2011.041472

[5] Altintas, Y., Brecher, C., Weck, M., Witt S., (2005). Virtual Machine Tool, CIRP Annals - Manufacturing, Technology, Vol. 54, Issue 2, pp. 115-138

[6] Lavernhe, S., Tournier, C., Lartigue, C, (2008). Kinematical performance prediction in multi-axis machining for process planning optimization', International Journal of Advanced Manufacturing Technology, Vol. 37, No. 5, pp. 534-544, DOI 10.1007/S00170-007-1001-4

[7] Y. Altintas, K. Erkorkmaz, W.-H. Zhu, (2000). Sliding Mode Controller Design for High Speed Feed Drives. Annals of the ClRP Vol. 49, Issue 1, pp. 265-270

[8] Min-Seok Kim a, Sung-Chong Chung, (2006). Integrated design methodology of ball-screw driven servomechanisms with discrete controllers. Part I: Modelling and performance analysis, Mechatronics 16, Elsevier, pp. 491-502, DOI:10.1016/J.Mechatronics.2006.01.008.

[9] H. B. Lacerda, Dr. Eng., E. M. Belo, (1999). Application of a PID+Fuzzy Controller on the Motion Control System in Machine Tools, Ph.D Thesis, J. Braz. Soc. Mech. Sci. Vol. 21, No. 3, Rio de Janeiro.

[10] Jin-Sung Kim, Jin-Hwan Kim, Ji-Mo Park, Sung-Man Park, Won-Yong Choe and Hoon Heo, (2008). Auto Tuning PID Controller based on Improved Genetic Algorithm for Reverse Osmosis Plant, World Academy of Science, Engineering and Technology 47., pp. 384-389.

[11] Isayyad Naimuddin, B. Warkad, Dr. G. M. Dhole, Mohammad Nasiruddin (2009). Enhancing The Operational Availability Of CNC Machine Using Fuzzy Logic Controller, Journal Of Theoretical And Applied Information Technology, pp. 145-154.

[12] Cus F., Zuperl U., Balič J., (2011). Combined feedforward and feedback control of end milling system, Journal of Achievements in Materials and Manufacturing Engineering. Vol. 45, ISSUE 1, pp. 79-88.

[13] Gayathri Monicka Dr. N.O.Guna Sekhar K. Ramash Kumar, (2011). Performance Evaluation of Membership Functions on Fuzzy Logic Controlled AC Voltage Controller for Speed Control of Induction Motor Drive, International Journal of Computer Applications, Vol. 13, No.5, pp. 8-12. ISSN 0975-8887

[14] Asllani, T. (2016). CNC Mechanical Device Design and Programming with Mach3, Master Thesis, Department of Construction and Design, University of Prishtina, Prishtina, Kosovo.

[15] Gjelaj, A., Balič, J. and Fičko, M., (2013). Intelligent optimal tool selections for CNC programming of machine tools. Transactions of FAMENA, Vol. 37, No. 3, pp.31-40, ISSN 1333-1124.

DOI: 10.2507/28th.daaam.proceedings.032

HIDDEN INFORMATION IN INFRARED SPECTRUM ON SAFETY CLOTHES

Jana Žiljak Gršić, Lidija Tepeš Golubić, Ula Leiner, Denis Jurečić

This Publication has to be referred as: Ziljak Grsic, J[ana]; Tepes Golubic, L[idija]; Ulla, L[einer] M[aksan] & Jurecic, D[enis] (2017). Hidden Information in Infrared Spectrum on Safety Clothes, Proceedings of the 28th DAAAM International Symposium, pp.0237-0243, B. Katalinic (Ed.), Published by DAAAM International, ISBN 978-3-902734-11-2, ISSN 1726-9679, Vienna, Austria
DOI: 10.2507/28th.daaam.proceedings.032

Abstract

We are observing duo images with double cameras that we have developed for the further development of infrared information VZ technology. The spectrum range goes from visual (V – RGB) to the near infrared spectrum (Z – NIR). We have developed procedures by reproducing a fine art work that contains a double image. The digital reproduction contains the same content from two images. The original fine work and the printing reproduction have been performed by different procedures, different dyes. Spectral analysis of painting and printing dyes has been performed. In this paper we are publishing a mathematical model of VZ separation of dyes in order to achieve reproduction of double images information on silk. The new model is based on twin dyes that have resulted from the experimental work with printing on silk and measuring their equality. We are using the spectral analysis of light absorption in the range from 400 to 1000 nm. The goal is to find the shortest way to fix the content of process dyes in the ensuing printing experiments. The results of development of hidden image information are being demonstrated on safety clothes.

Keywords: infrared painting; near infrared spectroscopy; NIR reproduction; hidden image; camouflage uniform

1. Introduction

We have built the hidden information in documents [1], clothes [1], bank notes and packaging [1]. Double graphic solutions are on these products. The first one; for direct observation with naked eye. The second solution is being observed through a digital infrared camera. Art paintings have a double content considering the way of observation and the manner of painting. It has been established by spectrographic methods that painting dyes have different properties of sunlight absorption in the visual and near infrared spectrum [4].

Painters Nada Žiljak and Dijana Nazor create two pictures at the same place. 'Infrared painting' has been developed that is nowadays being successfully presented in galleries and museums. Figure 1 (a and b) is Nada's picture in double edition. The painter uses dozens of different dyes. Opposite to that, the reproduction of art paintings is being performed with process dyes C, M, Y, K. Only four dyes from the printing 'color practice'. Their absorption properties have been published in the article named: „Near infrared spektroskopy in print tehnology“ [5]. The richness of various dye tones for fine art is being simulated with just four dyes.

We have developed the model of simulation of a big range of paints in the system of double images with cmyk printing dyes. We have asked the question: if the art painting has a duality, is it possible to simulate the painter's dyes with cmyk inks in digital print. The painter creates the Infrared painting in her own way. The innovation in this paper is based on the idea that there is that dual look into two light areas in the reproduction. The paintings have been taken from Sv. I. Zelina gallery where there is a large opus of modern infrared painting.

With this work we are expanding the topic of double information on textile, silk and artistic paintings. The first results of a forensic approach to art works has been demonstrated at the DAAAM symposium 2010 [9] with emphasis on "Hidden Information with Infraredesign". Products have been depicted as a pilot project during our second presentation 2014, (DAAAM Vienna 2014) [10]. The results of the INFRAREDESIGN procedure have been shown using the innovative design of topographic maps. An infrared method in the printing of special maps with integrated hidden information and recognition via IRD cameras has been introduced.

Fig. 1. Nada Žiljak. A double image, in a visible and near infrared Z-spectrum

2. Spectrograms of painting dyes

We have established spectrograms for painting dyes. Each dye has its own light absorption property in the visual (V) and near infrared spectrum (NIR-Z). Tube dyes (Burnt Umber, Cadmium Red Light, Ceruelan Blue, Phthalo Green) have been recorded with two cameras: V and Z.

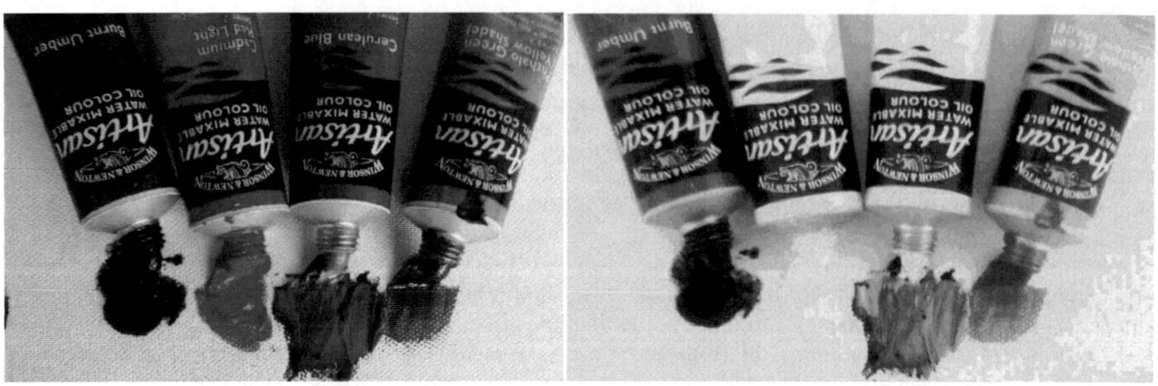

Fig. 2. (a and b) Tube dyes in V and Z spectrum

Each dye splits into two areas for which the light absorption control in the visual and NIR spectrum is being performed. The goal is to achieve two desired pictures at the same place that are mutually hidden. The pictures are called 'V' and 'Z' considering that they appear independently, reveal themselves in those light areas. Commercial dyes are being mixed according to the artists' plans and desires. New dyes with double VZ properties under the name of 'twin dyes' are being created.

On the graph, figure 3, we can see light absorption properties for a few commercial dyes in the field of fine art. Some dyes (brown and blue) have a significant response in the near infrared spectrum. Dyes: green and red, for example, do not have a response in the NIR spectrum.

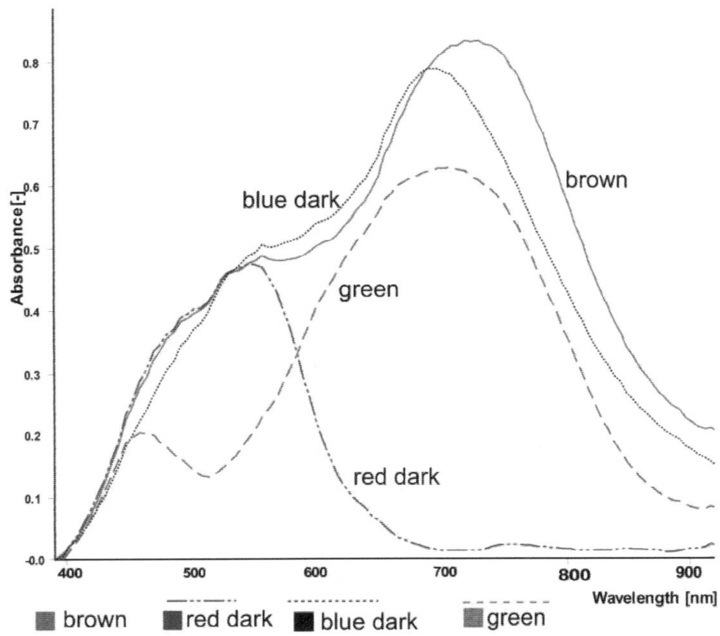

Fig. 3. Spectrograms of fine art dyes in the visual and near infrared spectrum

Dyes for color mixtures in the range from 400 to 900 nm have been studied through spectrograms. The analysis has introduced us to the field of 'twin dyes'. Those are pairs of dyes of identical colors in the visual spectrum but different in the near infrared Z spectrum. The painters have developed a new approach in art. The first step is development of 'twin dyes' by respecting the property of dye transparency in correlation with the thickness of in the layer 'color on color' coating. This innovation is the base for dual painting. Two dyes with an identical color tone in the visual spectrum and different near infrared spectrum absorption properties have enabled 'intimate painting' for a new way of observing a piece of art.

3. Twin dyes for printing raster painting technology

Table of twin dyes for printing: all the tones have an equal K value (carbon black) of 40%. This will ensure the management of applying inks (dyes) and the plan of content in CMYKIR printing application with an intention to provoke a controlled appearance of the hidden image [6]. The new printing methods, expansion of the "twin dyes" idea, are being continued in the papers and patents in the field of safety print [7]. The base of the twin idea starts with grey dyes testing and accompanying rasterization with small line screens [8].

In table 1 you can find the results of the experimental search for twin dyes for printing on silk. After calculating regressive parameters for the entire color area using an INFRAREDESIGNR method, we are accessing the calculation of the following twin dye pairs. For some materials (military canvas (2)) the targeted color tones that exist in nature are being calculated: Forest areas, area of camouflage in the field, area of hiding in a sandy desert.

We are providing a mathematical model with which painting tones of artistic creations are being simulated. The goal is to achieve a reproduction of original art paintings and for them to have two independent images at the same place that are being separately observed with RGB (V) and NIR – (Z) cameras.

Two dyes of the dual system, their light absorption in the range from 400 to 900 nm have been shown in the graph, figure 4. Each dye has been performed in two ways of mixing, dye pairs of identical tone.

Twins colours with: C0=78, M0=63; Y0=54%. Blue twins colour (print on silk) are composed of: 61; 25, 15 i 40% carbon black ink (HP 5000).
Red colour has values: X0: C0=43, M0= 95; Y0= 95 %. Her twin colour Z (print on silk) is: 0; 76; 90; i 40% K.
Red and blue colors have the same ΔZ values at 900 nm. The grey line at the bottom of the graph is the light absorption on pure silk on which the dress print has been performed.

X0, % C0, M0, Y0			X40, eksperimentalno Blizanac za K=40%			X40, C40, M40, Y40 REGRESIJA za K=40%		
73	91	94	45	65	84	41	62	74
55	74	78	32	47	65	22	45	56
89	96	50	67	64	11	76	67	23
78	86	89	50	55	64	50	54	64
42	55	56	10	25	26	11	24	28
39	55	66	4	25	38	4	24	40
40	44	43	11	11	9	13	29	6
39	59	71	2	33	34	2	6	48
35	40	40	1	6	10	7	31	6
44	63	99	0	21	85	1	0	77
95	50	95	60	0	40	72	5	44
40	40	40	12	5	4	14	0	4
87	48	81	67	1	33	66	0	32
79	45	71	59	2	17	59	6	23
67	47	47	52	2	2	49	6	3
75	45	95	49	0	42	46	0	50
85	48	48	77	0	0	73	3	0
89	48	48	80	0	8	78	2	0
75	48	48	68	9	8	59	6	2
65	48	48	45	8	7	45	8	6

Table 1. Twins dyes for silk

Twin dyes have an identical, similar light absorption in the visual area. Graphs of the same dye tone separate after 700 nm. We call them V and Z. Their difference is being manifested as a field in photographing with a NIR – Z camera. We added two sub-fields to INFRAREDESIGN® practice: Z1 as the field of separation that needs to be skipped in these technologies (The naked eye sees something in that area).

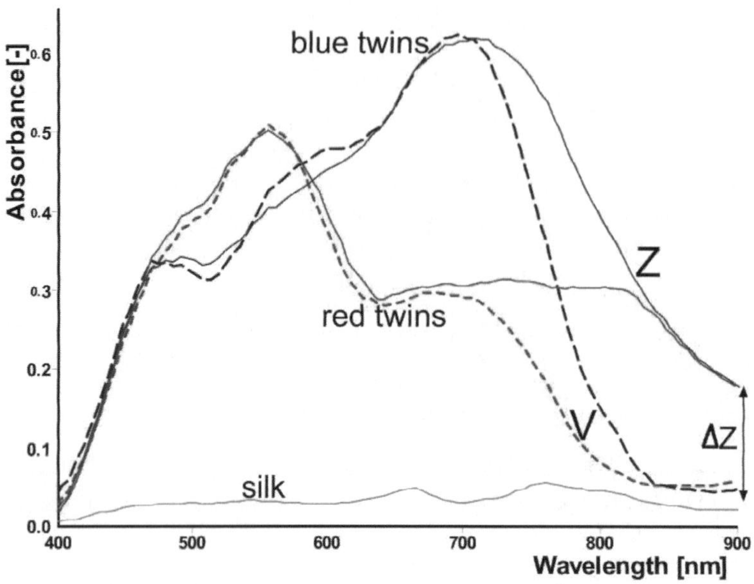

Fig. 4. Absorption of red and blue twins

Light blocking for IRD technology starts at 850 nm. We have set up a Z2 area that is blocked within 850 and 1000 nm. We have modified the photo cameras for recording at 1000 nm for which we use the name Z-camera. All the recordings are being performed with two (RGB and Z) cameras simultaneously for the purpose of a subsequent joining of two photographs [6].

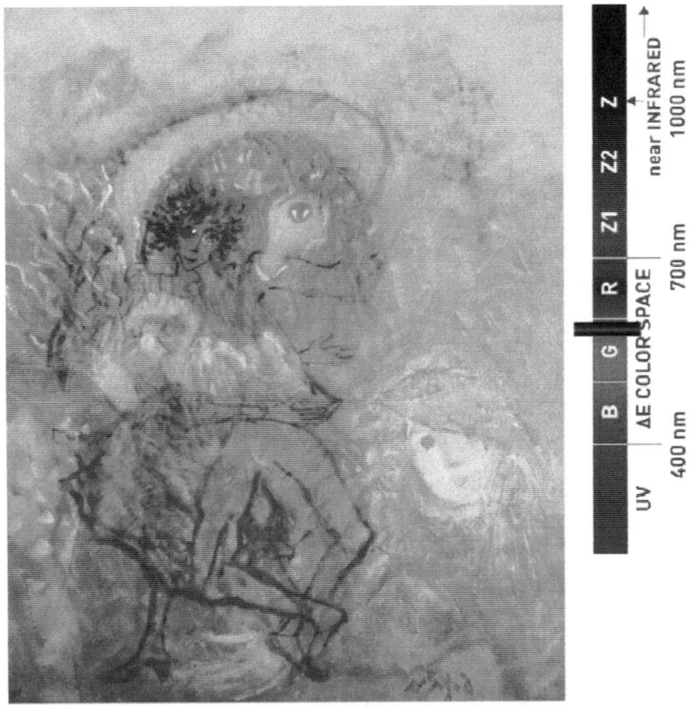

Fig. 5. InfraredArt; light with blocking from 400 to 1000 nm
http://jana.ziljak.hr/NadaDam17.html or: http://jana.ziljak.hr/NadaDam17.swf

Figure 5 shows a simulation of continuous transition of light blockages from 400 to 1000 nm. Animation of light blockage is being used for the control of vanishing and occurrence visible to the naked eye and invisible to the naked eye. The animation area includes V, Z1, Z2 areas. Digital control is being performed before printing with a model for printing on silk.

Model CMYKIR - v3, for silk that carries reproductions with process colors:

$$C40 = - 0.287243907*Y - 0.045801224*M + 1.367575664*C - 27.36785066$$
$$M40 = - 0.05868754*Y + 1.358549444*M - 0.277607606*C - 36.05324188$$
$$Y40 = 1.058889627*Y + 0.525560016*M - 0.42510473*C - 42.10206775$$

The experimental process of calculating twin dyes is lasting. Print iterations and measurements of L*a*b values are being performed. Spectroscopy controls of twin dyes after print are being added for the final iterations. In table 2 you can find twin dyes that represent the expansion of the color field. The goal is to reduce the number of iterations of printing experiments. New twin dyes are the base for determining corrections of parameter coefficients in dependency equations X40 o X0. Eight New Calculated Critical Colors are in table 2.

"X0, zadano C0, M0, Y0"			"Regresija X40, M40, Y40"		
60	60	94	25	23	63
42	95	45	13	79	18
44	80	77	7	56	63
44	50	50	16	17	18
82	91	87	56	60	63
36	36	36	10	1	0
99	53	48	92	6	0

Table 2. Examples of new "silk" twin colorants with 40% K

The painting, acrylic on canvas by Nada Žiljak, has been recorded with two cameras. In equal resolutions and equal number of pixels. The first camera has recorded RGB (3 x 8 bit) in the visual (V) spectrum, and the second camera is Z (8 bit) that records only one light absorption channel at 1000 nm. Projectina cameras are being used that give information in 12 light blockages in V and Z area [4].

Twin dyes and new rules in printing: it is measured with the print on the same material (silk) and the dyes that are going to be used for printing. Change of order of dye application is later not allowed. Raster technique, covering of dyes, has been based on mutual transparency. We take this into calculation of the composition of raster elements for all combinations of printing technique and associated dyes. Measurements on digital print, for example, are valid only for the given combination of silk, ink and toner type. Change of material means: repeated measurements for creation of their twin sets.

4. A dress with a reproduction of the art 'infrared painting'

The new mathematical model for reproduction of the art painting on silk has been applied on HP5000 plotter with the associated UV inks. This is an expansion of the safety area for safety uniforms, unique clothes and preservation of the artwork's authenticity.

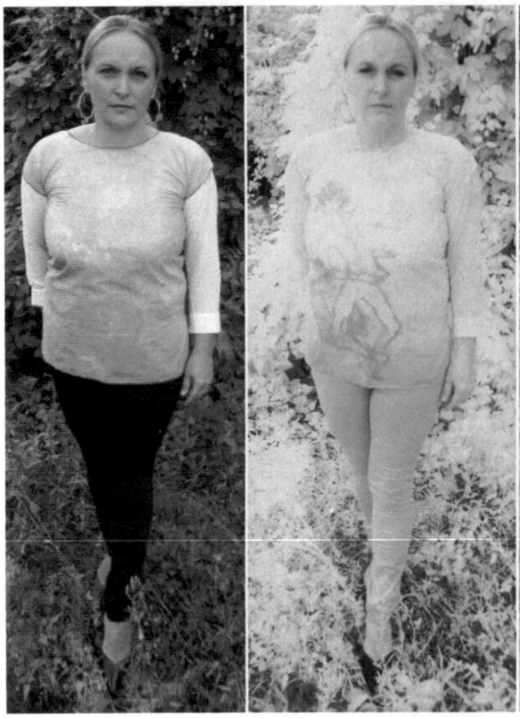

Fig. 6. (a and b). Infrared picture on a dress Fig. 7. (a and b). Infrared picture on black silk

The front part of the dress has been painted like the artist's painting. Visual and NIR pictures that are reproductions of the original are being observed with two cameras. Hidden data is the text about DAAAAM 2017 congress. On black canvas (back side of the dress) is a Z picture expanded with appropriate textual information: *28th DAAAM International Symposium Jana Žiljak, Lidija Tepeš Golubić, Ula Leiner, Denis Jurečić Zagreb, CROATIA Hidden information in infrared spectrum on safety clothes.* The canvas itself reflects the near infrared light. For Z camera that canvas is 'white'.

5. Conclusion

The painters are given an innovative method of creating a double image. Such artwork represents an insurmountable problem in the restoration technology. This paper demonstrates the future reproductive practice of art paintings. Each technique of graphic reproduction that relies on dual art painting is a new challenge in understanding dye properties in two light areas. This idea of 'hidden reproduction' or the new term 'infrared painting' will spread to other printing technologies: offset, screen printing, high printing and different digital print technologies. For each change of materials and dyes it is necessary to establish new color settings, a new mathematical model of separation of printing dyes. We are facing a new challenge in development of safety print and VZ reproduction. This paper is an example of a union between art and science.

These researches introduce new methods in the digital management of hidden information with printing processes. The use can be expanded to fine arts, a new form of painting that uses scientific methods of studying colour. The painters will start to use security principles to protect their works of art, as suggested in this article. Future research will be on the topic of colour spectroscopy for different applications. This will expand the procedures of understanding the properties of colour to an area beyond the reach of our eyes. New procedures will be opened in the forensic work of determining the origin of works of art.

6. References

[1] Aleksandra Bernašek, Ljilja Ivančević; Sigurnost u dizajnu dokumenata i vrijednosnicaSafety design of documents and securitie; Polytechnic & Design (.tvz.hr); Zagreb University of applied sciences; Vol. 4, No. 1, 2016. pp: 60 - 67; ISSN 2459-6302; ISSN ;1849-1995; DOI: 10.19279/TVZ.PD.2016-4-1-08

[2] Ivana Ž. Stanimirović, Jana Ž. Vujić, Nikolina Stanić Loknar; Marking of the camouflage uniform for visual and near infrared spectrum, Technics Technologies Education Management, Vol 8. No3. 2013, p: 920 - 026, ISSN 1840-1503

[3] Martina Friščić, Olivera Međugorac, Lidija Tepeš, Denis Jurečić, Invisible information on the transparent polymer food packaging with Infra V/Z technology, TTEM Vol 8/4,/ 2013; P: 1512 -1519, ISSN:1840-1503, e-ISSN 1986-809X

[4] Projectina AG Docucenter 4500, Switzerland, http://forensictechnology.com/projectina/

[5] Jana Žiljak; Near infrared spektroskopy in print tehnology; Polytechnic & design Vol. 5, No. 1, 2017. pp:32-36; DOI: 10.19279/TVZ.PD.2017-5-1-05

[6] Pap, Klaudio; Žiljak, Ivana; Žiljak-Vujić, Jana; Image Reproduction for Near Infrared Spectrum and the Infraredesign Theory, Journal of Imaging Science and Technology, vol. 54, no. 1, pp. 10502-1-10502-9(9), 2010

[7] Li, C.; Wang, C.; Wang, S.J. A Black Generation Method for Black Ink Hiding Infrared Security Image. // Applied Mechanics and Materials, Trans Tech Publications, Switzerland, Vol. 262 (2013), pp. 9-12.

[8] Ivan Pogarčić, Ana Agić, Maja Matas; Evaluation of the colorant twins for the neutral grey spectra in infrared graphic procedure; Tehnički vjesnik 23, 6(2016), p:1659-1664; ISSN 1330-3651, ISSN 1848-6339 ; DOI: 10.17559/TV-20150303132036 ; Hrcak ID: 169526

[9] Jana Žiljak-Vujić; Ivana Žiljak Stanimirović; Ivan Pogarčić: "Hidden Information with Infraredesign", Annals of DAAAM for 2010 & Proceedings of the 21st international DAAAM symposium intelligent manufacturing&automation: focus on interdisciplinary solutions, Zadar, Croatia (ed. Branko Katalinic), (2010.), p. 249-250, ISSN: 1726-9679, ISBN: 978-3-901509-73-5

[10] Jana Žiljak Vujić, Maja Matas, Mateo Pogarčić, Ivana Žiljak Stanimirović: " Topographic maps with infrared colors", 25th DAAAM International Symposium on Intelligent Manufacturing and Automation, 2014, Vienna, 2014-10-27, p. 1-8, ISSN 1877-7058 © 2015 The Authors. Published by Elsevier Ltd.Peer-review under responsibility of DAAAM International Vienna.

DOI: 10.2507/28th.daaam.proceedings.033

DESIGNING AN APPLICATION TO OPTIMIZE THE CALIBRATION OF OUTPUT DEVICES IN DIGITAL PRINT ENVIRONMENT

Milan Navrátil & Jaromír Polášek

This Publication has to be referred as: Navratil, M[ilan] & Polasek, J[aromir] (2017). Designing An Application To Optimize The Calibration Of Output Devices In A Digital Print Environment, Proceedings of the 28th DAAAM International Symposium, pp.0244-0252, B. Katalinic (Ed.), Published by DAAAM International, ISBN 978-3-902734-11-2, ISSN 1726-9679, Vienna, Austria
DOI: 10.2507/28th.daaam.proceedings.033

Abstract

Aim of this paper is analysis of measured data from spectrophotometric probe and design of methods applicable in printer operation for the optimization of linearization curves. The paper also deals with application design in Microsoft .NET platform to optimize the calibration of output devices in a digital print environment. The proposed application is based on both, the initial linearization requirements, i.e. before creating an ICC profile, as well as the possibilities of re-linearizing the printer to extend the life of the profiles. The statistical analysis, description of measured data, calculation of colour differences and possibilities that linearization for colour management offers are also described. The proposed application helps to eliminate the serious technological problems in the creation of output-ICC profiles just for the needs of digital printing machines and other output devices with CMYK output and with regard to the technology of sublimation of these colours on textile materials.

Keywords: Spectrophotometer; measurement; linearization; digital printing; .NET; application

1. Introduction

Current requirements for the operation of printers and printing plants generally place great emphasis on true colour processing as well as on output quality. Due to the increasing development of computer technology, the possibilities in the area of source data processing as well as the control of the output device as such are increased. Hand-in-hand with the development of new technologies, new challenges and related problems and obstacles are coming.

With the need for colourful fidelity, demands not only for accuracy but also for its own sustainability arise. The accuracy is assured with the well-established output profile. In this case, sustainability is meant to maintain the output profiles and to increase their lifetime. Materials and colours commonly available on the industrial processing market show some degree of variability in terms of their attributes within the range indicated by the manufacturer. Parameters may vary depending on the current production series, sub suppliers, regulatory or market requirements. Operation must be able to react flexibly to these changes and regularly re-calibrate the output profiles or create new ones.

Frequently neglected or mitigated technologies include linearization, which should be an integral part of creating and maintaining the output profile of the device. However, the current Raster Image Processor (RIP) tools and their equipment are only marginally devoted to this area and offer very little or no analysis or back testing. Some of them even encourage user in order to influence the measured values. This can lead to fatal consequences for back testing or re-linearization, as the same starting conditions are not maintained. The operator has modified the results, so they cannot agree with the same measurement in another time or by a measurement made by another person. Such a measurement does not reflect the real state of the machine but rather the ideal idea of the operator, based on his experience.

One of the problematic technologies for linearization is sublimation printing, which in the printing industry represents considerable potential in the field of printing of advertising articles and various textiles, both ordinary needs and specific designations. By introducing this technology, which significantly differs from the standard direct digital printing, at least in the way of its application, the tools used for maintaining and calibrating printers, possibly for output profiles, have appeared to be inadequate. Where existing tools have had rather shortcomings, sublimation has begun to show complete inability to make linearization at all. The measured values were noisy. After the introduction of this technology, even after several attempts and attempts to measure as accurately as possible and to observe the conditions set by the probe manufacturer, the linearization, especially the yellow channel, failed to be usable. This has fully demonstrated the need for an analytical tool to help reveal the problem or even to solve it. One of the solutions to this issue can be good analytical tools, which are already commercially available today, but with their educational requirements far exceed the knowledge and capabilities of the average machine operator. By repeating the procedure the operator chooses, the input and output conditions must be maintained. The entire operation must be repeatable, if necessary, with the same results or with differences caused by any changes in the parameters of the packing of the output device or of the print media itself.

Alternatively, the measuring device is a so-called spectrophotometric probe with a higher resolution or a smaller deviation. This solution, in addition to higher acquisition costs, brings further problems in terms of compatibility and widespread use. Top-class devices are designed and delivered with corresponding application-specific software especially for industrial use in production. From these reasons and due to their cost these devices cannot be often used. Operational requirements for these devices are also much higher than the requirements for an industrial print operator. The solution, therefore, should be to process data from a commonly used device that can be used with a wide range of RIPs and Profilers.

The last possible solution to this problem can be a simple analytical tool, capable of graphically displaying and archiving the measured values and the results of several measurements between them, or averaging the data so that the same input and output conditions are maintained for each linearization. This tool, if it retains the concept of simplicity and availability, could be a great help not only in large businesses, but also small in printers with limited portfolio that are interested in quality and sustainable production.

2. Digital print

The term digital printing includes a number of printing techniques. Its main feature is direct printing from digital data. This eliminates the preparation outside of the printing machine. In the case of digital printing, the machine only has the required print media. All that is needed for printing either arises directly in the machine or is a part of it. Digital printing works on a similar principle as desktop printers. The devices have their own RIP or a third party RIP that sorts print jobs and rasters print data. From the point of view of the principle, the most commonly used technologies are electrographic and jet or inkjet printing (so-called ink-jet). Powder or liquid toners or inks are used as the color. The advantage of this technology is minimal pre-printing and final output. Compared to other technologies, for example, the drying time is negligible [1], [2].

From a technical point of view, this is a very sensitive process that depends on many factors, such as the type and model of the output device, the type of media to which printing is applied, the ink setting, the resolution, the half toning method, print speed, output quality, preheat and media drying and more. Changing any of these attributes results in a change in appearance but also the color of the output. For this reason, it is necessary to check and record the environment in which the machine has been calibrated and, if any, change its recalibration. For example, simply changing the paperweight can completely change its water absorption and affect the overall output [1].

2.1. Calibration of digital print machines

Successful color reproduction across different materials and media requires calibration and color characterization of input and output devices. The calibration process will place the device in an up-to-date, unchanging condition that does not change or change over a certain range during the use of the device within a specified period. Characterization, on the other hand, de- fines the model of the input and output values of the device [3], [4].

Digital printing on textiles, compared to direct printing on a medium, requires much more production time. Printing, for example, on paper is relatively easy, but printing on textiles often requires further steps of the overall process such as sublimation or subsequent washing of textiles to remove remnants of non-dyed paint. The calibration of the output device for textile printing in one-step would be an effective way. However, this method is not suitable or feasible, since the overall calibration quality will be low. The two-step process is a good compromise between the quality of calibration and usability in real traffic on the other hand. The first step of the calibration is linearization and the second output of the output profile itself [1], [3], [5].

2.2. Linearization

Calibration of one color output channel is a basic and often neglected process of initial setting of the output device. The disadvantage of half-toning techniques used by modern printers is non-linear output. With regard to the most widespread techniques such as "dithering", "error diffusion" or "stochastic screening", the issue becomes even more complex. The use of modern headsets with variable droplet size and Light-Ink usage still contributes to the rise in non-linear output. The process consists in printing a calibration card with precisely defined fields, with the increment of each individual color and its subsequent measurement. The result is then displayed and can be manually adjusted if necessary [1], [6], [7].

The main aim of linearization is therefore the color characteristics of the output from the printer and the conversion of color information from the independent PCS color space into CMYK comprehensible to the printer. Since CIELAB and CMYK are not linearly dependent, there is a reason to create a linear approximation. Thus, linearization can be understood as the mathematical model of the output device as follows: [6]

$$(C, M, Y, K) = f(L^*, a^*, b^*) \quad (1)$$

This is one way to minimize the error between Original CMYK values and model values. Linearization of CMYK values into the CIELAB space will facilitate the colour characterization of the output device in many cases. Another reason for linearization is the often-used 3D Output Profiling Method called 3D LUT. Output profiles contain a 3D table defining a colour transformation from CIELAB to CMYK. Generally, it consists of a uniform 3D grid in CIELAB colour space. For each printed pixel, the colour transformation is calculated by linear interpolation between the values in the grid using interpolation patterns. This scheme can be a tri-linear or tetrahedral interpolation. Because conversion from CIELAB to CMYK is strongly nonlinear, this interpolation can lead to minor errors, especially if 3D LUT is defined on a very coarse grid. Linearizing CMYK values into CIELAB space using 1D LUT will convert more linearly using 3D LUT [6].

An important feature of linearization of printing machines is the resulting non-linearity. We are talking about getting non-completely linear output. Because the measured magnitude of most RIP and linearization tools is luminescent component L*, measured value 50 does not mean 50% coverage [1], [8].

The CIELAB color space is a 3D grid that has the L*, a* and b* axes. For measurements for linearization purposes, the following equation is offered:

$$D = \sqrt{(L')^2 + (a')^2 + (b')^2} \quad (2)$$

Where $L' = \max(L^*) - L^*$, $a' = \max(a^*) - a^*$, $b' = \max(b^*) - b^*$. The problem of this solution lies in the higher sensitivity of the human eye to the luminescent component (L*) than to the chromatic (a* and b*). The sum of all three components with the same weight will not result in a linear result. Another option is to use only the L* folder. For reasons of mesh sensitivity and positive values only, it seems to be the best solution. This solution will work with CMK channels, but for the Y channel, the yellow color is the dynamic range of the L* axis significantly smaller. Thus, the result on the L* axis will not yield satisfactory results. On the other hand, the b* axis is the largest in the yellow color.

2.3. Re-Linearization

Re-linearization is intended to recreate the output device model in order to correct errors resulting from its use. As time passes, the attributes of the output device change, reducing the quality of the output and its colour fidelity, i.e., match to the source. One option is to create a new device linearization. Another factor may be variations in the batch of fillings or print media. Generally, manufacturers recommend that they be linearized with each change of charge or a severe change in operating conditions. Re-linearization can bring the device with the output profile used so far to the state in which it was created when creating the first linearization and output profile without having to re-create the profile. Compared with profile formation, the linearization process is much less time consuming. Another possibility is a completely new linearization of the device. This operation is more time-consuming, but still incomparably less time consuming than creating a new profile. It is necessary to take into account that re-linearization or new basic linearization will remove only minor deviations. It cannot be expected that an error of magnitude ΔE of 3.2 will create an error ΔE = 1 [9], [10].

3. Measurement and data preparation

Textile materials with non-standard behaviour was selected. Additionally, the number of samples of each material was determined, and the measured data was subjected to statistical methods in StatSoft Statistica software. Measurements was accomplished using the third-party Roland WersaWorks Advanced RIP software supplied with the printer. The reasons for using this RIP are the possibility of direct communication with the printer and the export of data to *csv* format, which is easy to import into Statistica for subsequent data analysis.

3.1. Sample preparation

Print run on the Roland Versa Art RS-640 in default mode at 540 x 360 dpi, with three passes. Air conditioning in the printer room maintains a standard temperature of 18 ° C and a relative humidity in the range of 35 - 40 %. For our measurements, the temperature was raised to the desired 23 °C. Head feed rate was 964 $cm.min^{-1}$; feed rate was 30 $cm.min^{-1}$ (rates are specified by the machine manufacturer). Drops dropped onto transfer paper heated to 45 °C; the paper finished the drying at the same temperature. The machine was not equipped with airflow fans. Transfer to final materials was set after 24-hour technological break. Transfer was carried out on the machine from GMC Company. Next day, the machine was set to a temperature of 195 ° C and, after reaching this temperature; the prints were transferred to eight textile materials (see Table 1).

Mark / Stock weight (g/m^2)	**Using**	**Material, Composition (%)**
BLOCKOUT / 270	both-sided flags	PES 100
COOLMAX /200	thermos clothes	PES 100
DOUBLE / 160	sports dresses, t-shirts	PES 50, COOLMAX 50
LYCRA / 200	elastic clothes	PES 88, ELASTAN 12
SATIN / 135	pillows, dresses	PES 100
GARDEX / 172	tents, awnings	PES 100
UTEX / 69	inflatable items	PES, NYLON
POLYFLAG	flags, banners	PES 100

Table 1. List of measured fabrics and materials

3.2. Measuring process

The spectrophotometer *X-rite i1pro2* from Gretag MacBeth Company was used for colour measurement of the selected samples. This instrument is supported by many third party applications and from the point of view of the scan quality, is a good choice between quality and price and it is suitable for our purposes. The documentation lists the maximum and average differences between the measurements of twelve BCRA (British Ceramic Research Association) tiles. Differences are reported in ΔE94* at 23 °C for D50 lighting and 2° angle as follows: $\Delta E94*_{max} = 1.0$ and $\Delta E94*_{avg} = 0.4$ Documentation also provides a procedure for short-term repeatability where ΔE94* = 0.1 on a white surface (not specified) for D50 mode at an angle of 2 °, at ten repetitions every 3 seconds [11].

Before starting the measurement process it is necessary to determine the optimal number of measurements. If we proceed from the repeat measurement procedure that shows the smallest difference ΔE, the best adherent is ten repetitions for each material. It is customary to perform five measurements at the first linearization and only three measurements for control or re-linearization. By selecting ten measurements, we obtain a sufficient data reserve for extreme values. All measurements were made on lanes consisting of 21 samples with size 10 x 10 mm as shown in Fig. 1.

Fig. 1. Measurement of prepared CMYK testing target of a particular sample.

Each copy of the material was measured with the spectrophotometer and then the data was numbered and divided into folders by names for further analysis and visualization.

4. Design of user application

In order to create a usable application, it is necessary to define a list of requirements that will further guide the development of the application. This list serves as a guideline for designing individual components and logic of the entire application. It makes it possible to define clearly not only the framework in which the application will work, but also the means of support that it will require and the way, how it will be developed. Microsoft Visual Studio 2015 development environment was used.

4.1. Application requirements

Overview of non-functional requirement

The application must be capable of being usable in the operation for which it is designed. They have to meet their demands and requirements to fit existing equipment and at the same time maintain a certain degree of versatility. Its demands and requirements cannot exceed the standard equipment of a regular printer and its software tools. Most RIPs are running on Windows. On Mac platforms, it is more likely to prepare graphics as such than directly to print.

The first non-functional requirement, as mentioned in the previous paragraph, will be the MS Windows operating system in the currently still supported version 7. Advantage of maintaining the same platform as that of the RIPs is concurrency on a single PC. Therefore, you can switch between applications, you will not have to use another PC or run another operating system. The ".NET Framework" is available as an application runtime environment. This extensive software platform is designed to develop many kinds of different applications. It allows programming in several languages (e.g. C #, F #, J # VB). All these languages are translated into runtime itself (CLR) at start up. An enormous range of functions and libraries provides the necessary calculations or operations in development, the need for designing of own visual elements is avoided [12], [13].

The user application is written in the C# programming language using the *System.Windows.Form* libraries that are designed for application development in MS Windows graphical user interface.

Overview of functional requirements

Workspace:

- The application (hereinafter App.) must work in a clearly defined work environment. This environment must retain information about external links. Specifically, it relates to the device on which the linearization is created and the material for which it is processed. This environment is called "Workspace".
- This environment should also be a common collection for all linearization created on the same machine and material and must somehow archive the results of linearization operations.
- The environment should create linkages between linkages and allow for a general overview.

Linearization:

- App. must be able to display one or more linearization in a common context so that the user can visually compare their values and evaluate errors or deviations. If this requires manipulation of axis, zooming, or zooming data, the application must include these elements.
- App. must under no circumstances modify or otherwise modify the values from loaded linearization. For the results of different operations, it is necessary to create a new linearization. The original (measured) data must always be retained.

Calculations and modification:

- App. must include methods and elements for calculating the deviation ΔE by all methods specified in the theory. The user must be able to choose between these methods and, in the case of optional parameters, have the option to enter these parameters.
- App. must be able to transform the b* to L* axis on any channel to ensure the ability to adapt the application to future knowledge in the art.

File manipulation:

- App. must be equipped with standard file handling controls.
- App. should be able to manipulate disk data without having to wait for the result of this operation (background work).
- App. should periodically back up the work, so that the work can be restored in the event of a fall or a power outage. This operation should not restrict users while working with the application.
- App. should be able to save their work on termination.

Scenarios and diagrams

Use case diagrams help determine the scope of the application and present the user interface as the user accesses. It determines the mechanisms and operations that will be required after the application and the individual steps leading to the desired goal. This information is the starting point when designing the application design and ensuring that it fulfils the expected function.

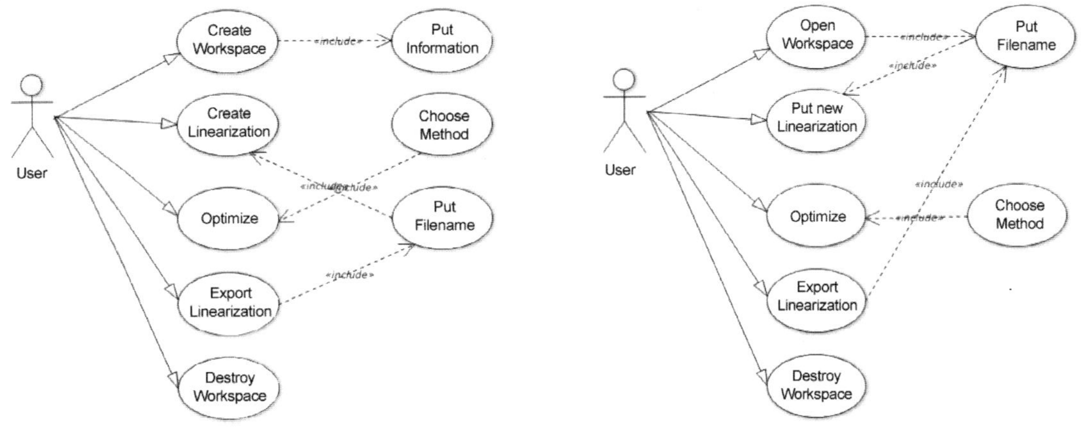

Fig. 2. Use case diagrams: Linearization (left) and Re-Linearization (right).

4.2. Structure

The application structure is based on a proven application-programming model for MS Windows. After launch, the main thread creates a custom application window, the *System.Windows.Form* descendant, which encapsulates all the classes and methods necessary for proper functionality.

Class Workspace

The instance of the workspace class is generated and extinct depending on the creation or termination of the work environment. It has methods to create custom instances based on user-entered data or data obtained by deserializing the stored instance on disk. It is also able to save to disk, and end its activity if necessary to release the resources for another instance.

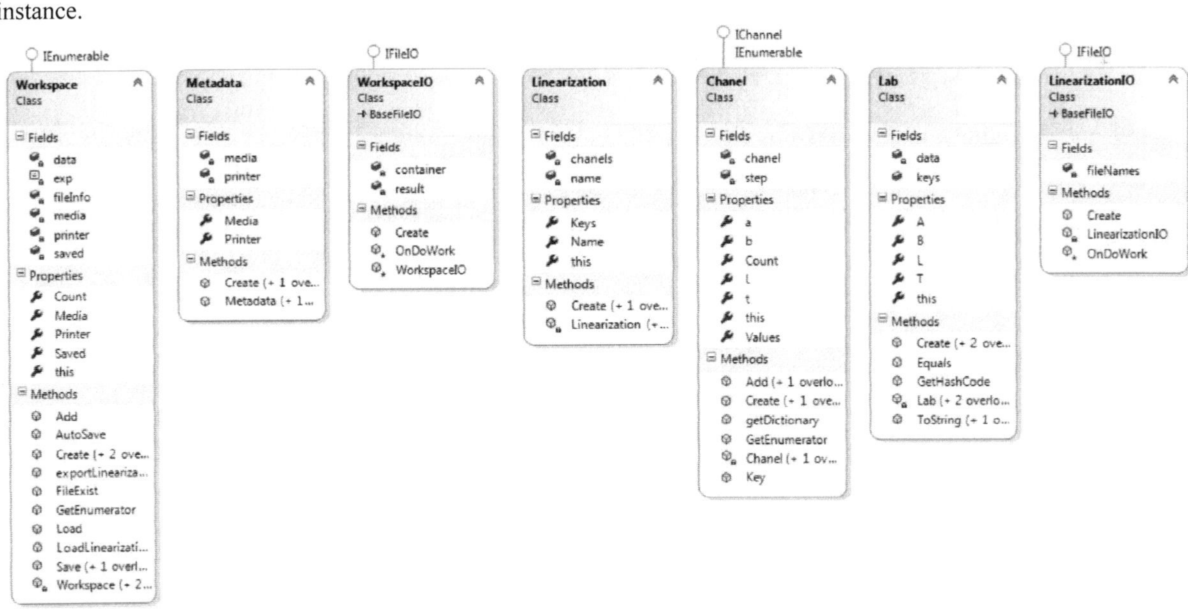

Fig. 3. Design of classes: Workspace, Metadata, WorkspaceIO, Linearization, Channel, Lab, and LinearizationIO.

The Workspace class requires a Metadata class to store printer and material information. It is serialized in parallel with the Workspace class. The Metadata class requires this information and is not able to change it during its existence, just as the workspace is not allowed the Metadata class to change. This enclosure guarantees that data in the work environment is always in relation to one specific machine and material.

The WorkspaceIO class implements the IFileIO interface, which defines methods for communication with a storage. It contains only the onDoWork method, which allows user to specify data to be written. The rest of the class is inherited by BaseFileIO, which includes methods for manipulating files and working with them. For its function, it needs a container class that encapsulates and structures the data before writing to the repository.

The Linearization class instance serves to preserve just one linearization. This class is the most important element of an application. It divides the linearization on and its data into three parts. The smallest element is the Lab class. This class holds the values of just one measurement, one specific field on a linearization target. The internal vocabulary is used for preservation and there are properties called after axes CIELAB L, A and B (see Fig. 4), these instances are grouped into the Chanel class, which represents one particular measured colour. The Chanel classes are then grouped into one Linearization class.

The LinearizationIO class, as well as its sister class, WorkspaceIO, implements the IFileIO interface and describes how to write to a disk using the OnDoWork method. This class will also inherit BaseFileIO. This procedure provides for the reusability of code for other linearization formats or files that the application might require in its further development.

Interface IDelta

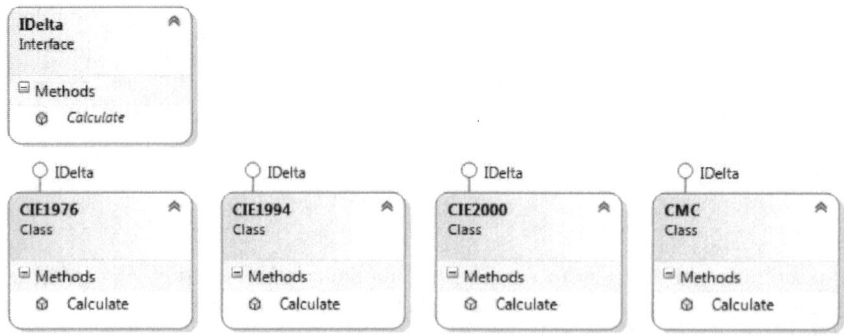

Fig. 4. Interface IDelta including implemented classes.

This interface implements four classes representing the calculation methods ΔE CIE1976, CIE1994, CIE2000 and CMC. The input parameters are instances of class *Lab* (at least two) in the form of the list, and the output is a variable of double type that represents the resulting ΔE.

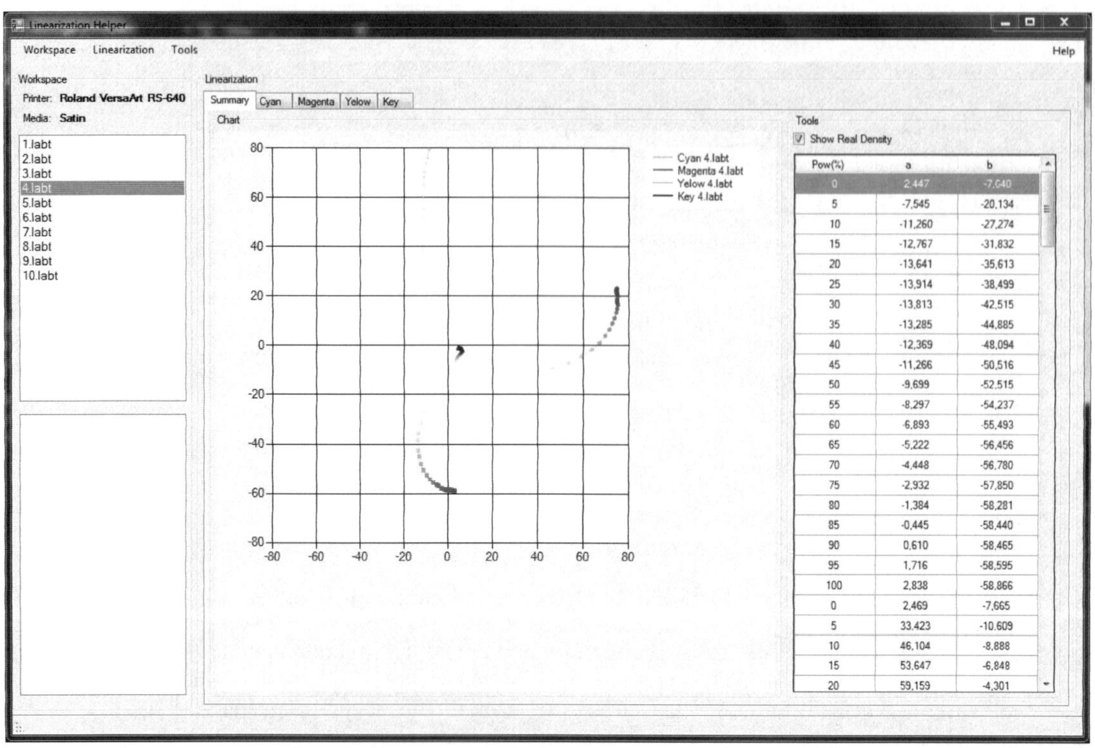

Fig. 5. Design of the main window of the application – Linearization panel - Summary.

The main application window is divided into the context menu at the top of the window, the status bar at the bottom; see Fig. 5. There are two vertical panels. The *Workspace panel* works with linearization files and the *Linearization panel* for visualizing and optimizing data.

Workspace panel contains the name of the printer and the material for which Workspace was created. This data has an informative function and cannot be changed so that Workspace cannot be confused. Two linearization lists are located below and belong to this Workspace. The linearization list at the top contains linearization retrieved from the files and the list at the bottom is created by this application. From these panels user can select individual linearization or groups of linearization, which they will then appear in the right side *Linearization panel.* It contains several tabs, their number depends on number of colors available for the printer to which linearization is managed. The first tab from left is *Summary* tab that shows the C_{ab} chromatic diagram in which the individual measurements in mutual relationship are placed.

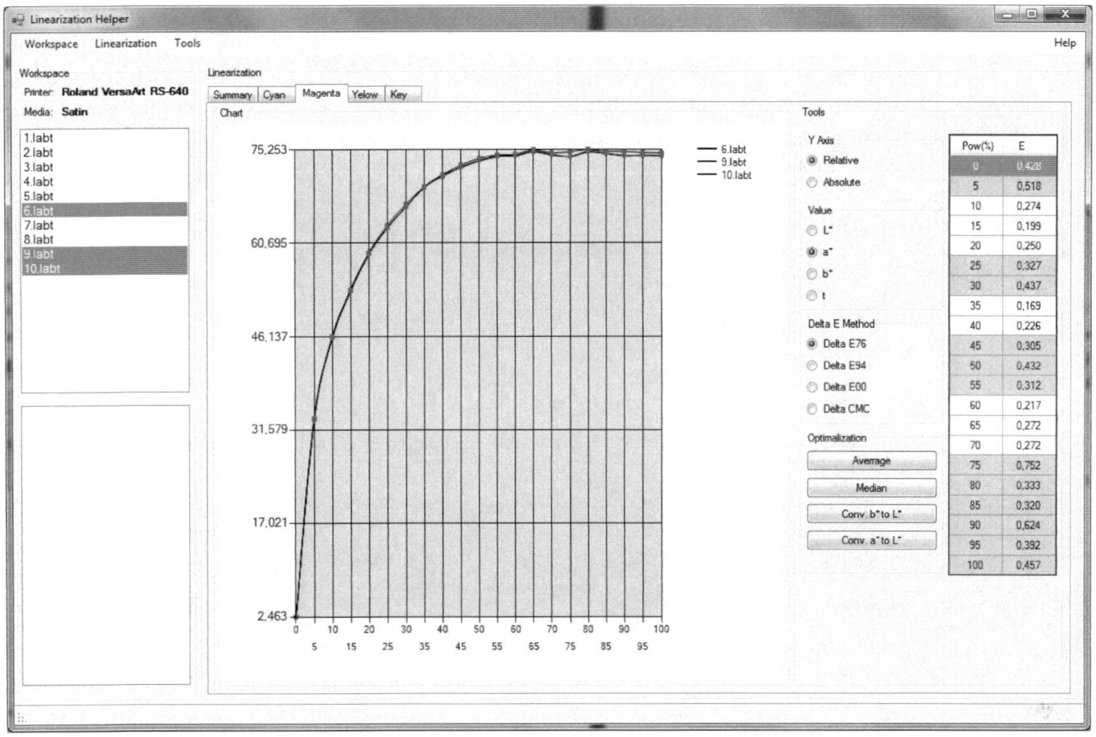

Fig. 6. Design of the main window of the application – Linearization panel – Magenta tab.

The color tabs are labeled with the names of the printer's direct colors for which the linearization was created (see Fig. 6). The tabs represent each color channel for the device. This view allows user to individually process each channel. The color tab content is divided into the *Chart* view and the *Tools* section. In the visualization section there is a graph showing one selected CIELAB axis or "Density" in one or more linearization in context. The legend to the right of the graph shows which linearization are currently displayed. On the color tab, there are shown optimization methods that allow correction of linearization or grouping of linearization.

5. Conclusion

The aim of this paper was the design of the user application to optimize the calibration of output devices in a digital print environment. It covered collection of spectrophotometric data, their subsequent analysis and design of suitable methods for optimization. The reason for its inception was the serious technological problems in the creation of output ICC profiles just for the needs of digital printing machines and other output devices with CMYK output and with regard to the technology of sublimation of these colors on textile materials.

The analysis of the measured data revealed numerous shortcomings in the resolution of the measuring device, which proved to be insufficient within the parameters set by the manufacturer. Furthermore, it has been found that this device exhibits, on some samples, several times greater variance of the measured data than the frame given by the average and maximum deviation set in the documentation by the manufacturer. Therefore, it is recommended for the above-described printing technologies and especially for sublimation printing to use devices capable of adhering to these tolerances also on textile materials.

During the inspection of the measuring device it was found that the range of data measured for the yellow channel (Y), namely the data for the CIELAB colour space axis L*, is less than the resolution of the measuring device. This feature has helped detect unpredictable results when measuring this colour channel. After further analysis, it was marked as the main cause. In addition, the dependence of error size on the spot colour used and the need to optimize each channel individually was determined. The ability to optimize linearization as a whole, i.e. all channels by the same method, proved inappropriate. On the contrary, from the theory and from the results of the statistical analyses, the method of gradual adjustments seems to be the most appropriate.

The main output of this work is, however, an application that we can describe as a librarian editor. This application is able to visualize data both directly in the table, graphs, and has tools to detect some problems that are difficult to detect by default. The application does not aim to compete with professional analytical tools, but only to enable digital printing machines to conduct timely detection of possible errors or undesirable properties of materials or technologies. It was designed so that without any more serious hardware or software entitlement, it could be used in existing operations and able to perform its function. Its main advantage is the ability to work with multiple linearization and to produce averages or medians from the measured values.

It is also capable of transforming axes always to take into account the most significant component of a particular colour channel whose changes will be most noticeable in the event of gradual degradation of equipment or filling or replacement.

The optimization features of the application, which, when called, create new linearization files, allow for easy archiving, the ability to compare each step with a predefined state, and always preserve the original measured values. On the other hand, they do not allow manual intervention in the values to prevent the user from influencing the results and thus the overall re-linearization process.

For future development, it is critical to the application that it has been written with the maximum use of programming methods over the interface. This property gives great potential for its expansion in the future, not only in terms of addition of new procedures and optimization methods, but it is also possible to add new file formats containing linearization and even in different colour spaces without major changes to the code.

6. References

[1] Ujiie, H. (2006) Digital Print of Textiles, Woodhead Publishing, ISBN 9781855739512, Cambridge (England).

[2] Dvořáková, Z. (2008) DTP and prepress: complete guide from graphic design to professional printing (in Czech). Computer Press, ISBN 9788025118818

[3] Fernandez-Maloigne, Ch. (2012) Advanced Color Image Processing and Analysis. Springer Science & Business Media, ISBN 1441961909, NY (USA)

[4] Homann, J (2008) Digital Color Management: Principles And Strategies For The Standardized Print Production, Springer Science & Business Media, ISBN 9783540693772, Berlin

[5] Schanda, J. (2007) Colorimetry: Understanding the CIE System, John Wiley & Sons, ISBN 9780470175620, USA

[6] Noyes, Y., Hardeberg, J. & Moskalev, A. (2000) Linearization Curve Generation for CcMmYK Printing. Proceedings of 8th IS&T/SID Color Imaging Conference 2000: Color Science and Engineering: Systems, Technologies, Applications, Scottsdale, Arizona (USA), ISSN 2169-2629, ISBN 9781622766284, pp. 247-251, Society for Imaging Science and Technology, Springfield (USA)

[7] http://support.rolanddga.com/docs/documents/departments/technical%20services/manuals%20and%20guides/support_doc_color%20calibrating%20a%20print%20mode.pdf, (2009). Pozorski, J., Roland DGA Corporation, Color Calibrating a Print Mode: Instructions on How to Linearize Your Printer, Accessed on: 2017-08-24

[8] https://www.xritephoto.com/documents/literature/en/L11-029_color_guide_EN.pdf, (2004). X-rite, The Color Guide and Glossary, Accessed on: 2017-08-24

[9] https://www.yumpu.com/en/document/view/30992971/kind-of-document-quentin, (2006), Kornetka, S., efi GmbH, Re-Linearization vs. L*a*b* Correction: Three Ways to Achieve Good Printer Adjustment, Accessed on: 2017-08-24

[10] http://docs.printfactory.cloud/, (2017), PrintFactory Knowledge Base, Accessed on: 2017-08-24

[11] https://my.xrite.com/documents/apps/public/Manuals/EO2UV-QSG_i1Pro_UVcut_User_Manual.pdf, (2016), X-rite, i1Pro2 User Manual, Accessed on: 2017-08-24

[12] https://msdn.microsoft.com/en-us/library/w0x726c2(v=vs.110).aspx, (2017), Microsoft, .NET Framework 4.7, 4.6, and 4.5. Microsoft Developer Network, Accessed on: 2017-08-24

[13] Djambic, G., Kucak, D. & Fulanovic, B. (2012) Empirical Comparison of Different View state approaches on Performance of ASP.Net Web Applications, Annals of DAAAM for 2012 & Proceedings of the 23rd International DAAAM Symposium, ISBN 978-3-901509-91-9, ISSN 2304-1382, pp 0733 - 0736, Editor B[ranko] Katalinic, Published by DAAAM International, Vienna, Austria

DOI: 10.2507/28th.daaam.proceedings.034

ANALYSIS OF MECHANICAL PROPERTIES OF ALTERNATIVE MATERIALS FOR PROCESS FUSED DEPOSIT MODELING

Ahmet Cekic, Murco Obucina, Derzija Begic-Hajdarevic & Miran Hebibovic

This Publication has to be referred as: Cekic, A[hmet]; Obucina, M[urco]; Begic-Hajdarevic, D[erzija] & Hebibovic, M[iran] (2017). Analysis of Mechanical Properties of Alternative Materials for Process Fused Deposit Modeling, Proceedings of the 28th DAAAM International Symposium, pp.0253-0257, B. Katalinic (Ed.), Published by DAAAM International, ISBN 978-3-902734-11-2, ISSN 1726-9679, Vienna, Austria
DOI: 10.2507/28th.daaam.proceedings.034

Abstract

For the greater use of additive technologies in the industry, research is needed to discover new materials for the rapid creation of more functional products. New materials for additive production should overcome, with their properties, all the shortcomings of the most common polymer materials. On the other hand, it is preferable for the materials to be environmentally friendly and biodegradable. This paper presents experimental research results of the additive process parameters influence of Fused Deposit Modeling (FDM) process on product quality, using different materials. Two polymeric materials, PLA and Laywoo-D3 were used for the production of the samples. The obtained results show that for certain uses of the product and certain exploitation conditions it is possible to replace the environmentally inferior materials with the new materials obtained from the wood waste with a certain percentage of the adhesive. This has a positive impact on the environment, because wood is a natural and ecologically pure material (CO_2 neutral).

Keywords: Additive Manufacturing; Fused deposit modeling; additive process parameters; mechanical properties of materials

1. Introduction

In order to meet the ever-demanding demands of the open world market and the earlier market entry of products as well as the faster return of invested funds, it is necessary, among other things, to reduce the development and production time to a minimum. An additional trend that is increasingly noticeable in certain segments of the market is the abandonment of mass production in favor of small-scale, and very often individual (personalized) production. To achieve these goals, Additive Manufacturing is increasingly being used today [1, 2]. The basic concept of Additive Technologies is that the products are made by adding and bonding the layer by layer. Additive Manufacturing processes do not require process flow planning, mold making, specific equipment for working with materials, transport between jobs, etc. It is important to note that to date on the world market, numerous of different Additive Manufacturing techniques (AM) have been implemented, and also devices with different technical characteristics are also available to users for all additive technologies. Namely, in addition to AM technologies, they differ in the way they generate the model, source of energy, starting material, etc. For each of these technologies, there are numerous devices produced by different manufacturers, and therefore different features (resolution, print speed, layer thickness, product dimensions, etc.). Although many different additive technologies have been implemented to date, they are not perfect and have certain limitations. The biggest drawback, at present, is the restriction on the use of certain materials [3 - 6]. One of the more frequent additive procedures is the *Fused Deposition Modeling* process - FDM, which is the subject of this paper. FDM is a process in which the wire-like material (polymer or composite) is melted and applied to the working surface by layer by layer in order to create a given model/product. According to the *Wohlers Report*, it is estimated that in the total market for additive technologies, FDM technology occupies more than 50% of the world market, [7]. In order to take advantage of this technology, it is necessary to analyze and define numerous influencing parameters. The most influential parameters of this process are:

- The temperature of the melted material for printing, as well as the temperature of the substrate and the chamber (quality and composition of the particular material matrix),
- Print speed and Layer thickness (device resolution),
- Filling density (from 0% to 100%) and types (rectangular, triangular, corrugated, compact, 3D) model
- Number of external shell models,
- Orientation of the model (vertical, horizontal or combined),
- Support structure, etc.

Additionally, given the availability of numerous different types of materials for making the product by FDM, the choice of the appropriate material is a challenge, [8, 9]. In addition to the model/product function, the selection of the appropriate material should take into account the following facts:

- Wire material diameter (type of device)
- Material color (model function)
- Class, type, condition and packaging materials, etc.

Materials for the FDM process can be polymeric (PLA, ABS, etc.) or composite, e.g. based on wood, metal or chalk powder or carbon fiber in combination with some of the polymers. The paper analyzes the possibilities of using alternative wood based material. Wood-based material (e.g. Laywoo-D3) is a mixture of recycled wood and a binding polymer. Wooden filaments from the PLA group contain an addition of wood fibers that give the finished products the appearance of wood. The percentage of these wooden accessories go up to 40%. The material has similar thermal endurance as a pure PLA. Using this material will produce products that will look and smell of wood, and can be further processed the same as wood.

2. Experimental research

Previous research and optimization of the process parameters for achieving maximum product quality was done at the Center for Additive Technologies (CATeh) in Zagreb. After analysing the impact of all parameters on product quality and determining the optimal parameters for producing better products, the tubes were manufactured according to EN ISO 527 2010 standard types, type 5. Tubes were manufactured with two different materials in order to test and compare their mechanical properties. The basic characteristics of the materials used for making the tubes are given in Table 1. It is important to note that the material Laywoo-D3 is a composite material consisting of a combination of polymeric material (PLA) and recycled wood (the share of wood in the mixture ~ 40%). The material has a very low vitrification and good adhesion to the substrate. It is very muddy (like a filament), but it's much more flexible after printing. After printing, the products are soft, but after 24 hours they become stiffer. The reason for the lack of information about the Laywoo-D3 material is that it is relatively new on the market ie. the necessary research for the purpose of its greater application for the FDM procedure has not been completed, and the authors of this work plan for future research.

Material	**PLA**	**Laywoo-D3**
Manufacturer	Flashforge	LAY Filaments
Accuracy of production	+/- 0.02 mm	+/- 0.02 mm
Diameter of the wire	1,75 mm	1,75 mm
Melting temperature	(160 – 190)°C	No information available
Print temperature	(180 - 230)°C	(175 - 250)°C
Tensile strength	(37 – 60) Mpa	No information available
Elongation	6%	No information available
Modulus of elasticity	4 Gpa	No information available
Density	1,3 g/cm3	No information available

Table 1. The basic characteristics of the materials used for making the tubes

The preparation of the tubes was performed on the *Flashforge Finder* and the process parameters are given in Table 2. The following software packages were used to create the tube:

a) **SolidWorks 2015** (used for three-dimensional modeling of the tubes/models in order to obtain 3D CAD models which will then be saved as *.stl files*),
b) **Autodesk Netfabb 2017** (used to validate and repair the *.stl f*ile, and can also be used to prepare a print model),
c) **FlashPrint** (used to set the model, parameters, and cut of the models for the print process, ie. for its preparation, this program is the device that follows the device used).

Material	PLA	Laywoo-D3
Thickness of the first layer	0,27 mm	0,27 mm
Thickness of other layers	0,18 mm	0,18 mm
Raft	YES	YES
Wall	YES	NO
Brim	YES	NO
Fill (Slow type)	100%	100%
Melting temperature	220°C	195°C
Number of layers	21	21
Print time	85 min	73 min
Predicted material consumption	11,9 m	9,91 m

Table 2. Process parameters in the preparation of the test tubes for tensile testing

In addition to the aforementioned tube designer, testing of the tensile material was carried out in the laboratories of the Department of Mechanical Production Engineering at the Faculty of Mechanical Engineering in Sarajevo. The aim of testing the tensile material is to compare the obtained values for PLA and Laywoo-D3 materials. Material testing on tension was done in two ways:

a) The classic procedure at the Zwick Materialprufung Tensile Testing Machine 1435. The gauge range of the Tensile Testing Machine is from 0 kN to 50 kN, and the displacement is 0.01 mm.
b) Using the same testing machine, measuring tape and digital signal amplifier for drawing diagram force-unit elongation (F-ε) for Laywoo-D3 material. The measuring strips for the 6/120LY11 steel of the manufacturer HBM (Figure 1a) with characteristics: resistance (120Ω ± 0.35%), the active length of the measuring strip (6mm), strip factor (k-factor 2.03 ± 1%) were used, thermal coefficient ($95 \cdot 10^{-6} K^{-1}$). Digital Measuring Amplifier System - DMC 9012 A manufacturer by HBM (Figure 1b) with built-in amplifier DMV 55 for connection of electro-resistive measuring strips with characteristics: accuracy class (0.05), measuring range (± 3mV / V), maximum sampling rate (9600Hz).

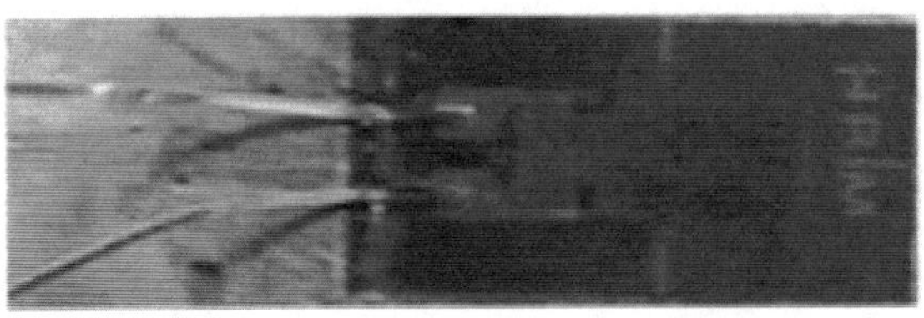

a) Electro-resistive measuring tape

b) Digital measurement booster system

Fig. 1. Testing with measuring strips

After testing the tubes made of PLA and Laywoo-D3 materials, it can be determined the brittle fracture on all the samples on the basis of fracture surfaces for both materials that it was performed. Figure 2 shows Hook diagrams for PLA and Laywoo-D3 materials obtained by testing 10 tubes for comparison. The general conclusion is that all tubes achieved roughly the same measurement results, which is confirmed by the validity of the obtained F- Δl diagram, and therefore, in Figure 2, is given one diagram (average value) for both materials for transparency. During testing tubes of PLA material, maximum force (load) of F = 1554N was recorded, while the maximum force (load) F = 377,3N was recorded by testing the tubes produced from the Layoo-D3 material. Based on the diagram (Figure 2), we can conclude that Laywoo-D3 is a ductile material, but significantly lower tensile strength. Based on the maximum measured force values, the tensile strength for PLA is σ_{PLA} = 38,86MPa, and for Laywoo-D3 it is $\sigma_{Laywoo\text{-}D3}$ = 9,55MPa. In addition to the above, the PLA has a lower contraction and elongation in yield on the Laywoo-D3 material. The contraction for PLA material on average is A_{PLA} = 4,71%, and elongation δ_{PLA} = 3,66%. For the Laywoo-D3 contraction, $A_{Laywoo\text{-}D3}$ = 5,86%, and elongation $\delta_{Laywoo\text{-}D3}$ = 4,93%.

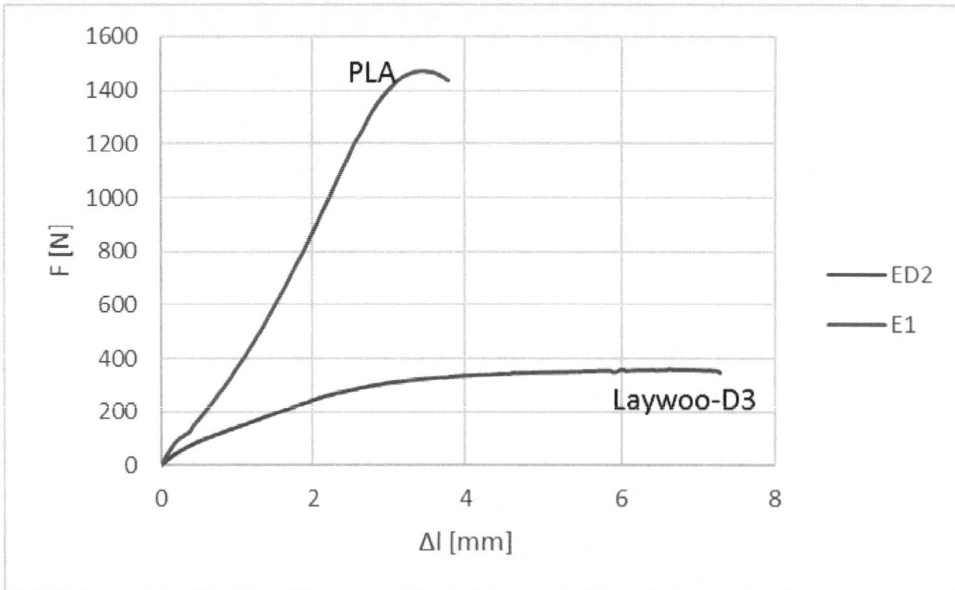

Fig. 2. Display of Hook Diagrams for PLA and Laywoo-D3 Material

Since it has been found that the Laywoo-D3 material is of a lower strength class, a measurement of the measured values and the use of measuring strip will be performed. Prior to the start of the measurement, the surface of the tubes was prepared for measuring the tape. In our case, two measuring strips were used (active and complementary) (Figure 3), where they used the Vilson semi-conductor for their connection. When connecting to the semi-conductor, it is essential that one contact from the active measuring strip is connected directly to the amplifier system. The second contact binds to one contact of the compensation measuring strip, so that these two contacts connect with one conductor to the measuring amplifier. The second contact, also from the compensation strip, is coupled to the measuring amplifier system via the conductor. Therefore, in the measuring reinforcement system there are three guides, one of which is common for two bands, and the other two go separately from each strip. In addition, the reinforcement system binds to ground.

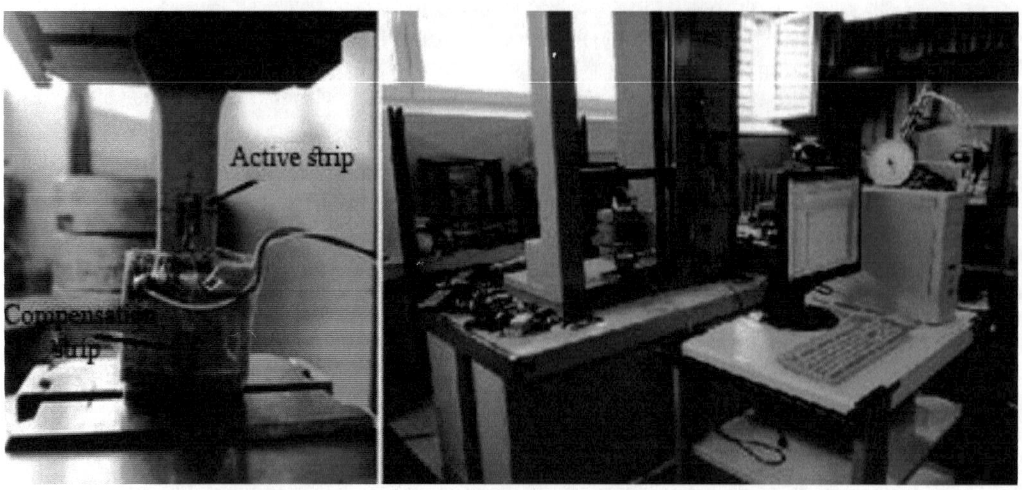

Fig. 3. Measuring equipment with the measuring strips

During the test with the measuring strips, due to their sensitivity, care should be taken to ensure that they do not get tangled. Testing of the material was made up to the 260N load. During measurement, unit elongation monitoring was performed, which was changed in the unit of time. The unit elongation values were recorded on the basis of the displayed load on the tensile testing machine. Measurement of unit elongation for load values from 0N to 260N was performed. After the measurement has been performed, in order to compare the results obtained by this method with the previously applied method, all elongation values are divided by the initial length of the measuring part of the tube, i.e. $L_0 = 30$mm, in order to obtain a unit elongation. In Figure 4, Hook Diagrams are shown, in order to compare the obtained data using the classical method and the use of measuring strips. On the basis of the obtained diagrams we can conclude that the measured values on the tensile testing machine are valid.

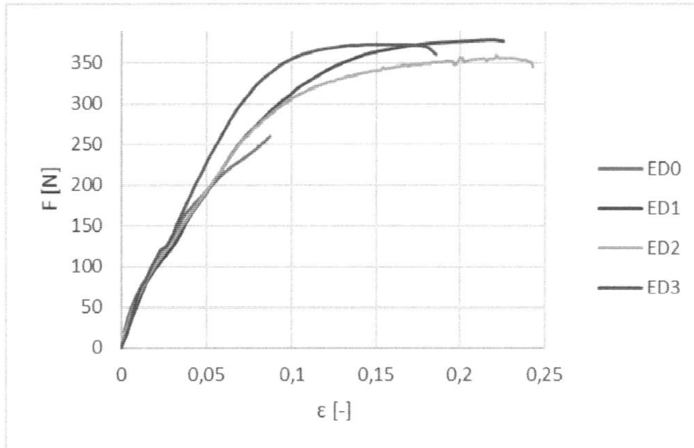

Fig. 4. Display and comparison of Hook diagrams for the Lajvoo-D3 material

3. Conslusions

Benefits of additive technologies, and therefore FDM (Fused Deposit Modeling) technology, are faster product development regardless of the complexity of the same. In this paper are presented the obtained experimental data on the research of the application of alternative materials with the additive process. The preceding testing, optimal parameters were defined for both materials: temperature and print speed, and resolution, occupancy, etc. parameters are defined for achieving the maximum quality of products on the given device. By testing the tensile material, it was found that the PLA material had about four times the tensile strength, but that the Laywoo-D3 was ductile. By further comparing the materials used, the Laywoo-D3 has an extremely good adhesion and no vitreousness occurs when printing, while an adequate substrate for better adhesion is required for the PLA material. Which material of the two listed will be used depends on the purpose of the product. It is recommended to use wood based materials, and if more solid products are required, the use of PLA materials is recommended.

For further research it is recommended to test the mechanical characteristics of (available and new) materials as well as other indicators of the quality of manufactured products with the added additive technology. It would also be desirable to make a test tube with the sameparameters, but on other similar FDM devices in order to compare the data obtained. The greatest emphasis on the development of this technology in the future is actually the development of new materials that can be used in the production of different products, which is a plan for further research by the authors of this paper.

4. Acknowledgments

The authors gratefully acknowledge the support of the Faculty of Mechanical Engineering University in Zagreb and Centre for Additive Technologies – CATeh Zagreb for this work.

5. References

[1] Valerga, A. P.; Batista, M.; Fernandez-Vidal, S. R.; Gomez-Parra, A. & Barcena, M. (2016). Preliminary Study of the Influence of Manufacturing Parameters in Fused Deposition Modeling, Proceedings of the 26th DAAAM International Symposium, ISBN 978-3-902734-07-5, ISSN 1726-9679B. Katalinic, B. (Ed.), pp.1004-1008, Published by DAAAM International, Vienna, DOI: 10.2507/26th.daaam.proceedings.141

[2] Cekic, A.; Rasovic, N.; Obad, M.; Kaljun, J.; Dolsak, B. & Begic-Hajdarevic, D. (2016). Production of optimized layered products using intelligent support, Proceedings of the 26th DAAAM International Symposium, ISBN 978-3-902734-07-5, ISSN 1726-9679, Katalinic, B. (Ed.), pp. 0271-0279, Published by DAAAM International, Vienna, City, DOI:10.2507/26th.daaam.proceedings.037

[3] Mohamed, O.; Masood, S. & Bhowmik, J. (2015). Optimization of fused deposition modeling process parameters: a review of current research and future prospects, Advances in Manufacturing

[4] Godec, D. & Šercer, M. Aditivna proizvodnja, ISBN 978-953-7738-26-6, Zagreb, 2015.

[5] Kalani, K. H. & Richard, H. 3D printing For Dummies A Wiley Brand, New Jersey 2014.

[6] Boschetto, A. & Bottini, L. (2015) Triangular mesh offset aiming to enhance Fused Deposition Modeling accuracy, International Journal of Advanced Manufacturing Technology, 80 (99 – 111).

[7] https://wohlersassociates.com

[8] Salam, N. K.; Syed, H. M. & Omar, A. M. An investigation on impact resistance of FDM processed Nylon-12 parts using response surface methodology, Australia 2016.

[9] Cupar, A.; Pogacar, V. & Stjepanovic, Z. (2015). Shape Verification of Fused Deposition Modelling 3D Prints, International Journal of Information and Computer Science, 4. pp 1-8.

DOI: 10.2507/28th.daaam.proceedings.035

METHOD OF SPATIAL PATH PLANNING FOR MOBILE ROBOT IN UNKNOWN ENVIRONMENT

Dmitry Yukhimets[1, 2], Alexander Zuev[1, 2] & Anton Gubankov[1, 2]

[1] Institute of Automation and Control Processes FEB RAS, Vladivostok, Russia
[2] Far Eastern Federal University, Vladivostok, Russia

This Publication has to be referred as: Yukhimets, D[mitry]; Zuev, A[lexander] & Gubankov, A[nton] (2017). Method of Spatial Path Planning for Mobile Robot in Unknown Environment, Proceedings of the 28th DAAAM International Symposium, pp.0258-0267, B. Katalinic (Ed.), Published by DAAAM International, ISBN 978-3-902734-11-2, ISSN 1726-9679, Vienna, Austria
DOI: 10.2507/28th.daaam.proceedings.035

Abstract

This paper deals with developing of method of spatial path planning for mobile robots of different types moving in unknown environment with obstacles. This method plans trajectories base on third-order Bezier splines using the data from onboard range sensors. It allows to automatically make decisions about the strategy of obstacles avoidance in horizontal or vertical planes considering the restrictions of mobile robots' propulsion system. Results of simulation confirm its high efficiency, low complexity and small computational capacity.

Keywords: mobile robot; path planning; obstacle avoidance; spline Bezier; smooth trajectory; unknown environment

1. Introduction

Today the application field of mobile robots (MR) of various types is continuously expanding. At the same time the tasks solved by the MR in the autonomous mode without control of an operator in conditions of an unknown in advance environment get special significance [1, 2]. These tasks include transportation of cargos of various function, inspection of structures, search operations and so on. In the process of performance of these tasks the information control system (ICS) of the MR must form movement trajectories on the basis of information obtaining from onboard sensors. These trajectories allow the robot to move quickly and safely in an unknown environment containing obstacles.

The formation task of the MR trajectories in an environment containing obstacles is one of the main problems in development of the ICS for MR. Many methods which can be divided into two groups is already developed for it solution. The first group methods include the methods forming trajectories based on use of various methods of finding the shortest path. Such methods include algorithms of finding an optimal path on a graph [3,4], genetic algorithms [5-7] and other methods [8-10], which are used in the presence of a priori information describing obstacles location. In the case when the MR moves in a partially unknown or dynamic environment the trajectory should be planned dynamically based on information coming from the onboard MR sensors and predefined movement patterns [11, 12]. The disadvantage of the approaches of this group of methods is the high computational complexity and a need to develop special heuristic rules for the trajectories formation. In addition, the trajectories formed by them are a sequence of base points, over which the MR must pass. It requires additional application of various interpolation algorithms to construct smoothed motion trajectories.

The second group includes approaches based on the method of potential fields [13-14], which allow the MR to move in an unknown in advance environment. The methods form control signals applied to MR actuators as a sum of the robot reactions to the signals obtaining from their sensors. However, such approaches do not allow to control the robots' trajectories during their movement. Also, they can lead to a sharp change in the movement direction of the MR, which are poorly carried out by feedback systems of the robots.

Despite the large number of methods of the MR trajectories formation almost all of them consider the problem of plane motion. It significantly reduces the functionality of the MR moving in space (underwater vehicles (UV) and unmanned aerial vehicles (UAV)). The possibility of spatial motion allows these robots in some cases to avoid obstacles from above without changing the movement direction in the horizontal plane. Also, the orientation of its onboard sensors continuously changes, when the MR is moving in the vertical plane. It can distort the obstacles picture around the MR and leads to the design of incorrect motion trajectories, when traditional methods are used.

In some papers the methods of trajectories formation for the MR moving in space are proposed. In [15] the hybrid method for trajectory formation of a UAV in a mountainous region is described. In this method the optimal flat motion trajectory is first formed based on a mesh description of the UAV working space. The resulting discrete trajectory is smoothed using the method of potential fields. The main disadvantage of the described method is needing to know in advance the map of the location of all obstacles, as well as the inability to fly over the encountered obstacles. In [16] the methods of obstacles avoiding in the vertical plane based on the use of information from their onboard sensors are described. However, in these papers it is assumed that the UV has no restrictions on the pitch angle and can always bypass the obstacle from above. This assumption is valid only for certain UV having vertical thrusters.

So, there is the task of developing a method of formation of spatial trajectories of the MR spatial motion with unknown obstacles, based on information obtaining from onboard sensors. This method will allow to use all the possibilities of spatial movement of the MR. In the article this problem is solved on the basis of the formation method of smooth motion trajectories in an unknown environment [17]. This method will be modified taking into account the spatial character of the MR motion.

2. Problem Statement

When developing the path planning method for MR we assume that their trajectories are set in advance. These trajectories pass through the sequence $\mathbf{W} = \{W_1, W_2, ..., W_n\}$ of base points (where $W_i = (x_{Wi}, y_{Wi})^{\mathrm{T}}$) and described by third order Bezier splines. Since the surface geometry is unknown in advance, the original trajectory of the MR is defined in a horizontal plane. Therefore, the z coordinate for each point of the source trajectory will be formed in the process of robot movement so that he was always at a specified height from the surface.

To get the information about surrounding objects the MR have a set of onboard sensors, which generate signals about the distance from the robot to objects in the direction of the respective sensors. These sensors include different sonars and Doppler velocity logs (for underwater robots), 3D sonar, and various distance sensors (laser or ultrasonic). These sensors are divided into two groups: sensors that measure distances to obstacles located around MR, and sensors measuring the distance to the surface. Based on information received from these sensors is continuously generated two vectors $\mathbf{D} = (d_1, d_2, ... d_i, ..., d_m)$ and $\mathbf{H} = (h_1, h_2, ... h_j, ..., h_n)$, where m and n is the number of relevant onboard sensors.

Further, we assume that MR has four sensors of distance to the surface, directed forward, backward, right and left relative to MR. (see Fig. 1). This situation is standard Doppler logs used onboard of UV.

Let's suppose that data obtained from onboard sensors satisfy the following conditions.

1. The zero-signal generated at the output of the i-th sensor indicates that the obstacle in the corresponding direction is not detected.
2. The range of detection of obstacles is limited by the value d_{max}, which can be different for different groups of sensors, but for simplicity is the same for all sensors.
3. Time of filling of vectors **D** and **H** by the data obtained from multiple independent onboard distance sensors does not exceed the sampling period ICS of the robot.
4. The orientation of all the sensors in the coordinate system of the MR is known and does not change during robot movement. In the article it is supposed that all sensors in each group are arranged symmetrically about the roll axis of MR.

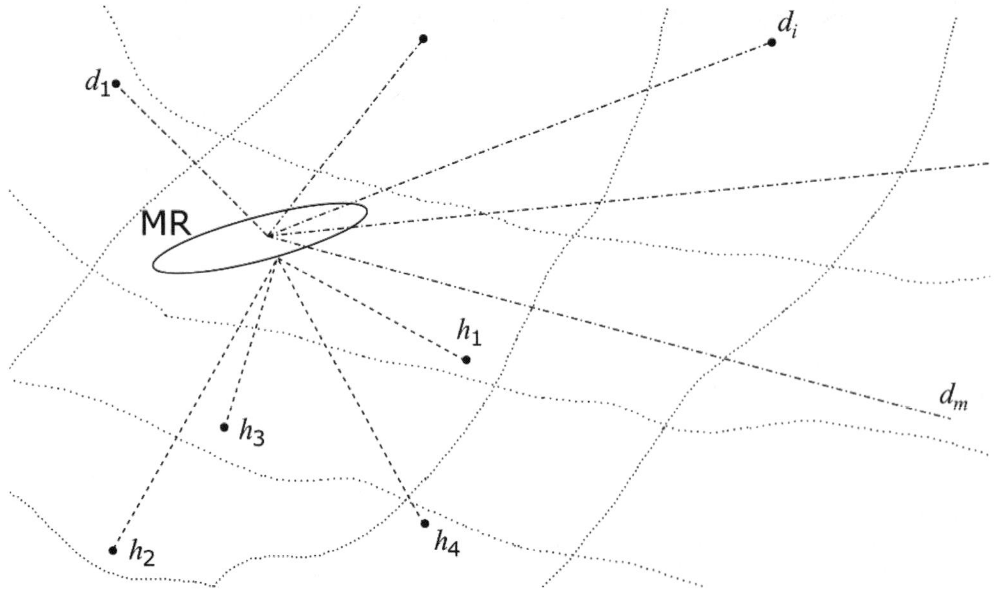

Fig. 1. Signal from onboard range sensors

Since the movement of the robot takes place in an unknown environment, the trajectory of this movement is defined in advance using a set of reference points. These points need to be adjusted in the process of movement based on data continuously received from onboard MP sensors. Thus, the process of generating of spatial trajectories have the following requirements.

1. The correction of these trajectories in the process of robot movement should not lead to an abrupt change of direction and speed of this movement.

2. Generated trajectory should allow the MR to going through the all base points in the specified sequence.

3. The generated trajectory should pass not less than the allowable distance $D_{\min}$ from the detected obstacle which is determined by the dynamic characteristics and the MR type.

4. All the parts of generated trajectories should be at a given height H_d from the surface of the robot movement.

5. The movement along the generated trajectories should not lead to the formation of the MR pitch modulo greater than the maximum allowed value $\varphi_{\max}$. This requirement is due to the features of construction and propulsion complexes of various types of MR.

As a basic method of forming the trajectory of the MR the method described in [17] will be used. This method generates and continuously tunes the smooth trajectory of the MR described by Bezier splines in the horizontal plane according to the data obtained from MR onboard sensors. This ensure the safe movement of the robot in a previously unknown environment containing obstacles. To use this method for MR performing spatial motion it is necessary to make the ability to avoid obstacles in the vertical plane and the always move at a given height from a surface with an unknown in advance relief.

Thus, the article solves the following problem. Let's set the initial smooth, flat trajectory of the robot movement, passing through a sequence of base points **W**. MR is moving in a previously unknown environment and equipped with sensors forming a vector **D** of distances to the detected obstacles and the vector **H** of the distances to the surface over which the movement of the robot is performing. It is necessary to correct the original trajectory that the resulting trajectory passes through the entire sequence of base points **W**, was located at the height H_d of the surface of the movement and the distance to the detected obstacles should not be smaller than the allowable $D_{\min}$.

The task of correcting the trajectory of the MR is solved in several steps.

Step 1. Analysis of environment based on data received from onboard MR sensors.
At this step there is evaluation of the surface relief and calculation of the coordinates of reference point X_d that specifies the movement of the robot. The specified coordinates X_d are calculated based on the current trajectory in the horizontal plane, the surface relief and the given speed of robot movement along the trajectory.

Step 2. Analysis of the location of obstacles relative to the MR and selecting of strategy of obstacle avoidance.
Based on information obtained from onboard sensors the MR evaluates magnitude of the pitch angle sufficient to avoid obstacles in the vertical plane. If this value is less than $\varphi_{\max}$, then MR can go around the obstacle in a vertical plane without correction of the trajectory in the horizontal plane. Otherwise, it makes decision of correction of the trajectory in the horizontal plane and avoid obstacles on the side saving the given height above the surface.

Step 3. The calculation of the parameters of the new trajectory using the algorithm described in [17], if MR makes the decision of correction.

Further, all these steps are described more detail.

3. Algorithm of forming of spatial trajectories of MR movement

Before beginning the work of the algorithm of formation of spatial trajectories ICS calculates the coordinates of points lying on obstacles and on surface of movement determined by the distance sensors. These coordinates considering the spatial orientation of MR is calculated as follows:

1) for sensors d_i:

$$X_{di} = X + R_G(\theta,\varphi,\psi)R_{MR}(a_{di},b_{di},c_{di})\begin{bmatrix}0\\0\\d_i\end{bmatrix}, i = \left(\overline{1,m}\right), \qquad (1)$$

where $X = (x, y, z)$ are coordinates of MR in world coordinate system; θ, φ, ψ are roll, pitch and yaw of MR respectively; a_{di}, b_{di}, c_{di} are angles of orientation of i-th sensor about the axis x, y and z in the MR body-fixed coordinate system; $R_G(.)$, $R_{MR}(.)$ are the rotation matrixes along axis x, y and z in world and body-fixed coordinate systems respectively;

2) for sensors h_j:

$$X_{hj} = X + R_G(\theta,\varphi,\psi)R_{MR}(a_{hj},b_{hj},c_{hj})\begin{bmatrix}0\\0\\h_j\end{bmatrix}, j = \left(\overline{1,4}\right), \qquad (2)$$

where a_{hj}, b_{hj}, c_{hj} are orientation angles of j-th surface distant sensor in in the MR body-fixed coordinate system. For definiteness, we assume that the point X_{h1} is ahead of MR, point X_{h2} is behind MR, point X_{h3} is on the left and point X_{h4} is on the right of the MR (see. fig. 1).

After determining the coordinates X_{hj} the coordinates of reference point X_d are calculated. They specify the desired movement of the robot along the trajectory. This calculation performs in two steps. At the first step the value of desired pitch angle φ_d is calculated. With φ_d the MR should move along the path to its height above the surface was equal to the given value H_d. On the second step the coordinates of the point X_d are calculated considering given trajectory in the horizontal plane, the φ_d and the value of v_d, which is reference speed of MR movement.
To calculate the φ_d the coordinates X_{h1} and X_{h2} are used. In the calculation it is considered that the profile of the surface relief under MR can be approximated by a straight line passing through these points (see fig. 2).

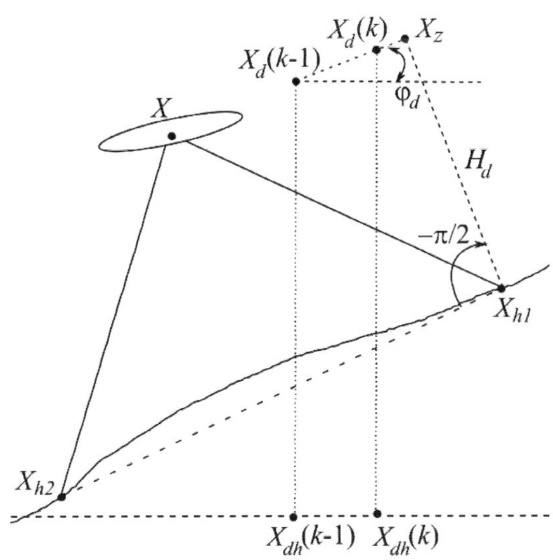

Fig. 2. Coordinate calculation of reference point X_d

In the beginning the coordinates of point X_z is calculated. This point is situated on distance H_d from point X_{h1} and on the plane passed through the points X, X_{h1} и X_{h2} (Fig.2). These coordinates are calculated by following expressions:

$$X_z = R_{\psi\theta\psi}\frac{X_{h1} - X_{h2}}{\| X_{h1} - X_{h2} \|}H_d + X_{h1}, \qquad (3)$$

where $R_{\psi\theta\psi} = R_G(0,0,\psi)R_G(0,-\pi/2,0)R_G(0,0,-\psi) = \begin{bmatrix} \sin^2\psi & -\cos\psi\sin\psi & -\cos\psi \\ -\cos\psi\sin\psi & \cos^2\psi & -\sin\psi \\ \cos\psi & \sin\psi & 0 \end{bmatrix}$. The matrix $R_{\psi\theta\psi}$ defines the rotation of vector (X_{h1}-X_{h2}) on angle π/2 relatively axis perpendicular this vector and lying in horizontal plan. If we know of point X_z coordinates we can calculate of value φ_d by following expression:

$$\varphi_d = \arctan\left(\frac{z_z - z_d(k-1)}{\| [x_z, y_z] - [x_d(k-1), y_d(k-1)] \|}\right), \tag{4}$$

where (x_z, y_z, z_z) и ($x_d(k$-1), $y_d(k$-1), $z_d(k$-1)) are the coordinates of points X_z and $X_d(k$-1), respectively. Index k-1 means coordinates X_d, calculated on previous sampling interval of MR control system.

Further we calculate the coordinates of point $X_d(k) = (x_d(k), y_d(k), z_d(k))$ with considering of value φ_d by following expressions:

$$\begin{aligned} X_{dh}(k) &= (1-\delta(k))^3 W_1 + 3\delta(k)(1-\delta(k))^2 P_1 + 3\delta(k)^2(1-\delta(k))P_2 + \delta(k)^3 W_2 = f_x(\delta(k)) \\ z_d(k) &= z_d(k-1) + \tan(\varphi_d) \| X_{dh}(k) - X_{dh}(k-1) \| \end{aligned}, \tag{5}$$

where $W_1=(x_{W1}, y_{W1})$, $W_2=(x_{W2}, x_{W2})$, $P_1 = (x_{P1}, y_{P1})$, $P_2 = (x_{P2}, x_{P2})$ are coordinates of starting, finishing and control points of spline describing current part of MR trajectory in horizontal plane; $X_{dh} = (x_d, y_d)$ is projection of point X_d on horizontal plane; δ=[0,1] is parameter set the point X_{dh} movement along spline from point W_1 to point W_2. The indexes k and k-1 mean values of different variables calculated for current and previous sampling interval of MR control system.

As we can see from expressions (5) the point X_d coordinates is defined by $\delta(k)$. Thus for calculation of X_d coordinates we have to find such value $\delta(k)$, which provide movement of point X_d with reference speed v_d. This task is solved by following iterative algorithm.

Step 1. $S_d = v_d \Delta\tau$, $p=0$, $\delta_t(p) = \delta(k-1)$, $S(p)=0$, $X_{dt}(p) = X_{dh}(k-1)$,

where S_d is the path of point X_d in a time Δτ when it moves with speed v_d; $\Delta\tau$ is the sampling interval of MR control system; p is the index of current iteration of algorithm.

Step 2. General calculation step.

$p = p+1$,

$\delta_t(p) = \delta_t(p-1) + \Delta$,

$X_{dt}(p) = f_x(\delta_t(p))$,

$S_x = \| X_{dt}(p) - X_{dt}(p-1) \|$,

$S_z = |\tan(\varphi_d)| S_x$,

$S(p) = S(p-1) + (S_x^{\,2} + S_z^{\,2})^{1/2}$,

where Δ is the small increment of δ_t.

Step 3. Checking of condition for calculation finish.

If $S(p) \geq S_d$, then $\delta(k) = \delta_t(p)$, else goto step 2.

The coordinates of point X_d for current sampling interval are calculated by expressions (5) after defining of $\delta(k)$. For increasing of dynamic accuracy v_d can be tuned according to [18, 19].

MR can meet obstacles during movement along trajectory ant it must avoid these obstacles on safely distance. For this purpose, it is necessary to define presence of obstacles near MR based on information about points $X_{di}, i=(\overline{1,m})$ and $X_{hj}, j=(\overline{1,n})$ and select the method of avoidance: in vertical plan without correction of trajectory in horizontal plane or sidewise with correction of trajectory in horizontal plan by algorithm described in [17]. This selection is made by estimation of value φ^* pitch angle needed for obstacle avoidance in vertical plane.

All range sensors d_i for calculation of φ^* are separated on three groups (Fig. 3): range sensors of MR left side **L** = $(d_1,\ldots, d_l)$, range sensors of MR forward side **F** = $(d_{l+1},\ldots, d_{l+f})$ and MR right side **R** = (d_{l+f+1}, d_{l+f+r}), where $l+f+r = m$. Each these groups associate with one of range sensor h_3, h_1, h_2 (Fig. 3). For each pair sensors d_i and h_j the estimation of pitch angle is calculated:

$$\varphi_i^* = \begin{cases} \arctan\left(\dfrac{z_{di} - z_{hj}}{\| [x_{di}, y_{di}] - [x_{hj}, y_{hj}] \|}\right), if\ d_i \neq 0 \\ 0, if\ d_i = 0 \end{cases}, i = \begin{cases} (\overline{1,l}), if\ j=2 \\ (\overline{l+1, l+f}), if\ j=1 \\ (\overline{l+f+1, l+f+r}), if\ j=3 \end{cases}, \tag{6}$$

where x_{di}, y_{di}, z_{di} are coordinates of point X_{di}; x_{hj}, y_{hj}, z_{hj} are the coordinates of point X_{hj}.

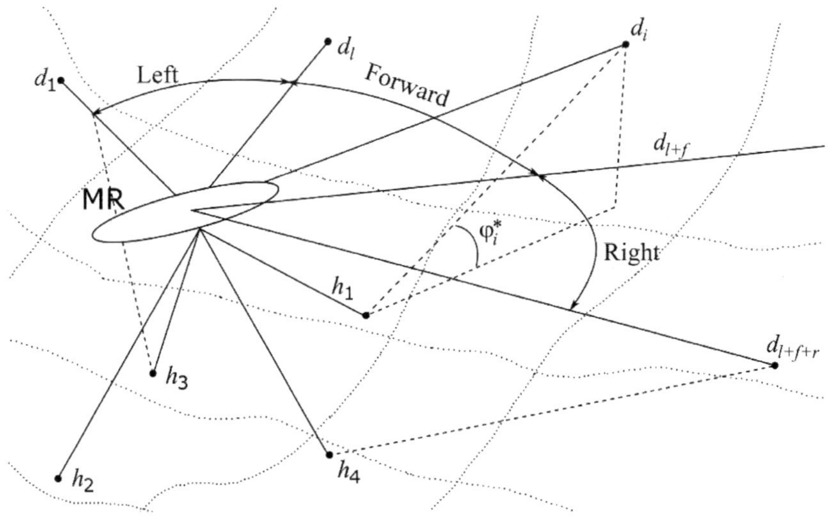

Fig 3. The selection of the method of obstacle avoidance

Finally, the value φ^* is defined as $\varphi^* = \max(\varphi_i^*), i = \overline{(1,m)}$. If $\varphi^* \le \varphi^*_{\max}$, then obstacle avoidance in vertical plan is possible and correction of MR trajectory in horizontal plane is not required. If $\varphi^* > \varphi^*_{\max}$, then obstacle avoidance must be in horizontal plane with correction MR trajectory. Herewith the value $\varphi^*_{\max}$ is defined by following expression:

$$\varphi^*_{\max} = \begin{cases} \min(\arctan(\frac{H_d}{D_{\min}}), \varphi_{\max}), if\ d_i \in \mathbf{L}\ or\ d_i \in \mathbf{R} \\ \varphi_{\max}, if\ d_i \in \mathbf{F} \end{cases}, i = \overline{(1,m)}. \quad (7)$$

As we can see from (7) if obstacle is left or right side of MR then value $\varphi^*_{\max}$ depend on relation of values H_d and D_{min} and can be less then value $\varphi_{\max}$. The explanation of this fact is shown in Fig. 4.

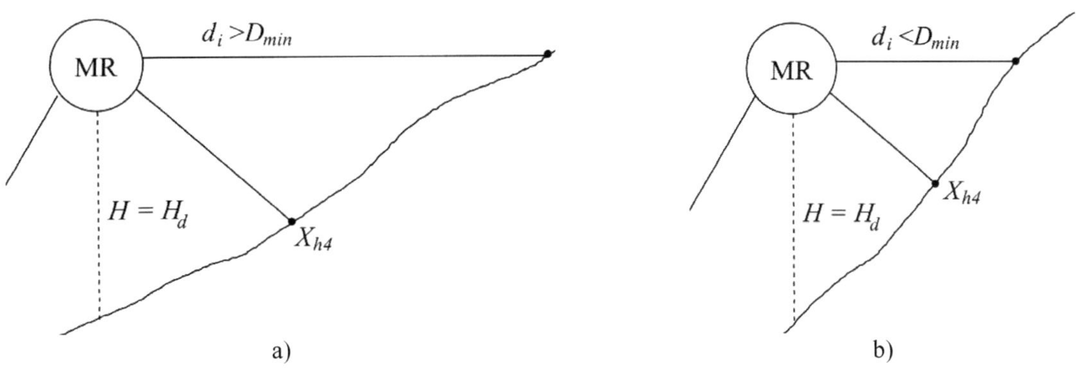

Fig 4. The features of MR movement along different obstacles

The MR movement along obstacle with smooth slope being from side of MR is shown in the Fig. 4a. In this case the MR can keep reference altitude from surface and simultaneously keep distance from obstacle more than minimal value $D_{\min}$. But if MR move along obstacle with stiff slope (Fig. 4b) then it became impossible simultaneously keep reference altitude from surface and minimal allowed distance from obstacle. If MR shifts from obstacle for increasing of side distance then altitude decrease too and MR come down. As result the side distance decrease again. So, for simultaneously keeping and altitude and minimal allowed side distance the value $\varphi^*_{\max}$ must consider of relation between values H_d and $D_{\min}$.

If obstacle must be avoiding by sidewise then trajectory setting in horizontal plane by Bezier spline of third degree is corrected by algorithm proposed in [17]. Herewith depend on obstacles position relatively MR there are two methods of trajectory correction: obstacle avoidance (from left or right side) or shifting trajectory for providing safely distance from obstacle. After correction the current part of MR trajectory is described by spline with new parameters which are used in expressions (5) for calculation of coordinates of point X_d.

4. The simulation result of proposed MR spatial path planning method

The mathematical simulation was carried out for investigation of proposed method. During this simulation the V-rep software is used for building of virtual training area (Fig. 5) containing different obstacles and MATLAB software is used for implementation this method of path planning.

Fig. 5. General view of virtual training area

During simulation it is proposed that MR has 7 range sensors of distance to obstacles and 4 range sensors of distance to surface. Sensors of distance to surface are directed to forward, backward, left side and right side relatively MR and have declination from yaw axis on angles $\pi/4$ rad in corresponding side. The reference altitude value was equal $H_d = 2$ m, minimal allowed distance from obstacles was equal $D_{\min} = 2$ m, maximal allowed value of pitch angle was equal $\varphi_{\max} = 0.6$ rad and reference MR velocity was equal $v_d = 1$ m/s. Starting point W1 has coordinates (24, 24), finishing point W2 has coordinates (-60, 30), and arrival angle to finishing point was equal π rad. Starting value of z coordinate was equal 4 m. During simulation it is supposed that MR control system provide its accurate movement along trajectory [20] and yaw and pitch angles are formed by direction of tangent lines to this trajectory in horizontal and vertical planes.

The simulation results of MR movement are presented in Fig. 6. In Fig. 6a the gray dashed line is the initial trajectory set in horizontal plane, and black solid line is the projection of resulting MR trajectory to horizontal plane. In Fig. 6b the changing of coordinate z along trajectory shown in Fig. 6a is presented. General view of resulting trajectory on virtual training area is shown in Fig. 6c. As we can see in these figures during MR movement its meet two obstacles: smooth and sharp hills. First obstacle MR avoids in vertical plan without correction of path in horizontal plane (gray and black line coincide in Fig. 6a). But MR cannot avoid second obstacle in vertical plane because of limitation of its pitch angle. In this case the path is corrected for providing avoidance of this obstacle from left side.

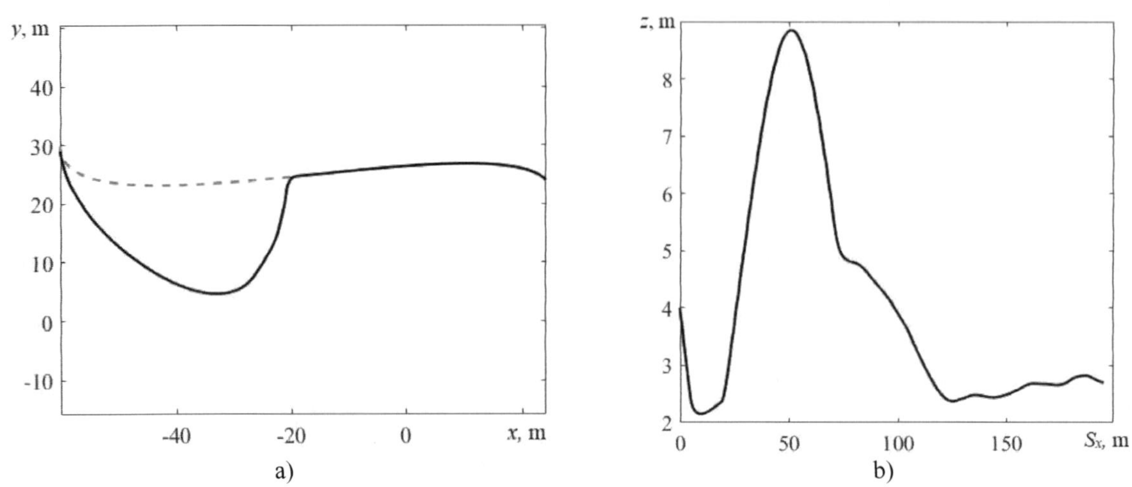

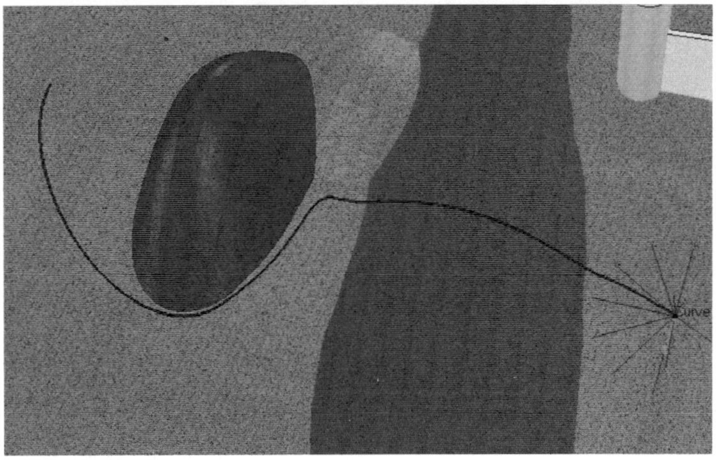

c)

Fig. 6. The resulting of MR trajectory

In the Fig. 7a the value φ^* changing is shown. As we can see when MR avoids first obstacle in vertical plane the value $\varphi^* < \varphi^*_{max}$ and when it avoids second obstacle $\varphi^* < \varphi^*_{max}$, and as result the initial trajectory is changed. In the Fig. 7b the changing of minimal distance from obstacles measured by range sensors is shown. In this figure zero distance means that there are no obstacles located near MR.

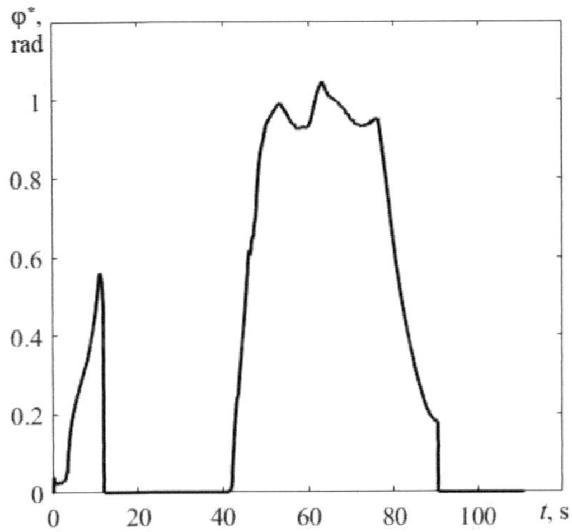

a)

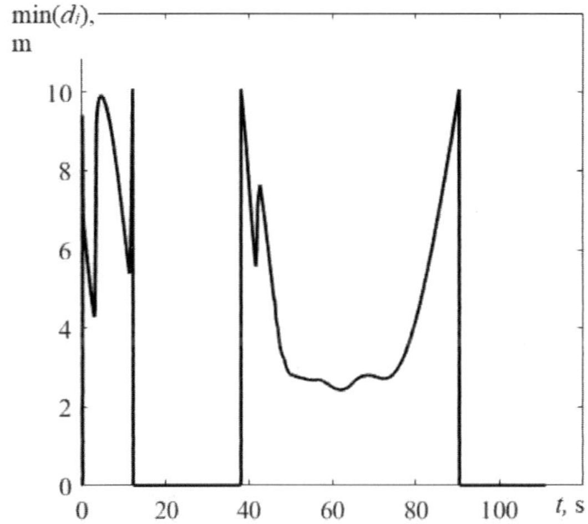

b)

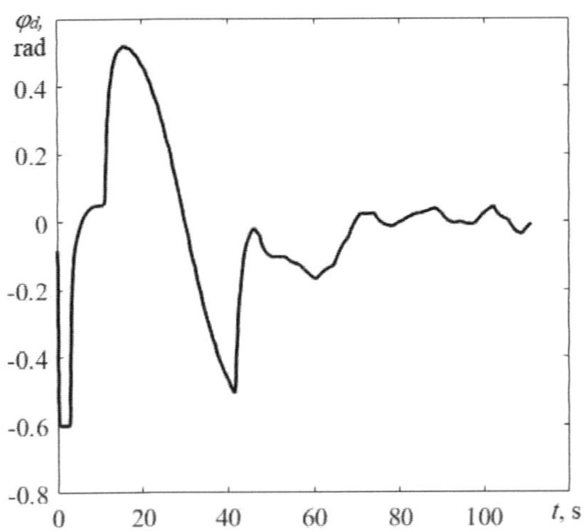

c)

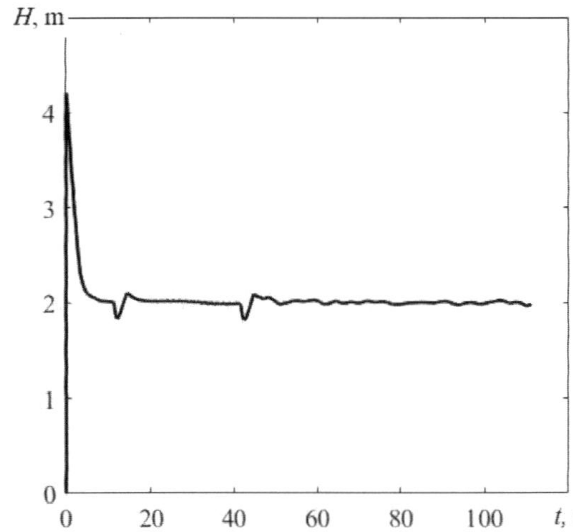

d)

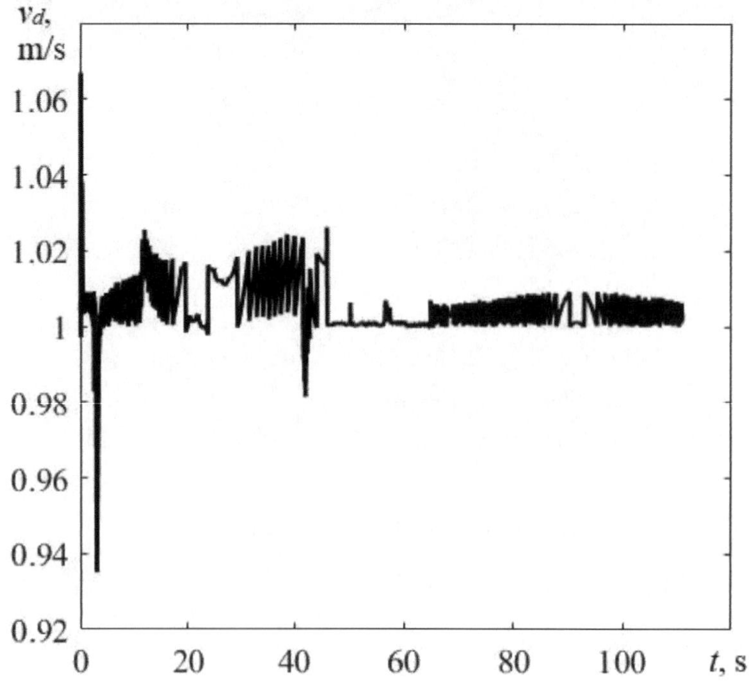

e)

Fig. 7. Changing φ^* (a), min(d_i), $i = (1,m)$, (b), φ_d (c), H (d), v_d (e) during MR movement

As we can see in this figure during MR movement minimal distance from obstacles always more than minimal allowed distance. Herewith this distance is minimal when MR moves along second obstacle. During movement the value of MR pitch angle not exceed $\varphi_{max} = 0.6$ rad (Fig. 7c). and deviation of altitude H under surface from reference value of altitude does not exceed 0.15 m (Fig. 7d). This deviation has maximal value when MR starts and finishes avoidance of first obstacle in vertical plane where relief of surface changes sharp. In other time this deviation does not exceed 0.07 m. The changing velocity of point X_d is presented in Fig. 7e. As we can see the deviation this velocity from its reference value does not exceed 0.02 m/s that provides performing any operations.

Thus, the simulation results confirm that proposed algorithm to allow planning spatial trajectories for MR in unknown environment and simultaneous keeping of reference altitude from surface and avoiding obstacles on safe distance. Also proposed algorithm has small computation complexity and allow automatically select mode of obstacle avoidance.

5. Conclusions

This paper proposes the new method of formation of spatial trajectories of the robot movement in an unknown environment containing obstacles, based on data obtained from onboard distance sensors. The proposed method is based on the previously developed method of formation of smooth trajectories. Feature of the proposed method is the simple mechanism of making decisions about bypass mode for detected obstacles: in the vertical plane without correction for the original trajectory, or side with the correction of this trajectory. The advantages of the proposed algorithm can indicate the formation of smooth trajectories, which allows to increase the speed of robot movement and its small computational complexity.

The future work will be continued according two ways. The first way is developing of intelligent approach to select the mode of obstacle avoidance in horizontal or vertical plans. For instance, on the base of fuzzy logic methods. The second way is developing method of selecting direction of obstacle avoidance on the base of analysis of covered distance.

6. Acknowledgments

This work is supported by RFBR (16-37-60025, 16-29-04195) and grant of President of Russia (MK-8536.2016.8).

7. References

[1] Fossen, T.I. (2011). Handbook of Marine Craft Hydrodynamics and Motion Control. Chichester: Wiley&Sons Ltd, ISBN: 978-1-119-99149-6

[2] Siciliano B. & Khatib O. Handbook of Robotics. (2008). Heidelberg: Springer-Verlag, Berlin.

[3] Wangy L.; Yongy L. & Marcelo H. (2002). Hybrid of global path planning and local navigation implemented on a mobile robot in indoor environment, Proceedings of the 2002 IEEE International Symposium on Intelligent Control (ISIC). Vancouver, British Columbia, Canada. pp. 821-826.

[4] Wu Zh. & Feng L. (2012). Obstacle prediction–based dynamic path planning for a mobile robot, International Journal of Advancements in Computing Technology, Vol. 4, No. 3, pp. 118-124.

[5] Jianping T. & Yang S. (2003). Genetic algorithm based path planning for a mobile robot, Proc. of the IEEE International Conference on Robotics and Automation, Taipei, Taiwan, Vol. 1, pp. 1221-1226.

[6] Shiltagh N. & Jalal L. (2013). Path Planning of Intelligent Mobile Robot Using Modified Genetic Algorithm, International Journal of Soft Computing and Engineering (IJSCE), Vol. 3, No. 2, pp. 31-36.

[7] Sedighi K.H.; Ashenayi K.; Manikas T.W.; Wainwright, R.L. & other. (2004). Autonomous Local Path-Planning for a Mobile Robot Using a Genetic Algorithm, Proc. of the Congress on Evolutionary Computation. Portland, USA, Vol. 2, pp.1338-1345.

[8] Shiltagh, N. & Jalal L. (2013). Optimal path planning for intelligent mobile robot navigation using modified particle swarm optimization, International Journal of Engineering and Advanced Technology, Vol. 2, No. 4, pp. 260-267.

[9] Garrido S.; Moreno L.; Blanco D. & Jurewicz P. (2011). Path planning for mobile robot navigation using Voronoi diagram and fast marching, International Journal of Robotics and Automation, Vol. 2, No. 1, pp. 42-64.

[10] Mohanraj T.; Arunkumar S.; Raghunath M. & Anand M. (2013). Mobile robot path planning using ant colony optimization, IJRET: International Journal of Research in Engineering and Technology, Vol. 213, No. 11, pp. 1-6.

[11] Saha O. & Dasgupta P. (2017). Real-time robot path planning around complex obstacle patterns through learning and transferring options, Proc. of the 2017 IEEE International Conference on Autonomous Robot Systems and Competitions (ICARSC), pp. 278-283.

[12] Jiang M.; Chen Y.; Zheng W.; Wu H. & Cheng L. (2017). Mobile robot path planning based on dynamic movement primitives, Proc. of the IEEE International Conference on Information and Automation, pp. 980-985.

[13] Ge S. & Cui Y. (2002). Dynamic motion planning for mobile robots using potential field method, Autonomous Robots, Vol. 13, pp. 207-222.

[14] Zhou J.-H. & Lin H.-Y. (2013). A self-localization and path planning technique for mobile robot navigation, Proceedings of the 8th World Congress on Intelligent Control and Automation, Taipei, Taiwan. pp. 694-699.

[15] Tan J.; Zhao L.; Wang Y.; Zhang Y. & Li L. (2016). The 3D Path Planning Based on A* Algorithm and Artificial Potential Field for the Rotary-Wing Flying Robot, Proc. of the 8th International Conference on Intelligent Human-Machine Systems and Cybernetics, pp. 551-556.

[16] Ageev M.D. (ed.). (2005). Autonomous underwater robots: systems and technology, Nauka, Moskow.

[17] Filaretov V.F. & Yukhimets D.A. (2017). Planning smooth paths for mobile robots in an unknown environment, Int. Journal of Computer and Systems Sciences, Vol. 56, No. 4, pp. 738–748.

[18] Filaretov V.F.; Gubankov A.S. & Gornostaev I.V. (2016). The formation of motion laws for mechatronics objects along the paths with the desired speed, Proc. of International Conference on Computer, Control, Informatics and Its Applications (IC3INA 2016), 03 - 05 October 2016, Jakarta, Indonesia. pp. 93-96.

[19] Filaretov, V. F. & Gubankov, A. S. (2012). Adaptive system forming extremely high speed of multilink manipulator gripper, Annals of DAAAM for 2012 & Proceedings of the 23rd International DAAAM Symposium, ISBN 978-3-901509-91-9, ISSN 2304-1382, pp. 0473 - 0476, B. Katalinic (Ed.), Published by DAAAM International, Vienna, Austria.

[20] Filaretov, V.F. (2007). Synthesis of Adaptive Control Systems for Electric Servo Actuators of Manipulators, Proceedings of the 28th DAAAM International Symposium, pp. 277- 279, B. Katalinic (Ed.), Published by DAAAM International, ISSN 1726-9679, ISBN 3-901509-58-5, Zadar, Croatia.

DOI: 10.2507/28th.daaam.proceedings.036

CALCULATION OF AIR-SLIDE CONVEYOR

Halima Hadziahmetovic & Ejub Dzaferovic

This Publication has to be referred as: Hadziahmetovic, H[alima] & Dzaferovic, E[jub] (2017). Calculation of Air-Slide Conveyor, Proceedings of the 28th DAAAM International Symposium, pp.0268-0273, B. Katalinic (Ed.), Published by DAAAM International, ISBN 978-3-902734-11-2, ISSN 1726-9679, Vienna, Austria
DOI: 10.2507/28th.daaam.proceedings.036

Abstract

Pneumatic conveying involves the transportation powder, granular and piece of material and is based on the phenomenon that at the appropriate speed of air in the pipeline, the solid particles are brought in the desired direction. If inlet air velocity is too high the material flow rate may be reduced, the power requirements will be excessive, and operating problems will be severe. Considering the advantages that air-slide conveyors can offer in relation to pneumatic conveying systems especially in terms of low power consumption and operating problems associated with abrasive particles, such as erosive wear of system components and degradation of friable particles. In this paper is presented design calculation of air- slide conveyor in the thermal power plant.

Keywords: particle; pneumatic conveying; air- slide conveyor; ventilator

1. Introduction

Pneumatic conveying is one of the key technologies in almost all branches of industry. Parallel to the development of science and technology, there was also the development of pneumatic components and systems, which contributed to the expansion of its application. Due to the many advantages, pneumatic conveying is used more than mechanically, mainly for the transport of dry, powdered and fine grained materials.

Pneumatic conveying is used to: unload and transfer dust and grain materials (cement, cereals, plant seeds and the like), transport vehicles by means of which they are transported in bulk (wagons, car tanks, cargo ships), transport of coal to powder in thermoelectric power plants and large boiler plants, transfer of electro-filter fly ash in thermal power plants, transport of cereals and their products in mills and silos, transport of sawdust and waste in wood industry and furniture factories, transfer of detergent to powder and grain materials in the chemical and process industry, sugar in the crystal, as well as other components in the food industry, the transfer of smoked tobacco in cigarette factories, the transportation and discharge of all kinds of dust collected by suction of contaminated premises and transport of administrative consignments. The area of application of pneumatic conveying to the industry is wide, especially in developed economy countries (Fig. 1.and Fig. 2.) [1,3,6].

Fig. 1. Transport of fly ash [8]

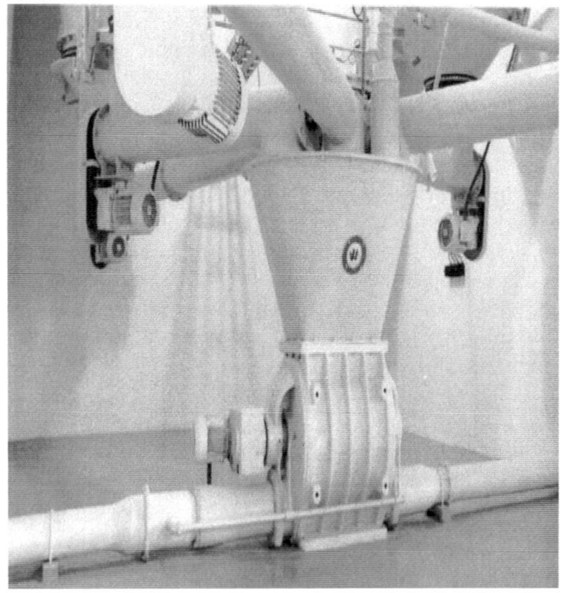

Fig. 2. Transport of flour [8]

Pneumatic conveying material is based on the principle of transferring solid particles at appropriate gas velocities in the pipeline. The required fluid pressure and flow depend on the quantity and type of material as well as the shape and length of the transport path. Each solid material provided in the dispersion form has different transport characteristics and when designing and modeling the transport system it is necessary to know these properties. The transport is realized by inserting the material into the flow of fluid into the pipeline through which it is transferred to a certain location. Subsequently, the material is separated from the material, which is discharged through the particle separator (cyclones, bag filters) into the atmosphere and the material is sent to the silo or to the further production [4,5,6].

With pneumatic conveying systems it is critical that the conveying line inlet air velocity is correctly specified. Because air is compressible, and very much higher air pressures are used in pneumatic conveying than in air-slide conveying, ensuring that the correct inlet velocity is achieved and maintained in a pneumatic conveying system is not a simple matter. If this inlet air velocity is too high the material flow rate may be reduced, the power requirements will be excessive, and operating problems will be severe. With the installation of air-slide conveyors, it will be provided in terms of low power consumption and operating problems associated with abrasive particles, such as erosive wear of system components are almost nonexistent [2,5]. In this paper is presented the design calculation of air-slide conveyor in the thermal power plant and the air-slide conveyor is in the process of being installed in the thermal power plant.

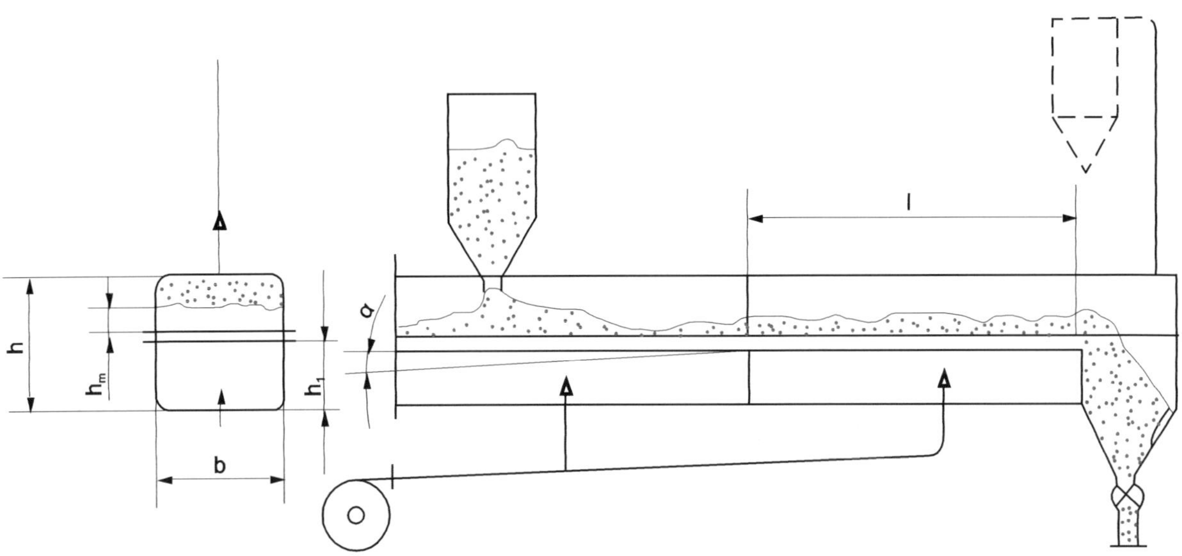

Fig. 3. Air- slide conveyor [4]

Different types of pneumatic conveying can be split into two groups: transporting the particles by means of air- slide conveyor and transporting the floating particles by means of a pipeline. Air- slide conveyor is a device used for the transport of powdered substances in a closed channel under the influence of gravitational force. The conveyor consists essentially of a channel, divided longitudinally by means of a suitable porous membrane on which the material is conveyed. A sketch of such a system is given in Fig. 3. If a small quantity of low pressure air is fed through the membrane, the inter-particle and particle/wall contact forces will be reduced and the material will behave like a liquid. If a slight slope is imparted to the conveyor, the material will flow. These conveyors are often referred to as 'air -slides'. Air gravity conveyors, ranging in width from 100 to 600 mm, can convey materials over distances of up to 100 m, and are suitable for material flow rates of up to about 3000 tonne/h. In general, most materials in the mean particle size and density ranges from 40 to 500 μm and 1400 to 5000 kg/m3, respectively, are the easiest to convey and will flow very well down shallow slopes [2,5].

2. Calculation of air -slide conveyor

For the known input data, the calculation of the air -slide conveyor was made. The input data for the calculation of the air -slide conveyor are: $d_{\varepsilon 50}$=40μm is middle particle diameter, ρ_m=2100 kg/m^3 is fly ash density, ρ_{ok} = 1,2 kg/m^3 is adopted air density during normal conditions, $\rho_{\varepsilon n}$ = 800 kg/m^3 is bulk density of material, η=17,3·10^{-6} Pas is dynamic viscosity of air during normal conditions and ε_f= 0,62.

Air velocity in phase of material aeration is defined by equation:

$$v_{vf} = \frac{d_{\varepsilon 50}^2 \cdot (\rho_m - \rho_{ok}) \cdot g}{178 \cdot \eta} \cdot \frac{\varepsilon_f^3}{1-\varepsilon_f}, \tag{1}$$

$$v_{vf} = \frac{(40 \cdot 10^{-6})^2 \cdot (2100 - 1{,}2) \cdot 9{,}81}{178 \cdot 17{,}3 \cdot 10^{-6}} \cdot \frac{0{,}62^3}{1 - 0{,}62} = 0{,}671 \cdot 10^{-2}\, m/s.$$

2.1. Ratio of heights of material layer during aeration

Ratio of heights of material layer during aeration is defined by equation:

$$\frac{h_m - h_f}{h_f} = \varphi_A \cdot \frac{\varepsilon_f^3}{1-\varepsilon_f} \cdot \left[\left(\frac{v}{v_{vf}}\right)^m - 1\right], \tag{2}$$

where h_m is height of material layer during aeration, h_f is height of fluidized bed at the minimum velocity, φ_A is coefficient depending on the porous membrane, $v = 10 \cdot v_{vf}$ is air speed calculation on the total surface of the air-slide conveyor (this ratio of velocity depends on the type of material being transported), $v = 10 \cdot 0{,}671 \cdot 10^{-2} = 0{,}0671$ m/s is air velocity and m = 0,5 is coefficient of velocity [4],

$$\varepsilon_f = 1 - \frac{\rho_{\varepsilon n}}{\rho_m}, \tag{3}$$

$$\varepsilon_f = 1 - \frac{800}{1200} = 0{,}62.$$

For ε_f=0,62 it obtains φ_A= 0,6.

Porous membrane	0,3	0,4	0,5	0,6	0,7	0,8
Coefficient	1,8	1,4	1,0	0,6	0,4	0,2

Table 1. Coefficient (φ_A) depending on the porous membrane [4]

Ratio of heights of material layer during aeration is defined by equation (2):

$$\frac{h_m - h_f}{h_f} = 0{,}6 \cdot \frac{0.62^3}{1 - 0.62} \cdot \left[\left(\frac{0{,}671 \cdot 10^{-1}}{0{,}671 \cdot 10^{-2}}\right)^{0{,}5} - 1\right] = 0{,}813,$$

$$\frac{h_m}{h_f} - 1 = 0{,}813 \ \ and \ \ \frac{h_m}{h_f} = 1{,}813.$$

Density of aerated layer in the channel be calculated according to the following:

$$\rho_\varepsilon = \rho_m \cdot (1 - \varepsilon_f) \cdot \frac{h_f}{h_m}, \tag{4}$$

$$\rho_\varepsilon = 2100 \cdot (1 - 0{,}62) \cdot \frac{1}{1{,}813} = 440{,}154 \ \frac{kg}{m^3}.$$

Dynamic viscosity of aerated layer in the channel should be calculated according to the following:

$$\eta_\varepsilon = \eta + K_\eta \cdot \frac{1-\varepsilon_f}{\varepsilon_f} \cdot \frac{v_{vf}\left(\frac{v}{v_{vf}}\right)^m}{d_{\check{c}50}^2 \left(\frac{h_m - h_f}{h_f}\right)^s}, \tag{5}$$

where $K_\eta = 3{,}7 \cdot 10^{-9}$ Pas is viscosity coefficient, $\eta = 17{,}3 \cdot 10^{-6}$ Pas is dynamic viscosity of air during normal conditions, $d_{\check{c}m}$=140 μm is arithmetic mean diameter of fly ash particles and dispersion coefficient is defined by equation

$$s = \left(\frac{d_{\check{c}m}}{d_{\check{c}50}}\right)^{0,6} = \left(\frac{140}{40}\right)^{0,6} = 2{,}12..$$

Dynamic viscosity of aerated layer in the channel should be calculated according to the following (5):

$$\eta_\varepsilon = 17{,}3 \cdot 10^{-6} + 3{,}7 \cdot 10^{-9} \cdot \frac{1-0{,}62}{0{,}62} \cdot \frac{0{,}671 \cdot 10^{-2}}{\left(40 \cdot 10^{-6}\right)^2} \cdot \frac{\left(\frac{0{,}671 \cdot 10^{-1}}{0{,}671 \cdot 10^{-2}}\right)^{0,5}}{(0{,}813)^{2,12}} = 0{,}046662 \text{ Pas}.$$

2.2. Channel width

Channel width (width of the conveyor) can be determined by:

$$b^4\left(1 + \frac{0{,}06}{b}\right) = \frac{M_m \cdot \eta_\varepsilon \cdot 64}{g \cdot \rho_\varepsilon^2 \cdot \sin\alpha}, \tag{6}$$

where b is channel width, M_m is capacity of the material is depending on the part of air -slide conveyor, M_{ml}=2· 10,55=21,1 t/h is capacity of materials for two drains on electrostatic precipitators and α= 2° is angle of inclination.
Inserting of obtained values into the equation for determining of channel width, it obtains:

$$b^4\left(1 + \frac{0{,}06}{b}\right) = \frac{21{,}1 \cdot \frac{1000}{3600} \cdot 0{,}046662 \cdot 64}{9{,}81 \cdot 440{,}154^2 \cdot 0{,}0349} = 2{,}64 \cdot 10^{-4}.$$

Or in the form of equation:

$$b^4 + 0{,}06b^3 - 2{,}64 \cdot 10^{-4} = 0.$$

By approximate solving of this equation it obtains channel width of b=115 mm and it adopts b=150 mm.

2.3. Necessary air quantity for aeration

Necessary air quantity for aeration be calculated according to the following:

$$Q_{vk} = A_{uk} \cdot \delta. \tag{7}$$

The known value of the specific loading on the porous membrane amounts to δ =2 m³/m²min and the total area of air -slide conveyor A=24,55 m², so necessary air quantity for aeration for the transport of materials equal to

$$Q_{vk} = 24{,}55 \cdot 2 = 49{,}1 \text{ m}^3/\text{min}.$$

2.4. Total pressure drop

Total pressure drop in air-slide conveyor when transporting bulk materials can be determined by:

$$\Delta p_{uk} = \Delta p_{dc} + \Delta p_{pp} + \Delta p_m + \Delta p_{oc}, \tag{8}$$

where Δp_{dc} is pressure drop across inlet pipeline, Δp_{pp} is pressure drop across the porous membrane, Δp_m is pressure drop across the fluidized bed and Δp_{oc} is pressure drop across the outlet pipeline, counting the end of the pipeline on the interlocker.

From practical experience and based on the recommendation of the manufacturer of air-slide conveyor, the company "Claudius Peters" from Hamburg [7], the sum of the pressure drop in the inlet pipeline and the pressure drop on the outlet pipeline is:

$$\Delta p_{dc} + \Delta p_{oc} = 3500 \text{ Pa},$$

where Δp_{pp}=1000 Pa is the value given by the producer of the porous membrane.

Pressure drop across the fluidized bed of material in air-slide conveyor be calculated according to the following:

$$\Delta p_m = h_m \cdot (1 - \varepsilon_f) \cdot (\rho_m - \rho_{ok}) \cdot g, \tag{9}$$

where $h_m = b_{max}/2$ is maximum height of the material layer is determined based on the maximum channel width,

$$\Delta p_m = 0{,}075 \cdot (1 - 0{,}62) \cdot (2100 - 1{,}2) \cdot 9{,}81 = 586{,}8 \text{ Pa}.$$

Total pressure drop is given by:

$$\Delta p_{uk} = 3500 + 1000 + 586{,}8 = 5086{,}8 \text{ Pa}.$$

2.5. The power of the ventilator motor

The power of the ventilator motor is calculated according to equation:

$$N_{em} = k \cdot \frac{\Delta p_{uk} \cdot Q_{vk} \cdot \dfrac{\rho_{rad}}{\rho_{ok}}}{\eta_v}, \tag{10}$$

where k=1,1 is safety coefficient, ρ_{rad}= 1,69 kg/m³ is operating air density in air -slide conveyor, ρ_{ok}=1,2 kg/m³ is adopted air density during normal conditions and η_v = 0,75 is coefficient of usefulness of ventilator.

Therefore, the power of the ventilator motor be calculated according to the following (10):

$$N_{em} = 1{,}1 \cdot \frac{5086{,}8 \cdot \dfrac{49{,}1}{60} \cdot \dfrac{1{,}69}{1{,}2}}{0{,}75} = 8{,}59 \text{ kW}.$$

2.6. Diameter pipeline

Diameter pipeline can be determined by:

$$D_{os} = \left(\frac{4 \cdot Q_{vk}}{\pi \cdot v_{cj}} \right)^{\frac{1}{2}}. \tag{11}$$

The value of the air velocity in the pipeline is adopted and is equal v_{cj}=20 m/s [4].

Diameter pipeline is given by:

$$D_{os} = \left(\frac{4 \cdot \frac{49{,}1}{60}}{3{,}14 \cdot 20} \right)^{\frac{1}{2}} = 0{,}228 \text{ m}.$$

The internal non-standard diameter of the pipeline Ø= 250 mm with a wall thickness δ = 2,5 mm is adopted.

3. Conclusion

With pneumatic conveying systems it is critical that the conveying line inlet air velocity is correctly specified. Because air is compressible, and very much higher air pressures are used in pneumatic conveying than in air-slide conveying, ensuring that the correct inlet velocity is achieved and maintained in a pneumatic conveying system is not a simple matter. If this inlet air velocity is too high the material flow rate may be reduced, the power requirements will be excessive, and operating problems will be severe. With the installation of air-slide conveyors, it will be provided in terms of low power consumption and operating problems associated with abrasive particles, such as erosive wear of system components are almost nonexistent.

In this paper is presented the design calculation of air-slide conveyor in the thermal power plant and the air-slide conveyor is in the process of being installed in the thermal power plant. Based on the known input data, the air velocity in phase of material aeration of the prepared transport material was calculated first, v_{vf} = 0.671 • 10-2 m / s. Then, the ratio of the layers of the fluid layer is calculated, which depends on the density of aerated layer in the channel ρ_ε = 440,154 kg / m3 and the dynamic viscosity of aerated layer in the channel η_ε = 0, 046662 Pas. To determine the channel width, it is necessary to know the capacity of materials for two drains on electro filters, which is M_{ml} = 21,1 t / h. By solving the equation of the fourth degree, the channel width is b = 115 mm, and it is adopted b = 150 mm. In order to determine the air flow required for the transport of the material, the total area of air -slide conveyor A= 24,55 m^2 and the specific loading on the porous membrane δ= 2 m^3/m^2min must be known. The air flow required for the transport of materials is Q_{vk} = 49,1 m^3/min. Total pressure drop is equal to Δp_{uk} = 5086,8 Pa. The air required for the transport of materials in the air troughs is prepared by means of ventilator, therefore it is necessary to determine , the power of the ventilator motor, which is N_{em} = 8,59 kW. Diameter pipeline is calculated on the basis of the air flow for the transport of materials and is equal to D_{os} = 0,228 m. The internal non-standard diameter of the pipeline Ø= 250 mm with a wall thickness δ = 2,5 mm is adopted.

4. References

[1] Bogović, I-N., Barišić, B., Katalinić, B., Kršulja, M., Car, Z.,(2011), Digitizing system ATOS - measuring turbo compressor housing, Annals of DAAAM for 2011 & PROCEEDINGS of the 22nd International DAAAM Symposium / Katalinić, Branko (ur.). - Vienna: DAAAM International Vienna, 2011. 1367-1368 (ISBN: 978- 3-901509- 83-4).

[2] Mills D., (2004), Pneumatic Conveying Design Guide, Second Edition, Butterworth-Heinemann, ISBN 0750654716, UK.

[3] Hadziahmetovic, H. & Dzaferovic, E., (2009), Ash pneumatic conveying from existing silos no. 4 to two new silos and ash loading in autocisterns - The 20th INTERNATIONAL DAAAM SYMPOSIUM "Intelligent Manufacturing & Automation: Theory, Practice & Education", 25-28th November 2009, Vienna, Austria.

[4] Barbalić N., Džaferović E., (2007), Transport čvrstih čestica fluidom, Faculty of Mechanical Engineering Sarajevo, Bosnia and Herzegovina.

[5] Mills D., Jones M.G., Agarwal V.K., (2004), Handbook of Pneumatic Conveying Engineering, CRC Pres.

[6] Hadžiahmetović, H., (2015), Numeričko modeliranje erozije u koljenu pri pneumatskom transportu, Ph.D. Dissertation, Faculty of Mechanical Engineering Sarajevo, Bosnia and Herzegovina.

[7] http://www.claudiuspeters.com/en-GB/documents/358/claudius-peters-pneumatic-conveying-brochure-en.pdf

[8] http://engineering.co.rs/cevni.htm, Accessed on: 2017-08-13.

DOI: 10.2507/28th.daaam.proceedings.037

CREATING CORPORATE CULTURE IN CONTEXT WITH STANDARDS ISO 9001: 2015

Jaromíra Vaňová, Iveta Paulová, Miroslav Rusko & Edita Hekelová

This Publication has to be referred as: Vanova, J[aromira]; Paulova, I[veta]; Rusko, M[iroslav] & Hekelova, E[dita] (2017). Creating Corporate Culture in Context with Standards ISO 9001: 2015, Proceedings of the 28th DAAAM International Symposium, pp.0274-0279, B. Katalinic (Ed.), Published by DAAAM International, ISBN 978-3-902734-11-2, ISSN 1726-9679, Vienna, Austria
DOI: 10.2507/28th.daaam.proceedings.037

Abstract

Corporate culture promoting the quality is focused on employees' behaviour concentrating on the continuous improvement of the processes' and products' quality. The final objective is to meet the changing requirements of customers at an appropriate management efficiency of the organization, which lead to a competitiveness. Any organization that is committed to the transformation of quality management or environs in accordance with the new requirements, has to create such environs in which the involvement of workers in the effective management of changes and new requirements is supported. The basic focus of the organization at the quality promotes a culture, which results in behaviour, opinions, activities and processes that add a value to the fulfilment of the needs and expectations of customers and other relevant stakeholders. The paper focuses on identifying changes in corporate culture by strengthening factors that are the result of a company's focus on fulfilling the requirements of standards ISO 9001:2015.

Keywords: Corporate Culture; Standards ISO 9001:2015; Quality System Management; Leadership

1. Introduction

Quality is ultimately defined by customers. There is significant literature available on how to measure quality and value as perceived by customers. In virtually every analysis, a major component of customer satisfaction is the ability of the company to provide a competitively priced product into which quality is designed, built, marketed, and maintained.

The aim of the organization is to meet the needs and expectations of customers as well as the other stakeholders and simultaneously to reach, maintain and improve the overall performance and the capacity of the organization. The surveys of economically successful enterprises show that the soft factor "corporate culture" contributes to their success. The change of the corporate culture belongs to some complex management tasks. Any organization that is committed to the transformation of quality management or environs in accordance with the new requirements (in context with ISO 9001:2015), has to create such environs in which the involvement of workers in the effective management of changes and new requirements is supported.

Total quality is the total systems approach (not a separate area or program) and an integral part of any high-level strategy; it works horizontally across functions and departments; involves all employees, top to bottom; and extends backwards and forwards to include the supplier and the customer chains. [1].

The social system has a significant impact on motivation and the ability to implement new ideas; it addresses corporate culture, structure, rewards, teamwork, values, and the creativity of individuals and groups. The social system is influenced by the values of the founders, leaders, families, peers, and supervisors, as well as group behaviours. Some authors are of the opinion that a certain type of corporate culture can help to maintain the long-term performance of the enterprise and be a source of competitive advantage.

That is why we also carried out secondary research of already existing studies available from subject area, in addition to the structured interviews and a questionnaire survey as a part of our project.

Our survey in the area was carried out from May 2016 to October 2016. For the purposes of the processing of this article only the questions that were related to the selected area of the present state of the corporate culture in organizations, which began with the transformation of the new standards ISO 9001:2015 were selected and used.

2. Corporate culture

Corporate culture, in the sense of a complicated and internally structured system is created and developed continuously in the company. Its creation in any particular company is timely compliant with the company foundations. As a company grows and develops gradually, so its culture is created, developed and continually changed (supplemented and filtered) also its culture 2010) [2].

The notion of corporate culture is defined by several authors. However, all definitions have the same characteristics and essence of: "...the shared philosophies, ideologies, (values, assumptions, beliefs, expectations, attitudes and standards ... how to make decisions and to solve problems…" [3]. In the case of a corporate culture that is essential for the transformation of new approaches to the management of quality, it is necessary for a top manager to be in the role of a leader, to be a personality that leads the way, to be a motivator and coach in the organization, to implement the creative management style and both acknowledges and objectively appreciates the achievements reached, the behaviour of individuals but also of the whole team [4].

At present, the transformation of ISO 9001:2015 standards becomes reality, that is affecting, in particular, some changes in approaches to the management of quality in areas such as the change of behaviour of the leaders, change of management (with respect to the risks) with an emphasis on continuous improvement, creativity and innovation. From all this it can be concluded that there is a change in the corporate culture of the organization. Continuous improvement is the basement of the corporate culture, which provides the organization of the management aimed at constantly improving, and educates employees to search for ways of possibilities for improving and solving some deficiencies. The corporate culture of a particular organization must be able to cope effectively with the changes in the organization. The characteristic features of a corporate culture that supports the transformation of requirements for quality management system and/or environmental management system are as follows:

- involved, active and competent employees identified with the organization objectives which they effectively perform,
- individual responsibility for performance of the tasks and resolving disagreements [5].

It cannot be automatically assumed that mere creation of a corporate culture is responsible for the success of the enterprise and its development. There must occur a mutual interaction and linking of corporate strategy - corporate culture - corporate structures – corporate communication. [6]

In this context, the theme of the corporate culture and related questions of quality of work, is getting into the centre of economic interest and is still more often associated with possibility that may increase competitive ability of company as well as satisfaction of employees.

3. Some of the requirements for the quality management system

The new standards ISO 9001: 2015 substantially extended the requirements to the top management– leaders. The top management has to perceive and quickly respond to changes in the business environs, to take an interest and make survey of stakeholders that can influence this business environs. New challenges and opportunities require 101 for the leaders to have a more sensitive approach to search and cope up with risks, the ability to respond to changes and to be creative enough or at least flexible [7]. It is further specified that the organizational changes must be implemented in a planned way. The most common changes can be those changes induced by the maintenance and improvement of the quality management system, the changing of some processes, internal and external aspects. Typical examples of organizational changes may be the introduction and take-up of new technologies, the changes in significant external provider, a change of the place of performance of services, major personnel changes, etc.

Employees of the organisation represent resources that may directly or indirectly influence their growth and the sustainability of their performance. The aim of each organization is to support its operation comprehensively. The management of the organization includes particularly the setting of the whole management system of the organization– setting the values and rules of the organization, setting up an organizational structure, resource management, processes and performance settings. For this reason, a basic requirement in the framework of the quality management system of the organization is to determine the responsibility of leaders for systematic and organized using of the knowledge, which is essential to the growth and wealth of the company. This is related to the requirement of continuous development of knowledge of all those employees who may need them [8]. In this case, we perceive the knowledge as a set of abilities, skills, experience, intuition, rules, and principles, personal notions of a specific person or a group of people. The requirement of the standards ISO 9001: 2015 is that the top management (leaders) have to assess how to primarily identify and continuously acquire the knowledge needed to meet the expected and anticipated future needs, to assess the knowledge requirements along with the current need of the staff knowledge, to assess the form of knowledge and the extent to which knowledge is identified and protected.

Communication in the framework of the quality management system increases the involvement of the people and improves the understanding of management and approach of the organization; it allows to clarify the needs and expectations of customers. Communication is a two-way process. Sharing information and subsequent feedback obtaining are the cornerstones of a good corporate culture [9].

4. Results of the survey dealing with current situation in fulfilment of the requirements of ISO 9001: 2015 standards in enterprises in Slovakia

As part of the project solution has been realized research through questionnaire about actual situation in applying principles of quality management in enterprises and comparison how are fulfilled some requirements in several enterprises in Slovakia. We carried out a survey on a sample of 125 enterprises in Slovakia. 40% of them was large and 60% medium sized enterprises.

Basic mathematical methods were used for questionnaires evaluation. We applied methods of analysis and synthesis for obtained information and various aspects of research problems solutions discussing. These methods were used to sort and summarize identified data. In this article we would like to present partial results of our survey focused on this issue.

The requirement for a successful and transparent application of the organization strategy is directed to the interlacing of the strategy at the lower levels through the objectives and quality policy, which can be fulfilled in the case, if there is a systematic approach created to the monitoring of the objectives achievement at various levels of governance, which must be all the key employees informed with. The level of corporate culture is determined in particular by leaders, who define responsibilities and powers for individual roles. It also includes a way of planning, managing, controlling and especially communicating. The survey revealed the level of leadership competencies for the successful transformation of the requirements of ISO 9001: 2015.

Figure 1 shows the difference between large and medium-sized organizations. Defined responsibilities and powers for effective organization management are better perceived by large organizations. Up to 77% of large organizations claim that organizational authorities and responsibilities are clearly, assigned, communicated and understood. Results from middle organizations point to the ambiguity of the approach to a clear definition of responsibilities and competences, and in particular communication. It is therefore necessary to draw attention to the unequivocal determination of competences for leaders at all levels of management as they are the carriers of the changes that need to be made in implementing ISO 9001: 2015 standards.

One of the key tasks of the leader in organizational changes, and thus in the transformation of the new requirements of ISO 9001: 2015, is to be an example to other employees, an authority that can directly present the company's goals of change. Creating trust between management and other employees is a guarantee of success. For this reason, we have looked at these aspects in large and medium-sized organizations.

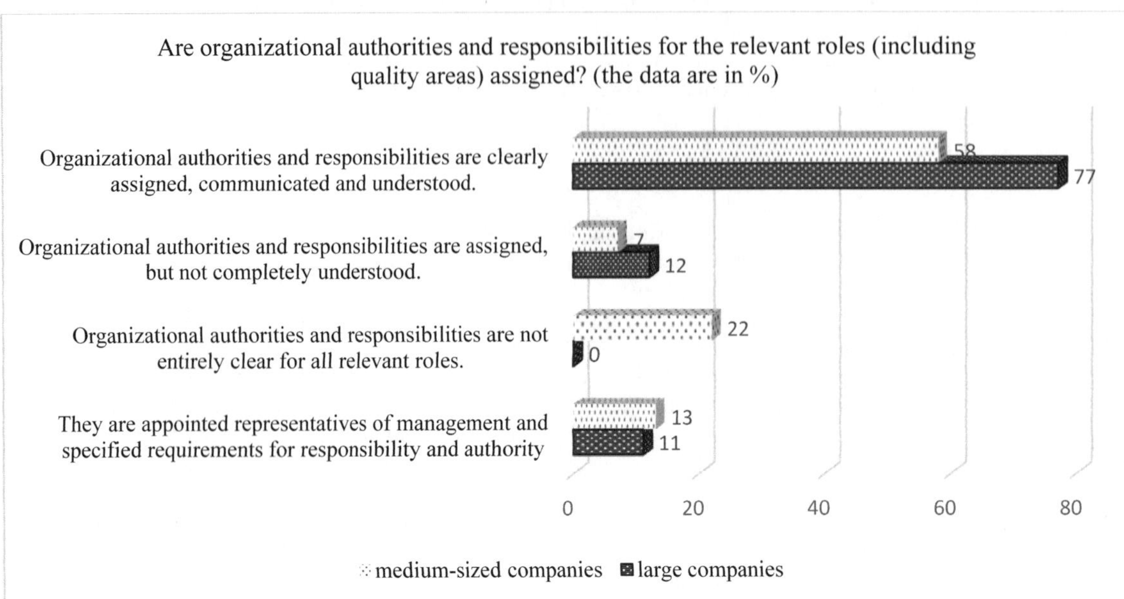

Fig. 1. Assigning responsibilities and authorities for relevant roles in the organization

Figure 2 presents the results of the survey and points to the fact that in 59% of the top management organizations has found its way to people and communicates with them openly. In middle organizations, 40% of top managers have found a way to people away and talk openly about the organization's intentions, but the same percentage is at the stage of looking for employees and not always communicating them.

Some organizations still have an authoritative style of governance (22% in medium-sized organizations and 11% in large organizations), which is not positive for changes in organizations. Employees are not motivated to change and managing changes in the organization is inefficient and cumbersome.

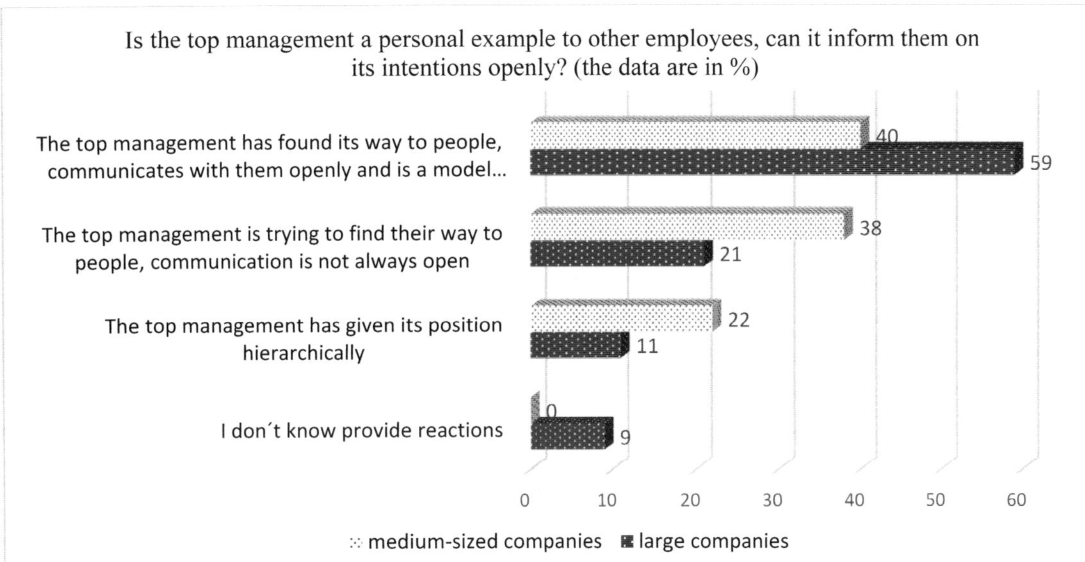

Fig. 2. Results of research about perceiving top managers in the enterprises (in %)

In addition to the need for effective communication of top managers with employees, it is essential to have effective communication with key stakeholders as well. Stakeholders are an essential part of an effective transformation of ISO 9001: 2015 standards in organizations. They share their insights into improving organizations' outcomes and perceiving the organization's external environment.

As it can be seen from Figure 3 in large organizations, levels of effective communication within the organization are equally balanced and 44% of stakeholders are interested in stakeholders and only 28% in medium-sized organizations. In medium-sized organizations, the flow of information in the organization is highest, feedback and verification of customer understanding is random. However, a third of large and medium-sized communication organizations take place internally with the management interface and outsourced employees with customers.

According to Sivri, Hennersdorf, & Krallmann (2016) "Research highlights that employees' awareness and involvement within the implementation are crucial success factors." [10]

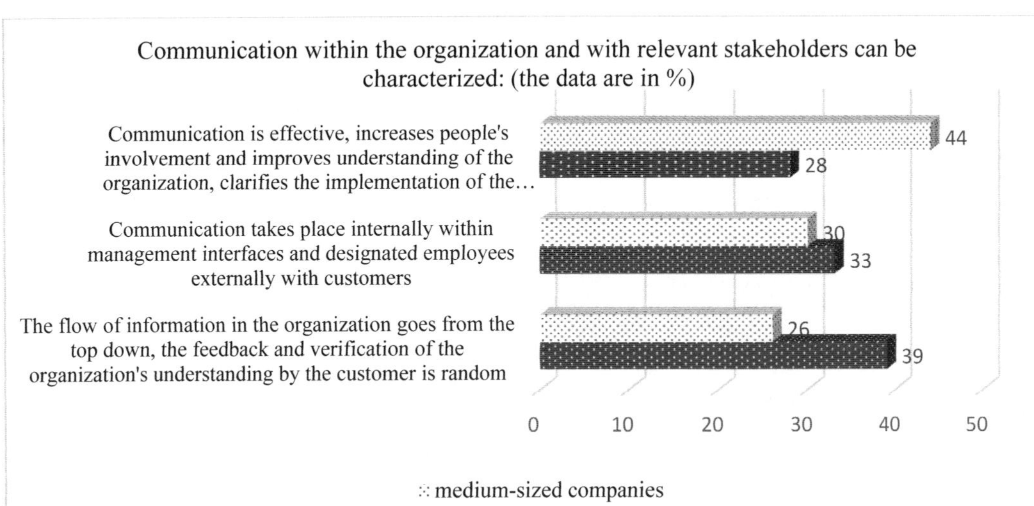

Fig. 3. The internal and external communications in organizations involved in research

5. Discussion

The organizations have to adapt to the requirements of the international market and, changes of the management methods and business philosophy in order to withstand any tough competitive struggles and to achieve a sustainable development. In all areas of economic and social life there are increased demands on corporate behaviour and there is a growing interest in maintaining and improving the product quality, the environment and safety at work.

Business performance is composed of the performance of the employees. Employees with their work, activities, work attitude and their performance should contribute to reaching company´s set goals and required business performance. Successful company performance at the market is influenced by potential of all company employees and managers [11]. Improvement of managerial work is one of the main tasks of human resource management in companies. It is due to the fact, that only a well-prepared manager can provide desired results. Main development requirement therefore is to educate the managers to be qualified leaders who can focus on strategically important issues and effectively engage employees to provide required performance. Recent trends bring a competency model approach to the management and human resource development

The organization identifies, develops and keeps competences of employees. If the organization creates a structure to allow its employees continuously develop their own competences, assume greater responsibility and show more initiative, then the employees contribute to the development of the workplace. Employee involvement is to create and environment in which employees have influence on the decisions and actions that affect their work. It involves creating a culture that supports the mission, vision, and values of the organization in practice. [12]

There are several factors that affect the performance of the company and one of them is a corporate culture. A suitable shaping corporate culture is achieved by working employee satisfaction, what inherently leads to the achievement of the business objectives and desired performance.

Through the corporate culture, company forms working environment and try to make the employees identify themselves with the preferred corporate values and ideas.

The survey shows that the challenge is to embrace and understand the soft factors and to define them as a concept that can be used at adapting the ISO 9001: 2015 standards. Because of this, it is necessary to deal with the question, what are the characteristics of corporate culture at the transformation of individual standards requirements.

The questions that can be asked, are e.g.: What characters and characteristics should a culture have and how they can be used as a tool of governance affecting the effective change of approaches, in accordance with the required standards? The exact rules and fixed and defined structures have a negative impact on creativity and motivation of employees, which are an important prerequisite for the first stage of the change at the transformation of standards.

Changes of the structure mean a change in the formal arrangements of the organization, modification in the extent of centralization or decentralization, authority, work descriptions, etc. Technological changes relate to the machinery and equipment used, and to the ways and methods of work performance. Changes in the field of people interfere with approaches of people to work, their expectations, their perception of work and organization, which is reflected in their behaviour.

In further research we will focus on how an organization enables its employees to use their full potential in accomplishing organizational objectives and how the organization maintains an environment that promotes performance excellence, full participation, and personal and organizational growth. Accomplishing this requires a close coupling between the leadership system and human resource system to enable defining and implementing the methods and structures needed to lead the organization in its strategic direction.

In the next step we want to find answers to questions:

- How do managers motivate employees to develop and utilize their full potential?
- How do managers ensure effective communication, cooperation, and knowledge/ skill sharing across work units, functions, and locations, as appropriate?
- How do managers identify characteristics and skills needed by potential employees?
- How do companies determine the key factors that affect employee well-being, satisfaction and motivation?

An organization must have approaches for enhancing employee well-being, satisfaction, and motivation based upon a holistic view of its entire workforce.

Measuring how well employees understand the connection between quality and customers can also show what an organization needs to do to achieve an organizational customer focus and competitiveness in the market

6. Conclusions

In addition to guiding and coordinating internal and external management and organisational processes in company, management of company has to nowadays meet also other important tasks. They have to harmonize profitability, humanity and legitimacy. If company management wants to be successful in markets with increasing competition, they should have set objectives to which it's heading, but also the values that will be respected for achievement of these objectives.

Analysis confirmed that humanely and effectively created and shaped corporate culture as well as model, exemplar behaviour of managers can be included among the crucial factors in the success of many companies.

Each organization that is committed to the implementation of new approaches of the organization to the requirements of the standards ISO 9001: 2015 must create such an environment in which the employees´ creativity is supported, which results in the possibility of different forms of improvement. Enterprises are trying to ensure higher productivity, greater employee satisfaction and the sustainability of the organization in relation to the environment through building a favourable corporate culture. A key role in this is met by leaders of the organization, who are changing also their roles in organization including the management style. The communication and its form and the way that should be efficient for the emerging environs are also important. It is essential to defuse an outdated management approach through commands and controls, and to look at people in what their strengths are and use it for the benefit of the organization. Satisfied and happy employees have a great impact on the organization's performance and profits. The human potential is a new competitive advantage [13].

The survey results indicate that there is a need to create a systemic approach in the sense of the current corporate culture analysis (including a selection of characteristics, marks and sphere of analysis), the proposal for a corporate culture that the way of communication within the organization, especially in medium-sized organizations. For this reason, it is necessary to focus on reconsidering the way of communication between the different levels of management, but especially the efficiency and the way of communication between the management and the employees. Sharing information on time and openly is a key aspect of the effective transformation of ISO 9001: 2015 standards. It is necessary to create a pro-innovative corporate culture where it will be possible to design impetus, improve and, in particular, discuss and improve at all levels of management.

For effective communication with stakeholders, it is necessary to create a database of key stakeholders and identify their relationships and the impacts of cooperation. The most important area is the change of managers' leadership competencies, where the personality features of the manager are especially important.

7. Acknowledgements

The paper is a part of research project VEGA 1/0990/15 "The readiness of industrial enterprises to implement the requirements of standards for quality management systems ISO 9001:2015 and environmental management systems ISO 14001:2014."

8. References

[1] Edosomwan, J. A. (1995), Organizational Transformation and Process Reengineering, St. Lucie Press, St. Lucie, FL.

[2] Blašková, M.: (2010). Systémové poňatie podnikovej kultúry. System concept of corporate culture. In: Ekonomika a manažment podnikov 2010 : Medzinárodná vedecká konferencia. Zvolen, 5.-6. 10.2010. - Zvolen : Technická univerzita vo Zvolene, 2010. - ISBN 978-80-228-2150-6. - pp. 168-175

[3] Killman, R., Saxton, M. J., Serpa, & all. (1985). Gaining Control of the Corporate Culture. San Francisco, London, JosseyBass Publishier, ISBN 0-87589-666-9.

[4] Zgodavova, K., Kisela, M. & Sutoova, A. (2016). Intelligent approaches for an organisation's management system change", The TQM Journal, Vol. 28 Issue: 5, pp.760-773, https://doi.org/10.1108/TQM-10-2015-0130

[5] Lukašová, R. & Nový, L (2004). Organizační kultura.Praha: Grada Publishing, ISBN 80-247-0648-2

[6] Gyurák Babeľová, Z. Effectiveness of investments into human resources development in Slovakia. Lódź : KSIEZY MLYN Dom Wydawniczy Michal Kolinski, 2011. 95 s. ISBN 978-83-7729-114-6.

[7] Composite authors of CAQ (2016). Komentované vydání ČSN EN ISO 9001:2016. Editor: Česká společnost pro jakost

[8] Zgodavova, K., Hudec, O. & Palfy, P. (2017). Culture of quality: insight into foreign organisations in Slovakia. Total Quality Management & Business Excellence Vol. 28, Iss. 9-10, 2017. DOI: 10.1080/14783363.2017.1309120.

[9] Paulová, I. & Vaňová, J. (2015). Change of the corporate culture in transforming of ISO 9001: 2015 and ISO 14001: 2014. In Journal of Environmental Protection, Safety, Education and Management. Vol. 4, No. 8 (2016), s. 100-104. ISSN 1339-5270.

[10] Sivri, S; Hennersdorf, S. & Krallmann, H. (2016).Enhanced Method for Quality-fit (ISO 9001) Operations within the Automotive Industry, Proceedings of the 26th DAAAM International Symposium, pp.0614-0619, B. Katalinic (Ed.), Published by DAAAM International, ISBN 978-3-902734-07-5, ISSN 1726-9679, Vienna, Austria DOI: 10.2507/26th.daaam.proceedings.083

[11] Vaňová, J., Gyurák Babeľová, Z. & Holková, A (2009). Company success dependency on management. In Annals of DAAAM and Proceedings of DAAAM Symposium 2009 : Vol. 20, No. 1. s.0741-0742. ISSN 1726-9679. V databáze: WOS ; SCOPUS.

[12] Paulová, I; Vaňová, J.& Dóza, P. (2016). Standard of fulfilment of requirements in the application of TQM in Slovak republic, Chapter 38 in DAAAM International Scientific Book 2016, pp.435- 446, B. Katalinic (Ed.), Published by DAAAM International, ISBN 978-3-902734-09-9, ISSN 1726 -9687, Austria, Vienna, Austria DOI: 10.2507/daaam.scibook.2016.38

[13] Muhlfeit, J. & Costi, M. (2017). Pozitívni leader. Albatros Media a.s. Praha. ISBN 978-80-265-0609-6

DOI: 10.2507/28th.daaam.proceedings.038

OPTIMIZATION OF THE IN-PROCESS CONTROL PROCESS USING SIX SIGMA METHODS AND TOOLS

Martin Januska& Adam Faifr

This Publication has to be referred as: Januska, M[artin] & Faifr, A[dam] (2017). Optimization of the In-Process Control Process Using Six Sigma Methods and Tools, Proceedings of the 28th DAAAM International Symposium, pp.0280-0288, B. Katalinic (Ed.), Published by DAAAM International, ISBN 978-3-902734-11-2, ISSN 1726-9679, Vienna, Austria
DOI: 10.2507/28th.daaam.proceedings.038

Abstract

Presented paper deals with optimization of technological processes, particularly it is focused at quality control. Quality measuring is fundamental for any quality or process improvement. "If is not possible to measure it, it is not possible to manage it." In general it is possible to say that quality is measured at the process input, during the process (in-process control) and at the process output. This paper is focused at the in-process control and its quality itself. Paper presents the case study of optimization of the in-process control by using Six Sigma methods and tools in company producing medical technology and medical products in Czech Republic. Firstly, the description and analysis of in-process control is carried out, while the problem-causing factors are examined. Based on these findings, the proposals for process optimization follow. According to methodology, the DMAIC cycle is completed by the evaluation of process performance, where partial process performance indicators are designed.

Keywords: in-process quality control; quality; Six Sigma; process improvement; process quality; quality improvement

1. Introduction

Goal of this paper is to describe the process of in-process quality control, its limitations and possibilities for improvement. Six sigma methods and tools are successfully utilized in the presented case study from Czech company dealing with production of medical supplies. Quality processes are interesting due to completely different way of their assessment in compliance with production processes.

The performance of quality process is determined by the ability to correctly detect deviations of manufactured parts during their production. However, it is not the only starting point for process performance. For the quality process is to be efficient, the efficiency of the resources used is also important in addition to the quality of the work done, which is derived from the speed of the quality checks and controls. Therefore, eventual changes of the process can improve its performance in one aspect and worsen it in the other one. The purpose of this study is to examine how changes of all process aspects affect each other and consequently propose the possibilities of the process performance improvement.

2. Process management approach

The process approach to organization management began to develop in the last two decades of the last century when it gradually began to replace the previous mainstream concept of corporate governance the functional approach. The main change over the functional approach is to look at the causes of the company's problems and the way they are solved where the processes are perceived purposefully in a relationship to the customer and not as standalone activities [12].

2.1. Measuring process performance

The systematic measurement of processes is an integral part of process control. Measuring process performance means activities designed to provide objective and accurate information about the progress of individual processes close to the real time so that processes can be continuously managed to meet all process requirements. There is huge difference between KPI (Key performance indicators) which are measured in real time or once a day or week and KRI (Key results indicators) measured once a month or year which are used for strategic planning.

From the business point of view measurement is necessary, among other things, in order to assess whether an enterprise fulfills defined objectives. In order to evaluate the performance of the process, it is also necessary to determine the base (etalon) with which the measured data will be confronted. Any changes to process outputs are conditioned by input changes and performance changes. If such changes in output or performance are not registered, no change can be expected. In order to control the change of output, the change of the inputs and the whole process of the process must also be controlled. Measurement of the process is a keystone of the process management. In other words if it is not possible to measure it, it is not possible to manage it [11].

2.2. Process improvement

In general, it can be argued that the need to improve processes arises at the moment when a certain shortcoming is discovered. From the point of view of keeping firms on the market, that is the time when is necessary to improve processes. The key factor in the need for change is the customer and his market position, which has changed significantly over the past two decades. Customers are demanding products and services higher and higher quality. Every business which wishes to stay on the market for a long time has to think the same way.

However, in order for processes to be improved, it is first necessary to identify their shortcomings. There are two basic approaches to improvement - continuous and discontinuous. Continuous approach is based on continuous improvement of processes based on continuous monitoring of process characteristics, ie process measurement.

The Six Sigma philosophy[2][8][9][14] used in process optimization is the DMAIC cycle consisting of five basic steps:

- Definition of process and problem
- Measurement of process variables (input, output, sources)
- Analysis of process deficiencies, finding causes of the problem
- Identifying opportunities and improving process performance
- Controlling proposed measures based on defined indicators

3. In-process control

For the presented case study the process of inter-operative control at Gerresheimer Horšovský Týn spol. s.r.o., a company operating in the Czech Republic, which is part of the global group of companies Gerresheimer. The Gerresheimer Group is a major producer of a wide range of medical technology and medical products.

The process of inter-operational control is part of corporate quality control, along with input and output control. Quality controls take place during ongoing production, where the checkpoint is always located behind each production step for better detection of nonconforming product but also to better determine the location and cause of the problem.

From the point of view of individual steps, the process can be divided into five basic phases, namely:

- Planning the test
- Selecting the sample
- Testing the sample
- Enter the result into the system
- Dismantling or archiving of the samples tested

The underlying issue of this case study is the reduced flexibility of the process leading to an increase in the cost of securing the process due to the increase in overtime. The following analysis is therefore carried out with regard to the problem examined in order to identify the options for solving the problem.

3.1. Input – Requirements for control

The product control plan is the main input for the control. The product control plan determines which tests and at what intervals will be performed at the product during the production cycle. This document is created separately for each component and determines the method of sampling, the type of tests to be performed, the interval of their execution and the location and equipment used for the testing.

On the basis of the control plan and data from current production, test requirements are generated. In the monitored period from April to June 2016, a total of 41,328 tests were performed in the department, which means a requirement to run the test every 3 minutes in 24/7 operation system.

The process is eligible if the measured values move between the acceptability limits and all fluctuations are caused by random causes. In this case, the temporary interruption of production on a smaller number of production facilities would be considered as such a random cause. If the process has only these random effects, the process is considered statistically mastered in the Six Sigma control system. The limit of the effects of these random causes is the deviation of values from the mean of the selection (UCL, LCL). The limit is usually determined as three times the standard deviation from the mean in both possible directions. In case of higher stability, the set tolerances may be lower [7] [9] [13].

The chart below shows a graph of the number of tests converted to a workplace in the two laboratories under study referred to as 4.7.I and 4.7.II. The graph shows marked variability in both directions from the CL border, especially in tests performed in the second control laboratory. Whilst during the month of April the number of tests performed exceeds the accepted upper limit (UCL), then in the second half of April this value decreases significantly so that in the following months the number of tests per worker is reduced to the lower acceptable limit (LCL).

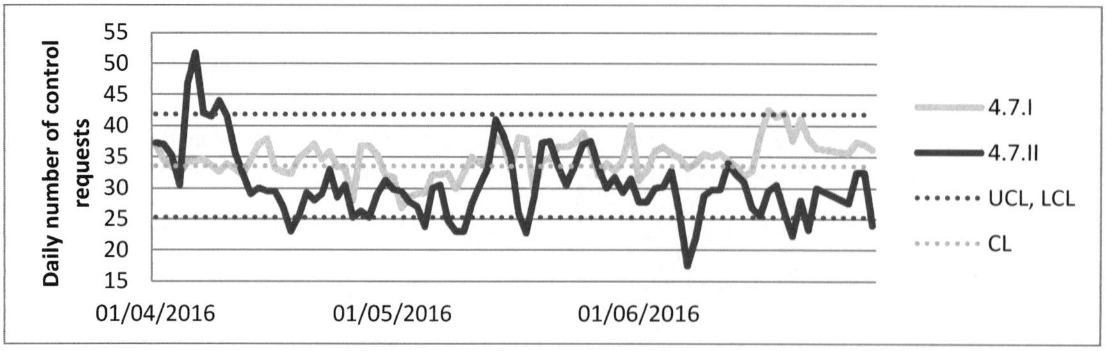

Fig. 1. Input variability progression

3.2. Output - Control results entering

The output of the process is that the results of the control are entered into the MES system. The number entered records should correspond to the number of requirements for control. Average necessary time since the requirement for a new test until entering the result into MES in the reference period lasted 49 minutes.

The following figure shows a graph describing the development of the daily average test time over the reference period. Trends were used to describe the trend over the last three days. It can be seen from the graph that, including fluctuations in average time, the test period are gradually extending.

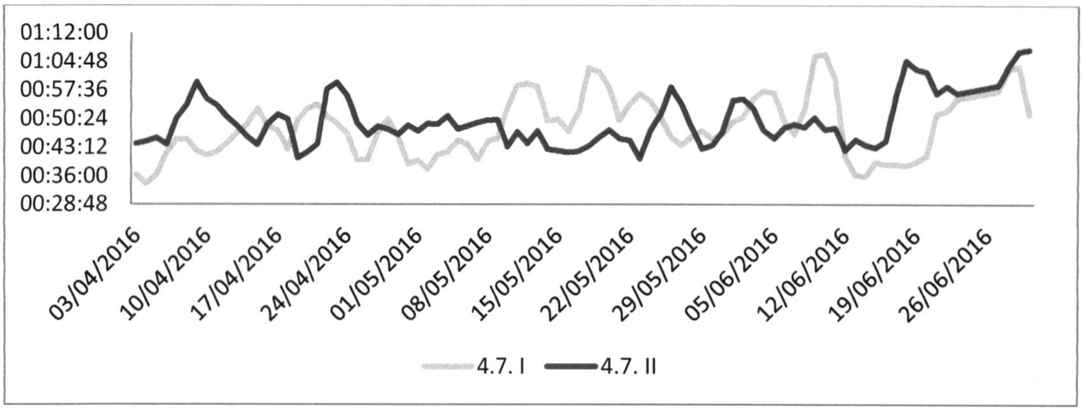

Fig. 2. Daily averages of inspection time

3.3. Material flow

The spaghetti diagram serves as a basis for visualizing the current flow of the process. This is a snapshot showing the current flow of material, information, or people in the process [3].

Control rooms are divided into two control laboratories. Non-control workers are also commonly employed in laboratories. The goal of material flow mapping is to place the process-related regulation into a real spatial framework. The current flow of material in IPC laboratories is described in the following figure. As shown in the blue interrupted rectangles, the spaces for the individual types of checks are indicated. The orange flow is then indicated by the material flow of the selected control type - combined visual and dimensional control.

The numbering of the control steps is assigned as follows:

1. Receiving the material for exam
2. 1st Control Point - Visual Exam
3. Preparations for the dimensional test
4. 2nd Inspection site - Dimensional test
5. Preparations for dimensional
6. Temporary storage - in the laboratory 4.7 I Transient sample retention has been observed after all tests have been performed. This storage is inconsistent with the process manuals.
7. Discard or archive samples

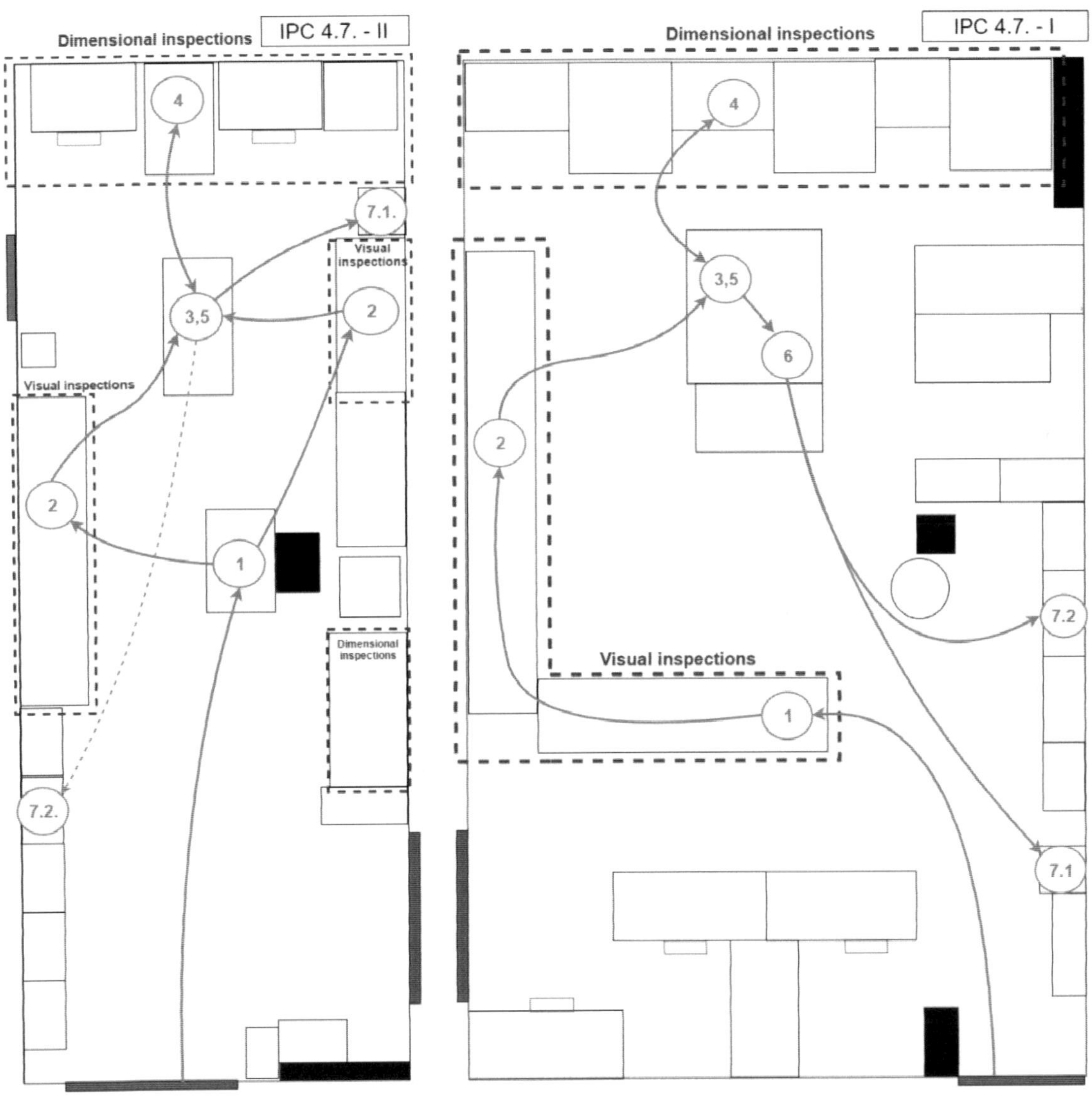

Fig. 3. Previous layout of control laboratories

4. Process analysis

Based on the description of the process characteristics, key process variables were described. From the above, the following factors are most important:

- Fluctuations in the number of tests per worker
- Different levels of laboratory testing
- The fluctuating average test duration
- Prolonged average test time
- There may be violations of process and work instructions
- Areas are shared with non-quality control staff

The aim of optimizing the process is therefore the elimination of individual risk factors as much as possible. The primary prerequisite for the process is that if the testing takes an average of longer time, then one worker performs fewer tests on average and increases the need to increase the number of workers on the shift, thereby increasing the cost of running the process accordingly.

Based on this assumption, an analysis of the dependence between input level and average test time will be performed. The dependency pattern is shown in the following figure where the trend is visible, but the low reliability of the measured data does not allow this hypothesis to be rejected.

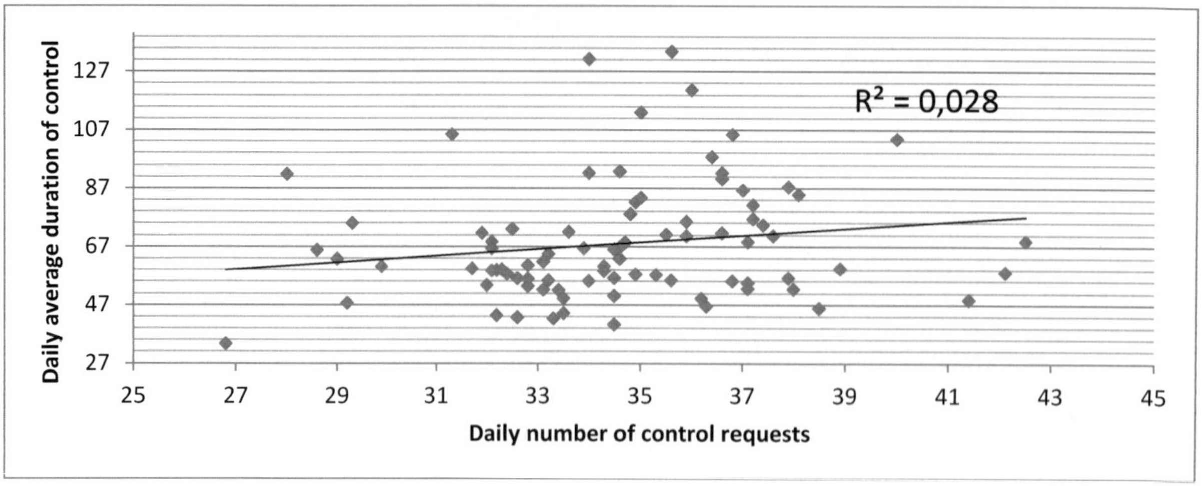

Fig. 4. Output regression analysis

In order to identify other possible causes of the prolongation of the test period, an Ishikawa diagram is compiled. The diagram defines, in an ordered form, the cause of the consequence. The diagram allows to find the real causes of the problem, not just the symptoms of the consequence [5].

The various potential causes can be divided according to defined areas:

Management

- Organization of control activities - refers to the division of the individual shifts and the assignment of tasks beyond the normal control activity

Workers

- Qualification of workers - Higher qualifications should also lead to higher speeds of testing
- Occupation of shift - temporary incomplete shift occupation (leave, incapacity for work, etc.)

Procedure

- Non-optimal follow-up of control steps - the control procedure is defined in cooperation with the customer (see the control plan)

Process

- Entry variability - workers may be overloaded with tests

Devices

- Device malfunctions - occupy 3.44% of control activity
- Insufficient equipment - may complain about the control activity, however the influence on the extension of the test processing time is minimal

Environment

- Spatial layout of laboratories - results from the current material flow in IPC laboratories. Some measuring devices are located outside of the lab, which can significantly increase transport time while complaining about handling control samples. There may also be a lack of clarity when handling samples.
- Disturbing elements - laboratories are shared with other workers, increased movement in laboratories may reduce the ability of workers to concentrate.

5. Process improvement proposals

The proposed measures should lead to increased process efficiency. Therefore, it is not possible to choose proposals that will achieve the desired effect (reducing the running time of the tests and reducing personnel costs) to the detriment of quality.

From the analyzed possible causes of the problem in the process, the proposal for a new organization of control laboratories and a change in the way control activities were organized were identified as potential opportunities for performance improvement. Another measure should aim to reduce the failures of used equipment.

5.1. Layout of control laboratories

In the cause and effect analysis, one of the possible potential causes of the extension of the test time was the current environment where the control is taking place. Specifically, this was a possible spatial arrangement that could prolong the transport time of the samples and thus extend the testing time as such. The second possible cause was the possible disturbance of control staff by non-control staff, which may, in addition to prolonging the testing, lead to a decrease in the number of workers and thus a decrease in the quality of the work performed.

The improvement proposal envisages a new spatial arrangement of control laboratories. The new plan envisages the centralization of control into a single room, where all control devices and places for workers will be moved from the IPC 4.7 II. laboratory to IPC 4.7 I. At the same time, all non-controllrealted devices will be moved to the IPC 4.7 II.laboratory.

This change should reduce the average transport distance of the controlled samples by 14.6%. Only direct manipulation of samples within the laboratories should save 36.78 km per month. Further improvements are expected from the viewpoint of the control environment where the movement of non-statutory personnel in laboratories should be significantly reduced, which should only enter the laboratory again with the benefit of parts for inspection and for the removal of empty containers back into production.

A new sampling site is set where the production staff will place the parts to be tested. Further, the objects for sample destruction are moved to the penultimate process step. This will, among other things, reduce the risk of mixing in the space for dimensional test preparations.

5.2. Increase of flexibility of the control team

Changing the organization of control laboratories and centralizing the process into one room will remove the physical barrier in the context of testing. Moving into one control laboratory can increase the substitutability of workers in the event of outages.

The limiting factor of change is the knowledge of inter-operative control workers in the field of checks on the remaining products. The knowledge of an inter-operative control worker can be structured into four layers in descending order as follows:

- Knowledge of the quality management system in the company
- Knowledge of capturing information
- Knowledge of the inter-operative control process
- Knowledge of the product, including how to perform specific controls

The first three levels of knowledge are common to all inter-operative control staff. Reorganization of the control system therefore requires the updating of the knowledge of the last level.

The first step of the reorganization is the implementation of staff training leading to the dissemination of knowledge in other processes. However, the training itself does not achieve the desired status. In order for a worker to carry out his / her own work in accordance with the knowledge gained, it is necessary to convert this explicit knowledge into tactile knowledge. This can only be achieved through practical activities, in this case by carrying out specific tests. The duration of this transfer depends on the entry level of the worker's knowledge and also the ability to acquire knowledge [1] [10].

The benefit of this measure is easier way to replace the individual IPC control staff in the event of a staff shortage, which should bring the desired effect of reducing the overtime hours by 69%. In total, this is a saving of 5,424 overtime hours. Another effect is a more even distribution of controls among individual workers, where the controls will not be allocated based on assignment to a specific project.

From the point of view of the development of the input, the variability of the variability (number of test requirements) will be significantly reduced, where the original fluctuations between the laboratories will be mutually attenuated.

6. Measuring process performance

There is no benchmark for any process that would ideally affect all aspects of the performance of each process. Therefore, the introduction of a new indicator is not a creation of a universal indicator, but one that best describes the activity under consideration and which is best suited to the needs of the user. Therefore, when introducing a new performance benchmark, it is important to ask: "How can this measure improve the performance of the activity?"

Cost of non-quality (IPC)
Measurement of costs due to inferiority can, in a reversed form, express the quality of the company's management or sub-process. Although this indicator is commonly used in the enterprise, it is not further analyzed. This provides information on the quality management system as a whole. The purpose of introducing this indicator is to allocate the costs incurred according to the place of inferiority. Its aim is therefore to measure only the costs that have arisen in a particular process. In this case, the process of inter-operative control.

The indicator calculation is defined by the following equation:

$$CNQ_{IPC} = \frac{\text{costs due to inferiority(in time period)}}{\text{unit turnover (in time period)}} * 100 \tag{1}$$

In relation to the CNQ business indicator, this indicator is given by the sum of the cost of the iferior partial processes to the unit's turnover.

Documentation Right First Time
The error caused by bad writing will normally lead to waste. This metric concerns misspellings in documents processed in the process. This is a list of internal complaints or mistakes when entering test results.

The pointer is defined by the following equation:

$$DRFT_{IPC} = (1 - \frac{\text{The number of documentation error reports}}{\text{total number of reports}}) * 100 \tag{2}$$

Controls processed in time
The purpose of introducing this indicator is to measure whether the process is capable of meeting its demands in terms of the speed of the tests being carried out. Two variables are defined for the calculation.

The calculation is given by the following equation:

$$CIT_{IPC} = \frac{\text{numer of tests performed in time}}{\text{total number of tests}} * 100 \tag{3}$$

Activity processed Index
In addition to control activities, IPC control officers are entrusted by their superiors with ad hoc control activities beyond the control plan. For example, it is a product review or additional measurement if the actual cause of the problem can be detected or in case of investigation of the influence of the newly used material etc. This indicator is inversely proportional to the CIT and QCI . The higher the amount of time consumed for conducting routine checks, the less time can be used for further testing.

The calculation is given by the following formula:

$$\boldsymbol{API}_{IPC} = \frac{\text{Number of finished activities in given time period}}{\text{Number of required activities in given period}} * 100 \tag{4}$$

Quality controls Index
Measurement of the level and variability of the input of the process provides information on the quantitative mode in which control is taking place. Although this indicator does not measure the ability to meet process requirements, it can analyze outputs based on these data.

A high entry rate can negatively affect either the quality of the tests performed or the speed of the tests being carried out. The calculation is defined by the following equation:

$$QCI_{IPC} = \frac{total\ number\ of\ required\ tests}{\text{monthly average of required tests}} \tag{5}$$

7. Evaluation of process performance

The previous section describes the possibilities of introducing sub-performance indicators in relation to the aspects of the inter-operational control process. As has been said above, improving one of the performance aspects can also mean a reduction in performance by the optics of the second aspect.

The aim of introducing a synthetic scale is to summarize all aspects of the process under review. In order to be able to evaluate the various aspects in a comprehensive way, it is necessary to build individual sub-indicators against each other and evaluate their impact on the overall performance of the process.

The so-called spider diagram (also referred to as the radar diagram) can be used to monitor process performance in individual aspects. The Radar Diagram is one of the tools used by Six Sigma philosophy. The graph shows the velocity data on its own axes in a spin-off manner. This method of recording results is appropriate when it is necessary to clearly evaluate several characteristics in one place. The characteristics are always based on the standard [4].

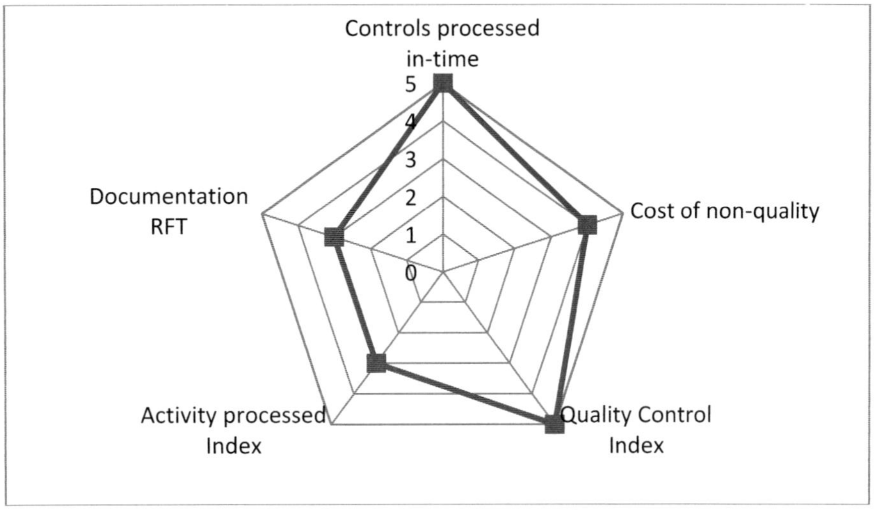

Fig. 5. Process performance diagram

The standard should be set to match the values of the characteristics. Values can be converted to spot scores. For example, 5 points means the best possible result, 3 points mean the average score, 0 means absolutely the worst possible result.

Besides the process performance determination, the proposed indicators can be further exploited in several other areas. The basic design calculates the aggregate indicators for the whole control process together. However, the method of processing the results can be tailored to the needs of the process management, both by time and by organizational units.

Unit rating permits a better chance of responding adequately to eventual performance reductions by locating the right site of potential shortage. All the information thus obtained can serve as a basis for better planning in all aspects of the process.

8. Conclusion

The aim of this paper was to investigate the process of interoperative control, its pitfalls and the possibilities of its improvement. In addition to the role of quality assurance in the enterprise, the choice of this process is a different way of evaluating its performance than that of value-creation (production) processes. The performance of this process is determined by the ability to correctly detect deviations of manufactured parts during their production. However, this is not the only starting condition for process performance. If the process is to be effective, it is also important for the quality of the work to be done, as well as the efficiency of the resources used, which is derived from the speed of the quality controls.

It is precisely the efficiency of resource utilization in the process and the possibility of increasing economic efficiency as the main objective of the investigation, while respecting all process constraints and process performance factors. The underlying issue of this case study was the reduced flexibility of the process leading to an increase in the cost of securing the process due to the increase in overtime. DMAIC Cycle, as a Six Sigma Continuous Improvement Project, was used to analyse this issue. In addition, tools that are commonly used in this process have been used.

Firstly, the process was defined and an analysis of the process was carried out, where important factors influencing the output were first identified, followed by the analysis of their causes with respect to the problem solved. Subsequent measures proposed could not offer a solution that would increase one of the monitored indicators at the expense of the other. To increase the flexibility of the process does not mean reducing the speed or quality of the controls performed.

Based on these findings, the optimization of the laboratory layout was proposed, as well as the reorganization of the control team. Due to these proposals the expected overtime costs reduced by approximately 69 %, where all related expenditure should be covered by increased process effectiveness within 130 days.

Moreover, continuous improvement is further driven by continuous measurement and monitoring of process developments. Therefore, partial performance indicators were designed to monitor both the development of inputs and, in particular, the development of output - the average duration of the controls and its quality.

As this work has shown, Six Sigma tools and methods can also be used for controlling processes.

9. References

[1] Dalkir, K. (2011). Knowledge management in theory and practice. 2nd ed.Mass.: MIT Press, ISBN 978-0-262-01508-0. Cambridge

[2] Furterer, S. (2009). Lean Six Sigma in Service: Applications and Case Studies, CRC Press, 9781420079104, Boca Raton

[3] George, N. (2010). Kapesnípříručka Lean Six Sigma: rychlý průvodce téměř 100 nástroji na zlepšování kvality procesů, rychlosti a complexity, SC&C Partner, 978-80-904099-2-7, Brno

[4] Gil, J., Howse, J., & Kent, S. (1999). Formalizing spider diagrams. In Visual Languages, 1999. Proceedings. 1999 IEEE Symposium on (pp. 130-137). IEEE.

[5] Ishikawa, K. (1985). What is total quality control? The Japanese way. Prentice Hall.

[6] Kolosowski, M.; Duda, J. & Tomasiak, J. (2016). Statistical process control in conditions of piece and small lot production, Proceedings of the 26th DAAAM International Symposium, pp.0147-0155, B. Katalinic (Ed.), Published by DAAAM International, ISBN 978-3-902734-07-5, ISSN 1726-9679, Vienna, Austria

[7] Kralikova, R., Rusko, M. & Wessely, E. (2012). Framework for Applying Six Sigma Methods within Environmental Management, Chapter 45 in DAAAM International Scientific Book 2012, pp. 545-552, B. Katalinic (Ed.), Published by DAAAM International, ISBN 978-3-901509-86-5, ISSN 1726-9687, Vienna, Austria

[8] Mahato, S.; Dixit, A. & Agrawal, R. (2017). Application of Lean Six Sigma for cost-optimised solution of a field quality problem: A case study. PROCEEDINGS OF THE INSTITUTION OF MECHANICAL ENGINEERS PART B-JOURNAL OF ENGINEERING MANUFACTURE, Vol. 231, (MAR 2017) page numbers (713-729), ISSN0954-4054

[9] Michalek, J. (2006). Statisticka regulace procesu (SPC), Ceska spolecnost pro jakost, 80-02-01810-9, Prague

[10] Mladkova, L. (2004). Management znalostí v praxi. Professional Publishing, 80-86419-51-7. Praha

[11] Nenadal, J. (2004). Mereni v systemech management jakosti, Management Press, 80-7261-110-0, Prague

[12] Repa, V. (2012). Procesne rizena organizace, Grada, 978-80-247-4128-4, Prague

[13] Tent, I.-D., Dumitrescu, C.-D. & Trandafir, N. (2010). Statistical quality control methods of products and services. In Proceedings of the 21st DAAAM International Symposium. Viena : DAAAM International, 2010. ISBN 978-3-901509-73-5.

[14] Wessel, G., & Burcher, P. (2004). Six Sigma for small and medium-sized enterprises. The TQM Magazine, 16(4), 264-272.7

DOI: 10.2507/28th.daaam.proceedings.039

STABILITY OF RETAINED AUSTENITE IN HIGH-STRENGTH MARTENSITIC STEELS DURING COLD DEFORMATION

Štěpán Jeníček, Dagmar Bublíková, Hana Jirková, Josef Káňa

This Publication has to be referred as: Jenicek, S[tepan]; Bublikova, D[agmar]; Jirkova, H[ana] & Kana, J[osef] (2017). Stability of Retained Austenite in High-Strength Martensitic Steels during Cold Deformation, Proceedings of the 28th DAAAM International Symposium, pp.0289-0294, B. Katalinic (Ed.), Published by DAAAM International, ISBN 978-3-902734-11-2, ISSN 1726-9679, Vienna, Austria
DOI: 10.2507/28th.daaam.proceedings.039

Abstract

The toughness of high-strength steels is improved when their martensitic matrix is combined with the presence of retained austenite. The stability of retained austenite is crucial, both upon cooling to room temperature and during subsequent cold deformation. One of the advanced heat treatment methods, which produces larger amounts of retained austenite in martensitic matrix thanks to an appropriate combination of parameters, is the Q&P process (Quenching and Partitioning). It produces microstructures consisting of martensite and retained austenite which exhibit strengths above 2000 MPa and elongation levels of more than 10%. The steel which was designed for this experiment was alloyed with manganese, silicon, molybdenum and chromium and had a low Ms temperature.
This experimental heat treatment led to a martensitic microstructure with retained austenite, with ultimate strengths in the range of 2200-2400 MPa and A_{5mm} elongations of 7-10%. The largest fraction of retained austenite, according to X-ray diffraction analysis, was 10%. The behaviour of retained austenite under mechanical load was studied by means of compression tests to various strain levels. The resulting microstructures were observed using optical and scanning electron microscopes..

Keywords: Q-P process; AHSS steels; retained austenite; X-ray diffraction

1. Introduction

Retained austenite in the final microstructure of advanced steels is an important phase which provides a combination of strength and ductility. An advanced processing technique for achieving the same combination is the Q&P (Quenching and Partitioning) process by which strengths in the vicinity of 2000 MPa and elongation levels of approximately 10% can be obtained [1], [2], [3]. Q&P processing relies on cooling from the fully austenitic region to a region between the M_s and M_f, whereby the martensite transformation is arrested, and subsequent reheating to a partitioning temperature. The stock is held at this temperature while retained austenite becomes stabilized by carbon which diffuses from the super-saturated martensite [4], [5]. For the purpose of application of high-strength steels, both chemical and mechanical stability of retained austenite must be known, as it is the behaviour of retained austenite during cold deformation which substantially dictates final mechanical properties. As the growing use of such steels also expands their potential for components operating at elevated temperatures, the stability of retained austenite under various thermal loading temperatures becomes relevant. The main factors determining whether retained austenite remains stable in martensitic matrix include its morphology, particle size and distribution. Its chemical stability also depends on alloying. Therefore, once appropriate heat treatment parameters are confirmed to provide a sufficient proportion of retained austenite, the behaviour of retained austenite under cold deformation must be ascertained as well.

2. Experimental programme

Since the Q&P process is characterized by interrupting the quench at a specific temperature between the M_s and M_f, its industrial use is technically demanding. For this reason, the proposed experimental steel had a manganese level of 2.45% (Table 1), which considerably depressed the M_s to 209°C and the M_f to 78°C. A very low M_f temperature is expected to simplify the entire heat treatment process. Manganese improves mechanical properties of steel through its effect on solid solution strengthening and retained austenite stabilization. The stability of retained austenite is also controlled by silicon which also prevents carbides from forming, and therefore provides sufficient super-saturation of martensite with carbon. The carbon content was 0.4%. Nickel was used for improving austenite stability and for better hardenability. Transformation temperatures were calculated by means of the JMatPro software [10].

C	Mn	Si	P	S	Cu	Cr	Ni	Al	Mo	Nb	M_s	M_f
0.419	2.45	2.09	0.005	0.002	0.06	1.34	0.56	0.005	0.04	0.03	209	78

Table 1. Chemical composition of experimental steel with reduced M_s temperature [wt %]

2.1. Q&P Process

Several different Q&P sequences were carried out on the experimental steel (Tab. 2). Their parameters were defined on the basis of earlier experiments [4]. Quenching temperature was chosen with reference to earlier experiments and a calculation using the Magee-Koistinen-Marburger formula (Eq. 1), [11], [12]. This equation yields an estimate of the martensite fraction dictated by the chosen quenching temperature (Fig. 1).

$$f_i^a = 1\text{-}\exp[-1.10\times10\text{-}2(Ms - Tq) \quad (1)$$

where f_i^a is the volume fraction of martensite, T_q is the quenching temperature (QT) and M_s corresponds to the M_s temperature calculated using JMatPro. This calculation shows that at a quenching temperature of 150°C, the material retains 53% of austenite. While the stock is held at the partitioning temperature (PT), a certain amount of retained austenite becomes stabilized by carbon which migrates from super-saturated martensite, whereas the remaining portion of retained austenite decomposes into martensite as the stock is cooled to room temperature.

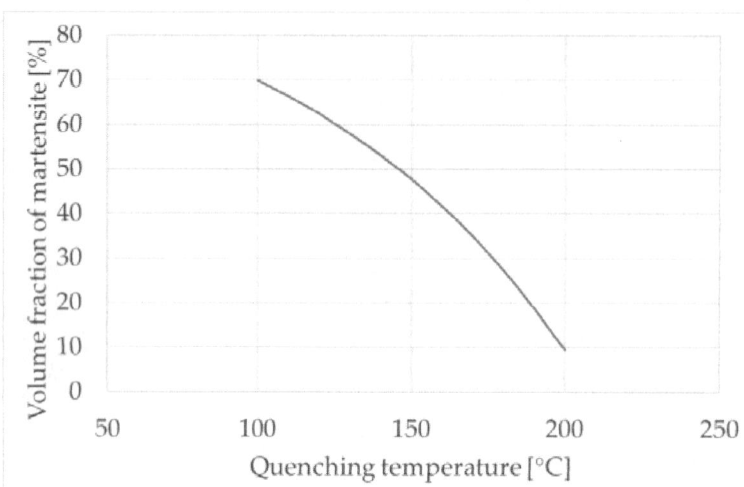

Fig. 1. Martensite volume fraction vs. quenching temperature (QT)

The Q&P process comprised heating to an austenitizing temperature (TA) of 850°C, soaking for 100 or 300 seconds (sequences 1, 2 and sequence 3) and quenching to QT at 16°C/s (sequence 1) or 1°C/s (sequences 2 and 3). The QT was 150°C in all sequences. This was followed by reheating to a partitioning temperature of 200°C, holding for 600 s and cooling to room temperature.
Since specific process parameters need to be used to obtain the desired mechanical properties, the heat treatment sequence was carried out in a thermomechanical simulator [6], [7], [8].

2.2. Stability of retained austenite during cold deformation

The stability of retained austenite and its transformation to strain-induced martensite during cold deformation is of key importance, which is why the behaviour of retained austenite under load needs to be characterized. Based on earlier results, sequence 2 was selected which had produced 12% retained austenite (RA) (Table 2). Specimens processed according to this sequence were compressed in the thermomechanical simulator to $\varphi = 0.13$ at room temperature and at various strain rates (10^{-3}, 10^{-1}, 10 s^{-1}). After this deformation, the retained austenite volume fraction was measured by X-ray diffraction again and compared with the value before deformation (Table 3).

Sequence No.	T_A [°C]/t_A [s]	Cooling rate [°C/s]	QT [°C]	PT [°C/s] /t_{PT} [s]	HV10 [-]	R_m (UTS) [MPa]	A_{5mm} [%]	RA [%]
1	850/100	16	150	200/600	716	2200	5	-
2	850/100	1	150	200/600	679	2308	10	12
3	850/300	1	150	200/600	708	2367	9	9

Table 2. Q&P sequences and results of mechanical tests

Strain rate [s^{-1}]	HV10 [-]	RA [%]
10^{-3}	452	6
10^{-1}	460	4
10	482	3

Table 3. Retained austenite volume fraction in relation to strain magnitude

The microstructures obtained by processing were examined using an optical microscope (Olympus XXX) and a scanning electron microscope (Tescan Vega 3).

Mechanical properties were determined via HV10 hardness and by means of tensile testing on miniature specimens with a gauge length of 5 mm.

The amount of retained austenite was measured by means of XRD phase analysis in the automatic powder diffractometer AXS Bruker D8 Discover with a position-sensitive area HI-STAR detector and a cobalt X-ray source ($\lambda K\alpha = 0.1790307$ nm).

3. Discussion of results

Three Q&P sequences were carried out on the experimental steel with a reduced martensite transformation temperature. They involved various soaking times and cooling rates of 16 and 1°C/s.

The first sequence, in which the austenitizing temperature was 850°C, the soaking time was 100 s and cooling took place at 16°C/s, ending at the quenching temperature of 150°C, produced a martensitic-bainitic microstructure with a hardness of 716 HV10 (Fig. 2). The ultimate strength reached 2200 MPa and elongation was A_{5mm} = 5%. High hardness and low elongation were due to high cooling rate. Slowing down the cooling rate from 16°C/s to 1 °C/s in sequence 2 has led to a larger bainite fraction and higher elongation: 10% (Fig. 3). Appreciable carbide precipitation and a small amount of ferrite were found by metallographic observation. Nevertheless, the volume fraction of retained austenite was high, reaching 12%. Extending the time at the austenitizing temperature from 100 s to 300 s in sequence 3 was not reflected in any way in microstructural evolution or mechanical properties (Table 2).

As the largest final fraction of retained austenite of 12% was found upon sequence 2, which involved a quenching temperature of 150°C and a cooling rate of 1°C/s, the specimens obtained with this sequence were selected for testing the stability of retained austenite upon after cold deformation.

Q&P-processed specimens were subjected to compressive deformation of φ= 0.13 applied at various rates: 10^{-3}, 10^{-1}, 10 s^{-1}. The resulting microstructure was a mixture of martensite and bainite with small amounts of pearlite and retained austenite and exhibited no appreciable effects of deformation. The highest retained austenite 6% austenite was preserved at the lowest strain rate (Table 3). Influence to the high strain rate, the residual austenite was transformed into martensite, which corresponded to the higher hardness of 482 HV10 (Table 3).

Fig 2. Sequence 1 – Martensitic microstructure with a smaller amount of bainite and retained austenite – scanning electron micrograph

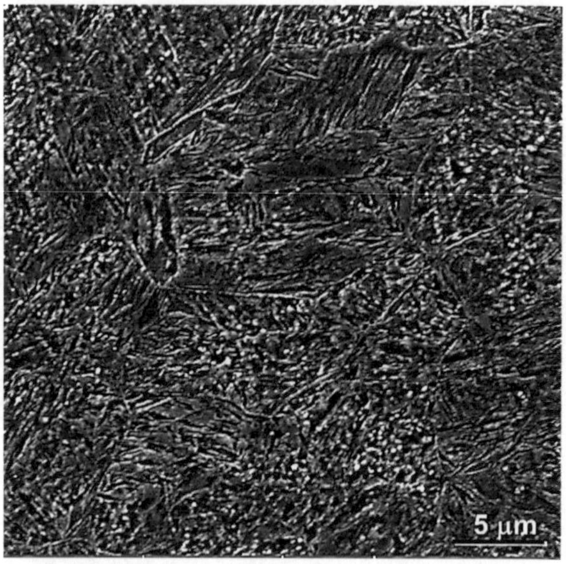

Fig 3. Sequence 2 – Martensitic microstructure with a larger amount of bainite and 12% of retained austenite – scanning electron micrograph

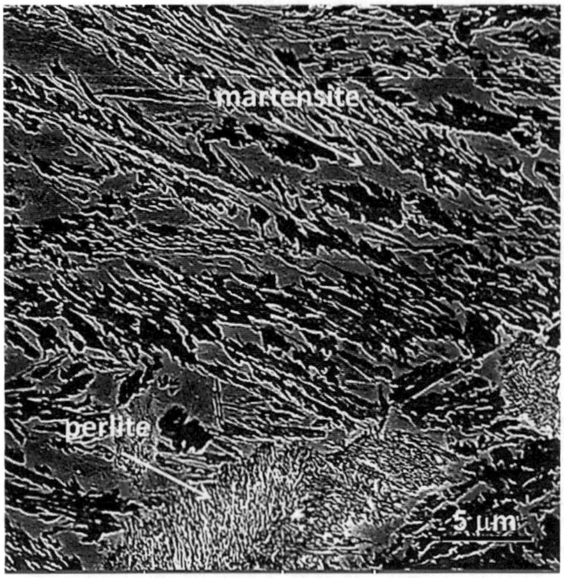

Fig 4. Sequence 2 – Post-compression condition, strain rate of 10^{-3} s^{-1}– martensitic-bainitic microstructure with a small volume fraction of pearlite and retained austenite – scanning electron micrograph

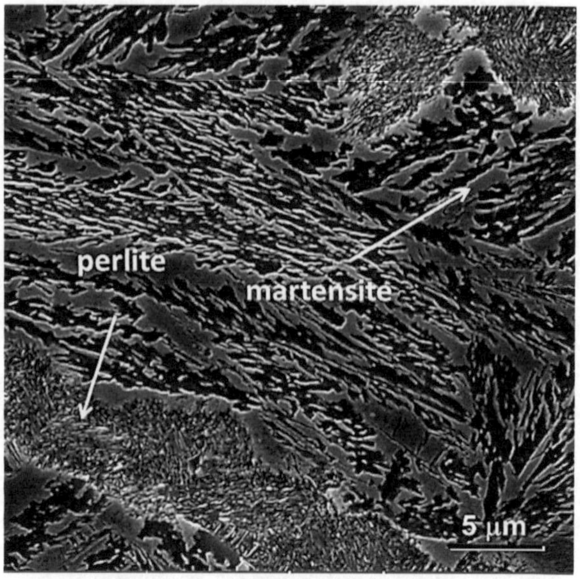

Fig. 5. Sequence 2 – Post-compression condition, strain rate of 10^{-1} s^{-1}– martensitic-bainitic microstructure with a small volume fraction of pearlite and retained austenite – scanning electron micrograph

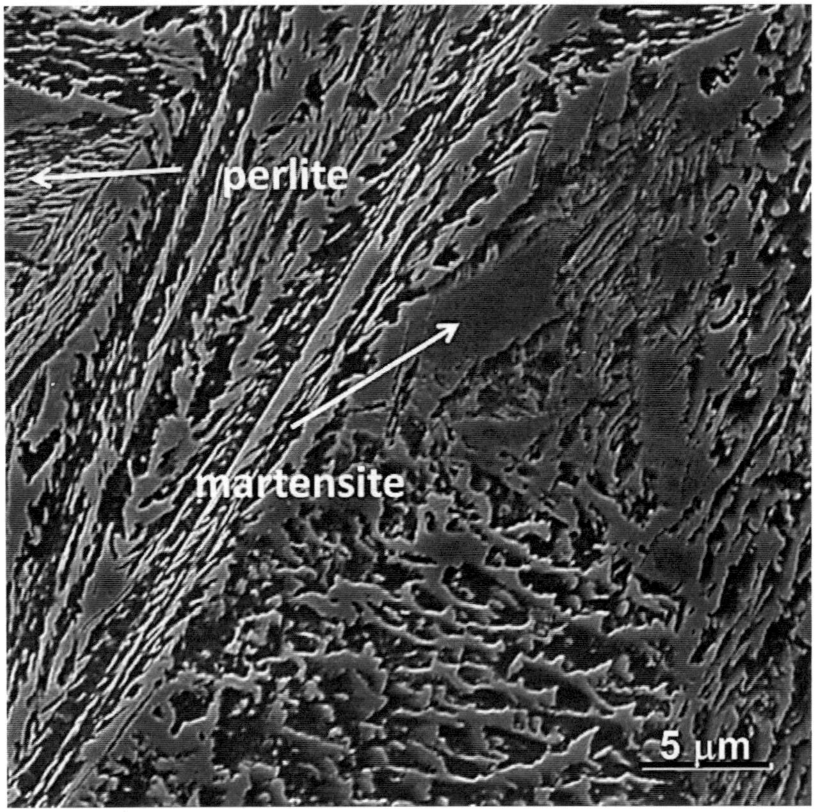

Fig 6. Sequence 2 – strain rate of 10^{-1} s-1– martensitic-bainitic microstructure with a small volume fraction of pearlite and retained austenite – scanning electron micrograph

4. Conclusion

An experimental AHS-type steel with a reduced M_s temperature and a carbon content of 0.4% and a manganese level of 2.5% was experimentally treated using the Q&P process with various times at the austenitizing temperature and various rates of cooling to the quenching temperature. Slowing down the cooling rate from 16°C/s to 1°C/s changed the character of the microstructure. A larger volume fraction of bainite was obtained, together with a small amount of pearlite. Despite that, XRD phase analysis revealed 12% of retained austenite in the material. The ultimate strength exceeded 2000 MPa and the elongation level was 10%.

Specimens processed in this manner were then compressed at room temperature and at various strain rates. Influence to the pressure deformation, retained austenite was transformed into martensite. The highest retained austenite rate was preserved at the lowest deformation rate of 6%.

5. Acknowledgments

This paper includes results achieved within the project SGS-2016-060 Research of Modern AHS Steels and Innovative Processing for their Manufacturing. The project is subsidised from specific resources of the state budget for research and development.

6. References

[1] Mašek, B.; Jirková, H.; Hauserová, D.; L. Kučerová, Klaubeová, D. (2010). TheEffect of Mn and Si on the Properties of Advanced High Strength Steels Processed by Quenching and Partitioning, Materials Science Forum, Vol. 654-656, pp. 94-97

[2] Jirková, H.; Kučerová, L.; B. Mašek, B. (2012). Effect of Quenching and Partitioning Temperatures in the Q-P Process on the Properties of AHSS with Various Amounts of Manganese and Silicon, Materials Science Forum, Vol. 706-709, pp. 2734-2739

[3] Edmondsa, D.V.; Hea, K.; Rizzo, F.C.; De Coomanc, B.C.; Matlock, D.K.; Speer, J.G. (2006). Quenching and partitioning martensite - A novel steel heat treatment, Materials Science and Engineering A, Vol. 438–440, pp. 25–34

[4] Ibrahim, K.; Bublíková, D.; Jirková, H.; Mašek, B. (2015) Stabilization of Retained Austenite in High-Strength Martensitic Steels with Reduced Ms Temperature, In METAL 2015. Ostrava: TANGER spol. s r. o., pp. 1-7. ISBN: 978-80-87294-58-1

[5] Qian, Z.; Lihe, Q.; Jun, T.; Jiangying, M.; Fucheng, Z. (2013) Inconsistent effects of mechanical stability of retained austenite on ductility and toughness of transformation-induced plasticity steels, Materials Science & Engineering A, Vol. 578, pp. 370–376,

[6] Jirková, H.; et al. (2014) Influence of metastable retained austenite on macro and micromechanical properties of steel processed by the Q-P proces, Journal of Alloys and Compounds, Vol. 615, pp. S163–S168

[7] Brushi, S.; Altan, T.; Banabic, D. (2014) Testing and modelling of material behaviour and formability in sheet metal forming, Cirp Annals In: Manufacturing Technology, Vol. 63, pp. 727-749,

[8] Mašek, B.; Jirková, H.; Malina, J.; Skálová, L.; Meyer, W. (2007) Physical Modelling of Microstructure Development During Technological Processes with Intensive Incremental Deformation In Key Engineering Materials, Vol. 345-346, pp. 934-946,

[9] J. Káňa, J.; I. Vorel, I.; Ronešová, A. (2015) Simulator of Thermomechanical Treatment of Metals, In Daaam 2015. Vienna: Daaam International Vienna, pp. 0513-05018, ISBN: 978-3-902734-07-5, ISSN: 1726-9679

[10] JMatPro, Release 9.0, Sente Software Ltd., 2016.

[11] Zhu, Y. et al. (2012) Stepping-quenching-partitioning treatment of 20SiMn2MoVA steel and effects of carbon and carbide forming elements. Sci China Tech Sci 2012, 55, 1838-1843

[12] Koistinen, D.P. (1959) Marburger, R.E. A general equation prescribing the extent of the austenite-martensite transformation in pure iron-carbon alloys and lain carbon steels. Acta Metallurgica, vol. 7, pp. 59-60.

DOI: 10.2507/28th.daaam.proceedings.040

INTERNET OF THINGS AND SOCIAL MEDIA: TOOLS OF A SUCCESSFUL INFORMATION ORGANIZATION

Vlado Majstorovic & Kresimir Rakic

This Publication has to be referred as: Majstorovic, V[lado] & Rakic, K[resimir] (2017). Internet of Things and Social Media: Tools of a Successful Information Organization, Proceedings of the 28th DAAAM International Symposium, pp.0295-0298, B. Katalinic (Ed.), Published by DAAAM International, ISBN 978-3-902734-11-2, ISSN 1726-9679, Vienna, Austria
DOI: 10.2507/28th.daaam.proceedings.040

Abstract

Information technology represents a conglomerate of microelectronics, computers, telecommunications and software, which enable input, processing and distribution of information. It is the fundamental tool of many organizations' strategies and key factor that enable their efficient work and their competitiveness in the market. Modern organizations are so dependent on information processes, flow of information and the application of modern technologies that we can legitimately call them "information organizations". These organizations arise from the traditional ones by application of business processes reengineering and they introduce the new principle of the communication range instead of the classic principle of the control range. Technology of the Internet of Things and communication through social media are two exciting new directions, that can contribute to more successful collection and management of data, important for the organization. Organizations need to adapt to these new technologies for an effective approach to creating, sharing, and analysing information. This paper emphasises the benefit that inclusion of these new technologies brings to the information organization and the potential need for organizational adaptation to the new reality.

Keywords: Modern technologies; Information organization; Social media; Internet of Things

1. Introduction

Proper functioning of the organization and achievement of the desired goals is possible only through the contribution of the information system that provides the raw material and the information necessary to establish the goals, tasks, powers and responsibilities of the management. Information represents essential support in decision-making and communication [1].

Most of the employees, at their workplace, are exposed to some form of information technology (IT). IT represents a conglomerate of microelectronics, computers, telecommunications and software, which enable input, processing and distribution of information. IT is the fundamental tool of many organizations' strategies and key factor that enable their efficient work and their competitiveness in the market. It helps in gathering data and creating information that is useful in making business decisions. New IT enables today's personal computers to use multiple media such as sound, text and video, as well as access to remote databases, leading to a revolution in the joint work of people with the aim to create a successful organization.

The most significant changes in work and business that are brought by the IT are: work in groups, the ability to work outside the office, the creation of non-hierarchical organizations and the outsourcing of a part of business to external organizations. IT today becomes a strategic element in the development of the organization and the society in general, and they are also transformed into one of basic management resources. Today, the organization from the point of view of modern management, becomes so dependent on information processes, flow of information and the application of modern technologies that it can rightly be called an "information organization".

The information organization arises from the traditional one by application of business processes reengineering. This organization introduces the new principle of the communication range instead of the classic principle of the control range. This means that unnecessary management levels are eliminated, i.e. workers who do not work but coordinate others work, and those who actually produce remain. The supervision function is partially performed by the IT. The information flow is circular - information is shared between subordinates and superiors. Also, they travel faster and easier, without unnecessary paperwork. Responsibility and initiative are brought down to "lower" levels.

Information-based organizations need a large number of expert staff who can work with an information system (transactional system, decision support system and/or expert system). The individual takes responsibility for his job and does it the way he knows, without interruption of his superior. The care for human capital is handled within the Human Resources Department.

The information organization is trying to respond as quickly as possible to the needs of its clients. They bind people for a particular client and always have to be available to them. This is helped by an information system for providing and exchanging information about the client and reporting to the supervisors about the performance of the business. Since information organizations manage knowledge, a new type of manager - Chief Knowledge Officer is provided to ensure knowledge development. The increasing reliance on technologies in contemporary organisations makes it vitally important to increase the capacity to plan for and manage technological change [2].

Two exciting new directions, that can contribute to more successful collection and management of data important for the organization, are the technology of the Internet of Things (IoT) and communication through social media. In the following chapters, we will discuss the features of these two innovations in the traditional way of using data in an organization.

2. Internet of Things in modern organization

The pace of technological progress in today's industrial society is essentially dictated by microelectronics and computer technology. Computers are becoming much smaller, inexpensive, last even longer, and part of everyday items: a table, our clothing, our environment and even our bodies can become computers and connected to Internet [3].

Using of sensors to monitor the condition of equipment and the work environment is not a novelty. However, connecting these sensors with the Internet and insights into the collected data revolutionizes almost all imaginable industries, from healthcare and retail to transportation and manufacturing. This new technology, technology of the IoT, even changes the way cities work. IoT supplies organizations with systematic insights to encourage smarter decisions and work processes, and to achieve better results.

The basic concept of the IoT is that everyday objects can be equipped with identifying, sensory, network and process capabilities that will enable them to communicate with each other, as well as to communicate with other devices and services over the Internet in order to achieve some useful purpose. IoT represents an evolution of the use of existing technologies, such as RFID technology, sensor networks, device communication, etc., in terms of number and type of device, as well as networking of these devices over the Internet.

The potential benefits of the IoT are almost unlimited and new applications are being developed every day. There are a number of different sensors that can be used in IoT applications, such as [4]:

- Sound sensors,
- Machine vision sensors and optical ambient light sensors,
- Light sensors,
- Acceleration and tilt sensors,
- Position and presence sensors,
- Movement, speed and shift sensors,
- Moisture, temperature and humidity sensors,
- Leakage and level sensors,
- Electric and magnetic sensors,
- Other sensors.

According to Cisco forecasts, IoT will make up to 500 billion devices connected to the Internet by 2030, which will ultimately improve productivity, create new business models, and generate new revenue streams [5].

Organizations that are willing to replicate the success of using the IoT must be aware of the fact that each initiative will consist of three steps [6]:

- In the first step, the organization should figure out the way to use IoT data about clients, associates, partners, and others. This step, commonly, consists of simple analysis and standard projects that are common within the industry as they bring clear added value.
- In the second step of implementation of the IoT, collected data begins to produce information that helps organizations launch projects that will differentiate them from competitors and provide them with unique features.
- In the final step, the organization has the opportunity to use its investment in IoT to develop new, exciting projects that change its business model. At this stage, organizations can really be separated from the competition.

In a modern company, employees can carry radio frequency identification (RFID) tags so they can be easily located in a case of emergency evacuation, and the management can identify potential bottlenecks in the work process.

For productivity analysis of employees, it is possible to use the cameras together with the software for tracking and analysing the video. The system can be linked to whiteboards, and the facility can be equipped with televisions that will provide work information about work team productivity to identify potential areas for improving the process.

Public security in the organization's facilities is another part in which IoT can play an important role. Capabilities of personal identification, audio and video analytics including face recognition and license plate readers, associated with cloud-based reporting systems will help prevent crime.

3. Potential benefit of using social media in organization communication

While there is still a certain caution on their application, the new media also affects the traditional media and the behaviour of information media users within organizational communications. As with traditional media channels, new media have their own advantages and disadvantages, and careful consideration is needed to optimally exploit their capabilities [7].

Using social media, public opinion communicates itself and reaches a significant targeted audience. In this way, the classical target audience of the organization is activated, ant it is not satisfied with just receiving and consuming information, but it also provides its contribution to data interpretation and commentary. Using social media also changes internal communication within the organization. By communicating through social media, employees contribute to the company public image [8] and create a feeling of belonging to the group. Outside the organization employees are perceived as real ambassadors of the company.

The challenge for the organization is to find ways to make their traditional communication structure more flexible and to enable successful integration of social media in it.

In order to determine the position of social media in communication within an organization, it is necessary to emphasize their particularities. Social networks represent a set of Internet applications based on the ideological and technological principles of Web 2.0 and enable the creation and exchange of user-generated content [9]. Web 2.0 is defined as a business revolution in the computer industry caused by moving to the Internet as a platform and attempting to understand the rules for success on this new platform [10]. We can say that Web 2.0 allows the transition from a reading web towards writing/reading web, a web filled with user-generated content which allows the usage of web application. Web 2.0 tools give communication professionals the opportunity to become more and more innovative and creative and add increasing value to the business [11].

Social media represents, so called, "me" communication, challenging communicators to use them to stimulate employee engagement, provide relevant information, and discover staff attitudes, insights and problems. Once students' informal tool has become a robust platform that connects billions of people around the world every day [12]. They represent new electronic communication channels based on the virtual environment: blogs, podcasts, wiki encyclopaedias, chat rooms, discussion forums, RSS feeds, web sites, social networks and other media that encourage dialogue. It is important for modern organization to find a way to fully exploit the vast potential of these modern communication media.

4. Conclusion

The amount of data, collected and shared within the organization, is getting bigger. In the planning of information infrastructure, it is necessary to look at it several years in advance, and it is almost impossible to estimate the future needs for the data space that organization will need. It is therefore important to find scalable and flexible storage solutions that can grow along with the deployment level of IoT. Cloud-based storage, whose size can easily be increased or reduced, is one of the possible solutions to this problem.

Social media has undergone substantial evolution in the new millennium. Its technology development will continue and soon it will include new technologies, such as artificial intelligence, virtual reality, and augmented reality. By developing social media opportunities, their potential for contributing to effective organizational knowledge management is growing. Organizations need to adapt to the new social media capabilities for an effective approach to creating, sharing, and analysing knowledge.

5. References

[1] Kurtuhuz, A. M.; Radu, D. L.; Anghel, A.; Zahiu, C. E.; Salisteanu, I. C.; Udroiu, I. N. & Dumitrica, M. A. (2013). The Information as the Essential Element in Communication Management, Chapter 34 in DAAAM International Scientific Book 2013, pp. 619-628, B. Katalinic & Z. Tekic (Eds.), Published by DAAAM International, ISBN 978-3-901509-94-0, ISSN 17269687, Vienna, Austria, DOI: 10.2507/daaam.scibook.2013.34

[2] More, E., & Laird, R. K. (1983). Modern technology and organisational communication implications for management, Journal of information science, Vol. 4-5, No. 7, pp 169-183.

[3] Buchmeister, B.; Leber, M.; Palcic, I. & Vujica-Hercog, N (2013). Future Development Trends and Challenges in Production and Social Systems, Chapter 04 in DAAAM International Scientific Book 2013, pp. 111-132, B. Katalinic & Z. Tekic (Eds.), Published by DAAAM International, ISBN 978-3-901509-94-0, ISSN 1726-9687, Vienna, Austria, DOI: 10.2507/daaam.scibook.2013.04

[4] Al-Fuqaha, A., et al. (2015). Internet of things: A survey on enabling technologies, protocols, and applications, IEEE Communications Surveys and Tutorials, No. 17, pp. 2347–2376.

[5] https://www.cisco.com/c/r/en/us/internet-of-everything-ioe/internet-of-things-iot/index.html, (2017). Cisco, Cisco Internet of Things, Accessed on: 2017-08-27

[6] IoT Whitepaper: Building a data-driven future, (2016). CDW, Available from: https://www.cdw.com/content/dam/CDW/resources/solutions/data-center/networking/iot-building-a-data-driven-future.pdf, Accessed on: 2017-08-27

[7] Berger, B. K. (2008). Employee/Organization communication, Institute for Public Relations, The science beneath the art of Public Relations, Available from: http://www.instituteforpr.org/topics/employee-organizational-communications, Accessed on: 2017-08-27

[8] Julich, T. (2012). Kommunikationskanal, aber mit anderen Spielregeln. BMV Group, Ein KommunikationsManager, PR Romania, Available from http://www.pr-romania.ro/articole/pr-20/1385-thorsten-julich-bmw-group-social-media-un-canal-de-comunicare-dar-cu-alte-reguli-de-joc.html, Accessed: 2017-08-27

[9] Kaplan, A. M. & Haenlein, M. (2010). Users of the world, unite!, The challenges and opportunities of Social Media, Business Horizons, Vol. 1, No. 53, pp. 59-68.

[10] O' Reilly, T. (2006). Web 2.0 Compact Definition: Traying Again, Radar. Available from http://radar.oreilly.com/2006/12/web-20-compact-definition-tryi.html, Accessed on: 2017-08-27

[11] Badea, M. (2014). Social media and organizational communication, Procedia - Social and behavioural sciences, No. 149, pp. 70-75.

[12] Kane, G. C. (2017). The evolutionary implications of social media for organizational knowledge management. Information and organization, Vol. 1, No. 27, pp. 37-46.

DOI: 10.2507/28th.daaam.proceedings.041

ANALYSIS OF ENERGY CONSUMPTION IN THE BUILDING AS THE BASIS FOR DEVELOPMENT OF THE MODULAR MODEL PREDICTIVE CONTROL SYSTEM

Boris Crnokić, Gordan Lješić, Željko Stojkić & Marijana Bandić Glavaš

University of Mostar, Faculty of Mechanical Engineering, Computing and Electrical engineering, Matice hrvatske bb, 88000, Mostar, BiH

This Publication has to be referred as: Crnokic, B[oris]; Ljesic, G[ordan]; Stojkic, Z[eljko] & Bandic Glavas, M[arijana] (2017). Analysis of Energy Consumption in the Building as the Basis for Development of the Modular Model Predictive Control System, Proceedings of the 28th DAAAM International Symposium, pp.0299-0308, B. Katalinic (Ed.), Published by DAAAM International, ISBN 978-3-902734-11-2, ISSN 1726-9679, Vienna, Austria
DOI: 10.2507/28th.daaam.proceedings.041

Abstract

This paper gives a comprehensive report about development and validation of the dynamic model of an exemplary building, and comparison of energy consumption from the dynamic model with real energy consumption, as the basis for development of the modular Model Predictive Control (MPC) system. The model is created in IDA ICE professional building modelling software and its performance verified against energy consumption data from energy bills. Analysis and comparison of energy consumption will serve to equip the building with sensors and data acquisition equipment so as to include the whole building into one common Building Energy Management System (BEMS). One of the key parts for development of BEMS is dynamic model of the building, which is derived in this paper.

Keywords: Dynamic Building Model; Construction Data; Simulation Scenarios; Heating/Cooling Building Demand; Zone Temperature Dynamics; Building Data

1. Introduction

Energy efficiency is a very important aspect of the development of modern society. Due to the non-optimal design of buildings, imperfect heating and cooling systems or insufficiently efficient energy management systems, energy consumption in practice is higher than the building itself demands. Approximately 40% of world energy consumption is caused by the operation of various systems in buildings, which includes heating, cooling and air conditioning systems. [1] Conventional energy management systems (EMS) did not give the desired results in terms of energy savings. Model predictive control (MPC) has been recognized as one of the essential solutions to achieve considerable energy savings in buildings. [2] In order to address the problem of energy efficiency, more and more scientific and professional projects are being launched all over the world, in the EU, including separate regions such as the Danube region. One of those projects is also "Smart Building – Smart Grid – Smart City (3Smart)" project. Project is co-funded by the European Union through Interreg Danube Transnational Programme, and partners from the following countries are involved: Croatia, Slovenia, Austria, Hungary, Bosnia and Herzegovina and Serbia.

Current tendencies in the Danube Region to integrate energy-efficiency measures and renewable energy are not followed by their adequate interactions to enable optimal operation of the building as a whole. Distribution system operators are reluctant in allowing renewable energy in the grid as it engages investments into the grid and increases costs for its correct and secure operation. The main objective of the 3Smart (Fig 1.) project is to provide a technological and legislative setup for cross-spanning energy management of buildings, grids and major city infrastructures in the Danube Region. It will provide optimal economical value to energy-efficiency and renewable energy investment in the building and at the same time it will result in optimized costs on the grid side whereas grid and buildings will also interact through exchanging energy and prices data. This is expected to motivate installation of distributed storages in both buildings and grids for improving energy security in the Danube Region. [3]

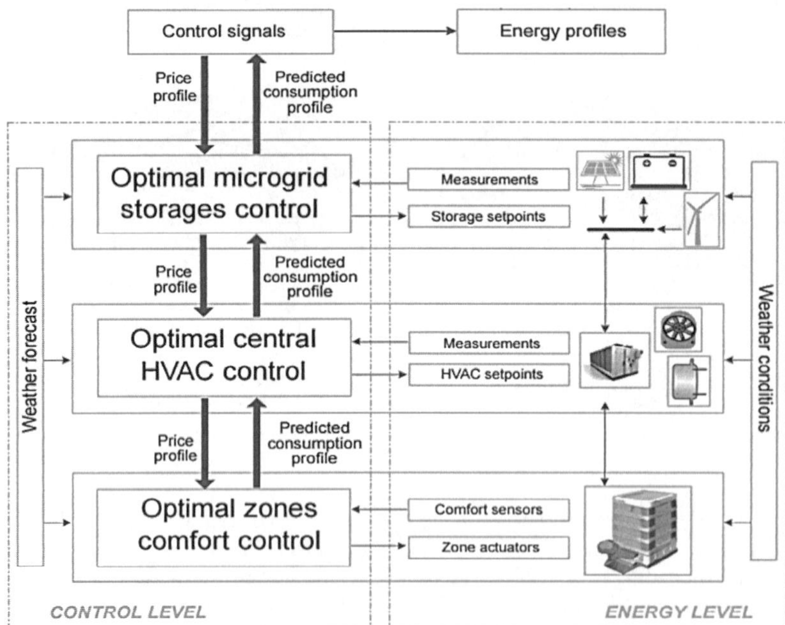

Fig. 1. The conceptual layout of 3Smart platform [3]

This paper presents one part of the project 3Smart, the thermodynamic model of the EPHZHB business building in Tomislavgrad (Bosnia and Herzegovina) which was used as one of the pilot buildings. The model is created in IDA ICE professional building modelling software and its performance verified against energy consumption data from energy bills. Building is consisted of one floor + ground floor and roof. The building is equipped with an advanced central control unit for heating/cooling, which enables the data acquisition from the building zone side to a central database. Based on these, the control commands for individual fan coils will be computed and transferred back through the existing communication network. This will result in a fully controllable building and in such way enable the zone-level smart predictive control for the entire building.

2. Building architecture

The building that is being considered in this paper is a pilot building of a project partner EPHZHB (JP Elektroprivreda Hrvatske Zajednice Herceg Bosne). Construction of the building in Tomislavgrad (Fig. 2) was finished in 2013. It has a useful area of 973,85 m^2 and is located in the industrial zone of "Vučilov brig". The building has a quadratic shape with ground floor dimensions 23,45 x 23,45 [m] which enables a rational arrangement of office space at the circumference of the building and the central position of the entrance hall and the meeting room. All offices have same illumination, but because of simple construction quick and easy modifications are possible if necessary. Illumination is the same in all working areas and comes laterally, and the corridors are illuminated through glazed door walls. Meeting rooms are illuminated and ventilated indirectly across the hallway and lobby, and these two problems are solved through special treatment in installation projects. In the central area are also facilities that do not require lighting (sanitary block, kitchen, elevator) so all the useful space on the facade is used for office lighting. In the entrance hall are vertical communications, staircase and elevator. [4]

Fig. 2. All sides of the considered building [4]

3. Building construction and materials

The technical-constructive solution consists of a proper raster of supporting walls with reinforced concrete horizontal and vertical cerclages, and if necessary with poles and beams. The structure between two floors is made of reinforced concrete slab (MB 30) with a thickness of 20 cm. The foundations were made of reinforced concrete slabs (MB 30) with a thickness of 60 cm. The walls are made of bricks with a thickness of 25 cm with thermal insulating facade system with a thickness of 15 cm. The roof structure consists of a wooden substructure on which a high quality panel is mounted. Thermal insulation is also mounted on reinforced concrete slab under the roof. External load-bearing walls are built with porotherm bricks (25 cm). On the inside of the wall is a layer of internal plaster (3 cm). Thermal insulation facade (15 cm) consists of: facade styrofoam (EPS), construction adhesive, polypropylene mesh and mineral facade plaster. All internal bearing walls are made of gypsum board panels with a thickness of 22 mm with aluminium substructure, 10 cm of stone wool insulation and perforated acoustic panels (12 mm). The wall height is about 3.60 m and the fire resistance class is 90. Total surface area of the internal bearing walls is 781,08 m2. [3] Detailed thermal properties of construction materials used in the building are given in Table 1.

Building element		**Layers (inside-outside) (top-bottom)**	**Thic-kness** *m*	**Thermal conductivity** *W/mK*	**Density** *kg/m³*	**Specific heat** *J/kgK*
External walls		Render	0,020	0,800	1 800	790
		Porotherm bricks	0,250	0,141	750	1000
		Construction adhesive	0,001	-	-	-
		Facade Styrofoam (EPS)	0.100	0.035	29	1213
		Construction adhesive	0,001	-	-	-
		Polypropylene mesh	0,003	0,100	900	1700
		Construction adhesive	0,001	-	-	-
		Mineral facade plaster	0,020	1.400	880	2300
Internal walls	Default walls	Gypsum board	0,022	0,160	950	840
		Perforated acoustic panel	0,012	0,370	117	-
		Stone wool	0,100	0.035	30	1000
		Aluminium substructure	-	-	-	-
		Perforated acoustic panel	0,012	0,370	117	-
		Gypsum board	0,022	0,160	950	840
	Sanitary blocks walls	Ceramic tiles	0,008	1,200	2000	1000
		Gypsum board	0,022	0,160	950	840
		Perforated acoustic panel	0,012	0,370	117	-
		Stone wool	0,100	0,035	30	1000
		Aluminium substructure	-	-	-	-
		Perforated acoustic panel	0,012	0,370	117	-
		Gypsum board	0,022	0,160	950	840

		Ceramic tiles	0,008	1,200	2000	1000
Floors	Ground floor	Floor coating:				
		polyamide carpet,	0,008	0,250	1100	1700
		ceramic tiles,	0,010	1,200	2000	1000
		decorative stone	0,025	1,700	2700	920
		Flooring screed MB 30	0,060	0,412	1200	840
		Separating layer	0,001	-	-	-
		Thermal layer-stone wool	0,050	0,035	30	1000
		Reinforced concrete slab MB 30	0,600	2,600	2500	1000
		Synthetic waterproof foil	0,002	0,190	87	840
		Tampon-gravel	0,200	1.441	1674	881
		Tampon-stone	0,200	1,700	2700	920
	First floor	Floor coating:				
		polyamide carpet,	0,008	0,250	1100	1700
		ceramic tiles,	0,010	1,200	2000	1000
		decorative stone	0,025	1,700	2700	920
		Flooring screed MB 30	0,060	0,412	1200	840
		Separating layer	0,001	-	-	-
		Thermal layer-stone wool	0,050	0,050	0,035	1000
		Reinforced concrete slab MB 30	0,200	2,600	2500	1000
Ceilings		Aluminium substructure	-	-	-	-
		Air gap	0,600	0,230	1,23	1006
		Gypsum board	0,022	0,160	950	840
Roof		Aluminium sheet metal plates	0,001	160,000	2800	896
		Synthetic (Wolfin) foil	0,024	0,190	87	840
		Geotextile layer	0,002	-	0,20	-
		Wooden boards	0,002	0,170	550	2500
		Stone wool	0,100	0,035	30	1000
		Wooden beams	0,100	0,170	550	2500
		Reinforced concrete slab MB 30	0,200	2.60	2500	1000
Glazing		Clear float glass	-	1,45	2500	910

Table 1. Detailed properties of construction material used in the building

3.1. Thermal bridges and infiltration

For the model of business building in Tomislavgrad the list of thermal bridges is shown in Figure 3.a. Values used in the IDA ICE software have been obtained by studying cases in the literature [5], [6], [7], [8], [9]. The determination of the coefficients from the above mentioned literature was made for similar materials and compounds as in the EPHZHB business building in Tomislavgrad. These are approximate values due to the complexity of the procedure for determining the thermal bridges. We did not have the ability to do a detailed examination of the building for that purpose.

Infiltration values are taken from the research of blower-door test from the other pilot building in the project (Faculty of Electrical Engineering and Computing building - UNIZGFER skyscraper building) [10]. We took the same data for "Air tightness": 3.7 ACH and "at pressure difference": 50 Pa. Infiltration for business building in Tomislavgrad defined in IDA-ICE is shown in Figure 3.b.

3.2. *Heating/cooling system installations*

Heating/cooling according to the main mechanical installation project, building is provided with following installations [11]:

- Heat pump
- Engine room
- Fan coil installation
- Radiators installation
- Air handling units
- Server room cooling
- Ventilation
- Automatic regulation system

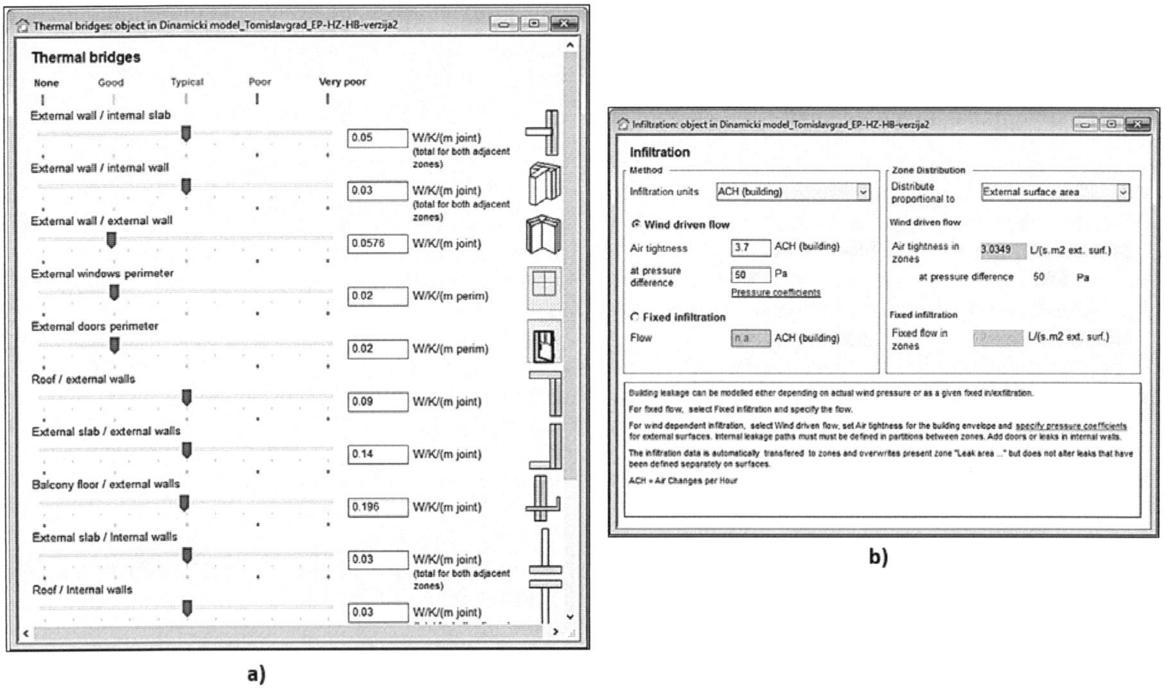

Fig. 3. a) Building thermal bridges defined in IDA-ICE; b) Infiltration for building defined in IDA-ICE

To ensure cold (hot) water 7/12°C (45/40°C), a heat pump with air-cooled condenser is installed. Water circulation is ensured via circulation pumps. Electricity is used as a basic energy source, and an electric boiler 88 kW is also placed in the engine room as an additional one, and at the temperatures below 7 °C as basic hot water source. Heating and cooling of the offices, conference rooms and halls are provided by parapet and ceiling fan coils. Steel panel radiators are provided in sanitary facilities, and they are connected to the fan coil pipeline. Ventilation is done via channel fans, suction valves, grilles and air ducts. Automatic regulation system and central monitoring and control system include management and control of HVAC. [11]

4. Building simulation model in IDA-ICE software

Based on all the information given, detailed building model is constructed in simulation software IDA-ICE [3]. Figure 4 shows 3D model of the building with: 3D view of the whole building, sectioned 3D view of first floor and 3D model of typical zones.

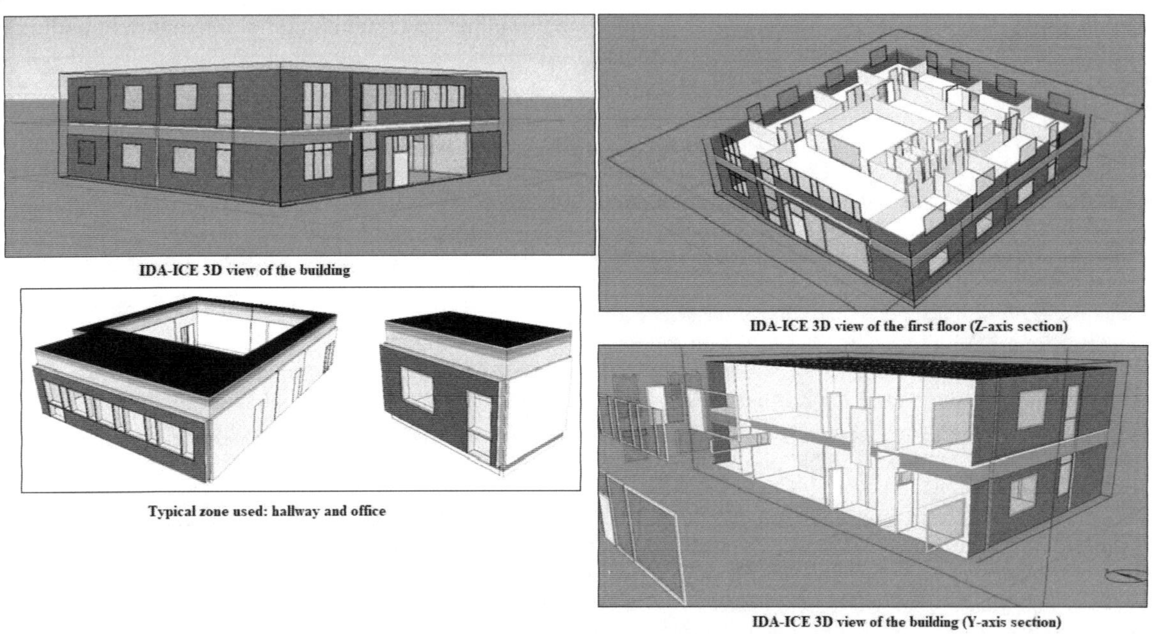

Fig. 4. IDA-ICE 3D model of the building

5. Simulation scenario and responses

Simulation scenario is chosen to fit the building occupancy profile, heating and cooling demands, window opening scenarios, etc.

Heating and cooling system in the building is a typical two-pipe system which implies seasonal heating and cooling. Cooling season covers approximately the period from May, 15th until October, 1st. The rest of the year only heating is available. On a daily basis heating/cooling system operates in two regimes, daily regime defined with working hours from 6:00-18:00 h and night regime outside working hours. During cooling season, temperature is regulated only in daily regime. In heating season, temperature is regulated in both regimes. In daily regime user can set arbitrary comfort conditions within some reasonable temperature range. Night regime is set automatically.

Occupancy schedules for the zones used as offices are generated in accordance with the zones occupancy from for EPHZHB business building technical documentation [3]. It is assumed that offices are not occupied during weekends and holidays. For all offices, it is assumed that working hours are from 08:00-17:00 h, with working break from 12:00-13:00. To simulate the occupancy of toilets, kitchenettes, hallways and stairways, cumulative occupancy is simulated such that each of these zones is occupied with 2 persons per hour. To speed up the simulations, door opening is not simulated due to the usually short duration of such interrupts.

Since building does not have air handling units installed for every room in the building, the fresh air supply depends on window opening. Air handling unit provides fresh air supply as well as ventilation for conference room on the ground floor, regulation room on first floor, hallways, restrooms and for all other rooms that do not have exterior walls. To ensure the fresh air, a ceiling air-chamber is located in lowered ceilings, for each room in particular. Air handling unit consists of:

- pressure fan
- air heater/air cooler with G3 filter
- regulatory blinds.

The air handling unit has the heater/cooler built-in regulator set, which maintains the temperature of the intake air in the winter time 22 °C, and in the summer time 24 °C. The heater has the freezing protection. In transition periods, the fresh air supply system can, apart from the ventilation, cover any heat loss or heat gain. For all other rooms/zones (mainly for offices) fresh air is obtained by occasionally opening the windows. Opened windows with active heating/cooling are major sources of unnecessary energy dissipation. Since window opening is inevitable for ensuring healthy working environment, in the simulation scenario windows are opened two times per 30 minutes during in one working day (from 08:00 – 17:00 h). The windows are opened from 08:00 – 08:30, and then again from 13:00 – 13:30. In all zones only one window is opened at the same time.

Lighting in the building is LED lighting so it does not contribute significantly to overall thermal energy consumption. Lighting is simulated such that the lights are on when zone is occupied and light intensity is below 500 Lux. The lights are turned off when daylight in office reaches 10000 Lux. Rated input per unit is 50 W. Electrical equipment is allocated across the model such that every person in every permanently occupied zone, has its own personal computer and monitor, and one printer per office with 150 W emitted heat per unit.

Weather data used for simulations are data gathered on a meteorological station close to the business building for year 2016. The data are provided by project partner EPHZHB. The data comprises outdoor temperature, direct and diffuse solar irradiance, wind speed and direction, and humidity.

6. Simulation results

Monthly energy demand for the simulation scenario in IDA-ICE software, defined in Chapter 5, for heating and cooling season in 2016 is given in Table 2.

Monthly energy consumption for heating and cooling seasons in 2016. is shown on graph depicted on Figure 5. Total floor area of the building is 973,85 m2 so thermal energy demand for heating in simulation scenario, for 2016., is **77,41 kWh/m^2**.

When running the simulation, one can choose an arbitrary sampling time for the zone level data such as mean air temperature, operative temperature, heat from heating and/or cooling room units, window thermal losses, heat from lighting, heat from occupants, etc. Such information can be of great importance for running a building model identification procedure when data from a real building are missing or are hard to reach. To capture the behaviour and impact of all disturbances affecting the building behaviour minute sampling time has been chosen.

Month		Thermal energy consumption (zone level) [kWh]		Season
		For heating	For cooling	
1.	January	15.179,0	3,8	Heating season
2.	February	10.198,0	24,2	
3.	March	9.157,0	181,3	
4.	April	4.452,0	231,5	
5.	May May	2.200,0	1.210,0	Heating Cooling
6.	June	1.063,0	1.875,0	Cooling season
7.	July	811,1	3.052,0	
8.	August	900,5	2.554,0	
9.	Sept. Sept.	2.181,0	1.085,0	Cooling Heating
10.	October	5.092,0	420,0	Heating season
11.	November	9.005,0	41,3	
12.	December	15.143,0	4,3	
Overall:		75.381,6 kWh	10.682,3 kWh	

Table 2. Monthly zone level thermal energy consumption during heating and cooling season in 2016.

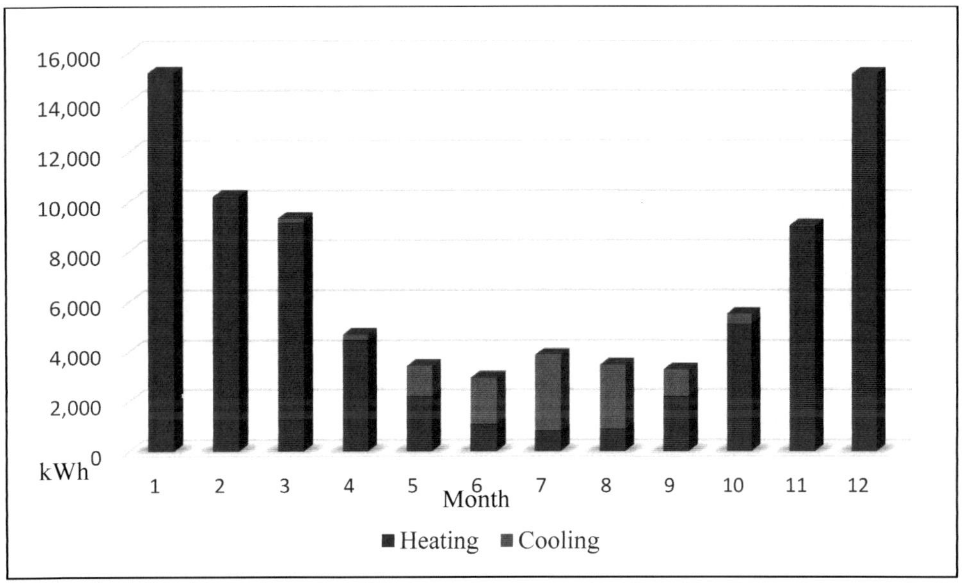

Fig. 5. Monthly energy consumption for heating/cooling in 2016 – IDA-ICE simulation

Simulation results for typical south oriented zone with the construction as shown on Figure 6.

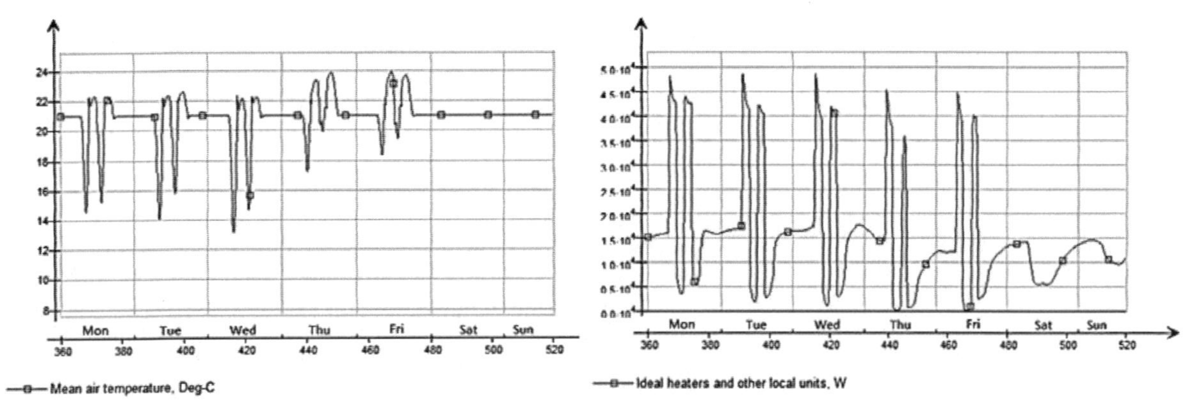

Fig. 6. Simulation results for typical south oriented office for the period from January 16. – 22.2016.

Daily zone temperature variations for an arbitrary selected north oriented zone are shown on Figure 7.

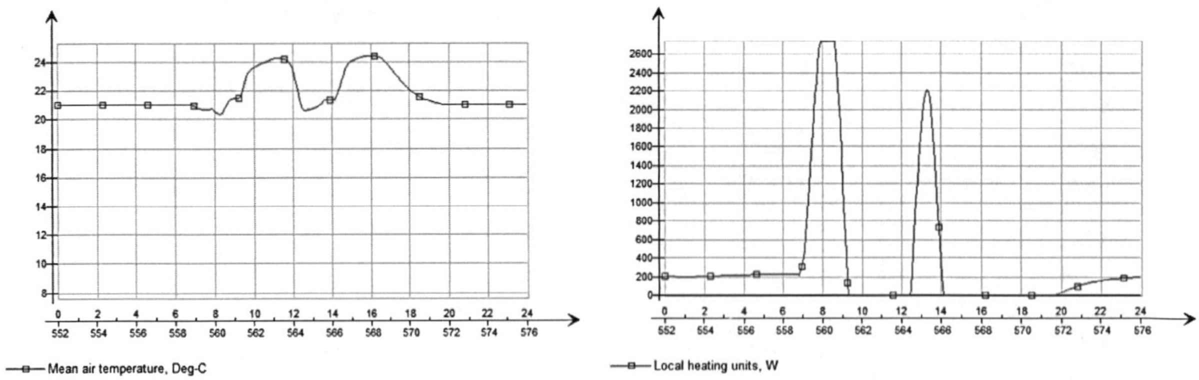

Fig. 7. Simulation results for an arbitrary selected north oriented zone on January 23. 2016.

7. Comparison with the real data

Total monthly energy demand for the EPHZHB business building during heating and cooling season in 2016. is given in Table 3.

	Month	Energy consumption for entire building [kWh]		Season
1.	January	17.370	-	Heating season
2.	February	15.028	-	
3.	March	17.503	-	
4.	April	8.517	-	
5.	May May	10.529		Heating Cooling
6.	June	-	2.930	Cooling season
7.	July	-	3.991	
8.	August	-	3.787	
9.	Sept. Sept.	3.505		Cooling Heating
10.	October	5.599	-	Heating season
11.	November	7.557	-	
12.	December	13.618	-	
Overall:		95.725 kWh	14.216 kWh	

Table 3. Monthly energy consumption during heating and cooling season in 2016.

Months are separated by the heating season and the cooling season. Only May and September fall in both seasons, so in that two months part of energy consumption is used for heating and part for cooling. By analysing the results of the simulation, and climate conditions in those months, we concluded that:

- In May approximately 80% of thermal energy was used for heating, and 20% of thermal energy was used for cooling.
- In September approximately 80% of thermal energy was used for cooling and 20% of thermal energy was used for heating.

Presented results are measured as the energy consumed for the entire building. These measurements are reduced by the energy consumption of lighting, computer equipment and other smaller consumers. Because we do not have precise results of separate energy consumption for heating/cooling, an estimation has been made that this group of electrical equipment has part in total energy consumption with approximately average of 1100 kWh/month. Here we have to consider the essential information from EPHZHB that in 2016. only the ground floor of the building was in use. The first floor was not used (the staff did not work in the offices), however, heating and cooling systems worked throughout the year. Taking this into account, Table 4 shows approximately monthly energy demand for heating and cooling season in 2016.

Month		Thermal energy consumption (zone level) [kWh]		Season
		For heating	For cooling	
1.	January	16.271	-	Heating season
2.	February	13.929	-	
3.	March	16.404	-	
4.	April	7.417	-	
5.	Ma May	7.543	1.886	Heating Cooling
6.	June	-	1.831	Cooling season
7.	July	-	2.892	
8.	August	-	2.678	
9.	Sept Sept.	481	1.925	Cooling Heating
10.	October	4.500	-	Heating season
11.	November	6.458	-	
12.	December	12.518	-	
Overall:		85.521 kWh	11.212 kWh	

Table 4. Monthly zone level approximately thermal energy consumption during heating and cooling season in 2016.

Total floor area of the building is 973,85 m^2 so thermal energy demand for heating in 2016 is **87,82 kWh/m²**, and for cooling is **11,51 kWh/m²**. Monthly energy consumption for heating and cooling in 2016. is shown on graph depicted on Fig. 8.

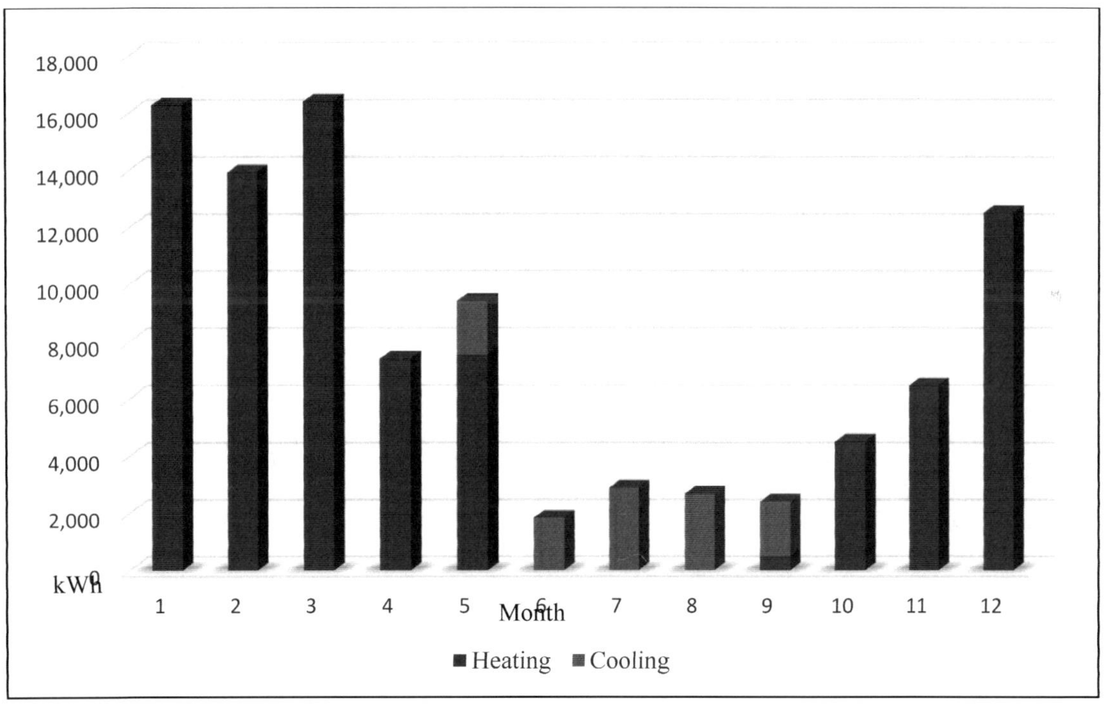

Fig. 8. Monthly energy consumption for heating and cooling in 2016.

In 2016, the total energy consumption in the building was 109.941 kWh, thereof thermal energy consumption was 96.733 kWh, which is 88 % of the total energy consumption in the building. When compared with the simulation results, the real thermal energy consumption in 2016. is in average **11% higher** than the expected building consumption defined with the expected use scenarios. This implies possible energy savings over 11%, i.e. approximately 5.000 KM (2.500 EUR) per year when the building usage would be aligned with the presumed simulation scenario.

Comparison of simulation scenario results in IDA-ICE and real energy consumption in 2016. is shown on graph depicted on Figure 9.

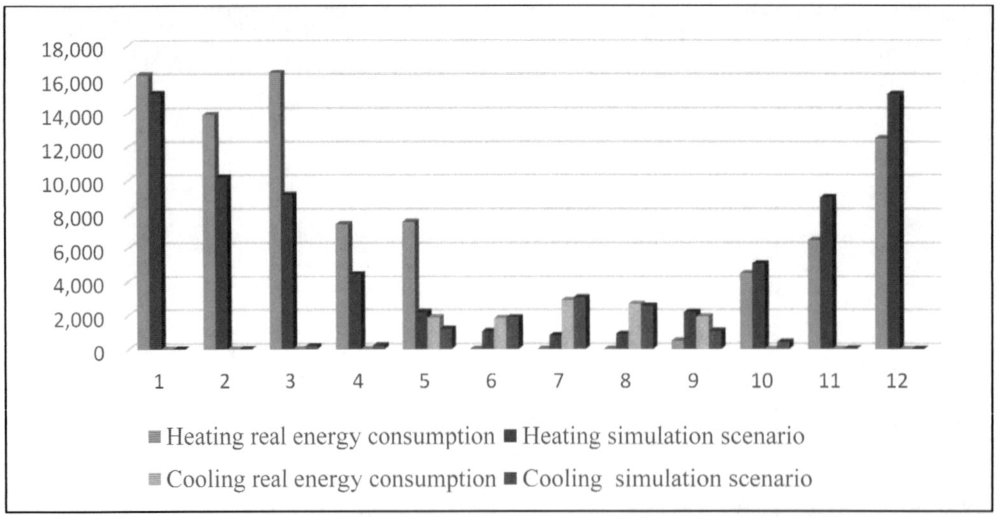

Fig. 9. Comparison of simulation scenario results in IDA-ICE and real energy consumption in 2016.

8. Conclusion

This paper presents a comparison of energy consumption parameters obtained from the energy dynamic model of the building and actual energy consumption parameters, as the basis for development of the future modular Model Predictive Control system. Comparison implies possible energy savings over 11%, i.e. per year when the building energy usage would be aligned with the presumed simulation scenario. Energy savings obtained through simulation in IDA-ICE takes into account weather data, comfort requirements and heat disturbance estimation to optimize energy consumption in each zone (room). This concept represents the basis for development of modular MPC system for pilot buildings in 3Smart project. Most of the buildings in the Danube region has analogous problem with building-side EMS. EMS is either non-existing or it is limited to particular ineffective subsystems. Such deficiencies result in excessive energy consumption, and ultimately with economical underperformance of a building. This knowledge provides the ability to create an optimized EMS in the building, and sets a good basis for creating modular software tool for energy management on building side. This concept will result significant economic gains and improved comfort compared to conventional EMS systems. The savings that would arise from this innovative EMS would not be negligible, especially when considering the application of such a system to larger buildings, even to entire cities.

9. References

[1] Birkeland J. (2002) "Design for Sustainability: A Sourcebook of Integrated, Eco-logical Solutions", Birkeland J., Eartscan Publications Ltd, 1-274, Sterling, ISBN: 1853838977, 9781853838972

[2] Martinčević A., Vašak M., Lešić V. (2016) "Model predictive control for energy-saving and comfortable temperature control in buildings", IEEE, Control and Automation (MED), Athens

[3] 3SMART project documentation (2017); [Online]. Available: http://www.interreg-danube.eu/approved-projects/3smart

[4] I.P.N. d.o.o. for design execution and supervision (2012), The main project of a business building in Tomislavgrad, Book I. Project Architecture, Posušje

[5] EQUA Simulation AB, IDA Indoor Climate and Energy, Sweden. [Online]. Available: http://www.equa-solutions.co.uk/en/software/idaice

[6] Little J., Arregi B., (2011) „Thermal Bridging-Understanding its critical role in energy efficiency", Construct Ireland, Issue 6., Vol. 5, 2011.

[7] Whale L. (2016), „An introductory guide to thermal bridging in homes", Zero Carbon Hub, London

[8] Schöck Isokorb®, (2014) „Design Guide-Solutions to Prevent Thermal Bridging", Kitchener, Ontario

[9] Constructive Details Ltd, (2014) „Thermal bridging solutions for external wall cavity details using Porotherm", Bucknalls Lane, Watford

[10] Martinčević A., Vašak M., (2017) "Dynamic building model for building simulation in a professional tool - UNIZGFER skyscraper building", 3Smart Deliverable D4.5.1, [Online]. Available: http://www.interreg-danube.eu/approved-projects/3smart/section/deliverables

[11] Starčić A., Vašak M., Mandrapa S., Međugorac M., et al. (2017) "Pilots conceptual projects with detailed analysis of pilot's current technical state and requirements and preliminary cost-benefit analysis of the planned investments", 3Smart Deliverable D6.2.1, [Online]. Available: http://www.interreg-danube.eu/approved-projects/3smart/section/deliverables

DOI: 10.2507/28th.daaam.proceedings.042

DEVELOPMENT OF TECHNOLOGY FOR IMPROVING PRODUCTIVITY AND QUALITY OF INJECTION MOLDING

Hong-Seok Park & Xuan-Phuong Dang

This Publication has to be referred as: Park, H[ong] S[eok] & Dang, X[uan-Phuong] (2017). Development of Technology for Improving Productivity and Quality of Injection Molding, Proceedings of the 28th DAAAM International Symposium, pp.0309-0314, B. Katalinic (Ed.), Published by DAAAM International, ISBN 978-3-902734-11-2, ISSN 1726-9679, Vienna, Austria
DOI: 10.2507/28th.daaam.proceedings.042

Abstract

This work presents a study on the application of 3D printing technology and the smart control for improving the productivity and quality in the injection molding. The cooling time and cycle time are greatly reduced using the conformal cooling channels made by metal 3D printing technology. The geometry of the cooling channels is optimized so that the cooling efficiency increases and the aesthetic of the molded part is also improved. A case study of molding of an automotive door module is applied to verified conformal cooling technology. The research result shows that the cooling time is reduced around 23%. Besides the application of 3D printing technology, a method that ensures the quality consistency of the molded parts regardless of the change of injection molding machine or the disturbance of uncertainty molding condition is proposed. The cavity sensors including pressure and temperature sensors are used to monitor the process condition inside the mold. A control strategy and an intelligent algorithms are figured out to compensate the variation of the molding condition in the mold cavity so that the quality consistency is always satisfied. This work is intended as a contribution to the development of the smart molding technology.

Keywords: smart injection molding; conformal cooling channels; cooling efficiency; 3D printing

1. Introduction

Injection molding is a popular manufacturing technology for making plastic products. The injection molding process includes three significant stages: filling and packing stage, cooling stage, and ejection stage. Among these stages, cooling stage is very important one because it mainly affects the productivity and molding quality. An appropriate design of cooling channel reduces cooling time, increases the productivity, and minimizes undesired defects such as sink marks, differential shrinkage, thermal residual stress and warpage. Simple and traditional cooling channels cannot result in an even and effective cooling performance when the geometry of the molded part is complex. Conformal cooling channels made by 3D printing method is one of the solutions for cooling complex part [1].

Conformal cooling channel has drawn great attention to mold designer and plastic molding researcher. Ferreira and Mateus [2] studied on rapid soft tooling with conformal cooling channels for plastic injection molding. Wang et al. [3] introduced an approach to generate spiral channels for conformal cooling system and concluded that the cooling channels derived from spiral curves introduce nearly no reduction of the rate of coolant flow. In literature, there are four types of conformal cooling channels: spiral conformal cooling channels [3, 4], zigzag-type conformal cooling channels [4], scaffold type conformal cooling channels [5], and Voronoi diagram type [6]. Each type of conformal channel is suitable for a specific case according to the geometry of the molded part and the decision of the mold designer.

Because of the advancement in metal 3D printing technology, the application of conformal cooling channels made by solid freeform fabrication has been becoming popular recently. The conformal cooling channels system is recognized as one of the best solutions for reducing cycle time, differential shrinkage, and warpage defects on molded parts [7, 8]. Jahan and El-Mounayri [9] studied on the way of optimization of cooling channels parameters for simple cylindrical shape using simulation and design of experiment. Various researches [9-13] developed the optimal design for cooling channels that can increase the effectiveness of cooling system in the injection mold; however, the optimal design of the cooling channels in the mold has been restricted by the relatively simple cooling channel configurations. The complex molded part requires a complicated cooling channel design than simple mold. In addition, how to design the best conformal cooling channel that meets the desired requirement has not been address intensively.

The quality of the molded part mainly depends on the skill of the molding technician. Plasticization, mold filling, and solidification cooling are invisible processes which take place inside electrical mechanical components such as the heat barrel, nozzle, and mold [14]. Commonly, injection molding parameter selection and adjustments were based primarily on the experience of the operator. This approach has the risk of unanticipated outcomes such as short shot, large warpage or sink mark. Also, the quality of molded part is inconsistent shot to shot. The technician only gets feedback about the molding quality after the molded part come out of the injection machine. This means that the only way for the technician to determine the cause of any molding faults is to check the molded part after they are produced and then adjust the molding parameters for the next cycle until an acceptable product is produced. The incorporation of statistics and the use of design of experiments can improve the process control setting, but we still didn't know what is happening in the mold cavity during injection. Therefore, it is necessary to adopt a real-time smart system that monitors, records, and controls temperature and pressure signals generated by in-cavity sensors. The system is integrated with the molding machine control system. Each molding cycle initiates an accumulation of data which is stored and refreshed at the start of the next cycle [15].

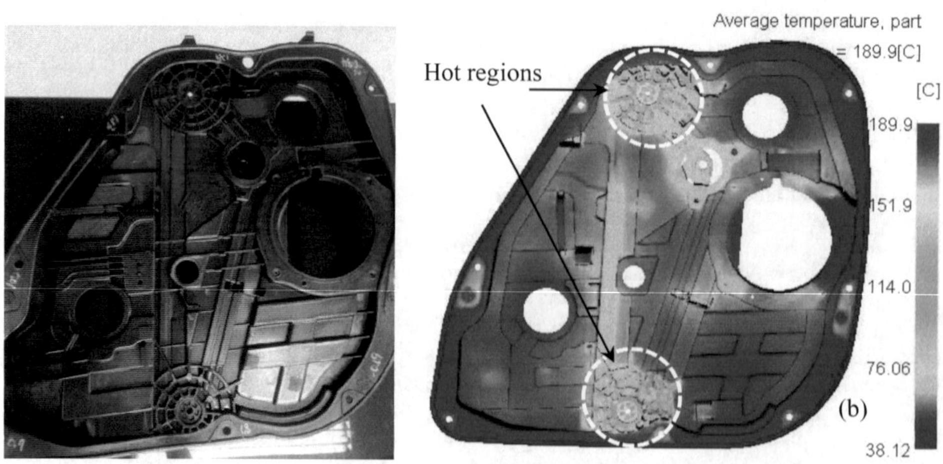

Fig. 1. The plastic door module (for automotive component) and the temperature distribution of the molded part

In this work, we solve a practical problem in injection molding industry. A complicated plastic car door module as shown in Figure 1 has been manufacturing. Due to the functionality of the molded part, the thickness of the molded part is varied. Especially, the thickness in the two regions with rib (web region) is largest. Therefore, the cooling time is longest compared to other regions. As the result, the total cycle time is long. Thus, the product producer wants to reduce the cycle time. In addition, they also want to ensure the quality and productivity regardless of the variation of process conditions. Therefore, it is necessary to develop a monitoring and the intelligent control system that can automatically find the optimal process parameters and produce a consistent product quality.

2. The systematic research procedure

To solve the previously mentioned problem, a systematic research procedure including six steps has been figured out as shown in Fig. 2. After analysing the problem and the development goal (reduce the cooling time for improving the productivity), we determined the appropriate cooling channel type and then designed and optimized of cooling channels by analytical method. Next, the numerical simulation was done for verification because the analytical method is based on semi-practical formulas. After design and simulation, we fabricated the cooling system with 3D printing technology and then performed the test on the real injection molding machine. The important steps in the research procedure for Improvement of cooling efficiency of injection molding are described in the next sections.

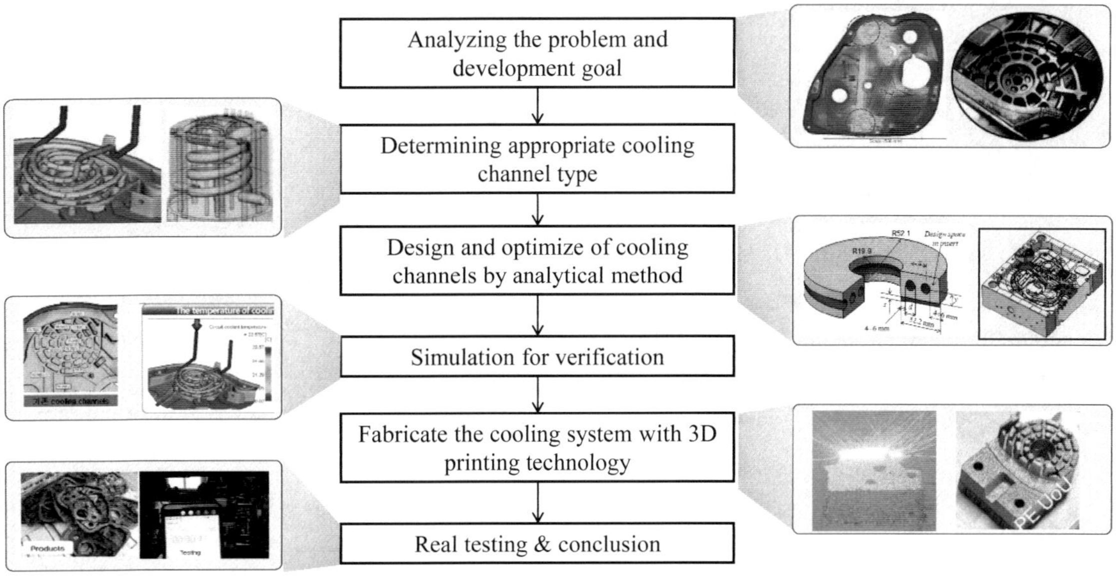

Fig. 2. The research procedure

3. Improvement of cooling efficiency of the injection mold by 3D printing technology

3.1 Analytical calculation

The conformal cooling channel with the spiral form was selected as the proper cooling channel for previously mentioned injection mold. It is difficult to determine the cooling channel configuration by intuition of the designer or by the iteration of simulation. The spiral cooling channel can be modelled as a simplified equivalent model that can use the analytical method. In this work, we use the semi-practical formula to determine the optimal cooling channel configuration. The configuration including pitch x, depth y and diameter d can be derived by the equation [16]:

$$\frac{[c_p(T_M - T_E) + i_m]\rho \frac{s}{2} x}{T_W - T_C}\left\{\frac{1}{2\pi k_{st}}\ln\left[\frac{2x\sinh(2\pi\frac{y}{x})}{\pi d}\right] + \frac{1}{0.03139\pi R_e^{0.8}}\right\} = \frac{s^2}{\pi^2 a}\ln\left[\frac{4}{\pi}\left(\frac{T_M - T_W}{T_E - T_W}\right)\right] \quad (1)$$

where $a = \frac{k_p}{\rho c_p}$ and Reynolds number $R_e = u\frac{d}{v}$. The nomenclature of other parameters is listed in Table 1.

T_E, T_M, T_W, T_C, k_{st}, s are ejection temperature, melt temperature , mold wall temperature, temperature of water (coolant), thermal conductivity of mold, molded part thickness, respectively.

It can be seen that the equation (1) shows the relation among cooling time and geometrical of cooling channel.

Mathematically, with pre-set T_M, T_E, T_W, T_C, predefined thermal properties of material, the equation (1) presents the relation between cooling time t_c and the variables related to cooling channels configuration including pitch x, depth y and diameter d.

Solving the equation (1) without constraints, multi solutions are obtained. However, due to the constraint $32.2 - 6\times2 \leq x + d \leq 32.2 - 5\times2$ of the design space, the diameter and the pitch of the cooling channel are selected as 7.0 and 13.9 mm, respectively. We used the analytical results to determine the geometry of cooling channel. This data was then used to model the cooling channels for numerical simulation. If the target time to reach the ejection temperature is not satisfied, the distance from the cooling channels to the mold surface is fine-tuned and additional simulations are carried out until the satisfaction is obtained.

3.2 Design and simulation of conformal cooling channels

The conformal cooling system has to be designed based on the current mold with conventional cooling channels. Because the mold is large and the molded part is mainly in flat shape, the main cooling system of the old mold is retained. Two inserts with conformal cooling channels are added into the mold core (Fig. 4).

To analyse the thermal performance of the original mold as well as to develop the new cooling system, we used the simulation tool (Moldflow software). The steady-state cool simulation was used to analyze both the temperature of the part and cycle averaged temperature distribution in the mold. The diameter, pitch, and depth of the cooling channels for building the simulation model were obtained from analytical method.

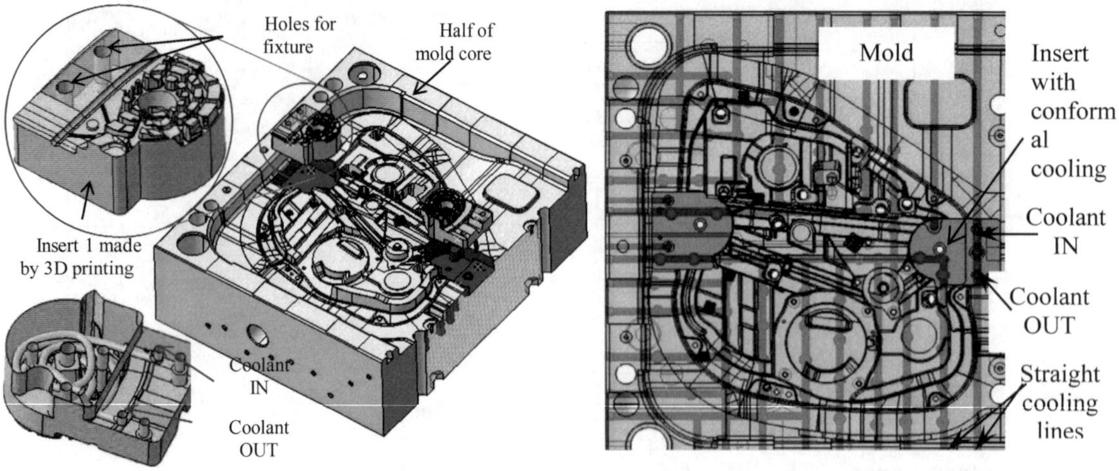

Fig. 3. The layout, structure and assembly of insert with conformal cooling channels in the mold

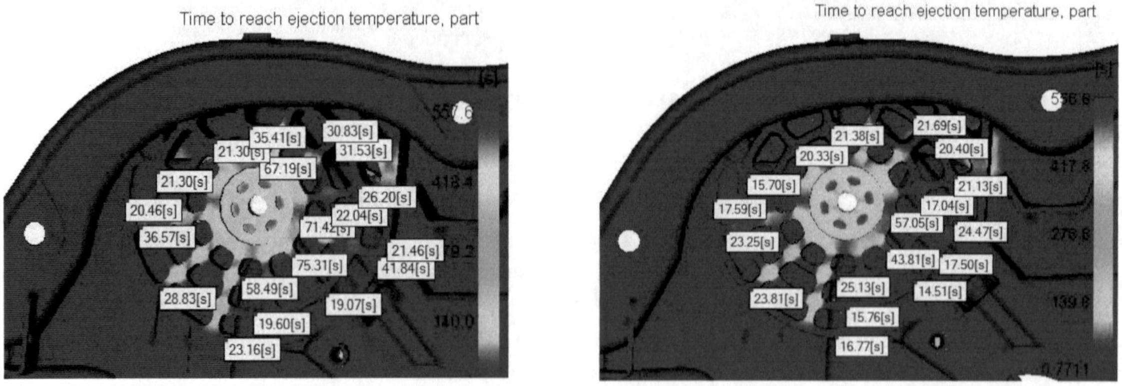

(a) Conventional cooling channel (b) Conformal cooling channel

Fig. 4. Comparison of time to reach the ejection temperature between conformal and conventional cooling channels

The simulation results show that the conformal cooling channels give more even temperature distribution. The temperature in the hot regions (Fig. 1) is reduced significantly (around 23÷28 ^{0}C depend on the position). The time to reach ejection temperature of molded par is reduced about (23.14-33.57)/33.57 = 31% (when 18 sample points are taken into account as shown in Fig. 5, the average time to reach ejection temperature of conventional and conformal cooling channels are 33.57 and 23.14 ^{0}C, respectively). This means that the cooling time can be reduced up to 31%.

3.3 Fabrication of inserts with conformal cooling channels using metal 3D printing technology

After design and simulation, the two insets were fabricated by selective laser melting method on MetalSys 150 3D printer with powder material that has composition as HP 4A(1). One of the insert is shown in Figure 6. The two inserts were machined and assembled to the mold for testing their cooling performance.

(a) Printing process (b) Insert after printing (c) Assembly to the mold

Fig. 5. Insert with conformal cooling channel inside was made by 3D printing technology

4. The smart mold system development

During the molding process, there may be undesired change of process parameters and uncontrolled disturbance. These variations can result in inconsistence of product quality such as sink marks, short shot, flash, or large warpage. Therefore, the sensor system has be installed in the mold cavity in order to monitors, records, and controls temperature and pressure signals. Base these data, smart molding system integrated with the molding machine control system will be develop. The diagram of the smart mold system has been figured out as shown in figure 7. Process monitoring criteria are based on maximum value of pressure and temperature, minimum value of pressure and temperature, integral, average value, threshold, and window of pressure and temperature. The control algorithm for the smart mold system can be fuzzy logic, PID control and neural network.

There are three issues that will be focused on smart mold system:

(1) Control the temperature of barrel and hot runner:

We will control the temperature of barrel and the temperature of hot runner to obtain the desired plastic's meld temperature inside the mold.

(2) Control the injection molding machine:

To make the smart mold in order to assure the quality consistency of the molded part regardless of the change of injection molding machine or the variation of environment, the system must be integrated with the molding machine control system (control injection speed and packing pressure).

(3) Control the mold temperature:

Mold temperature is very important to the cooling time and the quality of the molded part. To control the mold temperature, we have to control the coolant (water). There are two ways:

a) Control the inlet water temperature: lower the inlet temperature can effectively lower the mold temperature.

b) Control the flow rate: increase the flow rate will cool the mold better.

Method (a) is has large control threshold (large margin), but it is expensive and more complex than method (b). Method (b) uses a servo motor to control the speed of coolant pump. This method simple than method (a) but the threshold is narrow. For example, we can only change the mold temperature around $2 \div 3^0$C.

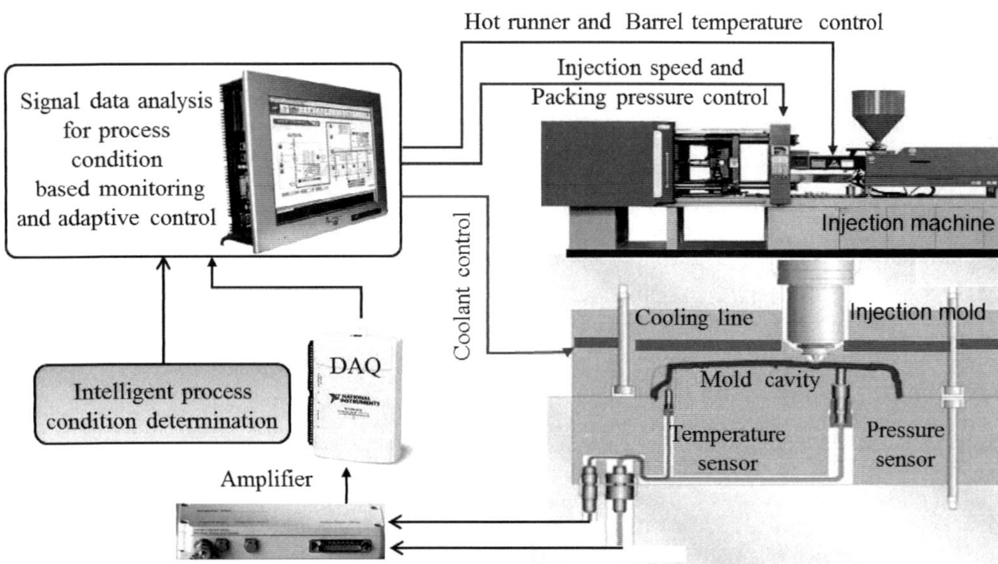

Fig. 6. The diagram of the smart mold system

5. Research results and discussion

After making the inserts, the testing process was done in order to verify the research results. The molding process was carried out on an industrial injection molding machine Toshiba IS2500DF with 2500 tons of clamping force. The result shows that's the real cooling time can be reduced from 39 sec to 30 sec (equivalent to 23 %) with the new cooling channel when using the coolant flow rate 2.25 litter/minute. There is error between simulation and real experiment due to the simulation error material properties. In addition, the microstructure and physical properties of 3D printing material is different from those of the same metal made by metallurgy. When using the conformal cooling channels, the appearance (a quality criterion) of the molded part also looks better than the one made with old cooling channels. The cycle time is 44 seconds. It is clear that the cooling time is reduced significantly. The error between simulation model and the real result is the limitation of this work. It is difficult to get the exact thermal conductivity data of material after 3D printing. I addition, the equivalent model is also an approximate approach. Therefore, error always occur.

6. Conclusion and future work

This work studies on a conformal cooling channels applied in a medium-size injection mold that makes an automotive part. We improved the cooling system of an existent mold in order to reduce the cycle time and improve the quality of molded part. In injection molding, cooling stage is very important because it account for more than two thirds of the molding cycle. For that reason, we used spiral cooling channels made by 3D printing method. It is very expensive to make a large mold by 3D printing method; therefore, local approach was adopted. As the results, inserts with conformal cooling channels that is used to cool extremely hot locations in the mold is an intelligent solution.

The design of conformal cooling channels was done using the combination of analytic and CAE simulation method. Analytic method was used to roughly identify the cooling channel parameters and the simulation method was employed to check and fine-tune the design and visualize the results. We analysed the current state and performance of the existent mold so that its drawback was drawn. The literature review method was also used to approach the state of the art and to decide the optimal solution. The spiral cooling channel was selected as the best choice. Selective laser melting method was used to fabricate the inserts. The real testing results show that the cooling time can be reduced to 23%. This is a significant improvement in the first phase of this project. In the next phase of this project, we develop the smart system for the mold as figured out in Section 4. The intelligent mold is still on the development process. By using the cavity sensor system on the second phase of our project, the temperature and pressure at considered (important) locations in the mold will be monitored; the intelligent molding control system will be developed and implemented based on the thermal and mechanical behaviour happen in the mold. The productivity and the quality of the molding process will be consistent regardless of disturbance and uncontrolled variations.

7. Acknowledgement

This work is supported by the Ministry of Trade, Industry & Energy (MOTIE, Korea) under Industrial Technology Innovation Program. No. 10062677. "For plastic injection molds with 50% improved cooling efficiency, development of 3D printing equipment & mold steel powder manufacturing technology having 30 μm grade in diameter".

8. References

[1] Balc N. C.; Berce P. & Pacurar R (2010). Comparison between SLM and SLS in producing complex metal parts, Annals of DAAAM for 2010, Vienna, Austria.

[2] Ferreira, J.C. & Mateus A. (2003). Studies of rapid soft tooling with conformal cooling channels for plastic injection moulding, Journal of Materials Processing Technology, 142(2): pp. 508-516.

[3] Wang, Y.; Yu K.-M.; & Wang C.C.L. (2015). Spiral and conformal cooling in plastic injection molding," Computer-Aided Design, 63: pp. 1-11

[4] Park, H.S. & Pham N.H. (2009). Design of conformal cooling channels for an automotive part, International Journal of Automotive Technology, 10(1): pp. 87-93.

[5] Au, K.M. & Yu K.M. (2007). A scaffolding architecture for conformal cooling design in rapid plastic injection moulding, The International Journal of Advanced Manufacturing Technology, 34(5): pp. 496-515.

[6] Wang, Y., et al. (2011). Automatic design of conformal cooling circuits for rapid tooling, Computer-Aided Design, 43(8): pp. 1001-1010.

[7] Shayfull, Z., et al. (2014). Potential of Conformal Cooling Channels in Rapid Heat Cycle Molding: A Review," Advances in Polymer Technology, 33(1), 2014.

[8] Rahim, S.Z.A., et al. (2016). Improving the Quality and Productivity of Molded Parts with a New Design of Conformal Cooling Channels for the Injection Molding Process," Advances in Polymer Technology, 35(1).

[9] Jahan, S.A. and Mounayri H. El (2016). Optimal Conformal Cooling Channels in 3D Printed Dies for Plastic Injection Molding, Procedia Manufacturing, 5: pp. 888-900.

[10] Ahn, D.-G., Park, S.-H. & Kim H.-S. (2010). Manufacture of an injection mould with rapid and uniform cooling characteristics for the fan parts using a DMT process, International Journal of Precision Engineering and Manufacturing, 11(6): pp. 915-924.

[11] Khan, M., et al. (2014). Cycle Time Reduction in Injection Molding Process by Selection of Robust Cooling Channel Design, ISRN Mechanical Engineering, 2014: pp. 8.

[12] Rännar, L.E.; Glad A.; & Gustafson C.G. (2007). Efficient cooling with tool inserts manufactured by electron beam melting, Rapid Prototyping Journal, 13(3): pp. 128-135.

[13] Park H. & Pham N. (2008). A new approach to conformal cooling channels in injection molding process. DAAAM International Scientific Book 2008, pp. 587-602

[14] Miyamoto, S. & Yukutomo H. (2011). Smart Molding Simplifies Molding Jobs - New Si-V Series Electric Injection Molding Machines, Hitachi Review, 60(5).

[15] Chrystal, S. (2013). Ensuring Injection Molded Part Consistency and Conformance through In-Mold Cavity Sensor Technology, in Precision Engineered Products 2013.

[16] Dang, X.-P. & H.-S. Park (2011), Design of U-shape milled groove conformal cooling channels for plastic injection mold, International Journal of Precision Engineering and Manufacturing, 12(1): pp. 73-84.

DOI: 10.2507/28th.daaam.proceedings.043

ENERGY EFFICIENCY ASSESSMENT AND IMPROVEMENT MEASURES FOR FURNITURE FACTORY

Dzana Kadric, Nijaz Delalic, Sanda Midzic-Kurtagic,
Berina Delalic, Kerima Medar

This Publication has to be referred as: Kadric, D[zana]; Delalic, N[ijaz]; Midzic-Kurtagic, S[anda]; Delalic, B[erina] & Medar, K[erima] (2017). Energy Efficiency Assessment And Improvement Measures For Furniture Factory, Proceedings of the 28th DAAAM International Symposium, pp.0315-0321, B. Katalinic (Ed.), Published by DAAAM International, ISBN 978-3-902734-11-2, ISSN 1726-9679, Vienna, Austria
DOI: 10.2507/28th.daaam.proceedings.043

Abstract

Energy consumption of a furniture factory is assessed through detail energy audit which was performed within a framework of the UNIDO National Cleaner Production Program in Bosnia and Herzegovina. Factory consist of three production plants: sawmills and kilns, furniture manufacturing facility and a pellets manufacturing facility. Energy audit was focused on analysing energy consumption of furniture manufacturing facility as well as detection of causes of excessed energy consumption. Following, several energy and resources saving measures are proposed with approximated values of energy and CO_2 savings. The paper presents detailed analysis of three improvement measures: replacement of the existing boiler, installation of frequency regulators at the engine of the dirt-blast system and introduction of alternative equipment for cleaning instead of the compressed air. Implementation of these measures will provide significant energy savings of 1.328.715 kWh/year and CO_2 savings of 85,74 t/year.

Keywords: energy efficiency; energy audit in industry; furniture factory; energy indicator; energy savings

1. Introduction

Share of industry energy consumption in Bosnia and Herzegovina in total energy consumption is approximately 23 %, which is much lower than in countries worldwide [1]. Furthermore, energy intensity of Bosnia and Herzegovina is 0,5 toe/1.000 USD of GDP, which is four times higher than average in the European Union (EU) [2].

Energy efficiency improvement and reduction of energy intensity are important targets for sustainable development worldwide [3]. Industry in Bosnia and Herzegovina, due to its high specific energy consumption and negligible percent of implementation of energy saving technologies offers a high energy saving potential which directly affect an exploitation of energy resources and thus diminishes the negative impact on the environment.

The furniture industry consumes electrical and thermal energy, wood and wood based board material as a basic raw material. Additionally, many auxiliary raw materials such as paints, adhesives, textiles, sponges, etc. are being used. Most of the annual energy consumption in furniture industry is related to the dirt-blast system (38%), wood processing machines (27%) and heating (10%). The lighting and the compressor station belong to a share of 7% therefore they are considered as less energy consumers [4].

Often, energy costs in these companies are seen as a fixed overhead, although it is actually one of the costs most easily managed and whose reduction can significantly increase their productivity. A recent study [5] has showed that this industry can reduce energy costs by up to 20% with relatively small investments and short payback periods. Energy savings are possible in following systems:10 – 15 % in dirt-blast system, 20 to 25 % in lighting system, 10 to 20 % in compressors and heating system from 10 to 40 % [6].

The KERN Schreinerei & Innenausbau furniture company has achieved a 34,5% reduction in energy consumption and 19,5% CO_2 emissions by replacing an existing boiler with a new boiler of higher efficiency. Investments in this improvement measure amounted to 60.000 € with a 15-year payback period [7]. Furniture factory in Germany, with the installation of a LED panel, achieved annual savings of 500 €, with a reduction in electricity consumption of 2.500 kWh/year and 1,4 t CO_2/year with a payback period of 2 years [8].

With the aim of determining the potential for energy savings, energy related cost savings and related reduction of environment impacts, preliminary assessment of resource efficiency and cleaner production (RECP) has been performed in this company within a framework of the UNIDO National Cleaner Production Program in Bosnia and Hercegovina. The RECP audit has addressed three production plants of the company: plant with sawmills and kilns, furniture manufacturing facility and a plant for the pellets manufacturing facility. The sawmills and kilns plant is located approx. 3 km far from the furniture manufacturing facility. Total number of employees is 109. The results of preliminary audit showed potential for improvement of resource efficiency in all of three facilities [9], however the company management decided to select furniture manufacturing facility for detailed analyses. The furniture manufacturing facility consists of three different production lines: manufacture of piece furniture, production of plate and production of taped furniture. Auxiliary systems in furniture manufacturing facility are: boiler room, dirt-blast system and compressor station. First stage of energy audit was focused on analysing energy consumption of furniture manufacturing facility as well as detection of causes of excessed energy consumption. Based on the collected data, and measurement of flue gas composition and measurement taken by the thermal imaging camera an analysis of the energy consumption has been performed. Also, energy indicators relevant to the furniture industry are calculated and compared with similar companies worldwide.

Following measures were selected for a detailed analysis: replacement of the existing boiler, installation of frequency regulators at the engine of the dirt-blast system and introduction of alternative equipment for cleaning instead of the compressed air. The evaluation of the measures from the technical, economic and environmental aspect was carried out.

2. Energy consumption and energy indicators

Cost structure analysis has shown that 62 % of energy costs in the plant are related to the furniture manufacturing facility while for plant with sawmills it is 21 % and for pellet production plant it is 17 %. Therefore, special attention and focus in the analysis was kept at the furniture manufacturing facility. In Fig. 1. building where production is placed and parts of production lines is shown.

Fig. 1. Production facility building and part of production lines

Installed power of all electrical consumers in the furniture manufacturing facility and their working hours was encountered for calculation and analysis of annual electric energy consumption of all subsystems. Dirt-blast system consumes 71 % of all electrical energy in the facility (Fig. 2.), which is much higher than average [4] and it shows that this system is not efficiently used. The dirt-blast system operates 17 hours a day and with a total engine power of 76 kW. Therefore, it makes the largest share in total electricity costs, followed by the machinery in the production plant (12%) and compressors (11%). Although the machinery in the production plant have large installed engine power their share in total power consumption of the drive is 12%. The reason for this is that the production process is carried out with minimal working hours of machines during the year and a large share of jobs are related to manual carpentry processing.

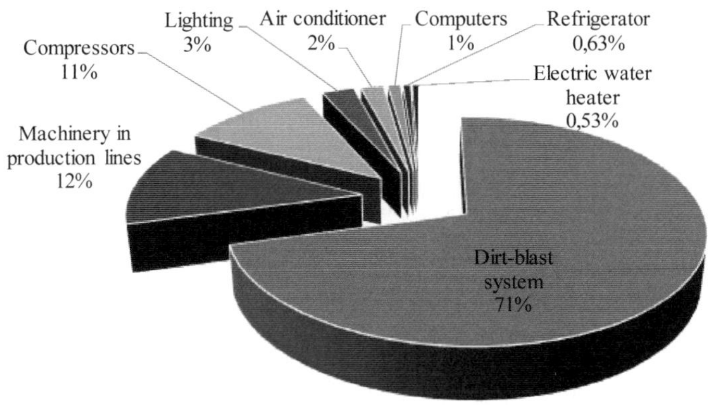

Fig. 2. Structure of electrical energy consumption in production plant

Thermal energy is provided to the facility from the boiler room, located close to the production plant and the dirt-blast system as it is shown in Fig. 3. Inside the boiler room, there is one boiler which serves to heat the water for heating system for administrative part and the production hall. Wood chips generated as a waste during the production process are used as fuel. Since overall efficiency of boiler is low (41,5 %) due to the poor state of boiler envelope (Fig. 4.) and high losses related to the flue gas, thermal comfort in the heated spaces is not satisfactory. As example, when outdoor temperature was -15 °C recorded indoor temperature in production hall was 8 °C. The existing boiler has the power of 700 kW and low efficiency. Additional load for the heating system is created by the process of dirt-blast when certain air volume is inserted in and extracted from the space. Instalment of any new boiler would not result with economic benefit if the current irregular thermal comfort situation is taken as baseline. Therefore, a hypothetical baseline and corresponding calculation of energy need is taken for scenario with 20 °C temperature of the office area of and 15 °C of the production area. The calculated boiler power, with the same boiler efficiency and existing system is 1,13 MW, therefore the existing one cannot meet the heat needs of the facility.

Fig. 3. Location of the boiler room and dirt-blast system

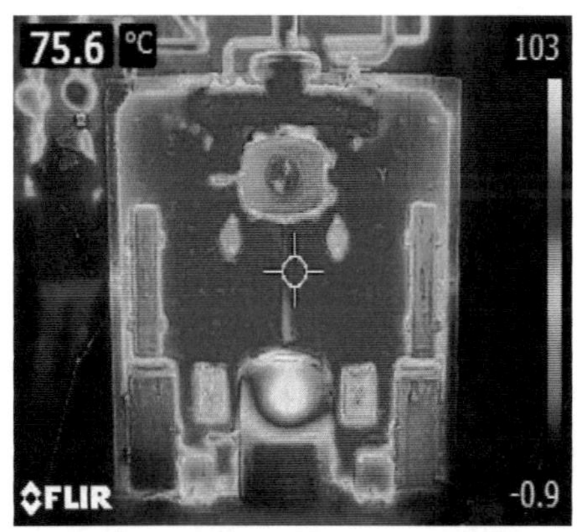

Fig. 4. Thermal image of the boiler

Energy need for heating is highest in the January due to the lowest ambience temperature (Fig. 5.) It is noticeable that energy need for heating of administrative part is smaller than for production hall, due to the smaller heating area off administrative part and high ventilation heat losses in the production hall due to the installed dirt-blast system. Annual energy need for heating for administrative part and production hall is 324.140 kWh/year.

Energy indicators represent a good basis for companies to plan more energy efficient production and allow benchmarking with similar facilities worldwide. Determination and continuous monitoring of indicators ensure detection of irregularities and faster implementation of improvement measures. The Energy and Economy Institute "Energieinstitut der Wirtschaft GmbH" published research results for 90 furniture manufacturing companies [10]. Within the scope of the research, 7 potential indicators for this type of industry are defined, namely: energy consumed per annual revenues, electricity consumed per square meter work surface, total energy consumed per square meter of work surface, electricity consumed per number of workers, total energy consumed per number of working hours, total energy consumed per cubic meters of treated wood and the ratio of total energy consumed and cubic meter of medium pressure and veneer.

Based on the data provided from the company, five energy indicators are calculated as it is shown in Table 1.

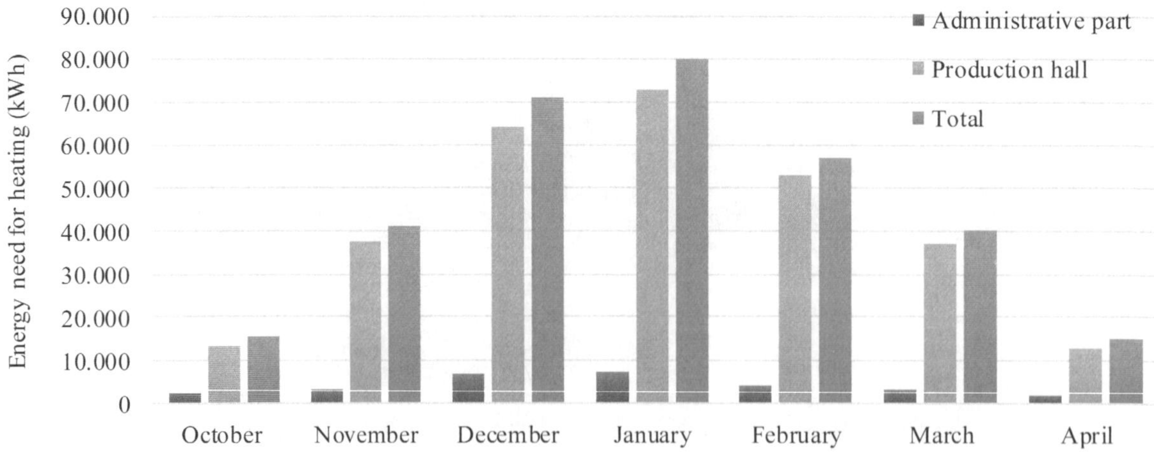

Fig. 5. Energy need for heating for production plant

No.	Energy indicator	Calculated value
1.	Electricity consumed per square meter work surface, EI_1	180 kWh/m^2
2.	Total energy consumed per square meter of work surface, EI_2	551 kWh/m^2
3.	Total energy consumed per number of working hours, EI_3	265 kWh/h
4.	Electricity consumed per number of workers, EI_4	6.113 kWh/no. work
5.	Total energy consumed per cubic meters of treated wood, EI_5	1.115 kWh/m^3

Table 1. Energy indicators of production plant

Fig. 6. shows values of energy indicators for plants whose annual energy consumption exceed 300 MWh [10] and values of the same indicators for considered furniture production plant which consumes annually 410 MWh of electrical energy. Green area represents highest production efficiency and good energy indicators while red area represents lowest energy efficiency.

It is noticeable that only one facility energy indicator EI_5 is located in the green area, while others are located in the area which represent low or very low energy efficiency. Low value of EI_5 could be explained with high share of manual carpentry processing in the production process. Values of all other indicators (EI_1 – EI_4) revealed that there is a large potential for energy efficiency improvement.

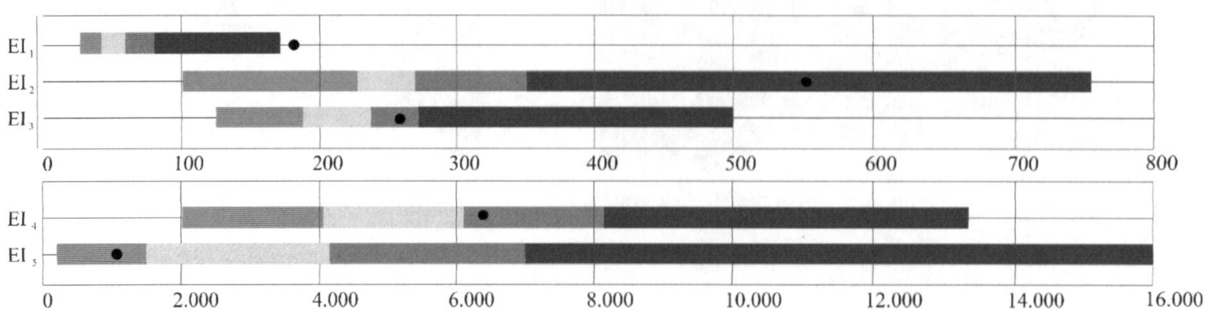

Fig. 6. Energy indicators for considered production plant (black dots) and for similar facilities (coloured bars)

3. Energy efficiency improvement measures

After completing a detail analysis of the company energy consumption, three measures are investigated in details: replacement of the existing boiler, installation of frequency regulators at the engine of the dirt-blast system and introduction of alternative equipment for cleaning instead of the compressed air.

Energy savings are calculated as a result of reduction in electricity and heating consumption, while CO_2 emission is calculated using the energy savings of energy carrier and related conversion factors.

3.1. Replacement of the existing boiler

Based on the calculated energy need for heating in the facility and a new boiler efficiency of 87%, which is widely available in the market, new boiler power would be 600 kW. Some additional investments should be made, such as new control system and several new fan coils since existing situation is not in line with the actual needs in production hall.

Baseline for the energy ad cost savings is theoretical energy need, boiler power of 1,13 MW and wood chips annual need of 634 m^3. Taking into account the available quantities of wood residues arising from production, the company should additionally procure 130 m^3 of wood residues at an annual cost of 4.600 €. New boiler of 600 kW power and 87% efficiency for space heating would need 225m^3 of wood chips, which would save 4.600 € and also can serve to generate additional earnings from selling the rest of the wood chips which remains from the production process.

Electrical consumption of existing fan coils is not shown in Fig. 2. since they are not in use. This also contributes to the low level of thermal comfort in production hall. Also, baseline for this measure is proper use of existing and new fan coils. Estimated investment for new boiler and 9 fan coils is approx. 45.000 €. With the total savings of 5.200 €, which include savings from the improved control system and boiler efficiency, as well as savings of electric energy, the payback period for the given measure would be 8 years and 7 months.

3.2. Installation of frequency regulators at the engine of the dirt-blast system

One of the most important benefit of using frequency regulators is energy savings and resulting electricity cost savings. Speed adjustment of the electric engine offers a number of benefits in terms of productivity increase and reduction of maintenance cost. With full speed control, it is possible to provide a slight acceleration and engine slowdown, avoiding strain and sudden stroke in the machine, while keeping the maintenance costs low and improving the working environment [11]. It is suggested to install frequency controllers on the engines of the dirt-blast system. Estimated investment for three frequency regulators is 9.970 € while reduction of electricity consumption will be 20%. As the dirt-blast system contributes with 71% of the total electricity consumption of the production plant, implementation of this measure will generate annual savings of electricity of 42.354 kWh/year, and 2.600 €/year. The payback period is 3 years and 9 months.

3.3. Replacement of the existing boiler joined with the installation of frequency regulators at the engine of the dirt-blast system

With installation of frequency regulators on the engines of the dirt-blast system it is reasonable to assume that ventilation losses will reduce up to a 20%. Therefore, calculated boiler power would be a 500 kW. The investment value for a new boiler, 3 frequency regulators and auxiliary equipment is estimated at 50.740 €. The payback period would be 6 years and 5 months with estimated annual savings of 7.845 € /year.

3.4. Introduction of alternative equipment for cleaning process instead of the compressed air

In order to calculate energy and cost savings for implementation of electric air blowers as an alternative to compressed air used for the cleaning process, it is necessary to determine the amount and cost of compressed air used in these processes. It includes the consumption of pneumatic tools used for machines, products and clothing cleaning. Taking in to a calculation the compressor power, volume flow rate and pressure of compressor and pneumatic tool, it is calculated that cost related to this activity is 13,15 € per month per one pneumatic tool. Electric air blower with power of 600 W consumes 7,5 € on a monthly basis for 8 hours use in a day. By implementing this measure, energy savings of 8.372 kWh/year and a decrease of carbon footprint of 12.74 t/year will be achieved. The value of the required investment is 640 €. Taking in to account a number of pneumatic tools (5), annual saving of 339 €/year could be achieved. Therefore, the payback period is 1 year and 9 months.

3.5. Energy and cost savings

Investment values, annual energy and cost savings with resulting CO_2 reduction and simple pay-back period (SPB) for all proposed measures are given in Table 2.

No.	Improvement measure	Investment (€)	Energy saving (kWh/year)	Cost savings (€/year)	SPB	CO_2 reduction (t/year)
1.	Replacement of the existing boiler	45.000	71.279	5.200	8 years, 7 months	8,33
2.	Installation of frequency regulators at the engine of the dirt-blast system	9.970	42.354	2.600	3 years, 9 months	64,46

No.	Improvement measure	Investment (€)	Energy saving (kWh/year)	Cost savings (€/year)	SPB	CO_2 reduction (t/year)
3.	Replacement of the existing boiler and installation of frequency regulators at the engine of the dirt-blast system	50.740	1.320.343	7.845	6 years, 6 months	73
4.	Introduction of alternative equipment for cleaning	640	8.372	339	1 year, 9 months	12,74

Table 2. Energy and cost savings after applying improvement measures

Replacement of the existing boiler joined with the installation of frequency regulators at the engine of the dirt-blast system contributes to largest proportion of energy savings and CO_2 reduction (Table 2.) In general, when applying all three proposed measures, energy savings would be 1.328.715 kWh/year with related CO_2 savings of 85,74 t/year and simple pay-back period 6 years and 3 months.

After applying all the proposed measures, energy indicators will be reduced (Table 3). In average, they are reduced by the 10 % therefore productivity and facility competitiveness and profit would increase.

No.	Energy indicator	Current value	Improved value
1.	Electricity consumed per square meter work surface, EI_1	180 kWh/m^2	151 kWh/m^2
2.	Total energy consumed per square meter of work surface, EI_2	551 kWh/m^2	528 kWh/m^2
3.	Total energy consumed per number of working hours, EI_3	265 kWh/h	254 kWh/h
4.	Electricity consumed per number of workers, EI_4	6.113 kWh/no. work	4.977 kWh/no. work
5.	Total energy consumed per cubic meters of treated wood, EI_5	1.115 kWh/m^3	1.068 kWh/m^3

Table 3. Energy indicators after implementation of improvement measures

4. Conclusion

Detail analysis of energy consumption of furniture manufacturing facility has shown that there are systems with exceeded energy use. One of the biggest energy users is dirt-blast system, which contributes to the electrical energy use for 71 %, which is much higher than in similar facilities. Reason for this are large installed engines power and long operating time.

Some parameters used for analysis are based on expert assessment, not on measured data. For example, structure of energy consumption of particular systems is calculated based on its installed power and working hours. More accurate results could be delivered by introducing the long-time measurements of electric energy consumption of particular electric energy users. The baseline of a thermal comfort in a production hall is assessed based on a single temperature measurement. More accurate baseline could be determined by long-time temperature measurements in the working area. In order to make accurate energy balance, there is need to register annual consumption of a fuel for the heating system.

Five energy indicators are calculated and compared with similar facilities worldwide. The indicator EI_5, total energy consumed per cubic meters of treated wood, indicated high energy efficiency of the company while other four indicated very low efficiency. Having in mind that the EI_5 is calculated as consumption of total energy per cubic meters of treated wood, it can not be relevant for companies that beside wood are using wood based board material, have carpentry and have significant proportion of manual work.

Detailed analysis of proposed improvement measures (replacement of the existing boiler, installation of frequency regulators at the engine of the dirt-blast system and introduction of alternative equipment for cleaning instead of the compressed air) has shown that their implementation will provide thermal comfort in production hall and also significant energy savings of 1.328.715 kWh/year and CO_2 savings of 85,74 t/year. After applying all the proposed measures, energy indicators will be improved by 10 %.

A particular challenge in the energy efficiency analysis of the wood industry lay in the selection and calculation of a relevant energy efficiency indicators in cases where the industry use a variety of wood raw materials and combines carpentry and manual work. In further research, the focus should be on selecting the appropriate indicators and benchmarking of such companies. The prerequisite for this kind of research is the establishment of a system of energy consumption measurement per production unit, resource and energy management system, in order to obtain a precise material balance.

5. References

[1] G. Granić, at all. (2008). Energy Sector Study in Bosnia and Herzegovina, Ministry of Foreign Trade and Economic Relations Bosnia and Herzegovina

[2] Energy Balances of non-OECD Countries (2014), International Energy Agency

[3] Palcic, I. & Buchmeister, B. (2016). Energy Efficiency in Slovenian Manufacturing Industry, Chapter 02 in DAAAM International Scientific Book 2016, pp.011-024, B. Katalinic (Ed.), Published by DAAAM International, ISBN 978-3-902734-09-9, ISSN 1726-9687

[4] Holz, F. & Schleswig-Holstein, K. (2002). Low-energy operation in carpentry, Hamburg

[5] D. Gordić, M. Babić, V. Šušteršič, D. Končalović, D. Jelić, (2010). Energy Saving Possibilities in the Furniture Industry, Energy / Economy / Ecology, Energy Saver List; Number 2; Academic editions, Belgrade.

[6] http://www.tischler-schreiner.de/energieeffiziente_werkstatt/, Accessed on: 20167-12-27.

[7] Rationalization and Innovation Center (RKW) of the German Economy e. V.RKW Competence Center, http://www.rkw-energieeffizienz.de, Accessed on: 2017-06-19.

[8] Institute for the Environmental Protection (2004), Efficient carpentry, Bavarian State Office for Environmental Protection

[9] Medar, K., Softić, A., Arnaut, S., Midžić-Kutragić, S. Kadrić, Dž., Delalić, N. (2017). Assessment of state and possible improvements resource efficiency in the furniture industry, IEEP, Society of thermal engineers Serbia

[10] Institute of energy GmbH | www.energieinstitut.net | In cooperation with ÖkoBusiness, (2014), Efficient carpentry, Accessed on 2016.12.27

[11] Monomentum automation. Frequency regulators. http://www.momentum-automation.com, Accessed on: 2017-03-15.

DOI: 10.2507/28th.daaam.proceedings.044

SYSTEM OF PROTECTION AND DIAGNOSTICS OF STRUCTURAL ELEMENTS OF SUPPORT OF HIGH-VOLTAGE POWER LINES

Iosif Breido, Vladimir Kaverin, Valeriy Ivanov,
Sofiya Voytkevich, Igor Levin

This Publication has to be referred as: Breido, I[ossif]; Kaverin, V[ladimir]; Ivanov, V[aleriy]; Voytkevich, S[ofiya] & Levin, I[gor] (2017). System of Protection and Diagnostics of Structural Elements of Support of High-Voltage Power Lines, Proceedings of the 28th DAAAM International Symposium, pp.0322-0330, B. Katalinic (Ed.), Published by DAAAM International, ISBN 978-3-902734-11-2, ISSN 1726-9679, Vienna, Austria
DOI: 10.2507/28th.daaam.proceedings.044

Abstract

A distributed system of protection and diagnostics of support structural elements of high-voltage power lines (HVPL) on 500 kV voltage was developed at Karaganda State technical university:
- current monitoring of leakage currents of bunches of suspension insulators;
- detection ice-rime depositions;
- cathode protection of underground elements of support constructions;
- information transfer from the sensors placed on supports on the control office on telemetric channels;
- creation of the passport of each support on SCADA – system basis.

The system contains set of local subsystems of protection, measurement, processing and information transfer on support, communication links and a subsystem processing and information storage on the control offices. Power supply of local subsystems is provided by the directed energy of electromagnetic fields. Detection of ice-rime deposits is carried out by the indirect methods based on the current monitoring of the support response of wind and ice loads and also temperatures and air humidity. Sensors information allows to control integral values of a leakage current of industrial frequency and surface currents of the partial discharges proceeding through strings of insulators and also to define the time-point optimum for drop-out ice-rime deposits.
Radio modem and cellular communication with an automatic choice of the communication channels in the conditions of the powerful electromagnetic fields have a significant effect on an opportunity and transmission quality of information are used for transmission the telemetric information to the control office. Tests with the positive results of the experimental samples of all components of system are carried out on the operating line and in laboratories. The developed system is a basis for methods creation and technical solutions on lowering the losses of the electric power in case of its transportation on 110 and 220 kV HVPL.

Keywords: High-voltage power lines, system of protection and diagnostics, telemetry, alternative channel.

1. Introduction

In the CIS countries power transmission on long distances is carried out by air high-voltage power lines (HVPL) [1].

Any accident, on HVPL of 110-500 kV, connected with arc overlapping suspended insulators, break of current carrying wires, supports falling, leads to interruptions of power supply of the whole regions.

Reliability of work and size of active losses of the HVPL electric power extremely depends on a condition of suspended insulators. Now remote contactless ways of preventive control of insulators with use of electron-optical defectoscopes [2-7] are used. Realization of these methods requires direct visual contact with supports therefore diagnostics can be made only selectively and with a big frequency. Therefore they are inapplicable for the supports established in boggy places, in mountainous and rocky areas. Big errors arise because of a corona effect on suspended insulators and influences of solar illumination at their use.

Considerable influence on reliability of power transmission on air-lines in the conditions of Kazakhstan and Russia is rendered by extreme climatic factors, including the ice-rime depositions leading to severe accidents (to 30% of emergency HVPL shutdowns) [8-10].

There are various ways of measurement control of the ice load of the current carrying elements of overhead power transmission lines [10-13] based on the analysis of various physical effects accompanying ice depositions. However these ways weren't widely adopted because of insufficient sensitivity, big mistakes, and also high complexity and insufficient reliability of ice detection on HVPL.

Essential decrease in a resource of support is caused by processes of electrochemical corrosion of the elements of fastening located underground, and protection systems against these processes on HVPL are absent.

Lack of the existing control methods and HVPL protection is their fragmentariness, they are not compatible among themselves and cannot be applied to continuous complex monitoring of a condition of each of the HVPL supports.

At the Karaganda State technical university the distributed system of protection and diagnostics of structural elements of the HVPL supports was developed [14-20]:

- current control of currents of bunches leakage of the suspended insulators;
- identification ice-rime depositions;
- cathodic protection of underground elements of the support constructions;
- information transfer from the sensors placed on support on control office on telemetric channels;
- creation of each support passport on the basis of SCADA – system.

Distinctive feature of system is its complexity that has allowed to solve within one project problems of protection and diagnostics of construction elements of HVPL in various parameters on the basis of the integrated telemetry and uniform approaches to design.

As a result, the experimental sample of system was developed on this basis, technical and scientific novelty of which was confirmed with inventions [16, 18, 19].

High reliability of the telemetry channel is provided with use of various communication channels between the local subsystems located on the HVPL support and a subsystem of processing and storage of information at control offices with automated variant choice (radio modem and cellular communication).

2. Description and characteristics of the protection system and diagnostics

The protection and diagnostics system represents set of the local of protection, measurement, processing and information transfer on supports, communication channels and a subsystem of processing and storage of information at control offices.

Separated phase sensors of leakage currents of strings of suspended high-voltage insulators are placed on each support.

There is insulator aging because of influence of powerful electromagnetic fields, surface discharges, a corona-formation, influences of the environment in use. Researches of currents statistics of leak on polymeric insulators have shown that the range of their change is less than 0.5 mA – to more than 25 mA depending on a condition of the insulators surface [21]. According to size of currents of leak it is possible to estimate a resource of isolation centers indirectly.

The combined sensor of leakage currents of suspended insulators strings which feature is a control of integrated values of leakage current of industrial frequency and current of partial discharge which design considers features of fastening in the grounded part of the suspended isolation center [22]. Sensors are placed between the top plates of strings on each phase and the top grounded traverses of supports.

About a condition of high-voltage polymeric insulators we can judge by currents of leak and gradual increase in quantity of impulses of partial surface discharges for a discrete phase interval of tension to serially going one by one partial discharges. It is a characteristic of a pre-breakdown situation of insulators [23]. Thus, a control of the surface discharges frequency and volume leak current allows to estimate an operational condition of a string of high-voltage insulators. Power supply is implemented on wires in the combined current sensor, and information from his outputs is transmitted through a radio modem that increases reliability and noise stability of the system. The structural diagram of the current sensor is showed in the figure 1.

The hybrid current sensor controls two parameters of leakage currents of suspension insulators HVPL [24]:
- an average value of the background leakage current module of the industrial frequency;
- an average value of the surface current module of the partial discharges.

The hybrid sensor consists of:
- a current sensor of the partial discharges – CSPD;
- a sensor of the background leakage current of industrial frequency – SBCIF;
- a current sensor controller – CSC;
- a radio modem – RM;
- a power supply unit – PSU.

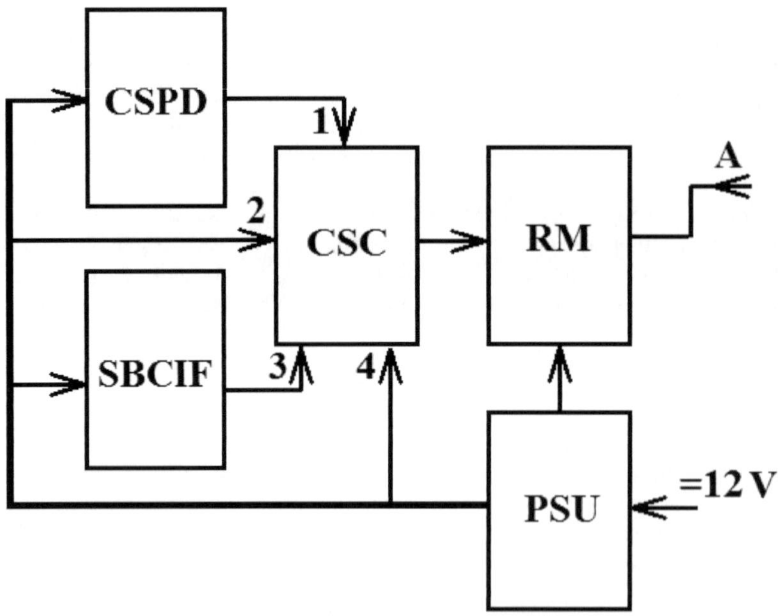

Fig. 1. The structural diagram of the sensor of leakage currents

A controller of the current sensor performs the following functions:
- constant inquiry of information analog outputs of CSPD and SBCIF, and also processing of the test signal with 1B tension, created by power supply unit and the discrete signals of synchronization in the cyclic mode;
- averaging of multiply repeated information signals from CSPD and SBCIF output;
- computation of admissible value of a zone of an output signals dispersion from CSPD and SBCIF sensors;
- synchronization of the controller current time of the control local subsystem (CLSS) and CSC in the cyclic mode.
- transfer of average signals of separated phase sensors of leak currents through a radio modem on a support in CLSS.

The following information signals are formed in memory of the sensor controller of current on the termination of the analyzed time interval:

1. test signal of 1V;
2. average signal of the background current size of the industrial frequency leakage;
3. average signal of current size of partial discharges;
4. value of admissible size of a zone of dispersion of an output signal of the sensor of background current of leakage of industrial frequency;
5. value of admissible amount of a zone of an output signal dispersion of the current sensor of partial discharges;
6. the current time of data recording in non-volatile memory;
7. code number with information number of a support and phase of the controlled insulator (record of a code with information about a support number and each of three phases is made right after completion of installation of three sensors of current on suspended insulators of current carrying wires of three phases in non-volatile memory of the current sensors controllers).

The physical form of an experimental sample of the sensor of leak current without the top cover is showed in the figure 2, where 1 - block of coordination of the current sensor of partial discharges, 2 - primary converter of the sensor of current of partial discharges, 3 - primary converter of the sensor of volume (background) leak current, 4 - block of coordination of the sensor of volume current, 5 - the converter of supply voltage of the combined current sensor, 6 - current sensor controller, 7 - radio modem.

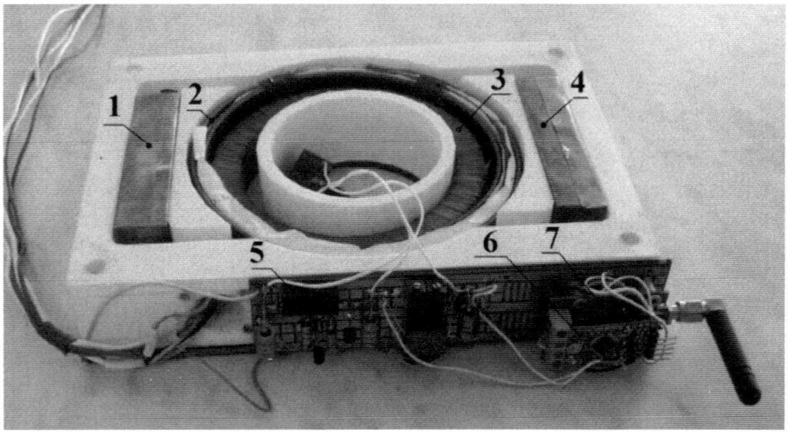

Fig. 2. An experimental sample of the combined current sensor

CSPD consists of primary converter performing functions of a galvanic outcome and current transformation of partial discharges to the electric signal coordinated on amplitude with input circuits of intensifying cascades and also the electronic block which performs functions of strengthening with the subsequent straightening and allocation of a constant component. In SBCIF functions of a galvanic outcome of primary converter with elements of fastening of the suspended insulator and also transformation of a harmonious component of the leakage current of industrial frequency proceeding via the insulator in an electric signal of constant tension are realized. Signals with CSPD and SBCIF enter the ADC current sensor controller (1 and 3 inputs of CSC in the figure 1 respectively). The basic tension of 1V enters on the CSC second input intended for control of ADC blocks operation, the controller and a radio channel. Supply voltage with PSU enters the 4th input controller. Also with PSU the stabilized tension enters the blocks – CSPD, SBCIF and RM. Information signal from the CSC output enters RM input. The telemetric signal from the output of the controller enters a radio modem input with the subsequent information transfer to the electronic block of collecting and preprocessing of information from the sensors located on a support.

CSC is executed on the Arduino-Micro platform.

The structural diagram of the local subsystem performing functions of collection and preprocessing with the subsequent transmission of the telemetric information to the control office is showed in a figure 3.

The local subsystem solves the following issues:

- collection of information from sensors;
- control of operability of the transfer channel of telemetric information from control and indicator sensors in the controller of local system;
- preprocessing information;
- saving the information in non-volatile memory;
- testing the information channels;
- the choice of information channel with the maximum level of a signal;
- transfer the saved information to the server of control office.

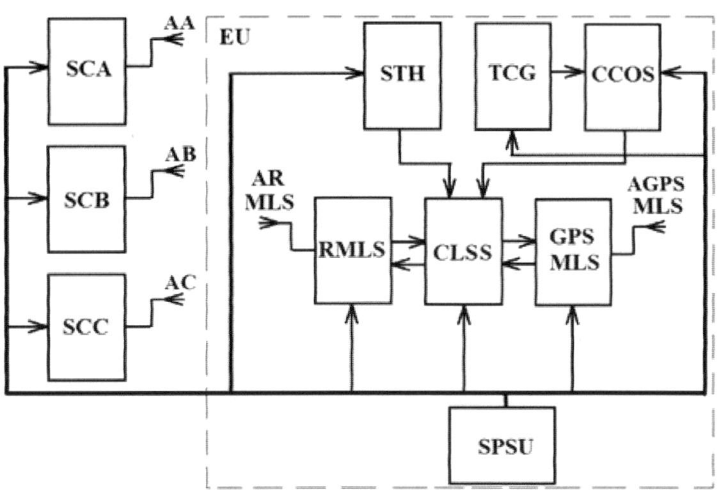

Fig. 3. Structural diagram of the local subsystem

The local subsystem consists of three identical, hybrid sensors of current and an electronic unit. Sensors of leakage currents on phases (SCA, SCB, SCC) transfer with the given periodicity on a radio channel the current information about value of the leakage current to the electronic unit (EU). A time interval of the inquiry of current sensors changes in value of a leakage current. The maximum time of an interval of the inquiry corresponds to the minimum value of the leakage current. The interval of time is reduced with increase in a leakage current. Control of the interval time inquiry of leak current sensors is executed by the controller of local subsystem (CLSS). CLSS is executed on the Arduino-Micro platform.

The Electronic Unit (EU) includes the hybrid sensor of temperature and humidity (STH), and also a three-coordinate gyroscope (TCG), a controller of conversion of output signals of TCG is CCOS, the controller of local subsystem (CLSS), two radio modems — cellular (GPSMLS) and a radio communication (RMLS), the stabilized power supply unit (SPSSU).

CLSS has bidirectional communication through channels of transferring information with control office. The sample frequency of these sensors increases at rising of leak currents to dangerous level or at gaining the frequency and signals amplitude of current sensors of the surface discharges.

Power supply of the local subsystems is situated on supports provided from the directed energy of electromagnetic fields, entering from a lightning-protective cable [18,19]. For protection against tension arising during lightning strokes, inputs disconnection of PSU power supply units and SPSSU from the converter of tension of a lightning-protective cable during its action was considered.

Detection ice-rime deposits is carried out by the indirect methods based on monitoring reaction of a support to wind and ice loads.

Formation of ice and snow sticking on HVPL wires happen at high humidity higher than 85% and air temperature in limits – (2÷5) °C. This process can be combined with big wind loading, followed by "dancing" and sagging of wires, it is shown by significant growth in mechanical dynamic loads of support. However, snow sticking can arise also at the windless weather. At the same time the tilt angle of support along HVPL changes, if it is absence of support accelerations in the horizontal plane, that is side and longitudinal accelerations.

Therefore the current control of acceleration of a support in three planes and also temperature and humidity of air by means of signals of the combined sensor of temperature and humidity is carried out (STH: measuring instrument of humidity and temperature ИВГ-1 МК-С-М) and also three-coordinate gyroscope (TKG: three-axial accelerometer MMA7361 of Freescale company) [19].

When comparing signals of the support acceleration in three planes and a support deviation corner on HVPL axis, and also temperature signals and air humidity with the preset values, the probability of icing process origin on current carrying wires is defined [19].

Increase of noise resistance from electromagnetic fields action and a corona-development is carried out on the basis of averaging signals of sensors in controllers of sensors of current and in the local subsystem controller. It is revealed that hindrances from corona-development processes and electromagnetic fields are concentrated in the range of frequencies 6-7 MHz. Radio modems of system provide signaling at frequencies more than 2 GHz. These measures provide noise resistance of the developed system.

On each HVPL support three sensors of leak currents according to phases are placed. Sets of temperature, humidity and accelerations sensors of a support are placed on HVPL support sites where there is a frosting and the drizzle on long-term observations.

Cathodic protection is a part of system. 2 options of cathodic protection of underground elements of the HVPL support design are developed:

- The autonomous cathodic protection which does not demand power supply source, based on connection of the semiconductor diode parallel to the isolating elements in backstay cables, that is equivalent to giving of negative potential on the protected element [16].
- Cathodic protection with power supply from the energy directed in a lightning-protective cable [18].

Radio modem, cellular communication with an automatic choice of channels of transferring information in the conditions of the powerful electromagnetic fields directed in the HVPL elements when transporting the electric power is used for telemetric information transmission to the control office.

Information transfer from support on the server is executed on cellular communication in the presence of stable cellular communication. At the same time the considerable sections of HVPL-500, because of remoteness of lines from settlements, and also because of a land features, are not enveloped by cellular communication. In this regard technical capability of information transfer on a chain from a support to a support is provided to the support which has a stable cellular communication with control office.

In that case if information transfer, for any reasons does not arise, CSC continues in the cyclic mode with the given interval of time to execute inquiry of CSPD and SBCIF information channels and to make computation of average values, and also the ranges of dispersion zones of output signals for each information channel. All information taking into account the current time is saved in the CLSS non-volatile memory.

The passport of each support of a HVPL controlled section which supports archive of information on technical condition of construction elements of HVPL support, trends of leakage about the current alert conditions with localization of the place, time are created in the control office.

Predicting mathematical development models of alert conditions, and also controlled elements aging of HVPL support construction were created on the basis of the analysis of the current values of leakage, technologic forecasting of frosting HVPL current elements is provided. Diagnostics of the equipment status of local systems and sensors is performed, recommendations, on changeover of elements of HVPL support are created.

3. The pilot laboratory studies of system

Due to the big organizational and technical problems of the research performance on the operating HVPL-500 line experiments have been conducted in laboratory conditions. The laboratory stand was manufactured; the structural diagram is showed in the figure 4.

In the figure 4 the following designations are entered: 1 - information transfer device; 1.1 - a controller (Arduino Pro Mini) forming the transferring file; 1.2 – HY RS485 transferring module on the radio modem basis; 2 – the device of receiving and visualization of parameters of the accepted information; 2.1 – the module of telemetric information receiving on the basis of HY RS485 radio modem; 2.2 – a controller (Arduino Pro Mini) with the program of the analysis of reliability degree of the accepted information; 2.3 – visualization device; 3 – The simulator of the electromagnetic field created by a corona of the HVPL high-voltage wires. L1 - distance between transferring and receiving devices; L2 – distance between the simulator of hindrances and the sending device.

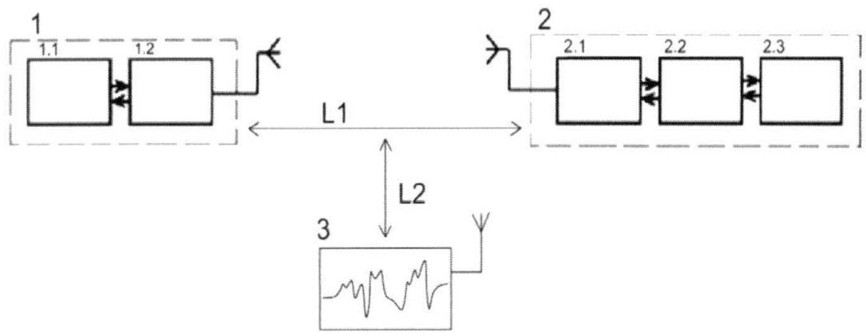

Fig. 4. Structural diagram of a subsystem of telemetric information transfer

The frequencies range of the hindrances generated by the simulator practically coincides with frequencies range of hindrances at a corona-development.

It was established that the optimum mode of data transmission was the data transmission mode with 4800 bit/sec speed. At this speed the greatest number of transfer cycles has excellent and good estimation of communication reliability. Communication reliability is characterized in number of transfer cycles as follows: bad - 20%, satisfactory - 12%, good – 32%; excellent - 35%. Data transmission deterioration is observed at increase speed of data transmission.

The average value of communication reliability at 4800 bit/sec speed for various distances between modems and radiator of hindrances is presented in the figure 5.

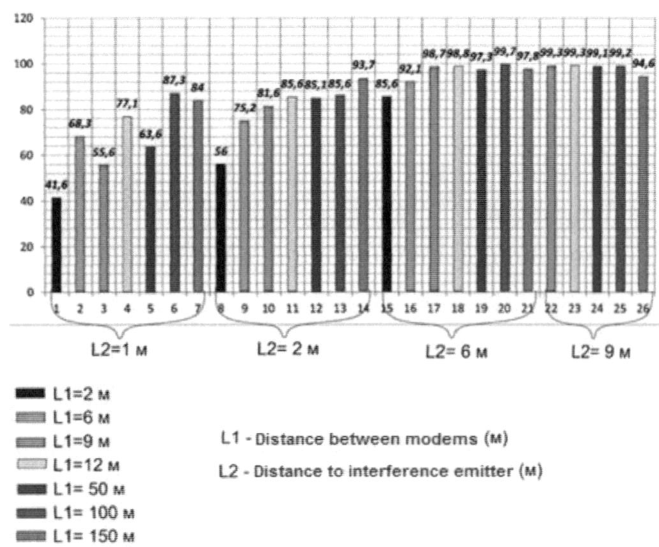

Fig. 5. Average value of communication reliability at speed 4800 % bit/sec

The number of the distorted files increase sharply at a distance less than 4 meters between the transferring module and the hindrances simulator.

Because the distance between the transferring modules, installed in leak current sensors, and current carrying wires on 500 kV HVPL makes not less than 6 m, it is possible to claim that the telemetric channel with high reliability will steadily transfer information.

Laboratory experiments on a research of the channel characteristics of telemetric information transfer while using cellular communication have been also executed.

The purpose of an experiment was assessment of influence of high-frequency hindrances on GSM signal level and also on quality of data transmission.

During the research the distance between the transferring GSM modem and the generator of hindrances changed and also the structure of the environment of transfer for obtaining various levels of a GSM signal modified.

It was established that at the level of weakening of a signal 101 dB data were transferred without loss at distance to the hindrances simulator more than 2 meters.

During the experiment packages of data were delivered without loss, or were not transferred at all if a signal was lost. It means that algorithms of sending data from the transferring GSM modem on control office provide their full transfer without loss as in the protocol of data, there are confirming bits about correctness of information transfer. Taking into account that on real HVPL-500 distance between current carrying wires and the system of telemetric data transmission will exceed by 6 meters, influence of current carrying wires on the level of a signal of the GSM channel will be insignificant and there will be enough single duplication of data for their correct receiving on control office.

4. Experiments on the operating HVPL-500

Tests of cathodic protection and pilot studies of the telemetry system were conducted on the operating HVPL-500 in the territory of NURA substation.

The model of the transferring block of telemetry containing the industrial controller of modular one-plate execution and the modem of cellular communication AZIMUTH have been developed and manufactured. Information via the channel of cellular communication BEELINE and Internet operator received to laboratory on the computer of the operator;

3 signals from sensors of current cathodic protection were transmitted in real time to the telemetry channel. The step of sampling the sensors of current has been established every 5 minutes.

The software established on the computer of the operator allowed to carry out the control in real time. Numerical values of telemetric information, including time and date of sampling the sensors of current cathodic protection were transferred.

The general time of tests was 10 months.

During experimental studies continuous record of the transmitted data for the purpose of carrying out the subsequent analysis of the revolting factors influence on reliability of the transmitted data was carried out. The reliability of the accepted information was estimated in comparison of information obtained at production tests with the similar information massifs received while testing control channels of cathodic protection a condition.

Diagrams of the cathodic protection current during one month are shown in figure 6.

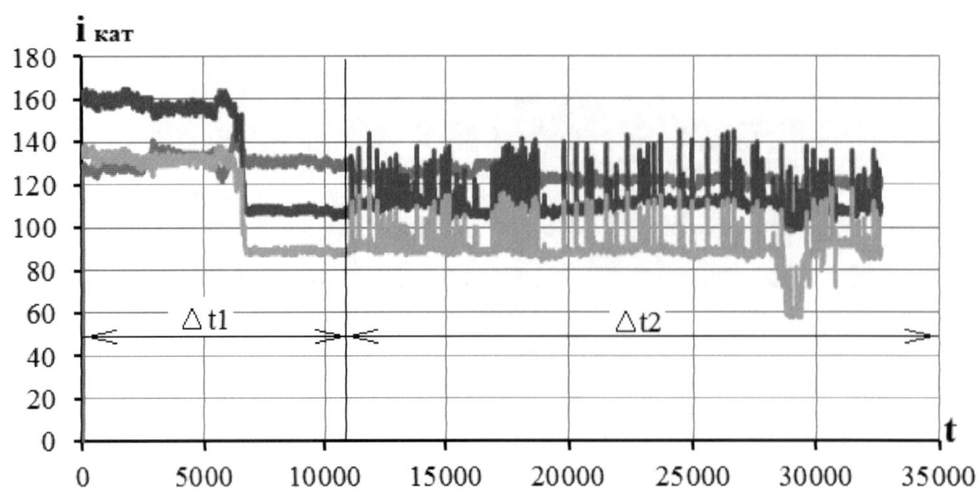

Fig. 6. The cathodic protection current during one month

Commissioning and testing of system of cathodic protection with the channel of telemetric information on the basis of cellular telephony was carried out on time interval Δ t1. The system operated normally on time interval Δ t2.

It is established that average loss of mass of anodes has made 11,7%, the gain of mass of cathodes has made 0,7% due to the surfaces of oxide formation. After carrying out tests of cathodic protection.

Quality of the accepted telemetry is quite satisfactory. Efficiency of the cathode protection of underground elements of construction support is confirmed. The minimum values of the cathodes currents are defined, and also developed constructions of cathodes are approved.

5. Conclusion

The system of protection and diagnostics of structural elements of support of high-voltage power lines (HVPL) created at Karaganda State Technical University on 500 kV voltage, it provides monitoring of currents of strings leakage of suspended insulators, identification of the ice-rime deposits, cathodic protection of the underground elements of the design support and creation of each support passport on the basis of SCADA – system.

The carried-out tests of experimental samples on the operating line and in laboratories have confirmed operability of all system components.

Its distinctive feature is a complexity that has allowed to solve within one project problems of protection and diagnostics of elements of HVPL design in various parameters on the basis of the integrated telemetry and uniform approaches to design.

Introduction of system will allow:

• to increase reliability of strategic high-voltage HVPL due to decrease in accident rate;

• to reduce losses of the electric power at its transportation by results of remote continuous monitoring due to timely replacement of the strings of high-voltage insulators at increase in currents of leak;

• to reduce losses of the electric power at its transportation due to the reduction of spreading currents size and currents induced in metal elements of the construction support.

The developed system is a basis for creation the methods and technical solutions on decreasing losses of the electric power at its transportation on 110 and 220 kV high-voltage HVPL.

The realized technical solutions will also be used for the complex analysis of losses on the operating high-voltage HVPL.

The components which are a part of the system have independent value and can be effectively applied to other applications. These are sensors of the leak currents, cathodic protection, the system of telemetry with use of various channels of communication.

So, sensors of the leak currents with telemetry will be used at the stands for tests of high-voltage insulators with control of the actual leak currents.

The main technical solutions created in the course of system development can be used during creation of software and hardware complexes for other applications where ensuring continuous control of the distributed electrical power systems is necessary.

6. References

[1] V. I. Chemodanov, N. V. Bobylyova., N. N. Utts. (2009). Development of the single national network in the existent economic conditions.//Electro. Russia, Moscow No. 5. pp. 2-7

[2] V. A. Oblozhin. (2000). Thermovision monitoring in case of the repairs organization of the electrotechnical equipment on its status. Power plants. No. 6. pp. 58-63.

[3] Methodical instructions for distant optical checking of insulation of air-lines of the electricity transmission and distributing devices of an alternating current 35 - 1150 kV. (2005). 56947007-29.240.003-2008 JSC "Federal network company of the unified power system". M. pp. 14-15.

[4] S.A. Bazhanov. (2001). Perspectives of infrared diagnostics use in power engineering//Power engineerings.. No. 8. pp. 31-32.

[5] M. N. Badretdinov, A. M. Gataullin, V. L. Matukhin, D.F. Gubayev. (2011). A method of non-contact and distant control of a bunches status of insulators of air high-voltage power lines, a patent No. 2359280 of 10.12.2011, Russia.

[6] N. M. Schmidt, I. M. Schmidt. (2001). A method of non-contact and distant control of electrodurability of bunches of insulators of air high-voltage power lines, the patent No. 2058559 of 17.09.2001, Russia.

[7] H.H. Tikhodeev. (1993). Test methods and reliability of the equipment for substations of high, super and ultrahigh tension;//News of RAS. Power engineering. - No. 3.1. pp. 42-60.

[8] E.V. Zakamsky, R. G. Minullin. (2002). The analysis of the experimental data obtained in case of diagnostics of air-lines of an electricity transmission//Materials of reports of VI Postgraduate and master scientific seminar of KSEU. Kazan: KSEU, pp. 45-46.

[9] A. L. Kulikov. (2010). A method of detection of ice formation on wires and lightning-protective ropes of power lines, a patent No. 2409882 of 27.01.2010, Russia.

[10] A.F. Dyakov, I.I. Levchenko, A. S. Zasypkin, A. A. Alliluyev, E. I. Satsuk, A. I. Bytkin, F. A. Dyakov. (2005). Information systems of monitoring of ice loadings on air-lines//Power engineering specialist. No. 11. pp. 20-25.

[11] I.F. Hasanoff, M.F. Latypov, M.I. Vafin. (2001). Classification of sensors of ice on air power transmission lines//Interuniversity scientific materials "Electrotechnical Complexes and Systems" Ufa: Ufa state.avia university. pp. 202-205.

[12] S.Voytkevich, I. Breido, V. Kaverin. (2014). Universal mathematical model of leakage currents and Currents Spread on Elements of High-Voltage Pillar. In Annals of DAAAM for 2014& Proceedings of 25th International DAAAM Symposium. Vienna: DAAAM International, 2014. ISBN 978-3-901509-73-5. ISSN 1726-9679.

[13] I. V. Breydo, V. V. Kaverin, A.U. Nenchenko. (2016) Application of smart grid concept in the control system and condition forecasting of HVPL supports. Materials of the international scientific and practical conference "Integration of Science, Education and Production — a Basis of Implementation of the Plan of the Nation". Part 4. June 22-24.

[15] I. V. Breydo, V. V. Kaverin, V. A. Ivanov, A. V. Sichkarenko, I. V. Levin, S. V. Voytkevich. (2016). The distributed noiseproof control complex of a condition of support of high-voltage power lines. Innovative solutions of economy problems of Belarus and Kazakhstan knowledge. Collection of materials of a scientific and practical conference, Minsk. October 13. pp. 94-95.

[16] I.V. Breydo, V.V. Kaverin, A. K. Beysenov, Yu.S.Laistsev, A.V. Voytkevich, S.V. Voytkevich. (2017). Patent of RK No. 98220. "Support of overhead power transmission lines", registered on 16.05.2017.

[17] Voytkevich S., Breido I., Kaverin V. (2014). Alternative sources of supply for stand-alone automation devices on supports high-voltage lines // In Annals of DAAAM for 2014 & Proceedings of 25th International DAAAM Symposium. – Vienna: DAAAM International.

[18] I.V. Breydo, V.V. Kaverin, S.V. Voytkevich. (2015). Innovative patent of RK No. 29977. "System of cathodic protection of elements of support of sites of high-voltage power lines", registered on 19.05.2015.

[19] I.V. Breydo, V.V. Kaverin, G.A. Em, A.V. Sichkarenko, V. A. Ivanov, S.V. Voytkevich. (2017). Application for the invention "The way of telemetry of ice loadings of high-voltage power lines and system for its implementation". Application for an invention, registered No. 2017/0424.1 from 19.05.2017.

[20] I. V. Breydo, V. V. Kaverin, A.V. Sichkarenko, S. V. Voytkevich, V. A. Ivanov. (2014). The distributed systems of protection and diagnostics of elements of high-voltage power lines. International scientific and practical conference. "Science and innovative developments - to the North". Collection of theses reports. Mirny city. pp. 89-90.

[21] Vlastos A. E., Sorqvist T. (1996). Field experience of ageing and performance of polymeric composite insulators. CIGRE.

[22] I. V. Breydo, V. V. Kaverin, V. A. Ivanov, G. A Em. (2016). Development and experimental study of the sensor of leakage current of high-voltage insulators. Works of university No. 3. Kazakhstan. pp. 105-108.

[23] A.M. Gataullin. (2015). The way of contactless remote diagnostics of a condition of high-voltage polymeric insulators. Patent of the Russian Federation No. 2566391, published on 27.10.15.

[24] V. A. Ivanov, V. V. Kaverin (2016) Diagnostics of the isolation condition of high-voltage insulators in leak current size. Works of the international scientific and practical conference "Integration of Science, Education and Production — a Basis of Implementation of the Nation Plan". Part 4. June 22-24, 2016.

DOI: 10.2507/28th.daaam.proceedings.045

SPECTROSCOPIC MEASUREMENT OF TEXTILE FIBRES

Hana Vaskova & Jan Hrano

This Publication has to be referred as: Vaskova, H[ana] & Hrano, J[an] (2017). Spectroscopic Measurement of Textile Fibres, Proceedings of the 28th DAAAM International Symposium, pp.0331-0333, B. Katalinic (Ed.), Published by DAAAM International, ISBN 978-3-902734-11-2, ISSN 1726-9679, Vienna, Austria
DOI: 10.2507/28th.daaam.proceedings.045

Abstract

Raman spectroscopy as an effective tool for material identification offers benefits that are important for the process control of the fibres in terms of quality control in production, in use, but also in the field of fibre analysis for forensic identification. In particular, it is specific information on chemical composition and structure of material; rapidity of measurement, non-destructiveness, independence on chemicals, portability, possibility of automation of measurements and ability to detect structural changes. In the study, samples of natural and synthetic polymer fibres were measured and analysed. For the evaluation of large spectral data sets multivariate analysis techniques were applied to find the patterns within the data. The spectral database was created on the basis of the acquired Raman spectra of textile fibres and was successfully tested on other samples. The results show a great potential of the used method for the fibre identification.

Keywords: Quality; control; textile fibre; Raman spectroscopy; multivariate analysis techniques; spectral database

1. Introduction

People have been using textile fibres of a different nature since the beginning of their existence. Plant sources were used for the production of textile fibres more than twenty thousand years ago [1]. Textile production is an old craft that has been greatly changed by technological advances. Many types of fabrics come from new technologies and raw materials. Fabrics used today are very often a mixture of multiple materials. The palette of modern use emphasizes the properties of fibres and fabrics, such as durability, strength, breathability, flexibility, bacteria resistance, etc. [2]. Many tests/methods, from simple to sophisticate, are used to study, measure, verify and improve [3] modern fibres.

One of the main motivations for this research was to find and study method offering rapid, non-destructive measurement with the ability to recognize and identify unknown samples. Although microscopy is often the most powerful method [4], Raman spectroscopic examination gives valuable information on the structure of material. Lately, this method finds more and more applications across scientific fields. Its potential allows studying natural fibres based on cellulose [5], carbon-based structures [6], as well as biological systems, artworks or food. Raman spectroscopy offers benefits that are important for the process control in terms of quality control in production, in use [7], but also for forensic identification [4]. Because of its many advantages, Raman spectroscopy was chosen for the study of textile fibres and creation spectral database. Another motivation is related with forensic application. Fibres are considered as a one form of trace evidence [8]. They are transferred between the perpetrator and the victim or left behind at the crime scene during committing a crime. A functional spectral library is then a useful and effective tool for the identification.

2. Materials, Methods and Instrumentation

The ability to obtain the relevant information on textile fibres is studied by a combination of Raman spectroscopic measurement with cluster analysis and principal component analysis.

2.1. Textile fibres

For the spectroscopic study of textile fibres, commonly used textile materials were chosen to cover various types of textiles. Twenty samples of natural and synthetic fibres, such as cotton, ramie, flax, wool, merino wool, silk, cellulose acetate, lyocell (rayon), viscose, nylon, polyamide and polyester were measured. A part of the mentioned materials included more samples. Therefore, duplicate samples were used to test the database. Colour of all used samples was natural/white.

2.2. Raman spectroscopy

Raman spectroscopy is a vibrational spectroscopic method providing specific chemical information of every single chemical substance and its modifications in the form of Raman spectra. A slight wavelength shifts occurring after irradiation facilitate the material identification and structural assessments. Substantial benefits arise from many advantages; method is rapid, non-destructive, contactless, applicable to all states of matter in different forms without special requirements for sample preparation, independent on chemicals, usable as *in situ* analysis, usable for measuring through transparent glass or polymeric covering layers. Since the Raman scattering is a weak effect, some adverse effects, such as luminescence, can influence the quality of spectral response and mask spectral information.

Raman spectra were measured using Renishaw InVia Basis Raman microscope with NIR diode laser (785 nm) with maximum output power 300 mW. Leica DM 2500 confocal microscope with the resolution 2 μm was coupled to the spectrometer. All measurements were collected with 4 s exposure time and 15 accumulations. The samples were firstly scanned in the range from 100 to 3200 cm^{-1}. A spectral range for the evaluation was reduced to 700 - 1800 cm^{-1}.

2.3. Multivariate analysis techniques

Raman spectroscopic data are multivariate, since they reflect the composition of material and its structure. Raman peaks are assigned to vibrations of chemical bonds in the analysed materials. For the evaluation it is appropriate to use multivariate mathematics. Methods that enable multivariate analysis such as principal component analysis and cluster analysis were used for data mining from the large spectral data sets to find the patterns within the data.

3. Results

Raman spectra of the textile fibres were measured on Raman microscope. A part of the obtained spectra is shown in Fig. 1a. The only spectrum that was excluded from the evaluation is the spectrum of viscose, because of its high luminescence. The linear baseline correction and unit normalization was applied on the spectra for further multivariate statistical evaluation. The data were analysed by principal component analysis, the output is displayed in Fig. 1b. The scores plot clearly shows the formation of three groups. Closer look at each of the groups obviously indicate the

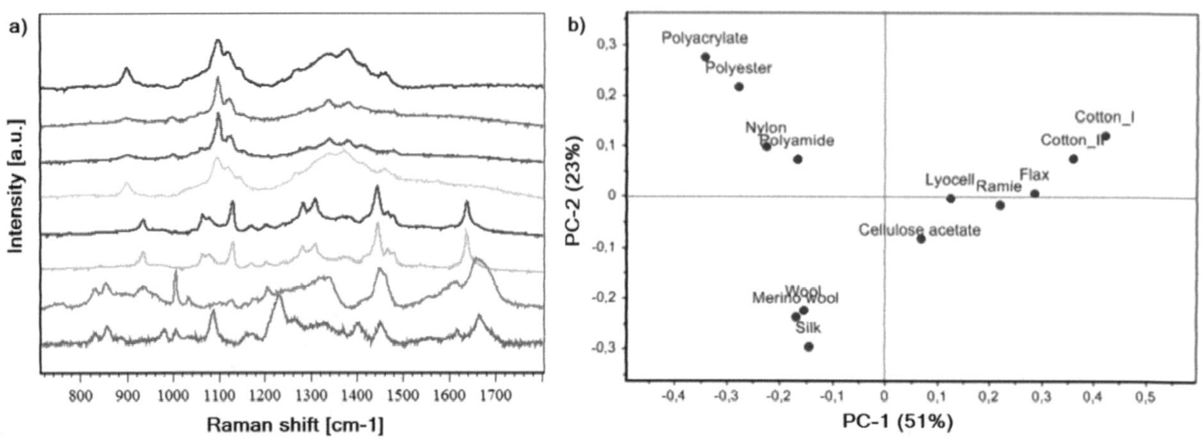

Fig. 1. a) Raman spectra of cotton, ramie, flax, lyocell, nylon, polyamide, merino wool, silk (top to bottom).
b) Principal component analysis of the Raman data of fibres

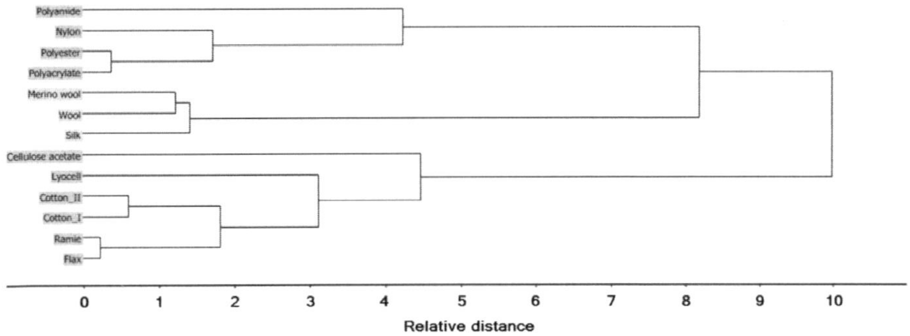

Fig. 2. Cluster analysis of the of the Raman data of fibres

differences associated with the source of the fibres – natural vegetable (based on cellulose), natural animal (based on protein) and man-made non-cellulosic polymers.

The hierarchical complete linkage based on correlation distance measure was used for the cluster analysis. There is the dendrogram in Fig. 2 which also shows the differentiation to three main sub-clusters according to nature of the fibre origin. An exception in the grouping could be lyocell and cellulose acetate. Although man-made fibres, these two samples are linked to the natural vegetable samples. The explanation arises from their cellulose source.

The database of Raman spectra of fibres was created on the bases of acquired data. The database was successfully tested on other samples of fibres with the known origin and composition. The database serves as a tool for identification of unknown samples. Samples can be due to advantages of Raman spectroscopy and coupled microscopy in the form of solitary pieces of fibre in dimensions in the order of micrometres. The database is helpful for identifying fibres as forensic traces.

4. Conclusion

Analytical method Raman spectroscopy was chosen for the examination of textile fibres. Data reflecting structural differences of thirteen textile fibres were acquired by this modern method. The evaluation of the datasets was performed using multivariate statistical methods principal component analysis and cluster analysis. Raman spectra show visible differentiation. The statistical results indicate agreement with the nature of the fibre origin. The database of Raman spectra was created for further identification of fibres using Spectral ID software. Forensic demonstration was done, however, the results will be published separately. Up to now only natural/white fibres were studied. The future perspective is to focus on enlarging the number of measured materials, on dyed fibres and another forensic application.

5. Acknowledgments

This work was supported by the Ministry of Education, Youth and Sports of the Czech Republic within the National Sustainability Programme project No. LO1303 (MSMT-7778/2014) and also by the European Regional Development Fund under the project CEBIA-Tech No. CZ.1.05/2.1.00/03.0089.

6. References

[1] Kvavadze, E., Bar-Yosef, O., et.al. (2009). 30,000-year-old wild flax fibers. Science, Vol. 325, pp. 1359-1359.

[2] Simoncic, B., & Tomsic, B. (2010). Structures of novel antimicrobial agents for textiles-a review. Textile Research Journal, Vol. 80, No. 16, 2010, pp. 1721-1737.

[3] Glogar, M. I., Parac-Osterman, Đ., Grundler, D., & Rolich, T. (2011). The rule-based approach in studying the influence of a different textile surfaces on colour similarity. In 22nd International DAAAM Symposium" Intelligent Manufacturing & Automation", pp 1207-1208.

[4] Muehlethaler, C., Leona, M., & Lombardi, J. R. (2015). Review of surface enhanced Raman scattering applications in forensic science. Analytical chemistry, Vol. 88, No.1, 2015, pp.152-169.

[5] Kavkler, K., & Demšar, A. (2011). Examination of cellulose textile fibres in historical objects by micro-Raman spectroscopy. Spectrochim Acta A Mol Biomol Spectrosc, Vol. 78, No. 2, 2011, pp. 740-746.

[6] Ivanov, R., Hussainova, I., Aghayan, M., & Petrov, M. (2014). Graphene coated alumina nanofibers as zirconia reinforcement. In 9th Int. DAAAM Balt. Conf, Vol. 348, 2014, pp. 348 – 353.

[7] Neumann, P., Houser, J., Pospíšilík, M., Skočík, P., & Adámek, M. (2015) Some methods for electronic component authenticity assessment. In Annals of DAAAM and Proceedings of the International DAAAM Symposium, pp.1-11.

[8] Chalmers, J. M., Howell G. E. & M. D. Hargreaves, Infrared and Raman spectroscopy in forensic science. 1st pub. Chichester, West Sussex, UK: Wiley, 2012

DOI: 10.2507/28th.daaam.proceedings.046

INFLUENCE OF SELECTED TECHNOLOGICAL FACTORS ON THE HOLE QUALITY DURING REAMING

Jaroslava Fulemova, Jan Kutlwaser, Miroslav Gombar & Jan Rehor

This Publication has to be referred as: Fulemova, J[aroslava]; Kutlwaser, J[an]; Gombar, M[iroslav] & Rehor, J[an] (2017). Influence of Selected Technological Factors on the Hole Quality During Reaming, Proceedings of the 28th DAAAM International Symposium, pp.0334-0340, B. Katalinic (Ed.), Published by DAAAM International, ISBN 978-3-902734-11-2, ISSN 1726-9679, Vienna, Austria
DOI: 10.2507/28th.daaam.proceedings.046

Abstract

Quality of reamed hole is influenced by many factors. Some of them have higher impact on accuracy and quality of reamed hole and the others have higher impact on tool life of a reamer. This article deals with influence of cutting speed, feed rate, reverse feed rate and eccentricity of the tool tip on cylindricity, surface roughness Ra and diameter of reamed hole. Design of experiment was carried out with CCD (Central Composite Design) with orthogonal blocking. As a tool was used cermet reamer with 6 brazed cutting edges. Results showed not only valuable data but also weaknesses of experiment preparation. The article also point out weaknesses and how to handle with them.

Keywords: Reaming; Design of Experiment; hole quality; machining parameters

1. Introduction

Reaming is well known technology for producing precise and very precise holes. The aim of reaming is increasing accuracy to shape and accuracy to size. Reamed holes are produced with very tight tolerance and with fine surface roughness. Generally, reamed hole can be produced in grade of accuracy from IT 5 to IT8 and the quality of machined surface up to Ra 0.4. Although reaming is finishing technology there can be demand for more precise holes (lower than IT 5 and Ra 0.4), then it is necessary to use the other technological operations like grinding, rolling, honing, etc. [1], [2].

To produce precise hole means a number of technological operations. First of all it is drilling. Accuracy grade of drilled hole is from IT9 to IT12, so it is roughing technology. After that it is boring or chamfering. At the end is reaming. In every step of the production chain is possible to make many mistakes which can influence accuracy and quality of reamed hole.

1.1. Problem statement

A part of results, which are presented in this paper, can be found in diploma thesis [3] and they belong to project Ro RTI A7. This project is focused on determination of all factors which can influence quality of reamed hole. The main idea of the project is based on basic research whose results will provide information for practical application. Results of literature research provide answer to a few specific problems and give general information about geometry and function of reamers. But nowhere is written which parameter influences the quality of reamed hole. This article tries to determine factors which can influence hole quality during reaming by using DoE (design of experiment) and at the same time draw attention to weaknesses of experiment preparation.

1.2. Literature research

Bhattacharyya et al. [4] designed mechanistic model to predict cutting forces for arbitrary reamer geometry. Input parameters for this model are reamer geometry, feed rate, cutting speed, initial hole geometry, process faults including parallel offset runout, spindle tilt, respective locating angles and tool/hole axis misalignment. Mechanistic model will predict torque, feed and radial force. The model valid over range of feed rate, cutting speed and varying reamer geometry. Designed model was validated by experiment. They found that parallel offset runout, spindle tilt, spindle tilt locating angle, and tool/hole axis misalignment have significant effects on the radial forces. These radial forces are shown to be correlated to hole quality. [4]

Hauer et al. [5] investigated the influence of a diagonal pre-drilled hole on hole quality during reaming using multi-blade tool. Presumption of their research ware following. The varying input parameters are influencing the width of cut and thus the quality of reamed hole. Interaction between positional tolerance of pre-drilled hole and the positioning accuracy of the machine tool lead to varying depth of cut and thereby higher radial force. Finally reaming tool is deflected and the hole quality worsens. They prepared experiment to find out radial deviation of pre-drilled hole made by two different types of drills. The results showed that smaller radial deviation has solid carbide drill than drill with replacable head. After that this pre-drilled holes were reamed to find out if the reamer follows pre-drilled hole. This theoretical presumption was not validated. [5]

Towfighian et al. [6] observed the influence of low speed reaming on vibrations of reamer with modified geometry. Designed FEM predict vibrations. The lobbed of multi-cornered holes are formed during low speed reaming. On the other hand tool chatter occurs at high speed reaming. They found out that irregular spacing of 6 cutting edged leads to perfect hole quality and the best combination of pitch angles also leads to the most stable condition. [6]

Bezerra et al. [7] investigated the influence of cutting parameters (depth of cut, cutting speed, feed rate, helix angle, number of blades, margin size and rake face) on real diameter, roundness, cylindricity and surface roughness during reaming aluminium-silicon alloy. They used reamer with brazed cutting edges from sintered carbide K10. Conclusion was following. Satisfactory hole quality can be achieved by smaller values depth of cut, lower cutting speed, higher feed rate, straight flute reamer with many blades and small margins. It is a paper which describes the whole factors very deeply. [7]

Müller et al. [8] used DoE for set up parameters of reaming to achieve parameters of machined surface and hole tolerance. They were focused on three factors (spindle speed, feed rate and lubrication) on two levels. As design of experiment was used full factorial experiment. The results show that at small spindle speed, feed rate and lubrication with oil the deviation of diameters is smaller than at high spindle speed. When they used lower spindle speed, feed rate and oil lubrication the surface roughness was lower than with higher parameters of cutting process and lubrication with cooling lubricant at 1% concentration. [8]

2. Experiment

Experiment was done on turn-mill complete machining centre CTX beta 1250 TC 4A. As a workpiece was used a bar stock with diameter 30 mm and length 550 mm. Final length of workpiece was 59 mm. The machine tool and workpiece are in the Fig. 1. The bar stock was clamped into three-jaw chuck and was made of 42CrMo4 steel. Chemical composition and mechanical properties are in Table 1. This material is very often used for tool holders.

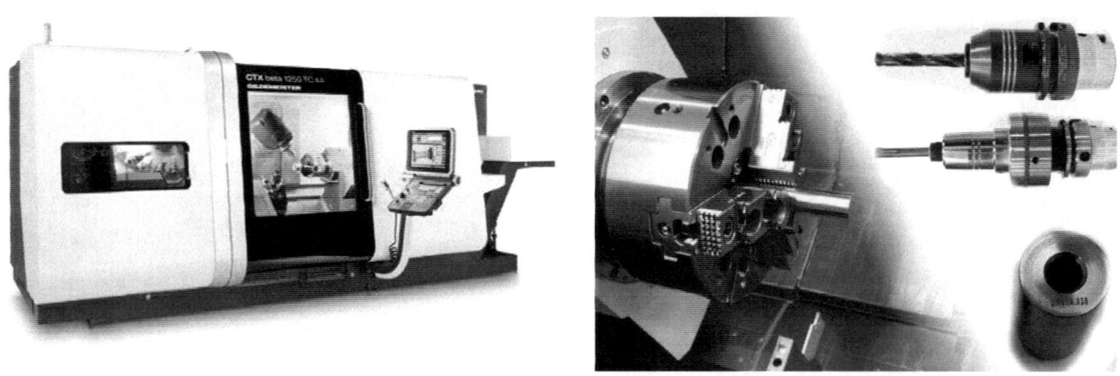

Fig. 1. Machining centre CTX beta 1250 TC 4A and workpiece with twist drill and reamer

42CrMo4 + QT + SH	Chemical element	**C**	**Mn**	**Si**	**P**	**S**	**Cr**	**Mo**	**Cu**	**Al**
	% representation	0.43	0.85	0.19	0.021	0.020	1.14	0.17	0.24	0.012
	Mechanical properties	**Rm [MPa]**		**Rp 0.2 [MPa]**		**A [%]**	**Z [%]**		**Kv [J] 20°C**	
		1036		880		15.1	58		82	

Table 1. Chemical composition and mechanical properties of steel 42CrMo4

2.1. Production process

There was used 6 blade reamer with brazed cutting edges from cermet. The reamer was clamped into tool holder with possibility to change eccentricity of the tool tip. This system is called Compensation tool holder. Production process is in the Table 2.

Nr. of operation	**1.**	**2.**	**3.**	**4.**
Technology operation	Spot-drilling	Drilling	Chamfering	Reaming
Dimension [mm]	ø3 x 1.5	ø11.8 x 60	0.5 x 45°	ø12H7 x 50
Material of the cutting tool	Sintered carbide coated TiAlN	Sintered carbide coated TiAlN	Sintered carbide coated TiN	Cermet
Number of blades	2	2	1	6
v_c [m/min]	50	80	100	*Variables*
f [mm/rev.]	0.05	0.19	0.1	

Table 2. Production process of the experiment

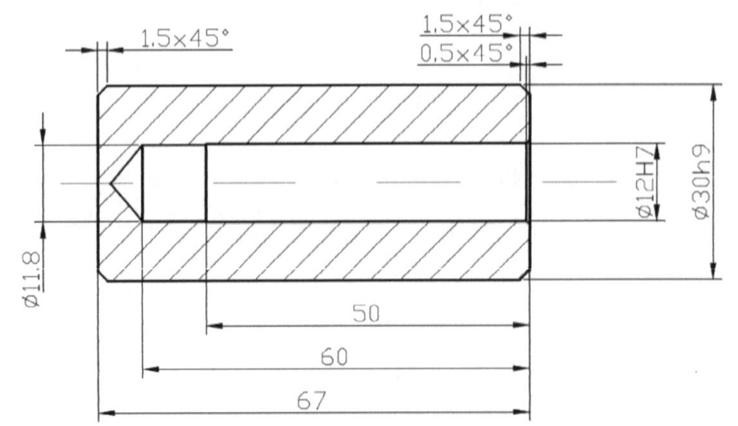

Fig. 2. Drawing of the sample

2.2. Design of Experiment

To evaluate the influence of single factors on quality of reamed hole there was chosen Central Composite Design with orthogonal blocking. This type of experimental plan is possible to detect nonlinear relation. The design of experiment takes into account 4 factors and every factor has 5 levels. The main advantage of CCD plan is small number of samples, to be specific 26 samples.

Code	**Labelling**	**Unit**	**Level of the factor**				
			-α	**-1**	**0**	**+1**	**+α**
x_1	v_c (cutting speed)	m/min	150	160	180	200	210
x_2	f (feed rate)	mm/rev.	0.6	0.64	0.72	0.8	0.84
x_3	v_f (reverse feed)	mm/min	15	20	30	40	45
x_4	eccentricity	µm	2	4	8	12	14

Table 3. Experimental plan

3. Experimental results

3.1. Cylindricity

Cylindricity of entire experiment were measured by Taylor Hobson TalyRond 585 roundness and form measurement instrument. The measured specimen was clamped in three-jaw chuck. Total measured height of the cylinder is a bit smaller than reamed hole height, because the measuring arm tip radius must be taken into calculation. Measuring arm with tip radius 2 mm was used for measurement. The measured height was divided into five levels and totally is the cylinder calculated from those five cross-sections using Least squares method.

Our goal was to approve the influence of selected parameters on the reamed hole cylindricity. We used a central composite design of the experiment that allows finding a mathematical model, which describes our response (cylindricity) in dependency on the process inputs, so called factors - cutting speed, tool tip eccentricity, feed. The first step in measured data evaluation, using mathematical model, is its creation by for example regression analysis and then its evaluation. Model suitability analysis is in Table 4.

Parameter	Value
RSquare	0.644137
RSquare Adj	0.555172
Root Mean Square Error	1.579146
Mean of Response	6.918407
Observations (or Sum Wgts)	26

Table 4. Regression model suitability analysis

We can see that the adjusted index of determination (Rsquare adj.) is only 0.55 that means that our mathematical model cannot explain approx. 44.5% of cylindricity variability. We can say that our selected reaming process inputs have only 55.5% influence on the cylindricity value. Let us have a look at model adequacy analysis, which was done by ANOVA. The results are in Table 5. The probability smaller than significance level (0.05) tells us that there is at least one mathematical model parameter that has an influence on investigated variable (in our case the cylindricity).

Source	DF	Sum of Squares	Mean Square	F Ratio	Prob > F
Model	5	90.27569	18.0551	7.2403	0.0005
Error	20	49.87404	2.4937		
C. Total	25	140.14974			

Table 5. Regression model ANOVA

The next step in the model analysis is the Lack of fit test. (See Table 6) This test proves us that the regression model perceives enough the observed dependency. It tests that the variability of residuals in comparison to variability of the values inside of groups. A null hypothesis tells that the residuals variability is smaller or equal to the variability of values inside of groups. The alternate hypothesis tells us that the variability of residuals is greater than inside of groups. The value of Fisher's test criterion is greater than significance level (0.05) we can assume that we do not have enough of evidence to reject null hypothesis. For us it means that our regression model is sufficient.

Source	DF	Sum of Squares	Mean Square	F Ratio	Prob > F	Max RSq
Lack of Fit	9	20.649730	2.29441	0.8636	0.5840	0.7915
Pure Error	11	29.224315	2.65676			
Total Error	20	49.874045				

Table 6. Lack of Fit test

We have sufficient and adequate regression model and now we can calculate model parameters estimation (Table 7).

Term	Estimate	Std Error	t Ratio	Prob > |t|	Lower 95 %	Upper 95 %	VIF
Intercept	6.9184069	0.309696	22.34	< 0.0001	6.2723923	7.5644215	.
x_1	0.3032367	0.349641	0.87	0.3961	-0.426102	1.0325751	1
x_2	-0.731483	0.349641	-2.09	0.0494	-1.460821	-0.002144	1
x_3	-0.343103	0.349641	-0.98	0.3382	-1.072441	0.3862352	1
$x_2 . x_1$	-1.767653	0.394787	-4.48	0.0002	-2.591163	-0.944143	1
$x_2 . x_1 . x_3$	1.2522606	0.394787	3.17	0.0048	0.4287504	2.0757709	1

Table 7. Regression model parameters estimation

From Table 7 we can see several facts. Firstly, at 5% significant level, the intercept has influence on the cylindricity value. This represents all other factors that were considered as fixed or neglected, or used factors intervals, other uncalculated errors etc. Next, other factors that have statistical influence on the cylindricity value are terms x_1 (cutting speed), x_2 (feed), x_3 (reverse feed), combination of cutting speed and feed and at the end is the combination of cutting speed, feed and reverse feed. What is interesting, the eccentricity of the reamer does not have influence on the cylindricity as independent factor neither as in combination with other factor. When we have a closer look at percentage influence of factors on the hole cylindricity, we can see that the intercept has 66%. The interaction of cutting speed and feed 13%, interaction of cutting speed, feed and reverse feed 9%, cutting speed has 3% of influence, feed influences the cylindricity from 6% and the reverse feed has 3% influence on the cylindricity.

At the end we have the prediction model of reamed hole cylindricity converted to the real factors value:

$$Cyl = 207.519 \times f + 0.865 \times v_c + 2.983 \times v_f - 1.187 \times f \times v_c - 4.174 \times f \times v_f - 1.667 \times 10^{-2} \times v_c \times v_f + 2.322 \times 10^{-2} \times f \times v_c \times v_f - 143.628 \quad (1)$$

For maximal complexity of regression analysis and selected regression model validation, it is necessary to evaluate residuals between measured and predicted values for their distribution and autocorrelation. The autocorrelation of residuals was tested using Durbin-Watson test and it showed us that there is no autocorrelation of residuals. The normal distribution of residuals we tested by Shapiro-Wilk test and it showed positive result for us, too. The model residuals have normal distribution. By those two tests, we approve that the model is statistically and numerically correct.

From analysis of regression model we can say that cutting speed as independent factor does not have significant influence (at statistical significance level 5%) on the reamed hole cylindricity. But, it is significantly involved in several combinations with other factors, see Fig. 3.

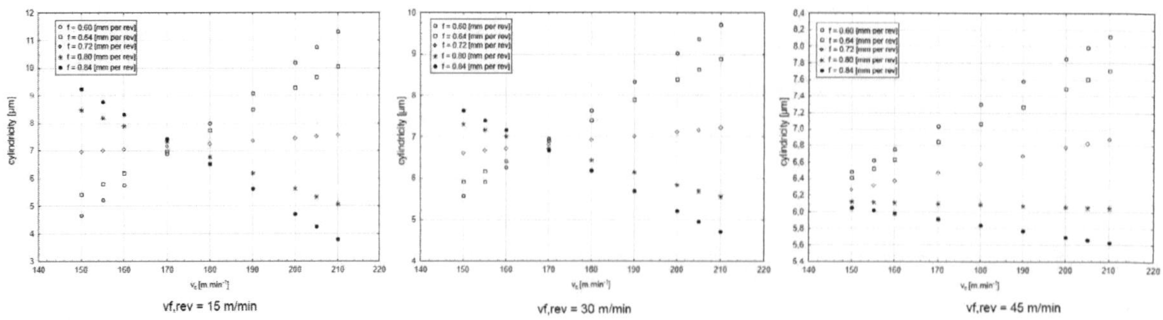

Fig. 3. Influence of cutting speed on cylindricity

The feed rate has its own influence as an independent factor with portion of 6%. It is also involved in interactions with cutting speed and reverse feed rate. Graphical representation of this influence is in Fig. 4.

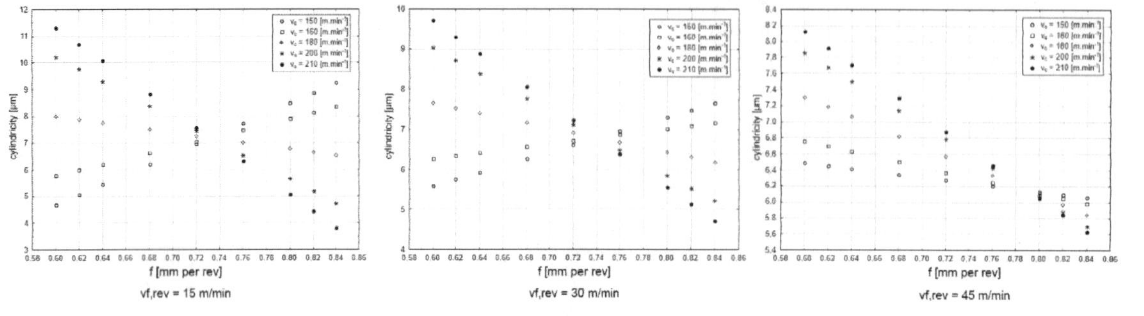

Fig. 4. Influence of feed rate on cylidricity

3.2. *Surface roughness Ra*

Measured values of surface roughness reached very low deviation for all 26 samples. Machined surface of reamed hole is possible to consider as homogeneous. Surface roughness Ra was from 0.2 to 0.6 µm. These phenomena can be explained by material of cutting tool. High cutting speeds, high feed rates and internal cooling are suitable for cermet reamer. Under these conditions the reamer works more reliably.

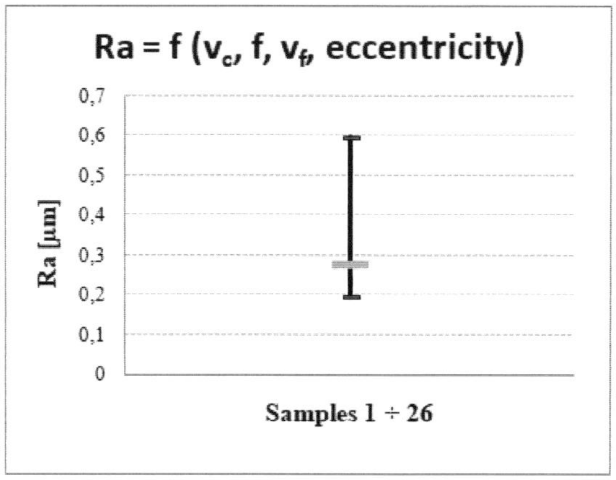

Fig. 5. Influence of tested factors on surface roughness Ra

3.3. Real diameter of reamed hole

Previous experiment did not prove the influence of eccentricity on cylindricity and surface roughness of reamed hole. So the new set of experiments was done and its conditions are in the Table 8. There are also values of real diameter which was measured at the top (D_up) and at the bottom (D_down) of the reamed hole. These values brightly show, that eccentricity of the tool tip does not influence quality of reamed hole. Eccentricity has influence mainly on tool wear and on position error. Clamping system is not perfectly stiff and overhanging of reamer is too big. If the tool was shorter (lower overhanging), eccentricity would influence measured parameters.

O.Nr.	Clamping system	Eccentricity [µm]	Measured diameter [mm]	
			D_up	D_down
1.	Compensation tool holder	4	12,0168	12,0179
2.	Compensation tool holder	12	12,0166	12,0178
3.	Compensation tool holder	26	12,0169	12,0177
4.	Hydraulic tool holder	26	12,0168	12,0179

Table 8. Influence of tool tip eccentricity on real diameter of reamed hole

4. Conclusion

This article dealt with influence of four factors (cutting speed, feed rate, reverse feed and eccentricity) on cylindricty, surface roughness and real diameter of reamed hole. The reamed hole was blind and was made by six blade reamer with brazed cermet cutting edges. The tool wear was controlled and was not take into account. The experiment proved that:

- the most significant component is constant. This constant includes everything what was not take into account. It means, for example, wrong range of cutting parameters; something what was changing during the experiment and we did not know about it; something what we could not influence, etc.
- main influence on cylindricity has feed, interaction of feed rate and cutting speed and interaction between cutting speed, feed rate and reverse feed. The higher feed rate and cutting speed the lower cylindricity.
- experimental factors had not any influence on surface roughness Ra. Cermet reamer proved produce homogeneous surface. It is because of cutting material. Cermet is called "queen" of finished surfaces.
- eccentricity did not influence the real diameter of reamed hole. The main reason is stiffness of clamping system and overhanging of the tool. At the first contact of the tool with the hole (chamfered part), the reamer was led by drilled hole. If the reamer was shorter (lower overhanging), we would suppose the bigger influence of eccentricity on real diameter. Eccentricity has influence mainly on tool wear. The higher eccentricity the lower tool life.

This experiment brightly proved that many factors has to be controlled before experiment. Statistical methods design of experiments can help us to find parameters which can influence the response and also save out time and money. On the other hand we have to know very well basic data, to understand the way of DoE, to know as much as possible about relations between dependent and independent variable, to set up right range of factors, etc. If we neglect something we will know it at value of constant. If the constant is significant we can start again. However, every mistake which we will do it, probably we will never repeat it and this mistake will move us forward.

Future plans are based on results presented in this paper. So it is necessary to choose right factors and their range. After that, we will repeat the experiment and evaluate the data. Hopefully, we will be able to express the influence of input parameters on the response.

5. Acknowledgement

The present contribution has been prepared under project LO1502 'Development of the Regional Technological Institute'under the auspices of the National Sustainability Programme I of the Ministry of Education of the Czech Republic aimed to support research, experimental development and innovation.

6. References

[1] Fulemova, J. & Rehor, J. (2016). Reaming of very precise and deep holes with cermet reamer. 27th DAAAM International Symposium on Intelligent Manufacturing and Automation 2016, Volume 27, Issue 1, 2016, Pages 275-282

[2] Klaic, M.; Staroveski, T. & Udiljak, T. (2014). Tool Wear Classification Using Decision Treesin Stone Drilling Applications: A Preliminary Study. 24th DAAAM International Symposium on Intelligent Manufacturing and Automation 2014. Volume 69, Pages 1326 - 1335. DOI: 10.1016/j.proeng.2014.03.125.

[3] Roub, J. (2017). Influence of selected technological factors of machining to surface integrity of precise holes - cutting conditions during reaming. Plzen. Diploma thesis. ZCU v Plzni.

[4] Bhattacharyya, O.; Kapoor, S.G. & Devor, R. E. (2006). Mechanistic model for the reaming process with emphasis on process faults. International Journal of Machine Tools and Manufacture. 2006, 46(7-8), 836-846. DOI: 10.1016/j.ijmachtools.2005.07.022

[5] Hauer, T.; Haydn, M. & Abele, M. (2012). Influence of a diagonal pre-drilled hole on hole quality during the reaming process using multiblade tools. Journal of the Brazilian Society of Mechanical Sciences and Engineering. Special issue 2, 2012, Vol. XXXIV, pp. 569 - 573

[6] Towfighian, S.; Behdian, K.; Papini, M.; Saghir, Z.; Zalzal, P. & Beer, J. (2006). Finite element modeling of low speed reaming vibrations with reamer geometry modifications. DOI: 10.1007/s10845-007-0038-4.

[7] Bezerra, A. A.; Machado, A. R.; Souza, A. M. Jr. & Ezugwu, E. O. (2000). Effects of machining parameters when reaming aluminium–silicon (SAE 322) alloy . DOI: 10.1016/S0924-0136(01)00561-1.

[8] Müller, P.; Genta, G.; Barbato, G.; Chiffre, L. & Levi, R. (2012) . Reaming process improvement and control: An application of statistical engineering. DOI: 10.1016/j.cirpj.2012.07.005.

DOI: 10.2507/28th.daaam.proceedings.047

EFFECT OF VENEER COMPOSITION ON THE STIFFNESS AND STRESS OF LAMINATED WOOD

Seid Hajdarevic, Murco Obucina & Elmedin Mesic

This Publication has to be referred as: Hajdarevic, S[eid]; Obucina, M[urco] & Mesic, E[lmedin] (2017). Effect of Veneer Composition on the Stiffness and Stress of Laminated Wood, Proceedings of the 28th DAAAM International Symposium, pp.0341-0350, B. Katalinic (Ed.), Published by DAAAM International, ISBN 978-3-902734-11-2, ISSN 1726-9679, Vienna, Austria
DOI: 10.2507/28th.daaam.proceedings.047

Abstract

The stiffness and stress analysis of beech wood laminated straight and curved cantilever beam is performed in the paper. Analysis was conducted layer by layer on the laminated beams with different veneer compositions. Finite element analysis is carried out with a '2D linear elastic model' for orthotropic materials. Analytic stiffness and stress evaluation is carried out by a 'transformed cross section method'. The influence of the veneer fiber orientation, veneer position, and thickness on the mechanical properties of laminated wood and the possibilities of both used methods for design of different shape of laminated wood have been analyzed. The results revealed that the laminated wood beam is stiffer when the outside veneers have longitudinal orientation in span direction or the thickness of outside veneers with longitudinal orientation is greater and that longitudinally oriented veneers sustain larger stress than veneers with tangential orientation. The results obtained by both used methods confirm that it can be used to achieve behaviour analysis of laminated wood in rectangular form. Curved forms require more complex analysis, but the methods can be used to achieve approximate data. The numerical in-plane stress analysis that applied to curved beam is not capable to calculating radial (interlaminar) stress.

Keywords: laminated wood; veneer composition; stiffness; stress; FEM; transformed cross section method

1. Introduction

Laminated wood composites have wide range of application in furniture industry due to their specific mechanical and physical properties. Today, plywood and curved veneer products e.g. plywood shell are the basic elements of a certain type of furniture constructions and there is considerable interest in the development of new complex forms. A better understanding of effect of veneer parameters on mechanical properties of laminated element can lead to improvement of quality of the laminated wood products.

Basic concepts of laminated wood analysis are introduced by analyzing a symmetric three-layer orthotropic laminate for the plane stress condition. Analysis of laminates is based on knowledge of the elastic parameters of individual laminae and component organization [1]. More extended analytic model can be finds in reference [2]. The model allows determining the stress condition of curved beam due bending. Numerical methods, such as the 'finite element method', are applicable and effective for the analysis of laminated wood. The limitations of numerical analysis depend on the complexity of numerical models and it is related to the simplifying assumptions that are introduced [3].

Plywood and curved veneer products are a balanced construction made of veneer sheets. The properties of the laminated product depend on parameters of production process and properties of individual veneers. Authors are focused on the determination of stresses and deformations develop during the manufacture veneer products [4], the analysis of shape stability with different fiber orientations and moisture content variations [5] and the impact of veneer layouts on strength of laminated wood [6]. Veneer properties are considered as design tool which has effect on the behavior of laminated wood structure [7, 8].

2. Research objective and methodology

The aim of this study was to determine stiffness and stress of the laminated wood structure, calculated by different methods and to compare the results in order to determine the effects of method, assumptions and simplifications used in the calculation of a structure's stiffness and stress. The objective was to make transparent what results are to be expected, depending on what approach to structural design was taken.

This study employed a numerical method (FEM) and the transformed cross section method for the analysis of the stiffness and normal stress of the straight and curved cantilever beams. The effects of the veneer composition on the behavior of laminate wood beams used in this study were determined in order to describe the influence of the veneer fiber orientation, veneer position, and thickness on the stiffness and stress distribution of laminated wood constructions.

2.1. Mathematical model

The equation of momentum balance, expressed in the Cartesian tensor notation [9]

$$\int_S \sigma_{ij} n_j dS + \int_V f_i dV = 0 \tag{1}$$

and of the constitutive relation for the elastic material

$$\sigma_{ij} = C_{ijkl}\varepsilon_{kl} = \frac{1}{2}C_{ijkl}\left(\frac{\partial u_k}{\partial x_l} + \frac{\partial u_l}{\partial x_k}\right) \tag{2}$$

describe the stress and strain of a loaded solid body in static equilibrium. In the equations above, x_j are Cartesian spatial coordinates, V is the volume of solution domain bounded by the surface S, σ_{ij} is the stress tensor, n_j is the outward unit normal to the surface S, f_i is the volume force, C_{ijkl} is the elastic constant tensor components, ε_{kl} is the strain tensor, and u_k represents the point displacement. Twelve non-zero orthotropic elastic constants A_{ij} are related to the Young's modulus E_i, the Poisson's ratio ν_{ij}, and the shear modulus G_{ij}.

In order to complete the mathematical model, the boundary conditions have to be specified. The surface traction f_{Si} and/or the displacement u_S at the domain boundaries are known, i.e.

$$\sigma_{ij} n_j = f_{S_i} \quad \text{and} \quad u_i = u_S. \tag{3}$$

Governing equations (1) combined with the constitutive relations (2) are solved a numerical method based on the finite element as a 2D problem with shell elements. Each layer is defined by a unique set of material properties and orientation. Calculations were performed by using the Catia software package.

2.2. Analytic evaluation - Transformed cross section method

Equations for analytic evaluation of a multilayer symmetric laminate subjected to bending are obtained by the method of transformed cross section. This method permits the conversion of the moduli of the various laminate to a single modulus of the beam and has the effect of reducing or increasing the width of laminae in relation to the reference width of the selected lamina [1]. The equation for transfer of width, Fig. 1, is:

$$b^i = b\left(\frac{E_z^i}{E_z^n}\right), \tag{4}$$

where: b – width of the face, reference lamina, E_z – modulus of elasticity in the direction of span (in the z direction). The general equation for the moment of inertia I'_x of the transformed cross section is:

$$I'_x = \frac{2b}{E_z^n} \sum_{i=1}^{n} \left[\frac{E_z^i (d^i)^3}{12} + E_z^i d^i \left(\sum_1^{i-1} d^{i-1} + \frac{d^i}{2} \right)^2 \right], \quad (5)$$

where: d – thickness of lamina.

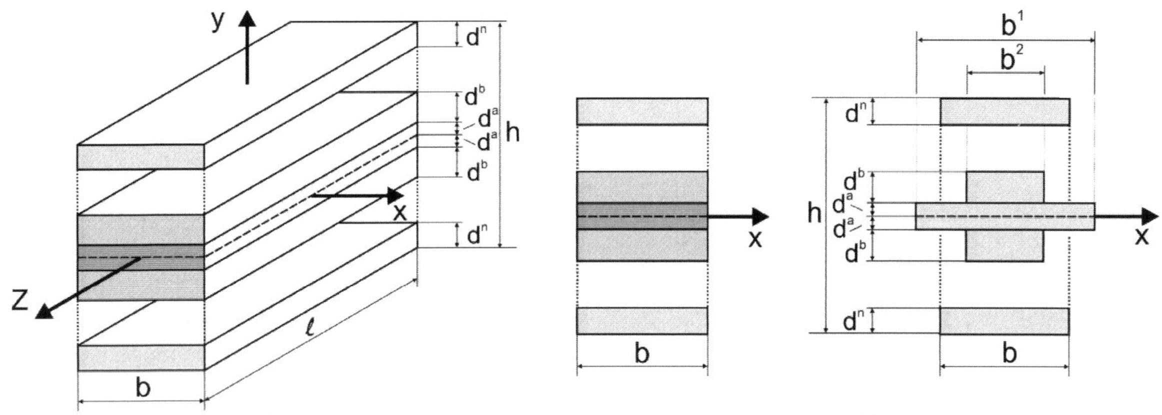

Fig. 1. a) symmetric multilayer orthotropic laminate, b) geometry and transformed cross section

2.3. Physical model

Two examples of laminated wood beam were analysed, Fig 2. The straight and curved beams are supported as a cantilever and are loaded at its end by a force of 10 N. Dimensions of rectangular beams cross-section are $\boldsymbol{b} = \mathbf{20}$ mm and $\boldsymbol{h} = \mathbf{8}$ mm. Length of the straight beams is $\boldsymbol{l_1} = \mathbf{200}$ mm and lengths of the curved beam straight sections are $\boldsymbol{l_2} = \mathbf{50}$ mm. The radius of centroidal axis of the curved cantilever beam is $\boldsymbol{R} = \mathbf{100}$ mm. The objective was to determine the vertical deflection at free end and the normal stress $\boldsymbol{\sigma_z}$ of the both beams.

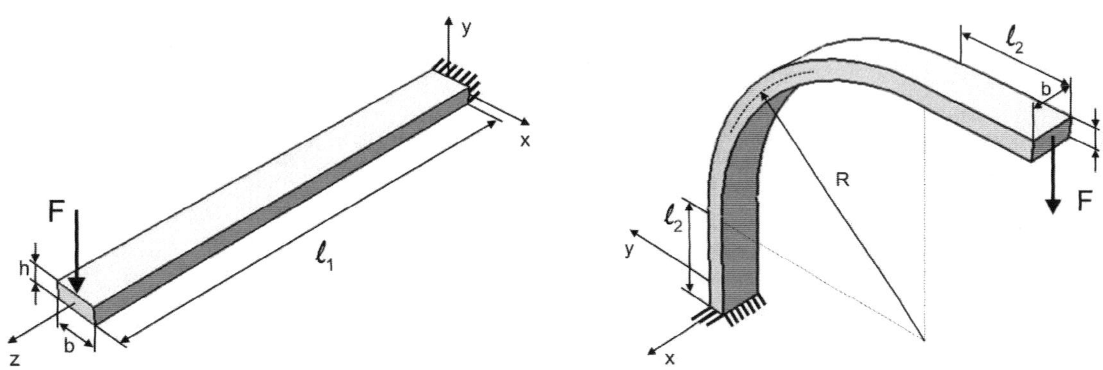

Fig. 2. Static scheme and dimensions: a) cantilever beam, b) curved cantilever beam

Calculation was carried out for beech wood (Fagus silvatica L.). Its elastic properties for wood density ρ=0,75 g/cm³ are presented in Tab. 1 [10]. Appropriate elastic properties for analysis in the LT plane were selected.

Modulus of elasticity, GPa			Rigidity modulus, GPa			Poisson´s ratio					
E_L	E_R	E_T	G_{LR}	G_{LT}	G_{RT}	ν_{LR}	ν_{LT}	ν_{RT}	ν_{TR}	ν_{RL}	ν_{TL}
13,969	2,284	1,160	1,645	1,082	0,471	0,450	0,510	0,750	0,360	0,075	0,044

Table 1. Elastic properties of beech (Fagus silvatica L.) [10]

Investigations were conducted on the laminated beams with six different veneer compositions. Orientation and thickness of veneers are shown in Fig. 3. Compositions are symmetric and consist of an even number of veneer.

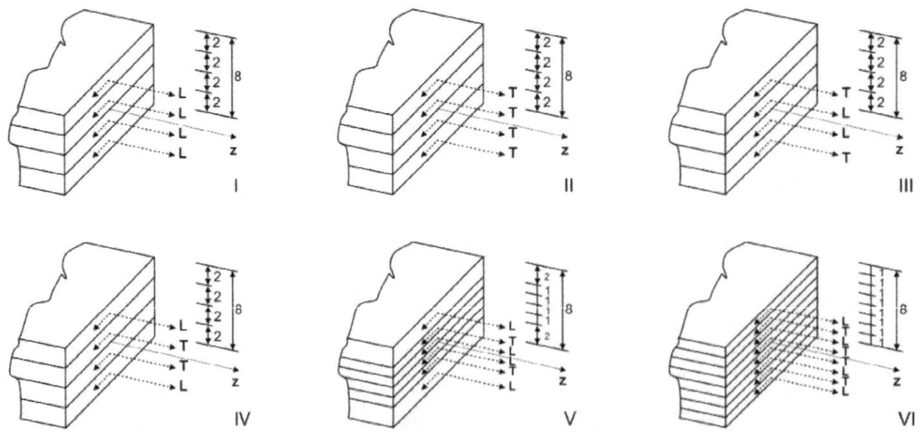

Fig. 3. Six different veneer compositions - orientation and thickness of veneers

3. Results

The results of the numerical calculation comprising vertical translation displacement magnitude of the straight cantilever beam and the curved cantilever beam are shown in Fig. 4 and Fig. 5.

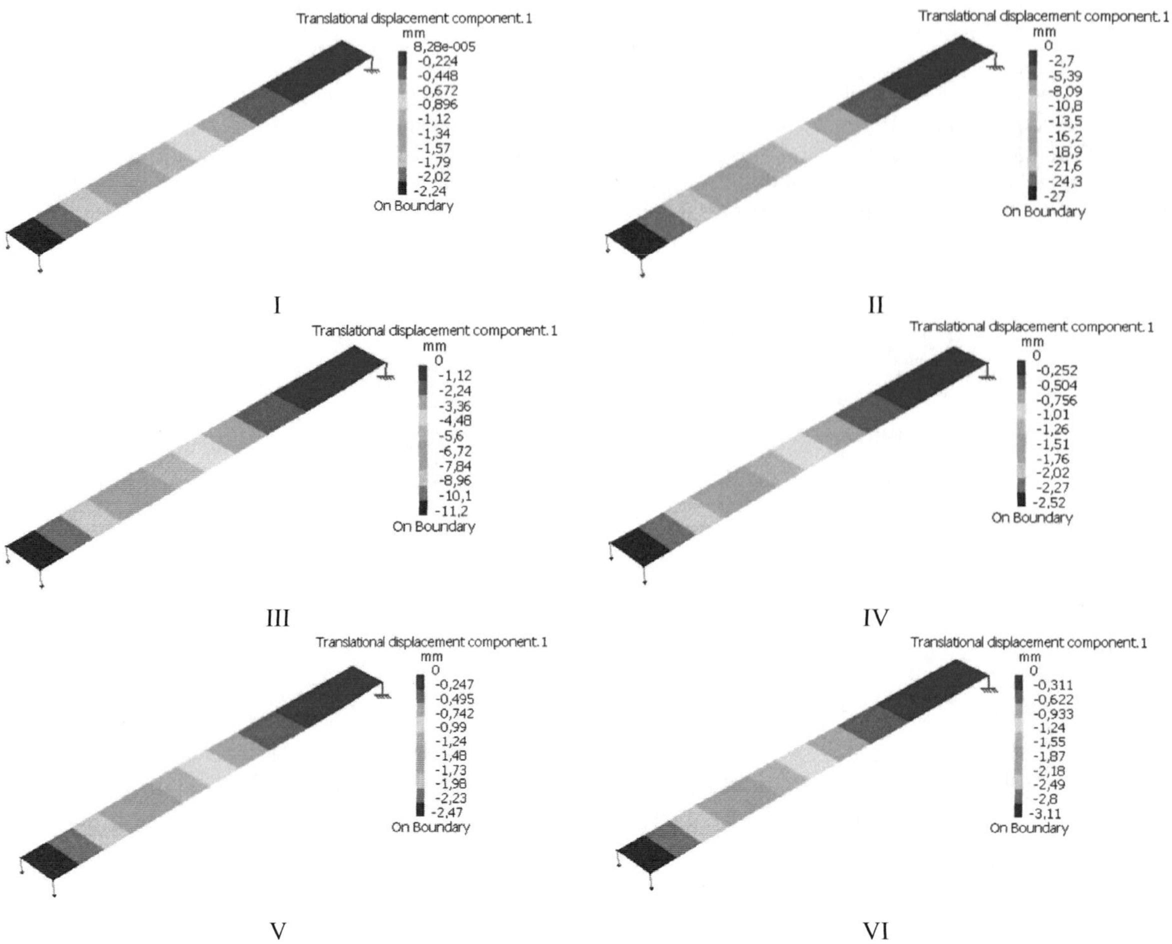

Fig. 4. Deflection straight cantilever beam for six veneer compositions

As was expected, the stiffness analysis revealed that the extreme values of displacement occur in cases of veneer composition with identical orientation of all veneers in span direction (case I and II). Effects of the elastic parameters and position of the individual veneer on the beam stiffness are evident in case III and IV. The outside veneers contribute more toward the bending stiffness than those close to the neutral plane. Beam stiffness is not significantly improved by placing veneers with longitudinally orientation in span direction near the neutral line (case V). By placing veneers with tangential orientation in span direction near the outside of a beam (case VI), beam stiffness is significantly decreased. The results of analysis of the curved beam displacement are shown identical effect of the veneer composition on the beam stiffness.

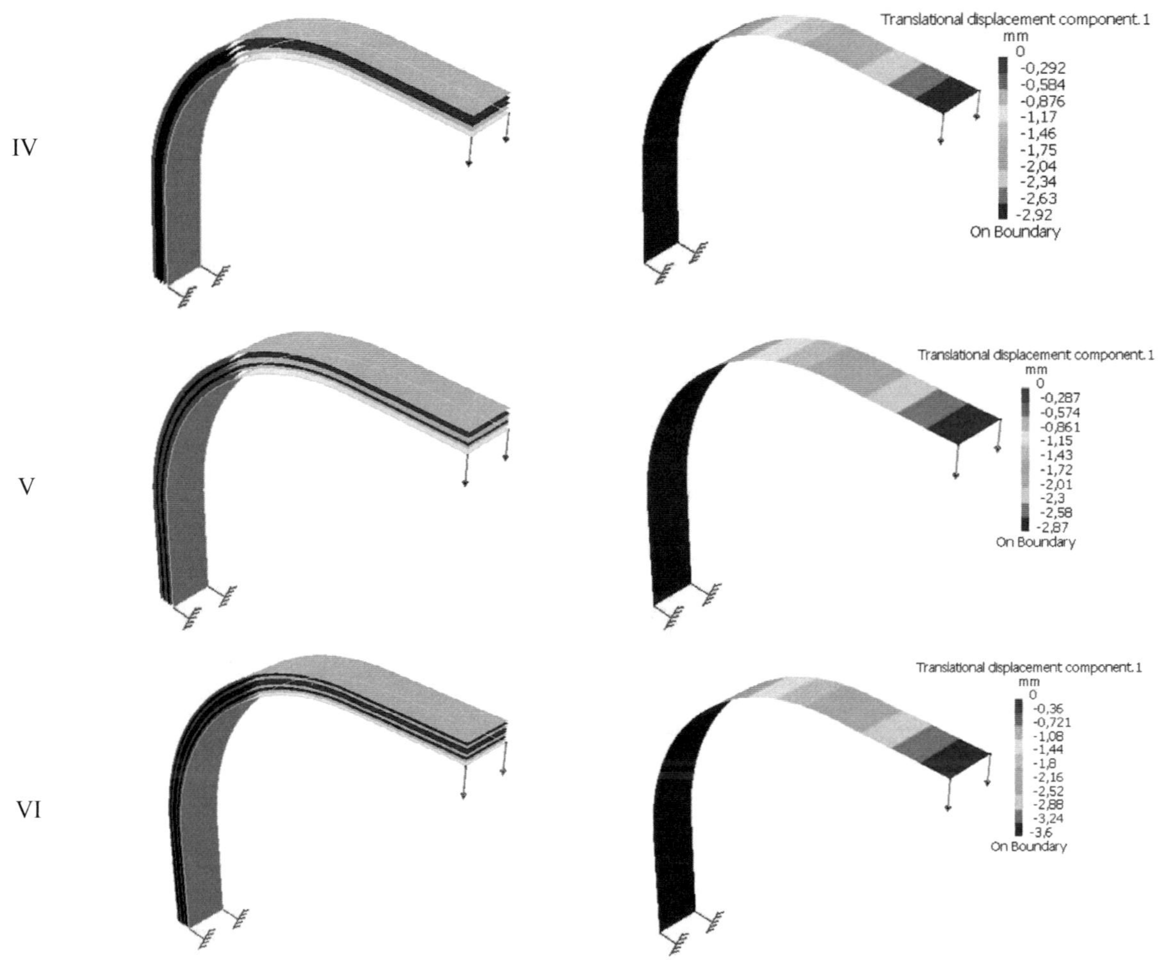

Fig. 5. Deflection of curved cantilever beam for three veneer compositions

Analitical results of the loaded end deflection of the straight cantilever beam are obtained from:

$$f_y = \frac{cFl_1^3}{E_z^n I'_x}, \tag{6}$$

where: F - concentrated load, $c = 1/3$ - the constant depends of the loading condition and l_1 - beam length. Vertical displacements of the curved cantilever beam at the loaded end are obtained using Castigliano's theorem. The strain energy due to axial and shear force are neglected and the strain energy due to bending moment in the curved section of the beam is approximated as:

$$W = \frac{1}{2}\int \frac{M_x^2 R}{E_z^n I'_x} d\theta \ \text{ for } \frac{R}{h} > 10, \tag{7}$$

where: R – radius of curvature measured to centroid of section and h - heigth of curved beam. The deflection of the curved beam in the direction of the force is found as:

$$\delta = \frac{F}{E_z^n I'_x}\left(\frac{4}{3}l_2^3 + l_2^2 R\frac{(\pi+4)}{2} + 3l_2R^2 + \frac{\pi}{4}R^3\right). \tag{8}$$

The analytical and numerical results comprising vertical displacement for six veneer compositions of the straight cantilever beam and for three veneer compositions of the curved cantilever beam are given in Tab. 2. Differences between displacement obtained by analytical and numerical calculation are <1% for straight beam and 8.6% for curved beam.

Beam cross section bxh=20x8 mm^2	Veneer composition	Analytical result		Numerical result
		Moment of inertia I'_x, m^4	Displacement, mm	Displacement, mm
Straight cantilever beam	I	$8{,}533 \cdot 10^{-10}$	2,237	2,24
	II	$8{,}533 \cdot 10^{-10}$	26,93	27,0
	III	$2{,}031 \cdot 10^{-9}$	11,31	11,2
	IV	$7{,}544 \cdot 10^{-10}$	2,530	2,52
	V	$7{,}679 \cdot 10^{-10}$	2,485	2,47
	VI	$6{,}117 \cdot 10^{-10}$	3,120	3,11
Curved cantilever beam	IV	$7{,}544 \cdot 10^{-10}$	3,173	2,92
	V	$7{,}679 \cdot 10^{-10}$	3,117	2,87
	VI	$6{,}117 \cdot 10^{-10}$	3,913	3,60

Table 2. Vertical displacements of straight and curved cantilever beams

The results of the numerical calculation comprising the normal stress $\boldsymbol{\sigma_z}$ in the individual layers of the straight cantilever beam are shown in Fig. 6.

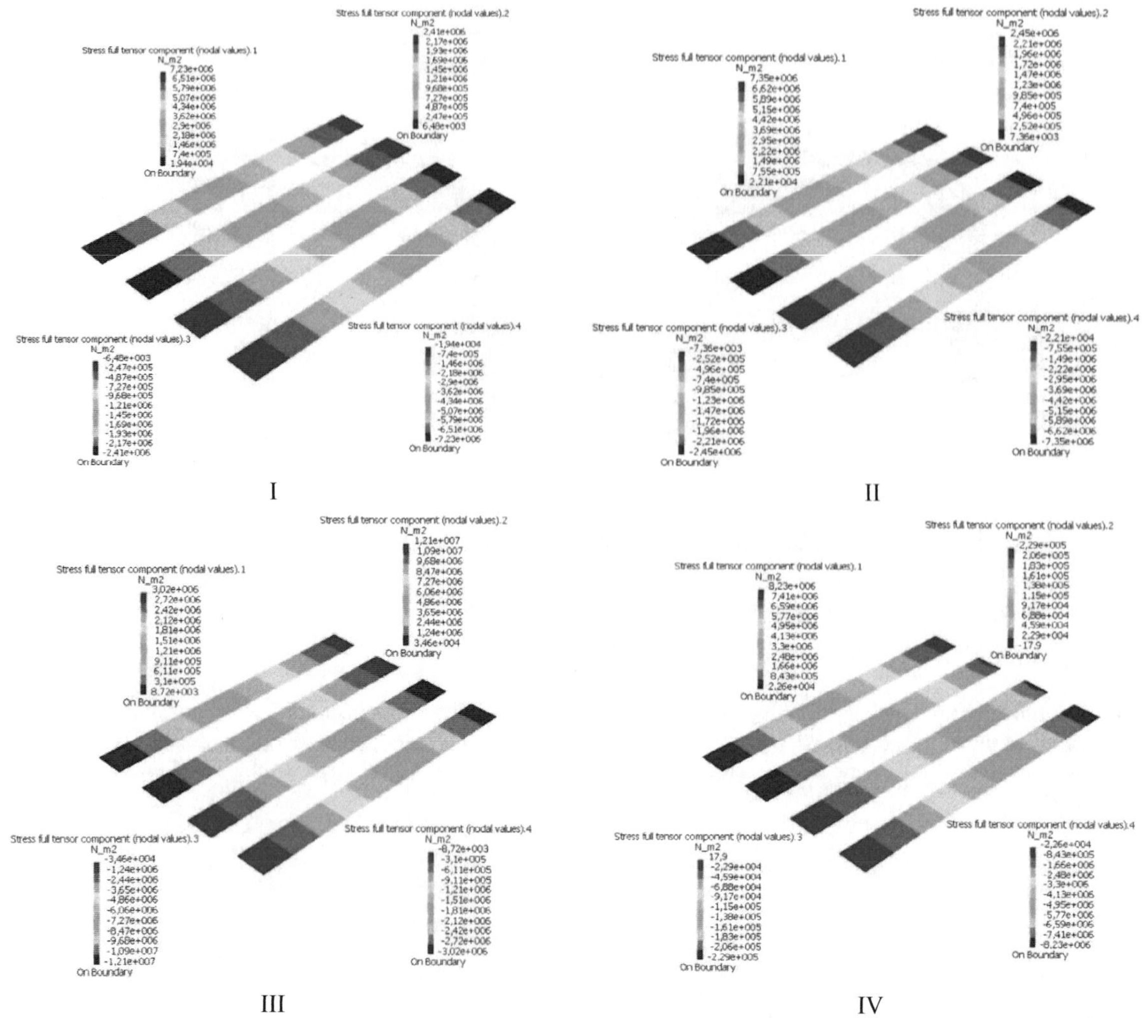

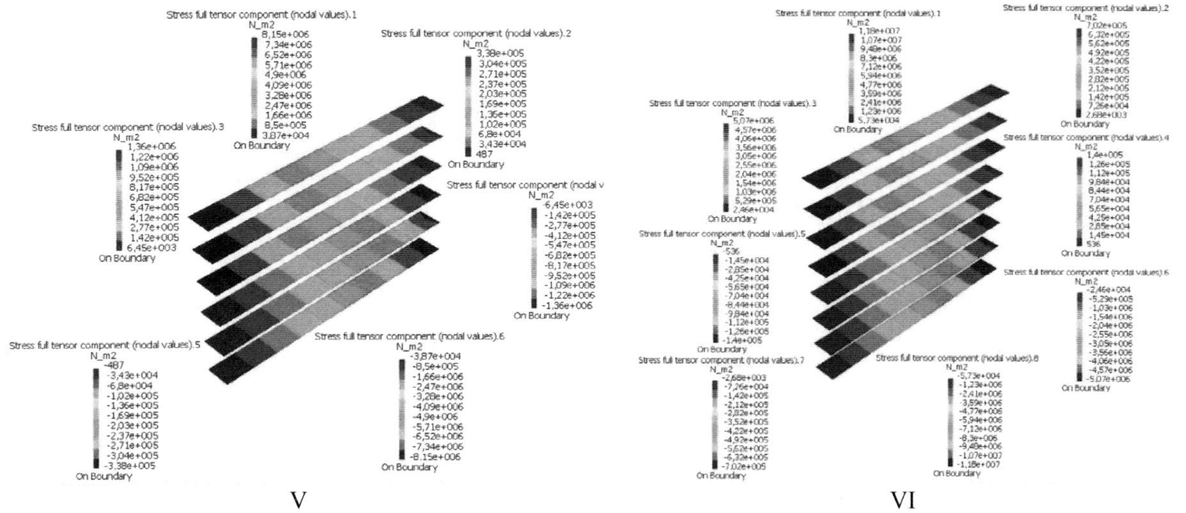

Fig. 6. Stress σ_z in centroidal plane of the individual layers of the straight cantilever beam

The analytical results of the normal stress of thestraight beam is obtained from:

$$\sigma_z = \frac{M_x \cdot y}{I'_x}\left(\frac{b'}{b}\right), \tag{9}$$

where: σ_z- normal stress, M_x - bending moment, y - predetermined distance, b' - the transformed width.

Normal stress distribution across the beam cross section ($z = l/2$) for six different veneer compositions are shown in Fig. 7. Values and distribution of the normal stress are identical for beams consisting of veneer of the same orientations (case I and II).

Veneers longitudinally orientated (high modulus of elasticity) in span direction are sustain larger stress than veneers with tangential orientation. Normal stress is significantly increased on the surfaces of the beam by reducing the thickness of the external veneers with longitudinal orientation.

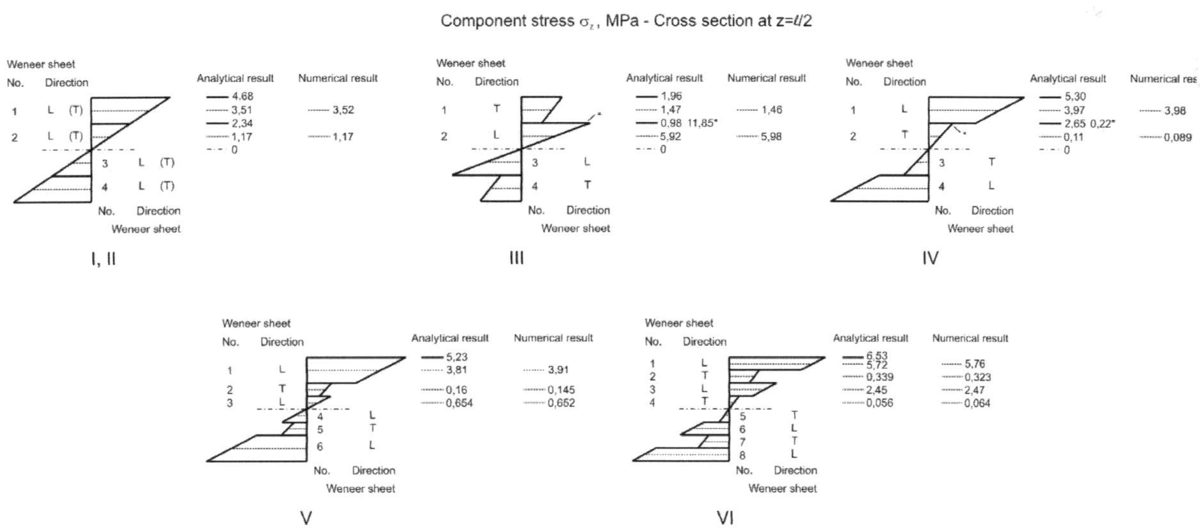

Fig. 7. Analytical and numerical results of the normal stress distribution across the beam cross section for six veneer compositions

The results of the numerical calculation comprising the normal stress $\boldsymbol{\sigma_z}$ in the individual layers of the curved cantilever beam are shown in Fig. 8.

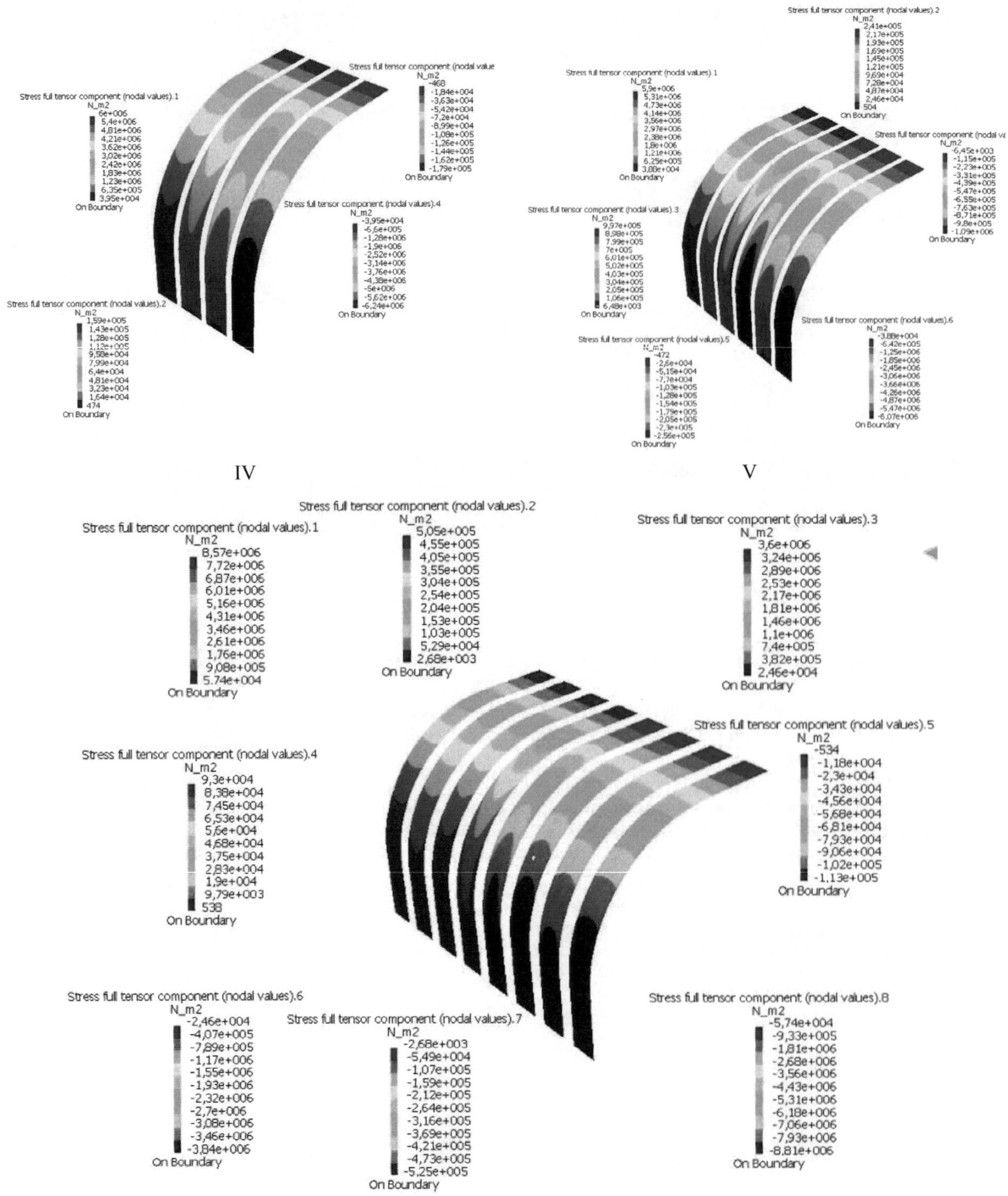

Fig. 8. Stress σ_z in centroidal curved plane of the individual layers of the curved cantilever beam for three veneer compositions

The neutral axis and the centroidal axis of a curved section of the beam are not coincident and also the stress does not vary linearly from the neutral axis [11]. An approximate estimate of the normal (circumferential) stress in curves section of beam (for $\frac{R}{h} > 8$) is found as:

$$e \approx \frac{I}{A r_c} \quad \rightarrow \sigma_\theta = \frac{M \cdot s}{I} \frac{r_c}{r} \tag{10}$$

where: r_c - radius of centroidal axis, r – radius of the elemental fiber from center of curvature, s - the distance from the centroidal axis. The sign convention used is: M is positive if it acts to straighten on the beam curved section and the distance s is positive inwards to the center of curvature.

The computation is based on a transformed section (I'_x). Approximate value of the actual normal stress in the curved section of the beam consisting of veneers of diferent properties is found as $\sigma_z = \sigma_\theta \left(\frac{b'}{b}\right)$. The analytical results comprising normal stress for three veneer compositions of the curved section cantilever beam are given in Tab. 3. The applied bending moment is $M = -F(l_2 + R \sin \theta)$ and $\theta = \pi/2$.

The analytical and numerical results show that the compressive stress at the inner layer (veneer 5, 6, 7 and 8) is greater that tensile stress at the symmetrical outer layer (veneer 4, 3, 2 and 1).

4 veneer sheets (case IV)			6 veneer sheets (case V)			8 veneer sheets (case VI)		
Veneer number	s, m	σ_z, MPa	Veneer number	s, m	σ_z, MPa	Veneer number	s, m	σ_z, MPa
1	-0,004	7,647	1	-0,004	7,513	1	-0,004	9,431
	-0,003	5,791		-0,003	5,689		-0,0035	8,292
2	-0,001	0,163	2	-0,0015	0,240	2	-0,0025	0,496
3	0,001	-0,167	3	-0,0005	0,972	3	-0,0015	3,624
4	0,003	-6,149	4	0,0005	-0,982	4	-0,0005	0,101
	0,004	-8,285	5	0,0015	-0,247	5	0,0005	-0,102
			6	0,003	-6,041	6	0,0015	-3,734
				0,004	-8,139	7	0,0025	-0,522
						8	0,0035	-8,894
							0,004	-10,217

Table 3. Normal stress for three veneer compositions of the curved section of cantilever beam

4. Conclusion

The influence of the veneer composition on the mechanical properties of rectangular and curved form of laminated wood has been investigated by finite element method and transformed cross section method. The possibilities of both used methods for design of different shape of laminated wood has been analysed.

The results show that the veneer fiber orientation, veneer position, and thickness have a considerable impact on the stiffness and normal stress distribution of laminated wood straight and curved beam. It is evident from the analysis that the laminated wood beam is stiffer when the outside veneers have longitudinal orientation in span direction or the thickness of outside veneers with longitudinal orientation is greater. The analysis of stress distribution shows that longitudinally oriented veneers sustain larger stress than veneers with tangential orientation. The normal (circumferential) stress of the inner zone of the curved beam is greater than the normal stress at outer zone.

Similarity between numerical and analytical results of straight beam analysis allows for the conclusion that the research models were designed correctly. The differences between results of curved beam are as a consequence of analytical method that was used, and assumptions and simplifications that were adopted. The results obtained by the transformed cross section method confirm that it can be used to achieve behavior analysis of laminated wood in rectangular form. Curved forms require more complex analysis, but the method can be used to achieve approximate data in early design phase. The research revealed that the numerical procedure used in the study, provides a convenient method of obtaining the information needed for determining basic mechanical properties of laminated wood. The in-plane stress analysis that was applied is not capable to calculating the radial (interlaminar) stresses that are significant factor and the main failure cause in curved forms of laminated wood e.g. laminated veneer products.

In the future investigation the numerical analysis will be extended on the real model of the laminated veneer product. Simulation results will be compared with experimentally obtained results of stiffness and stress.

5. References

[1] Boding, J. B.; Jayne, A. (1993). Mechanics of wood and wood composites, Krieger publishing Company, Malabar Florida

[2] Gonzalez-Cantero, J. M.; Graciani E.; Blazquez, F. & Paris, F. (2014). Analytic evaluation of radial stress in unfolding failure of composite materials Comparison with numerical solution, ECCM-16th European conference on composite materials, Seville, Spain, 22-26 June 2014

[3] Skovajsa, M.; Sedlacek, F. (2016). Numerical Simulation of Delamination in Laminated Composite Plates Using Cohesive Elements, Proceedings of the 27th DAAAM International Symposium, Vienna, Austria, ISSN 1726-9679, ISBN 978-3-902734-08-2, Katalinic, B. (Ed.), pp. 1726-9679, Published by DAAAM International, DOI: 10.2507/27th.daaam.proceedings.089

[4] Sandberg, D. & Ormarsson, S. (2007). Numerical simulation of hot-pressed veneer products, working paper no. 2007:19, Vaxjo University, School of Technology & Design, Sweden

[5] Blomqvist, L.; Johansson, J. & Sandberg, D. (2013). Basic knowledge of wood properties for improved performance of laminate veneer products, Pro Ligno, Vol.9 No 4 2013, pp. 549-556, ISSN-L 184-4737

[6] Jakimovska Popovska, V.; Iliev, B. & Zlateski, G. (2017). Impact of veneer Layouts on Plywood Tensile Strength, Drvna industrija, 68 (2) pp. 153-161, doi: 10.5552/drind.2017.1634

[7] Vratuša, S.; Kariž, M.; Ayrilmis, N. & Kitek Kuzman, M. (2017). Finite element simulations of the loading and deformation of plywood seat shells, Eur. J. Wood Prod. DOI 10.1007/s00107-017-1160-4

[8] Nestorović, B.; Skakić, D. & Grbac I. (2011). Determining the Characteristics of Composite Structure Laminae by Optical 3D Measurement of Deformation with Numerical Analysis, Drvna industrija, 62 (3) pp. 193-200, doi: 10.5552/drind.2011.1103

[9] Chung, T. J. (1996). Applied continuum mechanics, Cambridge University Press, Cambridge, 1996

[10] Smardzewski, J. (2008). Effect of wood species and glue type on contact stresses in a mortise and tenon joint, Journal of Mechanical Engineering Science, 222(12) 2008, pp. 2293-2299.

[11] Young, W. C. & Budynas, R. G., (2001). Roark's Formulas for Stress and Strain, McGraw-Hill Professional, ISBN 0-07-072542-X

DOI: 10.2507/28th.daaam.proceedings.048

EUROPEAN SWAPTION BY MONTE CARLO ESTIMATION ON MODERN INTEL ARCHITECTURES

Eugen Mudnic, Stipe Celar, Zeljko Seremet, Filipa Matic

This Publication has to be referred as: Mudnic, E[ugen]; Celar, S[tipe]; Seremet, Z[eljko] & Matic, F[ilipa] (2017). European Swaption by Monte Carlo Estimation on Modern Intel Architectures, Proceedings of the 28th DAAAM International Symposium, pp.0351-0356, B. Katalinic (Ed.), Published by DAAAM International, ISBN 978-3-902734-11-2, ISSN 1726-9679, Vienna, Austria
DOI: 10.2507/28th.daaam.proceedings.048

Abstract

Modern computing resources provide substantial processing power but require that the programmatic implementation of used numerical algorithms is fine-tuned to the target CPU architecture. The main performance gains could be obtained minimizing cache traffic and by efficient using of multiple levels of CPU parallel units. Monte Carlo estimation of LIBOR models is an example of challenging computational finance/computer programming problem. This paper compares the efficiency of multicore (Intel Core Haswell) and manycore platforms (Intel Xeon Phi Knight Corner and Intel Xeon Phi Knight Landing) for the calculation European LIBOR-based swaptions. Target optimized programmatic implementations of LIBOR calculation using the Intel Cilk Plus and OpenMP standards are presented and benchmarked. Results show that Intel Xeon Phi Knight Landing evaluate payoff European LIBOR-based swaptions faster, economically profitable and more energy efficient than the Intel Xeon Phi Knight Corner coprocessor and Intel Core Haswell.

Keywords: Libor; Monte Carlo simulation; Intel Cilk Plus; OpenMP; Intel Knights Landing

1. Introduction

The London interbank offered rate, also called LIBOR [1] is one of the main instruments in the debt market and it is the basis for many derived interest rate instruments. Along with numerous analytical methods, "brute force" Monte Carlo approach could be used to simulate prices of LIBOR-based swaptions [1]. Numerous papers [2][3][4][5][6][7][8][9][10][11] confirm that the Monte Carlo prices of LIBOR-based swaptions could not be calculated in a reasonable time without high level of parallelization. Currently Intel processors are evolving from multicore (Intel Core) toward manycore (Intel Xeon Phi) architectures, offering massive parallelization capabilities even on single CPU. Although the declared theoretical peak power (GFLOPs) show significant performance gain for manycore processors, the economic performance-per-price and performance-per-watt are still questionable. The purpose of the present work is therefore to first efficiently tune parallelization of the LIBOR Monte Carlo calculation both on multicore and manycore CPUs, and then to compare both their execution times and main economic performance.

There are many LIBOR test cases with different Monte Carlo settings [12][13][14]. In this paper we used algorithm [15] based on an original code by Paul Glasserman and Xiaoliang Zhao [12][16][18] with subsequent modifications by

Mike Giles [13][19]. The algorithm requires repeated simulations of chance that lend themselves well to parallel (thread) processing and vectorization on multicore and manycore Intel CPUs.

For writing the parallelization code, we used alternatively Intel Cilk Plus and OpenMP frameworks and Intel vectorization C++ language extensions as available using *the Intel Compiler v17.0*. The used algorithm was optimized to the target CPU architecture regarding to differences in available instruction sets (AVX 2.0, IMCI, and AVX-512) between target architectures: Intel Xeon Haswell (HSW), Intel Xeon Knight Corner (KNC), and Intel Xeon Knight Landing (KNL). We evaluated the economic profitability and the energy efficiency using computation time in representative benchmarks. On these benchmarks, we demonstrate best performance and also economic gains on the KNL.

The paper is organized as follows. Section 2 describes used algorithm. Section 3 describes implementation details. Section 4 evaluates showing performance results on real benchmarks. Section 5 presents our conclusion and future work.

2. Used algorithm

Monte Carlo simulation is very usefully technique in capacity planning [17] and to compute the correct prices for financial options in computational finance especially for the calculation of LIBOR. In mathematical finance, we have two classic methods used to compute LIBOR in a Monte Carlo setting: the pathwise sensitivities and the Likelihood Ratio Method. In this section, we brief review of LIBOR market models based on a finite set of maturities, as developed by Jamshidian [18].

The forward LIBOR rate at time t for the accrual period $[T_i; T_{i+1}]$, with $t \leq T_i$, is

$$L_i(t) = \frac{1}{\delta}\left(\frac{B_i(t)}{B_{i+1}(t)} - 1\right), i = 1, \dots, N. \quad (1)$$

It is at times notational convenient to extend the definition of L_i beyond the ith tenor date; we do so by setting $L_i(t) = L_i(T_i)$ for $t < T_i$. At a tenor date T_i, the price of any bond B_n, with n > i, that has not yet matured is given by

$$B_n(T_i) = \prod_{j=1}^{n-1} \frac{1}{1+\delta L_j(T_i)} \quad (2)$$

more generally, at an arbitrary time $t < T_n$, we have

$$B_n(T_i) = B_{\eta(t)}(t) \prod_{j=\eta(t)}^{n-1} \frac{1}{1+\delta L_j(t)} \quad (3)$$

The time-t value $C(t)$ and time-T value $C(T)$ of a derivative security (that can be replicated by trading in the basic bonds) are related by the pricing rule

$$C(t) = B^*(t)E\left[\frac{C(T)}{B^*(t)} | \mathcal{F}_t\right] \quad (4)$$

where $\{\mathcal{F}, t \geq 0\}$ is the filtration generated by the Brownian motion.
From (4) and the definition of B^*, it follows that its price at time 0 is

$$C_n(0) = B_1(0)\delta E\left[(L_n(T_n)) - K)^+ \prod_{i=1}^{n} \frac{C(T)}{1+\delta L_i(T_n)}\right] \quad (5)$$

See [16] for more details. The algorithm then proceeds as follows:

1. calculate Monte carlo paths,
2. calculate path for swaption kernel,
3. calculates the forward LIBOR rates over the time interval,
4. calculates the estimated strike price for the swaption,
5. discounts the forward LIBOR rate.

Each option in alghoritm step could be parellized using multiple threads or vectorization.

3. Implementation Details

Intel compiler v17.0 offers support for different multithreading parallelization frameworks. We have used alternatively Intel Cilk Plus and OpenMP since they use very similar code extensions. Each of the parallel programming frameworks has their own internal implementation and it is expected different fit to problem we are solving.

3.1. *Intel Cilk Plus*

Intel Cilk Plus [20] [21] is a powerful parallelization and vectorization framework that can effectively parallelize complex problems. In this framework, the programmer specifies the components of the application that can been run in parallel, and the runtime library takes care of assigning computing resources (cores) to the parallel tasks. This is done via an internal scheduling mechanism based on "work stealing" to distribute parallel work-items among "workers". Workers are a concept in Intel Cilk Plus similar to threads in other frameworks (for example, OpenMP). Because of the simplicity of the API and high degree of behind-the-scenes automation, Intel Cilk Plus can dramatically reduce the development workload and time while providing great performance. In the framework, we have only three keywords in Intel Cilk Plus: cilk_for, cilk_spawn and cilk_sync. Namely, in our study, we used cilk_for that enables programmers to parallelize C/C++ for loops.

However, note that this ease of use of Intel Cilk Plus functionality does not release the programmer from following the necessary precautions that apply to all parallelized loops. Avoiding issues such as race conditions is still the programmer's responsibility. In this case, Intel Cilk Plus provides C++ templates referred to as reducers in order to eliminate race conditions in parallel programs with certain patterns (see, e.g., [22]) [23]. In fact, Intel Cilk Plus is the vectorization and multi-threading capabilities are separable, and often the vectorization capabilities are useful in conjunction OpenMP. In our study, Intel Cilk Plus has particularly low cost "spawning", so it's a particularly good fit for highly recursive code where the base cases cannot be forced to be large chunks.

3.2. *OpenMP*

The OpenMP programming model provides an API with set of compiler directives, function calls, and environment variables that instruct the compiler how and where to use parallelism in the application. The directive based approach makes it possible to write sequentially consistent codes for easier maintenance. The well-known advantage of OpenMP is its global view of application memory address space that allows relatively fast development of parallel applications. OpenMP is supported by almost all major compilers [24].

The OpenMP can be used effectively for the parallelization of the originally sequential code because this requires only minimal code changes, and thus minimizing the logical mistakes of the programmer. The programmer does not create threads directly within the application code, but just inserts OpenMP directives for the compiler, which generates the code for threads during the translation. The advantage of this approach is obvious; in most cases, it is sufficient just to insert directives to indicate which section of code will running parallel. The original code does not need to be changed. The functionality of the originally sequenced code is clearly visible and the parallelization code is separable. Obviously, this approach is especially useful when it is necessary to parallelize the existing code.

In our study, we have used the following OpenMP directives valid for C/C++ programming languages:

- **#pragma omp parallel { }** – defines a section of the parallel calculation,
- **#pragma omp for schedule (dynamic)** – defines a parallel loop and each iteration will be assigned to threads dynamically.

The pragmas control how the program works. By the C/C++ standards, even if the compiler does not support pragmas, the program will run correctly, however, without parallelism. Therefore an adding the OpenMP directives can be done very safely. [25]

3.3. *Vectorization*

Used CPUs have different instruction vectorization sets. The HSW CPUs provide 256-bit vector registers and AVX/AVX2 (Advanced Vector Extensions) instruction sets. KNC provides 512-bit vector registers and IMCI instruction sets. The KNL nodes provide 512-bit vector registers and new AVX-512 instruction set. Intel compiler C/C++ language extension for array notations simplifies the code vectorization since it translates array notation any of targeted CPUs / instruction sets. However to get the most performance out of these processors, users need to take in account of CPU differences when applying vectorization instructions in their code, where the vector register size is the most significant parameter. In order to enable the compiler to generate efficient code for the assembly step, we have added *__restrict* and *const* keywords to our computational kernels.

This is rather straightforward to do as the computational kernels are implemented using simple data structures and abstractions are only build on top of that layer. We have used the vectorization report of the Intel performance profiling tool (Intel VTune Amplifier) to check that the compiler has indeed sufficient information to vectorize the time intensive portions of our algorithm. Note, however, that the code for the KNL is essentially identical to the optimized code for all platforms.

In this article, all tests are run on the three machines with the following specifications:

1) Intel multicore Platform Xeon Haswell-E (HSW): 6-core Intel Core i7-5930K series processor based on the Intel Xeon Haswell-E architecture with two thread per core; Frequency: 3.5 GHz; DDR4: 32GB; L1d cache: 32 kB; L1i cache: 32 kB; L2 Cache: 256 kB; L3 Cache: 15360 kB
2) Intel manycore Xeon Phi Knights Corner (KNC): a single 57-core Intel Xeon Phi coprocessor 3120A with four threads per core; Frequency: 1.1 GHz; GDDR5: 6 GB of RAM; L1d cache: 32 kB; L1i cache: 32 kB; L2 Cache: 28.5 MB
3) Intel manycore Xeon Phi Knights Landing (KNL): a single 68-core Intel(R) Xeon Phi(TM) CPU 7250 processor with four threads per core; Frequency: 1.4 GHz; DDR4: 96 GB of RAM with 16 GB MCDRAM memory at 1013.140 MHz; L1d cache: 32 kB; L1i cache: 32 kB; L2 Cache: 34 MB

The base scenario we used for most of our results sets δ=h=0.25 and N+1=40, corresponding to a ten-year term structure of quarterly rates. There are 3 possible swaptions with 5 possible maturities (lengths) per swaption and possible strike prices for the 3 swaptions. Amount price varies over time typically determined as a function of time to maturity in this example; however, it remains constant 0.2 values and number of simulations that Monte Carlo runs is 96000. The application of Libor Rate Model (LRM) to estimating delta entails knowledge of the transition density of the underlying state variables. No such density is available in forward LIBOR models, so we use a Gaussian approximation using Intel Math Kernel Library (MKL).

4. Performance Results

The algorithm requires repeated simulations of chance that lend themselves well to parallel processing and vectorization. The simulations in this example are run in the following cases:

1. Serially code without any parallelization,
2. Array Notation (AN) for vectorization using Intel specific C++ language extensions,
3. AN and Cilk implementation for task thread parallelism,
4. AN and OpenMP implementation for task thread parallelism.

To use target CPU architecture effectively, applications should effectively use VPU instructions on vector data, good locality of reference, and utilizes caches well in its core computations. However, the scheduling module (function cilkrts scheduler) of the Intel Cilk Plus runtime library is spent the majority of the CPU time on all platforms. It is beneficial to find a parallel scheduler with less overhead. For example, OpenMP framework is a better solution with less overhead and it is available for all processors (HSW, KNC, and KNL). The OpenMP standard provides similar functionality (tasking and dynamic number of threads in parallel regions), but in this application we were not able to achieve satisfactory performance results with OpenMP despite investing a greater development effort than we did with Intel Cilk Plus on HSW platform.

Table 1. shows the results for the single precision solution of the tiled. The performance is expressed in the computation time (run time). The first two trials for each data point were not taken into the statistical average.

	Intel Multi-core Platform	*Intel Many Integrated Core Xeon Phi Architecture*	
Test (Monte Carlo)	*Haswell (HSW)*	*Knights Corner (KNC)*	*Knights Landing (KNL)*
Performance time		**[ms]**	
1. Serial code	1973	25964	6837
2. AN	275	1640	439
3. AN + cilk for	47	21	6
4. AN + omp for	54	15	4
Recommended Customer Price (RCP)[1]	$594.00	$1,491.00	$2,436.00
Thermal Design Power (TDP)[1]	140 W	300 W	215 W
Theoretical Peak Performance (TPP)[1]	384 GFLOP/s	1.2 TFLOP/s	3 TFLOP/s

Table 1. Timing results for used algorithm on targets CPU architectures

On all architectures, we have achieved the same result (average discounted payoff is 49.439308) based on equation (5) and base scenario for an estimation of the valuation of a portfolio of European LIBOR-based swaptions using a Monte Carlo simulation. Timing results in table 1. show that the KNC times with serial code has 13 times more execution time, with 2.5 times more price and over 2 times more power than the HSW.

[1] *https://ark.intel.com/#@Processors*

In below formulas we have evaluated the economic profitability and the energy efficiency using executing time (run time) in representative benchmarks.

$$Cost_{HSW} = RCP*Time=594*47=27{,}918 \quad (6)$$

$$Cost_{KNC} = RCP*Time=1{,}491*15=22{,}365 \quad (7)$$

$$Cost_{KNL} = RCP*Time=2{,}436*4=9{,}774 \quad (8)$$

$$Dissipation_{HSW} = TDP*Time=140*47=6{,}580 \quad (9)$$

$$Dissipation_{KNC} = TDP*Time=300*15=4{,}500 \quad (10)$$

$$Dissipation_{KNL} = TDP*Time=215*4=860 \quad (11)$$

Results show that the KNL with OpenMP evaluates payoff European LIBOR-based swaptions significantly faster (small the computation time), more economically profitable (8) (small the Cost factor) and more energy efficient (11) (small the Dissipation factor) than nearly comparable KNC and HSW.

5. Conclusion

In this paper, we evaluate the performance of such an optimization algorithm on modern accelerators (more specifically, the Intel Multi-core Platform Xeon Haswell, the Intel Many-core Xeon Phi Knights Corner and the Intel Many-core Xeon Phi Knights Landing). We present timing results for all codes and discuss the similarities and differences between the four implementations. Optimization and parallelization for the HSW, KNC and KNL code is done using *the Intel Compiler v17.0* (vectorization) with Intel Cilk Plus and OpenMP frameworks. We've achieved same results (average discounted payoff is 49.439308) on all architectures.

This study shows that the KNL with OpenMP framework is significantly faster, more economic profitable and more energy efficient than KNC and the HSW. The results shows that the thread scheduling overhead in Intel Cilk Plus framework has a negative influence on the overall performance on all architectures, because this is done via an internal scheduling mechanism based on a "work stealing" to distribute parallel work-items among "workers". In OpenMP, we inserted directives, which defines a parallel loop and each iteration will be assigned to threads dynamically. Further, the study shows that KNC architecture has not significantly better results than HSW with regard to the price and the energy consumption.

In this way, KNL architecture have the best ways to compute, energy efficiency, economic profitability and achieve massive increases in performance with effectively used strong parallelization and vectorization (Cilk Plus and OpenMP) frameworks.

In further work, we are pursuing ways to further research/apply this, especially for newer targets that have different architectural characteristics.

6. References

[1] Ferreiro, A.M.; García, J.A.; López-Salas, J.G. & Vázquez, C. (2014). SABR/LIBOR market models: pricing and calibration for some interest rate derivatives, Appl. Math. Comput. 242, 2014, pp. 65–89.

[2] Dupire, B. (1994). "Pricing with a smile," Risk 7 (1), 1994, pp. 18–20.

[3] Derman, E. & Kani, I. (1994). Riding on a smile, Risk 7 (2), 1994, pp. 32–39.

[4] Andersen, L. B. G. & Andreasen, J. (2000). Volatility Skews and Extensions of the Libor Market Model, Journal Applied Mathematical Finance, Volume 7, 2000, pp. 1-32

[5] Merton, R. C. (1976). Option pricing when underlying stock returns are discontinuous, Journal of Financial Economics, Volume 3, Issues 1–2, January–March 1976, pp. 125-144

[6] Eberlein, E.; Keller, U. & Prause, K. (1998). New insights into smile, mispricing, and value at risk: the hyperbolic model, The Journal of Business, Volume 71, 1998, pp. 371-405

[7] Jamshidian, F. (2001). Libor market model with semimartigales, Available from: https://www.researchgate.net/publication/228750716_LIBOR_market_model_with_semimartingales, Accessed: 2017–06–04

[8] Glasserman, P. & Kou, S. (2003). The term structure of simple forward rates with jump risk, The Journal Mathematical Finance, Volume 13, Issue 3, 2003, pp. 345–444.
[9] Glasserman, P. & Merener, N. (2003). Numerical solution of jump-diffusion LIBOR market models, The Journal Finance and Stochastics, 2003, pp. 1–27.
[10] Belomestny, D. & Schoenmakers, J.G.M. (2011). A jump-diffusion Libor model and its robust calibration, Quantitative Finance, Volume 11, 2011, pp. 529–546.
[11] Eberlein, E. & Ozkan, F. (2005). The Lévy LIBOR model, The Journal Finance and Stochastics, 2005, pp. 327–348.
[12] Giles, M.B. & Glasserman, P. (2006). Smoking adjoints: fast Monte Carlo Greeks, RISK, 2006.
[13] Giles, M. (2007). Monte Carlo evaluation of sensitivities in computational finance, Tech. Rep. NA07/12, 2007.
[14] Glasserman, P. (2003). Monte Carlo Methods in Financial Engineering, Springer-Verlag, New York.
[15] Finance: Monte Carlo (2013). Available from: https://software.intel.com/en-us/code-samples/intel-c-compiler/application-domains/finance/Monte-Carlo, Accessed: 2017–06–04
[16] Glasserman, P. & Zhao, X. (2000). Arbitrage-free discretization of lognormal forward, Finance and Stochastics, Volume 4, Issue 1, 2000, pp. 35–68.
[17] Spicar, R. & Januska, M. (2015). Use of Monte Carlo Modified Markov Chains in Capacity Planning. Procedia Engineering, vol. 100, pp. 953–959.
[18] Jamshidian, F. (1997). Libor and swap rate models, Finance Stochast. 1, 1997, pp. 293–330.
[19] Keegan, S. (2008). Vibrato Monte Carlo and the calculation of greeks, Available from: http://eprints.maths.ox.ac.uk/716/1/Thesis_SK.pdf, Accessed: 2017–06–04
[20] Intel Cilk Plus, (2017). Available from: https://software.intel.com/en-us/intel-cilk-plus, Accessed: 2017–06–04
[21] Cilk home page, (2017). Available from: http://www.cilkplus.org/, Accessed: 2017–06–04
[22] McCool, M.; Reinders, J. & Robison, A. (2012). Structured parallel programming: patterns for efficient computation, Elsevier, 2012, ISBN: 978-0-12-415993-8
[23] Asai, R. & Vladimirov, A. (2015). Intel Cilk Plus for complex parallel algorithms: Enormous Fast Fourier Transforms (EFFT) library, Parallel Computing, Volume 48, 2015, pp. 125-142.
[24] Akhter, S. & Roberts, J., (2006). Multi-Core Programming: Increasing Performance through Software Multithreading. 1st Edn., Intel Press, ISBN: 13: 978-0976483243, pp: 360.
[25] Hofierka, J.; Lacko, M. & Zubal, S. (2016). Parallelization of interpolation, solar radiation and water flow simulation modules in GRASS GIS using OpenMP, Computers & Geosciences, Volume 107, October 2017, pp. 20-27

DOI: 10.2507/28th.daaam.proceedings.049

REGENERATIVE HEAT TREATMENT OF PROLONGED EXPLOITED HEAT RESISTANT STEEL

Damir Hodzic, Ismar Hajro & Petar Tasic

This Publication has to be referred as: Hodzic, D[amir]; Hajro, I[smar] & Tasic, P[etar] (2017). Regenerative Heat Treatment of Prolonged Exploited Heat Resistant Steel, Proceedings of the 28th DAAAM International Symposium, pp.0357-0360, B. Katalinic (Ed.), Published by DAAAM International, ISBN 978-3-902734-11-2, ISSN 1726-9679, Vienna, Austria

DOI: 10.2507/28th.daaam.proceedings.049

Abstract

Possibilities of regenerative heat treatment after prolonged exploitation under service conditions of low alloyed Cr-Mo-V cast steels are reported in published investigations. Considering this assumption that post service regenerative heat treatment could change microstructure and improve decreased mechanical properties after prolonged exploitation, heat resistant steamline steel 14MoV6-3 is investigated in this paper. Microstructure of post service regenerative heat treated steel 14MoV6-3 implicitly indicated increase of strength properties, but not toughness properties at same time. In order to investigate impact toughness properties, such as a crack initiation and propagation energy, following temperatures were selected: 20°C, 150°C, 400°C and 540°C (service temperature) for virgin, exploited and regenerative heat treated steamline steel 14MoV6-3. Further investigations related to possibilities of post service regeneration of exploited steel 14MoV6-3 should be directed to more adequate heat treatment cycle.

Keywords: Steel 14MoV6-3; Regenerative heat treatment; Crack initiation and propagation energy

1. Introduction

Power plants, which were originally intended to provide the base load, are frequently shut down and powered up. Variations in the steam temperature accompanying the power changes induce thermo-mechanical stresses in components, which lead to material degradation and consequently can cause failure, [1]. Results of some published investigations with low alloyed Cr-Mo-V cast steels reports about possibility of regenerative heat treatment after prolonged exploitation and influence of microstructure on properties, [2, 3]. Decrease of mechanical properties of exploited material is caused by changes in the steel microstructure due to long-lasting service operation. Reduction of toughness properties caused by long-term exploitation of the steel at elevated temperature depends to a large extent on the initial steel microstructure. Some published investigations reported that decrease of toughness properties caused by long-lasting operation is the least in the case of tempered bainite structure, [2]. It is important to emphasize that metallographic examinations of various steel grades after long-term service at elevated temperature revealed that transformations of carbides and morphological changes of phases have the most significant effect on service properties degradation, [4]. Considering assumption that post service regenerative heat treatment could change microstructure and improve decreased mechanical properties after prolonged exploitation, heat resistant steamline steel 14MoV6-3 is investigated in this paper. The low-alloyed steel 14MoV6-3 was chosen for this investigation, because it has been used for a long service period (194.207 service hours) with similar exploitation history and microstructure evolution as a previously mentioned Cr-Mo-V cast steels. Investigated material is taken from the Unit 5 main steamline (ø245×28mm) that operated at temperature 540 °C and pressure 13,5MPa in thermal power plant Kakanj, Bosnia and Herzegovina.

2. Regenerative heat treatment

Chemical composition testing of investigated steel 14MoV6-3 was accomplished in order to confirm that all delivered specimens of steamline are made from the same material, so the results of predicted investigation on virgin, exploited and regenerative heat treated material could be comparable. Method for determination of chemical composition was spectral analysis. Chemical composition of steel 14MoV6-3 was in accordance to normative DIN 17175/79 for all material conditions. The goal of regenerative heat treatment of exploited steel 14MoV6-3 was obtaining improved mechanical properties that were degraded, particularly toughness properties. Regenerative heat treatment (hardening and tempering) was done according to normative DIN 17175/79 for high-temperature seamless tube steel 14MoV6-3 on specimen of exploited steel. Regenerative heat treatment cycle of exploited material was consisted of:

- Hardening, heating to temperature 950 °C (1 hour) and accelerated cooling in oil because greater tube wall thickness, and
- Tempering, heating to temperature 700 °C (3 hours) with slow cooling together with furnace for heat treatment.

3. Hardness and microstructure

Brinell hardness test was done on outer surface, 1.5 mm under the outer surface, and at longitudinal and transversal cross section of steamline pipe. According to European normative EN 10216-2:2002, steel 14MoV6-3 are delivering as seamless steel tubes for elevated temperatures with acceptable value of hardness in range of 145 – 190 HB30 at room temperature. Results of measured hardness values (HB30) for virgin, exploited and regenerative heat treated steamline steel 14MoV6-3 are presented in Table 1.

Hardness of steel 14MoV6-3	Virgin, HB30	Exploited, HB30	Heat treated, HB30
Outer surface	135 – 143	134 – 147	270 – 288
1.5 mm under outer surface	155 – 161	146 – 159	252 – 265
Transversal cross section	159 – 162	150 – 153	288 – 293
Longitudinal cross section	161 – 164	149 – 151	290 – 293

Table 1. Hardness values for investigated material 14MoV6-3

Considering the highly increased hardness of regenerative heat treated steel 14MoV6-3, that is obviously consequence of large amount of insufficiently tempered constituents, it implicitly indicate increase of strength properties, but also significant decrease of toughness properties after regenerative heat treatment, [5].

Metallographic testing was accomplished for virgin, exploited and regenerative heat treated steel 14MoV6-3 by optical microscope with magnifications 500x. This was done in laboratory at IWS Institute TU Graz (Institute for materials and welding at Technical University Graz), Austria. Fig. 1 shows microstructure of investigated material 14MoV6-3 at transversal cross section of steamline pipe with same magnifications.

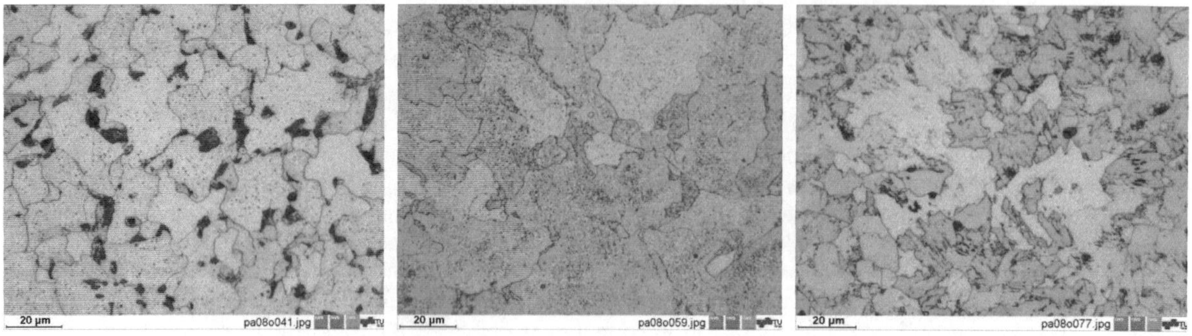

Fig. 1. Microstructure of virgin, exploited and regenerative heat treated steel 14MoV6-3 [5]

Results of microstructure investigation that are presented in this paper mainly can confirm facts that the initial microstructure of the 14MoV6-3 low-alloyed steel features the mixture of bainite with ferrite, sometimes with a small amount of pearlite and significant amount of the M_3C carbides and numerous, very fine MC type ones. The final structure image after long-term exploitation under service conditions is ferrite with rather homogeneously distributed precipitations inside grains and chains of the significant amount of precipitations on their boundaries, [6]. But in addition to mentioned microstructure evolution, there is also a significant growth of ferrite grain size after long-term operation at elevated temperature. As a consequence, grain growth has a significant influence on decrease of steel 14MoV6-3 impact toughness properties. Microstructure of regenerative heat treated steel 14MoV6-3 is ferrite structure with pearlite/bainite constituents on ferrite grain boundaries but also with insufficiently tempered constituents created by accelerated cooling. Comparing with virgin steel there is a less carbide precipitates in ferrite grains, less amount of ferrite at all and very small amount of pearlite and bainite.

4. Crack initiation and propagation energy

In order to investigate toughness properties of regenerative heat treated steamline steel 14MoV6-3, following temperatures were selected for impact toughness testing: 20 °C, 150 °C, 400 °C and 540 °C (service temperature). This was done by testing and comparison of crack initiation and propagation energy values of virgin steel, exploited steel and regenerative heat treated steel 14MoV6-3. For every testing temperature 3 Charpy V-notch specimens were used. In general, notch toughness is measured in terms of the absorbed impact energy needed to cause fracturing of the specimen. The change in potential energy of the impacting head (from before impact to after fracture) is determined with a calibrated dial that measures the total energy absorbed in breaking the specimen. Other quantitative parameters, such as fracture appearance and degree of ductility/deformation, are also often measured in addition to the fracture energy. Impact tests may also be instrumented to obtain load data as a function of time during the fracture event. The Charpy V-notch test continues to be the most utilized and accepted impact test in use in the industry, [6]. Results of average (3 specimens) impact crack initiation energy values per testing temperature for virgin, exploited and regenerative heat treated steel 14MoV6-3 are presented in Fig. 2.

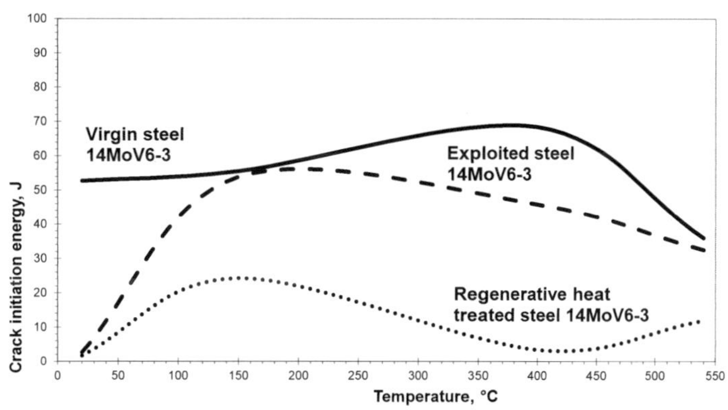

Fig. 2. Impact crack initiation energy

Results of average (3 specimens) impact crack propagation energy values per testing temperature for virgin, exploited and regenerative heat treated steel 14MoV6-3 are presented in Fig. 3.

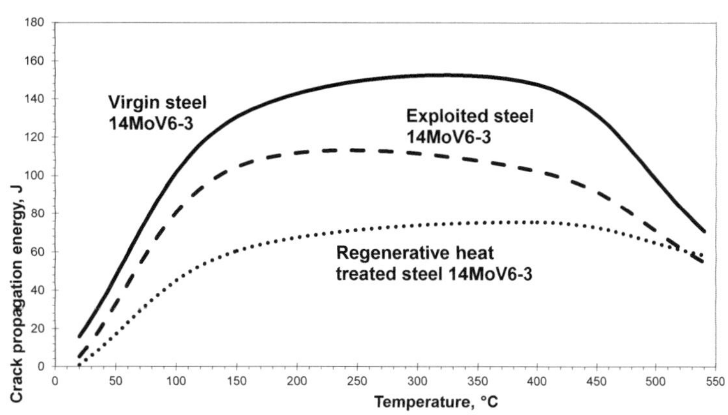

Fig. 3. Impact crack propagation energy

From the results of impact toughness testing it is notable that the impact crack initiation and propagation energy increases slightly up to 400 °C for virgin steel 14MoV6-3 and up to 150 °C for exploited steel 14MoV6-3. It is reduces significantly, but not drastically above 400 °C for virgin steel and above 150 °C for exploited steel, so that its values are still more than sufficient at steamline service temperature 540 °C. The most important result of exploited steamline steel impact toughness testing is extremely low value of impact crack initiation energy at room temperature 20°C, which is significantly beneath the allowed value, [6]. Considering regenerative heat treated steel, it is notable further decrease of toughness properties in overall testing temperature range. It should be mentioned that a large number of failures in engineering components occur due to preexisting defects, nonmetallic inclusions or other imperfections (casting, welding defects, etc.), [7]. However, it is of engineering interest to know how and why particular component has failed. The effort to extend designed lifetime of industrial plants operating for a long time at elevated temperatures is unthinkable without the knowledge of mechanical properties of materials prior to operation and mechanical properties after actual time of operation (actual mechanical properties), because the material properties can be reduced throughout the service life, [8].

5. Conclusion

Post service regenerative heat treatment of exploited material was accomplished with relative success. Toughness properties of steamline steel 14MoV6-3 depends mostly on development of the precipitation processes and also on development of the microstructure changes and structure discontinuities, as well as grain growth, originated during the long period of exploitation at elevated temperature. For this reason some changes are necessary in the microstructure degraded by long-term exploitation. These changes are:

- Grain size reduction allowing to increase the crack resistance, decrease the transition temperature and raise yield strength,
- Dissolving of carbides in austenite, particularly the carbides precipitated on grain boundaries, and
- Obtaining of tempered ferrite/bainite structure.

Further investigation related with possibility of exploited steel 14MoV6-3 post service regeneration by heat treatment should be directed to search for more adequate heat treatment cycle.

6. Acknowledgement

This investigation was partly supported by IWS Institute at Technical University Graz and OEAD Austrian Agency for International Cooperation in Education and Research.

7. References

[1] Linn S. & Scholz A. (2013). Creep-fatigue Lifetime Assessment with Phenomological and Constitutive Material Laws. Procedia Engineering (6th International Conference on Creep, Fatigue and Creep-Fatigue Interaction CF-6), Vol. 55, 2013, pp. 607-611, ISSN1877-7058, DOI 10.106/j.proeng.2013.03.302

[2] Golanski G., Stachura S., Gajda-Kucharska B. & Kupczyk J. (2007). Optimisation of Regenerative Heat Treatment Parameters of G21CrMoV4-6 Cast Steel. Archives of Materials Science and Engineering, Vol. 28, No. 6, (June 2007) page numbers (341-344), ISSN 1897-2764

[3] Golanski G. & Wieczorek P. (2008). Electron Microscopy Investigation of the Cr-Mo-V Cast Steel. Archives of Materials Science and Engineering, Vol. 30, No. 2, (April 2008) page numbers (73-76), ISSN 1897-2764

[4] Dobrzanski J., Zielinski A. & Krtzon H. (2007). Mechanical Properties and Structure of the Cr-Mo-V Low-alloyed Steel after Long-term Service in Creep Condition. Journal of Achievements in Materials and Manufacturing Engineering, Vol. 23, No. 1, (July 2007) page numbers (39-42), ISSN 1734-8412

[5] Hodzic D., Hajro I. & Tasic P. (2014). Regenerative Heat Treatment of Heat Resistant Steel 14MoV6-3. Proceedings of 18th International Research/Expert Conference "Trends in the Development of Machinery and Associated Technology", 10-12 September 2014, Budapest, Hungary, ISSN 1840-4944, Ekinović S., Yalcin S., Calvet J.V. (Ed.), pp. 117-120, Faculty of Mechanical Engineering in Zenica, Zenica

[6] Hodzic D. & Hajro I. (2016). Impact Crack Initiation and Propagation Energy of Prolonged Exploited Heat Resistant Steel. Proceedings of the 27th DAAAM International Symposium, pp.0308-0311, B. Katalinic (Ed.), Published by DAAAM International, ISBN 978-3-902734-08-2, ISSSN 1726-6979, Vienna, Austria, DOI: 10.2507/27th.daaam.proceedings.045

[7] Brnic J. & Brcic M. (2015). Comparison of Mechanical Properties and Resistance to Creep of 20MnCr5 Steel and X10CrAlSi25. Procedia Engineering (25th DAAAM International Symposium on Intelligent Manufacturing and Automation, 2014), Vol. 100, 2015, pp. 84-89, ISSN1877-7058, DOI 10.106/j.proeng.2015.01.345

[8] Matocha K. (2014). The Use of Small Punch Test for Determination of Fracture Behaviour of Ferritic Steels. Procedia Engineering (1st International Conference on Structural Integrity ICONS-2014), Vol. 86, 2014, pp. 885-891, ISSN1877-7058, DOI 10.106/j.proeng.2014.11.110

DOI: 10.2507/28th.daaam.proceedings.050

IMPLEMENTATION OF AUGMENTED AND VIRTUAL REALITY TECHNOLOGIES FOR AN EDUCATIONAL AND RESEARCH SITUATION CENTER

Konstantin Kharin, Sergey Kuvshinov, Valentin Pryanichnikov & Eleonora Anufrieva

This Publication has to be referred as: Kharin, K[onstantin]; Kuvshinov, S[ergey]; Pryanichnikov, V[alentin] E[.] & Anufrieva, E[leonora] (2017). Implementation of Augmented and Virtual Reality Technologies for an Educational and Research Situation Center, Proceedings of the 28th DAAAM International Symposium, pp.0361-0366, B. Katalinic (Ed.), Published by DAAAM International, ISBN 978-3-902734-11-2, ISSN 1726-9679, Vienna, Austria
DOI: 10.2507/28th.daaam.proceedings.050

Abstract

We consider some problems in the transformation of modern education in engineering fields and the Humanities. In addition, the need for improved clarity regarding the study and analysis of complex structures and objects is concerned. Augmented and virtual reality technologies are proposed for these purposes. Some examples of application of these technologies in various educational processes are given. Experience of such application allows to conclude that they can be used to implement a variety of educational scenarios for collective and individual activities under conditions of the educational and research situation center.

Keywords: engineering education; Humanities; 3D modeling; augmented reality; virtual reality; research situation center

1. Introduction

This study addresses a rather narrow problem (the process of creation and development of a specific training course at an educational situation center with a multichannel presentation of 3D multimedia data and augmented and virtual reality used to synthesize and analyze the structures of complex technical objects using the example of unmanned aerial vehicles) on the one hand and the conceptual problems of technical and humanitarian education, on the other hand. Therefore, the restructuring of a specific (in our opinion, any) training course cannot be adequately understood unless the situation of the global civilizational transition is taken into account, when the former European model of training specialists has been almost exhausted, while no visible prospective model is yet available [1]. In this situation, the education system as the main mechanism of social and cultural inheritance is certainly devoid of the key point—development ideals and values; in turn, this prevents the formation of a generally valid prospective model of educational development as a whole. Most often and most convincingly, many European countries speak about insufficient funding among "their" vital problems; however, although acute, this is not the only and, at present, not the most important problem of education.

The information, communication, and audiovisual technologies become the basis for constructing the structure of a new educational environment and organizing a new type of educational and research space: an educational and research situation center. This will allow the training of specialists in the synthesis of complex technical structures to go from the reproductive to a creative type of learning. The use of interactive information technologies as well as augmented [2] and virtual reality technologies in educational institutions provides a unique opportunity for teachers to find new ways of dealing with educational and scientific materials using the case study technology, particularly in a specially arranged audience that meets the criteria of an educational situation center [3, 4].

2. Project culture as an effective way to creative education

The term "project culture" is often used to characterize the culture of the entire 21st century as its pervasive feature. For a long time, education has been developing along two noncontiguous lines (humanitarian–artistic and scientific–technical), which currently exist as "two cultures". The conclusion about the existence of a "third culture" (project culture) is rather nontrivial; this was proposed to be called a "Design with a capital letter" and defined as "the cumulative experience of material culture and the cumulative array of experience, skills, and knowledge reflected in the art of planning, invention, creation, and implementation. Finally, it is reasonable to consider the design as a part of the general education system since the design involves fundamental cognition methods required in any activity. In other words, the design received a status of semantic activity. In the Russian professional technical literature, the concept of "design" is often identified with the implementation of a specific project, which, in turn, is focused on a final product. The practice of educational design—a kind of assembly of entire university learning—confirms, fixes, and retranslates this design ideology into the future. The transition in late 20th century to the organization of design (understood rather widely as both planning and forecasting) focused on function with the development and use of appropriate methods affects technical education quite inadequately. Discussing new approaches to engineering education and specifically the use of integrative university courses such as "Designing complex technical systems" on the example of aircraft design, one should note that all design problems can be conditionally described (simulated) by a form that governs the relationship between the research subject entities, attributes, properties, and behavior (characteristics) in an external environment. The problem has one unknown parameter (entity, environment, or characteristic), which should be determined from the remaining two components. Depending on what namely is unknown, all problems can be divided into three main classes.

The first class covers the problem of analysis when the entities are given and their characteristics (unknowns) under the conditions of a given external environment are investigated.

The second class involves synthesis problems when the required characteristics are created and one should determine the entities that provide the derivation of these characteristics under the conditions of the given environment.

The third class is called the problem of external environment assessment, when the entities and their characteristics are given and it is necessary to obtain an environment (yet unknown), such that the relevant entities manifest the given characteristics. There is no formal method that would make it possible to construct the most suitable (appropriate) model for a given problem. During the solution of the problem, the construction and subsequent correction of the model is put entirely on the man. A specific feature of the synthesis of complex object structures is the fact that the teams of people, who must interact and make decisions determining the fate of the project, create them. Therefore, it is so important now to train specialists under special conditions of an educational situation center, which develops not only skills of purely engineering activities, but also simulation and communication skills, and forms a sense of responsibility for making certain decisions.

Historically, the aviation education in many countries developed following the lead of industry together with scientific research institutes and design bureaus. The principle of professional knowledge transfer "from hand to hand" was rather efficient. In other words, the "knowledge orientation" of higher professional education as an absolute value recognized the content of appropriate scientific rather than even educational discipline, and the quality of teaching depended largely by the extent that the content of an educational subject approximates the corresponding scientific field. The structuring of the content of the subject studied and the educational process was determined by the logic of science rather than teaching. The methodological and especially psychological components of the pedagogical process were intuitive and one-sided (from teacher to student) character.

The current professional education based on both internal processes and the general course of scientific–technical and social and cultural development differs radically from the "classical" period. First of all, it became quite obvious that the activities in science and engineering and the relevant teaching are two fundamentally different types of activities. Their subjects and methods are not the same. These activities require completely different skills, dispositions, and motives. What they have in common is the information support. Here, the old division of knowledge into educational and scientific fields has lost its informational sense and acquired a new (methodical) sense. Secondly, it became obviously clear that the student couldn't be further regarded as a "black box" because the "pedagogical effect" cannot be adequately assessed. Thirdly, it is also clear that the teacher's role in the educational process changes both psychologically and conceptually. The teacher turns from the source (which often is almost unique) of knowledge into the organizer of the educational process, and his activities are directed to the learning of tools, methods, and the design culture in general, rather than to the design object (unlike the "designer"). It can be seen from the context that the design tools and methods likewise should not be absolutized in the educational process, which aims mainly to develop skills and abilities for independently choosing and justifying the design goals and finding the optimal ways for their achievement.

In this connection, one of the first tasks of formulating the design course and determining its structure should probably be resolved from the logic of teaching of "design rules" and the specific features of students' psychological perception, rather than from scientific and designer (mainly information–educative) logic of aircraft structure and operation. The teacher formulates (rather than creates, of course) and structures the game rules, teaches them (with individual control over their assimilation), and, if necessary, demonstrates their applicability. All this requires conditions for training specialists that are absolutely different from present-day computer classes and become increasingly more like situational centers. Fourthly, the reasoning (in a special sense of this word) begins only when consciousness is confronted with a contradiction that cannot be resolved by ready-made schemes, recipes, and algorithms. Therefore, the educational process (understood as the organization of students' reasoning) must be built as a specially organized system of possible contradictions (obstacles) in the implementation of a target (as opposed to a systematized description of ready-made (correct) solution sets). In other words, it is heuristically more important to study different errors and ways to overcome them than to study only the positive experience. Finally, the availability and remarkably rapid development of new information and computer technologies sharply (to a minimum) reduced the potential of the classical education model. Here, the question is not whether to use or not to use these technologies and not even how to use them, but how to find a new spatial and temporal model of the educational process such that any development of new technologies would only expand the capabilities of the pedagogical process.

3. Educational and research situation center based on AR technology

An example is the innovative Russian product that is based on the technology of augmented reality: the *EligoVision* interactive system [5]. The preparation of experts in the synthesis of complex technical structures requires a spatial representation of future objects and their systems, not to mention the complex processes that cannot be adequately understood in a planar representation. Under conditions of the educational situation center, the *EligoVision* interactive system is a set of audiovisual computer devices for data display, equipped with collective remote communication tools. This set allows the case runner, the teacher, and the entire group of students to be simultaneously involved in intensive activities, the voluminous graphical material to be handled in both 2D and 3D modes. The skills of collective communication for the synthesis of complex structures to be developed, and, which is especially important, a decision adequate to the given problem to be made.

The development of practical technologies for the educational process that emulates the situational center operation became possible in the International Institute of New Educational Technologies, which became a unique experimental platform to analyze and implement most advanced educational technologies in Russia.

The multifunctional audience *Visionarium* has a unique visualization system developed by the *EligoVision* company, and this new equipment is currently used for interactive lessons. There have been master classes for the teaching staff of both technical and humanitarian universities.

A new technique was proposed for the presentation and analysis of educational materials on the design and management of projects for creating complex technical systems, design and construction of aircrafts, etc.; conceptual teaching techniques were developed on the basis of collective and remote control of symbolic, graphical, and three-dimensional applications. For example, even all existing training courses and lectures prepared in traditional presentation programs with augmented and virtual reality technologies turn into an interactive discussion and a collaboration.

4. Implementations

The implementation of augmented reality technologies in education began a decade ago and has been developed with an explosive behavior. The elements of augmented reality are integrated into various educational and developmental tools, including children's coloring books, textbooks on school disciplines (usually, in natural sciences), and handouts for classes on specific topics. However, most applications of this technology have a "game" and fragmentary character and therefore, are rarely used in professional and higher education.

The experience of the authors in using digital technologies in general and augmented and virtual reality in particular allows us to state that it is these educational approaches that will govern the most prospective use of augmented reality in various fields of knowledge (both engineering and humanitarian). On the one hand, this is explained by the large number of topics, phenomena, processes, etc., which description by traditional teaching methods encounters problems in visualization.

On the other hand, the widespread use of 3D technologies, simulation, and design makes it possible to use the results of students themselves and the equipment used to increase the vividness of the given disciplines, including in the research and design activities. Often, these projects allow students to create their own 3D models and entire virtual worlds that can be used to create appropriate software. Thus, the learners themselves are engaged as "experts" and "knowledge engineers", which significantly increases their motivation in obtaining results and mastering new technologies [6].

An example of completed projects is the above-mentioned course of aircraft design and construction. The authors participated in all stages of this extensive course to develop a smart-scheme (a methodical teaching manual) that allows the student to focus on creative and personalized illustration of his own perception, existing supplementary schematics, diagrams, graphs, etc., rather than on accelerated and automatic recording of lecture fragments.

An important supplement to this manual was the augmented reality application that makes it possible to visualize most illustrations available in textbooks and smart-scheme with the help of animated 3D models that significantly increase the vividness of materials. In this case, a nonmarking technology was used that allows the illustrations themselves to be used as markers for optical tracking, including the supplements introduced during the smart-scheme completion. In the next step, the models were visualized by using modern 3D visualization systems, including *MotionParallax3D* technologies (virtual holograms) [7], rather than the flat screen of smartphones or tablets (see Fig. 1).

Fig. 1. Working with MotionParallax3D display

Another project was the preparation of materials for a professional course on oil production technologies. Here, the specific feature of the augmented reality technology was that the layout of drilling platforms, oil production and pumping equipment, etc. printed on a 3D printer were used as a marker for recognition in the application (see Fig. 2).

The superposition of additional information layers (geological, technological, etc.) together with the layout inspection from different angles significantly enhances the informative content of the educational material.

Fig. 2. AR application with a real object as a marker (drilling platform model)

The augmented reality technology can also be widely used for training in the fields of robotics, mechatronics, and sensorics. The combination of the complexity of visualization of many interacting devices and components with the study of programming, 3D modeling and design, and computer vision systems by most students [8] makes it possible to create integrated learning materials with the help of augmented reality by the students themselves during the completion of their qualification works or individual research projects. Here, in terms of the technology itself, the marking of individual components of devices, the pattern of conductor wiring or the arrangement of parts on the printed-circuit board, as well as specially designed markers (QR-codes) on the structural elements can be used as augmented reality markers.

Equally important is the use of the augmented reality technology in the humanitarian fields. An interesting example can be the learning in the conservative field of manual musical conducting. In some cases, the traditional visual method of teaching to conducting (*do as I do*) hinders the development of the musician creativeness. The learning process in the class of conducting is to form qualities that are professionally required for the musician and can allow him to become the leader of the performing team [9]. At the initial stage, both augmented and virtual reality technologies can be used as a developing approach to teaching to conducting. In the practice of teaching, the space of physical activities by the conductor is commonly represented as a cube since the graphics of the manual process is three-dimensional (width, height, and depth). The very concept of the conductor space is associated with a specific representation of the location of his hands in the low (the bottom of the cube), middle, and high (the upper plane of the cube) positions; in the near (front wall), middle (the cube center), and deep (rear wall) planes, and the horizontal movement of arms kind of connects the lateral planes of the cube. The gesture parameters and direction reflect the sound character, the degree of its saturation, the dynamic brightness, and specific features of the articulatory character of the movement (the "gait" of the sound stream). The initial possession of manual skills is associated with the formation of the ability to freely handle with conductor's three-dimensional space. It seems possible and prospective to use the traditional technique of mastering the conductor gesture on the basis of physical exercises seems in combination with audiovisual software and hardware tools recording diverse pieces of classical music. The use of augmented reality with markers as fragments of music score notation make it possible to read them with the help of a camera, process by a software application coupled with the reproduced music fragment, and interpret as a referenced real time video stream for displaying on a monitor or similar 2D or 3D devices. However, this is a somewhat static method that lacks to incorporate the emotional and tempo characteristics of musical works. The capabilities of the system can be expanded in the future by a rather complex system for automated analysis of professional conductor images with the translation of his gestures into a language of formal simplified description correlated to the phonogram. The comparison of the resulting generated texts will allow one to reveal the specific features of conducting or even to preliminary assess the trainee against the background of teacher's reference movements. The technical support for this can be realized, for example, on the basis of Kinekt-type contactless sensor controllers.

One should also note the use of augmented reality technologies in museum exposition and exhibition activities that has been actively developing in the last decade. Against the background of a large number of more or less successful examples, the authors developed and implemented the concept of multimedia expansion of the exhibition space, where the real exhibits are complemented by the demonstration of their 3D models on personal devices of visitors with the help of augmented reality and as virtual holograms in the stereo mode on devices with the *MotionParallax3D* technology. The modern technologies make these "hybrid" exhibitions additionally attractive for the young generation. Another factor of increased interest is (as, for example, in the exhibition project "Leonardo da Vinci in 3D" implemented by the authors) that the animated 3D models from sketches of the great Renaissance master were created by schoolchildren and students engaged in interdisciplinary projects, rather than by professional designers.

5. Conclusion

The augmented reality technologies allow one to look at conventional things with new eyes. The interactive augmented reality and virtual reality technologies support operation in a 3D mode: with special glasses, a team member observes voluminous objects and models and can deal with them in space with the help of special interactive manipulators. These manipulators can be used to intersect geometric objects by planes, consider sections, cut and join them, and make shifts and other operations necessary to solve the given problem. Even relatively simple applications of virtual reality make it possible to "inspect" the virtual object from different angles and "look inside" it simply by changing the position of sensors or a personal device (smartphone or tablet) in a quite natural way.

A shortcoming of this technology is the risk that the educational process can be easily turned into a game, which is especially challenging in the case of younger students. This problem can be solved by a careful elaboration of both the models used and the scenarios for their use in the educational process in order to adequately channel the "game" aspect into the right direction.

In prospective applications the development of intelligent modules is planned, which will process and analyze the queries and log data of students interaction with systems of augmented reality. It could result in constructing of new pedagogical models using the intelligent expert systems.

In a number of both engineering and humanitarian disciplines, this approach is not only justified but also very efficient. These technologies and devices open a new area for methodological developments. They can be used to implement a variety of scenarios of educational tasks for collective and individual activities under conditions of a new space—the educational and research situation center.

6. References

[1] Katalinic, B. (2010). Engineers for Knowledge Based Society, Annals of DAAAM for 2010 & Proceedings of the 21st International DAAAM Symposium, 20-23rd October 2010, Zadar, Croatia, ISSN 1726-9679, ISBN 978-3-901509-73-5, Katalinic, B. (Ed.), pp. 0001-0002, Published by DAAAM International Vienna, Vienna

[2] https://en.wikipedia.org/wiki/Augmented_reality. (2017). Augmented reality. Accessed on: 2017-10-16

[3] Andreev, V.; Karbanov, V.; Kharin, K.; Kuvshinov, S.; Poduraev, Y. and Pryanichnikov, V. (2015). Training Situation Center Based on Three-Dimensional Virtual Studio for Distributed Mobile Robotics Laboratory, in Proceedings of the 26th International DAAAM Symposium "Intelligent Manufacturing & Automation" 21-24th October 2015, Zadar, Croatia, EU, Ed. by B. Katalinic, 2015, Vol. 26, No. 1.

[4] Pryanichnikov, V.; Andreev, V.; Bobrov, P.; Biryukova, E.; Frolov, A.; Kharin, K.; Kirsanov, K.; Kostin, A.; Kuvshinov, S.; Marzanov, Y. and Prysev, E. (2012). Scientific-educational distributed laboratory — software and hardware means, in Annals of DAAAM for 2012 & Proceedings of the 23th international DAAAM Symposium "Intelligent Manufacturing & Automation" 24–27th October 2012 Zadar, Croatia, 2012, vol. 23, no. 1, pp. 1175–1178.

[5] http://www.eligovision.ru/en/. (2017). EligoVision company. Accessed on: 2017-10-05

[6] Andreev, V.; Pryanichnikov, V.; Poduraev, Y. & Kuvshinov, S. (2014). Education on the basis of virtual learning robotics laboratory and group-controlled robots, in 24th DAAAM Int. Symp. on Intelligent Manufacturing and Automation, 2013, Procedia Engineering, 2014, vol. 69, pp. 35–40.

[7] Kharin, K.; Kuvshinov, S. & Pryanichnikov, V. (2016). The Use of Three-dimensional Visualization on the Basis of the MotionParallax3D Technology in the Development of Complex Technical Devices and Training of Engineers, Proceedings of the 27th DAAAM International Symposium, pp.0338-0343, B. Katalinic (Ed.), Published by DAAAM International, ISBN 978-3-902734-08-2, ISSN 1726-9679, Vienna, Austria

[8] Pavlovsky, V.E. & Kirkov, A.Yu. (2012) Tonal acoustic communication of robots. Preprints of KIAM, No.85, 2012.

[9] Anufrieva, E.B. & Pichugina, L.N. (2012) The initial stage of training in choral conducting. Vuzovsky vestnik, Ekaterinburg, 2012.

DOI: 10.2507/28th.daaam.proceedings.051

MECHANICAL FORGING PRESS FOR PROGRESSIVE WORKING IN AUTOMATIC PROCESS

Zdenek Chval, Milan Cechura

This Publication has to be referred as: Chval, Z[denek] & Cechura, M[ilan] (2017). Mechanical Forging Press for Progressive Working in Automatic Process, Proceedings of the 28th DAAAM International Symposium, pp.0367-0371, B. Katalinic (Ed.), Published by DAAAM International, ISBN 978-3-902734-11-2, ISSN 1726-9679, Vienna, Austria
DOI: 10.2507/28th.daaam.proceedings.051

Abstract

This paper deals with variants of working force distribution in press dies in the working space of asymmetrical forging press with a nominal force 25 MN. The asymmetrical forging press was designed for a left-to-right forging in five dies with maximum forging force in the last die. This paper describes the possibilities the machine loading for progressive working in all dies during one working stroke. For this purpose, FEM calculations were used to determine maximum stresses and displacements in the whole structure of the forging press.

Keywords: forming machine; mechanical forging press; structural FEM analysis; press frame

1. Introduction

An asymmetric mechanical forging press was designed for progressive working in five dies with a maximal force 25 MN in last position (with maximal eccentricity).

This design solution was motivated mainly by the need for automation and mechanization. This need is resulting in one direction forging - from left to right. When the forging press works at fifth (side) die with a maximal force (Figure 1), this press is much more accurate comparing to a conventional press.

The press can also be used for forging in a conventional manner if the recommended forces shown in Fig. 1 are not exceeded. In this case, the maximum force at central die position is up to 30 MN.

Precision of the forgings in the central die position of forging on an asymmetrical press does not achieve the accuracy of the classical construction. This design [1, 2] is also applicable to a wide range of other technological operations.

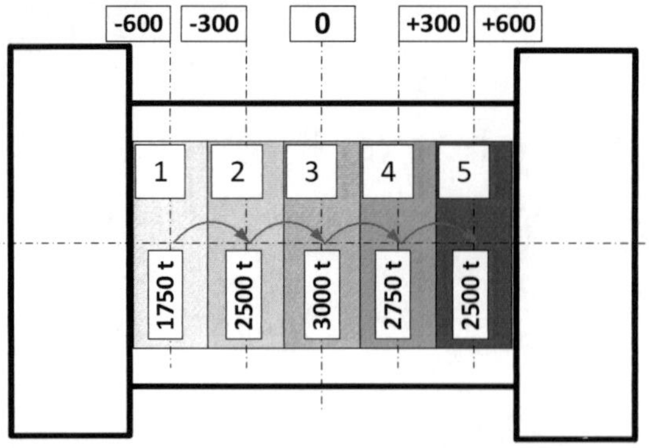

Fig. 1. Maximum work force for each die during the forging in one die [1, 2].

2. Analysis of the force distribution in the dies of multi-impression die

Current publications on an asymmetric forging press were focused on the ability of the press to work with maximum work force at the maximum possible eccentricity of the press. The design of the frame and all other components of the crank mechanism was made with respect to this requirement [6, 7].

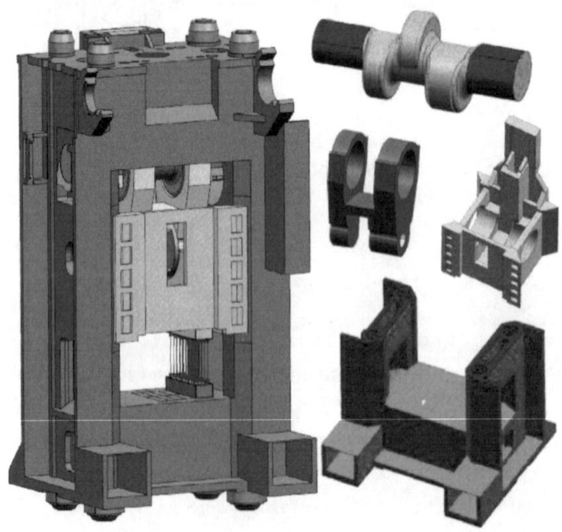

Fig. 2. An asymmetrical forging press with a working force 25 MN [1, 2].

The next step was to analyse possible variants of the working force distribution in the multi-impression die during one stroke of the machine. The variants are based on the results of the distribution of stress in the press stand from the proposed load forces. The following table shows the proposed forces distribution for which the asymmetrical press is sufficiently dimensioned.

		POSITION OF DIE				
		1	2	3	4	5
		FORGING FORCE (MN)				
VARIANT	A	4	6	7	9	4
	B	3,5	5	8	10	3,5
	C	8	5	10	3,5	3,5
	D	3,5	3,5	5	8	10
	E	0	10	5	10	0

Table 1. Variants of the forcing force distribution in the multi-impression die during one stroke.

The stress was evaluated for individual working force distributions (Fig.3) [3, 4, 5]. The simulation was performed using the Siemens NX 11 software [8, 9]. The stress do not exceed the allowable values and the forging press can therefore be used for all proposed variants of forging forces.

It can be seen in the figures that the largest stress is located under the anchor nut on the right side and below the work table near the anchors. A similar stress distribution in all these cases suggests that the proposed force distribution variants have been performed correctly.

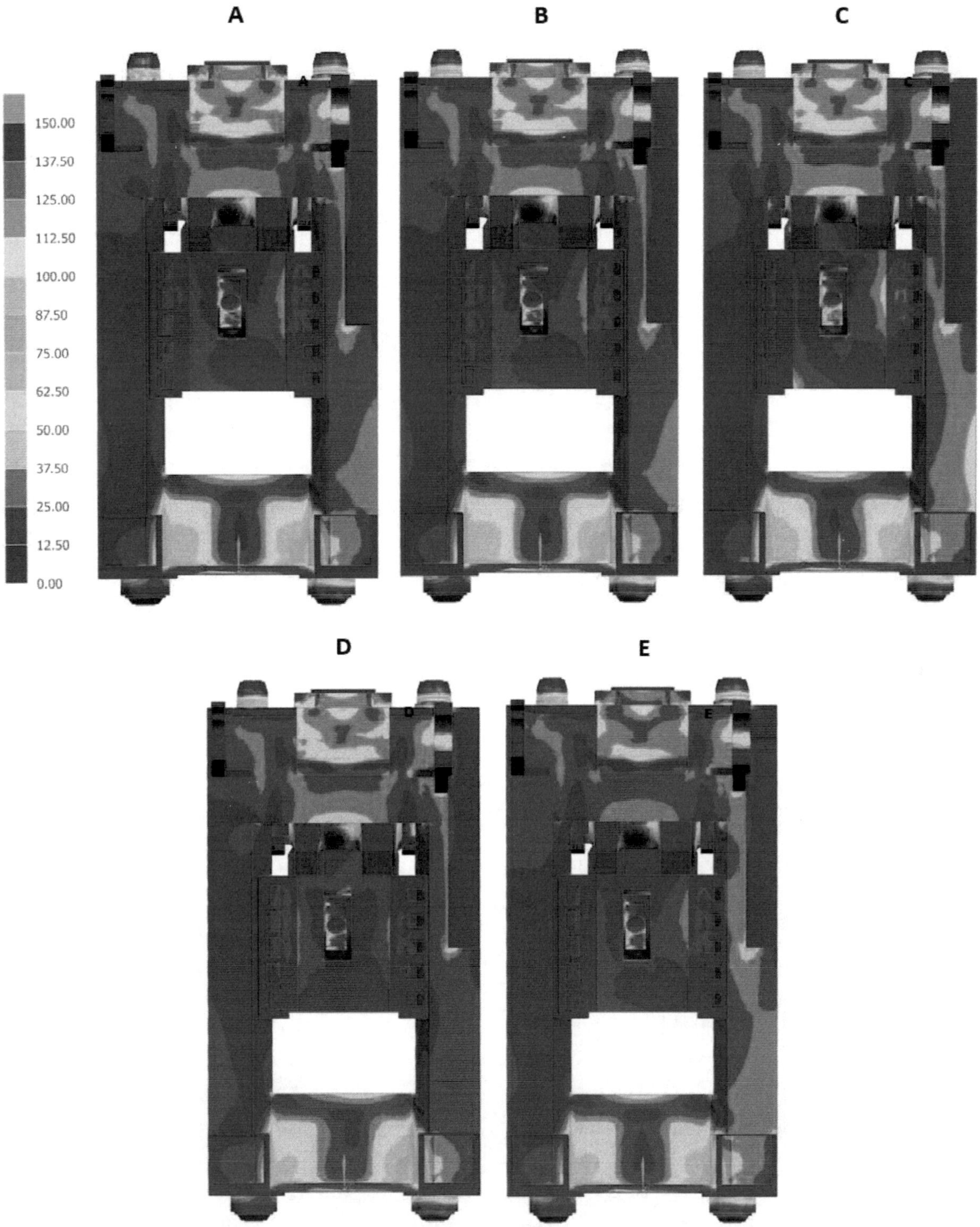

Fig. 3. Distribution of stresses on the forging press during one according to Table 1.

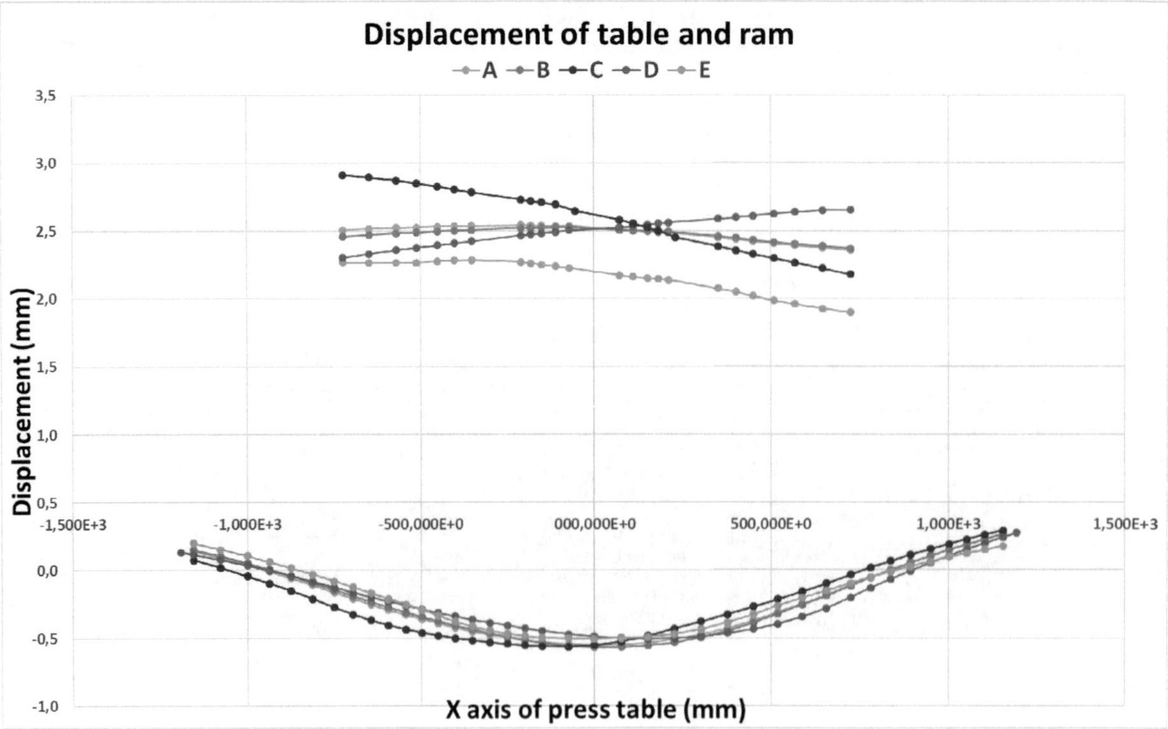

Fig. 4. Displacement of table (down) and ram (up) (mm)

For each variation of the forces distribution, the displacement of the work table and the ram was evaluated. It can be seen from Fig. 4 that the symmetrical loading of the press (variant E) results in a asymmetrical displacement of the table and the ram. Because of this, an asymmetrical press during a centric forging does not achieve such precision as a conventional forging press.

The proposed variants of the recommended load on the press work table can be modified as necessary so that the deformation of the table and ram is projected as little as possible in the final product. This is done by analyzing the curves shown in Figure 4.

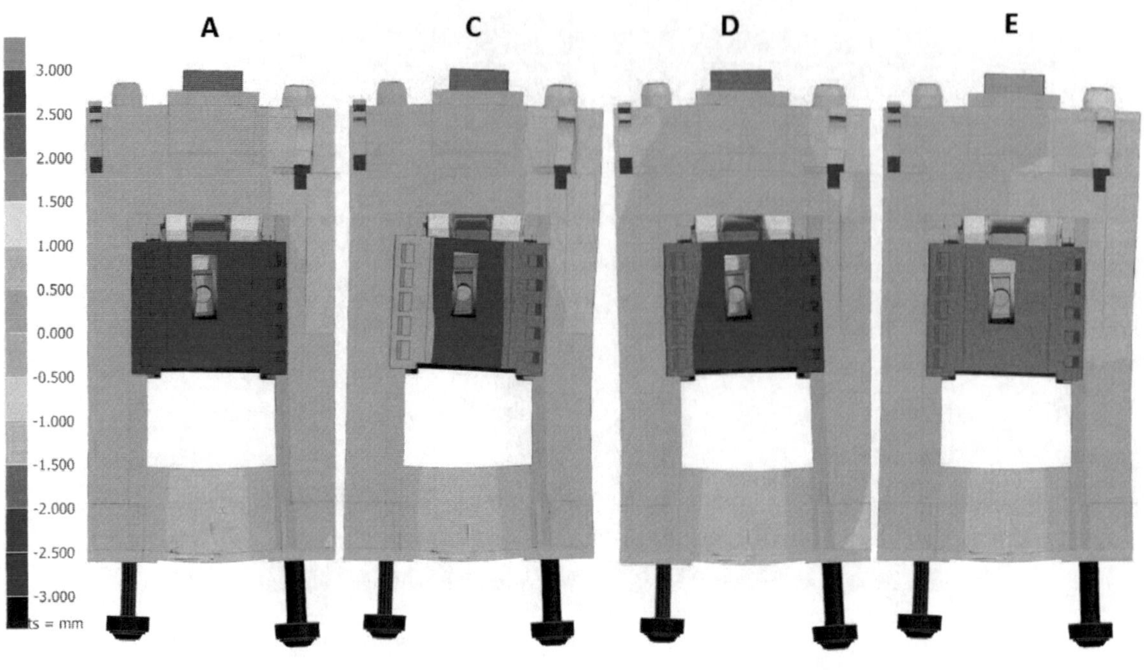

Fig. 5. Displacement of forging press in vertical direction.

3. Conclusion

Results described above were obtained with increasing of weight from 164t to 169t. The weight is considering frame, anchors, crankshaft, connecting rods and ram.

This solution is not single-purpose machine and can be used for various technological operations, when sorting operations from left to right - especially for automatic process. The press is primarily designed to place the maximum force centric or eccentric to the right of the center. As shown in Figures 4 and 5, when a dominant force is placed to the left side of the press table, a larger overall displacement occurs, a larger tilting of the die and it is resulting in lower forging accuracy. This results from the comparison of the curves C and D at the table and the ram of the press.

In the case of the forging in central position, it is more advantageous to use the press of a classic concept due to the tilting and machine precision. This is illustrated by the curves of variant E. Although it is a central forging, there is an asymmetrical displacement of the table and ram (see Figure 4).

4. Acknowledgments

This article was done by financial support of TA ČR – project TE01020075.

5. References

[1] Chval Z., Cechura, M. (2016). Design of mechanical forging press with maximal force in side cavity, MM Science Journal, November 2016, pp. 1358-1361, ISSN:1803-1269

[2] Chval, Z., Cechura M. (2015). Monitoring Extremly Stressed Points on Stands of Forging Presses, 25th DAAAM International Symposium on Intelligent Manufacturing and Automation 2014, Procedia Engineering, 2015, p. 841 – 846.

[3] Hawryluk, M., Kaszuba, M., Gronostajski, Z., Sadowski, P. (2016). Systems of supervision and analysis of industrial forging processes, Eksploatacja i Niezawodnosc, 18 (3), pp. 315-324.

[4] Hawryluk, M., Ziemba, J. (2017). Possibilities of application measurement techniques in hot die forging processes, Measurement: Journal of the International Measurement Confederation 110, pp. 284-295, ISSN:0263-2241

[5] Hawryluk, M. (2016). Review of selected methods of increasing the life of forging tools in hot die forging processes, Archives of Civil and Mechanical Engineering, 16 (4), pp. 845-866.

[6] Chval Z., Cechura, M., Raz, K. (2015). Analysis of stress distribution in the forging press, Annals of DAAAM and Proceedings of the International DAAAM Symposium, 2015-January, pp. 488-493

[7] Chval Z., Cechura, M. (2014). Optimization of power transmission on mechanical forging presses, Procedia Engineering 69, pp. 890-896, ISSN:1877-7058

[8] Raz, K.; Cechura, M., Chval, Z. (2015). Upper Crossbeam of Free Forging Press Optimization in Order to Improve Mechanical Properties and Durability, 26th DAAAM International Symposium on Intelligent Manufacturing and Automation, DAAAM 2015-January, pp. 524-529

[9] Zahalka, M.; Stanek, J., Raz, K. (2015) Cross Wedge Rolling and Forging Rolls as Additional Devices in Closed Die Forging, 26th DAAAM International Symposium on Intelligent Manufacturing and Automation, DAAAM 2015 - January, pp. 530-535

DOI: 10.2507/28th.daaam.proceedings.052

PREDICTION OF WEAR OF WOODWORKING TOOLS DEPENDING OF THE CHIP THICKNESS BY BOUNDARY ELEMENT METHOD SIMULATION

Ibrahim Busuladžić, Izet Horman, Seid Hajdarević

This Publication has to be referred as: Busuladzic, I[brahim]; Horman, I[zet] & Hajdarevic, S[eid] (2017). Prediction of Wear of Woodworking Tools Depending of the Chip Thickness by Boundary Element Method Simulation, Proceedings of the 28th DAAAM International Symposium, pp.0372-0378, B. Katalinic (Ed.), Published by DAAAM International, ISBN 978-3-902734-11-2, ISSN 1726-9679, Vienna, Austria
DOI: 10.2507/28th.daaam.proceedings.052

Abstract
A good quality of produced workpiece in wood industry requires the suitable machines and accompanying tools. Used tools must be constructed from adequate materials and with appropriate geometry depending of technological purpose. Using tools, cutting blade tend to blunt during the production cycle. One of the most important phenomena is the plastic deformation of cutting edge due to high temperature occurrence. In this study we analyse the influence of the chip thickness on temperature occurrence inside the blade. We use the Boundary Element Method (BEM) to analyse the temperature occurrence causing the plastic deformation in material and the blunting of the cutting edge. Different thickness of chips cause the different types of blunting whether the occurrence of craters or rounding of the blade.

Keywords: Wood cutting tools; tool wear; temperature occurrence; plastic deformation; boundary element method

1. Introduction

When processing wood by mechanical cutting, friction occurs between the separated particles and the front side of the tool, between the rake face of the tool and the cutting surface and internal friction between the particles of the treated material. In order to overcome the frictional forces, the mechanical energy should always be brought to the cutting place. Looking at the energy balance after cutting, it can be concluded that a very small portion of the generated energy is converted into the kinetic energy of the saw, while the bulk of the energy brought is transformed by friction into the heat energy.

Depending of the chip thickness the tool wear will be different. It can be described by the occurrence of craters or rounding of the blade. Plastic deformation caused by high temperature and frictional work between tool and workpiece is one of the most influential factors on tool wear. Experimental studies and different techniques can be used to determine the lifetime of cutting edges. Wear due to temperature is typical limitation of tool life. High temperature of the edge of a tool increases the wear of a tool as heat weakens and softens metals especially above 500–600 C^0. Sheikh-Ahmad et al. [1] have studied the effects of high temperature on tool wear of cemented tungsten carbide. Certain authors have employed different techniques to measure temperature on cutting tools. Chao et al. [2] have utilized infrared thermometer to measure the temperature on lathe tool. Csanady [3] has used sensors inside the tool in bisector line. To decrease the effect of tool heating Gazo et al. [4] have applied techniques of tool cooling using refrigerated air.

Sheikh-Ahmad et al. [5] have employed experimental measurement and Boundary Element Method to study influences of cutting speed on temperature distribution in a wood cutting tool. The Boundary Element Method will be used in this paper to represent the influence of chip thickness on temperature distribution inside the tool causing the plastic deformations. In this study, we use numerical technique to determine the type of tool wear without limitation on the outside or inside of the tool.

2. Blunting of cutting tools by plastic deformations

It can be important to determine temperature distribution by numerical means facilitating to develop predictive models of tool wear. When we investigate a process of cutting, we can see that on the edge and on the front surface of the knife, the pressure distribution develops such as shown in figure 1a, while the chip moves with speed v in regards to the knife. From the thermodynamic point of view, the material separation process is seen as a composition of four porters: particle, processing, tool and environment (figure 1b).

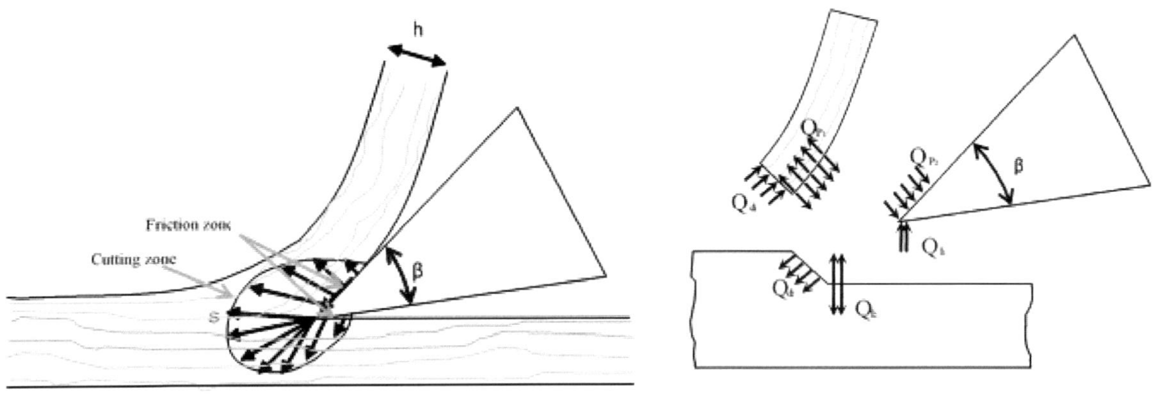

Fig. 1. a) Pressure distribution over cutting process, b) Material separation from thermodynamic - point of view

The theoretical value of the generated heat can be written as [6]:

$$Q = v \int_F \mu\sigma . df \tag{1}$$

The amount of heat produced is approximately equal to the work consumed (ignoring the kinetic energy of the chip). Therefore, the amount of heat that is generated in the unit of time and can be written as.

$$Q \approx F \cdot v \tag{2}$$

Q is the amount of heat generated in the unit of time, F is the force component on the blade in the direction of the cutting speed, and v the cutting speed.

Taking into account the energy contained in the movement of the chips, for total developed heat the term can be written as:

$$Q \approx F \cdot v \cdot \alpha_0 \tag{3}$$

α_0 is the factor that takes into account that part of the delivered mechanical energy that is transmitted as kinetic energy to the chip. Usually this part of the energy is neglected.

The resulting heat is the result of external and internal friction.

$$Q = Q_l + Q_P + Q_d \tag{4}$$

- Q_l is the amount of heat generated in the unit of time by friction between the rake face and cutting surface.
- Q_p is the amount of heat generated in the unit of time between the front surface and the chip.
- Q_d is the amount of heat generated in the unit of time by internal friction between the chips of the treated material.

On the front surface, in the unit of time it will be created heat equal to the product of these two quantities.

$$Q_p = F_{tp} \cdot v' \qquad (5)$$

F_{tp} is the frictional force between the separated particle and the front surface of the cutting blade and v ' is the velocity of the movement of the separated particle. On the front surface of the tool frictional force will be equal to:

$$F_{tp} = \mu_p \cdot F_N \qquad (6)$$

F_N is a normal (vertical) force on the tool's front surface, μ_p is the frictional factor between the separated particle and the front surface of the tool.

Comparing the experimental and theoretical values of temperature, it can be shown that only 60% of theoretical values of frictional forces appear such as effective heat on surface. The temperature generated on the tool body depends on several factors, mainly of; the thickness of the chip, the peripheral velocity and the sharpening angle.

The influence of thickness of the chip on friction and surface temperature causing the tool wear will be the parameter that will be analysed in this study. Numerical-experimental research approach will be adopted in this paper in order to determine the temperature distribution over the whole geometry of the tool in continuous process of cutting wood.

The target of this approach is to deliver a representative amount of the cutting power passing into thermal energy, which takes place in the tool. As this quantity of heat is not known, a priori the inverse technique is adopted in which the heat transfer in the tool is adapted to numerical model until the moment when the temperature in distant areas on the field coincide with the experimental measurements.

Measuring the cutting force one can determine friction coefficient. The previous questions can be answered partly from experimental results and partly from theoretical considerations. If the boundary conditions; the location of the heat flow and intensity, heat transfer on the surface are known, then the differential equation of the heat conductivity can be solved with a computer by numerical simulations. Boundary Element Method can be used to determine the whole temperature field.

The empirical equation describing the temperature can be written as follows [6]:

$$T = T_0 + (26.6\rho^{0.8} + 1053 \cdot x^{0.8}) \frac{v^{0.8}}{\beta^{0.95}} \qquad (7)$$

Csanady [2] has measured the temperature at bisector line of the tool at certain distance from the cutting edge. Extrapolation of the results at the known points in bisector line can give the value of temperature at the edge of tool.

Depending of the chip thickness we have different type of blunting whether craters or rounding of the blade. The equivalent length of the heat inflow on the front face depends primarily on the thickness of the chip and its value, from experimental observations and theoretical considerations, may be taken as:

$$L=(1,2-1,3)h \qquad (8)$$

In figure 2 we illustrate the process of cutting for thick and thin chip.

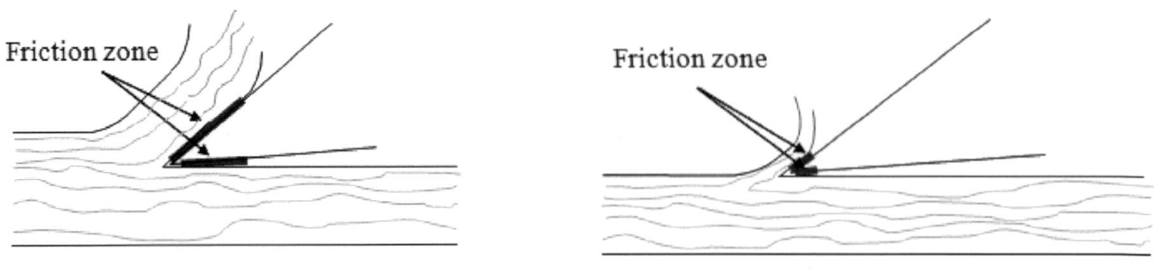

Fig. 2. Process of cutting with thick and thin chip

Increasing the thickness of the chip, the length of the heat inflow on the front face will increase. The chip takes a more curled shape if the chip is thinner and straighter if the chip is thicker. It is one of reasons that the length of heat inflow is longer when the chip is thicker. Also, the force exerted on cutting edge is higher with thicker chip resulting with larger heat generated on the front surface of the cutting tool.

With different thickness of the chip one can differentiate two types of cutting edge wear. In figure 3 we can see the different types of cutting edge wear in function of thickness of the chip and the cutting speed. Working with higher cutting speed the blunting is in form of craters already at a slight thickness.

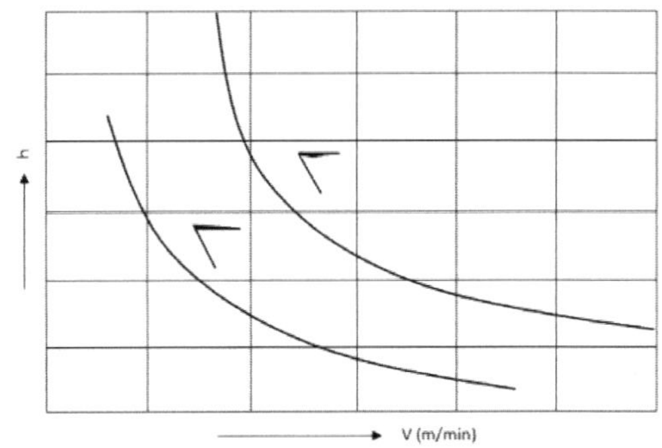

Fig. 3. Different types of tool blunting

Applying the boundary conditions in function of length of the chip in contact with the tool, respectively with cutting edge we change temperature field inside the tool. During the cutting process, thicker chip is longer in contact with tool then thinner ones. These boundary conditions will be applied during the numerical simulation via Boundary Element Method.

3. Boundary element method - BEM

Many numerical methods are available to solve thermal models and have been applied to solve thermal problems in cutting tools. In this study we have used the Boundary Element Method (BEM). BEM has the advantages that we profit in our study; the resolution in certain conditions is faster than with other numerical methods and it is easy to get solution in every point anywhere inside the domain.

Here below discretization is made for a simple two dimensional geometry with elements, nodes and internal points (figure 4).

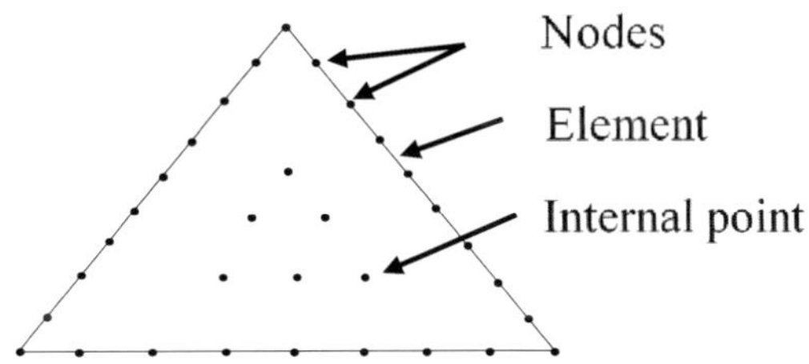

Fig. 4. Example of BEM meshing in 2D

Stationary heat conduction processes are governed by Laplace's equation. The problem is therefore written in general by the following equation:

$$\forall M \in \Omega,\ \vec{\nabla}^2 T(M) = 0 \qquad (9)$$

With following boundary conditions:

$$\forall M \in \Gamma_T, \qquad T(M) = T_0 \qquad (10)$$

$$\forall M \in \Gamma_\phi,\ -\lambda\vec{\nabla}T(M)\cdot\vec{n} = \phi_0 = -\lambda q_0 \qquad (11)$$

Note $q = \vec{\nabla}T\cdot\vec{n}$, the temperature gradient in the outward normal (and $q^* = \vec{\nabla}T^*\cdot\vec{n}$).

4. Fundamental Solution

The fundamental solution T* satisfies Laplace's equation and represents the field generated by a concentrated unit charge acting at a point 'i'. For an isotropic medium 2D, T* and q* are Green functions.

$$T^* = \frac{1}{2\pi}\ln\frac{1}{r} \qquad (12)$$

$$q^* = \vec{\nabla} T^* \cdot \vec{n} = \frac{d}{2\pi r^2} \qquad (13)$$

Where r is the distance from the point of collocation P to the point M of the considered border ($r = \|r\|$ and $r=PM$). "d" is the distance of the opposite projection of r on the outgoing normal M ($d = -\vec{r} \cdot \vec{n}$). The distances r and d are shown in figure 5.

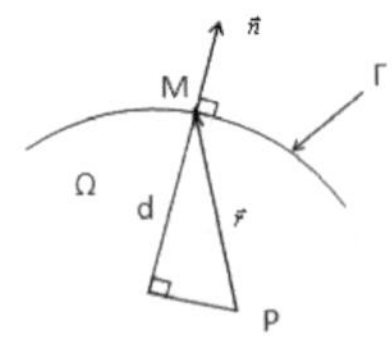

Fig. 5. Geometry definitions

5. Boundary integral equation

Using the definition of the Dirac function δp we obtain the integral formulation on the boundary. Applying this equation to a collocation point i (or the boundary area) [7], we have:

$$c_i T_i + \int_\Gamma T q^* d\Gamma = \int_\Gamma q T^* d\Gamma \qquad (14)$$

The most important features of boundary elements, for a 2D problem, however is that it only requires discretization of the contour rather than the surface. The calculation of integrals Γ of this equation is equivalent to a 1D integrals on the border. It is divided into elements which may be chosen from several types according to the degree of interpolation and continuity. In our simulations we have used the constant elements. The integral equation is initially applied to the nodes of the boundary, which allows to obtain solutions at these points. In a second step, the equation can be applied to points within domain called internal points.

If boundary integral equation is applied to the ''i'' collocation point located within the domain, the form c_i factor takes the value 1.

$$T_i = \int_\Gamma q T^* d\Gamma - \int_\Gamma T q^* d\Gamma \qquad (15)$$

6. Results and discussion

For first calculation we have validated our simulation on bisector line with experimental results. In this study we have taken a cutting tool to figure out the temperature distribution in this type of tools. The sharpening angle β of the tooth in the first simulation is 45^0. We take 28 elements and 9 internal points placed on bisector line. Comparing numerical with experimental results we have a good accordance (figure 6).

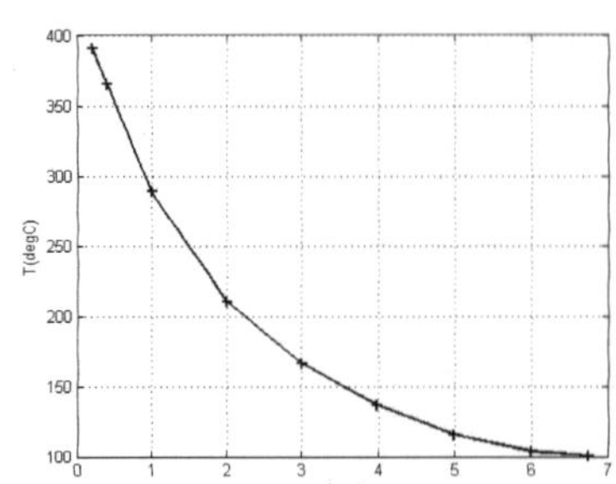

Fig. 6. Validation of BEM simulation

In next instance we have made a simulation for the case of simple blade applying two different thicknesses of chip. Those different thicknesses cause different types of blade blunting. Simulation is made with 28 elements and 18 internal points. The mesh of contour (1D) is finer near the cutting edge. In this section we place more internal points in denser spacing. First case with thin chip shows that we have occurrence of high temperature field close to the cutting edge causing the plastic deformations of the edge (figure 7.a.). This is the situation when we will have rounding of the cutting edge. The second case, with thicker chip shows that the plastic deformation is displaced with temperature field a little bit from cutting edge and that will cause deterioration of front surface forming shape of craters (figure 7.b.). This amount of displacement is correlated to the thickness of the chip and certainly does not just depend of it but also of other factors (material of tool, parameters of productions, main sped etc.) [8].

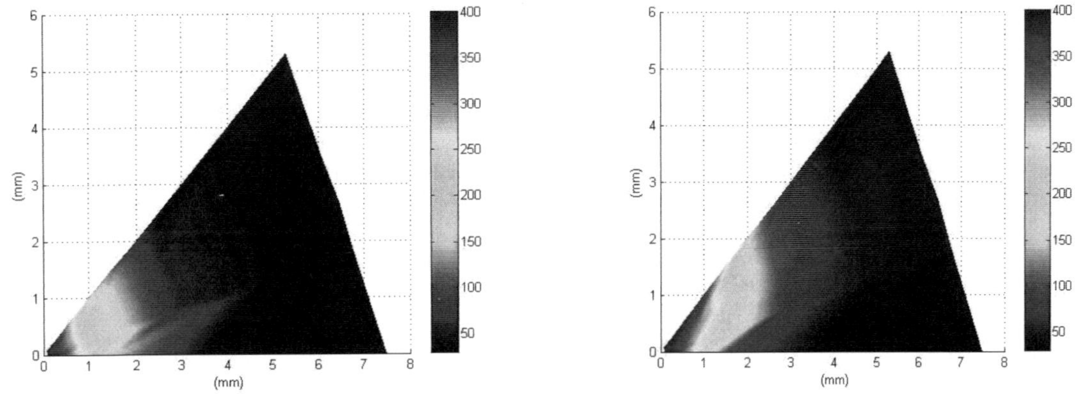

a) b)

Fig. 7. Temperature distribution for thin a) and thick b) chips

For the real case we take the geometry of the lathe tool on high speed turning. This case is also studied in 2D. The geometry of tool is shown in figure 8. The sharpening angle of this tool is 60°. To make this simulation by BEM we took 35 elements and 63 internal points, shown in figure 8 .In figure 9 we have shown the simulation for temperature distribution, processing the workpiece with thin a) and thick b) chip. Near the cutting edge we have generate finer mesh of the contour and we have concentrated dense number of internal points to have representative results in isographs. We applied the boundary conditions similar to the previous case.

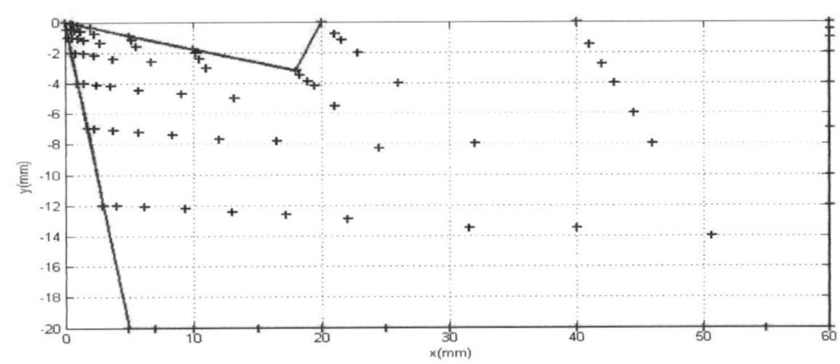

Fig. 8. Mesh of the contour of the lathe tool with internal points

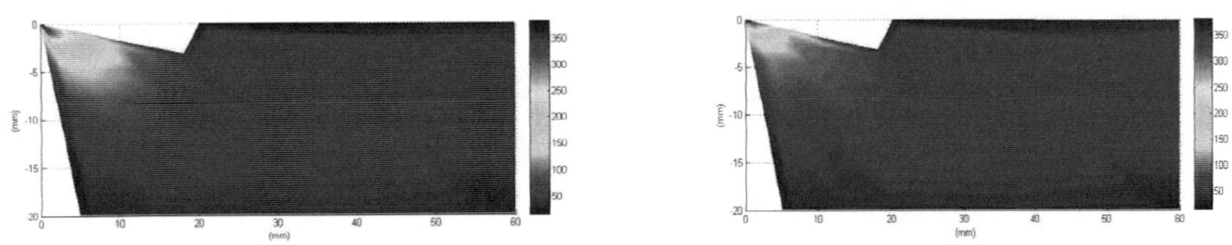

a) b)

Fig. 9. Temperature distribution for lathe tool for thin a) and thick b) chips

Knowing the thickness of the chip during the process, and helping us with the numerical simulations we can predict the way of cutting edge blunting, whether the occurrence of craters or rounding of the blade. In case of lathe tool we get the similar results such as in the case of simple cutting edge. The most interested regions to observe is near the cutting edge because the temperatures are quasi constants at the rest of geometry and quietly low to influence the tool behaviour. For that reason, the small exchangeable blades are usually used instead of complete tool.

7. Conclusion

Taking different technological conditions of cutting especially with different thickness we have different type of tool blunting. Experimental measurements inside the tool give us inputs to get the numerical solution over whole domain. This inverse technique can be useful to get rapid solution of temperature distribution inside the domain and on the surface of the cutting tool. Here, we use numerical tool via BEM simulation to have very close idea about the way of the tool blunting. Blunting of the tool is directly in relation with plastic deformations respectively with temperature increase. With thicker chip, the chip is straighter and longer in contact with front surface exerting higher force. This force produces immediately larger quantity of heat that is transmitted to the tool. Working with given main speed and thickness under certain value we get the blunting of the tool in the form of rounding of cutting edge. Above certain value of chip thickness we get the blunting in form of craters on front surface and rake face of the tool. The blunting will be pronounced on front surface of the tool. Of course, the technological conditions of cutting depend of other factors that can influence the cutting process such as ability of machines, material of tool, quality of product etc.

Mostly, the thickness of the chip depends of the required quality of wood cutting. Depending of the thickness of the chip, by BEM solution, we can determine the critical zone of tool wear. To decrease the tool wear, one solution is to cool this critical zone by refrigerated air and to use the coatings with good frictional characteristics.

8. References

[1] Sheikh-Ahmad, J. Y., Bailey, J. A., (1999). High-temperature Wear of Cemented Tungsten Carbide Tools While Machining Particleboard and Fiberboard, Journal of Wood Science, Vol. 45, No. 6, pp. 445-455, ISSN: 1435-0211

[2] Chao, L. & Xing, A., (2010). Study on Cutting Temperature and Wear Mechanism of Tool in High Speed Turning Superalloy GH2132, International Conference on Computing, Control and Industrial Engineering, Vol. 2, pp. 327-330, Wuhan, DOI:10.1109/CCIE.2010.199

[3] Csanady, E. (2007). Thermal loading in wood cutting tools, Proceedings of Third International Symposium on Wood Machining, ISBN 987-2-88074-725-1, pp. 151-154, Presses polytechniques et universitaire romandes, Lausanne

[4] Gazo, R., Gisip J. & Stewart H. A. (2011). Reducing tool wear by cryogenic treatment and cooling with refrigerated air when processing medium density fibreboard, In: Wood Machining, J. Paulo Davim), (Ed.), 28 (83-111), ISTE, ISBN 978-1-84821-315-9, London

[5] Sheikh-Ahmad, J. Y., Lewandowski, C. M., Bailey, J. A. & Stewart, J. S., (2003). Experimental and Numerical Method for Determining Temperature Distribution in a Wood Cutting Tool, Experimental Heat Transfer, Vol.16, No. 4, pp. 255-271, ISSN: 0891-6152, DOI: 10.1080/08916150390223092

[6] Csanady, E. & Magoss, E. (2013). Mechanics of Wood Machining, Springer, ISBN 978-3-642-29954-4, Berlin Heidelberg

[7] Brebbia, C.A. & Dominguez, J. (1992). Boundary Elements an Introductory Course, WIT Pess/Computational Mechanics Publications, ISBN 185312160 6, Southampton

[8] Horman, I., Busuladžić, I. & Azemović, E. (2014). Temperature Influence on Wear Characteristics and Blunting of the Tool in Continuous Wood Cutting Process, Procedia Engineering, Vol. 69, Katalinic, B., pp. 133-140, DOI.org/10.1016/j.proeng.2014.02.213.

DOI: 10.2507/28th.daaam.proceedings.053

IDENTIFICATION OF SPECIAL CAUSES

Katerina Bicova, Lukas Bebr

This Publication has to be referred as: Bicova, K[aterina] & Bebr, L[ukas] (2017). Identification of Special Causes, Proceedings of the 28th DAAAM International Symposium, pp.0379-0383, B. Katalinic (Ed.), Published by DAAAM International, ISBN 978-3-902734-11-2, ISSN 1726-9679, Vienna, Austria
DOI: 10.2507/28th.daaam.proceedings.053

Abstract
This paper deals with identifying the special causes of variability, which cause fluctuations in the monitored production process. SPC is used for uncovering these causes. The control diagrams can be described as a simple and effective tool for understanding the process under monitoring and therefore we can rank it as a commonly used tool in industrial practice. Critical criteria for determining the build-up of the control charts are the criteria for determining whether the process is stable or unstable. The aim of the paper is to focus attention on the case, where the process is unstable and the causes of fluctuations beyond the set limits, or the occurrence of other undesirable trends, need to be identified. Because the identification of special causes is essential for process control and further process monitoring and evaluation of the processes.

Keywords: SPC; control diagrams; special causes; manufacturing process; evaluation

1. Introduction

In today's industry, especially in the automotive industry, it is very important to monitor and follow up primary key production processes. For this purpose a very useful tool is the Statistical Process Control (SPC). It can be said, that SPC is a simple and effective instrument for understanding the process. For evaluation of the devised control charts determining criteria are valid, based on which we can ascertain whether the process is stable or unstable. The aim of this paper is to focus attention on the case where the process is unstable and where it is necessary to identify the causes of fluctuations beyond the established limits or the occurrence of other undesirable trends. The identification of special causes is essential for process control and further process monitoring and evaluation of the process.

Specific or definable causes relate to sources of fluctuation, that aren't permanently affecting the process and producing unexpected changes. These can be either harmful or beneficial.

Detecting the presence of special causes is the role of SPC. Elimination of special causes is then done in the form of:

- local actions (operating staff, operator within the scope of their authority)
- action within the system (these fall under the responsibility of the management).

The specific cause of variability needs to be identified, eliminated and safeguarded, so that it can no longer be repeated. [1].

As has been said, progressive SPC methods are used for detailed analysis of the production process. This ensures the identification of adverse effects and their correlation with the behaviour of the production process. It is a tool that can be used to monitor the long-term stability of the given system and to ascertain whether the monitored process behaves as expected. It is possible to assess whether the variability of the parameter is due to random fluctuations, or if it is a special cause. Use is possible everywhere, where the parameter is progressively monitored in time [2,3]

Control charts help by measuring to explain the results from the view of the process dispersion (variability) and also location (mean value of the process). Therefore the control charts are always presented in pairs. The most common pair of charts are (X, R) charts. These are the values of the sampling mean of the values obtained from the subgroups, which is the process position measure, and the range of values in each subgroup, which is the degree of process dispersion.

These diagrams are suitable tools for measuring process fluctuations the span can be easily calculated and is substantial for small ranges of subgroups. Graphic presentation is desirable in the SPC for its clearness and readability. [4]

2. Example - a specific solution

For a specific example of using SPC and evaluating the manufacturing process data from serial production of a simple machined part of a cylindrical shape are used here. The dimensions of the component are controlled by the innovative measurement technology system Equator. This innovative measurement system allows for the required 100% control, because it is a very precise and used for manufactured components marked in the automotive industry as critical. [9] The basic principle of the measurement process is based on the comparison of the measured part and the etalon. Upon customer agreement the diameter of the component has been determined as the critical feature.

From each day the relevant data are collected from machined parts inspections. In terms of cost-effectiveness and SPC effectiveness, it is important to correctly determine the extent of the logical subgroup and the length of the inspection interval. [5]

The measured data from one shift from 6:45 am to 2:15 pm are divided into subgroups after 15 minutes. Every 15 minutes 10 samples are taken. First, the basic statistical characteristics required for the construction of the control charts for the selected shift respectively given day are stipulated.

Calculated statistics for CD:	
n_j	10
k (amount of subgrups)	30
x bar bar	24.030180
Me bar	24.030300
R bar	0.005300
s bar	0.001668
s^2 bar	0.000003
s tot	0.001937
s xbar	0.000894

Table 1. Calculated statistics for control diagrams

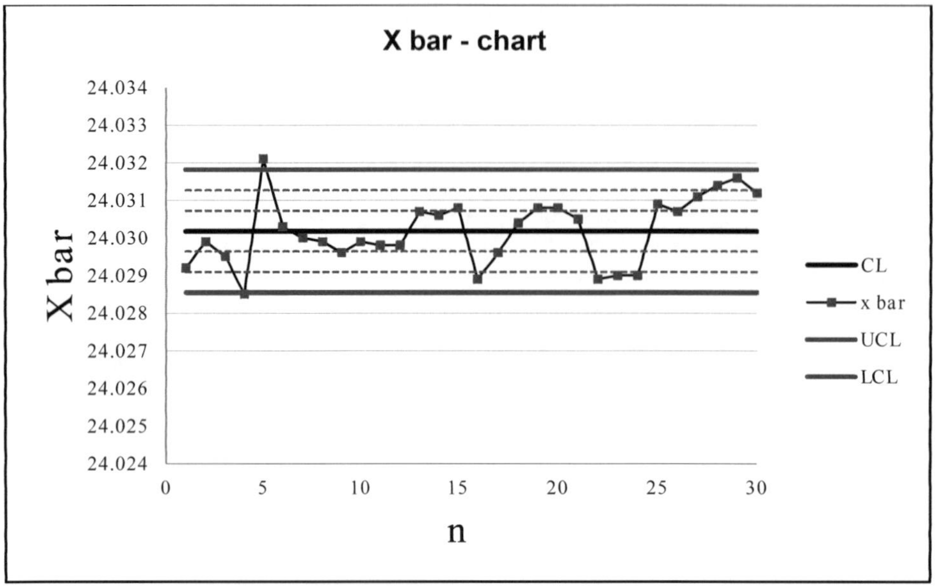

Fig. 1. Control chart for mean

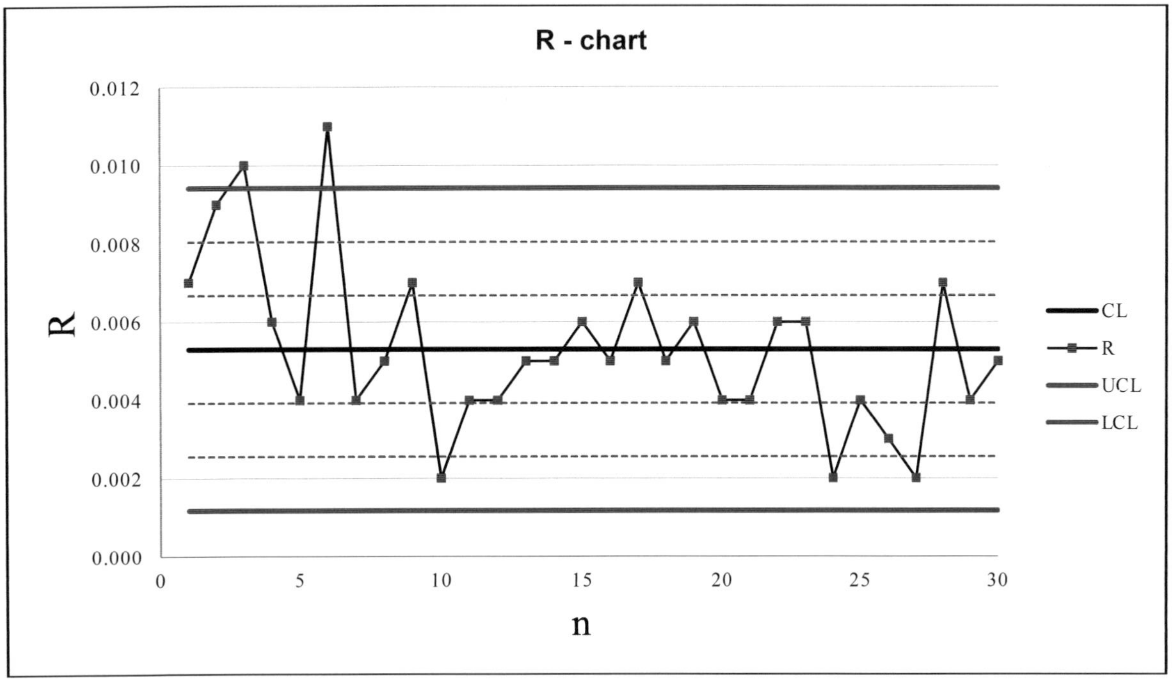

Fig. 2. Control chart for range

After the basic statistical characteristics are determined, control charts are construction. Figure 3 shows an example charts for the average value (mean) and figure 4 for the span.

3. Solution - Evaluation of the control charts

For regulation charts other decisive criteria also apply by which we can determine, whether the process is stable or unstable. In order to determine the stability of the process, i. e assessing the need to intervene in the process, e.g. the so-called non-random group tests or special groupings of points or the rules for determining special cases in the charts for the average from ISO 8258: 1998 [6, 7] are used. For example follow figure 3.

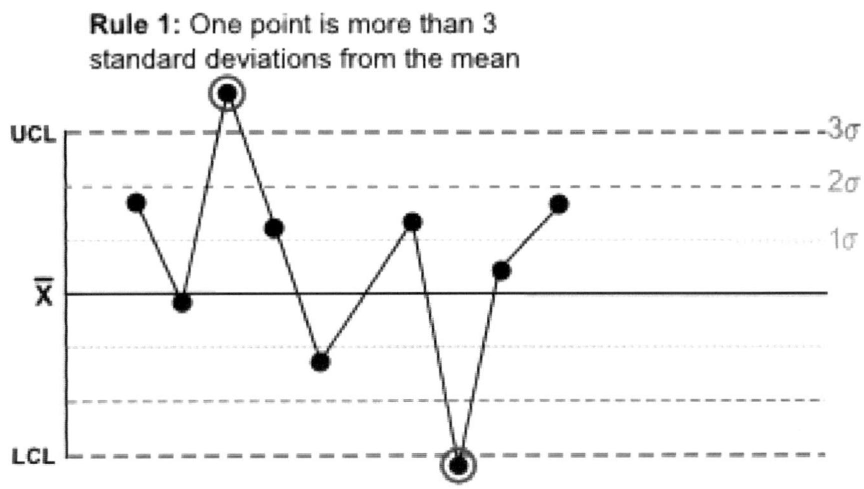

Fig. 3. Example of Nelson's rules [7]

Based on these sources, the built-in charts were evaluated and the following occurred (figure 4):
1. Points exist outside the regulation limits, namely the 4th subgroup under the UCL
2. Points exist outside the regulation limits, namely the 5th subgroup above the UCL.

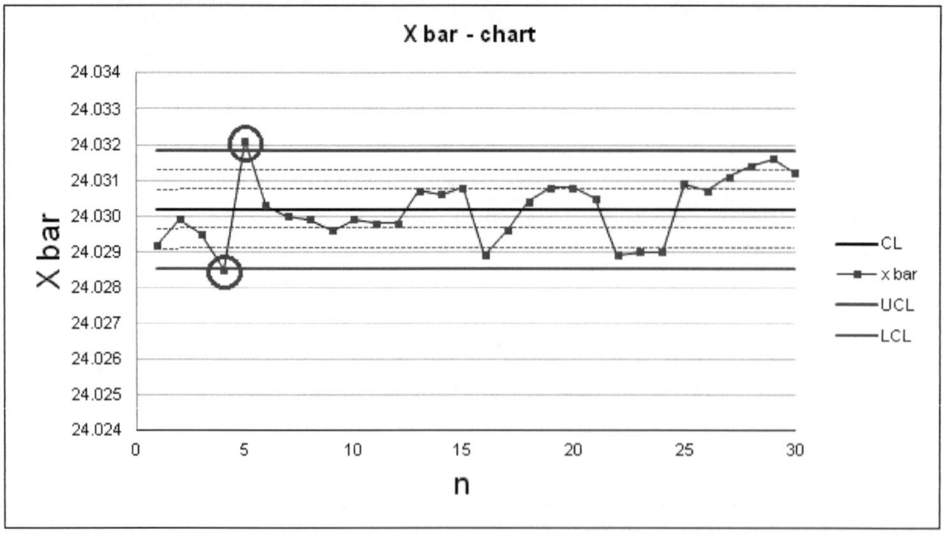

Fig. 4. Control chart with marked points outside the regulation limits

These phenomena can indicate that the process has changed at this point, the measuring system has changed, the control threshold is incorrectly calculated, or the point is poorly plotted in the graph etc.
On the basis of signaling the instability of the process, its causes were identified. One of the basic tools of quality management, namely the Ishikawa diagram (figure 5), was used for this purpose. The most probable causes, such as the used methods, the machine-tool, the cutting tools and the operators present in the production process, and others were analyzed.

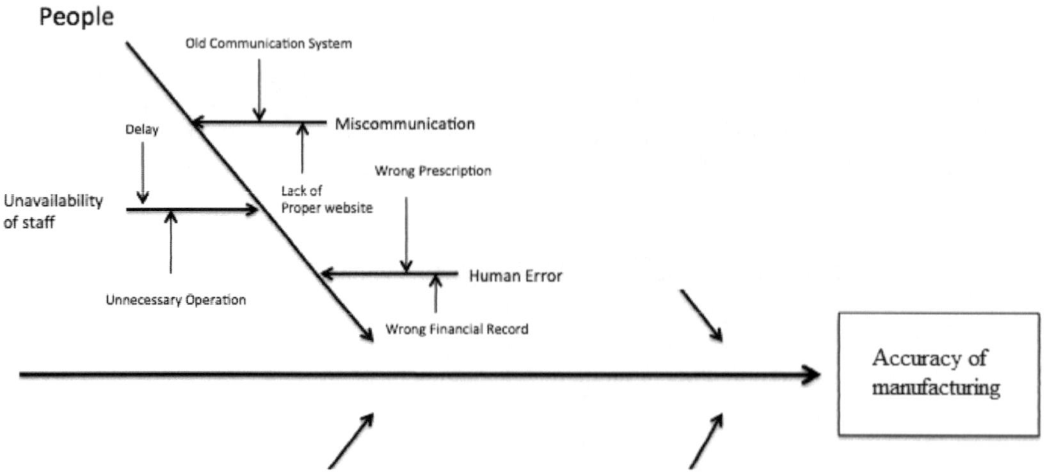

Fig. 5. A sample search for causes

As a result of this analysis, it was found that the cause of process fluctuation, i.e. points beyond the boundaries, was the lack of manufacturing precision of the machining process with respect to the influence of the human factor.
So this is a local cause where the operator made an unintentional error in changing of the cutting tool, caused by his inattention. The cutting tool has a prescribed replacement frequency in the working control procedure always after a specified number of machined pieces. This change was not performed by the operator and therefore a deviation was caused. The change took place later. This deviation was reflected in the accuracy of several manufactured pieces and is also evident in the control charts.
On the basis of the analysis and the identified cause, a corrective action was taken, namely re-training of the attending staff at the machine. If the cause were to repeat, it was proposed to further take another precaution in the form of setting the software of the machine so, that after a certain number of starts of the machining process the tool replacement is requested automatically from the operator.
Manufactured parts with insufficient manufacturing precision were blocked and therefore removed from the sample data set and the control charts were revised. In this case, the process was already stability and capable.

4. Conclusion

Please improve the conclusion. The conclusion must give clear answers on questions: what was the problem, how was it solved, what are the results/solutions, what is solved, what are future plans.
Human beings have a very important and indivisible share in the entire production process. Quality work in this area is very important. However, it is important to realize that high-quality work is not natural, but it is consistent and good-quality preparation, regular training and responsibility of each individual. In the future, the research team would like to concentrate on monitoring the manufacturing accuracy of the machining process with respect to the influence of the human factor. For example, shift work is common in many professions that directly affect the health and safety of others. The quality of life, health, safety and protection at work on shifts and commuting can affect workers in any field and thus their resulting work, e. i. total manufacturing process. [8]
This paper presented a way of evaluating the production process by using data from serial production, where was checked the critical character determined by the customer. From the measured data were created control charts. On the basis of this, the instability of the process was determined and its causes were determined by compiling the Ishikawa diagram. Based on the results was detected a local cause caused by a human factor. Corrective action has been taken and in case of repetition, local measures are proposed directly by setting the software on the machine. With further data collection, the process was already stability and capable.
The human factor has a significant share in the level of precision of production. It is a risk factor that must be remembered in the planning, production and control of machinery, especially in the automotive field, where the precision and quality of production is required on the first place.

5. Acknowledgments

This paper includes results created within the SGS-2016-005: Research and development for innovations in the field of mechanical engineering technology - machining technology.

6. References

[1] Jarošová E. a D. Noskievičov. (2015). Pokročilejší metody statistické regulace procesu. Grada Publishing a.s. ISBN 978-80-247-5884-8. Praha.
[2] Tabak, John. Probability and statistics. (2004). The science of uncertainty. Fact On File, Inc. ISBN 0-8160-4956-4. New York.
[3] Hrehova S. (2016). Predictive Model to Evaluation Quality of the Manufacturing Process Using Matlab Tools, In Procedia Engineering, Volume 149, 2016, Pages 149-154, ISSN 1877-7058, https://doi.org/10.1016/j.proeng.2016.06.649.
[4] Horálek, V. (1999). QS - 9000 SPC. ČSJ. ISBN 80-02-01293-3. Praha.
[5] Meran, R., John, A., Roenpage, O., Staudter, Ch. (2013). Six Sigma + Lean Toolset (Mindset for Successful Implementation of Improvement Projects). Springer. ISBN 978-3-642-35881-4. Berlin.
[6] ČSN ISO 8258. (1998). Shewhartovy regulační diagramy(Shewharts control charts). Český normalizační institut. Praha.
[7] Nelsonova pravidla pro regulační diagramy. [Online] [Cit.: 23. 5. 2017] http://upload.wikimedia.org/wikipedia/commons/7/72/Poster_-_Control_Charts_for_Nelson_Rules.svg
[8] Kenneth P. Wright Jr., Richard K. Boganb, James K. Wyatt (2013). Shift work and the assessment and management of shift work disorder (SWD), Sleep Medicine Reviews, Volume 17, Issue 1, February 2013, Pages 41–54 http://www.sciencedirect.com/science/article/pii/S1087079212000251
[9] Martin Melichar, Dana Kubátová. (2015). Processing Data from Automatic Measurement Device, Procedia Engineering, Volume 100, 2015, Pages 899-906, ISSN 1877-7058, http://dx.doi.org/10.1016/j.proeng.2015.01.447

DOI: 10.2507/28th.daaam.proceedings.054

CUTTING EDGE MICROGEOMETRY AND PREPARATION METHODS

Tomáš Vopát, Marcel Kuruc, Vladimír Šimna, Rudolf Zaujec, Jozef Peterka

This Publication has to be referred as: Vopat, T[omas]; Kuruc, M[arcel]; Simna, V[ladimir]; Zaujec, R[udolf] & Peterka, J[ozef] (2017). Cutting Edge Microgeometry and Preparation Methods, Proceedings of the 28th DAAAM International Symposium, pp.0384-0391, B. Katalinic (Ed.), Published by DAAAM International, ISBN 978-3-902734-11-2, ISSN 1726-9679, Vienna, Austria
DOI: 10.2507/28th.daaam.proceedings.054

Abstract

The paper deals with the issue of cutting edge preparation and tool treatment before the coating and their influence on the tool life of coated cutting tools. The aim was to examine the influence of cutting edge radius sizes on the tool life of cemented carbide turning inserts. Theoretical analysis was focussed on the issue of cutting edge microgeometry, discussing the classification and importance of cutting edge microgeometry. The most famous and used edge preparation methods are summarized in article along with the results of current research of famous tool companies and institutions. In this article, the tool lives of cemented carbide turning inserts which were prepared by several methods were compared. Cemented carbide turning inserts were deposited by AlCrSiN nanocomposite hard coating. Long-term tool live tests were carried out on DMG CTX alpha 500 turning centre. The tool wear of cemented carbide turning inserts was measured on Dino-Lite digital microscope during the turning. The tool lives of prepared cemented carbide turning inserts with required cutting edge radius and unprepared ones were compared. These results were investigated in the process of machining of austenitic stainless steel material X6CrNiTi18-10 by DIN EN (AISI 321), which shows higher strain hardening tendency.

Keywords: Cutting Edge Preparation; Treatment; Cemented Carbide, Cutting Tool; Cutting Edge Microgeometry

1. Introduction

In the past years research in cutting tools field focused on optimum cutting edge microgeometry development of the cutting edge geometry and also on the preparation of the cutting tool surface for the coating process. The cutting edge geometry and the applied edge preparation methods affect the tool wear of coated cutting tools [1-3]. A thorough study of the efficiency of various cutting edge rounding methods as well as the effects of these treatments on the wear behaviour of coated tools in milling various metallic materials is introduced in [1].

In terms of edge characterization, it has thus far been expedient in most instances to specify the geometry of honed edges by a simple edge radius parameter. This inherently assumes the edge profile that bridges the rake and flank faces to conform to an arc of a circle. This need not however be the case, and indeed tool performance may be enhanced by rendering the cutting edge to be appropriately asymmetric [4]. Given the critical influence of edge geometric attributes on process responses, and the evolving capability of aforementioned novel processes in the generation of tailored cutting edges, it is imperative to develop methodologies for the comprehensive geometric characterization of the cutting edge.

Publication [2] deals with the complexity of cutting edge microgeometry and evaluates its interactions with machining processes.

Publication [5] is concerned with the requirements and challenges in preparing and designing the cutting edge micro shape using wet abrasive jet machining. Wet abrasive jet machining with a robot guided system allows to prepare local tool areas and to realize a specific design of the cutting edge, as well as advantageous surface qualities.

Yussefian et al. [6] investigated sink electrical discharge machining as an optional method for controlled preparing of cutting edges by sinking the cutting edge into a counterface. Although the manufactured rounding variability is stated as minimal, the capability to prepare straight cutting edges is not limited by the tools' material hardness.

The application of laser machining for manufacturing the cutting edge roundness of cemented carbide inserts could have a beneficial effect on the wear behaviour of coated tools. The main advantage of using laser machining for manufacturing the cutting edge radii is the achievement of a reproducible micro-geometry [7, 8] as well as the capability of versatile application with respect to manufacturing of complex-shaped geometries and setting up high surface qualities for cemented carbide tools [9]. In [10], the feasibility of using laser machining for manufacturing the cutting edge roundness of cemented carbide inserts was investigated. In this context, diverse laser machining parameters such as of the laser-pulsing duration and –beam feed velocity were applied for shaping the cutting edge roundness. Hereupon, shorter laser pulsing durations in the order of pico-seconds and larger feed velocities eliminate the negative effects of a thermal affected zones associated with its thickness, material properties and adhesion on the coated tool life. Moreover, the conduct of micro-blasting on the already laser-treated cemented carbide substrates increases the effective coating adhesion, thus leading to a cutting performance improvement of the coated tools.

Experimental study on the surface roughness of tungsten alloy in Abrasive water jet cutting and laser cutting is discussed in [11, 12]. In [13, 14] authors investigated influence of the cutting edge preparation on tool life of cemented carbide inserts when milling. Publication [15] deals with testing cemented carbide cutting tools with different cutting edge qualities obtained by different grinding conditions during the experimental cutting of Inconel 718 super-alloy.

The role of the meso-geometry of the cutting edge on the mechanics of chip formation was first documented by Albrecht [16] in 1960. It has however taken several decades since, for this important aspect to be accorded its due consideration. As the meso-geometry of a tool edge profile comprises two tool faces that flank the cutting edge (Fig. 1.), unambiguous identification of the transition points that separate the edge from the tool faces is the first step that should precede edge characterization [17].

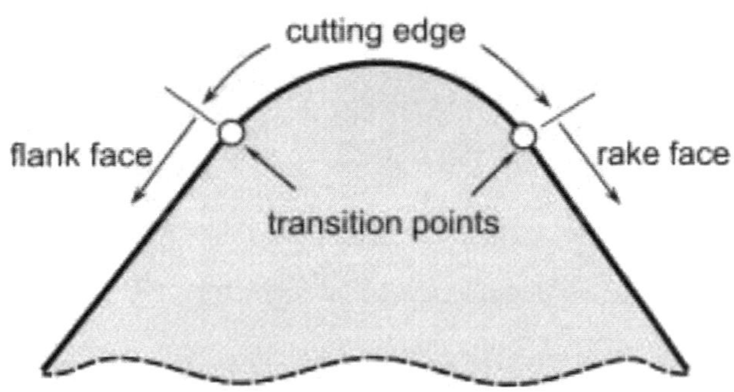

Fig. 1. Terminology of tool edge profile [17]

In terms of edge characterization, it has thus far been expedient in most instances to specify the geometry of honed edges by a simple edge radius parameter. This inherently assumes the edge profile that bridges the rake and flank faces to conform to an arc of a circle. This need not however be the case, and indeed tool performance may be enhanced by rendering the cutting edge to be appropriately asymmetric [4]. Given the critical influence of edge geometric attributes on process responses, and the evolving capability of aforementioned novel processes in the generation of tailored cutting edges, it is imperative to develop methodologies for the comprehensive geometric characterization of the cutting edge.

Denkena et al. [4] proposed that the edge geometry be characterized with reference to the virtual tool tip derived from the linear extension of flank and rake faces (Fig. 2). Any asymmetry in the edge geometry is signified by the ratio of distances S_α and S_γ from the tool tip to points 1 and 2, from where the edge profile diverges away from the flank and rake faces, respectively. The degree of edge flattening is specified by parameter Δr which is the distance from the virtual tool tip to the apex of the edge profile, and parameter φ locates the tool apex relative to the tool faces. Such a characterization is simple and facilitates easy visualization of the edge; however, the said parameters are evaluated based on just three points on the edge profile, which are not adequate to uniquely characterize the edge geometry. Furthermore, publication [4] did not specify a method to objectively determine the transition points 1 and 2 wherefrom the edge profile deviates off the tool faces.

Rodriguez [18] and Wyen et al. attempted to address the issue above by defining the different approaches as Denkena et al. described. Wyen et al. [19, 20] recently proposed a method for identifying the transition points that delineate the meso-geometry of the cutting edge from the flank and rake faces of the tool.

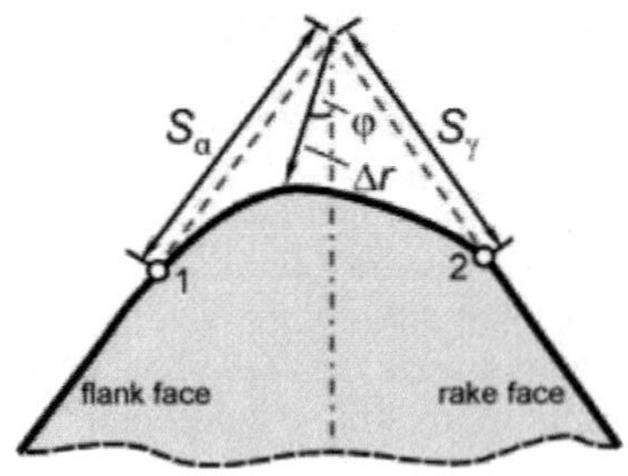

Fig. 2. Edge characterization methods due to Denkena et al. [4]

In publication [21], Denkena et al. presents a novel method for preparing of customized cutting edges by means of grinding. The reliability of the new method has been proven and the dependency of the obtained relative error on the chamfer size discussed. Moreover, an evaluation of the edge chipping has been carried out. It has been shown that preparing the edge as proposed by the method reduces the edge chipping by approximately 1 μm in comparison to a sharp edge. Varying the micro geometry does not lead to a variation of the edge chipping. Influence of used edge preparation method on the coating adhesion to substrate is discussed in [22].

2. Cutting edge preparation

Article deals with the issue of cutting edge preparation and tool treatment prior to coating and their influence on the tool life as mentioned before. It is a well-known fact that cutting tool wear is a function of basic process parameters (Fig. 3). The publications in introduction presented that cutting the edge microgeometry has a great influence on the tool life. Blue boxes in Fig. 3 show the factors that were tested and changed in the article.

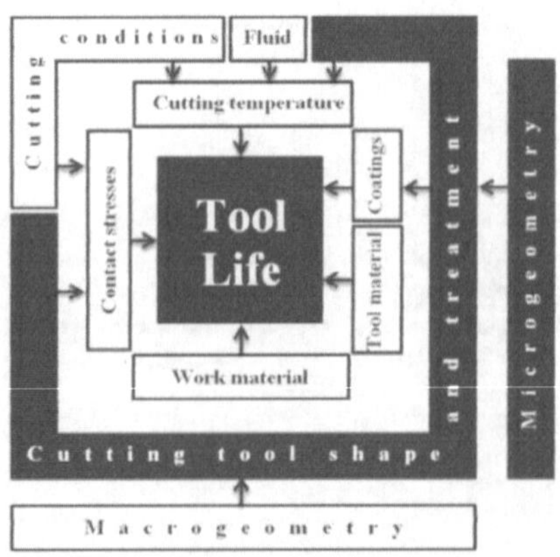

Fig. 3. Factors that were tested and changed

The aim of the article was to investigate influence of cutting edge radius r_n sizes on the tool life in turning of austenitic stainless steel material where authors expected the great importance for this material group. The other aim was to compare the various edge preparation methods in term of tool life of cemented carbide turning inserts. With support of the Pramet Tools, s.r.o. Company, two industrially usual edge preparation methods were selected in this article:

- brushing is a method uses a rotary brushing with SiC grains
- wet microblasting with Al_2O_3 medium

Furthermore, cemented carbide turning inserts were manufactured with the three cutting edge radius sizes: $r_n = 20$ μm, $r_n = 40$ μm, $r_n = 60$ μm. The fourth cemented carbide turning insert was manufactured with $r_n = 7.49$ μm, where dry microblasting was only used for deburring but cutting edges were not further prepared. Then, cemented carbide turning inserts were cleaned in Elmasonic P ultrasonic unit. After cleaning, all cemented carbide turning inserts were deposited by AlCrSiN nanocomposite hard coating when deposition conditions were constant. Since, the coating thickness was 4 μm, coated cemented carbide turning inserts were tested with the following cutting edge radius sizes in tool life test:

- $r_n = 11.49$ μm
- $r_n = 24$ μm
- $r_n = 44$ μm
- $r_n = 64$ μm

Zeiss LSM 700 laser confocal microscope and Accretech Surfcom measurement device were used to measure the cutting edge radius sizes.

3. Materials and methods

3.1. Tested cutting tools and workpiece material

The CNMG 120408E-SM turning inserts and PCLNL 2020K12 toolholder were selected for this research. Tested cutting tool material was cemented carbide. Specific grade of cemented carbide is HF10 by tool producer equivalent to K20 - K30 by ISO (89.6% of WC, 10% of Co and 0.4% of other carbides). The tool holder was cut due to reduce the effects of tool overhang on tool wear. There was selected very universal grade of cemented carbide with respect to semi-finishing strategy, PVD coating and testing of various workpiece materials.

The selected workpiece material was austenitic stainless steel material of DIN EN X6CrNiTi18-10 (AISI 321) grade in this paper. Chemical composition is shown in Table 1. We suggested a round bar with dimensions of ø 100×150 mm.

Element	C	Si	Mn	P	S	Cu	Cr	Ni	Mo	Ti
wt. %	0.064	0.83	1.44	0.026	0.025	0.66	17.5	9.89	0.46	0.47

Table 1. Chemical composition of machined material X6CrNiTi18-10

3.2. Tool life test and cutting parameters

For long-term tool life tests, DMG CTX alpha 500 turning centre was used. Turning without live centre support has to conform to the slenderness ratio $L/D \leq 4.5$ [23].

In cooperation with Pramet Tools Company, the feed and depth of cut were obtained from their research, which was oriented on chip forming of austenitic stainless steel material. For a long-term tool life test, cutting speed was investigated experimentally.

For studied coated cemented carbide turning inserts, the investigation of cutting speed corresponding to a tool life of 15 min was performed. Since, studied cemented carbide turning inserts had a several edge preparation methods, several cutting edge radii and modified own coating, there was necessary to determinate optimum cutting speed. This experiment was aimed at determining the cutting speed for the target function: tool life of 15 min. For this experimental investigation, the coated cemented carbide turning inserts with cutting edge radius r_n = 40 μm (as average value) were selected. The inserts were prepared by brushing.

On the basis of chip forming tests and long-term tool life tests, the cutting parameters (cutting speed, feed rate and depth of cut) of AlCrSiN coated cemented carbide turning inserts were established for turning austenitic stainless steel material and they are shown in Table 2. Experimental investigation of cutting parameters for this research in detail is published in [22].

cutting speed v_c (m/min)	feed f (mm)	depth of cut a_p (mm)
127	0.2	1

Table 2. Recommended cutting parameters for coated cemented carbide turning inserts

Tool wear of coated cemented carbide turning insert was measured on Dino-Lite digital microscope. Alusol ABF 10 coolant was used in the machining process. Since the flank wear of cemented carbide turning inserts was dominant, it was measured every 0.1 m length of cut for the first tests. After obtaining the flank wear behaviour, we measured width of flank wear after particular period of time considering evolution of flank wear. The machining tests were stopped when width of the flank wear reached over 0.25 mm because too large notch and edge chipping were observed over this value. Every type of cemented carbide turning insert was tested three times in order to exclude the impact of other factors involved in the process. After that, the average value was calculated.

4. Results and discussion

The graphs in Figs. 4 and 5 were plotted based on the width of flank wear value of particular measurement expresses the time dependence of width of flank wear. The time dependence of width of flank wear for prepared cemented carbides turning inserts with required cutting edge radii r_n = 24, 44, 64 μm and for unprepared cemented carbide turning inserts with cutting edge radii r_n = 11.49 μm were then compared.

For unprepared turning carbide inserts with r_n = 11.49 μm, the width of flank wear value reached over 0.25 mm after 26.1 min of cutting. For cemented carbide turning inserts prepared by specific edge preparation methods, the tool life was determined from the previous graphs (Fig. 4. and 5.) and inserted to Table 3.

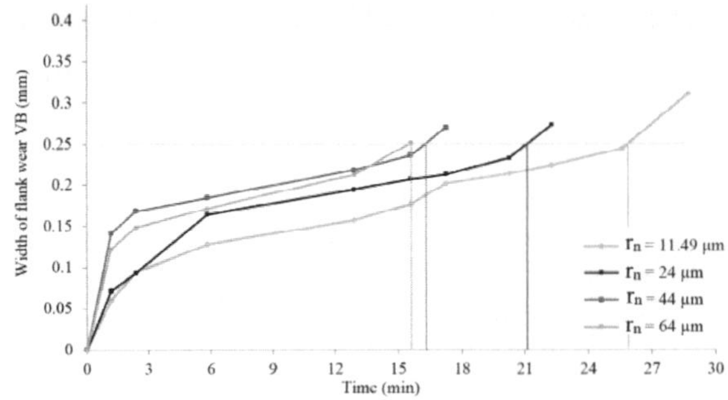

Fig. 4. Graph of time dependence of width of flank wear value VB (brushing)

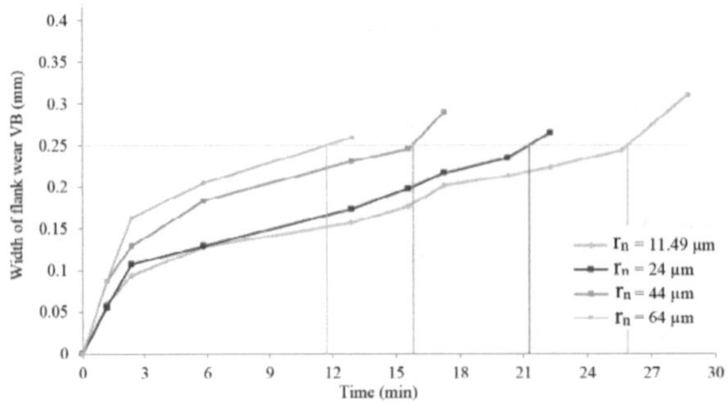

Fig. 5. Graph of time dependence of width of flank wear value VB (wet microblasting)

	Tool life (min)		
	r_n = 24 µm	r_n = 44 µm	r_n = 64 µm
Brushing	21.3	16.4	15.5
Wet microblasting	21.4	15.8	11.7

Table 3. The tool life of cemented carbide turning inserts prepared by brushing and wet microblasting

Fig. 6 records just some example of the flank wear evolution of unprepared cemented carbide turning insert during the turning. Observed were too large edge chipping and notch just over 0.25 mm of width of flank wear. Micrographs are magnified 230 times.

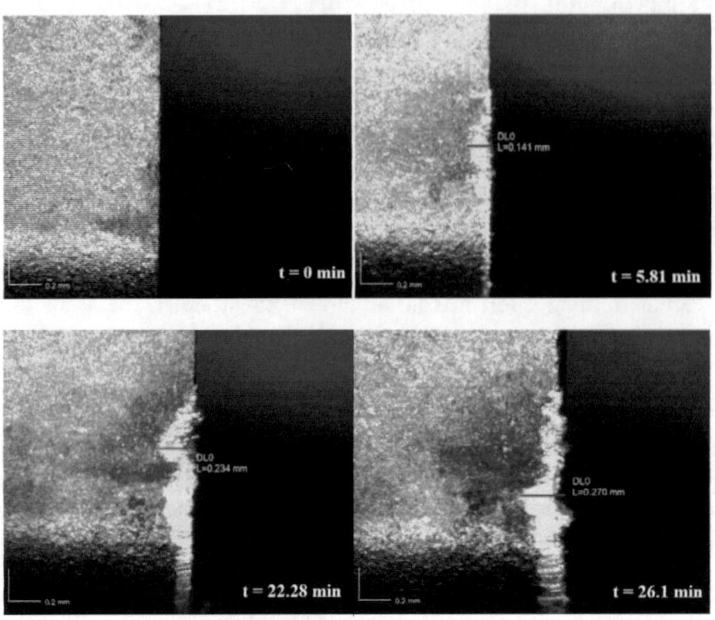

Fig. 6. Evolution of flank wear

Micrographs of flank wear for cemented carbide turning inserts prepared by brushing are shown in Fig. 7 in the first line and the second line records flank wear of cemented carbide turning inserts prepared by wet microblasting after attaining the value over 0.25 mm.

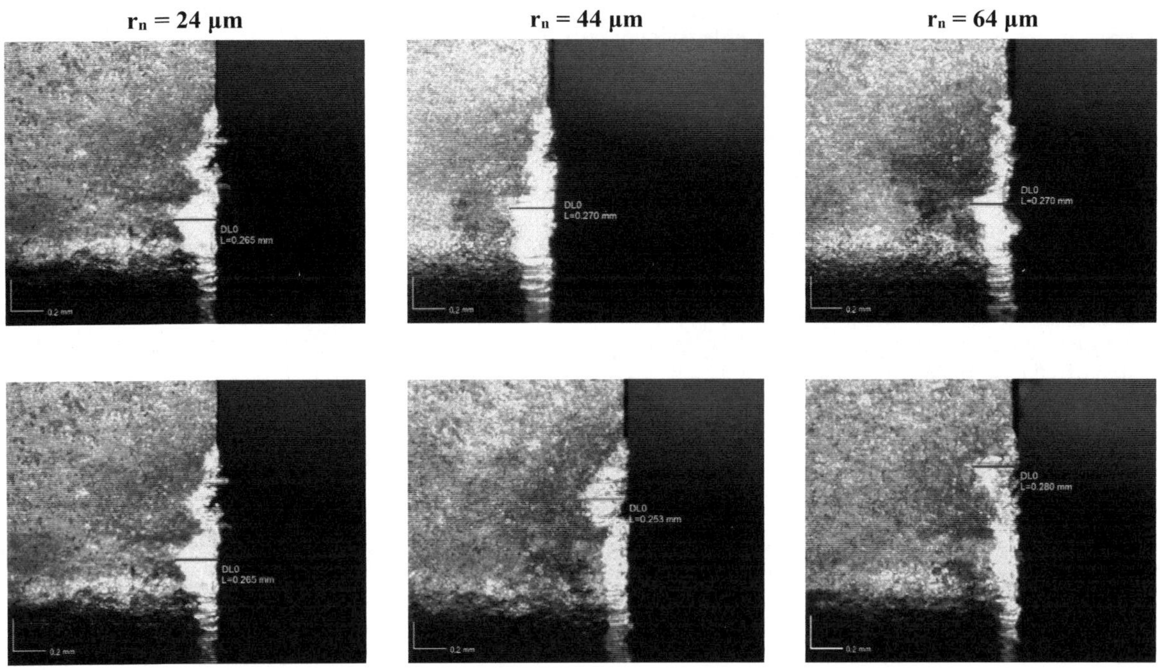

Fig. 7. Measured width of flank wear after attaining the value of 0.25 mm

The results (Fig. 4 and 5) (Table 3) show that the tool life of cemented carbide turning inserts increases when cutting edge radius is smaller. The tool life of prepared cemented carbide turning inserts with cutting edge radius $r_n = 44$ µm is approximately 18% longer than prepared cemented carbide turning inserts with cutting edge radius $r_n = 64$ µm. The tool life of prepared cemented carbide turning inserts with $r_n = 24$ µm is approximately 32% longer than prepared cemented carbide turning inserts with $r_n = 64$ µm. In addition, the tool life of unprepared cemented carbide turning inserts with $r_n = 11.49$ µm is approximately 22% longer than prepared cemented carbide turning inserts with $r_n = 24$ µm. These results were investigated in turning of austenitic stainless steel material, which shows higher strain hardening tendency.

Authors suppose that it is due to work hardening (also strain hardening) which is related to stagnation zone. Under stagnation (or separation) point (Fig. 8), workpiece material is pressed and therefore machined surface is significantly harder than work surface. Authors also suppose that bigger cutting edge radius made greater work hardening. Hence, if cutting edge radius is bigger than tool life will be shorter due to machining harder workpiece material.

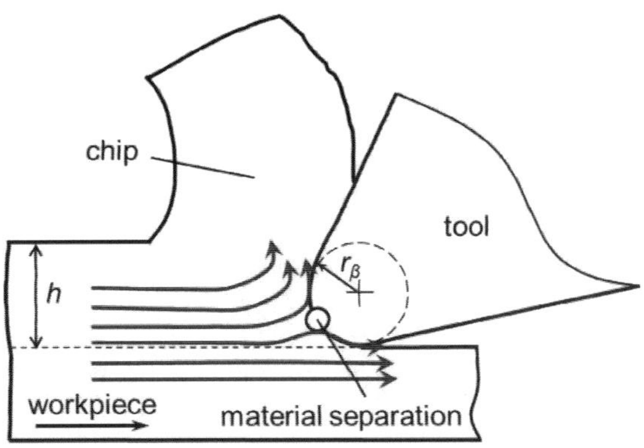

Fig. 8. Stagnation zone and stagnation (or separation) point [18]

The tool life of prepared cemented carbide turning inserts by brushing is approximately the same as tool life of prepared cemented carbide turning inserts by wet microblasting for cutting edge radii of 24 and 44 µm. The tool life of turning cemented carbide inserts prepared by brushing is approximately 32% longer than cemented carbide turning inserts prepared by wet microblasting for cutting edge radius $r_n = 64$ µm.

5. Conclusion

This paper presents importance of cutting edge microgeometry. The most famous and used edge preparation methods are summarized in article along with the results of current research of famous tool companies and institutions. In this article, the influence of cutting edge radius sizes on the tool life was investigated. Furthermore, the various edge preparation methods were compared in term of tool life of turning inserts.

Solution to this issue consists of the following steps: measurement of unprepared turning carbide inserts, cutting edge preparation and surface treatment, measurement of prepared turning carbide inserts, cleaning, PVD coating, long-term tool life tests and evaluation. Tested cutting tool material was cemented carbide and selected workpiece material was austenitic stainless steel material of DIN EN X6CrNiTi18-10 (AISI 321) grade. Cemented carbide turning inserts were deposited by AlCrSiN nanocomposite hard coating. Tool lives of cemented carbide turning inserts were determined during the machining on DMG CTX alpha 500 turning centre. Since the flank wear of cemented carbide turning inserts was dominant, evolution of flank wear was measured during the test. The machining tests were stopped when width of the flank wear reached over 0.25 mm because too large notch and edge chipping were observed over this value.

The tool life of prepared cemented carbide turning inserts by brushing is approximately the same as tool life of prepared cemented carbide turning inserts by wet microblasting for cutting edge radii smaller than 45 μm. The tool life of turning cemented carbide inserts prepared by brushing is approximately 32% longer than cemented carbide turning inserts prepared by wet microblasting for cutting edge radius $r_n = 64$ μm.

The results show that the tool life of cemented carbide turning inserts increases when cutting edge radius is smaller. In addition, the tool life of unprepared cemented carbide turning inserts is longer than prepared ones. This phenomenon occurs during the machining of austenitic stainless steel material when cutting parameters as depth of cut and feed are selected for semi-finishing and finishing operations.

In the future research, an experiments for roughing, that means higher feed and larger depth of cut, will be completed. In general, these results can't be used for other material for sure. Thus, if we want to determine the influence of cutting edge radius sizes on the tool life in machining then it will be necessary to study other workpiece materials. Authors are going to focus on investigation of issue of cutting edge microgeometry during the machining the difficult-to-materials material such as Inconel superalloy and hardened steel.

6. Acknowledgments

The article was written with the support of the Project of VEGA grant agency of the Ministry of Education, Science, Research and Sport of the Slovak Republic and Slovak Academy of Sciences, no. 1/0097/17: "The research of novel method for cutting edge preparation to increase the tool performance in machining of difficult-to-machine materials", and APVV Project of Slovak Research and development Agency of the Ministry of Education, Science, Research and Sport of the Slovak Republic, no. APVV-16-0057: "Research into the Unique Method for Treatment of Cutting Edge Microgeometry by Plasma Discharges in Electrolyte to Increase the Tool Life of Cutting Tools in Machining of Difficult-to-Machine Materials."

This article was supported by PRAMET TOOLS and therefore I am grateful to R&D department, which was involved in this research.

7. References

[1] Bouzakis, K.-D., Bouzakis, E., Kombogiannis, S., Makrimallakis, S., Skordaris, G., Michailidis, N., Charalampous, P., Paraskevopoulou, R., M'Saoubi, R., Aurich, J.C., Barthelmä, F., Biermann, D., Denkena, B., Dimitrov, D., Engin, S., Karpuschewski, B., Klocke, F., Özel, T., Poulachon, G., Rech, J., Schulze, V., Settineri, L., Srivastava, A., Wegener, K., Uhlmann, E., Zeman, P. (2014). Effect of Cutting Edge Preparation of Coated Tools on Their Performance in Milling Various Materials. CIRP Journal of Manufacturing Science and Technology, 7 (3), pp. 264-273

[2] Denkena, B., Biermann, D. Cutting Edge Geometries. (2014). CIRP Annals – Manufacturing Technology, Volume 63, pp. 631-653

[3] Uhlmann, E., Oberschmidt, D., Löwenstein A., Kuche, Y. (2016). Influence of Cutting Edge Preparation on the Performance of Micro Milling Tools. Procedia CIRP. Volume 46, Pages 214-217

[4] Denkena, B., Lucas, A., Bassett. (2011). E. Effects of the cutting edge microgeometry on tool wear and its thermo mechanical load. CIRP Annals – Manufacturing Technology, Volume 60, pp. 73-76

[5] Biermann, D., Aßmuth, R., Schumann, S., Rieger, M., Kuhlenkötter, B. (2016). Wet Abrasive Jet Machining to Prepare and Design the Cutting Edge Micro Shape. Procedia CIRP, Volume 45, pp.195-198

[6] Yussefian, N.Z., Koshy, P., Buchholz, S., Klocke F. (2010). Electro-erosion Edge Honing of Cutting Tools. CIRP Annals, Volume 59 (1), pp. 215-218

[7] Breidenstein, B., Gey, C., Denkena, B. (2013). Surface Integrity of Laser Machined Carbide Cutting Tools Within the PVD-coating Process Chain Materialwissenschaft und Werkstofftechnik, Volume 44 (8), pp. 704-709

[8] Aurich, J.C., Zimmermann, M., Leitz, L. (2011). The Preparation of Cutting Edges Using a Marking Laser Production Engineering –Research and Development, Volume 5 (1), pp. 17-24

[9] Uhlmann, E., Richarz, S., Mihotovic, V. (2009). Substrate Pre-treatment of Cemented Carbides using Abrasive Flow Machining and Laser Beam Ablation. Production Engineering, Volume 3 (1), pp. 81-86

[10] Bouzakis, K.-D., Charalampous, P., Kotsanis T., Skordaris, G., Bouzakis, E., Denkena, B., Breidenstein, B., Aurich, J.C., Zimmermann, M., Herrmann, T., M'Saoubi, R. (2017). Effect of HM substrates' cutting edge roundness manufactured by laser machining and micro-blasting on the coated tools' cutting performance. CIRP Journal of Manufacturing Science and Technology. Volume 18, pp. 188-197

[11] Begic-Hajdarevic, B., Cekic, A., Mehmedovic, M., Djelmic. A. (2015). Experimental Study on Surface Roughness in Abrasive Water Jet Cutting. Procedia Engineering, Volume 100, pp. 394-399

[12] Begic-Hajdarevic, D., Pasic, M., Cekic, A., Mehmedovic, M. (2016). Optimization of process parameters for cut quality in CO2 laser cutting using taguchi method. Annals of DAAAM and Proceedings of the International DAAAM Symposium. Volume 27, Issue 1, Pages 157-164

[13] Fulemová, J., Řehoř, J. (2015). Influence of form factor of the cutting edge on tool life during finishing milling. Procedia Engineering, Volume 100, pp. 682-688.

[14] Fulemova, J. Janda, Z. (2014). Influence of the cutting edge radius and the cutting edge preparation on tool life and cutting forces at inserts with wiper geometry. Procedia Eng., Volume 69, pp. 565-573.

[15] Baksa, T., Schornik, V., Adamek, P., Zetek, M. (2016). Machining of inconel 718 using uncoated cutting tools with different cutting EDGE quality. Annals of DAAAM and Proceedings of the International DAAAM Symposium, Volume 27, Issue 1, pp. 441-446.

[16] Albrecht P. (1960). New developments in the theory of metal cutting process. Part 1: The ploughing process in metal cutting. Journal of Engineering for Industry. Volume 82. pp. 348–57.

[17] Yussefian, N.Z., Koshy, P. (2013). Parametric characterization of the geometry of honed cutting edges. Precision Engineering. Volume 37, Issue 3, pp. 746-752.

[18] Rodriguez CJC. (2009). Cutting edge preparation of precision tools by applying microabrasive jet machining and brushing. Dissertation. Kassel University

[19] Wyen, C.-F., Knapp, W., Wegener. K. (2012). A new method for the characterisation of rounded cutting edges. International Journal of Advanced Manufacturing Technology, Volume 59, pp. 899-914

[20] Wyen, C.-F., Wegener, K. (2010). Influence of cutting edge radius on cutting forces in machining titanium. In: CIRP Annals – Manufacturing Technology. Vol. 59, pp. 93-96.

[21] Denkena, B. Kohler, J. Ventura, C.E.H. (2013). Customized cutting edge preparation by means of grinding. Volume 37, Issue 3, pp. 590-598.

[22] Vopát, T., Haršáni, M., Kuruc, M., Šimna, V., Zaujec, R., Peterka, J., Čaplovič, Ľ. (2017). Effect of Substrate Bias and Coating Thickness on the Properties of nc-AlCrN/a-SixNy Hard Coating and Determination of Cutting Parameters. Solid State Phenomena. Volume 261, pp. 229-236.

[23] Vopát, T., Beňo, M. et al. (2014). Errors in turning slender workpieces. In EQ-2014: In the framework of International Forum "Education Quality - 2014". Proceedings. Izhevsk, Russia. pp. 256-260.

DOI: 10.2507/28th.daaam.proceedings.055

INFLUENCE OF ULTRASONIC ASSISTANCE ON DELAMINATION DURING MACHINING OF DIFFERENT COMPOSITE MATERIALS

Marcel Kuruc*, Tomáš Vopát, Vladimír Šimna & Martin Necpal

This Publication has to be referred as: Kuruc, M[arcel]; Vopat, T[omas]; Simna, V[ladimir] & Necpal, M[artin] (2017). Influence of Ultrasonic Assistance on Delamination During Machining of Different Composite Materials, Proceedings of the 28th DAAAM International Symposium, pp.0392-0399, B. Katalinic (Ed.), Published by DAAAM International, ISBN 978-3-902734-11-2, ISSN 1726-9679, Vienna, Austria
DOI: 10.2507/28th.daaam.proceedings.055

Abstract

A lot of nowadays researches are focused on increasing strength of the components and decreasing their weight, or obtaining unique properties. Composite materials are one of the results of the researches. However, different material properties of matrix and reinforcing material usually cause certain difficulties during conventional machining, such as rapid tool wear and delamination of composite. Therefore, focusing on machinability of this kind of materials is necessary. Scientists have discovered several ways of how to satisfactorily machine different composite materials. Seco Tools Company starts developing a new design of the cutting tools. Researchers of the Stanford University have devised a diamond-like carbon (DLC) coating available to effectively cut matrix as well as reinforce fibres. DMG Mori Company has implemented ultrasonic assistance into machining process and reached enhanced machinability. This article compares conventional milling and ultrasonic assisted milling of the composite materials – glass fibre reinforced plastic (GFRP) and two types of carbon fibre reinforced plastics (CFRP) by using a special designed cutting tool. The results could be applied to determine when ultrasonic assistance is advantageous and what advantages it brings about. The purpose of this study is to determine ways to increase efficiency and productivity of machining different types of composite materials.

Keywords: Ultrasonic assisted milling; Composite materials; GFRP; CFRP; Delamination; Advanced machining.

1. Introduction

Composite materials are widely used in many applications in many fields, such as aerospace, aviation (Fig. 1.), marine, automotive, industry, medicine, sport and many others. Different mechanical properties of each compound in composite material cause challenges in its machining. Recommended machining parameters for hard and strong materials such as reinforcing fibre are different from the machining parameters for soft and tough materials such as matrix. That means, the machining parameters proper for one compound may be improper for other compounds. Improper machining parameters could cause melting of the matrix, tearing out the fibres, edge-chipping of the composite, rapid tool wear, formation of build-up-edges (BUE), etc. [1, 2]

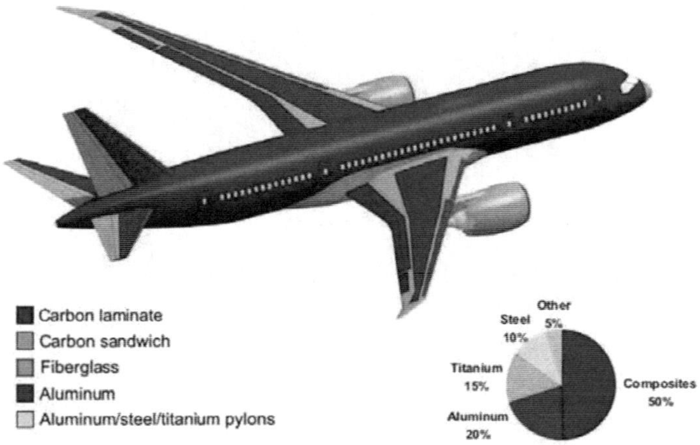

Fig. 1. Composite materials in Boeing 787 [2]

Different attitudes have been suggested to solve the problems, one of them being enchanting of the cutting tool geometry and design. Newly developed cutting tools for machining composite materials are typical by unusual shape. They could have additional notches on the teeth, and/or arrow shaped double screw, or pineapple-like shape, as is shown in Fig. 2. Another attitude is focused on advanced coatings, such as diamond or DLC (diamond like carbon) coated cutting tools. Another way to improve machining of composite materials is ultrasonic assistance. Ultrasound avoids creation of BUE, reduces cutting force and process heat. Resultant cutting parameters are still affected by the type of matrix and the type of reinforcing material. Also, different cutting parameters will be set for roughing and finishing [3, 4, 5, 6, 7, 8, 9, 10].

Fig. 2. Advanced design of milling cutters [11]

2. Machining method

A machine tool for rotary ultrasonic machining (RUM) was applied in this experiment. RUM usually utilize ultrasonic tool with diamond particles and undefined geometry. This tool rotates around the vertical axis and oscillates by ultrasonic frequency in vertical direction. Accordingly, very hard and brittle materials (such as ceramics) can be machined. When a tool with defined geometry (e.g. conventional milling cutter) is applied, the process is called ultrasonic assisted machining (UAM). RUM and UAM have a lot of common advantages, such as decrease of cutting force, reduction of heat generation, elimination of BUE formation, superior machined surface, increased tool life, etc. Generally, RUM is proper for the hard and brittle materials (such as optic glass), while UAM is proper for the soft and tough materials (such as metals and polymers). RUM and UAM execute similar movements, and therefore they could be usually performed on the same machine tool.

In this experiment, rotary ultrasonic milling machine DMG ULTRASONIC 20 linear (Fig. 3.) was used. This machine tool can operate continuously in five axes and it can operate as a conventional milling machine, high speed cutting machine, high feed milling machine, ultrasonic assisted milling machine and rotary ultrasonic milling machine. It is therefore able to machine almost every material [12, 13, 14].

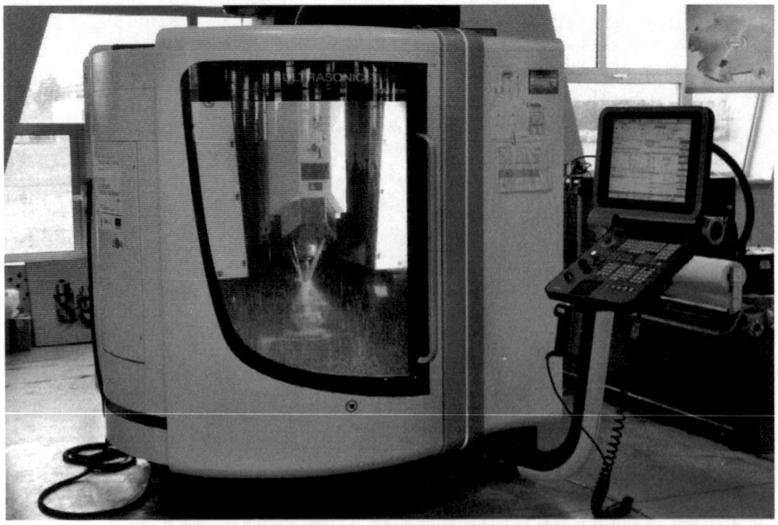

Fig. 3. Rotary ultrasonic machine tool DMG Ultrasonic 20 linear

As a cutting tool, a milling cutter for composite materials Fiber Line Honeycomb Hexacut – Fräse (Hexacut – End mill) with internal labelling 068HOA050 manufactured by Hufschmied Zerpanungssysteme was used. This tool has the cutting diameter of 5 mm, cutting length of 18 mm, shank diameter of 6 mm, overall length of 60 mm and 8 flutes (Fig. 4.). A 3D model of the workpiece was created by CAD software Autodesk PowerShape Ultimate 2017 and the NC program for machining (for control system Siemens Sinumerik 840D Solutionline) was generated by CAM software Autodesk PowerMill Ultimate 2017 [15, 16, 17, 18].

Fig. 4. Milling cutter Hexacut for composite materials

3. Description of the experiment

As workpiece materials, a glass fibre reinforced polymer (GFRP) and two types (M & S) of carbon fibre reinforced polymers (CFRP) were used. These composite materials have different mechanical properties, as shown Fig. 5.

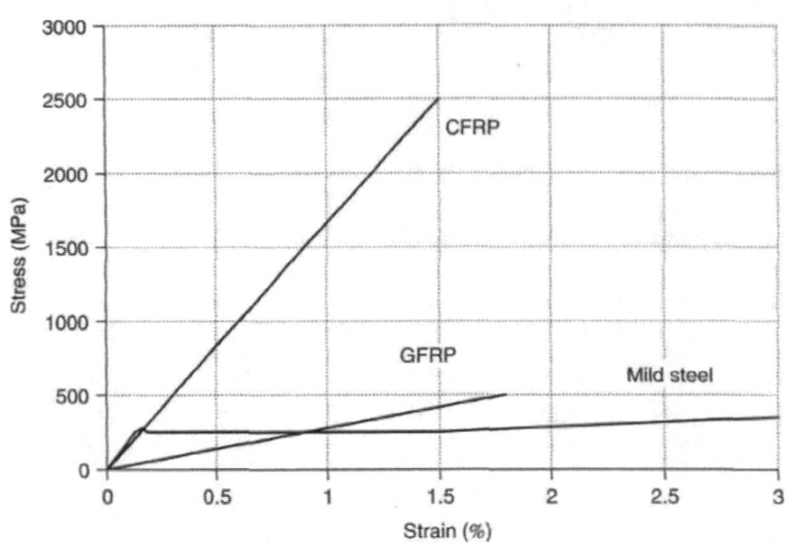

Fig. 5. Comparison of Yield stress for different materials [2]

GFRP material was provided by the Slovak company Armastek. Producer determine following mechanical properties: modulus of elasticity 21 GPa, tensile strength 190 MPa, yield strength 140 MPa, density 1850 kg.m^{-3}. The specimen was of a rod shape with rectangular section with filled edges with the dimensions 40 x 30 mm. Fibre orientation was parallel with the axis of the rod. The following cutting conditions were set up for every experiment, according to the SECO catalogue: cutting speed 50 m/min (spindle speed 3183 rpm), feed per tooth 0.015 mm (feed rate 382 mm/min) and depth of cut 0.7 mm. During ultrasonic machining, it is recommended to use harmonic frequency of the tool which was 23,500 Hz.

CFRP materials were provided by the Czech company Sanax (type S & M). Producer determine following mechanical properties: modulus of elasticity 170 GPa, tensile strength 3000 MPa, elongation 1.3 %, density 1550 kg.m^{-3} for type S and modulus of elasticity 210 GPa, tensile strength 2480 MPa, elongation 1.1 %, density 1550 kg.m^{-3} for type M. The specimens were a plate shape with the dimensions of 50 x 180 x 1.4 mm. The following cutting conditions were set up, according to the also SECO catalogue: cutting speed 100 m/min (spindle speed 6366 rpm), feed per tooth 0.015 mm (feed rate 764 mm/min) and depth of cut 0.7 mm. During ultrasonic machining, it is recommended to use harmonic frequency of the tool which was 23,500 Hz (the same cutting tool was used in all experiments) [19].

Eight notches were manufactured into GFRP. Two notches are parallel with the fibres, two notches were perpendicular to fibres. They were placed on the lateral surface of the rod. On the base surface four notches to the previous ones were placed. Half of the notches were machined without ultrasonic assistance and the rest of them with assistance of ultrasound. The notches were 1.4 mm deep. After machining, the specimen was digitised by GOM ATOS II TripleScan. Its precision is 0.02 mm. This device, as well as specimen prepared for 3D scanning are shown in Fig. 6.

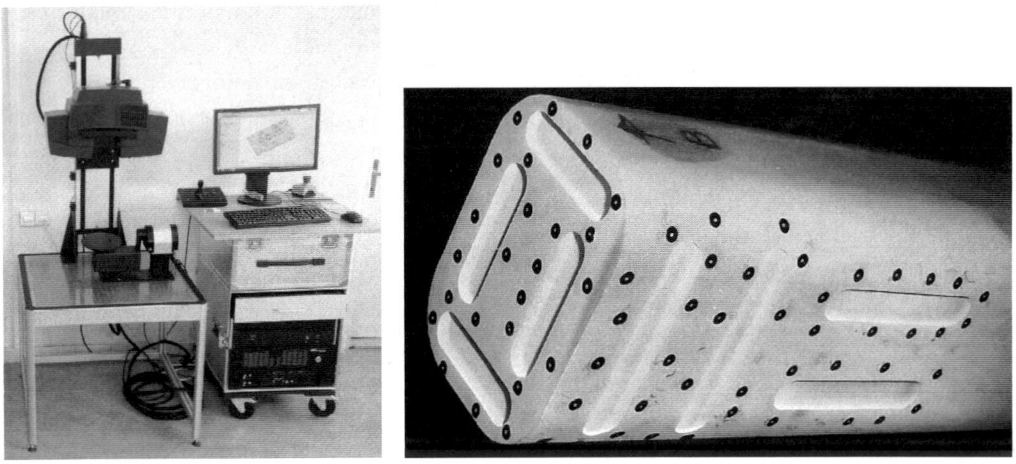

Fig. 6. Scanning device and prepared specimen

Eight notches were performed into each CFRP as well. Two notches were parallel with the fibres, two notches were perpendicular to the fibres. They were placed on the biggest surface of the rod. On the edges of CFRP, four notches to the previous ones were placed. Half of the notches were machined without ultrasonic assistance and the rest of them with the assistance of ultrasound. The notches were 0.7 mm deep. After machining, the specimens were digitized by Zeiss Metrotom 1500 computer tomography. The device, as well as its result of digitization are shown in Fig 7.

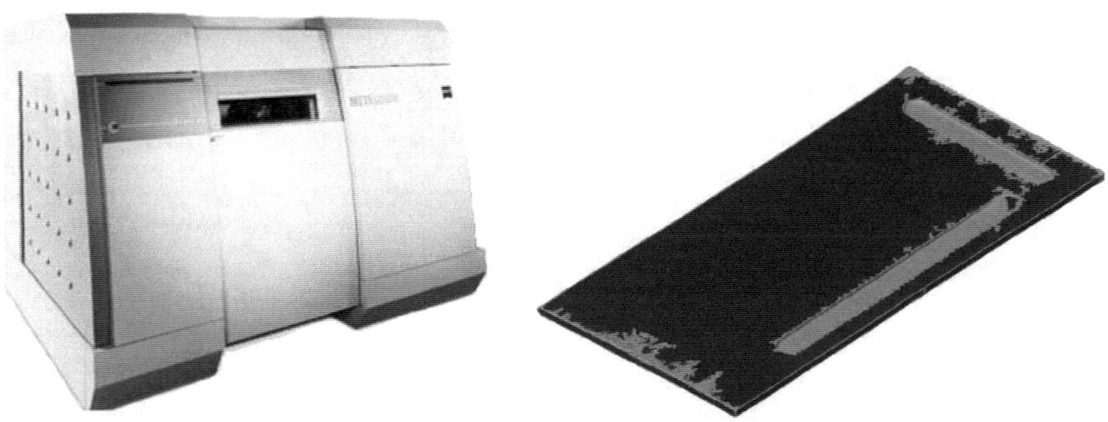

Fig. 7. Scanning device and digitised specimen [20]

Digitised models were compared with the original CAD model. Delamination was evaluated as the highest width of each notch (original width was 5 mm). Beside delamination, the machine load (in Z direction) and surface roughness were also evaluated (they are often connected with delamination itself). Machine load was obtained directly by the machine tool. Surface roughness was measured by ZEISS Surfcom 5000, which is shown in Fig. 8.

Fig. 8. Surface roughness measuring device [21]

4. Results of the experiment

Several factors of the delamination were obtained and evaluated, such as delamination width, machine tool load and surface roughness of composite materials. They were achieved for each machined notch. Labelling of the notches in GFRP is shown in Fig. 9. The green ones (letter "a") were machined without ultrasound. The blue ones (letter "b") were machined with ultrasonic assistance. Direction of the fibres is parallel with the groove numbered 2.

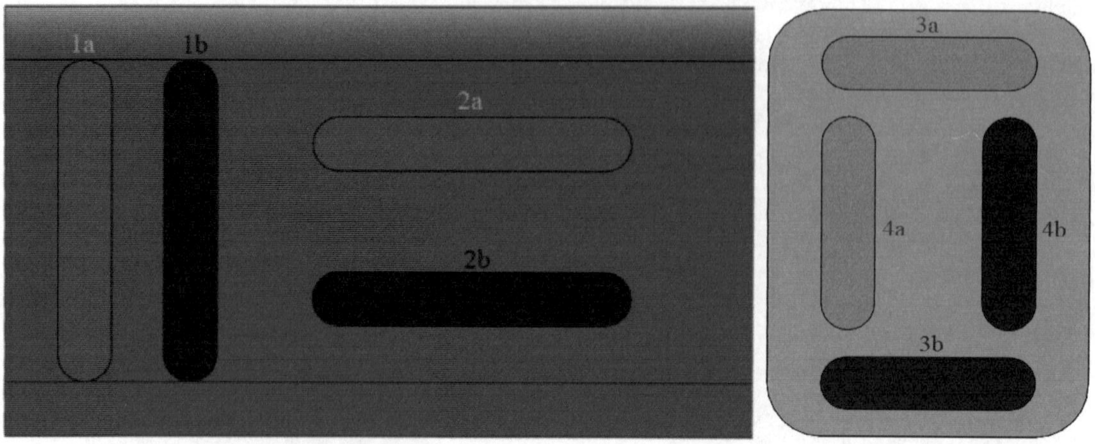

Fig. 9. Labeling of the notches

Labelling of the notches in CFRP is shown in Fig. 10. An extra notch (continuous one) was produced for easier determination of position. The green ones (letter "a") were machined without ultrasound. The blue ones (letter "b") were machined with ultrasonic assistance. Direction of the fibres is parallel with the grooves numbered 1 and 4.

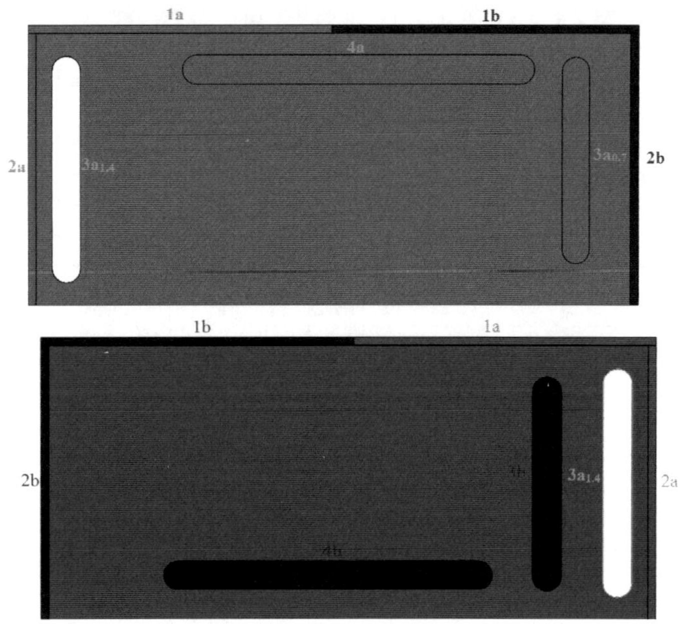

Fig. 10. Labelling of the notches (top and bottom side)

Results of the measurements are summarized in Tables 1, 2 and 3. Recorded are only average values (each measurement was repeated three times). In the Table, machine tool load (represent cutting force) in Z direction is labelled as *Load Z,* and it is the percentage of the maximum available load of the machine tool, measured directly by the machine tool. Delamination factor is labelled as *Df* and it is the percentage difference between the recommended width of the notch and the real width of the notch. It is expressed in Fig. 11. and by formula (1). As roughness parameters, its average (*Ra*) and maximum (*Rz*) values were chosen.

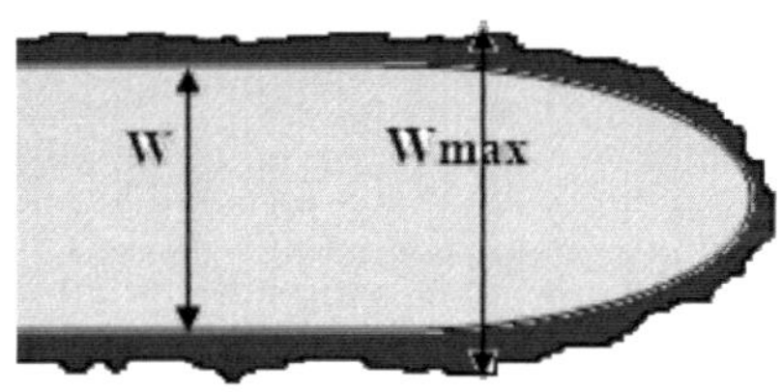

Fig. 11. Delamination of the notch

Turquoise colour represents recommended width of the notch (W) and brown colour represent the real (maximum) width of the notch (W_{max}). According Fig. 11. is possible to calculate delamination factor by following formula:

$$D_f = \left(\frac{W_{max} - W}{W}\right) \times 100 \quad [\%] \tag{1}$$

GFRP	Load Z [%]	Df [%]	Ra [μm]	Rz [μm]
1a	10	11.2	2.81	17.98
1b	5	14.8	2.86	17.25
2a	10	4.6	3.16	18.25
2b	5	3.6	3.18	17.35
3a	10	< 0.2	1.49	9.27
3b	5	< 0.2	1.56	10.59
4a	15	< 0.2	1.58	11.42
4b	10	< 0.2	1.59	11.42
avg. a	**11.25**	**7.9**	**2.26**	**14.23**
avg. b	**6.25**	**9.2**	**2.29**	**14.15**

Table 1. Results of the measurements for GFRP

Value of width of the delamination less than 0.2 % means, that there is present delamination lesser than accuracy of the digitization device. Therefore, it cannot be calculated (it looks like there is no delamination).

CFRP-S	Load Z [%]	Df [%]	Ra [µm]	Rz [µm]
1a	15	3.8	1.25	7.43
1b	5	1.9	1.15	7.34
2a	10	29.4	1.09	7.83
2b	5	38.1	0.92	6.92
3a	15	24.6	1.96	17.08
3b	5	17	2.15	18.60
4a	10	21.8	1.46	8.71
4b	5	14.2	1.45	9.42
avg. a	**12.5**	**19.9**	**1.44**	**10.26**
avg. b	**5**	**17.8**	**1.42**	**10.57**

Table 2. Results of the measurements for CFRP type S

CFRP-M	Load Z [%]	Df [%]	Ra [µm]	Rz [µm]
1a	20	2.5	0.89	5.47
1b	10	1.9	1.13	6.40
2a	20	10.6	0.81	6.17
2b	15	22.5	0.81	6.29
3a	20	31.8	1.64	10.30
3b	10	22.8	1.57	10.75
4a	20	21.2	1.39	7.84
4b	10	17.8	1.39	8.48
avg. a	**20**	**16.5**	**1.19**	**7.45**
avg. b	**11.25**	**16.2**	**1.23**	**7.98**

Table 3. Results of the measurements for CFRP type M

5. Conclusion

The obtained results allow us to conclude that ultrasonic assistance has a great influence on the reduction of machine tool load (decreasing by approx. 44 % for GFRP and decrease by approx. 60 % for CFRP-S and 44 % for CFRP-M). The surface roughness parameters were not significantly affected. Width of the delamination was also very similar (elevation by approx. 16 % for GFRP and reduction by approx. 11 % for CFRP-S and 2% for CFRP-M). Generally, during machining of GFRP was achieved lower machine load, lower delamination and higher surface roughness in comparison with machining of CFRP. During machining of CFRP-M (the one with lower tensile strength and higher modulus of elasticity) was achieved higher machine load, lower delamination and lower surface roughness in comparison with CFRP-S. Relatively low ultrasonic influence on delamination and surface roughness parameters could be caused by relatively low amplitude (tool was not directly manufactured for ultrasonic assistance – absence of concentrator). We also observed much lower delamination on the base surface of the GFRP rod and on the edge surface parallel to the fibres direction of the CFRP plate. On the lateral surface of GFRP was observed lower delamination at machining in direction of the fibres.

To sum up, ultrasonic assistance is proper especially in terms of decreasing of the machine tool load. Lower machine load allows usage of higher depth of cut and feed rate (increased material removal rate). It means, UAM allows increasing productivity even when common (cheap) cutting tool is used. If a tool will be designed for ultrasonic assistance, its improvement could be even greater. Results of unwanted delamination (edge-chipping of workpiece) could be applied in wanted cutting edge preparation (edge-chipping of tool). Further research will be focused on the determination of influence of ultrasonic assistance on different composite materials, such as metal matrix composite, or ceramic matrix composite, as well as sintered carbide for application as cutting tool with edge preparation.

6. Acknowledgments

This contribution is a part of the GA VEGA project of Ministry of Education, Science, Research and Sport of the Slovak Republic, No. 1/0477/14 "Research of influence of selected characteristics of machining process on achieved quality of machined surface and problem free assembly using high Technologies".

The article was written with the support of the Project of VEGA grant agency of the Ministry of Education, Science, Research and Sport of the Slovak Republic and Slovak Academy of Sciences, no. 1/0097/17: "The research of novel method for cutting edge preparation to increase the tool performance in machining of difficult-to-machine materials" and APVV Project of Slovak Research and development Agency of the Ministry of Education, Science, Research and Sport of the Slovak Republic, no. APVV-16-0057: "Research into the Unique Method for Treatment of Cutting Edge Microgeometry by Plasma Discharges in Electrolyte to Increase the Tool Life of Cutting Tools in Machining of Difficult-to-Machine Materials."

7. References

[1] Dai Gil Lee, et al. (2017). Adhesion characteristics of fiber-exposed glass composites. In Composite Structures. Vol. 165, pp. 9-14.
[2] Aviation. Available on the Internet: https://aviation.stackexchange.com/questions/22148/are-composite-materials-which-are-being-widely-used-in-newer-aircrafts-better
[3] Ning F., Wang H., Cong W., Fernando P.K.S.C. (2017). A mechanistic ultrasonic vibration amplitude model during rotary ultrasonic machining of CFRP. In Ultrasonics. Vol. 76, pp. 44-51.
[4] Su Z., Jia Z., Niu B., Bi G. (2017). Size effect of depth of cut on chip formation mechanism in machining of CFRP. In Composite Structures. Vol. 164, pp. 316-327.
[5] Khan M.A., Kumar A.S. (2011). Machinability of glass fibre reinforced plastic (GFRP) composite using alumina-based ceramic cutting tools. In Journal of Manufacturing Process. Vol. 13, Issue 1, pp. 67-73.
[6] Sonkar V., Abhishek K., Datta S., Mahapatra S.S. (2014). Multi-objective Optimization in Drilling of GFRP Composites: A Degree of Similarity Approach. In Procedia Materials Science. Vol. 6, pp. 538-543.
[7] Sreenivasaulu R. (2013). Optimization of Surface Roughness and Delamination Damage of GFRP Composite Material in End Milling using Taguchi Design Method and Artificial Neural Network. In Procedia Engineering. Vol. 64, pp. 785-794.
[8] Akbari J., et al. (2013). Applying Ultrasonic Vibration to Decrease Drilling-Induced Delamination in GFRP Laminates. In Procedia CIRP. Vol. 6, pp. 577-582.
[9] Palanikumar K. (2011). Experimental investigation and optimisation in drilling of GFRP composites. In Measurement. Vol. 44, Issue 10, pp. 2138-2148.
[10] Bosco M.A.J., Palanikumar K., Prasad B.D., Velayudham A. (2013). Influence of Machining Parameters on Delamination in Drilling of GFRP-armour Steel Sandwich Composites. In Procedia Engineering, pp. 758-763.
[11] Direct Industry – Product: Seco Tools. Available on the Internet: http://www.directindustry.com/prod/seco-tools-5699.html
[12] Kuruc M., Vopát T., Peterka J. (2015). Surface Roughness of Poly-crystalline Cubic Boron Nitride after Rotary Ultrasonic Machining. In Procedia Engineering. Vol. 100, pp. 877-884.
[13] Kuruc M., Zvončan M., Peterka J. (2014). Investigation of ultrasonic assisted milling of aluminum alloy AlMg4.5Mn. In Procedia Engineering. Vol. 69, pp. 1048-1053.
[14] Kuruc M. (2015). Machine tool loads in rotary ultrasonic machining of alumina, CBN and synthetic diamond. In Proceedings of the 26th DAAAM International Symposium, pp. 519-523. ISSN 1726-9679. ISBN 978-3-902734-07-5.
[15] Delcam – Advanced manufacturing solutions. PowerMILL. Available on the Internet: http://www.powermill.com/
[16] Zvončan M., Kováč M., Beňo M. (2012). Machine tool´s DMG Ultrasonic 20 linear simulation in Powermill CAM software. In: CA systems in production planning. ISSN 1335-3799. Vol. 12, No 1, p. 90-93.
[17] Siemens. Sinumerik 840D. Available on the Internet: http://w3.siemens.com/mcms/mc-systems/en/automation-systems/cnc-sinumerik/Pages/cnc-systems.aspx
[18] Hufschmied: CATALOGUE PLASTIC-MACHINING. Available on the Internet: http://www.hufschmied.net/tl_files/dokumente/HUF-Kuststoff-Low-080415.pdf
[19] SECO Tools. Monolithic Carbide Mills. Available on the Internet: https://www.secotools.com/CorpWeb/Czech%20Republic/katalogy/2015/CZ_Catalog_Solid%20end%20mills_2015_Inlay_LR.pdf
[20] Zeiss. Metrotom 1500. Available on the Internet: http://www.zeiss.com/industrial-metrology/en_de/products/systems/computedtomography/metrotom-1500.html
[21] Zeiss. SURFCOM 5000. Available on the Internet: http://www.zeiss.com/industrial-metrology/en_de/products/systems/form-andsurface/surface-and-contour/surfcom-5000.html

DOI: 10.2507/28th.daaam.proceedings.056

INFLUENCE OF CMM VELOCITY ON SCANNING PRECISION

Melichar Martin, Kubátová Dana, Kutlwašer Jan

This Publication has to be referred as: Melichar, M[artin]; Kubatova, D[ana] & Kutlwaser, J[an] (2017). Influence of CMM Velocity on Scanning Precision, Proceedings of the 28th DAAAM International Symposium, pp.0400-0409, B. Katalinic (Ed.), Published by DAAAM International, ISBN 978-3-902734-11-2, ISSN 1726-9679, Vienna, Austria
DOI: 10.2507/28th.daaam.proceedings.056

Abstract

CMM are one of the basics in today's industrial production based on the control activities of complex products. In particular, the area of parameter evaluation, such as shape and position deviations, point spacing from the 3-D model, as well as simple measurement of dimensions in the drawing documentation can't be imagined without the use of CMM. This article describes experiment of analysing influence of positioning velocity for the collecting of individual parameters. During testing in metrology lab of Regional technological institute in Pilsen had been etalon device tested for different speeds. Analysis was being concerned for basic elements of CMM measuring. Results were evaluated using estimation of measurement device capability.

Keywords: CMM; Scanning; Velocity; Precision

1. Introduction

In today's industrial production, coordinate measuring machines (CMM) are one of the cornerstones of inspection in most enterprises, being used throughout the manufacturing process. Without a CMM machine, it is difficult to imagine the evaluation of parameters related to simple dimension measurement based on drawings, not to mention more complicated and complex endeavours, such as evaluation of form, location and position deviations or the offset of certain points from a 3D model.

Essentially, 3D measuring machines (CMM) are sophisticated devices which are intended to be operated mainly in the controlled environment of a metrology laboratory. As defined in the ČSN 17025:2005 standard, controlled environment is characterized by a temperature of 20±2°C and a temperature gradient of 1°C/hour. Constant temperature of the machine, the environment and the specimen provided in this manner is a key underlying parameter of methodologically correct measurement. Under these controlled conditions, CMMs can ordinarily provide an accuracy down to 0.5+L/500µm, which is more than sufficient (considering a general-purpose machine) for a majority of technical applications in practice. The impacts of these factors, i.e. the environment temperature, were covered in the article

Another area which deserves attention involves the speeds used in the measuring process. The reason is that the continuously increasing pressure on shorter cycle times has reached the field metrology as well, leading to higher speeds to be used for measurement. It becomes necessary to study the practicability of the ever increasing measuring speeds with respect to not only measuring cycle time but mainly with regard to accuracy, repeatability and bias.

Expectations of the outcome were based on previous experience of working with earlier types of CMMs which were prone to vibration during travel and to a jolt upon the movement had stopped. Hence, measurement errors and resultant reduced repeatability and bias were expected to occur with increasing speeds.

2. Methodology or Experimental Procedure

2.1 Equipment

Testing was carried out using the Carl Zeiss Prismo Navigator 7 coordinate measuring machine which is able to be fitted with two variants of detection systems. One of them is a fixed scanning head (Fig. 1) (a scanning system for a single scanning direction; where scanning in other directions than the straight one is required, the probe tip must be rearranged in a star-shaped pattern or T-pattern).

The other system comprises a swivel head (Fig. 2)(a detection capable of rotating the probe tip in the direction of the point of interest). Each of these systems has its specific characteristics which were accepted in this experiment. This paper describes a test conducted exclusively with the indexed scanning system.

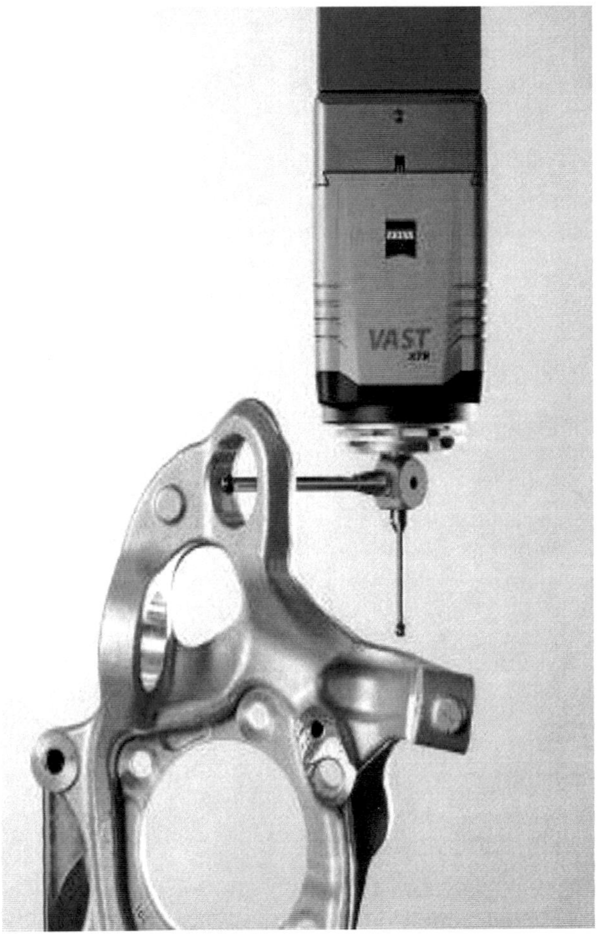

Fig. 1. Fixed head with probe tip[8]

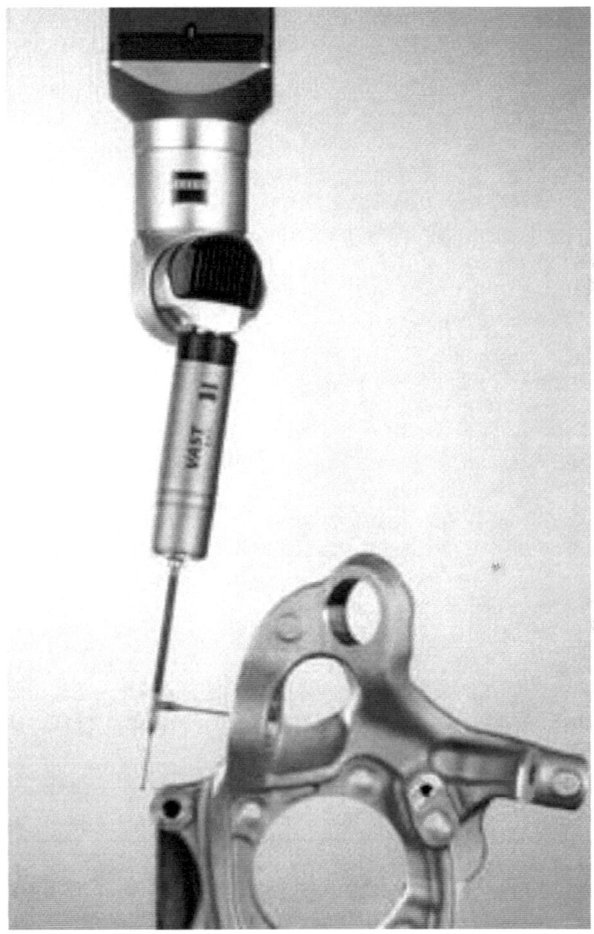

Fig. 2. Swivel head with probe tip[9]

The test item was a cube-shaped reference standard (referred to below as cube) of an aluminium alloy (Fig. 3). The cube was provided with a number of geometric features of known dimensions. These were suitable for making appropriate measurements in order to explore the effect of speed on readings during evaluation of basic dimensions such as length as well as more complex characteristics.[7] Therefore, the effect of speed could also be explored with respect to form, location and position deviations.

The measurements taken on the reference standard included length (large diameter of a cylinder) and form (concentricity of cylinders, flatness of the surface perpendicular to the cylinder axis and straightness). These characteristics were chosen on purpose because they represent typical measurement tasks on, say, engine blocks, gearboxes and pumps. [11]

Fig. 3. Tested etalon - Cube

Before the measurement was carried out, the cube had been kept at controlled conditions (20±2°C) for 48 hours. It was a sufficient period for thermal stabilisation to eliminate measurement uncertainty due to calculated thermal compensations of the machine. These play, as mentioned in 6;10 article, a major role in machine's accuracy.

2.2. Measurement parameters

a) Measuring speeds
- 20 mm/min
- 40 mm/min
- 70 mm/min

The CMM used also offers higher speeds in an "automatic" mode but these have not been ordinarily used for component inspection at the testing facility which is why they have not been employed for this experiment.

b) Point scanning systems

A whole range of characteristics and features can be evaluated using the machine. This experiment, however, only involved the most frequently measured characteristics. Table 1 summarizes selected characteristics of the measurement, including the scanning system and the numbers of points used.

Characteristic	Number of points	Scanning system
Mean	15	Scan
Flatness	45	Points
Concentricity	15/15	Scan / scan
Straightness	5	Points

Table 1. Summary of selected characteristics

2.3 Verification procedure

The practicability of the chosen speeds was tested on the above-described "cube" reference standard. For each of the speeds of 20, 40 and 70 mm/min, 11 values were measured which were evaluated subsequently. The first evaluation step only involved the time of measurement. After that, mean values for individual features were calculated, as well as their variance in relation to measurement speed. The next step involved the use of statistical methods where the measuring instrument capability was determined in the form of cg and cgk.

The original plan only concerned the time of measurement, the mean value and variance. That however proved insufficient. Consequently, the calculation of process repeatability (the Cg coefficient) was added. Yet, this parameter alone is insufficient as well. The issue here is that CMM technology exhibits a peculiar phenomenon: the variability between measurement runs does not increase with speed but may even diminish. This can be explained in terms of the control system, i.e. the static and, particularly, dynamic corrections applied to the data. Therefore, the information on precision alone (Cg) is not a reliable input for assessing the measuring process.[2,4]

The reason is that the measuring speed is reflected in the resultant value for the component, i.e. in its accuracy. Consequently, the Cgk coefficient must be used for process capability assessment, as it reflects this mechanism.

The indices Cg, Cgk were verified for both heads, using artificially defined tolerances Table 2. The limit capability value was chosen as follows: cg, cgk>1.67. The value for these parameters was intentionally set lower than the typical value of 1.33. It was motivated by the fact that the CMM is used for high-precision measurement and should therefore meet more stringent requirements. [5]

Characteristic	**Ref. value**
Mean	40.2
Flatness	0.01
Concentricity	0.013
Straightness	0.01

Table 2. Reference values of characteristics

Evaluation was based on reference values for the reference standard as well as on tolerance ranges and their positions, as indicated in Table 3.

Tolerance width	**0.1**	**0.05**	**0.01**
Mean	40.20/40.30	40.20/40.25	40.20/40.21
Flatness	0.01/0.11	0.01/0.06	0.01/0.02
Concentricity	0.013/0.113	0.013/0.063	0.013/0.023
Straightness	0.01/0.11	0.01/0.06	0.01/0.02

Table 3. Tolerance range of characteristics and their distribution (lower tolerance limit / upper tolerance limit)

Three procedures are available for calculating the capability of the measuring instrument cg and cgk. These are the procedures according to Škoda Auto a.s., FORD and GM. This experiment was based on the procedures used at the FORD company, Fig. 4 shows the relationships between capability indices and the tolerance range. The upper part of the figure shows the tolerance range T of the measurand and its limits LSL and USL. (T = USL - LSL). In evaluating the measuring instrument, only a part of this range is considered, represented in Fig. 4 as 20%, i.e. 0.2T, the same as in evaluating the test. The bottom of the figure shows an enlarged representation of this segment of the range. It contains the mean value from readings taken repeatedly on the reference standard $\overline{x}g$ and the reference value of the standard xr, which is sometimes referred to as xm (see Fig. 4). The 6Sg interval should be sufficiently smaller than the segment of the specification field under consideration. The value of 6Sg represents an interval, in which – assuming Gaussian distribution of the error of measurement – contains 99.73% of values. [1,3,4]

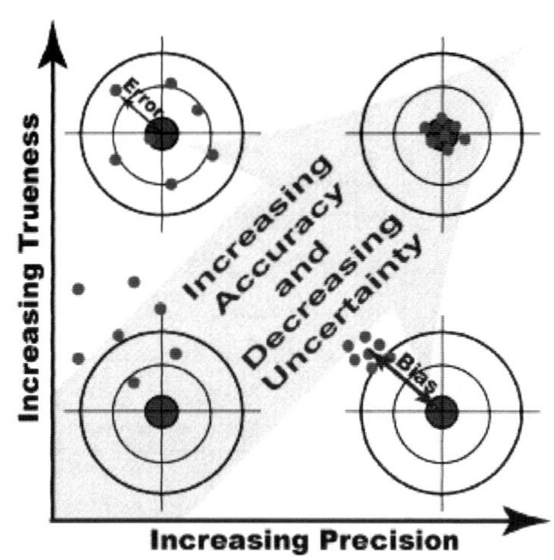

Fig. 4. Graphical representation of the statistical method of calculating the measuring instrument capability

Ford method formulae for calculation with the tolerance range T:

a) Sample standard deviation $S_g = \sqrt{\frac{1}{n-1}\Sigma(x_i - \bar{x})^2}$

b) Variance $S = s_g{}^2$

c) Arithmetic mean value $\bar{x} = \frac{1}{n}\Sigma x_i$

d) Repeatability $c_g = \frac{K*T}{6s_g}$

e) Bias $c_{gk} = \min\left\{\frac{(x_r+0{,}075T)-\bar{x}}{3s_g} \middle| \frac{\bar{x}-(x_r 0{,}075T)}{3s_g}\right\}$

n… number of measured values
xi…value of xth reading
x bar … arithmetic mean value
xr…reference value of the standard
T…tolerance range width
K…constant in the range of 10–30% defines the width of the tolerance range for calculation

3. Evaluation of the experiment

The evaluation sequence comprised the following steps. In each evaluation step, the actual values were converted into point scores on a scale from one to three. The higher the number of points, the poorer the result.

The first evaluation step for this test involved determining the time required for measurement at various speeds. The measuring times in Table 4 are times required for a single measuring cycle (program). Point scores were assigned in line with the above-described arrangement.

Speed	Measuring time	Result
20	42 min	3
40	17 min	2
70	13 min	1

Table 4. Measuring times and their point scores

This initial evaluation step was followed by ranking on the basis of mean values and variance for individual characteristics under measurement. Table 5 indicates the actual mean values and variances for individual characteristics and measuring speeds.

	20		40		70	
	mean value	variance	mean value	variance	mean value	variance
Mean	40.19373977	9.56918E-09	40.19323585	1.14424E-08	40.1935387	2.59506E-09
Flatness	0.010920845	3.87182E-09	0.010969845	1.0583E-09	0.010920473	6.62968E-09
Concentricity	0.008227182	3.26466E-07	0.008505027	1.92269E-07	0.008505027	1.92269E-07
Straightness	0.009515355	2.09533E-09	0.009514591	2.04718E-09	0.009514591	2.04718E-09

Table 5. Actual mean values and variances for the speeds of 20, 40 and 70 mm/min

The mean values and variance values were converted into point scores in such manner that higher values (poorer result) led to higher point scores. Table 6 this is the outcome of the conversion.

	20		**40**		**70**	
	mean value	variance	mean value	variance	mean value	Variance
Mean	3	3	1	1	2	2
Flatness	2	2	3	1	1	3
Concentricity	3	3	1	1	1	1
Straightness	3	3	1	1	1	1

Table 6. Point scores for mean values and variance values

The last table summarizes the results for the combination of the mean value and variance. Based on this calculation, Table 7 shows that if we only considered these two characteristics, our initial expectation would clearly be confirmed. i.e. the higher the speed, the higher the variance and the poorer the mean value. Not even this data. However, is in full agreement with our expectation: with straightness, higher speed leads to a better result. Nevertheless, it is important to consider the measuring experiment as a whole and identify the optimal speed for the entire measurement. This is why it was appropriate to also consider measurement precision (expressed by the cg parameter) and accuracy (expressed by cgk) and not only the readings.

	20	**40**	**70**
Mean	3	1	2
Flatness	2	2	2
Concentricity	3	1	1
Straightness	3	1	1
Result	**2,75**	**1,25**	**1,5**

Table 7. Summarized evaluation of mean value and variance

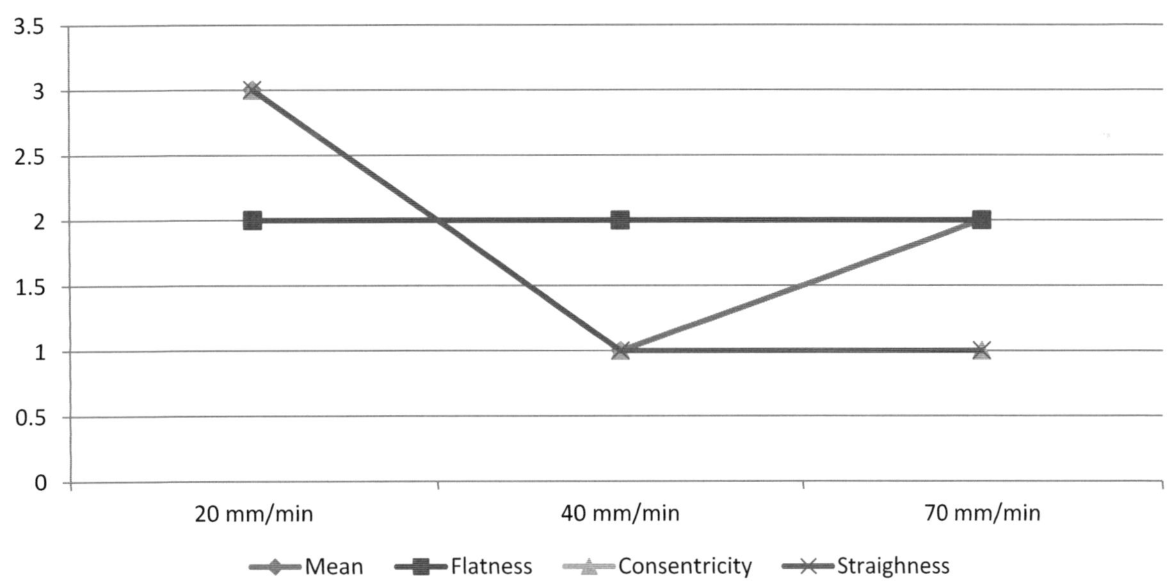

Graph 1. graphical representation of mean value and variance

The measuring instrument capability calculation took into account three widths of the measuring instrument tolerance ranges. Here, only the results for the lowest tolerance range value. 0.01 mm. are reported. With this width, the results began to be satisfactory. Furthermore, the values are essentially the same for the other 2 tolerance ranges but two and four times larger.

The evaluation used the same procedure as that of the mean value and variance. Table 8 again lists the actual calculated repeatability and reproducibility values according to the procedure outlined above.

	20		**40**		**70**	
	Repeatability	Bias	Repeatability	Bias	Repeatability	Bias
Mean	17,03771312	4,224578413	15,58081388	2,293014214	32,71711856	6,796654
Flatness	26,78496808	35,24448889	51,2324748	66,91119563	20,46927221	26,93563
Concentricity	2,916953882	4,242895045	3,800968745	5,317534541	3,800968745	5,317535
Straightness	36,41014189	51,08601088	36,83584585	51,67767789	36,83584585	51,67768

Table 8. Actual repeatability and bias values for the speeds of 20. 40 and 70 mm/min

They were also assigned point scores but it should be pointed out that in this case the procedure was different. Here. higher value did not receive more points. It was the opposite: the higher the repeatability or reproducibility. the better. and the lower the number of points.

	20		**40**		**70**	
	Repeatability	Bias	Repeatability	Bias	Repeatability	Bias
Mean	2	2	3	3	1	1
Flatness	2	2	1	1	3	3
Concentricity	3	3	1	1	1	1
Straightness	3	3	1	1	1	1

Table 9. Point conversions for repeatability and bias of measurement

The last table shows the aggregate results for the combination of repeatability and bias of measurement. This calculation reported in Table 10 shows that if we only took into account these two parameters, our expectations would have been disproved. The aggregate result indicates that the best speed was 40 mm/min.

	20	**40**	**70**
Mean	2	3	1
Flatness	2	1	3
Concentricity	3	1	1
Straightness	3	1	1
Result	**2,5**	**1,5**	**1,5**

Table 10. Summary of evaluation of repeatability and bias

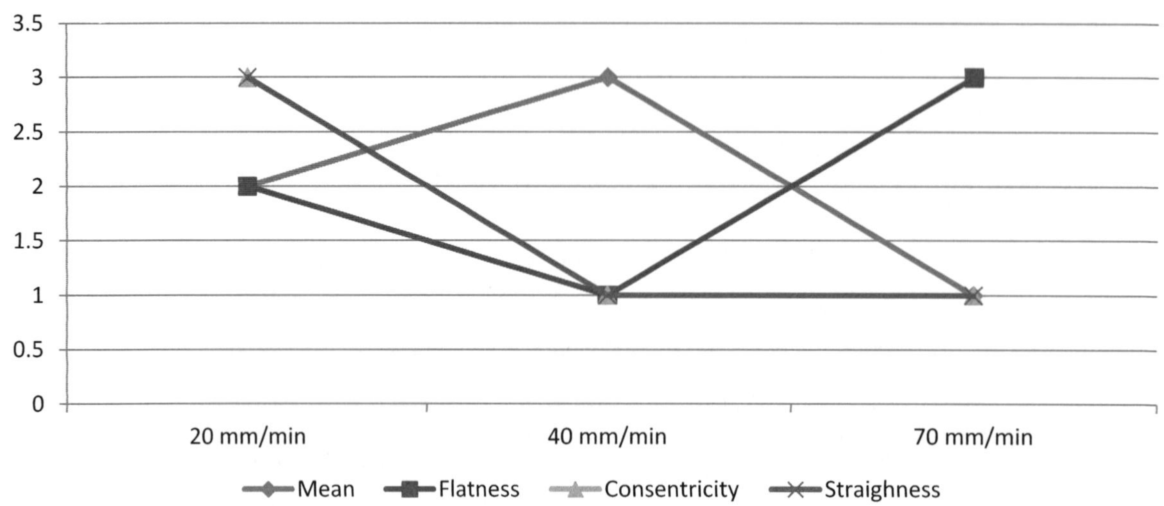

Graph 2. Graphical representation of results for repeatability and bias

The last two tables indicate overall results of the fixed head test. Table 11 shows point scores of the case where only accuracy, repeatability and bias of measurement were reviewed.

In this case, our hypothesis has been proved. Using lower speed provides better measurement values.

On the other hand, the difference between 20mm/min and 40 mm/min is not that striking. What is more, if this result is complemented with the point scores for the measuring time, as shown in Table 11,

Table 12 the picture becomes completely different.

	20	**40**	**70**
Result based on mean value and variance	2,75	1,25	1,5
Result based on repeatability and bias	2,5	1,5	1,5
Result	**2,625**	**1,375**	**1,5**

Table 11. Summary of the comparison of results I.

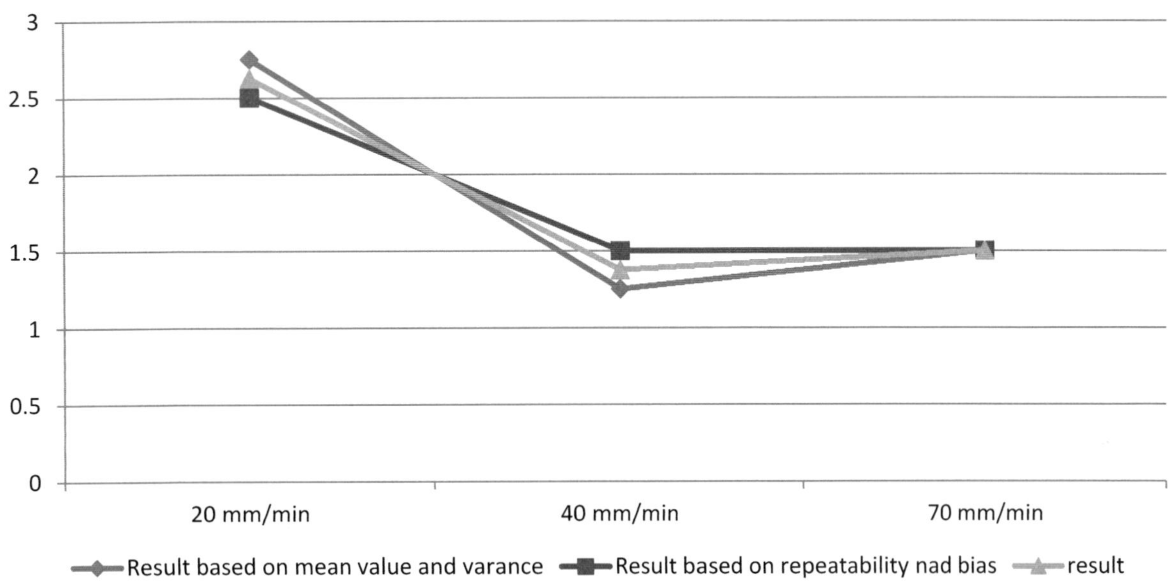

Graph 3. Summary of the comparison of results I

If this result is complemented with the point scores for the measuring time, as shown in Table 12, the picture becomes completely different.

The best one is now the measurement at the speed of 70 mm/min.

	20	**40**	**70**
Result based on measuring speed	3	2	1
Result based on mean value and variance	2,75	1,25	1,5
Result based on repeatability and bias	2,5	1,5	1,5
Result	**2,75**	**1,583333333**	**1,333333333**

Table 12. Summary of the comparison of results II.

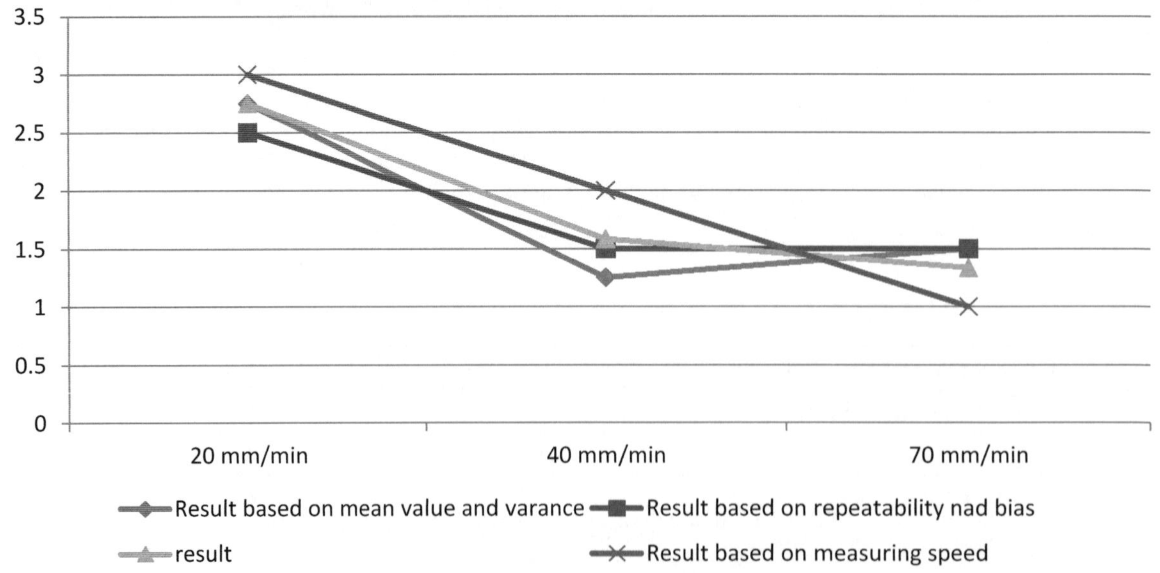

Graph 4. Summary of the comparison of results II.

4. Conclusions

Considering the comprehensive and time-consuming CMM-based inspections in production where machine parts are ordinarily checked with respect to hundreds of characteristics and the measuring times alone often reach tens of minutes of several hours on intricate components, opportunities for improvement are certain to exist in measuring centres, which can be brought to fruition by optimizing scanning speeds. Yet, such an effort must be undertaken with respect to the particular machine, for a particular configuration of the scanning system, specific characteristics of the part with prescribed tolerances and. where possible, by an experienced operator. It is therefore important to understand the capabilities and deficiencies of CMM settings.

This paper deals with an evaluation of the effects of measuring speed on reading accuracy was carried out as part of a research programme at the University of West Bohemia/Faculty of Mechanical Engineering/Regional Technological Institute. The actual evaluation of the test was based on point scores assigned to the data obtained, whether they represented the measuring times, mean values, variance, repeatability or bias and their combinations as well.

The test was based on an assumption that the measurement precision decreases with increasing measuring speeds (in our case. the speeds of 20. 40 and 70 mm/min were chosen).

Our assumption was confirmed in part. This was in the case when the test evaluation only considered the mean value and variance. i.e. when actual dimensions were taken into account. Here, when all measured characteristics were evaluated as a whole, the speed of 20 mm/min (i.e. the minimal measuring speed) was confirmed. Yet, a closer look at individual characteristics offers a less clear-cut picture: for instance, the evaluation of straightness shows that better results are obtained at a higher speed (70 mm/min).

This is why the test evaluation also used the processing based on measuring instrument capability (cg. cgk parameters). The minimum limit for these parameters was set at 1.67. All parameters calculated for all tolerance ranges met this value. However, when just repeatability and bias are considered, the result is not that definitive. Here, the point scores identified the speed of 40 mm/min as the best one.

Since the choice of measuring speed must be viewed in its entirety and one cannot choose different speeds for different characteristics, it is necessary to take a decision in each case and strike a compromise between the measuring speed, precision or repeatability.

Our test with the Carl Zeiss Prismo 7 Navigator machine fitted with a fixed head yields a recommendation that an appropriate compromise between these parameters and a chosen speed is the speed of 40 mm/min.

This experiment descripted in this paper is next step in larger study of measurement influencing factors.

5. Acknowledgements

The article contribution has been prepared under project LO1502 'Development of the Regional Technological Institute'under the auspices of the National Sustainability Programme I of the Ministry of Education of the Czech Republic aimed to support research, experimental development and innovation.

6. References

[1] Melichar M.. Kubátová D.. Kutlwaše J.: (2016). CMM measuring cycle and human factor. Proceedings of the 27th DAAAM International Symposium. ISBN 978-3-902734-08-

[2] Perníkář J.: Assessment of the competence of control means [online]. [cit. 2016-11-25]. Available from: http://gps.fme.vutbr.cz/STAH_IN FO/31_Pernikar_VUTBR.pdf

[3] Płowucha W.. Jakubiec W.. Wojtyła M.: Possibilities of CMM Software to Support Proper Geometrical Product Verification. Procedia CIRP. Volume 43. 2016. Pages 303-308. ISSN 2212-8271. http://dx.doi.org/10.1016/j.procir.2016.02.124

[4] Barini M. E.. Tosello G. Chiffre d L.: Uncertainty analysis of point-by-point sampling complex surfaces using touch probe CMMs: DOE for complex surfaces verification with CMM. Precision Engineering. Volume 34. Issue 1. January 2010. Pages 16-21. ISSN 0141-6359. http://dx.doi.org/10.1016/j.precisioneng.2009.06.009 .

[5] Jiroušek P.: Eligibility of measurement system in the gearboxes production of Škoda auto a.s [online]. 2012 [cit. 2017-03-08]. Available from:https://is.savs.cz/lide/clovek.pl?zalozka=13;id=148; studium=2015. Diplomová práce. ŠKODA AUTO a.s. vysoká škola.

[6] Kubátová D.. Melichar M.. Kutlwašer J: Evaluation of Repeatability and Reproducibility of CMM Equipment.

[7] Yu. C.-T & Ma. C.-M & Cheng. B.-W. (2009). An evaluation method of contact measurement system of the coordinate measuring machine (CMM). Journal of Quality. 16. 213-221.

[8] https://www.interempresas.net/Automotive/Companies-Products/Product-Measuring-probes-Zeiss-VAST-XTR-96591.html [online 18.9.2017]

[9] https://www.zeiss.cz/metrologie/produkty/senzory/vast-xxt.html [online 18.9.2017]

[10] Melichar M., Kubátová D., Kutlwašer J.: Influence of temperature compensation for evaluation using influence of temperature compensation for evaluation by continuous scanning.

[11] http://www.engineering.sk/clanky2/stroje-a-technologie/3188-vliv-rychlosti-snimani-na-zpusobilost-procesu-kontroly-geometrickych-specifikaci-s-vyuzitim-cmm [online 18.9.2017]

[12] https://www.nde-ed.org/GeneralResources/ErrorAnalysis/UncertaintyTerms.htm [online 22.9.2017]

DOI: 10.2507/28th.daaam.proceedings.057

EVALUATION OF METAL POWDER FOR ADDITIVE MANUFACTURING OF MARAGING STEEL

Ivana Zetková, Ludmila Kučerová, Miroslav Zetek, Jiří Česánek, Pavel Hanzl, Milan Daňa, Martin Nozar & Josef Káňa

Regional Technological Institute, University of West Bohemia – Faculty of Mechanical Engineering, Univerzitní 8, 306 14 Pilsen, Czech Republic

This Publication has to be referred as: Zetkova, I[vana]; Kucerova, L[udmila]; Zetek, M[iroslav]; Cesanek, J[iri]; Hanzl, P[avel] & Dana, M[ilan] (2017). Evaluation of Metal Powder for Additive Manufacturing of Margaring Steel, Proceedings of the 28th DAAAM International Symposium, pp.0410-0416, B. Katalinic (Ed.), Published by DAAAM International, ISBN 978-3-902734-11-2, ISSN 1726-9679, Vienna, Austria
DOI: 10.2507/28th.daaam.proceedings.057

Abstract

Additive technologies enable the creation of complex shapes and structures virtually unattainable using classical production processes. These technologies allow components to be produced not only from paper or plastic, but also from composite materials or metals, with the same or even better properties than their predecessors made by machining or forming technologies. However, for customers who are not specialists in these technologies it may often be difficult to orientate themselves in the wide range of options and take full advantage of the potential of these technologies. As a rule, production process parameters are not as decisive as final product characteristics, especially the price, quality and delivery date. Therefore, this article tries to answer elementary questions that customers might have when considering production using additive technology, and to describe its specifics and possibilities. Special attention is paid to metal 3D printing, whose specifics are illustrated using the example of printing a simple component in various ways, in different quantities and sizes. The article documents an experiment conducted under real conditions using DMLS (direct metal laser sintering) technology, which shows how different ways of printing affect the price, the processing time, and the quality of the final component.

Keywords: Additive manufacturing; Metal powder; metallography of the particles; shape of the particles

1. Introduction

The input material for additive manufacturing is the powder. There is a wide range of powders available. Each 3D print machine manufacturer recommends using their "own" powder. The most commonly used materials for laser sintering of metal powder are listed in Table 1. The choice of powder has the main influence on the mechanical properties, functionality and surface quality of the final part. The powders are transported to the users in closed barrels. The powder is taken from the barrels to the dispenser platform in the machine. A recoater applies the defined thickness of the powder on the building platform, where it is melted in the designated area. When the building process is finished, the powder around the sintered parts is sieved back to the dispenser platform for the next process.

	Material	DIN
Aluminium alloys	AlSi10Mg	3.2381
	AlSi7Mg	3.2371
	AlSi12	3.3581
Cobalt alloys	ASTM F75	2.4723
	CoCrWC	
Tool steels	AISI 420	1.2083
	Marage 300	1.2709
	H13	1.2344
	AISI D2	1.2379
	AISI A2	1.2363
	AISI S7	1.2357
Nickel alloys	Inconel 718	2.4668
	Inconel 625	2.4856
	Inconel 713	2.4670
	Inconel 738	
	Hastelloy X	2.4665
Stainless steel	SS 304	1.4301
	SS 316 L	1.4404
	SS 410	1.4006
	SS 440	1.4110
	15-5 PH	1.4540
	17-4 PH	1.4542
Titanium alloys	Titanium Grade 2	3.7035
	Ti6Al4V	3.7165
	Ti6Al4V ELI	3.7165 ELI
	TiAl6Nb7	
Precious Metal Alloys	gold	18 Carat
	silver	930 Sterling
Copper alloys	CC 480 K	2.1050

Table 1. Material for additive manufacturing [1]

The powder for additive manufacturing can be made in several different ways. The manufacturing process depends on the powder material and its final quality. Gas atomization, VIM gas atomization, plasma atomization, electrode melting gas atomization and hydride-dehydride methods are the most commonly used for powder manufacture. None of these methods have 100% yield. Grains of incorrect size are separated from the powder. In general, the size of particles for laser sintering is set in the range of 15 – 45 μm, and for electron beam sintering in the range 45–106 μm [2]. Results from some of these methods are shown in Fig.1.

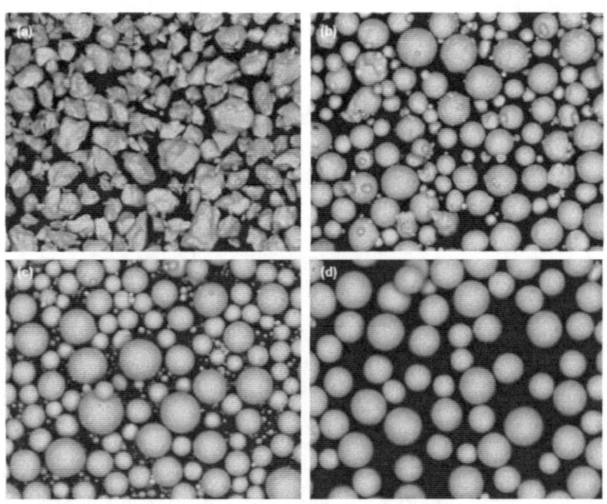

Fig. 1. SEM of typical particle shapes of Ti-6Al-4V powder for different manufacturing technologies - a) HDH; b) gas atomization; c) plasma atomization; d) rotating electrode atomization [3]

Mechanical properties for one type of material may be different depending on the manufacturing technology used. Fig. 2 shows the yield strength and hardness for materials which are made in different ways [1].

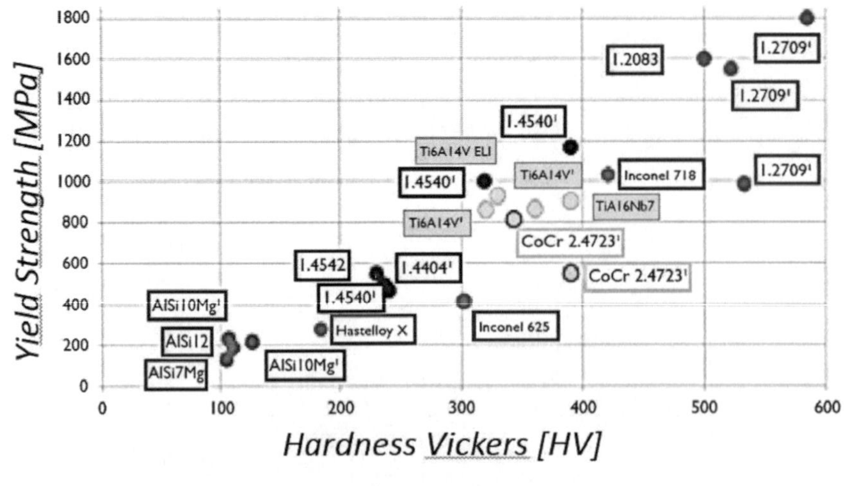

Fig. 2. Hardness and yield strength of materials used in additive technology [1, 2]

The rate of cooling and the gas used for spraying influence the diameter of the produced particles and their microstructure, according to reference [1], see Fig. 3.

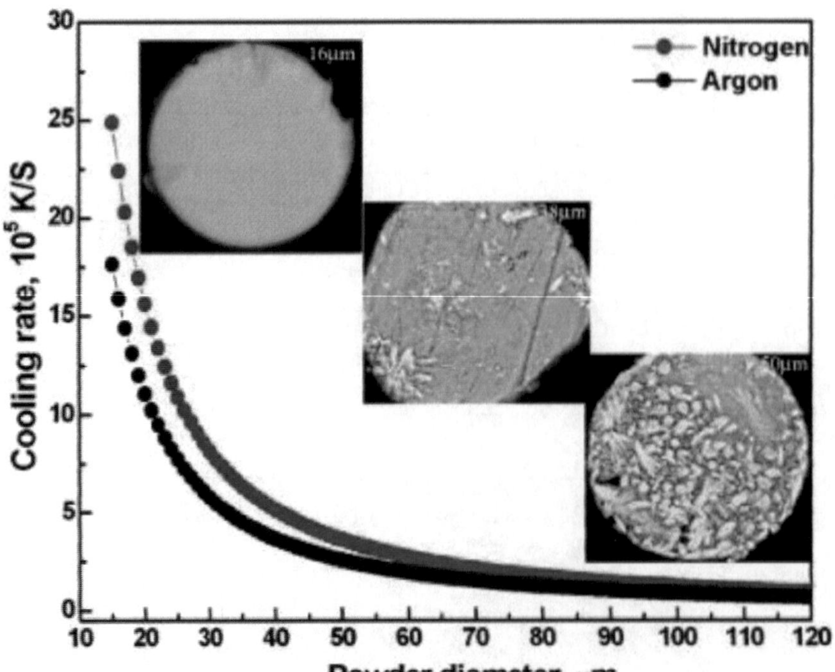

Fig. 3. Influence of cooling rate and type of atomizing medium on size and microstructure of powder Al 86 Ni 6 Y 4.5 CO 2 La 1.5 [1]

Additive technology is increasingly being used in mechanical engineering and the number of publications in this area is growing as well. They are usually focus on the mechanical properties of the printed parts [4,6] on to microstructure of the printed parts [5] but there is just basic information about the input material.
This paper is focused on the powder W-Nr.1.2709 (DIN X3NiCoMoTi18-9-5), on the evaluation of its shape and microstructure. It is usually used for plastic injection moulds, pressure casting of light metals, and cold pressing tools. The recommended chemical composition of printed and conventionally produced steels (hot formed) is shown in Tab. 2. Not much information has been found about this frequently-used powder.

	C	Si	Mn	P	S	Cr	Mo	Ni	V,W	Co	Ti	Cu	Al	Fe
MS1	≤ 0.03	≤ 0.1	≤ 0.1	≤ 0.01	≤ 0.01	≤ 0.5	4.5 - 5.2	17.0 - 19.0		8.5 - 9.5	0.6 - 0.8	≤ 0.5	0.05 - 0.15	balance
Conv.	≤ 0.03	≤ 0.1	≤ 0.15	≤ 0.01	≤ 0.01	≤ 0.25	4.5 - 5.2	17.0 - 19.0	-	8.5 - 10	0.8 - 1.2	-	-	balance

Table 2. Chemical composition of printed (MS1) and conventionally (Conv.) produced steel W-Nr. 1.2709

2. Evaluation of the shape and microstructure

The powder of steel W-Nr.1.2709 is supplied in 10 kilogram closed barrels. In general, it is necessary to observe the principles of safety at work, because some types of powders or their condensates may be self-inflammatory or explosive in combination with air [1]. A spherical shape is preferred due to good fluidity, but the manufacturer states that irregular particles do not reduce the quality of the final printed part [4].

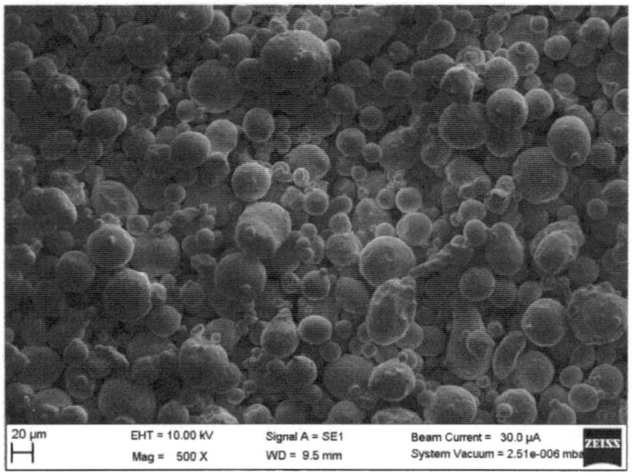

Fig. 4. SEM of powder W-Nr.1.2709

Firstly, it was necessary to determine the shape of the particles. Most of the grains are spherical and some are stuck together, see Fig. 5 This is the case for particles with similar and different sizes. This is because they were not divided before solid. This theory is confirmed below. It is possible to see two or more particles stuck together (Fig. 5 b), or a grain inside another grain. There is a minimal number of irregularly shaped particles in the powder. They probably did not arise during the production process, but during further manipulation. The peeling of layers (see below) or mutual contact of the grains cause the irregular shapes of the particles.

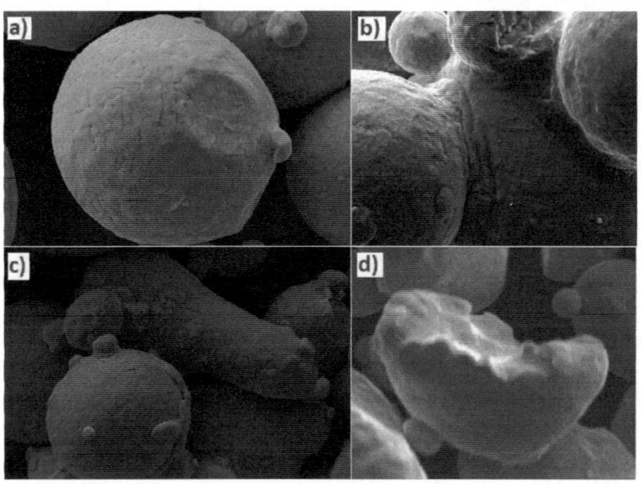

Fig. 5. Type of particles of the metal powder W-Nr. 1. a) spherical, b) several particles stuck together, c) elongated grain, d) irregular

A unique flaky structure, or "elephant skin" can be seen on the surface of the grains. It is a question how this structure was created. A metallographic section provides the answer, shown in Fig. 6. The etched grains have a cellular/dendritic microstructure from the soft, tough solution. In general, dendrites are formed from the solution of crystallization with a negative temperature gradient. This arises during the manufacturing process when the metal alloy is melted and the gas is atomized to fine droplets. They are formed after solidification in individual grains of powders with a relatively stable chemical composition. Smaller grains contain slightly more Ni and Co. The dendrite borders contain slightly more Mo and Ti in comparison with the middle of the grain.

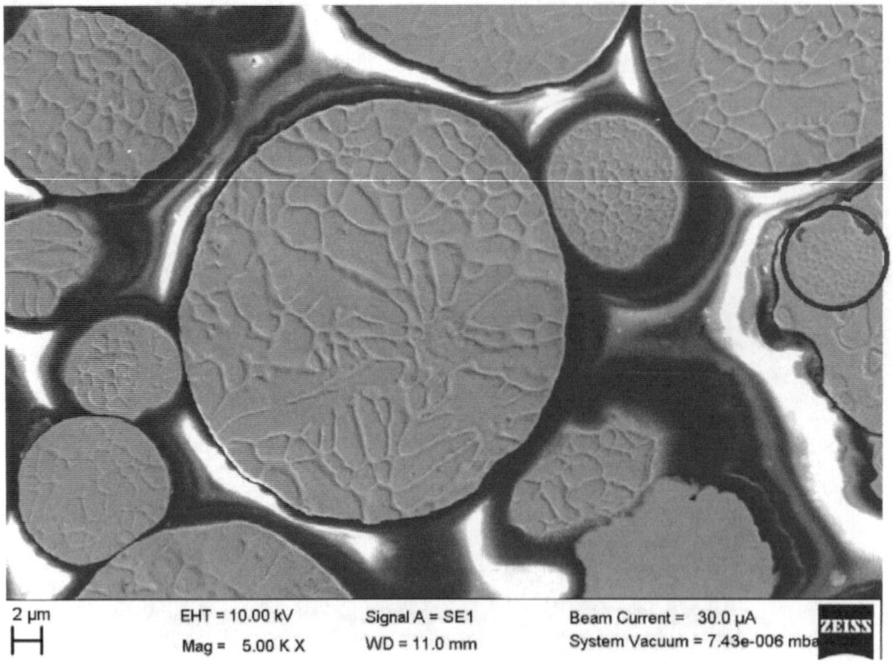

Fig. 6. Microstructure of powder W-Nr. 1.2709

Dendrites grow on the grain surface and create the characteristic "elephant skin" texture. The irregular growth of these grains stems from their production by gas atomization. In some cases, a grain grows inside another grain – Fig. 6 marked in red.

Fig. 7 shows particles stuck together. They contain more separate dendrites in one grain of irregular shape. This clarifies the theory of the formation of interconnected particles. Particles with a high portion of Fe and Ti were found in some grains (see Fig. 7, right)

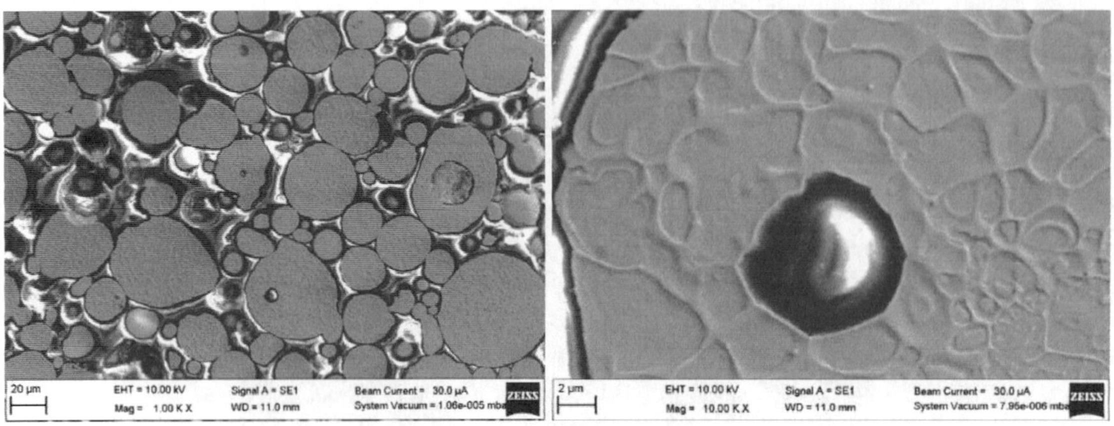

Fig.7. SEM of the microstructure

In the next step the powder which was used in the 3D print process was examined. This powder was not melted. It surrounded the printed component and was sieved back in the dispenser platform. A scan of this powder sample (see Fig 8) was compared with the new powder.

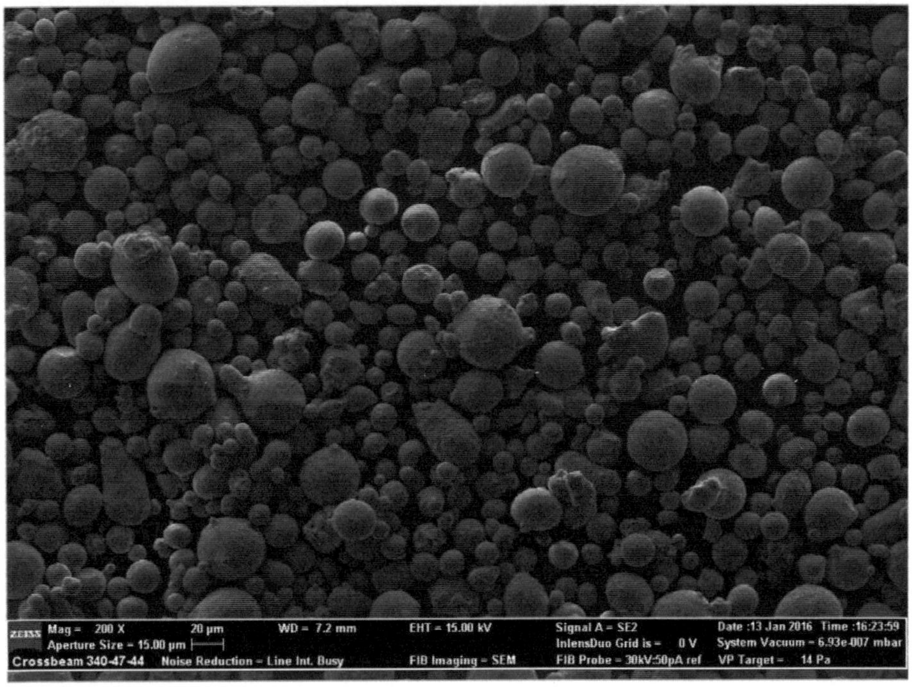

Fig. 8. Sieved powder

There were no visible differences between the new and the sieved powder. A very thorough investigation was made and a small percentage of powder particles without “elephant skin” was found. The surface of these particles was very smooth. They were not found in the new powder from any production batch, so this had to occur during the printing process.

Fig. 9. SEM of the particles in the sieved powder

A metallographic section was made to compare this type of particle and particles from the new powder. The results are shown in Fig. 10. The grain has smooth surface. The dendrites do not grow to the surface, so the "elephant skin" cannot be seen. The chemical composition is comparable with the particles from the new powder. The question is, how these particles influence the mechanical properties of the printed parts.

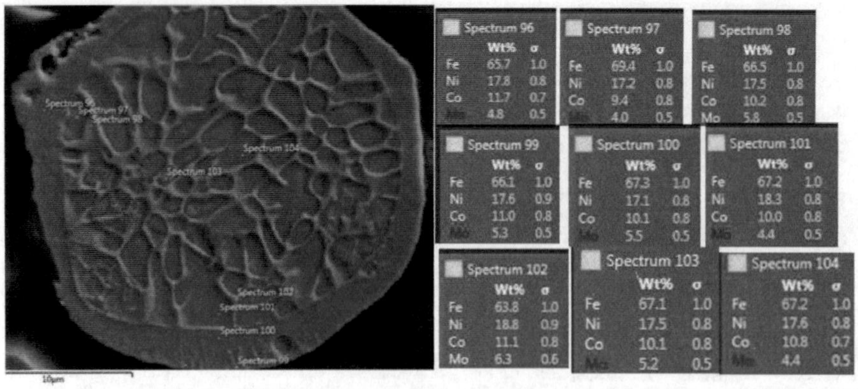

Fig. 10. Metallography of the particles in the sieved powder

The process of 3D printing was monitored by a high speed camera and a lot of particles were observed which are formed during the process. The metallographic section in Fig. 10 is completely unique. It has not been found in any other published scientific sources.

3. Conclusion

This paper was focused on the evaluation of metal particles of maraging steel W-Nr. 1.2709 according to their shape and structure. The new and used powder was observed after printing. The sieved powder was found to contain particles with the same chemical properties but with different structures on the grain surface. The dendrites form the basis for both types of particles. It was proved by using a high speed camera that smooth particles originate during the printing process. The next paper will describe how these particles influence the mechanical properties of the parts. There are scientific papers about metal powders for 3D printing, but there are none about the new particles which are formed in the printing process.

4. Acknowledgments

The article has been prepared under project LO1502 'Development of the Regional Technological Institute' under the auspices of the National Sustainability Programme I of the Ministry of Education of the Czech Republic aimed to support research, experimental development and innovation.

5. References

[1] Metal powders – the raw materials. Metal AM. (2015) . Available from: http://www.metal-am.com/introduction-to-metal-additive-manufacturing-and-3d-printing/metal-powders-the-raw-materials/
[2] Introduction to Additive Manufacturing Brochure. (2015). Available from: http://www.epma.com/additive-manufacturing
[3] Dawes, J., Bowerman, R., Trepleton, R., Introduction to the Additive Manufacturing Powder Metallurgy Supply Chain. Johnson Matthey Technol. Rev., 2015, 59, (3), 243-256
[4] Daňa M., Zetková I., Hanzl P. Need for support structures depending on overhang size. In MM (Modern Machinery) Science Journal, December 2016. [online]. [cit. 2017-03-15]. Available from: http://www.mmscience.eu/content/file/archives/MM_Science_2016192.pdf
[5] Kucerova, L. Zetkova, I. (2016) Metallography of 3D Printed 1.2709 Tool Steel. In MANUFACTURING TECHNOLOGY. February 2016, Vol. 16, No.1. s. 140 -144. ISSN 1213–2489.
[6] Hanzl, P., Zetková, I., Mach J. Optimization of the pressure porous sample and its manufacturability by selective laser melting. Manufacturing Technology, 2017, roč. 17, č. 1, s. 34-38. ISSN: 1213-2489
[7] Rubešová, K., Jirková, H., Mašek, B. Vorel, I. (2016). Effect of Mini-Thixoforming Parameters on Microstructural Evolution in Low-Alloy Steel, Proceedings of the 26th DAAAM International Symposium, pp.0803-0809, pp.0803-0809, B. Katalinic (Ed.), Published by DAAAM International, ISBN 978-3-902734-07-5, ISSN 1726-9679, Vienna, Austria
[8] Cubrova, J[ana]; Vorel, I[van]; Vancura, F[ilip] & Masek, B[ohuslav] (2016). Effect of Amount of deformation on Microstructure Evolution During Controlled Cooling of Forgings from FinishForging Temperature, Proceedings of the 26th DAAAM International Symposium, pp.0892-0896, B. Katalinic (Ed.), Published by DAAAM International, ISBN 978-3-902734-07-5, ISSN 1726-9679, Vienna, Austria DOI: 10.2507/26th.daaam.proceedings.124

DOI: 10.2507/28th.daaam.proceedings.058

INFLUENCE OF THE MILLING CUTTER DIAMETER ON THE CUTTING TOOL LIFE WHEN MACHINING INCONEL 718

Miroslav Zetek, Vojtěch Vozár, Tomáš Bakša & Ivana Zetková

This Publication has to be referred as: Zetek, M[iroslav]; Vozar, V[ojtech]; Baksa, T[omas] & Zetkova, I[vana] (2017). Influence of the Milling Cutter Diameter on the Cutting Tool Life when Machining Inconel 718, Proceedings of the 28th DAAAM International Symposium, pp.0417-0422, B. Katalinic (Ed.), Published by DAAAM International, ISBN 978-3-902734-11-2, ISSN 1726-9679, Vienna, Austria
DOI: 10.2507/28th.daaam.proceedings.058

Abstract

Indexable multi-insert milling cutters are the most economical and productive face milling tools. In case of the Inconel alloy machining there are some specific cutting conditions and strategies that influence cutting tool life and surface quality. When the indexable cutting inserts are used, it is recommended to use always down milling and avoid positioning cutter on–centre. For the tool entry is recommended reduce feed rate and use roll on strategy. When different tool dimeters are used there are different cutting conditions which is influenced by the principal cutting edge angle and it influence cutting tool life and tool wear character. This study is focused on the influence of the three different cutter diameter on the cutting tool life when the Inconel 718 is machined. For the tests different cutter diameters with same rounding cutting inserts are used. The trajectory of the movement are simulated for all diameters by the 3D view and on the base of it the real cutting conditions are calculate and setup for the practical experiment. During the tests, the cutting forces are monitored as well as cutting inserts tool wear (2D visualization and 3D visualization). For the comparing of the tool wear character the difference analyses is used and 2D cuts are made for the evaluation of the general tool wear parameters.

Keywords: Milling cutter; Tool diameter; Cutting tool life; Inconel 718; Tool wear

1. Introduction

Inconel 718 belongs to a group of difficult-to-machine materials. It offers excellent strength at low and high temperatures, corrosion resistance under demanding conditions and low thermal diffusivity. Therefore, it found use in components that operate under extreme loads in heavy-duty applications. Their examples include parts of combustion turbines, thin-walled aircraft components and parts of space shuttles. Its use in production has called for operation sequences which deliver optimum machining performance. Today, conventional machining of Inconel 718 poses a major difficulty in terms of the process economy. The motivation for this study arose from a search for appropriate machining parameters and from efforts to improve machining productivity of this material. For this purpose, information from worldwide databases was gathered. One of the key factors which govern the economy of cutting nickel alloys is the short durability of cutting tools. Numerous studies have been devoted to this aspect. One of them explored the effects of cutting conditions on tool durability. This study found that the dominant type of wear, regardless of cutting conditions, is the built-up edge which eventually leads to chipping of the cutting part of tool. Furthermore, if an inadequate thin film is chosen, the intensity of abrasive wear increases very rapidly. [1] This was confirmed by another study. Its authors were dealing with thin film optimization – which can improve overall durability by several per cent. [2] Cutting conditions affect the type and progress of wear very strongly. This issue was explored by authors of [3]. They monitored changes in the wear type depending on cutting speed, feed and cutting depth. They also correlated these aspects with the machined surface quality and monitored the impact of cutting conditions on changes in the material. It is clear that the greatest impact on durability results from changes in the cutting speed, whereas the quality of machined surface is dictated predominantly by changes in feed. Surface integrity in machining of Inconel 718 is the topic of additional studies. One of them [4] mapped defects in the machined surface in relation to the occurrence of wear and built-up edge. Similar results were reported by authors of [5] who correlated them with cryogenic machining. They found that better outcome is obtained with standard cooling. Another study describes industrial machining of turbine blades of this material. Its authors dealt with optimizing the production process while taking into account the occurrence of wear and maintaining the quality of machined surface and its accuracy. [6] They found that the profile shape affected the resultant roughness and accuracy. They used the findings in another publication [7] in which they added investigation of the influence of wear on roughness at various points along the turbine blade profile during machining with a ball mill. The dominant factor in the resultant quality of surface was the location of engagement of the ball mill along the blade profile. It was confirmed that roughness becomes less favourable with decreasing engagement radius, and therefore with decreasing cutting speed. The similar results gave authors which were focused on the turning with the different cutting tool radius [8]. One of the many available options for productive machining of Inconel 718 is High Speed Machining (HSC). Yet, the cutting part wear is a major factor in this process as well. This was the subject of the detailed study [9] which monitored the amount of wear of a cemented carbide cutting part vs. the cutting speed used. The results suggested that the amount of wear increased with cutting speed. The material, which is very appropriate for HSC machining of Inconel 718, is cubic boron nitride (PCB). The study [10] found that in this case, too, wear is governed by the cutting speed but also by the microstructure of the work and by the exact chemical composition of the cutting part's material. A very novel method of controlling the durability of CBN tools involves creating a defined surface texture on the cutting part of the tool. This is the focus of the study [11] which describes the texture shapes, cutting conditions and the benefits of this approach. Since Inconel 718 is difficult to machine, there are other studies which seek productive cutting and machining methods for this material. Authors of [12] explored the economy and environmental impacts of various waterjet cutting processes. The last article [13] deal with possibilities of evaluation oscillation absorption and damping factor is increased due to the large dissipative forces resulting from the resistance of the tool holder material that has texture with different metal deformation conditioned orientations. Clearly, this field offers many options for optimizing the machining process – and any improvement in tool durability counts.

On the base of these studies, the article will be deal with influence of the milling cutter diameter on the cutting tool life when machining Inconel 718. The main aim is to describe how the different cutting tool diameter influence the tool life, which is depending on side and rear plane and connected angels. For the evaluation, the different engaging conditions were used with the constant machining conditions.

2. Experiment

The objective of this experiment was to identify the relationship between the diameter and durability of a milling cutter. The diameters under test included 50, 100 and 125 (121) mm. The limit flank wear was set at 0.15 mm. During analysis of the static geometry and conversion into the kinematic geometry, it was found that the geometric set-up of these tools would differ. They were therefore adjusted to ensure that the resulting kinematic geometries of engagement were identical for all tool variants. Circular cemented-carbide exchangeable cutting inserts (below referred to as VBD) were used which had a thin film which had been specially developed for Inconel 718. No defects in this film were found upon a thorough inspection. All experimental runs were carried out at constant cutting speed v_c=40m/min, feed per tooth f_z=0.2mm and depth of cut a_p=2 mm. A constant width of engagement was chosen for the first variant, a_e=30% D of the milling cutter, as recommended by the manufacturer.

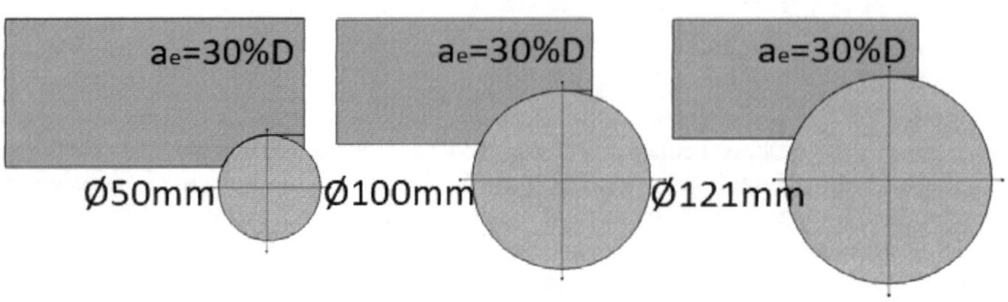

Fig. 1. Visualization of engagement conditions

In this case, the other parameters were set as follows:

	D = 50 mm	D = 100 mm	D = 121.76
vc [m/min]	40	40	40
vf [m/min]	51	25	21
ap [mm]	2	2	2
ae [mm]	15	30	36.53
fz [mm]	0.2	0.2	0.2
n [-]	255	127	105
Arc of engagement Ψ [°]	66.42	66.42	66.42
Q [cm3/min]	1.53	1.53	1.53

Table 1. Cutting conditions at constant ae

Only a single VBD was employed in the experiment in each case, which can lead to lower durability of the tool due to instability which occurs as a single VBD enters the work. To verify the findings, the experiment will be repeated with a fully-fitted milling head.

The entry into the workpiece was modified by pre-milling the work at 30% D for the tool diameter. Therefore, the VBD was not affected by the entry and exit variations, as illustrated in Fig. 2.

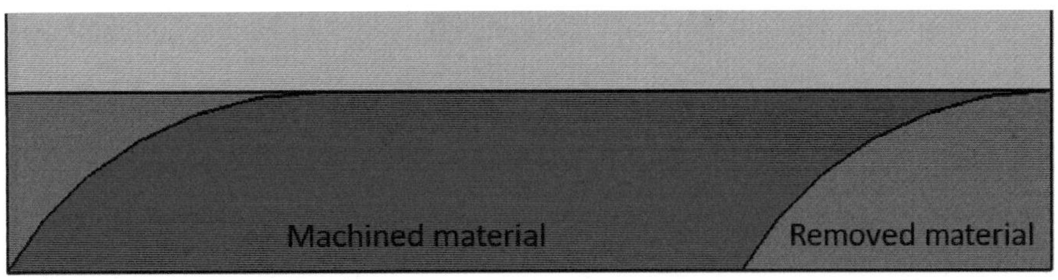

Fig. 2. Visualization of the pre-milled workpiece

It is important to check the material prior to testing: the chemical composition, microstructure, hardness and other mechanical properties if relevant. With regard to machinability and the effects on tool durability, the important aspects include the chemical composition, heat treatment history and the resultant microstructure and hardness.

C	Mn	Si	P	S	Ca	Cr	Al	Cu	Mo	Ni	N	Ti	Fe	Nb	Co
0,017	0,10	0,07	0,008	0,0005	<0,0003	18,23	0,51	0,04	2,98	53,3	0,006	0,98	18,22	5,09	0,31
B	Pb	Ta	O	Other											
0,0024	<0,0003	<0,020	<0,00003	Mg = 0,0010 ; Sn = 0,0011 ; Se = <0,0003											

Fig. 3. Chemical composition of workpiece

To check the homogeneity of microstructure of the workpiece, samples were taken from its surface, from a location 20 mm below the surface and from the workpiece centre.

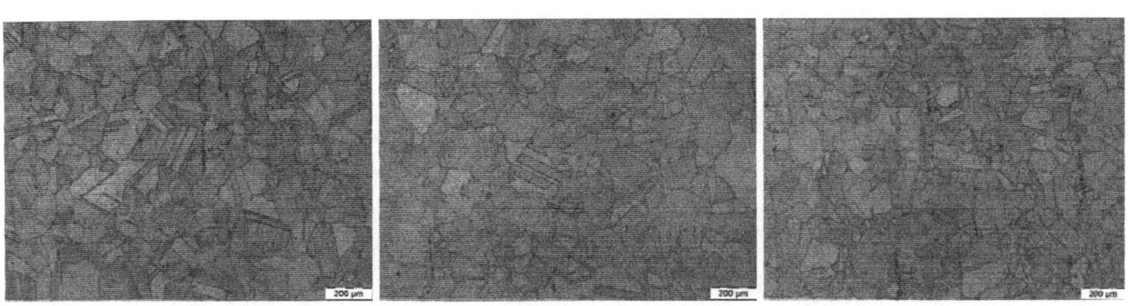

Fig. 4. Microstructure of the Inconel 718 workpiece (from left: surface, 20 mm below the surface, centre)

The images show that the grain shapes and sizes depend on the location within the workpiece. On the other hand, this is a common occurrence in the industrial practice. Therefore, no steps were taken to modify the microstructure. Instead, hardness was measured in these locations. Its values were in the range of 34–39 HRC, depending on the location.

Machining was carried out in a 5-axis milling centre. The workpiece was clamped in a hydraulic vice. In order to ensure that identical material (microstructure) is removed by tools of each diameter, they were all applied to a single layer of the workpiece in each case. The effect of the microstructure variation was thus eliminated.

The results are plotted in Graph 1.

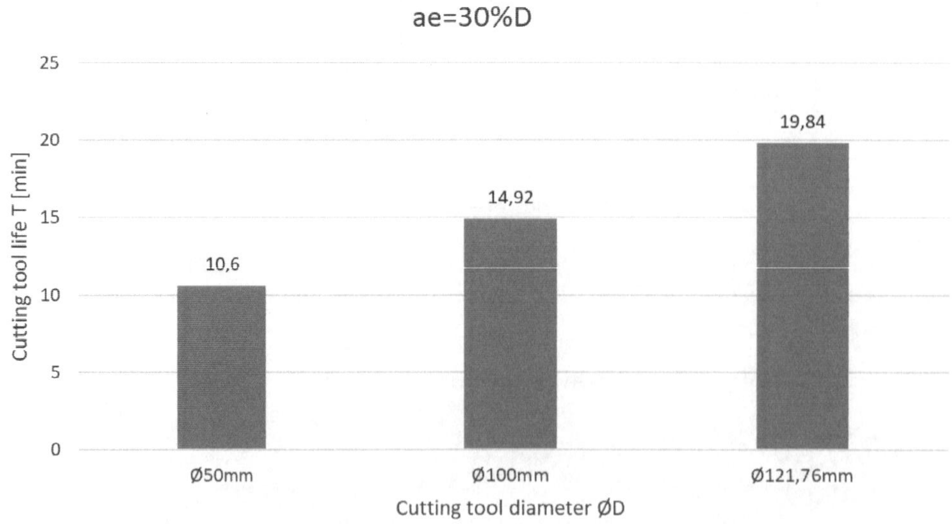

Graph 1. Tool durability vs. tool diameter in machining a pre-milled workpiece

The graph indicates that durability increases with the tool diameter. Constant a_e means that the arc of engagement is the same for all tools, although the width of engagement increases with the tool diameter. Hence, the smaller is the tool diameter, the more entries the tool must complete, which means that it sustains more impacts. Upon conversion per diameter basis, the following results were obtained:

	Ø50 mm	**Ø100 mm**	**Ø121.76 mm**
Alternative No. 2	2700	2375	2075

Table 2. Number of impacts of tool on the workpiece at the end of durability

One can therefore assume that with the number of impacts on the workpiece, the tool wear will increase more rapidly, reaching the limit value of 0.15 mm much earlier, as reflected in the durability.

For further comparison and verification, an experiment was carried out under identical cutting conditions on a workpiece which had not been pre-milled.

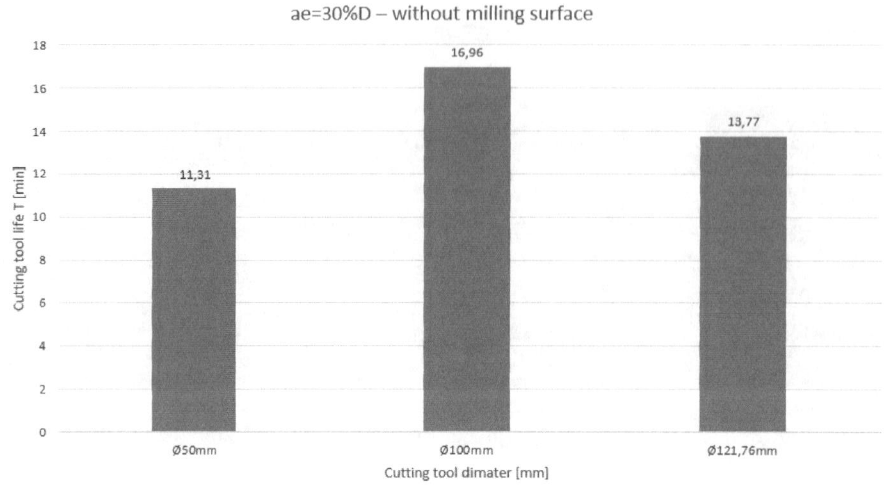

Graph 2. Tool durability vs. tool diameter in machining a non-pre-milled workpiece

Durability was higher than with the pre-milled workpiece but it no longer increased with the tool diameter. This set-up will therefore be analysed further because the results represent mean values from repeated runs.

The last experiment involved a comparison with a milling head with all VBDs fitted, under identical conditions. While the tools with 100 mm and 121 mm diameters had the same numbers of cutting inserts (VBDs) and an identical arc of engagement, the stability condition was fulfilled, i.e. more than one VBD being engaged at a time. For these reasons, only these two diameters were used for the comparison.

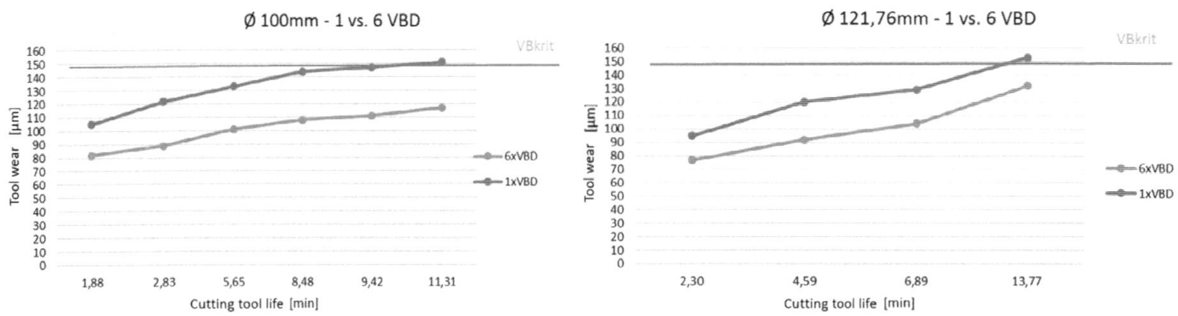

Graph 3. Effects of stability on tool durability in cutting with a single and multiple VBDs

The graphs clearly show that the use of a single VBD in the experiment affects the actual durability of the tool. It is mainly due to increased instability, as the tool suffers stronger vibration than a tool on which more than 1 VBD is engaged and which remains stable throughout the cutting operation. On the other hand, this has no impact on the plausibility of the results. Machining with a single VBD led to uniform abrasive wear, same as with the milling head with all positions occupied. No significant differences were found. The wear profile is shown in Fig. 5.

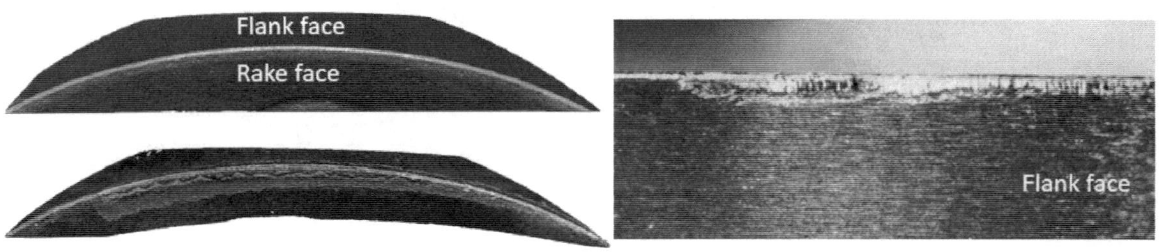

Fig. 5. Visualization of wear (left: 3D-scanned image of wear; right: 2D optical micrograph)

3. Conclusion

The results show the effects of the tool diameter and cutting conditions on tool durability. While the basic principle for defining cutting conditions was observed, i.e. a constant width of engagement, it was found that durability increases with the tool diameter. This may be of major importance in selecting the machining strategy and while considering the diameter for a particular size of machined surface. With Inconel 718, it is appropriate to use multiple tools of different diameters, instead of choosing a single diameter for removing material without the need for tool replacement. In many cases, this variant may prove very costly due to lower durability, and therefore the need for more frequent replacement of VBDs. These experiments also showed the effects of the stability of the cutting process on tool durability. When a single VBD is used, durability decreases. This means that in the choice of engagement conditions made in the industrial practice, it is important to ensure that more than a single point of the tool is engaged at a time. This may prove difficult or even unfeasible to meet with tools of small diameters. The above analyses provide underlying knowledge for future investigations which will focus on various engagement conditions.

4. Acknowledgments

This paper was supported by the programme of applied research, experimental development and innovation GAMA, No. TG02010011 - Support of UWB commercial opportunities.

5. References

[1] H. Abdul, J. A. Ghani, H. Che, and M. S. Kasim. (2016). Effect of cutting speed on the carbide cutting tool in milling Inconel 718 alloy, J. Mater. Res., Vol. 31, No. 13, pp. 1885–1892, DOI: 10.1557/jmr.2015.380.

[2] M. Sortino, S. Belfio, G. Totis, E. Kuljanic, and G. Fadelli. (2015). Innovative tool coatings for increasing tool life in milling Nickel-coated Nickel-Silver alloy, Energy Procedia, Vol. 100, pp. 946–952. DOI: 10.1016/j.proeng.2015.01.453

[3] M. A. Xavior, M. Manohar, P. Jeyapandiarajan, and P. M. Madhukar. (2017). Tool Wear Assessment During Machining of Inconel 718, Procedia Eng., Vol. 174, pp. 1000–1008, DOI: 10.1016/j.proeng.2017.01.252.

[4] C. Liu, C. Ren, G. Wang, Y. Yang, and L. Zhang. (2015). Study on surface defects in milling Inconel 718 super alloy, J. Mech. Sci. Technol., Vol. 29, No. 4, pp. 1723–1730, DOI: 10.1007/s12206-015-0345-1.

[5] A. Iturbe, E. Hormaetxe, A. Garay, and P. J. Arrazola. (2016). Surface Integrity Analysis when Machining Inconel 718 with Conventional and Cryogenic Cooling, Procedia CIRP, Vol. 45, pp. 67–70, DOI: 10.1016/j.procir.2016.02.095.

[6] J. Kyncl, L. Beránek, K. Kolarík, and Z. Pala. (2014). The research of the surface profile after profiling of inconel 738LC, Procedia Engineering, Vol. 69, pp. 974–979. DOI: 0.1016/j.proeng.2014.03.078.

[7] J. Kyncl and A. Molotovnik. (2015). The research of the surface profile after profiling of superalloys, Energy Procedia, Vol. 100, pp. 853–860. DOI: 10.1016/j.proeng.2015.01.441.

[8] N. Qehaja, K. Jakupi, A. Bunjaku, M. Bruçi, a H. Osmani. (2015). Effect of Machining Parameters and Machining Time on Surface Roughness in Dry Turning Process, Procedia Eng., roč. 100, s. 135–140, DOI: 10.1016/j.proeng.2015.01.351

[9] D. M. D'Addona, S. J. Raykar, and M. M. Narke. (2017). High Speed Machining of Inconel 718: Tool Wear and Surface Roughness Analysis, Procedia CIRP, Vol. 62, pp. 269–274, DOI: 10.1016/j.procir.2017.03.004.

[10] H. Tanaka, T. Sugihara, and T. Enomoto. (2016). High Speed Machining of Inconel 718 Focusing on Wear Behaviors of PCBN Cutting Tool, Procedia CIRP, Vol. 46, pp. 545–548, DOI: 10.1016/j.procir.2016.03.120.

[11] T. Sugihara, H. Tanaka, and T. Enomoto. (2017). Development of Novel CBN Cutting Tool for High Speed Machining of Inconel 718 Focusing on Coolant Behaviors, Procedia Manuf., Vol. 10, pp. 436–442, DOI: 10.1016/j.promfg.2017.07.021.

[12] G. S. Kadam and R. S. Pawade. (2017). Surface integrity and sustainability assessment in high-speed machining of Inconel 718 – An eco-friendly green approach, J. Clean. Prod., Vol. 147, pp. 273–283, DOI: 10.1016/j.jclepro.2017.01.104.

[13] J. Olt a V. V. Maksarov. (2015). Development of chatter-resistant system of cutting tool, Annals of DAAAM and Proceedings of the International DAAAM Symposium, 2015 p. 223–226. DOI: 10.2507/26th.daaam.proceedings.031

DOI: 10.2507/28th.daaam.proceedings.059

THE SIMULATION-BASED SMART MANAGEMENT APPROACH FOR CELLULAR NETWORK OPERATION AND PLANNING

Alexander Suleykin & Peter Panfilov

Higher School of Economics, Myasnitskaya Ulitsa, 20, Moscow, 101000, Russia

This Publication has to be referred as: Suleykin, A[lexander] & Panfilov, P[eter] (2017). The Simulation-Based Smart Management Approach for Cellular Network Operation and Planning, Proceedings of the 28th DAAAM International Symposium, pp.0423-0432, B. Katalinic (Ed.), Published by DAAAM International, ISBN 978-3-902734-11-2, ISSN 1726-9679, Vienna, Austria
DOI: 10.2507/28th.daaam.proceedings.059

Abstract

The number of mobile users in the world has been steadily increasing for last decades and is anticipated to continue its growth by the end to 2019 up to more than 5 billion users. This process brings new challenges to the telecom companies over the globe and to the cellular network infrastructures. The problem (dead) zones and dropped calls are now among the most critical issues that mobile providers tend to solve, while trying to increase the quality of service at the same time. The paper discusses modeling approaches to the cellular network's problem zones detection and analysis, on the basis of complaints, subscribers and base stations data aggregation and analysis. We describe algorithms to be used in building Heat Map geo-spatial visualizations of cellular network data, which show the distribution of problem zones within the network with their corresponding weights assigned according to the amount of complaints and population density in the area. This information is used in prioritization of distributed problem zones for the cellular network development and decision support on new base station deployment. It means that simulation-based approaches to problem zone detection and fixation in cellular network can help to increase the customer satisfaction of connection level and quality of mobile service. In addition, the paper presents the evaluation of economic effect of solving the problem zone issue that follows the subscriber traffic increase.

Keywords: cellular network; base station; problem zones; visual simulation; smart management

1. Introduction

In recent years, the popularity of mobile phones is dramatically increasing together with new technology development such as Internet of Things and Machine-to-Machine communications. The number of smartphone users has already reached 4.61 billion users in 2016, and upward trend is forecasted for the market with 5.07 billion users by the end of 2019 [1]. This trend might be explained by the new technologies appearance on the market with higher data speed transmission (for example 4G, 4G-Advanced, 5G creation etc.), the decrease of smartphone prices, the decrease of connection prices and others. As a result, smartphones are becoming more and more achievable for people with different income, and more users are appearing on the market. However, this process generates some new challenges for mobile operators.

As amount of smartphone and mobile users is growing and forecasted to continue growing at least until 2019, the load on the wireless mobile network is also increasing. This is a huge challenge that mobile operators are eventually facing and try to troubleshoot and overcome.

There is an obvious need for each mobile operator to provide the maximum coverage of the entire territory of the operator's presence with mobile network services. However, there is an issue of the planning optimal allocation of new base stations in problem zones of the network that needs to be done taking into account the following factors:

- The budget that is to be invested in construction of new stations. The construction of base stations is expensive and for mobile operators it is highly important to place new base stations in locations that cover problem zones and provide maximum user satisfaction at minimum cost.
- The minimization of the churn's average revenue per user (ARPU) value, which is supposed to decrease in cases when users decide to leave mobile operators for the bad connection in some locations. If user decides to alter the mobile provider, the ARPU of this user should be minimal. Thus, there is a case to find users with different ARPU values, prioritize them and try to satisfy the quality of connection for the most critical users for company from this list – the users with the highest ARPU.
- The decrease of number of such unsatisfied users who can face the dead connection problem.
- The current distribution of base stations.
- Subscriber's complaints registered on the web-site of the company, Contact Center or Speedtests (special bespoke company's measurements). In this paper, only these primary sources for complaints are considered. There are also many tools existing on the market, which allow to analyze the network quality in a more automated way (using special apps), but it will not be in scope of this study.

So the optimal planning of new base stations construction and deployment is extremely important for the telecom company, bringing better customer satisfaction, decreasing the churn losses and more effectively distributing connection power within the network.

Usually, telecom companies have their own private methods for problem zones detection and prioritization and not too much work has been done so far that is open for public. A lot of work were devoted to the femtocells usage [2, 3, 4] for dead zones coverage, mobile monitoring [5], some network delays caused by different factors [6], network optimization near wireless dead zones [7], architecture and its management in mobile networks [8, 9, 10].

One of the main challenges of operating a cellular network (CN) is associated with the fluctuations of user demand and the signal strength from the base station (BS), which may significantly depend on the environmental conditions. These issues make the management of the CN in terms of the operation and planning difficult. With this regard, a management strategy for CN operation and development that can mitigate these fluctuations and guarantee the CN operation in the most cost-effective manner is a necessary. In this research, a novel simulation-based data-driven smart management (DDSM) approach is proposed for determining problem zones in a CN, their prioritization for further use in planning new BS construction and deployment.

Our proposed approach is equipped with a database, a simulation model with visual simulation module, and an optimization module. In the database, data related to the CN's topology (i.e., subscriber demand and complaint, mobile traffic parameters, etc.), as well as those obtained from the measurements (i.e., signal strength from BS, environmental sensors, etc.) are stored. In the simulation, the operation of the considered CN is simulated under various system uncertainties and problem zones are detected and visualized. Meanwhile, the complexity of the optimization problem is determined and the optimization module is executed to obtain the optimal operation plan for the CN and plan for new BS construction and deployment in terms of the minimum operating/construction cost. In order to solve the optimization problem, a problem zone prioritization-based multi-objective optimization algorithm for operation planning of the CN is proposed. The results of the optimization module are finally sent back to the CN system for execution. The performance of the proposed approach is demonstrated via a synthetic CN. The approach has been constructed in a generic manner so that it can be employed by any CN system that has similar types of base stations by importing the necessary data capturing its characteristics (i.e. number of BSs installed), topology, and environmental conditions into the database.

The remainder of this paper is organized as follows. First we define the problem zones within the CN domain as those areas where quality of service parameters in mobile networks can be compromised due to bad subscriber's connection to base stations. We then discuss a related and important area – the protection of CNs from losing subscribers, followed by an idea of data-driven simulation and data visualization integration in a management process of the CN operation and planning to develop and model a reliable CN. Then we present our visual simulation-based approach to the problem zone detection and prioritization in the CN using examples of real-world geographical areas. Finally, we discuss methodological issues of the DDSM approach, especially in the context of economic impact of smart management of the CN. We will talk about the CN, its features and how problems in future can be overcome.

The architecture of proposed approach can be described as follows (Figure 1):

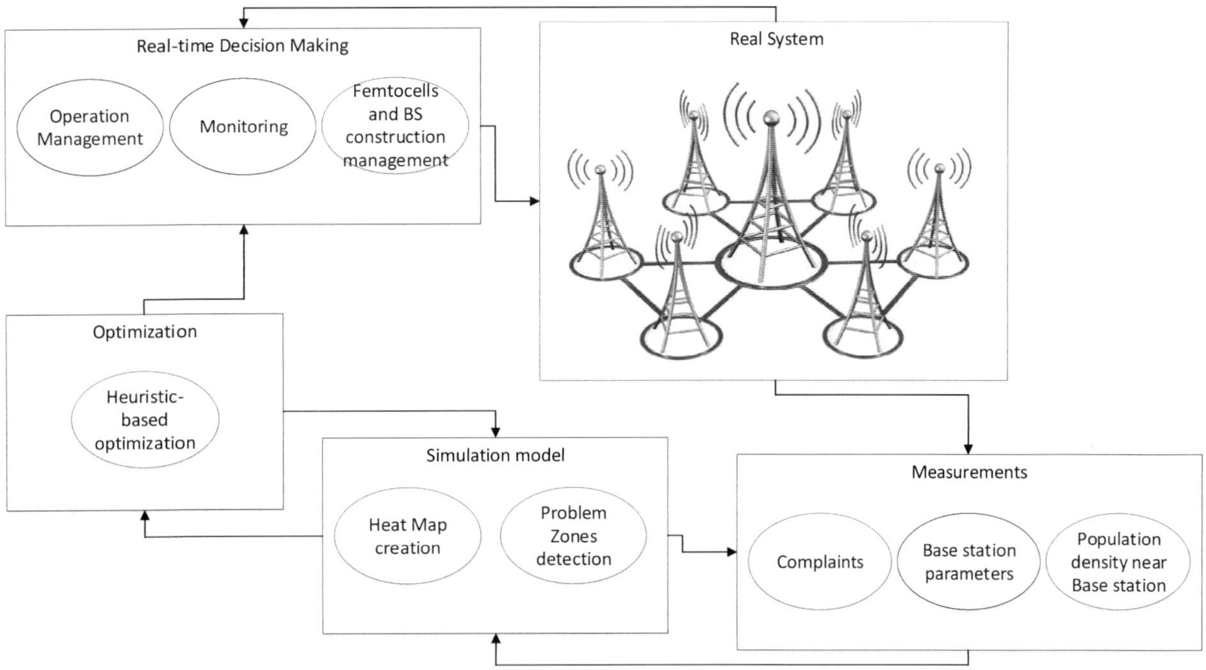

Fig. 1. Overview of DDSM approach for cellular network operation and planning

2. The domain peculiarities and problem description

2.1 Dead Zones

Basically, locations where mobile phones or User Equipment cannot transfer signal to the near base station or repeater are named *dead zones*. In these particular locations, the mobile connection is severely restricted or not available and there is no possibility for user to transmit data or have a voice call. Mostly it is caused by the fact that the signal in these zones is blocked by some buildings, hilly terrains or other obstacles that interfere with the correct signal transmission from User Equipment to base station.

There are many factors that can influence dead zones. They can be detected even in the locations, where a wireless carrier offers coverage with existing network technology and topology. Thus, it might be caused by the interference of other mobile sites, limited density of base stations in mobile network, or even the locations and capacity of antennas hanging on base station with particular height.

As a matter of fact, the transmission of data in cellular network is based on radio waves transmission, which can be dramatically reduced with distance and it makes mobile network transmission unreliable at times. Hilly terrains, high trees and buildings, even wind and seasonal changes can have an influence on data transmission and interfere, especially in urban areas, with its high density, clearly disturbing the unhindered and smooth radio waves distribution.

All these factors force mobile providers to continually work on the improvement of their network, upgrading existing ones and constructing new mobile sites. The purpose of this work is to reduce the amount of dropped calls, dead zones (or problem zones or coverage holes) and prevent from failures on the cellular network.

2.2 Dropped calls

Dropped calls are the calls that were terminated and caused by network issues, usually associated with "Dead Zones". It is highly important issue for mobile operators nowadays, even though the amount of such drops has been significantly reduced for last 20 years. This termination is considered as unexpected and technically calls "Abnormal release".

There are many reasons for dropped calls including the following:

- Mobile subscriber is located outside the coverage area after setting up a connection – this is simply the area, where radio towers can not receive the signal because of the distance. The standard base station in ideal conditions can provide a coverage radius of up to 35 km, and with additional software options - up to 120 km, which is in real cases rarely achievable because of interferences. The coverage radius of the micro-cell does not exceed 1 kilometer with good antenna.
- Drops during handovers between cells within one technology (2G/3G/4G) or drops during handovers from one technology to another.

- The lack of capacity of the network in some particular areas. Base stations and cells might be overloaded with traffic and amount of subscribers they serve, and in this cases become unavailable for other subscribers in this zone. Incorrect configuration or huge number of users being served by this base station might cause this issue.
- Transmission problems that might be caused by weather conditions or terrain structure. This issue is related to the radio waves interferences between User Equipment and base station.
- Loss of battery power at one of the ends of the call.

Dropped calls are considered by mobile providers as one of the most critical complaints from the user point of view. Many subscribers, which are experiencing dead zones or dropped calls very often tend to leave this particular mobile operator. Therefore, such companies are exceedingly interested in solving issues with dropped calls and dead zones. However, it is not possible to entirely solve all issues but the aim is to reduce amount of dropped calls and dead zones.

3. Proposed Approach

To address issues of problem zone detection and fixation in cellular network we propose a data-driven smart management (DDSM) approach which is based on modeling and analyzing particular business process of telecom company with a specific focus on the complaints management and planning new base station constructions and deployments. This business process is more or less similar in most of telecom companies (with regard to the user complaint processing and some network Speedtests), so the following workflow model of business process is applicable to any telecom company that acts as telecom operator and analyzes user's complaints and network performance:

1. At the end of each quarter the Department of Quality and Clients Experience (DOQCE) uploads the necessary complaints / measurements for the required period, collects them in one file in a certain format (CSV-format). The DOQCE downloads file with complaints from the Relational database, where all data from Speedtests and different complaints collected from the Contact Center and web-site are aggregated.
2. The DOQCE loads file for data processing, which is executed using simulation-based data analytics system.
3. The simulation based data analytics system performs the problem zone detection and analysis using data from existing and projected BSs. The reports on simulation results are generated for decisions on new BS deployment.
4. Based on reports and data visualizations received as an output of the simulation system, the Technical department at telecom company decides about where exactly it is critically to construct new base station or microcell so to increase the quality of service in the network.
5. After the construction of new base station or microcell, the effect of this station installation is evaluated.

In particular, the proposed DDSM approach is equipped with three modules: 1) a database including the CN's topology data (i.e., subscriber demand and complaints, mobile traffic parameters in the area, etc.), network measurements (i.e., Speedtest data), and (optionally) the data for environmental factors (i.e., solar radiation, wind speed, temperature, etc.); 2) a simulation system with data visualization modul, in which operation of the CN is simulated and problem zones are detected and visualized; and 3) a multi-objective optimization module which finds the near-optimal operation plan and new BS deployment plan in terms of minimum cost using a problem zone prioritization-based algorithm. The results obtained from the optimization module are sent back to the CN system to enhance its operation. The experiments conducted in this study demonstrate the power of the proposed approach in real-time assessment and control of operation and development management in CNs. The components of our proposed approach are presented in detail in this section.

3.1. Data Base

For easy adaptation of the proposed approach to varying types of CNs with different characteristics, the heterogeneous data associated with the considered CN's topology, operations, and environment are collected and stored in a database. The database contains the CN topology data (i.e., subscriber demand and complaint, mobile traffic parameters, BS locations and power, Speetest data, etc.) and the operational data for the BSs that characterize the load demand and operating cost of various stations. The environmental data (i.e., wind speed, solar irradiance, temperature, etc.) can also be collected and stored to estimate of the strength of signal that can be received from the base stations. For the data, both data from synthetic functions (i.e., probability distribution functions, pdf) and a real-world measurement data recorded from the sensors in specific geo-locations or obtained from special apps (e.g., Speedtest) can be used.

It is assumed that all collected data is stored in a single database in unified format in order to prevent any data manipulations before uploading it in a simulation.

There are three major sources of data in a database: subscribers (demand and complaints), base stations and mobile traffic in the area, and (optionally) environment.

The regarded primary sources for user complaints as have been listed above were data from company Contact Centers, web-site of the company and Speedtests data. Subscribers' data sectioned by stationary and stirring as well as data on the base stations for analyzed region is being received from relational database of a telecom company. The following base stations attributes are needed for a model: cell id, cell lac, standard network, base station name, launch date, latitude, longitude, area. The number of subscribers and the density of complaints are sectioned by squares of 500 x 500 meters (polygons). The proposed formula is as following:

$$Weight = (1 - (X * Subs\_Stay + Y * Subs\_Move) / 2 * Zone\_count\_claims), \quad (1)$$

Where
Weight – the average proportion of complaints per subscriber,
X – the proportion of stationary subscribers,
Subs_Stay – the amount of stationary subscribers,
Y – the proportion of stirred subscribers,
Subs_Move – the amount of stirred subscribers,
Zone_count_claims – the amount of complaints in this area.

The *Weight* metric shows the density of complaints per squares (tiles) of 500 x 500 meters, which can be easily visualized using Python. For the square of 500 x 500 meters, there is a distribution of complaints over time. The number of complaints with a stable problem is distributed according to Poisson's law and about 5% of complaints come out. In days, when the level of complaints exceeds 99% of the confidence interval, the number of complaints is reduced to an average of one quarter.

Radius to the nearest BS is a minimal distance to the nearest base station. The sense in determining the size of the problem zone is the following: the father is a zone from the base station - the wider and broader the problem area.

3.2. Simulation model for the problem zone detection in the CN

A simulation model provides an accurate representation of the considered CN (as a whole or a specific part of it) while simultaneously capturing the behavior of operating components as well as their interactions with each other. In this study, the operating components including demand (subscribers) and service providers (base stations) are designed as simulation models (e.g., the software agents in case of agent based modeling, ABM).

In the core of the simulation system, there is a data visualization system that provides interactive geo-spatial display of so called problem zones detected and identified as a result of real-time measurements and simulation.

Heat Map visualization is a Geo-spatial visualization of a map with imposing weight values as different colors per every square on this map.

The selected data visualization and analysis tool is Python with its powerful libraries for data processing and handling. The advantage of generating Heat Maps in Python is that there is the ability to fine-tune parameters (any nucleus, and not only monotonous), generating squares and polygons of any shape.

Hive is used as SQL-like database on top of Hadoop. But it is also possible to store polygons in any relational database.

GeoSpark is an open source in-memory cluster computing system for processing very large-scale spatial data. A fundamental data structure of Apache Spark is the Resilient Distributed Datasets (RDD) an immutable distributed collection of objects. Each dataset in RDD is divided into logical partitions, which may be computed on different nodes of the cluster. GeoSpark extends RDDs to form Spatial RDDs (SRDDs) and efficiently partitions SRDD data elements across machines and introduces novel parallelized spatial transformations and actions (for SRDD) that provide a more intuitive interface for users to write spatial data analytics programs [11].

The final visualization is performed in QGIS. QGIS is a professional GIS application that is built on top of and proud to be itself Free and Open Source Software (FOSS). QGIS is a user friendly Open Source Geographic Information System (GIS) licensed under the GNU General Public License. QGIS is an official project of the Open Source Geospatial Foundation (OSGeo). It runs on Linux, Unix, Mac OSX, Windows and Android and supports numerous vector, raster, and database formats and functionalities [12].

Table 1 shows the steps of the generalized procedure of visual simulation modeling of cellular network performance using network data collected and prepared as it was explained earlier.

The first step in visual simulation of a problem zones in cellular network is a Heat Map visual model generation in python with a bi-weight core with a radius and weight values taken from the attributes of the point of complaints.

Then separate tiles of a Heat Map are combined into zones and certain metrics are counted. According to the distance between the centers of tiles, tiles are united when having one common side. Upon completion of the union operation on tiles, a new zones are formed, the maximum density of the Heat Map is calculated which is exceeding the average level of complaints.

Next step involves splitting large zones into smaller ones and counting convex shapes, which is related to the model optimization strategy.

After model is optimized, special shape files are generated as an output from python and GeoSpark tools. These shapes files presents the layers in QGIS – BSs, complaints and Heat Maps. In QGIS this layers are imposed on the geographical map (physical, street, hybrid or satellite) and problem zones might be easily analyzed.

Step	Technology	Parameters
Convex shape of segments	GeoSpark	
Dynamics of complaints in zone	GeoSpark	
Combining the pieces (tiles) of the heat map into zones	GeoSpark	
Cutting the top of the heat map (filtering the tiles)	GeoSpark	cut-off percentage
Counting the final metric	GeoSpark	cell size of heat map kernel function for blurring a complaint blur radius
Building heat maps (generation of tiles)	GeoSpark	weight-of-complaint function method to find the nearest base station
Complaint attributes calculation	GeoSpark	upper area boundary
Segmentation of large zones into chunks	Python	
Removing intersections of convex hulls	GeoSpark	
Filtration of complaints on short-term accidents	Python	p-value - affects the proportion of cut-off
Storing polygons	Hive tables	number of last calculations for storage
Final visualizations	QGIS	

Table 1. Visual simulation workflow, used technology and parameters

3.3. Optimization strategy of the CN upgrade

In this study, the optimization of the CN operation and upgrade is executed using data from the identified problem zones in the CN to determine the new BS's construction plan in terms of the minimum cost of construction and maximum traffic increase in a particular area. These multi-scale problems are solved via heuristic-based optimization algorithm that uses problem zone prioritization technique.

The optimization is basically based on the level of the convex of map shapes. If the convex is more than the threshold, than we consider this zone a problem with particular weight. Using k-averaging technique tiles are divided into groups and convex hulls are counted. If the area of the convex hull is more than 4 square kilometers, the number of clusters in k-means increases by 1.

In QGIS, the zones obtained from the same broken large zone are checked for intersections. A random one of a pair of zones is replaced by its difference with the other. Thus, the model after optimization shows only convex geo-spatial zones with exceeds of threshold, which are the objects of this research – prioritized problem zones itself.

In future, we plan to explore the possibility of using both traditional and new "big data" approaches (for example, based on machine learning) to optimization problems in DDSM system for managing cellular networks.

4. Experiments and Results

4.1 Experimental setup

In order to demonstrate the capabilities of the proposed approach, experiments are designed and carried out over a synthetic CN. Each model execution is performed on particular region of a country or city, so the output of the model will show the prioritized problem zones distribution in a selected region.

The selected hardware is GeoSpark an open source in-memory cluster computing system with 30 cores and more than 250 GB RAM. All computations for processing very large-scale spatial data require a lot of RAM, processing cores and storage memory. The selected database is Hadoop with Hive for SQL-data structure as part of pre-installed Hortonworks Big Data application tools. Hortonworks is a leading innovator in the industry, creating, distributing and supporting enterprise-ready open data platforms and modern data applications [13]. So all chosen data for experiments are stored in Hadoop, and all data processing and algorithmic computations are performed on powerful cluster (edge nodes) using Hortonworks applications – Spark, Python.

For the experiments we have selected two geographical regions in the European part of Russian Federation, namely: the southwestern area of Moscow city as an example of huge (over 12 million inhabitants) city with a dense population (over 44 million mobile subscribers/SIM cards), and Stavropol – the medium-sized city (under 500,000 inhabitants) in the southern European region of Russian Federation with relatively low population density (2,7 million inhabitants in the territory of 66,500 sq. kilometers with some 4 million SIM cards).

4.2 Modeling results

The first selected region Stavropol has the following model input parameters (randomly generated numbers):

- Amount of complaints is 19668 (demand agents);
- Amount of cells is 9368 that is related to 1000 base stations (service agents).

The model output is QGIS geo-spatial visualization of problem zones. In our experiment we have received 339 Problem Zones. The output visualizations are the following: the Figure 2 shows the QGIS base stations and complaints distribution on geo-map for Stavropol city:

Fig. 2. The QGIS base stations and complaints distribution on geo-map for Stavropol city

The Heat Map visualization itself shows the density distribution of complaints per polygon (500x500 sq. meters):

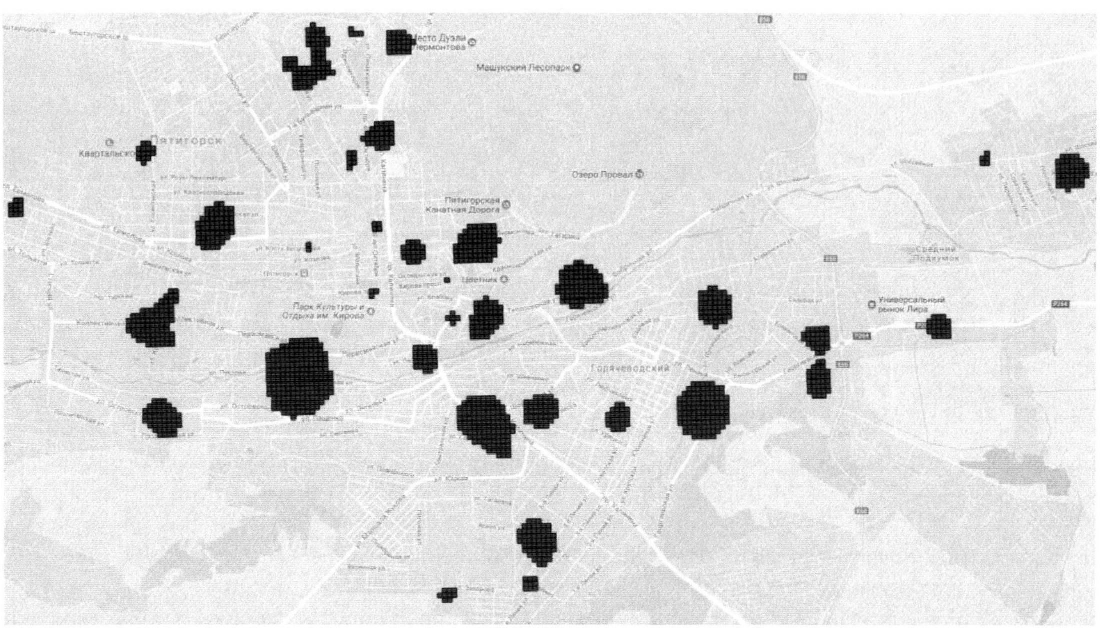

Fig. 3. The QGIS Heat Map for Stavropol city

And finally, filtering Heat Map by maximum complaints concentration gives the final Problem zones map (fig. 4):

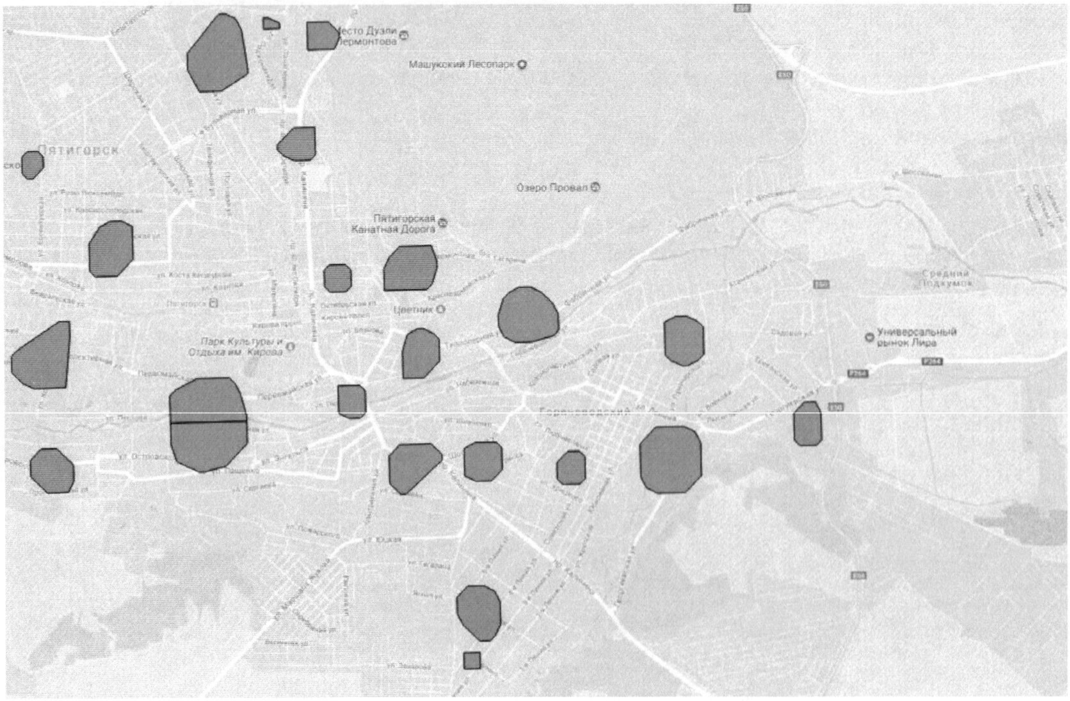

Fig. 4. The QGIS Problem Zones Map for Stavropol city

Additionally, the second experiment was performed for the geographic area in southwestern part of Moscow city. The Fig. 5(a) illustrates the basic final Heat Map visualization of network simulation data, Fig. 5(b) shows visualization of Problem Zones distribution in the area, and Fig. 5(c) depicts the potential effect from problem zone solution.

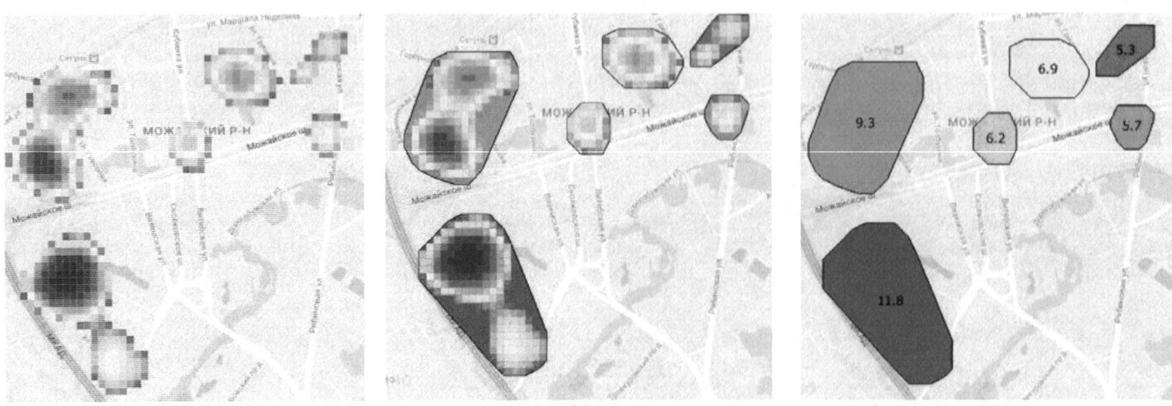

(a) (b) (c)

Fig. 5. The Heat Map visualization of problem zones: (a) the basic Heat Map; (b) Heat Map with problem zone borders; (c) Heat Map with projected complaint decrease numbers.

5. The approach to the economic effect calculation

Here we use an "economic effect" parameter as a major criteria for optimization strategy efficiency which is based on the following considerations:

- The problem zones are formed based on geo-analysis data of the feedback received from customers;
- In these zones there is an active unsatisfied demand for communication services;
- If we solve the problem in this zone, we satisfy this demand and the traffic growth in this zone is supposed to be higher than average traffic increase.

This kind of heuristics brings us to the following common sense approach to evaluating and managing economic efficiency of new base station installations in problem zones:

a. For each solved problem zone issue the base station(s) that serve(s) this zone are determined;
b. For this base station(s), the total data traffic for 6 months is calculated - preceding the issue solving and after 6 months;
c. The growth of traffic "before - after" in the problem zone is compared with the growth of traffic in the region as a whole over the same period;
d. The difference in the rate of growth is counted as an incremental effect from the elimination of the problem zone.

The following picture depicts data traffic dynamics in both problem zone and the average in region:

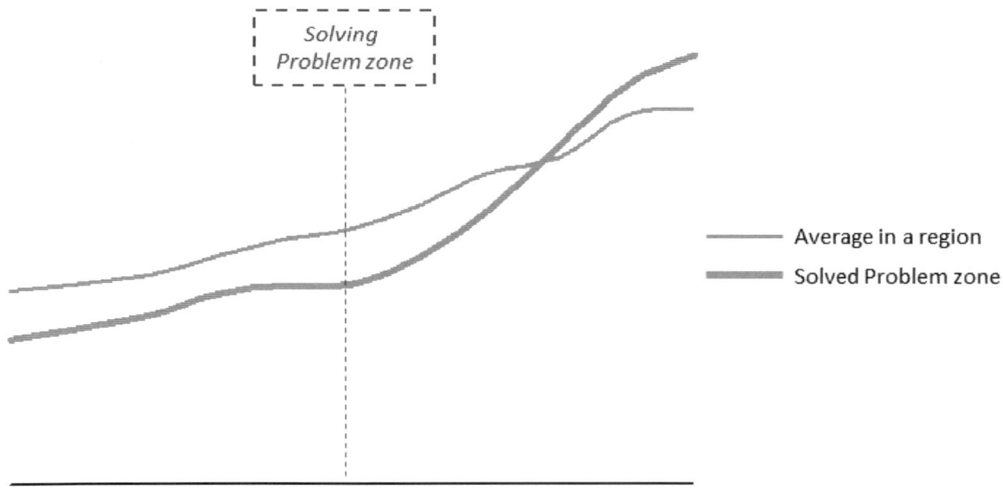

Fig. 6. Data traffic dynamics in both problem zone and the average in a region

6. Conclusion

In this work, a DDMS approach is proposed for the problem zone detection and solution within a cellular network. The proposed approach involves a database that stores the system-related operational, topological, and environmental data; a simulation system modeling the operation of the CN under uncertainties of the signal strength from the BSs and fluctuation of the subscriber demand; and an optimization module that solves the problem of allocation of new BSs in specific area according to the problem zones in the CN identified by the simulation. The performance of the proposed approach is demonstrated via a synthetic CN with simulated base stations and user demands. Results have shown that the proposed approach has the capability to provide optimal plan for the CN upgrade without compromising the quality of service requirements of the network.

The DDSM approach is developed in a very generic manner so that it could be adopted and implemented for any CN with similar types of BSs. The presence of one universal model prioritizes problem zones using Heat Maps taking into account the amount of complaints and the population density in particular zone. This model can be applied in any telecom company using proposed models, algorithms, GeoSpark and Python tools. Prioritized problem zones can be used for better distribution of new base stations construction or femtocells, which lead to increase of quality of service, mobile network and customer satisfaction.

In this study, problem zones in the CN are detected and visualized only on the basis of customer complaints data, but it is also important to estimate the quality of network using more automated solutions, taking into account only network quality parameters related to some particular geolocation. Thus, the automation of complaints collecting, architecture and altering the model accordingly are the further research topics. Supposedly, this will lead to more proper distribution of problem zones on the map and their weights. In the future, the use of new "big data" approaches such as machine learning and deep learning to the optimization problems in DDSM system will be explored. Also the environmental parameters in the CN will be taken into consideration in simulation when determining problem zones in the CN and the new BS construction and deployment plan.

7. References

[1] https://www.statista.com/statistics/274774/forecast-of-mobile-phone-users-worldwide (2017). Number of mobile phone users worldwide from 2013 to 2019 (in billions), Accessed on: 2017-10-11.
[2] M. Nagieb, M. Shokair (2014). Improvement of Coverage and Mobility in LTE-A Femto-Cell based on Cognitive Radio Network. International Journal of Computer Applications Vol. 86 No. 11 pp. 37-40 Jan 2014.

[3] J.M. Benita, V.M.A. Rajam (2017). Road Gateways for femtocell clusters to enable communication and handover in dead zones, Cluster Computing, 2017, ISSN 1386-7857.

[4] J. Benita, V. Rajam (2014). A novel architecture for enhancing communication in dead zones using femtocells. In: 2014 International Conference on Advances in Computing, Communications and Informatics (ICACCI), pp 2793–2796 (2014).

[5] M. Matysek, M. Adamek, T. Karafiat (2016). The Mobile Monitoring and Control of Real Systems, in: Annals of DAAAM for 2016 & Proceedings, DAAAM International, Viena, 2016, pp. 0104-0108.

[6] E.J. Hernandez-Valencia, M.C. Chuah (2000). Transport delays for umts voip. In: 2000 IEEE Wireless Communications and Networking Conference, vol 3, pp 1552–1556. WCNC (2000).

[7] G.J. Smith, G.W. Van Leeuwen (2004). Mobile communication optimization near wireless dead zone regions. US Patent 6,721,572 (2004).

[8] H. Kaaranen, et al. (2005). UMTS Networks: Architecture, Mobility and Services, John Wiley&Sons, Oxford, 2005.

[9] M. Rumney (2009). LTE and the Evolution to 4G Wireless, John Wiley & Sons, 2009.

[10] S. Fortes, A. Aguilar-García, R. Barco, F. Barba, J. Fernandez-luque, A. Fernández-Durán (2015). Management architecture for location-aware self-organizing lte/lte-a small cell networks. IEEE Commun. Mag. 53(1), 294–302, 2015.

[11] http://www.public.asu.edu/~jiayu2/geospark/ (2017). GeoSpark. A cluster computer framework for processing large-scale spatial data, Accessed on: 2017-10-21.

[12] http://www.qgis.org/en/site/about/index.html (2017). QGIS - The Leading Open Source Desktop GIS, Accessed on: 2017-10-21.

[13] https://hortonworks.com/about-us/ (2017). About Hortonworks, Accessed on: 2017-10-21.

DOI: 10.2507/28th.daaam.proceedings.060

EXTRACTING PATENT-RELATED INFORMATION FROM ONLINE SOCIAL NETWORKS: CASE OF FACEBOOK

Alexander Ivanov[1,2] & Zeljko Tekic[1]

[1]Skolkovo Institute of Science and Technology, Moscow, Russia [2]National Research University Higher School of Economics, Moscow, Russia

This Publication has to be referred as: Ivanov, A[lexander] & Tekic, Z[eljko] (2017). Extracting Patent-Related Information from Online Social Networks: Case of Facebook, Proceedings of the 28th DAAAM International Symposium, pp.0433-0438, B. Katalinic (Ed.), Published by DAAAM International, ISBN 978-3-902734-11-2, ISSN 1726-9679, Vienna, Austria
DOI: 10.2507/28th.daaam.proceedings.060

Abstract

In this study, we investigate how to identify and extract patent-related information from Facebook, the largest online social network. In the first step, we identified a list of trustworthy sources we started search from. Then, we developed algorithms for extracting and filtering information, and based on them software tool that is able to identify and deliver a patent, the Facebook post where it is mentioned and news / blog article which discuss it. We did a pilot test and collected more than 50 examples of articles (and Facebook posts) that add value to the patent they refer to. We classified collected articles and discussed how they can be used. Finally, we outlined where developed tool can be applied.

Keywords: Facebook; social networks; patent analysis; patents

1. Introduction

Since the beginning of this century social media and, especially online social networks (OSN) have exploded as platforms where users generate and share content and intensively engage with it through different actions. Online social networking sites like Facebook, Instagram, Twitter and LinkedIn attract more and more users every day. Being widespread, easy to use and available everywhere online social networks provide fast and powerful communication platform which sets trends and shapes public opinions in topics that range from politics and economy to technology and entertainment across societies.

OSNs have millions of members / users and these people possess different knowledge, expertise and skills sets, thus OSNs can be seen as a form of collective wisdom platform [1]. Bearing this in mind, we decided to investigate patents as a topic in OSNs. More specifically, we are interested in extracting patent-related information we are able to identify at Facebook, the largest OSN.

Why are we interested in patent-related information from Facebook? Patents are a powerful and unique source of data for innovation and technology analyses. However, extracting useful information from patents and freely available patent databases is not easy. Identification of relevant and valuable patents is still difficult, time- and manpower-consuming work, which requires special expertise [2]. Current research and developed software tools in the field of patent analytics [3, 4] try to respond to this challenge using exclusively patent data (i.e., bibliographic data, patent descriptions, abstracts, claims, etc.). This approach is limited in many ways – it does not solve the problem of decoding and interpreting patent language, it rarely allows the matching of patents with specific product features and/or specific products, and it requires strong expertise in the field of intellectual property law. To improve practical value of available patent information and democratize its usage ("patents for people, not only for experts"), we propose to use contextual information, information that is related to specific patent but is not part of patent (document) itself [5]. Instead of limiting ourselves to using only patent data, we will combine patent data with relevant information from context in which specific patent is mentioned by people intrinsically or professionally interested in the topic. Our idea is to identify and extract value from posts in which people with interest in patents and patented technology write about a certain patents, their features, potential application areas and importance. This information provide context needed to easier understand patent, its language and, possibly, its value. This source of information is unexplored so far in patent analytics, and we believe that it has potential to bring significant value to the field. However, before we start analysing contextual information as an input in patent analytics, we need to collect it. The objective of the paper is to describe the structure and functions of recently developed software tool that is able to identify and extract patent-related information from Facebook.

The remainder of the paper is organized as follows. In Section 2 we review relevant literature, while in Section 3 we describe our approach and algorithms used. Section 4 presents results and discusses implications for practice and future applications. Finally, in Section 5 we conclude with a summary of results, limitations and the future research directions.

2. Literature Review

OSNs, with billions of users, represent a very interesting source of data. It is recognised by many researchers who used them in different ways. We can differ them by the sources of data which is used and the application.

The first direction is the information distribution, where the speed of analysis matters, for example, in stock analysis. Such a direction includes Twitter analysis [6] and Facebook analysis [7]. The second one analysed millions of Facebook pages in order to predict stock behaviour. Correlations between the sentiment and volatility were found.

The next direction is sentiment analysis, when one would like to figure out people's attitude to a certain subject, for example, regarding mobile phone providers [8]. Another paper [9] studied the sentiment analysis in e-learning, and in the sentiment detection part they combined two approaches: machine learning approach and lexicon-based approach. The lexicon-based approach consisted of using a dictionary of keywords marked with the sentiment they represent and finding using the keyword search the sentiment of each phrase splitting it into tokens.

Social networks allow researchers to analyse text in order to find peculiarities. Such research may be restricted to a certain domain. For example, [10] studied the way how to alleviate the depressive symptoms. Instead of taking millions of pages and scanning large sets of data, they took 68 participants of the experiment and monitored only their texts. Apart from sentiment analysis, they also took into account additional features, e.g. the number of friends, number of comments, etc.

Finally, it is possible to analyse influencers [11]. Such methods represent social network as a graph with nodes and edges, and the task is to find the most influential nodes of the graph in terms of information distribution.

Next, we would like to discuss the methods and algorithms which are used in analysis. Some papers study not the specific applications of the Facebook data, but the algorithms and software for crawling data from Facebook. Rieder [12] introduces a Netvizz app which allows to grab data from specific Facebook groups and build friendship graphs. The software uses Facebook API and automatically downloads the data specified by the user. Several examples of applications are provided.

Additional line of research is studying the efficiency of different crawling algorithms. For example, Ye et al. [13] studied the efficiency of four search techniques (BFS, greedy, lottery and hypothetical greedy) applied to four different sources of data (Flickr, LiveJournal, Orkut and Youtube).

The next part of the literature review is the patent analysis. Patents are helpful for strategic planning purposes [14]. The main problem is that hundreds of thousand patents are published every year, and the task is to select the most valuable ones is hard and time-consuming. The most popular approach is to analyse patent parameters and the ways they indicate the patent value. The most influencing parameter is the number of forward patent citations [15] showing the industrial importance of the patent. Lee and Sohn [16] study the first patent citations as a faster indicator than the number of all citations, because some citations are received 5-10 years after the patent publication.

In [17], Ivanov and Tekic, instead of using patent parameters, used data from blogs. Websites are searched for the articles about specific patents in order to obtain expert insights about patents. As a result, it is said that such articles might give a clearer understanding of a patent.

In research presented here, we study Facebook as a primary social network and as a source of data, because it has more text content than Twitter, and it is the largest social network in the world with more than two billion users. We use Facebook application programming interface (API) for downloading and searching for patent-related data, the process will be described in details in next sections.

3. Approach and Implementation

The workflow of our approach is represented at the scheme at Figure 1.

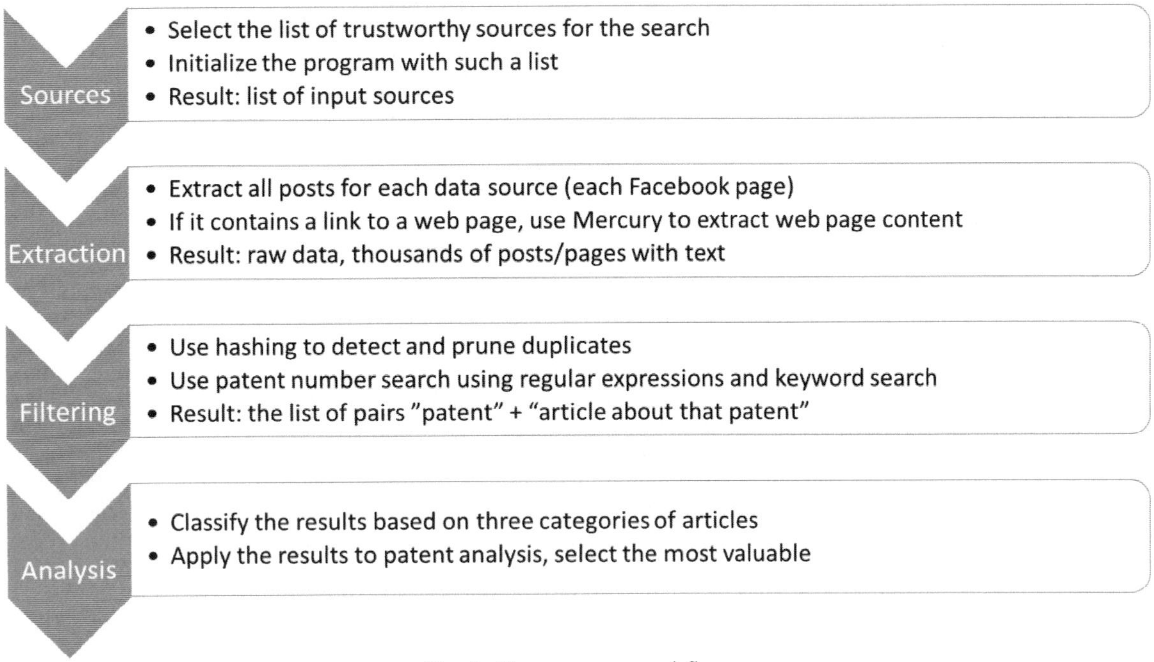

Fig. 1. The program workflow

Now we will discuss each of the steps from the workflow, one by one. Each data search must start from a list of entry points. It can be either the whole social network or a list of pages from a given social network. In the research connected to a specific topic, it might be a good idea to restrict the search domain. For example, in the patent research taking all Facebook pages may give a situation when not only expert opinions are taken into account, and the resulting dataset might include some irrelevant comments or even spam about patents. That is why it is better to take a list of trustworthy sources.

In the patent research we took only trustworthy sources by the following algorithms. First, we found several website catalogues about best websites devoted to intellectual property or technology. Second, for each website from the catalogue we looked for a Facebook page of this media. If it existed, we included that Facebook page into the initial list of search sources.

The next step is downloading data from social networks. Facebook provides a useful API [18] in order to access its data. Data which is available via API does not differ from the data one can see online on a certain page, but is represented in a structured way. The list of data which we extract includes the posts from a certain Facebook page, the number of likes, the number of shares and the number of comments. Facebook API might not be enough for the search in social networks. It turns out that many media projects do not include much information into their Facebook posts. In many cases a typical Facebook post of large and small media contains information about the title of the news and an external link to their website.

The next idea is to follow the link and to get information from the media website, but it is a complicated issue, because one will get an HTML web page which consists of not only text, but also the markup and a lot of sidebars and menus. Such information would be redundant for our research, we need only the content of the article which is mentioned in the Facebook post. For this reason we use additional tool for extracting content. Is it a library available for free use called Mercury [19]. For a given web page it returns the content of the webpage without sidebars, menus and most of the markup.

Next comes the text processing phase. It should select only relevant results from the whole set of found posts and pages. The phase consists of two steps: the keyword search step and removing duplicates step.

The keyword search step is aimed at finding the required keywords or word combinations in the text. In the case of patent search there are three situations when a certain piece of text is considered as a good result (by "good result" we mean an article about specific patent where it is possible to get the exact patent number the articles refers):

1. It contains a certain keyword (e.g. "patent") and the patent number (e.g. "US 1,234,567"). This search is performed by a combination of keyword search and regular expressions search. We selected a list of possible representation of patent numbers in the text including small and capital letters, spaces or commas as separators etc. and developed a list of regular expressions. If a given article matches at least one regular expression from our library, the article is considered as a good one.

2. It contains a link to the patent database to a specific patent. In this case we need to find all external hyperlinks in the text and compare them to the list of patent databases. If at least one hyperlink in the text matches with the patent databases by domain name, we consider the article as a good result.

The keyword search step takes every found article as input and checks whether it contains a patent number or a link to the patent database. The result of the keyword search step is the set of potential good results. We would like to underline the point that the keyword search step is the step which defines the application of the search algorithm. The program is universal and may be used in searching not only patent related data, but also data about brands or political preferences. In this case, the keyword search step should be fine-tuned if we would like to change the course of the reseach, while all other steps may remain intact.

The removing duplicates step is aimed at pruning pages which copy the content of previously found pages. It includes the following possible situations:

1. The same page with the same text is stored by different URLs
2. A certain media decided to republish the same text on their website by simple copying the content without any new comments or text

The program should not include duplicate results, for this reason it should contain a special module which deals with such situations.

The brute-force solution of such problem is to compare every new found piece of text with all previously found useful articles. The problem of such approach is that symbol-per-symbol comparison for each found article will give a computational complexity of O(length(text)*number_of_found_good_articles) for each article which might contain useful information. If we assume that our program should find thousands of good articles where each article consists of thousands symbols, the resulting computational complexity may significantly hurt the speed of the algorithm.

In order to solve this problem we use hashing to increase the speed of search. After removing all punctuation marks and other symbols except for the letters we calculate the hash value of a certain article using the formula:

$$\sum_{i=1..n} s_i * c^{i-1} \ modulo \ p$$

where s_i is the i-th symbol of the text, c and p are two hashing constants. In order to decrease the number of possible collisions, we use double-hashing with $c = 31, p = 2^{32}$ and $c = 53, p = 2^{32}$.

How the two strings are compared using hashes? If both hashes are different (first hash for string1 is not equal to first hash of string2 and second hash of string1 is not equal to second hash of string2), we assume that the two strings are also different. It means that we store the hashes for all found good articles and for each new candidate we simply calculate the two hash functions and look for those values in the current results, determining whether we already found such an article or it is a new result.

4. Results Overview

The program for finding contextual data about patent has been running for two months during summer 2017. As a result, more than 300,000 Facebook posts from 108 pages were analysed and 53 posts about specific patents were found. The average speed of the program is one parsed Facebook page with all its posts per one day if it is launched on one thread on one computer. This includes running all algorithms for finding the patent numbers, removing duplicates, hashing, etc. The run-time can be decreased if the program is launched concurrently on two or more computers using two or more threads on each computer, so the overall processing time can easily be reduced to several days.

In this Section we will classify the results. The 53 found pairs "article" + "patent" can divided into three groups:

1. Descriptive articles about patents. This category stands for articles which describe a recently issued patent (issued after 2004) and its possible applications.
2. Articles about historical / old patents. This category stands for patents which were published more than 13 years ago (before 2004). Why this threshold was chosen? The main reason is the scope of the research. We use Facebook as a source of acute data, Facebook was started in 2004, it means that before 2004 no media could publish a news article on Facebook about recently issued patent.
3. Articles about litigation. If Apple sues Samsung, the case is usually covered by a lot of websites and blogs and has many posts about it in social networks. If there is an article about a patent issued before 2004 and related to litigation, it will go to this category.

The most common case is the article about litigation. Approximately half (51%) of the results which we found are the articles about a litigation case with a patent involved. 21% of the results are articles about "old patents", i.e. patents issued before 2004. In this case, we cannot track the speed of media reaction to the publication and the value of the patent or we will get the biased results, because our source of data simply did not exist when the patent was published. Finally, 28% of found articles are the descriptive articles about specific patents. This is the most valuable part of the results. Why? Because instead of reading a patent for two hours a person without specific knowledge in the field may get the understanding of the substance of the patent.

The classification made based on the 53 found articles is presented at Figure 2.

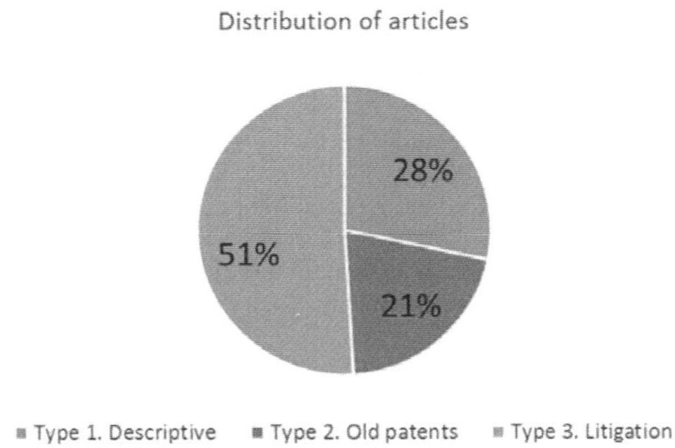

Fig. 2. Distribution of the found articles

Let us provide an example:
Patent number: US 8,253,639
Patent title: Wideband electromagnetic cloaking systems
Link to the patent: www.freepatentsonline.com/8253639.html
Link to the article: www.ipwatchdog.com/2012/09/07/uspto-issues-worlds-first-invisibility-cloak-patent/id=27841/

The article contains a comment of the inventor, a descriptive video and a prediction about its further development. Everything can be read within two minutes. Each out of 15 found articles of the category "descriptive articles about a specific patent" shows a more concise and easily readable description of a certain patent.

The use-case of such an approach may include a software which searches for experts' opinions for specific patents issued not so much time ago. Small and medium companies which do not have much budget for the intellectual property research may use it to monitor the competitive landscape. With thousands patents published every day, such a tool can decrease the amount of time needed for a person to understand each patent. With an easy understandable expert opinion it would take several minutes per patent instead of two hours.

5. Conclusion

Social networks represent a source of data which can be applied to different types of research. In this study, we investigated how to identify and extract patent-related information from Facebook, the largest online social network. We developed algorithms for extracting and filtering information, and based on them software tool that is able to identify and deliver a patent, the Facebook post where it is mentioned and news / blog article which discuss it. We did a pilot test and collected more than 50 examples of articles (and Facebook posts) that add value to the patent they refer to. We classified collected articles and discussed how they can be used.

The next steps of the research can be enhancing the list of data sources with more Facebook pages as well as adding websites to the list of sources. Additionally, the engagement numbers, e.g. the number of likes and comments should be studied as potential indicators of patent value.

The designed software tool is universal and may be applied not only to patent search, but also to all other types of keyword or regular expressions search. For example, the designed software can be easily switched to a brand-awareness search, when the task is to analyse how the crowd assesses the products of a certain brand.

What are the limitations of the approach? First, data from social networks represent data generated by all users of that social network. It means that if we need precision and accuracy, we need to restrict the list of sources or to check each source of information. In our case, we selected the ranking of media and took their pages on Facebook.

Second, the data on social networks related to specific topic is rather sparse. After processing 300,000 posts from the selected list of media, we found only 53 articles about specific patents. For example, if one would like to create a keyword monitor like a brand-loyalty monitor in social networks, he will need a lot of computational resources to get a representational set of opinions about his brand.

Third, most media put a link to their website and include only the title of the publication in the Facebook posts. It means that a direct web search, when not the pages on Facebook are analysed, but the media websites themselves, might give the same or even more results. In terms of applying contextual data analysis to patent intelligence, the next step might be the data search from the whole Internet or from the selected list of websites in order to find data about specific patents. Such an approach may give more than 53 received in this research results.

6. References

[1] S. Asur and B. A. Huberman, "Predicting the Future with Social Media," 2010 IEEE/WIC/ACM Int. Conf. Web Intell. Intell. Agent Technol., pp. 492–499, 2010.

[2] Z. Tekic and D. Kukolj, "Threat of Litigation and Patent Value: What Technology Managers Should Know," Res. Manag., vol. 56, no. 2, pp. 18–25, 2013.

[3] A. Abbas, L. Zhang, and S. U. Khan, "A Literature Review on the State-of-the-art in Patent Analysis." World Pat. Inf., vol 37, pp. 3–13, 2014.

[4] Z. Tekic, M. Drazic, D. Kukolj, and M. Vitas, "From Patent Data to Business Intelligence – PSALM Case Studies.". Procedia Eng. vol. 69, pp. 296 – 303, 2014

[5] A. Ivanov, and Z. Tekic, "Towards smart patent analytics - matching patent and contextual data", paper presented at R&D Management Conference 2016 "From Science to Society: Innovation and Value Creation" 3-6 July 2016, Cambridge, UK

[6] F. Corea, "Big Data Research Can Twitter Proxy the Investors ' Sentiment ? The Case for the Technology Sector," Big Data Res., vol. 1, pp. 1–5, 2016.

[7] A. Siganos, E. Vagenas-Nanos, and P. Verwijmeren, "Facebook's daily sentiment and international stock markets," J. Econ. Behav. Organ., 2014.

[8] N. A. Vidya, M. I. Fanany, and I. Budi, "Twitter Sentiment to Analyze Net Brand Reputation of Mobile Phone Providers," Procedia Comput. Sci., vol. 72, pp. 519–526, 2015.

[9] A. Ortigosa, J. M. Martín, and R. M. Carro, "Sentiment analysis in Facebook and its application to e-learning," Comput. Human Behav., vol. 31, no. 1, pp. 527–541, 2014.

[10] S. W. Lee, I. Kim, J. Yoo, S. Park, B. Jeong, and M. Cha, "Insights from an expressive writing intervention on Facebook to help alleviate depressive symptoms," Comput. Human Behav., vol. 62, pp. 613–619, 2016.

[11] E. Lahuerta-Otero and R. Cordero-Gutierrez, "Looking for the perfect tweet. The use of data mining techniques to find influencers on twitter," Comput. Human Behav., vol. 64, pp. 575–583, 2016.

[12] B. Rieder, "Studying Facebook via data extraction," Proc. 5th Annu. ACM Web Sci. Conf. - WebSci '13, pp. 346–355, 2013.

[13] S. Ye, J. Lang, and F. Wu, "Crawling online social graphs," Adv. Web Technol. Appl. - Proc. 12th Asia-Pacific Web Conf. APWeb 2010, no. February, pp. 236–242, 2010.

[14] H. Ernst, "Patent information for strategic technology management," World Pat. Inf., vol. 25, no. 3, pp. 233–242, 2003.

[15] M. B. Albert, D. Avery, F. Narin, and P. McAllister, "Direct validation of citation counts as indicators of industrially important patents," Res. Policy, vol. 20, no. 3, pp. 251–259, 1991.

[16] J. Lee and S. Y. Sohn, "What makes the first forward citation of a patent occur earlier?," Scientometrics, pp. 1–20, 2017.

[17] A. Ivanov and Z. Tekic, "Using Contextual Data for Smart Patent Analysis," 2016 IEEE Int. Conf. Cloud Comput. Technol. Sci., pp. 448–451, 2016.

[18] "Facebook API for Developers." [Online]. Available: https://developers.facebook.com/. [Accessed: 26-Sep-2017].

[19] "Mercury Web Parser." [Online]. Available: https://mercury.postlight.com/web-parser/. [Accessed: 26-Sep-2017].

DOI: 10.2507/28th.daaam.proceedings.061

FABLE REPRESENTATION IN FNOK AND DNOK FORMALISMS USING THE NOK CONCEPTUAL FRAMEWORK

Marina Rauker Koch, Martina Ašenbrener Katić & Mile Pavlić

This Publication has to be referred as: Rauker Koch, M[arina]; Asenbrener Katic, M[artina] & Pavlic, M[ile] (2017). Fable Representation in FNOK and DNOK Formalisms Using the NOK Conceptual Framework, Proceedings of the 28th DAAAM International Symposium, pp.0439-0445, B. Katalinic (Ed.), Published by DAAAM International, ISBN 978-3-902734-11-2, ISSN 1726-9679, Vienna, Austria
DOI: 10.2507/28th.daaam.proceedings.061

Abstract

The conceptual framework NOK (Node of Knowledge) consists of the NOK method, and DNOK and FNOK formalisms. Formalism DNOK enables a graphical representation and organization of knowledge in the form of a diagram as network nodes, and FNOK formalism represents knowledge in a textual form, which is suitable for entry into a relational database. This paper describes the representation of a text in DNOK form and formalized FNOK form, on the example of Aesop's fable "The Goose with the Golden Egg". The sentences from the fable consist of at least two actions, that is, they contain two verbs, and the paper shows that DNOK and FNOK formalisms can also be applied to such complex sentences.

Keywords: Node of knowledge (NOK); FNOK; DNOK; Knowledge Representation; Aesop

1. Introduction

Knowledge is a term that cannot be precisely determined and defined [1]. To convert knowledge into a form that can be coded using a computer, many methods have been developed applicable to different types of knowledge [2], [3]. The conceptual framework Node of Knowledge (NOK) is intended to represent sentences written in natural language [4].

In this paper, the conceptual framework NOK was applied to sentences from Aesop's fable "The Goose with the Golden Egg" [5], which are shown graphically in the form of DNOK (Diagram Nodes Of Knowledge) and in the formalized FNOK form [6]. Our aim is to show that the NOK conceptual framework, i.e. the NOK method, and the DNOK and FNOK formalisms can be applied to the so-called "complex sentences", i.e. sentences containing two or more verbs (multiple related actions). Examples of such complex sentences can be seen in this fable. The ultimate goal is to investigate data in the relational database, and it is therefore important to transform complex sentences into a form suitable for entering into a relational database. By using the conceptual framework NOK, we can transform complex sentences into a FNOK form that can be entered into a relational database, while DNOK serves as a more transparent and simplified representation, as shown in this paper. The second chapter gives an overview of relevant research in which the conceptual framework NOK was developed.

The third chapter describes the reasons for conducting this research, and the fourth chapter describes the applied methodology. The fifth chapter describes the application of DNOK and FNOK formalisms to sentences from the fable "The Goose with the Golden Egg". The sixth chapter describes the results, conclusions and possible problems of applying NOK to complex sentences, as well as plans for future research.

2. An Overview of Relevant Research

Knowledge Representation (KR) is one of the subtypes of artificial intelligence [7] and deals with methods of formal representation of knowledge in a computer. A lot of research is concerned with finding a method for knowledge representation [2], [8], [9], [10], [11], [12].

The NOK method belongs to a group of semantic networks that represent knowledge in a form of a graph. Its goal is to display textual knowledge as a knowledge network [4]. The formalization of the NOK method is presented in [6], [13]. This method analyses natural human language, sentences, words and their meanings, as well as related strings of words that form more complex phrases [14]. The NOK method translates knowledge into the model and then conducts a knowledge search in the model. It is then able to answer questions related to that knowledge.

Some of the problems of working with natural language are also homonyms and synonyms, and the model to resolve the problems is described in [15].

3. Research Motivation

The conceptual framework Nodes of Knowledge (NOK) consists of methods for modelling sentences written in a natural language. One fable in two languages translated by two different authors is shown in [16]. A knowledge network is created for these translations using the NOK method. For each of the fable versions, a diagram is created which contains all sentences from the fable. Then, the Croatian version of DNOK and the English version of DNOK are compared.

It is shown that the NOK method is applicable to modelling of sentences written in different natural human languages. Given the freedom of translation, some parts of the model remain similar, but there are some differences in the same story, in the same sentences.

This paper aims at graphically representing each sentence from the English version of the fable "The Goose with the Golden Egg" [5]in DNOK and formalized FNOK form. This will also show that complex sentences, with multiple related actions, can be represented in DNOK and FNOK forms. The resulting sentences are then suitable for entry into the relational database, which will be the subject of future research.

4. Methodology

The authors of the paper use the "one-way" NOK method in which the link between the nodes is realized by one node asking the question and the other node answering the question through the relationship in the model graphically represented in DNOK form. Fig. 1. shows an example of DNOK for the simple sentence "Ana is happy".

Fig. 1. DNOK for the sentence „Ana is happy."

FNOK for this sentence is:

is("who?" Ana, "what?" happy)

Complex human knowledge can be recorded in the form of a diagram that differs from the human language in such a way that one term exists only once in the DNOK but is linked as many times as necessary (unlimited number of links) [17], [14]. Node is the smallest element of knowledge that cannot be divided, and every new term or new meaning is a new node and it is indivisible [4]. Thus nodes can contain named persons, things, events, actions, ideas. The graphical representation of the basic concepts in the NOK method is given in Table 1.

Links connect nodes into a knowledge network. Links are represented by lines and have roles in the form of questions that connect two neighbouring nodes[18].

For this method, a formalized sentence representation was developed, i.e. the FNOK formalism [6], [13] is used to transform the sentences. The procedure for transforming sentences into FNOK form is given in [6], and it enables sentences with semantic link between nodes, using wh-questions, to be displayed textually [19].

Symbol	Meaning
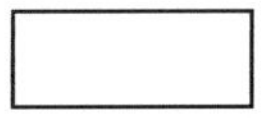	NODE – a part of knowledge in reality or thoughts
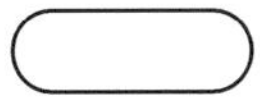	PROCESS NODE – action, event, activity
Role (question)	LINK – "one-way" link and a question with the link – linking nodes

Table 1. NOK symbols applied to modelling

5. Research Results

5.1. Translating into DNOK and FNOK

For the example of textual knowledge, the Aesop's fable "The Goose with the Golden Egg" [5] is used and separated into sentences as follows:

1. One day a countryman going to the nest of his Goose found there an egg all yellow and glittering.
2. When he took it up it was as heavy as lead and he was going to throw it away, because he thought a trick had been played upon him.
3. But he took it home on second thoughts, and soon found to his delight that it was an egg of pure gold.
4. Every morning the same thing occurred, and he soon became rich by selling his eggs.
5. As he grew rich he grew greedy; and thinking to get at once all the gold the Goose could give, he killed it and opened it only to find nothing.

Each sentence is translated into DNOK and FNOK forms separately. The first sentence was taken and each word in it was analysed and categorised according to word type. Verbs were separated from all the other word types. Each sentence in the fable contains more than one verb, so the sentences are complex. In simple sentences with only one verb, the verb is placed in the first position.

Here it was necessary to determine the exact semantic affiliation and the hierarchy between the verbs, i.e. the sequence of actions, so the first verb from the hierarchy is placed in the first position. If two actions, two verbs, in the sentence are mutually connected by conjunctions, then we place the conjunction in the first position, like in the second, third, fourth and fifth sentence. We link other verbs and words using questions with verbs on the higher level they belong to.

5.2. Fable Representation in DNOK

Each word in the sentence, as the smallest unit of knowledge, represents one node. Verbs, including conjunctions in complex sentences which link two clauses, are represented as process nodes in the ellipsis symbol. All the other nodes are represented in rectangular symbols. Semantically linked nodes are connected using links that contain a role, i.e. the question to which the other node in the representation gives the answer. In Fig. 2, the second node *egg* answers the question *found what?*.

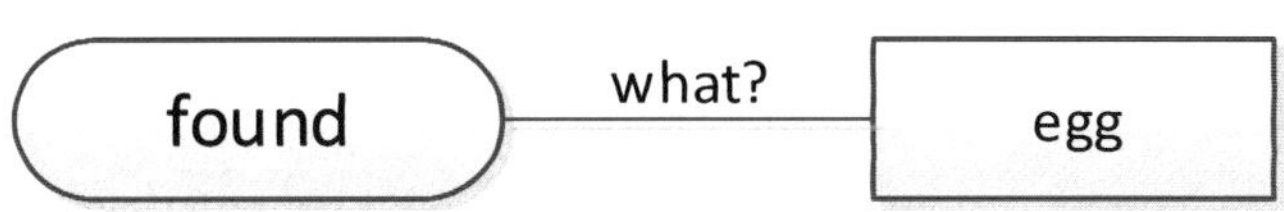

Fig. 1. Link between two nodes

For a more detailed description of the procedure, the fourth sentence is selected from the fable because, besides being complex, it contains three verbs, two of which are linked by a conjunction. The fourth sentence reads: *Every morning the same thing occurred, and he soon became rich by selling his eggs*. Sentence analysis resulted in these verbs: *occurred*, *became* and *selling*. Two simple clauses are connected into one complex sentence and interconnected using the conjunction *and*.

In this case, the conjunction *and* is observed as a process node and placed at the top of the hierarchy under which we then place verbs *occurred* and *became*. The verb *selling* is placed at a new level below *became* because it comes after *became*. At the next level, we put words related to verbs (e.g., *morning, thing*) and we continue to build the hierarchy depending on the interrelationship between the words (Fig. 4).

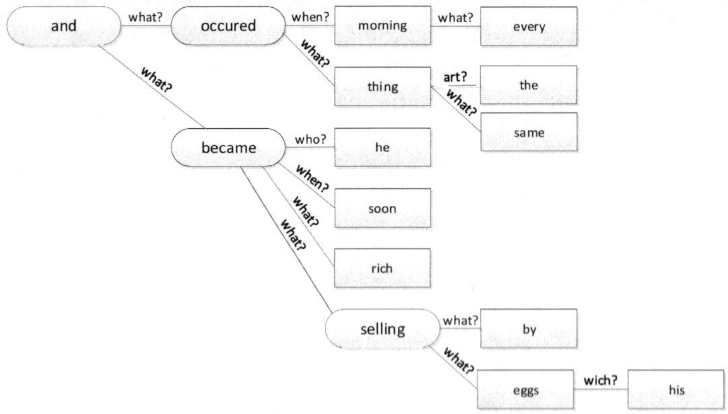

Fig. 2. DNOK of the fourth sentence

For the remaining sentences, the corresponding graphical representations in the form of DNOK are shown in Fig. 4 to Fig. 7.

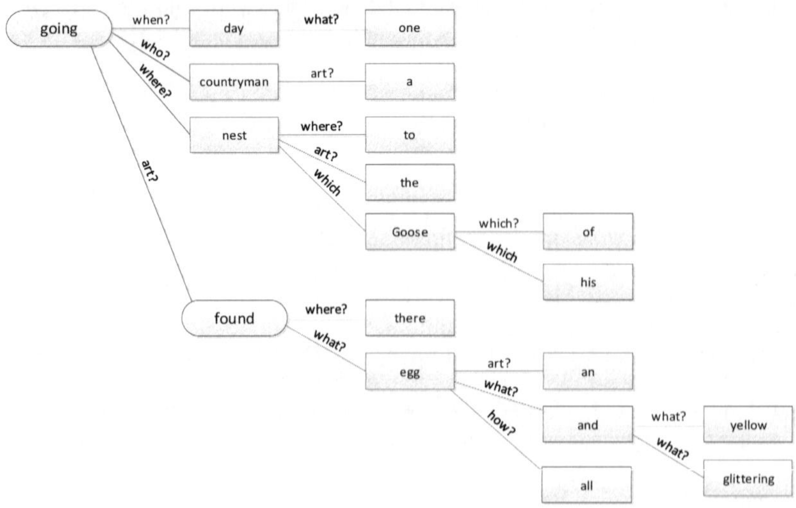

Fig. 3. DNOK of the first sentence

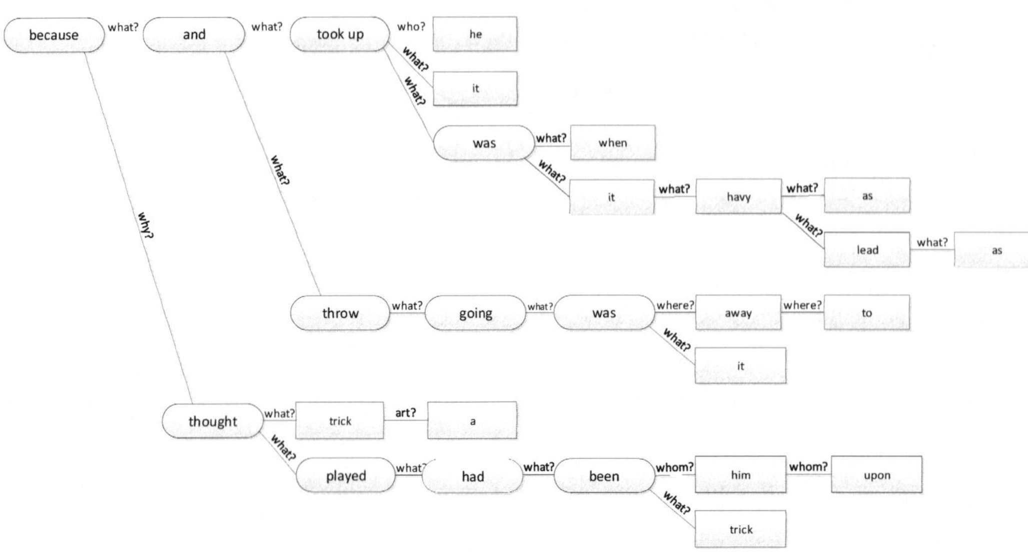

Fig. 4. DNOK of the second sentence

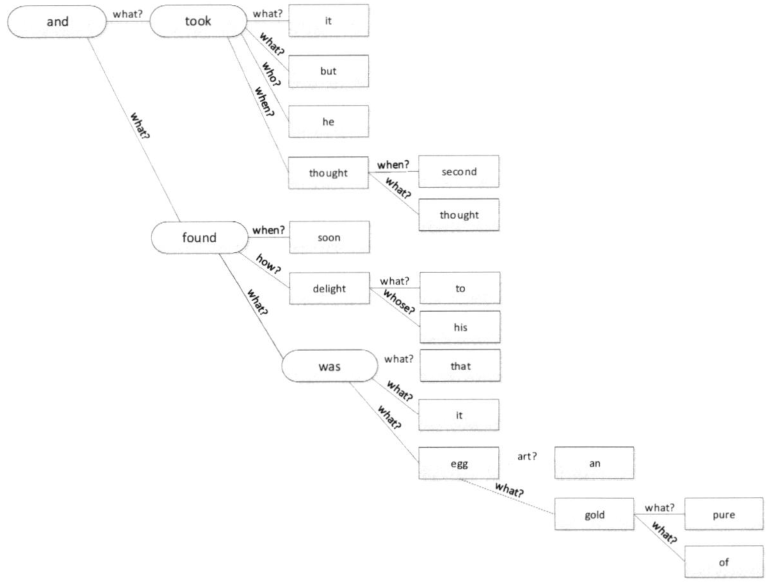

Fig. 5. DNOK of the third sentence

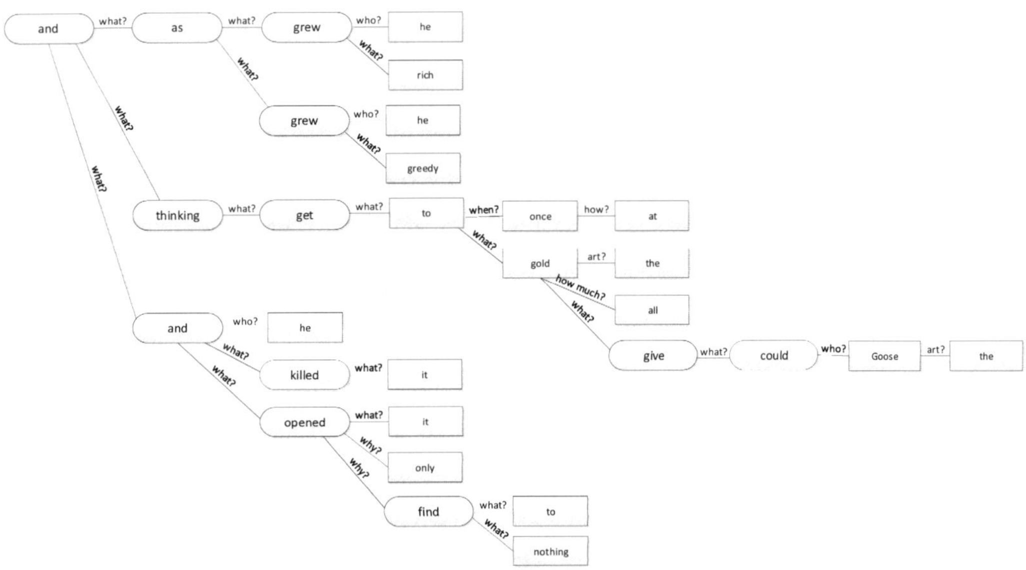

Fig. 6. DNOK of the fifth sentence

5.3. FNOK Representation of Fable Sentences

In order to prepare the sentences from the fable to be entered into the relational database, they need to be represented in a formalized FNOK form. When creating FNOK forms, a hierarchy visible in DNOK is followed, by opening parentheses for each new level, while the words from sentences with corresponding questions on the same level are separated by a comma.

The fable sentences in FNOK form:

1. going („when?" day („what?" one), „who?" countryman („art?" a), „where?" nest („where?" to, „art?" the, „which?" Goose ("which?" of, "which?" his)), "what?" found ("where?" there, "what?" egg ("art?" an, "how?" all, "what?" and ("what?" yellow , "what?" glittering))))
2. because („what?" and („what?" took up („who?" he, „what?" it, „when?" when, „what?" was („what?" heavy („what?" as, „what?" lead („what?" as)))), „what?" throw („what?" was („what?" going („where?" away („where?" to), „what?" it)))), „why?" thought („what?" trick („art?" a), „what?" played („what?" had („what?" been („whom?" him („whom?" upon), „what?" trick))))

3. and („what?“ took („what?“ it, „what?“ but, „who?“ he, „what?“ thought (“when?” on, „when?“ second)), „what?“ found („when?“ soon, „how?“ delight („what?“ to, „whose?“ his), „what?“ was („what?“ that, „what?“ it, „what?“ egg („art? an, „what?“ gold („what?“ of, „what?“ pure))))))
4. and („what?“ occurred („when?“ morning („what?“ every), „what?“ thing („art?“ the, „what?“ same), „what?“ became („who?“ he, „when?“ soon, „what?“ rich, „what?“ selling („what?“ by, „what?” eggs („which?“ his)))))
5. and (“what?” as (“what?” grew („who?“ he, „what?“ rich), „what?“ grew („who?“ he, „what?“ greedy)), “what?” thinking („what?“ get („what?“ to, „when?“ once („how?“ at), „what?“ gold („art?“ the, „how much?“ all, „what?“ give („what?“ could („who?“ Goose („art?“ the))))))), “what?” and (“who?” he, “what?” killed (“what?” it), “what?” opened („what?“ it, „why?“ only, „why?“ find („what?“ to, “what?“ nothing))))

6. Conclusion

This paper presents the application of the conceptual framework NOK and DNOK formalism, which is used to present the sentences graphically in the form of a diagram, and FNOK formalism, where the sentences are presented in a formalized form. So far, these formalisms have only been applied to simple sentences (sentences containing a verb in the present tense, i.e. a sentence represents one action). This paper went a step further; the research was extended to complex sentences and it was shown that complex sentences can be represented in DNOK and FNOK forms. The conceptual framework NOK was applied to sentences from the fable "The Goose with the Golden Egg", which are complex and each of them consists of at least two interdependent actions (verbs). By applying these methods, the interdependence is not lost and the semantics of the sentence is preserved, because every word is linked through a role (question) to the word it is related to. Complex sentences often contain conjunctions that link the two actions, and in that case their role is different from merely connecting two words. Therefore, they are placed at the top of the hierarchy. In simple sentences the words at the top of the hierarchy are verbs, and in the graphical representation they are represented as process nodes. Likewise, the conjunction at the top of the hierarchy is also represented as a process node. Types of conjunctions and their role in the sentence will be the subject of future research.

This research has shown that the conceptual framework NOK can be applied to complex sentences, because, in order to transform the fable into a relational database, it is necessary to represent it in a formalized form (FNOK), as well as graphically in the form of a NOK diagram (DNOK) so that the relationships between the nodes are more clearly presented. Transforming complex sentences into the relational database will be the subject of future research.

Future research will also concentrate on: comma analysis, different tenses, interrogative and reflexive pronouns, developing models and algorithms that will enable querying the relational database, the problem of synonyms and homonyms and etc.

7. Acknowledgments

The research has been conducted under the project "Extending the information system development methodology with artificial intelligence methods" (reference number 13.13.1.2.01.) supported by University of Rijeka (Croatia).

8. References

[1] Brachman, R. & Levesque, H. (2004). Knowledge representation and reasoning. The Morgan Kaufmann Series in Artificial Intelligence, Morgan Kaufmann.
[2] Harmelen, F., Lifschitz, V. & Porter, B. (2008). Handbook of Knowledge Representation. Elsevier, Oxford.
[3] Russell, S. & Norvig, P. (2010). Artificial Intelligence: A Modern Approach, Third edition. Prentice Hall.
[4] Pavlić, M., Jakupović, A. & Meštrović, A. (2013). Nodes of knowledge method for knowledge representation. Informatologia, 46, pp. 206–214.
[5] Aesop. The Goose with the golden egg, Available from: www.bartleby.com/17/1/57.html, Accessed: 2017-06-10
[6] Jakupović, A., Pavlić, M. & Dovedan Han, Z. (2014). Formalisation method for the text expressed knowledge. Expert Systems with Applications, 41, pp. 5308–5322. https://doi.org/10.1016/j.eswa.2014.03.006
[7] Morgenstern, L. & Thomason, R.H. (2000). Teaching Knowledge representation: Challenges and Proposals. KR2000,.
[8] Sowa, J.F. (2000). Knowledge representation: logical, philosophical, and computational foundations. Pacific Grove: Brooks/Cole.
[9] Chein, M. & Mugnier, M.-L. (2009). Graph-based Knowledge Representation, Computational Foundations of Conceptual Graphs. Springer.
[10] King, T.H. & Helbig, H. (2009). Knowledge Representation and the Semantics of Natural Language. Natural Language Engineering, 15, 310.
[11] Sowa, J.F. (2014). Principles of semantic networks: Explorations in the representation of knowledge. Morgan Kaufmann.
[12] Kamp, H. & Reyle, U. (2013). From discourse to logic: Introduction to modeltheoretic semantics of natural language, formal logic and discourse representation theory. Springer Science & Business Media.

[13] Pavlić, M., Dovedan Han, Z. & Jakupović, A. (2015). Question answering with a conceptual framework for knowledge-based system development "Node of Knowledge." Expert Systems with Applications, 42, 5264–5286. https://doi.org/10.1016/j.eswa.2015.02.024

[14] Asenbrener Katic, M., Pavlic, M. & Candrlic, S. (2015). The Representation of Database Content and Structure Using the NOK Method. 25th Daaam International Symposium on Intelligent Manufacturing and Automation 2014, Procedia Engineering, 100, pp. 1075–1081. https://doi.org/10.1016/j.proeng.2015.01.469

[15] Rauker Koch, M., Pavlic, M. & Asenbrener Katic, M. (2015). Homonyms and Synonyms in NOK Method. 25th Daaam International Symposium on Intelligent Manufacturing and Automation 2014, Procedia Engineering, 100, pp. 1055–1061. https://doi.org/10.1016/j.proeng.2015.01.466

[16] Rauker Koch, M., Pavlić, M. & Jakupović, A. (2014). Application of the NOK method in sentence modelling. Proceedings of the 37th Internation Convention MIPRO 2014, Croatian Society for Information and Communication Technology, Electronics - MIPRO. pp. 1426–1431.

[17] Pavlic, M., Dovedan Han, Z., Jakupovic, A., Asenbrener Katic, M. & Candrlic, S. (2017). Adjective representation with the method Nodes of Knowledge. 2017 40th International Convention on Information and Communication Technology, Electronics and Microelectronics (MIPRO), IEEE. pp. 1221–1226. https://doi.org/10.23919/MIPRO.2017.7973610

[18] Pavlić, M., Mestrović, A. & Jakupović, A. (2013). Graph-based formalisms for knowledge representation. Proceedings of the 17th World Multi- Conference on Systemic Cybernetics and Informatics (WMSCI 2013), Orlando. pp. 200–204.

[19] Asenbrener Katic, M., Candrlic, S. & Pavlic, M. (2017). Comparison of Two Versions of Formalization Method for Text Expressed Knowledge. Beyond Databases, Architectures and Structures (BDAS) 2017, CCIS, pp. 55–66. https://doi.org/10.1007/978-3-319-58274-0_5

DOI: 10.2507/28th.daaam.proceedings.062

DYNAMIC CHARACTERISTICS OF THE ELECTRIC VEHICLE

Adnan Masic, Boran Pikula & Dzevad Bibic

This Publication has to be referred as: Masic, A[dnan]; Pikula, B[oran] & Bibic, D[zevad] (2017). Dynamic Characteristics of the Electric Vehicle, Proceedings of the 28th DAAAM International Symposium, pp.0446-0451, B. Katalinic (Ed.), Published by DAAAM International, ISBN 978-3-902734-11-2, ISSN 1726-9679, Vienna, Austria
DOI: 10.2507/28th.daaam.proceedings.062

Abstract

In this paper, a method for analysis of dynamic characteristics of an electric vehicle is presented. The developed method is completely independent from vehicle's on-board systems and there is no interaction with vehicle's hardware or software. The test vehicle was a semi-custom electric light truck. Dynamic characteristics from realistic test scenario are presented and discussed. Special attention is dedicated to the energy consumption, kinetic energy recuperation and battery behaviour. Some suggestions for future research are proposed as well.

Keywords: Electric vehicle; Data acquisition system; Dynamic characteristics; Energy recuperation

1. Introduction

There is little doubt that the future of the road transportation belongs to the electric vehicles. Yet, it is not known how long will the transition take, nor what are the consequences. Therefore, it is very important to collect and analyse the relevant data, before the decisions are made.

Hybrid vehicles are intermediate step, and they are already on the roads in millions. Croitorescu and Jiga [1] analysed the model of a hybrid vehicle consisting of an internal combustion engine and an electric motor, while Abouelkheir and Vu [2] performed the simulation of the dual clutch transmission in the hybrid vehicle using the MATLAB and Simulink. Pikula [3] discussed the justification of electric and hybrid vehicles use in urban areas.

The aim of this research is twofold:

1. to develop an independent and reliable method for analysis of the dynamic characteristics of the electric vehicles,
2. to collect locally relevant data using the developed method.

Main research question is: *what kind of the relevant data can be obtained using this method and to what extent?*

For this purpose, we have acquired an electric vehicle: small truck powered by the three-phase asynchronous motor, EcoCarrier. Dimension of the vehicle are given in the Fig. 1, and some basic specification are:

- Power of the electric motor: 15 kW
- Max. velocity: 80 km/h
- Mass of the vehicle (including the battery, excluding the driver): 1140 kg
- Tyres: 195/65 R 15

- Brakes: ventilated disc brakes (front) / drum brakes (rear)

The mass of the vehicle is surprisingly low for its dimensions, because its chassis is made of aluminium alloy and body parts like the engine hood, doors and roof are made of the thermoplastic polymers. The vehicle is equipped with all necessary systems, making it legal to drive on public roads. Furthermore, it has the (electric) power steering, the vacuum servo brakes and the air suspension (rear)!

We have assembled battery for the vehicle using the individual components. Keeping the costs within the budget, AGM (Absorbent Glass Mat) type of the lead acid chemistry was selected. Some specifications of the battery are:

- Nominal voltage: 84 V
- Capacity: 120 Ah (10.08 kWh)
- Mass of the battery: 200 kg

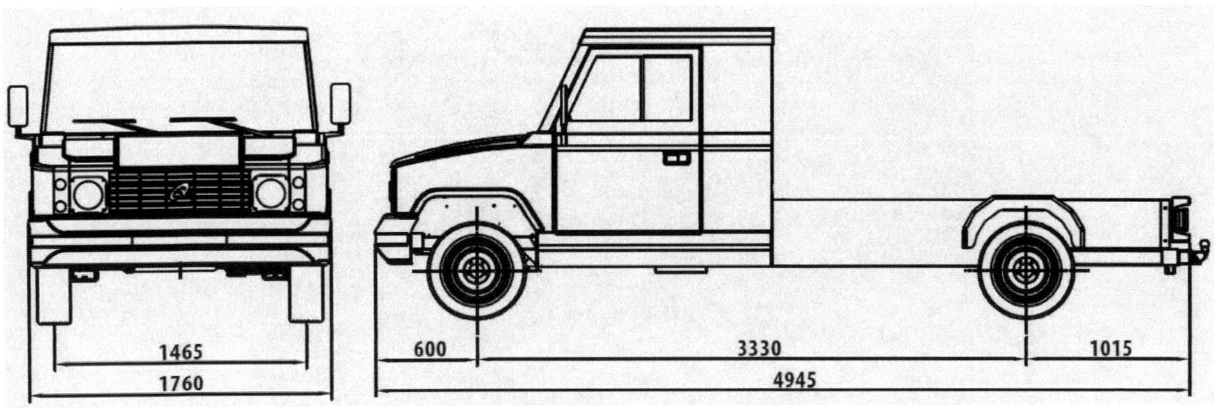

Fig. 1. Dimensions of the vehicle (all values are given in mm)

2. Method of data acquisition

Our goal was to develop a method of data acquisition which is independent from the vehicle's on-board system. The software in the vehicle is not aware that the test is underway, and all possible "software optimizations" (not to say cheats) are impossible. Detailed explanation of the methodology for testing the electric vehicles will be published in another manuscript soon. The platform for data acquisition has been developed in-house. It has its roots in the applications related to unmanned aerial vehicles and atmospheric measurements [4], [5].

The platform for data acquisition is based on open projects, such as Arduino [6]. The most challenging part was accurate and reliable measurement of the electric current, which is typically very high in case of electric vehicles. A suitable sensor is successfully adapted and installed on the vehicles main battery. The voltage is measured by means of custom voltage divider and a high performance ADC (Analog to Digital Converter). The list of important sensors is given in the Table 1.

Sensor	**Measured Quantity**
Allegro ACS770 [7]	Electric current
Bosch BMP085 [8]	Air pressure
Bosch BMA180 [9]	Acceleration
DHT22 [10]	Air temperature and humidity
Ublox NEO-6M [11]	GNSS coordinates

Table 1. Sensors

Simultaneously with the in-house developed data acquisition system (in-house DAS in further text), we have used a proprietary logger, Racelogic VBOX Sport (VBOX in further test), which is based on the high quality, 20 Hz GPS receiver. Basic specifications of the VBOX are listed in the Table 2.

VBOX Sport:	Velocity	Acceleration
Accuracy	0.1 km/h	0.5 %
Resolution	0.01 km/h	0.01 g

Table 2. VBOX specifications

3. Test track

The test track (Figure 2) was carefully chosen such that it simulates real driving scenarios, as much as possible. It is 15.5 km long with three characteristic segments:

- long straight path (1.8 km),
- cruising around the village (6.6 km) and
- mixed conditions with frequent stop-and-go cycles (7.1 km).

During the hole test, the vehicle consumed 2.48 kWh from the battery, which means that the average consumption was 16.0 kWh/100km. That is not surprising, taking into account mass and type of the vehicle. Taking into account this result, the fuel consumption of the similar vehicles with the internal combustion engine and domestic prices of the electric energy and the diesel fuel, we conclude that the electric version is more affordable by the factor of five, at least!

Fig. 2. The test track visualisation in Google Earth

4. Test results

In order to check for the consistency between two data loggers, we compare velocity of the vehicle recorded with two systems, on a drive segment throughout the village (Figure 3). From the Figure 3 we can see near perfect agreement of two independent data loggers! Proprietary DAS was extensively tested in various scenarios and we wanted to check if our in-house developed DAS can show at least similar performance. It is important to note that the in-house developed DAS offers almost endless possibilities (we can change the hardware, the software and generally any parameter), which is not the case with the proprietary solution.

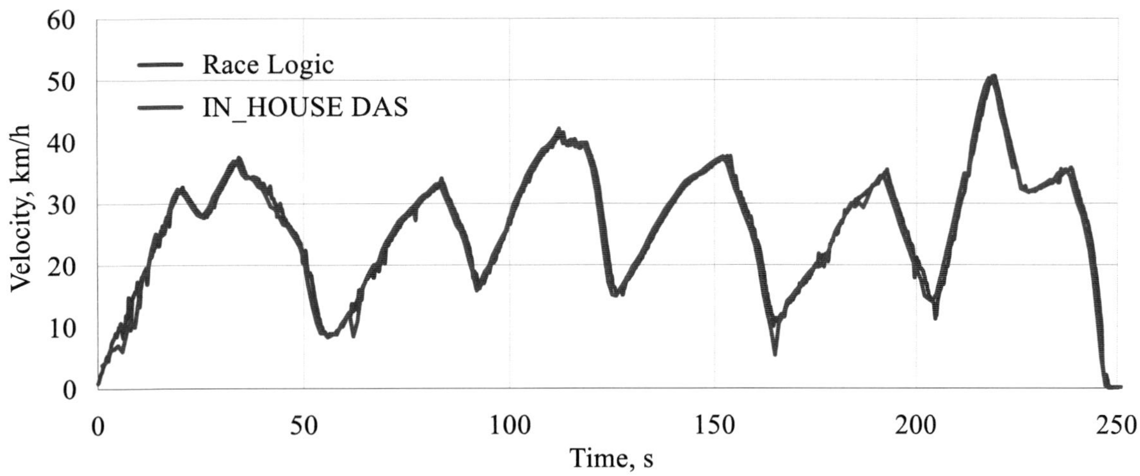

Fig. 3. Comparison of the results from two data loggers

In the Figure 4 voltage and current values are presented for the same drive segment. By convention, negative current corresponds to the battery drain, while positive current corresponds to the battery charge. From the Figure 4 we can see that the voltage of the main battery oscillates between 65V and 90V depending on the current intensity and the mode (draining or charging the battery).

This wide interval of voltage values is not desirable and indicates significant internal resistance of the battery. In order to evaluate the internal resistance of the battery, we need to record the voltage of the battery under the strong current drain, i.e. full scale volt-ampere characteristic of the battery.

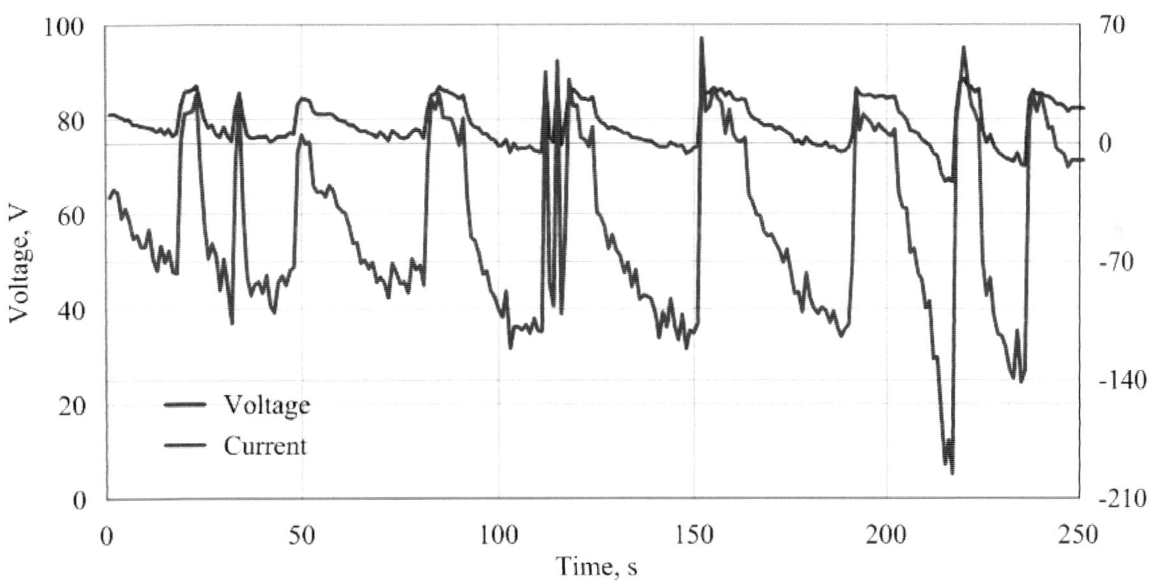

Fig. 4. Voltage and current readings from the in-house DAS

Volt-ampere characteristic of the vehicle's battery is presented in the Figure 5. The range of current intensity goes from about 200A drain to 60A charge, which is practically full scale test (taking into account specifications of the vehicle and the current sensor). No-load voltage was about 84V, just as expected. Under the load of 200A, however, the voltage drops by almost 20V! Using the linear regression fit from the Figure 5, and the Ohm's law, we find the internal resistance of the battery to be 90.5 mΩ (i.e. slope of the linear fit).

Clearly, this is not a high-performance battery, and another one, with lower internal resistance, would represent the improvement. Lithium-ion batteries would be the best choice, but their price exceeded the budget of this research project significantly. We hope that we will have an opportunity to test the performance of this electric vehicle equipped with the lithium-ion battery in the future.

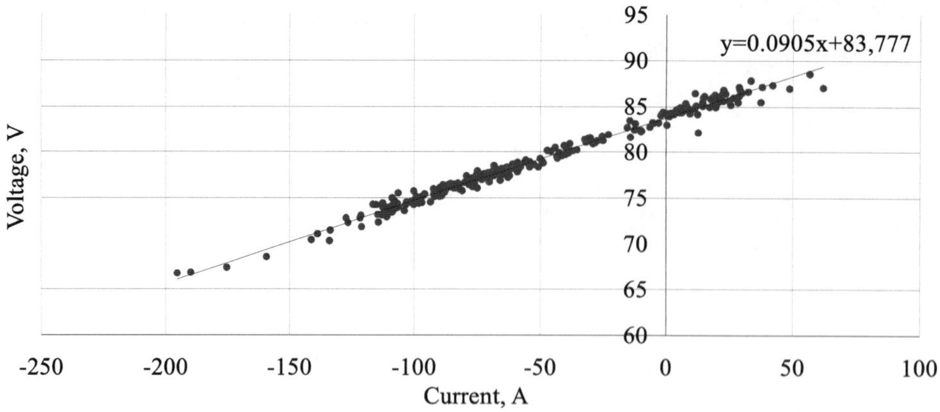

Fig. 5. Volt-ampere characteristic of the battery

Figure 6 shows the power-time dependence for the same track segment. Part of the graph "above zero" corresponds to the energy recuperation. By integrating power over time we get 300 Wh in the negative part of the graph and 26.6 Wh in the positive part, which means that 8.9 % of spent energy was recuperated back to the battery during the deceleration of the vehicle.

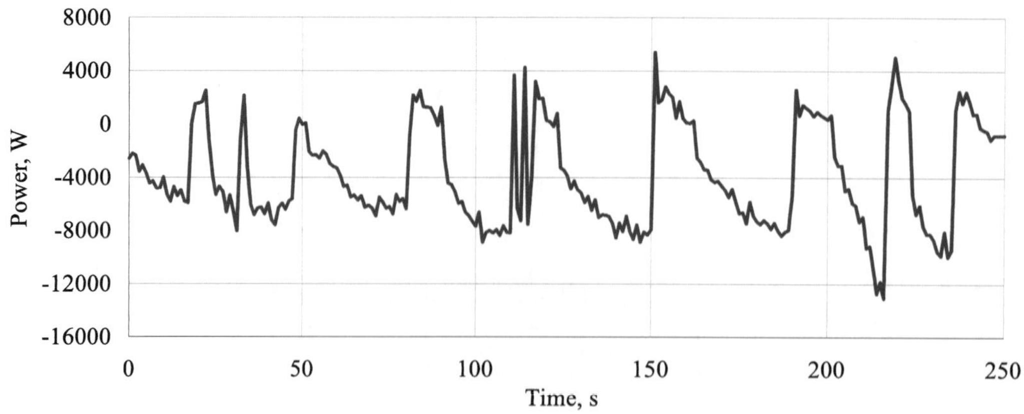

Fig. 6. Power consumption during drive on the test track

Finally, in the Figure 7 typical acceleration of the vehicle from zero to about 45 km/h is presented. The velocity profile was measured with the Race Logic VBOX. Maximum acceleration of the vehicle was 2.3 m/s^2, which is very good result, taking into account small declared power of the electric motor (15 kW only). This was achieved thanks to the single speed gearbox, specifically constructed for the electric vehicle. The velocity profile in the Figure 7 is very similar to the dynamic characteristic of the vehicle of the same size that is powered by the internal combustion engine.

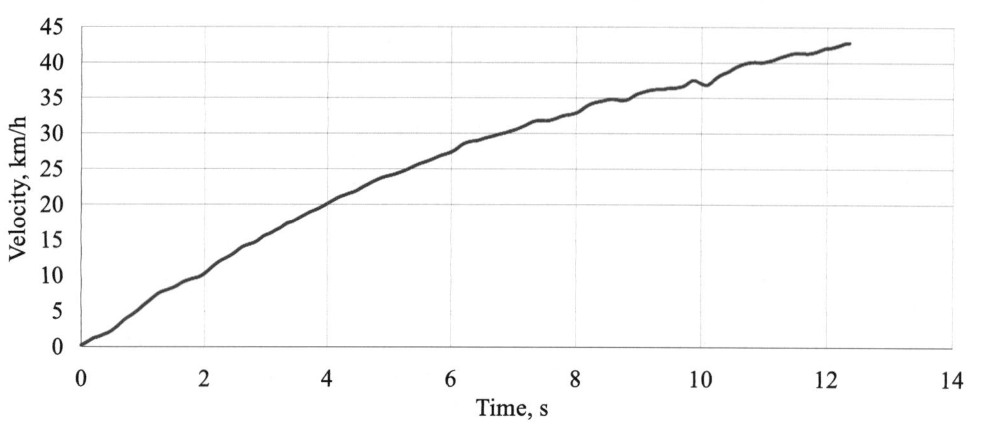

Fig. 7. Velocity profile during acceleration on the test track

5. Conclusion

Comprehensive method for testing dynamic and electric characteristics of the electric vehicle has been developed and successfully tested. The method was able to collect all relevant data (position of the vehicle, speed, acceleration, voltage and current) independently from vehicle's on-board electronic systems. The test track was chosen so that it reflects realistic driving scenarios. Special attention has been devoted to the performance of the battery and the energy recuperation. Internal resistance of the battery has been determined (90.5 mΩ). The amount of energy recuperated from the motor during the test was about 8 %, which fits to the range of expectancy. The average consumption of electric energy of the vehicle during the entire test was 16.0 kWh/100km. Taking into account local electricity price, this is at least five times cheaper than comparable diesel vehicle. In all, general dynamic characteristics of the electric vehicle are found to be similar to the characteristics of the comparable vehicles with the internal combustion engine. It is necessary, however, to educate drivers and encourage them to adapt the style of driving, so that maximal possible amount of energy can be recuperated.

The main limitation of the in-house developed DAS is the current sensor (limited to ±200A). This was sufficient for our particular vehicle, but may not suffice for some other vehicles. However, the in-house developed platform is very flexible and integration of new sensors is straight forward (once they are available).

Further tests are expected in the near future. They should include more test tracks, more driving scenarios, and (most important) different climate conditions. The test was performed during the summer, with very favourable weather. It would be interesting to re-do the test during the cold winter period. Battery behaviour should be carefully analysed under the extremely low temperatures (common for this geographical area).

Finally, we can give the answer to the main research question: this platform can provide all relevant data for determination of dynamic characteristics of the electric vehicle in realistic driving scenarios and can be easily adapted for tests of other electric vehicles (significantly smaller or bigger than the particular vehicle in this test).

6. Acknowledgments

Donation of the electric vehicle from Volkswagen Sarajevo Ltd is greatly acknowledged. So is the material support from the Faculty of Mechanical Engineering (Department for IC Engines), University of Sarajevo. Last, but not least, we would like to thank to Mr. Nijaz Delalic, for his support and collaboration on this project!

7. References

[1] Croitorescu, V. & Jiga, G. (2010). Influence of Using a Hybrid Electric Powertrain Combined with an ICE Variable Displacement in Automotive Architecture. Annals of DAAAM for 2010 & Proceedings of the 21st International DAAAM Symposium, Volume 21, No. 1, ISSN 1726-9679 ISBN 978-3-901509-73-5, Editor B. Katalinic, Published by DAAAM International, Vienna, Austria, EU, 2010

[2] Abouelkheir M. & Vu T. M. (2016). Modeling and Simulation of Dual Clutch Transmission and Hybrid Electric Vehicles. 11th International DAAAM Baltic Conference "INDUSTRIAL ENGINEERING" 20-22nd April 2016, Tallinn, Estonia

[3] Pikula B. (2011). Justification of Electric and Hybrid Vehicles Use in Urban Areas. Sarajevo Green Design Conference, Sarajevo, September 2011, ISBN 978-90-365-3451-2, pp 156-160

[4] Masic, A.; Musemic, R. & Dzaferovic-Masic, E. (2016). Temperature Inversion Measurements in Sarajevo Valley Using Unmanned Aerial Vehicles, Proceedings of the 27th DAAAM International Symposium, pp.0423-0427, B. Katalinic (Ed.), Published by DAAAM International, ISBN 978-3-902734-08-2, ISSN 1726-9679, Vienna, Austria, DOI: 10.2507/27th.daaam.proceedings.062

[5] Masic, A. (2015).Unmanned Aerial Vehicle as Data Acquisition System. Journal of Trends in the Development of Machinery and Associated Technology. Vol. 19, No. 1, 2015, ISSN 2303-4009 (online), p.p. 181-184.

[6] Bujdei, C. & Moraru S. A. (2011). A Design for a Network End Node, Based on Arduino and Xbee. Annals of DAAAM for 2011 & Proceedings of the 22nd International DAAAM Symposium, Volume 22, No. 1, ISSN 1726-9679, ISBN 978-3-901509-83-4, Editor B. Katalinic, Published by DAAAM International, Vienna, Austria, EU, 2011

[7] http://www.allegromicro.com/~/media/Files/Datasheets/ACS770-Datasheet.ashx?la=en?sc_camp=64EB2DD6B3FE4C088C07DB87D5D9B6EF, (2017). Accessed on: 2017-08-24

[8] https://ae-bst.resource.bosch.com/media/_tech/media/datasheets/BST-BMP180-DS000-121.pdf, (2017). Accessed on: 2017-08-24

[9] http://www.geeetech.com/Documents/BST-BMA180-FL000-03.pdf, (2017). Accessed on: 2017-08-24

[10] https://www.sparkfun.com/datasheets/Sensors/Temperature/DHT22.pdf, (2017). Accessed on: 2017-08-24

[11] https://www.u-blox.com/sites/default/files/products/documents/NEO-6_DataSheet_(GPS.G6-HW-09005).pdf, (2017). Accessed on: 2017-08-24

DOI: 10.2507/28th.daaam.proceedings.063

MOBILE MEASUREMENTS OF PARTICULATE MATTER CONCENTRATIONS IN URBAN AREA

Adnan Masic, Boran Pikula & Dzevad Bibic

This Publication has to be referred as: Masic, A[dnan]; Pikula, B[oran] & Bibic, D[zevad] (2017). Mobile Measurements of Particulate Matter Concentrations in Urban Area, Proceedings of the 28th DAAAM International Symposium, pp.0452-0456, B. Katalinic (Ed.), Published by DAAAM International, ISBN 978-3-902734-11-2, ISSN 1726-9679, Vienna, Austria
DOI: 10.2507/28th.daaam.proceedings.063

Abstract

The great attention is dedicated nowadays to the measurements of air pollutants' concentrations, and the air quality index (AQI), especially the particulate matter (PM) concentrations. However, measurements of PM concentrations are usually performed on fixed locations. Such data are very valuable, but not sufficient, because there is no information on spatial distribution of PM concentrations. For this purpose, a special low-cost device, based on laser diffraction method, for measurement of PM concentrations (PM1, PM2.5, PM10), has been developed and tested: both horizontal and vertical (using unmanned aerial vehicle) distributions are analysed over the urban area – city of Sarajevo.

Keywords: particle size analysis; laser diffraction methods; unmanned aerial vehicles; air quality index

1. Introduction

According to the World Health Organisation (WHO), the air pollution represents major environmental risk for human health [1]. Among all pollutants, the particulate matter suspended in the air (aerosol), have dominant effect on human health in Europe [2]. City of Sarajevo, Bosnia-Herzegovina, has characteristic and very strong air pollution during the winter, since it is located in the valley and affected by the natural phenomenon called temperature inversion [3]. Similar episodes can be observed in many other towns and cities with similar topology of terrain, for example Grenoble, France [4]. Therefore, monitoring of the PM concentrations in urban environment is essential. It is usually performed on fixed measuring locations operated by the Governmental institutions (if at all). Such measuring stations are very expensive, and their availability is limited. Data from such stations are very valuable, but there is no information about spatial distribution of air pollutants.

Another approach is presented here, as an alternative to the fixed monitors: development of the low-cost and highly mobile devices (air quality data loggers), which are suitable for recording of spatial distribution of PM concentration, both horizontally and vertically. These requirements (low-cost and high mobility) can be met using optical sensors, based on the laser diffraction method (instead of gravimetric analysis). Principal scheme of work of such optical sensors is presented in the Figure 1.

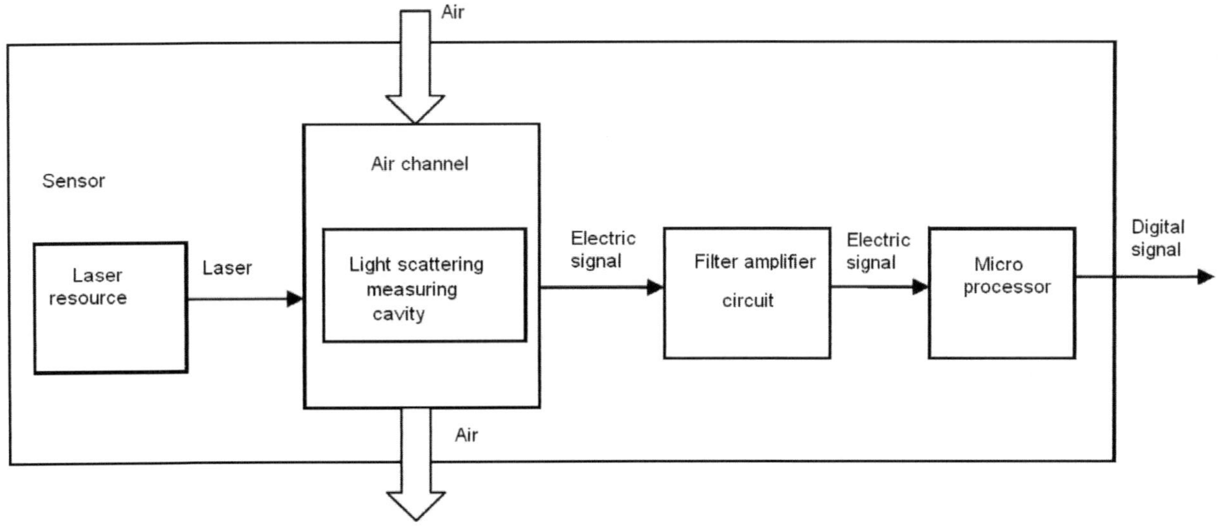

Fig. 1. Functional block diagram of the optical PM sensor

Laser light is scattered on the particulate matter in the air, producing certain effect on the photo-detector in the sensor. Electric signal from the photo-detector is amplified and processed internally using the Mie theory [5]. In short, the Mie theory is the solution of Maxwell's equations in the case of scattering of an electromagnetic plane wave by a homogeneous sphere.

The theory assumes that
1. all particles are spherical, optically homogenous and isotropic,
2. particles are illuminated by a plane wave of known wavelength,
3. the refractive index of both particle and medium (air) is known,
4. there is neither surface charge nor surface current on particles.

These assumptions should be carefully examined, since they can lead to significant uncertainty in the measuring process. Furthermore, the density of the particles must be assumed in order to calculate result in desired units ($\mu g/m^3$). An excellent guide to the reliable measurements of PM using the optical sensors is the ISO standard 13320:2009 [6].

2. The mobile particulate matter optical counter

Primary instrument for this research is the mobile particulate matter optical counter (mPMOC in further test), which is based on our previous, in-house developed, data acquisition system, which was used to investigate vertical temperature profiles using unmanned aerial vehicles [7]. It was redesigned for this project and major modification was integration of the optical PM sensor. Special enclosure for the new data acquisition system was developed and optimized using computational fluid dynamics (CFD) simulations. The enclosures are manufactured using 3D printers.

The optical particulate sensor for the mPMOC was carefully chosen using the aforementioned criteria. It is the Plantower PMS5003 [8], with following specifications:

- Minimum distinguishable particle diameter: 0.3 μm
- Range of measurement: 0-1000 $\mu g/m^3$
- Resolution: 1 $\mu g/m^3$
- Maximum consistency error: ±10%
- Single response time: <1 s
- Standard volume: 0.1 dm^3
- Total response time: <10 s
- Working temperature range: -10 ^{0}C – 60 ^{0}C
- Working humidity range: 0-99%
- Mean Time Before Failure: >3 years

Together with the PMS5003 sensor, mPMOC is equipped with: SHT31-D humidity and temperature sensor from ClosedCube, Measurement Specialties MS5611 barometer and u-blox NEO-M8N GNSS receiver. Communications with all sensors and logging of data was implemented by means of an Arduino Mega 2560 microcontroller. The SHT31-D and MS5611 sensors communicate with the microcontroller over the I^2C bus, while the PMS5003 and NEO-M8N sensors use two serial ports (UART).

3. Results

The mPMOC was designed in such a way that it can be easily adapted and installed on various mobile platforms (motorbike, bicycle, unmanned aerial vehicle etc.). First case, presented in the Figure 2, is the application of the mPMOC for use on the motorbike, so the city roads can be "scanned" for the PM10 air pollution. The city of Sarajevo has characteristic main longitudinal road of about 10 km. The map background in the Figure 2 is Google Terrain Map, the track is shown using GNSS coordinates from mPMOC, and the color of the track line corresponds to the PM10 concentration. The image is generated by means of GpsVisualizer tool [9].

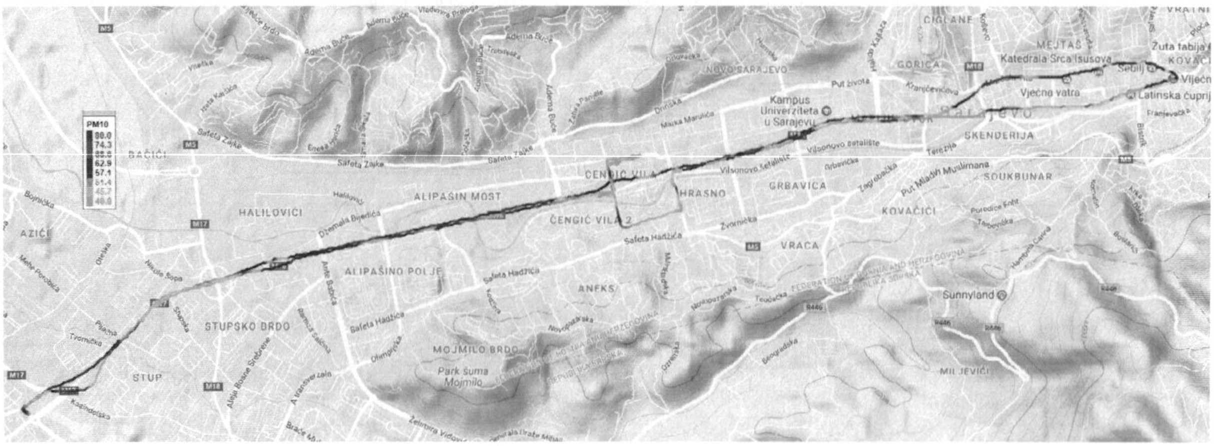

Fig. 2. Distribution of PM10 pollution over the main longitudinal road through the city

We can see red spots (increased PM10 pollution) in the Figure 2 on several road segments, but it is not clear from the figure if there is any specific reason for that. However, if we plot the combined graph of speed of the motorbike and PM10 concentration (Figure 3), we can notice an interesting correlation: whenever the speed of the motorbike drops to zero and keeps that value for certain amount of time, the corresponding PM10 concentrations are raising sharply! This happens on big intersections, where the traffic is slowed down, so there is increased air pollution from the vehicles' emission. In the left part of the Figure 2, we have the main interchange in the city (Stup interchange), and we can see that PM10 concentrations are below the average value from the graph, which looks surprising at first glance, taking into account high intensity of traffic on the interchange. Few years ago, that interchange was known for long traffic jams. However, the Stup interchange has been reconstructed recently. It is a modern and wide interchange now, and traffic jams are almost nonexistent. This is clear indicator of strong implications of urban planning to the air pollution levels, the PM10 in this case.

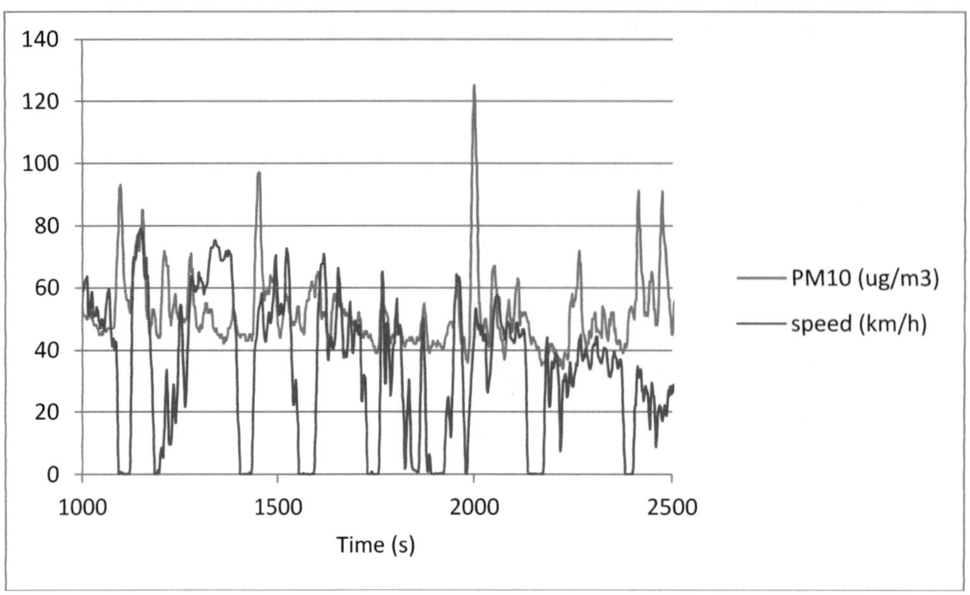

Fig. 3. Correlation between traffic speed and PM10 pollution

Second test case is performed with the mPMOC mounted to the bicycle. The test track was different know, with segmenets where motor vehicles were not allowed during the test duration (Vilsonovo promenade). The track included both urban and suburban areas (Figure 4). We can clearly see from the Figure 4 that the air pollution with PM10 particles is much weaker in the area with limited traffic and lot of trees (Vilsonovo promenade) in comparison to walk paths near the major roads. Especially strong air pollution was recorded over (many) city hills (Koševsko brdo, Kobilja Glava, etc.) due to the strong emissions from vehicles driving over the aformentioned inclinations.

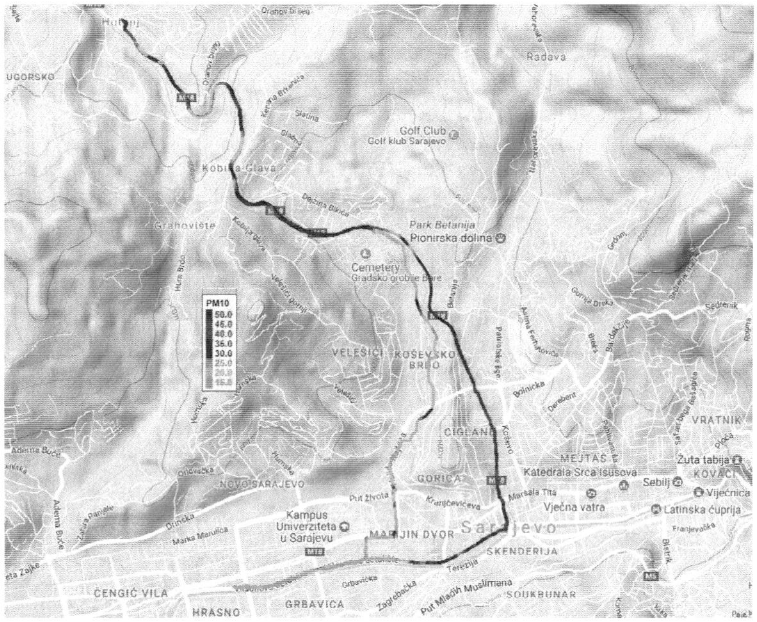

Fig. 4. PM10 pollution in urban and suburban section

Finally, the mPMOC was mounted to the in-house developed unmanned aerial vehicle (UAV) [3], to evaluate the possibility of measurements of vertical profiles of PM10 concentration. The test location was city quart Otoka, where the Governmental automatic station for measurements of air pollution is installed. Altitude of test location is about 500 meters, while the UAV lifted mPMOC approximately 700 meters above the city. The date of measurement was 19-03-2017 i.e. at the end of the winter. A very interesting vertical profile of PM10 concentration was observed (Figure 5): PM10 concentrations were initially decreasing with the altitude for several hundred meters, and then suddenly reversing the trend! The temperature profile indicates the interesting layer of temperature inversion some 300 meters above the city. This layer obstructed natural convection of the air, preventing the ventilation of major pollutants.

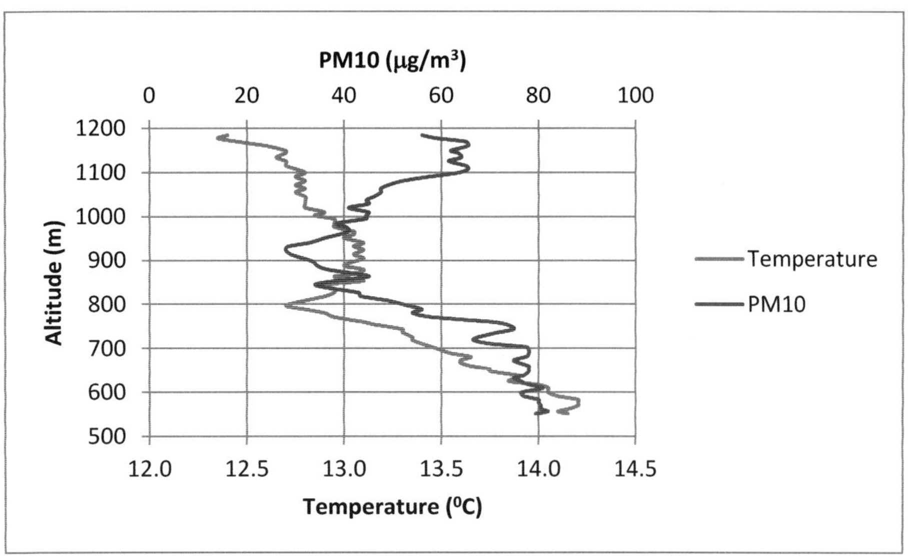

Fig. 5. Vertical profiles of temperature and PM10 concentration

4. Conclusion

Major goal of present research is the evaluation of the possibility of using the mPMOC on various mobile platforms. The mPMOC was tested for mobile measurements of the particulate matter concentrations in the air using the following carriers: the motorbike, the bicycle and the unmanned aerial vehicle. In these three applications, the mPMOC surpassed all expectations: it produced meaningful results, it had good resolution and fast reaction time. The obtained results indicated interesting correlations, such as:

- horizontal distribution of PM concentrations vs. traffic intensity,
- horizontal distribution of PM concentrations vs. orography of the terrain,
- vertical distribution of PM concentrations vs. vertical temperature profile etc.

In comparison to the fixed monitoring stations operated by the Governmental institutions, the mPMOC demonstrated clear advantages, most important of all: low cost and high mobility. Does it make the fixed and expansive monitoring stations obsolete? Actually no, it doesn't. The proposed method is more like the extension to the public network of monitors, rather than the replacement for them. It tries to overcome major disadvantages of the fixed stations (high costs and low mobility). However, properly calibrated and maintained fixed stations offer superior reliability of the measured pollutants concentrations, since they are based on gravimetric methods. All optical counters used for measurements of the air pollution use certain assumptions. These assumptions are mentioned in the introduction of this paper, and can have significant influence on the accuracy of measurements (especially assumptions 1-3).

Thus, we propose careful approach when using optical counters. They should be used as the supplement to the other techniques, such as: gravimetric methods, source appointment studies, computer simulations (turbulence modeling and dispersion programs) etc.

For the future work, we propose combination of different methods and evaluation of the mPMOC performance under various scenarios, such as the strong air pollution during the winter (due to the temperature inversion). More mPMOC devices should be constructed and used for simultaneous measurements of PM concentrations across the city. The application of unmanned aerial vehicles looks very promising, more measurements under different conditions would be very welcome. Finally, concentrations of other air pollutants (SO_2, NO_x, CO, O_3) should be measured as well. In other to keep the key advantages of mPMOC (low-cost and high mobility), the only viable option for mentioned pollutants is to use electrochemical sensors for measurements of their concentrations. We will try to implement more sensors in the future, and test them together with the optical PM sensor.

5. References

[1] WHO Air quality guidelines for particulate matter, ozone, nitrogen dioxide and sulfur dioxide (2006). World Health Organisaton

[2] Andrejiova, M[iriam]; Kralikova, R[uzena] & Wessely, E[mil] (2016). The Air Quality Assessment from the Stationary Source Air Pollution, Proceedings of the 26th DAAAM International Symposium, pp.0082-0088, B. Katalinic (Ed.), Published by DAAAM International, ISBN 978-3-902734-07-5, ISSN 1726-9679, Vienna, Austria DOI:10.2507/26th.daaam.proceedings.012

[3] Masic, A.; Musemic, R. & Dzaferovic-Masic, E. (2016). Temperature Inversion Measurements in Sarajevo Valley Using Unmanned Aerial Vehicles, Proceedings of the 27th DAAAM International Symposium, pp.0423-0427, B. Katalinic (Ed.), Published by DAAAM International, ISBN 978-3-902734-08-2, ISSN 1726-9679, Vienna, Austria, DOI: 10.2507/27th.daaam.proceedings.062

[4] Largeron Y. & Staquet C. (2016). Persistent inversion dynamics and wintertime PM10 air pollution in Alpine valleys. Atmospheric Environment, 135 (2016) 92–108, DOI: http://dx.doi.org/10.1016/j.atmosenv.2016.03.045

[5] Hergert W. & Wriedt T. (2012). The Mie Theory. Springer, ISBN 978-3-642-28737-4, ISSN 0342-4111, DOI: https://doi.org/10.1007/978-3-642-28738-1, Berlin, Heidelberg

[6] ISO 13320:2009 (2009). Particle size analysis – Laser diffraction methods. International Organization for Standardization, Geneva, Switzerland.

[7] Masic, A. (2015). Unmanned Aerial Vehicle as Data Acquisition System. Journal of Trends in the Development of Machinery and Associated Technology. Vol. 19, No. 1, 2015, ISSN 2303-4009 (online), p.p. 181-184.

[8] http://www.aqmd.gov/docs/default-source/aq-spec/resources-page/plantower-pms5003-manual_v2-3.pdf, (2016). Accessed on: 2017-09-20

[9] http://www.gpsvisualizer.com, (2017). Accessed on: 2017-09-20

DOI: 10.2507/28th.daaam.proceedings.064

IMPACT OF STYLUS SIZE IN ROUGHNESS MEASUREMENT

Dana Kubátová*, Martin Melichar, Jan Kutlwašer

This Publication has to be referred as: Kubatova, D[ana]; Melichar, M[artin] & Kutlwaser, J[an] (2017). Impact of Stylus Size in Roughness Measurement, Proceedings of the 28th DAAAM International Symposium, pp.0457-0466, B. Katalinic (Ed.), Published by DAAAM International, ISBN 978-3-902734-11-2, ISSN 1726-9679, Vienna, Austria
DOI: 10.2507/28th.daaam.proceedings.064

Abstract

In technical practice, the concept of surface roughness plays an irreplaceable role. This property has been evaluated by various techniques and mathematical approaches since the 1920s. However, evaluation is often underestimated and reduced to find the Ra or Rz parameter. This occurs regardless of methodology. It represents a totally unacceptable simplification of the issue. The following article will clarify a "small" part of this vast issue. Nowadays, there are two basic options for measuring roughness. The first option is the contact method; the second option is the non-contact measurement method. Each method has its advantages and disadvantages and is typically used only for a specific application (2D or 3D roughness measurement).This paper focuses on the contact measurement of 2D surface area, specifically the choice of the size and shape of the stylus tip. This article maps the problem of choosing a stylus tip in the interval where the standard ISO 4288: 1999 permits usage of both stylus tip sizes. This concerns the roughness range Ra = 0.1 ÷ 2 μm (aperiodic surfaces) and the range of the parameter Rsm = 0.13 ÷ 0.4 mm (periodic surfaces). The article further compares the use of a new stylus and an old (worn) one.

Keywords: Roughness; Contact and Non-contact measurement; Stylus size and shape; Standard ČSN EN ISO 4288:1999

1. Introduction

Measurement and evaluation of surface texture have seen major qualitative advances in recent years. Leading producers of measuring instruments (Hommel, Carl Zeiss, and others) have responded actively to new requirements. Surface texture is frequently checked by means of single-purpose measuring instruments. For this reason, some of the key players who put pressure on developing standards related to evaluating the quality of measurement of machined surfaces are the manufacturers of such instruments themselves. [2]

This has ultimately led to improved technologies of existing tools for surface texture measurement and evaluation, as well as to better methods, measuring systems and the system of assessment and evaluation of surface texture that are still under development. [6; 7] The system of assessment and evaluation of surface texture is defined by a body of standards which describe designations, measurement, and evaluation of surface texture, calibration of measuring instruments, and other aspects. They are the GPS (Geometrical Product Specification) standards. [5]

Generally, the measurement and assessment of surface texture represents a separate field of metrology. Using special techniques, the data required for characterizing the quality of surface can be obtained.

In order to assess the surface quality in an objective manner, relevant information on the surface in question must be obtained by measuring. First, the primary profile must be scanned using a stylus tip. From this profile, individual sets of irregularities are then filtered out (roughness, waviness, form of the surface) which comprise the actual surface texture. These irregularities differ predominantly in their spacing and their effects on the surface performance. This is why they must be separated for analysis. [3;8] Components of the surface texture are separated by filtering. In order to determine specific roughness parameters (Ra, Rz and others) from the measured profile (primary profile) of the surface, the roughness component must be separated from other types of irregularities found in the surface. However, when roughness is measured by a contact method, such as in this case, data distortion (filtering) by the probe arm. Just the size of the value distortion from the used tip was detected in the test described in this article.

2. Measurement of surface roughness by contact method

The ČSN EN ISO 3274:1998 standard defines a contact instrument as a measuring instrument which explores surfaces with a stylus and acquires deviations in the form of a surface profile, calculates parameters and can record this profile. One of the important components of the instrument is the measurement loop. It is a closed chain which comprises all mechanical components that connect the work piece and the stylus tip.

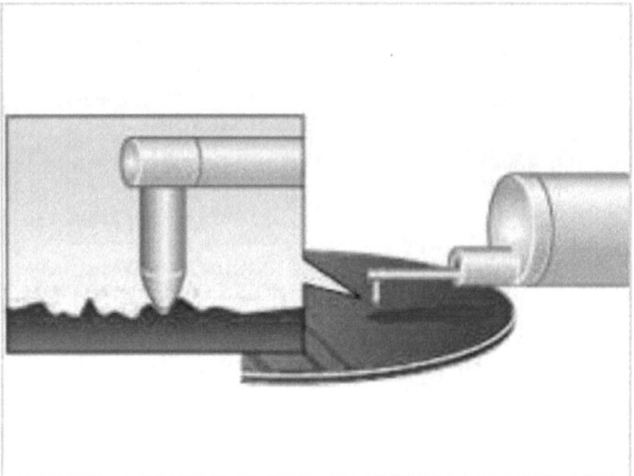

Fig. 1. Stylus [9]

The accuracy of the measurement reading is influenced by the following:

- stylus tip radius
- stylus tip apex angle
- measuring (loading) force
- rate of change of measuring force

In this article, we only focus on the effects of the stylus tip radius. Hence, changes in surface roughness readings obtained with styli of various sizes will be explored. The stylus tip should be pressed against the surface with such a force that it remains in contact with the surface under measurement during the transducer movement. According to the ČSN EN ISO 3274:1988 standard, the ideal stylus shape is a cone with a spherical tip. The nominal radius of the tip is 2 μm, 5 μm, or 10 μm; and the cone apex angle is 60°, or 90°.

3. Experimental

Testing was carried out in the Hommel Etamic T8000 machine. Styli of 2 and 5 μm with the apex angle of 90° were fitted to the machine. The tests also included investigation of the impact of using an old (worn) stylus; with both stylus tip sizes. Fig. 3 – 6 show electron micrographs of the actual stylus tip shapes.

Fig. 2. Old tip 2

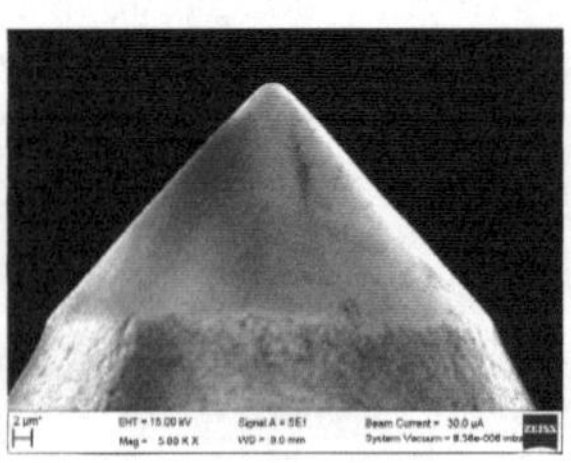

Fig. 3. New tip 2

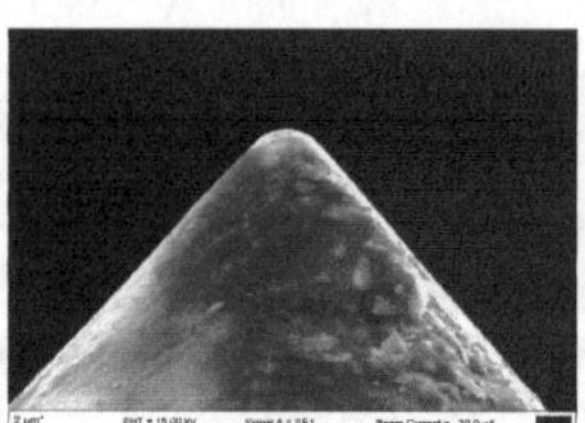

Fig. 4. Old Tip 5

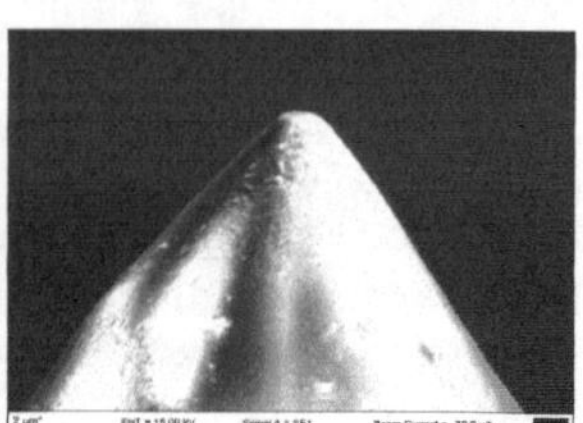

Fig. 5. New tip 5

The test pieces were roughness standards. Their roughness values were verified through calibration by an independent contractor. [12] These roughness values were chosen so that the experiment takes place within the Rsm parameter (periodic surface) interval in which the use of both sizes of stylus tips is permitted, as indicated in Tab. 1.

Periodic profiles
e.g. turning, milling

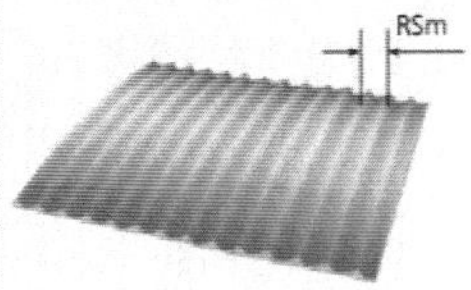

Measuring conditions

- lr sampling length
- ln evaluation length
- lt traverse length
- λc cut-off
- λs shortwave profile filter
- r_{tip} stylus tip radius
- ΔX digitization distance [1)]

1) The digitization distance is also standardized. This is set automatically by most roughness measuring instruments.

Aperiodic profiles
e.g. grinding, eroding

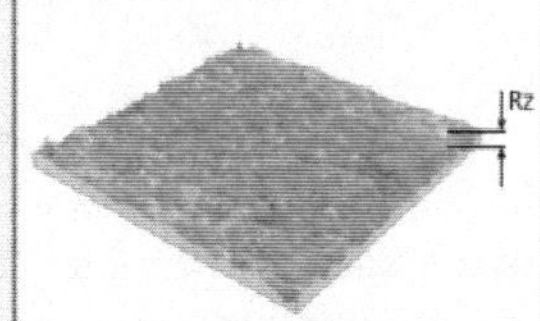

RSm (mm)	λc = lr (mm)	ln (mm)	lt (mm)	r_{tip} (µm)	λs (µm)	Ra (µm)	Rz (µm)
> 0.013 ...0.04	0.08	0.4	0.48	2	2.5	> (0.006) ...0.02	> (0.025) ...0.1
> 0.04 ...0.13	0.25	1.25	1.5	2	2.5	> 0.02 ...0.1	> 0.1 ...0.5
> 0.13 ...0.4	0.8	4	4.8	2 or 5*	2.5	> 0.1 ...2	> 0.5 ...10
> 0.4 ...1.3	2.5	12.5	15	5	8	> 2 ...10	> 10 ...50
> 1.3 ...4	8	40	48	10	25	> 10 ...80	> 50 ...200

Table 1. Table of values for roughness measurement [10]

Prior to measurement, each reference standard was divided into fields of 10×6 mm size (Fig. 6). Measurement was then carried out in each of these fields. The paths were arranged not to overlap which was made possible by the machine's precision measuring table. After the standard had been installed in the required position, its movement was effected exclusively by the measuring table. All measurement runs were repeated twice. The pattern of measuring paths started 0.5 mm from the field's boundary. Each subsequent path was spaced 1 mm from the previous one (0.5; 1.5; 2.5; …).

	A	B	C
1	IIII	IIII	IIII
2	IIII	IIII	IIII
3	IIII	IIII	IIII
4	IIII	IIII	IIII

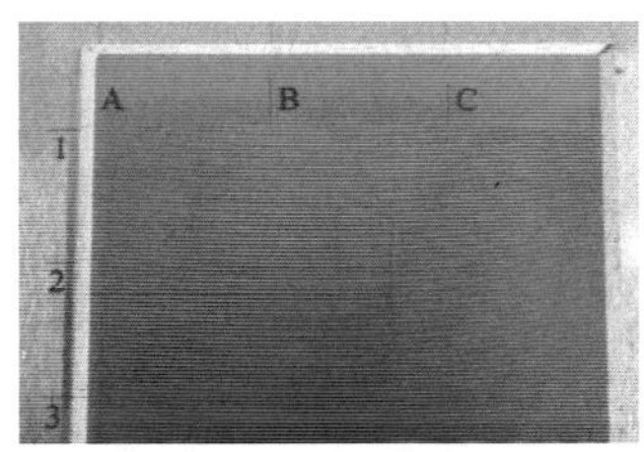

Fig. 2. Segmentation of the standard

In this test, a wide range of roughness parameters relating to all fields were evaluated, including all kinds of profile parameters, roughness parameters and waviness parameters. This article only describes those which are most often used in the automotive industry and aircraft engineering. Their values were calculated using the robust Gaussian filter which leads to the most faithful representation of the acquired profile. More information on filters can be found in articles 13 and 11.

4. Processing of results

The test machine was set up according to information obtained from preliminary tests conducted on individual standards. The set-up parameters were in agreement with the area indicated in red in Table 1. A detailed list is given in Table 2.

	Rsm [µm]	Lt [mm]	Lc [mm]	r_{tip} [µm]
Ra 0.5 µm/Rz	0.1301	4.8	0.8	2 or 5
Ra 1 µm/Rz	0.254	4.8	0.8	2 or 5
Ra 3.2 µm/Rz	0.408	4.8	0.8	2 or 5

Table 2. Surface roughness tester set-up parameters

Results were evaluated in several steps:

- In the first step, the parameter values were averaged across the entire standard and for the particular stylus tip. These values are plotted in graphs here.

- In the second step, the percent difference between values of selected parameters were calculated. These were values measured with stylus tips of the same size but different wear levels. The values measured with the new stylus tip were taken as the reference values.
- In the third step, the percent difference between stylus tips with the same wear level but different tip radii was calculated. The reference value was the value measured with the stylus tip of 2 μm size.

4.1 Ra 0.5 standard

The roughness value Ra 0.5 μm of this standard is the lowest roughness value used in this test. The surface of the standard is strictly periodic. Prior to measurement, the machine is set up on the basis of the roughness parameter Rsm ("Mean value of the profile element widths within a sampling length." [4]). During preliminary testing, the value Rsm = 0.1301 μm was very close to the lower limit of the marked area in Table 1. Nevertheless, it still met the condition of staying within the interval in which either of the 2 μm and 5 μm styli can be used.

4.1.1 Profile parameters

Since profile parameter results are not distorted by filters, they are good for exploring the effects of mechanical characteristics in the roughness measurement process. The following parameters were chosen:

- Pt – "Sum of the height of the largest profile peak height and the largest profile valley depth within the sampling length." [4]

	2 μm	5 μm
difference in %	21.9752347	18.43596647

Table 3. Percent difference between the new and the old stylus tip

	New	Old
difference in %	-3.916264416	-4.973875115

Table 4. Percent difference between the 2 μm and 5 μm stylus tips

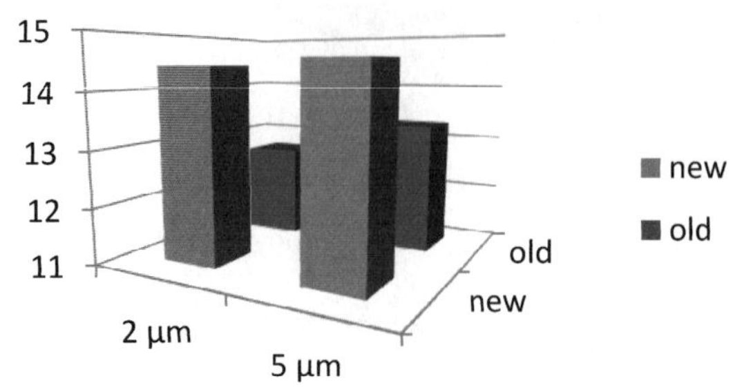

Graph 1. Graphic evaluation of Pt parameter

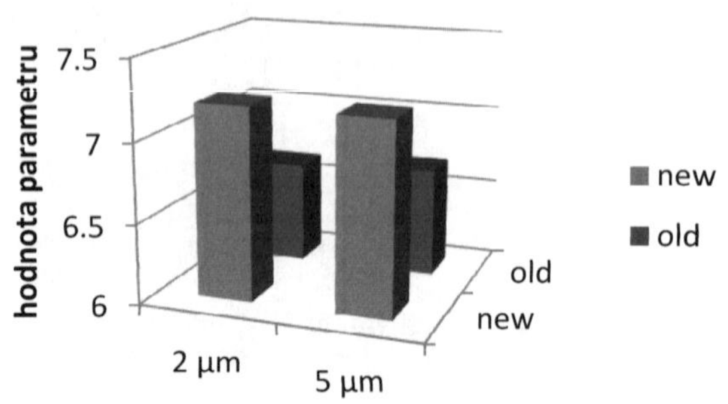

Graph 2. Graphic evaluation of Pa parameter

- Pa - "Arithmetic mean of absolute ordinate values within a sampling length." [4]

	2 μm	5 μm
difference in %	17.846048905	17.242688875

Table 5. Percent difference between the new and the old stylus tip

	New	Old
difference in %	0.046245448	-0.608182275

Table 6. Percent difference between the 2 μm and 5 μm stylus tips

4.1.2 Roughness parameters

Unlike profile parameters, roughness parameters are subject to distortion by the filter used for their calculation. For this reason, greater emphasis was placed on profile parameters during the processing of results. Despite that, the article reports the effects on two most common roughness parameters.

- Ra - "Arithmetic mean of absolute ordinate values within a sampling length." [4]

	2 µm	5 µm
difference in %	5.262520458	4.940866367

Table 7. Percent difference between the new and the old stylus tip

	New	Old
difference in %	-0.122749591	-0.345259897

Table 8. Percent difference between the 2 µm and 5 µm stylus tips

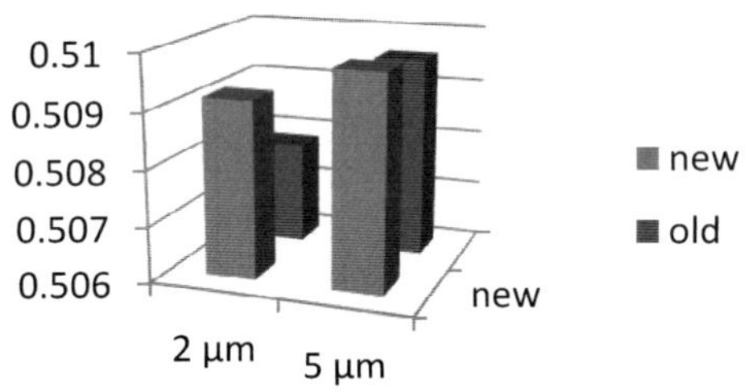

Graph 3. Graphic evaluation of Ra parameter

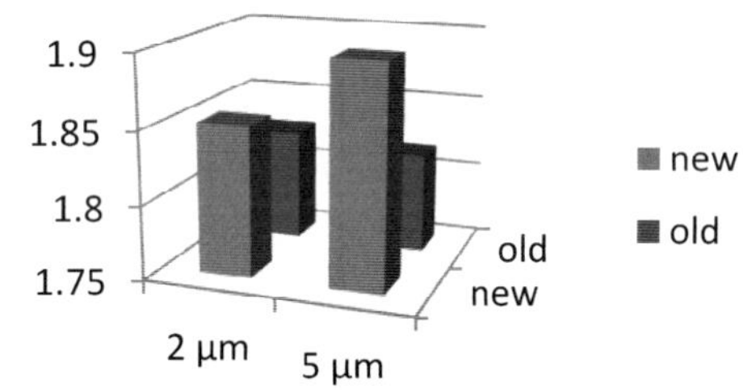

Graph 4. Graphic evaluation of Rz parameter

- Rz - "Sum of the height of the largest profile peak height and the largest profile valley depth within a sampling length." [4]

	2 µm	5 µm
difference in %	5.338913264	5.658038097

Table 9. Percent difference between the new and the old stylus tip

	New	Old
difference in %	-2.564967938	-1.19364718

Table 10. Percent difference between the 2 µm and 5 µm stylus tips

4.1.3 Interim summary

The differences between values of surface roughness parameters (i.e. R parameters) measured by 2 µm and 5 µm styli are neither large nor significant. With all the parameters, including those not listed here, the percent differences were under 1%. It is not so when one evaluates surface profile parameters (i.e. P parameters). Here, the differences become more notable. Ordinarily, they reach 5% but there are exceptions, such as with the Pa parameter in this article.

The differences between values measured by styli of different sizes are less than the differences between values from old and new styli. Although manufacturers of the equipment claim the stylus tips do not wear, some differences were expected even before starting this experiment. However, the differences found were larger than expected. With profile parameters, they reach as much as 20%. With roughness parameters, they are lower. Yet, the resulting difference of around 5% is appreciable, given the importance of each micrometre.

4.2 Ra 1 standard

The roughness of this standard represents a value which is ordinarily prescribed and used in practice. This standard, too, has an exclusively periodic surface. The value of the Rsm parameter which governed this test was in the middle of the interval required for this test. The value was Rsm = 0.254 μm, which was ideal for this test.

4.2.1 Profile parameters

- Pt – "Sum of the height of the largest profile peak height and the largest profile valley depth within the sampling length." [12]

	2 μm	5 μm
difference in %	2.429639298	3.273100033

Table 1. Percent difference between the new and the old stylus tip

	New	Old
difference in %	-2.71021078	-2.64078042

Table 2. Percent difference between the 2 μm and 5 μm stylus tips

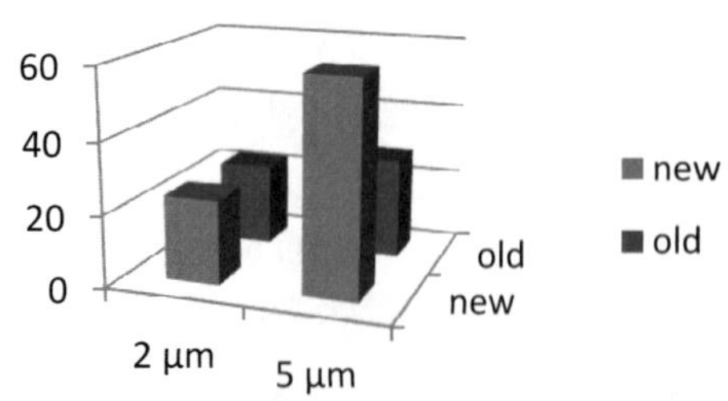

Graph 5. Graphic evaluation of Pt parameter

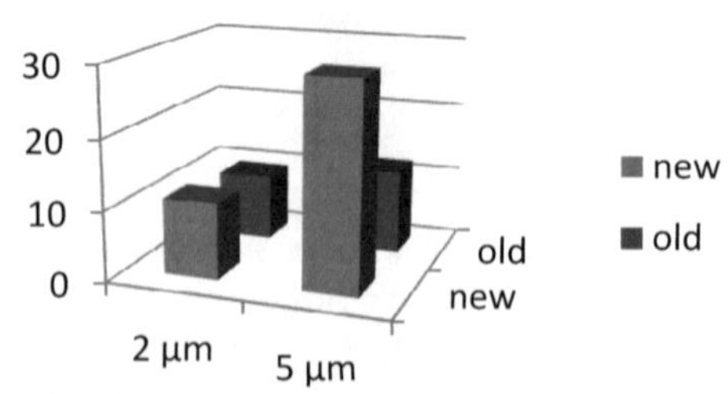

Graph 6. Graphic evaluation of Pa parameter

- Pa - "Arithmetic mean of absolute ordinate values within a sampling length." [12]

	2 μm	5 μm
difference in %	0.82355674	2.22645922

Table 3. Percent difference between the new and the old stylus tip

	New	Old
difference in %	-1.89526753	-1.80516959

Table 4. Percent difference between the 2 μm and 5 μm stylus tips

4.2.2 Roughness parameters

The following parameters were chosen:

- Ra - "Arithmetic mean of absolute ordinate values within a sampling length." [12]

	2 μm	5 μm
difference in %	1.414392181	0.42639593

Table 5. Percent difference between the new and the old stylus tip

	New	Old
difference in %	-0.285074323	-1.290102123

Table 6. Percent difference between the 2 μm and 5 μm stylus tips

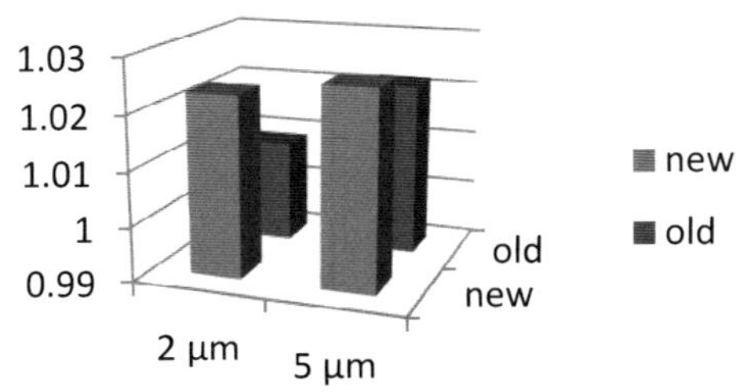

Graph 7. Graphic evaluation of Ra parameter

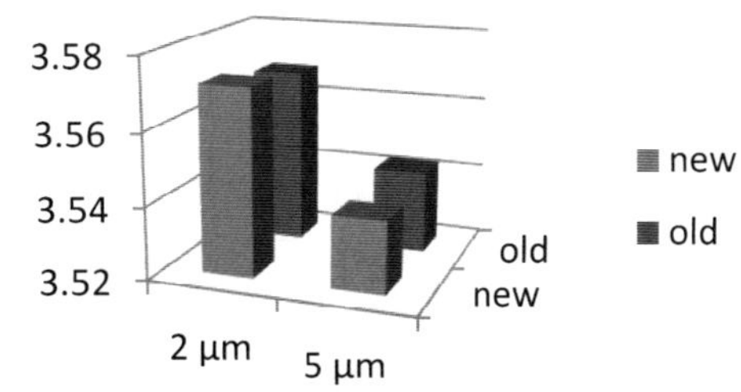

Graph 8. Graphic evaluation of Rz parameter

- Rz - "Sum of the height of the largest profile peak height and the largest profile valley depth within a sampling length." [12]

	2 μm	5 μm
difference in %	0.084792627	-0.082396563

Table 7. Percent difference between the new and the old stylus tip

	New	Old
difference in %	0.886659278	0.72081186

Table 8. Percent difference between the 2 μm and 5 μm stylus tips

4.2.3 Interim summary

The values collected with stylus tips of different sizes were compared. The differences in surface roughness readings (R parameters) from stylus tips of 2 and 5 μm size were negligible. With all the parameters, including those not listed here, the percent differences were under 1%. The differences in profile parameters (i.e. P parameters) are smaller for the Ra 1 μm standard than for the Ra 0.5 μm one. Here, the typical difference is 2-3%.

In this case, the difference between the old and new stylus tips disappears, somewhat unexpectedly after the findings from the standard with a lower value. The differences found here are several tenths of percent which represents rather negligible distortion of measurement readings.

4.3 Ra 3.2 standard

This standard had the highest roughness value in this test. Its surface was periodic, as in the previous cases. Its Rsm value was outside the desired interval by a small margin. Nevertheless, the findings were used for showing how the need for correct stylus tip identification rises with increasing roughness value. The parameter value was Rsm = 0.408 μm.

4.3.1 Profile parameters

- Pt – "Sum of the height of the largest profile peak height and the largest profile valley depth within the sampling length." [12]

	2 μm	5 μm
difference in %	32.931900778	31.908741288

Table 9. Percent difference between the new and the old stylus tip

	New	Old
difference in %	0.666839942	-0.380194729

Table 10. Percent difference between the 2 μm and 5 μm stylus tips

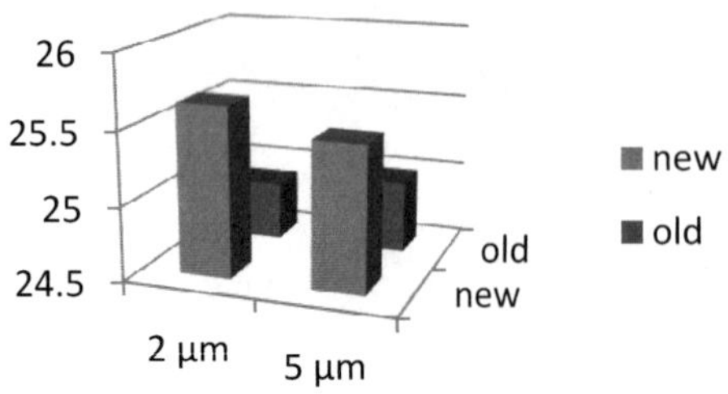

Graph 9. Graphic evaluation of Pt parameter

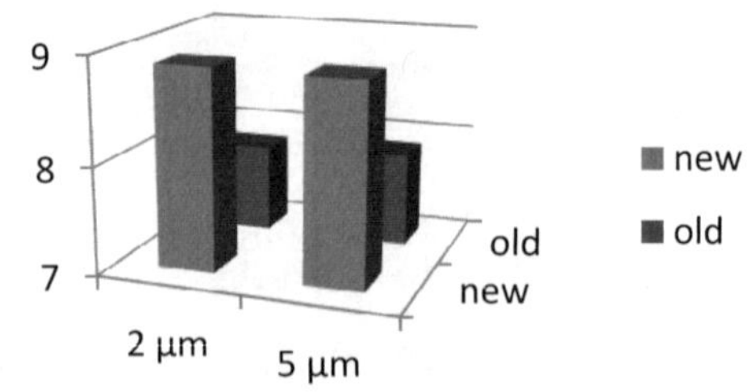

Graph 10. Graphic evaluation of Pa parameter

- Pa - "Arithmetic mean of absolute ordinate values within a sampling length." [12]

	2 μm	5 μm
difference in %	31.90680923	31.18939696

Table 11. Percent difference between the new and the old stylus tip

	New	Old
difference in %	0.161789533	-0.651271682

Table 12. Percent difference between the 2 μm and 5 μm stylus tips

4.3.2 Roughness parameters

The following parameters were chosen:

- Ra - "Arithmetic mean of absolute ordinate values within a sampling length." [12]

	2 μm	5 μm
difference in %	31.057187017	30.643301276

Table 13. Percent difference between the new and the old stylus tip

	New	Old
difference in %	0.225399279	-0.191965897

Table 14. Percent difference between the 2 μm and 5 μm stylus tips

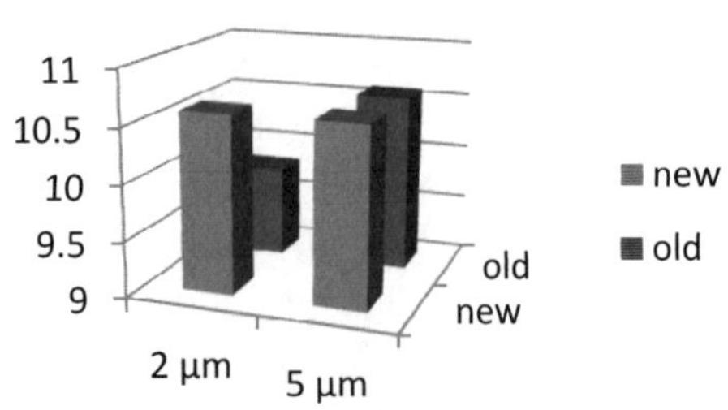

Graph 11. Graphic evaluation of Ra parameter

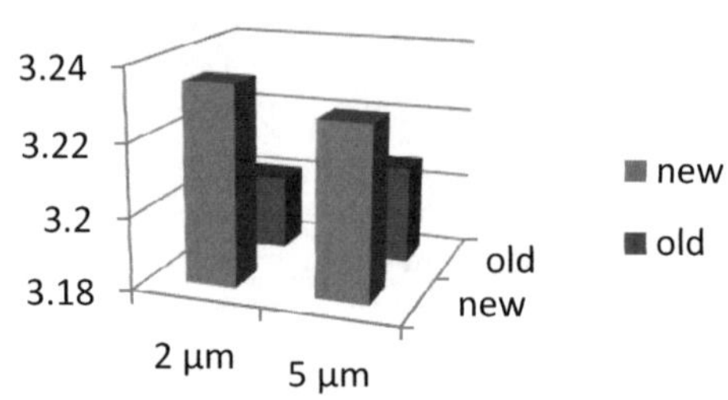

Graph 12. Graphic evaluation of Rz parameter

- Rz - "Sum of the height of the largest profile peak height and the largest profile valley depth within a sampling length." [4]

	2 µm	5 µm
difference in %	7.275296338	3.140746414

Table 15. Percent difference between the new and the old stylus tip

	New	Old
difference in %	-0.094356313	-7.795952047

Table 16. Percent difference between the 2 µm and 5 µm stylus tips

4.3.3 Interim summary

The graphs clearly show that values obtained with both styli are nearly identical. This holds for both roughness parameters (R parameters) and profile parameters (P parameters). With all the parameters, the percent differences were under 1%.

However, those between old and new styli were rising sharply. In this test, they even exceeded 30%. This is a value which is unthinkable in precision measurement.

5. Conclusion

Today's demands on high-speed components are increasing, giving rise to a need for measuring geometric parameters and deviations and even surface integrity parameters in such components. This is one of the reasons why measurement and evaluation of surface texture have seen major qualitative advances in recent years. Leading producers of measuring instruments (Hommel, Carl Zeiss, and others) have responded actively to new requirements. They did so despite the fact that surface texture is frequently measured and evaluated by means of single-purpose measuring machines – and developed new machines, equipment, methods, standards and guidelines.

The purpose of this article was to map at least a small part of this field. It describes an analysis of the effects that the size and shape of a stylus tip has on the readings that characterize surface texture elements. The results of this research will be further used and implemented in the design of methods for the selection of software filters for roughness measurement.

The tests were carried out using the Hommel Etamic T8000 machine housed at the Regional Technological Institute affiliated with the University of West Bohemia. Styli with two tip radii (2 and 5 µm) with an apex angle of 90° were employed. The styli tested were part of new and used probe arms. Test pieces were roughness standards of nominal values of Ra = 0.5; 1; 3.2 µm. Surface texture parameters were calculated with a robust Gaussian filter. Each standard was divided into 12 fields of 10×6 mm size. Four primary profiles were measured in each field. These measurement runs were repeated twice. This test is planned to be repeated in the future on higher and lower-roughness surfaces.

The data were evaluated in several steps. In the first step, the parameter values were averaged across the entire standard and for the particular stylus tip. These values are plotted in graphs here. In the second step, the percent differences between values of selected parameters were calculated. These were values measured with stylus tips of the same size but different wear levels. The values measured with the new stylus tip were taken as the reference values. In the third step, the percent difference between stylus tips with the same wear level but different tip radii was calculated. The reference value was the value measured with the stylus tip of 2 µm size.

Results that characterize the effect of the stylus tip radius can be divided into two groups. The first group comprises the profile parameters (P parameters). Their calculation is not affected by any filter. They are an effective indicator of the impact of the stylus tip radius. With all three standards, the differences between these parameters were under 3%. The second group consists of roughness parameters (R parameters). These are demanded by customers much more often than those of the former group. To calculate them, one has to use mathematical formulae to separate (filter) them from the primary profile. It is a source of a certain percent error. It also distorts any errors introduced by an inadequate choice of a stylus tip radius. Here, the percent difference was under 1%. This is a very favourable finding but caution should be exercised, considering the filter used. With the 2RC filter, these values in the second group reach 5-8%.

Nevertheless, these values, too, are negligible when compared to the differences between data from used and new styli. Some difference was expected there, despite manufacturers' claims that styli do not suffer any wear. However, most of the differences found were larger than expected. With profile parameters, they reached as much as 20%. With roughness parameters, the value was less extreme. Still, the final difference of about 5% is appreciable in the field where every micrometre matters.

6. Acknowledgments

The article has been prepared in the project LO1502 'Development of the Regional Technological Institute' under the auspices of the National Sustainability Programme I of the Ministry of Education of the Czech Republic aimed to support research, experimental development and innovation.

7. References

[1] ISO/TS 16610-1 Technical specification ISO/TS 16610. Geometrical product specifications (GPS) - Filtration, 2015.

[2] ČSN EN ISO 16610-20. Geometrical product specifications (GPS) - Filtration - Part 20: Linear profile filters: Basic concepts. Brussels: ÚNMZ, 2015.

[3] ČSN ISO/TS 16610-21 Geometrical product specifications (GPS) - Filtration - Part 21: Linear profile filters: Gauss filters. Brussels: ÚNMZ, May 2012

[4] ČSN EN ISO 16610-40. Geometrical product specification (GPS) – Filtration – Part 40: Morphological profile filter: Basic concepts, 2016

[5] ČSN EN ISO 4287. Geometrical product specification (GPS) - Surface structure: Profile method - Terms, definitions and surface texture parameters - Part 1; Brussels: CEN, 1999.

[6] http://www.hommel-etamic.cz/cz/technicke-informace/drsnost-povrchu-dle-din-en-iso/ [online]. [2016-08-27]

[7] http://www.techno-mat.cz/data/katedry/kom/KOM_MM_PR_10_CZE_Karasek _Geometricke_vlastnosti_povrchu.pdf [online]. [cit. 2015-02-01].

[8] https://www.olympus-ims.com/en/knowledge/metrology/roughness/[online]. [cit. 2015-09-01].

[9] https://www.hommel-etamic.cz/files/2009-13_en_roughness_poster.pdf[online]. [cit. 2015-09-01].

[10] Melichar M., Kutlwašer J.,Kubátová D.: (2016). Effect of Sweat Aggressiveness on Parameters of Surface Integrity. , 0536. DOI: 10.2507/26th.daaam.proceedings.073. Available at: http://www.daaam.info/Downloads/Pdfs/proceedings/proceedings_2015/073.pdf

[11] E.S. Gadelmawla, M.M. Koura, T.M.A. Maksoud, I.M. Elewa, H.H. Soliman,: (2002) Roughness parameters, In Journal of Materials Processing Technology, Volume 123, Issue 1, , Pages 133-145, ISSN 0924-0136, https://doi.org/10.1016/S0924-0136(02)00060-2. Dostupné také z:http://www.sciencedirect.com/science/article/pii/S0924013602000602

[12] E.Clayton Teague, Fredric E. Scire, Saul M. Baker, Stephen W. Jensen, (1982)Three-dimensional stylus profilometry, In Wear, Volume 83, Issue 1, , Pages 1-12, ISSN 0043-1648, https://doi.org/10.1016/0043-1648(82)90335-0. Available at:http://www.sciencedirect.com/science/article/pii/0043164882903350

[13] M. Shah Mohammadi, M. Ghani, M. Komeili, B. Crawford, A.S. Milani,(2017) The effect of manufacturing parameters on the surface roughness of glass fibre reinforced polymer moulds, In Composites Part B: Engineering, Volume 125, Pages 39-48, ISSN 1359-8368, https://doi.org/10.1016/j.compositesb.2017.05.028. Available at: http://www.sciencedirect.com/science/article/pii/S1359836816317887)

DOI: 10.2507/28th.daaam.proceedings.065

ACCURACY OF HOLES CREATED BY 3D PRINTING (DMLS)

Daňa Milan, Zetková Ivana, Hanzl Pavel, Hronek Ondřej

This Publication has to be referred as: Dana, M[ilan]; Zetkova, I[vana]; Hanzl, P[avel] & Hronek, O[ndrej] (2017). Accuracy of Holes Created by 3D Printing (DMLS), Proceedings of the 28th DAAAM International Symposium, pp.0467-0473, B. Katalinic (Ed.), Published by DAAAM International, ISBN 978-3-902734-11-2, ISSN 1726-9679, Vienna, Austria

DOI: 10.2507/28th.daaam.proceedings.065

Abstract

Additive manufacturing (AM) is a very important topic nowadays, therefore this paper is focused on 3D printing of metal. AM for metal covers many methods that are divided into two basic principles: powder bed technologies and powder deposition technologies. This paper is limited to Direct Metal Laser Sintering (DMLS) that works on the principle of a powder bed. This method is based on sintering metal powder in thin layers, layer by layer. It is able to print different kinds of metal, for example stainless steel, maraging steel, Inconel, aluminium, titanium, etc. An EOS M290 machine with maraging steel MS1 powder was used for the experimental part of this article. Almost every mechanical component contains at least one or several holes, therefore it is important to describe how accurately 3D printing can create these holes. The effects of different holes positions on accuracy and geometry are compared. The influence of the size of the hole on the geometry change at the bottom of an arc is investigated. The goal of the experiment is to find the size of a hole which can be printed without support structures and which does not show signs of degradation of its geometry.

Keywords: 3D print; Additive manufacturing; DMLS; maraging steel MS1; Holes

1. Introduction

Additive manufacturing (AM) is a very extensive area with many methods and several basic principles. One of the most effective AM techniques especially for metal is Direct Metal Laser Sintering (DMLS). This principle is based on the application of a thin layer of powder which is subsequently sintered using a laser beam with a diameter of approximately 80µm. The parts are built layer-by-layer. The printed parts are produced from very fine metal powder with spherical grain shape. This technology is suitable for producing components with very complex shapes and for parts that are impossible to produce by conventional technologies. For example, mould parts with conformal cooling, parts with internal cavities, etc.

Many studies have been carried out on Direct Metal Laser Sintering, but most of them are focused on changing the process parameters. This paper is focused on another of area of research in 3D metal printing. This area deals with the geometric limits on this technology from the point of view of internal holes. [10]

Reference [12] is focused on 3D printing, and especially on the geometrical limits of this method. The author has conducted several experiments regarding the printing of thin walls, overhangs and holes and states: '*The holes up to diameter 7 mm can be printed without support structures while maintaining relatively good geometry and good accuracy.*

These holes are built safely but will have an average of 0.5mm of sag at the top of the hole, which is uncontrollable even with supports, and prevents the overall hole from being perfectly circular.' [12] Reference [7] gives a comprehensive set of information about 3D printing rules. The author claims that holes up to a diameter of 10mm can be printed without support structures. Reference [6] is also focuses on 3D metal printing and especially on printing several basic features such as thin walls, narrow gaps and especially holes. The author measures the accuracy of holes using a microscope. The range of holes was chosen from 0.1 mm to 6 mm. Holes smaller than 0.45 mm were not printed. From the results, it can be stated that the printed holes are smaller than the model.

Holes with a horizontal axis are printed without problems and with very good geometry. The holes in this position can be printed without limitations to their size. However, printing holes with vertical axes has several limitations. The vast majority of publications deal with the geometry of printed holes, especially the poor vaulting of holes. This claim will be experimentally verified. The geometry of holes is not good either at the bottom of hole. This problem will be described further in this article.

2. Material and experimental devices

2.1 Material MS1

The material used for additive manufacturing is maraging steel with the commercial marking MS1. The conventional European standard designation is 1.2709 and, according to DIN, is X3NiCoMoTi 18-9-5. The material composition is shown in Table 1. [1],[2],[3],[4]

Elements	**Fe**	**Ni**	**Co**	**Mo**	**Ti**	**Al**	**Cr, Cu**	**Mn, Si**	**C**	**S, P**
in %	balance	17-19	8.5-9.5	4.5-5.2	0.6-0.8	0.05-0.15	≤0.5	≤0.1	≤0.03	≤0.01

Table 1. Material composition [2]

This kind of steel is characterized by having very good mechanical properties and is suitable for welding, machining, polishing and coating. It is easily heat-treatable. The steel is supplied in powder form. The powder particles have different diameters, which leads to a better volume fraction of the metal. The diameters of particles range from 10 to 40 micrometres. [2],[3],[4]

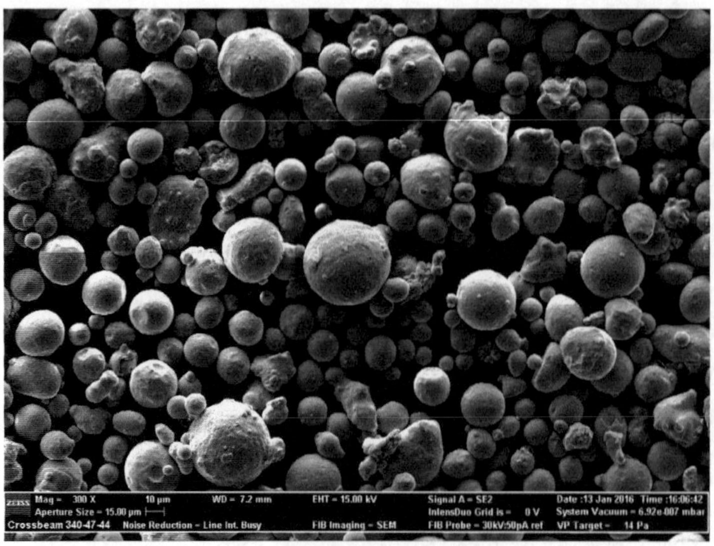

Fig. 1. The powder particles

2.2 Experimental devices

The EOS M290 was used for printing the experimental samples with holes. This printer is a very sophisticated device that enables fast, flexible and cost-effective production from various metals, for example titanium, aluminium, stainless steel, maraging steel, etc. The building volume is 250 x 250 x 325 mm. A 400-watt fibre laser provides an exceptionally high beam quality combined with stable performance. The scanning speed can be up to 7m/s. The experimental print started with the process parameters as recommended by the manufacturer. Individual parameters such as laser power, scan speed and so on are included under these designations. For example, the parameters for printing the part are: laser power 285W, Laser speed 960mm/s, Hatch 0.11mm, etc. The skin of the part has different parameters to the core of the part. The layer thickness was 40 micrometres. [2],[8]

An IFM G4 optical-scanning microscope from Alicona was used to measure the samples. This device allows scanning of the holes and high resolution measurements. The main advantage of this microscope is its high scanning speed and simplicity of measuring. [11]

2.3 Experimental setting

The samples were printed using support structures. The height of the support structures was 5mm to allow removal of the samples from the building platform using a bandsaw. The structure types 'Volume' and 'Block' were used together. The support structure of the block was set as follows: Hatching 0.65 x 0.65mm, Hatching teeth – Height 1.0mm, Top length 0.35mm, Base length 0.8mm and Base interval 0.15mm, Fragmentation 5.2 x 5.2mm and separation width 0.65mm.

The cones were situated in each corner, another two cones were situated in the middle of the part and the last two cones were situated in the middle, between them on both sides The support structure of the Cone was set as follows: Contact to part (r1) 1mm, Contact to platform (r2) 2mm. The cones ensure better strength of the support structures and also prevent breaking off of the part from the support structure.

The test samples were oriented in two positions relative to the recoater blade. The first orientation was that the axis of the holes was parallel to the recoater blade. The second orientation was with the axis of the holes perpendicular to the blade.

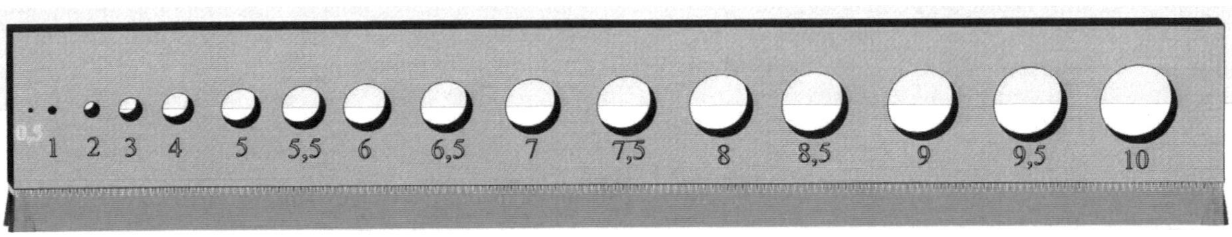

Fig. 2. Test sample

The test samples were printed, and heat treatment was not used after printing. Subsequently, the samples were cut off from the building platform to measure the printed holes on the IFM G4 optical scanning device. Not all the printed holes were scanned, but only representatives were selected and scanned to save time. Selected diameters were: 0.5, 1, 2, 3, 5, 6.5, 7 and 10 mm. These representatives were scanned from both sides. Fig. 3 shows some of the scanned holes.

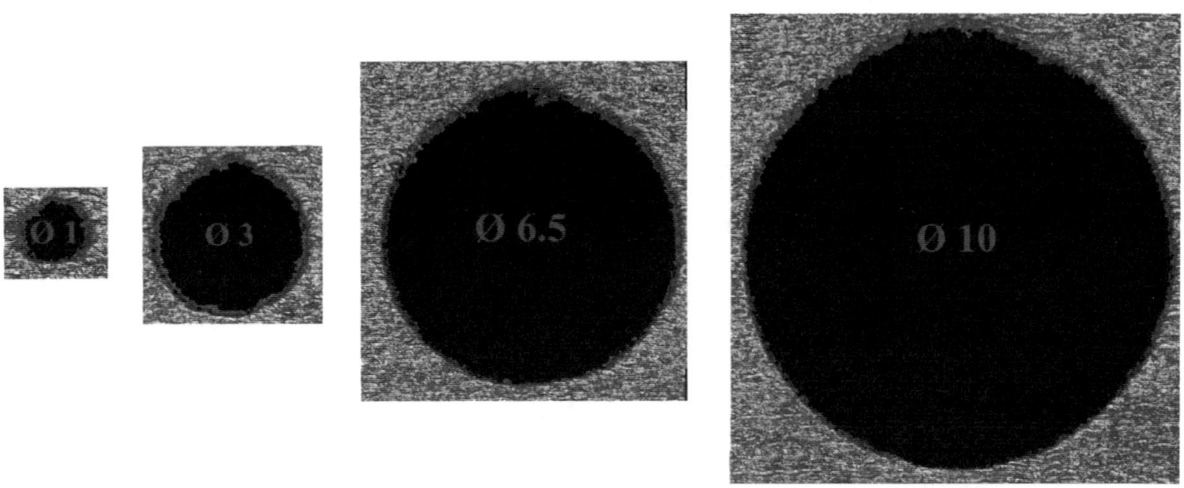

Fig. 3. Printed holes

The scanned holes were measured at 0.2 mm under the rake plane and the cut views are created in this position. These holes are evaluated using inscribed circles. The shape of the printed holes was not exactly circular, mainly for the small holes. This trend was reflected in holes with diameters 0.5 mm, 1 mm and 2 mm. This can be seen in Fig. 4.

Fig. 4. The holes from 0.5 to 3 mm

Holes larger than 3mm showed better circularity, but two areas were identified here where there is a problem with the circular shape. The first area is at the top of the hole. This problem has been described in many studies. A self-supporting angle is more than 45 angles upward from a platform. *'When the building process reaches to the top of the circular cross section the angle of the building direction is below the mentioned critical angle.'* [5]

This means that overhangs are created here, resulting in bad geometry and bad surface roughness. Sagging, burning or cavities are created in this area. Taking this into account, the best way to measure these holes is using inscribed circles. The diameter of the inscribed circle has a very good corresponding value. All measured holes have smaller diameters than the model. The measured values of the inscribed holes are listed in Table 2. [5],[9]

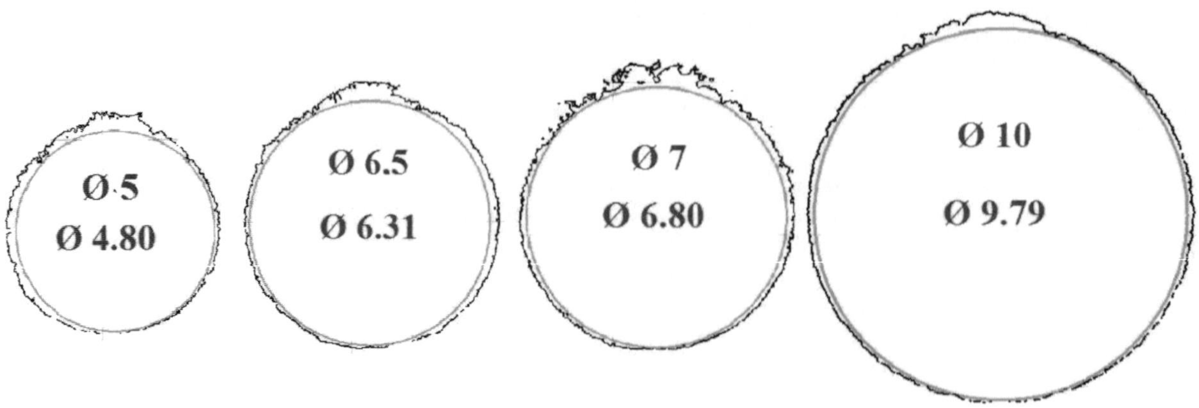

Fig. 5. The holes from 5 to 10 mm

Diameter	Sample oriented in Y		Sample oriented in X	
	Right side	Left side	Right side	Left side
0.5	0.458	0.37	0.398	0.358
1	0.932	0.862	0.926	0.81
2	1.858	1.848	1.858	1.802
3	2.862	2.81	2.916	2.826
5	4.778	4.694	4.802	4.842
6.5	6.306	6.232	6.308	6.314
7	6.696	6.79	6.748	6.824
10	9.792	9.758	9.828	9.72

Table 2. Measured values of inscribed circles

A 'staircase effect' results from the principle of 3D print technology that the parts are built up layer by layer. This defect is most evident when printing cylindrical bodies with a horizontal axis. For this reason, it is not possible to create a part with a smooth continuous surface that has been created by a very complex curve. The 'staircase effect' is shown in Fig. 6.

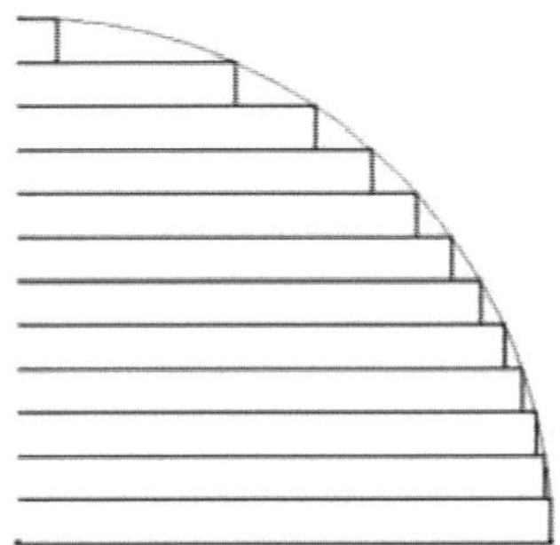

Fig. 6. 'Staircase effect' [13]

This effect causes poor geometry of a hole, even at two locations in one hole. The first location is situated at the beginning of the hole when it is printed. A flat surface (groove) is created in this place and stairs are formed at the ends of this flat surface. The same situation is at the top of the hole, but the experiment is not targeted at this area. It is possible to see the laser paths (hatching) on this flat surface. The width of this flat surface was measured and evaluated using a scanning microscope. The flat surface widths were measured in three locations and the average values are given in Table 3.

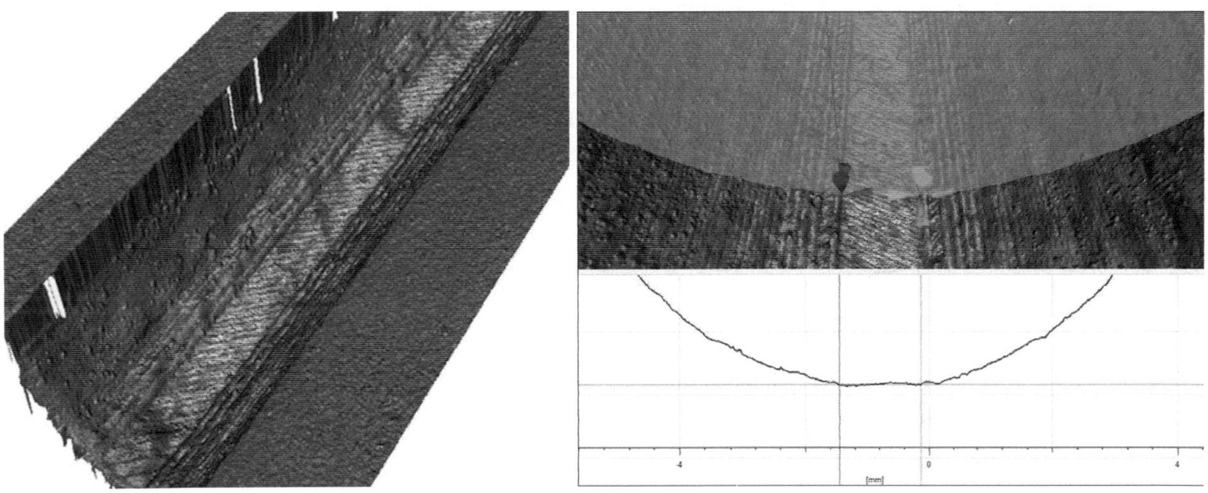

Fig. 7. Flat surface at the bottom of the hole

Hole diameter [mm]	Sample oriented in Y [μm]	Sample oriented in X [μm]
1	362	386
2	345	337
3	642	641
5	734	721
6.5	737	728
7	824	830
10	1335	1330

Table 3. Groove widths at bottom of holes

The values of the flat surface are dependent on the layer thickness. If the layer thickness is lower, then the width of the flat surface will be thinner. There is no other way to reduce the 'staircase effect'. However, the layer thickness cannot be changed during the printing.

The values in the table correspond approximately to the size of individual layers, but the holes with diameters of 2 mm and 6.5 mm do not correspond. This is due to the different start positions depending on the printed layer. If the part is sliced into individual layers with a height 0.04 mm, then a situation may occur as in Fig. 8. The start position of the circle may be between slices. This affects the width of the flat surface and explains the different values of the widths of the flat surfaces for holes with diameters of 2 mm and 6.5 mm.

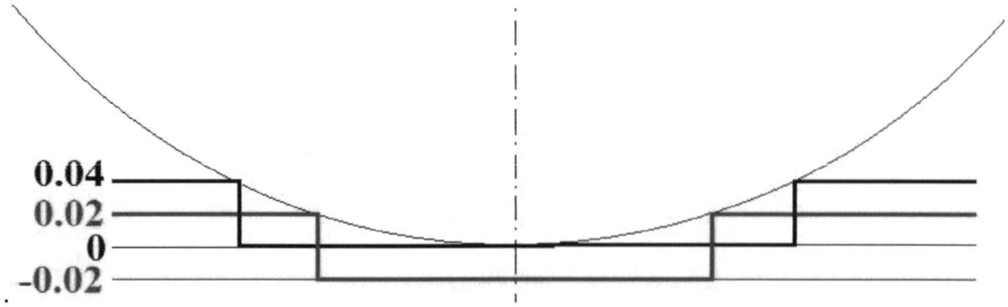

Fig. 8. Offset of start position

3. Conclusion

The main goal of the research presented in this paper is to understand the possibilities and limitations of the laser additive manufacturing process in terms of printing internal holes. The experiment confirms that the holes from 0.5 mm up to a diameter of 10 mm can be printed without support structures. However, the geometry of the holes is not so perfect; for example, a flat surface is created at the bottom of the holes and in some cases sagging, burning or cavities are created on the upper side. The holes were evaluated using inscribed circles that were measured by an optical-scanning microscope. All printed holes are smaller in diameter than the model. The maximum deviation from the basic size of the hole and the inscribed circle was -0.306 mm. The deviation was smaller than 0.2 mm in the majority of cases. The influence of changing the position of the test sample does not show any significant differences.

A flat surface at the bottom of the holes was detected. This problem is generated based on the 'staircase effect'. The 'staircase effect' results from the principle of 3D print technology that the parts are built up layer by layer. The values of the flat surface are dependent on the thickness of the layer and the diameter of the hole. Almost all flat surfaces correspond to the size of the hole and the layer thickness, but the holes with diameters 2 mm and 6.5 mm show different values. These differences were caused by the start position of the hole in relation to the printed layer, because the hole may be started between slices. This situation is clearly illustrated in Fig. 8. The geometry of holes is slightly changed during the slicing. It is not possible to create a part with a smooth continuous surface that has been created by a very complex curve.

These holes will be evaluated using differential analysis and further research will be focused on the upper side of the hole. The surface quality and surface roughness will be assessed in this area. It would be appropriate to choose holes with smaller steps between the diameters of the holes for further research.

4. Acknowledgments

This contribution has been prepared under the project LO1502 'Development of the Regional Technological Institute' under the auspices of the National Sustainability Programme I of the Ministry of Education of the Czech Republic aimed at supporting research, experimental development and innovation.

5. References

[1] Rubesova, K; Jenicek, S; Kana, J& Zetkova, I; (2016). Microstructure of MS1 Maraging Steel in 3D-Printed Products After Semi-Solid Processing, Proceedings of the 27th DAAAM International Symposium, pp.0467-0472, B. Katalinic (Ed.), Published by DAAAM International, ISBN 978-3-902734-08-2, ISSN 1726-9679, Vienna, Austria

[2] EOS MaragingSteel MS1. EOS. [online]. [cit. 2017-08-10]. Available from: http://ip-saas-eos-cms.s3.amazonaws.com/public/1af123af9a636e61/042696652ecc69142c8518dc772dc113/EOS_MaragingSteel_MS1_en.pdf

[3] Kucerova, L. Zetkova, I; (2016). Metallography of 3D Printed 1.2709 Tool Steel. In MANUFACTURING TECHNOLOGY. February 2016, Vol. 16, No.1. s. 140 -144. ISSN 1213–2489

[4] Dana, M. Zetkova, I. Hanzl, P; (2016). Prining of Thin Walls using DMLS. In MANUFACTURING TECHNOLOGY. October, Vol. 16, No.4. s. 883 -896. ISSN 1213–2489

[5] Additive Metal Manufacturing Inc. [cit. 2017-02-10]. Design for Additive Metal Manufacturing Part I Design for Minimum Impact of DMLS Geometrical Constraints. Canada, [online]. Available from: https://www.additivemet.com/wp-content/uploads/2016/07/design.pdf

[6] Matilainen, V.(2012.) Benchmarking of laser additive manufacturing process. Bachelor's thesis. Finland Available from: http://www.doria.fi/bitstream/handle/10024/84863/BSc%20Thesis%20Matilainen%20fv.pdf? sequence=1

[7] EPMA European Powder Metallurgy Association. Introduction to additive manufacturing technology - A guide for Designers and Engineers. Available from: http://www.epma.com/doc_details/427-introduction-to-additive-manufacturing-technology

[8] Hanzl, P; Zetek, M& Zetkova, I (2016). Cellular Lattice Structure Produced by Selective Laser Melting and its Mechanical Properties, Proceedings of the 26th DAAAM International Symposium, pp.0748-0752, B. Katalinic (Ed.), Published by DAAAM International, ISBN 978-3-902734- 07-5, ISSN 1726-9679, Vienna, Austria

[9] Hanzl, P., Zetek, M., Baksa, T., Kroupa, T. (2015). The influence of processing parameters on the mechanical properties of SLM parts. In Procedia Engineering. Vienna: Elsevier, s. 1405-1413. ISBN: 978-3-901509-99-5 ISSN: 1877-7058

[10] N. C. Balc, P. Berce, R. Pacurar, (2010) Comparison between SLM and SLS in producing complex metal parts, Annals of DAAAM for 2010 & Proceeding of the 21st International DAAAM Symposium, Volume 21, No. 1, ISSN 1726-9679, Vienna, Austria.

[11] Hronek, O., Zetek, M., Baksa, T., Adamek, P. (2016). Influences Of Holders Speed On The Cutting Edge During Drag Finishing. Manufacturing Technology, Vol. 16, No. 5, s. 933-939. ISSN: 1213-2489

[12] Thomas Daniel. The Development of Design Rules for Selective Laser Melting. Ph.D. Thesis National Centre for Product Design & Development Research University of Wales Institute. Available from: https://repository.cardiffmet.ac.uk/dspace/handle/10369/913

[13] Zetkova, I., Zetek, M., Dana, M., Kutlwaser, J. (2016). Critical Parameters of Metal Laser Sintering with its Impact on Geometrical Accuracy of a Component. Manufacturing Technology, Vol. 16, No. 5, s. 1216-1222. ISSN: 1213-2489

DOI: 10.2507/28th.daaam.proceedings.066

ORGANIZING INTERMODULAR COMMUNICATION FOR HETEROGENEOUS MODULAR MOBILE ROBOT

Victor Andreev & Pavel Pletenev

This Publication has to be referred as: Andreev, V[ictor] & Pletenev, P[avel] (2017). Organizing Intermodular Communication for Heterogeneous Modular Mobile Robot, Proceedings of the 28th DAAAM International Symposium, pp.0474-0480, B. Katalinic (Ed.), Published by DAAAM International, ISBN 978-3-902734-11-2, ISSN 1726-9679, Vienna, Austria

DOI: 10.2507/28th.daaam.proceedings.066

Abstract

The paper proposes an approach to intermodular communication method's specification for heterogeneous modular mobile robot based on principles of ROS (Robotic Operation System). One of key features of this specification is accounting for structure of modular robot's control system built on full functionality principle. The full functionality of a mechatronic device is the ability to perform its goal function using only its own facilities for executing instructions from an external control system. Consequently, every module is a fully-functional electronic or mechatronic device with its own control system, including all required components: sensors, computing modules and drives. As modules' computing devices are usually microcontrollers or other embedded computers which are very restricted in computing power, using ROS in every module is complicated and is not efficient. Proposed specification use ZeroMQ library combined with UDP protocol, which made it possible to implement specification for embedded devices. Rationale behind the choice of type and topology of communication networks is presented. Requirements for APIs (application programming interfaces) of modules of different types are formulated.

The results of an experimental study of the developed specification of the method of intermodular information interaction on the operating model of a heterogeneous modular mobile robot are presented.

Keywords: modular robot; heterogeneous robot; control system; local area network; network protocol

1. Introduction

This paper is the part of the research work on creation of a heterogeneous modular mobile robot (MR) for a wide range of purposes. The heterogeneous modular MR is a system of individual functional devices that, when assembled in one of a finite number of configurations, constitute a synergetic system – a mobile robot with executive, tactical and strategic control levels. Each individual device (a module) in such a robot is a mechatronic or an electronic device. Such individual devices should have universal mechanical, electrical and programming interfaces. This ensures the ability of the whole structure to be reconfigurable on site of works directly and on any time of operation. Even better interchangeability and flexibility of modules can be achieved if they are designed to carry out one specific function and to carry it out well i.e. to be fully functional. ***The full functionality*** of a mechatronic device is the ability to perform its goal function using only its own facilities for executing instructions from an external control system [1]. The full functionality of robot modules is achieved by designing each module as ***a full-featured electronic or mechatronic device*** with its own the information-measuring and control system (IMCS) containing all necessary components: sensors, computing devices and actuators. But here lies a contradiction. For example, the transport module carries out robot motion function. If such module will blindly make its way towards destination without considering environment's obstacles, it will eventually bump in some obstacle which was not expected by an external control system. But this does not mean that such a module must be equipped with its own distance sensors. To avoid collision a modular MR should be equipped with one or more distance sensors modules. Sensors modules' and transport modules' control systems are separated, therefore a necessity of synchronization between them becomes apparent. Also further increase in number of functional modules leads to organization of the general distributed control system which unites all control systems of each individual module. In such distributed control system every module should be able to exchange any information (including control commands) with any other module directly. Thus the whole system needs a specific communication method. Such method exists already and it is open-sourced project called Robot Operating System. It is considered state-of-the-art for mobile robotic systems at the moment. But also at the very same moment it is unusable in purposes of modular robotics as it cannot be used on low-power embedded computers directly.

The purpose of the study is to develop a specification of the intermodular communication method, built on principles of ROS (Robot Operating System), but able to run on embedded computers. As a basis, such specification should use the ***full functionality*** principle.

In order to achieve this goal it was necessary to complete the following tasks:

1. Select the type and topology of the communication network.
2. Develop the method of the information interaction of modules.
3. Define the hardware and software requirements for the modules.
4. Develop modules' systems of commands.
5. Develop new network protocols and modify well-known network protocols to implement the inter-module communication application programming interfaces (API).

2. State-of-the-art modular mobile robotics

Let's consider existing modular robots from the point of view of their intermodular communication system correspondence to the principle of full functionality and distributed control system. In 1987, Japanese specialists introduced the CEBOT modular robot [2]. This is the first of the known mobile robots, which is heterogeneous and fully modular, implying the unity of interfaces. Modular robots are divided into homogeneous and heterogeneous.

Homogeneous robots consist of many identical modules. Each module contains all the components necessary for it to function – motors, propellers, batteries, control systems, etc. In fact, a homogeneous modular robot is a superposition of a set of individual identical robots. An example of such robots is the modular robot M-TRAN [3]. Despite the simplicity of the movements carried out by one module, many modules, built into a certain structure, can perform complex coordinated movements. M-TRAN, as well as other homogeneous robots, for example, ATRON [4] and TRANSMOTE [5], have a common feature – all of them, even having united in the general structure, do not form a common full-fledged distributed control system. Each of the modules is independent, but all of the modules are controlled by one module-supervisor via execution level commands. The difficulty for a supervisor to control such modules directly is worth noting as it is necessary to develop additional software that translates the supervisor commands into a sequence of motion commands for individual modules, and, in general, this software runs on an external computer, rather than being embedded in the modules themselves. Also it's worth noting, that modules of all presented homogeneous robots are not full-featured. Modules of one type should be communicating with each other to carry out their primary function to move the robot.

Heterogeneous robots consist of a finite set of modules of different functionality that are connected to each other through unified mechanical and electrical interfaces. Unlike homogeneous, the modules of heterogeneous robots, in general, cannot function normally without each other.

SMART [6] and Thor [7] are heterogeneous robots. Each of these robots is a set of different modules and submodules (separate mechanical components of modules). The Thor robot is the direct successor of the homogeneous ODIN modular robot [8], including intermodular communication – both robots use RS 485 bus. The SMART robot is slightly different – one of its modules, connected via a Bluetooth wireless channel to the supervisor, becomes the master on the CAN bus through which it manages other modules, relaying commands received from the supervisor and the data from the other modules back to the supervisor via Bluetooth. In other words, all modules are driven by one module-supervisor on executive level only. Like homogeneous robots, they do not implement a full-fledged distributed control system, which makes them use a central high-power computer. This also leads to a lack of full functionality of modules of such modular MRs.

The most advanced in the field of intermodular interaction are the previously mentioned CEBOT and a modular robot for space research [9]. For them it is stated that all modules communicate with every other module – the modules have additional horizontal connections, which, for example, allowed the automatic configuration in the CEBOT modules

For each of the listed robots, developers have created special software that allows building the CS intended only for this design, which does not allow the use of third-party modules in the design. This, in turn, makes it very difficult to implement such a modular robot property as **operational reconfiguration**, which is especially important in cases of maximum uncertainty of the conditions of forthcoming works, i.e. in case of extreme situations, accidents, disasters and active counteractions, when it is not possible to determine in advance the composition and configuration of the necessary robotics [10]. Therefore, the task of creating a full-fledged distributed control system for modular robots that is capable of dynamically changing its structure and incorporating third-party modules into the robot design is *topical*.

Development of general APIs for full functional modules will enable realization of operational reconfiguration property of modular mobile robot.

This property can used successfully to develop robots for other planets' and Moon exploration. A review [11] of robotics needed to create a lunar base provides a large selection of robots for a variety of purposes: from robot bulldozers to robot builders and robot repairers. Such a wide range of robots is economically unprofitable to bring to the moon. It is sufficient to create a set of full functional modules and assemble needed robots on the Moon.

The intermodular communication method is being developed on top of functional-modular principle of modular mobile robot construction [12]:

1. Each module is a functionally completed electronic or mechatronic device with its own information measuring and control system (IMCS) the principle of distributed control.
2. IMCS of a robot is a local area network with modules as nodes.
3. Incorporation of any mechatronic device in the network structure of a robot is done using special software – drivers.

Functional completeness is a property of every module to be able to execute on its own the task that is given to it by a module supervisor (some general control system) [13]. Then a distributed control system is a system with mostly independent control processes, distributed between modules' computers, interacting with each other via intermodular communication method.

3. Communication networks: choosing type and topology

Modern modular robots use different serial busses, both homogeneous and heterogeneous, to achieve better modules' miniaturization and minimization of the number of mechanical interfaces' contacts:

- CAN – robots M-TRAN and SMART;
- IrDA (in fact RS-232) – robot ATRON;
- RS-485 – robots Thor and Odin;
- TRANSMOTE robot uses ZigBee interface – a serial bus in a wider meaning.

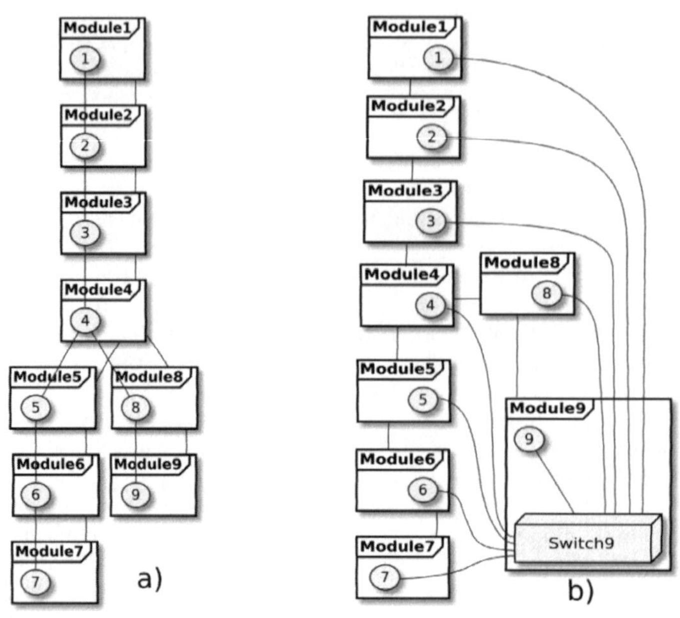

Fig. 1. "Bus" (a) and "Star" (b) topology for Ethernet

From the review above it is clear that communication in known modular robots is most often based on a topology of the "bus" type with local branches (Fig. 1a), rarely – on a topology of a "star" (Fig. 1b). Topology "bus" most fully reflects the structure of the modular robot - each module has several "neighbors" through which it interacts with the rest of the system. The disadvantage of such buses is their low throughput. But in the above constructions, the bandwidth is sufficient, since such robots have, in fact, two ways to function: (1) execute a set of simple rules that the authors placed in the microcontroller ROM in each of the modules, and (2) execute the commands from an external computer, supplied through a module to the bus of which (and, hence, to the common bus of the all connected modules) this computer is connected. This means that the role of the intermodular interaction system is small and, in fact, reduces only to the most rapid and efficient transmission of executive or tactical level commands and the readings of the sensors. Then the information flow in the process of intermodular interaction is small, since each module have either none or just a few of sensors and, in fact, it is necessary to transmit only executive level commands.

Many companies are now using the Ethernet standard. Its expansion for real-time operation EtherCAT has become the new standard among industrial field busses due to the versatility of the Ethernet standard [14]. A comparison of the various existing networks is given in Table 1.

Features	RS-485	CAN	Ethernet	EtherCAT
Real time	Hard(1)	Hard	Soft	Hard
Main topologies	Bus	Bus	Point-to-pont, various(2)	Point-to-pont, bus, various (3)
Top bandwidth (on certain connection length)	10 Mbit/s at 10 m	1 Mbit/s at 40 m	100 Mbit/s, 1 Gbit/s at 100 m	100 Mbit/s at 100 m
Collisions handling	Not defined	CR(4)	CD(4)	Master-Slave(5)
Service data length, bit	80(6)	44 or 64(7)	464 or 336(8)	160
Max data length, bit	2088(9)	64	11536 or 11664(8)	11888
Prevalence of equipment	High	Average	Very high	Low
Equipment cost	From low to average	From average to high	From low to average	High

Table 1. Different networks' properties

Notes for table 1: **(1)** – implementation dependent, standard does not defines this variable; **(2)** –topology can be both bus and star, depending on devices used; **(3)** – can have any topology, but most devices support bus topology; **(4)** - CAN is a synchronous bus with Collision Resolving (CR), (not Collision Detect (CD) as in Ethernet), it transmits messages deterministically; **(5)** – use a special frame format made by master device, which are read from and written to by slave devices, this ensures synchronicity of bus control at all times; **(6)** – for MODBUS implementation; **(7)** – regular and extended frame format; **(8)** –TCP and UDP protocols respectively; **(9)** –MODBUS implementation strictly regulates length of a message.

The principle of functional completeness of the modules eliminates the need to send executive level commands – only upper (tactical and strategic) level commands are transmitted. In this case, the "softer" requirements to the delivery time of messages and the use of interfaces with "soft real time" are acceptable. In our case, we are going to use in the robot's sensor system a large number of different sensors and a multi camera system of technical vision. Therefore, the intermodule communication channel must be broadband. According to the data in table 1, Ethernet is the most high-speed interface, and the wide prevalence of the corresponding equipment ensures its low cost. In the case of EtherCAT, all intermodule communication must occur via the "master" of the bus, which reduces the efficiency of direct inter-module communication. Although Ethernet cannot be considered as a hard real-time protocol, but, as noted above, there is no need.

Ethernet is not a bus by definition – it uses "one to one" connections. One can make it as one by installing switches between nodes enabling "one to many" connections. This design enables both bus and star topologies.

Fig. 1a and Fig. 1b show the topologies "bus" and "star" superimposed on the topology of a modular robot. The choice of topology must be made on the basis of the requirements for the desired frequency of messaging, requirements for assembly and economic indicators, as well as taking into account the limitations of the network functioning of a particular topology. A comparison of these topologies is presented in table 2.

Features	Bus	Star
Centralization	Decentralized	Centralized
Switch count	$\boldsymbol{n}$ switches with $\boldsymbol{s} + \boldsymbol{m_i}$ interfaces	1 switch with $\sum_{i=1}^{n} \boldsymbol{m_i}$ interfaces
Ethernet lines, passing though each module	$\boldsymbol{s} + \boldsymbol{m_i}$ lines from switch to other modules	In the worst case$(\sum_{i=1, i \neq i_c}^{n} \boldsymbol{m_i}) - \boldsymbol{m_k}$ lines, where $\boldsymbol{i_c}$ – module with a switch
Additional restrictions	Max 4 switches on the path of "point-to-point" connection to minimize collisions	None

Table 2. Different topologies' properties.

Variables, used in table 2: $\boldsymbol{n}$ – count of modules in a robot; $\boldsymbol{m_i}$ – count of Ethernet connections per module (in most cases $\boldsymbol{m_i}=1$); $\boldsymbol{s}$ – count of potential module neighbors (in most cases either 2 or 6).

As can be seen from the table, both topologies have drawbacks – The "bus" requires either to use a large number of switches, or to integrate several Ethernet interfaces into each device (similar to EtherCAT), whereas the "star" causes a much larger number of wires to be laid. The latter, in our case, does not have a significant value, since the wired system of the local computer network is located within the small dimensions of the robot design. *Therefore, it is proposed to use the topology "star".*

4. Communication networks: intermodular communication method

One of the most known intermodular communication method is ROS (Robotics Operation System) framework [15]. To use ROS, one needs a computer with a Linux-like operating system (OS), for which there is a port of ROS. Also, a large number of visualization, simulation and debugging tools were developed for it. At the moment ROS has several disadvantages - it was not designed to work directly on microcontrollers and it requires installation of a quite complex and voluminous software. Therefore, it is necessary to develop such a system so that it would be, on one hand, compatible with ROS and, on the other hand, convenient to implement on embedded systems.

Previous approach, proposed in [16], was based in Internet laboratory library (zmq_robot)[17]. It is built on top of ZMQ library [18] and allows creation of both modules' control systems and supervisor client software. This library has the same disadvantages as ROS – it requires installation of complex software, which only works on full-fledged OS, but not on embedded systems.

Intermodular communication method specification was created to solve this problem. It takes into account full functionality property of modules. For communications it is proposed to use ZMQ library instead of zmq_robot. In addition to ZMQ it is proposed to use UDP protocol with broadcast messages. This extension enables the usage on most basic embedded devices. Unlike ROS, each module has one interface common for all modules, through which it provides general information about itself – the name of the module, its description, functional interfaces, etc.

5. Hardware and software requirements for control systems' implementation

Hardware and software of module's control system, in a minimum implementation, mush have a computing device capable of working in the Ethernet through UDP protocol and with computing power sufficient to realize functional purpose of the module. It should have minimal possible cost, sizes and mass. As it was noted earlier, a distributed robot's CS as a whole is formed from the hardware and software of all modules. Due to the distribution of calculations between the microprocessors of the CS modules, the robot has a higher performance compared to conventional control systems that are built on a single computer.

The analysis of microcomputers of various manufacturers was carried out, among which a multicriteria Pareto optimization was carried out. The optimal candidates for hardware are Arduino Uno with Ethernet Shield, Arduino Mega with Ethernet Shield, Taijiuino Due Pro R3, Seeeduino Cloud, 86duino ZERO, 86duino One and Raspberry Pi B Plus.

6. Modules' command systems

Based on the principle of full functionality of the modules, it is possible to distinguish **transport**, **energy**, **information**, **communication**, **technological**, **auxiliary** functions and the **general control** function. Within the developed specification of the intermodule communication system, command systems have been developed for the following functional modules of the modular MR:

1. Transport module – provides movement of the robot (transport function).
2. Power module – provides power supply to electronic and electromechanical components of the robot, and their safe switching on and off (energy function).
3. Sensor system: short-range and long-range sensor modules - modules provide search for obstacles and objects of manipulation at different distances and provide sensory information to other modules (information function).
4. The intellectual control module – provides control of the robot as a whole, synchronizes information interaction of all modules, performs information processing and setting of tasks coming from the external supervisor of the robot, generates and distributes tasks between modules, being thus a supervisor for them, creates and controls The operation of the radio channel (communication function and the function of the general control system).

7. Experimentation

To study the efficiency of the developed method of intermodular communication, a working model of a modular MR was created, consisting of: a wheeled transport module (Arduino Yun running OpenWRT), a power module (Arduino Mega with an Ethernet Shield module) and a network management module (Raspberry Pi B+ running Arch Linux ARM).

Control programs for the IMCS of modules were created on top of the above specification. For the power module, as the least productive part of the system, it was possible to achieve a message exchange rate of up to 500 Hz. The experiments showed the effectiveness of the proposed solutions.

8. Conclusion

A specification of the intermodular communication method, built on principles of ROS, but able to run on embedded computers, was created. This specification is using the full functionality principle as a basis.

These results were achieved:

1. Ethernet with "star" topology was chosen to be communication network for heterogeneous modular robot due to its versatility and wide prevalence of devices implementing it.
2. The intermodular communication method specification was created. It is based on ideas of ROS. The specification is built on top of ZeroMQ TCP messaging library with additional UDP sockets as a mean to enable direct communication with embedded microcontrollers via the Ethernet. The latter cannot be achieved with ROS due to its design.
3. Embedded computers of different models and manufacturers were analyzed and Pareto optimization was carried out among them. Out of 10 selected, 7 computers are, by different criteria, efficient as processing units for modules of a modular robot.
4. In the first approximation, the main function and system of commands for transport, power, sensor and intellectual control modules have been defined.
5. The specification was implemented for 3 embedded devices (Arduino Yun, Arduino Mega, Raspberry Pi B+). Each device was used in a different module respectively – transport, power and supervisor control modules. Control systems for these modules were created using implementations mentioned above with respect to their systems of commands developed earlier.

In the future it is necessary to implement the specification for other architectures of embedded devices (such as ARM or Xtensa). It is necessary to develop systems of commands for other types of modules, such as manipulation module.

9. Acknowledgments

The authors are expressing their sincere gratitude to S. Kuvshinov, Director of International Institute of the New Educational Technologies of Russian State University for the Humanities, for assistance in construction of the experimental equipment.

This work was supported by the Russian Foundation for Basic Research, grants no. 16-07-00811a & 16-07-01264a.

10. References

[1] Andreev, V; Kim, V. & Pletenev, P. (2017). The principle of full functionality – the basis for rapid reconfiguration in heterogeneous modular mobile robots, Proceedings of the 28th DAAAM International Symposium, pp.xxxx-xxxx, B. Katalinic (Ed.), Published by DAAAM International, ISBN 978-3-902734-xx-x, ISSN 1726-9679, Vienna, Austria. DOI: 10.2507/28th.daaam.proceedings.xxx, in press.

[2] Toshio, F.; Tsuyoshi, U.; Yoshio, K. & Fumihito, A. (1992). Concept of cellular robotic system (CEBOT) and basic strategies for its realization. Computers & Electrical Engineering, Volume 18, Issue 1, January 1992, Pages 11-39, ISSN 0045-7906, DOI: 10.1016/0045-7906(92)90029-D.

[3] Kohji, T.; Satoshi, M.; Eiichi, Yo.; Haruhisa, K.; Akiya, K. & Shigeru, K. (2000). Development of a Self-Reconfigurable Modular Robotic System. Sensor Fusion and Decentralized Control in Robotic Systems III, Proceedings of SPIE, Vol. 4196, pp. 469-476.

[4] Jorgensen, M. W.; Ostergaard, E.H. & Lund, H.H. (2004) Modular ATRON: Modules for a self-reconfigurable robot. Proceedings 01 2004 IEEEIRSJ International Conference on Intelligent Robots and Systems – P. 2068-2073.

[5] Qiao, G.; Song, G.; Zhang; J. H.; Wang, W.; & Song, A. (2012) Design of Transmote: a Modular Self-Reconfigurable Robot with Versatile Transformation Capabilities. Proceedings of the 2012 IEEE International Conference on Robotics and Biomimetics. P.1331-1336.

[6] Baca, J.; Ferre, M. & Aracil, R. (2012) A heterogeneous modular robotic design for fast response to a diversity of tasks. Robotics and Autonomous Systems. Vol. 60. No. 4, P. 522–531, 2012.

[7] Lyder, A.H.; Stoy, K.; Mendoza-Garcia, R.-F.; Larsen, J.C. & Hermansen, P. (2013) On sub-modularization and morphological heterogeneity in modular robotics. Intelligent Autonomous Systems 12, volume 193 of Advances in Intelligent Systems and Computing, P. 649-661. Springer Berlin Heidelberg.

[8] Mendoza-Garcia, R.-F.; Lyder, A.; Christensen, D. J. & Stoy, K. (2009) Reusable Electronics and Adaptable Communication as Implemented in the Odin Modular Robot. 2009 IEEE International Conference on Robotics and Automation – P. 1152-1158.

[9] Hancher, M. D. & Hornby, G. S. (2006) A modular robotic system with applications to space exploration. 2nd IEEE International Conference on Space Mission Challenges for Information Technology (SMC-IT'06) Pasadena, CA, P. 8-132.

[10] Lopota, A. V. & Yurevich, E. I. (2013) Stages and development prospects of robotic systems design modular principle. St. Petersburg State Polytechnical University Journal. Computer Science. Telecommunication and Control Systems, Issue 1 2013. P.98 – 103 ISSN: 1994-2354.

[11] Platonov, A.K. (2010) Robotics of a moon base (Робототехника лунной базы). XXXIV Readings on Cosmonautics (XXXIV Чтения по космонавтике) Available from: http://www.keldysh.ru/section5/report.xhtm?src=section5.xml&filter=3. Accessed at 10.09.2017.

[12] Andreev, V.P.; Kim, V.L. & Poduraev, Yu.V. (2016) network–based design of heterogeneous modular mobile robotic systems. Robotics and Technical Cybernetics. Issue 3(12) 2016. P.23-29. ISSN 2310 5305.

[13] Andreev, V.P. (2017) The principle of the full functionality of modules in heterogeneous modular mobile robots. Extreme robotics. Proceedings of the International Scientific and Technological Conference, 2017. In press.

[14] EtherCAT Technology Group. Industrial Ethernet Technologies. Available from: https://www.ethercat.org/download/documents/Industrial_Ethernet_Technologies.pdf Accessed on: 08.09.2017.

[15] Quigley, M.; Gerkey, B.; Conley, K.; Faust, J.; Foote, T.; Leibs, J.; Berger, E.; Wheeler, R. & Ng A. (2009) ROS: an open-source Robot Operating System. ICRA workshop on open source software. № 3 2009 p. 5.

[16] Andreev, V.P. & Pletenev, P.F. (2016) Developing inter-modular communication for heterogeneous mobile robot. Extreme robotics. Proceedings of the International Scientific and Technological Conference, 2016. P.245 – 250.

[17] Kirsanov, K. (2015) Software architecture of control system for heterogeneous group of mobile robots. 25th DAAAM Int. Symp. on Intelligent Manufacturing and Automation, 2014. Procedia Engineering V.100. 2015 pp. 216-221 ISSN 1877-7058.

[18] Hintjens, P. (2011) "0MQ - The Guide." Available from: http://zguide.zeromq.org/page:all. Accessed on: 10.09.2017.

DOI: 10.2507/28th.daaam.proceedings.067

A COMPARISON OF LATTICE STRUCTURES IN METAL ADDITIVE MANUFACTURING

Pavel Hanzl, Ivana Zetková & Milan Daňa

This Publication has to be referred as: Hanzl, P[avel]; Zetkova, I[vana] & Dana, M[ilan] (2017). A Comparison of Lattice Structures in Metal Additive Manufacturing, Proceedings of the 28th DAAAM International Symposium, pp.0481-0485, B. Katalinic (Ed.), Published by DAAAM International, ISBN 978-3-902734-11-2, ISSN 1726-9679, Vienna, Austria
DOI: 10.2507/28th.daaam.proceedings.067

Abstract

The use of metal additive manufacturing is on the rise. One way to achieve significant weight reductions of components is by deploying periodic lattice structures. Selective laser melting enables the production of complex lattice structures in metallic components. In order to realize the full potential of these structures, their basic properties must be known. There are countless variations of topologies, but not all are suitable for SLM manufacturing. In this context, a comparison of commonly used structures is made in this article. The structures are evaluated by maximum capacity load, and the load directions (x, z) are taken into consideration. The results show significant anisotropy of some topologies. However, these topologies reached higher rigidity than topologies which have lower variations of capacity load depending on the direction.

Keywords: Schoen Gyroid; Lattice Structures; Rigid Constructions; Additive Manufacturing; Selective Laser Melting

1. Introduction

In recent years, certain branches of the mechanical engineering industry have been replacing solid materials with light porous materials. Metal Additive Manufacturing (MAM) opens new possibilities in the production of complicated components like lattice structures. Since the manufacturing is less constrained by the limits of traditional techniques, printed components can be smaller and more complicated. This is one of the reasons why the interest in additive manufacturing continues to grow. [2]

However, MAM has limits as well. Complex components with overhangs usually require support to hold the critical areas during the process. These support structures are built together with the component, and help prevent distortion from thermal stresses and anchor the overhang to the building platform. The support structures are removed after the production is complete. [1]

In some cases, the removal of the support structures can be very difficult, especially where the topology is very complex. This means that the choice of suitable lattice structures for MAM is narrow. Lattice structures with irremovable support structures would be counterproductive. A change to the orientation of the lattice structure can reduce the need for support structures. For these reasons, lattice structures based on self-supporting unit cells are in high demand and are used in additive manufacturing. [3]

Cellular lattice structures find wide range of usage due to their unique properties. They are used in applications where high strength accompanied by a relatively low mass is required. Porous structures also have other advantages such as thermal insulation properties and suppressed vibration. [7][11]

The interest in cellular lattice structures continues to grow, and a number of scientific articles have already been published dealing with this area of research and development. The authors in [9] and [10] focus on FEM analyses of different kinds of cellular lattices and the results were verified by practical experiments.

This article deals with a comparison of structures which are commonly used in additive manufacturing due to their self-supporting properties. The structures are evaluated by maximum capacity load, and the load directions (x, z) are taken into consideration.

2. Lattice structures with self-supporting cell units [2]

The experiment includes seven topologies with self-supporting unit cells, see Figure 1. All of the samples have 5 mm unit size and volume fraction is set at 13.5%. Cubic units such as Body Centred Cubic (BCC) and Face Centred Cubic (FCC) form the basis for the other variations such as PFCC and BCCz, where struts are added to gain stiffness along the z axis. F2BCC is created by combining BCC and FCC units.

Since the struts of BCC enclose a 45° angle with the main axes, BCC topology respects the rule of MAM and is self-supported along the main axes. The other topologies meanwhile are self-supported only in one z direction due to the added struts. While these lattices build along x or y direction, the MAM could fail on the horizontal struts. A typical representative of a fully self-supporting topology is a single gyroid, which does not require supports in a minimum range of cell size 3-8 mm. [7] [8]

The lattice structure marked as "Rhombic" does not respect the general rules of MAM and is therefore not self-supported. However, it is known from theory that AM is able to bridge an overhang of short distance without support structures. [4] Therefore its topology could also be built using this technology with the following parameters set.

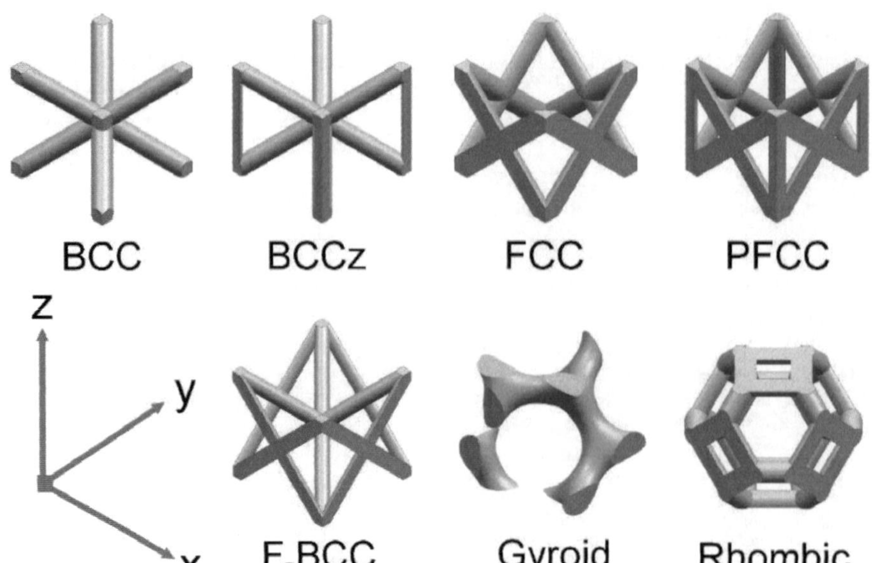

Fig. 1. Compared unit cells topologies

3. Production of Lattice Structure

The samples were made using an EOS M290. The construction material was EOS MaragingSteel MS1 and the layer thickness was set to 40 μm. Default process parameters were used from EOS. The set consisted of 11 sample types. Samples with a rotated porous core of about 90° were included. Some samples without a rotated core had a less suitable orientation for additive manufacturing of lattice structure topologies due to the existence of horizontal struts without support structures. All of the samples had rigid plates on the contact surfaces with the test machine. To ensure reliable production, the rigid plates were built vertically and therefore the axis z was oriented horizontally during AM, whereas the axis x was vertical. Some samples are shown in Figure 2.

None of the samples exhibits any defects. Only the horizontal struts in the samples with load direction z show an impaired quality of the lower surfaces and their diameters are slightly increased. This is attributed to overheating of these areas due to the absence of a support structure, resulting in the partial welding of powder particles to the bottom.

Fig. 2. Samples of lattice structures

4. Experiment

The porous samples were subjected to compression tests on the Zwick/Roell Z 250 testing device. The results of the load capacity for uniaxial loading along direction z and x are shown in Figure 3. Values of the load capacities have been determined for a 0.2% permanent deformation.

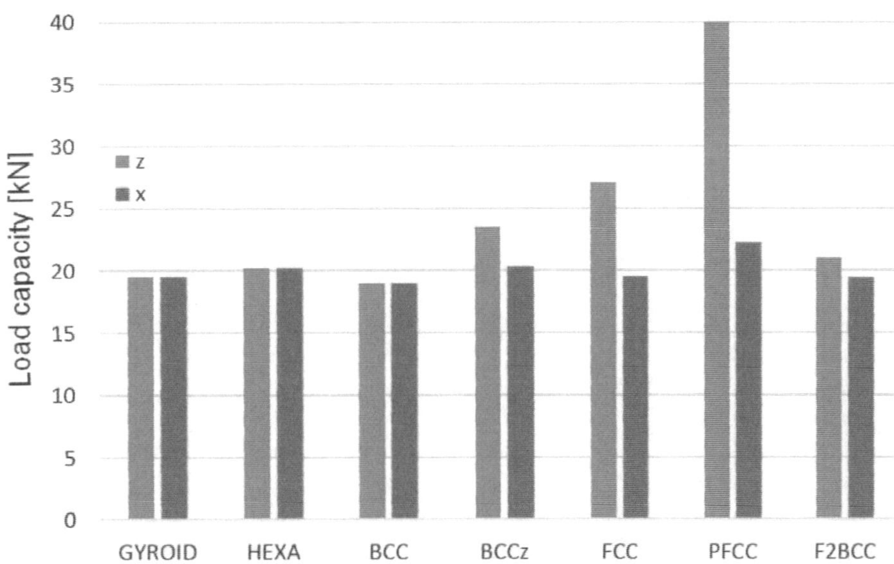

Fig. 3. Load capacity of the different topologies

The lattice structure with the PFCC unit achieved the highest rigidity in both load directions, but also exhibits the highest anisotropy along the investigated direction. Deformation curves which were obtained from the tests differed from each other significantly. The load along direction z increased until the yield point, then came a slow load reduction. The sample with rotated PFCC topology (load direction x) showed a different course of load. After attaining load peak, the load force fell sharply due to the collapse of one cell row. The peak of maximum local load was contained five times in the deformation curve since the sample had five rows of units. All the other topologies have deformation curves similar to the first one. A closer look at the deformation curves of the PFCC topology is shown in Figure 4.

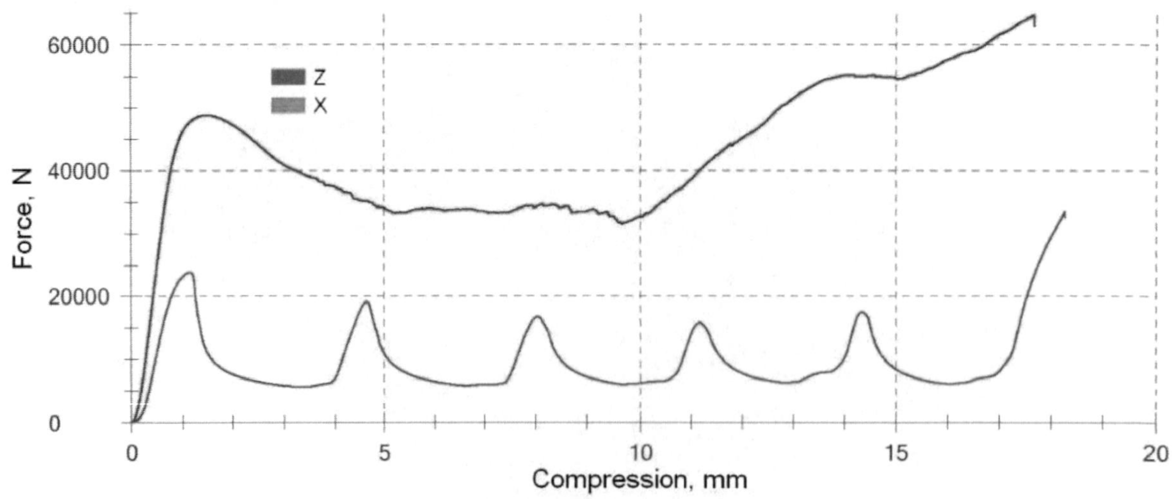

Fig. 4. Deformation curves of lattice with PFCC unit (a) direction z (b) direction x

5. Conclusion

This article deals with a comparison of self-supported lattice structures. The structures are evaluated by maximum capacity load, and the load directions (x, z) are taken into consideration. SLM is evidently very useful in the production of complex lattice structures. A set of porous samples with different topologies were all manufactured without defects. Since volume fraction and unit size were set as constant, the strut diameters were not constant and ranged from 0.64 mm to 0.96 mm in the different 3D models of the samples.

In general, the horizontal straight struts can distort self-supporting properties of cell unit because these struts do not respect the minimum building angle (40°) without supporting elements, and production may become risky. It was proved on the sample set that additive manufacturing could bridge a distance of 4.3 mm in the case of melting of horizontal struts with a minimum impact on the surface quality. The situation may be different where struts are slightly inclined from the horizontal position due to the absence of constrains on both ends of the struts during metal additive manufacturing. A consequence of this may be a deviation of the unconstrained end above the level of the melted layer under the influence of internal tensions. A collision of the recoater and strut end could follow. In this respect, the Gyroid is the most suitable for MAM, because its manufacturability does not depend on its orientation. [6]

The BCC, Rhombic and Gyroid structures have consistent mechanical properties along three orthogonal axes (x, y, z).[5] Therefore, the samples with these topologies were tested only in one z direction. These types of structures, including F2BCC can be used in applications where homogeneity of rigidity is required and could be useful in applications with combined stresses.

A higher load capacity was reached by BCCz and FCC topologies. Adding struts to BCCz gave it a higher load capacity in both directions than BCC. The PFCC topology was the stiffest with the least aligned results. This may suggest that its mechanical properties will degrade faster compared to other topologies, when the load is combined as pressure-torsion. However, some components are stressed along one direction and there the PFCC could be beneficial.

This conclusion corresponds with the results in article [3]. The authors focus on FEM analyses of different kinds of self-supported cellular lattices. It was confirmed that a double gyroid is particularly suitable for applications with multiaxial loading, since it provides identical rigidity of the porous sample in the x and z load axis. This also applies to a single gyroid, although achieving less rigidity. Geometrically simpler cellular lattices (such as BCCz, FCC, PFCC) have significantly increased stiffness in one of the main direction.

Additional mechanical tests, such as bending strength or cyclic fatigue, will be added in the following research activities.

6. Acknowledgments

The present contribution has been prepared under project LO1502 'Development of the Regional Technological Institute' under the auspices of the National Sustainability Programme I of the Ministry of Education of the Czech Republic aimed at supporting research, experimental development and innovation.

7. References

[1] Hussein, A., Hao, L., Yan, C., Everson, R., Young, P. (2013) Advanced lattice support structures for metal additive manufacturing, Published by Elsevier Ltd, Journal of Materials Processing Technology 213 (2013) 1019-1026

[2] Contuzzi, N., Campanelli, S., Casavola, G., Ludovico, A. D. (2010) Effect of heat treatment on selective laser melted steel parts, Proceedings of the 21th DAAAM International Symposium, Published by DAAAM International, ISBN 978-3-901509-73-5, ISSN 1726-9679, Vienna, Austria

[3] Aremu, A. O. all et. A Comparative finite element study of cubic unit cells for selective laser melting, EPSRC Centre for Innovative Manufacturing in Additive Manufacturing, Faculty of Engineering, University of Nottingham, Nottingham, NG7 2RD, UK.

[4] Thomas, D. (2009) The Development of Design of Design Rules for Selective Laser Melting, Ph.D. Thesis, University of Wales Institute, Cardiff, UK.

[5] Challis, V. J. all et. (2014) High specific strength and stiffness structures produced using selective laser melting, Published by Elsevier Ltd, Materials and Design 63 (2014) 783-788

[6] Yan, C., Hao, L., Hussein, A., Young, P., Raymont, D. (2014) Advanced lightweight 316L stainless steel cellular lattice structures fabricated via selective laser melting, Published by Elsevier Ltd, Materials and Design 55 (2014) 533-541

[7] Hao, L., Raymont, D., Yan, C., Hussein, A., Young, P. (2011) Design and Additive Manufacturing of Cellular Lattice Structures, College of Engineering, Mathematics and Physical Sciences, University of Exeter, Exeter EX4 4QF, Devon, United Kingdom

[8] Hanzl, P., Zetek, M., Zetková, I. (2015) Cellular Lattice Structure Produced by Selective Laser Melting and its Mechanical Properties, Proceedings of the 26th DAAAM International Symposium, pp.0748-0752, B. Katalinic (Ed.), Published by DAAAM International, ISBN 978-3-902734-07-5, ISSN 1726-9679, Vienna, Austria

[9] Li, P. (2015) Constitutive and failure behaviour in selective laser melted stainless steel for microlattice structures, Published by Elsevier Ltd, Materials Scince and Engineering A 622 (2015) 114-120

[10] Contuzzi, N., Campanelli, S. L., Casavola, C., Lamberti, L., (2013) Manufacturing and Characterization of 18Ni Marage 300 Lattice Components by Selective Laser Melting, Materials 2013, 6, 3451-3468; doi:10.3390/ma608345

[11] Marcian, P., Rehorek, Z. Florian, Z., Dlouhý, I, (2011) Estimation of the Properties Porous Structures by Experiment and Modelling, Chapter 47 in DAAAM International Scientific Book 2011, pp. 573-584, B. Katalinic (Ed.), Published by DAAAM International, ISBN 978-3-901509-84-1, ISSN 1726-9687, Vienna, Austria, DOI: 10.2507/daaam.scibook.2011.47.

DOI: 10.2507/27th.daaam.proceedings.068

USING OF WATERJET TECHNOLOGY FOR CUTTING ALUMINUM ALLOY

Tomáš Kroupa, Milan Daňa & Miroslav Zetek

This Publication has to be referred as: Kroupa, T[omas]; Dana, M[ilan] & Zetek, M[iroslav] (2017). Using of Waterjet Technology for Cutting Aluminum Alloy, Proceedings of the 28th DAAAM International Symposium, pp.0486-0492, B. Katalinic (Ed.), Published by DAAAM International, ISBN 978-3-902734-11-2, ISSN 1726-9679, Vienna, Austria
DOI: 10.2507/28th.daaam.proceedings.068

Abstract

This paper presents an analysis of the impact of AWJ Machining on the surface of Aluminium (AlMg3) parts. Samples of Aluminium are subjected to AWJ surface machining. The essence of this cutting method is the grinding of a material by using pressure from a waterjet. The jet is generated in the cutting head and terminates in the nozzle. A range of machining parameters is considered, including feed rate, cut quality, speed, and static shot. These samples are then subject to surface analysis. This analysis allows conclusions to be drawn with regard to optimal machining conditions for cut quality improvement. This experiment provided a practical verification of the theoretical assumption. The practical part with the cut samples should serve mainly to provide a better idea and easier understanding of this technology. Samples which were cut using AWJ were documented and compared.This experiment also shows that it is necessary to pay attention to ensuring the correct cutting parameters for a specific application, because even here can be seen the eternal duel between quantity versus quality.

Keywords: Unconventional machining technology; cutting material by waterjet; high pressure waterjet; abrasive material cutting

1. Introduction

Waterjet cutting technology uses the kinetic energy of a high pressure and high speed water jet for cutting material. Abrasive methods also use the kinetic energy of abrasive particles, which makes a significant difference. Material is cut by high speed erosion by abrasive particles which are concentrated into a thin ray over a small surface area. The speed of this beam is 600 – 900 m/s. It is called "hydroabrasive erosion" which is induced by a high speed hydroabrasive beam. The beam is the cutting tool in this type of technology. The particle removal of the cut material is micro-dimensional. This erosive removal goes on at practically non-elevated temperatures and therefore there is no thermal deformation around the cut. This means it is possible to cut materials that could not be divided by other methods.

Waterjet technology uses a mechanical-physical method of material removal. A very thin stream of water which has a high speed acts on the workpiece. This energy causes the beam to pass through the material, thereby machining it.

2. Methods of water jet cutting

Waterjet cutting is divided in two main methods. One method is waterjet machining, which is pure waterjet cutting. The second method is abrasive waterjet cutting (abbreviated AWJ) – which uses the addition of an abrasive medium. Both methods are similar, but they differ in the added abrasive material. In both of these methods, the cut material can be placed underwater and the device is similar. In both methods there is a minimal influence on the structure at the cut and a minimal temperature increase. The temperature increase is around 20°C and a small amount of water evaporation occurs, and sparks form around the beam. For some materials even this small increase in temperature can also mean a deterioration of the properties and possible destruction of the material.

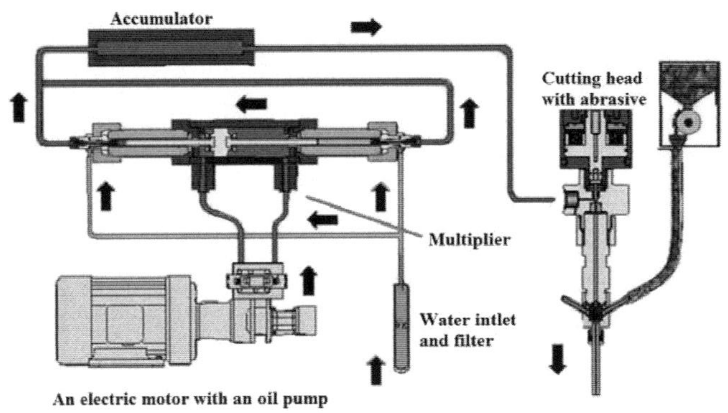

Fig. 1. Schematic of water jet device

3. Water jet machining

Cutting with pure water without adding abrasives involves the removal of material by the mechanical effect of the waterjet, which falls on the surface of the material at a high speed and high kinetic energy over a small surface area. Due to the absence of abrasive, there are higher demands on the working fluid, which is made up of water with certain required properties. Water seems to be the best choice for the working fluid, because it offers an acceptable compromise between the required properties.

In this method the beam is created in the cutting head. Water, which is compressed, passes through a nozzle with a small hole and is transformed into a thin waterjet with high kinetic energy and high speed. This beam impacts the machined material. After impact, the beam is decelerated due to the friction between the beam and the surface of the material. The pressure can reach values up to 690 MPa. The feed rate depends on the hardness, type and thickness of the material. The feed rate can be in the range of 5 to 400 m/min. The diameter of the nozzle is about 0.3 mm, the beam has a diameter of 0.1 to 1.5 mm. The width of the cut is slightly larger than the diameter of the nozzle (about 0.3 mm). The distance of the cutting head from the material is from 2.5 to 6.35 mm but can also be 10 to 25 mm. Pure waterjet technology is suitable for cutting soft, thin, non-rigid materials. It is possible to cut various plastic materials, rubber, plasterboard, foam materials, wood and plywood. It can also be used to remove corrosion or old coatings from products and devices. But a pure waterjet is not suitable for cutting metallic materials, because there is small removal of the material and it is more advantageous to use one of the abrasive methods where efficiency is higher. [1]

4. Abrasive waterjet

AWJ stands for abrasive waterjet and it is a method with a direct supply of abrasive. This technology works on the same principle as WJM with the material removed by the effects of a thin concentrated beam acting on a very small area. Because of the abrasive, the impact from AWJ has a much more mechanical effect. In this method, the abrasive is guided from the storage, which is outside the cutting device, straight into the mixing chamber in the cutting head. The flow of the beam causes mixing under-pressure in the chamber. This under-pressure causes suction, capture and acceleration of each abrasive particle. The dosage of the abrasive in this method is about 1 to 20 kg/min. Mixing power depends on the geometry of the focusing nozzle. The working pressure is in the range of 250 to 400 MPa. The diameter of the mixed beam is higher than the pure water one because of the abrasive. [2]

The feed rate depends on the cutting material. On hard materials, the feed rate is from 10 mm/min. Hard materials include cemented carbide, titanium and cobalt. Softer materials (aluminium, glass) can be cut with speeds up to 2500 mm/min. Mixing of the waterjet and the abrasive causes a deceleration of the beam by up to half of the original value, because it causes reduced performance. The selected abrasive must have a smaller particle size than the diameter of the hole in the focusing nozzle. The size of the abrasive should be at least five times smaller than the nozzle diameter. The smaller the hole in the nozzle, the more energy of the beam can be concentrated on one place. [3]

5. The experiment

Samples which were cut using AWJ were documented and compared. Ten different conditions and modes were created, in which identical parts were cut. Only the feed rates were changed. The abrasive flow rate was not changed because the dosing is selected by software which chooses the appropriate conditions according to the thickness and type of the material to be cut. Additionally, these changes would be hard to observe on the final surface. During the experiment the feed rate, type of bullet holes and other support technologies were changed. [4]

5.1. Samples

The samples were rectangular, with dimensions 100 x 40 mm. A 12 mm thick sheet was selected as a blank. At this thickness the beam delays and surface quality changes could be easily observed. The tested material was an aluminium-magnesium alloy EN AW-5754 (AlMg3). Sheets of this material are available in a wide range of thicknesses, accessible and have good mechanical properties and good weldability. This means it is widely used in industry, especially for lightweight aluminium constructions. For this reason, EN AW-5754 was the ideal choice for the cut samples. A hole with a radius R10 was made on one edge of the sample, and on the other edge there was a bevel 10 x 45°, to observe the surface when the trajectory is changed, other than 90°. For these reasons the sample also included a series of holes with diameter 10 mm. The theoretical weight of the sample is 115 grams. [5]

The experiment took place on a device which consists of an X-Y CNC table with working dimensions 3000 x 1500 mm, with abrasive dispensers and a JETS high pressure pump. The most important parameters of this pump are shown in the following table 1.

Parameter	Value
Ambient temperature	+5°C to 35°C
Device dimensions	1690 x 1350 x 1550 mm
Weight	1800 kg
Max. outlet water pressure	4100 bar
Max. amount of water for cutting	3.8 l/min
Max. total water consumption	3.8 l/min
Inlet water pressure	3.5 – 6 bar
Max total input power	39.3 kW
Electric motor	37 kW; 1500 rpm
Hydraulic oil filling	160 l
Max. working oil pressure	215 bar
Max. working oil temperature	+52 °C

Table 1. Parameters of high pressure pump

5.2. The state of each sample

Sample 1

Parameters: static shot, medium cut quality, speed 458.84 mm/min, DRC control off and Progress Jet on (Figure 2)

Fig. 2. Sample 1

Sample 2

Parameters: direct shot, medium cut quality, speed 458.84 mm/min, DRC control off and Progress Jet on (Figure 3)

Fig. 3. Sample 2

Sample 3

Parameters: circular shot, medium cut quality, speed 458.84 mm/min, DRC control off and Progress Jet on (Figure 4)

Fig. 4. Sample 3

Sample 4

Parameters: circular shot, the roughest cut quality, speed 458.84 mm/min, DRC control off and Progress Jet on (Figure 5)

Fig. 5. Sample 4

Sample 5

Parameters: circular shot, rough cut quality, speed 731.43 mm/min, DRC control off and Progress Jet on (Figure 6)

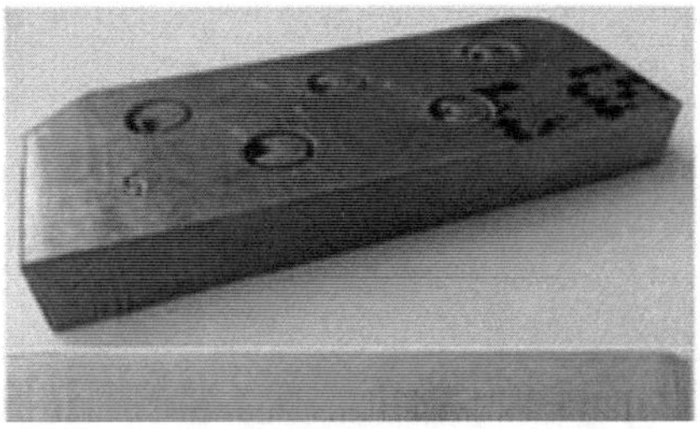

Fig. 6. Sample 5

Sample 6

Parameters: circular shot, medium cut quality, speed 458.84 mm/min, DRC control off and Progress Jet on (Figure 7)

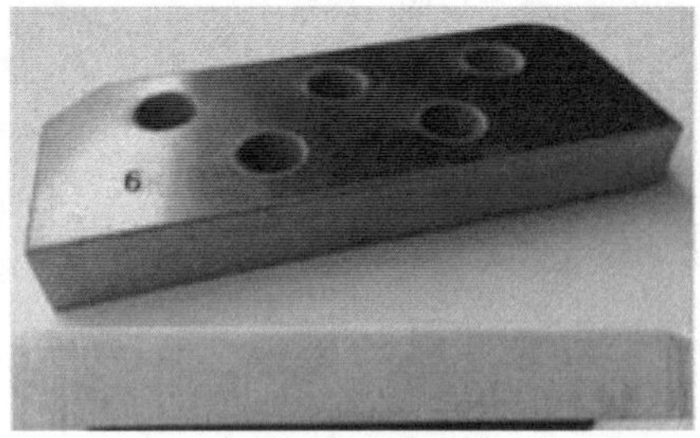

Fig. 7. Sample 6

Sample 7

Parameters: circular shot, smooth cut quality, speed 329.6 mm/min, DRC control off and Progress Jet on (Figure 8)

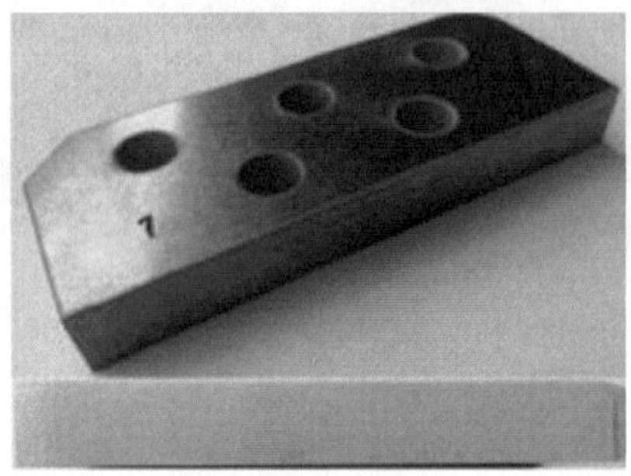

Fig. 8. Sample 7

Sample 8

Parameters: circular shot, the smoothest cut quality, speed 255 mm/min, DRC control off and Progress Jet on (Figure 9)

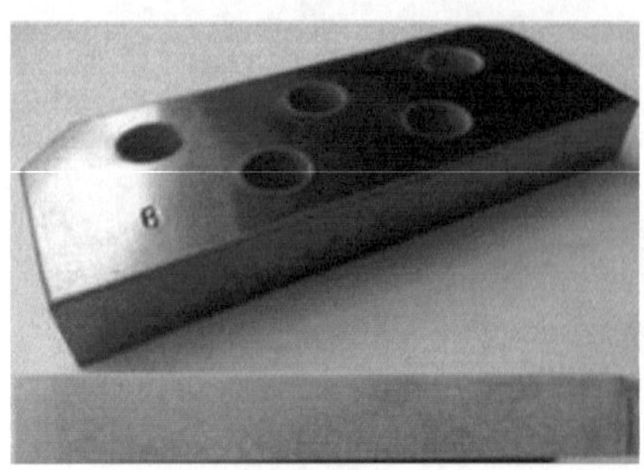

Fig. 9. Sample 8

Sample 9

Parameters: circular shot, smooth cut quality, speed 329.6 mm/min, DRC control on and Progress Jet on (Figure 10)

Fig. 10. Sample 9

Sample 10

Parameters: circular shot, smooth cut quality, speed 329.6 mm/min, DRC control off and Progress Jet on (Figure 11)

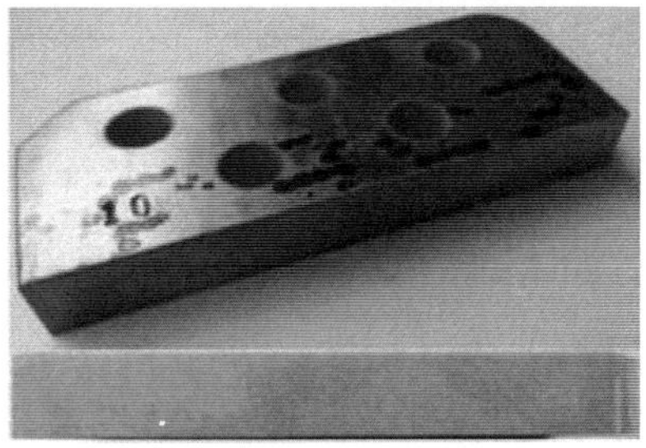

Fig. 11. Sample 10

5.3. Cutting parameters

Different setting properties were used for cutting the samples. Abrasive dosing settings were kept at 350 g/min for all samples. The same setting was used for the first three samples, the only difference was the type of the bullet hole: static, straight and circular. The abrasive dosing was set at 200 g/min for each bullet hole. Other samples remained with a circular bullet hole, which is the most suitable for most applications and the cutting speed was changed (samples 4 to 8). Speeds used are assigned to a particular cutting mode in the control system, the roughest, rough, medium, smooth and the smoothest. DRC controlling (radius correction) was used on Sample 9, and on sample 10 Progress Jet technology was turned off. Both the latest samples were cut in "smooth mode". A summary is shown in the following table (Table 2.)

Sample	Bullet hole	Quality of cut	Speed rate [mm/min]	DRC controlling	Progress Jet
1	static	medium	458.84	no	yes
2	direct	medium	458.84	no	yes
3	circular	medium	458.84	no	yes
4	circular	coarsest	1102.32	no	yes
5	circular	coarse	731.43	no	yes
6	circular	medium	458.84	no	yes
7	circular	smooth	329.6	no	yes
8	circular	smoothest	255	no	yes
9	circular	smooth	329.6	yes	yes
10	circular	smooth	329.6	no	no

Table 2. Parameters of cutting

5.4. Comparison in terms of the bullet hole

Different types of bullet hole through material were used for samples 1 to 3. The cutting mode was the same for samples with medium quality and therefore all the edges look practically the same. The first sample (static shot) has a distinct edge (mesh) across the whole depth. A substantially smaller edge was created on the second sample, which is only at the bottom of the sample depth. The third sample also has a small edge, but the area around the shot was cleaner. Therefore it appeared that the best and in practice the most used type of shot for thick walled sheet metal is the circular one. [4]

For samples 4 to 8, the circular type of shot was kept. The Progress Jet function was also enabled and DRC controlling was turned off. Only the speed mode was changed on these samples. On sample 4 the setting for the fastest cut was used, and therefore the worst quality of the cut surface was achieved. Exactly the opposite setting was used on sample 8, the slowest cut with best quality. The differences between each mode are very striking. The differences with changing speed were very noticeable on the samples. At higher speeds, in the rough and roughest modes, beam delays associated with the groove phenomenon can be observed. Sample 6 was cut with the middle mode, where the beam delay was less pronounced. [6]

The assumption was confirmed that with increasing feed rate there is a delay to the waterjet beam, which is subsequently shown on the grooves on the cut edge of the material. The result is that a higher feed rate means worse surface quality. The theoretical assumption was verified experimentally. It is interesting to note that on sample 4 the speed was so high that at several places on the cutting edge, complete cutting through the material did not occur. The sample had to be released from the blank using a hammer. [7]

6. Conclusion

It has been concluded from these experiments that waterjet feed rate may have an impact on cut quality surface. After evaluating the cut samples, the circular shot appears to the best type of shot material, because it had the least effect on the relief of its surroundings. On the contrary, the worst type of shot was static shot, where a significant edge remained throughout the whole depth of the sample. From a qualitative standpoint, direct shot is located between these shots. In practice, the most commonly used material is circular shot because of its better properties. While cutting samples on which the feed rate was changed, it could easily be seen how with increasing speed, the quality of the cut surfaces decreases. Delaying of the waterjet beam was seen at the edges, which increases with increasing feed rate. According to expectations, higher speeds mean more beam delay and, at the end, worse cut quality. This bad trait of the waterjet had a great impact on the bevel, where the cutting head did not slow down as on the sharp edges and the result is poor surface quality and incomplete cutting of the material. On the sample with the highest feed rate (sample 4), there were several uncut locations, which actually caused great problems when removing the sample from the blank. Slightly uncut parts often occur even at slower speeds, but cut samples can usually be removed without any problems or much effort from the blank. This experiment provided a practical verification of the theoretical assumption. The practical part with the cut samples should serve mainly to provide a better idea and easier understanding of this technology. This experiment also shows that it is necessary to pay attention to ensuring the correct cutting parameters for a specific application, because even here can be seen the eternal duel between quantity versus quality. [8,9]

In the future we will focus on optimizing cutting conditions, different process parameters and their influence on the process.

7. Acknowledgements

This paper is based upon work sponsored by project SGS-2016-005

8. References

[1] Aišman, D. Staňková, H. Skálová , L. Mašek, B. (2008). Testing of the parameters of the Q-P process in high strength low-alloyed steel. In Danube Adria Association for Automation and Manufacturing. Vienna Austria: DAAAM International, Vienna Austria, 2008. ISBN: 978-3-901509-68-1, Vienna, Austria

[2] Jenicek, S. Kotesovec, V. Kalina, T. & Masek, B. Use of Waterjet in Manufacturing Test Bars of High-Strength Steels, Annals of DAAAM for 2016 & Proceedings of the 27th DAAAM International Symposium, pp.0219-0224, B. Katalinic (Ed.), Published by DAAAM International, ISBN 978-3-902734-08-2, ISSN 1726-9679, Vienna, Austria

[3] L.M. Hlavac, I.M. Hlavacova, L. Gembalova, J. Kalicinsky, S. Fabian, J. Mestanek, J. Kmec, V. Madr, Experimental method investigation of the abrasive water jet cutting quality, Journal of Materials Processing Technology 209 (2009) 6190–6195.

[4] Fabian Lissek, Michal Kaufeld, Jacqueline Tegas, Sergej Hloch, Online-monitoring for Abrasive Waterjet Cutting of CFRP via Acoustic Emission: Evaluation of Machining Parameters and Work Piece Quality Due to Burst Analysis, Procedia Engineering, Volume 149, 2016, Pages 67-76

[5] Kana, J. Vorel, I & Ronesova, A. (2016). Simulator of Thermomechanical Treatment of Metals, Annals of DAAAM for 2015 &Proceedings of the 26th DAAAM International Symposium, pp.0513-0518, B.Katalinic (Ed.), Published by DAAAM International, ISBN 978-3-902734-07-5, ISSN 1726-9679, Vienna, Austria

[6] Loeser, C. Duerr, H. & Pilz, R. Application of the Micro Water Abrasive Injector Fine Jet for Precision Machining, Annals of DAAAM for 2010 & Proceedings of the 21st International DAAAM Symposium, Volume 21, No. 1, ISSN 1726-9679 ISBN 978-3-901509-73-5, Editor B. Katalinic, Published by DAAAM International, Vienna, Austria, EU, 2010

[7] Yuvraj K. Madhukar, Suvradip Mullick, Ashish K. Nath. An investigation on co-axial water-jet assisted fiber laser cutting of metal sheets. Optics and Lasers in Engineering, Volume 77, February 2016, Pages 203–218

[8] D.S. Srinivasu, D.A. Axinte, Surface Integrity Analysis of Plain Waterjet Milled Advanced Engineering Composite Materials, Procedia CIRP, Volume 13, 2014, Pages 371-376

[9] Begic-Hajdarevic. D. Cekic. A. Mehmedovic. M. Djelmic. Experimental Study on Surface Roughness in Abrasive Water Jet Cutting 25th DAAAM International Symposium on Intelligent Manufacturing and Automation, DAAAM 2014 Experimental Study on Surface Roughness in Abrasive Water Jet Cutting, Procedia Engineering 100 (2015) 394 – 399

DOI: 10.2507/27th.daaam.proceedings.069

ABRASIVE WATERJET CUTTING

Tomáš Kroupa, Milan Daňa & Miroslav Zetek

This Publication has to be referred as: Kroupa, T[omas]; Dana, M[ilan] & Zetek, M[iroslav] (2017). Abrasive Waterjet Cutting, Proceedings of the 28th DAAAM International Symposium, pp.0493-0499, B. Katalinic (Ed.), Published by DAAAM International, ISBN 978-3-902734-11-2, ISSN 1726-9679, Vienna, Austria
DOI: 10.2507/28th.daaam.proceedings.069

Abstract

Waterjet cutting is being used more and more frequently, for several reasons. One of its major advantages is the possibility to cut almost all materials and thicknesses. This technology can be used for very hard, tough materials (steel, aluminum, dural, mineral materials, plastic) and also soft materials such as rubber. The principle of waterjet cutting is the removal of material by the mechanical impact of a narrow waterjet at high speed. The waterjet is generated by the high pressure of the water as it passes through the nozzle. The cut width is about 0.3 mm, which ensures little loss of material. Deflection (curving) and deceleration of flow occurs during cutting, which is a consequence of the friction between the surface of the waterjet and the surface of the workpiece material. The trajectory of the waterjet is rounded and its radius changes according to the previous arc length. These properties affect the final shape of the surface and topography of the machined parts. Another significant advantage of this cutting method is that the workpiece is not affected by heat and relatively high accuracy is achieved. The use of modern devices often eliminates the need for further machining.

Keywords: Abrasive waterjet; cutting quality; cutting angle; eliminating bevels; high pressure pump

1. Introduction

The experiment is based on the comparison of results from cutting machines using AWJ technology and machines using DWJ technology. Machines with DWJ technology should achieve an overall higher quality of cutting and should eliminate chamfers on the machined material.

Below is shown the comparison of the cutting methods. For each sample is shown a graph of the dependence of the average feed rate speed of the cutting head on the width of the samples. For comparison, the values for cutting methods AWJ and DWJ are always shown in one graph for a given material and thickness. A digital caliper was used for detecting the width of the samples at two positions (estimating the size of the chamfer on the samples). The values in the tables are the averages of four measurements. The upper position was measured about 1 mm to 2 mm from the upper edge, the lower position was measured 1 mm to 2 mm from the bottom edge.

Another graph shows the dependence of feed rate speed on the average arithmetic deviation of the profile Ra cut area, which was measured on the length of the sample (in the central part) and in two positions - upper (1-2 mm from the top) and lower (1-2 mm from the lower edge). The values for comparing (AWJ, DWJ) are placed in one graph.

Type of waterjet cutting	AWJ	DWJ
Pressure [bar]	3800	5500
Abrasives	Indian garnet	Indian garnet
Grain abrasives	MESH 80	MESH 80
Abrasive flow [g.min-1]	540	540
Waterjet nozzle diameter (ruby) [mm]	0.3302	0.3556
Abrasive nozzle diameter [mm]	1.02	1.02

Table 1. Waterjet parameters

2. Figures Technological possibilities of the 3D abrasive waterjet

2.1. Technology of Dynamic Waterjet (DWJ)

Bending of the waterjet and chamfers can be eliminated by reducing the cutting speed. However, this means longer cycle times, thereby increasing production costs.

Dynamic Waterjet technology corrects the bending waterjet and chamfer of the cutting edge at maximum cutting speeds with high accuracy. A sample created without using DWJ is shown in Fig. 1a. The same sample made on a machine using DWJ is shown in Fig 1b. At first sight there is a noticeable difference, primarily in the incidence of the chamfers and the quality of the cut surface.

Fig. 1. Waterjet cutting (a) without DWJ (b) with DWJ

2.2. Abrasive Waterjet (AWJ)

When using AWJ, water contributes to the introduction of abrasives at higher speeds when it enters the mixing chamber (the abrasive obtains a high kinetic energy and momentum) and for removal of the abrasives and microparticles of the workpiece. It also acts as an opening wedge for increasing the rate of crack spreading in the erosion process. [7]

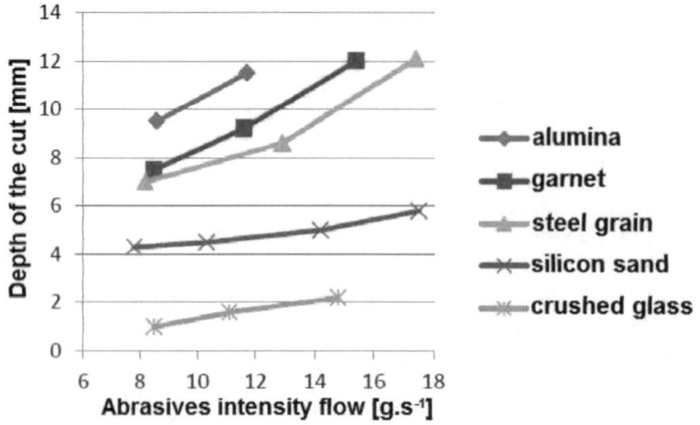

Fig. 2. Dependence of the depth of the cut on the intensity of the abrasive flow

Pressure [MPa]	speed [mm.s-1]	Type of steel
207	4,1	DIN 35CrNiMo6

Table 2. Cutting conditions

Abrasive factors which have an impact on the depth of the cut are the kind of abrasive used, its sharpness, hardness and grain. The most suitable shapes are abrasive particles with high roundness and low circularity. Fig. 2 shows the dependence of the depth of the cut on the intensity of the flow of various kinds of abrasives when cutting low alloy structural steel.

The graph shows that with an increase in the mass flow of the abrasive to the waterjet, the depth of the cut increases. However, this can only be applied until the critical mass flow. When it is exceeded, the depth of the cut does not increase, but starts to decline. This is because part of the energy is consumed during collisions between the abrasive particles.

Other factors affecting AWJ include the parameters of the abrasive mix. The shape and dimensions of the mixing chamber and the abrasive of the nozzle play an important role. The dimensions of the mixing chamber and the fluid pressure relate to the maximum mass flow of abrasive. In this chamber the character of the flow layers of fluid changes from laminar to turbulent, a vacuum is formed which results in the suction of the abrasive from the tray. Here direct proportion applies – the higher the vacuum, the greater is the maximum mass flow of the abrasive. The initial mixing of the abrasive with the liquid stream continues in the mixing chamber. [3, 7]

3. The experiment

In the following part, two tables are shown first (one for AWJ technology and the second for DWJ technology), which relate to one of the cut materials with a certain thickness. Then there are two graphs, which are based on the preceding tables.

The tables include information from a particular cut sample. The type of cut material, its thickness, the technology used for the cut, feed rate of cutting head and time of cutting. The measured values are also shown. First is given the sample width measured at the top and bottom positions. The Ra values of the cutting area (measured along the length of the sample) are shown, which were again measured at two positions, top and bottom. The values of Rz in two positions are also noted.

Below the two tables there are two graphs. The first shows the dependence of the width of the sample on the speed of the cutting head. The second graph shows the relation of the arithmetic deviations of profile Ra to the speed of the cutting head. [2]

3.1. Abrasive Waterjet (AWJ) Cutting of aluminium

Aluminium 5083AW 15mm Technology: AWJ		**Average feed rate speed [mm.min-1]**				
		150	200	250	300	350
Cutting time [s]		480	36	29	24	21
Measurement sample position width	**Upper [mm]**	19.81	19.82	19.82	19.85	19.86
	Lower [mm]	20	20.05	20.13	20.21	20.23
Measuring position Ra	**Upper [mm]**	4.41	4.76	4.28	4.99	4.51
	Lower [mm]	5.10	6.34	5.31	5.28	6.34
Measuring position Rz	**Upper [mm]**	27.09	28.60	25.42	30.70	29.71
	Lower [mm]	30.67	36.41	31.18	30.83	35.58

Table 3. Values of the sample from aluminium with thickness 15 mm, cutting technology AWJ

Aluminium 5083AW 15mm Technology: DWJ		**Average feed rate speed [mm.min-1]**				
		277	343	400	450	522
		60%	80%	100%	120%	140%
Cutting time [s]		26	21	18	16	14
Measurement sample position	**Upper [mm]**	19.94	19.98	19.92	19.90	20.03
	Lower [mm]	20	20.02	20.02	20.02	20.11
Measuring position Ra	**Upper [mm]**	4.31	4.06	4.56	4.29	4.35
	Lower [mm]	6.22	6.29	7.84	7.03	10.03
Measuring position Rz	**Upper [mm]**	24.74	24.83	29.46	26.07	26.05
	Lower [mm]	33.48	37.21	41.83	38.06	46.87

Table 4. Values of the sample from aluminium with thickness 15 mm, cutting technology DWJ

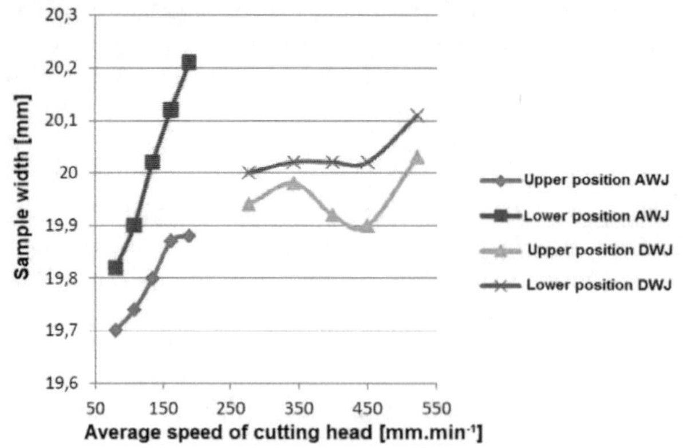

Fig. 3. Dependence of the sample width on the speed of cutting head (aluminium, width 15 mm)

Fig. 3 confirms that when cutting aluminum with thickness 15 mm using DWJ technology, the values of the chamfer are at a minimum. Increasing the average feed rate speed when cutting this thickness has a minimal effect on the size of the chamfer. However, this is different when cutting with AWJ technology, where the size of the chamfer increases with the increased average speed of the cutting head. [1]

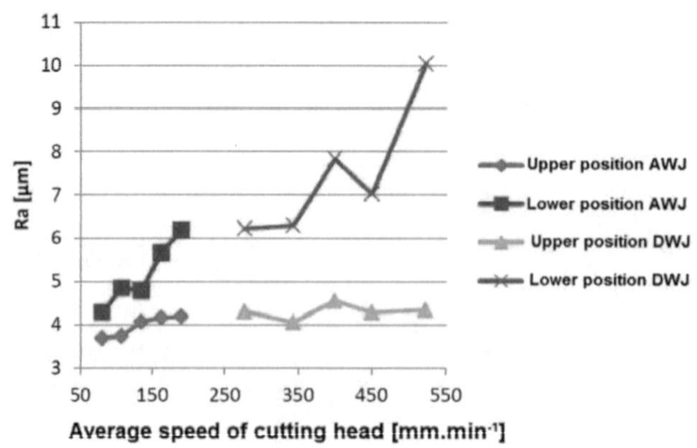

Fig. 4. Dependence of the Ra cut surface on the speed of cutting head (aluminium, width 15 mm)

Fig. 4 shows that when cutting aluminum with a thickness of 15 mm, higher values of Ra are obtained using DWJ cutting technology. There is also a significant difference between the values of the upper and lower positions because of the higher feed rate of the cutting head. The value of Ra is almost constant in the lower position with DWJ cutting technology. [4]

3.2. Cutting of steel

Steel 11 523 15mm Technology: AWJ		Average feed rate speed [mm.min-1]				
		60	80	100	120	140
Cutting time [s]		120	90	72	60	51
Measurement sample position width	Upper [mm]	19.90	19.93	19.98	19.99	20.00
	Lower [mm]	20.41	20.44	20.61	20.68	20.82
Measuring position Ra	Upper [mm]	3.15	2.89	2.8	3.08	3.63
	Lower [mm]	4.29	5.24	4.59	5.93	7.79
Measuring position Rz	Upper [mm]	20.51	18.74	17.72	20.81	22.29
	Lower [mm]	22.78	27.57	24.71	30.27	37.20

Table 5. Values of the sample from steel with thickness 15 mm, cutting technology AWJ

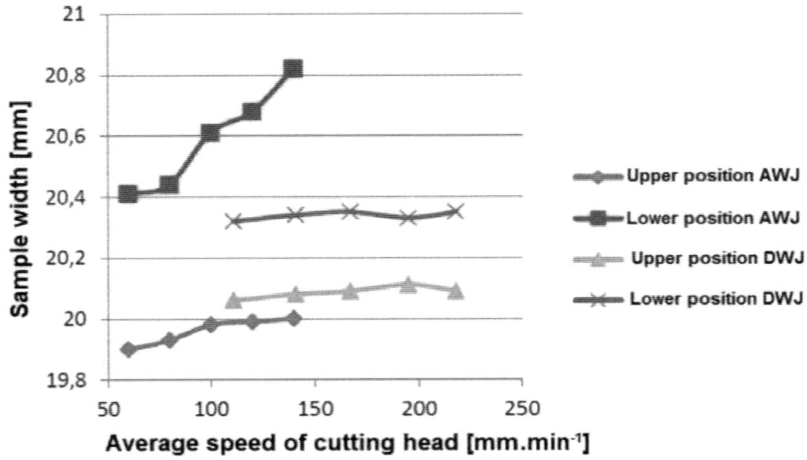

Fig. 5. Dependence of the sample width on the speed of cutting head (aluminium, width 15 mm)

In Fig. 5 for cutting steel with thickness 15 mm it is obvious that the chamfer when cutting with DWJ technology is significantly lower than with AWJ technology. Also, the value of the chamfer when using DWJ technology is almost constant when increasing the average speed of the cutting head. There is also a minimal increase in the width dimensions of the sample when increasing the average speed of the cutting head. However, with AWJ cutting technology, the value of the lower measurement position rises dramatically with the increasing speed of the cutting head. Therefore, the width of the sample in the lower position (chamfer) increases. [5]

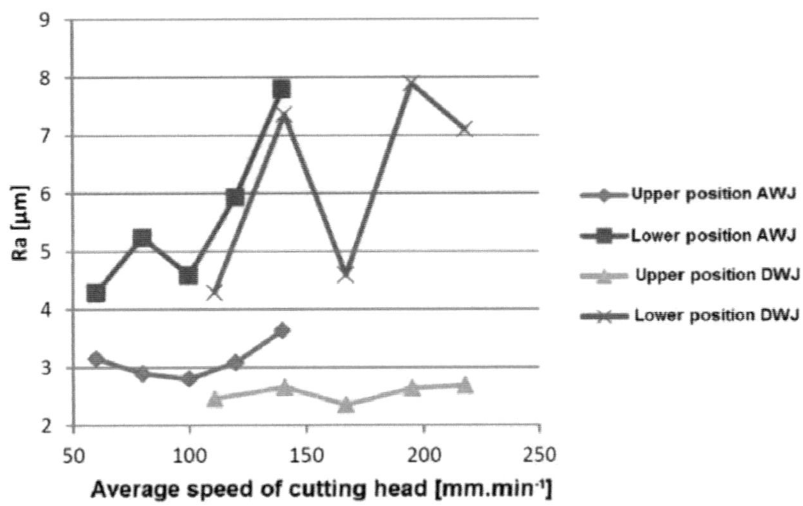

Fig. 6. Dependence of the Ra cut surface on the speed of cutting head (steel, width 15 mm)

Fig. 6 shows that when cutting steel with thickness 15 mm using DWJ technology, lower values of Ra (at speeds comparable with AWJ technology) are achieved. The growth of Ra values when using AWJ technology is small, if the notional boundary of the standard speed is not exceeded. Then comes a sharp increase in values in the lower position. Deviations of Ra (sharp increase of the value) when using DWJ cutting technology were observed. This deviation could be caused for example by placing the measuring tip into a place with distinct striations. Measured values of Ra in the upper positions cut with DWJ technology are very low. In standard operating conditions, the values of Ra measured on samples cut with DWJ technology were lower than with AWJ technology. [1]

The figures below are samples of AWJ and DWJ cutting technologies (steel, 15 mm thickness). Two samples are placed side by side on one flat surface. One was cut with the lowest average feed rate of the cutting head (on the highest quality of the cut – 60%), the second with the highest average feed rate of the cutting head (at the lowest quality – 140 %). Fig. 7a shows examples of cutting steel using AWJ technology. The large chamfer is noticeable in the figure, which is produced when cutting with AWJ technology.

Fig. 7b shows examples of cutting steel using DWJ technology. In this case the chamfer is minimal, and hardly recognizable by eye.

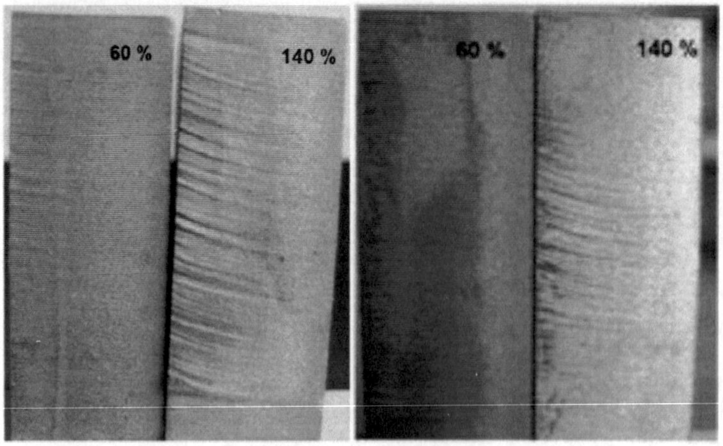

Fig. 7. Waterjet cutting (a) using technology AWJ (b) using technology DWJ

In both figures the deteriorating quality of the surface is obvious as the speed of the cutting head increases. Visual inspection reveals the lower quality of the sample surfaces (in this material) which are cut using DWJ technology.

4. Evaluation of the experiment

A total of 60 samples were cut in the experiment. Half the samples were cut by AWJ technology and the second half by DWJ technology. Evaluation shows that cutting using DWJ technology fulfils expectations regarding the chamfer of the samples. A minimal chamfer was measured on all the samples (two kinds of material) in comparison with the measured values from the AWJ technology. In most cases with cutting technology DWJ the chamfer increased minimally with increasing average speed of the cutting head. When comparing the size of the chamfers of each kind of material it was found that there are minor differences in the upper and lower positions on the aluminum samples. This is probably because it is a material with a lower specific density, meaning that the waterjet penetrates it more easily than steel without changing direction. Using AWJ technology there was a visible change in roughness when cutting corners. This was caused by the deceleration of the waterjet when it entered the corners. [2],[3]

5. Conclusion

For the second comparison, comparing the average arithmetic deviations of the profile Ra, the assumption has not been confirmed. Only when cutting steel with thickness 15 mm using DWJ technology were lower Ra values achieved than when cutting with AWJ technology. The comparison was made in terms of feed rate speed, which is used on each machine to achieve the quality of cut corresponding to the standard.

When measuring the arithmetical deviations of the profiles on the samples from aluminum and steel it was found that the expected results were not achieved. The samples from aluminum and steel cut by AWJ technology have significantly better Ra values than the samples using DWJ technology. Also, the differences between the upper and lower Ra values were significantly higher when cutting with DWJ technology. The results using the various methods from cutting at varying average speeds of the cutting head were compared. The cutting heads had enough velocity to achieve the optimal quality of the cut on the machine. [8]

At first sight, visual comparison of the samples reveals smaller chamfers on both materials when using DWJ technology. This method does not generate a change of roughness on the corners, which arises when using AWJ technology due to the slowing of the cutting head.

Regarding the quality of the surface and depth of the striations in the aluminum samples, the samples cut using DWJ technology seem visually worse. This finding refutes the assumption that there would be higher quality cutting for all the sample materials. The differences in values of the quality for aluminum are probably due to its low specific density. The waterjet with abrasive passes through the material easier (due to a pressure of 5500 bar and high average cutting speed) and leaves significant grooves and striations. On steel samples with a thickness of 15 mm, the samples cut with DWJ technology appear visually better, which confirms the assumption. [9]

For better results would be necessary to measure samples on the multiple levels. Sample measure also at the beginning, where the speed of the head increases, in the middle where the feed rate is highest and at the end, where cutting head slows down.

6. Acknowledgements

This paper is based upon work sponsored by project SGS-2016-005

7. References

[1] Loeser, C. Duerr, H. & Pilz, R. Application of the Micro Water Abrasive Injector Fine Jet for Precision Machining, Annals of DAAAM for 2010 & Proceedings of the 21st International DAAAM Symposium, Volume 21, No. 1, ISSN 1726-9679 ISBN 978-3-901509-73-5, Editor B. Katalinic, Published by DAAAM International, Vienna, Austria, EU, 2010

[2] Begic-Hajdarevic. D. Cekic. A. Mehmedovic. M. Djelmic. Experimental Study on Surface Roughness in Abrasive Water Jet Cutting 25th DAAAM International Symposium on Intelligent Manufacturing and Automation, DAAAM 2014 Experimental Study on Surface Roughness in Abrasive Water Jet Cutting, Procedia Engineering 100 (2015) 394 – 399

[3] D. Krajcarz, Comparison metal water jet cutting with laser and plasma cutting, in: Proceeding of 24th DAAAM International Symposium on Intelligent Manufacturing and Automation, Procedia Engineering 69 (2014) 838-843.

[4] F. Boud, L. F. Loo, P. K. Kinnell, The Impact of Plain Waterjet Machining on the Surface Integrity of Aluminium 74752nd CIRP Conference on Surface Integrity (CSI), Procedia CIRP 13 (2014) 382 – 386

[5] G. Holmqvist, U. Honsberg, Sensitivity analysis of abrasive waterjet cutting economy, in: Proceeding of 19th International Conference on Water Jetting, Nottingham University, UK, 2008, pp. 273–287 [6] http://www.insse.ro, (Year). The National Statistics Institute, Statistics Year Book, Accessed on: 2009-08-13

[7] M.Gent, M.Menédez, S.Torno, J.Toraño, Experimental evaluation of the physical properties required of abrasives for optimizing waterjet cutting of ductile materials, Wear, Volumes 284–285, 25 April 2012, Pages 43–51

[8] A. Akkurt, M.K. Kulekci, U. Seker, F. Ercan, Effect of feed rate on surface roughness in abrasive waterjet cutting applications, Journal of Materials Processing Technology 147 (2004) 389–396.

[9] Borkowski J., Sutowska M. (2007). The quality of surface cut after high-pressure abrasive-water cutting, Journal of Machine Engineering,Vol.7(3), pp.19-28

DOI: 10.2507/28th.daaam.proceedings.070

INFLUENCE OF CUTTING CONDITIONS ON THE SURFACE QUALITY AND GRINDING WHEEL WEAR DURING CYLINDRICAL GRINDING OF INCONEL 718

Tomas Baksa, Ondrej Hronek, Jindrich Farsky & Miroslav Zetek

This Publication has to be referred as: Baksa, T[omas]; Hronek, O[ndrej]; Farsky, J[indrich] & Zetek, M[iroslav] (2017). Influence of Cutting Conditions on the Surface Quality and Grinding Wheel Wear During Cylindrical Grinding of Inconel 718, Proceedings of the 28th DAAAM International Symposium, pp.0500-0505, B. Katalinic (Ed.), Published by DAAAM International, ISBN 978-3-902734-11-2, ISSN 1726-9679, Vienna, Austria
DOI: 10.2507/28th.daaam.proceedings.070

Abstract

This article deals with the experimental grinding of Inconel 718. This material is used for highly stressed components in the nuclear industry, such as combustion turbines. The material is very problematic for machining due to its rapid hardening and high heat generation. Two types of Inconel 718 in the form of testing bars were finished by cylindrical grinding using different cutting conditions and strategies. The main aim is to determine the impact of grinding conditions on the surface quality in terms of surface roughness after grinding. The resulting surface was scanned using an optical-scanning device to observe the topography of the ground surface. The grinding wheel was measured after grinding each bar to observe the wheel wear and the changes of the grinding wheel shape. The spindle load was monitored during grinding to see the effect of different strategies on the grinding process. The results from this work will be used in further studies and contribute to a better understanding of grinding Inconel 718.

Keywords: Grinding; Inconel 718; Roughness; Wheel wear; Cutting conditions

1. Introduction

Nickel-based alloys such as Inconel 718 are high-strength materials which are useful in many applications. It is used for example in the automotive, aerospace and nuclear industries for heat stressed components due to its excellent mechanical properties at high temperatures. Higher utilization of this material leads to higher demands on the productivity of component processing and production. However, the machining of Inconel 718 is very problematic due to its high cutting resistance. High strength at high temperatures and hardening during machining causes extreme cutting tool wear. Carbide cutting tools are widely used for machining Inconel 718 in the form of solid cutting tools or cutting inserts. Several studies [1, 2] deal with the lengthening of the cutting tool life by modification of the cutting edge. It was found that using an optimal value of the edge radius is important for good adhesion of the thin layer and longer tool life. For high performance cutting, CBN cutting tools in the form of cutting inserts are used. However, CBN cutting tools are characterized by higher cost. CBN and carbide cutting inserts are used for example for turning cylindrical components. However, the surface quality after turning is limited by the technology and it is very difficult to achieve low roughness.

Mechanical properties of components predominantly depend on the surface quality after machining. Therefore it is very important to ensure the surface quality of components for applications in areas such as the nuclear and aerospace industries. Low roughness and high surface quality can be achieved by using grinding technology where the material is removed by geometrically undefined cutting edges. Compared to other machining processes, grinding is characterized by higher specific energy. Most of this energy is changed into heat which has harmful effects on the surface quality as well as the tool wear [3]. During grinding Inconel 718, there is an accumulation of heat in the grinding zone. This is caused by the lower thermal conductivity of the material. The accumulation of the heat results in severe thermal damage on the ground products. An appropriate choice of grinding wheel is necessary to ensure a stable and effective grinding process.

Some conventional abrasives such as diamonds cause a chemical affinity with Inconel, which leads to sticking of the grinding wheel chips on the grinding wheel. Corundum grinding wheels are suitable for grinding materials such as Inconel 718. An advantage of corundum grinding wheels is their low cost. However, the wheel wear during grinding is relatively high, which leads to frequent truing of the grinding wheel. Study [4] deals with an assessment of the embodied energy of a common corundum grinding wheel and a coated cemented carbide end mill. It was found that higher sustainability is achieved by using the grinding wheel rather than the end mill, despite a much higher embodied energy. This is because the grinding wheel is capable of machining a considerably higher volume of material.

The grindability of Inconel 718 using a corundum and a silicon carbide (SiC) grinding wheel is described in [5]. It was found that the corundum grinding wheel is more effective than the SiC wheel for grinding Inconel 718. The higher grinding temperature leads to a chemical reaction between SiC grit and the workpiece material which causes severe attrition wear. One way to reduce wheel wear is to use a suitable grinding wheel material. CBN grinding wheels are characterized by high wear resistance, which increases grinding wheel durability. This is important especially for profile wheels such as radius shape wheels where it is desirable to keep the shape of the grinding wheel for a long time.

Grinding Inconel 718 with different grinding wheels was investigated by Tso [6], where the effect of various process parameters on the surface roughness, dimension accuracy and the cutting force were observed. The experiments led to the conclusion that a CBN grinding wheel is more suitable for grinding Inconel 718 than the SiC and corundum wheels. The grinding process creates heat which can lead to residual tensile stress on the surface after grinding. It may cause micro-cracks on the surface which reduce the fatigue life of the components. To reduce the residual stress after grinding we must first ensure a suitable and stable grinding process by choosing the appropriate cutting parameters and environment.

According to [7], residual tensile stress can be transformed to compressive stress by embedding an added heat source inside the workpiece. The grinding environment plays a significant role in grinding nickel based superalloys such as Inconel. The effects of different grinding environments such as dry, minimum quantity lubrication (MQL) and Cryo-MQL on grinding and surface performance are described in [8]. It was found that a Cryo-MQL environment achieved better results in terms of grinding forces, temperature and roughness compared to other environments. Research [9] deals with the application of nanofluids in small quantity lubrication mode during grinding Inconel 718.

The main aim of this research is to determine the influence of several different cutting parameters on surface roughness after grinding and the wheel wear on a CBN grinding wheel.

2. Grinding details

Several testing bars of Inconel 718 were ground with the same CBN grinding wheel under different grinding conditions. All the test bars have the same profile shape and the same dimensions. Roughing of the profile was carried out on a turning machine with carbide cutting inserts. Grinding was performed on a CNC 5-axis tool grinding machine with high precision.

The dimensions of the test bar are shown in Fig. 1. Testing bars were turned to diameter 5.25 mm which gives an allowance of 0.25 mm on the diameter for grinding. The profile of the test bar is composed of a linear part and a radius on each side. Therefore the radius grinding wheel was used to ensure the smooth transition and point contact between the grinding wheel and the surface. A small grain size of the wheel was chosen to achieve low surface roughness. The parameters of the grinding wheel are given in Tab. 1.

Wheel shape	Diameter [mm]	Abrasive rim height [mm]	Radius [mm]	Grain size	Clamping diameter [mm]
3F1	150	3	2	B33	32

Table 1. Grinding wheel parameters

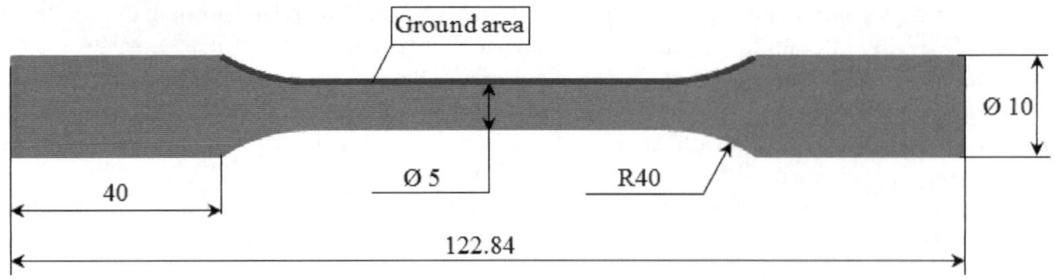

Fig. 1. Dimensions of the test bar

High-performance grinding fluid SintoGrind TTK was used to reduce temperature and friction during grinding. This mineral oil is designed for use in all grinding applications and for different materials. Coolant pipes were set to deliver the fluid to the grinding point. Coolant and the setting were constant for all cutting conditions. One cylindrical side of the testing bar was clamped into a collet of the machine and the second cylindrical side was supported by a steady bed to ensure a fixed position without bending. The steady bed of the support is V-shaped for a precise fit of the cylindrical part. Cutting conditions were changed during testing in terms of cutting speed, workpiece speed, feed rate, depth of cut and direction of rotation. The surface after cylindrical grinding was analysed using an optical-scanning device to measure the surface roughness. The grinding wheel profile was scanned before and during the grinding tests using an optical measuring device to observe the progress of the wheel wear.

Two types of Inconel 718 were tested during the experiment. Inconel 718 produced by conventional technology was used for grinding tests with different cutting conditions. Selected grinding conditions were used for grinding the second type of Inconel 718 which was produced by additive technology. The purpose of this is to observe the differences in grinding Inconel 718 produced by conventional and additive technology. Grinding was compared in terms of the surface roughness and the spindle load during grinding. Chemical composition of both materials is shown in Tab. 2.

[%]	**C**	**Mn, Cu**	**Si**	**P, S**	**Cr**	**Al**	**Mo**	**Ni**	**Ti**	**Fe**	**Nb**	**Co**	**B**
Conventional	0.01	0.02	0.05	<0.004	18.5	0.51	3.04	54.2	0.95	R17.64	4.95	0.03	0.004
Additive technology	<0.08	<0.35	<0.35	<0.015	17 - 21	0.2 - 0.8	2.8 - 3.3	50 - 55	0.65 - 1.15	balance	4.75 - 5.5	<1	<0.06

Table 2. Chemical composition of two types of Inconel 718

3. Results and discussion

Several different grinding conditions were used for grinding test bars made of Inconel 718 to find the best results in terms of surface roughness. The strategy for cylindrical grinding is shown in Fig. 2. Table 3 shows an overview of all the grinding conditions used.

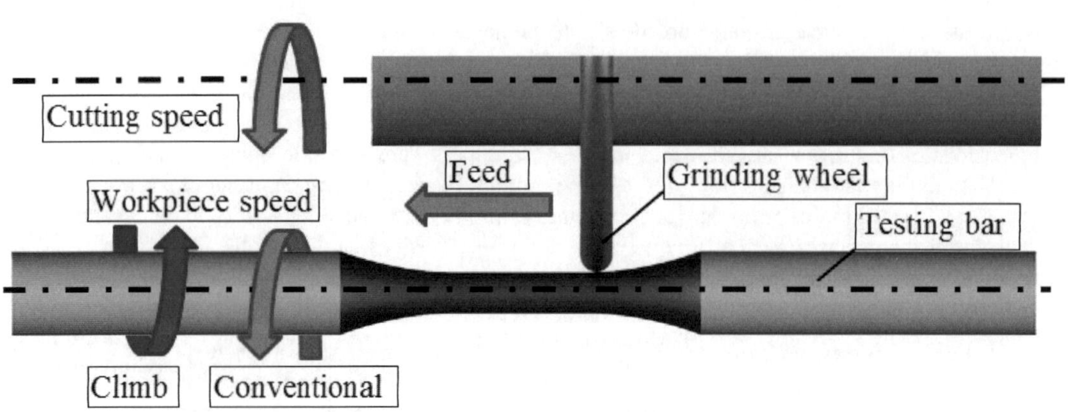

Fig. 2. Strategy of grinding

Test	Depth of cut a_p [mm]	Feed rate v_f [mm/min]	Workpiece speed n_w [rpm]	Cutting speed v_c [m/s]	Direction of rotation
T01	0.005	5	100	25	Conventional
T02	0.02	20	100	25	Conventional
T03	0.01	5	200	25	Conventional
T04	0.005	5	200	25	Climb
T05	0.005	5	200	25	Conventional
T06	0.005	3	200	25	Conventional
T07	0.005	5	100	30	Conventional
T08	0.005	5	100	35	Conventional
T09	0.005	5	150	35	Conventional
T10	0.005	5	200	35	Conventional
T11	0.003	5	100	25	Conventional
T12	0.005	5	100	25	Conventional

Table 3. Grinding conditions

Grinding conditions in test T01 were chosen as the reference grinding conditions with small depth of cut, slow feed rate and slow workpiece speed. The cutting speed was determined based on the grinding wheel limitations and recommendations. Conditions in T01 were also used in T12 where the testing bar produced by additive technology was ground. Surface roughness after each test is shown in Fig. 3. Test T02 was characterized by deeper cut and higher feed rate compared to other tests. A deeper cut allows more material to be removed during one pass. A higher feed rate allows faster passes during grinding. This combination of both is suitable for achieving low grinding times. However, this test is characterized by high surface roughness due to high scallop height left after grinding. Higher scallop height is caused by low workpiece speed which leads to high feed rate per workpiece revolution. Cutting conditions in T02 were evaluated as suitable for roughing, but not for finishing of Inconel 718. In test T03, feed rate was rapidly decreased and the workpiece speed was increased to reduce scallop height after grinding. The depth of cut was also reduced by half. This resulted in a significantly better surface than for T02. Several different combinations of workpiece speed and cutting speed were tested in T07 - T10. It was found that all these tests achieve very similar results for surface roughness to the reference conditions (T01). The best results were achieved by T09 and T10 where a high cutting speed and high workpiece speed were used. This combination proved to be the best in terms of surface roughness after grinding. However, it is important to remember that high cutting speed of the grinding wheel and the high speed of the workpiece, whose vectors go against each other, cause higher friction which can lead to higher temperature and thermal damage of the ground surface. Reducing the depth of the cut in test T11 shows no difference in surface roughness compared to T01. Test T04 was characterized by a different rotation of the workpiece than in the other tests. Also the speed of the workpiece was increased to reduce scallop height. No significant difference between conventional and climb grinding was found in terms of surface roughness. In test T12, the test bar produced by additive technology was ground with the same cutting conditions as T01 (reference). The roughness and the grinding behaviour was comparable with T01.

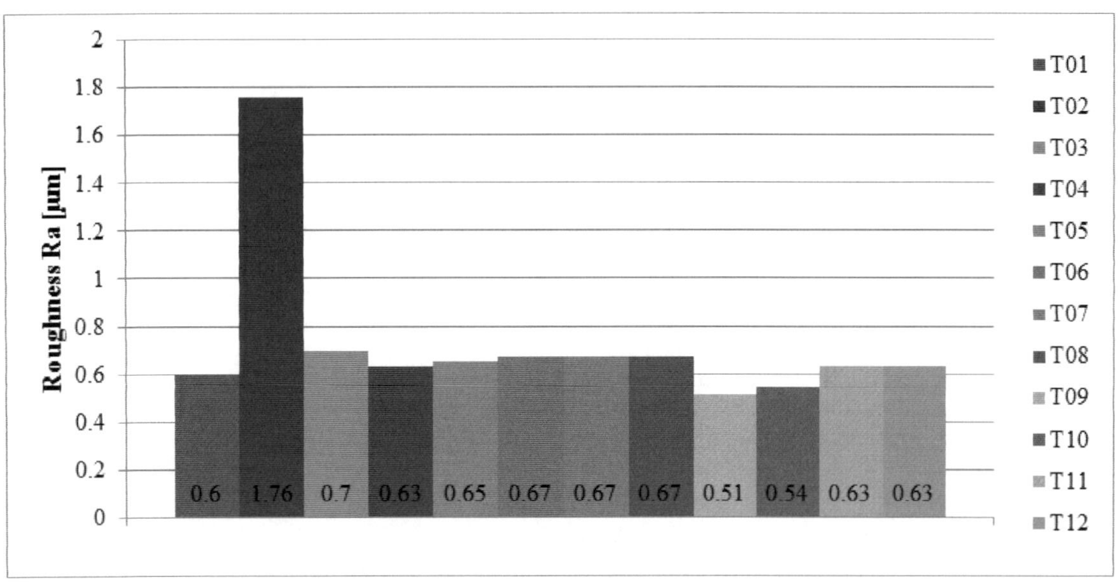

Fig. 3. Roughness after grinding Inconel 718

The profile of the radius grinding wheel was measured during grinding using an optical measuring device. Fig. 4a shows the wear of the grinding wheel after grinding eight test bars. All test bars have the same allowance after turning, so the volume of material removed during grinding was constant for all the bars. The red zone shows that grinding was performed on the top of the wheel profile. It seems that the wheel wear is slightly unsymmetrical and greater wear occurred at the bottom of the profile. This may be caused by inaccurate roughing of the profile during turning, which leads to a shifting profile. The wheel wear is shown in Fig. 4b. The greatest increase in wheel wear occurred after grinding the first test bar, when the new grinding wheel became stable. This phenomenon was expected. After that, the grinding wheel wear linearly increases with the number of test bars. There was no significant difference between conventionally produced Inconel 718 and Inconel 718 produced by additive technology in terms of grinding wheel wear.

The spindle load was measured during the grinding test to see the influence of the cutting conditions on the grinding process. The spindle load was around 3.3 % for all tests except test T02 where the spindle load was 3.5 %. This low spindle load was due to the small depth of cut, which was 0.005 mm for most tests. Test T02 achieved a slightly higher spindle load due to the greater depth of cut and higher feed rate. It can be said that grinding parameters such as cutting speed, workpiece speed and the feed rate have no significant influence on the spindle load if the depth of cut is small.

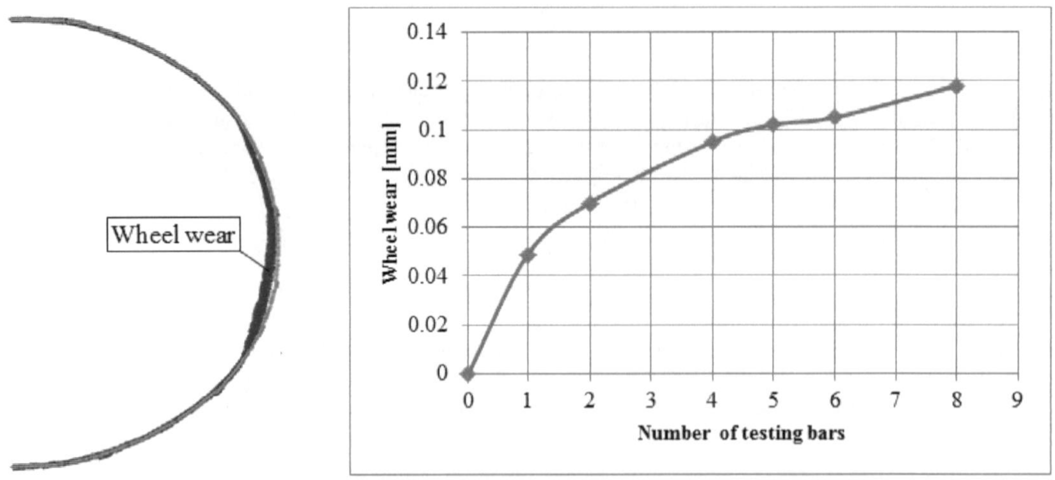

a) b)

Fig. 4. Profile of the grinding wheel (a) and the wheel wear (b)

4. Conclusion

This article deals with the experimental cylindrical grinding of Inconel 718 where several test bars with constant geometry were ground under different grinding conditions to see the effect on the surface quality in terms of the surface roughness. Inconel test bars were roughed using turning technology and then were finished using a radius CBN grinding wheel. The surface after grinding was scanned and measured to observe the surface roughness. It was found that tests T09 and T10 with a low feed rate, higher cutting speed (35 m/s) and higher workpiece speed (150-200 rev/min) achieved the best results in terms of surface roughness. The low feed rate, respectively the ratio between feed rate and the workpiece speed, is very important for achieving a low scallop height and thus the low surface roughness. Test T02 with high feed rate and a deeper cut achieved the worst result in terms of roughness and it was evaluated as suitable for roughing. Changing the direction of rotation of the workpiece during test T04 makes no difference to surface roughness. The profile of the grinding wheel was measured during grinding to see the progress of the wheel wear. Monitoring of the spindle load showed that all tests gave a spindle load of lower than 4 % due to the small depth of the cut during grinding. During the tests no less roughness than Ra = 0.5 μm was achieved. To achieve better surface roughness it would probably be necessary to use a grinding wheel with a smaller grain size. Two types of Inconel 718 were used for the test bars. The first was produced by conventional technology and the second by additive technology. It was confirmed that Inconel 718 produced by additive technology has the same grinding behaviour as conventional Inconel 718 in terms of achieved roughness, wheel wear and spindle load. The results from this research will be used for optimizing the cylindrical grinding of Inconel 718 and for better understanding of the behaviour during grinding. Further steps will be to determine the effect of different grinding wheels on grinding Inconel 718 to find out the most appropriate grinding wheel parameters.

5. Acknowledgements

This paper was supported by the programme of applied research, experimental development and innovation GAMA, No. TG02010011 - Support of UWB commercial opportunities.

6. References

[1] Zetek, M.; Česáková, I.; & Švarc, V. (2014). Increasing Cutting Tool Life when Machining Inconel 718, Procedia Engineering, vol. 69, pp. 1115–1124, ISSN: 1877-7058, Elsevier.

[2] Schorník, V.; Zetek, M.; & Bakša, T. (2015). Durability of the cutting tool with different thin layers when Inconel 718 is machined, Annals of DAAAM and Proceedings of the International DAAAM Symposium, pp. 678–682, ISBN 978-3-902734-07-5, ISSN 1726-9679, Vienna, Austria.

[3] Fritsche A. & Bleicher, F. (2015). Experimental Investigation of the Heat Flux Distribution in Grinding of Titanium Alloy, Procedia Eng., vol. 100, pp. 987–993, Elsevier.

[4] Kirsch, B.; Effgen, C.; Büchel, M. & Aurich, J. C. (2014). Comparison of the Embodied Energy of a Grinding Wheel and an End Mill, Procedia CIRP, vol. 15, pp. 74–79, Elsevier.

[5] Sinha, M. K.; Setti, D.; Ghosh, S. & Venkateswara Rao, P. (2016). An investigation on surface burn during grinding of Inconel 718, Journal of Manufacturing Processes, vol. 21, pp. 124–133, Elsevier.

[6] Tso, P.-L. (1995). Study on the grinding of Inconel 718, Journal of Materials Processing Technology, vol. 55, no. 3, pp. 421–426, Elsevier.

[7] Pei-zhuo, W.; Zhan-shu, H.; Yuan-xi, Z. & Shu-sen, Z. (2017). Control of Grinding Surface Residual Stress of Inconel 718, Procedia Engineering, vol. 174, pp. 504–511, Elsevier.

[8] Balan, A. S. S.; Vijayaraghavan, L.; Krishnamurthy, R.; Kuppan, P. & Oyyaravelu, R. (2016). An experimental assessment on the performance of different lubrication techniques in grinding of Inconel 751, Journal of Advanced Research, vol. 7, no. 5, pp. 709–718, Cairo University, Elsevier.

[9] Sinha, M. K.; Madarkar, R.; Ghosh, S. & Rao, P. V. (2017). Application of eco-friendly nanofluids during grinding of Inconel 718 through small quantity lubrication, Journal of Cleaner Production, vol. 141, pp. 1359–1375, Elsevier.

DOI: 10.2507/28th.daaam.proceedings.071

METHOD FOR SUPERVISORY IMPLEMENTATION OF MANIPULATION OPERATIONS BY UNDERWATER VEHICLES

Vladimir Filaretov, Alexander Konoplin, Nikita Konoplin & Bidyadhar Subudhi

This Publication has to be referred as: Filaretov, V[ladimir] F[edorovich]; Konoplin, A[lexander]; Konoplin, N[ikita] & Subudhi, B[idyadhar] (2017). Method for Supervisory Implementation of Manipulation Operations by Underwater Vehicles, Proceedings of the 28th DAAAM International Symposium, pp.0506-0512, B. Katalinic (Ed.), Published by DAAAM International, ISBN 978-3-902734-11-2, ISSN 1726-9679, Vienna, Austria
DOI: 10.2507/28th.daaam.proceedings.071

Abstract

At present, the underwater vehicles equipped with multilink manipulators are used to perform a wide range of survey, technological and research operations in the depths of the world ocean. But in most cases, said manipulation operations are performed with help of specially trained operators. Based on their experience and knowledge, these operators plan the trajectories of working tools of underwater manipulators. However, it is very difficult to quickly and accurately solve manipulation tasks, identifying the location of the object of work with help of video images. This leads to a decrease in the productivity and an increase in the probability of errors. The work proposes the new method for the supervisory implementation of manipulative operations by means of multilink manipulators mounted on underwater vehicles. This method involves the construction of mathematical models of objects of work with the help of on-board sonars. Herewith, the formation of target points and spatial trajectories of the manipulator's working tools is carried out by means of the targeting of the optical axis of the camera taking into account the requirements for the implementation of specific technological operations.

Keywords: supervisory implementation; underwater vehicle; multilink manipulator; underwater operations; spatial trajectory; control system

1. Introduction

The Analysis of current trends in the design and operation of underwater vehicles (UV) equipped with multi-link underwater manipulators (MMs) shows that the functionality of these vehicles is expanded due to the creation of the high-quality navigation systems and devices [1]. In addition, the analysis shows that new methods for the environment recognition and the motion trajectories planning of the UVs and their MMs [2, 3] also are created. At the same time, the majority of the underwater technological and research manipulation operations are still performed in a manual mode by the specially trained operators of the UVs. These operators basing on their experience and knowledge plan the trajectories of the movement of the MM working tools. However, the quickly and accurately performing of the manipulation operations without direct contact with the working object and with the information about its location from the video image only is very difficult for the operator. This leads to the performance decreasing of his work and to the probability increasing of the errors.

In particular, the indicated problem was confirmed in the deep-sea scientific research expeditions in the Japan, Bering and Okhotsk Seas when we have worked with the UV Sub-Atlantic Comanche 18 equipped with the Schilling Orion 7P MM. The manual taking of samples of the soil, the geological rocks and the bacterial mats (see Fig. 1) leads to the fast tiring of operators, to the significant increasing in the time of the operations performing, to the errors appearing in work and even to the working tools damage.

To solve this problem the work proposed and considered a new method of supervisory performing of manipulation operations using the MM installed on the UV.

Fig. 1. The performing of the manipulation operation by the Schilling Orion 7P MM

2. Description of the method of supervisory control of the MM installed on the UV

The proposed method firstly defines the geometric shape and the location of the objects of work with the help of the onboard sonars. Then the operator directs the camera installed on the UV to the working object and specifies one or more operating points belonging to this object. The camera can change the spatial orientation of its optical axis. The spatial trajectory of motion and the orientation of the MM working tool are automatically formed from the coordinates of the given points belonging to the surface of the working object and depends on the manipulation task.

The implementation of the supervisory control of a soil sampling box on the example of the taking of a surface layer of soil is shown in Fig. 2. This box is moved by the MM *1* installed on the UV *2*. The point cloud is formed via the multibeam hydroacoustic sonar *3* in a real time. The mathematical model *4* of the bottom surface is formed using the point cloud. This model is used to the further trajectory formation of the MM working tool. In particular, for the soil selection in the places of accumulation of bacterial mats, the operator specifies the s target points $A_i\,(i=\overline{1,s})$ via the moving of the optical axis *5* of the camera. These points are the intersection points of this axis with the bottom surface. The trajectory *6* of the movement of the MM working tool *2* is formed at the given target points. When the taking of surface layer of soil is performing, the vector $\vec{a}$ of the soil sampling box must always be perpendicular to the bottom surface, and the vector $\vec{n}$ must be directed to the next target point A_{i+1}, while the MM moves along the trajectory (see Fig. 2). Furthermore, the vector $\vec{n}$ is perpendicular to the vector $\vec{a}$.

In many cases, the surface of the bottom without significant unevenness can be represented in the form of an averaged plane [4]. This plane can be described by the equation in the normal form [5] in the rectangular coordinate system (CS) *xyz* rigidly connected to the UV body:

$$ax + by + cz = d, \tag{1}$$

where x, y and z are the axes of the connected CS. The origin of this CS is located in the UV center of buoyancy O, x axis coincides with the horizontal-longitudinal axis of the UV, z axis coincides with the vertical axis of the UV, y axis is right-handed. a, b and c are the direction cosines of the unit vector $\vec{m}=[a, b, c]$ of the normal to the plane (1), which coincides with the vector $\vec{a}$. The value d is the distance from the origin of the CS *xyz* to this plane. The elements of vector $\vec{m}$ and point d are calculated using the formed cloud of points belonging to the bottom surface, according to the method described in [6].

The coordinates of the target points A_i (see Fig. 2) must be calculated in the CS *xyz* for the forming of the MM working tool motion trajectories. If the camera is directed to the point A_i, then the position of this camera is determined in the CS *xyz* by the predetermined point $P_i=[P_{ix}, P_{iy}, P_{iz}]$ located on the optical axis. The orientation of the optical axis of the camera is determined by the unit vector $\vec{d}_i = [d_{ix}, d_{iy}, d_{iz}]$. In this case, the optical axis of the camera, which coincides with the vector $\vec{d}_i$, emerges from the point P_i to the direction of the bottom plane (1) and described by the equation:

$$A_i = P_i + t_i\vec{d}_i, \tag{2}$$

where t_i is the parameter varies within range $[0,\infty)$. The value of the parameter t_i at the point of intersection of the axis (2) with the plane (1) can be obtained by the substitution of the equation (2) in (1) [7]:

$$a(P_{ix} + t_i d_{ix}) + b(P_{iy} + t_i d_{iy}) + c(P_{iz} + t_i d_{iz}) - d = 0,$$

$$t_i = \frac{-(aP_{ix} + bP_{iy} + cP_{iz} - d)}{ad_{ix} + bd_{iy} + cd_{iz}}. \tag{3}$$

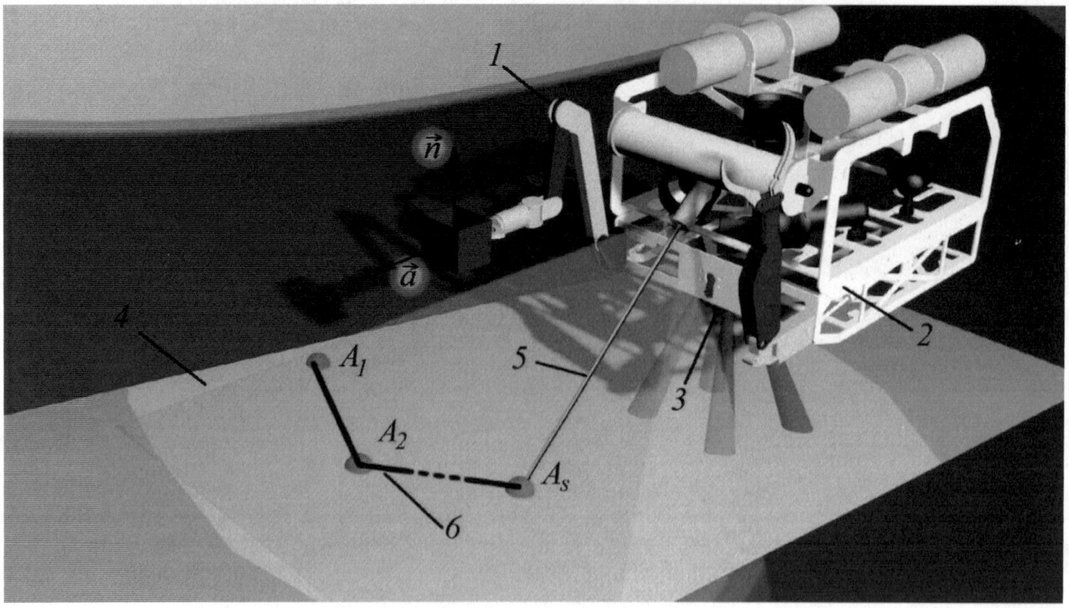

Fig. 2. The performing of an operation in supervisory mode

The coordinates of the point A_i of the intersection of the optical axis of the camera with the bottom plane are determined by the substitution of the found value of the parameter t_i into the equation (2):

$$A_i = P_i + \frac{-(aP_{ix} + bP_{iy} + cP_{iz} - d)}{ad_{ix} + bd_{iy} + cd_{iz}}\vec{d}_i.$$

After the operator directs the optical axis of the camera to the next target point A_{i+1}, the value of the parameter t_{i+1} is calculated using the equation (3) for the new point P_{i+1} and the vector $\vec{d}_{i+1}$. The coordinates of the remaining target points are determined in a similar way. As a result, the trajectory of the MM working tool is represented as its successive transition across the all target points.

3. Formation of the motion trajectory of the MM working tool in a complex bottom relief

If the bottom surface has a complex shape, then it is necessary to use the multi-beam 3D sonars [8, 9]. These sonars allow to accurately form the clouds of points, which belong to the bottom surface, in the work area of the UV. The triangulation model (see Fig. 3) is formed using these clouds and the known methods [10, 11]. This model is the set of triangular plates connected together. The motion trajectory of the MM working tool is formed on the basis of this model. The trajectory passes through the A_i target points, which specified by the operator, on the complex bottom relief. Each k-th triangle in the triangulation is given in the CS xyz by the coordinates of three vertices V_{k0}, V_{k1}, $V_{k2} \in R^3$, where $k = \overline{1, g}$, g – number of the triangles.

To find the coordinates of each target point A_i at the intersection of the optical axis of the camera and the resulting triangulation surface, it is necessary to determine which of the g triangles are intersected by the axis (2) at the current time. For this, the Moller-Trumbor algorithm [12] is used, which allows to determine the k-th triangle which is intersected by the axis (2). This algorithm uses the information about the coordinates of the point P_i, the elements of the vector $\vec{d}_i$, and also the coordinates of the vertices V_{k0}, V_{k1}, V_{k2} of the all g triangles. The coordinates of the point A_i in the CS xyz with respect to the vertices of the k-th triangle can be defined in vector form as.

$$A_{ik} = w_{ik}V_{k0} + u_{ik}V_{k1} + v_{ik}V_{k2}\,, \tag{4}$$

where u_{ik}, v_{ik}, w_{ik} are the barycentric coordinates of the point A_i with respect to the vertices of the k-th triangle ($u_{ik} + v_{ik} + w_{ik} = 1$).

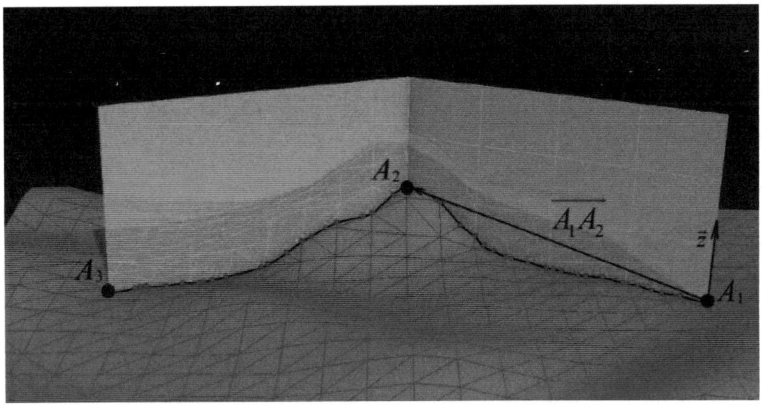

Fig. 3. Triangulation model of the bottom surface

After the calculation of $w_{ik} = 1 - (u_{ik} + v_{ik})$ and the substituting of the axis equation (2) into the equation (4) we obtain a system of equations with the parameters t_{ik}, u_{ik}, v_{ik}

$$P_i + t_{ik}\vec{d}_i = (1-(u_{ik}+v_{ik}))V_{k0} + u_{ik}V_{k1} + v_{ik}V_{k2}\,,$$

which can be written in a matrix form

$$\begin{bmatrix} -\vec{d}_i & V_{k1}-V_{k0} & V_{k2}-V_{k0} \end{bmatrix} \begin{bmatrix} t_{ik} \\ u_{ik} \\ v_{ik} \end{bmatrix} = \begin{bmatrix} P_i - V_{k0} \end{bmatrix}. \tag{5}$$

The values of the parameters t_{ik}, u_{ik}, v_{ik} are found by the solving of the system of equations (5) [12]:

$$\begin{bmatrix} t_{ik} \\ u_{ik} \\ v_{ik} \end{bmatrix} = \frac{1}{(\vec{d}_i \times (V_{k2}-V_{k0}))\cdot(V_{k1}-V_{k0})} \begin{bmatrix} ((P_i - V_{k0})\times(V_{k1}-V_{k0}))\cdot(V_{k2}-V_{k0}) \\ (\vec{d}_i \times (V_{k2}-V_{k0}))\cdot(P_i - V_{k0}) \\ ((P_i - V_{k0})\times(V_{k1}-V_{k0}))\cdot\vec{d}_i \end{bmatrix}, \tag{6}$$

where ($\cdot$) is the dot product of vectors; ($\times$) is the cross product of vectors.

As a result, the triangle k with the vertices V_{k0}, V_{k1} and V_{k2}, which intersects the axis (2), is found from the equations [12]

$$0 \le u_{ik} \le 1, \tag{7}$$

$$0 \le v_{ik} \le 1, \tag{8}$$

$$u_{ik} + v_{ik} \le 1. \tag{9}$$

The coordinates of the point A_i of the intersection of the optical axis of the camera with the found triangle k are determined by the substituting of the found parameter t_{ik} from the formula (6) into the equation (2):

$$A_i = P_i + \frac{((P_i - V_{k0})\times(V_{k1}-V_{k0}))\cdot(V_{k2}-V_{k0})}{(\vec{d}_i \times (V_{k2}-V_{k0}))\cdot(V_{k1}-V_{k0})}\vec{d}_i\,.$$

In order to create the MM working tool trajectory passing through the all the obtained points A_i this trajectory is represented as sequence of the triangulation surface cuts. These cuts are the vertical profiles of parts of this surface between the neighboring target points. These profiles can be represented in the form of a set of intersection points of the edges of the triangles in the triangulation and the plane formed by the vector $\overrightarrow{A_iA_{i+1}}$, which connects the neighboring target points, and the unit vector $\vec{z}$, which is parallel to the z axis of the CS xyz (see Fig. 3).

For the creation of the MM working tool trajectory, which has to pass through the intersection points of this plane and triangles, it is necessary to calculate which of the all g triangles are intersected by the plane. If the k-th triangle and the plane formed by the vectors $\overrightarrow{A_iA_{i+1}}$ and $\vec{z}$ intersect, then one triangle vertex will be on the opposite side of the plane than the other two. According to algorithm [7], the computing of the signed distance p_{ikh} between each of h-th vertex ($h = 0$, 1, 2) and this plane and comparing their signs can immediately determine if an intersection exists. The values p_{ikh} are calculated by the formula $p_{ikh} = \vec{r}_i \cdot \overrightarrow{A_iV_{kh}}$, $\vec{r}_i = \dfrac{\overrightarrow{A_iA_{i+1}} \times \vec{z}}{\left\| \overrightarrow{A_iA_{i+1}} \times \vec{z} \right\|}$, where $\overrightarrow{A_iV_{kh}}$ is the vector connecting the target point A_i and the h-th vertex V_{kh} of the k-th triangle.

If for any pair of the vertices of the triangle k the corresponding values p_{ikh} have the same signs ($p_{ik0}p_{ik1} > 0$ and $p_{ik0}p_{ik2} > 0$ or $p_{ik0}p_{ik1} < 0$ and $p_{ik0}p_{ik2} < 0$), then the plane does not intersect the k-th triangle. If p_{ikh} has different signs ($p_{ik0}p_{ik1} > 0$ and $p_{ik0}p_{ik2} < 0$ or $p_{ik0}p_{ik1} < 0$ and $p_{ik0}p_{ik2} > 0$), then the point T_{ike} lies on the edge of the triangle k, which is formed by this pair of vertices ($e = 1,2$, because in the general case the edges of each triangle can't have more than two points of intersection with the plane). The MM working tool must pass through this point. The coordinates of the point T_{ike} for the edge formed by the vertices V_{k0} and V_{k1} are determined by the formula [7]: $T_{ike} = V_{k0} + t_{ike} \dfrac{\overrightarrow{V_{k0}V_{k1}}}{\left\| \overrightarrow{V_{k0}V_{k1}} \right\|}$, where $t_{ike} = \dfrac{-(\vec{r}_i \cdot \overrightarrow{OV_{k0}} + \left| \overrightarrow{OA_i} \cdot \vec{r}_i \right|)}{\vec{r}_i \cdot \overrightarrow{V_{k0}V_{k1}}}$, $\overrightarrow{V_{k0}V_{k1}}$ is the vector connecting the vertices V_{k0} and V_{k1} of the triangle k [7], $\left| \overrightarrow{OA_i} \cdot \vec{r}_i \right|$ - the distance from the plane, which formed by the vectors $\overrightarrow{A_iA_{i+1}}$ and $\vec{z}$, to the initial point O of the CS xyz.

The coordinates of the second point T_{ike}, which lies on the one of the remaining two edges formed by the vertices V_{k0} and V_{k2} or V_{k1} and V_{k2} of the k-th triangle, are calculated similarly. If $p_{ikh} = 0$, then the h-th vertex of the k-th triangle lies on the plane. In this case $T_{ike} = V_{kh}$.

As a result, the set of points T_{ike} through which the MM needs to move the working tool (in particular, the soil sampling box) are formed. Moreover, the vector $\vec{a}_{ik}$ in the points T_{ik1} and T_{ik2} will always be perpendicular to the plane of the k-th triangle, and also when the MM working tool is moving between these points. Therefore, the vector $\vec{a}_{ik}$ in the CS xyz is always perpendicular to the plane defined by the three vertices of the k-th triangle.

The resulting set of points T_{ike} with the corresponding vectors $\vec{a}_{ik}$ is sorted into the sequence of points T_j with the vectors $\vec{a}_j$ for the creation of the MM motion trajectory. The points T_j are located between the neighboring target points A_i and A_{i+1}, where j is the sequence number of the point in the moving sequence of the MM working tool. The vector $\vec{n}_j$ of the working tool is always parallel to the vector $\overrightarrow{T_jT_{j+1}}$ connecting the passed and the next points of the sequence.

As a result, the obtained sequence of points T_j and corresponding vectors $\vec{a}_j$ and $\vec{n}_j$ forms the trajectory of the working tool movement, and also can set the different modes of this motion [13].

4. Program implementation and investigation of the supervisory control method

The developed method of supervisory control of the MM is implemented in the C ++ programming language. The input data for the created program is the set of points belonging to the bottom surface and obtained via hydroacoustic sonars. These points are determined in three-dimensional space (see Fig. 4a). This set is loaded into the program as a text file. The coordinates of the points P_i and the vectors $\vec{d}_i$ which determine the location of the optical axis of the camera in the CS xyz are recorded in this program.

The developed program builds a triangulation model of the bottom surface (see Fig. 4b). This program makes it using the algorithm [10] implemented in the Point Cloud Library [11]. After the operator sets the points P_i and the vectors $\vec{d}_i$, the program calculates the coordinates of the target points A_i, and also forms the sequence of points T_j (see Fig. 4b, 4c), and their corresponding vectors $\vec{a}_j$ and $\vec{n}_j$ of the MM working tool. The visualization of the results of the created program was performed using the 3ds Max software product and the ParaView graphic software package.

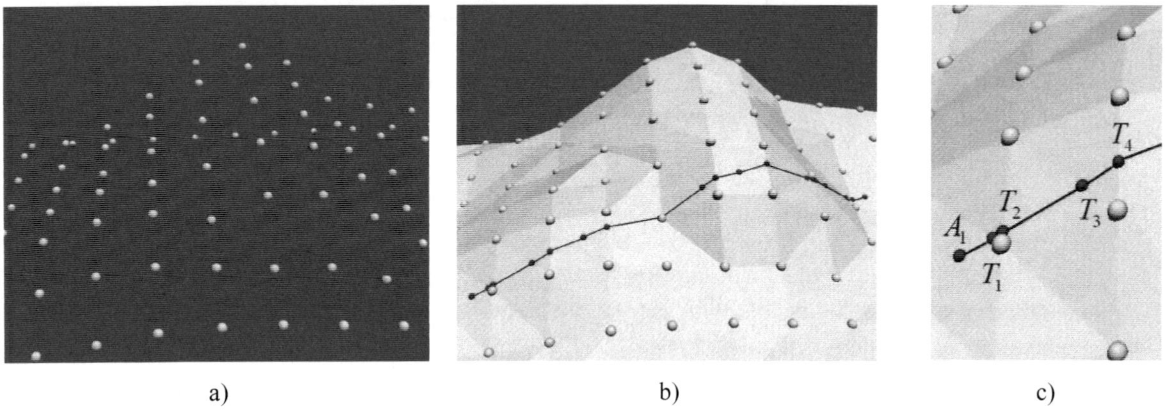

Fig. 4. a) set of points of the bottom surface in the CS *xyz*, b) triangulation model of the bottom and the formed trajectory of the MM working tool, c) detailed description (presentation) of the trajectory

5. Conclusion

The underwater technological and research manipulation operations are still performed only in a manual mode by the specially trained operators of the UVs. But quick and accurate performing of the manipulation operations is very difficult for the operator. This leads to the performance decreasing of his work and to the probability increasing of errors.

The developed method can be used to perform many underwater manipulation operations in the supervisory mode. The trajectories and orientation of the MM working tools in the process of motion are formed on the basis of the mathematical model of the work object and taking into account the requirements for the performing of specific technological operations. The developed method is implemented like a program that allows to build triangulation models of the seabed surface and calculate the spatial motion trajectories of the MM working tools.

The future research will include the creation of models of working objects based on the information from the technical vision systems, and the formation of smooth spatial trajectories of the MM working tools. The control systems will be created based on the developed method. These systems will be introduced into the underwater robotic complex, which will be used to perform real technological and research underwater operations.

6. Acknowledgments

This work was performed in Institute of Marine Technology Problems FEB RAS, Institute for Automation and Control Processes FEB RAS and also in Far Eastern Federal University (Vladivostok, Russia).

The development and software implementation of the method for the supervisory performing of manipulation operations was carried out with the financial support of the Russian Science Foundation (RSF) grant (project №17-79-10064). The theoretical and experimental works on the construction of mathematical models of the bottom surface was carried out with the financial support of the Russian Foundation for Basic Research (RFBR) grants (projects 16-29-04195 Ofi_m, 17-57-45055 Ind_a).

7. References

[1] Dubrovin F.S., Scherbatyuk A.F. (2015). Development of Algorithms for an Autonomous Underwater Vehicle Navigation with a Single Mobile Beacon: The Results of Simulations and Marine Trials // Proc. of the XXIIth International Conference on Integrated Navigation Systems, ICINS 2015, Saint Petersburg, Russia. P. 144-152. ISBN 978–5-91995-023-3.

[2] Filaretov V.F., Konoplin A.Yu. (2015). System of Automatically Correction of Program Trajectory of Motion of Multilink Manipulator Installed on Underwater Vehicle. Procedia Engineering, Vol. 100, 2015. pp. 1441-1449.

[3] Penalver A., Perez J., Fernandez J.J., Sales J., Sanz P.J., Garcia J.C., Fornas D., Marin R. (2015). Visually-guided manipulation techniques for robotic autonomous underwater panel interventions. Annual Reviews in Control 40, pp. 201–211, 2015.

[4] Filaretov V.F., Konoplin A.Yu., Konoplin N.Yu., Gorbachev G.V. (2016). Control system for underwater vehicle with multilink manipulator for automatic manipulation operations, Proceedings of the 27th DAAAM International Symposium, B. Katalinic (Ed.), Published by DAAAM International, Vienna, Austria, 2016, – pp. 714-720. DOI: 10.2507/27th.daaam.proceedings.103.

[5] Korn G.A., Korn T.M. (2000). Mathematical Handbook for Scientists and Engineers: Definitions, Theorems, and Formulas for Reference and Review. General Publishing Company, 2000. - 1151 p.

[6] Filaretov V.F., Konoplin A.Yu., Konoplin N.Yu. (2017). System for automatically performing of manipulation operations with help of underwater robot. Mechatronics, automation, control. 2017. Iss. 8. - pp. 543-549. (Филаретов В. Ф., Коноплин А.Ю., Коноплин Н.Ю. Система для автоматического выполнения манипуляционных операций с помощью подводного робота // Мехатроника, автоматизация, управление. 2017. №8. Т. 18. С. 543–549.)
[7] Schneider, P.J., and Eberly, D.H. (2002) Geometric Tools for Computer Graphics, Elsevier Science Inc., New York.
[8] Tritech Eclipse. Multibeam Sonar for 3D Model View of Sonar Imagery http://www.tritech.co.uk/
[9] BlueView 3D Multibeam Scanning Sonar http://www.teledynemarine.com/
[10] Marton Z., Rusu R., Beetz M. (2009). On Fast Surface Reconstruction Methods for Large and Noisy Datasets" Proceedings of the IEEE International Conference on Robotics and Automation (ICRA), 2009, Kobe, Japan, pp. 3218-3223.
[11] Point Cloud Library: Fast triangulation of unordered point clouds http://ns50.pointclouds.org/
[12] Möller T. and Trumbore B. (1997). Fast, Minimum Storage Ray-Triangle Intersection. Journal of Graphics Tools (JGT), 2(1):21–28, October 1997. 254.
[13] Filaretov V.F., Yukhimets D.A., Konoplin A.Yu. (2014). Synthesis of System for Automatic Formation of Multilink Manipulator Velocity// The Second RSI International Conference on Robotics and Mechatronics (ICRoM 2014). Tehran IRAN. International IEEE Conference. 2014. pp. 785-790. DOI: 10.1109/ICRoM.2014.6990999.

DOI: 10.2507/28th.daaam.proceedings.072

SYSTEM FOR COOPERATIVE MOVEMENTS OF ROV AND ITS SUPPORTING VESSEL

Vladimir Filaretov, Alexander Konoplin & Nikita Konoplin

This Publication has to be referred as: Filaretov, V[ladimir] F[edorovich]; Konoplin, A[lexander] & Konoplin, N[ikita] (2017). System for Cooperative Movements of ROV and its Supporting Vessel, Proceedings of the 28th DAAAM International Symposium, pp.0513-0516, B. Katalinic (Ed.), Published by DAAAM International, ISBN 978-3-902734-11-2, ISSN 1726-9679, Vienna, Austria
DOI: 10.2507/28th.daaam.proceedings.072

Abstract

The work proposes an intelligent system that automatically generates warnings and recommendations for the operators implementing coordinated movements of the remote operated vehicle, the supporting vessel and the lifting equipment in the process of underwater operations. The recommendations are formed on the basis of the continuous analysis of the current relative position of the supporting vessel, the remote operated vehicle and its depressor-weight, the seabed relief and the desired route of this vehicle. The proposed system provides information and intellectual support for the operators, this allows to significantly improve the accuracy and speed of the performing of complex and expensive underwater operations, excluding the emergency situations.

Keywords: intelligent system; remote operated vehicle; supporting system; underwater operations; expert system

1. Introduction

Remote operated underwater vehicles (ROV) are indispensable means of effectively solving of research, technological and rescue tasks in the depths of the World Ocean. At the present time, the high-precision navigation systems have been created, as well as the control systems for ROVs and supporting vessels [1, 2]. The intermediate depressor-weight is used in deep water, it connects the thick cable of the supporting vessel with the flexible cable of the ROV. The flexible cable provides the high mobility of the ROV. However, the tension of the flexible floating cable can occur under unknown currents unexpectedly for the operator [3, 4]. This tension limits the movement and working area of the ROV equipped with even a powerful propulsion-steering complex. In addition, the depressor-weight can often collide with the ground creating an emergency situation, which can lead to the breakage of the flexible cable. All this complicates the operation of the ROV at great depths and reduces the efficiency of the working operations performing.

It is necessary to simultaneously and consistently control the ROV and the supporting vessel to eliminate these emergency situations when the ROV moving along a given route. It is also necessary to control the length of the thick cable, constantly monitoring the relative position of the ROV and the depressor-weight and completely eliminating the collision of the depressor-weight with the ground.

The works and experiments with ROV Sub-Atlantic Comanche 18 in deep-sea expeditions in the Japanese Sea, the Okhotsk Sea and the Bering Sea have shown that the realization of relatively accurate coordinated movements of the supporting vessel, ROV and its depressor-weight in manual mode is almost impossible. In addition, this work has a great psychological burden to the operator and often leads to long delays in the performance of the scientific research and technological underwater operations.

The work deals to the developing a method for synthesizing an intelligent system to support the activities of operators to successfully solve the above problem. This system will be used for the precise coordinated movement of the supporting vessel, ROV and its depressor-weight in order to ensure the trouble-free movement of these objects during the performing of the underwater works.

2. The algorithm for control of coordinated movements of the ROV and the depressor-weight

The beacons of the sonar navigation system with an ultrashort base are installed on the ROV and its depressor-weight to determine their relative position. The antenna of sonar navigation system is fixed on a supporting vessel equipped with a dynamic positioning system [5]. The ROV work is possible only when the working object or the route target point is in the working area of the ROV, the dimensions of the working area are determined by the length of the flexible floating cable (see Fig. 1). The space location of the working area of ROV is determined by the location of the depressor-weight held by the thick cable of the supporting vessel. When the ROV, which is moving to the target point, approaches to the boundaries of its working area by the distance k, the developed algorithm calculates the direction and value l of the supporting vessel displacement for the movement of the depressor-weight towards the ROV. This displacement allows for the greatest freedom of maneuvering of the ROV.

At the same time, using the developed algorithm, the program signals are calculated for controlling the lifting device for holding the depressor-weight above the ROV on the desired value h (see Fig. 1). The calculations are made taking into account the relative location and direction of the movements of the supporting vessel, positions of the ROV and its depressor-weight and also the complexity of the bottom surface relief determined by the absolute Doppler lag using the method described in the works [6, 7]. The variable value h allows to efficient operate the ROV, excepting situations in which the depressor-weight and the flexible communication cable collides with the bottom.

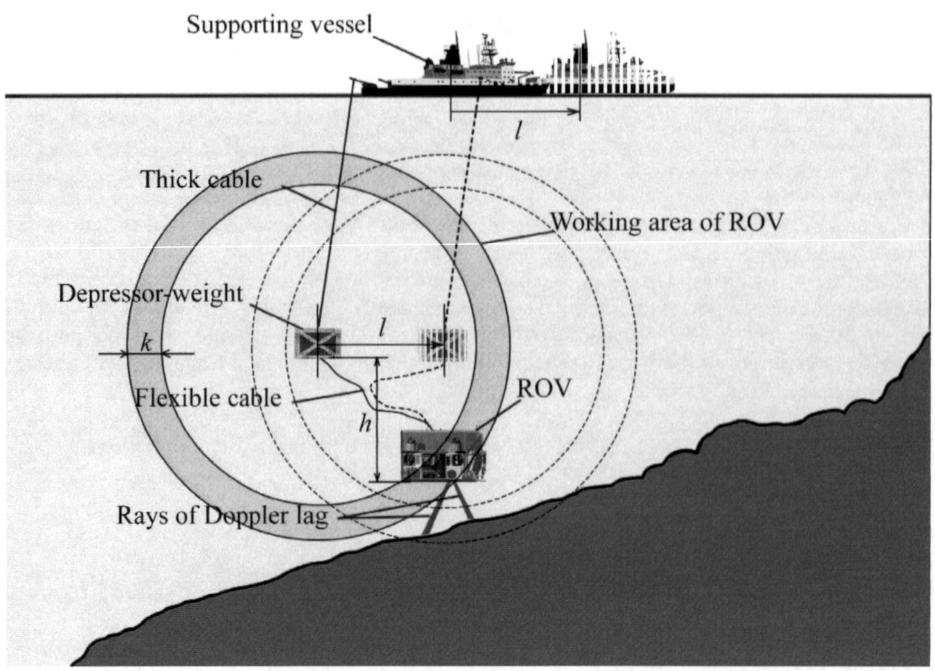

Fig. 1. Controlled underwater complex

3. Features of the developed system architecture

The architecture of the proposed intelligent system is shown on Fig. 2. This architecture allows to provide the operators with visual recommendations and warnings generated in real time on the basis of the expert evaluation of the information obtained from various sensors and navigation systems from different manufacturers.

The developed algorithm for the calculation of the values l and h, which determines the required location of the depressor-weight of the ROV when it approaches at the distance k to the boundaries of the working area, is realized in the decision-making system. Moreover, the variable k is given by this system on the basis of the formalizes estimates of underwater operation conditions stored in the expert knowledge database. These estimates are set by the operator or determined automatically based on the information from the sensors and also using the algorithm for determining the complexity of the bottom surface relief [7]. Wherein, the accumulated experience of the consequences of the choice of the particular value k is taken into account.

In addition, according to the scheme (see Fig. 2), the decision-making system compares the current data received from the sensors and external devices with the data of the base of formalized safety representations, such as: emergency tension of the flexible cable; complex relief of the bottom surface; possibility of the collision of the depressor-weight to the bottom or ROV. In this case, the possibility of an emergency situation is identified and the warnings and alarms for operators are implemented.

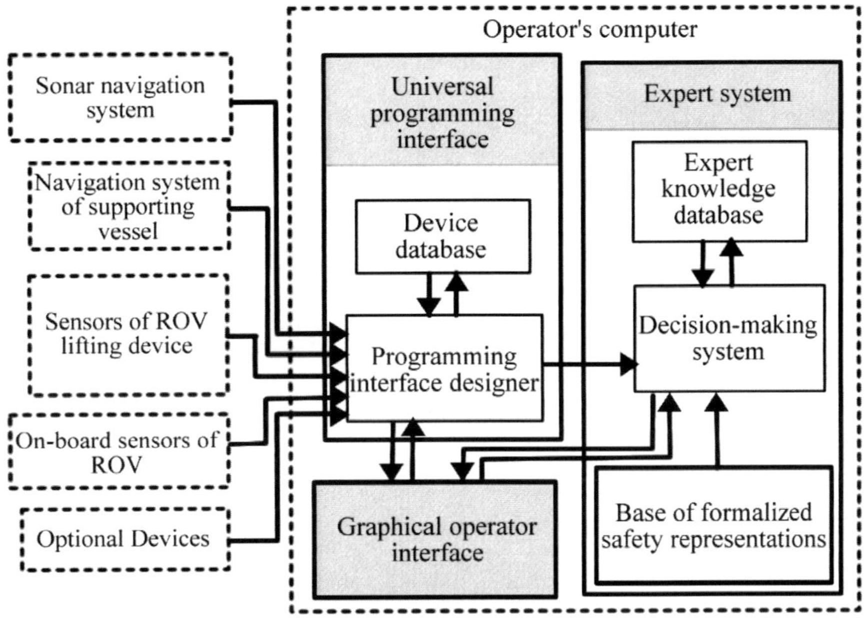

Fig. 2. Structural diagram of the intelligent system

The recommendations and warnings generated by the decision-making system are graphically displayed by the graphical operator interface. This interface has all the properties of the interface of the typical information-control system [8] of the ROV and provides the formation of the maps, as well as the routes planning for the ROV and the supporting vessel.

Since the navigation systems, ROV sensors and sensors of the lifting device require the using of various algorithms for processing data from these devices, then the universal programming interface is proposed to use for the devices communication with the expert system and the graphical operator interface. The programming interface designer using the device database converts the information received from various devices into a universal form, which used by both the expert system and the graphical operator interface. For this, the software implementations of data exchange interfaces for each device, as well as the software implementations of functional algorithms and data processing mechanisms are recorded in the device database.

As a result, the developed architecture will allow to create the software implementation of the intelligent system for the supporting of the activities of operators of ROVs. This system will not require the changes of the expert system and the graphical operator interface in case of the changing of the configuration and composition of devices and sensors, as well as the changing of the processing algorithms of the received data.

4. Conclusion

The paper suggests the system that automatically generates recommendations and warnings to the operators of ROVs for the most efficient control of a supporting vessel, ROV, as well as a lifting device. The system uses the expert estimates of the current relative position of the supporting vessel, the ROV and its depressor-weight, the bottom surface relief, the desired route of the vehicle, and the accumulated knowledge of the prerequisites, causes and consequences of emergency situations. The using of the created system will improve the accuracy and speed of the performing of complex and expensive underwater operations, as well as allows to eliminate the occurrence of emergency situations.

Currently, work is underway to establish an intellectual system for the supporting of the operators of the ROV Sub-Atlantic Comanche 18, which is based on the supporting vessel "Akademik M.A. Lavrentyev". The system will be tested in the deep-sea scientific research expedition of the National Scientific Center of Marine Biology FEB RAS in the Bering Sea in 2018.

5. Acknowledgments

This work was performed in Institute of Marine Technology Problems FEB RAS, Institute for Automation and Control Processes FEB RAS, Far Eastern Federal University and also in National Scientific Center of Marine Biology FEB RAS (Vladivostok, Russia).

The development and software implementation of the intelligent system for supporting of activities of ROV operators was carried out with the financial support of the Russian Science Foundation (RSF) grant (project №17-79-10064). The algorithm for control of coordinated movements of the ROV and the depressor-weight was developed with the financial support of the Russian Foundation for Basic Research (RFBR) grant (project 16-29-04195 Ofi_m).

6. References

[1] Scherbatyuk, A.F., Dubrovin F.S., Rodionov A.Yu., Unru P.P. (2016). Group Navigation and Control for Marine Autonomous Robotic Complex Based on Hydroacoustic Communication // Proceedings of the 2016 IEEE/RSJ International Conference on Intelligent Robots and Systems (IROS), Deajeon, - pp. 1388 - 1393.

[2] Conte G., Scaradozzi D., Mannocchi D., Raspa P., Panebianco L., Screpanti L. (2016). Experimental testing of a cooperative ASV-ROV multi-agent system // IFAC-PapersOnLine, Toulouse, France, Vol. 49, Iss. 23, - pp. 347-354.

[3] Kostenko V.V., Mokeeva I.G. (2009). Research of Tie Cable Influence on ROV's Manoeuvrability. Underwater Research and Robotics, № 1(7). - pp. 22-27. (Костенко В.В., Мокеева И.Г. Исследование влияния кабеля связи на маневренность телеуправляемого подводного аппарата // Подводные исследования и робототехника. 2009. № 1(7). - С. 22-27.)

[4] Rulevsky V.M., Yudintsev A.G. (2013). Power Supply Systems for Modern Remote-controlled Unmanned Underwater Vehicles // Proc. of 5th All-Russian Scientific and Technical Conf. "Technical problems of development of the World Ocean", Vladivostok, - pp. 81-86. (Рулевский В.М., Юдинцев А.Г. Системы электропитания современных телеуправляемых необитаемых подводных аппаратов // Материалы 5-й Всероссийской научно-технической конф. «Технические проблемы освоения мирового океана», Владивосток, 2013. – С. 81–86.)

[5] Zanin V.Yu. (2010). Using a Underwater Remote Operating Vehicle (ROV) with Specialized and Non-specialized Ships. News of SFedU. Technical science, №3(104). - pp. 179-186. (Занин В.Ю. Использование телеуправляемых подводных аппаратов (ТПА) со специализированных и неспециализированных плавсредств // Известия ЮФУ. Технические науки. 2010. №3(104). – С. 179-186.)

[6] Filaretov V.F., Konoplin A.Yu., Konoplin N.Yu., Gorbachev G.V. (2016). Control system for underwater vehicle with multilink manipulator for automatic manipulation operations, Proceedings of the 27th DAAAM International Symposium, B. Katalinic (Ed.), Published by DAAAM International, Vienna, Austria, 2016, – pp. 714-720. DOI: 10.2507/27th.daaam.proceedings.103.

[7] Filaretov V.F., Konoplin A.Yu., Konoplin N.Yu. (2017). System for automatically performing of manipulation operations with help of underwater robot. Mechatronics, automation, control. 2017. Iss. 8. - pp. 543-549. (Филаретов В. Ф., Коноплин А.Ю., Коноплин Н.Ю. Система для автоматического выполнения манипуляционных операций с помощью подводного робота // Мехатроника, автоматизация, управление. 2017. №8. Т. 18. С. 543–549.)

[8] Filaretov V.F., Konoplin N.Yu., Konoplin A.Yu. (2017). Approach to Creation of Information Control System of Underwater Vehicles // Proceedings of IEEE 2017 International Conference on Industrial Engineering, Applications and Manufacturing (ICIEAM), S.-Petersburg, Russia. - pp. 1-5.

DOI: 10.2507/28th.daaam.proceedings.073

USING UNITY3D AS AN ELEVATOR SIMULATION TOOL

Jiri Polcar, Petr Horejsi, Pavel Kopecek & Muhammad Latif

This Publication has to be referred as: Polcar, J[iri]; Horejsi, P[etr]; Kopecek, P[avel] & Latif, M[uhammad] (2017). Using Unity3D as an Elevator Simulation Tool, Proceedings of the 28th DAAAM International Symposium, pp.0517-0522, B. Katalinic (Ed.), Published by DAAAM International, ISBN 978-3-902734-11-2, ISSN 1726-9679, Vienna, Austria
DOI: 10.2507/28th.daaam.proceedings.073

Abstract

Simulations are used for finding answers to what-if scenarios prior to making decisions. Discrete event simulation (DES) tools are usually used for industrial processes. However, these tools require the creators of the simulation model to have highly specialized knowledge and they only rarely provide easily understandable graphic representations of the modelled situation. An elevator simulation model was created using an unusual approach: the simulation model was developed in Unity3D, an IDE intended for making computer games. The model was evaluated in terms of the accuracy of the results and of the suitability of using such a tool for creating an elevator simulation with a sequence dispatcher. The paper includes validation results and discusses the advantages, disadvantages and limits of such an approach. Although a DES tool would be better for the elevator itself, separate details of the simulation can be much more easily modelled in Unity3D. It can take into account 3D space (as opposed to 1D, which is usual for DESs), for example for finding probable paths of persons, calculating volumes or finding the right door opening times.

Keywords: 3D Visualization; Discrete event simulation; Elevator simulation; Unity3D

1. Introduction

Discrete-event simulations (DES) are a very good way to optimize many kinds of processes, but in some cases, the simplification of particular events can make the results of the whole simulation inaccurate. There are a lot of DES modelling tools, such as Anylogic, Rockwell Arena, Witness Simulation or Siemens Tecnomatix Plant Simulation. All of these tools share the concept that an entity representing a product moves through the pipeline of the simulation model, which is built of blocks or coded in a scripting language.

The specifics of such an approach are that the creators of the simulations must be trained in the use of special tools, which is usually very expensive – such tools are aimed at investments to make more profit, hence the high licence costs. Other than that, these tools are based on abstract graphics, be it 2D or 3D. Although not influencing the results, the graphics can influence the ability to understand the modelled situation and the gained results. Also, such models can only be used and run in the tool where they were created, with very limited or no possibilities to connect with other systems to use the results of the simulation. This makes the model useless for other purposes.

The lack of high definition graphics is discussed in [1]. The authors propose the use of gaming engines for presentation of the simulation because of their high optimization and very realistic graphics. Based on a survey, they claim that a good-looking model is good for its validation – especially when discussing the behaviour of the models with people who are experts in the simulated fields. Lower importance was given to the analysis phase, but a high importance for marketing purposes. Also, if the model is to be used for training purposes, the level of realism is important. The same authors also claim that simulation packages and gaming engines have a lot in common, implicating that game engines can be used as simulation platforms [2]. Another positive aspect is that the simulation itself is computed on a CPU, whereas the graphics layer is handled by a GPU – which leaves more CPU time for the actual simulation.

The situation, however, is that usually the animation engine of the simulation tool and the computational core are closely connected together, causing the animation to slow down the progress of the simulation, which equals an increase in the costs of the simulation. 'Loose coupling' of these two systems is proposed in [3]. The authors claim that loose coupling is necessary so that the simulation can run very fast and the graphic element of the simulation can be easily modified. This approach is compatible with the previous statements, where game engines could be used as a visualization layer, or a front-end, with the simulation layer as the back-end.

Use of a gaming engine is described in [4]. A large scale traffic simulation is populated by actors with their own decisions and behaviour. The whole platform runs as a distributed simulation and is powered by a Delta3D engine. The graphics are loosely coupled with the simulation model, as proposed in [3], allowing the option of turning the graphics off to speed up the simulation. The use of a game engine has the advantage of allowing the software to be modified. For example, the described simulation model allows a user to control one car directly, which means it can be used for driver training. Also, it enables non-expert users to create what-if simulation scenarios.

Another gaming engine, the Unity3D, was used in [5] for a shopping mall simulation. The multi-agent simulation was made in this tool, probably because of the ability to write in a full featured programming language (C#) rather than scripting in a specialized language or in a block diagram. Also, the ability of the agents, who act as entities, are easier to implement with such an approach.

Physics engines included in game engines can also be used for mechanism simulations, as described in [6]. The authors even claim that development of such a simulation model in Unity3D is simpler than using traditional methods. Plus, it enhances the possibilities of interaction with the model.

2. Simulation Model Development

Unity3D is a powerful tool used in various industrial and scientific applications. As mentioned above, common tools for DES are very specialized, requiring a lot of experience and they are very expensive. That is why the goal of this research is to discuss whether it is possible to use Unity3D for creating a simulation model that would normally be made using DES tools. Basically, anybody who can code in C# can make such a model in Unity3D using its API. Other than that, Unity3D provides possibilities for enhancing the realism of the simulation or reducing the need to simplify certain processes. These possibilities include:

- Collision detection
- Calculating navigational paths
- Implementing animations

Unity3D models can be parametrized and controlled using runtime arguments, file reading or named pipes communications. Such approaches are suitable for combination with various optimization methods, as described in [7] or in [8].

For further evaluation, an elevator simulation with a simple sequence dispatcher was created for this research. Elevators have been the subject of much research on optimization and simulation, such as [9].

The question is, whether the accuracy of a simply made Unity3D simulation model can give results comparable to simulation models made using DES tools. The persons in the simulation model are self-controlled and they interact with the virtual world like a real human would. Using a regular DES tool, they would go through the model on abstract paths and usually the objects of the model would drive them. As such, modifications to the behaviour of any objects could be done directly in their class, without worrying about modifications to other objects.

The modelled elevator has an interface for collecting requests, ascending, descending, opening and closing the door. The controller is in a separate class, which makes it easy to change the instance to another kind of controller. The implemented controller is a simple sequence elevator dispatcher, collecting requests in one direction and switching to the other direction when there are no requests left in the current direction. As for the persons, after instancing it is decided whether to use the stair or the elevator based on the fitness level. If the elevator is chosen, the person comes to the elevator where he waits in the queue. The first person in the queue calls the elevator if it has not arrived. If the elevator arrives, but the person fails to board, the person calls the elevator again. After boarding, the person makes an inside request and waits until the door opens and the elevator is on the desired floor. This is the real behaviour of an elevator and a person as opposed to standard DES models, where the decision of what the person (entity) will do next is usually up to other simulation objects. An algorithm flow chart is shown in Fig 1.

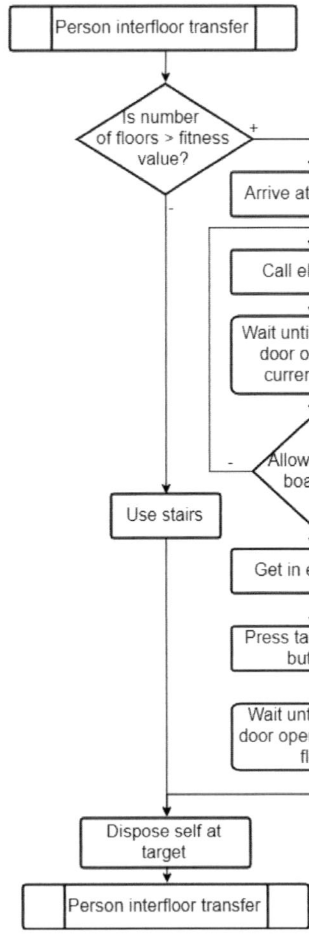

Fig. 1. Algorithm of the Person entity.

The model is designed to have as many floors as desired with a simple value changing. The floor consists of a flat 3D model, the positions of the spawn points and elevator queues. After changing the count of the floors value, the navigational mesh is computed and the model is ready to start. The model can be easily enhanced graphically, a spectator controller can be implemented or the whole model can be put into an existing scene. Also, Unity allows, after some tweaking, the persons to be controlled directly or to add some special events into the model, such as one time arrival of a lot of people and so on. The basic model can be seen in Fig 2.

Fig. 2. Overview of the simulation model.

This approach proved to have some disadvantages as well. Because Unity3D is not prepared for discrete events, but is strictly continuous, there could be issues with increasing the simulation speed. For every rendered frame, the engine decides what will be done in the next step. The more the simulation time scale increases, the higher is the delta time, which will result for example in unnecessarily prolonging the waiting time for the lift if it arrives at the beginning of the delta time interval. This means that the maximum speed of the simulation is ordinarily more limited than the speed of a regular DES.

The value of the time scale in the Unity engine can be adjusted directly. Usually, the engine runs at 60 FPS (frames per second). The higher this value, the higher the simulation accuracy should be. The time scale directly influences how much simulation time passes between the frames. Experiments were conducted on how the combination of average FPS and time scale influences the results of the simulation.

For this case, the simulation model consisted of five floors, where each floor had an independent person source, creating one at a time at a uniform interval from 5 to 25 seconds. Every person had a uniform distribution of the target floor and disappeared after reaching the target floor, contributing to the overall statistics. The elevator interfloor travel time was 5 seconds, door opening time 1 second, and door waiting time 1 second. Simulation time was 30 minutes.

The scene was not rendered between experiments, only GUI cameras displayed the current statistics. The screen size was always 640 x 400 pixels to ensure the frame rate was always the same. The frame rates were either capped at 60 FPS using vertical synchronization or reached values around 150 FPS with V-Sync off.

The results can be seen in Fig 3:

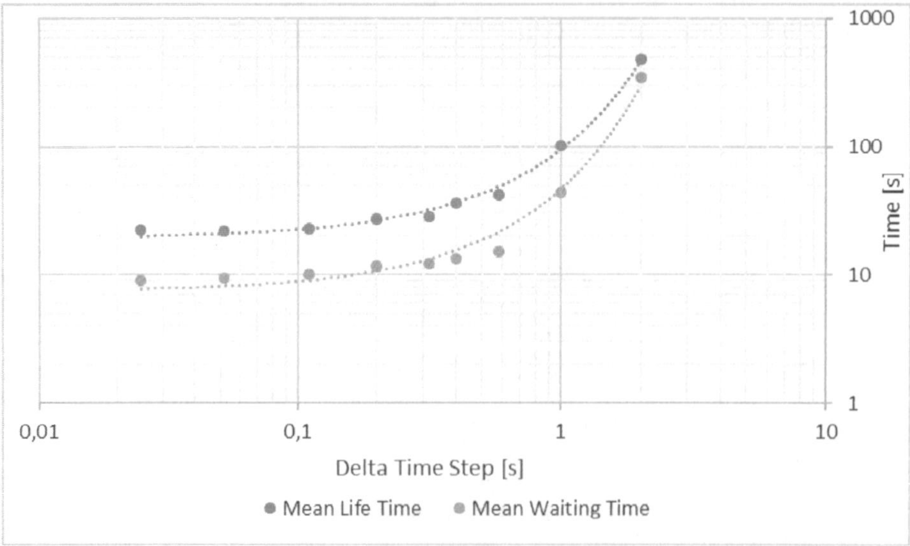

Fig. 3. Influence of the simulation time step on entities' life and waiting times.

As can be seen from Figure 3, the delta time has a big influence on the simulation results. Unfortunately, even relatively low values of time scale (which is the main factor of the delta time size), have effects on the results, meaning that the simulation cannot be sped up very much – in this case, a safe value of speeding up the simulation was less than 10 times, which is unsuitable for long term simulation runs.

In comparison with different elevator DES models, a former case study [10] was replicated in this experiment. The results are quite similar, although they are not exactly same. A model with an elevator speed between floors of 10 seconds, door opening time of 20 seconds and elevator capacity of 8 people was tested in three scenarios for 12 hours of simulation:

- Triangular distribution of arriving people of (1.8, 2.4, 2.0), (20, 60, 40) and (4, 8, 6) minutes at every storey
- Uniform distribution of target floor
- Number of storeys 6, 6 and 14 respectively

Of the published results, only the number of throughputs (number of people arrived at desired levels) was published, meaning only these values are compared. Other values in the described Unity model are also available in Table 1.

Experiment number	Described model	Case study [8]
1	2076	2421
2	116	87
3	1451	1796

Table 1. Total throughputs of elevator model described in this paper compared with the model described in [10]

3. Discussion

Making a simulation model in Unity3D which would be suitable for DES tools has both advantages and disadvantages. One of the main disadvantages is that the model cannot be sped up very significantly. This is due to the continuous computations of the objects' decisions and a too large delta time step can be very disturbing. The marginal value of the step is a function of the decision making frequency in the model. There are several ways to lower the delta time steps:

- Increasing CPU and GPU power
- Adjusting the camera to look away from complex scenery
- Lowering resolution and detail
- Reducing time scale (slowing down the simulation run)

One of the main advantages of the model is that it can easily be made to be agent based and have compliance with object oriented programming – any class can be implemented as long as it uses the model's interfaces. This means disturbing events and transitions can be simulated and the aim of the simulation can be changed quite easily. For example, if the people are not deleted on the target floor, they can stay there and move through different floors according to a stated schedule. Another thing that can be modelled is for example the capacity of the building's facilities merely by adding a decision whether to go or not to go to the toilet by evaluating some criterial functions, for example in school buildings during breaks. These are factors that can sometimes be implemented much more easily in a full-featured programming language than in DES tools. Last but not least, the model can be implemented for presentations using photorealistic graphics. There can even be multiple cameras in the model that can focus on different parts of the model, as shown in Fig 4 which can portray the same moment of the simulation as Fig 2 from different angles. Or a user controllable camera can be inserted, allowing users to see areas which they are interested in and better support their understanding of the model.

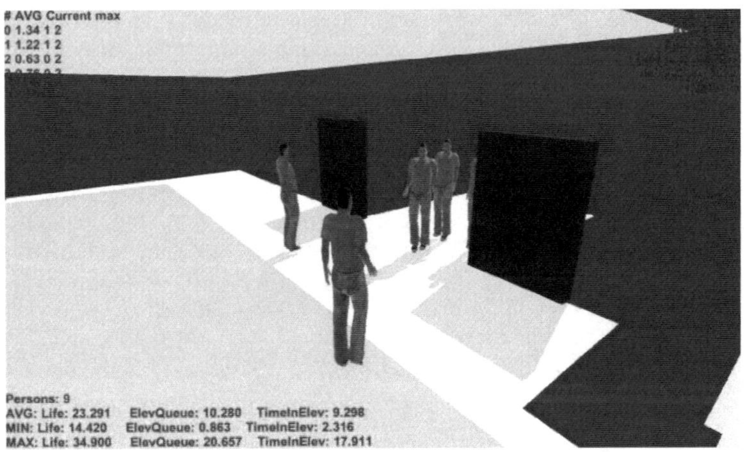

Fig. 4. Example of a custom aligned camera in the model.

Another advantage of using a completely different approach from DES tools is that the persons actually use the space to move. This allows for more complex and realistic computations when finding an optimum door opening time. If the door opening time is too short, it has to be opened again and waits for another interval to close. The walking speed of the persons can be randomly distributed, as well as the distance from the elevator door when it starts to open and the reaction time of the persons. Also, it is easy to simulate which elevator will be used if the elevators are placed in different corridors.

4. Conclusion

The basic idea was to prove whether DES simulation models need to be created strictly in DES simulation tools. For various reasons, we focused on using a different kind of tool. A simulation model was created using non-traditional tools – the Unity3D game engine and IDE. An elevator simulation model was chosen to compare using such an approach with using standard simulation tools. The development of the model proved that knowledge of C# and a basic overview of Unity3D API is enough to create such models. The tool proved to be usable for developing and evaluating such models, although, for an elevator simulation, a traditional tool would be recommended if available.

Unity3D can, however, simulate details that are usually neglected or abstracted to a very simplified level, such as boarding and exiting of the elevator with people walking at different paces and of given sizes when the elevator door has a given width. Unity3D can be used as a cheaper alternative with the possibility of using traditional programming in a full-featured programming language, whereas specialized discrete event simulation (DES) tools require a lot of special knowledge. Generally speaking, Unity3D could be used for more specialized tasks, like the behaviour of people in the system, such as which path will be chosen, and it can take into account the whole 3D space as opposed to the usual 1D paths in DES tools. Such ideas will be the subject of further research.

5. Acknowledgement

This article was prepared with the support of the Internal Science Foundation of the University of West Bohemia SGS–2015-065.

6. References

[1] Bijl, J. L. & Boer, C. A. (2011). Advanced 3D visualization for simulation using game technology, Proceedings of the 2011 Winter Simulation Conference (WSC), Phoenix, AZ, 2011, pp. 2810-2821. DOI: 10.1109/WSC.2011.6147985

[2] Bijl, J. L. & Van Nieuwenhuizen, P.R. (2009). Improving the visualization of 3D simulations using Computer Game technology, Master Thesis, Department of Computer Graphics, Faculty of Electrical Engineering, Mathematics and Computer Science, TU Delft, Netherlands.

[3] Fumarola, M.; Seck, M. & Verbraeck, A. (2010). An approach for loosely coupled discrete event simulation models and animation components, Proceedings of the 2010 Winter Simulation Conference, Baltimore, MD, 2010, pp. 2161-2170, DOI: 10.1109/WSC.2010.5678857

[4] Miao, Q.; Zhu, F.; Lv, Y.; Cheng, C. & Qiu, X. (2011). A Game-Engine-Based Platform for Modeling and Computing Artificial Transportation Systems, Proceedings of IEEE Transactions on Intelligent Transportation Systems, vol. 12, no. 2, pp. 343-353, June 2011, DOI: 10.1109/TITS.2010.2103400

[5] Christian, J. & Hansun, S. (2014). Implementation of multi-agent system using fuzzy logic towards shopping center simulation, 014 International Conference on Intelligent Autonomous Agents, Networks and Systems, Bandung, 2014, pp. 19-23, DOI: 10.1109/INAGENTSYS.2014.7005719

[6] Hu, W.; Qu, Z. & Zhang, X. (2012). A New Approach of Mechanics Simulation Based on Game Engine, 2012 Fifth International Joint Conference on Computational Sciences and Optimization, Harbin, 2012, pp. 619-622, DOI: 10.1109/CSO.2012.141

[7] Raska, P & Ulrych, Z. (2013) Testing Optimization Methods on Discrete Event Simulation Models and Testing Functions, Procedia Engineering, Volume 69, 2014, Pages 768-777, ISSN 1877-7058, DOI: 10.1016/j.proeng.2014.03.053

[8] Peschl, M.; Roening, J. & Link, N.: Distributed Simulation for Task-Driven Flexible Manufacturing Systems, Annals of DAAAM for 2012 & Proceedings of the 23rd International DAAAM Symposium, ISBN 978-3-901509-91-9, ISSN 2304-1382, pp 0171 - 0174, Editor B[ranko] Katalinic, Published by DAAAM International, Vienna, Austria, 2012

[9] Horejsi, P.; Horejsi, J.; Latif, M. & Ulrych, Z. (2011). Automatic Generator of Lift Dispatcher System Model, Annals of DAAAM for 2011 & Proceedings of the 22nd International DAAAM Symposium, 23-26th November 2011, Vienna, Austria, Volume 22, No. 1, ISSN 1726-9679, ISBN 978-3-901509-83-4, Katalinic, B. (Ed.), pp. 1217-1218, Published by DAAAM International Vienna, Vienna

[10] Horejsi, J. (2009). Use of ARENA Simulation Software for Lift Control Strategies Development in High-Level Buildings, Master Thesis, Department of Industrial Engineering and Management, Faculty of Mechanical Engineering, University of West Bohemia.

DOI: 10.2507/28th.daaam.proceedings.074

MAIN PRINCIPALS AND ISSUES OF DIGITAL TWIN DEVELOPMENT FOR COMPLEX TECHNOLOGICAL PROCESSES

Kirill Ponomarev, Nikita Kudryashov, Nikita Popelnukha & Vyacheslav Potekhin

This Publication has to be referred as: Ponomarev, K[irill]; Kudryashov, N[ikita] & Popelnukha, N[ikita] (2017). Main Principals and Issues of Digital Twin Development for Complex Technological Processes, Proceedings of the 28th DAAAM International Symposium, pp.0523-0528, B. Katalinic (Ed.), Published by DAAAM International, ISBN 978-3-902734-11-2, ISSN 1726-9679, Vienna, Austria
DOI: 10.2507/28th.daaam.proceedings.074

Abstract

To provide increasing of effectiveness of industrial systems with achievement of the guaranteed product quality, the manufacturing digital transformation paradigm is applied. According to Industry 4.0 concept, adaptive, client-oriented automatic control of product life cycle is the basic approach in global manufacturing digitalization. Such technologies as Big Data, Cloud Services, advanced analytic algorithms, have to be implemented to provide an effective control for complex technological processes. Therefore, different technologies operating information from Web as well as from field level and top level manufacture information systems should offer high degree of interoperability.
This article represents the, so-called, digital twin concept description. The main development issues and paradigms are described, applying to oil refining industry.

Keywords: Industry 4.0; Digital Twin; Intelligent Control System; Automation; Global Digitalisation

1. Introduction

In the modern world, due to the high speed of IT technology growth, it appears more opportunities to improve the efficiency of traditional industrial systems and the quality of distributed products. Applying the paradigm of digital production, with the help of modern information technologies, it is possible to create digital twin of the whole manufactory, and specify its pieces as detached systems [1]. This is especially actual for industries distributed in space and time. The main problem is to build a digital twin for a separate control object, it is necessary to take into account the nature of the object and a number of basic properties such as:

- discreteness / continuity;
- linearity / nonlinearity;
- stationarity / nonstationarity;
- distribution / concentration of parameters;
- the presence of uncertainties in the description of the structure / parameters;
- stochasticity / determinism;
- multiple connectivity and so on.

In this article, the following questions are considered:

- What does the term "Digital Twin" include?
- What parts do Digital Twins consist of?
- How to implement management based on a Digital Twin?
- How to build a Digital Twin using modern IT technologies such as Big Data, Cloud Services, and Advanced Analytic Algorithms?

2. What is a Digital Twin

The concept of digital twin envisages connection between the physical and digital world over the analytical program and machinery. The twin preimage of real model is generated at the design stage. Including the basic properties of all produced objects. The digital model and the physical object interact in real time [2]. This information exchange is characterized by a Big Data which generated by a multitude of sensors. The digital model is constantly updated. That is, it changes its parameters to better match the current working mode of the physical object. Thus, there is a real opportunity to identify emerging anomalies in the early stages, to predict the behaviour of the object and to ensure implementation algorithms of dynamic optimisation, which ultimately allows significantly improve the reliability and efficiency of the equipment [1, 3-5].

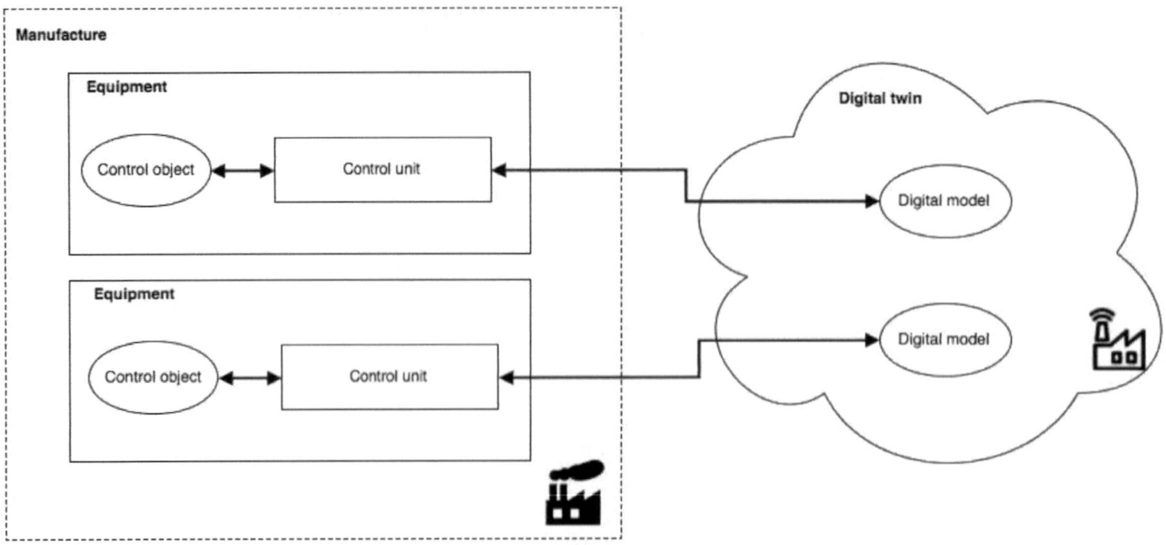

Fig. 1. Data exchange between a physical object and a digital model

3. Digital Twin Units

Digital twin – is the complex information and technical systems that cover practically all levels of the automated process control system (ACS): from the level of terminal equipment to the level of enterprise, resource planning (ERP). These levels are illustrated on Fig. 2.

In this article, we will consider digital twin as a multilayer system, consisting of five layers. Each of these levels represents itself as separate subsystem of Digital Twin. Development of each of them can be made independently, so this makes the process of development less complicated and provide multi-user access for developers [6]. Let us consider each level more precisely, Fig. 3.

3.1. Cyber-physical layer

At this level, the stabilization of the mode parameters and the program-logical control of the object are performed, which can be represented by separate installations or a technological process.

3.2. Primary processing/store data layer

This level, data are collected from devices represented by the cyber physical layer and their primary processing (for example: conversion of values) and temporary storage (for example: until the next layer is transferred). This layer can be represented by the implementation of OPC (Object Linking and Embedding for Process Control) [7].

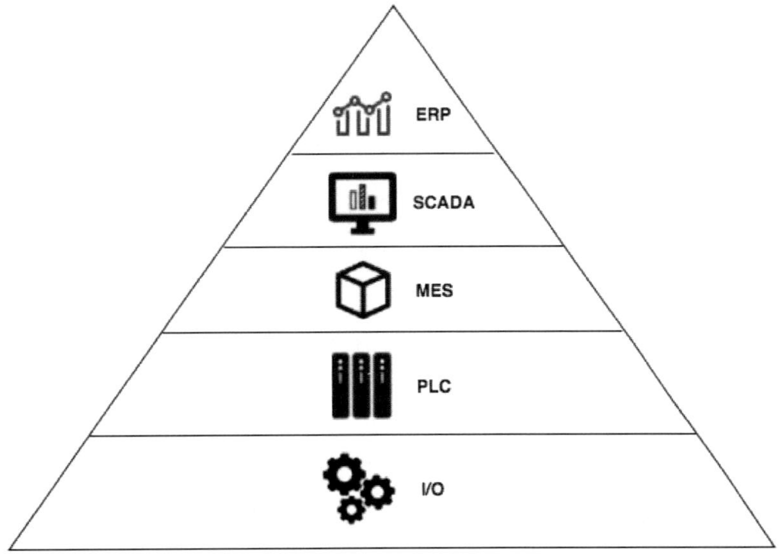

Fig. 2. Layers of the ACS

3.3. Distributed computing and storage layer

This layer is the kernel of the digital twin. It contains such units as:

- API (Application program interface). Through this interface we can communicate with multiple layers in the same time, using one of specified protocols (TCP, HTTP, WebSocket).
- Distributed computing management system parses tasks on huge number of subtasks and sends them to specified evaluation nodes [7]. Then it collects the results. So it is the way of representation of parallel evaluation approach.
- Distributed storage management system, provides the storage of technical data, master data, analysis results, etc. in on DDB. It can be represented by any DBMS(SQL, NoSQL, TSDB) for any type of data. (As an Example: for data from sensor – the Apache Cassandra can be used, but for project documentation- MongoDB, which is document-orientated) [7, 8]

3.4. Models and algorithms layer

On this layer, mathematical, statistical, neural-network models and CAD models can be stored.

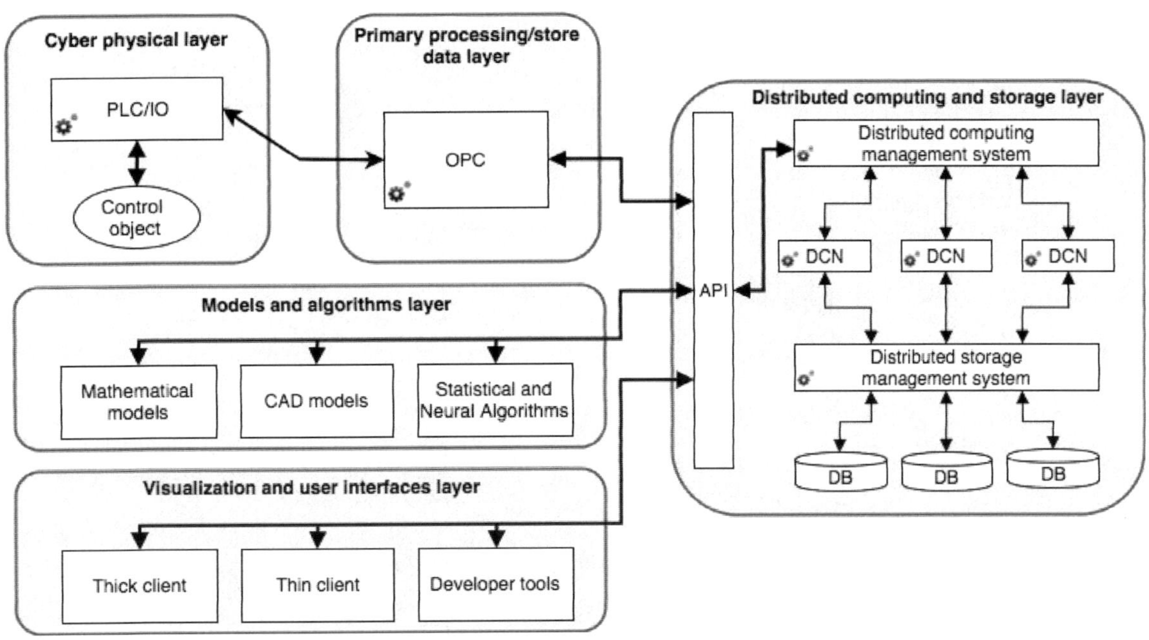

Fig. 3. Digital twin structure.

3.5. Visualisation and user interfaces layer

This layer provides the access for users to digital twin, using graphical interface. It contains:

- Rich-client provides AWP interface as technical schemes, screens. Can be viewed as:
    - SCADA (Supervisory Control And Data Acquisition);
    - HMI (Human-machine interface);
- Thin-client (for example WEB – application) provides analytic screens interface for experts:
    - Statistical model analysis
    - Object monitoring, according to dynamic of key parameter changes.
- Developer tools provides the tool for model and algorithm development. Can be represented with standalone specific application.

4. Control with a digital twin

A lot of digital twins functions are aimed at optimising control and making decision process:

- Identifying deviations in the current operation of equipment from the optimal mode;
- Identifying emergency situations;
- Alerting staff about events and situations;
- Analysing the causes of deviations

This functionality allows:

- Identify at an early stage of minor changes in the equipment, which in consequence can lead to serious damage and unscheduled downtime;
- Reduce the risk of failure of nodes and aggregates due to proactive diagnostics;
- Pre-alert operational staff of a potential threat to critical equipment failure;
- Reduce the time of unplanned downtime of equipment associated with its sudden breakage due to early warning and timely ordering of spare parts;
- Reduce the burden on operational personnel by attracting the attention of the operator only to statistically significant deviations in the parameters.

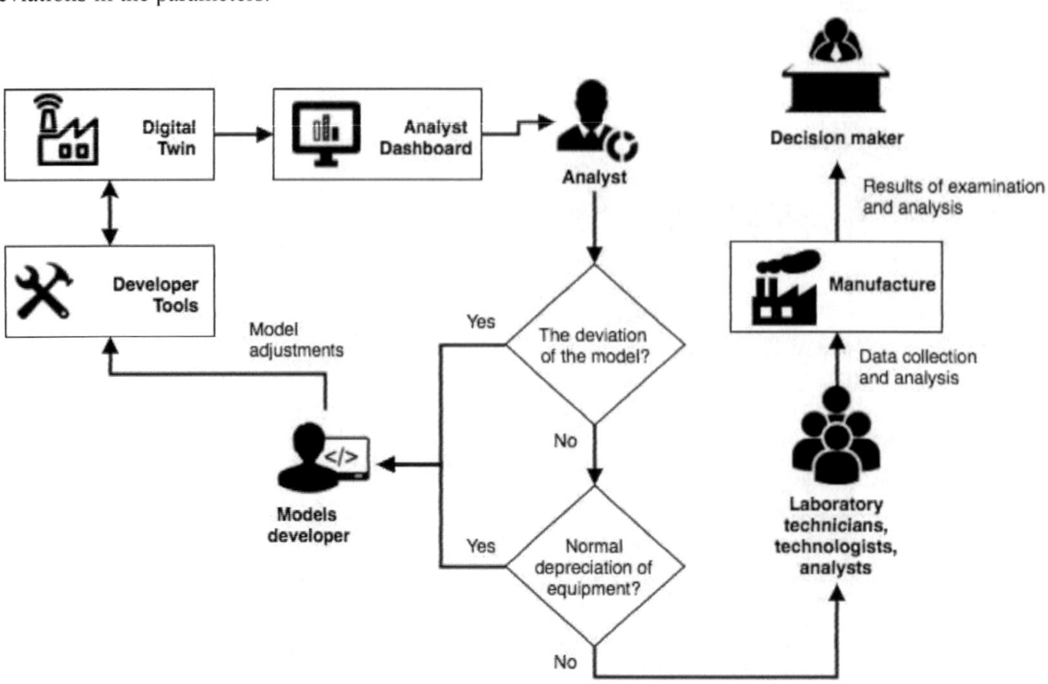

Fig. 4. Business decision-making process based on digital twin

The business decision-making process, when using the digital twin, can be represented, for example, in the same way as in Fig. 4. Thus, the process of equipment monitoring and diagnostics includes various production units from the operator and the instrumentation service to analysts and experts at the level of the company's main specialists [9].

5. Building a Digital Twin

As a basis for building a Digital Twin, we used the basic UML language and two profiles that define the object-oriented languages SysML and BPTL, proposed in the framework of OMG - Object Management Group. [10] SysML provides the ability to use 9 interrelated types of diagrams to describe the structure, behavior and system requirements:

- diagram of requirements;
- diagrams of activity (activities);
- diagrams of sequences;
- diagrams of states;
- diagrams of use cases (use cases);
- block definition diagrams;
- diagrams of internal block descriptions;
- diagrams of parameters;
- diagrams of packages.

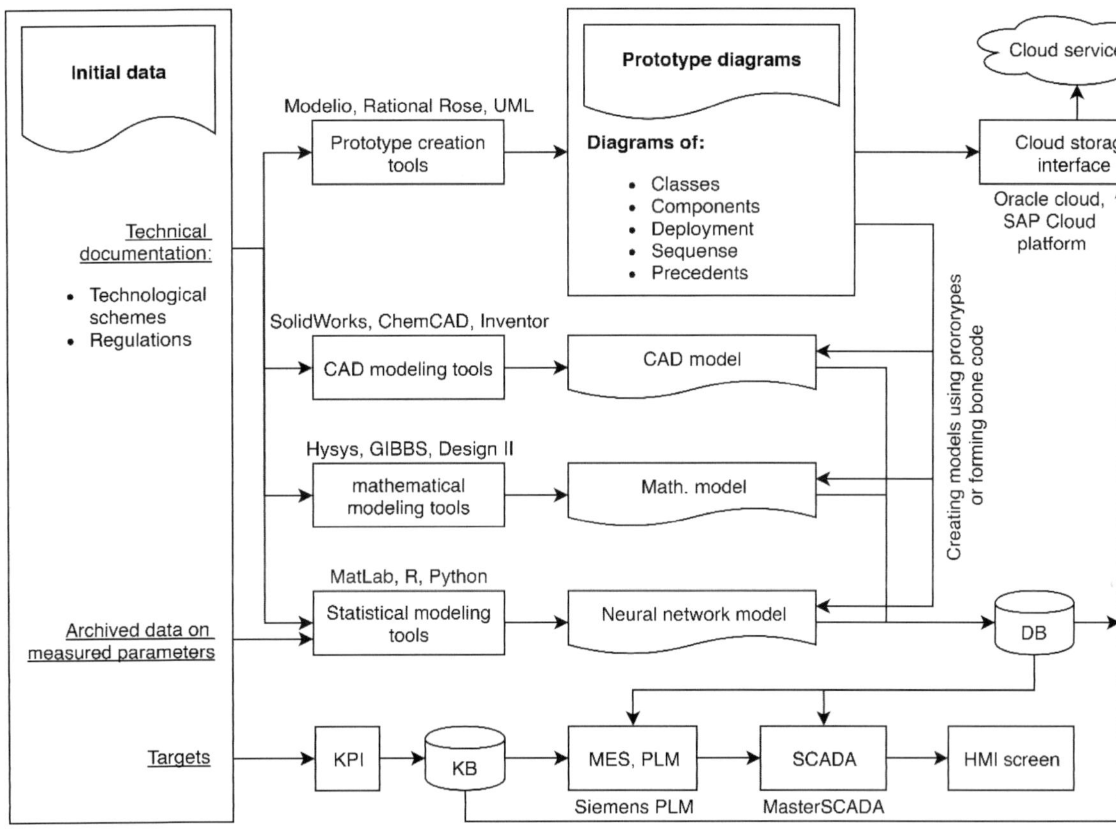

Fig. 5. The process of constructing the Digital Twin.

6. Conclusion

Digital Twin technology, which provides for the construction of "live" digital models and virtual simulators for local use or as an implementation of the Industrial Internet concept, is one of the key modern areas of operational analysis and increasing the efficiency of industrial equipment.

Expert and advisory diagnostic systems based on best practices can significantly reduce the costs of equipment maintenance, reduce the risks of their breakdowns, downtime and related material losses, and improve the efficiency and quality of technological processes. As a result of the development and verification of the architecture of Digital Twins, the problem of intelligent control system developing for distributed industrial object was solved. The results of the development of the architecture of digital twins are planned to be applied in one of the forming branches of Russia.

7. Acknowledgments

The article is published in the framework of the project Erasmus+ 573545-EPP-1-2016- 1-DE-EPPKA2-CBHE-JP and describes the part of the project conducted by SPbPU.

8. References

[1] B. Katalinic, I. Kukushkin, V. Pryanichnikov, D. Haskovic, Cloud Communication Concept for Bionic Assembly System, Procedia Engineering, Volume 69, 2014, Pages 1562-1568, ISSN 1877-7058, doi:10.1016/j.proeng.2014.03.156

[2] Turner, C. J., Hutabarat, W., Oyekan, J. and Tiwari, A. (2016) 'Discrete Event Simulation and Virtual Reality Use in Industry: New Opportunities and Future Trends', IEEE Transactions on Human-Machine Systems, 46(6). doi: 10.1109/THMS.2016.2596099

[3] Scheifele, S., Friedrich, J., Lechler, A. and Verl, A. (2014) 'Flexible, Self-configuring Control System for a Modular Production System', Procedia Technology. Elsevier B.V., 15, pp. 398–405. doi: 10.1016/j.protcy.2014.09.094.

[4] Kudriashov N., Protasov I., Markov S., Potekhin V., Yadgarova Y., Taratukhin V. (2016). Implementation of Cloud Services for Advance Management of Steel Transport for Continuous Casting Production, Proceedings of the 27th DAAAM International Symposium, pp.0457-0462, B. Katalinic (Ed.), Published by DAAAM International, ISBN 978-3-902734-08-2, ISSN 1726-9679, Vienna, Austria

[5] Gastermann, B., Stopper, M., Kossik, A., Katalinic, B. Secure Implementation of an On-Premises Cloud Storage Service for Small and Medium-sized Enterprises, Proceedings of DAAAM International Symposium on Intelligent Manufacturing and Automation, DAAAM 2014; Vienna; Austria; 26 November 2014 through 29 November 2014, Volume 100, Issue C, 2015, Pages 574-583

[6] Haskovic D., Katalinic B., Zec I., Kukushkin I., Zavrazhina A. (2016). Structure and Working Modes of the Intelligent Adviser Module, Proceedings of the 27th DAAAM International Symposium, pp.0866-0875, B. Katalinic (Ed.), Published by DAAAM International, ISBN 978- 3-902734-08-2, ISSN 1726-9679, Vienna, Austria

[7] Yerofeyev S.A., Ipatov O.A., Markov S.A., Potekhin V.V., Sulerova A.S. & Shkodyrev V.P. (2015) Adaptive Intelligent Manufacturing Control System, Proceedings; DAAAM International Symposium of Intelligent Manufacturing and Automation, DAAAM 2015; Vienna; Austria; Volume 26, No. 1, ISSN 2304-1382

[8] Haskovic, D., Katalinic, B., Kukushkin, I. Role o. The Adviser Module i. The hybrid assembly subordinating control structure, Proceedings of DAAAM International Symposium on Intelligent Manufacturing and Automation, DAAAM 2014; Vienna; Austria; 26 November 2014 through 29 November 2014, Volume 100, Issue C, 2015, Pages 1706-1713

[9] Yerofeyev S., Ipatov O., Markov S., Potekhin V., Sulerova A., Shkodyrev V. (2016). Adaptive Intelligent Manufacturing Control Systems, Proceedings of the 26th DAAAM International Symposium, pp.1016-1024, B. Katalinic (Ed.), Published by DAAAM International, ISBN 978-3-902734-07-5, ISSN 1726-9679, Vienna, Austria

[10] Rosen, R., Von Wichert, G., Lo, G. and Bettenhausen, K. D. (2015) 'About the importance of autonomy and digital twins for the future of manufacturing', IFACPapersOnLine. Elsevier Ltd., 28(3), pp. 567–572. doi: 10.1016/j.ifacol.2015.06.141.

DOI: 10.2507/28th.daaam.proceedings.075

INDUSTRIALLY USABLE NUMERICAL CALCULATION-BASED METHOD FOR ESTIMATING RESIDUAL STRESS DUE TO LASER CLADDING REFURBISHMENT

Stepans Sklariks, Toms Torims

Riga Technical University, Faculty of Mechanical Engineering, Transport and Aeronautics, Department of Material Processing Technology, 36A Viskalu Street, Riga

This Publication has to be referred as: Sklariks, S[tepans] & Torims, T[oms] (2017). Industrially Usable Numerical Calculation-Based Method for Estimating Residual Stress due to Laser Cladding Refurbishment, Proceedings of the 28th DAAAM International Symposium, pp.0529-0536, B. Katalinic (Ed.), Published by DAAAM International, ISBN 978-3-902734-11-2, ISSN 1726-9679, Vienna, Austria
DOI: 10.2507/28th.daaam.proceedings.075

Abstract

The current paper forms part of a series of research activities investigating the potential to implement a novel mobile device for marine diesel engine crankshaft journal surface refurbishment using laser cladding technology. Currently the laser cladding renovation process is not yet certified and is not an approved repair method for parts with highly stressed zones such as the fillets and bearing surfaces of marine diesel engine crankshafts, due to data deficiency or the absence of a validated method to evaluate high residual stress formation. In the current work, a method is proposed based on 3D transient temperature, calculating the residual stress field in the laser cladding model. The industrial application advantages of such a method are discussed and demonstrated in a case study. Further research campaigns and actions for demonstrating the industrial suitability of this method are discussed.

Keywords: Laser cladding; Marine Diesel Engine Crankshaft; Refurbishment Technology; Numerical Calculation; Transient Temperature Field; Transient Residual Stress Field.

1. Introduction

Laser cladding can provide a permanent structural repair and refurbishment of high-value components made of alloys that are generally considered unweldable, due to the small heat-affected zone, rapid solidification, improved cleanness and lower dilution, as well as greater control over the depth of the heat-affected zone. [1] [2]

The use of such technology would prove highly beneficial when integrated within a mobile device described in [3] and applied in the renovation of marine diesel engine crankshaft journals. According to [4] and [5], considerable economic benefits and saving of valuable time can be achieved by utilizing this kind of technological arrangement for marine diesel engine repairs. In [6] a prototype of such device was made and one of the biggest challenges for the implementation of this kind of device was established. In the view of several well-known classification societies the utilization of laser cladding technology as a repair method for crankshafts, especially in highly stressed zones such as fillets and bearing surfaces is not acceptable due to a high risk that the heat-affected zone could produce excessive hardness resulting in detrimental residual stress. The authors of the current work believe that such a view is based upon purely precautionary measures due to the thermomechanical nature of the process, the use of which indeed carries the risk of detrimental residual stress formation, however no quantifiable data is provided as justification.

It has been established that a numerical simulation method is a cost-effective technique to evaluate thermal history, evolution of residual stresses and distortions during the laser cladding process, which therefore make it possible to optimize process parameters without physically utilizing the technology. [7] Placing this conclusion in the context of marine diesel engine in-situ laser cladding technology development, numerical simulation therefore becomes an appropriate quantifiable means of establishing the level of residual stresses in the renovated section of the repaired part. Providing quantifiably justified means of risk evaluation for the certification of such technology would allow its further industrial implementation Hence, our current work is aimed at choosing the most appropriate laser cladding numerical simulation approach that would be efficient in terms of provision of necessary data on the levels of residual stress, while being industrially acceptable in terms of operational complexity and execution time.

2. Method Description

The method developed within this work is based on numerical simulation techniques, the most notable from the perspective of the present paper's authors being the laser cladding residual stress simulation model developed by Kovacevic et al. in [7], Bruckner et al. in [8] as well as Morville S. in [9] with negligible experimentally evaluated error where data on such evaluation was available. The current work utilizes all the fundamental laser cladding numerical simulation principles given in the aforementioned sources.

It is the authors' opinion that the main issue in all of the aforementioned works as well as other numerical studies is the level of complexity of the model setup and recreation of the results, requiring in-depth knowledge of the physical process that is being simulated and the method to simulate it. This considerably limits the use of such predictive tools to real-life scenarios, due to the need for qualified personnel to perform the operator's work, i.e. setting up the proposed mobile laser cladding device.

Another limitation is the significant time cost of performing such a laser cladding numerical simulation, where some complex simulation tools like the one developed in [9] can even take up to 3 months to execute.

Thus, despite the valuable and precise insight they provide on the modelled process, they cannot be considered industrially usable because of the two underlying factors explained above. This paper proposes an automated, simplified laser cladding model solution, in an attempt to eliminate these disadvantages.

For the successful development of the numerical simulation model, all the governing equations of the process first have to be established.

2.1. Thermal equilibrium equation

In the case of laser cladding process which can be modelled as thermo-mechanical coupled system, the thermal equilibrium equation can be written as:

$$\rho c\left(\frac{\partial T}{\partial t}+v_x\frac{\partial T}{\partial x}+v_y\frac{\partial T}{\partial y}+v_z\frac{\partial T}{\partial z}\right)=\dddot{q}+\frac{\partial}{\partial x}\left(\frac{K_x\partial T}{\partial x}\right)+\frac{\partial}{\partial y}\left(K_y\frac{\partial T}{\partial y}\right)+\frac{\partial}{\partial z}\left(K_z\frac{\partial T}{\partial z}\right) \quad (1)$$

Where:

K_{xx}, K_{yy}, K_{zz} – Conductivity in the element x, y and z directions respectively;
ρ - Density of material;
c- Specific heat;
T – Temperature of the material;
v_x, v_y, v_z – Mass transport of heat;
$\dddot{q}$ - heat generation rate per unit volume.

2.2. Stress – strain relationship

The stress is related to strains by:

$$\{\sigma\} = [D]\{\varepsilon^{el}\} \tag{2}$$

Where:

$\{\sigma\} - [\sigma_x \sigma_y \sigma_z \sigma_{xy} \sigma_{yz} \sigma_{xz}]^T$ -stress vector;
$[D]$ – elasticity or elastic stiffness matrix or stress-strain matrix;
$\{\varepsilon^{el}\} = \{\varepsilon\} - \{\varepsilon^{th}\}$ – elastic strain vector;
$\{\varepsilon\} = [\varepsilon_x \varepsilon_y \varepsilon_{xy} \varepsilon_{yz} \varepsilon_{xz}]^T$ total strain vector;
$\{\varepsilon^{th}\}$ – total strain vector.

For the 3-dimensional case, the thermal strain vector is:

$$\{\varepsilon^{th}\} = \Delta T[\alpha_X^{SE} \alpha_y^{SE} \alpha_z^{SE} 000]^T \tag{3}$$

Where:

$\alpha_X^{SE}; \alpha_Y^{SE}; \alpha_Z^{SE}$ – secant coefficient of thermal expansion in the x, y, z direction;
$\Delta T = T - T_{ref}$;
T – current temperature at the point in question;
T_{ref} – reference (strain-free) temperature.

2.3. Boundary Conditions

Four types of boundary conditions are considered. It is presumed that these cover the entire element.

1. Specified temperatures acting over the surface:

$$T = T^* \tag{4}$$

2. Specified heat flows acting over the surface are modelled as a travelling 2-dimensional distributed heat source with a Gaussian distribution:

$$q^* = \frac{3q_l}{\pi R^2} \exp(-\frac{3(X^2 + (Y - vt)^2)^2}{R^2}) \tag{5}$$

3. Specified convection acting over the surface (Newton's law of cooling):

$$\{q\}^T\{n\} = h_f(T_S - T_B) \tag{6}$$

4. Radiant energy loss:

$$Q_i = A_i \varepsilon_i F'_{ij} \sigma (T_i^4 - T_j^4) \tag{7}$$

$$F'_{ij} = \frac{F_{ij}}{F_{ij}(1 - \varepsilon) + \varepsilon_i} \tag{8}$$

Where:

T^* - specified temperature;
$\{n\}$ - unit outward vector;
q^*- specified heat flow;
T_B – bulk temperature of the adjacent fluid;
T_S - temperature at the surface of the model;
h_f – film coefficient evaluated at $\frac{T_B+T_S}{2}$; $[\frac{W}{m^2K}]$
ε_i – effective emissivity of surface i;
F_{ij} – Radiation view factors;

A_i – Area of surface i;
Q_i – Energy loss of surface i;
σ – Stefan-Boltzmann constant;
T_i – Absolute temperature of surface i;
R – Radius of the heat source;
q_l - final power output of the laser onto the substrate;
v – laser scanning speed;
t– laser cladding simulation time.

2.4. Modelling procedure and assumptions

The governing equations are solved with the FEM transient analysis procedure using commercial FEM software ANSYS APDL software.

The unique feature of the current model is that it provides a good indication of the melt pool shape, as well as the temperature-stress-strain history, considering material phase transformation and transformation plasticity during 1 cladding pass heating, whilst requiring a relatively low computation time and resources. The temperature field and stress-strain time-history calculations are obtained through an automated process, where the operator of the model enters the parameters of the workpiece geometry, laser heat source power, laser spot diameter and laser scanning speed, along with the selected workpiece material. Automated calculation is ensured by a script written in ANSYS APDL software which can be executed by opening the script with the appropriate parameters for the process being modelled.

However, the current model does not consider the influence of bead geometry on the temperature-strain distribution, which inarguably is very important when the aim is to obtain a precise mechanical response for the thermal contraction behavior of the consumable process materials. Additionally the current model does not take into account the gas/melt pool interaction, laser attenuation by powder, Bénard-Marangoni convection or a more precise laser beam mode heat source for the CO2, Nd:YAG and HPDL laser. The inaccuracy of the results in the model thus offer a subject for further improvement. However, the low time and computational resource cost will be gradually eroded as further improvements of the model are made.

3. Case Study

3.1 Boundary Condition Data Entry

Fig. 1. gives a schematic representation of the geometry model used within the current work.

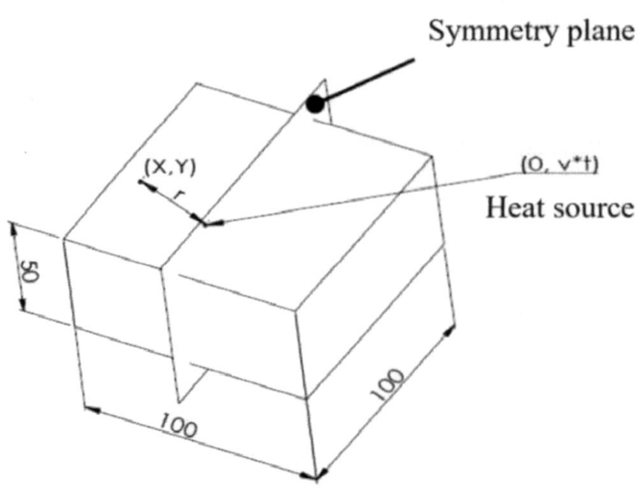

Fig. 1. Geometry of a modelled workpiece

To reduce the real-time simulation of the current process model, a symmetry plane for the symmetry boundary condition is introduced, as illustrated in Fig. 1.

The substrate material was modelled as C45. Given the very wide range of temperatures involved in the laser cladding process, the accuracy of a model strongly correlates with the precise definition of all non-linearities associated with the material mechanical property changes that are caused by temperature change. According to the standard DIN EN 10083-2 – technical delivery conditions for non-alloy steels [10] as well as the technical material specifications for C45 provided by some of the best-known steel suppliers [11] [12], all the material properties as time functions are defined accordingly.

To successfully consider the phase transformation phenomenon, various temperature-dependent thermo-physical material properties must be defined, such as density, Young's module, thermal expansion thermal conductivity, specific heat and Poisson's ratio. The defined model considers transformation plasticity. A simple method to define a material model with transformation plasticity is to define the bilinear stress-strain curve, which assumes the isotropic hardening of the material.

For the definition of the current model, two basic elements that were defined with "ANSYS" software are used: PLANE55 and SOLID 70. [13]

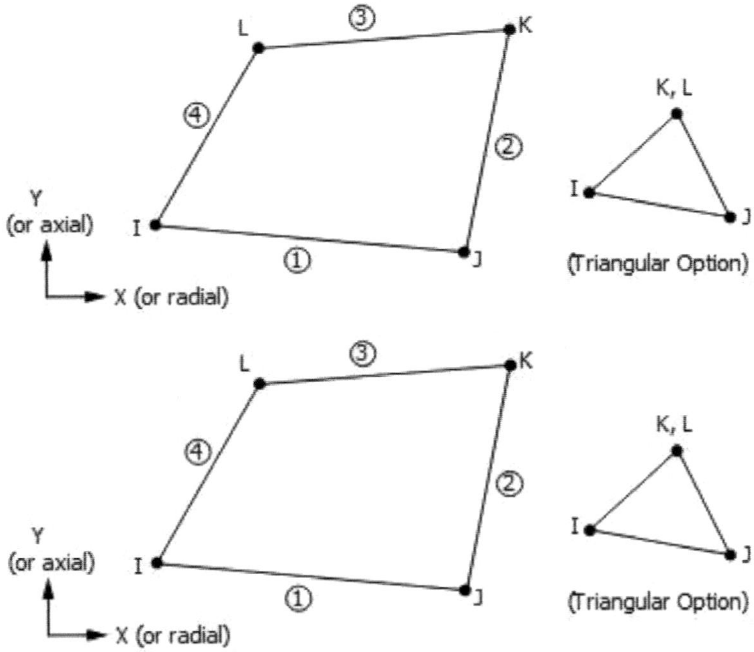

Fig. 2. PLANE55 and SOLID70 element geometry [13]

The initial temperature for the boundary conditions defined in (4) was chosen as per NTP (Normal temperature and pressure), which is $20°C$. Table 1 defines the technological parameters of the process to be modelled for the boundary condition given in (5).

P – final power output of the laser onto the substrate, $[W]$	700
R – radius of the laser spot, $[mm]$	1
v – laser scanning speed, $[\frac{m}{s}]$	10

Table 1. Technological equipment parameter chosen for the current numerical study

The convective heat transfer coefficient for all the surfaces of the modelled workpiece was chosen to be $h_f = 30 \frac{W}{m^2K}$, which is acceptable for normal ventilated in-laboratory conditions according to [14] and [15] which is later used for the boundary condition given in (6).

And finally, based on the information regarding the emissivity of various materials found in [16], the effective emissivity of the material C45 is chosen to be $\varepsilon_i = 0.94$.

3.2 Results

The numerical calculations performed within the scope of this work as mentioned previously were achieved using "ANSYS r16" FEM software. The numerical calculation consists of 2 major parts:

1. Obtaining the time-dependent temperature fields of the current study;
2. Reading the results of the first parts of this study and iteratively entering them within the second part of the study, to obtain the mechanical stress-strain response of the material model.

Given the iterative nature of the calculations described above, the scriptable "ANSYS r16 Mechanical APDL" part of the software had to be used, thus developing an automated numerical calculation tool to be used by the device operator mentioned in the introductory part of this paper.

The simulation was run on a regular office laptop.

The temperature field results obtained at different time intervals are given in fig. 3. The results obtained in the current sample numerical simulation indicate that the material is locally heated up to 2,451$°C@t = 0.2s$, 3,870 $°C@t = 5s$ and even reaches the temperature value of 4,132$°C@t = 10s$. But the most important information that our simulation can provide regarding the temperature field is an estimation of the weld pool size during the process, which gives weld pool measurements: depth = 1.56 mm and radius = 1.15 mm (diameter = 2.3 mm). This information can offer an insight into the various output parameters of the clad, such as its width and depth, as well as the height and width of the heat-affected area and dilution zone.

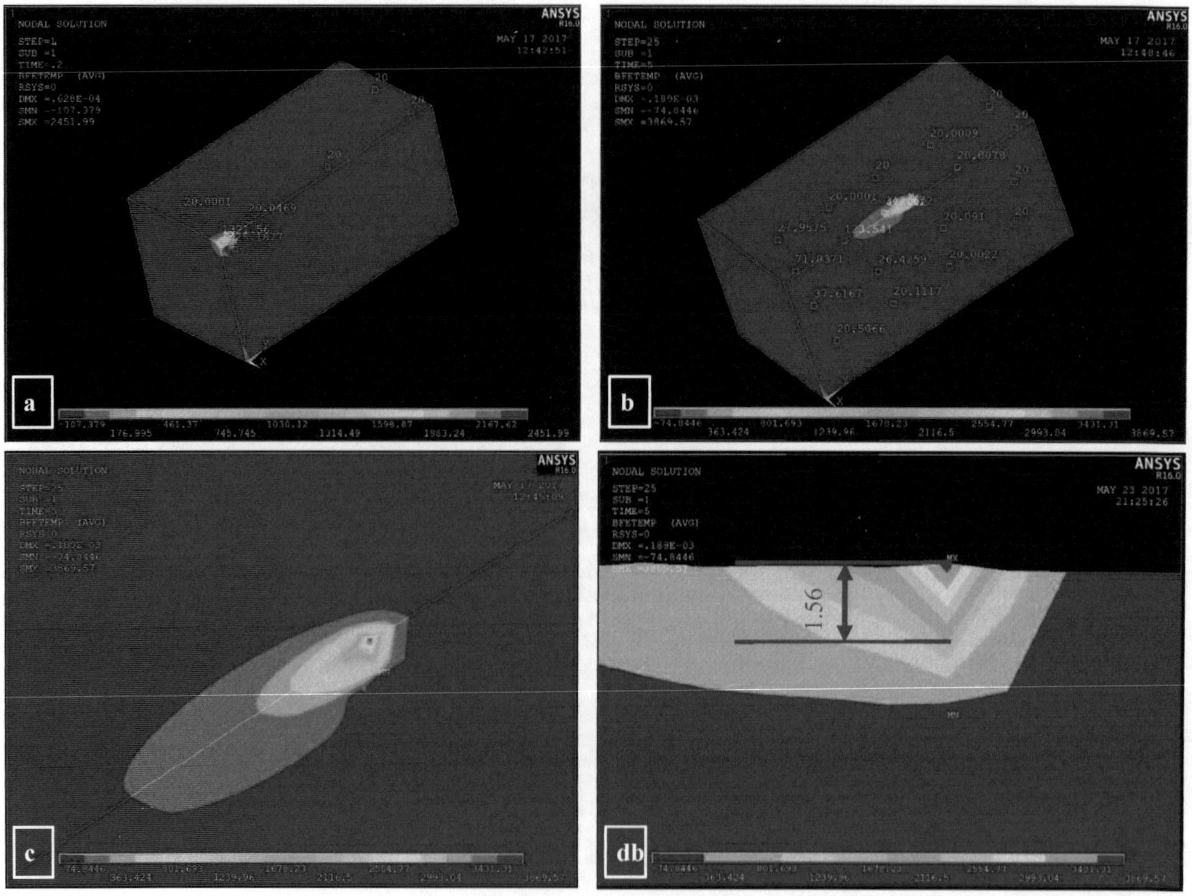

Fig. 3. Temperature field results at (a) t=0.2s; (b) t=5s; (c) Expanded view of the heat source at t=5s; (d) Expanded view of the heat source at t=5 s with depth measurement

In fig. 4, the residual stress field is shown at different time intervals. The results obtained here are particularly important for the specific aim of this work, which is to implement the laser cladding technology for the renovation of marine diesel-engine crankshafts. The von Mises stress field obtained here can be used to identify the safety margin between the operational stresses and ultimate stresses of the C45, which according to [11] is 620 MPa. Comparing this value with the values obtained in the study, the maximum residual stresses with the technological parameters specified within this study do not exceed 220 MPa, thus it can be deduced that there is a substantial safety margin of $\Delta = 620 - 220 = 400MPa$ which indicates that the risk of rupture or cracking of the material is almost non-existent.

It is acknowledged that the geometry of the clad as well as the material properties which are used for the coating are important parameters for determining a more realistic situation of cracking due to residual stresses. However, with only limited computer and time resources available for the evaluation of residual stresses, which is the case in typical operative field situations, the current model can be used to obtain results separately for the substrate and for the clad geometry, defining separately all the material and boundary conditions. In this way, the current model provides operational simplicity, speed and efficiency, which are very important in operational field conditions.

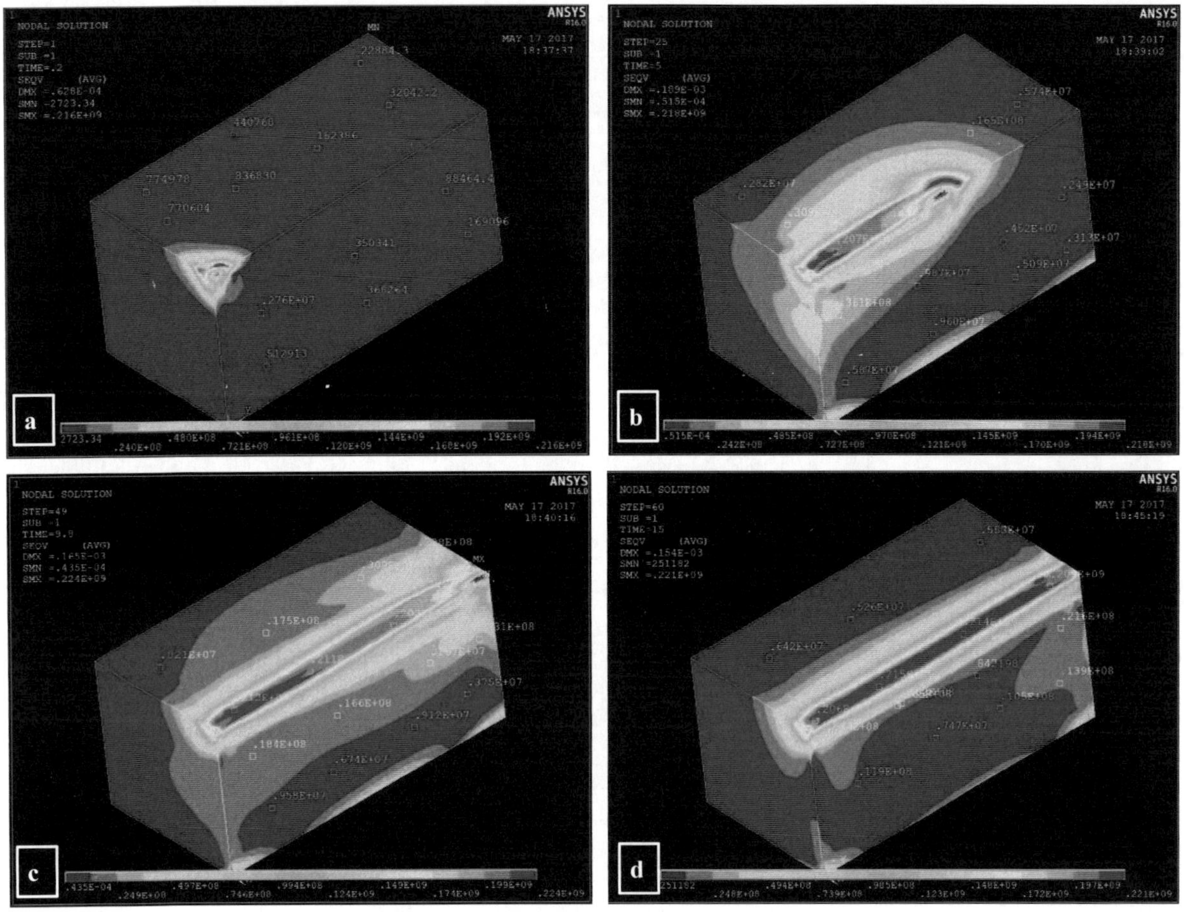

Fig. 4. Residual stress field results at (a) t=0.2s; (b) t=5s; (c) t=9.8 s; (d) t=15s.

4. Conclusions

Within the previous research efforts, it has been established that laser cladding technology is highly suitable for the application of marine diesel-engine crankshaft journal surface renovation in terms of the technological requirements of the resultant product, technology precision, repeatability and in-situ capability. However, a significant limitation on its real-life application exists, due to the complex nature of the laser cladding process, which has a great number of input and process parameters, along with the high sensitivity of these process parameters, which give rise to a detrimental residual stress within the renovated work.

To address the problem above, a novel script-written automated simulation tool was developed using FEM ANSYS software, which would enable a relatively quick and simple main process parameter study to be performed providing an insight for the in-situ laser cladding device operator into the levels of residual stress in renovated work. The scripted FEM tool limits the user input to such a level, that only the most important material properties, laser cladding technology and technological conditions are to be entered, while the rest of calculations are performed automatically, thus decreasing workload time-consumption and qualification requirements imposed on the intended user.

Experimental validation of the model utilized within the current work will be performed in future research campaigns to determine the accuracy of calculations made. Additional effort is required to achieve gas/melt pool interaction, laser attenuation by powder, Bénard-Marangoni convection and a more precise laser beam mode heat source integration within the tool developed in current work, while keeping the calculation time-cost and user workload to the same or even lower level.

5. References

[1] Toyserkani, E.; Khajepour A. and Corbin, S. (2005). Laser Cladding., CRC Press LLC, ISBN: 0-8493-2172-7. Florida

[2] Torims, T. (2013). The Application of Laser Cladding to Mechanical Component Repair, Renovation and Regeneration. Riga : DAAAM International, ISBN: 978-3-901509-94-0.

[3] Torims T. (2013). Laser cladding device for in-situ repairs of marine crankshafts. Advanced Materials Research Vols. 712-715., pp 709-714., DOI:10.4028/www.scientific.net/AMR.712-715.709

[4] Torims, T.; Vilcans, J.; Zarins, M.; & Ratkus, A. (2012). Development of the In-Situ Laser Renovation Technology for Marine Diesel Engine Crankshafts. ISBN 978-3-901509-91-9, ISSN 2304-1382, pp 0009 - 0014, Published by DAAAM International,Vienna, Austria.

[5] Torims, T.; Bruckner F.; Ratkus, A.; Fokejevs A.; Logins, A. (2014). The Application of Laser Cladding to Marine Crankshaft Journal Repair and Renovation, Proceedings of the ASME 2014 12th Biennial Conference on Engineering Systems Design and Analysis., Copenhagen

[6] Torims, T; Pikurs, G; Ratkus A.; Logins A.; Vilcans J.; Sklariks S. (2014). Development of Technological Equipment to Laboratory Test In-Situ Laser Cladding for Marine Engine Crankshaft Renovation, 25th DAAAM International Symposium on Intelligent Manufacturing and Automation, DAAAM, Elsevier Ltd., pp. 559-568., DOI: 10.1016/j.proeng.2015.01.405

[7] Farahmand, P.; Kovacevic R. (2014). An experimental–numerical investigation of heat distribution and stress field in single and multi-track laser cladding by a high-power direct diode laser. Dallas : Elsevier, Vol. 63. 0030-3992.

[8] Bruckner, F.; Lepski, D. and Beyer, E. (2006). Modelling the Influence of Process Parameters and Additional Heat Sources on Residual Stresses in Laser Cladding. Dresden, ASM International.

[9] Morville, S. (2012). Modelisation multiphysique du procede de Fabrication Rapide par Projection Laser en vue d'ameliorer l'etat de surface. Universite de Bretagne Sud, NNT : 2012LORIS280>.

[10] Deutsche Norm. (2006). Steels for quenching and tempering - Part 2: Technical delivery condition for non-alloy steels English version of DIN EN 10083-2:2006-10. DIN EN 10083-2.

[11] Group. (2017). Quality of C45E. s.l. : Lucefin Group.

[12] Saarstahl (2017). Material Specification sheet of C45. s.l..

[13] ANSYS INC. (2017). ANSYS Mechanical APDL Help Theory reference Heat flow fundamentals section. Southpointe.

[14] http://www.engineeringtoolbox.com/convective-heat-transfer-d_430.html (2017) The engineering toolbox. Convective Heat Transfer. The engineering toolbox. The engineering toolbox. Accessed on: 05 14, 2017.

[15] http://www.engineersedge.com/heat_transfer/convective_heat_transfer_coefficients__13378.htm, (2017) .Engineer's Edge. Convective Heat transfer coefficients table chart. Accessed on: 05 14, 2017.]

[16] MIKRON Vertretung Schweiz. (2012) Table of Emissivity of various surfaces.

DOI: 10.2507/28th.daaam.proceedings.076

METROLOGICAL SET-UP FOR CALIBRATING 1D LINE SCALES WITH SUB-MICROMETRE PRECISION

Rok Klobucar, Branko Strbac, Miodrag Hadzistevic, Bojan Acko

This Publication has to be referred as: Klobucar, R[ok]; Strbac, B[ranko] M[ile]; Hadzistevic, M[iodrag] & Acko, B[ojan] (2017). Metrological set-up for Calibrating 1D Line Scales with Sub-Micrometre Precision, Proceedings of the 28th DAAAM International Symposium, pp.0537-0544, B. Katalinic (Ed.), Published by DAAAM International, ISBN 978-3-902734-11-2, ISSN 1726-9679, Vienna, Austria
DOI: 10.2507/28th.daaam.proceedings.076

Abstract

The use of the optical standards in the industry is constantly growing. A high resolution 1D measurement instrument developed for calibrating precision line scales with sub-micrometre measurement uncertainty is presented in this paper. The mechanical foundation of this instrument is based around a custom designed, numerically controlled multi-axis granite stage manufactured by Newport Micro-Controle Spectra-Physics, but specifically designed to meet the metrological requirements of the Laboratory for Production Measurement (LTM) at the University of Maribor. The system integrates a numerically controlled multi-axis stage, a laser interferometer, and a vision system for detecting line position. The measurement and the analysis processes are completely automated in order to minimize manual labour during the calibration process, but also increase the calibration accuracy. Increasing calibration accuracy leads up to better quality of industrial measurements which is required by modern precision industry. The metrological capabilities of the presented measurement set-up were verified by some practical test measurements. Indirect users of the results of this research will be all manufacturers of precise products such as automotive and other industries.

Keywords: Measurement; high resolution measurements; multi-axis stage; metrological set-up; line scales; measurement uncertainty

1. Introduction

Industry is demanding more and more rapid and accurate dimensional measurements on diverse mechanical parts. In recent years, optical coordinate measuring machines (CMMs) having made substantial progress and are now often used for such applications. Especially CMMs equipped with imaging capabilities are frequently used for fast, non-contact measurements. Strong competition among manufacturers of such instruments and the demand for sub-micron accuracy has led to standardised tests, which are aimed for comparing and validating instrument performance [1]. Calibration, verification and also error correction of optical CMMs are mainly based on measurements using reference line scales or two-dimensional grid plates [2, 3, 4].

There are very many commercially available line scales made of different materials (steel, brass, invar, glass, quartz, zerodur). They can vary in length from below 1 mm to more than 1 m, and have resolutions (pitch) from below 10 nm to 1 cm. They can be calibrated by using different measurement set-ups, depending on their length and precision. Set-ups for calibrating high-precision line scales normally involve a microscope with an optical sensor for capturing and analysing the image of a line marker and a laser interferometer as a traceable measurement standard [5-7].

This paper introduces a measuring system developed for calibrating line scales with measurement uncertainty of less than 10 nm over a total length of 500 mm. The system integrates a numerically controlled multi-axis stage, a laser interferometer, and a vision system for detecting line position. With this measurement set-up, Abbe errors can be reduced to negligible levels. Spatial separation between the measured point and reference line is known as an Abbe offset [8].

Since it is possible to put optical components very close to each other, the air dead path error of the laser interferometer is also negligible. Software for detecting of the lines is based on earlier design from 2009 [9, 10]. In order to improve performance and to achieve better uncertainty, the software was improved and automatized, mechanics and optics have been redesigned and several uncertainty components better characterized. This facility will replace the old measurement facility that is manually operated. A model for evaluating uncertainty and the uncertainty budget for the demonstrated measuring system is presented in the paper.

2. Measurement system

Measurement set-up for calibrating line scales is shown in Fig. 1., while its schematic diagram is shown in Fig. 2. The system integrates a numerically controlled multi-axis stage, a laser interferometer, and a vision system for detecting line position [11].

Fig. 1. Measurement set-up for calibrating line scales

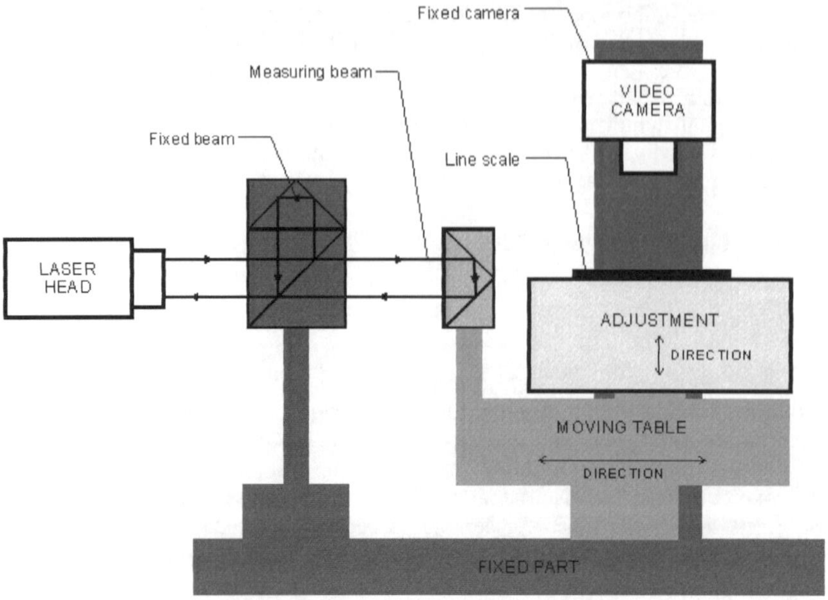

Fig. 2. Schematic diagram of measurement set-up for calibrating line scales

2.1. *Numerically controlled multi-axis stage*

Numerically controlled multi-axis stage is shown in Fig. 3. It was designed and manufactured by Newport-micro controle [12] for the Laboratory for Production Measurement (LTM), University of Maribor, Slovenia. The measuring system designed by the LTM is intended to perform 2D measurements on various objects, which will be fixed on a measuring table. This research is focused in 1D measurements. Newport's solution is based on a HybrYX-G5 stage featuring a ceramic carriage which freely slides in X and Y axes on a precision lapped granite reference plate, using proprietary pressure-vacuum air bearing design. The carriage is guided along Y axis by a rigid ceramic beam. The beam is supported and guided at each end with the ball bearing carriages of X axis. Both X and Y axes are motorized with linear actuators and include linear glass encoders. Both linear scale glass encoders are LIDA403 made by Heidenhain, length 440 mm for X axis and length 1250 mm for Y axis. This positioning system is equipped with the optional Z-Tip-Tilt and Theta stage. It is motorized by precision actuators equipped with miniature DC servo motors. The position system is built on a heavy granite table, which features a precision reference plane for the air bearing carriage. In addition, the granite structure includes a granite gantry allowing the fixation of a vertical motorized translation stage, which accommodates the measuring sensor, depending on an application. Two rails mounted on the gantry equipped with sliding carriages provide manual rough positioning capability. The granite structure is set on a heavy welded frame equipped with ND40 passive isolators. The position system has one XPS motion controller including dedicated drivers for the five DC motors and the three linear motors.

The X axis is dual axis composed of X1 and X2 actuators mounted on the flanks at both ends of the granite table. X1 and X2 carriages move the ceramic beam of the Y axes. The X axis travel range is +/- 175 mm. It has a linear encoder with 5 nm resolution. The maximal speed is 300 mm/s., while the repeatability is 500 nm.

The Y axis is a horizontal translation stage using air bearing technology and linear motor. The guiding beam of Y axes sub-assembly is attached to the carriages of the X1 and X2 short travel translation actuators. The guideways involve two parts: a large ceramic carriage and a Y-axis guide ceramic L-shape beam. The ceramic carriage freely slides in X and Y axes using pressure-vacuum air bearing design. The reference planes for the air bearing carriage are the precision lapped surface of the granite table and the vertical slide of the Y rigid ceramic beam. The Y axis travel range is +/- 500 mm. It contains a linear encoder with 5 nm resolution. Maximum speed is 600 mm/s., while the repeatability is 100 nm.

The ZTT Theta stage is mounted on the air bearing carriage; it is equipped at the top with the measuring platform, which allows installing measured objects. All axes are driven by the high precision motorized linear actuators. Three actuators are set in upright position with a travel range of 10 mm. One actuator is set horizontally to achieve the theta Z movement with a travel range of +/- 1°. All four actuators are equipped with a miniature DC motor made by Fulhaber.

Z axis is a vertical motorized translation stage on the middle carrier. It is a Newport catalogue stage reference M-IMS100V. The Z axis travel range is 100 mm, the minimum incremental motion is 0,3 μm, the encoder resolution is 0,1 μm, the maximum speed is 20 mm/s, while the repeatability is ± 0,5 μm.

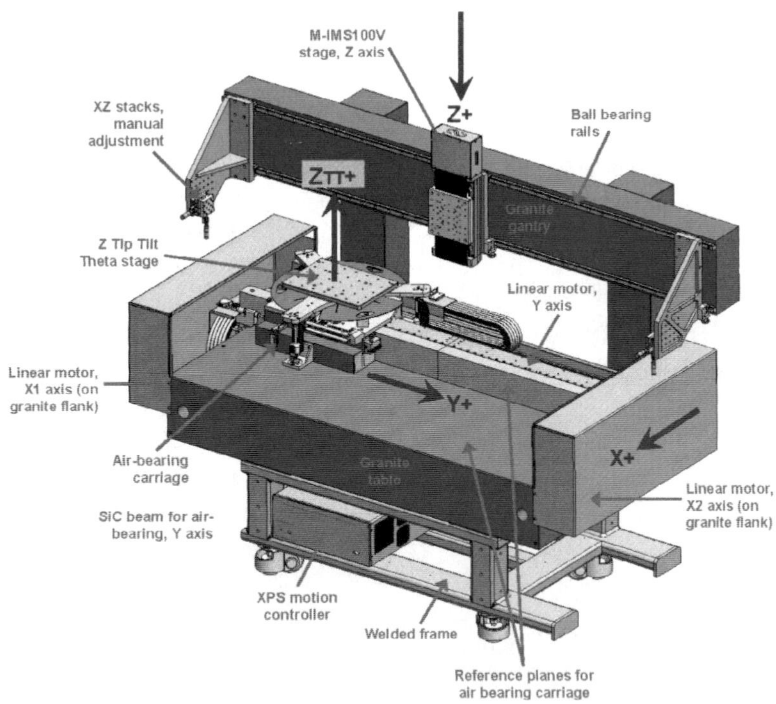

Fig. 3. Numerically controlled multi-axis stage

2.2. *Laser interferometer*

Laser interferometer position measurement systems provide very precise position or distance information [15]. Our system consists of a laser head HP 5528A, Agilent module 55292A, and a variety of optical components and accessories such as material sensors and air sensor. The basic system measures linear displacement. The system uses the wavelength of light from a low-power helium-neon laser as a length standard. We normally set the resolution of the laser interferometer to 10 nm.

Special mounting elements were constructed for the optics used for calibration. A schematic diagram of the laser interferometer and the position of the optics are shown in Fig. 2. The position of the optical elements and the moving parts are set in such way, that the measured object axis is set in the line with the centre of the laser linear retroreflector. It allows the Abbe errors to be reduced to negligible levels even for the most demanding dimensional metrology tasks [13]. The linear interferometer is fixed on the fixed part of the stage, while the retroreflector is placed on the moving table in the horizontal direction. Only the measured object is placed on the moving table that can move in the vertical direction.

2.3. *Vision system for detecting line position*

The vision system for detecting line position consists of a zoom microscope and a CMOS digital camera. The camera is connected to the computer via USB 3.0 port. The CMOS camera gets the images of the line scale and sends them to the computer software.

The software analyses the images and determinates the middle of the line., which is defended by the operator. The software calculates the distance in pixels. The CMOS digital camera takes 15 monochrome images per second in resolution 2592 x 1944 pixels. The software analyses the images in real time. The distance calculated in pixels is transformed into micrometers [14]. The software for calculation the distance between lines is more detailed presented by the authors in paper [10].

3. Uncertainty of measurement

Line scales are material measures made of glass, steel or other material, on which dimensions are marked with line marks. Since the materials have quite different temperature expansion coefficients, which are in many cases not exactly known, they are stabilized before calibration in the climatic room at 20 °C for 24 hours. The measurement system must be adjusted and initialized before the measurement. The line scale should be positioned under the camera by moving the measurement table. The measuring system, laser interferometer, vision system for detecting line position and line scale should be adjusted according to the measuring direction, which is defended by the movement of the horizontal direction of the table. The line scale should be fixed on the adjustment table under the camera.

The camera is fixed. It is adjusted in such a way, that the focus of the camera intersects the centre of the laser linear retroreflector. The camera is focused on the lines of the line scale by moving the adjustment table (ZTT Theta stage) in the vertical direction. With this procedure, the line scale axis, the images of the lines and the centre of the laser linear retroreflector with the laser beam are positioned in the measuring direction.

The image processing software should be initialized [10]. With the improved software, connection to the numerically controlled multi-axis stage and connection to the laser head, automatic measurements of distances are possible. The measurement goal is to measure distances between the reference line and the chosen measuring lines on the line scale automatically. Measured data are saved into the file where they are ready for further processing.

3.1. *Mathematical model of measurement*

The measured value in the calibration of a line scale is a deviation from a nominal distance between two line centres. The distance between two lines is calculated as a sum of laser interferometer indication and vision system for detecting line position indication. The vision system for detecting line position measures the distance between the measurement point (scale mark) and the reference line. Deviation e (measurement result) is given by the expression:

$$e = (L_{\mathrm{LI}} + L_{\mathrm{V}} - L_{\mathrm{LIref}}) \cdot (1 + \alpha_{\mathrm{m}} \cdot \theta_{\mathrm{m}}) - N + e_{\mathrm{cos}} + e_{\mathrm{mp}} + e_{\mathrm{ms}} + e_{\mathrm{a}} \quad (1)$$

where:

e deviation (measurement result) at 20 °C
L_{LI} corrected length shown by laser interferometer
L_{V} distance between the measurement point (scale mark) and reference line in the image window
L_{LIref} indication on the laser in the reference (origin) point
α_{m} linear temperature expansion coefficient of the scale

θ_m temperature deviation of the scale from 20 °C
N nominal value (without uncertainty)
e_{cos} cosine error of measurement (supposed to be 0)
e_{mp} dead path error
e_{ms} random error caused by uncontrolled mechanical changes
e_a error caused by the measuring table inclination

3.2. *Standard and expanded uncertainty*

For uncorrelated input quantities the square of the standard uncertainty associated with the output estimate y is given by (2) [14]:

$$u^2(y) = \sum_{\mathrm{i}=1}^{\mathrm{N}} u_{\mathrm{i}}^2(y) \tag{2}$$

The quantity $u_{\mathrm{i}}(y)$ $(i = 1,2, \dots, N)$ is the contribution to the standard uncertainty associated with the output estimate y resulting from the standard uncertainty associated with input estimate x_{i} [14]:

$$u_i(y) = c_i \cdot u(x_i) \tag{3}$$

where c_i is the sensitivity coefficient associated with the input estimate x_i, i.e. the partial derivative of the model function f with respect to X_i, evaluated at the input estimates x_i [14].

$$c_i = \frac{\partial f}{\partial x_i} = \frac{\partial f}{\partial X_i}\bigg|_{X_1=x_1 \dots X_N=x_N} \tag{4}$$

Value X_i	Estimated value	Standard Uncertainty	Distribution	Sensitivity coefficient	Uncertainty contribution
L_{LI}	0 mm	13 nm + $0{,}2 \cdot 10^{-6} \cdot L$	normal	1	13 nm + $2{,}5 \cdot 10^{-7} \cdot L$
L_V	<5 µm	25 nm	normal	1	25 nm
L_{LIref}	0 mm	9 nm	normal	-1	9 nm
α_m	10^{-5} °C^{-1}	$1{,}15 \cdot 10^{-6}$ °C^{-1}	rectangular	0,05 °C$\cdot L$	$0{,}06 \cdot 10^{-6} \cdot L$
θ_m	0 °C	0,05 °C	normal	10^{-5} °C$^{-1} \cdot L$	$0{,}05 \cdot 10^{-5} \cdot L$
e_{cos}	0	0 nm	normal	1	0 nm
e_{mp}	0	9 nm	rectangular	1	9 nm
e_{ms}	0	30 nm	normal	1	30 nm
e_a	0	13 nm	rectangular	1	13 nm
				Total:	$\sqrt{(45\ \mathrm{nm})^2 + (5{,}6 \cdot 10^{-7} \cdot L)^2}$

Table 1. Uncertainty budget for calibration of line scales

The expanded uncertainty for the coverage factor k=2 is then:

$$U = \sqrt{(90\ \mathrm{nm})^2 + (1{,}1 \cdot 10^{-6} \cdot L)^2} \tag{5}$$

4. Experimental results

The measurement accuracy of the numerically controlled multi-axis stage was verified with the laser interferometer [10] in Y direction over the distance (0 to 500) mm by twenty repeated measurements in eleven positions. The absolute difference between the reference value shown by the laser interferometer and the encoder value in the Y axis of the measuring system is shown in Fig. 4.

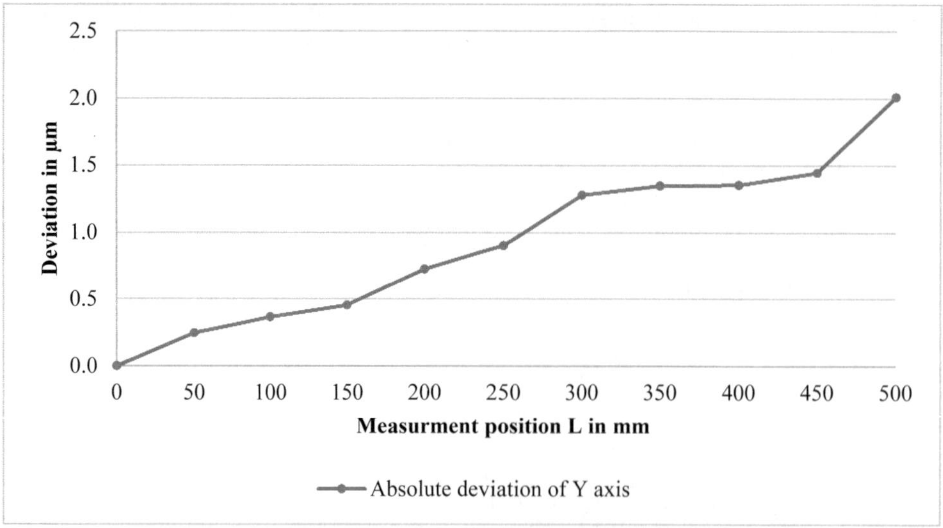

Fig. 4. Absolute deviation between laser values and encoder value

Experimental standard deviation that reflects random influences is shown in the diagram in Fig. 5. Random influences are caused by the multi-axis stage instability, vibrations and random changes of environmental conditions. Presented standard deviations were used in uncertainty budged of the calibration procedure.

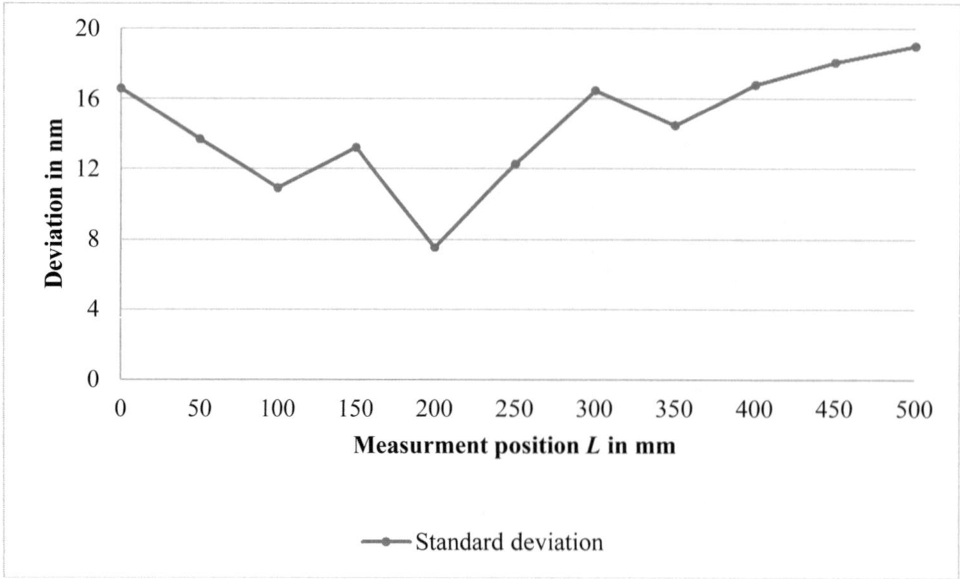

Fig. 5. Standard deviation on measurement positions

Measurement accuracy of the measuring system was verified on a calibrated line scale over the distance (0 to 100) mm. Experimental standard deviation $s(L)$ (6):

$$s(L) = \sqrt{\frac{1}{n-1}\sum_{j=1}^{n}\left(L_j - \bar{L}\right)^2} \tag{6}$$

that reflects random influences is shown in the diagram in Fig. 6. The diagram also represents the deviation of measured values from the reference and estimated standard uncertainty. These results characterize the measuring system which will allow calibrations of length measurements without laser interferometer. Measuring results represent a linear characteristic. It is possible to compensate the error with the appropriate error mapping. From the results, we can see that it is possible to improve the measurement uncertainty.

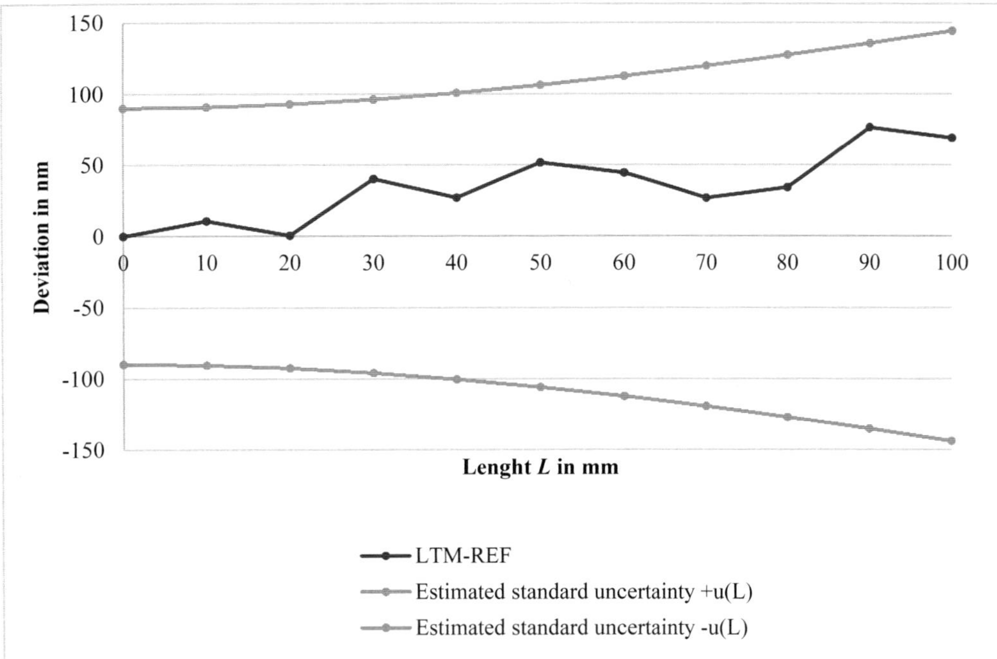

Fig. 6. Deviation of measured values from reference and estimated standard uncertainty

5. Conclusion

The main result of the presented research is a verified measuring set-up for calibrating precise line scales. The system integrates a numerically controlled multi-axis stage, a laser interferometer, and a vision system for detecting line position.

The presented procedure for calibrating line-scales with lengths up to 500 mm, with the measuring uncertainty expressed by (5) has already been accredited by the national accreditation body. In respect to the previous measurement set-up, we have improved the measuring system with new automatic numerical controlled multi-axis stage, better camera and better environmental conditions. Better calibration and measurement capability (CMC) was achieved and approved. The calibration, verification and also error correction of optical CMMs is mainly based on measurements using reference line scales or two-dimensional grid plates. Our further work will focus on calibrating 2D optical grids and step gauges. One of the final goals of this validation phase was to determine uncertainty of measurement in calibration and verification of new automatic high resolution measuring set-up for calibrating precise line scales.

6. Acknowledgment

The authors acknowledge the financial support from the Slovenian Research Agency (research core funding No. P2-0190), as well as from Metrology Institute of the Republic of Slovenia (funding of national standard of length; contract No. C3212-10-000072). The research was performed by using equipment financed from the European Structural and Investment funds (Measuring instrument for length measurement in two coordinates with sub-micrometer resolution; contract with MIRS No. C2132-13-000033).

7. References

[1] Meli, F. (2013). Calibration of photomasks for optical coordinate metrology, Physikalisch-Technische Bundesanstalt (PTB), doi: 10.7795/810.20130620C

[2] Mariko, K.; Tsukasa, W.; Makoto, A., & Toshiyuki, T. (2015). Calibrator for 2D Grid Plate Using Imaging Coordinate Measuring Machine with Laser Interferometers, International Journal of Automation Technology, Vol. 9, No. 5, pp. 541-545, doi: 10.20965/ijat.2015.p0541

[3] Ačko, B.; Brezovnik, S.; Črepinšek-Lipuš, L. & Klobučar, R. (2015). Verification of statistical calculations in interlaboratory comparisons by simulating input datasets, International journal of simulation modelling, Vol. 14, No. 2, pp. 227-237, doi: 10.2507/IJSIMM14(2)4.288

[4] Ačko, B.; Sluban, B.; Tasič, T. & Brezovnik, S. (2014). Performance metrics for testing statistical calculations in interlaboratory comparisons, Advances in Production Engineering & Management, Vol. 9, No. 1, pp. 44-52, doi: 10.14743/apem2014.1.175

[5] Flügge, J.; Köning, R.; Weichert, Ch.; Häßler-Grohne, W.; Geckeler, R. D.; Wiegmann, A.; Schulz, M.; Elster, C. & Bosse, H. (2009). Development of a 1.5D reference comparator for position and straightness metrology on photomasks, Proc. SPIE 7122, Photomask Technology 2008, doi:10.1117/12.801251

[6] Ačko, B. (2012). Final report on EUROMET Key Comparison EUROMET.L-K7: Calibration of line scales, Metrologia, Vol. 49, doi: 10.1088/0026-1394/49/1A/04006

[7] Lassila, A. (2012). MIKES fibre-coupled differential dynamic line scale interferometer, Measurement Science and Technology, Vol. 23, No. 9, doi: 10.1088/0957-0233/23/9/094011

[8] Leach, R. (2015). Abbe Error/Offset, CIRP Encyclopaedia of Production Engineering, Springer, Berlin, DE, 1-4, doi: 10.1007/978-3-642-35950-7_16793-1

[9] Klobučar, R. & Ačko, B. (2016). Experimental evaluation of ball bar standard thermal properties by simulating real shop floor conditions, International journal of simulation modelling, Vol. 15, No. 3, pp. 511-521, doi:10.2507/ijsimm15(3)10.356

[10] Družovec, M.; Ačko, B.; Godina, A. & Welzer, T. (2009). Robust algorithm for determining line centre in video position measuring system, Optics and Lasers in Engineering, Vol.47, pp. 1131-1138, doi: 10.1016/j.optlaseng.2009.06.017

[11] Klobucar, R. & Acko, B. (2017). Automatic high resolution measurement set-up for calibrating precise line scales, Advances in Production Engineering & Management, Vol. 12, No. 1, pp. 88-96, doi: 10.14743/apem2017.1.242

[12] Newport Corporation. Air Bearing Solution Guide, from https://www.newport.com/g/air-bearing-solution-selection-guide, Accessed on: 2017-06-15

[13] Köning , R.; Weichert, C.; Köchert, P.; Guan, J. & Flügge, J. (2013). Redetermination of the Abbe Errors' Uncertainty Contributions at the Nanometer Comparator, In: Proceedings of the 9th International Conference, Smolenice, Slovakia, pp. 171-174

[14] EA-4/02. Evaluation of the uncertainty of measurement in calibration, from http://www.european-accreditation.org/publication/ea-4-02-m-rev01-september-2013, Accessed on: 2017-06-18

[15] Lipus, L.C.; Budzyn, G.; Rzepka, J. & Acko, B. (2016). Calibration capability with laser frequency standard, DAAAM International Scientific Book 2016, pp. 197-206, doi: 10.2507/daaam.scibook.2016.18

DOI: 10.2507/28th.daaam.proceedings.077

DETERMINING THE STRENGTH OF COMPOSITE INTEGRATED JOINTS USING NUMERICAL SIMULATIONS

Frantisek Sedlacek, Vaclava Lasova

This Publication has to be referred as: Sedlacek, F[rantisek] & Lasova, V[aclava] (2017). Determining the Strength of Composite Integrated Joints Using Numerical Simulations, Proceedings of the 28th DAAAM International Symposium, pp.0545-0554, B. Katalinic (Ed.), Published by DAAAM International, ISBN 978-3-902734-11-2, ISSN 1726-9679, Vienna, Austria
DOI: 10.2507/28th.daaam.proceedings.077

Abstract

The paper deals with determining the strength of integrated high performance composite joints using numerical simulations. The solution of integrated joints made from composite materials is very problematic because it involves multi-axial stress. Wrapping loops were chosen as simplified specimens for validation of the numerical simulations and experimental tests. The wrapping loops were fabricated from unidirectional E-Glass roving in combination with an epoxy matrix. The experimental tensile tests were carried out on a Zwick/Roell Z050 machine. A ply-based FEM model was used for structural analysis. Four failure criteria for composite materials were used to determine the strength of the loops. There were favourable comparisons between the numerical simulations and the experimental tests, especially when using the Tsai-Wu failure criterion.

Keywords: Composite materials; Integrated High Performance Joints; numerical simulation; FEM

1. Introduction

Industrial development is becoming more and more rapid, thus increasing the demands on the products themselves. Especially in fields such as aerospace, automotive, railway, yachting or motorsports there are high requirements for low weight, high fatigue life, damping and vibro-acoustic properties of the materials. This is why using composite materials is advantageous. These materials are increasingly used today. Significant advances in virtual prototyping, which allow design and optimization of the complex composite parts using advanced numerical simulations is one reason for the growth in the use of composite materials.

The joints in composite parts are one of the most difficult areas when designing highly-stressed components. The joining of composite materials with composite materials or metals can be divided into five main groups: connections using additional materials (adhesive bonded joints or welded joints), friction joints (pressed joints), connections with an auxiliary element (bolted joints or riveted joints), joints that use the shape of the connected parts, and integrated high performance joints (IHPJ). The last group is preferred for composite materials, because this type of connection allows easy disassembly of the components and the reinforcement fibres are not disturbed by machining.

But this type of connection has many complications associated with it during the design phase and subsequently there are technological problems in fabrication. Determination of the strength of an IHPJ is most problematic because there is multi-axial stress across the whole joint. Much research has been done dealing with this issue [1], [2]. This paper deals with the determination of the strength of an integrated high performance joint using numerical simulation and a ply-based finite element model. Four three-dimensional failure criteria for composite materials were used to determine the strength of the joints.

2. Failure criteria for composite materials

Failure criteria for composite materials are still in the beginning against isotropic materials, whose has margin in an attempt to describe their violation more than two centuries (such as Coulomb, Mohr, etc. [3], [4]). Predicting the failure of these materials is very complex. Failure is influenced by many factors (such as type of load, layout of the plies of the laminate, the type of the constituents of the composite (fibres and matrix), their volume fraction, etc. The failure criteria for composite materials are based on the failure criteria for homogenous (isotropic) materials.

Most of these criteria assume that the material is homogenous (is not differentiated between fibres and matrix) and the stress is linear until failure occurs. There are several macro-mechanical failure criteria for composite materials (with transverse-isotropic or orthotropic properties).

In general, these failure criteria can be divided into three main groups: *non-interactive failure criteria, interactive failure criteria* and *direct-mode failure criteria.* The non-interactive theories do not consider the interaction between different stress components (normal and shear stresses in the main directions), whereas the interactive theories do [5]. The *direct-mode* criteria uses so-called failure modes which are described by independent conditions.

In this paper, four failure criteria for determining the strength of the wrapped loops are used. One non-interactive failure criterion – *maximum stress,* and three interactive failure criteria – *Hoffman, Hill* and *Tsai-Wu,* are used.

2.1. Maximum stress failure criterion

This failure criterion is one of the most frequently used theories for predicting the strength of composite materials. This criterion was first used by C. F. Jenkins in 1920 [6], and was created based on W. J. M. Rankin's theory *Maximum normal stresses for isotropic materials* [7]. According to this theory, failure occurs if any stress component has reached the ultimate strength of the material [8], [9]. Failure indexes for individual directions of this strength criterion are listed in equation (1).

Failure	Maximum stress criterion
F_{11}	σ_1/X^T if $\sigma_1 > 0$; σ_1/X^C if $\sigma_1 < 0$
F_{22}	σ_2/Y^T if $\sigma_2 > 0$; σ_2/Y^C if $\sigma_2 < 0$
F_{33}	σ_3/Z^T if $\sigma_3 > 0$; σ_3/Z^C if $\sigma_3 < 0$
F_{12}	$\lvert\tau_{12}/S_{12}\rvert$
F_{23}	$\lvert\tau_{23}/S_{23}\rvert$
F_{13}	$\lvert\tau_{13}/S_{13}\rvert$

(1)

where σ_1 (σ_2, σ_3) is normal stress in the longitudinal (transverse) direction; X^T (Y^T, Z^T) is tensile strength in the longitudinal (transverse) direction; X^C (Y^C, Z^C) is compressive strength in the longitudinal (transverse) direction; τ_{12} (τ_{23}, τ_{13}) is shear stress in plane 12 (23 or 13) and S_{12} (S_{23}, S_{13}) is shear strength in plane 12 (23 or 13) [10].

2.2. Hill failure criterion

The Hill failure criterion is based on the von Mises theory for isotropic materials. Hill extended this condition for orthotropic materials. This failure criterion assumes the same tensile and compressive strengths of the material (X^T=X^C) [11]. Assuming the plane stress (ie . $\sigma_{33} = \sigma_{13} = \sigma_{23} = 0$), Hill's failure criterion can be written in the form:

$$\left(\frac{\sigma_1}{X^T}\right)^2 + \left(\frac{\sigma_2}{Y^T}\right)^2 - \left(\frac{1}{(X^T)^2} + \frac{1}{(Y^T)^2} - \frac{1}{(Z^T)^2}\right)\sigma_1\sigma_2 + \left(\frac{\sigma_{12}}{S^T}\right)^2 = 1, \quad (2)$$

2.3. Hoffman failure criterion

Hoffman modified Hill's strength criterion for different tensile and compressive strengths [12]. Hoffman's failure criterion can be written (for plane stress) in the form:

$$\frac{\sigma_1{}^2}{X^T X^C} + \frac{\sigma_2{}^2}{Y^T Y^C} - \frac{\sigma_1 \sigma_2}{X^T X^C} + \frac{X^C - X^T}{X^T X^C}\sigma_1 + \frac{Y^C - Y^T}{Y^T Y^C}\sigma_2 + \frac{\sigma_{12}{}^2}{(S^L)^2} = 1. \tag{3}$$

2.4. Tsai-Wu failure criterion

The first attempt to develop a general theory of damage of anisotropic materials was made by Gol'denblat and Kopnov [13]. Their theory was an effort to predict the failure of a material under a general state of stress. The concept of a tensor strength is introduced, which takes into account the transformation from one coordinate system to another. It has the form of an invariant (formed by the stress and deformation tensor components), and it is able to explain the difference between the tensile and compressive strength. Based on this, Tsai and Wu created a failure criterion in the form of the polynomial

$$f_i \sigma_i + f_{ij}\sigma_i\sigma_j = 1; \quad i,j = 1,2,\dots,6. \tag{4}$$

Where f_i and f_{ij} are tensors of the first and second order of the strength, which can only be determined using experimental tests [14], [15]. For the plane stress condition, the Tsai-Wu failure criterion can be written in the form

$$\left(\frac{1}{X^T} - \frac{1}{X^C}\right)\sigma_1 + \left(\frac{1}{Y^T} - \frac{1}{Y^C}\right)\sigma_2 + \frac{\sigma_1{}^2}{X^T X^C} + \frac{\sigma_2{}^2}{Y^T Y^C} + \frac{\sigma_{12}{}^2}{(S^L)^2} + 2F_{12}^* \frac{\sigma_1 \sigma_2}{X^T X^C} = 1, \tag{5}$$

where the coefficient of the interaction F_{12}^* can be written as

$$F_{12}^* = \frac{1}{2\sigma^2}\left\{1 - \left[X^C - X^T + \frac{X^T X^C}{Y^T Y^C}(Y^C - Y^T)\right]\sigma + \left(1 + \frac{X^T X^C}{Y^T Y^C}\right)\sigma^2\right\}, \tag{6}$$

where σ is the stress at which failure occurs during a biaxial experimental test. The interaction coefficient F_{12}^* cannot be determined without this experimental test. Several papers deal with this coefficient, and most of them consider that the value of this coefficient is usually equal to zero [12], [16] , meaning that $F_{12}^* = 0$ provides more than enough accuracy for engineering applications.

3. Samples for experimental measurements

The simplified samples in the shape of loops were created to validate the numerical simulations. The loops were fabricated in a special mould. The wrapped loops were made from unidirectional E-glass roving (Aeroglass TEX 2400) and high-strength epoxy matrix LH289. The mechanical properties of the laminate are given in

Table 1. A total of sixteen samples in two geometrical variants were created (with four threads and seven threads of glass roving). The geometry of the specimen is shown in Fig. 1. The geometrical parameters of the samples (wrapping loops). Fig. 1.

V_f	(-)	0.60	Volume fraction of the fibres
V_m	(-)	0.40	Volume fraction of the matrix
E_1	(MPa)	46400	Young's module in direction 11
E_2	(MPa)	12700	Young's module in direction 22
G_{12}	(MPa)	3490	Shear module of elasticity in direction 12
G_{23}	(MPa)	3770	Shear module of elasticity in direction 23
G_{13}	(MPa)	3460	Shear module of elasticity in direction 13
ν_{12}	(-)	0.262	Poisson's ratio in plane 12
ν_{23}	(-)	0.350	Poisson's ratio in plane 23
X^T	(MPa)	1240	Ultimate tensile stress in direction 11
X^C	(MPa)	774	Ultimate compressive stress in direction 11
Y^T	(MPa)	43.9	Ultimate tensile stress in direction 22
Y^C	(MPa)	82	Ultimate compressive stress in direction 22
S_{12}	(MPa)	55.8	Ultimate shear stress in direction 12

Table 1. Mechanical properties of the composite material of the samples.

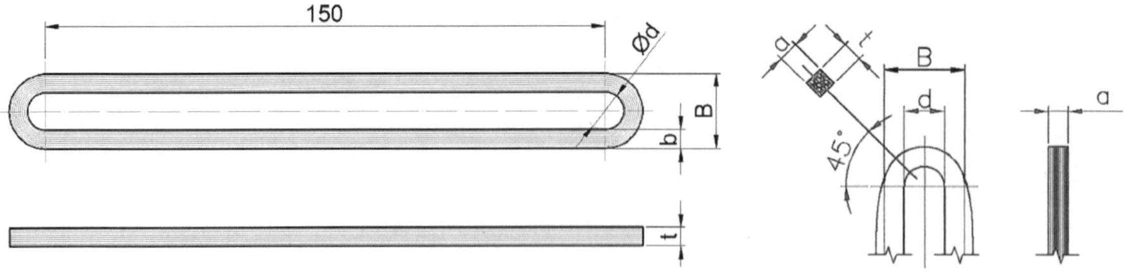

Fig. 1. The geometrical parameters of the samples (wrapping loops).

4. Experimental measurement of the samples

The experimental tensile tests of the loops were carried out using a quasi-static load (0.5 mm/sec) in a Zwick/Roell machine (Z050). The specimens were attached using pins in special jaws (without axial contact). The process of the failure of the loops was recorded using a DSLR camera (with full synchronization with the experimental output data) see Fig. 2.

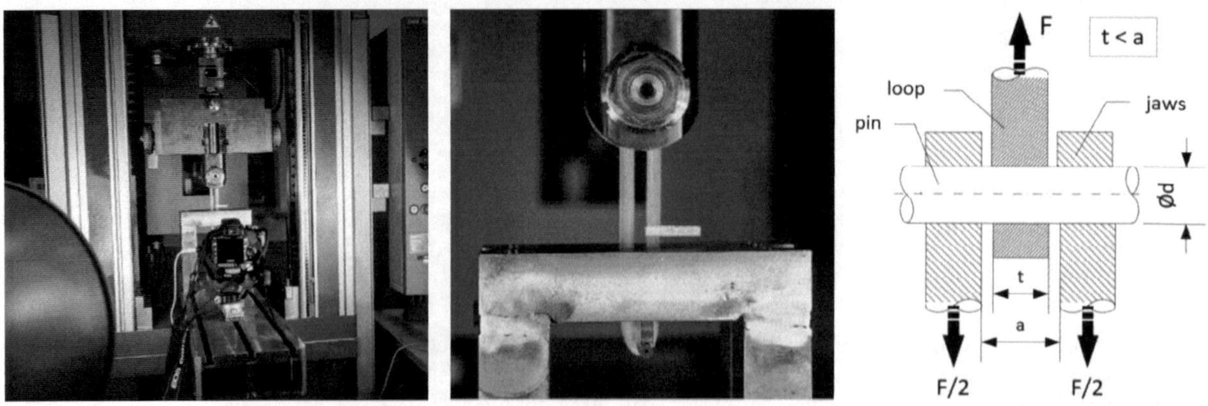

Fig. 2. Experimental tensile test of the specimen: (left) tensile test, (centre) detail of the attachment of the specimen, (right) diagram of the attachment of the specimen.

The test specimens were broken during the tensile tests in the following order: first, the matrix of the laminate in the contact cylindrical face of the loop was damaged. After reaching the maximum load the first (inner) thread of the fibre was broken.

Subsequently the remaining fibres were broken (fibre after fibre from inner to outer diameter until full failure of the loop). The loops were most often broken in the contact cylindrical face approximately at an angle 45°±10° relative to the longitudinal plane of the loop. The progression of the failure of the loop is shown in Fig. 3.

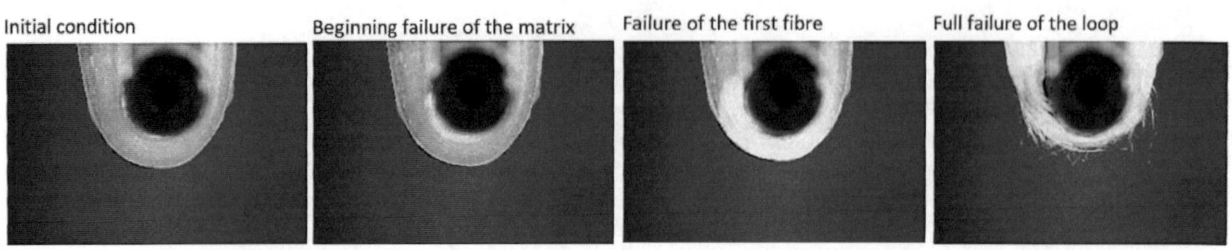

Fig. 3. Progression of the failure of the loop with seven threads of fibres.

The results of the tensile tests of the loops with four and seven threads of fibres are given in Fig. 4. Where the failure of the loops with four threads of fibres occurred on average around 5008 N, and for seven threads of fibres occurred on average around 5480 N.

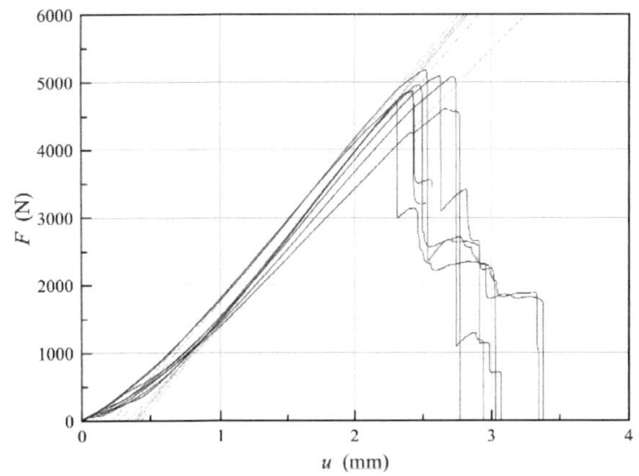

(a) Four threads of fibres

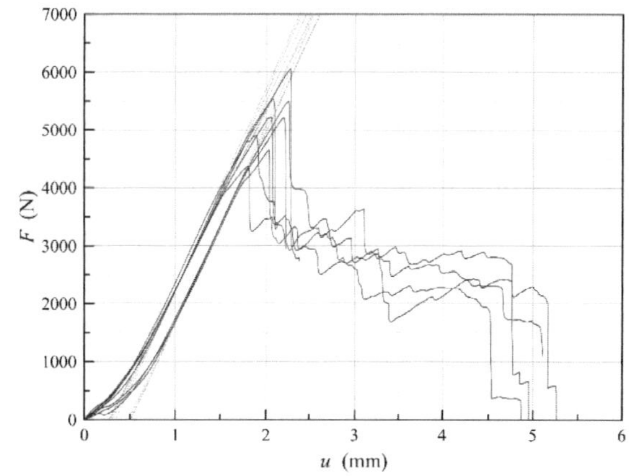

(b) Seven threads of fibres

Fig. 4. Results of the tensile tests of the loops (specimens).

5. Numerical simulation of the wrapping loops

5.1. FEM model of the wrapping loop

The numerical simulation was done using the NX Nastran 11 solver (based on FEM) and the Siemens Simcenter 11 pre/post-processor. The ply-zone process of meshing was used for this solution. A 2D mesh with CQUAD8 isoparametric hexahedron elements was applied on the model of the loop. The layout of the individual plies and the physical properties were defined.

Subsequently, this 2D mesh was extruded into a full three-dimensional FEM model (with CHEXA20 elements). The 3D mesh respects the layers, the thicknesses and the material orientation of the individual plies of the laminate (see Fig. 5). The possibility of determining the stress, strain and failure in individual planes (bottom/middle and top) of the plies is the main advantage of this type of meshing.

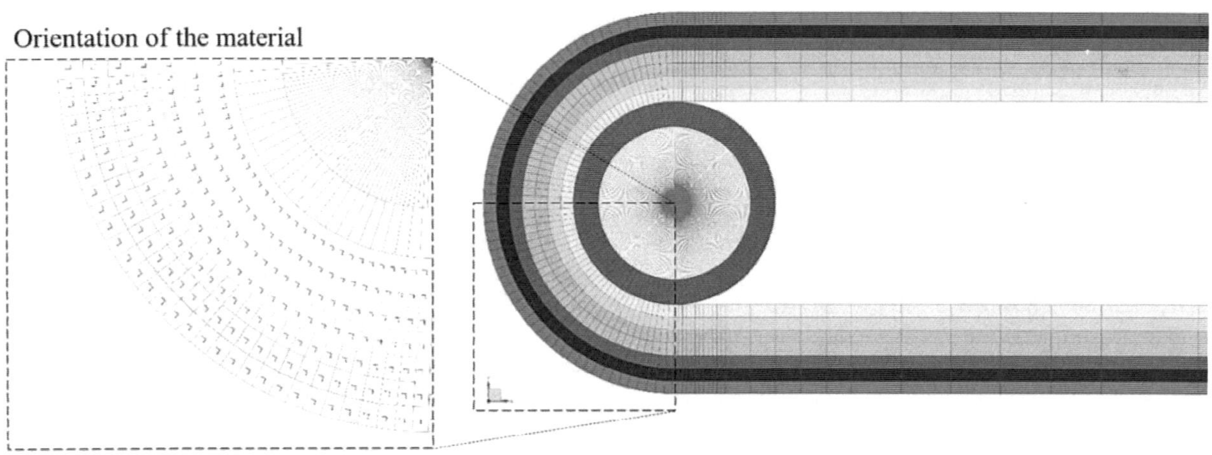

Fig. 5. FEM mesh of the wrapping loop with seven threads of fibres.

5.2. Results of the simulation

The structural analysis was done using the non-linear solver NX Nastran 401 and the NX Laminate Composite module. Four failure criteria were used: Maximum stress, Hill's, Hoffman's and Tsai-Wu failure criterion (see section 2). The joint was solved until failure index $F_I = 1$ was reached.

The results of the main normal stress of the loop (for average critical loading of the specimens during the experimental tests $F_{FF4\_AVG} = 5008$ N) are shown in

Fig. 6.

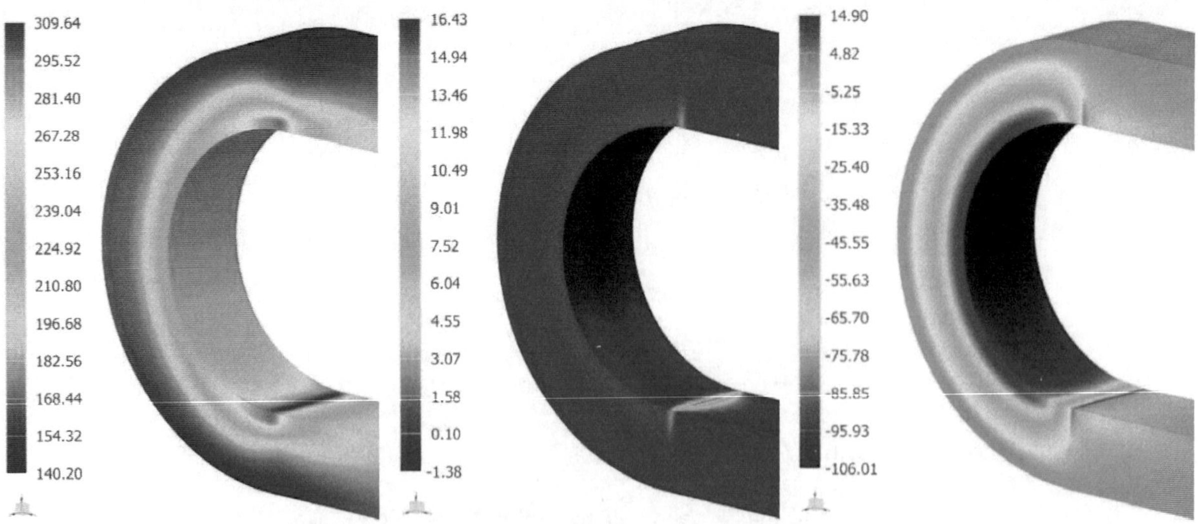

σ_{11} (MPa) σ_{22} (MPa) σ_{33} (MPa)

Fig. 6. The main normal stress of the loop with four threads of fibres.

The results of the failure criteria were compared with the experimental tensile tests of the specimens. The results of the bottom plane of the first ply of the wrapping loop (the inner contact face of the loop) were compared with the results from the beginning of the failure of the loop (failure matrix).

The results in the middle plane of the first ply of the loop were compared with the maximum loading condition from the experimental tests (= failure of the first fibre).

The results of the maximum stress failure criterion are given in Fig. 7. A third failure index F_{33} was determined as critical. This means that the most critical stress is in the third main normal direction of the laminate (radial stress). The critical area of the failure was found at the contact cylindrical face of the loop approximately at an angle 60°±10° relative to the longitudinal plane of the loop.

Compliance with a deviation up to 10% was determined in comparison with the experimental tests results (up to 7% for beginning of failure and up to 10% for full failure of the loop).

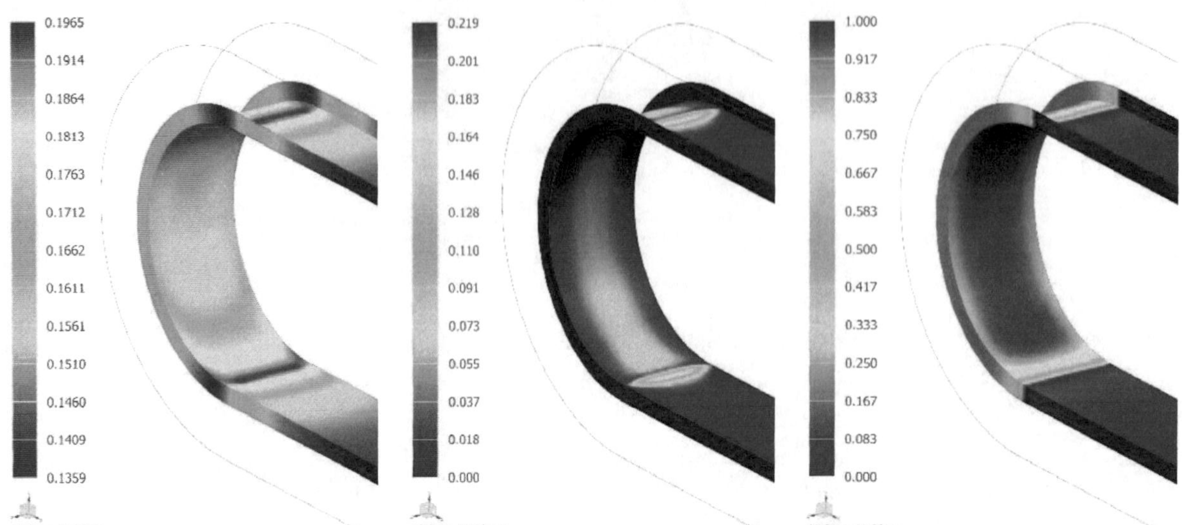

F_{11} (-) F_{22} (-) F_{33} (-)

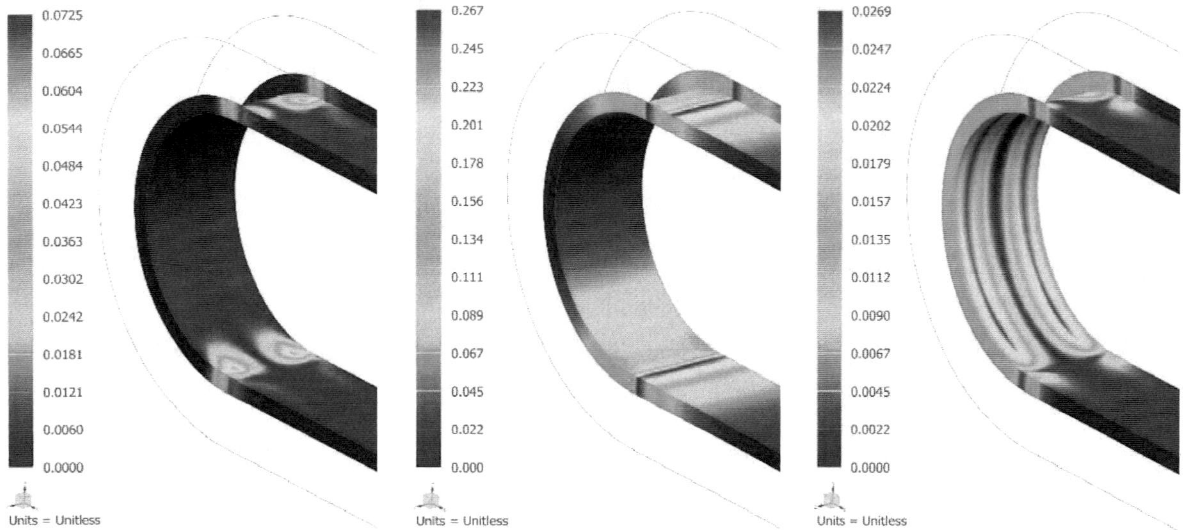

F_{12} (-) F_{23} (-) F_{31} (-)

Fig. 7. Results of maximum stress failure criterion in middle plane of first (critical) ply of loop with four threads of fibres.

The results of Hill's failure criterion are given in Fig. 8. This failure criterion has only one failure index. The critical area of the failure was found on the contact cylindrical face of the loop at an angle between 15° to 70° relative to the longitudinal plane of the loop. Compliance with deviation up to 8% was determined in comparison with the experimental tests results (up to 4% for beginning of failure and up to 8% for full failure of the loop).

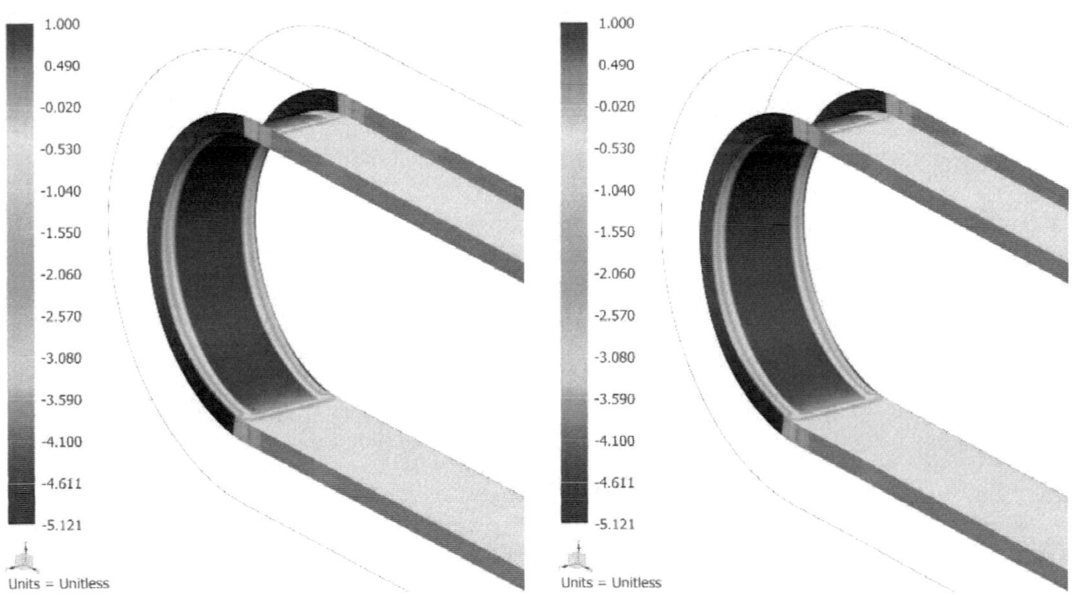

First ply - Bottom plane (BT) **First ply - Middle plane (MID)**

Fig. 8. Results of Hill's failure criterion in first (critical) ply of loop with four threads of fibres.

The results of the Hoffman's failure criterion are given in Fig. 9. The critical area of the failure was found on the contact cylindrical face of the loop approximately at an angle 60°±5° relative to the longitudinal plane of the loop. Compliance with deviation up to 9% was determined in comparison with the experimental tests results (up to 9% for beginning of failure and up to 3% for full failure of the loop).

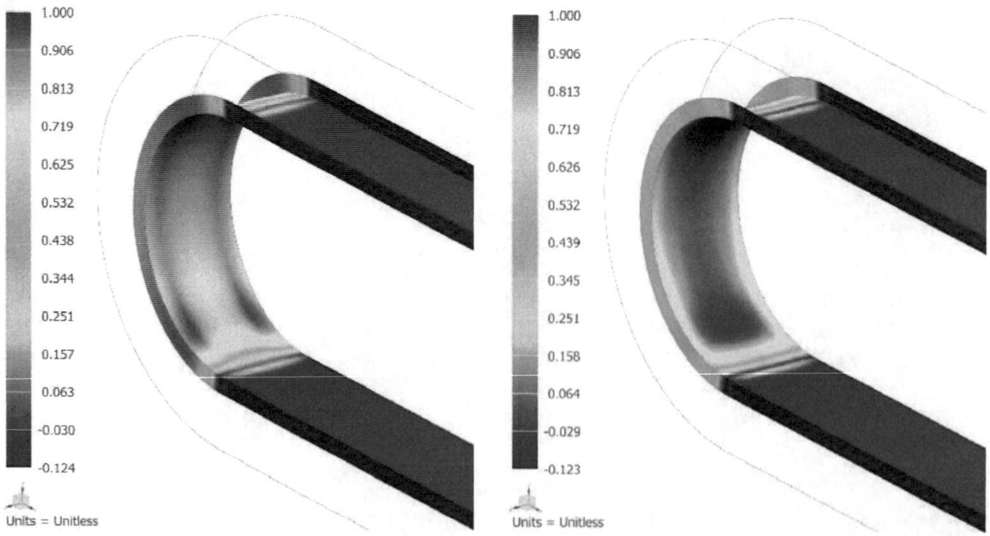

First ply - Bottom plane (BT) **First ply - Middle plane (MID)**

Fig. 9. Results of Hoffman's failure criterion in first (critical) ply of loop with four threads of fibres.

The results of the Tsai-Wu failure criterion are given in Fig. 10. The critical area of the failure was found on the contact cylindrical face of the loop approximately at an angle 55°±5° relative to the longitudinal plane of the loop. The results most similar to the experimental test were obtained using this failure criterion. Compliance with deviation up to 7% was determined in comparison with the experimental tests results (up to 5.5% for beginning of failure and up to 7% for full failure of the loop).

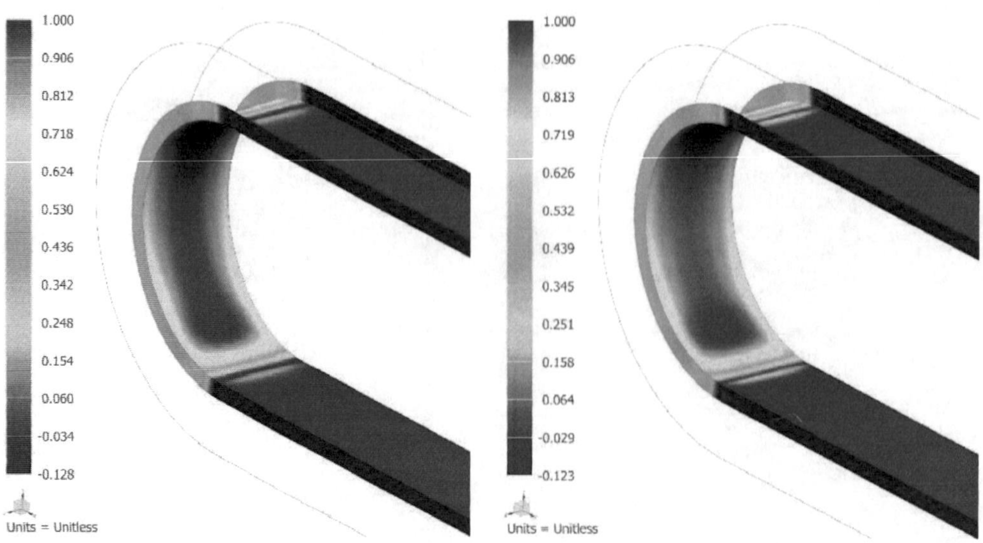

First ply - Bottom plane (BT) **First ply - Middle plane (MID)**

Fig. 10. Results of Tsai-Wu failure criterion in first (critical) ply of loop with four threads of fibres.

6. Discussion

In Fig. 11, the comparison of the results of the numerical simulation of the experimental tests for loops with four and seven threads of fibres can be seen. The 60% volume fraction of the fibres was used for the numerical simulation (that corresponds to the thickness of the loop Q = 2.3 mm for four threads of fibres and thickness of the loop Q = 3.5 mm for seven threads of fibres). The results show that the total strength of the wrapping loops increases with the thickness of the loop, but the strength of the matrix decreases. From a comparison of these results it was determined that the Tsai-Wu failure criterion has the closest compliance.

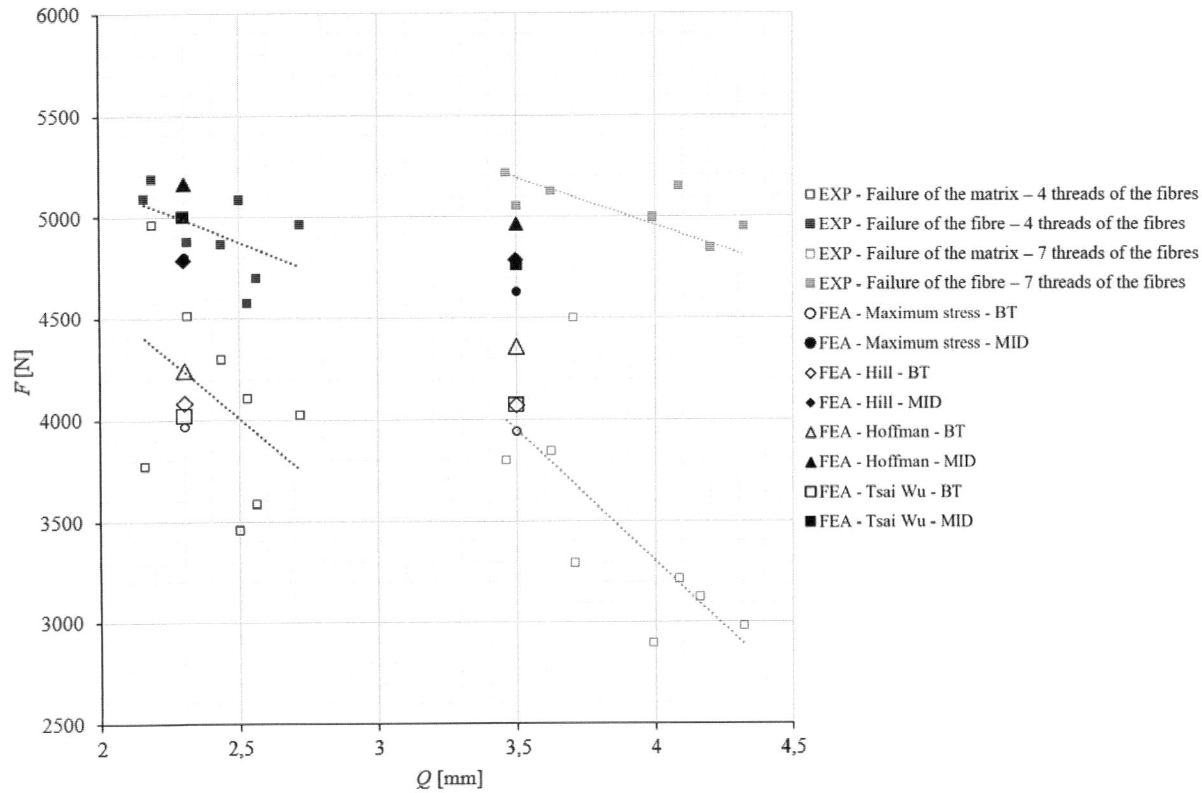

Fig. 11. Comparison of the results of the numerical simulations and experimental tests.

7. Conclusion

This paper deals with the determination of the strength of integrated high performance joints using ply-based numerical simulation. Simplified specimens (wrapping loops) made from unidirectional E-Glass roving were chosen for validation of the numerical simulation. Experimental tensile tests were done on sixteen specimens with two types of geometry. A ply-based FEM model of the loop was created. This type of FEM model allows the stress in individual plies of the laminate and their main planes to be evaluated separately (but suitability of used geometry is requirement/limitation of ply-based FEM method).

Four failure criteria for composite materials were used for determination of the strength of the wrapping loops. The failure in the middle plane in the most critical ply (first inner ply) was used for evaluating the strength of the loops. When comparing the numerical simulations and experimental tests, compliance with deviation up to 11% was found. The most accurate results (max. deviation up to 7%) were obtained using the Tsai-Wu failure criteria. This research will continue by testing other materials (carbon fibres) and using *direct mode* failure criteria.

8. Acknowledgments

This contribution has been prepared under the project LO1502 ‘Development of the Regional Technological Institute’ under the auspices of the National Sustainability Programme I of the Ministry of Education of the Czech Republic aimed at supporting research, experimental development and innovation.

9. References

[1] J. Krystek and R. Kottner. (2015). Load capacity prediction of carbon or glass fibre reinforced plastic part of wrapped pin joint. Materiali in Tehnologije, vol. roč. 49, pp. 957–960.

[2] R. Kottner, J. Krystek, R. Zemcik, J. Lombersky, and R. Hynek. (2014). Strength Analysis of Carbon Fiber-reinforced Plastic Coupling for Tensile and Compressive Loading Transmission’, in 52nd AIAA/ASME/ASCE/AHS/ASC Structures, Structural Dynamics and Materials Conference, American Institute of Aeronautics and Astronautics.

[3] C. A. Coulomb, ‘In memories de mathematique et de physique. (1773). Academie Royal des Sciences par divers sans, vol. 7, pp. 343–382.

[4] O. Mohr, 'Welche Umstände bedingen die Elastizitätsgrenze und den Bruch eines Materials. (1900). Zeitschrift des Vereins Deutscher Ingenieure, vol. 46, no. 1524–1530, pp. 1572–1577.
[5] V. Laš, Mechanika kompozitních materiálu. (2008). Pilsen. University of West Bohemia
[6] A. C. Long, Ed., Design and Manufacture of Textile Composites. (2005). 1 edition. Boca Raton, Fla.: Woodhead Publishing.
[7] R. C. Burk. (1983). Standard Failure Criteria Needed for Advanced Composites, AIAA, no. Vol. 21, p. Sv. pp. 58-62.
[8] V. Valery V. and E. Morozov V. (2013). Advanced Mechanics of Composite Materials and Structural elements, 3rd Edition., Oxford: Elsevier.
[9] T. Ramesh and S. Chandra V. (2012). Damage and Failure of Composite Materials. Cambridge: Cambridge University Press.
[10] E. J. Barbero. (2010). Introduction to Composite Materials Design, Second Edition, 2 edition. Boca Raton: CRC Press.
[11] R. Hill. (1998). The Mathematical Theory of Plasticity, New York.
[12] R. Narayanaswami and H. M. Adelman. (1997). Evaluation of the Tensor Polynomial and Hoffman Strength Theories for Composite Materials, Journal of Composite materials, no. Vol. 11, pp. 366–377.
[13] I. I. Gol'denblat and V. A. Kopnov. (1965). Strength of glass-reinforced plastics in the complex stress state', Polymer Mechanics, vol. 1, no. 2, pp. 54–59.
[14] W. Van Paepegem and J. Degrieck. (2003). Calculation of damage-dependent directional failure indices from the Tsai–Wu static failure criterion', Composites Science and Technology, vol. 63, no. 2, pp. 305–310.
[15] S. W. Tsai and E. M. Wu. (1972). A General theory of strength for anisotropic materials', in Technical Report, Ohio.
[16] W. C. Cui, M. R. Wisnom, and M. Jones. (1992). A Comparison of Failure Criteria to Predict Delamination of Unidirectional Glass/Epoxy Specimens Waisted Through the Thickness', Composites, no. Vol. 23, pp. 158–166.

DOI: 10.2507/28th.daaam.proceedings.078

SIMULATION OF VIRTUAL MACHINE TOOL DURING THE MACHINE DESIGN

Zdenek Hajicek, Vaclav Marek

This Publication has to be referred as: Hajicek, Z[denek] & Marek, V[aclav] (2017). Simulation of Virtual Machine Tool During the Machine Design, Proceedings of the 28th DAAAM International Symposium, pp.0555-0560, B. Katalinic (Ed.), Published by DAAAM International, ISBN 978-3-902734-11-2, ISSN 1726-9679, Vienna, Austria
DOI: 10.2507/28th.daaam.proceedings.078

Abstract

This paper deals with the simulation of CNC machine tools. The main focus is to develop a Case Study for a new virtual concept of Horizontal CNC Machine Tool. A virtual Machine Tool uses the main properties of a real machine. The goal is to simulate the functionality of the machine in virtual phases. This method is intended for complex and very expensive machines like Milling and Boring Machine Tools with numerous accessories. The current trend is to sell Machine Tools and equipment with complete technology of machining. Customers prefer a virtual Case Study of CNC technology. The idea of virtual prototyping is realized if the behaviour of the virtual Machine Tool is in compliance with real machine.

Keywords: Virtual Machine Tool; CNC Machine Tools; CSE Driver; Postprocessor

1. Introduction

The current trend in machine manufacturing is to deliver Machine Tools with a turnkey solution. This solution is based on a typical customer part and CNC technology. This solution provides the framework for the Case and Time Study. This Case Study consists of the Machine Tool, technical equipment and CNC technology in one complex 3D layout.

The main purpose of the Case Study is to guarantee reliable information about the future behaviour of the Machine Tool. The Time Study is the analysis of time in CNC technology for typical customer part. The result of this study is the timing of each CNC operation, usage of tool and machine accessories in one document. These studies are very important elements for a lot of tenders in industrial practice.

The main problem is to simulate behaviour and functionality of CNC Machine tool in virtual phases. CAD/CAM system NX with PLM Teamcenter is used for the methodology of Creating Machine Tool Simulator and virtual verification like one complex solution. This Machine Tool Simulator uses main property axis and real kinematics of a machine. The solution has been demonstrated for the Skoda Machine HCW 1000 and the application Automatic Tool Change with Industrial Robot.

1.1 Literature overview

Modern machine tools are very complex Mechatronical Systems. The capability and efficiency of a machine tool are mainly determined by its kinematics, structural dynamics, computer numerical control system and the machining process [1]. The CNC system consists of a computer, power electronic components, such as motor amplifiers and electronic circuits, and servo actuators. The computer control unit receives ISO standard NC-programs which describe the tool path geometry, tool number, feed and spindle speed at each path segment [2], [3]. The simulation of the CNC system involves virtual modelling of the machine tool kinematics and feed drive dynamics, update of the workpiece geometry as the material is removed and motions of the drives and auxiliary units, such as tool and pallet changes. In short, the rigid body motion of the machine tool and the CNC functions must be predicted as the workpiece is produced in order to realize a Virtual CNC system [1].

Once the NC Program is generated in a CAD/CAM environment, the present Virtual CNC technology allows the geometric update of the workpiece as the tool cuts the material at each NC block. In addition, the solid model of the machine tool, its multi-axis kinematics and the location of fixtures can be displayed in the CAD environment [4], [5].

The present technology of NC-path simulation allows the prediction of tool collision spots and correctness of the NC program by checking path errors and gauging on the workpiece surface graphically. Lauwers et al. [6], [7], [8], [9] take the CL file from the CAD system and simulate the machine motion by modelling the kinematics of the machine tool for collision detection and avoidance. In some commercial controllers machining simulation systems are integrated. During machining the simulation system runs a number of blocks ahead, and if there is a danger of a collision, the controller stops the machine immediately [1].

CSE Simulator consists of:

- Simplified 3D Layout of Machine Tool with all accessories and equipment
- Machine Tool Kinematic Structure model
- MCS Driver programmed for current Machine Controller
- NX Postprocessor programmed for current Machine Controller and simulation
- CSE Simulator implemented in NX CAM environment

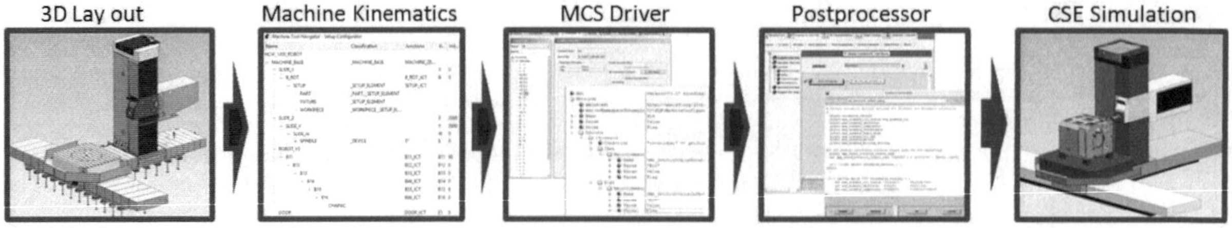

Fig. 1. Segments of Virtual Machine Tool (CSE Simulator)

2. Current process

2.1 Current status of Layout Design and simulation of Machine Tools

In the beginning of this project the company used one 3D Layout in three departments: Sales, Design and Customer Technology. These departments have different requirements. In the Sales Department, users need a very quick Layout of Machine for new tender and possibility of making revise a version. The Customer Technology Department needs a simple 3D assembly of Machine Tool for the preparation of Time and Case Study. This Case Study may consist of CAM Technology with all Machine equipment and accessories. This technology is supported by CSE Simulation for virtual verification of developed solution.

2.2 Current problems

The Machine Layout came from Design department in a non-optimized stated without simplified parts of a machine. If Application Engineer prepares a Collision Analysis of CNC technology the calculation will take long because the collision parts are not simplified. The CSE Collision Analysis builds a special mesh from collision parts automatically in the background. The analysis checks the penetration of these two meshes in real time of machine simulation. If we use non-simplified machine parts in this simulation (like Column with inside Ribs – Fig. 2) the mesh will be unnecessary difficult. For faster calculation of CSE analysis, it is necessary to use a simple solid model without inside parts and complicated shapes. After that, the simulation will be more efficient.

Disadvantages of current Layout Design and Technologic Studies:

- Not optimized 3D parts of Machine Tool Layout
- Bad quality of imported CAD data from Solid Edge ST2 to NX 10
- CAD data without Data Management
- Long time for the preparation of Machine Layouts
- Long time for the preparation of Technological Studies without any simulation
- Way to avoid mistakes in the preparation of CSE Simulation

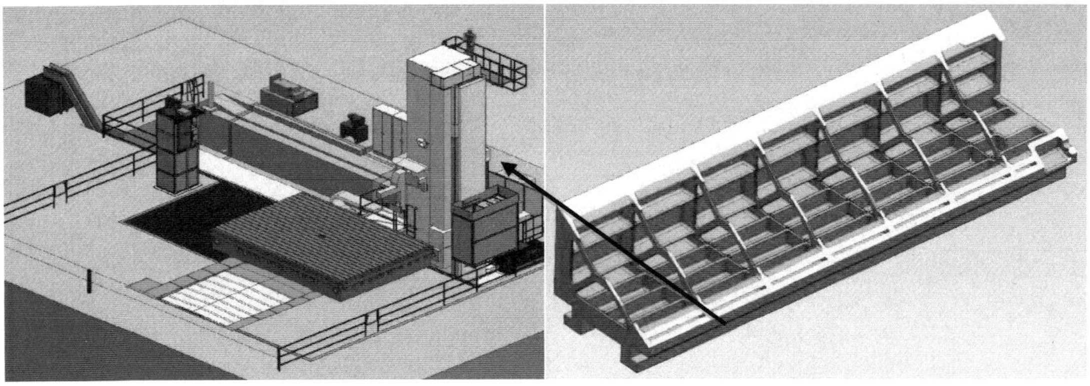

Fig. 1. Not optimized Machine 3D Layout and Column with Ribs

2.3 Research in using Machine Tool Simulation in the company

The main goal of this research is to create an application of ISV Simulation to design and develop of Heavy Machine Tools. The PLM System Teamcenter, NX CAD System, NX CAM System, NX ISV Simulation module are used to problem solving and the software the Post Builder is used for programming Postprocessors. PLM system Teamcenter is used for data administration and Approval Processes. NX CAD is used for the design of Machine Tool Layouts. NX CAM System is used for programming of CNC Technology and NX ISV module for CSE Simulation and verification of solution.

All these kinds of software are connected by one PLM platform in the company who is a producer of Heavy Machine Tools. To obtaining a contract is very complicated and lengthy.

The typical new contract starts by signing a new customer tender in the Heavy Engineering. This signing starts the workflow process in the company. The workflow is without CSE verification of solution and the time for the preparation of Machine Layout and Technology Studies is very long. These critical elements are eliminated by a more developed methodology of Automated Design 3D Layouts as well as Creating Machine Tool Simulator.

Main steps of new workflow for the contract:

- Design of Machine Tool Layout by new automated solution
- Customers revisions of the Machine Tool concept in PLM System
- Time and Case Study in CAM System
- CSE Verification of prepared solution
- Approved solution by customer and contract confirmation
- Design and production of new Horizontal Boring and Milling Machine Tool
- Assembly and tests of new Machine Tool
- Application Machine Tool in the customers company

3. New methodology layout design and machine simulation

3.1. Methodology of Automated Design Machine Tool Layout

This methodology is based on the principle of using simplified Machine Tool part from one shared library. This library is shaved in PLM database and is automatically integrated in NX CAD/CAM environment. The simplified Machine parts are designed as parametric model and this model can be driven by Excel table. The table contains only standard parameters of the machine dimension. The designer can put the machine part from the library. He can choose a current value of machine diameters from NX window and the model is automatically transformed to the recommended dimensions. The designer can assemble the Machine Layout very fast without using any Assembly Constrains. The component (Simplified Machine part) is automatically put on the correct position in the CAD assembly. This methodology reduces the design of Machine Layout from 30-40 hours to 4-8 hours. The time of preparation depends on the complexity of the Machine Tool.

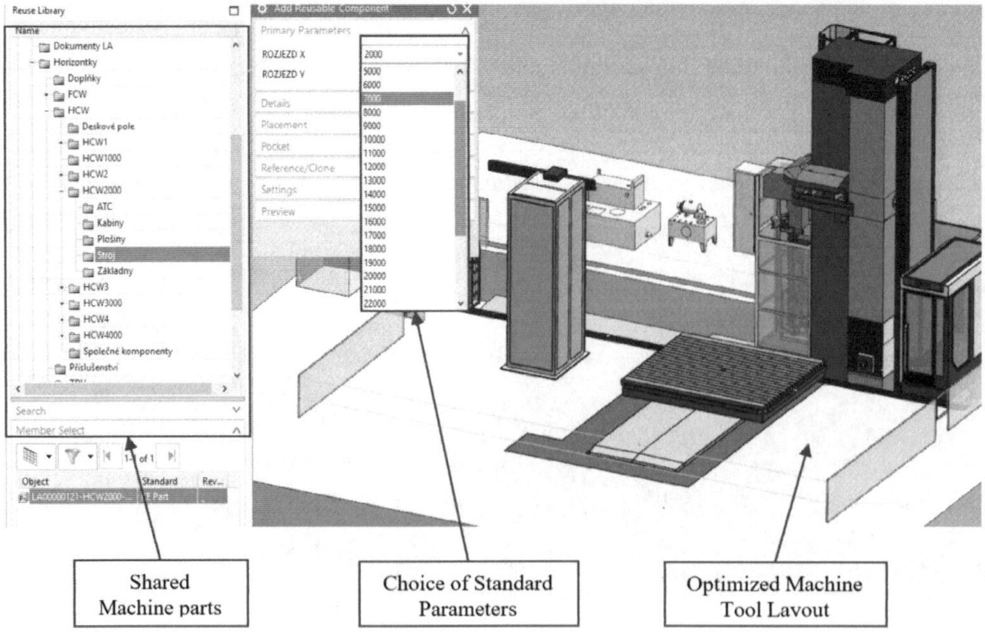

Fig. 3. Optimized Machine Tool Layout

3.2. Methodology of Creating Machine Tool CSE Simulator

This methodology is based on the principle of using Optimized Machine Tool Layout and CSE Template. The CSE Template consists of Machine Tool Kinematic structure (MTB), Postprocessor, MCS Driver and a service Library for each machine type. The Application Engineer can prepare Machine Kinematic in NX ISV module Machine Tool Builder. When creating the CSE Machine Tool Simulator, Application Engineer puts together CSE Template segments and new MTB Machine Tool. After that it is possible to implement a new CSE Simulator in NX CAM structure. The Workflow is described in the picture (Fig. 4).

If the new CSE Simulator is implemented in the NX CAM, the Application Engineer can prepare the customer CNC/CAM technology. The result of this methodology is a quick preparation of Time Study, Case Study and a verification of new Machine Tool.

All of these CSE Template elements are developed exactly for a producer of Heavy Machine Tools like Horizontal Boring and Milling Machines. These machines are very specific which makes the application of CNC technology difficult to implement. The new CSE Simulation is driven by the real CNC program and real CAM technology. This debugging, CNC program is possible to implement in real production line very quickly and safely.

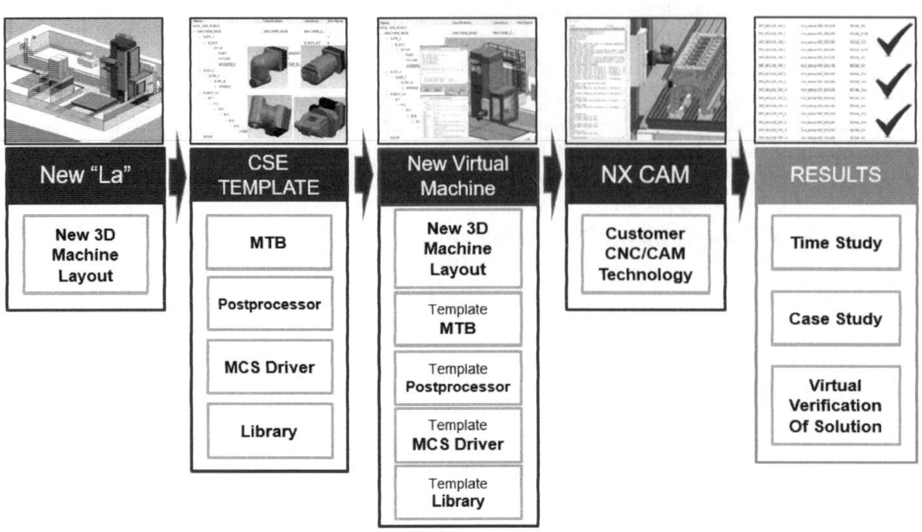

Fig. 4. Methodology of Creating Machine Tool Simulator

3.3. Sample calculation of the cost of the Machine Tool Layouts

A designer needs 40 hours for the preparation of a standard type of Machine Tool Layout by using an original methodology. If the designer uses the methodology of Automated Design Machine Tool, the time for preparing the standard type machine is eight hours. For this sample, the calculation of the internal rate per hour for one designer is 20 Euros. If the designer has to prepare ten layouts in one month, the price will be very different. By using the original methodology the price will be 8,000 Euros for ten layouts. By using the methodology of Automated Design, the price will be 1,600 Euros (for ten Layouts). The new solution will save 640 Euros per one layout and 6,400 Euros per ten layouts. In general, the methodology of Automated Design Machine Layout reduces the time and the price in preparing the machine layout by 80 %.

4. Application of CSE virtual machine simulation

This application is focused on the development of the virtual Horizontal Milling Machine for machining a block of headstock for the Turning Machine SR1000. The concept of the 3D layout is designed for high usage of tools. This solution uses tools up to 432. The machine tool is designed for the continuous cycle of production. It consists of a Pallet System for Workpiece changing and the Automatic Tool Change with the Industrial Robot KR 90. This system consists of the three pallets. Only one pallet can be in the workplace at each time. The other two pallets are used for clamping workpiece because this part must be machined to all faces. The first pallet is used for main clamping workpiece position and the second is used for repositioning for bottom milling. This technology uses two CNC programs which are changed by the current Workpiece position.

The customer requirement is a continous production and all of the tools must be delivered twice. Before inserting the tool to the spindle or to the Carousel Tool Housing Pocket, the robot checks the tool by the laser measuring system. If the tool shows any signs of damage, the robot forwards the tool to the machine operator repairs or changes it completely.

For the design, the Tool Housing has been selected: Representative Tools depending on size, Base Tool (Length 200 mm, Diameter 22 mm), Middle Tool (Length 320 mm, Diameter 80 mm) and Large Tool (Length 475 mm, Diameter 300 mm).

The industrial Robot uses the Tool Hydraulic Double Gripper with two pockets for tools. This Gripper can grasp two tools together due to the rapid tool change into the spindle. The robot with the Gripper must move to a changing position before the spindle, the first jaw mechanism removes the tool from the spindle, then it needs to rotate ninety degrees and the second jaw mechanism inserts new tool into the spindle. This movement takes a maximum of five seconds.

The Tool Housing is designed by three Carousels with three rotating shelves. This case study uses 144 Base Tools, 144 Middle Tools and 45 Large Tools. Altogether there are 333 mixed tools in one Tool Housing. The total time of this tool change simulation is 21 seconds which is also sufficient. The price of the solution is 190,000 Euros.

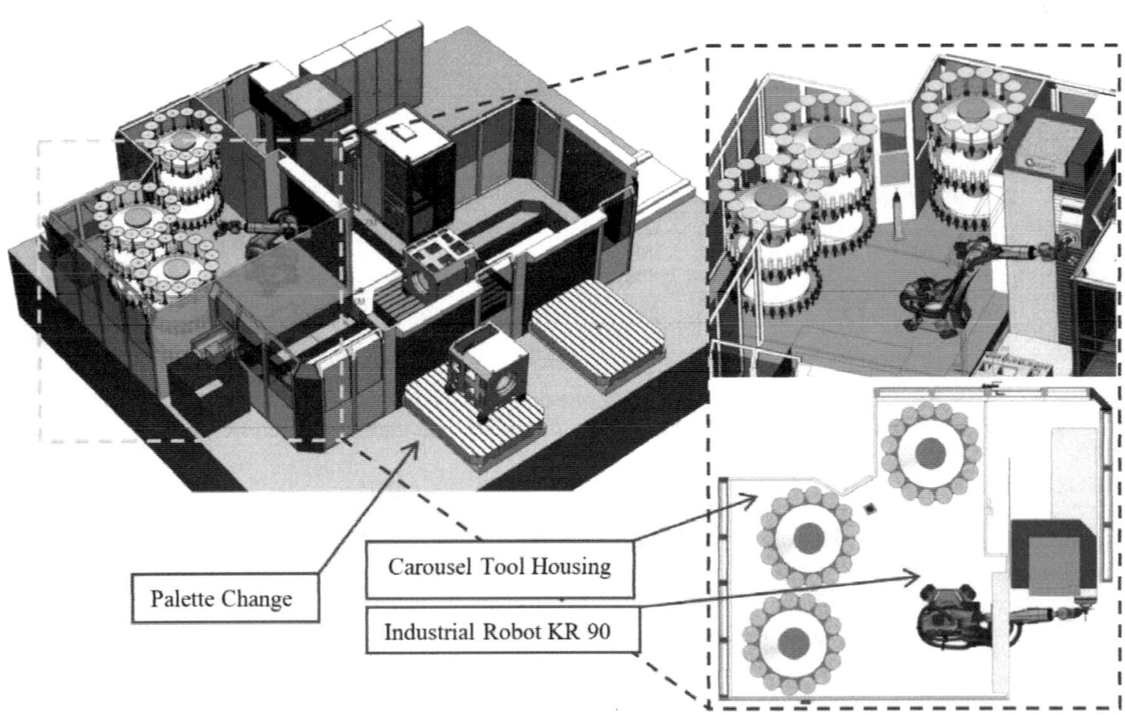

Fig. 5. CSE simulation of HCW 1000 and Robotic Tool Change with Carousel Arena

5. Conclusion and results

This thesis describes the problem of the simulation of Machine Tools and the possibility of using in virtual prototyping. The thesis shows the current status about the Machine Tool Layouts design and what the main problem is in the verification of new Machine concepts. A current trend in the machine manufacturing is to deliver Machine tools with a turnkey solution. It is important to prepare solution which is based on a typical customer part and CNC technology.

The main research is focused on development of methodology Layout Design and Machine Tool simulation. The methodology of Layout Design allows the use of automated design of new Machine Tool concepts. This solution is based on the principle of using simplified Machine Tool parts from one shared library. The solution becomes the possibility of making a revised version of PLM system Teamcenter which guarantees a fast interaction of different departments in the company. This methodology reduces the time of the preparation and the price of the layout by 80 %.

The CSE Templates, MTB Machine Kinematics, Postprocessors and MCS Drivers for all kinematic types of Horizontal Boring and Milling Machines Tool are developed for Creating Machine Tool CSE Simulator methodology. The main principle of this methodology is used for the application of Horizontal Machine HCW 1000 and Robotic Tool Change with Carousel Arena. The application verifies a complicated arrangement of the machine in virtual phase. The result of the simulation is reflected the Case Study, Time Study and price calculation of Robotic Tool Change. The CSE Simulator is driven by the real CNC program and the real CAM Technology. This debugged CNC program is possible to implement in the real production very quickly and safely.

5.1 Further research

The main principles will be applied in next research of another type of Machine Tools like Heavy Lathe Machines, Heavy Grind Machines or CNC Mill and Turn Machines for Automotive Industry. This research will be focus on Virtual Drive and Virtual Control Systems for highest precision of Virtual Machine Tool Simulation.

6. Acknowledgements

The paper was supported by " University of West Bohemia, Department of Machine Design, SGS 2017"

7. References

[1] Y. Altintas, C. Brecher, M. Weck, S. Witt, (2005), Virtual Machine Tool, CIRP Annals, 50/2: 115-138.,

[2] Pritschow, G., Altintas, Y., Javone, F., Koren, Y., Mitsuishi, M., Takata, S., Van Brussel, H., Weck, M., K.Yamazaki, 2001,Open Controller Architecture-past, present and future, Annals of the CIRP, 50/2: 446-463.

[3] Pritschow, G., Berkemer, J., Bürger, T., Croon, N., Korajda, B., Röck, S., (2003), Die simulierteWerkzeugmaschine, TagungsbandFertigungstechnischesKolloquium Stuttgart, 219-246

[4] Yeung, C.Ho, Altintas, Y., Erkorkmaz, K., (2004), Virtual CNC System, Part I: Architecture and Auto tuning of CNC Systems, Trans. ASME, J. Machine Tool Manuf. Eng.Manufacturing Science and Engineering.

[5] Yeung, C.Ho, Erkorkmaz, K., Altintas, Y., (2004), Virtual CNC System, Part II: Virtual Part Machining and auto correction of contouring errors, Trans. ASME, J. Manufacturing Science and Engineering.

[6] Kruth, J.P., Lauwers, B., Klewais, P., Dejonghe, P., (1999), "NC-postprocessing and NC-simulation for five-axis milling operations with automatic collision avoidance", International Journal for Manufacturing Science and Technology, 1(1), 1999, 12-18.

[7] Lauwers, B., Kruth, J.P., Dejonghe, P., Vreys, R., (2000), Efficient NC-programming of multi-axes milling machines through the integration of tool path generation and NC-simulation, Annals of the CIRP, 49/1/2000: 367-370.

[8] Lauwers, B., Van der Poorten, E., De Baerdemaeker, H., (2003), CAD/CAM based generation of collision free robot programs through the integration of a virtual machining environment, European Journal of Mechanical and Environmental Engineering,47/4: 215-220.

[9] Lauwers, B., Dejonghe, P., Kruth, J.P., (2003), Optimal and collision free tool posture in five-axis machining through the tight integration of tool path generation and machine simulation, Computer- Aided Design, 35 (5): 421-432.

[10] Ming, C.; Leu a Joshi A; (2008). Nx5 for engineering design. U.S. state of Missouri: Missouri Uneversity of Scieneceand Technology

[11] M. Wilson, Implementation of Robots Systems, United Kingdom, 2014, ISBN: 978-0-12-404733-4.

[12] F. Koenigsberger, J. Tlusty, Machine Tool Structures, Elsevier, 2013, 148316053X, 9781483160535.

[13] P. Janda, R. Polak, (2016). Virtual Prototyping and Optimization of Heavy Machine Tools, Published by DAAAM International, ISBN 978-3-902734-07-5, ISSN 1726-9679, Vienna, Austria

DOI: 10.2507/28th.daaam.proceedings.079

DETERMINATION OF FDM PRINTER SETTINGS WITH REGARD TO GEOMETRICAL ACCURACY

Roman Polak, Frantisek Sedlacek, Karel Raz

This Publication has to be referred as: Polak, R[oman]; Sedlacek, F[rantisek] & Raz, K[arel] (2017). Determination of FDM Printer Settings with Regard to Geometrical Accuracy, Proceedings of the 28th DAAAM International Symposium, pp.0561-0566, B. Katalinic (Ed.), Published by DAAAM International, ISBN 978-3-902734-11-2, ISSN 1726-9679, Vienna, Austria
DOI: 10.2507/28th.daaam.proceedings.079

Abstract

Material extrusion is one of the most used additive technologies. The most common application of this technology is in the production of prototypes, preparations and small serial parts. This article deals with relationship between different model geometries and parameters such as temperature, speed of printing and height of layer. Typical features of this technology are ease of printing, but it depends on the type of material used and the particular device. Printers with Fused Deposition Modelling (FDM) technology have no feedback about printed material, such as printing accuracy. This paper aims to easily find ideal parameters for FDM printing technology using Polylactic Acid (PLA) material.

Keywords: Additive Manufacturing; FDM technology; Polylactic-Acid; 3D print parameters; Material extrusion

1. Introduction

Additive manufacturing technology, known as 3D print, is more widespread in some sectors. Producing parts via this method offers many advantages over traditional manufacturing technologies. The most frequently used type of technology is Fused Filament Fabrication (FFF) [1]. This term is equivalent to Fused Deposition Modeling (FDM) which is used as trademark by Stratasys Inc. Objects printed with this technology are layered and software (slicer) allows change many settings. Producers provide information about their printers such as minimum layer, accuracy, print bed size and other technical parameters [2]. But it may not apply to all printed parts with different geometries.

Most printers have recommended settings provided through the profile in slicers. Profiles depend only on used material, type of printer and does make a provision for type and properties of printed geometry. Many articles deal with dimensional accuracy in last phase of additive manufacturing. They solve printing inaccuracies with post processing [13]-[14]. The technology used for surface finalizing is vapor smoothing, but at the expense of dimensional accuracy [10]-[11]. This article is focused on dependency between model geometry and parameters such as temperature, speed of printing and height of layer.

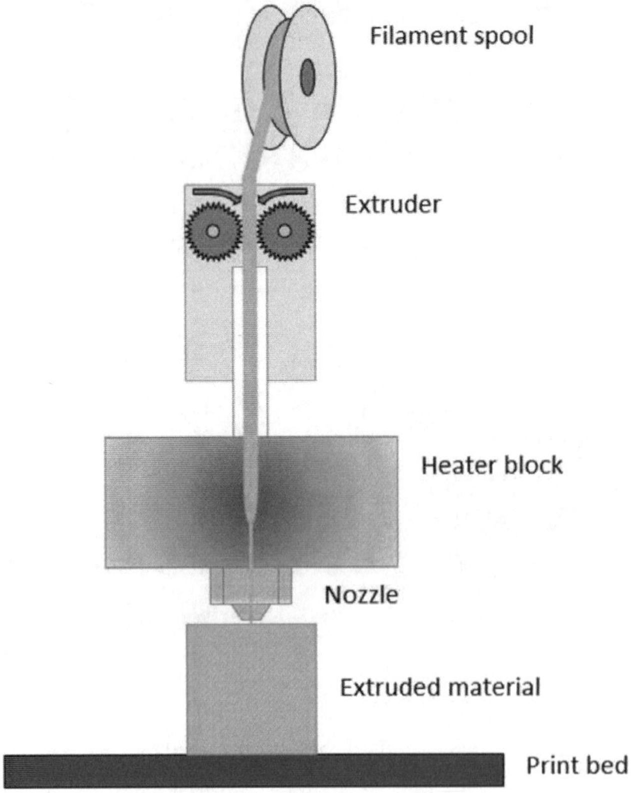

Fig. 1. FDM Technology

This technology is also known as "Material Extrusion". The material is loaded by an extruder (typically using a stepper motor) to the printer system [9]. Then is heated to a semi-liquid state by heater block and pressed through the nozzle to print bed. After the entire layer is applied, the pad is reduced by the thickness of the layer in the vertical axis and the gradual deposition continues again until the whole product is formed.

The abbreviations used below are trademarks of different manufacturers using the same technology [3]-[5]:

- FDM (Fused Deposition Modelling)
- PJP (Plastic Jet Printing)
- FFM (Fused Filament Modelling)
- MEM (Melted and Extruded Modelling)
- FFF (Fused Filament Fabrication)
- FLM (Fused Layer Modelling / Manufacturing)

2. Material description

The chosen material was Polylactic Acid (PLA). It is biodegradable material and is derived from renewable resources like corn starch or sugar cane. PLA with Acrylonitrile Butadiene Styrene (ABS) are the most used materials with material extrusion technology. PLA compared with ABS is easier and faster to print with the same initial conditions, but the products are significantly less resistant to higher temperatures. From the material point of view, PLA products are less flexible and glossier than ABS.

Property	**Value**	**Unit**
Melt Temperature	173 - 178	°C
Glass transition temperature	60	°C
Heat Deflection Temperature at 455 kPa	65	°C
Density	1.3	g/cm^3
Filament size	1.75	mm
Chemical formula	$(C_3H_4O_2)_n$	

Table 1. Physical properties of the PLA

3. Testing procedure

3.1 Data preparation

The FDM sample on following figure (**Error! Reference source not found.**) was used to verify the geometry with different settings of printer. This item was prepared in CAD software Siemens NX 11 [6]. In the drawing on figure (*Fig. 2*) are main parameters to verify with different printer settings. Source CAD data is [7].

Because software's for printers need stereo lithography file format (STL) [8], it was exported with following options. Chordal tolerance was set up to 0.05mm and 5° as angular tolerance, which is fully sufficient with regard to geometric shapes and print parameters. Smaller tolerance increase the detail of the export, but is unnecessary for printing with these specified parameters and only increases the size of the STL file.

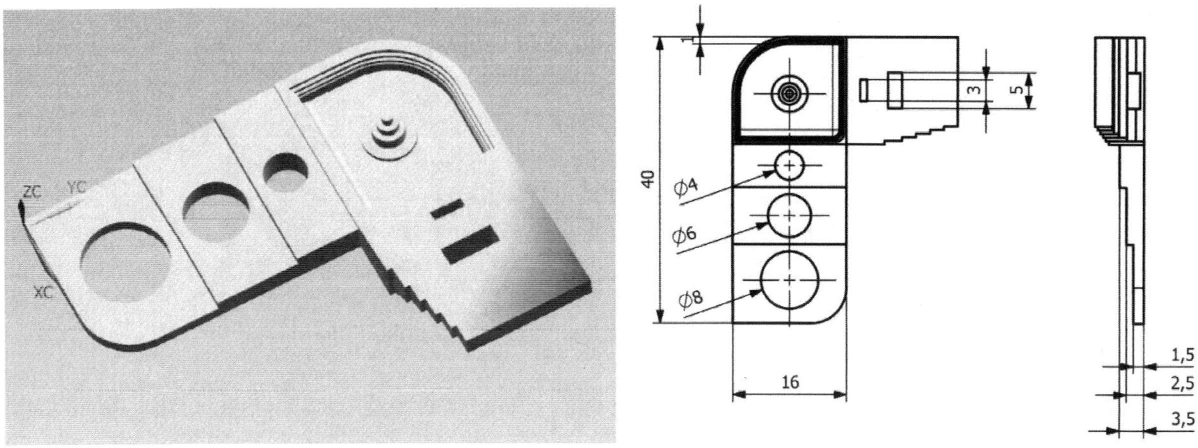

Fig. 2. CAD data with main drawing parameters

An STL file from Siemens NX was used as the input file for slicer software Cura 3.0.3. All printing parameters was set up in this slicer. Main parameters (speed, temperature and layer height) was changed in each print, but other parameters, such as printed line, cooling fan, bed heating, skirt etc. were the same for all printed samples. Printed line depends on nozzle diameter and must be the same as nozzle. Combination of bed heating and enabled cooling fan is the best for PLA bed adhesion with printing accuracy. Skirt ensures that the material is loaded into the entire printer device up to the print head.

All prints has the same settings. The main parameters are full infill, 0.5 mm size of nozzle, four walls, 40°C heated bed, print cooling from second layer etc. Tested parameters are temperature (180°C, 200°C, 220°C), speed (20 mm/s, 30 mm/s, 40mm/s) and layer height (0.1 mm, 0.2 mm) with all combinations. The printer working with G-code which is the output of this slicer software. The G-code is programming language with instructions for 3D printer. These instructions contain all the information's about print settings and model geometry converted to the print path.

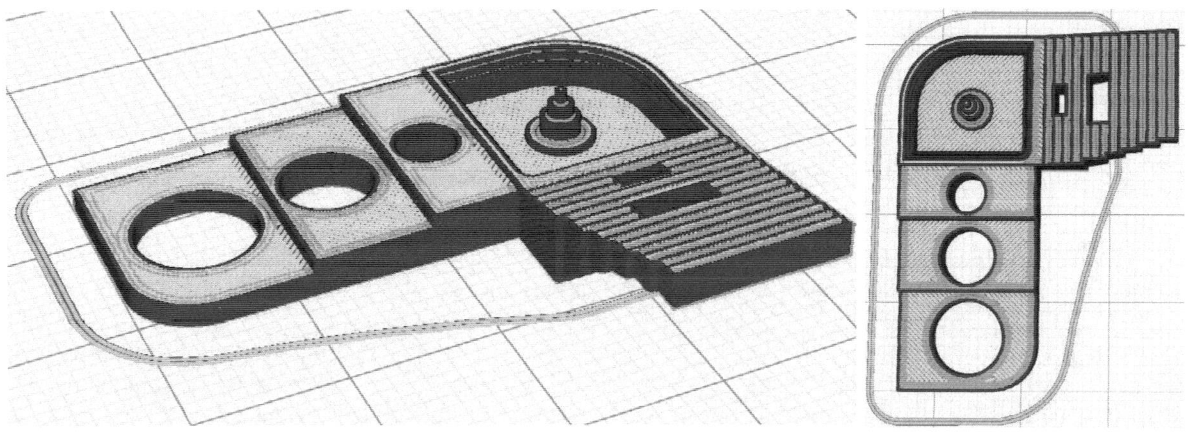

Fig. 3. Sample preparation in slicer Cura

3.2 Printing samples

All samples were printed on FDM printer based on Prusa i3. The manufacturer provided axis accuracy 0.012 mm for X and Y and 0.0004 mm for Z axis. Heated bed can be heated up to 120°C, but for PLA printing purposes is 60°C enough. Theoretical maximum printing speed is 150 mm/s on printing area 200 x 280 x 200 mm.

Nozzle diameter is 0.4 mm, but is customized by user from 0.2 mm to 0.5 mm. The values bellow are parameters which was used for printing. A combinations of prints is compiled from these parameters and each of these prints has 3 copies to minimize errors.

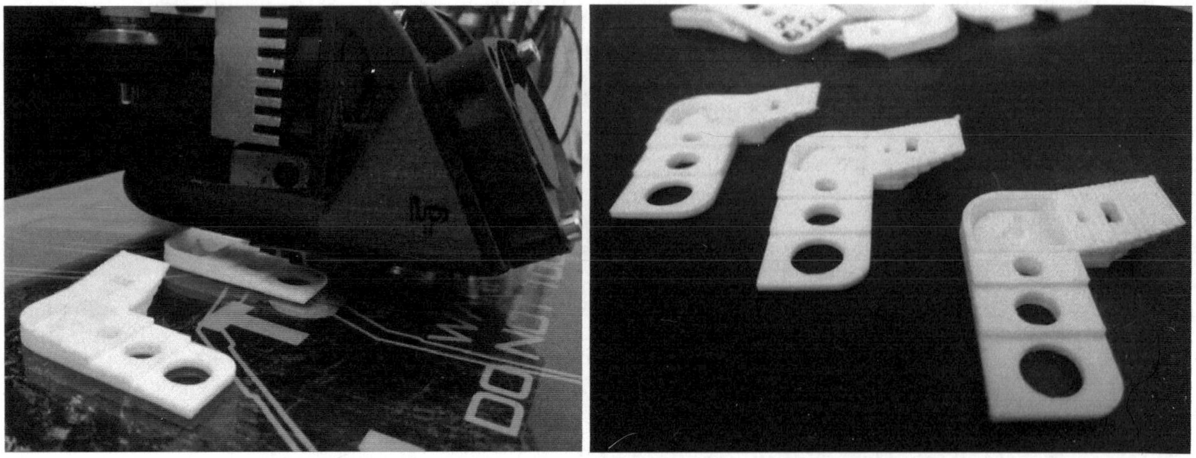

Fig. 4. Printed PLA samples

3.3 Data analysis

After the printing phase measurements and analysis are carried out. The FDM sample has many monitored and measure elements. All elements were measured and compared with the printer settings (speed, layer, temperature). There are elements circle 8, circle 6, circle 4, rectangle 3 and rectangle 5.

The shape (circle, rectangle) is the type of geometry and the numbers are diameters in millimetres. Than was monitored main values in all axis. The one diameter in X and Y axes. The Z diameter represent stair 1, stair 2 and stair 3. Where each stair has different height in the Z axis. The last parameter is minimum wall size. It means how accurate the thin wall is.

Data collected from all printed objects. Each print setting has 3 copies to minimize the errors rate. All prints have the same settings. The main parameters are full infill, 0.5 mm size of nozzle, four walls, 40°C heated bed, print cooling from second layer etc.

The tested parameters are temperature (180°C, 200°C, 220°C), speed (20 mm/s, 30 mm/s, 40mm/s) and layer height (0.1 mm, 0.2 mm) with all combinations. Values in the chart above (**Error! Reference source not found.**) show percentage accuracy according to print tested parameters.

The parameter for Z-axis accuracy has similar balanced results (Stairs in schema). It cannot be said that any tested parameter affects the precision in the Z-axis. The samples with higher temperature show signs of greater fluidity and less accuracy. This problem is typically shown at minimum wall size parameter.

The circle geometries have bigger accuracy with bigger diameter. The same rule applies to the rectangular shape, bigger diameter has better accuracy. The best results of samples, it means with minimum deviation from the required dimensions, are a complex combination of the lower layer height (0.1mm) and the lower temperature (180°C).

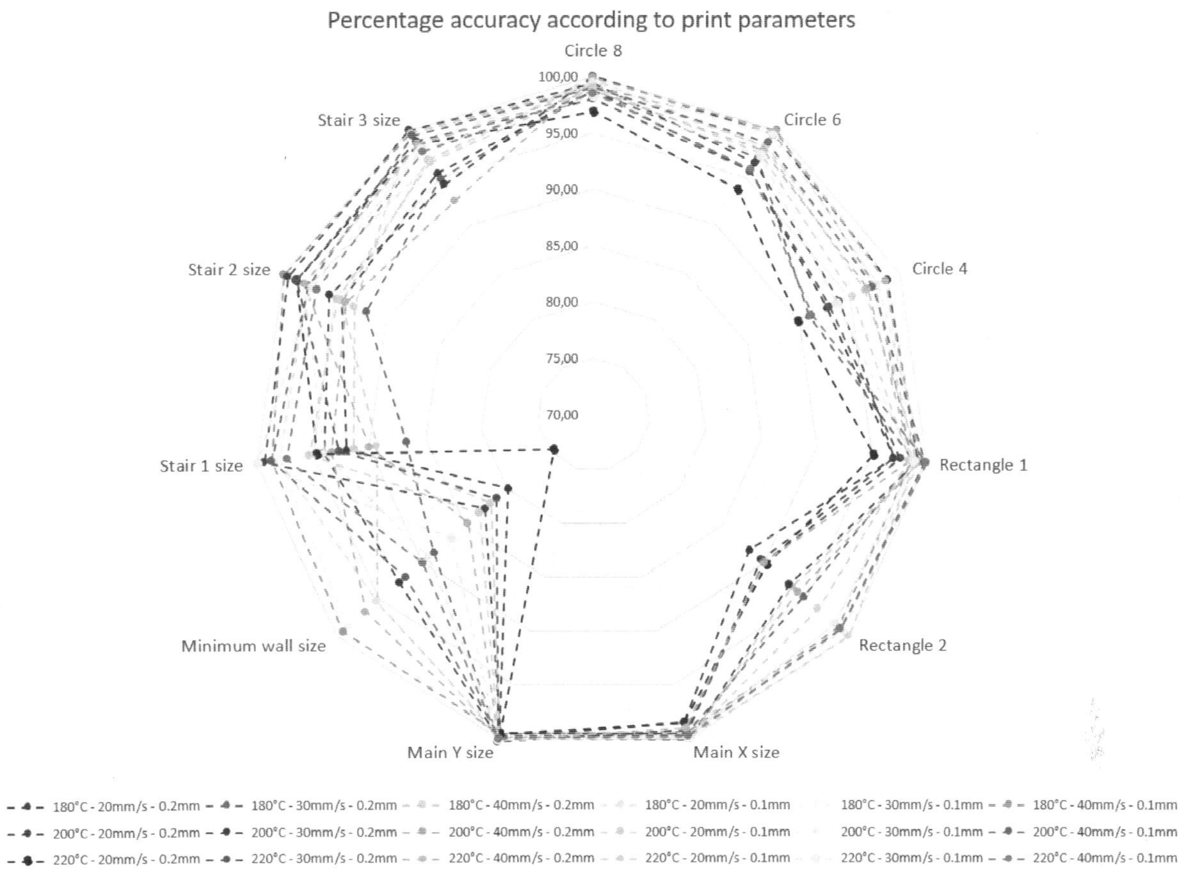

Fig. 5. Schema of data analysis

4. Conclusion

This paper describes the ideal settings of a Fused Deposition Modelling (FDM) printer. It is hard to find these settings during real production, therefore this experimental analysis was carried out. The collected data shows the dependence of the geometry on the print properties. The main tested parameters were speed, temperature and layer height. The results show the accuracy mainly depends on the temperature and layer height. The best results were obtained using lower temperature and thinner layer. Many articles deal with dimensional accuracy and model post processing with regard to the surface finish. The information contained in this article can be used for verification of these properties and finding the ideal parameters on other FDM devices.

This test has many possibilities for continuing and expanding the testing information for ideal FDM printer settings. Further research will continue with testing other print parameters, such as different nozzle sizes, other materials (ABS, Nylon, PETG etc.) and combining them with the parameters tested in this article.

5. Acknowledgments

This paper has been prepared under project LO1502 'Development of the Regional Technological Institute' under the auspices of the National Sustainability Programme I of the Ministry of Education of the Czech Republic aimed to the support research, experimental development and innovation.

6. References

[1] I. Gibson, D. Rosen, and B. Stucker. (2015). Introduction and Basic Principles in Additive Manufacturing Technologies, Springer New York, pp. 1–18, New York.

[2] C. Barnatt. 3D Printing: Third Edition. CreateSpace Independent Publishing Platform.

[3] ISO 17296:2015 - Additive manufacturing - General principles - Part 2: Overview of process categories and feedstock. ISO, 2015.

[4] ISO/ASTM 52915:2013 - Standard specification for additive manufacturing file format. International Organization for Standardization, Jun-2013.

[5] ISO/ASTM 52900:2015 - Additive manufacturing - General principles - Terminology'. International Organization for Standardization, Dec-2015.

[6] D. I. Wimpenny, P. M. Pandey, and L. J. Kumar, Eds. (2017). Advances in 3D Printing & Additive Manufacturing Technologies. Singapore: Springer Singapore, 2017.

[7] Siemens PLM. (2016) Siemens NX 11 User's Guide. Siemens AG..

[8] C. K. Chua and K. F. Leong. (2014) 3D Printing and Additive Manufacturing: Principles and Applications (with Companion Media Pack). Fourth Edition of Rapid Prototyping Fourth Edition. World Scientific Publishing Company.

[9] K. Kun. (2016). Reconstruction and Development of a 3D Printer Using FDM Technology. In Procedia Engineering, Volume 149, 2016, Pages 203-211, ISSN 1877-7058.

[10] Kozior T., Kundera C. (2017). Evaluation of the Influence of Parameters of FDM Technology on the Selected Mechanical Properties of Models. Procedia Engineering, Volume 192, 2017, Pages 463-468, ISSN 1877-7058.

[11] P.J. Nuñez, A. Rivas, E. García-Plaza, E. Beamud, A. Sanz-Lobera. (2015). Dimensional and Surface Texture Characterization in Fused Deposition Modelling (FDM) with ABS plus. Procedia Engineering, Volume 132, 2015, Pages 856-863, ISSN 1877-7058.

[12] Singh R., Singh J., Singh S. (2017). Investigation for surface finish improvement of FDM parts by vapor smoothing process. ISSN 1359-8368.

[13] Singh R., Singh J., Singh S. (2016). Investigation for dimensional accuracy of AMC prepared by FDM assisted investment casting using nylon-6 waste based reinforced filament. ISSN 0263-2241.

DOI: 10.2507/28th.daaam.proceedings.080

SPECIFIC DESIGN OF CRITICAL DRIVETRAIN COMPONENT

Jan Hnatik & Lubos Kroft*

This Publication has to be referred as: Hnatík, J[an] & Kroft, L[ubos] (2017). Specific Design of Critical Drivetrain Component, Proceedings of the 28th DAAAM International Symposium, pp.0567-0574, B. Katalinic (Ed.), Published by DAAAM International, ISBN 978-3-902734-11-2, ISSN 1726-9679, Vienna, Austria
DOI: 10.2507/28th.daaam.proceedings.080

Abstract

The automotive industry can be classified among one of the leading sectors of production engineering. From the perspective of the automotive parts is then the design and their production standard matter. The present contribution focuses on the management area and more specifically on the design and production of steering pinion. This is a common component, but in case of the historic cars, it is issue quite complicated. The aim of the project is just a design new steering pinion so as to be shortened step length and increased its fluidity, thus improving the current state of management over serial. It is therefore an upgrading of the existing type of steering pinion, which is based on the original assembly. On the basis of the input data will be calculated, designed and constructed a new type of steering pinion with steep transfer. This modified pinion steering will be made and implemented to control the assembly of the historic car, where they will test directly in the operation.

Keywords: Automotive industry; Historic cars; Steering pinion; Constructed; Test new part

1. Introduction

The automotive industry is one of the fastest-growing sectors with a major influence on economy in many countries. The trends of recent years include not only production of brand new cars but also restoration of older vehicles referred to as classic cars. Those active in this field are not only individuals but companies as well, particularly in the recent period.

Their motivations are varied but the restoration activities generally fall into three groups. The first one is restoration to the vehicle's original condition, the second one involves restoration and conversion of the vehicle into a special classic car for racing and the last comprises adaptation to modern trends – tuning.

The latter two types of restoration also involve efforts to improve the properties of the vehicle. In this respect, there is a difference between restoration of old cars and modification of volume-produced modern cars. Parts which can further improve vehicle properties are not generally available for classic cars.

Fig. 1. Škoda 130 RS - Rally Monte Carlo replica [8]

Design, production and testing of a new steering pinion for a Škoda 130 RS, which should have better features than the standard component, was the aim of project that is described in following article. New steering pinion, must be compatible with the original steering system while achieving a shorter steering ratio.

Component pinion with involute teeth is, complicated component from several aspects. An essential consideration is the service life of the pinion and especially the teeth that are the most stressed. For the new type of pinion, it is necessary to choose default material, heat treatment, and finishing operation. The issue of the life service of the gear deals in work with the sources [9] and [10]. Paper with filename "Failure analysis of a pinion" deals with the reason for the rupture of the teeth of the influence of fatigue. Article entitled "Engineering Failure Analysis" on the contrary, addresses the failure of the teeth from the point of view of material and its heat treatment.

For the calculation, design and construction of the new pinion information and formulas from sources [3] and [4] were used. Another important aspect is the machining. The chosen strategy has significant impact on the functional characteristics of the new component. For this reason, the results are quoted in the article "The Influence of Different Types of Copy Milling on the Surface Roughness and Tool Life of End Mills" [2], a very beneficial. Based on them it's possible to design new machining strategy.

2. Proposal for a new pinion

As mentioned above, the production of parts for classic vehicles is by far less common than for modern cars. The first parts are not generally available in today's market. Difficulties also arise from the fact that when a part is designed for today's cars, the entire assembly is de-signed at the same time, thanks to which the vehicle properties are improved. By contrast, when a part is being made for a classic vehicle, the resultant assembly comprises original parts and only the part in question is adapted. This is another reason why the opportunities for improvement are considerably limited here.

Therefore, when a new part is to be designed, one has to collect as much information as possible on the original assembly. It becomes yet another challenge because documentation on classic carts is very difficult to find. One often has to turn to reverse engineering to such obtain information.

2.1 Specifications

The controllability of a vehicle is of key importance. Unfortunately, a car as a whole is a complex system of interacting elements. Controllability is therefore governed by numerous components, such as wheel suspension, shock absorbers, springs, wheels as such and many others. The steering per se is just one of the many factors that play a role in controllability. However, its design determines the steering ratio. This parameter can be expressed as the number of steering wheel revolutions required for lock-to-lock movement of wheels. Steering ratio is a very important factor for manoeuvring at high speeds and through sharp turns in rallies and track races.

Steering ratio can be reduced by several methods. However, this pinion design is not intended for standard cars, which means that the options are rather limited here. The steering assembly is simple in principle. It consists of the steering housing in which the rotational movement of the steering wheel is transmitted by the steering pinion to the rack. The rack has at its ends tie rods which control the vehicle's axle. The entire system is shown in detail in Fig. 2. [1]

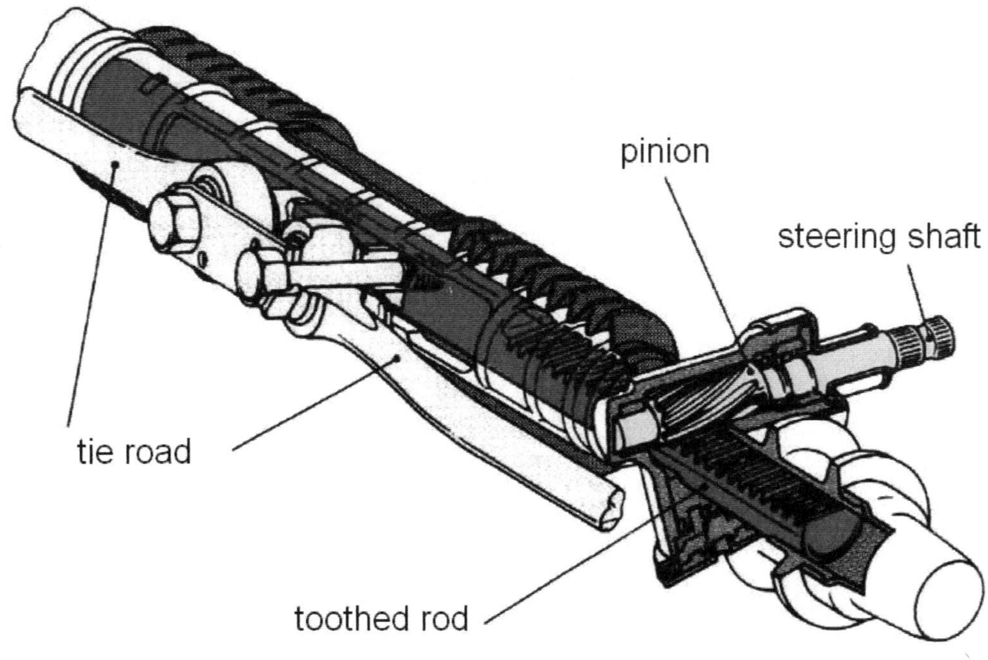

Fig. 2. Rack and pinion steering assembly [1]

The conclusion from the analysis is that the steering pinion represents the only acceptable route for dealing with the issue. This is because if the number of teeth on the steering pinion is increased, the gear ratio increases and the steering ratio becomes lower. For this to be accomplished, the part must be compatible with the original steering system and able to engage correctly with the original rack.

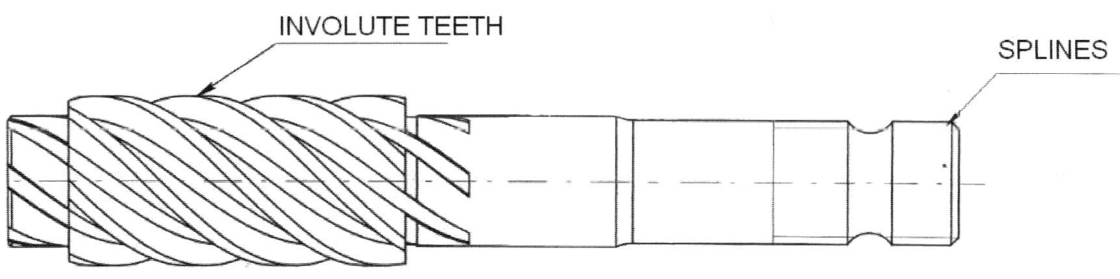

Fig. 3. Steering pinion

The steering pinion (Fig. 3.) is a rotation-symmetric part provided with teeth and splines. Splines are features which provide the connection between the pinion and the steering shaft. The splines are plain rectangular features with a recess at mid-length, at which the pinion is secured to the steering shaft. The teeth are the key functional part of the entire pinion. They are involute teeth in order to provide smooth action of the steering wheel. Besides highly-accurate dimensions and superior finished surface quality, the teeth also need a relatively high hardness. All these aspects must be taken into account in the design process and during manufacture.

2.2 Investigation of existing pinion

The purpose of this exercise was to propose, design and manufacture a new steering pin-ion. In order to be able to develop a design for a new steering pinion, it was essential to examine the existing one. This investigation was carried out at the Metrology Laboratory (ML) of the Regional Technological Institute (RTI). As the rack had straight teeth, it was measured using standard methods. The module was measured with a slide calliper. The remaining values were derived from engineering tables. The pinion was measured by means of a 3D measuring machine which offers laser scanning of profiles of parts.

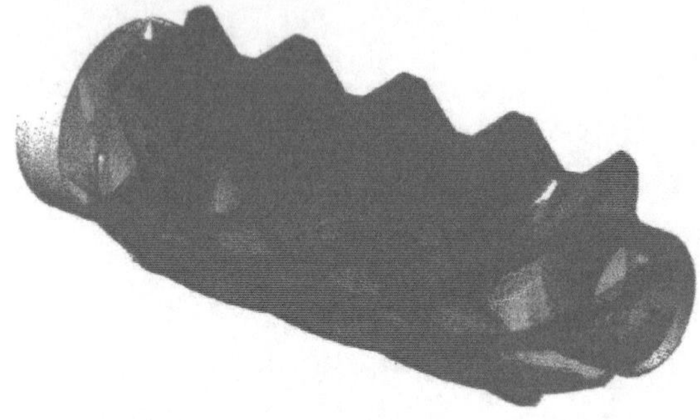

Fig. 4. Pinion model obtained by scanning

Although the new pinion would differ substantially from the existing one, the data that characterize the existing assembly were still important. It was necessary to determine the width of space, the helix angle and many other parameters which would be difficult to obtain by other means.

2.3 Design of a new pinion

The new pinion was proposed and designed on the basis of data obtained by examination of the rack and the existing pinion at ML. Since the new pinion is to have one tooth more than the existing one, the tooth profile, the width of space and other parameters had to be modified.

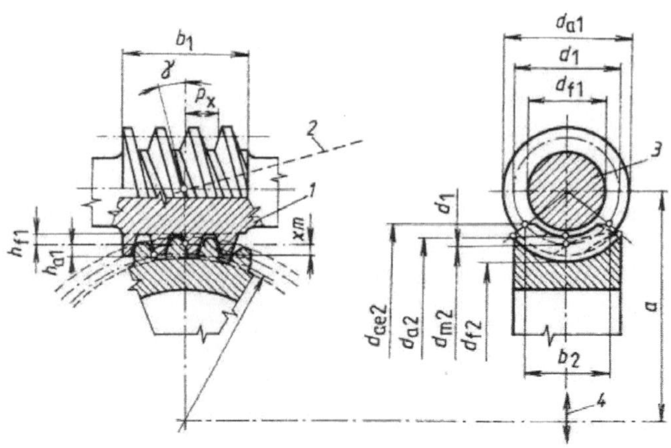

Fig. 5. General drawing of worm gear set and identification of parameters [4]

In the first phase, calculations using basic formulae were completed. As the rack parameters are fixed, only the module value is important:

Verification of the helix angle:

$$\tan\gamma = \frac{z_1}{q} \tag{3}$$

z1 – number of pinion teeth – to be decided
q – pinion radius (found by measurement)
γ – helix angle
Axial module:

$$m_x = \frac{m_n}{\cos\gamma} \tag{3}$$

Axial pitch:

$$P_x = \pi \times m_n \tag{3}$$

m_n – normal module (found by measurement)
m_x – axial module
Pitch circle diameter:

$$D_1 = \frac{z_1 \times m_x}{\tan \gamma} \tag{3}$$

D_1 – pinion pitch circle

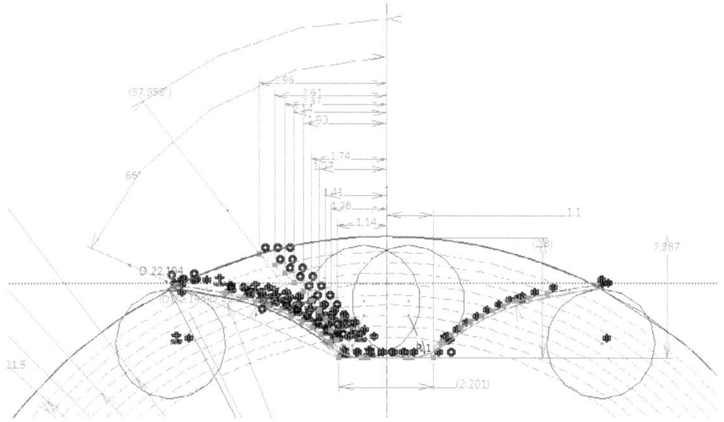

Fig. 6. Profile of tooth

Based on calculated and tabulated values, a new pinion with involute teeth was designed. Unfortunately, the correction of the existing gear was impossible to identify by measurement. Yet, even if this value had been available, the conversion for a pinion with a different number of teeth would have been complicated. Consequently, the solution was to actually manufacture and test the pinion. Fig. 7. shows the final solution which was tested in practice. [5] [6]

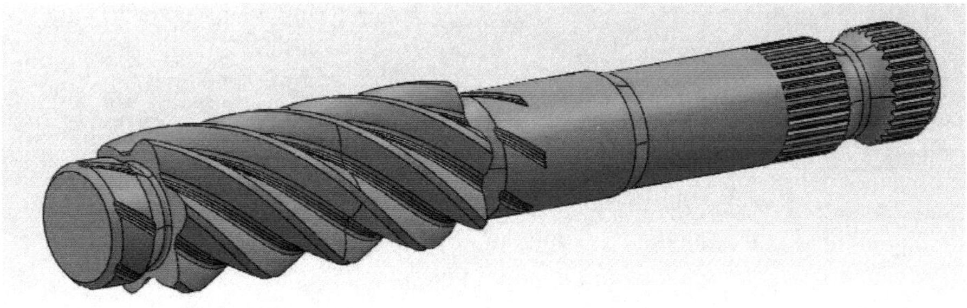

Fig. 7. New steering pinion

2.4 Selection of pinion material

As mentioned above, the proper function of the steering system depends not only on dimensional accuracy and good surface finish but also on certain material properties. The reason is that this part is intended not just for ordinary use but also for heavy-duty service in racing cars where loads, both static and dynamic, are significant.

As with any functional component of the transmission system, the part's surface must exhibit sufficient hardness. The value which is appropriate for a steering pinion is 58±2 HRC. This will guarantee that the component is not too soft and that the teeth do not suffer denting or seizure. At the same time, this hardness is not excessively high. Otherwise it would threaten to damage the steering rack or the steering shaft.

The prescribed hardness is required only on the surface and immediately below. The core of the pinion and the interior of the teeth should retain sufficient toughness. This is an essential requirement because of dynamic service loads. Had the part been hardened throughout, it could become brittle. Given the intricate shapes of the teeth, this would have compromised notch toughness and lead to teeth failure.

Upon considering these aspects, W. Nr. 16MnCr5, Czech Standard ČSN 41 4220.1 was selected as the initial material. It is a low-alloy manganese-chrome steel for case hardening. Its chemical composition is listed in Tab.1. When annealed, this material shows good machinability and is suitable for case hardening. By these processes, one can obtain the properties which are required in a steering pinion.

C	Si (max)	Mn	P (max)	S (max)	Cr	Mo	Ni	Al
0.14 - 0.19	0.40 1.00	1.30	0.035	0.035	0.80 – 1.10	-	-	At controlled austenitic grain size 0.015-0.050

Table 1. Chemical composition of 16MnCr5 steel [7]

3. Manufacture of prototype pinion

Ordinary volume production of gears is a highly specialized effort which requires particular machines and tools. It was impossible to use those because the goal of the entire project was to design, manufacture and test a new type of pinion. It was not cost-efficient to make an investment in special tools for prototype production of several pieces.

The production sequence and the operations were identical for all variants. The only difference in the designs concerned the shapes of the teeth.

Since the basic shape of the semi-finished product was rotation-symmetric, the first operation was the turning of a round rod. The Machining Technology Laboratory (MTL) at the RTI houses a multifunctional turning-milling machine from DMG MORI designated as CTX BE-TA 1250 TC 4A. In this machine, turning can be followed by other operations. This offers a major advantage in the reduction of operation times as transfers to other production machines are eliminated. Another advantage lies in that the machine offers automatic transfers and clamping of the work between spindles which benefits the concentricity of the product.

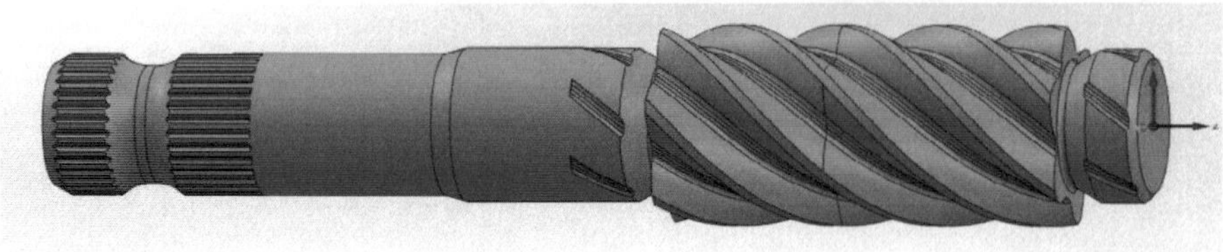

Fig. 8. Design model of the pinion

Fig. 9. Semi-finished pinion before finishing operations

Fig. 10. Finished pinion upon heat treatment and the finishing of all surfaces

The entire basic shape is thus produced with a single clamping operation, including the splines and involute gear teeth. The teeth were rough-cut by cemented carbide (CC) end mills of 2 and 4 mm diameters, as shown in Fig. 11. A) and Fig. 11. B). Finishing operations were completed with a CC ball mill of 1.8 mm diameter, as illustrated in Fig. 11. C). In order to achieve the desired quality, a relatively small step over of 0.12 mm was employed. This provided a sufficient quality of the finished surface, Fig. 11. D. It is not all. The next important parameter (also mentioned in article [2]) direction of the machining has influence influenced of tool life too. In our case strategy of contouring was used instead of copy-milling. As shown in the summary, for a constant lifetime is more appropriate to use the strategy from top to down. From this reason was also selected this option. Nevertheless, an additional polishing operation was carried out in OTEC DF 3 machine.

Polishing was included into the technological process due to the prevention of fatigue cracks. These cracks most often occur in places, notches as indicated in the research reports from the source [10] and [9].

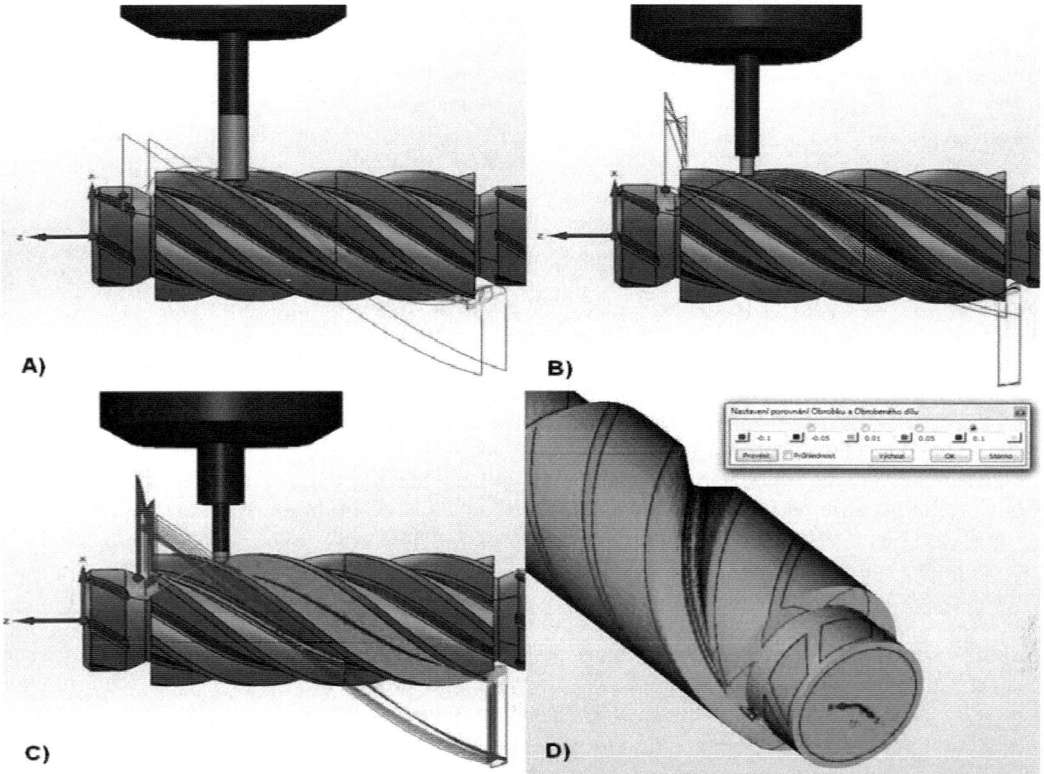

Fig. 11. Production sequence of a pinion tooth: A) roughing, B) pre-finishing, C) finishing, D) CAM analysis of the surface

The production accuracy was verified by the ML. This time, however, a 3D model was al-ready available thanks to which the inspection was completed by contact scanning of surfaces of individual teeth. Deviations of several hundredths of millimetre were detected. Such values are acceptable.

In order to be able to check the functioning of the pinion before subsequent production operations, one of the steering systems was adapted, as shown in Fig. 12., for testing of the pin-ion in the annealed condition. This dummy steering was assembled and areas of contact were checked using engineer's blue in order to select the appropriate variant.

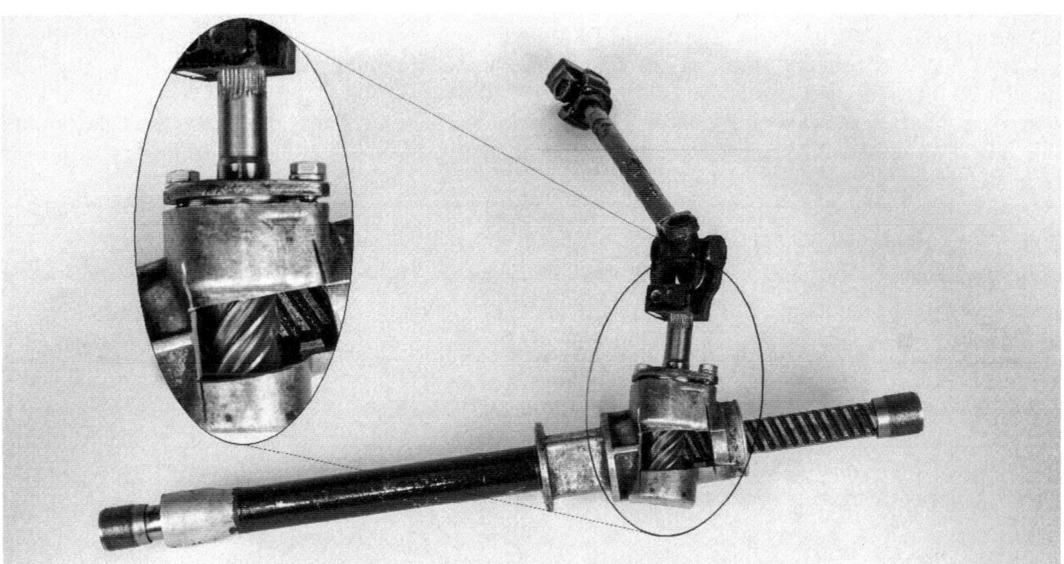

Fig. 12. Dummy steering assembly

The next production step involved carburizing. This operation was sourced from an authorized external partner. The carburized layer thickness was specified as no less than 0.3 mm. With this appropriate material, it was straightforward to obtain the desired depth and finish the part.

Carburizing in a vacuum furnace produced no scale on the surface and there was no need to regrind the pinion teeth afterwards. This made the production much simpler. The last operation was the grinding of the bearing seats.

The finished pinion was successfully tested in the dummy steering assembly and then installed in an actual car. Testing indicated very good properties of the entire assembly and good engagement of the newly-proposed tooth profile with older and worn racks, which was an un-foreseen extra benefit. Overall, the new pinion proved very successful.

4. Conclusion

Components for special classic cars for racing are in short supply. The problem with that led to a decision to upgrade an existing steering pinion type by making a new part which would better meet customer requirements. The project was undertaken jointly with the Regional Technological Institute of the University of West Bohemia.

The objective of the project was to reduce the steering ratio and thus improve controllability of the vehicle. At the start, there was an existing steering assembly. The upgrade was limited to the pinion, whereas the rest of the assembly remained without changes. After the existing pinion had been examined, a new steering pinion for a lower steering ratio was proposed, designed and manufactured to better meet market requirements. This new type of pinion is now being used in both special classic cars for racing and replicas of original vehicles and de-livers a more than satisfactory performance.

Further developments of the pinion were not yet realized, because these testing are so far too short and it is necessary to wait for more data from the real operation and races. If testing goes according to the assumptions of the aim for the future is to apply these outcomes to other types of pinions, or upgrade an already newly designed pinion.

5. Acknowledgments

The present contribution has been prepared under project LO1502 'Development of the Regional Technological Institute' under the auspices of the National Sustainability Programme I of the Ministry of Education of the Czech Republic aimed to support research, experimental development and innovation..

6. References

[1] Electronic textbook – Řízení / Driving. [online]. [accessed on 2017-08-17]. Available from: https://eluc.kr-olomoucky.cz/verejne/lekce/1491

[2] Vopat, T[omáš]; Peterka, J[ozef]; Simna, V[ladimír] & Kuruc, M[arcel] 2014. The Influence of Different Types of Copy Milling on the Surface Roughness and Tool Life of End Mills, Proceedings of the 25th DAAAM International Symposium, pp.0108-0116, B. Katalinic (Ed.), Published by DAAAM International, ISSN 1726-9679, Vienna, Austria

[3] Hradilova, P. 2013. Zabezpečovací pohon výrobního zařízení se šnekovou převodovkou a pojistnou spojkou / Security drive of the device with the worm gear and the locking clutch. Tomas Bata University in Zlín, Faculty of Technology.

[4] Taborsky, V. 2009. Návrh pohonu zařízení přes šnekovou převodovku a pojistnou spojku / Design of the drive device through the worm gear and the locking clutch. Tomas Bata University in Zlín, Faculty of Technology.

[5] Bolek, A[lfred] & K[ochman],. Části strojů / Parts of machines. 5th edition, revised, Praha: SNTL - Publishing house of technical literature, 1990. Technical guide. ISBN 80-030-0426-8.

[6] Leinveber, J[an] & Vavra P[avel]. Strojnické tabulky: pomocná učebnice pro školy technického zaměření / Mechanical table: an auxiliary textbook for schools technical focus. 4th edition, expanded. Úvaly: Albra, 2008. Technical guide. ISBN 978-80-7361-051-7.

[7] Bolzano – Technical handbook [online]. 2017 [accessed on 2017-08-21]. Available from: http://www.bolzano.cz/assets/files/TP/MOP_%20Tycova_ocel/EN_10084/MOP_16MnCr5.pdf

[8] Tym-motorsport - stavba závodních automobile / construction of the racing cars [online]. [accessed on 2017-08-22]. Avail-able from: http://www.tym-motorsport.cz/#utm_source=firmy.cz&utm_medium=ppd&utm_campaign=firmy.cz-2373099

[9] Kosec, G[orazd]; Nagode, A[leš]; Antic I[gor] & Kosec, B[orut] 2010. Failure of the pinion from the drive of a cement mill. Elsevier- Science direct [online].[cit. 2017-09-06]. Available from: http://www.sciencedirect.com/science/article/pii/S1350630710001858

[10] Das, C.R; Albert, S.K.; Bhaduri, A.K. & Ray, S.K. 2005. Failure analysis of a pinion. Elsevier- Science direct [online]. [cit. 2017-09-06]. DOI: Engineering Failure Analysis. Available from: http://www.sciencedirect.com/science/article/pii/S1350630704000834

DOI: 10.2507/28th.daaam.proceedings.081

THE EFFECT OF THE STRATEGY OF FINISHING ON DIMENSIONAL ACCURACY

Luboš Kroft*, Jan Hnatík & Kateřina Bícová

This Publication has to be referred as: Kroft, L[ubos]; Hnatík, J[an] & Bicova, K[aterina] (2017). The Effect of The Strategy of Finishing on Dimensional Accuracy, Proceedings of the 28th DAAAM International Symposium, pp.0575-0581, B. Katalinic (Ed.), Published by DAAAM International, ISBN 978-3-902734-11-2, ISSN 1726-9679, Vienna, Austria
DOI: 10.2507/28th.daaam.proceedings.081

Abstract

The trend of today is to manufacture even highly-intricate parts in the least number of technological steps. With this strategy, it is necessary to finish complex 3D surfaces with the best qualitative results. Where mill finishing is the last operation (e.g. in making a forging die), it is essential that the quality of the finished surface is acceptable without the need for other finishing technologies. This paper deals with the influence of the chosen finishing strategy on dimensional accuracy. In many types of production, dimensional accuracy is more fundamental than, for example, the quality of the finished surface. Also, efforts to meet very narrow tolerance limits for complex 3D surfaces with milling are becoming more and more common. Another aspect analysed here is the use of constant cutting speed in mill finishing with the aim to increase the efficiency of finishing operations.

Keywords: Machining; Finishing operations; Constant cutting speed; Qualitative results; Strategy

1. Introduction

Operations planning is being increasingly dominated by CA (computer-aided) systems. Advanced CAD/CAM systems offer a number of functionalities with a broad range of operations and upgrades which contribute to machining quality. [6] Their strengths are best manifested in tasks that involve complex tool paths. Such paths, which are typically used in machining three-dimensional intricate-shaped surfaces, are mostly associated with milling.

Milling has seen major developments in the recent decade. A majority of innovation efforts, whether involving improvement in the cutting performance of tools or upgrading the functional properties of machines and clamping equipment, aim to enhance the productivity and reduce the production costs of milled surfaces.CAM systems and their milling strategies are a potent tool for increasing the efficiency of preparation of both NC programs and the actual production. This applies particularly to roughing and semi-finishing strategies.

CAM systems offer superior solutions for these machining stages. The main reasons are the adaptive control of the tool and the growing use of trochoidal tool path generation. As a result, near-constant tool loads can be achieved. More aggressive cutting conditions can be applied with these strategies to improve productivity.

This article describes an investigation which focused on the finishing stage. The finishing strategies available in today's CAM systems are relatively powerful and of reasonable quality. One disadvantage is that tools for aggressive cutting conditions are not available for finishing strategies, unlike in the previous two stages.

Today, efficiency can only be improved by selecting a different machine, cutting tool, clamping method or by other alterations to the machine-tool-workpiece-fixture system. Finishing thus becomes rather time-consuming and costly, particularly with intricate-shaped surfaces.

Several scientific studies were therefore devoted to investigating the potential for improving the finishing efficiency. An example of how finishing strategies compare is presented in [10]. The paper compared finishing strategies in 5-axis operations but mainly from the productivity perspective.

More interesting information can be found in article [11] which dealt with machined surface accuracy but only in terms of interpolation of the surface and not as a result of a particular strategy.

Nevertheless, there are studies which are closer to the issue of our interest. The paper "The Influence of Different Types of Copy Milling on the Surface Roughness and Tool Life of End Mills" [7] focused on the impact of the finishing strategy on tool wear and on the dimensional accuracy of the finished surface.

In terms of certain parameters, these studies correspond with the topic discussed here. However, none of them explores the relationship between a machining strategy and the use of a constant cutting speed, the dimensional accuracy, and the geometric accuracy of the finished surface. The purpose of this article is to present the current results of research into these aspects.

CA	Computer Aided
CAM	Computer Aided Manufacturing
CAD	Computer Aided Design
CNC	Computer Numerical Control
CMM	Coordinate-measuring machine
Z	Linear Z axis
a_p	Stepover
f_z	Feed per tooth
v_c	Cutting Speed
DIN	Deutsche Industrie-Norm (German Industrial Standard)
D_{tool}	Tool diameter

Table 1. Symbols

2. Finishing strategy

The key distinction between finishing operations and roughing and semi-finishing is in the requirement for achieving specified dimensional accuracy and quality of surface, ideally within the shortest time. These requirements are reflected in the operations offered by CAM systems.

In finishing, the main difference is in the choice between parallel machining and flow-line machining. For illustration, Figure 1 shows the results of the research report presented in [9] which covered this topic. They clearly indicate that the choice of strategy has an impact on the dimensional accuracy of the finished surface. [8]

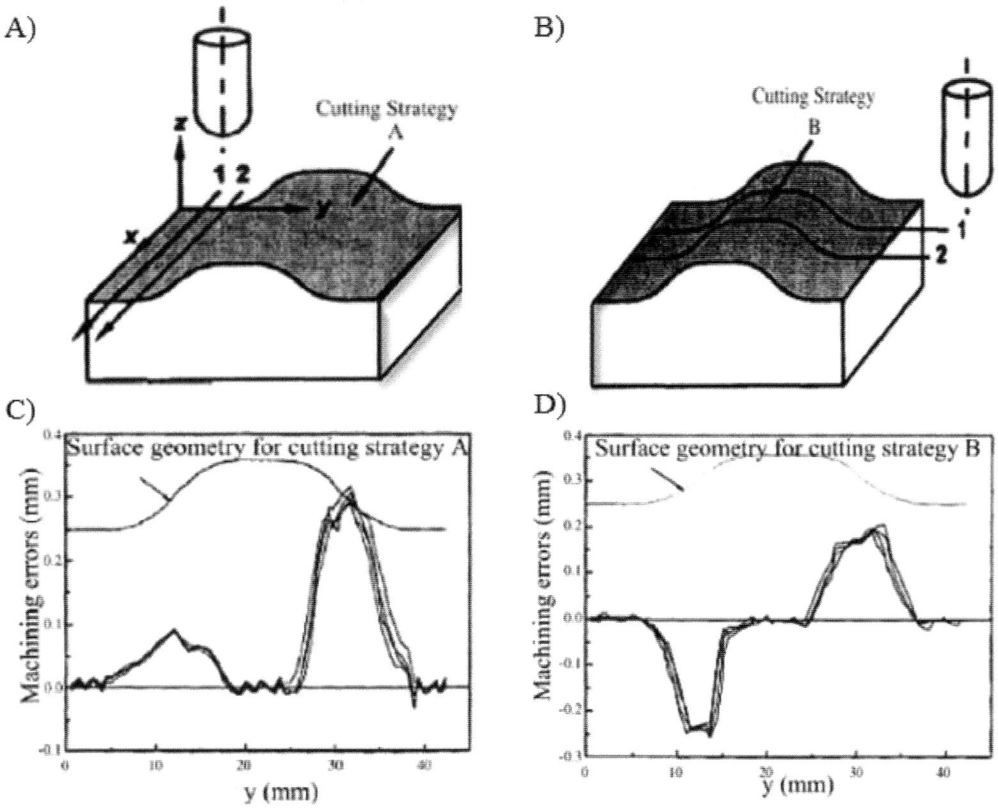

Fig. 1. Examples of stepover calculation strategies: A) Parallel machining strategy, B) Flow-line machining strategy, C) Results of parallel machining strategy, D) Results of flow-line machining strategy. [9]

However, this research did not deal with curve-based stepover calculation, and therefore it was not quite clear whether this parameter had an impact on the finished surface.

Machining methods	NC [KB]	Simulation diagram	Feed Path Length [mm]	Rapid Path Length [mm]	Cycle Time	Efficiency	quality
Parallel	515.8	3-a	2539.828	454.349	5m43.68s	③	great
Flowline	258.8	3-b	2127.253	17.000	4m43.83s	①	perfect
Contour	517.6	3-d	2095.197	820.479	4m48.43s	②	bad
Constant Scallop	742.4	3-e	1703.711	63.03	3m47.69s	/	worst
Constant Scallop&Shallow	1131.7	3-f	2640.632	941.057	6m00.01s	④	good

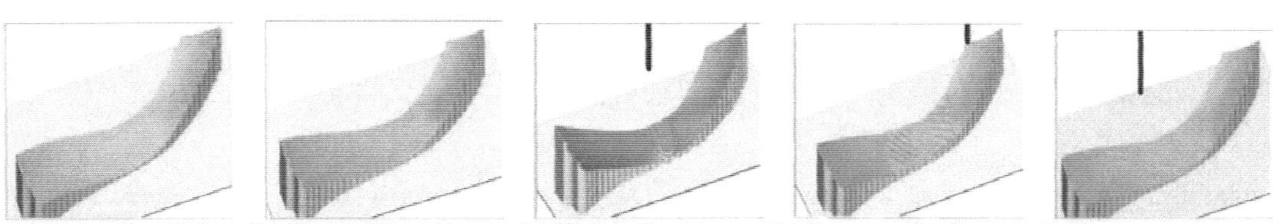

Fig. 2. Comparison between strategies applied to a shaped surface [3]

The topic is, however, described in the article "Tool-Path Scheduling for Free-Form Surface Based on MasterCAM" [3]. This study focused on the effects of stepover calculation but not on the impact on the accuracy of the finished surface, see Fig. 2. Nevertheless, it was clear that the type of stepover calculation had a fundamental bearing on the quality of the finished surface. In finishing of shaped surfaces, the choice of strategy is one of important aspects.

2.1. Role of rough-machined surface in the choice of finishing strategy

Dimensional accuracy, however, depends on many other variables and factors as well. One of them is the preceding manufacturing operation. There are smooth transitions between various parts of 3D surfaces, exhibiting different slopes and characters, depending on the surface complexity. The uniformity of semi-finished surface is manifested in the variation of the volume removed by the finishing tool. This is why the parameter represented by the distance between roughing or semi-finishing passes is so important, as shown in Fig. 3

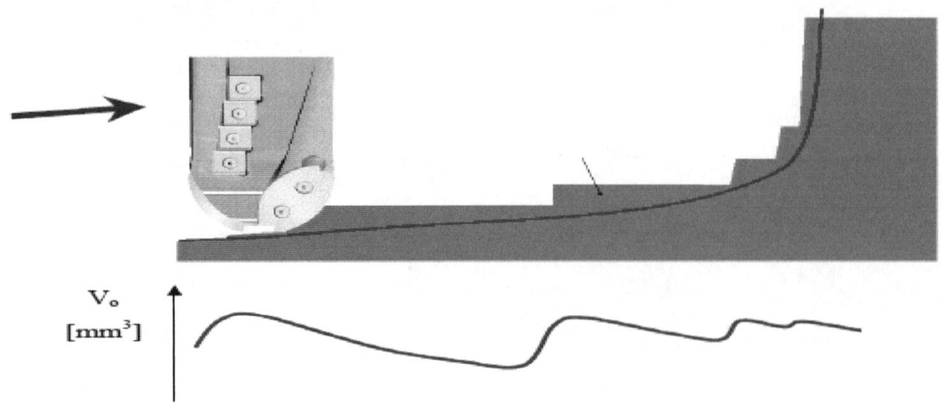

Fig. 3. Plot of volume removed profile [2]

This fact is well-known. Despite that, variation in the volume removed in finishing operations are, to a great extent, unavoidable. At the same time, the roughing stepover and/or semi-finishing stepover should be chosen with respect to the cutting process stability. Variation in the volume removed can lead to instabilities in the cutting process, as reflected, for instance, in the cutting force profile. Where cutting forces steeply increase, undesirable consequences may occur, such as out-of-tolerance and poor-quality finished surface and others.

These consequences can be addressed in several ways. One of them involves improved quality of semi-finished surface. However, that entails longer machining times, and therefore lower efficiency. Another one is based on using finishing tool paths with variable feed rates. This reduces cutting force fluctuations and improves the stability of the cutting process. [2, 4] The third option is to make finishing operations similar to adaptive roughing. The algorithm would evaluate the volume removed and if a certain limit was exceeded, it would generate such paths which prevent tool vibration.

3. Testing – effect of the choice of finishing strategy upon dimensional accuracy of finished surface

The purpose of finishing is to impart the workpiece its final form as specified on drawings. Besides the machined surface quality, another fundamental aspect is the dimensional accuracy of the finished surface. Dimensional accuracy is governed by numerous factors, including machine-tool stiffness, cutting tool type, workpiece clamping method and cutting conditions. [1, 5]

As the above sections demonstrate, the finishing strategy has a major impact on dimensional accuracy. For this reason, an experiment was designed in which curve-based stepover calculation is analysed, namely constant Z stepover calculation and constant stepover calculation along a curve. Furthermore, differences arising from the use of constant cutting speed and constant rotation speed were examined. The choice of constant rotation speed in cutting enables productivity for finishing operations to be increased.

3.1. Choice of test part

An experimental part was designed, as shown in Fig. 4. The part has several flat surfaces which slope at angles from 0° to 90°. It is a representation of a shaped surface and is suitable for straightforward evaluation of roughness and verification of dimensional accuracy. [1]

Experimental part	
Material [W.Nr.]	C45
Workpiece dimensions [mm]	60×90×100
Slope angles	0°,15°,30°,45°,60°,75°,90°

Table 2. Details of the experimental part

Details of the workpiece dimensions are given in Table 2.

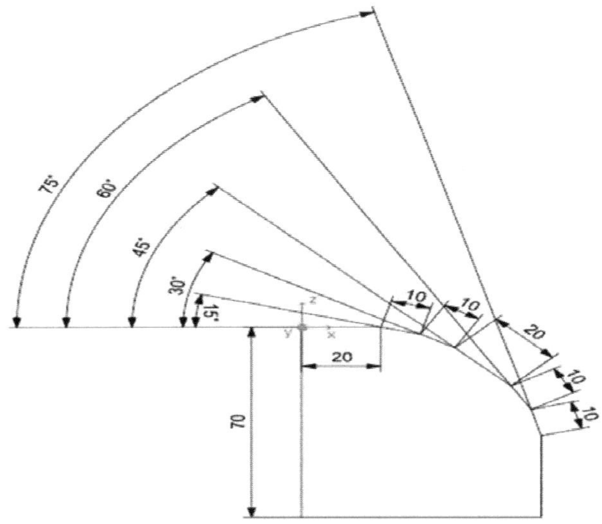

Fig. 4. Tested shaped surface [1]

3.2. Tool selection

A ball end mill was employed, which is used predominantly for finishing sculptured 3D surfaces. With such tools, the effective cutting diameter changes with the slope of the original surface. Therefore, the effective cutting speed varies when standard strategies are used in which rotation speed is based on the maximum diameter of the tool.

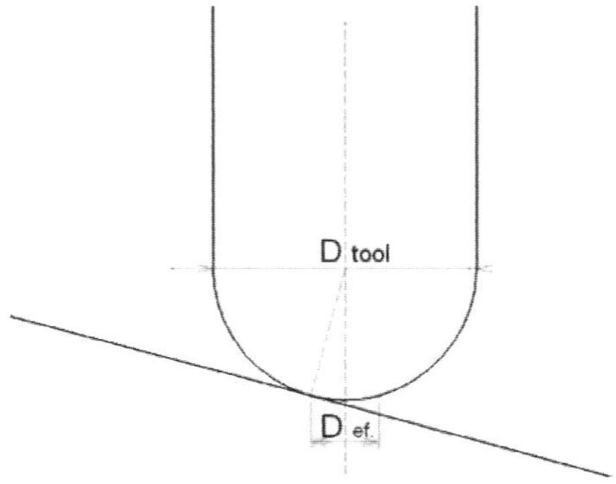

Fig. 5. Effective cutting diameter in cutting a sloping surface

Tool	
Supplier	ISCAR
Designation	HCM D16-A-L130-C16
D_{tool} [mm]	16
Number of teeth	2
Cutting conditions	
v_c [m/min]	150
a_p [mm]	0,2
f_z [mm]	0,2

Table 3. Tool and cutting conditions

3.3. Choice of strategy

In order to verify the assumptions, two tool path stepover calculation strategies were chosen:

- constant Z
- curve-based constant stepover calculation

Both strategies were implemented by manual programming. Therefore, any potential influence of a CAM system was eliminated and accurate stepover calculation for the tool path was obtained. Details of the experiment are given in Tables 3 and 4.

Cubes	Strategy
1	Stepover calculation along a curve – constant n
2	Constant Z stepover calculation – constant n
3	Stepover calculation along a curve – constant vc
4	Constant Z stepover calculation – constant vc

Table 4. Strategies chosen for individual cubes

Cutting forces were measured using a Kistler load cell. Dimensional accuracy was evaluated using a CMM instrument (Carl Zeiss Prismo 7 Navigator).

Evaluation of the experiment

In order to obtain objective data on dimensional accuracy, readings along three section planes were collected from each surface under test. This translates into 9 points for each surface under test. This is sufficient for subsequent evaluation. A mean value was calculated from these readings. Hence, seven values were obtained for each cube, i.e. one value for each machined surface.

The mean values from measured data for individual surfaces are plotted in the graphs. The graphs show that the total variance in cubes 2 and 4 is smaller than in cubes 1 a 3.

In cubes 1 and 3, the amount of rest material above the nominal surface increases uniformly up to the 4th surface, beyond which it decreases uniformly. By contrast, the amount of rest material above the nominal surface varies considerably in cubes 2 and 4.

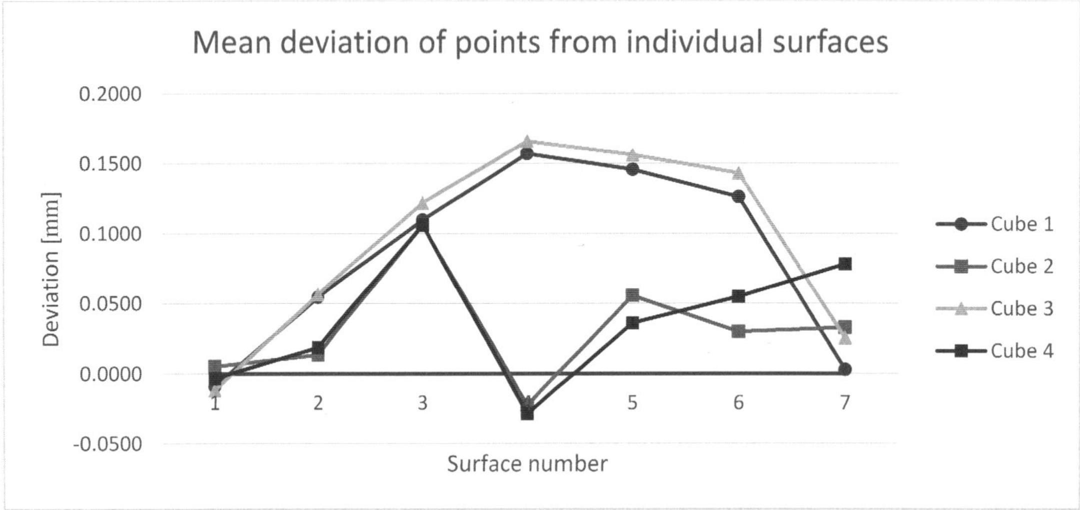

Graph 1. Average distance of points from individual surfaces

The cutting force values indicate that using a constant cutting speed can reduce the interval of cutting forces by up to several tens of newton's [N]. This is a relatively significant advantage from the perspective of finishing, where the maximal cutting forces are not of key importance because the purpose is to machine the surface as uniformly as possible.

Strategy	Cutting force interval [N]	Machine time for machining [min]
Stepover calculation along a curve – constant n	48.439	77
Constant Z stepover calculation – constant n	65.16	67
Stepover calculation along a curve – constant vc	28.85	47
Constant Z stepover calculation – constant vc	54.401	44

Table 5. Cutting force values and machining times for individual strategies

4. Summary

The measured data show that from the perspective of dimensional accuracy of the finished surface, the choice of strategy or tool path stepover calculation has a relatively significant impact. Both strategies used here, however, result in considerable deviations from the nominal surface. The research team considers the stepover calculation along a curve to be more favourable. This stepover calculation exhibits a uniform increase in the amount of rest material. In addition, it is easier to meet the prescribed dimensional tolerance with the stepover calculation along a curve. Better inaccuracy compensation is achieved with this strategy than with the other one where the profile varies considerably.

The use of constant cutting speed leads to a narrower interval of measured cutting forces. In addition, machining times are significantly shorter at constant cutting speed. One can therefore conclude that this option improves the efficiency of finishing.

The findings provide insight into the effects of finishing strategies which can lead to shorter machining times, better surface and, last but not least, control over dimensional accuracy. This article describes a pre-experiment which can be useful for developing a relevant methodology. The research team continues to explore this issue. Over a long term, further relationships between the parameters of finishing of shaped surfaces will be explored. On this basis, a methodology could be developed in future to simplify the choice of finishing parameters for specific requirements.

5. Acknowledgments

This paper includes results created within the SGS-2016-005: Research and development for innovations in the field of mechanical engineering technology - machining technology.

6. References

[1] Andrlik, J[iří]. (2017). Návrh strategie dokončování tvarových ploch - Draft of the strategy for the finishing contoured surfaces. Plzeň,. Diplomová práce. Západočeská univerzita v Plzni.

[2] Sova, F.; Hamouz, L. (2014). Softwarová optimalizace dráhy frézovacího nástroje při práci na čisto - Software optimization of the trajectory of the milling tool at work to finishing. In Strojírenská technologie - Plzeň. Plzeň: Západočeská univerzita. s. 1-15. ISBN: 80-7043-304-3

[3] Ren, W. J.; He, J. Y.; Yu, Y. (2014). Tool-Path Scheduling for Free-Form Surface Based on MasterCAM, Applied Mechanics and Materials, Vols. 556-562, pp. 1400-1403, Available from: https://www.scientific.net/AMM.556-562.1400

[4] Zuperl U.; Cus F.; Reibenschuh M. (2010). Neural control strategy of constant cutting force system in end milling [online]. [cit. 2017-5-29]. Available from: http://www.sciencedirect.com/science/article/pii/S0736584510001468

[5] Czerech, Ł[ukasz].(2013). Selection of Optimal Machining Strategy in the Manufacture of Elements Bounded by Curvilinear Surfaces. acta mechanica et automatica, 7.1: 5-10. Available from: http://www.degruyter.com/view/j/ama.2013.7.issue-1/ama-2013-0001/ama-2013-0001.xmla

[6] Sadílek, M. (2004). Vyspělé strategie ve 3D frézování - Advanced strategies in 3D milling. MM Spektrum. [online]., [cit. 2017-9-22]. Available from: http://www.mmspektrum.com/clanek/vyspele-strategie-ve-3d-frezovani.html

[7] Vopat, T[omáš]; Peterka, J[ozef]; Simna, V[ladimír] & Kuruc, M[arcel] (2014). The Influence of Different Types of Copy Milling on the Surface Roughness and Tool Life of End Mills, Proceedings of the 25th DAAAM International Symposium, pp.0108-0116, B. Katalinic (Ed.), Published by DAAAM International, ISSN 1726-9679, Vienna, Austria

[8] Vopat. T.; Peterka. J.; Kovac. M. & Buransky (2013). I.:The Wear Measurement Process of Ball Nose End Mill in the Copy Milling Operations. Proceedings of the 24th DAAAM International Symposium, B. Katalinic (Ed.), Published by DAAAM International, ISSN 1726-9679, Vienna, Austria.

[9] Lim, E.M.; Feng, H .Y.; Menq, C.H. & Z.H. Lin (1995). The prediction of dimensional error for sculptured surface productions using the ball end milling process. Part 1. Chip geometry analysis and cutting force prediction Int. J. Mach. Tools Manufact, 35 (8) (), pp. 1149–1169

[10] Tunc, L[ütfi]; Ozkirimli, O[mer]; Budak, E[rhan] (2015). Machining strategy development and parameter selection in 5-axis milling based on process simulations. The International Journal of Advanced Manufacturing Technology, 1¬18. Available from: http://link.springer.com/article/10.1007%2Fs00170¬015¬8001¬6

[11] Nittler, K[evin] M. & Faroukil, R[ida] T. (2016). A real-time surface interpolator methodology for precision CNC machining of swept surfaces. The International Journal of Advanced Manufacturing Technology, 83.1¬4: 561¬574. Available from: http://link.springer.com/article/10.1007%2Fs00170¬015¬7552¬x

DOI: 10.2507/28th.daaam.proceedings.082

DETERMINATION AND ANALYSIS OF ENERGY EFFICIENCY POTENTIAL IN SOCKS MANUFACTURING SYSTEM

Milovan Medojevic, Milana Medojevic, Ilija Cosic,
Milovan Lazarevic & Dusanka Dakic

This Publication has to be referred as: Medojevic, M[ilovan]; Medojevic, M[ilana]; Cosic, I[lija]; Lazarevic, M[ilovan] & Dakic, D[usanka] (2017). Determination and Analysis of Energy Efficiency Potential in Socks Manufacturing System, Proceedings of the 28th DAAAM International Symposium, pp.0582-0591, B. Katalinic (Ed.), Published by DAAAM International, ISBN 978-3-902734-11-2, ISSN 1726-9679, Vienna, Austria
DOI: 10.2507/28th.daaam.proceedings.082

Abstract

In this study a systematic approach for determination and analysis of energy efficiency potential in socks manufacturing system was applied in order stimulate rational energy consumption. Moreover, an overview of socks manufacturing process was provided upon which significant energy uses in form of systems/processes/operation were identified. Although, all significant energy users should have appropriate energy sub-meters which can then be simply used to quantify the consumption of each use, in reality, few or none of energy uses are sub-metered, indicating that their consumption should be estimated. Consequently, a variety of models and statistical methods which can be used to describe energy use in manufacturing systems were recognized, while a general multivariable regression method for measuring industrial energy efficiency potential that takes into account changes in weather and production volume as major drivers of energy use in majority of manufacturing systems was applied in order to reveal opportunities for improving productivity through energy efficiency.

Keywords: Energy efficiency; Socks manufacturing; Multivariable regression

1. Introduction

It is generally known that textile industry is among the most complicated manufacturing industries mainly because it is a fragmented and heterogeneous sector dominated by small and medium enterprises (SMEs). However, characterizing the textile manufacturing industry is complex because of the wide variety of substrates, processes, machinery and components used, as well as finishing steps undertaken, where different types of fibres or yarns, methods of fabric production and finishing processes (preparation, printing, dyeing, chemical/mechanical finishing, and coating) all interrelate in producing a finished product [1].

On the other hand, it was confirmed by numerous studies and research projects, summarized by Hasanbeigi & Price [1] that energy is one of the main cost factors in the textile industry. Dealing with energy issues is complex due to the fact that majority of energy forms are intangible or insensitive most times, while always representing the ability to perform

work. Having this in mind, determination of energy efficiency of a system or process is an essential step towards the control of the energy consumption and energy costs.
Consequently, research on industrial energy use and energy efficiency potential in various manufacturing sectors [2-6], as well as analysis concerning definition and determination of energy efficiency potential pertaining to a process [7-11], led to divulgation of variety energy efficiency indicators, such as thermal efficiency, specific energy consumption, energy intensity index, etc.
Here, the most widely used indicator in industry is the specific energy consumption (SEC) of a given output (or input) [7], in which case despite the simple definition, several complicating factors arise in its implementation [7-11], such as:

- Complex manufacturing facilities with multi production lines;
- Variety of products distinguished by their energy intensity factor;
- Relevant variables that affect energy efficiency and consumption;
- Suitability for SEC indicator application;
- Unlike thermal efficiency, the SEC indicator lacks information on whether energy is used efficiently;
- Definition of suitable system boundaries to ensure that all energy users are considered equally and precisely.

Moreover, majority of textile manufacturing companies still lack appropriate methods to effectively address energy consumption in a comprehensive and practical manner [11].

Although seemingly obvious, energy savings represent energy that is not used. However, it is quite complicated to measure them directly, except in some cases, such as a straightforward energy conversion process where savings are calculated from improvement of the ratio between measured output and input [12]. Having this in mind, this paper is based on a systematic approach with aim to highlight the methods and analysis which foster energy efficiency prioritization on the example of one socks manufacturing system.

2. Overview of socks manufacturing process

Sock manufacturing is intensive and widespread branch of textile industry. Socks of various types and designs can be produced by using cotton, acrylic, polyester, and nylon yarns of different counts, quality and shades. In general, socks manufacturing process could be divided into five steps. The process begins with knitting, then seaming, wet finish, board pairing and packaging respectively.

Knitting is the basic section from where the production of the socks starts. Here, the high tech circular knitting machines use a series of knitting needles in a cylinder formation. The yarn is fed to the needles row after row through the stands, where horizontal rows are called courses, while the vertical rows of stitches are called wales. The knitting section follows the size specifications and yarn description provided by the sampling section. Maintaining the quality of the socks is parallel activity of the knitting section controlled by the quality manager of the factory. The knitting section works according to a shift wise schedule and issues the production to both, batch section and greige store. The production which is to be batched immediately is sent to batching section and production which is to be batched later is directed to the greige store.

After the socks are knit, the toe seams must be closed. In order to achieve this, socks move to the next operation known as seaming. Once it leaves the knitting section and after its first quality inspection, the socks are funnel to the seaming area. Here the toe opening is closed. Although there are many ways to seam a sock, the selection is often based on two factors. These factors are the quality of the seam itself and the comfort specified by the wearer. Subsequently, the fibers and the sock's intended use are considered to determine the appropriate seaming process upon which seaming operators align the socks in the machine for seaming. The socks are usually placed inside out for the process where the seam will be sown. In addition, extra fabric must be removed for a comfortable toe seam, after which the machine turns the sock right side out as the final step.

Once seaming is completed, the work items now look like socks and they are ready for the next step which is Wet Finish Process. Wet finish process can involve several operations where some of them are common for all socks, while specific socks acquire additional operations and treatment. In addition, socks are washed and dried after the knitting and seaming due to many reasons. Washing removes knitting oils and residue from yarns. It allows adding softeners and conditioners to soften them. Also, drying helps certain fibres to adapt for further treatments and ensures bulk increase to some socks depending on given requirements. The cleaning, conditioning, and softening products and amounts used vary depending on the type of socks where, the temperature of the water and drying are important drivers in providing the optimum quality. Wet Finish area also includes dyeing, scouring, and bleaching operations. Socks that are to be white are either bleached or scoured. Bleaching is an oxidizing process that removes colour from a sock, leaving it white. On the other hand, scouring is a soap bath that removes tints and impurities. Lastly, coloured socks are either dyed after knitting or knit with previously dyed yarn. Already dyed yarn is also known as in-grain yarn. By the use of in-grain yarns, many attractive colour blends and combinations are possible.

After the Wet Finish, socks are ready for the next step which is Boarding and Pairing. This process consists of three sub processes: boarding, pairing, and comprehensive quality check. During boarding process, the socks are pulled on a flat metal foot forms. The forms represent the desired shape and size of the socks. Then, the forms are steam pressed between two heated surfaces in order to shape the sock final appearance. Thereafter, the freshly boarded socks are paired. Having in mind that even socks knit with the same yarns, same machines and under the same settings vary slightly they must be adequately paired before they proceed to next operation. Therefore, pairing process matches socks with the same size variation. The Board/Pair operation is the last phase in the socks manufacturing. Quality issues that are found in this process are traced back through the manufacturing procedure to the source, documented, and corrected. Once the socks are boarded and paired, the next step is packaging. The paired socks are either sent to the packaging line for immediate shipment or to fulfil future orders. In practice, packaging is driven by requirements of customers.

Above mentioned processes and operations represent the basic activities in the socks manufacturing. However, manufacturing process may vary depending on variety of factors, such as materials used, available equipment, specific requirements from customers, etc. Therefore, for the purpose of this study, socks manufacturing process given in the form of block diagram shown in the figure 1 was a subjected to further analysis. Even though the analysis could be deeper and more complex it stimulates logical identification for potential optimization spots, system effectiveness and process energy efficiency improvements [13].

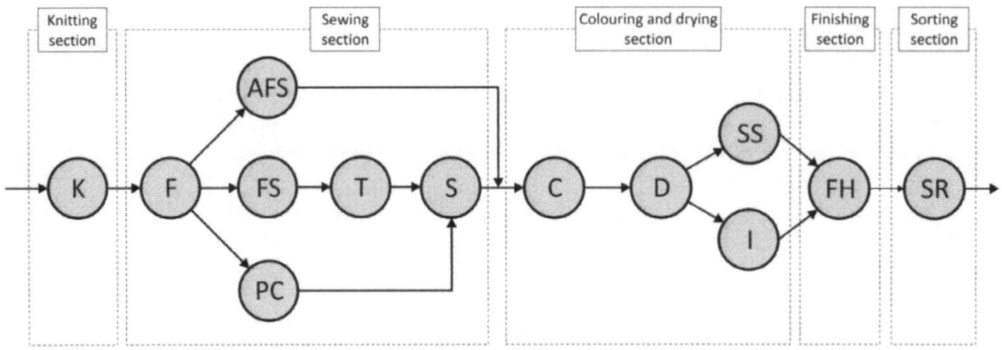

Fig. 1. Socks manufacturing operations flow diagram by relevant sections

In order to better understand the position of the operations identified by the process flow diagram (Fig. 1), they are shown on the factory layout given in the figure 2.

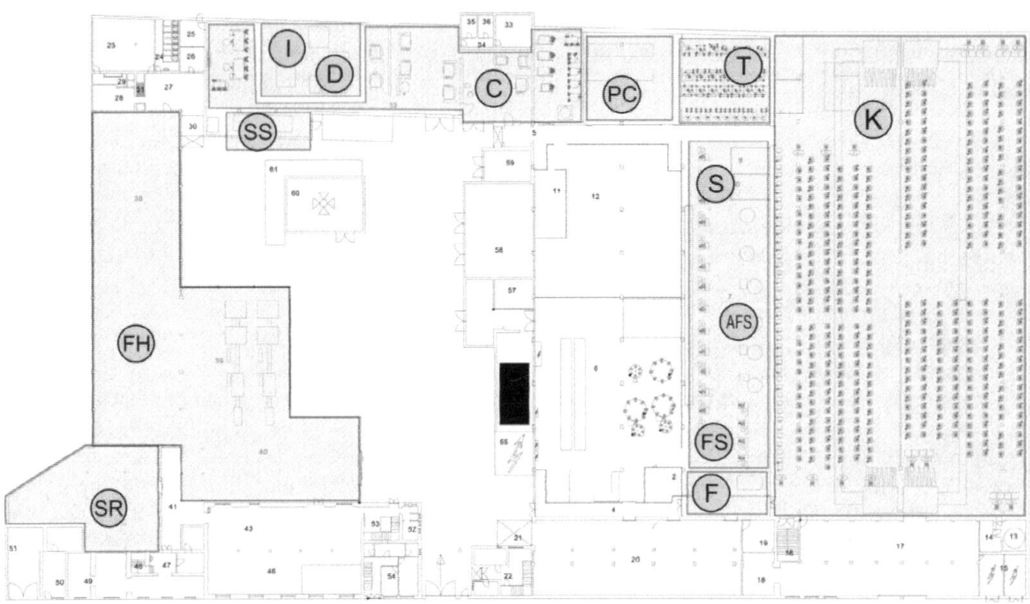

Fig. 2. Factory layout with the position of operations

Here the process starts form knitting (K), after which the socks are directed to Fixing machine (F) in order to improve the quality of socks and pads by thermal treatment. Afterwards, work items (pads and socks) continue to Fingers Sewing/Seaming (FS), Pad Cutting (PC) and Automated Fingers Sewing and sock fusing (AFS). PC is performed according to previously adopted specification (saddle, small, large, etc.) and after quality check pads are forwarded to sewing operation (S) where they are being attached to socks and sent for colouring (C). Simultaneously, AFS operation is fully automated and upon its completion work items are directly sent to C. Bearing in mind that FS differs from AFS, the tailoring operation (T) is introduced in order to perform sock cutting and altering. After T, this group of work items is sent to S and then to C. Colouring machines are divided into two basic groups. The first group consists of machines in which the fleet is stationary and work items move (these machines are used to paint thicker socks), while the second group consists of machines in which the work items are stationary and the fleet moves (On these machines more demanding items are coloured, such as liquor, compression, stretch, etc.). After C, it is necessary to dry coloured work items in the drying machines associated to drying operation (D). Some of dried socks are then directed to sock shaping (SS) in order to form a final shape of wrinkled socks by exposing them to microwaves, while others (for which SS is not necessary) are forwarded to steam and electric ironing (I). Thereafter, formed socks are subjected to quality control for scrap elimination and packaging in the finishing operation (FH). Lastly, socks which have passed quality check are being sorted (SR) according to the specification.

Due to the variety and complexity of the processes involved in the socks manufacturing process, there are too many operations to be explained. Therefore, brief descriptions of the major processes and operations relevant from the energy consumption point of view are considered hereafter.

3. Energy in the socks manufacturing

In the observed manufacturing process, large quantities of both electricity and fuels are used. More precisely, in knitting, seaming and sewing processes electricity is being used by variety of machines in order to perform desired work, as well as to produce compressed air which is necessary for the machines operation. Moreover, electricity is also used to ensure cooling during summer periods. On the other hand, in wet processing the major energy source is natural gas, which is mainly used for the steam production, required by the machines associated to operations such as colouring and steam ironing, as well as to provide heating of factory in winter periods.

Therefore, before any action taken towards increasing energy efficiency level of observed system, it is essential to conduct an overall site condition survey where attention should be given to identification where energy is obviously being used, or in other words, to identify spots of significant energy uses (SEUs). In order to identify SEUs, it is necessary to determine how much energy each process or system uses. In an ideal case, all large energy users will have appropriate energy sub-meters which can then be simply used to quantify the consumption of each use. In reality, few or none of energy uses will be sub-metered. In that case, their consumption should be estimated. This routine should be carried out and regularly updated for each energy source, i.e. electricity and each fuel type. However, in most cases it may be more appropriate to think in terms of processes or systems rather than pieces of equipment. Moreover, grouping equipment by energy systems (e.g. process heating, compressed air, steam systems, etc.) represents an important best practice, while understanding the dynamics of energy use in a system will lead to optimal energy savings. To achieve this, inventory and technical documentation analysis should be performed for each operation identified in Fig. 1. In addition to identified equipment energy requirements, it is also necessary to take into account other energy consuming equipment/process/system which has indirect influence on energy consumption besides manufacturing process.

3.1. Identification of SEUs

Given the aforementioned, a total of 490 circular knitting machines were installed in the knitting section. These machines can be divided into 9 groups while their relevant energy-related characteristics are briefly summarized in Table 1.

	Equipment Name	Quantity	Installed power		Compressed air consumption	
		[pcs]	[kW/pcs]	[kW]	[l/min/pcs]	[l/min]
	Knitting machines					
1	Santoni Mirabella 2	40	1.37	54.8	195	7800
2	Lonati L – 301 GE	118	1	118	40	4720
3	Lonati L – 412	146	1	146	41	5986
4	Lonati L10P5	11	0.55	6.05	33	363
5	Lonati La04MJ	10	0.55	5.5	33	330
6	Lonati L04MJ	11	0.55	6.05	33	363
7	Lonati La10P6	31	0.55	17.05	33	1023
8	Lonati La10P7	113	0.55	62.15	33	3729
9	Lonati La24E7D	10	0.55	5.5	33	330

Table 1. Energy-related characteristics of machines associated to knitting section

Based on the data given in Table 1, total installed power of knitting machines amounts 415.05 kW, while the maximum compressed air consumption amounts 24644 l/min. Furthermore, the sewing section include several operations such as F, PC, FS, AFS, T and S, while the typical machines and their relevant energy related characteristics are given in the Table 2.

	Equipment Name	Quantity	Installed power		Compressed air consumption	
		[pcs]	[kW/pcs]	[kW]	[l/min/pcs]	[l/min]
	Sewing machines					
1	Speedomatic	18	10	180	Data are not reliable	Data are not reliable
2	Flattlock sewing head	20	0.55	11		
3	Union sewing head	8	0.55	4.4		
4	Santoni lc320	2	3.5	7		
5	Santoni lc320 + TCR	3	12	36		
6	Takatori lc320	1	3.7	3.7		
7	Takatori lc320 + TCR	1	4.5	4.5		
8	Solis Turbo 11	1	5	5		
9	Solis 4C - George	3	11	33		
	Fixing machines					
1	Grandis 300 (Fixer)	1	7	7		
2	Low capacity fixer for pads	1	3	3		

Table 2. Energy-related characteristics of machines associated to sewing section

Sewing section with identified machines and associated operations, engages 294.6 kW of power. Subsequently, colouring and drying section refers to operations C, D, I and SS. Characteristic machines necessary to carry out these operations were identified and their relevant energy related properties are summarized in the Table 3.

	Equipment Name	Quantity	Installed power		Water consumption	
		[pcs]	[kW/pcs]	[kW]	[l/day/pcs]	[l/day]
	Coloring machines					
1	Borac 1 (900 l)	2	2	8	5400	10800
2	Borac 2 (1100 l)	2	5	10	6600	13200
3	Colori (40-45 kg)	7	15	105	3000	21000
4	Colori (1-5 kg)	1	6	6	300	300
5	Colori (5-10 kg)	1	7	7	1200	1200
6	TEN 1 (1000 l)	4	5	20	12000	60000
7	TEN 2 (2000 l)	2	7	14	12000	24000
8	TEN 3 (6000 l)	2	12	24	18000	36000
	Centrifuging machines					
1	Centrifuge	2	15	30	/	/
	Drying machines					
	Equipment Name	**Quantity**	**Installed power**		**Steam consumption**	
		[pcs]	[kW/pcs]	[kW]	[t/day/pcs]	[t/day]
1	Mielle 1 (electric driven)	1	15	30	/	/
2	Mielle 2 (steam driven)	1	3	3	Data are not reliable	Data are not reliable
3	Pasat (steam driven)	2	3	9		
4	RF System	1	50	50	/	/
	Ironing machines					
1	Technopea SD12	4	30	120	No data	No data
2	Cortesse	2	10	20	/	/

Table 3. Energy-related characteristics of machines associated to coloring and drying section

Overall installed power of colouring and drying section amounts 456 kW, which indicates that this section is the most intensive in the use of electricity. Having in mind that finishing section does not include any machine work, a more detailed analysis was not performed. However, a certain, indirect energy consumption related to FH and SR operations comes from lighting fixtures. Therefore, in the FH operation 135 fluorescent lamps are identified while their overall installed power amounts 4.86 kW. Similarly, 90 fluorescent lamps are identified in the SR with overall installed power of 3.24 kW.

Due to the fact that there are no reliable data regarding steam and compressed air consumption of specific operations in sewing and colouring sections, it can be assumed that the energy of steam and compressed air are not used rationally. The basis for this assumption lies in fact that the intensity of their use is not quantified or analyzed.

Beside SEUs identified in the manufacturing process, relevant spots of considerable energy uses are cooling and air conditioning system, compressor station, boiler station, lightning system, locksmith workshop, as well as office equipment. Therefore, their relevant energy-related properties are summarized in the table 4.

	Equipment Name	Quantity [pcs]	Installed power [kW/pcs]	Installed power [kW]
	Cooling and air conditioning system			
1	Vacuum fan (1)	7	15	105
2	Vacuum fan (2)	8	35	280
3	Chiller compressor	1	730	730
4	Chiller pump	1	19	19
5	Chiller fan	4	15	60
6	Industrial fan	9	0.5	4.5
	Cooling and air conditioning system in total			**1198.5**
	Compressor and Boiler station			
1	Boiler burner	2	1	2
2	Boiler pump	2	8	16
3	Waste press	1	4	4
4	Compressor (1)	2	132	264
5	Compressor (2)	1	110	110
6	Air dryer	1	20	20
7	Well pump (1)	2	5.5	11
8	Well pump (2)	2	1.5	3
9	Circulation pump (1)	2	4	8
10	Circulation pump (2)	1	5.5	5.5
11	Industrial fan	2	0.5	1
	Compressor and Boiler station in total			**445.5**
	Lighting fixtures			
1	Lighting fixtures (K)	1266	0.036	45.5
2	Lighting fixtures (S)	480	0.036	17.28
3	Lighting fixtures (C)	130	0.036	4.68
4	Lighting fixtures (FH)	135	0.036	4.86
5	Lighting fixtures (SR)	90	0.036	3.24
6	Lighting fixtures (Office)	50	0.036	1.8
7	Lighting fixtures (Other)	30	0.036	1.08
	Lighting fixtures in total			**78.44**
	Locksmith workshop			
1	Milling machine	1	1.1	1.1
2	Lathe (1)	1	2.5	2.5
3	Lathe (2)	1	4	4
4	Drill	2	1.5	3
5	Welding machine (1)	1	2.5	2.5
6	Welding machine (2)	1	2	2
7	Rendering machine	1	2.5	2.5
8	Grinding machine	1	2	2
9	Grinding stone	1	1	1
10	Frame saw	1	1	1
	Locksmith workshop in total			**23.6**
	Office equipment			
1	Air conditioner (1)	6	2.8	16.8
2	Air conditioner (2)	2	3.4	6.8
3	Water heater (1)	8	2.5	20
4	Water heater (2)	2	10	20
5	Desktop computers	37	0.4	14.8
	Office equipment in total			**78.4**

Table 4. Relevant energy-related properties of processes and systems indirectly linked to manufacturing

Bearing in mind the given data (Tables 1-4), SEUs and their share in overall electricity consumption depending on the engaged power were identified and illustratively given in the form of pie chart (Fig. 3). Here, the cooling and air condition systems represent most intensive electricity consumers with a share of 40%. In addition, manufacturing participates with a share of 39%, where most energy demanding sections are colouring and drying (39%), followed by knitting (36%) and sewing (25%) section respectively. Subsequently, compressor and boiler station are characterized by a share of 15%, while the least energy demanding SEUs identified are lightning fixtures (3%), office equipment (2%) and locksmith workshop (1%).

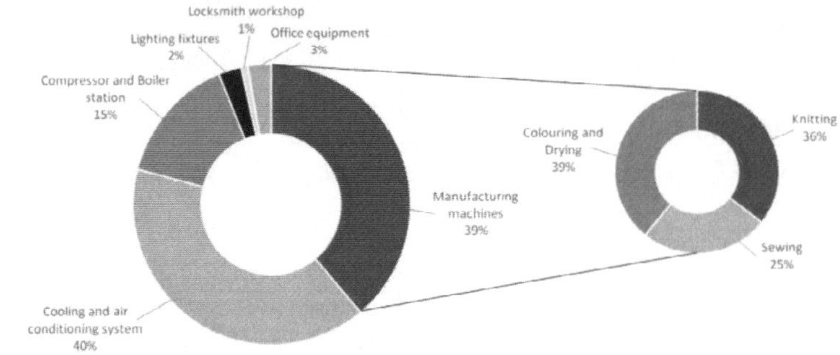

Fig. 3. Identified SEUs and their share in overall electricity consumption depending on the engaged power

3.2. Determination of energy efficiency potential in socks manufacturing

Accurate determination or measurement of energy consumption and industrial energy efficiency projects as well, can reduce uncertainty and stimulate development and implementation of future projects, while generating more reliable and more accurate estimations of expected savings, which finally lead to improved utilization of capital resources. Moreover, many efforts to determine industrial energy saving potential, or simply track progress toward efficiency goals, have had difficulty incorporating weather and production changes, which are frequently major drivers of energy use in majority of manufacturing systems. As the energy use often represents a function of weather and/or production, which frequently changes between the pre and post retrofit period, it is more difficult to measure energy savings and, as a consequence, savings are verified sparsely. This lack of verification hurts the effort to maximize industrial energy efficiency. In some cases, retrofit measures which would realize the expected savings are not implemented since there is no history of successful verification. In other cases, retrofits that do not achieve the expected savings get implemented, which wastes resources that may have been allocated to more effective measures. Both of these problems could be minimized by measuring savings systematically in order to compare expected and measured savings, where the adequate information could guide the selection of future retrofits, improvement of methods to calculate expected savings, promotion of financing energy efficiency through shared-savings agreements and utilization of resources [12].

Having this and all previously mentioned in mind, in this paper a general method for industrial energy efficiency potential determination was presented. At the very beginning, it is of high importance to establish the crucial physical components or subsystems for both energy and production that play a key role in the supply and demand side of overall process operation. In other words, when energy and process flow charts are put together, valuable information are provided on where, why and what type of energy is used. This represents the basis for decisions on identifying and determination of relevant measuring spots in order to monitor the process towards to energy efficiency and rational energy use [13]. Having this in mind, figure 4 illustrates the scope of this research as correlation between energy consumption and manufacturing process or other relevant energy consuming system in the observed factory. Thereafter, beside to the only 2, legally binding, measuring points, 30 potential measuring spots were identified. More precisely, 17 power, 5 steam flow, 7 air flow and a water flow data logging spots were suggested in order to monitor relevant parameters causing energy consumption. Subsequently, it is necessary to understand trends in energy consumption and in potentially relevant variables. After visibly assessing aforementioned, the significance of the relationship could be accessed by correlation of specific variables against energy consumption using a simple X-Y diagram. Here, if the variable is relevant, one expects to see evidence of a relationship in the scatter of points. In other words, if the points appear to be scattered around a mathematical function, shown as a trend line then this is indicative of the presence of relevant variables. Otherwise, if the points appear as a random cloud with no evident relationship, the variable is likely not relevant. Based on the above, the figure 5 illustrates correlation between total energy consumption reduced to kWh and production volume (left), as well as correlation between solely electricity consumption and production volume (right) in the observed factory. Similarly, figure 6 provides an overview regarding the correlation strength between natural gas consumption and production volume. Having in mind that the coefficient of determination (R^2) explains how many points fall on the regression line, it represents a measure of the goodness how trend line fits to the data, where a value of 1 indicates an ideal case, or a perfect fit respectively. In all three cases given in the figures 5 and 6, R^2 value does not exceed 0.31 indicating a very week correlation between energy consumption and production volume. More precisely, R^2 values of 0.31, 0.29 and 0.21 indicate that only 31%, 29% and 21% fits to the data from which it can be concluded that energy consumption does not depend on production. However, it is important to accentuate that when production volume is equalled to 0, baseload total energy consumption amounts 697.2 MWh/month. Of this 697.2 MWh/month, 393.64 MWh/month represent electricity consumption, while 303.55 MWh/month (41411.13 m^3/month) is consumed as natural gas. The ways the points appear as a random cloud with no evident relationship indicate a potential for energy efficiency improvement of observed manufacturing system.

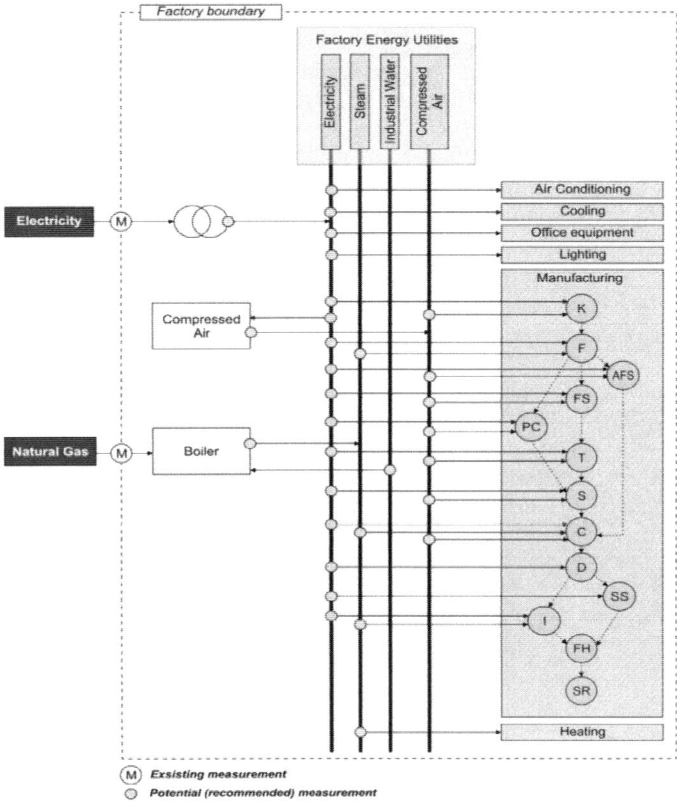

Fig. 4. Process and energy flow chart of a socks manufacturing factory with present and recommended measuring spots

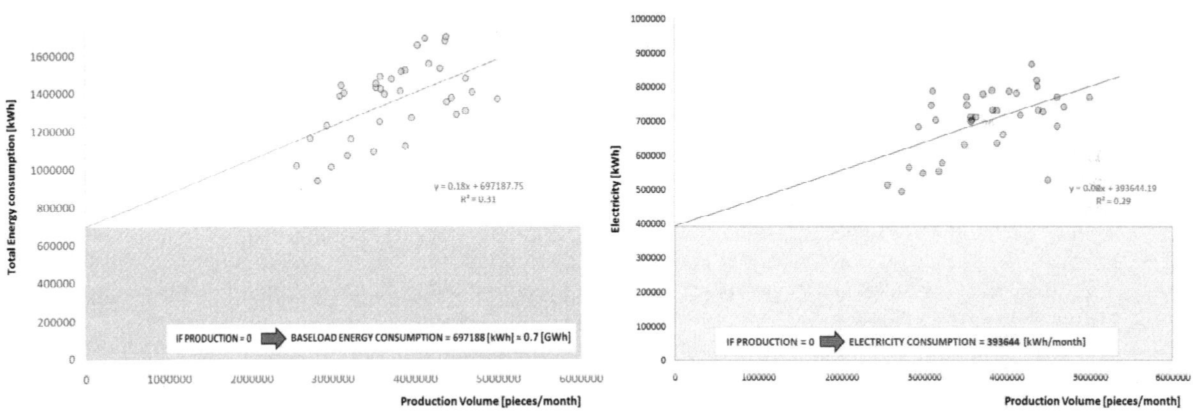

Fig. 5. Correlation between energy consumption and production volume in the observed factory

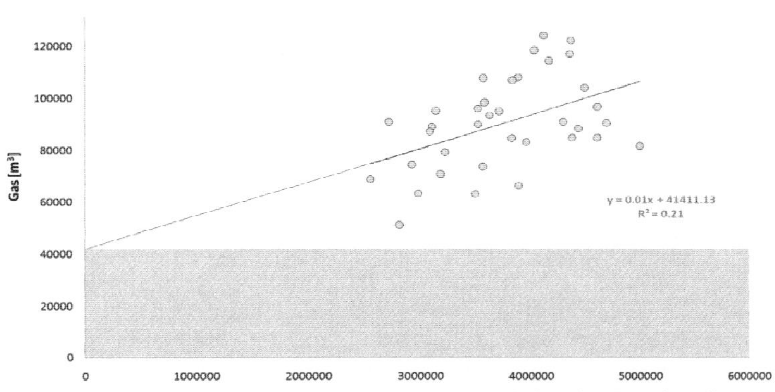

Fig. 6. Correlation between natural gas consumption and production volume in the observed factory

Subsequently, a variety of models and statistical methods can be used to describe energy use in manufacturing systems. For example, neural network models can accurately capture non-linear relationships and cross correlation among multiple independent variables, while Principal component analysis could be used to handle multi co-linearity associated with time series data. In addition, other examples of empirical modelling of industrial energy use introduced a productivity index to understand productivity, efficiency and environmental performance [12]. Although these methods all have appropriate applications, a multivariable regression modelling technique is chosen to identify segments for best practices application. Here, a general multivariable regression method for measuring industrial energy efficiency potential that takes into account changes in weather (through Heating degree days (HDD) and Cooling Degree Days (CDD)) and production was applied while the obtained results are given in the table 5.

Regression Statistics		**ANOVA Regression**				
Multiple R	0.75	*df*	3			
R Square	0.57	*SS*	$5.74\ e^{11}$			
Adjusted R Square	0.50	*MS*	$1.91\ e^{11}$			
Standard Error	148213	*F*	8.71			
Observations	24	*Significance F*	0.000675			
Interpret regression coefficients						
	Coefficients	*Standard Error*	*t Stat*	*P-value*	*Lower 95%*	*Upper 95%*
Intercept	447266.99	209036.81	2.14	0.045	11223.84	883310.14
Production volume	0.19	0.049	3.86	0.00098	0.088	0.29
HDD	661.3	299.16	2.21	0.039	37.27	1285.32
CDD	511.11	523.51	0.98	0.34	-580.92	1603.14

Table 5. Results of multivariable regression analysis

4. Discussion and results interpretation

The Regression analysis is carried out in order to determine the coefficients that yield the smallest residual sum of squares, which is equivalent to the greatest correlation coefficient squared. Moreover, the ANOVA probability (Significance F) that generated equation does not explain the variation in y, i.e. that any fit is purely by chance amounts 0.000675 indicating a quite meaningful correlation based on the F probability distribution. Also, the probability that the true value of the coefficient has the opposite sign to that found (P-value) suggests that variables in terms of production volume and HDD really influence energy consumption in the observed factory, which is not the case for CDD. In other words, this means that significant potential for energy efficiency improvement lies in the cooling and air condition systems retrofitting, which at the same time represent the most significant SEU. Thereafter, a 95% probability, that the true value of the coefficient, lies between the Lower 95% and Upper 95% values in the section of interpret regression coefficients (Table 5.). This probability is 2.5% that it lies in below the lower value, and 2.5% that it lies above. In addition to statistically verified data, a reduction of 30% in energy consumption is typical and achievable by improving operation control in managing compressed air system (CAS) in the observed factory by identifying and repairing air leaks, measuring and baselining CAS to determine the operating costs and efficiency or operating compressed air system at the lowest practical pressure. Also, other highly effective practical recommendations suggest installation, adjustment and maintenance of automatic system controls to coordinate operations of air compressors, cutting of compressed air supply to zones, equipment and applications that are not operating, using a blower rather than a compressor where and when appropriate, implementation of air reservoirs to reduce compressor cycling and respond to peak air demands, etc [14]. Beside CAS optimization, huge potential for improving energy efficiency was found in colouring and drying section of manufacturing where on daily basis 166.5 tons of hot water (98°C) is being released into the sink. By implementing an adequate system of heat exchangers in order to recover heat from the fleet which is being released to sink to preheat the water entering the steam generation boiler to the temperature of only 40°C, additional 30% reduction could be achieved in the natural gas consumption for steam generation. Lastly, although this method seeks to extract as much information about savings as possible from easily obtainable utility billing, production and temperature data, the extractable information is limited by the information in the data set, which is sparse in the both the system and time domain. This emphasizes the need for data logging at recommended measuring spots identified in the figure 4.

5. Conclusion

In this study, a straightforward structured framework has been proposed and applied to determine energy efficiency potential of a socks manufacturing system. The methodology has been illustrated through a concrete case in the example of an energy intensive branch of textile industry. The energy efficiency potential in whole energy system of a factory has been assessed through a statistical and deterministic approach.

Given the aforementioned, several limitations regarding the applied approach should be taken into account. Firstly, due to a variety and complexity of the processes involved in the socks manufacturing process, there are too many operations to be explained. Therefore, only the major processes and operations relevant from the energy consumption point of view were subjected to analysis. In addition, although in most cases it may be more appropriate to think in terms of processes or systems rather than pieces of equipment, grouping equipment by energy systems (e.g. process heating, compressed air, steam systems, etc.) represents an important best practice, where understanding the dynamics of energy use in a system could lead to optimal energy savings. Here, it is hard to understand the dynamics of energy use if not tracking the relevant changes over time, which points to the next important limitation, which is time awareness. It is important to simultaneously be aware when, where, why and what type of energy is used in order to make the best possible decision regarding improving productivity through fostering energy efficiency. Consequently, ensuring basic measurements in production segments where activities or production volume are quantifiable and where a significant amount of energy is used, valuable information are provided to understand the dynamics of energy use. Subsequently, the last limitation refers to statistical approach based on general multivariable regression method itself, due to the fact that, although the main drivers of energy consumption, only changes in weather and production were variables considered in this study. Also, even though this statistical approach indicates energy efficiency potential, it does not provide explanation what should be done to improve efficiency. However, it is believed that integration of other variables could provide clearer overview on how energy is actually being used up on which relevant energy flows could be identified and monitored carefully. Lastly, the methodology presented in this paper consists of a wide set of tools for energy management, which can be implemented within the systematic framework in order to reveal the energy efficiency potential necessary to foster the rational energy utilization in the observed manufacturing system, while plans for future research in this field are aimed to complement this methodology by revealing and understanding potential limitations, which are not identified at present. On top of that, the case study presented in this paper refers to single product processes, while the same methodology could be applied for multiple product processes, where it is more complex to comprehend how energy is provided and finally how energy consumptions allocation is stipulated among them.

6. References

[1] Hasanbeigi, A., Price, L. (2012). "A review of energy use and energy efficiency technologies for the textile industry", Renewable & Sustainable Energy Reviews., Vol. 16, No. 6, pp. 3648–3665, ISSN: 1364-0321.

[2] Gielen, D., Taylor, P. (2009). "Indicators for industrial energy efficiency in India", Energy, Vol. 34, No. 8, pp. 962–969, ISSN: 0360-5442.

[3] Palamutcu, S. (2010). "Electric energy consumption in the cotton textile processing stages," Energy, Vol. 35, No. 7, pp. 2945–2952, ISSN: 0360-5442.

[4] Ramirez, C., Patel, M., Blok, K. (2006). "From fluid milk to milk powder: Energy use and energy efficiency in the European dairy industry", Energy, Vol. 31, No. 12, pp. 1984–2004, ISSN: 0360-5442.

[5] Xu, T., Flapper, J., Kramer, K. J. (2009). "Characterization of energy use and performance of global cheese processing", Energy, Vol. 34, No. 11, pp. 1993–2000, ISSN: 0360-5442.

[6] Salta, M., Polatidis, H., Haralambopoulos, D. (2009). "Energy use in the Greek manufacturing sector: A methodological framework based on physical indicators with aggregation and decomposition analysis", Energy, Vol. 34, No. 1, pp. 90–111, ISSN: 0360-5442.

[7] "Reference Document on Best Available Techniques for Energy Efficiency," 2009. Available at: http://eippcb.jrc.ec.europa.eu/reference/BREF/ENE_Adopted_02-2009.pdf

[8] Siitonen, S., Tuomaala, M., Ahtila, P. (2010). "Variables affecting energy efficiency and CO_2 emissions in the steel industry," Energy Policy, Vol. 38, No. 5, pp. 2477–2485, ISSN: 0301-4215.

[9] Patterson, M. G. (1996). "What is energy efficiency?", Energy Policy, Vol. 24, No. 5, pp. 377–390, ISSN: 0301-4215.

[10] Tanaka, K. (2008)."Assessment of energy efficiency performance measures in industry and their application for policy," Energy Policy, Vol. 36, No. 8, pp. 2887–2902, ISSN: 0301-4215.

[11] Bunse, K., Vodicka, M., Schönsleben, P., Brülhart, M., Ernst, F. O. (2011). "Integrating energy efficiency performance in production management – gap analysis between industrial needs and scientific literature," Journal of Cleaner Production, Vol. 19, No. 6–7, pp. 667–679, ISSN: 0959-6526.

[12] Kissock, J.K., Eger, C. (2008). "Measuring industrial energy savings", Applied Energy, Vol. 85, pp. 347–361, ISSN: 0306-2619.

[13] Medojevic, M., Cosic, I., Sremcev, N., & Lazarevic, M. (2016). "Conceptual Theoretical Model for Life Cycle Energy Analysis of Photovoltaic Modules", Proceedings of the 27th DAAAM International Symposium, pp.0534 - 0543, B. Katalinic (Ed.), Published by DAAAM International, ISBN 978-3-902734-08-2, ISSN 1726-9679, Vienna, Austria DOI: 10.2507/27th.daaam.proceedings.079

[14] Medojevic, M. (2016). "Analysis of Current Automation Level in Specific Compressed Air System with Model for Optimization", Proceedings of the 26th DAAAM International Symposium, pp.1082-1090, B. Katalinic (Ed.), Published by DAAAM International, ISBN 978-3-902734-07-5, ISSN 1726-9679, Vienna, Austria DOI: 10.2507/26th.daaam.proceedings.152

DOI: 10.2507/28th.daaam.proceedings.083

DIFFERENTIAL EVOLUTION AS CALIBRATION TECHNIQUE FOR THREE AXIS GYROSCOPE

Ales Kuncar, Martin Sysel & Tomas Urbanek

This Publication has to be referred as: Kuncar, A[les]; Sysel, M[artin] & Urbanek, T[omas] (2017). Differential Evolution as Calibration Technique for Three Axis Gyroscope, Proceedings of the 28th DAAAM International Symposium, pp.0592-0596, B. Katalinic (Ed.), Published by DAAAM International, ISBN 978-3-902734-11-2, ISSN 1726-9679, Vienna, Austria
DOI: 10.2507/28th.daaam.proceedings.083

Abstract

The MEMS (Micro-Electro-Mechanical Systems) gyroscopes are widely used in many engineering applications and areas. These types of sensors are small, cheap, and have low power consumption; however, these sensors are not entirely accurate. They are influenced by many factors such as sensor errors (scale factor and bias) and manufacturing errors (misalignment); therefore, the calibration of the gyroscope is necessary before its application in systems. Thus, this research paper describes calibration technique for three axis gyroscope. The differential evolution (DE) algorithm is used for the determination of the scale factor, misalignment, and bias. The performance of this method is analysed in experiment on three axis low-cost gyroscope L3GD20.

Keywords: Calibration; Differential Evolution; Gyroscope; MEMS; Sensors

1. Introduction

Three axis gyroscopes are sensors that measure angular velocity (rotations in general) around each sensitive axis. MEMS gyroscopes use the principle of Coriolis force of the moving object. Coriolis force can be formulated as

$$F = 2 \cdot m \cdot \vec{\omega} \times \vec{v} \tag{1}$$

where m is the mass of the object, $\vec{\omega}$ is the angular velocity vector and $\vec{v}$ is the velocity with respect to the rotating system.

These sensors, along with accelerometers, are one of the most used sensors for inertial navigation systems [1][2] and motion sensing [3][4]. Gyroscopes are furthermore used in many areas and applications such as guidance systems, assistance of stability (bicycles, segway, etc.). However, these sensors based on MEMS technology are vulnerable to systematic errors (scale factor, bias and misalignment). Therefore, there is more emphasis on the calibration of the sensor.

The biggest error source, that has an impact on the accuracy of the measurements, is bias. This error can be simply minimized by stationary readings and then averaging these values [5]. To minimize the scale factor and misalignment angles, special equipment (turntable) is required. Olivares et al. [6] proposed a calibration technique using accelerometer to aid calibrate the gyroscope because the accelerometers are easier to calibrate and more accurate. On the other hand,

Wu et al. [7] used magnetometer instead of accelerometer. This method requires a sufficient number of rotations to cover the local magnetic field. Then these measurements can be used as a reference for calibration of gyroscope.

The reminder of this paper is organized as follows. In section 2, the differential evolution is introduced. The gyroscope error model and used equipment are briefly described in section 3 and section 4, respectively. The gyroscope calibration and its details are mentioned in section 5.

2. Differential Evolution

Differential evolution is an optimization heuristic algorithm which search the global minimum or maximum introduced by R. Storn and K. Price in 1995 [8]. This optimization method is an evolutionary algorithm based on population, mutation, and recombination.

The algorithm uses a mutation and recombination as a search mechanism and a selection that leads the search to perspective areas in the search space. The DE algorithm uses a recombination operator which can choose child parameters from one parent more often than from others. A recombination effectively shuffles information about successful combinations by the use of members from existing population allowing the search for a better solution space.

The optimization tasks consist of D parameters that represent a D-dimensional vector. The initial population of NP solutions is created randomly. Then the population is going to be adjusted using mutation, recombination and selection operators.

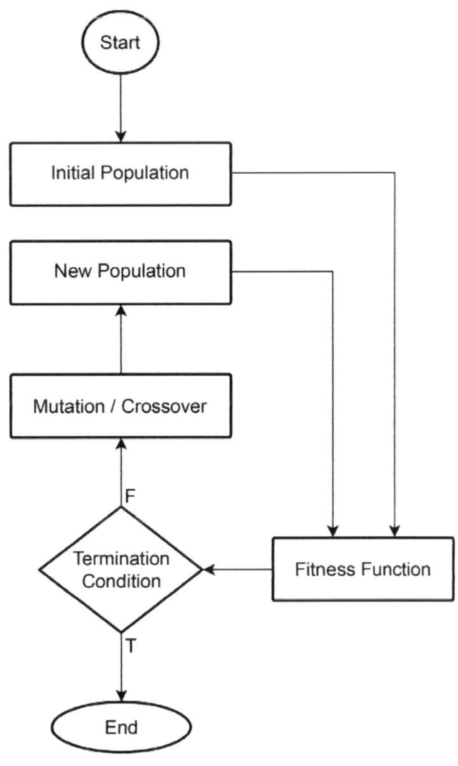

Fig. 1. Flow-chart of DE for parameters estimation.

Differential evolution uses only four parameters; therefore, it is easy to implement. Parameters are Generations, NP, F, and CR [9].

- Generations (Number of iterations) specifies the number of evolutionary cycles (generations) during which the entire population develops.
- NP (Number of population members) is a parameter which gives the size of the population. The value of this parameter cannot be lower than 4 because it is the minimum size at which the differential evolution algorithm still works. The optimal set up of this parameter is $5 \cdot D <= NP <= 10 \cdot D$ [8], where D is dimension of the problem.
- CR (Crossover probability). This parameter is a small value in range from 0 to 1. In case of separable function, this value is set close to 0 (clean copy of the fourth parent). Otherwise, it is set to the values close to 1 (random search).
- F (Mutation constant) is the last control parameter for differential evolution and its value ranges from 0 to 2. This parameter produces variability of the population.

The flow chart of differential evolution algorithm for the calibration parameters estimation is shown in Fig. 1.

3. Gyroscope Error Model

Sensor errors negatively influence the measurement results, thereby causing ambiguity results. Errors are caused by influences of the external environment, as well as internal influences. These errors can be divided up onto systematic and stochastic. Systematic errors can be reduced by calibration to remove their impact on measurements.

Measured values of three axis gyroscopes are affected by many error sources. These errors lead to a continuous increase in the position errors due to integration.

Mathematical model, that describes the measurements of the gyroscope, is expressed

$$\begin{vmatrix} \omega_X \\ \omega_Y \\ \omega_Z \end{vmatrix} = \begin{vmatrix} S_X & 0 & 0 \\ \alpha & S_Y & 0 \\ \beta & \gamma & S_Z \end{vmatrix} \cdot \begin{vmatrix} M_X \\ M_Y \\ M_Z \end{vmatrix} + \begin{vmatrix} O_X \\ O_Y \\ O_Z \end{vmatrix} \quad (2)$$

where ω_X, ω_Y, and ω_Z are calibrated gyroscope values. The coefficients S_X, S_Y, and S_Z and O_X, O_Y, and O_Z represent scale factor error, and bias, respectively. The angles α, β, and γ compensates the misalignment of the sensor axis. The M_X, M_Y, and M_Z coefficients in the matrix are the measured values.

- Scale Factor - Scale factor is defined as the ratio between digital output and mechanical input of the gyroscope, i.e. it represents the slope of the function of the sensor. Ideally, the scale factor should be 1, so any difference from 1 is affecting the sensor. The difference between the ideal and the actual value is a sensitivity error, which is referred to as the scale factor error.

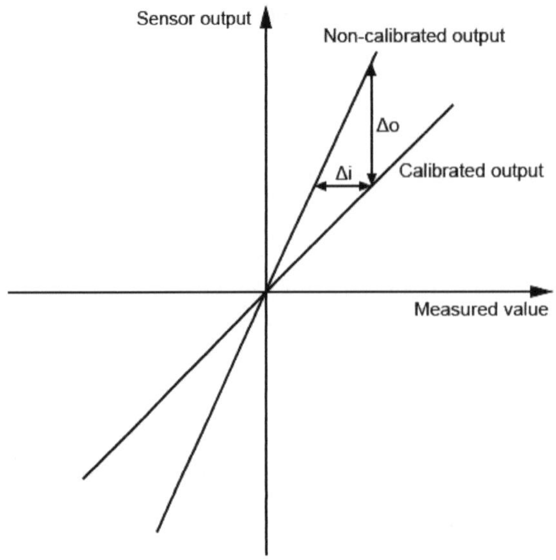

Fig. 2. Scale factor error.

- Misalignment error - This is an error in slight rotation of the sensor relative to the reference position. It is represented as angles between the sensor axis and the device body axis. This caused by imperfect mounting of sensor on the PCB (printed circuit board).
- Bias - If no external forces are applied on the gyroscope, its output should ideally show a zero value. However, there is an offset (zero shift), referred to as bias.

4. Equipment

The experimental measuring set (Fig. **3**. Measurement chain.Fig. 3) includes a control unit, an inertial measurement unit (IMU) and software for data collection.

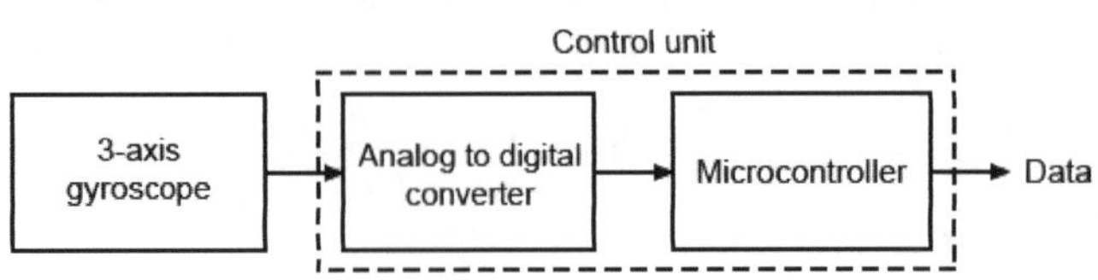

Fig. 3. Measurement chain.

The STEVAL-MKI109V2 motherboard/control unit is designed to provide a platform for the evaluation of MEMS inertial modules. These modules can be connected via a 24-pin expansion socket.

The unit also includes a 512kB high performance ARM-based 32bit STM32F103RET6 microcontroller, which acts as a bridge between MEMS and GUI, or dedicated software routines for customized applications.

For measurements, the STEVAL-MKI124V1 10-axis inertial measurement unit is connected to the control unit. The IMU includes a three-axis gyroscope with an internal thermometer (L3GD20), a three-axis accelerometer and a three-axis magnetometer (LSM303DLHC) and a barometer (LPS331AP). All these sensors are based on MEMS technology and are factory tested and trimmed. However, this factory calibration is only suitable for basic applications. Advanced calibration has to be provided for applications such as navigation systems.

Several different configurations allow specific usage settings. Sensor specifications are given in Table 1. and [10].

Parameters	Values
Full scale	250 – 2000 dps
Sensitivity	8.75 – 70 mdps/LSB
Min. zero level	±10 – ±75 dps
Zero level vs. temp	±0.03 – ±0.04 dps/°C
Noise density	0.03 dps/$\sqrt{Hz}$
Linearity	0.2 %FS

Table 1. Gyroscope Characteristics

To collect the measured data, the PC is connected to the control unit via a virtual serial port. The drivers for interaction and configuration of sensors are installed on the computer. This software is called Unico STSW-MKI109.

The collected data are pre-processed in Wolfram Mathematica 10 and then the differential evolution algorithm, programmed in Lua, is applied.

5. Gyroscope Calibration

The main objective of calibration is to reduce the errors, especially systematic errors (scale factor, bias). For this purpose, several measurements are needed in known conditions, in other words, the necessary data to determine the correction parameters.

5.1. Fitness Function

The fitness function is based on the sum of the squared errors where the error is the difference between the calculated output from current parameters and the true value. Therefore, the fitness function of the DE algorithm is modelled as

$$F = \sum_{i=1}^{n} (X + Y + Z - \Omega)^2 \qquad (3)$$

where *X*, *Y*, and *Z* are calibrated values and Ω is the true value of angular velocity.

5.2. Set-up of Differential Evolution

The set-up of differential evolution is shown in Table **2**. The parameter NP is set to the upper limitation suggested by [8]. That means to 10·D where D represents the dimension of the problem (the number of unknown calibration parameters). In this case the dimension is 9. The number of generations is given by the experiment that showed the optimal value around 300 (slow convergence rate to minimum). The parameters F and CR were also determined empirically.

Parameter	Value
NP	90
Generations	300
F	0.5
CR	0.7

Table 2. Set-up of differential evolution

6. Conclusion

The low-cost MEMS gyroscopes suffer from various sources of errors. In order to improve the accuracy of the measurements, the calibration of the sensor is necessary. Therefore, we presented calibration method for three axis gyroscope using differential evolution algorithm which compensates scale factor error, misalignment angles, and bias. For this method, simple turntable was used. The calibration was tested and analysed on gyroscope L3GD20. In the future work, we are going focus on sensor fusion using Kalman filters.

7. Acknowledgments

This work was supported by Internal Grant Agency of Tomas Bata University in Zlin under the project No. IGA/FAI/2017/007.

8. References

[1] Marinushkin, P.; Levitskiy, A. & Zograf, F. (2017) Improving pedestrian navigation system performance through the use of non-orthogonal redundant inertial measurement units, in 2017 International Siberian Conference on Control and Communications (SIBCON), pp. 1–5.

[2] Abdulrahim, K.; Hide, C.; Moore, T. & Hill, C. (2010). Aiding MEMS IMU with building heading for indoor pedestrian navigation, in 2010 Ubiquitous Positioning Indoor Navigation and Location Based Service, UPINLBS 2010, 2010.

[3] Galka, J.; Masior, M.; Zaborski, M. & Barczewska, K. (2016). Inertial Motion Sensing Glove for Sign Language Gesture Acquisition and Recognition, IEEE Sens. J., vol. 16, no. 16, pp. 6310–6316, Aug. 2016.

[4] Bardyn, F.; Savary, M.; Grassi, S.; Farine, P.-A.; Fasel, B. & Aminian, K. (2016). MEMS Inertial Motion Sensing Watch for Measuring Walking and Running Activities, in 2016 IEEE International Workshop on Signal Processing Systems (SiPS), 2016, pp. 104–109.

[5] Looney, M. (2010). A simple calibration for MEMS gyroscopes, Analog Devices, pp. 28–31, Jul-2010.

[6] Olivares, A.; Olivares, G.; Gorriz, J. M. & Ramirez, J. (2009). High-efficiency low-cost accelerometer-aided gyroscope calibration, in 2009 International Conference on Test and Measurement, 2009, pp. 354–360.

[7] Wu, Y. & Pei, L. (2017), Gyroscope Calibration via Magnetometer, IEEE Sens. J., vol. 17, no. 16, pp. 5269–5275, Aug. 2017.

[8] Storn, R. & Price, K. (1995). Differential Evolution - A simple and efficient adaptive scheme for global optimization over continuous spaces.

[9] Storn, R. (1996). On the usage of differential evolution for function optimization, in Proceedings of North American Fuzzy Information Processing, pp. 519–523.

[10] STMicroelectronics (2013), Data brief: STEVAL-MKI124V1. p. 4.

DOI: 10.2507/28th.daaam.proceedings.084

INFERRING PRODUCTIVITY FACTOR FOR USE CASE POINT METHOD

Tomas Urbanek, Alena Kolcavova & Ales Kuncar

This Publication has to be referred as: Urbanek, T[omas]; Kolcavova, A[lena] & Kuncar, A[les] (2017). Inferring Productivity Factor for Use Case Point Method, Proceedings of the 28th DAAAM International Symposium, pp.0597-0601, B. Katalinic (Ed.), Published by DAAAM International, ISBN 978-3-902734-11-2, ISSN 1726-9679, Vienna, Austria
DOI: 10.2507/28th.daaam.proceedings.084

Abstract

This paper aims to investigate and infer new range for productivity factor for Use Case Points method. Productivity factor is a ratio of man hours per use case point. In this paper, we use linear regression as an inference tool for productivity factor. This value is inferred from three different datasets. Then, we compared these inferred values with standard value for productivity factor. This study use MMRE (Mean Magnitude of Relative Error) measure and standard statistical methods for reliability assessment. The MMRE is also used for all comparisons. The experimental results show that the range for productivity factor is much narrow than recommended values with the mean of 15.

Keywords: Effort Estimation; Optimization; Productivity Factor; Software Engineering; Use Case Points

1. Introduction

Effort estimation is defined as the activity of predicting the amount of effort required to complete a development of software project [1]. Despite of a lot of effort of scientists and software engineers, there is still no optimal and effective method for every software project. The common way to improve effort estimation is to enhance the algorithmic methods. The algorithmic methods use mathematical formula for prediction.

It is very common that this group is also depending on the historical data. The most famous example of algorithmic methods are COCOMO (Constructive Cost Model) [2], FP (Function Points) [3] and last but not least UCP (Use Case Points) [4]. However, there are a many other algorithmic methods. It is essential that the calculation of effort estimation should be completed in early stage of software development cycle. The best case is if these calculations are known during the requirement analysis [4].

The accurate and reliable effort estimates are the crucial factor for the proper development cycle. These estimates are used for effective planning, monitoring and controlling the process of the software development. The prediction of effort estimations in software engineering is complex and complicated process. The main reason is that there are a lot of factors which influences the final prediction.

In this article, we investigated the properties of productivity factor for Use Case Points (UCP) method. We utilize linear regression model and t-test for inferring the confidence interval and the statistically significant value for productivity factor. In our best knowledge, no previous study has investigated the comparison of such models especially on different datasets when Use Case Points method and MMRE (Mean Magnitude of Relative Error) was used. Therefore, this study makes a major contribution to research of Use Case Points method.

1.1. Use Case Points Method

This effort estimation method was presented in 1993 by Gustav Karner [4]. It is based on a similar principle to the function point method. Project managers have to estimate the project parameters to four tables. Due to the aims of this paper, the detailed description of well-known Use Case Points method basic principles is insignificant and hence omitted. Please refer to [4], [5] for more detailed description of the Use Case Points method. The most basic equation for Use Case Points method is equation (1). The effort estimation is determined by multiplying the number of Use Case Points by the productivity factor.

$$UCP = (UUCW + UAW) \cdot TCF \cdot ECF \quad (1)$$

where UUCW is Unadjusted Use Case Weight, UAW is Unadjusted Actor Weight, TCF is Technical Complexity Factor and ECF is Environmental Complexity Factor.

1.2. Productivity factor

The Productivity Factor (PF) is a ratio of the number of man hours per use case point [6]. The setting of productivity factor is one of the most difficult tasks in an accurate estimation. According to industry experts, if no historical data has been collected, an interval between 10 and 30 hours per Use Case point can be used. The typical value for productivity factor proposed by Karner was 20 and this is also a suggested value for brand new development team. Schneider and Winters [7] proposed a method based on counting number of environmental factors. Additionally, the work of Silhavy et al. [8] propose a new algorithm for calibration of productivity factor based on historical data. The best way to estimate this value is through analysis of previous completed projects for each software organization. This value will be more accurate than multi-organizational dataset [9]. For the calculation of estimated effort (EE), we used equation (2).

$$EE = UCP \cdot PF \qquad [man/hour] \quad (2)$$

2. Research Objectives

This section presents the design of the research questions. The research questions of our study could be outlined as follows:

- RQ-1: What is the mean value of productivity factor of software projects?
- RQ-2: Is appropriate to recommend a standard value of productivity factor as 20?
- RQ-3: Is there a better value for setting up a productivity factor?

The first research question (RQ-1) aims to get an insight on the datasets of this research. We examine the datasets and then the productivity factor of each software project was calculated. This data will be then statistically summarized.

The second research question (RQ-2) aims to produce an evidence that the value for productivity factor (20) is statistically sufficient. Therefore we will perform one sample t-test of this claim.

To address research question (RQ-3), we experimented with simple linear model used for calibrating the productivity factor. To assess the evidences of statistical properties, we used exploratory analysis and hypothesis testing. For all comparisons, the MMRE measure will be used.

3. Experiment

For all models in this study, we used 10-fold cross validation method to assess the reliability of our research. The MMRE is chosen as criteria for all model comparison. Datasets used in this research are described in this section.

There are three datasets for comparison of productivity factor. And we record 5 values for each software project: UUCW, UAW, TCF, ECF and actual effort.

- Dataset from Poznan University of Technology [10] (referred to hereafter as Dataset1)
- Dataset from Subriadi's paper [11] (referred to hereafter as Dataset2)
- Our own dataset collected using document reviews and contributions from software companies (referred to hereafter as Dataset3)

	Dataset1 Value	Dataset2 Value	Dataset3 Value
n	14	10	143
Min.	4.22	11.18	2.18
1st Qu	11.30	12.00	10.39
Median	14.00	15.80	15.33
Mean	15.33	14.29	16.89
3rd Qu.	17.80	15.96	21.29
Max.	35.06	17.30	86.85

Table 1. Statistical comparison of PF in each dataset

Table 1. shows the descriptive statistical comparison of PF in each dataset. The most interesting part of this table is the mean of PF in each dataset. This number is about 15. We can also see some outliers especially for dataset3 (min and max values). We can also note that the median of PF in each dataset is also about 15.

3.1. Simple Linear Regression

In this research, we will utilize a simple linear regression for resolving productivity factor of Use Case Points method. The equation for simple linear regression can be seen on equation (3).

$$\hat{y} = \beta_0 + x \cdot \beta_1 \tag{3}$$

where $\hat{y}$ is prediction (dependent variable), β_0 is intercept, x can be seen as a number of Use Case Points (UCP) and β_1 is productivity factor for Use Case Points method. If we omit the intercept from the equation we get fallowing equation (4)

$$\hat{y} = x \cdot \beta_1 \Rightarrow \hat{y} = UCP \cdot PF \tag{4}$$

where $\hat{y}$ is prediction (dependent variable), x is number of Use Case Points (UCP) and β_1 is productivity factor for Use Case Points method.

4. Results

	CI low	CI high	p-value
Dataset1	10.87	19.78	4.96e-06
Dataset2	12.70	15.88	7.82e-09
Dataset3	15.11	18.67	2.2e-16

Table 2. Summary table for one sample t-test, if the PF is set to 20

In Table 2., the summary for one sample t-test can be seen, if the PF is set to 20. As can be seen in this table, the confidence interval ranges from 10-18. The most important is a fact that on each dataset the PF value of 20 was rejected (p-value). We also noted that the intersection of these entire interval ranges from 15.11 to 15.88.

The confidence interval for dataset1 can be seen in comparison with another as exceptionally wide. This is probably due to wide variety of software projects complexity.

The actual effort for each project in all dataset with linear models can be seen in Fig. 1.. As can be seen on this figure, all three models have very similar slope (the value of this slope can be seen in Table 3.). We also note that the intercept for each model was set to 0.

	PF (slope)	Std. Error	R^2
Dataset1	13.42	1.55	0.85
Dataset2	12.65	0.64	0.97
Dataset3	15.33	0.54	0.85

Table 3. Summary of linear models for each dataset

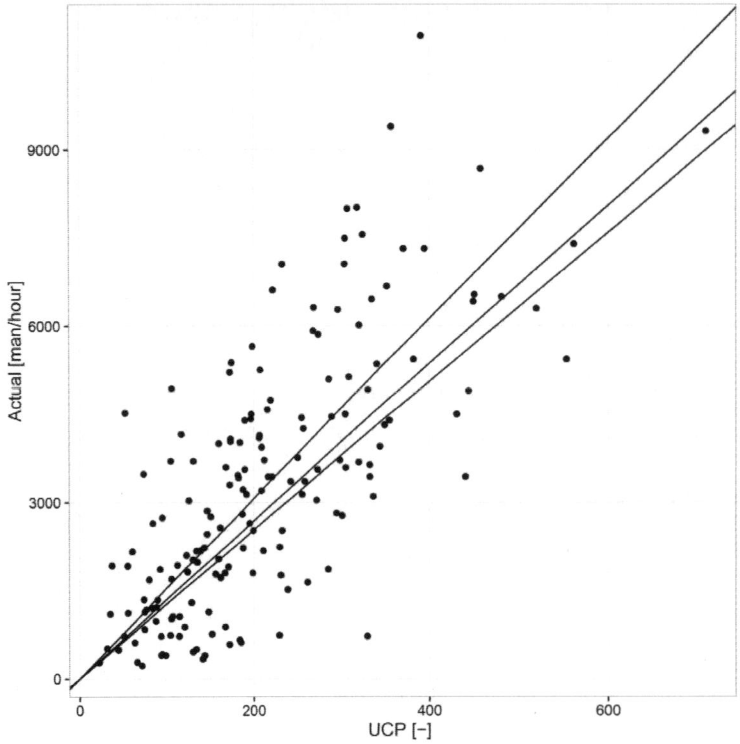

Fig. 1. Linear models on all datasets

In Table 3., the summary of linear models for each dataset is presented. As can be seen, all linear models have an exceptionally good value of R^2. The mean number of slope is about 13.8.

PF	20 MMRE [%]	15.5 MMRE [%]	15 MMRE [%]	Fit MMRE [%]
Dataset1	78	50	48	43
Dataset2	43	15	14	16
Dataset3	96	70	67	69

Table 4. Summary table for calculation of MMRE on chosen PF

As can be seen in Table 4., the summary for calculation of MMRE on chosen PF. We can see from that the standard value 20 for PF yield about 30% MMRE worse result than for value 15. The value of fit means, that we use a value (slope) from linear regression. Also we take note that the fit value is not the best MMRE for all datasets.

5. Discussion

The study started out with goals of answering three research questions outlined in research objective section. These questions are answered in the result section of this paper.

RQ-1: What is the mean value of productivity factor of software projects? This question is answered in result section respectively in Table 1.. The mean value for productivity factor is about 15. What is also important in this table is the values of min and max. These values are probably outliers. Nevertheless, we think that in this case these values are important. Nearly 75% values of PF are below 20. This is very surprising if we take a in account that the 20 is recommended value for productivity factor.

RQ-2: Is appropriate to recommend a standard value of productivity factor as 20? This question is answered in result section. In Table 2., we can see that all PF in our datasets did not exceed the 5% confidence level. Therefore, there is no evidence for supporting PF 20 as a good value. We can see also in Table 4. that the MMRE reached by PF 20 is not as good as for number 15. We can also see in Table 2. that the intersection of confidence interval ranges from 15.11 to 15.88. The number 20 is a relatively distant value from this range.

RQ-3: Is there a better value for setting up a productivity factor? To answer this question, we use 4 values of productivity factor. These numbers are 20, 15.5, 15 and fit value from linear regression. Table 4. shows the comparison of these numbers. The number 20 shows the worst results of MMRE error. Then there are two numbers 15.5 and 15. The

15.5 was chosen because it is an intersection for all confidence intervals. The number 15 was chosen because we would like to know the difference between 15.5 and the nearest natural number. Both values 15.5 and 15 show a difference in MMRE from 20 to about 30%. This can be seen as exceptional improvement from productivity factor set to 20. The fit value shows the slope for linear regression, which is different for each dataset. This value is calculated by least square method and this is also a reason why MMRE does not show the minimal value. We can also see that the number 15 is good choice if we consider only the value of MMRE.

6. Conclusion

Nowadays, 20 is a standard starting value for productivity factor. Also it is recommended use values from 10-30 for productivity factor. The current study found that, we must reconsider these recommendations. In this study nearly 75% of all projects in our dataset have PF value under 20. Moreover, the mean value is about 15. Also our study presented evidences that for all datasets is statistically significant to not set the PF value to 20. In our study, figures prominently range from 10-18. Hence, this is an exceptional difference to recommendation of 10-30. Nevertheless, there is a extensive threat of validity in the use of these datasets especially sample sizes and unknown collection methods. However, the findings of this study have a number of important implications for future research of the productivity factor and for Use Case Point method in general.

7. Acknowledgments

This work was supported by Internal Grant Agency of Tomas Bata University in Zlin under the project No. IGA/FAI/2017/007.

8. References

[1] Keung, J. W. (2008). Theoretical Maximum Prediction Accuracy for Analogy-Based Software Cost Estimation, Software Engineering Conference, 2008. APSEC '08. 15th Asia-Pacific, pp. 495–502, 2008.

[2] Boehm, W. (1984). Software Engineering Economics, IEEE Transactions on Software Engineering, vol. SE-10, pp. 4–21, jan 1984.

[3] Atkinson, K. & Shepperd, M. (1994). Using Function Points to Find Cost Analogies, 5th European Software Cost Modelling Meeting, Ivrea, Italy, pp. 1–5, 1994.

[4] Karner, G. (1993). Resource estimation for objectory projects, Objective Systems SF AB, pp. 1–9, 1993.

[5] Urbanek, T.; Prokopova, Z.; Silhavy, R. & Vesela, V. (2015). Prediction accuracy measurements as a fitness function for software effort estimation, SpringerPlus, 2015.

[6] Clemmons, R. K. (2006). Project estimation with Use Case Points, CrossTalk, vol. 19, no. 2, pp. 18–22, 2006.

[7] Schneider, G. & Winters, J. (2001). Applying Use Cases: A Practical Guide. Addison-Wesley Professional, 2001.

[8] Silhavy, R.; Silhavy, P. & Prokopova, Z. (2015). Algorithmic Optimisation Method for Improving Use Case Points Estimation, PLOS ONE, vol. 10, nov 2015.

[9] Jeffery, R.; Ruhe, M. & Wieczorek, I. (2000). A comparative study of two software development cost modeling techniques using multi-organizational and company-specific data, Information and Software Technology, vol. 42, pp. 1009–1016, 2000.

[10] Ochodek, M.; Nawrocki, J. & Kwarciak, K. (2011). Simplifying effort estimation based on Use Case Points, Information and Software Technology, vol. 53, pp. 200–213, mar 2011.

[11] Subriadi, A. P. & Ningrum, P. A. (2014). Critical review of the effort rate value in use case point method for estimating software development effort, Journal of Theoretical and Applied Information Technology, vol. 59, no. 3, pp. 735–744, 2014.

DOI: 10.2507/28th.daaam.proceedings.085

MODEL FOR CRITICAL INFRASTRUCTURE SAFETY MANAGEMENT

Prochazkova, D.; Prochazka, J.; Rusko, M.; Mikulova, M. & Ilko, J.

This Publication has to be referred as: Prochazkova, D[ana]; Prochazka, J[.]; Rusko, M[iroslav]; Mikulova, M[onika] & Ilko, J[an] (2017). Model for Critical Infrastructure Safety Management, Proceedings of the 28th DAAAM International Symposium, pp.0602-0610, B. Katalinic (Ed.), Published by DAAAM International, ISBN 978-3-902734-11-2, ISSN 1726-9679, Vienna, Austria
DOI: 10.2507/28th.daaam.proceedings.085

Abstract

The safe community is now at time of globalisation very dependent on a safety level of critical infrastructure ensuring the territory by basic services necessary for humans´ live as there are energy, water, food, information etc. Series of events from recent years connected with critical infrastructure failures showed its high importance. The critical infrastructures represent multistage mutually overlapping systems, i.e. big complex systems, the type of which is a systems system. The paper presents the model for critical infrastructure safety management based on this reality and it shows the way how simply to determine the criticality of individual infrastructures and the whole critical infrastructure.

Keywords: critical infrastructure; provision of territory services; security; safety; model for safety management

1. Introduction

For ensuring the human security and development, there is necessary the safe human system [1-3]. Ensuring the safe human system is not easy, because the human system is a system of systems [4], i.e. system of several mutually interconnected systems of a different nature. Consequences of interconnections (interfaces) are mutual dependences, the character of which is physical, cyber, territorial and organisational [4-6]. Mentioned interdependences are the sources of further vulnerabilities of human system that magnify the integral risk of a given system by increase of cross-section risks in the system of systems [4-6]. As a consequence of growing globalisation the new sources of disasters take on force, they cause critical infrastructure failures [7-9].
Corporate sustainability assigns firms the key role of integrating and pursuing economic, environmental and social goals. Thus, firms struggle to link corporate sustainability practices and organizational performance [10-13]. Therefore, using the systems approach may be a solution to realizing further necessary steps in the area of critical infrastructure protection.

The paper deals with problems of critical infrastructure in the broadest concept, i.e. not only from the viewpoint of critical infrastructure itself, i.e. from the viewpoint of its structure and co-operation of its individual parts, but also from the viewpoint of its impacts and profits for a given locality in it is in operation, i.e. for public assets in locality and region. By this feature the paper concept differs from the most of current works, and it is reality that its concept includes the public protection.

From the reasons of fulfilment of targets of humans that may be realised only if human communities are in safe territory, the object of present paper is the critical infrastructure safety that ensures the safe infrastructures that do not threaten their vicinities, i.e. also another system with which they are mutually interconnected or which they influence. The result of study, by help of methodology processed in the frame of project FOCUS [14-16], is the creation of model of infrastructure chains safety management.

2. Critical infrastructure and associated terms

The critical infrastructure includes the infrastructures that are parts of different technological systems that ensure the human society needs [5]. Each of considered systems consists of the control system and controlled systems [16], which are for company processes, social system (humans, organisational structures, assets and values, knowledge), and for own technological system (tools, equipment, procedures, technologies). It means that they are multistage systems at which among the individual stages in both directions they run flows of materials, finances, information and decisions. From these reasons the systems need to be also analysed from the viewpoint of interactions and interdependences among technical, human, social and organisational aspects of a system. The exception is the analysis of human survival that is either active or passive. The capability of passive survival is included in the system properties, there are based on knowledge on defects in environs; the defects are illustrated by causal chain. The capability of active survival manifests by system behaviour, it considers uncertainty in projection of future defects and failures.
From the methodological viewpoint the critical infrastructure and each its partial infrastructure is a system of systems [4, 5]. In engineering disciplines directed to risk at present we use two disciplines for trade-off with the risk [5]: a set of disciplines the target of which is the infrastructure security, i.e. security of infrastructure without regard to infrastructure vicinity (security management); and a set of disciplines the target of which is the infrastructure safety, i.e. security and development of both, the infrastructure and its vicinity. Many professional works deal with ensuring the first target, which has been pursued in engineering disciplines since the beginning of 80s [5]. The other discipline target is more ambitious on understanding, accessible data and methods of engineering disciplines. It has been pursued since a half of 80s but from reasons of big demands on: data (there are necessary data on: system, system vicinity, linkages and flows between system and its vicinity); comprehension of problems and their connections in a case of open system of systems; methods of problem structuring, analysis and solving the problems, it is only enforced in domain of nuclear technologies and astronautics [5], namely in spite of it solves interconnection of targets of humans in domains social, environmental and technological [3].

Regarding to present way of problem solving given above, we use two concepts for ensuring the safe entity [4, 5]; i.e. security management and safety management. The first mentioned concept being simpler is more often used in practice; i.e. the target is the critical infrastructure security and impacts of critical infrastructure on its vicinity are out of interest. The other ensures both, the critical infrastructure security and the security of vicinity of critical infrastructure.
With regards to works [3-5,16] the definitions of terms connected with security and safety are:

1. Each infrastructure belonging to the critical infrastructure and it alone is a multistage system in which among individual stages in both directions they run material, finance, information and decision flows.
2. The disasters for partial infrastructures and critical infrastructure are the phenomena that caused damages and losses. They include phenomena belonging to the category „All Hazards Approach" [17] and specific phenomena connected with humans and their behaviour that do harm to the critical infrastructure owners, operators and that influence the fulfilment of tasks for which they were established (insufficient co-ordination of activities – organising accidents, failure of outsourcing activities, intent attacks etc.).
3. The infrastructure vulnerability is a predisposition of infrastructure (its protected assets) to harm / damage origination.
4. The infrastructure resilience is an infrastructure capability to overcome impacts of a given disaster.
5. The infrastructure risk is a probable size of losses, harms and detriment caused by a disaster with size of normative hazard (mostly design disaster) on infrastructure and public assets or subsystems rescheduled on selected time unit (e.g.1 year), site unit (e.g. 1 km^2) and on basic assets of owners and operators of infrastructure.
6. The infrastructure security is a situation / condition at which the probability of infrastructure assets´ harms, damages and losses is acceptable (it is almost sure that harms, damages and losses cannot origin).
7. The infrastructure safety is a set of measures and activities for ensuring the security and sustainable development of infrastructure, its assets and public assets.
8. The infrastructure security management is a planning, organisation, allocation of resources, humans and tasks with aim to reach demanded security level of a supply chain.
9. The infrastructure safety management is a planning, organisation, allocation of resources, humans and tasks with aim to reach demanded safety level of infrastructure and its vicinity.
10. The infrastructure safety engineering is a set of engineering measures and activities by which the infrastructure safety is ensured in real conditions of a given site.

With regard to results from analyses of critical infrastructure safety and historical experiences, performed on the data given in the professional literature [1,5,16] and in sources quoted in given works, it is necessary to follow infrastructures for: energy supply, water supply, sewer handling, transport system, communication and information systems, bank and finance system, emergency services (police, fire rescue service, medical rescue service), basic services (food supply, waste liquidation, social services, funereal services), industry, agriculture, state and regional administrations, that are usually supported by the national legislative. To them there is necessary to join the infrastructures for both, the education and the research, which is supported by the EU legislation.

The safety and risk are not complementary quantities even though they together relate by a certain way. In each system both quantities depend on processes, acts and phenomena being under way in a given system and in its vicinity. In advanced concept the concentration to safety has higher targets than concentration to risk because it follows system security, system development, system existence, system vicinity existence and co-existence of different systems [4].

The risk sources are all phenomena included in the term: All Hazards Approach [17], the phenomena specified in work [4] and further fulfilled during the FOCUS project (from 77 disasters followed now in 2035 the number of disasters increases to 105) [18]. The risks connected with infrastructures are: partial that include risks connected with individual protected assets; integrated that include risks connected with several assets aggregated by a defined way; and integral that include risks connected with all protected assets, with linkages and flows among assets that cause couplings among assets, partial systems and with vicinity. It is clear that to be able to ensure the system safety, the system integral risk needs to be considered, managed and traded-off.

3. Method of criticality judgement and method of infrastructure safety management model building

For criticality judgement and for building the model for of management of critical infrastructure safety we use the method of risk engineering. With regard to the present knowledge it is necessary to give that for infrastructure safety management fundament, it is the risk analysis, risk assessment and trade-off with risks connected with mutual interconnections in infrastructure sectors and in whole infrastructure (i.e. in agreement with [4] it is necessary to consider interdependences in a system of systems; i.e. at risk identification it is necessary also to use cross-sectional criterions). The procedure of work with risk is shown in Figure 1. It starts with definition of concept of work with risk (system characteristics, determination of assets, specification of aims), on the basis of which risks are identified, analysed, assessed, judged, managed, traded-off and monitored. Feedbacks denoted in this Figure 1 are used if risk level is not on required level [16].

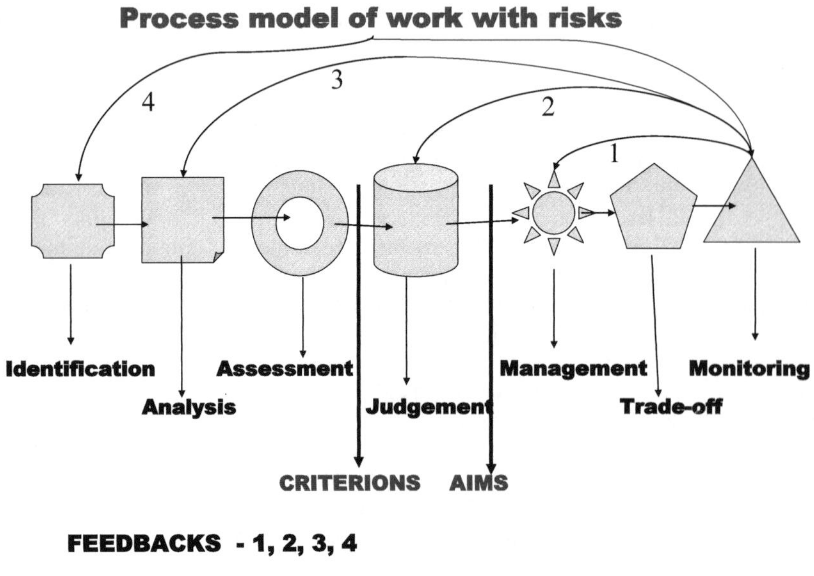

Fig. 1. Process model of work with risks, numbers 1, 2, 3 and 4 denote feedbacks.

In present practice we distinguish five different concepts for work with system risks: close system, technical faults as risk sources, aim – risk reduction; close system, human factor and technical faults as risk sources, aim – risk reduction; open system, external disasters, human factor and technical faults as risk sources, aim – secure system; open system, external disasters, human factor and technical faults as risk sources, aim – safe system; open system of systems, external disasters, human factor, technical faults and interdependences among the systems as risk sources, aim – safe system of systems. The details are summarized and described in work [5].

For human safety and for human system safety (i.e. territory, organisation, plant) we need to manage the integral risk including the human factor, i.e. to find the way of cross-section risks management and to concentrate the investigation on interdependences and critical spots with a potential to start the system cascade failures, domino effects, strange behaviour etc., and on the basis of such site knowledge to prepare measures and activities ensuring the continuity of limited infrastructure operation and of the human survival.

From results summarised in work [3] it follows that in each correctly strategically managed territory it is several disasters that can evoke the critical conditions in territory and in its facilities and infrastructures. Individual infrastructures that comprise the critical infrastructure possess the comparable and quite specific items that are mutually incommensurable; the critical infrastructure in practice contain: 1 - energy system; 2 - water supply system; 3 – sewage system; 4 - transport system; 5 - communication and information systems; 6 - bank and finance system; 7 - emergency services; 8 - basic services; 9 - state and regional administrations. For its capability to cope with abnormal and critical conditions it is necessary to create the tool by help of which it is possible by a simple way to control the critical infrastructure safety.

At infrastructure safety management and whole critical infrastructure safety management we need to concentrate to critical items, and therefore, it is necessary to judge the criticality of individual items. The assessment of criticality of individual systems (sectors) of infrastructures and the whole critical infrastructure is not trivial matter because under different conditions the sectors and the whole have a different role - active, reactive, critical or damping (not additive); e.g. the existence of several variants of electricity supply to one site decreases the energy infrastructure criticality but it increases expenses etc. The method for judgement of criticality of individual infrastructures and of whole critical infrastructure is described in [19]. According to this method, we:

- use factors that are targeted to protection of protected assets of human system: 1-rate of capability of protection; 2-rate of vulnerability; 3-rate of hazard for human lives and health; 4-rate of impact on environment; 5-rate of expensiveness of exchange or repair; 6-rate of time necessary for exchange or repair; 7-rate of relevance for ensuring the rescue and emergency functions in territory; 8-rate of relevance for ensuring the functions of government on levels local, regional and state; 9-rate of relevance for ensuring the functions of army and police; 10-rate of redundancy or replaceable service; 11-rate of relevance for ensuring the communication functions; 12-rate of impact of supply failure on economy of region (state); 13-rate of relevance of operability and interoperability; 14-rate of relevance in domain of symbols and culture,
- determine the scale for criticality rate 0 to 5 (0 - factor contributes to criticality little, …5 - factor contributes to criticality fundamentally); e.g. 0 – losses and damages are lower than 50 EUR, 1 - losses and damages are between 50 – 500 EUR, 2 - losses and damages are between 500 – 5000 EUR, etc. OR sum of maximum values for all 14 factors is SS = 14 x 5 = 70; if real value is more than 95% SS – criticality is extreme high, if real value is between 70 - 95% SS – criticality is very high; if real value is between 45 - 70% SS – criticality is high; if real value is more between 25 - 45% SS – criticality is medium; if real value is more between 5 - 25% SS – criticality is low; if real value is lower than 5% SS – criticality is very low (i.e. negligible),
- select five or six experts so that all important aspects connected with the human protection and the human system safety are covered who appreciate the real situation with help of all 14 factors for each critical disaster in a given territory and for each concept from five concepts of work with system risks (criterions for expert selection are in [16]),
- norm the criticality rate values to values between 0 and 1,
- determine the optimum values of criticality rate for each and each disaster on the basis of assumption that safety rate = 1 – criticality rate and application of Maximum Utility Theory [20] on safety rate,
- perform the same for the whole critical infrastructure. But here it is problem with lack of knowledge at evaluation of interdependences among individual infrastructures, and therefore, we often use integrated criticality rate (sum of criticality rates of individual infrastructures normed to value between 0 and 1or weighted sum of criticality rates of individual infrastructures normed to value between 0 and 1).

By the evaluation of real values criticality rates we can determine critical spots to which we need to concentrate attention at territory safety upgrade.

The purpose of model for infrastructure safety management is to show basic steps by which it is possible to ensure infrastructure security and infrastructure vicinity security. The model building method goes out from a system concept of infrastructures; it considers them as system of systems (several overlapping systems) [4], which means that their complex behaviour, function and development depend on both, the number and properties of partial systems and the diversities of their interconnections, i.e. their linkages and flows among them and also across them. The linkages and flows going across the partial systems are the originators of internal dependences (interdependences). The presented model is created by method of analogy to existing safety management models [3-5].

4. Assessment of capabilities of concepts of risk management and of engineering trade-off with risks applied to critical infrastructure to ensure the territory safety.

The results of research [21], based on the application of the Maximum Utility Theory [20], which dealt with the evaluation of the criticality rates of concepts of risk management and trade-off with risks, show that none of the concepts, used today for the management and trade-off with risks, has not a negligible rate of criticality, taking into account the assets of the human system (i.e. human lives, health and security; property and welfare; environment; critical infrastructures and technologies), i.e. the rate of criticality in the application:

- the classical concept of risk management and engineering trade-off with risk (considering the technical faults as risk sources) is extremely high,
- the classical concept of risk management and engineering trade-off with risk (considering the human factor and technical faults as risk sources) is very high,
- the concept of management and engineering trade-off with risk focused on secure system is high,
- the concept of management and engineering trade-off with risk focused on safe system is the medium,
- the concept of management and engineering trade-off with risk focused on the safe system of systems is low.

From safety reasons the above results means that tool for ensuring safety of important facilities' including the critical infrastructure is the concept system of systems.

5. Model for infrastructure safety management

With regard to: data and knowledge in [3-5,16,18,19,22-26]; the concept promoted by the OECD [27]; the method described in works [5,7]; and the assumption that each infrastructure is an open system (i.e. risk sources are internal and external disasters and human factor [3-5]), it is created a model for safety management having ten processes, i.e.:

1. ***Process 1*** that ensures the risk management of disasters, the sources of which are inside and outside of infrastructure plus human factor; i.e. it follows infrastructure and parameters of vicinity in which infrastructure operates. It is composed of: assessment of expected disaster size; determination of occurrence probability of important disasters; judgement of infrastructure vulnerabilities at important disasters; determination of impacts of important disasters on infrastructure. It creates a base for ensuring the safe infrastructure.
2. ***Process 2*** that ensures designing and planning the measures and activities for ensuring the infrastructure security at considering all important disasters [3,17]; i.e.: infrastructure layout (structure, function, sitting, buildings, equipment); performing the measures and activities for ensuring the infrastructure security; plan of renovation of infrastructure after disaster; plan of training the personnel performing the infrastructure; infrastructure activities´ monitoring; and correcting measures and activities for a case of important deviations in infrastructure operation.
3. ***Process 3*** that ensures designing and planning the measures and activities for ensuring the infrastructure vicinity security at considering all important disasters [3,17]; i.e.: infrastructure layout by a way that it may not threaten vicinity, i.e. all public assets; performing the measures and activities for ensuring the infrastructure vicinity security; plan of renovation of infrastructure vicinity after disaster; plan of training the personnel performing the infrastructure; infrastructure activities´ monitoring; and correcting measures and activities for a case of important deviations in infrastructure operation.
4. ***Process 4*** that ensures the harmony among the main activities connected with infrastructure commodities, i.e.: subject of supply (its manufacture, transport and distribution); following the deviations in a process of commodity management; and operating loops. It goes on ensuring the stabilities of processes, the minimisation of delays, the quality and the other critical aspects connected with the operation.
5. ***Process 5*** that ensures the safe assets of infrastructure, i.e. problems connected with: facilities, equipment or services; vehicles; shipping; products; and data systems. It also goes on averting of insiders' activities.
6. ***Process 6*** that ensures the safe human sources, i.e. problems connected with: acceptation of employee; understanding the employee behaviour features important for infrastructure operation; employee training; employee self-control; implementation of procedures that ensure correct employee behaviour; and employee stimulation.
7. ***Process 7*** that ensures good business partners, i.e. problems connected with: screening the possible partners; authentication of possible partners; producing the ways of negotiation with partners regarding to their behaviour; monitoring the partners' behaviours; and audits of partners.
8. ***Process 8*** that generates the capabilities for overcoming the impacts of extreme disasters that affect infrastructure, i.e. problems connected with: business continuity; specific response training; investigation of causes of extreme impacts; assembling the evidences; reparation of harms; and court settlement.
9. ***Process 9*** that ensures the dislocation of criminal and illegal infrastructures and chains, i.e. problems connected with: formation of base for disruption (ensuring the sources, determination of means, logistics, transport of means, distribution of means); and with support of governments and customers.
10. ***Process 10*** that ensures the integral safety of infrastructure, i.e. the coordination of all pillars, i.e. processes directing to infrastructure safety (PSM – process safety management).

The infrastructure safety management model is shown in Figure 2. The base constitutes the concept at which there are determined processes that are important for all infrastructures and the critical infrastructure. On this figure it is evident the principal role of concept on the basis of which the important internal and external processes and phenomena are determined. It is followed by: processes´ monitoring; judgement of impacts of all disasters (i.e. internal and external processes and phenomena) on infrastructure; and determination of optimal measures and activities directed to security of both, the infrastructure and its vicinity. Demands on determination of optimal solution for all processes and phenomena are fundamental [3,4] because there are under way frequent conflicts among the most suitable measures for some processes [28]. Because the implementation of measures and activities needs sources, forces and means and time for realisation, it is necessary in harmony with [3]: to process program for increase of safety of infrastructure; to determine measures for judgement of safety level in the sense of effectiveness of measures and activities for ensuring the infrastructure safety (indicators); and to fill program by projects that are interconnected and contain processes realising the individual measures and activities.

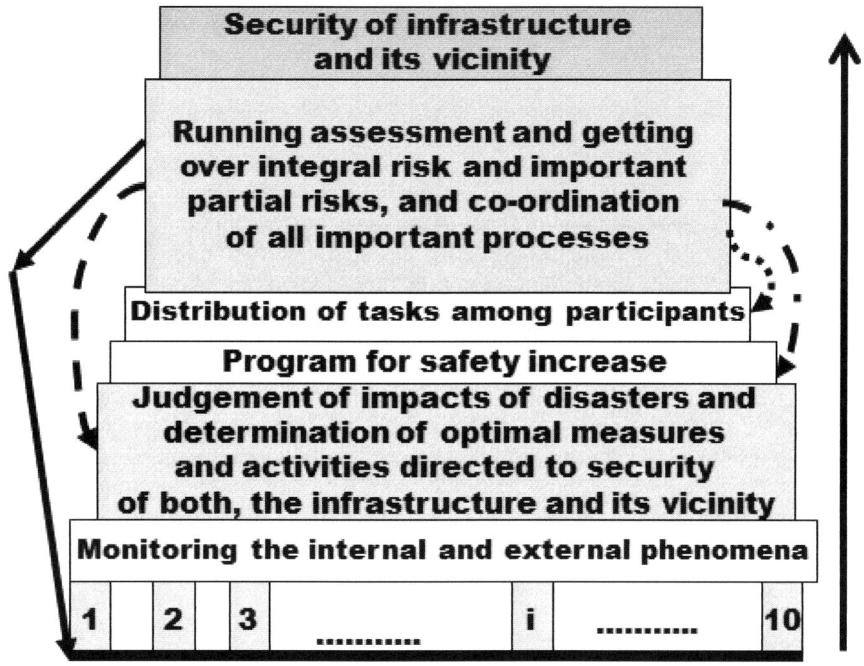

Fig. 2. Model of management of infrastructure safety; black block – concept for specification of important processes of infrastructure; dotted line – feedback 1; broken line – feedback 2; dashed line – feedback 3; full line – feedback 4.

The safety management system (SMS) of infrastructure operators includes the organisation structure, responsibilities, practices, rules, procedures and sources for determination and invoking the prevention for disasters that are results of processes inside and outside of infrastructure or at least mitigation of their unacceptable impacts. As a rule, it is connected with many aspects, apart from the organisation of employees, identification and assessment of hazard size, risk size, organising system, management of changes, emergency and crisis planning, safety monitoring, audits and scrutiny processes.

With regard to data in works [3,27] the program for increase of infrastructure chain safety has the following steps:

1. Determination of tasks (partial targets) and strategic goals for infrastructure with regard to safety directed to security of both, the infrastructure and the infrastructure vicinity.
2. For each process that is connected with infrastructure to determine suitable target and running indicators for safety level judgement.
3. To process dictionary for needs connected with integral safety management.
4. To harmonize standards, good practice methods and local procedures.
5. To determine set of target indicators.
6. To determine set of running indicators.
7. To determine way of assessment of target indicators specific for a given supply chain.
8. To determine way of assessment of running indicators specific for a given supply chain.
9. To determine way of assessment of all indicators together and marginal limits for a given infrastructure.

In practice it means that for each sector of selected authority the target and running indicators are determined and they have form of limits and checklists [3,27]. To them there are assigned criteria for assessment and scales by which it is determined if target is reached or is not reached. For creation of an effective safety management system the basic principle is that all participants play certain roles and at safety realization they must fulfil these roles (see stage in Figure 3 „distribution of tasks among participants ").
Because the world dynamically changes it is necessary to follow continuously the safety level, i.e. the size of integral risk that includes also the cross-sectional risks connected with interdependences and important partial risks of infrastructure. In case that limits and conditions are not kept, it is necessary to perform changes as shown feedbacks in Figure 2.

Because changes require sources, forces and needs, firstly it is realised feedback 1 and only if it does not ensure expected result the feedback 2 is realised etc. Only in the case of occurrence of extreme phenomena with catastrophic impacts, the feedback 4 is immediately realised.
Safety management system for infrastructure is lean on the concept of disaster prevention or at least of mitigation of severe disaster impacts that include the obligation to introduce and keep the safety management system [3,27] in which the following problems are taking into account:

- roles and responsibilities of persons participating in important hazards management on all organising levels and in ensuring the training,
- plans for systematic identification of important hazards and risks connected with them that are connected with normal, abnormal and critical conditions, and for assessment of their occurrence probability and severity,
- plans and procedures for ensuring the safety of all components and functions, namely including the object and facilities maintenance,
- plans for implementation of changes in territory, objects and facilities,
- plans for identification of foreseeable emergency situations by a systematic analysis including the preparation, tests and judgement of emergency plans for response to such emergency situations,
- plans for continuous evaluation of harmony with targets given in safety concept and in the SMS, and mechanisms for examination and performance of corrective activities in case of failure with aim to reach determined targets,
- plans for periodic systematic assessment of safety concept, effectiveness and convenience of the SMS and of criterions for judgement of safety level by top workers group.

6. Real result for the transport infrastructure in the Czech Republic

Transportation infrastructure belongs to the basic systems that make up the critical infrastructure in the European Union, in developed countries and also in the Czech Republic. According to the detailed discussion given in [16] ***the criticality of item means that item has simultaneously great importance and great vulnerability***. Detail investigation for transport infrastructure is in detail described in [19]; its results are in Table 1.

Factor	Rail transport	Road transport	Air transport	Water transport	Entire transport
1	5	4	5	2	16
2	5	4	5	3	17
3	4	3	4	3	14
4	4	3	3	2	12
5	4	3	5	2	14
6	3	2	4	3	12
7	5	5	4	1	15
8	3	3	2	2	10
9	5	5	5	2	17
10	3	2	3	1	9
11	3	2	3	1	9
12	5	4	4	4	17
13	4	4	4	2	14
14	3	2	3	2	10
All factors	56	46	54	30	186

Table 1. Rates of criticality of the transportation infrastructure in the Czech Republic

Table 1 shows that the greatest criticality has rail transport, which is followed by air transport, road transport and water transport. According to the chosen concept, the chosen scale and evaluation provided by the experts the criticality rates of both, the rail transport and the air transport are very high, the criticality rate of road traffic is high and the criticality rate of water transport is medium, and the criticality rate of whole transport system is high.

From Table 1 it is clear that to the size of criticality rate mostly contributes the vulnerability to attacks, little capability of protection (the protection of long linear networks in the territory is always a big problem), the importance for ensuring the functions of the army and the police, a big impact of transport failure on the economy of region (State), and the big importance for ensuring the emergency and rescue functions in the territory. The above facts show that it is urgent to pay increased attention to safety issues of transportation systems. It needs to be processed as a separate concept of transportation system safety, so the concept of critical infrastructure safety, as it shows the work [29].

From works [30,31] it follows that at reaching the threshold value for criticality rate "high", it is necessary from reason of protection of inhabitants and economic sphere to carry out preventive, mitigation and reactive measures and activities. Because measures against one disaster may be disserviceable against another one [28], it is necessary to seek for optimum for all possible disasters in a given territory that belong to critical ones as the strategic management principles stipulate [3].

7. Conclusion

The paper solves the problems of critical infrastructure safety. It shows that the critical infrastructure safety depends on concept of work with risks and on concept by which we model the critical infrastructure (close system, open system, open system of systems). The results given above shows that for ensuring the critical infrastructure safety it is necessary to use the concept of work with system risks which is directed to system of systems safety. This proves ideas of authors as Rinaldi, Moteff etc. who are cited in text. The paper shows the real plan how to solve the complex problem as the critical infrastructure safety management is. On the present knowledge basis, the model for safety management of infrastructures is compiled. This model is constructed as the process model in which they are represented the both, the individual important elements of process of safety management, and the feedbacks by which it is possible to correct cases in which demands of safety are not fulfilled. For application in practice the model for critical infrastructure safety management is supplemented by mechanism for ensuring the capability to be effective at abnormal and critical conditions. The authors verify this model in practice step by step and give the real recommendations to critical infrastructures entrepreneurs and operators.

8. Acknowledgement

Authors thanks to the Czech Technical University in Prague for support (grant SGS2015-17).

9. References

[1] UN (1994): Human Development Report. New York 1994, www.un.org.

[2] EU (2004): Safe Community. PASR projects, Brussels 2004.

[3] Prochazkova, D. (2011): Strategic Management of Territory and Organisation (in Czech). ISBN: 978-80-01-04844. Praha: ČVUT 2011, 483p.

[4] Prochazkova, D. (2011): Analysis and Management of Risks (in Czech). ISBN: 978-80-01-04841-2. Praha: ČVUT, 405p.

[5] Prochazkova, D. (2012): Safety of Critical Infrastructure (in Czech). ISBN: 978-80-01-05103-0. Praha: ČVUT, 318p.

[6] Prochazkova, D. (2009): Critical Infrastructure Safety Management. In: Reliability, Risk and Safety. Theory and Applications. ISBN 978-0-415-55509-8. Leiden: CRC Press / Balkema, pp 1875-1882.

[7] Necesal, L., Lukas, L. & Jasek, R. (2011): Measures for critical infrastructure protection in the Czech Republic. In: Annals of DAAAM for 2011 & Proceedings of the 22nd International DAAAM Symposium, Volume 22, No. 1, ISSN 1726-9679, Katalinic, B. [ed.] Published by DAAAM International, Vienna, pp. 0843-0844

[8] Lukas, L. & Hromada, M. (2011): Risk analysis in context of kritical infrastructure protection. In: Annals of DAAAM for 2011 & Proceedings of the 22nd International DAAAM Symposium, Volume 22, No. 1, ISSN 1726-9679, Katalinic, B. [ed.] Published by DAAAM International, Vienna

[9] Yusta, J. M., Correa, G. J. & Lacal-Arantegui, R. (2011): Methodologies and applications for critical infrastructure protection: State-of-the-art. Energy Policy, 39, Issue 10, October 2011, pp. 6100-6119

[10] Annunziata, E., Pucci, T., Frey, M. & Zanni, L. (2017): The role of organizational capabilities in attaining corporate sustainability practices and economic performance: evidence from Italian wine industry. Journal of Cleaner Production, Elsevier, DOI: 10.1016/j.jclepro.2017.10.035

[11] Varga, G., Szigeti, F., Deszö, G. (2015): Examination of surface roughness of burnished workpieces. Scientific Bulletin : serie C : Fascicle: Mechanics, Tribology, Machine Manufacturing Technology Vol. 29, 94-98 ISSN: 1224-3264

[12] Rusko, M. & Kralikova, R. (2012): Implementation of environmental management systems in the Slovak enterprises. In: DAAAM International Scientific Book 2012. Vol. 11, no. 31 (2012), p. 369-378. ISSN 1726-9687.

[13] Moldovan, L. (2015): Sustainability Assessment Framework for VET Organizations. Sustainability 2015, 7, 7156-7174. ISSN 2071-105,0 doi:10.3390/su7067156

[14] Prochazkova, D. (2012): Identification and Management of Risks of System of Systems. In: Proceedings. ISBN: 978-1-62276-436-5. IPSAM & ESRA, Helsinki, pp 6542-6551

[15] Prochazkova, D. (2013): Identification and Management of Risks of System of Systems. International Journal of Computer an Information Technology, ISSN: 2279-0764, 2, No 2, 232-239. http://ijcit.com/current.php

[16] Prochazkova, D. (2013): Principles of Management of Safety of Critical Infrastructure (in Czech). ISBN: 978-80-01-05245-7. Praha: ČVUT 2013, 213p.
[17] FEMA (1996): Guide for All-Hazard Emergency Operations Planning. State and Local Guide (SLG) 101. Washinton: FEMA
[18] EU: FOCUS Project. Brussels: EU. www.focusproject.eu
[19] Prochazkova, D. (2013): Criticality of Transport Infrastructure (in Czech). Periodica Academica, ISSN 1802-2626, VIII/2013, No. 2, pp 112-128.
[20] Keeny, R.L. & Raiffa, H. (1976): Decision Analysis with Multiple Conflicting Objectives. New York: J. Wilex
[21] Prochazkova, D. (2015): Optimum Concept of Management and Trade-off with Risks. Safety and Reliability: Methodology and Application. ISBN 978-1-138-02681-0. London: Taylor & Francis Group, pp 1463-1471.
[22] Stein, W., Hammerli, B., Pohl, H. & Posch, R. [Eds] (2003): Critical Infrastructure Protection – Status and Perspectives. Workshop on CIP, Frankfurt am Main, www.informatik2003.de
[23] Moteff, J., Copeland, C. & Fischer, J. (2003): Critical Infrastructures: What Makes an Infrastrucuture Critical? Report for Congress, CRS Web, Order Code RL31556.
[24] CISP (2005): Workshop on Critical Infrastructure Protection and Civil Emergency Planning-Dependable Structures, Cybersecurity, Commnon Standard. Zurich: Centre for International Security Policy
[25] Rinaldi, S. M. (2004): Modelling and Simulating Critical Infrastructures and Their Interdependencies. In: Proceedings of the 37th Hawaii International Conference on System Sciences–2004. Sandia: Sandia National Laboratories.
[26] Rinaldi, S. M., Peerenboom, J. P. & Kelly, T. K. (2001): Critical Infrastructure Interdependencies (Identifying, Understanding, and Analysing). IEEE Control Systems Magazine, Vol. 21, pp.12-25.
[27] OECD (2002): Guidance on Safety Performance Indicators. Guidance for Industry, Public Authorities and Communities for Developing SPI Programmes Related to Chemical Accident Prevention, Preparedness and Response. Paris: OECD 2002, 191p.
[28] Prochazkova, D. (2007): Methodology for Estimation of Costs for Renovation of Property at Territories Affected by Disaster (in Czech). SPBI Spektrum XI Ostrava, ISBN 978-80-86634-98-2, 251p.
[29] US (2001) US Critical Infrastructure Conception. Washington: US Federal Government,
[30] US (2005) Guide for Critical Infrastructure Protection. Washington: US Federal Government
[31] EMA (2003): Critical Infrastructure Emergency Risk Management and Assurance. Handbook Emergency Management Australia

DOI: 10.2507/28th.daaam.proceedings.086

MECHANICAL PROPERTIES OF HEAT-AFFECTED ZONE OF HIGH-STRENGTH STEEL WELDS

Ismar Hajro, Damir Hodžić & Petar Tasić

This Publication has to be referred as: Hajro, I[smar]; Hodzic, D[amir] & Tasic, P[etar] (2017). Mechanical Properties of Heat-Affected Zone of High-Strength Steel Welds, Proceedings of the 28th DAAAM International Symposium, pp.0611-0616, B. Katalinic (Ed.), Published by DAAAM International, ISBN 978-3-902734-11-2, ISSN 1726-9679, Vienna, Austria
DOI: 10.2507/28th.daaam.proceedings.086

Abstract

Mechanical properties testing of structural steel welds are well known and standardized. Results of those testing are important information regarding evaluation and qualification of welding technologies. However, using only standardized testing it is not possible to evaluate the weakest part of welds; e.g. coarse-grained heat affected zone, where peak temperature over 1300 °C is reached during fusion welding. Such high temperatures have for consequence significant change and drop of some mechanical properties. Thus, it is important to perform some more sophisticated evaluation and prior preparation of test specimens. Actually, such prior preparation include welding thermo-cycle simulation on test specimen ready for standardized testing.
The paper present one combination of testing results, both standardised and sophisticated, for the purpose of more detailed review of mechanical properties of high-strength steel welds with nominal yield strength in the range of 690-890 MPa.
Detailed distribution across welds is presented including the following generalised mechanical properties: hardness, strength, toughness.
Significant drop of some mechanical properties is observed within coarse-grained heat affected zone, which may require particular attention within everyday engineering practice.

Keywords: mechanical properties; strength; hardness; toughness; high-strength steel welds; heat-affected zone; coarse-grained heat-affected zone

1. Introduction

Use of high-strength steels (HSS) in various structural application is beneficial due to the various reasons, while most of them are related generally to provision of lighter and energy saving structures [1, 2]. While there is no some precise definition of HSS, they can be considered as one having yield strength equal or higher than 420 MPa. For the purpose of this paper two HSS are considered, in the range of nominal yield strength from 690 to 890 MPa, delivered in quenched and tempered (QT) condition.

Beside challenges which may be faced with application of various manufacturing technologies to HSS, such as banding or cutting; an application of welding technologies may become a particular challenge. This is mainly due to the two reasons: (1) loss of toughness; and (2) loss of crack resistance. Both problems are well known and particularly related to cooling time, $t_{8/5}$ [s]; which is principally opposite to a cooling rate, v_c [°C/s]. Actually, if fusion welding of HSS is performed with fast cooling rates (e.g. low $t_{8/5}$), there is a risk of microstructure hardening and further loss of cold crack resistance (e.g. hydrogen induced crack). Contrary, if welding is characterised with high $t_{8/5}$, or slow cooling rates; there is a risk of microstructural grain coarsening within heat-affected zone (HAZ; Fig. 1a), following significant loss of toughness. Such generalised concept is well known, and it is one of primary concerns while welding HSS. Therefore, it is important to follow recommended $t_{8/5}$; which is mostly found within HSS manufacturer's specification. In addition, according to EN 1011-2, general recommendation is $t_{8/5}$=10-25 s [2, 3]; while manufacturers recommendations for S690QL and S890QL are $t_{8/5}$=5-15(20) s [7, 8].

Another concern may be related to real welds, which must be qualified by standard procedures (e.g. according to ISO 15614-1 or ASME BPVC Sec.IX; etc.); which consist of standardised testing of mechanical properties, including surface and volumetric non-destructive tests (NDT). The qualification means that predicted welding technology parameters are set well to finally provide sound and acceptable welded joints in production. Thus, a test parts of sufficient size are welded, following the sampling of test specimens for further mechanical and technological testing [4, 5].

Somehow, such requirements may be achieved; and overall welding technology may become qualified. However, such approach may not evaluate the well-known weakest zone of welded joint – coarse-grained heat affected zone (CG-HAZ). It is a zone, where, during fusion welding, maximum (peak) temperature(s) close and over 1300°C is reached (Fig. 1b). On such high temperatures heat-affected zone microstructure tends to overgrow [6, 9], following formation of a quite large grains, which finally has for consequence significant drop in toughness. Here, importance of sufficient toughness become particularly important, because of sufficient resistance to crack initiation and growth [2, 6, 10].

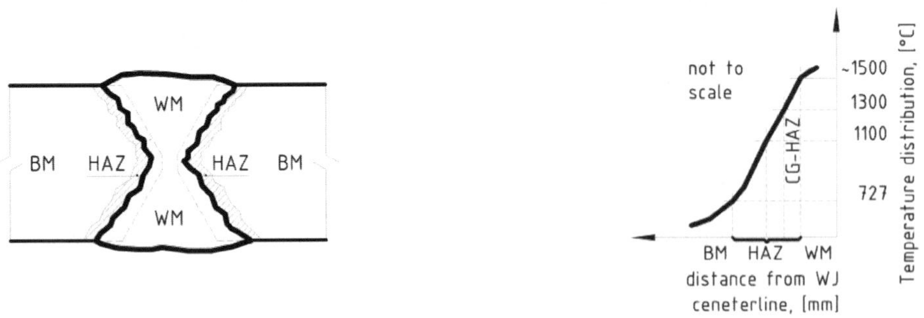

a) Welded joint zones: BM, HAZ, WM

b) Approximate location of CG-HAZ

Fig. 1. Welded joint zones, including CG-HAZ regarding peak temperature distribution

Therefore, even one welding technology (e.g. procedure) may become qualified in well-known and standardized manner [4, 5], the process of qualification cannot evaluate the weakest welded joint zone – CG-HAZ. Regarding its size, in comparison to complete welded joint, weakening within CG-HAZ should not be neglected easily. However, even everyday welding engineering practice may not relay on sophisticated approaches and techniques, such as simulation of welding - thermo cycles; a more detailed investigation is required, at least to provide general sense of a weakening effect.

The following experimental procedure, as well as provided analysis, is one [6] of a possible approaches [10] to gain reasonable sense about intensity and distribution of a welded joints weakening within HAZ / CG-HAZ.

2. Performed experiment

Set of welded joints were performed using gas metal arc welding (GMAW) on two HSS, S690QL and S890QL, both delivered in accordance to EN 10025-6. The general welding parameters are shown on Tab. 1, while the appearance of welded joints macro-sections are shown on Fig. 2.

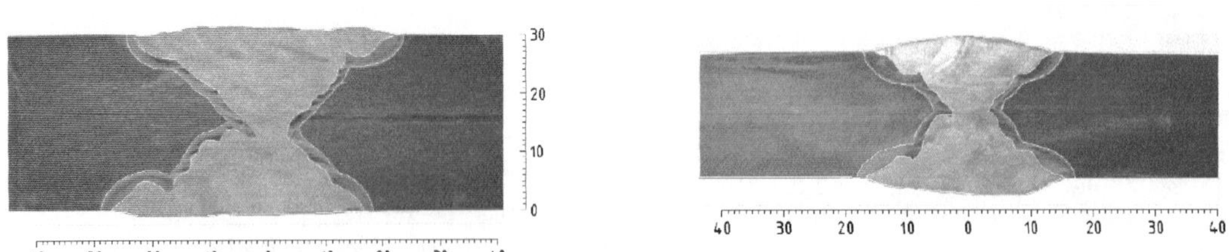

a) S690QL, 30mm

b) S890QL, 20mm

Fig. 2. Appearance of welded joints macro-sections [6]

HSS	Welding process	Thickness	Preheating	$t_{8/5}$	Filler Metal acc. to EN 12534
S690QL	GMAW	30 mm	200 °C	6-8 s	G 69 5 M Mn 3 Ni 1 Cr Mo
S890QL	GMAW	20 mm	150 °C	6-7 s	G 89 6 M Mn 4 Ni 2 Cr Mo

Table 1. General welding parameters

Beside real welds on both HSS, a sufficient number of specimens, 10x10x50 mm from both HSS were prepared for welding - thermo cycle simulation on thermo-mechanical simulator "SmithWeld".

Those specimens were simulated with previously calculated thermos-cycles, using application "Thermocycle t85" (according to welding parameters shown on Tab. 1; e.g. preheat temperature and $t_{8/5}$), which correspond to CG-HAZ, e.g. with peak temperature 1300 °C. Appearance of samples before, during and after welding thermo-cycle simulation are shown on Fig. 3.

The simulated samples were used for testing of hardness and impact toughness; while they actually represent the weakest zone of welded joint within heat-affected zone (HAZ).

a) before simulation b) during simulation c) after simulation

Fig. 3. Appearance of simulated specimens [6]

Sampling of test specimens for tensile and impact toughness are taken as shown on Fig. 4, including spots for hardness testing on welded joints macro-sections.

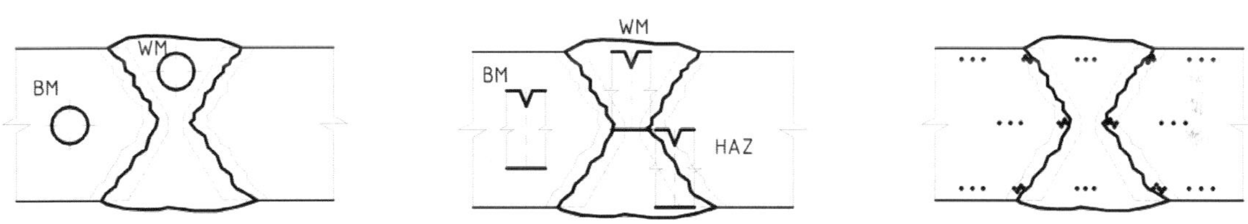

a) Tensile specimens for BM and WM; round transverse specimens were used.

b) Initial notch for BM, WM and HAZ for impact toughness specimens;

c) Hardness testing spots across welded joint macro-section (BM, HAZ, WM)

Fig. 4. Approximate sampling location for mechanical testing of real welded joints

All predicted mechanical testing were performed at room temperature, e.g. 20°C; and in accordance to:

- EN 10002-1 and EN 895, for tensile testing of base metal (BM) and weld metal (WM).
- EN 10045-1 for impact toughness (or general toughness) testing of base metal (BM), heat-affected zone (HAZ) and weld metal (WM).
- EN 1043-1 for hardness testing HV10 (Vickers method) of welded joints macro-section (including all three zones: BM, HAZ and WM).

3. Acquisition of mechanical properties

Tab. 2 and 3 shows acquired experimental results, i.e. general ranges, of tested mechanical properties for both HSS; while only the shaded cells values were calculated based on relationship provided in Fig. 5.

Obviously, from Tab. 2 and 3; high hardness has for consequence high strength and low toughness (as well as ductility); and vice-versa. Therefore, while considering generalised linear theoretical relationship [6] between hardness (HV10), strength (R_m) and toughness (KV); the following relationship (regression) analysis were performed (Fig. 5).

Mechanical property	Tested welded joint zone			
	BM	HAZ	CG-HAZ	WM
Strength / Yield stress, $R_{p0,2}$ [MPa]	745-780	-	-	695-720
Strength / Tensile strength, R_m [MPa]	857-898	1020	1193	770-779
Ductility, A [-]	0,133-0,159	-	-	0,144-0,164
Hardness, HV10	262-283	270-421	413	254-274
Impact toughness, KV [J] @ 20°C	184-212	184-187	85-94	114-171

Table 2. Acquired mechanical properties for HSS, S690QL [6]

Mechanical property	Tested welded joint zone			
	BM	HAZ	CG-HAZ	WM
Strength / Yield stress, $R_{p0,2}$ [MPa]	935-962	-	-	878-915
Strength / Tensile strength, R_m [MPa]	1007-1024	1129	1247	963-984
Ductility, A [-]	0,121-0,131	-	-	0,138-0,157
Hardness, HV10	327-351	322-455	434	317-351
Impact toughness, KV [J] @ 20°C	147-168	135-145	54-59	99-108

Table 3. Acquired mechanical properties for HSS, S890QL [6]

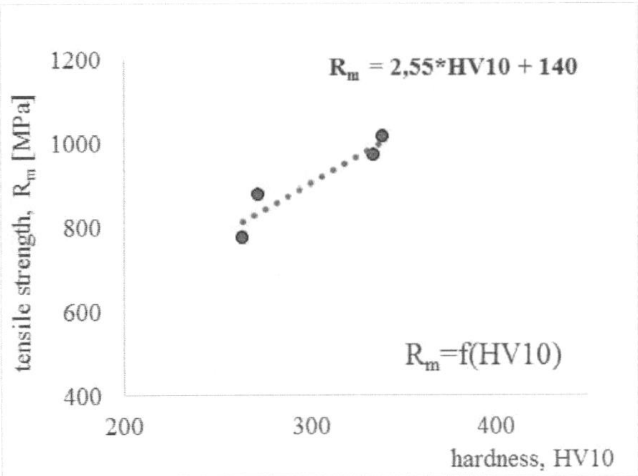

a) hardness vs. strength

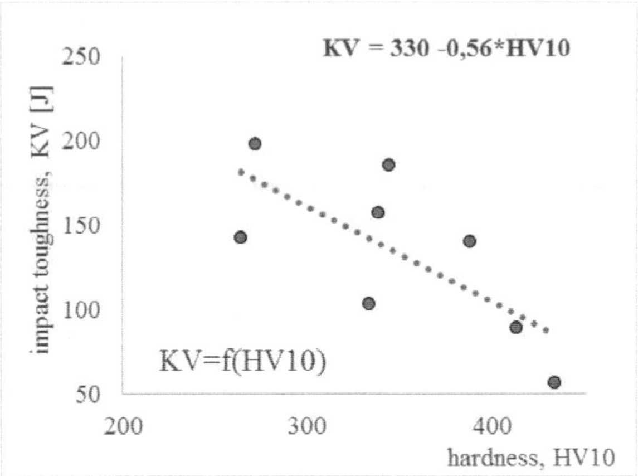

b) hardness vs. impact toughness

Fig. 5. Mechanical properties relationship(s)

4. Results analysis

Based on results, for R_m, HV10 and KV, provided in Tab. 2 and 3, the analysis (Fig. 6) of mechanical properties distribution is performed across one side of welded joint (e.g. right-hand of macro-section). However, it is reasonable to predict symmetry of mechanical properties distribution perpendicular to the butt welded joint axis.

Distance in [mm] of each zone (CG-HAZ, HAZ) from welded joint centre-line (e.g. c/l; which correspond to weld metal zone, WM) is determined from macro-section measurement (Fig. 2a and 2b). Actually, if considered on mid-thickness; the distances from WM (c/l) of CG-HAZ and HAZ are approximately 6mm and 8mm for S690QL; 5mm and 7mm for S890QL; respectively.

The distance between HAZ (rest of HAZ with peak temperatures less than 1100°C) and CG-HAZ is based on assumption that temperature distribution (according to Fig. 1b) is linear within HAZ. Actually, the width of CG-HAZ is 1,5-2,0mm on both HSS welded joints; based on theoretical assumption [6] that CG-HAZ is located between 1100 – 1500°C (Fig. 1b), with its peak on the 1300°C.

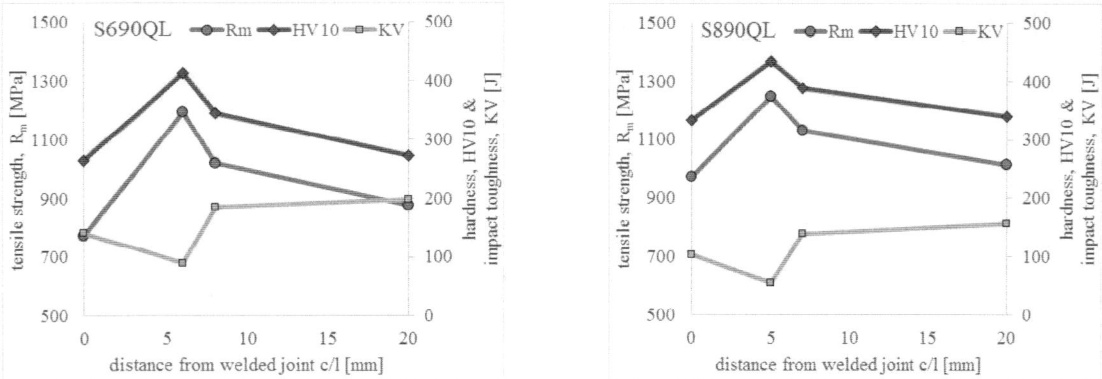

a) HSS, S690QL b) HSS, S890QL

Fig. 6. Distribution of mechanical properties across welded joints macro-sections

Finally, according to design requirements [4,5], that reference value for further comparison analysis are properties of a base metal (BM); than we can conclude mismatching of mechanical properties as shown in Tab. 4.

Mechanical property	Mismatch of mechanical properties in comparison to base metal (BM)			
	BM	HAZ	CG-HAZ	WM
Strength	1	(overmatching) 1,16 for S690QL 1,11 for S890QL	(overmatching) 1,36 for S690QL 1,23 for S890QL	(undermatching) 0,88 for S690QL 0,96 for S890QL
Hardness	1	(overmatching) 1,27 for S690QL 1,14 for S890QL	(overmatching) 1,52 for S690QL 1,28 for S890QL	(undermatching) 0,97 for S690QL 0,98 for S890QL
Toughness	1	(undermatching) 0,93 for S690QL 0,89 for S890QL	(undermatching) 0,45 for S690QL 0,37 for S890QL	(undermatching) 0,72 for S690QL 0,66 for S890QL

Table 4. Mismatch of mechanical properties across welded joint

Simply, undermatching (grey shaded cells in Tab. 4) may be considered as weakening of particular welded joint zone relative to the base metal; e.g. regarding strength, the weld metal (WM) is weaker in comparison to the base metal (BM) for 12% in the case of S690QL steel welded joint; or 4% weaker in the case of S890QL steel.

5. Conclusion

A problem of welded joint mechanical properties distribution, regarding level of mismatching effect, in real welds is present and it cannot be evaluated using standardised qualification procedures. Therefore, more detailed and sophisticated approach is required, at least by mean of preparation of additional specimens using simulation of welding - thermo cycle(s) of theoretically known weakest zone – coarse grained heat affected zone (CG-HAZ).

The presented investigation show one limited approach, mostly regarding selected high-strength steel(s) (its nominal strength values), and simulation used only for further evaluation of hardness and toughness. Also, it is possible to use presented simulation technique (as a mean of specimen's preparation) for any further mechanical testing; such as for strength testing which is not done within this research. For the purpose of this investigation, the strength of simulated CG-HAZ was recalculated (Tab. 2 and 3) based on the mechanical properties relationship analysis (Fig 5).

However, the provided investigation show clearly that the higher is strength of one high-strength steel (HSS) the lower ductility and toughness are. Beside that; welding may influence significantly a mechanical properties of welded joint.

Generally speaking, mismatching of properties are always present; but further important question arise: how intense mismatching effect is in major welded joint zones?! However, from a point of design requirements, a particular attention should be drawn to mismatching of hardness and toughness.

Thus, in the case of investigated high-strength steel's (HSS) GMAW welded joints, with the nominal strength in the range of 690-890 MPa; there is a significant undermatching of toughness; as well as overmatching of hardness; within heat-affected zone (HAZ).

The hardness overmatching goes from +(14-52)%, with its maximum value within coarse-grained heat-affected zone (CG-HAZ). Contrary, the toughness undermatching goes from –(7-63)%, with its maximum weakening within coarse-grained heat-affected zone (CG-HAZ).

Such degraded toughness may have for consequence notable loss of welded joint crack resistance. In addition, while so weakened zone of welded joint represent up to 40% of complete heat-affected zone (HAZ) width, it should not be neglected easily.

Finally, it is suggested that further studies and investigation(s) include wider range of structural steels; for example from 420 MPa, up to 1300 MPa; and more use of simulation technique. Actually, as long as simulation of welding - thermo cycles can be used for any peak temperature within HAZ (not only for CG-HAZ, 1300°C), it can provide more precise analysis of mechanical properties distribution.

6. References

[1] Janicek S.; Jirkova H.; Kucerova L. & Masek B. The behaviour of low-alloy high-strength steel after different types of processing for dynamic load, Annals of DAAAM for 2011 & Proceedings of the 22nd International DAAAM Symposium, Volume 22, No. 1, ISSN 1726-9679, ISBN 978-3-901509-83-4, Vienna, Austria, EU, 2011.

[2] Hajro I.; Burzic Z.; Kapor J. N.; Kokelj T. Experimental investigation of high-strength structural steel welds. Strojniski vestnik – Jounral of Mechanical Engineering 58(2012)6, p422-428, Slovenia, 2012.

[3] Kolarik L.; Kolarikova M.; Vondrous P. & Hrabina R. The choice of parameters for welding of steel S355NL. Annals of DAAAM for 2012 & Proceedings of the 23rd International DAAAM Symposium, Volume 23, No. 1, ISSN 2304-1382, ISBN 978-3-901509-91-9, Vienna, Austria, EU, 2012.

[4] ISO 15614-1, Specification and qualification of welding procedures for metallic materials - Welding procedure test - Part 1: Arc and gas welding of steels and arc welding of nickel and nickel alloys, ISO - International Organization for Standardization, 2017.

[5] ASME Boiler and Pressure Vessel Code - Section IX, Welding, Brazing, and Fusing Qualifications, ASME - American Society of Mechanical Engineers, 2017.

[6] Hajro I (2010). Research of ductile tearing and fracture resistance of quenched and tempered high-strength steel's welded joints, Ph.D. Dissertation, Department for manufacturing technologies, Faculty of Mechanical Engineering – University of Sarajevo, Bosnia and Herzegovina.

[7] Hot rolled Steel Plates, Sheets and Coils - Processing of material - Welding, Ruukki / Ruukki Metals Oy, Helsinki, Finland, 2017.

[8] High-strength heavy plates - Weight savings combined with excellent weldability, VoestAlpine Grobblech GmbH, Austria, 2008.

[9] Holub L.; Dunovsky J.; Kovanda K. & Kolarik L. SAW - Narrow Gap Welding CrMoV Heat-Resistant Steels Focusing to the Mechanical Properties Testing. Procedia Engineering - Conference Proceedings of the 25th International DAAAM Symposium on Intelligent Manufacturing and Automation, Volume 100-2015, ISSN 1877-7058, ScienceDirect.

[10] Inoslav R.; Gubeljak N. & Praunseis Z. The fracture behaviour of global / local mi-matched weld joints provided on HSLA steels. Materials and Technology 35(3-4)89(2001). ISSN 1580-2949. Slovenia, EU, 2001.

DOI: 10.2507/28th.daaam.proceedings.087

EXPERIMENTAL RESEARCH OF ROTATION TRANSMISSION DEVICE IN HIGH AMBIENT HYDROSTATIC PRESSURE CONDITIONS

Anton Kuzmin, Valeriy Popov & Sergey Stazhkov

This Publication has to be referred as: Kuzmin, A[nton]; Popov, V[alery] & Stazhkov, S[ergey] (2017). Experimental Research of Rotation Transmission Device in High Ambient Hydrostatic Pressure Conditions, Proceedings of the 28th DAAAM International Symposium, pp.0617-0620, B. Katalinic (Ed.), Published by DAAAM International, ISBN 978-3-902734-11-2, ISSN 1726-9679, Vienna, Austria
DOI: 10.2507/28th.daaam.proceedings.087

Abstract

In this article, the results of experimental research of rotation transmission device prototype characteristics are presented. The device can be used in different types of deep-water equipment. To confirm theoretical calculations of the device characteristics a prototype was designed and produced. The scheme of the prototype is explained in the article. The experiments were conducted for three different work fluids. Leakage volume dependencies on the operating pressure and shaft rotation frequency are examined. Friction torque and breakaway torque dependencies on operating pressure are examined. Efficiency coefficient and time of establishment of thermal regime of the rotation transmission device are found.

Keywords: machinery; deep-sea mining; power unit; hydraulics

1. Introduction

The World ocean is one of the most important sources of life sustain on Earth. The knowledge about its resources shows that in future resource depletion on Earth surface can be stopped by changing mining priority from land to ocean floor. Therefore, the World ocean research and development can significantly influence on solving a number of global problems.

Despite great prospects of using of bowels of the World Ocean, society at the modern stage of technical development is mostly focused on oil and gas production at easy accessible districts near continents. In most cases modern methods of World Ocean research requires application of universal underwater engineering. Recently increased interest in the development of the seabed and the ocean floor demands equipment that is capable of solving various tasks in scientific, ecological spheres as well as constructing of underwater objects etc.

A common problem that comes up during designing of underwater objects for different purposes is to transmit rotation through strong housing into high ambient hydrostatic pressure environment, and an essential question is to use proper shaft seal. In modern industry, it is mostly common to use translatory motion to transmit energy in deep sea environment. This fact is provided inter alia by different conditions of friction: when an interior element moves translatory relative to an outer element, contact area is changing during one operating cycle. When an interior element rotates in an outer element, there is a problem of contact stresses and constant friction area. Therefore, the problem of gap seal development is mostly solved for translatory motion, but this solutions can be not quite suitable for rotation. Wang et. al [1] decided to use a thin layer of polyetheretherketone (PEEK) as a seal material. However, while the inner element rotates with high speed relative to the outer element a covering on the surface of the inner element has a high probability to be erased. Wang, Zhi, Hu et. al [2] proved that gap seal has significant advantages over seal rings. As for application of hydraulic oil to transfer seawater pressure Selwyn and McCoy [3, 4] designed a new reciprocating hydrostatic motor.

An eccentricity of a gap seal has a large influence on leakage volume. Wieczorek and Ivantysynova examined this fact in their work [5]. To increase hydrodynamic forces that reduce the eccentricity of a piston that rotates and moves in axial direction in a bushing of axial piston pump Ivantysynova and Lasaar [6] proposed a design of a special barreled shape of the piston. This can also be not acceptable for a rotating shaft as it may cause some additional contact stresses, while variable section does not allow using minimal gap for the whole seal length. Hydrodynamic pressure can reach high values and handle external load in case of axial piston machines as it was shown in a work by Pelosi [7], though axial motion has a significant impact on the pressure that is built in the gap, therefore, this solution also may not be basic for advanced gap seal design.

To implement the task the rotation transmission device in High Ambient Hydrostatic Pressure Conditions was designed. Its constructive scheme and mathematical model for its characteristics calculation was examined in [8]. To verify the calculation methods of the rotation transmission device characteristics described in [8], to confirm its operability and to determine its characteristics experimentally its prototype was designed. A number of experiments was conducted under the following conditions: 1) ambient pressure 20 – 60 MPa, 2) rotation frequency 10 – 200 1/s, 3) torque 10 – 300 N m, 4) the research was conducted under actual conditions. The prototype was manufactured and comprehensively tested. The prototype design is shown on fig. 1.

2. The rotation transmission device prototype

The rotation transmission device prototype works in a following way:

Elastic elements 17 provide precompression of the floating bushing 7 through ground-in spherical washer 19 to the frontal surface of the partition 18 of the housing 1. Wherein tightness of the joint between the floating bushing 7 and the washer 19 and the face of the partition 18 of the housing 1.

At the impact of hydrostatic pressure the locking fluid under a pressure, equal to an ambient pressure, is supplied from a source to the primary locking chamber 8. Wherein further compression of the floating bushing 7 and spherical washer 19 to the partition 18 of the housing 1 is provided.

Face seal 5 serves for separation of the environment of the high pressure and the locking liquid. It operates at the pressure drop close to 0. The locking fluid is sealed on account of hydraulic resistance of an annular gap between the floating bushing 7 and the shaft 4.

The floating bushing 7 due to its mobility relative to the spherical surface and mobility of the spherical washer 19 in radial direction under an influence of hydrostatic and hydrodynamic forces is mounted coaxially to the shaft and traces its oscillation. Through the gap seal 7 leakages enter to the secondary locking chamber 9. From the chamber, the leakage is drained to a measuring capacitance. An axial force affecting the shaft 4, is perceived by a thrust bearing.

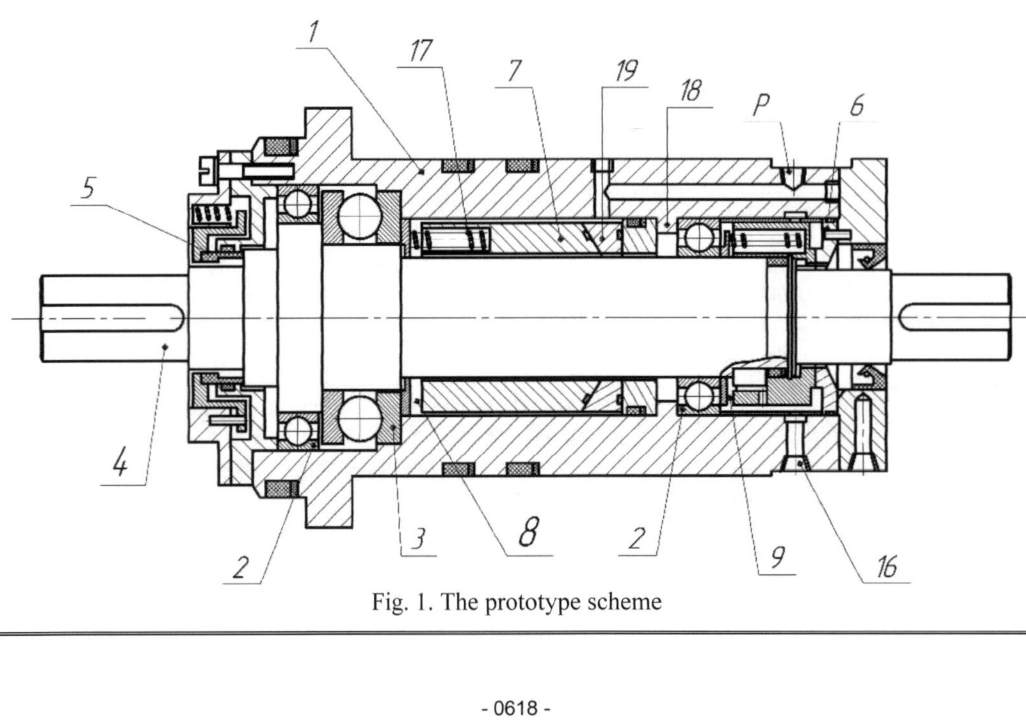

Fig. 1. The prototype scheme

This prototype is designed for moment transmission up to 20 kg·m at rotation frequency up to 3000 rpm and pressure value not exceeding 60 MPa. Its shaft diameter is 30 mm. The static characteristics of the rotation transmission device are volume losses (leakage), mechanical losses (friction torque, breakaway torque), and locking fluid temperature at the outlet of the device.

3. The experiment

The prototype of the rotation transmission device was tested for three types of locking fluid: spindle oil $\nu = 0{,}45\ cm^2/s$ vacuum oil 1 $\nu = 1{,}4\ cm^2/s$ and vacuum oil 2 $\nu = 3{,}2\ cm^2/s$. Due to the fact, that viscosity of the locking fluid has a major impact on amount of leakage, before the research a series of experiments were conducted to determine kinematic viscosity coefficient υ dependence on temperature.

Leakages were measured at two values of the radial gap between the shaft and the bushing, equal to 0.045 mm and 0.0175 mm. As thermal processes have a big impact on the amount of leakages for the small gaps and flow rates of the locking fluid, the leakages values for every pressure value were measured after establishment of stable thermal conditions – in about one or two hours after the rotation transmission device operation start. This time depends on the locking fluid pressure.

Leakage dependencies of the three oils on operating pressure for the gap of 0.045 mm is presented on Fig.2, and for the gap of 0.0175 mm – on Fig. 3. As it can be seen from the graphs, the dependencies are linear and most affected by the height of the gap between the shaft and the bushing. At the gap of 0.008 mm, vacuum oil 1 and operating pressure of 40 MPa a leakage rate of 10–15 cm^3/min was achieved.

At smaller gap height, values and pressure values of more than 7–8 MPa the bushing stopped its motion relative to the shaft, being pressed to it. The friction torque value dramatically increased. This fact can be explained by the bushing deformation, as its value becomes comparable to the gap height value and by a breakdown of the lubrication film between the bushing and the shaft.

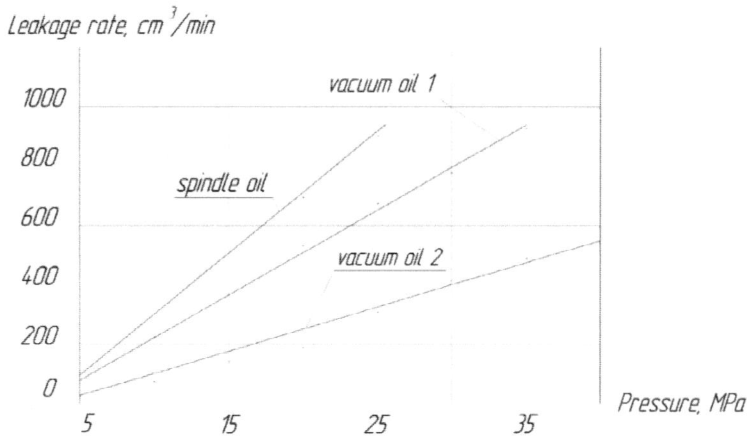

Fig. 2. Leakage rate at the gap height of 0.045 mm

The shaft rotation frequency, changing from 50 to 3000 rpm, showed no impact on the volume of leakage. The leakage volume when the shaft was not rotating at the same operating pressure was different in different experiments and varied from the leakage volume in the case of rotating shaft (Q_H) up to 1.5 Q_H. This fact can be explained by a different eccentricity of the bushing relative to the shaft, and minimal leakage in the case of rotating shaft indirectly shows a centering of the bushing when the shaft was rotating.

The friction torque of the rotation transmission device depends on pressure and viscosity of the locking fluid. It consists of several values: a friction torque in a liquid layer between the bushing and the shaft, a friction torque in the thrust bearing, friction torques in radial bearings, friction torques in the face seals. It is worth noting that only two first components depend on the locking fluid pressure, while the working fluid viscosity has an impact only on the first component.

As the locking fluid temperature grows at a fixed pressure value, the friction torque in the rotation transmission device decreases due to locking fluid viscosity decrease and becomes constant as it comes to stationary temperature conditions. With the pressure changing, the friction torque varies from 1.2–2.6 N m.

Comparing the friction torque function of the operating pressure to the friction torque in the thrust bearing, it is worth noting, that almost all the increase of the friction torque in the rotation transmission device with the pressure changing from 0 to 35 MPa is related to the increase of the friction torque in the thrust bearing.

The device breakaway torque dependence on the operating pressure value was also examined experimentally. Results of the experiment are shown on Fig. 4.

Comparing these experimental results to the experimental data of a friction torque at the spherical surface of the bushing and a friction torque at the face surface of the spherical washer, it can be concluded, that the breakaway torque value is determined by the bushing–shaft interface before the shaft starts its rotation, but not by the bushing turn along the face surface, as the breakaway torque value is much less than the value of the friction torque at the face surface.

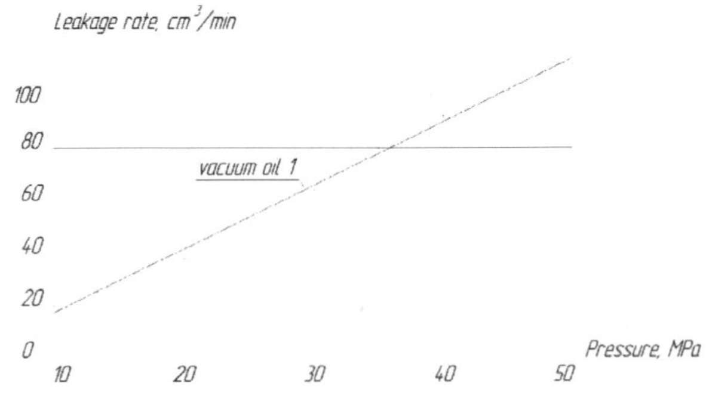

Fig. 3. Leakage rate at the gap height of 0.0175 mm

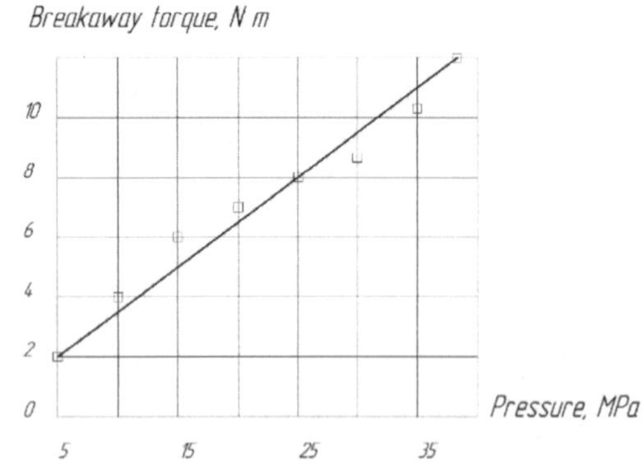

Fig. 4. Breakaway torque function of pressure

Experimental measurement of the locking fluid temperature at the outlet of the gap seal showed, that stationary temperature conditions is established in about an hour after the rotation transmission device starts. Wherein a liquid temperature increment at the outlet of the gap seal has a significant value and can reach up to a few dozens of degrees, which entails changes of the operating liquid viscosity 6–8 times.

The results of the experiments fully confirmed theoretical research [8]. Other investigations were done for short gaps [9], and in the present case, the gap is longer due to the spherical washer. Operability of this device was also confirmed, as well as its advanced characteristics.

4. Conclusion

The problem of transmission of rotary motion to an environment with high ambient pressure was solved. The constructive scheme and the device prototype were presented. The results of the experimental research confirmed the operability of the device and allowed to define its specifications.

The experimental static characteristics of the prototype confirmed an acceptability of the assumptions, introduced in theoretical calculations. It also showed that the difference between the values of theoretical and experimental data does not exceed 20%, what is acceptable for this kind of calculations. The experimental efficiency of the rotation transmission device at the pressure of 40 MPa was 98%.

The rotation transmission device was designed and examined theoretically and experimentally. In the future it is planned to conduct durability tests of the device.

5. References

[1] Jianjun Wang, Huawei Qin, Ying Chen, Energy performance evaluation of an innovative hydrostatic motor, In Renewable Energy, Volume 57, 2013, Pages 497-505, ISSN 0960-1481, https://doi.org/10.1016/j.renene.2013.01.059.

[2] Wang, Z. , Zhi, Y. , Hu, C. , Xi, J. , Hu, J. and An, J. (2013) Study of Seals on Long Stroke and Reciprocating Motion Condition. Journal of Power and Energy Engineering, 1, 40-44. doi: 10.4236/jpee.2013.17007.

[3] Selwyn S, McCoy FW. The hydrostatic motor: utilization of hydrostatic pressure differentials in the deep sea. Geo Mar Lett 1981;1:233e6.

[4] McCoy F, Selwyn S. The hydrostatic corer. Mar Geol 1984;54:M33e41.

[5] Uwe Wieczorek & Monika Ivantysynova (2002) Computer Aided Optimization of Bearing and Sealing Gaps in Hydrostatic Machines—The Simulation Tool Caspar, International Journal of Fluid Power, 3:1, 7-20, DOI: 10.1080/14399776.2002.10781124

[6] Monika Ivantysynova & Rolf Lasaar (2004) An Investigation into Micro- and Macrogeometric Design of Piston/Cylinder Assembly of Swash Plate Machines, International Journal of Fluid Power, 5:1, 23-36, DOI: 10.1080/14399776.2004.10781181

[7] Matteo Pelosi & Monika Ivantysynova (2012) A Geometric Multigrid Solver for the Piston–Cylinder Interface of Axial Piston Machines, Tribology Transactions, 55:2, 163-174, DOI: 10.1080/10402004.2011.639049

[8] Kuzmin, A[nton]; Popov, V[alery] &Stazhkov, S[ergey] (2016). Rotation Transmission Device in High Ambient Hydrostatic Pressure Conditions., Chapter 22 in DAAAM International Scientific Book 2016, pp.239-248, B. Katalinic (Ed.), Published by DAAAM International, ISBN 978-3-902734-09-9, ISSN 1726-9687, Vienna, Austria DOI: 10.2507/daaam.scibook.2016.22

[9] Slyozkin N.A. (1986) Sealing and compaction equipment, Mashinostroenie, pp. 464, USSR

DAAAM International Vienna
Bahnallee 10
A-2120 Wolkersdorf
Austria, EU

CURRAN ASSOCIATES INC.
proceedings.com

ISBN 978-1-5108-5327-0

28th International DAAAM Symposium on Intelligent Manufacturing and Automation 2017

Zadar, Croatia
8-11 November 2017

Volume 2 of 2

28th International DAAAM Symposium on Intelligent Manufacturing and Automation 2017

Zadar, Croatia
8-11 November 2017

Volume 2 of 2

Editors:

B. Katalinic

ISBN: 978-1-5108-5327-0

Printed from e-media with permission by:

Curran Associates, Inc.
57 Morehouse Lane
Red Hook, NY 12571

Some format issues inherent in the e-media version may also appear in this print version.

Printed by Curran Associates, Inc. (2018)

DAAAM International Vienna
Bahnallee 10
A-2120 Wolkersdorf
Austria, EU

Phone: +436646424398

secretary@daaam.info

Additional copies of this publication are available from:

Curran Associates, Inc.
57 Morehouse Lane
Red Hook, NY 12571 USA
Phone: 845-758-0400
Fax: 845-758-2633
Email: curran@proceedings.com
Web: www.proceedings.com

ISSN 1726-9679

DANUBE ADRIA ASSOCIATION FOR AUTOMATION & MANUFACTURING

DAAAM INTERNATIONAL VIENNA

www.daaam.info

PROCEEDINGS

OF THE 28TH INTERNATIONAL DAAAM SYMPOSIUM
"INTELLIGENT MANUFACTURING & AUTOMATION
08-11TH NOVEMBER 2017
ZADAR, CROATIA

ORGANIZED BY:
DAAAM INTERNATIONAL VIENNA

IN COOPERATION WITH:
ÖIAV 1848, VIENNA UNIVERSITY OF TECHNOLOGY, INTERNATIONAL ACADEMY OF ENGINEERING
AND UNIVERSITY OF APPLIED SCIENCES – TECHNIKUM WIEN

UNDER THE AUSPICES OF THE DANUBE RECTORS' CONFERENCE &
RECTORS' AND PRESIDENTS' HONOR COMMITTEE OF DAAAM INTERNATIONAL

EDITOR: B.[RANKO] KATALINIC

IMPRESSUM 2017

Proceedings of the 28th International DAAAM Symposium "Intelligent Manufacturing & Automation" 08-11th November 2017, Zadar, Croatia, Volume 28, No.1,
Abbreviated key title: ***Ann. DAAAM 2017***

ISSN 1726-9679

Ed. B. Katalinic,

Published by DAAAM International, Vienna, Austria, EU, 2017
Layout & Design: B[ranko] Katalinic/ D[amir] Haskovic
Technical Editors: D[amir] Haskovic / I[lija] Zec / R[oberta] Cvetkovska / I[lya] Kukushkin

Publisher:

DAAAM International Vienna
TU Wien, Getreidemarkt 9, 1060 Wien, Austria
E-mails: president@daaam.com / branko.katalinic@tuwien.ac.at
DAAAM homepage: http://www.daaam.info

CONTENTS

AUTHOR INDEX

A

B

Z

ORGANIZING COMMITTEE FOR 2017

COVIC, DRAGAN – PRESIDENT OF ORGANIZING COMMITTEE
KATALINIC, BRANKO – PRESIDENT OF DAAAM INTERNATIONAL
TOMIC, ZORAN – RECTOR OF THE UNIVERSITY OF MOSTAR
GUSEV, BORIS – PRESIDENT OF INTERNATIONAL ACADEMY OF ENGINEERING
UGLESIC, ANTE – COORDINATOR OF ORGANIZING COMMITTEE

BUCHMEISTER, BORUT
COLAK, IVO
COSIC, ILIJA
CVETKOVSKA, ROBERTA
EGOROV, SERGEY
FILARETOV, VLADIMIR
HASKOVIC, DAMIR (SECRETARY)
HOELLER, LIANE
KUKUSHKIN, ILYA
LEKO, TOMISLAV
MAJSTOROVIC, VLADO
NAUMBURGER, JOERG
PANFILOV, PETER
PARK, HONG-SEOK
PRYANICHNIKOV, VALENTIN
STAZHKOV, SERGEY
STOJKIC, INJA
STOPPER, MARKUS
TEKIC, ZELJKO
UDILJAK, TOMA
ZAVRAZHINA, ALINA
ZEC, ANTE
ZEC, ILIJA
ZEC, LUKA

RECTORS' AND PRESIDENTS' HONOR COMMITTEE

ADAMCZAK, STANISLAW	RECTOR, TECHNICAL UNIVERSITY OF KIELCE
AAVIKSOO, JAAK	RECTOR, TALLINN UNIVERSITY OF TECHNOLOGY
ABRUDAN, IOAN VASILE	RECTOR, TRANSILVANIA UNIVERSITY OF BRASOV
ALEXANDROV, ANATOLIY	RECTOR, BAUMAN MOSCOW STATE TECHNICAL UNIVERSITY
ANDJELINOVIC, SIMUN	RECTOR, UNIVERSITY OF SPLIT
ANISIMOV, NIKITA YU.	PRESIDENT, FAR EASTERN FEDERAL UNIVERSITY
ANTRANIKIAN, GARABED	PRESIDENT, HAMBURG UNIVERSITY OF TECHNOLOGY
BAAIJENS, FRANK	RECTOR, EINDHOVEN UNIVERSITY OF TECHNOLOGY
BALLUERKA LASA, NEKANE	RECTOR, UNIVERSITY OF THE BASQUE COUNTRY
BEIBST, GABRIELE	RECTOR, JENA UNIVERSITY OF APPLIED SCIENCE
BORAS, DAMIR	RECTOR, UNIVERSITY OF ZAGREB
BRUNGS, ATTILA	VICE-CHANCELLOR AND PRESIDENT, UNIVERSITY OF TECHNOLOGY, SYDNEY
BUCHVALD VIND, THOMAS	DIRECTOR, UNIVERSITY OF SOUTHERN DENMARK
CHEON, YEON	PRESIDENT, UNIVERSITY OF ULSAN
CISNEROS PÉREZ, GUILLERMO	RECTOR, UNIVERSIDAD POLITECNICA DE MADRID
COSTOIU, MIHNEA	RECTOR, UNIVERSITY POLITEHNICA OF BUCHAREST
DANIELS, MITCHELL	PRESIDENT, PURDUE UNIVERSITY
DEMA, MARJAN	RECTOR, UNIVERSITY OF PRISHTINA
DI SCIASCIO, EUGENIO	RECTOR, POLYTECHNIC OF BARI
FALKOV, VALERIY	RECTOR, TYUMEN STATE UNIVERSITY
FRUNZAVERDE, DOINA	RECTOR, "EFTIMIE MURGU" UNIVERSITY OF RESITA
GALIJEV, ILJHAM	PRESIDENT, OMSK STATE TRANSPORT UNIVERSITY
GARCÍA MARÍN, JUAN FRANCISCO	RECTOR, UNIVERSITY OF LEON
GELENCSER, ANDRAS	RECTOR, UNIVERSITY OF PANNONIA
GILLI, MARCO	RECTOR, POLITECNICO DI TORINO
GOLIAS, IOANNIS	RECTOR, NATIONAL TECHNICAL UNIVERSITY OF ATHENS

Gonokami, Makoto	President, University of Tokyo
Grigoryev, Sergey	President, Moscow State University of Technology "Stankin"
Guberac, Vlado	Rector, Strossmayer University of Osijek
Gusev, Boris V.	President, Russian Academy of Engineering, Moscow
Hannula, Mika	President, Tampere University of Technology
Hanselka, Holger	Rector, Karlsruhe Institute of Technology
Hanselka, Holger	President, Karlsruhe Institute of Technology
Herrmann, Wolfgang	President, Technical University Munich
Holecek, Miroslav	Rector, University of West Bohemia
Horatiu, Mihai	President, Lucian Blaga University of Sibiu
Ivanov, Konstantin	Rector, Baltic State Technical University
Jozsa, Janos	Rector, Budapest Univ. of Technology and Economics
Kainz, Harald	Rector, Graz University of Technology
Kanok-Nukulchai, Worsak	President, Asian Institute of Technology
Karlsson, Sigbritt	President, Royal Institute of Technology, Stockholm
Kazior, Jan	Rector, Cracow University of Technology
Kmet, Stanislav	Rector, Technical University of Kosice
Konvalinka, Petr	Rector, Czech Technical University in Prague
Koshkin, Valeriy	Rector, Sevastopol State University
Kosih, Anatoliy	Rector, Omsk State Technical University
Kosih, Anatoly	President, Omsk State Technical University
Lazovsky, Dmitry	Rector, Polotsk State University
Lodygowsk, Tomasz	Rector, Poznan University of Technology
Manz, Carsten	President, Konstanz University of Applied Sciences
Marat Kenesovich, Ibatov	Rector, Karaganda State Technical University
Mazo, Eduardo Gonzalez	Rector, University of Cadiz
McDowell, Jim	Chancellor, University of South Australia
Mezyk, Arkadiusz	Rector, Silesian University of Technology, Gliwice
Mihov, Georgi	Rector, Technical University of Sofia
Misra, Ashok	Chairman, Indian Institute of Technology Roorkee
Mueller-Steinhagen, Hans	Rector, Dresden University of Technology
Myners, Paul	Chancellor, The University of Exeter
Ovcharenko, Sergey	Rector, Omsk State Transport University
Prijić-Samaržija, Snježana	Rector, University of Rijeka
Radtke, Ulrich	Rector, University of Duisburg-Essen
Redhammer, Robert	Rector, Slovak University of Technology Bratislava
Ressel, Wolfram	Rector, University of Stuttgart
Ribickis, Leonids	Rector, Riga Technical University
Rittidech, Sampan	President, Mahasarakham University
Rogalev, Nikolai	Rector, Moscow Power Engineering Institute - TU
Rt Hon Lord Mandelson	Chancellor, Manchester Metropolitan University
Rudskoy, Andrey	Rector, Peter the Great St. Petersburg Polytechnic University
Saha, Petr	Rector, Tomas Bata University University Zlin
Sahin, Bahri	President, Yildiz Technical University
Schmoellebeck, Fritz	Rector, Univ. of Applied Sciences Technikum, Vienna
Seidler, Sabine	Rector, Vienna University of Technology
Sels, Luc	Rector, Catholic University of Leuven
Serban, Viorel-Aurel	Rector, Polytechnic University of Timisoara
Skrijelj, Rifat	Rector, University of Sarajevo
Smedescu, Ion	President, Romanian American University
Stepanek, Petr	Rector, Brno University of Technology
Stoll, Wilfried	Managing Partner- Festo
Strohmeier, Gerd	Rector, Chemnitz University of Technology
Sung-Chul, Shin	President, Korea Advanced Institute for Science and Technology
Szmidt, Jan	Rector, Warsaw University of Technology
Niemela, Ilkka	President, Helsinki University of Technology

INTERNATIONAL SCIENTIFIC COMMITTEE

INTERNATIONAL PROGRAM COMMITTEE

MOHORA, CRISTINA
MOK, H.
MORARU, SORIN AUREL
MORIWAKI, T.
MORTOIU, DOINA
MOTOMANCEA, ADRIAN
MRSA, ZORAN
MRVAC, NIKOLA
MUNTEAN, MIHAELA
MUNTEAN, SEBASTIAN
MUNTEANU, ANCA
MUSCA, GAVRIL
MUSTAFIZUL, KARIM
MYSHLYAVTSEV, ALEXANDR
NAGI, MIHAI
NAGLA, K. S.
NAKAMOTO, NOBUYUKI
NEUMANN, PETR
NICA COTET, GABRIELA BEATRICE
NIKOLIC, GOJKO
NKUMBWA, RODGER
NOH, SANG DO
NOVAC, OVIDIU
NOVAKOVIC, BRANKO M.
NOVOSELOVIC, DANIEL
NURUL AMIN A.K.M.
OANCEA, GHEORGHE I.
OBAD, MILENKO
OKAMOTO, J.
OLARU, ADRIAN, DUMITRU
OLARU, ONISIFOR
OLEKSIK, VALENTIN
OLTEANU, CIPRIAN
OMEROVIC, MIDHAT R.
OPRAN, CONSTANTIN
OPREAN, AUREL
OPRUTA, DAN
OSMANI, HYSNI
OSMERA, PAVEL
OTTOSSON, S.
PACAIOVA, HANA
PALCIC, IZTOK
PALOS, RAMONA
PAMINTAS, EUGEN
PANFILOV, PETER
PARASCHIV, NICOLAE
PARK, HONG-SEOK
PARPALA, LIDIA FLORENTINA
PASCUAL, R.COMPAMYS
PASIC, MUGDIM
PASTRAMA, STEFAN DAN
PATACHIA, SILVIA FLORICA CRISTINA
PAULIKOVA, ALENA
PAULOVA, IVETA
PAUNESCU, CARMEN MONICA
PAVIC, IVAN
PERJU, DAN
PERNIU, DANA
PERUTKA, KAREL
PETERS, JACQUES
PETKOVIC, DARKO
PETRE, CRISTIAN
PETRE, MARIAN
PETRESCU, VALENTIN
PINCA-BRETOTEAN, CAMELIA
PINTILIE, GHEORGHE
PIRTEA, GABRIEL MARILEN
PIVONKA, PETR
PJERO, ELENICA
PLESA, ANGELA
PODURAEV, JURY
POKORNY, PAVEL
POLJAK, KRUNOSLAV
POMAZAN, VALENTINA
POPA, CICERONE LAURENTIU
POPPEOVA, VIERA
PRODAN, DAN
PROKOP, ROMAN
PROSTREDNIK, DANIEL
PRYANICHNIKOV, VALENTIN
PYTKO, STANISLAW
QUINN, ROGER D.
RADIC, JURE
RAMAJ, VEHBI XH.
RAOS, PERO
RAVAS, RUDOLF
RAY, T. S.
RIZESCU, SABIN VASILE
ROGALE, DUBRAVKO
ROHDE, ULRICH L.
ROMERO, FERNANDO
ROS, OLIMPIA
ROSADO, V.O.G.
RUDAS, IMRE J.
RUGESCU, DRAGOS RADU DAN
RUJA, IOAN
RUSU ANGHEL, NICOLAE
RUSU, LIGIA
RUSU-CASANDRA, AURELIA
RUTEMI, GANI BRAHIM
RUZIC, DRAGO
SALIHU, AVDI HAJDIN
SAMEK, DAVID
SCHREIBER, PETER
SELVAKUMAR, ARUL LAWRENCE
SEONG, POONG HYUN
SERCER, MLADEN
SHIBASAKA, TOSHIROH
SHIH, MING CHANG
SIEMIATKOWSKI, M.
SIHN, WILFRIED
SINAY, JURAJ
SLUGA, ALOJZ
SMIRNOV, ALEXANDER
STEFANIC, NEDELJKO
STOPPER, MARKUS
STRGAR KURECIC, MAJA
SUBRAMANIAN, ARUNKUMAR
SUGAR, PETER
SYROVA, LIVIA
SYSALA, TOMAS
TADAMURA, KATSUMI
TAKALA, JOSU
TAMASILA, MATEI
TAN, KOK KIONG
THIERHEIMER, WALTER WILHELM
TORIMS, TOMS
TURKALJ, GORAN
ULRICH, GUENTHER
VAARIO, JARI
VASEK, VLADIMIR
VELISEK, KAROL
VEZA, IVICA
VISEKRUNA, VOJO
VRANJES, BOZO
VUJASINOVIC, EDITA
WANG, JYHWEN
WEIGANG, WU
WEIGL, K. HEINZ
WESSELY, EMIL
WIENER, PETER
WINKLER, T.
WOZNIAK, ADAM
XING, KE
XUE, D.
YAMAZAKI, K.
YORDANOV, RUMEN STOYANOV
YOSHIDA, SUSUMU
YU, Y.-C.
ZAJAC, JERZY
ZAJAC, JOZEF
ZAPCIU, MIRON
ZAVOIANU, ION
ZEQIRI, HAKIF
ZGODAVOVA, KRISTINA
ZHIRABOK, ALEXEY
ZICHIL, VALENTIN
ZIRRA, DANIELA
ZUEV, ALEXANDER
ZUPERL, UROS

REVIEWING COMMITTEE

DOI: 10.2507/28th.daaam.proceedings.088

HYDRODYNAMIC PROCESSES IN PISTON–BORE INTERFACE OF AXIAL PISTON SWASH PLATE MACHINE.

Anton Kuzmin, Valeriy Popov & Sergey Stazhkov

This Publication has to be referred as: Kuzmin, A[nton]; Popov, V[alery] & Stazhkov, S[ergey] (2017). Hydrodynamic Processes in Piston–Bore Interface of Axial Piston Swash Plate Machine, Proceedings of the 28th DAAAM International Symposium, pp.0621-0626, B. Katalinic (Ed.), Published by DAAAM International, ISBN 978-3-902734-11-2, ISSN 1726-9679, Vienna, Austria
DOI: 10.2507/28th.daaam.proceedings.088

Abstract

Kinematic analysis and numerical calculations for working fluid pressure field of piston – cylindrical bore interface is presented in this article. Possible types of piston motion relative to cylindrical bore are examined. Results of this analysis are the basis for Reynolds equation generation, that describes hydromechanical processes in piston – bore interface. In the present article, a simulation model of hydromechanical processes considering the studied kinematics is shown. The piston mechanism operation at low speed is examined and pressure fields of working liquid in the gap of the piston – bore gap are obtained. This simulation model is the basis for determination of constructive parameters and operating conditions that are leading to establishment of fluid friction regime in piston – cylindrical bore interface.

Keywords: axial piston machine; swash plate machine; hydrodynamics; friction; piston – cylinder interface

1. Introduction

Axial piston swash plate hydraulic machines (swash plate machines) are important components of different actuators. These hydraulic machines are remarkable for its design simplicity and manufacturability compared with other power hydraulic machines such as axial piston bent–axis hydraulic machines. Among other advantages it has less mass, smaller overall dimensions and high efficiency. The present machines are used as pumps and motors as well, and besides swash plate machines are often used in both modes by rotation. It is also frequently required for these machines to be able to reverse its shaft rotation.

To improve a quality of swash plate machines design and to meet the requirements been said, it is expediently to reduce friction forces between a piston and a cylindrical bore. Due to the construct of the swash plate machine, a location of a hydrostatic bearing of the piston slipper on a swash plate in particular, an axis of a cylindrical part of the piston is inclined relative to an axis of the cylindrical bore. Therefore, friction forces values can be reduced by providing an alignment of the axis of the elements. One of possible solutions of the present problem is in using hydrodynamic counteracting force that is caused by piston motion relative to the bore. This force establishes a moving cylinder concentric to a stable cylinder, as reduction of fluid layer thickness causes a local pressure peak.

This article is focused on obtaining of a mathematical model of hydrodynamic processes in the piston pair. This model will allow to describe a process of transition from mixed to liquid lubrication. The aim of the present work is to find numerical values of a pressure field in the gap between the piston and the cylindrical bore. Results of this investigation will be further confirmed experimentally.

2. Literature review

In his works Slyozkin [1] described hydrodynamical parameters and calculation dependencies of an operating fluid in a lubricating layer between two concentric cylinders. In these studies, the pressure field is calculated using Reynolds equations for the lubricating layer and solving the obtained equation system relative to velocities. In more detail, the case of concentric cylinders was examined in research by Korovchinsky [2] using an example of a sliding bearing concerning its real construct.

In his research N.D. Manring [3] presented a force analysis of the piston–bore kinematic pair and a dependency for minimum fluid-film thickness. U. Wieczorek and M. Ivantysynova [4] demonstrated a model of a gap flow in the piston pair. Y. Fang and M. Shirakashi [5] described conditions of metal contact in the piston–bore interface and a way to express characteristics of mixed lubrication. M. Pelosi [6] in his work solved a problem of piston–cylindrical bore interaction using elasthydrodynamic theory that allowed obtaining more accurate results of the fluid film in loaded conditions. However, this model is calculated for high machine shaft rotation frequency (100 rad/s). In this case, there is fully hydrodynamic lubrication and the piston rotates relative to the interior surface of the cylindrical bore with a frequency equal to the shaft rotation frequency. Investigations by B. Xu, J. Zhang, H. Yang and B. Zhang [7] included radial piston motion to show its influence on the whole carrying ability of the lubricating oil film. J. Jiang and K. Wang added pump flow to this model [8].

Pressure, causing transition from mixed to fully hydrodynamic lubrication, is built up through relative velocities of the piston and the bore. For hydraulic machine shaft rotation frequencies about 0,1–1 rad/s presented pressure is not enough for creating conditions for transition to liquid friction. In this regard, friction forces can take values, enough to define kinematics of the piston mechanism in the following way: rotation happens in spherical hinge between the piston and the slipper, while cylindrical part of the piston partly rolls along the interior surface of the bore. This situation should also be taken into account while describing hydrodynamic processes in the gap.

3. Mathematical model

For operating fluid pressure field in the piston–cylindrical bore interface calculation, Reynolds equations in the lubricating layer are applied.

$$\begin{cases} \frac{\partial p}{\partial x} = \mu \frac{\partial^2 U}{\partial y^2} \\ \frac{\partial p}{\partial y} = 0 \\ \frac{\partial p}{\partial z} = \mu \frac{\partial^2 W}{\partial y^2} \\ \frac{\partial U}{\partial x} + \frac{\partial V}{\partial y} + \frac{\partial W}{\partial z} = 0 \end{cases} \quad (1)$$

In works by N.A. Slyozkin [1], M. Pelosi [6] Y.Fang and M. Shirakashi [5] and other authors presented equation system was integrated without considering members of second order of smallness and members including the factor $\partial u/\partial x$. However, while formulating more accurate simulation models, considering deformations of mating surfaces of the elements of the piston mechanism and operating fluid viscosity change, it is also worth considering the following components. For high shaft rotation speed – a possible change of an angle between the axes of the piston and the slipper by the impact of counteracting force caused by operating liquid pressure. For low rotation speed – a circular motion, different from full sliding of the piston surface along the interior surface of the bore.

Integrating the system and using mathematical transformations, the following result is obtained:

$$\frac{\partial}{\partial x}\left(h^3 \frac{\partial p}{\partial x}\right) + \frac{\partial}{\partial z}\left(h^3 \frac{\partial p}{\partial z}\right) = 12\mu v + 6\mu u \frac{\partial h}{\partial x} + 6h\mu \frac{\partial u}{\partial x} + 18\mu v \left(\frac{\partial h}{\partial x}\right)^2 + 6\mu h \frac{\partial v}{\partial x}\frac{\partial h}{\partial x} + 6\mu h v \frac{\partial^2 h}{\partial x^2} + +6\mu w \frac{\partial h}{\partial z} + 18\mu v \left(\frac{\partial h}{\partial z}\right)^2 + 6\mu h v \frac{\partial^2 h}{\partial z^2} \quad (2)$$

p – pressure, μ – viscosity coefficient, U – velocity component, tangential to the bore surface, V – velocity component, perpendicular to the bore surface W – velocity of piston motion along the bore axis (considering the piston incline), h – fluid film thickness, v – velocity component, perpendicular to the piston surface, u – velocity component, tangential to the piston surface, w – velocity of piston motion along its axis (not considering the piston incline), ω – the machine shaft rotation frequency, φ – angular coordinate of a point on the piston surface.

Kinematics of the piston mechanism is defined by ratio of friction forces in kinematic pairs. For the piston operating throughout a duty cycle there are two pairs of rotary motion: the spherical hinge formed by the piston and the slipper, piston–cylindrical bore (Fig.1). This scheme has excess motion, as the rotary motion is caused only by a cylinder block rotation around its axis.

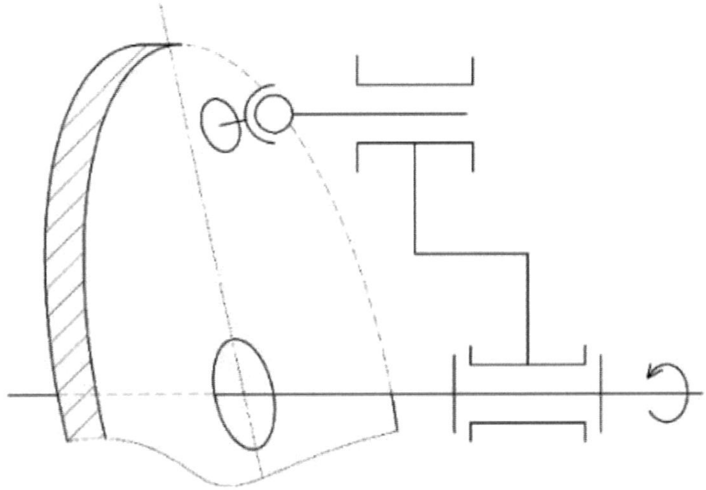

Fig. 1. Kinematic scheme of the piston mechanism.

At the impact of pressure, caused by the velocity of the relative piston to cylindrical bore motion, friction forces in piston–bore interface are significantly reducing, and rotation happens in this kinematic pair. At piston velocities that do not provide pressure, enough for significant friction forces reduction, rotation happens in the piston–slipper kinematic pair.

At the breakaway, friction forces in the piston–bore interface are so high, that in an artificial case of kinematics with rotation in piston–bore pair a joint opening between the slipper and the swash plate occurs [9]. Such a situation ruins the hydraulic machine operability.

In the swash plate machine progress, a case is possible, in which friction forces in the piston–bore prevent piston from rotating: either sliding or rolling along the bore surface (Fig.2). For this situation, the difference between two types of kinematics is visually shown: the piston sliding along the bore surface with a rotary speed, equal to the rotary speed of the shaft, and the piston local rolling along the bore.

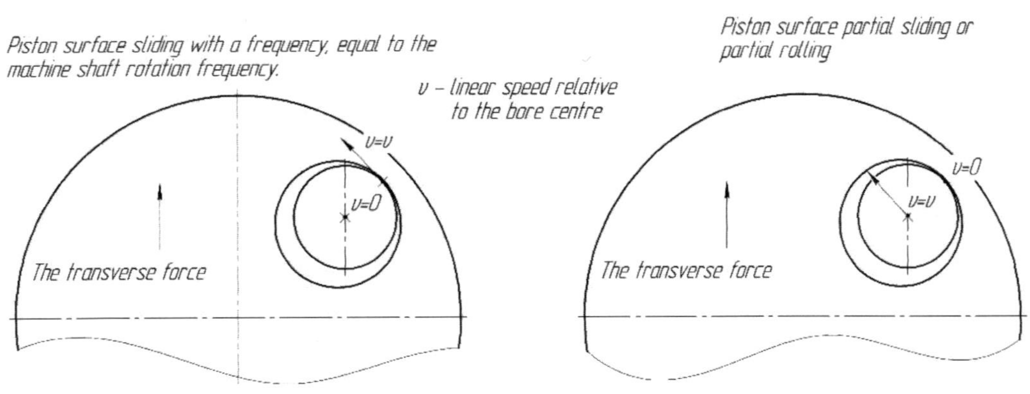

Fig. 2. The two types of piston surface rotation

In the present article, only these two cases are described. The intermediate cases will be examined in further works. The final forms of the Reynolds equation for the full piston sliding is:

$$\frac{\partial}{\partial x}\left(h^3\frac{\partial p}{\partial x}\right)+\frac{\partial}{\partial z}\left(h^3\frac{\partial p}{\partial z}\right)=6\mu u\frac{\partial h}{\partial x}+6\mu w\frac{\partial h}{\partial z} \tag{3}$$

Pelosi also examined this case in his simulation model [6]. For the other type of motion formula (2) should be applied, while

$$\begin{cases} v = 2\omega r \cdot \sin\left(\frac{\varphi}{2}\right) \cdot \cos\left(\frac{\varphi}{2}\right) \\ u = 2\omega r \cdot \left|\sin\left(\frac{\varphi}{2}\right) \cdot \sin\left(\frac{\varphi}{2}\right)\right| \end{cases} \quad (4)$$

For pressure field values calculation the tridiagonal matrix algorithm was used [10]. This method is enough for qualitative assessment of processes, occurring in the fluid film in the piston–cylindrical bore interface. Axes directions and schematic grid image (but not correct – the correct grid is defined for the fluid in the gap) are shown on Fig. 3. For presented problem, the method can be written in the following way:

$$\begin{cases} A \cdot P_{j+1} + B \cdot P_j + C \cdot P_{j-1} = F_j \\ A = \frac{h^3}{m_1^2} + \frac{\partial h^3}{\partial z} \cdot \frac{1}{2m_1} \\ C = \frac{h^3}{m_1^2} - \frac{\partial h^3}{\partial z} \cdot \frac{1}{2m_1} \\ B = 2h^3 \cdot \left(\frac{2}{m_2^2} + \frac{1}{m_1^2}\right) \end{cases} \quad (5)$$

For boundary i components the coefficient accepts the following values:

$$B = 2h^3 \cdot \left(\frac{1}{m_2^2} + \frac{1}{m_1^2}\right) + \frac{h^3}{m_2^2} \pm \frac{\partial h^3}{\partial \varphi} \frac{1}{2m_1} \quad (6)$$

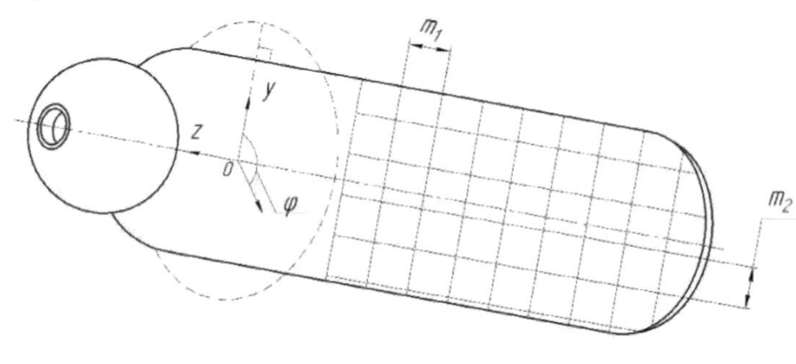

Fig. 3. Schematic grid representation

Coefficients α and β are calculated in the next way:

$$\begin{cases} \alpha_{j+1} = \frac{A_j}{B_j - C_j \cdot \alpha_j} \\ \beta_{j+1} = \frac{F_j + C_j \cdot \beta_j}{B_j - C_j \cdot \alpha_j} \end{cases} \quad (7)$$

For j=1: α=0, β=P_0 . Boundary conditions: on the inner edge of the cylindrical bore P_0 = 30 MPa, on the outer edge P = 0.

$$P_j = \alpha_{j+1} \cdot P_{j+1} + \beta_{j+1} \quad (8)$$

j – a number by axis 0Z, *i* – a number by angular coordinate *φ*, m_1 – step along axis 0Z, m_2 – step along angular coordinate *φ*, F – the right side of equations 2 and 3.

4. Pressure field

As a result, a pressure distribution of the operating fluid in the gap between the piston and the cylindrical bore was obtained. Parameters: piston diameter is 3,15 cm, cylinders centers circle diameter is 14 cm, the swash plate tilt angle is 18°, the gap height between the piston and the cylindrical bore in concentric state is 12,5 μm, the bronze bushing length is 6,15 cm. On Fig. 4-11 pressure fields on the edges of the cylindrical bore (bronze bushing) for shaft rotation speeds of 0.1 rad/s, 10 rad/s and 100 rad/s are presented.

From the values of the pressure field been obtained it can be concluded, that at maximal speed, coaxial with the bore, the rotation motion type has almost no impact on pressure peaks forming. Calculations can be more accurate while solving a complex system, M. Pelosi [6] model for example, with smaller steps values of pressure peaks reduce, while the order stays approximately the same. When the piston has no speed towards or backwards (it is situated at the dead centers), pressure peaks values on the outer edge of the bushing for sliding motion repeatedly exceed pressure that is formed by motion without sliding.

However, it is possible to conclude preliminarily, that this pressure is not enough to balance an applied to the machine shaft load. These facts give grounds to assert that no-sliding or rolling type of piston motion is more suitable for breakaway and at low speeds.

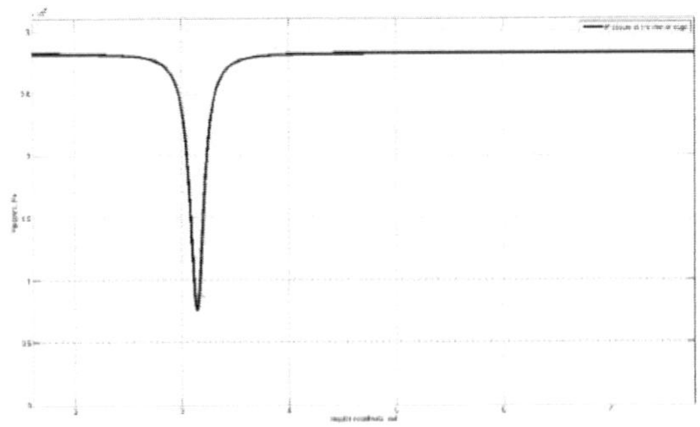
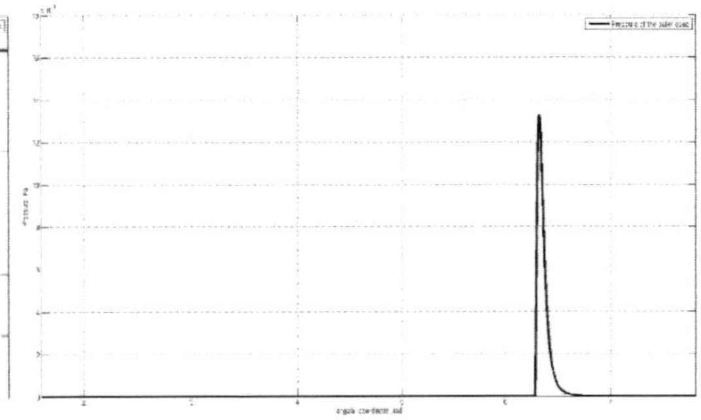

Fig. 4, 5. Pressure at piston dead centers and full surface sliding

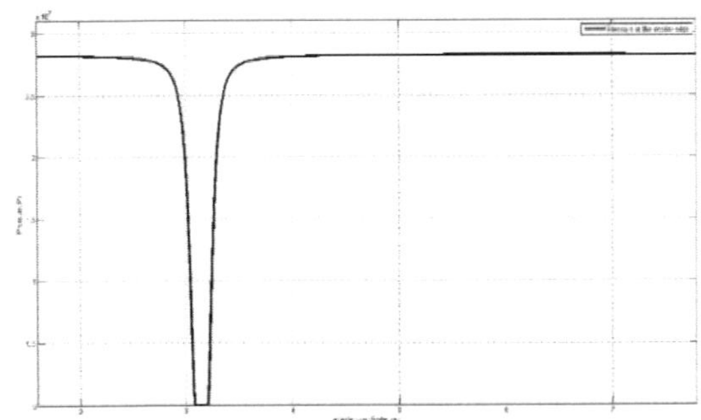
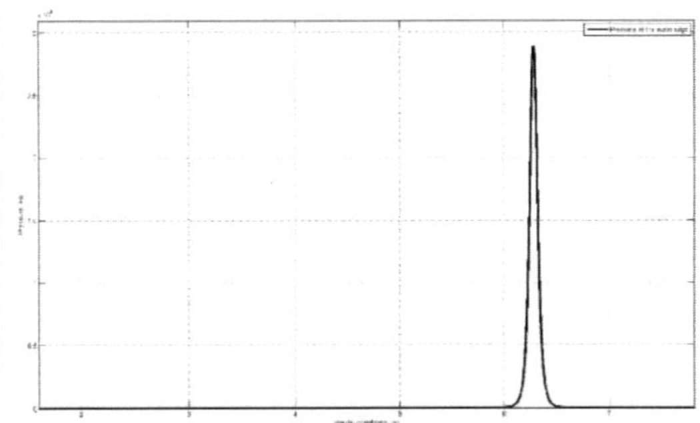

Fig. 6, 7. Pressure at piston maximal speed and full surface sliding

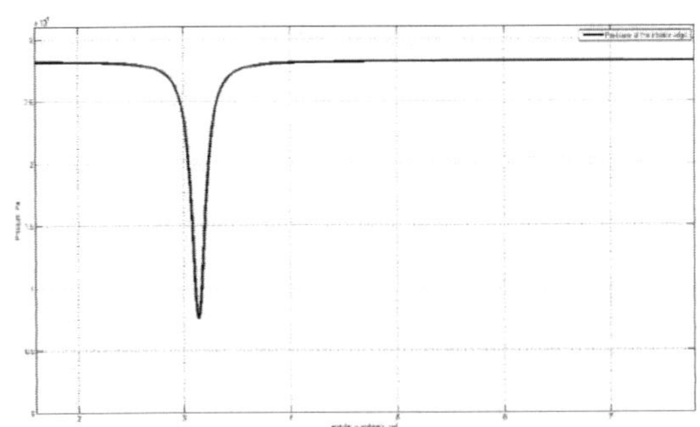
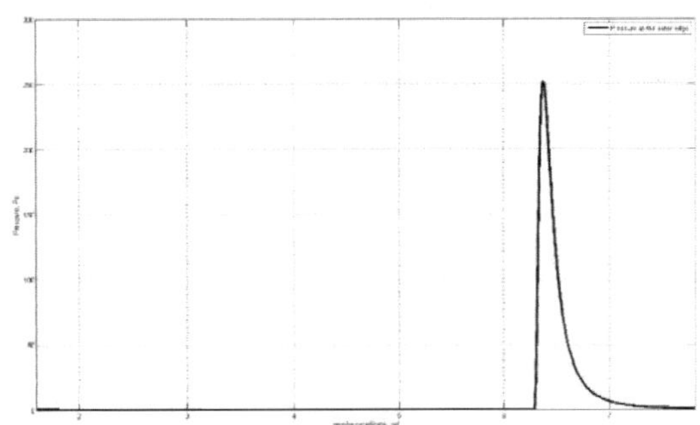

Fig. 8, 9. Pressure at piston dead centers and no surface sliding

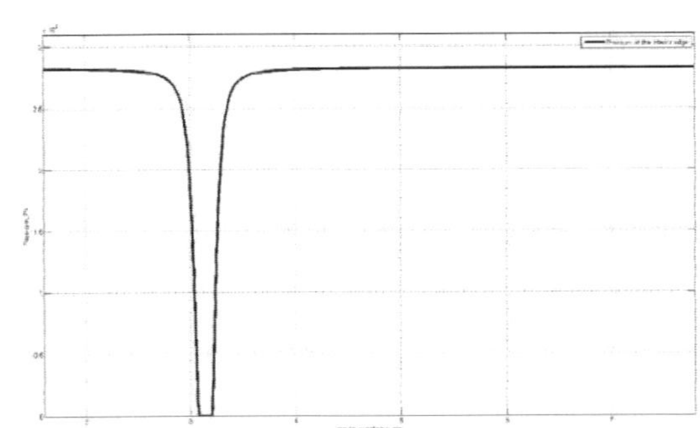
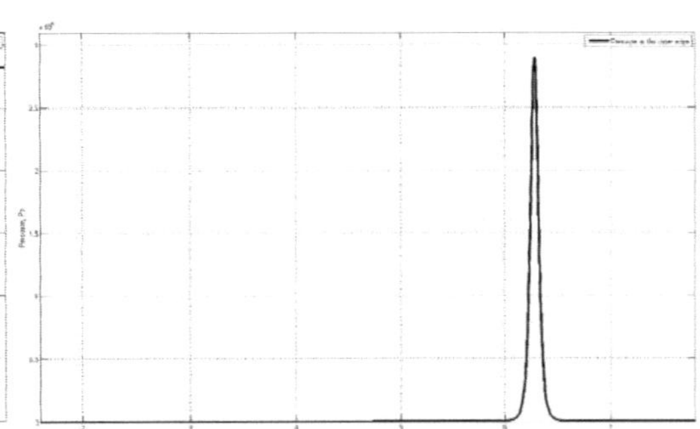

Fig. 10, 11. Pressure at piston maximal speed and no surface sliding

The pressure field values are mostly coincide with results of other researches [4] – [8] for the same kinematics of the mechanism, though the pressure field values for the kinematics that had not been described in these works were also found. These results are sufficient to find a speed value that will cause transition to liquid lubrication.

5. Conclusion and further work

The mathematical model of the hydrodynamic processes in the piston – cylindrical bore pair was designed. This model allows to solve the problem of transition from mixed to liquid friction.

The kinematics of the swash plate machine piston motion relative to the cylindrical bore was analyzed considering different types of relative motion. The first type was described by researchers and was examined in detail. The second type of piston motion relative to the cylindrical bore is the partial rolling along the bore (or the bronze bushing) outer edge. This type of motion had not been described in mathematical models of authors [4] – [8]. Presented kinematic dependencies give data that is necessary to solve the Reynolds equations. In future work transitional between these two types of kinematics will be examined. After a number of experiments a full model of the piston mechanism motion and correlations of the friction forces can be obtained.

For pressure field in the gap between the piston and the bore the tridiagonal matrix method was applied. This decision helped to solve the Reynolds equations with sufficient accuracy without designing a more complicated numerical model. Using this method will give more abilities for investigation of different parameters influence on the pressure built up in the gap and friction forces in future.

Numerical results of pressure values of the operating fluid in points that belong to the piston–cylindrical bore kinematical pair were obtained. Using these results it is possible to define constructive parameters that will cause earlier transition to liquid friction.

The results of this research can be a basis for recommendations on developing of the swash plate machine piston mechanism and reducing friction forces at low speed. This will cause less breakaway torque value for the axial piston swash plate machine and smaller dead band. It is also crucial for machines that work in a regime of frequent shaft rotation direction change.

Obtained values need confirming experimentally in further works. The first step is to conduct a number of experiments at a test rig, fixing a moment of transition from mixed to liquid lubrication. The next step is to apply the research results to a real hydraulic machine.

6. References

[1] Slyozkin N.A. (1955) Dynamics of viscous Incompressible flow, State publ. of Technical-theoretical literature, Moscow

[2] Korovchinsky M.V. (1959) Theoretical basis of sliding bearings work, Mashgiz, Moscow

[3] Manring N.D. (1999) The relative Motion between the Ball Guide and Slipper Retainer within an Axial-Piston Swash-Plate Type hydrostatic Pump, pp 531-537, Journal of Dynamic Systems, Measurement and Control, September 1999, Vol. 121

[4] Uwe Wieczorek & Monika Ivantysynova (2002) Computer Aided Optimization of Bearing and Sealing Gaps in Hydrostatic Machines—The Simulation Tool Caspar, International Journal of Fluid Power, 3:1, 7-20, DOI: 10.1080/14399776.2002.10781124

[5] Fang Y. and Shirakashi M. (1995) Mixed Lubrication Characteristics between the Piston and Cylinder inHydraulic Piston Pump-Motor. Journal of Tribology, Trans. ASME, Vol. 117, pp. 80-85.

[6] Matteo Pelosi & Monika Ivantysynova (2012) A Geometric Multigrid Solver for the Piston–Cylinder Interface of Axial Piston Machines, Tribology Transactions, 55:2, 163-174, DOI: 10.1080/10402004.2011.639049

[7] Xu, B., Zhang, J., Yang, H., Zhang, B. (2013) Investigation on the Radial Micro-Motion about Piston of Axial Piston Pump. Chinese Journal of Mechanical Engineering, 2013, Vol.26 (2), pp.325-333.

[8] J. Jiang and K. Wang, (2015) An integrated model of hydrodynamic lubricating for piston/cylinder interface, International Conference on Fluid Power and Mechatronics (FPM), Harbin, 2015, pp. 47-52. doi: 10.1109/FPM.2015.7337083

[9] Kuzmin, A[nton]; Popov, V[alery] & Stazhkov, S[ergey] (2016). Advanced Axial Piston Swash Plate Pump Parameters Recommendations, Proceedings of the 27th DAAAM International Symposium, pp.0556-0561, B. Katalinic (Ed.), Published by DAAAM International, ISBN 978-3-902734-08-2, ISSN 1726-9679, Vienna, Austria

[10] Samarsky A.A., Nikolaev E.S. (1978) Methods of grid equations solving, Nauka, Moscow

DOI: 10.2507/28th.daaam.proceedings.089

THERMAL SIMULATIONS BASED ON MACRO-MODELS

Václav Marek, Zdeněk Hájíček

This Publication has to be referred as: Marek, V[aclav] & Hajicek, Z[denek] (2017). Thermal Simulations Based on Macro-Models, Proceedings of the 28th DAAAM International Symposium, pp.0627-0634, B. Katalinic (Ed.), Published by DAAAM International, ISBN 978-3-902734-11-2, ISSN 1726-9679, Vienna, Austria
DOI: 10.2507/28th.daaam.proceedings.089

Abstract

Thermal transfer simulations, computed fluid dynamic solutions and finite element methods in general are vital sources of information about behavior of machine parts with heat loading. All these methods have common disadvantages, which include the requirements of a skilled engineer, hardware and software, and generally the high cost of computation. These problems are solved in this article. The reduction of costs is solved by a new tool, which uses simple macro-elements with predefined, parametric properties. These elements are able to describe cooling channels, heat transfers in bodies or identify the temperatures of heat sources. Computation by macro-element method takes a significantly shorter time with suitable accuracy. The macro-element method is suitable for customizing in a wide spectrum of thermal transfer cases. The article defines macro-elements, describes their implementation and the verification of the method.

Keywords: thermal simulation; liquid-cooled spindle unit; macro-element method, thermal displacements

1. Introduction

Machine tools are highly sophisticated mechatronic systems enabling manufacturing processes at a given precision. A machine tool's precision strongly depends on its thermo-elastic behaviour [8]: Internal and external heat sources lead to a non-uniform and non-nominal temperature distribution resulting in elongation and mechanical deformations of the machine tool structure. One of the main factors in mastering the thermal behavior of machine tools is effective cooling. It can be achieved by conduction through solid parts, by radiation, by free or enforced convection on outer surfaces or by forced convection with liquid cooling systems. The non-uniform distribution leads to areas with higher thermal load, such as the main drive, the bearings or the cutting tool. Direct cooling of cutting tools is used in almost every cutting process. It provides cooling of the tool and lubricates the cutting area. In order to reduce thermally induced displacements, fluid based cooling systems are used. Main drives of machines can be cooled by air or by liquid circuits, while bearings are often cooled by liquid circuits and oil mist. Thus, controlling the thermo-elastic behavior of machine tools is to a large extent depending on the design of their fluidic system. In order to speed up the development and minimize the necessity of physical prototypes, it is essential to verify the design before manufacturing. Simulation of a complex system is laborious task: It requires time to prepare computer aided design (CAD) models, skilled employees and computation time.

This paper follows up an idea of macro-models. Weber describes the approach for a heat-loaded motorized spindle unit [5]. It uses a network based computation model. It describes the transient, three-dimensional energy exchange processes in built-in motorized spindles – especially in a cooling sleeve with a single helical rectangular water-cooling channel. [1] A similar approach is used by Institute of Machine Tools and Manufacturing (IWF), at ETH Zurich (the Swiss Federal Institute of Technology in Zurich). It is characterized by the description of a cooling channel as a series of predefined, but parametric geometric features. The tool predicts the heat transfer properties and pressure loss to be expected for a given operational point and coolant. According to the authors, the main advantage is the simplicity of use, and significantly shorter computation times than with a CFD approach [8].

1.1 Current state

The macro-element method is based on simple elements. Macro-elements are distinct elements, defined by elementary equations. The method produces a description of the properties of a fluid in one dimensional dependency. The properties of the macro-element depend on the properties of the fluid, the duct geometry and the operational point [1, 2]. The properties of a wall like roughness also influence the results [9]. These dependencies determine pressure losses and heat transfer coefficients which are the two significant factors in cooling circuits. Macro-elements can be identified in Figure 1, which shows a CFD simulation of a cooling circuit. It is a solved case of CFD simulation. The simulation consists of $2x10^5$ elements, takes about 7 minutes and the solver uses a K-epsilon turbulent model of computation. This case is solved in NX10 simulation software. Examples of identified macro-elements are shown in table 1.

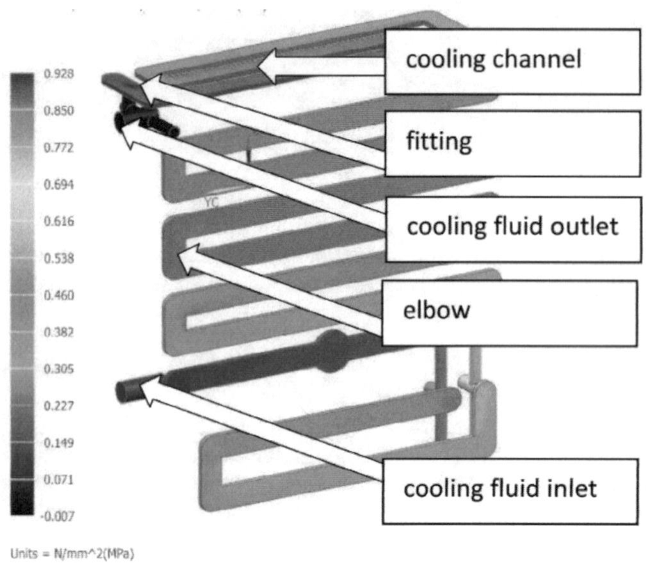

Fig. 1. CFD simulation of pressure losses in cooling circuit (NX10)

Samples of identified macro-elements

Channel	cooling channel, rectangular/circular
Elbow	elbow element, rectangular/circular
Helix channel	helix-shape cooling channel
Flow-Around	element used for cooling of a bearings
Fitting	connection, section change

Table 1. Basic macro-elements

2. Pressure loss

Macro-element definition consists of this pressure loss computation. Pressure loss is an important property of each macro-element. It is a necessary parameter for identifying the overall pressure loss in the cooling channel. The pressure loss Δp depends on velocity w, density ρ and friction coefficient ζ_u.

$$\Delta p = \zeta_u \frac{\rho w^2}{2} \; [Pa] \tag{1}$$

According to [4].
The friction coefficient is computed empirically:

$$\zeta_u = f(Re, D, r) \tag{2}$$

with the Reynolds number

$$Re = \frac{wD}{\nu} \tag{3}$$

and the pipe hydraulic diameter D of the pipe, the kinematic viscosity ν .

3. Verification of pressure losses

Calculations of properties are executed in order to define the basic shapes of each macro-element. The theoretical properties of the macro-elements presented in [2] were stated by using the equation for laminar and turbulent flow as presented in [7]. To verify the results of this theoretical equation, two commercially available CFD simulation software solutions are used: Ansys and Siemens NX. For each macro-element simulation, various operational and geometric conditions are considered [10]. These simulations provide the pressure loss and heat transfer coefficient (htc). Using these results in combination with the results of the macro-models, the macro-elements are verified. The result of CFD simulation of a macro-element can be seen in Figure 2. Verification was generally proved with flow $1x10^{-3}$[m3/s] with parametric variable dimensions. The graph in Figure 3 shows a comparison of CFD and macro-model computation. As can be seen, the results are very similar. The results of a specific case are shown in Table 1.

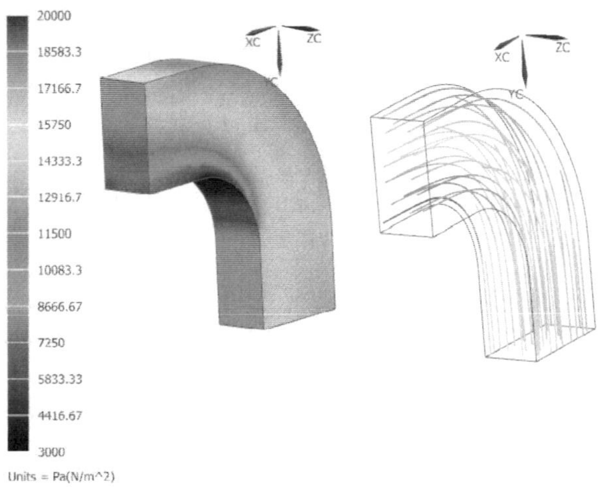

Fig. 2. CFD simulation of total pressure in flow in rectangular elbow

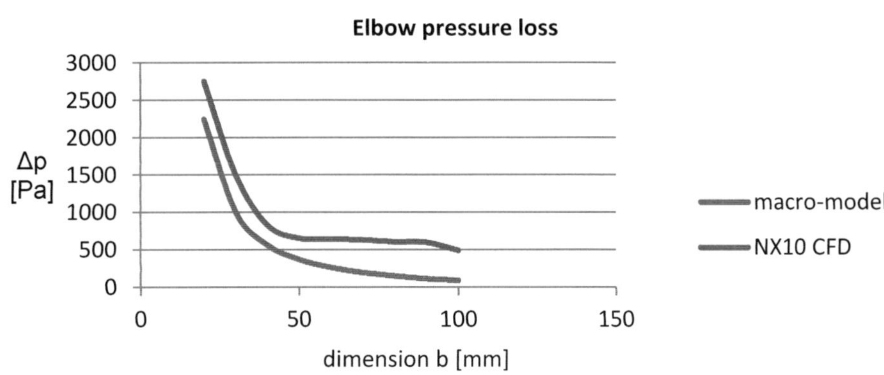

Fig. 3. Comparison of macro-element computation and CFD simulation

solver	section	dim.a [mm]	dim.b [mm]	Radius [mm]	Flow [m3/s]	Pressure loss [Pa]
NX10	Rectang.	10	20	25	$1x10^{-3}$	2874
DD	Rectang.	10	20	25	$1x10^{-3}$	2245
ANSYS	Rectang.	10	20	25	$1x10^{-3}$	2950

Table 2. Results of computation of elbow with rectangular section

4. Thermal field computation

The cooling characteristic is defined by thermal field computation. A similar approach as for pressure loss is performed. The element which performs heat transfer is examined. Heat transfer by convection is governed by the value of the heat transfer coefficient α or h [W/m²K]. The value of h depends on different quantities, including distinct physical characteristics of the fluid in question. Those are the isobaric specific heat (in J/kg K), the thermal conductivity (in W/m K) and the dynamic viscosity (in kg/ms) [6,7]:

$$htc_{convect} = Nu_D \frac{\lambda}{D} \tag{4}$$

$$Nu_D = Nu_D(Re_D, Pr) \tag{5}$$

$htc_{convect}$ – heat tranfer coefficient for heat convection in cooling channel
Nu – Nusselts number
Pr – Prandtls number
D – hydraulic diameter
λ - heat conductivity

4.1. Thermal phenomena in a motorized spindle

A spindle of a milling machine was chosen as an example of thermal phenomena. The motorized unit of a spindle of a milling machine consist of a few main parts which affect the thermal behavior of this assembly. [7]

- heat sources in the stator
- heat sources in the rotor
- heat sources and friction in bearing
- heat sources and friction in air gaps
- heat transfer between shaft and rotor
- heat transfer between housing and coolant

These effects are investigated.

4.2. Temperature field

Given an element with a constant wall temperature as boundary condition, the temperature at position and the wall heat flux can be calculated using the solution of the corresponding partial differential equation (PDE):

$$T(x) = T_w + \exp\left(-\frac{U \cdot x \cdot R_{th}}{c_p \cdot \dot{m}}\right) \cdot (T_{in} - T_w) \tag{6}$$

$$Q_w = (T(L) - T_{in}) \cdot c_p \cdot \dot{m} \tag{7}$$

T_w – Wall temperature
T_{in} – Inlet temperature (outlet temp. of the prev. element)
$\dot{Q}_w$ – Wall heat flux
U – Profile circumferential length
R_{th} - Thermal resistance between the wall and the fluid
c_p – Specific heat capacity of the fluid
$\dot{m}$ – Mass flow rate through the element
L – Element length

The conductive thermal resistance R_w of the wall and the surface A and htc α on the inside of the element:

$$R_{th} = \left(\frac{1}{R_w} + \frac{1}{\alpha \cdot A}\right)^{-1} \tag{8}$$

4.3. Thermal resistance of a wall

Thermal resistance is calculated as:

$$R_w = A \cdot \frac{\lambda}{\delta} \tag{9}$$

δ – Thickness of the wall
λ – Thermal conductivity of the wall
A – Surface of the wall

For a cylindrical element, the thermal resistance of the wall is calculated as follows [3]:

$$R_w = 2 \cdot \pi \cdot L \cdot \frac{\lambda}{\ln\frac{r_o}{r_i}} \tag{10}$$

r_o – Outside radius of the element
r_i – Inside radius of the element

5. Verification of thermal-field computation

Figure 4 shows the results of a CFD simulation of a macro-model. The CFD simulation in the picture shows temperatures in the housing and temperatures of the cooling fluid. In this comparison, the flow around the model was tested. The bearing unit simulation consists of a heat source which substitutes heat loss of bearing, in this case 800[W]. Cooling is provided by water with a flow of 0.4[kg/s].

The macro-model of the flow-around computes the heat transfer with good precision as can be seen in Figure 6. The temperature of output water is almost identical as computed by CFD, shown in Figure 5. Also, other computed temperatures of the wall and bearings are similar, given in Figure 6. The macro-model computation provides the correct information dependent on parametric dimensions, flow of coolant etc.

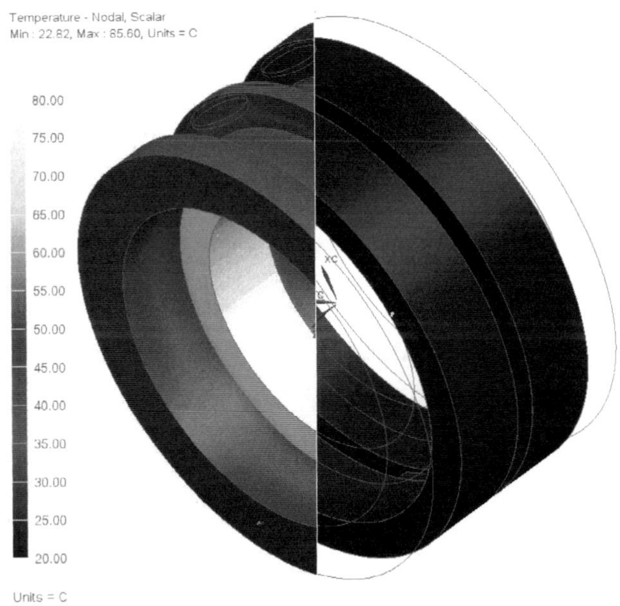

Fig. 4. CFD simulation (NX10) of temperatures of cooled bearing housing

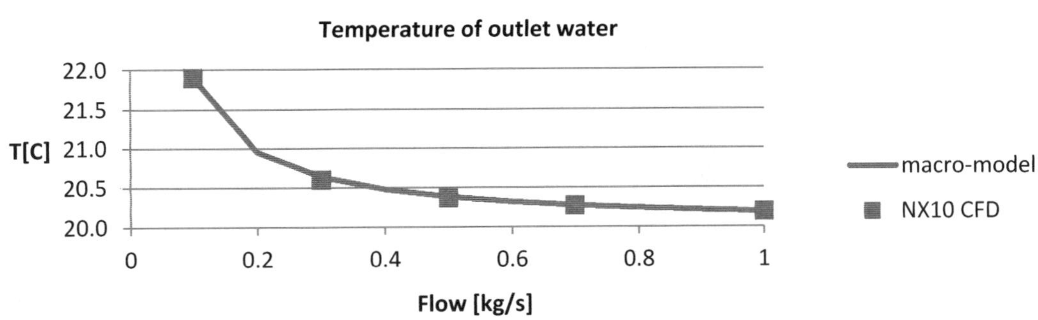

Fig. 5. Comparison of CFD and macro-element simulation - temperatures of outlet water

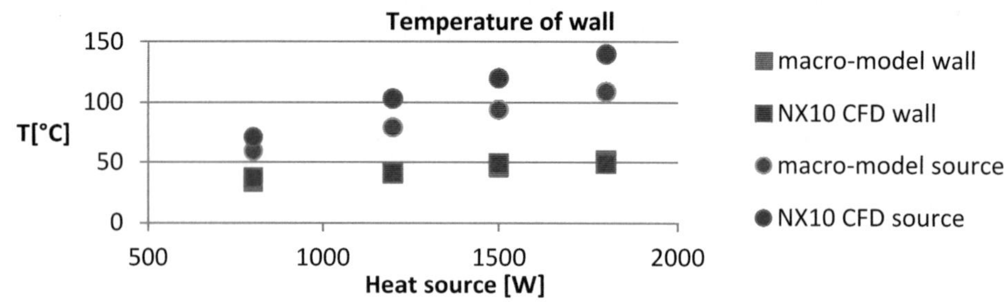

Fig. 6. Comparison of CFD and macro-element simulation - temperatures of wall and heat source (bearing)

6. New tool

The interface for the macro-element method computation is a tool called DuctDesigner. This tool was developed by the Institute of Machine Tools and Manufacturing (IWF), at ETH Zurich (the Swiss Federal Institute of Technology in Zurich). A Comparative calculation of the same cooling circuit performed by using the macro-element method provides results almost immediately. Results are shown in the graph in Figure 7. The simulation of a complex cooling circuit of a motorized spindle unit was performed. Figure 8 shows the results from a CFD simulation of the cooling circuit. Macro-element computation is shown in Figure 7. It shows pressure losses, temperatures and the heat transfer coefficient along the path of the cooling circuit. Pressure decrease can be seen in the graph, with an overall pressure loss of $3x10^5$ [Pa]. It also shows the results of temperature computation and heat transfer coefficient computation. Temperature results are given in Table 3.

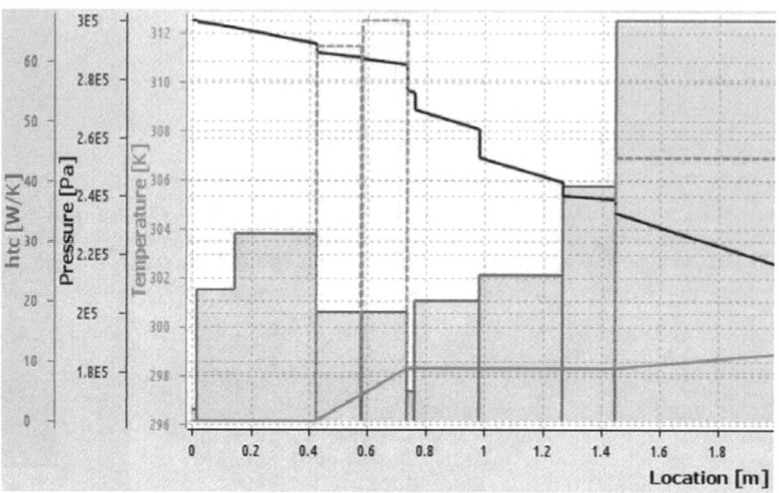

Fig. 7. Graphical output of temperatures, provided by DuctDesigner [1]

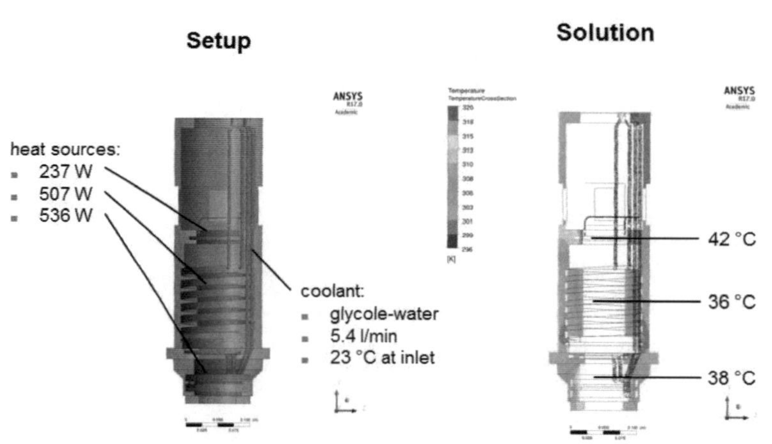

Fig. 8. Setup and solution of CFD simulation of motorized spindle unit, solved by ANSYS [1]

	Ansys CFX	DuctDesigner
Bearing back	42 °C	42 °C
Coil	36 °C	34 °C
Bearing front	38 °C	37 °C

Table 3. Results of CFD Ansys simulation and macro-element computation

7. Discussion

Temperature field computed by macro-elements is pretty same as CFD simulation. Temperature obtained by macro-elements is a function driven by a few input parameters (flow, fluid properties, channel dimensions etc.). It causes faster computation then CFD, which operates with many parameters of every single finite element.

Macro-element computation does not provide detail results, because of these facts. It provides required information, corresponding the design process. Comparison of method is shown in the Table 4. The table shows main differences between CFD calculation and macro-element prediction.

	Prepare time	Comp. time	Temp. Comp.	Result deviation	Costs
CFD	1hour	>600s	full	<5%	High
Macro-model	1/2hour	<1s	partial	<20%	Low

Table 4. Comparison of CFD and macro-element method

8. Conclusion

Overview of a design of a heat loaded machine was done in this paper. Macro-element method was integrated into a design and simulation process. Principles of the decomposition of a complex case to elementary case was described. The usability of the improved DuctDesigner software is demonstrated.

Its Macro-element calculation provides less calculation cost, shorter time configuration and no high skill requirements of users, as can be seen in Table 4. At this point, the concept of macro-elements is not a substitute of CFD or FEM, but it is an important addition to the existing design process. Similar results of CFD calculation and macro-element calculation verify precision of element computation. Results of this research should be verified by experiments. Experiments are included in plan.

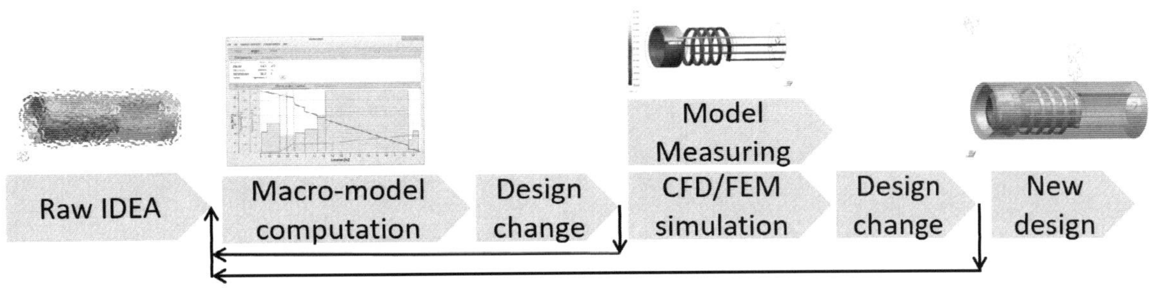

Fig. 9 Implementation of the macro-element prediction into design process

9. Outlook

A systematic verification of the existing macro models is thus required, to improve the performance of these models. Experimental measurement of single macro-elements and complex systems simulated by macro-elements is also required for verification. The main potential is identified in the possibility of using the method in a wide range of applications, in the prediction of parameters of a flow, thermal transfer effects and cases of heat loaded machine parts. Obtained data will be used for a research focused on compensation of thermal displacements of milling centres. Macro-element methodology will be optimize for prediction of thermal field in the spindles of large milling machines.

10. Acknowledgement

The project LO1502 'Development of the Regional Technological Institute' is carried out under the auspices of the National Sustainability Program I of the Ministry of Education of the Czech Republic

11. References

[1] Marek, V., Züst, S., Wegener, K., 2017, Thermal effects in fluidic cooling system of a spindle based on macro models, MMSciene Journal, in print

[2] Züst, S., et al. Model based prediction approach for internal machine tool heat sources on the level of subsystems, 3rd CIRP Global Web Conference, 2014, CIRP 28 (2015)28–33

[3] Bergman, T. L. a Frank P. Incropera. Fundamentals of heat and mass transfer. 7th ed. /. Hoboken, NJ: Wiley, c2011. ISBN 9780470501979

[4] Stephan, P., VDI-Wärmeatlas. Berechnungsunterlagen für Druckverlust, Wärme- und Stoffübertragung. Springer-Verlag, Berlin, Heidelberg, 2006. ISBN 3540255044.

[5] Weber, J., Weber, J., Thermo-energetic analysis and simulation of the fluidic cooling system of motorized high-speed spindles Institute of Fluid Power, Dresden University of Technology, 01062 Dresden, Germany

[6] Anaratone ,D., Engineering heat transfer. New York: Springer, 2010. ISBN9783642039324

[7] Zahedi, A. and Movahhedy ,M.R. Thermo-mechanical modeling of high speed spindles, Center of Excellence in Design, Robotics and Automation, Department of Mechanical Engineering, Sharif University of Technology, Tehran, Iran accepted 16 January 2012

[8] J. Mayr, J. Jedrzejewski, E. Uhlmann, M. A. Donmez, W. Knapp, F. Härtig, K. Wendt, T. Moriwaki, R. Schmitt, P. Shore, C. Brecher, T. Würz, K. Wegener, 2012, Thermal Issues in Machine Tools, CIRP Annals - Manufacturing Technology, 61: 771–791.

[9] Gerlich, V., Zálešák, M.(2010) Experimental validation of heat transfer model, Annals of DAAAM and Proceedings of the International DAAAM Symposium, 2010, pp. 969-970. ISSN 1726-9679, Zadar

[10] Marek, V.(2016) Basic Research Of Thermal Transfer Simulations, Annals of DAAAM and Proceedings of the International DAAAM Symposium, 2016, pp. 578-585. ISBN: 978-1-5108-3300-5, Mostar

DOI: 10.2507/28th.daaam.proceedings.090

THE APPLICATION OF ARTIFICIAL VECTOR FIELDS FOR MOTION CONTROL OF A HETEROGENEOUS MODULAR MOBILE ROBOT

Victor Andreev & Valerii Kim

This Publication has to be referred as: Andreev, V[ictor] & Kim, V[alerii] (2017). The Application of Artificial Vector Fields for Motion Control of a Heterogeneous Modular Mobile Robot, Proceedings of the 28th DAAAM International Symposium, pp.0635-0644, B. Katalinic (Ed.), Published by DAAAM International, ISBN 978-3-902734-11-2, ISSN 1726-9679, Vienna, Austria
DOI: 10.2507/28th.daaam.proceedings.090

Abstract

This paper presents a novel method for motion realization of differential drive mobile robot along a complex trajectory using two-dimensional vector fields (artificial forces). The method allows robot to move along trajectories that similar to B-splines without their preparatory computation. This is performed by using the given sequence of points (intermediate goal points) and manipulation of vector fields that defined by Gaussian functions. The feature of the method realization is as follows in its application in the transport module of a heterogeneous modular mobile robot which modules are functionally complete electronic or mechatronic devices. A functional completeness in this case means that every module must perform the function that built in its engineering implementation and control system by itself. The transport module realizes its functions upon receipt control goals (parameters of position vector) from module-supervisor that is part of modular mobile robot. In this case the transport module computes motion trajectory independently. Therefore a method of information interaction between the transport module and a module-supervisor is considered in this paper. The transport module performance under control of a module-supervisor is viewed by the example of mobile robotics typical tasks: motion in a known and unknown environment.

Keywords: mobile robot; heterogeneous modular robot; differential drive; motion planning; trajectory following

1. Introduction

At the present time mobile robots are used for different tasks and involved in various areas of activities [1-2]. Many researches consider the wheeled mobile robot control problems such as motion and path planning [3-6]. There are many fruitful results in these topics that are applied to different kinds of robots. One of the most challenging problems is a trajectory planning of mobile robot and path following [5-6]. In the most cases robot controllers for trajectory or path following guarantee the minimal tracking errors. But these controllers involve complex and sophisticated control laws that require more powerful on-board computers of mobile robots. Furthermore tracking controllers are even more complicated for nonholonomic wheeled mobile robots [7].

In this paper we present a novel motion planning method for a differential drive mobile robot using two-dimensional vector fields. This method requires that the given path for a mobile robot control system must be specified as a set of the reference points that we will call here the "nodal" points. Using the artificial vector fields mobile robot can move along trajectories like splines in proximity to nodal points. The robot motion in this case is smooth and continuous. Note that no trajectory planning is needed for this method. Our approach lies in a design of the such motion planning controller that allows a differential drive mobile robot to move along complex trajectories without classic tracking controllers.

The second objective of our research is the application of the proposed method to a heterogeneous modular mobile robot. This can be done efficiently because concerned modular robot control system can divide general task to several control goals for each module of the robot. This key feature considered in the fourth section.

2. Motion planning using vector fields

The essence of the method. Consider a given path as a set of the nodal points with respect to world coordinate system OXY: (X_1,Y_1), … , (X_i,Y_i), … , $(X_n,Y_n)=(X_G,Y_G)$, where (X_G,Y_G) - the goal position coordinates at the end of a path, n - number of points. The robot start position is in the point (X_0,Y_0). The robot control system takes this set of the nodal points from some robot path planner at the higher control level. Let us assign the vector field that defines the attractive force F_i to each nodal points of the given path (fig. 1).

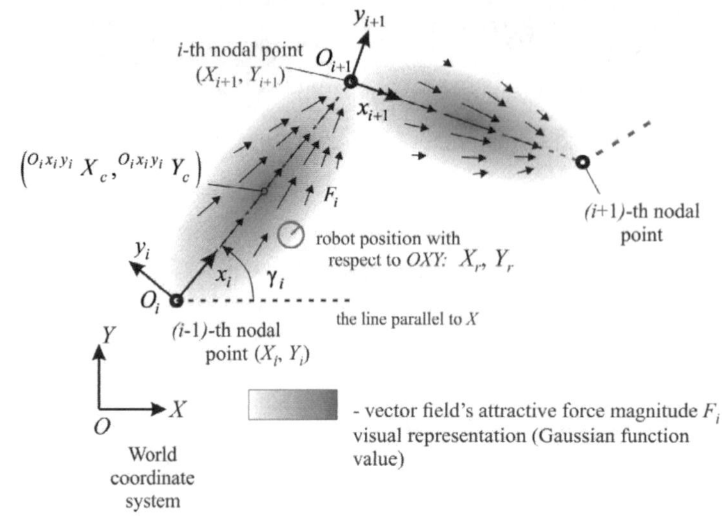

Fig. 1. Vector fields of attractive forces

The i-th $(i=\overline{1,n})$ attractive force magnitude F_i in each point of the vector field is a continuous function of distance between the robot and the centre of the line segment connecting previous $(i-1)$ and current i nodal points. For this function, we take two-dimensional Gaussian function. Each Gaussian function for the attractive force is rotated along line segment connecting previous $(i-1)$ and current i nodal points. Also a width of Gaussian function along this line segment is larger in comparison to transverse direction. It is needed because closely spaced Gaussian function can influence over robot motion. Considering all these properties we define the next Gaussian function:

$$F_i = F_0 e^{-\left[\left(w_{x,i}\left({}^{O_i x_i y_i}X_r - {}^{O_i x_i y_i}X_c\right)\right)^2 + \left(w_{y,i}\left({}^{O_i x_i y_i}Y_r - {}^{O_i x_i y_i}Y_c\right)\right)^2\right]},$$

where F_0 - the attractive force basis magnitude, ${}^{O_i x_i y_i}X_r$, ${}^{O_i x_i y_i}Y_r$ - the robot coordinates with respect to the turned Gaussian function coordinate system - $O_i x_i y_i$. Here we must mention that robot coordinates is the coordinates of its coordinate system $O'xy$ datum O'. Consider next parameters: ${}^{O_i x_i y_i}X_c$, ${}^{O_i x_i y_i}Y_c$ - coordinates of the Gaussian function centre with respect to the $O_i x_i y_i$ (see fig. 1), $w_{x,i}, w_{y,i}$ - real numbers that inversely related to the width of the Gaussian function along $O_i x_i$ and $O_i y_i$ axes of the $O_i x_i y_i$ respectively. Put it differently the larger the $w_{x,i}$, the smaller function width along $O_i x_i$. We can derive robot coordinates ${}^{O_i x_i y_i}X_r$, ${}^{O_i x_i y_i}Y_r$ using the appropriate homogeneous transform [8]:

$$\begin{pmatrix} {}^{O_i x_i y_i}X_r \\ {}^{O_i x_i y_i}Y_r \\ 1 \end{pmatrix} = \begin{pmatrix} \cos\gamma_i & \sin\gamma_i & -X_{i-1}\cos\gamma_i - Y_{i-1}\sin\gamma_i \\ -\sin\gamma_i & \cos\gamma_i & X_{i-1}\sin\gamma_i - Y_{i-1}\cos\gamma_i \\ 0 & 0 & 1 \end{pmatrix} \begin{pmatrix} X_r \\ Y_r \\ 1 \end{pmatrix},$$

where γ_i is the angle between the line segment connecting the $(i-1)$-th and i-th nodal points and OX axis, X_r, Y_r - the robot coordinates with respect to the world coordinate system OXY. The robot must move from one nodal point to another and this can be done if each Gaussian function that defines attractive force has appropriate width along axis $O_i x_i$ (along line segment). More precisely one vector field must intersect with another at the appropriate Gaussian function value. Thus parameter $w_{x,i}$ is defined as:

$$w_{x,i} = \frac{1.5\sqrt{2}}{\sqrt{(X_i - X_{i-1})^2 + (Y_i - Y_{i-1})^2}} = \frac{1.5\sqrt{2}}{d_i},$$

where d_i - the distance between $(i-1)$-th and i-th nodal points. As it was mentioned above width of the Gaussian function along line segment, i.e. along $O_i x_i$ axis must be greater than width along $O_i y_i$ axis (transverse direction). That can be done if we choose parameter $w_{y,i}$ greater than $w_{x,i}$:

$$w_{y,i} = k_w w_{x,i},$$

where $k_w > 1$ - constant of proportionality, that define how Gaussian function would be narrowed along $O_i y_i$ axis.

To smooth robot motion near with the nodal points the vector fields that define orbital (or rotating) and repulsive forces are used. These vector fields act around special points – barycentres. In this research barycentres are the nodal points except the final point (goal position) for which there is only an attractive force. The orbital and repulsive force magnitude is a function of the distance between robot and barycentre. For this function, we also take default Gaussian function with the same width along all directions. The j-th ($j = \overline{1,m}$) orbital force magnitude $F_{orb,j}$ then defined by the next equation:

$$F_{orb,j} = s_j F_{orb,0} \cdot e^{-\left[(w_{o,j}(X_r - X_{b,j}))^2 + (w_{o,j}(Y_r - Y_{b,j}))^2\right]},$$

where s_j - coefficient that specifies orbital force direction: for clockwise direction $s = 1$, otherwise $s = -1$, $F_{orb,0}$ - orbital force basis magnitude, $w_{o,j}$ - real number that define the orbital and the repulsive force area of action with the centre at the j-th barycentre, $X_{b,j}, Y_{b,j}$ - the j-th barycentre coordinates with respect to the world coordinate system OXY (see fig. 2). A number of the barycentres is given by $m = n-1$. Figure 2 shows the vector fields of orbital forces and necessary notations.

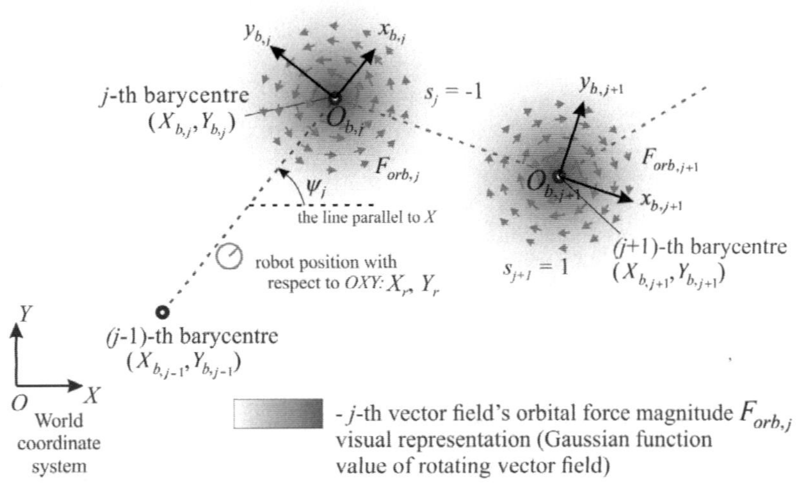

Fig. 2. Vector fields of orbital forces that acting around barycentres

The coefficient $w_{o,j}$ can be evaluated by this equation:

$$w_{o,j} = \frac{3}{r_j\sqrt{2}},$$

where r_j - the radius of action of the orbital force and the repulsive force. For r_j we use the next function:

$$r_j = g(d_j, d_{j+1}) = k_r \frac{d_j d_{j+1}}{d_j + d_{j+1}},$$

where k_r - constant of proportionality, d_j - the distance between $(j-1)$-th and j-th barycentres, d_{j+1} - the distance between j-th and $(j+1)$-th barycentres. To determine coefficient s_j it is necessary to find y coordinate of the $(j+1)$-th barycentre with respect to the j-th barycentre coordinate system $O_{b,j}x_{b,j}y_{b,j}$ (see fig. 2):

$$^{O_{b,j}x_{b,j}y_{b,j}}y_{j+1} = -(X_{b,j+1} - X_{b,j})\sin\psi_j + (Y_{b,j+1} - Y_{b,j})\cos\psi_j,$$

where $X_{b,j+1}, Y_{b,j+1}$ - coordinates of the $(j+1)$-th barycentre with respect to the OXY, ψ_j - the angle between line segment connecting j-th and the $(j+1)$-th barycentres and OX axis. If the coordinate $^{O_{b,j}x_{b,j}y_{b,j}}y_{j+1} < 0$, then $s_j = -1$, else $s_j = 1$. The j-th $(j = \overline{1,m})$ repulsive force magnitude $F_{rep,j}$ defined by the next equation:

$$F_{rep,j} = F_{rep,0} \cdot e^{-\left[(w_{o,j}(X_r - X_{b,j}))^2 + (w_{o,j}(Y_r - Y_{b,j}))^2\right]},$$

where $F_{rep,0}$ - the repulsive force basis magnitude. Fig. 3 shows the vector fields of repulsive forces and necessary notations.

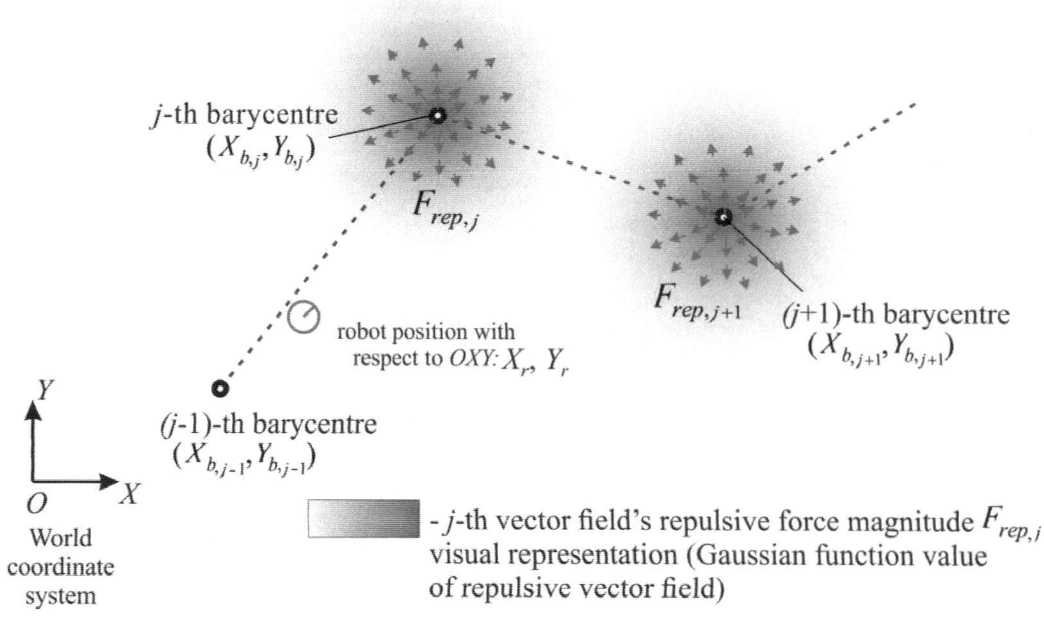

Fig. 3. Vector fields of repulsive forces that acting around barycentres

All forces act upon mobile robot but the most distant points from robot almost have no influence over a robot. Summarizing the above we can say that in each point of the two-dimensional space there are vector fields acting upon a mobile robot. The total force projection onto an $O'x$ axis of the robot proportional to its linear velocity V, the total force projection onto an $O'y$ axis of the robot proportional to its angular velocity ω (fig. 4).

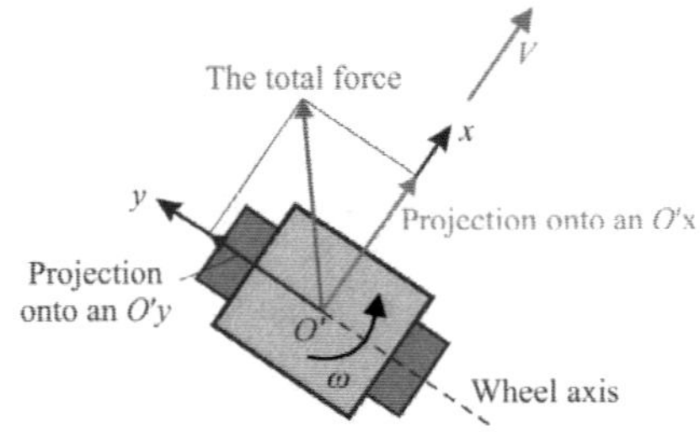

Fig. 4. The total force projections onto robot axes

To find total force projections onto axes $O'x$ and $O'y$ we should calculate projections of the attractive, orbital and repulsive forces onto these axes. The attractive force projections onto an axes of the robot coordinate system are defined by the next equation:

$$^{O'xy}F_{x,i} = F_i(\cos\alpha_i \cos\varphi + \sin\alpha_i \sin\varphi)$$
$$^{O'xy}F_{y,i} = F_i(\sin\alpha_i \cos\varphi - \cos\alpha_i \sin\varphi),$$

where $^{O'xy}F_{x,i}$, $^{O'xy}F_{y,i}$ - the attractive force projections onto an $O'x$ and $O'y$ axis respectively, α_i - the angle of i-th attractive force direction to the i-th nodal point, φ - the angle of the robot orientation with respect to the world coordinate system OXY. The repulsive force projections $^{O'xy}F_{repx,i}$, $^{O'xy}F_{repy,i}$ can be found similarly except for the signs:

$$^{O'xy}F_{repx,i} = -F_{rep,j}(\cos\beta_j \cos\varphi + \sin\beta_j \sin\varphi)$$
$$^{O'xy}F_{repy,i} = -F_{rep,j}(\sin\beta_j \cos\varphi - \cos\beta_j \sin\varphi),$$

where β_j - the angle of j-th repulsive force direction to the j-th barycentre. The orbital force projections onto axes of the robot coordinate system are defined by the equation:

$$^{O'xy}F_{orbx,j} = F_{orb,j}(\cos\beta_j \sin\varphi - \sin\beta_j \cos\varphi)$$
$$^{O'xy}F_{orby,j} = F_{orb,j}(\sin\beta_j \sin\varphi + \cos\beta_j \cos\varphi),$$

where $^{O'xy}F_{orbx,j}$, $^{O'xy}F_{orby,j}$ - the orbital force projections onto an $O'x$ and $O'y$ axis respectively.

Now we can evaluate the total force projections onto axes $O'x$ and $O'y$:

$$^{O'xy}F_{O'x} = \sum_{i=1}^{n} {}^{O'xy}F_{x,i} + \sum_{j=1}^{m}({}^{O'xy}F_{orbx,j} + {}^{O'xy}F_{repx,j})$$
$$^{O'xy}F_{O'y} = \sum_{i=1}^{n} {}^{O'xy}F_{y,i} + \sum_{j=1}^{m}({}^{O'xy}F_{orby,j} + {}^{O'xy}F_{repy,j}).$$

where $^{O'xy}F_{O'x}$, $^{O'xy}F_{O'y}$ - the total force projections onto axes $O'x$ and $O'y$. At the final step we define linear V and angular ω velocities of the mobile robot:

$$V = \frac{K_v}{M_r} {}^{O'xy}F_{O'x}, \quad \omega = \frac{K_\omega}{J_r} {}^{O'xy}F_{O'y},$$

where K_v - the linear velocity gain, M_r - the mass of the mobile robot, K_ω - the angular velocity gain, J_r - the moment of inertia of the mobile robot around vertical axis of rotation. Also for this control law we place usual limitations: if $V < V_{min}$, then $V = V_{min}$, if $V > V_{max}$, then $V = V_{max}$ and the same for the angular velocity. Here V_{min} - minimal linear velocity of the robot, V_{max} - maximum linear velocity of the robot.

The robot motion and its trajectory are depended on not only nodal points but also on a set of main (core) parameters that were defined above: $F_0, k_w, F_{orb,0}, k_r, F_{rep,0}, K_v, K_\omega$.

All these parameters can have different values but one condition must always hold:

$$F_{orb,0} \le F_{rep,0} < F_0 .$$

If this condition is not satisfied the robot can't reach any nodal point. In the next section the most significant simulation results would be considered.

3. Simulation results

We tested our method for various paths with different parameters using computer simulation (MATLAB Simulink). We assume that the mobile robot has a local map of the environment and the robot control system at the higher level can generate a path from the start position to the goal position. For example, the A* algorithm can be suitable for it.

1. Consider that the mobile robot has the given set of nodal points: (1, 1.5), (2, 0), (3.2, 1.7), (5.2, 1), (4.1, 0), (3.5, 0.5) and (4, 1) – coordinates of the goal position. Let $(X_0, Y_0) = (0, 0)$ be the start position of the robot and $\varphi_0 = 0$ - initial orientation angle of the robot. For this example we set $F_{orb,0} = 0$ and $F_{rep,0} = 0$, i.e. there are only attractive forces that act on the robot. We set other parameters like so: $F_0 = 3$, $k_w = 1.7$, $k_r = 2$, $K_v = 1.5$ and $K_\omega = 1.2$, $M_r = 10$ kg, $J_r = 0.487$ kg×m^2. Figure 5a shows the robot trajectory and the given path, figure 5b shows the actual linear and angular velocities of the robot during the motion.

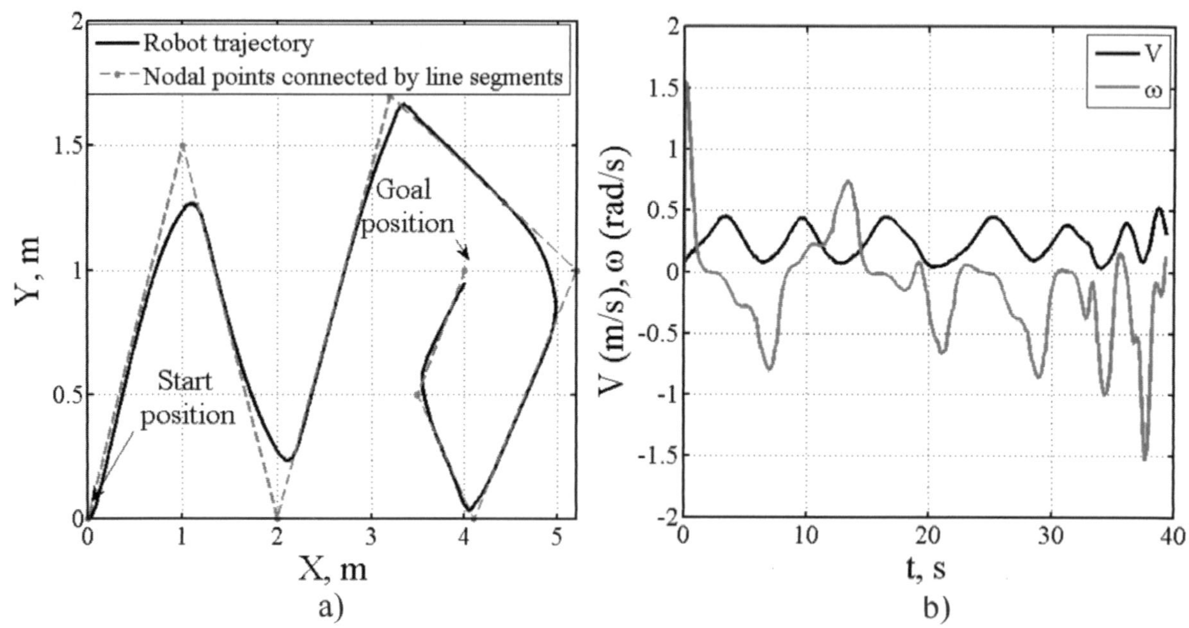

Fig. 5. The robot trajectory (a) and its velocities (b) during the motion without orbital and repulsive forces

From these simulation results it is clear that the robot motion nearby the nodal point (i.e. along the path) can be successfully performed without using the orbital and repulsive forces. The robot trajectory is smooth, linear and angular velocities change continuously (see fig. 5b). Note how the linear velocity drops when the angular velocity increases. It means that robot slows down at turns and accelerates at straight-line segments of the robot path.

2. Now consider the situation when all forces are in count for the same path. The orbital and repulsive forces provide an opportunity to make robot trajectories similar to for example B-splines. In this case we set parameters with the next values: $F_{orb,0} = F_{rep,0} = 1$, $k_r = 2$ and other are the same. Figure 6a shows the robot trajectory and the given path, figure 6b shows the linear and angular velocities of the robot during the motion.

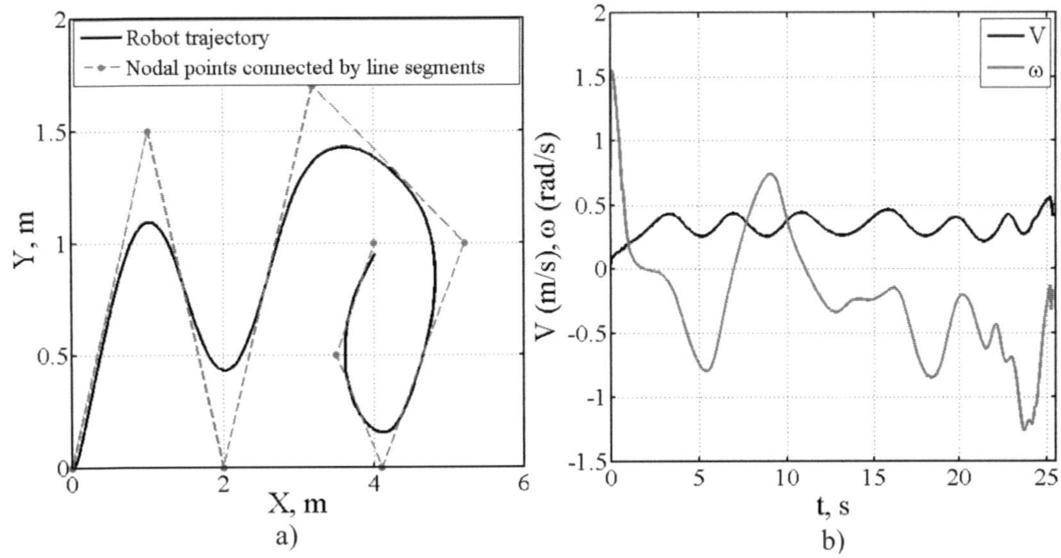

Fig. 6. The robot trajectory (a) that similar to B-spline and the robot velocities (b) during the motion

One can see from this plot that the robot trajectory is similar to B-spline. Also notice that the mobile robot moves faster in comparison to previous situation (in fig. 5b linear velocity decreases almost to zero, but in fig. 6b linear velocity is always larger than 0.25 m/s).

3. Consider the following example. The mobile robot must pass through the door by consequently visiting the set of nodal points: (1, 1.5), (2.3, 1.5), (2.3, 0.4) with the (1, 0.7) as the start position with the orientation angle $\varphi_0 = 0$. Coordinates of the goal position are (2.3, 0.4). We set orbital force basis magnitude in this case as $F_{orb,0} = 0.5$ and $k_r = 1.5$ while other parameters are same. Figure 7a shows the robot trajectory and the given path, figure 7b shows the linear and angular velocities of the robot during the motion.

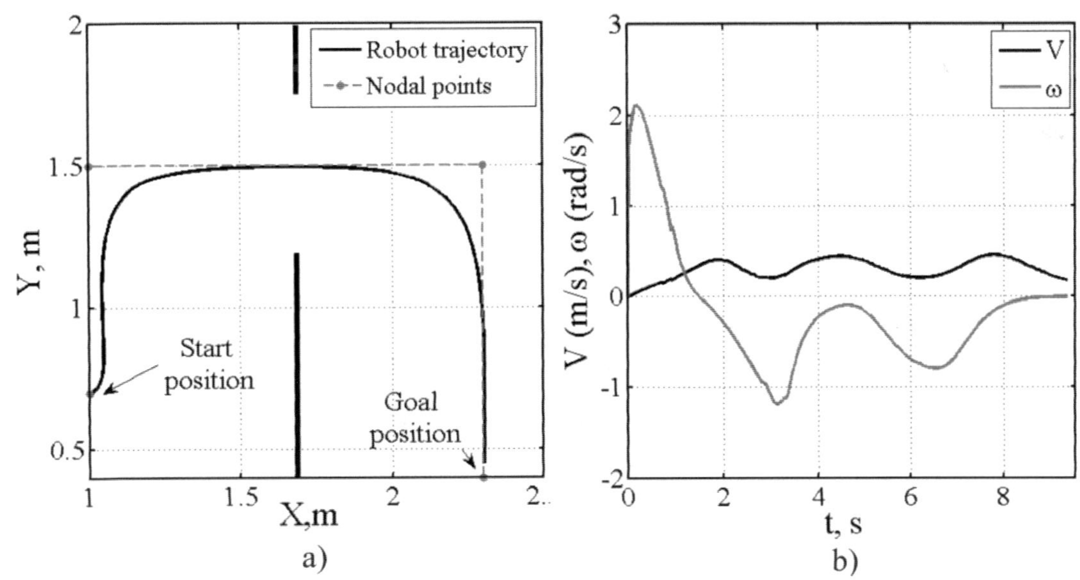

Fig. 7. The robot trajectory (a) as it goes through the door and corresponding velocities (b)

Like in other situations robot velocities change smoothly and continuously. Some other results are presented in the figure 8 with the same parameters. In the fig. 8a the robot initial position is $(X_0, Y_0, \varphi_0) = (1, 0, 0)$, in the fig. 8b - $(X_0, Y_0, \varphi_0) = (0, 0, 0)$, in the fig. 8c - $(X_0, Y_0, \varphi_0) = (0, 0, \pi/2)$. Coordinates of the goal positions are (1.1, 5), (-1, -3) and (1, 5) respectively.

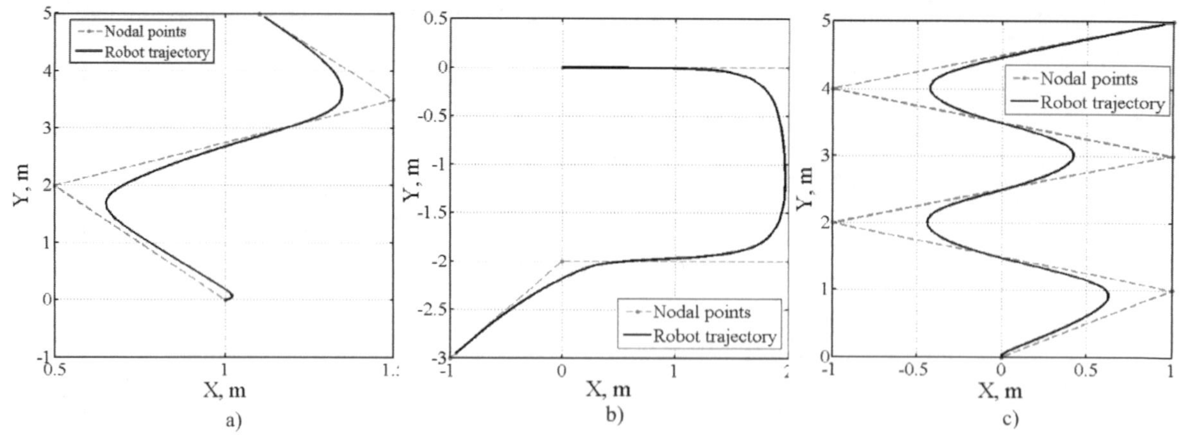

Fig. 8. The robot trajectories for different paths (sets of nodal points): a), b) and c)

Simulation results analysis. The robot motion planning using vector fields gives next advantages: a differential drive mobile robot can move along complex trajectories like B-splines *without* their planning, robot linear and angular velocities change smoothly and continuously, the variety of robot trajectories can be achieved by using different core parameters.

4. The application of the method to a heterogeneous modular mobile robot

This method of motion planning is well suited to a heterogeneous modular mobile robot which concept is considered in detail in [9]. The main feature of this modular robot is based on the principle of full functionality of its modules. According to the [9] the **full functionality** of a mechatronic device is *the ability to perform its goal function using only its own facilities for executing instructions from an external control system*. This heterogeneous modular mobile robot consists of the next basic modules: a transport module that performs locomotion of the whole robot, a module-supervisor that responsible for control goals definition, the power supply module and sensory modules. Some features of the transport module control system and design were considered in [10] in which old name was used for the transport module – the motion module.

In this research we will consider the wheeled transport module and its interaction with the module-supervisor. According to the definition of the full functionality the module-supervisor gives a goal position to the transport module but verifies only result of the module performance and not the working process (except for emergency situations). Therefore the transport module control system must plan its trajectory to the goal position and perform corresponding motion. In our heterogeneous modular robot this transport module can use different strategies for motion planning. The transport module also uses data structure from a sensor module for obstacle avoidance. Note that a sensor module sends to the transport module not the usual data as array of some measurements but data structures for example, positions of lines that approximates obstacle geometry.

We use simple transport module that technically is a differential drive mobile platform. Figure 9 shows the laboratory prototype of the transport module and the modular robot.

Fig. 9. The laboratory prototype of the transport module (without wheels) and modular mobile robot

Consider the interaction procedure between the transport module and the module-supervisor by the example of typical motion tasks. The modular robot must go to the goal position from the initial position in two cases: if the module-supervisor doesn't have a map of the environment (fig. 10a) and if the module-supervisor has a map (fig. 10b).

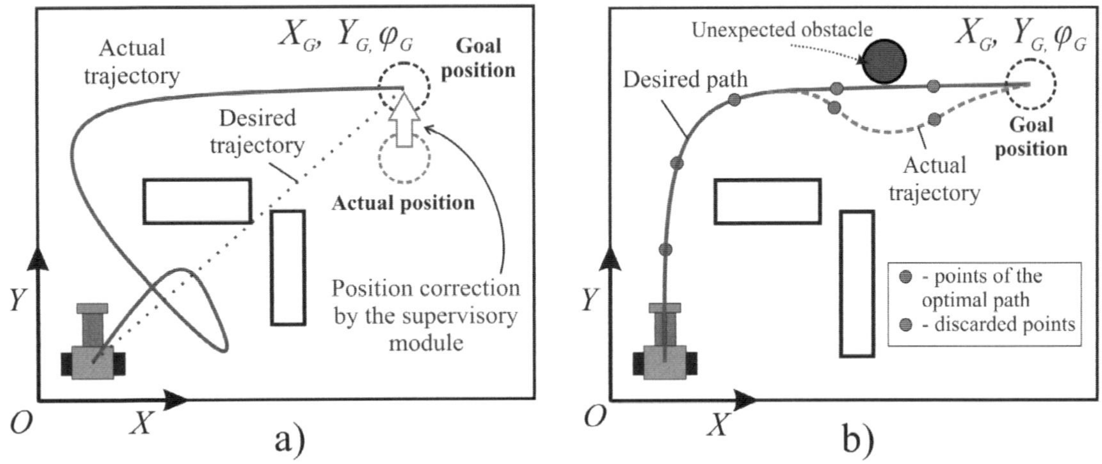

Fig. 10. Examples of typical motion tasks of the mobile robot: a) - the module-supervisor doesn't have a map of the environment; b) - the module-supervisor has a map

In both cases the interaction between the transport module and the module-supervisor is defined by next sequence of actions:
1. The module-supervisor and the transport module establish the connection and begin to exchange with initial data (IP addresses of the transport module and the module-supervisor in the shared robot network, used interfaces, data conversion).
2. The module-supervisor sends to the transport module the data about goal position as set of three numbers: X_G, Y_G - coordinates of the goal position in the world coordinate system, φ_G - orientation angle at the goal position.
3. The transport module goes to the goal position and avoids obstacles using the data from a sensor module.
4. If the goal position is reached (or not reached) the transport module stops and sends to the module-supervisor information about the robot actual position and the message that the task was completed (or not completed).

When the robot doesn't have information about an environment (fig. 10a) the robot motion is not optimal and the arrival to the goal position is not guaranteed. In this case the module-supervisor corrects a robot position error. If the robot motion occurs in the known environment the module-supervisor can define optimal path as the set of goal positions – nodal points (fig. 10b). In this case we can apply our method of motion planning using vector fields. The module-supervisor act as robot path planner at the higher control level and the transport module control system evaluates vector fields for nodal points. If the unexpected obstacle obstructs the robot motion then the transport module is trying to avoid it by itself requesting the new goal position from the module-supervisor.

As it was mentioned above the interaction between a sensor module and the transport module is a one-way communication. While the robot moves to a goal a sensor module perform pre-processing of the measurement data and sends it to the transport module. For the localization in the environment the transport module uses the sensor data from a sensor module and an encoder data (dead reckoning).

Note that the module-supervisor doesn't control motion planning of the robot because the transport module responsible for vector fields evaluating and computation of linear and angular velocities. That distributed control is the main advantage of the full functionality of the robot modules.

5. Conclusion

The motion planning of the differential drive mobile robot using artificial vector fields was presented. In this method we use vector fields that produce three kinds of forces: attractive, orbital and repulsive. These forces allow a mobile robot to move along complex trajectories without their preparatory definition. In this paper we demonstrated that robot trajectories can be similar to different splines (B-splines). The smooth and continuous motion of the robot in this case is the important advantage of the method. Note that for these results there is no need to use complex control laws for path following which require constant tracking of this path.

We approved that this method is well suited to a heterogeneous modular mobile robot which modules are full functionally devices. The task of motion planning can be efficiently divided between the module-supervisor, the transport module and the sensor module. The computation load from the central processing unit of the module-supervisor is distributed between other modules according to their functions.

Future work consists of implementation of this motion planning method in the laboratory prototype of the modular mobile robot. Also we will continue our work with this method for further modifications and improvements.

6. Acknowledgments

This work is supported by the Russian Foundation for Basic Research: grant 16-07-00811a, 16-07-00935a and 16-07-01264a.

7. References

[1]. Crnokic, B.; Rezic, S. & Pehar, S. (2016). Comparision of edge detection methods for obstacles detection in a mobile robot environment, Proceedings of the 27th DAAAM International Symposium, pp.0235-0244, B. Katalinic (Ed.), Published by DAAAM International, ISBN 978-3-902734- 08-2, ISSN 1726-9679, Vienna, Austria DOI: 10.2507/27th.daaam.proceedings.035.

[2]. Ilyukhin, Y. & Tatarintseva, A. (2016). Control mobile robot with hybrid drive, Proceedings of the 27th DAAAM International Symposium, pp.0405-0416, B. Katalinic (Ed.), Published by DAAAM International, ISBN 978-3-902734-08-2, ISSN 1726-9679, Vienna, Austria DOI: 10.2507/27th.daaam.proceedings.060.

[3]. Vinod, R.N & Abraham, T. M. Design, simulation and implementation of cascaded path tracking controller for a differential drive mobile robot, International Conference on Advances in Computing, Communications and Informatics (ICACCI), 2015, pp. 1085-1090.

[4]. Shojaei, K.; Shahri, A.M; Tarakameh, A. and Tabibian, B. (2010). Adaptive trajectory tracking control of a differential drive wheeled mobile robot. Robotica, Vol. 29, 2010, pp. 391-402, DOI:10.1017/S0263574710000202.

[5]. Lapiere, L.; Soetanto, D. & Pascoal A. (2006). Nonsingular path following control of unicycle in the presence of parametric modeling uncertainties. International Journal of Robust Nonlinear Control, No. 16, 2006, pp. 485-503, DOI: 10.1002/rnc.1075.

[6]. Fareh, R.; Saad, M.; Khadraoui, S. and Rabie, T. (2016). Lyapunov-based tracking control for nonholonomic wheeled mobile robot. International Journal of Electrical, Computer, Energetic, Electronic and Communication Engineering, Vol. 10, No. 8, 2016, pp. 1021-1026.

[7]. Samson, C. & Ait-Abderrahim, K. (1991). Feedback control of a nonholonomic wheeled cart in cartesian space, Proceedings of IEEE International Conference, 1991, pp. 1136-1141.

[8]. Kelly, A. (2013). Mobile Robotics, Mathematics, Models and Methods, Cambridge University Press, ISBN 978-1-107-03115-9, USA.

[9]. Andreev, V; Kim, V & Pletenev, P. (2017). The principle of full functionality – the basis for rapid reconfiguration in heterogeneous modular mobile robots, Proceedings of the 28th DAAAM International Symposium, pp.xxxx-xxxx, B. Katalinic (Ed.), Published by DAAAM International, ISBN 978-3-902734-xx-x, ISSN 1726-9679, Vienna, Austria, DOI: 10.2507/28th.daaam.proceedings.xxx, in press.

[10]. Andreev, V & Kim, V. (2016). Control system and design of the motion module of a heterogeneous modular mobile robot, Proceedings of the 27th DAAAM International Symposium, pp. 586-594, B. Katalinic (Ed.), Published by DAAAM International, ISBN 978-3-902734-08-2, ISSN 1726-9679, Vienna, Austria, DOI: 10.2507/27th.daaam.proceedings.086.

DOI: 10.2507/28th.daaam.proceedings.091

VACUUM GAUGE VERIFICATION METHODOLOGY FOR VACUUM COATING DEPOSITION PROCESSES

Antons Stekleins, Eriks Gerins & Artis Kromanis

This Publication has to be referred as: Stekleins, A[ntons]; Gerins, E[riks] & Kromanis, A[rtis] (2017). Vacuum Gauge Verification Methodology for Vacuum Coating Deposition Processes, Proceedings of the 28th DAAAM International Symposium, pp.0645-0653, B. Katalinic (Ed.), Published by DAAAM International, ISBN 978-3-902734-11-2, ISSN 1726-9679, Vienna, Austria
DOI: 10.2507/28th.daaam.proceedings.091

Abstract

The paper analyses vacuum gauge accuracy problem, which leads to the pressure deviation in vacuum nano-coating deposition process. Pressure deviation leaves significant impact on vacuum coating properties and quality. Therefore, new vacuum gauge verification methodology has been developed and the results are presented in this paper. Methodology consists of combination of several verification methods, including a method of constant pumping speed, method of comparison to the reference gauge, Student's t-distribution and other methods and approaches. In the following research, several experiments with ionization and diaphragm vacuum gauges were performed to obtain data on their accuracy and to validate thereof. Proposed methodology focuses on vacuum gauge multifunctional verification under manufacturing conditions, resulting in correction coefficient implementation for vacuum gauges. Developed methodology determines a necessity and a period of calibration to be needed for each vacuum gauge

Keywords: Verification methodology; vacuum gauge; calibration period; vacuum coating; deposition process

1. Introduction

Vacuum nano-coating deposition is performed using specialized vacuum systems. In order to ensure consistent vacuum nano-coating properties, technological process parameters must remain unchanged and controlled at certain level. Vacuum gauges are used to measure a pressure with the vacuum chamber in result of which proper vacuum level is controlled. Vacuum gauge accuracy may leave a significant impact on deposition technological process resulting in low nano-coating quality and properties. In order to minimize associated risks two gauges are used in vacuum systems as a standard. Nevertheless, after certain operation time pressure measurements of the vacuum gauges start to vary considerably. Hence a vacuum gauge calibration is commonly used when the pressure measurement accuracy starts to fail and it is impossible to secure necessary nano-coating quality and properties [5]. Some studies analyse vacuum gauge accuracy and calibration period, but there are no clear approaches for vacuum gauge verification under manufacturing conditions before calibration operation. [14] There are two particularly important considerations to be made to determine vacuum gauge calibration period [14]. First, has the device been calibrated at least twice before, without adjustment, with one calibration being quite recent? [14]. Second, what measurement uncertainty is needed? For the majority of instruments the recalibration interval is about a year, but it is important that the historical data relates to the device when used in an environment and in a way that is similar to the way it is normally used [14]. If no historical data is available it is not possible to answer the question [14]. In addition, users have to stop their manufacturing process for at least several weeks during the period of calibration [6]. This is inappropriate in manufacturing conditions.

Moreover, some users are worried about the damage of their vacuum gauge during transportation [6]. Vacuum gauge calibration decision must be made as final step based on reliable information. That's why it is so important to develop vacuum gauge verification methodology. Than vacuum gauges can be tested, adjusted, necessary coefficient implemented and historical data obtained. In the following paper, we are representing multifunctional vacuum gauge verification methodology using custom-made vacuum system. Experiments were conducted using various methods to obtain full-scale data. In the result of the proposed verification methodology it is possible to perform a vacuum gauge analysis and conclude whether a tested vacuum gauge can be used further or calibration must be performed in accredited metrology laboratory.

Main purpose of this paper is to present the new vacuum gauge verification methodology. The new approach allows to acquire a full spectre of vacuum gauge performance data and perform correction on tested vacuum gauge, thus enhancing deposited vacuum nano-coating quality and properties.

2. Discussion of Prior Art

Oxide coating process, such as SiO2 deposition is a reactive process using metallic targets because of cheapness of metallic target in comparison with conductive ceramic targets [2]. Therefore, vacuum deposited coatings of metal oxides have a wide use in various applications [3]. The main problem in the vacuum nano-coating deposition process is unreliable accuracy of vacuum gauge. Incorrect pressure measurements and even small pressure deviations can reduce nano-coating properties and quality. For example, SiO2 nano-coating is very sensitive to pressure deviation. In confirmation of said statement, the experiment was performed where a SiO2 nano-coating was deposited using laboratory vacuum system UV 80 and diaphragm vacuum gauge CDG 025D INFICON®.

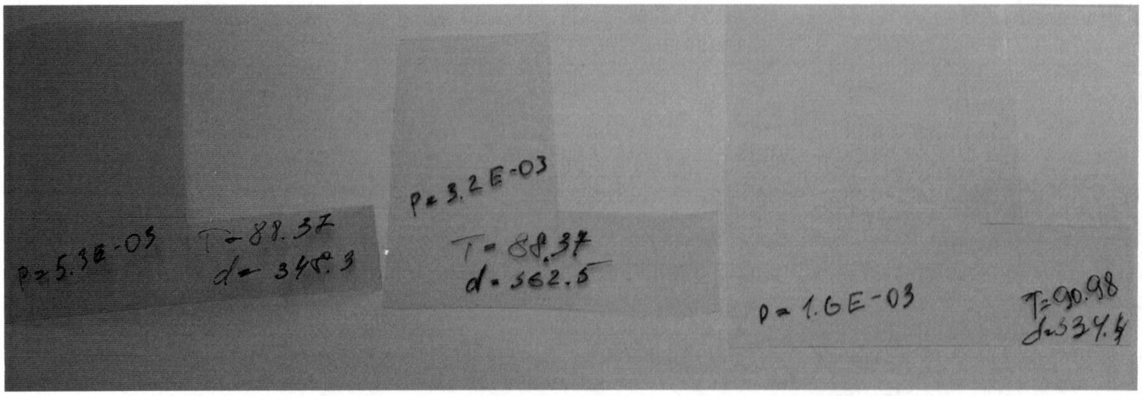

Fig. 1. SiO2 coating samples after vacuum coating deposition

Conducted experiment results are showed at figure 1 and table 1. Light transmission coefficients were measured using MC 122 spectrophotometer, but thicknesses were measured using FILMETRICS F20-UVspectroscopic reflectometer.

No.	Pressure, Torr	Transmission coefficient, %	Thickness, nm
1	5.3E-03	88.37	348.3
2	3.2E-03	88.37	362.5
3	1.6E-03	90.98	334.3

Table 1. Light transmission coefficients and thicknesses of SiO2 coatings under certain vacuum

After the preliminary analyses of experimental results it was concluded that transmission coefficient and coating properties changes depending on the pressure or vacuum value. In the following experiment the transmission coefficient values were above defined standard limit T = 87.5% for oxide coatings. Therefore, the coating quality at 3.2E-03 Torr and 5.3E-03 Torr was unacceptable and did not meet the necessary requirements.

Vacuum coating thickness also must be considered as one of the main parameters. It is well known that the technological process as such is combined from a set of factors and parameters, where a pressure or vacuum is very important. Vacuum gauge verification in certain periods of time should be performed, obtained data should be analysed and correction coefficient/-s should be implemented. The result should be in the form of a reliable and accurate vacuum gauge.

3. Vacuum gauge verification experimental approaches

Vacuum gauge verification using reference gauge. The first step of verification methodology was made using INFICION® diaphragm vacuum gauge CDG 025D and ionization vacuum gauge HPG 400. Any type of gas does not affect diaphragm vacuum gauge and it has high accuracy (inaccuracy is 0.2% of reading) [1, 4, 7]. Unlike CDG 025D, ionization vacuum gauge HPG 400 accuracy is ± 15% of reading [9]. Verification was performed in various pressure ranges, with argon correction coefficient option turned on for ionization vacuum gauge and without it. Experiment was performed using custom made vacuum system, where CDG 025D and two ionization HPG 400 vacuum gauges were verified. Ionization vacuum gauge coefficient was enabled. Argon flow was gradually reduced from 10 sccm to 1 sccm. In the same time a diaphragm vacuum gauge was considered as reference gauge. Results are showed at figure 2.

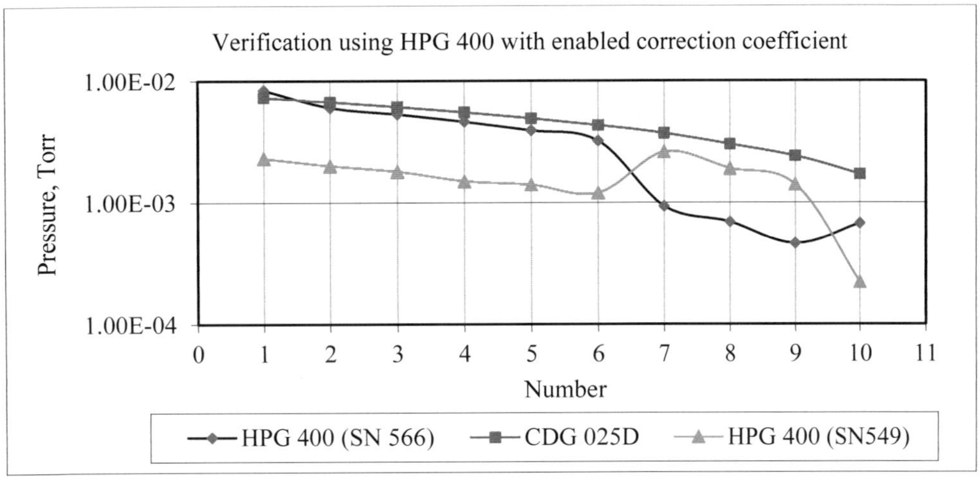

Fig. 2. Performance comparison of HPG 400 and CDG 205D vacuum gauges

After preliminary analysis of the results, it was concluded that the ionization vacuum gauge HPG 400 (SN 549) pressure measurement values differ significantly comparing to the reference gauge. Lack of accuracy exceeds acceptable boundary ±15% of reading, thus no further operations can be performed with this gauge. HPG 400 (SN 566) ionization vacuum gauge pressure measurements and graphic line is more similar to the reference gauge. Ionization vacuum gauge HPG 400 (SN 566) starts to show significant changes in pressure from seventh measuring point to ninth. Identical pressure deviation character was observed in several experiments with disabled argon correction coefficient. Overall ten experiments with enabled and disabled gas correction were conducted and necessary data obtained. This verification experiment combines several approaches such as: testing in different pressure ranges, gradual reduction of gas, and comparison to the reference gauge. All this approaches create vacuum gauge full-scale performance map necessary for the evaluation.

Ionization vacuum gauge correction coefficient test. Vacuum gauge built-in correction coefficient option may have its own inaccuracy. Accordingly, ionization pressure can be determined by following equation [9]:

$$p_{eff} = K \times pressure\ indicated \qquad (1)$$

No.	Argon flow, sccm	HPG 400 Measured pressure (correction off), Torr	HPG 400 Estimated pressure, Torr	HPG 400 Measured pressure (correction on), Torr	Uncertainty, %
1	10 sccm	3.10E-03	2.48E-03	2.32E-03	6.4%
2	9 sccm	2.70E-03	2.16E-03	2.10E-03	2.8%
3	8 sccm	2.40E-03	1.92E-03	1.84E-03	4.2%
4	7 sccm	2.15E-03	1.72E-03	1.55E-03	9.8%
5	6 sccm	1.80E-03	1.44E-03	1.37E-03	4.9%
6	5 sccm	1.50E-03	1.20E-03	1.10E-03	8.3%
7	4 sccm	1.30E-03	1.04E-03	1.02E-03	1.9%
8	3 sccm	9.5E-04	7.6E-04	7.4E-03	2.6%
9	2 sccm	8.0E-04	6.4E-03	6.25E-03	2.3%
10	1 sccm	6.8E-04	5.44E-04	5.35E-04	1.7%

Table 2. Measured and estimated pressures values of HPG 400 vacuum gauge

Ten identical experiments were conducted to check ionization vacuum gauge performance with enabled gas correction coefficient. Experiment results are shown in table 2. After preliminary analysis of experimental results, it was concluded that measured pressure values are lower than estimated pressure values. A maximum value of inaccuracy is 9.8%, which should be considered high because measuring device has inaccuracy ± 15% of reading. This approach shows that measuring device inaccuracy is not the only value to be considered.

Vacuum gauge verification using constant pumping speed. For this experiment, custom-made vacuum gauge verification system was developed and used together with turbo-molecular pump TURBO – V 3K – T for high vacuum and back up mechanical pump НВЗ-100Д for low vacuum. Prior art discusses that pumping speed can be constant at certain point, see figure 3 [8].

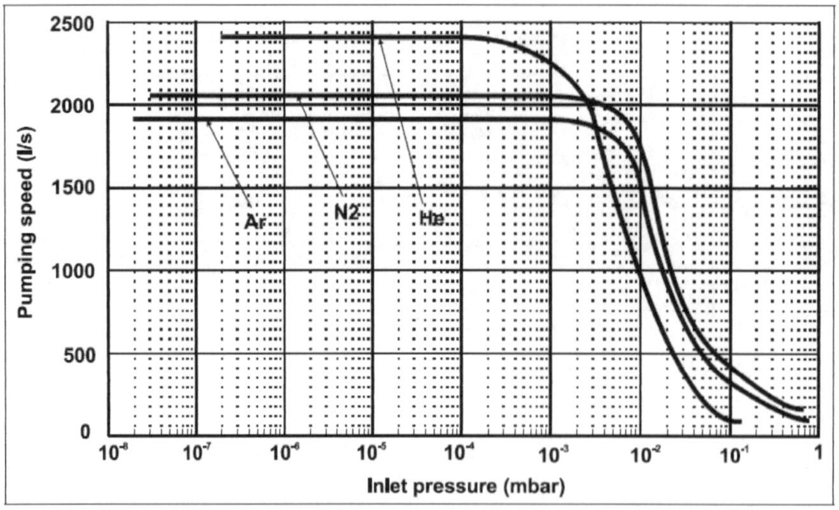

Fig. 3. Pumping speed curves

In this experiment diaphragm vacuum gauge CDG 025D was used to measure pressure within a vacuum chamber and to analyse its performance. Argon (Ar) and Oxygen (O2) gases were fed into the chamber in certain amounts. Argon pumping curve at figure 3 rapidly rises from atmosphere pressure till approx. 10^{-3} mbar, but after that point pumping speed is constant up to approx. $5x10^{-7}$ mbar. Therefore, the experiment was conducted in constant pumping speed range. Experiment results are shown in table 3.

No.	Ar, sccm	CDG 025D, Torr	Estimated pressure, Torr	Inaccuracy, %
1	1 sccm	1.70E-03	Input data	0%
2	2 sccm	2.40E-03	3.4E-03	29.41%
3	4 sccm	3.70E-03	6.8E-03	83.78%
4	6 sccm	4.90E-03	1.02E-02	108.16%
5	8 sccm	6.10E-03	1.36E-02	122.95%
6	10 sccm	7.30E-03	1.7E-02	132.88%

Table 3. Vacuum gauge estimated pressure and inaccuracy

The pressure can be determined by following equation [10]:

$$P = \frac{Q \times K}{S} \quad (2)$$

Where: P is pressure; Q is gas flow amount; S is pumping speed, K is unit conversion value.

Based on the experimental results, which are shown in table 3, it was possible to calculate pumping speed when a pressure and injected gas amount is known. Unit of throughput conversion is necessary for further calculations, thus $sccm = 1{,}27 \times 10^{-2} \frac{torr \cdot l}{s}$ [10].

First measurement in table 3 was taken as input data for pumping speed calculations [10].

$$S = \frac{Q \times K}{P} = \frac{1 \times 0.0127}{0.0017} = 7.47\frac{l}{s} \quad (3)$$

When pumping speed is known it is possible to estimate pressure in the remaining gas flow. Calculated pressure is acquired by using (2) and results are showed in Table 3.

After preliminary analysis of results, we concluded that diaphragm vacuum gauge indicated pressure and estimated pressure values are significantly different. Furthermore, the changes should be linear i.e., pressure value should change twice if gas flow amount is changed twice. Inaccuracy exceed 100% boundary and reached maximum value of 132.88%, which is unacceptable and high. Preforming vacuum coating deposition process using this vacuum gauge may result in poor coating quality.

This experiment proved that it can be performed for fast accuracy check of vacuum gauge when typical pumping curves are known.

Mass flow controller performance experiment. During technological process vacuum gauge measures pressure within vacuum chamber while gas is fed into it. Mass flow controller (MFC) is responsible for gas flow delivery and control. Therefore, it is possible that MFC inaccuracy will influence vacuum gauge indicated pressure. This vacuum system element check is necessary to ensure good deposition process and perform preliminary analysis of obtained data.

MFC verification experiments were performed by separately delivering argon (Ar) and oxygen (O2) in various amounts into vacuum chamber and obtaining pressure measurements. Experiment results are shown in table 4.

No.	Ar, sccm	O2, sccm	CDG 025D, Torr
1	50	0	3,40E-03
2	75	0	5,00E-03
3	100	0	6,68E-03
4	0	50	3,20E-03
5	0	75	4,67E-03
6	0	100	8,09E-03

Table 4. Measured pressure values depending on used gas type

Next step was to perform manipulations with gas delivery system by connecting argon flow channel though O2 mass flow controller. Schematic representation of gas connection establishing through O2 MFC is showed in figure 4.

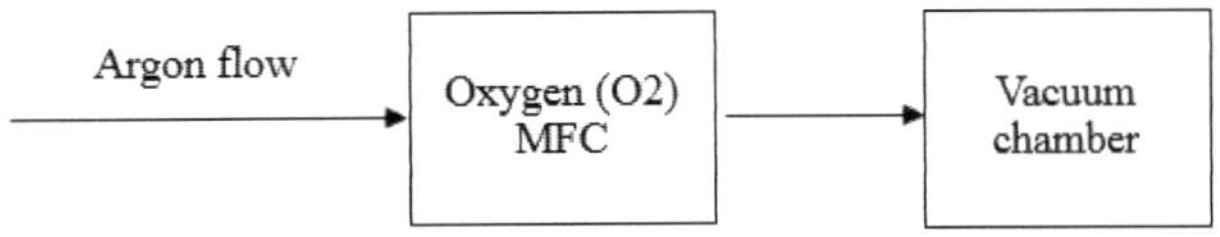

Fig. 4. Experimental gas connection scheme

Since argon gas is delivered through the O2 mass flow controller, gas correction factor must be implemented. For argon, a gas correction factor according to the @MKS data is 1.39 [11]. Obtained experimental results are showed in table 5.

No.	Argon flow value, sccm	O2 converted value, sccm	CDG 025D indicated pressure through O2 MFC, Torr	CDG 025D indicated pressure through Ar MFC, Torr	Inaccuracy, %
1	50	36	3.58E-03	3,40E-03	5,00%
2	75	54	5.26E-03	5,00E-03	4,94%
3	100	72	6.92E-03	6,68E-03	3,47%

Table 5. Vacuum indicated pressure values through different MFC's

After preliminary analysis of obtained data, it was concluded that measured pressure of the diaphragm vacuum gauge can be different delivering the same amount argon or oxygen into the chamber. Gas connection through Oxygen MFC did not result in accurate measurements. It was concluded that oxygen MFC has increased uncertainty. Pressure values should be identical, when argon is delivered into the chamber through Ar and O2 MFC.

Vacuum gauge verification using Student's t-distribution. T-distribution is used to obtain vacuum gauge confidence interval, thus its accuracy is checked. Experiment was performed using ionization HPG 400 vacuum gauge. Argon gas flow equal to 30 sccm was fed into the chamber, after pressure stabilization injection was stopped. When gas injection is stopped, pressure value should return to the initial value.

Pressure inside vacuum chamber was set up 4.05E-4 Torr. Experimental data and estimated values are shown at table 6.

The sample mean can be determined by following equation [12]:

$$\bar{x} = \frac{\sum x_i}{n} = 4{,}1E - 04\ (Torr) \tag{4}$$

Where: $\sum x_i$ – all pressure values, n – number of measurements.

No.	x_i HPG 400 pressure, Torr	$\Delta x = \bar{x} - x_i$, Torr	Δx^2, Torr
1	3.99E-04	+ 0.11E-04	0.0121E-04
2	4.11E-04	+ 0.001E-04	0.0001E-04
3	4.25E-04	+ 0.15E-04	0.0225E-04
4	3.80E-04	- 0.3E-04	0.09E-04
5	4.15E-04	+ 0.05E-04	0.0025E-04
6	4.22E-04	+ 0.12E-04	0.0144E-04
7	4.51E-04	+ 0.41E-04	0.1681E-04
8	4.35E-04	+ 0.25E-04	0.0625E-04
9	3.91E-04	- 0.19E-04	0.0361E-04
10	3.86E-04	- 0.24E-04	0.0576E-04

Table 6. Ionization vacuum gauge measured and estimated pressure values

When the sample mean estimation is complete, then a pressure standard deviation is calculated using following equation [12]:

$$S = \sqrt{\frac{\sum_i^n (x_i - \bar{x})^2}{n-1}} = 0.227E - 04\ (Torr) \tag{5}$$

After standard deviation, confidence interval of the mean is determined. We used 95% confidence level for (*f*) degrees of freedom:

$$f = n - 1 = 9, \tag{6}$$

Where: n is number of measurements. Multiplier $t = 2.26$ for 95% confidence level and 9 degrees of freedom [12]. Confidence interval equation is as follows [12]:

$$\Delta\bar{x} = \sqrt{\frac{S^2}{n}} = \frac{S}{\sqrt{n}} = \frac{2{,}26 \times 0{,}227}{\sqrt{10}} = 0.16223E - 04\ Torr \tag{7}$$

After calculations it was concluded that standard deviation confidence interval is 4.1±0.16E-04Torr. This approach shows vacuum gauge possible inaccuracy interval, which should be considered performing coating deposition process.

Vacuum gauge zero adjustment. As vacuum gauge verification final step, a zero adjustment must be performed. If the vacuum gauge accuracy if very low and exceeds accepted limits, then it is necessary to perform vacuum gauge zero adjustment.

For this experiment diaphragm vacuum gauge CDG025D was selected. The zero can be adjusted via the control button on the same gauge. Pressure in the vacuum chamber must be reduced to a pressure according to the manufacturers table, which is $<5 \times 10^{-6}$ Torr [13].

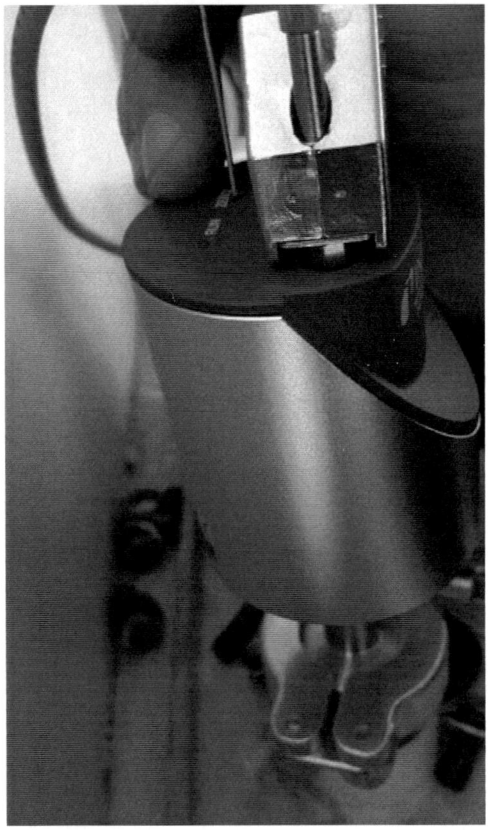

Fig. 5. Diaphragm vacuum gauge zero adjustment

The gauge was operated for at least 2 hours until signal was stable according to the instructions from manufacturer of the gauge [13]. The final step was to briefly press the zero button with a pin [13]. The zero adjustment runs automatically and LED flashed until the adjustment (≈ 8 sec.) was completed [13]. CDG 025D capacitance gauge experimental zero adjustment is shown at figure 5.

Diaphragm vacuum gauge accuracy test after zero adjustment. This operation is necessary if vacuum gauge accuracy is low after performing its verification by methods described in this paper. After zero adjustment vacuum gauge must be verified once again using all described methods. For example, based on the obtained experimental data CDG 025D diaphragm vacuum gauge pressure measurement accuracy was low. Therefore, zero adjustment was performed and re-verified. Adjusted vacuum gauge was checked using (3, 4) and calculating inaccuracy. Experimental results are shown in table 7.

No.	Ar, sccm	O2, sccm	CDG 025 D indicated pressure, Torr.	Estimated pressure	Inaccuracy, %
1	50	0	2.86E-03	Input data	0
2	75	0	4.33E-03	4.29E-03	0,93%
3	100	0	5.75E-03	5.72E-03	0,52%
4	150	0	8.52E-03	8.58E-03	0,70%

Table 7. Vacuum gauge estimated pressure and inaccuracy

After preliminary analysis of obtained experimental data, it was possible to conclude that diaphragm vacuum gauge pressure measurement accuracy was high. After the adjustment, the indicated pressure values were precise and uncertainty was low comparing to the diaphragm gauge experimental results before zero adjustment as seen in table 3.

4. Vacuum gauge verification methodology

The overall vacuum gauge verification methodology includes all the above described vacuum gauge verification approaches, which are represented as block scheme in figure 6.

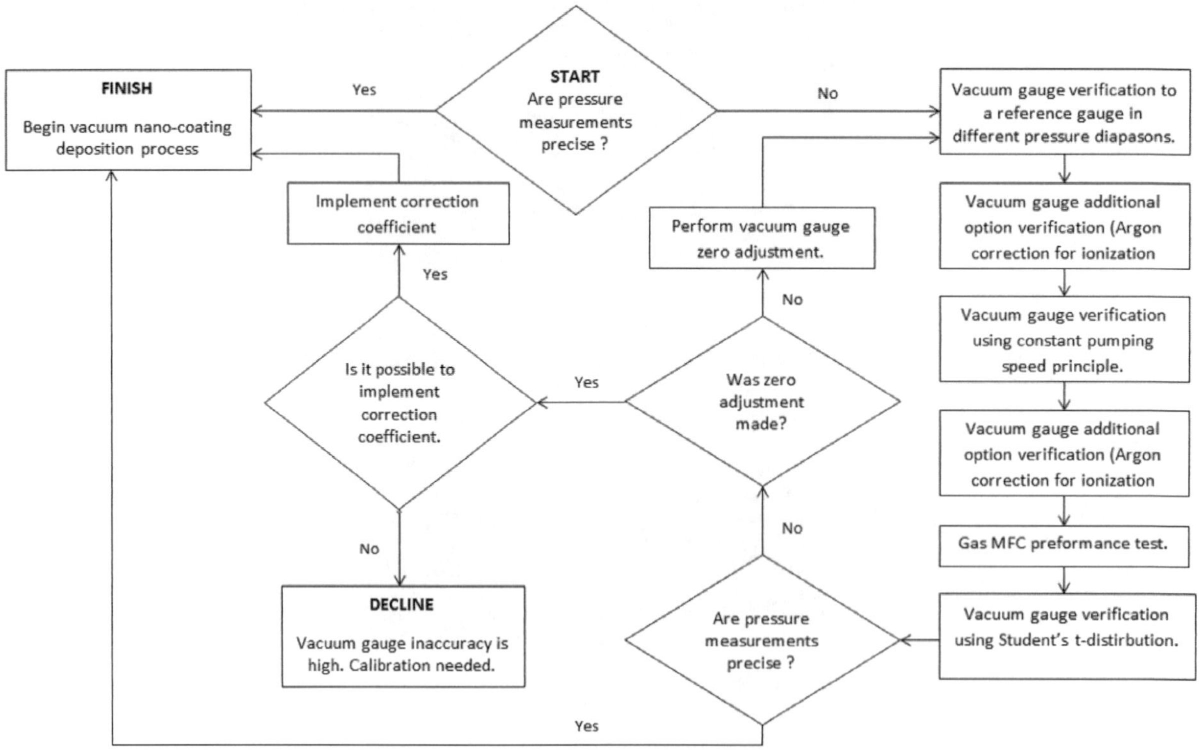

Fig. 6. Vacuum gauge verification methodology

Main goal of verification is to obtain reliable data under different conditions. Vacuum gauge performance will influence a quality of deposited nano-coating during manufacturing process, therefore it is necessary to secure reliable pressure indication. Each approach focuses on critical and possible vacuum gauge inaccuracies, which may take place or emerge. It was concluded that there is no need to perform expensive and long-term vacuum gauge calibration in accredited laboratories, because using the developed method it is possible to obtain full spectre data and act accordingly. If vacuum gauge verification was performed and even after adjustment pressure measurement uncertainty is high, then evaluation of current situation is needed. Evaluation includes technological process parameter and verification experimental data analysis to introduce a correction coefficient. If correction coefficient introduction is not possible, then vacuum gauge should be declined for further use and send to recalibration. Diaphragm vacuum gauge CDG 025D was verified using described methods. After adjustment vacuum gauge inaccuracy did not exceed 2% limit. In this situation correction coefficient implementation was not necessary.

5. Conclusion

The paper analyses vacuum gauge pressure deviation problems and its influence on coating quality. It was concluded that even small pressure changes will influence a visual look and parameters of the coating. To prevent following disadvantages a precise and reliable vacuum gauge shall be used for vacuum nano-coating deposition process. Vacuum gauge verification should be performed before deciding, whether a vacuum gauge recalibration is needed in special accredited laboratories. In a result of presented research a vacuum gauge verification methodology was designed. The method allows to obtain complete data on vacuum gauge pressure measurement accuracy and uncertainty thereof. Developed vacuum gauge verification methodology allows to perform all the verification operations under manufacturing conditions in short period of time in order to rapidly resume the coating deposition process. The methodology can be used as a guide in any company, which focuses on deposition of various vacuum nano-coatings. Further research should focus on correction coefficient implementation in vacuum gauges, its testing and observation.

6. References

[1] Barthélémy Daudé, Hadj Elandaloussi, Christof Janssen (2014). On the gas dependence of thermal transpiration and a critical appraisal of correction methods for capacitive diaphragm gauges, Vacuum, Vol. 104, 2014, pp. 77-87., DOI: http://doi.org/10.1016/j.vacuum.2014.01.002

[2] Kazuss J., Kozlovs V., Macevskis E. Control of reactive deposition process by stabilization of the power supply, Society of Vacuum Coaters 54th Annual Technical Conference Proceedings, USA, Chicago, 16.-21. April 2011. Santa Clara: Society of Vacuum Coaters, 2011, 419.-422.lpp. ISSN 0737-5921.

[3] Kazuss J., Kozlovs V., Macevskis E. Processes control for oxide layer deposition in roll-to-roll vacuum machines, Society of Vacuum Coaters 56th Annual Technical Conference Proceedings, USA, DOI: http://dx.doi.org/10.14332/svc13.proc.1038

[4] Kenta Arai, Hajime Yoshida, Tokihiko Kobata (2014). Measurement of volume change induced by a bulging of a diaphragm inside a differential type capacitance diaphragm gauge", Measurement, Vol. 48, 2014, pp. 149 –154., DOI: http://doi.org/10.1016/j.measurement.2013.10.042

[5] Stekleins A. Gerins E.; Kromanis A. (2016). Vacuum gauge performance verification system, Proceedings of the 27th DAAAM International Symposium, pp.0607-0614, B. Katalinic (Ed.), Published by DAAAM International, ISBN 978-3-902734-08-2, ISSN 1726-9679, Vienna, Austria DOI: 10.2507/27th.daaam.proceedings.089;

[6] Zhenhua X., Detian L., Yongjun C., Yongjun W., Huzhong Z., Jiatong L., Baoguo G., Kang J. (2016). Development of a transportable apparatus for vacuum gauge calibration", Vacuum, Vol. 127, 2016, pp. 30-37., https://doi.org/10.1016/j.vacuum.2016.01.025;[7] http://products.inficon.com/en-us/Product/Detail/CDG025D-X3?path=Markets_Applications%2FMarkets, (2017). INFICON, CDG025D Capacitance Diaphragm Gauge, Accessed on: 2017-04-19

[8] http://www.agilent.com/cs/library/usermanuals/public/TV-3KT%20Pumping%20System%20User%20Manual.pdf, (2017). Agilent Technologies, TV-3KT Pumping System User Manual, Accessed on: 2017-04-19

[9] http://products.inficon.com/GetAttachment.axd?attaName=BC0F2F3D-FCCF-458E-BB20-E00B043E630A, (2017). INFICON, HPG400 high pressure/ pirani gauge instruction sheet, Accessed on: 2017-04-19

[10] https://www.pfeiffer-vacuum.com/filepool/File/Vacuum-Technology-Book/Vacuum-Technology-Book-II-Part-2.pdf?referer=2012&request_locale=en_US, (2017). Pfeiffer, The vacuum technology book, Accessed on: 2017-04-18

[11] https://www.mksinst.com/docs/ur/MFCGasCorrection.aspx, (2017). MKS, Gas Correction Factors for Thermal-based Mass Flow Controllers, Accessed on: 2017-04-18

[12] http://www.himikatus.ru/art/modeling/ocwnka-errors.php, (2017). Chemical portal. Measurement uncertainty evaluation, Accessed on: 2017-04-19

[13] http://products.inficon.com/GetAttachment.axd?attaName=87f38c98-2d62-465d-99de-77a0e5d2124f, (2017). INFICON. CDG025D Capacitance diaphragm gauge operating manual, Accessed on: 2017-04-21

[14] http://www.npl.co.uk/upload/pdf/Guide_to_PV.pdf, (2017). National physical laboratory. Guide to the measurement of pressure and vacuum, Accessed on: 2017-04-16

DOI: 10.2507/28th.daaam.proceedings.092

PREDICTION OF SIO2 NANO COATING PROPERTIES USING FUZZY LOGIC

Antons Stekleins, Eriks Gerins & Artis Kromanis

This Publication has to be referred as: Stekleins, A[ntons]; Gerins, E[riks] & Kromanis, A[rtis] (2017). Prediction of SiO2 Nano Coating Properties Using Fuzzy Logic, Proceedings of the 28th DAAAM International Symposium, pp.0654-0660, B. Katalinic (Ed.), Published by DAAAM International, ISBN 978-3-902734-11-2, ISSN 1726-9679, Vienna, Austria
DOI: 10.2507/28th.daaam.proceedings.092

Abstract

The paper analyses SiO2 vacuum nano coating technological process and its influence on coating quality and properties. Use of fuzzy logic is proposed to develop a process prediction models for vacuum coating. Several experiments were designed and conducted using UV80 laboratory vacuum system, wherein SiO2 coating light transmission coefficient and thickness were measured. SiO2 nano coating light transmission and thickness predictions models were designed using fuzzy logic techniques. After preliminary analysis it was concluded that models developed by fuzzy logic have high accuracy and can be used to predict SiO2 vacuum nano coating properties if a leak in vacuum system appears during technological process, thus enabling continuation of the technological process as long as coating properties will match selected criteria.

Keywords: SiO2; vacuum coating; fuzzy logic; prediction model; coating deposition

1. Introduction

Nowadays nano coatings are one of the most commonly used and desired product in vacuum field, which can be applied onto webs, films and foils. For example, in SiO2 vacuum nano coating technology a magnetron sputtering is used and SiO2 coating itself is unique and demanded coating due to its properties. Thin layers having high insulation resistance are required for many applications in electronics and in sensor and photovoltaic technology [2]. SiO2 is one of these materials. These applications include gate oxide layers in microelectronics (which are only a few nanometers thick), insulating layers in sensor applications (for which higher dielectric strengths can be required) and insulating layers in photovoltaic technology (for which the requirements on dielectric strength are lower) [2]. Therefore, it is essential to maintain required coating technological process parameters to provide good product quality. System leakage is one of the main problems we encountered with during SiO2 coating process. Pressure deviations, process parameter changes and leakage can damage the coating, its quality and properties. Therefore, it is necessary to establish models, which can be used to predict vacuum nano coating properties when there is a leakage in vacuum system. Such models would help to evaluate SiO2 coating sputtering process condition and predict its output parameters (light transmission coefficient and thickness). In this study models were developed using fuzzy logic method. Fuzzy set theory which formed the basis of fuzzy logic, as an alternative to classical set theory, was presented by L.A. Zadeh [1]. In the fuzzy set theory, the transition between membership to non-membership is done progressively [1].

Unlike Boolean Logic fuzzy logic was designed to reproduce human thinking technique using mathematical approach [1, 5]. This particular approach helps to visualize and predict possible output data variables depending on the input data changes [3, 7, 6]. Several SiO2 experiments were designed and conducted, a light transmission coefficient and thickness were measured for each coating sample. Using fuzzy logic method thickness and light transmission coefficient models were developed and compared with measured values.

2. Design of SiO2 coating experiment

Experiments to develop prediction models were carried out at a laboratory vacuum machine UV80 for coating webs and films [8]. Silicon reactive sputtering was selected for the experiments, where SiO2 coating was applied onto polyethylene terephthalate (PET) film. For the process power supply (8kW), argon flow 75 sccm, variable oxygen and airflows were used. Main idea of the experiment was to change airflow quantity thus simulating leak condition, to verify coating quality, i.e. measure its thickness and a light transmission coefficient. After conducting SiO2 vacuum coating experiments samples were verified and measured. Measured results are given in table 1.

No.	O2, sccm	Air, sccm	d (thickness, nm)	T (transmission coefficient, %)
1	0	70	142.1	86.64
2	0	65	133	86.49
3	0	55	128.5	87.54
4	0	50	121.3	88.36
5	7	45	115.2	89.35
6	17	40	110.7	89.45
7	20	35	109.8	89.19
8	25	30	112.8	87.66
9	30	25	11.6	87.55
10	41	15	119.9	86.81
11	50	5	126	85.33
12	52	0	130.5	83.98

Table 1. Technological parameters and the results of SiO2 coating samples

Coating sample measurement was performed on spectrophotometer MC 122 for light transmission coefficient calculation and FILMETRICS F20-UV machine to determine coating thickness. SiO2 coating light transmission should not be less than 88% to match the requirements. Results showed that in leak situation under certain circumstances coating light transmission property does not match 88% boundary. After coating visual verification it was concluded that SiO2 coated samples No. 4, 5, 6, and 7 are pure, clean and doesn't have any browning signs, thus visual coating quality matches the desired requirements.

3. Design of prediction models using Fuzzy Logic

One of the main reasons of the research is to develop prediction models to control SiO2 coating light transmission coefficient and thickness under leak circumstances. Developed prediction models can be implemented in vacuum coating sputtering laboratories or factories. In the begining input values were defined, then the fuzzification was performed or in other words fuzzy logic was used [4]. The next step included generation of output values according to the defined fuzzy rules previously defined in the database [4]. After fuzzy model was developed, a defuzzification we performed in order to obtain 3D output parameter plot.

Fuzzy modelling was performed using fuzzyTECH 8.30b Professional Demo software. To develop fuzzy logic model two input variables were used: O2 flow rate and air flow rate, while output variables were coating light transmission coefficient (T) and thickness (d). The following model shows relationship between input data, output data and rule block (see figures 1 and 2).

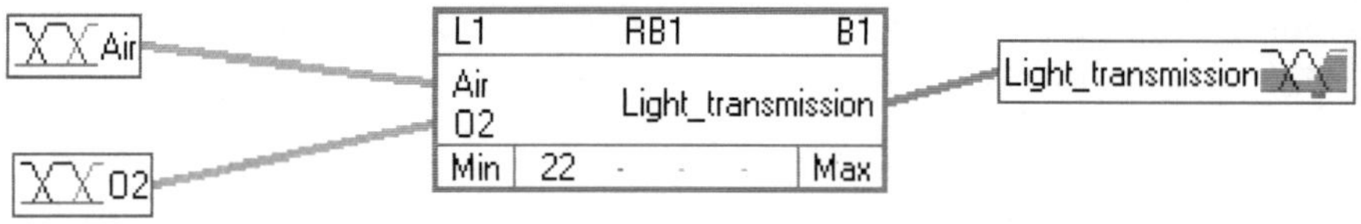

Fig. 1. Light transmission coefficient fuzzy model

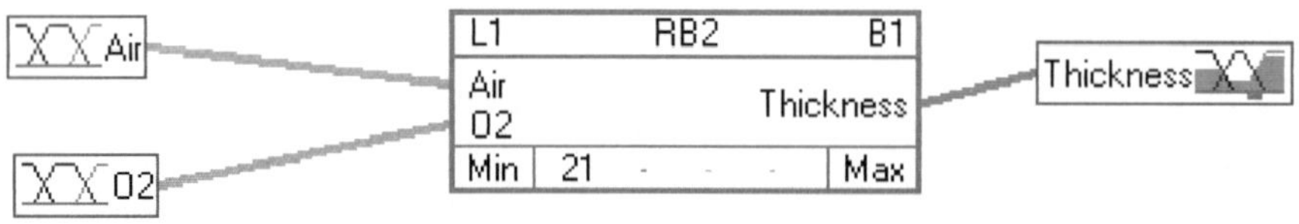

Fig. 2. SiO2 nano coating thickness fuzzy model

After fuzzy logic model development it was necessary to define membership functions to all input and output parameters and create rule block definitions between technological and output parameters. Logic variables have truth-values that lies in range between 0 and 1. Membership functions for input and output parameters were developed based on the results from experiment conducted and described, thus these results are considered as preliminary result database which is shown in table 1.

For the airflow five membership functions were selected that are as follows (shown at figure 3): very small (12 sccm); small (24 sccm); medium (36 scmm); big (48 sccm); very big (60 sccm).

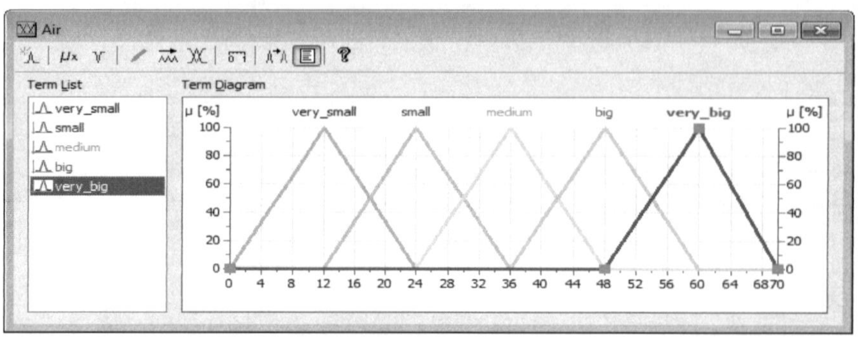

Fig. 3. Five membership functions for air flow rate

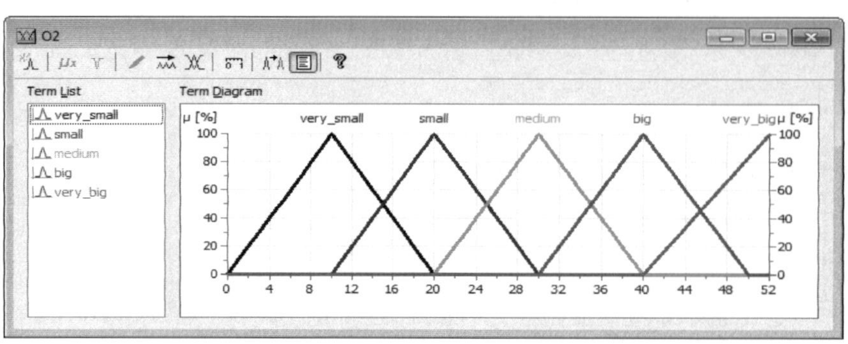

Fig. 4. Five membership functions for O2 flow rate

For oxygen flow rate (O2) five membership functions were selected that are as follows (shown at figure 4): very small (10 sccm); small (20 sccm); medium (30 sccm); big (40 sccm); very big (52 sccm).

Similar fuzzification procedure was made for thickness and light transmission coefficient (shown at figure 5). For light transmission coefficient five membership functions were selected that are as follows: very small (84, 5%); small (86%); medium (87%); big (88.5%); very big (90%).

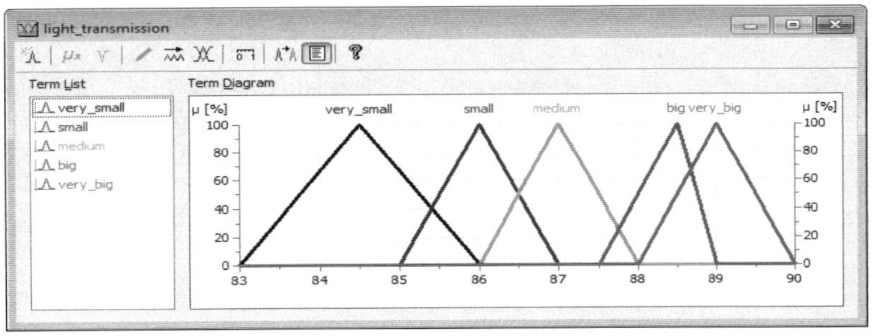

Fig. 5. Five light transmission membership functions

For thickness were selected five membership functions, as follows (shown at figure 6): very small (113 nm); small (119 nm); medium (126 nm); big (133 nm); very big (142 nm).

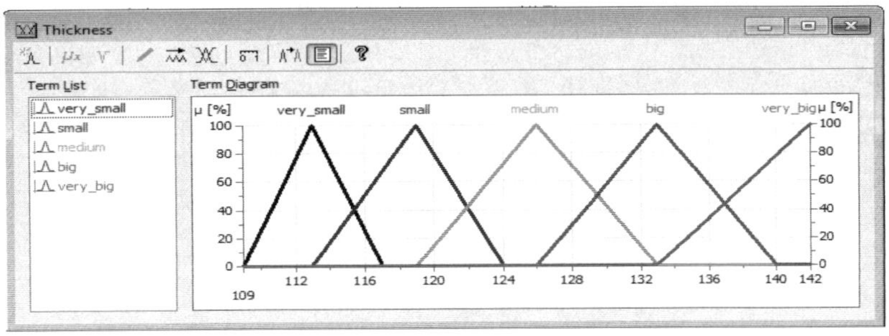

Fig. 6. Five thickness membership functions

The division created by membership function is sufficient to set up a full-fledged light transmission coefficient and thickness fuzzy logic model. Therefore, it is necessary to set a correlation between input and output data using fuzzy rule block, where a database of rules were formed. Since fuzzy logic is based on humanlike operating principles, thus the rule block operates by means of "IF", "THEN". In table 2 a rule block of light transmittance is shown. Rule block database is design based on the obtained and measured SiO2 coating sputtering experiment results, which are shown in table 1.

	IF		THEN
Nr.	**Air flow**	**O2 flow**	**Light transmission coefficient**
1	Very small	Big	Small
2	Very small	Big	Very small
3	Very small	Very big	Very small
4	Very small	Small	big
5	Very small	Medium	medium
6	Small	Small	Very big
7	Small	Medium	Big
8	Small	Very big	Big
9	Small	Very big	medium
10	Medium	Very small	Big
11	Medium	Very small	Very big
12	Medium	Very small	Medium
13	Medium	Small	Big
14	Medium	Small	Very big
15	Medium	Medium	Medium
16	Big	Very small	Big
17	Big	Very small	Medium
18	Big	Small	Very big
19	Big	Small	Big
20	Very big	Very small	Small
21	Very big	Very small	Medium
22	Very big	Very small	Big

Table 2. Fuzzy logic rule block of light transmission

4. Data defuzzification

Next step after development of the rule block was data defuzzification, where linguistic values are transferred into numerical values and prediction models are graphically displayed in 3D and 2D plots. Light transmission coefficient 3D prediction fuzzy model is shown in figure 7. Thickness 3D prediction fuzzy model is shown in figure 8. In additional window "Watch: Interactive debug mode" input values can be changed to obtain output values, thus oxygen and air flows can be changed and after data defuzzification thickness and light transmission coefficient values can be obtained.

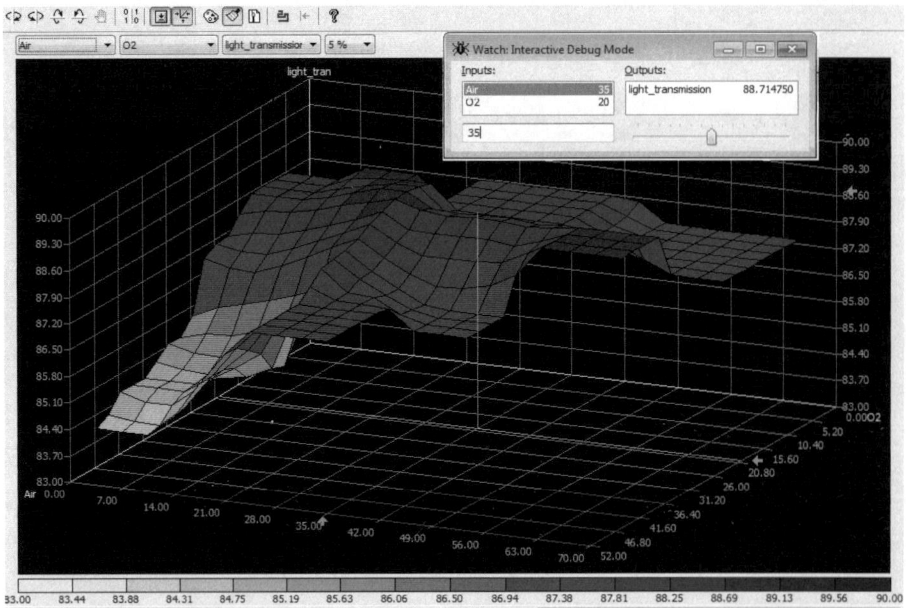

Fig. 7. Light transmission (T) prediction model 3D plot

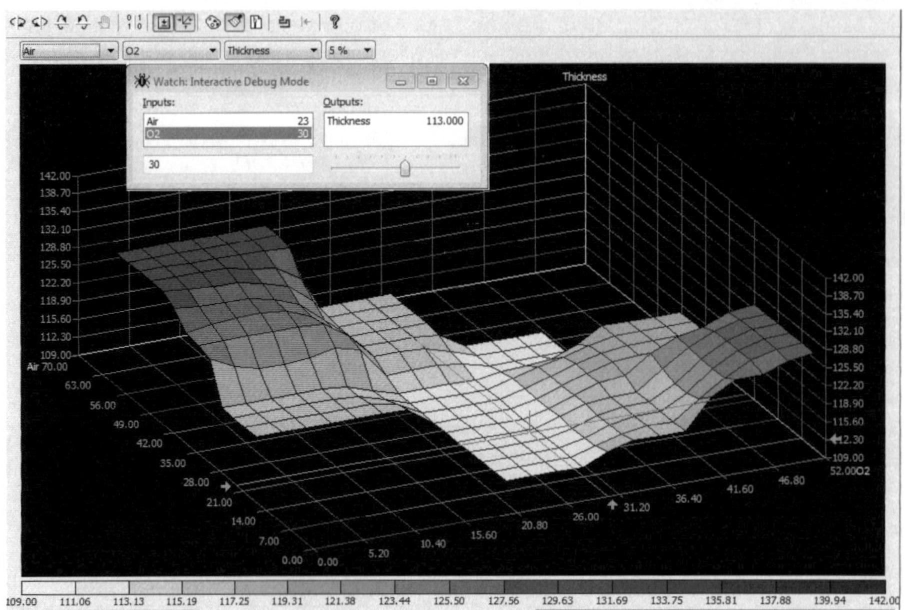

Fig. 8. Coating thickness (d) prediction model 3D plot

One particular example is showed in Fig. 8, where in the interactive debug mode the airflow is to 35 sccm and O2 flow is 20 sccm, thus data defuzzification calculates light transmission coefficient 88.71%. Same principle applies to nano coating thickness prediction model at figure 9, where airflow is 23 sccm and oxygen flow is 30 sccm and thickness prediction model result is 113 nm. Developed fuzzy logic models must be compared with experimental measured results. Therefore, several samples were selected from table 1 and combined with fuzzy logic generated predicted values changing gas flows (see figure 9, 10).

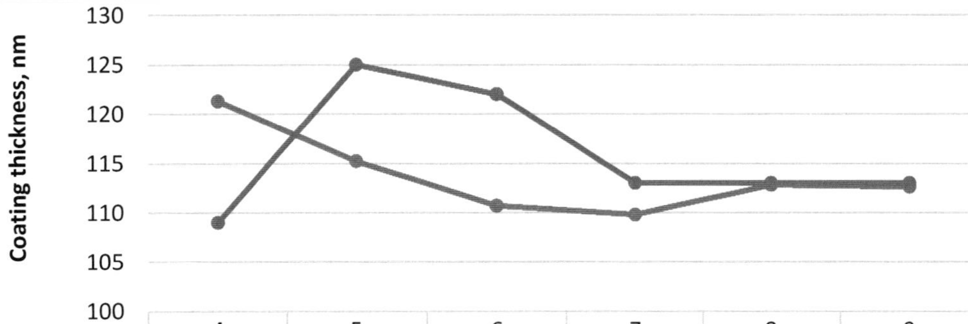

	4	5	6	7	8	9
d, nm (Measured)	121.3	115.2	110.7	109.8	112.8	112.6
d, nm (Predicted)	109	125	122	113	113	113

Sample Number

Fig. 9. Measured and predicted nano coating thickness (d)

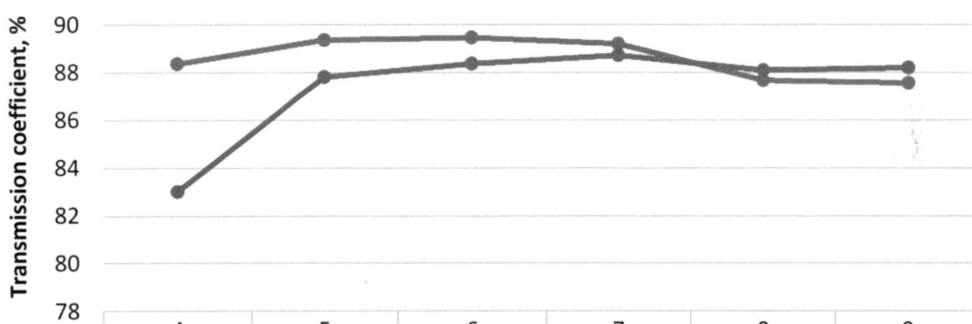

	4	5	6	7	8	9
T, % (Measured)	88.36	89.35	89.45	89.19	87.66	87.55
T, % (Predicted)	83	87.8	88.36	88.71	88.09	88.19

Sample Number

Fig. 10. Measured and predicted light transmission coefficient (T)

Vacuum SiO2 nano coating light transmission was evaluated as the main criteria, which it must meet. Therefore, if the leak appears it is necessary to follow light transmission prediction model in order to secure determined coating quality. Measured and fuzzy logic calculated light transmission and thickness results are shown in figures 9 and 10. Calculated light transmission results show high accuracy of 93.9% at fourth sample and highest value of 99.8% at eight sample. Calculated nano coating thickness results show high accuracy of 90% at fourth sample and highest value of 99.5% at eight sample. Light transmission model has high accuracy and can be used for vacuum nano coating properties prediction. Thickness model showed accuracy deviation is from 90% up to 99.5%. Thus, thickness model accuracy can be considered as decently high. Both models can be used to predict nano coating properties if a leak in vacuum system appears during technological process. As for the priority, the light transmission model remains as main.

5. Conclusion

This paper, investigate leak influence on coating quality and its properties during vacuum coating sputtering technological process. Experiment results and developed fuzzy logic models show that it is possible to predict vacuum nano coating light transmission coefficient and thickness under leak conditions with decently high accuracy. Experiments showed that light transmission accuracy diapason is from 93.9% up to 99.8% and thickness accuracy diapason is from 90% up to 99.5%. Fuzzy logic developed models present reliable accuracy, thus can be used in laboratory or vacuum coating sputtering facilities to predict coating properties and quality in leak conditions. Prediction models can save time, material and allow coating process to continue. Fuzzy model development is a complicated task, where precise definition of membership functions and rules are important.

6. References

[1] Ata, S. & Dincer, K. (2017). Fuzzy logic modeling of performance proton exchange membrane fuel cell with spin method coated with carbon nanotube", International Journal of Hydrogen Energy, Vol. 42, 2017, pp. 2626-2635., https://doi.org/10.1016/j.ijhydene.2016.04.134

[2] Frach, P., Bartzsch, H., Glöß, D., Fahland, M. & Händel, F. (2008). Electrically insulating Al2O3 and SiO2 films for sensor and photovoltaic applications deposited by reactive pulse magnetron sputtering, hollow cathode arc activated deposition and magnetron-PECVD", Surface and Coating Techonology, Vol. 202, 2008, pp. 5680–5683., https://doi.org/10.1016/j.surfcoat.2008.06.043

[3] Harun, T., Cemalettin, K., Özer, U. & Seher, A. (2006). FUZZYFCC: Fuzzy logic control of a fluid catalytic cracking unit (FCCU) to improve dynamic performance, Computers & Chemical Engineering, Vol. 30/5, 2006, pp. 850-863., https://doi.org/10.1016/j.compchemeng.2005.12.016

[4] Kromanis, A. & Krizbergs, J (2013). Prediction of 3D Surface Roughness Using Regression Analysis and Fuzzy Logic, and their Comparative Analysis, Proceedings of the 20th International DAAAM Symposium 2009, Vienna, Austria, 25-28 November 2009, ISSN: 17269679, ISBN: 978-390150970-4, Danube Adria Association for Automation and Manufacturing, DAAAMTop of Form, pp. 803-804.

[5] Mohammad Hossein Nadian., Mohammad Hossein Abbaspour-Fard., Alex Martynenko., Mahmood Reza Golzarian (2017). An intelligent integrated control of hybrid hot air-infrared dryer based on fuzzy logic and computer vision system, Computers and Electronics in Agriculture, Vol. 137, 2017, pp. 138-149., DOI: https://doi.org/10.1016/j.compag.2017.04.001

[6] Ooi, E., Sayuti, M. & Ahmed A.D.S (2015). Fuzzy logic-based approach to investigate the novel uses of nano suspended lubrication in precise machining of aerospace AL tempered grade 6061, Journal of Cleaner Production, Vol. 89, 2015, pp. 286-295., DOI: https://doi.org/10.1016/j.jclepro.2014.11.006.

[7] Zalnezhad, E. & Ahmed, A.D.S. (2014). A Fuzzy logic predictive model for better surface roughness of Ti-TiN coating on AL7075-T6 alloy for longer fretting fatigue life", Measurement. Vol. 49, 2014, pp. 256-265., https://doi.org/10.1016/j.measurement.2013.11.042

[8] http://www.sidrabe.com/assets/files/Processes%20Control%20for%20Oxide%20Layer%20Deposition%20in%20 Roll-to-Roll%20Vacuum%20Machines_2013.pdf, (2013). Sidrabe, Publications, Accessed on: 08.06.2017

DOI: 10.2507/28th.daaam.proceedings.093

ENHANCING THE AGILITY AND PERFORMANCES OF A PROJECT WITH LEAN MANUFACTURING PRACTICES

Nela Cvetković, Slobodan Morača, Miloš Jovanović,
Milovan Medojević, Bojan Lalić

This Publication has to be referred as: Cvetkovic, N[ela]; Moraca, S[lobodan]; Jovanovic, M[ilos]; Medojevic, M[ilovan] & Lalic, B[ojan] (2017). Enhancing the Agility and Performances of a Project with Lean Manufacturing Practices, Proceedings of the 28th DAAAM International Symposium, pp.0661-0670, B. Katalinic (Ed.), Published by DAAAM International, ISBN 978-3-902734-11-2, ISSN 1726-9679, Vienna, Austria
DOI: 10.2507/28th.daaam.proceedings.093

Abstract

This study presents the case study on implementation of Lean Kanban manufacturing principles into the existing Agile software development methodology. The aim of this study was to introduce the adjusted Lean principles which were seen as adequate for enhancing project agility (flexibility and adaptability) and to test if these changes consequently result in improved project team performances. In order to investigate possible effects of Lean approach on improved project agility (flexibility and capacity to adapt to changes), data from interviews, daily team meetings and sprint retrospective were examined. Furthermore, team velocity, defects rates and rework rate data were used for comparison of current Agile methodology success and proposed Lean Agile method. Results show that by adding adequate Lean Kanban principles based on needs of project and team, development process was improved. Defect rate notable decreased while velocity increased, mostly caused by Lean practices enabling more realistic effort estimation and optimized work flow. Moreover, team reported that improved visual management tools facilitated communication and made it more transparent, enabling better and faster problem solving and task management.

Keywords: Agility; Lean practices; Agile Project Management; Project Performance Improvement

1. Introduction

Term *agility* implies flexibility, proactivity [1] and reactivity [2]. Concisely, agility can be interpreted as "ability to proactively or reactively adapt to change"[2]. In this context, agility has significant relevance for software development projects, where high uncertainty and frequent change requirements are mostly inevitable.

Most of the software products are specifically tailored with intents to fulfil customers' requirements and needs as successfully as possible. However, only few of the projects intending to deliver these products are defined as successful at the end of its implementation; the deliverables of low percentage of software development projects are recognized as satisfactory by the customers or to have generated additional value to the users [3]. Alongside the major project constraints according to PMBOK [4]: scope, budget, schedule, risk, quality and resources, the most significant factors enhancing the complexity of managing IT projects are: continuous changes of underlying technologies, complexity of projects disabling realistic schedule and budget estimations, ineffective project management skills, lack of clients' involvement and clear requirements [2].These challenges could not be overpassed with so-called traditional project management methodologies since those entail extensive planning, up-front analyses and design [5], are seen as robust, bureaucratic and process-centric.

As a response to the abovementioned challenges, Agile methodologies have been introduced to the software project management. Agile approach has proved to be adequate and useful when it comes to the greatly important preconditions for successful projects: ability to respond and react to the changing needs of customers and to reduce delivery time [3]. Agile methodologies place the emphasis on stronger communication and creation of useful product, while strict frameworks of predefined project phases and extensive documentation are of the secondary importance [6]. Nevertheless, some Agile principles were inadequate for certain project types that require higher level of flexibility. The time-boxed processes [7], exhausting work scheduling, and non-applicability on large-scale projects, represent the most important possible demerits, causing the constant pressure on team to finish its sprint work on time in any circumstances, push system and in some cases wrong effort estimations and time waste. During the evolution of agile project methodologies, Lean approach has gained significant attention, especially caused by opinion that certain Lean manufacturing principles could improve agility and leanness of a project and redeem for abovementioned possible problems with the most popular Agile methodologies.

Considering that the practice implementation of lean manufacturing principles into the Agile development process is relatively new, the available literature and case studies are still limited. The aim of this study is to introduce the adjusted Lean principles, which were seen as adequate for enhancing project agility (flexibility and adaptability), to the Agile software development project and totestif these changes consequently result in improved project team performances.

2. Theoretical background

2.1. Lean manufacturing principles in Agile project management

Lean thinking has its roots in Toyota (then called Toyota Production System) where has been created in the mid-70s. The publication of the book "The Machine that Changed the World" [8] started the diffusion of lean manufacturing practices first of all in automotive industry and later in all other industries from agriculture to aerospace [9].
Lean has been recognized as the mainly operational management strategy in manufacturing sector [10].The basic principle of lean is responsiveness to change and waste minimization [11].

Recently the agile community has started to look toward Lean software development approaches [12], assaying the usage and benefits of Lean principles known for manufacturing and production industry on software development process. According to [13] there are claims that Lean software development provides the theory behind agile practices [12]. Many authors and practitioners see Lean software development as an instance of agile methods [14]. From case to case, there are different motives for implementation of Lean approach in software development. Since, on the one hand, many large companies using Lean techniques in their manufacturing operations have found it beneficial in many aspects, and on the other hand, have recognized software used in their company to be the substantial part of their products and business, they are seeking to transfer their successful lean experience to the software developers.

The main benefits of Lean development are improved efficiency in eliminating waste (including by that elimination of everything that is not creating value) and increased efficiency in delivering value to the customers. Additionally, pull system characteristic for Lean approach has proved to be essential for process flow optimization, where system can act as event-driven, producing and delivering what customer wants and where he wants, instead of producing in advance for inventory. Related to that, as the main overall goal of Lean development can be identified achievement of continuous and smooth flow of production with maximum flexibility and minimum waste in the process [15].

Lean concept involves three main elements: lean concepts, lean principles and lean practices [12]. Lean concept is providing organizations with possibility to "specify value, line-up value-creating actions in the best sequence, conduct these activities without interruption whenever someone requests them, and perform them more and more effectively". There are five guiding lean concepts:

- Value and clear understanding of what it is;
- Value stream;
- Process flow;
- Pull system and
- Perfection and constant strive for it [16].

Concerning software development, the relevant items that can be defined as waste that are tended to be eliminated are: "extra features, waiting, task switching, extra process, partially done work, movement, defects and unused employee creativity" [12], [13].

Poppendieck & Poppendieck [13] have adjusted lean principles into lean principles relevant to the software development:

- Eliminate waste;
- Build quality in;
- Create knowledge;

- Defer commitment;
- Deliver fast;
- Respect people;
- Optimize the whole.

Furthermore, other authors widened the list of lean practices relevant for software development [12]:
- Address bottlenecks;
- Defer decision making;
- Develop rewarding system;
- Workload levelling;
- Kaizen;
- Kano analysis;
- Use pull system;
- Kanban board;
- Limited WIP;
- Value stream mapping, etc.

One of the most popular instances of Lean approach is Kanban for which it is considered to offer a less prescriptive element as compared to Agile methods [17], [18].

2.2. Kanban

Kanban was developed by Taiichi Ohno, an industrial engineer at Toyota, as a system to improve and maintain a high level of production. It was one method to achieve Just In Time production. Kanban (kahn-bahn) is a Japanese word literally meaning “visible record” [19]. In manufacturing it refers to Kanban cards where one card is associated with each piece of work. It represents one of the manufacturing strategies and tools in Lean manufacturing systems where it may control the levels of buffer inventories in the system to regulate production. When a buffer reaches its preset maximum level, the upstream production activities are not being continued for that particular part type, that way achieving minimum inventory at any one time [20]. That is the main Kanban feature that practitioners have recognized as significant for software development. Furthermore, one of the most important premises of Kanban is that system requires production only when the demand of products is available. Kanban has proven to be useful and helpful tool for an organization, enabling advantages from improvement of productivity and minimization of waste [20], development of flexible work stations, to minimization of waiting time and logistics costs [19]. In other words, the idea of Kanban approach is “to use visual signals to synchronize the flow of work with process capacity, limit the waste of work interruption, minimize excess inventory or delay due to shortage, prevent unnecessary rework, and provide a means of tracking work progress”. Recently, with the application of Lean approach in software development, previously described Kanban principles have been tried to be “replicated" from manufacturing to software development.
In software development, the main focus of Kanban is “to accurately state what work needs to be done and when it needs to be done, by prioritizing tasks and defining workflow as well as lead-time to delivery"[18]. That means Kanban represents certain flow control mechanism for pull driven development. In brief, Kanban aims to provide visibility to the software development process, communicate priorities and highlight bottlenecks which results in a constant flow of releasing work items to the customers, as the developers focus only on those few items at a given time [21]. Regarding that, Kanban systems in software development have special significance from the aspect of scheduling work. Related, Kanban systems bring differences compared to Agile methodologies. They are focusing on continuous flow of work and disregard fixed iterations. Kanban, as in the case of Agile methodologies, sets the requirements and related work items and implements them incrementally. However, instead of time-boxed iterations, the Kanban team chooses those work items and starts working on them when that is needed and there is capacity. The team develops features one after the other and as soon as it is ready, the team works only on one or very few at a time.

In other words, the Kanban in the context of software development allows to the teams to visualize the workflow, limit WIP at each workflow stage and measure the cycle time [22].

In their study [23] the authors identify main motivational factors for adopting Kanban in software development projects:
- Improvement of team communication;
- Improvement of development flow;
- Reduced lead time;
- Increased productivity;
- Better transparency in organization and of work;
- Better control of flow; focus on flow and absence of fixed iterations.

3. Case study and results

3.1. Methodology

The aim of this study was to introduce the adjusted Lean principles which were seen as adequate for enhancing project agility (flexibility and adaptability) and to test if these changes consequently result in improved project team performances. In order to analyse and compare the project productivity and effectiveness before and after Lean principles were implemented in Agile software development process, available metrics were explored and adequate metrics have been chosen.

Based on extensive literature review and project methodology characteristics the most relevant and used metrics were selected for this study:

- *Team velocity*: Represents the amount of work accomplished in each Sprint expressed in story points [24], [25], [26]. This metric demonstrates how much work development team can accomplish within one iteration.
- *Defects rate (count)*: can be both, internal process measure of software testing and external measure as the number of defects experienced by the customer [26], [27], [28].
- *Rework items*: count of items demanding follow-on efforts.

Firstly, the existing Agile Scrum method used by teams was analysed based on previous 9 iterations. Following, the main challenges that team members are facing are identified and used as input to define the directions towards improved project agility and performances.

Based on the desired improvements, adequate modified Lean practices were suggested and integrated into the Agile Scrum methodology. New software development process was tested during 10 iterations and team performance in both cases was compared using chosen metrics. Additionally, data from sprint retrospectives, supporting programs and notes from interview were analysed in order to get insight into the project members' experience.

3.2. Analysis of applied Agile software development methodology

3.2.1. Agile Scrum software development process applied with the project team

The goal of development process analysis was to get insights into the software project management and identify improvement possibilities for this project. Development team has 7 collocated sub teams, counting all together 30 project team members. The final results of the projects was 3D computer game. The team was using adjusted Agile Scrum methodology, applying Scrum practices and artefacts presented in table 1.

Existing Scrum practices within project team	
Product backlog	In this case backlog product is "Game design document" which contains all game chapters with their rough description. The backlog items are prioritized from the beginning of the project.
Sprint and sprint backlog	There are two-week time-boxed iterations. Sprint backlog items are elements and objects of certain game scene. During the sprint duration, each member executes sprint items which go through different process phases.
Sprint planning and effort estimation	There is no traditional sprint planning. Since sprint backlog is made of elements and objects of one scene (small user stories), there is no need for sprint items planning. However, the team is revising the items together in order to discuss if the given effort estimation is realistic
Regular virtual board and burn down charts	For monitoring project progress Trello program is being used with integrated virtual table. Also, the program provides team members with burndown charts showing remained work within iteration. Overall, these artefacts are helping team members and product owner to track the progress of the sprint.
Scrum meetings	As the goal have discussion about necessary corrections, current problems and further instructions.
Sprint retrospective	Scrum retrospective helps team improve their further iterations. It is being conducted with aim to constantly improve following sprints.
Sprint demo	It represents the official representation to the product owner of functional sprint increment and official approval of sprint increment.

Table 1. Existing Agile Scrum practices and artefacts in case company

Following, the described development process with Scrum methodology is presented graphically (figure 1).

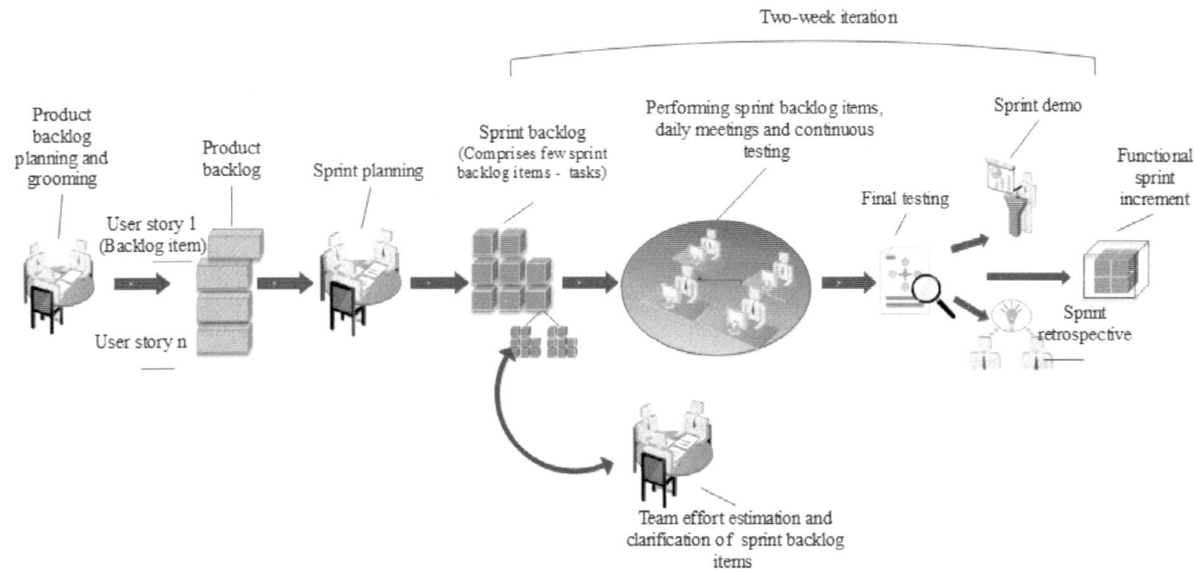

Fig. 1. Graphical presentation of Agile Scrum software development process applied with the project team

3.2.2. Team velocity analysis

Velocity was analysed and presented for each subteam and for the whole development team. In this paper the planned and real velocity of the whole development team is shown in chart 1.

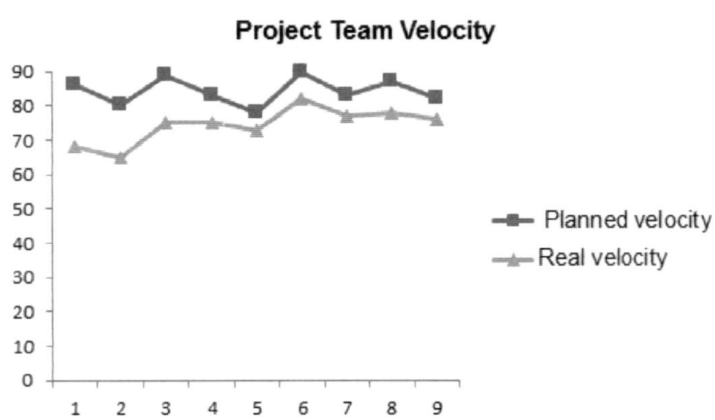

Chart 1. The planned and real team velocity

Results demonstrate high number of unfinished story points per iteration. With goal to identify more precisely potential problems within each sub team, planned and expected velocity was additionally analysed for each sub team separately. The major problem was in one sub team where members could not accomplish even 50 percentage of planned work. These events and delays in this particular sub team and similar problems in other sub teams caused serious bottlenecks, since all unfinished project tasks from one sprint need to be reallocated to the next iterations. Furthermore, one of the reasons for appearance of bottlenecks were interdependencies among tasks performed by different sub teams. Team members defined the unrealistic work load as important cause of poor velocity.

3.2.3. Analysis of defects and rework items

During the development process finished increments are regularly tested manually and automatically, in order to determine the quality of performed tasks and to detect necessary rework and possible improvements.
During the project duration project team has developed bug prediction practice. Team is forecasting possible defects that will demand additional human effort reducing the chances for bottlenecks. However, analysis show many of occurred

defects were not among predicted once, enhancing the lack of experience and appropriate methods for risk assessment regarding bugs and defects. Consequently, during iterations few more complex tasks were blocked. The execution of those tasks has been stopped in order to continue with the iteration which has resulted in accumulation of unfinished sprint items.

To get deeper insight on team's performance, defects count and rework items are demonstrated in chart 2. The number of defects and rework items is represented for each of the subteams.

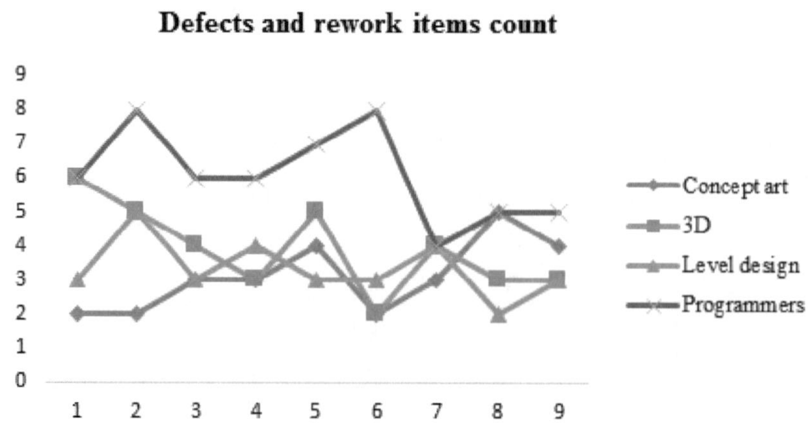

Chart 2. The average number of defects and rework items per each team

The highest number of bugs and needed rework is identified in Level design team. The main cause is the intensive communication Level design team had to have with 3D team members and Programming team, resulting in lack of time for execution of tasks and inefficient communication.

3.2.4. Analysis of sprint retrospective data

Alongside the examination of velocity and defects rate as indicators of project productivity and efficiency, data from sprint retrospectives, meetings and project management programs were analysed and the main challenges project team is facing during the Scrum development process are defined:

- *Incomplete tasks within a sprint* caused by work overload of team members, inflexible sprint duration ("time-boxed" iterations) and complex sprint items (tasks). Furthermore, since Scrum methodology do not allow changes in sprint backlog, change requirements occurred during the iteration remain unperformed regardless their priority, or, in case they are integrated in sprint backlog, other tasks are not being postponed, again resulting in incomplete tasks within sprint. Finally, there is lack of process transparency among sub teams which is essential in this case where tasks are highly interdependent. Even though time-boxed iterations can be motivational for team members to perform their tasks on time, in this case this practice showed lack of benefits for project agility.
- *Poor effort estimation for certain user stories or its parts* resulted in unrealistic scheduling, causing either idle time or bottlenecks.
- *Frequent change requirements during the sprint that team could not fulfil;*
- *Idle time for some team members:* Team reported that certain sprints were not optimized, respectively, work load among sub teams was not balanced enough. Consequently, not rarely some team members were required to wait for other team members to finish their task in order to continue their work. Since sprint backlog is locked, team member waiting for the upcoming task was not able to start new task, which resulted in idle time.
- *Bottlenecks:* During the iterations high number of sprint items were blocked in certain implementation phase. When certain problem occurred, team members were most commonly postponing blocked items and started new once without resolving issue, mostly because they were impeded to dedicated additional time to certain task due to the fixed iteration. In addition, poor effort estimation inflicted stoppage of certain work items.

3.3. *Identification of improvement directions*

In the next phase of this study, adequate directions for project performances improvement were defined. Observing identified challenges project team is facing and summarizing abovementioned problems, it was concluded that agility of the project needs to be improved in order to obtain better project results. Project needs more flexibility in order to successfully meet change requirements, better process optimization in order to mitigate bottlenecks and idle time and more efficient visual management. Main development processimprovement directions proposed were:

- *Improvement of work flow optimization*: Idle time and work overload could be mitigated by achieving smoother development process, respectively, by improving work flow optimization. In order to accomplish that, it is necessary to plan work more efficiently, to have more realistic effort and time estimation for each work item and clear instructions in case of blocked tasks.
- *Improvement of efficiency metrics based on experience from previous iterations:* Although it was observed that team members are not able to finish the backlog items within sprints, sprint planning method and velocity estimation was not changes during first 9 iterations. The major cause for delays on project were unrealistic effort estimation and complex work items. Thus, it is essential to dedicate attention to examination of previous iterations and performance results in order to obtain certain time norms and standards to tasks of approximate complexity.
- *Improvement of work flow transparency and team members' communication:* Blocked items and delays were also the result of inadequate cooperation and communication mechanisms. In addition, insufficient insight into the sprint progress of other sub teams caused inability for some team members to plan their related work properly.

3.4. Integration of Lean principles into the Agile project management methodology

Appropriately applied Lean development improves efficiency in eliminating waste and increases efficiency in delivering value to the customers [6]. In this case, it means Lean practices can provide project team with possibility to detect which change requirements are important, to prioritize them and include in their development process. Additionally, pull system characteristic for Lean approach has proved to be essential for process flow optimization, allowing the team to produce and deliver what customer wants while mitigating bottlenecks. The main overall goal of Lean development is to achieve continuous and smooth flow of production, maximizing flexibility [12].
According to outlined above and previous chapter, identified challenges and defined improvement directions, adequate Lean Kanban principles are suggested to be implemented and combined with existing (figure 2):

- Constant backlog planning and prioritization;
- Introduction of „pull" system;
- Introduction of „ready" queue;
- Work items of similar complexity;
- *WIP* (work in progress) limits:
- Improved work flow visualization;

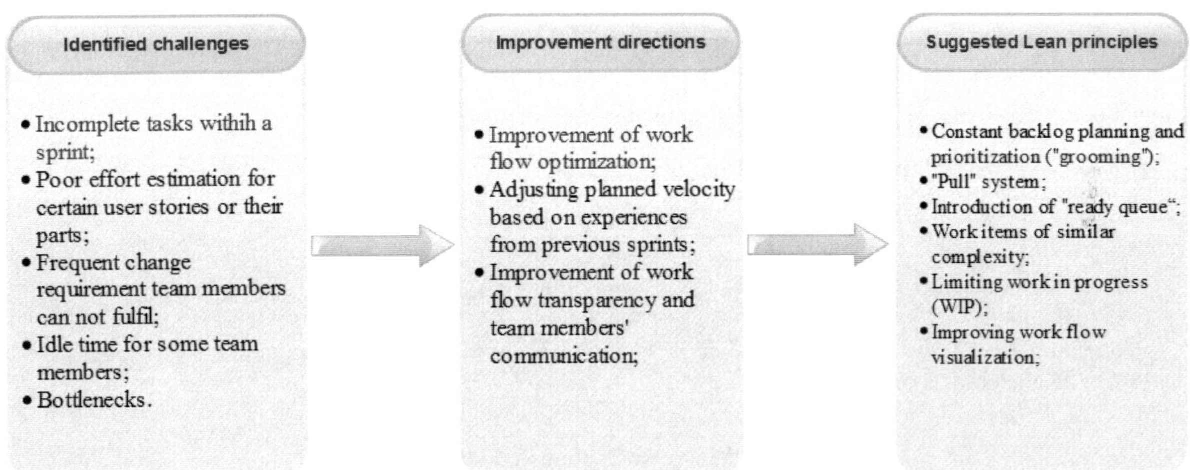

Fig. 2. Lean development principles suggested based on noted challenges and defined improvement directions

Constant backlog planning and prioritization: In Kanban Lean software development the sprint backlog is not locked. The backlog consists of items prioritized and chosen by creative director and game designer. This way, changes during the sprint are acceptable and idle time avoided, since as soon as a team member finishes one work, in case there is free capacity, employee pulls next items and works on it.
From "push" to "pull": The system implies that work is pulled as and when needed into a queue, unlike the approach used in a traditional Scrum where all the work to be completed within a sprint is assigned in the beginning of the sprint to the sprint backlog.
Introduction of „ready" queue: This ready queue contains items that are pending from the backlog, but have high priority and are chosen by product owner.
Work items of similar complexity: The suggestion is to decompose all of the work items on tasks of similar sizes as much as possible. This way during the following iterations it is possible to establish approximate time standards for familiar

tasks. Furthermore, team members will have more clear insight into the work scope and will be able to improve effort estimation.

WIP (work in progress) limits: This principle implies limiting the Product Backlog Items (PBIs) that are in progress at any point of time, including the sprint backlog. The idea is to keep team focused on completing work at hand rather than starting a new task. This means once a limit is reached within a particular stage of the workflow, rather than starting working on something new, it is time to help someone else within the team. This will ensure that the team's work flow becomes smoother and no stage becomes a bottleneck.

3.5. Results

The effects of proposed Lean Kanban principles on team performance were measured during 10 two-week iterations. During iterations before Lean principles were introduced, team was defining velocity as team capacity to perform certain number of story points within one sprint. Those story points weighted determined coefficient representing effort necessary for completing the task, while each item weighted certain number of those story points. In such a way,for every sub team expected velocity, i.e. number of story points is plannedin accordance with the number of sub team members and work complexity. However, sprint items could comprise significantly different story points. In the case of Lean development, work items contain same number of story points since one of the suggestions was to reduce work items to approximately same-size items.

It is noticed that modified agile methodology has resulted with the first two iterations with low team velocity. However, after the third sprint team accomplished significantly better velocity than with previous methodology. According to team members, the velocity of the first three iterations is the result of changes and new working practices which have required team members to adapt and learn. Planned and real velocity of the whole development team is shown in chart 3.

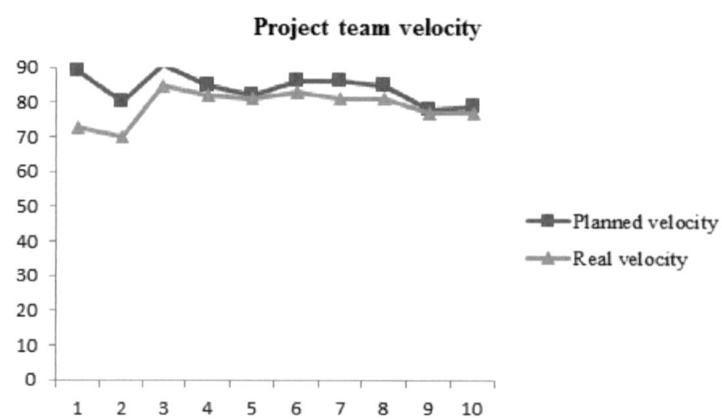

Chart 3. The planned and real team velocity after introduction of Lean principles

As in the case of initial software development process, defects and needed rework were monitored. The status of defects count after 10 iterations is shown in chart 4.

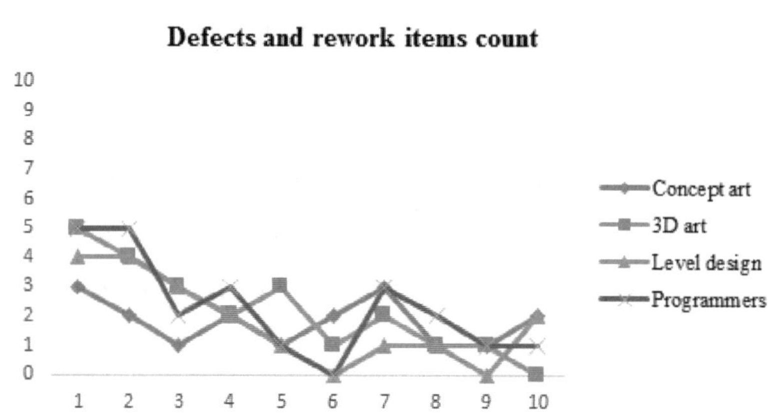

Chart 4. The average number of defects and rework items per each team after introduction of Lean practices

Although relatively high in first iterations, after the third sprint the number of defects and items demanded to be reworked has been significantly lower than in the first case. Team has shown more moticeable improvement regarding work quality compared to velocity melioration. Defect and rework rate was reduced for even 50 percentage. Team members emphasized that this results was mainly caused by decreased pressure due to the implementation of constant backlog grooimng practice and work in progress limitation.

Furthermore, during interviews and sprint retrospecives, team members pointed out that adeqaute usage of metrics and visual management tools facilitated communication and made it more transparent, enabling better and faster problem solving and task management. Rework was reduced, since the quality of work preformed was monitored and controlled during the whole sprint not only during the demo sessions. In addition, team was able to accomplish more user stories and improve its productivity, since metrics enabled more realistic estimation of work items planned for one sprint.

4. Discussion and conclusion

The aim of this study was to test the influence of Lean practices on project performances and project agility, i.e. flexibility and potential to adapt to changes.

The main challenges team using Agile Scrum approach is facing were identified and based on results suggestions for the improvement were created.

Scrum methodology, as other agile methods, is not being implemented in the same manner in every organization. In opposite, each organization adjusts methodology based on their specificities, project and team characteristics. Since the product of this project is alpha game, constant feedback from publisher and customers are included during the whole project duration. Thus, change requirements were highly frequent. Scrum methodology implies that sprint backlog could not be changed during the iteration, team was unable to fulfil new requirements on time. Consequently, the project milestones were often not accomplished, postponing the finish of the project. Furthermore, team consists of very different profiles of employees (artists, programmers, philosophers, etc.), which makes development process management more complicated and demands different techniques for effort management.

By adding adequate Lean Kanban principles based on needs of project and team, game development process was improved.

Pull system and constant backlog grooming enabled team to be responsive to the change requirements without causing delays and unfinished items within sprint.

WIP limits ensured that team members are focused on tasks they are executing at the moment and on solving potential problems and blocked items immediately, consequently decreasingthe occurrences of bottlenecks.

The practice of decomposing sprint items into similar-size tasks improved effort estimation process and facilitated the sprint progress monitoring. Additionally, it enabled team to establish average lead times for future iterations.

Considering the company, project nature, team structure, and obtaie results, further suggestion from authors is to completely eliminate sprints in order to optimize the flow additionally and improve capacity to react fast to the change requirements. Eliminating time-boxed iterations the optimization of work flow would be even more improved due to more flexible and event-driven work planning.Responsible project leader or product owner should continue grooming backlog and prioritizing items.

5. References

[1] Naylor, J. Ben, Naim, M. M., & Berry, D. (1999). Leagility: Integrating the lean and agile manufacturing paradigms in the total supply chain. Int. J. Production Economics, Vol.62, 107–118.

[2] Stratton, R., & Warburton, R. D. (2003). The strategic integration of agile and lean supply. International Journal of Production Economics, Vol. 85, No.2, 183–198

[3] The Standish Group International Inc. (2015). The Chaos Report. Available from https://www.projectsmart.co.uk/white-papers/chaos-report.pdf. Accessed: 01.09.2017.

[4] A Guide To The Project Management Body Of Knowledge (PMBOK Guide) (2004). Newtown Square, Project Management Institute, Inc., 2004. Print.

[5] Destefanis, G., Tonelli, R., Concas, G., &Marchesi, M. (2013). An Analysis of Anti-Micro-Patterns Effects on Fault-Proneness in Large Java Systems. In Proceedings of the 27th Annual ACM Symposium on Applied Computing (pp. 1251–1253).

[6] Majstorovic, V., Polajnar, A. & Bandic-Glavas, M., (2015). Innovative PM Approach to Produce an Environmentally Friendly Product, Chapter 36 in DAAAM International Scientific Book 2015, pp.423-430, B. Katalinic (Ed.), Published by DAAAM International, ISBN 978-3-902734-05-1, ISN 1726,Vienna, Austria, DOI: 10.2507/daaam.scibook.2015.36

[7] Aoyama, M. (1998). Web-based agile software development, IEEE Software, vol. 15, pp. 56-65.

[8] Herzog, V., Medic, M. & Kostanjevac, T., (2008). Value Stream mapping for Effective Lean Manufacturing, Annals of DAAAM & Proceedings, B. Katalinic (Ed.), Published by DAAAM International, ISN 1726-9679,Vienna, Austria

[9] Womack, J.P.; Jones, D.T. & Ross, D. (1990), The Machine that changed the World. Macmillan Publishing Company, Canada.

[10] Welo, T., & Ringen, G. (2015). Investigating Lean Development Practices in SE Companies: A Comparative Study Between Sectors. Procedia Computer Science, Vol. 44, 234–243

[11] Vujica Herzog, N & Buchmeister, B. (2012). Lean Manufacturing in Slovenian Companies, Chapter 11 in DAAAM International Scientific Book 2012, pp. 117-128, B. Katalinic (Ed.), Published by DAAAM International ISBN 978-3-901509- 86-5, ISSN 1726-9687.Vienna, Austria

[12] Wang, X., Conboy, K. &Cawley, O. (2012). "Leagile" software development: An experience report analysis of the application of lean approaches in agile software development," The Journal of Systems and Software, vol. 85, pp. 1287– 1299.

[13] Poppendieck, M. & Poppendieck, T. (2003) "Lean Software Development: An Agile Toolkit." Boston, Massachusetts, USA, Addison Wesley Professional.

[14] Dybĺ, T. & Dingsřyr, T. (2009) „What do we know about agile software development?" IEEE Software, vol. 26, pp. 6–9.

[15] Petersen, K. & Wohlin, C. (2010). "Software process improvement through the Lean Measurement (SPI-LEAM) method." The Journal of Systems and Software, vol. 83, pp. 1275–1287.

[16] Andersson, R., Eriksson, H. & Torstensson, H. (2006) "Similarities and differences between TQM, six sigma and lean", The TQM Magazine, Vol. 18, pp. 282-96.

[17] Hibbs, C., Jewett, S& Sullivan, M. (2009). The Art of Lean Software Development: A Practical and Incremental Approach, O'Reilly Media, Inc.

[18] Miller, J. & Al-Baik, O. (2015) "The kanban approach, between agility and leanness: a systematic review." Empir. Software Eng , vol. 20, pp. 1861-1897.

[19] Ohno, T. (1988). Toyota Production System - beyond large-scale production, Cambridge, Productivity Press.

[20] Rahman, N. A., Sharif S.M. & Esa, M. M. (2013) "Lean Manufacturing Case Study with Kanban System." in Procedia Economics and Finance, vol. 7:Proceedings of the international Conference on Economics and Business Research, Kedah, pp. 174-180.

[21] Oivo, M. & Ahmad, M. O. (2013) "Kanban in Software Development: A Systematic Literature Review." Proceedings of the 39th Euromicro Conference on Software Engineering and Advanced Applications, Santander, pp. 9-16.

[22] Kniberg, H. & Skarin, M. (2010). Kanban and scrum – Making the most of both, InfoQ.

[23] Ahmad, M. O., Oivo, M. & Kuvaja, P. (2014) "Usage of Kanban in Software Companies An empirical study on motivation, benefits and challenges. " Proceedings of the 9th International Conference on Software Engineering Advances, Nice, 2014, pp. 150-155.

[24] Mahnic, V., Zabkar, N. (2012) Measuring Progress of Scrum-based Software Projects. Elektronika ir Elektrotechnika, vol. 18, pp. 73-76.

[25] Nwakanma, C., Asiegbu, B., Ogbonna, C. and Njoku, P. (2013) Factors affecting successful implementation of information technology projects: experts' perception. European scientific journal, vol. 9, pp. 128-137.

[26] Kupianen E. et Al. (2015). Using metrics in Agile and Lean Software Development. Information and Software Technology, vol. 62, pp. 143-163.

[27] Talby, D., Hazzan, O., Dubinsky, Y. and Keren A. (2006). Reactions on reaction in agile software development. In Proceedings - AGILE Conference 2006, vol. 2006, pp. 100-110.

[28] Green P. (2011). Measuring the impact of scrum on product development at adobe systems. In Proceedings of the Annual Hawaii International Conference on System Sciences.

DOI: 10.2507/28th.daaam.proceedings.094

ANALYTIC DETERMINATION AND NUMERICAL SIMULATION OF TORSIONAL STIFFNESS OF A RACING CAR FRAME

Michal Skovajsa, Frantisek Sedlacek

This Publication has to be referred as: Skovajsa, M[ichal] & Sedlacek, F[rantisek] (2017). Analytic Determination and Numerical Simulation of Torsional Stiffness of a Racing Car Frame, Proceedings of the 28th DAAAM International Symposium, pp.0671-0678, B. Katalinic (Ed.), Published by DAAAM International, ISBN 978-3-902734-11-2, ISSN 1726-9679, Vienna, Austria
DOI: 10.2507/28th.daaam.proceedings.094

Abstract

The paper deals with the determination of the required torsional stiffness of the frame of a racing car. The required torsional stiffness of the frame was performed by analytic computation. The analytic computation deals with the transfer of the vertical load of the tyre between the front and rear axle for different roll stiffnesses of axles and torsional stiffnesses of the frame. The analytic computation determines the ratio between the roll stiffness of the axles and the torsional stiffness of frame. NX Nastran 11 software was used for the numerical simulation. The numerical simulation was performed by using four different numerical methods. Each numerical simulation was performed with different boundary conditions.

Keywords: Torsional stiffness; frame; numerical simulation; analytic solution

1. Introduction

Determining the correct ratio between the roll stiffness of the car and the torsion stiffness of the frame is covered by several publications. [1, 2, 3]. The numerical simulation of the torsion stiffness of the frame is covered by some publications [5,6]. These publications do not deal with the comparison of different numerical calculation methods. The real frame of the Formula Student car UWB05 of the University of West Bohemia was used for the simulation. The questions are: What is the required value of the torsional stiffness of the car? Which numerical simulation method should be chosen for computing the torsion stiffness of the frame?

The frame provides the transmission of force between the tyres and the centre of gravity of the car. The frame transfers four type of loads: Torsional load, Lateral load, Vertical load, and Longitudinal load [4]. The torsion stiffness is the most important parameter of a racing car frame. The frame ensures the lateral load transfers between the front and rear axles, thereby influencing the driving stability and even distribution of the vertical loads of individual tyres. This affects the driving stability and equally the distribution of the vertical loads of individual tyres.

Low values of torsional stiffness cause: incorrect distribution of lateral load, incorrect response of the car to chassis changes, big changes in suspension kinematics compared to the designed parameters, increases the dynamic strain on the car, reduces driving stability, increases the fatigue stress of the car. The torsion stiffness of the frame is defined as the ratio of the torsion stiffness of the frame and the roll stiffness of the axles. The torsion stiffness is defined by the static model of the vehicle (Figure 1), where the stiffness of the axles and the frame are replaced by springs [4].

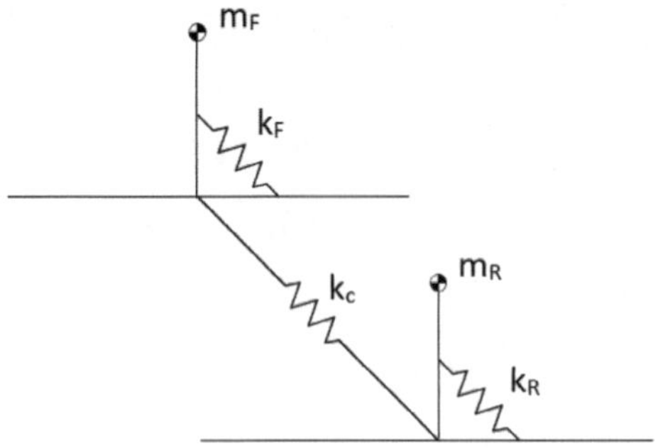

Fig. 1. Static model of the car [2]

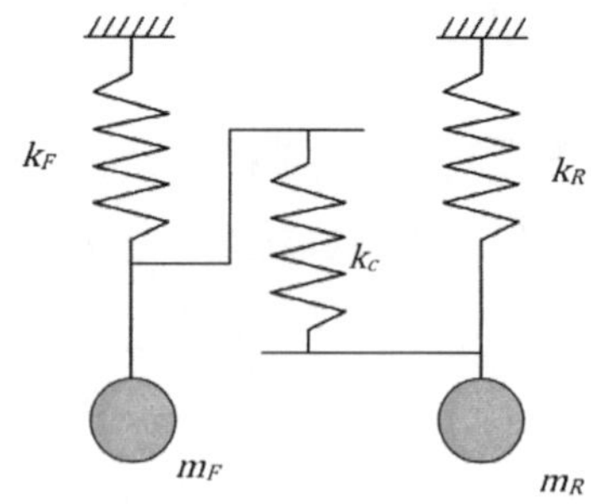

Fig. 2. Static model of the car as a classic spring model

Where m_F is load of front axle; m_R is load of rear axle; k_F is roll stiffness of front axle; k_R is roll stiffness of rear axle and k_c is torsion stiffness of frame. This static vehicle model is replaced by a classic spring system with two points of gravity. (Figure 2). The total torsional stiffness of the car is defined as the sum of the compliance of each element of the vehicle model.

2. Analytic computation

The vertical load of the wheels of one axle is identical in the static position of the car. The load on the inner wheel is reduced and the external wheel load is increased if the lateral force (Equation 1) is applied (turning) [1].

$$\Delta F_z = \frac{m h_{CG}}{t} a_y, \tag{1}$$

where ΔF_Z is the difference of the vertical load of the car; m is vehicle weight; h_{CG} is centre of gravity height; t is track; a_y is acceleration in lateral direction.

Setting assumptions for calculation: Lateral acceleration does not generate roll of the car, Rolling of front and rear axle is independent, The position of the centre of the vehicle and the centre of gravity lies in the longitudinal plane of the car, Unsupported masses are negligible, The front and rear axle track is the same, All norms to the roll axis are vertical.

2.1 Mathematical model of car with two masses and infinitely stiff frame

First was defined the mathematical model of a car with two masses and an infinitely stiff frame. This model is described by equations (2, 3, 4, 5, 6, 7) [1, 4, 7] and is shown in Figure 3

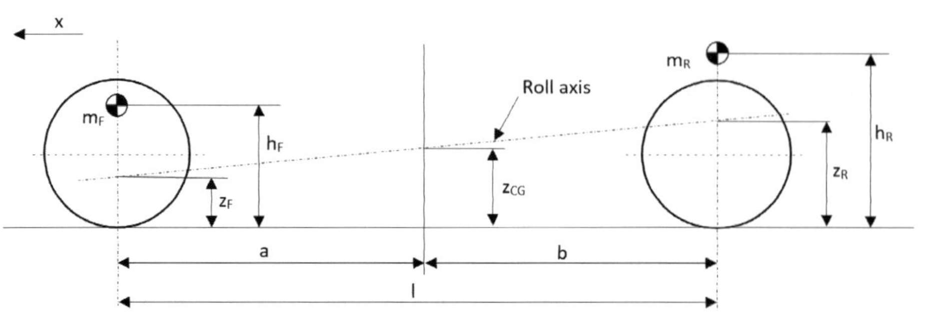

Fig. 3. Two mass model of a car [1]

$$m_F = \frac{b}{l} m \, ; \; m_R = \frac{b}{l} m, \tag{2; 3}$$

$$d_F = h_F - z_F; \; d_R = h_R - z_R, \tag{4; 5}$$

$$\Delta F_{zF0} = \frac{m}{t}\left(\frac{k_F}{k_F + k_R}\frac{b}{l} d_F + \frac{k_F}{k_F + k_R}\frac{a}{l} d_R + \frac{b}{l} z_F\right) a_y, \tag{6}$$

$$\Delta F_{zR0} = \frac{m}{t}\left(\frac{k_R}{k_F + k_R}\frac{b}{l}d_F + \frac{k_R}{k_F + k_R}\frac{a}{l}d_R + \frac{a}{l}z_R\right)a_y, \tag{7}$$

where ΔF_{ZF0} is the difference of the vertical load of the front axle with an infinitely stiff frame; ΔF_{RF0} is difference of vertical load of the rear axle with infinitely stiff frame; m is vehicle weight; m_F is load of front axle; m_R is load of rear axle; l is wheelbase; t is track; a is distance between centre of gravity and front axle axis; b is distance between centre of gravity and axis of rear axle; k_F is roll stiffness of front axle; k_R is roll stiffness of rear axle; a_y is acceleration in lateral direction; z_F is height of roll axis of front axle; z_R is height of roll axis of rear axle; d_F is distance between centre of gravity and roll axis of front axle; d_R is distance between centre of gravity and roll axis of rear axle; h_F is centre of gravity height of front axle; h_R is centre of gravity height of rear axle. This model was used for comparison with the model which is described below.

2.2 Mathematical model of car with two masses and finite stiffness of frame

The second step was to define the mathematical model of a car with two masses and finite stiffness of the frame. This model is described by equations (8, 9) [1, 4, 7] and is shown in Figure 3

$$\Delta F_{zF} = \frac{m}{t}\left(\frac{k_F}{k_F + \frac{k_R k_C}{k_R + k_C}}\frac{b}{l}d_F + \frac{\frac{k_F k_C}{k_F + k_C}}{k_R + \frac{k_F k_C}{k_F + k_C}}\frac{a}{l}d_R + \frac{b}{l}z_F\right)a_y \tag{8}$$

$$\Delta F_{zR} = \frac{m}{t}\left(\frac{\frac{k_R k_C}{k_R + k_C}}{k_F + \frac{k_R k_C}{k_R + k_C}}\frac{b}{l}d_F + \frac{k_R}{k_R + \frac{k_F k_C}{k_F + k_C}}\frac{a}{l}d_R + \frac{b}{l}z_R\right)a_y \tag{9}$$

Where ΔF_{ZF} is the difference of the vertical load of the front axle with finite stiffness of frame; ΔF_{RF} is difference of vertical load of the rear axle with finite stiffness of frame; m is vehicle weight; m_F is load of front axle; m_R is load of rear axle; l is wheelbase; t is track; a is distance between centre of gravity and front axle axis; b is distance between centre of gravity and axis of rear axle; k_F is roll stiffness of front axle; k_R is roll stiffness of rear axle; k_c is torsion stiffness of frame; a_y is acceleration in lateral direction; z_F is height of roll axis of front axle; z_R is height of roll axis of rear axle; d_F is distance between centre of gravity and roll axis of front axle; d_R is distance between centre of gravity and roll axis of rear axle; h_F is centre of gravity height of front axle; h_R is centre of gravity height of rear axle.

2.3 Comparison of mathematics with infinitely stiff and finitely stiff frame

The comparison was defined as the difference of the ratios of the vertical loads. The first ratio X_F was defined as the fraction of ΔF_{ZF} which is the difference of the vertical load of the front axle with a finitely stiff frame and ΔF_Z which is the difference of the vertical load of the car. The second ratio X_{F0} was defined as the fraction of ΔF_{ZF0} which is the difference of the vertical load of the front axle with an infinitely stiff frame and ΔF_Z which is the difference of the vertical load of the car.

$$X_F = \frac{\Delta F_{zF}}{\Delta F_z}, \tag{10}$$

$$X_{F0} = \frac{\Delta F_{zF0}}{\Delta F_z}, \tag{11}$$

where X_F is the ratio change in the vertical load of the front axle with a finitely stiff frame; X_{F0} is the ratio change in the vertical load of the front axle with an infinitely stiff frame.

Graphs 1, 2, 3, 4 were created based on equations (10,11) [1, 4, 7]. Graphs 1 and 2 show the difference of ratio 10 and 11 depending on the torsion stiffness of the frame for different roll stiffnesses of the front axle. Graphs 3 and 4 show the difference of ratios 10 and 11 depending on the roll stiffness of the front axle for different torsion stiffnesses of the frame. Graphs 1 and 3 were created by assuming equal distribution of the load on each axle ($m_F = m_R$). Graphs 2, 4 were created by assuming different distributions of the load on each axle in ratio 47:53 ($0.53m_F = 0.47m_R$). All graphs were created by assuming the same parameters for both axles ($z_F = z_R$; $h_F = h_R$; $d_F = d_R$)

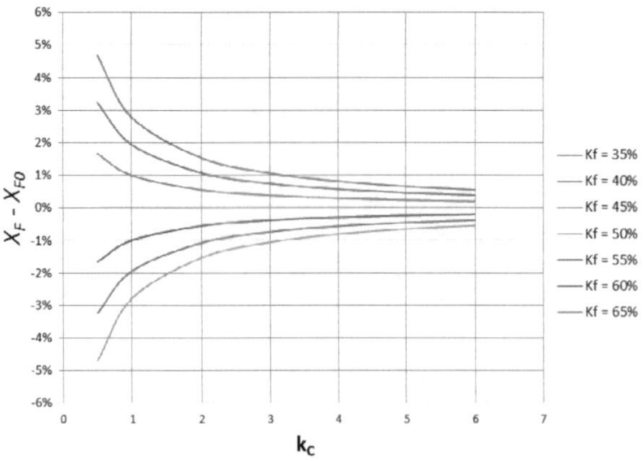

Graph 1. Difference of ratio X_F and X_{F0} depending on torsion stiffness of frame for different roll stiffnesses of front axle. For equally distributed load on each axle ($m_F = m_R$).

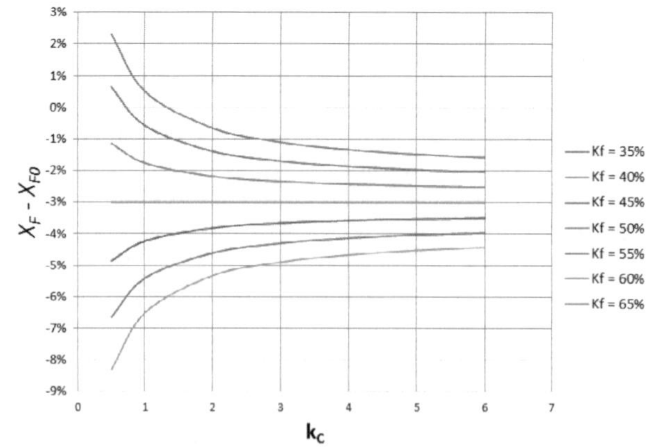

Graph 2. Difference of ratio X_F and X_{F0} depending on torsion stiffness of frame for different roll stiffnesses of front axle. For different distribution of load on each axle in ratio 47:53 ($0{,}53 m_F = 0{,}47 m_R$).

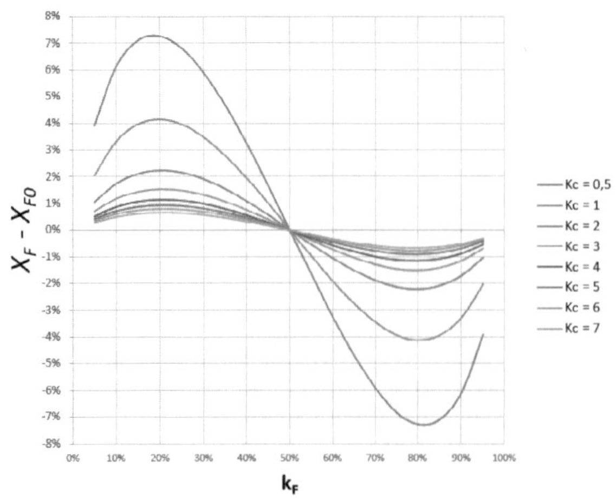

Graph 3. Difference of ratio X_F and X_{F0} depending on roll stiffness of front axle for different torsion stiffnesses of frame. For equally distributed load on each axle ($m_F = m_R$).

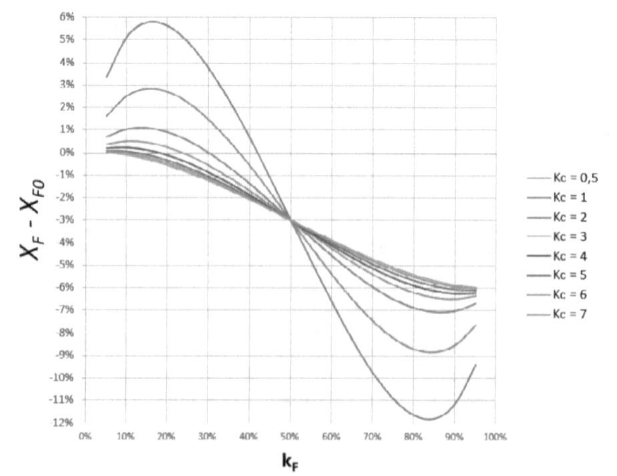

Graph 4. Difference of ratio X_F and X_{F0} depending on roll stiffness of front axle for different torsion stiffnesses of frame. For different distribution of load on each axle in ratio 47:53 ($0.53 m_F = 0.47 m_R$).

	k_C						
	0,5	1	2	3	4	5	6
$k_F = 35\%$	4.69%	2.78%	1.53%	1.06%	0.81%	0.65%	0.55%
$k_F = 40\%$	3.24%	1.94%	1.07%	0.74%	0.57%	0.46%	0.38%
$k_F = 45\%$	1.66%	0.99%	0.55%	0.38%	0.29%	0.24%	0.20%
$k_F = 50\%$	0.00%	0.00%	0.00%	0.00%	0.00%	0.00%	0.00%
$k_F = 55\%$	-1.66%	-0.99%	-0.55%	-0.38%	-0.29%	-0.24%	-0.20%
$k_F = 60\%$	-3.24%	-1.94%	-1.07%	-0.74%	-0.57%	-0.46%	-0.38%
$k_F = 65\%$	-4.69%	-2.78%	-1.53%	-1.06%	-0.81%	-0.65%	-0.55%
$(X_F\text{-}X_{F0})_{\Delta kF35\%}$ - $(X_F\text{-}X_{F0})_{\Delta kF65\%}$	9.38%	5.56%	3.06%	2.11%	1.61%	1.31%	1.10%
$((X_F\text{-}X_{F0})_{\Delta kF35\%}$ - $(X_F\text{-}X_{F0})_{\Delta kF65\%})_{i-1}$ - - $((X_F\text{-}X_{F0})_{\Delta kF35\%}$ - $(X_F\text{-}X_{F0})_{\Delta kF65\%})_i$		3.82%	2.50%	0.95%	0.50%	0.31%	0.21%

Table 1. Difference of ratio X_F and X_{F0} depending on torsion stiffness of frame for different roll stiffnesses of front axle. For equal distribution of load on each axle ($m_F = m_R$)

From all the graphs it was determined that a change of the torsional stiffness of the frame has no affect if the roll stiffness of the axles is the same ($k_F = k_R$). From all the graphs it was determined that a change in the ratio m_F/m_R does have an effect. The lateral load transfer is increased and the inflection point of the graphs is moved.

Because the car is loaded with dynamic action and the characteristics of the shock absorbers and progressive springs are nonlinear, then the roll stiffness characteristics of the individual axles are nonlinear too. The amplitude of roll stiffness of the axle was determined to be from 35% to 65%. Table 1 was created for this range. Table 1 contains the values of the differences and the ratios of the changes of differences of transfer of the lateral forces which is defined as X_F-X_{F0}.

A significant reduction of lateral force transfer was achieved with the value of k_C in a range from 0.5 to 4. An improvement of only about 0.5% was achieved for k_C over 4. For this reason, the final value of torsional stiffness of the frame was determined. The torsional stiffness of the frame is suitably set to a value 4 - 6 times the roll stiffness of the axles.

3. Numerical simulation of torsional stiffness of frame

The numerical solution was performed on the frame of the Formula Student racing car UWB05. Four solution variants of the numerical simulations of the torsional stiffness of the frame were carried out. The force and constrain for solutions one and two were applied to the frame. For example: The roll stiffness of the UWB05 car is 452 [Nm/deg]. Therefore, the required torsional stiffness of the frame is from 1808 – 2702 [Nm/deg].

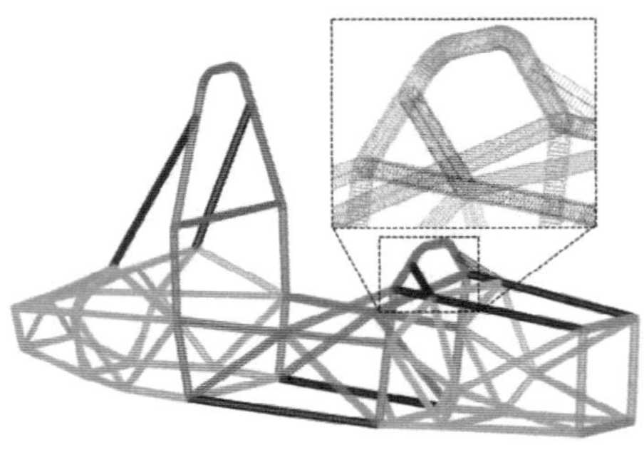

Fig. 4. 1D FE model of frame

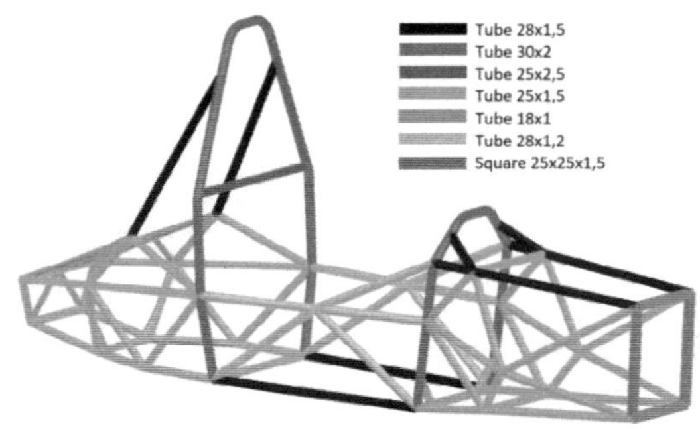

Fig. 5. Physical model of frame

Numerical simulations of the torsional stiffness of the frame were done using the linear structural solver NX Nastran 101 and pre-post-processor Siemens Simcenter 11 [8, 9]. A 1D model of the frame was created using 1D fine linear element mesh (cbeam elements) [10]. 5 mm FE mesh was used for the numeric simulation. The 1D FE model with mesh is shown in Figure 4. The boundary conditions of solution numbers 1 and 2 were applied directly to the frame. The numerical models for solutions 3 and 4 were extended using the model of the suspension, as shown in Figure 6. The connections between each link of the suspension and the suspension with the frame were formed by manual coupling. The manual coupling was set to transfer only the translation to ensure a rotary link. The forces and constrains for solutions 3 and 4 were applied to the suspension. In solution three the suspension was replaced by rigid links. In solution 4 the suspension was replaced by beams links, because in a real car the suspension is flexible. Therefore, they must be included in the simulation of the torsional stiffness of the frame.

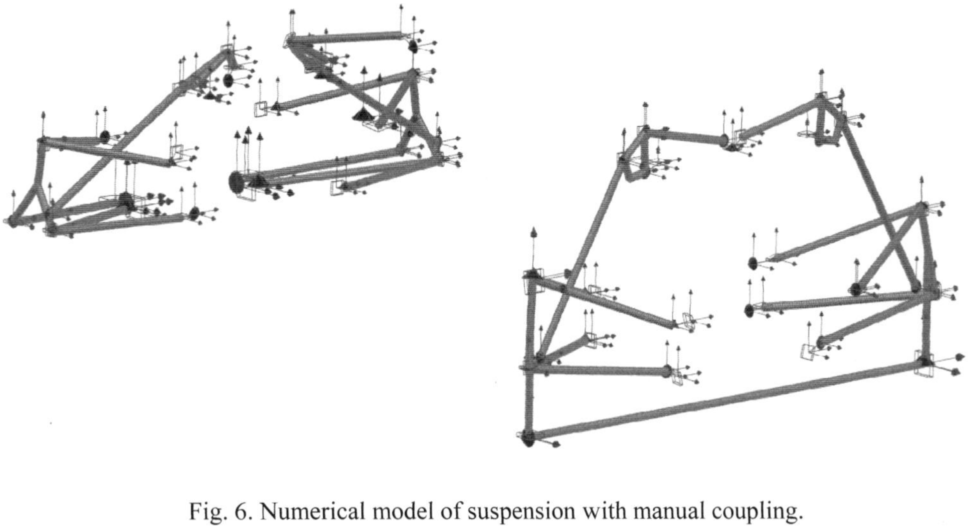

Fig. 6. Numerical model of suspension with manual coupling.

Solution 1

Setting:
Constrain:
Position: 1, 2, 3, 5 - Fixed Translation (x, y, z)
Forces:
Positon: 6 -1000N (z)
Measurement point:
Position: 7 (z)

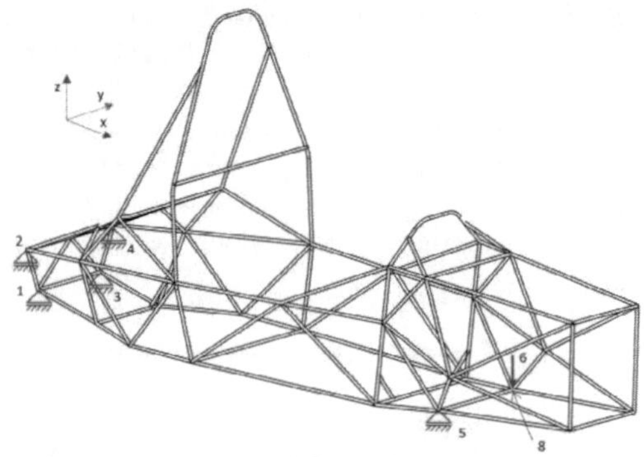

Fig. 7. Constrain and force for solution 1.

Solution 2

Setting:
Constrain:
Position: 1, 2, 3, 5 - Fixed Translation (x, y, z)
Forces:
Positon: 6 -500N (z)
Positon: 7 +500N (z)
Measurement point:
Position: 8 (z)

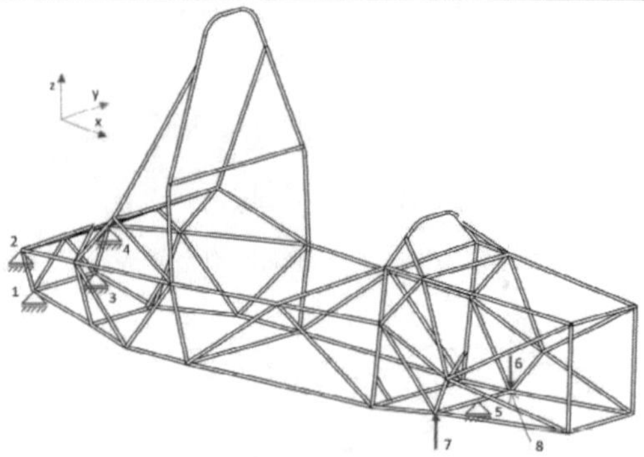

Fig. 8. Constrain and force for solution 2.

Solution 3

Setting:
Constrain:
Position: 1 - Fixed Translation (x, y, z)
2 - Fixed Translation (x, z)
3 - Fixed Translation (y, z)
Forces:
Positon: 4 -1000N (z)
Measurement point:
Position: 5 (z)
Suspension is rigid links

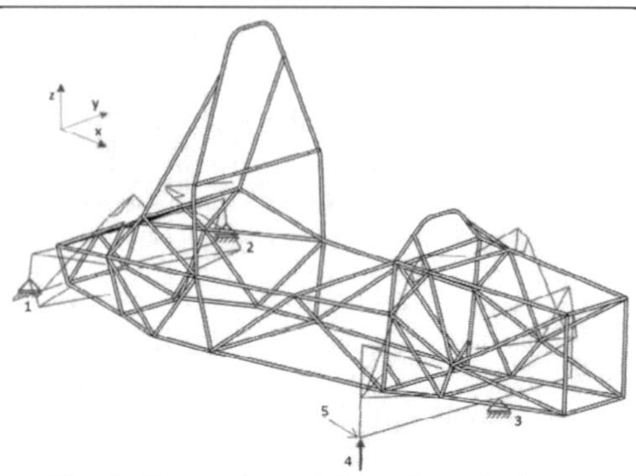

Fig. 9. Constrain and force for solution 3.

Solution 4

Setting:
Constrain:
Position: 1 - Fixed Translation (x, y, z)
2 - Fixed Translation (x, z)
3 - Fixed Translation (y, z)
Forces:
Positon: 4 -1000N (z)
Measurement point:
Position: 5 (z)
Suspension is flexible links

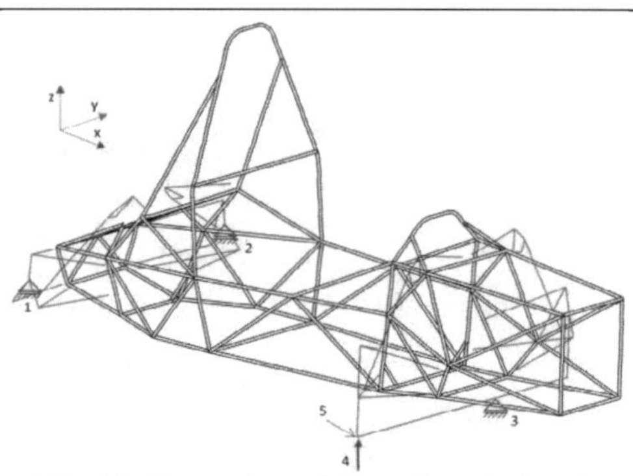

Fig. 10. Constrain and force for solution 4.

The results of the magnitude displacement of structural analysis in direction ‘z’ are shown for individual solutions in Fig. 10-13.

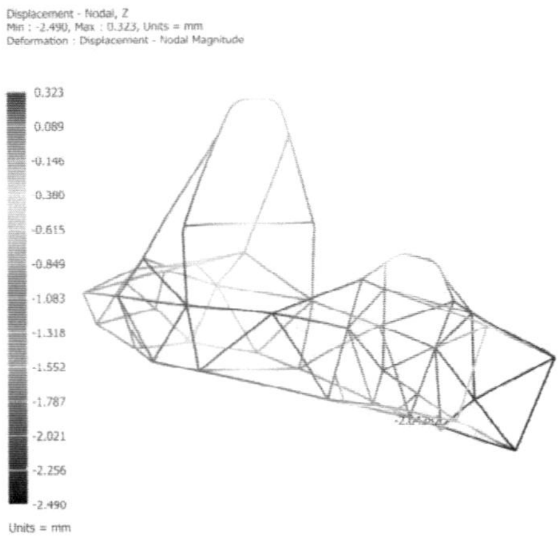

Fig. 11. Result of solution 1.

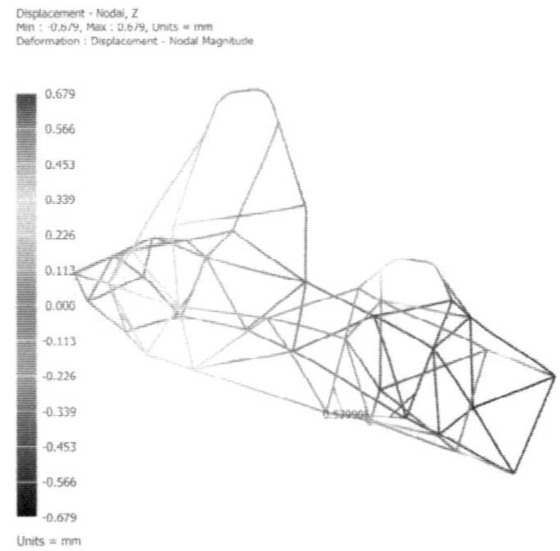

Fig. 12. Result of solution 2.

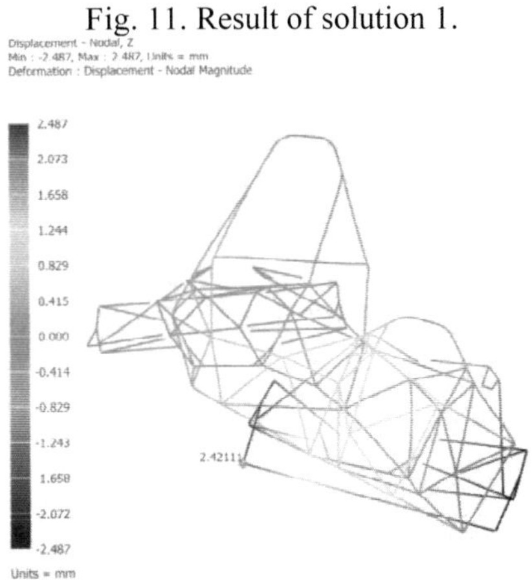

Fig. 13. Result of solution 3.

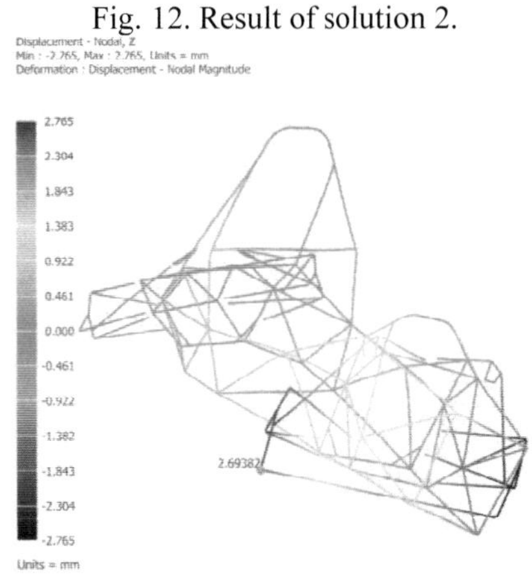

Fig. 14. Result of solution 4.

4. Comparison of the numerical simulation

The comparison of the numerical simulations is shown in Table 2. There are very significant differences between the numerical simulations. The maximum difference in torsional stiffness was determined to be 1 112.4 [Nm/deg]. Solutions number 1 and 2 are unrealistic in comparison to real car conditions. In solution number 3 the torsional stiffness of the frame was increased by the rigid links of the suspension. In this case the best result is number 4, where the suspension is connected by flexible links and the constrain and force are most consistent with reality.

Solution	Applied force			Deformation			Torsional stiffness
	Force	Arm of force	Torque	Deformation	Arm of deformation	Angle of deformation	
1	1000 [N]	460 [mm]	460 [Nm]	2.04282 [mm]	460 [mm]	0.25444 [deg]	1807.8 [Nm/deg]
2	2x 500 [N]	180 [mm]	180 [Nm]	0.5399 [mm]	180 [mm]	0.17185 [deg]	1047.4 [Nm/deg]
3	1000 [N]	547.37 [mm]	547.37 [Nm]	2.69382 [mm]	547.37 [mm]	0.25342 [deg]	2159.8 [Nm/deg]
4	1000 [N]	547.37 [mm]	547.37 [Nm]	2.42111 [mm]	547.37 [mm]	0.28197 [deg]	1941.2 [Nm/deg]

Table 2. The comparison of the numerical simulations

5. Conclusion

The paper deals with the determination of the required torsional stiffness of the frame of a racing car. The questions were: What is the torsional stiffness of the frame required for a specific roll stiffness of the car? Which type of boundary condition is it suitable to use for the numerical simulation of torsional stiffness in solver Nastran 101 in NX 11.

The torsion stiffness of the frame was defined as the ratio of the torsion stiffness of the frame and the roll stiffness of the axles. The torsion stiffness of the frame ensures transfer of the lateral forces between the front and rear axles. The difference of the vertical load of the front axle was defined for frames with infinite and finite stiffness (equations 6, 7, 8, 9). The difference of these vertical loads change for both types of frame, as shown in the Graphs (1, 2, 3, 4) and Table 1. From all the graphs it was determined that changing the torsional stiffness of the frame has no effect if the roll stiffnesses of the axles are the same ($k_F = k_R$). From all the graphs it was determined that changing the ratio of m_F/m_R does have an effect: the lateral load transfer is increased and the inflection point of the graphs is moved. When the racing car is loaded with dynamic action and the characteristics of the shock absorbers and progressive springs are nonlinear, then the roll stiffness characteristics of the individual axles are nonlinear too. The amplitude of the roll stiffness of the axle was determined to be from 35% to 65%. In this case, a significant reduction of lateral force transfer (about 3.82 - 0.95%) was achieved with a value of k_c in a range from 0.5 to 4. For k_c over 4 an improvement of only about 0.5% was achieved. For this reason, the final value of the torsional stiffness of the frame was determined. The torsional stiffness of the frame is suitably set to a multiple 4 - 6 times the roll stiffness of the axles. For example, if a vehicle with total weight 300kg is considered, then the lateral load transfers between the front and rear axle: for $k_c = 1$ it will be 8.34kg and the total distribution will change about 16.68kg; for $k_c = 5$ it will be 1.965kg and the total distribution will change only about 3.93kg.

The numerical simulations were performed in four solution variants with different boundary conditions which are described in section 3. The differences between the numerical simulations are very significant. The maximum difference of torsional stiffness was determined to be 1 112.4 [Nm/deg]. Solutions 1 and 2 are unrealistic in comparison to real car conditions. In solution 3 the torsional stiffness of the frame was increased by the rigid links of the suspension. In this case the best result is number 4, where the suspension is made of flexible links and the constrain and force are most consistent with reality. The numerical simulation will have to be verified by an experimental measurement of the real frame. We are currently working on experimental testing of the torsional stiffness of the frame of the UWB05 car, and verification of the numerical simulations to best-fit the numerical simulation to realistic values.

6. Acknowledgments

This paper is based on work sponsored by project SGS2016-012 (The complex support of designing technical equipment III.).

7. References

[1] Sampo E.; Sorniotti, A.; Crocombe, A. (2010). Chassis Torsional Stiffness: Analysis of tht Influence on Vehicle Dynamic, SAE International, ISSN 0148-7191, USA
[2] Deakin A.; Crolla, D.; Pablo J.; Hanley R. (2000). The Effect Chassis Stiffness of Race Car Handling Balnce, SAE International, 2000-01-3554, USA
[3] Riley W.; George A.; (2002). Design, Analysis and Tesing of a Formula SAE Car Chassis, SAE International, 2000-01-3300, USA
[4] Milleken F. W.; Millken L. D. (1995). Race Car Vehicle Dynamics, ISBN 1-56091-526-9
[5] Thompson L.; Lampert K.; Law H. (1998). Design of a Twist Fixture to Measure the Torsional Stiffness of a Winston Cup Chassis, SAE International, 983054, USA
[6] Thompson L.; Soni P.; Raju P.; Law H. (1998). The Effects of Chassis Flexibility on Roll Stiffness of a Winston Cup Race Car, SAE International, 983051, USA
[7] Giancarlo G.; Lorenzo M. (2009). The Automotive Chassis: Vol. 2: System Design, Springer, ISBN: 978-1-4020-8673-1
[8] Siemens PLM, 'NX Nastran 11 User's Guide'. Siemens AG., 2016.
[9] P. Goncharov, Engineering analysis with NX advanced simulation, vol. 2014. Raleigh: Lulu Press.
[10] O. C. Zienkiewicz, R. L. Taylor, and J. Z. Zhu, The Finite Element Method: Its Basis and Fundamentals, Seventh Edition, 7 edition. Amsterdam: Butterworth-Heinemann, 2013.

DOI: 10.2507/28th.daaam.proceedings.095

TECHNOLOGY FOR AUTOMATIC SOIL SAMPLING BY MANIPULATORS MOUNTED ON UNDERWATER VEHICLES

Vladimir Filaretov, Alexander Konoplin, Nikita Konoplin & Georgy Gorbachev

This Publication has to be referred as: Filaretov, V[ladimir] F[edorovich]; Konoplin, A[lexander]; Konoplin, N[ikita] & Gorbachev, G[eorgy] (2017). Technology for Automatic Soil Sampling by Manipulators Mounted on Underwater Vehicles, Proceedings of the 28th DAAAM International Symposium, pp.0679-0684, B. Katalinic (Ed.), Published by DAAAM International, ISBN 978-3-902734-11-2, ISSN 1726-9679, Vienna, Austria
DOI: 10.2507/28th.daaam.proceedings.095

Abstract

The work proposes the technology for an automatic implementation of the soil sampling by means of underwater multilink manipulators which are attached to submergible autonomous and remotely controlled vehicles. This technology includes the strategy of the safe approaching of a vehicle to a sea bottom surface, and also the formation of manipulator trajectories for the efficient soil sampling by hermetically sealed tubes and soil scoops. The results of sea tests are presented for the study of the working capacity of the proposed technology.

Keywords: underwater vehicle; multilink manipulator; automatic mode; soil sampling; trajectory planning

1. Introduction

Today, the world Ocean exploration is impossible without performing of research operations by means of underwater vehicles (UV). Up to now, the most of manipulation operations are performed with the help of remotely operated UV equipped with multilink manipulators (MM). Specially trained operators remotely operate the UV with MM from the laboratory on a carrier vessel in the process of the underwater operation performing. However, the efficiency of these operations is highly dependent on the weather conditions, since the presence of a flexible cable of a remotely operated UV does not allow to performs precisely manipulation operations under conditions of significant displacements of the carrier vessel. Moreover, the use of a remotely operated UV in the conditions of the northern seas for most of the year is impossible due to the presence of seas ice covers. Therefore, it is necessary to develop new modern control systems and technology for the UVs and its MMs to perform manipulation operations in an automatic mode. This will allow to use autonomous UVs equipped with MMs for the implementation of underwater operations independent of the weather conditions and even in conditions of seas ice covers.

The systems that provide accurately automatic movement of the UV along complex spatial trajectories [1, 2] are already developed. For the taking into account negative angular displacements of the UV under the influence of the grasped objects and probes the methods [3, 4] are developed. The technologies for high-precision automatic stabilization of the UV in the hang mode [5-8] are already created. These stabilization systems allow to perform manipulation operations without additional mechanical devices for the UV rigid fixation near work objects and exclude the introducing of disturbances into bottom layers of water.

There are the known cases of using the autonomous UV for the performing of the simplest manipulation operations in the automatic mode [9, 10]. In these cases, a stereo vision system and sophisticated image processing algorithms for detecting objects, which have a previously known form, are used to determine locations and spatial orientations of working objects with respect to the UV. Furthermore, a complexity of implementation of the MM automatic control is also due to high complexity of the MM working tools trajectories planning in conditions of variability and uncertainty of the environment.

The approach for automatic implementation of manipulative operations near the sea bottom in hang mode of UV was proposed in the previous paper [11] of the authors. The approach provides the definition of a sea bottom surface position relative to the UV using standard hydroacoustic onboard sensors, and allows to form trajectories of MM motions so that the MM provides minimum possible force and torque effects on the UV during the operations.

For the automatic performing of manipulation operations by UV it is necessary to develop a method that estimates a complexity of the sea bottom relief in a working area in a process of UV approaching to the bottom surface, as well as makes decisions about suitability of this sea bottom relief for the failsafe implementation of manipulation tasks. It is also necessary to rotate the UV towards the rise of the seabed surface [11]. This UV orientation will allow to take the most convenient position for the performing of the manipulation operation. Furthermore, a hovering position for the automatic UV stabilization near a bottom should be calculated automatically. In this place, a MM workspace should allow to properly take a sample of soil, avoiding of collision with a ground and resuspension of a bottom water layer. Thus, the trajectory of MM working tools should be formed taking into account dimensions of the sampler, characteristics of a MM workspace and with an information about continuously refines model of the sea bottom surface.

As a result, the process of automatic performing of soil sampling by UV has to be presented as sequence of the next actions (see Fig. 1):

- The submerging of the UV while the distance to the bottom will allow to estimate a complexity of the sea bottom relief in a working area.
- The estimation of the sea bottom relief complexity in a working area.
- The UV rotation toward the rise of the sea bottom surface.
- The approaching of the UV to the bottom surface, until the surface is in the working area of MM.
- The performing of manipulation operation.

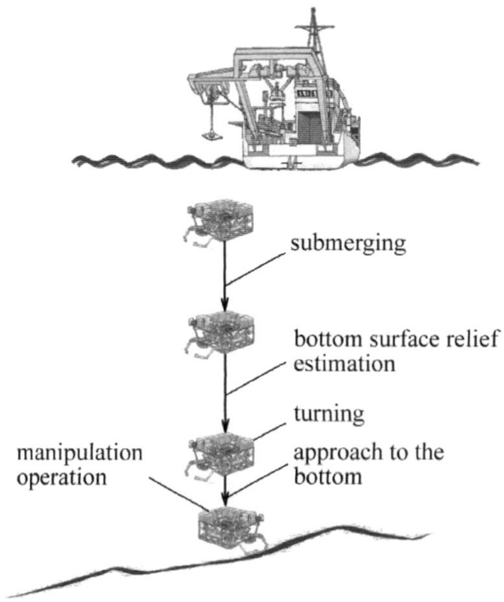

Fig. 1. The automatic performing of manipulation operation as sequence of the actions

2. Features of the UV approaching to the bottom surface

In the process of the UV submerging to the sea bottom surface, the determination of the space position of this surface relative of the vehicle can be obtained by the method [11]. According to this method, the bottom surface is described in the rigidly connected with the UV right rectangular coordinate system XYZ by the equation:

$$A^{*}x + B^{*}y - z + C^{*} = 0, \tag{1}$$

where x, y, z are the variables; A^*, B^*, C^* are the constant coefficients which are simultaneously not equal to zero. These coefficients, updated with each measurement of the Doppler lag, can be calculated by the equation:

$$A^* = \frac{n\sum_{i=1}^{n} z_i x_i \sum_{i=1}^{n} y_i^2 + \sum_{i=1}^{n} x_i y_i \sum_{i=1}^{n} y_i \sum_{i=1}^{n} z_i + \sum_{i=1}^{n} z_i y_i \sum_{i=1}^{n} y_i \sum_{i=1}^{n} x_i - \sum_{i=1}^{n} z_i \sum_{i=1}^{n} y_i^2 \sum_{i=1}^{n} x_i - n\sum_{i=1}^{n} z_i y_i \sum_{i=1}^{n} x_i y_i - \sum_{i=1}^{n} z_i x_i \left(\sum_{i=1}^{n} y_i\right)^2}{n\sum_{i=1}^{n} x_i^2 \sum_{i=1}^{n} y_i^2 + 2\sum_{i=1}^{n} x_i y_i \sum_{i=1}^{n} y_i \sum_{i=1}^{n} x_i - \left(\sum_{i=1}^{n} x_i\right)^2 \sum_{i=1}^{n} y_i^2 - n\left(\sum_{i=1}^{n} x_i y_i\right)^2 - \left(\sum_{i=1}^{n} y_i\right)^2 \sum_{i=1}^{n} x_i^2},$$

$$B^* = \frac{n\sum_{i=1}^{n} x_i^2 \sum_{i=1}^{n} z_i y_i + \sum_{i=1}^{n} z_i x_i \sum_{i=1}^{n} y_i \sum_{i=1}^{n} x_i + \sum_{i=1}^{n} x_i y_i \sum_{i=1}^{n} z_i \sum_{i=1}^{n} x_i - \left(\sum_{i=1}^{n} x_i\right)^2 \sum_{i=1}^{n} z_i y_i - n\sum_{i=1}^{n} x_i y_i \sum_{i=1}^{n} z_i x_i - \sum_{i=1}^{n} x_i^2 \sum_{i=1}^{n} y_i \sum_{i=1}^{n} z_i}{n\sum_{i=1}^{n} x_i^2 \sum_{i=1}^{n} y_i^2 + 2\sum_{i=1}^{n} x_i y_i \sum_{i=1}^{n} y_i \sum_{i=1}^{n} x_i - \left(\sum_{i=1}^{n} x_i\right)^2 \sum_{i=1}^{n} y_i^2 - n\left(\sum_{i=1}^{n} x_i y_i\right)^2 - \left(\sum_{i=1}^{n} y_i\right)^2 \sum_{i=1}^{n} x_i^2},$$

$$C^* = \frac{\sum_{i=1}^{n} x_i^2 \sum_{i=1}^{n} y_i^2 \sum_{i=1}^{n} z_i + \sum_{i=1}^{n} x_i y_i \sum_{i=1}^{n} z_i y_i \sum_{i=1}^{n} x_i + \sum_{i=1}^{n} x_i y_i \sum_{i=1}^{n} z_i x_i \sum_{i=1}^{n} y_i - \sum_{i=1}^{n} x_i \sum_{i=1}^{n} y_i^2 \sum_{i=1}^{n} z_i x_i - \left(\sum_{i=1}^{n} x_i y_i\right)^2 \sum_{i=1}^{n} z_i - \sum_{i=1}^{n} z_i y_i \sum_{i=1}^{n} y_i \sum_{i=1}^{n} x_i^2}{n\sum_{i=1}^{n} x_i^2 \sum_{i=1}^{n} y_i^2 + 2\sum_{i=1}^{n} x_i y_i \sum_{i=1}^{n} y_i \sum_{i=1}^{n} x_i - \left(\sum_{i=1}^{n} x_i\right)^2 \sum_{i=1}^{n} y_i^2 - n\left(\sum_{i=1}^{n} x_i y_i\right)^2 - \left(\sum_{i=1}^{n} y_i\right)^2 \sum_{i=1}^{n} x_i^2},$$

where x_i, y_i and z_i, $i = \overline{(1, n)}$ are coordinates of points of the array of n points which are formed in the coordinate system XYZ by the Doppler lag. These points belong to a sea bottom surface.

Equation (1) can be written in the normal form [12] by multiplying its left and right parts on the normalizing factor $\mathrm{sign}(C^*)\frac{1}{\sqrt{A^{*2}+B^{*2}+1}}$:

$$A_n x + B_n y + C_n z = p\,, \tag{2}$$

where $A_n = \mathrm{sign}(C^*)\frac{A^*}{\sqrt{A^{*2}+B^{*2}+1}}$, $B_n = \mathrm{sign}(C^*)\frac{B^*}{\sqrt{A^{*2}+B^{*2}+1}}$, $C_n = \mathrm{sign}(C^*)\frac{-1}{\sqrt{A^{*2}+B^{*2}+1}}$ are the direction cosines of the unit vector $\overline{n} = [A_n, B_n, C_n]^{\mathrm{T}} \in R^3$, which defines the shifting direction of the bottom surface plane from the origin of the coordinate system XYZ at a distance $p = \mathrm{sign}(C^*)\frac{C^*}{\sqrt{A^{*2}+B^{*2}+1}}$.

At the stage of the UV submerging, this UV has to evaluate the suitability of the surface relief in a working area. It is necessary for the failsafe performing of manipulative tasks. The sea bottom area, covered by Doppler lag rays, is decreasing in the process of the UV diving. And the coefficients of the equation (2) are also changed together with this area. Experimental investigation shown that the evaluation of a value of these changes in a limited working area should be performed with a fixed heading angle of the UV. The measurements of these changes start when the condition $p \le 10L$ is satisfied, where L is the maximum possible distance between the UV buoyancy center and the MM working tool.

The vector $\overline{n}_k = [A_{nk}, B_{nk}, C_{nk}]^{\mathrm{T}} \in R^3$ determines the bottom plane (2) spatial orientation in the coordinate system XYZ for the k-th measurements of all lag antennas, $k = \overline{(1, r)}$, where r is the number of measurements. Evaluation of the bottom relief complexity can be based on the measuring of changes of angles $\alpha_k = \arccos(A_{nk})$, $\beta_k = \arccos(B_{nk})$ and $\gamma_k = \arccos(C_{nk})$, which are formed by the vector $\overline{n}_k$ with the respective axes of the coordinate system XYZ. For all r measurements while the UV dives with the stabilized heading angle, the test results shown that for the suitable working area surface (without major irregularities) the following conditions are usually satisfied:

$$\alpha_{k\max} - \alpha_{k\min} \le 15°\,, \tag{3}$$

$$\beta_{k\max} - \beta_{k\min} \le 15°\,, \tag{4}$$

$$\gamma_{k\max} - \gamma_{k\min} \le 15°\,. \tag{5}$$

The UV submerging must be terminated if at least one of the conditions (3)-(5) are not satisfied.

As it is noted in [11], if the UV is near the bottom, the UV longitudinal axis should be directed to the rise of the plane (2), because a workspace of the MM is located in the front semisphere of this vehicle. This UV orientation allows to improve visibility of the working area and preserve the biggest workspace of the MM. As a result, when the conditions (3)-(5) are satisfied and the UV is approached to the bottom surface to a distance $p \leq 2L$, the UV should terminate its motion and turns to the rise of the bottom surface [11].

3. Features of performing of soil sampling

Place of the UV stabilization in the hovering mode [5 - 8] should be sufficiently close to a bottom surface for the soil sampling. But the UV has to be far from a bottom to avoid the collision with a ground and to avoid the resuspension of bottom water layers. For the automatically choosing of this place, the MM workspace can be presented as a finite numbers of s vertical segments inscribed within [11]. These segments have a parallelepiped shapes with identical bases. Axes of these segments are parallel to the vertical axis Z of the coordinate system XYZ. The value of s is chosen based on size of the MM workspace, soil samplers size and number of samples that you want to take.

The working MM is shown at Fig. 2. Which introduces the next labels: 1 is the bottom; 2 is the bottom surface plane (2); P_a is the approach point of the MM sampler 3 (its length is h); C_a is the crossing point of the vertical axis of the segment $a = \overline{1, s}$ with the plane 2; 4 is the MM workspace.

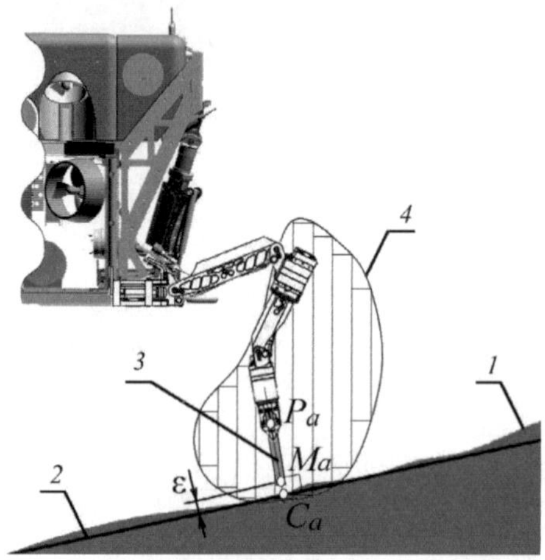

Fig. 2. Performing of the manipulation operation

For every segment a the coordinates x_{ca} and y_{ca} of the crossing point of the segment vertical axis and the plane XY are known. The coordinate z_{ca} of the crossing point C_a is defined by the formula:

$$z_{ca} = z(x_{ca}, y_{ca}) = A^* x_{ca} + B^* y_{ca} + C^*$$

The point C_a (x_{ca}, y_{ca}, z_{ca}) will belong to the MM workspace under the condition:

$$z_{ca} > z_{ca\min}, \tag{6}$$

where $z_{ca\min}$ is the coordinate of the lower base of the segment a. Until the point C_a satisfying the condition (6) appears, the UV should keep slow submerging to the bottom. When the point C_a appears, the UV stops and has to be fixed in this position with help of stabilization systems [5-8].

For the taking of q samples, the UV to dive until the points C_a satisfying the condition (6) appear in the q segments. The UV diving stops when q C_a points appear. When the UV is stabilized in the hovering mode the MM gripper trajectory can be described as a sequence of transitions from an initial position to the approach point P_a. Then the gripper transitions by the distance h along the vector $\overline{n}$ to the point M_a, in which the sampler is maximum immersed. The point P_a is formed based on the value ε (see. Fig. 2). The value ε allows to eliminate the possible contact of the sampler with the uneven bottom surface in the process of the MM gripper transition from the starting position to the point P_a:

$$P_a = C_a - \overline{n}(h + \varepsilon).$$

The point P_a must be located in the MM workspace. The coordinates of the point M_a are calculated by the formula:

$$M_a = P_a + \overline{n}h.$$

For the sampler removing the MM gripper moves to the point P_a without changing its orientation. Further, MM moves to the starting position and puts the sampler into a container. It should be noted, that the method [13] allows to perform the manipulation operation even under conditions of the UV unplanned displacements relative to the initial position of its stabilization under the influence of external factors.

4. Sea tests results

The sea tests of proposed technology were performed in the Bering Sea with the UV Sub-Atlantic Comanche 18 equipped with the MM Schilling Orion 7P. For this UV the value L is 3.4m. Coefficients of (2) had been updating with a frequency 5 Hz of Doppler lag Navigator Teledyne RD Instruments measurements in the process of the UV diving from a distance of $10L$ to a bottom. Fig. 3 shows the bottom surface plane 1 in the coordinate system XYZ. This plane was measured by the lag 2 at the distance of $2L$ to the sea bottom and described by (2) with the coefficients A_n= 0.072, B_n=0.055, C_n=–0.996, p=6.825. The vector $\bar{n}_k$ forms angles α_k=85.9°, β_k=86.8° and γ_k=174.8° with the respective axes of the coordinate system XYZ.

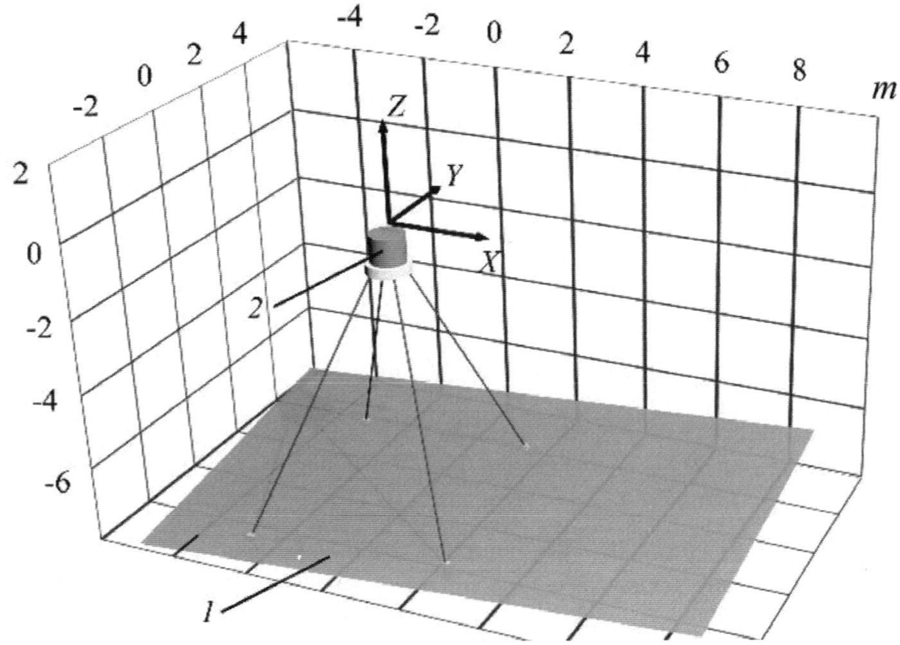

Fig. 3. The results of experimental tests

While the UV was submerged from a distance of $10L$ to $2L$ to the bottom, the maximal changes of angles α_k, β_k, γ_k were less than 5°. Therefore, the conditions (3)-(5) were not violated. The measurements have shown that the bottom surface had no significant irregularities and was suitable for performing of manipulation operations. Few dozen successful dives with the operations of taking of soil samples were performed. In all of these dives the conditions (3)-(5) were always satisfied.

5. Conclusion

In this paper, the technology for an automatic implementation of the soil sampling by means of underwater multilink manipulators was considered. This technology provides an automatic performing of the operations of taking soil samples by UV equipped with onboard MM. The position detecting of the sea bottom surface relative to the UV is based on information from a Doppler lag. In the process of UV moving to the bottom, this technology estimates the complexity of the sea bottom relief in the working area and decides on the suitability of this relief for the failsafe performing of desired underwater operations. Also, the algorithm for the automatic performing of operations of taking of soil samples from the bottom surface was proposed. The sea tests were performed for investigation of developed system and algorithm, and confirmed efficiency and simplicity of its practical implementation.

The future studies are aimed to perform more complex manipulation operations by UV in an automatic mode. Some of them are: taking the upper layer of a soil from the surface of the bottom with a complex shape, automatically detecting and taking of underwater objects, as well as performing technological operations, such as cleaning of underwater structures, underwater fouling, welding and carpentry. The technologies for the implementation of said operations will use the information about the object of work will be obtained using multi-beam 3D sonars and technical vision systems.

6. Acknowledgments

This work was performed in Institute of Marine Technology Problems FEB RAS, Institute for Automation and Control Processes FEB RAS and also in Far Eastern Federal University (Vladivostok, Russia).

The development of the technology for the autonomous soil sampling by UV was carried out with the financial support of the Russian Foundation for Basic Research (RFBR) grants (projects 16-29-04195 Ofi_m, 17-57-45055 Ind_a). The development of the method for the estimation of the complexity of a sea bottom relief was carried out with the financial support of the Russian Science Foundation (RSF) grant (project №17-79-10064).

7. References

[1] I.E. Tuphanov, A.F. Scherbatyuk. (2012) "Adaptive algorithm of AUV meander pattern trajectory planning for underwater sampling," Proceedings of the 10th ISOPE Pacific/Asia Offshore Mechanics Symposium, PACOMS 2012, Vladivostok, Russia. pp. 181 – 185.

[2] V.F. Filaretov, D.A. Yukhimets, A.F. Scherbatyuk, I.E. Tuphanov, E.A. Mursalimov. (2014). "Some Marine Trial Results of a New Method for AUV Trajectory Motion Control," Proceedings of the OCEANS 2014 MTS/IEEE Conference, St. John's, Newfoundland and Labrador, Canada. pp. 1-6.

[3] V.F. Filaretov, A.Yu. Konoplin, N.Yu. Konoplin. (2015). "Method of Synthesis of Automatic Correction Systems of Underwater Vehicles Linear Displacements" Procedia Engineering, Vol 100, pp. 1434-1440.

[4] K. Ishizu, H. Nakayama, N. Sakagami, M. Shibata, S. Kawamura, S. Matsuda, A. Mitsui. (2014). "Preliminary Experiments of a Human-Portable Underwater Gripper Robot for Dexterous Tasks," Proceedings of MTS/IEEE OCEANS 2014.

[5] V.F. Filaretov, A.Yu. Konoplin. (2015). "System of Automatic Stabilization of Underwater Vehicle in Hang Mode with Working Multilink Manipulator," International IEEE Conference on Computer, Control, Informatics and Its Applications (IC3INA), Bandung, Indonesia, pp. 132-137.

[6] V.F. Filaretov, A.Yu. Konoplin. (2016). "Development of Control Systems for Implementation of Manipulative Operations in Hovering Mode of Underwater Vehicle,"/ OCEANS 2016. Shanghai, China. International IEEE Conference.

[7] T.W. McLain, S.M. Rock, M.J. Lee. (1996). "Experiments in the coordinated control of an underwater arm/vehicle system," Autonomous Robots, vol. 3, no. 2-3, pp. 213–232.

[8] M. Santhakumar, K. Jinwhan. (2012). "Indirect adaptive control of an autonomous underwater vehicle-manipulator system for underwater manipulation tasks" Ocean Engineering 54, pp. 233–243.

[9] G. Marani, S.K. Choi, J. Yuh. (2009). "Underwater autonomous manipulation for intervention missions AUVs," Ocean Engineering, vol. 36, no. 1, pp. 15 – 23.

[10] D. Lodi Rizzini, F. Kallasi, J. Aleotti, F. Oleari, S. Caselli. (2016) "Integration of a stereo vision system into an autonomous underwater vehicle for pipe manipulation tasks," Computers & Electrical Engineering, pp. 1-12.

[11] V.F. Filaretov, A.Yu. Konoplin, N.Yu. Konoplin, G.V. Gorbachev. (2016). "Control system for underwater vehicle with multilink manipulator for automatic manipulation operations," Proceedings of the 27th DAAAM International Symposium, B. Katalinic (Ed.), Published by DAAAM International, Vienna, Austria, – pp. 714-720.

[12] G.A. Korn, T.M. Korn. (2000). "Mathematical Handbook for Scientists and Engineers: Definitions, Theorems, and Formulas for Reference and Review," General Publishing Company. - 1151 p.

[13] V.F. Filaretov, A.Yu. Konoplin. (2015). "System of Automatically Correction of Program Trajectory of Motion of Multilink Manipulator Installed on Underwater Vehicle," Procedia Engineering, Vol. 100, pp. 1441-144.

DOI: 10.2507/28th.daaam.proceedings.096

DYNAMIC BEHAVIOUR COMPARISON OF THREE DIFFERENT MATHEMATICAL MODEL COMPLEXITIES

Jiri Zatopek & Zdenek Urednicek

This Publication has to be referred as: Zatopek, J[iri] & Urednicek, Z[denek] (2017). Dynamic Behaviour Comparison of Three Different Mathematical Model Complexities, Proceedings of the 28th DAAAM International Symposium, pp.0685-0693, B. Katalinic (Ed.), Published by DAAAM International, ISBN 978-3-902734-11-2, ISSN 1726-9679, Vienna, Austria
DOI: 10.2507/28th.daaam.proceedings.096

Abstract

Dynamic behaviour comparison of three different mathematical descriptions complexity for first 3 joints of 6 degrees of freedom robotic structure is presented in this article. Firstly, 3D CAD model is designed in SolidWorks, which is used as the basis for a physical and mathematical model. The CAD model is exported directly from SolidWorks to SimMechanics as a physical model which is considered as the most accurate replacement for a real model in this work. The first type of a mathematical model is the most precise but also the most complex; it is based on SolidWorks inertia matrices and matrix form of Lagrange's motion equations of the second kind. The second type of a mathematical model is created by each part replacement with a suitable simplified shape; classical integration approach with Lagrange's motion equations of the second kind is used. The third type of a mathematical model is based on the same approach as the second type, but all the objects are replaced by mass points. At the end, all the results of dynamic behaviour are compared with the physical model, for utilization in controller design.

Keywords: Model-Based-Design; Mathematical model; Simulation; Motion equation; SimMechanics

1. Introduction

Nowadays, a wide range of different support software tools are available; it provides many possibilities and ways to determine mathematical or physical model [4, 5, 6, 7, 9, 10, 11, 13, 14, 15, 17, 18, 19, 20]. Accurate detection of a mathematical model is complicated in the case of a complex mechanical structure with many parts, and it should not be used for simulation purposes and controller determination at the same time. The chosen way is to split the problem between 2 models is: the most accurate model for a simulation as a real (physics) model substitution - physical model [4, 9, 14, 18] and appropriately simplified model purely for a controller determination - mathematical model [7, 19, 20].

For a mathematical and physical model, a CAD model is very helpful. SolidWorks is used for a CAD model design [12, 14, 15, 16, 17, 18]. Only the shape and the material properties of an object are needed to know for a full dynamic description; these parameters are included in the CAD model. This means that the CAD model should always be the first step in an accurate dynamic simulation of any system.

The physical model is exported directly from SolidWorks to the SimMechanics Second Generation [14, 15]. It includes all dimensions, joins and material properties of the CAD model. It is the most accurate model that includes all important shapes (inertia matrices) and mass details for dynamic simulation purposes [16, 17].

The first type of mathematical model is the most accurate but also the most complex. It is based on inertia matrices [4, 11], transformation matrices [1, 3, 14, 17, 20], centres of mass position vectors [2, 16] and model masses. All of these parameters are taken from the SolidWorks CAD model. In this case, the matrix form of Lagrange's motion equations of the second kind is used [2, 16].

The second type of mathematical model uses shape simplification. Every link is replaced by a mass rod, every actuator is superseded by a mass point and significant cylindrical part is replaced by a mass cylindrical plate. The whole assembly is divided into many essential parts whose contribution of energy sumis the energy of the entire assembly.

The third type of mathematical model is similar to the second type, but all stationary objects are replaced by mass points. This mathematical model is the simplest, but its accuracy is the lowest.

In the results of this work, the differences between the above mentioned mathematical descriptions during dynamic simulation should be visible. These results should help to understand when it is better to use a more precise mathematical model and when the simplest model is sufficient as a basis for design control law.

2. CAD model

The SolidWorks CAD model consists of 9 parts as shown in Fig.1. Each part has an associated local coordinate system, the material and joins each other by structural bonds to the final fully bonded assembly [12, 14, 15, 16].

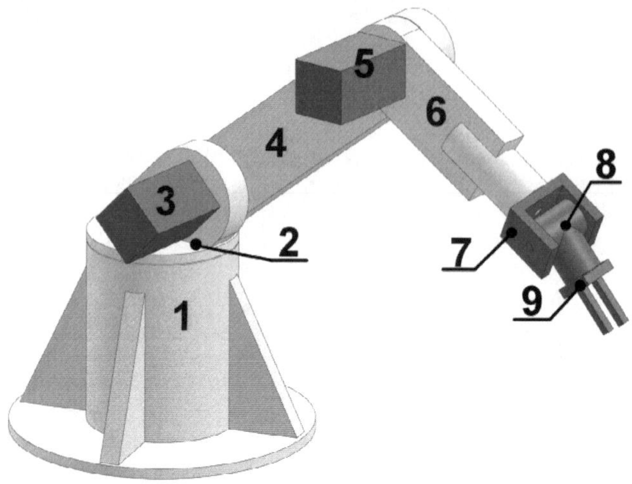

Fig. 1. The CAD model of the analysed system

The CAD model has 6 degrees of freedom (DOF) [3, 18] generally, but the first 3 joints have the most interesting dynamic behaviour and will be studied in this paper. Therefore, the last 3 DOF (parts 7, 8 and 9 in Fig. 1.) will be fixedly connected to each other and with the link 6.

3. Physical model

The SolidWorks model is exported directly into the SimMechanics Second Generation diagram along with all the linkages, dimensions and material properties [14, 15].

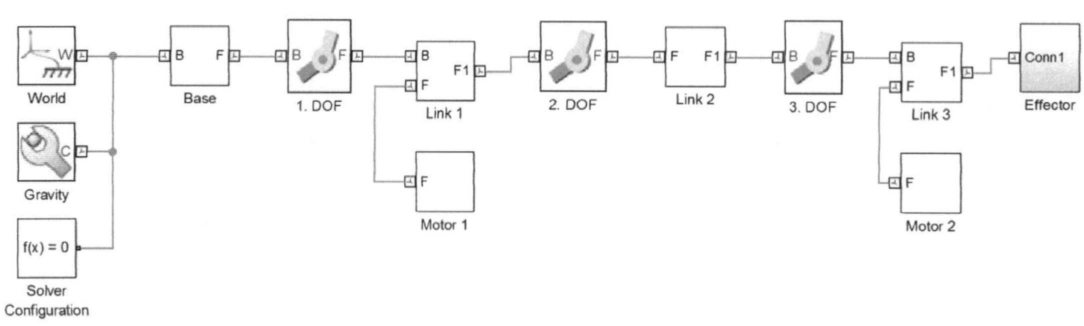

Fig. 2. SimMechanics Second Generation physical model

A physical model is considered to be the most accurate substitute for a real physics system in case of precise CAD model which is the SimMechanics model based on. All mathematical models dynamic behaviours are compared with the physical model to see how accurate the mathematical models are.

4. Mathematical models

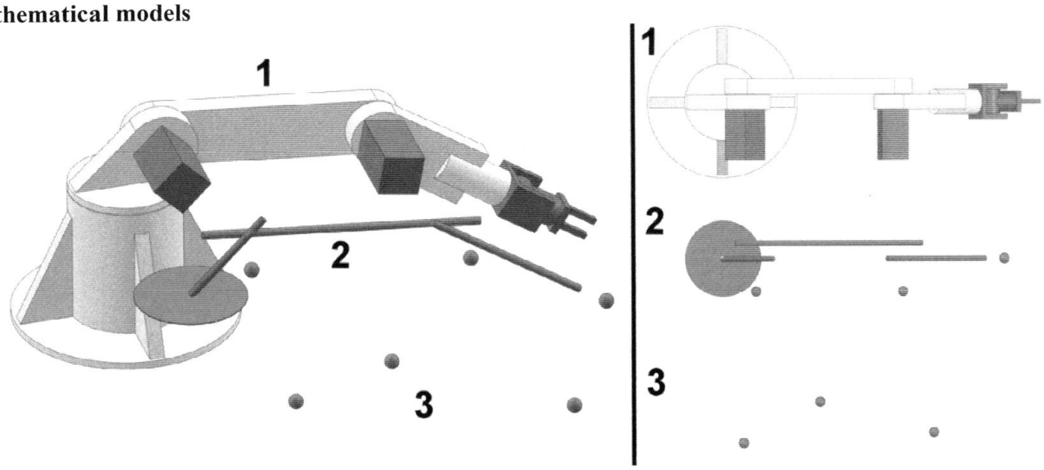

Fig. 3. Graphic preview of mathematical model simplifications

Three different CAD model simplifications are shown in Fig. 3. and their mathematical descriptions are derived below.

4.1 Transformation of coordinate systems

The transformation matrices are used to determine the position and orientation of a local coordinate system firmly connected with the component in the overall global coordinate system [8, 15]. They are an integral part of any complex mathematical model determination.

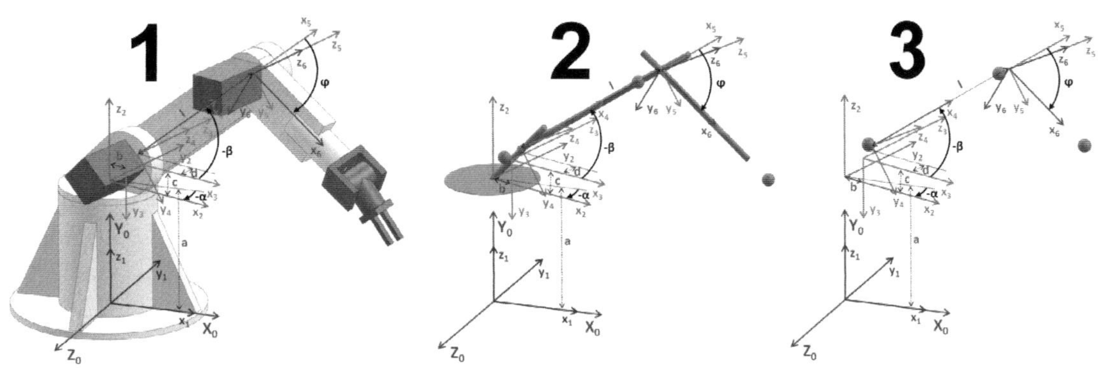

Fig. 4. Simplified models with the selected coordinate system placement

$$T_1 = {}^0T_2 = \begin{bmatrix} cos(\alpha) & -sin(\alpha) & 0 & 0 \\ 0 & 0 & 1 & a \\ -sin(\alpha) & -cos(\alpha) & 0 & 0 \\ 0 & 0 & 0 & 1 \end{bmatrix} \quad (1)$$

$$T_2 = {}^0T_4 = \begin{bmatrix} cos(\beta)cos(\alpha) & cos(\alpha)sin(\beta) & -sin(\alpha) & b \cdot cos(\alpha) - d \cdot sin(\alpha) \\ sin(\beta) & -cos(\beta) & 0 & a+c \\ -cos(\beta)sin(\alpha) & -sin(\alpha)sin(\beta) & -cos(\alpha) & -d \cdot cos(\alpha) - b \cdot sin(\alpha) \\ 0 & 0 & 0 & 1 \end{bmatrix} \quad (2)$$

$$T_3 = {}^0T_6 = \begin{bmatrix} cos(\beta+\varphi)cos(\alpha) & cos(\alpha)sin(\beta+\varphi) & -sin(\alpha) & b \cdot cos(\alpha) - d \cdot sin(\alpha) + l \cdot cos(\alpha)cos(\beta) \\ sin(\beta+\varphi) & -cos(\beta+\varphi) & 0 & a+c+l \cdot sin(\beta) \\ -cos(\beta+\varphi)sin(\alpha) & -sin(\alpha)sin(\beta+\varphi) & -cos(\alpha) & -d \cdot cos(\alpha) - b \cdot sin(\alpha) - l \cdot sin(\alpha)cos(\beta) \\ 0 & 0 & 0 & 1 \end{bmatrix} \quad (3)$$

The centres of mass position vectors for the above mentioned local coordinate systems are different for each mathematical model simplification. In general, they are listed as:

$$r_1 = {}^2r_2 = [r_{11} \quad r_{12} \quad r_{13} \quad 1]^T; \qquad r_2 = {}^4r_4 = [r_{21} \quad r_{22} \quad r_{23} \quad 1]^T \qquad r_3 = {}^6r_6 = [r_{31} \quad r_{32} \quad r_{33} \quad 1]^T \tag{4}$$

For the next steps, the real model parameters are:

$$\begin{matrix} a = 0{,}171\,m; & b = 0{,}065\,m; & c = 0{,}065\,m; & d = 0{,}031\,m;\ l = 0{,}3\,m \\ m_1 = 2{,}938\,kg; & m_2 = 1{,}008\,kg; & m_3 = 4{,}565\,kg; & g = 9{,}81\,m\cdot s^{-2} \end{matrix} \tag{5}$$

4.2 The first mathematical model - no simplification

The first model (Fig. 4. part 1) has a different approach to the motion equation derivation than the other two models. It uses the matrix form of Lagrange's motion equations of the second kind. This mathematical description is the most accurate of the three listed models because of inertia matrix using with a centre of mass position vectors from the CAD model, which are with transformation matrices all necessary parameters to need to know.

Motion equations matrix form has the general form [2, 16]:

$$D(q)\cdot\ddot{q} + H(q,\dot{q}) + G(q) = T \quad \Rightarrow \quad D(\alpha,\beta,\varphi)\cdot\begin{bmatrix}\ddot{\alpha}\\ \ddot{\beta}\\ \ddot{\varphi}\end{bmatrix} + H(\alpha,\dot{\alpha},\beta,\dot{\beta},\varphi,\dot{\varphi}) + G(\alpha,\beta,\varphi) = \begin{bmatrix}T_1\\ T_2\\ T_3\end{bmatrix} \tag{6}$$

in which:

$$\begin{aligned} D_{i,j} &= \sum_{r=\max\{i,j\}}^{n} tr\left[\frac{\partial^0 T_r}{\partial q_j}\cdot {}^r\bar{I}_r\left(\frac{\partial^0 T_r}{\partial q_i}\right)^T\right] \\ H_i &= \sum_{k=1}^{n}\sum_{m=1}^{n} H_{ikm}\dot{q}_k\dot{q}_m \\ H_{ikm} &= \sum_{r=\max\{i,k,m\}}^{n} tr\left[\frac{\partial^{2\,0} T_r}{\partial q_k \partial q_m}\cdot {}^r\bar{I}_r\left(\frac{\partial^0 T_r}{\partial q_i}\right)^T\right] \\ G_i &= -\sum_{r=i}^{n} m_r \left({}^0 g\right)^T\cdot\frac{\partial^0 T_r}{\partial q_i}\cdot {}^r\bar{r}_r \end{aligned} \tag{7}$$

As could be seen, the only two other variables need to be known. These parameters are found in the SolidWorks CAD model.

The centres of mass position vectors are:

$$r_1 = [0{,}057 \quad -0{,}053 \quad 0{,}058 \quad 1]^T; \quad r_2 = [-0{,}15 \quad 0 \quad -0{,}004 \quad 1]^T; \quad r_3 = [0{,}099 \quad 0 \quad -0{,}065 \quad 1]^T \tag{8}$$

The pseudo-inertia matrices are:

$$\bar{I}_1 = \begin{bmatrix} 0{,}0120 & -0{,}0101 & 0{,}0108 & 0{,}1675 \\ -0{,}0101 & 0{,}0124 & -0{,}0101 & -0{,}1557 \\ 0{,}0108 & -0{,}0101 & 0{,}0117 & 0{,}1704 \\ 0{,}1675 & -0{,}1557 & 0{,}1704 & 2{,}9380 \end{bmatrix}; \bar{I}_2 = \begin{bmatrix} 0{,}0352 & 0 & -0{,}0005 & 0{,}1512 \\ 0 & 0{,}0005 & 0 & 0 \\ -0{,}0005 & 0 & 0{,}0002 & -0{,}0040 \\ 0{,}1512 & 0 & -0{,}0040 & 1{,}0080 \end{bmatrix}; \bar{I}_3 = \begin{bmatrix} 0{,}1069 & 0 & -0{,}0140 & 0{,}4519 \\ 0 & 0{,}0011 & 0 & 0 \\ -0{,}0140 & 0 & 0{,}0268 & -0{,}2967 \\ 0{,}4519 & 0 & -0{,}2967 & 4{,}4650 \end{bmatrix} \tag{9}$$

The specific solution for the system shown in equation *(6)* is:

$$\mathbf{D1} = \begin{bmatrix} D_{11} & 5{,}008\cdot 10^{-5}\sin(\beta+\varphi) - 0{,}04242\sin(\beta) & 5{,}008\cdot 10^{-5}\sin(\beta+\varphi) \\ 5{,}008\cdot 10^{-5}\sin(\beta+\varphi) - 0{,}04242\sin(\beta) & 0{,}2712\cos(\varphi) + 0{,}5546 & 0{,}1356\cos(\varphi) + 0{,}108 \\ 5{,}008\cdot 10^{-5}\sin(\beta+\varphi) & 0{,}1356\cos(\varphi) + 0{,}108 & 0{,}108 \end{bmatrix} \tag{10}$$

$$D_{11} = 0{,}1356\cos(2\beta+\varphi) + 0{,}2228\cos(2\beta) + 0{,}05875\cos(2\beta) + 0{,}1977\cos(\beta) + 0{,}1356\cos(\varphi) + 0{,}05292\cos(2\beta+2\varphi) + 0{,}3389$$

$$\mathbf{H1} = \begin{bmatrix} \dot{\beta}^2\left[5,008\cdot 10^{-5}\cos(\beta+\varphi)-0,04242\cos(\beta)\right]+1,002\cdot 10^{-3}\cos(\beta+\varphi)\dot{\beta}\dot{\varphi}-2\dot{\alpha}\dot{\beta}\left[0,1356\sin(2\beta+\varphi)+0,2228\sin(2\beta)+\right. \\ \left.+0,02938\sin(\beta+\varphi)+0,09885\sin(\beta)+0,05292\sin(2\beta+2\varphi)\right]+5,008\cdot 10^{-5}\cos(\beta+\varphi)\dot{\varphi}-2\dot{\alpha}\dot{\varphi}\left[0,06779\sin(2\beta+\varphi)+\right. \\ \left.+0,02938\sin(\beta+\varphi)+0,06779\sin(\varphi)+0,05292\sin(2\beta+2\varphi)\right] \\ \dot{\alpha}^2\left[0,05292\sin(2\beta+2\varphi)+0,1356\sin(2\beta+\varphi)+0,2228\sin(2\beta)+0,02938\sin(\beta+\varphi)+0,09885\sin(\beta)\right]- \\ -0,1356\sin(\varphi)\dot{\varphi}^2-0,2712\sin(\varphi)\dot{\beta}\dot{\varphi} \\ \dot{\alpha}^2\left[0,06779\sin(2\beta+2\varphi)+0,02938\sin(\beta+\varphi)+0,06779\sin(\varphi)+0,05292\sin(2\beta+2\varphi)\right]+0,1356\sin(\varphi)\dot{\beta}^2 \end{bmatrix} \quad (11)$$

$$\mathbf{G1} = \begin{bmatrix} 0 \\ 4,433\cos(\beta+\varphi)+14,92\cos(\beta) \\ 4,433\cos(\beta+\varphi) \end{bmatrix} \quad (12)$$

4.3 The second mathematical model - shape simplification

The second model (Fig. 4. part 2) uses shape simplification as described in the introduction part. The Lagrange equations of the second kind were used to find the motion equations. In order to use Lagrange equations, it is necessary to define the kinetic and potential energy of the whole system. The transformation matrices and the centre of mass position vectors are used to determine the absolute position and velocity of each simplified part. This mathematical description is not as accurate as the previous model but should be sufficient even for simulation purposes because of the mass distribution cause - it is analyzed in results comparison.

Due to the motion equations derivation complexity (each part has similar but not the same approach to find kinetic and potential energy) will be described only one simplified part (Fig. 1. part 4) kinetic/potential energy determination and then the motion equations will be derived.

The position vector in the global coordinate system:

$$p_1 = \mathbf{T}_2 \cdot r_2 = \begin{bmatrix} p_{11} & p_{12} & p_{13} & 1 \end{bmatrix}^T \quad (13)$$

The velocity vector in the global coordinate system:

$$v_1 = \frac{dp_1}{dt} = \begin{bmatrix} v_{11} & v_{12} & v_{13} & 1 \end{bmatrix}^T \quad (14)$$

The kinetic energy:

$$dE_{K1} = \frac{1}{2}\cdot dm_1 \cdot \left(v_{11}^2 + v_{12}^2 + v_{13}^2\right) \qquad E_{K1} = \iint_D dE_{K1} \quad (15)$$

The potential energy (gravitation is in the global axis Y direction, therefore the p_{12} position is used):

$$E_{P1} = m_1 \cdot g \cdot p_{12} \quad (16)$$

The kinetic and potential energy of the other six parts (Fig. 1.) is determined in a similar way like for the above-mentioned part. The total kinetic and potential energy of the system is:

$$E_K = \sum_{i=1}^{7} E_{Ki} \qquad E_P = \sum_{i=1}^{7} E_{Pi} \quad (17)$$

The Lagrangian:

$$L = E_K - E_P \quad (18)$$

The final motion equations:

$$\frac{d}{dt}\left(\frac{\partial L}{\dot{\alpha}}\right) - \frac{\partial L}{\alpha} = T_1; \qquad \frac{d}{dt}\left(\frac{\partial L}{\dot{\beta}}\right) - \frac{\partial L}{\beta} = T_2; \qquad \frac{d}{dt}\left(\frac{\partial L}{\dot{\varphi}}\right) - \frac{\partial L}{\varphi} = T_3; \quad (19)$$

Equation (19) can be also rewritten into the motion equations matrix form shown in equation (6). For the parameters in equation (5) and the corresponding centres of mass position vectors, the motion equations for the second mathematical model are:

$$\mathbf{D2} = \begin{bmatrix} D_{11} & -0,04211sin(\beta) & 0 \\ -0,04211sin(\beta) & 0,2666cos(\varphi)+0,5497 & 0,1333cos(\varphi)+0,1045 \\ 0 & 0,1333cos(\varphi)+0,1045 & 0,1045 \end{bmatrix} \tag{20}$$

$$D_{11} = 0,1333cos(2\beta+\varphi) + 0,2226cos(2\beta) + 0,05776cos(\beta+\varphi) + 0,1975cos(\beta) + 0,1333cos(\varphi) + 0,05227cos(2\beta+2\varphi) + 0,3314$$

$$\mathbf{H2} = \begin{bmatrix} -\dot{\alpha}\dot{\varphi}[0,1333sin(2\beta+\varphi)+0,05776sin(\beta+\varphi)+0,1333sin(\varphi)+0,1045sin(2\beta+2\varphi)] - 0,04211cos(\beta)\dot{\beta}^2 - \\ -\dot{\alpha}\dot{\beta}[0,2666sin(2\beta+\varphi)+0,4451sin(2\beta)+0,05776sin(\beta+\varphi)+0,1975sin(\beta)+0,1045sin(2\beta+2\varphi)] \\ \\ \dot{\alpha}^2[0,05227sin(2\beta+2\varphi)+0,1333sin(2\beta+\varphi)+0,2226sin(2\beta)+0,02888sin(\beta+\varphi)+0,09873sin(\beta)] - \\ -\dot{\varphi}[0,2666sin(\varphi)\dot{\beta}+0,1333sin(\varphi)\dot{\varphi}] \\ \\ \dot{\alpha}^2[0,05227sin(2\beta+2\varphi)+0,06665sin(2\beta+\varphi)+0,02888sin(\beta+\varphi)+0,06665sin(\varphi)] + 0,1333sin(\varphi)\dot{\beta}^2 \end{bmatrix} \tag{21}$$

$$\mathbf{G2} = \begin{bmatrix} 0 \\ 4,359cos(\beta+\varphi)+14,9cos(\beta) \\ 4,359cos(\beta+\varphi) \end{bmatrix} \tag{22}$$

This mathematical description is not too complicated compared to the previous equations (10), (11) and (12). Nevertheless, there is one more solution with a simpler mathematical description determination.

4.4 The third mathematical model - mass point simplification

The third model (Fig. 4. part 3) uses mass point simplification which quantity is equal to the DOF number. The Lagrange equations of the second kind were also used to the motion equations determination. There is the same approach as in the previous chapter 4.3, but all the fixed parts are replaced by the mass points located in their centre of mass. Their number is three because the model has three DOF. This mathematical description is not as accurate as previous two models and should not be used for the simulation purposes because all the mass is concentrated in the centre of mass of each fixed body parts only. This means the mass distribution has been disturbed.

However, this model is easiest to derive and should be the most appropriate for the design control law purposes because of its expected simplicity. The differences between dynamic behaviour will be determined in the next chapter. The third model's motion equations matrix form for the parameters in equation (5) and appropriate centres of mass position vectors are:

$$\mathbf{D3} = \begin{bmatrix} D_{11} & -0,01537sin(\beta+\varphi)-0,04248sin(\beta) & -0,01537sin(\beta+\varphi) \\ -0,01537sin(\beta+\varphi)-0,04248sin(\beta) & 0,2712cos(\varphi)+0,4783 & 0,1356cos(\varphi)+0,04474 \\ -0,01537sin(\beta+\varphi)+0,04474 & 0,1356cos(\varphi)+0,04474 & 0,04474 \end{bmatrix} \tag{23}$$

$$D_{11} = 0,1356cos(2\beta+\varphi) + 0,2168cos(2\beta) + 0,05875cos(\beta+\varphi) + 0,1977cos(\beta) + 0,1356cos(\varphi) + 0,02237cos(2\beta+2\varphi) + 0,2865$$

$$\mathbf{H3} = \begin{bmatrix} -\dot{\alpha}\dot{\varphi}[0,1356sin(2\beta+\varphi)+0,05875sin(\beta+\varphi)+0,1356sin(\varphi)+0,04474sin(2\beta+2\varphi)] - 0,04248cos(\beta)\dot{\beta}^2 - \\ -\dot{\alpha}\dot{\beta}[0,2712sin(2\beta+\varphi)+0,4335sin(2\beta)+0,05875sin(\beta+\varphi)+0,1977sin(\beta)+0,04474sin(2\beta+2\varphi)] - \\ -0,01537cos(\beta+\varphi)\dot{\beta}^2 - 0,01537cos(\beta+\varphi)\dot{\varphi}^2 - 0,03073cos(\beta+\varphi)\dot{\beta}\dot{\varphi} \\ \\ \dot{\alpha}^2[0,02237sin(2\beta+2\varphi)+0,1356sin(2\beta+\varphi)+0,2168sin(2\beta)+0,02938sin(\beta+\varphi)+0,09885sin(\beta)] - \\ -\dot{\varphi}[0,2712sin(\varphi)\dot{\beta}+0,1356sin(\varphi)\dot{\varphi}] \\ \\ \dot{\alpha}^2[0,02237sin(2\beta+2\varphi)+0,06779sin(2\beta+\varphi)+0,02938sin(\beta+\varphi)+0,06779sin(\varphi)] + 0,1356sin(\varphi)\dot{\beta}^2 \end{bmatrix} \tag{24}$$

$$\mathbf{G3} = \begin{bmatrix} 0 \\ 4,433cos(\beta+\varphi)+14,92cos(\beta) \\ 4,433cos(\beta+\varphi) \end{bmatrix} \tag{25}$$

As could be seen, the simplest model with only three mass points does not mean simplest final motion equations. The second mathematical model is more accurate and not too complex, but its determination was hardest.

5. Comparison of results

The resulting motion equations are compared with the physical model torque outputs in Fig. 5. The SimMechanics physical model is considered to be the most accurate dynamic model here and therefore, any deviation indicates an inaccurate approximation.

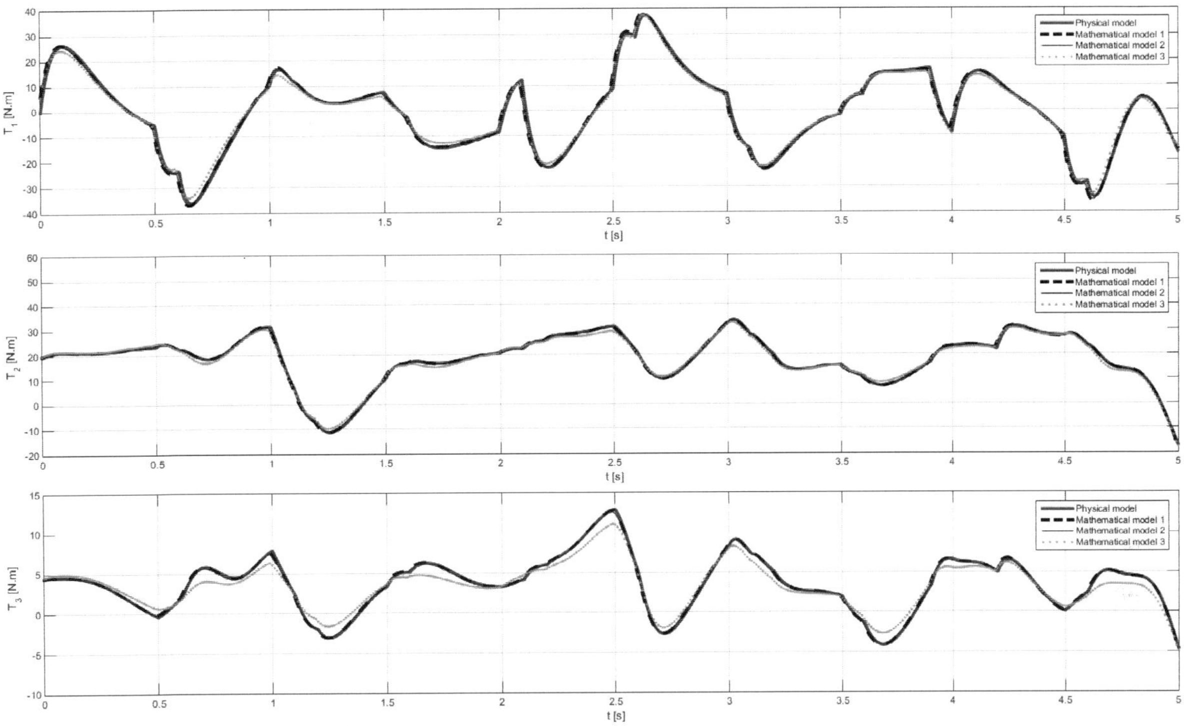

Fig. 5. The physical and the three different mathematical models comparison

As shown in Fig. 5., all the model's dynamic behaviour is very similar, only in the third DOF third mathematical model output is more obvious deflection. The motion, which the listed torques correspond to, was chosen as the most dynamic technological movement as could be required. For more detailed research, the Fig. 5. is enlarged in Fig. 6.

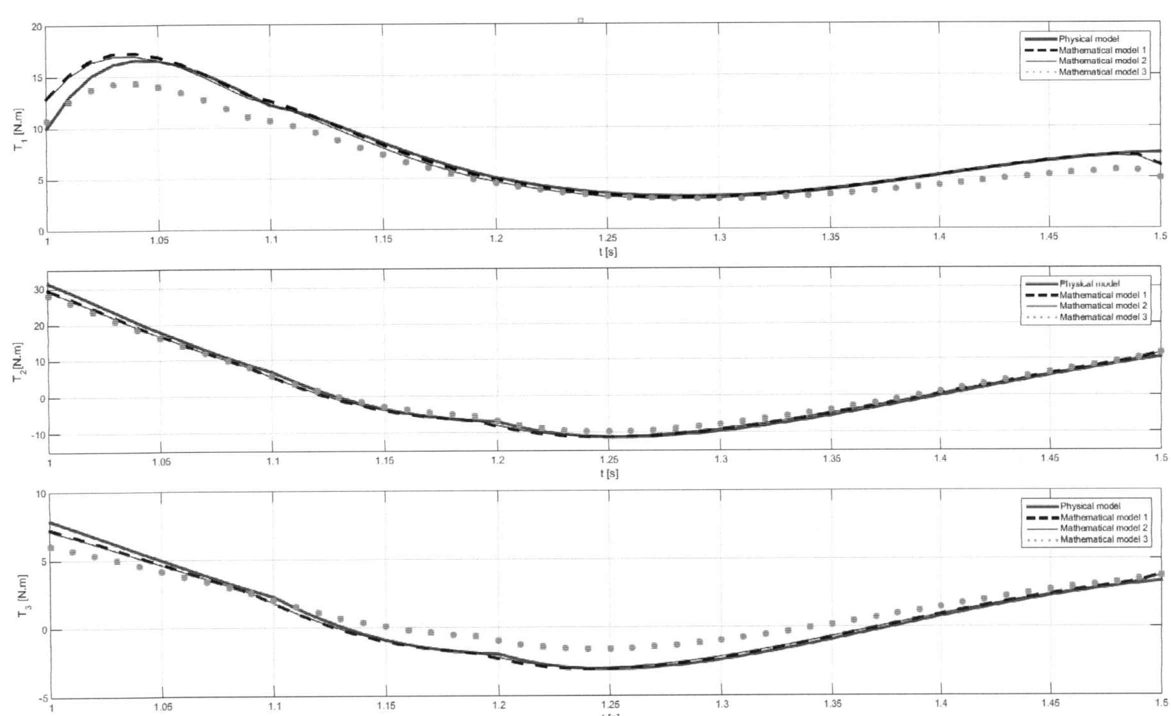

Fig. 6. The physical and the three different mathematical models comparison - detailed view

The differences between all the mathematical models and the physical model are clear now. The red thick solid line shows the physical model. The black dashed line associated with the first and the most accurate mathematical model almost overlaps the thin blue line related to the second model (shape simplification). All three of the above-mentioned lines are highly similar and the dynamic behaviour for all of them is essentially the same. The last green dot line associated with the third mathematical model (mass point simplification) is also noticeably similar, but this model should not be used for a simulation purpose, thus the real model substitution, because of its dynamic deviation compares to the physical model. The dynamic deflection is not crucial, but the other two mathematical models are better for this purpose for certain.

6. Conclusion

The effort in this work was devoted to the dynamic behaviour comparison of three different mathematical models. As could be found, everything starts with a CAD model. It is an essential part of every accurate dynamic model determination. The CAD model was exported to the SimMechanics physical model which is considered to be the most accurate replacement of a real physics system. All the dynamic behaviours of the mathematical models have been compared with the physical model to see how accurate the mathematical models are.

The first mathematical model derivation uses matrices of inertia from CAD model. That means an accurate description of mass distribution directly in motion equations. This mathematical model is the most accurate, but a physical interpretation is not clear during a derivation. Also, calculation errors are very difficult to find.

The second mathematical model uses an integration approach with every part simplification. This approach is the most complicated in terms of derivation, but it is intuitive and accuracy can be continuously explored. Its accuracy is, surprisingly, almost as precise as the first mathematical model. It means that the shape of any part is not decisive in the case of an appropriate approximation selection.

The third mathematical model derivation is similar to the second type, but all stationary objects are replaced with mass points. The accuracy of a mass distribution description is lowest and the shape of motion equations is not as simple as expected. However, the own motion equations derivation is the simplest.

Ultimate dynamic behaviour is highly similar to the first and second mathematical model and based on their accuracy; both mathematical models can be even used as simulation models. The third mathematical model has also satisfactory accuracy which is more than enough for the control law design but should not be used as a real model substitution for simulation purposes.

All things considered, the shape precision directly in motion equations has not a significant influence in a dynamic behaviour description as could be expected if a centre of mass position is known. Further work could be focused on the use of derived motion equations to the design control law and comparison of on that basis designed regulators accuracy.

7. References

[1] Denavit, J. & Hartenberg, R. S. (1955). ASME Journal of Applied Mechanics", A kinematic notation for lower-pair mechanisms based on matrices, Vol. 77, No. 22, June 1955, pp. 215-221.

[2] R. N. Jazar. (2010). Theory of Applied Robotics, Springer US, ISBN: 978-1-4419-1750-8, Boston, MA

[3] Damic, V. & Cohodar, M. (2016). Dynamic Analysis and Visualization of Spatial Manipulators with Closed Structure, Proceedings of the 26th DAAAM International Symposium, pp.0109-0117, B. Katalinic (Ed.), Published by DAAAM International, ISBN 978-3-902734-07-5, ISSN 1726-9679, Vienna, Austria, DOI: 10.2507/26th.daaam.proceedings.016

[4] Shala, A., Likaj, R., Bruqi, M. & Bajrami, X. (2014). Propulsion Effect Analysis of 3Dof Robot under Gravity, Proceedings of the 25th DAAAM International Symposium, pp.206-212, B. Katalinic (Ed.), Published by Elsevier Ltd, ISSN 1877-7058, Vienna, Austria, DOI: 10.1016/j.proeng.2015.01.359

[5] Dragomirescu, C. G. & Iliescu, V. (2010). About the Stability of the Motion of a Dynamic System in a Particular Case, Annals of DAAAM for 2010 & Proceedings of the 21st International DAAAM Symposium, 20-23rd October 2010, Zadar, Croatia, ISSN 1726-9679, ISBN 978-3-901509-73-5, Katalinic, B. (Ed.), pp. 0197-0198, Published by DAAAM International Vienna, Vienna

[6] Copilusi, C. P.; Marin, M. I. & Rusu, L. D. (2010). Design of a New Human Knee Prosthesis Based on CAM Mechanism, Annals of DAAAM for 2010 & Proceedings of the 21st International DAAAM Symposium, 20-23rd October 2010, Zadar, Croatia, ISSN 1726-9679, ISBN 978-3-901509-73-5, Katalinic, B. (Ed.), pp. 0709-0710, Published by DAAAM International Vienna, Vienna

[7] Shala, A. & Likaj, R. (2010). Fuzzy Logic Control and 3D Simulation of Road Vehicle, Annals of DAAAM for 2010 & Proceedings of the 21st International DAAAM Symposium, 20-23rd October 2010, Zadar, Croatia, ISSN 1726-9679, ISBN 978-3-901509-73-5, Katalinic, B. (Ed.), pp. 1457-1458, Published by DAAAM International Vienna, Vienna

[8] Ciortan, M. (2011). Kinematic Analysis of Jaw Movements, Annals of DAAAM for 2011 & Proceedings of the 22nd International DAAAM Symposium, 23-26th November 2011, Vienna, Austria, Volume 22, No. 1, ISSN 1726-9679, ISBN 978-3-901509-83-4, Katalinic, B. (Ed.), pp. 1599-1600, Published by DAAAM International Vienna, Vienna

[9] Damic, V. & Cohodar, M. (2014): Dynamic Analysis of Stewart Platform by Bond Graphs, 25th DAAAM International Symposium on Intelligent Manufacturing and Automation, 23-30th November 2014, Vienna, Austria, ISSN: 1877-7058, Procedia Engineering (2015), Volume 100, Katalinic, B. (Ed.), pp. 226 – 233, Published by DAAAM International Vienna, Vienna

[10] Vu, T. M.: Vehicle Steering Dynamic Calculation and Simulation, Annals of DAAAM for 2012 & Proceedings of the 23rd International DAAAM Symposium, ISBN 978-3-901509-91-9, ISSN 2304-1382, pp 0237 - 0242, Editor Branko Katalinic, Published by DAAAM International, Vienna, Austria, 2012

[11] Gebel, E. S. (2014): Dynamic Mathematical Modeling of Dynamics of Multi-Lever Linkages, 25th DAAAM International Symposium on Intelligent Manufacturing and Automation, 23-30th November 2014, Vienna, Austria, ISSN: 1877-7058, Procedia Engineering (2015), Volume 100, Katalinic, B. (Ed.), pp. 1562 – 1571, Published by DAAAM International Vienna, Vienna

[12] Aburaia, M.; Markl, E. & Stuja, K. (2014): New Concept for Design and Control of 4 Axis Robot Using the Additive Manufacturing Technology, 25th DAAAM International Symposium on Intelligent Manufacturing and Automation, 23-30th November 2014, Vienna, Austria, ISSN: 1877-7058, Procedia Engineering (2015), Volume 100, Katalinic, B. (Ed.), pp. 1364 – 1369, Published by DAAAM International Vienna, Vienna

[13] Isaenko, S.; Sochivko, O.; Shardyko, I. & Titov, V. (2016). Simulation of Manipulator as a Part of Space Robotic System Weightlessness Device, Proceedings of the 26th DAAAM International Symposium, pp.0554-0561, B. Katalinic (Ed.), Published by DAAAM International, ISBN 978-3-902734- 07-5, ISSN 1726-9679, Vienna, Austria DOI: 10.2507/26th.daaam.proceedings.076

[14] Damic, V.; Cohodar, M. & Tvrtkovic, M. (2016). Inverse Dynamic Analysis of Hobby Robot uArm by Matlab/Simulink, Proceedings of the 27th DAAAM International Symposium, pp.0095-0101, B. Katalinic (Ed.), Published by DAAAM International, ISBN 978-3-902734-08-2, ISSN 1726-9679, Vienna, Austria DOI: 10.2507/27th.daaam.proceedings.014

[15] Zatopek, J. (2016). Simulation and Visualisation of a Laboratory Model by Support Software Tools Connection, Proceedings of the 27th DAAAM International Symposium, pp.0749-0755, B. Katalinic (Ed.), Published by DAAAM International, ISBN 978-3-902734-08-2, ISSN 1726-9679, Vienna, Austria DOI: 10.2507/27th.daaam.proceedings.108

[16] Urednicek, Z. (2012). Robotika, T. Bata university in Zlin, ISBN: 978-80-7454-223-7, Zlin

[17] Gouasmi, M.; Ouali, M.; Fernini, B. & Meghatria, M. (2012): Kinematic Modelling and Simulation of a 2-R Robot Using SolidWorks and Verification by MATLAB/Simulink, International Journal of Advanced Robotic Systems, pp. 245-257, Volume 9, Issue 6, DOI: 10.5772/50203

[18] Udai, A. D.; Rajeevlochana, CG & Saha, S. K. (2011): Dynamic Simulation of a KUKA KR5 Industrial Robot using MATLAB SimMechanics, 15th National Conference on Machines and Mechanisms, pp. 1-8, 30th November to 2nd December 2011, Chennai, Tamilnadu, India, NaCoMM2011-96

[19] TLALE, N. S. & ZHANG P. (2005): Teaching the Design of Parallel Manipulators and Their Controllers Implementing MATLAB, Simulink, SimMechanics and CAD*, Int. J. Engng Ed. Volume 21, No. 5, pp. 838-845, 0949-149X/91

[20] Aldoom, S.; Adam, A.; Ji-Pin, Z. & Yi-hua, Z. (2017): Modeling and Simulation of 5DOF Robot Manipulator and Trajectory using Matlab and CATIA, 3rd International Conference on Control, Automation and Robotics, pp. 1-5, ISBN: 978-1-5090-6088-7, DOI: 10.1109/ICCAR.2017.7942657

DOI: 10.2507/28th.daaam.proceedings.097

EVALUATION OF THE ONTOLOGICAL METHOD OF INTEGRABLE DATA INCONSISTENCY IDENTIFICATION

Alexander A. Kropotin, Yuliya V. Bidulya, Alexander G. Ivashko

This Publication has to be referred as: Kropotin, A[lexander]; Bidulya V., Y[uliya] & Ivashko G., A[lexander] (2017). Evaluation of the Ontological Method of Integrable Data Inconsistency Identification, Proceedings of the 28th DAAAM International Symposium, pp.0694-0697, B. Katalinic (Ed.), Published by DAAAM International, ISBN 978-3-902734-11-2, ISSN 1726-9679, Vienna, Austria
DOI: 10.2507/28th.daaam.proceedings.097

Abstract

This paper evaluates the performances of the method of description logics implementation to identify semantic conflicts and inconsistency of integrable data in the context of relational databases integration presented by entity-relationship diagrams during information systems integration.
At first, we present a developed method method of description logics implementation to identify semantic conflicts and inconsistency of integrable data which based on original mathematical model and algorithms of description logics calculus applying.
We split evaluating into two sets. In the first set of experiments we compared how the classification and consistency checking of the ontology terminology could be applied for the identify semantic conflicts of integrable data. In the second set of experiments the consistency checking of the ontology was estimated for the identify semantic conflicts and inconsistency of integrable data.

Keywords: Ontology; description logics; database; ER-model

1. Introduction

The development of information technology has led to the emergence of a huge number of diverse information systems. As a rule these systems are designed to automate a business processes, to store the personal and reference data, and to generate different reports based on stored data [1]. It often occurs that the data in these systems duplicate each other, which creates big problems in their integration. That is why ensuring interoperability and the integration of information systems are the most urgent tasks [2].

Due to the fact that the volume of information processed is constantly growing, information technology specialists need to provide an effective data exchange between information systems avoiding the duplication and loss of information [1; 2; 3]. Increasingly, information systems developers are faced with many weak linked and disparate sources of information and they have to constantly perform routine actions to perform low-level operations with data in a heterogeneous environment. A special place is occupied by questions of data integration, which may not be consistent and lead to the emergence of structural and semantic conflicts. Traditional methods for solving this problem mainly based on linking between the elements of systems constructed manually, and are aimed at solving the problem for each specific case [1; 2; 4]. In this case, a change in the structure of one of the systems may cause the disruption of the program complex and requires re-integration [1; 3]. Thus, traditional methods do not provide a sufficiently flexible solution for creating the data integration system [2; 4].

At the present time, there are many methods for solving this problem, but most effectively is to apply the technologies of the semantic web and related data [1; 2; 3; 4; 8; 9; 10; 14].

These technologies have special standards for machine representation of knowledge bases and software products that implement inference algorithms [5]. To solve this problem, we developed the method of applying the descriptive logics formalism to identify automatically conflicts of entity-relationship conceptual schemes of relational databases [7; 14; 15]. The purpose of this paper is to evaluate the developed method using computational experiments.

2. Method description

The developed method of applying the formalism of descriptive logics [13; 14; 15] is based on the previously developed mathematical model of the ontological representation of relational databases based on their semantic description and the method of applying the calculus of descriptive logics to identify the conflicts of these schemes.

As a mathematical model of the ontological representation of relational databases based on their semantic description, the ontological model of conceptual objects was previously proposed [7; 14; 15], which provides rules for describing of conceptual schemes of the entity-relationship relational databases as axioms and statements of descriptive logic. These rules are based on the syntactic similarity of relational databases and ontologies, are substantiated by their semantic difference and are aimed at ensuring the possibility of identifying semantic conflicts of metadata and inconsistencies of integrable data through checking the feasibility of their ontologies. In accordance with the ontological model rules for representing conceptual objects, the elements of the entity-relationship schemes of databases are represented as follows [7; 14; 15]:

- each set of entities and each set of values are represented as the atomic concept;
- each set of links and attribute are represented atomically as role;
- specific elements of such sets are represented as corresponding individuals.

The method of applying the calculus of descriptive logics for identifying conflicts of entity-relationship conceptual schemes of relational databases allows us to reduce the process of identifying of semantic conflicts of metadata and the inconsistency of data which occurs when their integration, to solving of algorithmic problems of consistency and classification of terminology, and the consistency of the ontology of these data [7; 14; 15]. It should be noted that under the consistency of ontology or terminology is understood as the existence of the given ontology or terminology model in which all concepts and roles are not empty.

3. Evaluating

The applicability of the ontological model of conceptual objects and the method for identifying semantic conflicts and the inconsistency of the integrable data were experimentally evaluated of the using 861 conceptual database schemas that represent 368 possible cases of conflicts: attribute-attribute, attribute-entity, entity-relationship, dependency, behavior, inheritance, key, data types, acceptable values, permitted values, and the relevance of values.

This evaluation included three stages:

1. Preparation of experimental data, represented by sets of conceptual schemas of the entity-relationship of relational databases, each of which represents different cases of occurrence of a corresponding conflict of two such schemes. Based on the proposed ontological model for representing conceptual objects and the algorithm for constructing terminological axioms and statements based on the semantic representation of the relational databases of each of such a set, ontologies are constructed according to the OWL 2.0 standard.
2. Conducting computational experiments, focused on the tabular algorithm tasks using the experimental data and revealing the conflicts that they simulate in accordance with the proposed way of applying the formalism of descriptive logics to identify semantic conflicts and data discrepancy of their integration. There are tasks performing:
   - the consistency checking and terminology classification to identify conflicts: attribute-attribute, attribute-entity, entity-relationship, behavior, inheritance and key;
   - the consistency of the ontology entirely to detect conflicts of the format of values, the presentation of values and the relevance of values;
   - the consistency of terminology and ontology consistency to identify dependency conflicts, data types, acceptable values, permitted values.
3. Evaluation of the computational experiment results, which are based on a comparison of the number of actual semantic conflicts of entity-relationship conceptual schemas, and ones revealed during the computational experiments provided for corresponding inconsistencies in terminological axioms and / or statements.

The essence of the computational experiments consists in solving algorithmic problems of ontology consistency, consistency and terminology classification on sets of experimental data, and in identified inconsistencies calculation.

In accordance with the conflict classification of the conceptual schemes, the experimental data sets for the experimental evaluation were divided into two groups: the metadata conflict group and the data conflict group.

As a result of the experimental evaluation, on the sets of experimental data of the group of metadata conflicts, in all, 813 databases were processed and 328 cases of conflicts of their metadata from 350 possible were revealed. The results of the experimental evaluation on the sets of experimental data of the group of metadata conflicts are given in Table 1.

Experimental data		**Number of semantic conflicts**	
Conflict type	**Database number**	**Actual**	**Revealed**
Attribute-attribute	68	26	24
Attribute-entity	40	17	17
Entity-relationship	132	56	40
Dependencies	20	8	8
Behavior	14	7	5
Inheritance	12	6	4
Key	24	8	8
Data types	424	210	210
Acceptable values	16	6	6
Permitted values	16	6	6
Total	**813**	**350**	**328**

Table 1. The results of the experiment on the sets of experimental data of the group of metadata conflicts.

As a result of the experimental evaluation, on the sets of experimental data of the data conflict group, in all, 48 databases were processed and 18 cases of conflicts emergence out of 18 possible. The results of the experimental evaluation on the sets of experimental data of the data conflict group are given in Table 2.

Experimental data		**Number of semantic conflicts**	
Conflict type	**Database number**	**Actual**	**Revealed**
Format of values	16	6	6
Presentation of values	16	6	6
Relevance of values	16	6	6
Всего	**48**	**18**	**18**

Table 2. The results of the experiment on sets of experimental data of the data conflict group.

An experimental evaluation of the developed method of applying the descriptive log formalism to automate the process of identifying the conflicts of conceptual schemes of the entity-relationship of relational databases clearly showed that:

1. The developed model of the ontological representation of conceptual relational databases is oriented to ontological modeling of only the semantic component of such bases, excluding their structural representation.
2. The proposed method of using the descriptive calculus to identify the semantic conflicts of database schemas and data inconsistencies allows us to identify only the semantic inconsistency of such schemes, ignoring their structural differences. Such applicability of the method is due to the tasks and results of research into integration problems, which showed that in view of the semantic coherence, structural heterogeneity is solvable by standard data integration tools.
3. The proposed method cannot be applied for rigid typing of data and does not extend to conflicts formed by differences only in the cardinality constraints of equivalent link sets. This limitation of the method is due to various assumptions about the openness and closure of the world, which is accepted in ontologies and databases, respectively.

4. Conclusion

The paper presents the results of an experimental evaluation of the applicability of the developed method of using the descriptive log formalism to automate the process of identifying conflicts of conceptual schemes of the essence-relationship of relational databases.

The experimental estimation was carried out on 861 database on conflict detection: attribute-attribute, attribute-essence, entity-relationship, dependencies, behavior, data types, allowed and allowed values, value format, representation of values and relevance of values.

The results of the experiments showed that the proposed method of identifying semantic conflicts and inconsistency of integrable data, based on solving algorithmic problems of consistency and classification of terminology, and the consistency of ontology, makes it possible to reveal 95% of cases of different occurrence of semantic inconsistency.

The applicability of the proposed method for identifying semantic conflicts and the inconsistency of integrable data that is impossible with rigid data typing is determined experimentally and does not extend to conflicts formed by differences in only the cardinality constraints of equivalent sets of constraints.

In the future work it is planned to develop a model for the ontological representation of business processes and its further comparison with the ontological representation of databases in order to identify the correspondence of the database to the business process and vice versa. This approach will avoid the occurrence of many conflicts and inaccuracies when designing a database for each particular business process.

5. References

[1] Bubareva, O. A. (2014). Model. algorithms and software for data integration of information systems based on ontologies (on the example of the university) [Text]: the thesis for the degree of Candidate of Technical Sciences: 05.13.11: protected 24.03.2015 / Bubareva Olesya. - Biysk. - 137 p.

[2] Semarkhanov, I. A. (2014). Methods and algorithms for the automated integration of information resources on the basis of the ontological approach [Text]: the thesis for obtaining the scientific degree of Cand.Tech.Sci .: 05.13.12: protected 18.06.2014 / Semerkhanov Ilya Aleksandrovich. - St. Petersburg. - 140 p.

[3] Kogalovsky, M. R. (2003). Methods of Data Integration in Information Systems [Text] / M. R. Kogalovsky // Collected Works of the Third All-Russian Conference "Standards in Modern Information Systems Projects". Moscow, Russia. - 23-24 April. - P. 1-8.

[4] Semerkhanov, I. A. (2013). Integration of Information Systems with Related Data [Text] / IA Semerkhanov, DI Muromtsev // Scientific and Technical Herald of Information Technologies, Mechanics and Optics. - No5 (87). - P. 123-128.

[5] Grigoriev, A. V. (2013). Mathematical methods and algorithms for determining the consistency of knowledge bases [Text]: the thesis for the degree of Candidate of Technical Sciences: 05.13.18: protected 20.12.2013 / Grigoriev Andrey Viktorovich. - Tyumen. - 115 p.

[6] Zolin, E. E. Chapter 3 [Electronic resource]: Facts and ABox / EE Zolin // Descriptive logic (lectures). - Access mode: http://lpcs.math.msu.su/~zolin/ dl / pdf / DL_03_ABox.pdf. - 11/05/2016.

[7] Zolin, E. E. Chapter 6 [Electronic resource]: Extensions of logic / EE Zolin // Descriptive logic (lectures). - Access mode: http: // lpcs.math.msu.su/~zolin/dl/pdf/DL_06_ALCOIQ.pdf. - 11/05/2016.

[8] Kropotin, A. A. (2013) Application of the table algorithm for verification of business processes [Text] / A. A. Kropotin, A. G. Ivashko, A. V. Grigoriev, E. O. Ovsyannikova // Bulletin of the Tyumen State University. - No. 7. P. 202-213.

[9] Semerkhanov, I. A. (2013). Integration of relational databases using rdf \ owl [Text] / IA Semerkhanov, GV Wargin // Scientific and Technical Herald of Information Technologies, Mechanics and Optics. - No 1. St. Petersburg: NIUITMO. - P. 117-118.

[10] Kropotin, A. A. (2016). Database Schema Method for Automatic Semantic Errors Resolving During Information Systems [Текст] / A. A. Kropotin, A. V. Grigoryev, A. G. Ivashko // Integration, Informal Proceedings of the 2016 10th IEEE International Conference on Application of Information and Communication Technologies, AICT2016. Azerbaijan, Baku. – 12-14 October.

[11] Kropotin, A. A. (2016). Realization of the Ontologically Based Method for Checking Structural Inconsistences of Relational Databases [Текст] / A. A. Kropotin, A. V. Grigoryev, Y. V. Bidulya, A. G. Ivashko, N. S. Durynin // Proceedings of the 27th International DAAAM Symposium on Intelligent Manufacturing and Automation 2016. – vol. 27. Mostar, Bosnia and Herzegovina. – 26-29 October. – P. 762-767. – DOI 10.2507/27th.daaam.proceedings.110.

[12] Kropotin, A. A. (2016). Application of formalism of description logic for revealing the semantic conflicts of conceptual schemes of essence-connection [Text] / AA Kropotin // Journal: INTELLIGENCE. INNOVATION. INVESTMENTS. Orenburg State University. - Issue 7. - P. 93-98. - ISSN 2077-7175.

[13] Connolly, T. (2003). Databases. [Text]: Design, implementation and maintenance. Theory and practice / Thomas Connolly, Carolyn Begg; translated from English and edited by R. Imamutdinov, K. Ptitsyn. - 3 edition. - M .: Publishing house "Villamé". - 1436 p. - Translated from: Database Systems: A Practical Approach to Design, Implementation, and Management Third Edition / T. Connolly, C. Begg. - 3 edition. - ISBN 5-8459-0527-3, 0-201-70857-4 (in translation).

[14] Chen, P. P-S. (2002). The Entity Relationship Model — Toward a Unified View of Data [Текст] / P. P-S. Chen // Software Pioneers. Contributions to Software Engineering. – Springer Berlin Heidelberg. – P. 311-339. – ISBN 978-3-642-59412-0. – DOI 10.1007/978-3-642-59412-0_18.

[15] Codd, E. F. (1970). Extending the Database Relational Model to Capture More Meaning [Текст] / E. F. Codd. // ACM Transactions on Database Systems. – vol. 4. – issue 4. – P. 397-434. – DOI 10.1145/320107.320109.

DOI: 10.2507/28th.daaam.proceedings.098

YARN UNWINDING FROM PACKAGES

Predovič Aleksander

This Publication has to be referred as: Predovic, A[leksander] (2017). Yarn Unwinding from Packages, Proceedings of the 28th DAAAM International Symposium, pp.0698-0703, B. Katalinic (Ed.), Published by DAAAM International, ISBN 978-3-902734-11-2, ISSN 1726-9679, Vienna, Austria
DOI: 10.2507/28th.daaam.proceedings.098

Abstract

The Yarn unwinding from a package is important in many textile processes. The stability of the unwinding process has a direct influence on the efficiency of the process and on the quality of the end product. During the unwinding, the tension is oscillating. This is especially noticeable in over-end unwinding from a static package, where the yarn is being withdrawn with a high velocity in the direction of the package axis.
The optimal form of the package allows an optimal shape of the yarn balloon and low and steady tension even at very high unwinding velocities. Using experimental data as an input, we will calculate oscillations of yarn tension for different package geometries and winding angles. These results will be used to suggest a design new packages.

Keywords: yarn unwinding; balloon theory; simulations; design new packages

1. Introduction

Oscillations in the yarn during the yarn unwinding from stationary packages have a direct influence on the quality of the fabric. The characteristics of the unwinding process are thus important for production of high quality garments and should therefore be optimized. All authors (since 1950 [1-4] to 2001 [4-12]) who developed theory of unwinding yarn from packages, used a few assumptions, which help them to set the winding angle of the packages on value 0° ($\Phi = 0°$). In practical way, the winding angle in winding machines has values from 12° to 22°. For this reason, it is almost importance to determine the package geometry and the winding angle which allow to maximize the unwinding velocity given the allowed highest oscillations in the yarn.

2. Theoretical part

Angle ϕ is the winding angle of the yarn on the package. Its magnitude is the angle between the yarn at the unwinding point and the tangential direction at the package surface, while its signs determines whether the yarn is being withdrawn in the direction from the front end toward the rear end of the package ($\phi>0$) or from the rear end toward the front end ($\phi<0$) (Fig. 1).

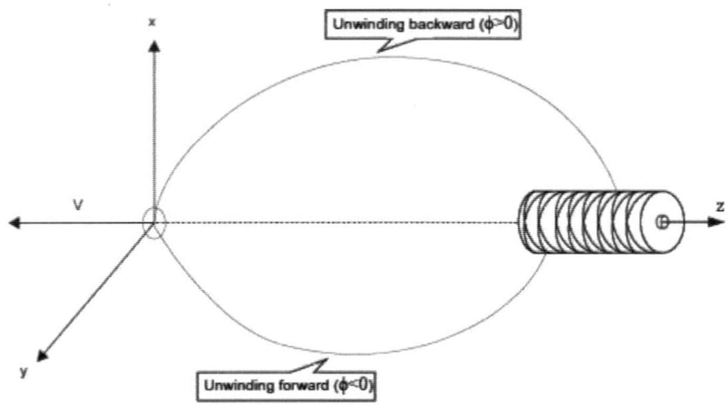

Fig. 1. Winding angle during the yarn unwinding from package

The general equation of motion which govern the motion of the yarn are known [13-40]:

$$\rho\left(D^2\mathbf{r}+2\boldsymbol{\omega}\times D\mathbf{r}+\boldsymbol{\omega}\times\left(\boldsymbol{\omega}\times\mathbf{r}\right)+\dot{\boldsymbol{\omega}}\times\mathbf{r}\right)=\frac{\partial}{\partial s}\left(\mathrm{T}\frac{\partial\mathbf{r}}{\partial s}\right)+\mathbf{f}. \tag{1}$$

Here ρ is the mass per unit length of the yarn, **r** is the radius vector to a given point on the yarn, D is the co-moving time derivative operator, T is the yarn tension and **f** is the density of external forces acting on the yarn.

Equation of motion expressed in terms of the dimensionless quantities [7].

$$\begin{aligned}&\bar{\mathbf{r}}=\mathbf{r}=/c,\bar{r}=r/c,\bar{z}=z/c,\ \bar{s}=s/c,\\&\bar{t}=t/\tau=\omega t,\bar{\mathbf{v}}=\mathbf{v}/V,\bar{\mathbf{v}}_n=\mathbf{v}_n/V,\\&\bar{\mathbf{f}}=\frac{\mathbf{f}c}{\rho V^2},\bar{\mathbf{n}}=\frac{\mathbf{n}c}{\rho V^2},\bar{T}=\frac{T}{\rho V^2}.\end{aligned} \tag{2}$$

When we rewrite the equation of motion with these dimensionless quantities we obtain

$$\bar{D}^2\bar{\mathbf{r}}+2\Omega\times\bar{D}\bar{\mathbf{r}}+\Omega\times\left(\Omega\times\bar{\mathbf{r}}\right)+\Omega\frac{\partial\Omega}{\partial\bar{t}}\times\bar{\mathbf{r}}=\frac{\partial}{\partial\bar{s}}\left(\bar{T}\frac{\partial\bar{\mathbf{r}}}{\partial\bar{s}}\right)+\bar{\mathbf{f}} \tag{3}$$

Only one parameter remains in this equation it is the dimensionless angular velocity $\Omega=\omega c/V$.

On cylindrical packages the angular velocity depends on the winding angle Φ according to relation [5,6]:

$$\omega=\frac{2\pi}{t}=\frac{V\cos\Phi}{c\,(1-\sin\Phi)} \tag{4}$$

The dimensionless angular velocity can obviously be expressed as:

$$\Omega=\frac{\cos\Phi}{1-\sin\Phi} \tag{5}$$

According to our simple model the dimensionless angular velocity thus only depends on the winding angle which will change with time because this angle is different for layers that are unwinding from front towards rear edge and those that are unwinding as the unwinding point moves from the rear towards front edge. The dependence Ω on the winding angle is shown in Fig. 2.

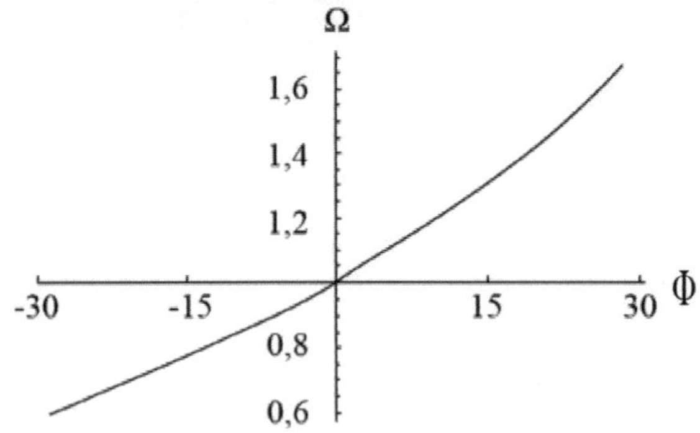

Fig. 2. Dependence of Ω on the winding angle [27-31].

During unwinding the lift-off point moves up and down the package. We can presume that the winding angle is approximately constant in the middle of the package and it changes at the edges of the package where its sign is reversed. To describe the time dependence of the winding angle we must look for a periodic function, because motion of the point is periodic to a good approximation.
The most known periodic functions are trigonometric function, such as sine function. This function should be modified so that it will change only slightly when the point moves up or down the packages. We can achieve this by raising the sine to a low fractional power, say 1/40 (we have to be careful about the signs, so we take absolute value of sine function and restore the sign using the signum function [26]:

$$f(t) = sign(\sin t)\left|\sin t\right|^{\frac{1}{40}} \tag{6}$$

3. Practical part

We measured how the tension at the eyelet depends on the angular velocity of unwinding. We measured the unwinding speed, package radius and tension of yarn as we unwound yarn from parallel cylindrical packages. We used a balloon limiter, that limits the balloon radius.

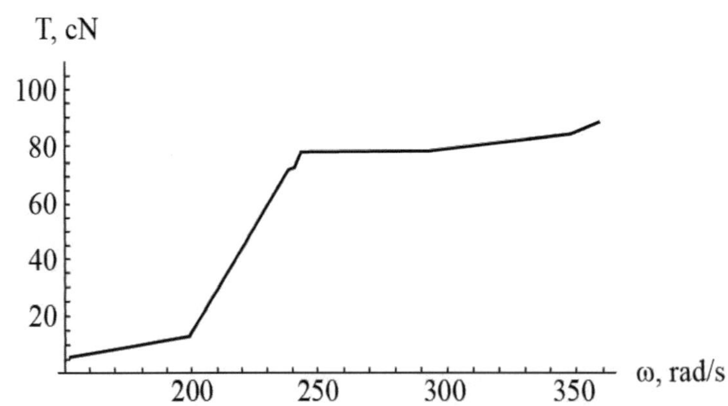

Fig. 3. Dependence of tension T on angular velocity ω of the yarn [37].

We must emphasize that this dependence was obtained by measurements on a simple, parallel wound package. In split of this we will use these date in our simulations of cross-wound packages. This approximation has little influence on the qualitative features of our results.

In our simulation, we calculate the winding angle using function (6), then we determine the corresponding angular velocity ω and finally we obtain an approximation for the tension using data from Fig. 3.

In our calculations, we considered unwinding for two consecutive layers of yarn, so that the package radius remains approximately constant during this time.

Figure 4 presents the changing tension in the yarn as we unwind yarn from a cylindrical package. The time is expressed in units of phase: 2π corresponds to one cycle of unwinding point up and down the package.

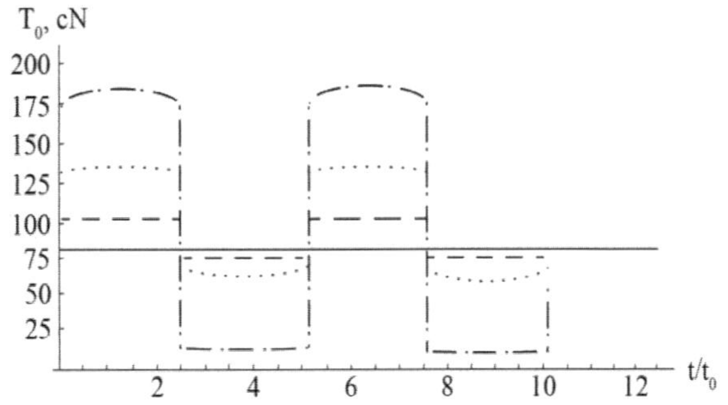

Fig. 4. Variation of the tension during unwinding, V=1400 m/min, c=70 mm. Φ~0° ,10° ,20° ,30°.

In a, cross-wound package with a very small Φ=0 the tension is constant, since in such packages the angular velocity doesn't change with time. In packages with higher winding angle Φ, the tension does change with time. This happens because the angular velocity is higher when the unwinding point moves backwards as it is when the unwinding point moves forward. The tension is a function of angular velocity, so it is oscillating. When the direction of unwinding changes near the edges of the package, the yarn tension undergoes a rapid change. Such sudden jumps lead to strong strain in the yarn, the yarn can be damaged or even broken in two parts.

Figure 5 depicts how the amplitude of oscillations depends on the package radius and winding angle:

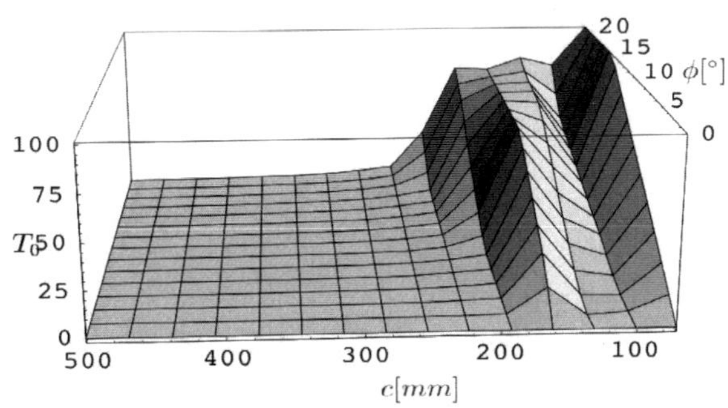

Fig. 5. Amplitude of the tension oscillation, V=2000 m/min.

3.1. New cross-wound packages

To reduce the tension oscillations, we devised new cross-wound packages. The package should be prepared so that the parallel wound layers are unwound when the lift-off point is moving backwards, and the interstitial cross wound layers are unwound when the lift-off point is moving forwards. This choice is motivated by the observation that cross wound packages lead to very high tensions when the yarn is unwound backwards. In addition, cross wound layers between two parallel wound layers are necessary to reduce the possibility of slips during unwinding.

In Fig. 6 we compare the amplitude of oscillations in regular and new packages. This amplitude is higher for old packages at all velocities and winding angles. The amplitude attains very high values in old packages while it is substantially lower in the new cross-wound packages.

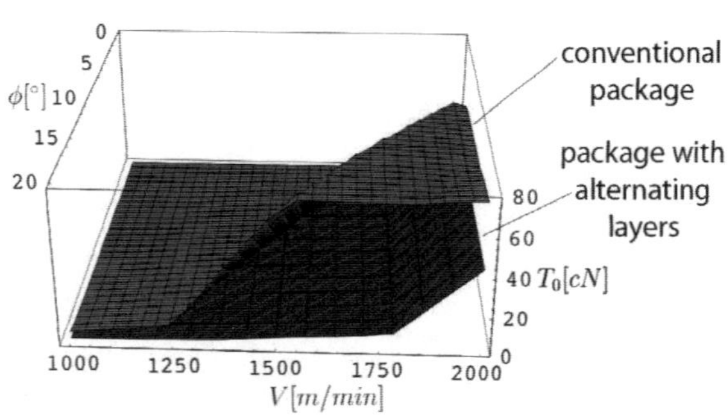

Fig. 6. Comparison of tension oscillation in regular (blue) and new packages (red), c=120 mm.

4. Conclusion

We have derived the expressions for the time dependence of the angular velocity of the balloon rotation around the package axis, of the unwinding velocity and the winding angle during the yarn unwinding from cylindrical packages. The expressions allow to study the motion of the lift-off point on the package surface as we have shown. This method makes it possible to consider more general package designs where the package radius and winding angle are arbitrary functions which describe the package form and the type of winding of yarn in individual layers. This procedure allows to simulate yarn unwinding from general packages and to determine the package construction with the desired properties. We have thereby established that the parallel and the cross-wound packages differ by the absence or presence of the characteristic variations of the angular velocity and consequently the yarn tension during the unwinding. The problem of high yarn tension and its high oscillations can be avoided by constructing packages, where two types of layers are alternating, parallel layers and layer with higher winding angle. The alternating design helps to reduce sudden changes of tension, and it leads to higher stability of the unwinding process. In the near future, we would like and hopefully will implement our knowledge in the textile production and try to find out, how all research fields affect quality.

5. References

[1] Mack, C. (1953). Theoretical study of ring and cap spinning balloon curves. J.Text.Inst., 44, pp. 483-498, ISSN 00405000

[2] Padfielf, D.G (1956). A note on fluctuations of tension during unwinding. J.Text.Inst., 47, pp. 301-308, ISSN 00405000

[3] Padfield, D. G. (1958). The Motion and tension of an unwinding thread. Proc. R. Soc., A 245, pp. 38 2–407, ISSN 1364-5021

[4] De Barr, A. E., Catling, H. (1965). Manual of cotton spinning. Vol.5, The principles and theory of ring spinning. Textile Institut. Manchester, ISSN 00405000

[5] Kothari, V. K., Leaf, G. A. V. (1979). The unwinding of yarns from packages. Part I, The theory of yarn-unwinding. J. Text. Inst., 3, pp. 89-95, ISSN 00405000

[6] Kothari, V. K., Leaf, G. A. V. (1979). The unwinding of yarns from packages, Part II: Unwinding from cylindrical packages. J. Text. Inst., 3, pp. 96-104, ISSN 00405000

[7] Fraser, W. B., Ghosh, T. K., Batra, S. K. (1992). On unwinding yarn from cylindrical package. Proc. R. Soc., A, 436, pp. 479–498, ISSN 1364-5021

[8] Fraser, W. B. (1992). The effect of yarn elasticity on an unwinding ballon. J. Tex. Inst., 83, pp. 603-613, ISSN 00405000

[9] Fraser WB. (1993). On the theory of ring spinning. Phil. Trans. R. Soc. London, A342, pp. 439-468, ISSN 1364-5021

[10] Clark, J.D., Fraser, W.B., Sharma, R. Rahn, C.D. (1998). The dynamic response of a ballooning yarn: theory and experiment. Proc. R. Soc. Lond. A, 454, pp. 2767-2789, ISSN 1364-5021

[11] Fraser WB and Stump DM. (1998). Yarn twist in the ring-spinning balloon. Proc R Soc London, A454, pp. 707-723, ISSN 1364-5021

[12] Clark JD, Fraser WB and Stump DM. (2001). Modelling of tension in yarn package unwinding. J Engi Mathe, 40, pp. 59-75, ISSN 0022-0833

[13] Praček, S. Jakšić, D. (2002). Theory of yarn unwinding off a package and derivation of differential equations.1, Tekstilec, 45, 5/6, pp. 119-123, ISSN 0351-3386

[14] Praček, S, Jakšić, D. (2002). Theory of yam unwinding off a package. 2, Tekstilec, 45, 7/8, pp. 175-178, ISSN 0351-3386

[15] Praček, S. (2004). Mathematical model for simulating yam unwinding from packages.1, Tekstilec, 47, 9/12, pp. 289-291, ISSN 0351-3386

[16] Praček, S. (2004). Sliding motion of yam on the package during unwinding.1, Derivation of equations. 47, 1/2, pp. 26-28, ISSN 0351-3386

[17] Praček, S. (2004). Sliding motion of yam on the package during unwinding. 2, Tekstilec, 47, 7/8, pp. 229-231, ISSN 0351-3386

[18] Praček, S. (2005). Mathematical model for simulating yam unwinding from packages. 2, Tekstilec, 48, 7/9, pp. 173-176, ISSN 0351-3386

[19] Praček, S. (2005). Virtual forces in the mathematical model of yam unwinding. 1, Tekstilec, 48, 10/11, pp. 252-254, ISSN 0351-3386

[20] Praček S, Jakšić D. (2005). Yarn unwinding from package s - discussion of kinematic and dynamic properties of yarn. Strojniški Vestnik, 51, 2, pp. 74-89, ISSN 0039-2480

[21] Praček, S. (2006). Virtual forces in the mathematical model of yam unwinding. 2, Tekstilec, 49, 1/3, pp. 5-7, ISSN 0351-3386

[22] Praček S. (2007). Theory of string motion in the textile process of yarn unwinding. International journal of nonlinear sciences and numerical simulation, 8. 3, pp. 451-460, ISSN 1565-1339

[23] He, Ji-Huan; Zhang, Li-Na (2008). On Convergence Point of the Two-strand Yarn Spinning Textile Research Journal. vol. 78,11, pp.1022-1024, ISSN 0040-517
[24] Yang, Rui-Huan; Wang, Shan-Yuan. (2009). Determination of the Convergent Point in the Rotor-spun Composite Yarn Spinning Process Textile Research Journal. vol. 79, 6, pp.555-557, ISSN 0040-517
[25] Praček, S. and Sluga, F. (2010). Numerical simulations of yarn unwinding from packages. Mathematical and computational applications in science and engineering, vol. 15, no. 5, pp.846-852, ISSSN 1300-686X
[26] Praček, S. Sluga, F. (2010). Matematical model for yarn unwinding from packages. Mathematical and computational applications in science and engineering, vol. 15, no. 5, pp.853-858, ISSN 1300-686X
[27] Praček, S. (2010). Theoretical model of unwinding process from packages, Annals of DAAAM for 2010 & Proceedings of the 21 st International DAAAM Symposium, 20-23rd October 2010, Zadar, Croatia, ISSN 1726-9679, ISBN 978-3-901509-73-5, Katalinic, B. (Ed), pp. 1469-1470, Published by DAAAM International Vienna, Vienna
[28] Praček, S. (2010). High-speed yarn transport systems simulation, Annals of DAAAM for 2010 & Proceedings of the 21 st International DAAAM Symposium, 20-23rd October 2010, Zadar, Croatia, ISSN 1726-9679, ISBN 978-3-901509-73-5, Katalinic, B. (Ed), pp. 1471-1472, Published by DAAAM International Vienna, Vienna
[29] Praček, S. Sluga, F. Možina, K. Franken, G. (2011). Unwinding from cylindrical packages, Annals of DAAAM for 2011 & Proceedings of the 22 st International DAAAM Symposium, 23-26th November 2011, Vienna, Austria, ISSN 1726-9679, ISBN 978-3-901509-83-4, Katalinic, B. (Ed), pp. 0917-0918, Published by DAAAM International Vienna, Vienna
[30] Praček, S. Sluga, F. Možina, K. Franken, G. (2011). Cylindrical packages simulations, Annals of DAAAM for 2011 & Proceedings of the 22 st International DAAAM Symposium, 23-26th November 2011, Vienna, Austria, ISSN 1726-9679, ISBN 978-3-901509-83-4, Katalinic, B. (Ed), pp. 0919-0920, Published by DAAAM International Vienna, Vienna
[31] Praček, S. Sluga, F. Možina, K. Franken, G. (2011). Conic packages oscillations, Annals of DAAAM for 2011 & Proceedings of the 22 st International DAAAM Symposium, 23-26th November 2011, Vienna, Austria, ISSN 1726-9679, ISBN 978-3-901509-83-4, Katalinic, B. (Ed), pp. 0921-0922, Published by DAAAM International Vienna, Vienna
[32] Praček, S. Možina, K. Sluga, F. (2012). Yarn motion during unwinding from packages. Math. comput. model. dyn. syst., vol. 18, 6, pp. 553-569, ISSN 1387-3954
[33] Praček, S. (2013). Theory of yarn dybamics during unwinding from packages. Adv. mat. res. vol. 601, pp. 275-279, ISSN 1662-8985
[34] Praček, S, Sluga, F. Možina, K. (2012). Mathematical model for yarn unwinding.1, Adv. mat. Res., vol. 403-408, pp. 5131-5135, ISSN 1662-8985
[35] Praček, S. Sluga, F. Možina, K. (2012). Mathematical model for yarn unwinding. 2, Adv. mat. res., vol. 403-408, pp. 5136-5141, ISSN 1662-8985
[36] Praček, S. (2013). Model for yarn transport systems. Adv. mat. res., vol. 601, pp. 237-241, ISSN 1662-8985
[37] Praček, S. Nace, P. Simončič, B. Petra, Eva, F.T. (2015). Model for simulating yarn unwinding from packages. Fibres &Textiles in Eastern Europe.,23.2(110), pp.25-32, ISSN 1230-3666
[38] Praček S, Pušnik N, Franken G, and Simončič B. (2016). Balloon theory of yarn during unwinding from packages. Text. Res. J., 86, 14, pp. 1522-1532, ISSN 0040-5175
[39] Praček, S. Pušnik, N. (2016). Fictitious forces in yarn during unwinding from packages. Textile research journal, ISSN 0040-5175
[40] Pušnik, N. Praček, S. (2016). The effect of winding angle on unwinding yarn. Transactions of famena, vol. 40, iss. 3, pp. 29-42, ISSN 1333-1124

DOI: 10.2507/28th.daaam.proceedings.099

STRATEGIC MANAGEMENT AND MULTIPLIER EFFECT

Milan Jemelka, Bronislav Chramcov

This Publication has to be referred as: Jemelka, M[ilan] & Chramcov, B[ronislav] (2017). Strategic Management and Multiplier Effect, Proceedings of the 28th DAAAM International Symposium, pp.0704-0707, B. Katalinic (Ed.), Published by DAAAM International, ISBN 978-3-902734-11-2, ISSN 1726-9679, Vienna, Austria
DOI: 10.2507/28th.daaam.proceedings.099

Abstract

Many companies are addressing the low performance of individual employees. Management of multiple companies intend to prepare individual coaching programs for employee development. Before going into this work, on one side it is the development and improvement of the employees' results, but the other side is the overall strategic management of the company and the related set of rules within the company that often influences this performance. The company's overall analysis should be conducted and the results subsequently confirm these assumptions. Human mistakes account for a maximum of 15% on the causes of problems in the company and unsatisfactory results. 85% of mistakes come from a poorly configured system, a lack of strategic management. This thesis is properly simulated in the program Witness from Lanner group. In this paper, we are concerned with discovering bottlenecks in production and with the multiplier effect aimed at increasing production. All simulation results are presented and explained.

Keywords: Simulation tool; Simulation; Witness program; Employee performance; Strategic management; Multiplier effect

1. Introduction

Qualified employees have recently become scarce goods. The economy is doing well, and one who's smart usually has a job or has so many offers to choose from, rather than choosing a company to accept or not. Personality politics can no longer be approached by a style in which simply those who do not suit, the company will throw away and accept new powers instead [1, 2, 3]. Looking for new talents is currently difficult. Companies need to work better with what they have, but how do we work to improve the performance of existing workers? Many companies deal with low performance of individual workers [4, 5, 6].

Companies want to prepare individual coaching programs for employee development. Before going into this work, on one side it is the development and improvement of the employees' results, but on the other side is the overall strategic management of the company and the related set of rules within the company that often influences this performance [7, 8, 9].

The company's overall analysis should be conducted and the results subsequently confirm these assumptions. Human mistakes account for a maximum of 15% on the causes of problems in the company and unsatisfactory results. 85% of mistakes come from a poorly configured system, a lack of strategic management [10, 11, 12].

It is recommended in companies dealing primarily business model, how are the company's overall settings such as a strategy to win the market. Only then it is advisable to address the performance of people. If the strategy and business model are improved, which are undersized, it will bring multiplier effect in all aspects. If an undersized area of low employee performance is developed, it brings only some minor partial successes, but at the expense of disproportionately great effort [13, 14, 15].

The price-to-performance ratio is completely unmatched. That's why it is recommend working in the first place on the strategy and setting of the whole company. If the system is well set up, it will automatically lead to better performances of existing workers even though they are not a source of improvement [16, 17, 18].

In this paper, we are concerned with discovering bottlenecks in production process and with the multiplier effect aimed at increasing production [19, 20, 21].

2. Problem formulation

An inefficient overall strategic management of the company can have a negative effect on business performance. The two main variables, which every company's performance is based are the production machines and employees. It is very common that the company's management puts a lot of pressure on the employees to make the company more successful.

However, this pressure brings only little success. The greatest success is the identification of critical locations in production, production processes and the consequent streamlining of these processes. This approach brings a multiplier effect in the manufacturing company. The limitations of the study are those characteristics of design or methodology that impacted or influenced the interpretation of the findings from our research.

2.1. Description of current manufacturing company

The factory produces rollers in bearings. At the input of the factory, the purchased raw material is delivered. This purchased raw material is a steel rod. This material is stored in the warehouse. From this warehouse, it is transported to two saws, where it is cut into several pieces. The subsequent cut pieces are transferred to the coater for further processing.

This is the subject of an inspection. Good production pieces continue to the hardener. Unsuitable pieces are sent back for rework. All the hardener pieces then continue to the grinder. The grinder output goes into the intermediate storage from which the material is transferred to the final workplace cleaner.

- The simulation time is always 432000 minutes. The warm up period is 550 minutes.
- The initial production of the factory is 8056 units.
- The initial workload of a machine operator is 22% of time busy and the rest idle. Number of jobs is 6066.

Statistics in %	Blocked	Busy	Idle	Broken
Saw 1	97,47	1,12	0,00	1,41
Saw 2	97,46	1,13	0,00	1,41
Coater	76,11	23,32	0,37	0,20
Inspect	43,36	49,77	4,45	2,42
Hardener	64,37	35,39	0,24	0,00
Grinder 1	88,89	8,20	0,38	2,53
Grinder 2	87,26	9,36	0,40	2,98
Cleaner	0,00	84,01	0,00	15,99

Table 1. Overall statistics of default production machines

2.2. Employee pressure strategy

Most companies erroneously increase the pressure on employees instead of looking for and removing bottlenecks in production. This step simulates an increase in the performance of the operator and measure the final production of the factory.

As proof of our thesis of employers' unnecessary pressure on individual workers, we increase the workforce and performance by 20 %. The new production of the factory is 8322 units. The measured impact on total production growth is only 3,3 %. From these figures is obvious, this approach does not produce the expected result.

2.3. Improving the efficiency of processes and strategies

It is very important to thoroughly examine all workplaces and then improve workflow. In the first step, it is important to find narrow places in the production process. From the table above, you can see how the workplaces follow each other. First, it is important to focus on the workplaces that are the busiest. The busiest workplace is the cleaner. In this case, it is recommended to purchase another cleaner.

In the second step, it is important to solve the blocked manufacturing machines. The least used manufacturing machines are the saws. They are blocked most of the time. According to the small use of the saws, we recommend using only one production saw device, and the other device either not to use or sell.

The last place of our focus is the inspection. There is blocking between coaters and inspections. Therefore, we recommend increasing the storage space three times between these two workplaces.

All these solutions are an attempt to find a harmonious tuning between the workplace and the achievement of a better production solution.

As part of these changes, we have increased the production of the entire plant to 15466 units. All these manufacturing changes bring up a better solution by 91,98 %.

The new workload of a machine operator is 42,69 % of time busy and the rest idle. Number of jobs is 11666. The multiplier effect is almost doubled compared to the default status for the machine operator. Measured improvement in percent for the operator, new value is 92,32 %.

Statistics in %	Blocked	Busy	Idle	Broken
Saw 1	88,27	4,30	0,00	7,43
Saw 2	NA	NA	NA	NA
Coater	52,66	44,77	2,20	0,37
Inspect	0,05	95,50	0,00	4,45
Hardener	0,24	97,14	2,62	0,00
Grinder 1	3,40	15,68	76,04	4,88
Grinder 2	3,25	17,92	73,17	5,66
Cleaner 1	0,00	80,62	1,36	18,02
Cleaner 2	0,00	80,68	1,42	17,90

Table 2. Overall statistics of production machines with new configuration

2.4. Improving the efficiency of processes and strategies and Employee pressure strategy

Following the improved model, we have once again applied a strategy of increased pressure on employees. Manufacturing changes and increase in employee's performance about 10% brought up production of 15961 units. The improvement is only 3,20 % better, then the previous one.

3. Analyses and results

In the first step, it is explained that the pressure on staff alone does not produce the expected outcome. Profit in this case is not proportional to the energy consumed.

In the second step, improvements are made to individual processes and machine settings. This approach brings great improvement to the production process.

In the third step, there is an increase in pressure on employees with an efficient production configuration. Another benefit is very small.

Scenario	Production in units
Default model	8056
1. Employee pressure strategy	8322
2. Improving the efficiency of processes and strategies	15466
3. Improving the efficiency of processes and strategies and Employee pressure strategy	15961

Table 3. Overall review of analyses and results

It's more than clear from the model that the right strategy is to focus on improving processes at the factory.

4. Conclusion

The aim of the article is to demonstrate on a simple functional model, that explain the problem of the pressure on employees is not bringing any significant improvements. In the subsequent simulations, we solved it by the computer simulations, that the correct way is to focus on individual processes in the production company and harmonize them. This is the result of the article. As proof of our thesis of employers' unnecessary pressure on individual workers, we increase the workforce and performance by 20 %. The new production of the factory is 8322 units. The measured impact on total production growth is only 3,3 %. From these figures is obvious, this approach does not produce the expected result.The change of the factory strategy changed the results vigorously. The production was increased in the entire plant up to 15466 units. All these manufacturing changes bring up a better solution by 91,98 %. The multiplier effect is almost doubled compared to the default status for the machine operator. This is the correct strategy to use. In the last step, there is an increase in pressure on employees with an efficient production configuration. Another benefit is very small. If somebody requires a fast recipe to increase employee performance, one way exists. It often helps to precisely define the results and tasks that are required. To know what is expected of them and then check their performance results on a weekly basis. Let people just to make regular accounts of their activities. We would like to focus on the multiplier effect in the future.

5. Acknowledgments

This work was supported by the Ministry of Education, Youth and Sports of the Czech Republic within the National Sustainability Programme project No. LO1303 (MSMT-7778/2014) and also by the European Regional Development Fund under the project CEBIA-Tech No. CZ.1.05/2.1.00/03.0089 and also by the Internal Grant Agency of Tomas Bata University under the project No. IGA/FAI(CEBIA)/2016/033.

6. References

[1] Cecil-Wright, J. (1986). How the boardroom can influence warehousing costs. International Journal of Retail & Distribution Management, 14 (3), 67-69.

[2] Segerstedt, A., Pettersson, A.I. (2012). Measurements of excellence in a supply chain. International Journal of Logistics Systems and Management, 13 (1), 65-80.

[3] Hausman, W.H., Schwarz, L.B. and Graves, S.C. (1976). Optimal storage assignment in automatic warehousing system. Management Science, 22 (6), 629–638.

[4] Muppani, V.R., Adil, G.K. and Bandyopadhyay, A. (2010). A review of methodologies for class-based storage location assignment in a warehouse. International Journal of Advanced Operations Management, 2 (3-4), 274-291.

[5] Kovacs, A. (2011). Optimizing the Storage Assignment in a Warehouse Served by Milkrun Logistics. International Journal of Production Economics, 133 (1), 312-318.

[6] Glock, C.H. and Grosse, E.H. (2012). Storage policies and order picking strategies in U-shaped order picking systems with a movable base. International Journal of Production Research, 50 (16), 4344-4357.

[7] De Koster, R.B.M., Le-Duc, T. and Zaerpour, N. (2012). Determining the number of zones in a pick and-sort order picking system. International Journal of Production Research, 50 (3), 757-771.

[8] Bragg S. M. (2013) Inventory Management, Accounting Tools, Colorado.

[9] Waters D. (2003) Inventory Control and Management (Business), John Wiley and Sons, 2nd edition.

[10] Piasecki D. J. (2009)., Inventory Management Explained, Ops Publishing.

[11] Granville D. (2007). Excellence in Inventory Management: How to Minimise Costs and Maximise Services, Cambridge Academic; First edition.

[12] Milner C. (2015). Inventory Management: Advanced Methods for Managing Inventory within Business Systems, Kogan Page; 1 edition.

[13] Grinsted S. (2013). The Logistics and Supply Chain Toolkit: Over 90 Tools for Transport, Warehousing and Inventory Management, Kogan Page; 1 edition.

[14] Muller M. (2011). Essentials of Inventory Management, Amacom, 2nd edition.

[15] Bottani E., R. Montanari, M. Rinaldi and G. Vignali (2015). Intelligent algorithms for warehouse management. Intelligent Systems Reference Library, 87, pp. 645-667.

[16] López J. A, Mendoza A. and J. Masini (2013). A classic and effective approach to inventory management. International Journal of Industrial Engineering: Theory Applications and Practice, 20 (56), pp. 372-386.

[17] Xiao Y., Zhang R. and I. Kaku (2011). A new approach of inventory classification based on loss profit. Expert Systems with Applications, 38 (8), pp. 9382-9391.

[18] Paweł P., Marek F. and G. Paulina (2008). Using ABC Classification to determine production sequence in automotive industry, In Proceedings of 2008 World Automation Congress, WAC 2008, IEEE, pp. 1-6.

[19] Smith A.D. (2011). Inventory management and ABC analysis practices in competitive environments. International Journal of Procurement Management, 4 (4), pp. 433-454.

[20] Miculescu, M. N.; Lut, D. M. & Miculescu, C. (2011). Current Trends of Production Cost Accounting, Annals of DAAAM for 2011 & Proceedings of the 22nd International DAAAM Symposium, 23-26th November 2011, Vienna, Austria, Volume 22, No. 1, ISSN 1726-9679, ISBN 978-3-901509-83-4, Katalinic, B. (Ed.), pp. 0941-0942, Published by DAAAM International Vienna, Vienna

[21] Pasic, M.; Kadric, E. R. & Bajric, H. (2010). Relationship between Inventory Investment and Forecasting and Inventory Control, Annals of DAAAM for 2010 & Proceedings of the 21st International DAAAM Symposium, 20-23rd October 2010, Zadar, Croatia, ISSN 1726-9679, ISBN 978-3-901509-73-5, Katalinic, B. (Ed.), pp. 0511-0512, Published by DAAAM International Vienna, Vienna

DOI: 10.2507/28th.daaam.proceedings.100

VIRTUAL REALITY AND ITS INFLUENCE ON TRAINING AND EDUCATION - LITERATURE REVIEW

Sergo Martirosov, Pavel Kopecek

This Publication has to be referred as: Martirosov, S[ergo] & Kopecek, P[avel] (2017). Virtual Reality and its Influence on Training and Education - Literature Review, Proceedings of the 28th DAAAM International Symposium, pp.0708-0717, B. Katalinic (Ed.), Published by DAAAM International, ISBN 978-3-902734-11-2, ISSN 1726-9679, Vienna, Austria

DOI: 10.2507/28th.daaam.proceedings.100

Abstract

This paper presents a robust literature review on the subject of Virtual Reality and its use in training and educational systems. Firstly, the beginnings of Virtual Reality are presented, then, advantages and disadvantages of virtual reality environments are described. Various characteristics of Virtual Reality are highlighted, which have positive or negative influence on learning or training ability of individuals. It is shown how various organizations could benefit from using virtual reality in their daily activities. Such aspect as sense of presence is also described, which has tremendous effect on how subjects perceive information in virtual world.

Keywords: Virtual Reality; Sense of Presence; Training; Education

1. Introduction

Previous research on this new technology is very helpful as it shows potential benefits and answers many questions, thus allowing to continue new studies based on the accumulated knowledge. Nowadays, Virtual Reality (VR) technology is used in various areas for various purposes. It is used in training, education, psychology, etc. Studies carried out in this field suggest that training in virtual environments can replace the hours spent in the real world. This comes very handy when training in real environments is impossible due to risk or cost. Although this technology is still in its infancy, it is confidently gaining momentum and in the near future many people will be dealing with VR on a frequent basis, be it watching movies in VR, playing games or improving skills for their profession.

The first traces of VR came from the world of science fiction. Stanley G. Weinbaum's 1935 short story "Pygmalion's Spectacles" is recognized as one of the first works of science fiction that explores VR. It describes a google-based VR system with holographic recording of fictional experiences including smell and touch [1]. Furthermore, if we go back to the early 1950s, it was the time when Morton Heilig developed the Sensorama, a multisensory simulator with wind and scent production, vibratory sensation and 3D display [2][3].

Fig. 1. Morton Heilig's Sensorama

It was an arcade-style theatre cabinet that could simulate all the senses, not just sight and sound. It included stereo speakers, stereoscopic 3D display, fans, vibrating chair and smell generators, that would allow nearly full immersion inside the short films that Morton Heilig edited himself. Next Heilig's invention was the Telesphere Mask in 1960, which was the first example of a head-mounted display. The headset provided stereoscopic 3D and wide vision with stereo sound but was non-interactive and without motion tracking. In 1987 Jaron Lanier, founder of the visual programming lab through his company developed a range of VR devices, which include the Dataglove and the EyePhone head mounted display. These inventions led to major breakthroughs in developments in this area.

Around that time, in 1968, Ivan Sutherland, a leader in computer graphics developed the "Sword of Damocles", one of the first Head Mounted Displays (HMDs), which overlaid computer graphics onto the real world, today known as augmented/mixed reality systems [2][3].

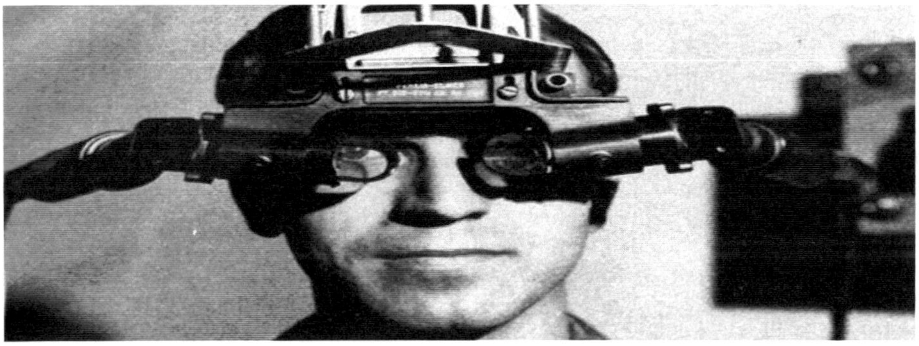

Fig. 2. The Sword of Damocles by Ivan Sutherland

NASA also had big influence on the evolution of VR technology as they worked on developing HMDs for their astronauts [3]. Throughout 1990s, we began to see VR devices to which the public had access. In 1993, SEGA announced the Sega VR headset for the Sega Genesis Console, which prototype glasses supported head tracking, stereo sound and LCD screens in the visor. Next was Nintendo Virtual Boy in 1995, a 3D gaming console that was promised to be the first ever-portable console that would display true 3D graphics. Despite low prices, it was a commercial failure due to lack of colors in graphics and lack of software support. As to the modern times, in 2010, the first Oculus Rift prototype saw the light, in 2014, Sony announced PlayStation VR, a VR headset for PlayStation 4 video game console, and Google announced Cardboard, a do-it-yourself stereoscopic viewer for smartphones. In 2015, HTC and Valve Corporation announced the VR headset with controllers.

2. Virtual Reality and Augmented Reality, the Difference

There is also another technology called Augmented Reality (AR), which is not the same as VR. The difference between those is that AR simulates artificial objects in the real environment, while VR creates a simulated environment to inhabit. VR and AR technology can be categorized as immersive VR, non-immersive VR, AR and mixed reality [4].

Both technologies have great potential in almost every industry, and each of them has different objectives. Although in some cases they can be used for the same cause/idea, the results will always be different, for example, assembly of objects can be shown with both technologies, but in different ways. Both will bring benefits if used as intended.

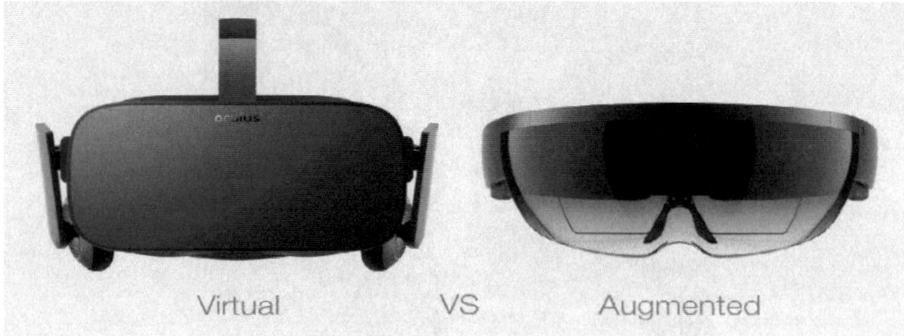

Fig. 3. Virtual Reality vs Augmented Reality

In immersive VR, the real world is completely occluded from the field of view and the user has a stronger feeling of presence in that virtual environment [4]. Examples of immersive VR are Oculus Rift and HTC Vive. In non-immersive VR, the user remains aware that they are viewing a virtual environment [4]. The examples of non-immersive VR systems are Samsung Gear VR and Google Cardboard. In AR, the virtual objects are like holograms, transparent in the real world in daylight, whereas in mixed reality, virtual images appear solid to the user in daylight [4]. Examples of AR are Meta and DAQRI. Examples of mixed reality are Microsoft HoloLens and Magic Leap. The integration of these technologies in different practices (i.e., clinical practice) and commercial success is dependent on various variables, which are: affordability, flexibility, wear ability, usability, immersion, mobility and vision [4][5]. VR and AR have the potential to enhance the learning environment in comparison to traditional learning. It is mentioned that 3D learning can increase learner's motivation/engagement, improve spatial knowledge representation, improve contextualization of learning and develop superior technical skills [6].

3. Virtual Reality and Sense of Presence

In the modern time VR, technology became very popular and is used in many different areas such as entertainment, military, healthcare, education, engineering etc. VR is a simulated environment that is created with computer technology and presented to the user in such a way, that user starts to feel like in real environment. A Simulation is a model of a real world where user has the ability to interact with the environment [7]. Simulations are helpful and useful as they provide realistic context in which individuals can explore, experiment and see immediate results as they create models of their own or try theories on the modelled concept [8]. Depending on the amount of senses simulated in VR, such as vision, hearing, touch, balance, even smell, the immersion level in the artificial world can vary.

Thanks to modern technological advance of computer hardware and software, it became possible to incorporate 3D VR in innovative applications of teaching, training, and learning [9][10][11]. Virtual environment or virtual world is a computer generated 3D representation of real or fictional environments. A user can interact in such an environment independently in the same pace, one would experience events in the real world [12]. VR has the ability to support realistic and immersive simulation and enables transfer of taught skills in VR into real contexts as well as provide multi-user, embodied and interactive active learning [13]. VR is believed to be a promising tool for training and complex problem solving, which requires weighing multiple variables and situational decision making [13][14][15][16].

Sense of presence is defined as "a state of consciousness", the psychological sense of being in the virtual environment [17][18][19][20][21]. It is noted by various authors that sense of presence reduces social distance between learners and improves skill acquisition and knowledge transfer by permitting multiple perspectives and situated performance [16][20][22][23]. It was noted that sense of presence in virtual learning environments increases learner's motivation and engagement in learning process and potentially learning outcomes, by enabling focused and realistic interactions with learning materials and activities [24][25][26].

A similar concept related to sense of presence is immersion – degree of "the subjective impression that one is participating in a comprehensive, realistic experience" and "the semi-voluntary experience of being transported into an alternate context for an extended duration" [15][27]. Previous research suggests that higher levels of immersion are associated with higher levels of presence [20][28][29]. The psychological perception of being inside a virtual world is known as presence [30]. Among many factors affecting user is feeling of presence, one is the content of the virtual world [31]. Immersive experience can be created via the sensory and environmental fidelity in 3D virtual environment, engagement with virtual objects and activities, and emotional experience activated through realistic scenarios in virtual world [16]. VR has been implemented as a promising learning tool both for formal and informal learning contexts in a variety of educational activities [11][32]. It was also reported that immersive VR could provide a virtual environment to simulate teaching challenges and thus act as a pedagogical tool for the collaborative/cooperative teaching/training program [33].

It has been noted that haptic feedback increases the sense of presence in VEs [34]. Haptic feedback is usually described as incorporating both kinesthetic (coming from muscles, joints and tendons) and tactile (coming from nerve receptors in the skin) information [35][36][37]. Haptic feedback can be very useful and have positive effect on task performance. For instance, it is said that teleoperators benefit from haptic feedback when manipulating remote objects [38][39]. Various studies have been conducted within collaborative virtual environments (CVE) to find out how haptic feedback influences performance. In one study a person was asked to hand over an object to another person within a CVE, which resulted in faster task completion, higher sense of presence and participant's belief in superiority of such actions, compared to those without haptic feedback [40][41]. Similar study that also used CVE, where two people collaboratively had to move a virtual ring along a virtual wire without touching a wire, showed that performance was significantly better using haptic feedback and sense of presence was increased as well [42][43][44]. Lastly, another study within a CVE in which two participants had to lift a virtual box together, suggests that haptic feedback evokes a greater sense of co-presence [45].

VR is characterized with representational fidelity, learner interaction, support of psychological sense of presence, immersion and multi-user environment for embodied interactions [6]. Prior studies note that immersive VR learning environment also enhances situated learning through simulation of realistic contexts and providing contextualized learning activities [46][47][48], thus improving performance transfer from learning environment to the real-world setting [49][50]. Nowadays flight simulators are used widely both for entertainment and training purposes, as they provide realistic environments to the extent that flight training hours can be replaced by corresponding simulator hours [51]. Several unique affordances of an immersive VR learning environment include: (I) virtual agents/avatars that act like real personalized and interactive learning partners that sometimes cannot be easily arranged in a real-world, (II) unlimited variations of 3D VR environments, and (III) possibility to transfer sensory representations to improve virtual learning actions [16][46].

4. Virtual Reality in Training and Education

VR simulations prove to be effective for not only educational purposes but also in situations, where human lives can be at risk or it could be hard and costly to check with a high precision, weather some changes are worthwhile to introduce to the real environment.

Interactivity is the most fundamental aspect in learning. It is mentioned that both, providing a learning environment and interaction with PCs may promote a better understanding of things [52]. Despite many studies in this area, there is still controversy about the real effects of VR on learning and cognition, but preliminary results indicate that the use of VR can increase learning performance [53], facilitate usability and enhance interaction [54], help to reconstruct and navigate through non-existent environments [55], help disabled learners with knowledge accessibility [56], etc. Serious games tend to be used often in formal education and with sufficient support are shown to be highly motivational and effective in learning complex tasks [57][58].

VR Technology can be of great use for emergency service personnel such as police, ambulance, fire fighters and army, as they have to be prepared for various operations at all times, thus continuous training is crucial for their success and security. But training in real environment is sometimes impossible, due to cost or complexity and the good alternative that comes handy is virtual training environment, which allows training for such complex collaborative tasks [59][60]. Thus, if the situation does not predispose for real life training practice due to danger, cost or effort, emergency service personnel can profit from virtual training to train for scenarios in real environment [61]. VR is also widely used in ergonomic applications, where workplaces are constructed virtually with the help of simulator softwares such as Tecnomatix Jack or DELMIA5 [62], than they can analyzed to make specific changes in the real workplace environment. Often VR training is the only option that is possible [63]. Some studies suggest that virtual training environments can be used successfully for training specific skills, such as spatial perception, route learning or the reactions in certain emergency situations [61][64][65][66]. There is also evidence that 3D presentation of visual information in virtual environment promotes better collaborative decision-making compared to 2D presentation [67].

Virtual training environments (VTEs) keep learners motivated and engaged in learning process [68] and have also been proven to be very useful in pedagogical contexts [69]. VTEs allow wide variety of learning styles and support collaborative knowledge building and reflective thinking [69] as well as experiential learning [70]. To enhance the effects of training, adequate instructions can be adapted to pre-training, during-training and post-training phase [71]. Some authors suggest debriefings during and/or after a training session, where individuals reflect on the training, to foster transfer of such complex team tasks [72]. It is also mentioned that virtual team performance resembles team performance in reality [73]. Team performance is dependent on whether procedures are established within a team, how conflicts are resolved and if collaboration occurs [73].

Many authors highlight the need for realistic virtual training, where training simulation holds information overload and pressure during an operation, especially when the training is meant for crisis management [74]. It is important for the emotional state to be invoked during training and it should be as close to emotional state trainees would confront during and actual operation as possible, because of the strong effect of emotions on decision-making, learning and problem-solving [74]. Similar research suggests that virtual training for action forces such as military or fire fighters should provide first-person experiences and situated learning without time for conscious thoughts, but with the possibility of gaining knowledge [63]. The levels of immersion or presence in the virtual environment seem to be crucial for the development of first-person experiences and the knowledge gained is then "direct, personal, subjective and often tacit" and can be transferred to the real world [75]. It is noted that first-person perspective can improve learning in virtual environment [76]. In addition, it is suggested that mental representations acquired in a VTE might be similar or close to representations of real objects [64]. Experiences that individuals build in virtual environments can then become recallable and used successfully in the real world [77]. Some authors mention that in order to maximize the transfer of skills from virtual training to reality, the context of the virtual training task should be as close and similar to the task in reality as possible [78]. In addition, structuring virtual learning according to psychological factors along with fidelity to real equipment plays big role in knowledge transfer [79]. As virtual simulations or games allow exploration of realistic scenarios, learning and performance can be enhanced [72]. There is a promising potential of VTEs in educational curriculum [80].

It is important to note that assessing the level of relevance and value of training to trainees to find out the usefulness of training content to them is crucial [81], as this kind of information can help predict the transfer of knowledge and the impact of training on job performance [82]. Learning measures alone cannot be used to show whether gained knowledge will be applied, but a positive attitude towards the training makes it more probable that transfer will be successful [82]. If trainees accept training and see personal relevance for their job, they will have higher motivation to learn [83].

One of the uses of VR can be helping those with autism to lessen social anxiety and frustration and improve academic and social skills. Individuals with autism suffer impairment in social functions which leads to difficulties in social interactions, communication and emotion recognition [84][85]. Despite their good academic performance, social communication barrier can lead to frustration [85][86] and this in turn can worsen academic performance due to low self-esteem, even though their intellectual abilities are average or above that [87]. VR can be used for social training and provide safe and unlimited contexts to practice social scenarios that are most commonly encountered in our everyday life [88][89][90] and as a result, social anxiety can be reduced [91]. Additionally, it is noted that computer technology is often highly motivating and rewarding for those suffering with ASD (Autism Spectrum Disorder) and especially children with HFA (High Functioning Autism) [92]. Various studies suggest that VR environments can help individuals with autism disorders improve social problem solving and social understanding (i. e., knowing when to initiate conversation, what other person feels, choosing appropriate seat) [93][94][95].

The fact that VR reality systems can allow users to explore immersive 3D environments from any location at any time, could have a profound impact on science education [96]. VR allows exploration of hidden phenomena, distant locations and manipulation of otherwise immutable objects [97]. For instance, VR gives the possibility for medical students to explore delicate internal organs that would otherwise require cadaver (corpse) dissection [98]. VR can be a promising replacement to cadaver dissection for learning anatomy and practicing surgical procedures [99]. Direct manipulation of objects in virtual environment may promote development of "embodied", multi-modal mental representations of represented objects [100]. Embodied learning allows students to practice via simulations or mental imagery when there is no physical opportunity [100]. The ability to imagine and mentally manipulate anatomical structures is a crucial skill that medical students must have [101]. VR and computer models represent a good alternative for medical students and other professionals to practice their skills when access to patients or cadavers is limited [102]. It is also noted that to create a realistic and complex simulation/serious game with interactive objects requires some methodology and as the size and complexity of a game/simulation increases, it's becoming harder to control the environment and objects, thus a methodology is required to steadily construct a 3D Virtual World[103].

Many games have been developed using 3D images with which users need to interact in order to learn certain tasks. Interaction and immersion are the key elements when it comes to making learning process more enjoyable. 3D animations can be used to teach students various procedures and mechanisms for carrying out specific tasks [104]. Furthermore, some VR environments can support multiple user collaboration/cooperation, where students learn together and from each other [105]. This is a good way to increase their motivation for studying. It was noted, that text-based learning content often leads to boredom and can prevent students from understanding the subject [106]. It is also noted that in many different contexts video is a more effective form of learning compared to books [107].

Student motivation is vital as it has a direct influence on his/her desire and dedication to accomplish specific task [108]. Motivation is a major cognitive factor influencing learning and thus the better-motivated students can learn with higher efficiency [109]. There is a research suggesting that 3D full immersive VR Reality Learning Environments (VRLE) will increase learner‘s interest and motivation towards learning process more compared to 2D animated environments [110].

It has been argued by some authors that emotions play big role in learning. Some suggest that different emotional states induce different kinds of thinking [111], and others mention that a slight positive mood effects thinking in a positive way as well, thus leading towards greater creativity and flexibility in problem solving [112]. Thanks to today's knowledge and technologies, VR environments can be created and filled with various interesting elements, making it motivative and fun for students to explore, thus making the learning process more fruitful.

5. Conclusion

In this paper, we showed a multitude of studies on the subject of Virtual Reality. This is a new technology which has to be tested in order to find the potential of use in different areas of our lives, be it enjoyment, education, training, etc. As there are some flaws in this technology, still, most of the studies on Virtual Reality emphasize its substantial benefits. Previous research is very helpful as it displays what has been done in this field as well as what new could be done. Furthermore, it answers some questions in regards if Virtual Reality could be used in specific scenarios and what outcomes to await.

This knowledge is not final, as there are more studies to come, some will be controversial as well, but it is already proved, that there is great significance in Virtual Reality applications and that Virtual Reality technology will take big part of our lives in the near future.

6. Acknowledgement

This paper is based upon work sponsored by project LO1502 Development of Regional Technological Institute.

7. References

[1] History of Virtual Reality. In Wikipedia. Retrieved August 10, 2016 Available from https://en.wikipedia.org/wiki/Virtual_reality#History

[2] Drummond K.H, Houston T, Irvine T, et al. (2014). The rise and fall and rise of virtual reality. Vox Media.

[3] Virtual reality systems. San Diego, CA: Academic Press Limited; 1993, ISBN: 0-12-227748-1

[4] Merel, T. (2015), The 7 drivers of the $150 billion AR/VR industry. Aol Tech; Available from https://techcrunch.com/2015/07/08/the-7-drivers-of-the-150-billion-arvr-industry

[5] Sekhar LN, Tariq F, Kim LJ, et al. (2013). Commentary: virtual reality and robotics in neurosurgery. Neurosurgery; 72:1–6. DOI:10.122/NEU.0b013e31827db647

[6] Dalgarno BL, Mark JW, (2010). What are the learning affordances of 3-D virtual environments? British Journal of Education Technology; 41: pp.10–32. DOI:10.1111/j.1467-8535.2009.01038.x

[7] Thurman, R. A., (1993). Instructional simulation from a cognitive psychology viewpoint. Educational technology research and development, 41(4), pp.75-89. DOI:10.1007/bf02297513

[8] Rieber, L.P. (1992). Computer-based microworlds: a bridge between constructivism and direct instruction. Educational Technology Research and Development, 40(1), pp.93-106. DOI:10.1007/BF02296709

[9] Abulrub, A. G., Attridge, A. N., & Williams, M. A. (2011). Virtual reality in engineering education: the future of creative learning. In Global engineering education conference (EDUCON), 2011 IEEE (pp. 751-757). DOI:10.1109/EDUCON.2011.5773223

[10] Gregory, S., Gregory, B., Reiners, T., Fardinpour, A., Hillier, M., Lee, M., Basu, A. (2013). Virtual worlds in Australian and New Zealand higher education: remembering the past, understanding the present and imagining the future. In ASCILITE 2013 "Electric dreams", 1-4 december 2013. Sydney, NSW: Macquarie University. ISBN 9781741384031

[11] Jou, M., & Wang, J. (2013). Investigation of effects of virtual reality environments on learning performance of technical skills. Computers in Human Behavior, 29(2), pp.433-438. DOI: 10.1016/j.chb.2012.04.020

[12] Mitchell, P., Parsons, S., & Leonard, A. (2007). Using virtual environments for teaching social understanding to 6 adolescents with autistic spectrum disorders. Journal of autism and developmental disorders, 37(3), pp.589-600. DOI: 10.1007/S10803-006-0189-8

[13] Cheng, Y., & Wang, S. H. (2011). Applying a 3D virtual learning environment to facilitate student's application ability-the case of marketing. Computers in Human Behavior, 27(1), pp.576-584. DOI: 10.1016/j.chb.2010.10.008

[14] Bertram, J., Moskaliuk, J., & Cress, U. (2015). Virtual training: making reality work? Computers in Human Behavior, 43, pp.284-292. DOI: 10.1016/j.chb.2014.10.032

[15] Dede, C. (2005). Millennial learning styles. Educause Quarterly, 28(1), pp.7-12.

[16] Dede, C. (2009). Immersive interfaces for engagement and learning. Science, 323(5910), pp.66-69. DOI: 10.1126/science.1167311

[17] Barfield, W., & Hendrix, C. (1995). The effect of update rate on the sense of presence within virtual environments. Virtual Reality, 1(1), pp.3-15. DOI: 10.1007/BF02009709

[18] Heeter, C. (1992). Being there: the subjective experience of presence. Presence: Teleoperators and virtual environments, 1(2), pp.262-271. DOI: 10.1162/pres.1992.1.2.262

[19] Lee, K. M. (2004). Presence, explicated. Communication theory, 14(1), pp.27-50. DOI: 10.1111/j.1468-2885.2004.tb00302.x

[20] Slater, M., & Wilbur, S. (1997). A framework for immersive virtual environments (FIVE): speculations on the role of presence in virtual environments. Presence: Teleoperators and Virtual Environments, 6(6), pp.603-616. DOI: 10.1162/pres.1997.6.6.603

[21] Witmer, B. G. B. G., Jerome, C., & Singer, M. (2005). The factor structure of the presence questionnaire. Presence, 14(3), pp.298-312. DOI: 10.1162/105474605323384654

[22] Mantovani, F., & Castelnuovo, G. (2003). The sense of presence in virtual training: enhancing skills acquisition and transfer of knowledge through learning experience in virtual environments. In G. D. F. Riva (Ed.), Being there: concepts, effects and measurement of user presence in synthetic environments (pp. 167-182). Amsterdam: IOS Press.

[23] Palloff, R. M., & Pratt, K. (1999). Building learning communities in cyberspace (vol. 12). San Francisco: Jossey-Bass.

[24] Mikropoulos, T. A. (2006). Presence: a unique characteristic in educational virtual environments. Virtual Reality, 10(3-4), pp.197-206. DOI: 10.1007/s10055-006-0039-1

[25] Persky, S., Kaphingst, K. A., McCall, C., Lachance, C., Beall, A. C., & Blascovich, J. (2009). Presence relates to distinct outcomes in two virtual environments employing different learning modalities. CyberPsychology and Behavior, 12(3), pp.263-268. DOI: 10.1089/cpb.2008.0262

[26] Selverian, M. M., & Hwang, H. S. (2003). In search of presence: a systematic evaluation of evolving VLEs. PRESENCE: Teleoperators and Virtual Environments, 12(5), pp.512-522. DOI: 10.1162/105474603322761306

[27] Buchanan, K. (2006). Beyond attention-getters: Designing for deep engagement (Unpublished doctoral dissertation). Michigan: Michigan State University. Retrieved from http://education.uwsp.edu/publications/buchanan_2006_beyond_attention.pdf.

[28] Bystrom, K. E., Barfield,W., & Hendrix, C. (1999). A conceptual model of the sense of presence in virtual environments. Presence: Teleoperators and Virtual Environments, 8(2), pp.241-244. DOI: 10.1162/105474699566107

[29] Faiola, A., Newlon, C., Pfaff, M., & Smyslova, O. (2013). Correlating the effects of flow and telepresence in virtual worlds: enhancing our understanding of user behavior in game-based learning. Computers in Human Behavior, 29(3), pp.1113-1121. DOI: 10.1016/j.chb.2012.10.003

[30] Witmer B. and Singer M. (2000). "Measuring presence in virtual environments: A presence questionnaire". Presence: Teleoperators and Virtual Environments, vol. 9, issue 5, pp.497-503. DOI: 10.1162/105474698565686

[31] Ijsselsteijn W.A, H. de Ridder, 1. Freeman and S.E. Avons, (2000). "Piesence: Concept, determinants and measurement", Proceedings of the SPIE, Human Vision and Electronic Imaging V, San Jose, CA, pp.3959-3976. ISBN 0-8194-4577-5

[32] Hew, K. F., & Cheung,W. S. (2010). Use of three-dimensional (3-D) immersive virtual worlds in K-12 and higher education settings: a review of the research. British journal of educational technology, 41(1), pp.33-55. DOI: 10.1111/j.1467-8535.2008.00900.x

[33] Quintana, M. G. B., & Fernandez, S. M. (2015). A pedagogical model to develop teaching skills. The collaborative learning experience in the Immersive Virtual World TYMMI. Computers in Human Behavior, 51, pp.594-603. DOI: 10.1016/j.chb.2015.03.016

[34] Hoffman H., GroeN J. Rousseau S., Hollander A., Winn W., Wells M. and Furness T. (1996). "Tactile augmentation: Enhancing presence in virtual reality with tactile feedback from real objects". Proceedings of the 1996 Convention of the American Psychological Society.

[35] Bowman, D., Kruijff E., La Viola J. and Poupyrev I. (2000). "The art and science of 3D interaction". Tutorial notes from the IEEE International Virtual Reality 2000 Conference, New Brunswick. NJ.

[36] Durlach, N.I. and Mavor A.S., (1995). Vinual Reality: Scientific and Technological Challenges. Academic Press. Washington, D.C. DOI: 10.17226/4761

[37] Hoffman, H.G., Prothero J.D., Wells M.J. and Groen I. (1998). "Virtual chess: Meaning enhances users' sense of presence in virtual environments", International Journal of Human Computer Interaction, vol. 10 issue 2, pp.251-263. DOI: 10.1207/S15327590ijhc1003_3

[38] Das, H., Zak, H., Kim, W.S., Bejczy, A.K. and Schenker, P.S. (1992). "Operator performance with alternative manual control modes in teleoperation", Presence: Teleoperators and Virtual Environments, vol. I issue 2., pp.201-218. DOI: 10.1162/pres.1992.1.2.201

[39] Petzold, B., Zach, M.F., Faerber, B., Deml, B., Egermeier, H., Schilp, J. and Clarke, S. (2004). "A study on visual. auditory and haptic feedback for assembly tasks". Presence: Teleoperators and Virtual Environments. vol. 13 issue 1, pp.16-21. DOI: 10.1162/105474604774048207

[40] Sallnäs, E.-L. (1999). "Presence in multimodal interfaces", Second International Workshop on Presence, University of Essex.

[41] Sallnäs, E.-L., Rassmus-Gröhn and Sjöström, C. (2000). "Supporting Presence in collaborative environments by haptic force feedback ", ACM Transactions on Computer- Human Interaction", vol. 7 issue 4,461476. DOI: 10.1145/365058.365086

[42] Basdogan, C., Ho, C., Slater, M. and Srinivasan, M.A. (1998). "The role of haptic communication in shared virtual environments", Proceedings of the Third Workshop on Phantom User Group, Dedham. MA.

[43] Basdogan, C., Ho, C., Srinivasan, M.A. and Slater, M., (2000). “An experimental study on the role of touch in shared virtual environments”, ACM Transactions on Computer-Human Interaction. vol. 7 issue 4.443460. DOI: 10.1145/365058.365082
[44] Basdogan, C. Ha. C., Slater, M., Durlach, N. and Srinivasan, M.A. (1998). “An experiment on the influence of haptic communication on the sense of being together”. Proceedings of the British Telecom Workshop on Presence in Shared Virtual Environments, Ipswich, England.
[45] Jordan, I., Mortensen, J., Oliveira, M., Slater, M., Tay, B.K., Kim, J. and Srinivasan, M.A. (2002). “Collaboration in a mediated haptic environment”. Proceedings of the 5th Annual lnremarional Workshop on Presence, 2002.
[46] Bailenson, J. N., Yee, N., Blascovich, J., Beall, A. C., Lundblad, N., & Jin, M. (2008). The use of immersive virtual reality in the learning sciences: digital transformations of teachers, students, and social context. The Journal of the Learning Sciences, 17(1), pp.102-141. DOI: 10.1080/10508400701793141
[47] Barab, S., Sadler, T., Heiselt, C., Hickey, D., & Zuiker, S. (2007). Relating narrative, inquiry, and inscriptions: a framework for socio-scientific inquiry. Journal of Science Education and Technology, 16(1), pp.59-82. DOI: 10.1007/s10956-006-9033-3
[48] Ketelhut, D. J., Dede, C., Clarke, J., Nelson, B., & Bowman. (2007). Studying situated learning in a multi-user virtual environment. In E. Baker, J. Dickieson, W. Wulfeck, & H. O'Neil (Eds.), Assessment of problem solving using simulations (pp. 37-58). Mahwah, NJ: Lawrence Erlbaum Associates.
[49] Bossard, C., Kermarrec, G., Buche, C., & Tisseau, J. (2008). Transfer of learning in virtual environments: a new challenge? Virtual Reality, 12(3), pp.151-161. DOI 10.1007/s10055-008-0093-y
[50] Park, J., MacRae, H., Musselman, L. J., Rossos, P., Hamstra, S. J., Wolman, S., et al. (2007). Randomized controlled trial of virtual reality simulator training: transfer to live patients. The American journal of surgery, 194(2), pp.205-211. DOI: 10.1016/j.amjsurg.2006.11.032
[51] Federal Aviation Administration, Qualification and Approval of Personal Computer-Based Aviation Training Devices (Advisory Circular No. AC 61–126, 5/12/97), U.S. Department of Transportation, Washington D.C., 1997.
[52] Seymour, P., Nicholas, N., The Connected Family: Bridging the Digital Generation Gap, Longstreet Press, (1996). ISBN - 1563523353
[53] Carpenter, T.& Anderson, L. (1996). Using virtual reality technology for learning design skills, Proceedings of ED-MEDIA 96, Boston, USA, pp. 752
[54] Mikropoulos, T.& Nikolou E. (1996). A virtual hand with tactile feedback for virtual learning environments,Proceedings of ED-MEDIA 96, Boston, USA, pp. 792
[55] Littman, M. (1996). Enhancing instruction through virtual reality, Proceedings of ICTE, New Orleans, USA,pp. 31-33
[56] Lumbreras, M., Sánchez J., Barcia M. (1996), A 3D sound hypermedial system for the blind. In Proceedings of the First European Conference on Disability, Virtual Reality and Associated Technologies, pp.187-191, Maidenhead, UK.
[57] Chatti, Mohamed Amine, et al. (2012). "A reference model for learning analytics." International Journal of Technology Enhanced Learning 4.5- 6: pp.318-331. DOI: 10.1504/IJTEL.2012.051815
[58] Eseryel, D., Law, V., Ifenthaler, D., Ge, X., & Miller, R. (2014). An investigation of the interrelationships between motivation, engagement, and complex problem solving in game-based learning. Journal of Educational Technology & Society, 17(1), pp.42-53. ISSN: 11763647
[59] Moskaliuk, Bertram & Cress. (2013a). Impact of virtual training environments on the acquisition and transfer of knowledge. Cyberpsychology, Behavior, and Social Networking, 16(3), 210–214. DOI: 10.1089/cyber.2012.0416
[60] Moskaliuk, Bertram & Cress. (2013b). Training in virtual environments: Putting theory into practice. Ergonomics, 56(2), pp.195–204. DOI: 10.1080/00140139.2012.745623
[61] Rose, F. D., Attree, E. A., Brooks, B. M., Parslow, D. M., Penn, P. R., & Amibihaipahan, N. (2000). Training in virtual environments: Transfer to real world tasks and equivalence to real task training. Ergonomics, 43(4), pp.494–511. DOI: 10.1080/001401300184378
[62] Bures, M. (2016). Time Demands of Virtual Ergonomic Modelling – Experimental Study, Proceedings of the 26th DAAAM International Symposium, pp.0286-0291, B. Katalinic (Ed.), Published by DAAAM International, ISBN 978-3-902734-07-5, ISSN 1726-9679, Vienna, Austria, DOI:10.2507/26th.daaam.proceedings.039
[63] Romano, D. M., & Brna, P. (2001). Presence and reflection in training: Support for learning to improve quality decision-making skills under time limitations. CyberPsychology & Behavior, 4(2), pp.265–277. DOI: 10.1089/109493101300117947
[64] Arthur, E. J., Hancock, P. A., & Chrysler, S. T. (1997). The perception of spatial layout in real and virtual worlds. Ergonomics, 40(1), pp.69–77. DOI: 10.1080/001401397188387
[65] Gamberini, L., Cottone, P., Spagnolli, A., Varotto, D., & Mantovani, G. (2003). Responding to a fire emergency in a virtual environment: Different patterns of action for different situations. Ergonomics, 46(8), pp.842–858. DOI: 10.1080/0014013031000111266
[66] Lloyd, J., Persaud, N. V., & Powell, T. E. (2009). Equivalence of real-world and virtualreality route learning: A pilot study. Cyberpsychology & Behavior, 12(4), pp.423–427. DOI: 10.1089/cpb.2008.0326

[67] Van Der Land, S., Schouten, A. P., Feldberg, F., Van Den Hooff, B., & Huysman, M. (2013). Lost in space? Cognitive fit and cognitive load in 3D virtual environments. Computers in Human Behavior, 29(3), pp.1054–1064. DOI: 10.1016/j.chb.2012.09.006

[68] Monahan, T., McArdle, G., & Bertolotto, M. (2008). Virtual reality for collaborative elearning. Computers & Education, 50(4), pp.1339–1353. DOI: 10.1016/j.compedu.2006.12.008

[69] Mikropoulos, T. A., & Natsis, A. (2011). Educational virtual environments: A ten-year review of empirical research (1999–2009). Computers & Education, 56(3), pp.769–780. DOI: 10.1016/j.compedu.2010.10.020

[70] Jarmon, L., Traphagan, T., Mayrath, M., & Trivedi, A. (2009). Virtual world teaching, experiential learning, and assessment: An interdisciplinary communication course in Second Life. Computers & Education, 53(1), pp.169–182. DOI: 10.1016/j.compedu.2009.01.010

[71] Vogel-Walcutt, J. J., Fiorella, L., & Malone, N. (2013). Instructional strategies framework for military training systems. Computers in Human Behavior, 29(4), pp.1490–1498. DOI: 10.1016/j.chb.2013.01.038

[72] Salas, E., Tannenbaum, S. I., Kraiger, K., & Smith-Jentsch, K. A. (2012). The science of training and development in organizations: What matters in practice. Psychological Science in the Public Interest, 13(2), pp.74–101. DOI: 10.1177/1529100612436661

[73] Johnson, S. D., Suriya, C., Won Yoon, S., Berrett, J. V., & La Fleur, J. (2002). Team development and group processes of virtual learning teams. Computers & Education, 39(4), pp.379–393. ISSN 0360-1315

[74] Bacon, L., Windall, G., & MacKinnon, L. (2012). The development of a rich multimedia training environment for crisis management: Using emotional affect to enhance learning. Research in Learning Technology, 19. DOI: 10.3402/rlt.v19s1/7780

[75] Winn, W. (1993). A conceptual basis for educational applications of virtual reality. Human interface technology laboratory of the washington technology center. Seattle: University of Washington. Technical Publication R-93-9.

[76] Lindgren, R. (2012). Generating a learning stance through perspective-taking in a virtual environment. Computers in Human Behavior, 28(4), pp.1130–1139. DOI: 10.1016/j.chb.2012.01.021

[77] Tichon, J. G. (2007). Using presence to improve a virtual training environment. CyberPsychology & Behavior, 10(6), 781–787. DOI: 10.1089/cpb.2007.0005

[78] Kozak, J. J., Hancock, P. A., Arthur, E. J., & Chrysler, S. T. (1993). Transfer of training from virtual reality. Ergonomics, 36(7), pp.777–784. DOI: 10.1080/00140139308967941

[79] Dorsey, D., Campbell, G., & Russell, S. (2009). Adopting the instructional science paradigm to encompass training in virtual environments. Theoretical Issues in Ergonomics Science, 10(3), pp.197–215. DOI: 10.1080/14639220802151427

[80] Chittaro, L., & Ranon, R. (2007). Web3D technologies in learning, education and training: Motivations, issues, opportunities. Computers & Education, 49(1), pp.3–18. DOI: 10.1016/j.compedu.2005.06.002

[81] Salas, E., Milham, L. M., & Bowers, C. A. (2003). Training evaluation in the military: Misconceptions, opportunities, and challenges. Military Psychology, 15(1), pp.3–16. DOI: 10.1207/S15327876MP1501_01

[82] Alliger, G. M., Tannenbaum, S. I., Bennett, W., Jr., Traver, H., & Shotland, A. (1997). A meta-analysis of the relations among training criteria. Personnel Psychology, 50(2), pp.341–358. DOI: 10.1111/j.1744-6570.1997.tb00911.x

[83] Ryan, R. M., & Deci, E. L. (2000). Self-determination theory and the facilitation of intrinsic motivation, social development, and well-being. American Psychologist, 55(1), pp.68–78. DOI: 10.1037110003-066X.55.1.68

[84] American Psychiatric Association. (2000). Diagnostic and statistical manual of mental disorders (4th ed.). Washington, DC: American Psychiatric Association. ISBN 0-89042-062-9

[85] Hooper, S. R., Poon, K. K., Marcus, L., & Fine, C. (2006). Neuropsychological characteristics of school-age children with high-functioning autism: performance on the NEPSY. Child Neuropsychology, 12, pp.299-305. DOI : 10.1080/09297040600737984.

[86] Rinehart, N. J., Bradshaw, J. L., Brereton, A. V., & Tonge, B. J. (2001). Movement preparation in high-functioning autism and Asperger disorder: a serial choice reaction time task involving motor reprogramming. Journal of Autism and Developmental Disorders, 31(1), pp.79-88. DOI: 10.1023/A:1005617831035

[87] Welsh, M., Parke, R. D., Widaman, K., & O'Neil, R. (2001). Linkages between children's social and academic competence: a longitudinal analysis. Journal of School Psychology, 39(6), pp.463-482. DOI: 10.1016/S0022-4405(01)00084-X

[88] Kandalaft, M. R., Didehbani, N., Krawczyk, D. C., Allen, T. T., & Chapman, S. B. (2013). Virtual Reality Social Cognition Training for young adults with high-functioning autism. Journal of Autism and Developmental Disorders, 43, pp.34-44. DOI: 10.1007/s10803-012-1544-6.

[89] Parsons, S., Mitchell, P., & Leonard, A. (2005). Do adolescents with autistic spectrum disorders adhere to social conventions in virtual environments? Autism, 9, pp.95-117. DOI: 10.1177/1362361305049032.

[90] Wallace, S., Parsons, S., Westbury, A., White, K., & Bailey, A. (2010). Sense of presence and atypical social judgments in immersive virtual environments: responses of adolescents with Autism Spectrum Disorders. Autism, 14, pp.199-213. DOI: 10.1177/1362361310363283

[91] Maskey, M., Lowry, J., Rodgers, J., McConachie, H., & Parr, J. R. (2014). Reducing specific phobia/fear in young people with Autism Spectrum Disorders (ASDs) through a virtual reality environment intervention. PLoS One, 9(7), DOI: 10.1371/journal.pone.0100374

[92] Parsons, S., & Mitchell, P. (2002). The potential of virtual reality in social skills training for people with autistic spectrum disorders. Journal of Intellectual Disability Research, 46(5), pp.430-443. DOI: 10.1046/j.1365-2788.2002.00425.x

[93] Parsons, S., Mitchell, P., & Leonard, A. (2004). The use and understanding of virtual environments by adolescents with autistic spectrum disorders. Journal of Autism and Developmental Disorders, 34(4), pp.449-466. DOI: 10.1023/B:JADD.0000037421.98517.8d

[94] Cheng, Y., & Ye, J. (2010). Exploring the social competence of students with autism spectrum conditions in a collaborative virtual learning environment: the pilot study. Computers and Education, 54, pp.1068-1077. http://dx.doi.org/10.1016/ j.compedu.2009.10.011.

[95] Herrera, G., Alcantud, F., Jordan, R., Blanquer, A., Labajo, G., & De Pablo, C. (2008). Development of symbolic play through the use of virtual reality tools in children with autistic spectrum disorders: two case studies. Autism, 12, 143-157. DOI: 10.1177/1362361307086657.

[96] Merchant, Z., Goetz, E. T., Cifuentes, L., Keeney-Kennicutt, W., & Davis, T. J. (2014). Effectiveness of virtual reality-based instruction on students' learning outcomes in K-12 and higher education: A meta-analysis. Computers and Education, 70, pp.29-40. DOI: 10.1016/j.compedu.2013.07.033

[97] Lee, E. A. L., & Wong, K. W. (2014). Learning with desktop virtual reality: Low spatial ability learners are more positively affected. Computers and Education, 79, pp.49-58. DOI: 10.1016/j.compedu.2014.07.010

[98] Nicholson, D. T., Chalk, C., Funnell, W. R. J., & Daniel, S. J. (2006). Can virtual reality improve anatomy education? A randomised controlled study of a computer-generated three-dimensional anatomical ear model. Medical Education, 40(11), pp.1081-1087. DOI: 10.1111/j.1365-2929.2006.02611.x

[99] Lee, E. A. L., & Wong, K. W. (2014). Learning with desktop virtual reality: Low spatial ability learners are more positively affected. Computers and Education, 79, pp.49-58. DOI: 10.1016/j.compedu.2014.07.010

[100] Barsalou, L. W. (1999). Perceptual symbol systems. Behavioral and Brain Sciences, 22, pp.577-609. DOI: 10.1017/S0140525X99532147

[101] Stull, A. T., Hegarty, M., & Mayer, R. E. (2009). Getting a handle on learning anatomy with interactive three-dimensional graphics. Journal of Educational Psychology, 101(4), pp.803-816. ISSN 0022-0663

[102] Hessinger, M., Holzinger, A., Leitner, D., & Wassertheurer, S. (2008). Hemodynamic models for education in physiology. Mathematics and Computers in Simulation, 79(4), pp.1039-1047. DOI: 10.1016/j.matcom.2008.02.015

[103] Polcar, J, Gregor M, Horejsi P, Kopecek P. (2016). Methodology for Designing Virtual Reality Applications, Proceedings of the 26th DAAAM International Symposium, pp.0768-0774, B. Katalinic (Ed.), Published by DAAAM International, ISBN 978-3-902734-07-5, ISSN 1726-9679, Vienna, Austria, DOI:10.2507/26th.daaam.proceedings.107

[104] Nijholt, A. (2000). Agent-supported cooperative learning environments. In Proceedings of the international workshop on advanced learning technologies. DOI: 10.1109/IWALT.2000.890554

[105] Kitchen, D., & McDougall, D. (1998). Collaborative learning on the Internet. Journal of Educational Technology Systems, 27, pp.245–258. ISSN-0047-2395

[106] Zhang, D., Zhao, J. L., Zhou, L., & Nunamaker, J. F. (2004). Can E-learning replace classroom learning? Communications of the ACM,47(5), pp.74–79. DOI: 10.1145/986213.986216

[107] McNeal B. J. and Nelson K. R., (1991). "Meta-analysis of interactive video instruction: A 10 year review of achievement effects,"J. COMP. BASED INSTR, vol. 18, no. 1, pp. 1–6. ISSN-0098-597X

[108] Keller J. M. and Litchfield B. C. (2002) "Motivation and Performance." Trends and Issues in Instructional Design and Technology. (R. A. Reiser & J. V. Dempsey, Eds.). Upper Saddle River, NJ: Pearson Education. ISBN-13: 9780132999441

[109] Sutcliffe, A. (2003). Multimedia and virtual reality. New York: Lawrence Erlbaum Associates. ISBN: 080583950X

[110] Limniou, M., Roberts, D., & Papadopoulos, N. (2008). Full immersive virtual environment CAVETM in chemistry education. Computer & Education, 51, pp.584-593. DOI: 10.1016/j.compedu.2007.06.014

[111] Minsky, M (2003): The Emotional Machine, http://web.media.mit.edu/~minsky/

[112] Picard, R.W., Papert, S., Bender, W., Blumberg, B., Breazeal, C.,Cavallo, D., Machover, T., Resnick, M., Roy, D. and Strohecker, C (2004): Affective learning — a manifesto. BT TechnologyJournal 22(4). DOI: 10.1023/B:BTTJ.0000047603.37042.33

[113] Fig. 1. Available from www.mortonheilig.com/InventorVR.html

[114] Fig. 2. Available from www.geek.com/news/the-history-of-virtual-reality-games-1652225

[115] Fig. 3. Available from www.cl3ver.com/blog/vr-vs-ar-in-architecture

DOI: 10.2507/28th.daaam.proceedings.101

CYBER SICKNESS IN VIRTUAL REALITY - LITERATURE REVIEW

Sergo Martirosov, Pavel Kopecek

This Publication has to be referred as: Martirosov, S[ergo] & Kopecek, P[avel] (2017). Cyber Sickness in Virtual Reality - Literature Review, Proceedings of the 28th DAAAM International Symposium, pp.0718-0726, B. Katalinic (Ed.), Published by DAAAM International, ISBN 978-3-902734-11-2, ISSN 1726-9679, Vienna, Austria
DOI: 10.2507/28th.daaam.proceedings.101

Abstract

In this paper, we present a review of the major issue of Virtual Reality - Cyber Sickness. As Virtual Reality becomes more and more popular and advanced technology, it is being used in a variety of areas, such as military, surgery, schools, etc. for training and education. Although at this point, Virtual Reality technology has multiple problems to overcome, Cyber Sickness issue is one of the most difficult ones, as it is closely linked to our physiology and Virtual Reality technology as well. Cyber Sickness symptoms are very similar to those of Motion Sickness, but are caused by Virtual Reality Environments. Due to its potential harm to human health, it is important to know the causes of this issue and possible ways to combat it. In this paper, we will highlight what aspects of Virtual Reality technology cause these symptoms and their outcomes.

Keywords: Virtual Reality; Cyber Sickness; Motion Sickness; Postural Stability

1. Introduction

In the modern time Virtual Reality (VR) technology became very popular and is used in many different areas such as entertainment, military, healthcare, education, engineering etc. Thanks to modern technological advance of computer hardware and software, it became possible to incorporate 3D VR in innovative applications of teaching, training, and learning [1][2][3]. Virtual environment or virtual world is a computer generated 3D representation of real or fictional environments. A user can interact in such an environment independently in the same pace, one would experience events in the real world [4]. VR has the ability to support realistic and immersive simulation and enables transfer of taught skills in VR into real contexts as well as provide multi-user, embodied and interactive active learning [5]. VR is believed to be a promising tool for training and complex problem solving, which requires weighing multiple variables and situational decision making [5][6][7][8]. As an example, VR can be used in ergonomics in order to find out time requirements for specific tasks, as well as help minimize the risks related to physical injuries [9].

Despite many studies in this area, there is still controversy about the real effects of VR on learning and cognition, but preliminary results indicate that the use of VR can increase learning performance [10], facilitate usability and enhance interaction [11], help to reconstruct and navigate through non-existent environments [12], help disabled learners with knowledge accessibility [13], etc. Serious games tend to be used often in formal education and with sufficient support are shown to be highly motivational and effective in learning complex tasks [14][15].

VR Technology can be of great use for emergency service personnel such as police, ambulance, fire fighters and army, as they have to be prepared for various operations at all times, thus continuous training is crucial for their success and security. But training in real environment is sometimes impossible, due to cost or complexity and the good alternative that comes handy is virtual training environment, which allows training for such complex collaborative tasks [16][17]. Thus, if the situation doesn't predispose for real life training practice due to danger, cost or effort, emergency service personnel can profit from virtual training to train for scenarios in real environment [18]. Often VR training is the only option that is possible [19]. And in order to make complex VR applications/simulations, some authors suggest methodologies for designing VR applications [20].

One of the topics that frequently comes up about VR is Cyber Sickness (CS). This is something similar to Motion Sickness (MS), as the symptoms are similar, but one is caused in virtual world, another in the real one. There are various applications and suggestions that try to measure and combat those symptoms of CS as they can potentially affect health in a negative way, which could lead to fatal accidents, such as car crash due to feeling sick. Previous research on this topic helps immensely, as it is important to understand what to expect and find the way to deal with that issue. The symptoms include: headache, nausea, vomiting, fatigue, postural instability, disorientation, etc. This sickness is a natural reaction to an unnatural environment and it happens due to the mismatch between sensory inputs. When those inputs don't agree, that's when you start feeling sick. In traditional MS the mismatch occurs because you feel movement in your muscles and your inner ear, but you don't see it, like it happens on the boat. With CS it's the opposite, you see movement on the screen, but you don't feel it.

Most of the experiments on evaluating CS in subjects are carried out in some kind of simulators, such as flying simulator, driving simulator, surgical simulator, etc. Or other variant of such environment is maze. A Simulation is a model of a real world where user has the ability to interact with the environment [21]. Simulations are helpful and useful as they provide realistic context in which individuals can explore, experiment and see immediate results as they create models of their own or try theories on the modeled concept [22]. Depending on the amount of senses simulated in VR, such as vision, hearing, touch, balance, even smell, the immersion level in the artificial world can vary. As every person is susceptible to CS in different ways, various environments and scenarios can cause divergent sickness levels in users as well. Thus, it would be wise to conduct more studies in different VR simulations to establish more precise link between training in that particular sphere and the levels of CS caused by it. Narrowing research to specific areas, would allow better understanding of things.

2. Cyber Sickness – Symptoms, Causes and Outcomes

There are a wide variety of research conducted in the field of VR on simulator sickness: simulator sickness in different types of moving images [23], comparison of one-screened and three-screened display types [24][25], relationship between age and sickness level [26], and gender effect on simulator sickness [27][28]. Research on virtual environments has provided converging evidence that being placed in a virtual environment with an HMD can lead to MS [29]. Previous studies suggest that a twenty-minute exposure to VREs can increase CS symptoms in over 60% of participants [30]. It has been found that exposure to VEs and simulators caused 30% of exposures result in nausea, up to 40% in eyestrain, as well as the experience of numerous other symptoms of MS including headache, dizziness/vertigo, loss of postural stability, drowsiness, salivation and sweating [31]. Numerous factors are related to the susceptibility and severity of visually induced motion sickness (VIMS) [32][33][34], including the sensation of illusory self-motion [35][36][37], gender [38], age [39], or personality factors like neuroticism and anxiety [40]. Due to many factors contributing to the cause of simulator sickness both from technological and individual standpoints, it is very difficult to predict [41]. It has been noted that psychological conditions such as anxiety can significantly increase when simulated motion increases [42]. Based on the theories of sensory conflict [43]and neural storage [44], changes in visual stimulus could cause sensory conflicts between different modalities (vision, touch, etc.) which in turn could evoke different levels of virtually induced MS.

Little is known on the subject of virtual environment effects on symptoms of CS, thus, a new system describing VE systems, the Virtual Environment Description and Classification System (VEDACS [45]) was proposed. VEDACS describes VEs in terms of 42 variables which are then divided into seven categories: (1) environmental dimensions, (2) visual complexity, (3) the motion of objects, (4) the tasks that users perform, (5) the provision of help/assistance, (6) the provision of sound, and (7) global measures such as the amount of time that each person spent in a VE.

It has been suggested that the size of the monitor and the resulting display angle can also influence the severity of VIMS [46][47]. Often head-mounted devices are used for displaying virtual environments and they are known to cause MS [48][49]. A research conducted to analyze simulator sickness score using a video game in VR with a HMD, notes that sickness occurred in all participants who played the game for a maximum of 50 min [50]. Similar research suggests that sickness occurred both, in users who played a game on a standard TV screen and those with a HMD [51]. One study, targeted at finding out effects of immersion in VRE using a 20 m experiment, concluded that remarkable effects of immersion could not be found under such a short duration condition [52]. Other research suggests that the duration of exposure might be a causal factor in postural disequilibrium and that exposure of less than 3 h would not be sufficient to induce an outstanding detriment in postural steadiness [53]. According to the sensory conflict theory [43][44][54], anything that strengthens the non-veridical visual stimulus should increase MS.

According to various studies, in virtual environments, high-latency that can result from low frame rates can lead to reduced sense of presence [55], impaired task performance, reduced user response [56][57] or simulator sickness [58]. It has been determined by some authors [59] that the critical end-to-end latency is at 17ms, but other authors [60] found this threshold to be even lower. When rotating head in virtual environment with head mounted display on, adaptation to a newly displayed scene has to happen and due to the delay in visual image, during the process of adaptations, some users may develop MS and show postural disequilibrium [61][62].

It is noted that VR environment influences automatic nervous system and equilibrium [63]. Sensory conflict theory states that when there are conflicting visual and vestibular inputs disequilibrium occurs, which results in MS [64]. When sensory conflict happens, the instability of persons standing positions occurs [65] and the influence of this instability may appear in humans walk after the righting reflex occurs [66]. It is a reflex that corrects the orientation of the body when it is taken out of its normal upright position. Various studies note that CS can effect 50-80 % of people, depending on the quality of the digital content and how it's presented [67]. Also, some studies show that women are more susceptible to MS than men, and those with a history of migraines or concussion also suffer a higher chance of getting motion sick [68]. Also, the older subjects are more susceptible to Simulator Sickness than younger subjects [26][69].

Some authors suggest that major contributor to MS is the duration of motion with visuals [70]. Others argue, that consumer HMDs require ideal frame rates in order to reduce/prevent the discomfort of MS [71]. Various researchers believe that decreasing Field of View (FOV) tends to decrease VR sickness, though at the cost of sense of presence [72]. A few authors came to a conclusion that a wide FOV display can maximize "immersion" in VE, whereas narrow one may hinder sense of presence [73][74]. As a result of increased FOV, users exhibited more postural instability, which has been suggested to be a surrogate measure for simulator sickness [75][76]. Differences in FOV have been related to navigation and memory differences [77][78][79]. Larger FOV can improve spatial awareness as well as increase sense of presence and consequently may help exhibit better memory structure of that VE [80]. Various methods, such as brain wave readings [81], behavioral metrics [82], and self-reported measures [83][84][85], have been employed over the years in order to measure users sense of presence in VEs.

Sense of presence is defined as "a state of consciousness", the psychological sense of being in the virtual environment [86][87][88][89][90]. A similar concept related to sense of presence is immersion – degree of "the subjective impression that one is participating in a comprehensive, realistic experience" and "the semi-voluntary experience of being transported into an alternate context for an extended duration" [7][91]. Previous research suggests that higher levels of immersion are associated with higher levels of presence [89][92][93]. Furthermore, the psychological perception of being inside a virtual world is known as presence [94]. Among many factors affecting user's feeling of presence, one is the content of the virtual world [95]. Immersive experience can be created via the sensory and environmental fidelity in 3D virtual environment, engagement with virtual objects and activities, and emotional experience activated through realistic scenarios in virtual world [8].

In some research it is noted that 3D video yields higher discomfort levels than 2D video [96]. There is an interesting conclusion made by some authors, stating that the highest value of sickness occurs during rotation along roll axis and that the results were highly consistent [97]. A study was conducted to analyze Visually Induced Sickness during day-night driving sessions, which showed that sickness level in night sessions was much higher than in morning sessions. In addition to that it was proved that sleep deprivation may strengthen the symptoms [98]. It's interesting to notice that in traditional MS studies, younger subjects were found to be more prone to sickness than older subjects [43], whereas in others, many older visitors appeared to suffer from effects of CS more than younger subjects [99].

It has been noted that some users might exhibit symptoms of CS, both during and after VR experience [100][101]. CS is different from MS in the way that the user is stationary, but has a compelling sense of motion as the visual imagery changes [102][103]. Vection or otherwise illusions of self-motion and sickness can be caused by watching changing moving scenes in VE [104]. Also, it is found that users spending more than 20 min in VE experience progressively more sickness symptoms [105]. Studies suggest that both exposure duration and repeated exposure are significantly linearly related to sickness outcomes [106]. The repetitive watching of the same video image reduced subjective score of MS in 8 of the 14 participants [107]. Some authors found that scene oscillations produce vection and promote CS significantly [108]. Simulation Sickness Questionnaire (SSQ) has been found to be valid and reliable subjective measure tool for both simulator sickness and side effects experienced in VEs [109][110]. It's interesting to note, that MS after immersion in VRE is much greater when both pre and post questionnaire are given rather than when only post test questionnaire is used [111].

Some previous studies suggest that the discomfort of MS can cause self-control ability decline and lead to serious traffic accident fatalities [43]. And as it was mentioned that the discomfort of all motion related sicknesses may even last for the entire day [112]. This adds some risk to using VR, as many people are very susceptible to VR Sickness and have to cope with that discomfort after their experience with it through the rest of the day, during which fatal accidents may occur and not only during driving a vehicle. For these reasons some studies are conducted to determine a set of MS indicators that would predict the occurrence of a person's MS as soon as possible [113]. Specifically they use electroencephalography (EEG), as among other imaging technologies, it outperforms the other methods in terms of portability and temporal resolution, which refers to the precision of a measurement with respect to time.

Currently there is a debate concerning VR technology and VR sickness. Some argue that the problem of VR sickness is inherent from this technology itself and as long as there is mismatch between what is visually perceived and what is physically sensed, fixing the VR display properties does not address the root cause [114]. Particular study indicates that the weight of the HMD devices themselves play big part in causing VR sickness [114]. The 1994 US Navy study reported that techniques such as biofeedback, relaxation training, and counseling had an estimated 40 to 85 % success rate in helping reduce susceptibility to MS [114]. MS drugs for the most part claim to prevent, not treat, MS [114]. It is suggested that a visible horizon has a mitigating effect on MS [115][116].

Researches were conducted to find out the effect of virtual and real walking in virtual environment and the results seem controversial. A study found that during the task requiring a navigationally complex environment and for a long period of time, it is preferable to use simulated walking [117]. On the other hand, another study reported that natural walking caused less MS than simulated walking in one out of two experiments involving a virtual maze [118]. During a similar research no differences were found in simulator sickness between real walking and several virtual travel techniques in a small virtual room [119]. Furthermore, no differences were found between real walking, gaze-directed and pointing-directed travel techniques in a complex 3D maze [120].

3. Conclusion

In this paper we have showed a multitude of studies conducted on the subject of Cyber Sickness. This issue, which is similar to Motion Sickness, is directly linked to Virtual Reality technology and as research suggests, could potentially harm human health and life. As benefits of Virtual Reality are tremendous, and there is no doubt that it will be used widely in the future, this means that dealing with the issue of Cyber Sickness is inevitable, thus it is important to understand what are the causes of it and what could be done to combat it. Previous research and experiments help with this immensely, as they provide useful information that could be used for future studies. It could be used to allow understanding of what to await while working with Virtual Reality applications and simulations, take every precaution and be prepared for most scenarios.

As every person is susceptible to Cyber Sickness in different ways and every scenario in Virtual Reality is different, in order to have precise knowledge on this topic, it is necessary to produce more research and narrowing the scope to specific areas and scenarios would allow better understanding of this issue.

4. Acknowledgment

This article was created with the subsidy of University of West Bohemia internal grant number SGS-2015-065 titled "Development of a sustainable production system parameters".

5. References

[1] Abulrub, A. G., Attridge, A. N., & Williams, M. A. (2011). Virtual reality in engineering education: the future of creative learning. In Global engineering education conference (EDUCON), 2011 IEEE (pp.751-757). IEEE. DOI:10.1109/EDUCON.2011.5773223

[2] Gregory, S., Gregory, B., Reiners, T., Fardinpour, A., Hillier, M., Lee, M., Basu, A. (2013). Virtual worlds in Australian and New Zealand higher education: remembering the past, understanding the present and imagining the future. In ASCILITE 2013 "Electric dreams", 1-4 december 2013. Sydney, NSW: Macquarie University. ISBN 9781741384031

[3] Jou, M., & Wang, J. (2013). Investigation of effects of virtual reality environments on learning performance of technical skills. Computers in Human Behavior, 29(2), pp.433-438. DOI: 10.1016/j.chb.2012.04.020

[4] Mitchell, P., Parsons, S., & Leonard, A. (2007). Using virtual environments for teaching social understanding to 6 adolescents with autistic spectrum disorders. Journal of autism and developmental disorders, 37(3), pp.589-600. DOI: 10.1007/S10803-006-0189-8

[5] Cheng, Y., & Wang, S. H. (2011). Applying a 3D virtual learning environment to facilitate student's application ability-the case of marketing. Computers in Human Behavior, 27(1), pp.576-584. DOI: 10.1016/j.chb.2010.10.008

[6] Bertram, J., Moskaliuk, J., & Cress, U. (2015). Virtual training: making reality work? Computers in Human Behavior, 43, pp.284-292. DOI: 10.1016/j.chb.2014.10.032

[7] Dede, C. (2005). Millennial learning styles. Educause Quarterly, 28(1), 7-12.

[8] Dede, C. (2009). Immersive interfaces for engagement and learning. Science, 323(5910), pp.66-69. DOI: 10.1126/science.1167311

[9] Bures, M. (2016). Time Demands of Virtual Ergonomic Modelling – Experimental Study, Proceedings of the 26th DAAAM International Symposium, pp.0286-0291, B. Katalinic (Ed.), Published by DAAAM International, ISBN 978-3-902734-07-5, ISSN 1726-9679, Vienna, Austria, DOI:10.2507/26th.daaam.proceedings.039

[10] Carpenter T. & Anderson L. (1996). Using virtual reality technology for learning design skills, Proceedings of ED-MEDIA 96, Boston, USA, pp.752

[11] Mikropoulos T. & Nikolou E. (1996). A virtual hand with tactile feedback for virtual learning environments,Proceedings of ED-MEDIA 96, Boston, USA, pp.792

[12] Littman M.(1996). Enhancing instruction through virtual reality, Proceedings of ICTE, New Orleans, USA,pp.31-33

[13] Lumbreras M., Sánchez J., Barcia M. (1996), A 3D sound hypermedial system for the blind. In Proceedings of the First European Conference on Disability, Virtual Reality and Associated Technologies, pp.187-191, Maidenhead, UK. DOI: 10.1145/302979.303101

[14] Chatti, Mohamed Amine, et al. (2012). "A reference model for learning analytics." International Journal of Technology Enhanced Learning 4.5- 6: pp.318-331. DOI: 10.1504/IJTEL.2012.051815

[15] Eseryel, D., Law, V., Ifenthaler, D., Ge, X., & Miller, R. (2014). An investigation of the interrelationships between motivation, engagement, and complex problem solving in game-based learning. Journal of Educational Technology & Society, 17(1), pp.42-53.

[16] Moskaliuk, Bertram & Cress (2013a). Impact of virtual training environments on the acquisition and transfer of knowledge. Cyberpsychology, Behavior, and Social Networking, 16(3), pp.210–214. DOI: 10.1089/cyber.2012.0416

[17] Moskaliuk, Bertram & Cress (2013b). Training in virtual environments: Putting theory into practice. Ergonomics, 56(2), pp.195–204. DOI: 10.1080/00140139.2012.745623

[18] Rose, F. D., Attree, E. A., Brooks, B. M., Parslow, D. M., Penn, P. R., & Amibihaipahan, N. (2000). Training in virtual environments: Transfer to real world tasks and equivalence to real task training. Ergonomics, 43(4), pp.494–511. DOI: 10.1080/001401300184378

[19] Romano, D. M., & Brna, P. (2001). Presence and reflection in training: Support for learning to improve quality decision-making skills under time limitations. CyberPsychology & Behavior, 4(2), pp.265–277. DOI: 10.1089/109493101300117947

[20] Polcar, J, Gregor M, Horejsi P, Kopecek P. (2016). Methodology for Designing Virtual Reality Applications, Proceedings of the 26th DAAAM International Symposium, pp.0768-0774, B. Katalinic (Ed.), Published by DAAAM International, ISBN 978-3-902734-07-5, ISSN 1726-9679, Vienna, Austria, DOI:10.2507/26th.daaam.proceedings.107

[21] Thurman, R. A. (1993). Instructional simulation from a cognitive psychology viewpoint. Educational technology research and development, 41(4), pp.75-89. DOI: 10.1007/BF02297513

[22] Rieber, L.P. (1992). Computer-based microworlds: a bridge between constructivism and direct instruction. Educational Technology Research and Development, 40(1), pp.93-106. DOI: 10.1007/BF02296709

[23] Kuze, J. and Ukai, K. (2008) : "Evaluation of Visual Fatigue Caused by Motion Images", Elsevier, Display, Vol.29, pp.159-166. DOI: 10.1016/j.displa.2007.09.007

[24] Seay A.F., Krum, D.M., Hodges, L. and Ribarsky, W., (2001). "Simulator Sickness and Presence in a High FOV Virtual Environment," IEEE Virtual Reality Conference 2001 (VR2001), Japan, March 13-17, pp.299-300. DOI: 10.1109/VR.2001.913806

[25] Chompoonuch J. and Kazuhiko H., (2011) "Study on Parallax Affect on Simulator Sickness in One-Screen and Three-screen Immersive Virtual Environment", Proceeding of The School of Information and Telecommunication Engineering, Tokai University, Vol.4, No.1, pp.34-39.

[26] Laura L. Arns and Melinda M. Cerney (2005). "The Relationship Between Age and Incidence of Cyber sickness Among Immersive Environment Users", Proceedings of the IEEE Virtual Reality, pp.267-268. DOI: 10.1109/VR.2005.1492788

[27] Thomas D. Parsons, Peter Larson, Kris Kratz, Marcus Thiebaux, Brendon Bluestein, J. Galen Buckwalter, and Albert A. Rizzo (2004): "Sex Differences in Mental Rotation and Spatial Rotation in a Virtual Environment", Elsevier Ltd., Neuropsychologia 42, pp.555-562. DOI: 10.1016/j.neuropsychologia.2003.08.014

[28] Evan A. Suma, Samantha L. Finkelstein, Seth Clark, Paula Goolkasian, and Larry F. Hodges (2010). "Effects of Travel Technique and Gender on a Divided Attention Task in a Virtual Environment", IEEE Symposium on 3D User Interface, March 20-21, USA, pp.27-34. DOI: 10.1109/3DUI.2010.5444726

[29] DiZio, P. and Lackner, J.R. (1998). "Alleviation of motion sickness and postural instability during and after virtual environment exposure" (Technical Report). Prepared for Naval Air Warfare Center Training Systems Division, Orlando, FL.

[30] Regan, E. and Price, K. (1994). "The Frequency and Occurrence and Severity of Side-Effects of Immersion in Virtual Reality,". Aviation Space and Environmental Medicine, vol. 65, pp. 527-530.

[31] Kennedy, R. S., Fowlkes, J. E. and Lilienthal, M. G., (1993). "Postural and performance changes in Navy and Marine Corps pilots following flight simulators," Aviation, Space, and Environmental Medicine, Vol. 64, pp.912-920, 1993.

[32] Golding, J.F. (2006). Predicting individual differences in motion sickness susceptibility by questionnaire, Personality and Individual Differences 41 (2), pp. 237– 248. DOI: 10.1016/j.paid.2006.01.012
[33] Shupak, A. and Gordon, C.R. (2006). Motion sickness: advances in pathogenesis, prediction, prevention, and treatment, Aviation Space and Environmental Medicine 77 (12), pp.1213–1223.
[34] Stanney, K.M, Hale, K.S., Nahmens, I., Kennedy, R.S. (2003). What to expect from immersive virtual environment exposure: Influences of gender, body mass index, and past experience, Human Factors 45 (3), pp.504–520. DOI: 10.1518/hfes.45.3.504.27254
[35] Flanagan, M.B., May, J.G. and Dobie, T.G. (2004). The role of vection eye movements and postural instability in the etiology of motion sickness, Journal of Vestibular Research-Equilibrium and Orientation 14 (4), pp.335–346.
[36] Hettinger, L.J., Berbaum, K.S., Kennedy, R.S., Dunlap, W.P. and Nolan, M.D. (1990). Vection and simulator sickness, Military Psychology 2 (3), pp.171–181. DOI: 10.1207/s15327876mp0203_4
[37] Smart, L.J., Stoffregen, T.A. and Bardy, B.G. (2002). Visually induced motion sickness predicted by postural instability, Human Factors 44 (3), pp.451–465. DOI: 10.1518/0018720024497745
[38] Flanagan, M.B., May, J.G. and Dobie, T.G. (2005). Sex differences in tolerance to visuallyinduced motion sickness, Aviation Space and Environmental Medicine 76 (7), pp.642–646.
[39] Turner, M. and Griffin, M.J. (1999). Motion sickness in public road transport: passenger behavior and susceptibility, Ergonomics 42 (3), pp.444–461. DOI: 10.1080/001401399185586
[40] Bick, P.A. (1983). Physiological and psychological correlates of motion sickness, The British Journal of Medical Psychology 56 (Pt 2), pp.189–196. DOI: 10.1111/j.2044-8341.1983.tb01546.x
[41] LaViola Jr., J.J. (2000). "A Discussion of Cybersickness in Virtual Environment", SIGCHI Bulletin, Vol.32, No.1, pp.47-56. DOI: 10.1145/333329.333344
[42] Bruck, S. and Watters, P.A. (2009). "Cybersickness and Anxiety During Simulated Motion: Implications for VRET." Annual Review of CyberTherapy and Telemedicine (ARCTT). in press. DOI: 10.3233/978-1-60750-017-9-169
[43] Reason, J.T. and Brand, J.J. (1975). Motion Sickness, Academic Press, London. ISBN-13: 978-0125840507
[44] Reason, J.T. (1978). Motion sickness adaptation: a neural mismatch model, J. R. Soc. Med. 71 (11), pp.819.
[45] Nichols, S., Ramsey, A.D., Cobb, S., Neale, H., D'Cruz, M. and Wilson, J.R. (2000). Incidence of virtual reality induced symptoms and effects (VRISE) in desktop and projection screen displays, HSE Books: Sudbury, Suffolk, UK. ISBN 0 7176 1810 2
[46] Duh, H.B.L., Parker, D.E. and Furness, T.E. (2004). An independent visual background reduced simulator sickness in a driving simulator, Presenence: Teleoperators and Virtual Environments 13 (5), pp.578–588. DOI: 10.1162/1054746042545283
[47] Emoto, M., Sugawara, M., and Nojiri, Y. (2008). Viewing angle dependency of visuallyinduced motion sickness in viewing wide-field images by subjective and autonomic nervous indices, Displays 29 (2), pp.90–99. DOI: 10.1016/j.displa.2007.09.010
[48] Mon-Williams, M., Wann, J.P. and Rushton, S. (1993). Binocular vision in a virtual world: visual deficits following the wearing of a head-mounted display, Ophthalmic & Physiological Optics 13 (4), pp.387–391. DOI: 10.1111/j.1475-1313.1993.tb00496.x
[49] Patterson, R., Winterbottom, M.D. and Pierce, B.J. (2006). Perceptual issues in the use of head-mounted visual displays, Human Factors 48 (3), pp.555–573. DOI: 10.1518/001872006778606877
[50] Merhi, O., Faugloire, E., Flanagan, M. and Stoffregen, T.A. (2007). Motion sickness, console videogames, and head-mounted displays,Human Factors 49 (5), pp.920–934. DOI: 10.1518/001872007X230262
[51] Stoffregen, T.A., Faugloire, E., Yoshida, K., Flanagan, M.B. and Merhi, O. (2008). Motion sickness and postural sway in console video games, Human Factors 50 (2), pp.322–331. DOI: 10.1518/001872008X250755
[52] Cobb, S.V.G. (1999). "Measurement of postural stability before and after immersion in a virtual environmlent," Applied Ergonomics, vo1.30, pp.45-57. DOI: 10.1016/S0003-6870(98)00038-6
[53] Kennedy, R.S. and Lilienthal, M.G. (1995). "Implications of balance disturbances following exposure to virtual reality systems," Proceedings of Virtual Reality Annual Intemational Symposium'gfi, IEEE Computer Society Press, Los Alamitos : CA, pp.35-39. DOI: 10.1109/VRAIS.1995.512477
[54] Oman, C.M. (1990). Motion sickness: a synthesis and evaluation of the sensory conflict theory, Canadian Journal of Physiology and Pharmacology 68 (2), pp.294– 303. DOI: 10.1139/y90-044
[55] Slater, M., Lotto, B., Arnold, M.M. and Sanchez-Vives, M.V. (2009). How we experience immersive virtual environments: the concept of presence and its measurement. Anuario de Psicolog´ıa, 40(2): pp.193–210.
[56] Ellis, S.R., Adelstein, B.D., Baumeler, S., Jense, G.J. and Jacoby, R.H. (1999). Sensor Spatial Distortion, Visual Latency, and Update Rate Effects on 3D Tracking in Virtual Environments. In Proceedings of IEEE Virtual Reality, pp.218–221. DOI: 10.1109/VR.1999.756954
[57] MacKenzie I.S. and Ware, C. (1993). Lag As a Determinant of Human Performance in Interactive Systems. In Proceedings of the INTERACT '93 and CHI '93 Conference on Human Factors in Computing Systems, pp.488–493. DOI: 10.1145/169059.169431
[58] Buker, T.J., Vincenzi, D.A. and Deaton, J.E. (2012). The Effect of Apparent Latency on Simulator Sickness While Using a See-Through HelmetMounted Display: Reducing Apparent Latency With Predictive Compensation. Human Factors, 54(2): pp.235–249. DOI: 10.1177/0018720811428734

[59] Adelstein, B.D., Lee, T.G. and Ellis, S.R. (2003). Head Tracking Latency in Virtual Environments: Psychophysics and a Model. In Proceedings of the Human Factor Ergonomics Society, pp.2083–2087. DOI: 10.1177/154193120304702001

[60] Jerald, J.J. (2010). Scene-Motion- and Latency-Perception Thresholds for HeadMounted Displays. PhD thesis, University of North Carolina at Chapel Hill.

[61] Kolasinski, E.M. and Gilson, R.D. (1999) Ataxia following exposure to a virtual environment, Aviat. Space Environ. Med. 70, pp.264–269.

[62] Regan, E.C. and Price, K.R. (1994). The frequency of occurrence and severity of side-effects of immersion virtual reality, Aviat. Space Environ. Med. 65, pp.527–530.

[63] Nakagawa, C., Ohsuga, M., Takeda, T., (2000). Basic study toward VE-sickness assesment, The Japanese Journal of Ergonomics,vol.36(3), pp.130-138. DOI: 10.5100/jje.36.131

[64] Money KE (1970) Motion Sickness. Am. Psychol. Soc. 50: pp.1–39

[65] Murata. A. and Miyoshi, Y., (2000). Effects of Duration of Immersion in a Virtual Environment on Postual Stability, Proc of IEEE-SMC2000, pp.961- 966. DOI: 10.1109/ICSMC.2000.885974

[66] Hreljac, A. and Marshall, N., (2000). Algorithms to determine event timing during normal waking using kinematic data, Journal of Biomechanics,Vol.33, pp.783-786. DOI: 10.1016/S0021-9290(00)00014-2

[67] Chang, C., Pan, W., Tseng, L. et al. Experimental Brain Research (2012) 217: pp.299-309. DOI: 10.1007/s00221-011-2993-4

[68] Koslucher, F., Haaland, E., TA Stoffregen. (2016). Experimental Brain Besearch 234 (1), pp.313-322. DOI: 10.1007/s00221-015-4462-y

[69] Johnell O. Brooks, Richard R. Goodenough, Matthew C. Crisler, Nathan D. Klein, Rebecca L. Alley, Beatrice L. Koon, William C. Logan Jr., Jennifer H. Ogle, Richard A. Tyrrell and Rebekkah F. Wills, (2010). "Simulator Sickness during Driving Simulation Studies," Elsevier Accident Analysis and Prevention 42, pp.788-796. DOI: 10.1016/j.aap.2009.04.013

[70] Mohsen, S., Nika, Z., Kristina, M., (2013). Abdominal Muscle Behavior and Motion Sickness During Paired Visual Input with Roll Motion, Poland. DOI: 10.1109/HSI.2013.6577831

[71] Edward, P., Christoff, L., Burkhard, W., (2013). More for Less: Fast Image Warping for Improving the Appearance of Head Tracking on HMDs, 28th Intern ational Conference on Image and Vision Computing, New Zealand. DOI: 10.1109/IVCNZ.2013.6726990

[72] Ajoy, F., Steven, F., (2016). Combating VR Sickness through Subtle Dynamic Field-Of-View Modification. DOI: 10.1109/3DUI.2016.7460053

[73] Prothero, J.D. and Hoffman, H.D. (1995). "Widening the field-ofview increase the sense of presence within immersive virtual environments", Human Interface Technology Laboratory Tech. Rep. R-95-4, Seattle: University of Washington.

[74] Stanney, K. and Salvendy, G. (1998). "Aftereffects and sense of presence in virtual environments: formulation of a research and development agenda", International Journal of Human- Computer Interaction, 10(2), pp.135-187. DOI: 10.1207/s15327590ijhc1002_3

[75] Duh, H.B.L., Lin, J.J.W., Kenyon, R.V., Parker, D.E. and Furness, T.A. (2001). "Effects of field of view on balance in an immersive environment", Proceedings of IEEE Virtual Reality 2001, Yokohama, Japan. DOI: 10.1109/VR.2001.913791

[76] Duh, H.B.L., Lin, J.J.W., Kenyon, R.V., Parker, D.E. and Furness, T.A. (2002). "Effects of characteristics of image quality on balance in an immersive environment", Presence: Teleoperators and Virtual Environments. DOI: 10.1162/105474602317473259

[77] Alfano, P. and Michel, G. (1990). "Restricting the field-of-view: perceptual and performance effects", Perceptual and Motor Skills, 70, pp.35-45. DOI: 10.2466/pms.1990.70.1.35

[78] McCreary, F.A. and Williges, R.C. (1998). "Effects of age and field-of-view on spatial learning in an immersive virtual environment", Proceedings of the Human Factors and Ergonomics Society 42nd Annual Meeting, 1998, pp. 1491-1495.

[79] Witmer, B.G., Bailey, J.H. and Knerr, B.W. (1994). "Training dismounted soldiers in virtual environments: Route learning and transfer", ARI Technical Report 1022, Alexandria, VA: U.S. Army Research Institute for the Behavioral and Social Sciences. From http://www.dtic.mil/get-tr-doc/pdf?AD=ADA292900

[80] Usoh, M., Catena, E., Arman, S. and Slater, M. (2000). "Using Presence Questionnaires in Reality", Presence: Teleoperators and Virtual Environment, 9(5), pp.497-503. DOI: 10.1162/105474600566989

[81] Pugnetti, L., Mendozzi, L., Cattaneo, A.M., Guzzetti, S., Mezzetti, C., Cglati, C., Brancotti, A. and Venanzi, D. (1995). "Psychophysiological activity during virtual reality (VR) testing", Journal of Psychophysiology, Vol. 8, No. 4, pp.361-362.

[82] Usoh, M., Arthur, K., Whitton, M., Bastos, R., Steed, A., Slater, M. and Brooks, F. (1999). "Walking > Walking-in-Place > Flying in Virtual Environments", Proc. of ACM SIGGRAPH 99, pp.359-364. DOI: 10.1145/311535.311589

[83] Kennedy, R., Lane, N., Berbaum, K. and Lilienthal, M. (1993). "A Simulator Sickness Questionnaire (SSQ): A New Method for Quantifying Simulator Sickness", International Journal of Aviation Psychology, Vol. 3, No. 3, pp.203-220. DOI: 10.1207/s15327108ijap0303_3

[84] Slater, M. and Steed, A. (2000). "A Virtual Presence Counter", Presence, Vol. 9, No. 5, pp.413-434. DOI: 10.1162/105474600566925
[85] Slater, M., Usoh, M. and Steed, A. (1994). "Depth of Presence in Virtual Environments", Presence, Vol. 3, No. 2, pp. 145-157. DOI: 10.1162/pres.1994.3.2.130
[86] Barfield, W., & Hendrix, C. (1995). The effect of update rate on the sense of presence within virtual environments. Virtual Reality, 1(1), 3-15. DOI: 10.1007/BF02009709
[87] Heeter, C. (1992). Being there: the subjective experience of presence. Presence: Teleoperators and virtual environments, 1(2), pp.262-271. DOI: 10.1162/pres.1992.1.2.262
[88] Lee, K. M. (2004). Presence, explicated. Communication theory, 14(1), pp.27-50. DOI: 10.1111/j.1468-2885.2004.tb00302.x
[89] Slater, M., & Wilbur, S. (1997). A framework for immersive virtual environments (FIVE): speculations on the role of presence in virtual environments. Presence: Teleoperators and Virtual Environments, 6(6), pp.603-616. DOI: 10.1162/pres.1997.6.6.603
[90] Witmer, B. G. B. G., Jerome, C., & Singer, M. (2005). The factor structure of the presence questionnaire. Presence, 14(3), pp.298-312. DOI: 10.1162/105474605323384654
[91] Buchanan, K. (2006). Beyond attention-getters: Designing for deep engagement (Unpublished doctoral dissertation). Michigan: Michigan State University. Retrieved from http://education.uwsp.edu/publications/buchanan_2006_beyond_attention.pdf
[92] Bystrom, K. E., Barfield,W., & Hendrix, C. (1999). A conceptual model of the sense of presence in virtual environments. Presence: Teleoperators and Virtual Environments, 8(2), pp.241-244. DOI: 10.1162/105474699566107
[93] Faiola, A., Newlon, C., Pfaff, M., & Smyslova, O. (2013). Correlating the effects of flow and telepresence in virtual worlds: enhancing our understanding of user behavior in game-based learning. Computers in Human Behavior, 29(3), pp.1113-1121. DOI: 10.1016/j.chb.2012.10.003
[94] Witmer B. and Singer, M. (2000). "Measuring presence in virtual environments: A presence questionnaire". Presence: Teleoperators and Virtual Environments, vol. 9, issue 5, 2000, pp.497-503. DOI: 10.1162/105474698565686
[95] Ijsselsteijn, W.A., H. de Ridder, Freeman., J. and Avons, S.E. (2000). "Presence: Concept, determinants and measurement", Proceedings of the SPIE, Human Vision and Electronic Imaging V, San Jose, CA,3959-3976. ISBN 0-8194-3577-5
[96] Naoki, K., Hiroki, Y., Masahiro, I., Yutetsu, M., (2015). Effects of Visual Induced Motion Sickness of Stereoscopic 3D Interactive Video, IEEE 4th Global Conference on Consumer Electronics (GCCE). DOI: 10.1109/GCCE.2015.7398678
[97] Ujike, H., Yokoi, T. and Saida, S. (2004). Effects of Virtual Body Motion on Visually-Induced Motion Sickness, Proceedings of the 26th Annual International Conference of the IEEE EMBS. DOI: 10.1109/IEMBS.2004.1403694
[98] Sunu, W., Titis, W., Hanung, A., Muhhamad, B., Mumtaz, N., (2015). Quantifying visual attention and visually induced motion sickness during day-night driving and sleep deprivation, International Conference on Data and Software Engineering. DOI: 10.1109/ICODSE.2015.7436996
[99] Pausch, R., Crea, T., & Conway, M. (1992). A literature survey for virtual environments: Military Flight Simulators Visual Systems and Simulator Sickness. Presence,1(3): pp.344-363. DOI: 10.1162/pres.1992.1.3.344
[100] Bowman, D., Kruijff, E., LaViola J.J. and Poupyrev, I. (2001). "An introduction to 3-D user interface design," Teleoperators and Virtual Environments, 2001, vol. 10, pp. 96-108. DOI: 10.1162/105474601750182342
[101] Rich, C.J. and Braun, C.C. "Assessing the impact of control and sensory compatibility on sickness in virtual environments," Proceedings of Human Factors and Ergonomics Society 40th Annual Meeting 1996, 2-6 September, Philadelphia, USA. DOI: 10.1177/154193129604002203
[102] Hale, K.S., Stanney, K.M. (2002). Handbook of Virtual Environments, New York: Earlbaum. ISBN 9780805832709
[103] Lathan, R. (2001). "Tutorial: a brief introduction to simulation sickness and motion programming," Real Time Graphics, vol. 9, pp.3-5.
[104] So, H.Y., Lo, W.T. and Ho, T.K. (2001). "Effects of navigation speed on motion sickness caused by an immersive virtual environment," Human Factors, 2001, vol. 43, pp.452-461. DOI: 10.1518/001872001775898223
[105] Regan, C. (1995). "An investigation into nausea and other side-effects of headcoupled immersive virtual reality," Virtual Reality, 1995, vol. 1, pp.17- 32. DOI: 10.1007/BF02009710
[106] Kennedy, R. S., Stanney, K. M. and Dunlap, W. (2000). "Duration and exposure to virtual environments: sickness curves during and across sessions," Presence: Teleoperators and Virtual Environments, vol. 9, pp.463-472. DOI: 10.1162/105474600566952
[107] Norihiro Sugita, Makoto Yoshizawa and Makoto Abe, (2007). "Evaluation of Adaptation to Visually Induced Motion Sickness by Using Physiological Index Associated with Baroreflex Function," in Proceedings of the 29th Annual International Conference of the IEEE EMBS, France, pp.303-306, , 23-26 August. DOI: 10.1109/IEMBS.2007.4352284
[108] Lo, W. T. and So, R.-H. Y. (2001). "Cybersickness in the presence of scene rotational movements along different axes," Applied Ergonomics, vol. 32, pp.1-14. DOI: 10.1016/S0003-6870(00)00059-4

[109]Kiryu, T. and So, R. H. Y. (2007). "Sensation of presence and cybersickness in applications of virtual reality for advanced rehabilitation," Journal of NeuroEngineering and Rehabilitation, vol. 4, pp.34. DOI: 10.1186/1743-0003-4-34

[110]Hale K.S. and Stanney, K.M. (2006)."Effects of low stereo acuity on performance, presence and sickness within a virtual environment," Applied Ergonomics, vol. 37, pp.329-339. DOI: 10.1016/j.apergo.2005.06.009

[111]Sean D. Y., Bernard D. A. and Stephen R. E., (2006). "Demand Characteristics of a Questionnaire Used to Assess Motion Sickness in a Virtual Environment," in Proceeding of the IEEE Virtual Reality Conference, pp.97-102, 25-29. DOI: 10.1109/VR.2006.44

[112]Gower, D.W. and Fowlkes, J.E. (1989). "Simulator Sickness in the UH-60 (Black Hawk) Flight Simulator", U.S. Army Aeromedical Research Laboratory, September.

[113]Chin-Teng, L., Shu-Fang, T., Li-Wei, K., (2013). EEG-Based Learning System for Online Motion Sickness Level Estimation in a Dynamic Vehicle Environment, IEEE Trans. on Neural Networks and Learning Systems, Vol. 24, no. 10. DOI: 10.1109/TNNLS.2013.2275003

[114]Michael, P., (1998). Motion Sick in Cyberspace, IEEE Computer Graphics and Applications. DOI: 10.1109/38.637264

[115]Steele, J.E., (1961). Motion Sickness and Spatial Perception. A Theoretical Study. from http://www.dtic.mil/get-tr-doc/pdf?AD=AD0273602

[116]Tal, D., Wiener, G. and Shupak, A. (2014). Mal de debarquement, motion sickness and the effect of an artificial horizon, J. Vestibular Res.: Equilibr. Orient. 24 (1), pp.17–23, DOI: 10.3233/VES-130505.

[117]Evan A. Suma, Samantha L. Finkelstein, Myra Reid, Amy Ulinski and Larry F. Hodges, (2009). "Real Walking Increases Simulator Sickness in Navigationally Complex Virtual Environments, " in IEEE Virtual Reality, Louisiana, USA, pp.245-246. DOI: 10.1109/VR.2009.4811037

[118]Chance, S.S., Gaunet, F., Beall, A.C and Loomis, J.M. (1998). Locomotion mode affects the updating of objects encountered during travel: The contribution of vestibular and proprioceptive inputs to path integration. Presence, 7(2): pp.168–178. DOI: 10.1162/105474698565659

[119]Zanbaka, C.A., Lok, B.C., Babu, S.V., Ulinski, A.C. and Hodges, L.F. (2005). Comparison of path visualizations and cognitive measures relative to travel technique in a virtual environment. IEEE TVCG, 11(6): pp.694–705. DOI: 10.1109/TVCG.2005.92

[120]Suma, E.A., Babu, S. and Hodges, L.F. (2007). Comparison of travel techniques in a complex, multi-level 3d environment. In IEEE Symposium on 3D User Interfaces, pp.147–153. DOI: 10.1109/3DUI.2007.340788

DOI: 10.2507/28th.daaam.proceedings.102

USING AUGMENTED REALITY IN SMART MANUFACTURING

Crina Narcisa Deac, Gicu Calin Deac, Cicerone Laurentiu Popa,
Mihalache Ghinea & Costel Emil Cotet

This Publication has to be referred as: Deac, C[rina] N[arcisa]; Deac, C[rina] N[arcisa]; Popa, C[icerone] L[aurentiu]; Ghinea, M[ihalache] & Cotet, C[ostel] E[mil] (2017). Using Augmented Reality in Smart Manufacturing, Proceedings of the 28th DAAAM International Symposium, pp.0727-0732, B. Katalinic (Ed.), Published by DAAAM International, ISBN 978-3-902734-11-2, ISSN 1726-9679, Vienna, Austria
DOI: 10.2507/28th.daaam.proceedings.102

Abstract

Augmented Reality (AR) is a technology widely used in smart manufacturing, which makes data visible in real time, while the user is interacting with the real world. This paper describes a new web based AR application developed by the authors for smart manufacturing architectures, providing portability to any device, free from any existing AR platforms based on reading and interpreting markers placed on different components of industrial machines. The application provides useful real-time information to the operators: technical manuals, operating diagrams, maintenance history, components availability in the warehouse or in supplier's stock with the possibility of direct order, training videos for maintenance and work safety. Also, being connected to the smart manufacturing software and performing easier data analysis, the application is especially useful for time crucial predictive maintenance.

Keywords: smart manufacturing; predictive maintenance; augmented reality

1. Introduction

Recently, thanks to the emergence of the Industry 4.0, the MAS (Multi Agent Systems) combined with the new technologies (Internet of Things - IoT, wireless sensor networks, big data, cloud computing and mobile Internet, made possible the appearance of the "smart manufacturing" concept [1].

In simple terms, smart manufacturing can be considered the pursuit of data-driven manufacturing, where real-time data from sensors in the factory can be analyzed to inform decision-making. More generally, smart manufacturing can be considered a specialization of big data, whereby big data technologies and methods are extended to meet the needs of manufacturing. [2]

Industry 4.0 proposes the predictive manufacturing in the future industry. The machines are connected as a collaborative community. Such evolution requires the utilization of advance prediction tools, so that data can be systematically processed into information that can explain the uncertainties and thereby make more "informed" decisions [3].

The augmented reality has a broad applicability in the industry, from real-time viewing of product information and production flow[10], viewing kinematic simulations and displaying deformations and loads [9], to product design, maintenance and logistics.

Several empirical studies were conducted on the field in order to research the effectiveness of AR applied to industrial operations: J. Saaski et al. and S. J. Henderson et al. results show a performance improvement up to 50%, which means that AR enables workers to perform a given task faster and with a lower possibility of error [4,5].

Following the investigations of IIoT (Industrial Internet of Things) platforms, we noticed that:

- There is no integration of predictive analytical maintenance applications with actual physical operation maintenance processes, for direct transmission and field registration of information through the operator team or autonomous robots that perform component replacement.
- There are no horizontal implementations for the availability of necessary maintenance information (components, accessories, supplies etc., or their availability to suppliers and delivery times)
- There are no integrated technical bookstores for maintenance processes (technical catalogues, procedures, etc.) that are immediately available depending on the required maintenance.
- There is no universal open bookstore for gathering and querying concrete cases of malfunctions and process parameters that led to them for easy consultation and modelling. (Through this bookstore, the cases experienced by a company can be analysed and taken into account by other companies.)

The PTC ThingWorx platform offers users the opportunity to develop VUFORIA-based AR applications, but there are no complete predefined implementations [6].

2. Open-source mobile application for maintenance

The research theme was proposed by FESTO Didactic and consists in building an open-source mobile application for maintenance based on WEB technologies and AR.

The purpose of this application is to provide useful information for intervention teams, such us:

- Identification of machines and their components
- State of the machinery components and subassemblies
- Information on the history of previous maintenance processes (date and time of the maintenance operation, cause, technical team etc.)
- Information on inventory of components, suppliers, prices
- Technical library (user manuals, technical manuals, labour safety etc.)
- Media library with video clips on maintenance
- Enhanced videoconference application for remote support

To implement the application, we only used web technologies, so the application can run directly into the web browser, on any device or operating system, and server-type architecture, ensuring centralization of information and ease of maintenance for libraries. The major challenge was to identify existing web technologies that could support this deployment.

We started from the basic principle of building an augmented reality application, namely:

- Taking a video stream from a webcam
- Analysis of AR identification markers
- Generating 2D or 3D augmented content
- View video stream and augmented content

For the video stream coming from the camera, we selected WebRTC / getUserMedia, which is a new HTML5 technology that does not require the installation of any additional driver or plugin.

The interface of the application was made using HTML5, CSS and javascript and for the server application we have used PHP and MySQL.

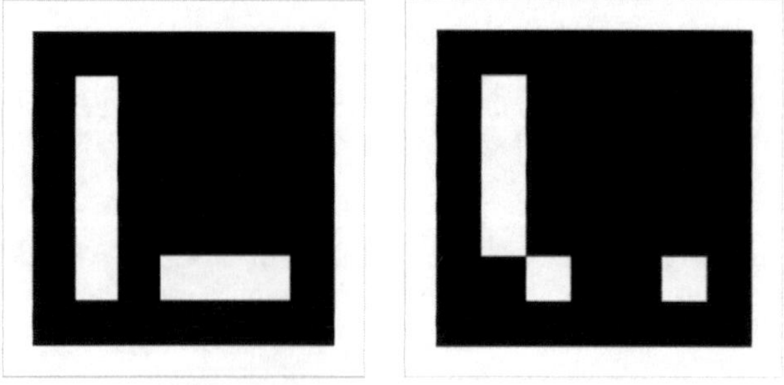

Fig. 1. Example of ArUco Markers

For AR markers, we chose the Aruco model (figure 1.), for which there is already implementation in javascript (Aruco libraries based on OpenCV technology) [7].
For augmented 2D or 3D content we chose WebGL 2.0 technology (Web graphics library) implemented in the THREE.js library, which allows creating, importing and displaying 3D animations using Javascript without importing plugins [8].

The operating principle is as follows:

a. The operator scans by means of tablet, phone or laptop using the video camera, the markers on each machine and its components.

b. When identifying markers, the component identification information and the minimum description (image, name, serial) overlap with the image provided by the camera.

c. The operator is approaching the target component, and at that point after it is identified in the web interface, all useful information about that component is loaded.

Fig. 2. Application starting screen

Once the application is accessed, the home screen is displayed and here the camcorder is selected (Camera Settings, front or rear - Figure 2). By clicking the red button, the user gets inside the Scanning screen (SCAN) where the AR application is deployed (Figure 3). The interface also contains the main menu of the application (Scan, Library, Stocks, Maintenance). The user starts the scanning. After the scan is done the software identifies the component Id and brings more information about the scanned component that can be interrogated by clicking the other items from the menu.

Fig. 3. Scanning page

By selecting Stocks from the main menu, an info screen (Figure 4) containing different useful information related to the scanned component is presented: general information about the component including inventory, condition, warranty and model data, purchase information, issues, existent opened tickets, info attached files and admin interface for asset edit.

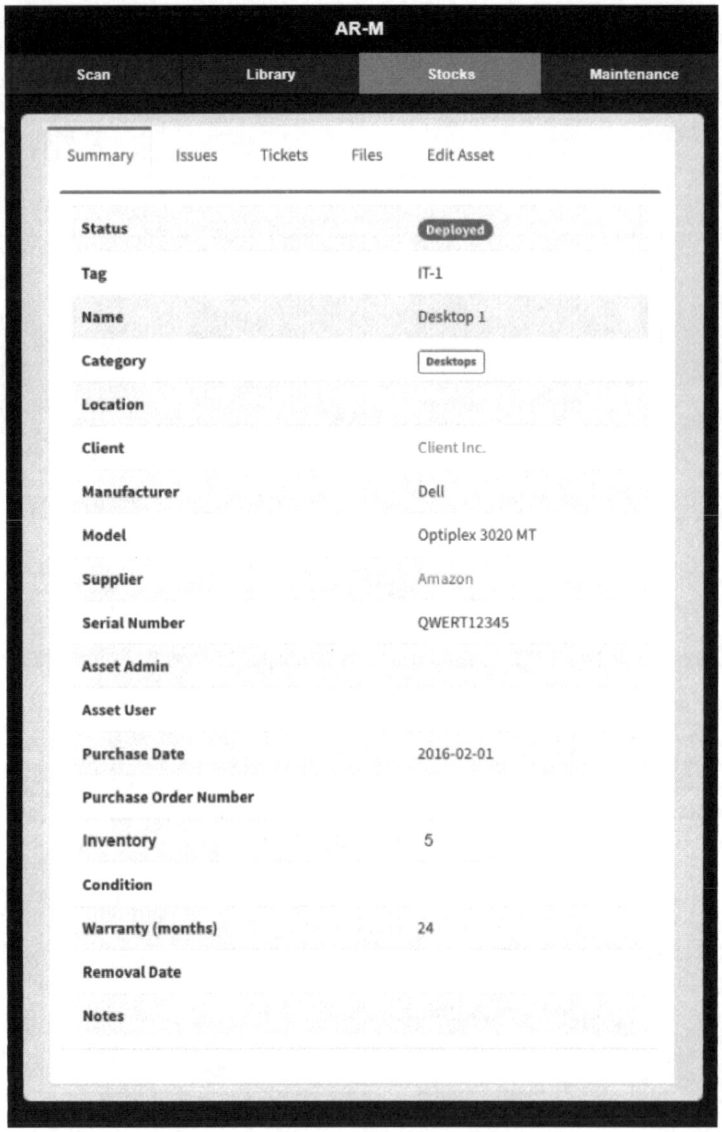

Fig. 4. Info screen

The Library menu item loads selectively, depending on the scanned AR marker, the corresponding technical library organised by visual tabs with pdf links to different articles regarding installation and maintenance or technical information about the scanned component (Figure 5).

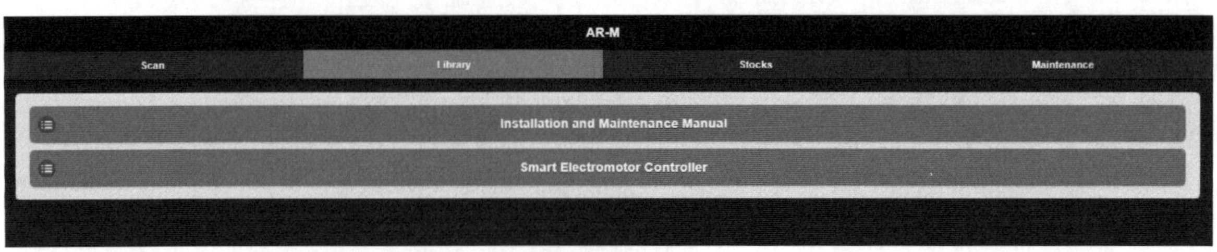

Fig. 5. Technical Library

By selecting Maintenance, the user can obtain information about the maintenance processes and history of that component, including maintenance tasks assigned to the team, statuses of the tasks and due dates, also a video library about maintenance processes and useful instructions offered by a video guide. (Figure 6)

Fig. 6. Maintenance

In case the operator needs support, a Videoconference tab can be accessed and the logged user can enter a videoconference for being assisted during the maintenance process or for asking other information. Audio, video, chat, file sharing and screen sharing are the main features available inside this module (Figure 7).

The video conferencing allows simultaneous connection of multiple users and includes the following: High Definition video, audio, screen sharing, file transfer, text chat, video sharing from Youtube, SIP connection with other proprietary conference systems and allows a great remote support.

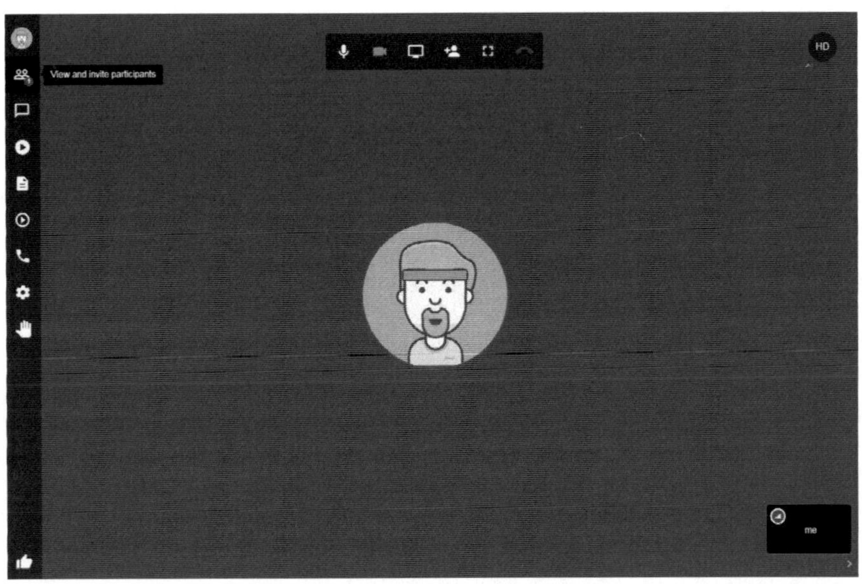

Fig. 7. AR Videoconference

3. Conclusion

In our research, we have succeeded to implement an open source web AR application for maintenance which does not depend on any other proprietary cloud platform, the processing (scanning, marker identification, tracking, augmentation) is performed directly in the client's web browser using javascript, unlike other solutions requiring the installation of proprietary applications and the payment of licenses or subscriptions.

This software extends the IIoT platforms capabilities, providing useful real-time information to the operators who are on the ground and need support or immediate information: technical manuals, operating diagrams, maintenance history, components availability in the warehouse or in suppliers' stock with the possibility of direct order, training videos for maintenance and work safety.

Limitations: The only limitations of this application are that Web RTC technology used for scanning and video conferencing is not yet implemented in web browsers running on the iOS platform (iPhone, iPad) being still under implementation.

Future work: our future work will be focused on extending the mobile application for maintenance with new modules and functionalities, based on client's requests. It will contain also data processing analyze for predictive maintenance. We intend to publish online the application as an open source project.

4. Acknowledgments

We want to thank to our partner FESTO Didactic for collaboration and support.

5. References

[1] Wang, S., Wan, J., Li, D., & Zhang, C. (2016). Implementing smart factory of industrie 4.0: An outlook. International Journal of Distributed Sensor Networks, pp. 1-10, ISSN: 15501477

[2] O'Donovan, P., Leahy, K., Bruton, K. & O'Sullivan, D. T. J. (2015). An industrial big data pipeline for data-driven analytics maintenance applications in large-scale smart manufacturing facilities, Journal of Big Data, pp 1-26, ISSN 2196-1115

[3] Lee, J., Kao, H. A., & Yang, S. (2014). Service innovation and smart analytics for industry 4.0 and big data environment. Procedia CIRP, pp. 3-8, ISSN: 2212-8271

[4] Sääski, J., Salonen, T., Liinasuo, M., Pakkanen, J., Vanhatalo, M., Riitahuhta, A. (2008). Augmented Reality Efficiency in Manufacturing Industry: A Case Study, Proceedings of NordDesign 2008 Conference, ISBN 978-9985-59-840-5, pp. 99-109

[5] Henderson, S. J. & Feiner, S. (2009). Evaluating the benefits of augmented reality for task localization in maintenance of an armored personnel carrier turret, 8th IEEE International Symposium on Mixed and Augmented Reality, ISBN: 978-1-4244-5390-0, pp. 135-144

[6] https://www.ptc.com/en/internet-of-things/technology-platform-thingworx, 2017, Accessed on: 2017-08-10

[7] https://sourceforge.net/projects/aruco/files, Aruco Markers for Augmented Reality applications based on OpenCV, Accessed on: 2017-08-10

[8] https://threejs.org/, Accessed on: 2017-08-10

[9] Park, H. S. & Park, J. W.: Application Method of Augmented Reality Including FEM to Manufacturing System, Annals of DAAAM for 2011 & Proceedings of the 22nd International DAAAM Symposium, Volume 22, No. 1, ISSN 1726-9679 ISBN 978-3-901509-83-4, Editor B. Katalinic, Published by DAAAM International, Vienna, Austria, EU, 2011

[10] Novak – Marcincin, J.; Barna, J.; Fecova, V. & Novakova – Marcincinova, L.: Augmented Reality Applications in Manufacturing Engineering, Annals of DAAAM for 2012 & Proceedings of the 23rd International DAAAM Symposium, Volume 23, No.1, ISSN 2304-1382, ISBN 978-3-901509-91-9, CDROM version, Ed. B. Katalinic, Published by DAAAM International, Vienna, Austria, EU, 2012

DOI: 10.2507/28th.daaam.proceedings.103

MACHINE VISION IN MANUFACTURING PROCESSES AND THE DIGITAL TWIN OF MANUFACTURING ARCHITECTURES

Gicu Calin Deac, Crina Narcisa Deac, Cicerone Laurentiu Popa,
Mihalache Ghinea & Costel Emil Cotet

This Publication has to be referred as: Deac, C[rina] N[arcisa]; Deac, C[rina] N[arcisa]; Popa, C[icerone] L[aurentiu]; Ghinea, M[ihalache] & Cotet, C[ostel] E[mil] (2017). Machine Vision in Manufacturing Processes and the Digital Twin of Manufacturing Architectures, Proceedings of the 28th DAAAM International Symposium, pp.0733-0736, B. Katalinic (Ed.), Published by DAAAM International, ISBN 978-3-902734-11-2, ISSN 1726-9679, Vienna, Austria
DOI: 10.2507/28th.daaam.proceedings.103

Abstract

This paper describes some results of authors' research in machine vision, as a key technology, essential in connecting the manufacturing processes with the digital twin of the manufacturing architecture. The focus is on designing a system that uses machine vision in combination with deep learning algorithms. The result is a human-like vision system for real-time analysis and interpretation of all video streams of cameras to continuously detect and count the various products on the production line. All detected features and counting information are available in real-time by augmenting on a video screen and streamed to the cloud where they can be easily processed and stored.

Keywords: machine vision; Industry 4.0; augmented reality

1. Introduction

In automated manufacturing, vision systems have been essential in many industries such as: electronics, medical devices, consumer goods, semiconductor, packaging, automotive [1]. Machine visions are implemented mainly for inspecting and identifying parts, measuring dimensions [7], or guiding machines or robots during assembly operations, for pick-and-place [8]. It is not at all clear what the term "computer vision" means. Most people in the field would agree that the use of a single photocell detector to count bottles on a manufacturing line, for example, does not constitute computer vision [2]. They would also agree that the new powerful systems that inspect components and aid robots in assembly most definitely exemplify computer vision. But there appears to be almost a continuum of systems and applications in between.

Research on machine vision has been in progress for over 25 years, yet practical accomplishments have been modest. To date most installed machine vision systems are relatively simple and involve only simple computer processing. In most cases, only silhouette images are utilized and frequently the illumination is specially structured to reduce burden on the pattern recognition computation. Vision is by far the most powerful and versatile sensing mechanism, and once we achieve machine vision with a capability approaching that of human vision at a competitive cost, the installation of machine vision systems will virtually explode [3].

To define our research, we started from the following questions:

1. Is it possible to integrate a distributed vision system across the entire production line?
2. What is the solution for real-time analysis of video streams to identify and count different objects?
3. Is such a solution feasible in terms of deployment cost?

2. Method - Tool

Our research is focused on developing a vision system close to the human vision, using video streams and image processing based on a neural network. The vision system should be able to obtain a real-time classification and to count different products from different stages of processing on an assembly line and send the information to a cloud IIoT application.

After analysing the technical solutions existing on the market at this moment, we designed the following system, whose diagram is presented in Figure 1. The system is composed by several video cameras placed in different points on the production line. The video signal from all cameras is processed by a video multiplexer, resulting a single video stream with all images combined. This video stream is captured by a video capture board and processed by the classification application running on the vision computer. To obtain real-time classification, we used YOLO9000 [4], [5], compiled with CUDA support. The parallel CUDA architecture, gives extra processing power at a low cost. The processing is done in real time (using 4 Nvidia GeForce Titan X video cards, we could analyse the video stream at 60 frames per second). In the classification process the combined image is augmented and tracked.

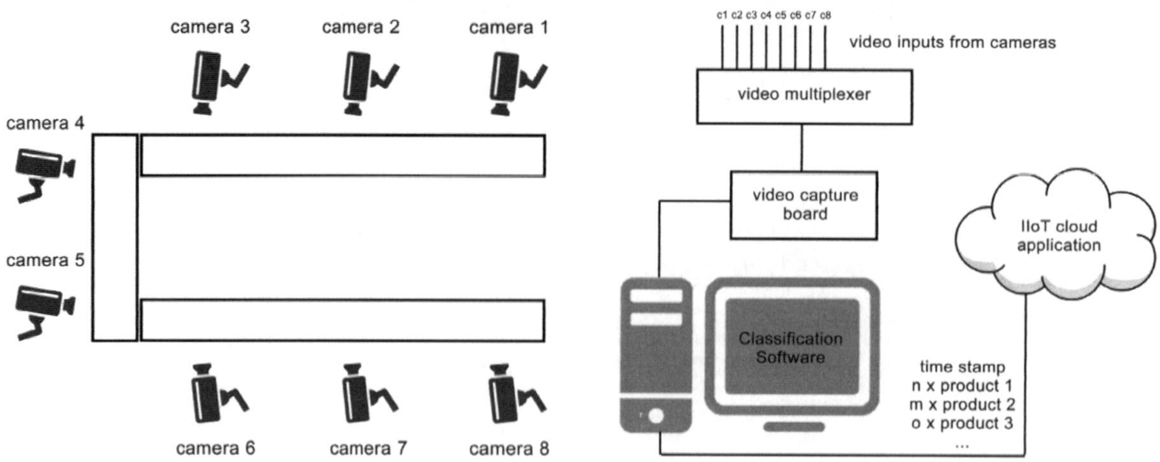

Fig. 1. Vision system diagram

The video cameras used in this application are colour SD 4:3 aspect ratio with resolution of 720x576 and with composite video output.

The video multiplexer stitches the images provided by the cameras, side by side, top/bottom in a matrix resulting a single video stream with a higher resolution. The video capture card digitises the video output of the video multiplexer and feed the stream to the application.

The application for classification is based on YOLO9000 [4], [5], a convolutional network with anchor boxes which predicts the coordinates of bounding boxes directly using fully connected layers on top of the convolutional feature extractor. YOLO9000 is a real-time framework for detection of more than 9000 object categories by jointly optimizing detection and classification. It uses WordTree (figure 2) to combine data from various sources and a joint optimization technique to train simultaneously on ImageNet and COCO. YOLO9000 is a strong step towards closing the dataset size gap between detection and classification.

The WordTree representation of ImageNet offers a richer, more detailed output space for image classification. Dataset combination using hierarchical classification would be useful in the classification and segmentation domains. Training techniques like multi-scale training could provide benefit across a variety of visual tasks. [5]

Starting with YOLO 9000 pre-trained on ImageNet and COCO we have performed a re-training based on images of different products in different stages of production.

Processing images with YOLO is simple and straightforward. The system resizes the input image to 448 x 448, runs a single convolutional network on the image, and thresholds the resulting detections by the model's confidence (figure 3). A single convolutional network simultaneously predicts multiple bounding boxes and class probabilities for those boxes. YOLO trains on full images and directly optimizes detection performance [5].

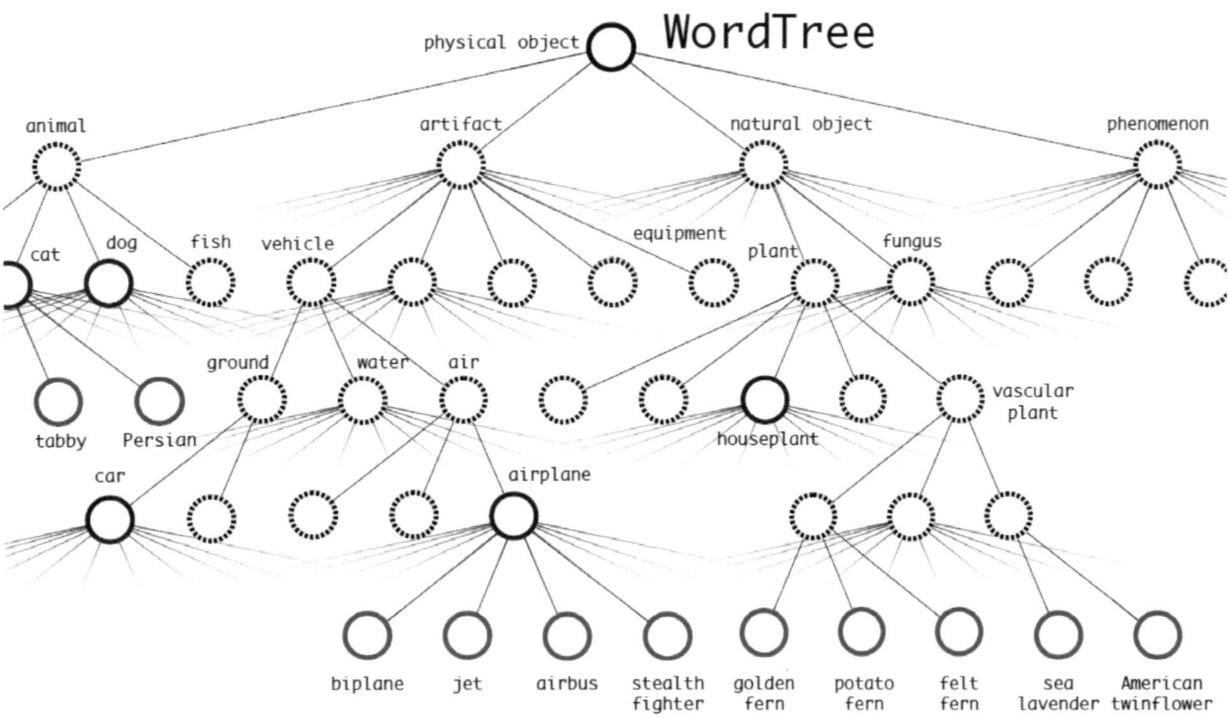

Fig. 2. Combining datasets using WordTree hierarchy. [5]

This unified model has several benefits over traditional methods of object detection, being faster and unlike sliding window and region proposal-based techniques, YOLO sees the entire image during training and test time so it implicitly encodes contextual information about classes as well as their appearance [5].

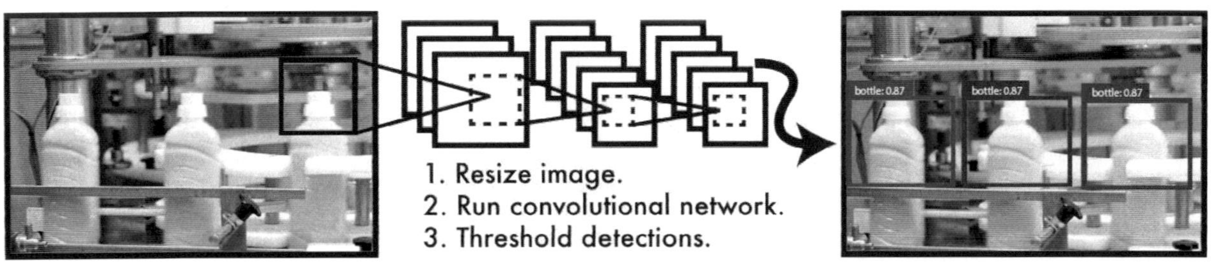

Fig. 3. The YOLO detection system (You Only Look Once) [5]

The detection network has 24 convolutional layers followed by 2 fully connected layers. Alternating 1 x 1 convolutional layers reduce the features space from preceding layers. The convolutional layers are pretrained on the ImageNet classification task at half the resolution (224 x 224 input image) and then double the resolution for detection (figure 4) [5].

YOLO predicts multiple bounding boxes per grid cell. At training time, we only want one bounding box predictor to be responsible for each object. This leads to specialization between the bounding box predictors. Each predictor gets better at predicting certain sizes, aspect ratios, or classes of object, improving overall recall [6].

YOLO saves the detection and classification information in a text log file. Using a log parser script, we get the latest information and create a Json file containing time stamp in milliseconds, a list of detected objects and their counts. The Json file is uploaded using a secured socked connection to the cloud server, where data is stored in a database and analysed.

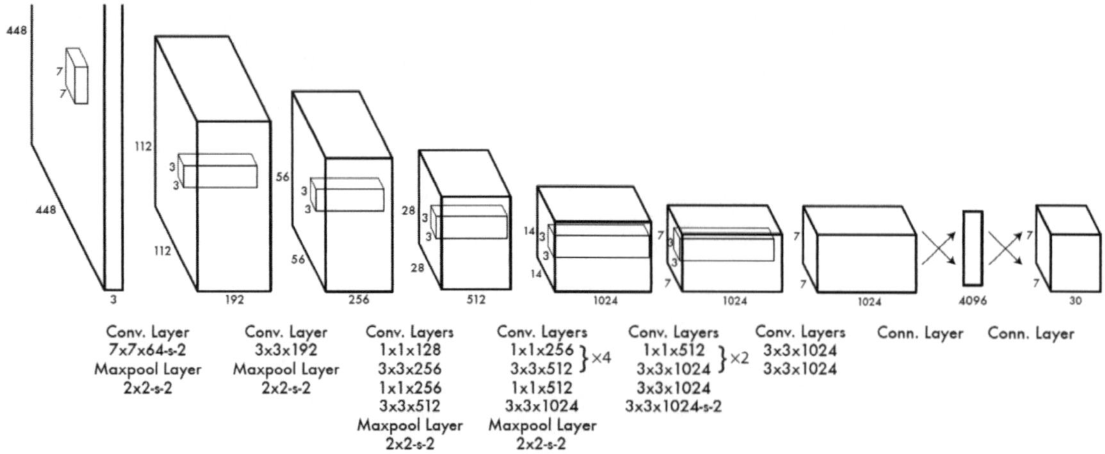

Fig. 4. Yolo architecture [5]

Our future work will be focused on testing and implementing the new Nvidia Jetson TK1 developer kit and Vision Works in the same workflow, to cut the prices and create a more simplified and robust architecture. Also, we will extend the system with new functionalities such as filtering damaged items, alarming on work safety incidents (when a person passes in a dangerous zone).

3. Conclusion

By choosing to combine multiple images from different cameras into a single high-resolution stream we have succeeded to integrate a distributed system across the entire production line, that allows the use of a single computer and an application to detect, classify and count all products in different manufacturing stages from the production line. By using YOLO compiled with CUDA support we benefit from fastest detection platform which allows real time classification at a small price.

Limitations: Due to the use of a computer and an application requiring real-time video processing, which results in intense system loading, an efficient cooling system is required to not compromise the stability of the application. A dedicated hardware-based approach would probably be more efficient in terms of long-term exploitation.

4. Acknowledgments

This work was supported by a grant of the Romanian National Authority for Scientific Research and Innovation, CNCS/CCCDI – UEFISCDI, project number PN-III-P2-2.1-BG-2016-0437, within PNCDI III.

5. References

[1] Abdulrahman, W. (2014). Overview on Industrial Vision Systems, International Journal of Engineering Research & Technology, Vol. 3, Issue 5, pp. 383-386, ISSN: 2278-0181

[2] Agin G. J. (1980). Computer Vision Systems for Industrial Inspection and Assembly, Computer, Vol. 13, Issue: 5, pp. 11 – 20, ISSN: 0018-9162

[3] Freeman, H. Machine Vision (1988): Algorithms, Architectures, and Systems, Academic Press, ISBN 0-12-266720-4, USA

[4] Redmon, J., Farhadi, A. (2016). YOLO9000: Better, Faster, Stronger; Available from: https://arxiv.org/abs/1612.08242, Accessed: 2017-08-09

[5] Redmon, J., Divvala, S., Girshick, R., Farhadi, A. (2016). You only look once: Unified, real-time object detection, pp. 779-788, Proceedings of the IEEE Conference on Computer Vision and Pattern Recognition, ISBN 978-1-4673-8851-1

[6] R. Girshick, J. Donahue, T. Darrell, and J. Malik. (2014) Rich feature hierarchies for accurate object detection and semantic segmentation. Available from: https://arxiv.org/abs/1311.2524 Accessed: 2017-08-09

[7] Hruz, M.; Siroky, J. &Manas, D. (2010). Knoop Hardness Measurement Using Computer Vision, Annals of DAAAM for 2010 & Proceedings of the 21st International DAAAM Symposium, Volume 21, No. 1, ISSN 1726-9679 ISBN 978-3-901509-73-5, Published by DAAAM International, Vienna, Austria, EU, 2010

[8] Feriancik, M. & Liska, O. (2011). Autonomous Positioning Based on Machine Vision, Annals of DAAAM for 2011, Proceedings of the 22nd International DAAAM Symposium, Volume 22, No. 1, ISSN 1726-9679 ISBN 978-3-901509-83-4, Published by DAAAM International, Vienna, Austria, EU, 2011

DOI: 10.2507/28th.daaam.proceedings.104

MITIGATING FRICTION IN MULTICULTURAL VIRTUAL ORGANIZATIONS / TEAMS

Ionut Savu, Cicerone Laurentiu Popa & Costel Emil Cotet

This Publication has to be referred as: Savu, I[onut]; Popa, C[icerone] L[aurentiu] & Cotet, C[ostel] E[mil] (2017). Mitigating Friction in Multicultural Virtual Organizations / Teams, Proceedings of the 28th DAAAM International Symposium, pp.0737-0742, B. Katalinic (Ed.), Published by DAAAM International, ISBN 978-3-902734-11-2, ISSN 1726-9679, Vienna, Austria
DOI: 10.2507/28th.daaam.proceedings.104

Abstract

In an increasingly global marketplace, the rise and fall of actors that want to compete on the world stage has become dependent on their ability to search for and find creative solutions to a fast paced and changing market. This search for fast problem solving, plasticity of ideas and increased efficiency has more often than not lead to creation of virtual teams and organizations, some spanning several countries, mindsets and cultures. By describing the problems arisen in working in a multinational company that tries to apply local values to a global workspace and the corresponding solutions found, this paper aims to find general guidelines by which friction in communication could be reduced to better facilitate efficiency in multicultural virtual teams as well as discover what principles should be applied for better remote management. This article leverages research done in anthropological studies of intercultural cooperation to contextualize the issues at hand in the behaviour of team members and managers in a virtual space.

Keywords: Virtual organizations; Multiculturalism; Virtual teams; Communication; Efficiency

1. Introduction

Organizations are currently facing important and unprecedented challenges in an ever dynamic, constantly changing and complex environment [1]. Economic activity of all types is moving in the direction of globalization [2][3]. With the rapid development of electronic information and communication media in the last decades, distributed work has become much easier, faster and more efficient [4]. To respond to the increasing de-centralization and globalization of work processes, many organizations have adapted to these dynamic environments by introducing virtual teams that collaborate by communication technologies across geographical, temporal, cultural and organizational boundaries to achieve common goal in their organization's outputs. Virtual teams are growing in popularity and the rapid development of new communication technologies such as the internet has accelerated this trend so that today, most of the larger organizations employ virtual teams to some degree [4]. Information technology is providing the infrastructure necessary to support the development of new forms of organizations. Virtual teams represent one such form, one that could revolutionize the workplace and provide organizations with unprecedented level of flexibility and responsiveness [5]. Virtual teams are important mechanisms for those seeking to leverage scarce resources across geographic and other boundaries [6] and represent a large pool of new product know-how which seems to be a promising source of innovation. With the rise of globalization, virtual teams are more distributed than ever, across time, space, and culture. These dimensions correspond to challenges that can significantly impact the productivity of virtual teams and the quality of the inter-personal relationships of their members. By describing the problems arisen in working in a multinational company that tries to apply local values to a global workspace and the corresponding solutions found, this paper aims to find general guidelines by which friction in communication could be reduced to better facilitate efficiency in multicultural virtual teams as well as discover what principles should be applied for better remote management.

2. Historical Context

For a complete understanding of virtual teams we have to take into account the historical context of their development as a result of globalization. Hertel et al. viewed it as the free movement of goods, capital, services, people, technology and information across the world, the action or procedure of international integration of countries arising from the convergence of world views, products, ideas, and other aspects of culture [7]. Advances in means of transport (such as the steam locomotive, steamship, jet engine, and container ships) and in telecommunications infrastructure have been major factors in globalization, generating further interdependence of economic and cultural activities [8][9]. Large-scale globalization began in the 1820s [10]. In the late 19th century and early 20th century, the connectivity of the world's economies and cultures grew rapidly. During the 19th century, globalization approached its form as a direct result of the Industrial Revolution which allowed standardized production of household items using economies of scale while rapid population growth created sustained demand for commodities. Also, steamships reduced the cost of international transport significantly and railroads made inland transport cheaper. More nations embraced international trade [10] and the invention of shipping containers in 1956 also helped advance the globalization of commerce [11]. Student exchange programs became popular after World War II, and were intended to increase the participants' understanding and tolerance of other cultures, as well as improving their language skills and broadening their social horizons. Between 1963 and 2006 the number of students studying in a foreign country increased 9 times [12]. The emergence of the virtual workplace has been tightly linked to the fast increase of PC sales in the 1960s, the popularity of cellular phones in the 1970s, the internet and the World Wide Web in the 1990s. Thus necessity, economic and technological viability had merged to make the concept of virtual teams a reality.

3. Virtual Teams

Virtual teams are the group of individuals united by a common goal but spread geographically, culturally, linguistically or, ethnically. As defined by Powell, Piccoli and Ives, virtual teams are described 'as groups of geographically, organizationally and/or time dispersed workers brought together by information and telecommunication technologies to accomplish one or more organizational tasks'. Virtual teams mainly meet the demand for flexibility and agility in the delivery of services or products in a marketplace with increased competition and a decentralized and globalized work processes. The need for talent and competency has lead organizations to look to structures which leverage available talent across boundaries. More and more organizations are adopting virtual team approach to reduce their operating costs, encourage knowledge sharing among their employees to promote organizational learning and expand their business hours to 24/7 by utilizing the different time zones of virtual team members [13]. Human relationships draw their power from the proximity of the individuals but due to the fact that virtual team members have limited personal interaction this creates a myriad of issues. For a virtual team to be productive it is important to overcome the cultural differences, communication barriers, power struggles and conflict to build trust, collaboration and commitment among the individuals. By using the right strategies, tools and processes, organizations can benefit greatly from this new-age trend of virtual teams [13].

4. Types of Virtual Teams

Virtual teams can be classified into 6 categories depending on lifespan, temporal distribution, role of members and objectives.

Networked Teams:

They are short-lived and serve the function of resolving a short-term objective. Team members are sometimes brought from outside the organization to share their expertise as cross-functional participants and for their broader perspective. Networked teams, upon completion of their objectives are dismissed thus, their lifespan varies on how complex the issue at hand is.

Parallel Teams:

They are highly task oriented teams that usually consist of specialized professionals. While they are generally only required for very short span of time, unlike networked teams, they are not dissolved after completion of the tasks. The team may be either internal or external to the organization [14].

Project or Product Development Teams:

These were actually the pioneer in the development of virtual teams. They work on complex sets of activities, are composed of subject matter experts brought together from different parts of the globe, over a long time period and may be formed to develop new products, deliver a new technology system, or redesign operational processes.

Work, Production or Functional Teams:

They are formed when members with specific skill sets come together to perform a single type of ongoing activity. These teams are totally function specific as they only work on a particular area within an organization. Operating virtually from different geographical locations, these teams exist to perform regular or ongoing tasks.

Service Teams:

Service teams are geographically located in different time-zones and are assigned to a particular service such as customer support to provide 24/7 continuity. Though each member works independently, they together perform work in continuation. This kind of team also has value in technical support activities.

Offshore ISD Teams:

They are independent service provider teams that a company can subcontract portions of work to. This is the model that is most used in outsourcing organizations, especially software development or R&D. The team is usually based in a low-cost location and coordinates with the onshore team to deliver the expected results.

5. Advantages and Disadvantages of Virtual Teams

Regarding the advantages and disadvantages of virtual teams we can contextualize them from three perspectives, individual, organizational and societal. The table below summarizes the attribute, positive and negative, of virtual teams so that a clear pathway to identifying constructive solutions can be uncovered.

Advantages	Disadvantages
Cost savings Organizations can optimize real estate costs by using virtual teams as well as utility or travel expenses. Most outsource operations in low-cost regions to also take advantage of lower wages of the employees.	**Cost of Technology** One of the main costs in running virtual teams comes from the necessary technology required to run them. Without the use of multiple communication technologies it is either impossible or impractical to run such teams.
Global Talent By leveraging human capital from around the world, organizations can bring in the fold greater understanding of local markets as well as benefit from specialist and experts that increase innovation and knowledge sharing.	**Conflicts, Lack of Trust & Collaboration** One of the main reasons for conflict in virtual teams is the cultural divide as well as the different styles of communication which give rise to mistrust and difficulties in collaboration that are intrinsic to their success. Other challenges come from the lack of direct interaction as most communication is done through non-verbal interactions..
Increased Productivity As opposed to local teams, virtual teams tend to have a greater focus on the objectives dealt with as they benefit from a flatter structure of organization. This translates into increased productivity and greater profits for the company.	**Social Isolation** Although collaboration in the virtual space has many advantages, on a personal level, the lack of direct contact with coworkers may lead to an adverse affect on the psychological welfare of team members and consequently on productivity.
Decreased Time to Market Virtual teams can leverage the different time zones that the members span and run the project continuously 24/7. This greatly speeds up development as well as response time.	
New Opportunities Any tasks that do not require physical presence are now obtainable by candidates that formerly were restricted by availability thus opening the doors to new potential talent.	

Table 1. Advantages & Disadvantages of Virtual Teams [15]

Although challenges clearly exist in implementing a successful virtual team, these can be overcome by following more efficient communication strategies that take into account the cultural makeup of the team as well as following different leadership strategies.

6. Case Study and Anthropological Context

Through research it has been documented that the success of multicultural teams is vital for the performance of organizations [16][17]. In the context of virtual teams and increased contact with multiculturalism, interest has been renewed in understanding and developing communication skills that are needed for high performing cross cultural teams. In a study by Matveev and Nelson (2004) [18] there was found a significant effect of national culture on cross-cultural communication competence, suggesting the need to train managers to become more effective in culturally complex workplaces. National culture of team members plays a significant role in determining whether communicative behavior is perceived as competent, thus emphasizing the central role that national culture and ethnicity play in assessing communication competence already documented in the literature [19][20][21]. For companies that leverage virtual teams, their performance becomes vulnerable to cross-cultural problems [18]. Managing cultural diversity, cultural differences, and cross-cultural conflicts have surfaced as frequent challenges for cross-cultural teams [22].

Matveev and Nelson (2004) emphasize five cultural orientations that could influence cross-cultural communication competence: long term orientation, power distance, individualism, uncertainty avoidance, and performance orientations [22].

The intent of this study was to document the points of failure and success in the studied virtual team and draw a correlation with the cultural orientation of the members involved, drawing from prior research done in anthropological studies of cross-cultural interactions. To do so, a secondary analysis was done, using data from Hofstede's 6-D Model of cultural overview. The team was composed of 13 members from Finland, Serbia, Switzerland, Denmark, Sweden, Poland and Romania with management and IT support done by israeli members. It can be characterized as a functional team as it came together to provide account development services for the business clients of the company.
The table below presents the characteristics of each of the team members' native cultures.

	Long term orientation	Power Distance	Individualism	Uncertainty Avoidance	Performance Orientations
Finland	38	33	63	59	26
Serbia	52	86	25	92	43
Switzerland	74	34	68	58	70
Denmark	35	18	74	23	16
Sweden	53	31	71	29	5
Poland	38	68	60	93	64
Romania	52	90	30	90	42
Israel	38	13	54	81	47

Table 2. Cultural Orientation Score [23]

Long term orientation describes how every society has to maintain some links with its own past while dealing with the challenges of the present and future, and societies prioritise these two existential goals differently. Normative societies which score low on this dimension, for example, prefer to maintain time-honoured traditions and norms while viewing societal change with suspicion. Those with a culture which scores high, on the other hand, take a more pragmatic approach: they encourage thrift and efforts in modern education as a way to prepare for the future [23].

Power distance deals with the fact that all individuals in societies are not equal – it expresses the attitude of the culture towards these inequalities amongst us. Power Distance is defined as the extent to which the less powerful members of institutions and organisations within a country expect and accept that power is distributed unequally [23].

Individualism is the degree of interdependence a society maintains among its members. It has to do with whether people's self-image is defined in terms of "I" or "We". In Individualist societies people are supposed to look after themselves and their direct family only. In Collectivist societies people belong to 'in groups' that take care of them in exchange for loyalty [23].

Uncertainty Avoidance has to do with the way that a society deals with the fact that the future can never be known: should we try to control the future or just let it happen? This ambiguity brings with it anxiety and different cultures have learnt to deal with this anxiety in different ways. The extent to which the members of a culture feel threatened by ambiguous or unknown situations and have created beliefs and institutions that try to avoid these is reflected in the score on Uncertainty Avoidance [23].

Performance Orientations A high score on this dimension indicates that the society will be driven by competition, achievement and success, with success being defined by the winner/best in field – a value system that starts in school and continues throughout organisational life. A low score on the dimension means that the dominant values in society are caring for others and quality of life. The fundamental issue here is what motivates people, wanting to be the best or liking what you do [23].

Going forward we report on the successful interactions and some of the main problems observed in the studied virtual team and interpret through the lens of the stated cultural dimensions.

Lack of Differential Treatment to the Complexity of Virtual Teams

Organizations have a tendency to not differentiate between virtual and traditional teams thus not taking into account the complexity of virtual teams in regard to time, distance and cultural differences. There is no specialized training imparted to either team members nor team leaders. The particular case studied manifested friction especially between management and team members from Romania, Serbia and to some extent Poland. A sentiment of weak management style was developed. This is evidence of the difference in the power distance index between the parties. In Israel power is decentralized, managers count on the experience of their team members and employees expect to be consulted whereas the others accept a hierarchical order in which everybody has a place and which needs no further justification. Hierarchy in an organization is seen as reflecting inherent inequalities, centralization is popular, subordinates expect to be told what to do and the ideal boss is a benevolent autocrat [23].

Ambiguity towards Responsibility and Task

Due to the reliance on virtual communication and lack of face to face contact, a lack of clarity emerges towards the specific role a member has. Moreover since most communication within the team is process oriented working relationships are difficult to develop which affects information sharing and trust. Again this issue manifested mainly in team members from countries with high power distance index but less in polish members. We speculate this is because of their high level of individualism.

Leadership

As stated formerly, virtual teams rely heavily on modern communication technology thus the absence of face to face contact needs to be addressed [24]. To compensate for the particularities of managing a virtual team, leaders need to incorporate technical abilities as well as interpersonal and team building aptitudes into their skillset. In the studied virtual team, the leaders managed to compensate the downfalls of the virtual communication medium by having a more involved attitude towards the team members [26][27]. Low power distance and a high uncertainty avoidance score is indicative of this cultural inclination. There is an emotional need for rules, people have an inner urge to be busy and work hard, precision and punctuality are the norm, security is an important element in individual motivation [23]. Cultures with a high score on this dimension are often very expressive.

Lack of Trust

The socialization process intrinsic to the development of trust in a team is absent in the virtual team environment [25]. Lack of trust at the level of either the team member or the team leader results in low relational commitment and conflicts evident in low productivity and performance of the team. The main issues observed were in cases where close collaboration was needed between members of Denmark/Sweden and Romania/Serbia/ Poland. The high delta of uncertainty avoidance between northern and eastern european cultures could be a factor. Whereas countries exhibiting high uncertainty avoidance index (UAI) maintain rigid codes of belief and behaviour and are intolerant of unorthodox behaviour and ideas, low UAI societies maintain a more relaxed attitude [23].

Lack of Effective Communication

Since there is a lack of nonverbal cues which are central to trust building, issues might arise that affect performance. Although modern communication tools as emails, video conferencing, instant messaging, phone are being used they do not convey information at all levels. Even though eventually mitigated by team building exercises, communication between team members from cultures with polarised individualism indexes was lacking a personal level that might have improved efficiency from the onset [23].

7. Conclusion

The need for faster time to market, lower costs, greater innovation and talent has brought forth the concept of virtual teams as a solution but for organizations to fully take advantage of them and reap the benefits of this competitive edge potential failure points must be identified. The paper found there is evidence of a link between performance/productivity and cultural dimensions that each team member possess. As stated in the introduction, challenges appear in virtual teams in regard to cultural dimensions. However, these dimensions should not be taken individually but comparatively and as prior research shows they don't always apply to all members of a particular culture(hofstede). Furthermore, the authors acknowledge the limitations of the research due to the small number of subjects in the monitored virtual team, but we consider it representative of a multicultural virtual team as defined in the paper. When building and managing a virtual team consideration should be taken of the diverse background of the members so as to facilitate a more efficient collaboration between members. Further and more comprehensive research should be done on the topic so as to create a more complete guideline for cross-cultural virtual team management as well as for communication improvements, as such we intend to extend our study into more examples of multicultural virtual teams to strengthen our conclusions as well as to test our hypothesis, that the cultural makeup should be taken into account when forming and managing a virtual team.

8. References

[1] Rezgui, Y. (2007). Exploring virtual team-working effectiveness in the construction sector. Interacting with computers, 19(1), 96-112.

[2] Acs, Z. J., & Preston, L. (1997). Small and medium-sized enterprises, technology, and globalization: Introduction to a special issue on small and medium-sized enterprises in the global economy. Small Business Economics, 9/1, 1-6.

[3] Ale Ebrahim, N., Ahmed, S., & Taha, Z. (2009). Virtual teams: A literature review.

[4] Hertel, G., Geister, S., & Konradt, U. (2005). Managing virtual teams: A review of current empirical research. Human resource management review, 15(1), 69-95.

[5] Powell, A., Piccoli, G., & Ives, B. (2004). Virtual teams: a review of current literature and directions for future research. ACM Sigmis Database, 35(1), 6-36.

[6] Munkvold, B. E., & Zigurs, I. (2007). Process and technology challenges in swift-starting virtual teams. Information & Management, 44(3), 287-299.

[7] Albrow, M. Elizabeth King, eds. 1990. Globalization, Knowledge, and Society.

[8] (n.d.). Imagining the Internet - Elon University. Retrieved August 1, 2017, from http://www.elon.edu/e-web/predictions/150/1830.xhtml

[9] Stever, H. "Guyford.(1972). Science, Systems, and Society." Journal of Cybernetics 2, no. 3: 1-3.

[10] O'rourke, K. H., & Williamson, J. G. (2002). When did globalisation begin?. European Review of Economic History, 6(1), 23-50.

[11] Levinson, M. (2016). The box: how the shipping container made the world smaller and the world economy bigger. Princeton University Press.

[12] Varghese, N. V. (2008). Globalization of higher education and cross-border student mobility (pp. 1-34). Paris: Unesco, International Institute for Educational Planning.

[13] (n.d.). Virtual Team - Management Study Guide. Retrieved August 1, 2017, from http://www.managementstudyguide.com/virtual-team.htm

[14] Duarte, D. L., Snyder, N. T. (2006) Mastering virtual teams: Strategies, tools, and techniques that succeed. Wiley

[15] (n.d.). Advantages and Disadvantages of Virtual Teams. Retrieved August 1, 2017, from http://www.managementstudyguide.com/virtual-teams-advantages-and-disadvantages.htm

[16] Jackson, S. E., May, K. E., & Whitney, K. (1995). Understanding the dynamics of diversity in decision-making teams. Team effectiveness and decision making in organizations, 204, 261.

[17] Snow, C. C., Snell, S. A., Davison, S. C., & Hambrick, D. C. (1996). Use transnational teams to globalize your company. Organizational dynamics, 24(4), 50-67.

[18] Matveev, A. V., & Nelson, P. E. (2004). Cross cultural communication competence and multicultural team performance: Perceptions of American and Russian managers. International Journal of Cross Cultural Management, 4(2), 253-270.

[19] Hughes, Everett Cherrington. The sociological eye: Selected papers. Transaction publishers, 1971.

[20] Jarvenpaa, S. L., Shaw, T. R., & Staples, D. S. (2004). Toward contextualized theories of trust: The role of trust in global virtual teams. Information systems research, 15(3), 250-267.

[21] Watson, W. E., Kumar, K., Michaelsen, L. K. (1993). Cultural diversity's impact on interaction process and performance: Comparing homogeneous and diverse task groups. Academy of management journal, 36, 590-602.

[22] Geert, H., & Jan, H. G. (1991). Cultures and organizations: Software of the mind. McGaw-Hill, London.

[23] (n.d.). Countries - Geert Hofstede. Retrieved August 1, 2017, from https://geert-hofstede.com/countries.html

[24] Harangus, Daniela, and Dana Codruta Duda-Daianu. "The teamwork abilities versus leadership." Annals of DAAAM for 2011 & Proceedings of the 22nd International DAAAM Symposium, Volume 22, No. 1, ISSN 1726-9679 ISBN 978-3-901509-83-4, Editor B. Katalinic, Published by DAAAM International, Vienna, Austria, 2011.

[25] Cotet, Nica. "Trust and Reputation Systems in Two Online Auction and Shopping Websites from Romania and the Netherlands." Annals of DAAAM for 2012 & Proceedings of the 23rd International DAAAM Symposium, Volume 23, No.1, ISSN 2304-1382 ISBN 978-3-901509-91-9, CDROM version, Ed. B. Katalinic, Published by DAAAM International, Vienna, Austria, EU, 2012.

[26] G.B. COTET & A.B. BALGIU, (2017). Using a constellation of personal capabilities in defining Industry 4.0. Soft Skills, The 4th International Multidisciplinary Scientific Conference on Social Sciences and Arts SGEM 2017, www.sgemsocial.org, SGEM2017 Conference Proceedings, ISBN 978-619-7408-22-5 / ISSN 2367-5659, 24 - 30 August, 2017, Book 3, Vol 5, 649-656 pp, DOI: 10.5593/sgemsocial2017/35/S13.085

[27] Gabriela Beatrice Cotet, Beatrice Adriana Balgiu and Violeta – Carmen Zaleschi (Negrea) (2017), Assessmentprocedure for the soft skills requested by Industry 4.0, The 8th International Conference on Manufacturing Science and Education – MSE 2017 "Trends in New Industrial Revolution", Volume 121, Article Number 07005, https://doi.org/10.1051/matecconf/201712107005, MATEC Web of Conferences 121, 07005 (2017), eISSN: 2261-236X

DOI: 10.2507/28th.daaam.proceedings.105

THE GAP BETWEEN THE KNOWLEDGE OF VIRTUAL ENTERPRISE ACTOR AND KNOWLEDGE DEMAND OF INDUSTRY 4.0

Anca Georgiana Costache (Stochitoiu), Cicerone Laurentiu Popa,
Tiberiu Dobrescu, Costel Emil Cotet

This Publication has to be referred as: Costache (Stochitoiu), A[nca] G[eorgiana]; Popa, C[icerone] L[aurentiu]; Dobrescu, T[iberiu] G[abriel] & Cotet, C[ostel] E[mil] (2017). The Gap Between the Knowledge of Virtual Enterprise Actor and Knowledge Demand of Industry 4.0, Proceedings of the 28th DAAAM International Symposium, pp.0743-0749, B. Katalinic (Ed.), Published by DAAAM International, ISBN 978-3-902734-11-2, ISSN 1726-9679, Vienna, Austria
DOI: 10.2507/28th.daaam.proceedings.105

Abstract

The alert steps of industrial digitization makes now credible the use of "Industrie 4.0" term launched by Henning Kagermann in 2011 and by most appreciations, interconnected digital technology is the Fourth Industrial Revolution. All the major innovations that occurred in digital technology require from the virtual enterprise actors a level of knowledge not so easy to be achieved. Knowledge comes in multiple forms and types (tacit, explicit, learned individually or through collective learning) and is in close link with the management of virtual enterprises. When it is absent the activity of the company will be faulty. It is very important to find the reason for the existing gap between demand and offer of knowledge and establish in what way it will affect the management of virtual enterprise. This study aims to present the level of existing knowledge versus the level of knowledge demand, how we can help increase competence, in which areas organisations can benefit from it and what type of competences we will need in the future.

Keywords: Knowledge; Management; Virtual Enterprise; Competences; Fourth Industrial Revolution

1. Introduction

Over the last few years there have been gradual changes towards the new technology, but from now on, there is already a difference in the peace at which things need to change. Progress awaits no one, and all that is new must be absorbed as fast as possible. Virtual organizations are at the point in which all the strategies must be reorganized in order to keep up with the new demands coming from Industry 4.0: better knowledge, better and new specializations, soft skills, management and communication skills.

The intensive development of computer technology and industry or telecommunication services over the past decades has led to a better outreach of the information society. A society of this type is clearly dependent on artificial intelligence, virtual environment and the evolution of industrial society, to the informational, through computerization, is a trend of contemporarily clearly characterized by global expansion, rapid change and the social system revolution, while the social and economic environment, with the help of cross-border information, tends towards a globalization process that will have a strong positive impact on information technology and communications. [7] Gradually, the way of access to information and public information resources will be very diversified in all areas nowadays, from government to culture or education, will create new jobs and a significant increase in occupation.

Information resources and knowledge will take on a very large and high degree of importance. Research and technological innovation activities will play a key role in economic and social development. The economic sector that deals with the production and dissemination of information assets will feel an increased dynamics, and computer-assisted activities, interactive work, remote work within virtual organizations (where occupants of various geographic territories will overcome geographical barriers) will become predominant. A major share will be held by virtual enterprises and distance learning. [8]

An informational society is based on an intensive technology-information economy and provides information-based businesses with the utmost use of material and human resources through a high intelligence investment. This type of economy leads to a wide range of IT products and easy citizen access to information system facilities, both in workplace and private life. The infrastructure of this society is represented by computer networks, telecommunication, intelligent workstations, expert teams and systems or multimedia systems that become operational, integrating information (be it text, sound or image) to assist the process of access to knowledge, designing new ideas and making decisions.

Technologically advanced development must be sustainable and qualitative, therefore limited natural resources, environment and society, all need to be protected by applying appropriate industrial policies and implementing new strategies to support the education and training system research. We are witnessing the development of Industry 4.0 and all changes and forms of organization or reorganization imposed by it.

The current innovation potential of the companies involved in the production will ensure competitiveness, maintenance and development in the future. Post-industrial society is progressing from a technical and technological point of view, and highly developed countries tend to deindustrialize, transferring the areas of labor-intensive activity to emerging states. But there is a problem: can the service industry provide enough work to balance the growth of industrial strength and productivity as a result of automation and robotics? In the last decades, the new information technology has had a huge impact and decentralization, reorganization requires more and more accessible information to the management of organizations and enterprises, and to top or middle managers. The service market is over-saturated and competition in the field is getting stronger.

There have been changes in many areas such as e-government, e-commerce, education (e-learning), culture (multimedia and virtual libraries), setting the transition to a new industrial revolution based on the progress of knowledge, innovation, discoveries and new technologies. As the gradual evolution of science, the emergence of important inventions, and the desire of the people to evolve have generated major changes and have always led to industrial revolutions, the development and evolution of technology, technology in general, information and communication technology in particular the beginning of the Fourth Industrial Revolution we are currently witnessing: Revolution 4.0.

This Industrial Revolution is characterized by the development and use of green energies, the strong impact of the Internet, the spread of robotics, biotechnology, nanotechnology, microelectronics, the emergence of advanced materials and additive production (3D printing).

Internet development has led to a change in the way goods and services are offered. It has moved from direct trade to online commerce, thus facilitating access to information and communication by both companies and consumers.

As a result of all changes in the field of information and communication technology, there is a need for a different form of business organization at the functional level, namely the one we currently call a virtual organization. As a result of the changes in the economic and industrial environment, the possibility of a new form of organizing business and production at the institutional level was identified: the virtual enterprise.

At the same pace as society, technology and industry change, the way in which businesses operate and is organized is changing. In order to be present in the current climate of technical and technological progress, we see the emergence of virtual organizations, virtual teams and virtual organization of enterprises and the need for a better level of knowledge and knowledge management.

2. Literature review

Every few years, a new technological development or managerial philosophy attracts the attention of thinker's strategic organizations and business practice. Not long ago, it was conceptualized and developed "Total quality management", then "redesigning Business Process Reengineering". More recently, a new interest emerged for "Knowledge Management", although as early as 1950 Peter Drucker introduced the concept of "knowledge Workers" for workers capable of using knowledge in the organization to produce intangible products. Because the market has become more competitive and less predictable, organizations realized that their main values are what they know and using the knowledge on technologies, organizational processes, in order to obtain the competitive advantage [2]. If the industrial society has different types of organization, coordination, planning of technological resources, products and financial resources in the informational company or knowledge-based approach is pursued finding principles, methods and techniques of investigation, planning, organizing the critical resource-knowledge [12]. There is no standard definition regarding all that stands for knowledge. It generally refers to a set of organizational practices and ways of generating, capture and disseminate the necessary knowledge to achieve objectives of the organization.

There is a lot of confusion about differences between time, information and knowledge. While the data reflects the numerical description of some actions, processes, facts, phenomena, information bring an increase of knowledge reflecting a set of data grouped in some models and forms, and knowledge groups a set of information with a strong human and contextual determinant. In this sense, the knowledge describes an ensemble of information acquired or applied to a particular context through human thinking. A significant difference between information and knowledge is determined by the transfer them. While the information can be easily transferred from one person to another, knowledge has a lower degree of transferability, they also have a psycho-social content, contextually reflecting intuition, creativity and experience the person who possesses this knowledge. [13]

Nonaka and Takeuchi identify the following types of knowledge [9], [10], [11]:

- Explicit knowledge that can be expressed through words and numbers. It can be easily communicated and distributed under form of data, scientific formulas, coded procedures or universal principles.
- Tactical knowledge that is very personal and difficult to understand by formalized, experience-dependent and feature-rich individuals.
- Implied knowledge stored in entities that depend on the practical context (products, technologies, processes etc). The value of knowledge is reflected in products and physical services of the organization in intellectual products (patents and licenses), processes (structural capital) and people (intellectual capital).

In the view of the authors these are the main characteristics of knowledge:

- Is an abundant resource that develops continuously. While world evolves, human knowledge develops. With the help of information and communication technologies, methods innovation and generation of ideas, this resource can have a special momentum [1].
- The benefits of a good knowledge management can be very high. In an organization with 1000 people we don't actually have 1000 people who develop and share their knowledge possessed, but the ones that do are a great asset.
- The transfer of knowledge and information is not subject to laws, taxes and trade barriers. It assumes a lower cost of this resource.
- Knowledge is contained in information packets. They cannot be assimilated to information that is perceived in seconds. Knowledge is assimilated in days and weeks by reading, listening, experience or human interaction.
- Knowledge about work can be learned through work - during practical exercise, behind reflections on mistakes and the success of the process.
- Knowledge is generative. Knowledge about a domain involves not just reproduction of information which characterizes that area, but the enunciation of some Knowledge that requires correlations, abstractions and syntheses. Human knowledge thus becomes the most important production factor that can provide organizations and nations, competitive advantage in the new millennium [3].

Human knowledge practically embraces all structures and its processes: organizations, technologies, products, human resources, being the foundation structure of an organization.

Translating human knowledge into manageable entities, which can be effectively acted through different actions and practical tools have spurred the emergence of a theory regarding the management of knowledge, which can be defined as the art of predicting, manage and develop the knowledge of a socio-human system in order to achieve the objectives.

3. Existing knowledge and knowledge demand

To offset the rapid pace with which industry and society in general are moving towards adopting Industry 4.0, we need to balance between knowledge needs and current knowledge. The limitations of this research and the authors approach are related to the knowledge and the knowledge gap between the existing level and the level required.

Managing knowledge is the process by which the starting knowledge that exists in a company is organized and coordinated. This process involves:

- Grouping knowledge by type from the main sources of the organization, through processes, to building internal memory. The group is considering different criteria set by knowledge system designers.
- Storing the knowledge that is being pursued in knowledge infrastructure organization in terms of efficiency.
- Classification of knowledge for evaluation and award priorities that reflect the quality of knowledge, respectively the degree of suitability to process requirements.
- Selection of knowledge by which the user is allowed using criteria to identify the most appropriate ones.
- Dissemination of knowledge through which knowledge is stored in various forms (tacit, explicit or implicit) and accessed by all members of the organization and even by its stakeholders - customers, suppliers, etc. The knowledge management process has developed mostly based on technology systems information and communication.

Knowledge development refers to the accumulation of new knowledge for innovation, adaptation and achievement objectives of the organization. Forms of knowledge accumulation vary depending on the entity that accumulates them. Human resources accumulate knowledge through socialization (human interaction), explicit internalization (learning in formal or informal framework) and implicit internalization (carrying out work processes, or by doing). Artificial intelligence systems accumulate knowledge by processing new facts in the context determined by an initial knowledge base and a model of thinking. The organization, viewed as a learning entity organization, modifies its potential actions in a given context. The rhythm with which an entity learns to anticipate and adapt to environmental developments is a source of competitive advantage [4]. The organization accumulates knowledge from outside by renting (hiring a consultant and using his expertise for acquiring new acquaintance) or by buying (purchasing a society for the knowledge it holds) or from inside, through the dissemination of it from a particular source component and the development of existing sources. An important aspect of development is the process of converting implicit knowledge (from the practical environment) into tacit knowledge (related to the resource human) through the exercise of work processes. Efficiency of this process is determined by the mental model or the way of thinking of the individual, his previous knowledge and the practical context [14]. The role of each of these factors can be deduced from the following analogies: the context is practically analogous to the raw material supplier; mental model is analogous to the technology of transformation the raw material, and previous knowledge has the role of the catalyst which allows the raw material to be transformed. The mental model is the result of the accumulation and is considered the most important factor in the process development. In the theories of knowledge management there have been delimitated two types of knowledge mental models: linear thinking and systemic thinking. Studies on the two models of thinking reveals that systemic thinking allows efficiency and high efficacy of human decisions in professional or social life. By systemic approach the system is building decisions that can balance the negative effects of decisions that are taken in the case of linear thinking. Previous knowledge may be characterized by volume, structure and quality. The higher the volume of the platform on which new knowledge can be built and the better the structure or diversity of knowledge is, it contributes to building deeper knowledge. Quality acquaintance reflects the ability to respond more effective to practical needs; the more knowledge more appropriate to practical needs we have, both accumulation is intensifying. The practical context has the role of providing its information the knowledge that represents the substance necessary for the accumulation of new; through its dynamics, the practical context determines new connections and new knowledge, respectively.

It is in our belief that new specialist are needed for managing different tasks: knowledge transfer experts (people who extract knowledge from different sources, organize and them so that anyone can use it and update them periodically); strategies experts for knowledge management (people that develop the strategies related to the knowledge base - auditing knowledge sources, determining what requirements derives from the mission, purpose and objectives assumed, the strategic planning of the necessary knowledge etc.); knowledge designers (people with concerns similar to the specialists who have developed for two decades rules and facts for expert systems; in this new profession, designers of knowledge follow design rules and the knowledge base at the whole organization); knowledge management Officers (responsible with the creation of knowledge infrastructure, structures and related processes and organized organizational culture for learning and acquiring knowledge) [5].

We need knowledge-based technologies. Intranet represents the Internet concept within an organizations : each entity, individual, compartment, office, working groups, other formal or informal structures have its own web page thus allowing access via the Internet to the information and knowledge dispersed within organization. The intranet also contains a page web with access to the central knowledge of the organization. This form of network uses Internet infrastructure and leads to minimal resources needed for its design. Navigation can be done with Netscape navigator or other specialized browsers. Employees of the company can thus be in permanent contact with organization and with its knowledge base from any point on Globe through the Internet. Groupware is the information technology that drives of group words and software, which means software for group work or in other words, knowledge and software accessed at the same time by multiple users. Most relevant construction of the software market is Lotus Notes, often used for brainstorming and electronic teleconferences in the process of generating and disseminating new knowledge. Document management software is a product computer system that allow the integration of entities within the documents as well as on-line access to them. Open distributed hypertext systems are software programs that allow embedding of components information technology - text, graphics, audio and video - which are also distributed via virtual network connections computers. The Internet is such a hypertext system distributed but are read-only, not open as webpages are.

Technologies and techniques of representation of knowledge through which knowledge is considered object between which different connections are established are in full swing development due to the emphasis on explicit knowledge. It is appreciated that the hypertext distribution through virtual connections between the components of an entity is the most used representation technique and will be more and more used.

Already exist in the world processes that help increase knowledge competence modeling. One of those is the process of analysis for organizational identification of specialists and their competencies, profile design (knowledge, values, aspirations, behaviors and attitudes) and development competencies in the directions needed to support the areas critical of the organization. The memory of the organization is the process that has as an objective the transformation of silent knowledge from within the organization in explicit knowledge, easily transferable. The existence of internal memory allows the application of contestant practices engineering, rapid adaptation to environmental changes outside and the formation of virtual organizations in which components are horizontally integrated. Communities of knowledge practice are communities of experts, practitioners, politicians, teachers and researchers etc. which share a common interest in certain themes. They use in general an electronic platform, to develop concepts and debate issues in discussion. Benchmarking is the managerial practice with the role of investigate best practices from different industries and organizations and report the organization's internal processes to the new standards.

We have to develop national design policies for a knowledge-based economy. Most of the countries OECD have begun building the human infrastructure and informational technologies that will translate economies to the new knowledge-based society. The main competencies on which these policies are focused: production of knowledge, transmission, transfer and measurement of knowledge.

Organizations are increasingly focusing on their intellectual assets, and less on the material goods they have administered. Knowledge solutions become the key used to build and support capital goods and they are used to create economic value and enable individuals, teams, and communities to achieve much improved performance in creating, capturing, sharing, and exploiting knowledge [6].

4. Conclusion

It is necessary to clearly delimit the concepts of data, information and knowledge. Data is a series of directly observable properties of things, information is the accumulation of data in a higher-level beam with meaning filtered through a person's perceptions and thoughts (cognition), information stimulates one wide range of actions and activities and knowledge is extracted from data and built on information. The two distinct forms of knowledge : explicit knowledge, consisting of forms of information, expertise, or experience that can be expressed in detail, archived, encoded, and can be distributed with IT and can take the form of a database, a document, a drawing, formula, patent, video, or presentations and silent knowledge, or intrinsic science, the ability of an individual to possess, is a system Common, unexpressed values, visions, goals, and behaviors that direct the activities of an organizations; the tacit knowledge derives from the accumulated experience.

The main obstacle for most organizations is identifying this tacit knowledge system. Challenge raised by this consists of finding out how to recognize, generate, distribute, and administer the system. Tactical knowledge can only be deduced from human actions, it is are not observable directly. It is a human attribute that predisposes people to act in certain ways depending on circumstances. An accumulation process can use and re-use information, explicit knowledge, experience, and expertise to obtain a special business benefit, meeting a goal, or to increase the profitability and competitiveness of the organization. Through this process, an organization generates value from intellectual and goods based knowledge. Knowledge management is a good catalyst for innovation and learning. Solutions in this area have the duty to fill the gap between explicit and tacit knowledge and implement content technologies that support the capture and management of explicit information. Collaborative technologies allow individuals and communities to create, distribute, and publish information to meet the specific business objectives.

The gap between existent knowledge and demanded knowledge must be filled. In order to do that, organisations need a good knowledge management. We think that in today's economic climate, where the focus is on information, organizations get greatest value from their intellectual goods rather than from physical ones. They have to maintain the sharing of knowledge to lay the foundations for collaboration. In addition, a good management of it narrows the gap and helps organizations in the following areas:

- Supporting innovation - provides a network infrastructure electronic and social, so that they can develop new products or services; it supports and provide access to sources of ideas so that they can be transformed into capital.
- Supporting collaboration - enhances opportunities for collaboration; this enriches the exchange of tacit and explicit knowledge among people; and encourages free flow of ideas.
- Encouraging and using learning - facilitates and accelerates the learning process; it creates for individuals and groups the possibilities to implement new knowledge; it exploits organizational knowledge and brings the right knowledge to the right people, in an intelligible context that explains the new challenges; it capitalizes on individual learning through rewarding and speculating on the benefit of the entire organization.
- Developing social capital - enhances the transfer of individual knowledge within the organization; supports the exchange outside the organizational boundaries, and puts together the people who have the necessary tacit and explicit knowledge to carry out their work.
- Attracting and Maintaining Human Capital - increases storage rates employees by recognizing the value of each employee's knowledge and rewarding it for this luggage; capture and capitalize people knowledge; promotes career development.
- Creating and using structural capital - transforms intellectual capital in structural capital; and directs strategic thinking towards capitalization of knowledge above focus on budget issues.
- Creating the opportunity for online government - provides a basis IT-enabled information technology to enable citizens and customers to access the information and services they need; it establishes a network of well-trained government employees which can increase value for citizens and customers.
- Increasing productivity - fluidizes operations and reduces costs, risks, learning curves and initialization times by eliminating redundant or unnecessary processes; it contributes to the fundamental goals / mission.
- Exchange of good practices / processes - distributes good practice throughout within the organization; it learns from the failed efforts; provides a platform for innovation and innovation reuse of knowledge; sets the benchmarks for internal and external performance for individuals and teams and helps new employees become more familiar with the culture of the organization.
- Increasing customer satisfaction - improves customer service through cutting response times; it focuses on the knowledge of the needs of the customer in viewing the efforts of an organization; and improves customer experiences and results With regard to the services that are available to them.
- Establishing a distinction between competitive advantage and the business market - helps to ensure the superiority of knowledge; when it takes the form of a proposal business, products or services, it changes the proposal to capitalize on

the supply of goods and services to provide knowledge and expertise for these goods and services and concentrates on customer needs to meet business / mission goals.

In creating innovative products and services, some organizations may not have the ability to plan in the way effectively and strategically, access the operation, and be able to successfully implement the systems needed for the business. These can become obstacles to implementing a good knowledge management.

Some of the other problems regarding the gap between demand and offer in terms of knowledge include:

• Recruitment of employees - a culture that recognizes silent knowledge and encourages employees to make exchanges is critical for good management. Employees are encouraged to give up their knowledge and experience - the features that define their value at individual level. In addition, employees who participate have the opportunity to become aware of the quality or relevance of the information they contribute with.

• Using technology to deliver optimal output. Knowledge is not focused on technology. That's why an organization should not be attracted to the trap of establishing a knowledge management program exclusively on the premises designed by a collaborative tool. Technology can support knowledge management, but this is not the starting point.

• Lack of a specific business goal - managers must incorporate a specific business goal.

• Ensure updating in the knowledge process. Content of a program must be permanently updated, rectified, or deleted. In addition, the relevance of knowledge at some point changes, just as skills employees do.

• Changing information into knowledge. Organizations must be vigilant about overloading with information. The goal of is to filter and spread the most valuable knowledge from a true ocean of information.

If we think about solving problems and about the future needs here are 10 competencies a virtual enterprise actor will need at work after 2020 in order to meet any knowledge demand:

• Flexible thinking - the ability to give up ineffective actions and solve problems in a way that will lead to the desired results, the ability to approach new strategies easily, to see the situation in development and to predict the outcome of different variants, the ability to select information and find solutions by using a variety of approaches;

• Negotiation - the capacity to find the solutions in which both parties gain, the ability to create a compromise that both parties benefits from, excluding imposing the view on others;

• Service-oriented - that competence that facilitates the provision of the best service, using existing resources in a shorter time. This means acting proactively and having the ability to observe both the details and the ensemble, that is, to work in such a way as to avoid the occurrence of problems, to keep the person informed about how to process the problem solving process, to have the ability to use the information you have in the most efficient way and to ensure that the service you offer has the impact you have planned;

• The ability to discern and make a decision - a complex package of interconnected competences, such as the ability to perceive and distinguish the details of a situation, the ability to follow a cognitive process of analysis and selection of conclusions, the ability to formulate an opinion or make a critical distinction on ideas, people or situations, the ability to evaluate, compare and deliberate;

• Emotional Intelligence - the capacity to manage your own emotions and correctly identify the emotions of others, the awareness of inner states and effective social interactivity;

• Coordination with others - the ability to organize your own activity by linking it to the whole package as part of it, the ability to cope with obstacles and to keep moving towards reaching the goal, separating tasks and correlating them with work, the ability to prioritize activities so as to ensure logical, fluid performance in relation to the entire package to which it belongs;

• Ability to manage relationships with people - the ability to maximize the efficiency and productivity of interaction with others for the benefit of all;

• Creativity - the ability of man to realize the new, in different forms, to reveal special, unknown aspects of reality, to develop original paths and solutions to solve the problems and to express them in new personal forms;

• Critical Thinking - an amount of competencies that support a mental process of analyzing and evaluating an idea or situation, the ability to reflect and find a logical connection between ideas and arguments, to understand the relevance of information or knowledge, and to build new arguments, as well as the ability to distinguish between facts, opinions and judgments;

• Solving complex problems - that is, a combined use of all the above competencies.

This article represents only a beginning of the study regarding the problem of knowledge. In the following period, the authors will focus their attention on further research and completion of this study.

5. References

[1] Anand, N.; Gardner, H.K. & Morris, T. (2007). Knowledge based Innovation: Emergence and Embedding of New Practice Areas in Management Consulting Firms. Academy of Management Journal, Vol. 50, Vol. 2, pp. 406-428. ISSN 0001 – 4273

[2] Drucker, P. (1998). Peter Drucker on the profession of management. Mass.: Harvard Business School Press, ISBN 0-87854-836-2, Boston, US

[3] Foss, N. J. & Pedersen, T. (2004). Organizing knowledge processes in the multinational corporation, Journal of International Business Studies, Vol. 35, No. 5, pp. 340-349, ISSN 0047-2506

[4] Maleki, E. J. (2003). Knowledge, industry and environment: Institutions and innovation in territorial perspective. Annals of the Association of American Geographers, Vol. 93, No. 2, pp. 523-524, ISBN 0-75463-052-8, DOI : 10.1111/1464-8306.930201412

[5] Hurt, U./, Otto, T./, Kaare, K. K. & Koppel, O. (2014). New Approach to Knowledge Transfer Environment Development. Procedia Engineering, 24th DAAAM International Symposium on Intelligent Manufacturing and Automation, 2014, Zadar, Croatia, ISSN 1877-7058, Katalinic, B. (Ed.), Vol. 69, pp. 273-281, Elsevier Ltd., DOI : 10.1016/j.proeng.2014.02.232

[6] Kogut, B. & Zander, U. (1993). Knowledge of the firm and the evolutionary theory of the multinational corporation, Journal of International Business Studies, Vol. 24, No. 4, pp. 625-645, ISSN 1478-6990, DOI : 10.1057/Palgrave.jibs.8490248

[7] Kurtuhuz, A.; Salisteanu, C.; Udroiu, I.; Petrescu, M.; Issa, C. & Udroiu, F. (2014). The new organization based on knowledge and communication. Chapter 17 in DAAAM International Scientific Book 2014 Vienna, Austria, ISSN 1726-9687, ISBN 978-3-901509-98-8, Katalinic, B. (Ed.), pp. 189-196, Elsevier Ltd., DOI : 10.2507/daaam.scibook.2014.16

[8] Leber, M.; Buchmeister, B. & Ivanisevic, A. (2015). Impact of knowledge on innovation process. Chapter 21 in DAAAM International Scientific Book 2015, Vienna, Austria, ISSN 1726-9687, ISBN 978-3-902734-05-1, Katalinic, B. (Ed.), pp. 235-248, Elsevier Ltd., DOI : 10.2507/daaam.scibook.2015.21

[9] Nonaka, L. & Takeuchi, H. (1995). The knowledge creation company: How Japanese companies create the dynamics of innovation. Oxford University Press, Oxford, New York, ISBN 0-19-509269-4

[10] Nonaka, L. &Takeuchi, H.; & Umemoto, K. (1996). A theory of organizational knowledge creation. International Journal of Technology Management, Vol. 11, No. 7-8, pp. 833-845, DOI : 10.1504/IJTM.1996.025472

[11]Nonaka, L. & Takeuchi, H. (2011). The wise leader. Harvard business review, Vol. 89, No. 5, pp. 58-67, ISSN 0017-8012

[12] Rodgers, P.A.; Caldwell, N.H.M.; Hoxor, A.P. & Clarkson, P.J., (1999). WEBCADET: a knowledge management support system for new product development, In : 6th International Product Development Management Conference, Cambridge University Press, pp. 937-948, Cambridge/UK

[13] Welbourne, M. (2002). Knowledge. Taylor and Francis, Chesham, Acumen

[14] Wouters, P.; Beaulieu, A.; Scharnhorst, A. & Wyatt, S. (Eds.). (2013). Virtual Knowledge: Experimenting in the Humanities and the Social Sciences. MIT Press. Available from http://www.jstor.org/stable/j.ctt5vjrxn, Accessed : 2016-10-15

DOI: 10.2507/28th.daaam.proceedings.106

USE OF SIMULATION SOFTWARE ENVIRONMENTS FOR THE PURPOSE OF PRODUCTION OPTIMIZATION

Robert Ojstersek & Borut Buchmeister

This Publication has to be referred as: Ojstersek, R[obert] & Buchmeister, B[orut] (2017). Use of Simulation Software Environments for the Purpose of Production Optimization, Proceedings of the 28th DAAAM International Symposium, pp.0750-0758, B. Katalinic (Ed.), Published by DAAAM International, ISBN 978-3-902734-11-2, ISSN 1726-9679, Vienna, Austria
DOI: 10.2507/28th.daaam.proceedings.106

Abstract

This paper presents an introduction of two simulation environments for the purpose of optimizing production line. The time in which we are implementing the concept of Industry 4.0, a part of which is based on automated and robotic production systems, we want to pre-check financial and production efficiency of the proposed new updated concept of our existing production line. A comparison of two different programming environments for discrete systems simulation will be presented in this paper on the applied industrial example. Simulation model enables calculation of machine utilization, throughput time, number of finished products and minimum needed workers (MNW). If we want to justify the use of simulation environments for the purpose of optimizing production, we must firstly show the real world example as in presented paper. The result will justify the redesign of the existing production line and justify the use of simulation software tools for carrying out preliminary tests to create new more productive production line, which is designed on concept of Industry 4.0.

Keywords: Simulation; Simio; Autodesk Process Analysis 360°; Industry 4.0; Optimization

1. Introduction

Modern concepts of production are based on the introduction of automated production systems, this raises the questions; firstly, the financial justification of new investments in the existing manufacturing process, secondly, an appropriate allocation of labour force. Thirdly, if we had reached satisfying optimization of our production line which will be able to follow new trends and demands for a satisfactory period of time. Modern industry is turning to get the help from simulation software environments through which we can precisely define, optimize and simulate new manufacturing process. Throe the simulation results we can accurately determine production line specifications and parameters. In the final phase, operation of the system is simulated in a virtual environment to test real world parameters. The results of the simulations present the basic data for the comparative calculation of the eligibility of the existing system and proposed updates. With the introduction of the concept of Industry 4.0 [1] we must be also careful to a social aspect of labour. We must properly allocate the existing workforce. We must provide professional fulfilment in a new working area while we should not resort to unnecessary firing of the workers.

The paper is organized as follows: in section 2 the literature review of simulation software is provided, in section 3 some basic approaches of simulations model are introduced. The two discrete system simulation environments Autodesk Process Analysis 360° and Simio are presented in sections 4 and 5. Section 6 provides the details of designing, developing and implementing of simulation software's on applied example for the purpose of production optimization. Final section provides some conclusion about benefits and future advancements of simulation environments.

2. Literature review

Simulation environments are more and more powerful tools for precise execution of the new proposed concepts, models and production lines in field of scheduling and optimizing production environments. For this purpose, we can use a lot of different available simulation software packages. The problems are appearing when we want to combine our theoretical knowledge and mathematics concepts to new future proposed concepts in simulation environments. Many different authors have tried to integrate and compare multiple diverse simulation software and theoretical-mathematical models to get near optimal results for optimization of production environments.

Shirazi et al. [2] presented the use of simulation tools for solving flexible manufacturing systems, where the machines and tools are flexible; they can allocate the position and time of variables. Dehghanimohammadabadi et al. [3] present the concept of the intelligent simulation in which they integrate two different powerful simulation tools Simio and MatLab to create decision support system for optimization. We can see that many authors use simulation tools for: solving job shop scheduling problem (JSSP) with multiple-objects such as Ripon et al. [4] presents a use of mathematical simulation tool for solving crossover in JSSP. Wang et al. [5] present the simulation approach for process planning, operation determining and operational sequencing. Shundar et al. [6] represented the use of bee colony algorithm for solving JSSP in which they use some data analyser software to compare the mathematical data received from the simulation environment to real world data. The same problem is solved by the Wisittipanich and Kachitvichyanukil [7] which are using Microsoft Visual Studio and C# programming language to solve this problem. Most of big factories around the world use simulation tools to simulate different approaches, methods and new concepts. Haider and Mirza [8] represent the manufacturing in Toyota production system, the methods of lean manufacturing and job shop production is presented on one-piece flow simulation. Another way to use the simulation environments is optimizing the layout and material flow analysis, Centobelli et al. [9] present the concept of digital factory which is founded on the simulation environments.

Many authors are still using priority rules for solving different non-deterministic polynominal-time hard (NP-hard) problems, but we can see that they use simulation environments instead of mathematical calculations. Zupan et al. [10] present the hybrid algorithm based on priority rules for simulation of workshop production, the results show the advantages of using simulation environments. We can also see that a lot of researchers use simulation environment to create and test new metaheuristic approaches such as Shuffled frog-leaping algorithm [11, 12], Intelligent water drops algorithm [13 - 15], Cloud-Entropy enhanced genetic algorithm [16] and so on [17]. Zgang et al. [18] present new hybrid multi-neighborhood based shuffled frog leaping algorithm with path relinking (RMN-SFLA-PR) for solving JSSP and MNW. RMN-SFLA-PR algorithm has a 100 % success rate in the low-dimensional cases of the 4 benchmark (dj38) and converged to optimum ten times faster than individual algorithms in JSSP to solve MNW for the real world production line. They use MatLab environment for creating mathematical model and discrete system simulation environment Simio for testing the newly proposed algorithm on real world example. The mentioned things above show how useful are simulation environments and in how many diversified fields we can used them. But in some cases the usage of the simulation environments it is not so easy, programmes or researcher must have extended computational knowledge to create reliable and working simulation models. This paper presents the use of two different simulation environments, Autodesk Process Analysis 360° and discrete system simulation tool Simio, to enable testing the productivity and the utilization of the new production line in comparison with the existing one.

3. Simulation

Simulation presents the transferred real system into computer programming language [19]. Designer build a computer model, which has the characteristics of the real world system. Created model represents a true picture of real industrial production line, to which we can add real world variables as mathematical variables or as a constant value in a specified period of time. In the simulation optimization of the production system, the world's leading companies use Autodesk software package Process Analysis 360° and discrete system programming environment Simio. In the following chapters we will present this two simulation environments, their features and characteristics, which will be presented in the applied example of real world production line. In our simulation of production line we are focused on optimizing three main criteria; maximum makespan C_{max}, defined by (3.1), the total flow time F (3.2) and the utilisation of the machine centres UR (3.3) [12].

$$C_{max} = \max\{C_i\}, 1 \leq i \leq n \tag{3.1}$$

C_i is the completion time of job i, $i = 1, \ldots, n$. n is number of jobs.

$$F = \sum (C_i - S_i),\ i = 1, \dots, n \tag{3.2}$$

C_i is the competition time and S_i is the starting time of job i, $i = 1, \dots, n$.

$$UR = \frac{\sum_{i=1}^{n} \sum_{j=1}^{m} p_{ij}}{m\, C_{max}},\ m \text{ is number of machines.} \tag{3.3}$$

The processing time p_{ij} represent the time job j has to spend on machine i [21].

4. Autodesk Process Analisys 360°

Simulation of the production system in the programming environment Autodesk Process Analysis 360° [22] enables as direct transfer of three-dimensional model from the software environment Autodesk Inventor to the software simulation environment Autodesk Process Analysis 360°. The advantage of the interconnectivity is represented in the entire software environment of Autodesk Factory Design Suite. With transferring the production system model in which are included all the data defined in the three-dimensional model created in Autodesk Inventor software environment, we transfer also mutual length, data of processing machines and the number of workers. In our case, we want to determinate the MNW. Real world example shows that the company already uses software from the manufacturer Autodesk environment, for example, AutoCAD, which is often used software environment for two-dimensional designs of the layout processing lines, and synchronization is possible between the two-dimensional model in AutoCAD and a three-dimensional model in Autodesk Inventor. Synchronization between the two-dimensional model in AutoCAD and the three-dimensional model is shown in Figure 1.

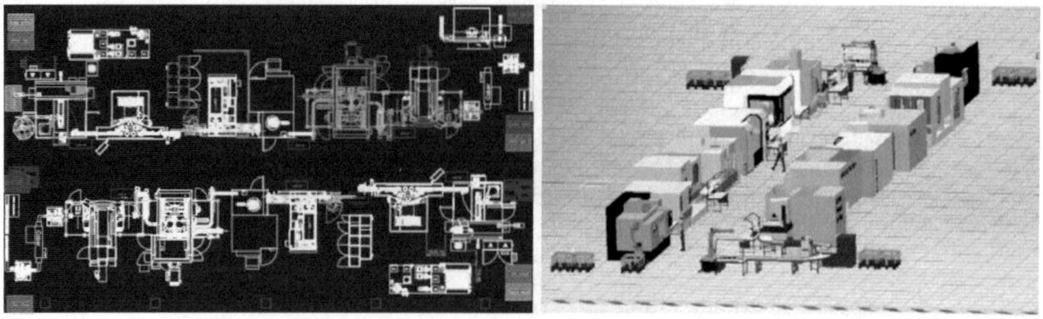

Fig. 1. A two-dimension model (left) and a three-dimensional model (right) of the production process

In the simulation environment Autodesk Process Analysis 360° is required to enter real information about the vehicle speed, the speed of workers and conveyors, as well as time data: time to wait before processing, wait times on means of transport, downtime of production line and the duration of the simulation. The integrity of the software environment is reflected in the ease of use, versatility of existing information and interconnectivity of the entire software package Autodesk Factory Design Suite.

5. Software environment Simio

Simio is a simulation environment based on the use of intelligent objects [23]. Intelligent objects can be built by ourselves or we can use an existing object from the library. The use of existing objects is often the initial choice of the software users. Firstly, the object can be assigned as symbolic three-dimensional animation models and secondly, maybe even more important, is the possibility of transferring the real data captured from files of Microsoft Office Excel and SAP systems from production floor to the Simio software environment. For the mathematical assignment of characteristics and functions we used object-oriented programming languages C ++, C #. For the simplified implementation of object-oriented programming language Simio provides already configured graphical programming framework. The pre-existing objects are divided into the following groups:

- Fixed object, representing objects that are stationary relative to the location.
- Agents, freely moving objects in three-dimensional space.
- Entity, subsystem of moving objects, often presented as a moving product.
- Link and Node, connect the movement of moving objects.
- Transporter, intended to define routes, positions, and the final destinations.

Proper implementation of the objects is very important in order to obtain credible results in created simulation model. The model is well-defined with real world application characteristics and properties, which can be determined by the fixed values, or mathematical functions. The simulation model with real world characteristics and properties is presented in Figure 2. The advantage of the 3D graphical simulation model are simulation results, which can be used as a basis for a comparative simulation study between the existing and the proposed model of the new production line.

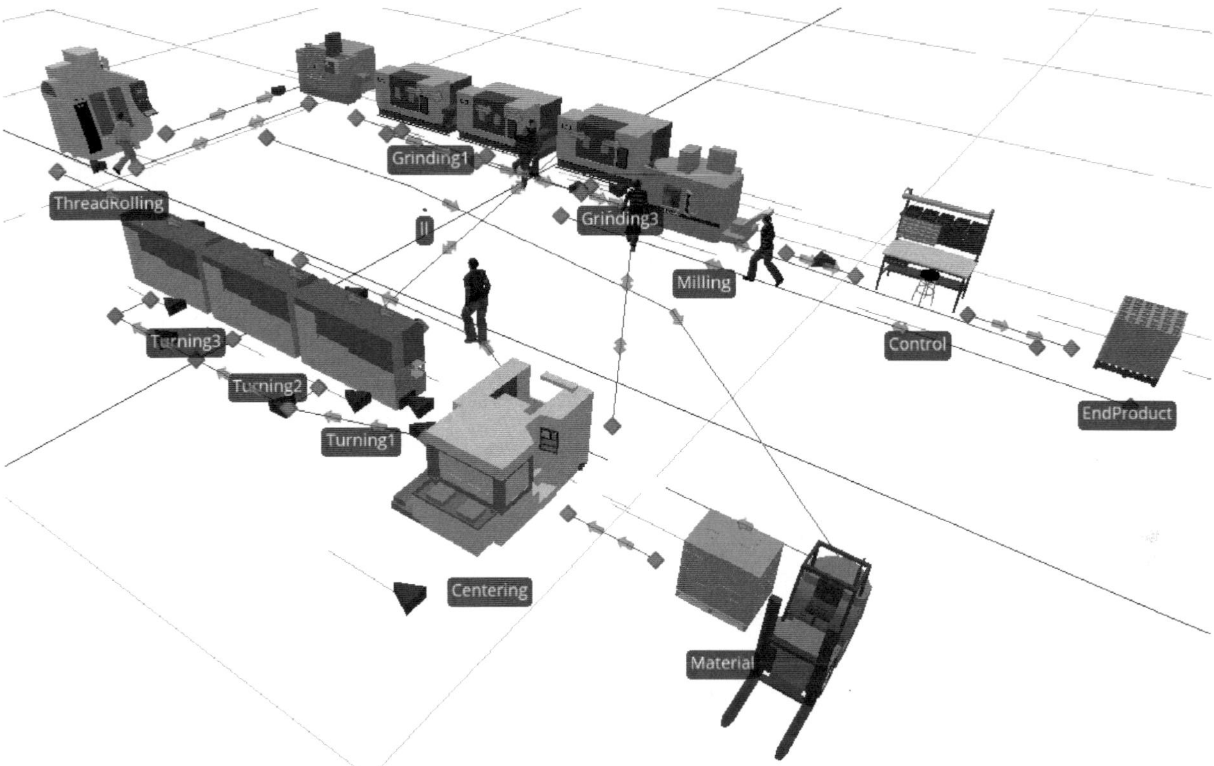

Fig. 2. Discrete system simulation in Simio

6. Example of the use of simulation software environments for the purpose of production optimization

Usage of simulation environment is very effective in the introducing of new concepts into an existing manufacturing process. Figure 3 shows the existing production line, which is not optimized. It consists of older types of machines with longer processing times and short mean time between failures (*MTBF*). For the smooth functioning of the production line care six workers. As we know, the human factor reduces the reliability, robustness and productivity of the production line. Therefore, we propose new concepts of automated and robotized production lines (fundament on concept of Industry 4.0). The existing system consists of ten individual machining centers, including transportation systems, partly managed over the conveyor belt, partly manually transported using transport carts. Due to the influence of the human factor we can assume the loss of production time occurring during manual transporting and handling of the work pieces. Older un-optimized machining centre have longer production time per work piece. Ten machining centre perform operations listed below (in brackets are the operational times per work piece in seconds):

1. Centering (20 s),
2. External turning (67 s),
3. External turning (64 s),
4. External turning (66 s),
5. Rolling (18 s),
6. Hardening (24 s),
7. Grinding (65 s),
8. Grinding (68 s),
9. Grinding (66 s),
10. Groove milling (17 s).

It should be noted that the operations of external turning and grinding are carried on only one of the three machining centre. Transport of the work pieces from one machine centre to another is partly implemented via a conveyor belt, which travels at a speed of 0.1 m/s, and partly by carts pushed by workers which are working at the production line. Their speed of movement is between 70 m/min and 75 m/min, depending on the weight of the transport carts. Delivery of semi-finished products on the production line and the initial filling of the first machine center (centering machine) and also the removal of finished product from the production line is provided with a forklift. Speed of the forklift is 5 km/h or 83.3 m/min and it is electronically limited.

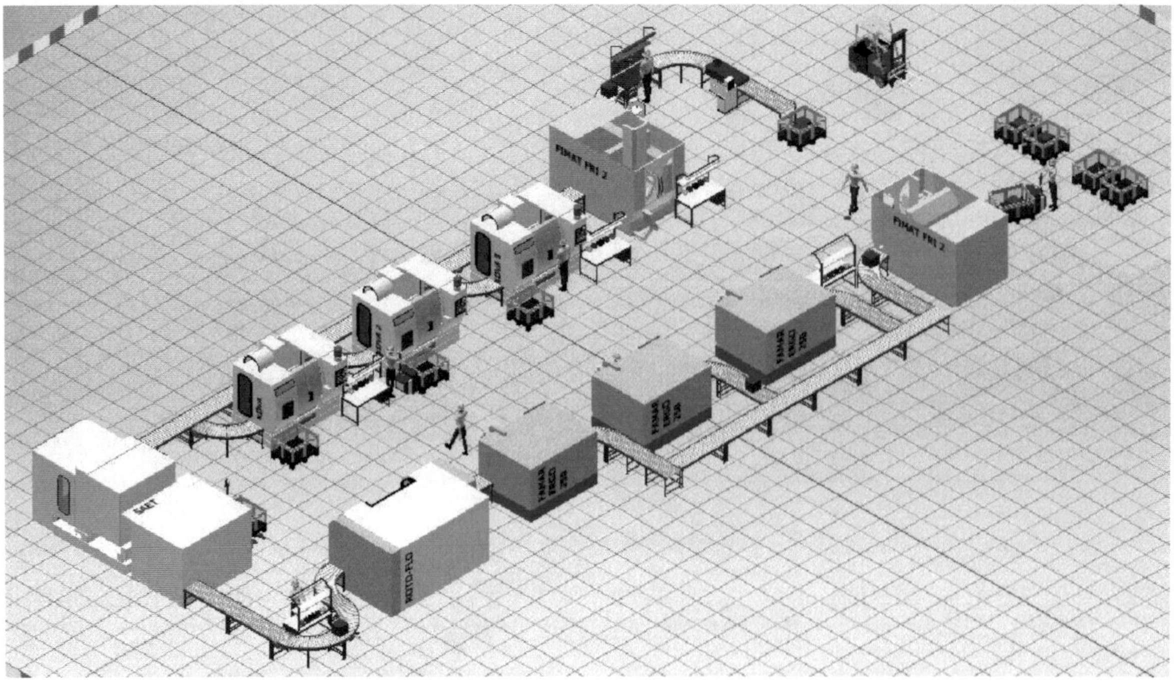

Fig. 3. A three-dimensional model of the existing production

The simulation results in Figure 4 show that the total utilization of the existing production line is 33.82 %. The simulation time is represented by one eight hour shift or 480 min. During 480 min shift a worker has one 30 min lunch break, two 10 min break and one 5 min break to get some rest between hard work. That's mean that the total production time of the simulation is 425 min. In this time production line produced 440 finished products.

The simulation results are:

- Simulation time: 07:05:00 (h:m:s)
- Total Uptime of all machine centre: 1:02:21:18:203 (d:h:m:s:ms)
- Total Downtime of all machine centre: 03:57:11:000 (h:m:s:ms)
- Utilization: 33,82 %

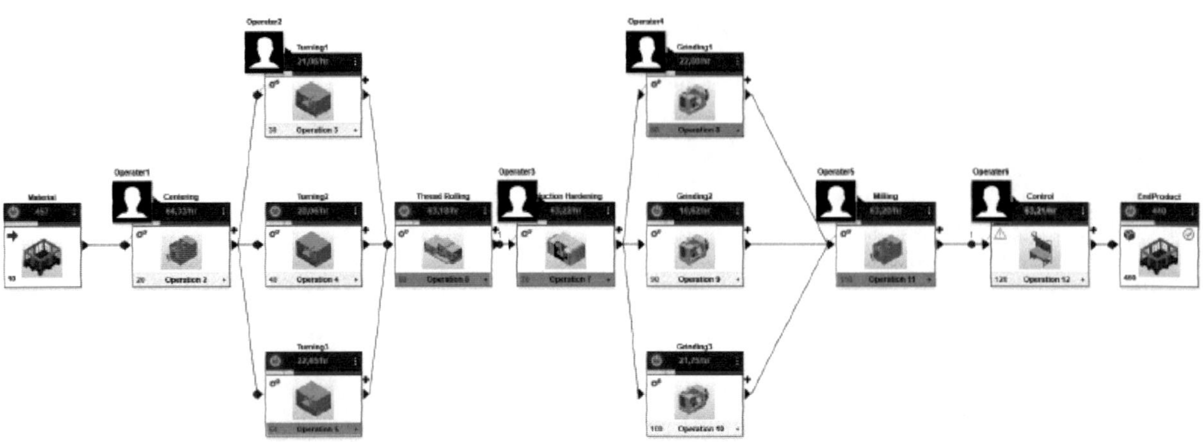

Fig. 4. Simulation model in Autodesk Process Analysis 360°

Figure 5 represent the graphical representation of the simulation results for the utilization of each individual machine centre, through which we can see single utilization, waiting times and potential bottlenecks in the production process.

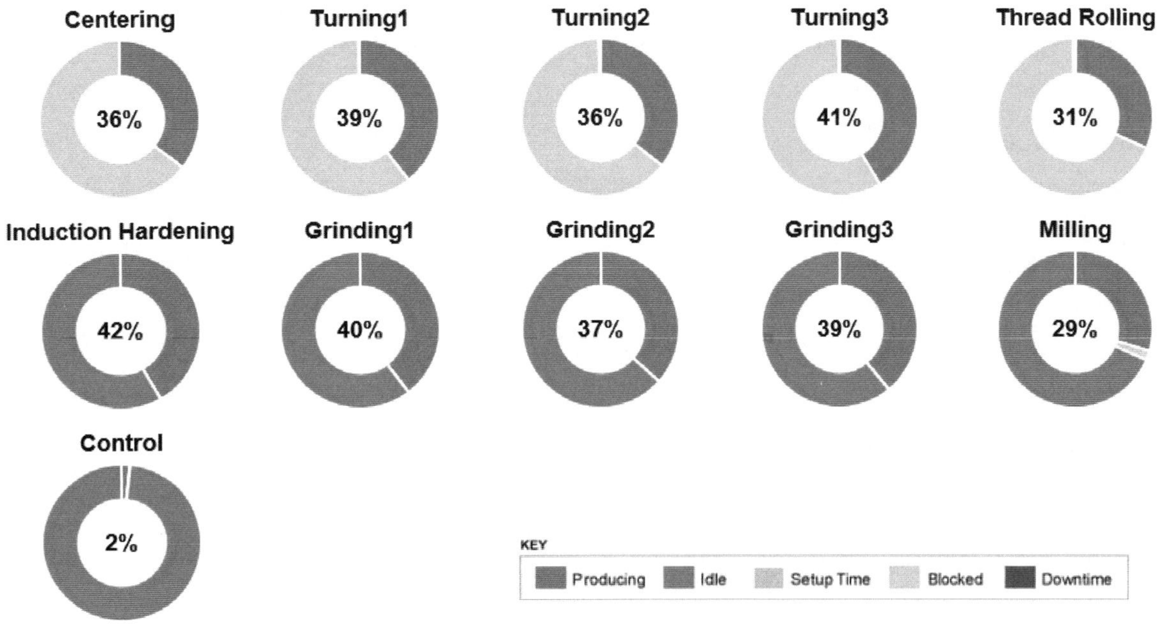

Fig. 5. Graphical presentation of the simulation results for the existing system

For better reliability of the simulation results we have also tested the simulation in simulation environment Simio. Simulation results in Simio shows that the total number of finished products is 448 parts within the simulation time period of 425 min, which is the time of one work shift. Results shows that the amount of finished products in simulation environment is higher for 8 finished product compared to the simulation carried out in the software environment Autodesk Process Analysis 360 °. Minimum difference between this two simulation environment can be attributed to real data entered into the simulation model and the impact of production downtime. We can also predict that the number of the completed products is related to the reliability factor of the workers. The simulation model has been tested several times, we tested with several applications and two durations:

- Short simulation time of one work shift, 425 min.
- Long simulation time of the eighteen work shifts, 7650 min.

From the obtained results we can see that the number of completed products differs just for two in positive or negative way, which demonstrate the reliability of the simulation. Then we started with creating of new proposed concept of production line. New production line has more productive machine centre with higher production rate lower *MTBF* and totally automated machining process. New production line consists of ten machine centers that perform operations listed below (in brackets are the operational times per work pieces in seconds):

1. Thread rolling (15 s),
2. Induction annealing (17 s),
3. Labelling (15 s),
4. Induction slackening (17 s),
5. Tooth controlling (17 s),
6. Hard turning (24 s),
7. Hard turning / milling (55 s),
8. Groove milling (28 s),
9. Groove controlling (17 s),
10. Product washing (17 s).

In new production line the transport of the product is carried out with conveyor belts, their speed is 0.1 m/s. Functioning of the production line is provided by three workers, who are responsible for the smooth operation of the machining centre and carrying out control checks on the work pieces. Delivery of semi-finished products on the production line is provided by a forklift. Speed of the forklift is 5 km/h or 83.3 m/min (electronically limited). Semi-finished products, which arrive in the production line from the external supplier are pre-treaded, also finished products are sent for corrosion protection, provided by external supplier. Feeding semi-finished products to the production line is carried out by using the robot to load the first machine centre (thread rolling). Also at the end of the production line a robot move finished products from the conveyor belt to the box. The implementation of robotic systems in the production process contributes to a more efficient, more reliable and faster production of the products. New concept of the production line is represented in Figure 6, in which we can see new machine centre and the robots at the beginning and at the end of the production line.

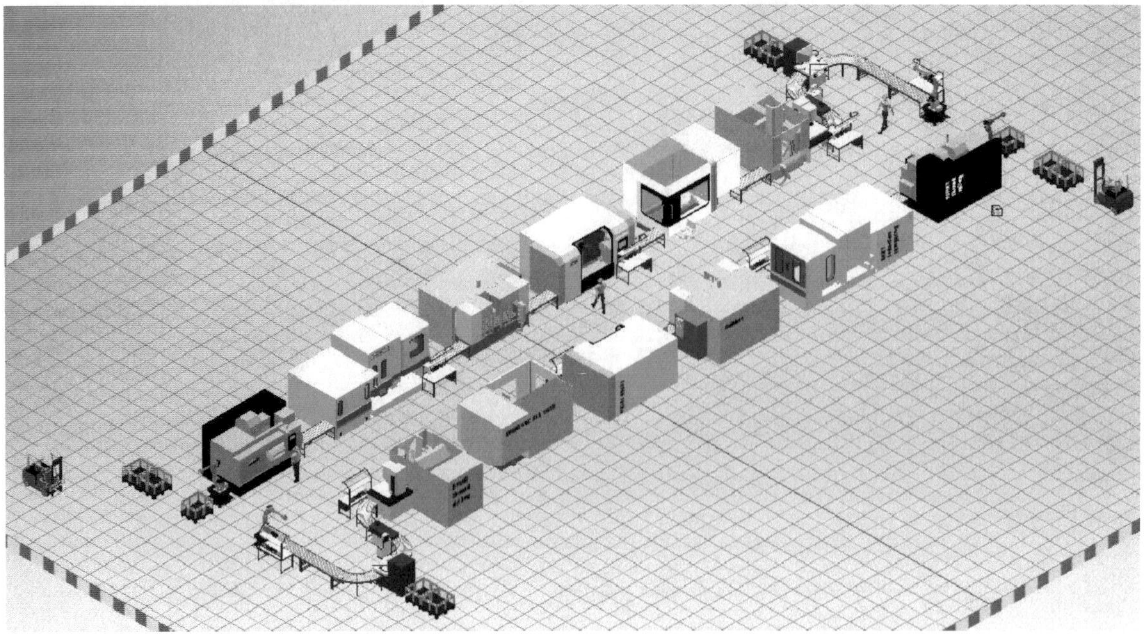

Fig. 6. Three-dimensional simulation model of the proposed new machining line

From the following Figure 7 can be seen an increase in the utilization of the production line from 33.82 % (old production line) to 40.58 % (new production line), utilization is increased by 6,76 % or relatively by 20 %. It should be emphasized that the number of workers is decreased from the six workers at the old production line to three workers at the new production line. Smaller number of workers and higher productivity of the new machine centers are responsible for the smoother, faster, more repayable operations at the new production line. As we said the implementation of the automation and the robotisation in the new production line contributes to higher reliability, efficiency and robustness of the operating process.

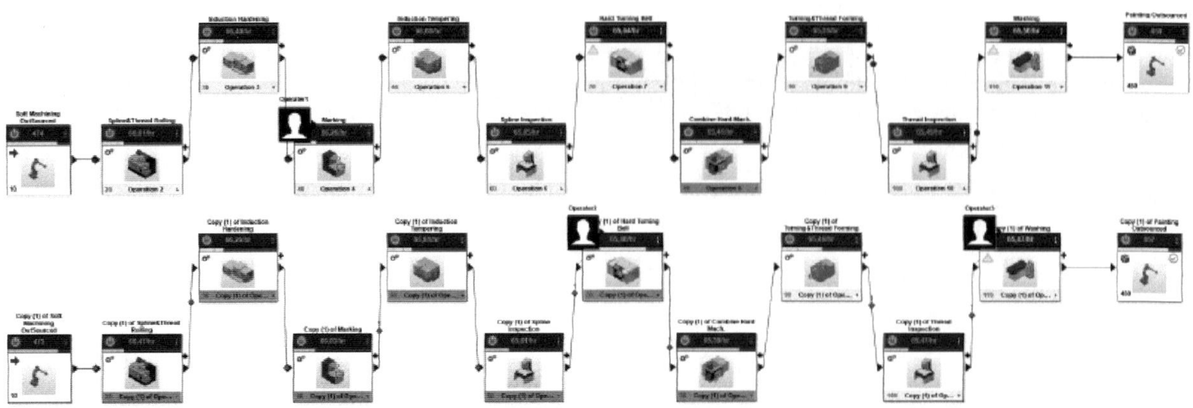

Fig. 7. Simulation model of the proposed concept

The results predict that the productivity of the new production line with implementation of a new concept will improve the finished number of products form 440 products to 457 products, manufactured on an individual branch of the newly proposed concept of the production line. Total number of finished product in the new line is 914 products, over a period of 425 minutes.

The simulation results are:

- Simulation time: 07:05:00 (h:m:s)
- Total Uptime of all machine centre: 2:09:29:09:228 (d:h:m:s:ms)
- Total Downtime of all machine centre: 08:37:42:150 (h:m:s:ms)
- Utilization: 40,58 %

From the graphical representation in Figure 8, we can see the increase of each machine centre utilization. The increase of the utilization can be attributed to the automation and robotisation of the production process.

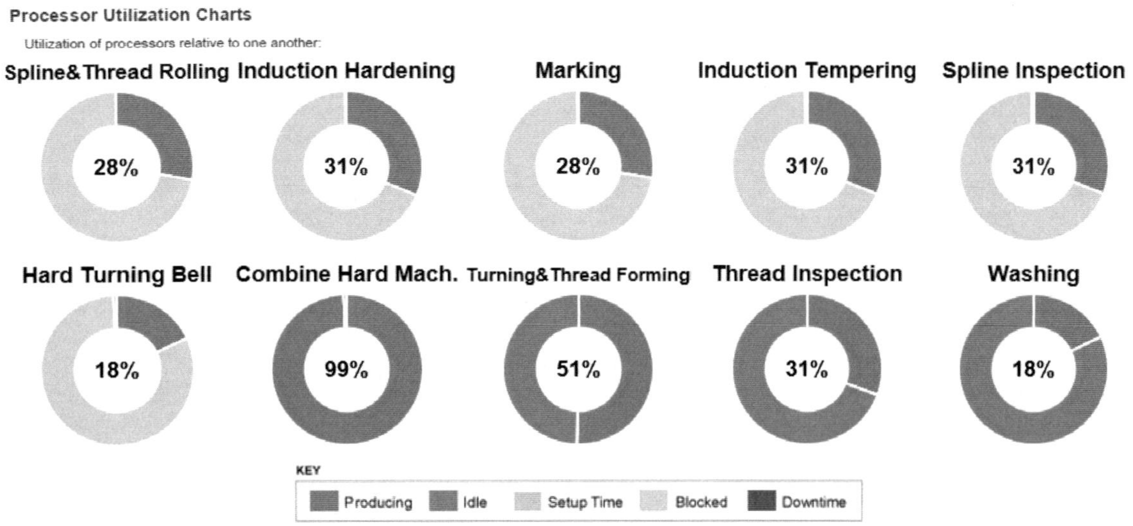

Fig. 8. Graphical display of occupancy of the machining line

Simulation study implemented in the programming environment Simio showed that the number of completed products of two central mirrored production lines (two branches) is higher than number of completed products of existing production line (Figure 9). From the obtained results we can see that the number of finished products differs just for four products in positive or negative way, which demonstrate the reliability of the simulation made in simulation programs Autodesk Process Analysis 360° and Simio.

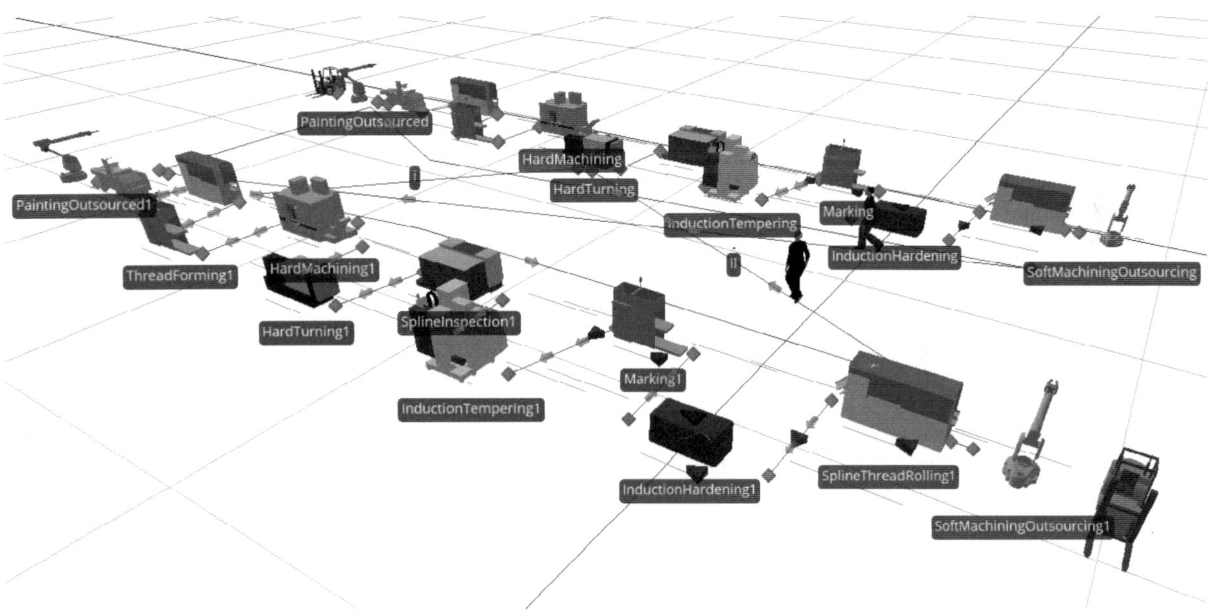

Fig. 9. Three-dimensional simulation model of the proposed new concept

7. Conclusion

The use of simulation packages, designed for simulation studies, indicates eligibility performance simulation, both for the purposes of comparing and of introducing the new concept of the production process. It provides a preliminary analysis of the financial justification of introducing new concepts into an existing manufacturing process. From the above, case it can be seen that the introduction of a new concept increases the productivity of the production line for 20 %, with a 10 % raise the occupancy of machining centre. By reducing the number of workers (MNW), from six at the existing production line to two at the new proposed concept. Introduction of automated production systems increases the robustness of the

system, the higher the reliability, the less the number of downtime, which leads to the financial justification of introducing automated systems in the production process. It proves the correctness of the simulation results matching final value of the two separate software environments. We can say that the introduction of simulation environments has a positive impact on the implementation of the concept of Industry 4.0 in the manufacturing process.
In the advancement, we will add some new metaheuristic methods for solving JSSP and MNW problem; we will combine MatLab code with discrete system simulation and metaheuristic methods. Newly proposed metaheuristic models in MatLab will be the fundament for creating near optimal multi-criteria optimization of JSSP. The focus will be on calculation of MNW at the production line.

8. Acknowledgement

The authors acknowledge the financial support from the Slovenian Research Agency (ARRS).

9. References

[1] Schwab, K. (2016). The Fourth Industrial Revolution, Word Economic Forum, Cologny/Geneva, Switzerland

[2] Sivakumar, A.I. (2001). Multiobjective dynamic scheduling using discrete event simulation. International Journal of Computer Integrated Manufacturing, Vol. 14, No. 2, pp. 154-167.

[3] Dehghanimohammadabadi, M. & Keyser, T.K. (2017). Intelligent Simulation: Integration of SIMIO and MATLAB to deploy decision support systems to simualtion environmnet. Simulation Modelling Practice and Theory, Vol. 71, pp. 45-60.

[4] Ripon, K.S.N.; Siddique, N.H. & Torresen, J. (2011). Improved precedence preservations crossover for multi-objective job shop scheduling problem, Evolving Systems, Vol. 2, No. 2, pp. 119-129.

[5] Wang, J.F.; Kang, W.L.; Zhao, J.L. & Chu, K.Y. (2016). A simulation approach to the process planning problem using a modified particle swarm optimization. Advances in Production Engineering & Management, Vol. 11, No. 2, pp. 77-92.

[6] Sundar, S.; Suganthan, P. N.; Jin, C. T.; Xiang, C. T. & Soon, C. C. (2015). A hybride artificial bee colony algorithm for job-shop scheduling problem with no-wait constraint. Soft Computing, pp. 1-10.

[7] Wisittipanich, W. & Kachitvichyanukul, V. (2013). An Efficient PSO Algorithm for Finding Pareto-Frontier in Multi-Objective Job Shop Scheduling Problems. Industrial Engineering & Management Systems, Vol. 12, No. 2, pp. 151-160.

[8] Haider, A. & Mirza, J. (2015). An implementation of lean scheduling in a job shop environment. Advances in Production Engineering & Management, Vol. 10, No. 1, pp. 5-17.

[9] Centrobelli, P.; Cerchione, R.; Murino, T. & Gallo, M. (2016). Layout and material flow optimization in digital factory. International Journal of Simulation Modelling, Vol. 15, No. 2, pp. 223-235.

[10] Zupan, H.; Herakovic, N.; Starbek, M. & Kusar, J. (2016). Hybride algorithm based on priority rules for simulation of workshop production. International Journal of Simulation Modelling, Vol. 15, No. 1, pp. 29-41.

[11] Xu, Y.; Wang, L. & Wang, S. (2013). An effective shuffled frog leaping algorithm for the flexible job shop scheduling problem. Proceedings of IEEE Computational Intelligence in Control and Automation, Published in 2013 IEEE Symposium on 16-19 April.

[12] Eusuff, M.; Lansey, K. & Pasha, F. (2006). Shuffled forg-leaping algorithm: a memetic meta-heuristic for doscrete optimization. Engineering Optimization, Vol. 38, No. 2, pp. 129-154.

[13] Shah-Hosseini, H. (2008). Intelligent water drops algorithm: A new optimization method for solving the multiple knapsack problem. International Journal of Intelligent Computing and Cybernetics, Vol. 1, No. 2, pp.193-212.

[14] Shah-Hosseini, H. (2009). The intelligent water drops algorithm: a nature-inspired swarm-based optimization algorithm. International Journal of Bio-Inspired Computation, Vol. 1, No. 1/2, pp. 71-79.

[15] Shah-Hosseini, H. (2009). Evoltionary Computation: Optimization with the Nature-Inspired Intelligent Water Drops Algorithm, I-Tech, Vienna, Austria.

[16] Yongxiang, L.; Xifan, Y. & Jifeng, Z. (2016). Multi-objective Optimization of Cloud Manufacturing Service Composition with Cloud-Entropy Enhanced Genetic Algorithm. Strojniški vestnik – Journal of Mechanical Engineering, Vol. 62, No. 10, pp. 577-590.

[17] Supsomboon, S. & Vajasuvimon, A. (2016). Simulation Model for Job Shop Production Process Improvement in Machine Parts Manufacturing. International Journal of Simulation Modelling, Vol. 15, No. 4, pp. 611-622.

[18] Zhang, H.; Liu, S. & Ojstersek, R. (2017). An effective use of hybride metaheuristic algorithm for job shop scheduling problem, International Journal of Simulation Modelling, in press

[19] Woolfson M. M.; Pert G. J. (1999). An Introduction to Computer Simulation. Oxford University Press, New York, USA.

[20] Buchmeister, B. (2013). Advanced job shop scheduling, DAAAM International, Vienna, Austria.

[21] Pinedo, M. L. (2005). Planning and Scheduling in Manufacturing and Services, Springer Science+Business Media, LLC, New York, USA.

[22] Strijker P. (2014). Hands-on: Factory Design Suite 2014, Autodesk Inc.

[23] Pegden C. D. & Sturrock D.T. Intruduction to Simio. Winter Simulation Conference (2010), pp. 1-10

DOI: 10.2507/28th.daaam.proceedings.107

AUTONOMOUS CONTROL SYSTEM FOR A VEHICLE WITH ACTIVELY ARTICULATED FRAME

Kharuzin Sergei Vadimovich, Ivanov Aleksandr Aleksandrovich
& Shmakov Oleg Aleksandrovich

This Publication has to be referred as: Kharuzin, S[ergei]; Ivanov, A[lexander] & Shmakov, O[leg] (2017). Autonomous Control System for a Vehicle with Actively Articulated Frame, Proceedings of the 28th DAAAM International Symposium, pp.0759-0766, B. Katalinic (Ed.), Published by DAAAM International, ISBN 978-3-902734-11-2, ISSN 1726-9679, Vienna, Austria
DOI: 10.2507/28th.daaam.proceedings.107

Abstract

In this paper we suggest a structure of an autonomous control system for a six-wheeled vehicle with actively articulated frame that is traversing through cluttered spaces with undetermined placement of specified-type obstacles. Description of active frame transformation control strategies and algorithms also is given. Control algorithms utilize information that is provided by proprioceptive sensors (used in direct estimation of current vehicle configuration) and exterioceptive sensors (used in indirect estimation of the obstacle height and distance from it). These algorithms are active while vehicle overcomes random consequence of obstacles belonged to three classes: step (overcoming), step (descending) and ditch. For each obstacle type we suggest specific vehicle movement mode. Patterns of each feasible switching between modes and dependency between desired frame configuration and obstacle height are then discussed. Suggested strategies and algorithms are tested via complex virtual modelling of control system and vehicle dynamics.

Keywords: Unmanned ground vehicle; mobile robot; all-terrain vehicle; autonomous movement

1. Introduction

Unmanned ground vehicles that have actively articulated mechanisms as a part of their frames can effectively traverse through uneven terrain with randomly positioned obstacles. Height of an obstacle that can be overcome by these robots highly depends on reconfiguration manner and may exceed one of vehicle's nominal dimensions. However, adaptive reconfiguration or transformation of robot's frame under control of human operator that is observing both obstacle and robot is executable but challenging task. In real working conditions only visual information captured by robotic platform onboard cameras is provided to operator. Such information may be insufficient for effective control. In order to increase quality of control process, operator may be supplied with information captured by exterioceptive and proprioceptive sensors. But in some cases it can be not enough for efficient control of robotic platform. Moreover, complexity of incoming information paired with redundancy of robotic platform (for it has many degrees of freedom) can cause sensitive overload and decrease operator performance. Full or partial automation of the control process may be performed to establish robotic platform effective movement in unstructured environments.

In this paper we suggest a structure of an autonomous control system for a six-wheeled vehicle with actively articulated frame that is traversing through cluttered spaces with undetermined placement of specified-type obstacles. Description of active frame transformation control strategies and algorithms also is given. Control algorithms utilize information that is provided by proprioceptive sensors (used in direct estimation of current vehicle configuration) and exterioceptive sensors (used in indirect estimation of the obstacle height and distance from it). These algorithms are active while vehicle overcomes random consequence of obstacles belonged to three classes: step (overcoming), step (descending) and ditch. For each obstacle type we suggest specific vehicle movement mode. Patterns of each feasible switching between modes and dependency between desired frame configuration and obstacle height are then discussed. Suggested strategies and algorithms are tested via complex virtual modelling of control system and vehicle dynamics.

2. Problem statement

In this paper we raise a problem of establishing robust control for articulated six-wheeled vehicle. Robotic platform is minimally equipped with exterioceptive sensors and features uncommon mechanical structure. Concerned device must efficiently traverse through cluttered spaces or move on uneven terrain. Furthermore, all platform active frame transformations must be performed in autonomous mode – without overriding from human-operator. This means that operator is only able to choose movement direction for platform. For method verification we use complex virtual model of robotic platform kinematics and dynamics.

3. Related work

The main part of research works in the area of developing obstacle negotiation algorithms for articulated devices with continuous ground trace of propulsion mechanisms was performed with wheeled and tracked robots.

While overcoming obstacles wheeled devices often rely on additional mechanisms and structural improvements. One of the common ways to increase passing ability of such robots is utilizing leg-wheel structures. The way of efficient usage of leg-wheel mechanisms in a role of adaptive suspension is presented in paper [1]. Each mechanism of a suggested four-wheeled robot consists of two linkages one of which is connected with robot's base and other with motorized wheel. Vertical loads and inclinometers are proposed by authors to be implemented as control algorithm inputs. Having been tested via virtual modelling this method has shown its efficiency. Another way to improve obstacle negotiation capability is to gather more information about robot's environment. Researchers [2] suggested torque control algorithms for minimizing wheel slip of the six-wheeled robot equipped with so-called tactile wheels and active suspension mechanism. Tactile wheels allow robot's computing unit to estimate direction and scale of wheel-terrain normal forces. Method has shown its efficiency with two different sets of tires used to achieve different friction force values. In paper [3] another approach was demonstrated. Authors have developed six-wheeled device with semi-passive compliant suspension mechanism. It consists of parallel linkages that hold side wheels and a front four-linked suspension front wheel. Rear wheel is fixed on the individual link. Experiments both with physical and virtual model of the robot have shown that it can overcome obstacles that is higher that device's wheel. Adaptive leg gait for four-wheeled robot with 5 actuated axes (4 for wheels and 1 for robot's frame twisting) has been shown in [4]. Considered adaptive semiautonomous movement pattern allows vehicle to negotiate singular step-like obstacles.

Another subject of discussion presented in recent scientific publications is developing situational control strategies for wheeled robots with common structures. Different strategies for overcoming and descending from step or moving over a ditch for six-wheeled vehicles were described in [5-7]. Fuzzy algorithm for obstacle classification under information gathered with range imaging device was developed in [8]. It allows robot to autonomously classify obstacles and choose desired movement pattern for their negotiation. Situational control and strategies for overcoming obstacles by six-wheeled robot was considered in [9]. Robot's computing unit monitors energy consumption in active joints that connect robots frame and wheel-legs. When consumption rises over preliminary stated threshold it considered that step-like obstacle has been met.

Some auxiliary methods suitable for many wheeled devices also were recently discussed by several authors. Wheel anti-slip method for robotic platform has been discussed in [10]. Main principles of this method may be used as recommendations for developing of effective anti-slip controller. Paper [11] describes tip-over algorithms for movement of wheeled vehicle on rough terrain.

Some studies have examined obstacle negotiation for tracked robots. Authors of [12] have presented algorithm for morphology control of the tracked vehicle with active tracked levers. Suggested algorithm utilizes information provided by an onboard camera and 3-axial gyroscope. The main issue of this work is that robot adapts to its environment without prior knowledge of the stair geometry. Paper [13] represents control system for tracked robot with same structure. Suggested control system was built based on fuzzy logic algorithms for control of track movement speeds. Another approach for establishing autonomous obstacle negotiation was presented in [14]. Authors have considered algorithms that utilize incomplete information about robot's environment gathered via probing its surroundings with sensible manipulator. Stages of development of wheel-tracked robot with reconfigurable tracks and control algorithms for such device have been discovered in [15]. This vehicle has unfordable tail that acts as a third footing for movement on flat terrain when both tracks are fully folded and act as wheel. In [16] two-phased motion algorithm for movement on ledges is suggested. The main issue of the proposed method is that it is not belong to the group of the classification-based methods. Therefore, robot desirable condition is generated based on the obstacle parameters.

We propose structure of control system that has reduced pre-requisites for robot's sensor equipment. The only exterioceptive information that is used by our system for task generation is information provided by optical range sensors and information about each vehicle's module orientation. Our approach is quite simple from the view of technical realization and can be implemented in already existing physical models without changing their mechanical structure. Suggested below strategies describes robotic platform behavior in three common scenarios reviewed by other authors: step (overcoming), step (descending) and ditch negotiation.

4. Robotic platform

3D CAD-model of articulated six-wheeled vehicle with 9 active drives is depicted at Fig. 1 (a). Robotic platform consists of 3 two-wheeled modules (pt. 1-3). Rotation axes of motorized wheels (pt. 4-9) are fixed to modules. Middle module is separated into two similar pieces connected via active rotational joint. It includes rotational drive that performs rotation of first and third modules in relation to middle module. The axis of such rotation is identical to middle motorized wheels axis. Achievable rotation angle belong to range of ±90°. For their part, first and third modules each include one lateral and one rotational active joint. Lateral joint allows modules (pt. 1, 2) to move towards and away from middle module (pt. 3). Rotational joint allows them to rotate around lateral movement axis. All considered active frame transformations performs under lateral drives movement and rotation of active joint that placed in middle module. Distance between axes of first and middle or third and middle modules may be expanded for up to 33% of nominal dimensions.

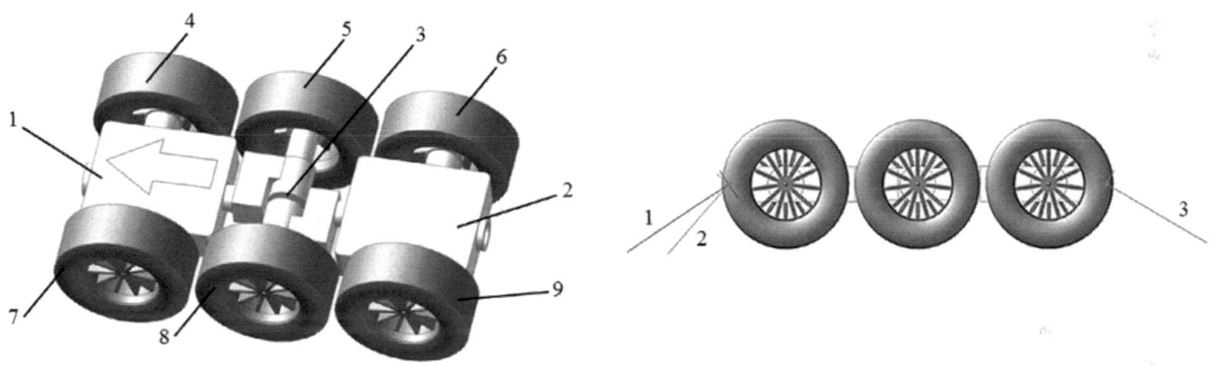

a) b)

Fig. 1. 3D CAD-model of transformable six-wheeled vehicle (a) and optical range sensor beams (b)

Information for obstacle discovery and classification is provided by three optical range sensors and 2 orientation sensors, each with two active axes. First two range sensors are placed on the frame of first module and one range sensor is placed on the rear module. Orientation sensors placed in both rear and front modules. Optical range sensors beams are depicted at Fig. 1 (b) (pt. 1) and (pt. 2-3). Angles between beams and horizontal plane are 30° and 45°. With considered approach minimal number of proprioceptive sensors corresponds to amount of active drives. Number of external sensors, therefore, must be sufficient for obstacle classification and determination of vehicle condition.

5. Structure of the control system

Autonomous control system contains nodes as depicted at Fig. 2. Target generation node produces time-dependent expressions that characterize movement of active drives between update intervals. This process is based on discrete information that is periodically provided by exterioceptive/proprioceptive sensors and rules of obstacle classification. Regulator in this particular case plays standard role, it forms control signals based on error between target and current condition.

We suggest three obstacle classes: (1) obstacles that can be overcome by simple movement with passive rotation of all joints; (2) obstacles that can be overcome by performing a sequence of active frame transformations; (3) obstacles that can't be overcome in any affordable vehicle conditions and configurations. Rules for overcoming typical obstacles are realized for overcoming obstacles outlined in first two classes. Discovery of third class obstacles causes initiation of emergency stop algorithms.

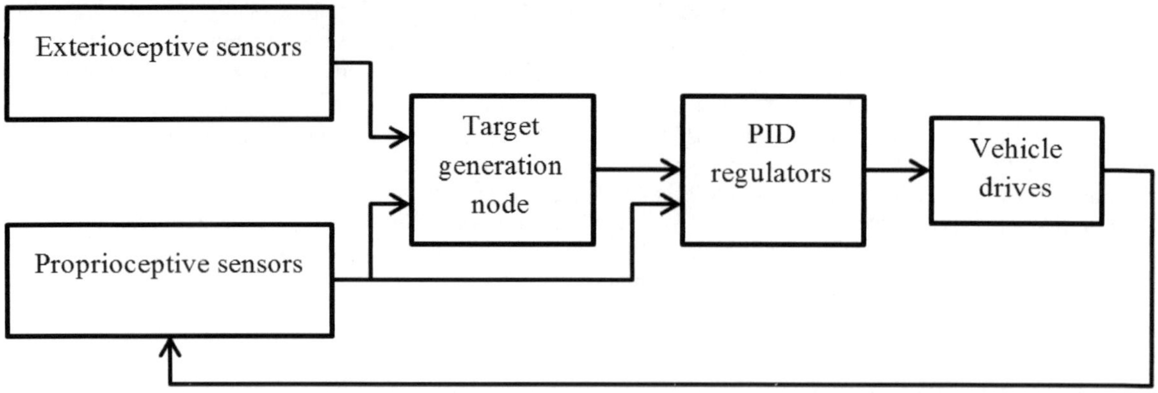

Fig. 2. Generalized structure of control system

6. Strategies overview

In current subsection we suggest behavioral strategies for five typical cases. These cases are: overcoming a step, descending from a step, ditch negotiation, simple forward movement and emergency stop in case of detection of obstacle that can't be overcome. Every typical case corresponds to specially designed movement mode.

Diagram of executable switches between modes is depicted at Fig. 3. It is possible to switch nominal mode (simple forward movement) to four different modes: overcoming a step, descending from a step, ditch negotiation mode and emergency stop mode. First three modes can be characterized by consequently executable stages of vehicle frame transformations. After finishing sequence of stages in each mode nominal mode become active.

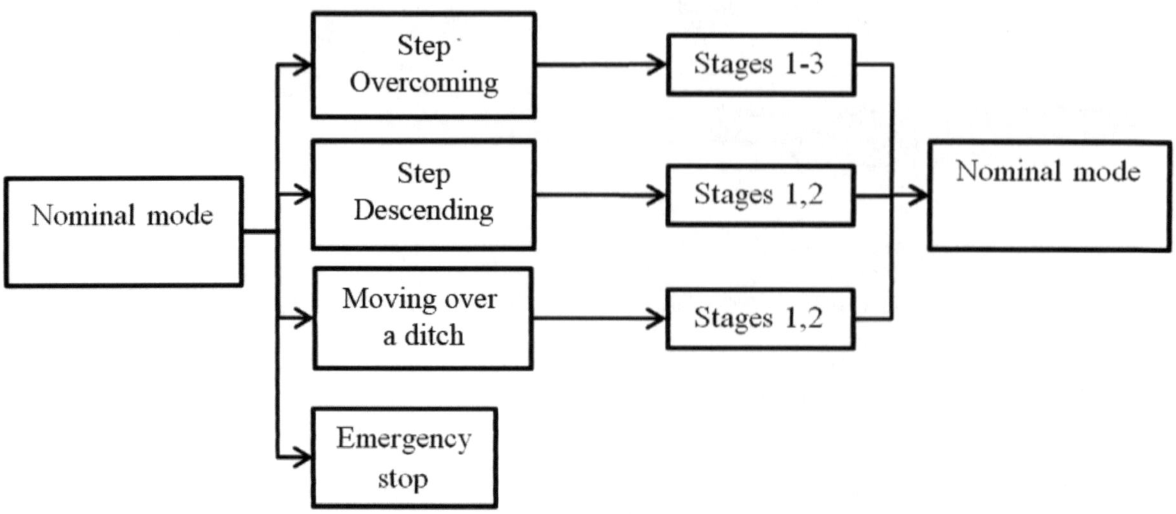

Fig. 3. Diagram of executable switches between movement modes

Obstacle discovery and classification performs under the values of estimated height of the obstacle and estimated distance to it. By estimated height we assume the length of vertical projection of interval point of between range sensor beam and obstacle surface intersection and point of front or middle wheels and terrain contact. In this particular case we assume that front and/or middle wheels are in contact with terrain.

For estimation of rear sensor group parameters we assume that rear and/or middle wheels are in contact with terrain. By estimated distance between vehicle and obstacle we assume horizontal projection of interval between beam origin and point where it intersects with an obstacle. Below we suggest expressions for calculating estimated parameters for range sensor that is placed on front module. Calculation diagram is depicted at Fig. 4.

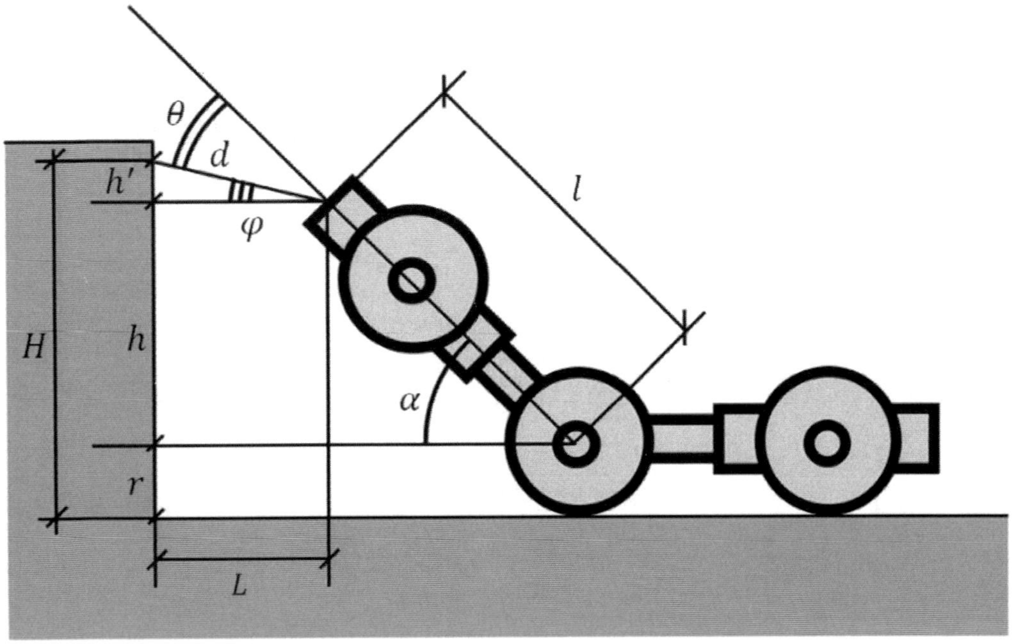

Fig. 4. Diagram for calculation of estimated parameters. Range sensor is placed on front module

Calculations of estimated height H and distance L are performed in accordance with equation (1) and (2):

$$H = r + l * \sin(\alpha) + d * \sin(\alpha - \theta) \quad (1)$$

$$L = d * \cos(\alpha - \theta) \quad (2)$$

here r is a wheel radius; l – distance between front wheels axis and sensor beam starting point; d – interval between beam starting point and point of intersection between beam and obstacle; α – angle between rear and front modules; θ – angle that characterizes direction of sensor beam towards nominally horizontal plane.

First and second range sensors provide information for obstacle discovery and classification. If provided by second sensor estimated height corresponds to a cavity with depth that exceed wheel radius and provided by first range sensor estimated length is less or equal to maximum distance between platform adjacent pairs of wheels then ditch negotiation mode activates. If the ditch width exceeds distance between wheels of maximally exerted front module and wheels of middle module then mode of descending from escarp activates. Obstacle that classified as escarp corresponds to estimated height of second sensor that exceeds ¼ of wheel radius. This rule guarantees obstacle negotiation under a small dynamic friction. If estimated height of the obstacle exceeds preliminary established limits then emergency stop mode activates. If estimated height belong interval between –r/2 and r/4 then simple forward movement mode activates.

Movement of platform in environment with undetermined positions of typical obstacles can be represented as set of consequently carried stages that corresponds to movement modes. Simplified images of platform configurations at different stages of movement modes depicted at Fig. 5. If there are no obstacles on the way of platform then it moves in nominal mode (pt. 1). In this mode both front and rear modules are retracted and only rear pair of wheels is actuated. If nearest environment irregularity was classified as an escarp then nominal mode switches to the first stage of the escarp negotiation mode. Configuration of the platform when the first stage is finished depicted as (pt. 2). It corresponds to leap in estimated height and a length values due to range sensor beam begins to intersect with escarp floor. Configuration (pt. 3) can be achieved by rotating the front module in opposite direction with simultaneous correction of wheels speed. Finishing stage of escarp negotiation mode corresponds to (pt. 4). At this stage vehicle performs active frame transformation until its middle wheels get in contact with escarp floor. Then it lifts rear module and switches current mode to simple forward movement (pt. 5). Detection of a deep and sufficiently wide cavity leads to switching nominal mode to escarp descending mode (pt. 6). At this stage both front and rear modules extracts on the length that provides safe descend. This configuration corresponds to the case when simultaneous contact of frontal wheels and lower floor and middle wheels and upper floor is achievable. During this process only rear wheels are actuated. When achieving stage (pt. 8) through stage (pt. 7) front and middle wheels are actuated and PID regulator coefficients for middle transformation drive is set to be 10 times lower than its nominal condition. At the first stage of a ditch negotiation mode (pt. 9) front module extracts so distance between frontal and middle wheels become equal to ditch width. If range sensors have detected a floor the second stage (pt. 10) initiates. Stage consists of transferring middle module towards to the front module with simultaneous extraction of rear module (pt. 11). Platform current mode switches to nominal when rear range sensor detects floor.

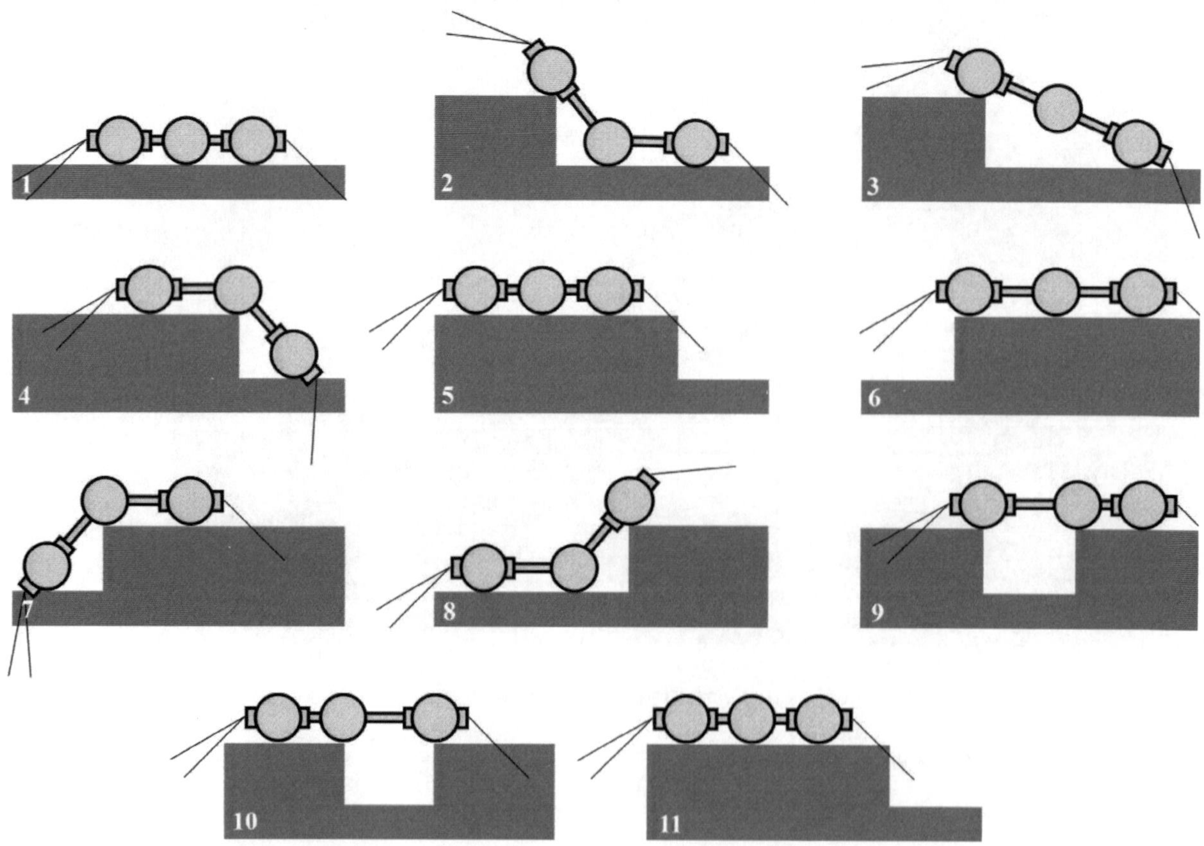

Fig. 5. Simplified images of different platform configurations at different stages of movement

7. Control algorithm

Control algorithm consists of 6 stages that are consequently executed on each virtual modelling step. In the first stage proprioceptive and exterioceptive information are gathered and obstacle height and distance to it are estimated. This information then is utilized to choose movement mode. Then based on current platform condition and current movement stage target drive conditions for robot transformation are generated. This target further is modified in correspondence to current targets for wheel rotation speeds and modules extraction. Equation (3) is suggested to be used for wheels speed correction in accordance to module extraction speed:

$$\omega = \omega 0 \pm (v/r\_w) \tag{3}$$

here ω is wheel speed; v is module extraction speed; r_w is wheel radius.

At the final stage errors between current and target platform configurations are calculated and send to PID regulators. Torques and forces that are formed this way then applied to mechanical parts of the actuators.

8. Virtual experiment

Suggested strategies and algorithm have been tested on complex virtual model of dynamics and kinematics for six-wheeled articulated platform. During virtual testing each step of the suggested algorithm was consequently performed with frequency of 100 Hz. Torques and forces formed by control system model were applied to mechanical parts of platform actuators. Testing was performed under the various parameters of wheel and ground contacts. At each testing iteration we were selecting dynamic friction from the range of 0,3 to 0,7.

We discuss algorithm efficiency under the example of step overcoming process. Dynamic friction coefficient during considered virtual test was chosen to be 0,3. Obstacle negotiation process depicted at Fig. 6. While approaching to the obstacle platform moves in first mode (pt. 1). Then lifting process of the first module begins (pt. 2). After achieving preliminary assigned angle platform start to rotate front module in opposite direction (pt.3, 4). At third stage rear module are lifted to get in contact with escarp floor (pt. 5). Platform returns in nominal mode at final stage (pt. 6).

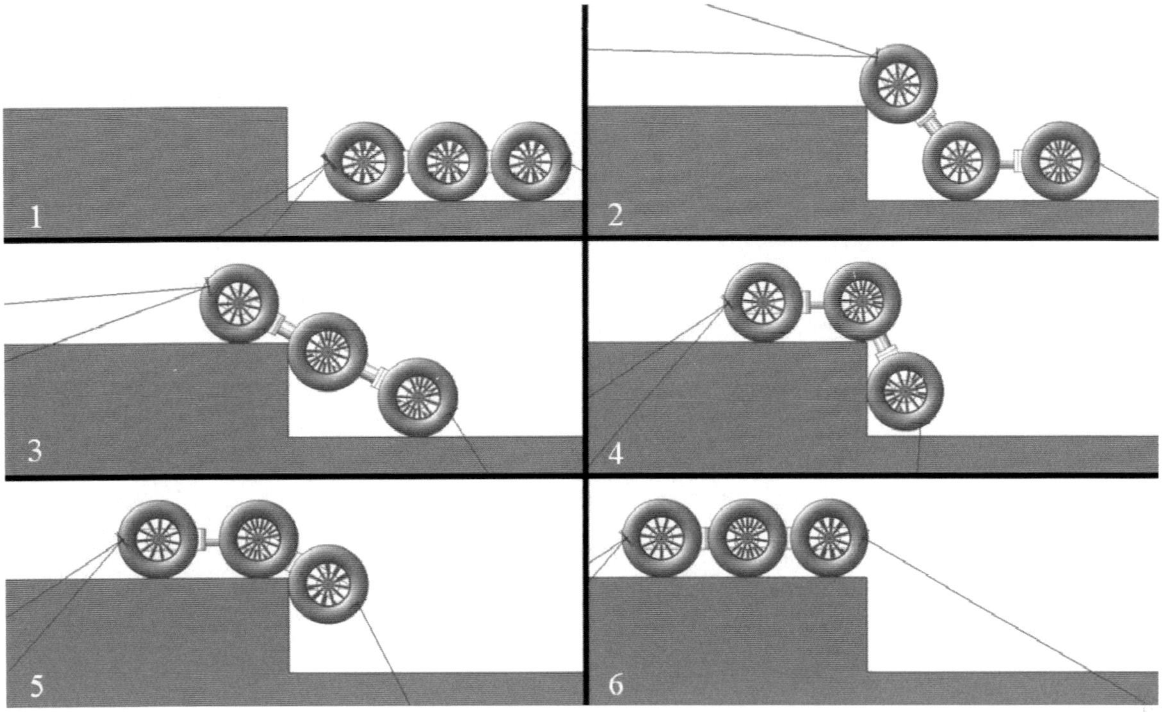

Fig. 6. Obstacle negotiation process

9. Conclusion

The approach for solving a problem of establishing robust control for articulated six-wheeled vehicle that is moving through cluttered spaces has been discussed. During concerned research new strategies for overcoming relatively large obstacles have been developed. Rules for obstacles classification were suggested. Continuously preformed platform transformation stages allows articulated vehicle to overcome a set of specified obstacles belonged to classes of step (overcoming), step (descending) and ditch. Preliminary efficiency approval of suggested control scheme and strategies was performed via complex virtual modelling of platform kinematics and dynamics. Virtual experiments showed that suggested strategies provide effective negotiation of specified obstacles. Including it in form of algorithms in the control system of transformable robotic platform provides it with capability to overcome relatively large obstacles with dynamic friction coefficient between wheels and ground higher than 0,3.

The main disadvantage of the suggested control system is that it has highly determined structure. Therefore switching between movements modes is not possible before exit condition for current mode is achieved. Furthermore described virtual modelling was performed with simplified models of idealized obstacles. For real world conditions additional rules for transformation and classification might need to be developed. Nevertheless we consider that stated above problem is solved. Based on achieved results we suggest that further implication of suggested algorithms will allow physical model of the platform to effectively traverse through undetermined environments with minimal aid of human-operator.

Future work is focused on further algorithm refining, extending the set of classified obstacles and expanding control scheme capabilities on the case of three-dimensional movement.

10. Acknowledgements

This research is supported by the Ministry of Education and Science of Russia (an agreement № 14.578.21.0124 on granting for the implementation of applied scientific research and experimental development). Unique identificator of ASRED – RFMEFI57815X0124.

11. References

[1] Alexey, B.; Dmitriy K. & Eugeny, L. (2016). Use of walking mechanisms as an adaptive suspension of a planet rover, Proceedings of the 26th DAAAM International Symposium, ISSN 1726-9679, ISBN 978-3-902734-07-5, Katalinic, B. (Ed.), pp. 0989-0996, Published by DAAAM International, Vienna, Austria, DOI 10.2507/26th.daaam.proceedings.139

[2] Bin X.; Cedric P.; Ambroise K.; Roland S. & Fuchun S. (2011). Composite control based on optimal torque control and adaptive kriging control for the crab rover, Proceedings of the 2011 IEEE International Conference on Robotics

and Automation, Shanghai International Conference Center, ISSN 1050-4729, ISBN 978-1-61284-386-5, Published by IEEE, Shanghai, China, DOI 10.1109/ICRA.2011.5979800

[3] Min X.; Runhuai Y.; Yong C. & Hongcheng X. (2011). Kinematics modeling and step climbing study of an all-terrain wheeled mobile robot on uneven terrains, Proceedings of the 2011 International Conference on Electronic & Mechanical Engineering and Information Technology, ISBN 978-1-61284-086-4, Published by IEEE, Harbin, China, DOI 10.1109/EMEIT.2011.6023597

[4] Shuro N. & Kazuhisa I. (2012). Concept of adaptive gait for leg-wheel robot, RT-mover, Proceedings of the 2012 IEEE International Conference on Robotics and Biomimetics, ISBN 978-1-4673-2127-3, Published by IEEE, Guangzhou, China, DOI 10.1109/ROBIO.2012.6490982

[5] Kyeong B. L.; Sim J. K. & Yong-San Y. (2010). Deliberative planner for ugv with actively articulated suspension to negotiate geometric obstacles by using centipede locomotion pattern, Proceedings of the International Conference on Control, Automation and Systems 2010, KINTEX, ISBN 978-89-93215-02-1, Published by IEEE, Gyeonggi-do, Korea, DOI 10.1109/ICCAS.2010.5670132

[6] Jingli L. & Chunguang B. (2009). Study on the mobile robot reconfiguration control methods, Proceedings of the IEEE International Conference on Automation and Logistics, ISSN 2161-8151, ISBN 978-1-4244-4794-7, Published by IEEE, Shenyang, China, DOI 10.1109/ICAL.2009.5262605

[7] Weidong W.; Dongmei W.; Wei D.; Chun X. & Pengfei S. (2012). The optimization of obstacle-crossing and the simulation in ADAMS of the composite six-wheeled-legged robot, Proceedings of the 2012 9th International Conference on Fuzzy Systems and Knowledge Discovery (FSKD 2012), ISBN 978-1-4673-0024-7, Published by IEEE, Sichuan, China, DOI 10.1109/FSKD.2012.6233733

[8] Kyeong B. L.; Jeong-hoon K.; Yong-San Y. & Sang H. L. (2008). Obstacle-overcoming algorithm for unmanned ground vehicle with actively articulated suspensions on unstructured terrain, Proceedings of the International Conference on Control, Automation and Systems 2008, ISBN 978-89-950038-9-3, Published by IEEE, Seoul, South Korea, DOI 10.1109/ICCAS.2008.4694665

[9] Kyeong B. L.; Sukhoon P.; Suengwoo K.; Jae M. J. & Yong-San Y. (2009). Behavior planning of an unmanned ground vehicle with actively articulated suspension to negotiate geometric obstacles, Proceedings of the 2009 IEEE/RSJ International Conference on Intelligent Robots and Systems, ISSN 2153-0858, ISBN 978-1-4244-3803-7, Published by IEEE, St. Louis, MO, USA, DOI 10.1109/IROS.2009.5354137

[10] Pierre L.; Ambroise K.; Michel L. & Roland S. (2004). Wheel torque control for a rough terrain rover, Proceedings of the 2004 IEEE international Conference on Robotics and Automation, ISBN 0-7803-8232-3, Published by IEEE, New Orleans, LA, USA, DOI 10.1109/ROBOT.2004.1302456

[11] Leah K., Kurt T.; Patrick L. & Garret C. (2015). Tip-over prevention: adaptive control development and experimentation, Proceedings of the 2015 IEEE International Conference on Robotics and Automation (ICRA), ISSN 1050-4729, ISBN 978-1-4799-6923-4, Published by IEEE, Seattle, WA, USA, DOI 10.1109/ICRA.2015.7139802

[12] Anastasios I. M.; Nikolas T.; Stergios L. R.; Daniel M. H. & Larry M. (2007). Autonomous stair climbing for tracked vehicles. The International Journal of Robotics Research, Vol. 26, No. 7, (July 2007) pp. 737-758, DOI 10.1177/0278364907080423

[13] Mihankhah E.; Kalantari A.; Aboosaeedan E.; Taghirad H.D. & Moosavian S.Ali.A. (2009). Autonomous staircase detection and stair climbing for a tracked mobile robot using fuzzy controller, Proceedings of the 2008 IEEE International Conference on Robotics and Biomimetics, ISBN 978-1-4244-2678-2, Published by IEEE, Bangkok, Thailand, DOI 10.1109/ROBIO.2009.4913304

[14] Martin P. & Karel Z.(2017). Controlling robot morphology from incomplete measurements. IEEE Transactions on Industrial Electronics, Vol. 64, No. 2, (Feb. 2017) pp. 1773-1782, ISSN 0278-0046

[15] Wenzeng G.; Shigong J.; Chengguo Z. & Xueshan G. (2014). Development of a transformable wheel-track mobile robot and obstacle-crossing mode selection, Proceedings of the 2014 IEEE International Conference on Mechatronics and Automation, ISSN 2152-7431, ISBN 978-1-4799-3979-4, Published by IEEE, Tianjin, China, DOI 10.1109/ICMA.2014.6885957

[16] Michael B.; Bernd B. & Dirk S. (2013). Hierarchical rough terrain motion planning using an optimal sampling-based method, Proceedings of the 2013 IEEE International Conference on Robotics and Automation (ICRA), ISSN 1050-4729, ISBN 978-1-4673-5643-5, Published by IEEE, Karlsruhe, Germany, DOI 10.1109/ICRA.2013.6631372

DOI: 10.2507/28th.daaam.proceedings.108

MECHANICAL FUNCTIONALISATION OF CUTTING INSERTS BY MACHINE HAMMER PEENING

Manuel Reiter*, Stephan Krall & Friedrich Bleicher

This Publication has to be referred as: Reiter, M[anuel]; Krall, S[tephan] & Bleicher, F[riedrich] (2017). Mechanical Functionalisation of Cutting Inserts by Machine Hammer Peening, Proceedings of the 28th DAAAM International Symposium, pp.0767-0772, B. Katalinic (Ed.), Published by DAAAM International, ISBN 978-3-902734-11-2, ISSN 1726-9679, Vienna, Austria
DOI: 10.2507/28th.daaam.proceedings.108

Abstract

Modern developments of functional components result in increasing demands on new materials and their characteristics. Therefore, the complexity of machining applications is also rising, and the challenges concerning wear resistance and longer lifetime of cutting tools continue to grow. The focus of this work is on the mechanical functionalisation of cutting inserts made of cemented tungsten carbide with a Ti-Al-N coating. The functionalisation of the cutting tools is caused by machine hammer peening of the cutting insert's rake face in order to influence surface-near layers, which results in material compaction of the treated areas and an inducement of residual compressive stresses.

Keywords: machine hammer peening; cutting insert; cemented carbide; Ti-Al-N coating; residual stress

1. Introduction

Over the last years, the process of machine hammer peening (MHP) has grown to an important method for influencing the surface properties and the mechanical state of the subsurface material layers (see e.g. [1]). By the use of MHP, the treated surfaces may show increased surface hardness and residual compressive stresses in near-surface layers [2]. These characteristics may be advantageous when it comes to cutting tools that are exposed to extremely abrasive wear. In the present work, cutting inserts (insert code: TCMW 16T308; cemented tungsten carbide base body with Ti-Al-N coating) that are usually applied for the simultaneous machining of a material combination (a casted aluminium-alloy with a wire-arc sprayed coating; for further relevant information regarding material combinations and machining types, see [3]) get mechanically functionalised.

During this simultaneous machining process, the cutting insert's rake face is exposed to significant normal stress gradients between the cutting areas of the malleable aluminium and the abrasive, wear-resistant spray-coating. For this reason, the rake faces of the mentioned cutting inserts are treated by means of the machine hammer peening technology to induce residual compressive stresses that should prevent tearing up of the coating (initially caused by the occurring tensile stresses in the coating due to the normal stress discrepancy on the rake face).

2. Machine hammer peening

Machine hammer peening is based on the oscillating movement of an axially guided ram with a tool tip on its end (mostly spherical in shape), which periodically comes in contact with the workpiece. The actuator, by the company *accurapuls,* used for this research work functions using an electromagnetic principle. A magnetic reaction force is generated by an alternating sinusoidal current applied on a plunger coil within a magnetic field in the MHP-system. The therefore resulting Lorentz force accelerates the ram over the stroke distance towards the surface, where the impact of the tool tip at the surface transforms its kinetic energy into elastoplastic deformation of the surface [2]. The ram may be equipped with different spherical tool tips, in order to use appropriate materials for the various applications of the machine hammer peening process. The NC-control of the machining centre, in which the actuator is mounted, provides the relative movements between tool (the oscillating MHP-actuator) and workpiece.

The composition of the actuator and the relevant process parameters are depicted in Fig. 1.

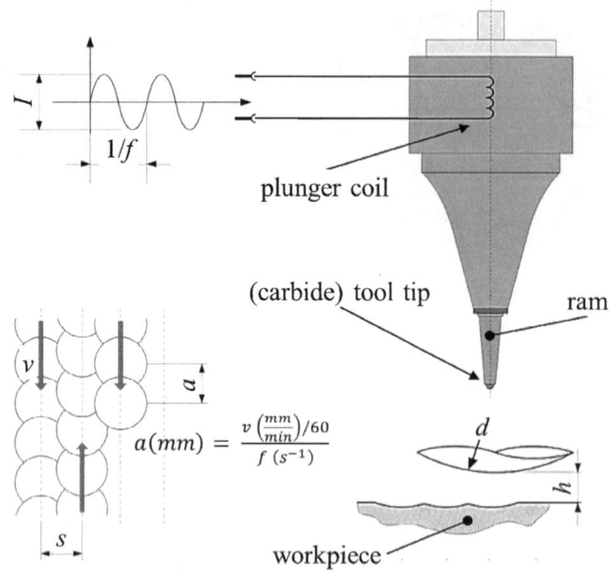

Fig. 1. Schematic overview of the electromagnetic MHP-actuator and the relevant process parameters: *I*...intensity (percentage of voltage amplitude); *f*...frequency (Hz); *v*...feed rate (mm/min); *a*...distance of indentation (mm); *s*...stepover distance (mm); *d*...tool tip diameter (mm); *h*...stroke (mm) (cf. [1]).

Previous works like [4],[5] have shown that due to the preceding MHP-treatment on metallic materials, significant residual compressive stresses have been detected in the near-surface layers of the covered areas. These are higher in orthogonal direction to the tool path than parallel to it [5]. Moreover, it is obvious that the smaller the distance of indentation and stepover distance as well as the lower the feed rate are chosen, the more overlap and therefore higher energy is induced [6]. The distance of indentation is influenced by the choice of feed rate and the applied frequency of the actuator's oscillation. In general, for ductile materials the surface roughness will be decreased by increasing the tool tip diameter. Apart from that, increasing stepover distance, hardness of the tool tip and feed rate will lead to greater surface roughness. For most applications, the main cause for an increase in work hardening of the surface layer is a decreasing tool tip diameter [6]. In addition, the microstructure may be influenced by MHP, which is further investigated in [7].

3. Machining setup for the mechanical functionalisation of cutting inserts

To ensure the machining of cutting inserts by the use of machine hammer peening, a clamping system was designed and manufactured (see Fig. 2a). It should provide appropriate fixation and orientation in the machining centre of both the clamping and the cutting tools to be processed. This clamping system consists of a substructure out of conventional hardened steel with a plate of carbide metal mounted on its top. The top plate has four recesses, which represent the precise negative form of the cutting inserts. Moreover, the substructure may be used for additional top plates according to other types of cutting inserts.

In Fig. 2b, the setup for MHP-treating of cutting inserts in the machining centre is shown. Herein, the MHP-actuator as well as the designed clamping system is depicted. The research work was carried out on a 5-axis machining centre (*Hermle C20 U*).

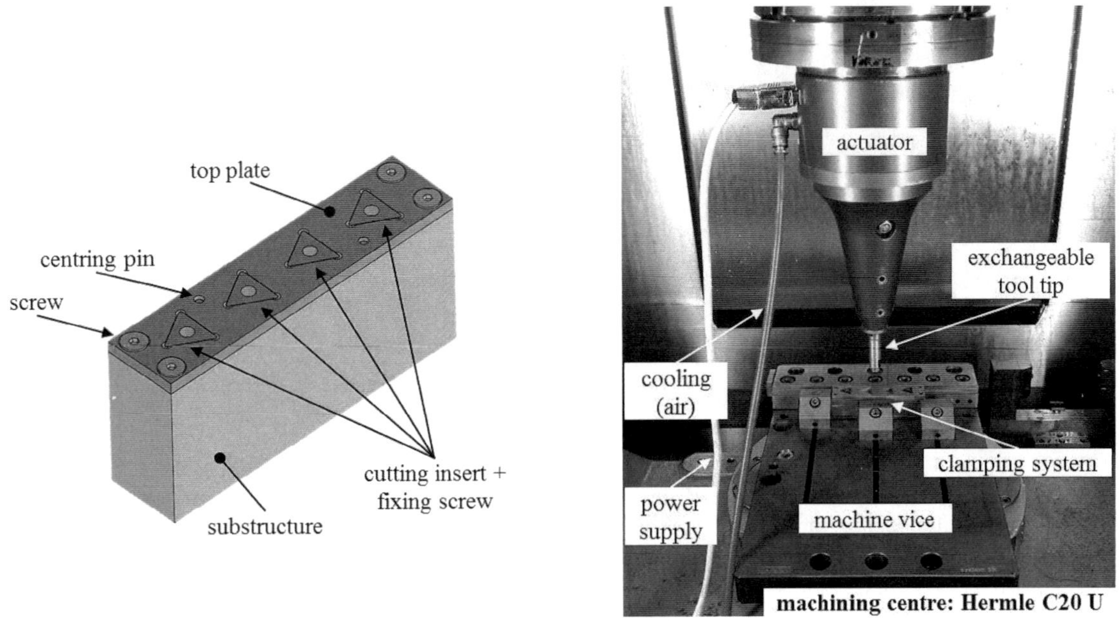

(a) (b)

Fig. 2. (a) CAD-model of designed clamping system for the mechanical functionalisation of cutting inserts; (b) experimental setup in the machining centre.

4. Configuration of the machining process

The main advantage of the MHP-process in combination with NC-machines is the precise machining of workpieces. Therefore, different NC-strategies for the exact movement of the oscillating hammer peening device across the cutting inserts were investigated. The final form of the tool paths on the cutting insert's rake face is shown in Fig. 3. This strategy was chosen due to optimized results compared to earlier attempts that led to damaged cutting inserts, as used tool path arrangements were placed too close to the cutting edge or even across the cutting edges. The finally chosen strategy could be placed at the required distance to the cutting edge without damaging it.

Furthermore, the process parameters have been investigated and tested. As already discussed in section 2, the stepover distance and the distance of indentation were set very low to create a high percentage of overlapping tool paths on the processed areas. This is also the reason for a relatively high stroke value that delivers a significant amount of kinetic energy to the ram, which is transformed into mechanical deformation of the cutting insert. The subsequently applied process parameters are listed in Table 1.

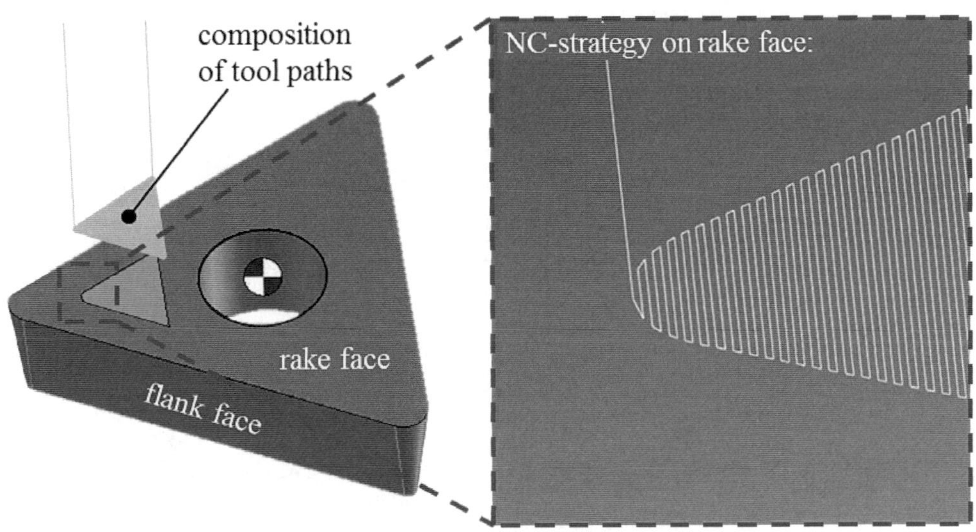

Fig. 3. CAD-model of cutting insert with final NC-strategy.

Intensity (*I*)	100 %
Frequency (*f*)	200 Hz
Feed rate (*v*)	1.200 mm/min
Distance of indentation (*a*)	0,1 mm
Stepover distance (*s*)	0,02 mm
Tool tip diameter (*d*)	6 mm
Stroke (*h*)	0,5 mm

Table 1. Process parameters.

Various materials of the tool tip were investigated, with the best results reached using the carbide metal tool tip. Other tip materials that have particularly higher hardness, e.g. ceramic materials, showed poor wear behaviour like flat spots after just a single machining sequence, fine cracks on the spherical surface, etc. Moreover, the amounts of induced residual compressive stresses on the cutting insert resulted in considerably smaller values (see also section 5.2).

5. Effects on the processed surface layers

To verify the orthogonal distance of the MHP-treated area to the cutting edge and to gain knowledge of the reached depths due to mechanical deformation on the cutting insert, optical measurements were carried out using a digital microscope as well as a 3D surface measuring device. The detection of residual stresses in near-surface regions of the MHP-treated cutting inserts were carried out by means of X-ray diffraction measurements.

5.1. Optical measurements

In the upper section of Fig. 4, the corner of a cutting insert right after MHP-treatment is shown, where no postprocess cleaning had taken place, and therefore marks of abrasion are still visible. For this specific insert, that was processed with the above presented NC-strategy (Fig. 3) and process parameters (Table 1), the orthogonal distance of the MHP-treated area to the cutting edge was just 0,7 mm. The image was taken using a *Keyence VW-9000* digital microscope with a 100x magnification. Further measurements, especially to get quantitative knowledge of the mechanical deformation of the MHP-treated areas, were carried out using an *Alicona InfiniteFocus G4* 3D surface measuring device, with a vertical resolution of 250 nm and a lateral resolution of 5 µm regarding a 10x magnification. Concerning this, the lower section of Fig. 4 shows the false-colour plot and the belonging depth profile of the marked area on the cutting insert.

The orthogonal distance between MHP-treated area (composed of the strung together tool paths) and cutting edge of 0,7 mm was defined as a default value. Former experiments using this distance and the favoured process parameters (Table 1) showed positive results, thus the stable processing of the cutting inserts have been ensured.

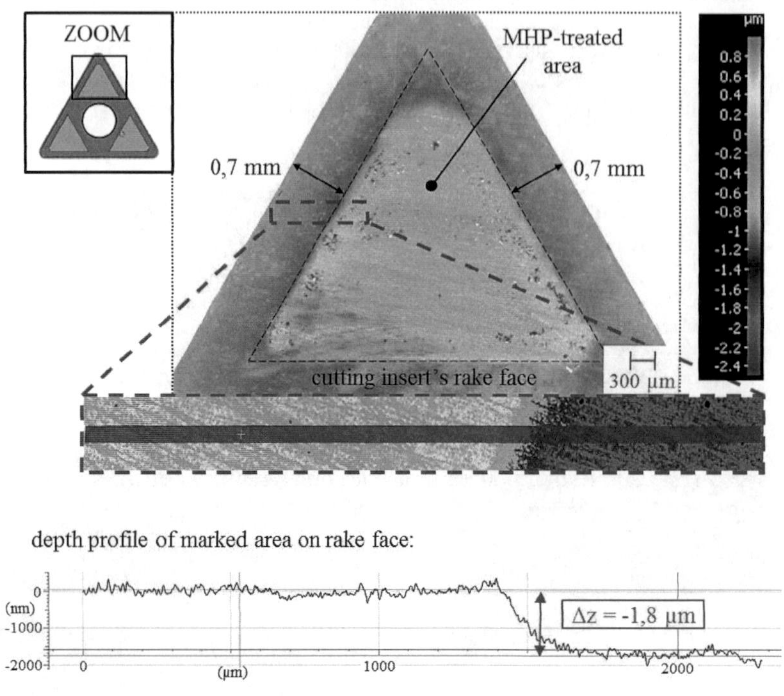

Fig. 4. Microscopic image of the MHP-treated cutting insert and depth profile of marked area on the rake face.

As visible in Fig. 4, an overall depth difference (from the untreated to the treated surface) of 1,8 µm was measured for the marked area of the cutting insert's rake face. Due to this significant mechanical deformation respectively material compaction, residual compressive stresses should be induced as well as the hardness in the subsurface layers may be increased (see section 5.2).

5.2. X-ray diffraction measurements

X-ray diffraction (XRD) belongs to the group of non-destructive testing methods for residual stresses on material surfaces. It bases, according to its indication, on the diffraction of ordered structures like in crystalline metal. As a result of mechanical stress, the spacing between atomic planes in the structure change, which leads to different diffraction patterns before and after applying the load. The resulting values deliver information on the strains in the material, and therefore the material's residual stresses can be calculated. The underlying principle for this kind of diffraction is the so-called Bragg equation (1). For further relevant information regarding Bragg's law and X-ray diffraction, see referred literature, e.g. [8],[9].

$$n * \lambda = 2 * d * \sin\theta \tag{1}$$

n ... integer (typically 1)
λ ... X-ray wavelength
d ... lattice spacing
θ ... diffraction angle (so-called Bragg-angle)

The X-ray diffraction measurements were carried out using a *Bruker D8 Discover* device. As the coating has a thickness of approximately 10 µm and the depth of the X-ray diffraction is under 10 µm, the induced residual stresses in the coating could be measured. The measurement point was placed in the centre of the machined triangular area, but the XRD-measurements were performed in orthogonal direction to the cutting edge (see schematical drawing in Fig. 5b). This is the crucial direction for the cutting tool, because the material cutting occurs also in orthogonal direction to the cutting edge. In Fig. 5a, the resulting residual stress values in the subsurface layers of the cutting insert's rake face are shown. This specific cutting insert was machined as follows: Corner 1 functioned as a measuring reference (no MHP-treatment, cutting insert respectively coated material in as-delivered condition). Corner 2 was MHP-treated with the presented process parameters (Table 1) and NC-strategy (Fig. 3; with the default distance of 0,7 mm to the cutting edge). Corner 3 was machined in the same way as corner 2, but was MHP-treated twice.

After the machining, corner 2 showed an overall depth difference from the untreated to the treated surface on the rake face of about 1,4 µm, as it was measured over the entire width of the cutting edge (from the main cutting edge to the minor cutting edge). This is the reason for the discrepancy in depth to the example in section 1, which was measured just in a small area on the outside of the MHP-treated area, where the turnaround of the tool paths happened and therefore more indentations took place. The overall depth difference to corner 3 was measured approximately 2,2 µm, in the same way as corner 2.

It can be seen that the reference point shows minimal residual tensile stresses. Corner 2 shows significant induced residual compressive stresses, which was already expected due to the mechanical deformation of the processed area. Corner 3 shows higher induced residual stresses than corner 2, but as also mentioned in [6] and considering the given confidence intervals, double treatment does not change the amount of residual stresses in a neither significant nor proportional manner.

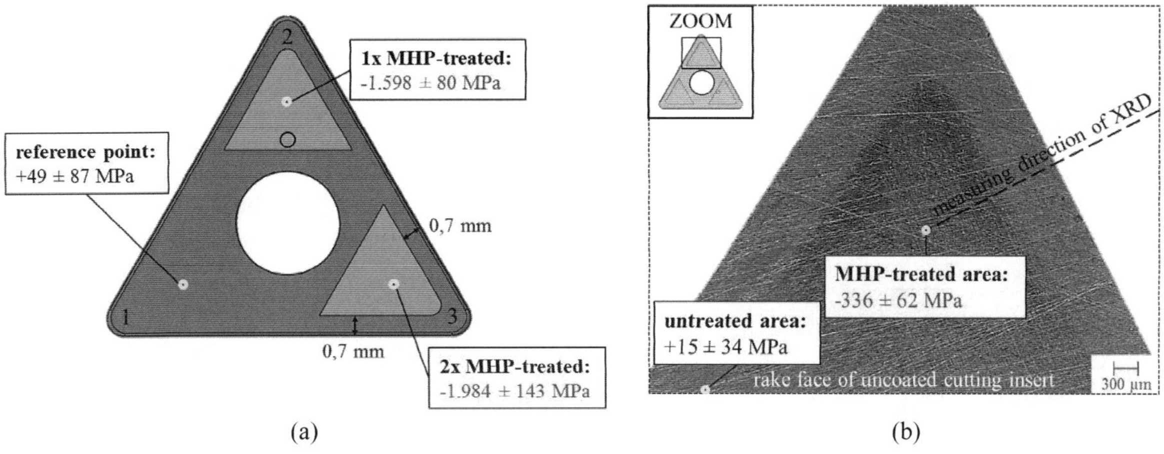

Fig. 5. (a) Schematic overview of MHP-treated areas on the cutting insert's rake face and measured residual stresses; (b) image of the MHP-treated uncoated cutting insert and the measured residual stresses.

The MHP-treatment was also performed on uncoated inserts (with the same process parameters as in Table 1 and NC-strategy as in Fig. 3) to prove the inducement of residual compressive stresses not only into the coating, but also in the subjacent substrate. The X-ray diffraction measurements therefore also showed an inducement of compressive residual stresses in the substrate, although the outcomes were at a lower level than for the coating (see Fig. 5b). The processed area on the uncoated cutting insert can be seen as a shaded, triangular field (image taken using the *Keyence VW-9000* digital microscope).

As already mentioned, residual stress measurements of the cutting inserts that were MHP-treated with ceramic tool tip materials (aluminium oxide Al_2O_3 and silicon nitride Si_3N_4) instead of carbide metal have been performed. The generated findings pointed out that these tool tip materials had too high wear susceptibility. Therefore, the resulting residual compressive stresses of the machined areas on the cutting insert's rake face were significantly lower than for the carbide metal tip – less than half the stress value of the ones processed with the carbide metal tip was reached (subject to the use of the identical cutting inserts, the same process parameters as in Table 1 and the same NC-strategy as depicted in Fig. 3).

6. Conclusion and outlook

The present work's focus was on the mechanical functionalisation of cutting inserts that are usually applied for the simultaneous machining of a material combination. Due to this functionalisation, caused by the use of machine hammer peening, the processed surface area on the cutting insert's rake face should show induced residual compressive stresses.

In conclusion, the positive effects of machine hammer peening for the application on cutting inserts have been successfully demonstrated. Due to the previous investigations and setting of appropriate processing parameters as well as suitable arrangement of tool paths, a significant material compaction has been reached on the cutting insert's rake face by means of machine hammer peening. Hence, a major amount of residual compressive stresses has been induced into the coating and also the sintered substrate beneath.

The aim of further investigations should analyse the effects on the hardening of subjacent layers of the MHP-treated areas. In addition, greater insight into the impacts of MHP-treatment on the coating-substrate bond, as it was a cemented tungsten carbide substrate with Ti-Al-N coating for this present research, have to be gained. The consequences on the performance of the cutting tool, especially the reachable tool life, must be further investigated in experimental cutting tests. Subject to the success of these mentioned and further future developments, this kind of mechanical functionalisation of cutting tools may be established in the field of production engineering to consequently lower production costs and improve sustainability.

7. References

[1] C. Lechner, F. Bleicher, C. Habersohn, C. Bauer & S. Gössinger (2012). "The use of machine hammer peening technology for smoothening and structuring of surfaces," Proc. 23rd Int. DAAAM Symp., vol. 23, pp. 331–336.

[2] D. H. Trauth (2016). Tribology of Machine Hammer Peened Tool Surface for Deep Drawing. Aachen: Apprimus Verlag.

[3] D. Boehnke (2007). Qualitätsorientierte Zerspanung von Parallelverbunden im kontinuierlichen Schnitt. Hannover: PZH, Produktionstechn. Zentrum.

[4] C. Lechner (2014). "Oberflächenmodifikation unter Einsatz der Technologie des Schlagverdichtens (Machine Hammer Peenings)," Technische Universität Wien.

[5] F. Bleicher, C. Lechner, C. Habersohn, E. Kozeschnik, B. Adjassoho & H. Kaminski (2012). "Mechanism of surface modification using machine hammer peening technology," CIRP Ann. - Manuf. Technol.

[6] V. Schulze, F. Bleicher, P. Groche, Y. B. Guo & Y. S. Pyun (2016). "Surface modification by machine hammer peening and burnishing," CIRP Ann. - Manuf. Technol., vol. 65, no. 2, pp. 809–832.

[7] B. Adjassoho et al. (2013). "Controlled surface treatment with machine hammer peening," Met. 2013, Brno, vol. 4, pp. 13–18.

[8] N. S. Rossini, M. Dassisti, K. Y. Benyounis & A. G. Olabi (2012). "Methods of measuring residual stresses in components," Mater. Des., vol. 35, pp. 572–588.

[9] P. S. Prevéy (1996). "Current applications of X-ray diffraction residual stress measurement," Dev. Mater. Charact. Technol., no. 513, pp. 103–110.

DOI: 10.2507/28th.daaam.proceedings.109

CONTROL DEPLOYMENT OF MOBILE UNITS OF LARGE-SIZED SPACECRAFT

Fedor Mitin, Alexey Krivushov

This Publication has to be referred as: Mitin, F[edor] & Krivushov, A[lexey] (2017). Control Deployment of Mobile Units of Large-Sized Spacecraft, Proceedings of the 28th DAAAM International Symposium, pp.0773-0779, B. Katalinic (Ed.), Published by DAAAM International, ISBN 978-3-902734-11-2, ISSN 1726-9679, Vienna, Austria
DOI: 10.2507/28th.daaam.proceedings.109

Abstract

The variants of automatic deployment of mobile units of a large-sized spacecraft in orbit are considered. In view of the large-sized of the apparatus, its deployment from the folded state to the working one occurs in stages. The main task at each stage, in addition to reliable deployment, is to reduce the vibrations and deflections of the structure. At present, after the completion of the regular stage, it takes a considerable time to wait for the damping of the fluctuations and only then proceed to the next stage. This article explores possible options for deploying mobile units of a spacecraft: independent for each stage; joint for several stages, allowing to reduce fluctuations, deflections and, therefore, the time of opening the structure. Mathematical models of rotational and translational movements are presented. As an executive element, the application of an electric machine is considered, since it allows to control the deployment process. The results of numerical modeling allowing to evaluate the effectiveness of various variants of deployment are presented.

Keywords: large-sized spacecraft; deployment of a spacecraft; mathematical model

1. Introduction

At present, antennas are an integral part of modern spacecraft. The history of their creation is inextricably linked with the history of the development of space and rocket technology. Active research and development aimed at the creation in space of designs of various classes, having large size in one dimension or another, are introduced throughout the world.

Despite significant advances in the design of large-sized transformable space-based structures being deployment, the important task of their smooth and reliable deployment remains [1-5].

A distinctive feature of large-size spacecraft (large-size antennas, space telescopes, etc.) is the presence in their composition of a multitude of mobile units. Of particular interest are transformable antennas, the profile of which can be changed both at the stage of preparation for work and in the process of functioning. Due to the large size, the translation of the reflector from the folded state to the workers is divided into stages that can be regarded as independent of each other.

Deployment of various types of large-size reflectors (trusses, umbrellas, rings, etc.) is done at the expense of the energy of springs or deformed elastic elements [6-8].

The disadvantages of this method of deployment include the inability to control the deployment process, the presence of fluctuations and deflections of the structure after completion of the deployment stage, the need to create a layout and perform additional calculations to determine the degree of compression of springs or twisting of elements.

An alternative to classical systems of deployment can be automatic systems in which the energy for the deployment of reflector elements is produced by drives, in particular an electric machine. The application of this type of actuator will allow to control the process of opening, to reduce the fluctuations and deflections of the design of a large-sized spacecraft.

2. Mathematical model of mobile units

The process of deployment of a large-sized spacecraft can be divide into several separate stages. Let us consider in more detail the most frequently encountered types of motion describing these stages: rotational and translational.

In the course of the study, it was proposed to simultaneously open several elements of a large-sized spacecraft in order to shorten the time taken for the apparatus to move to the working position. Consider the problem of opening the root element and the simultaneous extension of the intermediate element of the end elements of the power frame, shown in Figure 1.

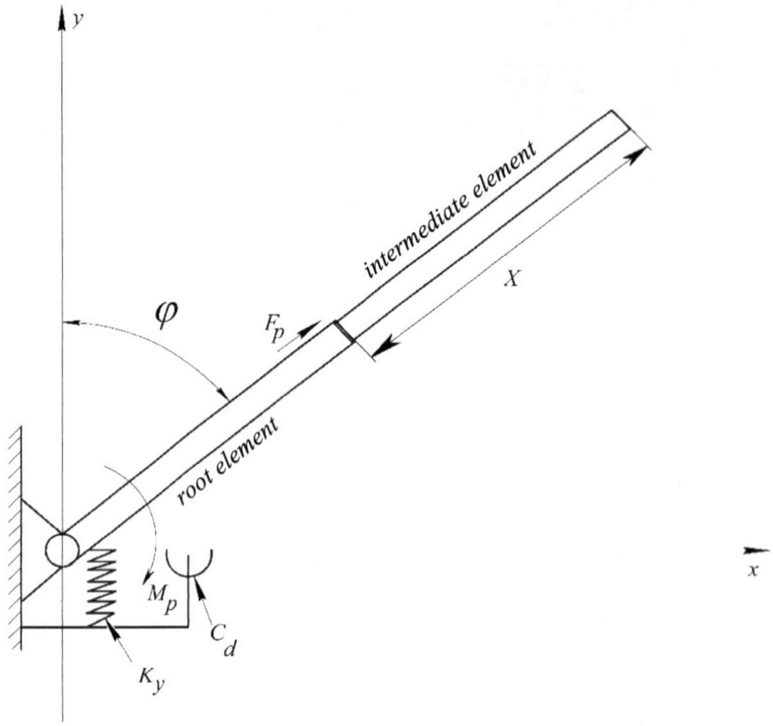

Fig. 1. Scheme of mutual arrangement of the elements of the base of the power frame

The system describing the joint deployment is a combination of two systems: angular motion and translational motion, the mathematical models of which are presented in [9, 10]. We obtain the system:

$$\dot{\varphi} = w - h, \tag{1}$$

$$\dot{w} = \frac{M_p - M_{tr} - (k_y \cdot \Delta\varphi + C_d \cdot \dot{\varphi}) \cdot H(\Delta\varphi) - \mu \cdot \dot{\varphi}}{I}, \tag{2}$$

$$\dot{x} = V \pm \Delta x_{sz}, \tag{3}$$

$$\dot{V} = \frac{F_p - F_{tr} - (k_y \cdot \Delta x + C_d \cdot \dot{x}) \cdot H(\Delta x) - \mu \cdot \dot{x}}{m}; \tag{4}$$

where φ is the angle of rotation of the root element; ω is the angular velocity of rotation; l is the length of the rotating part; ρ is the material density; E is the Young's modulus for the material; I is the moment of inertia; k_y is coefficient of rigidity of the elastic component of the stop (lock) model; C_d is the coefficient of damping of the viscous component of the stopping device; $\Delta\varphi = (\varphi - \varphi_{\text{к}})$ is the relative angle corresponding to the moment of installation at the stop (lock); μ is the coefficient of dissipativity; $H(\Delta\varphi)$ is a function that takes the value 0 ($\Delta\varphi > 0$) or 1 ($\Delta\varphi < 0$); M is the engine torque; $M_{tr} = \varphi_{ish}\, \varphi_{rs}\, M_{rr} + M_{sl} + M_{seal} + M_{drag}$, M_{rr} is the rolling friction moment; M_{sl} is the sliding friction moment; M_{seal}

is the friction moment of the seals; M_{drag} is the friction moment due to the resistance of lubrication, agitation, sprays, etc., φ_{ish} is the reduction factor due to the heating of the lubricant; φ_{rs} is the reduction factor in the kinematic fasting mode; x ist he displacement; V is the linear velocity; $\Delta x_{sz} = \frac{F_p \cdot l}{E \cdot S}$; $F_p = k_f \cdot M_p$ is the useful force, push intermediate element; $F_{tr} = \mu_{tr} \cdot N_i$ is the frictional force; μ_{tr} is the coefficient of sliding friction of the guide sections; N_i is the normal forces in the sliding supports of the i-th section; $\Delta x = (x - x_\kappa)$ is the relative movement corresponding to the moment of installation at the stop (lock); $H(\Delta x)$ is a function that takes the value 0 ($\Delta x > 0$) or 1 ($\Delta x < 0$).

During the operation of the system, it may be necessary to take into account the bending of the loose end of the element. Starting from [11] we consider the applicability of the approximate theory. The approximate equation for calculating the deflection value is calculated as:

$$h = \frac{M(l+\Delta l)}{2EI}, \tag{5}$$

where Δl is the elongation of the element; $M = M_p - M_{tr} - (k_y \cdot \Delta\varphi + C_d \cdot \dot{\varphi}) \cdot H(\Delta\varphi) - \mu \cdot \dot{\varphi}$ is the total moment acting on the base of the power frame.

To create a useful torque M_p, we choose an electric motor that is most suitable for the problem, that is, Taking into account the field of application, environmental conditions, service life, power source, operation cycle, heat generation, efficiency, controllability, torque, speed, weight. Analyzing current trends in the application of electric motors for similar purposes [12-14], one can come to the conclusion that the most suitable are brushless (gate) DC motors. Since they have a long service life, they provide high moments, high efficiency and low heat dissipation.
The torque is calculated by the equation [14]

$$M_p = \frac{mpE_0U_1}{w_1X_c} sin\vartheta, \tag{6}$$

where m is the number of phases of the rotor; p is the number of pairs of poles of the magnetic field, E_0 is the effective value of the emf on the stator winding; U_1 is the supply voltage, x_c is the synchronous resistance ($x_c = x_\sigma + x_a$, where x_σ and x_a are the inductive scattering resistances and reactions respectively), ϑ is the angle of error (between U_1 and E_0, for the motor is within [0, π/2]).

3. Modeling

The first method of joint deployment of the stages is the use of the engine for each of the components of the motion. In modeling the dynamics of the joint opening of the reflector, the damping factors and the elastic component of the stop varied, the actuators for both types of movements had the same parameters and were switched on simultaneously. The following system parameters were selected during simulation: ABS material plastic QHF 0140: ρ = 1040 kg · m³, E = 9900 · 10⁶ Pa, U_1 = 5 V, retractable part with length x_k = 2 m, angular movement stop φ_k = 1 rad. The behavior of the system is shown in Figures 2-3.

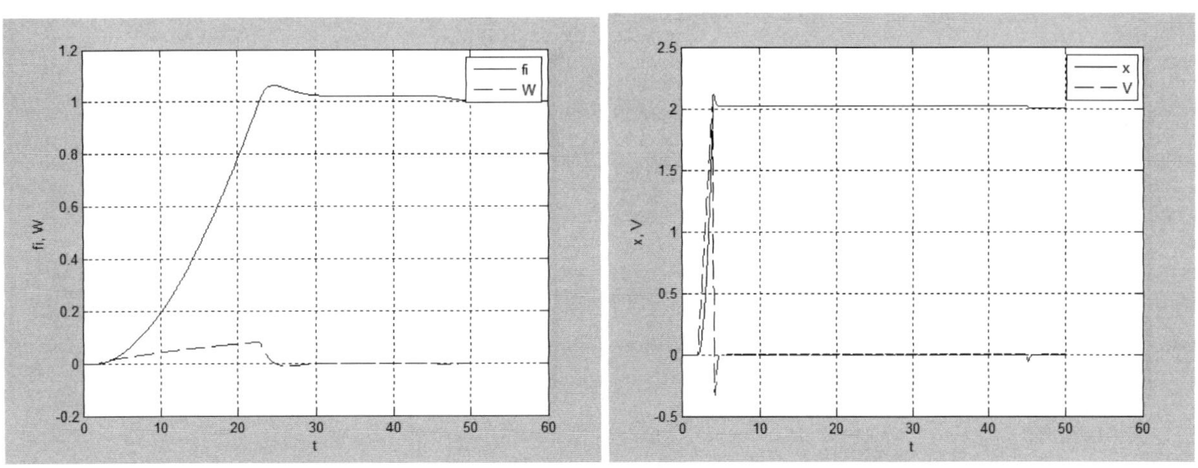

a) b)

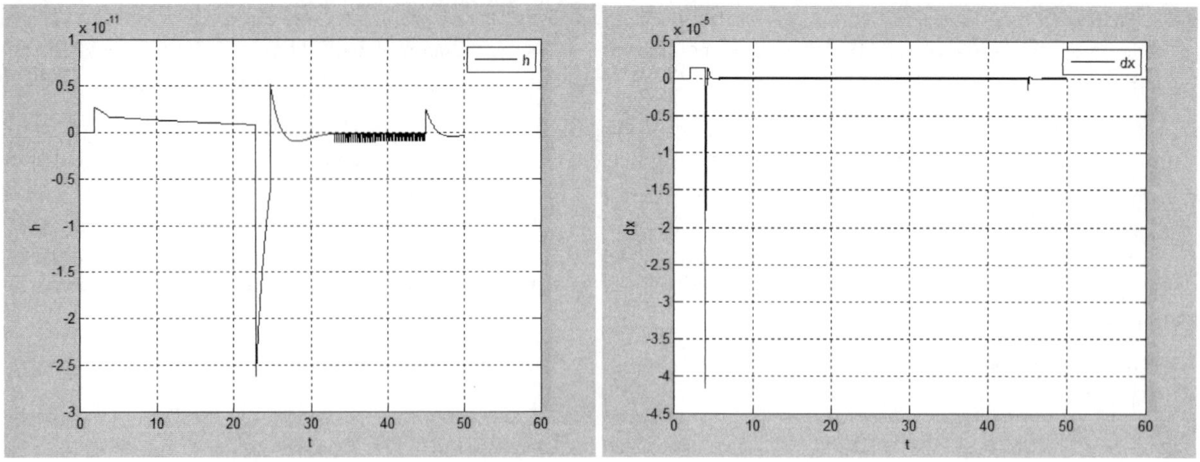

c) d)

Fig. 2. The characteristic of the dynamics of joint deploymentof the elements: a) the graph of $\varphi(t)$ and $\omega(t)$ at $k_y = 10$, C_d =25; b) the graph of $x(t)$ и $V(t)$ at $k_y = 100$, C_d =25; c) the graph of $h(t)$; d) the graph of $\Delta \boldsymbol{x}_{sz}(t)$

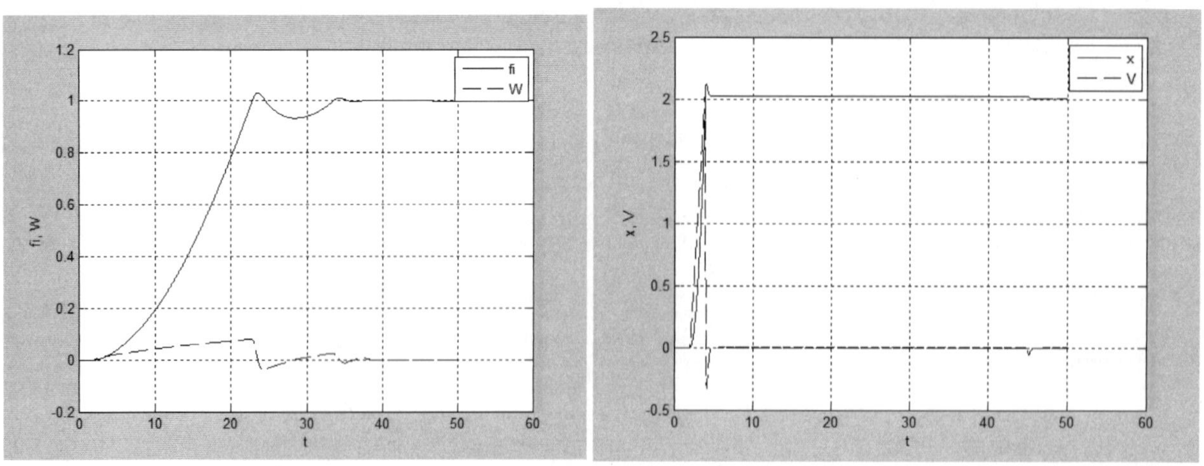

a) b)

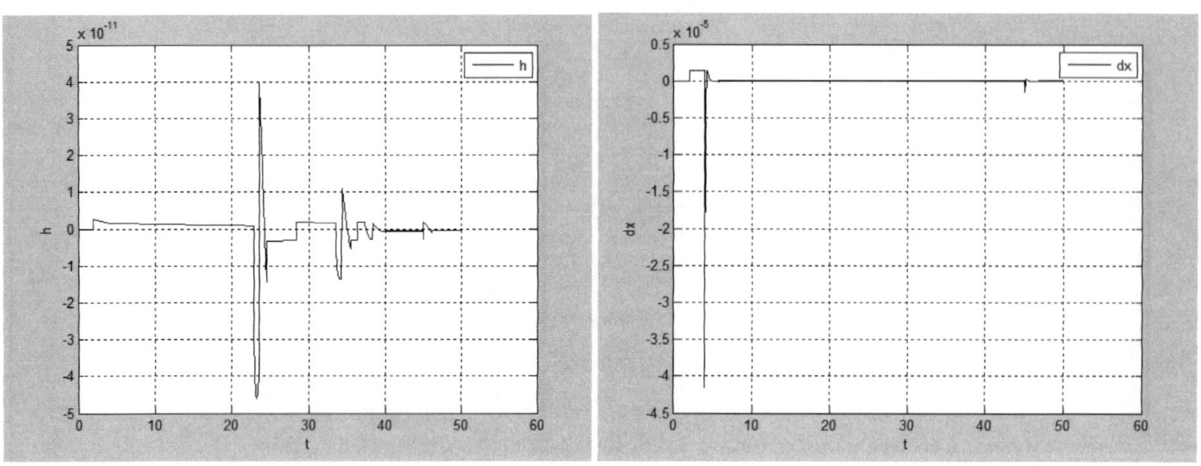

c) d)

Fig. 3. The characteristic of the dynamics of joint deploymentof the elements: a)) the graph of $\varphi(t)$ and $\omega(t)$ at $k_y = 100$, C_d =25; b) the graph of $x(t)$ и $V(t)$ at $k_y = 100$, C_d =25; c) the graph of $h(t)$; d) the graph of $\Delta \boldsymbol{x}_{sz}(t)$

Based on the above, we can make the following conclusions. The use of several executive devices allows implementing various programs for the deployment of design elements. The disadvantages include the fact that with the increase in the number of executive machines, the mass of the entire design, the inertia of the elements in the design, will also grow.

The second way is to use centrifugal force to deployment intermediate element. In this method, the useful force F_p is created by the centrifugal force. The system describing the joint expansion of the structural elements has the same form as the system (1-4), except that F_p=0,5·m·ω^2·(x+0.5·l).

During the simulation, the same system parameters were used as for the first method. From figure 4 it can be seen that the actuator of the angular motion creates insufficient centrifugal force for the full extension of the intermediate element at the initial parameters. Based on this, the control signal U_l was increased from 5 V to 15 V. Due to the increase in the control signal, the torque also increased. Figure 5 shows the process of expansion and extension at U_l=15 V. From the obtained characteristic, it can be concluded that it was possible to achieve setting on the stops of both the root element and the intermediate element of the reflector.

This embodiment is devoid of the disadvantage of the first method in the form of additional weight of the actuators. The problem of voltage control is complicated, since it becomes necessary to vary U_l the actuator based on the desired characteristics of not only the angular motion but also the translational component.

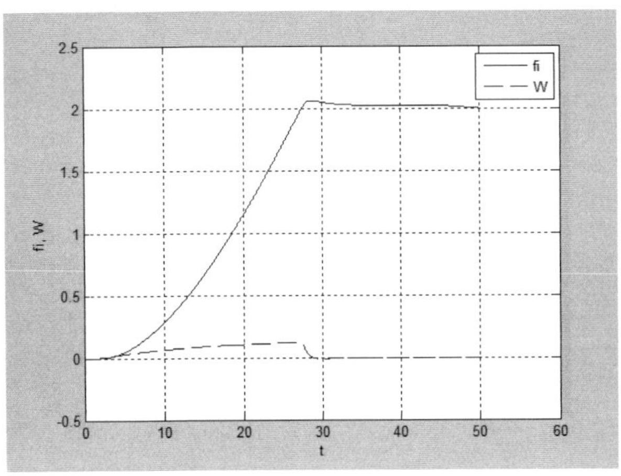

a)

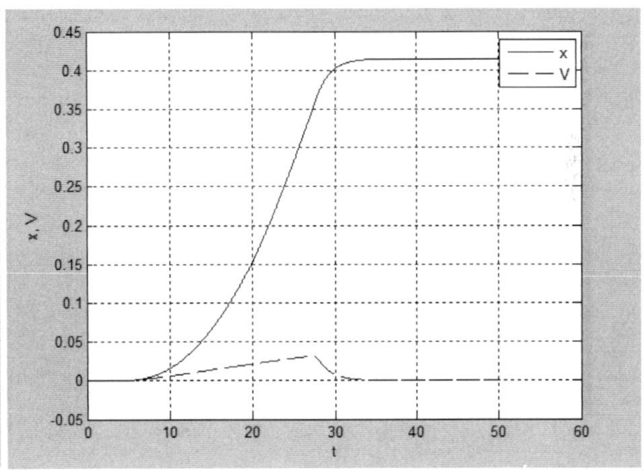

b)

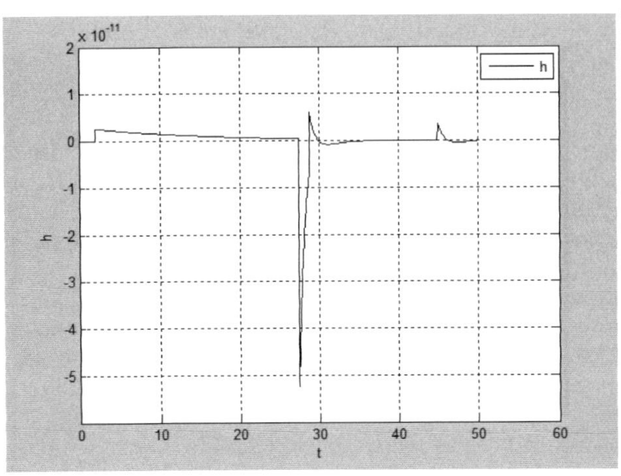

c)

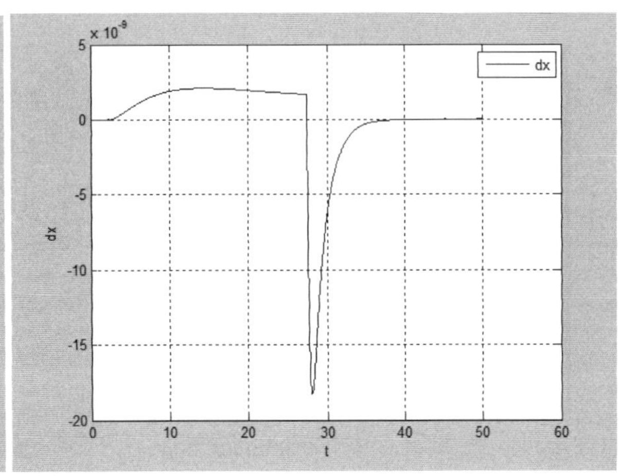

d)

Fig. 4. The characteristic of the dynamics of joint deployment of the elements: a) the graph of $\varphi(t)$ and $\omega(t)$ at U_l = 5 V, k_y = 10, C_d =25; b) the graph of $x(t)$ and $V(t)$ at k_y = 100, C_d =25; c) the graph of $h(t)$; d) the graph of $\Delta x_{sz}(t)$

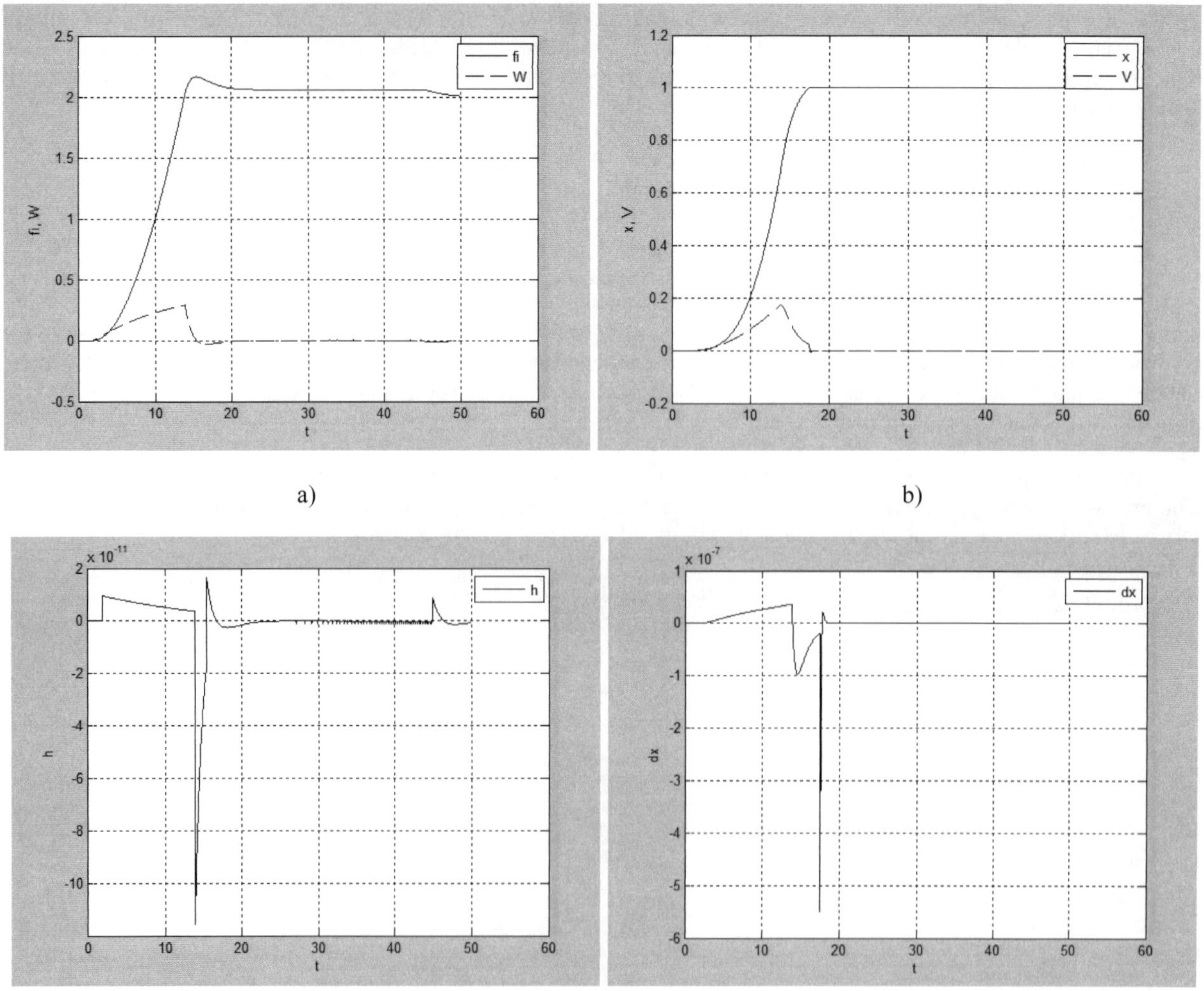

a) b)

c) d)

Fig. 5. The characteristic of the dynamics of joint deploymentof the elements: a) the graph of $\varphi(t)$ and $\omega(t)$ at $U_I = 15$ V, $k_y = 10$, C_d=25; б) the graph of $x(t)$ and $V(t)$ at $k_y = 100$, C_d=25; c) the graph of $h(t)$; d) the graph of $\Delta \boldsymbol{x}_{sz}(t)$

4. Conclusion

In the course of the study, two universal models of simultaneous realization of several stages of the reflector opening were obtained. Versatility lies in the possibility of using them for structures with different parameters: from changing the material, mass, dimensions to selecting the control body (motor). Accounting for such important parameters as the magnitude of the bending of the structural elements, dissipation, dynamic elongation, the influence of the properties of the stop on the angular motion of the body when framed on the stop, made it possible to describe the behavior of the system with a sufficiently high accuracy.

The main interest in choosing the control option was to find a version of the deployment, which reduces the vibrations and deflections of the structure. The two options considered allow us to manage the deployment process, which will allow us to further implement the control algorithm solving this problem. In order to implement the first variant, an additional actuator is required, the second variant is deprived of this drawback and is preferred. When using both methods, it is necessary to take into account the limitations on maximum loads and design moments, since the forces created by the system can exceed critical forces.

The results of these studies can be used to select the executive body, assess the necessary efforts. This mathematical model is approximate. To improve the quality of the system and the accuracy of the results, it is necessary to consider options for solving the direct and inverse problems of dynamics for physical systems, including the account of the energy components of the motion presented in [15, 16].

In the future, to improve the characteristics of the deployment of the reflector, it is supposed to solve the problem of a smooth approach of the structural elements to the stops and setting them. To solve this problem, it is proposed to use the methods of optimal control.

5. Acknowledgments

The research was carried out within the framework of the implementation of the Federal Target Program "Research and Development in Priority Areas for the Development of the Russian Science and Technology Complex for 2014-2020", the Ministry of Education of Russia, agreement No. 14.577.21.0201 (unique identifier RFMEFI 57715X0201): "Creating a highly efficient wireless system Directional transmission of energy and information for managing the state of space-based objects based on laser and fiber-optic technologies. " The work was carried out on the equipment of the Center for Innovative Technologies, the Center for Additive Technologies and Volume Scan, the Center for Radio-Technical and Optical Measurements, the Center for IPI Technologies, the Center for Materials Research, the Center for Educational Research Center of expert testing and quality control of new functional materials of innovative and nanotechnology. "

6. References

[1] Bei N.A., Zimin V.N. (2005). Transformable antennas of large dimensions for geostationary spacecraft, Antennas, No. 10(101), 2005, pp. 24-27.
[2] Lopatin A.V., Rutkovskaya M.A. (200)7. Overview of the constructions of modern transformed space antennas (Part 1), Vestnik of Siberian State University of Automation, No. 2, 2007, pp. 51-57.
[3] Seefeldt Patric, Spietz Peter, Sproewitz Tom, Grundmann Jan Thimo, Hillebrandt Martin, Hobbie Catherin, Ruffer Michael, Straubel Marco, Three Norbert, Zander, Martin. (2017). Gossamer-1: Mission concept and technology for a controlled deployment of gossamer spacecraft, Advances in Space Research, Vol. 59, No. 1, 2017, pp. 434-456.
[4] Khoroshilov, V. S., Zakrzhevskii, A. E. (2012). Dynamics of spacecraft due to elastic ring antenna deployment, Vibrations in Physical Systems, Vol. 25, 2012, pp. 229-234.
[5] Isaenko, Sergei; Sochivko Oleg & Dalyaev Igor. (2016). Analysis of Vibration Modelling Validity of Space-Borne Robotic System, Proceedings of the 26th DAAAM International Symposium, ISSN 1726-9679, Vienna, Austria, 2016, pp. 0548-0553
[6] Grinevich D.V. (2013). Investigation of the dynamics of unfolding extended structures, Questions of electromechanics, Proceedings of VNIIEM, Vol. 134, 2013, pp. 37-42.
[7] Kabdulin G.V., Komkov V.A., Melnikov V.M., Harlov B.N. (2009). Dynamics of controlled opening by centrifugal forces of space structures with compensation of the kinetic moment, Journal of Cosmonautics and Rocket Science, No. 1 (54), 2009, pp. 189-198.
[8] Bagheri Ghaleh, P. and Malaek, S.M. (2015). On dynamic stiffness of spacecraft flexible appendages in deployment phase, Aerospace Science and Technology, Vol. 47, 2015, pp. 1-9.
[9] Krivushov A.I. (2017). Creation and investigation of the mathematical model of the rotational motion of the spokes of a large-sized space-based reflector, Questions of radio electronics, No. 7, 2017, pp. 122-126.
[10] Mitin F.V. (2017). Development and analysis of the mathematical model of translational motion of transformed space structures, Questions of radio electronics, No. 7, 2017, pp. 117-121.
[11] Rabotonov Yu. N. (1962). Resistance of materials, Fizmatgiz, Moscow, 1962.
[12] Woldek A.I., Popov V.V. (2007). Electric cars. Introduction to electromechanics. DC machines and transformers: A textbook for high schools, Peter, St. Petersburg, 2007.
[13] Ivashov E.N., Panfilova E.S., Shikhov A.I. (2014). The choice of the electric motor for vacuum switching equipment, Advances in modern natural science, No. 12 (Part 4), 2014, pp. 409-411.
[14] Usoltsev A.A. (2009). General electrical engineering: Textbook, St. Petersburg State University ITMO, St. Petersburg, 2009.
[15] Kostin G.V., Saurin V.V. (2009). The method of integro-differential relations for the analysis of natural oscillations of membranes, PMM, Vol. 73. No. 3, 2009, pp. 459-473.
[16] Kostin G.V., Saurin V.V. (2010). Modeling of spatial motions of an elastic beam on the basis of the method of integro-differential ratios, Modern problems of continuum mechanics, Proceedings of the XIV International Conference, Vol. 2, Rostov - on - Don, Azov, 2010. pp. 165-169.

DOI: 10.2507/28th.daaam.proceedings.110

DESIGN CONCEPT OF LUNAR ROVER FOR THE MOON GEOLOGICAL EXPLORATION

Andrey Vasiliev[a], Igor Dalyaev[a], Evgeny Slyuta[b]

[a]Russian State Scientific Center for Robotics and Technical Cybernetics (RTC), Saint-Petersburg, Russia
[b]Vernadsky Institute of Geochemistry and Analytical Chemistry of Russian Academy of Sciences, Moscow, Russia

This Publication has to be referred as: Vasiliev, A[ndrey]; Dalyaev, I[gor] & Slyuta, E[vgeny] (2017). Design Concept Lunar Rover for the Moon Geological Exploration, Proceedings of the 28th DAAAM International Symposium, pp.0780-0786, B. Katalinic (Ed.), Published by DAAAM International, ISBN 978-3-902734-11-2, ISSN 1726-9679, Vienna, Austria
DOI: 10.2507/28th.daaam.proceedings.110

Abstract

The paper discusses the current results of mobile robotic system «Robot-Geologist» development. The robot is intended to maintain geological exploration on the surface of the Moon. Distinctive features of the project in comparison with the known planetary rover projects comprise: superior length of the estimated research route up to 500 km; an autonomous control system capable of moving from one point of scientific interest to the other; a wide range of scientific equipment and an extensive program of research, including lunar surface sampling along with drilling and investigation of wells. The key feature of the "Robot-Geologist" is the ability to transport samples in the course of the expedition and upload them to returnable spacecraft. The paper presents a description of the proposed robot concept as a result of the project development.

Keywords: Mobile robotic system; geological exploration; Moon research; moon (lunar) rover; drilling rig; manipulation system; ground sampling; robot-geologist

1. Introduction

Research of the Moon is one of the immediate tasks for space exploration. The moon is the largest celestial body close to Earth, a source of important natural resources, a unique base for astrophysical observations and other scientific research, and in the long term it can be considered as a reference point for long-range manned space missions.

The first stage of the development of the moon should be research on its surface involving a network of stationary and self-propelled scientific stations [1]. The thematic geological survey and preliminary geological exploration constitute together an important and necessary stage in the development of the Moon, which should provide not only extensive scientific information, but also provide the information necessary for the optimal location determination of a long-term or periodically inhabited lunar base, the deployment of a primary industrial infrastructure and the preparation for future manned lunar missions.

The use of a stationary research apparatus on the lunar surface is advisable in areas characterized by relatively stable geological properties. However, full-scale studies in areas where there are a variety of complexes of rocks of different ages, genesis, composition and spectral classes are practically impossible without the use of a mobile apparatus. The estimated route variations for such studies, the length of which is up to 400-500 km, were developed by scientists of Vernadsky Institute of Geochemistry and Analytical Chemistry (Moscow) [2]. According to its geological structure, the region of Mons Rümker in the northwestern part of the Oceanus Procellarum is one of the primary landing sites and research areas for the mobile research apparatus [1, 2]. The scheme of this route is shown in Fig.1. The main geological studies are expected to be carried out by the methods of well drilling with sampling of stratified cores of regolith, and by collecting small samples (up to 50-100 g) of the surface layer of regolith and stones.

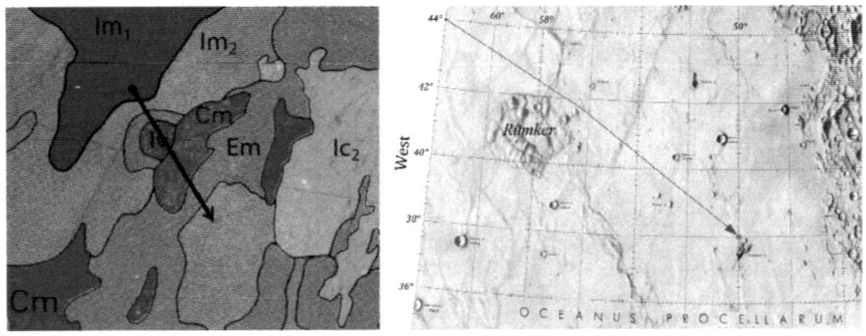

Fig. 1. Geologic map of Mons Rümker region [3] and schematic representation of the route (the arrow length is about 400 km): Im1 – Lower Imbrian Series basalts; Im2 – Upper Imbrian Series basalts; Em – Eratosthenian System basalts; Cm – Copernican System basalts; Iv – volcanic province; Ic2 – crater material from Mare Imbrium basin

In connection with the foregoing, the first and foremost task is to create a new generation of automatic research apparatus – the Mobile Robotic System (MRS) for geological survey, equipped with a drilling rig, a manipulation system and a set of scientific tools.

The purpose of this project is the preliminary development of the design appearance of such an apparatus under the conventional name "Robot-Geologist". The term "design appearance" here means not only the overall technical appearance of the whole system and basic technical solutions for its composition and design, but also scenarios of the MRS activity.

The tasks for the MRS, set by the skilled geochemists, are as follows:

- locomotion along the route up to 400-500 km;
- drilling of several wells up to 3-6 m deep with sampling of continuous cores of lunar regolith while maintaining their complete stratification;
- surface soil sampling with a certain periodicity;
- carrying out a wide range of scientific research (gravimetry, magnetometry, spectrometry, active and passive seismic studies, etc.);
- deployment of a long-term monitoring automatic scientific research station (ASRS) near one of the wells;
- delivering and re-laying of collected samples with a total mass of up to 100 kg into the automatic lunar transport spacecraft, waiting for the MRS at one of the intermediate points of the route or at its final point.

The project is being developed by the Russian State Scientific Center for Robotics and Technical Cybernetics (RTC) in St.-Petersburg with the support of Russian Space Agency in accordance with the initiative and recommendations of the Vernadsky Institute of Geochemistry and Analytical Chemistry of Russian Academy of Sciences.

2. Prototypes

To meet the set of requirements, the MRS has a number of unique features. The review of the stationary and self-propelled apparatus equipped with drilling rigs for conducting studies on the surface of other celestial bodies that were created earlier or are being developed at present [4-8], shows that there are no analogues of the MRS. Current research in the world in similar areas usually involve drilling of shallow wells without maintaining stratification of the core samples, short-term missions that cover geologically small areas (i.e., limited in length and duration of research), do not involve the delivery of collected samples to the Earth [6, 7]. Fig. 2 shows some of the examples of already employed and currently developed systems for drilling in space.

The developed concept of geological research on the Moon's surface by means of the lumokhod "Robot-Geologist" takes into account the experience of all the previous lunar missions, such as studies by means of Soviet scientific stations "Luna" and american "Apollo" missions.

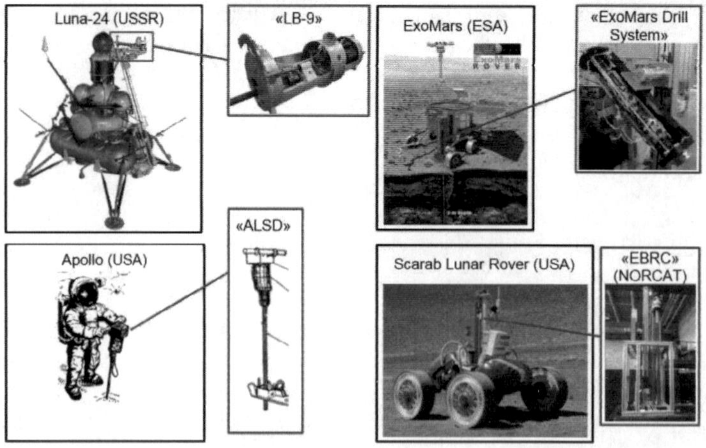

Fig. 2. Examples of drilling systems for the Moon and the Mars

3. MRS concept

The MRS concept aggregates the development of basic technical solutions for all the rover subsystems, the general technical design of the MRS and the development of scenarios for scientific tasks performing.

The fundamental difference between MRS and all previously created planet rovers is the required route length which is more than an order of magnitude greater than the maximum distance traveled by terrestrial mobile vehicles along the surface of another celestial body [5, 9]. The length of the proposed MRS route is not less than 400 km. This fact, combined with the general specificity of the MRS, justifies the desire to simplify the chassis design as much as possible in order to ensure its maximum reliability.

The length of the mission and the unprecedented travel distance also impose special requirements to the MRS control system, which should provide not only remote control mode, but also autonomous locomotion with real-time environment assessment and insurmountable obstacles avoiding [10].

In the course of the project, the MRS design appearance was accomplished. As a result, a three-dimensional computer model of the MRS was developed, that is shown in Fig. 3. The MRS can be represented as two main parts: service systems and payload.

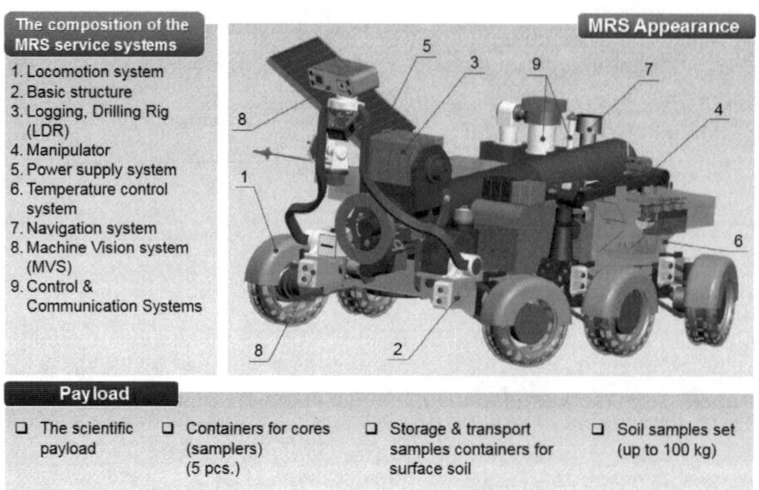

Fig. 3. The Lunokhod «Robot-Geologist» appearance, payload and service systems composition

Service systems comprises (see Fig.3): locomotion system (1); basic structure (2); logging and drilling rig (LDR) (3); manipulation system (4); power supply system (5); thermal control system (6); navigation system (7); machine vision system (MVS) (8); integrated on-board control system and communication system (9).

The payload includes a set of scientific tools, equipment for storing and transporting cores from wells and samples of the surface layer of regolith, as well as the samples themselves with a total mass of up to 100 kg.

The following scientific equipment is situated onboard: 1) scientific navigation devices, including: radio beacon, TV-spectrometer, IR-spectrometer, TV-camera for manipulator service area; 2) geophysical research tools, including: set for active and passive seismic survey, geological radar, logging device, magnetometer and gravimeter; 3) tools for studying the composition of the lunar soil and gases, including: gas analyzer for the study of weakly bound volatiles in lunar regolith and for the study of the lunar atmosphere, gamma-spectrometer and neutron detector; 4) the sampler devices, including: drilling rig and manipulator with cassettes for the collection and storage of core samples and soil samples.

The basic structure of the MRS consists of three main parts: a frame, a hermetical equipment bay and an open body platform for placing autonomously operating scientific equipment to be unloaded during the mission to the lunar surface by means of a manipulator [11, 12].

The universal chassis is built on the basis of six identical modules of the motor-wheels. Each module is attached to the MRS frame and contains a wheel, a steering drive, suspension levers and mounting shield.

The MRS design specification is presented in Table 1.

Parameter	Unit	Value
Mass (without soil samples set)	kg	1350
Load capacity of soil samples set	kg	100
Chassis mass	kg	542
Scientific equipment mass	kg	200
Autonomy (time), not less than	hour	20
Autonomy (distance), not less than	km	40
Maximal obstacle height	mm	250
Angle of surface inclination	grades	20
Maximal speed on a flat surface - in autonomous mode - in remote control mode	 km/h km/h	 3 5
Range	km	500
Active lifetime	year	2

Table 1. The MRS design specification

4. Logging-Drilling Rig (LDR) composition

The main and most massive part of the MRS is the LDR. It has a sufficiently large length (up to 4 m) and is placed on a lunar rover in a horizontal position.

Such LDR placement was chosen after comparing two ways: a vertical fixed placement (Fig.4a) and a horizontal deployable one (Fig.4b). Both of them have their own advantages and drawbacks. The analytical comparison is presented in Table 2.

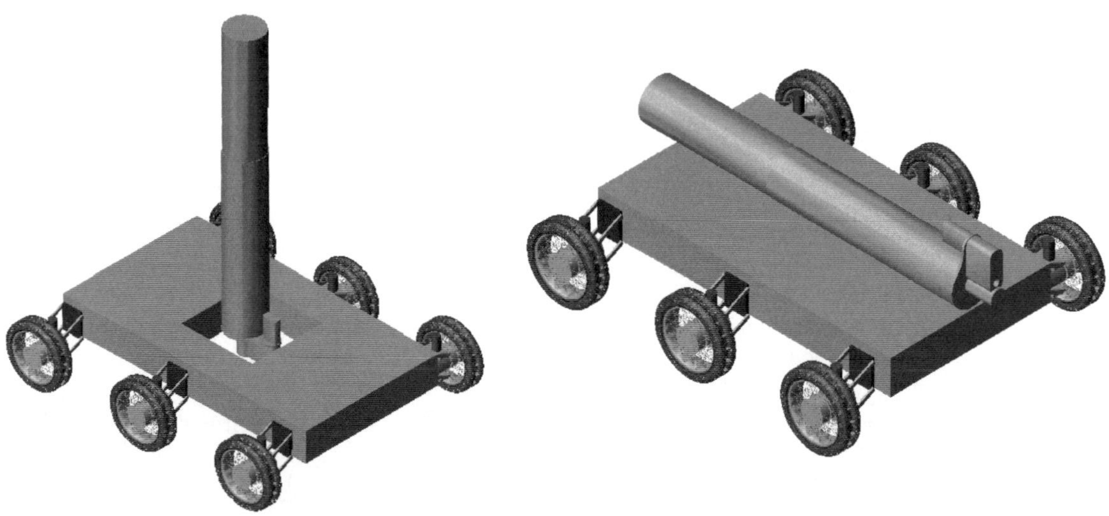

a) b)

Fig. 4. The schemes of the MRS on-board LDR placement

Estimated characteristic	**Analysis results**			
	Scheme on Fig.4a		Scheme on Fig.4b	
Overall dimensions	Significant overall height dimension	–	Minimal overall dimensions	+
Layout	The minimal space LDR occupied on the MRS platform	+	A significant space LDR occupied on the MRS platform	–
Design advantages	Design symmetry. No need for additional outriggers	+	The minimal height of the center of gravity of the MRS	+
Design drawbacks	High location of the MRS center of gravity. Large required service area of the manipulator (if there is a requirement to reach the upper part of the LDR);	–	The MRS center of gravity shifts forward when the LDR is in the working position. Additional outriggers are necessary.	–
Operational specificities	Less stable driving over typical obstacles	–	More stable driving through typical obstacles	+
Possibility of emergency LDR undocking from the MRS	Impossible	–	Possible	+
Placement of MRS in the transport volume of the launch spacecraft vehicle	Non-rational	–	Relatively compact	+

Table 2. The LDR on-board placement comparison

Large dimensions of the LDR predetermined the overall layout of the MRS: the platform is divided into three areas, as shown in Fig.5a: area (I) is reserved for the placement of the equipment bay and the solar battery with its orientation drive; area (II) in the center is for the placement of the LDR; area (III) comprises an open body platform for the scientific equipment along with the manipulation system and other equipment.

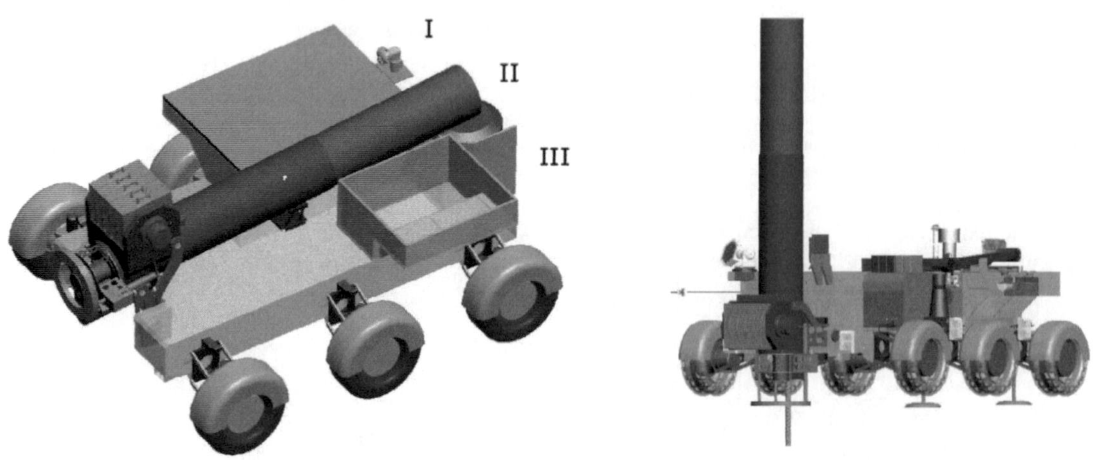

a) b)

Fig. 5. The LDR on-board placement (a) and the MRS drilling configuration (b)

When drilling, the LDR is being vertically moved, as shown in Fig.5b. At the same time, the MRS rises on three pull-out supports: one is located along the axis of the LDR and the other two are located at the rear of the MRS.

The main part of the LDR is an automatic logging-drilling machine, the prototype of which is the LB-10 machine [13]. The LB-10 machine shown in Fig.6a, with a potential drilling depth of up to 15 m, was developed in the USSR as the next generation of the LB-9 machine that was successfully used on the "Luna-24" station.

As a result of considering the MRS drilling scenarios, the design of the LDR was substantially modified. The LDR, as shown in Fig.6b, in addition to the logging-drilling machine (1) includes: a frame (2) with a deployment drive, a clamping mechanism (3), sample containers (4).

The deployment drive serves to transfer the LDR from the transport position to the vertical working one. The clamping mechanism ensures the unloading of the drilling machine tool and acts in conjunction with two retractable supports (outriggers) in the rear part of the MRS.

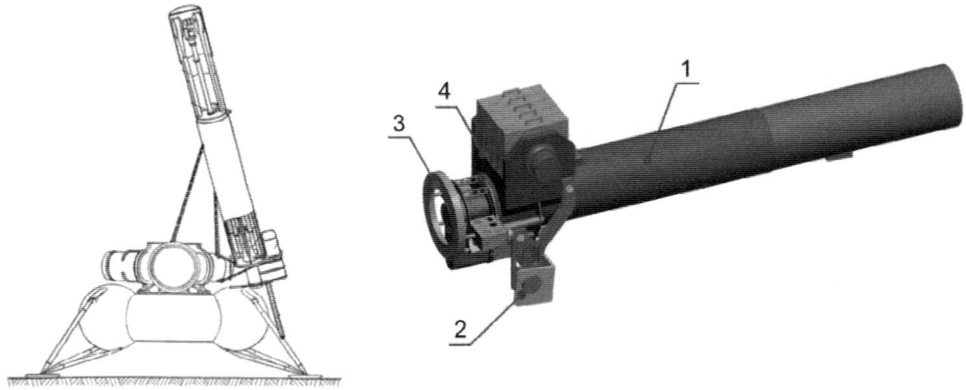

a) b)

Fig. 6. Drilling rigs: a) LB-10 automatic logging-drilling machine; b) the MRS logging-drilling rig (LDR)

The LDR specification is given in Table 3.

Parameter	Unit	Value
Drilling depth	mm	3000 - 6000
Well diameter	mm	57
Core diameter	mm	32
The drilling tool maximum force	N	200
Nominal force of LDR to surface clamping	N	1200
Nominal power consumption	W	1000
Overall dimensions	mm	4000 x 720 x 492
Mass	kg	120

Table 3. The LDR specification

The LDR is equipped with detachable containers with the possibility of their capture by the manipulator and transfer onto the returnable automatic transport spacecraft.

5. Conclusion

In connection with the developed concept of geological research on the surface of the Moon the first and foremost task is to create a new generation of automatic research apparatus – the Mobile Robotic System (MRS) for geological survey, equipped with a drilling rig, a manipulation system and a set of scientific tools.

To meet the set of requirements, the MRS has a number of unique features. Distinctive features in comparison with the known planetary rovers comprise: superior length of the estimated research route up to 500 km; an autonomous control system capable of moving from one point of scientific interest to the other; a wide range of scientific equipment and an extensive program of research, including lunar surface sampling along with drilling and investigation of wells. The key feature of the MRS is the ability to transport samples in the course of the expedition and upload them to returnable spacecraft.

The project is aimed at the preliminary development of the design appearance of such the MRS under the conventional name "Robot-Geologist". The term "design appearance" here means not only the overall technical appearance of the whole system and basic technical solutions for its composition and design, but also scenarios of the MRS activity.

In the course of the project, the MRS design appearance was accomplished. The project development involved the necessary design work, the calculation of all the main MRS subsystems, MRS activity scenarios development, determination of the design specification and the main technical requirements for the MRS and its subsystems.

The article provides a brief description of the design concept development of a MRS "Robot-Geologist" intended for conducting geological studies on the lunar surface. The crucial and most massive part of the MRS is a logging-drilling rig LDR, the description of which has been given special attention.

LDR composition and MRS on-board placement was accomplished. The main part of the LDR is an automatic logging-drilling machine. Its prototype is the LB-10 machine, which was developed in the USSR as the next generation of the LB-9 machine successfully used on the "Luna-24" scientific research station. As a result of considering the MRS drilling scenarios, the design of the LDR was substantially modified.

Further research is aimed at refining the design of the key subsystems of the MRS, namely: locomotion system, LDR and manipulation system.

6. Acknowledgments

The research is performed in the Russian State Scientific Center for Robotics and Technical Cybernetics (RTC) with the support of Russian Space Agency. Scientific tasks and technologies of geological research on the lunar surface are developed in Vernadsky Institute of Geochemistry and Analytical Chemistry of Russian Academy of Sciences.

7. References

[1] Fundamental Space Research (2014), Raykunov, G.G. (Ed.), Book 2, Fismatlit, ISBN: 978-5-9221-1559-9, Moscow, 457 p.

[2] Slyuta, E.N.; Vasilev, A.V. & Dalyaev, I.Yu. (2017). Lunokhod "Robot-Geologist": scientific tasks and technical configuration, Proceedings of 48th Lunar and Planetary Science Conference, March 20-24, 2017, Houston, USA, Available from: http://www.hou.usra.edu/meetings/lpsc2017/pdf/1929.pdf Accessed on: 2017-03-09

[3] Wilhelms, D. E. (1987). The Geologic History of the Moon, United States Government Printing Office, ISBN: 1495919854, Washington, 329 p.

[4] Drilling in Extreme Environments. Penetration and Sampling on Earth and other Planets. (2008). Bar-Cohen, J. & Zacny, K. (Ed.), WILEY-VCH, ISBN: 978-3-527-40852-8, Germany, 767 p.

[5] Young, A.H. (2007). Lunar and Planetary Rovers. The wheels of Apollo and the Quest for Mars, Springer, Praxis Publishing, ISBN 10: 0-387-30774-5, Germany, 393 p.

[6] Bartlett, P.; Wettergreen, D.; Whittaker, W.L.; et al. (2008). Design of the Scarab Rover for Mobility and Drilling in the Lunar Cold Traps, Proceedings of International Symposium on Artificial Intelligence, Robotics and Automation in Space, Available from: http://www.ri.cmu.edu/pub_files/2009/1/09aiaa.scarab.pdf Accessed: 2017-06-23

[7] Sanders, G.B.; Barid, R.S.; Rogers K.N.; et al. (2012). RESOLVE. Lunar Ice/Volatile Payload Development and Field Test Status, Presentation to LEAG, October, 24, 2012, Available from: http://www.lpi.usra.edu/meetings/leag2012/presentations/Sanders.pdf Accessed: 2017-06-23

[8] Krylov, I. (2015). "Yutu" told about the Moon, Available from: http://www.gazeta.ru/science/2015/12/23_a_7985801.shtml Accessed: 2017-06-23

[9] Planetokhody (Planet Rovers) (1993), Kemurdzhian, A.L. (ed.), Mashinostroenie, Moscow, 400 p.

[10] Smirnova, E.; Stepanov, D. & Goryunov, V. (2016). A Technique of Natural Visual Landmarks Detection and Description for Mobile Robot Cognitive Navigation, Proceedings of the 26th DAAAM International Symposium, pp.0905-0911, B. Katalinic (Ed.), Published by DAAAM International, ISBN 978-3-902734-07-5, ISSN 1726-9679, Vienna, Austria, DOI: 10.2507/26th.daaam.proceedings.126, Available from: http://www.daaam.info/Downloads/Pdfs/proceedings/proceedings_2015/126.pdf Accessed: 2017-06-23

[11] Lopota, A.; Dalyaev, I.; Shardyko, I.; Kuznetcova, E. & Belezyakov, I. (2016). Means of Robotic Support for On-Orbit Servicing, Proceedings of the 26th DAAAM International Symposium, pp.0865-0870, B. Katalinic (Ed.), Published by DAAAM International, ISBN 978-3-902734-07-5, ISSN 1726-9679, Vienna, Austria, DOI:10.2507/26th.daaam.proceedings.120, Available from: http://www.daaam.info/Downloads/Pdfs/proceedings/proceedings_2015/120.pdf Accessed: 2017-06-23

[12] Isaenko, S.; Sochivko, O.; Shardyko, I. & Titov, V. (2016). Simulation of Manipulator as a Part of Space Robotic System Weightlessness Device, Proceedings of the 26th DAAAM International Symposium, pp.0554-0561, B. Katalinic (Ed.), Published by DAAAM International, ISBN 978-3-902734-07-5, ISSN 1726-9679, Vienna, Austria, DOI: 10.2507/26th.daaam.proceedings.076, Available from: http://www.daaam.info/Downloads/Pdfs/proceedings/proceedings_2015/076.pdf Accessed: 2017-06-23

[13] The Moon – a step towards the technologies of the solar system development (2011), Legostaev, V.P. & Lopota V.A. (ed.), RSC Energia, ISBN: 978-5-91820-046-9, Moscow, 584 p.

DOI: 10.2507/28th.daaam.proceedings.111

DEVELOPMENT OF INNOVATION IN FIELD OF PRECISION AGRICULTURE

Aleksandar Vekic[a], Jelena Borocki[a], Stevan Stankovski[a] & Gordana Ostojic[a]

[a]University of Novi Sad, Faculty of Technical Sciences,
Trg Dositeja Obradovica 6, 21000 Novi Sad, Republic of Serbia

This Publication has to be referred as: Vekic, A[leksandar]; Borocki, J[elena]; Stankovski, S[tevan] & Ostojic, G[ordana] (2017). Development of Innovation in Field of Precision Agriculture, Proceedings of the 28th DAAAM International Symposium, pp.0787-0795, B. Katalinic (Ed.), Published by DAAAM International, ISBN 978-3-902734-11-2, ISSN 1726-9679, Vienna, Austria
DOI: 10.2507/28th.daaam.proceedings.111

Abstract

Important changes in technology have brought great progress both in the development of precision and future agriculture. Modern systems have facilitated functioning of the whole process and led to increased yields and benefits for all participants in the chain. Innovations in the field of precision agriculture are recognized as an important lever of development, and as such supported various programs by the EU funds. The importance of this support is reflected in very good results, and constant progress of the sector since its inception in the 80's of the 20th century. The aim of this paper is to show the importance of innovation in precision agriculture to the participants in the process, and for society on the whole.

Keywords: Precision Agriculture; Growth; Innovations; EU support

1. Introduction

This paper focuses on innovation in the area of Precision Agriculture and give reviews of modern solutions which are implemented for this purpose. Dynamic changes in all areas of human life have changed the views and ways of doing certain jobs, especially those of the traditional ones. The effort to reduce the amount of time spent and investing in a particular area always leads to the rapid development of innovation. Innovative activity is always inherent in human society, so many authors try to explain the notion of innovation. Some definitions of innovation are [1]: Schumpeter (1930) explains innovation" as process of introducing a new product or modifications brought to an existing product". Drucker (1954) considers that innovation is „one of two basic functions of an organization". Simmonds (1986) theoretical view of innovation is: „Innovations are new ideas that consist of: new products and services, new use of existing products, new markets for existing products or new marketing methods, and that is basic creative process". For Devenport (1991) innovations are: „Complete a task development in a radically new way". Authors: Covin şi Slevin (1991), Lumpkin and

Dess (1996), Knox (2002) defined innovation as „A process that provides added value and a degree of novelty to the organization, suppliers and customers, developing new procedures, solutions, products and services and new ways of marketing". For Rogers (1998):" innovation involves both knowledge creation and diffusion of existing knowledge". For European Union innovations are the basis of future development so main institution such as The European Commission gave definition (1999) of innovation as: „Successful production, assimilation and exploitation of novelty in the economic or social environment".
According to Thrott (2005):" An innovation is an idea that has been transformed into practical reality" [2]. Authors Tidd &Bessant state that „ Innovation is driven by the ability to see connections, to spot opportunities and to take advantage of them" [3]. Many of huge multinational companies all potential for their development and growth can find in innovation. For example, Siemens company state: Innovation is our lifeblood"[4], the founder of Apple company Steve Jobs, is quoted as saying „innovation distinguishes between a leader and a follower" [5]. Richard Brenson, the founder of Virgin Group, said, that: „An innovative business is one which live breathes „outside the box". It is not just good ideas, it is a combination of good ideas, motivated staff and an instinctive understanding of what your customers want" [6].
Strong growth and development of the ICT sector (including software and hardware) has enabled a high degree of change in all spheres, even in the field of agriculture. One of the reasons for higher development of agriculture segment is growing of the global population, and that causes a changed perspective of development. The concept of technological innovation is particularly important for the development of Precision Agriculture. OECD gave the next definition of technological innovations: "Technological innovations comprise new products and processes and significant technological changes of products and processes. An innovation has been implemented if it has been introduced on the market (product innovation) or used within a production process (process innovation). Innovations therefore involve a series of scientific, technological, organizational, financial and commercial activities" [7]. So product and process innovation are associated with the concept of the technological innovation.

According to Oslo Manual, these two types of innovation are defined as [8]:

- A product innovation is the introduction of a good or service that is new or significantly improved with respect to its characteristics or intended use. This includes significant improvements in technical specifications, components and materials, incorporated software, user friendliness or other functional characteristics.
- A process innovation is the implementation of a new or significantly improved production or delivery method. This includes significant changes in techniques, equipment and/or software.

The future brings an opportunity for unimaginable development of Precision Agriculture, using modern technologies and increasing improvements of mobile platforms which support this kind of managing in agriculture production and storage process. It is clear that innovation is the basis for development of Precision Agriculture, therefore it is very important to study the possibilities that their implementation can bring.

2. Methodology

This paper presents insight into the area of Precision Agriculture with modern solutions and systems, and possibilities for innovation for this purpose. We collected a data through this research by using relevant literature and other open sources. We used our own experience too, collected through development of different solutions and applications in this field. Data which used for this purpose are from European Parlament Study about Agriculture and Rural development from 2014. This study is the largest European report about Precison Agriculture, and the name of the study is: Precision Agriculture: An opportunity for EU farmers, potential support with the CAP 2014-2020. The study was developed within Directorate - General for internal policies, Policy department B – structural and cohesion policies.

3. Precision Agriculture – A new era of agriculture innovation

In the middle of 80s, the notion of Precision Agriculture emerged and began with sensors for soil matter, and have quickly diversified to include satellites, aerial, and hand held or tractor mounted sensors [9]. The world's population has reached 7,5 billion people, therefore food needs are growing. Limited availability of agricultural resources and capability of supply chain result in a search for new ways in the production of agricultural products.
Bearing in mind the digital revolution we are witnessing, it is clear that most of the answers to some important questions can be found right in this domain. The technological development in Precision Agriculture has been rapid since the 90s, and it continues through the development of the ICT sector. This kind of development has significantly changed the perspectives of agriculture's future.
Precision Agriculture can be understood as: „An integrated information- and production-based farming system that is designed to increase long-term, site-specific and whole-farm production efficiency, productivity and profitability while minimizing unintended impacts on wildlife and the environment" [10]. This definition understood PA in a context of „whole farm" management strategy, not in terms of individual farms or crop fields. It's also common definition: „PA is form whereby decisions on resource application and agronomic practices are improved to better match soil and crop requirements as they vary in the field" [10]. According to this definition, the main idea is that PA involving management strategy and focus is on the decision making. The main goal of the concept is to improve production

management and to limited impact of environment, especially in changeable climatic conditions which are typical of today's life. This kind of agriculture requests using of modern information technology and technical solutions, with a goal of providing benefits, such as: economic, environmental and social.

The objectives of PA can be illustrated:

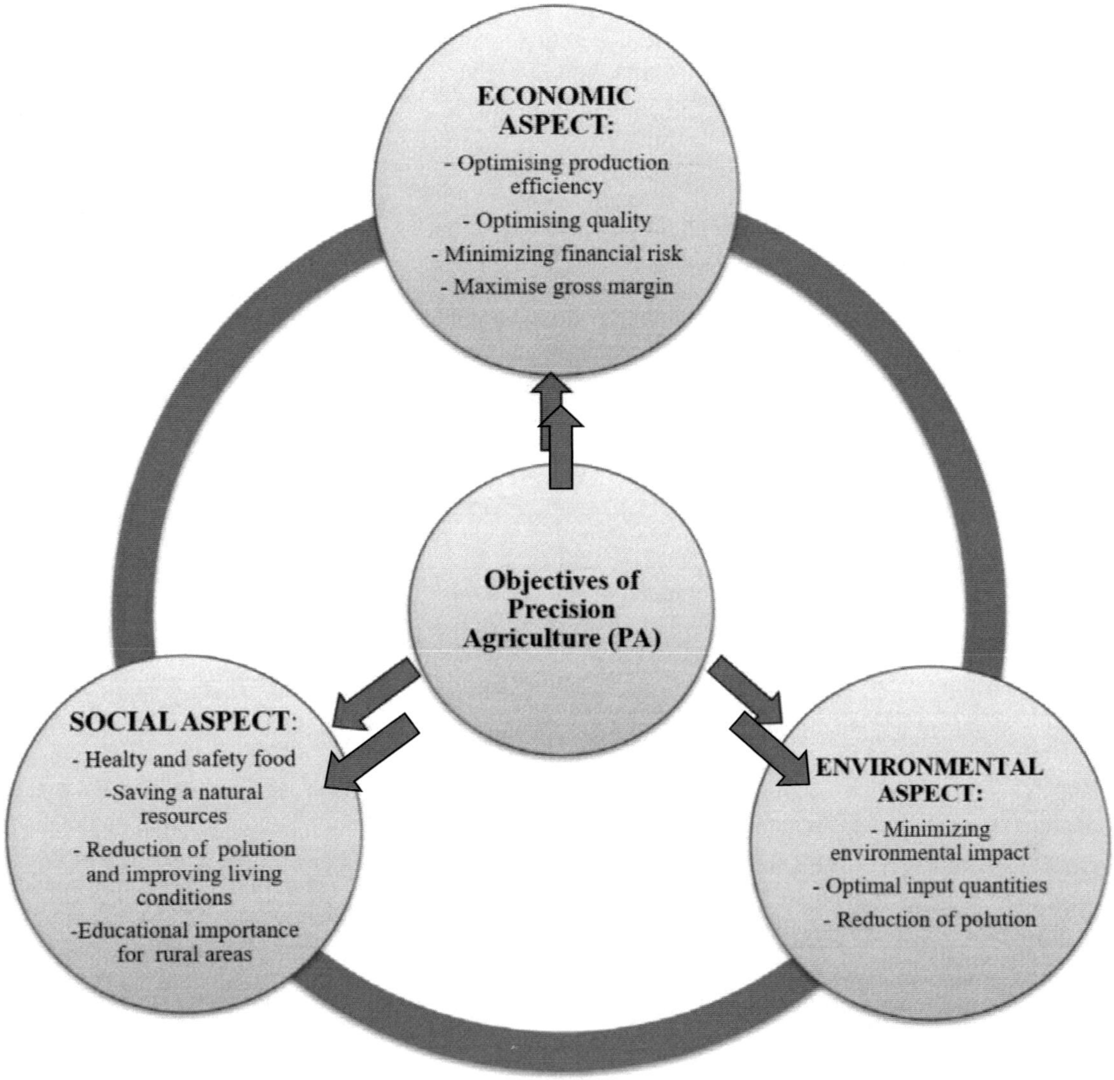

Fig. 1. Objectives of Precision Agriculture (PA) [10]

The primary goal for the development of Precision Agriculture is to make measurements of important parameters such as: temperature, humidity, soil moisture and acidity, reduction of pesticides and fertilizers use, in order to get higher yields and a higher level of quality. It's very important to satisfy requests of economic and environmental aspects. For farmers are very important to take care about basic components of PA [11]:

- ***Background data*** – means data about crop production fields and help to make important decisions. The sources of this data can be: grid soil samples, detailed soil mapping, aerial photography, topographic maps, yield maps, soil texture maps, environmentally sensitive areas.
- ***A recordkeeping system*** - allows the storage of large amounts of data that are later used in the process and decision making. Usually, special software or external data warehouses such as clouds are used for this purpose.
- ***Analysis and decision making process*** – this is a crucial step in developing Precision Agriculture. It is often necessary to analyze huge amounts of data and do their systematization. With the development of IT sector, special web and mobile platforms are developed to perform this complex task and help in making the decision as simple as possible.

- ***Specialized implementation equipment*** – this is precondition to perform tasks within precision agriculture. Development and innovation of the modern equipment have been always in the progress, and that providing more precision data with a higher usage value. Equipment usually includes: GPS guidance systems, equipment for yield monitoring, variable rate application equipment, special types of sensors, different cameras and photo equipment, drones, etc.
- ***Evaluation and revision-*** Continuing learning and gaining experience are inseparable part of the process. This is very important, especially after harvest season, and that help farmers to create precision and achievable agriculture plan for the next season.

Within these process components, there are continuous improvement and innovation. This is essential for maintaining the vitality of the entire concept of Precision Agriculture. This implies monitoring of the development of modern and applied technologies and analyzing current trends. Most innovations in precision agriculture are led by big agriculture companies, smaller innovative agricultural technology companies and top agriculture universities and institutes [12]. Innovation in this sphere requires a higher level of investment, but they are very much in demand until their commercialization and return of investing funds.

Most of the funds are spent in the development and implementation of adequate equipment and modern services for its functioning. Their development plays a key role in achieving innovation goal. In the following table, we will show the most important technologies used in the field of Precision Agriculture, their purpose and benefits:

Type of technology	Purpose and benefits
Computational decision tools	Use data to develop recommendations for management and optimize multitudes of farm tasks.
The cloud	Provide efficient, inexpensive, and centralized data storage, computation, and communication to support farm management.
Sensors	Gather information on the functioning of equipment and farm resources to support management decisions
Robots	Implement tasks with efficiency and minimal human labour.
Digital communication tools (mobile, broadband, LPWAN)	Allow frequent, real-time communication between farm resources, workers, managers, and computational resources in support of management
Geo-locationing (GPS, RTK)	Provide precise location of farm resources (field equipment, animals, etc.), often combined with measurements (yield, etc.), or used to steer equipment to locations.
Geographic information systems	Use computerized mapping to aid inventory management and to make geographical crop input prescriptions (fertilizer, etc.).
Yield monitors	Employ sensors and GPS on harvesters to continually measure harvest rate and make yield maps that allow for identification of local yield variability.
Precision soil sampling	Sample soil at high spatial resolution (in zones) to detect and manage fertility patterns in fields.

Spectral reflectance sensing (proximal and remote)	Measure light reflectance of soil or crop using satellite, airplane, or UAS, imaging, or field equipment–mounted sensors, to make determinations on soil patterns, crop, or Animal performance, or on nutrient/pest problems.
Auto-steering and guidance	Reduce labour or fatigue with self-driving technology for farm equipment (including robots); can also precisely guide equipment in fields to enable highly accurate crop input placement and management.
Variable rate technology	Allow continuous adjustment of application rates to precisely match localized crop needs in field areas with field applicators for crop inputs (chemicals, seed, etc.).
On-board computers	Collect and process field data with specialized computer hardware and software on tractors, harvesters, etc., often connected to sensors or controllers.
Radio frequency ID	Transmit identity data with tags attached to production units (mostly animals) that allow data collection on performance as well as individualized management.
Automated milking, feeding, and monitoring systems	Perform milking or feeding operations automatically with robotic systems, often combined with sensors that collect basic biometric data on animals, thereby reducing labour needs and facilitating individualized animal management.

Table 1. Enabling technologies for Precision Agriculture [12]

Very important part of the Precision Agriculture concept is the Remote Sensing Technology. This modern technology provides a huge range of data which are important for agriculture, such as data about: temperature, soil, moisture analysis, monitoring of crop growth, yield forecasting, etc.

This data can be used by national governments, in order to make important decisions and implement them into their agricultural policy. Individual farmers can also use this valuable data which are collected by remote sensing technology, when dealing with their individual crops, about their health status and how to deal with any problems. It is especially important that this data allows farmers to make rough estimates to the final crop yield just by looking at the plant and health status half-way through the season and taking into account the climatic conditions up until that point.

Introduction of this technology always requests detailed planning of agricultural business and making decisions for the next period. Remote sensing can be used for Precision Agriculture in a number of ways by providing input on soil and plant condition and variability in the overall management and decision support system [13]. Remote sensing can significantly contribute to providing a timely and accurate picture of the agricultural sector, as it is very suitable for gathering information over large areas with high revisit frequency [14].

This equipment can provide recorded data in real time. Prior to seeding, air photos, multi-spectral scanners, or high resolution satellite imagery can be used to translate results from point soil test samples, acquired using a grid or stratified sampling scheme, to a spatial coverage for the whole field [13].

Some possibilities of modern solutions of Precision Agriculture are illustrated on following pictures:

Fig. 2. The use of sensors and drones (with spectral camera) on the crop filed for Geothermal analysis [15] [16]

Optimal combination of shown technologies and services leads to the positive growth of all the most important parameters that are observed. Some researchers predict that digital data will be getting collected at the rate of 40 zettabytes (1 ZB = 40 trillion GB) per year by 2020 [17]. This is a huge potential for the Precision Agriculture, and it is to be expected that these opportunities will be exploited in the best possible way. The primary goal of Precision agriculture is to enable the production of enough food for the growing world population, according to UN data world population will reach 9,6 billion by 2050 [18]. Development of agricultural technologies requires knowledge (ranging from broad to specific) and talent in science and entrepreneurship [12].

4. The main benefits of agricultural innovation and EU support

PA nowadays is seen as an "environment friendly system solution that optimizes product quality and quantity while minimizing cost, human intervention and the variation caused by the unpredictable nature" [19]. It's very important to understand the applicability technologies in Precision Agriculture and to gain knowledge of how to use them in the best way. Their implementation leads to more efficient: decision-making process, monitoring soil and plant parameters, collecting real time data, reducing waste and the chemicals in the production process, automated field management, increase profit, protection of the environment, use of labor and resources, adapting to regulations, etc. For a successful business, farms must increase their arable land, reduce risk of crop failure, minimize operating costs, reach the highest price for their products, work on expanding their market, etc.

The precision farming market is expected to grow, at a CAGR (Compound annual growth rate), of 13.47% between 2016 and 2022 [20]. Some research shows that implementation of digital technology in agriculture can increase overall profitability by 55$ to 110$ per acre (1 acre = 0,40 ha) [21]. The global market for Precision Agriculture is expected to be around 3.7 billion euros in 2018 [22]. By using the modern applications and management tools, the farmer can easily obtain data on profitability and the impact that the application of modern solutions has on his business and the expected returns. This allows farmers to gain a higher level of competitive advantage and become more dominant in the market. It's estimated that over 50 percent of growers currently engage in some form of Precision Agriculture Technology [23]. Agriculture retailers have the opportunity to help their customers by providing them with local expertise on how the various technologies available today, might best be implemented within their geography and more specifically within their individual operation to help them improve their return on investment in the most appropriate ways [23].
The following table shown relationship between costs and benefits:

	Costs	Benefits
Background data	Time is required to collect data or pay a consultant for the service. Soil sampling generally costs $3–7/a, but can be considerably more for the more intensive collection. Other services from a crop consultant are generally based on an hourly charge.	Data may identify problems and opportunities the farmer was unaware of. This can be particularly true of low-yielding areas or environmentally sensitive areas where application of crop inputs may cause problems.
Recordkeeping	Software can range from very little for shareware or public domain packages to several thousand dollars for commercial mapping or geographic information system programs.	Organized data allow the farmer to make logical decisions.
Analysis and decision-making	Crop consultants generally charge $40–80/ha.	The analysis allows the farmer to make the right economic and environmental decisions.
Equipment	Auto-steer: $30,000–40,000 Light bar guidance: $3,000–5,000 Yield monitors: $4,000–7,000 Variable rate spray/fertilizer controller: $2,000–3,000	Crop inputs can be reduced and the environment protected.

Table 2. Relationship between costs and benefits, according to basic components of PA [11]

The cost types associated with the implementation of PA are [19]:

1) Information costs, related to the necessary investments in the technology, including rental fees for specific hardware or machinery;
2) Costs which involving data processing, specific license fees, software and hardware products for data analysis;
3) Learning costs, mainly due to the additional time required for the farmer to develop management schemes, calibration of the machinery, as well as 'lost' opportunity costs due to inefficient use of the PA technology

The potential benefits from PA can be [19]:

1) Crop yield improvement
2) Optimization of inputs
3) Improvement of management and quality of the work.

One of the most important factors which affects the profitability of a farm is its size. In Europe, there are 4,500 manufacturers with a mix of large multinational companies and numerous SMEs producing 450 different machine types with an annual turnover of €26 billion and employing 135,000 people directly and a further 125,000 in the distribution and service network.

The European Union continually develops support programs for the development of agriculture as the main pillar of its development. Support programs for development of PA and innovation in this area are contained in Regulation (EU) No 1305/2013 of the European Parliament and of the Council of 17 December 2013. Several of them are available for Member States to support PA development through their R&D programs [19]:

- Article 17 (Investments in physical assets)
- Article 28 (Agri-environment-climate)
- Article 35 (Co-operation)
- Article 14 (Knowledge transfer and information actions)
- Article 15 (Advisory services, farm management and farm relief services)

The European Innovation Partnership for Agricultural Productivity and Sustainability (EIP-AGRI) can play an important role in both developing and mainstreaming precision farming in the EU. This program includes: Article 55 (Aims), Article 56 (Operational groups), Article 57 (Tasks). The stated EU report confirmed that Precision Agriculture could play a substantial role in meeting the increasing demand for food while ensuring sustainable use of natural resources and the environment. The Common Agricultural Policy (CAP) is Europe's answer to the need for a decent standard of living for 22 million farmers and agricultural workers and a stable, varied and safe food supply for its 500 million citizens [24]. As a common policy for all 28 EU countries, the CAP strengthens the competitiveness and sustainability of EU agriculture by providing Direct Payments aimed at stabilizing farm revenues and finances projects responding to country-specific needs through national (or regional) Rural Development Programs, which also cover the wider rural economy [24]. The CAP also provides a range of market measures, including tools to address market difficulties, and other additional elements such as quality logos, promotion for EU farm products which complete CAP action to support farmers [24]. The CAP budget fixed for the period from 2014-2020 provides a total of EUR 408.31 billion, with EUR 308.73 billion intended for direct payments and market measures (the so-called First Pillar) and EUR 99.58 billion for Rural Development (the so-called Second Pillar) [23]. The European Commission promotes precision farming, mostly via its EUR 78 billion through Horizon 2020 program for research.

5. Conclusion

Innovation potentials in precision agriculture are enormous, therefore all countries in the world will have to work on strengthening them, in order to achieve greater competitive advantage in this area, as one of the main promoters of growth and development. If we take into account the rapid growth of the population on the planet and the increased need for food, it is clear that any investment in agricultural production will be very profitable but also necessary. Accelerated changes lead to an increase in the number of innovations in all areas of life, even in agriculture that we know today, as it becomes increasingly digital. It is very important to manage these innovations effectively in order to achieve appropriate positive effects in terms of social, economic and environmental aspects. Innovation usually relates to the development of new technologies and equipment, development of new seed types, decision support systems, automation of the business process, introduction of new methods, education of process participants, development of management and new business models, etc. A special subject of research can be to determine the relationship between investment and yielded benefits, based on detailed analysis and creation of cause-consequence models. The specific role in the development of Precision

Agriculture has innovation in Remote Sensing Technology. This technology provides a huge range of data, which can be used for improvement of agriculture business. These technologies are becoming more precise, and the quality of the data obtained is increasing. The growing use of these technologies (especially sensor systems with base stations and drones) in agriculture is noticeable in all developed market economies as well as in emerging countries. In further research, it will be necessary to determine the impact of the application of these technologies on the observed parameters in the multi-annual framework, taking into account the changing conditions. Our predictions are that the development of this area will be very dynamic, with almost automatic effects on the economy and society, at the level of one country, region or the whole world. The European Union is a good example of how much attention needs to be given to the development of agriculture. This process is followed by the number of regulations, development and support programs without which the entire system would not be functional and that requires the investment of enormous assets in order to achieve the set goals. All Member States have an important role in this project. It is necessary to establish a clear mechanism for overcoming the obstacles that accompany this complex process. It will be very interesting to see which further moves will be in terms of defining agriculture policy and their effects in the future. We expect to see in the future: the emergence of new and improved software and hardware systems to support all operations within precision agriculture, the emergence of new partnerships, the development of new agricultural crops, new harvesting approaches, etc.

6. Acknowledgments

The authors acknowledge the financial support of the Ministry of Education and Science of the Republic of Serbia, within the Project No. 47028.

7. References

[1] Popa, L. I.; Preda, G. & Boldea, M. (2010). A theoretical approach of the concept of innovation, Managerial Challenges of the Contemporary Society, Vol.1, pp. 151-156, ISSN: 2069-4229.

[2] Trott, P. (2008). Innovation management and new product development, Prentice Hall, ISBN: 9780273713159, London.

[3] Tidd, J. Bessant, J. (2016). Managing innovation: Integrating technological, market and organizational change, Wiley, ISBN: 978-1-118-36063-7, London.

[4] http://www.siemens.dz/en/about/innovation.htm, (2017). Accessed on: 2017-07-07.

[5] http://www.denverleadership.org/wp-content/uploads/2015/02/AR10_11.pdf, (2011). Accessed on: 2017-07-07.

[6] Fitzgerald, B. Wynn, E. (2004). IT innovation for adaptability and competitiveness, Kluwer Academic Publisher, ISBN: 978-1-4020-7999-3, Boston.

[7] OECD, Oslo manual. (1993). The Measurement of Scientific and Technical Activities, ISSN: 1990-0414, Available from: http://dx.doi.org/10.1787/19900414, Accessed on: 2017-07-07.

[8] Palcic, I.; Koren, R & Buchmeister, B. (2014). Technical Innovation Concepts is Slovenia Manufacturing Companies, 25th DAAAM International Symposium on Intelligent Manufacturing and Automation, DAAAM 2014, Procedia Engineering 100 (2015), Katalinic, B. (Ed), Published by DAAAM International, pp.141-149, Vienna, Austria.

[9] Mulla, J. D. (2013). Twenty five years of remote sensing in precision agriculture: Key advances and remaining knowledge gaps, Biosystems engineering, Vol. 114, Issue 4, pp. 358-371, ISSN: 1537- 5110.

[10] Whelan, B. Taylor, J. (2013). Precision Agriculture for Grain production systems, CSIRO PUBLISHING, ISBN: 978-0643107472, Collingwood VIC.

[11] USDA, NRCS, East National Support Center, Soil Quality National Technology Development team. (2007), Precision Agriculture: NRCS Support for Emerging Technologies, Agronomy Technical Note No.1, Available from https://www.nrcs.usda.gov/Internet/FSE_DOCUMENTS/stelprdb1043474.pdf, Accessed on: 2017-09-07.

[12] Es Van, H. Woodard, J. (2017). Innovation in Agriculture and Food Systems in the Digital Age, In: The Global Innovation Index 2017- Innovation Feeding the World, pp. 97-104. Cornell University, INSEAD, and the World Intellectual Property Organization (WIPO), ISBN: 979-10-95870-04-3.

[13] Brisco, B. Brown, R.J. Hirose, T. McNairn, H. & Staenz, K. (2014). Precision Agriculture and the Role of Remote Sensing: A Review, Canadian Journal of Remote Sensing, pp. 315-327 http://dx.doi.org/10.1080/07038992.1998.10855254

[14] Atzberger, C. (2013). Advances in Remote Sensing of Agriculture: Context Description, Existing Operational Monitoring Systems and Major Information Needs, Remote Sensing, pp. 949-981, ISSN 2072-4292.

[15] http://www.homburg-holland.com/nieuws/homburg-holland-on-tour-2016, (2016). Accessed on: 2017-09-27.

[16] http://panamaadvisoryinternationalgroup.com/blog/news-from-panama/drones-in-agriculture/, (2017). Accessed on: 2017-09-27.

[17] Tien, J. M. (2013). Big Data: Unleashing Information, Journal of Systems Science and Systems Engineering, Vol. 22, No.2, pp. 127–151, ISSN: 1004-3756.

[18] Roser, M. Ortiz-Ospina, E. (2017). World population growth, Our world in data, Available from: https://ourworldindata.org/world-population-growth/, Accessed on: 2017-09-27.

[19] European Parlament (2014). Precision agriculture: An opportunity for EU farmers – potential support with the CAP 2014-2020, Publication office, ISBN: 978-92-823-5575-6.

[20] http://www.marketsandmarkets.com/Market-Reports/precision-farming-market-1243.html, (2017). Accessed on: 2017-09-29.

[21] https://www.accenture.com/_acnmedia/Accenture/ConversionAssets/DotCom/Documents/Global/PDF/Digital_3/Accenture-Digital-Agriculture-Point-of-View.pdf, (2017). Accessed on: 2017-09-29.

[22] https://www.statista.com/statistics/720062/market-value-smart-agriculture-worldwide/, (2017). Accessed on: 2017-30-09.

[23] http://inthefurrow.com/benefits-precision-agriculture/ , (2017). Accessed on : 2017-09-30.

[24] https://ec.europa.eu/agriculture/sites/agriculture/files/cap-in-your-country/pdf/cz_en.pdf , (2016). Accessed on: 2017-09-30.

DOI: 10.2507/28th.daaam.proceedings.112

PREDICTIVE MANUFACTURING SYSTEMS IN INDUSTRY 4.0: TRENDS, BENEFITS AND CHALLENGES

Bojana Nikolic, Jelena Ignjatic, Nikola Suzic,
Branislav Stevanov, Aleksandar Rikalovic

This Publication has to be referred as: Nikolic, B[ojana]; Ignjatic, J[elena]; Suzic, N[ikola]; Stevanov, B[ranislav] & Rikalovic, A[leksandar] (2017). Predictive Manufacturing Systems in Industry 4.0: Trends, Benefits and Challenges, Proceedings of the 28th DAAAM International Symposium, pp.0796-0802, B. Katalinic (Ed.), Published by DAAAM International, ISBN 978-3-902734-11-2, ISSN 1726-9679, Vienna, Austria
DOI: 10.2507/28th.daaam.proceedings.112

Abstract

The fourth industrial revolution, known as Industry 4.0, has tendency to push the boundaries of science and technology. This is especially true for the manufacturing industry. One of the biggest challenges facing the manufacturing industry today is how to make intelligent systems for production with "self-aware", "self-predict and "self-maintain" abilities. Predictive manufacturing systems (PMS) are new intelligent systems that provide these abilities in the production, processes and machines. The PMS combines different technologies and techniques: statistics, data mining, modelling and artificial intelligence methods. These technologies and techniques are used to convert data into information and make predictions about the observed system. This paper provides an overview of the various challenges, existing solutions and benefits of PMS, with a focus on success factors in Industry 4.0.

Keywords: Big Data; Data mining; Industry 4.0; Predictive analytics; Predictive Manufacturing Systems (PMS)

1. Introduction

So far, we have witnessed the development of three industrial revolutions [1]. Constant evolution of industry as well as industrial processes, is now based on cyber-physical systems (CPS), and represents an integrated computational and a physical subsystem [2]. This and the Internet of Things (IoT) leads us to think about what would be needed for the future development of manufacturing systems. The fourth industrial revolution (i.e. Industry 4.0) came as a result of the use of intelligent systems and machines that are able to predict future events and states and to suggest the optimal solutions to possible future problems [1,2,3]. In manufacturing environment, these intelligent systems are named predictive manufacturing systems (PMS).

On the one hand, PMS rely on the development of computer science, information and communication technologies, statistics, modelling and simulation [4]. On the other hand, PMS depend on the development of manufacturing science and technology [4]. PMS is an insufficiently researched emerging field that deals with issues of uncertainty and invisibles related to productivity, efficiency, flexibility and safety [5,6]. The main goal of PMS is to create manufacturing environment that enables the implementation of all new technologies based on capabilities such as "self-awareness", "self-predicting", "self-maintaining" and "self-learning". According to Lee et al. [5] those capabilities allow that predictive manufacturing systems have a greater transparency in order to avoid the potential issues inside the system.

Transparency is the ability of a system to eliminate and quantify the uncertainties in order to determine an objective estimation of its manufacturing capability [6]. In order to achieve transparency, manufacturing enterprise has to utilize different advanced predictive analytics along with IoT, data mining, intelligent sensor networks and big data to convert a large amount of collected data into useful information and to store it in the cloud, where it can be available at anytime and anywhere. According to [5,6,7], this represents the key effort in order to discover the invisible uncertainties in the manufacturing systems.

The trends in predictive manufacturing systems which are related with development of new technologies may vary in different geographic areas. The results and benefits achieved by implementation of PMS, are already noticeable [7] and reflected in increasing increased efficiency, enabling the transparency of the entire manufacturing system, thus reducing the production time and costs. Also, there are challenges that engineers and computer scientists are faced with, such as a large amount of generated data and the proper way of understanding, using and storing it.

The objective of this paper is to review trends, benefits and challenges for predictive manufacturing systems in Industry 4.0 environment. This paper gives the preliminary literature review of conference papers and journal articles in order to focus on further research of PMS.

The paper is organised as follows: Section 2 provides the background for the Industry 4.0, and also presents its relation with PMS, and outlines the main technologies which are used for implementation of PMS (i.e. big data, statistics, data mining and intelligent systems); Section 3 outlines the current trends, long term benefits and disparate challenges in Industry 4.0 with focus on PMS; Section 4 presents conclusions and provides directions for the future research.

2. Background

The social, economic and technological changes are the cause of all of the industrial revolutions [7]. The rapid growth of technologies interconnected with electronics and internet of things enables the development of manufacturing which has led to the paradigm shift and has become a fourth industrial revolution [8]. Nowadays, Industry 4.0 became a global trend which started in Germany as a high-tech manufacturing concept [8,9].

The vision of Industry 4.0 is not limited to automation of a single manufacturing facility. Its goal is to incorporate vertical and horizontal integrations of all core functions, from manufacturing, procurement and warehousing, all the way to sales of the final product which leads to more rapid product development, customized production, improved management of manufacturing environments, faster supply chains, and so on. This high level of integration allows the high level of transparency across business processes, combined with new technologies which will enable a greater efficiency inside the responsive and predictive manufacturing systems that deal with various challenges including digitalization, delivery and clear decision recommendations for users [10,12].

The is implementation of cyber-physical systems combined with internet of things in Industry 4.0 which can provide predictive intelligent systems, digital representation and interoperable interfaces in order to support predictive manufacturing systems and networked production environments. The production becomes flexible and intelligent embedded devices begin to work together seamlessly [10].

In order to achieve intelligent and flexible systems, a large amount of data is required. In this large amount of data, the statistics and data mining play important role. These methods and technologies combined can transform data into useful information which can be further processed.

Figure 1 presents the PMS framework with technologies that support the development of the PMS and their interconnectedness. The big data represents the main data source without which it would not be possible to find the knowledge in the data. The process of finding the knowledge in the data, extracting it and transforming it into information is presented as knowledge discovery in databases (KDD) which uses all of the data mining techniques [11]. Other techniques that have the same goal as the data mining, but with different approaches, are statistics [12] and intelligent systems [22]. The statistical approaches are based on mathematical equations, while intelligent systems are part of artificial intelligence (AI), a subfield of computer science [2]. A separate technology that uses statistical and intelligent systems techniques in order to recognize the patterns in data regardless of its type is known as pattern recognition [30], where it could be recognized the patterns of number, picture or audio. The manufacturing systems are learning by using pattern recognition technique, where machines adopt present behaviour patterns, so that they know how to behave in the future situations. In this way, the manufacturing systems become intelligent systems with the features of prediction future events in the entire manufacturing environment.

In the rest of this section, the main technologies that enable implementation of predictive manufacturing systems in the Industry 4.0 are presented, namely: big data, statistics, and data mining and intelligent systems.

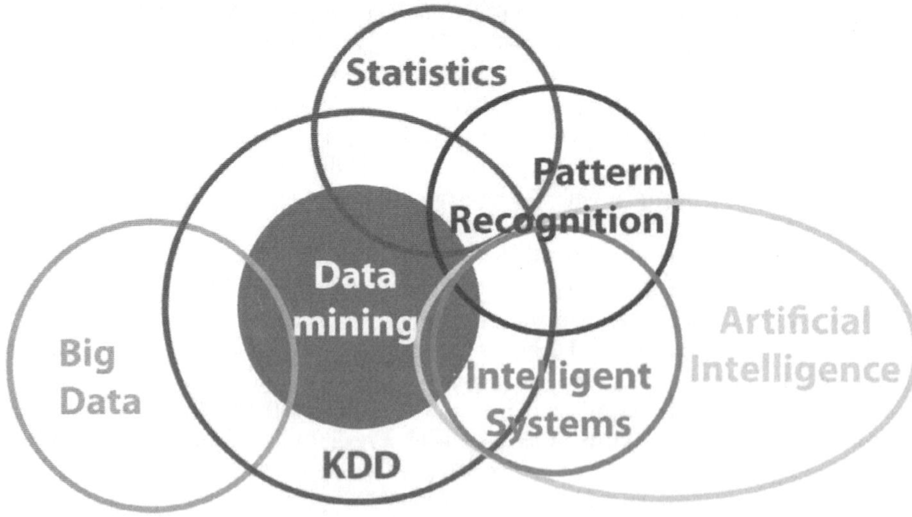

Fig. 1. The framework of predictive manufacturing systems (adopted from Srivastava, 2015 [34])

2.1. Big data

The Industry 4.0 includes different technologies that are joining together to discover a more effective way in order to improve manufacturing systems in every aspect of management [13]. With the implementation of new technologies in manufacturing environments, the amount of data approximately doubles every couple of years, which results in a large quantity of raw data [7]. Using the appropriate methods, algorithms and software tools, various types of data sets can be collected and extracted from the different layers in the manufacturing environment. Such extremely large and complex sets of data are named Big data [3,13,15].

Today, the big data cannot be processed by using existing technologies [15,16]. Traditional software application processing is not capable to preserve, acquire, manage and process large amount of upcoming signals [3,14]. Technological improvements in sensing and connectivity enable seamless aggregation of raw data from various production facilities [16]. According to [5,6,14,17] the actual processing of big data into meaningful information is performed by the prognostics and health management (PHM) predictive analytics which is chosen based on the following criteria: speed of computation, investment cost and ease of deployment.

The fact is that as more and more data will be generated by ICT equipment (intelligent devices, sensors and others from the environment). Big data has ability to provide various statistical samples and results to the users. Combining the big data technology with analytics, the result is big data analytics which is able to create one of the most profound trends in manufacturing and management [16]. The big data analytics have the power to dramatically modify the current manufacturing landscape, where it could be combined with manufacturing control through the IoT, which produces many opportunities in all areas of production [10].

One of the main challenges in Industry 4.0 is how to acquire huge amounts of data from the manufacturing environments due to different technologies, machines, sensors, CPS, IoT devices and communication networks. Another challenge is transformation of collected data into information [18]. Finally, after finding a proper way of collecting all required data and transforming it into information, a challenge is how to use Big data analytics in order to discover the invisible uncertainties in the PMS [5,17].

2.2. Statistics

A large number of new methods (e.g. advanced statistics, machine learning, computer science and data mining) aim at taking advantage of big data [12]. Statistics has always been the most used mathematical method, that presents the traditional data analysis for gathering, reviewing, analyzing and giving conclusions from data hidden in a batch of chaotic datasets with the main goal of maximizing data value [14,15,18].

According to Hoerl, Snee, and De Veaux [12], fundamentals of statistical way of thinking is crucial for the success of big data and data mining and it is based on three fundamental principles:

- All work occurs in a system of processes that are interconnected,
- Variation exists in all processes inside the system, and
- Understanding and reducing variation is key factor to success.

These principles have been known and applied for long period of time. Nowadays, it is common thing for statistics to be replaced with more innovative methods, but advanced statistical analysis still plays a huge guidance role in making development plans in manufacturing environments in Industry 4.0. Still, advanced statistical analysis represents the basis for new techniques such as big data and data mining in combination with classical statistical methods such as correlation, regression, factor analysis, clustering, classification, and so on [12,13].

Correlation analysis is a statistical method for determining the law of relations and mutual restriction among observed data and according to that making predictions and giving conclusions. Also, regression analysis represents a statistical tool for revealing correlations between one variable and several others and identifies dependence relationships among variables based on a group of experiments or observed data. Factor analysis is targeted at describing the relation among many elements with only a few factors where several closely related variables are grouped into a single factor, and the few of factors are used for revealing the most information of the original data. An unsupervised study method without training data represents the cluster analysis which is used for grouping, differentiating and classifying objects according to some particular features and divide them into clusters according to these features [14]. All of these classical statistical analysis methods have one thing in common, and that is the analysis of the big data.

Big data analytics can be inferred in manufacturing environments by means of advanced statistical analysis, where the inferential advanced statistical analysis is able to make conclusions from data subject to random variations [14]. Also, it is able to make predictions based on conclusions which represent a major step forward in the implementation of predictive manufacturing systems [15].

2.3. Data mining

All kinds of data can be collected from the environment of PMS and transformed into valuable information which can be meaningful and useful, but the question is what the appropriate data collection technique is considering the amount and size, but also the invisibleness of raw data. The most popular technique for collecting, searching, extracting and filtering data in a systematic way is data mining [18].

Data mining is defined as a method that uses a variety of the analytical and modelling techniques for excavation of raw, but sometimes hidden and unknown data [11,13]. Also, it parses large data sets in order to find unsuspected relationships and locate latent patterns that can make accurate predictions, ensuring that the result is potentially valuable and useful information and knowledge from massive, incomplete, noisy, fuzzy and random data [15,16,19].

Data mining is used in many different areas in manufacturing and plays an important role in predictive manufacturing, where its application is most reflected in decision making and making prediction for detecting defects inside the manufacturing environments by developing the suitable software. In order to improve the efficiency and quality of software development, data mining uses different statistical techniques, such as regression, clustering, classification, but also association mining to analyse and predict defects based on collected data [12,19,20].

However, there are many challenges which need the attention of researchers. They involve industrial data such as ability to handle different types of data which can be noisy and incomplete, representation of data mining requests and results, mining at different abstraction levels of data and different sources and protection of data security [21].

2.4. Intelligent Systems

With the constant expansion and development of new technologies as well as an enlargement of data, the complexity of problems is increasing in industry. With that issues in mind, the intelligent systems are being investigated with the aim of finding intelligent solutions for solving complex problems [22].

The intelligent systems (IS) present systems capable providing efficient decisions making in real applications. IS are based on two major approaches – artificial intelligence and soft computing which include a variety of different techniques, such as neural networks, expert systems, fuzzy logic and fuzzy systems, genetic algorithms and genetic programming, and others. All of this techniques operate together in order to improve strategic decision making and ensure flexible processing abilities of data and information for dealing with real world situations [21,22,23,24].

In comparison with the conventional systems, the intelligent systems often provide a competitive advantage [22]. The difference between intelligent systems and conventional systems is reflected in the range of techniques and number of applications, where IS emphasize the simulation of intelligent behaviours using computing devices or software systems, and conventional systems execute the operations that people have set with the lack of any signs of intelligent behaviour [21,25]. Another advantage of intelligent systems over conventional systems is that IS recognizes the available data which might be incomplete, uncertain or fuzzy, where they can still make a reasonable solutions, while conventional systems can manipulate only with data which are complete or exact [26]. Moreover, automatic decision support, artificial evolution, parallel execution, virtual integration, intelligent search and optimization present unique abilities of intelligent systems which are not available in conventional information system [23,24].

The intelligent systems represent the pillar in predictive manufacturing systems in Industry 4.0. The emphasis is placed on predicting the future state of the system to reduce the unnecessary issues in early stage and tackling that issue from problem solving to problem avoidance using the machine learning techniques. With the help of adequate machine learning technique, software and tools of the intelligent systems, the PMS is able to detect disturbance and predict the occurrence of abnormality and quality of the products in the manufacturing environments where it reduces the impact of uncertainties and prevent performance loss of manufacturing systems [5,26].

Nowadays, machine learning, as a part of IS, is broadly applied in various areas of manufacturing (e.g. optimization, control, quality and troubleshooting) where its techniques are designed to extract knowledge out of existing data. The new knowledge supports the process of decision-making or it is used to improve the system directly. The end goal of machine learning techniques is detection of the patterns or regularities that describe the relationships and structure between the elements [27].

Different researches have different approaches to structure the field of machine learning, but the structure that is most widely used is classified by learning processes is reinforcement, supervised and unsupervised learning [28]. The reinforcement learning refers to the kind of learning where less feedback is given, because knowledgeable expert has to evaluate which of the chosen action gives the best results [27,28]. The supervised learning is machine learning where the correct response is provided by the knowledgeable expert, while the unsupervised learning is intelligent learning where evaluation of the action is not provided, because there is no knowledgeable expert [26,27]. According to Wuest et al. [27], the supervised learning is the most commonly used, because it is the best fit for most manufacturing systems due to the fact that the majority of applications can provide labelled data.

3. Trends, benefits and challenges in predictive manufacturing systems

As a new concept, PMS brings many opportunities and improvements inside the manufacturing environments. The conducted review led to identification of trends, benefits and challenges of PMS systems which are discussed in the following subsections.

3.1. Trends

The trends of PMS vary in different geographic area, but according to Liao et al. [1] and Kang et al. [3] the leading countries that are paying the most attention to new technology are Germany, USA and South Korea which allocate great resources for developing new technologies. Germany focuses on creating new values that have not been seen before and resolving various social problems by linking the things inside and outside the manufacturing environments, but also on construction of a new business models for predictive manufacturing systems [29]. In the USA the research and development programs are related to predictive manufacturing with accent on key technology assignments, including data mining, big data analytics, intelligent systems along with IoT and CPS [3]. The trends in South Korean manufacturing are similar with the trends in Germany with small change on the main manufacturing strategy from the existing assembly and device-oriented production to the innovation of development and implementation a new predictive environments [3].

The current trends in using social media networks and their development have influenced the demands and perception of consumers on product's quality, variety, innovation and speed of delivery, which lead to shorter life-cycle of a product. Such demands established the predictive manufacturing approach where the manufacturing systems in combination with corresponding software could be changed in a short period of time, which results with increase of production capacity and rapid adaption of system functionality [5]. With these demands on mind, there is no space for failure of some part inside the machine, so trends of predictive manufacturing systems are going to the next level, where they are able to predict the unsuspected events. Prediction for supply chain is also the big trend because of the possibility of prediction the need and demands of the market.

Also, according to [5,6,15] there is a need for systematic integration, managing and analyzing machinery and raw data during different stages of machine life-cycle. In order to transform data into information more efficiently and to achieve better transparency of a machine's condition in the PMS, the prognostics and health management represent a critical research area that plays a huge role the development of advanced predictive tools.

3.2. Benefits

The technological benefits that are currently noticeable in manufacturing environments are the base of Industry 4.0. The entire present manufacturing systems is transformed into predictive manufacturing systems with the "self-aware", "self-predict", "self-maintain" and "self-learning" abilities which allow that decisions are going to be made autonomously [29]. According to Lee et al. [5] and Ferreiro et al. [30], in contrary to the conventional systems which are in use today, predictive manufacturing systems improve the performance of the equipment and strengthen the business model of industry, with the help of a set of intelligent sensing, predictive analytics and network techniques. Such an improvement of the performance leads to the prediction of the critical equipment failures, visibility of overall manufacturing process and the quality of a product, which will be guaranteed before it is produced [5].

The other advantage of the PMS is visibility of the entire manufacturing processes that leads to transparency which results the overall health condition of production systems [5]. Also, as Lee and Lapira [16] and Lee et al. [5] state that the increase of transparency in manufacturing environments has a number of benefits:

- **Cost reduction** – is reflected in just-in-time maintenance equipment, which maximizes the life and utilization of machine components.
- **Operational efficiency** – represents the knowledge about failure of the equipment that can be inferred by degradation patterns which allows better scheduling of maintenance and production. That leads to maximizing the equipment availability and uptime. Also, other associated advantages include reducing the mean time between failures and mean time to repair.

- **Product quality improvement** – implies the minimization of product quality deviations in production process control, where the aim is avoidance of unnecessary rework.

PMS also have effects in the economy through cost reductions and productivity increases with constant creation of new values [3].

No one mentioned the benefits concerning the increased flexibility of the PMS in the researched papers, which allows for faster adaptation to the required changes in the entire manufacturing environments, but only as a vision of Industry 4.0 in following papers [2,3,10,31].

3.3. Challenges

Despite the consistent development of the related technologies for the realization of PMS, there are still many issues to be considered. Transformation from present systems into more intelligent ones requires further research, development, and commercialization [3,13].

In the majority of the researched papers, the first mentioned and biggest issues are related to data. According to [13,17,29], the major challenge that has an effect in machine's abilities of sensoring, monitoring and gathering the big data from the manufacturing environments in order to achieve "self-aware", "self-predict", "self-maintain" and "self-learning" machines by adopting algorithms of the PHM. The development of the algorithms that are able to learn from and to handle different situations is still a great issue which can be considered as a general challenge not only for the PMS, but for entire Industry 4.0 [3,13]. Also, the collected set of data can contain an irrelevant and redundant information which can have an impact on the performance of the learning capabilities of PMS, and today's generation of PMS can only operate with continuous and nominal data values [27]. Another major concern is data security in PMS since the big data are stored in virtual cloud platforms [15]. Thus, additional researches are needed in order to improve the storage of big data.

As Liao et al. [1] and Wuest et al. [27] pointed out, the PMS represents a very complex system for managing that requires development of appropriate planning and explanatory algorithms and development of a single set of common standards which are still the big issues for the scientists. The planning and explanatory models will provide a basis for managing complex manufacturing systems, and standardisation will support a collaboration and technical description of standards. Also, standardisation can act together as a framework of standards for improving overall information exchange, sharing and integration, especially during the implementation process of Industry 4.0 [32,33].

The question about an uncertainty is still a challenge in PMS because of its invisibleness. This topic has been processed in the following papers [5,15,26].

None of papers were found dealing with the issue of finding human talents for development of complex algorithms for PMS as well as other related technologies, and only one paper [10] was found that is dealing with the problems of their training for usage of intelligent devices in predictive manufacturing systems.

4. Conclusion

This paper provides the overview of the main technologies such as big data, statistic, data mining and intelligent systems that are used for development and implementation of predictive manufacturing systems within the Industry 4.0.

The trends are related to the development of the algorithms with accent on the key technologies. Those algorithms enable devices to work seamlessly within the PMS that will have an impact on productivity, efficiency, flexibility and safety. Other trends are related to predictive manufacturing approaches where manufacturing systems in combination with corresponding software could be changed in a short period of time. The result is production capacity incensement and functionality can adapt more rapidly.

The most challenging task is gathering the big data from the manufacturing environments in order to have enough information for developing the algorithms that are able to transform the conventional systems into the intelligent predictive systems, as well as the security of the data. Standardisation, also, represents a big issue that need to be improved in order to achieve exchange, sharing and integration of the information. Without the support of standards, the process of implementing PMS can be more difficult, which has reflections in overwhelming waste of cost, manpower and material resources.

We also found that the numerous advantages are achieved by implementation of predictive manufacturing systems such as cost reduction, operational efficiency, product quality improvement and reduction of production time. The transparency of the entire manufacturing environment represents another benefit of PMS that results an overall health of the manufacturing system.

This paper is the initial part of the research, so in future its continuity will contribute to the further exploration and development of PMS and its implementation inside intelligent manufacturing environments with focus on Industry 4.0.

5. References

[1] Y. Liao, F. Deschamps, E. de F. R. Loures, and L. F. Ramos. (2016). "Past, Present and Future of Industry 4.0 - A Systematic Literature Review and Research Agenda Proposal," Int. J. Prod. Res., vol. 55, no. 12, p. 0.

[2] M. Dopico, A. Gomez, D. De la Fuente, N. Garcia, R. Rosillo, and J. Puche. (2016). "A vision of industry 4.0 from an artificial intelligence point of view," in Proceedings on the International Conference on Artificial Intelligence (ICAI), p. 407.

[3] H. S. Kang et al. (2016). "Smart manufacturing: Past research, present findings, and future directions," Int. J. Precis. Eng. Manuf. - Green Technol., vol. 3, no. 1, pp. 111–128.

[4] L. Monostori. (2014). "Cyber-physical production systems: Roots, expectations and R&D challenges," in Procedia CIRP, vol. 17, pp. 9–13.

[5] J. Lee, E. Lapira, S. Yang, and A. Kao. (2013). Predictive manufacturing system - Trends of next-generation production systems, vol. 46, no. 7. IFAC.

[6] J. Lee, E. Lapira, B. Bagheri, and H. an Kao. (2013). "Recent advances and trends in predictive manufacturing systems in big data environment," in Manufacturing Letters, vol. 1, no. 1, pp. 38–41.

[7] P. Tamas and B. Illes. (2017). "Process Improvement Trends for Manufacturing Systems in Industry 4.0," Acad. J. Manuf. Eng., vol. 14, no. February.

[8] J. Qin, Y. Liu, and R. Grosvenor. (2016). "A Categorical Framework of Manufacturing for Industry 4.0 and beyond," Procedia CIRP, vol. 52, pp. 173–178.

[9] S. Weyer, M. Schmitt, M. Ohmer, and D. Gorecky. (2015). "Towards industry 4.0 - Standardization as the crucial challenge for highly modular, multi-vendor production systems," IFAC-PapersOnLine, vol. 28, no. 3, pp. 579–584.

[10] International Electrotechnical Commission. (2015). "Factory of the future ®".

[11] S. S. Ghuman. (2014). "A review of data mining techniques," Ind. Manag. Data Syst., vol. 3, no. 4, pp. 1401 – 1406.

[12] R. W. Hoerl, R. D. Snee, and R. D. De Veaux. (2014). "Applying statistical thinking to 'Big Data' problems," Wiley Interdiscip. Rev. Comput. Stat., vol. 6, no. 4, pp. 222–232.

[13] J. Lee, H. A. Kao, and S. Yang. (2014). "Service innovation and smart analytics for Industry 4.0 and big data environment," in Procedia CIRP, vol. 16, pp. 3–8.

[14] M. Chen, S. Mao, and Y. Liu. (2014). "Big data: A survey," Mob. Networks Appl., vol. 19, no. 2, pp. 171–209.

[15] N. Khan et al. (2014). "Big Data: Survey, Technologies, Opportunities, and Challenges," Sci. World J., vol. 2014, pp. 1–18.

[16] J. Lee and E. Lapira. (2013). "Predictive Factories: The Next Transformation," Manuf. Leadersh. Journal. Frost Sullivan, pp. 2–9.

[17] J. Lee, E. Lapira, S. Yang, and A. Kao. (2013). "Predictive manufacturing system - Trends of next-generation production systems," in IFAC Proceedings Volumes (IFAC-PapersOnline), vol. 46, no. 7, pp. 150–156.

[18] M. Khan, X. Wu, X. Xu, and W. Dou. (2017). "Big Data Challenges and Opportunities in the Big Data Challenges and Opportunities in the Hype," in IEEE ICC 2017 SAC Symposium Big Data Networking Track Big, no. May.

[19] M. Chandamona and A. R. Pon Periasamy. (2016). "An Improved Analysis of Data Mining Techniques on Medical Data," Int. J. Mod. Educ. Comput. Sci., vol. 8, no. 2, pp. 42–48.

[20] A. R. Pon Periasamy and A. Mishbahulhuda. (2017). "Data Mining Techniques in Software Defect Prediction," Int. J. Adv. Res. Comput. Sci. Softw. Eng., vol. 7, no. 3, pp. 301–303.

[21] R. Ismail, Z. Othman, and A. A. Bakar. (2009). "Data Mining In Production Planning and Scheduling: A Review," in 2009 2nd Conference on Data Mining and Optimization, no. October, pp. 154–159.

[22] Z. Irani and M. M. Kamal. (2014). "Intelligent systems research in the construction industry," Expert Syst. Appl., vol. 41, no. 4 PART 1, pp. 934–950.

[23] K. A. Pupkov. (2017). "Intelligent Systems: Development and Issues," in Procedia Computer Science, vol. 103, no. October 2016, pp. 581–583.

[24] M. K. Bharatkumar. (2015). "Intelligent System ' s Design Approaches : A Review," Int. J. Eng. Manag. Res., vol. 5, no. 4, pp. 208–213.

[25] M. Gregor, M. Haluska, M. Fusko, and P. Grznar. (2016). "Model of Intelligent Maintenance Systems," in 26TH DAAAM Internacional Symposium on Intelligent Manufacturing and Automation, no. 2016, pp. 1097–1101.

[26] Z. Sun. (2016). "A Framework for Developing Management Intelligent Systems," Int. J. Syst. Serv. Eng., vol. 6, no. 1, pp. 37–53.

[27] T. Wuest, D. Weimer, C. Irgens, and K.-D. Thoben. (2016). "Machine learning in manufacturing: advantages, challenges, and applications," Prod. Manuf. Res., vol. 4, no. 1, pp. 23–45.

[28] L. Monostori. (2002). "Ai and Machine Learning Techniques for Managing Complexity, Changes and Uncertainties in Manufacturing," in IFAC Proceedings Volumes, vol. 35, no. 1, pp. 119–130.

[29] J. A. Saucedo-Martinez, M. Perez-Lara, J. A. Marmolejo-Saucedo, T. E. Salais-Fierro, and P. Vasant. (2017) "Industry 4.0 framework for management and operations: a review," J. Ambient Intell. Humaniz. Comput., pp. 1–13.

[30] S. Ferreiro, E. Konde, S. Fernández, and A. Prado. (2016). "Industry 4.0 : Predictive Intelligent Maintenance for Production Equipment," in European Conference of the Prognostics and Health Management Society, no. June 2016, pp. 1–8.

[31] H. Foidl and M. Felderer. (2016). "Research Challenges of Industry 4.0 for Quality Management," in Innovations in Enterprise Information Systems Management and Engineering, vol. 245, no. April, pp. 73–80.

[32] M. Freitag and M. Zelm. (2015). "Standardisation connecting the initiative 'industry 4.0' and service life cycle," CEUR Workshop Proc., vol. 1414, no. acatech 2013.

[33] L. Qing, J. Hongzhen, T. Qianlin, C. Yaotang, L. Jun, and Z. Jian. (2017). "Smart Manufacturing Standardization: Reference Model and Standards Framework," Springer Int. Publ., vol. 10034, pp. 16–25.

DOI: 10.2507/28th.daaam.proceedings.113

THE INFLUENCE OF CAM STRATEGIES ON THE TOOL WEAR AND SURFACE ROUGHNESS

Rudolf Zaujec, Tomáš Vopát, Vladimír Šimna & Peter Pokorný

This Publication has to be referred as: Zaujec, R[udolf]; Vopat, T[omas]; Simna, V[ladimir] & Pokorny, P[eter] (2017). The Influence of CAM Strategies on the Tool Wear and Surface Roughness, Proceedings of the 28th DAAAM International Symposium, pp.0803-0811, B. Katalinic (Ed.), Published by DAAAM International, ISBN 978-3-902734-11-2, ISSN 1726-9679, Vienna, Austria
DOI: 10.2507/28th.daaam.proceedings.113

Abstract

The article deals with the tool life of ball nose end mills and surface roughness of steel C45 depending on contouring and copy milling. The aim was to determine and compare the wear of ball nose end mill and surface roughness parameter for different types of ball end milling strategies, as well as to specify particular steps of the measurement process. For tool life test, DMG DMU 85 monoBLOCK 5-axis CNC milling machine was used. In the experiment, the cutting speeds, feed rates, axial and radial depth of cut were constant. The coated cemented carbide was used as tool material. The cutting tool wear was measured on Zoller Genius 3s and laser Blum Micro Compact NT. Surtronic 3+ (Taylor Hobson) was used for surface roughness measurement. The results show different achieved surface roughness parameter of machined material C45 and tool life of ball nose end mills depending on milling strategy.

Keywords: Contouring; Copy Milling; Wear; Machining; Tool

1. Introduction

Tools wear is generally considered as negative factor that accompanies each of machining process. This affects the cutting forces, cutting temperature and surface quality. The complete elimination of wear is not possible, but with a well-selected material of tool, coating and machining conditions it can be minimized. Taking into account these general conditions research was focused on the impact of CAM strategies for wear and tool life of ball nose end mills. Research will be focused on contouring and copy milling strategies. In this case climb milling during up and down-copying and up and down-contouring were used. Nowadays research of wear of milling tools isn't focused in one field. The position of the tool in relation to the machined surface (inclination angle) has a strong impact on the cutting forces [1-3]. In [4] is an influence of up and down-copying on tool wear and [5] on the surface roughness. Tian et al. [6] studied effect of cutting force to wear mechanisms for down and up copying. Influence of inclination angles of cutting tool on wear of cutter after machining process are in [7] and [8]. Impact of different types of hard machining materials on the tool wear concretely Ti-6Al-4V [9], Hastelloy C-22HS [10], 3Cr13Cu [11], compacted graphite iron and graphite iron [12]. Zetek et al. [13] observed flank wear and measured cutting forces during the face milling of Inconel 718. They optimized size of edge radius and increased tool life about 20%. Kasim et al. [14] found that notch wear was the predominant failure mode during end milling of Inconel 718. Begic-Hajdarevic et al. investigated the effect of cutting parameters on surface roughness in up- and down milling. In major part of these publications was flank wear, the main criterion for evaluation of experiments. In our experiment surface roughness and accuracy as a criterion were used.

2. Experiment

The experiment was focused on the basic principles of finish milling strategies (contouring, copy milling) with ball nose end mill see. Fig.1. Finish milling strategies during and down-copying and up and down-contouring were compared. The material was medium carbon steel ISO C45 (AISI 1045) grade. Block material had dimensions of 200×100×100 mm. The tested cutting tools material was coated cemented carbide. Diameter of tool was 8 mm. Model of workpiece was carried out in CAD system PowerShape 2017 and CAM strategies were created in PowerMILL 2017 software.

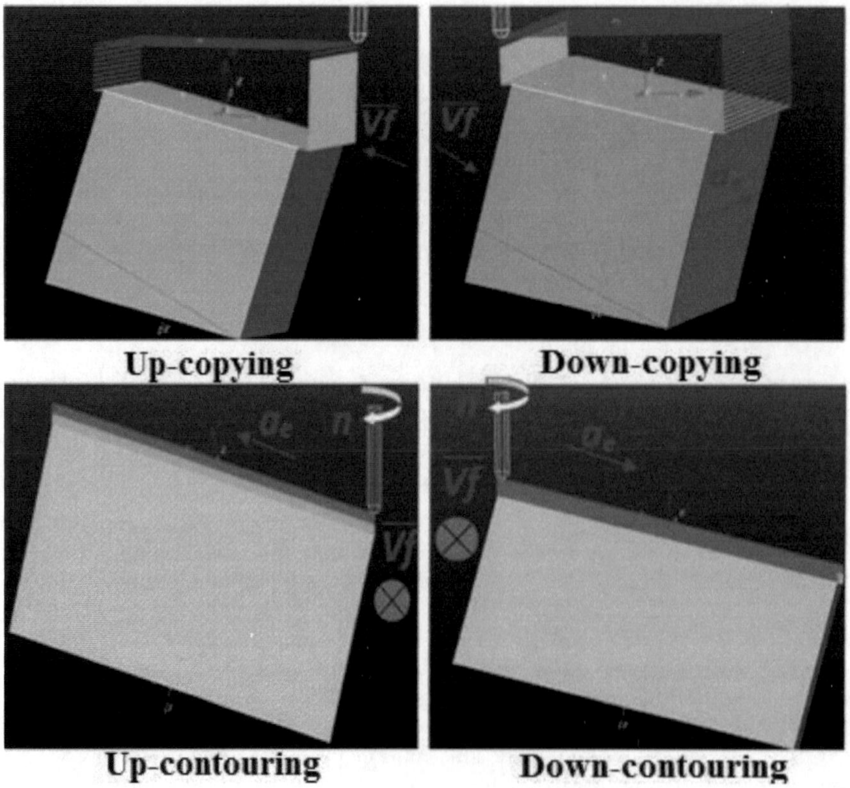

Fig. 1. Milling strategies used in experiment

For milling, DMG DMU 85 monoBLOCK 5-axis CNC milling machine was used. Air coolant in the machining process was used. For cutting tests, we set the same cutting conditions (cutting speed, feed rate, axial and radial depth of cut). Hence, we intended to investigate just the influence of cutter contact areas on the cutter plane on wear of ball nose end mill for the same cutting parameters. The inclination angle of the workpiece was 15°. Tonshoff et al. [15] found that the optimum inclined angle is 15° for ball end-milling of block materials.

Since the flank wear of ball nose end mill was dominant, it was measured on Zoller Genius 3s universal measuring machine. For nominal tool radius lost of Blum Micro Compat NT was used. Tool nominal radius lost (Nom-DR2) had influence on accuracy of machined surface. Blum measuring cycle 588 was used for measure 20 radiuses on the cutting edge from 0° (tool tip) to 45°. These settings were carried out because up to 45° tool wear did not occur. Surtronic 3+ was used for measurement of surface roughness.

For tool wear criteria Nom-DR2 = 0.025 mm and surface roughness parameter Ra = 1.6 μm were established. Flank wear VB = 0.15 mm was also evaluated in conjunction with Nom-DR2 and Ra.

All tool life tests were carried out with the following parameters: cutting speed v_c = 452 m/min, spindle speed n = 18000 rpm, feed rate v_f = 2160 mm/min, axial depth of cut a_p = 0.3 mm, radial depth of cut a_e = 0.1 mm, feed per tooth f_z = 0.06 mm. Effective cutting radius and effective cutting speed in up and down (copying, contouring) were different. In upward milling, tangential curve is placed on the one side from the axis of rotation of ball nose end mill. In downward milling, tangential curve is placed around the axis of rotation of ball nose end mill on the both sides, and crosses the centre of rotation of ball nose end mill. The scheme with symbols for copy milling is shown in Fig. 2.

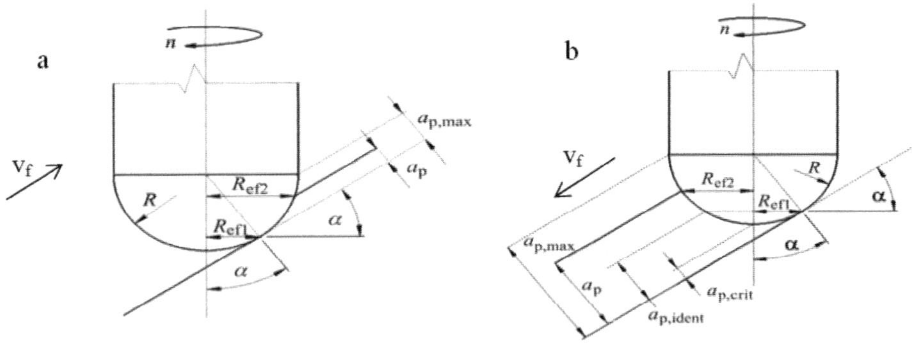

Fig. 2. (a) Scheme of up-copying; (b) scheme of down-copying [16].

The symbols in Fig. 2:

R - radius of the cutter (mm)
v_f - feed rate (mm/min)
α - slope angle of milling surface (°)
n - spindle speed (1/min)
R_{ef1} -effective radius of the cutter on machined surface (mm)

a_p -depth of cut (mm)
$a_{p,max}$ - maximum depth of cut (mm)
$a_{p,iden}$ - identical depth of cut (mm)
$a_{p,crit}$ - critical depth of cut (mm)
R_{ef2} - effective radius of the cutter on work surface (mm)

Determination of effective radius is very important for obtaining the effective cutting speed. Equations for effective radius and effective cutting speed have the following forms [17]:

$$R_{ef1} = R \cdot \sin\alpha$$
$$R_{ef1} = 4 \cdot \sin 15° = 1.035\, mm \quad (1)$$

$$v_{c1} = \frac{2\pi \cdot R_{efi} \cdot n}{1000} = 117.06\, m/\min \quad (2)$$

- The situations for up-contouring, copy milling:

$$R_{ef2} = R \times \sin\left(\alpha + \arccos\frac{R - a_p}{R}\right) = 2.4257mm \quad (3)$$

$$v_{c2} = \frac{2\pi \cdot R_{efi} \cdot n}{1000} = 274.34m/\min \quad (4)$$

- The situations for down-contouring, copy milling and provided that $a_p > a_{p,crit} \wedge a_p < R$:

$$R_{ef2} = R \times \sin\left(-\alpha + \arccos\frac{R - a_p}{R}\right) = 0.5104mm \quad (5)$$

From the equation for effective cutting speed results that the highest cutting speed is calculated for the larger of the effective radius R_{ef1} *and* R_{ef2}. Therefore, in up-contouring, copying (equation 3), $R_{ef2} > R_{ef1}$, the effective cutting speed was calculated for R_{ef2} (equation 4). Since in down-contouring, copying (equation 5), $R_{ef2} < R_{ef1}$, the effective cutting speed was calculated for R_{ef1} (equation 2).

3. Results

In the Fig.2 you can see time dependence of VB flank wear value (first evaluated criteria). Micrographs of flank wear of carbide ball nose end mills are shown in Fig.3. During contouring was flank wear bigger because constant (higher) cutting speeds were achieved. In down-contouring cutting tool had zero cutting speed (machining with of tool centre), but in down-copy milling not. In the copying milling deform layer of C45 could caused in a certain way lower cutting speed and removing of material layer with the tool back. Cutting tool in down-contouring and copying had milling with the tool centre according effective radius formulas in these cases. But in copy milling did not have. Theoretical calculations based on Blum measuring in our conditions were confirmed only for contouring.

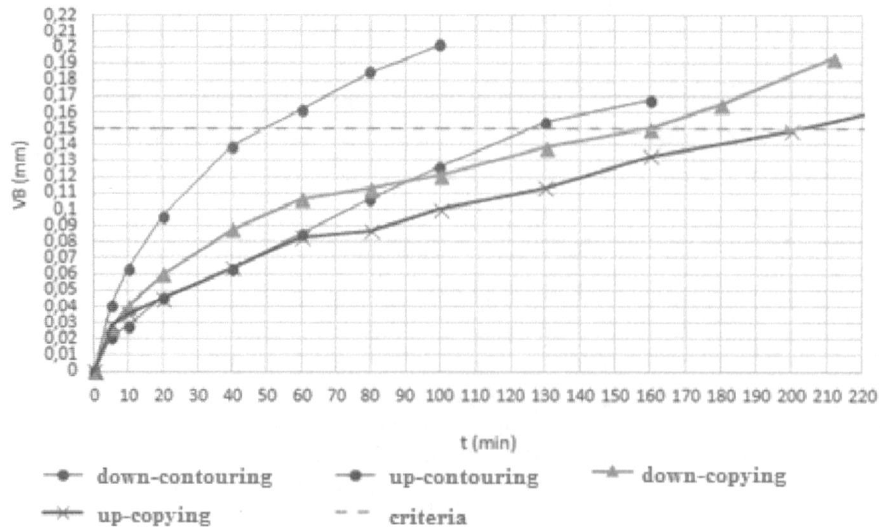

Fig. 3. Time dependence of average values of flank wear during climb milling

So if we had a set criterion VB = 0.15 mm, then copy milling strategies were the most appropriate. For the Ra criterion showed below, the most appropriate was different. In the figures you can see that in running-in stage of wear were little differences for all strategies.

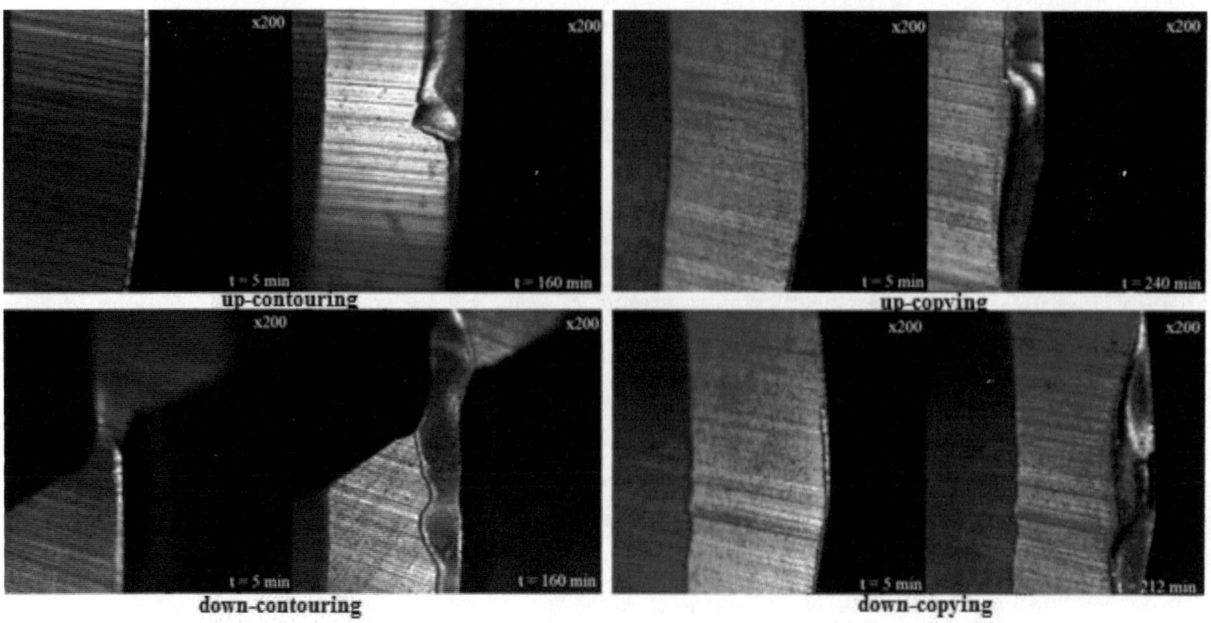

Fig. 4. Flank wear on tooth 1 measured in Zoller Genius 3

When the surface roughness parameter Ra was evaluated, we determine that impact of flank wear was less than influence of CAM strategy. Used milling strategy had significant affect of the Ra parameter of machined surface and also on the tool life. For example in up-contouring was Ra = 1.6 μm obtained for VB = 0,154 mm (VB = 0.162 mm for down-contouring), while in up-copying was VB = 0.064 and VB = 0.074 during down-copying. It follows that VB criteria must be different for copy milling and contouring. This fact was probably closely related to the constant Z axis value in contouring to the machining area. Positioning of the tool in Z axis during copy milling has a significant negative effect on the quality of the machined surface. It is closely related to the material removal stability, which was reflected by the worse Ra parameter see Fig. 5.

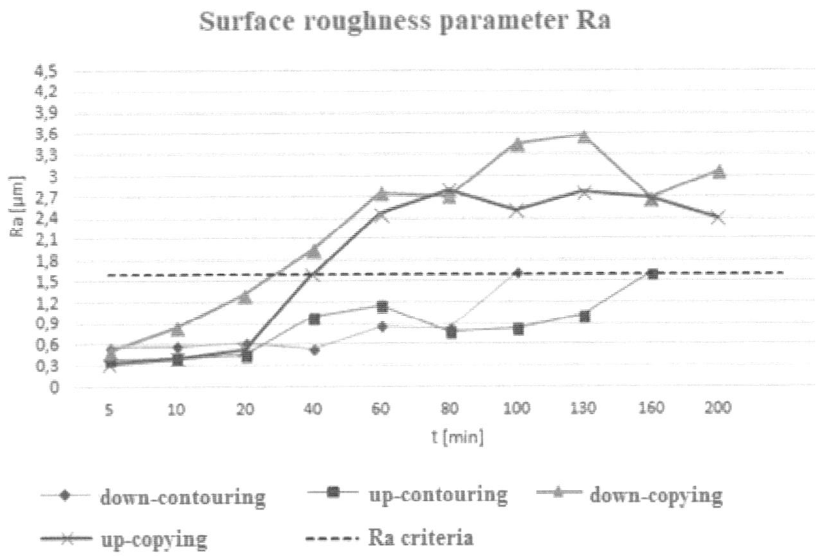

Fig. 5. Average values of Ra measured in cross direction on the tool paths

The last evaluated parameter obtained in experiment was nominal tool radius lost (Nom-DR2). Evaluating of this parameter in the research was focused on the common correlation between VB and Nom-DR2. The main reason for finding of correlation was aimed to the measuring of tool wear directly on the CNC machine. Second one was directional on the surface accuracy.

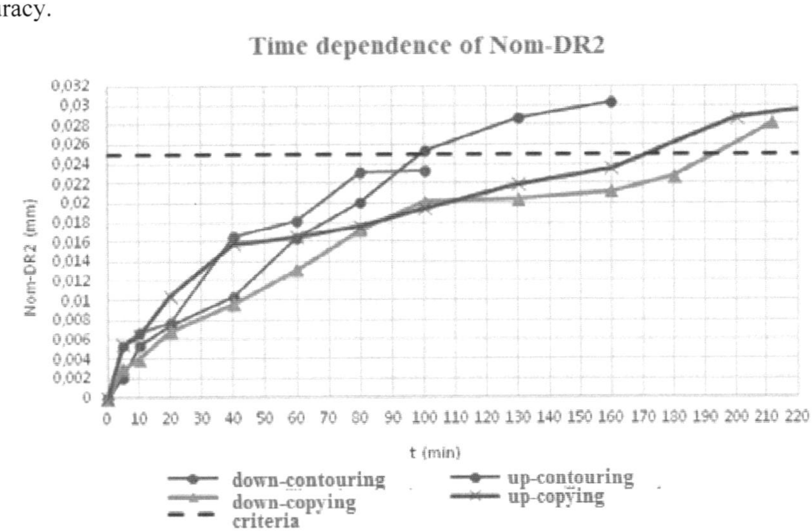

Fig. 6. Time dependence of Nom-DR2

In the Fig. 6 is showed time dependence of Nom-DR2. Results indicate that values from VB and Nom-DR2 were very similar. Little differences between results in copy milling strategies were obtained. During up-copying was the most suitable for VB criteria, but for Nom-DR2 was the best strategy of down-copying. We must realise that differences in copy milling strategies was abreast approximately 0,002 mm. Copy milling strategies deviation arose in running-in stage of wear. Reason of the less nominal tool radius lost in the copy milling strategies are in lower values of effective cutting speeds. Specific situation in down-contouring arose (Tab. 1 and Fig. 6). As has been said tool machining with the centre. This fact was reflected in the Nom-DR2 results, that values were in positively numbers. According Tab.1 down-contouring had the lower values to the VB. But In this specific case, the bigger nominal tool lost was in the tool length. In begin of experiment was tool L = 129.165 mm and in the end (100 min) L = 129.1275 mm (approximately deviation 0.04 mm). Positively numbers in Blum measuring device indicates to us machining with the tool centre, because ellipse shape is created. Blum device needs to measure of Nom-DR2 with the 588 cycle position of the tool tip and centre of ball nose shape, which represented nominal radius (4 mm in our case). Ratio of length to the radius can be reflected only in positively numbers – ellipse shape. Based on the results we can say that downward contouring had the worst influence on the surface accuracy. Influence of the Nom-DR2 to the surface accuracy is closer explained in the Fig. 7 and Tab. 2.

Copy milling					
down	**up**	**down**	**up**	**down**	**up**
VB (mm)	VB (mm)	Nom-DR2(mm)	Nom-DR2(mm)	coefficient	coefficient
0,0275	0,0285	-0,003	-0,0055	9,16667	5,18182
0,04	0,0355	-0,004	-0,0064	10	5,54688
0,0605	0,045	-0,0068	-0,0104	8,89706	4,32692
0,088	0,064	-0,0096	-0,0158	9,16667	4,05063
0,1065	0,083	-0,0131	-0,0165	8,12977	5,0303
0,113	0,087	-0,0172	-0,0176	6,56977	4,94318
0,121	0,1	-0,0201	-0,0194	6,0199	5,15464
0,1385	0,1135	-0,0204	-0,0219	6,78922	5,18265
0,1505	0,133	-0,0212	-0,0235	7,09906	5,65957
0,165	0,1485	-0,0227	-0,0287	7,26872	5,17422
0,194	0,1685	-0,0282	-0,0302	6,87943	5,57947
Average value				**7,81693**	**5,07548**
Contouring					
down	**up**	**down**	**up**	**down**	**up**
VB (mm)	VB (mm)	Nom-DR2(mm)	Nom-DR2(mm)	coefficient	coefficient
0,0405	0,0215	0,0054	-0,0021	7,5	10,2381
0,0635	0,0285	0,0067	-0,0054	6,04478	5,27778
0,096	0,046	0,0078	-0,0075	8,14103	6,13333
0,139	0,064	0,0166	-0,0104	5,78313	6,15385
0,162	0,085	0,0182	-0,0164	7,63736	5,18293
0,185	0,107	0,0232	-0,02	6,98276	5,35
0,2015	0,1265	0,0233	-0,0253	7,93991	5
	0,154		-0,02865		5,37522
	0,1675		-0,0303		5,52805
Average value				**7,1474**	**6,02658**

Table 1. Measured values of VB and Nom-DR2 with the calculated coefficients of their correlations

In Tab. 1 are values of Nom-DR2 and their coefficients of correlation with VB calculated. Is obvious that lower coefficient in up-copying and contouring were obtained. Lower coefficient means higher values of tool nominal radius lost to the VB. The highest effective cutting speed caused, that the biggest Nom-DR2 was in up-copying and contouring. In down-contouring was length coefficient 5,512. Downward copy milling was based on the accuracy criteria the most applicable strategy for finish milling. During the down-copying was layer of material distributed over the biggest part of cutting edge with the lowest values of effective cutting speed. In the Tab. 2 are added values from measuring of surface accuracy with the touch probe Heidenhain TS649. Measuring with the probe was carried out in end position of time cycle in four points along Y axis.

Accuracy for up-contouring							
Time[min]	**1**	**2**	**3**	**4**	**average value**	**Nom-DR2**	**Deviation**
5	-0,3408	-0,3418	-0,3427	-0,341	-0,34158	0,0021	-
10	-0,34	-0,3398	-0,336	-0,3356	-0,33785	0,0054	**0,0037**
20	-0,3361	-0,336	-0,336	-0,3356	-0,33593	0,0075	**0,00565**
40	-0,333	-0,3331	-0,333	-0,3322	-0,33283	0,0104	**0,00875**
60	-0,3272	-0,3281	-0,3277	-0,3282	-0,3278	0,0164	**0,01378**
80	-0,321	-0,322	-0,323	-0,322	-0,322	0,02	**0,01958**
100	-0,3169	-0,3169	-0,3199	-0,3189	-0,31815	0,0253	**0,02343**
130	-0,3137	-0,314	-0,316	-0,3151	-0,3147	0,02865	**0,02688**
160	-0,3108	-0,313	-0,3117	-0,312	-0,31188	0,0303	**0,0297**

Table 2. Measured values of surface accuracy with touch probe in CNC machine and Nom-DR2

$$k = \frac{VB}{NomDR2} \tag{6}$$

Dependence of Nom-DR2 to the final accuracy of machined surface according the measured values in the Tab. 2 and Fig.7 were obvious. It is very hard to say how exactly depends Nom-DR2 wear or another causes of inaccuracies to the final accuracy. In general, we would suggest taking into account the values from Blum device for accuracy criteria.
In Fig. 7 shows the measured deviation of surface and Nom-DR2. During the measurements we have not been made automatic correction of tool. If we use auto-correction, the value of the correction from Nom-DR2 = -0.02 is written to the tool table in positive values, e.g. DR2 = 0.02 mm. In principle, it is the same as without correction. Basic our experiments we recommend to done correction manually, e.g. Nom-DR2 = -0.02 correction will be DR2 = -0.02. Then the cutting tool should machine of surface accurately.

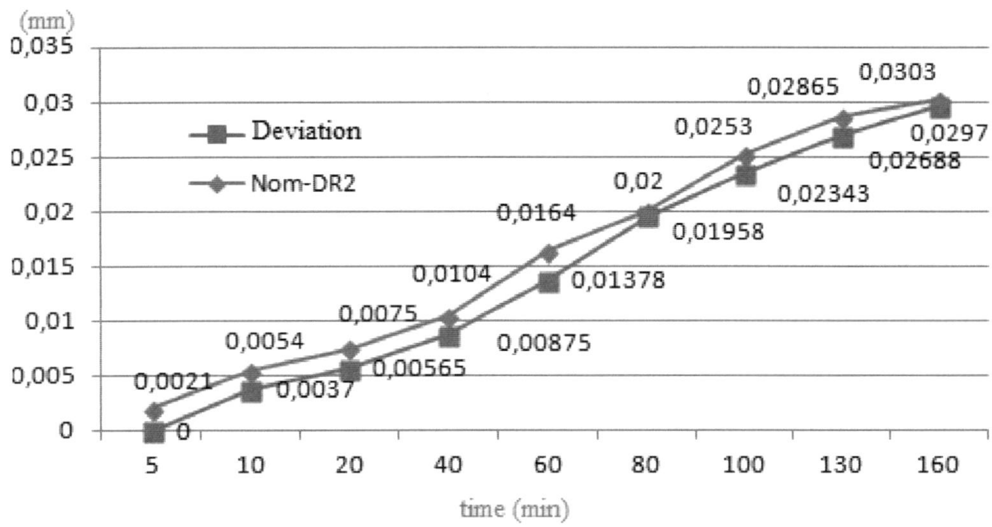

Fig. 7. Example of demand Nom-DR2 to surface accuracy during up-contouring

Based on experiments in terms of accuracy the copy milling strategies were the best. Mutual correlating between VB and Nom-DR2 was obtained. All copy milling strategies for Nom-DR2 criterion, which were related to accuracy, have been attained later. Opposite effect based on Ra criterion for contouring was reached. Smaller VB and Nom-DR2 values did not significantly affect the achieved surface roughness parameter Ra. The impact of VB on achieved surface roughness compared to use CAM strategy was secondary. This argument is also based on the fact that for contouring the values of VB were doubled and tool life for Ra was 2-3 times better see Tab. 3

1. Up-contouring		**2. Down-contouring**	
Criteria	**Time [min]**	**Criteria**	**Time [min]**
Ra = 1,6 [μm]	160	Ra = 1,6 [μm]	100
Nom-DR2 = 0,025 [mm]	100	Nom-DR2 = 0,025 [mm]	60
VB = 0,15 [mm]	130	VB = 0,15 [mm]	50
Tool life	**100 [min]**	**Tool life**	**60 [min]**
VB value to criteria [mm]	0,154	VB value to criteria [mm]	0,162
3. Up-copying		**4. Down-copying**	
Criteria	**Time [min]**	**Criteria**	**Time [min]**
Ra = 1,6 [μm]	40	Ra = 1,6 [μm]	30
Nom-DR2 = 0,025 [mm]	170	Nom-DR2 = 0,025 [mm]	195
VB = 0,15 [mm]	220	VB = 0,15 [mm]	160
Tool life	**40 [min]**	**Tool life**	**30 [min]**
VB value to criteria [mm]	0,064	VB value to criteria [mm]	0,083

Table 3. Established tool wear criteria for Ball nose end mill D8.

4. Conclusion

The article deals with the issue of CAM end milling strategies and their influence on the surface roughness parameter Ra and tool life of ball nose end mills. The aim was to investigate the wear of ball nose end mill for copy milling and contouring. Furthermore, surface roughness parameters Ra were determined and compared for each end milling strategy. For tool life test, DMG DMU 85 monoBLOCK 5-axis CNC milling machine was used. There were specified particular steps of the measurement process. In the experiment, the cutting speeds, feed rates, axial and radial depth of cut were constant. The coated cemented carbide was used as tool material. The cutting tool wear was measured on Zoller Genius 3s and laser Blum Micro Compact NT. Surtronic 3+ (Taylor Hobson) was used for surface roughness measurement.

The results show differences in use of CAM end milling strategies. Mutual correlating between VB and Nom-DR2 was obtained. All copy milling strategies for Nom-DR2 criterion, which were related to accuracy, have been attained later. Opposite effect based on Ra criterion for contouring was reached. Smaller VB and Nom-DR2 values did not significantly affect the achieved surface roughness parameter Ra. The impact of VB on achieved surface roughness compared to use CAM strategy was secondary. The CAM strategy has the most pronounced impact on tool life to the selected criteria (Contouring vs. Copy milling). Differences in CAM strategies are also reflected in different values of effective cutting speed. Another fact was arisen during small depths of cut of C45, because this kind of material in certain way was ductile and deformed. This was particularly evident in copy milling.

Based on experiment results, we can measure and quantify the amount of wear directly on the CNC machine. This is possible with a device that can measure the dimension of the shaped tool. For defined criteria is possible automatically control of cutting tools during or after machining of surface. For Blum device is possible use of R_BREAK or L_BREAK function. An example is possible to choose a value in the tool table R_BREAK = 0.025. Cutting tool will not machine after reaching this value. CNC machine can then automatically change the used tool for a new one. The second more complicated option is to change the tool tilting to the workpiece after tool wear, for example from 15° to 45°. With this setting, we would get a new unused cutting edge on the used tool. This would significantly increase the tool life.

Results from this research will be used for machining die and moulds when cutting tool microgeometry will be varied. Then, cutting tool microgeometry will be optimized to increase the tool life during the machining of Difficult-to-Machine materials. In the future work these strategies for conventional milling will be also investigated.

5. Acknowledgment

The article was written with the support of the project of VEGA grant agency of the Ministry of Education, Science, Research and Sport of the Slovak Republic and Slovak Academy of Sciences, no. 1/0097/17: "The research of novel method for cutting edge preparation to increase the tool performance in machining of difficult-to-machine materials", project and APVV Project of Slovak Research and development Agency of the Ministry of Education, Science, Research and Sport of the Slovak Republic, no. APVV-16-0057: "Research into the Unique Method for Treatment of Cutting Edge Microgeometry by Plasma Discharges in Electrolyte to Increase the Tool Life of Cutting Tools in Machining of Difficult-to-Machine Materials." The authors would like to thank for financial contribution from the STU Grant scheme for Support of Young Researchers project no. 1375 with acronym "SKOPF" and the project VEGA 1/0477/14. Research of influence of selected characteristics of machining process on achieved quality of machined surface and problem free assembly using high Technologies.

6. References

[1] M. Fontaine, A. Devillez, A. Moufki, D. Dudzinski. (2007). Modeling of cutting forces in ball end milling with tool surface inclination. Part II. Influence of cutting conditions, run-out, ploughing and inclination angle. J. Mater. Process. Technol., 189 (1–3), pp. 85–89.

[2] A. Daymi, M. Boujelbene, J.M. Linares, E. Bayraktar, A. Ben Amara. (2009). Influence of workpiece inclination angle on the surface roughness in ball end milling of the titanium alloy Ti–6Al–4V. J. Achiev. Mater. Manuf. Eng., 35 (1), pp. 79–86.

[3] C.E.H. Ventura, A. Hassui. (2013). Evaluation of static cutting forces and tool wear in HSM process applied to pocket milling, Int. J. Adv. Manuf. Tech. 65, pp. 1681-1689.

[4] VOPÁT, Tomáš - PETERKA, Jozef - ŠIMNA, Vladimír - KURUC, Marcel. (2015). The influence of different types of copy milling on the surface roughness and tool life of end mills. In Procedia Engineering. Vol. 100, pp. 868–876. ISSN 1877-7058.

[5] D. Begic-Hajdarevic, M. Kulenovic, A. Cekic. (2014). Experimental study on the high speed machining of hardened steel. Procedia Engineering, Vol. 69, pp. 291–295.

[6] T. Xianhua, Z. Jun. (2013). Effect of cutting speed on cutting forces and wear mechanisms in high-speed face milling of Inconel 718 with Sialon ceramic tools, Int. J. Adv. Manuf. Tech. 69, pp. 2669-2678

[7] C. Xiaoxiao, Z. Jun, H. Shiguo. (2013). Effects of inclination angles on geometrical features of machined surface in five-axis milling, Int. J. Adv. Manuf. Tech. 65, pp. 1721-1733,

[8] H. Schulz, S. Hock. (1995). High-speed milling of die and moulds – cutting conditions and technology. Annals of the CIRP 44, pp. 35-38

[9] Z. Song, L. Jian-feng. (2010). Tool wear criterion, tool life, and surface roughness during high speed steel end milling Ti-6Al-4V, J. Zhejiang Univ. Sci. A. 8, pp. 587-595

[10] K.A. Kadirgamaa, A. Abou-el-hosseinb. (2011). Tool life and wear mechanism when machining Hastelloy C-22HS, Elsevier Wear. 270, pp. 258–268

[11] Q. Xinyi, L. Pengnan. (2014). Study on serrated chip formation and tool wear of cerment tools for milling stainless steel 3Cr13Cu. Int. J. Adv. Manuf. Tech. 77, pp. 461-467

[12] M.B. da Silva, V.T.G. (2011). Naves, Analysis of wear of cemented carbide cutting tools during milling operation of gray iron and compacted graphite iron. Wear, 271, pp. 2426-2432

[13] M. Zetek. I. Cesakova, V. Svarc. (2014). Increasing Cutting Tool Life when Machining Inconel 718. Procedia Engineering, Vol. 69, pp. pp. 1171–1179.

[14] M.S. Kasima. C.H. Che Haronb, J.A. Ghanib et al. (2013). Wear mechanism and notch wear location prediction model in ball nose end milling of Inconel 718 Wear. Vol. 302, Issues 1–2

[15] H.K. Tonshoff, J. Hernandez-Camacho. (1989). Die manufacturing by 5 and 3 axes milling, Journal of Mechanical Working Technology 20, pp. 105-119.

[16] C.C. Tai, K.H. Fhu. (1994). A predictive force model in ball-end milling including eccentricity effects, Int. J. Mach. Tools Manufact. 34, pp. 959–979

[17] J. Peterka. (1997). Analysis of the geometry and kinematics of copy milling, in Vedecké práce (Science Theses) MTF STU in Trnava, 5, pp. 53- 58.

DOI: 10.2507/28th.daaam.proceedings.114

RESEARCH OF ABS AND PLA MATERIALS IN THE PROCESS OF FUSED DEPOSITION MODELING METHOD

Ján Milde, Róbert Hrušecký, Rudolf Zaujec, Ladislav Morovič & Augustín Görög

This Publication has to be referred as: Milde, J[an]; Hrusecky, R[obert]; Zaujec, R[udolf]; Morovic, L[adislav] & Gorog, A[ugustin] (2017). Research of ABS and PLA Materials in the Process of Fused Deposition Modeling Method, Proceedings of the 28th DAAAM International Symposium, pp.0812-0820, B. Katalinic (Ed.), Published by DAAAM International, ISBN 978-3-902734-11-2, ISSN 1726-9679, Vienna, Austria
DOI: 10.2507/28th.daaam.proceedings.114

Abstract

This article focused on comparing of the properties of the two most used materials, Acrylonitrile-Butadiene-Styrene (ABS) and Polylactic Acid (PLA), in the Rapid Prototyping (RP) process using the Fused Deposition Modeling (FDM) method. The thin-walled components were manufactured by Zortrax M200 and DeeGreen 3D printer. Print settings and printer parameters are also mentioned in the article. The printed models were 3D digitized using the GOM ATOS Triple Scan II optical 3D scanner. Main section is focused on the analysis of individual components in the GOM Inspect software. Authors used this software to measure the individual distances. The last section of the paper focuses especially on the evaluation of obtained results in the form of tables and a chart.

Keywords: Thin-wall; FDM; ABS; PLA; 3D scanning; inspection

1. Introduction

While traditionally plastic material ABS has been used in FDM for industrial applications, desktop printers started with ABS but gradually shifted to PLA due to its green reputation, bio-compatibility, bio-degradability, low shrinkage and good printability. PLA based materials used in FDM printers don't have the best properties (mechanical, material). Many print defect such as curvature (especially at the corners) and warpage of printed parts are commonly observed, which become more visible as the size of printed parts increase. PLA material also suffer from weak temperature resistance, which may influence in the deformation of printed objects under elevated temperatures experienced during storage and shipping or even during usage, e.g. when the objects are placed under direct sun [1,2].

In this study we established a research which compares the dimensional accuracy of the materials in consideration and performed an experiment to establish the superiority of a material in terms of accuracy over each other. The overall inaccuracy of the printed parts being built by FDM has been one of the major challenges that need to be overcome. Errors due to warpage and shrinking dominate to the inaccuracy of the part [3].

2. Statement of the problem

The purpose of this research is comparing of dimensional accuracy of thin-walled 3D printed parts PLA and ABS when compared with dimensions of the CAD models from which these parts are printed. The ABS parts were printed using the 3D printer Zortrax M200 built by Zortrax. The PLA parts were printed using the 3D printer DeeGreen built by Be3D [4, 5].

The goal of this research is to evaluate the limitations of the printing material, to rule out factors that do not contribute significantly to print accuracy, and to provide a practical, quantitative guide for accuracy measurement as an engineering tool. A methodology or a practical approach for calculating accuracy of a material will be developed which can be used in future to test accuracy. It will establish which material is more accurate and what is the reason behind it.
Part, which was fabricated and evaluated, was a thin-walled extruder cover for a 3D printer.

3. ABS and PLA materials

Acrylonitrile-Butadiene-Styrene (ABS) is a terpolymer made by polymerizing styrene and acrylonitrile in the presence of polybutadiene. The proportions can vary from 15 to 35 % acrylonitrile, 5 to 30 % butadiene and 40 to 60 % styrene. ABS is an engineering plastic that has butadiene part common distributed over the acrylonitrile-styrene matrix. It possesses excellent toughness, good dimensional and geometrical stability, easy processing ability, chemical resistance, and cheapness. Typically, ABS is a product of systematic polymerization of monomers, namely, acrylonitrile, butadiene, and styrene as shown in Fig. 1. [6]. Main properties of ABS plastic are following: density (ρ) 0,9$\frac{g}{cm^3}$ - 1.53$\frac{g}{cm^3}$, glass transformation ~105 °C. ABS is amorphous and therefore has no true melting point, however 230 °C is the standard for printing.

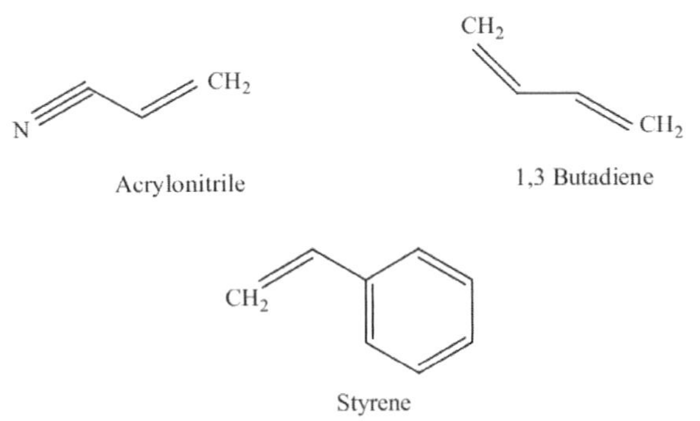

Fig. 1. Monomer units of ABS [7]

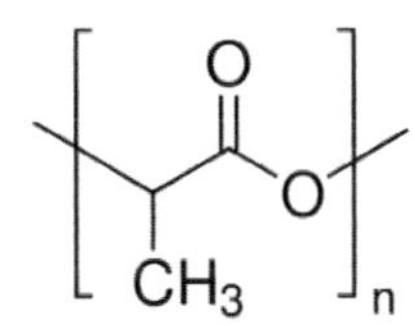

Fig. 2. Scheme of PLA [8]

Polylactic acid or polylactide (PLA) is a biodegradable and bioactive thermoplastic aliphatic polyester derived from renewable resources, such as corn starch, cassava roots, chips or starch, or sugarcane. PLA is very brittle, with less than 10 % elongation at break, thus it is not suitable for demanding mechanical performance applications unless it is suitably modified. This material is also strongly hydrophobic and can elicit an inflammatory response from the tissues of living hosts, because of its low affinity with cells when it is used as a tissue engineering material. Another limitation of PLA material towards its wider industrial application which is limited by gas barrier properties which prevent its complete access to industrial sectors such as packaging [8]. Main properties of PLA plastic are following: density (ρ): 1.210$\frac{g}{cm^3}$ - 1.430$\frac{g}{cm^3}$, melting point: from 150 to 160 °C.

Authors Olivera S. et al. [7], focused on the properties of ABS, processes and mechanisms of plating, and studies of plating on ABS involving mainly eco-friendly methods of plating by discussing the literature published in recent years. The details of electroplating of ABS carried out in the authors' laboratory are also presented.

Authors Dunne R. et al. [9] aims on the investigates the mechanical characterization of various natural fiber composites with an ABS. The tensile strength, impact strength and air flow resistivity of samples with different densities have been measured, and have been found.

Authors Xiao L. et al. [10] focused on the properties of PLA material, biodegradability and industrial application.

Mudassir A. [3] focused on the compare the dimensional accuracy of 3D printed parts fabricated from PLA and ABS when compared with dimensions of the CAD models. The goal of this research was evaluate the limitations of the printing material, to rule out factors that do not contribute significantly to print accuracy, and to provide a practical, quantitative guide for accuracy measurement as an engineering tool. A methodology or a practical approach for calculating accuracy of a material will be developed which can be used in future to test accuracy. It will establish which material is more accurate and what is the reason behind it.

4. Thin-walled 3D model used for the experiment

The model used in this experiment is thin-walled cover of extruder of 3D printer DeeGreen. This model was chosen due to its importance for manufacturing of thin-walled parts in the process of FDM. This kind of models are used in also in architecture, automotive industry, medicine, and electrotechnical industry. Buranský [12] defined thin-walled part as a part that is built by walls that are very thin. According to Trebuňa, a thin-walled profile is a profile where the wall thickness t_i is low when considering its other dimensions. Its deformations are slight when considering the thickness of walls b_i and that the rigidity of laying is infinitely large [13]. Aijun defined a thin-walled part as a part, where the thickness h is lower than height b, i.e. $(1/80 \sim 1/100)\ b < < h << (1/8 \sim 1/5)\ b$, (where h is the wall thickness (mm), and b is the wall height (mm)) [14].

Thickness of wall of this part is 3.00 mm. This thin-walled model (Fig. 3) was created in Autodesk Inventor Professional 2017 software [4].

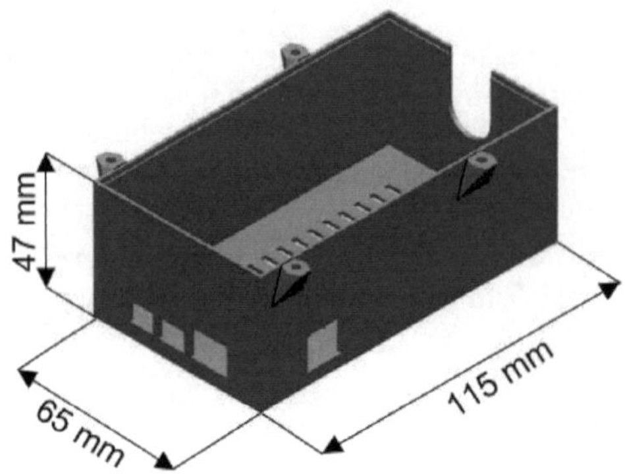

Fig. 3. Thin-walled part used in the experiment

5. Fused Deposition Modeling method

Fused Deposition Modeling (FDM) is a process in which a part is produced using layer-by-layer deposition of material. It is an important technology as it has potential to reduce the manufacturing lead time of the product up to 30–50 % even the relative part complexity is very high. Rapid Prototyping (RP) cycle consists of creation of geometric model using a solid CAD modeler (SolidWorks, PowerSHAPE, CATIA, etc.) or 3D digitization of existing part, determination of suitable deposition orientation, slicing, generation of material deposition paths, layer thickness, internal structure and then post processing operations. Layer thickness in FDM process has significant effect on many key characteristics that determines the final quality and cost of the prototype [11].

In the FDM method, used material is in a plastic wire form, unwound from a coil. This material fiber is fed to the printing head where it is heated, fused and extruded through the nozzle in the form of fine fibers. There are several kinds of kinematic structures used in Fused Deposition Modeling.

The movements of the head and 3D printer can be divided into three most used groups:

- head is moved in axis X, Y, Z,
- head is moved in axis X, Y and build platform is moved in Z axis,
- head is moved in axis X, Z and build platform is moved in Y axis [11].

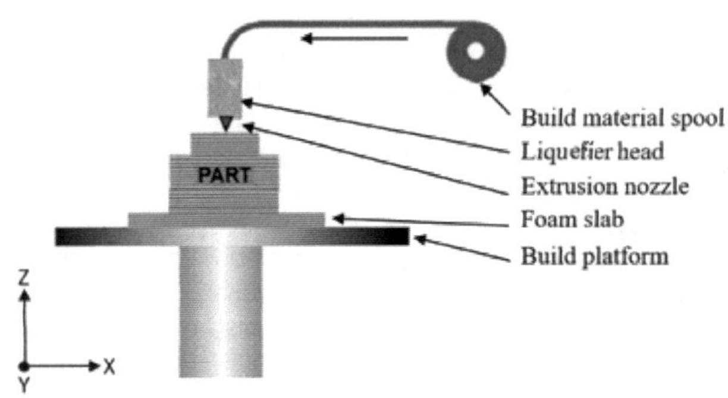

Fig. 4. Scheme of Fused Deposition Modeling [11]

6. Parameters settings of additive manufacturing

In this experiment were used two FDM printer in total. The first one, Zortrax M200 3D printer was used for manufacturing of thin-walled part from ABS material. Second one, DeeGreen 3D printer was used for manufacturing of thin-walled part from PLA material.

Manufacturing parameters used for ABS and PLA parts are in the Tab. 1.

Manufacturing parameters	ABS part	PLA part
Layer thickness [mm]	0.10	0.10
Orientation [°]	0	0
Infill [- , %]	Medium	50
Support [%]	30	30
Quality	High	High

Table 1. Parameters used for manufacturing of thin-walled parts

7. Process of measuring

For part measuring, GOM ATOS II Triple Scan optical 3D scanner was used (Fig. 5) with measuring volume MV 170 (170 x 130 x 130 mm).

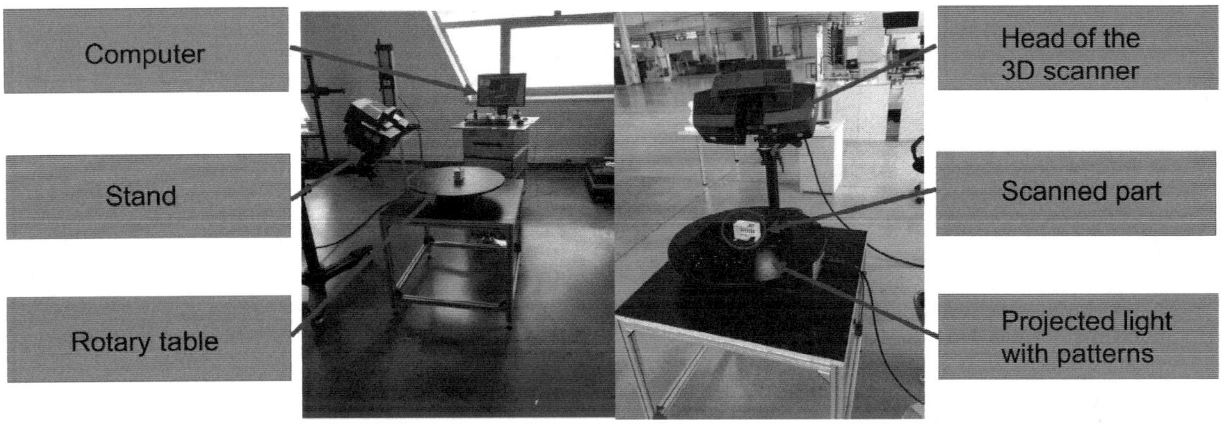

Fig. 5. GOM ATOS II Triple Scan optical 3D scanner

Before scanning, it is necessary prepare the parts for digitization (sticking the uncoded reference points to the parts 1.5 mm in diameter).

After 3D scanning, component can be shown in the software. Was found, that the scanner also scanned the unwanted parts. These object were removed in software. After scanning all model, it is necessary to make a comparison of parts. Parts will compare and analyze in the software GOM Inspect V8 with nominal model.

If model is imported, next step is the following: *Operations - Alignment - Initial Alignment - Prealignment* and both models will be set up. Then, go to the *Operations - Alignment - Main Alignment - Local Best-Fit* menu, mark the two models and set up. Colour deviation map was created, which serves to graphically display deviations [4].

In the second step, sections on the part were created. The sections were performed on all three axes X, Y and Z. Sections in the X-axis, were performed at a distance of 67.4 mm, 110 mm and 139.4 mm. Sections in the Y-axis, were performed at a distance of 23 mm and 73 mm. Sections in the Z-axis were conducted of 10 mm, 25 mm and 40 mm. Numerical values of deviations have been displayed for the sections thus created.

In next steps, inner and outer distances were measured in the Y-axis and X-axis for both thin-walled components.

For a measurement of outer distances, function *Outer Disc Caliper* was used. For a measurement of inner distances, function *Inner Disc Caliper* was used. This functions can found in GOM Inspect software in the menu *Contstruct – Distance.* Then, all distances on nominal and actual models were created. After creation of all measured distances, we connected this distances on nominal model with distances on actual model with the function *Link to Actual Element.* Measured distances was shown using the function *Distance,* where are also shown deviations. For better orientation in the software GOM Inspect, for a single distances we created tags. The tags was created with the marking of single elements in the menu *Edit – New Tag Scene* [4].

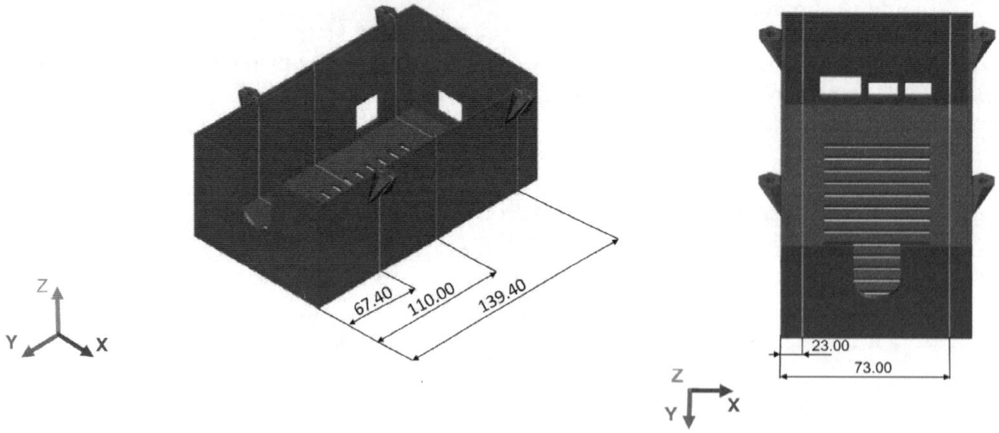

a) b)

Fig. 6. Sections on thin-walled component: a) section in Y-axis, b) section in X-axis

8. Analysis of results

The thin waled-part was printed six times in total. Two different printers were used for manufacturing of the parts. The model from ABS material was printed three times and model from PLA material was printed three times too. After digitizing printed models using an optical 3D scanner, the distances between two point on outer and inner surfaces was measured. The analysis of the measured values is summarized in the chart (Fig. 10 according to measured values listed in the tables 2 – 9. The measuring of distances was done in the software GOM Inspect V8.

Measurements are divided into:

- measurement of Y-axis distance Y,
- measurement of outer surfaces in X, Y axis,
- measurement of inner surfaces in the X, Y axis.

Measurement of the component was according to a certain principle. Measurement points were chosen at intersections of the cuts. Distance was measured on the Y-axis so that the distance created at the distance X +67.4 mm and points where they cross the cut at Z +10, 25, 40 mm, were created. After measuring these points, arithmetic mean was created of the entire section and entered the measured values table. This is principle, how to measure the other distances in the respective sections and their intersections.

After calculating of the arithmetic mean of distance, values were imported into one common chart (Fig. 10). Measurement of Y-axis is shown on Fig. 7

Measurement of outer distances in Y-axis – ABS material						
Section X [mm]	**Nominal value [mm]**	**Print 1 [mm]**	**Print 2 [mm]**	**Print 3 [mm]**	**Arithmetic means [mm]**	**Deviation [mm]**
[67.40]	65.00	64.43	64.52	64.71	64.55	- 0.44
[110.00]	65.00	64.73	64.66	64.92	64.77	- 0.23
[139.40]	65.00	64.67	64.59	64.64	64.63	- 0.36
Arithmetic mean of ABS prints in Y-section [mm]						**- 0.34**

Table 2. Measurement of outer distances in Y-axis – ABS material

Measurement of outer distances in Y-axis – PLA material						
Section X [mm]	**Nominal value [mm]**	**Print 1 [mm]**	**Print 2 [mm]**	**Print 3 [mm]**	**Arithmetic means [mm]**	**Deviation [mm]**
[67.40]	65.00	64.66	64.68	64.80	64.71	- 0.28
[110.00]	65.00	64.83	64.82	64.85	64.83	- 0.16
[139.40]	65.00	64.76	64.61	64.89	64.75	- 0.24
Arithmetic mean of PLA prints in Y-section [mm]						**- 0.23**

Table 3. Measurement of outer distances in Y-axis – PLA material

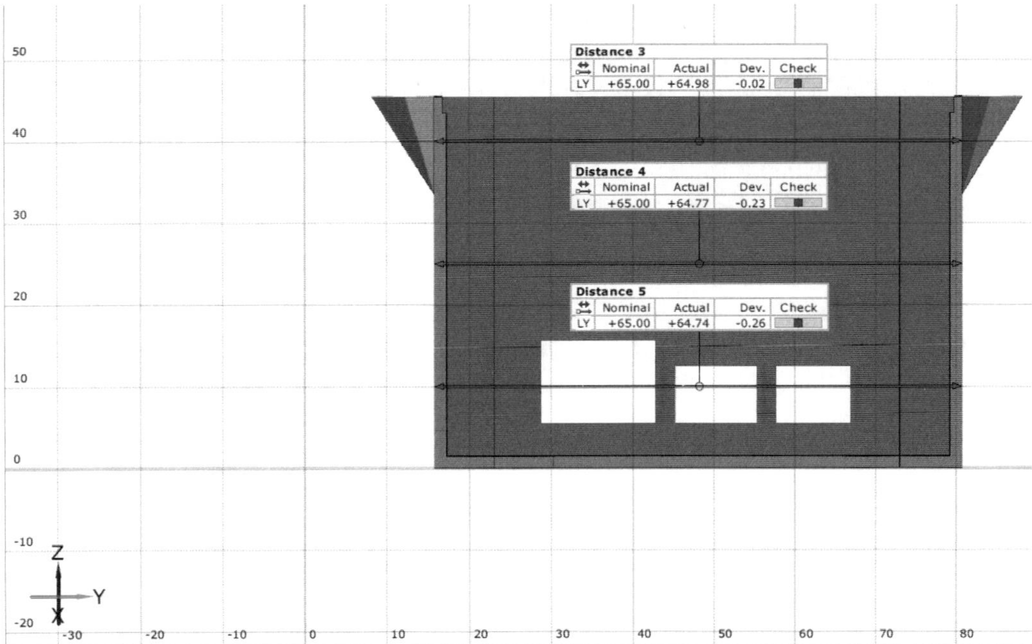

Fig. 7. Cross-section view - measurement of outer Y-axis distance Y – PLA material

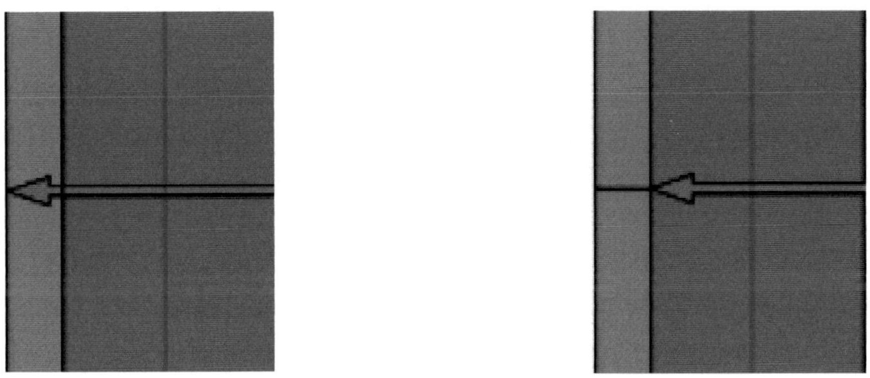

a) b)

Fig. 8. Difference between outer and inner distances: a) outer distance, b) inner distance

Measurement of inner distances in X-axis – ABS material						
Section Y [mm]	**Nominal value [mm]**	**Print 1 [mm]**	**Print 2 [mm]**	**Print 3 [mm]**	**Arithmetic means [mm]**	**Deviation [mm]**
[23.00]	115.00	114.31	114.37	114.28	114.32	- 0.68
[73.00]	115.00	114.29	114.31	114.26	114.28	- 0.71
Arithmetic mean of ABS prints in X-section [mm]						**- 0.69**

Table 4. Measurement of outer distances in X axis – ABS material

Measurement of inner distances in X-axis – PLA material						
Section Y [mm]	**Nominal value [mm]**	**Print 1 [mm]**	**Print 2 [mm]**	**Print 3 [mm]**	**Arithmetic means [mm]**	**Deviation [mm]**
[23.00]	115.00	114.78	114.63	114.43	114.61	- 0.38
[73.00]	115.00	114.78	114.67	114.39	114.61	- 0.38
Arithmetic mean of PLA prints in X-section [mm]						**- 0.38**

Table 5. Measurement of outer distances in X axis – ABS material

Measurement of inner distances in Y-axis – ABS material						
Section X [mm]	**Nominal value [mm]**	**Print 1 [mm]**	**Print 2 [mm]**	**Print 3 [mm]**	**Arithmetic means [mm]**	**Deviation [mm]**
[67.40]	62.00	61.76	61.92	61.77	61.81	- 0.18
[110.00]	62.00	61.95	61.88	61.82	61.88	- 0.11
[139.40]	62.00	61.92	61.93	61.93	61.84	- 0.15
Arithmetic mean of ABS prints in Y-section [mm]						**- 0.15**

Table 6. Measurement of inner distances in Y axis – ABS material

Measurement of inner distances in Y-axis – PLA material						
Section X [mm]	**Nominal value [mm]**	**Print 1 [mm]**	**Print 2 [mm]**	**Print 3 [mm]**	**Arithmetic means [mm]**	**Deviation [mm]**
[67.40]	62.00	61.77	61.65	61.68	61.70	- 0.30
[110.00]	62.00	61.72	61.77	61.75	61.74	- 0.25
[139.40]	62.00	61.76	61.74	61.71	61.73	- 0.26
Arithmetic mean of ABS prints in Y-section [mm]						**- 0.27**

Table 7. Measurement of inner distances in Y axis – PLA material

Measurement of inner distances in X-axis – ABS material						
Section Y [mm]	**Nominal value [mm]**	**Print 1 [mm]**	**Print 2 [mm]**	**Print 3 [mm]**	**Arithmetic means [mm]**	**Deviation [mm]**
[23.00]	112.00	111.53	111.61	111.52	111.55	- 0.44
[73.00]	112.00	111.51	111.64	111.48	111.54	- 0.45
Arithmetic mean of ABS prints in X-section [mm]						**- 0.45**

Table 8. Measurement of inner distances in X axis – ABS material

Measurement of inner distances in X-axis – PLA material						
Section Y [mm]	**Nominal value [mm]**	**Print 1 [mm]**	**Print 2 [mm]**	**Print 3 [mm]**	**Arithmetic means [mm]**	**Deviation [mm]**
[23.00]	112.00	111.55	111.73	111.68	111.65	- 0.34
[73.00]	112.00	111.51	111.77	111.70	111.66	- 0.34
Arithmetic mean of ABS prints in X-section [mm]						**- 0.34**

Table 9. Measurement of inner distances in X axis – PLA material

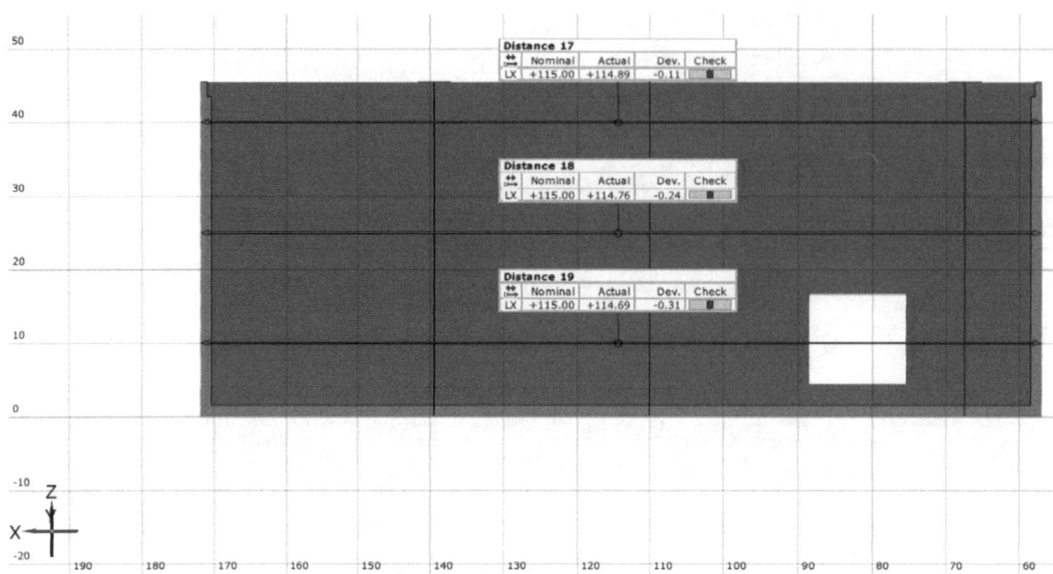

Fig. 9. Measurement of distances in X axis

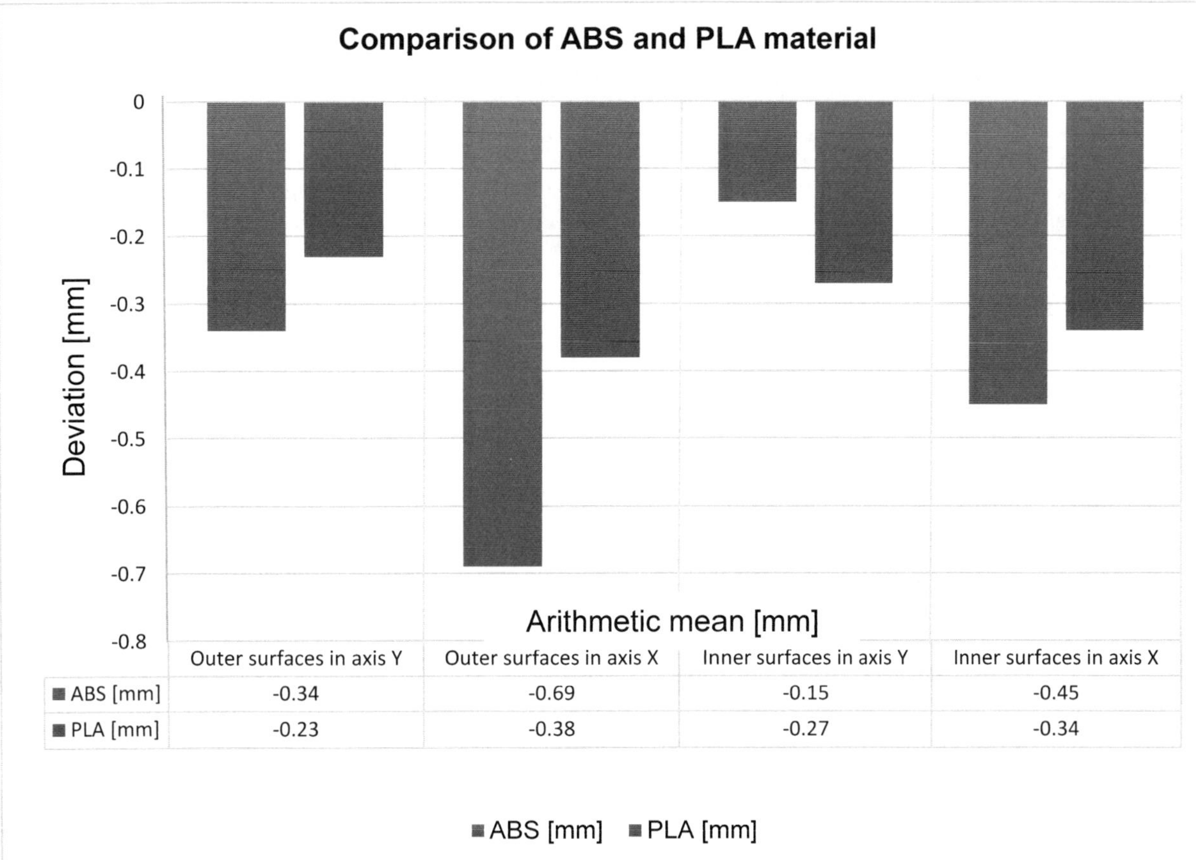

Fig. 10. Comparison of deviation of ABS and PLA thin-walled components

9. Conclusion

The shrinkage of plastics designates the volume contraction of polymers during the cooling step of the processing of polymers. This contraction is partly due to the difference of density of polymers from the melt state and the cooled, rigid state. In the process of molding is known, that ABS plastic is susceptible to shrinkage. Problem was, to know how is the dimensional differences between two most used materials in the additive manufacturing method known as Fused Deposition Modeling.

Component was designed in the Autodesk Inventor Professional 2017 CAD software. Thin-walled parts were manufactured by additive manufacturing with the technology, common known as Fused Deposition Modeling. Zortrax M200 3D printer was used for ABS components and for PLA part was used Be3D DeeGreen 3D printer. Before fabrication, converting the CAD model on to .stl format was necessary. After import the .stl file to the printer, the print process was set. Z-Suite and DeeControl software offer following properties: layer thickness 0.09 mm, high quality, medium infill (50 %), 9 full top layers, 4 full bottom layers and automatic fan speed. Orientation of model was set as 0°. Created CAD model was printed on two different printers, so we had to repeat the same procedure on the DeeGreen printer. Six parts was made in total. Three for ABS and three for PLA material. At the third point, it was necessary to digitize the printed components. Digitization was realized at the Centre of Excellence of the 5-Axis Machining of STU MTF in Trnava. Measure volume was used MV 170. Before scanning, printed models had to be marked for scanning by attaching 1.5 mm reference points. After scans of the first series of scanning, component was turned to scan all the faces, and we combined these series of scans into one. We have been able to export the scanned model to the .stl format.

The data was divided according to the measurement method. After evaluating the experiment, was found that the actual impact on the resulting quality and speed of production was due to the used printer itself. The time of manufacturing of ABS components was 3 hours and 54 minutes longer than the PLA material. In the overall assessment, for ABS components was needed 8 hours 42 minutes more time to manufacture. According to Fig. 10 the biggest difference between components manufactured for ABS and PLA materials was measurement on outer surfaces in X-axis. The value was more than 0.3 mm higher for components from ABS material. Because is known, that ABS material is more prone to retraction than PLA. The measurement of X-axis of outer surfaces was the highest distance and the reason for the largest deviation was retraction. In just one case, the value of the ABS material was less than the PLA material. It was in case measurement of inner surfaces in Y- axis. Subsequent digitization and component inspections were found: the PLA material is less susceptible to shrinkage at the X and Y-axis outer surfaces. The ABS material exhibited smaller mean deviations on the Y-axis inner surfaces. The arithmetic mean deviation of the Y-axis contour of the outer surfaces is less for the PLA material (0.23 mm) compared to the ABS material (0.34 mm). Comparison of this measured distances are shown on Fig. 10. The biggest deviation was on ABS part on outer surfaces in X - axis with the value 0.6967 mm. That the PLA material was less susceptible to shrinking the plastic on the X and Y-axis outer surfaces measured. The ABS material was less prone to the Y-axis inner surfaces.

Suggestions for future advanced research in this area are following: print the same model from other available plastic materials as High Impact Polystyrene (HIPS), Polycarbonate combined with Acrylonitrile-Butadiene-Styrene (PCABS) and Polyethylene terephthalate glycol (PET-G). Manufacturing of circular components by the FDM technology and measuring of them in the second step will be also good suggestion for future research.

10. Acknowledgments

The authors would like to thank for financial contribution from the STU Grant scheme for Support of Young Researchers project no. 1375 with acronym "SKOPF" and the project VEGA 1/0477/14. Research of influence of selected characteristics of machining process on achieved quality of machined surface and problem free assembly using high Technologies.

11. References

[1] Kochesfahani S. H. (2016). Plastics Engineering, Imerys Filtration and Performance Additives, ISSN 123-0-1234567, ANTEC 2016 Indianapolis, USA

[2] Kochesfahani S. H., Abler, C., Crepin-Leblond, J., Jouffret, F. (2010). Enhancing Biopolymers with High Performance Talc Products, Available from: http://www.4spe.org/Resources/resource.aspx?ItemNumber=11876

[3] Mudassir A. (2016) Measuring Accuracy of Two 3D Printing Materials, Department of Engineering Technologies, Bowling Green State University, Bowling Green, Ohio, USA

[4] Jančuš L. (2017). Research of ABS a PLA materials in the process of Fused Deposition Modeling, Department of (Bachelor´s thesis). Slovak University of Technology in Bratislava, Faculty of Materials Science and Technology in Trnava, Institute of Production Technologies. Supervisor: Ing. Ján Milde

[5] Vittek, M. (2016). Influence of CAD orientation in the process of Rapid Prototyping on geometrical and dimensional accuracy of components (Bachelor´s thesis). Slovak University of Technology in Bratislava, Faculty of Materials Science and Technology in Trnava, Institute of Production Technologies. Supervisor: doc. Ladislav Morovič, PhD.

[6] Olivera S., Muralidhara B. H., Venkatesh K., Gopalakrishna K., Vivek CH. S. (2016). Plating on acrylonitrile–butadiene–styrene (ABS) plastic: a review", Journal of Material Science, Vol. 51, pp. 3657-3674, ISSN 1573-4803, https://doi.org/10.1007/s1085

[7] Plastic Properties of Acrylonitrile Butadiene Styrene (2010). Dynalab. Internet archive wayback machine. http://www.dynalabcorp.com/technical_info_abs.asp

[8] Cuiffo, M.A.; Snyder, J., Elliott, A.M.; Romero, N.; Kannan, S.; Halada, G.P. (2017). Impact of the Fused Deposition (FDM) Printing Process on Polylactic Acid (PLA) Chemistry and Structure. Appl. Sci. 2017, 7, 579.

[9] Dunne R., Desai N., and Sadiku R. (2017). Material characterization of blended sisal-kenaf composites with an ABS matrix", Applied Acoustics, Vol. 125, pp. 184-193, 003-682X, https://doi.org/10.1016/j.apacoust.2017.03.0

[10] Xiao L., Wang B., Yang G., Gauthier M. (2012). Poly (Lactic Acid)-Based Biomaterials: Synthesis, Modification and Applications, 978-953-307-471-9, INTECH, Shanghai, China.

[11] Milde J., Morovič L. (2016). The Influence of Internal Structures in Fused Deposition Modeling Method on Dimensional Accuracy of Components. Slovak University of Technology, Faculty of Materials Science and Technology in Trnava, Institute of Production Technologies. Research Papers, Vol. 24, pp. 73-80. ISSN 1336-1589

[12] Buranský I. (2011). Thin-walled parts machining. Scientific Monographs. Hochschule Anhalt. Kolten, Germany. ISBN 978-3-86011-043-0.

[13] Trebuňa, F., Šimčák, F. (1999). Tenkostenné nosné prvky a konštrukcie. (Thin wall bearing elements and constructions). Košice: Vienala. ISBN 80-7099-444-4

[14] Aijun, T., Zhanqiang L. (2008). Deformation of thin-walled plate due to static end milling force. In Journal of Materials Processing Technology. Volume 206. http://www.sciencedirect.com

DOI: 10.2507/28th.daaam.proceedings.115

STRUCTURAL OPTIMIZATION OF SPACE COMPONENTS ADAPTED FOR 3D PRINTING

Camelia Elena Munteanu, Alexandru-Mihai Cismilianu,
Alina-Ioana Chira & Daniela Baran

This Publication has to be referred as: Munteanu, C[amelia] E[lena]; Cismilianu, A[lexandru] -[Mihai]; Chira, A[lina-Ioana] & Baran, D[aniela] (2017). Structural Optimization of Space Components Adapted for 3D Printing, Proceedings of the 28th DAAAM International Symposium, pp.0821-0825, B. Katalinic (Ed.), Published by DAAAM International, ISBN 978-3-902734-11-2, ISSN 1726-9679, Vienna, Austria
DOI: 10.2507/28th.daaam.proceedings.115

Abstract

In the space sector reducing the cost through innovative designs, can be achieved through mass reduction and shorter development time. Additive layer manufacturing (ALM) is a process which enables, not only a new cycle of optimization in terms of mass and performance, but also a minimum lead time for small series production of complex parts.
The objective of this paper is to design a part as light as possible while respecting the specific requirements. Specialized optimization tools were used in order to significantly reduce the number of design iterations. In order to cut down the mass furthermore, while increasing the stiffness, a hollow structure was considered. Internal cavities may raise a problem of powder evacuation in case of powder fusion processes. Concerning contamination problems of the spacecraft's (S/C) components, a powder removal procedure was closely developed with the manufacturer. As an outcome, a lighter space component with a lower production cost and fewer points of potential failure was obtained.

Keywords: Structural optimization; 3D Printing; internal cavities; powder evacuation; space applications

1. Introduction

The quest for light and stiff structures by industries such as the medical industry, the aerospace industry and other industries drives the progress in the additive manufacturing technology[1].

Cost reducing has become the focus of most researches and activities developed in the space sector. Although it is extremely expensive to manufacture and launch a space vehicle, the most expensive part is the development of all the sub-processes and the optimization of all the components. An important approach of cost cutback is weight reduction and shorter development time.

ALM is an emerging manufacturing technology which provides the potential for significant weight savings through optimization due to the relatively relaxed design constraints imposed. The part cost for ALM is independent of complexity, and so there is a 'virtuous circle' whereby weight savings through optimization also result in cost savings as the amount of material used to make the part reduces.[2]

The process of building an object by layering material instead of subtracting it from a larger block of material is often cited as having the potential to revolutionize the manufacturing industry.

Aerospace and Space industry has taken a leading role in the development, implementation and industrialization of ALM. The main benefits of the ALM process are design flexibility, low material waste and low cost of producing parts from hard materials that are otherwise difficult to manufacture [3].

ALM is known to provide more design freedom than conventional manufacturing methods, which encourages the implementation of numerical optimization methods in the design process in order to reduce weight by eliminating unneeded material [4]. Topology optimization offers a faster way to create load specific structures thus, enabling the industry to unlock enormous lightweight design potential by using a powerful design tool in combination with the use of ALM.

Using topology optimization, the entire structure can be modified. This type of analysis reduces drastically the number of exchanges between the design and stress departments which leads to a downsize in time and cost.

The key to realizing metal parts using ALM is in understanding that 3D printed metal parts differ in properties from machined ones just like aluminium cast parts differ from aluminium machined parts.

2. Input

The aim of the project was to demonstrate the potential weight savings achievable using the design freedom offered by ALM, while respecting the conditions imposed by the customer.

The utility of an optimization software is presented by determining the optimum material distribution of a support bracket for space use. For the design of the bracket the following input data was set (Fig. 1.). In the input data figure there are two groups of thrusters disposed in the front of the envelope. We also have the attachment points which are disposed in the same plane.

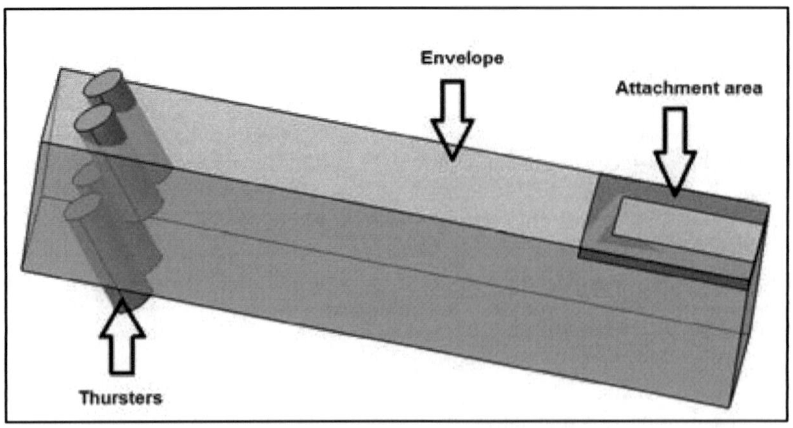

Fig. 1. Envelope & boundary conditions

The main requirements are to develop a bracket which can hold the thrusters in the given positions and fit the envelope. A first frequency value of 90 Hz or above has to be provided and the resulted structure must withstand a 30 g load on all principal directions (X, Y, and Z). The aim is to obtain a structure by minimizing the mass while maximizing the frequency with respect of the constraints. Eight attachment points on the orange area, were considered to be enough. The thruster attachment points to the structure was developed closely with the customer.

3. Design approach

By carefully analysing the given input, eight attachment points on the orange area, were considered to be enough. In order to generate a conceptual design, topological optimization is used. The software used to perform the analysis is INSPIRE [5].

The design space is presented in Fig. 2.a). The objective function considered was to minimize the weighted compliance. The result of the optimization analysis is presented in Fig. 2.b). It can be observed that are no internal elements and most of the material is distributed on the sides. An interesting and expected fact are the rounded shape of the elements that connect the last screws with the rest of the structure.

Using this structure, a new geometry presented in Fig. 2.c). was developed and used for a second iteration of optimization. In order to give more material and space for the optimization analysis, all the constructive elements were considered thicker. The upper and lower sides were not fully defined from the first analysis, so in those areas, for the second iteration of optimization, thin plates were considered. The second optimization offered a better understanding of the distribution of the material.

The result is presented in Fig. 2.d). The elements starting from the last screws are even more rounded. Also in the upper and lower part, some elements are significantly more visible.

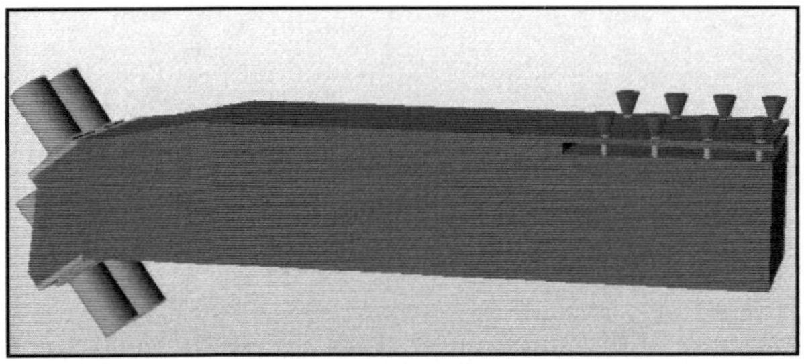

a) Input design space

b) Optimization result

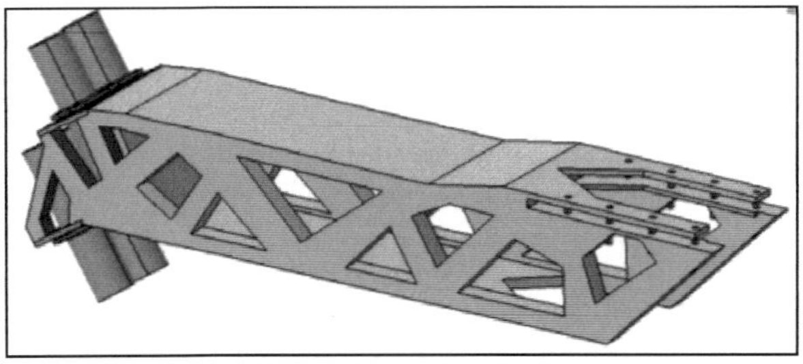

c) Second iteration design

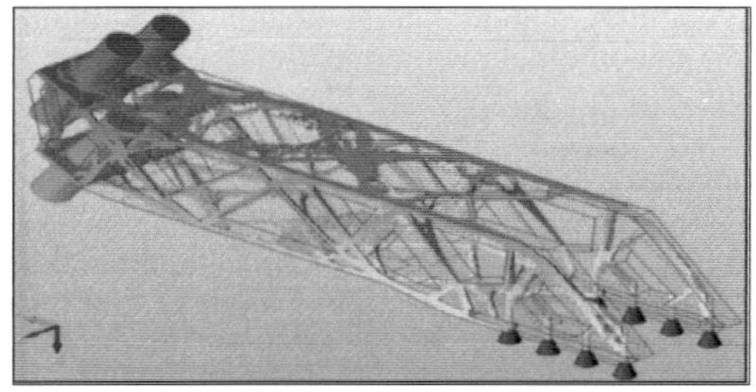

d) Second iteration of optimization

Fig. 2. Design iterations

4. Final design and manufacturing

By carefully analysing the distribution of the material given by the optimizer and respecting some conditions, like access in the interior of the bracket, ease of installation, providing connections with external elements, a new design was considered (Fig. 3.).

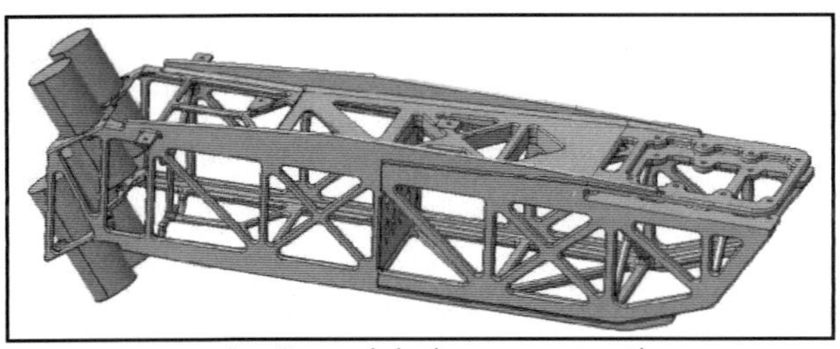

a) Second design reconstruction

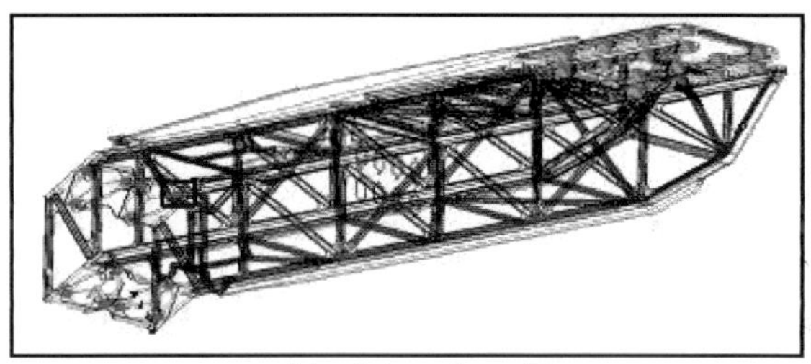

b) Internal cavities

Fig. 3. Final design

The aim was to provide a part as lightweight as possible while respecting a stiffness requirement of over 90 Hz. Taking the advantages offered by ALM and its fewer restrictions regarding the geometry complexity, it was considered a part with internal cavities Fig. 3.b). which lead to a structure 30% lighter.

The new design was analysed using finite element method and proven to withstand the loads considered and with the first frequency of 95 Hz. Topology optimization allowed selecting the best elements in the given design space to maximize the use of material.

The next important step is manufacturing. As said before, ALM was chosen to manufacture the part. Although this process gives enormous shape freedom of the part, which can't be easily obtained with traditional manufacturing techniques, there are some aspects to consider when using this method. The parts that are intended to be made via ALM need to fit as a piece, or as many components in the largest printer available for the design team. For this article, Concept Laser Xline 1000R [6] was considered because of the large overall dimensions of the bracket.

For an evaluation of the structure, LAAM made the following estimation regarding the volume of the part, the estimated support volume and the estimated build time, presented in Fig. 4.

- Machine: Xline1000r
- Build volume: 630x400x500 mm
- Material: AlSi7Mg0,6
- Volume: 672 cm3
- Support volume: 1100 cm3
- Time estimation: 10 days

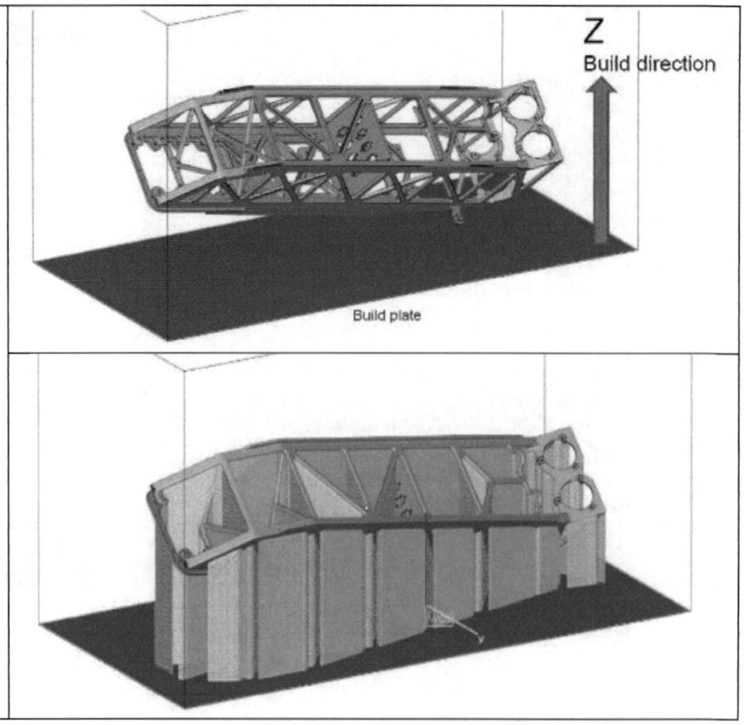

Fig. 4. Part orientation on build plate and support disposal courtesy of LAAM

As mentioned before, the bracket is intended to be manufactured hollow. For this purpose, evacuation holes are needed in order to safely evacuate all the powder.

Powder evacuation holes are directly dependent with the oriented part on the manufacturing build plate. The evacuation is intended to be made by shaking the part and by using pressured air. The powder must be evacuated before the elimination of supports, because if the support is removed without a heat treatment for stress relieve, the part will deform. If the heat treatment is made before the support removal there is a high risk that the powder will weld with the bracket making impossible the evacuation.

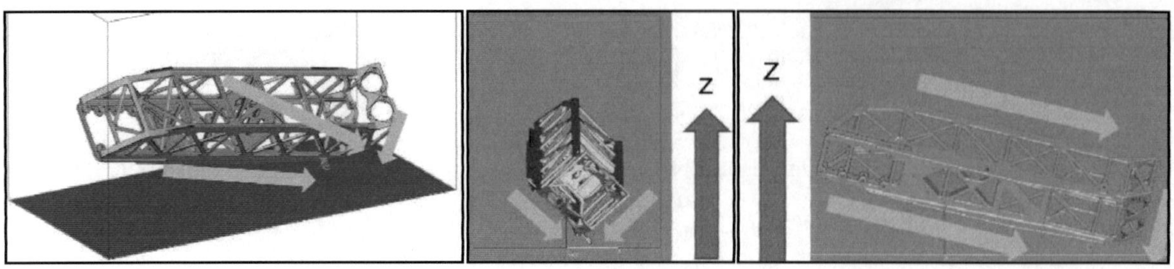

Fig. 5. Evacuation holes placement indications courtesy of LAAM

Before placing evacuation holes on the part, all the internal cavities must be interconnected, where possible, for an efficient removal of the powder. Where the cavities can't be interconnected, special evacuation holes must be made. All holes have to be placed then, as much as possible at the edge of the cavity taking in account the evacuation holes orientation directions presented in Fig. 5.

.

Taking into account the fact that the powder evacuation will be made by shaking the part and pressurized air, a powder flow circulation is presented. On the side, the cavities have evacuation holes only on the far ends (one of them represented in lower right of Fig. 6. The powder flow circulation is made because, if the red and blue pipe where connected as one, the pressurized air blown through the end will dissipate and the powder evacuation will be inefficient.

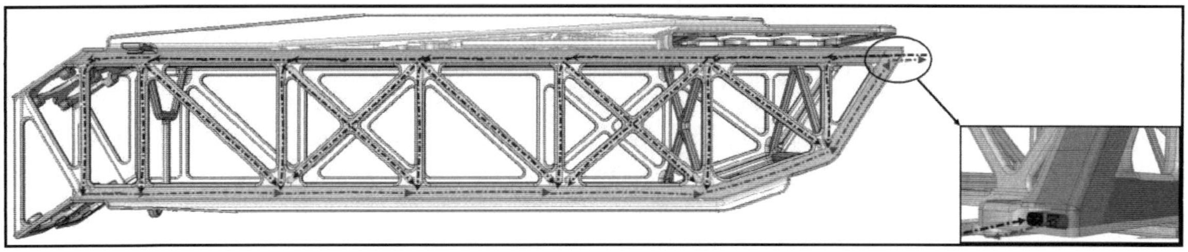

Fig. 6. Proposed powder-flow circulation

Because contaminating nearby components is a high-risk issue, and any remaining powder in the cavities prove to be a problem. For a designer, the essential question is how much contamination from all sources can be tolerated without causing a given spacecraft system to degrade below a critical performance level, or fail altogether [7]. In order to mitigate the risk CT scan has to be made in order to see if there is any remaining powder.

Choosing and properly implementing the best ALM process, material and post-processing combination for intended application is critical for success [8].

5. Conclusions

The aim of this paper was to obtain a part with an optimum distribution of material capable to withstand the loads applied while respecting the boundary conditions. Another important aspect was to provide access to some area inside the structure and to supply connections with external elements which imposed some limitation on the geometry obtained.

Using INSPIRE a load sufficient structure was obtained and analysed in a much shorter time of development. A detailed FEM analysis was performed only at the end of the process.

The support brackets evolution from the design space given as input to a final design ready to be printed was presented.

It was proven that INSPIRE is a powerful tool which if properly used, leads to advanced structures, with a dramatically reduced number of iterations between the design and stress departments. Being able to achieve faster an optimized structure enables a downsize in lead time and qualified personnel needed. Applying the steps presented in this paper to a larger scale can provide significant cost reductions.

6. Acknowledgement

The manufacturing part of the paper was possible through the know-how developed from the iterations that we had with LAAM - LISI AEROSPACE ADDITIVE MANUFACTURING Powered by POLY-SHAPE. We want to thank LAAM for providing us pictures for the manufacturing chapter. We also want to thank Elisabeth REY and Sébastien EYRIGNOUX for their cooperation.

7. References

[1] Hanzl, P[avel]; Zetek, M[iroslav] & Zetkova, I[vana] (2016). Cellular Lattice Structure Produced by Selective Laser Melting and its Mechanical Properties, Proceedings of the 26th DAAAM International Symposium, pp.0748-0752, B. Katalinic (Ed.), Published by DAAAM International, ISBN 978-3-902734-07-5, ISSN 1726-9679, Vienna, Austria; DOI:10.2507/26th.daaam.proceedings.104

[2] Topology Optimisation of an Aerospace Part to be Produced by Additive Layer Manufacturing (ALM),Case Study, ALTAIR HyperWorks

[3] Maximizing the Potential of Additive Manufacturing with Design Optimization, Altair ProductDesign library of „Success Stories", www.altairproductdesign.com

[4] Joona Seppälä, Andreas Hupfer, Topology Optimization in Structural Design of a LP Turbine Guide Vane: Potential of Additive Manufacturing for Weight Reduction, ASME Turbo Expo 2014: Turbine Technical Conference and Exposition, ISBN: 978-0-7918-4576-9

[5] http://www.solidthinking.com/Inspire2017.html

[6] https://www.concept-laser.de/en/products/machines.html

[7] Gary Pippin, Russ Crutcher, Spacecraft contamination issues from LDEF: Issues for design

[8] I. Gibson, D. W. Rosen, B. Stucker, Additive Manufacturing Technologies - Rapid Prototyping to Direct Digital Manufacturing, Springer Science+Bussiness Media, LLC 2010, ISBN 978-1-4419-1119-3

DOI: 10.2507/28th.daaam.proceedings.116

METHODS OF TASK CLINCHES PREVENTION IN SOFTWARE APPLICATIONS FOR REAL-TIME SYSTEMS

Viktor Nikiforov, Vladimir Pavlov

This Publication has to be referred as: Nikiforov, V[iktor] & Pavlov, V[ladimir] (2017). Methods of Task Clinches Prevention in Software Applications for Real-Time Systems, Proceedings of the 28th DAAAM International Symposium, pp.0826-0834, B. Katalinic (Ed.), Published by DAAAM International, ISBN 978-3-902734-11-2, ISSN 1726-9679, Vienna, Austria
DOI: 10.2507/28th.daaam.proceedings.116

Abstract

Real-time software applications are built as a set of tasks sharing hardware and information resources. A consistency of shared information resources is achieved by the usage of mutex-type synchronizing means. The usage of mutexes may lead into task clinches. A traditional mean to prevent task clinches is the Priority Ceiling Protocol (PCP). But PCP utilization excludes the using of efficient scheduling modes with task priorities reassigned dynamically. Therefore the implementation efficiency may be decreased significantly for a wide range of real-time software applications. Conducted experiments showed that such decrease can be especially significant when software applications are realized in a multi-core processor environment. The paper represents methods of the task clinch prevention which ensure that a wide range of applications can be scheduled with efficient scheduling modes permitting dynamic task priorities by using a specialized multi-partite oriented graph reflecting features of inter-task synchronizing links appearing when using mutexes. The absence of inter-partite contours means that task clinches are impossible. If such contours exist, it is possible to determine the code section that shall be modified to break these contours. If such modification is impossible then the new developed protocol can be utilized.

Keywords: real-time systems; multitasking software applications; task synchronization; scheduling modes; resource access protocols

1. Introduction

The study concerns a feasibility of software applications for real-time systems that relates to the problems of the efficient real-time operating systems development which provide an effective task scheduling without processor resource usage productivity decreasing. Real-time systems shall ensure a well-timed information exchange with external processes. The key property of real-time software applications is that they run in a "time structure" that is defined by the course of external processes [1]. Such correspondence between the course of the real-time system software application components execution and the course of external processes is caused by the presence of more or less hard time frames for information exchanges with external processes. One of the main consequences of this fact is a requirement to build a real-time software application as a set of cooperative tasks $\tau_1, \tau_2, \ldots, \tau_n$.

During execution of the multitasking software application its cooperative tasks share common hardware resources: executive resources (first of all, processors and processor cores) and information resources (global data files, interface registers of peripheral devices, man-machine interface elements, etc.). To ensure an execution consistency the cooperative tasks exchange data and synchronizing (signal) information messages. A logical correctness of the task interaction structure consists in an assurance of the following: a) consistency of the shared information resources; b) absence of the risk of clinches when two or more tasks are in the state of mutual blocking. An additional condition of the dynamic correctness exists for the real-time systems: it is necessary to ensure a timeliness of the task executions.

A traditional approach to an assurance of the shared resources consistency is a usage of mutexes [2]. For each of the shared resources g_k a specific program object (mutex *mut_k*) is created that is a synchronizing element indicating occupancy of the resource g_k. The code section in which the task τ_i has an access to the resource g_k (critical interval on the access to g_k) is restricted by the synchronizing operators *lock(mut_k)* and *unlock(mut_k)* which restrict a capture and a release of the resource g_k, respectively.

The purpose of this study is to find new approaches to such computational process organization that ensure a task clinch prevention as well as a retention of the possibility to use highly efficient scheduling modes with dynamic task priorities.

2. Multitasking application profile

To fulfill a static analysis of the multitasking application with shared information resources or a simulation of the execution progress of such application the models are built in which a code of each task is represented as a chain of segments. Each segment contains a program code fragment that is terminated by the operator associated with the system event:

- segment terminated by the information resource request (by the *lock* operator);
- segment terminated by the *unlock* operator (resource release);
- final segment (task completion).

Tasks that don't contain resource access operators consist of the single (final) segment. A set of data about segment types sequence (with an indication of the specific information resource for which *lock* or *unlock* operator is executed) presents the task scheme. A set of schemes of all tasks represents the application scheme.

An example of the application scheme consisting of two tasks is presented in fig. 1*a* (using route net means [3]). The code of task τ_1 as well as the code of task τ_2 contains overlapping critical intervals on the access to resources g_1 and g_2. The schemes of tasks τ_1 and τ_2 differ in the overlapping kind of critical intervals: the code of task τ_1 contains the intersection of critical intervals but the code of task τ_2 contains the nesting of critical intervals, i.e. the critical interval on the resource g_2 is nested into the critical interval on the resource g_1. Application schemes contain all necessary and sufficient information to check a logical correctness of the application, in particular, to check an absence of the risk of clinches during the application execution.

To check a dynamic correctness of software applications used in real-time systems the scheme of each task shall be complemented with the following parameter values:

- weight w of each segment (i.e. an amount of computations performed in the frame of the segment);
- period T_i and deadline D_i of each task τ_i.

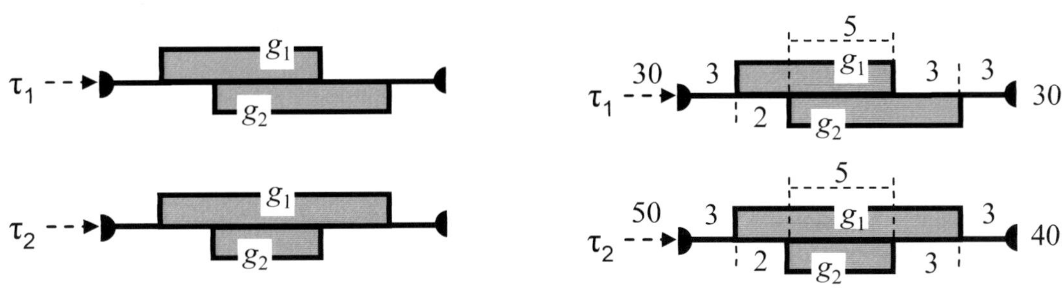

a) Application scheme b) Application profile

Fig. 1. Presentation of data about software application consisting of two tasks using route net means.

The weight w of the segment corresponds to the amount of computations executed within the segment (in a form of the equivalent amount of reference operations, e.g. arithmetic operations with floating point). The sum of weights of all segments contained in the code of task τ_i is the weight W_i of the complete task τ_i. If a processor throughput P is known (as a number of reference operations per second) then the task weight can be expressed in seconds of the processor time: $C_i = W_i/P$. The ratio $u_i = C_i/T_i$ is a processor load from the task τ_i (the load corresponds to the processor time quota that shall be allocated for the execution of instances of the task τ_i). The total load $U=\Sigma u_i$ indicates a measure of the processor time usage by the application with the given profile and with the given processor throughput value P.

The period T_i represents a minimum acceptable duration of the time interval between adjacent activations of the task τ_i; the deadline D_i represents a maximum acceptable duration of the time interval between a moment of the j-th activation of the task τ_i (i.e. a moment of the j-th task instance generation) and a moment of this instance execution termination.

The profile of the task τ_i is a set of data about a sequence of the segment kinds forming the task model with an indication of each segment weight as well as T_i and D_i values corresponding to this task. The application profile is a set of profiles of all its tasks.

The profile of the application with the scheme represented in fig. 1*a* is shown in fig. 1*b*. The first, second, third, fourth, and fifth segments of both tasks τ_1 and τ_2 have values of weights 3, 2, 5, 3, and 3 respectively. The weight C_1 of the task τ_1 (as well as the weight C_2 of the task τ_2) is equal to 16 units of the processor time measure (e.g. milliseconds). The value of the period T_i of the task τ_i is pointed out above a dotted arrow in front of the first segment of the task code. The deadline value is pointed out after the last task segment ($T_1 = 30$, $T_2 = 50$, $D_1 = 30$, $D_2 = 40$).

The ratio $H_i = T_i/D_i$ indicates a hardness of requirements for the task τ_i deadline. The hardness H_1=1 of the task τ_1 in fig. 1*b* means that the time interval of the τ_1 current execution shall not be overlapped with the time interval of its next execution. The hardness H_2=1.25 of the task τ_2 means that the time interval of the τ_2 current execution shall be terminated in advance so that it shall be not less than 20% of the period T_2 before the next τ_2 execution.

3. Processor throughput and execution configuration

An application execution is carried out in specific conditions defined by the known processor having known throughput P, known number m of cores, known scheduling mode and known access protocol to shared resources. If these conditions are defined and the application profile is known then the question about feasibility of this application with the given profile in the given conditions becomes actual one. Feasibility means a guaranteed deadline of all tasks in any acceptable scenarios of system events. If it is provided to be that in the given defined conditions one can find such task τ_i and such acceptable scenario of system events in the application with the given profile that some j-th instance of the task has no time to complete its execution in the frame of the time interval of the length D_i then this application is considered to be unfeasible in the given conditions.

Theoretically it is possible to get the application feasibility via increasing the value P of the processor throughput under invariability of the application profile and invariability of the rest application execution condition except P. Let's call these rest conditions the application execution configuration. The configuration specification includes the number m of processor cores, used scheduling mode, and used access protocol to the shared information resources.

When the value P increases, the values u_i of loads from each task τ_i and the value U of the total processor load from the whole application will decrease. Varying values P it is possible to find (with some prescribed accuracy) such maximum value U_{max} when the application still remains feasible. The value $Dens = U_{max}/m$ (where m — number of processor cores) is called a density of the application in regard to the specific execution configuration [4].

The value of $Dens$ is no more than 1. This value reflects an efficiency of the specific execution configuration of the application with specific profile. Here efficiency is a maximum attainable quota of the processor time used effectively. For example, the value $Dens = 0,5$ means that an application with given profile cannot use effectively more that 50% of the processor time under guarantees of the each task deadline in the given execution configuration (the remaining time a processor will either stand idle or make calculations not relevant to an execution of tasks $\tau_1, \tau_2, \ldots, \tau_n$). This circumstance allows to consider the value $Dens$ as a criterion of the comparative efficiency of various execution configurations usage in an application with the specific profile.

In particular, such comparison can be used to choose a scheduling mode in the real-time system functioning under control of the software application with the specific profile when it is necessary to resolve collisions between requirements to decrease a volume of allocated computational resources and requirements to keep a feasibility of the real-time software application.

Studies devoting to a development of efficient scheduling modes for real-time systems based on individual single-core processors have been started more than 40 years ago. In [5] the authors proved the optimality of the Rate Monotonic (RM) scheduling mode in the class of scheduling modes with static task priorities as well as the Earliest Deadline First (EDF) scheduling mode in the class of scheduling modes with task priorities assigned dynamically. They found a critical profile with the following features shown in fig. 2*a*:

- total load U is distributed evenly between application tasks (for each task $u_i = U/n$);
- values of periods T_i are increased exponentially with the increasing of index i.

A comparison of RM and EDF scheduling modes for the application consisting of 5 independent tasks executed on a single-core processor is presented in fig. 2*b* for hardness values $H_i = 1$ and $H_i = 1.25$ that shows an advantage of the EDF scheduling mode relative to the RM scheduling mode.

The application profile with parameter values such as in fig. 2*a* is the critical one to the effect that applications with such profile have the minimum density value $Dens$ for the RM scheduling mode. Below we will call application profiles with such correlations between periods and loads as Liu-Layland profiles.

Density values presented in fig. 2*b* have been derived as analytically as via simulation. To check a validity of the measurement results obtained during simulation these results have been compared with results obtained via analytical methods. A coincidence of density values obtained via these two methods confirms a validity of the results obtained via simulation. The density values presented in fig. 2*b* indicate that EDF scheduling is more efficient then RM scheduling under standard and more hard requirements concerning the task deadline.

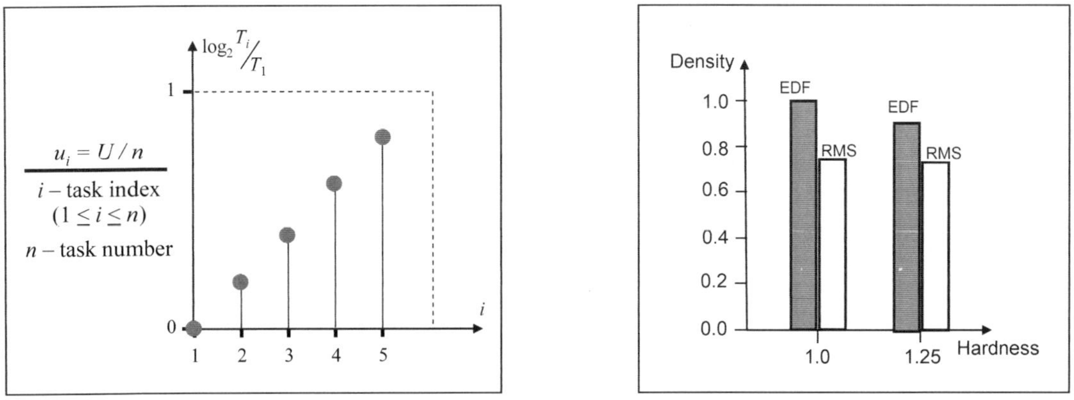

a) Parameters of the application with Liu-Layland profile

b) Density values for RM and EDF scheduling modes with various hardness value

Fig. 2. Density values for the application consisting of five independent tasks with Liu-Layland profile

Thus the simulation confirmed that RM and EDF scheduling modes are the optimal ones in their classes when the applications are run on single-core processor. These scheduling modes win with respect to any other scheduling modes in their classes not only when applications with Liu-Layland profiles are executed but also when applications with any other profiles are executed.

Nowadays multitasking software applications are developed to be implemented on multiprocessor hardware or on multi-core processors. For example, scheduling modes for multiprocessor hardware are considered in [6]. But below we will discuss the scheduling modes for multitasking software applications implemented on multi-core processors.

In the case of multi-core processors RM and EDF scheduling modes are not optimal ones. They become very inefficient ones when the total load has essentially irregular distribution between tasks).

In the application with profile shown in fig. 3*a* 60 % of the total load carries the task τ_5 therefore it can be called "heavy", then the tasks τ_1, τ_2, τ_3, τ_4 can be called "light". It is reasonably to use the modified RM scheduling mode (ModifRM) for applications with the shifted load: a heavy task will have the highest priority while the light tasks scheduling is carried out with usual RM mode. The histograms presented in fig. 3*b* and 3*c* show that RM and EDF scheduling modes lose essentially with respect to ModifRM scheduling mode as for two-core processor as for four-core processor.

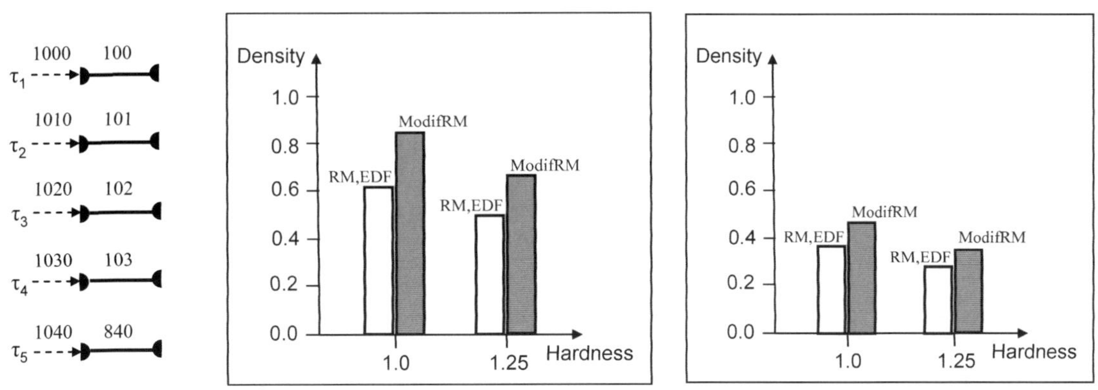

a) Application profile

b) Execution on two-core processor

c) Execution on four-core processor

Fig. 3. Density values for the application consisting of five independent tasks with the shifted load

An efficiency of the scheduling mode with fixed task priorities is decreasing when the number of processor cores is increasing. When an application is run on the multi-core processor the best of the known scheduling modes with fixed task priorities ensure a dynamic application execution correctness if total load values do not exceed $0.4m$ (where m – number of processor cores) [7].

In the late 1990s it was proved the existence of Pfair scheduling modes that ensure the feasibility of applications consisting of independent tasks on multi-core processors when *Dens* = 1 [8]. Later the algorithms were developed that realize Pfair scheduling modes in systems with time slicing (e.g. [9]) as well as in systems without time slicing [10].

Thus, the scheduling modes with dynamic task priorities can ensure an essential gain in the processor resource usage efficiency in real-time systems especially when software applications are realized on multi-core processors.

The examples considered above relate to applications with independent tasks. In regard to multitasking software applications with interdependent tasks the specialized protocols of task requests processing on an access to information resources have been developed. These protocols differ in conditions and/or actions checking/performing in the order of *lock* and *unlock* operations execution [11].

When the simplest access protocol (SP) is applied the single condition of the requested resource allocation is that the resource is free at the current moment [12]. If the resource is occupied then the task requesting an access to it is converted into the state of this resource release waiting. No additional actions are performed in this case. A disadvantage of this protocol is that it allows:

- a clinch appearance and
- a priority inversion that can lead to the value *Dens* significant decreasing.

The Priority Inheritance Protocol (PIP) requires a performance of the following additional (relative to SP) actions: if high priority task τ_i requests a resource g that is used by low priority task τ_j, then task τ_i (temporally, until τ_j will release g) gives its own priority to τ_j. In the case of PIP usage a priority inversion is prevented but a possibility of the clinch appearance is preserved [13].

The Priority Ceiling Protocol (PCP) prevents as priority inversion as clinch appearance possibility. When this protocol is used it is necessary to make a specific static processing of the scheme (on the stage of software application design) reflecting a structure of inter-task links in a multitasking software application of the real-time system [14]: each resource g is provided with the special scheduling parameter $pc(g)$ that is called a priority ceiling of g . The value $pc(g)$ is equal to the highest static priority $prio(\tau_i)$ among the tasks which have critical interval on g. The Priority Inheritance Protocol requires a check of the more strong condition (relative to SP and PIP) when task τ_i requests a resource g: if a resource g with $pc(g) \geq prio(\tau_i)$ is occupied by τ_j (j is not equal to i) then task τ_i is converted into the waiting state until all resources g with such high priority ceiling will be released.

New protocol of Inter-Partite Contours (IPC protocol) not only prevents a priority inversion and a clinch appearance but retains a possibility also to use efficient scheduling modes with dynamic priorities [15]. When IPC protocol is used as well as when PCP is used a specific static processing of the application scheme is required. This processing is based on construction and analysis of the bundle graph that is a specialized multi-partite oriented graph reflecting features of inter-task synchronizing links.

4. Bundle graph

A scheme of the multitasking software application containing interdependent tasks first of all should be checked on the clinch appearance possibility. If this check indicates that the clinch appearance is impossible then to ensure a consistency of the shared resources it is sufficiently to apply the SP protocol which allows to use any scheduling modes including (if necessary) the most efficient scheduling modes with dynamic task priorities.

The simplest method of this check is a search of the critical interval bundles. If the critical interval bundles are not discovered then it is sufficiently to apply the SP or PIP protocols which allow a usage of the scheduling modes with dynamic task priorities.

If the critical interval bundles exist then developers apply the PCP protocol. A difference between PCP and SP or PIP is that in addition to the check whether the requested resource is unoccupied it is necessary to make an additional check of the current state of other shared resources. To ensure a mutual blocking prevention by the PCP usage it is necessary to pay with a decreasing of the value *Dens*. The application density can decrease because of PCP does not allow the usage of the highly efficient scheduling modes with a dynamic reallocation of the task priorities.

An availability of the critical interval bundles is a necessary condition of the clinch appearance possibility. However this fact in itself does not mean that the mutual blocking is really possible. For example, the mutual blocking is impossible in the application with the scheme shown in fig. 1*a* despite the presence of the critical interval bundles but the application with the structure shown in fig. 2 can get into the state of the mutual blocking of the tasks τ_2, τ_3, τ_4. Therefore when the critical interval bundles are available it is still unclear is it necessary to apply PCP which although protects an application from the clinch appearances but excludes a usage of the efficient scheduling modes. Real-time software application developers are obliged to apply PCP accompanying with the appropriate efficiency decreasing though it is not excluded that such precaution is unnecessary in fact.

To apply PCP in those cases only when it is really essential a developer shall have a necessary and sufficient condition of the clinch appearance possibility as well as an efficient tool to check a fulfillment of this condition. Such tool is a construction and analysis of the critical interval bundle graph (or simply the bundle graph) [16].

In the application consisting of two tasks τ_1 and τ_2 that is represented by the scheme in fig. 1*a* each task contains overlapping critical intervals on the access to resources g_1 and g_2: critical intervals of the task τ_1 are concatenated ones and critical intervals of the task τ_2 are nested ones. Let's call two critical intervals of the task τ on resources g and g^* bundled ones (i.e. they form a bundle $L = <\tau, g, g^*>$) if they overlap, i.e. they contain common segments of the task code.

In the starting (leading) section of the bundle $L = <\tau, g, g^*>$ the task τ has an access to the head resource g of the bundle. In the central section the task τ has an access to the both resources of the bundle: head g and additional g^*. In the closing section one of the bundle resources is released already by the task τ.

The bundle graph is the multi-partite oriented graph that reflects dependence relations of the critical interval bundles. The bundle L_x=< g_a, g_b> depends on the bundle L_y=< g_c, g_d > if τ_i and τ_k are different tasks and $g_b = g_c$. In other words, the bundle L_x depends on L_y if L_x and L_y belong to different tasks and the head resource of the bundle L_y coincides with the additional resource of the bundle L_x. A node of the bundle graph has a one-to-one correspondence with a bundle existing in the application scheme.

If two graph nodes L_x и L_y are bundled by the arc directed form L_x to L_y then the bundle L_x depends on the bundle L_y. The profile and the bundle graph of the application consisting of five tasks which share four resources are represented in fig. 4*a*. Figure 4*b* represents the profile and the bundle graph of the application that differs from the application shown in fig. 4*a* in that the request operator of the resource g_2 in the task τ_4 is shifted so that the resource g_2 is captured before the resource g_4.

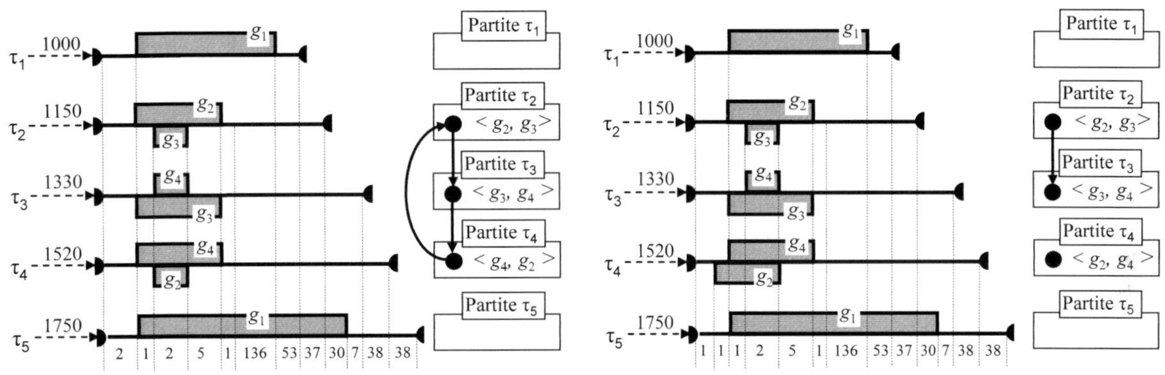

a) Application profile allowing a clinch appearance possibility

b) Modification of the profile eliminating a clinch appearance possibility

Fig. 4. An example of the application profile modification eliminating a clinch appearance possibility

The bundle graph is a multi-partite graph (with one partition per each task) because dependent bundles cannot belong to the same task. The contour in the bundle graph shown in fig. 4*a* is an inter-partite contour in the sense that all its nodes belong to different partitions of the graph. Just the presence of such contour indicates a possibility of the appearance of the mutual blocking of tasks. It is necessary and sufficient for the mutual blocking appearance possibility during application execution that an inter-partite contour exists in the corresponding bundle graph.

The application with the structure shown in fig. 1 has no dependent bundles and the corresponding bundle graph does not contain arcs (and therefore inter-partite contours). The protocols SP or PIP allowing the efficient scheduling modes usage can be utilized during such application execution. In the case of the software application shown in fig. 4*a* the usage of these protocols could lead to the clinch appearance.

A single way exists to prevent a clinch without the bundle graph usage, namely, the PCP usage with the corresponding efficiency loss. When the bundle graph is used there are two ways of SP and PIP usage with the corresponding maintenance of the possibility to use highly efficient scheduling modes.

The first way is connected with a local modification of the application scheme that ensures a break of the contour in the bundle graph. So the contour corresponding to the application in fig. 4*a* will be broken if an order of the capture of resources g_2 and g_4 by the task τ_4 will be changed. In the scheme shown in fig. 4*a* the request of the resource g_4 precedes the request of the resource g_2. If to relocate a request operator of the resource g_2 so that it precedes the request of the resource g_4 then the contour in the bundle graph will be broken (see fig. 4*b*), hence, the PCP usage becomes unnecessary and the possibility of the highly efficient scheduling modes usage is maintained.

The second way is connected with the possibility check of the ICP protocol usage. This protocol can be applied if the bundle graph does not contain overlapping inter-partite contours. If the required modification of the application scheme is unallowable on some reasons and overlapping inter-partite contours are absent (as in application represented in fig. 5) then the ICP protocol usage is allowed which, from the one hand, prevents clinches and, from the other hand, allows the usage of the scheduling modes with dynamic task priorities.

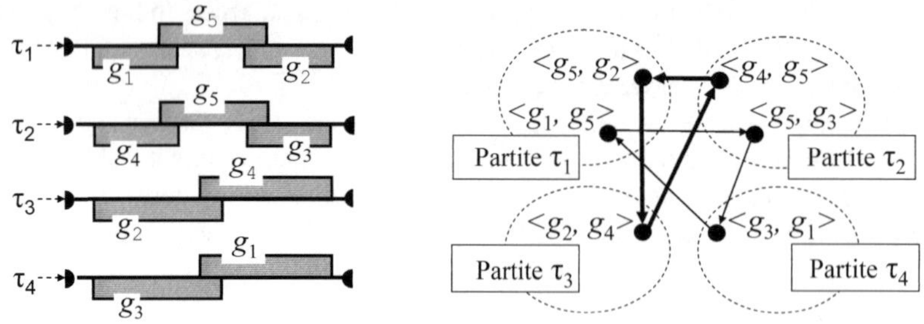

Fig. 5. Application with the bundle graph containing two non-overlapping inter-partite contours

When the IPC protocol is applied as well as the PCP is applied it is required a static processing of the application scheme. Values *pc*(*g*) of the resource ceiling priorities are calculated during such processing for the PCP, but as for IPC protocol the list of inter-partite contours $\{C_1, C_2, .., C_k\}$ existing in the bundle graph is built. The IPC protocol usage is possible only when the bundle graph corresponding to the application does not contain overlapping inter-partite contours. In this case each critical interval bundle *L* existing in the application either is included in the specific inter-partite contour *C*(*L*) or is not related to any inter-partite contour.

To execute an application with the IPC protocol usage each inter-partite contour *C* is provided with a counter *cnt*(*C*) of the number of tasks which currently execute a code segment corresponding to any leading section of the bundle containing in the contour *C*. Besides the length *lng*(*C*) of each contour is fixed.

When the task τ requests an access to the code segment being a leading section of the bundle *L* that belongs to C the additional check of the condition $cnt(C) < (lng(C) - 1)$ is performed besides conditions and actions corresponding to PIP. If this condition is satisfied then the task τ obtains an access to the requested resource and the value *cnt*(*C*) is increased by one. If this additional condition is not satisfied then an execution of the task τ is suspended while it will be satisfied.

The bundle graph in fig. 5 contains two non-overlapping inter-partite contours. The length of each of these contours is equal to 3. If the application scheme can't be modified so that each contour will be broken then the IPC protocol may be used. In this case clinches will be impossible and efficient scheduling modes become available.

5. Density of applications with interdependent tasks

The means for a simulation of the application execution with interdependent tasks were included into the simulation software ensuring an acquisition of density values to compare a relative efficiency of the RM and EDF scheduling modes.

Histograms in fig. 6 represent a dependence of density on hardness for applications with the profile shown in fig. 4*b* when using EDF and RM scheduling modes. Note that this application profile contains critical interval bundles. Therefore a developer who has no a checking tool to detect a clinch appearance possibility during such application execution with SP or PIP protocols need to apply PCP with RM scheduling mode. In this case, when a real-time system is realized on a single-core processor and a hardness has a standard value $H = 1$, it will be received approximately 25 % loss of efficiency in comparison with the case when a developer knows about the absence of inter-partite contours and can apply EDF scheduling mode without fear of clinch appearances (fig. 6*a*). When a real-time system is realized on a two-core processor this loss will be 40 % (fig. 6*b*). When inter-partite contours exist in a bundle graph a developer can find such modification of the application scheme which provides inter-partite contour breaks (just as it occurs when the application scheme shown in fig. 4*a* is replaced by the application scheme shown in fig. 4*b*) and it becomes possible to increase the application implementation efficiency in the tens of percent.

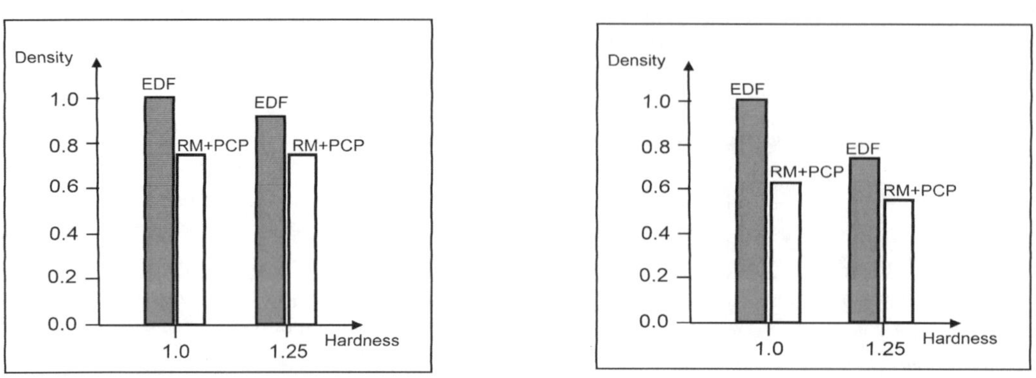

a) Single-core processor b) Two-core processor

Fig. 6. Execution of the application with interdependent tasks on a single-core processor

A modification of the application scheme by some reason can appears to be unwanted or impossible. In this case it is necessary to check a fitness for use of the new IPC protocol that prevents clinch appearances and at the same time allows applying scheduling modes with task priorities assigned dynamically. Fig. 7 represents density values obtained as a result of the simulation of the application with profile shown in fig. 4*a* realized on a single-core processor.

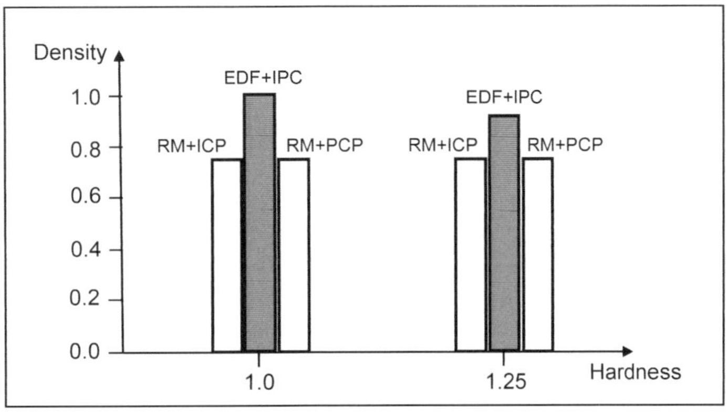

Fig. 7. Density values when IPC protocol is applied on the single-core processor

The data represented in fig. 7 indicate that IPC protocol usage together with EDF scheduling mode ensures the same gain in the tens of percent that is obtained due to application scheme modification eliminating inter-partite contours in the bundle graph.

6. Conclusion

The problem was that the usage of existing access protocols to shared information resources ensuring a task clinch prevention is connected with the necessity to reject the usage of highly efficient scheduling modes with dynamic task priorities that leads to processor resource usage productivity decreasing The necessary condition for clinch appearances during multitasking applications execution is a presence of critical interval bundles in task codes. If an analysis of the application scheme detects an absence of bundles then a developer has no limitations in the configuration choice of such application execution (i.e. in the choice of appropriate scheduling modes and access protocols to shared resources). The traditional way to prevent clinches when critical interval bundles are present is a usage of the Priority Ceiling Protocol. But this protocol usage is inconsistent with the usage of highly efficient scheduling modes with task priorities assigned dynamically.

A suggested approach to solve this problem consists in the construction and analysis of the multi-partite graph of critical interval bundles. An application has no risk of clinch appearances if this graph does not contain inter-partite contours. In this case the most efficient execution configurations can be used for application implementation. The inter-partite contour existing in this graph indicates those sections of task codes a local modification of which leads to the contour break. Such modifications transfer the application into a class of applications which are free from the risk of clinch appearances.

The results obtained consist in a specific static processing of the application scheme based on the multi-partite graph analysis. This analysis can help a software application developer to find the task code sections, modifications of which will break inter-partite contours and prevent clinch appearance. If such task code modifications by some reason are unwanted or impossible then the new developed Inter-Partite Contours protocol can be used to prevent clinch appearance and to use highly efficient scheduling modes if there are no overlapping inter-partite contours in the multi-partite graph.

The future plans concern the Inter-Partite Contours protocol improvement that will give a possibility to use this protocol for software applications with overlapping inter-partite contours in the graph of critical interval bundles.

7. References

[1] Davidenko, K. (1985). A Programming Technology for Automated Control Systems of Manufacturing Processes, Energoatomizdat, Moscow (in Russian)

[2] Laplante, P. (2004). Real-Time Systems Design and Analysis, John Wiley & Sons, Inc., ISBN 0-471-22855-9, NJ, USA

[3] Nikiforov, V. & Shkirtil, V. (2010). Route nets – graphical form for structural representation of real-time software applications. SPIIRAS Proceedings, No. 3(14), pp. 7–28, ISSN 2078-9599 (in Russian)

[4] Baranov, S. & Nikiforov, V. (2015). Density of multi-task real-time applications, Proceedings of the 17th Conference of Open Innovations Association (FRUCT 2015), 20-24 April 2015, Yaroslavl, Russia, ISSN 2305-

7254, ISBN 978-5-7577-0493-7, Balandin, S. (Ed.), pp. 9-15, Publishing house of Reseach University of Information Technologies, Mechanics and Optics, Saint-Petersburg, Russia, DOI: 10.1109/FRUCT.2015.7117964

[5] Liu, C. & Layland, J. (1973). Scheduling algorithms for multiprogramming in a hard-real-time environment. Journal of the ACM, Vol. 20, No.1, January 1973, pp. 46-61, ISSN 0004-5411

[6] Amol, A. S. (2016). Multicruteria dynamic scheduling by topsis and goal programming, Proceedings of the 26th DAAAM International Symposium, 21-24 October 2015, Zadar, Croatia, ISSN 1726-9679, ISBN 978-3-902734-07-5, Katalinic, B. (Ed.), pp. 0426-0434, Published by DAAAM International, Vienna, Austria, DOI: 10.2507/26th.daaam.proceedings.057

[7] Anderson, B. (2008). Global static-priority preempted multiprocessor scheduling with utilization bound 38%, Proceedings of the 12th International Conference on Principles of Distributed Systems, 15-18 December 2008, Luxor, Egipt, ISSN 0302-9743, ISBN 3-540-92220-9, Baker, T. P., Bui, A. & Tixeuil, S. (Eds.), pp. 73-88, Springer-Verlag Berlin Heidelberg, DOI: 10.1007/978-3-540-92221-6_7

[8] Baruah, S. (1995). Fairness in periodic real-time scheduling, Proceedings of 16th IEEE Real-Time Systems Symposium, 5-7 December 1995, ISSN 1052-8725, ISBN 0-7803-3199-0, Pisa, Italy, pp. 200-209, IEEE Computer Society Press Los Alamos, California, DOI: 10.1109/REAL.1995.495210

[9] Anderson, J. & Srinivasan, A. (2000). Early-release fair scheduling, Proceedings 12th Euromicro Conference on Real-Time Systems, 19-21 June 2000, ISSN 1068-3070, ISBN 0-7695-0734-4, Stockholm, Sweden, pp. 35-43, IEEE Computer Society Washington, DC, USA, DOI: 10.1109/EMRTS.2000.853990

[10] Cho, H.; Ravindran, B. & Jensen, D. (2006). An optimal real-time scheduling algorithm for multiprocessors, Proceedings of 27th IEEE Real-Time Systems Symposium, 5-8 December 2006, ISSN 1052-8725, ISBN 0-7695-2761-2, Rio de Janeiro, Brazil, pp.101-110, IEEE Computer Society Washington, DC, USA, DOI: 10.1109/RTSS.2006.10

[11] Sha, L.; Rajkumar, R. & Lehoczky, J. P. (1990). Priority inheritance protocols: an approach to real-time synchronization. IEEE Transactions on Computers, Vol. 39, No 9, September 1990, pp. 1175–1185, ISSN 0018-9340

[12] Danilov, M. (2001). Methods of task scheduling in real-time systems. Software Products and Systems, No. 4, December 2001, pp. 28-35, ISSN 0236-235x (in Russian)

[13] Baranov, S. & Nikiforov, V. (2015). Transitive task priority inheritance in real-time multi-task applications. SPIIRAS Proceedings, 2015, No. 6(43), pp. 114-134, ISSN 2078-9599 (in Russian)

[14] Nikiforov, V. & Danilov, M. (2000). Static processing of real-time software system specifications. Software Products and Systems, No. 4, December 2000, pp. 13–19, ISSN 2036-235x (in Russian)

[15] Nikiforov, V. (2014). A protocol for prevention of task mutual blocking in real-time systems. Journal of Instrument Engineering, Vol. 57, No. 12, December 2014, pp.21-27, ISSN 0021-3454 (in Russian)

[16] Nikiforov, V. & Pavlov, V. (2011). Structured models for analysis of multitasking software systems. Information-Measuring and Control Systems, Vol. 9, No. 9, pp.19-29, ISSN 2070-0814 (in Russian)

DOI: 10.2507/28th.daaam.proceedings.117

COMPARISON OF METHODS FOR MODELLING PRODUCTION FLOOR LAYOUT

Strapek Martin, Hořejší Petr & Polcar Jiří

This Publication has to be referred as: Strapek, M[artin]; Horejsi, P[etr] & Polcar, J[iri] (2017). Comparison of Methods for Modelling Production Floor Layout, Proceedings of the 28th DAAAM International Symposium, pp.0835-0844, B. Katalinic (Ed.), Published by DAAAM International, ISBN 978-3-902734-11-2, ISSN 1726-9679, Vienna, Austria
DOI: 10.2507/28th.daaam.proceedings.117

Abstract

This paper deals with comparison of two methods for production floor layout design or reengineering – conventional capturing method and 3D laser scanning. Firstly there is an analysis of the current state, where possible uses of scanning technology for production floor design or reengineering are mentioned. Then the processes of both methods are shortly described. The main part of the article is comparison of these two method. For the comparison we have used data acquired during capture of the real factory hall with both methods. The comparison is based on determined criteria – time consumption, user and hardware requirements and quality of outputs. In the conclusion, the suitability of the methods is evaluated. It also shows time consumption differences between methods and difficulty and level of requirements. In the end there are recommendations for scanning method use and for the future work.

Keywords: laser scanning; production floor modelling; factory point cloud; factory modelling

1. Introduction

3D laser scanning technology is no longer a new technology. This fast and precise data acquisition method is used in several areas: architecture, geodesy, archeology, criminology, show business, etc. It also penetrates into engineering, which this paper is focused on, specifically on the use of the technology for production floor planning.

The work continues on work [1], where possible methods of production floor layout modelling were described. Production floor layout [2] is one of the most important tools for production planning.

In this paper these methods are compared with the common capturing method for production floor modelling. Both methods are briefly explained. Then there are described three criteria for comparison. These were selected based on the type of expected outputs. Next, the evaluating criteria and their weights are set and then the methods are compared. The conclusion describes which method is best for each type of output.

2. Production floor scanning

The first company to use 3D laser scanning for layout planning was Volvo. At that time (1996-1997), the available technology was limited. But since then, many experiments using the technology in different areas have been made. Production floor layout modelling [3], visualization and problem solving in production [4] and connecting scanned models into a simulated environment [5] have been developed for the engineering field.

Data acquisition and data processing are described by [6], and for production floors specifically by [1]. During the scanning process, the spatial coordinates of a scanned object are acquired and these are then converted into digital form, a 'point cloud'. An example of a point cloud is given in Fig. 1. Due to the very high performance of scanning devices, it is more important to focus on processing the acquired data. Work with these data is sometimes very complicated, especially when we want to generate a full 3D model from a point cloud model. This issue is dealt with for example by [7] and [8]. Other possible uses of algorithmization are being worked on [9]. This algorithm is used in the data processing method [1] for rapid creation of a 2D production floor layout from a point cloud model. Parts of scanned production floors can be also used as models for augmented reality systems [10].

Fig. 1. Example of building point cloud

When acquiring data with laser scanning it is necessary to use a scanning device. As was mentioned, performance of these devices is now at a much higher level than older models. The main limiting parameter of a scanner is its scanning speed. The highest scanning speeds these days are about 1 000 000 points per second. It is probably the highest possible speed. Even higher speeds are not necessary, because they would require new technologies, and the time savings would be insignificant compared to the present devices.

Instead of increasing speed, the producers of these devices are focused on more tailor made scanners. A great example is the BLK 360 [11]. This scanner has 360 000 pts/s speed, which is definitely enough for high quality point cloud acquisition in a very short time. This speed is more than enough for production floor scanning, so it would be useless to increase the cost of a faster scanner. On top of that, BLK360 excels with its dimensions, which reduce the problems with scanner positioning in a limited production floor space.

3. Methodology

The following sections describe how the use of scanners effects production floor layout modelling. The times of individual parts of the process will be shown, as will other criteria. These values will be compared with the basic method for production floor modelling and recommendations will be made for when to use a specific method.

3.1 Conventional capturing method

For classic capturing and modelling, common measuring tools (laser meter, camera, machine and equipment sketches) are used. For the classic method, employees must capture the dimensions of the basic factory hall, machines, equipment and objects in the factory and their relative positions. Based on these acquired data and use of special software, a layout model can be made. It is necessary to create models of machines, equipment and objects, then the factory hall itself and place all the objects exactly as in the real state.

3.2 Laser scanning method

For the scanning method we used a 3D laser scanner (in our case Leica C5). In this method, the whole factory hall (including all objects) was scanned. Then the acquired data were loaded into special software, where a complete 3D model was created. This model can be used for further processing as described in section **Production floor scanning**. The technical parameters of our scanning devices are on [12]. The most important parameter is scanning speed, which is 25 000 points per second (pts/s) in our case.

3.3 Reference model and outputs

A real factory was chosen to compare the methods. The factory was first measured and modelled using the classic method. Next we tried to acquire the same (or similar) output using 3D laser scanning. The time between both captures was about one year. Each capture was done by two different teams of employees. For both methods, two employees were used. In each group, we used experts in the specific method.
Both of these methods were tested in three main categories, which were determined based on the required output:

- 2D block scheme layout (layout modelled on the level of workplaces or parts of the production)

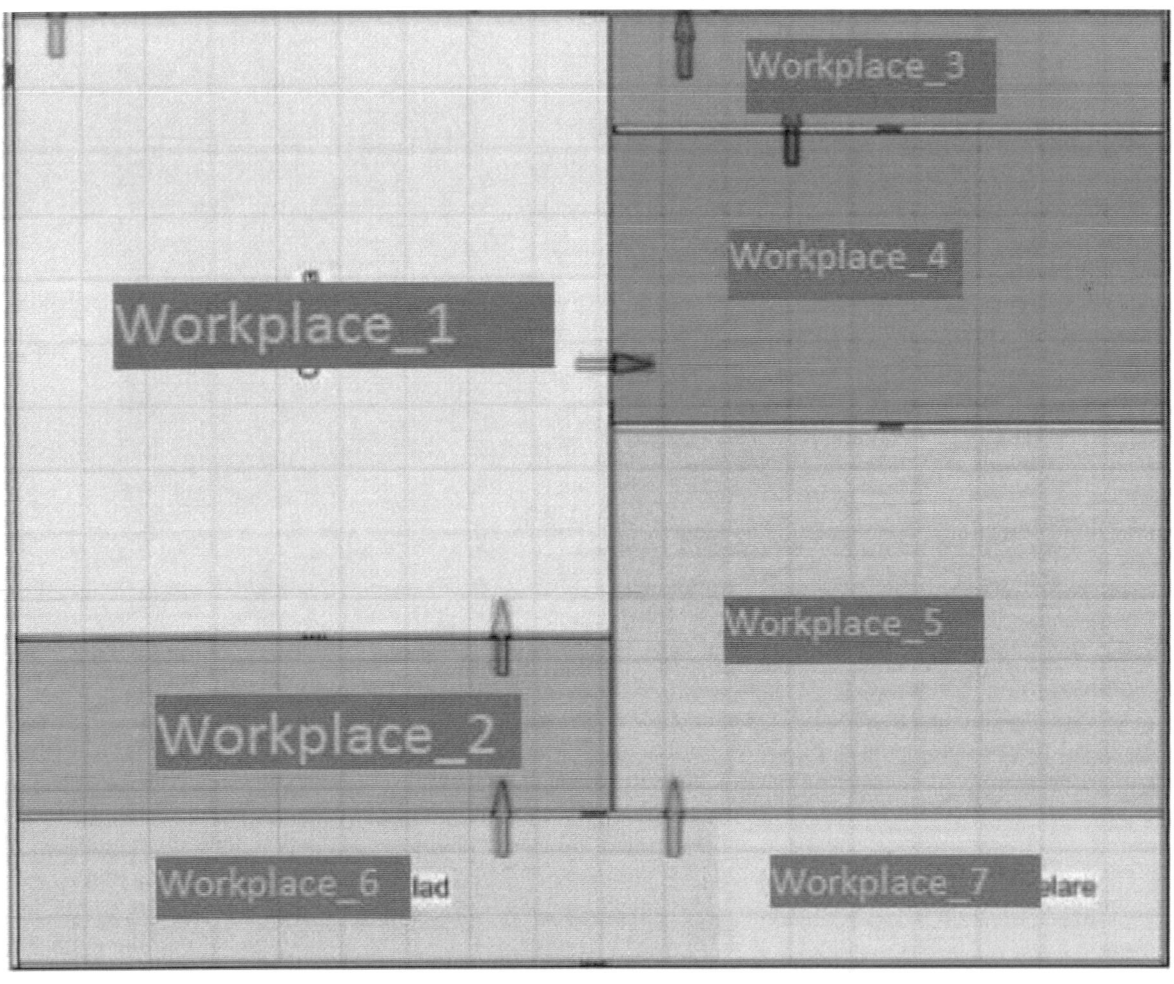

Fig. 2. Example of 2D block scheme layout

- 2D machine layout (layout modelled on the level of individual machines and equipment)

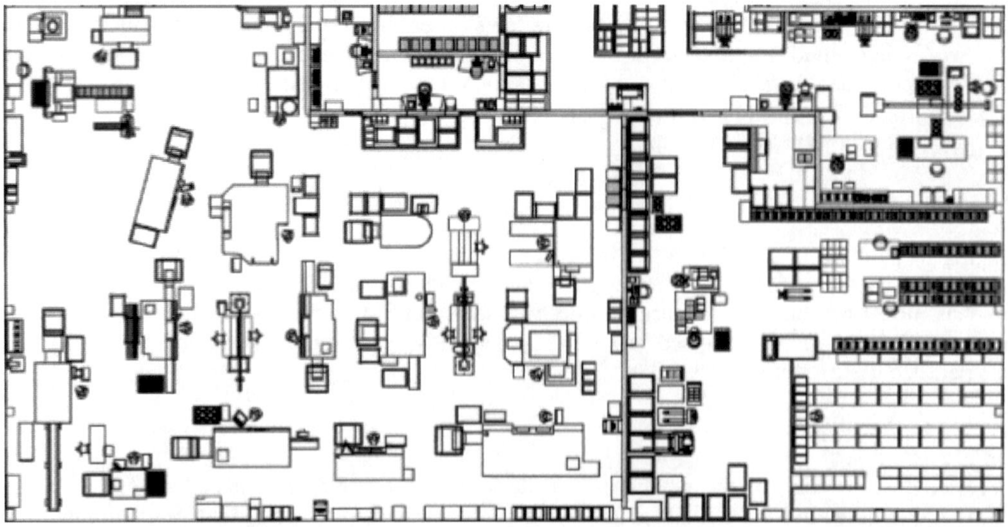

Fig. 3. Example of 2D machine layout

- 3D machine layout (layout modelled on the level of individual machines and equipment and visualized in 3D)

Fig. 4. Example of 3D machine layout

The effort was to create models that have a similar use for further processing. Currently, our most used software for layout modelling is VisTable software [13]. This software is very user friendly, cheap and it contains the analyzing tools which are required for layouts. We wanted to create outputs from the scanned models that have the same possibilities for analysis as the outputs from the classic method.

3.4 Criteria

The evaluation criteria were selected for each category. These criteria evaluate the whole process of output creation and its quality:

a) Data acquisition time

These times were taken during data acquisition in both methods. As already mentioned, the selected factory was first captured using the classic method, and then it was scanned using the Leica C5 laser scanner.
For comparison, we estimated the final times for a more powerful scanning device (for example, Leica BLK360). The time duration is considered from the start of measurement to the acquisition of the last dimension. Times needed for acquiring missing dimensions are also counted in this time.

b) Output development time

This time was measured from the first processing of the acquired data to the end of creation of the complete model. In the classic method this is the total of:

- Sum of times needed for creating models of machines, equipment and objects in CAD modelling software (times are converted to one man work)
- Time needed for creating layout in visTable software or Autodesk Factory Design (times for creating model in both types of software are similar)

The classic method output was a layout in visTable format. The output is a layout which is completely ready for subsequent analysis.
In the scanning method, the output creation time is the total of:

- Time needed for transferring scanned data into computer (because scanned data has very big volume, it was necessary to count this time)
- Time needed for cleaning data (removal of unnecessary point clouds)
- Time needed for data registration (registration = assembling individual scans into one complete model)

c) User requirements

User qualifications and skills must be at a level such that s/he can produce the required output at the required quality using the appropriate method. This criterion was set based on the quantity of operations, software and equipment which the user must handle during the process of data acquisition and output creation.
In the classic method, the user must handle a laser meter and sketching. These are activities which do not require any special skills. For layout creation, the user must handle visTable software and modelling CAD software. These activities usually go in parallel, so only knowledge of one particular type of software is needed for the user.
In the scanning method the user must handle the scanning device. Basic training for using a scanning device can take 1 – 2 days. The user must handle work with software for scanned data processing (Leica Cyclone in our case). Basic training and mastering skills with the software can take 1 day. For further data processing (AutoCAD, visTable, Inventor) the user must also know how to work with this software. Due to the connection of all activities during data processing, it is appropriate for one user to master them.
The classic method requires the knowledge of two types of software for a given methodology. The scanning method requires working with point clouds in this software and also requires the knowledge of scanned data processing software.
The scale of evaluating user requirements was set based on these criteria. The scale is from low requirements to very high requirements.

d) Hardware requirements

In hardware requirements, the computer equipment for creating smooth outputs was evaluated. The classic method does not require any special equipment. The recommended requirements for visTable software [13] and for AutoCAD software [14] meet the performance of commonly used computers.
For the scanning method, even for meeting the Leica Cyclone requirements [15], there is a need for a high performance computer. When working with large projects, like factory halls, there is a need for the computer to process a big volume of data (sometimes GBs files).

e) Quality

Match with the real object, visualization and clarity of model is evaluated in the quality criterion. In the classic method, we only model things which affect the production system- the layout model is clear for understanding the process.
The scanning method captures the whole production system including objects unusable in production floor analyses. However, these objects can be used for other analyses (pipeline, distribution of electricity, air conditioning, etc.).

4. Method comparison

The multi-criteria decision making method was used to compare the methods. For the above-mentioned criteria, we set criteria weights according to the Table 1. We used the pair comparison method for determination of weights.

The most important criteria are time values, then hardware and the user requirements. Quality has the lowest weight because this is mainly the visualization of the output. Dimension matches with real objects are within the available norm for all methods.

Criterion	Data acquisition time (h)	Output creation time (h)	User requirements	Hardware requirements	Quality	SUM
Weight	0.3	0.3	0.15	0.175	0.075	1

Table 1. Criteria and its weigths

Input values for both methods are given in the Table 2., 3. and 4. Numerical values were set based on the measured data, others were set according to 3.4 Criteria. We compared the classic method, the scanning method using a Leica C5 and a method using a BLK360 device (time values were converted using the scanning speed parameter).

Output/Criterion	Data acquisition time (h)	Output creation time (h)	User requirements	Hardware requirements	Quality
2D block scheme	14	2	Low	Low	High
2D machine layout	31	63	Medium	Low	High
3D machine layout	34	63	Medium	Medium	High

Table 2. Input data for classic method

Output/Criterion	Data acquisition time (h)	Output creation time (h)	User requirements	Hardware requirements	Quality
2D block scheme	8	4.5	High	Very high	Very high
2D machine layout	15	5	High	Very high	Very high
3D machine layout	15	6	High	Very high	Very high

Table 3. Input data for laser scanning method using C5 device

Output/Criterion	Data acquisition time (h)	Output creation time (h)	User requirements	Hardware requirements	Quality
2D block scheme	2	4.5	High	Very high	Very high
2D machine layout	3	5	High	Very high	Very high
3D machine layout	3	6	High	Very high	Very high

Table 4. Input data for laser scanning method using BLK360 device

Next, the criteria values were scored 1 – 5 (1 – the worst, 5 – the best), followed by multiplying by criteria weights. The weighted criterion sum represents the suitability of using individual methods.

4.1 2D block scheme layout

According to Table 5. And Fig. 5. we can see that for creating a simple block scheme layout, it is most appropriate to use the classic method. Even though Fig. *6* shows that data acquisition time and output creation time are a little shorter for the scanning method, according Fig. 7, using the scanning method is unnecessarily complicated and difficult. Because the graphs for other criteria comparison are almost the same, we have merged them into one graph.

Method/Criterion	Data acquisition time (h)	Output creation time (h)	User requirements	Hardware requirements	Quality	SUM
Classic	1*0.3	5*.03	5*0.15	5*0.175	3*0.075	3.65
Leica C5 scanner	3*0.3	3*.03	1*.015	1*0.175	5*0.075	2.5
BLK360 scanner	5*0.3	3*.03	1*0.15	1*0.175	5*0.075	3.1

Table 5. 2D block scheme layout evaluating

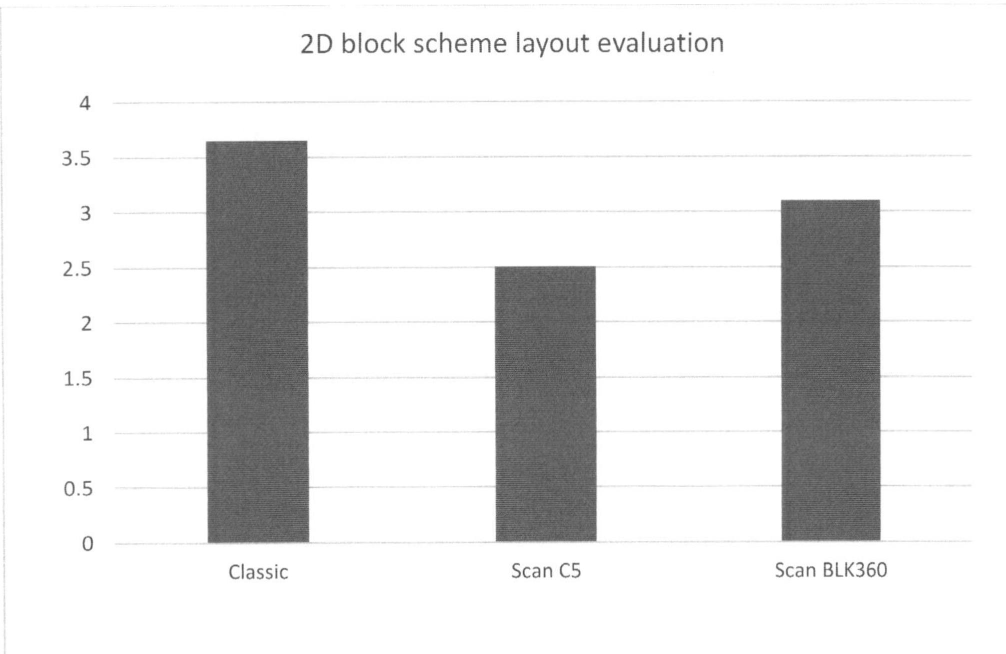

Fig. 5. Comparison of methods for 2D block scheme layout modelling

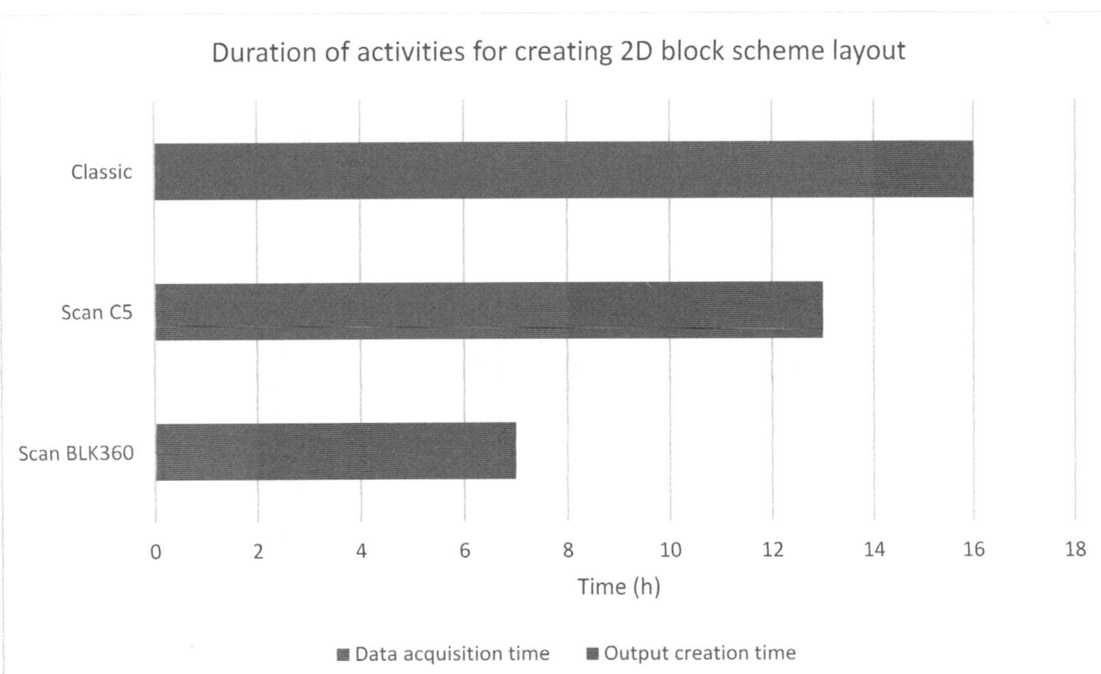

Fig. 6. Activities duration comparison for 2D block scheme layout

Fig. 7. Criteria comparison for 2D/3D layout

4.2 2D/3D machine layout

Also, because time values and other criteria values for both of these outputs are very similar, we have merged them into one category. For these outputs according to Table 6.and Fig. 8. it is preferable to use scanning methods. The best choice by far is the device BLK 360. Time savings are approximately 5 times shorter, which compensates for the higher requirements of the method.

Method/Criterion	Data acquisition time (h)	Output creation time (h)	User requirements	Hardware requirements	Quality	SUM
Classic	1*0.3	1*.03	5*0.15	5*0.175	4*0.075	2.525
Leica C5 scanner	3*0.3	5*.03	1*.015	1*0.175	5*0.075	3.1
BLK360 scanner	5*0.3	5*.03	1*0.15	1*0.175	5*0.075	3.7

Table 6. 2D/3D machine layout evaluating

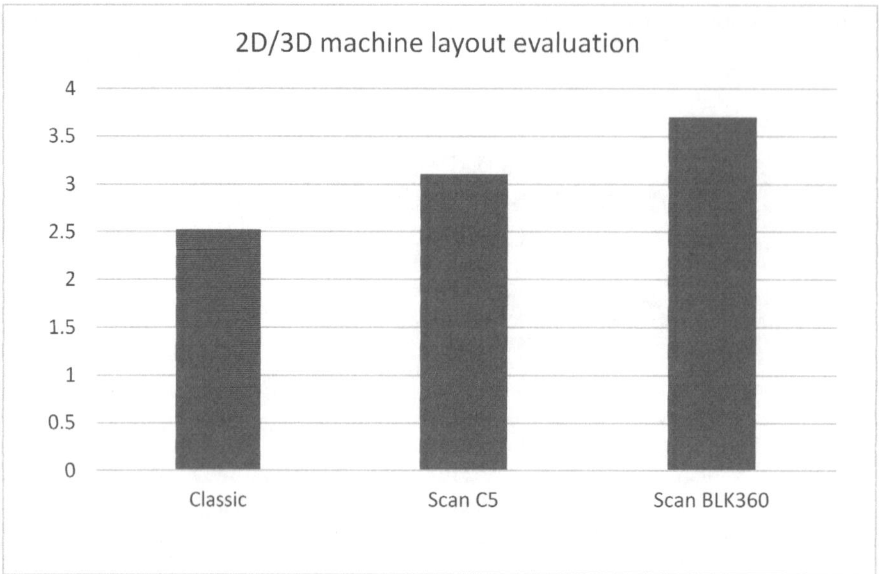

Fig. 8. Comparison of methods for 2D/3D machine layout modelling

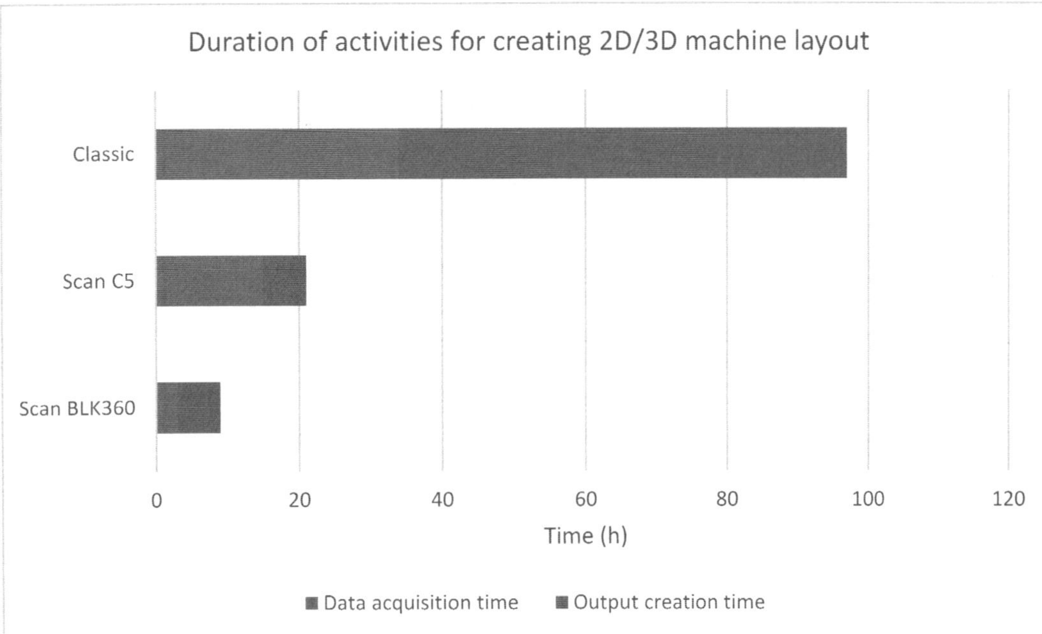

Fig. 9. Activities duration comparison for 2D/3D machine layout

5. Evaluation of results, recommendations, future work

The main purpose of this work was to compare efficiency of two methods for production floor layout modelling. For the comparison, the real factory was chosen and both methods were tested. During testing, we were acquiring data for comparison – time values, requirements for methods and quality of outputs.
Based on the results we can say that for creating a block scheme it is more suitable to use the classic measuring method.

Even though creating a layout using this method takes longer, using the scanning method would be too complicated. We would have to use complicated processes (scanning, data transformation, clearing data and creating output) and the time savings are not that big. When using the scanning method for machine layout capture, especially 3D, the time saving is already big enough that the method is more suitable.

Economically, purchase of scanning equipment and necessary hardware is recommended for repeated use of the scanning device. It is more suitable for bigger companies with a large number of factories, or for companies which offer scanning as their service. For scanning a production floor only once it is better to outsource the services of a company which already has the needed equipment.

Due to the falling price of scanners with parameters suitable for production floor measurement, it is probable that more companies will use this method for capturing their layouts. With the increasing amount of scanned data used, new opportunities in algorithmization can arise, for example, an algorithm for automatically generating proper primitives and library models from point clouds.

Because we have found scanning technology useful for production floor modelling, we will focus on developing algorithms for modelling simplification in the future work. Also we research possibilities for processing scanned data with the use of virtual and augmented reality devices like motion tracking systems.

6. Acknowledgement

This paper was prepared with the support of the Internal Science Foundation of the University of West Bohemia SGS–2015-065.

7. References

[1] Strapek, M., Hořejší, P., Polcar, J. (2016) 3D Laser Scanned Data Processing Possibilities for Production Floors Models. In Proceedings of the 28th International Business Information Management Association Conference. Norristown: International Business Information Management Association, 2016. pp. 2920-2930. ISBN: 978-0-9860419-8-3

[2] Chen, D., Shariatzadeh, N. (2012), „Software Eraluation Criteria for Rapid Factory Layout Planning, Design and Simulation“ at 45 CIRP Conference on Manufacturing Systems, ISBN 978-1-627-48512-8 Stockholm.

[3] E. Shellshear, R. Berlin, J. S. Carlson (2015), "Maximizing Smart Factory Systems by Incrementally Updating Point Clouds", IEEE Computer Graphic and Applications, sv. XXXV, č. 2 ISSN 0272-1716, pp. 62-69.

[4] Lindskog E., Berglund J., Vallhagen J., Johansson B. (2014), „Lean Based Problem Solving using 3D Laser Scanned Visualizations of Production Systems" at International Journal of Engineering Science and Innovative Technology, ISSN: 2319-5967.

[5] Lindskog E., Berglund J., Vallhagen J., Berlin R., Johansson B. (2012), „Combining Point Cloud Technologies with Discrete Event Simulation" at 2012 Winter Simulation Conference, ISBN: 978-1-467-34781-5 Berlin

[6] Bi, Z.M., Wang, L. (2010), Advances in 3D data acquisition and processing for industrial applications. From: Robotics and Computer-Integrated Manufacturing 26. March 2010. ISSN: 0736-5845

[7] Xiong, X., Adan, A., Akinci, B., Huber, B. „Automatic creation of semantically rich 3D building models from laser scanner data" at Automation in Construction. ISSN: 0926-5805. [2013]

[8] Tang, P., Huber, D., Akinci, B., Lipman, R., Lytle, A. „Automatic reconstruction of as-built building information models from laser-scanned point clouds: A review of related techniques" at Automation in Construction. ISSN: 0926-5805. [2010].

[9] Polcar, J., Gregor, M., Hořejší, P., Kopeček, P. Projection of LiDAR Point Cloud Slices to Raster Images as 3D Modelling Underlays. In Innovation Management and Sustainable Economic Competitive Advantage: From Regional Development to Global Growth. Madrid: International Business Information Management Association, 2015. s. 2220-2226. ISBN: 978-0-9860419-5-2

[10] Hořejší, P., (2015), Augmented Reality Systems for Virtual Training of Parts Assembly in Procedia Engineering, Volume 100 – 2015, ISSN 1877-7058, edit Katalinic, P., pp. 699-706, Zadar, Croatia.

[11] https://lasers.leica-geosystems.com (2017). Leica BLK360 Product specification, Accessed on: 2017-06-21

[12] https://gefos-leica.cz (2017). Leica C5 Product specifications, Accessed on: 2017-06-21

[13] http://cloud.vistable.de (2017). Systemvoraussetzungen-Requirements, Accessed on: 2017-06-20

[14] https://knowledge.autodesk.com (2017). System requirements for AutoCAD 2017, Accessed on: 2017-06-20

[15] https://hds.leica-geosystems.com (2017). Leica Cyclone datasheet, Accessed on: 2017-06-19

DOI: 10.2507/28th.daaam.proceedings.118

GAIT PHASE RECOGNITION FOR EXOSKELETON CONTROL USING ADAPTIVE NEURO FUZZY INFERENCE SYSTEM

Broniszewski Michał, Dąbroś Jakub, Iwaniec Marek,
Patyk Mateusz & Wesół Jacek

This Publication has to be referred as: Broniszewski, M[ichal]; Dabros, J[akub]; Iwaniec, M[arek]; Mateusz, P[atyk] & Wesol, J[acek] (2017). Gait Phase Recognition for Exoskeleton Control Using Adaptive Neuro Fuzzy Inference System, Proceedings of the 28th DAAAM International Symposium, pp.0845-0853, B. Katalinic (Ed.), Published by DAAAM International, ISBN 978-3-902734-11-2, ISSN 1726-9679, Vienna, Austria
DOI: 10.2507/28th.daaam.proceedings.118

Abstract

Pressure measurement at the foot-ground contact zone provides necessary information for detecting the phases of the human gait. The human gait is a complicated process and, therefore, it is not possible to distinguish its stages simply by comparing the foot-ground surface pressure to threshold values. In this paper we propose a new method for human gait phases recognition using an adaptive network-based fuzzy inference system which processes data measured by pressure sensors located underneath footbed inserts. This method is intended to bring some additional information to the lower limb exoskeleton control system as well as for gait diagnostic purposes. The device uses only three pressure sensors for each foot. In order to verify the results of the study, measurements with sensors placed under the footbed inserts has been performed and the pressure at the foot-ground contact points during the walk has been investigated. Twelve models of gait phase classifiers were created. The performance of each classifier has been tested for a previously unknown pressure pattern that corresponds to the natural gait. Based on the quality of gait matching the test run, the best model of the classifier was selected.

Keywords: gait-phase detection; adaptive neuro fuzzy; pressure sensors; force sensitive resistor (FSR)

1. Introduction

Walking is an execution of repetitive sequences composed of the movements of the lower limbs in order to proceed forward with the simultaneous keeping of proper balance [1]. It is also one of the most convenient forms of locomotion [1] which enables people to perform basic daily activities [2]. A single, non-recurring sequence of movement performed by one leg is called a gait cycle [1]. Gait cycle considered in the extended perspective can be divided into eight unique walking phases [3].

Recognition of user current walking phase is a key element of the control algorithms of the exoskeleton of the lower extremities [3][4][5]. Very often, in order to simplify the control system in these devices, designers reduce the amount of reproduced walking phases, in spite of the awareness of seven or more gait phases [3][4][5].

In most cases, signals such as threshold values of pressure sensors, angles observed in knee and hip joints, or readings from electromyographic sensors, are used to determine the state of the structure and its operator [3][4][6].

While the angular trajectories should be very similar for most people, the ground reaction forces (GRF) will vary according to the weight of the tested person. For this reason, we believe that gait phase recognition based on the input value comparison with a set of predetermined threshold values is an inadequate and strongly restrictive approach to the functionality of an exoskeleton.

Modeling systems based on traditional mathematical tools is not suitable for working with vaguely formulated or difficult to define models that are better suited for use with fuzzy reasoning systems simulating human decision-making [7].

In this paper, we propose a solution to the problem of gait phase detection for diagnosis or use in the control system of the lower limb exoskeleton in form of adaptive neuro-fuzzy inference systems. The proposed adaptive system is based on an artificial neural network, in which neurons are defined and linked together in such a way as to simulate a fuzzy reasoning system [7]. The great advantage of this solution is that there is no need to know the exact model of the object or process being analyzed. In order to determine the appropriate number of fuzzy sets to which the reading at any given time may belong, it is sufficient to provide input data and targeted output data with an addition of an appropriate input data conceptual analysis.

2. Review

Recent advances in a wearables have contributed to the development of techniques for a gait phase automatic detection and recognition [8]. These techniques are based on the measurement of muscle activity (EMG) [9], joint angles by using encoders or goniometers [10][11], but also accelerometers and gyroscopes [12] or the fusion of them both [13].

Walking aids such as exoskeletons and active orthoses use built-in sensors to control and to detect a user's intentions [4][9]. However, methods based on EMG will not necessarily be the best solution for supporting devices due to their individual variability [2]. There are also problems with electrodes placement on the body and difficulties with EMG signal processing. Exoskeleton constructions frequently use encoders or potentiometers for measuring joint angles, but also inertial measurement units (IMUs) as well as pressure sensors [14].

Foot-to-ground contact detection signals are directly translated into information about the current walking phase, hence the common use of pressure sensors in exoskeleton construction [HAL]. However, GRF signals do not provide information for the device control but give the basis for human movement detection which allows the control systems to adjust the control algorithms to the current walking phase [2]. However, it is often a problem to adapt pressure sensor systems to accommodate to different footprints and working conditions [14].

Research shows that the use of artificial neural networks based on force sensitive resistors (FSRs) located under the foot are the most precise methods of gait phase recognition, even for cases where the network has not been taught before. Similar neural network, which instead used accelerometer data, achieved significantly higher error values calculated with respect to gait pattern. Additionally, the accelerometers introduced recognition delays of over 120ms [15].

Much previous work uses at least four FSR sensors in their systems, which are typically located in four distinct places: under the heel, 1st, 4th or 5th metatarsal bones, and under the hallux [2][13][15]. Jung et. al. in their work, used eight pressure sensors located between the upper and lower shoe inserts [14]. Thanks to the use of velcro materials, the position of each sensor can be changed at any time to match the size of the measurement system to different foot sizes. This approach requires that the location of sensors individually adjusted for every patient before use. The commonly used method for determining gait phases is a method that compares readings with predetermined threshold values. The method is called the threshold method or the discrete event method [2]. Jung et. al. use the collected data from all eight sensors of the exoskeleton foot to determine the stance and swing phase. If the data from each sensor exceeds the predetermined threshold, it means that the exoskeleton is in the stance phase [15]. The use of eight pressure sensors to determine two gait phases is definitely uneconomical, as such a system can be replaced with one sensor on each foot.

Significantly better results are obtained by using artificial neural networks using pressure sensor or accelerometer data [15]. The authors used four pressure sensors that became inputs for the multi-layer perceptron (MLP). The structure of the used network consisted of four neurons (for each input), one output neuron and fifteen neurons in each of the two hidden layers. The network was taught to detect the stance phase and swing phase. Its results showed a 2.8% error [15]. Despite the high efficiency of the method, it was possible to recognize only two phases of walking, which in the case of the support device control is insufficient.

The use of gait phases in control algorithms requires the changes between each phase to be more fluid rather than rapid. Similarly, during the walk, walking phases are very smooth, since the human gait is extremely energy efficient and to a large extent, allows suppression of the reaction forces of the substrate [1]. The smooth transitions between the walking phases were obtained by Kong [2], who used the fuzzy logic method to recognize the walking phases using four pressure sensors placed under the shoe insert. The approach proposed by the author allowed the detection of six phases of walking [16]. However, the use of fuzzy logic requires an understanding of the problem to formulate a rule base [2]. Our adaptive neuro-fuzzy inference system (ANFIS), unlike fuzzy models, does not require previous knowledge of the object being analyzed and does not require the creation of a rule base. The rule base is created by the system based on the input data provided, thus excluding the possibility of error.

3. Materials and methods

The simplest walk cycle can be divided into two stages: *stance* (or support) and *swing* [1]. In the extended view of the gait cycle, it has been divided into eight unique phases [1]. Each phase has its own footprint pattern, that is, the surface that contacts the ground. The distinction between the stance and swing period is possible with one pressure sensor on each foot (pressure is present, or no pressure is present). On the other hand, the distinction between all eight walking phases requires the use of multiple measuring systems.

3.1. Design concept

Human gait is an extremely complex operation, so it is difficult to determine when the gait cycle begins and ends. The Rancho Los Amigos Committee from the American National Rehabilitation Center has established that the heel strike (*Initial Contact, IC*) phase is the initiating phase of the gait cycle [1]. Contrary to this theoretical consideration, this study considered a first phase of the gait (zero index) to be the swing phase, secondary to the heel strike and the following phases.

During walking, body mass swings across the arch along the surface of the foot from the heel toward the fourth and fifth metatarsal bones, and then follows a straight line toward the large toe. Body weight is transferred through the closed kinematic chain created by the skeleton to the ground. The places of greatest pressure at the foot-ground contact zone correspond to the anatomical structure of the skeleton of the foot [17]. Although tissues act as depreciation, the greatest pressure values occur at the areas where the foot bones are closest to the ground. Therefore, it is reasonable to place the pressure sensors in these places.

In the presented solution, only three pressure sensors were used for each foot. Sensors were located at the locations of the largest reaction forces of the substrate. Fig. 1 shows the location of the sensors placed under the footbed inserts. The first sensor is located at the center of the heel (Fig 1, label 3), the second is located near the fifth metatarsal bone (Fig. 1, label 2), and the third is placed under the hallux (Fig. 1 label 3).

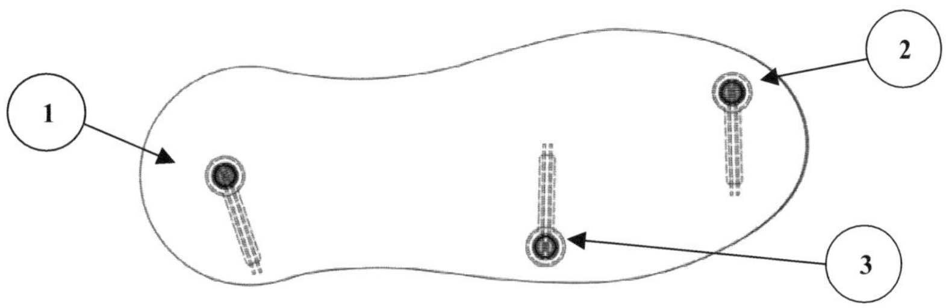

Fig. 1. Location of the pressure sensors: 1 - heel, 2 - 5th metatarsal bone, 3 - hallux.

Each gait phase can be identified by a unique support surface. Fig. 2a shows the sequence of changes in the foot support surface that contacts the ground during the support period. The Initial Contact (IC) phase begins with the first contact of the heel with the ground. It takes a very short time (about 2% of the total gait cycle, GC), so it is often analyzed along with the *Loading Response* (LR) phase [1].

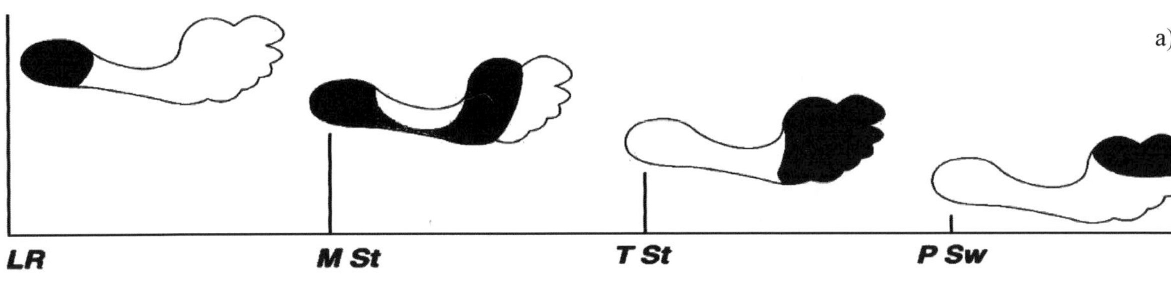

Fig. 2a. Sequence of changes in the foot support areas during stance. Based on [18].

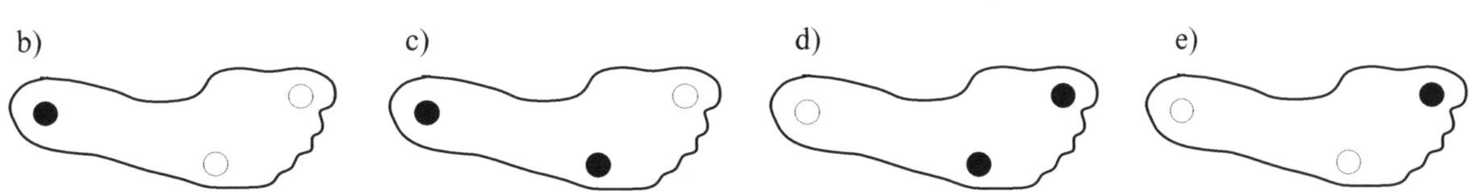

Fig. 2b-e. Representation of pressure exerted by the foot on the sensors at specific gait phases.

At this time only the back of the foot, specifically the heel, contacts the ground, while a body weight is gradually transferred to the lower limb. During the initial two phases of the walk, a large amount of pressure is recorded from the sensor located on the heel. Fig. 2b shows the pressure measured by the sensor, which has been represented by the black circle. The other two sensors record no pressure or very small pressure, as shown in Fig. 2b - circles in short dash lines. During *Mid Stance* (*M St*) phase, the foot remains straight, so pressure is measured on the heel and the fifth metatarsal bone (Fig. 2c). During *Terminal Stance* (*T St*) phase, the heel pressure disappears while the pressure on the forefoot increases. At this time pressures are recorded by sensors located under the fifth metatarsal bone and under the hallux (Fig. 2d). The last recordable phase during the support period is the *Pre-Swing* (*PSw*) phase, in which the pressure is only recorded by the sensor placed under the hallux (Fig. 2e).

During the swing period, the foot does not touch the ground, so each sensor on the transferred leg will measure the pressure close to zero. Therefore, the swing phases will be interpreted as one phase.

Despite the fact that only three sensors were used, all of the subphases during the stance period could have been distinguished. Sensor placement points were selected based on analysis of the changes in the foot support areas during stance (Fig. 2a). Placing sensors under the heel and hallux is necessary. On the other hand, placing the sensor under the fifth metatarsal bone allows the distinction between the Terminal Stance phase and the Pre-Swing phase. Such sensor placement ensures that each phase is uniquely distinguished on the basis of a readout from at least one sensor.

3.2. Hardware, software and data acquisition

The measuring system consists of three pressure sensors, resistance-voltage converters, voltage followers and microcontroller with analogue-to-digital converters (Fig. 3). The acquisition system was based on an Arduino Due developer kit with an Atmel ATSAM3X8E processor (84 MHz, 32bit ARM Cortex-M3 architecture). The microcontroller is equipped with 12-bit analog-to-digital converters, resulting in measurement accuracy of ≈0.8 mV with a measuring range of 3.3V.

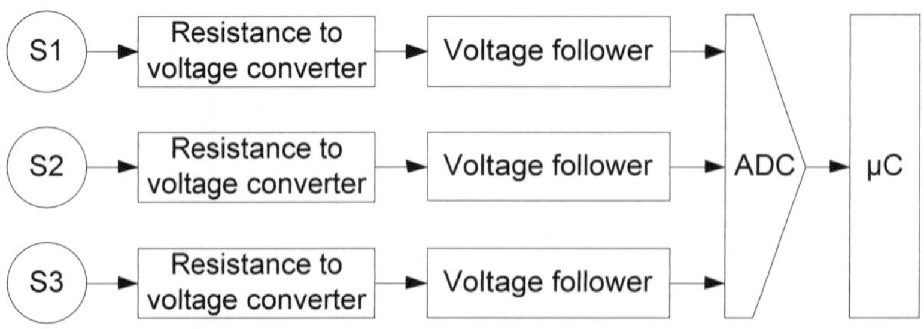

Fig. 3. Block diagram of the measuring system.

Three CP151NS sensors from IEE International Electronics & Engineering were used for one foot. The used sensors are the Force Sensitive Resistor (FSR), which shows a decrease in resistance due to the increasing pressure exerted on them. Each of these sensors is plugged into a voltage divider that comes with the potentiometer. It allows the voltage range to be adjusted at the output of the V_{OUT} divider (1) by controlling the resistance of the R_{POT} potentiometer.

$$V_{OUT} = V_{+} \frac{R_{POT}}{R_{POT} + R_{FSR}} \tag{1}$$

In such an arrangement, the voltage at the output of the divider increases as the force applied to the sensor surface increases, which is equivalent to the decrease in the resistance of the R_{FSR} sensor. Potentiometer resistance R_{POT} was set to value 10 kΩ and 3.3V supply voltage V_{+} was applied to the voltage dividers. The output of each voltage divider was connected to the voltage follower input and then the output of the divider was connected to the corresponding ADC input.

The transmission and pressure sensors program was written in C ++ using Atmel Studio 7, then compiled using Arduino toolchain and finally loaded into the microcontroller memory.

The subject whose gait data was collected was 185 cm tall and weighed 77 kg. The measurements were made for three sensors located under the footbed of the right foot. The sensors have been positioned as shown in Fig. 1. The recorded data was sampled with 1.3 kHz sampling rate, filtered in real time with a digital lowpass filter (2) implemented in the program structure and sent to a PC via a serial port.

$$y[i] = \alpha * x[i] + (1 - \alpha) * y[i - 1] \tag{2}$$

$y[i]$, $x[i]$ and $x[i - 1]$ denote the filter output in i-th sample respectively, the output from the previous sample filter, and the current readout value from the ADC. Parameter α is a smoothing agent and is in the range (0,1), where the value close to zero means a strong smoothing of the waveform. The value of α was set at 0.1.

3.3. ANFIS

Angular variations in the joints of a subject's lower limbs during a walk in a given cycle can be described as non-linear in time [1]. Taking into account the differences between individuals, such as varying body weight or foot geometry or the possibility of deformation during the development of anatomical arches of the foot (e.g. flat foot), it is impossible to clearly describe the human gait with an exact mathematical model that can be applied to population at large. Pressure distributions in feet with lowered physiological arches due to the transfer of part of the human weight to a healthy support area that is not available in healthy people foot will appear to be completely different than those with properly developed foot geometry. It is therefore difficult to create an accurate, commonly used, gait classifier based on standardized sizes, in which case they are usually affiliated to collections limited by the predefined thresholds. This requires the use of membership to collections defined by the use of traditional logic used by exoskeleton creators. For example, if the detection of the passage between the heel strike and the loading response is defined as exceeding the threshold pressure value by the sensor readings on the fifth metatarsal bone, it is essential to provide information on the body weight of the threshold value used. If the calculated hypothetical value is assumed to be equal to half of the measuring range, based on the pressure-to-weight ratio of the 50 kg person-to-weight contact, the 100 kg subject will exceed the threshold value at an earlier stage of analogically considered walking cycle [20]. In addition, surface contact patterns for different patients may be different, hence threshold-based methods may not be able to correctly interpret the current walking phase in such diverse cases [2]. Due to the strong nonlinearity of a human gait dynamics in exoskeleton control systems, the control models vary depending on the phase detected at any given moment [4][14]. Despite problems with the manufacture of systems suitable for different footprints and conditions, classifiers based on pressure sensors are commonly used in active lower leg orthoses [14].

The proposed solution to the above problems is the adaptive neural-fuzzy inference system (ANFIS), first described by Jang with his example of systems that allow the modeling of nonlinear functions [7]. The ANFIS structure is based on an artificial neural network and consists of five layers in which the neurons have different structures and perform different functions [19]. The proposed structure is designed to respond to the analyzed problem. Each layer corresponds to another element of the fuzzy inference system. For the purpose of explaining the principle of operation of the network, a model based on three fuzzy sets describing each of the inputs was adopted.

The first, in order of the signal flow direction, is the input layer with the number of $n = 3$ perceptrons carrying information about the filtered data collected by the pressure sensors. As to the values, the individual nodes correspond to the values of the sensors on the heel, the fifth of a metatarsal, and the toe. The second layer present in the model is the layer of the input membership functions. At this level of the algorithm, each of the inputs is described by $m = 3$ membership functions, symbolizing respectively: no pressure or low pressure, medium pressure, high or total pressure. In the case of the second set we cannot speak of the mean as the value describing the signal, but as the subjective conclusion of the network. The value stored by each node in this layer is the result calculated by substituting the neuron response of the input layer with the pattern for the applied membership function. The third layer is the layer of inference rules, which contains $m^n = 3^3 = 27$ rules describing the inference system with logical expressions. The scheme of operation of this network layer can be presented in linguistic form "if $\mu_{0,i}$ and $\mu_{1,j}$ and $\mu_{2,k}$ then μ_{out}", or in mathematical form (3), where $\mu_{0,i}$, $\mu_{1,j}$ and $\mu_{2,k}$, are the results of the input membership functions, where the indices $\{i, j, k\}$ tell us which of the sets is considered for each input, as μ_{out} analogous value for the layer output.

$$\mu_{0,i} \wedge \mu_{1,j} \wedge \mu_{2,k} \Rightarrow \mu_{out} \quad (3)$$

The penultimate in the model, the layer of the function of the membership of the outputs is composed of the same number of fuzzy sets as the number of rules in the preceding layer $m^n = 3^3 = 27$. The values received by each neuron are subjected to the input of membership function of the layer output and they are dependent directly on the inference of the previous layer. This layer's nodes' output values have been normalized. The output value of the model is provided by the layer of aggregation and defuzzyfication. It is a layer composed of only one neuron with $m^n = 3^3 = 27$ inputs, each of a specific weight, which, when collected and defuzzyfied, is represented as a specific crisp output value.

The parameters of the model that can be changed (apart from the number of inputs already mentioned and the number of fuzzy sets belonging to the given input) include: the shapes of the membership functions, the choice between the AND and OR functions and the corresponding *t-norm* implemented in the rules base layer, learning vectors, *t-norm* for implication, output aggregation definition, and output defuzzification method. Our study involved the preparation and teaching of a set of 12 networks, two of which served as "master" networks, the starting point for modifying the parameters, which differed in learning time between them. These are networks c) and d), with the learning of the first being interrupted in about 80% of the second learning period. For a detailed description of the variability of the models, see Table 1.

4. Results

Neural network learning results were derived from a mean squared error between the pattern and the neural network result and presented in Table 1. The highest of the errors noted was the model with the linear function of the outputs membership and it was an error of 2 orders of magnitude larger than the others.

On the other hand, the smallest error characterized the network with an increased number of fuzzy sets - 5 for each input. The standard parameters for the study were: 19991 samples in teaching vectors, Gaussian input functions, the function of membership of outputs in the form of one-element sets, using the t-standard operation in the form of the product of sets, and finally using the method of teaching with the use of reverse propagation error.

Model	Length of learning matrix [samples]	Fuzzy sets number	Input membership function	Output membership function	*t-norm* operation for AND	Network learning method	Network Error
a)	10000	[3 3 3]	Gauss	singletons	prod	back propagation	0,2700
b)	15000	[3 3 3]	Gauss	singletons	prod	back propagation	0,2740
c)	19991	[3 3 3]	Gauss	singletons	prod	back propagation	0,2875
d)	19991	[3 3 3]	Gauss	singletons	prod	back propagation	0,2803
e)	19991	[3 3 3]	triangular	singletons	prod	back propagation	0,2688
f)	19991	[3 3 3]	trapezoid	singletons	prod	back propagation	0,2989
g)	19991	[3 3 3]	Gauss	linear	prod	back propagation	12,0755
h)	19991	[3 3 3]	Gauss	singletons	min	back propagation	0,3040
i)	19991	[3 3 3]	Gauss	singletons	prod	hybrid	0,2649
j)	19991	[4 4 4]	Gauss	singletons	prod	back propagation	0,2759
k)	19991	[5 5 5]	Gauss	singletons	prod	back propagation	**0,2460**
l)	19991	[7 7 7]	Gauss	singletons	prod	back propagation	0,5992

Table 1. Examined models of gait classifiers. The column representing the number of fuzzy sets presents the vector of the number of sets for each input.

Fig. 4 shows the comparison of the determined gait phase obtained from the operation of a given model with gait phase patterns, which were manually defined in the same way (using the same phase separation criteria) as for generating the resultant pattern vectors for the data provided to the learning network.

Waveform observation allows to provides a reference for the reliability of the numerical results describing the network's mean squared error. Single samples with high error values relative to the corresponding standard samples may lead to an increase in RMSE value. Detecting and counteracting such phenomena can lead to obtaining better classifiers than would be suggested by learning outcomes.

In addition to the network comparison chart, the pattern of absolute error values (Fig. 5) was also determined, allowing to confront the network operation for each phase of the walk. The highest values of errors were observed at the end of each walk cycle (assuming the order of gait phases used in the study). The shape of gait pattern should be examined carefully. The model is characterized by low error rates and relatively small derivatives, but it contains single peaks with higher values, which can be considered as a good match for the network.

This means that for most of the walk cycle, the fit to the pattern was high, but there were moments when the gait phase obtained from the classifier differed significantly from the one observed in reality. On the other hand, patterns characterized by multiple local maxima or relatively high values with minor variations may be understood as inferior network matching and greater inconsistency in real and classified walking phases.

Fig. 6 shows a direct comparison of the mean error of the classifier's performance against the standard shown in the average error and maximum error. The top line shows a comparison of all the discussed models, but due to significant differences (2 orders of magnitude) between the models, it was decided to reject two of them and display the crop data again in the following line, zeroing values for skipped models. The smallest RMSE model with five fuzzy sets for each of the inputs did not turn out to also be the model with the lowest maximum error, which was characterized by model e) from Table 1 which was using the triangular function of the inputs.

On both Fig. 4 and Fig. 5, only portions of the whole 40 seconds duration measurement are visualized. Fig. 6 shows the results from the analysis of the whole chart. For this reason, the maximum error values may not be visible in previous waveforms.

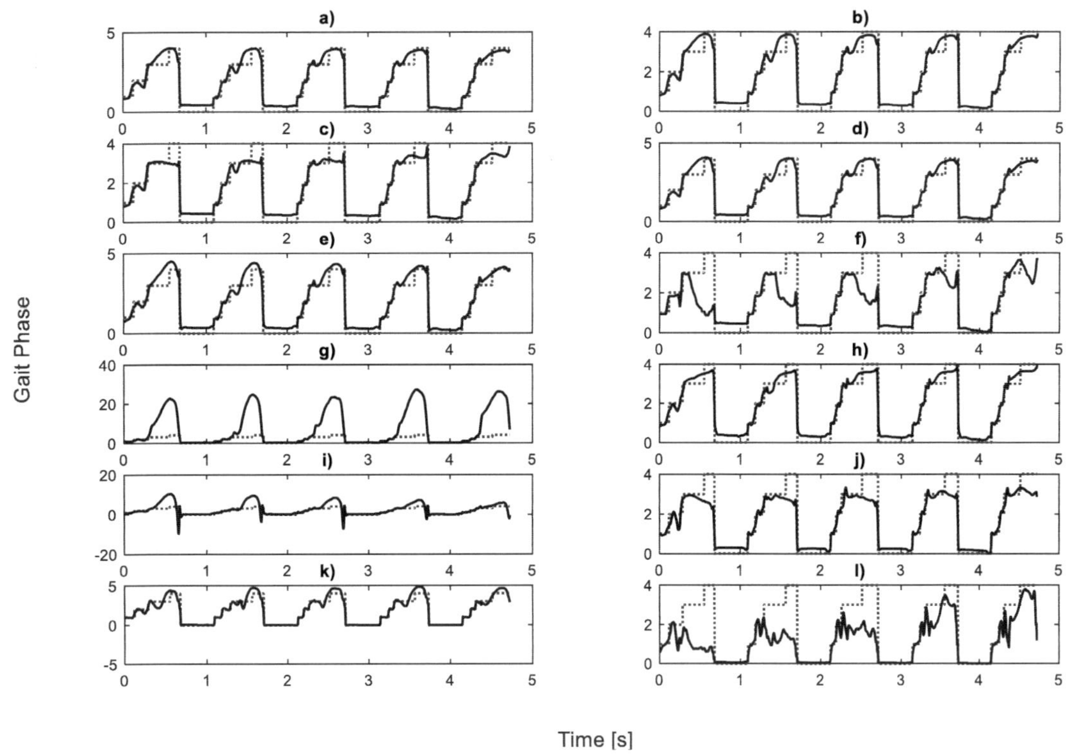

Fig. 4. Comparison of ANFIS results with manually prepared pattern. Only a fragment of the entire chart is presented. The individual graphs are described in the following lines in Table 1 showing the parameters of the system.

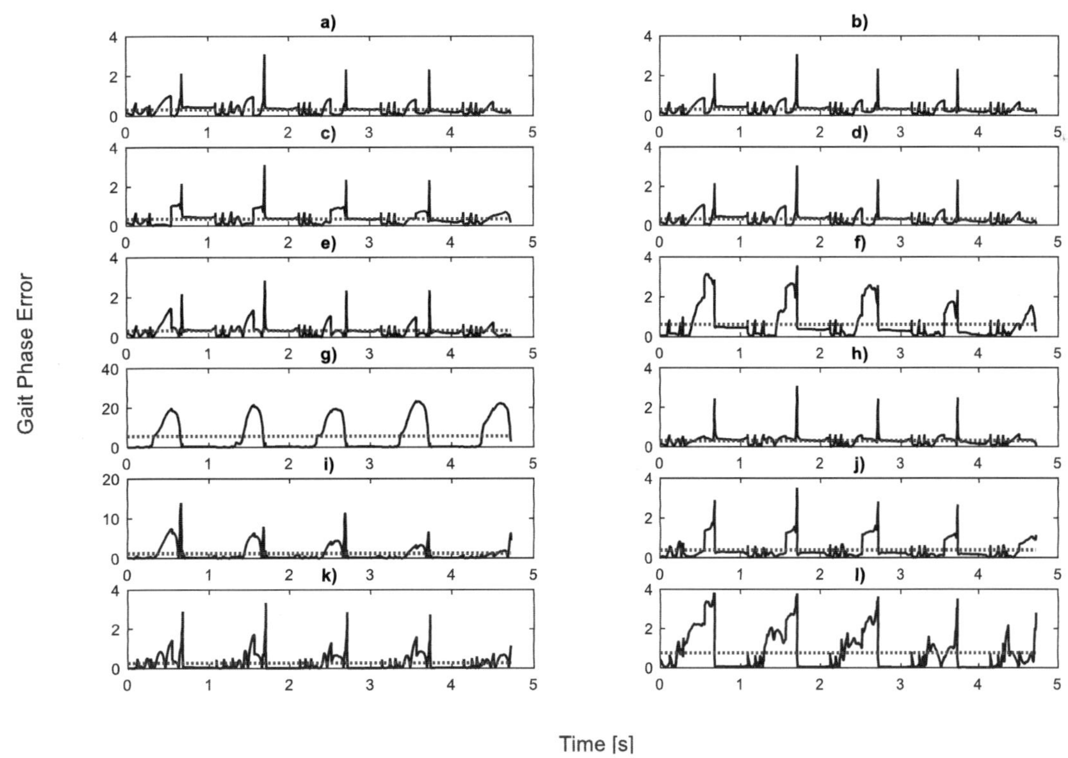

Fig. 5. Comparison of absolute error values of ANFIS results. Only a fragment of the entire waveform is presented. The dashed line symbolizes the average error value of the model. The individual graphs are described in the following lines in Table 1 showing the parameters of the system.

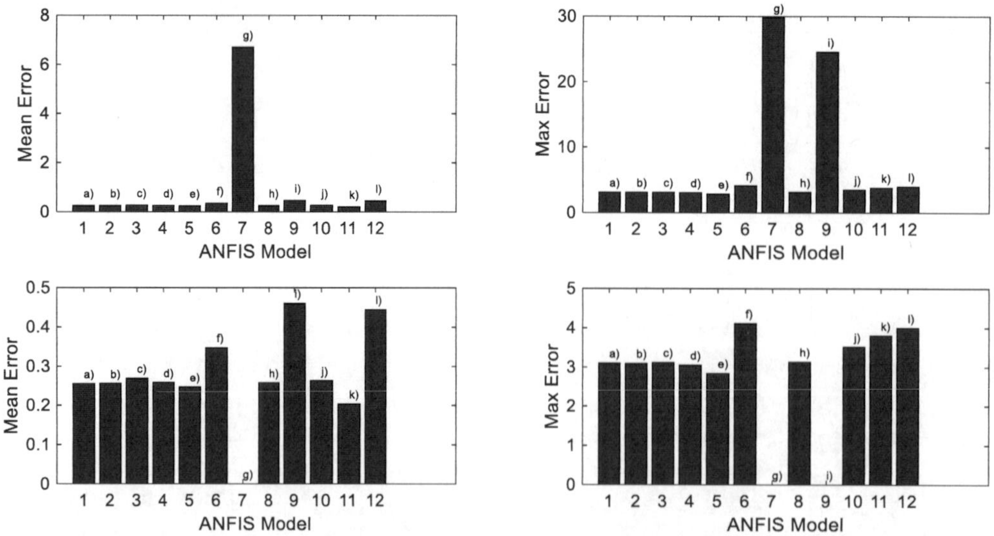

Fig. 6. Comparison of mean and maximum errors for each model. The top row contains graphs showing all tested models. The bottom line excludes results for models with much higher error values than the rest of the models.

5. Conclusion

This article presents a proposal for classifiers based on adaptive neuro-fuzzy inference systems that enable gait phase recognition. Proposed neural network was designed for use in control systems of bipedal robots including lower limb exoskeletons for rehabilitation or enhancement of physical capabilities of the operator.

The characteristic feature of the proposed classifiers is the use of only three inputs corresponding to the pressure sensors mounted at the foot-ground contact point, which considerably reduces the number of operations required to be performed by the central processing unit of the control system. This solution provides much shorter response time of the system, thereby allowing better control of the device stability, which increases the safety of the exoskeleton operator.

Twelve classifier models differing in their parameters were tested. The best results were achieved by model which was provided in full trajectories as learning vectors with a structure of five fuzzy sets, each described by the Gaussian membership function in sets. Outputs of the mentioned function were described by the membership function to one-element sets which used the *t-norm* operation in form of the product of fuzzy sets and finally using the method of output defuzzification using a weighted average. Despite higher than other models' values of maximum errors, the overall alignment of the classifier's output pattern to the input pattern was the best, and with proper data manipulation, it should lead to ultra-precise classifiers with low specificity, allowing work with a wide range of people.

Despite theoretically greater possibilities to differentiate the input states for the seven fuzzy sets model, it turned out that the results were unsatisfactory. Improvement of performance against classifiers using less fuzzy sets was expected. It is concluded that the low value of membership for a sample value centered between the center of gravity of two adjacent sets is more advantageous to the system than the ability to qualify a given sample to an additional set. This is a desired feature for implementation on microprocessor systems - fewer sets mean less demand for computing power. It would be useful, however, to investigate the reproducibility of the mentioned feature using different network parameters.

Further work should focus on carrying out measurements on a larger group of subjects with varying body weight. It is also proposed to research a problem of handicapped patients with walking difficulties in order to assess the ANFIS use for gait diagnostics.

6. References

[1] Perry, J. (1992). Gait analysis: normal and pathological function, SLACK, 978-1-55642-192-1, Thorofare, NJ

[2] Kong, K. & Tomizuka, M. (2008). Smooth and continuous human gait phase detection based on foot pressure patterns, 2008 IEEE International Conference on Robotics and Automation, Pasadena, CA

[3] Vaughan, C. L.; Davis, B. L. & O'connor, J. C. (1992). Dynamics of human gait, Human Kinetics Publishers, 978-0-620-23558-6, Champaign, Illinois

[4] Kazerooni, H.; Racine, J. L.; Huang, L. & Steger, R. (2005). On the Control of the Berkeley Lower Extremity Exoskeleton (BLEEX), Proceedings of the 2005 IEEE International Conference on Robotics and Automation, Barcelona, Spain, pp. 4353-4360, doi: 10.1109/ROBOT.2005.1570790

[5] Herr H. et al. Exoskeleton for running and walking, U.S. Patent No. 2007/0123997 A1, May 31, 2007

[6] Kenta, S.; Gouji, M.; Hiroaki, K.; Yasuhisa, H. & Yoshiyuki, S. (2007). Intention-based walking support for paraplegia patients with Robot Suit HAL, Advanced Robotics, Vol. 21, No. 12, pp. 1441-1469

[7] Jang, J.-S. R. (1993). ANFIS: Adaptive-Network-Based Fuzzy Inference System, IEEE Transactions on Systems, Man, and Cybernetics, Vol. 23, No. 3, pp. 665-685

[8] Tao, W.; Liu, T.; Zheng, R. & Feng, H. (2012). Gait analysis using wearable sensors, Sensors, Vol. 12, No. 2, 2012, pp. 2255–2283

[9] Sankai, Y. (2010). HAL: Hybrid assistive limb based on cybernics, Robotics Research, Springer, 2010, pp. 25-34

[10] Zoss, A. B.; Kazerooni, H. & Chu, A. (2006). Biomechanical design of the Berkeley lower extremity exoskeleton (BLEEX). IEEE/ASME Transactions On Mechatronics, Vol. 11, No. 2, pp. 128-138.

[11] Donno, M.; Palange, E.; Di Nicola, F.; Bucci, G. & Ciancetta, F. (2008). A new flexible optical fiber goniometer for dynamic angular measurements: Application to human joint movement monitoring. IEEE Transactions on Instrumentation and Measurement, Vol. 57, No. 8, 2008, pp. 1614-1620

[12] Zeng, H, & Zhao, Y. (2011). Sensing movement: Microsensors for body motion measurement, Sensors, Vol. 11, No. 1, 2011, pp. 638-660

[13] Parikesit, E.; Mengko, T. L. R. & Zakaria, H. (2011). Wearable gait measurement system based on accelerometer and pressure sensor, Proceedings of 2011 2nd International Conference on Instrumentation, Communications, Information Technology, and Biomedical Engineering (ICICI-BME), Bandung, Indonesia, pp. 395-398

[14] Mijailovic, N.; Gavrilovic, M.; Rafajlovic, S.; Đuric-Jovicic, M. & Popovic, D. (2009). Gait phases recognition from accelerations and ground reaction forces: Application of neural networks. Telfor Journal, Vol. 1, No. 1, pp. 34-36

[15] Jung, J. Y.; Heo, W.; Yang, H. & Park, H. (2015). A neural network-based gait phase classification method using sensors equipped on lower limb exoskeleton robots. Sensors, Vol. 15, No. 11, pp. 27738-27759

[16] Kong, K. & Tomizuka, M. (2009). A gait monitoring system based on air pressure sensors embedded in a shoe. IEEE/ASME Transactions on mechatronics, Vol. 14, No. 3, pp. 358-370

[17] Cavanagh, P. R. & Michiyoshi Ae. (1980). A technique for the display of pressure distributions beneath the foot. Journal of biomechanics, Vol. 13, No. 2, pp.69-75

[18] Barnett, C. H. (1956). The phases of human gait. The Lancet, Vol. 268, No. 6943, pp. 617-621

[19] Nyrkov, A.; Chernyi, S.; Zhilenkov, A. & Sokolov, S. (2016). The use of Fuzzy Neural Structures to Increase the Reliability of Drilling Platforms, Proceedings of the 26th DAAAM International Symposium, pp. 0672-0677, B. Katalinic (Ed.), Published by DAAAM International, 978- 3-902734-07-5, 1726-9679, Vienna, Austria, DOI: 10.2507/26th.daaam.proceedings.091

[20] Birtane, M. et al. The evaluation of plantar pressure distribution in obese and non-obese adults, Clinical Biomechanics, Vol. 19, No. 10, pp. 1055 - 1059

DOI: 10.2507/28th.daaam.proceedings.119

TECHNOLOGICAL AND ECONOMICAL ASPECTS OF ARC WELDING OF AL 5754

Petar Tasic*, Ismar Hajro & Damir Hodžić

This Publication has to be referred as: Tasic, P[etar]; Hajro, I[smar] & Hodzic, D[amir] (2017). Technological And Economical Aspects Of Arc Welding Of AL 5754, Proceedings of the 28th DAAAM International Symposium, pp.0854-0858, B. Katalinic (Ed.), Published by DAAAM International, ISBN 978-3-902734-11-2, ISSN 1726-9679, Vienna, Austria

DOI: 10.2507/28th.daaam.proceedings.119

Abstract

Aluminium and its alloys represent very important group of structural materials. They have many applications in mechanical and civil engineering, and welding is considered to be one of the most important joining techniques. However, welding of aluminium has several issues, like high thermal conductivity and easy formation of oxide layer with high melting temperature. Recently, solid state friction stir welding process has been developed to overcome such issues, but it is not easily applicable in every situation. Therefore, welding of aluminium still greatly relies on arc welding. Among commonly used TIG and MIG processes, pulsed MIG (and its variants) are developed to fulfil requests regarding heat input and oxide layer removal during welding. This paper gives brief general overview of technical and technological aspects of these processes, and then more details regarding welding of widely used aluminium 5754 alloy with thickness of 4 mm. Comparison of relevant costs is given as well. All welds were made using digitally controlled equipment, while voltage and current were additionally measured. Based on evaluation of welds and calculated costs, conclusion regarding feasibility of each process is given.

Keywords: aluminium; TIG; MIG; pulsed MIG

1. Introduction

Aluminium and its alloys have numerous applications for mechanical and civil engineering. There are few reasons for this, but most important ones are excellent corrosion resistance to various environments, high specific strength, and great technological properties. [1] Aluminium alloys are generally easily cast, formed and machined, while joining can be done by nuts and bolts, adhesive bonding, and welding. In particular, welding is often used in automotive industry and vessel manufacturing. [2] However, welding of aluminium and its alloys is more complex than welding of most of the steels [3]. Among all reasons for this, main ones are:

- Aluminium and its alloys have excellent heat conductivity,
- In general, aluminium alloys are prone to hot cracking,
- Aluminium and its alloys have relatively low melting point, generally between 480 and 660 °C,
- Aluminium forms oxide layer on surface, with melting temperature slightly above 2,000 °C.

Therefore, aluminium welding requires more detailed knowledge about alloy (in particular its chemical composition), specific equipment, precisely controlled welding conditions (i.e. welding parameters and composition of shielding gas), and well-trained welding personnel.

Many issues related to welding of aluminium and its alloys usually can be solved by precise control of heat input. If heat input is as low as possible, and as high as required, many adverse effects of welding can be avoided. That enabled solid state friction stir welding process, and its variants, to become often used to weld some aluminium alloys that previously were not easily weldable. [4,5] However, this process still has certain disadvantages, leaving significant part of aluminium welding industry to fusion welding processes. Two main fusion welding processes are in use nowadays for aluminium and its alloys, tungsten inert gas welding (TIG or GTAW) and metal inert gas welding (MIG). [2] MIG process has evolved into several variants, each offering some advantages. The idea is to reduce heat input and avoid problems induced by it. [6] This is achieved by using digitally controlled heat sources, automatization, and welder/operator training. However, it requires significant investment in equipment and training. [7] Thus, it is necessary to carefully evaluate feasibility of such investment of each case of welding, independently and ubiasedly. The problem is to conclude when, and under which conditions, it is justified to use such unconventional equipment to make welds.

2. TIG and MIG welding processes

There are two main fusion welding processes used nowadays for welding of aluminium and its alloys, TIG and MIG. Both are using inert gases (usually argon, helium and their combinations) to shield the weld pool. Beside these, there are few other fusion processes applicable to aluminium, but they are used in specific situations and/or configurations. [3]

TIG process is still predominantly manual welding process, highly dependent on welder's skills. It is often used where no other option is available, mostly in individual manufacturing, on-site and in shop floor conditions. Despite certain advantages, this process is slow, having very low deposition rate.

MIG process is mechanized by default, and used mostly in serial or mass production. Nevertheless, it can be used in individual production. It has significantly higher deposition rate than TIG. Several modifications of MIG process have been developed, but important one is pulsed MIG. [6] It is a variant of the conventional MIG welding process in which the current is pulsed. Pulsing was introduced to control metal transfer at low mean current levels by imposing short-duration high-current pulses. [8] The cycle consists of applying the repeated pulse current over a constant background current, as illustrates Figure 1. Background current can be as low as 15 A.

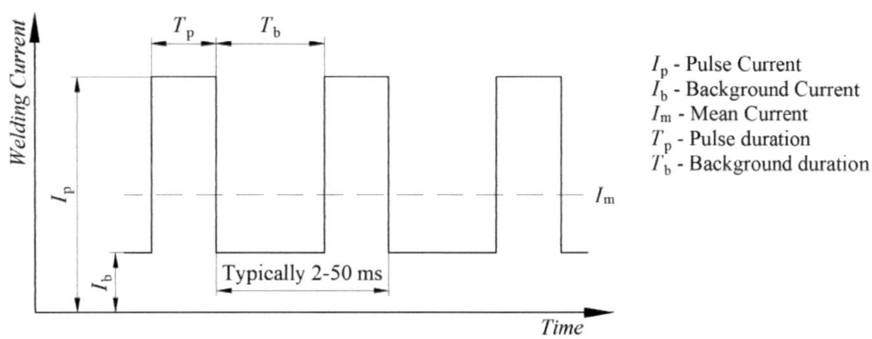

Fig. 1. General principle of pulsed MIG [8]

Modern welding sets permit the use of a wide range of pulse amplitudes, durations and waveforms at frequencies from a few Hertz to a few hundred Hertz. Pulse amplitude and duration are best combined to melt and detach a single droplet of the same/slightly smaller diameter as the electrode wire. Selection of pulse parameters for a given wire feed speed is a complex operation. Pulse height and duration are a function of wire composition, diameter and, to a lesser extent, shielding gas composition. Pulsing must provide enough heat to melt filler material (wire) and to detach exactly one drop of molten metal per pulse.

Generally, advantages of pulsed MIG welding are [8]:

- It allows the use of smooth, spatter-free welding at mean currents, which would otherwise be too low for all except dip transfer with its irregular transfer and associated spatter.
- Pulsing can extend spray operation below and through the natural transition from dip to spray where globular transfer would normally occur.
- Pulsed transfer is midway between spray transfer and dip transfer mechanism, which can be too 'cold' (due to non-continuous arcing, the arc effectively 'goes out' between each melting cycle). This makes it ideal for welding of thicker sections where more heat is needed but for which spray transfer is still too 'hot'.
- Pulsed MIG allows welding at higher deposition rates for all positions where dip or spray transfer is not applicable.

Pulsed MIG is used mainly for welding of aluminium and stainless steel [8]. Since this process uses advanced electronics to control process, it is not economical to use where cheaper alternatives are available. Different manufacturers of welding equipment use various brand name for pulsed MIG, but the principle is the same [6,9].

As possible to see, pulsed MIG offers advantages when welding aluminium. It yields the lowest possible heat input, and most of the issues regarding unnecessarily high heat input are avoided. However, it has also few disadvantages. Production rate (meters of weld per minute) is lower than with dip transfer (i.e. MIG) due to decrease of wire feed rate as heat input increases, and there are limitations regarding usage of shielding gases in comparison with dip transfer. [10]

3. Experiment

To compare feasibility of TIG, MIG and pulsed MIG welding, an experiment has been conducted. Samples of aluminium plates were welded, and visually examined afterwards. This ensures that all welds have comparable quality. Based on measured parameters, costs are calculated.

Although primary intention was to compare only MIG and pulsed MIG, it is decided to include TIG as well, since it is commonly used process for these base and filler materials. The investment in welding equipment and welder training is not considered here, because it can vary significantly from one manufacturer to other.

3.1. Welding of samples

Sheets of aluminium alloy 5754 (chemical composition given in Table 1) with thickness of 4 mm were welded.

	% wt.								
	Mg	**Mn+Cr**	**Mn**	**Si**	**Fe**	**Cr**	**Zn**	**Ti**	**Cu**
5754	2.60-3.60	0.10-0.60	0.50	0.40	0.40	0.30	0.20	0.15	0.10

Table 1. Chemical composition of Al 5754 [11]

For each process several horizontal fillet welds were made, each 120 mm long, without weld preparation. All welds are made as single-pass. Table 2 gives data about filler materials for welding processes. They have been chosen according to recommendations of welding equipment or filler material manufacturers. Filler material for TIG process is Al 4043 wire (diameter 2.0 mm), and for MIG and pulsed MIG it is Al 5356 wire (dimeter 1.2 mm).

	% wt.								
	Mg	**Mn+Cr**	**Mn**	**Si**	**Fe**	**Cr**	**Zn**	**Ti**	**Cu**
TIG	0.05	-	0.05	4.50-6.00	0.80	-	0.10	0.20	0.30
MIG pMIG	4.50-5.50	-	0.05-0.20	0.25	0.40	0.05-0.20	0.10	0.06-0.20	0.10

Table 2. Chemical composition of filler materials [12,13]

All welds were made in the laboratory, simulating shop floor conditions. Welding parameters were taken from welding procedure specifications where available, while other were chosen according to recommendations from either welding equipment or filler material manufacturer. They are given in Table 3. Shielding gas for all processes is argon.

	Current, A	**Voltage, V**	**Wire dimeter, mm**	**Shielding gas, l/min**
TIG	90	12.0	2.0	18
MIG	150	21.5	1.2	20
pMIG	105	19.5	1.2	20

Table 3. Welding parameters [11-13]

Prior to welding, aluminium sheets were cleaned with trichloroethylene, rinsed with distilled water, dried, and then cleaned with stainless steel brush. [14]

In this way, parameters for all evaluated processes are as close to each other as possible, yet within recommended boundaries. Therefore, it is possible to make fair comparison of weld quality, efficiency and price.

3.2. Examination of samples

This step is required to ensure that all welds have comparable quality level in accordance with ISO 10042 (Welding — Arc-welded joints in aluminium and its alloys — Quality levels for imperfections). Welds were evaluated visually after welding and, additionally, cross-sections were cut from each weld to evaluate weld geometry and porosity. Samples were visually examined in accordance with ISO 17637 (Non-destructive testing of welds — Visual testing of fusion-welded joints). Macro sections have been prepared and examined in accordance with ISO 17639 (Destructive tests on welds in metallic materials — Macroscopic and microscopic examination of welds).

Most importantly, no cracks have been detected in welds. After cutting and polishing of cross-section specimens, porosity has been found only in MIG welds, with dimensions that correspond to quality level B and coverage of approximately 0.5 %. Geometry of welds has been examined as well. All welds are symmetrical, with properly made root and full penetration. No root gap or excessive convexity has been detected.

It should be noted that ISO 10042 governs significantly more imperfections than examined in this study, but not all of them are always required to be analysed. Also, there are some imperfections not described or defined, e.g. spattering. Only MIG welds had some spattering, while no other welds had other imperfections described by ISO 17637.

Aesthetically, all welds look acceptable, but it should be noted that this is rather subjective evaluation. TIG welds are superior. Welds made with pulsed MIG are slightly behind, and welds made with MIG are last (due to spattering).

3.3. Calculation of welding costs

Welding costs are one of the most important and limiting factors of application of certain welding process. A rather simple method has been employed to calculate welding costs of three chosen processes. [15,16] Simple method has been chosen since it is illustrative enough, yet avoiding relatively complicated cost calculation with amortization rates and additional expenses. Later, costs will be shown as relative to each other and mutually compared.

Costs calculated in this study include only direct costs of filler material, shielding gas, electricity and welder. As mentioned before, investments required for equipment and welder's training are not included. Costs of pre-heating and post-weld heat treatment are not included as well.

All costs were calculated using parameters measured during and after welding. Current and voltage were measured by independent devices. Weight of deposited metal has been determined by weighing samples before and after welding. Shielding gas flow was measured at reduction valve (TIG) and welding apparatus (MIG and pulsed MIG).

4. Results and discussion

As demonstrated, it is possible to successfully utilize all three analysed fusion welding processes of aluminium alloy 5754 in given conditions. Regarding filler metal, it is enough to follow recommendations of their manufacturers. There is enough sources that recommend welding parameters (available welding procedure specifications, equipment and filler material manufacturers). Generally, this should not be concern when welding easily weldable and widely used aluminium alloy. Problems are likely to occur when welding rather obscure and rare aluminium alloy.

All three processes gave similar weld quality. After visual examination of welds and cross-sections, no unacceptable imperfections have been found. Based on examination done, all welds could fit into class B (highest quality class) according to ISO 10042. The only difference in welds is subjective, and it considers aesthetical appearance of welds. Considering this as criteria, TIG has the best look, while MIG, due to spattering, has the worst. However, all of them are acceptable. It can be concluded that all three analysed processes give adequate weld quality under described conditions.

Analysing welding costs shown that values of MIG and pulsed MIG are similar, while those for TIG are generally higher. To illustrate that, Figure 2 shows total costs of welding and costs of welder, while Figure 3 shows costs of filler material and shielding gas. Relative scaling is applied, where the most expensive process represents 100 %.

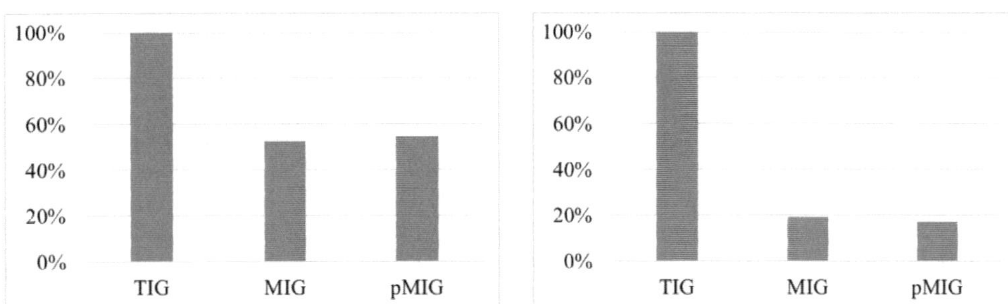

Fig. 2. Total costs of welding and costs of welder

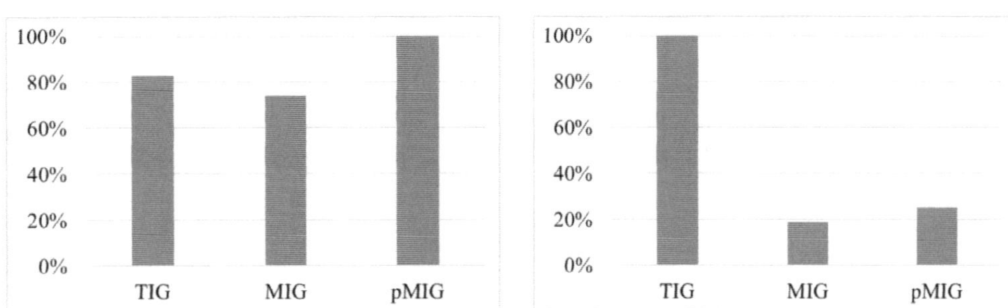

Fig. 3. Costs of filler material and shielding gas

As noticeable, TIG has the highest welding costs. That is mainly because it is fully manual process, with very low deposition rate. That significantly increases welding time (arc time) leading to increased costs of welder and shielding gas. However, it has slightly lower costs of filler material. This is because there is no spattering and because it is partly autogenous welding process (i.e. filler material is not absolutely necessary all time). However, TIG would not be used for this weld anyway. Here is included to show more superior characteristics of MIG and pulsed MIG when applied for considered conditions. Deposition rates, which can be obtained through welder's costs (Figure 2), of MIG and pulsed MIG are close. So are the total welding costs.

It is necessary to bear in mind that this evaluation does not include expenses required for equipment and training. If that is taken into account, pulsed MIG costs would go significantly higher.

5. Conclusion

There are several fusion welding processes applicable for welding of aluminium alloy 5754. Among others, there are MIG, TIG and pulsed MIG, latter offering significant advantages regarding heat input. However, more factors should be considered when evaluating feasibility of certain welding process (e.g. costs and production time). To evaluate abovementioned processes, an experiment has been conducted. Plates of aluminium alloy 5754 with thickness of 4.0 mm have been welded and welds examined. Based on information obtained through this experiment, feasibility in given conditions has been estimated. All three processes have similar output, with welds of acceptable quality and look. Subjectively, TIG welds have the best appearance, with pulsed MIG slightly behind. However, pulsed MIG requires significantly higher investment in equipment and training than other two processes. MIG and pulsed MIG shown similar deposition rates and costs for sheets of average thickness (e.g. 4 mm, as in this study), so it can be concluded that the most feasible process in this case would be MIG. Advantages offered by pulsed MIG could be visible in other conditions. That includes thin aluminium sheets (e.g. 1 mm), specific welding positions, or specific products where heat input is strictly limited (e.g. measuring equipment). Further research, including welding of other aluminium alloys, could confirm this by conducting similar experiment.

6. References

[1] Kuruc, M., Zvončan, M., Peterka, J. (2014). Investigation of Ultrasonic Assisted Milling of Aluminum Alloy AlMg4.5Mn, in: Proceeding of 24th DAAAM International Symposium on Intelligent Manufacturing and Automation, Procedia Engineering, Vol. 69, 2014, pp. 1048–1053. doi: 10.1016/j.proeng.2014.03.089

[2] Müller, U. (2011). Introduction to Structural Aluminium Design, Whittles Publishing, ISBN 978-184-99-5007-7, Dunbeath

[3] Zaki A. [Ed.] (2012). Aluminium Alloys - New Trends in Fabrication and Applications, InTech, ISBN 978-953-51-0861-0, Rijeka, doi: 10.5772/3354

[4] Casalino, G., Campanelli, S., Mortello, M. (2014). Influence of Shoulder Geometry and Coating of the Tool on the Friction Stir Welding of Aluminium Alloy Plates, in: Proceeding of 24th DAAAM International Symposium on Intelligent Manufacturing and Automation, Procedia Engineering, Vol. 69, 2014, pp. 1541–1548. doi: 10.1016/j.proeng.2014.03.153

[5] Mishra, R. S., Ma Z. Y. (2005). Friction stir welding and processing, Materials Science and Engineering R, Vol. 50, 2005, pp. 1–78.

[6] Horvat M., Kondić V. (2012). The examples of modified procedures of MIG/MAG welding, Tehnički glasnik, Vol. 6, No. 2, 2012, pp. 137–140.

[7] Rosado T., Almeida I., Pires I., Miranda R., Quintino L. (2008). Innovations in arc welding, Proceedings of 5. Congresso Luso-Moçambicano de Engenharia, Maputo, 2-4 September 2008.

[8] http://www.twi-global.com (2017). What is Pulsed MIG/MAG welding and what are its advantages over conventional MIG/MAG processes?, Accesed on 2017-07-16

[9] Brown, E. (2017). An Introduction to Pulsed GMAW, Welding Journal, Vol. 96, No. 2, 2017, pp. 76–78.

[10] Lincoln Electric (2017). AC Aluminum Pulse (GMAW) Weld Process Guide, Identification code TE12.010

[11] AlcoTec (2015). AlcoTec Aluminum Technical Guide, Identification code ALC-10029B

[12] AlcoTec (2017). Alloy 4043 weld data sheet

[13] AlcoTec (2017). Alloy 5356 weld data sheet

[14] Carriere, M. D. (2017). Preparing Aluminum before Welding, Welding Journal, Vol. 96, No. 2, 2017, pp. 32–35.

[15] http://www.twi-global.com (2017). Welding costs, Accesed on 2017-07-18

[16] O'Brien, A. [Ed.] (2004). Welding Handobbok: Welding processes, Part 1, AWS, ISBN 0-87171-729-8, Doral

DOI: 10.2507/28th.daaam.proceedings.120

POSSIBLE APPLICATION OF HETEROGENEOUS ROBOT GROUP TO SEARCH AND LOCALIZE IONIZING RADIATION SOURCES

Artem Kulichenko[a], Ekaterina Smirnova[a]

[a]Russian State Scientific Center for Robotics and Technical Cybernetics (RTC), Saint-Petersburg, Russia

This Publication has to be referred as: Kulichenko, A[rtem] & Smirnova, E[katerina] (2017). Possible Application of Heterogeneous Robot Group to Search and Localize Ionizing Radiation Sources, Proceedings of the 28th DAAAM International Symposium, pp.0859-0866, B. Katalinic (Ed.), Published by DAAAM International, ISBN 978-3-902734-11-2, ISSN 1726-9679, Vienna, Austria
DOI: 10.2507/28th.daaam.proceedings.120

Abstract

Let us define heterogeneous robot group as the main subject of research. The aim of this work is to investigate the possibilities when heterogeneous robot group can be used to search ionizing radiation sources. To compare the duration of operations, fulfilled by single robot and by robot group, experimental studies have been made. On the basis of the results, gained from the experiments, the strategy and procedures for ionizing radiation sources detection with given accuracy, are determined. Justification for efficiency increase is also given.

Keywords: Heterogeneous robot group; group robotics; complexes for radiation reconnaissance; UAV (unmanned aerial vehicle); group robots application possibilities

1. Introduction

Robots application appears to be the most appropriate for extreme environments, where both human work and activity are rather difficult, impossible or can be a real hazard for human life and health[1]. Such environments include working within ionizing radiation fields, for example, to mitigate consequences of technological disasters. The advantages of robot group application for the tasks of extreme robotics are quite obvious [2]. In general they are: greater range of operation – that is due to the robots distribution within entire working area; extended operation abilities – that is due to mounting on each robot its individual actuator; and the higher success probability while fulfilling the task – that is due to the possibility of goals redistribution between the robots within group.

Let us consider the problem of robot group formation for the task of searching ionizing radiation source (IRS) and task planning for its work. IRS search and localization is a kind of very specific task, resulting in case of accidents or emergencies in IRS applying industries (of course, restricted IRS application is meant, for example in chemical industry, food industry, etc.), relating to operations and service of interim spent-fuel or industrial IRS storage facilities, or to the threats of terrorist attacks when the so-called "dirty bombs" can be used, and some other local accidents.

On the one hand, such situations can be described as having relatively small and "known" accident zone. On the other hand, there can be no special team for consequences mitigation in the area of emergency, and this may lead to the necessity of delivering special equipment from local emergency centers nearby. Therefore, for the robot group the mobility is needed, and its transportation to the accident zone or emergency area should be carried out with cost-effective and relatively inexpensive means (vehicles) to also serve the base point for monitoring and control.

Let us define the optimal robot group configuration and the structure of robotic complexes (RC) needed for functional tasks accomplishment. To make a decision about robot group configuration for the tasks of IRS detection and localization, the following key considerations should be kept in mind:

• to organize data transmission between the group members in conditions of unstable communication as well as to significantly increase the communication range with ground-based robots, use of unmanned aerial vehicles (UAV) can provide best possibilities;

• to detect and define direction to IRS at the same time being at the safest possible source-distance, each member of the robot group should be equipped with special devices, providing remote sensing of the radiation field;

• to make it possible to solve localization and navigation tasks, members of the robot-group should also have some gridding equipment for geo-location (e.g. GPS/GLONASS). However, there should also be provided scenarios of performing operations within freak zones, when communication of ground-based robots with satellites can be lost.

2. Group operation planning

Factor of time is crucially important especially for the tasks of extreme robotics, often responsible for operation in zones/areas of heavy and rugged terrain.

Robot group operation planning should provide:

• reducing the time for operation accomplishment;
• saving energy due to optimal scheduling;
• making it possible to preliminary evaluate hazard level (e.g., level of radiation danger) and detail the maps for pass-ability within area of operation with the help of UAV.

One more advantage of robot group application could be minimizing RC dose loads when working within the radiation fields. To achieve the advantages above, it is necessary to elaborate logic of functioning and develop typical scenario of robot group operation.

Despite the above mentioned, the following tasks should also be taken into consideration when designing the prototype:

• providing communication and interaction of members in the group;
• ensuring the group navigation using all available navigational information in general;
• creating the point of centralized management overall group control with corresponding hardware and software.

However, these general tasks are not covered by the subject area of this paper, focusing on the tasks of IRS search and localization, that can be solved on the basis of well-known theoretical approaches [2].

The paper describes and analyzes existing approaches and techniques of solving the tasks for IRS search and localization. On the basis of analysis given, the possibility of improving is stated - to be used for heterogeneous robot group operation tasks.

Technologies and devices of IRS search and localization are generally based on analyzing parameters of radiation field, that can be measured remotely, at IRS safe distant. For the task of IRS detection, gamma-ray direction finders are widely used. They are able of measuring course angle on target (that is ionizing radiation source) with the accuracy of 1-3 degrees [3].

While operating, the ground-based robot sequentially fulfils the tasks of:

• IRS detection;
• IRS localization in robot system of coordinates;
• IRS reaching for the robot to be capable of fulfilling operations with manipulator equipped.

3. General description of the algorithm for IRS detection and localization by ground-based robot

First and foremost, to start procedure of IRS detection, measurement of the intensity of the background radiation is performed. Then the detecting procedure begins: the mobile RC studies the area moving across and making stops from time to time. While searching, IRS location in relation to RC is considered to have two parameters: course angle to target **a** and distance **d** (see Figure 1).

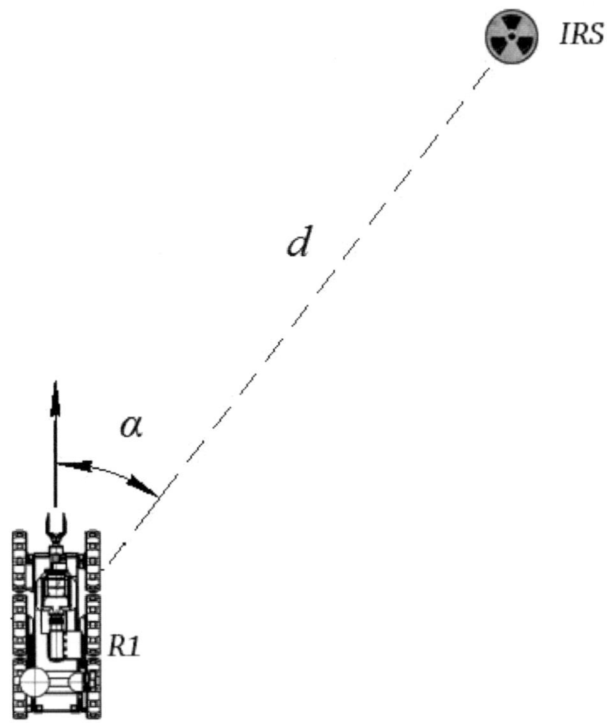

Fig.1. Finding Direction to IRS by the Robot (R)

Main measurements and statistical analysis based on their results for the robot to find IRS location, are performed during the stops. The procedures at points of stop suppose the following:

• within exposure period the robot registers the radiation via gamma-finder, estimates direction to the IRS and calculates direction estimation error;

• in case the robot has fulfilled controlled transfer (that is measuring the path covered) from the previous stop to the current one, the distance between RC and IRS is estimated by means of comparing radiation level at these two stops, and so the distance estimation error is calculated.

Taking into account the angle measured, RC then makes a turn in IRS direction and moves to it – with regard to measured distance. While moving, gradient of radiation level is controlled, and approximate estimation of distance to the IRS is calculated.

It is possible for the ground-based RC to define IRS location with accuracy enough to take IRS and put it into container with the help of manipulator. The main disadvantage of technology described is the following: when accomplished by the only (single) ground-based RC, the task of IRS localization cannot be solved with needed accuracy until the robot comes closer to the IRS location, e.g. enters higher radiation fields. The farther is IRS location, the weaker is radiation field intensity and the more is estimation error of distance ***d***, since radiation field gradient value is used for this very parameter evaluation. The closer is IRS location, the more is either risk of RC electronic components destruction or necessity of applying shielding alloys and radiation resistant elements, what makes the robot much more expensive. Improving accuracy of distance-parameter estimation is possible through triangulation method use, that is due to concurrent operation of the two ground-based RCs in our case.

4. Simulation of the task-mission for a single RC. Selection of the robot group

Due to statistical modeling program, it has become possible to simulate set of experiments for single robot to detect and localize IRS. The goal of simulation has united two main tasks: the first one – to calculate IRS average search time and the second one – to estimate the percentage of successful operations fulfilled by single RC in terms of minimal radiation exposure for both robot and RC operator. In accordance with the radiation safety standards, the exposure dose rate (EDR) of 50 mcR/h has been taken as minimal at the start point, and EDR of 250 mcR/h – as maximal acceptable exposure limit. The distance of 500 meter is considered as distance to the radiation source, that is standard radius for light-class robot operation zone. The value of radiation background is accepted as equal to 14 mcR/h (natural background, NB) and to 28 mcR/h (natural increased background, NIB).

Based on simulation results, there is SD chart obtained, where we can see that angle estimation at robot starting point depends on exposure period. Please, see the results of simulation in Table 1. SD angle dependency graph on single element searching time within each simulation held is presented in Figure 2.

EDR at starting point, mcR/h			
50			
NB		**NIB**	
initial conditions, option I		initial conditions, option II	
Average mean time of exposure,s	RCS angle, °	Average mean time of exposure,s	SD angle, °
5989,74	0	28156,7	0,86
3070,62	2,22	6481,61	2,22
1148,63	4,19	2131,46	4,59
345,79	6,06	608,28	6,82
174,56	7,7	316,36	9,64
134,42	8,8	242,94	9,9
EDR at starting point, mcR/h			
250			
NB		**NIB**	
initial conditions, option III		initial conditions, option IV	
Average mean time of stop,s	SD angle, °	Average mean time of stop,s	SD angle, °
1841,94	0,86	2267,49	0,83
416,67	2,2	499,51	2,31
152,34	3,94	186,09	4,02
46,1	5,98	54,06	6,7
27,34	7,01	29,07	7,89
22,03	7,9	24,08	8,5

Table 1. Simulation results for SD dependencies of IRS angle estimation on exposure time at starting point

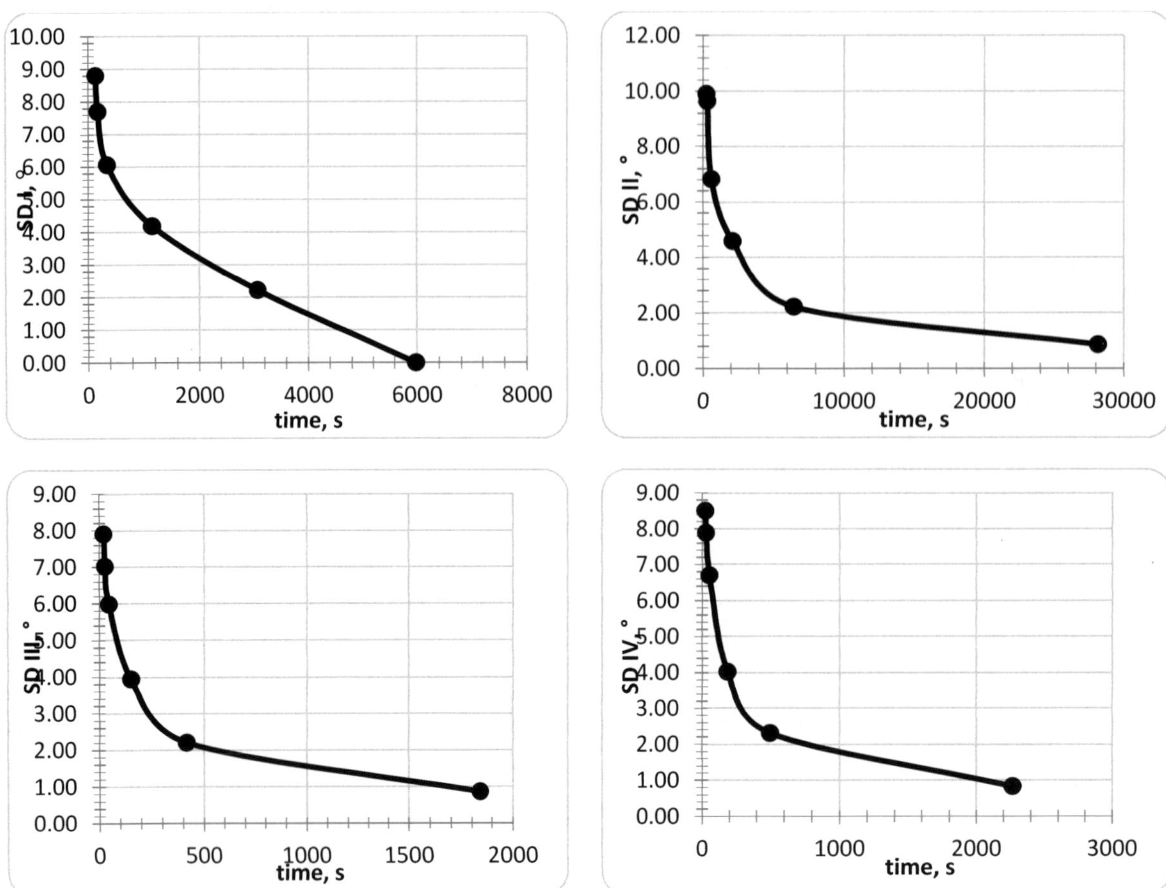

Fig. 2. SD dependencies graphs of IRS angle estimation on exposure time at starting point

The data show that when starting from the safe zone (calculation options I and II), errors under 6° are provided by increasing the duration of the first stop.

Then search-mission simulation has been made, that is – simulating automatic movements of the robot to IRS-source for each initial conditions option I-IV provided that the time of stop is 50 minutes. The search missions have been considered successful in case the robot reaches IRS location.

Simulation results have allowed to calculate the percentage of successful operations and average time for mission operation. When determining the average operation time, all missions have been taken into consideration, including the failed ones. Please, see simulation results presented in Table 2.

Characteristics	Value			
	Option			
	I	II	III	IV
Average mission operation time for IRS searching, s	3611,5	12291,33	4159,5	4074
Successful missions percentage, %	40	30	90	90

Table 2. Simulation Results for IRS search-mission operation by single robot in the automatic mode

The simulations have shown that time of mission-operation for single robot takes up to 3.5 h at average, that is nearly the limit of operation-time for light-class robot without recharging or replacing the batteries. In terms of minimum radiation burden for the operator at the starting point, it is impossible to fulfill the task successfully at 100 %.

To reduce time-period for searching without increasing the radiation hazard for the operator becomes possible in case of coordinated IRS direction-finding at maximum possible safe distance by means of two and more RCs.

Equipped with device of radiation search, UAV is able to carry on radiation reconnaissance from the height, without being exposed. UAV application gives possibility for a number of ground-based robots to stay in relatively safe zone at the same time defining approximately the area of IRS location, to estimate possible obstacles within RCs operation zone, to provide reliable communication as well as to keep observation in the process of mission. However, when doing aerial radiation reconnaissance, there is one main technological disadvantage –significant IRS localization error due to the field spatial smoothing.

Thus the best result can be achieved when using the group of robots, consisting of:

- Ultra-light class ground-based RC equipped with gamma finder;
- Light-class ground-based RC equipped with gamma-finder, manipulator and container;
- Two UAVs of the helicopter type: one - for preliminary radiation reconnaissance, the other - for video surveillance tracking the progress of ground-based RCs operation and communication with them.

5. Simulation of the task-mission for robot group. Results and analysis

Let us analyze the IRS search-mission operation for two ground-based robots with UAV application. We assume that UAV use has allowed to define the following:

- border of the safe zone; and
- approximate area of IRS localization taking in mind the error of ±75 meters (due to field spatial smoothing and dynamics of UAV);

At the same time the following reasonable assumptions can be accepted:

- Placing the ground-based robots at the border of the safe zone is initially performed with the help of vehicle;
- On-Board GPS/GLONASS receiver system defines the coordinates of robots within absolute coordinate system with negligible error;
- Robots are identically oriented in azimuth.

First and foremost it is necessary to find the best place for starting points of the robots in correspondence to each other, having in mind, that the best ones are such initial positions where mission successful operation probability is equal to 100% and the time spent for the mission-task is the least. You can find the task illustrated in Figure 3, where: a prior preliminary IRS location is for IRS point, and the initial starting positions of robots are shown as R1 and R2.

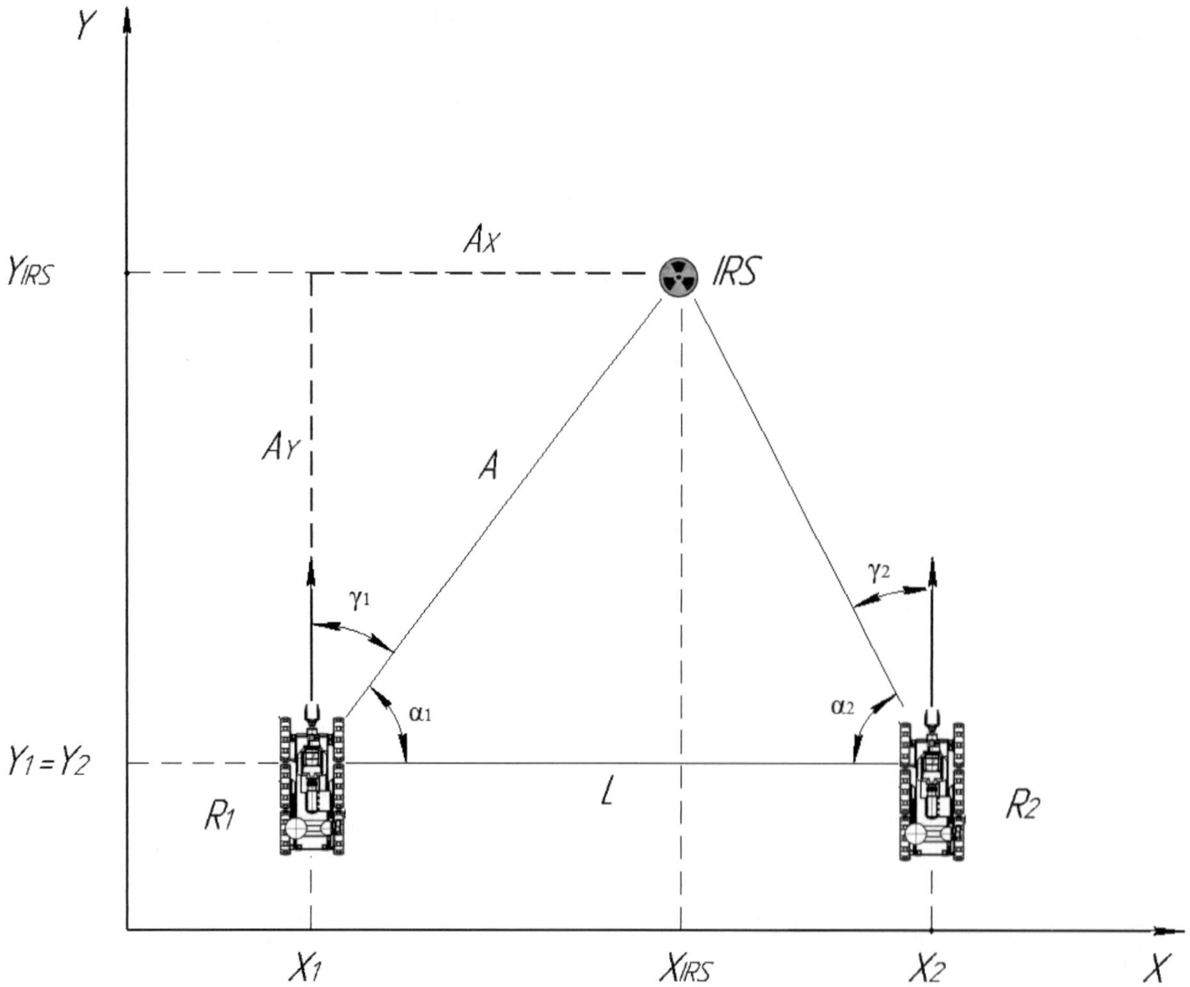

Fig. 3. Example of two robots positioning in the starting points at the beginning of IRS search-mission

It is quite obvious, that the coordinates of this source X_{IRS} и Y_{IRS} can be found from the system of equations:

$$\begin{cases} X_{IRS} = X_1 + L\dfrac{\cos(\gamma_2)}{sin(\gamma_1+\gamma_2)}\cdot\sin(\gamma_1); \\ Y_{IRS} = Y_1 + L\dfrac{\cos(\gamma_2)}{sin(\gamma_1+\gamma_2)}\cdot\cos(\gamma_1) \end{cases}$$

Values X_{IRS} and Y_{IRS} do depend on uncorrelated random variables – estimates of the angles γ_1 and γ_2 , and therefore are also random values. To calculate SD estimates of these variables, it is necessary to find their dispersions using method of linearization [4]:

$$\begin{cases} б^2_{X_{IRS}} = \sum_{i=1}^{n}\left(\dfrac{\partial X_{IRS}}{\partial \gamma_i}\right)^2 \cdot б^2_{\gamma_i}; \\ б^2_{Y_{IRS}} = \sum_{i=1}^{n}\left(\dfrac{\partial Y_{IRS}}{\partial \gamma_i}\right)^2 \cdot б^2_{\gamma_i}; \end{cases}$$

where $б\gamma_1$ и $б\gamma_2$ – are SD angles estimates for IRS direction, γ_1 and γ_2 correspondingly. Setting the range of errors for the coordinates determining, restrictions on the values $б\gamma_1$ и $б\gamma_2$ can be got.

Figure 4 shows a graphical interpretation of $б\gamma_1$ and $б\gamma_2$ correlation and failures of IRS finding. Rhombus (blue line) defines the boundaries of IRS localization area, and the projection of rhombus diagonals are $3•бX_{IRS}$ and $3•бY_{IRS}$ correspondingly.

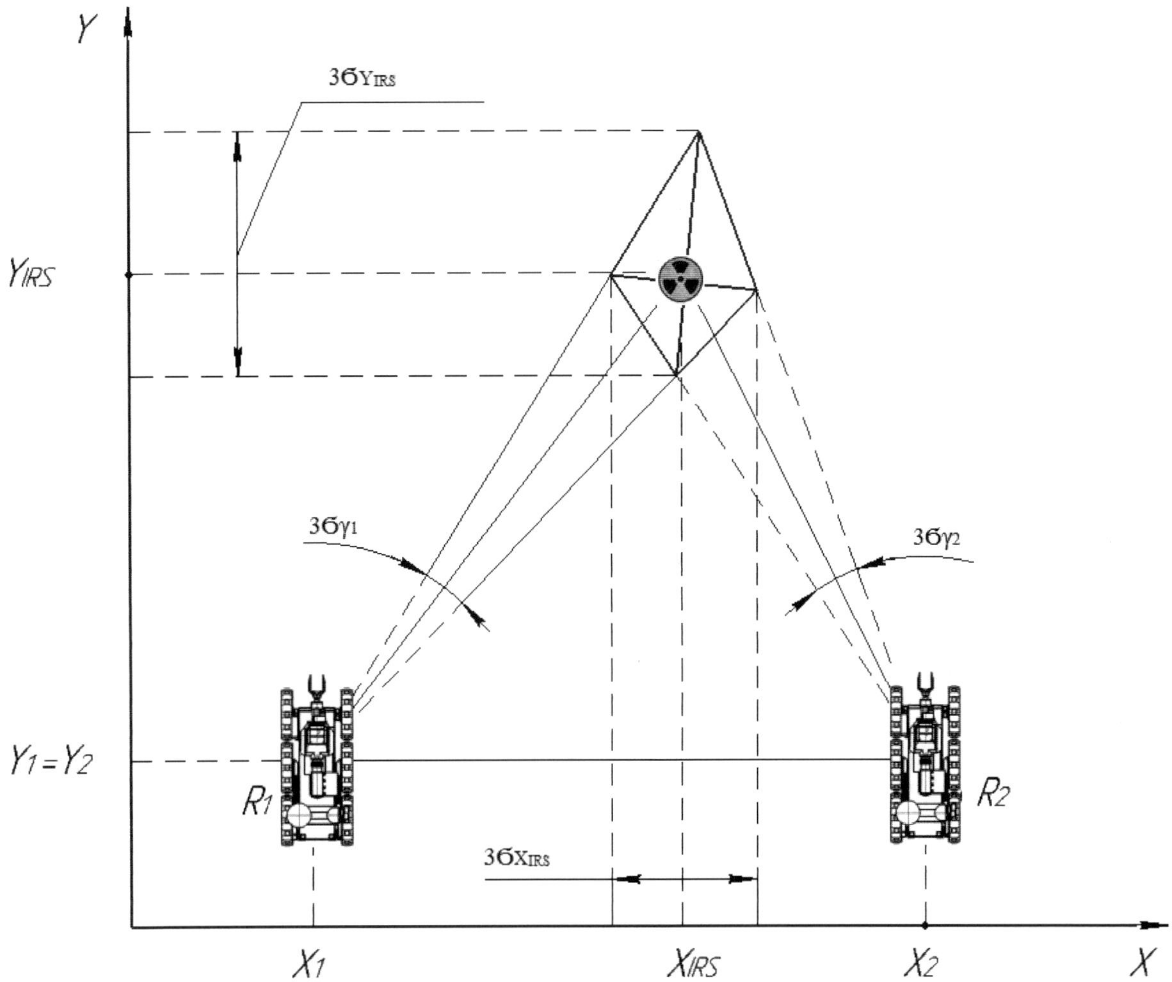

Fig. 4. IRS localization area

Results of calculations show that the minimum values of RMS for X_{IRS} and Y_{IRS} functions can be achieved on condition that angles γ_1 and γ_2 are in the range of 32° to 48°. Thus, the best initial positioning of robots is believed to be found.

For further simulation the following algorithm of search-mission organization has been chosen:

- estimating IRS angular directions by both robots simultaneously with accuracy error of 6;
- calculating of refined (in comparison to a priori) coordinates of IRS and its localization area (see Figure 4);
- non-stop moving of robot equipped with manipulator according to the shortest possible distance to the borders of

IRS localization area;

- defining refined direction to IRS and the refined IRS coordinates from calculated localization area border;
- entering into the IRS operation zone for robot equipped with manipulator in the way described for single robot.

The results of the search-mission simulation for options I-IV of initial conditions are presented in Table 3.

Characteristics	Value			
	Option			
	I	II	III	IV
Average mission operation time for IRS searching, s	1645	2300	1111	1165
Successful missions percentage, %	100	100	100	100

Table 3. Results of calculation of IRS search time for two robots operating in automatic mode

6. Conclusion

In the frames of this work, disadvantages of existing methods and technologies for IRS search and localization have been defined. Analysis of current robotic complexes for radiation reconnaissance has proven robot group practicability application. Current technologies for IRS search and localization missions are not intended for use in group of robots. Defining the main procedures for operation fulfillment makes it possible to determine capabilities of improving to make them possible to be applied to heterogeneous robot group. Based on existing algorithms realized for search-mission application task by single robot, a number of mathematic simulation models have been made for heterogeneous robot group.

Simulation results show that time of operation for IRS search and localization task reduces by 2-5 times in correlation with initial conditions if a group of robots is applied. This means that integral absorbed radiation dose for every group-member decreases as well. The percentage of successful missions in the case of robot group application reaches maximum of 100%. Task-mission procedures described in this paper are meant for automatic mode operation to minimize participation of the human-operator in the mission, and this leads to efficiency effect increase [7]. On the basis of results of the research, robot group members are justified. Heterogeneous robot group should mainly include light class ground-based robot (up to 300 kg) equipped with manipulator, ultra-light class ground-based robot (up to 50 kg) and two small-size helicopter type UAVs. Time of continuous operation for the members of such group corresponds to 2-3 hours, ensuring solution for the task of IRS search, localization and evacuation in correspondence with proposed technology. Simulation results also show that integral absorbed radiation dose is approximately by 1.4-2.0 times decrease (if compared to single robot operation process). When combined with corresponding services (decontamination, mitigation of consequences), there can be provided RC life and durability, speaking of both maintenance and midlife, full/complete and overhaul repairs. Dimensions and mass for both ground-based robots and UAVs make it possible to place them within single small-sized vehicle for transportation – for example, inside the cargo-passenger off-road vehicle [8]. Thus prompt response in conditions of emergencies can be granted.

The investigations held have proven the main disadvantage of aerial radiation reconnaissance – that is significant error of IRS localization due to field spatial smoothing. This means that elaboration of algorithms for data processing from UAV detection units, is necessary.

The results of work done will be further used in operating procedures for radiation reconnaissance complex being under development of RTC. They can also be applied for other robotic complexes and systems to solve detection and localization tasks.

7. References

[1] Yurevich E.I. Avariya na Chernobylskoi AES i ekstremalnaya robototekhnika [Chernobyl NPP Accident and Extreme Robotics]. Mechatronics, Automation, Control, 2004, vol.3, pp. 22-24 (in Russian).

[2] Boyko A.Yu., Vasiliev A.V., Denisienya Yu.A., Lipovckiy D.D. Opredeleniye prioritetnykh napravlenii razvitiya tekhnologii robototekchnicheskikh kompleksov voennogo naznacheniya voisk radiatsyonnoi, khimicheckoi I biologichesloi zashchity Vooruzhennykh Sil Rossiyskoi Federatsyi [Determination of the Priority Directions of Technology Development of Military Robotics of NBC Defense Troops of Russian Federation Armed Forces]. Robotics and Technical Cybernetics, 2017, vol.1(14), pp. 21-31 (in Russian)

[3] Kalyaev I. A., Gaiduk A. R., Kapustyan S. G. Modeli i algoritmy kollektivnogo upravleniya v gruppakh robotov [Models and Algorithms of Coordinated Control in Groups of Robots] Moscow, Fizmatlit, 2009, 280 p. (in Rssian)

[4] Russian State Scientific Center for Robotics and Technical Cybernetics. Available at: https://www.rtc.ru/ru/sredstva-radiatsionnogo-kontrolya/imd-24. (access date: 20.05.2017) (in Russian)

[5] Spassky B.A. Sovmestnoye primeneniye bespilotnykh apparatov razlichnogo bazirovaniya, Obzor sostoyaniya razvitiya [Application of Heterogenous Robotic Systems. State-of-the-Art]. Robotics and Technical Cybernetics, 2016, vol.2(11), pp. 8-19 (in Russian)

[6] Ventzel E. S., Ovcharov L. A. Teoriya veroyatnosti i yeyo inzhenernye prilozheniya [Probability Theory and Its Engineering Applications] Moscow, KNORUS, 2016. 480 p. (in Russian)

[7] Spassky B.A. Sovmestnoye upravleniye robotami, avtonomnoye i ot cheloveka-operatora [Robot Control: from Assisted Teleoperation and Mixed Initiative to Full Automation] Robotics and Technical Cybernetics, 2016, vol.2(11), pp. 69-76 (in Russian)

[8] Andreev V., Kirsanov K., Pryanichnikov V. (2016),Geographically distributed "multi-operator" Control for Mechatronic Devices via the Internet, In Proceedings of the 26th DAAAM International Symposium., B. Katalinic (Ed.), Published by DAAAM International, ISBN: 978-1-5108-1839-2, ISSN 1726-9679, Vienna, Austria. 2015. Curran Proceedings, New York,.pp. 0004-0011.

DOI: 10.2507/28th.daaam.proceedings.121

THERMAL REGIME ASPECTS WHEN GRINDING RUBBER ROLLS FOR PAPER INDUSTRY

Petre Valea & Eugen Străjescu

This Publication has to be referred as: Valea, P[etre] & Strajescu, E[ugen] R[adu] (2017). Thermal Regime Aspects when Grinding Rubber Rolls for Paper Industry, Proceedings of the 28th DAAAM International Symposium, pp.0867-0874, B. Katalinic (Ed.), Published by DAAAM International, ISBN 978-3-902734-11-2, ISSN 1726-9679, Vienna, Austria

DOI: 10.2507/28th.daaam.proceedings.121

Abstract

The paper investigates the thermal regime when grinding largerubber parts that are used like coat for rollers in paper industry. In order to analyse the variation of workpiece temperature when finishing grinding of the rollers rubber coat a full factorial design of experiments was used considering the cutting parameters as independent variables. The proper selection of cutting conditions for the grinding of large rubber rolls should take into consideration the thermal regime and the workpiece temperature variation because it is well known that this material is extremely sensitive to high temperature exposure and that build-up of rubber on the face of the grinding wheel may occur. In the experiments, abundant flow of cutting fluids is used.

Keywords: Grinding rubber rolls; abundant cutting fluid flow rate; thermal regime; cutting parameters

1. Introduction

The heat generated by the cutting processes is a consequence of the material deformation, chips segmentation and removal, as well as of the friction between the chips and the tool and between the tool and the machined surface. The heat resulting from this main is shared by the tool, chips, blank and the environment.

Thermal phenomena, which inevitably accompany the grinding process of the rubber parts, has a significant influence on the workpiece accuracy and a smaller one on the cutting tool wear and the deformation process. Increased heat in the contact area between the grinding wheel and the machined surface can cause errors that can lead to grinding surface degradation (pinching, burns, cracks, cuts).

In printing and textile industry a wide variety of rollers are used with a steel core and an overlying rubber coating. The main industrial application of this rollers are printing and deflection processes. The surface quality, roundness and concentricity accuracy that are required are in the range of a few hundredths of a millimetre. This level of precision is typically achieved by external cylindrical grinding. processes.

In the case of the rubber grinding processes, it is recommended that the temperature in the cutting area to not exceed a limit of 15°C and that an abundant cutting fluid flow rate (80-100 l / min) to be used. Also, the scientific literature [1, 2, 3] recommends that for every 10 mm of the abrasive wheel height is useful to provide a cooling fluid flow rate of 5..8 l/min. The higher the hardness of the abrasive material used is and the denser the wheel structure is, a higher cutting fluid flow rate must be used. In order to improve the quality of the coolant it is necessary to introduce certain additives, as well as the elimination of the impurities using the filters and separators.

Taking into account the above considerations, the rigorous control of the temperature developed in the cutting area is required for the grinding of the hard rubber parts. The cutting parameters must be carefully chosen as to ensure during the cutting process, in the contact area between the grinding wheel and the rubber workpiece, temperatures lower than the rubber thermal degradation temperature.

2. Literature review

The grinding process is the high temperature characterized by arising in zone of the cutting, promoting appearance heat defect on processed surface, reducing working factors of the details [18]. There is poor scientific literature on the grinding processes of the rubber rollers used in the paper industry. Most of the scientific work in this field[4, 5] deals with the machinability by grinding of the parts used in metallurgy and paper industry. The rubber covered rollers used in the paper industry are subjects of special inspections due to the demanding precision required in terms of surface finishing, roundness, concentricity, static and dynamic balance [4, 20]

The literature review of the cutting processes of rubber parts deals mainly with the turning process of small dimension parts [6]. The grinding processes of the large rollers covered by hard rubber (usually 85-92 Shore) are poor documented in the scientific literature. Ştețiu [3] presents aspects of the choice of grinding wheels, abrasive liquids, especially for difficult grinding materials such as rubber, cork, felt, skin, wood, asbestos, and heat-sensitive materials - such as stainless steels, high-alloy steels. Also, firms that are specialized in machining parts for the paper industry give some technical recommendations and manufacturing solutions for grinding rubber rollers [4, 20, 21]

3. Identify the problem

The purpose of the research is to identify the problems that occur in the case of hard rubber grinding, in order to issue reliable technical solutions for avoiding thermal machining errors when grinding rubber covered rollers for the paper industry. In order to avoid thermal deformations an adequate thermal regime in the cutting area during the grinding process must be maintained.

The main technological objectives that must be considered when rough grinding rubber parts are obtaining the prescribed roughness, assuring the integrity of the machined surfaces (avoiding burns, cracks, pinching, scratches, and residual stresses) and the prescribed geometrical and shape precision. Rubber is being considered a highly thermal sensitive material for mechanical machining.Precision grinding processes for rubber rollers depends on a number of factors like machine stiftness, genmetrical and cinematical precision, tool type and precision, cutting conditions, etc One of the main factors that can aaffect the grinding process of rubber rollers precision is the thermal regime durring cutting.

Grinding of rubber is considered a difficult process, because of the build-up of rubber chips on the face of the grinding wheel and because of the thermal deformations of the part as a result of the cutting heat generated. Also, the friction generated by the abrasive wheel- rubber part contact, as the wheel cuts, combined with the poor thermal conductivity of rubber, leads to excessive heat generation. This can lead to wheel cracking and sometimes even breakage. For this reasons, the abrasive industry has often been hesitant to supply grinding wheels for this particular application.

The paper deals with optimizing the cutting parameters of the grinding process of large dimension hard rubber covered rolls with different or unburned sleeves with porous abrasive wheels. If the rubber covered rollers have not been properly grinding (resulting a crown deflections of k = 1,420 mm on the rubber sheath rollers, Φ = 800 mm, see Figures 1) then the laminate strip breaks and the whole technological process is affected.

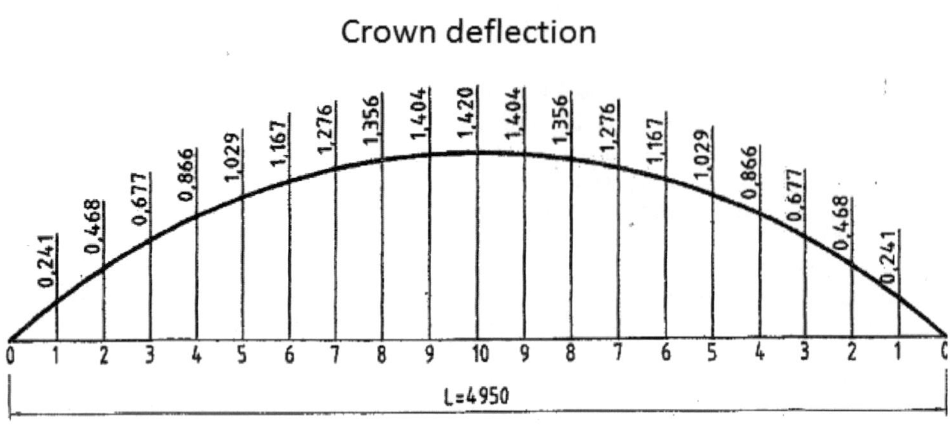

Fig. 1. The rubber rollers crown deflection, k=1,420 mm

4. The thermal regime of rubber grinding

The manufacturing process grinding is characterized by high thermal loads caused by large contact areas and a high number of grain cuts as well as negative chip angle [19]. For the study of the thermal regime of the rubber grinding process, the infrared thermography method was used. This method consists in the remote detection of infrared radiation emitted by a product under examination and the transformation of the information gathered by point-to-point scanning into a visible image on which temperatures can be highlighted at any point. (See Figures 2, 3 and 4).
A thermovision camera, model FLIR type E-45 from Vrancart Adjud company from Romania was used.

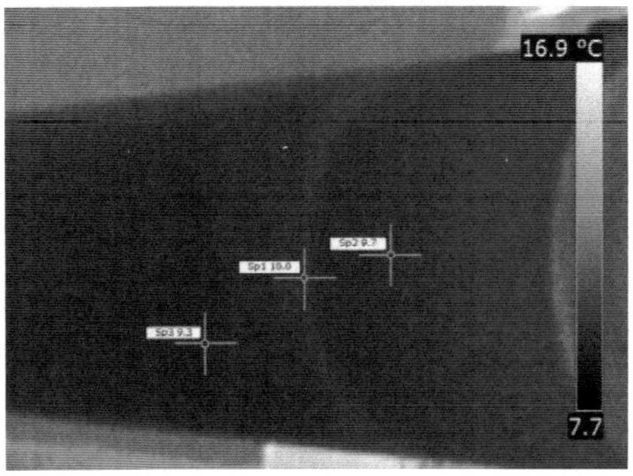

Fig. 2. Workpiece temperature measurement with FLIR - E 45

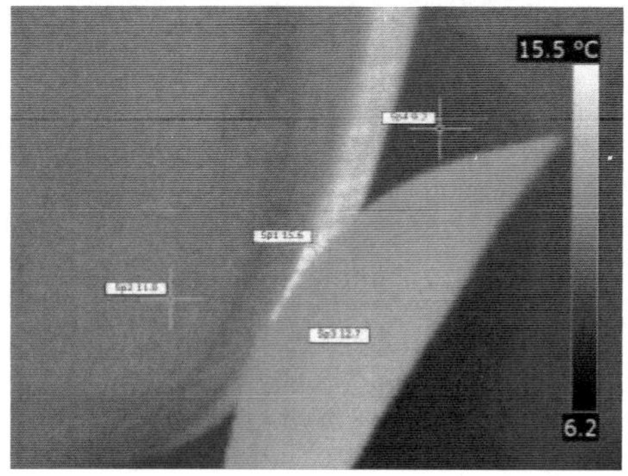

Fig. 3. Temperature measured with the FLIR – E 45 thermographic camera at short cooling fluid interruption

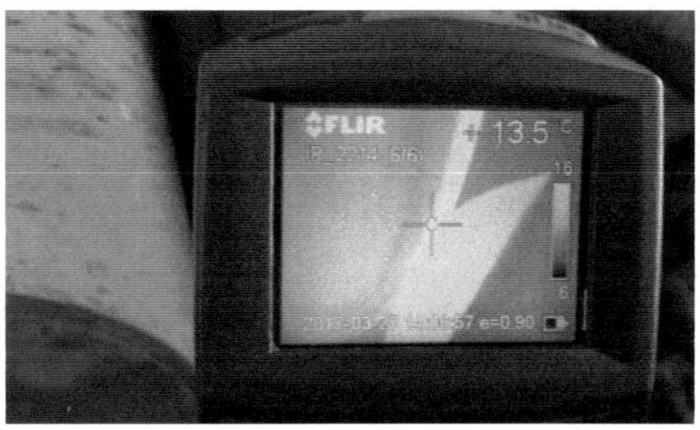

Fig. 4. Temperature measuring with the FLIR-E45 thermographic camera

5. Experimental method

Taking into account the recommendations from technical literature [7-11] and the cinematic possibilities of the Mockba Model XH-193H8 Grinding Roll Machine from the Mechanical Workshop ofVrancart Company, Adjud, Vrancea, (See Figures 6 and 7) the experiments were carried out by modifying the depth of cuta$_p$ [mm], the longitudinal feed rate f_l [mm/min] and the workpiece speed (roller speed) v_p [m/s]. The chosen values for these machining parameters are shown in Table 1.

The experiments were carried out with the purpose to draw some pertinent conclusions about the optimal cutting parameters for grinding large rubber rollers in terms of thermal cutting regime.

Machine-tools: the Mockba model XH-193H8 grinding roll machine [10] Abrasive wheel: Super-porous stone 600 x 304,5 x 80 (D x d x H), H [12]		
Cutting parameters	Values	U.M.
Depth of processing a_p	0,025; 0,100	[mm]
Longitudinal feed f_l	111; 250	[mm/min]
The speed of workpiece, v_p	0,33; 0,55	[m/s]
Observations: - the grinder operations were done with coolant (See Figures 6 and 7)		

Table 1. The grinding conditiones

Fig. 5. Rubber rollers grinding process using abundant cooling liquid at SC Vrancart SA Adjud, Vrancea

Fig. 6. Front view of the rubber rollers grinding process

Super porous abrasives wheels are successfully used for the process of grinding rubber rollers used for the paper industry [4, 12]. The super porous abrasive wheels with the following dimensions: 600 x 304,5 x 80 (D x d x H) and the hardness H were used on the grinding machine XШ-193H8 Mosckba for grinding of large rubber rollers. The grinding wheels used were: 33A46J10V for roughing and 22C80J10V for finishing.
We mention that the same types of abrasive wheels are used for the grinding of rubber parts in other paper factories than the one where the research have been done, for example SC Ambro SA Suceava and SC Ecopaper SRL Zărneşti (See Figures 8 and 9).

Fig. 7. Grinding of rubber roller at SC Ambro SA Suceava

Fig. 8. Grinding machine at SC Ecopaper SRL Zarnesti

6. Design of experiments - DOE

Design of experiments (DOE) is an important statistical technique for improving product quality and solving production problems.
The plan chosen for the experiments is shown in Table 3.

Experiment number	Factors		
	The speed of workpiece v_p [m/s]	Longitudinal cutting feed f_l [mm]	Depth of cut a_p [mm]
1.	+ 1	+1	+1
2.	+1	+1	-1
3.	+1	-1	+1
4.	+1	-1	-1
5.	-1	+1	+1
6.	-1	+1	-1
7.	-1	-1	+1
8.	-1	-1	-1

Table 2. Standard 2^3full factorial design matrix

In this paper, three input independent variables and two levels (-1) and (+1) were chosen for the variance of the parameters and the prediction of the optimal values, as shown in Table 2. The symbolization used was: (-1) for the lower level of a factor; (+1) for the upper level of a factor. According to this convention, the matrix of experiences can be written, Table 3 and 4.

Factor / NivelLevel	v_p [m/s]	f_l[mm/min]	a_p [mm]
-1	0,33	111	0,025
+1	0,55	250	0,100

Table 3. Selection of the independent variables and their levels

7. Workpiece temperature analysis during grinding

In the development of the present research it was considered necessity to determine the correlation between the cutting parameters: the workpiece speed v_p [m / s], the longitudinal feed f_l [mm / min] and the depth of cut a_p [mm] over the part temperature. For the workpiece temperature, a regression function was determined with a form of polytrophic equation [13]:

$$T = C_T v_p^{\alpha} f_l^{\beta} a_p^{\gamma} \tag{1}$$

Another indicator of a good cooling in the grinding process is the temperature of the liquid after the process: the lower the liquid temperature is, the cooling is considered to be optimal [3].
The cutting liquid that is most commonly used is the aqueous solution with small amount of soda ash, soap. When finishing grinding, the cutting fluid chemical composition may influence the surface quality. It is recommended to keep a clean coolant during machining. The washing capacity of the coolant influences the finishing grinding; in this sense it is recommended to use liquids having a high washing capacity. [14-17]
$2^3 = 8$ full factorial experiments plan was performed.

Exp. num.r	Factors			Temperature T [^{0}C]
	The speed of workpiece v_p [m/s]	Longitudinal cutting feed f_l [mm/min]	Depth of cut a_p [mm]	
1.	0,50	250	0,100	12
2.	0,50	250	0,025	11
3.	0,50	111	0,100	10,5
4.	0,50	111	0,025	10
5.	0,33	250	0,100	9,8
6.	0,33	250	0,025	9,6
7.	0,33	111	0,100	9,5
8.	0,33	111	0,025	9

Table 4. Experimental plan with assign parameters and response values

For the considered regression equation: $T = f(v_p, f_l, a_p)$**,** after the experiments were carried out the experimental data was processed with a specialized software, MINITAB 17 free version. For the experimental values obtained, the software returned the following temperature variation law (regression equation):

$$T = 8{,}3903 \times v_p^{0{,}27352} \times f_l^{0{,}105955} \times a_p^{0{,}03992} \tag{2}$$

As we can see, the influence of the workpiece speed had manifested a higher influence on the workpiece temperature compared to the longitudinal cutting feed and the cutting depth.

With the software MINITAB 17 free version, the workpiece temperature variation with the longitudinal cutting feed f [mm/min] and the cutting depth [mm] graph was plotted (See Figure 10). There can be observed a rise in temperature with the increase of the cutting depth and the longitudinal cutting feed.

Figure 11 shows the variation of the temperature in relation with the workpiece speed v_p [m / s] and the longitudinal cutting feed f_l [mm / min]. An increase in workpiece temperature can be observed with the increase of the workpiece speed v_p [m / s] and the longitudinal feed rate f_l [mm / min]. As we can see the cutting speed exercises a higher influence over the workpiece temperature.

Figure 12 shows the temperature variation according to the workpiece speed v_p [m / s] and the cutting depth [mm]. There can be observed an increase in temperature with the workpiece speed increase and the workpiece depth [mm] increase.

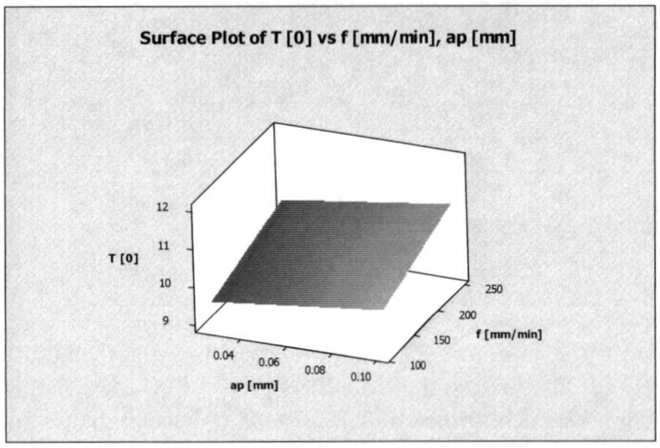

Fig. 9. 3D surface plot for effect of longitudinal cutting feed f_l [mm/min] and depth of cut a_p [mm]on workpiece temperature T [^{0}C]

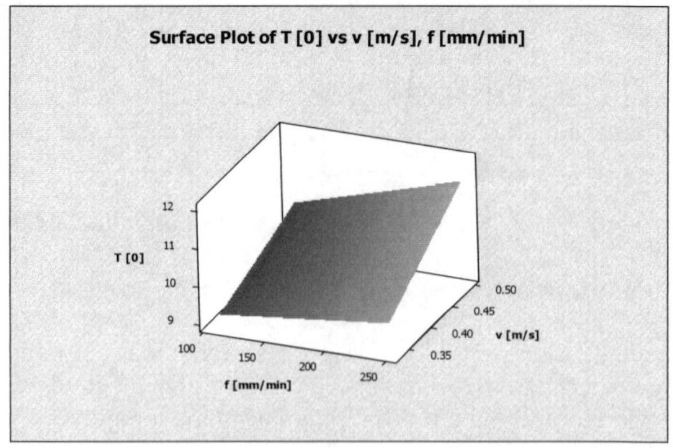

Fig. 10. Surface plot of T [^{0}C] versus workpiece speed vp [m/s], longitudinal advance fl [mm/min]

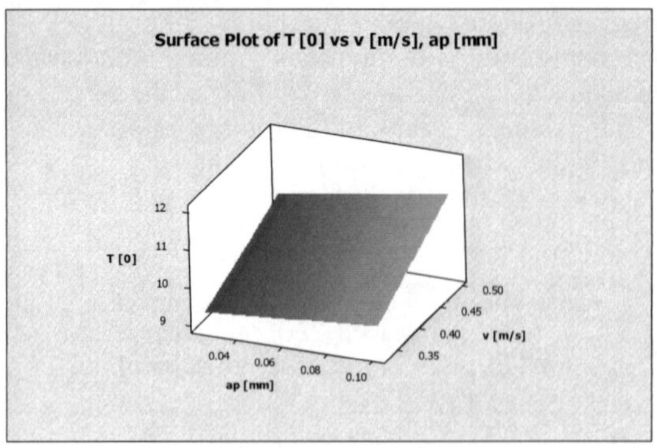

Fig. 11. 3D surface plot for effect of workpiece speedv_p [m/s] and depth of cut a_p [mm]on workpiecetemperature T [^{0}C]

Figure 13 and 14 show the factors main effects and factors interactions on the workpiece temperature variation. The three input parameters chosen for the study have a negative influence on the workpiece temperature during grinding. The main effects plot gives the optimal combination of grinding process parameters for minimal workpiece temperature during machining. The optimal combination of parameters is found to be $(v_p - (-1), f_l - (-1), a_p - (-1))$, lower-levels of workpiece speed, feed rate and depth of cut. Also, the main effect plot throws some light on the significance of the parameters on the system response. The slope of the main effect plot for each parameter reveals the intensity of the influence of the parameter. The plot having higher inclination will have higher influence. From Figure 13, it can be observed that the factor workpiece speed is more significant while the factor depth of cut has the lowest significance.

Interactions between the parameters were studied by using interaction plot drawn and shown in Figure 14. As expected, the interaction between worpiece cutting speed and the longitudinal cutting feed has the greatest influence over the workpiece temperature during grinding.

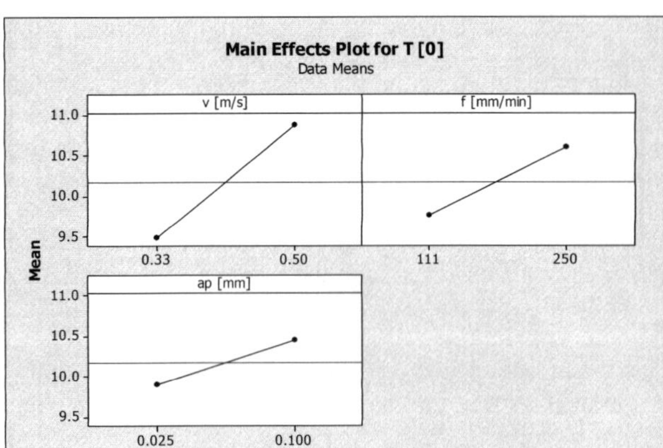

Fig.12. Main effects plot for T ^{0}C

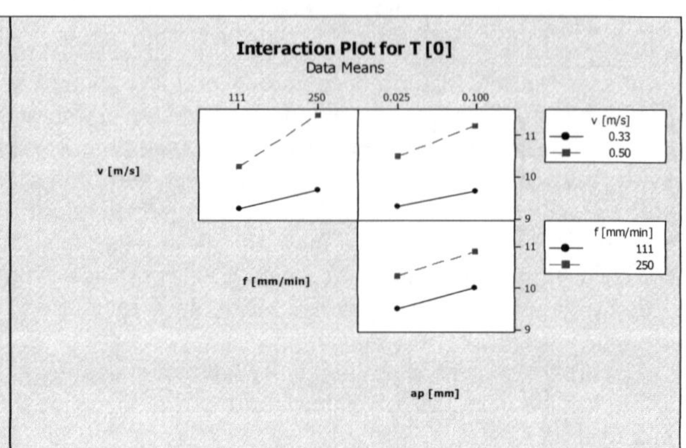

Fig. 13. Interaction plot for T ^{0}C.

8. Regresion analysis

The Minitab out put for the regression is presented below:

Equation

T [C] = 8.3903 * 'vp [m/s]' ^ 0.27352 * 'fl [mm/min]' ^ 0.105955 * 'ap [mm]' ^ 0.0399203

Parameter Estimates

Parameter	Estimate	SE Estimate	95% CI
c_T	8.39030	1.21337	(5.61102, 12.4962)
α	0.27352	0.04128	(0.15928, 0.3887)
β	0.10596	0.02581	(0.03440, 0.1779)
γ	0.03992	0.01509	(-0.00193, 0.0819)

T [C] = c_T * 'vp [m/s]' ^ α * 'fl [mm/min]' ^ β * 'ap [mm]' ^ γ

Summary

Iterations	6
Final SSE	0.363791
DFE	4
MSE	0.0909479
S	0.301576

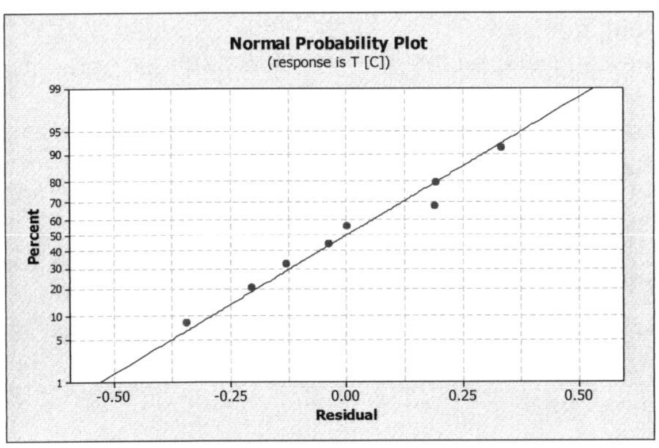

Fig. 14. Normal probability plot for T [C]

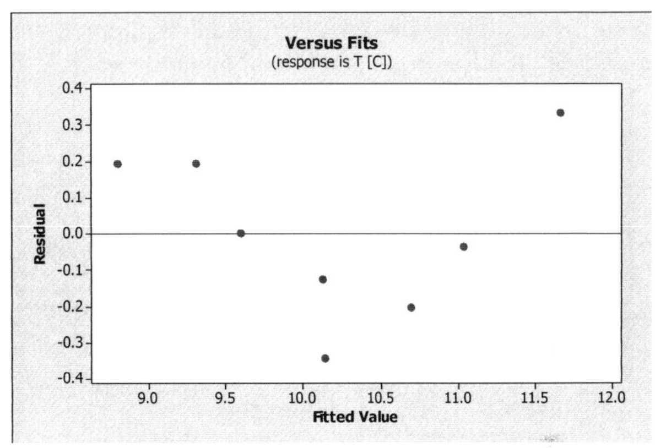

Fig. 15. Residuals versus the fits plot for T [C]

In the section Parameter Estimates there are given the points estimated for the parameters and the 95% confidence intervals. We consider this to be a reasonable range that implies significance.

The value 0.301576 for the Standard error of the regression S indicates that the observed temperature values fall a standard distance (roughly an average absolute distance) of 0.301576 units from the fitted mobility values determined with the regression equation.

The normal probability plot shows an approximately linear pattern consistent with a normal distribution. (See Figure 15)

The plot of residuals versus the fitted values shows a random pattern, which suggests that the residuals have constant variance. (See Figure 16)

Therefore, the regression equation obtained ''(2)'' can be consider adequate for analyzing the thermal regime of the rubber workpiece during grinding with abundant cutting fluid flow rate. The regression model specify correctly the relationship between the response-workpiece temperature and the predictors - workpiece speed, longitudinal cutting feed f [mm / min] and the cutting depth [mm].

According to the regression equation obtained, the optimal conditions that can assure a minimal workpiece temperature during grinding and therefore smaller thermal deflection of the rubber rollers machined with abundant lubricant flow rate (100 l / min), imply using smaller values for the cutting parameters, especially for the workpiece speed v_p [m/s].

9. Conclusion

This paper purpose was to investigate the influence of cutting conditions over the workpiece temperature durring rubber rollers grinding process. The process thermal regim and esspecialy the workpiece temperature has a significant influence over the grinding process ability to produce the round, concentric rubber rolls with tight tolerances and free from surface inconsistencies and variations. High cutting temperature can lead to thermal deflections that result in crown shaped external diameter rollers.

That is why it is very important to analyse and prevent temperaturevariations on the roll face durring machining. By analyzing the measured temperatures of the workpiece (that were in the range of 9-12^{0}C) during the grinding tests that were carried out it was found that the abundant lubricant flow rate (100 l / min) can limit and eliminate much of the amount of heat released in the cutting zone.

For the specific cutting conditions used in the experiments, the surface quality obtainedwas the one prescribed and no burn, cracks or other defects could be observed on the machined surfaces. It has also been found that the influence of the workpiece (the roller) speed manifests a greater influence over the workpiece temperature during grinding, compared to the longitudinal cutting feed and the cutting depth.

Since the grinding process of rubber parts is a difficult industrial application (because of the thermal sensitivity of the machined material), it is necessary to continue the researches with higher cutting speeds and even high flow rates for the cutting coolant, or using different cooling methods in order to establish the proper conditions necessary for grinding rubber covered rollers with high productivity and accuracy .

Also, as a further research direction, it is proposed to continue the grinding of hard rubberin a cryogenic medium (liquid nitrogen) with higher cutting speeds with the purpose of increasing process productivity.

10. References

[1] Vlase, A. &Gheorghiu, S.,(1995). Processing technologies on grinding machines, Technical Publishing House,ISBN 973-31-0629-1, Bucharest, Romania

[2] Vlase, A., Bălan, I.R &Vlase, B.G., (2011). Design of technologies on grinding machines, Printech PublishingHouse,ISBN 978-606-521-729-4, Bucharest, Romania

[3] Ştețiu, M., Lăzărescu, I.D., Oprean, C., &Ştețiu, G., (1994). Theory and practice of cutting tools, vol. III, Design of cutting tools (II), University of Sibiu Publishing House, Sibiu, Romania

[4] Negrea, I &Kaima, A., (1987) Abrasive bodies for grinding cylinders, Edited by Carbochim Cluj-Napoca, Cluj-Napoca, Romania

[5] Axinte, D.A., (1996), The correlation between the grinding process and the rolling mill grinding machine, PhDDissertation, University of Galati, Faculty of Mechanics, Galati, Romania

[6] Popescu, I., (1998). Research on cutting of non-ferrous and non-metallic materials), Reprography of the Universityof Craiova, Craiova, Romania

[7] Alexis, J., (1999). Taguchi Method in Industrial Practice - Experience Plans, Technical Publishing House, ISBN973-31-1352-2, Bucharest, Romania

[8] Muscă, G., Amarandei, D. & Ionescu, R., (1998), Experimenting, modeling and optimizing products and processes,Technical Publishing House, Chisinau, Moldavia

[9] Ionescu, R. &Amarandei, D., (2004). Experiment Planning - Efficiency and Quality,A.G.I.R.PublishingHouse,ISBN 973-8466-43-1, Bucharest, Romania

[10] ***Roller Grinding Machine Technical Paper model XШ-193H8, (1982). Mockba, URSS

[11] ***Grinding machines, (1982).Stankoimport, Moscou, URSS, 1982

[12] ***Catalog of abrasive products Carbochim, (2015). Cluj-Napoca, Romania

[13] Străjescu, E., (2004). The roughness of cutting tools, BrenPublishing House, ISBN 973-648-366-5, Bucharest, Romania

[14] Savin, I.M., (2006). Cooling - spray lubrication in metal cutting, PerformanticaPublishing House, ISBN 978-973-730-321-9, Iasi, Romania,

[15] Bărbulescu, V., (1973). Rectifiers - Quality and quality control of machined parts by grinding, no.3, Ministry ofHeavy Machinery Construction, Office of Documentation and Technical Publications, Bucharest, Romania

[16] Ciolacu, F.G., (2006). Managing the technological fluids for cutting, SitechPublishing House, ISBN 973-746-338-2, Craiova, Romania

[17] Ştefănescu, I., (2011). Improved machinability by using fluids for cutting, SitechPublishing House, Craiova,ISBN 978-606-11-1873-1, Craiova, Romania

[18] Starshev, D. (2008), Influence of the mode of the cooling under grinding on quality of surface layer.(Report)" Annals of DAAAM & Proceedings. DAAAM International Vienna.

[19] Schneider, M.; Klein, F. & Abt, C. (2010), Reductionof coolinglubrificationsin the grinding process, 7th International DAAAM Baltic Conference

[20] Practical Roll Techniques: Grinding & Balancing, Valmet technical paper series, september 2013, http://www.valmet.com/globalassets/media/downloads/white-papers/roll-services/wpr_grindandbalance.pdf, Accessed on: 2017-06-10

[21] The polishing of rubber covered rollers, Roller Manufacturing Articles, Rollertech, http://www.rollertech.co.uk/articles/polishing.html Accessed on: 2017-06-10

DOI: 10.2507/28th.daaam.proceedings.122

ANALYSIS OF STUDENTS' AND TRAM DRIVERS' BODY RATIOS IN ORDER TO SIMPLIFY THE CONTROL PANEL DESIGN

Sandro Tokić, Davor Sumpor, Srećko Đuranović & Nedžad Musabašić

This Publication has to be referred as: Tokic, S[andro]; Sumpor, D[avor]; Duranovic, S[recko] & Musabasic, N[edzad] (2017). Analysis of Students' and Tram Drivers' Body Ratios in Order to Simplify the Control Panel Design, Proceedings of the 28th DAAAM International Symposium, pp.0875-0884, B. Katalinic (Ed.), Published by DAAAM International, ISBN 978-3-902734-11-2, ISSN 1726-9679, Vienna, Austria
DOI: 10.2507/28th.daaam.proceedings.122

Abstract

A total of 10 anthropometric measures were taken from a sample of 21 male tram drivers of ZET Zagreb, from a sample of 39 male students of the University of Zagreb, as well as from a sample of 48 female students of the University of Zagreb, out of which the mean arithmetic values of body segment ratios were determined. According to the scientific findings that body height decreases with age, male tram drivers were divided by age into three groups. Some of the calculated mean arithmetic values of body segment ratios h_i/h for individual anthropometric measures h_i in relation to the body height h have been compared with theoretical values of body ratios calculated by harmonic analysis by Zerderbauer and Muftić and the golden section, and it is obvious that there are no big differences in values. Body ratios h_{mdr}/h and h_{ndr}/h can be used for simplified calculation of anthropometric measures h_{mdr} and h_{ndr} only by knowing the value of body height h, during the control panel design in tram cab or train cab, because body ratios h_{mdr}/h and h_{ndr}/h do not depend on gender, age and occupation for adult respondents from Croatia.

Keywords: anthropometric measures; theoretic harmonic analysis; golden section; body segment ratios; simplified calculation

1. Introduction

The main objective of all the human body harmonic analyses, such as harmonic analysis by Zederbauer and Muftić [1], is to ensure a more accurate and more simplified way to calculate other anthropometric measures, based only on knowing one measured anthropometric measure, which is the most common standing body height h. Generally considered, the widely known are the following significant five groups of factors of the human body sizes [2, 3, 4, 5, 6]: gender, age, ethnic differences, socioeconomic factors and demographic factors. In accordance with the proved scientific claims that the gender, age and socioeconomic factors of participants significantly affect body height, the researchers have wondered: "Is it possible that the gender, age and socioeconomic factors of respondents do not significantly affect body ratios h_i/h in relation to the current standing body height h for anthropological measures h_i important for cabin design?" In the smaller part of the study whose results are published in this paper the participants were male and female students of the University of Zagreb and male tram drivers of ZET Zagreb.

The presented results represent only a smaller part of the larger scale research which includes several separate studies of male engine drivers from all regions of Croatia [7], male and female tram drivers of ZET Zagreb and male and female students of the University of Zagreb.

2. Harmonic circle by Zederbauer and Muftić

The construction of the harmonic circle with radius *R* was given by Zederbauer at the beginning of the last century, in the way that the harmonic circle is a geometrical structure of an isosceles triangle, on whose sides there are lifted squares. The relations between radius of the circle *R* and the length of the sides of isosceles triangle *a* and *b* are always one and the same harmonic numbers, regardless of the change in size of the circle radius. Muftić chose the diameter of the harmonic circle *2R* for the human body height *h*, so the network of canon of eight head heights h_g is associated with the diameter of the harmonic circle and the human body height, so in this way are established the connection between harmonic numbers and anthropometric measures of the human, according to equation (1) and Figure 1 [1].

$$2 \cdot R = h = 8 \cdot h_g \tag{1}$$

The circle in Figure 1 is known in literature under the name of harmonic circle by Zerderbauer and Muftić [1], from whose constructions the sizes called harmonic numbers are obtained, according to Table 1.

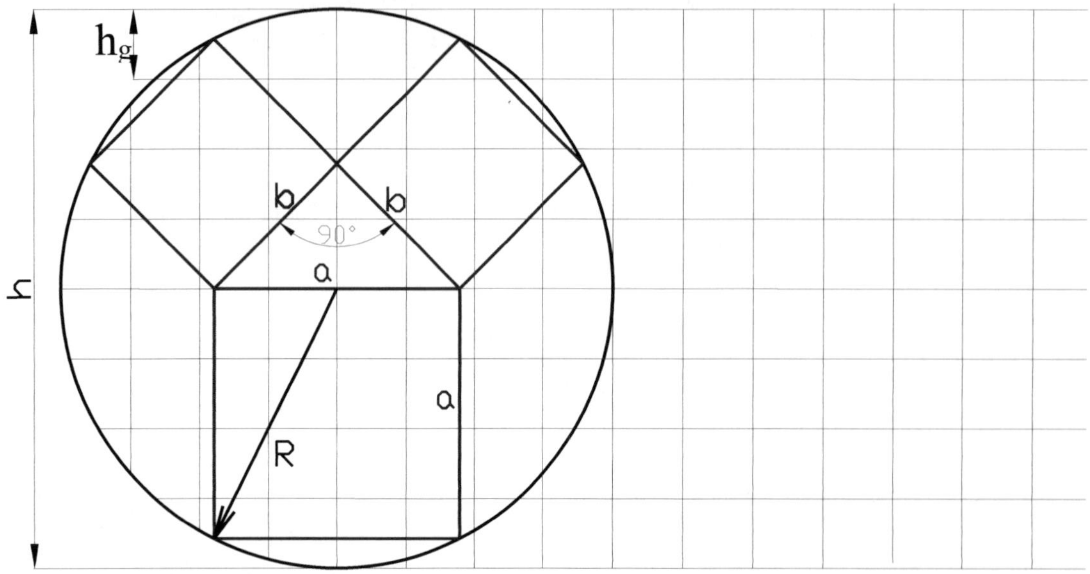

Fig. 1. Harmonic circle by Zederbauer and Muftić with associated canon of eight head heights as a measure for the standing body height
Source: Modified taken from Muftić, O. et al., Zagreb, 2001 [1])

Basic harmonic sizes:			Derived harmonic sizes:		
Label:	Connection with *a:*	Amount for *a=1*:	Label:	Connection with basic harmonic sizes:	Amount for *a=1*:
a	1a	1	r_h	$=b-\frac{a}{2}=\frac{\sqrt{2}-1}{2}\cdot a$	0,207
b	$\frac{\sqrt{2}}{2}\cdot a$	0,707	d_h	$=R-\frac{a}{2}=\frac{\sqrt{5}-1}{2}\cdot a$	0,618
R	$\frac{\sqrt{5}}{2}\cdot a$	1,118	$b+r_h$	$=\frac{\sqrt{2}}{2}\cdot a+\left(b-\frac{a}{2}\right)=\frac{2\sqrt{2}-1}{2}\cdot a$	0,914

Table 1. Harmonic sizes as harmonic numbers
Source: Modified taken from Muftić, O. et al., Zagreb, 2001 [1]

Length of body segment:	Label of length	Function $hi=hi(h)$:	Length of body segment	Label of length	Function $hi=hi(h)$:
Length of arm	h_r	$=\frac{25}{64}h=0.39\cdot h$	length of upper arm	h_{ndl}	$=\frac{5}{32}h=0.156\cdot h$
Length of forearm	h_{pdl}	$=\frac{h}{8}=0.125\cdot h$	length of hand	$h_š$	$=\frac{7}{64}h=0.109\cdot h$
Leg length	h_n	$=\frac{17}{32}h$	length of upper leg	h_3	$=\frac{9}{32}h$
Length of lower leg	h_2	$=\frac{7}{32}h$	foot length	h_1	$=\frac{h}{8}$
Length of the mobile part of the spine in the standing position	h_k	$=\frac{1}{3}h$	height of foot	h_{11}	$=\frac{h}{32}$

Table 2. Anthropometric measures of a human as a function of a standing body height *h*
Source: Modified, taken from Muftić, O. et al., Zagreb, 2001 [1]

Harmonic analysis of the human shows that functions of the anthropometric sizes i.e. segmental lengths depend on the standing body height of the human, on the basis of the relationship according to equation (1). All functions $hi=hi(h)$ which are shown in Table 2 should be universally valid for the young and healthy subjects of both genders.

According to many authors [3, 5, 6], age and gender are significant factors of body stature. The whole development of body dimensions reaches its peak towards the end of teenage or 20s in men, while women reach this development a few years earlier. After maturity, the body dimensions of both genders begin to decrease with age which is illustrated in Figure 2.

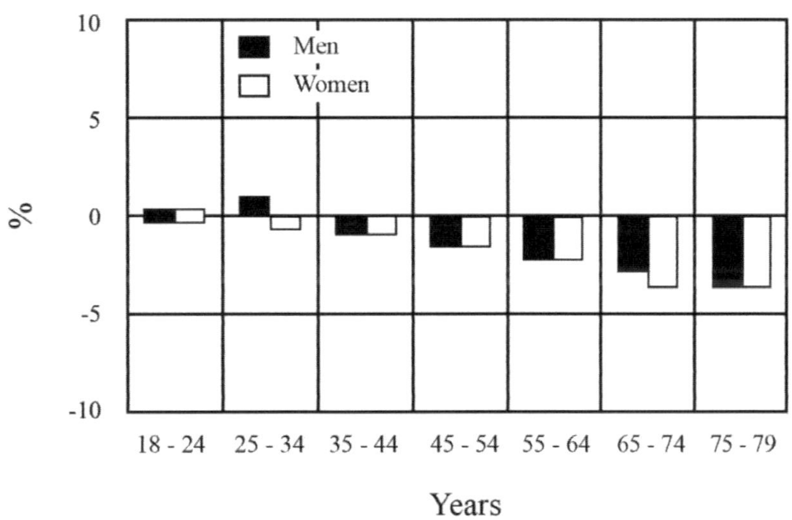

Fig. 2. Relative changes in body stature depending on age and gender, for men and women at the age of 18 to 79
Source: Taken from the National Center for Health Statistics, 1965 [6]

However, in reality, there are minor deviations of functional dependencies i.e. body segment ratios h_i/h of anthropometric sizes h_i about the standing body height h between men and women, which have been analysed in this paper on male and female students. Also, possible deviations of the real body segment ratios have been analysed in relation to theoretical values shown in Table 2 which were calculated using the harmonic analysis by Zederbaur and Muftić.

Relative changes of body segment ratios h_i/h with age have been analysed on the male tram drivers from Zagreb, who were divided into three age groups.

The anthropomeasures variations in adult men and women compared to the dimensional canon of eight head heights are within the range of one module, i.e. it is considered that the total standing body height of a human can vary within an interval of 7.5 to 8.5 head heights; however, it can be even 9 head heights for the so-called heroically built people [2]. This scientific claim for the value of the body ratio h/h_g between standing body height h and head height h_g has been verified and confirmed in this paper on the male and female students of the University of Zagreb until the age of 29, as well as on the male tram drivers in Zagreb, divided into three age groups.

3. Analysis of the human body based on the golden section

The geometric construction of the golden ratio [3] is illustrated on the rectangular triangle shown in Figure 3. The standing body height h is divided in two parts, the larger section i.e. height to the navel h_B, and the smaller section i.e. height above the navel h_C, which is shown in Figure 4.

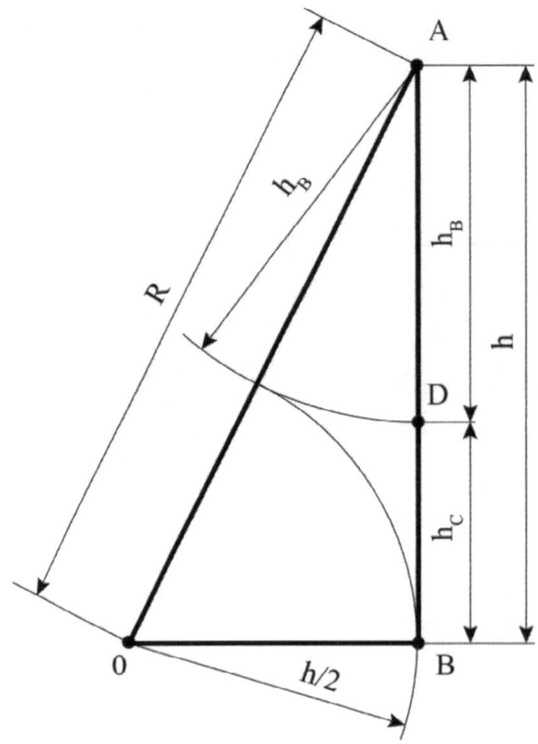

Fig. 3. Analysis of the human body based on the golden section
Source: Modified, taken from Ujević, D. et al., Zagreb, 2009 [3]

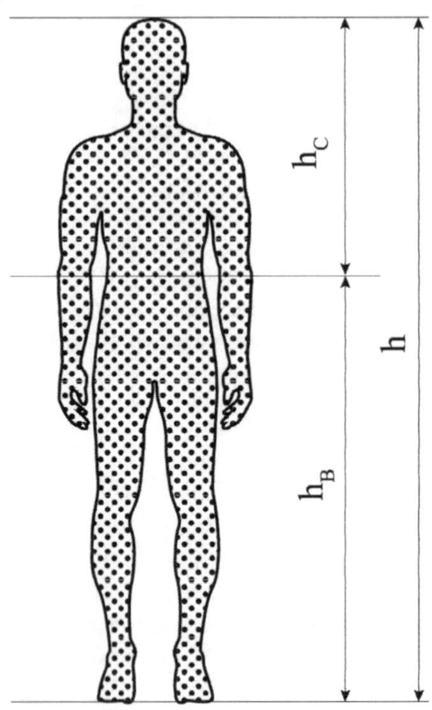

Fig. 4. Human body and the golden section

From the rectangular triangle ΔOAB shown in Figure 3 there follows:

$$R = h/2 + h_B \tag{2}$$

$$R^2 = h^2 + (h/2)^2 = 5/4h^2 \tag{3.1}$$

$$R = \sqrt{\frac{5h^2}{4}} = \frac{\sqrt{5}}{2}h = 1{,}1180 \cdot h \tag{3.2}$$

$$h_B = R - h/2 = \frac{\sqrt{5}}{2}h - h/2 = 0.618 \cdot h \tag{4}$$

$$h_B / h = 0.618 \tag{4.1}$$

$$h_C = h - h_B = h - 0.6180 \cdot h = 0.3819 \cdot h \tag{5}$$

$$h_C / h = 0.3819 \tag{5.1}$$

$$h_C / h_B = h_B / h = 0.618 \tag{6}$$

It is obvious that the theoretical value of the golden section is 0.618 in accordance with equation (6) and model shown in Figures 3 and 4. The paper presents a deviation of the real values of the body ratio h_B/h in all groups of respondents in relation to the theoretical value of the golden section 0.618.

4. Results

For all groups of respondents (male and female students of the University of Zagreb and male tram drivers in Zagreb) only a few of the most important anthropometric measures were measured, shown in Table 4 for male tram drivers in Zagreb (for the entire sample, and for sample divided into three age groups). Dynamic anthropometric measures maximum arm reach h_{mdr} and normal arm reach h_{ndr} together with static anthropometric measure bi-acromial range (shoulder width) $h_{šr}$ are the most important anthropometric measures for control panel design in the tram cabin.

The static anthropometric measures of the length of upper arm h_{ndl} and the length of forearm h_{pdl} were calculated using the measured values of the next three static anthropometric measures: arm length h_r, length of forearm and hand h_{10} and hand length $h_š$. The distance from the floor to the navel h_B was measured for the purpose of calculating the real value of ratio h/h_B for the golden section, for all groups of respondents.

If the arithmetic mean M and sample standard deviation SD are known, then 5 centile and 95 centile for all anthropometric measures can be calculated, according to equations (7) and (8) taken from Kroemer and Grandjean [4], because 5% of the tallest and 5% of the shortest individuals of the entire sample of respondents should be excluded (in the physical dimension to which the analysis applies).

$$5{,}0 \cdot c = M - 1{,}65 \cdot SD \tag{7}$$

$$95{,}0 \cdot c = M + 1{,}65 \cdot SD \tag{8}$$

If we have a mixed population of drivers (as a mixed male-female population of tram drivers of ZET operator in Zagreb), the range of anthropometric measures for central 90% should be calculated as a range between 5 centile for females (for the age group with the smallest values of anthropometric measures) and 95 centile for males (for the age group with the biggest values of anthropometric measures). But this is not the main objective of this research. The static and dynamic anthropometric measures were used for the calculation of body segment ratios h_i/h for anthropometric measure h_i in relation to the standing body height h.

Figure 5 shows typical anthropometric measures in the sagittal plane with labels by Kroemer and Muftić [1, 4]. Label 19 was added by co-authors.

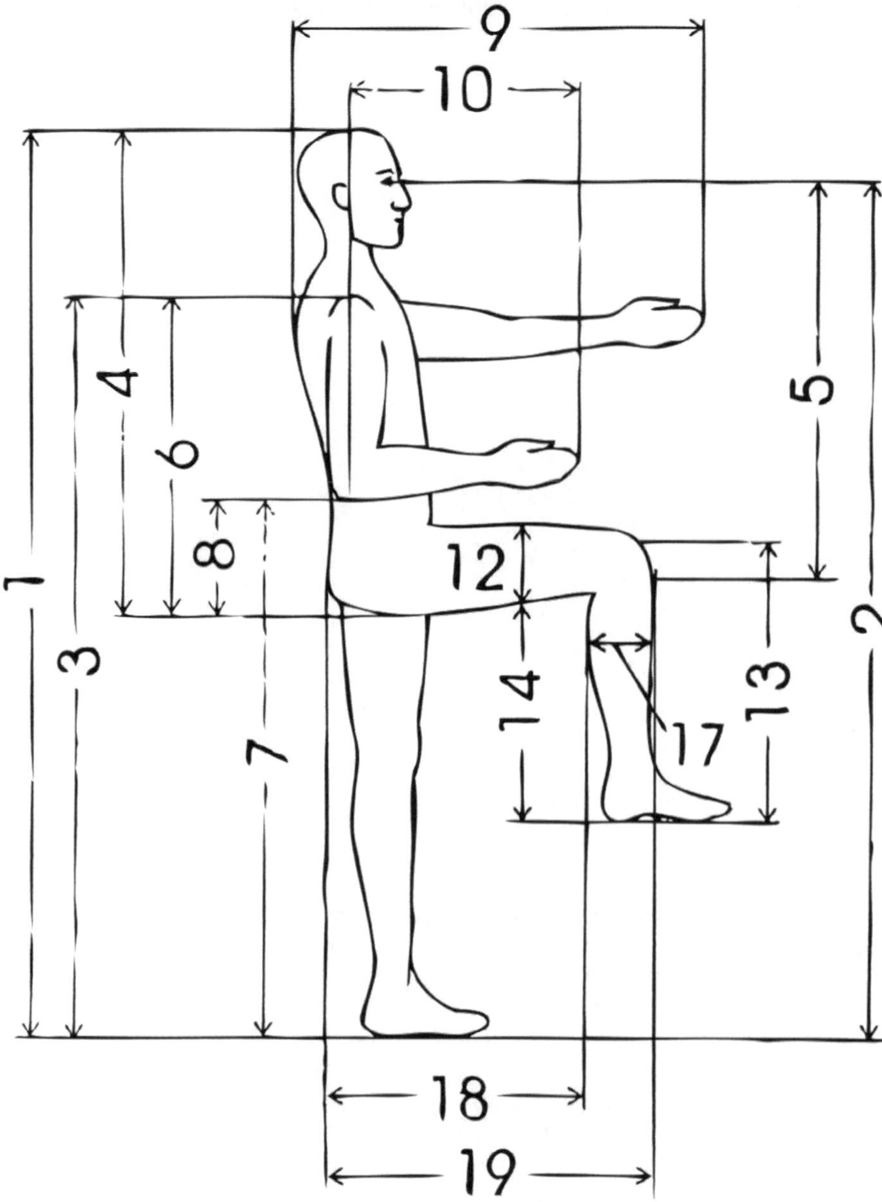

Fig. 5. Showing typical anthropometric measures in the sagittal plane by Kroemer
Source: Complemented, taken from Muftić, O., Milčić, D. 2001 [1]

The results of a few own studies confirm that gender is a significant factor of body stature. The age of fifty-one (51) researched engine drivers from Croatia, which are in the range from 27 to 56 years of age, significantly affects body height h and other anthropometric measures h_i that are functionally dependent on the body height h [7], which is for body height h confirmed by the results shown in Table 3.

Anthropometric measure	Symbol (unit)	Age groups									
		Total sample (*n = 51)*		Up to 29 years *(n = 9)*		From 30 up 39 years *(n = 13)*		From 40 up 49 years *(n = 18)*		From 50 up 59 years *(n = 11)*	
		M	*SD*	*M*	*SD*	*M*	*SD*	*M*	*SD*	*M*	*SD*
Body height	*h (cm)*	178.9	5.7	177.2	6.0	180.6	6.2	178.6	6.1	178.5	4.0

Table 3. Body height of male engine drivers in Croatia depending on age groups
Source: Taken from Mikulčić, M. et al., 2015 [7]

The values of anthropometric measures for male tram drivers in Zagreb shown in Table 4 partially confirm that age is a significant factor of body stature, because the largest value of arithmetic mean M of body height h is in the age group from 30 up to 39 years of age.

But the arithmetic mean of body height h in the age group from 50 up to 59 years is not the smallest value, and possible reasons for this deviation of results shown in Table 4 can be: insufficient total number of tram drivers in the sample, unequal and insufficient number of tram drivers in three age groups, as well as possible impacts of demographic factors within a particular age group (the origin of tram drivers from different parts of Croatia was not analysed and not taken into account during the sampling).

Name of the anthropometric measures or body segments	Symbol	Label by Fig. 5	Total sample		From 30 up to 39 years		From 40 up to 49 years		From 50 up to 59 years	
			n=21		n=5		n=8		n=8	
			M	SD	M	SD	M	SD	M	SD
			cm (kg*)	cm (kg*)	cm (kg*)	cm (kg*)	cm (kg*)	cm (kg*)	cm (kg*)	cm (kg*)
Standing body height	h	1	179.6	6.00	182.0	5.10	177.8	6.52	180.0	6.02
Mass*	m		95.1	17.4	94.1	16.07	91.8	16.53	98.9	20.27
Arm length (from acromion to the tip of the middle finger in vertical position)	h_r		78.2	3.49	79.0	4.36	76.6	3.16	79.3	3.06
Length of forearm and hand (from rare side of the elbow to the tip of the middle finger in a horizontal position)	h_{10}	10	48.6	2.11	49.2	2.17	48.0	2.51	48.9	1.73
Length of upper arm	h_{ndl}		29.6	1.99	29.8	2.49	28.6	0.92	30.4	2.26
Length of forearm	h_{pdl}		28.6	1.43	29.0	1.73	28.4	1.69	28.5	1.07
Hand length (distance between tip of the middle finger and the first crease in the wrist)	$h_š$		20.0	1.02	20.2	0.84	19.6	1.06	20.4	1.06
Distance from the navel to the floor	h_B		105.0	3.99	108.6	2.30	104.3	3.11	103.6	4.57
Normal arm reach (from the rear side of the elbow to the middle of a clenched fist)	h_{ndr}		35.6	1.91	36.4	1.67	34.9	1.96	35.8	1.98
Maximum arm reach (from the rear side of the acromion to the middle of a clenched fist)	h_{mdr}		64.6	2.94	65.4	3.85	63.9	3.56	64.8	1.58
Bi-acromial range (shoulder width)	$h_{šr}$	15	39.5	2.68	38.0	3.54	40.3	2.43	39.8	2.25

Table 4. Static and dynamic anthropometric measures of male tram drivers in Zagreb depending on age

Body segment ratios or golden section ratios	Male students up to 29 years of age				Female students up to 29 years of age				Male tram drivers in Zagreb from 30 up to 59 years of age			
	n=39				*n=48*				*n=21*			
	M_r	SD_r	*min.*	*max.*	M_r	SD_r	*min.*	*max.*	M_r	SD_r	*min.*	*max.*
h_B/h	**0.61**	0.015	0.58	0.66	**0.60**	0.013	0.57	0.62	**0.59**	0.015	0.55	0.61
h_C/h_B	**0.63**	0.040	0.52	0.72	**0.66**	0.036	0.61	0.76	**0.71**	0.045	0.64	0.83
h_C/h	**0.39**	0.015	0.34	0.42	**0.40**	0.013	0.38	0.43	**0.42**	0.016	0.39	0.45
h/h_g	**8.10**	0.660	6.73	9.37	**8.38**	0.481	7.43	9.26	**8.06**	0.340	7.50	8.75
h_r/h	**0.44**	0.017	0.40	0.47	**0.43**	0.013	0.41	0.47	**0.43**	0.012	0.41	0.46
h_{ndl}/h	**0.17**	0.015	0.13	0.20	**0.17**	0.012	0.14	0.20	**0.16**	0.009	0.15	0.18
h_{pdl}/h	**0.15**	0.008	0.14	0.17	**0.15**	0.008	0.13	0.16	**0.16**	0.006	0.15	0.17
$h_{š}/h$	**0.12**	0.006	0.10	0.13	**0.11**	0.005	0.11	0.12	**0.11**	0.005	0.10	0.12
h_{mdr}/h	**0.36**	0.020	0.33	0.40	**0.36**	0.016	0.32	0.40	**0.36**	0.015	0.33	0.39
h_{ndr}/h	**0.20**	0.014	0.16	0.24	**0.20**	0.010	0.16	0.22	**0.20**	0.008	0.18	0.21
$h_{šr}/h$	**0.20**	0.019	0.16	0.23	**0.20**	0.018	0.17	0.24	**0.22**	0.015	0.20	0.24

Table 5. Body segment ratios and the golden section ratios of male and female students of the University of Zagreb and male tram drivers in Zagreb

Results in Tables 5 and 6 show that all mean values M_r of the golden section ratios h_B/h for all groups of respondents (for both genders and for all age groups of tram drivers) deviate very little from the theoretical value of the golden section which is 0.618. The largest range of variations of the golden section ratios h_B/h around the mean arithmetic value M_r is for the male students of up to 29 years of age, from the minimal 0.58 up to the maximal 0.66.

Body segment ratios or golden section ratios	Male tram drivers from 30 up to 39 years				Male tram drivers from 40 up to 49 years				Male tram drivers from 50 up to 59 years			
	n=5				*n=8*				*n=8*			
	M_r	SD_r	*min.*	*max.*	M_r	SD_r	*min.*	*max.*	M_r	*SDr*	*min.*	*max.*
h_B/h	**0.60**	0.009	0.59	0.61	**0.59**	0.009	0.57	0.60	**0.58**	0.019	0.55	0.60
h_C/h_B	**0.68**	0.026	0.64	0.71	**0.70**	0.025	0.66	0.74	**0.74**	0.058	0.66	0.83
h_C/h	**0.40**	0.009	0.39	0.41	**0.41**	0.009	0.40	0.43	**0.42**	4.50	0.40	0.45
h/h_g	**8.13**	0.354	7.78	8.71	**8.06**	0.423	7.50	8.75	**8.01**	0.271	7.57	8.38
h_r/h	**0.43**	0.015	0.42	0.46	**0.43**	0.012	0.41	0.45	**0.44**	0.008	0.43	0.45
h_{ndl}/h	**0.16**	0.010	0.15	0.18	**0.16**	0.004	0.15	0.17	**0.17**	0.011	0.15	0.18
h_{pdl}/h	**0.16**	0.007	0.15	0.17	**0.16**	0.008	0.15	0.17	**0.16**	0.005	0.15	0.17
$h_{š}/h$	**0.11**	0.005	0.10	0.12	**0.11**	0.004	0.10	0.12	**0.11**	0.003	0.11	0.12
h_{mdr}/h	**0.36**	0.016	0.34	0.38	**0.36**	0.018	0.33	0.39	**0.36**	0.014	0.34	0.39
h_{ndr}/h	**0.20**	0.010	0.19	0.21	**0.20**	0.009	0.18	0.21	**0.20**	0.007	0.19	0.21
$h_{šr}/h$	**0.21**	0.016	0.19	0.23	**0.23**	0.010	0.22	0.24	**0.22**	0.015	0.20	0.24

Table 6. Body segment ratios and golden section ratios of male tram drivers in Zagreb depending on age

Results in Tables 5 and 6 show that all mean values M_r of the body ratios h/h_g for all groups of respondents (for both genders and for all age groups of tram drivers) deviate very little from the theoretical value based on the canon of eight head heights h_g which is 8. There is a small difference between the mean arithmetic value of the body ratio for male students ($h/h_g = 8.1$) and the mean arithmetic value of the body ratio for female students ($h/h_g = 8.38$).

The largest range of variations of body ratios h/h_g is around the mean arithmetic value M_r for male students up to 29 years, from minimal 6.73 up to the maximum of 9.37 (variations gain value of 2.64 head heights). Differences of the mean arithmetic values of body ratios h/h_g for tram drivers in relation to the theoretical value $h/h_g = 8$ decrease depending on age; for male tram drivers from 50 up to 59 years of age, the body ratio h/h_g is 8.01.

Results shown in Tables 5 and 6 confirm that there is no big difference between the mean arithmetic values of body ratios h_r/h (ratio of arm length and body height), h_{ndl}/h (ratio of length of upper arm and body height), h_{pdl}/h (ratio of length of forearm and body height) and $h_š/h$ (ratio of hand length and body height) depending on gender (male and female students), age (three age groups of male tram drivers) and occupation (students and tram drivers).

Mean arithmetic values of body ratios h_r/h and $h_š/h$ of all groups of respondents shown in Tables 5 and 6 are almost same with the mean arithmetic values of body ratios h_r/h and $h_š/h$ of all four age groups for fifty one (51) researched engine drivers from Croatia, which were in the range from 27 to 56 years of age in 2015 [7]; and also those are almost the same with the mean arithmetic values of body ratios h_r/h and $h_š/h$ of 68 female student from the University of Zagreb in 2016 (another sample from the same population) [8].

Mean arithmetic values of body ratios $h_š/h$ and h_{ndl}/h of all groups of respondents are almost the same with the theoretical values shown in Table 2 which are based on the harmonic analysis and body height h defined by the canon of eight head heights h_g.

Mean arithmetic values of body ratios h_r/h and h_{pdl}/h of all groups of respondents are different in relation to the theoretical values of body ratios shown in Table 2 which are based on the harmonic analysis, because real mean arithmetic values of body ratio h_r/h are bigger in relation to the theoretical value $h_r/h = 0.39$ for ca. 13 %, and the real mean arithmetic values of body ratio h_{pdl}/h are bigger in relation to the theoretical value $h_{pdl}/h = 0.125$ for ca. 28 %.

Mean arithmetic values of body ratios $h_{mdr}/h = 0.36$ (ratio of maximum arm reach and body height) and $h_{ndr}/h = 0.2$ (ratio of normal arm reach and body height) of all groups of respondents shown in Tables 5 and 6 are completely the same with the mean arithmetic values of body ratios h_{mdr}/h and h_{ndr}/h of all four age groups for fifty one (51) researched engine drivers from Croatia [7], and also those are completely the same with the mean arithmetic values of body ratios h_{mdr}/h and h_{ndr}/h of 68 female students from the University of Zagreb in 2016 (another sample from the same population) [8]. Generally considering, dynamic anthropometric measures, maximum arm reach h_{mdr} and normal arm reach h_{ndr} are linear functionalities depending on the standing body height h [7], and this fact can be the reason why those body ratios do not depend on the changes of body height depending on age.

Body ratios h_{mdr}/h and h_{ndr}/h can be used for simplified calculation of values of anthropometric measures h_{mdr} and h_{ndr} only by knowing the value of the body height h, during the control panel design in tram cab or train cab, because real mean arithmetic values of body ratios $h_{mdr}/h = 0.36$ and $h_{ndr}/h = 0.20$ do not depend on gender, age and occupation for adult respondents from Croatia. Quick and simplified calculation of normal arm reach h_{ndr} is very important, because frequently used commands on the locomotive, railcar or tram control panel need to be arranged mainly within the normal reach of the arm, using multi-purpose controllers for serving several important and frequently used functions by one hand, whenever possible [9].

Mean arithmetic values of body ratios $h_{šr}/h$ of all groups of respondents shown in Tables 5 and 6 are not the same. The range of variations of mean arithmetic values of body ratio $h_{šr}/h$ is from minimal 0.20 (male and female students up to 29 years) up to the maximum 0.23 (male tram drivers from 40 up to 49 years). Static anthropometric measure bi-acromial range (shoulder width) $h_{šr}$ have to be measured for all respondents divided into age groups from target population of drivers, and the mean arithmetic value of body ratio $h_{šr}/h$ cannot be used for simplified calculation of the value of anthropometric measure $h_{šr}$ only by knowing the value of body height h. The available studies [10] also indicate very weak correlations ($r = 0.42$) between bi-acromial range (shoulder width) and body height h in males (r = 0.42), which means that there are no linear functional dependences $hšr=hšr(h)$.

5. Discussion and conclusion

The largest part of the measured participants are not harmonic beings. Just a few respondents have body dimensions in accordance to the golden section and harmonic analysis by Zederbauer and Muftić based on the canon of eight head heights h_g.

But the calculated mean arithmetic values of body segment ratios h_i/h for individual anthropometric measures h_i in relation to the body height h have not deviated significantly from the theoretical values of the body ratios calculated by the harmonic analysis by Zerderbauer and Muftić and the golden section. Between the analysed body ratios the biggest deviation is for body ratio h_{pdl}/h (ratio of the length of forearm and body height), and real mean arithmetic values of body ratio h_{pdl}/h are ca. 28% bigger in relation to the theoretical value calculated by Zerderbauer and Muftić.

Much more important results of this study are that the real mean arithmetic values of body ratios $h_{mdr}/h = 0.36$ (ratio of maximum arm reach and body height) and $h_{ndr}/h = 0.2$ (ratio of normal arm reach and body height) can be used for simplified and quick calculation of the values of anthropometric measures h_{mdr} and h_{ndr} only by knowing the value of the body height h, during the control panel design in tram cab or train cab, because of real mean arithmetic values of body ratios h_{mdr}/h and h_{ndr}/h do not depend on gender, age, and occupation (socioeconomic factors) for adult respondents from Croatia.

This partial study is not finished, and will be continued with improvements during the sampling as follows.

There was an insufficient total number of tram drivers in the sample; the number of tram drivers in three age groups was unequal and insufficient; there were no tram driver respondents of up to 29 years of age; there were no tram driver respondents over 60, and possible impacts of demographic factors within a particular age group (the origin of tram drivers from different parts of Croatia was not analysed and was not taken into account during the sampling). Since the population of tram drivers of the ZET operator in Zagreb is mixed male-female, anthropometric measures should also be measured for female tram drivers, for all age groups, for the purpose of comparing the results of the mean arithmetic values of body ratios h_i/h with the results presented in this paper.

Of course, for a final study conclusion and final results of the larger scale research the results published in this paper will be compared with the results of other planned participants. If it is necessary due to mathematical and methodological reasons, some of the partial research will be repeated with a bigger number of respondents.

The final results of the larger scale research should not be limited only to tram and train drivers in Croatia.

6. Acknowledgments

Authors are grateful to female and male students of the University of Zagreb and tram drivers of ZET Zagreb, who were participants of this study. Without their cooperation, this study could not have been carried out. Also, special thanks to the representative of the Association of Croatian Trade Unions of Drivers and Transport Workers Mr. Anto Jelić for his patience in coordinating the research in ZET Zagreb. This study is supported by the Faculty of Transport and Traffic Sciences, University of Zagreb, under the name "Program of support for establishing of the research activities and groups PROM-PRO 995-12 (2017)".

7. References

[1] Muftić, O., Milčić, D. (2001). Ergonomija u sigurnosti, Visoka škola za sigurnost na radu, Iproz, Zagreb

[2] Jurum-Kipke , J., Baksa, S., Kavran, Z. (2007). Anthropometric Relations of Human Body in the Function of Traffic Environment Analysis, Proceedings of 3rd International Ergonomics Conference "Ergonomics 2007", June 13-16, 1007, Stubičke toplice, ISBN: 978-953-98741-4-6, Mijović, B. et al. (ed.), pp. 239-247, Croatian Ergonomics Society, Zagreb

[3] Ujević, D. et al. (2009). Theoretical Aspects and Application of Croatian Anthropometric System, Faculty of Textile Technology, University of Zagreb, ISBN: 978-953-7105-28-0, Zagreb

[4] Kroemer, K.H.E.; Grandjean, E. (1997). Fitting the Task to the Human, A Textbook of Occupational Ergonomics, Fifth Edition, Published by Taylor & Francis Ltd., ISBN: 0748406654, London, (K.H.E. Kroemer, E. Grandjean, Prilagođavanje rada čovjeku, Naklada Slap, Jastrebarsko, 2000., ISBN: 953-191-096-0).

[5] Panero, J.; Zelnik, M. (1979). Human dimension & Interior space, Watson Guptill Publications, ISBN: 0-8230-7271-1, New York

[6] https://www.cdc.gov/nchs/data/series/sr_11/sr11_008.pdf, (1965). U.S. Department of Health, Education and Weffare, National Center for Health Statistic, Weight, Height, and Selected Body Dimensions of Adults, Series 11, Number 8, Rockwille, Maryland, USA, Accessed on: 2016-10-23

[7] Mikulčić, M.; Modrić, M.; Sumpor, D. (2015). Application Possibility of Engine Drivers' Body Segments Ratios in Designing the Cab' Working Environment in Croatia, Proceedings of the 26th DAAAM International Symposium on Intelligent Manufacturing and Automation, DAAAM 2015, Zadar, 21st – 24th October 2015, ISSN: 1726-9679, Katalinić, B. (ed.), pp. 842-848, Published by DAAAM International, Vienna

[8] Sumpor, D.; Mikulčić, M.; Modrić, M. (2016). Female Students' Body Segments Ratios, Book of Proceedings of the 6th International Ergonomics Conference „Ergonomics 2016 – Focus on synergy", June15-18, 2016, Zadar, ISSN: 1848-9699, Sušić, A. et al. (ed.), pp. 323-330, Croatian Ergonomics Society, Zagreb

[9] Sumpor, D.; Taboršak, D.; Jurum-Kipke, J. (2015). Anthropometric Measures Important for Control Panel Design in Locomotive or Railcar, Technical Gazette, Scientific proffesional journal of technical faculties of the Josip Juraj Strossmayer University of Osijek, Vol.22, No.1, Slavonski Brod, 2015, pp. 1-10, ISSN: 1330-3651

[10] Ozaslan, A.; Karadayi, B.; Kolusayin, M.O.; Kaya, A. (2011). Stature Estimation from Bi-acromial and Bi-iliocristal Measurements, Romanian Journal of Legal Medicine, Romanian Society of Legal Medicine, Vol.19, No.3, Bucharest, 2011, pp. 171-176, ISSN: 1844-8585

DOI: 10.2507/28th.daaam.proceedings.123

ROBOTIC MACHINING SIMULATION USING A SIMPLIFIED MULTIBODY MODEL

Hoai Nam Huynh, Olivier Verlinden & Edouard Rivière-Lorphèvre

This Publication has to be referred as: Huynh, H[oai] N[am]; Verlinden, O[livier] & Riviere-Lorphevre, E[douard] (2017). Robotic Machining Simulation using a Simplified Multibody Model, Proceedings of the 28th DAAAM International Symposium, pp.0885-0894, B. Katalinic (Ed.), Published by DAAAM International, ISBN 978-3-902734-11-2, ISSN 1726-9679, Vienna, Austria
DOI: 10.2507/28th.daaam.proceedings.123

Abstract

Robotic machining is a promising manufacturing technology combining the agility of an industrial robot with the potential of a net shape process. Nowadays, robotic machining is springing up everywhere. A large number of companies attempt the conversion of their process using this technology, but without great success. Indeed, the material removal using a robot is very different from the conventional milling. Even though it is cheaper for a larger working envelope, the understanding of robotic process must be widened in order to overcome technological hurdles. Among them, the lack of stiffness at joints and the inherent low frequency structure are mainly accountable for the quality depletion of the machined workpiece. The simulation of an industrial process can be an interesting approach to better perceive the on-going phenomena during the machining phase. For this reason, a multibody modelling of a robot coupled with the simulation of milling was implemented. The robot was modelled as an anthropomorphic arm comprising 4 degrees of freedom and including the joint stiffness. Results were finally confronted with experimental data in terms of cutting forces, vibrations and roughness for which a good accordance could be observed.

Keywords: chatter; multibody; robotic machining; simulation; validation

1. Introduction

Robotic machining, which is the use of an industrial robot as a machine tool, is a manufacturing technique in demand these days. Many industrials are trying this relatively new technology by transposing their current process of material removal to the potential of robotics, but without great success. Although offering many advantages such as a lower cost for a larger workspace and a greater flexibility in terms of part accessibility, its development is held back due to several issues leading to the quality depletion of the final product. Limiting factors comprise the low stiffness of robot joints and their lack of accuracy. Furthermore, due to their low frequency structure, typically around 10 [Hz], industrial robots are more exposed to vibratory instabilities such as the chatter phenomenon. While there are two types of chatter named regenerative chatter and mode coupling chatter, Pan Z. et al. [1] found that the latter was the dominant source of vibration in robotic machining. Above all, chatter phenomenon must be avoided as it leads to a sharp increase in cutting forces and vibrations wrecking the surface finish of the workpiece and sometimes conducting to tool breakage.

Barnfather J.D. et al. [2] also stated that a barrier to its deployment is the dimensional errors coming from the low dynamic stiffness and their low resistance to machining forces. Dumas et al. draw attention on the fact that the low joint stiffness affects the surface finish [3]. Geometrical errors in robot links and joints cause assembly misalignments and affect the positioning accuracy of the end-effector [4]. In other words, issues commonly found in conventional machining such as the tool deflection, the backlash at joints and wear are magnified due to the less rigid structure [5].

Yet robotic machining is consequently bound to operations requiring low cutting forces in order to avoid the aforementioned issues. Its field of applications especially concerns finishing operations such as grinding, polishing, deburring, drilling, sanding, cutting and the milling of soft materials, for example, plastic, wood, foam, aluminium and even steel when small layers are removed. Different industrial sectors benefit from robotic machining keeping in mind that they will take full advantage of the technology when machining large workpieces. It is certainly the case of aerospace for the polishing of turbine blades and the drilling of wing segments and in foundries for the deburring of cast parts (Fig. 1) [6].

a) Plastics b) Carbon-fiber c) Aluminium

Fig. 1. Applications in robotic machining

Even though industrial robots will not replace conventional machine tools in all machining operations because of their lack of stiffness and precision, research is conducted in order to better understand such a process. On-going research focusses on the development of indicators pointing stiffer robot configurations, the optimal placement of the workpiece with respect to the robot, the implementation of calibration procedure to improve part quality, the installation of extra actuator to compensate robot positioning error, the improvement of existing controllers, the creation of off-line compensation methods, etc.

For instance, Ilyukhin Y.V. proposed a new control structure ensuring high-performance machining [7]. Their algorithm generated corrected values for the velocity path taking into account the currents in the windings of all servomotors. As a result, the milling tool followed continuous path without overheating and overloading the drives. Klimchik A. et al. ranked five industrial robots according to a developed approach based on the circularity norm that evaluates the contortion degree of a benchmark circle [8]. This industry-oriented methodology allows its user identifying which robot will offer the best machining accuracy on the basis of a model. The method was validated experimentally and proved to be in good agreement with the theoretical model.

Finally, Bauer J. et al aimed at developing a strategy to improve the accuracy in robotic machining by prediction and offline compensation of path deviations on the basis of the ideal cutting forces and offset mirroring [9]. For this purpose, they adapted a cutting force model and applied it for industrial robots in order to perform simulations. The robot motion was computed using the Newton-Euler formulation. Both sub-models were first tested separately before their coupling to simulate the machine process interaction. Milling experiments were eventually achieved to confirm the compensation concept.

This paper is also related to the simulation of robotic machining with process interaction but rather targeted to the model validation. A simplified multibody model of an existing robot was built and coupled with a milling routine in order to simulate the whole process. Section 2 is dedicated to the presentation of the robot and the cutting force models, followed by their coupling. Section 3 describes the two milling operations studied in this paper as well as the cutting parameters employed to replicate milling experiments. The comparison of the results is based on the analysis of the cutting force amplitudes, the vibrations and the lateral roughness of the machined workpiece in section 4. The last section will draw the final summary and future work.

2. Robotic machining modelling

In order to develop a simulation environment of robotic machining, the industrial robot and the milling process should first be modelled separately and then coupled together. The robot was modelled using the multibody approach while cutting forces were computed according to a macroscopic representation of the material removal. Both models were programmed in C++ and were coupled in one single simulation environment of robotic machining.

2.1. Robot modelling

As milling tests were performed using a serial 6-dof industrial robot from Stäubli, the robot model was naturally inspired by its geometrical dimensions. The TX200 model from Stäubli is a heavy payload robot featuring a nominal load capacity of 1000 [N], a reach of 2194 [mm] and a repeatability of ±0.06 [mm]. Its stiffness is more than sufficient to withstand the cutting forces as the machined part quality can attest. The aim was to simulate the dynamic behaviour of the robotic arm in a simplified way.

The anthropomorphic arm was simplified into a 4-axis manipulator whose motion can span a 3D envelope. The simplification was considered to alleviate the computational load of the inverse kinematic algorithm and to keep a simple model while being apt to represent the main trend in robotic machining. Figure 2 presents the real industrial robot comprising its 6 degrees of freedom (DOF) and its simplified model with 4 DOFs and 5 bodies. As the robot was intended to machine a workpiece following a $2D^{1/2}$ path, the motion of the last link belonging to the simplified model was such that the latter remains in vertical position.

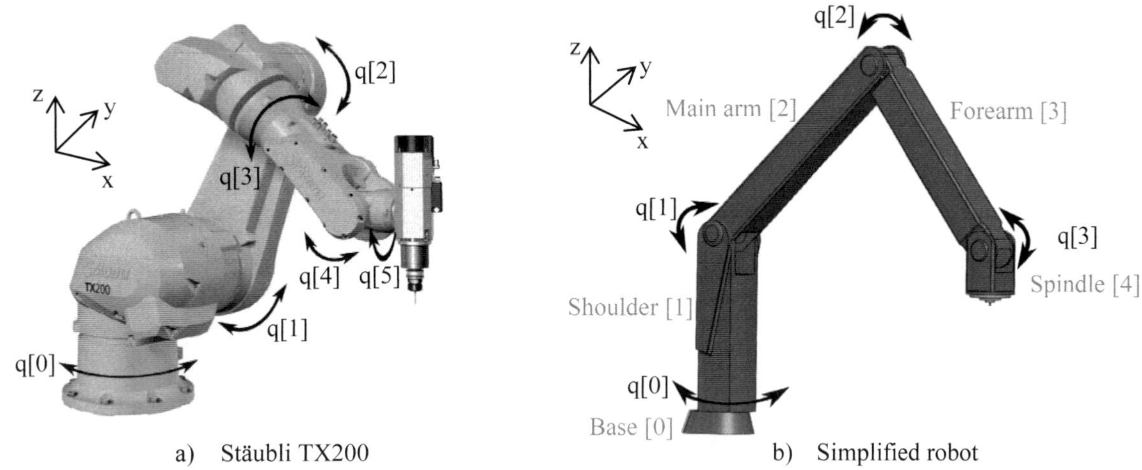

Fig. 2. Simplified robot modelling

In order to define the mass and inertia properties of each body pertaining to the simplified model, it was necessary to know the mass distribution of the real robot components. Nevertheless, such data are not directly available from the robot manufacturer. Only the total mass of the Stäubli robot can be retrieved and equals 981 [kg]. To better reproduce the behaviour of the industrial robot with the simplified model, it was therefore essential to distribute the total mass amongst the five bodies, respectively named the base, the shoulder, the main arm, the forearm and the spindle. For that purpose, two considerations had to be taken into account:

- The mass distribution of the real robot is similar to its volume distribution. CAD designs of the robotic arm are provided by the robot manufacturer but the mass of each component cannot be inferred as the virtual shapes are not hollowed. Consequently, shapes can only give information about their volume.
- The shoulder of the robot (body [1]) is naturally heavier than its other components as it embeds the two main motors for joints 1 and 2.

As a result, it was decided to distribute the mass of the robot according to the volume of each of its component. Concerning the identification of their moment of inertia, each robot part was simplified into a rectangular parallelepiped with an equivalent volume. Table 1. displays a synopsis of the mass and moment of inertia for each part of the simplified robot. It was not necessary to define the moments of inertia for the base (body [0]) as it does not intervene in the motion of the robot.

Body	**Mass [kg]**	**Ixx [kg.m²]**	**Iyy [kg.m²]**	**Izz [kg.m²]**	**Length [m]**
Base [0]	183	/	/	/	Lg: 0.1
Shoulder [1]	342	18.044	29.213	30.014	L1: 0.65
Main arm [2]	243	35.033	37.463	4.050	L2: 0.95
Forearm [3]	202	5.507	5.507	4.547	L3: 0.85
Spindle [4]	11	0.103	0.103	0.041	L4: 0.2

Table 1. Mass and moments of inertia of the simplified robot components

On the basis of the inertia properties and the positions of each centre of mass, it was then possible to model the simplified robot as a multibody system. To help with this modelling, the Theoretical Mechanics, Dynamics and Vibration Department of the University of Mons (Belgium) developed a multibody framework called EasyDyn allowing the dynamic simulation of mechanical systems [10]. Initially created for teaching purposes, it was successfully used in several Master's and PhD theses and is available for free on the Internet.

EasyDyn offers its user the possibility to solve problems represented by second-order differential equations and, more particularly, the dynamic behaviour of mechanical multibody systems. The approach currently implemented is based on the minimal coordinates for the choice of the configuration parameters q defining the motion of the bodies. The kinematics of all bodies is provided by the user through homogeneous transformation matrices T function of the chosen configuration parameters as far as they are independent. The number of configuration parameters is consequently equivalent to the number of degrees of freedom. The expressions of position and orientation of each body i comes from its transformation homogeneous matrix $T_{0,i}$ considered at their centre of mass. The latter is then symbolically derived in order to get the expressions of velocities and accelerations ($T_{0,i}$ gives the position and orientation of the frame attached to centre of mass of body i with respect to a global reference frame 0) (Fig. 3). The kinematics of each body i is accordingly a function of q and its time-derivatives $\dot{q}$ and $\ddot{q}$. Aside from the kinematics, the user may also apply forces on each body.

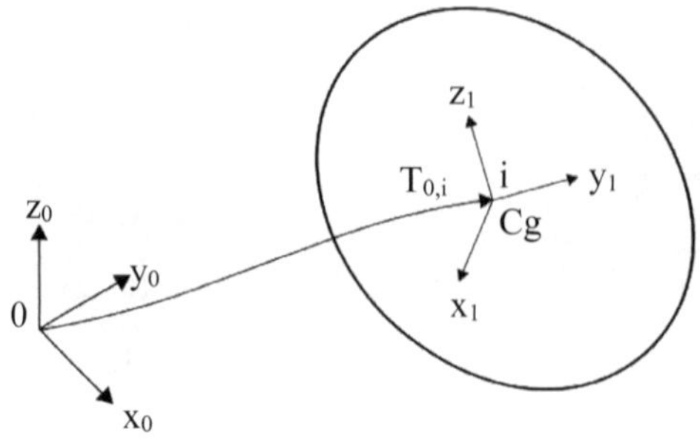

Fig. 3. Frame situation of body i with respect to the global reference frame 0

As the simplified robot model comprised five bodies, its kinematic model involved the definition of five homogeneous transformation matrices expressed as follows:

T_0 = Tdisp(0, 0 , -Lg/2) (1)
$T_{0,1}$= Trotz(q[0]) x Tdisp(0, 0, L1/2) (2)
$T_{1,2}$= Tdisp(0, 0, L1/2) x Troty(q[1]) x Tdisp(0, 0, L2/2) (3)
$T_{2,3}$= Tdisp(0, 0, L2/2) x Troty(q[2]) x Tdisp(0, 0, L3/2) (4)
$T_{3,4}$= Tdisp(0, 0, L3/2) x Troty(q[3]) x Tdisp(0, 0, L4/2) (5)

where

- Tdisp(x, y, z) represents a homogeneous transformation matrix corresponding to a displacement without rotation by coordinates x, y and z;
- Troty(γ) or Trotz(γ) represents a homogeneous transformation matrix corresponding to a rotation around y- or z-axis respectively of an angle γ.

The n_{cp} equations of motion were then built according to the given kinematics involving the n_{cp} configuration parameters and the applied forces on the n_B bodies. Their form derives from the application of the d'Alembert's principle:

$$\sum_{i=1}^{n_B}[d^{i,j} \cdot (\vec{R_i} - m_i\vec{a_i}) + \theta^{i,j} \cdot (\vec{M}_{G_i} - \Phi_{G_i}\dot{\vec{\omega}}_i - \vec{\omega}_i \times \Phi_{G_i}\vec{\omega}_i)] = 0 \qquad j=1, n_{cp} \qquad (6)$$

- m_i is the mass of body i;
- Φ_{G_i} is the inertia tensor of body i;
- $\vec{R_i}$ is the resultant force of all applied forces acting on body i;
- $\vec{M}_{G_i}$ is the resultant moment at the centre of mass G_i of all applied forces acting on body i;
- $\vec{v_i}$ is the velocity of the center of mass of body i;
- $\vec{\omega_i}$ is the rotational velocity of body i;
- $d^{i,j}$ is the partial contribution of $\vec{\dot{q}_j}$ in the translational velocity of body i: $\vec{v_i} = \sum_{j=1}^{n_{cp}} d^{i,j} \cdot \vec{\dot{q}_j}$;
- $\theta^{i,j}$ is the partial contribution of $\vec{\dot{q}_j}$ in the rotational velocity of body i: $\vec{\omega_i} = \sum_{j=1}^{n_{cp}} \theta^{i,j} \cdot \vec{\dot{q}_j}$.

Once built, the equations of motion were integrated according to the Newmark integration method, the parameters corresponding to the standard Newmark 1/4.

The robot modelling would not have been complete without the introduction of joint stiffness, being the main weakness of industrial robots. The component bringing such flexibility is mainly the gearbox of each motor which was modelled as a rotational spring along with a damper located between each link (Fig. 4).

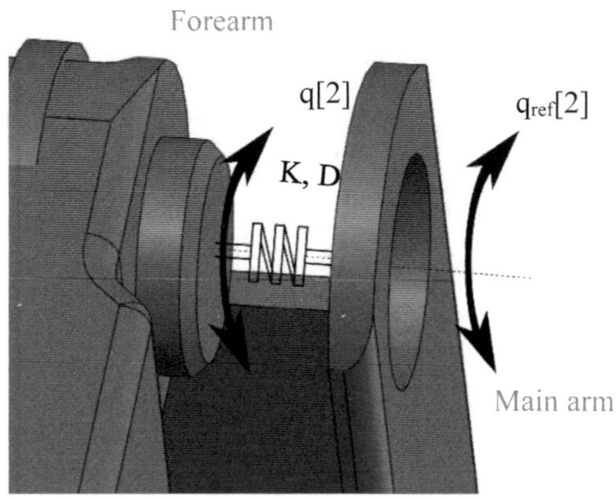

Fig. 4. Joint flexibility modelled as a spring K and a damper D

The torque applied at joint i+1 simply results from the difference between the reference position $q_{ref}[i]$ at joint i, computed by the inverse kinematic algorithm, and the real joint position q[i] after the spring and the damper (7). The reaction torque is applied at joint i (8).

$$M_{i+1} = K_i \cdot (q_{ref}[i] - q[i]) - D_i \cdot \dot{q}_i \quad (7)$$
$$M_i = K_i \cdot (q_{ref}[i] - q[i]) - D_i \cdot \dot{q}_i \quad (8)$$

The values of stiffness were picked up from an article presenting a methodology to identify the joint stiffness experimentally. The studied industrial robot (Kuka KR240-2) was also very similar in terms of size [3]. Joint damping coefficients were harder to find. In the PhD thesis of Oueslati M., the damping coefficient of joint 2 was experimentally identified for the Stäubli RX170B robot [11]. It was deduced from the exponential decay of the vibration signal measured on the position of the second link when subjected to position step. However, the estimated value seemed quite high (D_{ref}=4960 [N.m.s/rad]) and lead to purely real poles characterised by a damping coefficient of 100 %. In order to find more realistic values, the damping coefficient was rescaled as a function of the squared root of the ratio involving K_{ref} (provided in [14]) and the stiffness to associate, as indicated by equation (9). Table 2 finally provides the stiffness and damping values used for the simplified robot model. Additionally, a torque compensation for the gravity was applied at joints 1 and 2 as a function of the mass of the links following them.

$$\frac{D_i}{D_{ref}} = \sqrt{\frac{K_i}{K_{ref}}} \quad (9)$$

	Joint 1	Joint 2	Joint 3	Joint 4
Stiffness [N.m/rad] K	$3.8 \cdot 10^6$	$6.6 \cdot 10^6$	$3.9 \cdot 10^6$	$0.66 \cdot 10^6$
Damping [N.m.s/rad] D	1057.2	1393.3	1071.0	440.6

Table 2. Joint stiffness and damping

2.2. Machining model

In order to simulate the whole process of robotic machining, the machining process has to be modelled accurately to correctly predict the interaction between the workpiece and the mechanical system. Although microscopic and mesoscopic models exist, these methods require large simulation time that industrials can rarely afford. A fair compromise is to rely on macroscopic approaches combining a mechanistic model of the cutting forces, a numerical model of the dynamic response of the mechanical system and a geometrical representation of the machined workpiece.

The machining modelling was handled by an in-house milling routine, called DyStaMill for "Dynamic Stability of Milling operations" that implements a macroscopic model of material removal. It was also developed at the University of Mons by Rivière-Lorphèvre E. during his PhD thesis about the stability of high-speed milling [12]. It was first created to study the stability of machine tool for operations generating $2D^{1/2}$ shapes such as slotting, contouring, pocketing or face milling. In a practical way, dynamic simulation of milling involves three different features:

1. The prediction of the cutting forces;
2. The modelling of the workpiece;
3. A representation of the mechanical system which is in this case the aforementioned robot model.

The representation of the machined surface has to be as accurate as possible for being able to predict the milling instabilities. Indeed, it is of great importance when chatter phenomenon occurs as it is a regenerative process appearing when the tool removes material from a previously machined surface [13]. In DyStaMill, the update of the workpiece is inspired from the "eraser of matter" model proposed by G. Peigné [14]. The part is divided into superimposed slices and each 2D contour is approximated by a succession of straight segments. The end mill is also discretised into superimposed discs along its revolving axis in such a way that each of them interacts with one 2D contour. The motion of the cutting edges is also considered in the model and an intersection procedure is set up to compute the chip thickness. It is assumed that the displacements of the tool lie in a plane thus yielding to a $2D^{1/2}$ workpiece representation (Fig. 5).

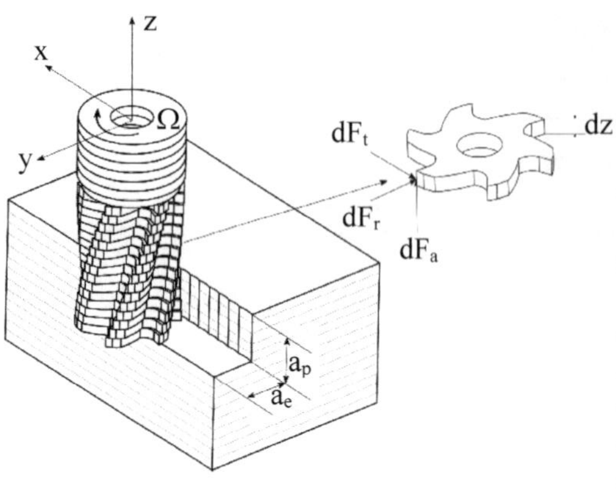

Fig. 5. Machining model

$$\vec{dF_i} = K_{i,c} \cdot h \cdot dz \tag{10}$$

$$\begin{Bmatrix} dF_x \\ dF_y \\ dF_z \end{Bmatrix} = \begin{bmatrix} -cos\theta & -sin\theta \cdot sin\kappa & -sin\theta \cdot sin\kappa \\ sin\theta & -cos\theta \cdot sin\kappa & -cos\theta \cdot cos\kappa \\ 0 & -cos\kappa & -sin\kappa \end{bmatrix} \begin{Bmatrix} dF_t \\ dF_r \\ dF_a \end{Bmatrix} \tag{11}$$

- a_e and a_p are the radial and axial depths of cut respectively [mm];
- $\vec{dF}_{i:a,r,t}$ is the local cutting force element in the axial, radial or tangential direction [N];
- $K_{i,c}$ is the cutting force coefficient [N/m²];
- h is the chip thickness [mm];
- dz is the height of an infinitesimal disc [mm];
- Ω is the angular speed of the spindle [RPM];
- θ and κ are the rotation angle of the cutting edge and the angle orienting the radial direction [rad];
- $\vec{dF}_{x,y,z}$ is the cutting force element projected into the global reference frame [N].

On the other hand, the computation of the cutting forces is based on a mechanistic model. This kind of model allows the prediction of the cutting forces from a given tool/ material couple using simple analytical laws. The latter assumes that the cutting force is proportional to the undeformed chip cross section. In DyStaMill, the popular model from Altintas Y. is used (10) to compute the cutting forces locally along the axial dF_a, radial dF_r and tangential dF_t directions at the cutting edges from the undeformed chip thickness h [15]. The local cutting forces are then projected into the global reference frame on the basis of cutting angles (11). Cutting force coefficients $K_{i,c}$ can be determined experimentally from slotting operations whose measured signals are processed by an inverse analysis [16].

2.3. Coupling of the two sub-models

With the objective of simulating the whole process of robotic machining, the DyStaMill routine was coupled and integrated into the multibody framework EasyDyn. The coupled environment was validated on machine tool test cases [17], [18]. Figure 6 presents the interactions between the milling routine and the multibody framework. The first step consists in entering the inertia properties of each body and the simulation parameters for the milling phase such as the tool diameter $Ø_{tool}$, the number of edges, the radial and axial depths of cut, the spindle angular speed and the cutting coefficients. The complete kinematics of the mechanical system can then be derived from the homogeneous transformation matrices. On the basis of the prediction of the tool centre point position at the end of the time step, it is possible for DyStaMill to predict the size of the chip thanks to intersection functions looking for the meeting point between the workpiece and the cutting edge. From then, the global cutting forces can be inferred for being considered into the integration procedure as force elements. Once the configuration parameters have been recomputed, the workpiece geometry is updated before resuming the procedure.

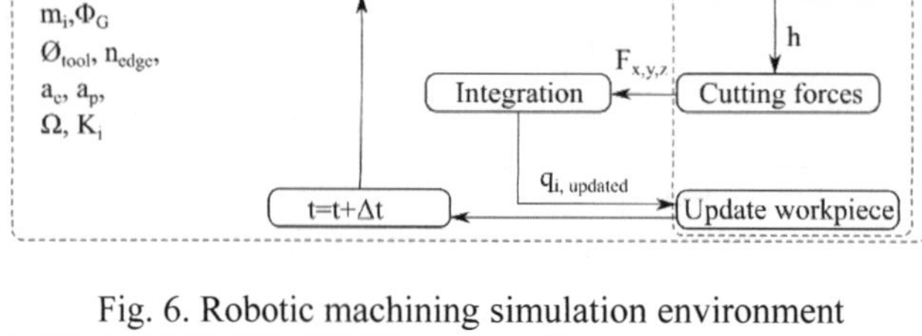

Fig. 6. Robotic machining simulation environment

3. Definition of the milling operations

This brief section lays out the two considered milling operations that the robot achieved within the robotic machining simulation environment. The cutting parameters adopted in this simulation context are also presented.

3.1 Longitudinal and transversal milling pass

As announced, the robot performed two milling operations in such a way that its end-effector remained in a horizontal plane, ending up with two shouldering operations. One shoulder was machined in the expansion direction of the arm (longitudinal pass, x-axis) while the other one was made perpendicular to the former one (transversal pass, y-axis) (Fig. 7). The reference position of each joint $q_{ref}[i]$ was computed on the basis of an inverse kinematic algorithm simply yielding in the resolution of two triangles. As a result, the end effector position set point was given in terms of x and y coordinates and translated into joint coordinates, assuming that the z coordinate was fixed. The motion of the end effector was settled so that its velocity increased smoothly from a null speed to the desired feed rate. Figure 8 shows the followed velocity profile by the end effector, it amounted to successively apply a constant jerk for 0.4 [s], a constant acceleration for 0.2 [s] and the reverse jerk for 0.4 [s] before reaching the desired feed rate in 1.0 [s].

It should also be mentioned that to speed up the computation, two integration time steps, in terms of duration, were chosen. The first one was set to $\Delta t=0.001$ [s] while the end effector stabilises around the desired feed rate. Then, it was necessary to strongly reduce the integration time step to take into account the machining phase. As a matter of fact, the workpiece geometry has to be updated at least 30 times per tool revolution. For the present simulation, the machined surface was kept up-to-date 100 times per tool revolution which turned out to be $\Delta t=3.21\times10^{-5}$ [s]. To sum up, the end-effector accelerated outside the workpiece before stabilising at the desired speed and entering into the workpiece.

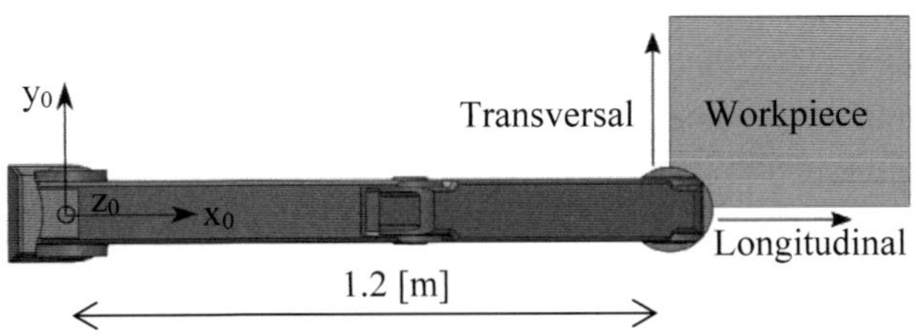

Fig. 7. Top view of the milling operations

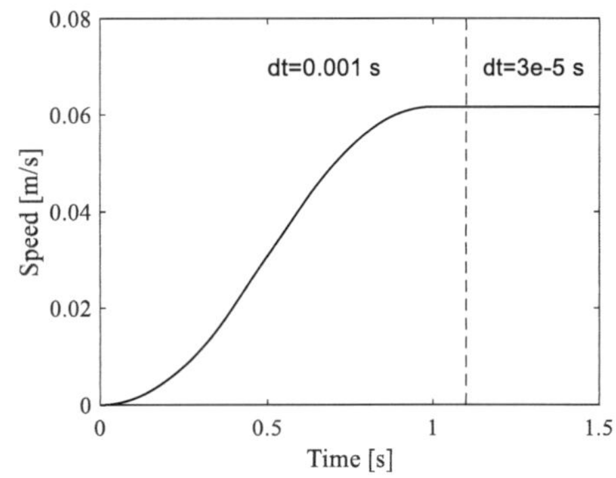

Fig. 8. End effector velocity profile

3.2 Cutting parameters

Cutting parameters were chosen according to real robotic machining tests which were performed with a Stäubli TX200 robot (Table 3). Four different milling experiments were accomplished in an aluminium plate 6060 T6:

1. Longitudinal pass in down-milling;
2. Longitudinal pass in up-milling;
3. Transversal pass in down-milling (Fig. 9);
4. Transversal pass in up-milling.

Solid Tool	**Diameter**	10 [mm]
	Nb. of teeth	2
	Helix angle	30°
	Variable tooth pitch	170° - 190°
Milling operation	**Radial depth of cut (a_e)**	4 [mm]
	Axial depth of cut (a_p)	1.6 [mm]
	Spindle speed	18700 [RPM]
	Feed rate	3700 mm/min
	Cutting coefficients (K_{tc}, K_{rc})	661.513, 253.458 [MPa]

Table 3. Cutting parameters

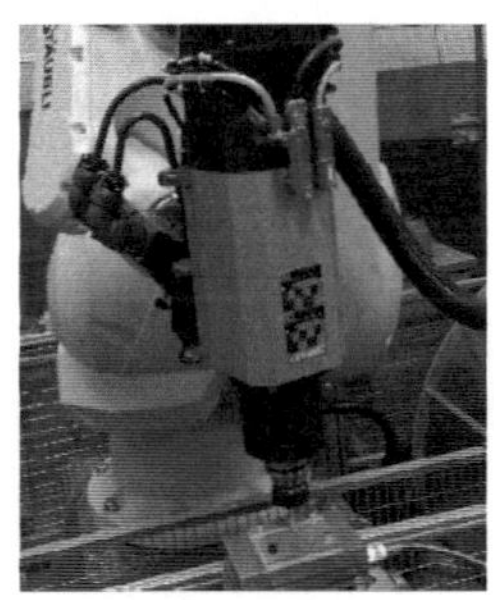

Fig. 9. Robotic machining tests

Cutting forces F_x, F_y and F_z were recorded thanks to a force sensor located under the workpiece and a triaxial accelerometer was in charge of the vibration signal acquisition. Data were sampled at 10,000 [Hz]. Cutting force data were already exploited in order to extract the cutting coefficients K_{tc} and K_{rc} from an inverse analysis. K_{ac} was not identified as the axial signal is often of a lesser quality.

4. Comparison with experiment

With the robotic machining simulation environment on the one hand and experimental data available on the other hand, it is now possible to compare the obtained results and to comment on the modelling quality to replicate real data. Results are compared in terms of cutting forces, vibrations and lateral roughness of the machined part.

4.1 Cutting forces

First of all, cutting forces were compared with experimental data within the context of the four experiments described in section 3.2. Figures 10 to 13 superimpose simulation (plain lines) and experimental (dashed lines) data over two tool revolutions for cutting forces F_x (in blue) and F_y (in red) measured at tooltip and projected in the global reference frame (F_z signal is not shown since K_{ac} couldn't be identified properly).

Overall, the simulated cutting forces followed the experimental signals in steady state. They were well correlated in terms of amplitude which meant that the cutting coefficients were well identified. Their periodicity was also in concordance with the set spindle speed of 18700 [RPM]. As expected, it can be observed that the signal perpendicular to the feed direction (F_y) is always higher since the tooth enters normally to the material surface. Furthermore, given that the end mill featured a variable tooth pitch (170°|190°), cutting force amplitudes alternated between high and lower peaks as it was the case for experimental data.

As shown in Fig. 10 and 11, F_y signal was slightly more noisy but respected the theoretical trend. Lastly, experimental peaks along the x-axis were slightly higher than predicted, thus resulting in a falling edge occurring a bit later.

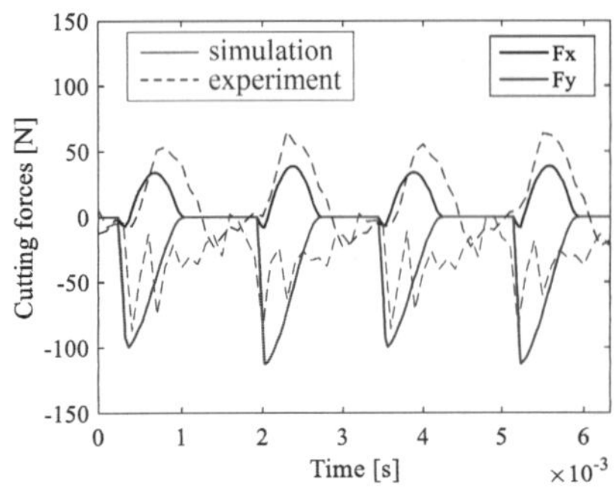

Fig. 10. Longitudinal pass in down-milling

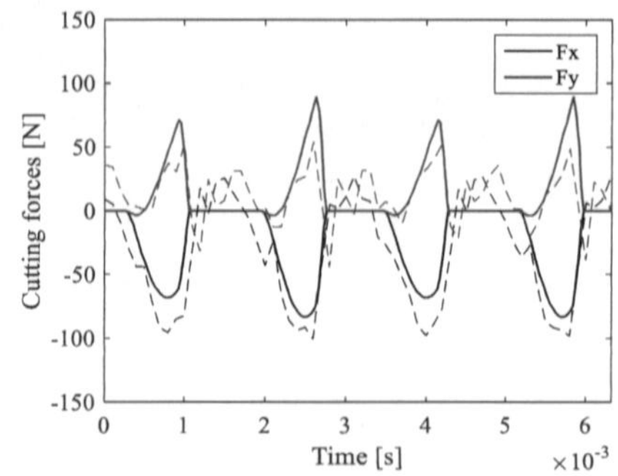

Fig. 11. Longitudinal pass in up-milling

As for transversal tests (Fig. 12 and 13), simulated cutting force F_y correlated quite well its experimental counterpart in terms of amplitude and periodicity. Nevertheless, its falling edge still appeared before the experimental fall. Concerning cutting force F_x, experimental signal was very noisy and it was difficult to notice the analogy. Amplitudes were still coherent.

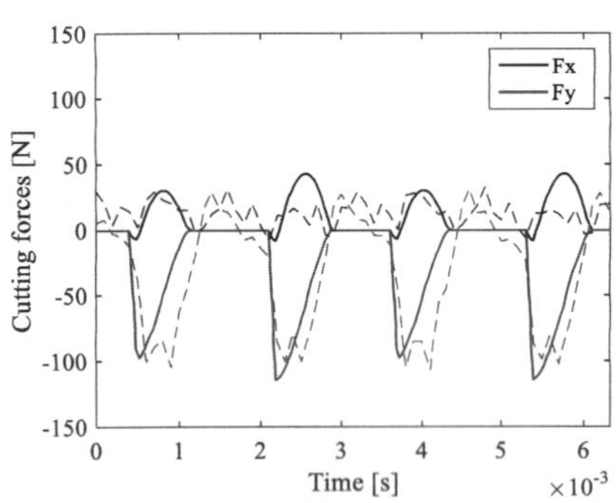

Fig. 12. Transversal pass in down-milling

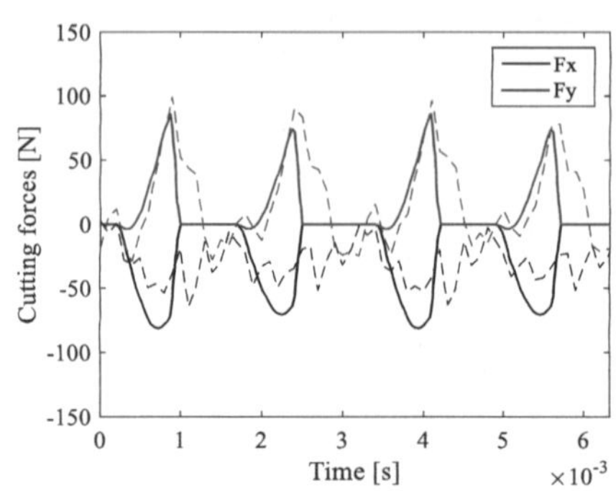

Fig. 13. Transversal pass in up-milling

4.2 Vibrations

In terms of vibrations, if signals along the feed direction were compared, again the steady state amplitudes correlated experimental data (Fig. 14). However, such statement could not be drawn for vibration signals perpendicular to the feed direction: the simulated vibration signals were much lower. This effect might be partially related to the fact that the simplified robot model did not take into account that the links of the real robot are not in the same plane (Fig. 2.a).

A frequential analysis on the cutting force signals was also carried out and revealed peaks and harmonics at frequencies corresponding to the spindle rotational speed (311.67 [Hz]) and to the tooth passing frequency (623.33 [Hz]) (Fig. 15).

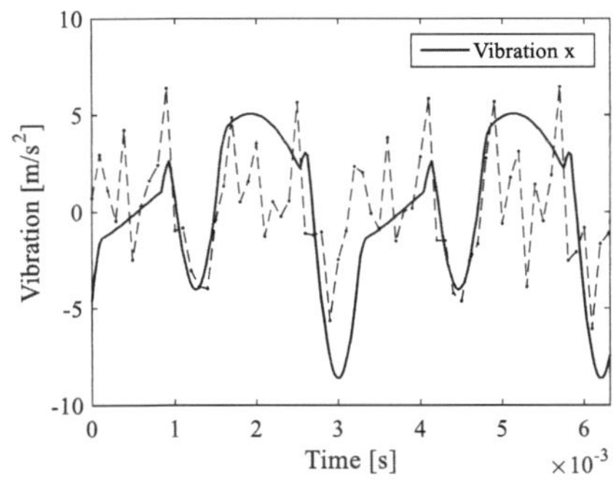

Fig. 14. Longitudinal pass in down-milling

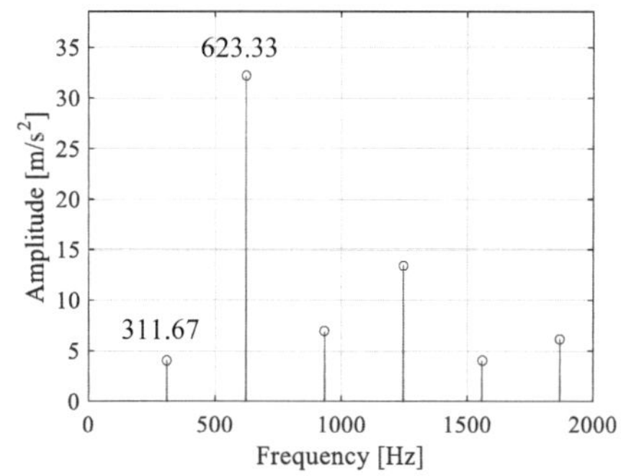

Fig. 15: Spectral analysis on the simulated signal

4.3 Lateral roughness

The virtual machined workpiece was compared with the real finished product in terms of lateral roughness. Before presenting the roughness results, it should be noted that the tool deviation in relation to the nominal value (a_e=4 [mm]) was around 10 [μm] in any simulation. Although this deviation is difficult to evaluate on the real part, no large offsets were detected which tend to support that the order of magnitude was well estimated (Fig. 16).

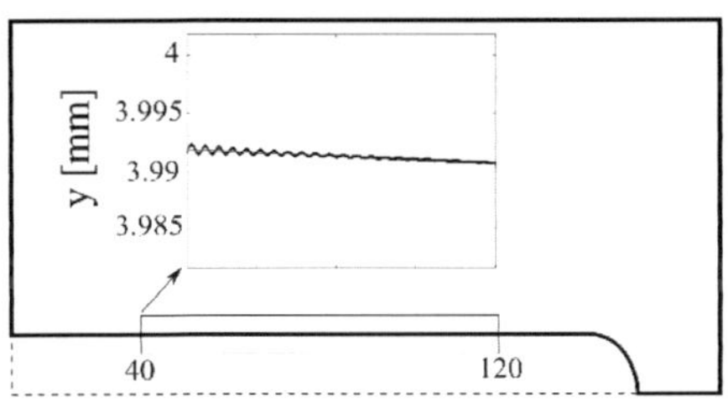

Fig. 16. End mill deviation (down-milling)

Simulation: R_a:0.32, R_t: 1.98, R_q: 0.39 [μm] (Down-milling)

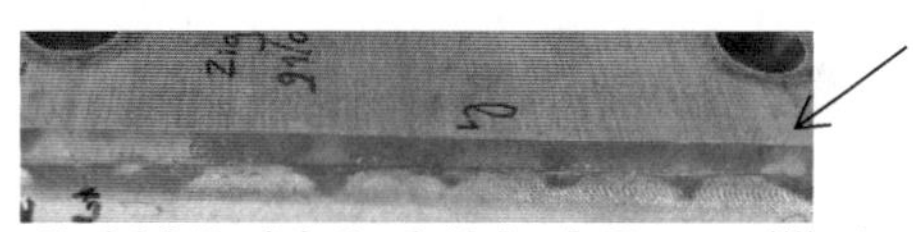

Experiment: R_a:0.36, R_t: 2.0, R_q: 0.43 [μm] (Down-milling)

Fig. 17. Roughness comparison

At a first glance, according to Fig. 17, roughness evaluated on experimental and simulated data is almost equivalent. Arithmetic roughness R_a, quadratic roughness R_q and total roughness R_t were computed on the basis of their classical formulae. Both the simulation values and the real data shown on the figure originate from the longitudinal pass in down-milling. In robotic machining, the latter constitutes the preferential cutting condition. An inspection of real machined parts revealed that up-milling tests should be prohibited in robotic machining as it leads to higher cutting forces and the pull-out of the surface material resulting in a sharp increase of roughness. The experimental roughness values retrieved for the transversal pass in down-milling were slightly higher but the surface finish was still acceptable. As the simplified modelling was not able to capture the roughness quality depletion in up-milling, the model should be refined in future work.

5. Conclusion

Robotic machining is a growing technology but investigations still need to be conducted to better understand the on-going phenomena while milling. The proof of concept subsists in the simulation of the whole robotic machining process. Once validated, the model should be able to predict in advance for which set of cutting parameters the machining will reach an optimum in terms of part quality, productivity and stability.

Even though simplistic, this work presented a simplified modelling of robotic machining. A 6-axis robot was transposed and modelled as a 4-axis anthropomorphic arm using the multibody approach and coupled with a milling routine. Four different machining tests were accomplished and compared with experimental data. Results showed a good accordance in terms of cutting force and vibration amplitudes, roughness and tool deviation. More specifically, cutting forces were well correlated. Vibration amplitudes along the feed direction were correctly approximated whereas signals found in the perpendicular direction were too low probably because of the geometrical approximations of the robot model. Roughness and tool deviation were only in accordance with real data for a longitudinal pass in down-milling. The simulation model should therefore be revised appropriately even though it was already able to capture the overall trends.

Before developing a simulation environment comprising a 6-axis robot, it would be interesting to check whether the current model is able to predict unstable milling conditions. Future work will therefore focus on the possibility to define stability lobes commonly used in machining to identify optimal cutting parameters.

6. Acknowledgments

The authors would like to thank the Belgian National Fund for Scientific Research (NFSR-FRIA) for the financial support and the CNC Solutions - VDS company for the access granting to their machining robot.

7. References

[1] Pan, Z.; Zhang, H.; Zhu, Z. & Wang, J. (2006). Chatter analysis of robotic machining process, Journal of Materials Processing Technology, Elsevier, vol. 173, pp. 301-309, doi:10.1016/j.jmatprotec.2005.11.033.

[2] Barnfather, J. D.; Goodfellow M. J. & Abram T. (2016). A performance evaluation methodology for robotic machine tools used in large volume manufacturing, Robotics and Computer-Integrated Manufacturing, Vol. 37, pp. 49-56 from: http://dx.doi.org/10.1016/j.rcim.2015.06.002.

[3] Dumas, C.; Caro, S.; Garnier, S. & Furet, B. (2011). Joint stiffness identification of six-revolute industrial serial robots, Robot. Comput. Integr. Manuf., Vol. 27(4), pp. 881-888, http://dx.doi.org/10.1016/j.rcim.2011.02.003.

[4] Weill, R. & Shani, B. (1991). Assessment of accuracy in relation with geometrical tolerances in robot links. CIRP Ann. Manuf. Technol, Vol. 40, pp. 395-399, http://dx.doi.org/10.1016/S0007-8506(07)62015-0.

[5] Turek, P.; Jedrzejewski, J & Modrzycki, W. (2010). Methods of machine tool error compensation, J. mach. Eng. Vol. 10 (4), pp.5-25, doi: 10.1016/j.procir.2013.06.078.

[6] Iglesias, I.; Sebastian, M. A., Ares, J. E. (2015). Overview of the state of robotic machining: current situation and future potential, Procedia Engineering, Vol. 132, pp. 911-917, doi: 10.1016/j.proeng.2015.12.577.

[7] Ilyukhin, Yu. V.; Poduraev, Yu. V.; Tatarintseva, A. V. (2014). Nonlinear adaptive correction of continuous path speed of the tool for high efficiency robotic machining, 25th DAAAM International Symposium on Intelligent Manufacturing and Automation, Procedia Engineering: Vol. 100, pp. 994-1002.

[8] Klimchik, A.; Ambiehl, A.; Garnier, S.; Furet, B. & Pashkevich A. (2017). Efficiency evaluation of robots in machining applications using industrial performance measure, Robotics and Computer – Integrated Manufacturing, Vol. 48, pp. 12-29.

[9] Bauer, J.; Friedmann, M.; Hemker, T.; Pischan, M.; Reinl, C.; Abele & von Stryk, O. (2013). Analysis of industrial robot structure and milling process Interaction for path manipulation (chapter 11), Process Machine Interactions, Springer, ISSN 2194-0525, DOI 10.1007/978-3-642-32448-2.

[10] Verlinden, O.; Ben Fékih, L. & Kouroussis, G. (2013). Symbolic generation of the kinematics of multibody systems in EasyDyn: From MuPAD to Xcas/Giac, Theoretical and Applied Mechanics Letters, Vol. 3 (1), pp. 013012, doi: 10.1063/2.13013012.

[11] Oueslati, M. (2013). Contribution à la modélisation dynamique, l'identification et la synthèse de lois de commande adaptées aux axes flexibles d'un robot industriel (translated: contribution to the dynamic modelling, the identification and the synthesis of command laws adapted for industrial robot flexible joints) (PhD thesis), L'école National Supérieure d'Arts et Métiers, Arts et Métiers ParisTech – Centre de Lille.

[12] Rivière-Lorphèvre, E.; Filippi, E. & Dehombreux, P. (2007). Chatter prediction using dynamic simulation, International review of mechanical engineering (I.R.E.M.E.), Vol. 1, pp. 78-86.

[13] Tlusty, J. & Polacek, M. (1963). The stability of the machine tool against self-excited vibration in machining, ASME International Research in Production Engineering, pp. 465-474.

[14] Peigné, G.; Paris, H. & Brissaud, D. (2003). A model of milled surface generation for time domain simulation of high-speed cutting, Proceedings of the Institution of Mechanical Engineers, Vol. 217, pp. 919-930.

[15] Altintas, Y. & Lee, P. (1996). A general mechanics and dynamics model for helical end mills, CIRP Annals – Manufacturing Technology, Vol. 45, pp. 59-64, doi: 10.1016/S0007–8506(07)63017–0.

[16] Rivière-Lorphèvre, E.; Letot, C.; Ducobu, F.; Dehombreux, P. & Filippi, E. (2017). Dynamic simulation of milling operations with small diameter milling cutters: effect of material heterogeneity on the cutting force model, Meccanica, Vol. 52, pp. 35-44.

[17] Huynh, H.N.; Rivière-Lorphèvre, E.; Verlinden, O. (2016). Integration of machining simulation within a multibody framework: application to milling, IMSD: The 4th Joint International Conference on Multibody System Dynamics.

[18] Huynh, H.N.; Rivière-Lorphèvre, E. & Verlinden, O. (2016). Milling simulations with a 3-DOF flexible planar robot, World Academy of Science, Engineering and Technology, International Journal of Mechanical, Aerospace, Industrial, Mechatronic and Manufacturing Engineering, Vol. 10 (9), pp. 1647-1656.

DOI: 10.2507/28th.daaam.proceedings.124

CHALLENGES OF THE CHALLENGERS: AN INSIGHT INTO THE INTERNATIONALIZATION PATHWAY OF CROATIAN DIGITAL AGENCIES

Nikola Draskovic, Milivoj Markovic & Domagoj Hruskar

This Publication has to be referred as: Draskovic, N[ikola]; Markovic, M[ilivoj] & Hruskar, D[omagoj] (2017). Challenges of the Challengers: An Insight into the Internationalization Pathway of Croatian Digital Agencies, Proceedings of the 28th DAAAM International Symposium, pp.0895-0901, B. Katalinic (Ed.), Published by DAAAM International, ISBN 978-3-902734-11-2, ISSN 1726-9679, Vienna, Austria
DOI: 10.2507/28th.daaam.proceedings.124

Abstract

Development of internet-based technologies provided companies with an ideal infrastructure for rapid internationalization. The internet era facilitated the emergence of a new type of companies that rapidly develop global presence through the utilisation of various internet-based technologies. According to the traditional internationalization theories, the process of internationalization requires vast amounts of funds, significant time, competent international workforce and network support by suppliers. The so cold born global companies rely far less on these elements (prerequisites). For them, the development of internet infrastructure and the overall market penetration of internet are much more essential. Their process of internationalization (or globalization) is usually rapid, without any preceding long-term domestic or internationalization phase. This paper analyses theoretical framework behind the born global companies' internationalization process in order to shed some light on the recent developments among Croatian digital agencies and their internationalization pathways. Although born global companies are challenging traditional theories, there exists a scope for potential complementarity between these essentially opposite pathways of internationalization.

Keywords: internet-based technologies; digital agencies; internationalization; born global companies.

1. Introduction

With around 4.2 million inhabitants and GDP per capita at EUR 10,162 [1], Croatia is fairly small and emerging market. Typically, local companies focused on growth are forced to internationalise their business activities in relatively short time span. In other words, Croatian companies cannot maintain sustainable growth focused solely on the local market. During the latest economic downturn, Croatian companies focused on exporting activities performed above the average [2]. Traditionally, a Croatian company will initiate its internationalization process by entering one or more markets within the region (i.e. the area of former Yugoslavia), which is fairly close geographically and with modest cultural distance. Encouraged by regional success, a company will probably consider entering markets outside the region. The whole process will take time and it requires vast amounts of capital, often times attainable only at a high cost.

Therefore, companies are highly incentivised to seek alternative approaches. With the development of internet and internet-based technologies, a completely new sales and promotion channel emerged, giving small and medium enterprises (SMEs) and opportunity to rapidly scale their business internationally.

The aim of this paper is to analyse and contrast traditional internationalization theories and the born global phenomenon, or the process of rapid globalization, in the context of the Croatian digital agencies' internationalization process. Furthermore, this paper also serves a purpose of a preliminary study with the intention to plot a path for further research of this topic. From the methodological standpoint, this paper is based on the literature review and research of secondary sources addressing the phenomenon of recent Croatian digital agencies internationalization efforts.

2. An overview of internationalization theories: the organic versus born global

Figure 1 represents a concise comparison between two opposite pathways of internationalization. In the context of a company's internationalization, the organic pathway (or the Uppsala model) represents more traditional and time-consuming approach. On the opposite side, born global pathway represents an accelerated process of internationalization that takes less time, but is also riskier from the company's perspective.

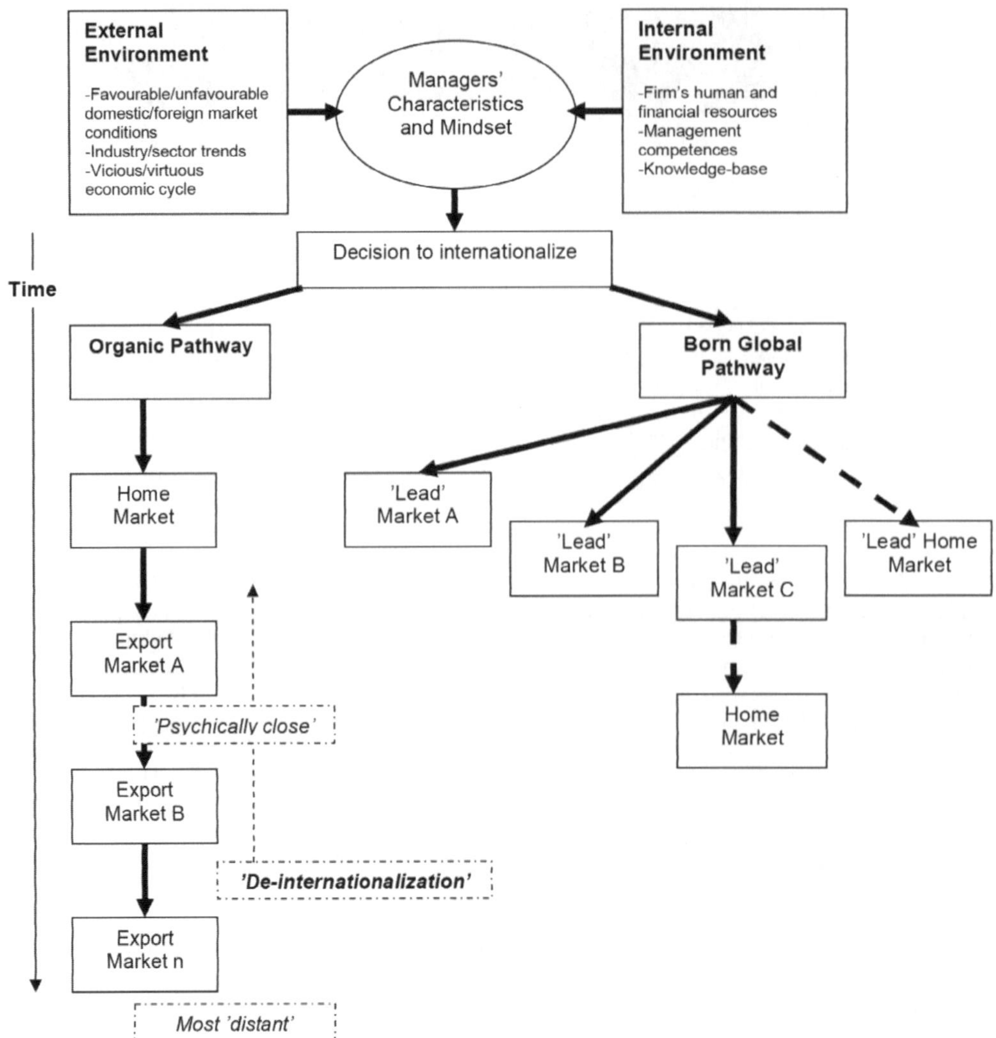

Fig. 1. Two extreme pathways of software firm internationalization: the organic pathway (Uppsala model) versus born global [3]

2.1. The Uppsala internationalization model

Researchers at the University of Uppsala in Sweden developed the Uppsala internationalization model in late 1970s in order to develop a better understanding of internationalization pathways. The theoretical framework was built around observations and analysis of Swedish manufacturing companies and their internationalization processes [4].

According to the research findings, Swedish manufacturing companies usually started their international operations in countries that were relatively close geographically and further internationalization will continue in small incremental steps. Commonly, companies will enter new markets through exports. However, after several years a company will increase its foreign market commitment.

As Figure 2 implies, the Uppsala model has two main interacting and mutually influencing parts – state factors and change factors. As the company cumulates experience, its market knowledge increases, which influences the commitment decisions. With the increasing commitment, company increases the level of activities, which leads towards cumulating new experience and gaining new knowledge.

The concept of market commitment recognizes just two factors – the amount of resources (i.e. investment) committed and the degree of commitment (i.e. the complexity of discovering an alternative use for the resources and relocating them to the alternative use) [5]. Besides the aforementioned geographical proximity, another important factor influencing the degree of commitment and new market entry is the psychic distance. [6]

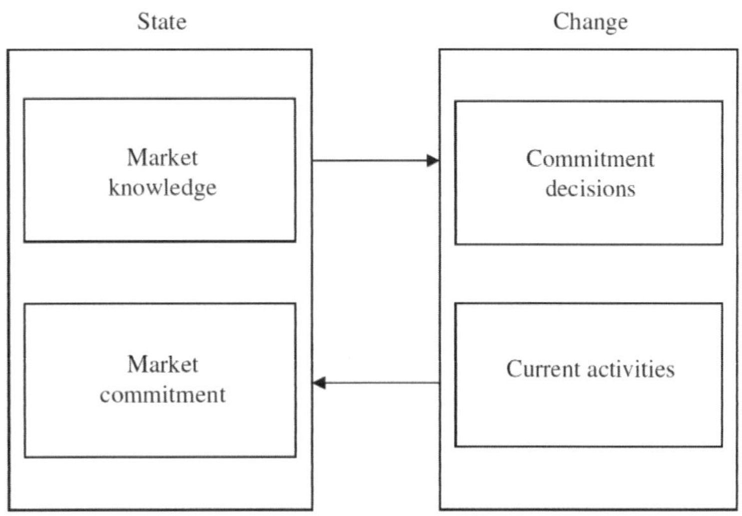

Fig. 2. The original Uppsala model [7]

Johanson and Wiedersheim-Paul [4] identified four foreign market entry modes or stages with different levels of international involvement:

- Stage 1: no regular export activities or just sporadic export
- Stage 2: export via independent representatives
- Stage 3: establishment of a foreign sales subsidiary
- Stage 4: foreign manufacturing units

According to the Uppsala model, a company will move through successive stages increasing its international market commitment. Obviously, it is a slow and iterative process, but with a reason. As more recent research suggests [8], market's high complexity and overall uncertainty call for a slow, thorough approach to internationalization. Business risk is often too high for bold and quick decisions about new market entries.

Due to the increasing complexity and challenges of the global market, a need for further improvement of the original Uppsala model emerged. Consequently, the importance of company's business networks has been recognized as a success factor [9], which lead to the revision of the original Uppsala model [10].

According to the study of the internationalization process of SMEs [11], companies tend to rely on their relations with upstream and downstream networks while selecting the market and the mode of entry. In other words, the relationships that a company develops and maintains in a domestic network can be used as bridges to similar networks abroad [6]. The revised model (Figure 3) emphasises the fact that the internationalization process takes place within the network context [12]. The company's network position is influenced by the activities such as learning, creating and trust-building.

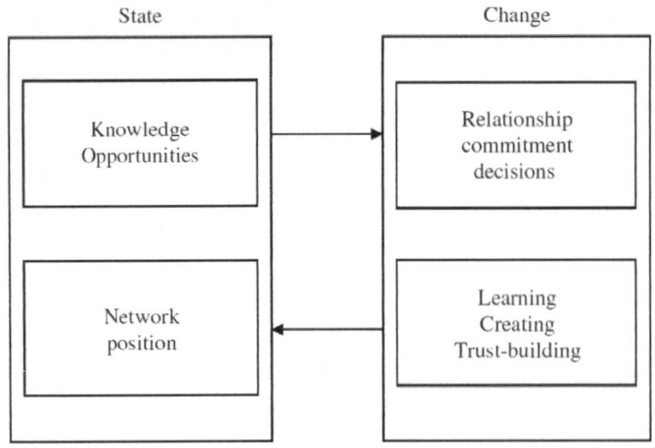

Fig. 3. The revised Uppsala model with the application of a basic networking prospect [10]

Subsequent research has examined complementary variables to the ones originally considered by Uppsala model which added to the understanding of the internationalization pathways of companies. Research has examined cultural distance and found support for the proposition that when internationalizing, companies tend to choose markets and countries that exhibit closer cultural distance to their current markets [13] [14]. Institutional distance was also found to be a significant factor when businesses chose a host country for internationalization. In general, evidence support the fact that companies tend to choose host countries for foreign entry which exhibit smaller institutional distance vs. countries which exhibit larger institutional distance [15] [16].

Common to these studies is that, similarly to Uppsala model, they look on internationalization pathways of companies as gradual processes that companies engage into by analysing similarities (different measurements of "distance") between the markets/countries they are present in and the new ones they are targeting for entrance. As is analysed next, there exist a number of companies that do not fit any of the internationalization pathways presented so far.

2.2. Born global

Globalization processes, fuelled by advances in technology and communication, and removal of trade barriers, have resulted with dramatic changes and challenges within the companies' environment. However, with these disruptive changes, new business opportunities emerged. Noticed by researchers in the recent years, there is an increasing number of companies riding the wave of globalization that do not follow traditional internationalization stages pattern [6]. Instead, their globalization process is accelerated, without any preceding time-consuming step-by-step internationalization.

Born global can be defined as companies that, from the start-up, follow a vision of becoming global and usually rapidly globalize their business activity, skipping long-term domestic or internationalization phase [17] [18]. Born globals are typically SMEs with rapid globalization strategy from the moment of their inception or soon after. More strict definition suggests that the born global company has to rapidly internationalise its business within three years from the inception and its international activities should account for at least 25% of the total revenue [19]. Many researchers contributed to the identification of the reasons behind the emergence of the born globals. Some of the most common are the following [20] [21] [22]:

- Small domestic niche markets,
- Uniqueness of the product/service
- Globalization of sourcing activities and financial markets,
- Technological development (especially in the field of internet-based technologies, communication and logistics),
- Founders and employees are nowadays more capable of doing international business thanks to the possibilities of modern technology,
- Quick acquisition of international experience and knowledge,
- Existence of international networks that support global business activities.

In the traditional model, the risk related to the uncertainty of the foreign market entry is typically hindering companies' internationalization efforts. However, the increased level of globalization in some industries contributes to the lowering of the perceived risk of entering foreign markets, creating a kind of me-too effect among SMEs [23]. Thus, the born global approach requires certain level of entrepreneurship and risk-taking among founder(s) [24]. Due to a different level of internationalization, there are different types of born global companies [19] [25]:

- Born regionals – international companies focused on markets within their region,
- Born internationals – companies with focus on low-distance markets,
- True born globals – companies focusing on both low- and high-distance markets
- Born-again globals – companies previously focused only on domestic markets, which at some point embrace rapid and dedicated internationalization

It is often argued that internet capabilities have strongly impacted emergence of born global companies by offering a plethora of opportunities for SME's to quickly expand their global reach and market presence [26]. Most common case analysed, for which research has provided positive evidence, is the pervasive use of internet [27] as a marketing channel in accessing global markets. While positive effects have been found, researchers note that internet cannot be considered as a substitute to physical market presence, it does however contribute positively to company performance [28] [29].

3. Croatian digital marketing internationalization efforts

3.1. An overview of the Croatian digital marketing industry

For decades, Croatian marketing industry was tightly linked with the political and economic developments. After the period of socialism, Croatia started its transition towards democracy and free market in early 1990s. That was the era when a majority of established local marketing agencies either bankrupted or were acquired by some large international advertising network [30]. Additionally, a large number of micro and small agencies emerged, usually founded by former employees of the bankrupted agencies. Some of them managed to survive harsh economic environment of the post-war Croatia, while majority simply disappeared over the years. Soon, the marketing industry was dominated by international networks (e.g. McCann-Ericksson, BBDO, Grey), together with a number of smaller local agencies. In the mid 2000s, marketing industry bloomed reaching yearly revenues at EUR 500 million [31]. The global economic downturn in 2007/2008 had heavy impact on the Croatian economy. The marketing communications budgets of companies in Croatia deflate, resulting with the sinking of the whole marketing industry and new bankruptcies among local agencies.

Unfortunately, Croatian marketing industry did not manage to recover in the past decade. According to the available reports [32], Croatian marketing communications industry is currently worth around EUR 200 million. Similar to the global trends [33], digital side of marketing is continuously growing, while traditional marketing communications and traditional media are stagnating. More or less, all marketing communications agencies in Croatia offer digital marketing service, with no exception among the biggest traditional agencies [34]. On average, digital marketing contributes around 26% to the traditional agencies' revenues [35].

The growing number of newly founded digital agencies initiated the transition of traditional agencies towards digital marketing. While traditional agencies simply added a digital department to their organization structure, digital agencies were usually micro or small start-ups exclusively focused on the digital aspect of marketing. Since the founding of a digital agency does not require some large-scale investment, entry barriers for new players are quite low, resulting with an inflation of digital agencies in Croatia right now [36]. While the overall marketing industry and media expenditure is stagnating, the increasing number of digital agencies escalated competition, resulting with the decline of service fees.

3.2. Internationalization specifics of Croatian digital marketing agencies

Faced with international competition within their domestic market, internationalization through an organic pathway is a common growth strategy for the traditional Croatian marketing agencies. Consequently, many of them entered neighbouring markets, such as Bosnia or Serbia, mostly due to small geographic and cultural distance. The common entry modes are usually joint ventures with local agencies or acquisitions. Direct investments are not that common. Although, there are also some exceptions. For example, Bruketa & Zinic agency expanded its business to Belgrade and Vienna (which counts as a region), and, in 2011, the agency established a subsidiary in Baku (Azerbaijan) after winning a local telecom account [37]. However, the agency withdraw from both Baku and Belgrade this year, after selling its share to the WPP group and merging with Grey Croatia [38]. Obviously, the agency did not have sufficient funds to continue with its internationalization process and the owners switched to de-internationalization strategy.

While for the traditional agencies, establishing a foreign subsidiary is a time consuming process that requires vast amounts of money and significant risk-taking, the internationalization pathway of Croatian digital agencies seems to be less complex. Typically, a digital agency will, at the beginning, look for some additional business opportunities within the region using internet and social media for communication with prospects. This way, an agency does not have to establishing a subsidiary abroad, which reduces risk and investment. If the business would not develop as expected, an agency will simply change its focus to another market.

Besides regional strategies, digital agencies are also developing selective international and global strategies. Usually, they target just a few developed markets, such as UK, US or Germany. Although a majority of prospecting activities and communication is conducted online, agencies sometimes rely on the assistance of local agents. Local representation is especially needed in distant markets, such as US, due to both geographic and psychic distance.

It is a job of local representative to bridge this distance and to close a deal with the prospect found either by him/her or by a domestic sales representative.

The main question that arises at this point is how a digital agency from Croatia can compete on these highly competitive markets? The business model of Croatian digital agencies is very simple. The only really internationalized part of the company is the sales department, while all the job is done locally, in Croatia. This way, a Croatian digital agency utilises home country cheaper labour and offers same services as the host country agencies, but at lower price. Potentially, this looks like a very sustainable born global strategy that benefits from both the development of internet based technologies and increasing budgets for online advertising. The only potentially negative development that may threat this business model is the emergence of similar agencies from countries with even cheaper highly skilled labour.

4. Conclusion

This paper contrasts two leading and competing approaches to internationalization by revisiting their key tenants and contrasting them with development paths of Croatian digital marketing agencies. It is evident that support exist for the proposition that Croatian digital marketing agencies have bypassed traditional internationalization pathway (stage-by-stage approach) as described by Uppsala model (and subsequent extensions) and opted for a fast and far-flung internationalization approach (born global).

Using internet marketing channels, well-educated and cost competitive workforce, and building on their domestic experience, they have been able to successfully enter regional markets as well as more distant ones. However, it would seem that greater challenge from entering foreign markets (winning initial accounts) is establishing firm market position and scaling the business on those markets in order to sustain pressures from competition. It would also seem that Croatian digital marketing agencies have successfully used the possibilities provided by internet to establish international presence (born global approach) with a combination of traditional stage models of internationalization (short "distanced" regional expansion).

Future research endeavours should explore internationalization pathways among the regional digital marketing agencies which could support the finding here on multicountry basis. From the methodological standpoint, instead of secondary data analysis, researchers should opt for primary data collecting, ideally through series of in-depth interviews with managers from digital marketing agencies. Also, it would be worth exploring the validity of born global approach in other industries impacted heavily be digital technology. Although internet provides opportunities for fast and distant internationalization the key challenge apparently revolves around how to scale the business once the market has been entered.

5. References

[1] Ostroski, L. (Ed.) (2015). Hrvatska u brojkama/Croatia in Figures – 2015, State Bureau of Statistics of Republic of Croatia, Zagreb, ISSN 1848-6673 Ostroski, L. (Ed.) (2015). Hrvatska u brojkama/Croatia in Figures – 2015, Državni zavod za statistiku Republike Hrvatske, Zagreb, ISSN 1848-6673

[2] Sector of International Affairs/EEN (2015). Internationalization and Financing: Committee of Republic of Croatia for Internationalization of Croatian Economy, Croatian Chamber of Commerce, Zagreb. Sektor za medjunarodne poslove/EEN (2015). Internacionalizacija i mogućnosti financiranja: Povjerenstvo Vlade Republike Hrvatske za internacionalizaciju hrvatskog gospodarstva, Hrvatska gospodarska komora, Zagreb

[3] Âijö, T., Kuivalainen, O., Saarenketo, S., Lindqvist, J. & Hanninen, H. (2005). Internationalization Handbook for the Software Business, Centre of Expertise for Software Product Business, ISBN: 952-99529-1-0, Espoo

[4] Johanson, J. & Wiedersheim-Paul, F. (1975). The internationalization of the firm — four Swedish cases. Journal of Management Studies, Vol. 12, No. 3, 1975, pp. 305-323, ISSN: 1467-6486

[5] Andersen, O. (1993). On the internationalization process of firms: a critical analysis. Journal of International Business Studies, Vol. 24, No. 2, 1993, pp. 209-231, ISSN: 0047-2506

[6] Hollensen, S. (2017). Global Marketing, Pearson Education, ISBN: 978-1-292-10011-1, Harlow

[7] Johanson, J. & Vahlne, J.E. (1977). The internationalization process of the firm: A model of knowledge development and increasing foreign market commitments. Journal of International Business Studies, Vol. 8, No. 1, 1977, pp. 23-32, ISSN: 0047-2506

[8] Vahlne, J.E., Ivarsson, I. & Johanson, J. (2011). The tortuous road to globalization for Volvo's heavy truck business: extending the scope of the Uppsala model. International Business Review, Vol. 20, No. 1, 2011, pp. 1-14, ISSN: 0969-5931

[9] Johanson, J. & Vahlne, J.E. (1990). The mechanisms of internationalization. International Marketing Review, Vol. 7, No. 4, pp. 11-24, 1990, ISSN: 0265-1335

[10] Johanson, J. & Vahlne, J.E. (2009). The Uppsala internationalization process model revisited: From liability of foreignness to liability of outsidership. Journal of International Business Studies, Vol. 40, No. 9, 2009, pp. 1411-1431, ISSN: 0047-2506

[11] Coviello, N.E. & Munro, H.J. (1997). Network relationships and the internationalization process of smaller software firms. International Business Review, Vol. 6, No. 4, 1997, pp. 361-384, ISSN: 0969-5931

[12] Hosseini, M. & Dadfar, H. (2012). Network-based theories and internationalization of firms: applications to empirical studies. The Business & Management Review, Vol. 3, No. 1, November 2012, pp. 182-191, ISSN: 2047-0398

[13] Goerzen, A., & Beamish, P.W. (2003). Geographic scope and multinational enterprise performance. Strategic Management Journal, 24, 1289–1306.

[14] Filippaios, F., & Rama, R. (2011). Cultural distance and internationalization: The world's largest food and drink multinationals. Agribusiness, 27(4), 399-419. doi:10.1002/agr.20283

[15] Gama, M. A. B., Lana, J., Vilasboas Calixto, C., & Bandeira-de-Mello, R. (2016). Business group internationalization: Choosing a host country according to institutional distance. Review of Business Management, 18(61), 327-347. doi:10.7819/rbgn.v18i61.2894

[16] Pogrebnyakov, N., & Maitland, C. F. (2011). Institutional distance and the internationalization process: The case of mobile operators. Journal of International Management, 17(1), 68-82. doi:10.1016/j.intman.2010.12.003

[17] Luostarinen, R. & Gabrielsson, M. (2004). Finnish perspectives of international entrepreneurship, In: Handbook of research on international entrepreneurship, Dana, L.P. (Ed.), pp. 383–403, Edward Elgar Publishing, ISBN: 184376069X, Cheltenham

[18] Oviatt, B. M., & McDougall, P. P. (1994). Toward a theory of international new ventures. Journal of International Business Studies, Vol. 25, No. 1, 1994, pp. 45–64, ISSN: 0047-2506

[19] Kuivalainen, O., Sundqvist, S. and Servais, P. (2007). Firms' degree of born-globalness, international entrepreneurial orientation and export performance. Journal of World Business, Vol. 42, No. 3, 2007, pp. 253–267, ISSN: 1090-9516

[20] Knight, G.A. & Cavusgil, S.T. (2004). Innovation, organization capabilities, and the born-global firm. Journal of International Business Studies, Vol. 35, No. 2, 2004, pp. 124–141, ISSN: 00472506

[21] Andersson, S. & Wictor, I. (2003). Innovative internationalisation in new firms: born globals – the Swedish case. Journal of International Entrepreneurship, Vol. 1, No. 3, 2003, pp. 249-275, ISSN: 1570-7385

[22] Madsen, T. K. & Servais, P. (1997). The internationalization of born globals: an evolutionary process?, International Business Review, Vol. 6, No. 6, 1997, pp. 561-583, ISSN: 0969-5931

[23] Verdier, S., Prange, C., Atamer, T. & Monin, P. (2010). Internationalization performance revisited: the impact of age and speed on sales growth. Management International, Vol. 15, No. 1, 2010, pp. 19-31, ISSN: 1206-1697

[24] Andersson, S. & Florén, H. (2009). Exploring differences in the work of owner-managers in small international and non-international firms. Competitive paper presented at the 12th McGill International Entrepreneurship Conference Hanken School of Economics, Vaasa, Finland September 18-20.

[25] Bell, J., McNaughton, R. & Young, S. (2001). "Born-Again Global" Firms: An Extension to the "Born Global" Phenomenon. Journal of International Management, Vol. 7, No. 3, 2001, pp. 1–17, ISSN: 1075-4253

[26] Marcus Vinicius de Oliveira Brasil, Ogasavara, M. H., Francisco Correia de Oliveira, Tassigny, M. M., & Raimundo Eduardo Silveira Fontenele. (2013). the role of internet in the born global companies. Revista De Administração Da UFSM, 6(2), 431-442. doi:10.5902/198346599061

[27] Simicevic, V.; Zoroja, J. & Pejic Bach, M.: Cluster Analysis of Internet Usage in European Countries, Annals of DAAAM for 2012 & Proceedings of the 23rd International DAAAM Symposium, ISBN 978-3-901509-91-9, ISSN 2304-1382, pp 0525 - 0528, Editor B[ranko] Katalinic, Published by DAAAM International, Vienna, Austria, 2012.

[28] Sinkovics, N., Sinkovics, R. R., & "Bryan" Jean, R. (2013). The internet as an alternative path to internationalization? International Marketing Review, 30(2), 130-155. doi:10.1108/02651331311314556

[29] Gabrielsson, M., & Gabrielsson, P. (2011). Internet-based sales channel strategies of born global firms. International Business Review, 20(1), 88-99. doi:10.1016/j.ibusrev.2010.05.001

[30] Draskovic, N., Pavicic, J., Vignali, G. & Baric, Z. (2006). Advertising in Croatia: Historical and Environmental Change Framework. International Journal of Management Cases, Vol. 8, No. 3, 2006, pp. 21-32, ISSN: 1741-6264

[31] Kupinic Guscic, D. (2005). Market of 426 million Euros. Economic Bulletin, No. 3399, May 2005, pp. 16-23, ISSN 2459-8968 Kupinic Guscic, D. (2005). Trziste tesko 426 milijuna eura. Privredni vjesnik, No. 3399, May 2005, pp. 16-23, ISSN 2459-8968

[32] Antolovic, K. (2017). Advertising in Croatia from its beginnings till today, Available from: http://www.medijskapismenost.hr/oglasavanje-hrvatskoj-svojih-pocetaka-do-danas/ Accessed: 2017-07-19 Antolovic, K. (2017). Oglašavanje u Hrvatskoj od svojih početaka do danas, Available from: http://www.medijskapismenost.hr/oglasavanje-hrvatskoj-svojih-pocetaka-do-danas/ Accessed: 2017-07-19

[33] Letang, V. & Stillman, L. (2016). Global advertising forecast: Winter update. MAGNA, New York.

[34] Macan, K. (2011). 7 TOP agencies for digital advertising in Croatia, Available from: http://www.broj1.hr/2011/08/18/7-top-agencija-za-digitalno-ogla%C5%A1avanje-u-hrvatskoj Accessed: 2017-07-19 Macan, K. (2011). 7 TOP agencija za digitalno oglašavanje u Hrvatskoj, Available from: http://www.broj1.hr/2011/08/18/7-top-agencija-za-digitalno-ogla%C5%A1avanje-u-hrvatskoj Accessed: 2017-07-19

[35] Ipsos Connect (2014). Agency Census 2014, Available from: http://hura.hr/wp-content/uploads/2016/11/HURA-agencijski-cenzus-2014-1.pdf Accessed: 2017-07-19 Ipsos Connect (2014). Agencijski census 2014, Available from: http://hura.hr/wp-content/uploads/2016/11/HURA-agencijski-cenzus-2014-1.pdf Accessed: 2017-07-19

[36] Dupanovic, E. (2015). Ilija Brajković: Most digital agencies are doing social network campaigns at unrealistically low prices. Media Marketing, Available from: http://www.media-marketing.com/intervju-dana/ilija-brajkovic-vecina-digitalnih-agencija-rade-kampanje-na-drustvenim-mrezama-po-nerealno-niskim-cijenama/ Accessed: 2017-07-19, Dupanovic, E. (2015). Ilija Brajković: Većina digitalnih agencija rade kampanje na društvenim mrežama po nerealno niskim cijenama. Media Marketing, Available from: http://www.media-marketing.com/intervju-dana/ilija-brajkovic-vecina-digitalnih-agencija-rade-kampanje-na-drustvenim-mrezama-po-nerealno-niskim-cijenama/ Accessed: 2017-07-19

[37] Kranjec, R. (2015). Letter from Azerbaijan: Bruketa&Žinić OM in "fiery" country of incredibly fast changes, Available from: http://www.media-marketing.com/kolumna/pismo-iz-azerbajdzana-bruketazinic-om-u-vatrenoj-zemlji-nevjerojatno-brzih-promjena/ Accessed: 2017-07-20, Kranjec, R. (2015). Pismo iz Azerbajdžana: Bruketa&Žinić OM u "vatrenoj" zemlji nevjerojatno brzih promjena, Available from: http://www.media-marketing.com/kolumna/pismo-iz-azerbajdzana-bruketazinic-om-u-vatrenoj-zemlji-nevjerojatno-brzih-promjena/ Accessed: 2017-07-20

[38] Bruketa&Zinic OM (2017). Bruketa&Žinić OM sold stakes in Baku, Belgrade and Brokuli&Ž, Available from: http://bruketa-zinic.com/hr/2017/04/27/bruketazinic-om-prodali-udjele-u-bakuu-beogradu-i-brokuliz/ Accessed: 2017-07-20, Bruketa&Zinic OM (2017). Bruketa&Žinić OM prodali udjele u Bakuu, Beogradu i Brokuli&Ž, Available from: http://bruketa-zinic.com/hr/2017/04/27/bruketazinic-om-prodali-udjele-u-bakuu-beogradu-i-brokuliz/ Accessed: 2017-07-20

DOI: 10.2507/28th.daaam.proceedings.125

TRACKING SYSTEM FOR MOVING OBJECT WITH FORECASTING

Alexander Kobzev, Natalia Novikova & Anastasia Lekareva

This Publication has to be referred as: Kobzev, A[lexander]; Novikova, N[atalia] & Lekareva, A[nastasia] (2017). Tracking System for Moving Object with Forecasting, Proceedings of the 28th DAAAM International Symposium, pp.0902-0910, B. Katalinic (Ed.), Published by DAAAM International, ISBN 978-3-902734-11-2, ISSN 1726-9679, Vienna, Austria
DOI: 10.2507/28th.daaam.proceedings.125

Abstract

A characteristic feature of tracking systems for movable objects is the change of trajectory according to various laws. Under the uncertainty of these trajectories, the question of ensuring of a given tracking accuracy is topical. One of the possible options for constructing a rational structure of the automatic control system that enhances the control characteristics is the introduction of a parallel model that generates a signal for «additional control». Hence, it is possible to use two structures of a parallel model: 1) a model similar to the original ACS, but with the main feedback on the adjustable coordinate; 2) a model that forms a given («reference») transient process under various external influences. Studies have shown that the introduction of a parallel model having a structure similar to the original ACS, but with a control object completely covered by feedback, allows to stabilize the response characteristics only in a limited range of external influences. ACSs with a parallel model of the second type have higher response performance. Therefore, both in the first and in the second case it is necessary to introduce an element with a variable transmission coefficient into the circuit of "additional control" signal synthesis, which significantly complicates the structure of the automatic control system and the control problem, since its value depends on many factors.
The introduction of the predictive model allows to improve the response characteristics of the system significantly. The results of the ACS simulation with the parallel predictive model indicate that the maximum absolute error of the controlled quantity does not exceed 1% of the maximum value of the control input.

Keywords: Dynamic objects with the incompletely observed regulated coordinate; predictive control; pattern model; correction algorithms of the controlling action; forecasting of the external effects

1. Introduction

When designing automatic control systems (ACS) for various dynamic objects, a careful analysis of the features of their design is required. This analysis is dictated, inter alia, by the requirement of optimal use of all available resources (energy, information, computing, etc.) at each operation stage.

An important feature of a large class of systems and processes is the incomplete observability of the controlled coordinate. This means that part of the control object is not covered by feedback, and therefore, with its typical management structure, its dynamic properties are not taken into account. Such situations arise when certain characteristics of an object are either not available to direct observation, or cannot be automatically measured due to the absence of the required sensors or are measured with a long delay. In addition, cases of inability to use instrumentation due to the presence of an aggressive environment, shaking, vibration, and also because of the lack of space to accommodate them during the process, should be explicitly noted [1].

The absence of the feedback on the controlled coordinate (i.e. presence of the unaccounted dynamic of the object of the control) results in the significant decrease of the effectiveness of the control, inability to provide required quality characteristics of the control process which includes non-compliance with the requirements of the stability. [2].

In technical terms, the problem under consideration is the most acute for three types of technical systems: 1) tracking objects on mobile vehicles (land, sea, aviation); 2) industrial robots with angular and anthropomorphic kinematics with «long» arms; 3) metalworking machines (turning, milling and other classes of machines).

One of the possible options for designing a rational structure of the ACS that enhances the control characteristics of the class of objects described earlier is the introduction of a parallel predictive model (PPM). The advantage of this approach, which allowed it to develop successfully in the practice of building and operating control systems, is the relative simplicity of the basic feedback generation loop, combined with the high adaptability. The latter makes it possible to control multidimensional objects with a complex structure and multiple connections including non-linearities, to optimize real-time processes within the constraints on control and controlled variables and to take into account the uncertainties in the descriptions of objects and perturbations [3]

Among the most common methods of predictive control special mention should go to: 1) Dynamic matrix control (DMC). Application area: linear open stationary processes. [5]; 2) Model algorithmic control (MAC). MAC is applicable only for the stable processes with an open scheme [6]; 3) Predictive functional control (PFC). PFC algorithm requires real time optimization method [7]; 4) Extended prediction self-adaptive control (EPSAC). It is characterized by simplicity of the formulated law of control. Possible settings are horizon of the prediction, weighting factor and polynomial filter. Though the real time setting of the predictive horizon influences on the structure of the multi-level predictor and control structure. [8]; 5) Extended horizon adaptive control (EHAC). It doesn't require knowledge of the matrix analysis systems and bears the effect of the output disturbances. However, multimetrical regulators based on one-step criteria of the prediction are sensitive to the choice of the delay structure. [9]; 6) Generalized predictive control (GPC). GPC is applicable for the nonminimum phase systems, with the unstable open loop and changing delay time. It can consider both constant and changing future values of the control points. [10]; 7) Neural inverted predictive control [11].

The aim of this work is an adaption of the forecasting control methods to the objects with the incompletely observed regulated coordinate for the increase of their control characteristics.

2. Automatic control systems with an incompletely observed controlled coordinate

The concept of distinguishing a class of automatic control systems with an incompletely observed controlled coordinate can be represented as the following sequence of procedures.

- Analysis of the process or mode of operation of the control object.
- Analysis of the media on which the location of the actuator or object control subsystem.
- Analysis of the structure of the original system. Allocation of subsystems.
- Establishment of the structural interconnection of subsystems in the hierarchy and the formation of control signals and perturbations.
- Isolation of links that generate the non-observability of the coordinates of the state space. The definition of zones and time intervals is not complete observability.
- Determination which links of the system are not covered by the main feedback.
- Designing a model of links not covered by the main feedback [14].

The block diagram of a tracking automatic control system with an incompletely observed controlled coordinate is shown in Fig. 1.

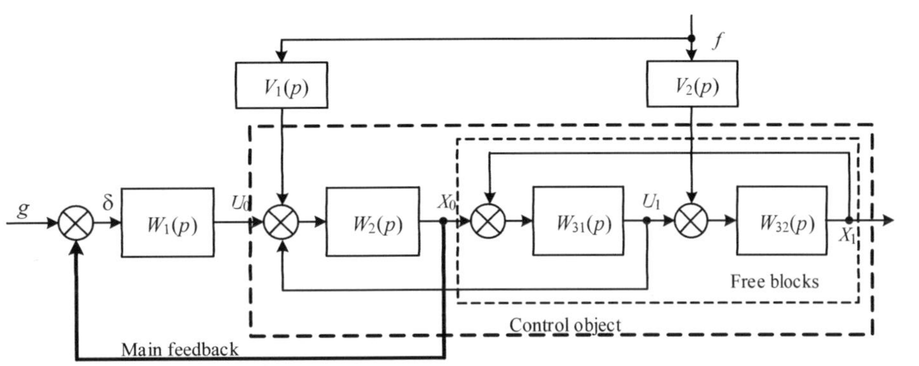

Fig.1. Block diagram of a tracking ACS with an incompletely observed controlled coordinate

Three blocks of the forward contour are marked on the figure 1: an actuator with kinematic transmission, described by the transfer functions $W_1(p)$, $W_2(p)$; the control object, partially covered by the main feedback, with the transfer function $W_{31}(p)$; the control object, partially not covered by the main feedback, with the transfer function $W_{32}(p)$. The block diagram also denotes: $V_1(p)$, $V_2(p)$ - transfer functions of the actuator for perturbation for the covered and not covered parts of the object by the main feedback; g, f - control signal and perturbations; X_1, X_0 - directly adjustable coordinate and its intermediate value, covered by the main feedback of the ACS; δ - is the error of the system.

Consider the features of the ACS for control of the tracking objects located on mobile vehicles:

- the ACS represents two sequentially linked (i.e. in series) interconnected systems. The first (mobile object) forms perturbations, which are subject to mitigation and stabilization. Coordinate oscillations of the hull are generating control signals and perturbations for the second system that performs direct tracking of mobile objects;
- the law governing the change in the control effect is random and represents a subharmonic function;
- the explicit control signal for the system is not synthesized;
- the signal of the direct path of the system is the error signal obtained from the gyro sensor.
- «platform» represents a beam, fixed in the hinge, with one arm l_1 at least an order of magnitude shorter than the other l_2;
- an adjustable coordinate determining the «quality of the technological process» is the extreme point of the longitudinal axis of the long arm beam l_2;
- the gyroscopic main feedback sensor is located on the short arm of the beam in the immediate vicinity of the hinge axis;
- if the beam is considered to be a nonrigid link, the main feedback sensor does not provide reliable information about the adjustable coordinate under perturbations with large accelerations and with a considerable mass of the beam (platform);
- for constructive and «technological» factors it is impossible to install a gyroscopic feedback sensor at the end of a long arm of a beam (platform).

3. The concept of predictive control

Predictive control refers to a class of control algorithms that use models to predict future system reactions. Fig. 2 shows a typical block diagram of the ACS with predictive control.

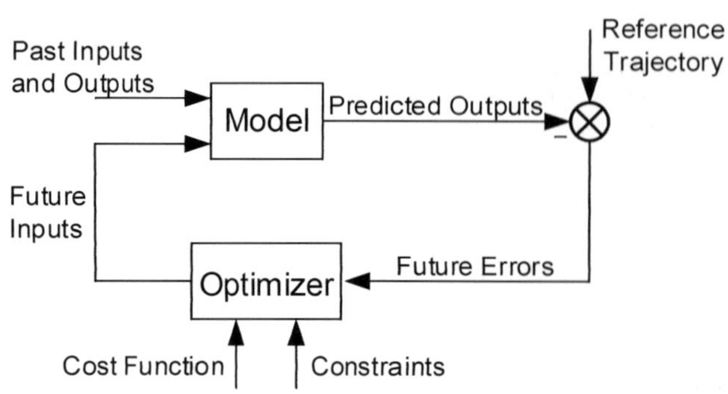

Fig. 2. Block diagram of ACS with predictive control

The implementation of the control with the predictive model can be described as the sequential execution of the following actions [3]:

1. Estimation of the state vector of the control object:

$$\hat{x}[k+1] = \hat{A}\hat{x}[k] + \hat{B}\hat{u}[k] + \hat{H}d[k], \quad \hat{x}[0] = \hat{x}_0,$$
$$\hat{y}[k] = \hat{C}\hat{x}[k] + w[k],$$

where, $k = 0,1,2, \ldots$ - is the clock cycle counter that determines the discrete time moment $t_k = kT$, Δt – is the discretization step, $\hat{x}[k] \in E^n$ - is the state vector of the control object at time t_k, $\hat{u}[k] \in E^m$ - is the control vector at time t_k, and $\hat{y}[k] \in E^r$ - is the measurement vector, $d[k] \in E^{nd}$ - external perturbation, $w[k] \in E^r$ - noise in the measurements.

2. Selection of control on the forecast (prediction) horizon optimizing the motion of the predictive model (approximating the dynamics of the control object):

$$x[i+1] = f(x[i],u[i]),\quad i = k+j,\quad j = 0,1,2,...,\quad x[k] = \hat{x}[k],$$
$$y[i] = Cx[i],$$

where, $x[i] \in E^n$ - is the model state vector, $u[i] \in E^m$ - the control vector, $y[i] \in E^r$ - the output vector. The predictive model is initialized at the clock cycle $j = 0$ by the current state of the real object $\hat{x}[k]$.

As a functional of the control quality of the predictive model, we consider the forecast horizon P in the following form:

$$J_k = J_k(y,\Delta u) = \sum_{j=1}^{P} \left[(y[k+j] - r[k+j]^T R[k+j](y[k+j] - r[k+j] + \Delta u[k+j-1]^T Q[k+j]\Delta u[k+j-1] \right]$$

where, $R[k+j]$, $Q[k+j]$ - given positive-definite matrices, $y = (y[k+j]\, y[k+j+1]...y[k+P])^T \in E^{rP}$, $\Delta u = (\Delta u[k+j]\, \Delta u[k+j+1]...\Delta u[k+P-1])^T \in E^{mP}$ - vectors of adjustable and control sequences on the forecast horizon, respectively.

3. Realization of the calculated optimal control for the object on the current clock cycle (from moment k to moment $k+1$).

4. Repetition of the steps 1 to 3 for the next cycle, starting from the time moment $k+1$.

The traditional scheme of predictive control has a number of shortcomings that limit its application in real conditions:

- a continuous evaluation of the model state vector is necessary in order to implement the forecast [15], [16]. However, the Wiener or Kalman filters require knowledge of the statistics of the signals, which are not always available. In addition, active uncontrolled perturbations introduce a significant error in the estimation process, and the problem of joint identification of the state vector and perturbations is considered as the problem of mathematical optimisation without a proven global extremum [17];
- for the purpose of designing a robust controller, for example, using optimization $H\infty$, one resorts to approximating of the delay element using Pad' series, which leads to inadequate results.

In this article authors propose a modification of the traditional predictive control scheme. In the ACS, a control correction circuit with the predictive model is introduced. The predictive analyzer included in the channel estimates the external influences a step forward, which allows to evaluate the system's responses. In this case, the parallel model generates an additional component of the ACS mismatch signal ahead of time, determined by the extrapolation step. For a model, two types of structures can be considered: 1) a model having a structure similar to the original ACS, but with a control object completely covered by feedback; 2) the model that forms the «reference transient process». In the second case, the structure of the model does not matter, since a system which is built into the parallel system is not the one with a certain scheme and parameters, but an independent structure that forms the required dynamic process.

In this case, we can distinguish three basic structures of the system with a PPM:

1) systems with a parallel predictive management model;
2) systems with a parallel predictable model by perturbation;
3) ACSs for stabilization with a predictive model in the error circuit (here: two predictive models, both for control and for perturbation). Figures 3a and b show block diagrams of the ACS with PPMs for control and perturbation.

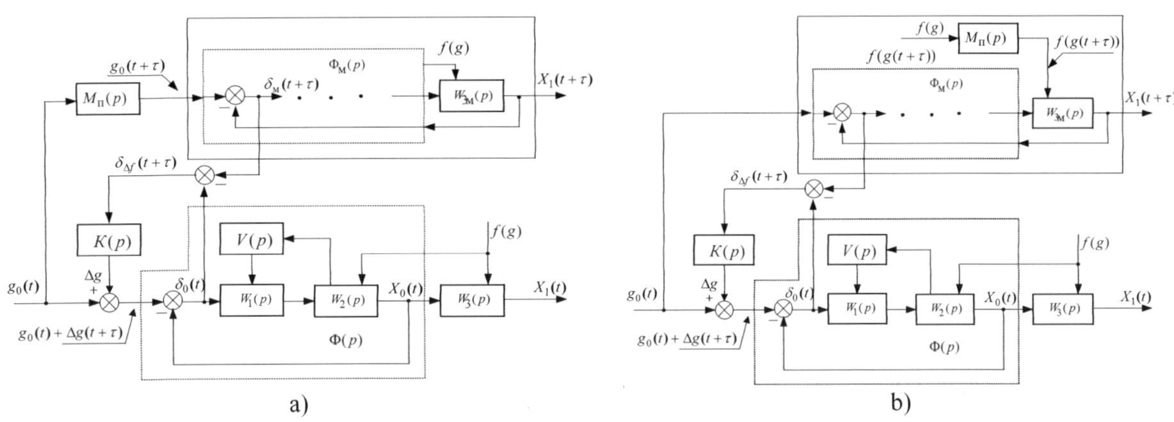

Fig. 3 - block diagrams of ACS with parallel predictive model: a) for control and b) for disturbance

Features of structures:

1) at the input of the model there is a leading block with the transfer function $M(p) = e^{p\tau}$; where τ - is the lookahead time;

2) the model of the main circuit includes all links of the direct path, including the transport system; its transfer functions

$$\Phi_{м}(p) = \frac{W_{м}(p)}{1+W_{м}(p)} \qquad \Phi_{fм}(p) = \frac{V_1(p)}{1+W_{м}(p)};$$

3) an additional component in the control action is formed on the basis of the difference between the model and system error signals

$$\delta_{\Delta f}(t+\tau) = \delta_{\Delta f}(t+\tau) - \delta_0(t);$$

4) additional control eventually forms a link with a transfer function $K(p)$.

To synthesize algorithms for additional control, we use the technique proposed in [18]-[21]. Consider the structural diagrams of the formation of an additional component with a forecast in control. There are two possible options: 1) the formation of an autonomous component in the current control signal; 2) the formation of an additional component at the stage of decomposition of the predetermined path in the device for generating control actions.

Thus, depending on the type of control action unit $K(p)$, which is supplied to an error signal should have a transfer function:

1) the first variant of additional management implementation:

$$g(p) = g_{м} = X = \text{const}: \qquad K(p) = \frac{\nu_0}{p};$$

$$g(p) = g_{м} = \frac{V}{p},\ V = \text{const}: \quad K(p) = \frac{\nu_1}{p^2};$$

$$g(p) = g_{м} = \frac{\varepsilon}{p^2},\ \varepsilon = \text{const}: \quad K(p) = \frac{\nu_2}{p^3},$$

where: X - adjustable coordinate; V - velocity input signal; ε - acceleration of the reference signal; ν_0, ν_1, ν_2 - transmission coefficients in the channel for generating additional control. This method is not related to the methods of control signal synthesis. It can be effectively applied in tracking systems for mobile objects.

2) the second variant of additional control implementation:

$$g(p) = g_{м} = X = \text{const}: \qquad K(p) = \frac{\gamma_0}{p};$$

$$g(p) = g_{м} = \frac{V}{p},\ V = \text{const}: \quad K(p) = \frac{\gamma_1}{p};$$

$$g(p) = g_{м} = \frac{\varepsilon}{p^2},\ \varepsilon = \text{const}: \quad K(p) = \frac{\gamma_2}{p},$$

where γ_0, γ_1, γ_2 - transmission coefficients in the additional control signal synthesis channel.

4. The choice of the algorithm for the predictive analyser

This part of the article is dedicated explicitly to the choice of predictive functions or polynomials and the analysis of the accuracy of their work in real time.

When selecting a method for predicting functions in dynamical systems, it is required to obtain the best approximation within the minimum number of measures. Authors propose an approach to the choice of the algorithm of the predictive model, which includes the following components [14]:

- Analysis of characteristic operation modes of the control object for a reasonable choice of the approximating function. When choosing a predictive function, it is necessary to consider the frequency spectrum of the control signals and perturbations. It is necessary to take into account the own transmission frequency of the servo actuators, the duration of the impulse response of the actuator, the precision of tracking and/or stabilization and the frequency properties of the links not covered by the main feedback.

- The choice of the method of calculating of the approximating function. The prediction problem can be solved using interpolation methods, and therefore on this step the problem of «forward interpolation» - extrapolation or approximation of the received data - is formulated.
- Determination of the amount of initial data for prediction with the required accuracy - the number of nodes and their fixation (equidistant or with arbitrary intervals).
- Selection of the extrapolation step based on theoretical data and simulation results, while taking into account the statements above and prediction errors.

The results of the study of the predictive model algorithms are presented in [22]. The analysis of the obtained results testifies to the advantage of the prediction using the Newton interpolation polynomials. High accuracy of calculations with a relatively small computing time, the simplicity of the algorithm makes it most preferable in the construction of automatic control systems with predictive models.

5. Results of the study

The investigated movement actuator with the main feedback on its position (DC motor with independent excitation) includes two internal circuits: the voltage and the speed of the motor and the PID controller, and has stable characteristics (no over-adjustment (overshoot) and a specified response rate of 0.2 s) which in practice do not depend on external influences. The control object is not covered by the feedback and has a finite rigidity c and the damping factor b, its design scheme shown in Fig. 4.

The system of differential equations of the dynamics of the control object has the form

$$M_1 - M_2 = J_1\varphi_1;$$
$$J_2\varphi_2 = c(\varphi_1 - \varphi_2) + b(\phi_1 - \phi_2) + M_n,$$

where M_1, J_1 - dynamic moment and the moment of inertia of the actuator, M_2, J_2- moment of the elastic forces and reduced moment of inertia of the load, φ_1, φ_2 - angular displacement between the actuator and the load, M_n - external static moment. The moment of the elastic forces M_2 is a major perturbation and determines the impact not covered by the feedback control of the object during the functioning of ACS. The system of differential equations describing the dynamics of the ACS with an incompletely observable control object is of the sixth order and considers the limitations on the motor current and voltage and nonlinearity of the control object. The ACS research was carried out by simulation of the dynamic modes in the MatLab environment (Simulink). The qualitative parameters of transient processes and the accuracy of adjustment were determined under different laws of the input signal and load parameters (static moment and moment of inertia of the load).

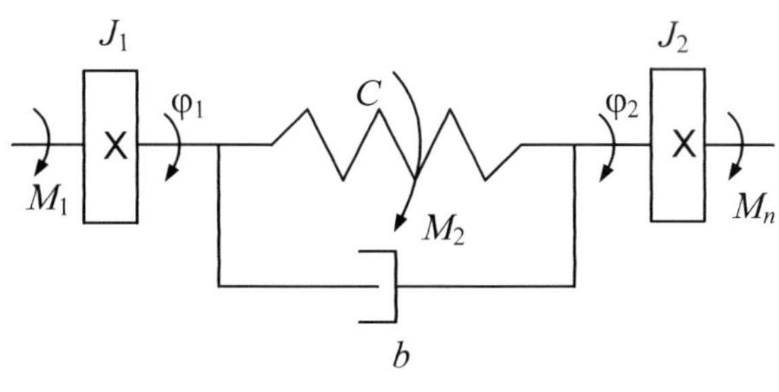

Fig. 4. The design scheme of the control object

The simulation results showed that the lack of complete information on the adjustable coordinate sharply reduces the performance of the ACS, especially with variable external influences. The increase in the control signal causes an increase in the oscillation index in such systems and can lead to instability. An increase in the static moment of the load $M_н$ results in an increase in overshoot during transient conditions and reduce the marginal stability of the system. A similar effect on the dynamics of ACS has a change in load inertia J_2. With harmonic input signals, increasing their frequency causes a decrease in accuracy, the steady-state dynamic error increases to an unacceptable value.

Studies have shown that the introduction of a parallel model having a structure similar to the original ACS, but with a control object completely covered by feedback, allows to stabilize the dynamic characteristics (eliminate overshoot, shorten the transient process, reduce the error) only under a limited range of external perturbations. With significant changes in the driving force, the model can become a source of additional perturbation. This will often lead to a deterioration in the dynamic characteristics and stability of the ACS.

ACS PPMs of the second type have high dynamic performance. These models tend to include the synthesis of the «additional control» element with variable transmission coefficient $K(g)$ in their circuits.

The value of this coefficient depends on the value of the reference variable *g*. This inclusion ensures invariance of the ACS to external perturbations and minimize the negative influence of the actuator and load nonlinearities

$$\Delta g = K(g)(\varphi_2^* - \varphi_1) \quad (1)$$

Here Δg - «additional control», a difference between the output coordinates of the model – φ_2^* and ACS actuator - φ_1.

Fig. 5 shows the dependency of coefficient of transfer (1) from the magnitude of the rapidly changing reference variable. Results were achieved from simulations under conditions of monotonic transient lasting no more than 0.2 s.

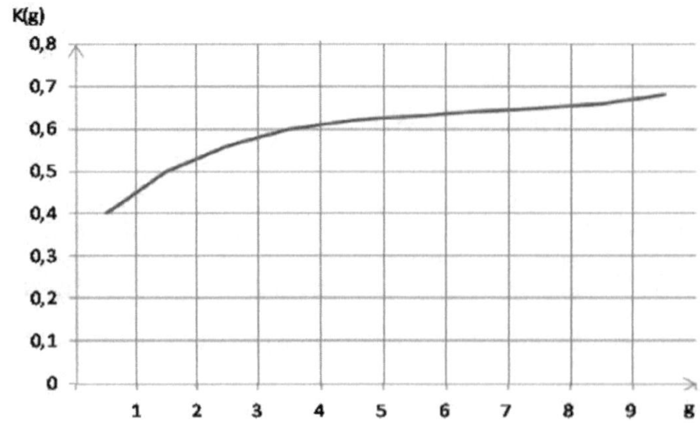

Fig. 5. Dependency of additional transmission coefficient *K*(*g*) from the magnitude of the reference variable

A significant influence on the qualitative parameters of the ACS is caused by the nonlinearity of the control object not covered by the main feedback. For example, object dead zone in the control system without a parallel model increases the dynamic error to 20% of the steady-state value of the output quantity. The introduction of parallel model can significantly reduce the steady-state error, but doesn't completely eliminate the effect of nonlinearity. To ensure the stability of dynamic characteristics in automatic control with a nonlinear control object it is expedient to use an element with a variable transmission coefficient, as in the additional control channel, and in the PPM. Mathematical function of this nonlinear operation correction unit depends on the amount of control. Dependencies of the control channel transmission factors and the model to ensure stable performance of ACS (under conditions of monotonic transient lasting no more than 0.2 s) from the magnitude of the rapidly changing input signal are shown in Fig. 6. Similar results are obtained with harmonic input signal with varying amplitude.

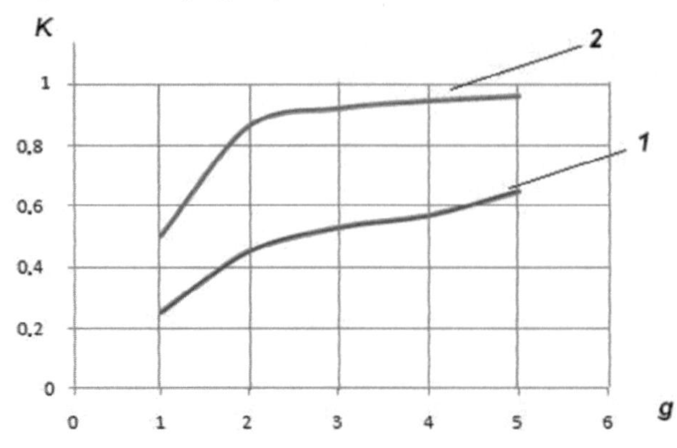

Fig. 6. Dependencies of the transmission coefficients of the nonlinear element in a control channel (1) and parallel model (2) from the magnitude of the input signal

Introduction of the external influence prediction element to the signal synthesis channel of the ACS with PPM allows to improve the dynamic characteristics of the system without resorting to the use of the nonlinear element with a variable transmission ratio, which would significantly complicate the structure of ACS and control problems, because its value depends on many factors.

We investigated the movement modes with different parameters of external influences (static moment and the moment of inertia of the load) using operation algorithm for the predictive element that implements the Newton's second interpolation formula.

The greatest influence on performance of ACS under harmonic and polyharmonic perturbations is caused by the frequency of their change. Ensuring the stability of ACS characteristics is possible via the variation of prediction time. Fig. 7 shows the dependency of the interpolation step (τ- prediction time) from the fundamental frequency (ω) of external perturbations.

The simulation results demonstrated that the maximum absolute error of the controlled variable does not exceed 1% of the maximum value of the reference variable.

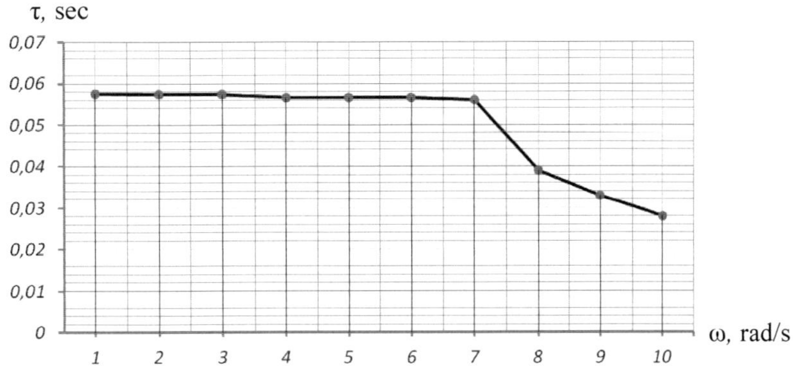

Fig. 7. Dependencies of the prediction time from the input signal

6. Conclusion

The absence of the feedback on the controlled coordinate in the objects of the control (i.e. presence of the unaccounted dynamics of the control object) results into the significant decrease of the control effectiveness, inability to provide required quality characteristics of the control process which includes noncompliance of the requirements of the stability. The introduction of the circuit into ACS forming an additional component on the basis of the predictive model is an effective method of the increase of the dynamic characteristics of the objects considered. Herewith it is rational to use the model forming desired transient process in the correction circuit. The algorithm of the work of the predictive analyser is based on the extrapolation Newton's polynom. The results of the simulation showed that maximum absolute uncertainty of the controlled condition doesn't exceed 1% from the maximum of the reference signal.

The researches made with the introduction of the parallel model without using predictive analyzer showed that for the provision of the desired dynamic characteristics it is necessary to add a link with variable transition factor into the circuit of the formation additional control, and that significantly perplex the structure of the system and control problems.

It is possible to move the development of the control predictive method suggested on the other dynamic objects. It is necessary to make a research in choosing an appropriate time of prediction and synthesis of the algorithms of its determination.

7. Acknowledgments

Research is executed with financial support of the Russian Fund of basic researches in the scientific project No. 16-38-00638 mol_a.

8. References

[1] Gudzenko, L.I. (1970). Analysis of an unregulated object using the method «black box without entry», In: the collection Identification and equipment for statistical research, Ovsepian F.A. (Ed.), Publishing house Science, Moscow, Russia.

[2] Kobzev A.A.; Mishulin Yu.E.; Novikova, N.A. & Noskov, E.V. (2010). Analysis of automatic control systems with an incompletely observed controlled coordinate. Proceedings of the Institute of Engineering Physics, Vol. 4, No. 18 (december, 2010), pp. 16-19, ISSN: 2073-8110.

[3] Veremey, E.I.; Eremeev, V.V. (2004). Introduction to management problems based on predictions, Proceedings of the All-Russian Scientific Conference «Design of Scientific and Engineering Applications in the MATLAB Environment», may 25-26, 2004, Moscow, Russia, ISBN 5-201-14971-5200-4, Prangishvili, I.V. (Ed.), pp. 98-115, Publisher house of Institute of Management Problems, Moscow, Russia.

[4] Holkar, K.S. (2010). An Overview of Model Predictive Control. International Journal of Control and Automation, Vol. 3, No. 4, 2010, pp. 47–63, ISSN: 1598-6446.

[5] Cutler, C.R. & Ramaker, B. (1980). Dynamic Matrix Control – A computer control Algorithm. Proceedings of the Joint Automatic Control Conference, 1980, San Francisco: American Automatic Control Council, Vol. 3, pp. 5- 13.

[6] Mehra, R.K. & Eterno, J.S. (1980). Model Algorithmic Control for Electric Power Plants. IEEE, 1980, pp. 280–285.

[7] Camacho, E.F. & Bordons, C. (2007). Model Predictive Control, Springer edition, ISBN: 978-1-85233-694-3, London.

[8] De Keyser, R.M.C.; Van De Velde, PH.G.A. & Dumortier, F.A.G. (1988). A comparative Study of Self-adaptive Long range Predictive Control Methods. Automatica, Vol. 24 (Issue 2), pp. 149–163, ISSN: 0005-1098.

[9] Golden, M.P.; Liu, L.K. & Ydstie, B.E. (2005). Multivariable Control using Extended Prediction Horizons, Proceedings of 25th Conference on Decision & Control, December, 1986, Athens, Greece, pp. 1550–1551.

[10] Clarke, D.W. (1987). Generalized Predictive Control: A Robust Self-Tuning Algorithm, Proceedings of American Control Conference, June 10–12, 1987, Minneapolis, Minnesota, pp. 990–995.

[11] Markov, N.A. (2012). Predictive inverse neurocontrol by experimental aerodynamic plant. Vestnik Engineering School of the far Eastern Federal University, Vol. 4 (13), pp. 58-63.

[12] Antos, J[an] & Kubalcik, M[arek] (2016). Alternative Approach to Optimization in Model Predictive Control Using Hill Climbing Algorithm, Proceedings of the 26th DAAAM International Symposium, B. Katalinic (Ed.), pp. 0856-0864, Published by DAAAM International, ISBN 978-3-902734-07-5, ISSN 1726-9679, Vienna, Austria, DOI:10.2507/26th.daaam.proceedings.119.

[13] Macku, L[ubomir] & Samek, D[avid] (2016). Self - Organizing Migrating Algorithm in Model Predictive Control: Case study on Semi-Batch Chemical Reactor, Proceedings of the 26th DAAAM International Symposium, B. Katalinic (Ed.), pp. 0238-0246, Published by DAAAM International, ISBN 978-3-902734-07-5, ISSN 1726-9679, Vienna, Austria, DOI:10.2507/26th.daaam.proceedings.033.

[14] Kobzev, A.A.; Mishulin, Yu.E.; Novikova, N.A. & Lekareva, A.V. (2014). Automatic control systems with parallel predictive model: monograph. ISBN: 978-5-9984-0507-5, Publishing house of the Vladimir State University named after A.G. and N.G. Stoletovykh, ISBN: 978-5-9984-0507-5, Vladimir, Russia.

[15] Liu, Ch.-Sh.; Peng, H. (2002). Inverse-dynamics based state and disturbance observers for linear time-invariant systems. J. of Dynamic Systems, Measurement, and Control, Vol. 124 (July 23, 2002), pp. 375-381, DOI: 10.1115/1.1485748.

[16] Yan, J. & Bitmead, R.R. (2005). Incorporating state estimation into model predicative control and its application to network traffic control. Automatica, Vol. 41, pp. 595-604, ISSN: 0005-1098.

[17] Camacho, E.F. & Bordons, C. (2007). Model Predictive Control, Springer edition, ISBN: 978-1-85233-694-3, London.

[18] Novoselov, B.V.; Gorokhov, Yu.S.; Kobzev, A.A. & Shchitov A.I. (1975). Automatic machines-adjusters tracking systems, Publishing house Energy, Moscow, Russia.

[19] Kobzev, A.A. (1991). Adaptation of the control action in the drives of the assembly robot. News of Higher Educational Establishments. Electromechanics, No. 12, pp. 73-79, ISSN: 0136-3360.

[20] Kobzev, A.A.; Novikova, N.A. & Lekareva, A.V. (2015). Investigation of adaptation control algorithms for drives of robotic systems using the intercoordinate displacement simulator. News of Higher Educational Establishments. Electromechanics, No. 3, pp. 50-55, ISSN: 0136-3360.

[21] Lekareva, A.V. (2016). Ensuring the invariance of error by perturbation in systems of automatic control of trajectory movements of technological objects. Scientific and Technical Herald of Information Technologies, Mechanics and Optics, Vol. 16, No 5, pp. 787–795, DOI: 10.17586/2226-1494-2016-16-5-787-795.

[22] Kobzev, A.A., Novikova, N.A. & Lekareva, A.V. (2017). Algorithm selection for the predictive model in control systems with incompletely observable control coordinate. Procedia Computer Science, Vol. 103, pp. 426–431, Available from: http://dx.doi.org/10.1016/j.procs.2017.01.009, Accessed: 2017-08-30.

DOI: 10.2507/28th.daaam.proceedings.126

AN OPTIMIZATION OF PRODUCT DESIGN AND MANUFACTURING BASED ON SUSTAINABILITY

Halla Atiya & Luma Adnan Alkindi

This Publication has to be referred as: Atiya, H[alla] & Alkindi, L[uma] (2017). An Optimization of Product Design and Manufacturing Based on Sustainability, Proceedings of the 28th DAAAM International Symposium, pp.0911-0920, B. Katalinic (Ed.), Published by DAAAM International, ISBN 978-3-902734-11-2, ISSN 1726-9679, Vienna, Austria DOI: 10.2507/28th.daaam.proceedings.126

Abstract

In the 21st century, manufacturing companies are facing the challenge to remain competitive in global markets by finding a more sustainable approach to product design and manufacturing. This is associated with the main features of companies' competitive landscapes namely, simultaneous cost reduction and quality improvement of the product. It may appear that reaching both these goals together is prohibitively problematic as they are considered by "manufacturing leaders" as opposing objectives. However, much recent research has been done on introducing approaches to delivering low cost, high quality sustainable products based on finding a link between the design or the manufacturing processes and other key elements of sustainability; economic, environmental, and social. Indeed, these approaches often times include limitations such as the lack of integration between the design and manufacturing stages of the product and do not adequately address three key parts of sustainability; economic, environmental, and social. Consequently, these approaches, have a low applicability, and hence few opportunities for success. However, for delivering a sustainable product, the key parts of sustainability, as well as the integration of product design and manufacturing processes must be optimised in terms of one effective measure. This project aims to propose an approach for design and manufacturing sustainability. The combined sustainability index (CSI) is developed as a comprehensive evaluation metric for alternative scenarios of the process and also as a tool that enables decision makers to run optimal processes. The applicability of the CSI has to be tested and verified in a real industrial setting with the aim to improve the sustainability performance by designing a product and the corresponding manufacturing processes for that product. Process optimisation is carried out in terms of the highest CSI score achieved.

Keywords: Sustainable Manufacturing and Design; combined sustainability index (CSI); model, multi-objective

1. Introduction

The need to implement sustainability concept in designing and manufacturing products has become, more than ever, a matter of crucial importance in today's economy. From the descent of raw materials through to the final use and disposal, sustainable or "green" products outweigh conventional products on their environmental and social intrinsic worth. Manufacturers need to follow activities of production internalize negative environmental externalities while profits maximize and social economic benefits. Most of the manufacturing organizations expend large quantities of resources at the same time generate wastes and pollution. However with recent increase in sustainability issues, organizations should consider taking suitable measures to enhance their sustainability aspects. It has thus become the need of the time for manufacturers to pursue manufacturing activities, which helps in minimizing environmental impacts while maintaining social and economic benefits (Joung et al. 2013) [1]. One of definitions of the sustainability is Brundtland description. It explains sustainability as "meeting the needs of the present generation without compromising the ability of future generations to meet their own needs [2]. The purpose of this paper is to determine the sustainable product and to set major operational parameters to maximize a proposed manufacturing sustainability index.

2. Literature Review

Siddhartha Kushwaha (2005), proposed index term as "A Comprehensive Sustainability Index (CSI) Balanced Social, Economic & Environmental Approach" to measure sustainability of countries based on the triangular tradeoff among environment, economy and society. The comprehensive sustainability index (CSI) is based on genuine saving, ESI, EPI and HDI. Collected hierarchal clustering method is used to group 101 countries with their peers into two groups :(1) More Sustainable Nations (63 countries) and

(2) Less Sustainable Nations (38 countries).The low standard deviation in CSI represents the clustering of nations about mean CSI value. The clustering is an indicator of mediocre sustainability of nations. Hence severe measures are needed to be taken by governments, policy makers, industry and residents to increase the sustainable environment [3].

Schmidt and Butt (2006) developed method term as product sustainability index (PSI), was based on LCA impact assessment categories. It includes eight key environmental indictors across three major aspects of the product sustainability. The economic indicators are from Life-cycle Costing (LCC) assessment method. The societal indicators are aimed at evaluating the safety and mobile capability of a product. This methodology was further applied to two of their vehicle models [4].

de Silva et al.(2009) , Ungureanu et al.(2007) developed sustainability scoring method to quantitatively assess the potential benefits of using aluminum alloys for manufacturing of an auto body. This method involves six major elements of the Design for Sustainability (DfS): environmental impact, functionality, manufacturability, recyclability and re-manufacturability, resource utilization/economy and societal impact. Each of these elements was sub-divided into several corresponding sub-elements. Different levels of influencing factors were categorized based on their significance to the product. The level of importance of each sub-element was assigned with high, medium, or low weights. Finally, sustainability scores of two different materials (steel alloy and aluminum alloy) were computed and their levels of sustainability performance were compared [5] [6].

Nomeda Dobrovolskiiené and Rima Tamošiūnienė (2015) proposed a tool term as composite sustainability index of a project (CSIP) which to measure the sustainability of a business project in the construction industry, take 15 criteria (four economic, six environmental and five social criteria) of construction sustainability selected, then group of professionals had to rank the importance of each criterion from 1 to 15, where the highest score was 15 points and the lowest 1 point. The next step is normalization by applying a distance to a reference method, the normalized value is calculated as the ratio between the indicator and an external benchmark (or target value). The final step is aggregation. At this stage, an aggregate index, called a composite sustainability index of a project is developed. The most common form of aggregation is the summing-up of the weighted and normalized individual indicators [7].

Peihao Li and Mingjia Liu (2015) develop a mathematical model term as Comprehensive Sustainability Index (CSI) that can be used to comprehensively evaluate and clearly define a country's sustainability based on factor analysis. Sustainability index was determined by a linear combination of a set of indicators .They selected 10 representative indicators coming from three aspects: social development, economic development and environmental protection. They used factor analysis to convert observations of correlated variables into values of linearly uncorrelated variables called factors thus determining the weights of indicators. In the evaluation section, they select 15 countries to form the sample, 5 least developed countries, 5 developing countries and 5 developed countries. By calculating the value of CSI of each country, they found out that there are clear differences among the CSIs of the three kinds of countries. And based on such differences, they reached classification of CSI [8].

Zhang et al. (2013) Presenting methodology involving the use of total life-cycle approach, including the life-cycle assessment (LCA) method, for improving the product sustainability performance of metallic automotive components sustainability aspects (economic, environmental and social) [9].

C. Vilaa, J.V. Abellán-Nebota , J.C. Albiñanaa and G. Hernándezb proposed framework developed the key stages for a sustainable product lifecycle strategy. Proposed model establishes all the processes carried out through the product lifecycle and indicates the application and interaction of specific methodologies, tools and knowledge at each stage that will help to achieve sustainability. The processes applied are controlled by means of the ecodesign and sustainable manufacturing approach throughout the lifecycle of the product. This makes it possible to increase the efficiency in product development and sustainable lifecycle management. [10]

The contribution of this paper is to determine the sustainable product capturing the totality of sustainability aspects (economic, environmental and social). The current work differs from previous work in that it offers a mathematical model considering the totality of aspects of sustainability, not only environmental, to help systematically plan and improve sustainable manufacturing activities.

3. Mathematical Model Development

3.1. Problem Definition

Manufacturing companies aims to increase profits by reducing manufacturing cost of product, but given to the increase in taxes paid versus each amount of CO2 emitted through manufacturing of product .Therefore the world towards to manufacture "green product".Consider a manufacturing system, such as that described in Figure 1, producing a variety of products. The system transforms a set of natural resources with the aid of labor, technology and financial resources to outputs. These outputs are the required products to be sold in the market in addition to scrap resulting from inefficiencies in the transformation process and some recyclable material. The operation of this manufacturing system has economic, environmental, as well as social impacts on its surrounding environment. Most of researchers introducing approaches to reducing manufacturing cost and others introducing approaches reducing CO2 emission, often include limitations such as the lack of integration between the design and manufacturing stages of the product and three key parts of sustainability; economic, environmental, social. Quantity and variety decisions have to be made that shape the efficiency of the manufacturing operation. These decisions are Production process method, Skill levels of workers, number of workers, Number of machine, and type of material. These decisions heavily affect the economic, environmental and social aspects of the facility and, thus, determine the manufacturing sustainability. The target is to arrive at decisions concerning the sustainable product and the aforementioned operating parameters, so as to maximize system sustainability.

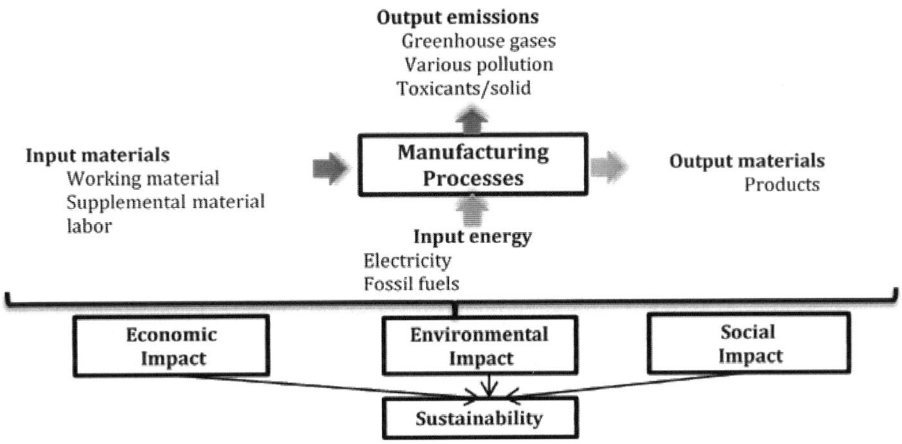

Fig. 1. Elements of a manufacturing system affecting its sustainability [11]

3.2. Mathematical Model

The aim of this paper is to devise a mathematical model as a comprehensive evaluation metric for alternative scenarios of the process and also as a tool that enables decision makers to setting up optimal vital operating parameters, so as to maximize manufacturing sustainability. Thus, the objective function should reflect sustainability, while the constraints reflect the limited available resources. Since the sustainability consists of three pillars, economic, environmental and social, the objective function has to include these elements. An integration of these incommensurable aspects is challenging. It addresses the three pillars of sustainability: environmental, economic and social. Each of the pillars is evaluated by a set of sub-indicators; economic evaluated by total manufacturing cost of product (C_{TM}); environment and social evaluated by carbon dioxide emission from machines during manufacturing process (C_E). All of the indicators are so devised such that they have a value ranging from zero to one. The two performance measures discussed in this paper C_{TM} and C_E belong to the smaller-the-better characteristic. In order to find the best decision variable set values satisfying the two performance measures applying linear programming method (LP), it is a requirement that these multiple measures are converted into a single measure. This single measure is referred to as comprehensive sustainability index (CSI). The CSI is an overall representative measure and needs to be finally optimised. The desirability function approach is used to convert from multiple measures to a single measure as described in the following sub-sections.

3.2.1 Model Assumptions

The following assumptions

- Only one type of emissions is considered, which is CO2.
- Demand is deterministic and constant.
- Exact amounts of raw material required for producing each product.
- Ability of machine to produce exact amount of product each day.

3.2.2 Model Nomenclature

Parameters

C_{Tm}: Total manufacturing cost required to produce specified production quantity each year.
C_o: Ordering cost of raw materials monetary unit (MU).
C_H: Holding cost of raw materials within the production.
C_{tr}: Transportation cost for delivering raw materials to the production system (MU).
C_P: The purchasing cost of raw materials required to produce the product (MU).
C_U: Utilities cost of the final product (MU).
C_D: Duties cost arising from procuring raw material from an external supplier (MU).
T_{Pi}: Transfer price required for procuring raw material i from an external supplier i (MU).
S: Origin of ordered raw materials.
v: Destination of required raw materials.
m_i: Transportation mode for transporting raw material i to its customer.
N_T: Number of transportation modes used for shipping raw materials to the production system (unit).
M_i: Raw material types required in producing one unit of product i.
N_P: Number of parts required to produce one product (unit).
C_{Uo}: Ordering cost of raw materials for the final product (MU).
C_{UH}: Holding cost of raw materials of each final product in the production system warehouse (MU/ day).
C_{UMi} Unit cost of the raw material i at the beginning of each cycle (MU);
d_{RM}: Daily demand of raw materials required to meet customer satisfaction (unit);
$\%d_{RM}$: Daily demand percentage of raw materials required to meet customer satisfaction (unit).
C_{TOmang}: Total managerial ordering cost per order (MU).
C_{THmang}Total managerial holding cost for keeping raw materials to produce final product during the lead time duration.
SF: Storage factor for keeping raw materials in the warehouse.
OF: Ordering factor for procuring each order from the supplier.
%Util: Utilities cost percentage of the final product (MU).
t_p: Raw material cost percentage incurred for procuring raw material i from an external supplier j (MU).
$T_{Si,v,m}$: Tensor for transportation cost per critical measurement (MU).
t_m: Critical transportation measurement of raw materials shipped using transportation mode m.
%V: Volume percentage value required for transporting raw materials to their customer.
D_i: Duty rate (%) per price of raw material i supplied by external supplier j (MU).
%sale: Sales percentage value offered from selling extra raw materials during the same order.
N_{RM}: Number of raw materials supplied to the production system (unit).
C_E: Carbon emission.
C_{En}: Energy consumption kwh/year by machines.
C_{OP}: Operating cost of machine
C_E: Energy cost, the cost of energy used to operate the machines
$C_{L_{i,\gamma}}$: Cost of labour by time of one operation.
h: Hourly operation time required to produce the product (hr.)
AU: Annual use of power (kwh) per year.
$E_{(C)}$: Energy consumption of power (Kwh/hr.) used per day.
P : power (kWh).
P_E : Price of one kWh used to operate machine.
Sk: Skill level of worker.
P_r : Price of the machine.
C_{Ei}: Carbon dioxide (CO_2) emission for the quantity of raw material i ordered in each patch required to produce the final product per week (unit)
Q_{Mi} : i Raw material weekly batch quantity ordered to produce final required product (unit).
Mi: Raw material type required in producing one unit of product i.
C_{EC}: Carbon dioxide (CO_2) emission from machines c in each department.
TEU: Twenty Foot Equivalent Unit. The two most common international standardized container types are twenty and forty foot. A twenty foot unit measures about 6 meters, a forty foot about 12 meters (external dimensions).
C_{CP} : The cost of waste collection per CP per year
c_{vy} : is the annual cost of collection per vehicle
N_{cp}: is the number of served CPs.
b_v: is the backup rate for the collection vehicle (backup vehicle per collection vehicle).
C_{cv} : The capital cost of vehicle amortized over its economic life.
N_{wb}: The number of waste bins that a vehicle can service during one collection trip.
C_{cwb}: The annual capital cost per waste bin.

C_{ov}: is the annual operating cost of vehicle.
e: is the administrative rate.
C_v: is the unit price of vehicle.
CRF_{wb}: is the capital recovery factor.
U_c : is the vehicle utilization factor (usable per total vehicle capacity).
V_c : is the vehicle capacity.
F_c : is the frequency of waste collection per week.
W_{mix}: is the weekly waste generation rate per CP.
P_{mix}: is the overall density of mixed waste. Overall density of mixed waste.
P_i: is the weight fraction for waste component i.
p_{ci} : is the compacted density of waste component i.
C_{vwb}: is the unit price of waste bin and CRF_{wb} is the capital recovery factor.
i: The yearly discount rate.
L: is the economic life of collection vehicle.
a: is the fringe benefit rates.
b_w: is the backup rate for collection workers.
S_w: is the salary rate per hour per worker.
N_w: is the number of collection workers per vehicle.
S_d: is the driver salary rate per hour.
MH_D Man-hours/Day

Decision Variables:

γ Skill levels of workers
N_W No. of workers
N_c No. of machine
T_{RM} Type of material

3.2.3 Objective Function

The objective function for the proposed model is the comprehensive sustainability index, which is to be maximized is by minimizing the total manufacturing cost C_{Tm}, carbon emission C_E, in production system. This objective can be formulated as following:

$$Minimise \sum (C_{Tm} + C_E) \tag{1}$$

I. Total manufacturing cost C_{Tm}

Various costs associated with total manufacturing costs, in this paper total manufacturing cost includes the operating cost "C_{op}", , raw material cost "C_{RM}", Utilities Cost "C_U", , and cost of waste collection "C_{cp}" therefor it can be calculate as :

$$C_{TM} = C_{op} + C_{RM} + C_U + C_{cp} \tag{2}$$

Each component of the total cost of the final product is described as follows:

A. Operating cost (C_{OP})

Operating cost is the expenses related to operating machines to produce final product, involves cost of workers operate on the machine to produce product and cost of energy to operate machines, it can be calculated as:

$$C_{OP} = \sum (C_E + C_W) \tag{3}$$

C_W worker cost is considered as the rate resulting from multiplying the cost of working time by the time required for producing the final product (hour), which can be calculated as follows Equation (4):

$$C_{OP} = \sum (C_E + C_W) \tag{4}$$

Skill of i = (1 or 0.9, 0.8, 0.7)

$$C_E = AU \times P_E \tag{5}$$

B. Raw Material Cost (C_{RM})

Raw material cost is the cost of raw materials used to produce the final product in the production system. In this research, raw material cost includes the ordering cost C_O, holding cost C_H, purchasing cost C_P, transportation cost C_{TR}, duties cost C_D and transfer price cost TP. Therefore, it can be calculated as equation (6) [12]:

$$C_{RM} = C_O + C_H + C_P + C_{TR} + C_D \tag{6}$$

- *Ordering Cost (C_O)*

Ordering costs are the cost of ordering and receiving an amount of raw materials each order. These costs include determining how much is required, preparing invoices, inspecting goods upon arrival for quality and quantity against shipping lists and moving the goods to temporary storage. Generally, they are expressed as a fixed value per order, regardless of the order batch size (Chopra and Meindl, 2010).

$$C_{UOj} = \frac{C_{Tomang}}{\%d_{RM} \times L_T} \tag{7}$$

$$\%d_{RM} = \frac{d_P}{d_{RM}} \tag{8}$$

$$C_O = C_{UO} \times OF \tag{9}$$

- *Holding Cost (C_H)*

Holding cost is the cost of keeping one unit of raw materials in warehouses for a specified time period (Chopra and Meindl, 2010). The rate of C_H if the production system procures its raw materials from an external supplier equals Equation (10):

$$C_{UHj} = \frac{C_{THmang}}{N_{RM}} \tag{10}$$

$$C_H = \sum_{i=1}^{N_P} (C_{UHi})_s \times \%d_{RM} \times (LT + SF) \tag{11}$$

$i = 1, 2, 3, \ldots, N_P$.

- *Purchasing Cost (C_P)*

Purchasing cost is the cost of goods acquired from suppliers required to produce the final product in the production facility.

In this research, it is considered as the unit cost of raw material at the beginning of that cycle (CUM). $\boldsymbol{C_P}$ Can be calculated as Equation (12):

$$C_{UMSEj} = \%sale \times C_{UPi} \tag{12}$$

$$C_P = \sum_{i=1}^{N_P} (C_{UMSE_i}) \tag{13}$$

- *Duties Cost (C_D)*

This is the tax incurred by importing goods from supplier in one country to a customer in another country. It is based on the value of goods or upon some criteria of the item such as weight and dimensions.

This work considers C_D as the cost arises from supplying raw materials by a regular external supplier SEj to the production system. It can be calculated as Equation (14):

$$C_D = \sum_{i=1}^{N_P} C_{UPi} (1 - IF) \times D \tag{14}$$

IFj = 0 if $SE_j = 1$, otherwise IF = 1.

C. Utilities Cost (C_U)

Utilities cost is the cost arising from using the required utilities such as electricity, gas, heat, water and maintenance to produce final product in the production facility. This research considers C_U as a raw material cost percentage of the final product. It equals Equation (15):

$$C_{UMSEj} = \sum_{i=1}^{N_P} \%C_{RMi} \qquad (15)$$

D. The cost of waste collection (C_{CP})

The cost of waste collection per CP (collection point) per year is calculated according to [13]:

$$C_{CP} = \frac{c_{vy}}{N_{cp}} \qquad (16)$$

$$c_{vy} = [(1+b_v) \times C_{cv}] + N_{wb} \times C_{cwb} + C_{ov} \qquad (17)$$

$$C_{cv} = (1+e) \times C_v \times CRF \qquad (18)$$

$$N_{wp} = \frac{N_{cp}}{a_{wp}} \qquad (19)$$

$$N_{cp} = \frac{U_c \times V_c \times F_c}{\frac{W_{mix}}{P_{mix}}} \qquad (20)$$

$$P_{mix} = \frac{1}{\sum \frac{P_i}{P_{ci}}} \qquad (21)$$

$$C_{cwb} = (1+e) \times C_{vwb} \times CRF_{wb} \qquad (22)$$

$$CRF_{wb} = \frac{i.(1+i)^L}{(1+i)^L - 1} \qquad (23)$$

$$C_{ov} = (1+e).\,(1+a)\,(1+b_w).\,(S_w.\,N_w + S_d) \qquad (24)$$

II. CO2 emissions (C_E):

In regard to the carbon emissions from machines, formula (24) computes the total carbon emissions occurring during the operating of machines

$$C_{EC} = E_c \times E_{fi} \qquad (25)$$

3.3 Constraints

Three type of constraints considered in the proposed model, as described next. Production quantity: The amount of product that produces each month should be less than or equal to the maximum demand of that product as in Equation (26),

$$S_K \leq D_K \qquad (26)$$

Manufacturing cost: the cost of manufacturing product that comprises, energy cost, labour cost, raw material cost, and utilities cost should be not excessed working capital as in Equation (27),

$$C_{TM} \leq B_{TM} \qquad (27)$$

CO2 emission: Ensures that the amount of CO2 cannot exceed the maximum technically feasible emission amount during manufacturing processes as in Equation (28),

$$C_E \leq \lambda_E \qquad (28)$$

Finally Type of material, Number of machines, number of labor and skill: should be between the minimum and maximum target values set by the authority or the organization as in equation (29), (30), (31), and (32) respectively:

$$RM_{min} \leq RM_t \leq RM_{max} \quad (29)$$

$$C_{min} \leq C_n \leq C_{max} \quad (30)$$

$$W_{min} \leq W_n \leq W_{max} \quad (31)$$

$$\gamma_{min} \leq \gamma \leq \gamma_{max} \quad (32)$$

4. Numerical Example Illustration

In this section, an example problem is tested to evaluate the performance of the proposed approach. The objective is to demonstrate how the proposed approach works in optimizing the product design and its manufacturing processes based on maximum sustainability. For this purpose, alternative configurations of the product are performed to obtain highest level of sustainability. The proposal considers a bottle opener design and its manufacturing processes (see table 1) in local small-sized manufacturing company. It is beneficial to reiterate that the fixed data are variables that are set as constant values during the numerical experimentation in implementing the modeling product design and manufacturing approach (see table 2)

	Alternative Design	Material	Manufacturing Processes	No. of stations
1	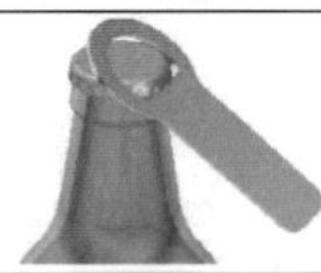	Steel	• Cutting • Punching • Quality control	3
2		Aluminium	• Cutting • Extrusion • Sawing • Remove burrs • Drilling • Quality control	6
4		Thermoplastics	• cutting • Punching • Insert molding • Quality control	4

Table 1. Bottle opener design and manufacturing alternatives

Parameter	Value	units	Parameter	Value	unit	Parameter	Value	unit
MH_D	7	hour	N_{RM}	500.5 100.5 60.5	kg	I	2	
$C_{L_{i,\gamma}}$	7 8 9 10	$	C_{UPi}	3.42 1.92 2.4	$	L S_w N_w S_d	12 15 2 15	
AU	21 36.4 176.4	kwh	D	0.15 0.05 0.05	$	RM_{max}	3	
P_E	0.2	$	E_{fi}	0.812045	(kg CO2-e/kWh)	RM_{min}	1	—
N_P	0.05 0.01 0.006	kg	C_v	9000	$	C_{max} C_{min}	4 2	—

C_{TOmang}	10 8 9	\$	CRF_{wb} U_c V_c	7 8 2	W_{max}	5	—
C_{THmang}	100 70 50	\$	F_c W_{mix}	1 1	W_{min}	2	—
d_{RM}	25 5 3	kg	P_i p_{ci}	5 5	γ_{max}	100%	—
D_K	500	part	C_{vwb}	2	γ_{min}	70%	—

Table 2. Input data

The input data are based on the following assumptions: number of hour's work is 8 hour/day; the production ability of machines is 500 parts each day; and scrap and recyclable percentages is constant product.

5. Results and Discussion

The proposed model has been solved using linear programming optimization (Microsoft excel 2010), and a global optimum solution has been obtained. Table 3 summarizes results. The optimum sustainability achieve for the manufacturing system under study is 0.8. The results call, the best type of material is 2 Aluminium, Number of worker is 2, Number of machine 2, and skill level of worker is 70%.

variable	value
T_{RM}	2 Aluminium
N_W	2
N_c	2
γ	70%

Table 3. Summary of the result

Performance measures values and individual desirability's				Sustainability index (CSI)
Environmental Impact		Economic Impact		
C_E	$d_q(C_E)$	C_{Tm}	$d_q(C_{Tm})$	D_a
0.0008535	1.00	1.001904	0.778	0.8

Table 4. Performance measure value and sustainability index

6. Conclusion

This proposed model introduced a formulation for a sustainable product problem in the manufacturing industry. The contribution of the proposed formulation is its ability to addresses pillars of sustainability at the planning level.The essential feature of the proposed model for measuring sustainability is the least amount of effort required in data collection.it mainly depends on data generally collected in all plants for cost analysis. This fact makes the model applicable in facilities introducing sustainability concepts.

It thus contributes to encouraging the implementation of sustainable practices in manufacturing, especially in emerging economies, where there is still a lack of sustainability awareness and related legislation. The inclusion of Microsoft excel (2010) to apply optimization in the form of alternative scenario for manufacturing product, make model suitable to small-scale MTO manufacturers because it can meet the required specifications according to the customer order. The result provides the least total cost and at the same time maintain environment by decreasing emission of carbon dioxide.

7. Acknowledgment

The authors acknowledge the time and effort of the unknown person referee whose useful comments have helped get better the quality of this manuscript.

8. References

[1] Joung, Che B.; John, C.; Prabir,S.; and Shaw C. (2013), Categorization of indicators for sustainable, Ecological Indicators, pp. 148-157.
[2] "World Commission on Environment and Development (WCED). Our Common Future, Oxford, Oxford, UK, New York, NY, USA.
[3] Kushwaha, S. (2015), A Comprehensive Sustainability Index (CSI):ABalanced Social, Economic & Enviromental approach, International Journal of Management and Social Sciences Research (IJMSSR), Vols. Volume 4, No. 5.
[4] Schmidt, W.-P.; Butt, F. (2006), Life Cycle Tools within Ford of Europe's Product Sustainability Index: Case Study Ford S-MAX & Ford Galaxy, International Journal of Life Cycle Assessment, vol. 11, pp. 315-322, No.5.
[5] Ungureanu, C. A.; Das, S.; & Jawahir, I. S.(2007), Development of a sustainability scoring method for manufactured automotive products: a case study of auto body panels, In the proceedings of ASME International Mechanical Engineering Congress and Exposition.
[6] De Silva, N.; Jawahir, I. S.; Dillon, O., & Russell, M. (2009), A New Comprehensive methodology fo rthe evaluation of product sustainability at the design and development stage of consumer electronic products, International Journal of Sustainable Manufacturing, pp. 251-264, 2009.
[7] Nomeda, D.; Rima, T. (2015), An Index to Measure Sustainability of a Business Project in the Construction Industry.
[8] Peihao L. ; Mingjia L. (2015), Statistical Methods Application in Comprehensive Sustainability Index and Its Application in Regional Sustainability Measurement, Information Technology, Computer and Electrical Engineering (ICITACEE), Indonesia.
[9] Zhang, X. (2012), Anew metric-based LCA method for assessing the sustainability performance of metallic automotive components.
[10] Vilaa, C.; Abellán-Nebota,J.; Albiñana J.; Hernández,G.(2015), An approach to Sustainable Product Lifecycle Management (Green PLM), In 25th DAAAM International of DAAAM Symposium on intelligent Manufacturing Engineering Society International Conference, p. 585 – 592.
[11] Ashrafi, N.(2014), A Review of Current Trend in Design for Sustainable Manufacturing, IOSR Journal of Mechanical and Civil Engineering (IOSR-JMCE), vol. 11.
[12] Faraj, E. D.,Romeo,M., Yousef, A.(2013), A novel optimizatiion model for simultanous cost-risk eduction multi-sppliers just-in-time systems, Journal of computer science.
[13] Goran, B., Nebojsa, J., Sasa, J.(2016), Calculating the costs of waste collection:A methodological proposal, Waste Management & Research (WM&R).

DOI: 10.2507/28th.daaam.proceedings.127

CHANGES OF PVC NANOCOMPOSITE PROPERTIES CONNECTED WITH CLAY CONTENT

Alena Kalendova*[1], Lenka Brazdilova (Sabadasova)[1],
Dagmar Merinska[1], Simona Mrkvickova[1]

[1] Tomas Bata University in Zlin, Faculty of Technology, Vavreckova 275,
760 01 Zlin, Czech Republic, email:kalendova@utb.cz

This Publication has to be referred as: Kalendova, A[lena]; Brazdilova Sabadasova, L[enka]; Merinska, D[agmar] & Mrkvickova, S[imona] (2017). Changes of PVC Nanocomposite Properties Connected with Clay Content, Proceedings of the 28th DAAAM International Symposium, pp.0921-0925, B. Katalinic (Ed.), Published by DAAAM International, ISBN 978-3-902734-11-2, ISSN 1726-9679, Vienna, Austria
DOI: 10.2507/28th.daaam.proceedings.127

Abstract

The *polyvinylchloride (PVC)* is an extensively used thermoplastic material, therefore the research were focused on this polymer. The aim of this work was to determinate the current development of nanocomposite materials based the polyvinylchloride and the influence of filler loading on their thermal and electrical properties. As nanofillers were used clays derived from montmorillonite (MMT): *MMTNa+* and *MMT30B*. *MMTNa+* presented sodium type of montmorillonite and *MMT30* is organically modified type of MMT. Both of the clays were intercalated or cointercalated with plasticizer *bis(2-ethylhexyl) phthalate (DEHP)***.** As the compounder the Buss KO-kneader was selected. This type of compounder is suitable for processing temperature-sensitive polymers. The properties as thermal and electrical conductivity were evaluated in comparison to the structure.

Keywords: PVC; nanocomposite; clay; conductivity

1. Introduction

Among all types of plastics, polyvinyl chloride (PVC) is one of the most important but also potential environmentally hazardous polymer materials for the content of chlorine [1] and large amount of processing additives. But it is still the world´s third-most widely produced synthetic polymer after polyethylene and polypropylene. The reason for its huge utilization derivate from this facts: low flammability, high chemical resistance, barrier properties, low cost, and formulating versatility. Next PVC is one of the typical insulating materials, and electric charges can be easily generated and accumulated on PVC surfaces, resulting in static-electric spark and even fire disasters [2]. In some circumstances such as chemical plants, gas stations, and coal mines, anti-static properties of PVC products are extremely important for ensurance of safety and reliability [3]. In that case, developing anti-static PVC materials has aroused much attention from academic and industrial fields for several decades.

Next, there has been a rapid development of solid-phase polymeric nanocomposites over the past 20 years. In particular, organically modified montmorillonites have been widely studied from fundamental and practical points of view in the area of organic–inorganic hybrid nanocomposites [4-10]. The structure of polymer–clay mineral nanocomposites mostly depends on the thermodynamic compatibility between the polymer, clay mineral and organic modifier.

Montmorillonite (MMT) is an aluminosilicate type of smectite clay mineral having a 2:1 expanding crystal lattice. Its crystalline structure is composed of an alumina octahedral sheet between two tetrahedral sheets of silica. Montmorillonite (MMT) s usually organically modified by ion-exchange reactions with primary, secondary, tertiary and quaternary alkylammonium or alkylphosphonium salts [11]. These types of MMT have, when exfoliated, a number of improved properties including mechanical strength, barrier properties, flammability resistance, electrical/electronic properties and polymer blend compatibilization with most polymer matrices [12,13].

The aim of this work was established the influence of filler content on the polymer composite properties especially thermal and electrical conductivity in correlation to the structure of material.

2. Material and methods

2.1. Materials

The polymer matrix was presented by the suspension type of *PVC (Neralit 652, K = 65).* Two types of montmorillonite, *MMT30B* and *MMTNa+,* from Southern Clay Products, were used as the nanofillers. *MMTNa+* represents a sodium montmorillonite (d-spacing = 1.2 nm) with no organic surface treatment. *MMT30B* contains methyl tallow bis-2-hydroxyethyl ammonium cation at a loading of 90 meq/100g, (d-spacing = 1.8 nm). The level of loading varied from 0.5, 1, 3, 5, 10, 15 and 20 %. Plasticizer *bis(2-ethylhexyl) phthalate (DEHP)* plays role of intercalant or cointercalant. As the compounder the Buss KO-kneader MKS 30 was selected. This single-screw compounder is suitable for processing temperature-sensitive polymers. The barrel diameter was 30mm and speed 50 rpm. The operating temperature changed from 130 °C at the entrance zone to 160 °C in the exit zone. After the pelletization of obtained strings the pellets were pressed into 1 mm sheets. The changes from 20-36 % in case of thermal conductivity and more than 400 % for el. conductivity were observed.

2.2. *Experimental methods*

Thermal conductivity is a property of the environment and depends generally on the temperature, pressure, composition of the system and also on the density of the substance being investigated. Its values greatly depend on the type of material and its state. For this study was employed non-stationary heat conduction measurement based on Fitche instrument [14]. This measurement has many advantages including, in particular, the short measurement time, the simplicity of the process and the measuring device, as well as the possibility of determining the important influence of the moisture content of the insulating material samples on their thermal conductivity. Measurements were performed from 45 °C to 25 °C. Software LabView Signal Express was used for data collection. Tested samples have dimension 40x40x1 mm.

The electric conductivity were measured using a two-point method with a Keithley 6517A programmable electrometer, which at the same time served as a source of the DC voltage. The range of the measured voltage was 0 - 100 V. Based on these measurements, it was calculated the specific electrical conductivity according to the equation (1):

$$\sigma = \frac{I}{U} * \frac{l}{S} \tag{1}$$

where U-voltage (V), I-current (A), S-electrode area (cm^2), l-sample thickness (cm), σ-electric. conductivity (S/cm^{-1}).

Further the structure was monitored using X-ray diffraction (PAN analytical X'Pert PRO with a cathode CuKα in a reflective arrangement from angles 2Θ: 1-30 ° and SEM (SEM Vega Tescan). Samples were placed in liquid nitrogen and afterwards freeze-fractured.

3. Results

With the low-cost polymer as a matrix material, the composite can often be formed into products with higher thermal conductivity than the original polymer. In this study, the thermal conductivity was measured by a non-stationary method on a type of Fitche device. The average values of thermal conductivity of prepared samples are given in Table 1,2 and Fig. 1. The thermal conductivity is growing with filler loading for both filler. The bigger influence was recorded for *MMTNa+*. The increase 24% was observed for loading 3 and 5 % *MMTNa+*. The improvement 36 % show sample with 20 % of natural MMT. The mixture *MMT30B* show lower conductivity growth comparing *MMTNa+*. The increase was 9 and 19 % for mixture *PVC/3%MMT30B* and *PVC/5%MMT30*. The highest change of 25 % was recorded for *PVC/20%MMT30B*. The lower changes by lower filler loading may be connected with the morphology of sample. The mixtures with loading to 5% show very fine structure. The filler is well dispersed in the matrix with the particles in nm range. In case of higher loading the morphology study show the mixture of nanoparticles and microparticles as can be seen in Fig. 2.

Next the DC electrical conductivity was measured. The values, obtained from the measurements, are shown in the Table 3, 4. There can be no free charge carriers in pure PVC, therefore we consider the polyvinyl chloride matrix as an insulator. The significant increase was observed in case *MMT/15%30B* and *MMT/20%30B* (more than 400 %). The *MMTNa+* cause the drop of electrical conductivity comparing pure PVC. This facts are in good correlation with the morphology of tested samples. *MMT30B* is modified clay and this modifier may create the charge carrier. Next the *MMT30B* created, due to the good dispersion of filler in matrix a "filler network, which can be helped to lead the el. charge. In case of *MMTNa+* the microparticles are separated (isolated) in polymer matrix, which is dielectric.

Sample	**λ ($W.m^{-1}.K^{-1}$)**	**Δλ ($W.m^{-1}.K^{-1}$)**
Pure PVC	0.127	0.002
PVC/0.5 MMT Na+	0.129	0.001
PVC/1 MMT Na+	0.135	0.001
PVC/3 MMT Na+	0.157	0.001
PVC/5 MMT Na+	0.158	0.001
PVC/10 MMT Na+	0.168	0.002
PVC/15 MMT Na+	0.172	0.002
PVC/20 MMT Na+	0.163	0.002

Table 1 Thermal conductivity *MMTNa+*

Sample	**λ ($W.m^{-1}.K^{-1}$)**	**Δλ ($W.m^{-1}.K^{-1}$)**
Pure PVC	0.127	0.002
PVC/0.5 MMT30B	0.128	0.001
PVC/1 MMT30B	0.136	0.001
PVC/3 MMT30B	0.139	0.002
PVC/5 MMT30B	0.151	0.001
PVC/10 MMT30B	0.152	0.001
PVC/15 MMT30B	0.158	0.001
PVC/20 MMT30B	0.159	0.001

Table 2 Thermal conductivity *MMT30B*

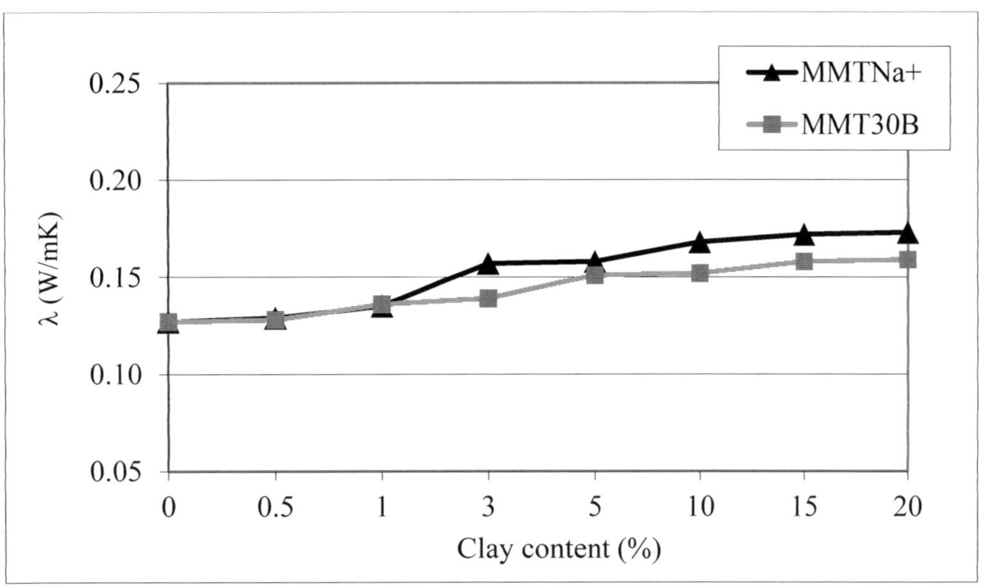

Fig.1. Thermal conductivity

Sample	σ (S/cm^{-1}*10^{-13}) for U=80V	$\Delta\sigma$ (S/cm^{-1}*10^{-14}) for U=80V
Pure PVC	2.48	2.01
PVC/0.5 MMT Na+	3.47	2.48
PVC/1 MMT Na+	2.01	1.00
PVC/3 MMT Na+	1.80	1.27
PVC/5 MMT Na+	2.24	1.37
PVC/10 MMT Na+	2.52	1.32
PVC/15 MMT Na+	2.15	1.99
PVC/20 MMT Na+	1.85	1.15

Table 3 DC Electrical conductivity for samples with *MMTNa+*

Sample	σ (S/cm^{-1}*10^{-13}) for U=80V	$\Delta\sigma$ (S/cm^{-1}*10^{-14}) for U=80V
Pure PVC	2.48	2.01
PVC/0.5 MMT30B	5.05	2.30
PVC/1 MMT30B	6.38	1.94
PVC/3 MMT30B	3.75	1.83
PVC/5 MMT30B	5.25	0.51
PVC/10 MMT30B	6.83	0.45
PVC/15 MMT30B	12.17	2.27
PVC/20 MMT30B	17.90	1.98

Table 4 DC Electrical conductivity for samples with *MMT30B*

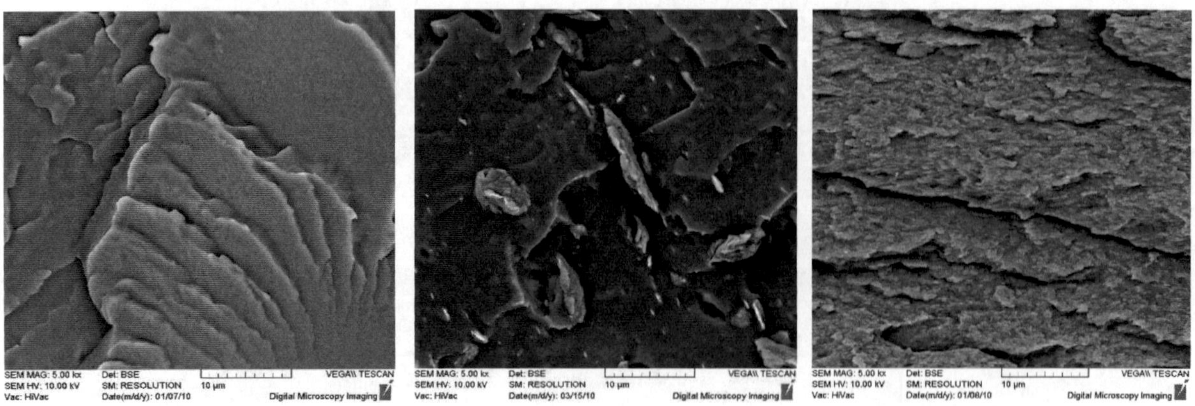

Fig. 2. SEM: *Pure PVC, PVC/10% MMTNa+, PVC/10% MMT30B* (from left)

4. Conclusion

The polyvinylchloride (PVC) is an extensively used thermoplastic material, therefore the research were focused on this polymer. Next developing anti-static PVC materials has aroused much attention from academic and industrial fields. The aim of this work was to determinate the influence of filler loading to the thermal and electrical conductivity in correlation to the structure of prepared mixtures. In accordance with the obtained results, the thermal conductivity is growing with filler loading for both filler.

The bigger influence was recorded for *MMTNa+*, where improvement 36 % for 20 % loading was observed. While, it was only 25 % for 20% loading of *MMT30B*. This fact may be connected with the morphology of the sample. The filler *MMT30B* seems to be well dispersed in the matrix and it is restricted the polymer chain motion. This may create a barrier for thermal conduction.

In case of DC el. conductivity, the significant increase was observe for *MMT/15%30B* and *MMT/20%30B* (more than 400 %). This fact is probably connected with the filler modification which may create the charge carriers in the filler surface and the "filler network" in polymer matrix. The future plan is testing other types of intercalated or cointercalated fillers especially for el. conductivity.

5. Acknowledgments

This project was supported by internal grant of TBU in Zlin RVOE. Next I wish to thanks Mr. Antonin Blaha for the help with the thermal conductivity measurement.

6. References

[1] Han Wang; Guiyuan Xie; Minghe Fang; Zhe Ying; Yu Tong; You Zeng, (2015). Electrical and mechanical properties of antistatic PVC films containing multi-layer graphene. Composites Part B. 79. 444-450.

[2] Saad ALG; Sayed WM; Ahmed MGM; Hassan AM. (1999). Preparation and properties of some filled poly(vinyl chloride) compositions. J Appl Polym Sci. 73(13). 2657-70.

[3] Moulay S. (2010). Chemical modification of poly(vinyl chloride)- still on the run. ProgPolym Sci. 35(3). 303-31.

[4] Giannelis E.P. (1996). Polymer layered silicate nanocomposites. Adv. Mater. 8. 29–35.

[5] Alexandre M.; Dubois P. (2000). Polymer-layered silicate nanocomposites: preparation, properties and uses of a new class of materials. Mater. Sci. Eng. R. Rep. 28. 1–63.

[6] Fornes T.D.; Yoon P.J.; Hunter D.L.; Keskkula H.; Paul D.R. (2002). Effect of organoclay structure on nylon 6 nanocomposite morphology and properties. Polymer. 43. 5915–5933.

[7] Stretz H.A.; Paul D.R.; Li R.; Keskkula H.; Cassidy P.E. (2005). Intercalation and exfoliation relationships in melt-processed poly(styrene-co-acrylonitrile)/montmorillonite nanocomposites. Polymer. 46. 2621–2637.

[8] Peprnicek T.; Kalendova A.; Pavlova E.; Simonik J.; Duchet J.; Gerard J.F. (2006). Poly(vinyl chloride)-paste/clay nanocomposites: investigation of thermal and morphological characteristics. Polym. Degrad. Stab. 91. 3322–3329.

[9] Shah R.K.; Kim D.H.; Paul D.R. (2007). Morphology and properties of nanocomposites formed from ethylene/methacrylic acid copolymers and organoclays. Polymer. 48. 1047–1057.

[10] Merinska D.; Kubisova H.; Kalendova A.; Svoboda P.; Hromadkova J. (2012). Processing and properties of polyethylene/montmorillonite nanocomposites. Journal of Thermplastic Composite Materials. 25(1). 115-131.

[11] McAtee J.J. (1956). Heterogeneity in montmorillonite. Clay Minerals. 5. 279–288.

[12] Paul D.R.; Robeson L.M. (2008). Polymer nanotechnology: nanocomposites. Polymer. 49. 3187–3204.

[13] Petersen H.; Jakubowicz I.; Enebro J.; Yarahmadi J. (2015) Organic modification of montmorillonite for application in plasticized PVC nanocomposites. Applied Clay Science. 107. 78–84

[14] Blaha A.; Pavlica. R. Thermal conductivity measurement by non-stationary method. Mereni tepelne vodivosti nestacionarni metodou. (2001). CHISA.15. - 18. října 2001. Srni. Sumava. Czech Republic. Czech Society of Chemical Engineering . ISBN 80-86059-32-4.

DOI: 10.2507/28th.daaam.proceedings.128

METHODS OF EVALUATING NOISE POLLUTION IN THE MEGALOPOLIS

Andrei Ermolaev

This Publication has to be referred as: Ermolaev, A[ndrei] (2017). Methods Of Evaluating Noise Pollution In The Megalopolis, Proceedings of the 28th DAAAM International Symposium, pp.0926-0929, B. Katalinic (Ed.), Published by DAAAM International, ISBN 978-3-902734-11-2, ISSN 1726-9679, Vienna, Austria
DOI: 10.2507/28th.daaam.proceedings.128

Abstract

The purpose of the research is to build an effective system for assessing noise pollution during sports events in residential areas of the megalopolis. A combination of methods is proposed - measuring the level of noise pollution using a sound meter and analyzing the complaints of local residents. The influence of noise on the humans is estimated. Investigation of noise pollution near two sports grounds in residential areas of Moscow has been carried out. Methods for reducing noise are proposed. Conclusions about the effectiveness of the proposed system of assessments are made and ways of its improvement are proposed.

Keywords: environment; urbanization; noise pollution; assessment system

1. Introduction

The state of environment in the modern world is directly related to the degree of development of social consciousness as well as to the understanding of relationship between the nature and a human being.

Ecological monitoring is the system designed for observing, gathering and storing of information related to the state of environment, well-targeted environmental assessment of such information, forecasting the state of environment taken in conjunction with natural and anthropogenic factors. The purpose of ecological monitoring can be described as obtaining reliable information concerning the state of environment, data processing, forecasting changes in the state of environment, and assessment of mutual influence between the environment and a man [1]. The most observable (whether in a negative or in a positive way) mutual influence can be assessed using the example of living in megalopolis, where large gatherings of people structure their interactions with the environment.

Moscow is the largest city of Russia (as of 2017, it has a population of more than 12 300 000 citizens), the center of the economic life [Federal State Statistics Service – www.moscow.gks.ru]. Moscow is the center of the ringed structure of transport routes in the European part of Russia, which provides for additional increase of traffic in the city. These factors always result in the increased noise pollution level in Moscow. The additional factor contributing to the noise in residential areas is the holding of sports events.

The topicality of the research lies in the fact that noise pollution problem is oftentimes ignored, although it has an influence on the second most significant sensory receptor of humans – the sense of hearing.
The purpose of this study is to build an effective system for evaluation of the influence on noise pollution caused by placing sports fields in the city's residential districts. In this study, the following methods are discussed:

1. Measuring noise pollution level;
2. Analyzing the complaints of local residents posted on municipal information portals.

2. Effect of noise pollution on humans

Noise pollution is the excess of natural background noise level or abnormal change of sound characteristics: rate of recurrence, sound volume, etc. Noise pollution is among the three most significant ecological disturbances in the world. Along with the growth of urbanization, noise became a permanent part of human life and one of the major parametric pollutants of the urban environment. The problem of protection of population against the excessive noise is primarily a healthcare issue [2].

Today, it is acknowledged that noise is a common biological irritant, i.e. it has an influence not only on the organs of hearing, but on the entire organism as well. First of all, noise affects cerebral structures, which leads to adverse changes in the functioning of various organs and systems. The central nervous system is the most sensitive to the impact of such noise levels. Its first reaction to the impact of noise is the change of its functional status indicators. People affected by noise most typically complain of headaches, which may vary in terms of amount and distribution, dizziness after a postural change, hypomnesia, undue fatigability, somnolence, sleep disturbances, emotional instability, loss of appetite, hyperhidrosis, chest pain [2][7].

Noise is one of the strongest factors of stress. Noise has an impact on the functions of endocrine and immune systems of the body, which may manifest itself, in particular, as three main biological effects [3]:

- Decrease in immunity to infections;
- Decrease in immunity to the development of neoplastic processes;
- Presence of a conducive environment for the emergence and development of allergic and autoimmune processes.

The impact of noise on the organism is quite complex and should be avoided. This is the reason why measures for improvement of personal environment with the purpose of protecting it against the noise factor are essential for the modern society of technology and urbanization.

3. Methods of studying noise pollution in residential areas of Moscow

According to Russian standards, the following noise requirements are set for residential premises:
— maximum sound level during the daytime (from 7 AM to 11 PM) – no more than 55 dB;
— maximum sound level during the nighttime (from 23 PM to 7 AM) – no more than 40 dB.

Therefore, the level of noise in residential premises must not exceed the volume level of a calm conversation [4].

3.1 Method of measurement of noise pollution level

In September 2017, measurements of the noise level were performed in the districts of Moscow. The measurements were performed on 17th and 18th of September in the area of Geroev–Panfilovtsev str. 1-5 (Point **A**):

Fig. 1. Point A

and on 24th and 25th of September in the area of Shirokaya str. 9 (Point **B**):

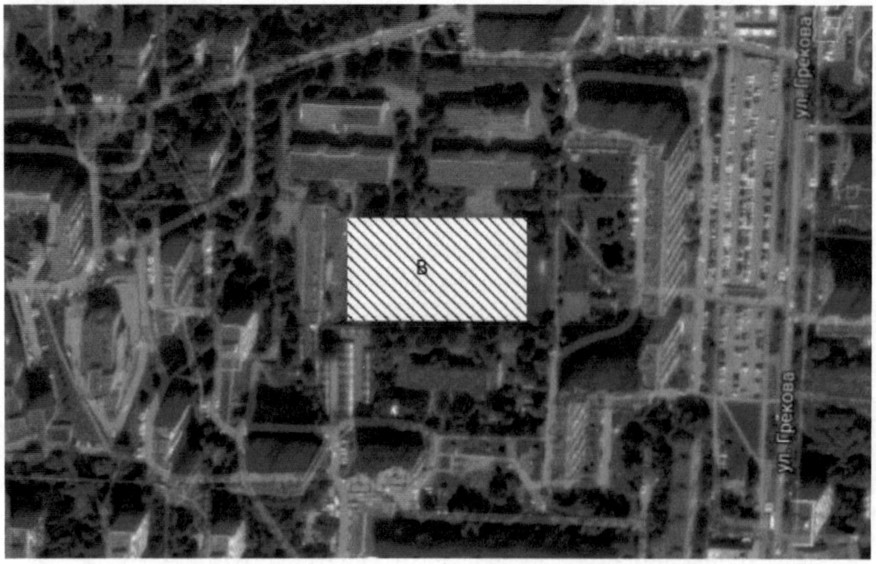

Fig. 2. Point B

In point A, the distance from the sports field to the nearest residential building is 18 meters, in point B – 22 meters. The measurements were taken from 3 to 11 PM under normal conditions and during sports events. "Digital Sound Meter AR854" device was used. The measurement results are given in Table 1 and Figure 1:

Point	Maximum permitted value, dB	Maximum values when measuring under normal conditions, dB	The value of violation of sanitary standards under normal conditions, dB	Maximum values when measuring during sports events, dB	The value of violation of sanitary standards during sports events, dB
A	**55**	74	19	87	32
B		78	23	90	35

Table 1. Measurement results.

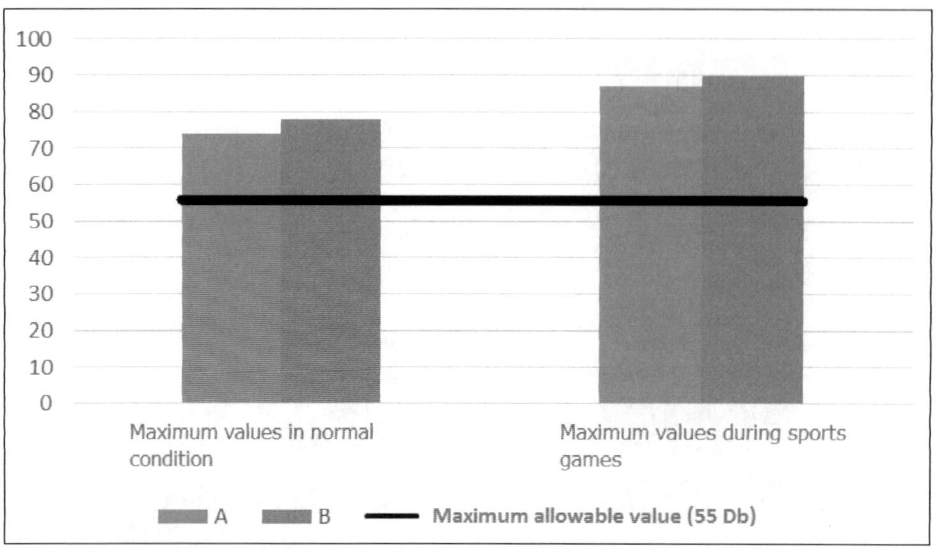

Fig. 3. Correlation of noise level measurements in points A and B.

3.2 Method of analyzing complaints of local residents

This method involved the analysis of municipal public information portal "Active Citizen", the purposes of which include submission of complaints by, and control over administration of complaints given by, the citizens. Search of complaints submitted with the indication of a residential address adjacent to the sports fields in point A or B has shown low efficiency of this method. Thus, there is only one complaint per 700 citizens who live in the houses located near points A and B [5]. Among the factors reducing the efficiency of this method, we can mention that the user must have a certain level of computer skills in order to register and submit a complaint. We must note that citizens don't have no way of objection evaluation of the degree of impact of noise pollution on their health, and taking account of the passivity in solving secondary problems which is peculiar to the most of people, they usually don't give due consideration to it.

4. Methods of noise level reduction

Currently, most of sports fields are built without the use of sound insulation materials, which leads to high values of impact noise during sports events (from ball slams, etc.). Thus, the level of impact noise from the ball slamming against the banking board of the sports field in points A and B at the 20 meters distance is equal to 82 dB. This value significantly exceeds the permitted value, so it is necessary to use sound insulation materials and structures. Modern industry offers a variety of options. For example, cork rolls are able to reduce the impact noise level by 12 dB [6]. Multilayered structures are the most efficient: using combinations of various materials, it is possible to reduce the impact noise level during sports events to the permitted values. For protection of the upper floors, it is reasonable to use noise screens.

The alternative way is the use of vegetation plantations around the sports fields as a natural noise-absorbing material. In addition, this would allow improving the ecological and aesthetical aspects of the residential area.

The other method of reducing noise pollution is to place the sports fields at such a distance from residential buildings that the sound wave would disperse before reaching them. Oftentimes, one can observe violations of design and planning standards for placement of sports fields against residential buildings, which lead to the increase of noise pollution.

5. Conclusion

As a result of the study performed, the following conclusion should be made about the effectiveness of the proposed system for assessing noise pollution during sporting events in the residential neighborhoods of the megalopolis. The method of estimation by a sound meter proved to be effective, but requires a systematic approach and regular implementation with the maintenance of a citywide measurement base, that allows estimating changes in indicators over a long period. The assessment method of analyzing complaints of local residents proved ineffective due to the poor awareness of the residents about the negative impact of the increased noise level, which indicates that the problem is undervalued. The main problem is the lack of public objective data and discussions in the society.

To improve the effectiveness of the evaluation system I propose to supplement it by statistical information from health organizations. Thus, it will be possible to monitor the effect of noise pollution on local residents without regular measurements of the level of noise pollution. According to the statistics of residents' complaints, which are a consequence of the influence of the increased noise level, it is possible to identify areas with increased noise level and to carry out the necessary work to eliminate noise pollution by constructing sound insulation structures or reconstructing objects that are sources of noise.

6. References

[1] A.V. Shamraev (2014). Environmental monitoring and assessment: teaching medium, Federal State Budgetary Institution of Higher Education "Orenburg State University". - Orenburg: OSU, 2014, ISBN 978-5-905383-37-3

[2] V.V. Minaeva; A.V. Gaponenko (2015). "The impact of noise on a human organism", International students' science bulletin, 2015 – No. 3 (part 1), pp. 56-58, ISSN 2409-529X

[3] S.V. Alekseev, Yu.P. Pivovarov, O.I. Yanushanets. (2002) Human ecology: Textbook. – M.: Ikar,. ISBN 5-7974-0054-5

[4] GOST 12.1.003-2014. Occupational safety standards system "Noise. General safety requirements", SN 2.2.4/2.1.8.562-96 "Noise at Workplaces, in Residential and Public Spaces, and in Areas of Residential Development"

[5] Portal "Active Citizen", https://ag.mos.ru/info

[6] Gruppa7 website, http://gruppa7.ru/articles/27/

[7] Hoffmann, Barbara; Moebus, Susanne; Stang, Andreas; Beck, Eva-Maria; Dragano, Nico; Möhlenkamp, Stephan; Schmermund, Axel; Memmesheimer, Michael; Mann, Klaus (2006-11-01). "Residence close to high traffic and prevalence of coronary heart disease". European Heart Journal. 27 (22): 2696–2702. ISSN 0195-668X

DOI: 10.2507/28th.daaam.proceedings.129

INFLUENCE OF GRINDING CONDITIONS ON SURFACE QUALITY AND ACCURACY WHEN A TOOL GRINDER IS USED

Jindřich Farský, Miroslav Zetek, Tomáš Bakša, Pavel Adámek

This Publication has to be referred as: Farsky, J[indrich]; Zetek, M[iroslav]; Baksa, T[omas] & Adamek, P[avel] (2017). Influence of Grinding Conditions on Surface Quality and Accuracy when a Tool Grinder is Used, Proceedings of the 28th DAAAM International Symposium, pp.0930-0936, B. Katalinic (Ed.), Published by DAAAM International, ISBN 978-3-902734-11-2, ISSN 1726-9679, Vienna, Austria
DOI: 10.2507/28th.daaam.proceedings.129

Abstract

This work deals with the investigation of the influence of grinding conditions on the surface quality and accuracy of dimensions when grinding complex shaped surfaces on an ANCA MX7 5-axis tool grinder. The upper part of a knee implant was selected as a representative example of a complex shaped surface. NC programs in NX CAM software with a postprocessor from ANCA were created for this work. The aim of this work was to design and implement an experiment to study the influence of cutting conditions during grinding. To carry out this experiment it was necessary to design the technology for grinding the knee implant, create an NC program for grinding, and construct a clamp for the knee implant.

Keywords: Grinding; 5-axis tool grinder; surface quality; accuracy; grinding conditions; NX

1. Introduction

It is necessary to achieve accurate dimensions and high surface quality on components with complex shapes. Therefore attention must be paid to grinding complex shaped surfaces on multi axis-grinding machines. One of these grinders is the ANCA MX7 tool-grinding machine, on which it is possible to grind complex shaped surfaces. Because tool-grinding machines are mainly designed for grinding rotary cutting tools or general rotary components, the use of these grinders has many pitfalls for grinding components which have complex shapes.

One of the problems is creating NC data using CAM software with a special postprocessor. Other issues are designing and creating special clamping fixtures for these complex shaped surfaces, choosing the right grinding strategy, selecting the right cutting conditions, etc.

As already mentioned, accurate dimensions and high surface quality of components with complex shaped surfaces are required. [1] These two conditions are most affected by the selection of the correct grinding conditions. By changing one of these parameters we can achieve different surface quality and different dimensions, because changing the grinding conditions influences both the surface quality and the temperature at the cutting point. These influences are shown in Fig. 1., 2., 3., 4.

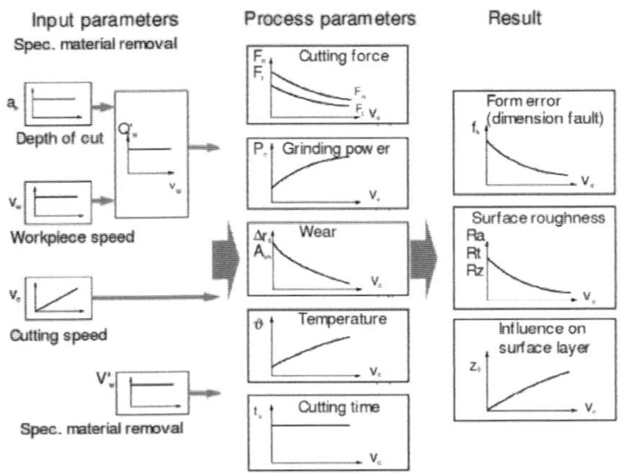

Fig. 1. The influence of cutting speed on process parameters and output [2]

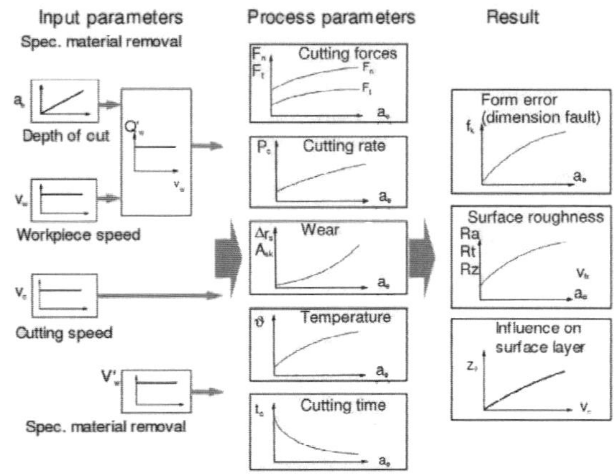

Fig. 2. The influence on feed rate and depth of cut on process parameters and output [2]

The aim of this work is to investigate the influence of grinding conditions on the surface quality and accuracy of dimensions when grinding complex shaped surfaces on the upper part of a knee implant which is not intended for medical purposes. Surface quality in this work means surface roughness with parameters Ra, Rq and Rz, and accuracy dimensions are derived from the amount of deviation away from the surface of the component. To minimize the errors in the experiment caused by wear of the grinding wheel, the wear of the grinding wheel was measured after each grinding and the diameter correction was entered in the grinder control software. [5], [6]

2. The selected component

The component selected for this experiment was the upper part of a knee implant, shown in Fig. 3. The semi-finished product for this experiment was printed on a 3D printer using martensitic steel, which is designated by the producer as EOS MaragingSteel MS1 to DIN classification X3NiCoMoTi 18-9-5. This material is used in cutting tools and high performance industrial and engineering parts. It is characterized by a hardness of around 35HRC in its natural state and after age hardening its hardness rises to 50 - 56 HRC. [7] [8]

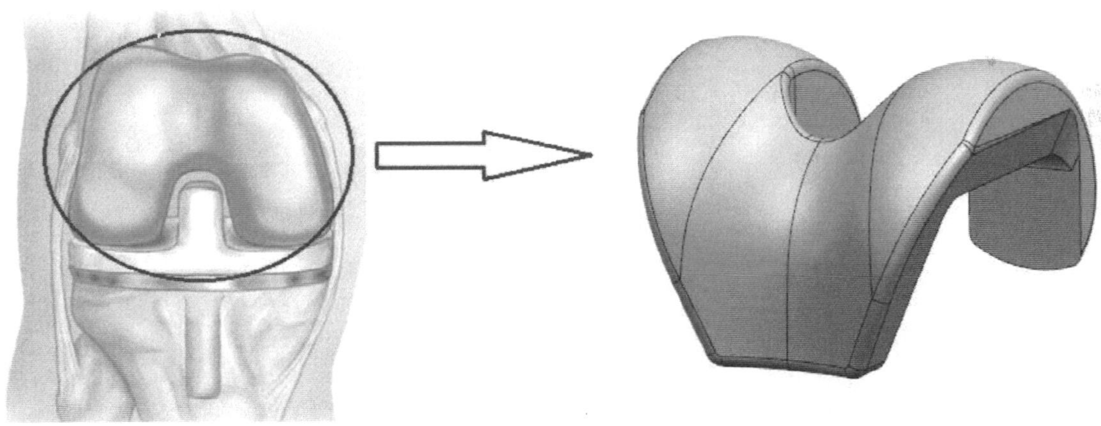

Fig. 3. Knee implant [8]

3. Introducing the experiment

As has already been mentioned, the aim of this work is to study the influence of grinding conditions on surface quality and the accuracy of dimensions. We needed to limit the potentially very large scope of the experiment. These limitations were in terms of the equipment and the selected input parameters. The tool grinding machine ANCA MX7 is limited by two things. The first limit is the maximum revolutions on the spindle, which is 10 000 rpm. The second limit of the machine is the spindle clamping system, which only allows the clamping of cylindrical components. The grinding wheel 1FF1 CBN was the next limit, due to its maximum grinding speed of 63 $m.s^{-1}$, and its composition, which is marked as B64C75B54. The limits from the input parameters are based on research from other articles that show which grinding parameters most influence the quality and dimensions of the ground surfaces. In order to correctly design the experiment, it is necessary to select those grinding parameters which most influence the evaluated parameters. Three grinding parameters are selected: grinding speed, axial feed and feed rate. Another variable parameter is the tilt of the tool spindle by 30° degrees to test whether tilting has an effect on the parameters being evaluated. The axial feed in this case is set by the maximum scallop height, because CAM software NX only allows set up of this parameter, from which can be calculated the approximate rate of the axial feed. The equation for calculating axial feed from maximum scallop height is

$$sc = R - \sqrt{R^2 - (\frac{a_p}{2})^2} \tag{1}$$

$$a_p = 2 \cdot \sqrt{R^2 - (R - sc)^2} \tag{2}$$

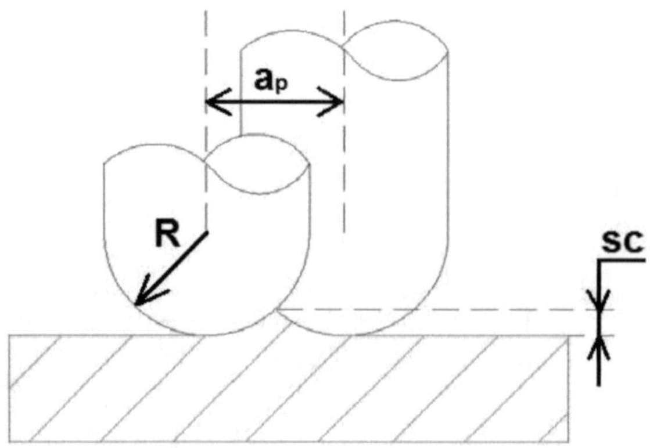

Fig. 4. Diagram of maximum scallop height [9]

The cutting speed is limited by the maximum spindle speed, which is 10 000 rpm, and the maximum cutting speed of the grinding wheel, which is 63 m/s. The value of the feed speed was selected from the lower values, because this is the first grinding of this type on the ANCA MX7 grinder using NX CAM software with the new postprocessor. Parameters for the experiments are given in the following tables.

Table 1 shows the values for preparation of the component for the main experiment. The preparation grinding with parameter values HR00 was performed several times until the knee implant surface was ready for the experiment. Values in the last operation in the pre-grinding were the same as the first value in the experiment so as not to affect the experiment.

Table 2 shows the values for individual experiments. Parameter "*sc*" is the maximum scallop height and the next column gives the approximate values of axial feed which are calculated from the formula.

	sc [mm]	*a_p [mm]*	*v_c [m.s^{-1}]*	*v_f [mm.min^{-1}]*	*a_e [mm]*	*Distance [mm]*
HR00	0.1	1.536	20	50	0.1	From 0 to -1
HR01	0.005	0.346	30	150	0.05	-1.05

Table 1. Preparatory grinding

	sc [mm]	*a_p [mm]*	*v_c [m.s^{-1}]*	*v_f [mm.min^{-1}]*	*a_e [mm]*	*Distance [mm]*
EXP01	0.005	0.346	30	150	0.05	-1.10
EXP02	0.005	0.346	40	150	0.05	-1.15
EXP03	0.005	0.346	30	200	0.05	-1.20
EXP04	0.005	0.346	40	200	0.05	-1.25
EXP05	0.001	0.155	30	150	0.05	-1.30
EXP06	0.001	0.155	40	150	0.05	-1.35
EXP07	0.001	0.155	30	200	0.05	-1.40
EXP08	0.001	0.155	40	200	0.05	-1.45

Table 2. The main experiments

4. Preparation of experiment

Before the experiment it is necessary to design the grinding technology so that the experiment is not affected by errors which could affect the results. The following section gives some basic information about the design of the grinding technology, construction of the clamping jig, creation of NC data in the NX CAM software and checking the NC data on the grinder which was generated from the NX software.

4.1. Design of the grinding technology

Designing the grinding technology includes the definition of the surface which is to be ground, the grinding strategy, definition of the grinding wheel, and basic kinematics of the grinding machine. This design is very important, because errors may be introduced in to the experiment which affect the evaluated parameters. First of all it is necessary to define which surfaces are to be ground. Fig. 5. highlights the surfaces to be ground in orange. The second step is to choose a grinding strategy. In this case a "zig-zag" strategy is selected, where the grinding wheel moves backwards and forwards, and moves the axial feed value after each step. A radius-grinding wheel (Fig. 6) was selected to grind the surface, which has a basic designation of 1FF1 100 x 6 x 20 with a maximum cutting speed is 63 m/s. The cutting speed is limited by the kinematics of the grinding machine, in particular the maximum revolutions per minute on the spindle of the machine, of 10 000 rpm. The basic kinematics of the ANCA MX7 tool grinder is shown in Fig. 7.

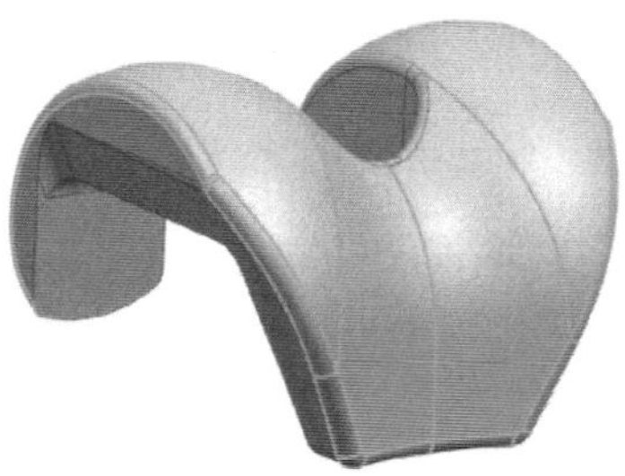

Fig. 5. Highlighted surface to be ground

Fig. 6. The grinding wheel

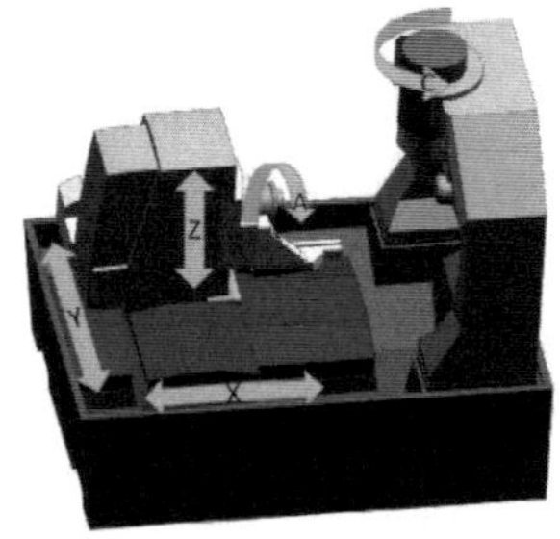

Fig. 7. Kinematics of ANCA MX7

4.2. The clamp

In order to grind the knee implant, it was necessary to design and manufacture a clamp, as the clamping collet on the machine only allows cylindrical components to be clamped. The clamp (Fig. 8) is divided into three main sections, the clamping part for the collet, the clamping part for the knee-implant and the reference part. There is a flat section on the reference part which is for setting the zero position axis A on the grinder and to ensure the repeatability of clamping in the same position both on the grinder and on the measuring machine. This surface is also parallel to the clamping surface for the knee-implant, which is in the middle part of the clamp.

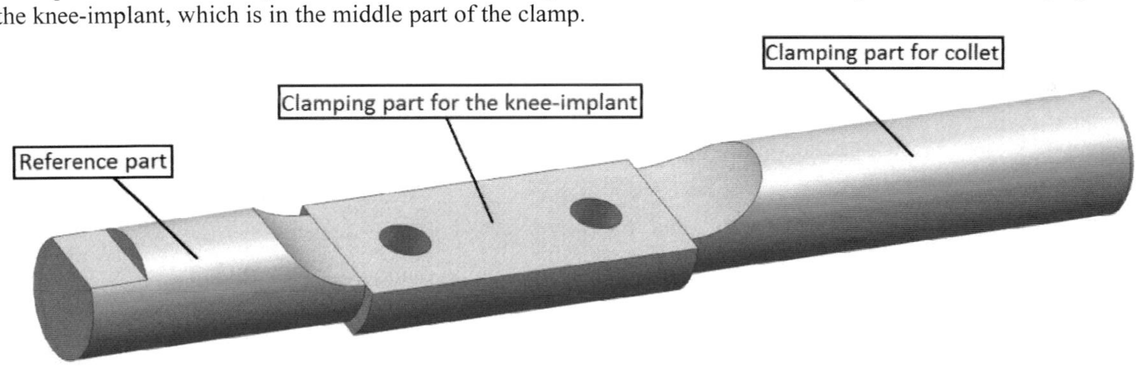

Fig. 8. Clamp for the knee implant

4.3. Creating NC data in NX software

To carry out the experiment it was necessary to create the NC data in NX software. This software also provides collision checking between the machine, the grinding wheel, the component and the clamp. This collision checking discovered two collisions between the clamp and the spindle. As mentioned earlier, one variable is the tilting of the tool spindle by 30° degrees. This tilt is located in the back half of the knee-implant. Therefore the grinding technology is divided into two parts.

The first part is grinding without tilting the spindle and is located on the front half of the knee-implant (Fig. 9). The second part, which is located on the back half the knee-implant, is performed with the spindle tilted by 30° degrees (Fig. 10). This spindle tilt prevents collision between the spindle and the clamp, as shown in Fig. 11. After creating the grinding technology in NX software, it is necessary to generate the NC data in the special postprocessor which was designed for the ANCA MX7 tool grinder.

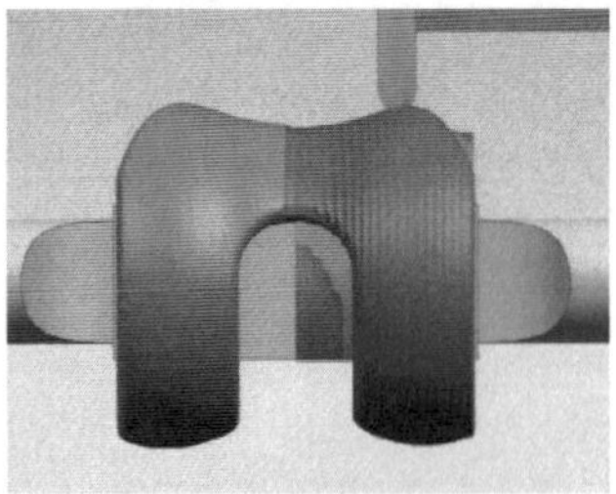

Fig. 9. Grinding without tilting the spindle

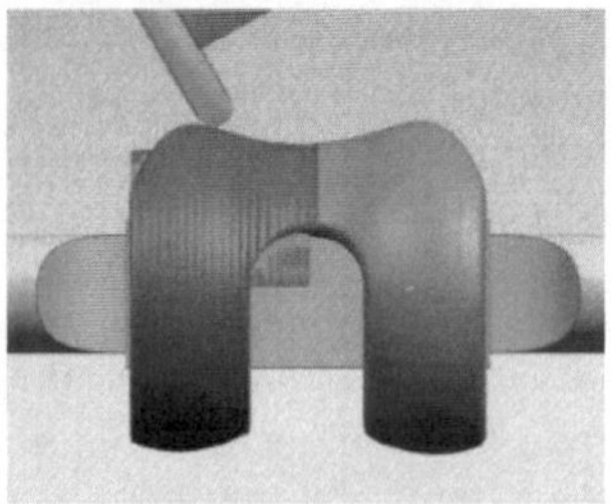

Fig. 10. Grinding with spindle tilted by 30° degrees

Fig. 11. Collision between spindle and clamp

5. Rating experimental results

As mentioned in the introduction, this work deals with the influence of the cutting conditions on the surface quality and accuracy of dimensions. Parameters of surface roughness Ra, Rq and Rz are selected for evaluation of surface quality. The accuracy of dimensions after grinding were evaluated according to the size of the deviations from the desired dimensions. The size of the wear of the grinding wheel was also monitored, which was subsequently corrected in the control software on the grinder.

5.1. Grinding wheel wear

To minimize the errors in the experiment caused by wear of the grinding wheel, the wear of grinding wheel was measured after each grinding, because grinding releases worn abrasive grains from the binder of the grinding wheel. Because a new grinding wheel was used, this wear is considerable at the start, because the abrasive grains are not so tightly bound to the binder in the surface layer. The wear process is shown in Fig. 12, and on the left the high wear of the grinding wheel can be seen, which was recorded during the roughing process. This wear was corrected using the average correction of the grinding wheel in the control software. In the experiment, the wear was minimal and reached values of 0.01 mm per diameter. The wear of the grinding wheel during the experiment is shown on the right side of the graph.

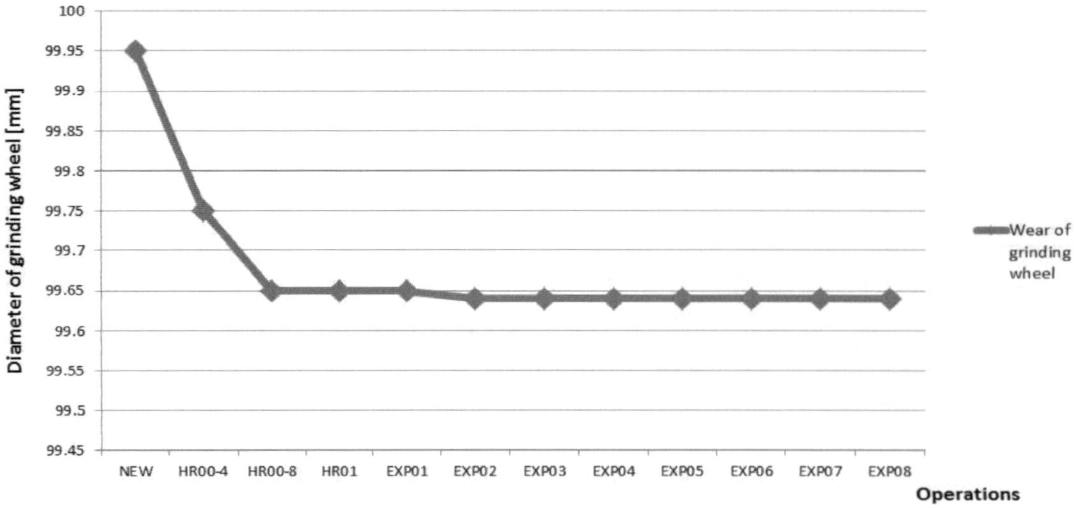

Fig. 12. Wear of the grinding wheel

5.2. Surface roughness

Roughness was measured as it is an excellent indicator of surface quality. The change in the surface roughness is evaluated in relation to the change in the cutting conditions and on the change of the tilt of the spindle. For each experiment at least ten surface roughness measurements were made on eight sections of the component. The average value of the surface roughness for each experiment was calculated from the measured values. These values are compared with each other depending on the input parameters. The greatest influence on the surface roughness was the axial feed value, which is here called the maximum scallop height "sc". The second value which influences the surface roughness is the cutting speed. The third value is the feed rate. These values are shown in Fig. 13. The tilt of the spindle also influences the surface roughness, but only up to a certain value of maximum scallop height. As shown in Fig. 14, the difference of values for sc = 0.005 is 0.35 μm, but for sc = 0.001 it is 0.025 μm.

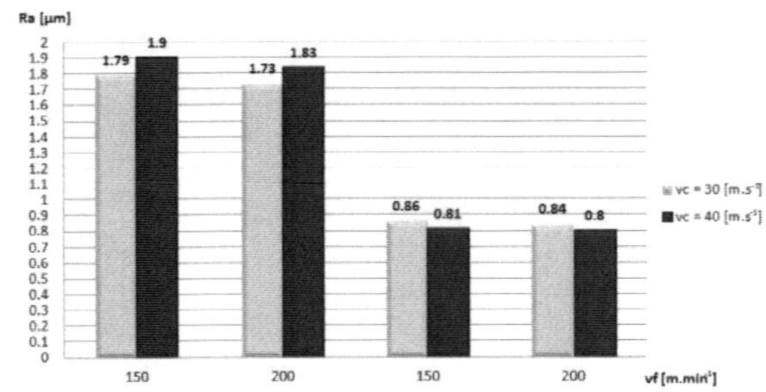

Fig. 13. Surface roughness in relation to cutting conditions

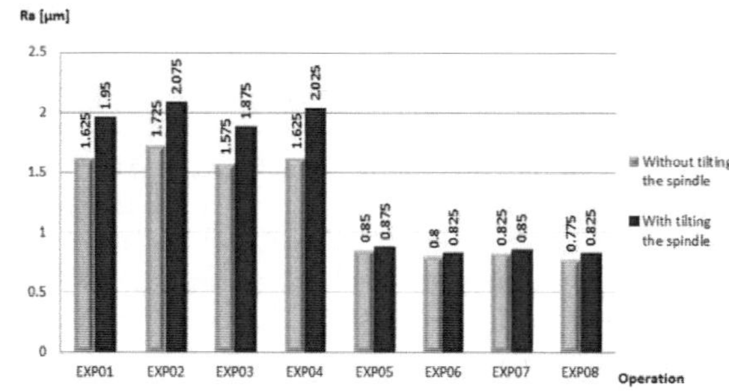

Fig. 14. Surface roughness in relation to spindle tilt

5.3. Accuracy of dimensions

The next parameter which determines the accuracy of the dimensions is the value of the deviation from the desired dimension. This deviation was measured on a coordinate measuring machine using the contact method. After each experiment the component was measured at eight sections and at least ten points were measured in each section. The average value of the deviance in each section is calculated from the measured values. Subsequently, the mean variance for each experiment is calculated from these values. These values are included in the graph, where the average deviance from the desired dimension is 0.0067 mm without tilting the spindle (Fig. 15) and 0.2476 mm with tilting the spindle (Fig. 16). Accuracy achieved here does not match the required accuracy, and therefore it was necessary to verify the accuracy of the displacement of the tracks of the experiments which have the same input variables. These are the last preparatory roughing "HR01" and the first experimental grinding "EXP01", because these grindings have the same input variables. From these values were calculated deviation displacements of the tracks of the grinding and these deviations are shown in Fig. 17, where the average deviation displacement of the track is 0.003 mm without tilting the spindle and 0.011 mm with tilting the spindle. Only the value without tilting can be considered correct. These geometrical inaccuracies may be caused by changes to the grinding wheel profile by wear, or errors in the conversion of 3D correction in the postprocessor.

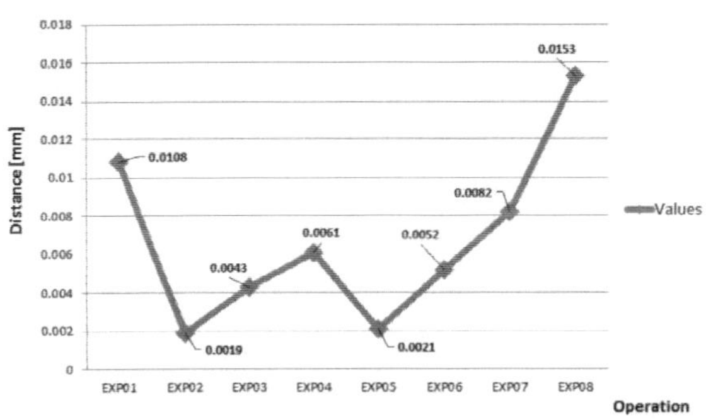

Fig. 15. Accuracy of dimensions without tilting

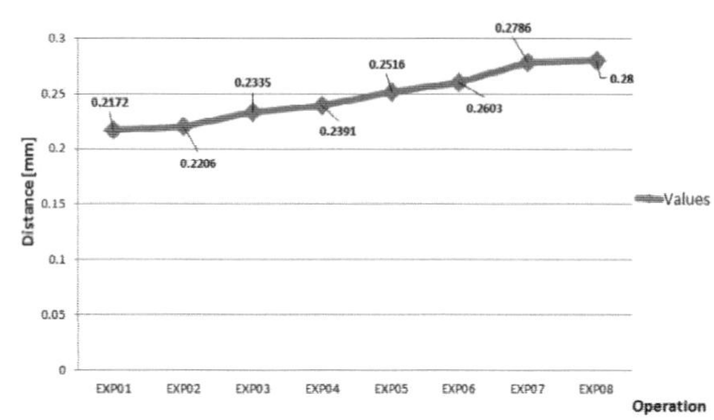

Fig. 16. Accuracy of dimensions with tilting

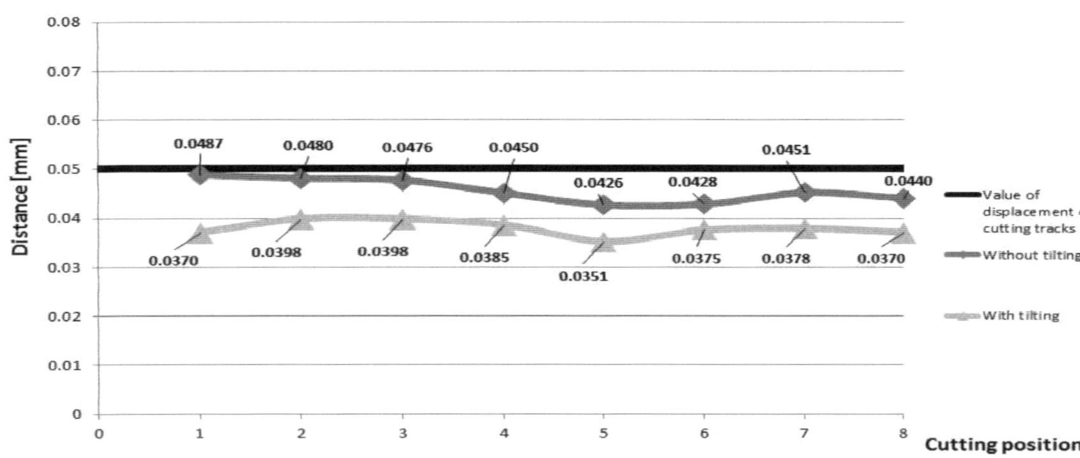

Fig. 17. Geometric accuracy of displacement of cutting tracks

6. Conclusion

The issue of grinding complex shaped surfaces is very complicated and extensive due to the large number of changes that affect the resulting accuracy and surface quality. Therefore, this work demonstrates only parts of this issue, because the main idea of this work is the investigation of the influence of the cutting conditions on the surface quality and the accuracy of dimensions. This investigation proved that the required roughness parameters can be achieved by using the right combination of cutting parameters, axial feed rate and cutting speed. The biggest problem for the accuracy of dimensions is the tilting of the spindle, because the tilting of the spindle changes the cutting point and this change leads to the deterioration of the accuracy of dimensions by uneven wear of the grinding wheel. This problem can been removed by keeping the profile curve of the grinding wheel radius.

Because this is a very extensive issue, it is necessary to focus more on the preparation of the experiments, and above all, on the wear of the grinding wheel, dressing the grinding wheel and the corrections of the tools. Also it is necessary to pay attention to the manufacturing accuracy of the clamp, because inaccuracies from manufacturing the clamp can introduce errors into the experiments.

This experiment provides the basis for future research dealing with grinding complex shaped surfaces. This experiment shows the influence of the grinding conditions on the surface quality and accuracy of dimensions on complex shaped surfaces. From the results, it is evident that it is necessary to focus in the future on the accuracy of dimensions when the spindle is tilted. Therefore further research will be focus on the accuracy of dimensions during grinding with a tilted spindle and during continuous 5-axis grinding. This article also serves as a default article for future research in this field.

7. Acknowledgments

The present contribution has been prepared under project LO1502 'Development of the Regional Technological Institute' under the auspices of the National Sustainability Programme I of the Ministry of Education of the Czech Republic aimed at supporting research, experimental development and innovation.

8. References

[1] Denkena, B.; Grove, T. & Lucas, H. (2016). Influences of grinding with Toric CBN grinding tools on surface and subsurface of 1.3344PM steel. Journal of Materials Processing Technology, vol. 229, pp. 541–548, ISSN: 0924-0136.

[2] Klocke, F. (2009). Manufacturing Processes 2: Grinding, Honing, Lapping. Springer Science & Business Media, ISBN: 978-3-540-92259-9.

[3] Davim, J. P. (2011). Modern Machining Technology: A Practical Guide. Elsevier, ISBN: 978-0-85709-494-0.

[4] Rokyta, L. (2011). Testing of grinding wheel for tools manufacturing. Proceedings of the 22nd International DAAAM Symposium, Volume 22, No. 1, pp 1137–1138, B. Katalinic (Ed.), Published by DAAAM International ISBN 978-3-901509-83-4, ISSN 1726-9679, Vienna, Austria.

[5] Novak, M. (2012). Surfaces with high precision of roughness after grinding. Manufacturing Technology, vol. 12, pp. 66–70, Published by Institute of Technology and Production Management University of J.E. Purkyne ISSN 1213-2489, Usti nad Labem, Europe, Czech Republic.

[6] LUKOVICS, I. (2006) 'HIGH-SPEED GRINDING OF TOOLS AND TOOLING'. A shortened version of habilitation work. Brno 2016. ISSN 1213-418X, ISBN 80-214-3305-1. [cit. 2017-08-20]. Available: http://www.vutium.vutbr.cz/tituly/pdf/ukazka/80-214-3305-1.pdf.

[7] MaragingSteel MS1 Material data sheet, [Accessed: 2017-08-21] © 2007 EOS GmbH – Electro Optical Systems. Available: https://www.axisproto.com/pdf/MaragingSteel_MS1_Material_data_sheet_07-07_en.pdf

[8] 'Unicompartmental Knee Replacement-OrthoInfo - AAOS'. [Accessed: 2017-08-21]. Copyright ©1995-2017 by the American Academy of Orthopaedic Surgeons. Available: http://orthoinfo.aaos.org/topic.cfm?topic=a00585.

[9] Router-CIM 2016, Constant Scallop Height. [Accessed: 2017-08-21]. © 2016 CIM-Tech.com, Inc. Available: http://www.cimtechsoftware.com/RouterCIM_Online_Help/constant_scallop_height.htm.

DOI: 10.2507/28th.daaam.proceedings.130

ABRADED SURFACE PROPERTIES OF SINTERED CARBIDE SPECIMENS BEFORE AND AFTER DEPOSITION OF A THIN LAYER USING PVD

Bricín David, Průcha Vojtěch & Kříž Antonín

This Publication has to be referred as: Bricin, D[avid]; Prucha, V[ojtech] & Kriz, A[ntonin] (2017). Abraded Surface Properties of Sintered Carbide Specimens Before and After Deposition of a Thin Layer Using PVD, Proceedings of the 28th DAAAM International Symposium, pp.0937-0943, B. Katalinic (Ed.), Published by DAAAM International, ISBN 978-3-902734-11-2, ISSN 1726-9679, Vienna, Austria
DOI: 10.2507/28th.daaam.proceedings.130

Abstract

Two series of samples were produced from sintered carbide WC-Co by using a pulsed laser. Due to the different parameters used for the ablation process, the ablation textures which were created have significantly different properties. From each series of samples, some were selected for deposition of a thin layer by PVD (physical vapour deposition). Metallographic analysis was used to evaluate the properties of the laser treated surface, supplemented by analysis of the chemical composition, hardness, fracture toughness of KIC, tribological properties, and adhesion properties of the prepared samples. The results from the analyses of the ablated surface samples were then compared with the results obtained from the ground surface samples.
By comparing the results, it was found that with the correct parameters set for laser ablation, it is possible to obtain a surface with better abrasion properties for the deposition of the PVD layer on the substrate than with the ground surface.

Keywords: Ablation texture; surface treatment; adhesion; laser power; indentation; wear

1. Introduction

Sintered carbides are widely used for machining tools because of their mechanical properties, such as hardness and fracture toughness.[6] For some applications, these tools are further deposited with thin PVD or CVD layers. These layers provide a longer tool life because the layer takes over the function of the contact surface between the leaving chips and the face of the tool. In order for the thin layer to fulfil its proper function, it must have good adhesion to the substrate and good tribological properties. These properties are derived both from the chemical composition of the deposited layer and the substrate, and from the surface treatment of the cutting tool by grinding, baking, brushing, and the like. Grinding is mainly used as the technology for surface treatment before the deposition process because residual pressure stresses are initiated in the surface layer of the cutting tool, which is better for prolonging the life of the cutting tool.[1-2] This paper presents some of the results from a study to assess the suitability of surface treatment of cemented carbide by laser and to determine how this surface treatment affects the adhesion and tribological properties of the deposited PVD film. Two series of samples with a laser-treated surface were created, each one made using different parameters.

For one series of samples, the lowest laser power was used so that the input energy density (Ev) was as low as possible. For the second series of samples, a completely opposite approach was chosen, using the highest possible laser power, and the input value of energy density was approximately 100x higher than in the first series. A comparative series of samples with ground surfaces was then made.

Metallographic analysis using a TESCAN VECTRA electron microscope, supplemented by hardness, fracture toughness and chemical composition analysis using an EDX analyser was used to evaluate the properties of the laser-treated surfaces. The Scratch Test, Pin-on-Disc Test and Mercedes Test were used to evaluate the properties of the deposited thin films.

2. Results and discusion

2.1. Metallographic evaluation of the samples

The first analysis was the visual assessment of the condition of the treated surface. This evaluation was performed using a TESCAN VECTRA electron microscope, as shown in the photographs below (Fig. 1), which document the samples' surface conditions after laser ablation and grinding.

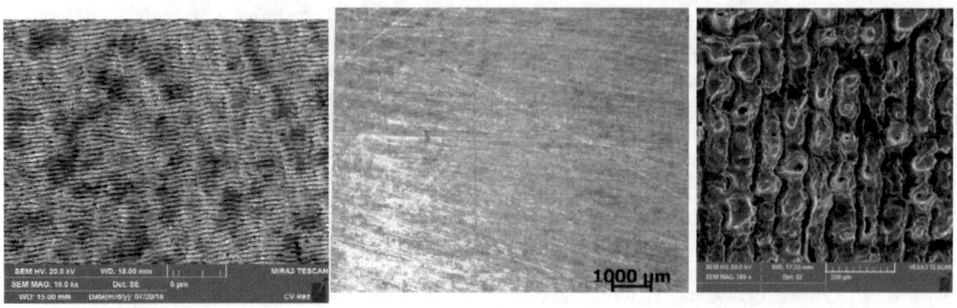

Fig. 1. Samples of treated surfaces. Left: sample surface treated with low energy density of laser. Centre: sample surface with ground surface. Right: sample surface treated with high energy density.

From Fig. 1 it can be seen that the modified surfaces of the samples differ significantly due to the different production parameters. The centre photograph shows typical wear of the surface caused by the grinding wheel. Since the tracks produced go in different directions, it is likely that multi-directional grinding was used to grind the sample surface. By comparing the left and the right images, it can be seen that the high and low energy densities produced a change in the surface layer of the treated cemented carbide. The structure created using low energy density was considerably finer than the structure created by the high energy density. It also shows that the resulting texture has a certain spatial orientation. This orientation was dependent on the direction of the laser movement.

Other differences detected by the electron microscope were differences in chemical composition. Before surface treatment, the chemical composition of each sample was determined by EDX analysis. This chemical composition corresponded to the material supplier's datasheets. This analysis was repeated after the surface treatment of the samples. Comparing the results of the EDX analysis before and after the surface treatment, it was found that all of the samples tested showed a decrease in (Co) in the surface layer of the samples, see Table 1.

	W+C	Co	O	Surface modification	W+C	Co	O
Sample Before surface treatment	94±0,076	6±0,076	-	Laser with low Ev	91.4	5.6	3
				Grinding	92.2	5	2.8
				Laser with high Ev	95.5	2.4	2.1

Table 1. Chemical composition of the samples with treated surfaces detected by EDX analysis.

All samples showed slight surface oxidation during surface treatment due to the fact that no inert protective atmosphere was used in the process. Binder Co losses were caused by the different sets of parameters and technologies which were used for surface treatment of the samples. The use of higher energy density Ev led to a more pronounced loss of binder, resulting in the formation of a brittle surface layer with a thickness of more than 50 microns. The low toughness of the layer was also manifested by the cracking of the layer due to large residual heat stresses. The low tenacity of this layer was also demonstrated in further experiments. There was a loss of binder on the ground surface due to leaching, which occurred during grinding of the sintered carbide. The lowest loss of the binder was detected on the laser-treated samples with low energy density Ev. This meant that the layer was tougher than the other samples with laser modified surfaces. The thickness of this affected layer was only a few microns.

The Vickers hardness test at a load of 30 kgf (HV30) was used to show the fragile behaviour of the formed surface layers. See Fig. 2.

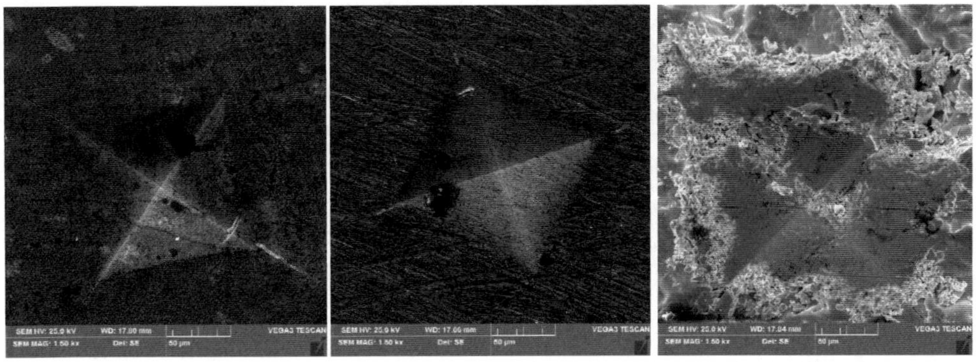

Fig. 2. Indentations created during hardness measurement. For samples with ground surfaces (centre) and samples with surfaces treated by laser with low energy density (left), cracks were found in the corners of the imprints. The shape of indentations was regular. When measuring surface hardness of samples with surface ablation using high energy density Ev (right), the substrate dropped and the shape of the imprints was uneven.

It can be seen from Fig. 2 that when applying a high energy density, the surface layer of the sample was brittle. This embrittlement is caused by a significant loss of binder from this layer by evaporation. This occurs because its evaporation temperature was reached during the ablation process. Due to the decay of the brittle surface layer into fragments, it was not possible to precisely determine its hardness. The hardness measurement was further supplemented by the Palmquist fracture toughness calculation, see Equation 1 and Table 2. However, this measurement was only possible for samples of ground and laser-treated samples with low energy density Ev. Other types of samples did not initiate cracks at the corners of the imprints. This was probably because the applied load caused dropping and breakage of the brittle layer under the indenter so that the energy to create these cracks was used for disruption of the brittle layer, meaning that cracks did not initiate.

[4] $$W_k = A \cdot \sqrt{HV} \cdot \sqrt{W_g}\left[MN \cdot m^{-\frac{3}{2}}\right] \quad (1)$$

[4] $$W_g = \frac{P}{T}[N \cdot mm^{-1}] \quad (2)$$

Where:

A ... constant, HV ... hardness according to Vickers HV (30), P ... imprinting load (kgf), T, total length of cracks produced (mm)

Vickers HV30 surface hardness and calculated fracture toughness values Kic		
Type of sample	Hardness HV (30)	Kic (Palmquist)
Samples with ground surface	1 700 ± 38.5	12.4 ± 0.52
Samples with laser treated surface using low energy density Ev	1 690 ± 19.3	9.7 ± 0.32
Samples with laser treated surface using high energy density Ev	-	-

Table 2. Calculated fracture toughness and HV30 hardness

The behaviour of the treated sample surfaces was further evaluated using a Pin on Disc tribological test, where the volume loss of material (mm3) at the point of contact where the indenter compressed the rotating surface of the test sample was evaluated. [3, 8] This test was further used to evaluate the properties of the deposited PVD layer. For samples which were deposited by a PVD thin layer, the Mercedes Test and Scratch test were used to evaluate the cohesive-adhesive behaviour of the deposited PVD thin film on the modified surfaces of the samples.

2.2. Tribological behaviour of the treated specimens

For the Pin on Disc Test, a 3 mm diameter carbide ball was forced into the rotating surface of the specimens with a constant force of 10 Newtons. After 20 000 cycles, tribological traces were formed on the sample surfaces, see Fig. 3.

Fig. 3. Differences between tribological traces. Left: tribological trace created on the sample surface treated by laser with low density of input energy (E_v). Centre: tribological trace created on sample surface treated by grinding. Right: tribological trace created on sample surface treated by laser with high density of input energy (E_v).

Tribological traces were then measured at four different locations. The volume loss of the material was then calculated according to Equation 2 below. Fig. 4 shows a comparison of the volume of material removed from the treated samples surface layers before and after deposition by thin layer PVD.

[7] $$V = \frac{\pi(wear\ track\ radius, mm)(track\ width, mm^3)}{6(sphere\ radius, mm)}\ [mm^3] \qquad (3)$$

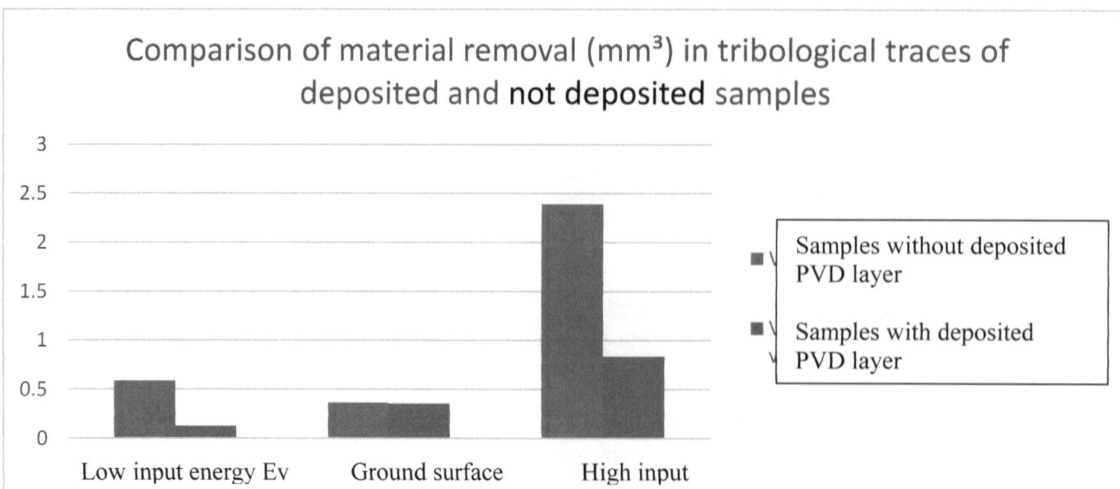

Fig. 4. Comparison of the volume of material removed during Pin on Disc tribological test.

From Fig. 4 it can be seen that the smallest material volume removed was recorded in the samples with ground surfaces prior to the deposition of the thin film PVD on the treated sample surface. This is probably due to the fact that residual pressure stresses were initiated during grinding. These stresses are better for the treated surface layer because they have a tendency to close any cracks which may form at places where the surface of the ground samples is damaged. For samples with a laser-ablated surface treatment, residual stresses are created in the sample surfaces. These types of residual stresses tend to increase cracking which may form at the places where the surface of the laser ablated samples is damaged. Therefore the volume of material removed from the treated surface layer of the ground samples is lower than for the samples with laser ablated surfaces. Another possible reason why the material removal rate was lower for samples with ground surfaces may be the lower surface roughness of these samples. Due to the smoother surface, without extreme rises and drops in the roughness profile, there is no intense tearing of the material from the sample surfaces during their contact with the indenter during the Pin on Disc test. This is another possible reason why the amount of the removed material was smaller for the ground samples than for the samples with laser ablated surfaces. The largest volume loss of material was recorded in samples with laser treated surfaces using high energy density (Ev). This is associated with the brittleness of the layer thus created, which was confirmed in previous experiments.

After deposition of a thin layer by PVD, there was a decrease in material removal from the surface of the samples. The most noticeable decrease of removed material was observed in samples with surfaces treated by laser. This is probably because surface wrinkling occurred during this technological operation which increases the contact surface between the thin layer and the laser treated substrate. This meant that the adhesion of the deposited thin layer was higher and the material removal was less than for samples with ground surfaces. A reduction in the volume of removed material was calculated for samples with surface treatment by laser using high energy density, but due to the brittleness of this layer, the amount of removed material was still higher than measured on the other samples. In the case of the ground samples, after the deposition of a thin layer by PVD, there was also a reduction of the material removal in the tribological trace, but this decrease is minimal compared with the other samples. The reduction of the volume of material removed for samples with laser treated surfaces using low energy density after thin layer PVD is interesting from a technological point of view, as described above.

2.3. Adhesive-cohesive behaviour of samples deposited using thin layer PVD

The adhesive-cohesive behaviour of the deposited thin films was evaluated using the Mercedes test.[5-9] In this test, a Rockwell Cone is pressed into the surface of a specimen with a force of 1500 N. The indenter induces stresses at the edges of the indentation to create tension, which then leads to cracking and peeling of the PVD layer. A Zeiss Observer Z1m light microscope and a comparison table (Table 3) were used to evaluate the state of the indentation interfaces.

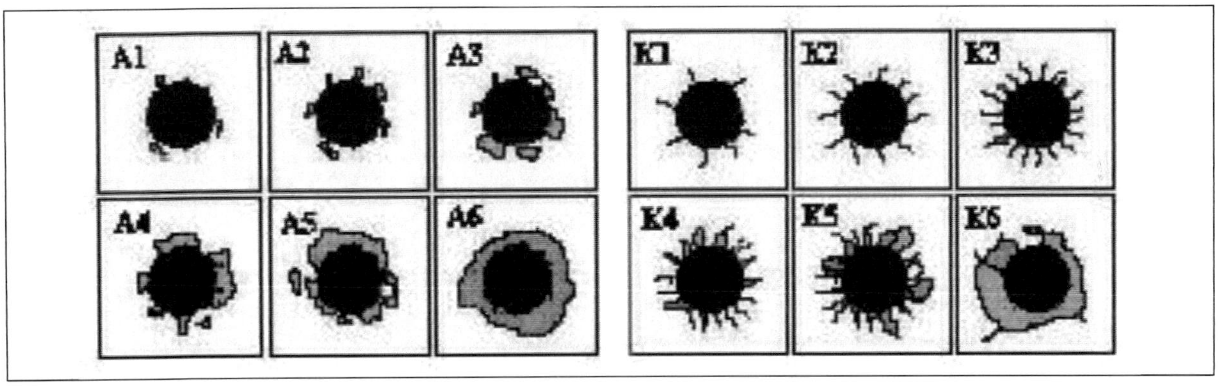

Table 3. Comparison card used to evaluate the cohesive-adhesive behaviour of a pair of thin-layer substrates.[5]

A deposited layer is considered to have satisfactory adhesive properties (A) and cohesive properties (K) with values 1-3. At higher values, the adhesion or cohesion of the layer to the substrate is considered unsuitable, see Table 3.

Fig. 5 shows photographs of the interface of the imprint.

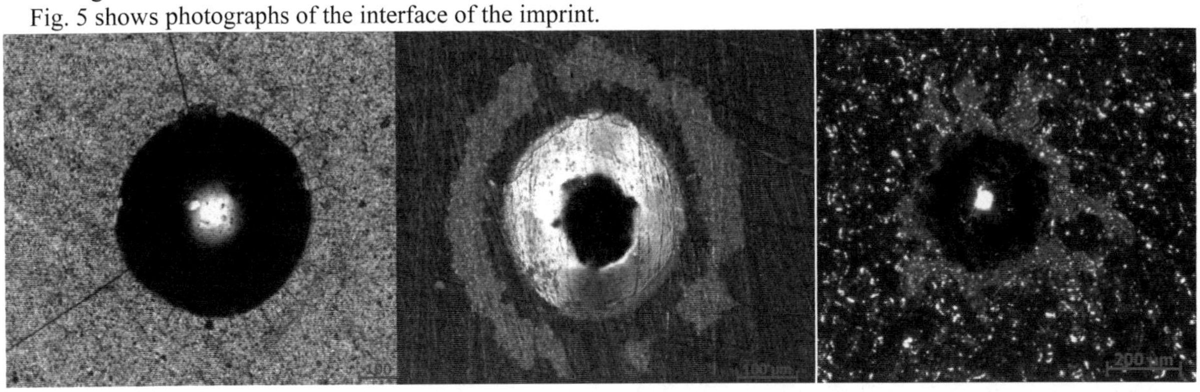

Fig. 5. Sample surface condition after Mercedes test. Left: indentation edge on sample with laser surface treatment using low density input energy (Ev). Centre: indentation edge on sample with ground surface. Right: indentation edge on sample with laser surface treatment using high density input energy (Ev).

It can be seen from Fig. 5 that for samples with ground surfaces and samples with laser treated surfaces using high density input energy there was a tearing off of the deposited thin layer from the substrate surface. This indicates that the adhesion of the layer to the substrate is inadequate. For samples with surfaces treated by laser using low density input energy (Ev) cohesive cracks were found (see Fig. 5, left) at the edge of the indentation. This type of cohesive damage was not as significant as that observed in the other samples. Therefore the cohesive behaviour of all the series of all the samples was evaluated as similar and satisfactory.

For sample surfaces with a deposited thin layer, a Scratch test was also performed.[5, 8] The results of this test are, however, only indicative, because the surfaces of the samples were too rough after their laser treatment. This caused the indenter, in its straight forward motion, to copy the surface roughness of the specimens and jump from them, thus disturbing the acoustic emission signal.

The traces created during this test could then only be evaluated by using an electron scanning microscope, which allows EDX analysis of the chemical composition of the treated surface layer. This made it possible to find the places where the tearing of the thin layer out of the substrate occurred.

Critical loads Ls (N) were then deducted at the places where the tearing started, which signify the adhesion of the PVD layer against the substrate. In this case it was possible to find the place where tearing started due to the increased percentage of tungsten detected by the EDX analyser. Traces which were created during the Scratch Test are shown in Fig. 6.

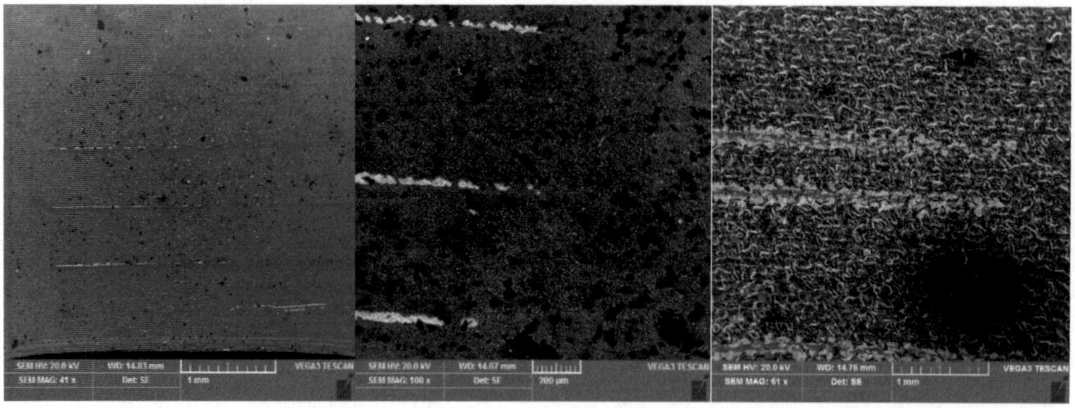

Fig. 6. Indentation tracks created during the scratch test. Left: indentation tracks on sample with laser surface treatment using low density input energy (Ev). Centre: indentation tracks on sample with ground surface. Right: indentation tracks on sample with laser surface treatment using high density input energy (Ev).

Sample type	Critical load value Ls (N)
Low energy Ev	46
Ground surface	9
High energy EV	12

Table 4. Measured Critical Load Values Ls.

From the results given above, it was found that the highest resistance to tearing of the layer was shown by the thin layer deposited on the surface which was treated with a laser using low energy density Ev. For samples with ground surfaces and surfaces treated with a laser using high density input energy, lower resistance to tearing of the deposited thin layer from the substrate was measured. These results confirm the results of the Mercedes test.

3. Conclusion

Our results show that we can achieve better adhesion of a PVD layer to a sintered carbide substrate with the correct setting of laser ablation parameters, than for the deposition of a PVD layer on a ground surface. With the correct parameters of laser treatment for sample surfaces with low input energy density, it was possible to achieve the appropriate roughness, i.e. the wrinkling of the surfaces of the samples.

The sample surfaces had good toughness, which differed from the sample surfaces prepared by laser using high density energy input, which was brittle. After deposition by thin layer PVD, other experiments showed that samples with surfaces treated with low input energy density had better tribological properties and adhesion to the substrate.

In this case, the abrasion wear of the laser treated surface of the samples was reduced by about 35 % when using low input energy density, compared to samples treated by grinding. Further experiments will focus on practical applications of the wrinkling created in the machining process.

4. Acknowledgement

This article was made possible by the funding for the SGS-2015-016 project "Analysis of Surfaces of Structural Details and Tools by Surface Integrity Method, and Impacts on End-Use Properties".

5. References

[1] Denkena B. B et al., Residual Stress Distribution in PVD-Coated Carbide, Leibniz Universität Hannover, Germany, Institute of Production Engineering and Machine Tools

[2] Hegeman J. et al., Grinding of WC–Co hard-metals, University of Groningen, Laboratory of Applied Physics, Materials Science Center and The Netherlands Institute for Metals Research.

[3] Hájek J. (2007).Laboratory and practical tests on systems with thin layers used in cutting applications. Ph.D. thesis. UWB in Pilsen. Faculty of Mechanical Engineering, Department of Material Science and Technology. Pilsen 2007.

[4] Janoušek A. (2014). Properties and Cutting Tool Applications of Sintered Carbides. Master thesis. UWB in Pilsen. Faculty of Mechanical Engineering, Department of Material Science and Technology. Pilsen 2014.

[5] Sosnová M. (2007). Relationship of methods, which evaluate the adhesive cohesive behaviour of systems with thin layers. Ph.D. thesis. UWB in Pilsen. Faculty of Mechanical Engineering, Department of Material Science and Technology. Pilsen 2007.

[6] Špirit Z. (2013). Selected Properties and cutting tools applications of sintered caribides. Master thesis. UWB in Pilsen. Faculty of Mechanical Engineering, Department of Material Science and Technology. Pilsen 2013.

[7] ASTM G99 – 17, Standard Test Method for Wear Testing with a Pin-on-Disk Apparatus, (2017) ASTM International, 1. 1. 2017, Available from: https://www.astm.org/Standards/G99. Cited: 7. 8. 2017.

[8] Valova, M. & Scuhanek, J. (2011). Evaluation of Tribilogical Characteristics of Duplex Coated 31CrMoV9 Steel, Annals of DAAAM for 2011 & Proceedings of the 22nd International DAAAM Symposium, 23-26th November 2011, Vienna, Austria, Volume 22, No. 1, ISSN 1726-9679, ISBN 978-3-901509-83-4, Katalinic, B. (Ed.), pp. 0635-0636, Published by DAAAM International Vienna, Vienna.

[9] Lind, L., Peetsalu, P., Põdra, P., Adoberg, E., Veinthal, R. and Kulu, P. (2010). Description of punch wear mechanism during fine blanking process. In Proc. 7th International Conference DAAAM Baltic Industrial Engineering. Tallinn.

DOI: 10.2507/28th.daaam.proceedings.131

PARAMETRIC VERSUS NONPARAMETRIC STATISTICAL PROCESS CONTROL

Tereza Smajdorova & Darja Noskievicova

This Publication has to be referred as: Smajdorova, T[ereza] & Noskievicova, D[arja] (2017). Parametric Versus Nonparametric Statistical Process Control, Proceedings of the 28th DAAAM International Symposium, pp.0944-0949, B. Katalinic (Ed.), Published by DAAAM International, ISBN 978-3-902734-11-2, ISSN 1726-9679, Vienna, Austria DOI: 10.2507/28th.daaam.proceedings.131

Abstract

This paper represents the limitations of commonly used parametric methods of statistical process control, especially problems associated with the use of classical Shewhart control charts, and shows the benefits of non-parametric and robust control charts. The aim of this article is to define the various deviations from normality that cause problems in the use of classical parametric methods of statistical process control. Another goal is to define the differences between nonparametric and robust methods and to present these methods as one of the possibilities of solving the problems of normality violations. Finally, there is introduced the development and the future direction and robust nonparametric statistical process control.

Keywords: Statistical process control; parametric methods; nonparametric methods; basic data assumptions

1. Introduction

One of the main aims of statistical process control is to distinguish between two sources of variability in a given process. These are sources of variability that cannot be economically identified and corrected (common causes) and those that can be identified and subsequently removed (assignable causes). [3,9,21]

- Common causes: The causes that are inherently connected with the process, affect it every day. [4]
- Assignable causes: The causes that do not affect the process continuously, but they arise under various circumstances. [4]

If the process is affected only by the random causes of variability, it is said to be in a state of statistical stability (in-control). [3,9]. The main tool of statistical process control is the control chart. In case, the process is affected by a assignable cause of variability, the control chart should signal as soon as possible that the process is out of control. Obviously, the faster the effect of the assignable cause of variability is identified, the more effective the control chart is. [3,9] One of the most prominent pioneers in quality management was Walter A. Shewhart, since 1918 the Bell Telephone Laboratory employee.

In May 1924, he designed the first control chart to assess whether the variability of the monitored process parameter was due to random fluctuations or special causes (machine alignment, change of raw materials, etc.) and laid the foundations for modern industry. [8]

To use these classical Shewhart control charts, the data must meet certain basic assumptions (normality, independence, constant mean and variance). In practice, however, these assumptions often cannot be met. Even so, companies use Shewhart control charts, but it does not have to deliver relevant results. Therefore, their attention should be redirected to non-parametric control charts that are a suitable alternative to the classical Shewhart control charts.

However, nonparametric or robust control charts are not available in current publications or in teaching, whether in schools or courses. There is not enough software support for these alternative methods as well. The aim of this paper is to clarify the problems that may arise when using Shewhart classical control diagrams and to present the advantages of nonparametric methods.

2. Classical Shewhart control charts

The classical Shewhart control charts count among the parametric methods of statistical process control. For the use of classical Shewhart control charts, some basic data assumptions must be met. These assumptions include: [5]

1. A general assumption
   - compliant capability of the measurement system

2. The basic statistical assumptions
   - normal distribution of the quality characteristics,
   - constant mean and variance,
   - mutual independence of quality characteristics values.

3. The other assumptions
   - a sufficient amount of data,
   - sensitivity to greater changes in a process,
   - monitoring one quality characteristics per unit of a product.

These basic assumptions need to be verified and a number of tests and graphical tools can be used:

- Verification of normality:
  - Shapiro - Wilk test [5]
  - Shapiro –Francia test [1]
  - Royston modification [7]
  - Anderson - Darling test [6;10]

- Verification of independence:
  - Autocorrelation test [7]
  - Test of iterations up and down [5]
  - Runs Above and Below the Median [7]

- Verification of Homogeneity of means and variances:
  - Analysis of variance ANOVA [5; 7]
  - Brown - Forsyth test [5; 7]
  - Levene test [5; 7]

In manufacturing practice, however, these basic assumptions cannot always be fulfilled (Table 1).

Industry/Technology/Quantity	Normality	Independence	Constant Mean	Constant Variance
Mechanical Engineering, Automotive Industry (dimension)	-	-	-	-
Mechanical Tests (strength, flexibility,...)	x	-	-	x
Chemistry, Metallurgy (concentration, contents,...)	-	x	x	x

Environment (different concentrations)	x	x	x	x
Electrical Quantities	-	-	-	x
Energy	x	x	x	x
Plastics, Polymers, Textiles, Physico-mechanical Quantities	x	-	x	-
Biochemistry, Pharmacy, Food Industry	x	x	-	-
Economic and Financial Indicators	x	x	x	-
Sociology, Human Resources	x	x	x	x

Table 1. Typical breaches of assumptions [8]

The "x" denotes a violation of the assumption, and the "-" designation then determines the predominant fulfillment of the assumption.

3. Violation of basic data assumptions

This chapter lists the possible ways of violating the underlying assumptions about the data.

- Violation of the assumption of normality

Unlike the normal probability distribution, the data file may be sloping right or left, or may have different spikes. These violations of normality can be clearly seen on the Q-Q graph. The following 4 images (Figure 1) show the relationship between the histogram (upper image) and the Q-Q graph (bottom image):

a) the concave curve of the Q-Q graph shows a slope to the right and a greater variability of higher values
b) the convex curvature indicates a left slope and a greater variability of lower values
c) the convex concave curve indicates a division with less tails (heavy tails)
d) the concave convex curve shows a division with higher tails (light tails) [20]
e) this image indicates the occurence of an outlier value

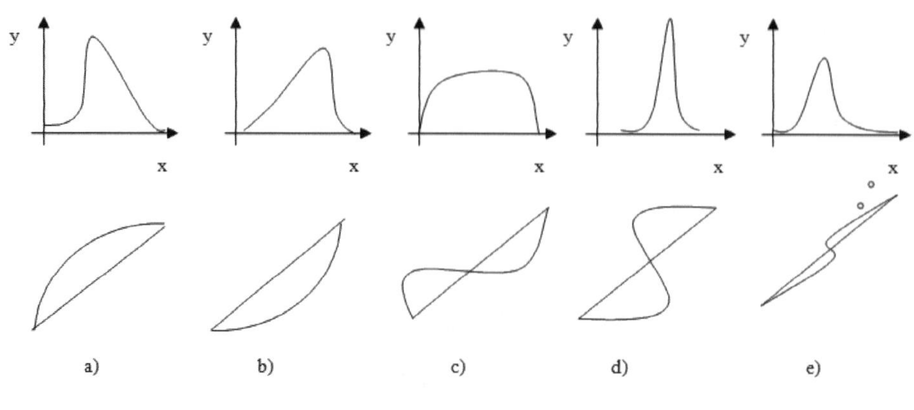

Fig. 1. Relationship of histogram and Q-Q graph

- Violation of the data independence assumption

Independence of data can be verified by the autocorrelation test. Measurement dependency is usually caused by instability of the measuring device, inconstant measurement conditions, neglecting factors that significantly affect measurement results or improper (non-random) sampling. There is a negative and a positive independence. Figure 2 shows the autocorrelation chart:

a) Negative autocorrelation
b) No dependence
c) Positive autocorrelation [8]

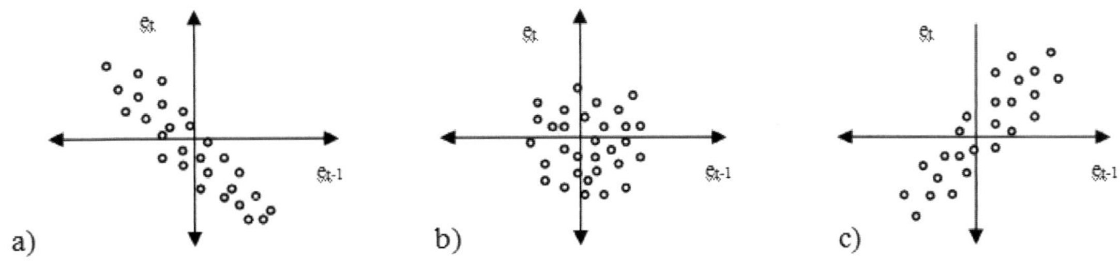

Fig. 2. Chart of autocorrelation

- Violation of the assumption of constant scattering

Constant scatter can be verified using a residual graph. With increasing x, the value of the residue should not increase or decrease, it should still be equidistant from the x-axis. Figure 3 shows:

a) Constant dispersion
b) Non-constant dispersion [8]

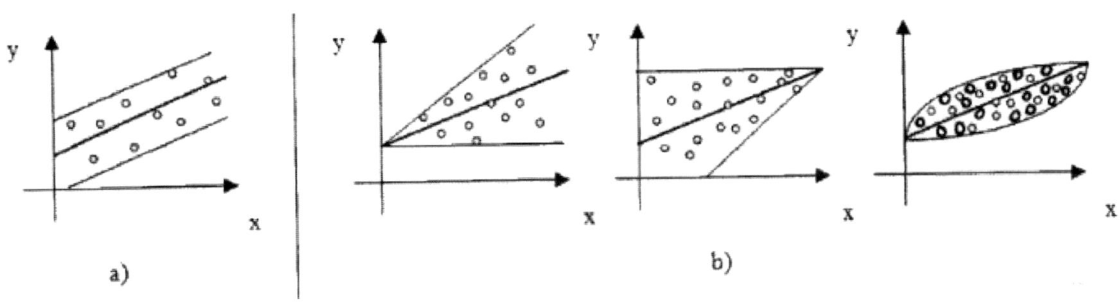

Fig. 3. Non-constant dispersion

- Violation of the assumed constant mean value

Non-compliance with the assumed constant mean value can occur, for example, when changing inputs or technology. The t-test can be used to assess the significance of the difference between the mean values. How the change in the mean value is reflected in the individual value control chart is shown in Figure 4. [8]

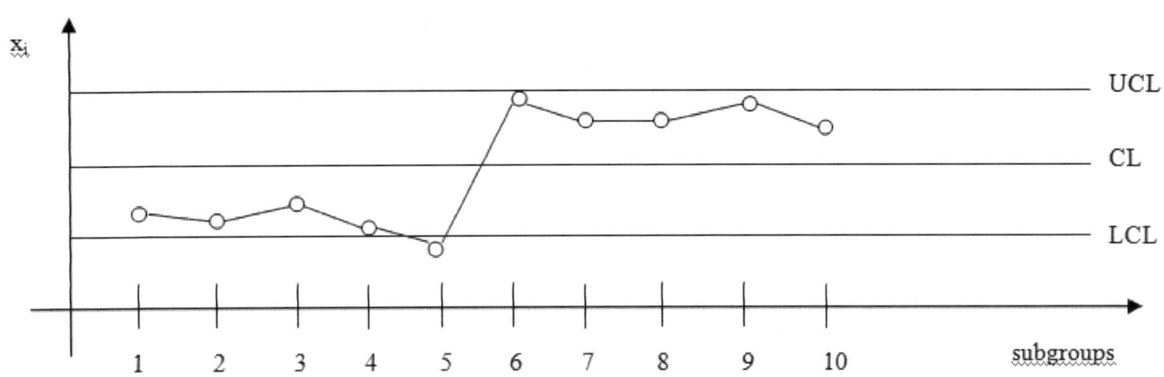

Fig. 4. Change of mean value in individual value control chart

4. Nonparametric or robust statistical methods

Robust statistics attempt to provide appropriate methods that replace commonly used parametric statistical methods but are not affected by remote values or other small deviations from model assumptions. In statistics, classical parametric estimation methods rely on data assumptions (normality, independence, constant mean and scattering) that often are not met in practice. Unfortunately, if data do not meet these assumptions, classical parametric methods often do not provide the right results. For these cases, robust statistical methods have been developed that are not affected by the failure to meet the basic assumptions about the data.

Nonparametric methods are not based on any particular probability distribution and can thus be used when the assumption of normality is not met. Non-parametric methods are more robust than parametric methods. Nonparametric methods can thus be classified as robust.

Roughly speaking, a nonparametric procedure is a statistical procedure that has certain desirable properties. The rapid and continuous development of nonparametric statistical procedures over the past 7 1/2 decades is due to the following advantages enjoyed by nonparametric techniques:

- Nonparametric methods require few assumptions about the underlying populations from which the data are obtained. In particular, nonparametric procedures forgo the traditional assumption that the underlying populations are normal.
- Nonparametric methods enable the user to obtain exact P-values for tests, exact coverage probability for confidence intervals, exact experimentwise error rates for multiplate comparison procedures, and exact coverage probabilities for confidence bands without relying on assumptions that underlying populations are normal.
- Nonparametric methods are often (although not always) easier to apply than their normal theory counterparts.
- Nonparametric methods are often quite easy to understand.
- Although at first glance most nonparametric methods seem to sacrifice too much of the basic information in the sample, theoretical efficiency investigations have shown that this is not the case. Usually, the nonparametric methods are only slightly less efficient that their normal theory competitors, when the underlying populations are normal (the home court of normal theory methods), and they can be mildly or wildly more efficient than these competitors when the underlying populations are not normal.
- Nonparametric methods are relatively insensitive to outlying observations.
- Nonparametric methods are applicable in many situations where normal theory procedures cannot be utilized. Many nonparametric methods require just the ranks of the observations, rather than the actual magnitude of the observations, whereas the parametric methods require the magnitudes. [2]

With regards to the use of these methods at statistical process control, non-parametric and robust control charts have been developed. These control charts are not only suitable for processes that do not meet normality and independence of the data, but especially in the beginning of the SPC implementation, when there are not enough data available. [3]

5. The future of nonparametric statistical process control

In recent years, non-parametric and robust control charts have received more attention than it used to be in the past. It is indisputable that they deserve this attention because they form an important part of statistical process control. Already since the 1960s, alternatives to the classical Shewhart control charts have been developed based on a non-normal distribution. In 1958, Ferrell [11] defined a control chart for the lognormal distribution. Nealson [12] obtained in 1979 control limits for median, range scale and location scale for the Weibull distribution. Unlike these control diagrams, the nonparametric ones are more flexible. Since the 1990s, various nonparametric and scrupulous control patterns have been proposed. Alloway and Raghavachary [13] in 1991, and then Pappanastos and Adams [14] in 1996 proposed a control chart for Location based on the Hodgest – Lehmann estimator. Amin, Reynolds and Bakir [15] presented in 1995 a control chart for the process mean and process variability based on the sign statistics.

In 1995, Bai and Choi [16] proposed a regulatory diagram for the sloping distribution. In 2008, Nandini [17] dealt with control charts based on the two sample rank – sum test of Ansari and Bradley. Chakraborti, Mark and Wiel [18] in 2008 engaged in a nonparametric control chart based on the Mann – Whitney statistics. In 2015 Zhang, Chen and Zou [19] suggested a robust multivariate control chart based on the goodness-of-fit test. Unfortunately, this area of statistical process control is not used in practice, it is not a part of courses and publications on statistical process control. There is not even the necessary software support, which could change in the future and these simple methods could become a match for the classical parametric methods.

6. Conclusion

This article summarizes shortcomings of classical Shewhart control charts (such as the need for normal data distribution, mutual data independence, and more). It offers the possibility to use nonparametric methods to eliminate these deficiencies. It presents the advantages of nonparametric methods. The results will contribute to the further development of statistical process control and process capability analysis. The proposed methodology could help in the decision-making processes in practice. In the future, I would like to test different non-parametric control charts on data that violate in different ways basic assumptions for using classical Shewhart control charts. And based on the results, I would elaborate their overview, including a software support.

7. Acknowledgments

The work was supported by the specific university research of the Ministry of Education, Youth and Sports of the Czech Republic No.SP 2017/63.

8. References

[1] D'Agostino, R. B., et al. (1986). Goodness-of-fit techniques. New York: M. Dekker, ISBN 0824774876

[2] Hollander, M., et al. Nonparametric statistical methods. Third edition. ISBN 978-0-470-38737-5.

[3] Chakraborti, S., et al. (1999). Nonparametric statistical process control : an overview and some results. (Memorandum COSOR; Vol. 9908). Eindhoven: Technische niversiteit Eindhoven.

[4] Ion, R. A. (2001). Nonparametric statistical process control, Ph.D. Dissertation, Korteweg-de Vries Institute for Mathematics, Amsterdam, Netherlands,

[5] Jarosova, E., & Noskievicova, D. (2015). More advanced methods of statistical process control. 1st ed .. Prague: Grada Publishing. ISBN 978-80-247-5355-3.

Jarošová, E., & Noskievičová, D. (2015). Pokročilejší metody statistické regulace procesu. 1st ed.. Praha: Grada Publishing. ISBN 978-80-247-5355-3.

[6] Kotlorz, L .. (2012). Normal tests. Prague,. Bc. Dissertation. Charles University in Prague, Prague, Czech republic Kotlorz, L..(2012). Testy normalty. Praha,. Bakalářská práce. Karlova Univerzita v Praze, Praha, Česká republika

[7] Madansky, A.. (1988). Prescriptions for working statisticians. New York: Springer-Verlag,. ISBN 0387966277.

[8] Meloun, M., et al. (2004). Statistical analysis of experimental data. Ed. 2., eds. Prague: Academia ,. ISBN 80-200-1254-0.

Meloun, M., et al.(2004). Statistická analýza experimentálních dat. Vyd. 2., upr. a rozš. Praha: Academia,. ISBN 80-200-1254-0.

[9] Nenadal, J.; Petrikova, R.; Plura, J.; Noskievicova, D. & Tosenovsky, J. (2008). Modern Quality Management: Principles, Procedures, Methods. Prague: Management Press ,. ISBN 978-80-7261-186-7.

Nenadál, J.; Petříková, R.; Plura, J.; Noskievičová, D. & Tošenovský, J. (2008). Moderní management jakosti: principy, postupy, metody. Praha: Management Press,. ISBN 978-80-7261-186-7.

[10] Salda, Z. (2010). Measurement uncertainties and statistical models. Bc. Dissertation, Brno University of Technology, Brno, Czech Republic

Šalda, Z. (2010). Nejistoty měření a statistické modely. Bakalářská práce, VUT v Brně, brno, Česká Republika

[11] Ferrell, E. B. (1958). Control charts for Lognormal Universe. Industrial Quality Control 15, pp 4-6.

[12] Nelson, P. R. (1979). Control Charts for Weibull Processes with Standardss Given. IEEE Transactions on reliability 28, pp. 283-287.

[13] Alloway, J. A., et al. (1991). Control chart based onn the Hodges-Lehmann estimator. Journal of Quality Technology 23, pp. 336-347.

[14] Pappanastos, E. A., et al. (1996). Alternative designs off the Hodges-Lehmann control chart. Journal of Quality Technology 28,, pp. 213-223.

[15] Amin, R. W., et al. (1995). Nonprametric quality control based on the sign statistic. Communication Statistical Theory and Methods 24, pp. 1597-1623

[16] Bai, D. S., et al. (1995). X and R Control Charts for Skewedd Populations. Journal of Quality Technology 27, pp. 120-131.

[17] Nandini, D., (2008). A Note of the Efficiency of Nonparametric Control Chart for Monitoring Process Variability. Economic Quality Control, vol. 23, No. 1, ISSN 0910-5151.

[18] Chakraborti, S., et al. (2008) A nonparametric control chart based on the Mann-Whitney statistic. Beyond Parametrics in Interdisciplinary Research: Festschrift in Honor of Professor Pranab K. Sen [online]. Beachwood, Ohio, USA: Institute of Mathematical Statistics, pp. 156 [cit. 2017-09-01]. DOI: 10.1214/193940307000000112. ISBN 978-0-940600-73-7.

[19] Zhang, Ch., et al. (2015), Robust multivariate control chart based on goodness-of-fit test.

[20] Sebera, M. Multidimensional Statistical Methods [online]. [feeling. 2017-08-28]. Available from: http://www.fsps.muni.cz/~sebera/vicerozmerna_statistika/pojmy.html

Sebera, M. Vícerozměrné statistické metody [online]. [cit. 2017-08-28]. Dostupné z: http://www.fsps.muni.cz/~sebera/vicerozmerna_statistika/pojmy.html

[21] Kolosowski, M[ariusz]; Duda, J[erzy] & Tomasiak, J[acek] (2016). Statistical process control in conditions of piece and small lot production, Proceedings of the 26th DAAAM International Symposium, pp.0147-0155, B. Katalinic (Ed.), Published by DAAAM International, ISBN 978-3-902734-07-5, ISSN 1726-9679, Vienna, Austria DOI:10.2507/26th.daaam.proceedings.021

DOI: 10.2507/28th.daaam.proceedings.132

MODULAR MOBILE ROBOTIC KIT FOR PROTOTYPING AND DEBUGGING OF CONTROL ALGORITHMS

Oleg Shmakov, Dmitrii Korolev, Dmitrii Popov, Nikolai Kitaev & Alexei Korotkov

This Publication has to be referred as: Shmakov, O[leg]; Korolev, D[mitrii]; Popov, D[mitrii]; Kitaev, N[ikolai] & Korotkov, A[lexei] (2017). Modular Mobile Robotic Kit for Prototyping and Debugging of Control Algorithms, Proceedings of the 28th DAAAM International Symposium, pp.0950-0956, B. Katalinic (Ed.), Published by DAAAM International, ISBN 978-3-902734-11-2, ISSN 1726-9679, Vienna, Austria
DOI: 10.2507/28th.daaam.proceedings.132

Abstract

Rising popularity of robotics in education leads to necessity for modern training programs to have high-quality tool – robotic kit. Authors propose a new kit of mobile robotic modules. A modular construction will reduce the time needed for prototyping robotic systems, and will allow to focus directly on tuning control algorithms. In addition, authors focus on the circuit design of the future kit and suggest possible software. The models and structural configurations of wheeled and tracked platforms and manipulator made as basic robotic systems using proposed kit are shown in the article.

Keywords: Modular Kit; Mobile Robotic; Prototyping; Algorithmic Debugging

1. Introduction

Robotics is an advanced course of science and technology and includes a variety of fields such as construction, modeling, programming, etc. One of the major problems of modern robotics appears to be the development of education programs [1-3]. Today, "sport" approach to the education is generally used; its intention is to prepare projects for national or international competitions and festivals. The teaching process in that case includes learning element base and basic constructions, basics of programming language, solving simple tasks such as path following, obstacle avoidance, etc., iterative preparations of robots due to the regulations of specific competition. However, drawbacks of "sport" robotics are obvious – quick results do not guarantee getting high quality knowledge, as well as, vide variety of competitions do not often allow to make an effective learning process. Opposite to the mentioned approach is a popular paradigm STEM (Science Technology Engineering Mathematics), which presents robotics as an interactive tool, which helps to practice theoretical knowledge about Exact and Natural Sciences [4]. An issue during STEM appliance is to create qualitative and reliable training tools, which could provide essential hardware and software to solve robotics tasks [3, 5].

Available on the market robotics kits (LEGO Mindstorms, Fischertechnik, Makeblock, etc. [6-10]) despite their popularity usually have the element base which is insufficient to execute complex algorithms, besides, since they consist of small parts, the major part of learning process is occupied by prototyping instead of studying control tasks. To solve mentioned problems authors of the article offer a kit of mobile robotic modules (MRM). Simple and fast assembly using complete modules will reduce the time needed for prototyping robotic systems and will allow to focus directly on control algorithms implementation.

2. Constructions and configurations of a mobile modular robotic kit

A large modules assembly is used for the realization of the modular kit. According to that method, all modules should be the same size (as Moss and Cubelets have [11, 12]) or connectors should be scalable. Joining of modules and installation on the frame (if there is one) could be done without fasteners, such as screws, nuts, etc., for that purpose modules have quick-detached connectors, such as:

- the universal coupler to fasten drive modules and sensors. That coupler is a combination of rail type mountings on the frame. A module is placed on a rail, and there are options whether to let it to move forward-backward along the rail or to fix it with vices;
- fasteners (thread bushing) to connect drive modules and robot power supply, such as propulsion engines, mechanical transmissions. That joint needs knurled screws. Using that connection type, kit modules could be installed on different frames besides one from MRM kit;
- the universal coupler to transmit torque between drive and mechanical transmissions modules, as well as to connect with locomotion actuators such as wheels, caterpillar tracks, etc. Those connectors are represented as a separate category in order to assure strong connection and the absence of mechanical gaps, using inner system of shaft misalignment elimination. Bodies of modules are connected with nuts on the output and threads on the input of transmissions modules, while an elastomer coupling transmits the moment.

According to suggested kit design, each module is a completely closed system. Regarding the necessary functionality which MRM model should have in order to perform different control algorithms, the following modules were designed:

- CSM – control system module, which the user directly works with and upgrades its software;
- APS – autonomous power supply to support other modules;
- PDM – power drive module, which is motor with control electronics to move around the whole robotic system;
- IM – instrumentation modules, which have different sensors, such as distance, touch, light and other sensors;
- CM – communication module, which connects a robot with a console (a personal computer);
- MTM – mechanical transmissions modules (connectors between motors, grippers, rotators and different locomotion actuators as wheels modules, track modules and caterpillars themselves), which transmit torque from the drive modules and allow the user to increase or reduce full speed of rotation and the moment of motor output shaft and also to put shaft angularly to the motor or to move it.

Here are the most usual control tasks [13-15] for robotic systems (RS):

- controlling robotic system which moves on the high speed and orients itself using variety of markers (along solid and dashed lines) or obstacles (static and moving) in order to automatically correct its path according to the task (avoidance, collision, stop to avoid collision, etc.);
- centralized or decentralized control of RS group, such as an allocation of the robots; modification of group behavior and priorities of individual assignments, if one of the robots has malfunctions; area reconnaissance. In all that cases (and also for the tasks mentioned before), wheeled robots are most suitable, due to their ability to adapt for changes in the surface;
- RS control in the conditions of rough terrain: overcoming various obstacles, both natural and artificial, using running advantages of a particular device, in that case, tracked robots are suitable because of their passing abilities;
- RS control in order to detect and deliver a particular object: specifying the object's location, selecting among other objects according to the input, grasping the object by the most appropriate way from available options or prediction of a failure, delivery in the particular area, using given reference points, in those cases, usually a manipulator is required.

Regarding mentioned control tasks and in order to show all functions of MRM kit (using relevant sensors and other equipment), three basic RS could be made using the kit – wheeled, tracked and four-step manipulator (see Fig. 1).

Besides, other RS models could be designed based on the mentioned MRM. Actually, variety of constructions, which could be assembled using the kit, is only limited by a number of modules. For example, using several MRM kits, which are intended for several basic wheeled RS, there could be made one platform with six wheels, each of which has two degree of freedom. RS like that not only has good passing abilities but also makes it possible to perform turning maneuvers using each wheel. In addition, PDM can be installed on MTM, using body element, so that a gap between the plane with MTM fastening and the one of wheel contact could be modified. Four assemblies like that and a location sensor allow the platform to keep the same level while moving along rough terrain. If you put two PDM on the short base and equip the third one with "passive" leg, two-wheeled RS could be made. The advantages of that model are the facts that few modules are required and prototyping takes a little time. Finally, two short platforms, connected by rotators, allow RS to rotate about the axis, so that all four wheels always remain in connection with rough terrain.

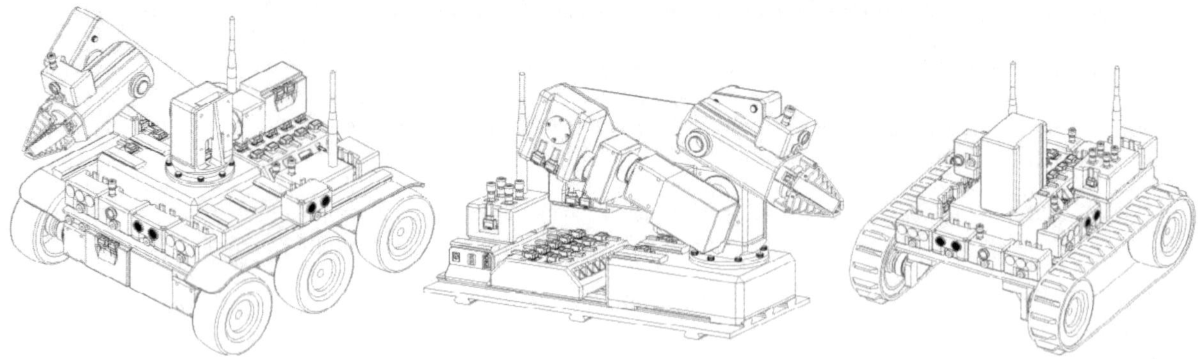

Fig. 1. Basic RS using MRM kit

3. Circuit design of MRM kit

Following design options are suitable for any MRM kit, regardless of the RS configuration or model. All modules that have electronics are connected with the set of RS-485 wires or directly with CSM. To communicate and control the modules, connected with bus, address formats for data exchange are used. Modules get their addresses automatically according to command of CSM. The order of address assignment is defined in advance and depends on the succession of modules. If places of modules from different MRM kits coincide, the same addresses are guaranteed.

The central control element is CSM. It is suggested that one completed system has no more than one CSM. In the base of a module there is STM32F469 microcontroller with ARM Cortex-M4F core. CSM control is performed by user's programs, loaded using USB port in the body of the module. For that purpose, there are the JTAG debugger and USB-to-UART Bridge Controller in the module. CSM supplies connected modules with power while controlling current flow and protecting against overpower, for that purpose it is connected with APS. Operating voltage range of APS is 12 V. Moreover, CSM could be connected with other devices, such as sensors and low-power actuators. For that purpose, modules have some PLT pin connectors with power line of 5 V and 3.3 V, as well as digital and analog inputs/outputs, protected against polarity reversal and overload

All power drive modules are based on gear motors IG-32PGM and IG-16GM with incremental transducers. Electronics, built in modules, provide three-circuit system, controlling motors by winding current, speed and position of a shaft, so that it is possible to perform control algorithms, based on holding the moment, speed and position.

To perform and implement specific algorithms, installation on your RS of different sensors and payloads may be required, depending on the control task you work with. The minimum set of modules, included in MRM kit, consists of four modules, three of which contain several devices and connect with CSM (which connected with CM) by the bus:

- module1: light, temperature sensors and a microphone;
- module 2: a touch sensor, ultrasonic and infrared distance sensors;
- module 3: a dynamic speaker and a lamp;
- module 4: a camera.

In MRM kit there are two communication modules – one is to process and transmit video from cameras, another – to receive control commands and the telemetry data.

To analyze circuit design of each basic RS model from Fig. 1, we will study each module of the RS separately as an assembly scheme. Four-wheeled basic model (see Fig. 2, 3) has battery (APS), control system module (CSM), communication modules (CM) for radio control and video transmission, instrumentation modules (IM), such as a camera, distance sensors and lamps and a manipulator base.

On the base there are power drives (PDM), a gripper (MTM) and a camera (IM). On the lower part of platform there are four propulsion engines (on the PDM and MTM base). There are two fenders, to which you can fasten IM, above the wheels. Thus, using MRM kit, RS with four-wheeled drive and expandable functionality could be designed.

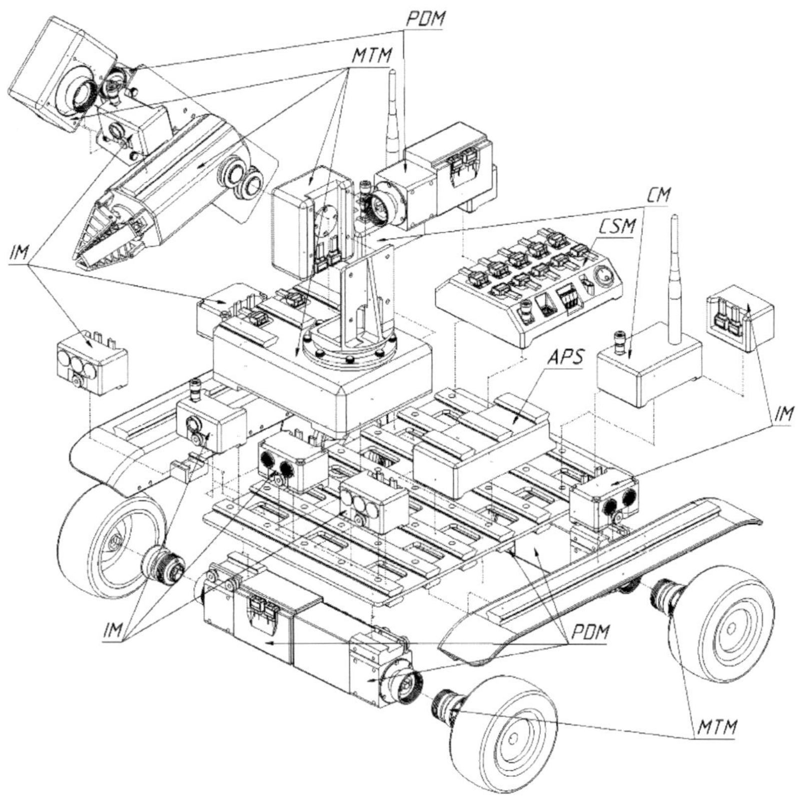

Fig. 2. The assembly scheme of basic four-wheeled RS

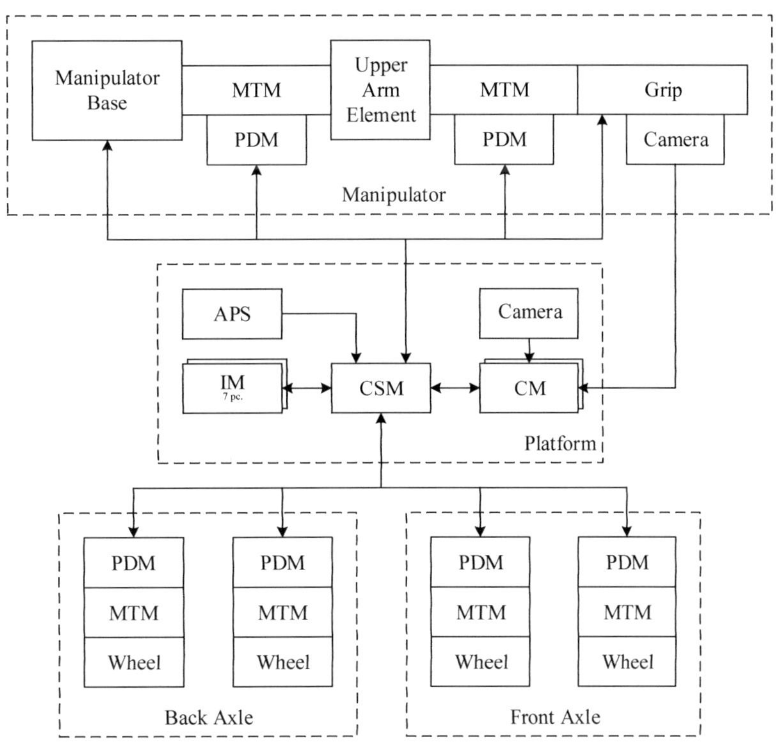

Fig. 3. The structure of four-wheeled RS

Considering applied modules, there is a slight difference between tracked model (see Fig. 4, 5) and the wheeled RS: locomotion actuators are tracks; there are only two propulsion engines (PDM), and each has one cylindrical transmission (MTM); instead of manipulator there is a two-step camera orientation module (MTM).

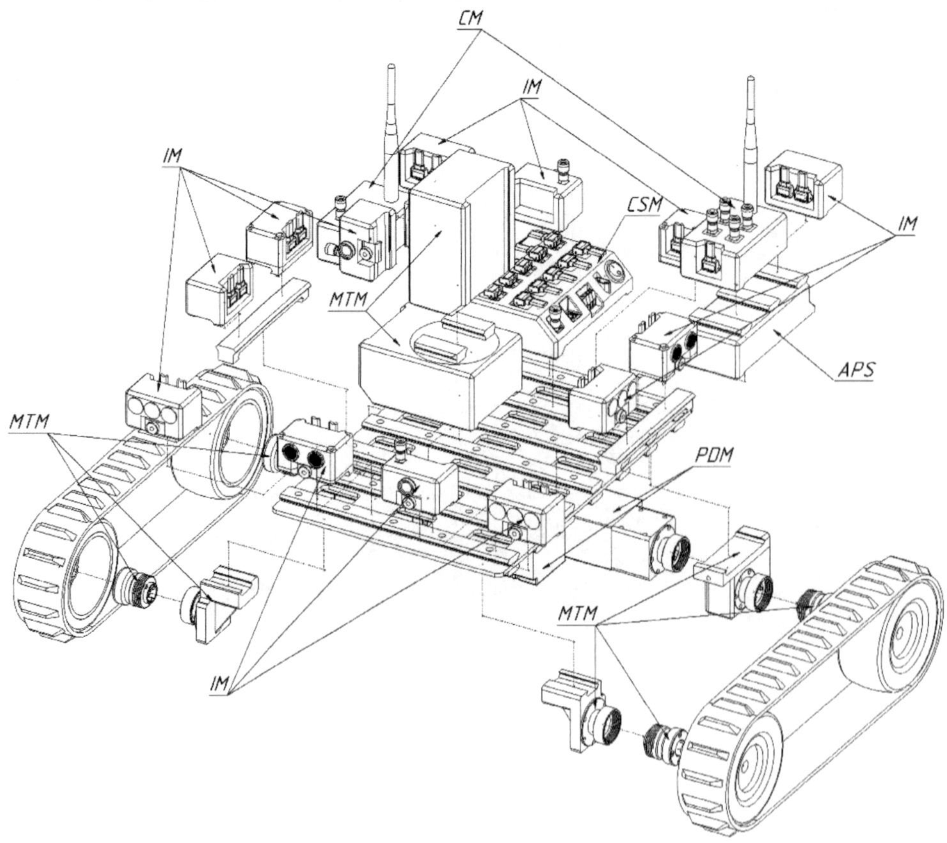

Fig. 4. The assembly scheme of basic tracked RS

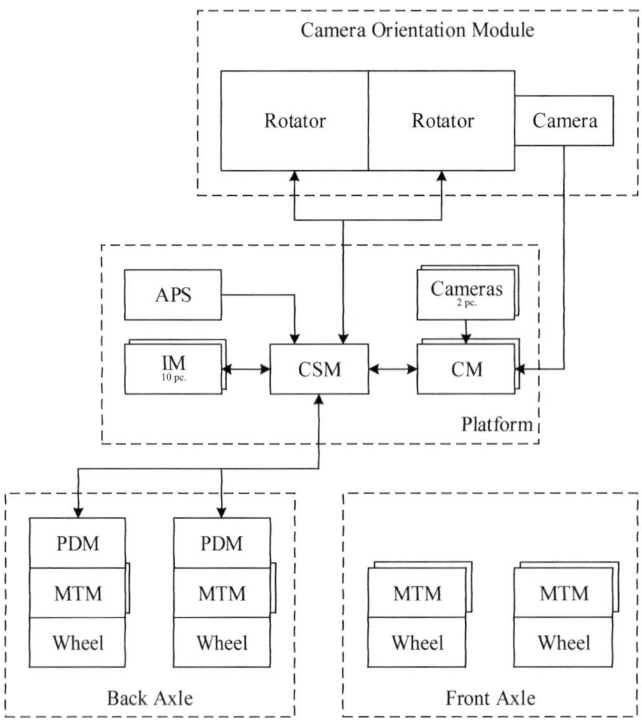

Fig. 5. The structure of tracked RS

Basic model of four-step manipulator with a grip (see Fig. 6, 7) consists of the same modules as the former two, but it also has special transitional body parts.

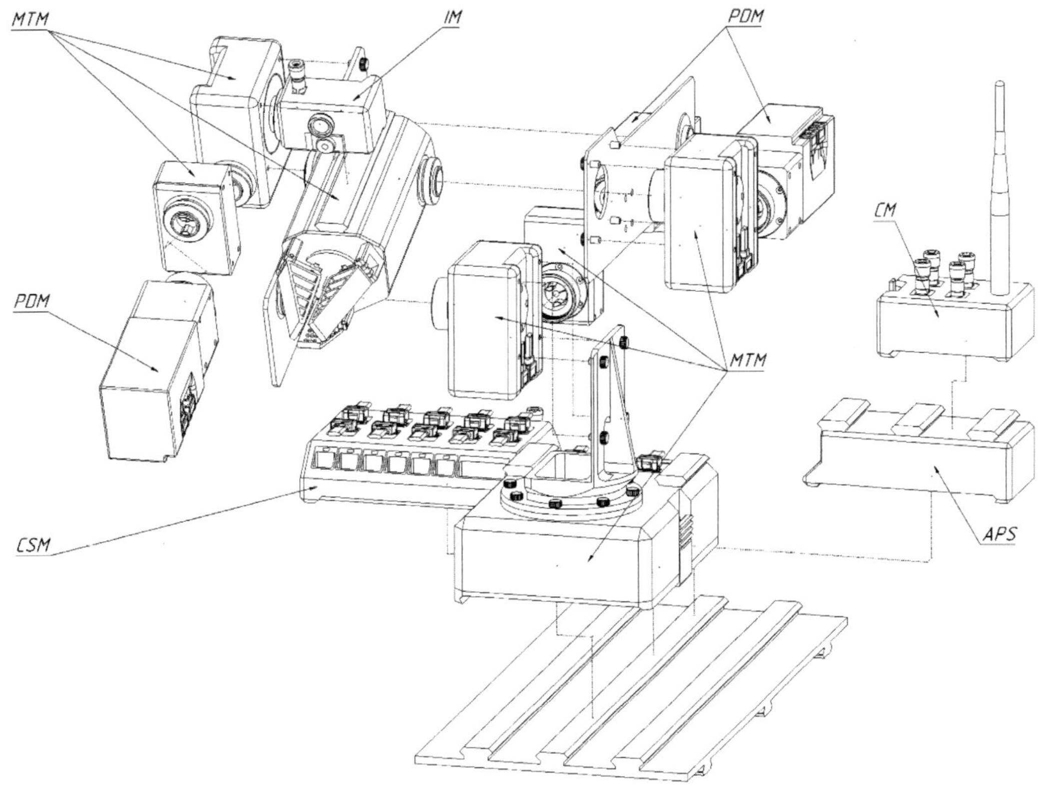

Fig. 6. The assembly scheme of basic RS with four-step manipulator

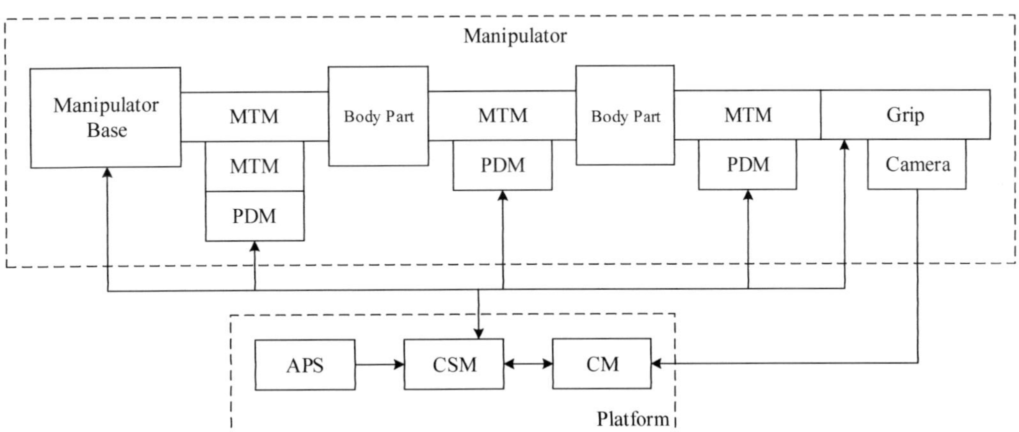

Fig. 7. The structure of RS with four-step manipulator

4. Software solution of MRM kit

Now, the authors do not have final decision on the development environment to debug control algorithms. Still, for convenience and complete understanding of MRM kit there are key aspects for further development below. So that the user can easily begin software development, there are two courses of the development environment:

- programming using textual modeling language;
- programming using graphical modeling language.

Textual modeling involves writing source code of a program using keyboard and presumably in C and C++ languages. For graphical modeling, instead of source code of a program, there are graphical annotations. They could represent whole blocks (lines) of source code and higher abstractions (behavior, module). In view of the successful implementation of a visual development environment for the cybernetic controller TRIK [16], the authors consider the possibility of TRIK STUDIO integration in the software.

5. Conclusion

The main problem of modern educational robotics is to create such kit, which will be useful both in elementary and in middle schools to teach children the basics of engineering and robotics and in high schools and universities for executing and implementing control algorithms. The authors successfully solved this problem by developing MRM kit, the main concept of which was based on the combination of the advantages of a large modular assembly (simplicity and quickness) and the equipment like a powerful controller and a variety of high-quality sensors, which are essential for testing modern algorithms. Moreover, suggested solution could be an effective tool to design and model mobile robotic platforms and industrial robotic complexes for qualified specialists in the research centers and enterprises. Future plans include the production of MRM kit and development of software solutions.

6. Acknowledgements

This research is supported by the Ministry of Education and Science of Russia (an agreement № 14.578.21.0124 on granting for the implementation of applied scientific research and experimental development). Unique identificator of ASRED – RFMEFI57815X0124.

7. References

[1] http://raor.ru, (2016). Official web site of Russian association of educational robotics, Accessed on: 2016-10-31
[2] Pryanichnikov, V.; Andreev, V.; Bobrov, P.; Biryukova, E.; Frolov, A.; Kharin, K.; Kirsanov, K.; Kostin, A.; Kuvshinov, S.; Marzanov, Y. & Prysev, E. (2012). Scientific-educational distributed laboratory-software and hardware means. Annals of DAAAM for 2012 & Proceedings of the 23rd International DAAAM Symposium, Vol. 23, No. 1, pp. 1175-1178, ISSN 2304-1382
[3] Seiler, S.; Ptasik, D. & Sell, R. (2011). Remote and virtual labs in a distance learning environment. Annals of DAAAM for 2011 & Proceedings of the 22nd International DAAAM Symposium, Vol. 22, No. 1, pp. 611-612, ISSN 1726-9679
[4] Alvarez, I. (2011). VEX Robotics: STEM Program and Robotics Competition Expansion into Europe, In: Research and Education in Robotics – EUROBOT 2011, Obdrzalek, D. & Gottscheber, A., pp. 10-16, Springer, ISBN 978-3-642-21975-7
[5] Filippov, S.A. (2013). Robotics for children and parents, rus: Robototekhnika dlya detey i roditeley, Nauka, ISBN 978-5-02-038-200-8, St. Petersburg, Russia
[6] Baichtal, J. (2015). Hacking Your LEGO Mindstorms EV3 Kit, Que Publishing, ISBN-13: 978-0-7897-5538-4
[7] Ferrari, M. & Ferrari, G. (2011). Building Robots with LEGO Mindstorms NXT, Syngress, ISBN-13: 978-1597491525
[8] Cuéllar, M.P. & Pegalajar, M.C. (2014). Design and implementation of intelligent systems with LEGO Mindstorms for undergraduate computer engineers. Computer Applications in Engineering Education, No. 22, pp. 153-166, ISSN 1099-0542
[9] Beliovskaya, L.G. & Beliovskiy, A.Ye. (2013). Programming microcomputer NXT with LabVIEW, rus: Programmiruyem mikrokompyuter NXT v LabVIEW, DMK Press, ISBN 978-5-94074-594-5, Moscow, Russia
[10] Bagnall, B. (2012). Maximum LEGO NXT: Building Robots with Java Brains 3rd, Variant Press, ISBN-13: 978-0986832222
[11] Thai, C. N. & Paulishen, M. (2011). Using robotis bioloid systems for instructional robotics, Proceedings of IEEE Southeastcon, Nashville, USA, 17-20 March 2011, ISSN: 1558-058X, ISBN: 978-1-61284-738-2, pp. 300-306, IEEE, DOI 10.1109/SECON.2011.5752954
[12] http://www.modrobotics.com, (2016). Official web site of kits MOSS and Cubelets, Accessed on: 2016-10-31
[13] Kee, D. (2011). Educational Robotics Primary and Secondary Education. IEEE Robotics & Automation Magazine, No. 18(4), pp. 16-19, ISSN 1070-9932
[14] Khalamov, V.N. & Sagritdinova N.A. (2012). Fischertechnik – basics of educational robotics, rus: Fischertechnik – osnovy obrazovatelnoy robototekhniki, Vzglyad, ISBN 978-5-93946-191-7, Chelyabinsk, Russia
[15] Correll, N.; Wailes, C. & Slaby, S. (2014). A One-Hour Curriculum to Engage Middle School Students in Robotics and Computer Science Using, In: Distributed Autonomous Robotic Systems, Ani Hsieh, M. & Chirikjian, G., pp. 165-176, Springer, ISBN 978-3-642-55146-8
[16] http://www. trikset.com, (2016). Official web site of TRIK project, Accessed on: 2016-10-31

DOI: 10.2507/28th.daaam.proceedings.133

A CUSTOMER'S VIEW ON KEY ASPECTS OF METAL ADDITIVE MANUFACTURING

Martin Nozar, Ivana Zetková, Pavel Hanzl & Milan Daňa

Regional Technological Institute, University of West Bohemia – Faculty of Mechanical Engineering, Univerzitní 8, 306 14 Pilsen, Czech Republic

This Publication has to be referred as: Nozar, M[artin]; Zetkova, I[vana]; Hanzl, P[avel] & Dana, M[ilan] (2017). A Customer's View on Key Aspects of Metal Additive Manufacturing, Proceedings of the 28th DAAAM International Symposium, pp.0957-0966, B. Katalinic (Ed.), Published by DAAAM International, ISBN 978-3-902734-11-2, ISSN 1726-9679, Vienna, Austria
DOI: 10.2507/28th.daaam.proceedings.133

Abstract

Additive technologies enable the creation of complex shapes and structures virtually unattainable using classical production processes. These technologies allow components to be produced not only from paper or plastic, but also from composite materials or metals, with the same or even better properties than their predecessors made by machining or forming technologies. However, for customers who are not specialists in these technologies it may often be difficult to orientate themselves in the wide range of options and take full advantage of the potential of these technologies. As a rule, production process parameters are not as decisive as final product characteristics, especially the price, quality and delivery date. Therefore, this article tries to answer elementary questions that customers might have when considering production using additive technology, and to describe its specifics and possibilities. Special attention is paid to metal 3D printing, whose specifics are illustrated using the example of printing a simple component in various ways, in different quantities and sizes. The article documents an experiment conducted under real conditions using DMLS technology, which shows how differences in ways of printing affect the price, the processing time, and the quality of the final component.

Keywords: Additive manufacturing; build orientation; processing time; build cost

1. Introduction

1.1. Additive manufacturing

Additive manufacturing (AM), also known as 3D printing, rapid manufacturing, additive fabrication, additive processes, additive techniques, additive layer manufacturing, layer manufacturing, and freeform fabrication is defined by the American Society for Testing and Materials (ASTM) [10] as "The process of joining materials to make objects from 3D model data, usually layer upon layer, as opposed to subtractive manufacturing methodologies".

These technologies have witnessed intensive development in recent years and have found their applications not only in industrial and prototype manufacturing. Nonetheless, they are very specific and their production potential significantly differs from conventional production technologies. Although they can create very complicated shapes and structures which were previously almost impossible to manufacture, these technologies have their own limitations and are not capable of doing everything that a designer could invent. Specialists working with this technology are well aware of its capabilities and limitations, but for customers it could be quite difficult to properly consider its capabilities with respect to the required parameters and to correctly formulate their requirements. Because of this, they are often dependent on recommendations from specialists, who have to identify and formalize all the relevant data. However, in order to take complete advantage of the opportunities offered by the technology, it is necessary to know exactly what is manufacturable using this technology and what is not. For this purpose, this article tries to take a closer look at the options offered by additive manufacturing from a customer's point of view and to study the printing of one specific component, with emphasis on three important factors - price, quality and delivery date. Its motivation is to point out, with the help of a simple experiment, the differences from conventional manufacturing that may not be obvious but which could have a significant impact on the properties of the final product, and to answer questions that customers may have when considering the manufacture of a particular component using this technology. The experiment, in which several test prints and print simulations are carried out, clearly illustrates the differences in price, time and quality of identical components printed in different ways and in different sizes and quantities. In doing so, the key aspects of this technology are demonstrated here, and the question is answered whether higher quality always means higher costs in the additive production process, and to what extent cost-saving efforts could negatively affect the quality of the manufactured component.

1.2. Advantages, limitations and key aspects of additive technologies

Additive technologies in themselves have a number of advantages over conventional manufacturing methods. Among the most prominent is the possibility to design parts with unlimited complexity, allowing twisted and contorted shapes, variable wall thickness, blind holes and screws, and very high strength-to-weight ratio. And therefore, since the manufacturing of complex aesthetic shapes is no longer a problem, designers can focus on designing exceptional part functionality and assembly – additive technology allows several parts to be combined into an integrated assembly. [4] Moreover, additive technologies allow the true customization of every product – every single part, built at the same time, can have a different shape or size. And in doing so, there is no need for tooling development, so additive technologies make the production process faster by passing from design to production.

On the other hand, there are some drawbacks associated with these processes including high machine, material, and maintenance costs, slow production rate, limited build sizes, dimensional accuracy, and the small variety of materials available for processing. In comparison to conventional manufacturing, there are some unexpected downsides, which are generally based on a particular AM machine (i.e. its building method and parameters), and particular settings of the building process. A specific product's design, considering possible build problems and constraints, needs to be carefully analysed in order to achieve such optimal parameters for the production process [12].

In terms of quality, the most important aspects of a final product are its mechanical properties, dimensional and shape accuracy, and surface roughness. All of these aspects can be influenced by the correct setting of the processing parameters (for more details see e.g. [7]), above all with:

- Laser power and scanning speed on the structure of the material
- Hatch angle (the angle between laser scanning directions on individual layers)
- Building orientation (the acute angle between the longitudinal axis of a fabricated part and the vertical axis)
- Layer thickness
- Overlap rate (the amount of area which is influenced by repeated sintering with the energy beam)

While there is often no ideal combination of these parameters, it is always necessary to consider the following negative factors which decrease the quality of the final component:

- Staircase effect (a negative effect causing surface roughness to increase and degradation of dimensional accuracy due to orientation of the printed part on the platform)
- Volume shrinkage (risk of degradation of shape and dimensional accuracy caused by print settings)
- Heat dissipation (risk of possible deformation of printed parts due to laser heat transfer to surrounding material)
- Removability of support structures (hard to remove or impossible to remove supports, particularly of internal structures and cavities)
- Tension in material (presence of temperature gradient along the Z-axis of the part bed, which must be removed by subsequent heat treatment)

To a certain extent it is possible to influence all of these factors by suitable settings of the process parameters, but in the described experiment the setting of the layer thickness, the printed component orientation and creation of effective support structures are crucial (while maintaining the other parameters of the recommended settings). While the layer thickness could be set up as needed (according to the machine parameters), the support structures depend on the chosen orientation. However, the decision on a component's orientation has much more significant consequences. In addition to a component's mechanical properties (anisotropy), it influences the surface quality, processing time and costs. These aspects, which are particularly important and examined in this article, are influenced by the component orientation as follows [2]:

- The height of the part in the build direction, which is directly related to the total build time and hence final cost
- Total volume of support material used. The support structure does not contribute towards the finished part, wasting both build time and material
- Total area of contact of the part with the support structure. Reducing contact area decreases the time and the cost of removing supports and finishing surfaces
- The quality of selected faces (or total surface area) measured by surface accuracy. How a part is oriented determines which faces are subjected to the staircase effect and which are in contact with supports. Both these factors deteriorate the surface quality of the manufactured part

2. Specifications and parameters of the experiment

2.1. Building technology

The experiment described in this article was carried out using Direct Metal Laser Sintering (DMLS) technology. This technology is one of the major additive methods used to manufacture metal components. In a nutshell, this method operates as follows: A 3D model in "stl" format is redesigned as required (above all, any surface errors and inappropriate shapes are eliminated), support structures are generated, road paths are calculated and then the model is digitally cut into discrete slices. These slices are sent to the DMLS machine, which recombines them in a layer-by-layer sequence. The machine selectively scans the surface of the metal powder bed with a fibre laser, effectively creating a thin, planar slice of solid part geometry. Once the sintering of the layer is complete, a new specific increment of metal powder is deposited and the sintering of the next layer commences. This cycle is repeated until the build is complete [2].

2.2. AM machine technical data

An AM machine of type EOS M 290 was used for this experiment. This model has a building volume 250 x 250 x 325 mm and uses a Yb fibre laser with 400 W beam power, up to 7.0 m/s scanning speed and 100 μm focus diameter. EOS Maraging Steel MS1 (AKA 1.2709 or X3NiCoMoTi 18-9-5) with 8.0 g/cm³ density was used as build material, and Argon was used as the protective atmosphere.

2.3. Model component

The model part selected for printing can be seen in Figure 1. This component in various orientations and quantities was used for several trial prints and simulations. The orientation of the component on the platform has a significant effect on building time, quality and the cost of the final part. The aim of the experiment is to compare 4 orientations to find the lowest cost, processing time and the best quality of the component.

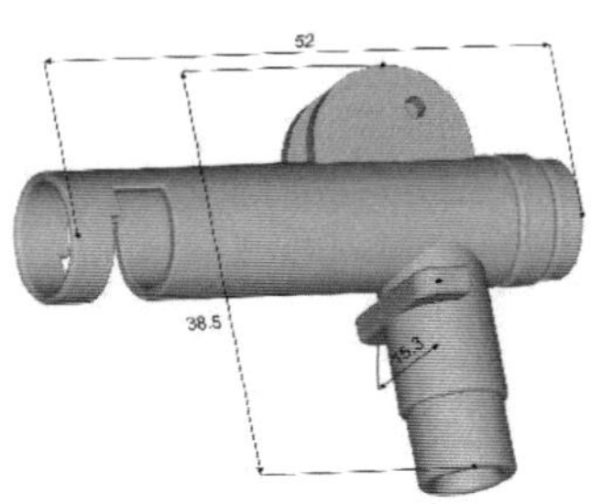

Fig. 1. Model component

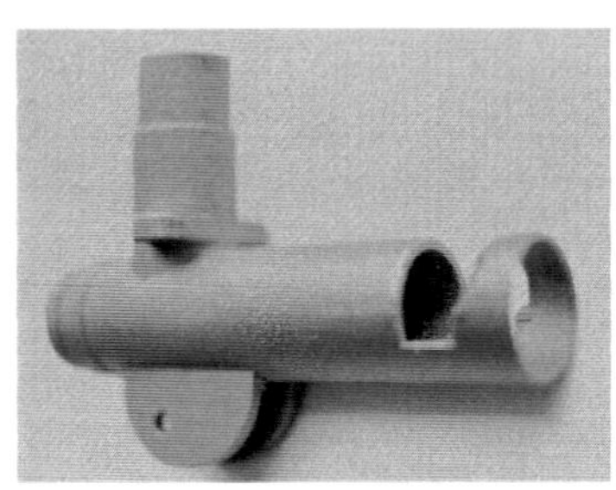

Fig. 2. Built component

For the comparison, the component was printed in 3 variants with 20 and 40 μm layers and in different orientations (see Fig. 3) which have specific advantages and disadvantages. The basic variant was a printing simulation of a single component on the platform. The next, in order to take into account the effect of the size on detected values, was the simulated component print in the given orientations but with four times larger dimensions. And the last, due to the obvious high cost of printing single components, represented a print simulation of the maximum possible number of components on the platform (Fig. 4).

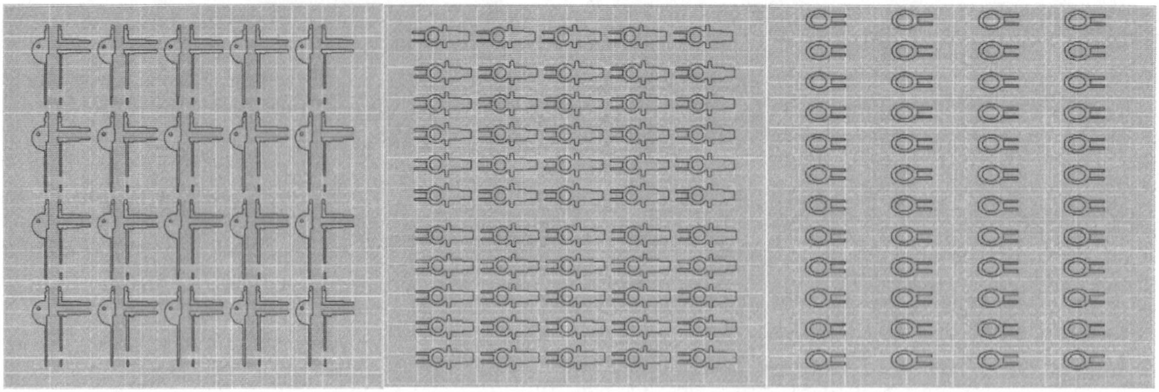

Fig. 3. - 5. Distribution of components on the platform

3. Qualitative aspects

3.1. Build orientation and support structures

Building orientation is a very important factor that affects the build process as well as the post-processing costs and mechanical properties. For instance, it causes anisotropy of the tensile properties of every fabricated part and therefore the placement and angle of rotation on the platform have to be chosen optimally according to the functionality of the part. Special attention must be paid to the fact that this direction also predetermines the number and volume of the support structures required.

The model component was printed in positions as shown in Fig. 6 - 9. The supports are shown in red. The volume of these components and their supports are given in Table 2. The supports were automatically generated by Materialize Magics software and adjusted as necessary.

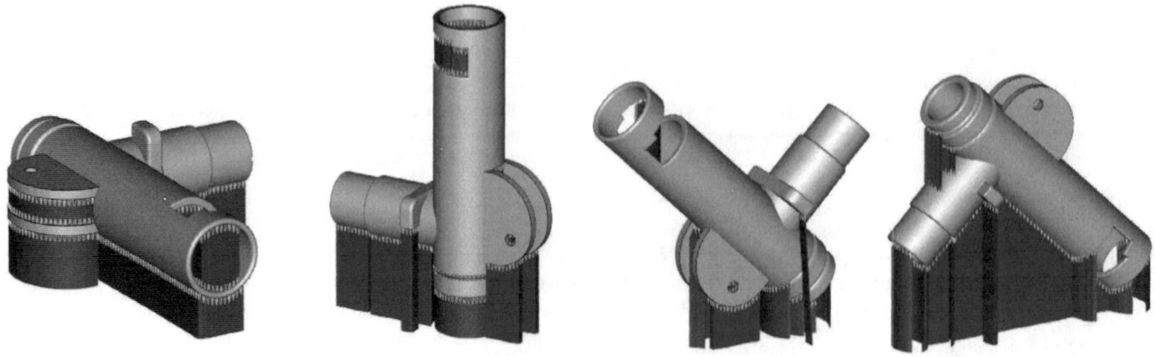

Fig. 6. – 9. Component orientations No. 1 - No. 4

The positons in Fig. 6. - 9. were chosen intentionally, above all with respect to the following principles of building orientation and support creation [2]:

- Minimizing the part's height (consistent with Orientation No. 1)
- Maximizing the stability of the object by selecting the largest base convex hull (consistent with Orientation No. 1)
- Minimizing the volume of the support to reduce excess time and cost of depositing and removing wasted material (consistent with Orientation No. 3)
- Minimizing total surface area of contact with the support to reduce support removal and finishing time (consistent with Orientation No. 3 and partly No. 4)
- Maximizing the surface accuracy of the part for better part quality, which will reduce the finishing time (consistent with Orientation No. 2).

3.2. Quality evaluation of individual components

As already mentioned, using the appropriate build orientation and support structures can significantly improve the quality of the part, because a major source of part inaccuracy and degraded surface properties are contacts with support structures and the staircase effect.

Therefore, it is wise to minimize these factors as much as possible to avoid post-processing (grinding, polishing, etc.) operations. Unfortunately, meeting these requirements can be quite complicated because on the one hand functional areas should be oriented either horizontally or vertically to maximize surface accuracy, and on the other hand horizontal down-facing surfaces need to be supported, while vertical ones do not. This is exactly the main problem of orientations No. 3 and 4 in our experiment.

In the experiment, the parts were printed in 40 μm and 20 μm thin layers. As expected, the thinner layers were always of higher quality, but the printing time was roughly double, as can be seen from Table 1. A partial solution to this could be the option of setting individual layers, which enables different parameters to be used locally and depositing thinner layers only where necessary. Modern AM machines already have this functionality, which can improve surface properties, without increasing costs too much. In the case of the EOS M 290 printer, this is enabled by the latest EOS Print 2.0 software [9].

Orientation No. 1: This orientation is not very recommendable due to the great volume of the supports. These supports negatively affect the quality by contacting the printed components across a too large surface, even in the internal parts, from which it is relatively hard to remove them. Thus, the printed component will still have to be machined.

Orientation No. 2: This orientation shows a relatively small volume of supports, but unfortunately, some of them are internal. The main advantage of this orientation is the shape and dimensional accuracy. However, some internal and functional areas in the lower part of the component need to be machined.

Orientation No. 3: This orientation has the minimum volume of supports, which is achieved by rotation of the component, meaning that there is no need for internal supports (except for two tiny and easily removable internal supports). The main disadvantage is the degraded surface quality due to the staircase effect. For a component printed in this way, it is suitable to create it in 20 μm layers or use some subsequent finishing operations.

Orientation No. 4: The component with this orientation has relatively extensive supports, but only external ones (except for one tiny and easily removable internal support), and provides the printed component with good stability and heat dissipation. Due to the large contact of the supports with the body of the component, subsequent finishing operations are required, as in Orientation No. 3.

4. Time aspects

4.1. Processing times

Many relevant studies [1], [3], [6] in accordance with observations from practice confirm that the manufacturing time per part is the key factor to optimizing the costs of additive manufacturing. The time spent on the whole building process consists of:

- Pre-processing time (model preparation, machine set-up, machine atmosphere generation, machine warm-up, etc.)
- Processing time (scanning time, Z-axis movement, levelling, non-manufacturing movement of the nozzle/laser, etc.)
- Post-processing time (cooling time, machine cleaning, unused powder sifting, sawing pieces from the platform, finishing operations such as heat treatment, milling, polishing, etc.)

In our experiment, we took into account only the processing time, since pre-processing times are practically the same for all positions, and the post-processing times differ primarily in the finishing operations required by a customer or by the function of the printed component. Generally, post-processing times can be reduced in various ways that are, however, different for each and every component, and therefore they have not been taken into consideration, neither have we considered time losses caused by possible failures or repairs of defective components.

A simulated print was performed to compare the print times of model components in different orientations - the prepared models were sent to the machine, which determined the exact printing times. The simulated print was made for variants with single component printing, enlarged component printing, and print of multiple components with original size. In this variant, the building platform was completely filled with components so the part in Orientation No. 1 was printed 20 times, 55 times in Orientation No. 2, and 48 times in Orientations No. 3 and 4. See Fig. 6. - 9.

Processing times of the model component in the different orientations and variants are given in Table 1.

Component type	Processing Times in different Orientations (min.)							
	No. 1		No. 2		No. 3		No. 4	
	layer$_{20\,\mu m}$	layer$_{40\,\mu m}$	layer$_{20\,\mu m}$	layer$_{40\,\mu m}$	layer$_{20\,\mu m}$	layer$_{40\,\mu m}$	layer$_{20\,\mu m}$	layer$_{40\,\mu m}$
Single	239	**120**	537	**241**	466	**218**	478	**224**
Multiple	1 666	**1 037**	3 969	**2 025**	2 887	**1 976**	3 661	**2 300**
Large	4 349	**2 346**	5 158	**2 206**	4 352	**2 195**	4 798	**2 440**

Table 1. Print Times (in minutes)

4.2 Evaluation of component processing times

As shown in Table 1., the fastest possible print of the component can be achieved in Orientation No. 1. when applying 40 µm thin layers. In contrast, when applying 20 µm thin layers in Orientation No. 2, the print is almost four times longer, due to the largest building size of the component in this orientation. This is also the reason why the time values for orientations No. 3 and 4 are quite similar, since the components have the same building height. Observable differences are only caused by the difference in support print times, and these are more massive in Orientation No. 4.

In general, printing using 20 µm thin layers is roughly two times longer than for 40 µm layers. A big difference is evident between the time of printing the original and the enlarged component, which is 4 times higher, but the print time is 6-10 times longer. This disproportion is caused not only by the printing of a greater volume, but above all by the significant building height, which means many more deposited layers, and hence more time spent on the recoating procedures.

5. Price aspects

5.1. Costing estimation for additive manufacturing

There are several approaches to determining the cost model of component production using AM. It is possible to use the model below as an example [1]. The highlighted costs are important for the following calculations and the printing cost evaluations.

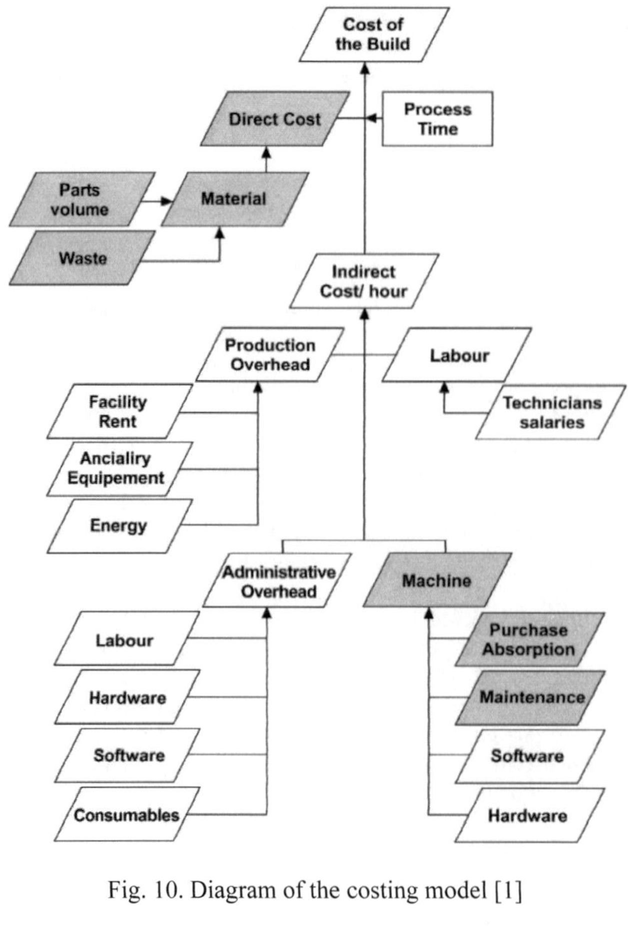

Fig. 10. Diagram of the costing model [1]

5.2. Structure of the processing costs

It is possible to find these with economic (or cost) models (e.g. the model above) and the prices of particular components in various publications (e.g. [1], [5]). However, this article is not concerned with determining the total cost or price of printing, just costs that are directly related to the printing of the model components in the given orientations. For the purpose of costs quantification real and current prices were used, which are not stated in detail in this article, because these calculated costs, no matter how thoroughly they were identified, have only relative informative value and they are stated here just for the possibility of making different options and comparing orientations.

It is not possible to compare these costs with the cost of manufacturing using a different AM machine as material prices depend on the time when the calculation was made, on the country where the survey is conducted and the current exchange rate (this calculation was made with prices valid in the Czech Republic in September 2017 and in Czech Crowns). Further differences in the calculation may be caused by the selected (or required) material, which may have different processing properties and parameters.

Using a different printer could also cause significant differences, as it may have different production capabilities. For example, a printer could have a significantly larger workspace (i.e. multiple components can be printed concurrently), it could be a hybrid AM machine (i.e. capable of machining printed shapes during a printing process) or a multi-laser machine (usually using 4 lasers to shorten the processing time), or a printer could be equipped with a sieving module (to shorten times of powder recycling and to reduce powder material losses). However, all these benefits mean increased investment costs for the acquisition of the AM machine and its appropriate equipment.

Very significant differences in values can also be obtained if we take into consideration the operator's capabilities and the equipment's operating conditions, in particular the service lifetime and worktime fund of a given AM machine. In the case of the AM machine used in this article, its user is a university research centre, which uses this AM machine for research and prototype development. This means a larger portion of pre-processing time, shorter working hours and potentially longer service lifetime than printers used for commercial production. All this is reflected in the hourly machine rate, which is based on the purchase price and the depreciation method of the AM machine. That is why there are almost no absolute sums here, as they virtually would not be predictive.

To complement this topic, we recommend an article [1] in which the authors made a cost calculation based on 60% utilization of the machine time fund in 8 years, i.e. at the limit recommended by the machine manufacturer. A study [8] from 2016 is also very interesting and relevant, in which its author compared three different AM systems (including the EOS M 290) and stated the hourly machine rates with proportions of their individual components. Considering a 5-year depreciation period, utilization 5 000 h p.a., 10% maintenance and downtime, the hourly machine rates (include only processing time, i.e. without material and direct labour costs), range from 35 - 45 EUR/h for a 'small single-laser' machine to 90 - 120 EUR/h for a 'large machine'.

5.3 Relevant costs calculation

Calculated costs considered in our experiment include material costs (energy and gas) and processing costs (which result from processing time, so they include machine depreciation and regular maintenance costs). The consumption of metal powder is calculated according to the volumes of the printed components and their respective support structures. Both these volumes are stated in Table 2.

Component type	Volumes in different Orientations							
	No. 1		No. 2		No. 3		No. 4	
	Component	Support	Component	Support	Component	Support	Component	Support
Single	**3 422**	7 665	**3 422**	3 146	**3 422**	903	**3 422**	3 840
Multiple	**68 440**	153 302	**188 210**	173 003	**164 256**	43 351	**164 256**	184 344
Large	**219 002**	348 825	**219 002**	144 872	**219 002**	29 642	**219 002**	158 356

Table 2. Volumes of printed components and their support structures (in mm^3)

It is necessary to add the volumes of metal power depreciated and lost during processing and post-processing operations and which cannot be reused, to the volumes stated in Table 2.. According to the producer, 2- 4 % of the powder material is considered waste.

However, according to our own measurements (accurate material weighing before and after printing), EOS MS1 metal powder losses reach up to 25% of the printed component's volume. Therefore, this amount of metal powder was added into the calculated consumption. The consumption of electric power (above all of the printer and compressor) and gas (e.g. Argon; to create a protective atmosphere) was also put into the calculation.

This calculation revealed (in accordance with some previous studies [1], [4]), the major part of the cost is found to be due to investment in the machine and regular maintenance costs. The costs of energy and materials only contribute to a small extent. The sum of costs (C_Σ) and share of processing cost (C_{pt}) for all simulated orientations and variants are shown in Table 3. The sums of costs are stated in Euros (recalculated with current exchange rate 26 EUR/CZE). Note: Q_{max} is the maximum number of components on the platform.

Orientation	Single				One of Q_{max}				Large			
	layer 20 µm		layer 40 µm		layer 20 µm		layer 40 µm		layer 20 µm		layer 40 µm	
	C_Σ	C_{pt}/C_Σ	C_Σ	C_{pt}/C_Σ	C_Σ	C_{pt}/C_Σ	C_Σ	C_{pt}/C_Σ	C_Σ	C_{pt}/C_Σ	C_Σ	C_{pt}/C_Σ
No. 1	172	92%	**94**	**85%**	69	80%	**48**	**71%**	3 615	80%	**2 289**	**68%**
No. 2	364	98%	**168**	**95%**	57	84%	**33**	**74%**	3 909	87%	**1 955**	**75%**
No. 3	315	98%	**150**	**96%**	46	87%	**33**	**82%**	3 239	89%	**1 811**	**80%**
No. 4	326	97%	**158**	**94%**	60	84%	**41**	**77%**	3 687	86%	**2 126**	**76%**

Table 3. Sum of costs (in EUR) and share of processing costs in them

5.4. Evaluation of variants

As outlined in Table 3., when printing single components, the cheapest component is printed in Orientation No. 1, with no difference when printing in 20 or 40 µm thin layers. In other cases, the cheapest printing is in orientation No. 3, with the exception of printing with a maximum number of components in 40 µm thin layers, where costs are even slightly lower. The values given above also indicate how beneficial the horizontal filling of an AM machine working space is. The greatest savings are thus obtained when printing in 20 µm thin layers (on average 78%), especially in orientations No.2 (84%) and No.3 (85%). As expected, printing in 20 µm layers is always more expensive; for single component print variants on average twice as expensive, for multiple component print variants 1.5 times as expensive, and for large print variants 1.7 times as expensive.

This table also shows that the decisive factor in the calculated costs is the processing cost (in another words machine investment cost), which reaches on average 94% (for all orientations and both layer thicknesses) during single component printing. This value does not fall below 80% on average, even in the cost calculation of the multiple component print when the total cost is distributed per piece, or in the cost calculation of the enlarged component print. Even when considering the maximum recommended machine utilization, as used in the article [4], the average percentages of processing costs do not fall below 88% in single print variants and below 64% in large and multiple print variants. In this case, the total lowest values were reached by printing on a full platform with components in orientations No. 1 and No. 2. However, the percentage values exceed 52% here as well. Therefore, it is clear that there is no print variant where the processing costs would not be the decisive part of print costs.

As a result, it is possible to state that in terms of price, the volume of the support structures is not a decisive factor (unless such support structures require complicated and costly removal), and filling the available AM machine build space is the prime determinant for the efficient operation of the technology.

6. Evaluation of experiment results

The purpose of the experiment was to evaluate the printing of a model component in various orientations and three different variants which were simulated by AM machine. The evaluation was carried out with regard to quality, processing time and cost. At the end, the component was printed in order to confirm all our conclusions.

As it turned out during the evaluation, it is not possible to find an ideal component orientation when taking into account the aspect of quality – each orientation has its own advantages and disadvantages, and it is a matter of individual assessment to determine which of these factors are less problematic for the customer and the function of the printed component. A qualitative aspect is therefore useful purely as a guide. A quite similar case is also the time aspect where the component orientation is not decisive, unlike the size of the printed volume. However, processing time is a determining factor for processing costs, which, as has been shown, creates the main part of the printing costs.

In terms of time, orientation No. 1, which has the smallest building height and therefore the smallest number of printed layers, would be the ideal case. Unfortunately, this variant is not the most cost-effective option (it is most cost-effective only when printing a single component, which is the least economical variant) and, moreover, a high volume of internal and external support structures require a large number of finishing operations which increase production time and costs. The disadvantage of this orientation is also the fact that when printing more components at the same time, this component can be placed in the smallest quantity (20 copies) on the building platform. Therefore, when printing a large number of components in this orientation, this printing option ceases to be favourable in both cost and time.

As expected, it reveals the key aspect to be the price, implicitly including the time aspect. Thus, if we prefer the price aspect, it seems orientation No. 3 is the most favourable one in most cases. However, in this case it is necessary to take into account the deteriorated surface quality due to the staircase effect. This can be partially avoided by printing this component in 20 µm thin layers, which means an increase in costs of 40%.

In this context, it is necessary to draw attention to the possible instability of this component during printing due to the small number of support structures that could cause printing defects. Because of this risk the component in orientation No. 4 is equipped with more massive support structures.

These support structures increase the processing costs on average by 16%, but provide the printed component with greater stability and better heat transfer from the laser beam into the surrounding material. However, a component oriented in this way is also liable to suffer a deterioration in surface quality due to the staircase effect.

This undesirable effect has almost no impact on the component in orientation No. 2, which is oriented perpendicularly to the building platform, and thus attains the highest quality in shape and size. A component printed in this orientation is one of the most expensive ones in most cases, but it is possible to print it in the largest number of copies (55 copies) concurrently, but when printing even with a full building platform, the price per component exceeds even the cheapest orientation No. 3. In contrast to this orientation, it has the advantage of having a good surface even when printed in 40 μm thin layers, but the disadvantage is that there are internal support structures at the bottom of the component, which need to be removed.

When considering time and cost, it is quite easy to identify the optimum orientation in given circumstances. Unfortunately, in the case of quality the situation is much more complicated because each orientation has its pros and cons. And insufficient quality could mean repeated printing, which doubles its original price. Therefore, for practical printing it is advisable (due to the significant disparity between material and processing costs) to print the component with a more massive but safer support structure (unless it could cause complications during its removal), to carry out a test print, e.g. add this component to a different print batch in order to verify its properties, or to print this component in several efficient orientations concurrently. There are several approaches to manufacturing a suitable component, but when seeking the best one it is always necessary to know whether quality (i.e. functional properties), time or costs matter the most.

7. Conclusion

This article is dedicated to metal additive manufacturing, which uses a laser beam to melt metal powder to form solid objects. Basic aspects of this technology are presented, its capabilities and limitations that need to be taken into account when considering manufacturing using laser sintering. The article explains that it is not enough just to solve whether the required piece or component can be printed at all, but also the importance of the orientation of this component during printing. From the customer´s point of view it is essential to what extent just this orientation directly affects the quality (surface and mechanical properties), processing time and processing cost. Therefore, this article attempts to give answers to the basic questions regarding which factors affect price, quality and processing time; how to save costs without compromising quality; and whether higher cost and longer processing time always result in better quality. At present, there is some software capable of optimising the arrangement and orientation of parts on the platform, but it only provides recommendations made with regard to appropriate criteria. But as demonstrated in the article, as a rule, every single orientation has its positive and negative aspects, so the final decision about the build orientation is always up to the specialist/customer who has to thoroughly assess the specific situation.

For the purpose of comparison of different building orientations on the platform, i.e. above all for finding the connection between quality, costs and building time, we prepared different printing models, created a calculation formula, quantified cost components and simulated printing of these models. The values we obtained enabled us to compare and evaluate all the orientations and variants from the customer's perspective.

The experiment was carried out on an EOS M 290 machine and based on printing a selected component in four different orientations, quantities and sizes; and the values obtained from each of the simulations were recorded into 3 tables for cross-referencing. As it turned out, when the results were evaluated with regards to quality, time and cost, every orientation and print variant has its own specific advantages and complications, so when deciding on individual print adjustments (not just the number and orientation of components), the key question is always whether to prefer quality (i.e. properties), time or price of these components. Sometimes the solution is unambiguous, but in a number of cases it is, unfortunately, a matter of compromise. However, since the processing time is the largest price factor, all measures to reduce the time also reduce the cost accordingly. As a result, additive technology users should fill the available build space with as many parts as possible because the manufacturing time per part is the key factor to optimizing the costs of additive manufacturing.

This article describes the issue from the perspective of the customer, who is primarily interested in quality, costs and printing time of the required component. However, this issue is more complicated and subsequent research will be conducted in two principal directions. On the one hand, the complete cost formula will be specified (i.e. all parts of the cost model on Fig. 10 will be considered) and a precise formula for building time calculation will be created (all the time values stated in this article were individually indicated by the machine). These formulas will enable us to create a special computer programme able to calculate and recommend optimum placements and building orientations of every piece by using advanced simulations and genetic algorithms. The orientations considered in this article are intentionally chosen to point out the differences between these de facto extremely different building directions, however, it is probable that the best option in many cases would be a combination of these orientations. Although it is possible to use a computer simulation to find such a combination, it requires a clearly and precisely determined criterion function, i.e. exact values of key parameters and their connections with processing parameters. Therefore, parallel research will investigate the impact of all the processing parameters on costs, quality and building time.

8. Acknowledgments

The article has been prepared under project LO1502 'Development of the Regional Technological Institute' under the auspices of the National Sustainability Programme I of the Ministry of Education of the Czech Republic aimed at supporting research, experimental development and innovation.

9. References

[1] Ruffo, M.; Tuck, C.; Hague, R. (2006). Cost estimation for rapid manufacturing – laser sintering production for low to medium volumes, Available from: https://www.researchgate.net/publication/48352741_Cost_estimation_for_rapid_ma-nufacturing_-_Laser_sintering_production_for_low_to_medium_volumes, Accessed: 2017-09-01

[2] Alexander, P.; Allen, S.; Dutta, D. (1998). Part orientation and build cost determination in layered manufacturing. Available from: http://www.sciencedirect.com/science/article/pii/S0010448597000833, Accessed: 2017-09-01

[3] Piili, H.; Happonen, A.; Väistö, T.; Venkataramanan, V.; Partanen, J.; Salminen, A.;(2015). Cost Estimation of Laser Additive Manufacturing of Stainless Steel, Available from: http://www.sciencedirect.com/science/article/pii/S1875-389215015436, Accessed: 2017-09-01

[4] Atzeni, E.; Salmi, A.; (2012). Economics of additive manufacturing for end-usable metal parts. Available from: https://link.springer.com/content/pdf/10.1007%2Fs00170-011-3878-1.pdf, Accessed: 2017-09-01

[5] Baumers, P.; Tuck, C.; Hague, R.; (2016). The cost of additive manufacturing: machine productivity, economies of scale and technology-push, Available from: http://www.sciencedirect.com/science/article/pii/S0040162515000530, Accessed: 2017-09-01

[6] Baumers, M.; Tuck, C.; Wildman, R.; Ashcroft, I.; Rosamond, E.; Hague, R.; (2012). Combined Build–Time, Energy Consumption and Cost Estimation for Direct Metal Laser Sintering, Available from: http://citeseerx.ist.psu.edu/viewdoc/download?doi=10.1.1.476.5195&rep=rep1&type=pdf, Accessed: 2017-09-01

[7] Hanzl, P.; Zetek, M.; Bakša, T.; Kroupa, T. (2015). The Influence of Processing Parameters on the Mechanical Properties of SLM Parts, Available from: http://www.sciencedirect.com/science/article/pii/S1877705815005378, Accessed: 2017-09-01

[8] Berger, R. (2010). Additive Manufacturing – next generation, Available from: https://www.rolandber-ger.com/publications/publication_pdf/roland_berger_additive_manufacturing_next_generation_amnx_study_2016 0412.pdf, Accessed: 2017-09-01

[9] https://www.eos.info/systems_solutions/software/eosprint, (2017). EOS, EOSPRINT 2.0, Accessed: 2017-09-23

[10] https://www.astm.org/Standards/F2792 (2017). ASTM, Standard terminology for additive manufacturing, Accessed: 2017-09-01

[11] Ruffo, M.; Hague, R. (2007). Cost estimation for rapid manufacturing – simultaneous production of mixed compo-nents using laser sintering, Available from: https://dspace.lboro.ac.uk/dspace-jspui/bitstream/2134/4808/1/ruffo2.pdf, Accessed: 2017-09-01

[12] Cekic, A.; Rasovic, N.; Obad, M.; Kaljun, J.; Dolsek, B.& Begic-Hajdarevic, D. (2016). Production of Optimized Layered Products using Intelligent Support, Proceedings of the 26th DAAAM International Symposium, pp.0271-0279, B. Katalinic (Ed.), Published by DAAAM International, ISBN 978-3-902734-07-5, ISSN 1726-9679, Vienna, Austria DOI:10.2507/26th.daaam.proceedings.037

DOI: 10.2507/28th.daaam.proceedings.134

CLASSIFICATION OF AUTOMATIC CONTROLLERS DIAGRAMS

Sviatoslav Klos, Roman Fedoryshyn, Volodymyr Savytskyi,
Yevhen Pistun & Fedir Matiko

This Publication has to be referred as: Klos, S[viatoslav]; Fedoryshyn, R[oman]; Savytskyi, V[olodymyr]; Pistun, Y[evhen] & Matiko, F[edir] (2017). Classification of Automatic Controllers Diagrams, Proceedings of the 28th DAAAM International Symposium, pp.0967-0973, B. Katalinic (Ed.), Published by DAAAM International, ISBN 978-3-902734-11-2, ISSN 1726-9679, Vienna, Austria
DOI: 10.2507/28th.daaam.proceedings.134

Abstract

The analysis of the existing classifications of automatic controllers being applied in the industrial automation systems is presented in the paper. The main classification features in the existing classifications of automatic controllers are as follows: purpose, principle of operation, construction, type of energy being used, type of controlling action etc. A new classification of automatic controllers diagrams with control channels is proposed. According to this classification the control channel is considered as a part of the control loop consisting of an automatic controller and an intermediate element between the controller and the controlled plant. It is proposed to classify the automatic controllers diagrams on the basis of the following three features: type of control law, type of controller output signal and type of controlling action at the input of the controlled plant. The new classification is presented in a graphical and a table form. Each diagram is analyzed in details. The new classification provides exact demonstration and clear understanding of the place of switch, continuous and step controllers among other controllers. The classification also provides the possibility to obtain new types of diagrams of automatic controllers with the control channels. These diagrams can be implemented in PLCs by programming.

Keywords: automatic controller; diagram; classification; industrial automation; control law; output signal

1. Introduction

The main field of automatic control instruments application is technological processes and objects. These instruments may consist of individual control devices or programmable logic controllers (PLCs). Automatic regulation technique is the most important part of a traditional system of automatic control and automation. And the automatic regulation itself is an action destined at maintaining the controlled parameter at a set value.

Nowadays automated control systems are being more and more widely applied. These are human machine systems providing automatic acquisition and processing of data needed for optimization of the control process. Optimization here means choosing such a way of control at which the extreme value of the control criterion is reached.

Automation is still one of the main tasks to be accomplished for industrial production and for social life. As time goes by, automation becomes a wider concept which includes some new tasks of its scientific and engineering progress, i.e. computerization, robotics and other specialized fields of science.

However, the main purpose of automation remains unchanged – it is making easier or complete replacement of human physical and brain work by means of automation instruments.

There is no field of industry where there would be no need to apply the automated systems for technological processes control. The main advantages of the automated control systems application are reduction (or even elimination) of the so called human factor influence on the technological process, minimization of raw materials consumption, improvement of quality of the output product and significant improvement of production efficiency. The main functions of such systems are automatic control, regulation, monitoring, data acquisition, exchange and storage, alarming, automatic protection and blocking as well as generating the curves and reports.

Automation engineers often meet the problem of choosing the instruments for solving the automation problems. There is one more important problem to be solved by the engineers in order to control the technological process – the problem of choosing the structure of the automatic controller (control law) and defining the numerical values of its tuning parameters. This problem can be solved on the basis of recommendations given in the specialized literature and in the technical documentation. To simplify solving of this problem, the classifications of automatic controllers were developed by various authors.

The goal of this work is to analyze the existing classifications of automatic controllers and to present the new developed classification of automatic controllers diagrams with the control channels.

2. Analysis of the existing classifications of automatic controllers

There are various classifications of automatic controllers in literature and electronic resources. Classifications of the following authors are known: Klyuyev A. S., Sharkov A. A., Kazakov A. V. and other.

Automatic controllers are classified by Klyuyev A. S. [1] on the basis of the following features: purpose of the controller, principle of operation, construction, type of energy being used and type of control law.

Based on the type of the controlled parameter the automatic controllers are divided into temperature controllers, pressure controllers, level controllers, flow rate controllers etc.

Based on the construction features the automatic controllers are divided into instrument controllers, device controllers, aggregate (unit) controllers and module (element) controllers. An instrument controller consists of the controller and the sensor (primary transducer). Instrument controllers operate independently (in parallel) to the other measurement instruments for a given technological parameter.

A device controller consists of the controller and the secondary transducer. There is no sensor (primary transducer) in the device controller.

The aggregate (unit) controllers consist of individual units with a specific function of each unit. The input and output signals of the units are uniform which provides the possibility of designing the controllers for various functional purposes.

The module (element) controllers consist of individual modules (elements) that accomplish simple operations. The input and output signals of the modules are uniform which provides the possibility of designing the controllers for various functional purposes (like in unit controllers).

Based on the source of energy being used the automatic controllers are divided into the direct action controllers and the indirect action controllers. In the direct action controllers part of the energy is taken from the controlled plant during measurement of the controlled value. This energy is used for controller operation and its action on the actuator. Thus, there is no external energy supply to the automatic system "controlled plant – automatic controller".

In the indirect action controllers the additional sources of energy are used to provide the controller operation and its action on the actuator.

Depending on the type of energy being used the indirect action controllers are divided into electric (electromechanical, electronic), pneumatic, hydraulic and combined (electro-pneumatic, electro-hydraulic etc.) controllers. The choice of the controller from the point of view of the type of energy being used is defined by the nature of the controlled plant and by the peculiarities of the automatic system.

Based on the type of the control law the automatic controllers are divided into controllers with linear and nonlinear control law. Switch controllers (two-point switch and three-point switch controllers) are the examples of the controllers with nonlinear control law.

According to Sharkov A. A. [2] the electric controllers being applied in the industry are divided into the following two types:

- device controllers consisting of the measurement devices with the built-in electric regulation devices for proportional or switch regulation;
- apparatus controllers consisting of autonomous control instruments which receive signals from the automatic sensors, implement a definite control law and transmit the controlling action signals to the actuators.

Classification of automatic controllers based on the purpose, principle of operation, construction, type of energy being used, type of controlling action etc. is given in [3]. This classification is presented as a diagram in Fig. 1.

Classification of automatic controllers based on a number of features is presented in [4] (Kazakov A. V.). This classification is similar to the one presented in [1]. However there is one more classification feature in [4], i.e. type of controlling action. Based on the type of the controlling action the controllers are divided into continuous and discrete controllers. The continuous controllers permanently act on the controlled plant during the control process.

The discrete controllers act on the controlled plant only at definite moments of time or at reaching some definite values by the controlled parameter. The most simple and the most widespread type of a discrete controller is the switch controller.

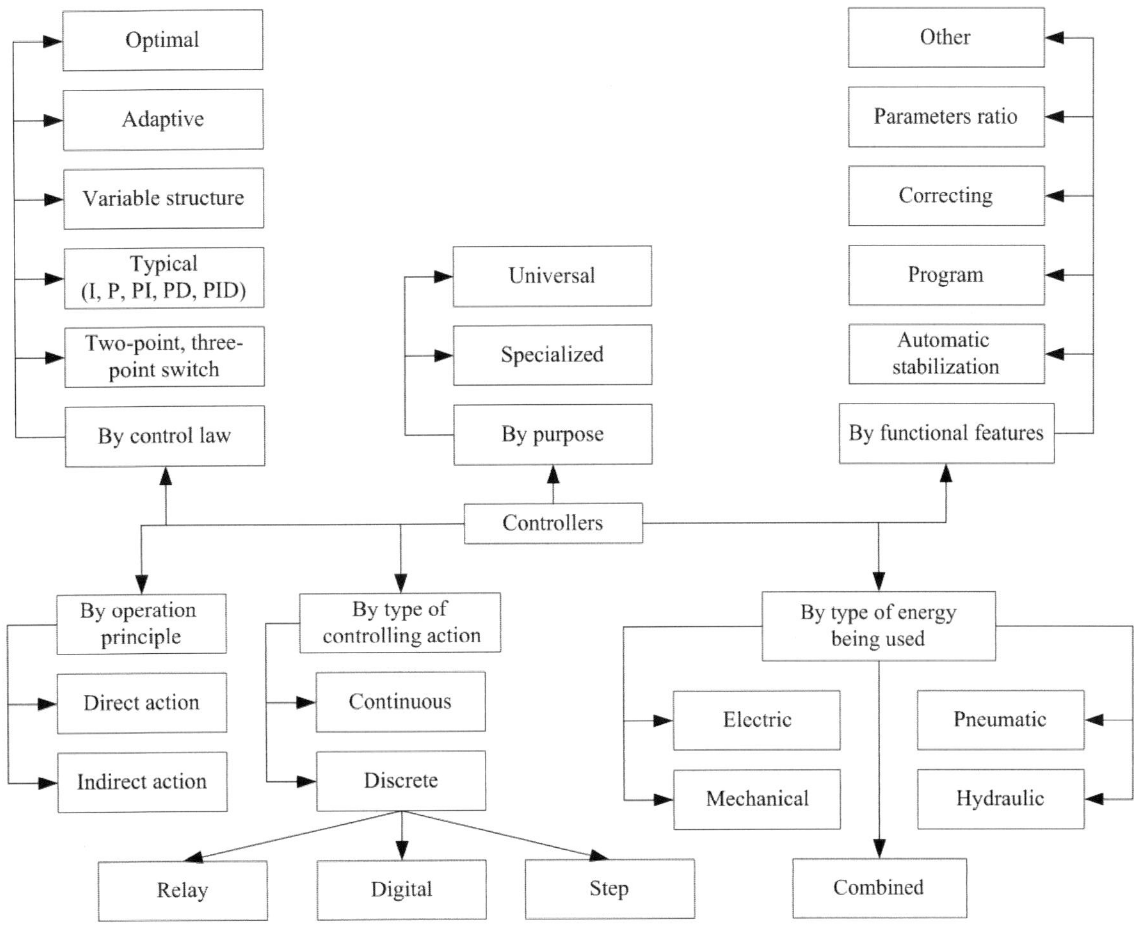

Fig. 1. Classification of automatic controllers according to [3]

One of the most important features for classification is the characteristics of the controller, i.e. the functional relation between the displacement of the valve and the deviation of the controlled parameter. The characteristics of the controller like of any other element in the automatic control loop (ACL) can be linear or non-linear [4].

Based on the type of the characteristics the controllers are divided into I-controllers (non-static), P-controllers (static), PI-controllers (isodromic) as well as controllers with differentiation, i.e. PD-controllers and PID-controllers [4].

The term "control law" came to the control systems theory from the classical theory. The "law" implies the mathematical relation based on which the controlling instrument would act on the controlled plant in case if it (instrument) possessed no inertia. Thus the "law" is an idealized algorithm of control. The actual algorithm of control would be quite different from the "law" due to the dynamics in the controlling instrument. However, since the inertia of most of the industrial controllers is small enough in comparison to the inertia of the controlled plant, the algorithm of control in the controllers is close to the "law" and this term is widely used. These laws are called typical and they are implemented by the controllers [3].

3. New classification of automatic controllers diagrams

Based on the accomplished review and analysis of the existing classifications of automatic controllers it was defined that there are a number of features for classifying the controllers (purpose of the controller, principle of operation, construction, type of energy being used, type of controlling action etc). However, little attention is paid to putting in order and classifying the diagrams implemented in the automatic controllers.

For better understanding of the distinction between the switch, continuous and step controllers as well as for understanding the place of these controllers among other controllers a new classification of automatic controllers diagrams with the control channels is proposed.

According to the new classification the control channel is considered as a part of the automatic control loop consisting of an automatic controller (AC) and an intermediate element (IE) between the controller and a controlled plant (CP), see Fig. 2.

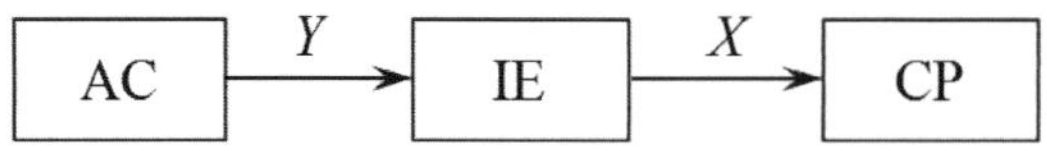

Fig. 2. Structure of the control channel: Y is automatic controller output signal;
X is controlled plant input signal (controlling action)

The diagrams of automatic controllers with control channels are classified based on the following three features:

- type of control law;
- type of automatic controller output signal;
- type of controlling action on the controlled plant.

The control law implemented in an automatic controller can be of the following two types: continuous and switch. PID control law with its subtypes (i.e. P, I, PI, PD, ID) belongs to the continuous control law. The switch control law can be implemented by a two-point switch or a three-point switch.

The automatic controller output signal can be analog or impulse. And the controlling action on the controlled plant can also be analog or impulse.

The developed classification of automatic controllers diagrams with control channels is presented in Fig. 3. Conventional numbering of the automatic controllers diagrams with control channels from 1 to 8 is introduced.

The peculiarity of the developed classification is that the classification features are located in a hierarchical manner (see Fig. 3). Different types of controlling action on the controlled plant are considered for different types of controller output signal. And different types of controller output signal are considered for different types of control law. All possible combinations between the types of the control law, types of the controller output signal and types of the controlling action on the controlled plant are considered in the developed classification with taking into account the hierarchy of the classification features.

The diagrams of automatic controllers with control channels (diagrams 1 – 8) according to the developed classification are presented in Fig. 4 – 11. The description and analysis of these diagrams is made with respect to a thermal controlled plant (electric oven with air flow through it). Air temperature in the oven is the controlled parameter. Electric power at the input of the oven is the controlling action on the controlled plant. The detailed description of this controlled plant is presented in [5].

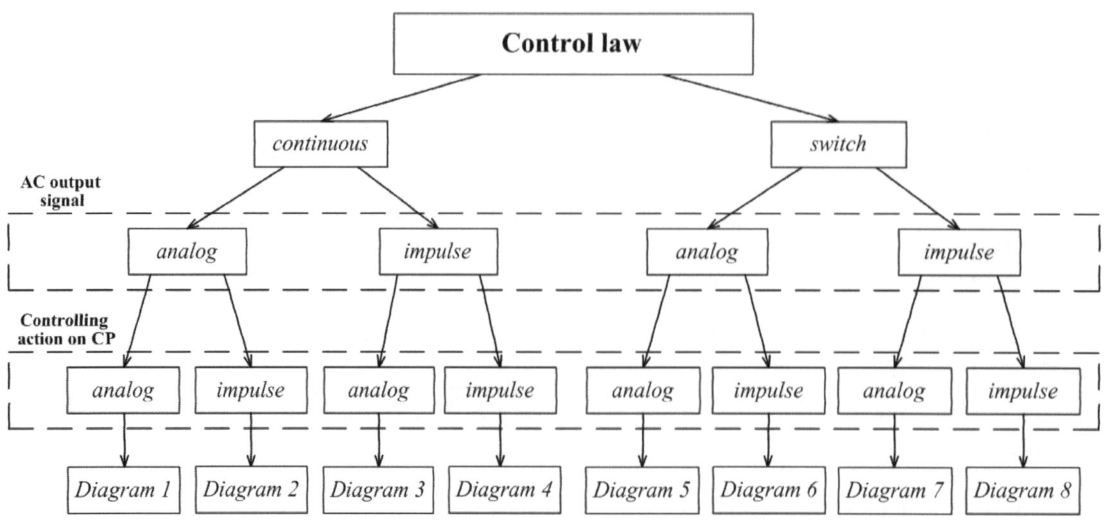

Fig. 3. New developed classification of automatic controllers diagrams with control channels

In diagram 1 according to the developed classification (see Fig. 4) the automatic controller is built on the basis of PID control law with analog output signal. This signal goes to the intermediate element in which proportional function (amplification with proportional gain coefficient K_P) is accomplished. The analog signal from the intermediate element is supplied to the controlled plant. In case of the thermal controlled plant the role of the intermediate element can be played by a thyristor amplifier. It increases the electric power at the input of the electric oven depending on the value of the analog output signal from the controller.

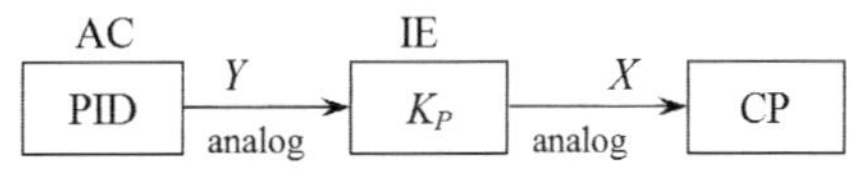

Fig. 4. Diagram 1 according to the developed classification (continuous PID controller and an amplifier)

In diagram 2 (see Fig. 5) the automatic controller like in diagram 1 is also built on the basis of PID control law with analog output signal. However, the intermediate element here is different. It consists of the pulse-width modulator (PWM) and an amplifier. The PWM impulse output signal is unipolar and it can be "ON" ("1") or "OFF" ("0"). The amplified impulse signal after the intermediate element is supplied to the controlled plant (electric oven). The amplifier here may consist of a relay or a thyristor amplifier working in the relay mode.

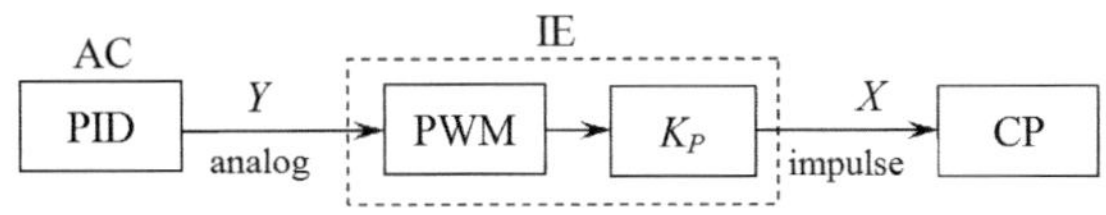

Fig. 5. Diagram 2 according to the developed classification (continuous PID controller and a pulse-width modulator with an amplifier)

In diagram 3 (see Fig. 6) the automatic controller is built on the basis of PID control law with impulse output signal. This type of the controller is called step PID controller. The purpose of the impulse signal is to control the electric actuator (motor) with constant speed of movement. The impulse output signal of the controller is bipolar and it can be "UP" ("+1"), "DOWN" ("-1") or "OFF" ("0"). The intermediate element in this diagram consists of the integral action actuator presented by the integral element and a valve presented by the proportional element (see Fig. 6). The intermediate element provides analog controlling action at the input of the controlled plant. In case of the thermal controlled plant the role of the valve can be played by a variable autotransformer. It provides analog change of electric power at the input of the electric oven depending on the position of the autotransformer secondary connection rotated by the actuator [5].

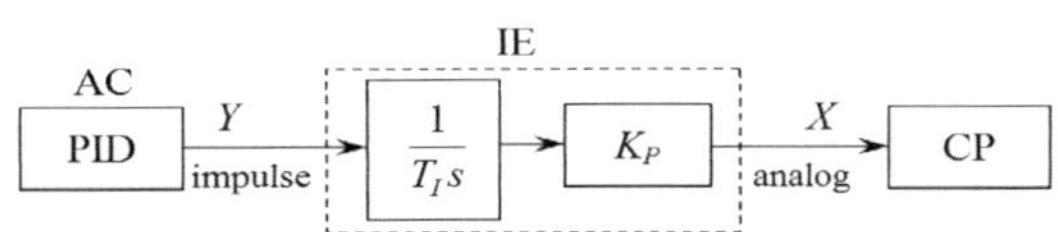

Fig. 6. Diagram 3 according to the developed classification (step PID controller and an integral action actuator with a valve)

Step PID controller for controlling the integral action electric actuator can be built on the basis of one of the following diagrams:

- step PID controller with double differentiation;
- continuous PID controller and a pulse-width modulator;
- continuous PID controller and a simulated positioner;
- step PI controller with a pulse generator.

The detailed description and analysis of step PID controllers diagrams is presented in [6]. In diagram 4 according to the developed classification (see Fig. 7) the automatic controller is built on the basis of PID control law with a pulse-width modulator (unipolar). The impulse output signal goes to the amplifier (relay). And the impulse controlling action is supplied to the controlled plant.

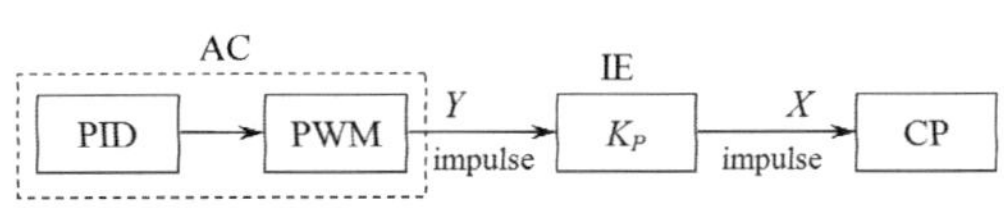

Fig. 7. Diagram 4 according to the developed classification (continuous PID controller with a pulse-width modulator and an amplifier)

In diagram 5 (see Fig. 8) the automatic controller is built on the basis of a three-point switch with an integrator. The analog output signal goes to the amplifier. And the analog controlling action is supplied to the controlled plant.

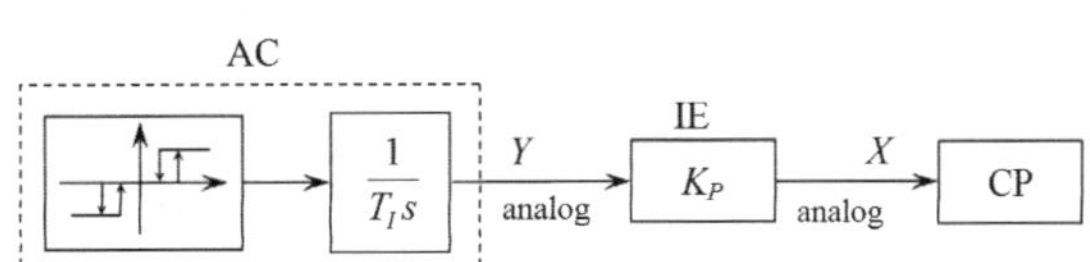

Fig. 8. Diagram 5 according to the developed classification (three-point switch with an integrator and an amplifier)

In diagram 6 (see Fig. 9) the automatic controller like in diagram 5 is built on the basis of a three-point switch with an integrator. The analog output signal goes to the intermediate element consisting of a pulse-width modulator and an amplifier (relay). Impulse controlling action from the intermediate element is supplied to the controlled plant.

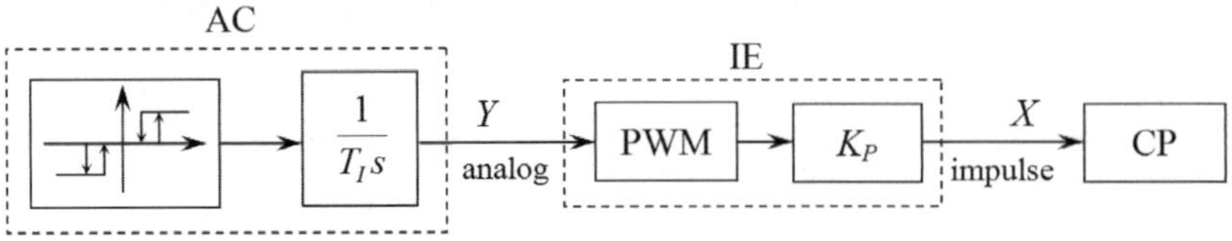

Fig. 9. Diagram 6 according to the developed classification (three-point switch with an integrator and a pulse-width modulator with an amplifier)

In diagram 7 (see Fig. 10) the automatic controller is built on the basis of a three-point switch. The impulse output signal (bipolar) goes to the intermediate element consisting of an integral action actuator presented by the integral element and a valve presented by the proportional element (see Fig. 10). Analog controlling action from the intermediate element is supplied to the controlled plant.

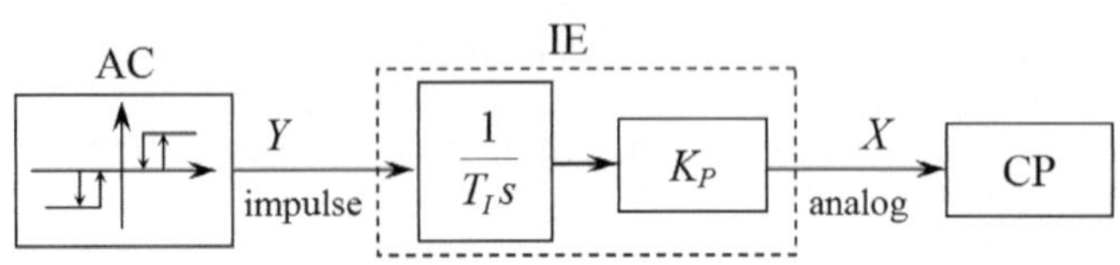

Fig. 10. Diagram 7 according to the developed classification (three-point switch and an integral action actuator with a valve)

In diagram 8 (see Fig. 11) the automatic controller is built on the basis of a two-point switch. The impulse output signal (unipolar) goes to the amplifier (relay). And the impulse controlling action is supplied to the controlled plant.

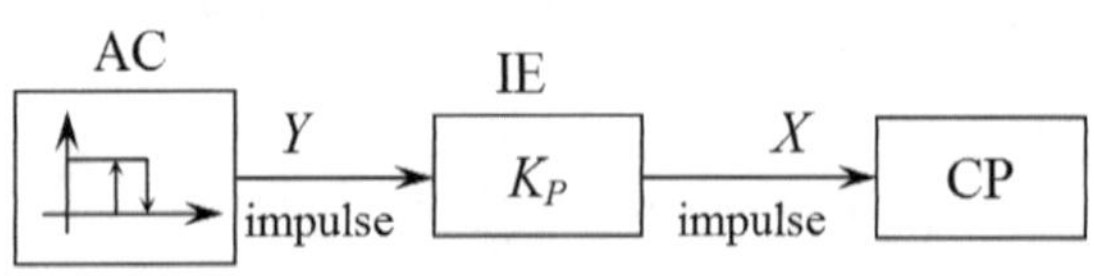

Fig. 11. Diagram 8 according to the developed classification (two-point switch and an amplifier)

The developed classification of automatic controllers diagrams with control channels is also presented in a table form (see Table 1).

Diagram number	**Control law**	**AC output signal**	**Controlling action on CP**	**Intermediate element**
Diagram 1	Continuous (PID)	Analog	Analog	Amplifier (K_P)
Diagram 2	Continuous (PID)	Analog	Impulse	PWM + Amplifier (K_P, relay)
Diagram 3	Continuous (PID)	Impulse	Analog	Actuator ($1/(T_I \cdot s)$) + Valve (K_P)
Diagram 4	Continuous (PID + PWM)	Impulse	Impulse	Amplifier (K_P, relay)
Diagram 5	Switch (3-point switch + Int.)	Analog	Analog	Amplifier (K_P)
Diagram 6	Switch (3-point switch + Int.)	Analog	Impulse	PWM + Amplifier (K_P, relay)
Diagram 7	Switch (3-point switch)	Impulse	Analog	Actuator ($1/(T_I \cdot s)$) + Valve (K_P)
Diagram 8	Switch (2-point switch)	Impulse	Impulse	Amplifier (K_P, relay)

Table 1. Diagrams of automatic controllers with control channels according to the developed classification

Diagram 8 is the most widespread case for controlling a thermal plant (electric oven in particular) since its adjustment and construction is simple. A shortcoming of this diagram is the self-oscillating transient process in the ACL. In order to remove the self-oscillations a PID controller with a pulse-width modulator is applied [7], i.e. diagram 2 or 4. These diagrams provide better quality of the transient process and absence of self-oscillations. However, adjustment of these diagrams and definition of numerical values of PID control law tuning parameters is a complicated process. This problem is present not only in diagrams 2 and 4 but in all the diagrams where PID control law is applied (i.e. diagrams 1 – 4). In order to simplify this problem a special algorithm for automatic definition of numerical values of PID control law tuning parameters (PID autotune) [8] is proposed by manufacturers of PLCs.

The main principle of PID autotune algorithm work can be described in general by the following two steps. Step one: generation of self-oscillating transient process in the ACL by applying the two-point switch. Step two: definition of numerical values of PID control law tuning parameters based on the parameters of the self-oscillating transient process. It means that diagram 8 is applied during the first step. And diagram 1 is applied during the second step.

It should be mentioned that there were no diagrams 5 and 6 when the classification was developed at first. However, in order to fill the gaps in the right part of the developed classification (Fig. 3) the three-point switch with an integrator was introduced to provide an analog output signal for a switch control law. This option (diagram 5) in some cases provides even better quality of transient processes than it is done by a continuous PID controller (diagram 1). The presented diagrams of automatic controllers can be implemented in PLCs by programming.

4. Conclusion

The existing classifications of automatic controllers being applied in the industrial automation systems were analyzed. Based on the accomplished analysis it was defined that automatic controllers today are classified on the basis of a number of features (such as purpose of the controller, its principle of operation, construction, type of energy being used, type of controlling action and other). However, little attention is paid to putting in order and classifying the diagrams implemented in the automatic controllers. For better understanding of the distinction between the switch, continuous and step controllers as well as for understanding the place of these controllers among other controllers a new classification of automatic controllers diagrams with the control channels is proposed. All possible combinations between the types of the control law, types of the controller output signal and types of the controlling action on the controlled plant were considered in the developed classification. Conventional numbering of the automatic controllers diagrams with control channels from 1 to 8 was introduced.

The developed classification simplifies the problem of choosing and analyzing the diagrams of automatic controllers for controlling a technological process since it provides a clear demonstration of all possible ways of automatic controller action on the controlled plant with taking into account the type of control law, type of controller output signal and type of controlling action on the controlled plant.

The authors' future plans are to carry our experimental study of automatic control loop behaviour for each type of the automatic controller diagram presented in this paper and to analyze the quality of the transient processes.

5. References

[1] Tuning of automation instrumentation and automatic control systems: reference book / A.S. Klyuyev, A.T. Lebedev, S.A. Klyuyev, A.G. Tovarnov. – M.: Energoatomizdat, 1989. – 368 p. (Naladka sredstv avtomatizatsii i avtomaticheskih sistem regulirovaniya: spravochnoe posobie/ A.S. Klyuev, A.T. Lebedev, S.A. Klyuev, A.G. Tovarnov. – M.: Energoatomizdat, 1989. – 368 s.)

[2] Automatic control and controllers in chemical industry / Sharkov A. A., Prityko G. M., Palyukh B. V. – M.: Chemistry, 1990. – 288 p. (Avtomaticheskoe regulirovanie i regulatory v khimicheskoi promyshlennosti / Sharkov A. A., Prityko G. M., Palyukh B. V. – M.: Khimia, 1990. – 288 s.)

[3] Electronic review portal. Virtual laboratory on energy saving and energy efficiency. http://eef.misis.ru/, Accessed on: 2017-08-21.

[4] Fundamentals of automatics and automation of chemical production. Textbook for higher educational institutions. Kazakov A. V., Kulakov M. V., Melyushev Y. K. – M., "Machine-building", 1970. – 376 p. (Osnovy avtomatiki i avtomatizatsii khimicheskikh proizvodstv. Uchebnoe posobie dlya vuzov. Kazakov A. V., Kulakov M. V., Melyushev Y. K. – M., "Mashinostroenie", 1970. – 376 s.)

[5] R. Fedoryshyn, S. Klos, V. Savytskyi, O. Masniak. Identification of Controlled Plant and Development of Its Model by Means of PLC. Energy Eng. Control Syst., 2016, Vol. 2, No. 2, pp. 69 – 78. https://doi.org/10.23939/jeecs2016.02.069

[6] S. Kril, R. Fedoryshyn, O. Kril, Y. Pistun. Investigation of Functional Diagrams of Step PID Controllers for Electric Actuators // Procedia Engineering, Volume 100, 2015, Pages 1338-1347. doi: 10.1016/j.proeng.2015.01.549.

[7] J.-K. Woo, D. Yang, K. Najafi, S. Lee, J. Mitchell. (2016). Miniaturized digital oven-control microsystem with high power efficiency and ±1.8ppm frequency drift. Frequency Control Symposium, 2016 IEEE International, pp. 1 – 4. DOI: 10.1109/FCS.2016.7563578.

[8] J. Korbel & R. Prokop. Aperiodic Principle for Automatic Controller Tuning // Annals of DAAAM for 2011 & Proceedings of the 22nd International DAAAM Symposium, Volume 22, No. 1, ISSN 1726-9679, ISBN 978-3-901509-83-4, Editor B. Katalinic, Published by DAAAM International, Vienna, Austria, EU, 2011, Pages 399-400.

DOI: 10.2507/28th.daaam.proceedings.135

SIMULATION AND MODELLING SYSTEM FOR THE PURPOSES OF CRISIS MANAGEMENT

Martin Ficek & Dušan Vičar

This Publication has to be referred as: Ficek, M[artin] & Vicar, D[usanka] (2017). Simulation and Modelling System for the Purposes of Crisis Management, Proceedings of the 28th DAAAM International Symposium, pp.0974-0980, B. Katalinic (Ed.), Published by DAAAM International, ISBN 978-3-902734-11-2, ISSN 1726-9679, Vienna, Austria
DOI: 10.2507/28th.daaam.proceedings.135

Abstract

Hazardous chemical substances are used virtually every day. When these are stored or transported, extraordinary events occur from time to time, and as a consequence, health or lives of people within the reach of the accident or in the surroundings through which a hazardous cloud of this substance may spread are in danger. For a more efficient and easier operation of the emergency services and of the workers in crisis management, software products are used that model and simulate leakages of these substances and this way, they provide valuable information. This article compares the selected software and evaluates its efficiency.

Keywords: software; Aloha; crisis management; modelling; simulation

1. Introduction

The today's world depends on a whole number of chemicals and these substances have a wide use in industry. In many cases, these are hazardous chemical substances. Hazardous chemical substances have a number of pitfalls. In this article, we will deal with the issues of a leakage of a hazardous chemical substance. Leakages happen mostly during the transport or storing of a substance. For modelling and simulation of extraordinary events with hazardous substances, a number of programs are used. These programs can be used for management and for the corresponding company workers, as well as for workers of the crisis management and units of the emergency services. There, this program can relatively quickly provide basic necessary information to solve the extraordinary event at hand.

These issues are broad and appropriate attention is given to them practically around the whole world. An example can be the article called Crisis Management Modelling and Simulation Laboratory [1]. This article deals with, for example, the danger of a leakage of a chemical substance during transport that is carried out in a densely populated area and with the impacts of such leakage.

In their article, Bernatik, A. et al. [2] deal with the very necessary addition to the regular software (the abbreviation SW) products such as Terex or Aloha in the form of models of computational fluid dynamics or physics modelling in aerodynamic tunnels, and they provide data that they subsequently compare with the outputs of regular SW products. In a similar fashion, the article Application of Computational Fluid Dynamics for Manufactured Gas Dispersion Modelling and Leakage Consequence Analysis by Sun, H. et al. [3] deals with the computation of fluid dynamics.

Since not only the transported substance has to be a hazardous substance, but also the fuel itself can be like, for example, the nowadays already relatively frequently used Liquefied Natural Gas (the abbreviation LNG), it is necessary to count even with these substances, and even in this case, it is possible to use appropriate SW products for the purposes of crisis management, as stated by the authors of the article called LNG as a potential alternative fuel – Safety and security of storage facilities. [4]

The article Using the SW modelling and simulating tools in transport of hazardous cargos by Ficek, M. et al. [5] is directly covers the comparison of two selected SW products. The article follows these publications and tries to further elaborate on them. The aim of this article is to provide background information for the evaluation of the selected software, both on the level of statistics, as well as based on a comparison with real-life situations and last but not least based on the SWOT analysis.

2. Used tools and methods

ALOHA – uses statistical mathematical models (SLAB and DEGEADIS). This software (there is an abbreviation in the article: SW) displays maximum concentrations and their reach. It does not display the development of substance leakage, but the overall maximum impacts. Modelling in the Aloha SW provides three options from the viewpoint of a leakage of hazardous substances and these are: gas leakage in the form of a cloud, whose output is only the final concentration (without relation to time). An instantaneous (one-time) leakage, which means diffusion of one gas or fumes and by dispersion with the wind – here, it's possible to get information about the time behaviour. The third type is an "accidental" leakage that is time-dependent, there is no specific statistical model and that's why the Aloha SW computes a series of five final durations in a steady state of concentrations, with the cloud not reacting with the environment being meant here. Every cloud contributes to the resulting concentration at any point and time. [material in German 15] Despite the fact that he Aloha SW does not display the course of the accident in time, it is possible to display the time behaviour of the concentration at a chosen point. [5, 6, 7]

The tool called Area Locations of Hazardous Atmospheres, abbreviated as ALOHA, created by the US Environmental Protection Agency in cooperation with the National Oceanic and Atmospheric Administration, (the United States of America) is a program designed for the modelling of leakages of hazardous chemical substances with leakage to the atmosphere. It has the advantage of a relatively large database of substances; its disadvantage is that it works only with the variant of non-reacting substances (it does not take into account reaction with the environment). It also does not take into account any other influences. An advantage is that it is possible to connect the program with measuring devices and to work with their data. The output is a figure of the threatened area (Threat Zone). However, these areas can be influenced by the setting of the level of the concern indicator (Level of Concentration – LOC). The result is tables and graphs of mathematical calculations that can be visualised using the MARLPLOT tool, which is, along with ALOHA, part of the package of the CAMEO Software Suite, or using the Arc Tool extension at Arc View and Arcmap.[5, 6, 7]

Key characteristics: Non-commercial, shareware, great range and possibilities of numerical results and calculations, calculation of the speed of substances leakage with a prognosis for the future, cooperation with other CAMEO tools. [5, 6, 7]

Works with the following input data:

- Data about the leaked substance
- Data about the state of the atmosphere
- Data about the source of the leakage
  - Direct source (type of the leakage – immediate or continuous, amount of the substance leaked or the speed of the leakage, height of the source above the ground)
  - Puddle (surface of the spilled liquid, volume, weight or depth of the puddle, type of the foundation, temperature of the foundation, temperature of the leaked substance)
  - Container (type of the container and its orientation – spherical, cylindrical, vertical or horizontal, diameter, height or volume of the container, state of the substance in the container, temperature of the stored substance, weight or volume of the substance, type and dimensions of the escape hatch – rectangular or circular, type of leakage – punctured hole in the shell or short piping, height of the opening above the bottom, type of foundation and its temperature, presence of a leakage sump, or its dimensions) [5, 6, 7]

The output data are:

- Maximum speed of the leakage, in liquids, this concerns the speed of evaporation, not the speed of the leakage.
- Maximum average of the leakage speed per minute (average speed of the leakage during the minimum time of 1 min), in liquids, this again concerns the speed of evaporation. In devices that work under pressure, overestimation may occur by influence of a swift initial leakage.
- The total amount of the leaked substance during the period of maximum 1 hour.
- Maximum reach of the threat zone, where the concentration will reach the entered value.
- Maximum concentration of the leaked substance in a randomly entered place.
- Maximum dose in the selected place that the organism took in after the leakage in the course of 1 hour. [5, 6, 7]

Emergency Response Planning Guidelines (ERPGs)

"ERPGs are exposure guidelines designed to anticipate health effects from exposure to certain airborne chemical concentrations." [8]

"ERPGs estimate the concentrations at which most people will begin to experience health effects if they are exposed to a hazardous airborne chemical for 1 hour. (Sensitive members of the public—such as old, sick, or very young people—aren't covered by these guidelines and they may experience adverse effects at concentrations below the ERPG values.)

A chemical may have up to three ERPG values, each of which corresponds to a specific tier of health effects." [8]

The three ERPG tiers are defined as follows:

"ERPG-3 is the maximum airborne concentration below which nearly all individuals could be exposed for up to 1 hour without experiencing or developing life-threatening health effects." [8]

"ERPG-2 is the maximum airborne concentration below which nearly all individuals could be exposed for up to 1 hour without experiencing or developing irreversible or other serious health effects or symptoms which could impair an individual's ability to take protective action." [8]

"ERPG-1 is the maximum airborne concentration below which nearly all individuals could be exposed for up to 1 hour without experiencing more than mild, transient adverse health effects or without perceiving a clearly defined objectionable odor." [8]

Paired two-sample t-test

In this article, the paired two sample t-test will be used for the statistical evaluation.

This test compares data that form "paired variance rows", that means that they come from entities that have been subjected to two measuring. So we are conducting 2 measuring in a single selection set: 1. measuring before the application of the experimental intervention (in our case real measured data / data acquired by CEI calculation), 2. after the application of the experimental intervention (simulated data). The values acquired in this way from pairs and in testing represent both the control and experimental group of the compared data. [9]

In the test, we are building on the differences of the measured pair values in the compared variance rows. We are testing a hypothesis that states that the median values of the measurement before the experiment and after it are equal (or: the difference of the median values of the pair measuring is zero). [9]

First, we calculate the differences of the pair values in the selection set (n – number of pairs) and from the established differences, we calculate the arithmetical average X and the standard deviation "s" (or dispersion s2). [9]
Then we calculate the test criterion (statistics) t:

$$t = \frac{|X|}{\sqrt{\frac{s^2}{n}}} \tag{1}$$

To find the table critical value, it is necessary to determine the number of degrees of discretion of the selection set: v = n-1 and to choose the significance level α. [9]

We compare the resulting statistics t with the table critical value $t1\text{-}\alpha/2(v)$, where v = n-1 and α is chosen to be 0.05. [9]

If $t \leq t1\text{-}\alpha/2(v)$, then the statistically insignificant difference is μ1 and μ2 with a chosen α. (We are not disapproving the null hypothesis H0 that is that the median value of the measuring before the experiment does not differ from the median value of the measuring after the experiment). [9]

If $t > t1\text{-}\alpha/2(v)$, then the statistically significant difference is μ1 and μ2 (α = 0.05) (We are disapproving the null hypothesis H0 that is that the median value of the measuring before the experiment differs from the median value of the measuring after the experiment). [9]

3. Results

Many pieces of information are necessary for the individual situations, however, these are not completely relevant for the purpose of this article and by reason of conciseness, only the most important will be included here and these are: the amount and the type of the leaking substance. In the following table, you can see both the acquired real values and the values (from a threat zone corresponding with ERPG-3) acquired using simulation in the appropriate SW.

Situation	Data		REAL (km)	ALOHA (km)
1.	Substance	Chlorine	2.0 /0	7.2
	Amt. of Sub. (kg)	188		
2.	Substance	Ammonia	0.220	2.8
	Amt. of Sub. (kg)	86		
3.	Substance	Hydrogen Chloride	0.50	0.390
	Amt. of Sub. (kg)	49		
4.	Substance	Methyl Chloride	0.15	0.056
	Amt. of Sub. (kg)	112		
5.	Substance	Acetaldehyde	0.17	1.2
	Amt. of Sub. (kg)	70		
6.	Substance	Benzene	0.23	0.6
	Amt. of Sub. (kg)	127		
7.	Substance	Hydrogen Fluoride	0.8	3.99
	Amt. of Sub. (kg)	76		
8.	Substance	Ammonia	0.16	2.7
	Amt. of Sub. (kg)	46		
9.	Substance	Formaldehyde	0.78	4.96
	Amt. of Sub. (kg)	58		
10.	Substance	Chlorine	1.4	7.6
	Amt. of Sub. (kg)	93		

Table 1. Resulting values

After the data from the table (real and ALOHA) have been applied on the statistical part mentioned above, we get the resulting decision on the hypothesis H0. H0 is a hypothesis that states that the median values are equal.

If we follow the procedure, we will find, that the arithmetical average of the difference between the actual measured reach of the substance and the simulated reach in the Aloha SW is -2.5086, the number of measuring is 10 and the dispersion s2 of the difference between the actual measured reach of the substance and the simulated reach in the Aloha SW is 2.224222621.

After substitution into the formula:

$$t = \frac{|X|}{\sqrt{\frac{s^2}{n}}} \quad (2)$$

The resulting value is t = 5.319151

The table critical value $t_{1-\alpha/2}(v)$ where v = n-1 and α is chosen as 0.05 is 3.25.

The conclusion is that t >t1-α/2(v). We have a statistically significant difference here. We're disapproving the null hypothesis H0, that means that the median value of the real measured values differs from the median value of the simulated values acquired using the Aloha SW.

SWOT-based analysis

The selected SW is evaluated using the SWOT analysis, which evaluates strengths, weaknesses, opportunities and threats. This analysis will also be quantified. It is necessary to realise, that the analysis will be related to the findings that have been stated above. That means that it will be influenced in comparison to a real situation.

SWOT - Aloha	Helpful	Harmful
Internal origin	Strengths: Models allow for a greater possibility of simulations. The option to select many simulations provides complexity. Possibility to expand the database of hazardous substances.	Weaknesses: The results are significantly overrated and they absolutely don't correspond with reality.
External origin	Opportunities: Specification of the resulting simulations. Simplification of operation with the possibility to use the SW out in the field.	Threats: If we're planning measures according to the results of this SW, we will use unnecessarily many powers and means and we will disrupt operation in an area that is unnecessarily large.

Table 2. SWOT analysis from the Aloha SW

Strengths	Evaluation	Weight	Resulting value
Models allow for a greater possibility of simulations.	4	0.3	1.2
The possibility to select many simulations provides complexity.	4	0.3	1.2
Possibility to expand the database of hazardous substances.	5	0.4	2
Total			4,4
Weaknesses	**Evaluation**	**Weight**	**Resulting value**
The results are significantly overrated and they absolutely don't correspond with reality.	-4	1	-4
Total			-4

Opportunities	Evaluation	Weight	Resulting value
Specification of the resulting simulations.	5	0.5	2.5
Simplification of operation with the possibility to use the SW out in the field.	5	0.5	2.5
Total			5
Threats	**Evaluation**	**Weight**	**Resulting value**
If we're planning measures according to the results of this SW, we will use unnecessarily many powers and meansand we will disrupt operation in an area that is unnecessarily large.	-4.5	1	-4.5
Total			-4.5

Table 3. Quantification of SWOT analysis of the Aloha SW.

- Internal value = strengths + weaknesses = 4.4 - 4 = 0.4
- External value = opportunities + threats = 5 - 4.5 = 0.5
- Resulting value = Internal value + External value = 0.4 + 0.5 = 0.9

The resulting number is not staggeringly high by any means, nevertheless, it is still positive and therefore, the SW can be considered a success. The fact that the results of the simulations are significantly overrated is at fault. If the information from the simulations would be used in a crisis management (there is an abbreviation in the article: CM), the measures would take place on an unnecessarily large area and they would restrict normal operation in this area, not speaking of a significantly increased price of the intervention. Moreover, the SW is relatively complicated and it can be used out in the field only with greater difficulties, which is another disadvantage. The Aloha Software is designed rather for more complicated simulations and its task is to show the worst possible scenario.

4. Conclusion

There is a great number of software products available that create models and simulations of hazardous substances leakage. Where is their place in the area of crisis management? Definitely in crisis management and preparation for emergencies, nevertheless, also in the course of a reaction to an emergency. These areas seem to be similar, but upon a closer look, you will be able to find that each of them requires a specific SW. For example, if we're using SW in crisis management, we need to know the worst possible scenario and conduct complicated simulations. To do that, we need a SW that provides a wide spectrum of settings (types of leakages, types of containers, types of buildings, weather settings, inversion settings, etc.) – during planning, the worker has enough time for a complicated simulation and settings. By contrast, in an intervention, it is necessary to react as quickly as possible and there's no time to enter much data. Here, the SW that needs only several basic information in a simple interface is used, which will provide relatively precise and reliable information necessary for an intervention.

In this article, we tested the Aloha SW. SW Aloha is more suitable for more complex situations, when it its necessary to know the worst possible variant even at a price of a relatively great distortion of the threatened areas. Both the finding from the statistical comparison and the SWOT analysis correspond to this.

It is necessary to realise, that a program is only an aid. The main responsibility remains on the crisis workers and authorised persons. They must treat the results from a SW critically and confront them with their experience and findings from practice. Only a decision taken based on such information is usually of quality and beneficial.

The future in this field will focus on making the results more precise and creating an interface that will be simple and that will provide immediately available information for an intervention, but at the same time, that will be possible to expand and used for more complicated and complex situations. Also, it will be more common for SW products to be connected to sensors and actuators in the future and security will significantly increase. This situation can be used as an example: A hazardous substance will start leaking from the piping, the sensor detects and localises the leakage. Actuators close the given part of the building and they prevent further leakage. Automatically, the relevant workers are alarmed, along with the appropriate organs of the CM and the emergency services and a warning is issued for the surroundings. Such systems are only in development, but this is exactly the way that you should take.

5. Acknowledgments

This work was supported by the European Regional Development Fund under the project CEBIA-Tech Instrumentation No. CZ.1.05/2.1.00/19.0376 and also by the Internal Grant Agency of Tomas Bata University under the project No. IGA/FAI/2017/015

6. References

[1] Björnham, O. (2013), Crisis Management Modeling and Simulation Laboratory. In: . pg. 759. DOI: 10.1007/978-94-007-5577-2_129. Also available at: http://link.springer.com/10.1007/978-94-007-5577-2_129

[2] Bernatik, A. (2008). Modelling accidental repases of dangerous Gates into the Loir troposphere from mobile sources. In: Process Safety and Environmental Protection. 86(3), pgs. 198-207. DOI: 10.1016/j.psep.2007.12.002. ISSN 09575820. Also available at: http://linkinghub.elsevier.com/retrieve/pii/S0957582008000037

[3] Sun, H. ; Ning, P. & Tang, L. (2010). Application of Computational Fluid Dynamics for Manufactured Gas Dispersion Modeling and Leakage Consequence Analysis. In: 7th International Symposium on Safety Science and Technology (ISSST). Hangzhou, PEOPLES R CHINA. ISBN 978-7-03-029089-2.

[4] Bernatik, A. (2011). LNG as a potential alternative fuel – Safety and security of storage facilities. In: Journal of Loss Prevention in the Process Industries. 24(1), pgs. 19-24. DOI: 10.1016/j.jlp.2010.08.003. ISSN 09504230. Also available at: http://linkinghub.elsevier.com/retrieve/pii/S0950423010001014

[5] Ficek, M.; Vičar, D.; Rak, J. & Svoboda, P. (2016) Using the SW modeling and simulating tools in transport of hazardous cargos, at the conference. In: TRANSPORT MEANS 2016. Juodkrante, Lithuania: Kaunas university of technology, 2016, pgs. 862-865. ISSN 2351-7034.

[6] Hrušková, V. (2010). Zhodnocení vybraných softwarových nástrojů určených pro modelování úniku nebezpečných látek při haváriích v chemickém průmyslu, in English: Evaluation of selected software tools for modeling leakage of hazardous substances in accidents in the chemical industry. Brno. Thesis. Vysoké učení technické v Brně (Technical University Brno). Head of thesis: doc. Ing. Vladimír Adamec, CSc.

[7] Kudlička, R. (2013). Postavení simulace a modelování v rámci záchranných a likvidačních prací, in English: Position of simulation and modeling in rescue and disposal work. Zlín. Thesis. Univerzita Tomáše Bati ve Zlíně (Tomas Bata University in Zlín). Head of thesis: Ing. Martin Hromada, Ph.D.

[8] Emergency Response Planning Guidelines (2017). Office of Response and Restoration [online]. [quoted on 2017-08-16]. Available at: https://response.restoration.noaa.gov/erpgs

[9] Parametrické testy - Studentův t-test, in English: Parametric tests - Student's t-test (2013). Veterinární a farmaceutická univerzita Brno - statistika a výpočetní technika, (Veterinary and Pharmaceutical University of Brno - Statistics and Computing). [online]. Brno [quoted on 2017-08-16]. Available at: https://cit.vfu.cz/statpotr/POTR/Teorie/Predn3/ttest.htm

[10] Blistanova, M. (2014). Data Preparation for Logistic Modeling of Flood Crisis Management. In: 24th DAAAM International Symposium on Intelligent Manufacturing and Automation, 2013. Vienna, 1529 – 1533. ISBN 1877-7058.

DOI: 10.2507/28th.daaam.proceedings.136

DEVELOPMENT OF A SAFETY INDUSTRY 4.0 PRODUCTION ENVIRONMENT

Maximilian Papa, David Kaselautzke, Thomas Radinger, Kemajl Stuja

This Publication has to be referred as: Papa, M[aximilian]; Kaselautzke, D[avid]; Radinger, T[homas] & Stuja, K[emajl] (2017). Development of a Safety Industry 4.0 Production Environment, Proceedings of the 28th DAAAM International Symposium, pp.0981-0987, B. Katalinic (Ed.), Published by DAAAM International, ISBN 978-3-902734-11-2, ISSN 1726-9679, Vienna, Austria
DOI: 10.2507/28th.daaam.proceedings.136

Abstract

In modern industrial environments, the usage of autonomous mobile robots is very common. Building up the mobile robot's prerequisites not only financial resources but interdisciplinary knowhow too. Therefore, these preconditions barriers small and medium-sized companies from the up to date industrial concepts. In order, to support small and medium-sized companies this paper shows, how to design and implement mobile robot systems capable to work safety into complex industrial 4.0 environment.

This research task can be accomplished in three steps. The first step deals with the conceptualization and creation of a test environment inspired by a small and medium-sized environment. The difficulty is to build a new suitable test factory, which fits in an office room. In the second step, a simple work station for different test cases were conceptualized. The third step deals with the creation of different mobile robots. Because of the limited space the mobile robots in this work were not bigger than 20x20x20cm.

A concept for a test industry 4.0 test factory with simple work stations was created. Different miniature mobile robots were bought and tested, whereby different risks of the standards were researched. For the next steps, an implementation and guideline of a solution in a small- or medium-sized company will be tested and written.

Keywords: safety; mobile robot; SIP4.0; ROS

1. Introduction

Nowadays the industry gets more often confronted with the terms of "industry 4.0" or "internet of things". These terms stand for an intelligent cooperation of devices in a factory which are communicating with each other. The great advantage for a company will be a flexible production line and that the products can be produced individual for the respective customer. [1] Therefor the change to industry 4.0 comes also with a competitive advantage against former industry companies. But there is a big difference between big-sized companies to small- and medium-sized companies. Namely the smaller companies normally don't have as much money to invest for this change. [2]

The goal of this paper will be to create and test a cost-efficient mobile robotic solution in the industry 4.0 environment. Mobile robots are used for the material transportation between different work stations and machines and in an industry

4.0 environment they have act full autonomy. That means to avoid obstacles and plan their path to the right work station. This goals obviously must confirm with the actual standards and guidelines. [3]

Further the safety and security are big parts of the Industry 4.0 and is often called as "SIP4.0 – Safety & Security in Industry 4.0 environments". Security is used for the secure connection and communication between the different devices. But the security will not be processed in this paper, instead the safety of the mobile robots in an industry 4.0 environment will be dealt with. Therefor different problems must be solved. First the safety aspects should be thought off. The mobile robot must be designed and programmed so it won't harm people or other machines under no circumstances. If it is not possible, the danger must be set to the lowest possible way. [4]

For these reasons a suitable miniature industry 4.0 test factory must be build. To work within an office room with more than one robot and different stations at the same time, the whole test factory within the stations and mobile robots must be very small. Furthermore, the mobile robots were chosen with a size not larger than 20x20x20cm for an optimal test place.

Another goal in this work is, that the whole hardware- and software-solution could be easily transformed to the real and bigger mobile robots, after the successful completion of this work. Through the limitation in financial resources some cost-efficient solutions should be created, because these solutions are very interesting for the small- and medium-sized companies. Once the creation of the test factory with different stations and at least two mobiles robots is done, it should be tested for a SIP4.0 Standard.

Obvious all build robots must be built within the newest safety standards and the state of the art robot controller. Additionally, all stations and mobile robots should have prepared a way for the network communication. This will be needed, if the mobile robot should be requested by the station, whereby the mobile robot must navigate locally to the right station through moveable and fixed obstacles. To avoid a harm from the mobile robot for humans or other machines, all standards will be influenced in this context.

2. State of the art and research

The first step for the mobile robots in the safety industrial environment is to design them properly. Per definition of ISO 13482:2014 (Robots and robotic devices – Safety requirements for personal care robots) [5] a mobile robot is a personal care robot. The ISO 13842:2014 shows that a risk assessment for a safety concept for a mobile robot in an industrial environment is necessary. The exact definition and concept of the risk assessment is shown in ISO 12100 (Safety of machinery -- General principles for design -- Risk assessment and risk reduction) [6], which contains the provision of a risk assessment and a risk analysis on basis of possible danger identifications.

For a risk assessment it must be clear, if a risk is acceptable or not. Therefore, in the first place an inherent construction of the mobile robot should be made. If this isn't practicable it is necessary to implement protective measures at the mobile robot. For last instance, if there is any dangerous possibility left caused due to the mobile robot, it is required to write user information's of them into the user manual. In other place, for dangerous risks which cannot get solved it is necessary to have warning signs directly mounted on the mobile robot or at the factory which can be directly recognized from every human.

Some danger identifications for a risk assessment are listed in chapter 4.2 or in Annex A in ISO:13482. However, it should be considered that this list is not a complete list of all risks with a mobile robot. Thus, for a risk assessment it is important to list every single risk which stands in connection with the mobile robot and should be supplemented for every other application.

After all possibilities for an inherent construction and protective measures for the unsolved risk were made a residual risk of the mobile robot must be calculated. However, an acceptable risk of the mobile robot should be audited. In ISO:13842 it is recommended to setting up a risk assessment as individual cases. Thereby an acceptable risk can be evaluated. For example, if a mobile robot is allowed to enter safety-relevant obstacles or not.

For a state-of-the-art solution regarded to the robot controller the robot operation system (ROS) will be used. ROS is an open-source framework especially design for the robotic use. Many programmers around the world are using this framework and extend the content with their own programs. This concept allows an easy and fast way to program a robot, without programming each function from the start on the own. [7]

3. Methods

In this chapter are the different concepts listed for the planed test factory in the office room. When this is successful and has proven, it can get transformed for a small or medium-sized company.

3.1 Factory concepts:

Three different test factory concepts have been created. In the first concept an ideal factory with perfect pasted lines on the floor was thought off. The mobile robot would use this lines for the navigation through the factory from station to station. The symbolic sketch of this concept can be seen in Fig. 1.

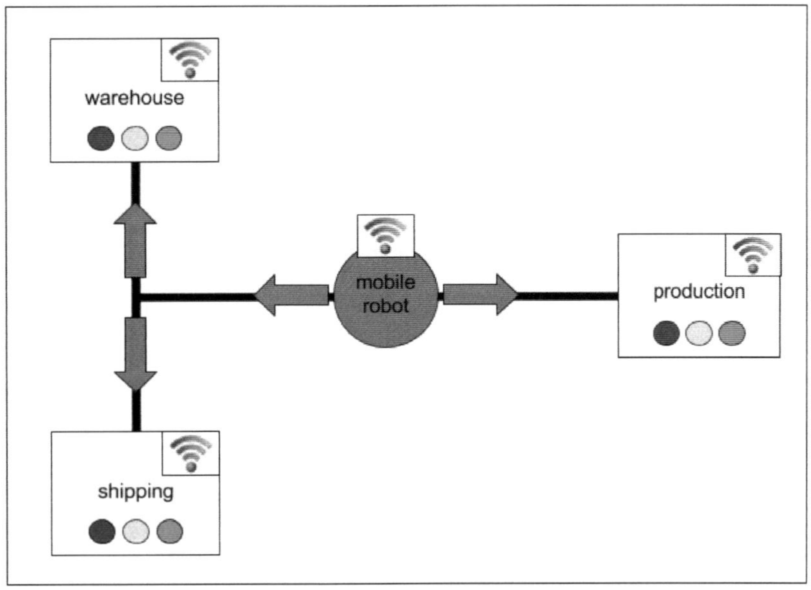

Fig. 1. Concept of an SIP4.0 factory with pasted lines

In an actual small- and medium-sized-factory it is nearly impossible to work with this kind of navigation. Possible errors in a small- and medium-sized company with pasted lines would be strong wear or pollution of the lines. Also, a perfect contrast between the lines and the ground is rare and if there is a high contrast, it is possible that the sun irritates the line following sensors of the robot. Additionally, sometimes people or machines would cross the lines of the robot, which have to avoid a collision.

Because of these reasons a second concept for the test factory was created. In this concept a two-dimensional plan of the factory would be generated. The factory would be divided in rectangles of the same height and width, preferably in the size of the stations or the mobile robot. This plan would be saved into the mobile robot and would be represented there through an array. With this array the mobile robot knows the factory structure and could plan the path global through the factory. For this path planning an algorithm named A* would be very common. [8]

On the other hand, for the local navigation sensors like a gyroscope, odometry or infrared could be used. The infrared would be used to detect obstacles, which are not recorded in the global array plan and for a communication with other mobile robots and working stations. Therefore, the work stations would have an infrared transmitter, so the robot have a redundancy to know which station is sending infrared data and knowing also, due to that where his position in the factory is. But a big disadvantage of this concept is the weakness in dynamic factory structures. Because each time the factory arrangement would change, the whole plan or array must be updated and saved into the robot again. The symbolic sketch of this second concept is shown on the Fig. 2.

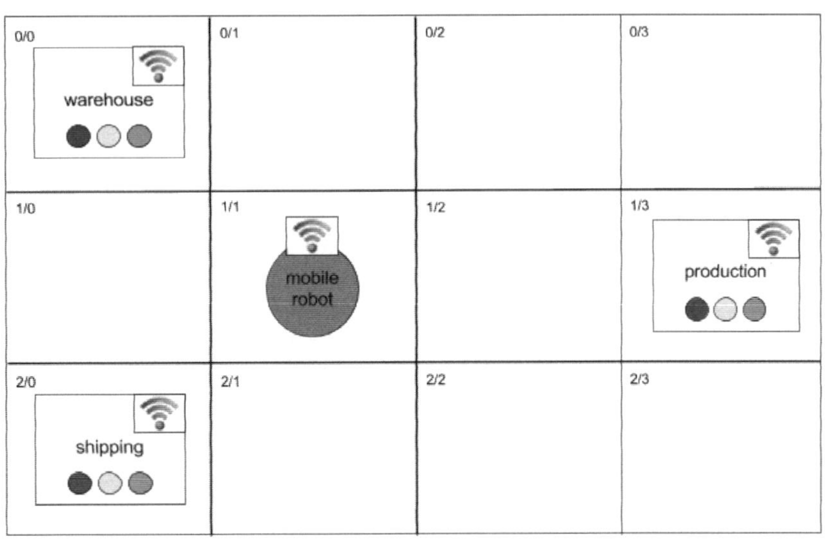

Fig. 2. Concept of a SIP4.0 factory with a two-dimensional plan

The second concept has how already mentioned a big disadvantage with big and variable stations which has a dynamic factory structure. However, the plan creation must be perfect and this is very exhausting for the user when it comes to rebuilding it. Therefore, a third concept for small- and medium-sized companies were created. These factories of this companies are in most cases chaotically formed because they have been growing over time. Thus, these factories didn't have perfect structed machines and working stations. In addition, obstacles like humans or lift trucks can get in the way of mobile robots.

Because of these points in this third concept all machines (stations) can communicate with the mobile robots which calculates its way to the stations completely dynamical with an additional local obstacles avoidance. With infrared and ultrasonic sensors, the local obstacle avoidance will be made. In addition, the infrared signals will be used for a station recognition like in the second concept. Optional the mobile robot could use a camera or a laser to generate a dynamic plan of the factory. Through appropriate placement of the sensors the robot can identify the height of the obstacles and can drive beneath tables or chairs for example. [3]. Also for this third concept a symbolic sketch was included as Fig. 3.

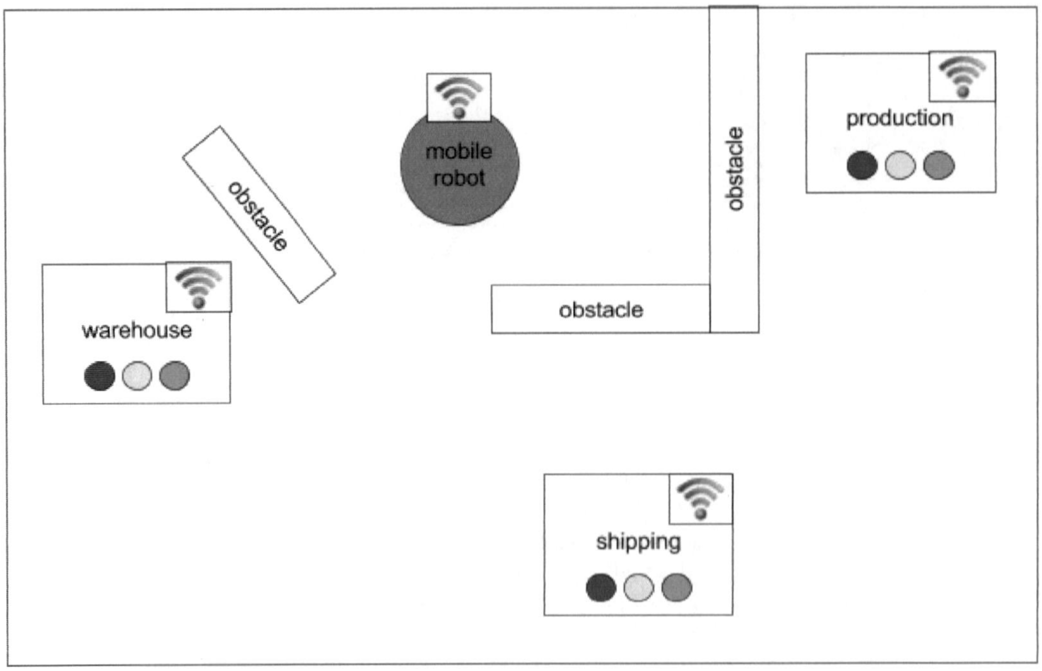

Fig. 3. Concept of a chaotically formed factory

This concept would be the best fitting concept to build in every small- and medium-sized company. With a laser sensor the mobile robot could generate a map of the factory for the self-localization in a dynamically created global map. In addition to that, the mobile robot could get information's where the allowed safety areas to drive though are. This should be used as redundancy for the safety driving along safety areas in the factory.

3.2 Station concepts:

After the concept of the own test factory was clear, the stations, which represent the machines in this test factory, and the mobile robots were conceptualized. First it was considered which states are necessary to display different test cases. For the communication between the control system and every station a wireless network access is needed. Therefore, a Raspberry Pi (RPi) will be used. However, two different Raspberry Pi's got analysed in greater detail [9]:

	Raspberry Pi 3	**Raspberry Pi Zero W**
Size	85,6mm x 56mm	65mm x 30mm
CPU cores	4	1
CPU frequency	4x 1200 MHz	1000 MHz
RAM	1024 MB	512 MB
WLAN	Yes	Yes

Table 1. Comparison of Raspberry Pi 3/Zero W

The Raspberry Pi Zero W and the Raspberry Pi 3 provide both a WLAN communication. But in combination with ROS the RPi Zero W would be too weak because of its lower specifications like the CPU und RAM. [10]

Thus, the Raspberry Pi 3 was chosen. The RPi was also chosen because it is possible to run Linux with ROS on it. If a ROS connection isn't required for the mobile robot, it is possible to use a combination of an Arduino and an WLAN module. The best module with the best value of money is the ESP8266. With this module it is possible to provide a WLAN connection to communicate with the control system of the test factory.

For the stations it is not necessary to supply them with big tasks in first place. To begin with different states of the stations will get visualized with LEDs. Therefore, every station has three LEDs for the state recognition. It can visualise that the station is free, is working or is ready to collect. Further an infrared transmitter is planned for a local communication between station and mobile robot. And of course, also a limit switch will be installed, to recognise the delivery of the mobile robot.

3.3 Mobile robot concepts:

As mentioned before it is necessary that at least more than two mobile robots are working simultaneous at the test factory. Because the test factory will be built in an office room with limited space, the robots should need lesser space than 20x20x20cm, Furthermore, the dodge scenarios and the communication could only be tested with more than two mobile robots. To build an own mobile robot the

Table 2 lists important components which were researched and needed for a simple tiny mobile robot.

Component	Costs per unit [€]	Units	Costs [€]
Gear motors	4,6	1	4,6
Encoder	14,65	1	14,65
Driving wheel	3,5	1	3,5
Support wheel	4,05	1	4,05
Infrared Sensor	10,6	2	21,2
Arduino	19,95	1	19,95
Raspberry Pi	18,99	1	18,99
SD card	11,49	1	11,49
Ultrasonic sensor	10,99	1	10,99
Battery holder	1,99	1	1,99
Battery clip	0,51	1	0,51
Accumulator	18,99	1	18,99
Motor driver	3,59	1	3,59
IC socket	0,26	1	0,26
Voltage regulator	0,26	1	0,26
Heat sink	1,09	1	1,09

Table 2. Components for an own tiny mobile robot

The total costs for the own built mobile robot would be 136.11€ without any circuit boards or the whole assembly. Simultaneous to the research of the components some existing complete miniature mobile robots were compared. The research showed, that the existing mobile robots where much cheaper than self-built one. The self-built one would be more flexible, but in order to the limited time and budget some on the market existing tiny mobile robots where bought based on the comparison in

Table 3:

Mobile Robot	Advantages	Disadvantages
AREXX AAR 04	+ Integrated Arduino + Very cheap + Optical Line following + Plug and play + Odometry sensors	- Rectangle shape - No Infrared or ultrasonic sensor available
Elegoo UNO	+ External Arduino programming + Ultrasonic sensor + Line following + Infrared and Bluetooth remote controlling	- No infrared sensors - Oval shape

FRANZIS	+ Integrated Arduino + Big round platform for expansions + Line following	- Discontinued by Arduino - No ultrasonic sensors
NIBO Burger	+ Integrated Arduino + Line following, Robot following, Obstacle avoidance, Colour detection + 4 infrared sensors, 2 infrared wheel encoders + 4 programable LEDs on top of the mobile robot + Documentation and libraries + I^2C, UART, SPI interface for Raspberry Pi + Expansions possible on top of the robot	- No ultrasonic sensors
Variobot	+ Line and light following + Interaction with obstacles and other robots + Obstacle avoidance (due to patented sensors)	- No programming possible
ZUMO Robot	+ Integrated 32U4 microcontroller + Line following, Robot following, Obstacle avoidance + 4 infrared sensors, Wheel encoders + Documentation and libraries + Expansions possible on top of the robot	- No ultrasonic sensors

Table 3. Comparison of different mobile robots

The different advantages and disadvantages of each existing mobile robot were compared and as a result some NIBO Burger and ZUMO Robots were bought. Both mobile robots offer many useful functions as line tracking or obstacle avoidance. The NIBO Burger is the more cost-efficient solution, but it uses a plastic gearbox instead of a metal one like the ZUMO robot. Also, both bought mobile robot systems are working with an Arduino processor, which unfortunately cannot be used directly with ROS. A Raspberry Pi 3 with the Ubuntu Mate operating system will be needed, where the ROS package could be installed for the robot controlling. With the combination of the Arduino processor and the Raspberry Pi with the ROS package some basic question about safety and the communication in the industry 4.0 environment can be solved.

If a Raspberry Pi will be used, an external power supply will be needed. A power bank would offer the needed portability and the output current of minimum two amperes. All used mobile robots are needing modifications for this work.

4. Results

To compare the different mobile robot solutions and the possible needed improvements the ZUMO Robot and the NIBO Burger were tested in some load tests.

The ZUMO Robot was supplied a 2700mAh accumulator, where the robot drove in an obstacle avoidance program full automatically. Under full motor speed the ZUMO mobile robot was capable to drive over 3 hours and 45 minutes. This long test drive was very overwhelming, but even if longer running time is required, the power supply can be extended by a power bank. Further, the robot hit often an obstacle, because the infrared sensors are not covering the whole robot body. The edge, the middle front and the top obstacle avoidance of the robot could be improved through more or better placed infrared and additional ultrasonic sensors. Therefor the sensors must be placed at the top of the robot so it can recognize, if the robot can drive beneath obstacles or not.

At the research of the standards some risk analyses for the mobile robot were done and listed in the next few sentences:

- The mobile robot may not squeeze persons or machines in a crash. To avoid this kind of injuries, the mobile robot must ensure different safety precautions, like bumpers.
- The gears are dangerous if they are reachable, in order they must be hidden under cases. Also, other harmful mechanical components must be protected by a case.
- The mobile robot must ensure precautions against unintentional contact with the power supplies. Again, a case around the dangerous components would help to prevent any harm.
- All outer sides of cases and the mobile robot should not be maintaining sharp edges, where persons or machines could injury in a collision.
- The mobile robot may not move spontaneous over a big distance after turning it on or after a warm restart.
- In the end may all risks though emission like noise emission, risks through vibrations or extreme temperature be prevented.

All above written risks must be ensured and have precautions to build a standardized mobile robot. But these are only a few very important risks for the final mobile robot concept which will have a new and more detailed risk analysis. This will be made for the finished test factory and is not already necessary along the standpoint of this project.

5. Discussion

This project has just started and is in the initial phase. In first place the research began about the topic SIP 4.0. Afterwards the goal was to develop concepts for a mobile robot for companies in small- and medium-sized companies. Due to that it was decided to build an own test factory in a miniature design in an office room to have infinite possibilities for expanding and designing a mobile robot and working stations.

In the next step the mobile robots, which were already tested, will get expanded with Raspberry Pi's for a communication between each mobile robot and the working station. Additionally, the robots will get ultrasonic sensors and replaced infrared sensor for a better obstacle avoidance. However, the next step is to build the working stations and equip them with Raspberry Pi's or ESP8266 for an WLAN communication in the test factory with the mobile robots.

Another improvement for the mobile robots would be the extension with a camera. Especially for the concept of a chaotically formed factory would the additional camera navigation bring benefits. For example, an improved obstacle avoidance through visual detection or for a tele-operated emergency function. Furthermore, the mobile robot needs a way to localize in an industry 4.0 indoor application, because GPS is not working there. Through the camera special natural landmarks could be detected. [11] And with the detected landmarks the path of the mobile robot could get planned. [12]

When this is done, the project goes into the next step which means to transform this test factory to a fully SIP 4.0 factory. Therefor the topics safety and security of the whole factory will get updated to the state of the art. If this is working together the project will begin to expand with bigger mobile robots, which will get equipped with the safety and security standards of the miniature mobile robots. Therefore, a risk assessment as it stands in the ISO 13482 will created.

The final goal will be to have a mobile robot for a small- or medium-sized company which get tested for an autonomous system which brings parts from Machine A to B. Therefor the mobile robot may get equipped with a robotic arm to grab parts from the machines and delivers them. Based on these results a guideline for the usage of cost-efficient mobile robots for companies in an industry 4.0 environment will be created.

This miniature test factory will help for the research of an optimal SIP4.0 production environment. With the created and tested smaller mobile robots all important questions, methods and standards were prepared and can easily adapted on bigger mobile robots. For different test cases the test factory and their stations, built in this work, can be used.

6. Acknowledgments

We would like to thank the "MA23" and also the UAS Technikum Wien for the opportunity to work on this paper and its very interesting topic. This accomplishment would not have been possible without them and therefor a big thank you.

7. References

[1] Stojkić, Ž.; Veža, I. & Bošnjak, I. (2016). „A concept of information system implementation (CRM and ERP) within Industry 4.0", Proceedings of the 26th DAAAM International Symposium, pp.0912-0919, Vienna, Austria.

[2] Bischoff, J.; Taphorn, C.; Wolter, D.; Braun, N.; Fellbaum, M.; Goloverov, A.; Ludwig, S.; Hegmanns, T.; Prasse, C.; Henke, M.; Hompel, M.; Döbbeler, F.; Fuss, E.; Kirsch, C.; Mättig, B.; Braun, S.; Guth, M.; Kaspers M. & Scheffler, D. (2015). „Erschließen der Potenziale der Anwendung von ‚Industrie 4.0' im Mittelstand", Mülheim an der Ruhr: agiplan GmbH.

[3] Siegwart, I. N. R. (2004). „Introduction to Autonomous Mobile Robots", MIT Press.

[4] Markis, A.; Montenegro, H.; Neuhold, M.; Oberweger, A.; Schlosser, C.; Schwald, C.; Sihn, W.; Ranz, F.; Edtmayr, T.; Hold P. & Reisinger, G. (2016). „Sicherheit in der Mensch-Roboter-Kollaboration", TÜV AUSTRIA Holding AG & Fraunhofer Austria Research GmbH, Wien.

[5] ISO 13482:2014, (2014). „Robots and robotic devices -- Safety requirements for personal care robots".

[6] DIN EN ISO 12100:2011-03, (2011). „Safety of machinery - General principles for design - Risk assessment and risk reduction".

[7] http://www.ros.org, (2017). „Robot Operating System", Accessed on: 2017-09-25.

[8] Hart, P. E.; Nilsson, N. J. & Raphael, B. (1968). „A Formal Basis for the Heuristic Determination of Minimum Cost Paths", IEEE Transactions on Systems Science and Cybernetics, Bd. 4, Nr. 2, pp. 100-107.

[9] https://www.datenreise.de/raspberry-pi-unterschiede-zwischen-den-modellen/, (2017). „Raspberry Pi - Modellvergleich", Accessed on: 2017-09-15.

[10] https://kofler.info/ubuntu-mate-16-04-raspberry-pi/, (2016). „Ubuntu Mate 16.04 für den Raspberry Pi", Accessed on: 2017-09-12

[11] Smirnova, E.; Stepanov, D. & Goryunov, V. (2016). „A technique of natural visual landmarks detection", Proceedings of the 26th DAAAM International Symposium, pp.0905-0911, Vienna, Austria.

[12] Katsurin, A. (2016). „Planning trajectory of the mobile robot with a camera", Proceedings of the 26th DAAAM International Symposium, pp.0407-0416, Vienna, Austria.

DOI: 10.2507/28th.daaam.proceedings.137

MATHEMATICAL MODELING OF THERMAL MANAGEMENT AT DIFFERENT ATMOSPHERIC PRESSURE CONDITIONS

Juan Carlos Beltrán-Prieto, Luis Antonio Beltrán-Prieto &
Long Huynh Bach Son Nguyen

This Publication has to be referred as: Beltran Prieto, J[uan] C[arlos]; Beltran Prieto, L[uis] A[ntonio] & Long Huynuh Bach Son, N[guyen] (2017). Mathematical Modeling of Thermal Management at Different Atmospheric Pressure Conditions, Proceedings of the 28th DAAAM International Symposium, pp.0988-0995, B. Katalinic (Ed.), Published by DAAAM International, ISBN 978-3-902734-11-2, ISSN 1726-9679, Vienna, Austria
DOI: 10.2507/28th.daaam.proceedings.137

Abstract

Cooling process is an important element in the correct functioning of all electronic devices. While some components can resist high temperatures, others are damaged during overheating conditions. We performed numerical simulation to study the temperature variation during cooling process of electronic devices that are subject to the effect of different pressures. Commonly the correct functioning of instruments is tested under specific conditions given by the location of the manufacturer but generally the consumer uses the device in a different region. Cooling process performed by convection uses the air provided by a fan. Fluid properties such as kinematic viscosity are influenced by pressure and the effect in cooling is demonstrated by analyzing the variation of surface temperature at different pressures. This study allows us to understand the importance of fluid flow speed in controlling the heating rate. We performed numerical simulation at different air speed (10 – 20 m/s) and pressure (77 – 100 kPa) which corresponds to the elevation of cities that are located between 31 m and 2240 m., additionally we studied the maximum amount of power dissipation as a function of air velocity with radiation contribution and without radiation effect.

Keywords: Air flow velocity; Nusselt parameter; atmospheric pressure; kinematic viscosity; power dissipation

1. Introduction

Forced convection is a process in which heat transfer occurs due to simultaneous conduction and fluid flow. During the process of cooling caused by high velocities of flow, laminar regime is present in the interphase of the surface and the fluid while turbulent flow is present far from the surface. The mathematical model of cooling process is frequently expressed by differential equations. In some cases, analytical solutions exist but frequently numerical methods are required. One of the most common cases of cooling is caused by air flowing in parallel direction of a hot solid plate [1]. As the air flows, the temperature profile changes at different cross sections normal to the plate surface. The direction of heat flows is always from regions of higher to regions of lower temperature. Energy is transferred through an area in the form of heat at a rate proportional to the temperature gradient normal to the area. [2]

Intensive research has been performed about cooling systems. Heating and cooling experiments were performed by Vogel and coworkers, who studied heat transfer in a rectangular enclosure with different widths and heights [3]. Earlier studies by Bareiss and Beer described the observation of two-regime models, a conduction regime was observed firstly, followed by a quasi-steady convection regime [4]. Studies on heat transfer of fluids in cylinders by convection at different heat generation rates and temperatures were performed by Shibahara et al., [5]. Experiments were also performed using different Rayleigh numbers according to the numerical value of heat flux input, rectangular box length and initial properties of the fluid and defined Prandtl number identifying three different stages, namely conduction through the fluid layer, Rayleigh-Bénard convection and formation of longitudinal rolls [6]. Other studies were aimed to validate theoretical correlations that take part in fluid flow and heat transfer characteristics in microchannels (i.e. Nusselt number) [7, 8]

Cooling system on computers is normally unnoticed by the user while using the device but it is a key element for reliable operation and correct functioning of the device. Without the presence of a cooling system, the heat generated while operating the instrument would cause overheating, damage and not only the breakdown of the system but also could represent a risk to users' health [9]. The proper design of the cooling system has to be compact and also efficient. During the operation of a computer, the temperature of some parts of the component could reach up to 100 °C. Heat transfer plays an important role in controlling the temperature by distributing the heat generated to fins located in the electronic device. A fan is used to circulate the air which adsorbs and release the heat. There exist several alternatives to cool a system, among the most common are by air and by water (watercooling). Tubes containing copper filaments and water in liquid state are in continuous contact with the electronic parts where heat is generated. Transfer of heat occurs from the computer microprocessors and heated components of the device to the tubes due to conduction. Inside the cavity of the tube, the temperature increases as water adsorbs the heat, evaporates and changes the gaseous state. Heated water vapor fills the cavity of the tube surrounding the filaments and the heat starts to be distributed to the fins due to the continuous air flow provided by a near fan. When the water vapor reaches the extreme of the tubes, the heat is adsorbed and the vapor condensates in the copper filaments and changes its state to liquid. The space between the filaments is filled gradually until reaching again the hot area of the device to repeat the cooling process. Therefore, the heated vapor is continuously replaced with cooled water and the cycle continues. A gradient of temperature exists due to the difference in temperature between the solid surface and the fluid surrounding it [10–12]. Recent publications on heat transfer states the importance of simulation and modeling of thermal processes in several research fields of study [13–16].Often, the influence of parameters such as atmospheric pressure on convection process is not taken into account. In the present paper, we aim to model the variation of kinematic viscosity of cooling air and surface temperature of a transistor at different pressures. The different values of corresponding pressures depend on the elevation of the cities which were randomly selected for numerical simulation purposes. Mathematical description of air flow variation is presented to account for the regulation of surface temperature. This information would be helpful to assist in the design of new cooling devices of electronics as the control of temperature in the surface of transistors is a key element in troubleshooting thermocouple failures. Furthermore, we studied the effect of air flow in the theoretical maximum value of power dissipation that could be achieved.

2. Description of the mathematical model

For the analysis of the problem, we consider steady state conditions, assuming that the surface of the transistor is isothermal. Air is used to cool the upper layer of the chip surface entering at 25°C and flowing in parallel over the surface of the chip with length (*l*). The remaining layers of the chip are well insulated from the surrounding. The average convection coefficient for the external flat surface of the chip located at distance (L) from the leading edge is related to the local value according to equation $\bar{h} = \frac{1}{L}\int_0^L h \cdot dx$ [17]. Additionally, we suppose that the system can be confined to two dimensions if the transistor dimensions of length are assumed to be considerably greater than the thickness. This system can be represented as described in Figure 1. Considering the Newton's law of cooling, the rate of convection ($\dot{q}$) is proportional to the temperature difference as expressed in (1). The heat transfer from the solid to the stagnant fluid layer is by conduction, following Fourier's law (2)

$$\dot{q} = h(T_S - T_\infty) \tag{1}$$

$$\dot{q} = -k_f \left.\frac{\partial(T - T_S)}{\partial y}\right|_{y=0} \tag{2}$$

Where h is the convective heat transfer coefficient ($W/m^2 \cdot K$), T_S is the chip surface temperature (K), T_∞ is the fluid temperature (K), k_f represents the thermal conductivity of the fluid, and T is the temperature at distance (y) [18]. Combining (1) and (2) leads to (3) and to determine the value of h we derive expression (4)

$$h(T_S - T_\infty) = -k_f \left.\frac{\partial T}{\partial y}\right|_{y=0} \tag{3}$$

$$h = \frac{k_f}{(T_S - T_\infty)} \cdot \frac{\partial (T_S - T)}{\partial y}\bigg|_{y=0} \tag{4}$$

Fluid properties are normally evaluated at the film temperature, which is given by expression (5) representing the average value between the transistor surface and the surrounding fluid temperatures [19].

$$T_F = \frac{(T_S + T_\infty)}{2} \tag{5}$$

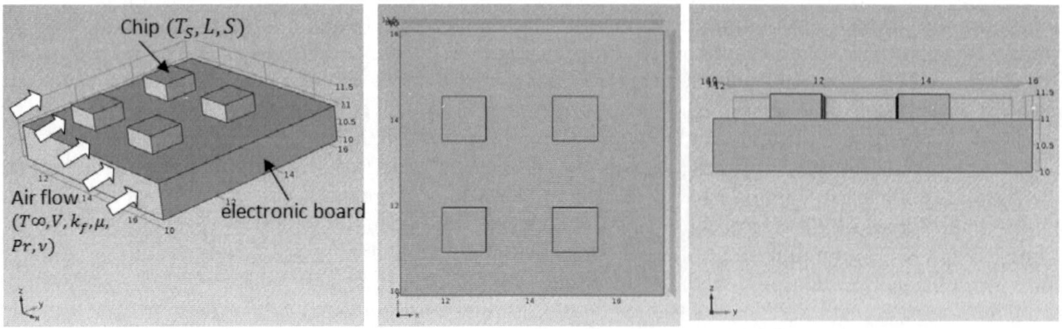

Fig. 1. Graphical representation of the system under study showing the electronic board, direction of air flow, location of transistor and geometrical parameters.

The main fluid properties that are important to consider are the thermal conductivity (k_f;W/m·K), the viscosity (μ; N·s/m^2), Prandtl number (Pr), kinematic viscosity (v;m^2/s) which can be obtained from Thermodynamic tables available in literature after linear regression as described through (6) to (9) for a temperature range of $100 \leq T,(K) \leq 550$

$$v = \left[-2.072 + 0.0276T_F + 1.078 \times 10^{-4}{T_F}^2\right] \times 10^{-6} \tag{6}$$

$$\mu = \left[6.109 + 0.6945T_F - 3.323 \times 10^{-4}{T_F}^2\right] \times 10^{-7} \tag{7}$$

$$k_f = \left[-0.0386 + 0.097T_F - 3.1 \times 10^{-5}{T_F}^2\right] \times 10^{-3} \tag{8}$$

$$Pr = \left[0.8396 - 6.2857 \times 10^{-4}T_F + 6.32 \times 10^{-7}{T_F}^2\right] \tag{9}$$

Then, for the case of air at 25°C at sea level (P=1 atm), the parameters are $v = 1.5726 \times 10^{-5}\ m^2/s$, $\mu = 1.8356 \times 10^{-5}\ N \cdot s/m^2$, $k = 0.0261\ W/m \cdot K$, $Pr = 0.7084$

Reordering equation and multiplying both sides by length (L) we obtain

$$\frac{hL}{k_f} = \frac{L}{(T_S - T_\infty)} \frac{\partial (T - T_S)}{\partial y}\bigg|_{y=0} \tag{10}$$

were the term hL/k_f represents the dimensionless Nusselt parameter which expresses the relationship between the heat transfer by convection and the heat transfer by conduction across the boundary[20]. This parameter is function of Reynolds and Prandtl numbers. For Reynolds number in the range $10^4 \leq Re \leq 5\times10^4$, the Nusselt number can be evaluated from the expression $Nu = aRe^bPr^c$, were a, b, c are defined constant values depending on the geometry of the solid and Reynolds number. Particularly, for turbulent flow, the Nusselt number takes the form of $Nu = 0.023Re^{4/5}Pr^{1/3}$ for $0.6 \leq Pr \leq 60$ based on Chilton-Colburn analogy [21]. Moreover, from the definition of Reynolds number: $Re=Vx/v$, and evaluating when $x=L$ leads to $Re = VL/v$. Thus

$$\frac{hL}{k_f} = 0.023\left[\frac{VL}{v}\right]^{4/5} Pr^{1/3} \tag{11}$$

Following from expression (11), we can obtain the value of the local heat transfer coefficient

$$h = \frac{k_f}{L^{1/5}} \cdot 0.023\left[\frac{V}{v}\right]^{4/5} Pr^{1/3} \tag{12}$$

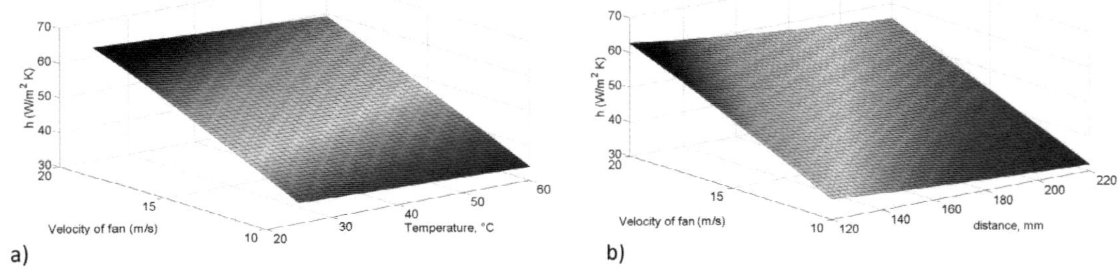

Fig. 2. Local heat transfer coefficient profile as function of velocity of input air (m/s), Temperature (°C) and distance (mm)

As described in figure 2, the value of convective heat transfer coefficient does not vary with the temperature. However, the distance of the transistor from the leading edge has an impact in increasing the value of h. From expression (1), it is clear that the local surface temperature (T_s) can be calculated as expressed in (13) and depicted in Figure 3 where the variation of chip temperature is described as function of chip dimensions and convective coefficient. Accordingly, Figure 3b) shows the influence of air velocity and chip dimensions in temperature profile.

$$T_S = \frac{\dot{q}}{h} + T_\infty = \frac{q}{A \cdot h} + T_\infty \qquad (13)$$

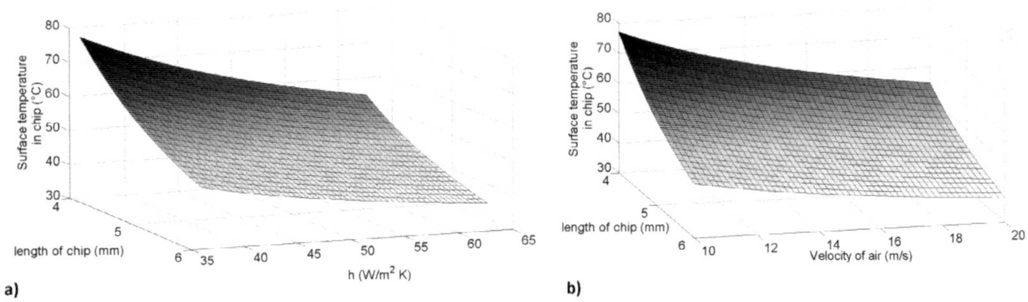

Fig. 3. Temperature on the surface of the chip variation (°C) as function of length (mm), h (W/m2·K) and input air velocity (m/s).

We proceed to calculate the influence of atmospheric pressure on the cooling performance of the electronic device. We select a diversity of cities with an altitude between 31m and 2240 m as described in Table 1. Pressure is determined as described in (14)

$$P = 101.325[1 - 2.25577 \times 10^{-5} \cdot h]^{5.25588} \qquad (14)$$

were h is the altitude of the city (m) and Pressure is given in kPa

Altitude (m)	Pressure (kPa)
2240	77.15
2147	78.05
1804	81.45
1537	84.17
1137	88.39
530	95.12
31	100.95

Table 1. Variation of pressure from selected cities around a country with corresponding values of altitude

In case of ideal gases we have $\rho = P/RT$, in which ρ is constant for incompressible liquids. As the kinematic viscosity is $\nu = \mu/\rho$, then $\mu = \nu P/RT$. Considering case 1 the conditions at atmospheric pressure (P_1) and case 2 the pressure at different city (P_2), the value of viscosity does not change with pressure, then

$$\mu_1 = \mu_2 \tag{15}$$

$$\frac{v_1 P_1}{\mu_1 R T_1} = \frac{v_2 P_2}{R T_2} \tag{16}$$

$$v_2 = \frac{v_1 P_1}{P_2} \tag{17}$$

$$v_2 = \frac{(1.5726 \times 10^{-5}\ m^2/s)(101.325\ kPa)}{P_2} = \frac{0.001593}{P_2} \tag{18}$$

Kinematic viscosity relates the ability to transmit momentum disturbance with the ability to sustain momentum disturbance. As a result, the value of convective heat transfer coefficient and the temperature at the chip surface at different pressure can be calculated from expressions (19) and (20) respectively

$$h_2 = \frac{k_f}{L^{1/5}} \cdot 0.023 \left[\frac{P_2 V}{0.001593}\right]^{4/5} Pr^{1/3} \tag{19}$$

$$T_S = \frac{q}{A \cdot h_2} + T_\infty \tag{20}$$

Pressure (kPa)	υ (m²/s)	Vf=10 m/s		Vf=20 m/s	
		h (W/m²·K)	T (°C)	h (W/m²·K)	T (°C)
77.15	21×10^{-6}	28.9	89.73	69.59	51.79
78.05	20.4×10^{-6}	29.16	89.13	70.24	51.54
81.45	19.5×10^{-6}	30.17	86.97	72.67	50.64
84.17	18.92×10^{-6}	30.98	85.36	74.61	49.97
88.39	18×10^{-6}	32.21	83.04	77.59	49
95.12	16.7×10^{-6}	34.16	79.72	82.28	47.63
100.95	15.78×10^{-6}	35.83	77.15	86.29	46.57

Table 2. Corresponding values of *h* and chip Temperature at given values of air flow speed and Pressure

This data is represented in Figure 4 for the limit of air velocity flow studied (10m/s - 20 m/s). As pressure decreases (higher altitude) the temperature on the transistor surface is higher. Using A speed of 10m/s the maximum chip temperature at normal pressure is 77 °C but in a city with different altitude, the temperature could reach 89.73 °C which is a significant variation and should not be considered irrelevant. In order to reduce the temperature, air speed should be increased. This can be calculated from expression (21)

$$V = v \left[\frac{h L^{1/5}}{k_f 0.023 Pr^{1/3}}\right]^{5/4} \tag{21}$$

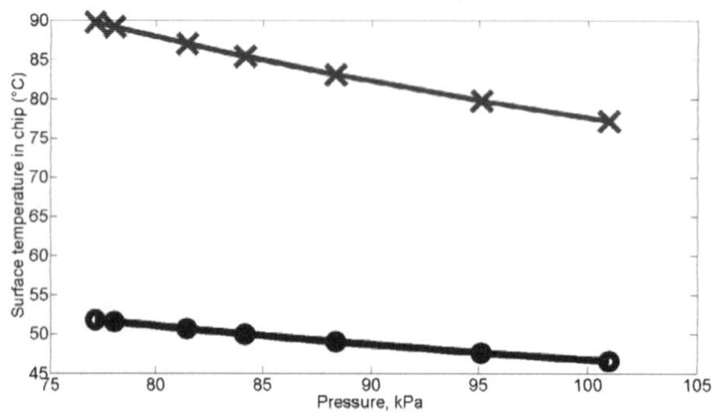

Fig. 4. Chip surface temperature range for selected values of input air flow (10 m/s ×; 20 m/s •).

Pressure (kPa)	υ (m²/s)	$v = 1.57 \times 10^{-5}\, m^2/s$, $h = 35.93\ W/m^2 \cdot K$		$v = 1.57 \times 10^{-5}\, m^2/s$, $h = 62.57\ W/m^2 \cdot K$	
		T	Vf	T	Vf
77.15	21×10^{-6}	89.73	13.3	51.79	26.26
78.05	20.4×10^{-6}	89.13	12.98	51.54	25.96
81.45	19.5×10^{-6}	86.97	12.44	50.64	24.88
84.17	18.92×10^{-6}	85.36	12.03	49.97	24.07
88.39	18×10^{-6}	83.04	11.46	49	22.92
95.12	16.7×10^{-6}	79.72	10.65	47.63	21.3
100.95	15.78×10^{-6}	77.15	10.03	46.57	20.07

Table 3. Readjustment of air velocity to cool the temperature on the surface of the transistor at same levels as tested at 1 atm

Now we proceed to determine the influence of air velocity at 25°C on the maximum amount of power dissipation as a function of air velocity. Considering that the maximum recommended temperature in the transistor surface should not be more than 85°C for reliable operation over prolonged periods of time, the properties of air at the film temperature, which is given by $T_F = (85 + 25)/2 = 55°C$ are $v = 1.86 \times 10^{-5}\ m^2/s$, $\mu = 1.98 \times 10^{-5}\ N \cdot s/m^2$, $k = 0.0285\ W/m \cdot K$, $Pr = 0.7014$. We considering an emissivity of 0.8 and performing an energy balance on the chip which is stagnant in a dielectric board with a surrounding temperature of 25°C. The graphical description of power dissipation can be observed in Figure 5.

$$P_c = \dot{q} + \dot{q}_r = hS(T_S - T_\infty) + h_r S(T_S - T_{sur}) \tag{22}$$

$$h_r = \varepsilon\sigma(T_S + T_{sur})(T_S{}^2 - T_{sur}{}^2) \tag{23}$$

$$P_c = hS(T_S - T_\infty) + \varepsilon\sigma S(T_S + T_{surrounding})(T_S{}^2 - T_{surrounding}{}^2)(T_S - T_{surrounding}) \tag{24}$$

$$P_c = S\left[\frac{k_f}{L^{1/5}} \cdot 0.023 \left[\frac{V}{v}\right]^{4/5} Pr^{1/3}\right](T_S - T_\infty) + \varepsilon\sigma S(T_S + T_{sur})(T_S{}^2 - T_{sur}{}^2)(T_S - T_{sur}) \tag{25}$$

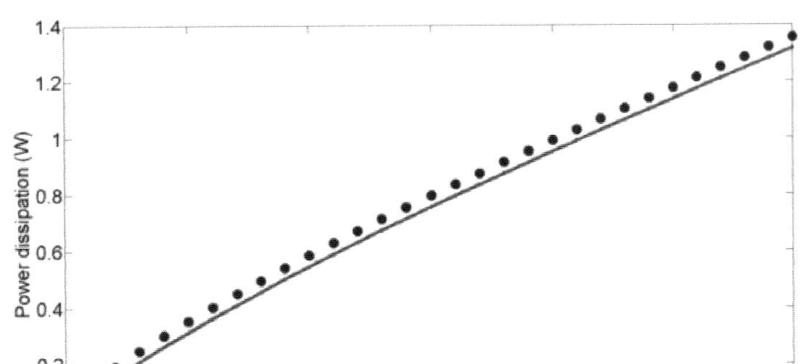

Fig. 5. Power dissipation profile as function of velocity of inlet air considering a maximum of 30 m/s with radiation contribution (dotted) and without radiation effect (continuous).

The convective heat transfer coefficient increases with the velocity of air in the same proportion as the power of dissipation. This can be seen in Figure 6 were it is plotted the value of h at different values of inlet air speed. Power dissipation increases with the velocity of air. The effect of radiation increases at a constant rate the dissipation of heat ($0.04W$) while the convective heat transfer coefficient reaches a maximum value of 82.2 at 30 m/s. Accordingly, it is also possible to evaluate an average convection coefficient according to the integral over a range of the leading edge distance.

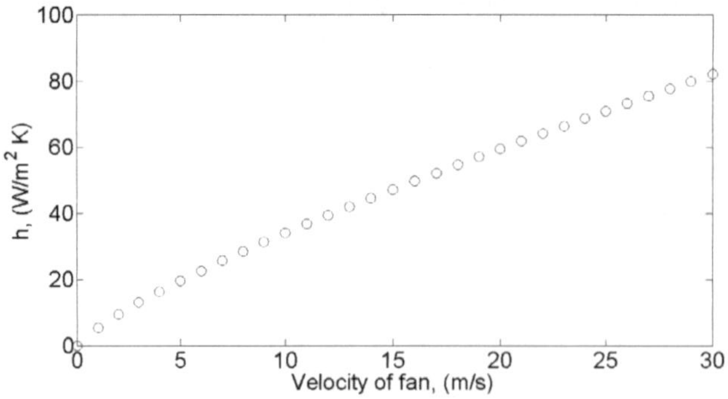

Fig. 6. Influence of air cooling velocity speed (0-30m/s) in heat transfer coefficient.

Variation of elevation affects several thermodynamic properties. The air is less dense at higher elevation, which decreases the atmospheric pressure value. Boiling point is other property that is affected as it decreases with the increment in elevation due to the effect of vapor pressure. In the case of convective heat transfer, there are several dimensional parameters that are important, namely Nusselt number, Reynolds number Prandtl number, and Rayleigh number. After Rayleigh number reaches certain values, the dominant mechanism is convection, rather than conduction. These values are strongly dependent on the characteristics of the system like the shape of the solid, fluid state (gas or liquid), viscosity and kinematic viscosity of the fluid, heat capacity of the air that is used for cooling, as well as the thermal conductivity, among others. Cooling process is affected considerably when the process is performed at different altitudes as it has been demonstrated by the different value of surface temperature (12.58 °C difference) achieved when the device is used in cities above 2240 m. Adjustments in the input air flow have to be made in order to compensate for this difference. This has a big impact in the design of new types of electronic devices which should take into consideration whether or not some changes have to me made according to the maximum allowed temperatures recommended for the components to maintain or improve their performance.

3. Conclusion

Mathematical modeling and simulation is an important element in several applications in many different fields of sciences such as chemistry, physics, mechanics, biology, and electronics among others. The study of fluid flow through a hot surface that allows the distribution of heat due to a temperature gradient is a practical problem that faces many challenging problems. Numerical simulation and modeling are tools that are frequently used to understand many processes and be applied to real cases to improve the design and its performance. In this paper we have used numerical simulation to study the variation of temperature on a transistor that is heated on an electronic device and is subject to the effect of different pressures. This is the case of using the device in different cities as the pressure changes with altitude and more important the density of the air since this parameter increases proportionally to the increment in pressure. Accordingly, at constant mass flow rate a decline in density causes a linear drop in the pressure. Many fans that are manufactured conventionally cool at constant speed even with the change of pressure or flow rate due to the design of blades which run at constant speed. The pressure output of the fan is strongly connected with the density of the air. Even at constant volumetric flow rate values a drop in mass flow rate with density is observed. Other type of fans is called slip fans, which allows the use of different speeds from the motor driving. This is extremely important from the point of view of electronic and mechanical cooling design for the control of temperature. In this type of fans, changes in altitude do not causes a drastic variation in the pressure flow rate curves in comparison to the use of conventional fans. One of the most important parameters to take into consideration for the effect of cooling through convection is the mass flow of fluid. When the air flow mass in the system is increased, the rise in temperature is lower. For example, using a flow of 10 m/s, a temperature near 77 °C can be achieved in the surface of the chip but this temperature decreases up to 46 °C when the speed is doubled. In the numerical study range of elevation (31-2240 m) the temperature values achieved using fans with 10 m/s leads to temperatures between 77.15 °C and 89.73 °C while in the case of doubling the air mass flow the temperature range is between 46.57 °C and 51.79. This decrement of temperature is a key element in the correct functioning and maintenance of devices that generates heat such as the case of electronics. Selection of fans appropriate for cooling of electronic devices should be base on the proper interpretation of fan supply curve and system pressure drop, particularly in the identification of stable operating points with enough flow rates. At higher altitudes, is important to take into account the influence of low density and temperature on pressure drop. Variation of volumetric flow rate is related to the type of fan used. This study allows us to evaluate the variation of temperature and to understand the importance of fluid flow speed in controlling the heating rate. The value of heat flux is restricted to the maximum temperature that the chip can resist after technical and thermal characteristics of design. Additionally, we studied the maximum amount of power dissipation as a function of air velocity considering a maximum of 30 m/s with radiation contribution and without radiation effect. Other consideration that could be incorporated in further studies is the influence of temperature in the surroundings (i.e. temperatures during winter and summer seasons), particularly the hot days in summer which would characterize the design for critical and worst conditions.

4. Acknowledgments

This work was supported by the Ministry of Education, Youth and Sports of the Czech Republic within the National Sustainability Programme project No. LO1303 (MSMT-7778/2014) and also by the European Regional Development Fund under the project CEBIA-Tech No. CZ.1.05/2.1.00/03.0089. Further it was supported by Grant Agency of the Czech Republic—GACR 588 P103/15/06700S and by Internal Grant Agency of Tomas Bata University in Zlin under the project No. IGA/CebiaTech/2017/004. L.A.B.P author also thanks the doctoral scholarship provided by the National Council for Science and Technology (CONACYT) and the Council for Science and Technology of the State of Guanajuato (CONCYTEG) in Mexico.

5. References

[1] Incropera, F. P., DeWitt, D. P., Bergman, T. L., & Lavine Adrienne S. (2007). Fundamentals of Heat and Mass Transfer, John Wiley & Sons Inc, ISBN: 978-1-119-32042-5, United States.

[2] Lienhard V, J. H., & Lienhard V, J. H. (2004). A Heat Transfer Textbook Dover, ISBN-10: 0486479315, United States

[3] Vogel, J., Felbinger, J., & Johnson, M. (2016). Natural convection in high temperature flat plate latent heat thermal energy storage systems. Applied Energy, Vol. 184, pp. 184–196. doi:10.1016/j.apenergy.2016.10.001

[4] Bareiss, M., & Beer, H. (1984). Experimental investigation of melting heat transfer with regard to different geometric arrangements. International Communications in Heat and Mass Transfer, Vol. 11, No. 4, pp. 323–333. doi:10.1016/0735-1933(84)90060-5

[5] Shibahara, M., Liu, Q., & Fukuda, K. (2016). Transient natural convection heat transfer of liquid D-mannitol on a horizontal cylinder. Renewable Energy, Vol. 99, pp. 971–977. doi:10.1016/j.renene.2016.07.076

[6] Sanmiguel Vila, C., Discetti, S., Carlomagno, G. M., Astarita, T., & Ianiro, A. (2016). On the onset of horizontal convection. International Journal of Thermal Sciences, Vol. 110, pp. 96–108. doi:10.1016/j.ijthermalsci.2016.06.019

[7] Yu, J., Kang, S.-W., Jeong, R.-G., & Banerjee, D. (2016). Experimental validation of numerical predictions for forced convective heat transfer of nanofluids in a microchannel. International Journal of Heat and Fluid Flow, Vol. 62, pp. 203–212. doi:10.1016/j.ijheatfluidflow.2016.11.001

[8] Kim, B. (2016). An experimental study on fully developed laminar flow and heat transfer in rectangular microchannels. International Journal of Heat and Fluid Flow, Vol. 62, pp. 224–232. doi:10.1016/j.ijheatfluidflow.2016.10.007

[9] Yeh, L. T. (1995). Review of Heat Transfer Technologies in Electronic Equipment. Journal of Electronic Packaging, Vol. 117, No. 4, pp. 333. doi:10.1115/1.2792113

[10] Jang, S. P., Kim, S. J., & Paik, K. W. (2003). Experimental investigation of thermal characteristics for a microchannel heat sink subject to an impinging jet, using a micro-thermal sensor array. Sensors and Actuators A, Vol. 105, pp. 211–224. doi:10.1016/S0924-4247(03)00103-1

[11] Baby, R., & Balaji, C. (2013). Thermal optimization of PCM based pin fin heat sinks: An experimental study. Applied Thermal Engineering, Vol. 54, No. 1, pp. 65–77. doi:10.1016/j.applthermaleng.2012.10.056

[12] Kalbasi, R., & Salimpour, M. R. (2015). Constructal design of phase change material enclosures used for cooling electronic devices. Applied Thermal Engineering, Vol. 84, pp. 339–349. doi:10.1016/j.applthermaleng.2015.03.031

[13] Nunez-Vega, A-M; Hugenschmidt, S & Hofacker, W (2012). Numerical Simulation of the Convective Drying of Apple Slices, Chapter 29 in DAAAM International Scientific Book 2012, pp. 339-356, B. Katalinic (Ed.), Published by DAAAM International, ISBN 978-3-901509-86-5, ISSN 1726-9687, Vienna, Austria DOI: 10.2507/daaam.scibook.2012.29

[14] Marek, V (2016). Basic Research of Thermal Transfer Simulations, Proceedings of the 27th DAAAM International Symposium, pp.0578-0585, B. Katalinic (Ed.), Published by DAAAM International, ISBN 978-3-902734-08-2, ISSN 1726-9679, Vienna, Austria DOI: 10.2507/27th.daaam.proceedings.085

[15] Gerlich, V., Zálešák, M.(2010) Experimental validation of heat transfer model, Annals of DAAAM and Proceedings of the International DAAAM Symposium, 2010, pp. 969-970. ISSN 1726-9679, Zadar

[16] Chval, Z & Raz, K (2016). Effect of Heat Load on a Mechanical Forging Press, Proceedings of the 27th DAAAM International Symposium, pp.0344-0348, B. Katalinic (Ed.), Published by DAAAM International, ISBN 978-3-902734-08-2, ISSN 1726-9679, Vienna, Austria DOI: 10.2507/27th.daaam.proceedings.050

[17] Holman, J. P. (2009). Heat transfer, McGraw-Hill, ISBN: 0073529362, London

[18] Cengel, Y., & Turner, R. (2001). Fundamentals of Thermal-Fluid Sciences, McGraw-Hill Higher Education.ISBN: 0078027683, New York.

[19] Welty, J., Wicks, C., & Wilson, R. (2008). Fundamentals of Momentum, heat, and Mass Transfer. Jonh Wiley and Sons, ISBN: 978-0470128688, Mexico.

[20] Bird, R. B., Stewart, W. E., & Lightfoot, E. N. (2002). Transport Phenomena. John Wiley & Sons, Inc. ISBN: 0470115394, New York

[21] Tsilingiris, P. T. (2010). Modeling heat and mass transport phenomena at higher temperatures in solar distillation systems – The Chilton–Colburn analogy. Solar Energy, Vol. 84, No. 2, pp. 308–317. doi:10.1016/j.solener.2009.11.012

DOI: 10.2507/28th.daaam.proceedings.138

COLLABORATIVE WORK BETWEEN HUMAN AND INDUSTRIAL ROBOT IN MANUFACTURING BY ADVANCED SAFETY MONITORING SYSTEM

Vladimir Kuts, Martins Sarkans, Tauno Otto, Toivo Tähemaa

Tallinn University of Technology, School Of Engineering, Department of Mechanical and Industrial Engineering, Ehitajate tee 5, 19086 Tallinn , Estonia

This Publication has to be referred as: Kuts, V[ladimir]; Sarkans, M[artins]; Otto, T[auno] & Tahemaa, T[oivo] (2017). Collaborative Work Between Human And Industrial Robot In Manufacturing By Advanced Safety Monitoring System, Proceedings of the 28th DAAAM International Symposium, pp.0996-1001, B. Katalinic (Ed.), Published by DAAAM International, ISBN 978-3-902734-11-2, ISSN 1726-9679, Vienna, Austria
DOI: 10.2507/28th.daaam.proceedings.138

Abstract

Collaborative work between human and industrial robot in different areas is developing rapidly. High payload industrial robots that can harm humans execute their tasks in separated protected areas and are regulated by safety standards. The main aim of this study was to design experiments, analyze, and model the collaborative work and safety monitoring systems by using small industrial robots with inbuilt power and force limiting. Different methods of collaboration are standardized by the ISO/TS 15066:2016. Human co-work with robots gives us a variety of options that can be used in manufacturing and other related areas. The advanced multi-level monitoring system of co-work processes can reach the highest level of the human safety with minimum robot downtime. Several methods were analyzed to find the optimal solution for the online monitoring processes. As a result, a safety multi-level monitoring system was designed, tested in simulation software with real-room conditions with an industrial robot. Safety multi-level system, as a part of the control and monitoring online system consist of different types of sensors and a microcontroller that can be attached to the Cyber Physical Production System (CPPS). Those control and monitoring methods and processes were implemented in experimental setup in Robot Operating System (ROS).

Keywords: collaborative industrial robots; Cyber Physical Production System; Robot Operating System

1. Introduction

The collaboration between industrial robots and humans is a widely-discussed topic nowadays. Many research papers have been published in this field, but there are still a few uncovered areas, mostly relating to the devices ranging between collaborative robots and small simple industrial robots. There is also lack of optimized safety systems, which could fulfill all safety requirements and manufacturing needs.

Different system components, software, and other tools were used in making proper methodologies for studying industrial robot and human collaboration [1], whereas ISO/TS 15066:2016 is the latest recognized standard for this kind of systems. As the development process in the field of collaborative robots is so rapid, we predict that this standard gets updates in the nearest years.

1.1. Problem statement

There are two main problems in small industrial robots and human worker collaboration systems:

- Human safety monitoring system control
- Downtime of the industrial robot, while safety system is activated

In the modern robot systems, every large manufacturer has developed its own controller and software where off-line programming (OLP) is done. In this kind of systems, industrial robots are executing their tasks and moving according to the pre-programmed path which is already done before the real work is performed. To gain better control of the movement of the industrial robot it must be possible to affect robot behavior online, while the task is being executed. In this paper, we demonstrate a method for online monitoring and control of the robot with the help of the robot operating system (ROS). In the experimental part of this article, the comparison of the off-line programming (OLP) vs online control, and monitoring method is shown, which was enabled with the new computer technologies and machine vision applications.

1.2. Background and current situation

Non-collaborative industrial robots must usually comply with safety requirements described in standards ISO 12100, ISO 13850, ISO 13855 etc. As there is no need for human intervention during the manufacturing process (i.e. painting, welding, deburring, etc.) strict rules exist for the safety of the human and for emergency halts of the industrial robot. Standards and guidelines describe the allowed speeds, distances, required safety equipment and safety procedures to establish the required safety. When human enters into the workspace of the robot, the process (and the robot) is stopped immediately. To restart the production process, the fault situation must be cleared (worker must leave from the workspace or remove the obstacle) by resetting emergency button outside the cell. As the production processes are growing more complex and interconnected, the human and robot collaboration principles development has become an important issue during the last years. Industrial robots have advantages in areas (processes) where speed, power, repetitiveness, and durability is needed. The human worker can be added into the process to increase agility and flexibility (if the process or product changes rapidly).

Different authors have researched human-robot collaboration topics such as:

- ROS based coordination - For coordination of assembly tasks between human and robot an ROS based software architecture is used. Human and robot tasks are simulated using OLP tool and tasks are recalled using graphical user interface (GUI) thus enabling to separate the tasks for a robot and human operators. [2]
- Human safety - Concept of kinetostatic safety for human-robot collaboration is introduced and its computational methodology is presented. [3]
- Assembly cell - Methodology for task assignment and scheduling for human-robot cooperation (HRC) assembly cell. By using ROS software platform the overall framework for HRC is developed. [4, 5]
- Assembly Factories of the Future (FoF) - Project ROBO-PARTNER is introduced. Its main goal is to develop the integration platform for safe human-robot collaboration (HRC). Different areas like safety, collaboration tasks planning, robot programming, and integration are considered. [6]
- Speed and separation monitoring – Developing a solution for human-robot collaboration in the automotive industry (PSA) for the assembly process. [7]
- Sensors – implementation and integration of different types of sensors to establish required safety for human-robot collaborations [8]

ISO 15066 standard gives advice for human-robot co-operation safety issues. These are recommended for the development of robot cell solutions. By standard, the workspace is divided into two parts: robot workspace (operating space); and collaborative workspace. Collaborative workspace is defined as "space within the operating space where the robot system (including the workpiece) and a human can perform tasks concurrently during production operation" [9]. As long as the industrial robot operates in its allowed operating space then the general rules of safety are applied (robot stops immediately when someone enters the robot workspace). As robot enters into collaboration workspace, the standard ISO 15066 must be applied. Our research paper concentrates mainly on collaboration workspace issues.

Usually, the robot is programmed by using the teach pendant or offline programming (OLP) method. After the program is simulated in the computer, it is uploaded to the robot controller storage memory. To execute this program on the robot the program is uploaded into program (RAM) memory. During this step, the program syntax is controlled to prevent any faulty code or program.

This kind of programming approach is quite rigid and usually does not include any human safety factors. The program can be stopped by using the external stop command, additional sensors, safety stop, or a multi-task option. However, in real-life conditions, these are not flexible and consume additional process time. Also, the response of the robot controller can be too slow to halt (or decelerate) the robot before the human gets harmed.

The proposed solution described in this article is to unload the program (connected to the production process) from the robot controller and to convert robot into the client mode for listening only "external" commands from the server. After the conversion, the program can be divided into smaller subprograms and the safety commands and parameters can be added.

Some of the authors have contributed to the robot communication, machine vision, and robot trajectory planning:

- Planning of the trajectory for the robot by using an external camera for this purpose. [10]
- Communication principles between networked robots and remote control solutions are described [11]
- Use of Robot Operating System (ROS) for implementation and programming of mobile robots for process [12]
- Use of computer vision for mobile robots trajectory planning and mapping [13]

Abovementioned concepts can be used as a base for further development also in this case. Although this case includes only one robot in future several robots will be included in this study.

2. Methodology

Both of the solutions - robot path planning, and preliminary program - are done in OLP software. The difference can be seen on control and execution level.

2.1. OLP and online programming comparison

Many different tools exist for OLP programming – originally all industrial robot manufacturers – ABB, Kuka, Fanuc, Universal Robot, and much more, have their own tools for giving the robot the ability to execute the exact task – welding, measurement, or grabbing depending from the production line/user needs. This way of making a program is nowadays mostly used by users and robotic cell system planners [14]. Generally, the system consists of the robot controller, industrial robot manipulator and a number of sensors. Additional specific components can be added to the system. (See. 1.).

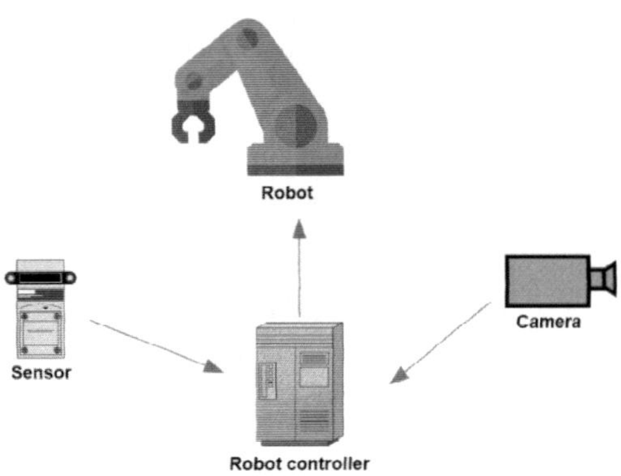

Fig. 1. Robot path executed from OLP

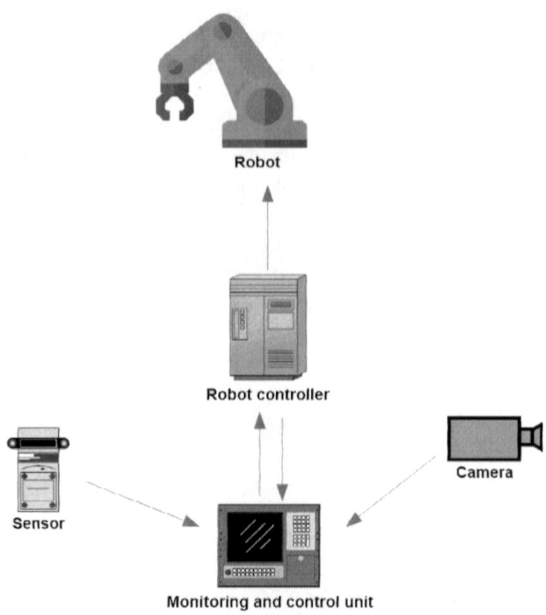

Fig. 2. System with online control

With the online control and monitoring method program for the industrial robot, the path is still programmed by manufacturer's software, but we add an additional control tool, as ROS for example which can be run on a separate controller, or server machine (See Fig. 2.).

This method gives the system unlimited abilities for:

- process monitoring and control;
- program online modifications;
- automated decision making;
- advanced safety system (security levels); and
- automatic path planning with the camera.

For example, if we have rapid code command which consists of:

- MoveL, which is linear movement command
- XYZ – coordinates in the system
- V – speed in mm/sec
- Z – curve on path while moving through coordinates

This is enough to execute the task from point A to point B (see Fig. 3.). However, to gain additional artificial intellect to our robot program we need to add an additional factor to the code. Fully effective usage of this can be done via online monitoring and control tool.

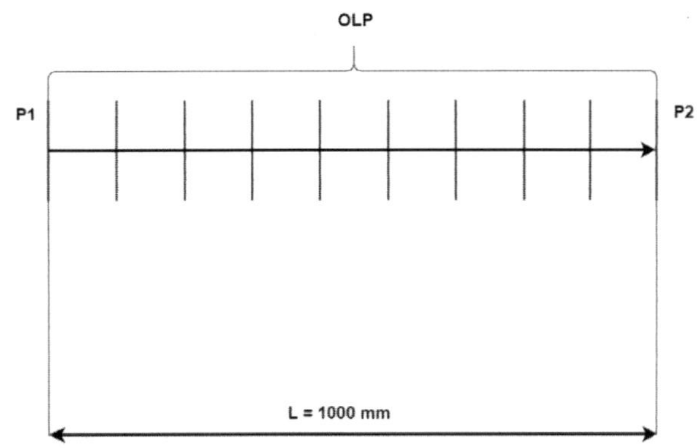

Fig. 3. Pre-programmed robot path

3. Research

3.1 Design of Experiment

An additional factor in our system is security level of each trajectory part. As in the previous study with OLP, we have 10 coordinates, which must be passed through by the industrial robot according to the pre-done program. By adding the additional factor, it is possible to give to every one of those points a security level on a scale from 1 to 10, where 10 means "highly secure". So, each step will be monitored with the refresh rate of 0.01 seconds online during the whole program execution progress. If sensor or camera will detect any object in the range of the industrial robot, additional control unit, will decide and give a command to shut down, go to sleep mode, change the path or jump over to the next task. So, as the safety status drops (somebody or something enters the area), the parameter gets a smaller value (0...9) depending on the distance between the object and the duration of the breach. And when the object/human leaves the zone, the program automatically gives permission to proceed with the same step or to go back to the not performed task. On the next round of the program, control unit already remembers where the fault appeared and moves with an exact level of additional notice into that coordinate, and reduces it with every cycle (See Fig.4).

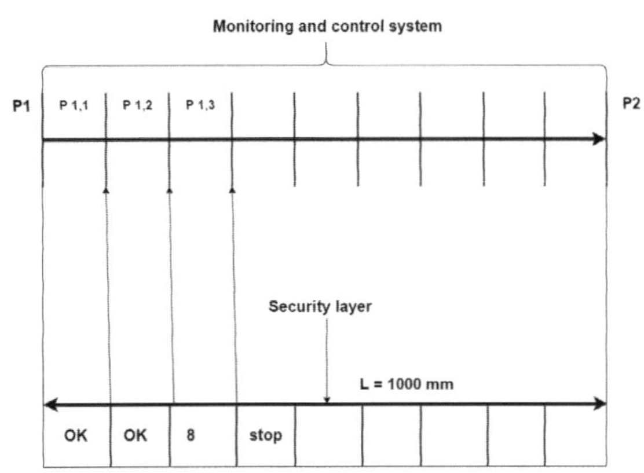

Fig. 4. Robot path with online monitoring

3.2. Case study – safety system example

The example was done using ROS [15] and ABB robot IRB120. Industrial robot and sonar scanner with data about objects, joints location coordinates and physical parameters (See Fig.5) were added to the system.

Fig. 5. Sonar scanner parameters

The robot is executing its task. After an object was put into the range of the sonar sensor in the simulation software – the robot stopped executing the task and automatically proceeded with work, when the object was removed (See Fig. 6)

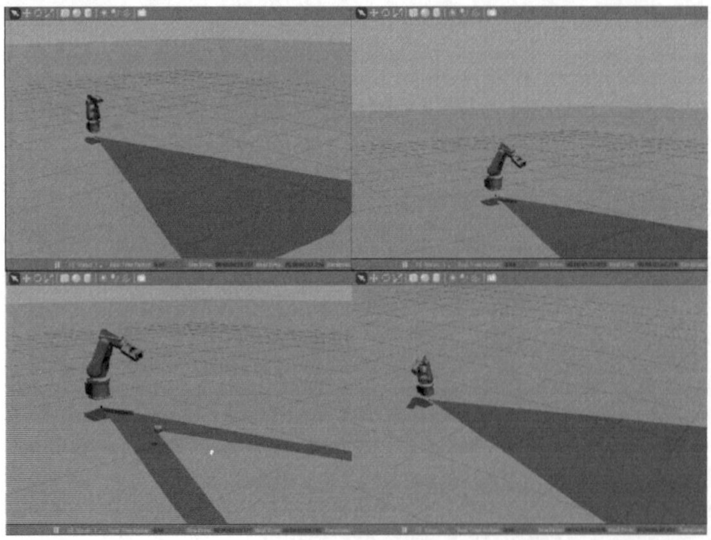

Fig. 6. Case study with safety system

3.3 Comparison of different methods

This experiment, comparing OLP and online monitoring methods, has shown that the side control of the robotic path leads to more effective results. While executing the path, additional factors can be added and in most cases, a pre-done program is not needed – according to the commands from the sensors, the robot path execution can be generated online.

This method makes industrial robots more intelligent and gives them the ability to decide about one or another task by themselves. Still, many factors must be considered like movement kinematics and maximum speeds of the robot joints. Considering this and implementing an online monitoring method into manufacturing process will give an unlimited amount of additional features to the manufacturing line. Moreover, first steps to create a tool for online programming in Virtual Reality (VR) were done with the development of real and virtual manufacturing line communication architecture [16], where methods described in this research can be tested.

4. Discussion

OLP and online monitoring and control methods were compared and discussed in order to find an optimal way for robot re-programming. Proposed solution based on ROS enables pre-check unattended human intervention and stop the robot cell work until danger of harming the human is de-activated. The methodology was designed for ABB small sized IRB series industrial robot, but the due openness of ROS it can be used also in other manufacturers' industrial robots. With online monitoring of the industrial robot process, there is no limit for the features of the robotic system. The method described in this paper and the case study is only the beginning of the future development of this topic.

Next steps are:

- Control of different manufacturers' robots with one tool
- Simulation improvements (reducing response time and increasing flexibility)
- Human presence simulation in the VR
- Flexible robot cell design using VR and simulations tools

In the future, this area should be further investigated and new tools for more precise simulations developed as the human precision simulation using VR tools.

5. Acknowledgements

The research was supported by I4MS project SmartIC Robotics - Regional Digital Innovation Hub in Robotics in Estonia. Authors are grateful to the Integrated Engineering students of Tallinn University of Technology – Mohammad Tavassolian, Aleksei Tanjuhhin, and Tengiz Pataraia helping in experiments and simulations.

6. References

[1] Vysocky A., Novak P. (2016) Human-Robot Collaboration in Industry. MM (Modern Machinery) Science Journal, June 2016

[2] Tsarouchia P., Makrisa S., MichalosaG., Matthaiakisa A.-S., Chatzigeorgioua X., Athanasatosa A., Stefosa M., Aivaliotisa P., Chryssolourisa G. (2015) ROS based coordination of human-robot cooperative assembly tasks - An industrial case study. Procedia CIRP 37, 2015, 254 – 259

[3] Polverini M.P., Zanchettin A.M., Rocco P. (2017) A computationally efficient safety assessment for collaborative robotics applications, Robotics and Computer-Integrated Manufacturing, 46, 2017, 25-37.

[4] Ore F., Vemula B.R., Hanson L., Wiktorsson M. (2016) Human – Industrial Robot Collaboration: Application of Simulation Software for Workstation Optimisation, Procedia CIRP, 44, 2016, 181-186.

[5] Tsarouchi P, Matthaiakis AS, Makris S, Chryssolouris G. (2016): On a human-robot collaboration in an assembly cell, International Journal of Computer Integrated Manufacturing

[6] Michalos G, Makris S, Spiliotopoulos J, Misios I, Tsarouchi P, Chryssolouris G. (2014) ROBO-PARTNER: Seamless Human-Robot Cooperation for Intelligent, Flexible and Safe Operations in the Assembly Factories of the Future, Procedia CIRP 23 (2014) 71-76

[7] Cherubini A, Passama R, Crosnier A, Lasnier A, Fraisse P. (2016) Collaborative manufacturing with physical human-robot interaction, Robotics and Computer-Integrated Manufacturing 40 (2016) 1-13

[8] Bdiwi M. (2014) Integrated sensors system for human safety during cooperating with industrial robots for handing-over and assembling task, Procedia CIRP 23 (2014) 65 – 70

[9] ISO/TS 15066:2016 Robots and robotic devices – Collaborative robots (2016)

[10] Katsurin, A. (2016). Planning Trajectory of the Mobile Robot with a Camera, Proceedings of the 26th DAAAM International Symposium, pp.0407-0416, B. Katalinic (Ed.), Published by DAAAM International, ISBN 978-3-902734-07-5, ISSN 1726-9679, Vienna, Austria, DOI:10.2507/26th.daaam.proceedings.055

[11] Andreev, V.; Pletenev, P.; Poduraev, Y. (2016). The Method of Network Control of Robotic Systems of Different Models and Manufacturers, Proceedings of the 26th DAAAM International Symposium, pp.0727-0730, B. Katalinic (Ed.), Published by DAAAM International, ISBN 978-3-902734- 07-5, ISSN 1726-9679, Vienna, Austria, DOI:10.2507/26th.daaam.proceedings.101

[12] Khassanov, A.; Krupenkin, A.; Borgul, A. (2014). Control of the Mobile Robots with ROS in Robotics Courses, Proceedings of the 25th DAAAM International Symposium, pp.1475-1484, B. Katalinic (Ed.), Published by DAAAM International, Published by Elsevier in Procedia Engineering 100 (2015) 1475 – 1484.

[13] Gryaznov, N.; Lopota, A. (2014). Computer Vision for Mobile On-Ground Robotics, Proceedings of the 25th DAAAM International Symposium, pp.1376-1380, B. Katalinic (Ed.), Published by DAAAM International, Published by Elsevier in Procedia Engineering 100 (2015) 1376 – 1380.

[14] Kuts, V.; Tähemaa, T.; Otto, T.; Sarkans, M.; Lend, H. (2016). Robot manipulator usage for measurement in production areas. Journal of the Machine Engineering, 16 (1), 57–67.

[15] ROS documentation, Available from: http://wiki.ros.org/ Accessed: 2016-11-25 25.11.2016)

[16] Kuts, V.; Modoni, G. E.; Terkaj, W.; Tähemaa, T.; Sacco, M.; Otto, T. (2017). Exploiting factory telemetry to support Virtual Reality simulation in robotics cell. Augmented Reality, Virtual Reality, and Computer Graphics, 1: 4th International Conference, AVR 2017, Ugento, Italy, June 12-15, 2017. Ed. L. Tommaso De Paolis, P. Bourdot, A. Mongelli. Springer, 212–221. (Lecture Notes in Computer Science; 10324).10.1007/978-3-319-60922-5_16.

DOI: 10.2507/28th.daaam.proceedings.139

CONTROL SYSTEM OF THE ARTICULATED ARM BRAKING MECHATRONIC MACHINE (AABMM)

Mikhail Solovyev, Vorotnikov Andrei, Daniil Klimov,
Kovalskii Vladislav, Yuri Poduraev

This Publication has to be referred as: Solovyev, M[ikhail]; Vorotnikov, A[ndrei]; Klimov, D[aniil]; Kovalskii, V[ladislav] & Poduraev, Y[uriy] V[.] (2017). Control System of the Articulated Arm Braking Mechatronic Machine (AABMM), Proceedings of the 28th DAAAM International Symposium, pp.1002-1009, B. Katalinic (Ed.), Published by DAAAM International, ISBN 978-3-902734-11-2, ISSN 1726-9679, Vienna, Austria
DOI: 10.2507/28th.daaam.proceedings.139

Abstract

An approach to solving the problem of fixing objects in space with a given position and orientation is considered in the article. The concept of the Articulated Arm Braking Mechatronic Machine (AABMM) is described, which allows solving this problem in various fields of science and technology. The synthesis of the control system based on modular approach has been carried out. A rig for testing and calibration of sensors, designed to improve the reliability of the system and the accuracy of the feedback loop elements was developed. The proposed control system will allow to operate the AABMM with different kinematic structures and function sets.

Keywords: control system; medical robotics; mechatronic; sensor calibration; articulated arm

1. Introduction

In various areas of human activity, there are tasks that require fixing objects in space with a given position and orientation, relative to other objects. Similar problems can be found in industry, where it is required to keep the workpiece or tool in a certain position during processing, or in cinematography, to position the imaging equipment during process shooting for alignment with 3D-graphics [1]. A wide class of similar problems can be found in medicine. For example, in neurosurgery, when performing operations in the region of the spinal cord or brain, it is necessary to fix the position of the medical instruments relative to the operating field for their further accurate surgical insertion into the living tissue of the patient. Precise positioning of the instruments at the time of the medical operation reduces the risks caused by the human factor, ensures a low invasiveness of the intervention, increases the speed of the operation. To effectively solve these problems, the concept of a mechatronic device – Articulated Arm Braking Mechatronic Machine (AABMM) is proposed, consisting of a series of links capable of fixing relative to each other in a certain position necessary for the user, with built-in brakes A similar approach to the solution of these problems in the field of medicine is considered in [2,3], however, despite the close solution in link fixing, there are no positioning devices provided, so the navigation of the medical instrument is done manually.

In the RSSS (Robotic Spinal Surgical System) system described in [4,5], it is proposed to use a manipulator capable of working in a collaborative mode, allowing the individual links to move manually, which makes it functionally close to the AABMM. Nevertheless, the problem of navigation in RSSS is solved on the basis of stereo photogrammetry, using an external stationary stereo camera and markers on the end effector of the manipulator. In AABMM, in turn, precise angular sensors are used to determine the position, thus allowing navigation without external measuring equipment.

The aim of the work is to cover the problems of design and development of the electronic subsystem of the AABMM, including low-level software. The paper considers the synthesis of the structure of the modular control system of the mechatronic device, the approaches to transferring and receiving information between the elements of this structure, as well as the development of the rig and experimental studies for testing and calibration of absolute position sensors - an essential element of AABMM.

2. Description of the kinematics and internal components of the AABMM

The developed mechatronic device is an articulated arm of the open kinematic model. Inside of the joints of the kinematic pairs, precision angular position sensors and brake devices are located. The structure of the kinematic scheme, the number and the size of the links can vary depending on the requirements of operation. The general view of the AABMM (Articulated Arm Braking Mechatronic Machine) is shown in Figure 1. The main function of this device is to limit the movements of the end effector to perform the tasks assigned. When performing an operation, the brake devices fixes the articulated arm and does not allow the end effector to deviate from the preset position.

The position and orientation of the end effector is specified to the AABMM in the control program, as a set of lines of code necessary for execution. After loading the control program, the control system, in which the kinematic model is set, reports the necessary moving commands to the user. The movement of the end effector from one position to another takes place manually and can be done in two ways: by moving around or along each selected axis to the specified position one by one, or by manually moving the articulated arm to the desired position, and then fixing all the joints. The braking devices can be switched on and off by means of control elements located on the AABMM body.

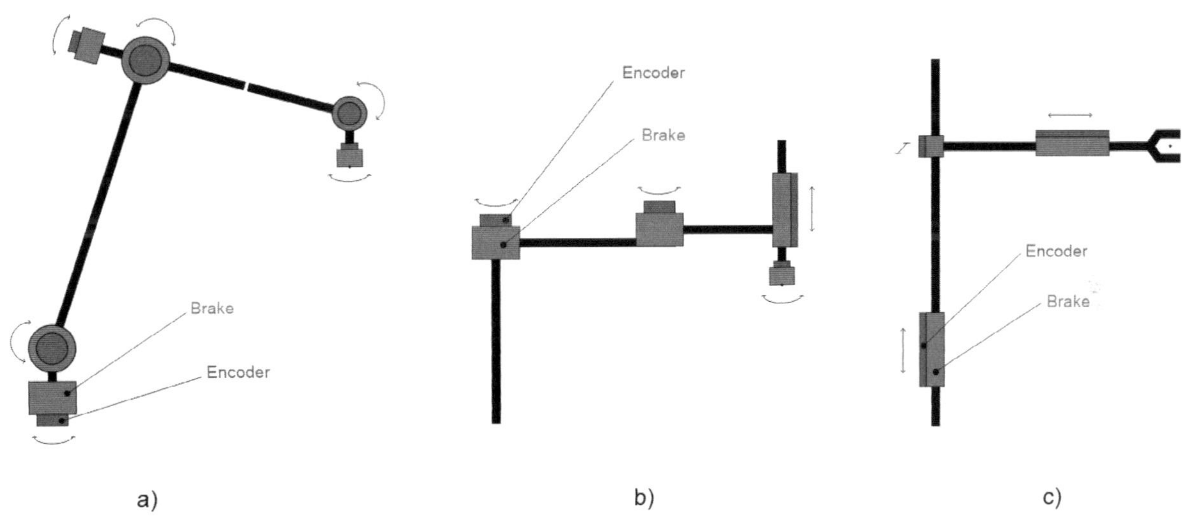

Fig. 1. Kinematic models of typical AABMM

The basis of the mathematical description of AABMM in the control system is the direct kinematic problem. The solution of the direct kinematics problem allows to transform the space of generalized coordinates of the position of the AABMM joints into the Cartesian space by means of several transitions from the previous link to the next one. Each transition is carried out in accordance with certain rules established by Denavite and Hartenberg [6] and is described by a homogeneous transformation matrix of 4 by 4 dimensions:

$$A_i^{i-1} = Trans(z_i, d_i) \cdot Rot(z_i, \theta_i) \cdot Trans(x_i, a_i) \cdot Rot(x_i, \alpha_i) \quad (1)$$

where $Trans(h_i, \chi_i)$ and $Rot(h_i, \chi_i)$ – homogeneous transfer and rotation transformation matrices, respectively, h_i - rotation or transfer axis (x_i , y_i or z_i) i coordinate systems, χ_i - angle of rotation or length of transfer i coordinate system to one of the geometric parameters: d_i , θ_i, a_i , α_i . The number of matrices corresponds to the number of joints in the mechanism.

For the developed kinematic model, it is necessary to perform a calibration by determining the actual values of the geometric parameters (d_i , θ_i, a_i , α_i): displacements of generalized coordinates (initial relative position of links), real lengths of links and displacements between them. Calibration of AABMM is possible to perform in a variety of approaches, for example, using:

- Identification of all geometric parameters of the kinematic model of the control system. For this purpose, the coordinates of the points of the position of the working element, measured throughout the entire working area of the AABMM, are approximated.
- Definition of individual geometric elements of the robot using CPA (circle point analysis) and further transformation of the entire kinematic model.

A more detailed description of these approaches was proposed in [**Error! Reference source not found.- Error! Reference source not found.**].

3. Modular control system of the AABMM.

When designing the control system for AABMM, a modular approach was used [10]. This approach consists in combining devices for various purposes into a constructively and functionally independent product. For AABMM, it is rational to combine the sensor and the brake in each joint into a separate module. All modules include a unified control board. In this board, feedback is created by means of a high-frequency survey of absolute angular position sensors and, in accordance with the specified program or an external command from other modules, interaction with the control object (brake) occurs.

The generalized AABMM block diagram, using the modular principle, is shown in Figure 2. The communication between modules is carried out using a single data bus (DATA BUS). All modules (Slave module) interact with control objects in the joints. This approach allows you to distribute the computational load by high-frequency data processing from the sensors parallel between each module, which leads to an increase in the processing speed. It also makes it possible to change the number of joints in AABMM without any additional changes in the control system. In addition, with this approach, the production of identical modules for AABMM becomes more economically feasible [10]. The control system uses a master module to interact with an external PC via Ethernet protocols, which manages the system by sending commands and monitoring the position of the links by surveying the slave modules.

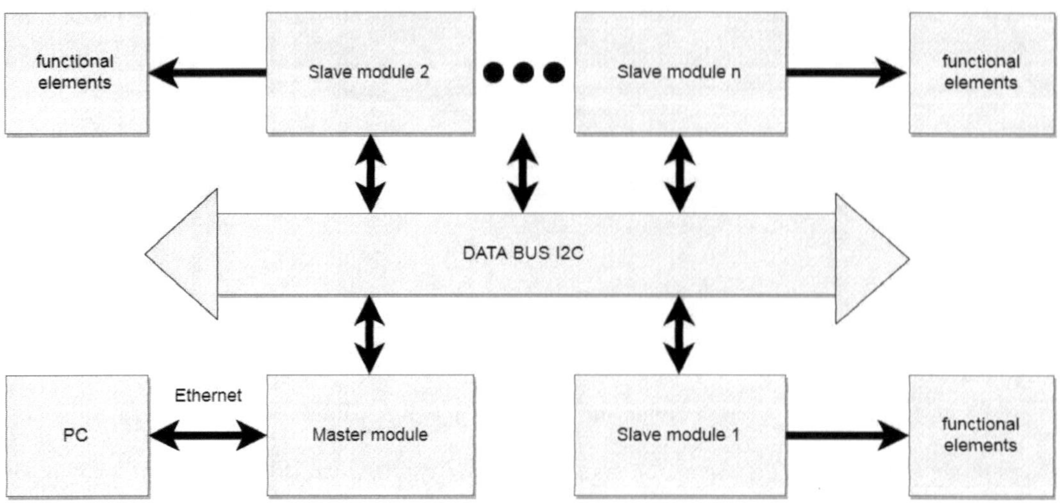

Fig. 2. Generalized block diagram of the AABMM control system

A data bus I2C was used to connect the modules. This choice is due to two criteria: a small number of wires and the simplicity of the data exchange protocol, compared to other bi-directional buses. A small number of wires allows the use of an angular position sensor with a hollow shaft, which has a small diameter, which makes it possible to reduce the dimensions of the structure. The choice of I2C as the data exchange protocol was made to simplify the development of additional modules. This bus is present in most controllers and processors, which allows you to create additional modules through software modification of the connected devices. The development of modules for finite devices is reduced to designing an adapter between the control interface and the data bus. As additional modules are considered specialized end effectors, fixed on the flange. It is also possible to use information devices that transmit patient data during the operation or the parameters of the working environment directly by the flange. There are 7 modules for joint management and 6 additional modules for connection of end effectors.

During the development of the control system, two changes were made to the data bus. The first change is to increase the voltage on the two power wires. The voltage value was increased from 5 volts to 12. This is due to the need for increased supply of brake components in the module. The second change is related to the data rate on the bus clock and data bus. From the "classic" standard value of 100 Kbits / s, it was increased to another speed value of 400 Kbits / s. This is due to the provision of a time reserve for data transmission in case of a critical situation, for example, stopping of the operation of all modules by the PC command.

In addition to the implemented method of instant brake locking in all modules, in response to a specialized signal, it is possible to transmit an emergency command sequentially to each module. The time for such a transmission should not exceed the button contact jitter processing time of 0.01 seconds. Although the maximum design response time of the system is 0.0026 seconds, under operating conditions, data and data processing errors are possible, so the speed has been increased with a safety factor of 4. Thus, a balance was observed between the number of wires and the speed of the system, depending on the data transfer rate on the bus.

Let us consider the structure and operation of the modules in more detail. A detailed block diagram is shown in Figure 4. The slave module (Figure 4a) produces high-frequency data reading from the position sensor. The data is written to the ring buffer of the device and, when requested via the data bus, it can return both the last recorded value and the entire buffer. Brake control in the module occurs in accordance with the program or commands received via the data bus. The module's functionality allows you to enable or disable the brake depending on the angle of relative position of the links in the joint, from the command on the data bus or from the command of the control on the housing, such as a button. As a controller, the AVR ATMEGA328P microcontroller is used.

The I2C interface module of the microcontroller is directly connected to the common data bus. The slave module has three control elements: a brake, a sensor and a control board mounted on the body. The brake is connected to the board via a galvanic isolation using a relay. This is due to the difference in the brake control voltage and the operating voltage of the microcontroller. A power converter is used to power the entire system, which provides 12 volts for controlling the magnet and 5 volts for powering the controller and sensor. The absolute position sensor is connected to the controller via the RS-422 / RS-485 signal converter.

The task of the master module (Figure 4 b) is to control the system by sending commands and monitoring the position of the links by low-frequency surveying of slave devices. In addition, it communicates with an external control device, a personal computer through the physical network interface rj45 over TCP / IP protocols.

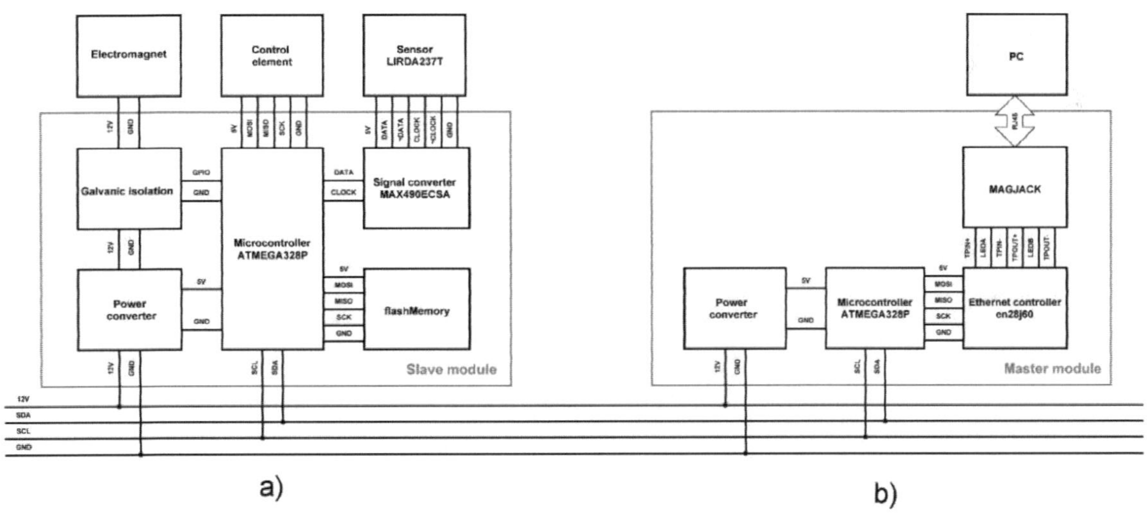

Fig. 4. Slave (a) and master (b) devices.

In the master module, the microcontroller as well as in the slave is directly connected to the data bus. The controller is connected to the PC by the ENC28J60 chip, which is an Ethernet controller designed to work with connections based on the 10BASE-T family interfaces. Control of this chip is carried out using the SPI protocol. For connection to a personal computer, a connector is used for the rj45 interface of the magjack type.

The master module is the master on the data bus. It initiates the reception and transmission of data with slave modules. Thus, there is no possibility of overlaying information from two different modules. This implementation allows the use of basic TCP / IP protocol tools to communicate with the device and interpret data in any certified software [11], which is designed to work with a medical device.

4. Testing and calibration rig for absolute position sensors.

It is possible to use AABMM in medical practice to perform operations on a patient. Therefore, the measuring devices used in medical AABMM should have increased requirements for accuracy and reliability. As measuring devices for AABMM, it is advisable to use absolute position sensors, which, unlike incremental sensors, save information about the current position when the power is turned off. Absolute position sensors are installed in the joints and allow to determine the position of the next link relative to the previous one. In connection with the increased requirements for absolute position sensors, it is necessary to evaluate their accuracy against the manufacturer's specified qualitative characteristics.

A specialized rig has been developed for the testing of angle sensors. Its principle of operation is to test a less accurate sensor using a different, more accurate one. Both sensors are fixed to a single shaft, which moves to an angular discrete smaller than the accuracy of the sensor being tested. For each discrete value is read from the two sensors, which are recorded in a look up table. Values of the look up table for full shaft rotation are used to determine the actual characteristics of the sensor being tested. With the help of such a specialized rig, it is also possible to calibrate the sensors, increasing their accuracy. The appearance of the stand is shown in Figure 5. It combines two sensors through a single shaft, the rotation of which provides a stepper motor through a worm gearbox. To reduce the initial value of the eccentricity and the relative inclination of the sensors, they are located at a minimum distance from each other. In this case, the design of the rig is designed in such a way, that it makes it possible to replace the tested absolute sensors without disassembly.

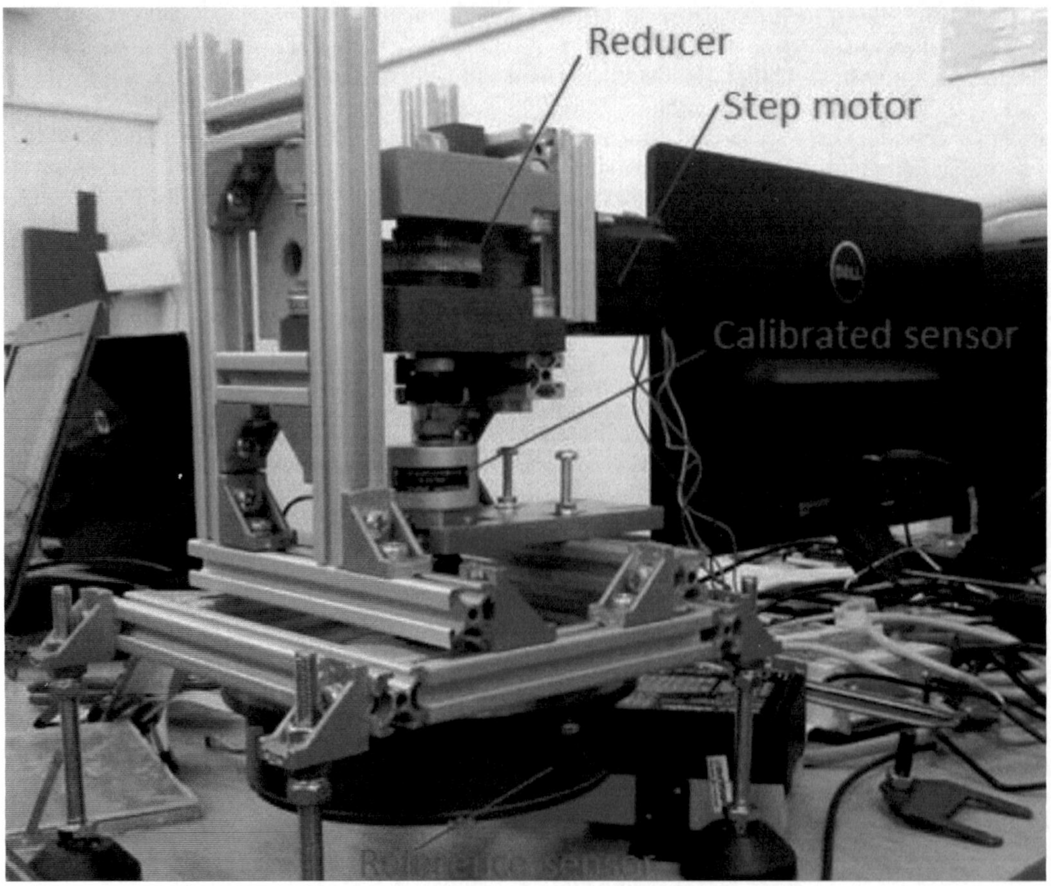

Fig. 5. Experimental rig for testing and calibration of sensors

The block diagram of the developed rig for testing absolute sensors is shown in Figure 6. The stepper motor with a worm gear is used as the device for creating the angular movement of the shaft. The stepper motor is controlled by a driver with the ability to generate a microstep. Microstepping allows you to "divide" the steps of the motor, thereby reducing the discretion of moving the common shaft. The microstep selection is carried out in accordance with the following formula:

$$k_{microstep} > \frac{360 * 60 * 60}{n_{step} * i_{reducer} * s_{sensor}} = 16.2 \qquad (2)$$

where: n_{step} - the number of discretes of the stepping motor, $i_{reducer}$ - gear ratio, s_{sensor} - the resolution value per second. Since the driver uses coefficients equal to powers of 2, the microstep coefficient is assumed equal to 32.

As a controller for the test rig and calibration of sensors, a single-board computer Tion-pro270 was used. This computer has a large number of GPIO ports. With their help, the driver of the stepper motor is controlled by the STEP/DIR/ENABLE protocol. The shaft rotation is measured using a calibrated (LIRDA237T) and a reference sensor (ROD800).

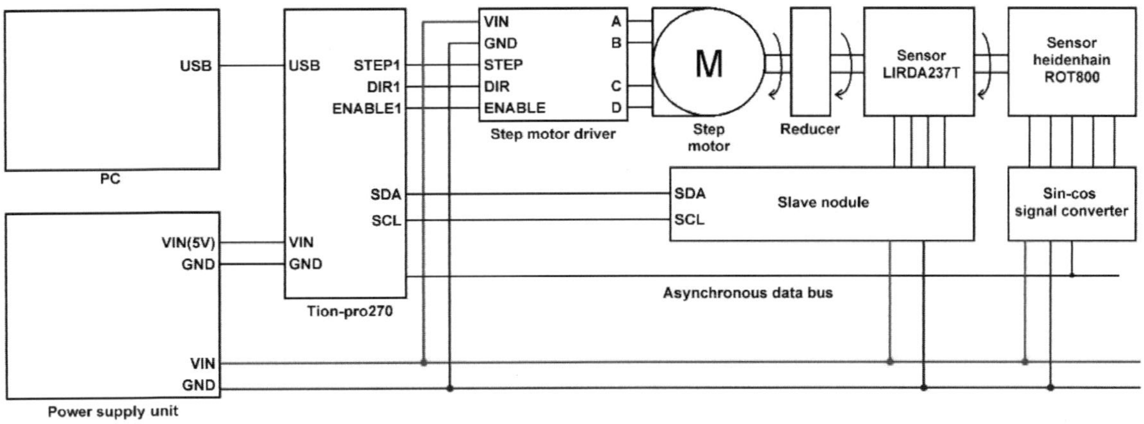

Fig. 6. Structural electrical diagram of the rig for testing and calibration of sensors

The sensors used differ not only in their accuracy and resolution, but also in the way they deliver data. Comparison of the characteristics of the sensors is presented in Table 1. LIRDA237T, is an absolute encoder with a hollow shaft and having 2097152 discrete per revolution. It is this model that is used in all slave modules AABMM. The data from the sensor is read from the slave module, which is connected to the control board of the single-board computer via the I2C data bus. The reference sensor has a higher resolving power and accuracy. Since it is an incremental encoder with a sinc-cosine signal, the position information is obtained by means of a sinus-cosine converter LIR960. This converter is connected with tion-pro270 via a dedicated asynchronous data bus.

	LIRDA237T	**ROD800**
Type	Absolute	Incremental
Resolution	5''	2''
Accuracy	30''	2''
Signal type	SSI	SINCOS

Table 1. Comparison of the characteristics of the sensors

For the manager of a single-board computer operating on the basis of the Linux operating system, a control program via the TFT screen was implemented. As a development environment, a cross-platform framework was used to develop Qt creator software. The complexity of the management program implementation is associated with the critical influence of each thread, on the duration of the calibration process. The solution to this problem is to divide the program into several threads. The number of streams was minimized, thereby facilitating the load on the computing power of the board. By setting thread priorities in a multithreaded system, all time delays were minimized. A simplified diagram of the distribution of threads in the program is shown in Figure 7.

Three threads are distinguished. Draw tread is designed to display the graphic interface on the TFT screen and to read the touch data to the capacitive matrix. Since the draw tread occupies a large resource of computing power, a low priority is set for the stability of the entire system and an increase in the calibration speed of a single sensor for this thread. The data thread from the asynchronous bus is only started when there is a command from the thread. The read function is based on writing the value to the ring buffer, which is located directly in the hardware of the bus. Thus, when the buffer overflows, a critical error will be created and the process of reading the data will need to be interrupted. Therefore, it is critically important to continuously parse the parcels from the asynchronous bus.

To stabilize the system, this thread was given the highest priority. For the main thread running linearly, high priority is set. In it, after the activation of the read command, the values coming from the two sensors are processed, the stepper motor is controlled via the galvanic isolation keys and the calculation of the traversed paths. Also, the function of saving data to the internal memory of the device or to a flash drive was realized.

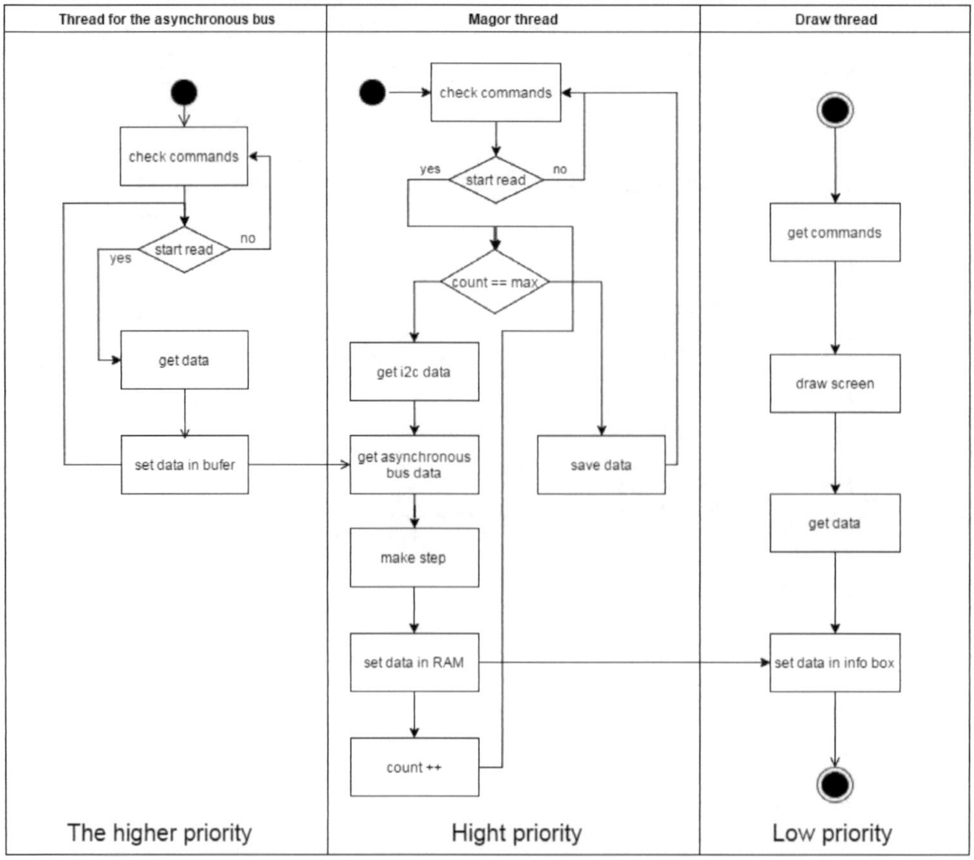

Fig. 7. Thread diagram.

To fully compensate the eccentricity and the mutual tilt of the two sensors, an algorithm for processing the look up table was implemented. After obtaining a look up table between the data coming from the calibrated sensor and the reference sensor data, the signal is processed. The processing consists in the formation of an error signal of the sensor readings (blue graph). Next, a direct Fourier transform for the error signal and a transition to the frequency domain are performed. The first two harmonics create a compensating error signal and perform its inverse transformation from the frequency domain to the time domain (red graph). By subtracting the compensating error signal from the original signal, the eccentricity and mutual tilt of the two sensors are compensated and a look up table is created (green graph). Calibration is carried out by transferring table values to the memory of the module working with the calibrated sensor. The signal graphs are shown in Figure 8.

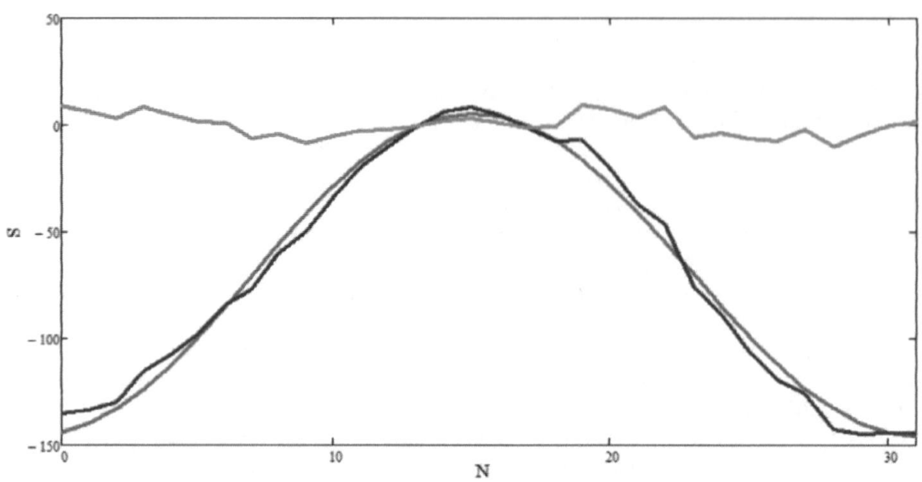

Fig. 8. Graphs of received sensor signals during calibration.

5. Conclusion

The article considered the problem of developing devices for fixing objects in space with a given position and orientation. In particular, a solution for the electronic control subsystem is proposed. The solution is based on the modular principle, which allows to combine control objects in the joints into functionally independent devices connected by a single data bus. In this way, it is possible to easily modify the lengths of the links, the number of joints and the device functionality by developing additional modules. This makes this type of device applicable for various classes of tasks, including operations in medical practice. Solutions for the electronic subsystem are described, structural and schematic diagrams and low-level software were developed. An approach to testing and calibrating the sensors of the modules is also described. The proposed solution in the form of a specialized sensor calibration rig is aimed at automating the processes of data acquisition, processing and analysis. Based on the described developments, it is planned to create a prototype of the AABMM, for use in medical practice.

6. Acknowledgments

The Ministry of Education and Science of Russia as part of State assignment № 9.3408.2017/4.6 supported the work.

7. References

[1] Klimov, D.; Poduraev, Y. (2014). Design and Experimental Testing of a Robotic System for High Speed Recording, Proceedings of the 25th International DAAAM Symposium, ISSN 1877-7058, ISBN 978-3-901509-97-1, Katalinic, B. (Ed.), pp.1332-1337, Published by DAAAM International, Vienna, Austria, doi: 10.1016/j.proeng.2015.01.501

[2] Hu, Y.; Zhang, J.; Li, C.; Cheng, S; Wang, L. (2008). Safety design of an Assisting Robotic Arm for minimally invasive thoracic surgery, Proceedings of the 2008. ROBIO 2008. IEEE International Conference on. – IEEE, 2009. , pp. 709-714, IEEE, Bangkok, Thailand, 10.1109/ROBIO.2009.4913088

[3] Wang, X; Tian, C; Duan, X.; Gu, Y.; Huang N. (2014). A medical manipulator system with lasers in photodynamic therapy of port wine stains BioMed research international T. 2014.

[4] Tian, W.; Han, X.; Liu, B.; Liu, Y.; Hu, Y.; Han, X. (2014) A robot-assisted surgical system using a force-image control method for pedicle screw insertion, PloS one, Vol 9,No 1, 01.2014, pp. 1-9

[5] Hu Y.; Jin H.; Zhang L.; Zhang P. & Zhang J, (2014) State recognition of pedicle drilling with force sensing in a robotic spinal surgical system, IEEE/ASME Transactions on Mechatronics, 05. 2014, Vol 19, No 1. P. 357-365, Shenzhen, China

[6] Denavit J. & Hartenberg R. (1955) A kinematic notation for lower-pair mechanisms based on matrices, Transactions of ASME, 09.1955, Vol 22, No 2, pp. 215-221, ISSN:0098-2202

[7] Mooring B.; Roth Z. & Driels M. Fundamentals of manipulator calibration, John Wiley & Sons, ISBN 0-471-50864-0, United States

[8] Vorotnicov, A.; Isaev, A.; Romash, E.; Ilyukhin Yu.; Bashevskaya, O. & Poduraev Y. (2015) Geometrical Approach for Industrial Robot Axis Calibration Using Laser Tracker., Proceedings of the 26th DAAAM International Symposium, ISSN 1877-7058, ISBN 978-3-901509-97-1, Katalinic, B. (Ed.), pp. 0897-0904, Published by DAAAM International, Vienna, Austria, DOI: 10.2507/26th.daaam.proceedings.125

[9] Vorotnicov, A.; Romash, E.; Isaev, A.; Bashevskaya, O.; Bianchi G. & Poduraev Y. (2016). Uncertainty Estimation of Axes Direction Determination of Industrial Robot Using an Ellipsoid Concentration Model, Proceedings of the 27th DAAAM International Symposium, , ISBN 978-3-902734-08-2, ISSN 1726-9679, B. Katalinic (Ed.), pp.0480-0486, Published by DAAAM International, Vienna, Austria. DOI: 10.2507/27th.daaam.proceedings.072

[10] Lopota, V. & Yurevich E. (2011). Stages and development prospects of robotic systems design modular principle. St. Petersburg Polytechnic University Journal of Engineering Science and Technology, Vol 2, 09.2011, pp. 98-103, ISSN: 1994-2354

[11] ISO/IEC 62304 The instrumental standard for medical device software

DOI: 10.2507/28th.daaam.proceedings.140

DEVELOPMENT OF 2-DOF ADAPTIVE MECHATRONIC DEVICE WITH CORRECTIVE ADJUSTMENT OF LASER TRACKER REFLECTOR FOR INDUSTRIAL ROBOT CALIBRATION

Dmitrii Kosterev, Andrei Vorotnikov, Simeon Nedelchev,
Elena Romash & Yuri Poduraev

This Publication has to be referred as: Kosterev, D[mitrii]; Vorotnikov, A[ndrei]; Nedelcev, S[imeon]; Romash, E[lena] & Poduraev, Y[uriy] V[.] (2017). Development of 2-DOF Adaptive Mechatronic Device with corrective adjustment of laser tracker reflector for Industrial Robot Calibration, Proceedings of the 28th DAAAM International Symposium, pp.1010-1014, B. Katalinic (Ed.), Published by DAAAM International, ISBN 978-3-902734-11-2, ISSN 1726-9679, Vienna, Austria
DOI: 10.2507/28th.daaam.proceedings.140

Abstract

In this paper, a design of adaptive mechatronic device for industrial robot calibration was proposed. An adaptive mechatronic module can be used to permanently preserve the orientation of the reflector with respect to the source of measuring tool. A testing setupfcor based on coordinate measuring machine - laser tracker LTD 800 was developed. The calibration method of one degree of freedom of proposed device was described and experimentally evaluated.

Keywords: industrial robot calibration; laser tracker; measurement; mechatronics; control system

1. Introduction

Currently, industrial robots are actively used in all areas of production. They perform a large number of different tasks, such as palletizing, welding, painting, assembling or metalworking. Each operation can be performed with an appropriate quality, which is provided by the performance characteristics of the robot. In accordance with the standard ISO 9283 [1], one of the main characteristics of an industrial robot is the pose accuracy, path accuracy, the path velocity accuracy and repeatability. Achievement of these characteristics is provided in two ways: the appropriate calculation at the design stage, and calibration, which is necessary as a procedure after the assembly of the manipulator. If the robot has a good repeatability, which can only be provided by design, calibration can be used to achieve the corresponding performance in the remaining characteristics. Thus, calibration is necessary to improve the performance and accuracy of an industrial robot.

To carry out the calibration procedure, a software approach is used, in which the parameters of the control model are calibrated. Parameter calibration involves 2 steps: measuring of the necessary positions or trajectories of the robot end effector depending on the calibrated characteristic, and processing of the obtained data. After estimation of parameters, it is necessary to perform validation of the calibration state, which allows to guarantee correctness of the procedure.

There are different schemes for measuring the position and orientation of end effector of the robot during calibration, different metrological equipment is used [2]. Thus, the researchers in [3] calibrate the robot using cameras. Robertson, Slocum and Willoughby [4] shows a specialized precision tooling with goniometers. Approach based on using virtual linear guidance for calibration presented in [5]. Slamani, Nubiola and Bonev [6] used the Renishaw XL-80 and Ball Bar systems to calibrate robots.

All the presented measurement approaches have a limitations on the possible range of the measured robot positions, therefore in this article the measurement of the positions or trajectories of the robot required for calibration and validation is carried out using a coordinate measuring machine - the laser tracker shown in Fig. 1. For the measurements, the reflector of the laser tracker is mounted on the robot flange. The laser tracker measures the coordinates of the reflector's during the movement of the robot. Unfortunately, the reflector of the laser tracker has a limited laser reception area and a high cost. In order to cover full robot work space, some researchers use at least eight expensive reflectors [7]. Leica laser trackers [8], provide solution of this problem, based on optical principles. However, practically the entire range of the working area can be measured using a specialized equipment - 2-DOF mechatronic module, presented in the patent API Radian [9]. But, unfortunately, the rotation angles in the joints of this module are related to the robot control system, which makes the user's work more difficult. Therefore, it is advisable to develop an adaptive mechatronic module that has its own control system capable of making decisions without a high-level control system, hence the connection between the mechatronic module and the robot control system will be absent. Such a control system can be built on the inertial sensors.

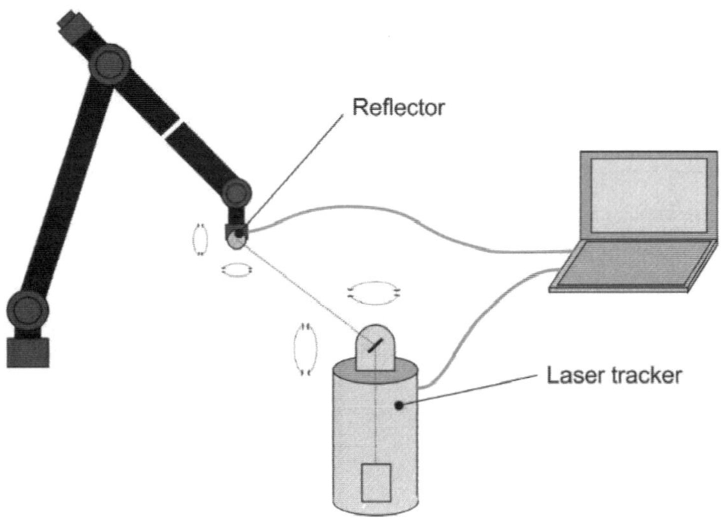

Fig. 1. Scheme of measurements of the manipulator calibration using a laser tracker.

This paper is aimed at describing the concept of 2-DOF adaptive mechatronic module that adjusts to the beam of a laser tracker regardless of the position of the industrial robot and experimental evaluation of 1-DOF of proposed module. A single adaptive mechatronic module will allow to obtain the coordinates of the positions and trajectories of the industrial robot necessary for calibration and validation procedures without additional manual rearrangement of the reflectors. Proposed system will help to maintain constant orientation of the reflector relative to the laser tracker during movement of robot, which leads to significant reducing a time required for calibration and validation of industrial robots. Since proposed module also required calibration, we perform experimental

2. The concept of a 2-DOF adaptive mechatronic module

The appearance of the 2-DOF adaptive mechatronic module used to calibrate industrial robots is shown in Figure 2.b. Its kinematic diagram, shown in Figure 2.a, is designed so that the reflector mounted on the adaptive mechatronic module is in the center intersection of the axes of the mechatronic module being developed. The direct kinematics problem was solved with using Denavit-Hetenberg approach [10], position and orientation of the reflector in space are calculated depending on the given angles of motors shafts. Coordinate systems $X_1Y_1Z_1$ (first axis) $X_2Y_2Z_2$ (second axis), are shown in Figure 2.a. Matrices of homogeneous transformations of the transition between local coordinate systems module include the following parameters: Θ_i – angle that characterizes the amount of rotation of the X_1 around the Z_1 axis. n the adaptive mechatronic module, Θ_1 is the generalized coordinate for the first joint; d_i – displacement between X_1 and X_2 along the mutual perpendicular along Z_1; a_i – the shortest distance between the beginning of the first coordinate system and the 2-nd coordinate system measured along the X_2 axis; α_i – angle, on which it is necessary to turn the axis Z_1 around the axis X_2, so that it becomes co-oriented with the axis Z_2. Each of the parameters defines a matrix of homogeneous transformations. The product of these matrices is a homogeneous matrix of the Denavite-Hetenberg transformations.

$$A_{(i-1,i)} = A^{d}_{(i-1,i)} A^{\theta}_{(i-1,i)} A^{a}_{(i-1,i)} A^{\alpha}_{(i-1,i)} = \begin{pmatrix} \cos\theta i & -\sin\theta i\cos\alpha i & \sin\theta i\sin\alpha i & ai\cos\theta i \\ \sin\theta i & \cos\theta i\cos\alpha i & -\cos\theta i\sin\alpha i & ai\sin\theta i \\ 0 & \sin\alpha i & \cos\alpha i & di \\ 0 & 0 & 0 & 1 \end{pmatrix} \quad (1)$$

Thus, the complete kinematic model of the mechatronic module can be represented by the following relation:

$$H = \begin{pmatrix} \cos\theta_1 & -\cos\alpha_1\sin\theta_1 & \sin\alpha_1\sin\theta_1 & a_1\cos\theta_1 \\ \sin\theta_1 & \cos\alpha_1\cos\theta_1 & -\sin\alpha_1\cos\theta_1 & a_1\sin\theta_1 \\ 0 & \sin\alpha_1 & \cos\alpha_1 & d_1 \\ 0 & 0 & 0 & 1 \end{pmatrix} \cdot \begin{pmatrix} \cos\theta_2 & -\sin\theta_2 & 0 & 0 \\ \sin\theta_2 & \cos\theta_2 & 0 & 0 \\ 0 & 0 & 1 & 0 \\ 0 & 0 & 0 & 1 \end{pmatrix} \quad (2)$$

Where, H- the resulting matrix of position and orientation of the adaptive mechatronic module. Using this matrix, the current orientation of the reflector is calculated, depending on the value of the motors shafts angles. A significant advantage over similar systems is the absence of mechanical transmission, which reduce the cost, improve controllability, accuracy and simplicity of system.

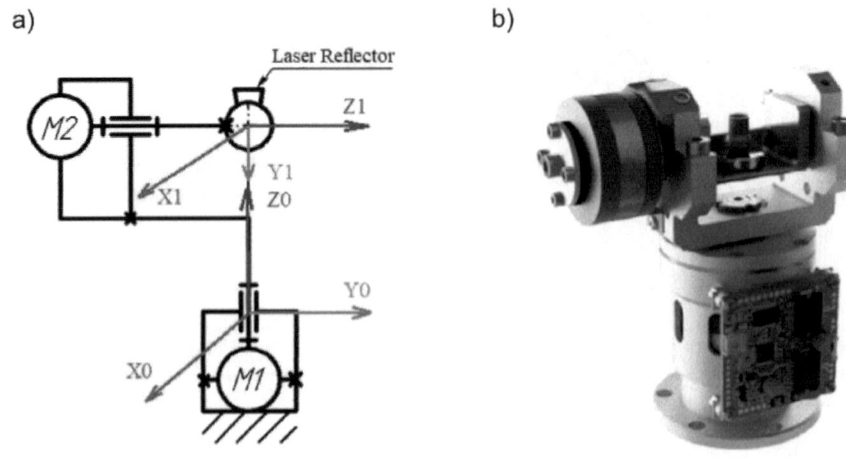

Fig. 2. a) Kinematic diagram of the mechatronic module with local coordinate systems, b) Appearance of mechatronic module.

Proposed module able to make decisions about the necessary change of the motor shaft angles preserve the orientation of the reflector, relative to the tracker beam source. Information about the current orientation of the reflector comes from the inertial sensors (accelerometer, gyro, compass). The data enters the BDPT microcontroller and, after filtering by the Kalman algorithm, converted into Euler orientation angles. Thus, inertial sensors operate as feedback elements of the mechatronic module servo drives.

The implementation of all the above solutions, should provide the accuracy of the reflector positioning in a sphere with a diameter of 15 μm, and the error in orientation of the reflector within an angle of 8 degrees; however, all this is achievable with precision manufacturing. For an affordable implementation, without using of high-precision CNCs, the concept of a specialized trimmer is proposed. It is able to adjust the reflector of the laser tracker to the point of intersection of the two axes of the mechatronic module. This is achieved by adding to the reflector 4 microscrews with micromotors, capable of alternately leading the reflector to the intersection point shown in Figure 3.

3. Experimental results.

In order to ensure the accurate operation of the adaptive mechatronic module, it is necessary to calibrate it. As it was described in previous section, it is necessary to perform a multi-stage adjustment, which will ensure the rotation of the reflector around the intersection point of the module axes. The main two parts of this procedure are the adjustment of the center of the reflector to the axes of rotation of the mechatronic module, first to the first axis, then to the second axis. The adjustment operation to the actual position of each axis of the mechatronic module includes a set of identical actions: the rotation of the reflector around the tuned axis and its linear displacement. And linear movement is carried out by means of two micro-screws.

To confirm the hypothesis of the possibility of adjusting the reflector to the actual position of one axis of rotation, an experimental setup was developed and an appropriate experiment was carried out. The experimental setup shown in Figure 3. consists of the following set of elements:

2-DOF micro stage (1) with 2 micrometer screws (2) Is mounted on the output shaft of the stepping motor with the aid of a rig (3), made on a 3D printer. A magnetic base (4) is mounted on the micro stage, which, in turn, is fitted with a reflector (5) of the laser tracker (6) Leica LTD800. The motor (7) is rigidly fixed by means of a vice (8). The control of the stepper motor is realized with the help of a controller (9) and a driver (10) of the stepper motor, which are powered by 12 V from a laboratory power source (11).

The elements and control system of the rig are located on the antivibration table (12). The motor shaft rotates with the reflector, while the laser tracker measures its position. With full rotation of the motor shaft and measurement of the position of the laser tracker, a circle with a center lying on the axis of rotation of the engine is formed.

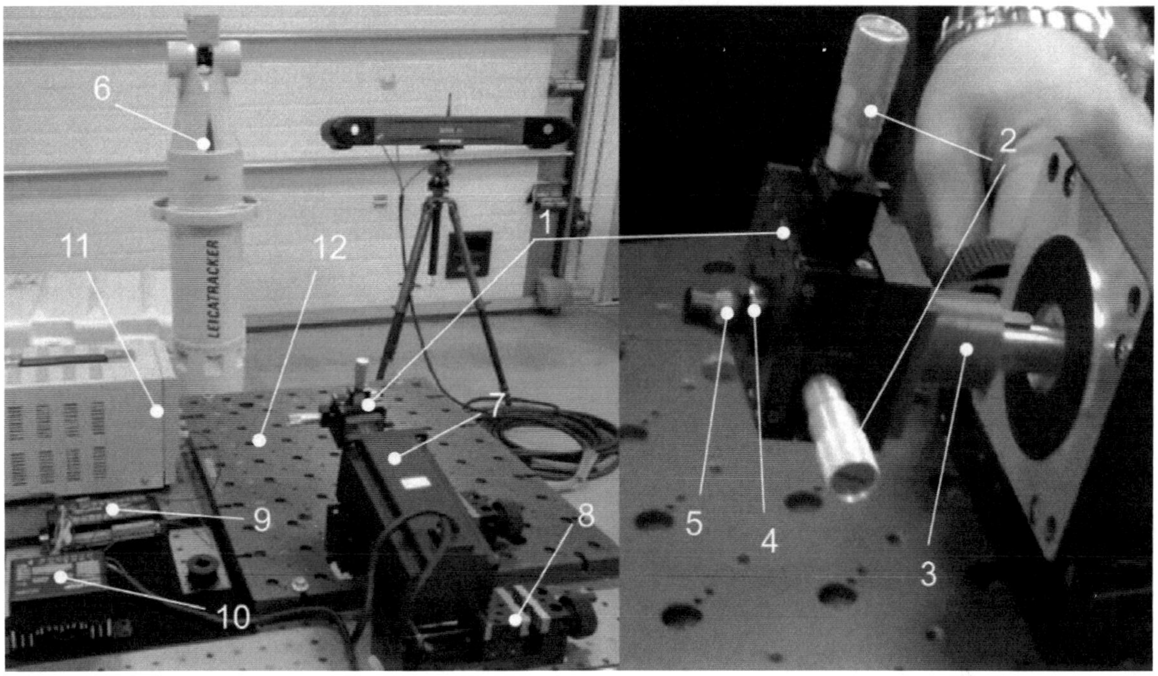

Fig. 3. Experimental setup.

The experiment for adjusting the reflector contained 7 iterations, each of which is a 360 ° rotation of the motor shaft. With each complete rotation, the reflector position was measured by a laser tracker. The measurements were carried out in dynamic mode. After receiving the data points (x_i, y_i, z_i) from each iteration, the circle center coordinates (x_0, y_0) - which lies on the motor shaft axis, were calculated. Further, one of the micro-screws is used, which moves the magnetic base together with the reflector closer to the center of the circle.

The next iteration involves another micro-screw. And, thus, in turn, the reflector moves to the axis of rotation of the motor shaft. Data processing is performed using the methods proposed in [11-13], in order to solve the following optimization:

$$\sum_{i=1}^{n}(Ax_i + By_i + Cz_i + 1)^2 \rightarrow \min \tag{3}$$

$$\sum_{i=1}^{n}\left[(x_i - x_0)^2 + (y_i - y_0)^2 - R_0^{\,2}\right]^2 \rightarrow \min \tag{4}$$

Where (A, B, C) - parameters of norm vector of plane where lies points (x_i, y_i, z_i), R_0 - radius of reflector in each iteration.

The data obtained from each rotation is illustrated in figure 4.a. The first 6 iterations give circles, using which it is necessary to determine the centers of circles. The last iteration gives a set of points with a maximum spread of 13 μm, and these are extreme points that are outside the range of 6σ of the normal distribution. 2σ make up a little more than 4 μm. A graph illustrating the decrease in the circle radius with respect to the iteration number is shown on the figure 4.b.

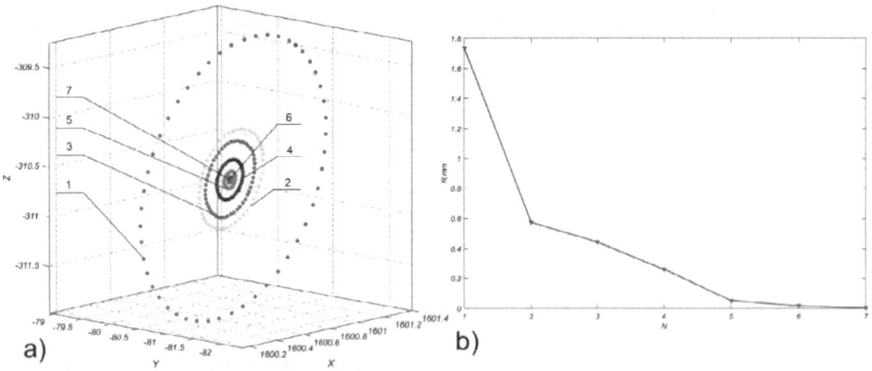

Fig. 4. a) A visual display of the data obtained, b) A decrease in the radius of the circle as the iteration of the corrective adjustment is increased.

Experiment showed that procedure described above let us to reduce error between center of the motor shaft and reflector from 1.8 mm to 13 μm after 7 iterations.

4. Conclusion

In this paper concept of 2-DOF adaptive mechatronic device has been proposed. Following device allow to improve volume of working space in which we can measure position of industrial robot without manual rearrangement of the reflectors. The control system of the adaptive mechatronic module will able to make decisions without a industrial robot control system. With using the proposed adaptive mechatronic module, the time of procedures for calibrating and validating mechanisms will be significantly reduced. The use of a specialized system for the installation of reflectors will reduce the requirements for the accuracy of the components of the adaptive mechatronic module, which was experimentally approved thru calibration of 1-DOF stepper motor based system. The maximum projected error of position determination on the basis of the performed experiments and after the appropriate calculation does not exceed 15 μm.

The next step will be to conduct an experiment for 2-DOF and prepare for the production of the device prototype, debug the control system and software to interact with the computer.

5. References

[1] Manipulating Industrial robots (1995). – Performance criteria and related test methods, ISO 9283:1998.

[2] Manipulating industrial robots (1998). – Informative guide on test equipment and metrology methods of operation for robot performance evaluation in accordance with ISO 9283, ISO/13309:1995.

[3] Petar Maric and Velibor Djalic, (2011). Choice of window size in calibrating the geometry of manipulators based on the regions correlation, ELECTRONICS, VOL. 15, NO. 1, JUNE 2011, pp.45-53.

[4] Robertson, A., Slocum, A., Willoughby, P. (2001). "Precision robot calibration using kinematically Placed inclinometers." Proceedings of 2001 ASPE Conference, p 229 - 232.

[5] Biqiang Du, Ning Xi, Erick Nieves, (2012). Industrial robot calibration using a virtual linear constraint, International Journal On Smart Sensing And Intelligent Systems, Vol. 5, No. 4, December 2012, pp. 987-1001.

[6] Mohamed Slamani, Albert Nubiola and Ilian Bonev, (2012). Assessment of the positioning performance of an industrial robot, Industrial Robot: An International Journal, 39/1, 2012, pp. 57–68, DOI 10.1108/01439911211192501.

[7] Albert Nubiola, Ilian A. Bonev, (2013). Absolute calibration of an ABB IRB 1600 robot using a laser traker, Robotics and Computer_Integrated Manufacturing, 2013, Vol. 29, pp. 236-245.

[8] Isolde Dressler, (2012). Modeling and control of stiff robots for flexible manufacturing, Department of Automatic Control Lund University Lund, Printed in Sweden by Media-Tryck, 2012.

[9] Kam C. Lau, Yuanqun Liu, Guixiu Qiao, Liangyun Xie, (2010). "Volumetric error compensation system with laser tracker and active target," U.S. Patent US20100176270 A1, July 15, 2010.

[10] Denavit J., Hartenberg R. (1955). A kinematic notation for lower-pair mechanisms based on matrices. // Transactions of ASME – Journal of Applied Mechanics, 22(2), pp. 215-221, June 1955.

[11] Mooring B., Roth Z., Driels M. Fundamentals of manipulator calibration// John Wiley & Sons, INC, 1991.

[12] Vorotnikov A., Bashevskaya. O., Ilyukhin Yu., Romash, E., Isaev A.V., Poduraev Yu. (2016). Geometrical Approach for Industrial Robot Axis Calibration Using Laser Tracker // Proceedings of the 26th DAAAM International Symposium. – Vienna, Austria, 2016. – P. 0897-0904

[13] Vorotnikov A[ndrei]; Romash E[lena]; Isaev,A[lexander]; Bashevskaya O[lga]; Bianchi G[iacomo] & Poduraev Y[uri] (2016). Uncertainty estimation of axes direction determination of industrial robot using an ellipsoid concentration model, Proceedings of the 27th DAAAM International Symposium, pp.0480-0486, B. Katalinic (Ed.), Published by DAAAM International, ISBN 978-3-902734-08-2, ISSN 1726-9679, Vienna

DOI: 10.2507/28th.daaam.proceedings.141

A WEARABLE FALL DETECTOR FOR ELDERLY PEOPLE

Miroslav Musalek

Tomas Bata University in Zlin, Faculty of Applied Informatics,
Nad Stranemi 4511, Zlin 760 05, Czech Republic

This Publication has to be referred as: Musalek, M[iroslav] (2017). A Wearable Fall Detector for Elderly People, Proceedings of the 28th DAAAM International Symposium, pp.1015-1020, B. Katalinic (Ed.), Published by DAAAM International, ISBN 978-3-902734-11-2, ISSN 1726-9679, Vienna, Austria
DOI: 10.2507/28th.daaam.proceedings.141

Abstract

Falls and their subsequent possible side effects are for elderly people and people with various illnesses, such as epilepsy, are present the main obstacle to an independent and more dignified life. Even for people living independently, falls are common occurrences. Proper use of a fall detector will not only add more security to the seniors, but will also allow for timely medical intervention that can be life-critical. One of the limitations of this wearable technology is its limited lifetime and the occurrence of possible errors during the data acquisition and processing process. A fall detector designed to be able to monitor the activity of elderly people and designed to make the necessary warnings to ensure their safety as a result of the fall. In addition, other options that are used for this purpose are compared. This paper summarizes the functionality, architecture and implementation of the system.

Keywords: Accelerometer; Gyroscope; Wearable; Fall detection

1. Introduction

Falls present a great risk of injury, especially for older people and people suffering from neurodegenerative diseases. Falls and related consequences have various definitions.

A fall can be defined as a change in position that ends with body contact with the ground, which may be accompanied by impaired consciousness and injury. Falls affects 20-30% of people aged 65-69 years, and up to 50% of people over 85 years of age in a given calendar year. There is a higher incidence of falls among women, acute and chronically ill patients, hospitalized and long-term institutional care patients [1].

In 1987 the Kellogg international working group on the prevention of falls in the elderly defined and fall as "unintentionally coming to ground, or some lower level not as a consequence of sustaining a violent blow, the loss of academic press journal consciousness, sudden onset of paralysis as in a stroke or an epileptic seizure " [2].

World health organization (WHO) defines a crash as-defined as an event which results in a person coming to rest inadvertently on the ground or floor or other lower level.

Research on the automatic detections of falls has been taking place since the 1990s. Lord a Colvin [3] studied the causes and consequences of falls in seniors, tried to prevent falls and proposed the use of a fall detection accelerometer. The first prototype fall detector was developed in 1998. There was use a piezoelectric shock sensor to detect abnormal falls and a mercury tilt switch to detect the user's orientation after falling.

Sixsmith et al. [4] in 2004 a range of cheap infrared cameras mounted on the wall was used. The alarm was triggered if activity was not recorded for too long or upon the detection of a fall. The attempts (20 falls + 10 attempts without falling) unfortunately showed that only 30% of falls were being correctly determined.

Nyan et al. [5] attempted fall detection based on a high-speed camera and 3 gyroscopes built within an undershirt. They were specifically placed on the chest, armpit and waist. The camera was used to study the body's position during the fall, while the angular velocity was a guide to the fall detection.

Miaou et al. [6] carried out a fall detection based on a panoramic camera and user information (body height to width ratio and BMI. This system has 70% accuracy without user information and 81% accuracy with this information.

Bianchia et al. [7] have developed a fall detection system based on a barometric pressure sensor and a triaxial accelerometer located on the waist. Based on the assumption that the atmospheric pressure between the waist and ground level is different, then the results of the experiments demonstrated that the obtained sensor information is useful for the detection of falls.

In recent years, interest in exploring the use of detectors in mobile phones has grown. Other studies [8], however demonstrated that a three-axis accelerometer built into a smartphone may be of relatively low quality and this means worse fall detection results compared to other accelerometers available on the market.

Current work on automatic fall detection methods can be classified into three main categories in terms of the sensors they use: video-based methods, acoustics-based methods and wearable sensor-based methods.

In the article [9] a comparison of the various fall detecting possibilities is performed and we can discover the most suitable detection methods and their effectiveness.

Currently methods using triaxial acceleration sensors are the main methods for motion and fall detection in older populations:

1) [10]
2) [11]
3) [12]

Nyan and colleagues have developed a fall detector using a triaxial accelerometer (MEMS) to collect data. To collect signals use dictated wavelet transformation (DWT). These data are processed to recognize daily activity and fall. The sensor is placed in a jacket at shoulder-height and other components, such as the microcontroller, are inside the pocket. This solution has a limited period of use [13].

Using the accelerometer and the gyroscope together allows for more accurate fall detection. The accelerometer can provide kinetic power while the gyroscope can help estimate the current position.

Li and colleagues used the accelerometer and gyroscope. By using two triaxial accelerometers located at various points in the body, they can recognize individual daytime activities. Transitions between these static positions are treated as dynamic transitions. If the transition before going to sleep is not intentional, a fall is detected. The recognition of whether movement gradients are intentional or not are determined by measuring linear acceleration and angular velocity [14].

Accelerometer data may sometimes be distorted. Kuncar [15] compares various filtration methods and calibration methods for MEMS sensors in order to remove measurement noise.

In the Czech Republic the Senior Inspect is being implemented, whose development began in 2009 and currently offers continuous assistance in the lives of seniors.

This paper introduces a proposed sensor for detecting the movement of older people.

2. Fall detection

The motion sensor is used here this purpose. The sensor is a wearable device capable of wireless communication. This device is capable of recognizing user activity and evaluating falls. Using a triaxial accelerometer, we obtain acceleration along three X, Y and Z coordinate axes. The angular acceleration and orientation of the sensor can be measured using a triaxial gyroscope.

In order to detect falls, the sensor must be able to sense the motion and the various measurable properties associated with motion.

2.1 Hardware Descriptions

The proposed fall detection system is shown in Fig. 1. Its individual elements are described in the text below.

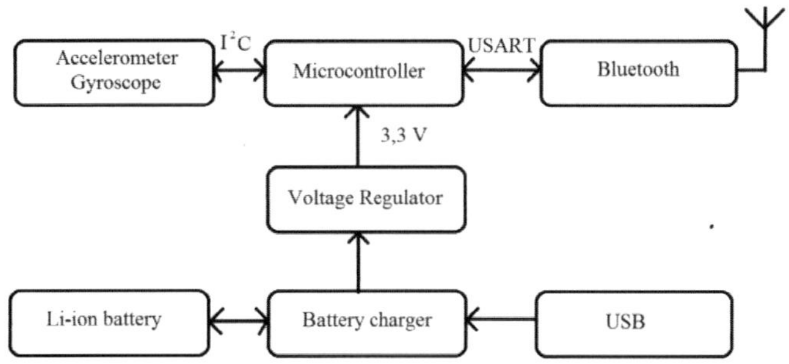

Fig. 1. Block diagram of the proposed fall detection system

I. Motion module

A MEMS sensor MPU-6050, which includes a triaxial accelerometer, a triaxial gyroscope and a thermometer. It also includes the Digital Motion Processor™ (DMP), which allows the performance of complex calculations between the accelerometer and the gyroscope. For both fast and slow movements, the gyroscope can programmatically set the measuring range to ± 250, ± 500, ± 1000 and ± 2000 ° / sec, and accelerometer within ± 2 g, ± 4 g, ± 8 g, and ± 16 g. The sensor operates at a supply voltage range of 2.375 V to 3.46 V. In normal operation, the consumption of the gyroscope, accelerometer, and DMP is 3.9 mA. Due to its small size and low power consumption it is suitable for wearable devices. Its power is selected at 3.3 V.

II. Interface for communication

Wireless detector communication uses the Bluetooth® standard. For the largest range of equipment, which is 100 m, the Bluetooth ®RN41 Class 1 is used. This module includes a built-in antenna. Communication takes place asynchronously through the USART interface (Universal Synchronous/Asynchronous Receiver and Transmitter).

III. Microcontroller

As the microcontroller for the detector, the STM32F051x model is used, which has a high performance ARM Cortex ™ -M0 32-bit RISC core operating at 48 MHz Its advantages include low consumption, dimensions and I2C and UART bus. The flash memory size of this model is 64 kB. The motion sensor is connected to it through the I2C bus.

IV. Detector power module

The power block of the human motion detection module is composed of a battery, a battery charging circuit, and a power circuit. The battery in the module can be charged through the USB interface.
A lithium ionic battery LP-443440-IS-3 is used at a capacity of 600mAh at 0.2 °C and nominal voltage of 3.7V The capacity of the battery will last for 24-hour operation. The maximum charging current is 560mA and the charging voltage is 4.2V. That is why the BQ24032A circuit from Texas Instruments was chosen as the integrated charging circuit for this battery through the USB interface. This circuit can be used to charge single-cell Li-Ion or Li-Pol batteries. This is a linear charger with an energy consumption management system designed for portable applications.

The MPU-6050 can be powered in the range 2,375 - 3,46 V, Bluetooth® module 3 - 3.3 V and microcontroller 2 - 3.6 V. Out of the voltage ranges, a uniform voltage of 3.3 V was chosen. To maintain a constant voltage, the step-up and step-down of the inverter must be combined because the voltage from the BQ24032A circuit output may be below the 3.3V threshold. These requirements are met by the TPS61201 circuit designed to power the circuitry from the battery. In order to maintain the output voltage of 3.3 V, the input voltage must be within the range of 0.3-5.5 V. The converter is based on a fixed frequency, a pulse (PWM) with the aid of synchronous repairs to achieve maximum efficiency. During low load currents, it enters standby mode and maintains high circuit efficiency

2.2 Fall detection algorithm

To determine the overall acceleration vector and angular velocity, similar parameters are used as in the [16] study. The total acceleration vector we can call the vector size signal (SVM) can be calculated from the relationship (1).

$$Acc = \sqrt{(ACC_x)^2 + (ACC_y)^2 + (ACC_z)^2} \quad (1)$$

The Accx, Accy, and Accz constants are gravitational accelerations in the x, y and accelerometer axes (g).

The equation (2) is designed to calculate the angular velocity of the connected gyroscope.

$$Gyro = \sqrt{(Gyro_x)^2 + (Gyro_y)^2 + (Gyro_z)^2} \quad (2)$$

In the upper equation we mean constants Gyrox, Gyro y and Gyro z angular velocity in the x, y and z axes.

To determine a fall, thresholds are set to make it possible to recognize the fall. This algorithm not only detects the fall, but also compares boundary values with daily activity. Daytime activity includes, in particular, normal walking or running. Prior to evaluating the detection, the potential fall is evaluated by a potential fall limit, which was determined by gradual measurement to 1 g (picture below). After this evaluation, it will check for orientation changes. If a change in orientation occurs, the fall information is sent. The course of the entire algorithm is represented by Fig. 2.

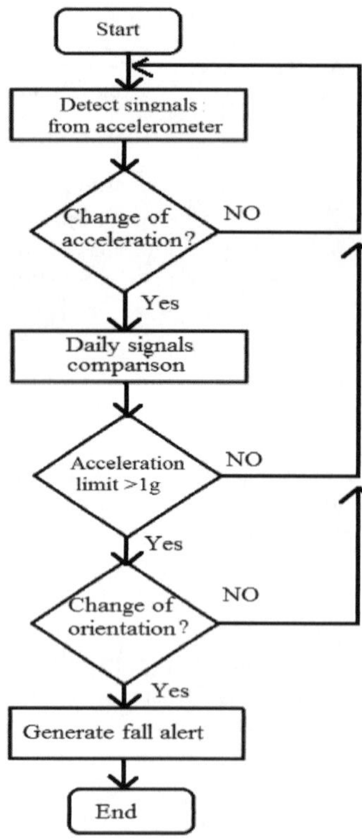

Fig. 2. Fall detection algorithm

Daily activity is determined by the diameter of the acceleration vector and the standard deviation. Values were determined by gradual measurement and effectively set as accurately as possible to identify daily movements such as running and walking. Using a gyroscope, we are able to determine the person's orientation, for example, whether it lies on the right or left side.
Furthermore, sensitivity and specificity measurements were performed to determine the reliability of this system. The above parameters can be calculated from equations (3) and (4).

$$Sensitivity = \frac{TP}{TP+FN} \quad (3)$$

$$Specificity = \frac{TN}{TN+FP} \quad (4)$$

True Positive (TP) is a status where the system can recognize the fall, when it actually occurred. False Positive (FP) is a status where the system detects a fall, although it did not occur. True Negative (TN) means a status in which the system does not detect a fall and a fall actually did not occur. False Negative (FN) is a system state where you cannot detect a fall when it occurs.

In Fig. 3, we see the dependency of the acceleration vector size over time. From the measured data, an accelerometer with a range of at least ± 4 g is required to detect falling, walking and running. The idle state of the sensor corresponds to an acceleration of approximately 1 g, walking to 2.5 g and running 2.5 g or more.

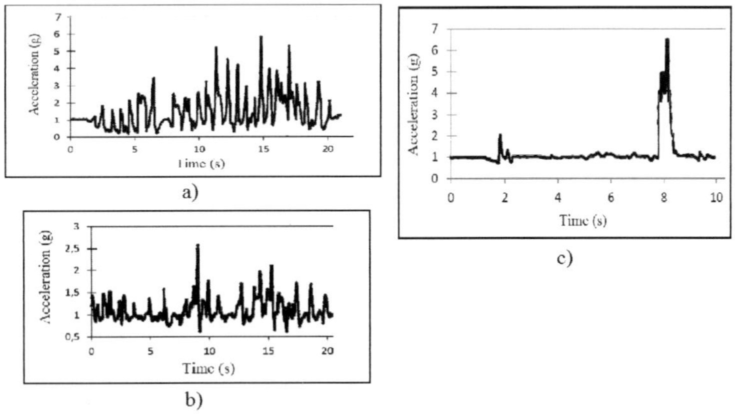

Fig. 3. Acceleration data a) running, b) walking and c) fall detection

The size of the acceleration of the fall ranges from 2 g and more. Fall detection only with maximum acceleration and subsequent idling is not possible since erroneous detection can occur with faster movement and stopping, as e.g. referred walking or running. To eliminate false detection, it is possible to evaluate the fall warning only when the orientation of the sensor is changed.

3. Conclusion and future work

In this article, a fall detector is designed and implemented. Fall detection is an important part in the care of the older population. Because falls are often underestimated, although they significantly complicate life. They often may lead to the lowering of the body's capabilities, such as disability and the impaired quality of life. These factors, the loss of certainty or fear can cause other disorders. For example, depression, powerlessness and social isolation. These measurements were performed with a lateral detector. The acceleration and angular velocity data obtained from the sensors was evaluated by the relevant algorithm. The most important function is to detect a fall that is evaluated when the acceleration greater than 1 g. Suppressing false falls is addressed by detecting daily activity and changing orientation. Measurement results demonstrate that this system can detect the sensitivity and specificity of 92.3% and 87.6% of each other.

For the recognition of either standing or sitting, it would be best to use on your thigh, but then it would prevent the detection of lying down. When using with hands, it is not possible to precisely determine the position of the measured object. In case of chest or hand use, the module could be equipped with a heart rate sensor. Further examination will have a physiological effect on the accuracy of the measurement itself. Furthermore, the system can be equipped with a GSM and GPS module.

In the future, portable sensors (e.g. body temperature, glucose, blood pressure, etc.) may be integrated into this system in order to ensure full health care. The fall detection algorithm itself can be adapted to detect traffic accidents that mathematically correspond to the energy impact of a fall. It is further necessary to distinguish between laboratory measurements that may not always correspond to reality. In the foreseeable future, it was necessary to create an extensive database for this purpose. However, this procedure may impact the area of the individual's privacy.

4. Acknowledgments

This work was supported by the European Regional Development Fund under the project CEBIA-Tech Instrumentation No. CZ.1.05/2.1.00/19.0376 and also by the Internal Grant Agency of Tomas Bata University under the project No. IGA/FAI/2017/015.

5. References

[1] Topinkova, E. (2005). Geriatrics for practice, Prague, ISBN 80-7262-365-6

[2] Gibson, M.J. et al. (1987). "The prevention of falls in later life. And report of the Kellogg International Work Group on the Prevention of Falls by the Elderly "Danish Medical Bulletin, Supplement 34 4:1-24, 1987.

[3] Brownsell, S. a M. S. Hawley (2004). Automatic fall detectors and the fear of falling. Journal of Telemedicine and Telecare of. 10 (5): 262-266 DOI: 10.1258/1357633042026251. ISSN 1357-633x.

[4] Pannurat, Natthapon, Surapa Thiemjarus a Ekawit Nantajeewarawat (2014). Automatic Fall Monitoring: A Review. Sensors. 14(7): 12900-12936. DOI:10.3390/s140712900. ISSN 1424-8220.

[5] Annual International Conference of the IEEE Engineering in Medicine and Biology Society: Lyon, France, 22-26 August 2007. 2007. Piscataway, NJ: IEEE Service Center [distributor], 16 v. (cx, 6759, 13, 46 p.). ISBN 14-244-0787-7.

[6] Miaou, S.-G., Pei-hsu Sung a Chia-Yuan Huang (2006). A Customized Human Fall Detection System Using Omni-Camera Images and Personal Information. In: 1st Transdisciplinary Conference on Distributed Diagnosis and Home Healthcare, 2006. D2H2. IEEE, s. 39-42. DOI: 10.1109/DDHH.2006.1624792. ISBN 1-4244-0058-9.

[7] Shane Kim, Jihoon Ryoo (2017). "WiSDom: A model-driven solitary death prevention system based on WiFi signals and real-time supervised training", Consumer Communications & Networking Conference (CCNC) 2017 14th IEEE Annual, pp. 131-136, 2017, ISSN 2331-9860

[8] Albert MV, Kording K, Herrmann M, Jayaraman A (2012). Fall Classification by Machine Learning Using Mobile Phones. PLoS ONE 7 (5): e36556. https://doi.org/10.1371/journal.pone.0036556

[9] Igual, Raul, Carlos Medrano and Inmaculada Plaza (2013). Challenges, issues and trends in fall detection systems. BioMedical Engineering OnLine. 12(1): DOI: 10.1186/1475-925X-12-66. ISSN 1475-925 x.

[10] Sriborrirux, W.; Leamsumran, P.; Dan-klang,P. (2014). Real-time system for monitoring activity among the elderly using an RF SoC device with triaxial accelerometer data over a wireless sensor network:2014 IEEE MTT-S International Microwave Workshop Series on RF and Wireless Technologies for Biomedical and Healthcare Applications(IMWSBio), Thailand,December 8-10,2014 [C].IEEE,2014.

[11] Yanjun Li,Gan Chen,Yueyun Shen,et al. (2012). Accelerometer- based fall detection sensor system for the elderly:2012 IEEE 2nd International Conference on Cloud Computer and Intelligent Systems (CCIS),Hangzhou, China, October 30- November 1,2012[C].IEEE,2012.

[12] Aguiar,B.;Rocha,T.;Silva,J., et al. (2014). Accelerometer- based fall detection for smartphones: 2014 IEEE International Symposium on Medical Measurements and Applications (MeMeA), June 11-12, 2014 [C].IEEE,2014.

[13] Nyan, M., et al. (2006). Garment-based detection of falls and activities of daily living using 3-axis MEMS accelerometer. in Journal of Physics: Conference Series. 2006. IOP Publishing.

[14] Li, Q., et al. (2009). Accurate, fast fall detection using gyroscopes and accelerometer-derived posture information. in Wearable and Implantable Body Sensor Networks, 2009. BSN 2009. Sixth International Workshop on. 2009. IEEE.

[15] Kuncar, A[les] (2016). Basic Techniques for Filtering Noise Out of Accelerometer Data, Proceedings of the 26th DAAAM International Symposium, pp.1122-1128, B. Katalinic (Ed.), Published by DAAAM International, ISBN 978-3-902734-07-5, ISSN 1726-9679, Vienna, Austria DOI: 10.2507/26th.daaam.proceedings.158

[16] Quoc T. Huynh, Uyen D. Nguyen, Lucia B. Irazabal, Nazanin Ghassemian, and Binh Q. Tran (2015). "Optimization of an Accelerometer and Gyroscope-Based Fall Detection Algorithm," Journal of Sensors, vol. 2015, Article ID 452078, 8 pages, 2015. doi:10.1155/2015/452078

DOI: 10.2507/28th.daaam.proceedings.142

EVALUATION OF PROCESS RISKS IN INDUSTRY 4.0 ENVIRONMENT

Michaela Švingerová, Martin Melichar

This Publication has to be referred as: Svingerova, M[ichaela] & Melichar, M[artin] (2017). Evaluation of Process Risks in Industry 4.0 Environment, Proceedings of the 28th DAAAM International Symposium, pp.1021-1029, B. Katalinic (Ed.), Published by DAAAM International, ISBN 978-3-902734-11-2, ISSN 1726-9679, Vienna, Austria
DOI: 10.2507/28th.daaam.proceedings.142

Abstract

Industry 4.0 has recently become a frequent topic at a number of expert conferences and even in the mass media. It is typically discussed in the context of process productivity improvement but also in relation to the fear of job losses. The situation is unprecedented where for the first time in history white-collar jobs are under threat and high-qualification professional class faces the prospect of economic decline. This paper reviews Industry 4.0 from the perspective of benefits and risks which are not related to societal aspects but exclusively to serial production data from the automotive industry. It deals with implementation and evaluation of the effect of substitution of human labour upon RPN (risk priority number) indicators which are used for process classification in the PFMEA (Process Failure Mode Effects Analysis) method as required by the IATF 16949 (International Automotive Task Force) international automotive standard. The article describes the PFMEA method, the inputs and analyses the inputs and outputs of the FMEA method in a real-world environment.

Keywords: Industry 4.0; PFMEA; Risk

1. Introduction

In today's industrial production, process risk assessment by means of PFMEA is one of the cornerstones of control processes in most companies. The earlier automotive standard version, ISO TS 16949, merely recommended FMEA as an appropriate tool for companies involved in the automotive supply chain to master their preventive measures. Its revision and transformation into IATF 16949 and linking to an equally updated version of ISO 9001:2015, the role of risk management was greatly strengthened. IATF-certified companies are required to identify and assess the risks of all company processes on a regular basis as part of their internal management system. Now, the FMEA or PFMEA tool is not just recommended but expressly required by this standard.

The history of FMEA goes back to the 1940s when the U.S. Army sought a technique for eliminating errors in the machines and equipment in use. 1960s saw the the first non-military use of this method when NASA (National Aeronautics and Space Administration) began to use FMEA to identify potential risks in Apollo and Gemini programmes. The truly civilian era of risk assessment began 10 years later when the method was adopted by the automotive industry, particularly the Ford company in an attempt to address the low quality of its Ford Pinto model.

PFMEA is a team-based approach to potential risk assessment where the multidisciplinary team is usually led by the engineer who is responsible for the product design. Active involvement of representatives of all areas concerned is expected, including assembly, production, design, testing and others [7]. It is advisable to begin risk mapping with a flow chart of the process in order to map it clearly and identify its boundaries. The customer is typically the end user but, depending on the definition, it may comprise the downstream operation or subsequent assembly operation.

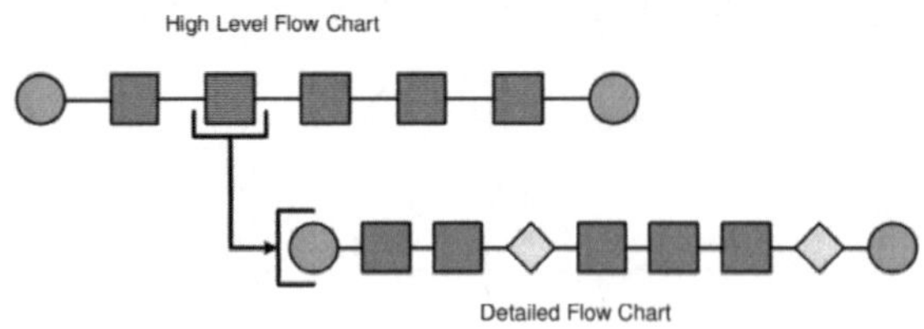

Fig. 1. Flow chart [10]

Once the process and its boundaries have been clearly defined, brainstorming, Ishikawa diagram or other tools are used to identify as many risks as possible which may threaten the process[2;6]. These risks are then evaluated according to a consistent method from three aspects: severity of consequences in respect of the customer, occurrence of consequences in respect of the customer and the chance of detection. Each of these aspects is assigned a score on a scale from 1 to 10 as indicated in the tables below.

Consequence	Severity in respect of the customer	Score
Failure to meet safety of regulatory requirements	Affects safe operation without warning	10
	Affects safe operation after warning	9
Impaired or lost primary function	Vehicle disabled	8
	Vehicle drivable, reduced power	7
Impaired or lost secondary function	Loss of comfort	6
	Reduced comfort	5
Inconvenience	Noticed by 75% of customers	4
	Noticed by 50% of customers	3
	Noticed by 25% of customers	2
No consequence	No perceptible consequence	1

Table 1. FMEA – severity [10]

Consequence	Occurrence in respect of the customer	Score
Major	100000ppm	10
Medium	50000ppm	9
	20000ppm	8
	10000ppm	7
Medium	2000ppm	6
	500ppm	5
	100ppm	4
Low	10ppm	3
	1ppm	2
Very low	Eliminated by prevention	1

Table 2. FMEA – occurrence [10]

Chance of detection	Probability of identification of the cause	Score
Detection impossible	Cannot be detected	10
Near impossible	Low probability of detection	9
Negligible	E.g. visual inspection	8
	Detection of the problem in the workplace	7
Very low		
Low	Measurement after the operation was completed by the operator	6
Medium	Detection by operator with the aid of automated tools	5
Medium to high	Inspection with automated tools and separation	4
High	Inspection with automated tools and separation	3
Very high	Automated tools for error detection	2
Fault prevention	Error prevention	1

Table 3. FMEA – identification [10]

After the assessment, risk priority numbers (RPN) are determined for individual risks by simple multiplication of values for all three aspects. Risks with RPN values higher than an internal or contractual limit and those with the greatest severity are addressed on a priority basis as part of prevention measures.

2. Experimental Procedure

2.1. Real case study

This paper describes an assessment of automation of production using real-world data from a company which operates in the automotive industry. The data comes from serial production of canned catalysts. Two production lines were compared. Both of them are currently used in production. The one which has not been automated yet has been in operation for several years with minor improvements gradually introduced. One year ago, an automated line was installed next to it[2;3;4]. In this new line, most operations which used to be performed by a worker/operator are done by a robot. Even this automated line, though, requires the operator to carry out basic infeed and outfeed tasks. The automated line is monitored continuously and its processes are being refined. Such preparations may eventually lead to full automation.

2.2. Finished product

The finished products of both lines are canned catalysts. They consist of the monolithic catalyst, a support mat and a steel tube. It should be pointed out that for the purposes of this paper, the canned catalyst is considered the final product, although this only applies to the production lines described here. The end use requires its incorporation into the exhaust system of a vehicle.

The monolithic catalyst is the single most important and most expensive part of the catalytic converter. It is a ceramic cylinder which contains elements such as platinum, rhodium and palladium. It is a very brittle part. In the catalytic converter, its role is to reduce the amount of harmful products in the exhaust gases. This is provided by its sophisticated inner structure which is not identical in all monolithic catalysts. Every car manufacturer and every car model has its specific monolithic catalyst structure and size. The figure below shows various cell structures as seen from the top or bottom side of the monolithic catalyst[1;5;7].

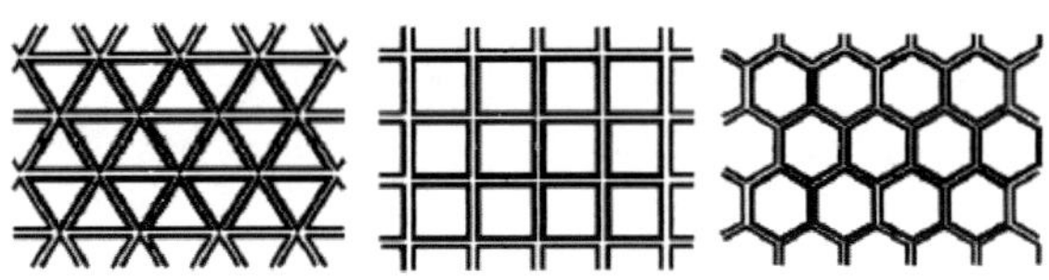

Fig. 2. Example of cell structures [9]

The sizes of monolithic catalysts differ as well, with diameters of approximately 15 cm and heights around 15 cm in passenger vehicles. The typical appearance of a monolithic catalyst is shown in the figure below.

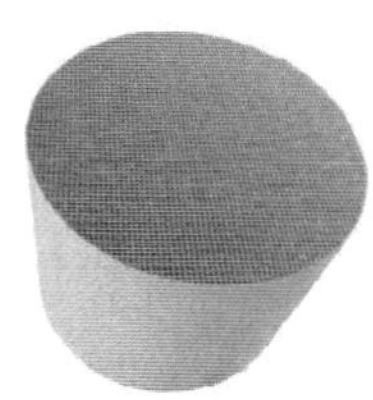

Fig. 3. Monolithic catalyst Source: Internal company documents

The support mat in the canned catalyst is made of glass wool. It provides both thermal and noise insulation and protects the brittle monolithic catalyst from damage. Shapes of the support mat can differ, as illustrated below, but the insulation effect must remain.

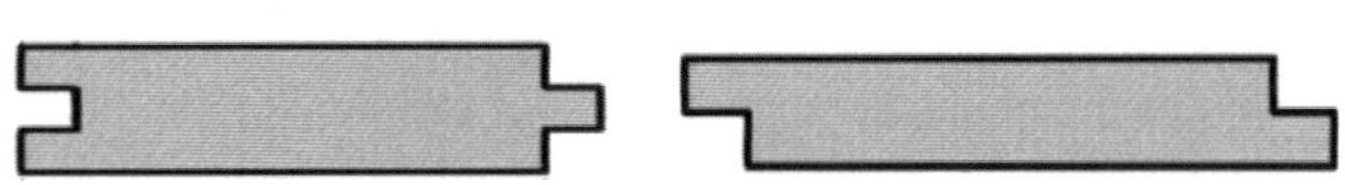

Fig. 4. Detail of isolation [9]

To give an idea of the glass wool texture, the diagram below shows the classification of fibres of which the insulation is composed.

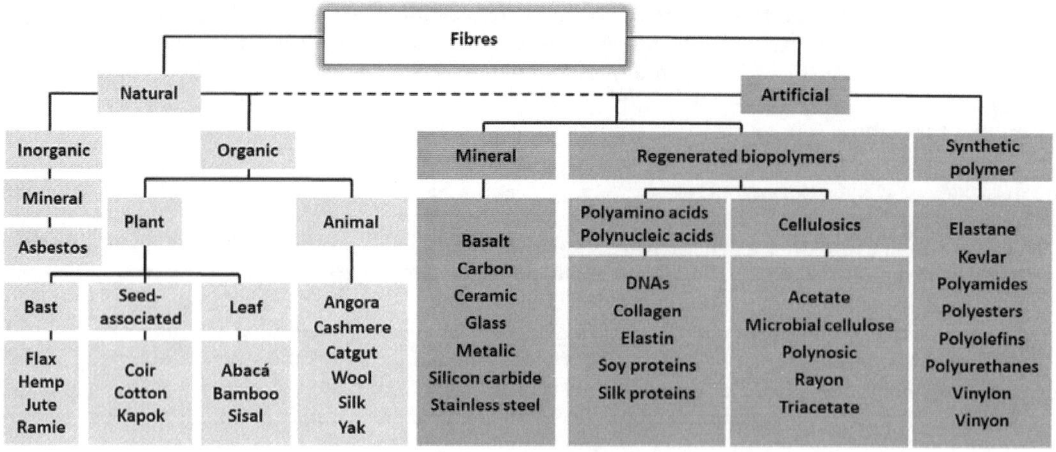

Fig. 5. Fibres classification [9]

The last part of the canned catalyst is its outer shell. It is in fact an iron tube. The first step involves rolling up sheet metal to a cylinder shape and laser welding it together in the machine. The resulting tube is cut by laser to the desired lengths to be used as "cans" for the catalytic converter. This part is shown in the figure below, on the left. The article in the centre is the monolithic catalyst and the one on the right is the rolled-up support mat. The figure below them shows a 3D rendering of the completed canned catalyst.

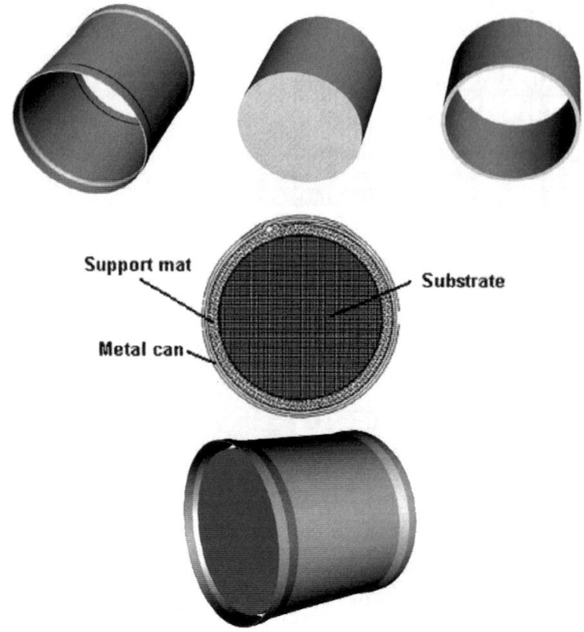

Fig. 6. Canned catalyst [9]

2.3. Production process

The sequence for making the canned catalyst is obvious from the parts used. The process flow below indicates the individual steps.

The automated line will be considered first. The boxes to the right of and below the dotted rectangle in the process flow chart indicate the operations which are done by human operator even in the automated line. They include feeding individual parts to the machine, final inspection, and transfer to downstream lines for final processing. Inside the dotted rectangle, there are the operations which are performed by the machine.

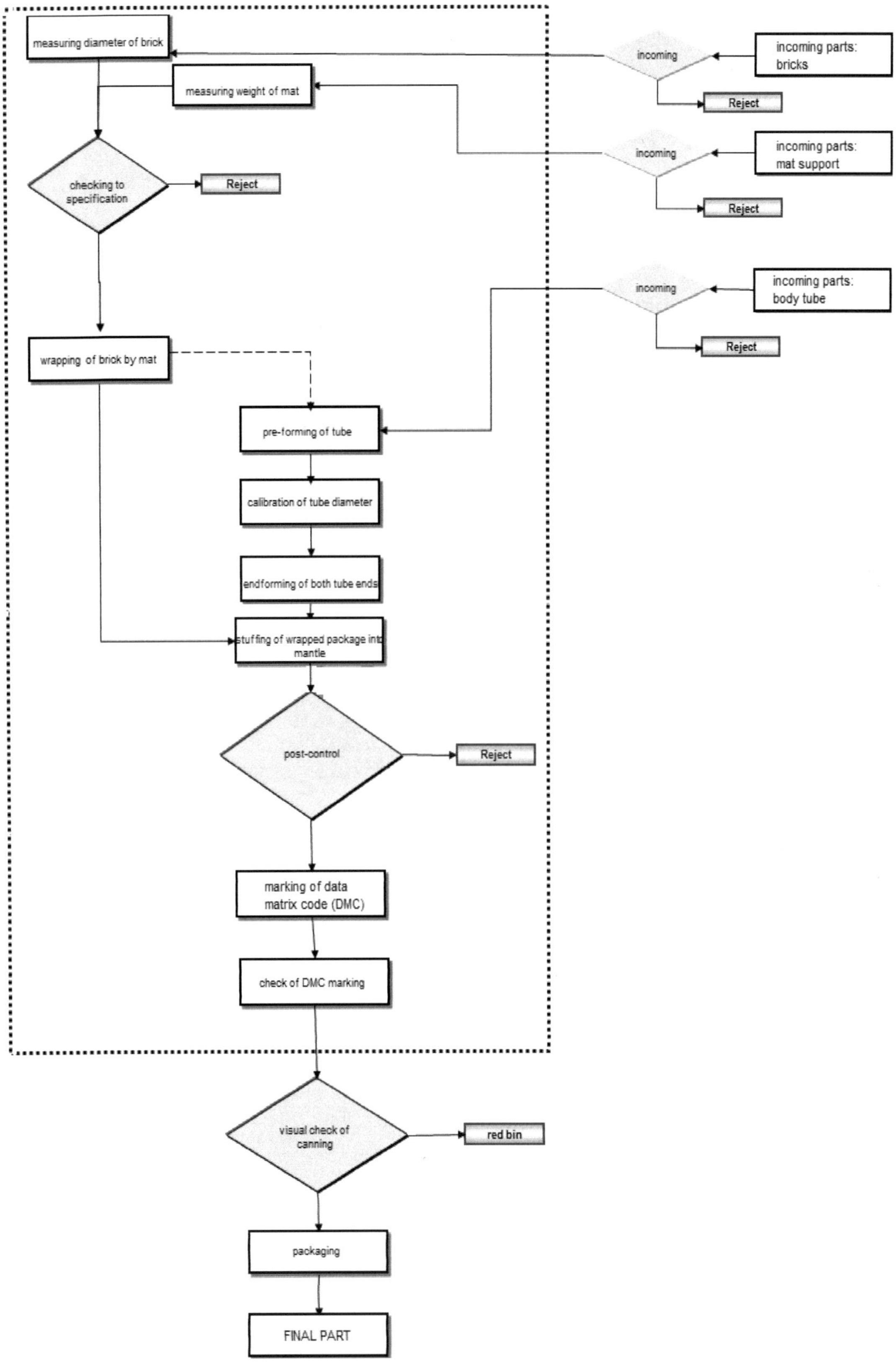

Fig. 7. Making catalyst schema [9]

The photograph (Fig. 9) below the flow chart shows the “Kirschenhofer Maschinen” automated canning line.

Fig. 8. Robotics work place [9]

In the conventional canning line, all the operations indicated in the flow chart are carried out by a human operator in several work stations with production machines and respective control computers [8]. Today, this conventional line is only used occasionally, on a customer request, typically for making small batches of canned catalysts. Due to confidentiality restrictions, the photograph of this conventional line – referred to as "shrinking line" – cannot be shown.

3. FMEA analysis

3.1. Manual operator

Step number	Process Function Operation	Requirements Item.	Potential Failure Mode	Potential Effect(s) of Failure	Severity	Class	Potential Cause(s)/ Mechanism(s) of Failure	Occurrence	Current Process Controls Prevention	Current Process Controls Detection	Detection	R.P.N
1	Components loading into the machine		Wrong part orientation	Improper fit to vehicle	7		Operator mistake, Standardize working instruction not followed	4	Process steps description in Standardize work instruction	sensors control correct orientation	2	56
			Insuficient part position	Tooling damage, time loss	8		Operator mistake, Standardize working instruction not followed	4	Process steps description in Standardize work instruction, pin is controling correct position (sensor on end stop)	pin is controling correct position (sensor on end stop)	3	96
			Insufficient part position (	Tooling damage, time loss	8		Operator mistake, Standardize working instruction not followed	4	Process steps description in Standardize work instruction	visual control, operator has to ensure position with small equipment	5	160
			missing component	Improper fit to vehicle	7		Operator mistake, Standardize working instruction not followed	4	Process steps description in Standardize work instruction	sensor checking right position	5	140
			extra component	Improper fit to vehicle	7		component inserted twice	3	Work instruction	install sensor to control right material thickness before clamping	3	63
					7		components sticked with oil	3	Work instruction	install sensor to control right material thickness before clamping	3	63

2	clamping	clamping parts	clamp quality	Part may not meet system design requirements	7		parts weren't clamped	4	operator training	sensor checking right position of clamps	4	112
					7		parts weren't clamped	4	operator training	sensors checking right position of clamps	4	112
					7		parts weren't clamped	4	operator training	sensors checking right position of clamps	4	112
				Part may not meet system design requirements	7		unsifficient clamping	4	operator training	sensor checking right position of clamps	4	112
					7		unsifficient clamping	4	operator training	sensors checking right position of clamps	4	112
3	part unloading from the machine	assy	Part damage	Improper fit to vehicle	6		Falling of the part	2	Design of the holders allowing easy part manipulation	Operator training	8	96
4	packaging	Box	Wrong Container type	Time loss, wrong number of parts	3		Material handler mistake	4	no control	Operator training, packaging instruction	8	96
			Wrong number of parts in container	Inventory discrapency	3		Operator mistake, packaging instruction not followed	4	Operator training, packaging instruction	Discrapancy for next station	8	96
			Missing label	Lost of tracebility	4		Operator did not print label or forget to place printed label on box	4	Operator training, automatic printing after last piece inspection	Reading lable on next process step	4	64
		part condition	skipped operation	Improper fit to vehicle	7		operator mistake	3	No control	100% Touchpoint inspection on another work station	7	147

Fig. 9. FMEA analysis – manual operator [10]

3.2. Automated process

Step number	Process Function Operation	Requirements Item.	Potential Failure Mode	Potential Effect(s) of Failure	Severity	Class	Potential Cause(s)/ Mechanism(s) of Failure	Occurrence	Current Process Controls Prevention	Current Process Controls Detection	Detection	R.P.N
1	Components loading into the machine	components (Stamping)	Wrong part orientation	Improper fit to vehicle	7		Operator mistake, Standardize working instruction not followed	2	Process steps description in Standardize work instruction	sensors control correct orientation	1	14
			Insuficient part position	Tooling damage, time loss	8		Operator mistake, Standardize working instruction not followed	2	Process steps description in Standardize work instruction, pin is controling correct position (sensor on end stop)	pin is controling correct position (sensor on end stop)	1	16
			Insufficient part position (	Tooling damage, time loss	8		Operator mistake, Standardize working instruction not followed	2	Process steps description in Standardize work instruction	visual control, operator has to ensure position with small equipment	2	32
			missing component	Improper fit to vehicle	7		Operator mistake, Standardize working instruction not followed	3	Process steps description in Standardize work instruction	sensor checking right position	2	42
			extra component	Improper fit to vehicle	7		component inserted twice	2	Work instruction	install sensor to control right material thickness before clamping	2	28
					7		components sticked with oil	2	Work instruction	install sensor to control right material thickness before clamping	2	28

2	clamping	clamping parts	clamping quality	Part may not meet system design requirements	7		parts weren't clamped	2	operator training	sensor checking right position of clamps	2	28
					7		parts weren't clamped	3	operator training	sensors checking right position of clamps	2	42
					7		parts weren't clamped	3	operator training	sensors checking right position of clamps	2	42
				Part may not meet system design requirements	7		unsifficient clamping	3	operator training	sensor checking right position of clamps	2	42
					7		unsifficient clamping	3	operator training	sensors checking right position of clamps	2	42
3	part unloading from the machine	assy	Part damage	Improper fit to vehicle	6		Falling of the part	2	Design of the holders allowing easy part manipulation	Operator training	3	36
4	packaging	Box	Wrong Container type	Time loss, wrong number of parts	3		Material handler mistake	4	no control	Operator training, packaging instruction	8	96
			Wrong number of parts in container	Inventory discrapency	3		Operator mistake, packaging instruction not followed	4	Operator training, packaging instruction	Discrapancy for next station	8	96
			Missing label	Lost of tracebility	4		Operator did not print label or forget to place printed label on box	4	Operator training, automatic printing after last piece inspection	Reading lable on next process step	4	64
		part condition	skipped operation	Improper fit to vehicle	7		operator mistake	3	No control	100% Touchpoint inspection on another work station	7	147

Fig. 1. FMEA analysis – robot [10]

3.3. Summary

An FMEA analysis based on a previous SPC analysis of process stability has clearly shown that substitution of manual operations with an automated process considerably reduced RPN values. The reduction was in the range from 33% to 84%. The only area where the risk priority number has not changed was the packaging of the part which remained a manual operation even after process automation. The most interesting item in the analysis is the number of critical processes, i.e. the potentially critical ones.

Like its many automotive counterparts, the company has defined its internal RPN acceptance level as 100. As the analysis shows, appropriately configured automation eliminated seven critical processes which are of major importance to the end user.

4. Conclusions

Research team of university of West Bohemia had the goal to analyse influence of risk management connected with replacement human work.

As mentioned in the introduction, today's automotive industry relies in its mandatory risk assessment activities exclusively on the FMEA method which is required by the IATF standard. The method was also used for experimental analysis of a production process, i.e. a substitution of a manual assembly workplace with a robotic cell, an upgrade which can undoubtedly be classified under Industry 4.0. Effects of a specific robotisation step were evaluated in collaboration with researchers from the University of West Bohemia/Regional Technological Institute.

Both processes were evaluated by a multidisciplinary team comprising members from all areas of the production process. Using brainstorming and data from real-world production (which were also utilized in the previous contribution [9]), risks were identified for each process. Each of the risks was classified by the FMEA team from three perspectives: severity/occurrence/detection. Since supply chain experience suggests that setting internal limit RPN values is not effective, this calculation was only use for statistical evaluation and for comparing both groups of results.

The results clearly show that despite the high initial investment costs for introducing an Industry 4.0 workplace, this investment pays off, at least from the risk management perspective. One general finding is that although the robotic cell comprises elements for active detection of errors and integrates a number of features of poka-yoke philosophy, the outcomes are virtually incomparable. RPN values become lower by an order of magnitude, on account of a drop in the detection and occurrence aspects.

The results achieved are in agreement with an earlier experiment, in which this process was assessed from the perspective of capability and stability. Next step in this analyse will be evaluation of risk during mass production inspection.

5. Acknowledgements

This paper was created due to the project GA ZCU v Plzni: SGS-2016-005 "Research and development for innovation in field of Manufacturing processes – Technology of metal cutting II.

6. References

[1] Melichar M., Kubátová D., Kutlwašer J.: (2016). CMM measuring cycle and human factor, Proceedings of the 27th DAAAM International Symposium, ISBN 978-3-902734-08-2

[2] Sahno J., Shevtshenko E., Zahharov R.:(2015) Framework for Continuous Improvement of Production Processes and Product Throughput, Procedia Engineering, Volume 100, Pages 511-519, ISSN 1877-7058, http://dx.doi.org/10.1016/j.proeng.2015.01.398.

[3] Hiltmann K.: Predicting Unknown Failures,(2015) Procedia Engineering, Volume 131, Pages 840-849, ISSN 1877-7058, http://dx.doi.org/10.1016/j.proeng.2015.12.392.

[4] Hak Soo Mok, Hyun Su Song, Deuk Jung Kim, Jin Eui Hong, Seung Min Lee, Jung Tae Ahn: Determination of Failure Cause in Remanufacturing,(2015) Procedia Engineering, Volume 100, Pages 14-23, ISSN 1877-7058, http://dx.doi.org/10.1016/j.proeng.2015.01.337.

[5] Chandrajit P., Anand S.: Correlating Failure Mode Effect Analysis (FMEA) & Overall Equipment Effectiveness (OEE), (2012) Procedia Engineering, Volume 38, Pages 3482-3486, ISSN 1877-7058, http://dx.doi.org/10.1016/j.proeng.2012.06.402.

[6] Xiaoli Zhang, Zhen He, Liangxing Shi, Process Quality Metrics for Mechanical and Electrical Production Lin, (2011), Procedia Engineering, Volume 24, Pages 6-11, ISSN 1877-7058, http://dx.doi.org/10.1016/j.proeng.2011.11.2592. (http://www.sciencedirect.com/science/article/pii/S1877705811054440)

[7] Vladimir Filaretov, Alexander Zuev, Anatoly Khvalchev, Development of Approach to Automatic Machining of Composite Parts without their Rigid Fixing by Means of Multilink Manipulators, (2014) Procedia Engineering, Volume 69, Pages 4-12, ISSN 1877-7058, http://dx.doi.org/10.1016/j.proeng.2014.02.196. (http://www.sciencedirect.com/science/article/pii/S1877705814001982)

[8] Qi Guo, Kuangjie Sheng, Zheng Wang, Xilin Zhang, hengyi Yang, Rui Miao, Research on Element Importance of Shafting Installation Based on QFD and FMEA,(2017), Procedia Engineering, Volume 174, Pages 677-685, ISSN 1877-7058, http://dx.doi.org/10.1016/j.proeng.2017.01.205.

[9] Faurecia's internal sources

[10] Švingerová M., Melichar M., Evaluation of automated production,(2017) Pro-tech-ma

DOI: 10.2507/28th.daaam.proceedings.143

THE IMMERSION DEPTH INFLUENCES ON CUTTING EDGE RADIUS DURING DRAG FINISHING

Ondřej Hronek, Miroslav Zetek & Tomáš Bakša

This Publication has to be referred as: Hronek, O[ndrej]; Zetek, M[iroslav] & Baksa, T[omas] (2017). The Immersion Depth Influences on Cutting Edge Radius During Drag Finishing, Proceedings of the 28th DAAAM International Symposium, pp.1030-1036, B. Katalinic (Ed.), Published by DAAAM International, ISBN 978-3-902734-11-2, ISSN 1726-9679, Vienna, Austria

DOI: 10.2507/28th.daaam.proceedings.143

Abstract

The article deals with the intensity of the preparation of a cutting edge during the drag finishing process. The cutting edge radius of a milling tool is investigated because it is a very important element of cutting tool behaviour during machining. The intensity of the process was evaluated relative to the immersion depths of samples during drag finishing. The aim of this experiment is to confirm the idea that a greater depth in the working container causes a higher intensity of the process. A DF – 3 drag finishing machine was used for the experiment. An Alicona IFM G4 microscope was used to measure the microgeometry of the cutting edges. Cutting edges were always measured before and after the modification. The main goal of this article is to find the dependence between cutting edge modification intensity and immersion depth.

Keywords: Drag finishing; Process conditions; Immersion depth; Cutting edge modification

1. Introduction

Drag finishing is one of four types of mass finishing. Rotors and holders make the main rotary movements. There are multiple combinations of rotary motions. The samples are attached in the holders for finishing. The holder and rotors can make counter-clockwise and clockwise motions. The holders also make a vertical movement in the working container. The container does not move. High circumferential speeds and great depth of immersion cause high pressure between the media and the parts. [1] [2]

In mechanical engineering this method is mainly used for finishing milling and drilling tools. It is especially used in the cutting edge preparation of cutting tools. Cutting edges do not have a high quality after grinding. [3] Defects can be seen on the cutting edge under a microscope. For example, the flank and rake surfaces are badly shaped, there are small areas on the main edges where the cutting edge is missing, or there is a variable size of the radius. Most defects on the cutting edges are removed by drag finishing, which creates a higher quality cutting edge. The cutting edge has a better surface roughness and a modified microgeometry with a precise value of the radius. [4] Also the cutting edge radius affects cutting tool wear. [8] The influence of cutting edge microgeometry was proved when milling super-alloy Inconel 718 during a practical experiment with different cutting edge radii (15 μm; 20 μm; 25 μm). [9]

In many cases other experiments are performed after drag finishing. These experiments monitor the cutting tools in the process of machining. [5] [6] Drag finishing is a common process prior to the deposition of thin coatings on cutting tools. This type of machining improves the adhesive properties of the tools. This technology can create a barrier on the surface which prevents the development of cracks or other defects, so this is a big advantage. Different radius sizes cause changes to the effect and the size of the cutting forces. In many cases, drag finishing causes longer tool life. [7]

1.1. Problem statement

Cutting tool microgeometry is a topical issue. As microgeometry affects cutting tool life and efficiency, research and development need to continue and look for new possibilities for improving cutting tool quality. Therefore, this article is concerned with cutting tool microgeometry modification. The effects of the drag finishing process parameters are investigated on the cutting edge radius. The cutting edge radius is one of the most important parameters on a cutting tool. This article investigates the effect of the immersion depth on the cutting edge radius. We investigated how the immersion depth affects the intensity of the microgeometry modification. The research is also focused on the positive or negative impact of immersion depth on the shape and quality of the cutting tool wedge microgeometry.

2. The apparatus used in the experiment

The IFM G4 microscope is a 3D optical / scanning measuring device which works on the principle of changing focal lengths. It is used for measuring surface characteristics, analysis and also for actual dimensions. The accuracy of the device is in nanometers. It is also suitable for cutting edge measurement.

Fig. 1. IFM G4 Microscope

A DF – 3 drag finishing machine is used for preparing the edges of the tools for machining. The main movement is a planetary rotary motion made by the holders and the rotors. This movement causes pressure between the media and the workpieces. 16 workpieces can be clamped into the DF – 3 at the same time. A big advantage is the simplicity of the drag finishing machine. The drag finishing machine has five main parts: the frame, the control panel, the process container, the rotor and the holders.

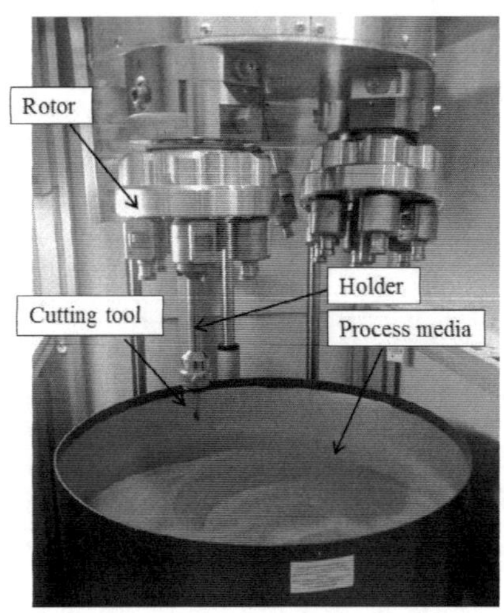

Fig. 2. Drag finishing machine

HSC 1/300 medium was used for experiment. The medium consists of two components: walnut shells and silicon carbide. It is a dry type of abrasive. It is used only for drag and stream finishing. HSC 1/300 is used mostly for finishing cutting tool microgeometry. For an efficient drag finishing process with a short time by using HSC 1/300, the maximum value of the cutting edge radius is 20 μm.

Fig. 3. Process media HSC 1/300

3. Experiment

The tool diameter is 8 mm and the length is 55 mm. Sintered carbide milling tools were used for the experiment and evaluating the cutting tool microgeometry. Twelve end mill tools were used and six tests were prepared. There were two samples for different immersion depths (Table 1.).

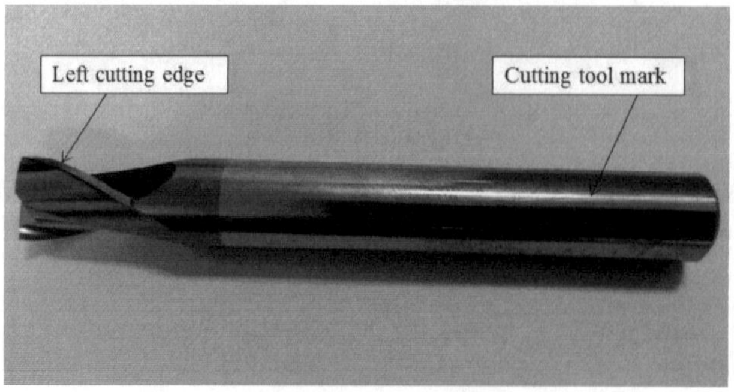

Fig. 4. Modified cutting tool

Before the experiment, it was necessary to select the appropriate finishing parameters. The DF – 3 drag finishing machine has six parameter settings. The aim is to find out the relationship between the intensity and the immersion depth of drag finishing. The Table 1. shows that the only variable parameter is the immersion depth. Other parameters are constant during the experiment.

Overall time [min]	CCW [min]	CW [min]	Rotor speed [min^{-1}]	Holder speed [min^{-1}]	Immersion depth [mm]
6	3	3	40/-40	65/-65	330
6	3	3	40/-40	65/-65	350
6	3	3	40/-40	65/-65	370
6	3	3	40/-40	65/-65	390
6	3	3	40/-40	65/-65	410
6	3	3	40/-40	65/-65	430

Table 1. Process parameters for drag finishing

All 12 samples were measured by IFM before finishing. The most important parameter was the radius of the cutting edge. The left and right edges were measured separately. The scanning area was 2 mm from the tip of the milling tool.

Sample number	Left radius edge [μm]	Right radius edge [μm]
1	1.31	1.73
2	1.55	1.39
3	1.66	1.27
4	1.33	1.61
5	1.26	1.83
6	1.66	1.23
7	1.47	1.51
8	1.30	1.36
9	1.90	1.39
10	1.24	1.26
11	1.66	1.54
12	1.48	1.36

Table 2. Values of cutting edge radius before drag finishing

Fig. 5. shows the cutting edge of the sample. This edge was created by grinding. It is magnified 20x by an Alicona IFM microscope. At this magnification it is easy to identify the marks created by the grindstone. Moreover, the radius of the cutting edge is also evident, even though it is only about 1.47 microns. Despite the fact, that the cutting tool is much sharper after grinding, the cutting edge is brittle and prone to chipping and also to deformation.

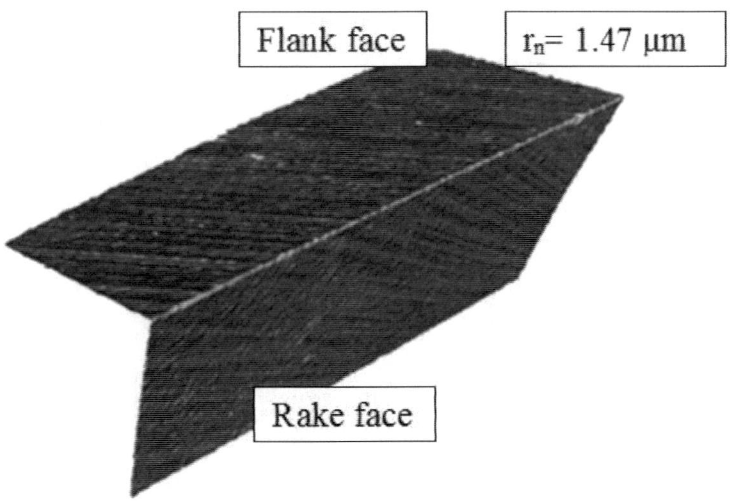

Fig. 5. Cutting edge microgeometry before drag finishing

4. Results

Twelve samples were finished at six different depths. The depth was changed after every second process. The immersion depth increased in 20 mm intervals. The difference between the initial and the final depth was 100 mm. Every cutting edge was measured by IFM after drag finishing. The left and right radii of the cutting edge were measured at the same location as before drag finishing. The measuring point was on the edge, 2 mm from the tip (Fig. 6.).

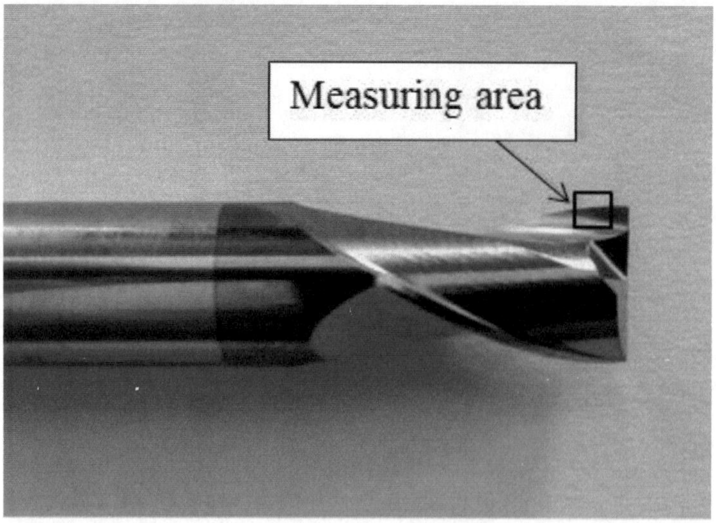

Fig. 6. Example of measuring area on cutting edge

The smallest measured radius was 8.1 microns on the sample with 330 mm immersion. The biggest radius was achieved at the bottom of the process container (430 mm). This was 14.6 microns. So, with a distance of 100 mm, the intensity of finishing the radius increases by 6.5 microns. At an immersion depth of 430 mm, the radius of the sample was 44% larger than the sample at a depth of 330mm with the same entry condition. Fig. 7. shows the increase in the size of the radius by drag finishing. The values of the radius increase gradually. The biggest increase of the intensity is between depths of 390 and 410 mm. The increase of the cutting edge was more than 2 microns at this point. The curves on the graph have relatively the same shape. The radius sizes for the left and right milling edges are approximately the same. The biggest difference was measured at a depth of 410 mm. The difference between the left and right cutting edges was 0.74 microns.

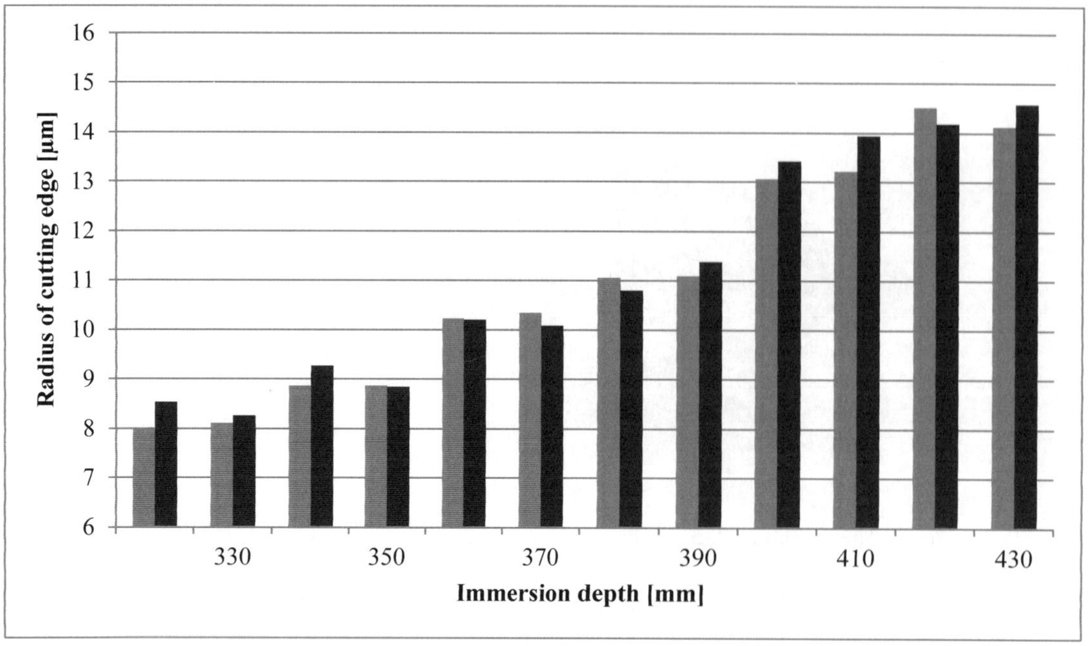

Fig. 7. Results of experiment

Fig. 8. shows the development of the cutting edge before and after finishing at depths of 330 mm and 430 mm. There is a gradual enlarging of the cutting edge radius. There is also a difference in the flank and rake surfaces. The quality of these surfaces improved with greater depth.

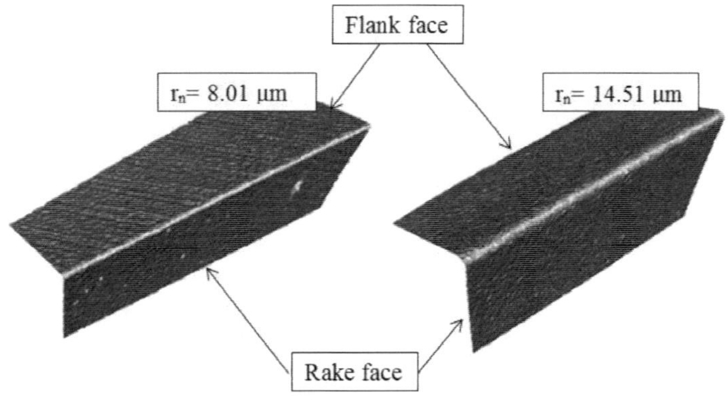

Fig. 8. Cutting edge microgeometry after modification; immersion depth 330 mm and 430 mm

Differential analysis was used to make an interesting comparison between the cutting edge before and after finishing. Differential analysis is based on the color spectrum and provides the user with quick and clear information. Fig. 9. shows the edge which was finished at 330 mm. The radius of the cutting edge is 8.1 microns.

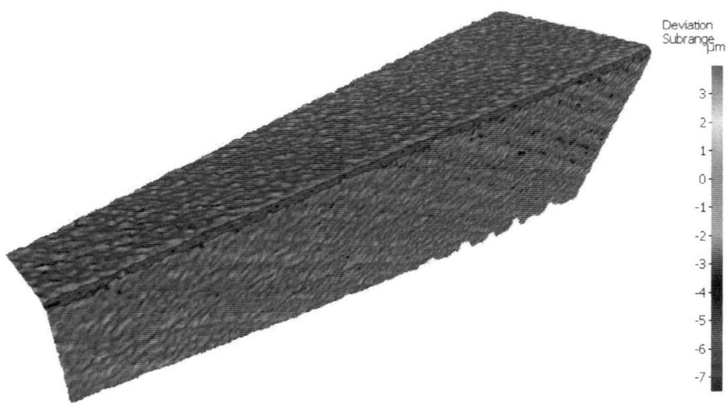

Fig. 9. Differential analysis; immersion depth 330 mm

The cutting edge in Fig. 10. is finished at an immersion depth of 430 mm. The radius of the edge is 14.1 microns. Compared with Fig. 9., it is evident that the radius has increased. Additionally, the finishing at 430 mm gives better surface quality, because there are fewer tracks from the grindstone.

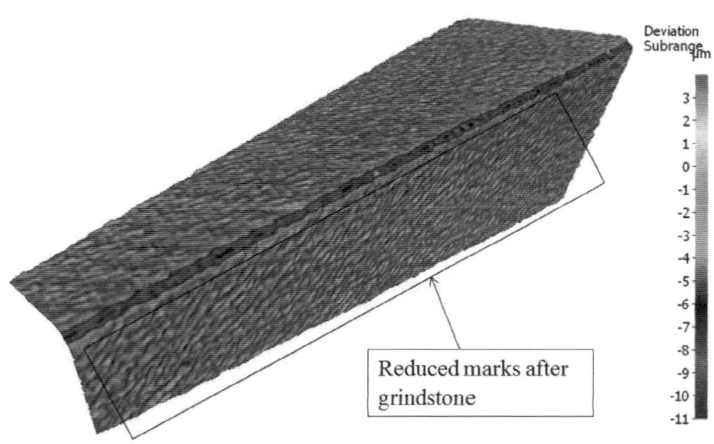

Fig. 10. Cutting edge microgeometry; immersion depth 430 mm

5. Conclusion

The main focus of this experiment was to verify the influences of the immersion depth on the cutting edge radius of an end mill tool. Therefore, an experiment was proposed in which the drag finishing immersion depth was changed. Walnut shells with SiC were used as the medium for the experiment. Immersion depths of 330 mm; 350 mm; 370 mm; 390 mm; 410 mm and 430 mm were selected. By evaluating the measured values, the relationship was found between the cutting edge radius and the immersion depth during the drag finishing process. The smallest radius, 8.1 microns, was measured at a depth of 330 mm, whereas at the depth of 430 mm, the value of the radius increased by 44%, to 14.6 microns.

Besides from the fact, that a deeper immersion depth causes a higher intensity of cutting edge modification, marks after grindstone disappear from flank and rake face. Another advantage is that the symmetry of cutting edge is achieved after drag finishing. Functional areas of the cutting tool with higher qualities are necessary not just for cutting process, but also for a coating process. It exists many cases, where cutting edge modification has brought a cutting tool life increasing, for example during machining stainless steel. But it is important to realize, that materials with toughness, such as Al and Cu alloys can't be machined by tools with high value of cutting edge radius. So it is necessary to understand a system, cutting tool – microgeometry – cutting material – workpiece material for proper cutting edge modification.

In the near future, these tools will be tested. This is necessary in order to obtain relevant information about the potential influence of microgeometry on the durability of the cutting tool. At the same time, cutting forces will be measured during the experiment. This will provide additional information and the relationships between the cutting tool microgeometry and the cutting process. This article is one of the tests that are designed to increase the quality of monolithic carbide end mill cutters.

6. Acknowledgments

This paper includes results created within the project SGS-2016-005.

7. References

[1] INFOCUBE.CZ [online]. [Accessed: 2017-09-15]. http://infocube.cz/wp-content/uploads/2017/04/Vlečné-omílání-zlepšuje-kvalitu-nástrojů.

[2] Pospíšilová, I., Vojtěch, D. & Kubásek, J. (2014). Structure and Properties of Zn-Mg Alloys for Medical Implants. Manufacturing Technology, Vol. 14, No. 3, pp. 422 – 427, ISSN: 1213 – 2489

[3] Čilliková, M., Mičúch, M., Neslušan, M. & Mičietová, A. (2013). Non-destructive micromagnetic evaluation of surface damage after grinding. Manufacturing Technology, Vol. 13, No. 2, pp. 152 – 157, ISSN: 1213 – 2489

[4] Mádl, J., Razek, V., Koutný, V. & Kafka, J. (2013). Surface Integrity in Notches Machining. Manufacturing Technology, Vol. 13, No. 2, pp. 188 – 193, ISSN: 1213 – 2489

[5] Mádl, J. & Martinovský, M. (2015). Cutting Tool Wear Monitoring. Manufacturing Technology, Vol. 15, No. 3, pp. 380-384, ISSN: 1213 – 2489

[6] Česáková, I., Zetek, M. & Švarc, V. (2013). Evaluation of Cutting Tool Parameters, Proceedings of the 24th DAAAM International Symposium, pp. 1105 – 1114, Published by Elsevier, ISBN 978-3-902734-07-5, ISSN 1877-7058, Zadar, Croatia

[7] Pilc, J. & Vasilko, K. (2013). Development and applications of a rotating turning tool. Manufacturing Technology, Vol. 13, No. 2, pp, 226-231, ISSN: 1213 – 2489.

[8] Lenz, J. & Westkaemper, E. (2017). Wear Prediction of Woodworking Cutting Tools based on History Data. Procedia CIRP, Vol.63, pp 675 – 679, ISSN: 2212 – 8271

[9] Hronek, O. (2017). The Influence of the Cutting Tool Surface Quality on Durability, Diploma thesis, Department of Machining Technology, University of West Bohemia, Pilsen, Czech Republic

[10] Baksa, T., Schornik, V., Adamek, P. & Zetek, M. (2016). Machining of Inconel 718 Using Uncoated Cutting Tools with Different Cutting Edge Quality, Proccedings of the 27th DAAAM International Symposium, pp. 0441 – 0446, B. Katalinic (Ed.), Published by DAAAM International, ISBN 978-3-902734-08-2, ISSN 1726-9679, Vienna, Austria

DOI: 10.2507/28th.daaam.proceedings.144

THE INFLUENCES OF CUTTING EDGE RADIUS ON SURFACE ROUGHNESS WHEN MILLING NICKEL ALLOY

Ondřej Hronek & Miroslav Zetek

This Publication has to be referred as: Hronek, O[ndrej] & Zetek, M[iroslav] (2017). The Influences of Cutting Edge Radius on Surface Roughness when Milling Nickel Alloy, Proceedings of the 28th DAAAM International Symposium, pp.1037-1043, B. Katalinic (Ed.), Published by DAAAM International, ISBN 978-3-902734-11-2, ISSN 1726-9679, Vienna, Austria
DOI: 10.2507/28th.daaam.proceedings.144

Abstract

Currently one of the most important tasks of industrial enterprises is fulfilling the required quality of the products. This article is focused just on surface roughness, because product quality is an extensive concept. To obtain the desired product shape and dimensions, a cutting tool must remove the necessary volume of chips. The cutting tool and process conditions also affect the surface roughness during the machining process. The effect of the cutting edge radius on the roughness is studied during the experiment. The cutting tools have modified microgeometries. There are six cutting tools with cutting edge radii of 15 μm; 20 μm and 25 μm. The roughness measurement was performed using the non-contact method on an IGM G4.

Keywords: Cutting edge modification; Drag finishing; Immersion depth; Cutting tool microgeometry

1. Introduction

The nickel alloy Inconel 718 is often used for extremely heavy-duty products. Its properties such as excellent mechanical properties, heat resistance and toughness are the main reasons why it is used for products that are thermally and mechanically stressed. However, because of its high heat resistance, the abrasive effects of the chemical composition of the alloy and the self-hardening ability [1], these alloys are incorporated into the ISO S group, the group of materials which are difficult to machine. These alloys exhibit a high cutting resistance, moreover, there is a high proportion of heat transfer to the cutting tool. This is the reason for the research and development of machining nickel alloys, such as Inconel 718.

When adjusting the cutting tool microgeometry mechanical, or thermal and chemical methods to increase the durability, it is necessary to examine the effect of microgeometry modification on the surface quality. The goal is to modify the cutting edge, so that the cutting tool achieves the required durability. At the same time, the cutting process cannot have excessive vibrations, and the required roughness of the workpiece surface and shape accuracy must be maintained. Because of the frequent use of this material for stressed products, it is important to measure the residual stress on the surface layers. It has been found [2] [8] that machining of materials which are difficult to machine (ISO S) leads to different orientations and sizes of residual stresses depending on the machining environment.

Therefore, it is necessary to know and examine how the combination of the cutting process and the machining environment affects the residual stresses in the layers on and under the surface.

The final roughness is affected by the machining method, the machined material and its heat treatment, the cutting tool, the cutting conditions, etc. The cutting angles of the cutting tool have one of the biggest influences on the quality of the machined surface. Also the choice of the cutting conditions is also related to this. However, the system of surface quality – the cutting tool – machined material also influences the machining environment. Kumar et al. [3] investigated the influences of the machining environment on the roughness of nickel alloy Inconel 718. It was found that machining without cutting fluid negatively affects the surface roughness compared to the MQL (minimum quantity lubrication) and processing with cutting fluid. Currently, there are other types of machining environment that affect workpiece surface quality. Shokani et al. [4] focus on using MQL, cryogenic cooling and their combinations. In these environment types, the combination of MQL and cryogenic cooling the lowest roughness was achieved (Ra= 0.3 μm – 0.4 μm).

1.1 Roughness measurement

Currently, two methods are used for measuring roughness: contact and non-contact measurement. Both of these methods have their own advantages and disadvantages. The principle of contact measurement consists in the contact of a stylus with the measured surface. The measuring stylus moves over the measured surface at a constant speed. The stylus copies the shape of the surface. The stylus movement is recorded on the measuring software. The results of the roughness measurement are evaluated from this software. [5] This type of measurement has many advantages. It is simpler, quicker, an cheaper than non-contact roughness measurement. However, the disadvantage of contact measurement is the limited capability for measuring stylus on the complex components. In some cases the measured values are distorted. The measuring stylus does not accurately copy the shape of the measured profile of the measured area. Also, the tip of the measuring stylus is worn during measurement. This wear causes errors in the on measurement evaluation. [5] The non-contact method is the second type of roughness measurement. Measuring devices can work on the focus variation principle. In this case, they are the optical scanning microscopes (e.g. IFM G4). Non-contact roughness measurement provides the ability to measure the roughness on complex-shaped surfaces. For example, measuring the roughness on the flank and rake face of a cutting edge. It is also possible to evaluate the roughness of curved surfaces. The software tools for these microscopes allow far greater capabilities for working with the measured data.

2. Cutting Edge Modification

Monolithic end mill tools were used for the experiment. Sintered carbide was selected as the cutting material. To confirm or disprove the effect of cutting edge microgeometry on the surface roughness it was necessary to modify the cutting edge microgeometry. After grinding the cutting tools, the microgeometry was measured. An optical/scanning microscope was used for microgeometry measurement (Fig. 1. a)). The microscope is based on the focus variation principle. It is a non-contact microscope, suitable for cutting tools microgeometry measurement. The cutting edge radius ranged from 2 to 3 microns after grinding. Cutting edge modification followed after measurement. A drag finishing machine was used for cutting edge modification (Fig. 2. b)). The theoretical values of the cutting edge radius were selected: 15 μm; 20 μm and 25 μm. Six tools were used in the experiment, so two end mill tools were selected for each cutting edge modification.

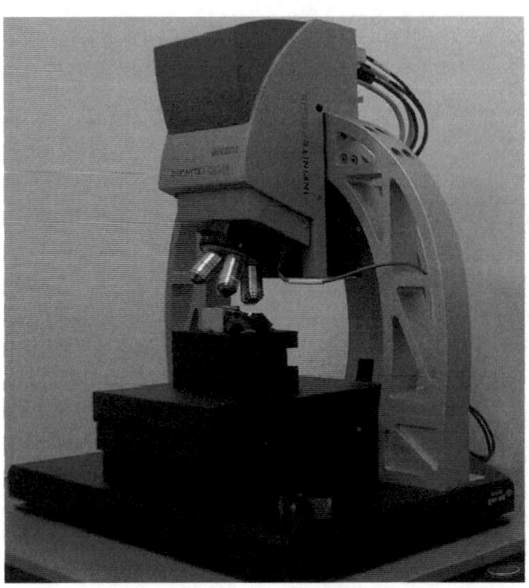

Fig. 1. a) IFM G4, b) Drag finishing machine

The drag finishing process parameters used to achieve a proper cutting edge radius are shown in Table 1. At the same time, the cutting edge before and after modification is compared using differential analysis software. In each analysis, the influence of different cutting edge radius values is made visible (by a colour spectrum). From these analyses, it can be seen that the cutting edges have no defects on the cutting edge.

After measurement, the coating process followed. Schornik V. [6] confirmed in a practical experiment that the chemical composition and the thickness of the thin layer influences the durability and cutting forces when milling nickel alloy Inconel 718. Therefore, it is important to pay attention to the choice of the proper thin layer for machining Inconel 718. It is necessary to select a thin layer with properties such as: hardness, heat resistance, thin, with a low friction coefficient, etc. A thin layer called Triple Coating Cr was selected for this experiment. The properties of this thin layer are suitable for materials which are difficult to machine.

Modification	Modification parameters	Differential analysis
Variant 1 (Cutting tool 1; 2)	Cutting edge radius: 15µm Process time: 6 min Media: HSC 1/300	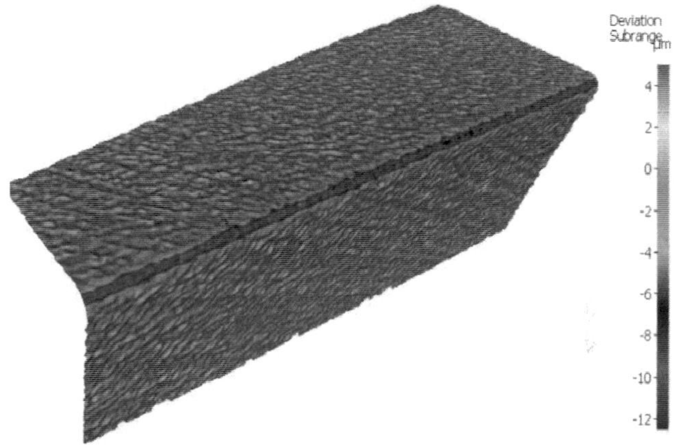
Variant 2 (Cutting tool 3; 4)	Cutting edge radius: 20µm Process time: 6 min Media: HSC 1/300	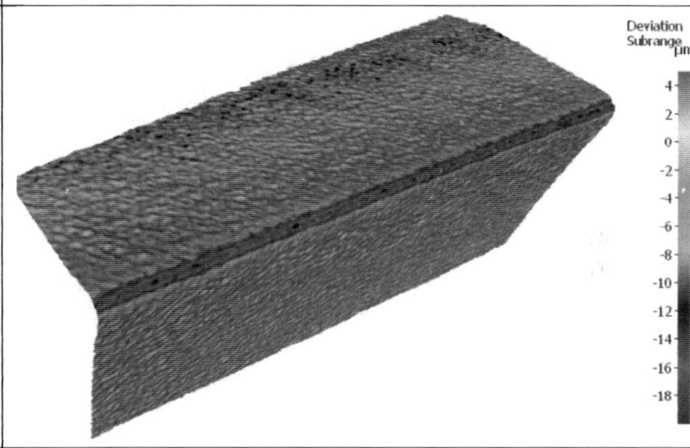
Variant 3 (Cutting tool 5; 6)	Cutting edge radius: 25µm Process time: 23 min Media: HSC 1/300	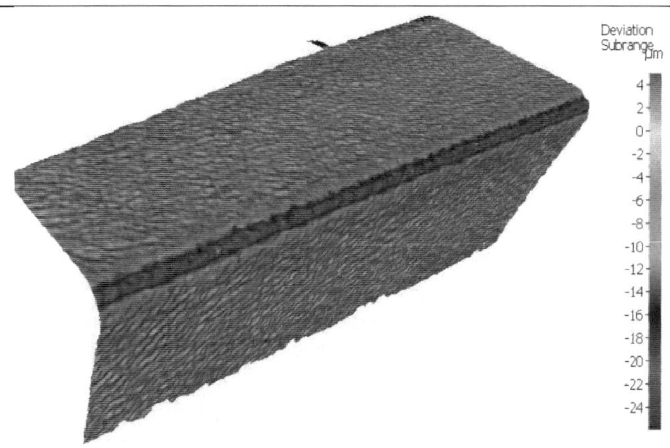

Table 1. Variants of cutting edge modification

3. Experiment

In the first step, the process parameters for milling were selected. The machined material had the biggest influence on the choice of these parameters (see Table. 2). The Inconel 718 has properties such as high temperature strength, self-hardening by plastic deformation, hardness and vulnerability to abrasive tool wear. It is necessary to select appropriate cutting conditions that will increase the cutting tool life and also stabilize the cutting process. [3] [4] [6] Process parameters are also based on experience gained from previous tests of machining Inconel 718. Table 2. Shows the process parameters used. Kroft L. [7] focused on the influence of step-over on the cutting force Fy . In this article it was found that the lower step-over [mm] causes a reduction of force Fy. In the case of milling, the step-over value is equal to a_e (cutting width). For this experiment, cutting width was 0.5 mm.

v_c [m/min]	f_z [mm]	a_p [mm]	a_e [mm]	Milling type	Machining environment
35	0.04	3	0.5	Down Milling	Cutting fluid

Table 2. Process parameters

The MCV 750A three-axis milling centre used for the experiment. Cutting tools were clamped into a polygonal tool holder which is characterized by appropriate stiffness and a solid radial runout. This will make the system Cutting tool-Holder more rigid than to the frequently used collet clamp. The workpiece was clamped into a chuck. Each cutting tool has its own workpiece. It was necessary to use six workpieces for surface roughness evaluation. The experiment consisted of several simple steps. Once the cutting tool and workpiece clamped are, the cutting tool makes one cross. After that, the cutting tool and workpiece are replaced by another tool and workpiece. By that simple principle, each of the six cutting tools were tested. As mentioned earlier, the cutting environment also has an impact on the surface roughness during machining. In this case, a process media was used during machining for cooling the cutting process.

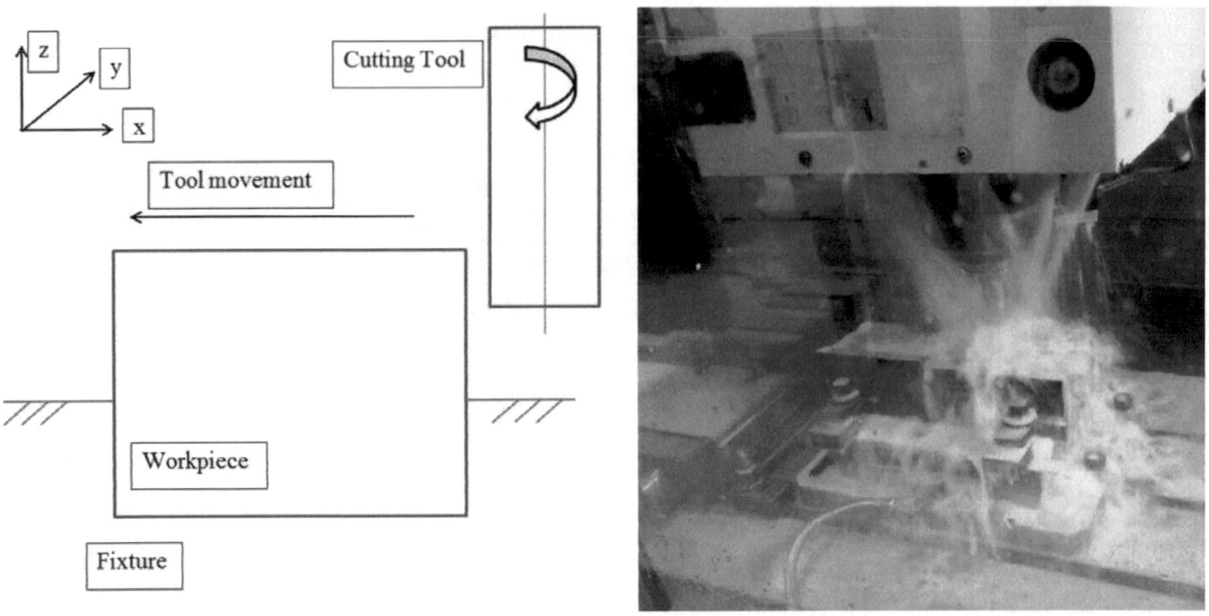

Fig. 2. a) Cutting process diagram, b) Real cutting process

4. Results and Discussion

After the experiment was completed, the machined surface was measured. An IFM G4 microscope was used for the roughness measurement. In addition to measuring shape and size, the advantage of this microscope is its ability to measure and evaluate surface roughness. The 50x objective was selected to measure the machined roughness. Also, it is necessary to select suitable lateral and vertical resolution, contrast and lighting settings. These set up parameters remains constant for each of the six workpiece surface measurements. Fig. 3. a) shows the surface on which the roughness was measured and evaluated. This area was created by the cylindrical part of the cutting tool.

The roughness measurement was carried out on an area 1.5 x 1.5 mm. However, it was not yet possible to measure the surface roughness of the measured surface. An area created by the microscope showed waviness, so it was not possible to measure roughness on this type of surface. Therefore, the next step was to remove the waviness, because the waviness affects the values of the roughness measurement.

The marks from the cutting tool can be seen on the measured surface (Fig. 3.b)). These marks are visible on all of the workpieces. These marks are created by elastic/plastic deformation, by cutting forces and material properties. During measurement it was found that the distance between the track approximately corresponds to the a feed value f_z= 0.04 mm.

Fig. 3. a) Machined surface, b) Detailed view of the machined surface by IFM G4

Roughness measurement and evaluation is efficiently carried out using an IFM G4 microscope. It is possible to measure linear and surface roughness. The main roughness parameters were selected for evaluation. Two of these three parameters are linear Ra and Rz parameters. The last parameter to be evaluated is the surface roughness Sa. Table 3. shows the measured values on the surface created by tool number 1. The same measurement method was used for each workpiece. The roughness was measured perpendicular to the tracks of the cutting tool. Five measurements were made on each workpiece (Table. 3). The arithmetic mean (x) was calculated from each of these measurements

Cutting tool 1 – roughness parameter Ra measurement		
Roughness measurement	**x_i [μm]**	**$\Delta_i = x_i - x$ [μm]**
1	0.278	-0.0756
2	0.322	-0.0316
3	0.288	-0.0656
4	0.490	0.1364
5	0.390	0.0364
x	**0.3536**	**$\Sigma\Delta_i = 0$**

Table 3. Evaluation of measured values

4.1 Results

This section deals with the evaluation of the measured values. Fig. 4. Shows the measured roughness values (Ra; Rz; Sa) depending on the cutting edge radius. It has been found, that the cutting edge radius affects the roughness of the machined surface during the milling process, with constant process parameters. The lowest linear and surface roughness of the workpiece were made by cutting tools with r_n= 15 μm. Although the cutting edge radius is created on these tools (Variant 1), the cutting edge is still sufficiently sharp for easy chip creation during the machining of Inconel 718. Workpiece properties such as high hardness and toughness negatively affect the workpiece roughness during milling when using cutting tools with r_n= 25 μm. Due to the higher cutting edge radius, there is an energetically more complex process of chip formation. In addition, higher r_n negatively affects plastic deformation and the final surface roughness.

The roughness parameter Ra ranges from 0.343 μm to 0.688 μm. It is possible to say that all of the tested cutting tools with different microgeometries have created high quality workpiece surfaces. However, modern quality evaluation uses more roughness parameters for the evaluation of surface roughness, and the results are more credible and also the conclusions are more accurate. Another evaluated parameter was the largest profile height (Rz). Even in this case, the highest roughness values of Rz were obtained with Variant 3, r_n= 25 μm. Cutting tools with cutting edges of 15 μm and 20 μm have similar roughness Rz. But compared to the tool with r_n= 25 μm, there is a significant difference in the surface roughness evaluated by parameter Rz. This is related to the fact that a cutting tool with a higher cutting edge uses higher energy when cutting the chip during the milling process. Also, this energy can cause increased vibration and cutting forces. These effects lead to a deterioration of the surface roughness.

The last and final measured parameter was the arithmetical mean height Sa. This is an extension of the Ra parameter on the surface area. The result of the evaluation of roughness Sa give us the same dependence as the Ra evaluation. So, cutting tools with lower cutting edge radius improve the workpiece surface. The surface roughness Sa ranges from 0.522 μm to 0.787 μm.

It is necessary to realize that many parameters influence non-contact measurement. These parameters could negatively affect the final measured values. Parameters which could affect the measurements are: objective selection, lateral and vertical distinction, contrast, brightness, waviness removal, etc.

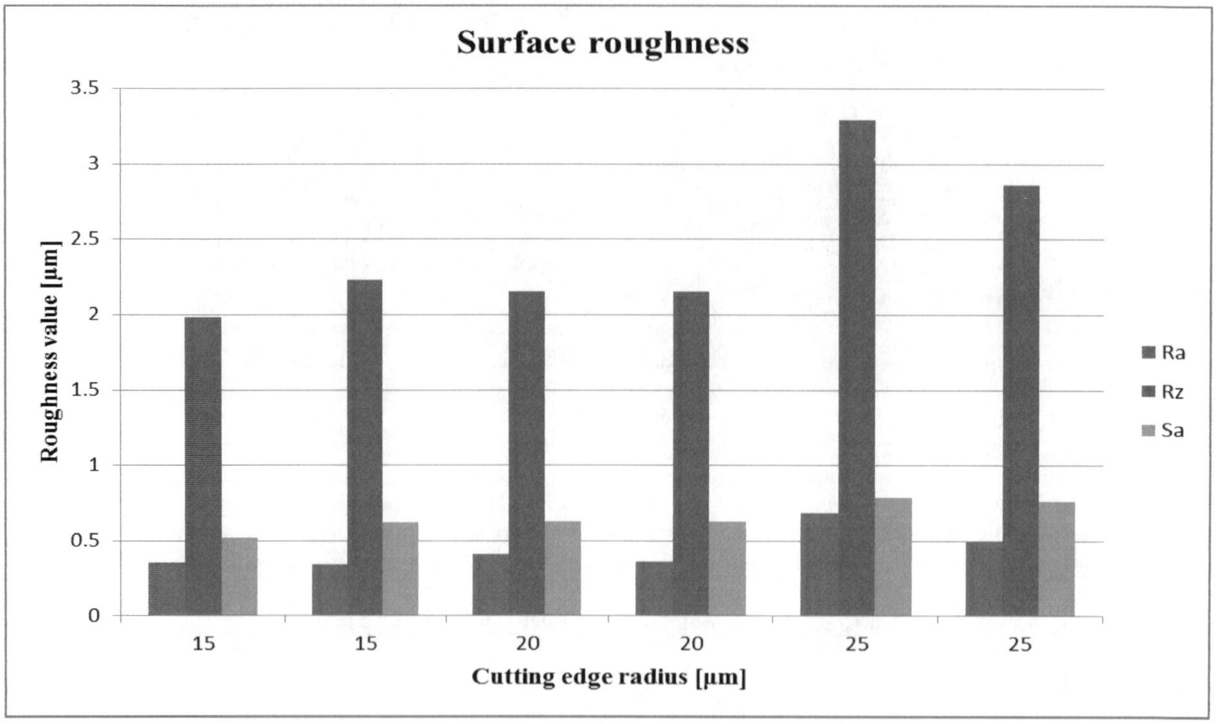

Fig. 4. Influences of r_n on workpiece roughness

5. Conclusion

This article deals with the influences of the cutting edge radius on the workpiece surface roughness. Three variants of cutting edge radius (15 μm; 20 μm and 25 μm) were tested during the experiment. Process parameters were constant during the milling process. Nickel alloy Inconel 718 was the machined material.

The experiment shows the dependence between the cutting tool radius and the roughness of the workpiece. After the practical part and evaluation, the following conclusions were reached:

a) Higher cutting edge radius causes an increase of roughness parameters Ra; Rz; Sa

b) The largest difference in Ra parameter was between the cutting tools of Variant 1 and Variant 3. The cutting tool with a cutting edge radius of 15 μm had a roughness Ra= 0.343 μm while the cutting tool with r_n= 25 μm had a roughness of Ra= 0.688 μm.

c) The largest profile height Rz was the second roughness parameter that was evaluated. In this case, the same influence between the cutting edge radius and workpiece roughness was detected. This parameter revealed greater differences between the highest profile height. The cutting tool with r_n= 15 μm had a roughness of Rz= 1.982 μm, while the cutting tool with r_n= 25 μm had a much higher roughness; Rz= 3.292 μm.

d) The surface roughness parameter Sa was also used for evaluation. The dependence between the cutting edge radius and the workpiece roughness was equal to the parameters Ra and Rz.

It is important to say that the impact of the cutting tool on the surface quality cannot be considered separately. Therefore, even in this experiment, it is necessary to realize that the machine, the cutting tool, the workpiece, and the workpiece holder influence the roughness and geometrical accuracy.

In the future, the influences of the microgeometry on surface quality will be investigated further. The aim of these experiments is to increase the quality of the workpiece surface. Moreover, these experiments will find the relationships between cutting tool microgeometry, process parameters, the machined material and the surface roughness. The aim is also to increase the quality of cutting tools.

6. Acknowledgments

The present contribution has been prepared under project LO1502 'Development of the Regional Technological Institute' under the auspices of the National Sustainability Programme I of the Ministry of Education of the Czech Republic aimed at supporting research, experimental development and innovation.

7. References

[1] Lopéz de lacalle, L. N., Peréz, J., Llorente, J. I., Sánchez, J. A. (2000). Advanced cutting conditions for the milling of aeronautical alloys, Journal of Materials Processing Technology, Vol. 100, 2000, pp. 1 – 11, ISSN: 0924-0136

[2] Malý, J., Zeman, P. & Mádl. J. (2009). Effect of Cutting Media and Cutting Conditions on Temperature and Residual Stress of Machined Surface at Milling of Difficult-to-cut Alloy Ti6Al4V, [online]. Available from: https://otik.uk.zcu.cz/xmlui/bitstream/handle/11025/16437/Maly.pdf?sequence=1&isAllowed=y, [Accessed: 2017-10-16]

[3] Kumar, S., Singh, D. & Nirmal, S. K. (2017). Experimental Investigations of Surface Roughness of Inconel 718 under different Machining Conditions, Materials today: PROCEEDING, Vol. 4, 2017, pp. 1179 – 1185, ISSN: 2214-7853

[4] Shokrani, A., Dhokia, V. & Newman, S. T. (2017). Hybrid Cooling and Lubricating Technology for CNC Milling of Inconel 718 Nickel Alloy. Procedia Manufacturing. Vol. 11, 2017, pp. 625 – 632, ISSN: 2351-9789

[5] Stout., K. J. & et al. (2004). Development of Method for the Characterisation of Roughness in Three Dimensions, Penton Press and contributors 2000, ISBN 1-8571-8023-2, London N1 9JN, UK

[6] Schornik, V., Zetek, M. & Baksa, T. (2016). Durability of the Cutting Tool with Different Thin Layers when Inconel 718 is Machined, Proccedings of the 27th DAAAM International Symposium, pp. 0678 – 0682, B. Katalinic (Ed.), Published by DAAAM International, ISBN 978-3-902734-08-2, ISSN 1726-9679, Vienna, Austria

[7] Kroft, L. (2016). The Influence of the Finishing Strategy on the Quality of the Surface, Proccedings of the 27th DAAAM International Symposium, pp. 0524 – 0533, B. Katalinic (Ed.), Published by DAAAM International, ISBN 978-3-902734-08-2, ISSN 1726-9679, Vienna, Austria

[8] Wang, J., Zhang, D., Wu, B., Luo, M. (2017). Numerical and Empirical Modelling of Machining-induced Residual Stresses in Ball end Milling of Inconel 718. Procedia CIRP. Vol. 58, 2017, pp. 7 – 12, ISSN: 2212-8271

DOI: 10.2507/28th.daaam.proceedings.145

DEMOGRAPHIC ENERGY-BALANCED FRAMEWORK FOR SUSTAINABLE URBAN ENVIRONMENTS

Tobias Teich, Mirko Bodach, Matthias Hoffmann, Tim Neumann, Philipp Werler, Tobias Härtel, Sebastian Junghans, Christian Blumhagen

This Publication has to be referred as: Teich, T[obias]; Bodach, M[irko]; Hoffman, M[atthias]; Neumann, T[im]; Werler, P[hilipp] & Hartel, T[obias] (2017). Demographic Energy-balanced Framework for Sustainable Urban Environments, Proceedings of the 28th DAAAM International Symposium, pp.1044-1050, B. Katalinic (Ed.), Published by DAAAM International, ISBN 978-3-902734-11-2, ISSN 1726-9679, Vienna, Austria
DOI: 10.2507/28th.daaam.proceedings.145

Abstract

The social-political tendency in Germany clearly shows that the integration of issues that ensure affordability and quality of life, such as the use of intelligent options, taking into account demographic aspects, are necessary for sustainable development in order to ensure the viability of housing. The starting point is defined in the problem of the hitherto separated view on the areas of human, health and energy with the demographic change of the people in a district.
The goal of the project is therefore to enable a demographically oriented operationalization of these concepts for the attainability of a "sustainable livelihood" through the integrated and multidisciplinary consideration of the existing developments. The main objectives are: to develop a generic networking model for the integration of the partial models of an original independent competence cluster; to use previously unused potentials from the currently missing integration of different perspectives. A demonstrable improvement of the energetic situation is to be achieved by a systemic integration. In particular, the impact of demographic effects on people as tenants as well as integratable solutions with regard to technological infrastructure and feedback possibilities on thermal, electrical and information technology components are analysed.

Keywords: networking; ubiquitous; sustainability; integration; demography

1. Introduction

To enable a demographic-oriented operationalization of concepts for the accessibility of a Sustainable Way of Living through the integrated and multidisciplinary consideration of previous developments identified gaps should be concluded with the project in the form of an operationalizable framework (figure 1). Living or "housing" in a building is a basic need of people in a modern civilisation and the providing of housing is a fundamental right. However, the financing of these essential factors must still be possible in 20 to 30 years. The urban environment, with its neighbourhoods of buildings and infrastructure, is one of the most important drivers of climate change. At the same time it forms the basis for a sustainable life. (see further in [1])

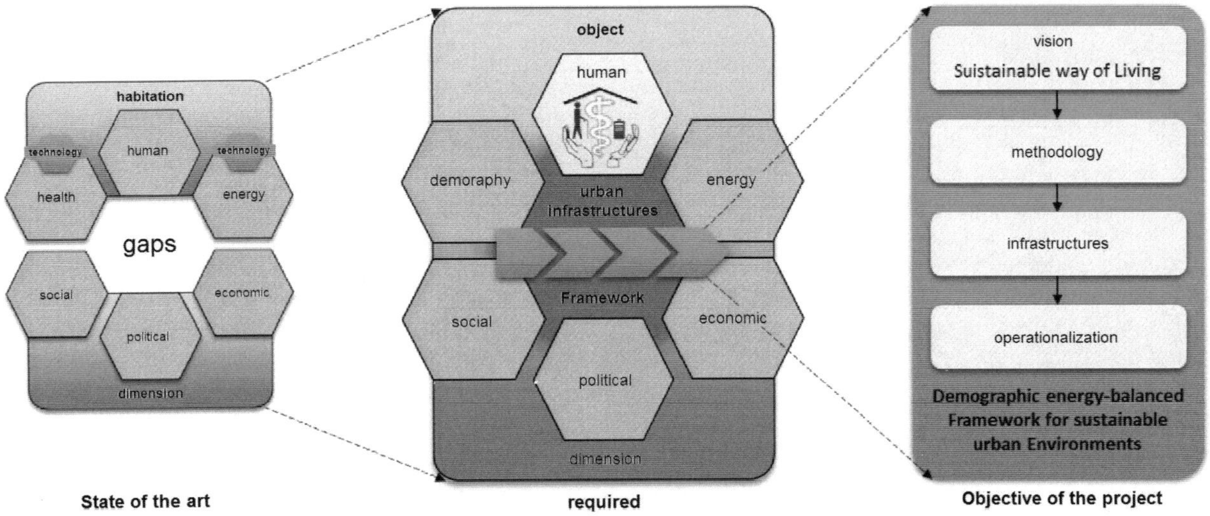

Fig. 1. Objective of the project

The demographic development and its serious impacts are of central importance for future, generational concepts of self-determined life, especially in priority regions such as Zwickau (see figure 2). The Zwickau region is a demographic focus region in nationwide comparison. Due to the ongoing demographic change and the aging society, resulting problems will also affect other municipalities. The resulting demographic stress field is increasingly aggravated by the negative development of the relationship between persons making the supply and benefit recipients in the socio-economic context (persons with a social security obligation / the recipient of social benefits, and in particular pension payments). The ratio is currently about three to one [2]. It will deteriorate drastically because of the demographic development. In less than 20 years it will be about one to one [3]. This imbalance consequently leads to a reduction in the income of pensioners.

In the same period there is a development of pensions in eastern Germany from an average of EUR 879 to 530 which shows how dramatically the situation of affordable rents, health and energy costs will develop [4]. Thus, the old age poverty is one of the greatest risks of the future for the region. This illustrates the enormous societal importance of the issue for Germany and manifests the demand situation for the problem beyond the region of Zwickau, too [5]. From the global climatic as well as the regional demographic and therefore necessary infrastructural change results not only the demand for partial and complete restructuring of the health service and in particular the outpatient care, but also the technical approaches in thermal and electrotechnical supply systems.

A Sustainable Way of Living can only be achieved with the help of novel concepts and systems, which are defined and planned at district level. The personal requirements of the residents for individual care cause an increasingly frequented form of media supply in the fields of energy and health care. These requirements must be countered with a redefinition of application concepts for technical building equipment and integration across hierarchical levels. At the moment, there is a steady development and optimization of partial models of individual competence clusters, e.g. care in medicine, medicine and energy supply and ambient-based life.

However, the transfer into a generic networking model with holistic methods for the integration of these competence clusters has been completely missing (figure 3). In this way, only isolated solutions result, although the objective of integration has been formulated in many applications. Roll-outs are practically not observable after the end of the funding period due to unproven profitability. The necessary realization of an interdisciplinary interaction and functional integration fails due to lack of integration or interoperability interest or restricted possibilities of the different branches, manufacturers or products. As a university of applied sciences, the connection between research and teaching is essential for the Westsächsische Hochschule Zwickau (WHZ).

The development of the student enrolments and, in particular, the notable negative trend of the MINT quota (Mathematics, computer science, natural sciences and technology) require new and innovative study contents for a successful continuation of the university landscape. Particular attention must be paid to the interdisciplinary combination of different technical fields currently required by the market. Energy efficiency and health as cross-cutting issues offer such a possibility.

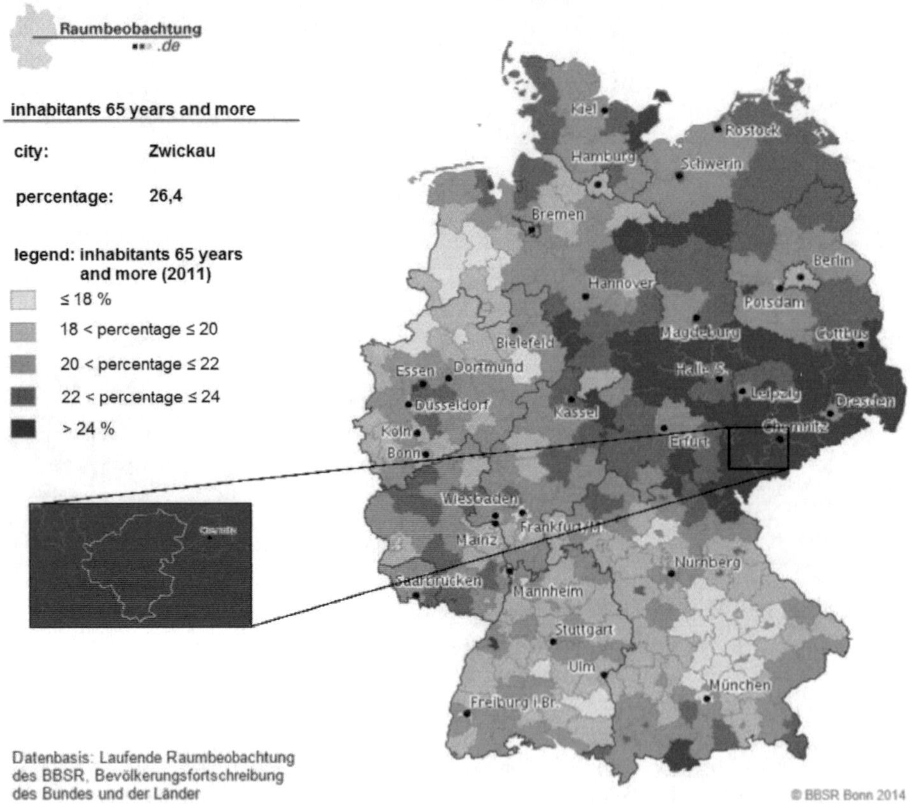

Fig. 2. age structure of Germany in 2011

2. Specific objectives of the project

A generic network model is to be developed for the integration of the partial models of an original, independent competence cluster in order to use previously unused potentials resulting from the currently missing integration of different perspectives. With the help of the project results a demonstrable improvement of the energetic situation is to be achieved by systemic integration. In particular, the influence of demographic effects on people as tenants as well as integrative solutions with regard to technological infrastructure and feedback possibilities on thermal, electrical and information technology components are analyzed. This approach is clearly distinguished from the widespread singular technical approaches by creating superior technical procedures and integrating networks on the infrastructural, energetic and organizational level. This results in the possibility to synthesise novel, cross-sectoral potentials for increasing the efficiency in networked systems - both for people as tenants in their homes, as well as for the lessors, such as thermal and electrical supply at neighbourhood levels. The project is supported by a competence center for energetic and health care services, which was financed by the city of Zwickau and the Sächsische Aufbaubank on the basis of the results of the previous research groups and was built in 2015 and 2016. It is used to transfer results in the direction of large-scale tenants, healthcare providers and, in particular, citizens as an information and counseling center.

3. Preliminary work

Various completed research projects formed the methodological and organizational basis for the intended project. The experiences of these projects, the cooperation of the project management in the panel on climate issues of the city of Zwickau as well as in the Advisory Board "AlterLeben" of the Association of Saxon Housing Cooperatives (Wohnungsbaugenossenschaften e.V. - VSWG) and numerous other preparatory works clearly show that in the future only cooperative structures as well as interoperability and intersectoral business processes including the required Systems can lead to comprehensive and sustainable solutions. (see in [6] and [7]) The choice of the cooperating partners involved in the project illustrates the regional strength and the user-friendliness of the applicant's past projects in the research area. In the project described here, this existing network is to be used and expanded accordingly. In the "Low Energy Living" project (sponsored with 1.7 million euro by the European Social Fund, ESF) the possibilities for sustainable energy saving were examined using intelligent building system technology in the practical environment of municipal housing construction. (see further in [8])

The main focus and the results to be expected are listed below:

main focus	objective / result
Framework zum Sustainable Way of Living	Scientific framework for the operationalization of a generic networking model for the realization of an adaptive network with the purpose of securing the Sustainable Way of Living
ecoBSC	Instrument for large-scale tenants to support the (investment) decisions of sustainable neighbourhood development within an integrated urban development concept as well as the assessment of the energy and demographic situation of jointly supplied neighbourhoods and the derivation of relevant measures
ADL in the district / neighbourhood	Quartier-oriented further development of the sensor-assisted detection of relevant activities of daily life (ADL) taking into account demographic aspects as well as billing event-driven health services
Multidisciplinary interoperability	Model for multidisciplinary information management as well as integration and enhancement of the interoperability of sensor networks in residential districts to the interdisciplinary and manufacturer-independent information exchange
energy accounting	New billing models for the active energy consumption awareness of tenants with a transparent representation of the tenant's potential for room regulation in the network
planning methods	Appropriate technical planning approaches to take account of the future requirements for an economic and sustainable development of districts
Modular swarm-intelligent network nodes	Involving and intelligently linking decentralized heat storage possibilities and heat generators as well as increasing energy efficiency with the help of novel heat distribution variants by using swarm-intelligent network nodes
network validation	Generation of forecast algorithms for efficient and domain-spanning energy management as a result of established requirements profiles taking into account demographic aspects in the district / neighbourhood
Transient simulation	Possibility of control energy participation (consumption and supply), more precise dimensioning of a modular energy storage device, use of the stationary network simulation program Cerberus for transient simulation
scenarios of memory usage	Efficient operation of the energy storage by taking into account the interactions of all disciplines and analysis of power-to-heat concepts, determination of the most economical energy storage, possible combinations and the application method
Qualification of young scientists	Transfer of knowledge and results into the teaching concept and content of a MINT Master's course "Intelligent Building Systems" to increase the MINT rate at the Westsächsische Hochschule Zwickau (MINT … Mathematics, computer science, natural sciences and technology)

Table 1. Focus and results

The project was successfully completed in October 2012. In Zwickau as well as numerous other cities in Saxony the concept was implemented and achieved a reduction of more than 20 percent primary energy. This led to the definition of further future-oriented research topics. On the one hand, this is the subject of the development and implementation of value-added services on the basis of ubiquitously available and intelligently networked information (ubiquitous infrastructures). For this purpose research projects focussed on health and care management, especially in the Ambient Assisted Living (AAL), are being carried out at the Westsächsische Hochschule Zwickau. On the other hand, energetic potentials were exploited with the use of intelligent building equipment combined with a holistic, inter-branch information infrastructure. The analysis of the energy supply of a district showed tremendous potential for optimization, which could be opened up by networking several different levels of information. Information levels can be, for example, rooms, apartments, entrances, buildings, quarters. Investigations in cooperation with the Brunata-Metrona Group (German market leader for the consumption-based heating and water cost accounting) provide a clear potential for savings through intelligent networking of ubiquitous information of more than 20% in multi-family houses. (see further in [9])

The project "Ambient Assisted Living in Intelligent Controlled Environments" (A^2LICE), which was also funded by the European Social Fund (ESF) with 1.1 million euros, represents the second major line of the project, starting with the results of the Low Energy Living project. The focus of this project was the development of a concept for closing the gap in the medical care chain between in-patient care and self-determined life in the usual environment, leading to a significant improvement in health and care management, self-determination of the elderly and / or handicapped people, in the field of housing construction as well as in commercial and public buildings. The results of the research work so far and the demands resulting from demographic change show that demography and energetics already form a closely interwoven paradigm, which necessarily requires innovative and holistic solution concepts (figure 1).

Within these two ESF projects as well as in further research work, essential foundations were created in the areas of demography and energy. This can also be demonstrated by further projects which were supported, for example, by the SMWK (Sächsisches Staatsministerium für Wissenschaft und Kunst, engl.: Saxon State Ministry for Science and the Arts): "Combined use of technical and economic instruments to increase energy efficiency in buildings" (March 2009 – December 2010), "Ambient Assisted Living in urban house building to overcome demographic problems in health care" (March 2011 – December 2011) or "Ubiquitous infrastructures" (January 2013 – December 2014). The future interdisciplinary consideration of the topic of energy efficiency at the level of energy-assisted shared districts is already taken into account when creating so-called district concepts (KfW432) for the cities of Zwickau, Borna, Riesa, Netzschkau or Oelsnitz/Erzgebirge.

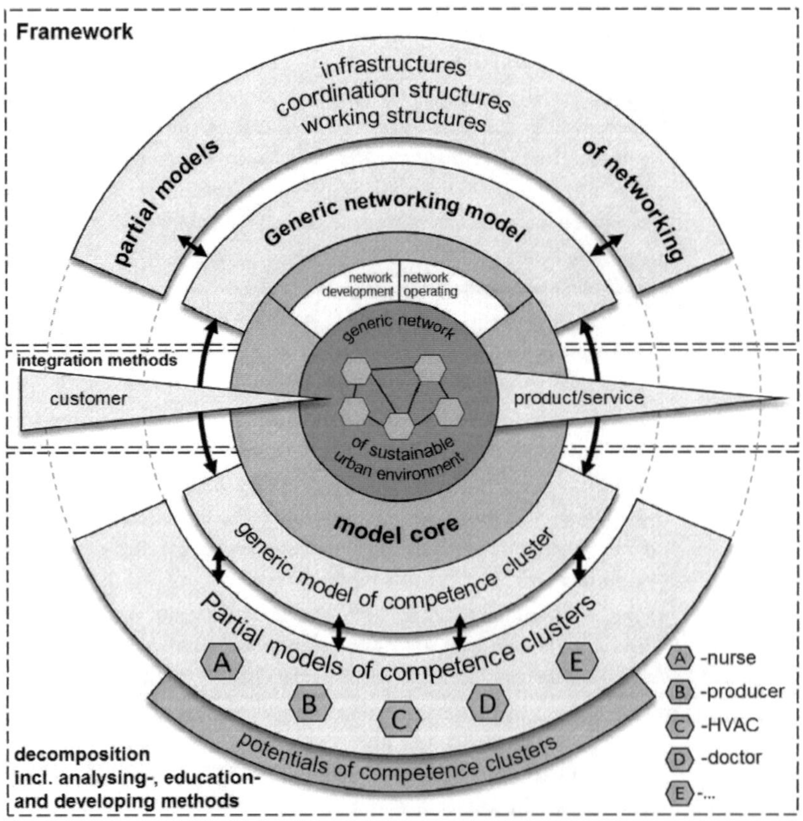

Fig. 3. Presentation of the project objective

Some research work resulted in the publication of patents, in order to provide a basis for the realization of future projects. These include: "Device arrangement for recording, evaluating and influencing heating energy distributions within a building envelope" and "Ubiquitous sensor networking with methods for controlling the heating energy output from local to regional living space structures". In addition to these research activities, which are mainly based on the content, the qualification of young scientists is also promoted. With the help of the co-operative doctoral procedure, scientists have already been successfully supported for their doctorate. These former scientific staff members of the WHZ are in leading positions in the Saxon economy today and strengthen the region of Saxony.

4. Achievement of objectives and applied operations

The topics of the project are accompanied by three professorships at the Westsächsische Hochschule Zwickau. The professorship for networked systems in business management is assumed by the project management as well as the content of the demographically motivated subproblems. The professorship for thermal engineering / computer-based planning takes over the part of the thermal energy efficiency of a holistic view. The use of potential synergy effects of electrical energy efficiency is pushed forward by the professorship for electrical energy technology / regenerative energies. All professorships are characterized by preparatory work on the respective subject areas and can thus fall back on the necessary technological basics. The transfer of knowledge and results takes place both through the teaching on the training of highly innovative graduates and on the qualification of service providers of the Saxon economy within the emerging competence center. The research profile of the Westsächsische Hochschule Zwickau is sustained and strengthened externally.

5. Conclusion

A central element in the publication and documentation of the results of the project, already mentioned in the first and second chapters, is the parallel construction of a new competence and demonstration center, which is economically independent from the project, but is interlinked with the project. To this end, an important step was taken in the Zwickau region four years ago with the founding of the Intelligente Infrastruktur Zwickau mbH. This created a political unique selling proposition feature. Large-scale tenants and the local energy provider put their interests back a bit, working together to find solutions to the major challenges faced by demographic trends in the future. With a competence center for "Affordable living in old age" as well as the development of new business processes for other economic stakeholders, the city of Zwickau is to become a leadership example for other cities in Germany, but especially in Saxony. The consideration of the topics of residential, health and energy supply costs happens interdisciplinary by different groups of interest. One of these interest groups is the Westsächsische Hochschule Zwickau, which is an important component of the transfer of knowledge and results in the research profile of energy efficiency. Thus, the sustainable use and documentation of the project results for the Zwickau region and its inhabitants can be ensured over the project duration. At the same time, the competence center plays an important role as a multiplier for supraregional actors in Saxony and beyond. Furthermore, the center of competence is the spatial basis for a possible spin-off of some of the scientists. It is also planned to publish the results of the project on a common internet platform in order to make the results available to a broad public [10]. At the end of the project, the project results are also planned to be presented in a book.

The reference to the principles of environmental protection and resource protection has already been clarified in the previous chapters of this paper. Furthermore, the integration of demographic aspects in the development of the social innovation content of the project is enhanced. In addition there is a close co-operation with the "Social Affairs and Projects" department of the Verband sächsische Wohnungsbaugenossenschaften e.V. (Association of Saxon Housing Co-operatives), which analyzes the compatibility of the results for tenants and landlords. The consideration of so-called "ethical, legal and social implications" - also called ELSI aspects - makes an important contribution to the safeguarding of basic questions of equality, equal opportunities and non-discrimination as well as in particular data protection and safeguarding the personality rights in this sensitive environment of the tenants. For this purpose, a subcontractor is provided for monitoring the project with regard to these questions after the second and last project year. This has already been done in a similar way in previous research projects. In addition to the aforementioned cooperation in the course of the project, the planned project is supported by other stakeholders. Both the panel on climate issues of the city of Zwickau as well as the headquarters urban development of the Lord Mayor of Zwickau are involved. Last but not least, the project is also supported by the local energy supplier, Zwickauer Energieversorgung GmbH (ZEV).

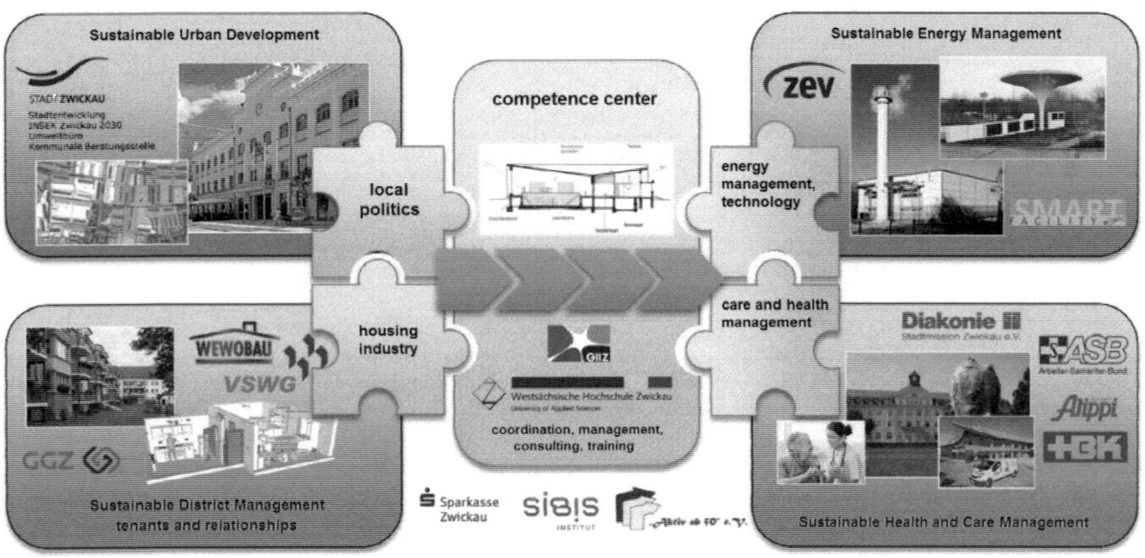

Fig. 4. Research and demonstration center

6. Acknowledgments

In the framework of the project, young scientists are primarily supported by the corresponding professors. During individual weekly workshops, current work progress is directly discussed and thus evaluated at all times during the project process. The professors ensure a correspondingly necessary cooperation, which is also reflected in the project calculation. In the regularly planned milestones, the entire research group meets for an interdisciplinary exchange of results together with representatives of the municipal workgroups of related topics.

A number of cooperating practice partners are also taking part in these meetings. In doing so, particular attention is paid to existing and long-term successful cooperations with companies in the commercial sector or associations. This network is to be expanded by relevant partners within the framework of the project. As a result, several partners can contribute to project evaluation and profit from research results as well as the socially competent and professionally advanced employees of the research group. The influence of the cooperating partners will evaluate the results directly in terms of practicality and thus strengthen the application-oriented research of the young scientists. At the same time, these milestones represent a platform for the presentation of young scientists. This is an important component of personal development. In addition to this, the young researchers are themselves encouraged to ensure quality in research by ensuring transparency, application orientation and sustainability in the research work. The inclusion of tenants 'and senior citizens' representations has already provided a basis for ensuring the quality of the research work of young scholars and is also to be used in joint discussions during this project. As a further result of the preparatory work, the research group can draw on the knowledge and experience as well as the network of the municipal advisory center "Better life in old age through technology", a collaborative project between the city of Zwickau and thus further ensure the quality of the research results. After all we would like to thank all the partners who are part of the success of the project results for their commitment.

7. References

[1] van Bueren, E. M.; van Bohemen, H.; Itard, L. (2012); Visscher, H. (Eds.): Sustainable Urban Environments; Springer Verlag; ISBN 978-94-007-1294-2

[2] Raffelhüschen, B.; Müller Ch. (July 2011): Demografischer Wandel: Künftige Handlungsoptionen für die deutsche Rentenpolitik; Forschungsinstitut zur Zukunft der Arbeit; p. 4

[3] https://www.iwkoeln.de/themen/finanz-und-sozialpolitik/rentenversicherung; Accessed on: 2017-09-07

[4] Geyer, J.; Steiner V. (November 2010): Künftige Altersrenten in Deutschland: Relative Stabilität im Westen, starker Rückgang im Osten; Deutsches Institut für Wirtschaftsforschung; Wochenbericht; pp. 3 – 11

[5] IfD-Allensbach (2014), Kapitel „Von Altersarmut bis Verkehrsunfall", p. 267; Forschung und Lehre; Deutscher Hochschulverband; Ausgabe 4, 21. Jahrgang; Bonn.

[6] T. Teich, S. Junghans, D. Kretz and M. Trommer (2016): "Non-Hierarchical Services Networks, steered by Competence Cells". B. Katalinic ed. DAAAM internati-onal Publishing, pp. 453.

[7] T. Teich, S. Wolf, S. Junghans, T. Neumann, S. Franke (2013): Concept for a service-oriented architecture in building automation systems. 24th DAAAM Interna-tional Symposium on Intelligent Manufacturing and Automation

[8] Teich, T.; Igel, W. (2010): Energieeffizienz in Wohngebäuden; GUC-Verlag; ISBN 978-3-934235-89-2

[9] Gentemann, R. B.; Schröder, F.; Teich T. (August 2012): Smart Metering belegt Energieeinsparung - Geniax System im Mietwohnungsbau; in: HLH Bd. 63 Nr. 8, Springer-VDI-Verlag, Düsseldorf, pp. 83-87

[10] http://ubineum.de/home.html; ; Accessed on: 2017-09-12

DOI: 10.2507/28th.daaam.proceedings.146

BLOCKCHAIN AND SUPPLY CHAIN MANAGEMENT: AIRCRAFTS' PARTS' BUSINESS CASE

Yash Madhwal & Peter B. Panfilov

This Publication has to be referred as: Madhwal, Y[ash] & Panfilov, P[eter] (2017). Blockchain And Supply Chain Management: Aircrafts' Parts' Business Case, Proceedings of the 28th DAAAM International Symposium, pp.1051-1056, B. Katalinic (Ed.), Published by DAAAM International, ISBN 978-3-902734-11-2, ISSN 1726-9679, Vienna, Austria
DOI: 10.2507/28th.daaam.proceedings.146

Abstract

To serve target customers better than their competitors, supply chain management (SCM) teams today look into new technologies such as Big Data, Internet of Things (IoT) and Blockchain. These new technologies allow managers to develop and provide complex supply chain services and products faster with improved reliabilities. With these technologies, SCM teams can build complex models of a supply chain or systems of supply chains using a data-driven approach. With the growth of aviation domain across the world, there has been increasing demand in aircraft for airlines and other customers. In this domain, SCM teams deal with complex networked supply chains for aircraft's spare part purchase and delivery for aircraft's maintenance and repair. Aircraft's spare parts are shipped to single assembly hubs, located globally. All parts come with certain life expectancy, specific requirements and maintenance attributes. With thousands of spare parts, hundreds of parameters, and number of manufactures distributed globally, SCM team need to deal with very large amount of data. In this paper, we use an industrial scenario of aviation industry SCM to demonstrate the necessity of having decentralized system based on distributed data-driven application technologies such as Blockchain, not only to assist in maintaining inventory of the aircraft's parts but also to monitor the performance, usage, etc. This will help to achieve a transparent network of supply chain for aircraft's parts and reduce the risk of availability of aircraft's parts in black market. These new data-driven technologies when embedded into SCM scenarios will help the SCM managers to analyse the supply, demands, source of availability of spare parts and provide methods to procure them from the right sources.

Keywords: Supply Chain Management; Blockchain; Aircraft's Parts; Segment; Distributed Data-Driven Application Systems

1. Introduction

Supply chain management is the backbone of any industrial sector. Different industrial organizations, distributed globally and specialized in manufacturing of particular types of products are operated as production houses for particular segments, which would be required by another manufacturer to produce another product or part of final product. Before reaching the final product, individual parts of the product go through series of supply chains called multi-tier supply chain. Aviation industry represent perfect example of how complex this process can be and how all this process works.

Individual parts or aircraft body segments are imported, a proper inventory is maintained and then assembled domestically. Aircraft may overhaul to another region depending on type of service required.

These supply chains are well regulated and properly monitored by individual companies at their individual levels. There is a possible risk of parts or segments replacement, bad quality replication, and decommissioned or preceding products being sold in black markets. There is no single supply chain of product manufacturing but a web on interdependent manufacturers, heavily depended on each other for segment production. There are certain pre requisites that should satisfy specific requirements before moving forward in this chain that would make product worth and efficient for future use. In most cases, when specific requirements are not met, they are either returned or remanufactured, further increasing the delay in obtaining the final products.

Traditional SCM models and approaches are useful when the amount of data to be processed is relatively small and the complexity of the problem is low. But what about larger problems with a lot more data, like in case of the aircraft's spare parts business? This challenge requires new data-driven techniques such as Big Data and Blockchain. This emerging technologies will push us into the next era of SCM design and Industrial Internet of Things (IIoT) applications.

In this paper we take industrial scenario for aviation's supply chain management, on which, we implement the concept of Blockchain technology where each transactions is monitored by individual tiers of this supply chain, later smart contracts platform between individual tiers can monitor quality and proper flow of products in this chain. Individual databases at individual level will record and update all transactions that would be flowing along different tiers. This concept and technology into supply chain management can really bring significant added value to business process through more easy flow, higher reliability and better quality in aircrafts parts or segments.

The remainder of this paper is organized as follows. First we discuss the exploitation of SCM at different levels of OEMs (Original Equipment Manufacturers), MRO (Maintenance, Repair and Overhaul) businesses in aviation industry. Then we discuss a related and important area – the protection of supply chains from counterfeiting, followed by an idea of Blockchain integration in a tier to develop and model a transparent and reliable chain. Section 5 discusses approaches to apply Blockchain to the SCM in industry as well as its application in aviation. Section 6 continues and focuses this discussion on Blockchain applications in aircraft's parts business. Finally, in Section 7 we discuss methodological issues in Blockchain, especially in the context of smart supply chain management. We will talk about supply chain, its features and how problems in future can be overcome.

2. Challenges in Smart Supply Chain Management

Protecting legitimate supply chains from counterfeiting has become a significant challenge. In 1998, the OECD's report on the Economic Impact of Counterfeiting [1] showed that counterfeit products were often made by the same manufacturer that was contracted to produce the authentic product. The counterfeiting strategies are increasingly sophisticated as in many cases counterfeiters apply the same technologies and use the same suppliers as legitimate brands [2]. Thus, a variety of counterfeiting strategies has emerged and affects the legitimate supply chain such as genuine parts or products are stolen from the legitimate supply chain (e.g.: disposed-of genuine products are recovered), factory over-runs or near copies are illegally produced by sub-contractors, counterfeit products and genuine products are bundled and distributed together.

Over the past years many industrial sectors have been reporting cases of infiltration of counterfeit products into their supply chains. For instance, the aviation sector has been expressing strong concerns about the detection of counterfeit aircraft 's parts in the legitimate supply chain. Such detections have been reported both in the civil and military aircraft supply chains in the US as well as in Europe.

Several factors contribute to this phenomenon. First, global supply chains constitute a complex and large network of actors involved in production and distribution processes. In such circumstances, many companies find it difficult to map and monitor their suppliers and sub-suppliers beyond the second tier of their supply chains. This limited visibility results in lower control over the supply chain and thereby increased exposure to a variety of risks such as counterfeiting. In some cases, these vulnerabilities allow illicit networks to penetrate legitimate supply chains and exploit the services provided by supply chain intermediaries.

Therefore, it becomes urgent to take the necessary measures such as smart SCM techniques to better protect legitimate supply chains from counterfeiting threats, including by enhancing cooperation between supply chain actors, improving information exchange and security, and actively promoting transparency and best practices in SCM.

The smart SCM vision relies on extensive use of sensors to monitor the supply chain status and ICT to adapt activities and operations of the supply chain to meet its objectives. Realization of the smart SCM vision requires solution of many technical challenges. To serve target customers better than their competitors, supply chain management (SCM) teams today look into new technologies such as Big Data, Internet of Things (IoT) and Blockchain. These new technologies allow managers to develop and provide complex supply chain services and products faster with improved reliabilities. With these technologies, SCM teams can build complex models of a supply chain or systems of supply chains using a data-driven approach.

Blockchain is a paradigm that involves dynamically incorporating real-time transactions into chains of blocks in order to steer the information exchange process of an application system. Information on a Blockchain exists as a shared database which isn't stored in any single location, meaning the records it keeps are truly public and easily verifiable. No centralized version of this database exists for a hacker to corrupt. Hosted by millions of computers simultaneously, its data is accessible to anyone on the internet.

The Blockchain network automatically checks in with itself every ten minutes. A kind of self-auditing ecosystem of a digital value, the network reconciles every transaction that happens in ten-minute intervals. Each group of these transactions is referred to as a "block". Two important properties result from this:

- **Transparency:** data is embedded within the network as a whole, by definition it is public.
- **It cannot be corrupted:** altering any unit of information on the Blockchain would mean using a huge amount of computing power to override the entire network.

In theory, this could be possible. In practice, it's unlikely to happen. Taking control of the system to capture Bitcoins, for instance, would also have the effect of destroying their value.

Blockchain technology initially was introduced in cryptocurrency where because of decentralized ledger, it was easy to record all the financial transaction among all the actors simultaneously avoiding any kind of error. However, following to Don and Alex Tapscott: "The blockchain is an incorruptible digital ledger of economic transactions that can be programmed to record not just financial transactions but virtually everything of value." [3]

The Blockchain concept of other domains has been lately proposed but yet to be implemented. In aviation industry there is increasing demand of aircraft, and logistics or supply chains play important role in support of the aviation industry. Introduction of the Blockchain technology into SCM practice of aviation industry will enhance the aviation industries' performance and increase the passengers' safety.

It is immediately apparent that the Blockchain paradigm is directly applicable to realizing smart supply chains. For example, Blockchain provides an approach to address issues such as creating more efficient and reliable supply chains and to mitigate counterfeiting. This paper discusses the applicability of this paradigm to create more resilient and sustainable supply chains in aviation industry.

3. Supply Chain in Aviation

By 2035, there will be a significant growth of passengers' traffic and consequential demand in aircraft, as a result, significant increase in aircraft production is expected [4]. This growth will seek huge demand for more flight carriers, therefore the supply chain should proceed in perfect sequence, without any error. Soaring demands also lead to pressure on the supply chain causing delay and failure. Example, the problem to unhandled this situation could be at any level where the supplier is not able to process demands due to unavailability of products at preceding level [5]. In this environment of supply chain, individual level competes instead of co-operating by not sharing the information. This leads to error as we proceed up the supply chain [6]. Behind every final product, there is a long chain or levels of assemblies, but some researches have combined these individual levels, which produce for OEMs into 3 main tiers supplies [7]:

- Tier 1: This level directly works for OEMs, for example, dealing with engines, brakes, etc. they are first suppliers to the assemblers.
- Tier 2: They are the suppliers to tier 1, they supply products, which are manufacture from their own productions. Example, assembling individual parts of Engine.
- Tier 3: These are individual small-scale component production companies, which usually provide small components like electrical components, raw material.

Figure 1 below is a sample tier representation of Aviation Turbine Company, though this company would be serving at some tier for another "Originating Buyers" i.e. each component of any supply chain at any tier would have its own tier of supplies.

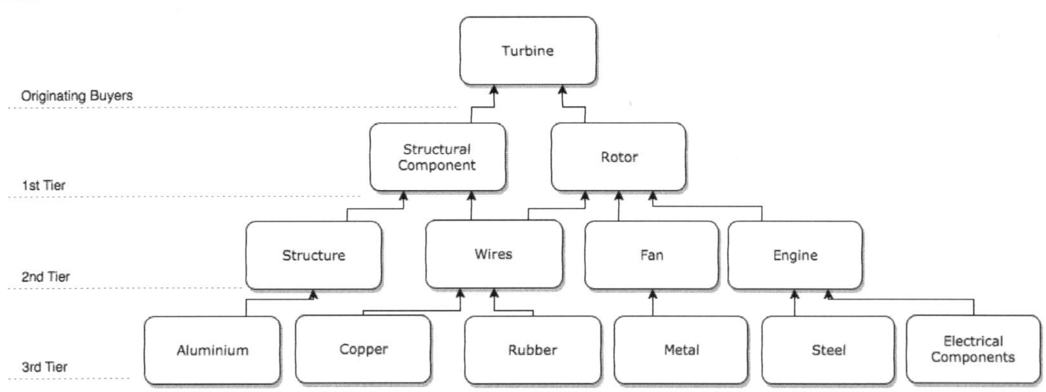

Fig. 1. The aircraft's Turbine's Supply Chain

Even after the proper path of supply chain, the risk and challenges like careful coordination among different bodies, challenges of managing the risk across global network and complex supply network, risks like guarantee, functioning, quality control, usage control, etc. still exist in OMEs. Necessity by the organizations to understand not just the direct suppliers they buy from, but also those who indirectly contribute component or service across the extended supply chain [7].

4. MRO Supply Chain Methodology

The purpose in any supply chain is to ensure that right product is produced at the right time and is at proper flow in the chain. It is highly dependent on the external factors like mode of transportation and internal factors like availability of inventories to produce and meet the required demands. These chains consist of exorbitant amount of flow information, product and money [8]. Components of supply chain management are complex but based on:

- Planning of Inventory i.e. from which part of manufacturing hub, it will be easy to procure the required product, etc.
- Development in making relation with supplier of raw material.
- Make i.e. manufacturing and measuring the quality of product.
- Delivering final product to required party
- Run i.e. finally testing or running the product as required.

There are instances when an aircraft is required service, part replacement, or sometime overhauling where it as a whole may be used for performance testing. Aircrafts move from one place to another for getting serviced as that specific service could be present at some particular place hub. This ensures the quality and performance marking and if there is any error of fault, then it is either repaired or replaced.

5. Designing Supply Chain Management with Blockchain Technology

Blockchain technology [9] plays a vital role in today's era of IT. It was initially introduced by Bitcoin, which focused to decentralize the common database used to record transaction between different users by enabling individual ledgers among the users that will update all the transactions anonymously and would be difficult for a single party to manipulate to interfere with. Blockchain calls for traceability, with open ledger we can know which product is transacted to which member of the chain, though two parties stand anonymous.

In aviation supply chain network, User will have the access of involving parties in the transaction. If there is some error or fault in the final product, then we can easily trace it back to the path of supply. We can monitor the progress and with it plan the future manufacturing. In supportive of design principles in Industry 4.0 [10], we propose the following:

- Interoperability: Each individual segment to be enabled with sensor or special id, helping different tiers bodies to connect and communicate among each other.
- Information Transparency: Ability to show the individual processes at each level, for example, if any changes made at any level in this chain all the users will be updated with the change with new details.
- Technical Assistance: This would enable to assist systems to support individual in chain by aggregating and visualizing information comprehensibly for making informed decisions, future planning and resolve problems quickly.
- Decentralized Decisions: Enabling individual companies with make their own decision and to perform their task.

Introduction of Smart Contracts, can help to achieve the desires, required and up to the mark product. This process will have all the details like, serial number, validity of products, i.e. all the information that would guarantee its authentication and uniqueness. Initially the supply chain was reordered on papers and where never digitalized where bundles of record books were piling and lot of time was wasted when details were requiring. Now many multinational companies provide with user the status updates with the location or time of arrival. This helps to attain secure transparency across global chain supply, which will help to reduce frauds, reduce delays, paper works, other wastes, etc.

6. Industrial Case Study

Now we will imply Blockchain to aviation industries' supply chain and monitor how, if details that are share across the platforms of individual users in this network, can actually benefits the aviation sectors in terms profit and safety. We will take the help of figure 2 below describing small scenario.

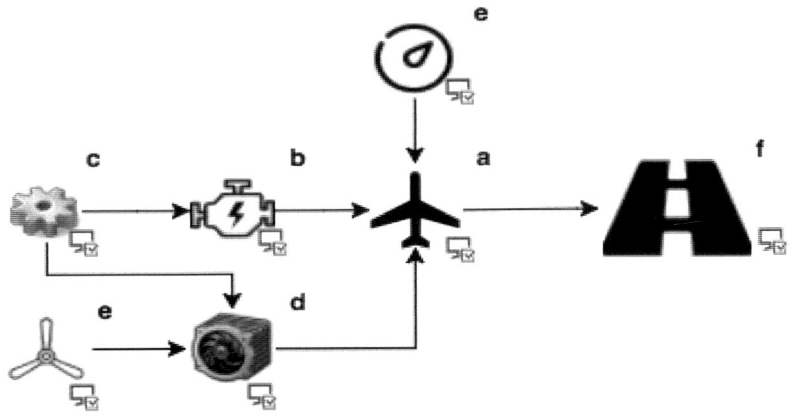

Fig. 2. Small Sample Supply Chain

Let us begin at an assembly hub 'a', these assembly hubs procure different parts from different places and are assembled into one individual standing aircraft. This hub is one of the individual bodies of the supply chain that procured parts like engine and turbine from different companied 'b' and 'd' respectively. When procuring and assembling one part for e.g. turbine into an aircraft at hub 'a', a transaction and assembly ledger will be updated across all the members of the network indicating that turbine of unique id along with other segments were assembled together, producing aircraft which itself will act like a new product of unique id. Assembly ledger will keep the record of all the new products generated at each level, which would be integrated with transaction ledger that will show the transacting details among the individual members. This whole network could also works as inventory registry for individual companies along with details of the products, which will show the production details.

Now at individual assembly hubs, will maintain their own supply chain path. At some tire level in this chain, if a product is replaced or tampered and the database of the hub, that will receive this product, is manipulated, due to the decentralized ledger, this process would not be validated as there would be mismatch, for example, may be in serial number and hence this can avoid and kind of future malfunctioning that could have happened.

Once the part is assembled into a plane and is serving for a purpose (commercial, private, etc.), during the overhaul at any maintenance hub, the engineers would have a proper and easy access to check for all the required maintenance, remaining life expectancy of products, hence, reducing the time and increasing the efficiency of work. If any part is close to expiration, it will indicate beforehand and the performance can also be monitored during the working and with parallel monitoring.

7. Conclusion

This work is devoted to the application of Blockchain technology to support supply chain system design and operation in aircraft parts manufacturing and distribution sector of aviation industry. Blockchain is inherently capable of providing significant advantages of timely, efficient and transparent transactions to the supply chain systems in the same way it has been proven for the value adding processes in financial/banking systems.

The approach used was to examine most recent works to establish the connection between Blockchain and supply chain taking aviation industry as an illustrative example. This study shows that Blockchain has well established link in supply chain research, and the use of Blockchain in supply chain management (SCM) has been increasing. In logistics, the Blockchain has mostly used as a digital distributed ledger to create a single digital place where all documents related to a shipment could be housed.

One of the key advantages of Blockchain is that it is more secure than traditional IT solutions. We suggest that Blockchain could be effective in enhancing aircraft parts traceability and establishing the authenticity of spare parts. In future, we plan to develop and test the technology to authenticate parts manufactured in Russia delivered to Indian airlines and government customers. The solution takes information from sensors or RFID tags and records it on the Blockchain to track the parts from factory to aircraft. We are also interested in aircraft parts traceability. With the availability of all information on transactions in the supply chain, anticipation for the reduction in counterfeit products would be achievable. Further, the introduction of the smart contract to this network will minimize the possibility of corrupt practices as it can eliminate the involvement of the third party and with the data information spread across, the buyers from different sectors like defence, private players will be self-capable to purchase.

There are some other interesting areas for future investigation of the Blockchain introduction to the SCM practice. In particular, we are also aiming to use Blockchain to lower the cost of doing business in global aircraft parts supply chains, where it is common for a big company to buy spare parts from small manufacturers. This usually increases costs across the end-to-end supply chain. Potentially, the Blockchain technology could greatly reduce the financial costs associated with strategic procurement.

8. Acknowledgments

This research was partially supported by the National Research University Higher School of Economics (HSE) of Moscow, Russia.

9. References

[1] OECD, (1998) "The Economic Impact of Counterfeiting," [Online]. Available: https://www.oecd.org/sti/ind/2090589.pdf.
[2] M. S. &. J. Busby, (2015), "An exploratory analysis of counterfeiting strategies," International Journal of Operations & Production Management, vol. 35, no. 1, pp. 110 -144.
[3] D. T. &. A. (2016) Tapscott, Blockchain Revolution: How the Technology Behind Bitcoin Is Changing Money, Business, and the World, Penguin Random House, Portfolio, p. 368, 2016-05-10.
[4] Y. M. &. Z. Avdeeva, (2017) "Planning in Aircraft Industry Based on Aircraft Prediction," Unpublished, Moscow.
[5] PricewaterhouseCoopers, "Price Waterhouse Coopers," [Online]. Available: www.pwc.com/aerospaceanddefence.
[6] B. &. P. I. Buchmeister, (2013) "Bullwhip Effect Simulation Of A Supply Chain With Level Constraints," Daaam International Scientific Book, Pp. 133-148.
[7] D. MOCENCO, "Supply Chain Features of the Aerospace Industry Particular Case Airbus And Boeing," Scientific Bulletin, pp. 17-25.
[8] B. Raza, "Supply Chain for MRO," Frankfurt University of Applied Sciences, Frankfurt.
[9] Blockchain, "Blockchain," [Online]. Available: blockchain.info.
[10] W.-I. 4.0, "Wikipedia," [Online]. Available: en.wikipedia.org/wiki/Industry_4.0.

DOI: 10.2507/28th.daaam.proceedings.147

EVALUATION OF THE ENERGY INTENSITY OF FORMING MACHINES COMPARED TO APPROACHES IN OTHER INDUSTRIES

Miroslav Běle & Jan Hlaváč

This Publication has to be referred as: Bele, M[iroslav] & Hlavac, J[an] (2017). Evaluation of the Energy Intensity of Forming Machines Compared to Approaches in Other Industries, Proceedings of the 28th DAAAM International Symposium, pp.1057-1064, B. Katalinic (Ed.), Published by DAAAM International, ISBN 978-3-902734-11-2, ISSN 1726-9679, Vienna, Austria
DOI: 10.2507/28th.daaam.proceedings.147

Abstract

Monitoring energy consumption of machines which are used in manufacture is a current trend. We can achieve the reduction of energy consumption by appropriate optimization. The goal of this article is to compare different approaches to energy balance in various industries. We focus on one working cycle of a forming machine, where we monitor the ratio between the used and consumed energy. We divide the consumed energy into losses caused by friction and losses due to the elastic deformation of the machine. We do not focus on energy losses during unproductive time. Different industrial sectors have different ways of dealing with energy balance we therefore decided to create a basic comparison in order to improve our existing approach to energy balance.

Keywords: Approach of energy balance; used energy; consumed energy; energy balance

1. Introduction

There are currently attempts in all scientific disciplines to reduce the energy demands of all products and equipment to achieve the smallest energy losses so as to minimize waste energy. These efforts not only focus on machine design and construction but also on losses during the operation of machines. Forming machines are no exception. The effort to reduce energy consumption of forming machines is very significant because there is considerable energy consumption. (For mechanical presses it is about 100kW, and for hydraulic presses, about 1MW). One needs to find the places in a machine where the biggest energy losses are during the design of the machine. This is why it is important to focus on the way energy losses are evaluated. The principle of energy balance is used for forming machines, especially for crank presses. Different approaches to energy balance are used for different industrial products (for example [7]). The goal of this article is to map the options for energy balance assessment.

The main objective is to verify whether our approach to evaluating the energy balance of forming machines is correct and complete. We want to achieve this by using research of other approaches in other industrial branches. If we find anything useful, we will adjust our existing methodology.

2. Energy balance of forming machine

Energy balance (the amount of energy which a machine must deliver to carry out its technical operation) is solved for one work cycle during continuous operation of the machine.

$$A_c=A_u \pm A_d+A_z \tag{1}$$

A_c... Total work
A_u... Useful work
A_d... Work of elastic deformations of the machine
A_z... Loss work of friction

Energy balance of forming machines can also be evaluated for a discontinuous work cycle. A discontinuous work cycle occurs when, after each technological operation, the crank shaft and connected ram stop in the upper dead centre of the mechanism. This means that energy is evaluated from the time when the front shaft and crank shaft of the mechanism connects with clutch until the time when the front shaft and crank shaft of the mechanism disconnects from the clutch and the crank shaft is stopped by the brake. Energy which is needed to start the machine is not considered.

3. Example of energy balance of crank press

Energy balance assessment of a crank press with press force 80MN considering these assumptions (see Table 1.). Crank press parameters:

Stroke	500	mm
Rod length	1700	mm
Crank radius	400	mm
Nominal angle	7,5	degree
Bearings diameter shaft/frame	980	mm
Bearings diameter shaft/rod	1530	mm
Bearings diameter rod/pin	750	mm
Coefficient of friction in the pins	0,03	
Coefficient of friction in the ram guide	0,05	
Nominal stroke	2,45	mm
Radius of pivot friction shaft/frame	14,7	mm
Radius of pivot friction shaft/rod	22,95	mm
Radius of pivot friction rod/pin	11,25	mm

Table 1. Parameters of crank press

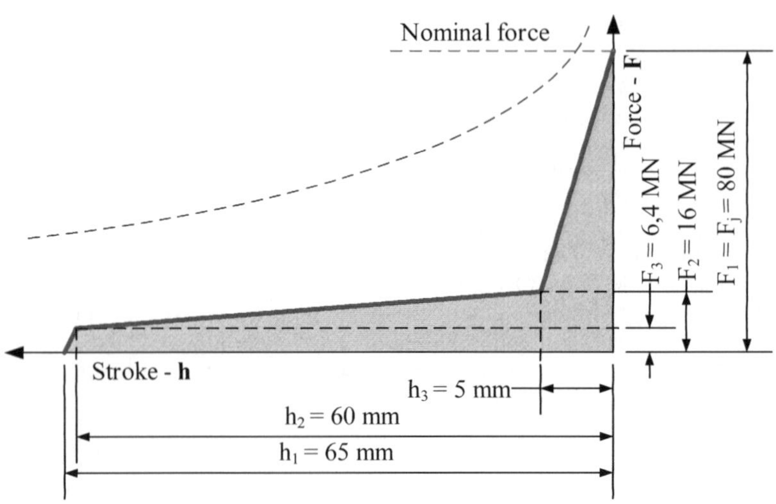

Fig. 1. Simplified graph of forming forces in relation to strokes in die forging [1]

Useful work A_u

Useful work A_u is work which is needed for implementation of the technical operation. For the purposes of this research, useful work size is described by an idealized course of forming characteristics (Fig. 1.).

The work is determined as the area under the curve of the shaping force. On a crank press with a nominal force F_j=80 MN and nominal stroke h_u=2.45 mm, A_u=872 kJ.

Work of elastic deformations of the machine A_d

Work of elastic deformations of a machine A_d is work which is accumulated into the elastic deformations of the machine caused by working force. Total deformation of the work area of the press is approx. y=4.2 mm, which is A_d=168 kJ. The specific distribution of the share of each component is shown in the following figure (Fig. 2.).

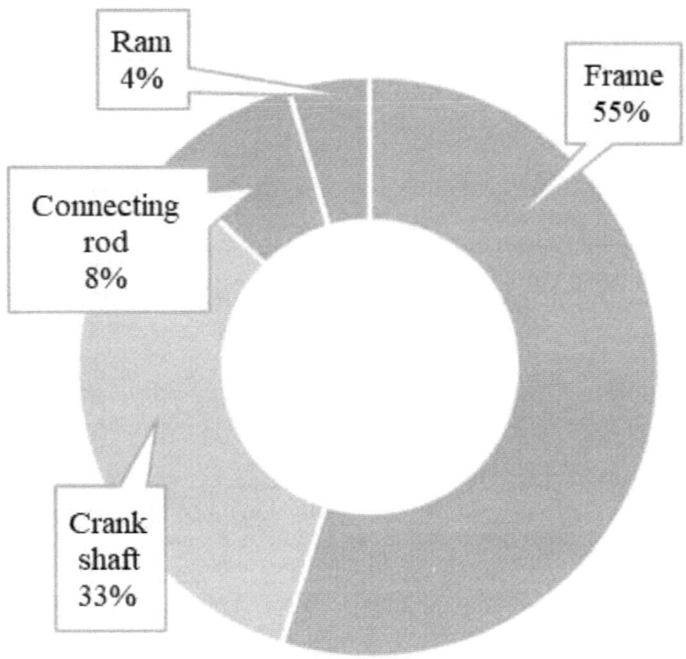

Fig. 2. Distribution of elastic deformation in press components

From the above it can be seen that when designing a machine it is necessary to focus on the frame construction, and in the drive mechanism it is necessary to focus on the parameters of the crank shaft, including the crank shaft bearing.

Loss work A_z

Loss work A_z is work which is lost due to friction.

Coefficient of friction in the planar surface of 0.05 and in the cylindrical surface of 0.03 are considered for a specific crank press.

The following results are based on these assumptions.

A_{zL}=320 kJ
A_{zLe}=500 kJ
A_{zLb}=58 kJ
A_{zvb}=3 kJ

Total loss work is A_{zL}=882 kJ. Percentage of the share of individual works is shown in the following graph (Fig. 3.).

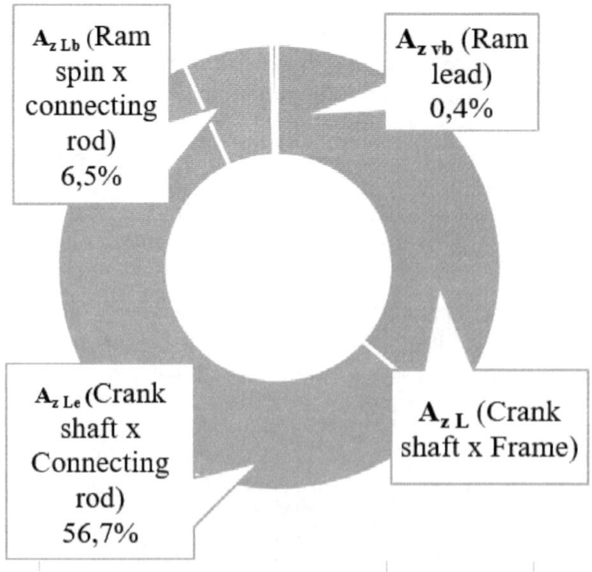

Fig. 3. Distribution of friction losses

A_{zL}... friction between the crankshaft and its placement in the rack
A_{zLe}... friction between the connecting rod and its bearing on the crankshaft
A_{zLb}... friction between the connecting rod and its mounting on the ram shaft
A_{zvb}... friction between the ram and its lead in frame

The friction losses are calculated for the position of the crank mechanism behind the lower dead centre because there is still force in the mechanism due to the machine springs. Losses in the ram are small and irrelevant in terms of the total energy balance.

As the analysis shows, the biggest energy losses are between the rod and the eccentric of the crank shaft (this can be more than 60% of the total A_z of the machine). The size of this friction moment increases with increasing eccentric diameter. A significant part of the energy is lost in the place of the crankshaft mounting in the frame (up to 30% of total A_z).

4. Evaluation of energy balance

The way that energy balance is calculated for forming machines (crank presses) is clear. But we are interested in how energy balance and its evaluation is used in other areas, especially for other machines and equipment in industry. We want to compare and evaluate whether this approach is correct, and whether it would be possible to use it in some of the other approaches which are applied in other areas.

Distribution was created according to the time taken for the evaluation of energy balance based on extensive research from foreign and Czech sources.

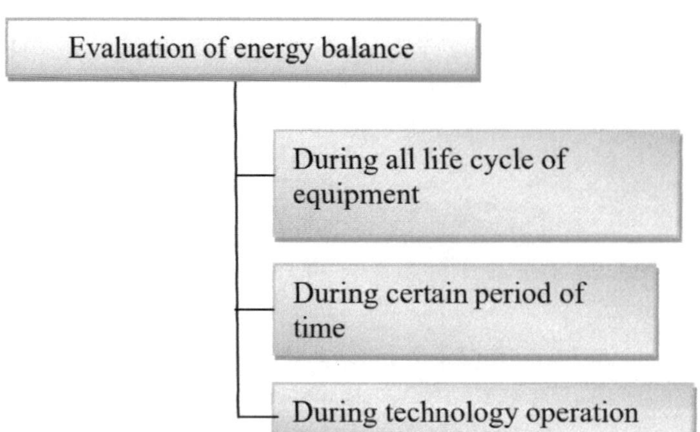

Energy balance can be evaluated:

- **For the whole life cycle of the equipment:** During use or during the whole life cycle. Costs are calculated for extracting raw materials and costs of liquidation in the second case.
- **For a certain period of time:** A certain period of time is selected for which the energy balance is evaluated.
- Or **during technological operation** (cutting, forming). This approach is used primarily in the field of machining technology, when energy savings of new machining technologies are evaluated.

4.1. Energy balance during all life cycle of equipment

The energy which is needed for the production of a machine (or equipment) is analysed. Energy for extraction of raw materials, energy for production, energy for transport to the customer, energy for lifetime use and energy for liquidation are also calculated. This is life cycle assessment (Fig. 4.).

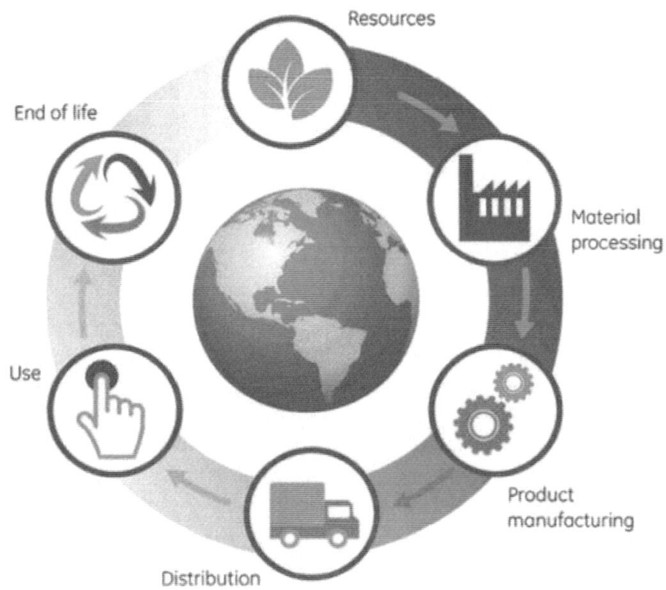

Fig. 4. Life Cycle [4]

This term is used in relation to the lifetime of equipment in some cases for example for solar panels (Fig. 5). The energy is calculated which is needed for production, and the energy which is produced by this equipment. The energy balance of the equipment for the whole lifecycle is based on these factors. This branch is marked in blue in the table. (see Table 2.)

Fig. 5. Solar panels [5]

4.2. Energy balance during certain period of time

The next approach is used for evaluation of machines and equipment. A time interval is considered which is described for example by one cycle of a machine or equipment (Fig. 6.). Energy balance is evaluated for this time interval. The result is the knowledge of the amount of energy which is used for a technical operation and the amount of energy which is lost by passive effects.

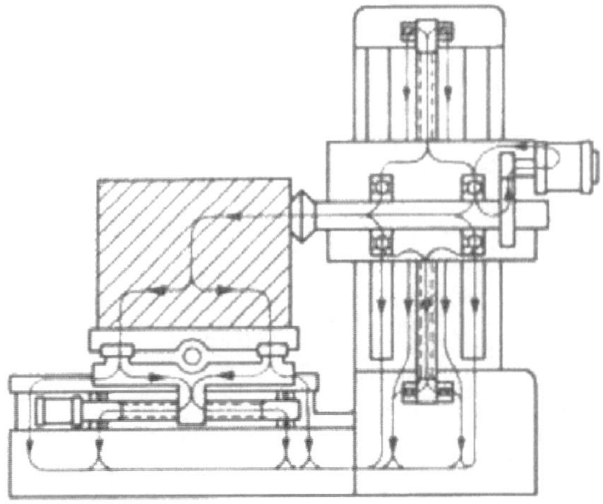

Fig. 6. Energy flow in a machine [2]

In other cases, for example in the energy balance of buildings, there is a different approach. A thermal energy source (central heating) is assumed to be installed in the building which supplies thermal energy. The amount of energy which escapes from the building is calculated. The energy class of the building is determined on the basis of the energy analysis. This approach does not evaluate the energy balance for the whole life cycle, but only for a certain period of time (Fig. 7.).

Buildings have properties which do not change over a long period of time, and therefore this approach is understood as an evaluation for the whole lifecycle. Energy which is needed for extraction of raw materials, construction and liquidation are not calculated here. This branch is marked red in the table. (see Table 2.)

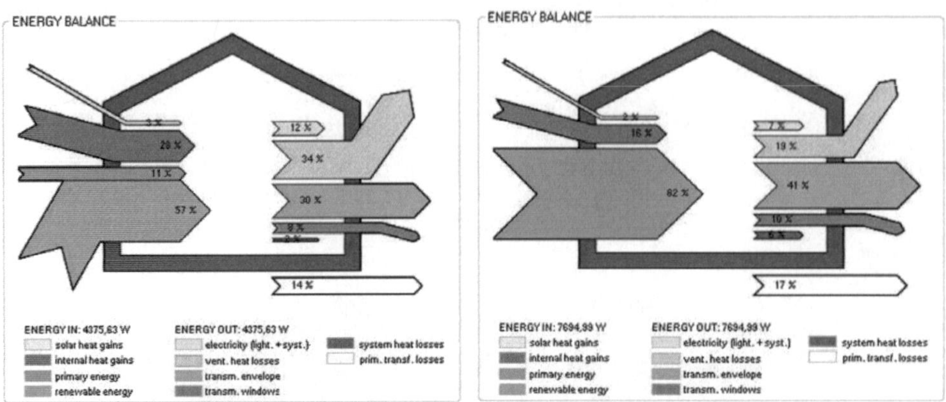

Fig. 7. Energy balance of buildings – Czech Republic [6]

4.3. Energy balance during technology operation

The last of approach to evaluating energy balance is used for cutting machines (for example [8]). Energy which enters the machine and energy which is used for cutting are measured. The electric motor is measured during the technological operation (milling), from which the energy for the technological operation is found. Energy which is used for cutting is measured by a dynamometer. The dynamometer is situated between the milling tool and the spindle. Cutting forces are recorded by the dynamometer during the cutting operation, from which the cutting performance and the amount of energy used for cutting is evaluated. The difference between these two energies provides information about energy balance during cutting. It describes the energy transmitted between the machine and the workpiece. This branch is marked in green in the table. (see Table 2.)

Assessment period	Branch of science	Some cases of specific use	Description
During all lifecycle of equipment	Electro technics	Solar panels	It is used to map the energy needed to produce traffic and dispose of the equipment.
		Light sources	
		Electric devices	
	Energetics	Power plants	It is used to map the energy needed to produce, transport and dispose of the equipment.
		Turbines	
		Boilers	
During certain period of time	Construction	Buildings	Short-term energy evaluation (eg: thermal energy leakage from buildings) is used. Due to the fact that the properties of the objects do not change for a long time, this evaluation is often valid throughout the life of the object.
		Houses	
	Engineering	Machines	The energy intensity of the machine (device) during the working cycle is determined. This cycle is constantly repeated over the service life of the machine.
		Equipment	
		Production lines	
	Electro technology	Electric machines and equipment	The energy intensity of the machine (device) during the working cycle is determined. This cycle is constantly repeated over the service life of the machine.
During technology operation	Machining technologies, other production technologies	Cutting	Determines the energy taken by the machine during a particular machining operation.

Table 2. Summary of evaluation times of the life cycle

5. Evaluation of achieved result and future

The survey shows that approaches to energy balance are different in different industrial branches. Approaches range from an entirely local view of one duty cycle to an overview of the energy performance of the entire life cycle. If we want to choose the right method for describing the energy balance we must set a target against which the comparison is to be conducted.

High energy intensity is typical for forming machines because the forming process is very energy intensive. This is why we are interested in performance during forming operations. Most forming machines work in a cyclic operation, so it is appropriate to describe just one cycle. Other types of forming machines work continuously (rolling), but we did not analyse these machines.

The approach used to evaluate energy balance which is actually used here describes one work cycle of the machine and seems to be the best method.

There are other options resulting from this research. For example, to focus on the consumption of a machine out of its working cycle. For example, a flywheel spins and revolutions are maintained outside the working cycle.

In the future it would be appropriate to expand the description of energy losses to include the drive unit of a machine. A drive unit consists of a gear box, clutch, brake and also electric motor. We can achieve further energy savings by optimising these components.

6. Conclusion

This article summarizes the current basic and theoretical knowledge in the field of energy balance. The work describes this concept of energy balance across various scientific branches and a map shows how it is covered in other sectors. The largest part is dedicated to energy balance as it is used in the field of forming machines. Based on our research, we will expand the existing description of energy losses to include the drive unit of a machine. We did not pay attention to drive unit of machine until now.

7. Acknowledgments

This article was done by financial support SGS-2016-012 of University of West Bohemia.

8. References

[1] Kováč, A. and Jenkut, M. (1978). Forming Machines. Bratislava: The publishing house of technical and economic literature
[2] Kamelandr, I. (1982). Forming machines I. Technical University Brno
[3] Kamelandr, I. (1982). Forming machines II. Technical University Brno
[4] BioProcess Matters life cycle, Available at: https://promo.gelifesciences.com/gl/BP/UP_art4.html, Accessed: 2017-09-25
[5] What Czechs, that solar panel, Available at: https://byznys.lidovky.cz/v-cr-pripada-temer-na-kazdeho-obyvatele-jeden-solarni-panel-pa6-/firmy-trhy.aspx?c=A121226_160056_firmy-trhy_vs, Accessed: 2017-09-25
[6] Building information model based energy performance assessment in early design stage, Available at: http://www.sciencedirect.com/science/article/pii/S0926580508001064, Accessed: 2017-10-03
[7] Baumann, H. and Tillman, A. (2004). The Hitch Hiker's Guide to LCA. An orientation in life cycle assessment methodology and application, External organization
[8] Rief, M. and Kalhofer, E. (2017). Evaluation and modelling of the energy demand during machining. Aalen University of Applied Sciences

DOI: 10.2507/28th.daaam.proceedings.148

METHODS OF ASSESSMENT AND TRAINING OF A COMPANY TOWARDS THE ENTERPRISE 4.0

Baicu Andrei Valentin

This Publication has to be referred as: Baicu, A[ndrei] V[alentin] (2017). Methods of Assessment and Training of a Company Towards the Enterprise 4.0, Proceedings of the 28th DAAAM International Symposium, pp.1065-1073, B. Katalinic (Ed.), Published by DAAAM International, ISBN 978-3-902734-11-2, ISSN 1726-9679, Vienna, Austria DOI: 10.2507/28th.daaam.proceedings.148

Abstract

This paper is about identifying an enterprise's analysis and evaluation tool, starting from the general characteristics of an organization and initial testing on two companies: a real company that has recently gone through a process of change and a theoretical enterprise type 4.0.
Finding an analytical tool leads to a method of working with this tool. Analysing a company allows us to identify vulnerabilities and imbalances. The method proposed by this paper consists in analysing the enterprise based on an instrument that determines the balance of the complementary factors that describe the enterprise. These complementary factors and relations of determination are detailed.

Keywords: Industry 4.0; Evaluation tool for organization; Complementarity; Complementarity factors; Determinant and determined

1. Introduction

Adaptation to the new requirements of the world economy, adaptation of new technologies, automation, led to a new industrial revolution, 4.0. Industry 4.0 turns the entire production cycle of a product from its design to its delivery, increasing production speed by 30%, efficiency by 25% (according to the Boston Consulting Group) and raising product personalization to previously unseen levels. According to BCG, industry transformation has as sources 9 elements [8].

It is needed to follow a series of steps to move a company to industry 4.0 whose steps are described in BCG [9]. One of them is organization preparation [3]. From the above, some observations are made:

- The proliferation of information systems, automation and interconnection will change the current business system.
- Using machines for analytical purposes, the explosion of information will lead to a new way of understanding systems and organizations. For such a radical change, you need preparation.

The method proposed by this paper consists in analysing the enterprise based on an instrument that determines the balance of the complementary factors that describe the enterprise. These complementary factors are detailed in chapters 2 and 3 of the paper. With the help of the relations of determination - described in chapter 4 - measures for balancing the enterprise with the purpose of its transition to the enterprise are established.

2. Complementarity

Analysing contemporary models of organizations - such as virtual enterprises and learning organizations - we have found that the differences between them are complementary. More precisely, their combined qualities form an ideal model.

As proof of the need to combine the two business models, it is noticed that more and more large companies develop common features with the two types of organizations mentioned above. The possibility of combining features proves that they are complementary, not opposite.

For example, the BEST Awards recognize organizations that demonstrate enterprise-wide success as a result of employee talent development [12]. Looking to the list of winners, we can see that many of the winners have the characteristics of the two types of organization.

One of the organization's permanent issues is getting new capabilities. One of the resources is permanent innovation. In order to be able to analyse these adaptive processes, we must observe the relationship between exploring new possibilities and exploiting the already existing certitudes.

Innovative exploitation generates procedures, methods, solutions or tools that are used in the production or sales processes and that enable the company to gain competitive advantage by increasing customer satisfaction.

Innovative exploration enables the company to position itself properly in the marketplace, find new opportunities and new business solutions, cooperate with partners in other production / service areas that produce new technologies.

Both methods are necessary in the activity of a company; they can support each other or alternate in the good work of a company. It should be remembered that the excessive use of one of the two methods could adversely affect the evolution of the company. It can be considered consider that the two methods are complementary.

Analysing models of organizational structure, we distinguish two representatives:

1. ***The hierarchy*** is an organizational structure in which ranks are classified according to the levels of importance. The image of this type of structure of an organization can be represented as a pyramid one.
2. ***A heterarchy*** is an organization system in which the elements of the organization are irregular (non-hierarchical) or possess the potential to be classified in several different ways. This system it is named by some authors as a non-hierarchical system [10][11].

The two ways of structuring (hierarchically and heterarchically) are not mutually exclusive, so they are complementary, and they can be found at departmental level within the same company. We have noticed that a company's model, the methods used by it and the structure can be analysed in terms of two complementary elements. Obviously, an organization cannot manifest only one of the complementary factors.

The question is whether we can build an analysis tool based on the complementarity of some organizational factors that can help us evaluate the company's status and provide clues about what the business should do in the future.

In order to be able to make this assessment, we need to find the right indicators that describe the organization and which we will call complementarity factors. The factors give us quantitative information about characteristics considered representative of an organization.

E.g.:	*Element:*	*Organizational Structure*
	Feature:	*Degree of formalization*
	Complementarity factors:	*Verticality - horizontality.*

The analysis of the organization cannot be done solely on the basis of finding the complementarity factors and their individual evaluation. Only together with the sum of all complementarity factors and their assessment they can provide us with information about the balance of the organization.

For evaluation, each complementarity factor will be given a score of 1 to 5, even if an organization can be described as missing one of the two complementary factors. I think that in fact, the second factor also manifests itself in some way within the organization so we will give at least one point.

The factors chosen in this work will be described and explained in the next chapter without claiming that they are the only ones. I have, however, sought the simplest approach of the study.

In the table below we included the most common elements and features founded in management textbooks and organization theory [2][4][10].I followed particularly those written by S. P. Robbins [5].

The complementary factors are entered in columns 3 and 4 as follows:

- Centric complementary factors are characterized by firm, strong, well-rounded, inward-facing features.
- Centrifugal complementary factors are distinguished by flexibility, low contour, horizontality, orientation towards the outside.

Elements of the organization	Characteristics	Complementary factor centripetal	Complementary factor centrifugal
Structure	1. Complexity	Vertical diferentiation	Orizontal diferentiation
	2. Formalization	Clearly defined	Permanent adaptable
	3. Centralization	Concentrated	Dispersed
Design		Functional	Divisional
Management		Omnipotent	Symbolic/Democratic
Culture		Existent	Wanted
Proceses	1. Changing processes	Internal	External
	2. Recruitment	Internal	External
	3. Learning	Formal	Informal
Strategies	1. Corporate	Growing	Renewal
	2. Competitive	Better	Different
Employees Relationships		Formal	Informal

Table 1. Descriptive complementary factors of an organization

If the total point in column 3 is equal to the total in column 4, we consider that we have a balanced organization. The score of the two columns gives us a representation of the company's "health condition". If the sum in column 3 is very different from column 4, we need to analyse where the maximum points of differentiation are and what is their significance.

The economic indicators, balance sheets, on-going contracts give us a snapshot of the situation of a company. Based on these, some forecasts may also be made over certain periods of time.

3. Complementarity factors

As the organization is an open system, the state of equilibrium is given by the result of permanent actions to and from the system on one hand, and by the internal balance of the system on the other. With complementarity factors, we seek to analyse this internal balance. These factors must present the qualities of the system at the structural level; the balance between them is representing the balance of the system.

To determine which complementarity factors can describe the status of an organization, we need to look at the elements that describe an organization in general.

The elements of an organization have several features. We will seek to attribute each of them two complementary factors. We have chosen the most common features in management textbooks and organization theory [6] [7].

What elements can we meet in an organization and which are the appropriate complementarity factors? The identified elements are: structure, design, management, organizational culture, processes, strategies and relationships between the members of the organization.

3.1. Organization Structure

The structure of the organization defines how the tasks are assigned, who reports to whom, what are the formal coordination mechanisms, and the patterns of interaction that will be followed.

The organizational structure has three components [5]:

1. Complexity
2. Formalization
3. Centralization

"Complexity refers to the degree of differentiation that exists within an organization.

In this situation, vertical differentiation and horizontal differentiation can be considered complementary factors that determine the complexity of the organization.

Formalization refers to the degree to which workplaces and procedures within the organization are standardized. If a job is very formal, those who carry out their work have a minimum freedom in what needs to be done, when it should be done and how to do it. From this point of view we can consider that we have two complementary factors describing formalization: clearly defined and permanently adaptable.

"Centralization is the most difficult definition of the three components. Most organizational theorists agree that the term refers to the degree to which the decision-making process is concentrated in a single point of the organization." - P. Robbins.

For the description of centralization we have two complementary factors: Concentrated and dispersed.

3.2. Design of Enterprises

Robbins and Coulter describes three traditional business design patterns [5].

- The simple structure is an organizational structure with poor divisions across departments, the authority being centralized to one person. It is typical of an organization with a small number of people.
- Functional structure is the organizational design that makes clustering of specialties based on similarity or association. The structure can be viewed as a split of functional departments across the organization.
- The divisional structure is that organizational structure made up of separate business units. In this type of structure, each department has limited autonomy and a manager responsible for its performance.

In the literature, we also encounter the matrix structure, a structure that represents practically the combination of the divisional structure and the functional one. Given that when an enterprise grows any simple structure, it becomes a functional or divisional structure, to describe the enterprise design, we have chosen two complementary factors: Functional and Divisional.

3.3. Management

Managerial competence has constraints from the organizational environment and organizational culture. According to S. P. Robbins, there are two perspectives in management theory [5].

- *The omnipotent vision of leadership*: managers are directly responsible for the success or failure of an organization.
- *The symbolic vision of leadership*: much of the success or failure of an organization is given by external forces controlled by managers. The complementary factors are: omnipotent and symbolic.

3.4. Organizational Culture

Organizational culture is another element that we encounter within an organization and can have a significant influence on its results. M. Armstrong summarized the most common classifications of Roger Harrison, Charles Handy and Edgar Schein. Most of the time, there are 4 dimensions [1]. To simplify and reach only 2 elements that can represent the two complementarity factors, we have made the following classification:

- Culture based on own members (can be a culture of power), on the existing procedures or competencies of the members of the organization.
- The culture of the results - is based on the expected results (culture of achievements) or on the personal motivation (can therefore be a culture of support).

The two complementary factors are represented by what they have/existent and what they want/ wanted.

3.5. Processes

The process of change

Changes in the company may be determined by internal or external factors. The internal ones refer to a new organizational strategy, changing the composition of the workforce, new equipment, or changing the attitude of the employees. External changes mean changing consumer needs and wishes, new governmental laws, new technologies, or economic changes. The two complementary factors are internal change and external change.

Recruitment

The two complementarity factors are internal and external recruitment.

Employee training

Employee training can be done informally or formally [10]. The informal learning is done permanently; the skills development takes place based on feedback from the workplace. Formal learning is planned, systematically and requires structured training and programs that contain training and practice.
The complementarity factors that define the training are formal and informal learning.

3.6. Strategies

According to S. P. Robbins there are three types of strategies: corporate, competitive and functional [5]. To identify the complementary factors, we considered the first two types because functional strategy belongs to managers in lower levels. It is the corporate strategy that determines where the business is or where it will be. There are 3 main types of corporate strategy: growth, stability and renewal.

Considering that the strategy which promotes stability, actual preserves what exists, it does not seem relevant to our study, so we have selected the two complementarity factors as: growth and renewal.
The competitive strategy determines how the company will behave in relation to its competitors. We can talk about a lower price strategy or a better product at the same price or the strategy of differentiation - when your product is different through its qualities. The complementarity factors that define the competitive strategy will be better or different.

3.7. Relations between members

Relationships between employees may be formal, established by rules, procedures, protocols, their degree of complexity being directly related to the size of the organization, or may be informal, outside the rules and procedures, of a more personal nature. The two factors that describe relationships between employees are: formal and informal relationships.

4. Determinant and determined

Practically, an organization can be described by analysing the structure and design, management, strategy, processes, organizational culture, and relationships among its members. A better understanding of the organization can be achieved through a representation of its elements and the determinations between the elements. For a simplified representation, we will consider the structure and design of the organization as one element - just as they are in some places, as well as the organizational culture and the relationships between its members.

Analysing the elements in the fig.1, establishing the determinations between the elements - who determines and who are determined, we can make an image of the organization that can be represented as follows:

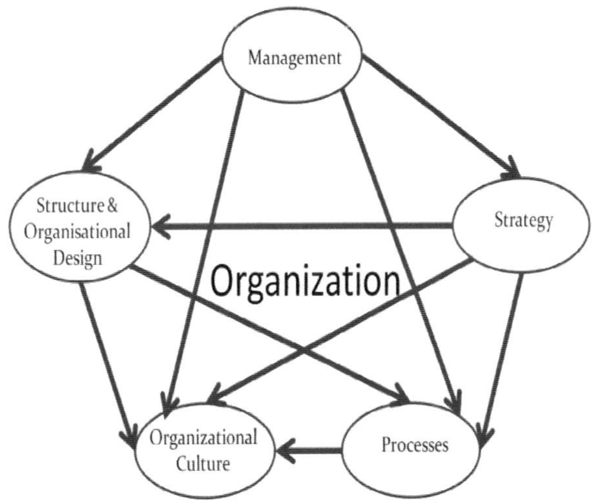

Fig. 1. Representation of an organization through the elements determined and determinant

If we find an imbalance of an element, we can predict which would be the next element affected. At the same time we can find out if the unbalanced is the result of the element that determines the unbalanced element and this will help to identify the remedial solutions. If we put together the structure and design of the organization with the strategy, being determined by management and determining the same elements (processes, culture and organizational relationships) and introducing the results of the organization as a determinant for its management, the result is the following scheme:

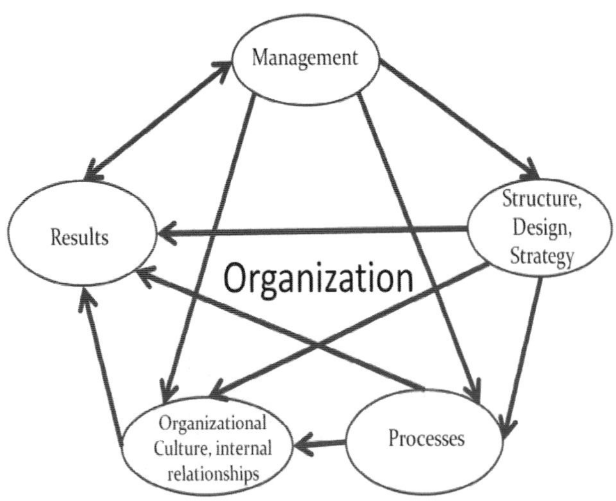

Fig. 2. Representation of an organization by including the results in the elements determined and determinant.

People who are part of the organization create organizational culture and relationships between members. Practically, the element in our schemes represents the members of the organization. Structure, organization design, and strategy are the result of thinking and planning made by the management. Processes are the action, the concrete activities, the work itself.

Fig. 3. Representation of a company by redefined elements

Redrawing the above scheme, by redefining the elements, we get figure 3. In this determination cycle, the ultimate goal of the enterprise is to achieve results. Higher performance results are conditioned by innovation and permanent adaptation. Under the conditions of the Fourth Industrial Revolution, exploitation innovation can be largely taken over by cars, computers and the Internet, all of which resolve many of the structures, strategies and processes, continuously improving them. What about innovative exploration? It is dependent on the human factor and will remain as it is at least until the moment of the development of an artificial intelligence superior to the one existing today.

4.1. Case Study – The application of the assessment tool on a real business

Elements of the organization	Characteristics	Complementary factor centripetal	Complementary factor centrifugal
Structure	Complexity	Vertical diferentiation	Orizontal diferentiation
		1	2
	Formalization	Clearly defined	Permanent adaptable
		1	4
	Centralization	Concentrated	Dispersed
		5	2
Design		Functional	Divizional
		5	1
Management		Omnipotent	Simbolic
		5	1
Culture		Existent	Wanted
		3	2
Processess	Changing processess	Internal	External
		1	1
	Recruitment	Internal	External
		2	5
	Learning	Formal	Informal
		1	1
Strategies	Corporate strategy	Growing	Renewal
		2	1
	Competitive strategy	Better	Different
		5	5
Employees Relationships		Formal	Informal
		5	5
Total		**36**	**30**

Table 2. The situation of a company in October 2016 evaluated through the complementarity factors

The study was done for a company in France, a subsidiary of the multinational company that has subsidiaries in the United States, Europe, China, Russia, India, and South Africa. The company had good results, in accordance with the forecasts. The company has between 50 and 100 employees. The field of activity is animal genetics. The information has been supplied in the framework of an interview with the regional director of human resources, responsible for France, Italy, Germany and Poland. On the basis of the information received we have evaluated and summarized the situation of the company in October 2016 and July 2017. By analysing the table 2 we find that there is a significant difference between the two complementary groups (16.7%) that indicates a possible imbalance. We find that the biggest differences come from management and structural design (4 points). Other differences we encounter in the structure of the organization, the degree of complexity, the degree of centralization and the recruitment process.

According to the principle stated above, for optimum results, a company should have the complementary factors in equilibrium and should reduce the differences through organizational changes applied in areas where the differences are greatest. This means that changes at management level, organizational design, structure and processes should be made.

In an absolutely unexpected way in November 2016, the general manager decides to leave and it is appointed a new general manager and a new human resources manager from within the global organization but from different countries.

They find that there are some procedural problems, imbalances in the structure and a high level of dissatisfaction among employees due to the failure to keep the promises made by the former manager. The new manager implements a series of changes with medium and short-term impact. In July 2017, the company's situation is described by those in the company according to the table below.

By implementing relatively simple changes, there is a decrease in the difference between the two groups of complementary factors to 11.2% within a few months. We have a single 4 point difference in organizational design - this can not be changed dramatically due to the small size of the company and we still have a 3 point difference in the recruitment process that is mainly external due to the small size of the company on one hand and the fact that the company it is growing, on the other hand. The management gap also dropped to three points due to the implemented changes: an additional level of management has been added, the number of individual trainings increased, the degree of delegation increased, few processes have been changed and a process of changing the employment contracts begins.

4.2. Results and Interpretation

The biggest imbalance has been found in October 2016 exactly where the first change appear through the resignation of the general manager. Scoring changes were noted between the two tables, generally where the differences were large in the first table (representing the initial diagnosis of company performance.) This shows that the imbalances of the initial situation are found in the weaknesses identified using the proposed analysis tool here.

Elements of the organization	Characteristics	Complementary factor centripetal	Complementary factor centrifugal
Structure	Complexity	Vertical diferentiation	Orizontal diferentiation
		2	2
	Formalization	Clearly defined	Permanent adaptable
		2	4
	Centralizatiom	Concentrated	Dispersed
		3	3
Design		Functional	Divisional
		5	1
Management		Omnipotent	Simbolic
		4	1
Culture		Existent	Wanted
		4	3
Procese	Changing processess	Internal	External
		2	1
	Recruitment	Internal	External
		2	5
	Learning	Formal	Informal
		3	3
Strategies	Corporate strategy	Growing	Renewal
		3	2
	Competitive strategy	Better	Different
		1	1
Employees Relationships		Formal	Informal
		4	5
Total		**35**	**31**

Table 3. The Company as described in July 2017

Imbalances or malfunctions could not be found by analysing the classical economic indicators. By strict reporting to them, the company's results were good. However, following global questionnaires made in 2013 and 2015, there may be some misunderstandings and discontents of the company's employees in France in 2015.

A very good correlation between the performance of the company as a result of changes made by the new management and the evolution of the complementarity factors in Table 3 was found. This concordance validates the analysis of complementarity factors as a pertinent diagnosis of a company's health and also the ability to suggest actions that can be taken to maximize the performance of the organization. It is possible to estimate the consequences of existing imbalances and propose solutions to remedy them.

The company is interested to progress toward the model 4.0 and will start with small steps by implementing first a proper ERP system. However before talking about the 4.0 version of this company, the management need to solve the current problems and this is facilitated by the usage of this instrument. The transition to 4.0 implies significant changes in current processes, and this will have to be forced by the external environment or the need to become more competitive.

5. Enterprise analysis 4.0 based on the evaluation tool

5.1. Simulation – The application of the assessment tool on a theoretical 4.0 enterprise

Doing a simulation for a 4.0 enterprise by using the complementarity factors, we will give a maximum score (5 points) to the complementary factor that is likely to be representative for a 4.0 enterprise and a minimum score (1 point) for the complementary factor that is unlikely to be representative or even defining the enterprise at all. The results are presented in table 4.

Elements of the organization	**Characteristics**	**Complementary factor centripetal**	**Complementary factor centrifugal**
Structure	Complexity	Vertical diferentiation	Orizontal diferentiation
		5	5
	Formalization	Clearly defined	Permanent adaptable
		5	5
	Centralizatiom	Concentrated	Dispersed
		5	5
Design		Functional	Divisional
		5	5
Management		Omnipotent	Simbolic
		5	5
Culture		Existent	Wanted
		5	5
Procese	Changing processess	Internal	External
		1	1
	Recruitment	Internal	External
		5	1
	Learning	Formal	Informal
		5	1
Strategy	Corporate strategy	Growing	Renewal
		5	5
	Competitive strategy	Better	Different
		5	1
Employees Relationships		Formal	Informal
		5	1
Total		**54**	**40**

Table 4. Enterprise 4.0 evaluated with the assessment tool presented previously

5.2. Interpretation

By doing detailed analysis, we notice that the differences between the complementary factors occur in the recruitment, training, competitive strategy and employee relations.

Considering the degree of complexity of specialization in the Enterprise 4.0, internal recruitment becomes a priority compare with the external recruitment. External recruitment brings, besides the new skills required by the job description, experiences from outside the system and new knowledge. Permanent recruitment from inside leads to recurrence and systematic errors.

Continuous automatization of processes determines the prioritization of formal training, standardization. Thus, the role of informal learning becomes smaller.

The competitive strategy for Enterprise 4.0 is based on sustained efficiency, continuous improvement of products and services, so the complementary factor of the 'better' competition strategy prevails over the 'different' one.

Email use, teleconferencing will reduce informal relationships, and the stimulation induced by direct contact between people tends to disappear.

It appears that there is a risk of diminishing creativity at the organization level, entering a recurrence that can cause problems for the organization. Competition can also appeal to the same technological solutions with similar results. Differentiation will be crucial in this case. An important element is the creative element, which is weakened by diminishing external recruitment, informal training and informal relationships among employees.

6. Conclusion

Although the instrument defined here is new and has been tested on one real enterprise, the results are encouraging and the recipient's feedback very good. Simulation on the model of an enterprise 4.0 has led to results and conclusions that outline the actions needed in such a company to achieve a good level of indicators of long-term performance. Both the analysis tool created and the schemes that show us the determinants and determinations can show the general situation in an enterprise. The diagnostic tool outlined in this paper aims to assist managerial decisions in this critical area for the company's long-term performance. It needs a refinement to highlight the problems that exist in a company. This can only be done by analysing several companies having different organizational cultures to obtain a general formula for application. These results are the premises of refining this tool by testing it on a larger number of interested companies.

7. References

[1] Armstrong, M; Taylor, S. (2006). Human Resource Management Practice, Kogan Page, ISBN 9780749469641, London, England

[2] Coteț, C. E.; Coteț, C. E. (2014). Managementul Resurselor Umane în Ingineria Industrială-Human Resources Management in Industrial Engineering, Printech, ISBN 978-606-23-0320-4,Bucharest, Romania

[3] Lorenz, M.; Rüßmann, M. & Strack, R. (2016) . Man and Machine in Industry 4.0, Boston Consulting Group, Available from: https://www.bcg.com Accessed: 2016-10-10

[4] McAuley, J.; Duberley, J. & Johnson, P. (2007). Organization Theory-Challenge and Perspectives, Pearson, ISBN 10: 0-273-68774-3,Harlow, England

[5] Robbins, S. P.; Coulter, M. (2012). Management, Pearson, ISBN 978-0-216384-2, Boston

[6] Robbins, S. P.; Judge, T. A. (2015). Essentials of Organizational Behavior, Pearson,ISBN:10 1292090073, Harlow England

[7] Robbins, S. P.; Judge, T. A. (2013). Organizational Behaviour, Pearson, ISBN 10: 0-13-283487-1, Boston

[8] Rüßmann, M.; Lorenz, M. & Gerbert, P. (2015). Industry 4.0 The future of Productivity and Growth in Manufacturing, Boston Consulting Group, Available from: https://www.bcg.com Accessed: 2016-10-10

[9] Sirkin, H. L.; Zinser, M. & Rose, J.R. (2015) – The Robotics Revolution- Next Great Leap in Manufacturing, Boston Consulting Group, Available from: https://www.bcg.com Accessed: 2016-10-10

[10] Teich, T.; Junghans, S. Kretz, D. & Trommer, M. (2016). Non-Hierarchical Services Networks, Steered by Competence Cells, Chapter 10 in DAAAM International Scientific Book 2016, pp.103-112, B. Katalinic (Ed.), Published by DAAAM International, ISBN 978-3- 902734-09-9, ISSN 1726-9687, Vienna, Austria

[11] Thietart, R.A. (2017) Le management-Management, Presses Universitaires de France, ISBN 9782130794721, Paris

[12] https://events.td.org/Events/ATD-Best/Winners, (2017). 2017 BEST Awards Winners, Accessed on: 2017-06-02

DOI: 10.2507/28th.daaam.proceedings.149

MULTI-RESPONSE OPTIMIZATION OF PLASMA CUTTING PARAMETERS USING GREY RELATIONAL ANALYSIS

Kenan Muhamedagic, Derzija Begic-Hajdarevic,
Cekic Ahmet & Muhamed Mehmedovic

This Publication has to be referred as: Muhamedagic, K[enan]; Begic-Hajdarevic, D[erzija]; Cekic, A[hmet] & Mehmedovic, M[uhamed] (2017). Multi-Response Optimization of Plasma Cutting Parameters using Grey Relational Analysis, Proceedings of the 28th DAAAM International Symposium, pp.1074-1082, B. Katalinic (Ed.), Published by DAAAM International, ISBN 978-3-902734-11-2, ISSN 1726-9679, Vienna, Austria
DOI: 10.2507/28th.daaam.proceedings.149

Abstract

In this paper was performed the optimization of the parameter of the plasma cutting process using Grey relational analysis. Design of experiments were carried out based on Taguchi L9 orthogonal array. Stainless steel X5CrNi18-10 plates with thickness of 5 mm were cut using different cutting speed and plasma gas pressure, while other process parameters were constant. The quality of the cut has been monitored by measuring surface roughness parameter *Rz*, cut perpendicularity and kerf width. Effect of each process parameter on quality characteristics of cut was shown. Optimal cutting parameters has been obtained by Grey relational analysis (GRA) method.

Keywords: plasma cutting; optimization; Grey relational analysis; Taguchi method; stainless steel

1. Introduction

Plasma arc cutting technology is one in which nitrogen, oxygen and compressed air are used to generate a plasma jet and then different type of materials such as nonferrous metal, stainless steel and others materials can be cut. During the plasma cutting process a source of heat is an electric arc generated by a plasma torch. The focus of manufacturers using plasma cutting process is optimization of productivity and the required quality of products made by plasma cutting. Both of these aspects are dependent on the appropriate choice of the process parameters. Plasma arc current, type gas, plasma gas pressure (plasma gas mass flow rate), cutting speed are parameters which have important effect on the quality characteristics. The effect of process parameters on the quality of the high tolerance plasma arc cutting of a 5 mm commercially pure titanium sheet was investigated in [1]. Authors concluded that the quality characteristics of the cutting edge of high tolerance plasma arc cutting of commercially pure titanium from a geometrical point of view need to be integrated with the considerations on the functional requirements for the cutting edge and on economical considerations that are partially correlated to microstructural modifications occurred at the cutting edge. A high tolerance plasma arc cutting system was used to cut plates of a 15 mm thick mild steel [2] and the arc voltage and cutting speed, plasma gas flow rate, shield gas flow rate and shield gas mixture are included in the analysis and their effects on kerf position and shape are evaluated.

It was shown that very good quality can be achieved for all the sides by varying the cutting speed and the arc voltage only. The effect of cutting speed and plasma gas pressure in plasma cutting of a 5 mm thick low alloy steel was analysed in [3]. It was shown that good quality cuts can be produced at the cutting speed from 400 to 700 mm/min and at the plasma gas pressure from 4 to 5 bar. Numerical method [4] and [5], genetic algorithms [6], statistical techniques [7] are common for the machining processes models. A new control strategy for plasma arc cutting was developed by [8] and this algorithm reduced the complexity of the nonlinear system modelling and achieved a real-time and online control for the cutting process by combining the advantages of fuzzy control and PID neural network control.

The regression analysis was used in [9] for the development of empirical models able to describe the effect of the process parameters on the cut quality in plasma arc cutting of a 15 mm thick mild steel sheets. It was found that the surface roughness and the kerf taper angle are mainly affected by the cutting height, whereas the heat affected zone is mainly influenced by the cutting current.

Investigation on the optimization and the effect of process parameters on material removal rate and surface roughness parameters during plasma arc cutting of EN31 steel using Taguchi method with grey relational analysis was presented in [10]. Also, analysis of variance was performed to get the contribution of each process parameters on the performance characteristics and it was observed that gas pressure is significant process parameters that affects the response.

The aim of this paper are to determine the effect and optimization of process parameters (cutting speed and plasma gas pressure) on surface roughness, kerf with and cut perpendicularity in plasma arc cutting of stainless steel a 5 mm thick using Grey relational analysis coupled with Taguchi method.

2. Experimental procedure

2.1. Test material, plasma cutting system and measuring equipment

The test material used in this experiment are X5CrNi18-10 steel plates. This steel belongs to austenitic stainless steel group. The chemical composition and mechanical properties of this steel are given in Tables 1 and 2.

Steel name	**Steel number**	**Chemical composition (percentages by mass)**							
		C	Si	Mn	P max.	S	N	Cr	Ni
X5CrNi18-10	1.4301	≤0,07	≤1,00	≤2,00	0,045	≤0,030	0,12-0,22	17-19,5	8,5-11,5

Table 1. Chemical composition of X5CrNi18-10 steel

Steel name	**Steel number**	**Maximum hardness HB**	**Minimum 0,2 % proof strength, $R_{p0,2}$ N/mm²**	**Tensile strength, R_m N/mm²**
X5CrNi18-10	1.4301	215	190	500-700

Table 2. Mehanical properties of X5CrNi18-10 steel

Cutting of test samples was performed using *the HiFocus 280i neo* plasma device. HiFocus technology is based on the plasma cutting principle with narrowed and stabilized plasma arc. Narrowing and stabilizing the plasma arc is achieved by using a smaller diameter nozzle, increased plasma gas rotation, and the additional application of the rotary gas to the plasma arc through a non-potential coaxial nozzle.

Measurement of surface roughness was performed on the *Mitutoyo SJ-210* measuring device. The surface roughness parameter *Rz* was measured at five different places along the cut edge. The *Mitutoyo TM-505* microscope was used to measure the cut perpendicularity and kerf width, which were measured at 9 different places along the edge of the cut.

2.2. Design of experiments

In this paper, a full experiment plan was used. Cutting speed and plasma gas pressure are varied for this experiment. Other process parameters were constant. A total of 9 samples were sampled for each combination of cutting speed and plasma gas pressure (Fig. 1).

Fig. 1. Test samples

In this experiment, there are two cutting parameters at three levels each. The cutting parameters levels are shown in the Table 3. Other constant parameters like current, voltage, nozzle distance from workpiece, types of plasma and secondary gas are shown in the Table 4.

Simbol	**Cutting parameter**	**Level**		
		1	2	3
A	Cutting speed V, mm/min	2000	2500	3000
B	Plasma gas pressure p, bar	6	8	10

Table 3. Cutting parameters and their levels

Constant parameter	**Value**
Current	60 A
Voltage	140 V
Nozzle distance	4 mm
Material thickness	5 mm
Plasma gas	N_2/H_2
Secondary gas	N_2

Table 4. Constant cutting parameters

The experimental layout for Taguchi L9 (3^2) orthogonal array is shown in Table 5.

Exp. No.	A Cutting speed	B Plasma gas pressure
1.	1	1
2.	1	2
3.	1	3
4.	2	1
5.	2	2
6.	2	3
7.	3	1
8.	3	2
9.	3	3

Table 5. Ortogonal array L9 (3^2) of the experimental run

3. Results and discussions

The aim of this paper is to determine the optimal plasma cutting parameters for minimizing surface roughness, kerf width and cut perpendicularity. The mean values of the surface roughness parameter *Rz*, kerf width and cut perpendicularity are given in Table 6.

Exp. No.	A Cutting speed *V*, mm/min	B Plasma gas pressure *p*, bar	Mean height of the profile *Rz*, µm	Cut perpendicularity *u*, µm	Kerf width *a*, µm
1.	2000	6	15,996	615,11	1991,1
2.	2000	8	13,756	514,05	2055,5
3.	2000	10	9,861	527,81	1971,4
4.	2500	6	8,192	424,82	1934,2
5.	2500	8	10,572	411,00	1990,7
6.	2500	10	11,851	421,59	1954,4
7.	3000	6	9,365	812,74	1871,4
8.	3000	8	10,272	426,77	1834,1
9.	3000	10	11,022	503,23	1962,8

Table 6. Experimental results

3.1 Effect of input parameters on the analyzed quality characteristics

In plasma cutting of termally cut materials, the most significant impact on the quality of the cut surface is the combination of input parameter. In this case, effect of cutting speed and plasma gas pressure on the roughness of the cut surface, cut perpendicularity and kerf width are analyzed.

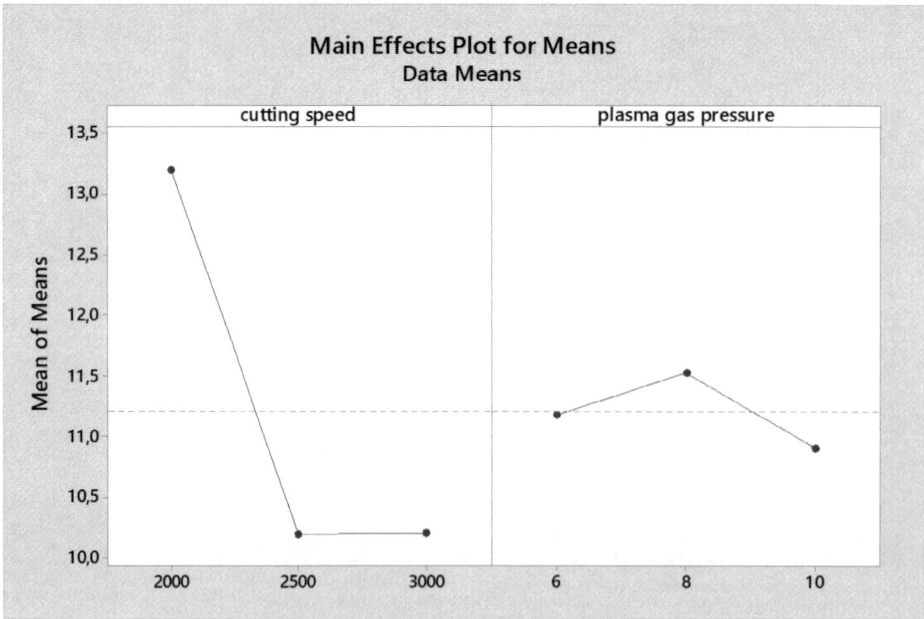

Fig. 2. Effect of cutting speed and plasma gas pressure on surface roughness

From Fig. 2. can be seen:

- When cutting speed increases from 2000 to 2500 mm/min, the surface roughness decreases rapidly. Increasing the speed from 2500 to 3000 mm / min does not have a significant impact on surface roughness.
- When plasma gas pressure increases from 6 to 8 bar, the surface roughness increases. Increasing plasma gas pressure from 8 to 10 bar, roughness decreases.

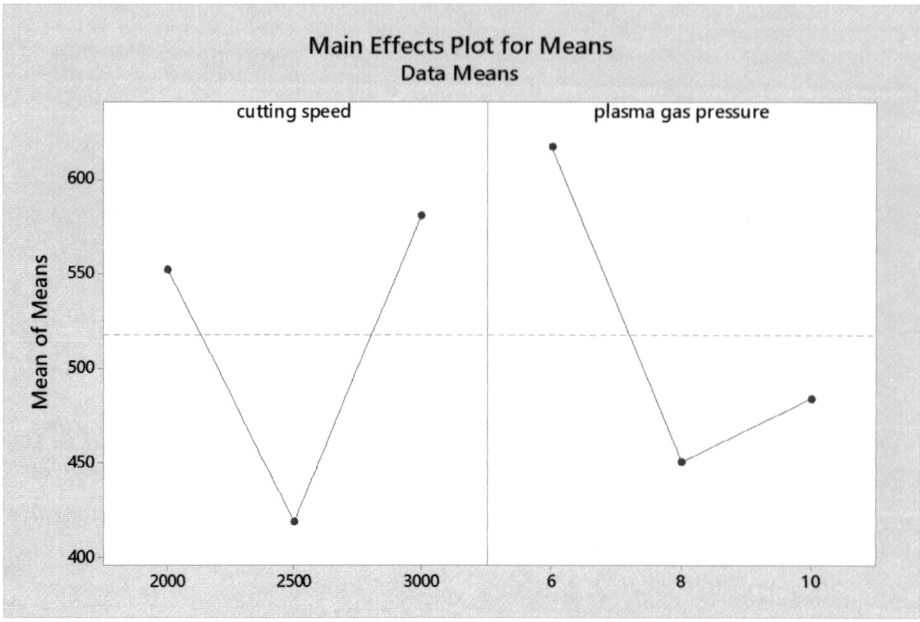

Fig. 3. Effect of cutting speed and plasma gas pressure on cut perpendicularity

From Fig. 3. can be seen:

- When cutting speed increases from 2000 to 2500 mm/min, the cut perpendicularity decreases. Increasing the speed from 2500 to 3000 mm / min, cut perpendicularity increases.
- When plasma gas pressure increases from 6 to 8 bar, the cut perpendicularity decreases. Increasing plasma gas pressure from 8 to 10 bar, cut perpendicularity increases.

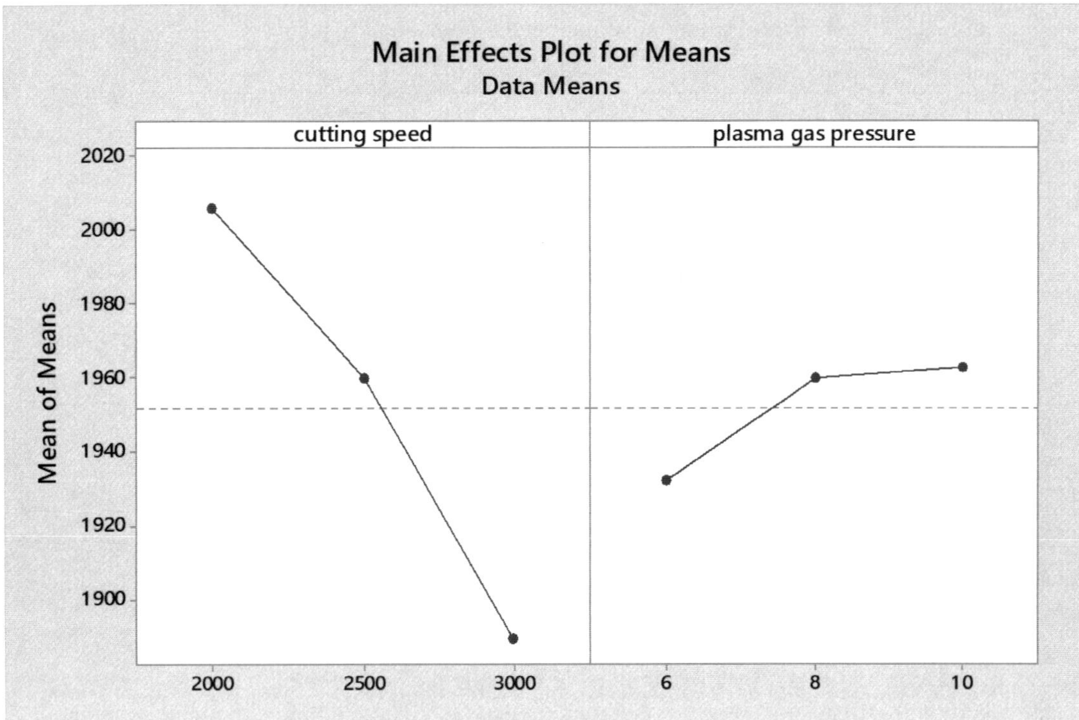

Fig. 4. Effect of cutting speed and plasma gas pressure on kerf width

From Fig. 4. can be seen:

- When cutting speed increases, kerf width decreases.
- When plasma gas pressure increases, kerf width also increases.

3.2 Grey Relational Analysis

Gray relational analysis is used for optimization multi-response characteristics. Optimization of such complicated characteristics can be converted into optimization single-response characteristic based on calculating Grey Relational Grade (GRG).

In this case, there are three response characteristics that should be minimzed. These are, the surface roughness parameter *Rz*, cut perpendicularity and kerf width.

The first step in the grey relational analysis is that the experimental results are normalized in the range between 0 and 1. All of three quality characteristics are required to be minimized, so *smaller-is-better* approach is used. This criterion can be expressed as:

$$x_i(k) = \frac{\max y_i(k) - y_i(k)}{\max y_i(k) - \min y_i(k)} \tag{1}$$

where, $y_i(k)$ is the *i*-th eksperimental result for the *k*-th process responses, $\max y_i(k)$ is the largest value of $y_i(k)$, $\min y_i(k)$ is the smallest value of $y_i(k)$, and $x_i(k)$ is the normalized value of $y_i(k)$.

The normalized values of surface roughness, cut perpendicularity and kerf width calculated by (1) are shown in Table 7. After normalization, the deviation sequence was calculated by (2) and given in Table 7.

$$\Delta_{0i}(k) = |x_0(k) - x_i(k)| \tag{2}$$

where, $\Delta_{0i}(k)$ is difference of the absolute value between $x_0(k)$ and $x_i(k)$. $x_0(k)$ is the reference sequence of the *k*-th process response.

Exp. No.	Normalized values of responses			Deviation sequences $\Delta_{0i}(k)$		
	Rz	*u*	*a*	*Rz*	*u*	*a*
1.	0	0,492	0,291	1	0,508	0,709
2.	0,287	0,743	0	0,713	0,257	1
3.	0,786	0,709	0,380	0,214	0,291	0,620
4.	1	0,966	0,548	0	0,034	0,452
5.	0,695	1	0,293	0,305	0	0,707
6.	0,531	0,974	0,457	0,469	0,026	0,543
7.	0,850	0	0,832	0,150	1	0,168
8.	0,733	0,961	1	0,267	0,039	0
9.	0,637	0,770	0,419	0,363	0,230	0,581

Table 7. Normalized values and deviation sequences of responses

The next step is to determine grey relational coefficient based on normalized values of responses. The grey relational coefficients (GRC) can be calculated as:

$$\xi_i(k) = \frac{\Delta_{min} + \varsigma\Delta_{max}}{\Delta_{0i}(k) + \varsigma\Delta_{max}} \tag{3}$$

where, Δ_{min} denotes smallest value of $\Delta_{0i}(k)$, Δ_{max} denotes largest value of $\Delta_{0i}(k)$, and ς is a distinguishing coefficient. Its value can be chosen in the range of 0 to 1. Generally, this coefficient has value of 0,5.

Average value of the Grey relational coefficients for all responses determine the Grey relational grade (GRG). Grey relational grade can be calculated by (4).

$$\gamma_i = \frac{1}{n}\sum_{k=1}^{n} \xi_i(k) \tag{4}$$

where, $\xi_i(k)$ is the Grey relational coefficient of *k*-th process response in *i*-th eksperiment and n is the number of process responses.

The calculated values of Grey relational coefficient and Grey relational grade are given in Table 8.

Exp. No.	Grey relational coefficient			Grey relational grade γ_i	
	Rz	*u*	*a*	γ_i	Rank
1.	0,333	0,496	0,414	0,414	9
2.	0,412	0,661	0,333	0,469	8
3.	0,700	0,632	0,446	0,593	6
4.	1	0,936	0,525	0,820	2
5.	0,621	1	0,414	0,678	3
6.	0,516	0,950	0,479	0,648	4
7.	0,769	0,333	0,748	0,617	5
8.	0,652	0,927	1	**0,860**	**1**
9.	0,580	0,685	0,462	0,576	7

Table 8. Grey relational coefficients and Grey relational grades

The higher value of Grey relational grade represents that the corresponding process reponses are close to the optimal condition. In this case, it is experiment number 8 with the best combination of process parameters for surface roughness, cut perpendicularity and kerf width among 9 experiments.

The means of the Grey relational grade for each level of cutting speed and plasma gas pressure are given in Table 9. According to the results shown in Table 9, the cutting speed has the most significant effect in reducing surface roughness, cut perpendicularity and kerf width.

Level	**Cutting parameters**	
	A: Cutting speed	**B: Plasma gas pressure**
1	0,492	0,617
2	**0,715**	**0,669**
3	0,684	0,606
Δ	0,223	0,063
Rank	1	2
Total mean of Grey relational grade = 0,6305		

Table 9. Response table for the mean Grey relational grade

Fig. 5. shows the response graph for mean Grey relational grade from which the optimal combination of cutting parameters can be determined. This is the combination of parameters for which the highest mean Grey relational grade are obtained. Therefore, optimal plasma cutting parameters for X5CrNi18-10 stainless steel plates with respect to surface roughness, cut perpendicularity and kerf width are, cutting speed of $V = 2500$ mm/min and plasma gas pressure of $p = 8$ bar.

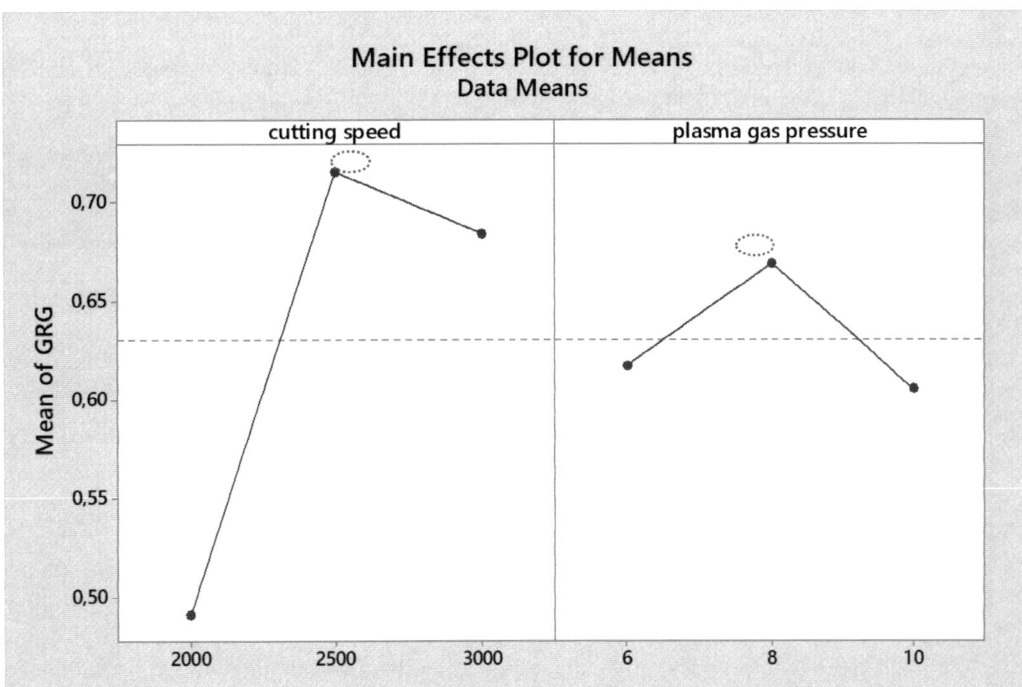

Fig. 5. Response graph for mean Grey relational grades with selected optimal cutting parameters

4. Conclusion

In this study, the Grey relational analysis was applied for optimization of the parameters of the plasma cutting process. Stainless steel X5CrNi18-10 plates with thickness of 5 mm were cut using different cutting speed and plasma gas pressure. Three output process parameters, surface roughness, cut perpendicularity and kerf width, were monitored. The optimal plasma cutting parameters are: cutting speed $V = 2500$ mm/min and plasma gas pressure $p = 8$ bar. The same parameters are recommended by the manufacturer of cutting equipment.

However, the author's recommendation is that a lower plasma gas pressure ($p = 6$ bar) can be used at the same cutting speed of 2500 mm/min, as the cut quality remains the same and lower production costs are achieved. In both cases (plasma gas pressure, $p = 6$ bar and p = 8 bar) the cuts belong to the same quality class according to EN ISO 9013.

The experimental result showed that the cutting speed has the most significant effect in reducing surface roughness, cut perpendicularity and kerf width.

The proposal for future research is that other input parameters (current, type of plasma gas, nozzle distance, etc.) are varied besides the cutting speed and the plasma gas pressure, and then to analyse the influence of these parameters on the cut quality for different thickness of material.

5. Acknowledgments

The authors acknowledge the project (Modern machining processes) was financially supported by the Ministry of Education, Science and Youth of Sarajevo Canton.

6. References

[1] Gariboldi, E. & Previtali, B. (2005). High tolerance plasma arc cutting of commercially pure titanium. Journal of Materials Processing Technology, 160, (2005) 77-89, ISSN 0924-0136

[2] Bini, R.; Colosimo, B.M.; Kutlu, A.E. & Monno, M. (2008). Experimental study of the features of the kerf generated by a 200A high tolerance plasma arc cutting system. Journal of Materials Processing Technology, 196, (2008) 345-355, ISSN 0924-0136

[3] Begic, D.; Kulenovic, M.; Cekic, A. & Dedic, E. (2012). Some experimental studies on plasma cutting quality of low alloy steel, Annals of DAAAM for 2012 & Proceedings of the 23rd International DAAAM Symposium, Vol.23, No.1, Vienna, Austria, ISSN 2304-1382, ISBN 978-3-901509-91-9, Katalinic, B. (Ed.), pp. 0183-0186, DAAAM International Vienna, Vienna

[4] Freton, P.; Gonzalez, J.J.; Gleizes, A.; Camy Peyret, F.; Caillibotte, G. & Delzenne, M. (2002). Numerical and experimental study of plasma cutting torch. Journal of Physics D: Applied Physics, 35, 2 (2002) 115-131, ISSN 1361-6463

[5] Begic-Hajdarevic, D. & Bijelonja, I. (2014). Experimental and numerical investigation of temperature distribution and hole geometry during laser drilling process. Procedia Engineering, 100, (2014) 384-393, ISSN 1877-7058

[6] Ficko, M. & Palcic, I. (2013). Designing a layout using the modified triangle method, and genetic algorithms. International Journal of Simulation Modelling, 12, 4 (2014), 237-251, ISSN 1726-4529

[7] Begic-Hajdarevic, D.; Vucijak, B.; Pasic, M. & Bijelonja, I. (2017). Analysis of the influence of cutting parameters on surface roughness in laser cutting process of tungsten alloy using control charts. Tehnicki vjesnik – Technical Gazette, 24, Suppl. 2 (2017) 339-344, ISSN 1848-6339

[8] Deli, J. & Bo, Y. (2011). An intelligent control strategy for plasma arc cutting technology. Journal of Manufacturing Processes 13 (2011) 1–7, ISSN 1526-6125

[9] Salonitis, K. & Vatousianos, S. (2012). Experimental Investigation of the Plasma Arc Cutting Process. Procedia CIRP 3 (2012) 287 – 292, ISSN 2212-8271

[10] Kumar Das, M.; Kumar, K.; Barman, T. & Sahoo, P. (2014). Optimization of process parameters in plasma arc cutting of EN 31 steel based on MRR and multiple roughness characteristics using grey relational analysis. Procedia Materials Science, 5 (2014), 1550-1559, ISSN 2211-8128.

DOI: 10.2507/28th.daaam.proceedings.150

ESTIMATION OF THE LASER CUT PERPENDICULARITY USING LINEAR REGRESSION ANALYSIS

Mirza Pasic, Derzija Begic-Hajdarevic & Izet Bijelonja

This Publication has to be referred as: Pasic, M[irza]; Begic-Hajdarevic, D[erzija] & Bijelonja, I[zet] (2017). Estimation of the Laser Cut Perpendicularity Using Linear Regression Analysis, Proceedings of the 28th DAAAM International Symposium, pp.1083-1089, B. Katalinic (Ed.), Published by DAAAM International, ISBN 978-3-902734-11-2, ISSN 1726-9679, Vienna, Austria
DOI: 10.2507/28th.daaam.proceedings.150

Abstract

The aim of this paper is to develop a mathematical model using a linear regression analysis for the laser cut perpendicularity depending on the cutting parameters and to compare the values of the cut perpendicularity according to ISO 9013:2002 (E) standard. Cutting parameters that are used as independent variables in this research are: laser power P, cutting speed v and assist gas pressure p. During the experiment 75 samples of 1.4307 DIN stainless steel and 1 mm of thickness were cut on HSG fiber laser using nitrogen N_2 as an assist gas. Each sample was cut using different combination of values of independent variables. Cut perpendicularity was measured on these samples with total of 225 measurements. Obtained results show that the cutting speed is only significant variable in the model and that the linear regression analysis is suitable for the analysis of the cut perpendicularity. Every experimental value of the cut perpendicularity is located in the first quality category according to ISO 9013:2002 (E) standard.

Keywords: laser cutting; fiber laser; regression analysis; cut perpendicularity; stainless steel

1. Introduction

Multiple regression analysis is shown to be useful technique for mathematical modeling of many industrial engineering processes including material cutting process. These processes happen as result of interaction of variables. Some of these variables are independent, while some are dependent. To explain these processes, it is important to determine relations between these variables and to create the model that will explain certain process and prediction of dependent variables [1].

During the laser cutting process there are many variables. Some of them are independent variables, known as cutting parameters, defined before laser cutting process, while some are dependent variables which determine cut quality and their values depend on variability of independent variables. Regression analysis is used to obtain mathematical model that explains variability of dependent variables based on variability of independent variables.

Nowadays in many fields like informatics, measurement technique, telecommunications and medicine there is increasing need for the high precision of small parts and for reduction of material and energy used in production processes. The use of laser is more and more present and necessary in industrial practice. Laser cutting is the modern cutting method and its use is in expansion because of many advantages in comparison to conventional cutting processes. Laser cutting is thermal cutting process and has advantages even to other thermal cutting processes [2, 3]. Advantages of the laser cutting are high quality of the cut surface, narrow kerf width, low heat affected zone and sharp and perpendicular edges. However, during the laser cutting process the cut quality can vary depending on many factors such as cutting parameters, type of assist gas and type of material [4].

Many papers examine influence of these factors on certain aspects of the cut quality. In [5] influence of cutting speed, focus position, assist gas pressure and stand-off on surface roughness R_a and heat affected zone using two different types of steel is investigated. It is concluded that R_a is smaller when the cutting speed is higher and with focus position above the surface, while the assist gas pressure does not affect R_a.

In [6] dependency of kerf width is analyzed based on variability of laser power and type of assist gas. The cutting is performed on 3 stainless steel plates of different thickness. It is concluded that with increase of energy per unit of length there is increase of the kerf width. With the same laser power kerf width is higher if nitrogen is used as an assist gas instead of oxygen.

In [7] influence of laser power and cutting speed on surface roughness R_a is examined during CO_2 laser cutting using oxygen as assist gas. Influence of increase of cutting speed on surface roughness and heat affected zone is analyzed in [8]. Research [9] uses regression analysis for development of model that describes influence of laser power on kerf width, surface roughness, heat affected zone and striation frequency.

Regression analysis is shown to be very reliable method for analysis of parameters and quality of laser cutting process. Analysis of research papers shows that general model of dependency of the laser cutting quality on cutting parameters has not be developed yet, but each experiment is special case depending of laser type, type of assist gas, thickness of material etc.

In this paper regression analysis is used to analyze cut perpendicularity with respect to cutting parameters: laser power, cutting speed and assist gas pressure during fiber laser cutting with N_2 as assist gas. Multiple regression analysis is performed to develop mathematical model to predict cut perpendicularity and to determine which cutting parameters are significant for cut perpendicularity.

2. Experimental setup

In this experiment 75 samples of 1.4307 DIN (X2CrNi18-9) stainless steel of 1 mm of thickness were cut on HSG fiber laser using nitrogen N_2 as an assist gas. HSG fiber laser with nominal power of 1000 W was used in this experiment. Cutting parameters that were varied in this experiment were laser power P, cutting speed v and assist gas pressure p. These cutting parameters were used as independent variables in the linear regression analysis. During cutting process five different values of laser power and cutting speed and three different values of assist gas pressure were used. Those values are shown in table 1, table 2 and table 3.

Laser Power P	Values
P_1	800 W
P_2	850 W
P_3	900 W
P_4	950 W
P_5	1000 W

Table 1. Laser power values

Cutting speed v	Values
v_1	320 mm/s
v_2	300 mm/s
v_3	280 mm/s
v_4	260 mm/s
v_5	240 mm/s

Table 2. Cutting speed values

Assist gas pressure p	Values
p_1	1 bar
p_2	1,3 bar
p_3	2 bar

Table 3. Assist gas pressure values

Cutting parameters that were constant during the cutting process are presented in table 4.

Constant cutting parameters	Values
Focus position	-1 mm
Distance of the nozzle	1 mm
Diameter of the nozzle	1,5 mm

Table 4. Values of cutting parameters that were constant during the cutting process

Each sample was made with different combination of values of cutting parameters. There are 75 different combinations of values of cutting parameters and thus 75 samples were cut. Kerf width was measured by microscope Mitutoyo TM 505 and camera Motic 10 MP. Camera was placed on the microscope and they both were connected to computer via software Motic Images Plus 3.0.

On each sample nine places were marked on both input and output side of the laser beam side of the sample. By microscope and camera, on each marked spot, kerf was captured and the picture was downloaded to the computer. After that kerf width was measured on each captured picture using software Motic Images Plus 3.0. Nine measurements of the kerf width were made on each side of the sample, leading to 675 measurements for input and 675 measurements for output side of laser beam.

Next step was to calculate average values of the kerf width for input and output side of laser beam for each sample. Difference between average values of the kerf width on the input and the output side of the laser beam was calculated for each sample. This difference is called cut perpendicularity. There were 75 values of cut perpendicularity. After the measurements, linear regression analysis was done using cut perpendicularity as dependent variable and laser power P, cutting speed v and assist gas pressure p as independent variables. These 75 values of cut perpendicularity were categorized according to the ISO 9013:2002 (E) standard.

3. Analysis of results

Based on the experiment, backward multiple regression model was developed using sample size $n = 75$ and number of independent variables $k = 3$. Regression analysis results are given in table 5.

Regression Statistics	
Multiple R	0,732
R Square	0,536
Adjusted R Square	0,517
Standard Error	6,961
Observations	75

ANOVA

	df	SS	MS	F	Significance F
Regression	3	3980,351	1326,784	27,379	0,000
Residual	71	3440,641	48,460		
Total	74	7420,992			

	Coefficients	Standard Error	t Stat	P-value	Lower 95%	Upper 95%
Intercept	25,469	13,274	1,919	0,059	-0,999	51,937
P	-0,078	0,011	-6,889	0,000	-0,101	-0,056
v	0,166	0,028	5,846	0,000	0,109	0,223
p	1,373	1,918	0,716	0,477	-2,452	5,198

Table 5. First results of regression analysis

Calculated F test value ($F_{stat} = 27{,}379$) is greater than the critical F value ($F_{crit} = 2{,}734$) for $\alpha = 0{,}05$ level of significance with 3 and 71 degrees of freedom which means that there is at least one regression coefficient b_i ($i = 1, 2, 3$) different from zero. Comparing statistical t values with critical t value, as well as comparing p values with $\alpha = 0{,}05$ it can be seen that b_1 and b_2 regression coefficients are statistically different from zero, while b_3 is zero as shown in (1), (2) and (3).

$$b_1: t_{stat}(P) = -6{,}889 < t_{crit} = -1{,}99;\ p - value(P) = 0 < \alpha = 0{,}05;\ \Rightarrow\ b_1 \neq 0 \quad (1)$$

$$b_2: t_{stat}(v) = 5{,}846 > t_{crit} = 1{,}99;\ p - value(v) = 0 < \alpha = 0{,}05;\ \Rightarrow\ b_2 \neq 0 \quad (2)$$

$$b_3: t_{crit} = -1{,}99 < t_{stat}(p) = 0{,}716 < t_{crit} = 1{,}99; p - value(p) = 0{,}477 > \alpha = 0{,}05; \Rightarrow\ b_3 = 0 \quad (3)$$

Since assist gas pressure is not significant variable, regression analysis was done again, this time with cutting speed and laser power as independent variables. Regression analysis results are presented in table 6.

Regression Statistics	
Multiple R	0,656
R Square	0,431
Adjusted R Square	0,415
Standard Error	4,163
Observations	75

ANOVA

	df	SS	MS	F	Significance F
Regression	2	943,428	471,714	27,219	0,000
Residual	72	1247,785	17,330		
Total	74	2191,213			

	Coefficients	Standard Error	t Stat	P-value	Lower 95%	Upper 95%
Intercept	-28,259	7,766	-3,639	0,001	-43,740	-12,778
P	0,002	0,007	0,365	0,716	-0,011	0,016
v	0,125	0,017	7,369	0,000	0,091	0,159

Table 6. Second results of regression analysis

For $\alpha = 0{,}05$ level of significance with 2 and 72 degrees of freedom $F_{crit} = 3{,}124$. From table 4 it can be seen that $F_{stat} = 27{,}219$. F test shows that at least one out of two regression coefficients is different from zero. Comparing t_{stat} values for each regression coefficient with critical t value ($t_{crit} = \pm 1{,}99$) it can be concluded that laser power is not significant variable in the model. Finally, regression analysis with cutting speed as only independent variable was performed. Results of regression analysis are presented in table 7.

Regression Statistics	
Multiple R	0,655
R Square	0,429
Adjusted R Square	0,422
Standard Error	4,138
Observations	75

ANOVA

	df	SS	MS	F	Significance F
Regression	1	941,113	941,113	54,957	0,000
Residual	73	1250,099	17,125		
Total	74	2191,213			

	Coefficients	Standard Error	t Stat	P-value	Lower 95%	Upper 95%
Intercept	-26,023	4,754	-5,474	0,000	-35,499	-16,548
v	0,125	0,017	7,413	0,000	0,092	0,159

Table 7. Third results of regression analysis

From table 5 it can be seen that cutting speed is significant variable since $t_{stat} = 7{,}413 > t_{crit} = 1{,}99$. Regression equation for the cut perpendicularity U is given by (4).

$$U = -26{,}023 + 0{,}125 \cdot v \tag{4}$$

Figures 1 and 2 depict diagrams of residuals and standardized residuals respectively.

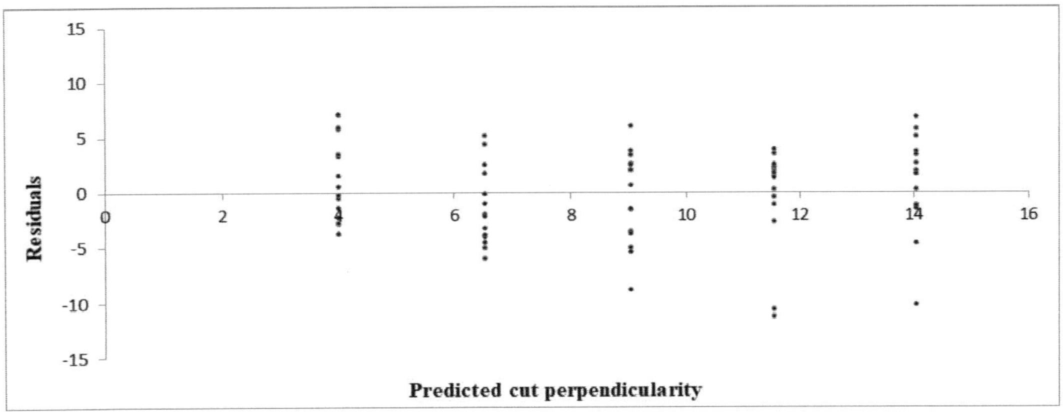

Fig. 1. Residuals vs. predicted cut perpendicularity

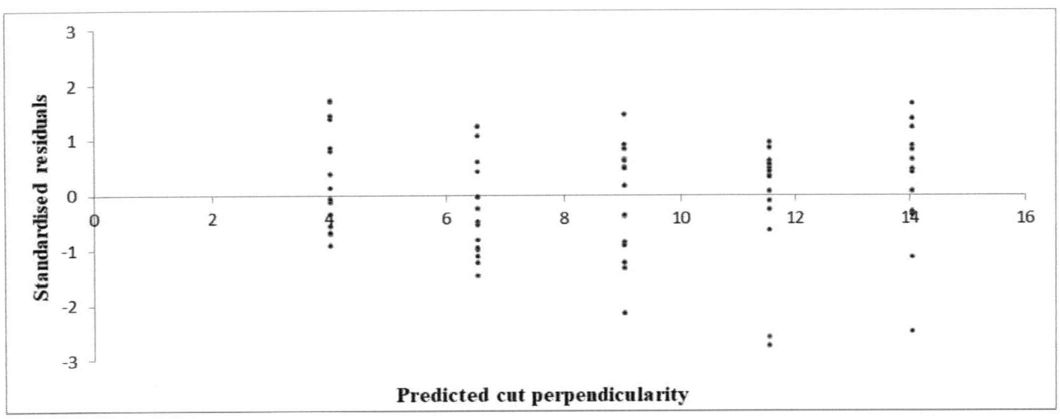

Fig. 2. Standardized residuals vs. predicted cut perpendicularity

From figures 1 and 2 it can be seen that the values of residuals are located around zero value without visible pattern, while standardized residuals are within three standard deviations. There are no extreme values of residuals and it can be concluded that regression equation is suitable to predict cut perpendicularity with respect to cutting speed. Histogram of residuals is depicted on figure 3.

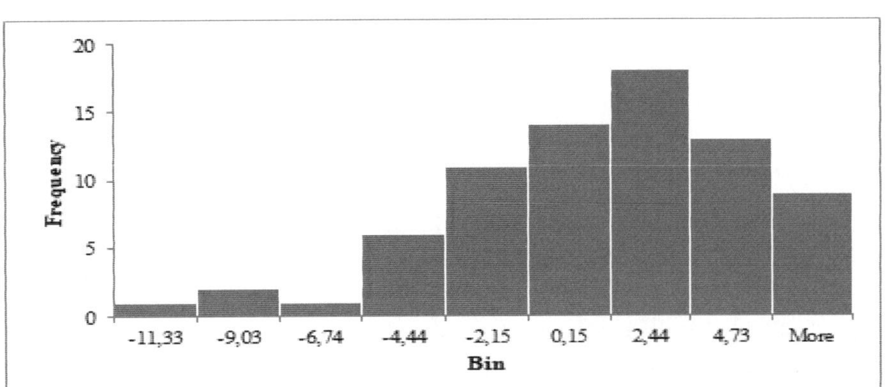

Fig. 3. Histogram of residuals

From figure 3 it can be seen that the residuals are approximately normally distributed and that the condition of normality of residuals is satisfied.

Coefficient value for the cutting speed is 0,125 which means that with the unit change of cutting speed the cut perpendicularity change is 0,125. Standardized coefficient value for cutting speed is 0,356 which means that with unit change of standard deviation of cutting speed, the change of the cut perpendicularity is 0,356 standard deviation. Since coefficient of determination is $R^2 = 0{,}429$, it can be concluded that 42,9% of variability of the cut perpendicularity is explained with variability of the cutting speed.

4. Verification of results according to ISO standard

During the experiment for the each sample average cut perpendicularity was computed, so 75 values of the variable cut perpendicularity was obtained. Since cut perpendicularity depends on the cutting speed, diagram of the cut perpendicularity vs. cutting speed was made as presented in figure 4. Diagram shows average values of cut perpendicularity, as well as horizontal line which presents the border of the first quality category of the cut perpendicularity according to ISO 9013:2002 (E) standard.

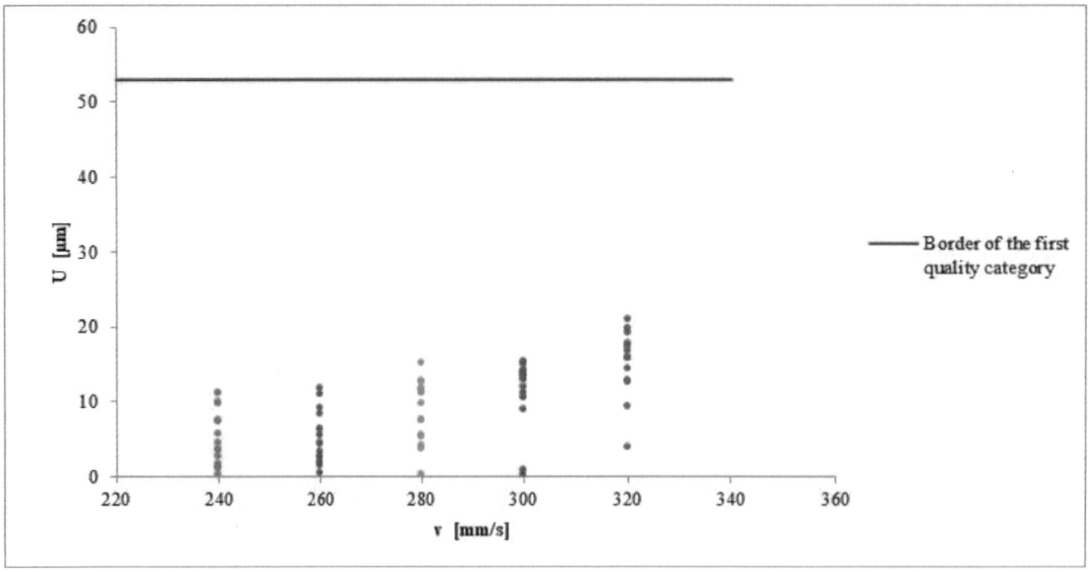

Fig. 4. Cut perpendicularity values

From figure 4 it can be seen that all average values of the cut perpendicularity are located well beyond the border of the first quality category of the cut perpendicularity according to ISO 9013:2002 (E) standard. It means that there are no large deviations of the kerf width at input and output side of the laser beam.

5. Conclusion

In this research the stainless steel 1.4307 DIN, 1 mm of thickness was cut by HSG fiber laser. The effect of laser power, cutting speed and assist gas pressure on the cut perpendicularity was examined. Backward elimination linear regression analysis was used to create mathematical model for the estimation of the laser cut perpendicularity. From the regression model, it can be concluded that cutting speed is the only significant independent variable and that 42,9% variability of the cut perpendicularity is explained by the cutting speed. According to the model, cut perpendicularity increases with the increase of the cutting speed.

Analysis of residuals shows that there is no visible pattern. Also, analysis of standardized residuals shows that all standardized residuals are within three standard deviations. Histogram of residuals shows that residuals are approximately normally distributed. It can be concluded that there are no extreme measured values and that the obtained regression model is suitable to predict values of the laser cut perpendicularity depending on the cutting speed.

Every experimental value of the cut perpendicularity is located in the first quality category according to ISO 9013:2002 (E) standard. The recommendation could be to use the highest experimental speed of 320 mm/s so to increase efficiency of the cutting process.

In the future research experiment with the same combination of the laser power and assist gas pressure values could be performed, but to use higher cutting speeds. In that way, it can be determined at which cutting speed the value of the cut perpendicularity moves into the second category of the quality according to ISO 9013:2002 (E) standard.

6. References

[1] Levine, D. M.; Stephan D. F.; Krehbiel, T. C. & Berenson, M. L. (2006). Business statistics, a first course, sixth edition, Pearson Education Inc., ISBN 0-13-611349-4, New Jersey

[2] Al-Sulaiman, F. A.; Yilbas, B. S. & Ahsan M. (2006). CO_2 laser cutting of a carbon/carbon multi-lamelled plain-weave structure, Journal of Material Processing Technology, Vol. 173, No. 3, (April 2006) page numbers (345-351), ISSN 0924-0136

[3] Yan, Y.; Lin, L; Kursad S.; Wei W.; David W.; Lingfei J.; Yong B. & Yijian J. (2011). CO2 laser underwater machining of deep cavities in alumina, Journal of the European Ceramic Society, Vol. 31, No. 15, (December 2011) page numbers (2793-2807), ISSN 0955-2219

[4] Avanish, K. D. & Vinod, Y. (2008). Laser beam machining – A review, International Journal of Machine Tools & Manufacture, Vol. 48, No. 6, (May 2008) page numbers (609-628), ISSN 0890-6955

[5] Cekic, A.; Begic-Hajdarevic, D; Kulenovic, M. & Omerspahic, A. (2013). CO_2 laser cutting of alloy steels using N_2 assist gas, Procedia Engineering, Vol. 69, page numbers (310-315), ISSN 1877-7058

[6] Baumeister, M.; Dickmann, K. & Hoult, T. (2006). Fiber laser micro-cutting of stainless steel sheets, Applied Physics A, Vol. 85, No. 2, (November 2006) page numbers (121-124), ISSN 0947-8396

[7] Begic-Hajdarevic, D.; Pasic, M.; Vucijak B. & Cekic A. (2016). Statistical process control of surface roughness during CO2 laser cutting using oxygen as assist gas, Proceedings of the 26th DAAAM International Symposium, ISSN 1726-9679, ISBN 978-3-902734-07-5, Katalinic, B. (Ed.), pp. 0247-0255, Published by DAAAM International, Vienna, Austria, DOI:10.2507/26th.proceedings.034

[8] Jarosz, K.; Löschner, P. & Nieslony, P. (2016). Effect of cutting speed on surface quality and heat-affected zone in laser cutting of 316L stainless steel, Procedia Engineering, Vol. 149, page numbers (155-162), ISSN 1877-7058

[9] Rajaram, N.; Sheikh-Ahmad, J. & Cheragi, S. H. (2003). CO_2 laser cut quality of 4130 steel, International Journal of Machine Tools & Manufacture, Vol. 43, No. 4, (March 2003) page numbers (351-358), ISSN 0890-6955

DOI: 10.2507/28th.daaam.proceedings.151

PERFORMANCE INDICATORS OF PRODUCTION PROCESSES

Ajdin Vatres*, Izet Bijelonja, Mugdim Pasic

This Publication has to be referred as: Vatres, A[jdin]; Bijelonja, I[zet] & Pasic, M[ugdim] (2017). Performance Indicators of Production Processes, Proceedings of the 28th DAAAM International Symposium, pp.1090-1095, B. Katalinic (Ed.), Published by DAAAM International, ISBN 978-3-902734-11-2, ISSN 1726-9679, Vienna, Austria
DOI: 10.2507/28th.daaam.proceedings.151

Abstract

In this paper results of the research of the application of methods for designing and managing production processes as well as the presence of different process performance indicators in companies in Bosnia Herzegovina are presented. A survey questionnaire was designed to obtain required data for the analysis. The questionnaire consisted of two parts. The first part of the questionnaire contained questions about general characteristics of the companies, while the other part was related to the applied methods and tools for production process design and methods of evaluating the performance of the process. According to the analysis it can be concluded that 56,3 % of companies use production management software solutions like Pantheon, F 300, MIS, SAP. At the same time 37,5 % of the analysed companies use software packages to design their production process. In 62,5 % of the cases the companies analyse the performance indicators of the production process on a monthly basis.

Keywords: business process; performance indicators measurement; process modelling tools; production procedures

1. Introduction

Performance can be viewed simply as a company's ability to consistently achieve its goals. However, if measurement is carried out too often, it uses more resources and becomes counterproductive. Therefore, every organization should focus on the critical areas of its processes in tailormade fashion [1].

Depending on the company's priorities, key performance indicators (KPI) can be financial and non-financial measures that organizations use to determine how successful they are in accomplishing long lasting goals [2]. Performance indicators can be implemented within one manufacturing company, a network of manufacturing companies, a complete supply chain, as well as the complete distribution channel of a product. Through the development and implementation of key performance indicators, the company aims to monitor product performance, supplier selection, distribution channel management, sales and production planning. This requires significant financial resources [3]. In the paper [4] an integrated approach is proposed in order to prioritize key performance indicators. It is noted that, for instance, the size and complexity of organizations can affect the type and number of selected KPIs. Therefore it is imperative to select and utilize the most appropriate indicators.

Research paper [5] shows that traditional performance measures have many limitations. They are based on outdated traditional cost management systems, lagging metrics, not related to corporate strategy, inflexible, expensive and contradict continuous improvement. The limitations of traditional performance measures resulted in research that confirmed the necessity of introducing a new set of operational efficiency measures that should provide the necessary and timely information for day-to-day decision making. These measures should be flexible, primarily non-financial and can be changed as needed. One of the possible alternatives is time-based performance measurement, with its main drawback in overstating the role of time and neglecting the impact of other performance performance measures. Another way to respond to traditional problems is the development of integrated performance measurement systems, which also have their own limitations.

In [6] process modeling techniques are classified as diagrammatic models, formal/ mathematical models, business process languages programming languages. When speaking of graphing techniques, it is pointed out that the unavoidable advantage of the graphical approach is its simplicity and ease of creation, as well as the fact that no technical expertise is required for its analysis, which is why it still has a wide application in business practice. However, modeling based on pure graphic notation lacks formal semantics and quantitative information which results in its inconsistency and inadvertence. Furthermore, every analysis based on graphical model types often comes down to a visual overview of a chart that depends on the skills of an analyst, and is highly subjective.

The need for formal semantics in business modeling has led to the development of the second generation of models. Analyzing formal modeling techniques, in [6] it is emphasized that formal processes model concepts are defined rigorously and precisely, enabling them to mathematically analyze the consistency of the process and its other characteristics. On the other hand, the lack of formal methods to support process design is that the elements and constraints of business processes are mostly of a qualitative nature and are difficult to characterize in a formal way that can be matched by mathematical analysis methods. In [6] it is presented that mathematical representation of the process does not allow to describe the complexity of the actual environment (decision making, feedback loop, parallel or hierarchical flow). Also, complex mathematical notations are required for maintenance, which acts discouraging at business analysts.

Programming languages as the third and the latest generation of business process modeling techniques are the result of the need to solve problems arising from the complexity of formal techniques, which would preserve the consistency and potential of formal analysis. Languages for process modeling include elements and formal modeling and visualization of the business process in terms of building standardized and repeatable models [6].

As indicated in the research paper [7] during the late nineties it became imperative for successful organization of a production company to do process oriented reorganization of all business flows, and it was a regular practice for large companies. Today that is true for small and medium sized companies too. That is why the paper [7] gives a concept of introducing information systems while using the moment of the information system implementation to change and improve business processes and to continuously improve processes. This implementation concept allows parallel improvement of the business processes of the companies and business process automation. It allows for transparent measurement of improvement of business process efficiency.

Article [8] presents a set of criteria and proposes guidelines for the selection of the most beneficial business process management system. The proposal has been developed on the basis of literature overview and the feedback from practical experience. The research done in this paper supports the use of a complex set of criteria which is intentionally quite narrow and general, so that it can be applied to different organizations from different industries. However, specific criteria must be added in order to assure the applicability of guidelines in specific cases.

Paper [9] presents the results of empirical research conducted in Croatian and Slovenian companies. The main objective of our research was to determine whether a higher level of business process orientation resulted in better organizational performance in companies in transitional economies. The analyzed data supported this claim.

Research [10] extends on the research carried out by Skrinjar and associates (2008). The findings show that managers should not focus on financial indicators alone. Monitoring of various nonfinancial performance indicators significantly contributes to making good decisions in real time.

In this paper methods of designing and managing production processes as well as the presence of different process indicators are examined in companies in Bosnia and Herzegovina.

2. Research methodology

For the purpose of research of the methods of designing and managing production processes as well as the presence of different process indicators in companies in Bosnia and Herzegovina a survey questionnaire was designed. The survey questionnaire was sent electronically to 100 companies. The company list was created from the data obtained from the Chamber of Commerce of the Federation of Bosnia and Herzegovina. The response rate was 16 %.

Regarding the type of work performed by the surveyed companies, it should be noted that this is a broad spectrum of manufacturing activities: production of chemicals; ammunition production; production of primary aluminum and aluminum alloys; metal machining, fabrication and machining of metal structures; manufacture of electrical equipment for motor vehicles; cement production; production of air and liquid filters; production of boilers and biomass boilers; production of PVC and ALU carpentry, thermal insulation glass, metal plasticizing; production of plastic packaging; manufacture of furniture; production of textile for the auto industry.

The questionnaire contained 26 questions with pre-offered answer options. It was designed to provide an objective view of the respondents' condition by trying to remove all possible preferences according to the alternatives offered and gain an objective insight into the way production process management was performed. To avoid further clarification, the questions were intentionally simple. With the intent that companies of different production orientations could be involved in the research, the offered answers were of a typical character. The first part of the questionnaire contained questions about the basic characteristics of the investigated companies, while the second part related to companies applying methods and tools for process design and ways of evaluating the performance of the process.

3. Data analysis

Table 1 shows the basic characteristics of the surveyed companies (average number of employees, utilization of business capacity, production orientation and number of clients for which the company produces). The size of the company was determined according to the number of employees and from table 1 it can be seen that 50 % of the surveyed companies employ between 50 and 249 employees. Of the companies that participated in the survey, 87,4 % of companies have a production capacity utilization of more than 70 %. Furthermore, none of the companies are oriented exclusively to the domestic market, and 31,3 % of the companies are selling their products exclusively on the foreign market. Also, 93,7 % of analyzed companies have a wide range of clients.

Average number of employees	**%**
Up to 50	6,3
50 - 249	50,0
250 - 499	18,7
More than 500	25,0
Utilization of production capacities	**%**
85 - 100%	50,0
70 - 85%	37,4
55 - 70%	6,3
Less than 50%	6,3
Orientation of production	**%**
Domestic market	0
Foreign market	31,3
Domestic and foreign market	68,7
Number of clients	**%**
One	0
2 - 5	6,3
More than 5	93,7

Table 1. Overview of the general characteristics of the surveyed companies

When asked about the use of software for production management, 43,7 % of surveyed companies gave a negative response. In contrast, 56,3 % of companies use production management software, some of which are Pantheon, F 300, MIS, SAP.

Table 2 gives information on the use of manufacturing process design tools per company. In the sample, 12,5 % of companies do not use tools. By contrast, in designing and defining their processes, in most cases, companies use traditional graphical flow chart and process map type methods: 37,5 % of companies use flowcharts and 18,7 % process maps. Of the analyzed companies, 37,5 % use software packages such as CORPUS, F300, MIS, CIMDB and APIS.

Company	**1**	**2**	**3**	**4**	**5**	**6**	**7**	**8**	**9**	**10**	**11**	**12**	**13**	**14**	**15**	**16**
Flow Charts			X					X	X		X			X		X
Process Charts			X										X			X
Software packages		X		X			X			X		X				X
Other tools	X														X	X
No use of tools					X	X										

Table 2. Use of tools for production process design

When it comes to the way companies make a daily production plan, it can be concluded from the data that 31,2 % of companies use only experience, 56,3 % of companies use software, 12,5 % of companies based on work order and 6,3 % of companies based on aggregate production plan.

Table 3 shows the ways in which the surveyed companies measure the production cycle time and set working standards. The most commonly used method of determining the production cycle is the time required for each operation, represented by 81,2 % of the company. When determining work standards, 37,5 % of the companies combines multiple methods for estimating the work time, 31,2 % of companies measure the work time of a prototype sample, 18,7 % of the companies determine working standard by experience, and 12,6 % of companies apply different standards.

The manner in which surveyed companies determine the working standards	**%**
By experience	18,7
Measuring the work on the prototype sample	31,2
MTM-, MOST-, MODAPTS-standard	6,3
REFA-standard	6,3
Combining multiple methods for estimating work time	37,5
The way the surveyed companies measure the time cycle	**%**
Measuring total production time	0
By measuring the time of each operation and collecting these times	81,2
Combination of total time measurement and time measurement obtained by the time of each operation	12,5
Unknown	6,3

Table 3. Determination of the working standards and measurement of the time cycle

In 93,7 % of cases, companies have prescribed procedures for transferring information in the production process. When manufacturing processes are performed, the most commonly used method of defining work tasks is in the written form (68,8 %). Also, companies do so via the information system (50 %).

The next set of questions related to the use of certain types of production planning files by companies. Table 4 is a representation of the utilization of work center master files, item master files and routing files. In the case of 62,5 % of companies, all three types of files are represented. In contrast, 18,6 % of companies record only information about each component that the company produces (item master file), while 6,3 % of companies have only written procedures for item delivery (routing file). When it comes to the control file, 50 % of companies track current progress for each order by software, 31,3 % of the companies do it through checklists, while 18,7 % of the companies decided to combine both methods.

Company	**1**	**2**	**3**	**4**	**5**	**6**	**7**	**8**	**9**	**10**	**11**	**12**	**13**	**14**	**15**	**16**
Work center master files	x	x	x	x			x	x	x	x	x		x			x
Item master files	x	x	x	x	x	x	x	x	x	x	x	x	x		x	x
Routing files		x	x	x			x	x	x	x	x	x	x	x		x

Table 4. Types of design files and their representation

In order to transport semi-finished goods companies use different equipment: cranes (50 %), lifters (37,5 %), conveyors (62,5 %), forklifts (25 %), pushcarts (6,3 %), pipelines (6,3 %), or do it manually (12,5 %).

For 62,5 % of the companies, control of the product in the production process implies control of the finished product and the control of the semi-finished products at each workplace after the completion of the operation. At the same time, 31,5 % of companies control only semi-products, while 6.3% of companies use other forms of control. In the event of a defective product, 62,5 % of companies have a written procedure for the product rework, while 6,3 % of companies rely only on oral communication or on the information system only. In 12,5 % cases it is a combination of oral and written method of defining the rework, in 6,3 % of cases this combination refers to written procedures and an information system or a combination of oral and defining procedures through an information system. Regarding the number of products returned for rework, 93,7 % of companies track (document) their number.

It was found that 75 % of respondents measure the utilization of specific resources (workforce, machines, etc.). Also, when it comes to downtime causes, which are shown in the table 5, the most common cause of downtime is a failure in 75 % of cases. For 56,2 % of companies downtime was caused by switching to a new product, scheduled maintenance was noted in 25 % cases and the human factor in 18,8 % of cases. Furthermore, downtime as a result of a bottleneck occurs in 12,5 % of cases and the lack of financial resources is listed in 6,3 % of companies. A resource which would represent a bottleneck, for 62,5 % of companies is changing depending on the current production program.

Company	1	2	3	4	5	6	7	8	9	10	11	12	13	14	15	16
Malfunction	x	x	x	x	x		x		x	x	x			x	x	x
Planned maintenance							x				x	x			x	
Moving to new product	x		x	x		x			x		x		x	x		x
Bottleneck						x								x		
Human factor											x			x		x
Insolvency								x								

Table 5. Causes for downtime

Table 6 represents the time frames for reviewing the performance indicators of a manufacturing process by companies. Analysis of the performance indicators of a company's manufacturing process most frequently occurs only on a monthly basis in 62,5 % of cases. In 12,5 % of cases, the company, in addition to reviewing performance indicators at the monthly level, also approaches quarterly, semi-annual and yearly.

Company	1	2	3	4	5	6	7	8	9	10	11	12	13	14	15	16
Weekly									x							
Monthly	x	x	x	x		x	x	x		x	x		x		x	x
Quarterly			x				x							x		
Semiannually			x				x					x				
Annually			x		x		x									

Table 6. Time frames for reviewing the performance indicators of manufacturing process

When it comes to how companies relate to key performance indicators (KPIs) and benchmarking, it can be stated that 56, 3% of companies are tracking KPIs and 62,5 % of companies are performing benchmarking. In 50 % of cases, the company simultaneously monitors KPIs and performs benchmarking, while 31,3 % of the companies do not track KPIs or conduct benchmarking. Table 7 shows the attitude of individual companies towards KPIs and benchmarking.

Company	1	2	3	4	5	6	7	8	9	10	11	12	13	14	15	16
KPI	x	x		x			x	x			x		x	x		x
Benchmarking		x	x	x			x	x		x	x		x	x		x

Table 7. The attitude of individual companies towards KPIs and benchmarking

4. Conclusion

Based on the questionnaire survey, it can be concluded that a relatively small number of companies are performing benchmarking and tracking KPIs. The number of companies that use information systems in their business practice is inadequate. It should also be noted that the production downtime caused by failures is much more frequent than by the scheduled maintenance. Production procedures in general are often not clearly defined, since they are defined in a large number of cases through personal oral communication. Analysis of performance indicators only on a monthly basis, the most commonly used, is insufficient and allows companies to use only lagging metrics.

Analysis shows that although there is no visible nonresponse pattern, the limitation of this research is relatively small response rate. Future research could include the use of more alternatives for companies to complete and return questionnaires in order to achieve higher response rate. In this way, companies could be grouped according to industry branches. The behaviour of companies in service sector should be explored as well.

5. References

[1] Stainer, L. (2006). Performance management and corporate social responsibility: the strategic connection. Strategic Change, Vol. 15, (August 2006) pp. 253-264, ISSN 1099-1697

[2] Velimirovic, D.; Velimirovic, M. & Stankovic, R. (2011). Role and importance of key performance indicators measurement. Serbian Journal of Management, Vol. 6, No. 1, (May 2011) pp. 63-72, ISSN 2217-7159

[3] Atanasov, N. & Lecic-Cvetkovic, D. (2009). Production key performance indicators in global crisis environment, Available from: http://spin.fon.bg.ac.rs/doc/Zbornik09.pdf Accessed: 2017-04-29

[4] Shahin, A. & Mahbod, M. A. (2007). Prioritization of key performance indicators: An integration of analytical hierarchy process and goal setting. International Journal of Productivity and Performance Management, Vol. 56, No. 3, (March 2007) pp. 226-240, ISSN 1741-0401

[5] Ghalayini, A. M. & Noble, J. S. (1996). The changing basis of performance measurement. International Journal of Operations & Production Management, Vol. 16, No. 8, (August 1996) pp. 63-80, ISSN 0144-3577

[6] Vergidis, K.; Tiwari, A. & Majeed, B. (2008). Business process analysis and optimization: Beyond reengineering. IEEE Transactions on Systems, Man, and Cybernetics, Part C (Applications and Reviews), Vol. 38, No. 6, (January 2008) pp. 69-82, ISSN 1094-6977, DOI 10.1109/TSMCC.2007.905812

[7] Stojkic, Z. & Majstorovic, I. (2012). Improving Business Organization through the Implementation of Integrated IS, Annals of DAAAM for 2012 & Proceedings of the 23rd International DAAAM Symposium, 24-27th October 2012, Zadar, Croatia, ISSN 2304-1382, ISBN 978-3-901509-91-9, Katalinic, B. (Ed.), pp. 0853-0856, Published by DAAAM International, Vienna, Austria, 2012

[8] Bosilj Vuksic, V.; Brkic, Lj. & Baranovic, M. (2016). Business Process Management Systems Selection Guidelines: Theory and Practice. 39th International Convention on Information and Communication Technology, Electronics and Microelectronics (MIPRO) 2016, May 30-June 3 2016, Opatija, Croatia, ISBN 978-953-233-088-5, Biljanovic, P. (Ed.), pp. 1476-1481, Croatian Society for Information and Communication Technology, Electronics and Microelectronics-MIPRO, Rijeka, Croatia, DOI 10.1109/MIPRO.2016.7522372

[9] Skrinjar, R.; Bosilj Vuksic, V. & Indihar Stemberger, M. (2008). The impact of business process orientation on financial and non-financial performance. Business Process Management Journal, Vol. 14, No. 5, (September 2008) pp. 738-54, ISSN 1463-7154

[10] Hernaus, T.; Pejic Bach, M. & Bosilj Vuksic, V. (2012). Influence of strategic approach to BPM on financial and non-financial performance. Baltic Journal of Management, Vol. 7, No. 4 (October 2012) pp. 376 -396, ISSN 1746-5265

DOI: 10.2507/28th.daaam.proceedings.152

AN ANALYSIS OF STRATEGIES OF MEASUREMENT OF 3D ROTARY ELEMENTS

Stanisław Adamczak, Krzysztof Stępień, Paweł Zmarzły*

This Publication has to be referred as: Adamczak, S[tanislaw]; Stepien, K[rzysztof] S[tanislaw] & Zmarzly, P[awel] (2017). An Analysis of Strategies of Measurement of 3D Rotary Elements, Proceedings of the 28th DAAAM International Symposium, pp.1096-1100, B. Katalinic (Ed.), Published by DAAAM International, ISBN 978-3-902734-11-2, ISSN 1726-9679, Vienna, Austria
DOI: 10.2507/28th.daaam.proceedings.152

Abstract

Measurements of form deviations of 3D elements can be conducted with the use of various strategies, differing in a number and distribution of sampling points located on an investigated surface. Low number of sampling points or measured sections can lead to the situation that some surface irregularities are not detected by a measuring system. An application of higher density of sampling points, in turn, results in significant lengthening of measurement time, which is undesirable, if one takes into account requirements of modern manufacturing processes. The main goal of this paper is to analyze major advantages and drawbacks of measurement strategies of 3D rotary elements that are described in the scientific literature. Additionally, the concept of a novel, adaptive measurement strategy is presented.

Keywords: form deviation; measurement strategy; rotary elements; uniform sampling; adaptive strategy

1. Introduction

Rotary element are significant and numerous group of machine parts. They are used in a number of branches of engineering industry, including bearing industry (components of rolling bearings) [1], automotive industry (shafts, engine pistons and cylinders), metallurgy (metallurgical cylinders), ship industry (drive shafts) and many more [2]. In industrial practice measurements of form deviations of such elements is simplified – usually roundness measurement is conducted in one or a few preselected cross-sections of investigated part.

Measurements of 3D parameters are practically limited to the problem of evaluation of cylindricity, which is performed with the use of special-purpose radial systems for measurements of roundness and cylindricity deviations (for example series of Talyrond instruments by Taylor Hobson). If tolerances of parts permit, such measurements are conducted with the use of coordinate measuring machines, too. Considering the problem of correct measurement of form deviations of 3D elements it is noteworthy that a very important issue is to apply relevant measuring strategy, i.e. scanning trajectory, which is basically a path along which the sensor moves on the surface to be measured. Selection appropriate measuring strategy is a matter of great importance and refers to the problem of obtaining satisfying measurement accuracy and measurement time. The aim of this paper is to discuss major advantages and limitations of the strategies described in the literature. In addition, outlines of the new strategy proposed by authors is discussed.

2. Strategies used to measure form deviations of 3D elements

An analysis of the problem of cylindricity measurements shows that there are three main groups of measuring strategies. The first group are uniform sampling strategies that are most common strategies usually applied in industrial practice. The second group constitute strategies fitted to measured surface on the basis of predicted pattern of irregularities. The last group of the strategies that are described in the literature are so-called adaptive strategies.

As it was previously mentioned, at present time the most popular are uniform sampling strategies. However, due to the efforts aiming at shortening measurement time intensive research work is conducted that is focused on applying adaptive strategies. These are relatively novel concepts and they are still under study by a number of scientists dealing with metrology of geometrical quantities.

2.1 Uniform sampling strategies

Standard ISO 12180 describes following uniform sampling strategies used to measure cylindricity: cross-sections, strategy, longitudinal sections strategy, 'bird-cage" strategy and single points strategy. Apart from mentioned above, also helical trace strategy is quite often applied in industry. The strategies are shown in Fig. 1.

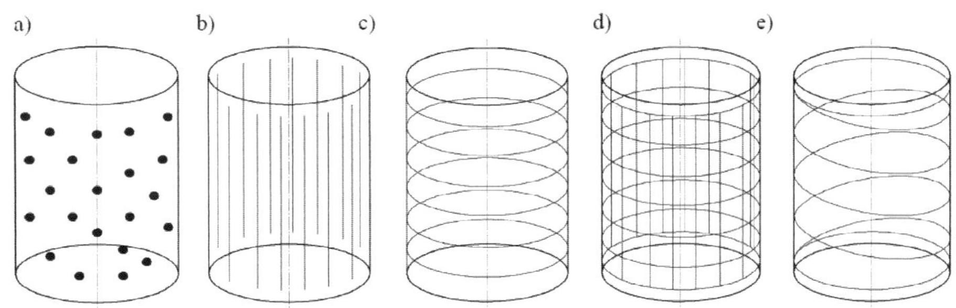

Fig. 1. Strategies used to measure cylindricity deviations: a) single points strategy, b) longitudinal sections strategy, c) cross-sections strategy, d) "bird-cage" strategy, e) helical trace strategy [3].

Usually measurements of cylindricity deviations are performed with the use of the cross-sections strategy and the longitudinal sections strategy. Single points strategy is less common as the limited number of sampling points allows a rough evaluation of form deviations only. The strategy that permits the most accurate measurement is the "bird-cage strategy. However, due to computational problems this strategy has not been available in the software of measuring systems for a long time. At the moment, there are new measuring systems that allow application this strategy. The most important advantage of the "bird-cage" strategy is that it permits dense covering of measured surface with a grid of sampling points. Therefore such results are most reliable. The major drawback of the "bird-cage" strategy is that it is very time consuming. Thus, its application under industrial conditions is limited. The strategy that was not described in the standard ISO 12180 is the helical trace strategy. This strategy is a compromise solution between the strategies of cross-sections and longitudinal sections and it is quite often applied under industrial conditions.

In general, it should be noted that the strategies shown in Fig. 1 allow uniform sampling. It means that the whole surface of the specimen is scanned in the same way. Thus, for the case of cross-section, longitudinal section and "bird-cage" strategy all the linear distance between all cross-sections and angular distance between longitudinal sections is the same. The spiral trace strategy can be regarded as a uniform sampling strategy, too since the parameter that describes sampling density, which is pitch angle, remains constant during the measurement. Application of the single points strategies also requires approximate uniform distribution of sampling points on the surface under investigation. It can be performed with the use of specific numerical methods, for example by applying so-called Hammersley sequence.

Uniform sampling strategies are very useful if surfaces under study are characterized by regular pattern of form errors. However, sometimes significant irregularities are located only in a certain area of the surface. Such area should be then sampled with the use of many more points than other part of the surface. This is the reason why a research activities are conducted that are focused on development of a strategy that would denser sampling in areas where significant surface defects are predicted.

2.2 Strategies to be fitted to predicted form errors pattern

These strategies are fitted to the surface under study on the basis of predicted distribution of surface irregularities. If we have information about the machining process of the surface, then we can predict most probable distribution of the irregularities of the surface. For example, cylindrical surface of workpieces machined by turning that were placed in a three-jaw chucks are very often characterized by a triangular error (i.e. the third harmonic component is dominant if we conduct its spectral analysis). Additionaly, workpieces that were placed in lathe centers are usually characterized by a saddle-shaped error.

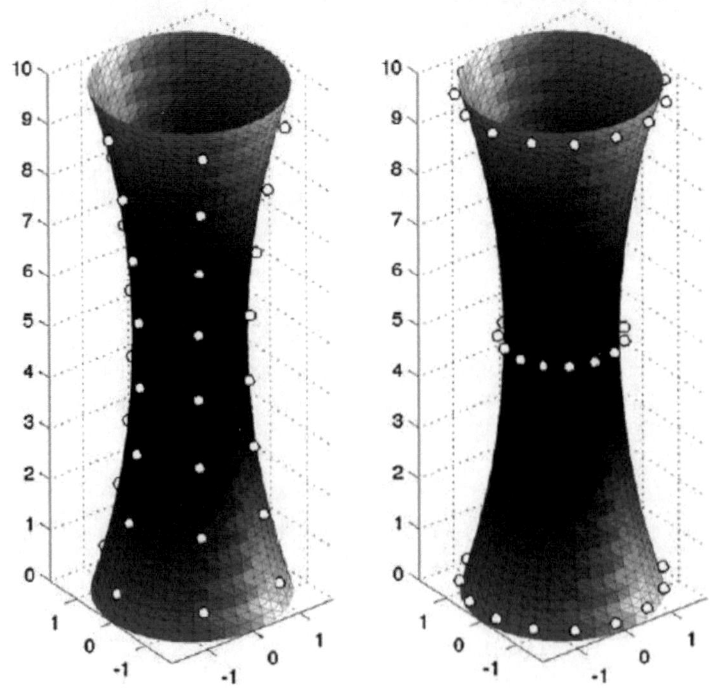

Fig. 2. Strategies of form error measurement of saddle-shaped element: on the left – the strategy applied usually – not fitted to the predicted form error of the part; on the right – the strategy designed on the basis of predicted form error of the element [4].

Strategies fitted to predicted model of irregularities can be designed on the basis of data of preliminary measurements or on the basis of assumed pattern of form errors. An example of the former approach is a method of harmonics fitting described in work [5]. This method applies statistical evaluation of characteristics of individual Fourier components of the profile in order to establish as small set of sampling points data as possible that allows reliable evaluation of form errors of the surface under study.

The latter approach is given in work [6]. In this method, a preliminary model of the shape of the surface is assumed. The model is described mathematically by the linear combination of the set of base functions (for example, polynomials, Fourier components or eigenfunctions). The model is then used to design the grid of sampling points. After the measurement on the basis of the values in preselected sampling points coefficients of assumed model of the linear combination are calculated (taking into account uncertainty of fitting). The coefficients are then used to reconstruct the image of the whole surface under study.

2.3 Adaptive strategies

Adaptive strategies are iterative ones. In the case of application of adaptive strategies, the sampling is conducted in a few stages. The first stage is a preliminary measurement. The preliminary measurement is usually performed with the use of uniform sampling. After the preliminary measurement Next, on the basis of obtained measurement data, a relevant algorithm is applied to find the areas where the risk of occurring of large local disturbances is high. After conducting measurements in these areas the value of presumed coefficient is calculated.

The value of this coefficient indicates if the measurements should be finished or it is necessary to conduct the next series of measurements applying denser sampling. Such approach uses so-called Krige models quite often. The name of the method comes from the name South African engineer who has the first researcher to apply this method to predict changes of the signal in geostatics in the sixties of the XX. century.

At present, these models are more and more common to predict localization of subsequent sampling points. In Krige models the signal is modified with the use of the set of base functions and their coefficients supplemented by a random component whose expected value is equal to zero and for whom the covariance between the input and output remains constant. Fig. 3 shows an algorithm that presents an application of Krige model to evaluate form deviations by an adaptive strategy proposed in work [7].

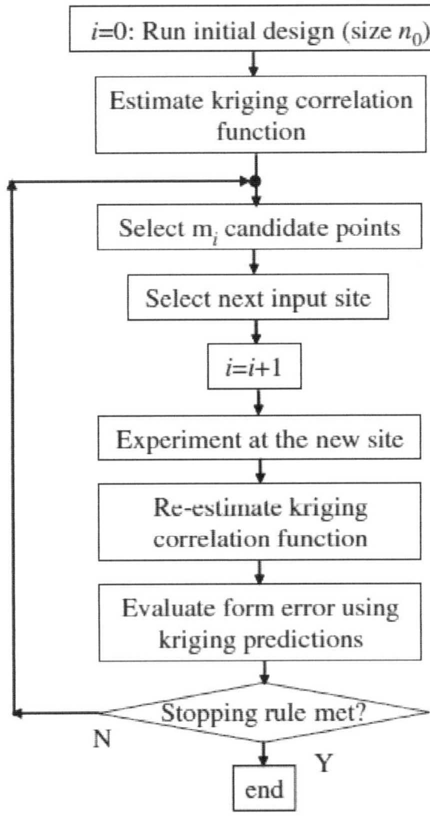

Fig. 3. Block diagram showing an example of the application of Krige model to evaluate form deviations of elements [7]

2.4 An approach to adaptive strategy proposed by authors

Taking into account limitations of the strategies presented above authors have developed a novel concept of applying the adaptive strategy to evaluate form deviations of 3D rotary elements. The strategy is to be applied with the use of coordinate measuring machines [8, 9].

The strategy consists of two stages: the former is a preliminary measurement and the latter are additional measurements. Preliminary measurement will be conducted along preselected scanning path. Authors propose to apply following uniform sampling strategies for the preliminary measurement: helical trace strategy, cross-sections strategy and longitudinal sections strategy. Let us assume that we conduct preliminary measurement by the helical trace strategy. If measurement results show that in a certain area of the surface occurrs a large change of the signal, then we conduct additional measurements, performed for example in the cross-section and the longitudinal section lying in analyzed fragment of the surface. Additionally, we can also carry out a new measurement along the helical trace applying lower value of the pitch angle, which will contribute to denser covering the surface with the grid of sampling points. For the preliminary measurements with the use of cross-section and longitudinal sections strategy the procedure can be as follows: we calculate correlation coefficients between neighboring profiles. If the value of the coefficient is lower than presumed permitted value then we conduct additional measurements in the area between these profiles.

3. Conclusions

Strategies based on predicted model of irregularities and adaptive strategies described in the section 2.2 and 2.3. are still under investigation and they have not been widely applied under industrial conditions. They have lots of advantages, however also severe drawbacks. The strategy based on predicted model of irregularities allows accurate sampling only in preselected areas. However, real technological processes are disturbed by many random factors that can contribute to arising irregularities in areas different from those that were predicted before the process initialization (in particular if the process becomes statistically unregulated). Referring to the problem of adaptive strategies described in section 2.3 it should be noted that these are iterative methods, thus they are quite complex computationally. Moreover, their application with the use of coordinate measuring machine can be troublesome as there is a risk of the collision of the measuring tip with the surface of the specimen. The risk results from the fact that sampling points can be generally different for elements whose nominal dimensions are the same.

In addition, it is noteworthy that at present time form deviations are usually measured by coordinate measuring machines equipped with scanning probe heads. Such probe heads allow obtaining very dense sampling in a relatively short time. Thus, methods for whom areas requiring denser sampling are to be calculated become out-of-date very quickly. These methods are still relevant if touch-triggering probe heads are used but such type of probe heads is dynamically replaced with scanning ones at present time.

The concept of two-stage adaptive strategy of measurements of 3D machine parts seems to be interesting from the technological point of view. Such strategy will allow obtaining accurate information about surfaces under study and will be less time-consuming than other adaptive sampling strategies. However, in order to evaluate the applicability of this strategy under industrial conditions it will be necessary to conduct a series of experiments, which is planned in the nearest future.

4. Acknowledgments

The paper has been elaborated within the framework of the research project entitled "Theoretical and experimental problems of integrated 3D measurements of elements' surfaces", reg. no.: 2015/19/B/ST8/02643, ID: 317012, financed by National Science Centre, Poland.

5. References

[1] Adamczak, S. & Zmarzły, P. (2017). Influence of raceway waviness on the level of vibration in rolling-element bearings. Bull. Pol. Ac.: Tech. 65(4), pp. 541-551.

[2] Adamczak, S.; Zmarzły, P.; Kozior, T.; Gogolewski, D. (2017). Assessment of roundness and waviness deviations of elements produced by selective laser sintering technology. Engineering Mechanics 2017, pp. 70-73.

[3] Nowakowski, P. & Gapiński, B. (2013). Analiza niepewności pomiaru odchyłki walcowości w zależności od stosowanej strategii pomiaru. Inżynieria Maszyn, Vol. 18, No. 3, pp. 35-47, (in Polish).

[4] Colosimo, B. M.; Moroni, G. & Petro, S. (2010). A tolerance interwal based criterion for optimizing discrete point sampling. Precision Engineering 34, pp. 745-754.

[5] Capello, E. & Semeraro, Q. (2001). The harmonic fitting method for the assessment of the substitute geometry estimate error. Part I: 2D and 3D theory. Int J Mach Tools Manuf. 41(8), pp. 1071-1102.

[6] Summerhays, K. D.; Henke, R. P.; Baldwin, J. M.; Cassou, R. M. & Brown, C. W. (2001). Optimizing discrete point sample patterns and measurement data analysis on internal cylindrical surfaces with systematic form deviations. Precis Eng 26, pp. 105-121.

[7] Pedone, P.; Vicario, G. & Romano, D. (2009). Kriging-based sequential inspection plans for coordinate measuring machines. Appl. Stochastic Models Bus. Ind. 25(2), pp. 133–49

[8] Beranek, L.; Volf, L. & Mikes, P. (2011). Coordinate Metrology Education Using Virtual CMM, Annals of DAAAM for 2011 & Proceedings of the 22nd International DAAAM Symposium, 23-26th November 2011, Vienna, Austria, Volume 22, No. 1, ISSN 1726-9679, ISBN 978-3-901509-83-4, Katalinic, B. (Ed.), pp. 1323-1324, Published by DAAAM International Vienna, Vienna

[9] Runje, B. & Medic, S. (2011). Measurement Uncertainty in Process of Line Scales Calibrating, Annals of DAAAM for 2011 & Proceedings of the 22nd International DAAAM Symposium, 23-26th November 2011, Vienna, Austria, Volume 22, No. 1, ISSN 1726-9679, ISBN 978-3-901509-83-4, Katalinic, B. (Ed.), pp. 0093-0094, Published by DAAAM International Vienna, Vienna

DOI: 10.2507/28th.daaam.proceedings.153

AN ANALYSIS OF CONTEMPORARY METHODS FOR MEASUREMENT OF FORM ERRORS OF SPHERICAL MACHINE PARTS

Stanisław Adamczak, Dariusz Janecki, Krzysztof Stępień, Mateusz Wrzochal

This Publication has to be referred as: Adamczak, S[tanislaw]; Janecki, D[arius]; Stepien, K[rzysztof] S[tanislaw] & Wrzochal, M[ateusz] (2017). An analysis Of Contemporary Methods For Measurement Of Form Errors Of Spherical Machine Parts, Proceedings of the 28th DAAAM International Symposium, pp.1101-1107, B. Katalinic (Ed.), Published by DAAAM International, ISBN 978-3-902734-11-2, ISSN 1726-9679, Vienna, Austria
DOI: 10.2507/28th.daaam.proceedings.153

Abstract

Spherical elements constitute significant group of machine parts. In particular, they are commonly applied in bearing industry. Spherical elements should be usually manufactured very precisely and their form deviations are relatively small. Deviations of balls used in rolling bearings usually do not exceed 1 micrometer. The problem of accurate measurement of spherical parts is complicated and standards do not provide too many recommendations referring to this problem. The paper presents up-to-date solutions of the problem of measurements of spherical parts. It also presents the method of sphericity measurements developed at Kielce University of Technology. In addition, measurement strategies are presented that are based on developed method.

Keywords: measurement; methods; sphericity; form deviations; measurement systems

1. Introduction

A key issue in contemporary manufacturing technology is to control dimensional accuracy of machine elements. Apart from the dimensions, quality of surfaces of machine parts is very important, too. This is why accurate surface measurements are nowadays a matter of great importance [1-3]. Considering measurements of rotary elements it is noteworthy that problems of measurements of roundness and cylindricity are well described in the scientific literature.

However, measurements of roundness seem to be insufficient way to evaluate form deviations of spherical elements. This is why in a number of research centres studies focused on this problem are conducted.

The main aim of this study is to critically evaluate major advantages and drawbacks of sphericity measurement methods that are presented in the scientific literature. Additionally, the concept of sphericity measurement developed by authors is presented. The concept is based on applying the radius-change instrument equipped with special-purpose unit for controlled rotation of spheres

2. Sphericity measurements by the contact method

Generally methods of measurements of sphericity can be divided into two main groups: contact and optical methods. Contact methods can be classified as industrial and laboratory ones. Industrial methods apply simple instruments and provide quick information if measured element fulfils its technological requirements. Laboratory methods are more complex and they aim at as accurate measurement of the whole surface of the specimen as possible. In the laboratory methods the process of measurement is usually much longer and complicated than in the industrial methods.

2.1 Industrial methods

Standardization documents used in the industrial practice describe following strategy of measurements of sphericity of machine elements: roundness profiles to be measured in three mutually perpendicular cross-sections of the specimen (see Fig. 1). Results of the measurements are presented in the form of polar diagrams and the diagrams are the basis to evaluate approximate sphericity deviations. Such approach allows rough evaluation of sphericity only, as significant part of the surface is not sampled. Moreover, it should be noted that in industrial practice evaluation of form deviations of spherical elements is conducted on the basis of an analysis of only one roundness profile measured in a random cross-section of the sphere.

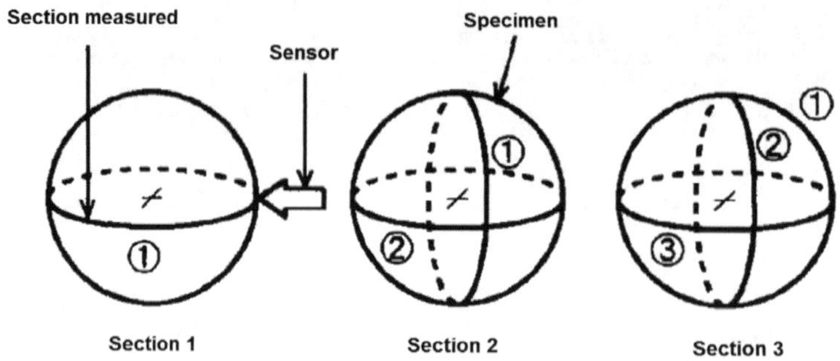

Fig. 1. Strategy of measurements of spherical elements described in industrial standards [4]

Industrial documents state usually that the best reference feature to evaluate form deviations of spherical elements is the least circle circumscribed on the measured profile. It is noteworthy that industrial approach does not require to apply any special-purpose systems for sphericity measurements, but recommends using a typical roundness measurement instruments. Usually such instruments are based on the radius-change method.

2.2 Laboratory measurements

Apart from the radius-change method also a 3-point method can be used to measure sphericity, which was described in [5]. Generally, in this concept the sphere is placed on two-point support [see Fig. 2]. The signal is obtained from the sensor whose axis intersects the nominal centre of the sphere. During the measurement the sphere rotates and due to irregularities of the surface the sensor readings change. This concept requires mathematical transformation of the sensor readings since the signal is affected by irregularities occurring in all three contact points of the sphere. It can be easily conducted with the use of Fourier transform [6, 7].

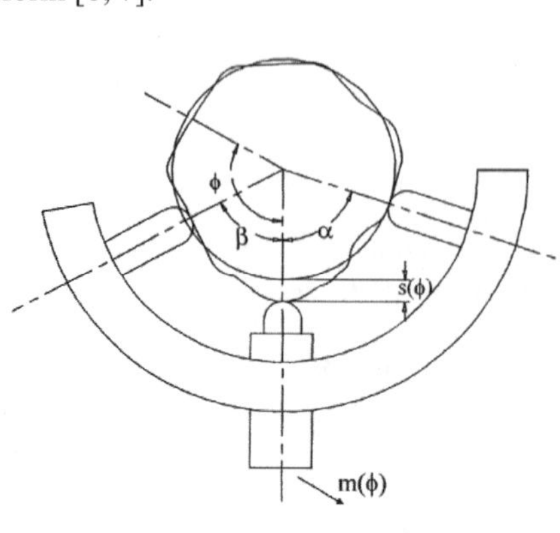

Fig. 2. Sphericity measurements by the 3-point method . φ- rotation angle, s(φ)- deviation, m(φ) – sensor readings, α, β angles between supports and the sensor axis [5]

Sensor readings in a given point of the profile is a function of the angles between supports and the change of the displacement of the sensor tip in this point. The major drawback of this method is that there are some harmonic components of the profile that cannot be detected by the measuring system.

Other concept of sphericity measurements has been developed at Kielce University of Technology [8]. This concept is based on using typical roundness measuring instrument additionally equipped with e special-purpose unit allowing accurate rotation of measured spheres (see Fig. 3). Owing to that it is possible to measure roundness profiles of the sphere in a number of strictly defined cross-sections.

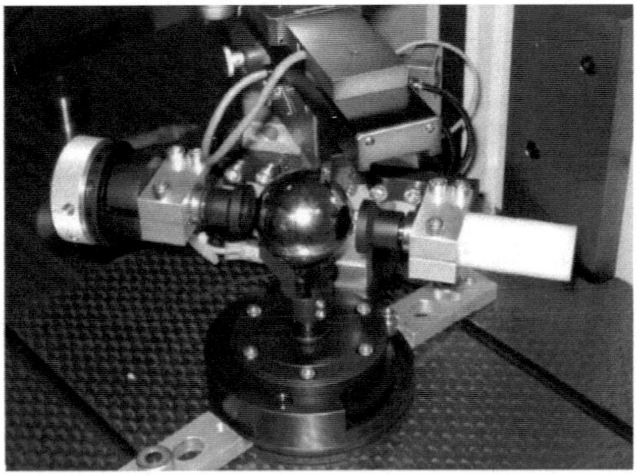

Fig. 3. The unit for accurate positioning of spheres

3. Non-contact methods of sphericity measurements

Except for contact methods, a number of researchers focus on studying methods of sphericity measurements with the use of optical systems. Such methods can be divided into two main groups: measurements applying the light beam and measurements based on image processing.

3.1 Light beam measurements

It should be noted that already in the sixties of the XX. century in former Czechoslovakia an automated system for quality control of balls of rolling bearing has been developed that applied an optical system. The system operated on the basis of observing changes of luminosity of the light reflected from the sphere surface. The spheres during measurements were rotated by the rolls. Focused light beam falls on the surface of the sphere and the reflects to a photodiode. If the surface is damaged then the luminosity of the light reflected changes [9].

At present, intensive research work is carried out in the field of measurements of surface topography of spherical elements with the use of laser interferometers, which was described for example in work [10]. In the method presented in Fig. 4 small fragments of the sphere are measured with the use of Fizeau interferometer and then the procedure of stitching is applied to obtain the image of the whole surface.

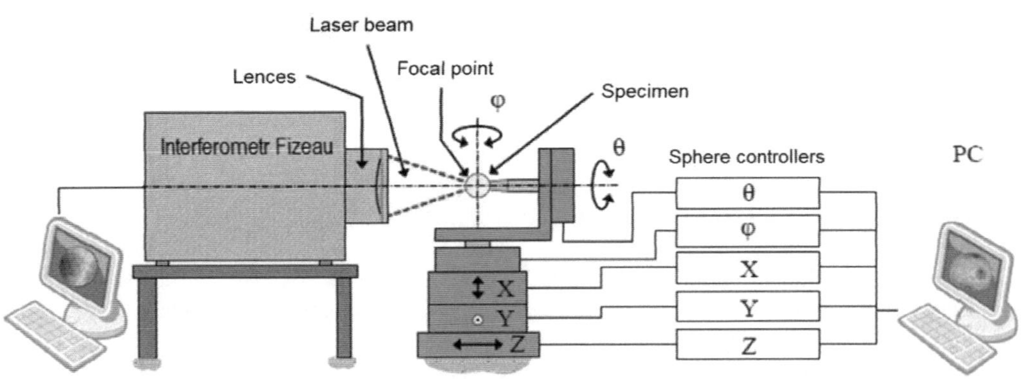

Fig. 4. Measurements of sphericity by Fizeau interferometer presented in work [10]

Method presented in Fig. 4 requires to determine accurate position of the sphere during subsequent measurements. If the position of the sphere were not determined accurately then it would not be possible to stitch the fragments. Thus, rotation of the sphere is performed with a set of controllers. Relatively similar system is described in work [11]. This system applies a modified Fizeau interferometer, too. Similarly to the method presented in [11] the image of the whole sphere is reconstructed from measured fragments, which is shown in Fig. 5.

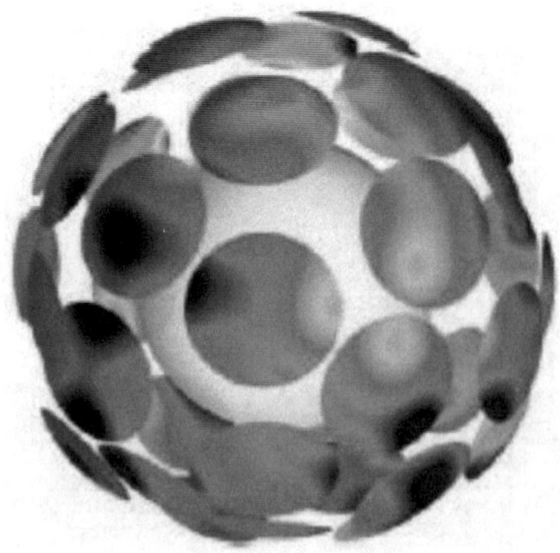

Fig. 5. Reconstruction of the image of the sphere from measured fragments of the surface [11]

3.2 Measurements applying image processing methods

In work [12] the method for determination of sphericity deviation of the electrodischarge machining tool with the use of an image processing techniques. The instrument is here properly calibrated camera that observes an image of the tool tip. Next, relevant software is used to determine coordinates of points lying on the edge of the tool, which is shown in Fig. 6.

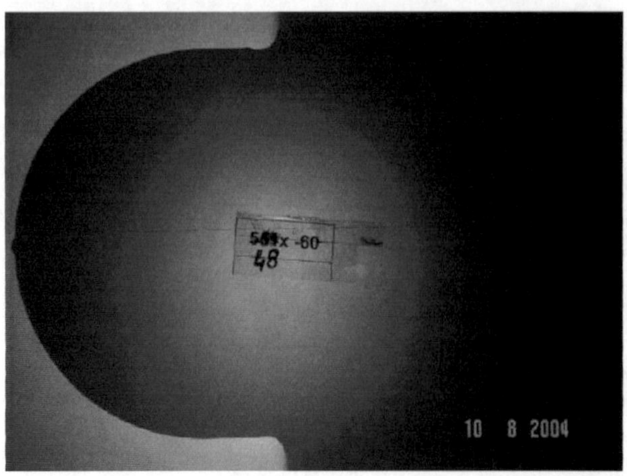

Fig. 6. One of the images used to evaluate sphericity deviations of the tool [10]

After determining the coordinates in a certain position the specimen is rotated to the next position and the procedure repeats then. This way a matrix containing values of coordinates of the sphere is obtained. The values from the matrix are then used to evaluate relevant roundness and sphericity parameters.

An interesting method allowing sphericity measurements by the optical method is presented in work [13]. Authors of this work propose to apply digital image processing technique to reconstruct a 3D view of the surface from two-dimensional images. In the presented method a semisphere can be reconstructed from 2D images. The method applies properly calibrated camera to get 2D images. As distances of various points on the sphere are different then the intensity of shading scale is different, too. The changes of this intensity is the basis to generate 3D view of the surface. One of the 2D images and reconstructed 3D view of the surface are shown in Fig. 7.

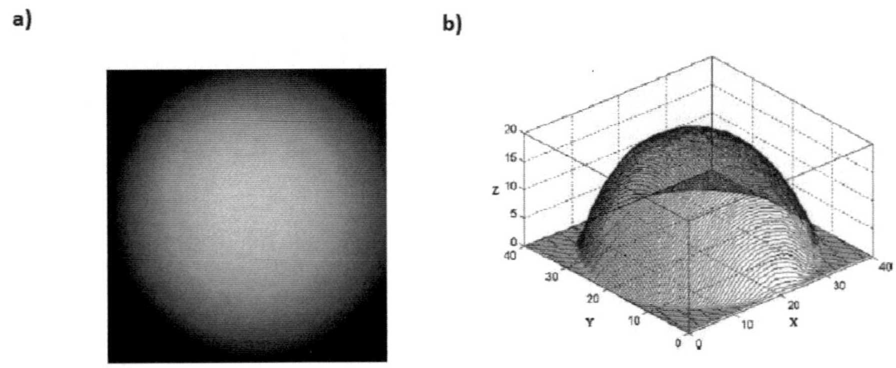

Fig. 7. Reconstruction of the 3D view of the surface from 2D images of the sphere: a) an image captured by the camera. b) reconstructed 3D view [13]

4. Sphericity measurement strategy in contact measurements

At present most common instruments used to measure form deviations of rotary elements are radius-change systems, described for example in work [14]. Equipping such systems with a unit for accurate positioning of the spheres, shown in Fig. 3 allows application of various measurement strategies that permit covering the sphere with a dense grid of sampling points. Authors have conducted a series of preliminary measurements to find most useful strategies that could be used to evaluate form deviations of spherical elements. Two of these strategies are shown in Fig. 8 and Fig. 9. Fig. 8 shows a so-called equator&meridians strategy.

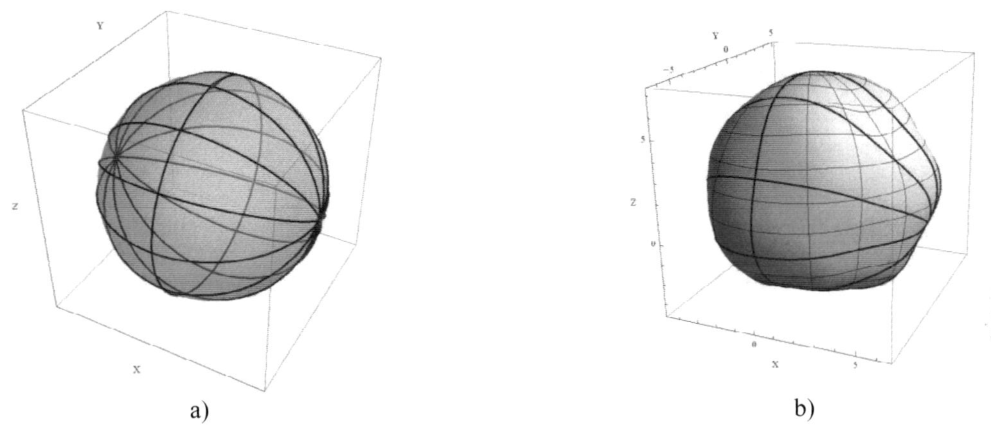

Fig. 8. Equator&meridians strategy: a) scanning trajectory, b) distribution of measured sections on a real element

Another analyzed strategy was a "bird-cage" strategy shown in Fig. 9.

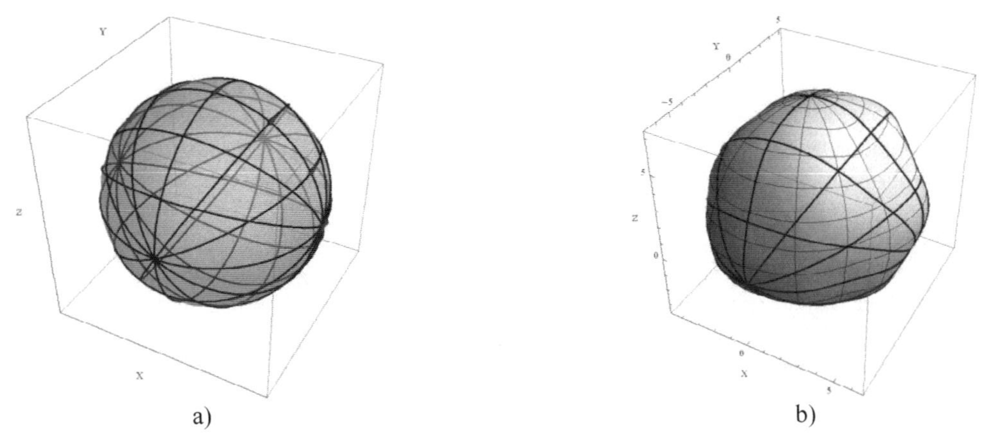

Fig. 9. "Bird-cage" strategy: a) scanning trajectory, b) distribution of measured sections on a real element.

The unit shown in Fig. 3 allows applying many more measurement strategies that are more complicated that those presented above. However, experience of authors indicates that in practice "bird-cage strategy" provides high quality results [15, 16]. It should be noted that accurate evaluation of sphericity deviations with such strategy requires performing a number of mathematical operations used to process measurement data, such as: profiles matching, filtration, etc. [17-19].

5. Conclusions

The problem of accurate measurements of sphericity devations is very complex. In industrial practice contact methods are dominant, however an industrial approach to evaluate form errors of spheres is very simplified. A number of researchers deals also with studying non-contact methods to measure out-of-sphericity. Apart from that coordinate measuring machines are investigated to apply in such measurements. Considering dynamic development of coordinate metrology it is possible that in the near future this technique will be one of the leading when measuring spherical elements. However, at present time, coordinate measuring machine do not still offer accuracy sufficient to satisfy manufacturers of precise spheres used for example in bearing industry.

Generally, it is noteworthy that laboratory methods presented in sections 2.2 and 3 are quite complicated from practical point of view, as they usually require complicated measurement data processing procedures. This is why it would be very difficult to apply them under industrial conditions. The solution can be the method developed by authors which is based on applying a typical radius-change measuring instrument equipped with the unit for controlled rotation of spheres. Owing to such system it is possible to conduct a series of measurements of roundness profiles that can be further used to reconstruct a 3D view of the surface of measured element.

Additionally, it must be underlined that no matter what method is used to measure spherical parts, accurate evaluation of form errors of sphericity deviations is generally very complex and difficult. It requires applying complicated mathematical apparatus using spherical harmonic components and complex measurement data procedures.

6. Acknowledgment

The paper has been elaborated within the framework of the research project entitled "Theoretical and experimental problems of integrated 3D measurements of elements' surfaces", reg. no.: 2015/19/B/ST8/02643, ID: 317012, financed by National Science Centre, Poland.

7. References

[1] Jermak C. J., Rucki M. (2016), Static Characteristics of Air Gauges Applied in the Roundness Assessment, Metrology and Measurement Systems, 23 (1), 85-96.

[2] Zorawski W. et al. (2015), Microstructure and tribological properties of nanostructured and conventional plasma sprayed alumina-titania coatings. Surface & Coatings Technology, 268 , 190-197.

[3] Kundera C., Kozior T. (2014), Research of the elastic properties of bellows made in SLS technology. Advanced Materials Research, 874, 77-81.

[4] Kanada T. (1997), Estimation of sphericity by means of statistical processing for roundness of spherical parts. „Precision Engineering", Vol. 20 (2), pp. 117-122.

[5] Gleason E., Schwenke H. (1998), A spindless instrument for the roundness measurement of precision spheres. „Precision Engineering", Vol. 22 (1), pp. 37-42.

[6] Głowacz A., (2014), Diagnostics of DC and Induction Motors Based on the Analysis of Acoustic Signals. Measurement Science Review, 14 (4), 257-262.

[7] Głowacz A., Głowacz Z., (2016), Diagnostics of stator faults of the single-phase induction motor using thermal images, MoASoS and selected classifiers. Measurement, 93, 86-93.

[8] Janecki D., Stępień K., Adamczak S. (2016) Sphericity measurements by the radial method: II. Experimental verification, Measurement Science and Technology, Vol. 27, 1, pp. 015006

[9] Ratajczyk E. (1968) Kontrola jakości powierzchni łożysk tocznych automatami AVIKO. „Mechanik", Vol. 4/68, s. 203-207.

[10] Hagino T. et al., (2012), Sphericity Measurement Using Stitching Interferometry, Key Engineering Materials, Vols. 523-524, pp. 883-888.

[11] Guido Bartl et al., (2010), Interferometric determination of the topographies of absolute sphere radii using the sphere interferometer of PTB. "Meas. Sci. Technol", Vol. 21, 115101 (8pp).

[12] Halkaci H.S., Mavi Ö., Yigit O., (2007) Evaluation of form error at semi-spherical tools by use of image processing. „Measurement", Vol. 40 (9-10), pp. 860-867.

[13] Song L.M. et al., (2005) Novel 3D sphericity evaluation based on SFS-NDT. "NDT&E International", Vol. 38, pp. 442-447.

[14] Stępień K. (2014), In situ measurement of cylindricity—Problems and solutions, Precision Engineering, Vol. 38 No. 3, pp. 697-701

[15] Janecki D., Stępień K., Adamczak S. (2016) Sphericity measurements by the radial method: I. Mathematical fundamentals, Measurement Science and Technology, Vol. 27, 1, pp. 015005

[16] Janecki D., Stępień K., Adamczak S. (2012), Problem of profile matching in sphericity mesurements by the radial method, Metrology and Measurement Systems, Vol. XIX, No. 4, pp. 703–714.

[17] Beranek, L.; Volf, L. & Mikes, P. (2011). Coordinate Metrology Education Using Virtual CMM, Annals of DAAAM for 2011 & Proceedings of the 22nd International DAAAM Symposium, 23-26th November 2011, Vienna, Austria, Volume 22, No. 1, ISSN 1726-9679, ISBN 978-3-901509-83-4, Katalinic, B. (Ed.), pp. 1323-1324, Published by DAAAM International Vienna, Vienna

[18] Runje, B. & Medic, S. (2011). Measurement Uncertainty in Process of Line Scales Calibrating, Annals of DAAAM for 2011 & Proceedings of the 22nd International DAAAM Symposium, 23-26th November 2011, Vienna, Austria, Volume 22, No. 1, ISSN 1726-9679, ISBN 978-3-901509-83-4, Katalinic, B. (Ed.), pp. 0093-0094, Published by DAAAM International Vienna, Vienna

[19] Poniatowska M. (2009), Research on spatial interrelations of geometric deviations determined in coordinate measurements of free-form surfaces. Metrology and Measurement Systems, 16 (3), 501-510.

DOI: 10.2507/28th.daaam.proceedings.154

POSTNONCLASSICAL PARADIGM OF ERGONOMIC DESIGN OF COMPLEX SYSTEMS

Oleg Ipatov & Sergey Sergeev

This Publication has to be referred as: Ipatov, O[leg] & Sergeev, S[ergey] (2017). Postnonclassical Paradigm of Ergonomic Design of Complex Systems, Proceedings of the 28th DAAAM International Symposium, pp.1108-1110, B. Katalinic (Ed.), Published by DAAAM International, ISBN 978-3-902734-11-2, ISSN 1726-9679, Vienna, Austria
DOI: 10.2507/28th.daaam.proceedings.154

Abstract

The article discusses the problems of engineering-psychological design, automated systems design, including virtual prototyping and polyparametric optimization. We have considered the prospects of applying post-non-classical models of complex systems for the analysis of interactions between the operator and the design system with artificial intelligence.

Keywords: complex systems; self-organization; communication; environment and engineering design environment

1. Introduction

Methods and technologies of designing complex technical systems, their maintenance, optimization and automation of project management are classical objects not only in engineering sciences, but they also define the key problems of engineering psychology [1]. Their study is devoted to the works of prominent representatives of the technical direction the engineering and psychological knowledge (Galaktionov, A.I., Gubinskiy, A.I., Koroteev, G.L., Krylov A.A., Lvov V.M., Lomov B.F., Naftaliev A.I., Paderno P.I., Chernyshev A.P., Shadrikov V.D., Shchedrovitsky L.P., etc.). The general conclusion obtained in their work is that individual human consciousness is able to create objects and systems of high complexity, their implementations do not meet the criteria of optimality in the power of biologically determined limitations of the human cognitive system that performs on the associative principle, precluding the exact knowledge. The possibility of collective action in implementing complex projects is much higher, but takes a long time to form the team and its constitutive communication. On average, the creation of the effective team for technical complexes and systems takes at least 10-15 years. The only reasonable way to deal with this situation is automation of design activities that reduce the cognitive load on a person. However, there is a risk of a new problem – ensuring an efficient symbiotic interaction between a designer (design group) and a virtual simulation environment, transforming product ideas into physical product and its optimization. This problem cannot be solved only by creating touch interfaces; it is important and effective to combine the human cognitive system with the design environment. Unfortunately, the scientific solution to this problem is, currently unavailable.

The fact that scientists turn to new forms of scientific rationality, which operates in a complex and evolving systems, is not accidental. This is quite a reasonable reaction to the crisis of modern methodology of designing complex technical systems and complexes that have exhausted their possibilities within the framework of classical mechanistic ideas, which do not allow creating an artificial system of organized complexity [2, 3]. The main problem is the necessity to work with high-precision polyparametric models of physical systems.

Turning to cyber-physical systems, "industry 4.0" technology and bionic design technologies (bionic design) is a consequence of dealing with complexity in creating modern engineering products [4].

Developing a new class of engineering instruments provide computer-aided design in a virtual environment along, with the translation of virtual products in the real physical world using technologies and platforms CAD/CAE/CAM/PLM [5]. Their emergence at the forefront of science and technology is determined by the needs of practice and the evolution of the man-made environment of human civilization.

Development processes, technological environment, tools and principles of its generation in the last decade have become global [6]. We were completely unarmed in front of the new reality related to the emergence of complex network systems, implementing functions of artificial intelligence, interacting with a natural intelligence. Complex systems are a special class of systems. The purpose of this study is to develop a paradigm of ergonomic design of complex systems on a new stage of development, technological environment. It will require a paradigm shift for the technical design, the use of the new categorical apparatus, including interdisciplinary discourses generated in humanities, biology, neurophysiology, genetics, etc. But this is not enough. More profoundly we mean changing the scientific picture of the world, emergence of the new philosophy that reflects the laws of forming a complex world beyond the limits of ordinary consciousness, operating within the framework of classical models of rationality [7].

2. The problem of design complexity in developing systems

The problem of designing complex systems cannot be solved without the introduction of categorical forms, reflecting the significant features of this class. In the modern scientific and psychological knowledge on mental and personality regulation of constituent entities of the project activity (Abulkhanova-Slavske K.A., Antsiferova L.I., Bodrov V.A., Zavalova N.D., Zhuravlev A.L., Lomov B.F., Oboznov A.A., Oshanin D.A., Ponomarenko V.A., Sergienko E.A., etc.) the personal, the subjective component is crucial in forming a specialist and it opportunities in the ergatic system. According to the subject-activity approach (Rubinstein S.L., Anan'ev B.G., Leontiev A.N., Lomov B.F, Brushlinsky A.V., Klimov E.A., etc.), human involvement in the labor activity forms a specific property and a quality of self-organization, self-regulation, self-affirmation and self-control, coordination of external and internal conditions and activities.

Cyclical interaction of developers with the virtual environment while solving a design problem is not a guarantee of order in a complex system of disordered, chaotically mixed potential ideas of the future product. The fact of "advance notice" is long known in psychology; it is formulated in the famous aphorism of Marx that the worst architect from the best bees is innately characterized by the ability to build the wax cell in the head, before building it in reality. In order words, the result of labor is pre-imagined, or pre-planned in the mind.

Designing in virtual reality a human with freedom to choose options for a project (product) and supercomputing and modeling of the design system provide physical implementation.

3. Communication in smart design environment

Communication processes in the system design are considered within the classical approach as the exchange of information between human and machine parts of the system, but the modern post-non-classical views on communication emphasize its self-organizing nature, subordinating "subject-quasisubjectivity" relations [8]. In the design system there are subjects of design and means of interaction between them and simulation environments. The designer is aware of a problematic situation in developing a complex simulation environment and then work with the intelligence of the symbiotic environment to make a decision on its settlement. There is used an interactive method of in which the environment responds flexibly to actions of the designer shaping the design discourse in the form of optimized solutions, which results into the final version. The next stage is to form the "integration platform" that contains the recognized and shared (by all the designers) knowledge about the problem, which needs solving. The Analysis of the processes occurring in the supercomputing system design, in our opinion, is possible with the introduction of the description of the intra-system communication in terms of the paradigm of second order Cybernetics (Heinz von Foerster, Humberto Maturana, Francisco Varela, Sokolov B.V., and Yusupov R.M.) [9-14]. This approach allows solving the classical problem of searching sources of order in self-organizing systems (including in ergonomics design). It is logical, in our opinion, to use reflective models in developing network systems and group design (Lepsky, V.E., and Lefevre V.A.) [15-17]. In the system aspect, reflection is implied as the ability of some systems to build models themselves and other systems, and simultaneously to see themselves building such models. In this way, it is possible to draw a line between structural differences of human knowledge about himself and awareness of being a bearers of such knowledge, which is important in transforming subjective models into their physical embodiment.

4. Conclusion

We set out the basic content of the non-classical paradigm of ergonomic design of complex systems. Automation project activities focus the attention of researchers on using non-conventional methods for interdisciplinary synthesis and close cooperation of representatives of humanitarian and technical knowledge, when creating computer-aided design of.

This is especially important when creating complex ergonomics and socio-technical systems with integrated artificial intelligence that gets into a symbiotic relationship with the intelligence and experience of the designer. The following work will be devoted to communication in complex systems.

5. Acknowledgments

The work is fulfilled within the implementation of the State Assignment from the Ministry of Education and Science of the Russian Federation No. 25.8444.2017/БЧ for the project "Engineering and psychological design of simulation training systems based on symbiotic polymodal technogenic environments".

6. References

[1] Sergeev, S.F. (2011). Introduction to engineering psychology and ergonomics immersive environments, Saint-Petersburg, Russia

[2] Sergeev, S.F. (2012). The Problem of difficulty in ergatic systems. In: materials of the conference "Control in technical, ergatic, organizational and network systems" (UTEOS-2012), pp. 802-805, "Concern "CRI "Electropribor" SPb., Saint-Petersburg, Russia

[3] Sergeev, S.F. (2014). Methodological problems in the design of complex ergatic systems. Problems of control and modeling in complex systems: proceedings of the XVI International conference (30 June - 03 July 2014 Samara, Russia) / edited by: Acad. E.A. Fedosov, Acad. N.A. Kuznetsov, Professor V.A. Wittikh, pp. 386-392, Samara scientific centre of RAS, ISBN 978-5-93424-703-5, Samara, Russia

[4] Sergeev, S.F. (2013). Global, technological environment in the evolution of human civilization. Bulletin of Moscow University Witte. Series 1: Economics and management, No. 1, pp. 80-86.

[5] Sergeev, S.F. (2015). Ergonomics and engineering psychology, man-made world: issues of methodology, theory and practice. In: Modern trends in psychology of labor and organizational psychology, L.G. Dikaja, A.L. Zhuravlev, A.N. Zankovskiy, (Ed.), pp. 47-56, Publishing house "Institute of psychology RAS", ISBN 978-5-9270-0303-7, Moscow, Russia

[6] Sergeev, S.F. (2013). Science and Technology of the 21st Century. Communications and NBICS-Convergence. In.: Global Future 2045/ Convergence Technologies (NBICS) and Trans-Humanistic Evolution. Edited by Prof. D.I. Dubrovsky, pp. 158-168, OOO "Izdatelstvo MBA", ISBN 978-5-906325-26-6, Moscow

[7] Stepin, V.S. (2009). Classics, Non-Classics, Post Non-Classics: Criteria for Differentiation, In: Post Non-Classics: Philosophy, Science, Culture, pp. 249-295, Publishing House "Mir", ISBN 5988460372, 9785988460374, St. Petersburg

[8] Sergeev, S.F. & Sergeeva, A.S. (2014). Educational communication and interface in computer education systems and media, Open education, Vol. 106, No. 5, pp. 41-48, ISSN 1818-4243, Moscow, Rusia

[9] Maturana, H. & Varela, F. (2001). The Tree of Knowledge, Progress-Traditsia, 2001. ISBN 5-89826-103-6, Moscow, Russia

[10] Knyazeva, E.N. (2006). Epistemological Constructivism, Philosophy of Science. Issue 12, pp. 133-152, Moscow: IF RAS, ISSN 1560-7488, Moscow, Russia

[11] Tsokolov, S. (2000). Discourse of Radical Constructivism. Traditions of Scepticism in Modern Philosophy and Theory of Cognition, PHREN, ISBN 3-9805074-2-4, Munich

[12] Sergeev, S.F. (2009). Teaching and Professional Immersive Environments, Narodnoie Obrazovanie, ISBN 978-5-87953-256-2, Moscow, Russia

[13] Grgurevic, I; Perakovic, D; Forenbacher, I & Milinovic, T. (2015). Application of the Internet of Things Concept in Carsharing System, Chapter 34 in DAAAM International Scientific Book 2015, pp.401-414, B. Katalinic (Ed.), Published by DAAAM International, ISBN 978-3-902734-05-1, ISSN 1726-9687, Vienna, Austria

[14] Vaupotic, B.; Brezocnik, M. & Balic, J. (2007). Use of Self-Organizing Neural Network in Modern Manufacturing Systems, Chapter 12 in DAAAM International Scientific Book 2007, B. Katalinic (Ed.), Published by DAAAM International, ISBN 3-901509-60-7, ISSN 1726-9687, Vienna, Austria

[15] Arshinov, V.I. (2015). The reflexivity paradigm of complexity. Reflexive processes and control. The collection of materials of the X International Symposium on October 15-16, 2015, pp. 86-88. ISBN 978-5-89353-465-8, Saint-Petersburg, Russia

[16] Lepsky, V.E. From monodisciplinary to transdisciplinary approach in the development of scientific knowledge (perspective control). The collection of materials of the X International Symposium on October 15-16, 2015, pp. 7-13. ISBN 978-5-89353-465-8, Saint-Petersburg, Russia

[17] Buchmeister, B.; Leber, M.; Palcic, I. & Vujica - Herzog, N. (2013) Future Development Trends and Challenges in Production and Social Systems, Chapter 04 in DAAAM International Scientific Book 2013, pp. 111-132, B. Katalinic & Z. Tekic (Eds.), Published by DAAAM International, ISBN 978-3-901509-94-0, ISSN1726-9687, Vienna, Austria

DOI: 10.2507/28th.daaam.proceedings.155

DIGITAL-ANALOG INPUT / OUTPUT DEVICE WITH THE MATLAB/SIMULINK ENVIRONMENT IN REAL TIME MODE

Peshko Michael, Ryabchevsky Vladislav

This Publication has to be referred as: Peshko, M[ichael] & Ryabchevskii, V[ladislav] (2017). Digital-analog Input / Output Device with the Matlab/Simulink Environment in Real Time Mode, Proceedings of the 28th DAAAM International Symposium, pp.1111-1117, B. Katalinic (Ed.), Published by DAAAM International, ISBN 978-3-902734-11-2, ISSN 1726-9679, Vienna, Austria
DOI: 10.2507/28th.daaam.proceedings.155

Abstract

Data capture about control object and its analysis in real time with minimal delays is an actual problem in the learning process and in many researches. In this article we suggest that the input/output can provide data exchange via digital analogue channels between the Matlab / Simulink field and the industrial control device / control object in two modes of operation. The device is developed with using open technologies and based on a cheap microcontroller AVR, provides a two-way communication channel with the top-level system through popular USB / Ethernet interfaces. There are shown technical solutions of the input and output channels of the input-output device compatible with industry standards.

Keywords: DAQ; data acquisition; input/output devices; industrial control systems; Matlab/Simulink; automatic control systems

1. Introduction

The development of automatic control systems of technological processes and production is closely connected with the collection and date analysis on the control object. The software MATLAB / Simulink is more preferable for data analysis, identification of the model control object and synthesis of control systems [1], [2], [3]. The software package allows to make researches of student scientific researches and serious scientific ones, allowing to develop your own solutions using the built-in programming language. It is necessary to make experiments on real equipment with obtaining empirical data for consolidate the theoretical knowledge and confirm the effectiveness of the developed control algorithms.

The most appropriate equipment for these tasks are DAQ-devices (Data acquisition), which are presented by Nation Instruments [4], [5], [6]. DAQ devices can work with a wide range of signals; they have a high-speed inner tire with convenient high-level system, but DAQ ones are expensive. Industrial input-output devices [7], [8], equipped with a network interface (RS485, Ethernet, RS232) can be used as alternative clients. These devices can work directly from the MATLAB / Simulink field via the OPC driver or via manually created network exchange code. This approach allows you to organize two-way data exchange, but with high delays, that can play a good role in the process of control dynamic objects in real time. The developments are known [9-10] with using low-budget equipment as DAQ devices, but these devices do not need an electromechanical compatibility with standards of industrial automation.

2. Problem statement

According to the work, the task is to develop the input-output device for industrial signals, which makes possible to obtain data in real-time conditions which are sufficient for solving the learning tasks of researching control systems and scientific research developing new and improve existing control algorithms. The device should be maintainable, cheap and developed with using transparent technologies for exchanging two-way data.

3. Field of application and tasks of the input-output device

The main propose of input/output devise is transferring data between the virtual implementation and the real object. Laboratory stand, industrial technology process and programmable logic controller can serve as a real object and a similar objects in a Matlab / Simulink field can be served as a virtual object. Thus, we can use input/output device in two base configurations. The first type configuration (fig. 1) is aimed to train identification skills of control object, to research its description, to synthesize and research the regulator of the control object in real time. This configuration is appropriate as for educational purposes and as well as for scientific work and for development new algorithms.

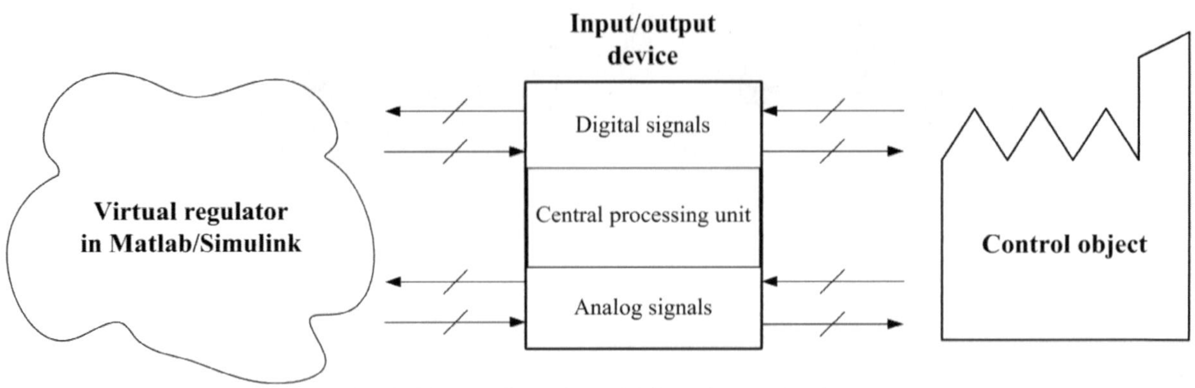

Fig. 1. First configuration with input/output device

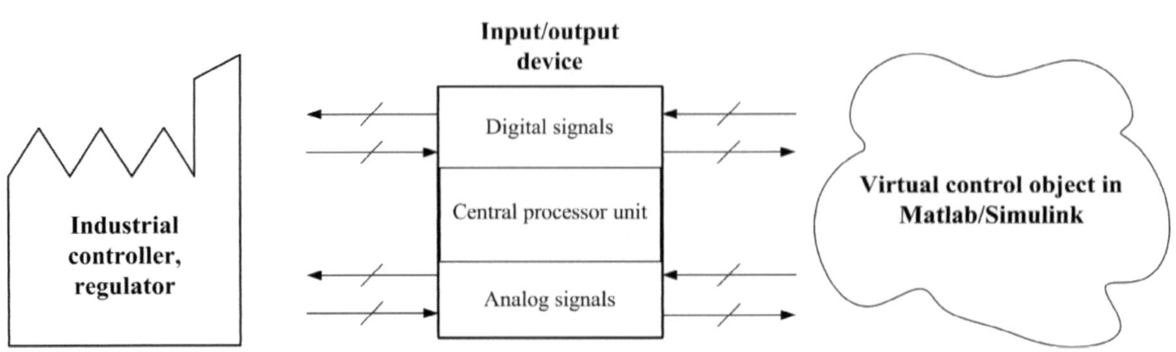

Fig. 2. Second configuration with input/output device

The second type configuration (fig. 2) is used to implement new modern algorithms, to control technological process, to train skills of programming control systems on industrial equipment with next training on virtual object. This approach provides an unlimited number of virtual control objects in MATLAB/Simulink with visualization of the process for education and organization laboratory or practical classes. It allows to get empirical data on the effectiveness of new control algorithms before their implementation and validation on real industrial objects.

4. Technical solutions of the input/output device

In accordance with requirements, the block-diagram of device is showed on fig. 3. 8-bit microcontroller ATMega2560 is chosen as a processing core, it has a sufficient processing power 16MIPS for signal processing, user programs and organization of necessary communication interfaces with external equipment. Using a specialized program loader in the microcontroller by open project «Arduino» will allow to convert the microcontroller to a DAQ device for two-way data exchange with the Simulink. Data exchange between the microcontroller and the top-level system is realized through USB or Ethernet interfaces inside the interface module. This configuration allow to make some researches, data capture and real time control of technological process.

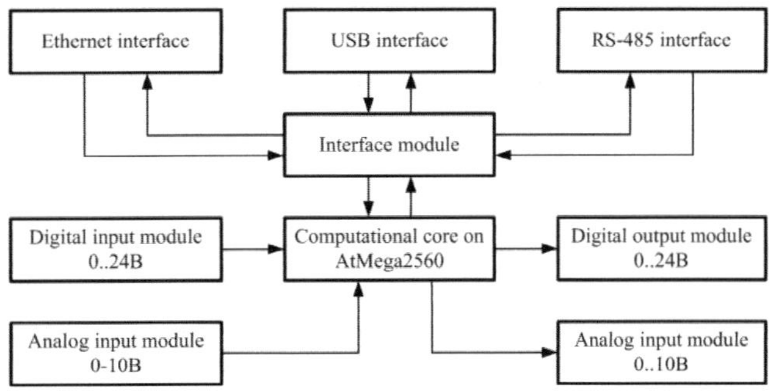

Fig. 3. Structure of Input-output device based on ATMega2560

For creation communication input-output device in real time with real control object or programmable logical controller, we developed the modules of digital input-output operations with appropriate industrial standard 0-24V according to the industrial standard [10] and also the models of analog input-output operations with appropriate industrial standard 0-10V [12]. On fig. 4 digital input channel is showed. Its contents are D1 and D2 LEDs, which are indicate about wrong connection and about logic "1" level of signal.

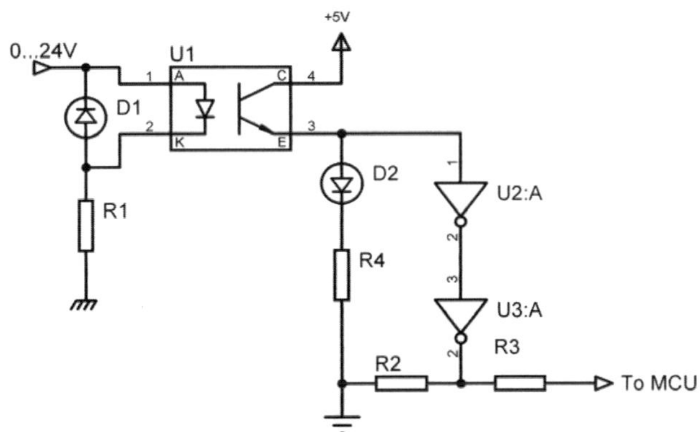

Fig. 4. Digital input channel

The U1 optocoupler is used to protect the input circuits of microcontrollers and matching signals level. Trigger Schmitt U2 is used for forming a logic level and for the input of the microcontroller. The U3 signal inverter changes the signal after the trigger transmits direct signal for ATMega2560 chip through the current limiting resistor R3.

Digital output channel (fig. 5) consists of optocoupler U4 which matches signal levels between microcontroller and MOSFET Q1, fuse FU1 which protects short circuit and LEDs D3 and D4 as indicators. Without signals on the input circuit of optocoupler, the gate of transistor Q1 has a voltage 24 V, which makes a difference between potentials Gate-Source channel to 0 V.

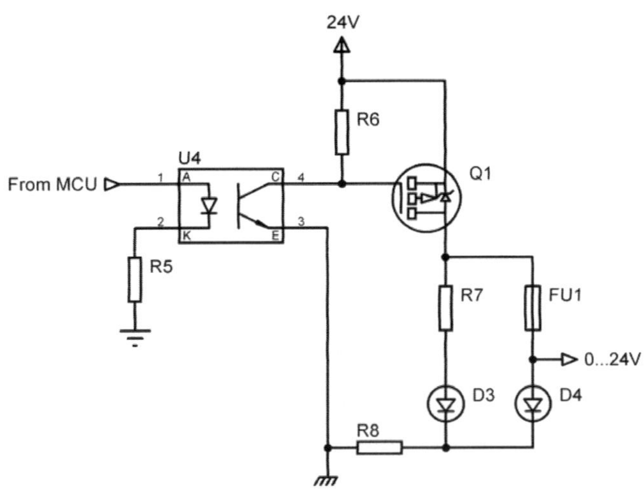

Fig. 5. Digital output channel

Microcontroller makes a logic «1» signal on the input circuit of optocoupler, and it opens the last one, which makes potential on transistor's gate close to zero. Also, difference between potentials Gate-Source channel is going to max, which is fully open transistor. Also, difference between potentials Gate-Source channel is going to max, which is fully open transistor. Power current from drain Q1 passes through the fuse FU1 to load. There is LED D4 with the limited resistor R8 in parallel with load here. If there is a short circuit or unacceptable load on the output channel here, the fuse opens power circuit. After that circuit with the LED D3 and the limited resistor R7 will signal about emergency operation.

For measuring operations in range 0-10V is used 10-bit analog-to-digital convertor in microcontroller ATMega2560 (on fig. 6).

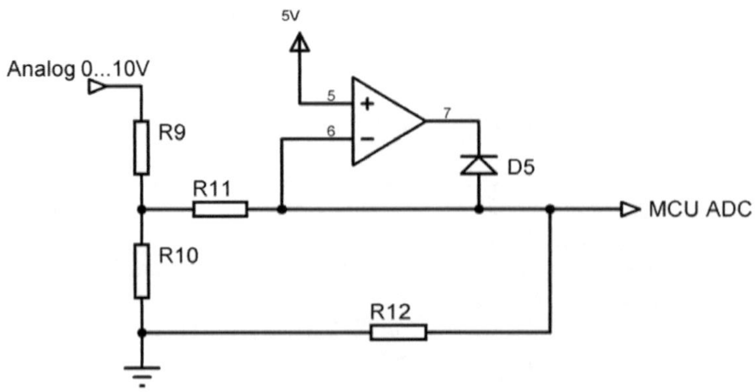

Fig. 6. Analog input channel

Voltage divider on resistors R9-R10 is installed on input circuit. The operation amplifier (AO) U5, which workes in voltage limit mode, has limiting input voltage on ADC in range allowed for microcontroller 0-5V. While input voltage is smaller then 5V, operation amplifier is in a positive SATURATION, which makes diode in REVERS SHIFT and doesn't work in circuit. Input voltage is gone to output through resistor R11. When input voltage is bigger than the limited, output voltage on AO changes its sing, diode opens, feedback circuit is closed, and on the output of circuit is setting limited voltage.

The analog output channel (fig. 7) consists of the microchip U4 (12-bit digital-to-analog converter), which receives digital value via I2C tire and generate voltage in range 0-5V. Convention of signal to standard industrial 0-10V takes place through operation amplifier, which works in amplifier mode. On this circuit, the signal 0-5V is applied to non-inverted amplifier's input. The amplifier circuit has series inverted feedback by voltage, which created by resistor R13 and applied on inverted amplifier's input.

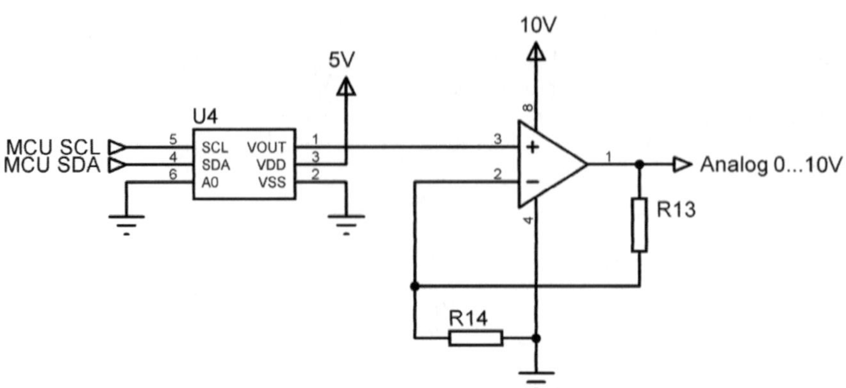

Fig. 7. Analog output channel

5. Experimental study of input-output devices

The investigation of the module was done at a compact station, which could meet the goals of holding liquid, spending and pressure level. Schematic structure of the compact station is shown in the fig. 8. The station operates with 3 analog output signals (level, spending and pressure) in the range of 0-10 v; with 1 analog input signal (pump speed) 0-10 v; with 2 digital output signals (extreme positions of the liquid level in the tank) 0-24 v; with 2 digital input signals 0-24 v (analog/discrete operating mode and discrete pump control).

The test is a permanent control for the pump to achieve the free motion of the controlled object. During the experiment, the input/output module continually reads the analog parameters of the station (level, spending and pressure) and transmits their values to the Simulink model, where they are recorded for further analysis. The Simulink model which was used during the experiment is shown in fig. 9.

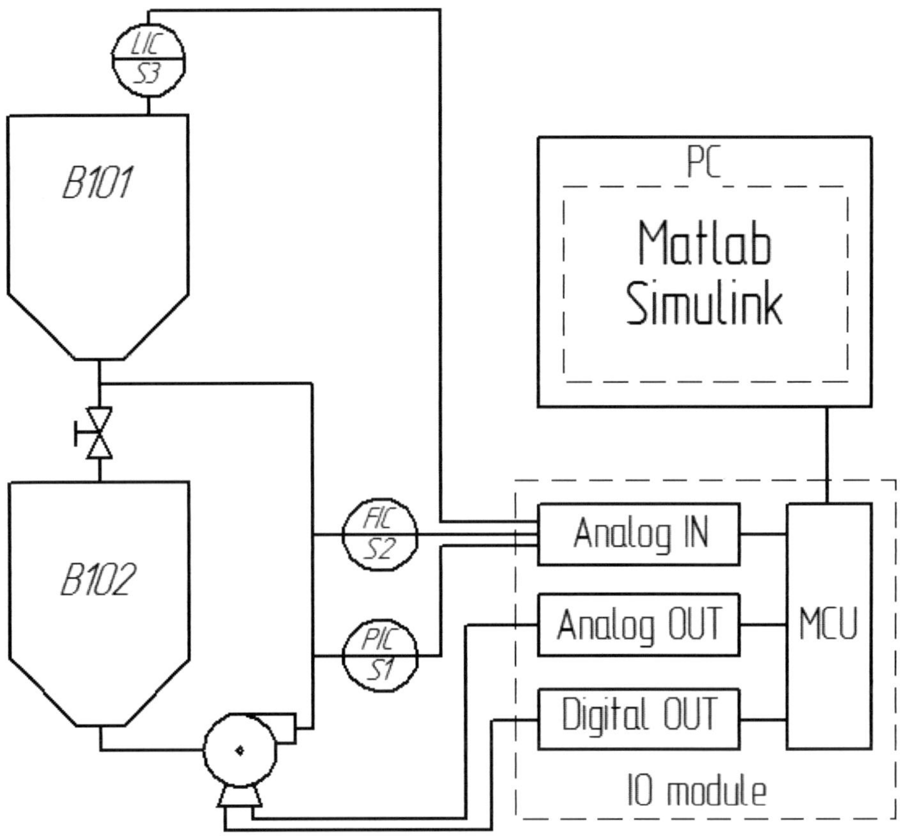

Fig. 8. Schematic structure of the test bench

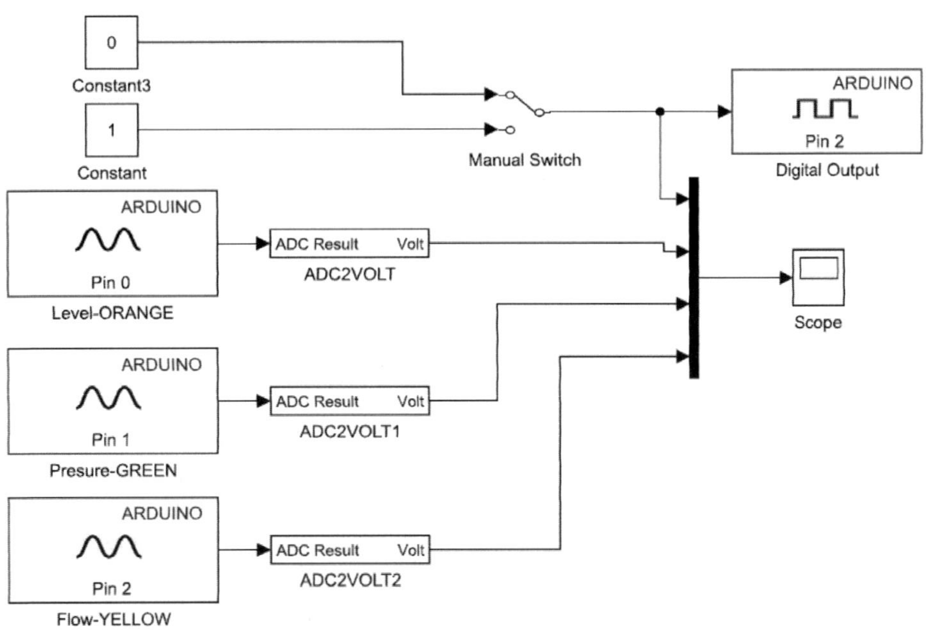

Fig. 9. Simulink model for the removal of analog data from the test bench

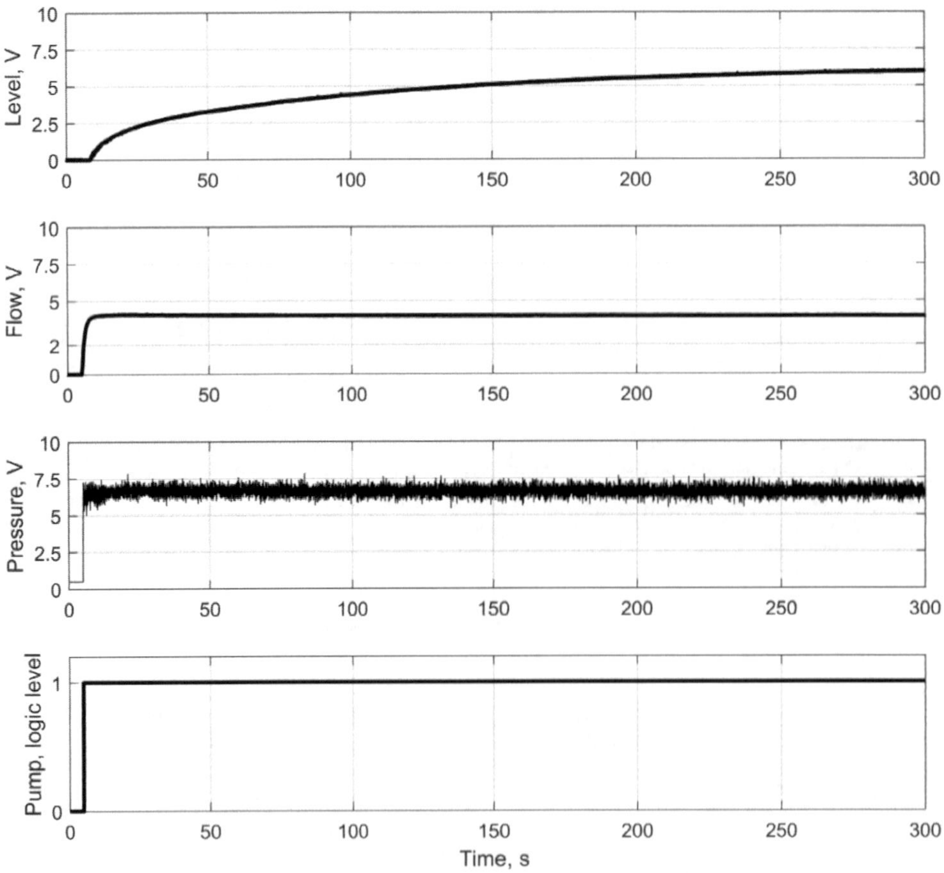

Fig. 10. Signal Data from the test bench

The data which are shown in the fig. 10 were obtained in the test results. The measuring frequency of each channel is 40 Hz (Sample time = 0.025), which is sufficient to track the channels of level and spending without losing their information. The pressure parameter requires a higher data density (Sample time < 0.025) per unit of time, but the module can't provide it without garbling and losing the information. The Simulink model complexity impacts significantly on value of frequency of input/output module poll program. In that way making simple tasks such as monitoring of system parameter values and the polling frequency value can be up to 40 Hz, as it is shown in the experimental schedule, and during the implementation of PID controller the upper threshold of an effective value of Silulink model poll frequency doesn't exceed 10 Hz. This study shows the limited processing power of the input/output module and suggests that in future a more efficient controller can be used for fast parameters such as pressure.

6. Conclusion

The issue in the cheap method of acquisition and processing data from industrial control objects in real-time conditions is solved by the proposed input / output module.

The input-output device providing two-way data transfer between the real object and Matlab / Simulink field are proposed in this work. The device allows to work in two base modes: using real industrial process or laboratory stand with industrial electrical signals and virtual control system inside MATLAB/Simulink field or using programmatic logical controller with virtual control object. The obtained information in the Matlab system allows to evaluate the nature of the control object and also to control the object in real time by setting the desired slider for a particular parameter.

The device is developed with cheap computing platform, which provides electrical compatibility with signals of industry standards (0-10V) and data transfer to the virtual environment of Matlab / Simulink. The device allows to work with a polling frequency up to 40 Hz without distortion and data loss in real time.

In the future, we plan to increase the performance of the MCU to obtain a higher signal polling frequency, to work with processes with a low time constant. A device prototype of the input / output module of the industrial signals having the extended nomenclature in accordance with the IEC 61131-5 standard is being developed.

7. References

[1] Martyanov A.S.; Solomin E.V., Korobatov D.V. (2015). "Development of control algorithms in Matlab/Simulink" Procedia Engineering 129. Elsevier, 2015. pp.922-926.

[2] Outeiro M.; Carvalho A. (2011). "MatLab/Simulink as design tool of PEM Fuel Cells as electrical generation systems", European Fuel Cell Forum 2011.

[3] Saghafinia A.; Hew W., Ping M., Uddin N., Amindoust A. (2013). "Teaching of Simulation an Adjustable Speed Drive of Induction Motor Using MATLAB/Simulink in Advanced Electrical Machine Laboratory" Procedia - Social and Behavioral Sciences 103 Elsevier, 2013. pp.912-921.

[4] Wang L.; Tan Y., Cui X. and Cui H. (2012). "The Application of LabVIEW in Data Acquisition System of Solar Absorption Refrigerator", 2012 International Conference on Future Energy, Environment, and Materials.

[5] Liao H. Q.; Qiu Z. R. and Feng G. H. (2011). "The Design of LDF Data Acquisition System Based on LabVIEW", 2011 3rd International Conference on Environmental Science and Information Application Technology

[6] Bruce-Boye C.; Rüdiger zum Beck, Kazakov D. (2006). "An approach to distributed remote control based on middleware technology, MATLAB/Simulink - LabMap/LabNet framework", Advances in Computer, Information, and Systems Sciences, and Engineering. Springer, 2006. pp.37–41.

[7] Owen Input output modules Mx110// Owen. – URL: http://www.owen.ru/catalog/99197162 (accessed 08 June 2017).

[8] Remote I / O Modules – Advantech// OWEN. – URL: http://www.advantech.ru/products/remote-i-o-modules/sub_1-2mlcjf (accessed 08 June 2017).

[9] Gula M.; Žáková K. (2015) "Matlab Adapter – Online Access to Matlab/Simulink Based on REST Web Services", Advances in Intelligent Systems and Computing 348. Springer International Publishing Switzerland 2015. pp.199-205.

[10] Barber R.; Horra M., Crespo J. (2013). "Control Practices using Simulink with Arduino as Low Cost Hardware", 10th IFAC Symposium Advances in Control Education The International Federation of Automatic Control. August 28-30, 2013. pp.250-255.

[11] IEC 61131-2-2012. Programmable controllers. Part 2. Equipment requirements and tests. Functional requirements. 5.2 Digital I/Os Introduced 2014-07-01.

[12] IEC 61131-2-2012. Programmable controllers. Part 2. Equipment requirements and tests. Functional requirements. 5.3. Analogue I/Os Introduced 2014-07-01

DOI: 10.2507/28th.daaam.proceedings.156

THE APPLICATION OF THERMOGRAPHIC DIAGNOSTIC IN INVESTMENT CASTING PROCESS

Ondřej Vrátný, Aleš Herman & Irena Kubelková

This Publication has to be referred as: Vratny, O[ndrej]; Herman, A[les] & Kubelkova, I[rena] (2017). The Application Of Thermographic Diagnostic In Investment Casting Process, Proceedings of the 28th DAAAM International Symposium, pp.1118-1125, B. Katalinic (Ed.), Published by DAAAM International, ISBN 978-3-902734-11-2, ISSN 1726-9679, Vienna, Austria
DOI: 10.2507/28th.daaam.proceedings.156

Abstract

The paper aims to describe the dependence of cooling rate of a ceramic shell on the number of refractory slurries and on the way of insulation in investment casting process. The infrared thermography was the basis for measurement of temperatures during experiments. This method has not been used during Investment Casting process yet according to scientific databases. From a plenty of experiments there are shown the top results in this paper. The point was to investigate the cooling behaviour of ceramic shell to find out the dependence of a decreasing temperature on time. This knowledge is essential in determining the maximum period of time to assure the minimum required preheating temperature of the ceramic shell.

Keywords: Thermographic analysis; Investment casting; Ceramic shell; A Cooling rate

1. Introduction

The goal of the paper is an investigation of behaviour of ceramic shell in investment casting process from a cooling point of view. There are not many articles dealing with the issue of thermographic analysis in the casting process. Some authors investigated the cooling behaviour of the mould after the liquid metal had been poured in. [1, 8]

Thermographic cameras based on infrared thermography method usually detect radiation in the long-infrared range of the electromagnetic spectrum (approximately 9,000 – 14,000 nm or 9–14 μm) and produce images of that radiation, called thermograms. Since infrared radiation is emitted by all objects with a temperature above absolute zero according to the black body radiation law, thermography makes it possible to see one's environment with or without visible illumination. [2, 7, 9, 10]. Investigation of cooling behaviour in casting process is nothing new. Eric Wulf and his colleagues were investigating the influence of different cooling rates in-situ and ex-situ. The temperatures of the geometric specimens were determined during the experiment with the aid of a near-infrared camera and thermocouples. The aim was to both determine a function between the specimen's cooling rate and the speed of the solid-liquid interface as well as establish a relationship between the thermographic images and the solidification's morphology. [1]

The new idea of the paper is to investigate the cooling behaviour using IRT in Investment casting process, especially to carry out temperature measurements of the ceramic shell after being removed from the preheating furnace.
It is a request from foundries to estimate the shell temperature after a certain period of time since pulling out of the preheating furnace.

This knowledge is very important in determining the maximum period of time to assure the minimum required preheating temperature of the ceramic shell. If the casting process is delayed, there should be a strict time limit after which the shell has to be return back to the preheating furnace. The minimum preheating temperature is determined by a technologist or by a metallurgist. [3, 4, 5, 6]

2. Experimental part

Following calibrated devices and equipment were used for measurement:

- Thermographic camera FLIR T640 + evaluation software TOOLS+
- Contact thermometer Ahlborn Therm 2420 with thermocouple FT 106
- Stopwatch
- Tape measure

In the case of uninsulated shell there were used only parameters related to ceramic material. It means especially the value of emissivity = 0,97. The rest of changeable values were set the same for all measurement (apparent reflected temperature = 30 °C, distance = 2 m).

To all points of the figures in this paper there are assigned the same following parameters:

Ceramic shell (Zr+)	Emissivity	0,97
	Apparent reflected temperature	30 °C
	Distance	2 m
Insulation (Sibral)	Emissivity	0,36
	Apparent reflected temperature	30 °C
	Distance	2 m

Table 1. Set parameters for thermographic analysis

2.1. The cooling process of ceramic shell A

The measurement of cooling rate of ceramic shell *A* was executed in two variants.

- First, the mould was as fast as possible transported from the preheating furnace to the bottom chamber of vacuum furnace to simulate real casting conditions. The bottom chamber was immediately closed and the space was evacuated. In this case, vacuum does not refer to a state where there is no atmosphere at all (absolute vacuum), but rather, is simply a specific space where the gas pressure is less than atmospheric pressure. As a result, the degree or level of vacuum is quantified in terms of pressure (this case $p = 10^{-6}$ bar). Three thermal images were captured. First one after taking the mould out of preheating furnace, second one just before place to the chamber of vacuum furnace and the last thermal image (as it is shown in the figure 1) is taken immediately after being removed from the chamber. It took 251 seconds since removing the ceramic mould from the preheating furnace to opening the vacuum furnace. The temperature decreased from 1085 °C to 668 °C, it means that the ceramic shell cooled down by 417°C.

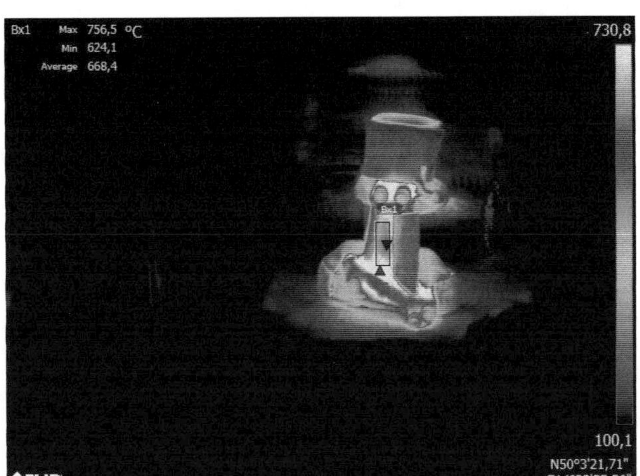

Fig. 1. The thermal image of ceramic shell *A*

- Second measurement was recorded using thermographic camera. The record was created in sequence after 2 seconds. The evaluated part of the shell was as in the vacuum furnace trailing edge of a blade. In the Table 2 there are temperature values related to this area depending on time.

Cooling rate of the ceramic shell *A* - 13 slurries			
Time [s]	**Temperature [°C]**	**Time period [s]**	**Average decrease [°C/s]**
30	900,58	2 to 30	4,76
60	843,87	31 to 60	1,89
90	802,88	61 to 90	1,37
120	766,84	91 to 120	1,20
150	740,38	121 to 150	0,88
180	722,41	151 to 180	0,60
210	701,49	181 to 210	0,70
240	683,71	211 to 240	0,59

Table 2. The cooling rate of the ceramic shell *A*

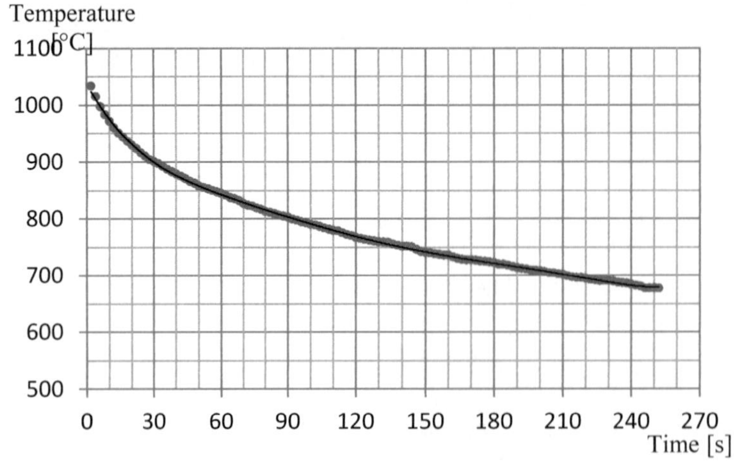

Fig. 2. The cooling rate of ceramic shell *A*

The following graph in Fig. 3 shows the comparison between cooling under vacuum and cooling in the surrounding atmosphere. Despite the different number of measurement we can make a conclusion, that the temperature values are almost the same at the time of 251 second after the ceramic shell is removed from the preheating furnace. In the case of cooling in the vacuum chamber there is the temperature value 688 °C and during the cooling in atmosphere the temperature decreased to 677 °C. The preheating temperature was the same in both cases. We may say, that the cooling process under vacuum and in the atmosphere have similar character.

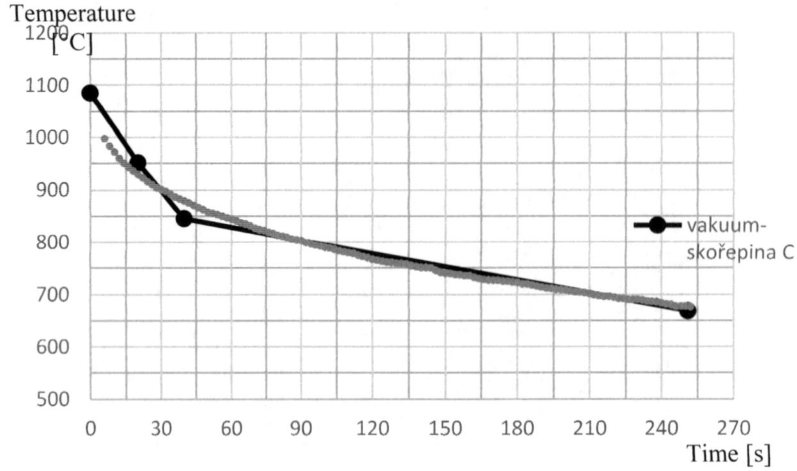

Fig. 3. The comparison of cooling process under vacuum and in the atmosphere

2.2. The cooling process of the ceramic shell B

The ceramic shell *B* was cooled down in the surrounding atmosphere from the temperature of 1100 °C and evaluated in two areas indicated in Fig. 4. First area was around root part, marked as *Bx1* and the area *Bx2* was placed near to pouring basin.

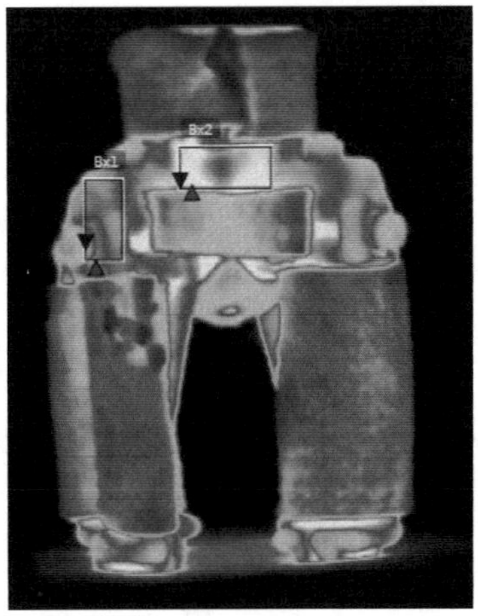

Fig. 4. The thermal image of ceramic shell *B*

In the Table 3 there is a list of values of temperature depending on time for cooling process of the ceramic shell *B*.

Cooling rate of the ceramic shell *B* - 12 slurries							
The area under the pouring basin				**The area around root**			
Time [s]	**Temperature [°C]**	**Time period [s]**	**Average decrease [°C/s]**	**Time [s]**	**Temperature [°C]**	**Time period [s]**	**Average decrease [°C/s]**
30	958,88	4 to 30	4,77	30	864,93	6 to 30	5,37
60	878,51	31 to 60	2,15	60	784,35	31 to 60	2,75
90	810,05	61 to 90	2,91	90	733,29	61 to 90	2,03
120	786,29	91 to 120	1,57	120	685,60	91 to 120	1,46
150	746,37	121 to 150	1,20	150	646,87	121 to 150	1,29
180	718,87	151 to 180	1,21	180	614,27	151 to 180	1,09
210	679,31	181 to 210	0,84	210	585,10	181 to 210	0,97

Table 3. The cooling rate of ceramic shell *B*

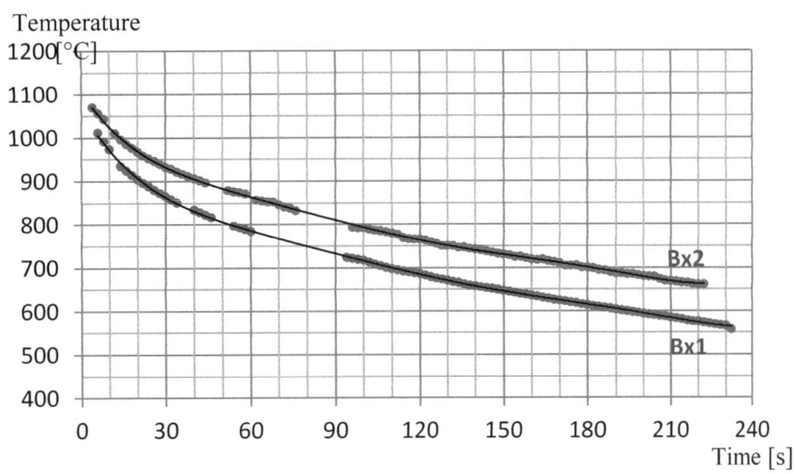

Fig. 5. The cooling rate of ceramic shell *B*

From the Fig. 4 is obvious, that area around root is cooled down faster than the area near to pouring basin. The reason is probably thinner shell around the external rounding and the fact, that this area is exhibited to surrounding relatively cold atmosphere. It is an opposite situation in compare to the second evaluated area *Bx2*, where the shape of the ceramic shell is characterized by internal rounding closed to environment and probably thicker shell.

2.3. The cooling process of the ceramic shell C

The ceramic shell *C* with 11 slurries was evaluated in area shown in the Fig. 6, this thermal image was captured 4 minutes after being removed from the preheating furnace and shows the differences in temperature between areas near to the warm centre and parts, which are exhibited to environment. The most obvious it is around the top and bottom of pouring basin.

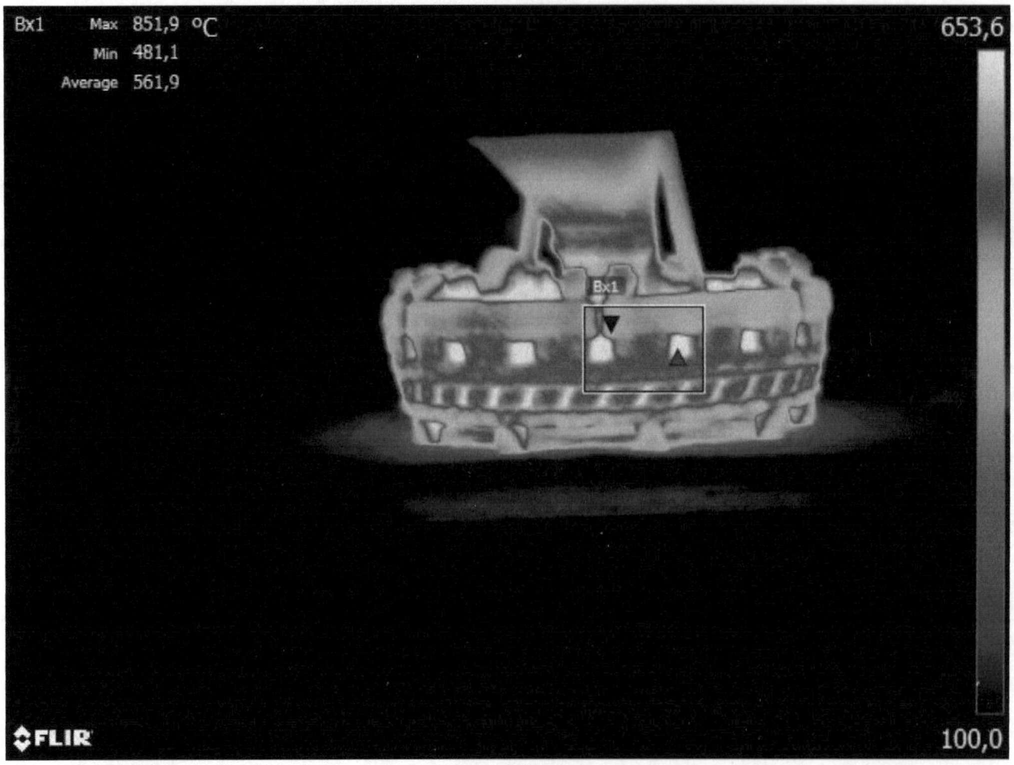

Fig. 6. The thermal image of ceramic shell *C*

In the Table 4 there is a list of values of temperature depending on time for cooling process of the ceramic shell *C*.

Cooling rate of the ceramic shell *C* - 11 slurries			
Time [s]	**Temperature [°C]**	**Time period [s]**	**Avarage decrease [°C/s]**
30	880,88	6 to 30	5,47
60	788,10	31 to 60	3,09
90	732,28	61 to 90	1,86
120	686,25	91 to 120	1,53
150	652,32	121 to 150	1,13
180	616,51	151 to 180	1,19
210	588,51	181 to 210	0,93
240	561,93	211 to 240	0,89
270	541,56	241 to 270	0,68
300	518,54	271 to 300	0,77

Table 4. The cooling rate of ceramic shell *C*

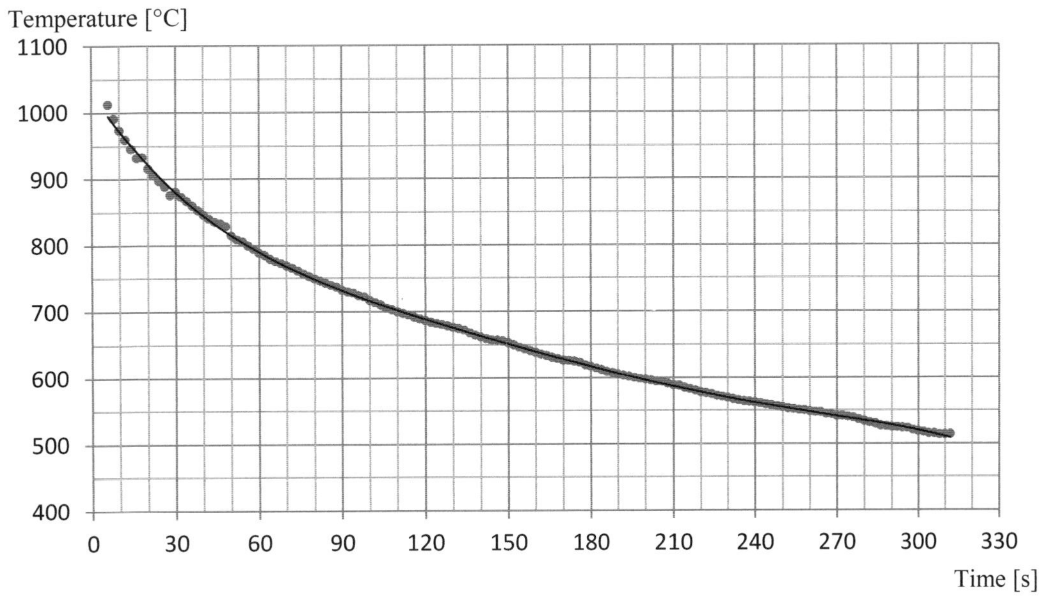

Fig. 7. The cooling rate of ceramic shell *C*

2.4. The effect of insulation

In the study of cooling behaviour effected by insulation there were used Sibral wraps with different thickness (13, 25 and 38 mm). To eliminate undesirable effect of the shape of ceramic shell there was prepared easy-shape ceramic cylinder which was preheated to the temperature of 1100 °C. This temperature was also the same for all measurements.

The Fig. 8 shows different areas to be evaluated. Based on the knowledge that the ceramic material has the same temperature as the Sibral wrap (through the ceramic shell and insulation should be the same temperature because of heating for more than 3 hours), we were able to match the value of emissivity to the isolation, which was approximately 0,36.

Fig. 8. The evaluation of areas with different value of emissivity

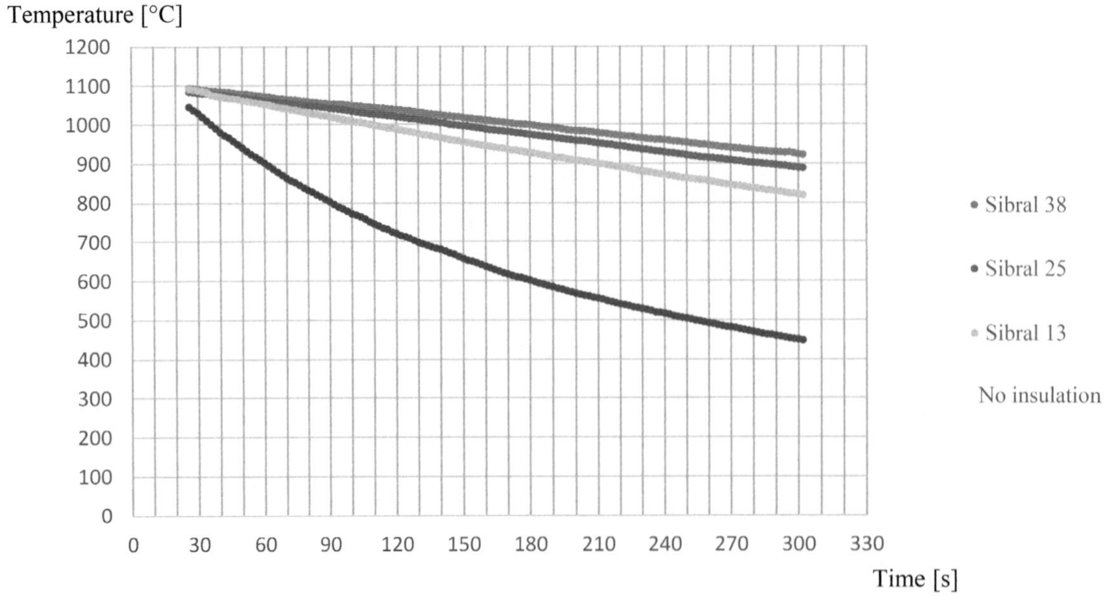

Fig. 9. The cooling rate for wraps of different thickness

2.5. A comparison of ceramic shells with different number of slurries

The summary of results related to measurement of cooling process for shells with different number of slurries is shown in the Fig. 10. The temperature goes down slower with increasing number of ceramic layers. After 210 seconds after removing the shell from the preheating furnace there was a temperature of the ceramic shell *D* with 8 slurries 403 °C, on the opposite side the ceramic shell *A* with 13 slurries had 701 °C. These 5 ceramic layer difference caused temperature difference of 298 °C. Nevertheless there is almost no difference between the ceramic shell *B* (12 slurries – 588 °C) and *C* (11 slurries – 585 °C). The reason could be in the evaluated areas, which are not comparable with each other. The root area in the case of the ceramic shell *B* cooled down faster than the area evaluated in the case of the ceramic shell *C*. Anyway the temperature is influenced not only because of number of layers but also because of the shape of shell and different areas, which are exhibited to the environment.

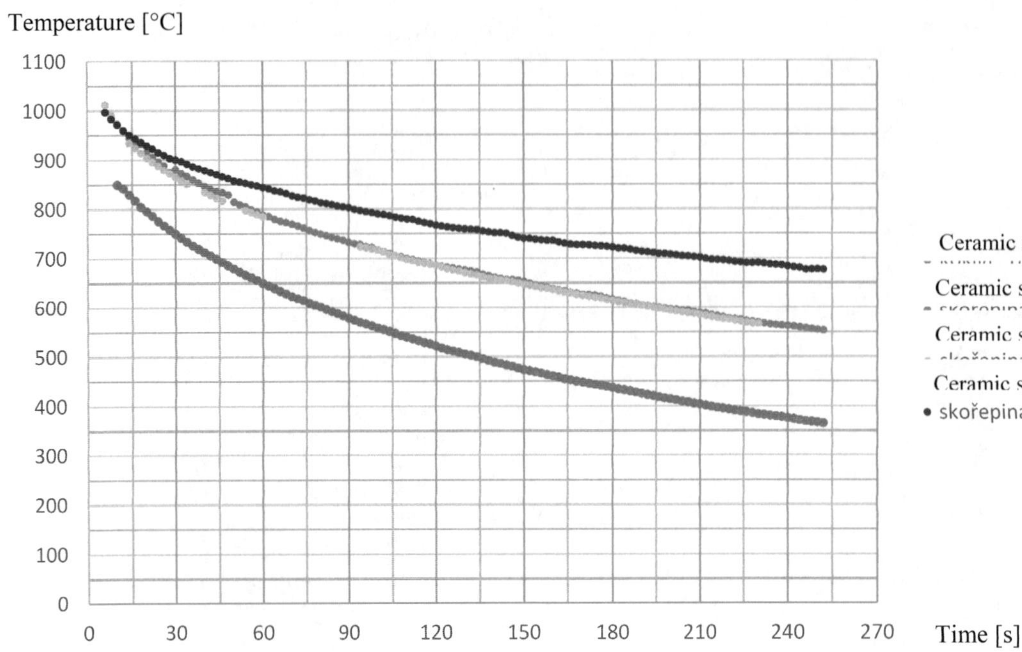

Fig. 10. The cooling rate for shells with different number of slurries

3. Conclusion

The aim of the paper was to investigate the thermal behaviour of ceramic shells used in Investment casting process. The main problem is from the site of foundries to know the maximum period of time the ceramic shell is still allowed to be used. If not, it must be removed back to the preheating furnace.

As a very useful method, as it turned out, was the infrared thermography. Testing was performed using the FLIR T640 thermographic camera, which allowed to take thermal images or videos at the moment of experiments and later evaluation of results. In this experiment the following conclusions were made:

- We may say, that the cooling process under vacuum and in the surrounding atmosphere have similar character.
- For comparison of different shells is very important to choose and evaluate proper area. Every point on the shell has different value of temperature.
- The effect of Sibral heat insulation on the cooling process was investigated. It can be seen from Fig. 9 that when using the smallest thickness of Sibral 13 mm, there is a significant reduction in the cooling rate compared to the uninsulated shell. It is also shown in the graph that the difference between the use of Sibral 13, 25 or 48 depending on temperature is not very important.
- In Fig. 10 is a comparison of the cooling rate depending on the number of slurries. The results show that the increasing number of slurries slow down the cooling rate.

Thermographic analysis is a very useful method in investment casting process. There is a plan to use it to analyse thermal influence on wax pattern deformation and to investigate the whole dipping process in the refractory slurries.

4. Acknowledgments

This publication is under the Centre advanced aerospace technologies, reg. No. CZ.02.1.01 / 0.0 / 0.0 / 16_019 / 0000826, which is co-financed by the European Regional Development Fund through the Operational Program Research, development and education.

5. References

[1] Wulf, Eric, Christian Krause, Dirk Bormann, Mirko Schaper, Jörg August Becker, Verena Becker a Friedrich-Wilhelm BACH. (2009) Thermographic analysis of AlSi12 during crystallisation as a function of cooling rate. International Journal of Materials Research [online]. 2009, 100(1), 97-103 [cit. 2017-10-24]. DOI: 10.3139/146.101781. ISSN 1862-5282. Dostupné z: http://www.hanser-elibrary.com/doi/abs/10.3139/146.101781

[2] Helen E. Kourous, Behrouz N. Shabestari, Spencer D. Luster, Jaroslaw P. Sacha, (1998) "Online industrial thermography of die casting tooling using dual-wavelength IR imaging", Proc. SPIE 3361, Thermosense XX, (26 March 1998); doi: 10.1117/12.304731; http://dx.doi.org/10.1117/12.304731

[3] Herman, A.; Vrátný, O.; Kubelková, I., (2016) Vyhodnocení kritických míst na voskových modelech lopatek, Velká Bíteš: PBS Velká Bíteš, 2016. U12133/2016/006.

[4] Herman, A.; Česal, M., (2012) Temperature Stability of the Process of Production of Wax Patterns for Investment Casting Technology, Manufacturing Technology. 2012, 12(12), 108-112. ISSN 1213-2489.

[5] Herman, A.; Česal, M., (2013) The Solving of a Wax Pattern Deformation for the Large Castings of Blades for Gas Turbines, Technolog. 2013,(2), 27-34. ISSN 1337-8996.

[6] Kovanda, K.; Rohan, P.; Herman, A., (2015) Study of Temperature Fields during FCAW Welding, In: Technological forum 2015, 6th International Technical Conference, Book of Proceedings. Jaroměř: Ing. Jan Kudláček, 2015, pp. 211-214. ISBN 978-80-87583-13-5.

[7] Greenberg, A.R., S.S. Shojaie, W.B. Krantz a S.B. Tantekin-Ersolmaz. (1995) Use of infrared thermography for temperature measurement during evaporative casting of thin polymeric films. Journal of Membrane Science [online]. 1995, 107(3), 249-261 [cit. 2017-10-26]. DOI: 10.1016/0376-7388(95)00123-2. ISSN 03767388. Dostupné z: http://linkinghub.elsevier.com/retrieve/pii/0376738895001232 [8] Filaretov, V. & Adamczak, S. (Year if known). Title of paper, unpublished.

[8] Solórzano, E., F. Garcia-Moreno, N. Babcsán a J. Banhart. (2009) Thermographic Monitoring of Aluminium Foaming Process. Journal of Nondestructive Evaluation [online]. 2009, 28(3-4), 141-148 [cit. 2017-10-26]. DOI: 10.1007/s10921-009-0056-6. ISSN 0195-9298. Dostupné z: http://link.springer.com/10.1007/s10921-009-0056-6

[9] Padalko, A. G., A. N. Veselov a G. V. Talanova. (2008) Thermographic investigation of the phase equilibria in an aluminum-based alloy at high pressures and temperatures. Russian Metallurgy (Metally)[online]. 2008, 2008(5), 413-418 [cit. 2017-10-26]. DOI: 10.1134/S0036029508050091. ISSN 0036-0295. Dostupné z: http://link.springer.com/10.1134/S0036029508050091

[10] Minkina, Waldemar. a Sebastian Dudzik. (2009) Infrared thermography: errors and uncertainties. Hoboken, NJ: J. Wiley, c2009. ISBN 9780470747186.

DOI: 10.2507/28th.daaam.proceedings.157

EVALUATION OF CRITICAL PLACES ON WAX PATTERNS OF BLADES

Irena Kubelková, Aleš Herman & Ondřej Vrátný

This Publication has to be referred as: Kubelkova, I[rena]; Herman, A[les] & Vratny, O[ndrej] (2017). Evaluation of Critical Places on Wax Patterns of Blades, Proceedings of the 28th DAAAM International Symposium, pp.1126-1135, B. Katalinic (Ed.), Published by DAAAM International, ISBN 978-3-902734-11-2, ISSN 1726-9679, Vienna, Austria
DOI: 10.2507/28th.daaam.proceedings.157

Abstract

The production of precision castings by investment casting is becoming an increasingly important manufacturing industry and many of the pitfalls of this method have to be addressed. This paper deals with evaluation of critical points on wax patterns of blades. After the casting of this type of part, the casting had a deviation from the required dimensions. Investigation revealed that the effect on the resulting dimensional deviation is not only the casting process but also the blade injection process itself. The turbine blade is one of the most important parts in turbine machinery. Casting deformation is an important index to evaluate the quality of the turbine blade. In order to control the deformation of the turbine blade during investment casting, a novel compensation method based on reverse deformation was proposed in this study. The article investigates and evaluates critical points for deformation of the blades after their production on wax-press. In addition, the effect of the pre-deformation preparation and the human factor influencing effect during assembly is evaluated.

Keywords: Investment casting; Wax patterns; Deformation; Thermography; Turbine blade

1. Introduction

The preparation of the wax pattern is a key procedure in investment casting, playing an important role with the solidification of the metal on the final dimensional accuracy of the casting part. This study considered the injection procedure and the metal solidification as a whole and tried to discover the individual influence of both factors on the dimensional accuracy of the part. [3]

Film cooling is one of most developed technologies which enhance the gas turbine blades operation in today's high-thrust-to-weight-ratio gas engines. The determination of the accurate cooling holes of turbine blades is of vital importance to improving the cooling performance. However, in all current methods for manufacturing drilled cooling holes on turbine blades, the complexity of the production processes will ineluctably cause unexpected consequences such as the deformation of the turbomachinery parts, the locating error caused by fixture layouts. These issues will cause deviations in the geometry and position of the drilled holes. In this paper, a methodology is proposed to analyse the deviation and to establish an improved geometric model of drilled cooling holes with accurate positional and geometrical parameters on turbine blades. [4]

2. Experiment

The experiment is divided into 3 parts: the evaluation of critical locations where deformations can occur on wax patterns of blade A, blade B and the third part evaluates the thermal impact of waxy trees during gluing.

Used calibrated equipment and devices:

- Thermographic camera FLIR T640 + evaluation software TOOLS+
- Contact thermometer Ahlborn Therm 2420 with thermocouple FT 106
- CAD software SolidWorks 2015
- Stopwatch
- Tape measure

2.1 Measurement procedure of blade A

The blade is pressed onto the injector and set into the rack in the position of the blade lock up.

Fig. 1. Placing the blade in the rack

Measuring parameters:

- Ambient temperature 23 ° C
- Reflected temperature 23 ° C
- Emissions of 0.96
- Distance: 2 m
- Relative Humidity 60%

On this blade the following critical points were evaluated - from side 1 (against the notch) Fig. 3:

- Lock in the strongest place
- The thickest place in the blade (where the blade / cooler is no longer)
- Attaching the lock and blade
- Bandage.

From Side 2 (Slot Link Side) - See Fig. 4:

- The bottom of the blade lock
- Attaching the lock and blade
- Maximum temperature in bandage
- The thickest place in the blade (where the blade / cooler is no longer)

From these pictures it can be seen that the temperature field after removal from the mold is in the functional part of the blade the temperature below the critical 28 ° C only in the bandage, otherwise the temperature is above 33 ° C.

Thickness range	Number of areas	Surface area	% of the analysed area
5,00 – 7,92 mm	103	65 637,45 mm^2	40,32%
7,92 – 10,84 mm	55	20 244,69 mm^2	12,44%
10,84 – 13,76 mm	18	6 865,77 mm^2	4,22%
13,76 – 16,68 mm	6	1 853,64 mm^2	1,14%

Table 1. Details of the analysis

Parameter	Value
Total surface area analysed	162 793,01 mm^2
Critical surface area (% of the area under analysis)	94 601,55 mm^2
Maximum deviation from target thickness	11,68 mm
Average weighted critical area thickness	7,32 mm
Weighted average thickness of the analyzed area	5,75 mm
Number of critical areas	182 areas
Number of critical elements	12
Minimum thickness of the analyzed area	0,00 mm
Maximum thickness of the analyzed area	16,68 mm
Surface area	162 943,37 mm^2
Volume	465 310,13 mm^3
Weight	465,31 g

Table 2. Details of the analysis and physical properties

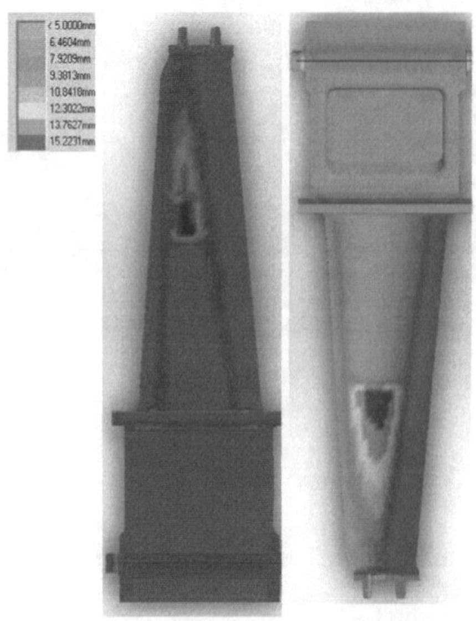

Fig. 2. Graphical thickness analysis SolidWorks

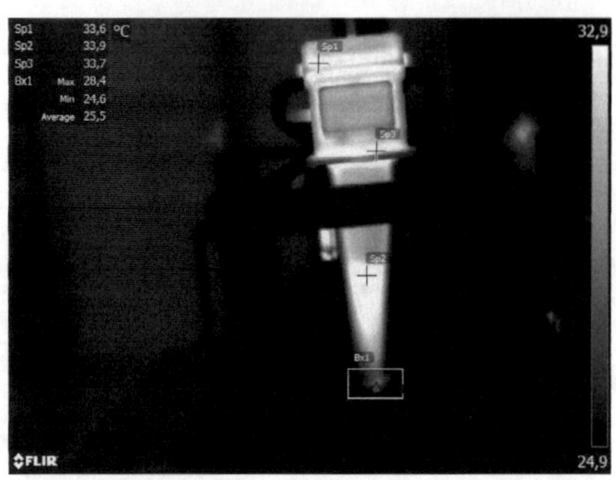

Fig. 3. Blade after removal from mould – side 1

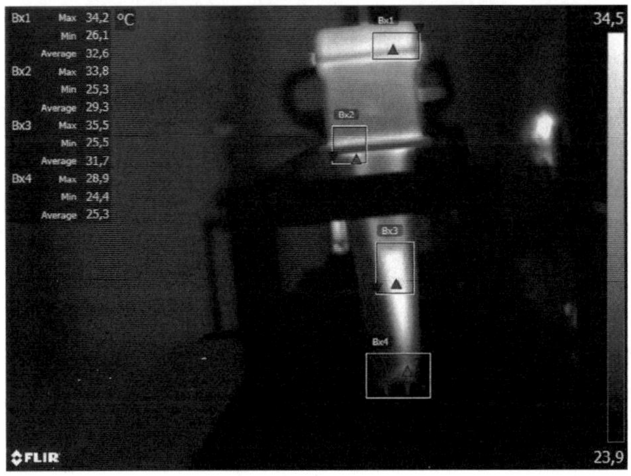

Fig. 4. Blade after removal from mould – side 2

Figure 3 shows a blade without a cooler, that is, just the injected blade space. It can be seen that the cooler is poorly designed - around the lock and the blade of the blade, the thickness of the moulded wax is about 5 - 7 mm, while the area above the cooler has a thickness of up to 17 mm. This is one of the factors that can affect the deformation of this type of model. As shown in Figure 4 the shape of the cooler must be extended to at least the yellow area - here the cooler would be 2 - 3 mm thick so that a gap of 5 mm would occur. The blade cooling is in the correct position - the highest core and surface alignment temperature is over 39 ° C. This is, in itself, very critical for deformation of wax models, and it is necessary to apply watering to the water bath after pressing. We recommend the procedure - after inserting it into a fixation device and using a cooling bath at a temperature of 10 - 15 ° C, in order to quickly rinse and fix the position.

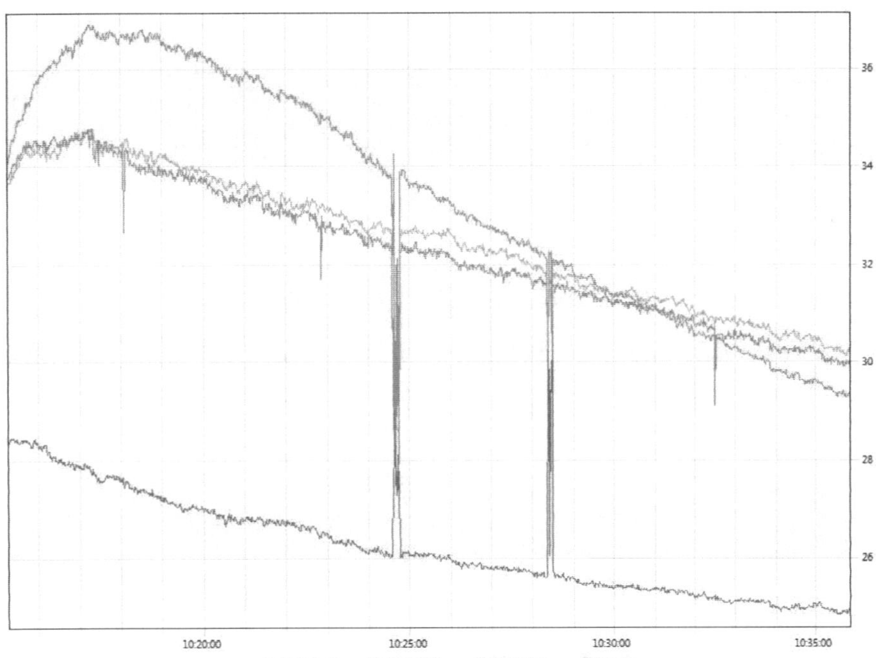

Fig. 5. Temperature range depending on thickness – side 1

From the course of these dependencies it can be seen that the temperature in the critical walls was 34-35 ° C (under the bandage below 28 ° C) after removal from the mould and standing on the stand. This is, however, a condition where the surface is torn off the injection mould, but the device is warmer. So, in the first few minutes, you can see a temperature increase of up to 39 ° C (side 1 of a large blade - at the point where there is no blade / chiller in the blade). It can be seen that at the initial temperature after removal from the mould, the surface of the blade returns for 10 to 11 minutes of standing in the upright position - there may be a relatively large deformation due to wax withdrawal during polymerization since the wax is a critical temperature of 28 ° C begins to soften.

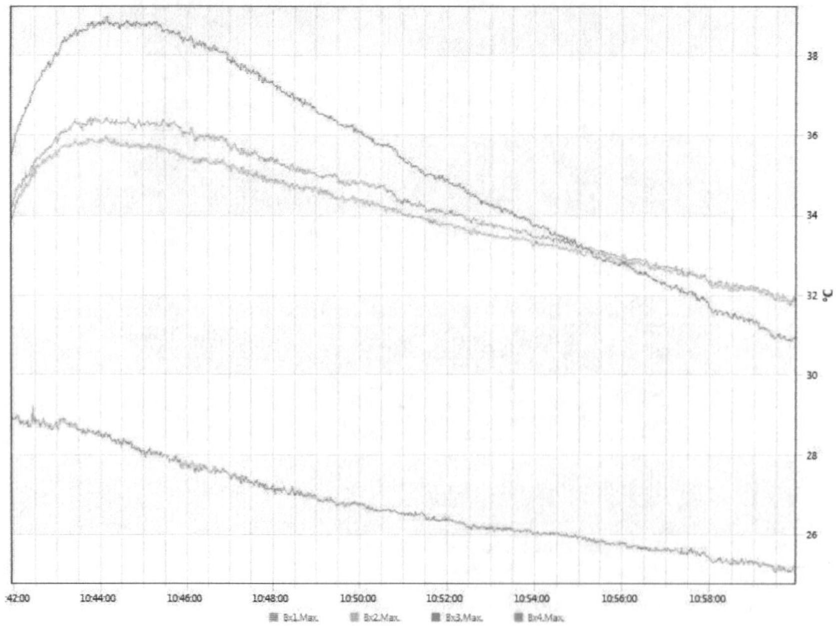

Fig. 6. Temperature range depending on thickness – side 2

2.2 Measurement procedure of blade B

The blade is pressed onto the injector and set into the rack in the position of the blade lock down.

Measuring parameters:

- Ambient temperature 25 ° C
- Reflected temperature 25 ° C
- Emissions of 0.96
- Distance: 1 m
- Relative Humidity 60%

On this blade the following critical points were evaluated - from side 1 (against the notch) Fig. 8:

- The thermal node in the bandage
- The thickest place in the castle
- Attaching the lock and blade
- The warmest place against the notch.

From Side 2 (Slot Link Side) - See Figure 9:

- Heat node - connection of notch
- The bottom of the blade lock
- Attaching the lock and blade
- Maximum temperature in bandage
- Temperature in different cuts of different blade thicknesses

Thickness range [mm]	**Number of areas**	**Surface area [mm^2]**	**% of the analysed area**
1,00 – 3,25	78	11 049,94	46,57 %
3,25 – 5,50	32	4 769,17	20,10 %
5,50 – 7,75	12	3 016,13	12,71 %
7,75 – 10,00	39	1 245,66	5,25 %

Table 3. Details of the analysis

Parameter	Value
Total surface area analysed	23 726,46 mm^2
Critical surface area (% of the area under analysis)	20 080,90 mm^2
Maximum deviation from target thickness	8,99 mm
Average weighted critical area thickness	3,65 mm
Weighted average thickness of the analyzed area	4,46 mm
Number of critical areas	161 areas
Number of critical elements	9
Minimum thickness of the analyzed area	0,00 mm
Maximum thickness of the analyzed area	17,05 mm
Surface area	23 741,54 mm^2
Volume	58 835,17 mm^3
Weight	58,84 g

Table 4. Details of the analysis

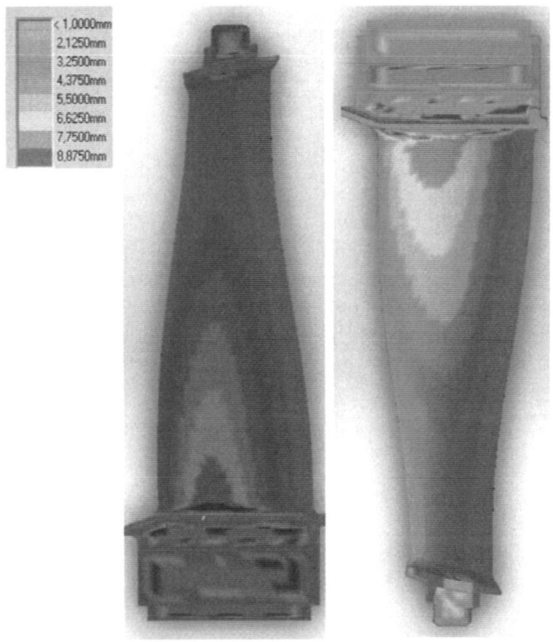

Fig. 7. Graphical thickness analysis SolidWorks

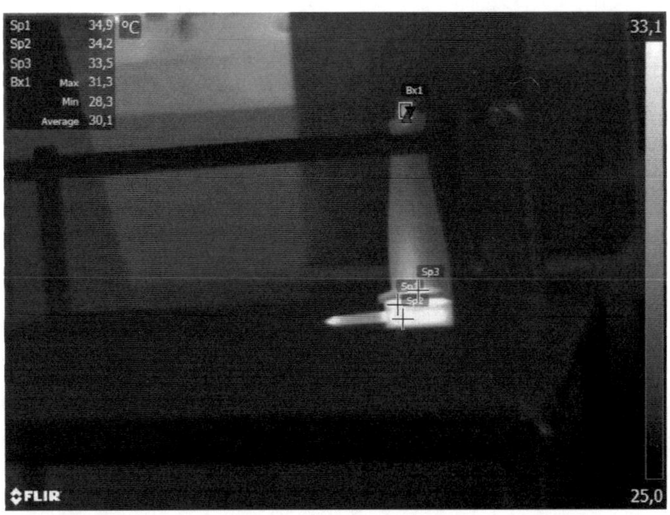

Fig. 8. Blade after removal from mould – side 1

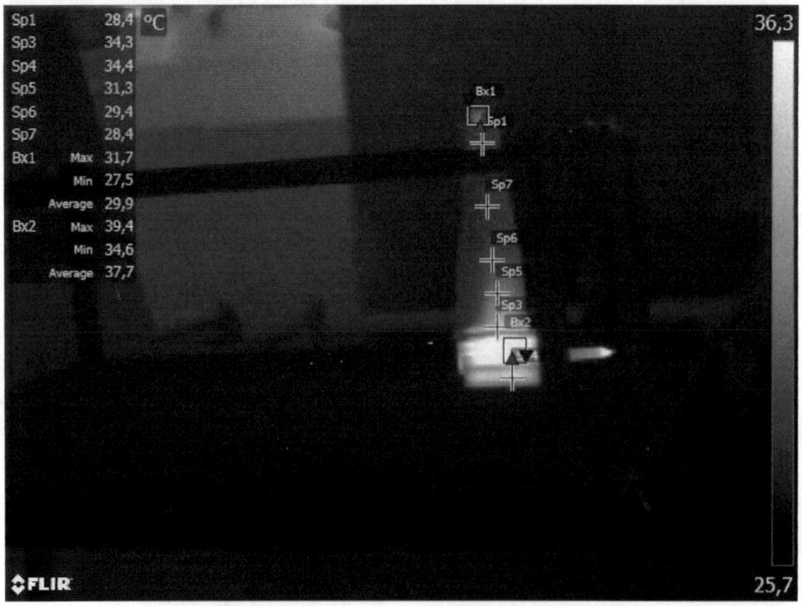

Fig. 9. Blade after removal from mould – side 2

From the course of following dependencies it can be seen that in the critical (strong) walls the temperature was in the range of 30-34 ° C after removal from the mold and standing on the stand. This is, however, a condition where the surface is torn off the injection mold, but the device is warmer. So, in the first few minutes, you can see a temperature increase up to 44 ° C (blade B side 2 - connecting the notch to the lock - fig. 8). It can be seen that at the initial temperature after removal from the mold, the blade surface returns for about 10 Minutes standing in the upright position - there may be a relatively large deformation due to the wax withdrawal during polymerization since the wax is a critical temperature of 28 ° C when it begins to soften.

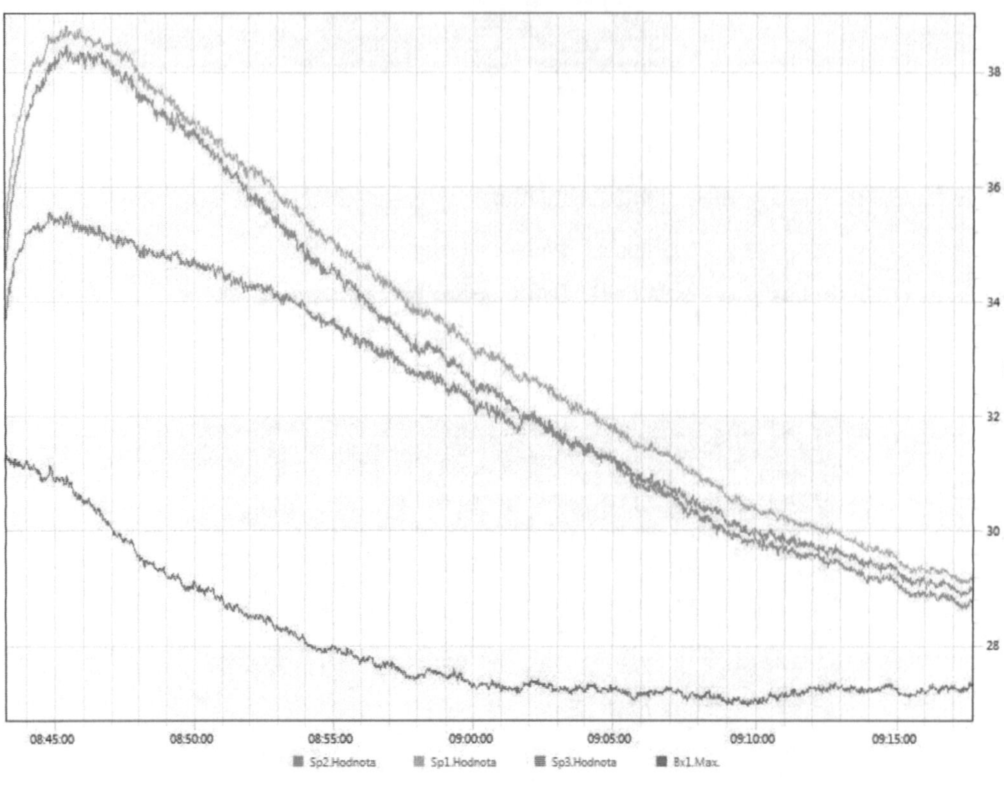

Fig. 10. Temperature range depending on thickness – side 1

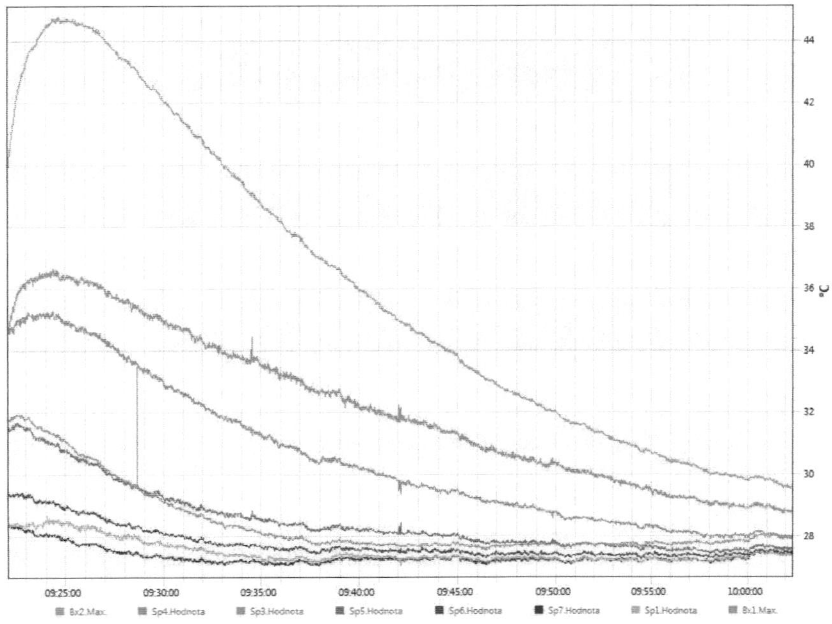

Fig. 11. Temperature range depending on thickness – side 2

Depending on the individual wall thicknesses, the greatest problem in the vane lock and the connection of the inlet system, which is compacted with the vane model, is evident. In this regard, it would be better to stick the inlet part to the model. If we take the shape of a blade, cooling the blade should take place in a hinged form behind the lock, the bandage down (exactly the opposite of what it was found).

2.3 The influence of cooling in the preparation

After being removed from the injection molding machine, a large blade is transferred to the stabilizing fixative, where it is allowed to cool for 10 minutes and then transferred to a straightening preparation, where it is cooled for another 10 minutes.

Measuring parameters:

- Ambient temperature 23 ° C
- Reflected temperature 23 ° C
- Emissions of 0.96
- Distance: 1 m
- Relative Humidity 60%

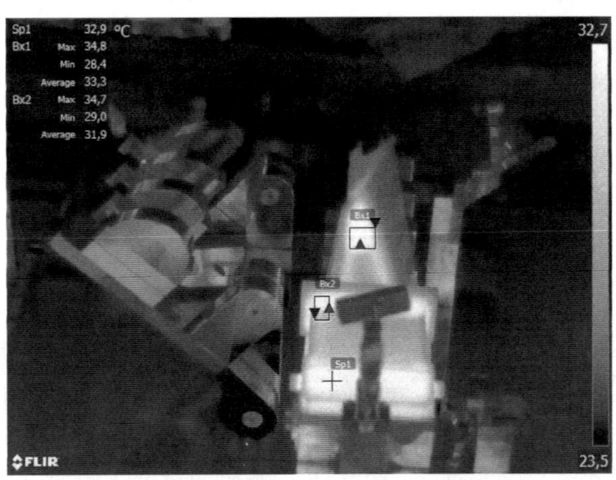

Fig. 12. Insert a blade into the fixative

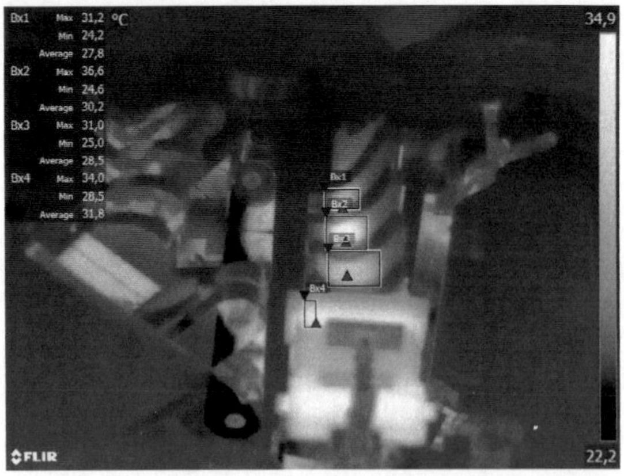

Fig. 13. Opening the fixative after 10 minutes

Figure 12 shows that the temperature distribution corresponds to the temperature field after the removal, recorded on the free-flowing blade in the stand. It can be seen from Figure 13 that despite the use of the fixation device, the temperature is practically at a maximum of 36 ° C - 2 ° C more than it was inserted. In the places where the fixation device touches the vanes, it can be seen that the heat dissipation was more pronounced, but the space between the fins of the fixative is still extremely high. It can be seen from Figure 12 and 13 that it is necessary to quench in a different and substantially faster manner, because the pressing cycle is about 4 minutes and the cooling in the preparation was tested 2.5 times longer than the production cycle of the new wax model of the blade. This results in the need for more products, which is extremely costly in nature (there must be at least 3 fixing and 3 straightening products). Therefore, it is advisable to perform a fixation of the preparation in the cooling water at 10 ° C after removal. This should compensate for the temperature field and the maximum temperature in the blade should not exceed the ambient temperature. This will then ensure dimensional stability of the process

2.4 Thermal influence of trees on the assembly site

In the framework of our experience, we have also addressed the possible thermal influence of glued trees on the model workplace. The amount of heat that may be around the workers (whether radiant heat from the plates, maintenance cups or light sources) can affect the temperature rise above 28 ° C and thus impair the dimensional stability of the wax patterns.

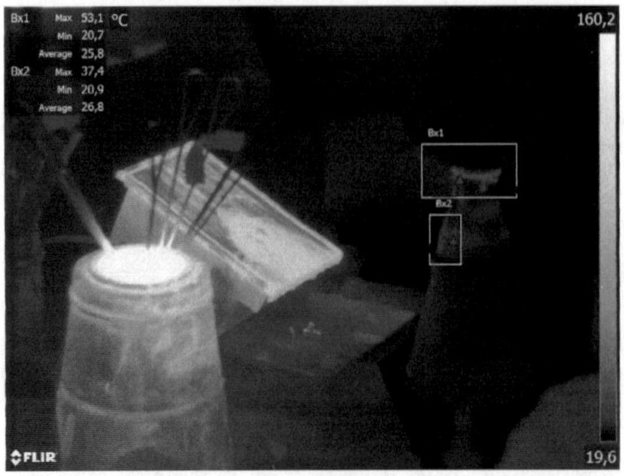

Fig. 14. Figure example

Thermal influences of the tree on Figures 14 show the structure of the workplace for gluing trees and the temperature of the tree placed. Here it is seen that the turbocharger rotor was stuck to the inlet hole in the Bx1 area, a bonded joint having a maximum temperature of about 53 ° C can be seen. While there was nothing glued to the surface of the gulley, a heat reflection of about 37 ° C is visible. From this it can be seen that there is a significant thermal influence on the adhesive workplace, and it will be necessary to reorganize the workplace - either by moving the glued trees to the other side of the table, or by thermal shading of the heating sources in order not to affect the temperature at the tree. In the next phase, it will be necessary to focus directly on influencing sticking of the blades to the tree and their temperature field at this workplace in order to find a suitable position for bonding that will not affect the dimensional stability of the wax models.

3. Conclusion

The main problem / task of the research were by thermography to determine when the blades deform and prevent this. Based on the results from the thermography image, it is clear that the blades are most deformed during compaction after compression due to the influence of residual heat inside the blades. After evaluating all the measurement results we can give the following recommendations:

• Do not inject the blade B together with the notch - it forms a distinct thermal knot.
• In the case of a blade B, it would be advisable either to use a chiller / chisel (to minimize the thermal knot in the lock) or to use a fuser with cold water
• In the case of a large blade, we have shown a poorly designed chiller / jaw - it should be considerably longer - the blade is a large thermal knot that affects deformation behavior and increases dimensional instability
• Thoroughly tighten the large blade straight after being pressed into the fuser (align it with production so that it requires up to 2 fixation products - the current situation seems to be very unprofitable)
• The workshop for the gluing of trees significantly influences the wax models - it will be necessary to measure the effect on sticking the blades to the tree and, if necessary, to adjust the ergonomics of the workplace so as to avoid any possible influence

The future plants are to analyze other types of blades and describe their production - it will take place in the subsequent period. We will focus especially on the stability of the process, avoiding unwanted deformations of the blades caused by inappropriate thermal control.

4. Acknowledgments

This report was created in cooperation with the company První brněnská strojírna Velká Bíteš, and in the form of an economic contract under the project TRIO No. FV10105, Research on the enhancement of the shape accuracy of wax models for turbocharger blade castings and strain gas turbines.

5. References

[1] Herman, A.; Vrátný, O.; Kubelková, I. (2016). Vyhodnocení kritických míst na voskových modelech lopatek, Velká Bíteš: PBS Velká Bíteš

[2] Jiang, R., Zhang, D., Bu, K. et al. Int J Adv Manuf Technol (2017) 88: 3195. https://doi.org/10.1007/s00170-016-9030-5

[3] Wang, D., He, B., Liu, S. et al. Int J Adv Manuf Technol (2016) 85: 201. https://doi.org/10.1007/s00170-015-7836-1

[4] Dong, YW.; Zhao,Q.; Li, XI.; Li, XJ. & Yang, J. (2017). Methodology to develop geometric modeling of accurate drilled cooling holes on turbine blades, proceedings of the ASME Turbo Expo: Turbine Technical Conference and Exposition, AMER Soc Mechanical Engineers, Three Park Avenue, New York, NY 10016-5990 USA, ISBN:978-0-7918-5091-6

[5] Hancock, Phil . Technical Report : Wax Evaluation. Cambridge, UK : [S.N.], (2011). 24 Quality Control Manual (Summary). [S.L.] : Blayson Olefines LTD, 200?. 28 S.

[6] Herman, A., Česal, M., A Píša, V. (2014). Měření Vlastností Voskových Směsí Jako Vstupních Dat Do Databáze Simulačního Software. Strojírenská Technologie., XIX(1), S. 8-12. ISSN 1211-4162.

[7] Herman, A., Česal, M., A Mikeš, P. (2012). Problematic Of Model And Castings Deformations In Investment Casting Technology. In: ULEWICZ, Robert A Beneš, L., Eds. New Trends In The Field Of Materials And Technologies Engineering. Czestochowa: Oficyna Wydawnicza Stowarzyszenia Menedzerów Jakości I Produkcji. S. 160-177. ISBN 978-83-934225-2-4.

[8] Bond, David; Nishikawa, Koji.(2013). Investigation Of Wax Expansion. Blayson Technical Lectures. 2002, S. 1-44.Volume 834-836, 2013, Pages 1575-1579,2013 3rd International Conference On Materials And Products Manufacturing Technology, ICMPMT 2013; Guangzhou; China; 25 September 2013 Through 26 September 2013; Code 100757

[9] Herman, A., Česal, M., A Mikeš, P. (2011). Problematic Of Model And Castings Deformations In Investment Casting Technology. In: Techmat 11. TECHMAT 2011. Svitavy, 10.11.2011. Pardubice: Univerzita Pardubice, Dopravní Fakulta Jana Pernera. 2011, ISBN 978-80-7395-431-4.

[12] Modukuru, S.C.-Ramakrishnan, N.-Sriramamurthy, A.M.(1996). Determination Of The Die Profile For The Investment Casting Of Aerofoil-Shaped Turbine Blades Using The Finite-Element Method. Journal Of Materials Processing Technology 58 (1996) 223-226

[13] Marek, V[aclav] (2016). Basic Research of Thermal Transfer Simulations, Proceedings of the 27th DAAAM International Symposium, pp.0578-0585, B. Katalinic (Ed.), Published by DAAAM International, ISBN 978-3-902734-08-2, ISSN 1726-9679, Vienna, Austria DOI: 10.2507/27th.daaam.proceedings.085

DOI: 10.2507/28th.daaam.proceedings.158

PERSONAL DATA IN CLOUD. RUSSIA EXPERIENCE

Anna Zharova, Vladimir Elin & Petr Panfilov

* Director of Centre for Cyberspace Research, Associate Professor, Docent,
National Research University Higher School of Economics, Moscow
** Department of Information Security, Associate Professor, National Research University,
Higher School of Economics, Moscow

This Publication has to be referred as: Zharova, A[nna]; Elin, V[ladimir] & Panfilov, P[eter] (2017). Personal Data in Cloud. Russia Experience, Proceedings of the 28th DAAAM International Symposium, pp.1136-1142, B. Katalinic (Ed.), Published by DAAAM International, ISBN 978-3-902734-11-2, ISSN 1726-9679, Vienna, Austria
DOI: 10.2507/28th.daaam.proceedings.158

Abstract

The use of cloud computing to ensure interaction between the state and citizens allows to speed up information interaction, to realize state services, to reduce the costs of providing such interaction, but at the same time this interaction raises important questions about the reliability of the cloud provider and security of interaction. Providers of the cloud can be both public authorities and private organizations. In the event that the cloud provider is a government agency, it can be assumed that all the requirements for security will be met. However, if the cloud provider is a private person, then we cannot be sure of security, if these requirements for security are not mandatory. It should be noted that Russian legislation does not require the mandatory application of information security standards. In this regard, the security of stored information in the clouds and its legislative support, the responsibility of providers providing cloud access services are very significant for the use of this technology in Russia.

Keywords: cloud computing; the cloud provider; security; personal data; Russian legislation; cloud access services

1. Introduction

1.1. Main issues of this article

The term «cloud computing» is often used in various spheres of public life. «Cloud computing» refers to technology, which is created to optimize business processes and improve the service quality, different service providers and to increase people's informatization level. Questions on people's interpersonal relationship are arising more often in the field of medicine, government, the banking and private sector. The application of cloud computing is a strategy important development for the computer technologies and provides different services.

The active introduction of cloud computing (CC) put questions to the legislator about the specifics of regulating these people's behavior online. These are questions of data security when it is stored in the cloud computing and question of whether the cloud computing is private or government owned. Question as to the requirements for the cloud Internet Service Provider's (ISP) effective functions as well as ISP's responsibility for it is own action and those of third parties, remain unanswered. Is it possible to on sure that technology which allows us access to the cloud computing is reliable and safe? What if public information and IT systems are hosted abroad?

Whether the government has determined the requirements for ensuring the security of information and emerging relationships the proper regulation and trust of citizens to the security of relationships depends.

The authors of this article will try to answer these questions.

1.2. Definition "cloud computing" in Russia

The State Program of the Russian Federation «Information Society (2011-2020) » [4] considers the development of cloud computing to be one of the main priorities. The creation of a national platform of cloud computing is provided in this government program. It is planned to connect all the public authorities of the Russian Federation in the period of 2011-2020. In 2010 to a Russian platform, the Deputy Minister of Ministry of Communications and Mass Communications of the RF said that «these kinds of proprietary technologies in Russia are practically absent therefore, this sector is actively engaging with foreign companies such as Google» [25].

Despite the active application of the term «cloud computing» in the Russian Federation, this term has no legislative definition. Despite this, «cloud computing» was taught at technical universities in 1990s and the further development of this technology at public institutes was predicted. Consequently, we used to definition which the Institute of Electrical and Electronics Engineers (IEEE) gave. According to the IEEE's definition published in 2008, «Cloudy data processing is a paradigm within which information is constantly stored on servers on the Internet and is temporarily cached with the client party, for example, on personal computers, game consoles, laptops and smartphones».[14] The IEEE, investigating the prospects of developing cloud computing drew a parallel between this technology and the Internet. In Russia, the following IEEE companied operate in Siberia, Northwest (St. Petersburg) and Central Russia (Moscow)[15].

The Strategy of the development of the information society in the Russian Federation for the years 2017-2030[10] defines the notion of "cloud computing" as an model of information technology, by use of the Internet, for ubiquitous and convenient access to a common set of configurable computing resources ("cloud"), data storage devices, applications and services which can be quickly provided and freed from the load with minimal operating costs or practically without the participation of the provider. Cloud computing is considered to be a platform, which provides various different services.

However, it is only classified in the technical sphere, and so the following classification of cloud computing is accepted:

1) SaaS (Software as a service) – in this case the user of cloud computing is granted the right to use and access the necessary software.

2) IaaS (Infrastructure as a Service) – the user of cloud computing is granted the right to use the provider's hardware-software complex for work and data storage.

3) PaaS (Platform as a Service) – the user of cloud computing is granted the right to use a hardware-software complex for to development, testing, expansion and support of web applications.

On 2013, the Russian Federation service for technical export control (FSTEC) set out project the State standard which defined the cloudy services such as:

- Hardware as a service (HaaS);
- Safety as a service (SecaaS);
- Business process as a service (BPaaS);
- Data as a service (DaaS);
- Trust as a service (TaaS);
- Infrastructure as a service (IaaS);
- Cloud development environment as a service (CDPaaS);
- Communication as a service (CaaS);
- Platform as a service (PaaS);
- The connection as a service (CnaaS);
- Transparency as a service (TraaS);
- Workplace as a service (WaaS).

In this article we will not consider the questions of processing of the state secrets or publicly accessible data by the CC. Of course, answers to the above questions are linked to category owner of cloud, whether they are a state or individuals , so cloud computing depending on the subject to whom they belong, can be public or private. Public clouds belong to the state authorities or the local council.

Furthermore, public cloud data may be better protected than the private cloud computing. The cloud technology can belong to foreign or Russian ISPs. Information can be public or have limited access (ch.2 Art. 5 of Federal law «Information, Informational Technologies and Protection of Information»). Limited access information can be personal data, «know-how» and secrets data importation from companies, and is regulated only by the Federal legislation.

2. Legal regulation on use of personal data in Russia

In 2014, a Russian Advisory council from the Ministry of Telecoms and Mass Communications was created to improve the development and implement a state policy and standard and legal regulation of cloud computing. The Advisory council aimed to develop suggestions for improving the legislation for cloud computing; ensure the regular interaction of the Ministry of Telecoms and Mass Communications with the expert community on cloud technologies.[2].

Experts in the technical sphere claim that it is impossible to ensure the absolute safety of these technologies. [18] Noriswadi Ismail has also written about these problems. [21], P.250]

We consider the difficulties that legislation encounters when trying to ensure the protection of limited access data, such as personal data[19]. As Angela Adrian remarks, "how should legal rules change to accommodate the new communication technology? If cloud computing does not alter our fundamental values, how should legal rules adapt and change in order to maintain our current values? What should the substance of our rules be in light of the changing environment for the actors in the cloud? [17], P.49]

According to Art. 7 of the Russian Federal law «On personal data» [5] «the operators of personal data and other persons who accessed to personal data are obliged not to disclose this information to third parties and not to impart personal data without consent of the subject of personal data . The other cases must be provided by the law».

What ought to be done if information is stored in a foreign cloud?

According to item 11. Art. 3 of the Federal law «On personal data» the cross-border transfer of personal data is the transfer of personal data to the territory of the foreign state, to the authority of the foreign state, to a foreign national person or the foreign legal entity.

Therefore, we can understand the cross-border transfer of personal data as the transmission of personal data of any person from the Russian Federation to a foreign person abroad. For example, when any person is filling in a form online to purchase goods from a foreign seller.

Art. 19 of the Law «On personal data» states that the Russian operator of personal data has to establish in the agreement with the foreign handler of the information, the duty to comply with the confidentiality of personal data, to ensure to safety of personal data, and also to comply with other requirements to protect the personal information. According to item 3. Art. 12 of the Law «On personal data» the operator of personal data must be convinced that the foreign country to which the data are transferred provides an adequate protection of the people's rights to their personal data. The Federal law «On personal data» specifies that states can be submitted as two groups of countries:

1) Member countries of the Convention for the Protection of Individuals with regard to the Automatic Processing of Personal Data (1981); (This Convention came into force in the Russian Federation on September 1, 2013) [11]

2) Countries which are not the party to the Convention for the Protection of Individuals with regard to the Automatic Processing of Personal Data (1981), but which are listed as foreign states which provide adequate protection of the people's rights to their personal data. This is such countries[1].

The Russian Ministry of Telecom and Mass Communications in unofficial letter on May 13, 2009 explained that «the adequacy of protection means that foreign state provides the level of security of people's rights to their personal data, not below the standards provided in Russia. One of the assessment criteria of the state in this aspect is ratification of the Convention for the Protection of Individuals with regard to the Automatic Processing of Personal Data (1981)» [12]

However, if the cross-border transfer of personal data is being carried, out in foreign states which are not providing adequate protection people's rights to their personal data, then the operator of personal data must receive, in written form, that consent person's for their personal data to be transferred. This relationship between foreign ISP (operator of personal data) and the owner of personal data is regulated by two the Russian laws. They are the Federal law «On personal data» and the Federal law «On electronic signatures». [7]

Accordance with part. 4 Art. 9 of the Federal law «On personal data» the participants in the relationship have the right to use an electronic signature to create a written form of their agreement. Art. 4 of the Federal law «On electronic signatures» defined the rights of parties in the electronic interaction to use any information technology and (or) the technical means, to apply an electronic signature. However, the Federal law «On electronic signatures» requires that participants (operator of personal data and the owner of personal data) must approve of these actions. This approval can be placed on the main web page of the cloud operator which handles personal data (ISP of personal data) on the conditions of using the type electronic signature and the technologies. The subject of personal data must agree with these conditions.

Although, there is another opinion, Kanashevsky V. writes that sending an e-mail message of the Russian's users to the foreign cloud ISP's of personal data no is electronic document signed the electronic signature. «The foreign operator potentially can be brought to civil liability in accordance with Russian law. In connection with this action, the claim for the damages and compensation for moral damages can be brought to the foreign operator. (Art. 17 of the Federal law «On Personal Data»). [20]

[1] Australia-Australian alliance; The Argentine Republic ; The State of Israel; Canada ; Kingdom of Morocco; Malaysia; United Mexican States ; Mongolia ; new Zealand; The Republic of Angola ; The Republic of Benin ; The Republic of Cape Verde ; The Republic of Korea ; The Republic of Peru ; The Republic of Senegal; Republic of Tunisia; Republic of Chile; Hong Kong Special Administrative Region of the People's Republic of China ; The Swiss Confederation // The Order of the Federal Service for Supervision of Communications, Information Technology and Communications of March 15, 2013 № 274 "On approval of the list of foreign countries that are not parties to the Convention of the Council of Europe Convention on the Protection of Individuals with regard to Automatic Processing of Personal Data, and provide adequate protection of human subjects personal data" // http://www.garant.ru/products/ipo/prime/doc/70268490/#ixzz3E8YwESIQ

This opinion is erroneous. Firstly, the Federal law «On electronic signatures» points directly to the recognition of foreign electronic signatures. Article 7 of this Federal law indicated that «if the electronic signature is created in accordance with the law of a foreign state and the international standards then it is being recognized in the Russian Federation as the electronic signatures of the type which correspond the Federal law «On electronic signatures». Also, according part 2 of Art. 7 of the Federal law «On electronic signatures» an electronic document signed by electronic signatures cannot be regarded as illegal only on the basis that the key's certificate of the electronic signature verification is issued in accordance with the rules of the foreign law. [7]

Secondly, part 2 of Art. 9 of the Federal law «On personal data» determines that consent of the subject of personal data can be given in any form which allows officially confirm receiving this consent.

Thus, the Federal law «On personal data» and the Federal law «On electronic signatures» define the legal conditions of recognition the consent.

However, be aware that Federal law «On personal data» allows to the ISP (operator) of personal data to not obtain the consent of the subject of personal data if cross-border transfer of personal data is carried out «for the execution of the contract. And party to this contract is the subject of personal data (person) » (part 4 Art. 12 of the Federal law «on Personal Data »). Thus, this Federal law indicates the presence of the preliminary agreement on cooperation.

In addition, there is one additional exception by which the telecom operator does not have to obtain the consent of the subject personal data. So, in 2013, the Russian Federal law «On communications» was changed [8] [Federal law «On Communications»]. Art. 53 of this law was added the right of the telecom operator This right allows to the telecom operator entrust the processing of personal data to third parties without received the consent of person on this. But only, if the telecom operator processes the parties' personal data in according with contract for the provision of telecommunications services. And (or) the telecom operator acts for the implementation of the rights and legitimate interests of the service provider or the citizen. However, the Federal law «on Communications» did not disclose whether the operator must warn of the fact that personal data was processed by a third party.

Thus, the opinion about the complexities of the processing of personal data by the foreign cloud ISP (or operator of personal data) is a fallacy in most cases.

In 2013-2014, the total transmission model (provision) of personal data to external operators was changed in Russia. Expanded the list of cases in which the consent is not required of the subject of personal data on processing his data by third parties.

3. Ensuring the technological security of the transmission and processing of personal data.

Of course, the cross-border transfer of data is associated with problems of the ensuring the technological security of the transmission and processing of personal data. However, here it is necessary to take into account the opinion of technical experts, who believe that there are no absolutely safe technologies. Furthermore, there is a problem of difficulty of selecting a required standard of the existing standards. For example, in the report of the Europe Commission are indicated a problem of the difficulty in the selection of standards: "a jungle of standards generates confusion by, on one hand, a proliferation of standards and on the other hand a lack of certainty as to which standards provide adequate levels of interoperability of data formats to permit portability; the extent to which safeguards are in place for the protection of personal data; or the problem of the data breaches and the protection against cyberattacks…» [27]

But here it is necessary to bear in mind that there is no legal prohibition on the choice of the standards on the territory of the Russian Federation countries. The Federal Service for Supervision of Communications, Information Technology, and Mass Media (Roskomnadzor) of the Russian Federation accepted the «Methodical recommendations on the choice of methods of depersonalization of personal data». [16] In this methodical recommendations are submitted various methods to anonymize personal data. In case the depersonalization of personal data, the complexity of the selection standards is eliminated.

Providers of cloud services must address the questions of data security. Technological complexity of security in the cloud is that the virus-infected file from one client allows to infect this virus other connected and non-performing clients. In connection with this, there is the danger of cracking data is stored in the cloud. The responsibility for protection lies on the provider. Users of cloud services need to be sure that the provider is using security tools that reduce the likelihood of offending. The provider of cloud services should allow the user to check the level of protection and should to record the actions of all actors who have gained access to the cloud.

If there are no special requirements of the law, an owner of cloud computing can voluntarily be audited and certified by the ISO / IEC 27001: 2005 and (or) SAS 70 Type II and Type I.

The European Network and Information Security Agency (ENISA) proposed using ISO27001. Moreover ENISA pointed out that «all suppliers must demonstrate compliance. They are NOT required to be accredited but compliance is verified through yearly submission of their information security management system and associated policy documents.

In Russia, according to the Order of the Government of the Russian Federation of 22.02.2012 №238-p OJSC «Rostelecom» appointed as the sole executor for the further creation and development of complex information technology and telecommunications infrastructure and e-government for 2012-2014.

«Rostelecom» must execute the list of activities of the Russian state program «Information Society (2011 - 2020 years)». This program include the development of the portal of public services, the development of a common space of trust in electronic signatures, the development of interagency electronic interaction and a single identification system. It is planned that «Rostelecom» will engage the development of mechanisms which enable the use of mobile devices for access to e-government services and to the state-mail address which using for the interaction between citizens and government agencies.

The «Rostelecom» must create the common directories and the qualifiers for the state information systems and it must create the projects such as «Electronic Registry Office», «E-Region» and «E-democracy». The basis for the implementation of these systems is cloud computing platform. The project offers infrastructure and software as a service (IaaS and SaaS).[16]

Strategy of development of branch of information technologies in the Russian Federation for 2014-2020 and on prospect till 2025 provides creation of a national platform of cloud computing.[9] For these purposes the State defines itself as the customer of these works in the Russian market of information technologies.

The «Rostelecom» has developed and has launched cloud platform in August 2011. This platform helps the "Rostelecom" in selling their computing resources and software to third-party. In the future, this platform can be an integrated into the national platform of cloud computing.

Currently, in the Russian Federation plans to create seven powerful data centers, on the basis of which will be provided the cloud services for government agencies and academia. Including, the public services platform «Gospriklad» was developed for the provision of public services. [23]

4. Conclusion

The Decree of the Government of the Russian Federation of July 20, 2011 N 1275 p approved the Concept creation and development of an integrated information system of state management of public finances «Electronic budget». This Concept is recommended to the executive authorities of the Russian Federation and local authorities for the development of systems of public (municipal) finance. The subsystems of the system «Electronic budget» will be available as service subsystems for use by the subjects of the Russian Federation, the municipalities or the government sector organizations on a pro bono basis. The subsystems of the system «Electronic budget» will be submitted as model of «software as a service» (SaaS).

Single portal of the budget system of the Russian Federation will consist of closed and of open parts.

The open part of the portal is being provided as a public information resource, providing free access to the regulatory, statistical and analytical information in the management of public finances.

The storage and the processing is being carried out in a centralized service of the system «Electronic budget» which is a source of information to be placed in the open part of the portal.

Closed part of the portal consists of a «private offices» available to users with the appropriate powers and which have keys of an electronic signature certificates. Closed part of the portal is the single point of user access to the functions of centralized and service subsystems «Electronic budget». This system includes a single, unified system of management of user access levels depending on their powers for the all functions of subsystems «Electronic budget».

The government services will be available to users cloud platform in the areas applications of the medicine, housing, energy, transport and others. In addition, the online service has created for work with documents.

Unfortunately, in the Russian Federation there are not many examples of the use of models of public services based on cloud computing. Although, with 2013 in Russia is being actively implemented the idea of cloud computing in government.

Thus, we need to understand the technology and their reliability before using cloud technology. The ISP of cloud computing should prescribe in the contract the level of service. Of course, the using the cloud computing has disadvantages which carry new risks. These risks are actually as Russia and other country.

Firstly, the user does not own or have not access to the inside the cloud. Safety of the user data is highly dependent from the company's ISP.

Secondly, there are no generally accepted standards in the direction of cloud computing security.

Originally, the owner of the «cloud» developed the standards of the safety. He is free to choose the models and mechanisms of protection.

Thirdly, there is no definite answer about the responsibility of the subject providing the service to the user.

Fourthly, ISP of personal data must ensure conditions of use of electronic signatures.

The results of this study will be used in further science researches that aimed at identifying approaches to organizing and securing the digital economy in Russia.

5. Acknowledgments

We thank the Russian Humanitarian Science Foundation for their assistance in the project "Comparative legal research methods of information security in the Russian Federation and EU Members (№ 16-03-00679)".

6. References

[1]. https://cloudsecurityalliance.org/media/news/csa-survey-reveals-emerging-international-data-privacy-challenges/ (2014) Cloud Security Alliance. Published 2014, 23, September. New Cloud Security Alliance Survey Reveals Emerging International Data Privacy Challenges. Accessed on: 2017-10-13.

[2]. The Order of the Ministry of Communications and Mass Communications of the Russian Federation N 178 of June 30, 2014 «On the Council of Experts on the use of cloud computing in the Ministry of Communications and Mass Communications of the Russian Federation»

[3]. The Strategy of the development of the information society in the Russian Federation for the years 2017-2030. approved by the Decree of the President of the Russian Federation of May 9, 2017 No. 203

[4]. The State Program of the Russian Federation "Information Society (2011-2020)" endorsed by Order of the Government of the Russian Federation No. 1815-r of October 20, 2010

[5]. Federal law "On Personal Data" No. 152-FZ of July 27, 2006

[6]. Federal law "On Information, Informational Technologies and Protection of Information" No. 149-FZ of July 27, 2006

[7]. Federal law "On Electronic Signature" No. 63-FZ of April 6, 2011

[8]. Federal law "On Communications" No. 126-FZ of July 7, 2003

[9]. The Strategy of development of branch of information technologies in the Russian Federation for 2014-2020 and on prospect till 2025 endorsed by Order of the Government of the Russian Federation of November 1, 2013. Implementation of the cross-border transfer of personal data

[10]. The Strategy of the development of the information society in the Russian Federation for the years 2017-2030. approved by the Decree of the President of the Russian Federation of May 9, 2017 No. 203

[11]. http://conventions.coe.int/Treaty/EN/Treaties/HTML/108.htm (1981) The Convention for the Protection of Individuals with regard to Automatic Processing of Personal Data. Strasbourg, 28.I. 1981 .Accessed on: 2017-10-13

[12]. Letter of the Ministry of Communications and Mass Communications of the Russian Federation of 13 May 2009 N CP-P11-2502 «On the implementation of the cross-border transfer of personal data»

[13]. http://www.enisa.europa.eu/media/news-items/security-issues-in-cross-border-e-authentication (2017) Security Issues in Cross-border e-Authentication. Accessed on: 2017-10-13

[14]. https://ec.europa.eu/digital-agenda/node/1322 (2017) About Cloud computing. Accessed on: 2017-10-13

[15]. http://ieee.tusur.ru/ru/index.htm (2016) The Tomsk group and student's office of Institute of engineers on electrical equipment and radio electronics. Accessed on: 2016-11-09

[16]. http://minsvyaz.ru/ru/monitoring/index.php?id_4=43034 (2017) Official website of the Ministry of Telecom and Mass Communications of the Russian Federation, Accessed on: 2017-12-12

[17]. Adrian A.. (2013) How much privacy do clouds provide? An Australian perspective, Computer law & Security review, Vol. 29 Issue 1, pp. 48-57 doi.org/10.1016/j.clsr.2012.11.010

[18]. Baranov A. (2012) Can we protect in the «cloud» of confidential information? .- High availability systems № 2, Vol. 8

[19]. Zharova A., Elin V. (2017), The use of Big Data: A Russian perspective of personal data security, Computer Law & Security Review, Volume 33, Issue 4,pp. 482-501, doi.org/10.1016/j.clsr.2017.03.025

[20]. Kanashevsky V.A. (2012) Legal regime of cross-border transfer of Personal Data .- Legislation

[21]. Noriswadi I. (2011) Cursing the Cloud (or) Controlling the Cloud?. - Computer law & Security review Vol. 27

[22]. Official sire of the Federal Service for Supervision of Communications, Information Technology, and Mass Media (Roskomnadzor) (Accessed December 12, 2017). Retrieved from: http://rkn.gov.ru/news/rsoc/news23181.htm

[23]. http://minsvyaz.ru/ru/monitoring/index.php?id_4=41944 (2011) In 2011 the Ministry of Telecom and Mass Communications will create a resource - a gospriklad.ru. Accessed on: 2017-12-12

[24]. The Internet interview to the head of the department of information technologies, communication and information security of the Ministry of Internal Affairs of the Russian Federation, lieutenant general of internal service Tyurkin M. L. «Development and introduction of modern information technologies in system of the Ministry of Internal Affairs of Russia

[25]. http://minsvyaz.ru/ru/monitoring/index.php?id_4=43034 (2016) Official website of the Ministry of Communications and Mass Communications of the Russian Federation. Accessed on: 2017-10-13

[26]. http://www.enisa.europa.eu/activities/risk-management/files/deliverables/cloud-computing-risk-assessment (2009) Cloud Computing. Benefits, risks and recommendations for information security. November 2009. Accessed on: 2017-10-13.

[27]. http ://eur-lex.europa.eu/LexUriServ/LexUriServ.do?uri=COM:2012:0529:FIN:EN:PDF (2012) Communication from the commission to the European parliament, the council, the European economic and social committee and the committee of the regions Unleashing the Potential of Cloud Computing in Europe. Brussels, 27.9.2012 . Accessed on: 2017-10-13.

[28]. Zharova A. K., Elin V. (2015) Legal Aspects of the Use of Educational and Scientific Information Posted on the Intern et for Organizing e-Learning According to the Russian Example, in: Strategic Innovative Marketing. 4th IC-SIM, Mykonos, Greece 2015. Switzerland : Springer International Publishing, 2016. doi P. 615-623.

[29]. Bahtovski A., Gusev M. (2014) Cloudlet Challenges, Procedia of 24th DAAAM International Symposium on Intelligent Manufacturing and Automation, pp. 704–711, doi: 10.1016/j.proeng.2014.03.045
[30]. Kostoska M..; Gusev M..;.Ristov S. (2014) New Cloud Services Portability Platform, Procedia of 24th DAAAM International Symposium on Intelligent Manufacturing and Automation, pp. 1268 – 1275, doi: 10.1016/j.proeng.2014.03.118
[31]. Katalinic. B.; Kukushkin I.; Pryanichnikov V.; Haskovic D. (2014) Cloud Communication Concept for Bionic Assembly System, Procedia of 24th DAAAM International Symposium, pp.1562 – 1568, doi: 10.1016/j.proeng.2014.03.156
[32]. Davidovic. V.; Ilijevic D.; Luk V.; Pogarcic I. (2015) Private Cloud Computing and Delegation of Control. Procedia Engineering, Vol.100, pp. 196-205. ISSN: 1877-7058

DOI: 10.2507/28th.daaam.proceedings.159

METHODOLOGY OF DESIGN AND VALIDATION OF THE SHOP FLOOR MILLRUN METHOD

Michal Šimon, Josef Babor, Michal Bochinský & Pavel Kábele

This Publication has to be referred as: Simon, M[ichal]; Babor, J[osef]; Bochinsky, M[ichal] & Kabele, P[avel] (2017). Methodology of Design and Validation of the Shop floor Millrun method, Proceedings of the 28th DAAAM International Symposium, pp.1143-1150, B. Katalinic (Ed.), Published by DAAAM International, ISBN 978-3-902734-11-2, ISSN 1726-9679, Vienna, Austria

DOI: 10.2507/28th.daaam.proceedings.159

Abstract

This article discusses a methodology that is used to set the appropriate material feed rate using the Millrun method. The methodology uses basic calculations and data about products, warehousing technology. Basic data used in this methodology are static and also the simulation of Millrun is static. This methodology using basic data as an input and provides four determined steps leading to successful Millrun implementation. Methodology uses only one type of vehicle for supplying. Methodology is made as easy as it can be to ensure applicability for everyone and requires personal approach to solutions based on common sense in rationalization of the process. Methodology aims to provide a sequence of steps leading to implementation of shop-floor Millrun.

Keywords: Millrun; shop-floor; implementation; lean logistics; pull system

1. Introduction

Current state of industrial companies is focusing on reduction of cost by rationalization. Achievement of cost reduction is conditional to implementation of more effective processes.

If the companies want to do so, they need to have easy to use, straightforward methodologies based on skilled people, methodologies robust enough to be easily and quickly implemented into production or logistics.

The Milk run system or method is not a novelty today. It originates in the early 20th century, where its essence consisted of regular sips of fresh milk from individual farmers, for example, to a Milkrunner. The family could rely on the fact that empty bottles would be replaced for full ones at the set time.

The correct setting of the Millrun system is a tricky task and a lot of effort is needed. [3]

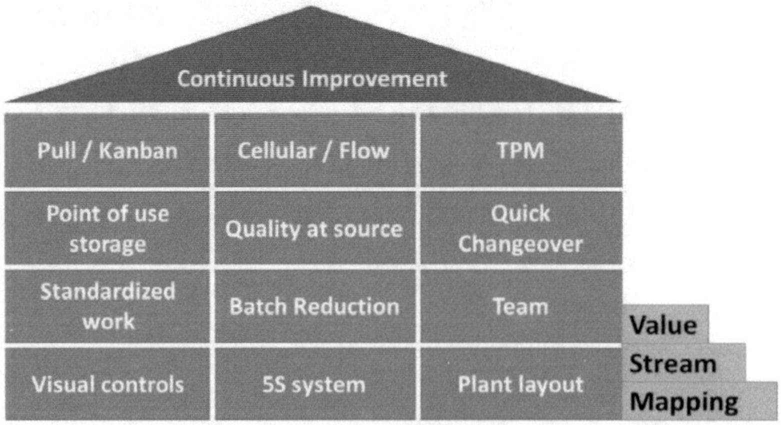

Fig. 1. Pillars of lean manufacturing [4]

In the case study described in this article, a focus was placed on the already established Milk run system, which supplied the production on one assembly line and the accompanying workplaces. The pull system as such is one of the basic pillars of lean manufacturing, and this fact is shown in Figure 1. Generally, setting up the efficient functioning of the Milk run is a complex process that is very demanding, and it is necessary to solve all the modifications together with the logistics, lean and purchasing departments. However, even the loading pull system with a Milk run can be problematic and have problems. In addition to problems such as the fact that the operator loses the Kanban card, takes the wrong amount, does not notice the Kanban cards, etc., the train for the Milk run circuit may be overloaded or unloaded. One of the issues that need to be tackled may be the case solved in this case study. There was a Milk run in the company that suffered from the problem that there was an irregular accumulation of a great deal of material loading requirements in such a way that the trainer would have to load much more material than he would be physically and physically capable. There are several possible causes of this problem, so it was necessary to carry out analyzes to identify the root cause of the problem. This article gives an insight into the methodology that leads to the supply system settings using a static simulation. [1], [2]

There is a lot of papers written about implementation of Millrun method but most of these papers focus on setting up the Millrun routing as vehicle routing problem (VRP). [5] Toshinori Nnemeto [6] write about JIT concept on the assembling line and also created mathematical model to help in station demand modelling. This paper issued the number of Kanban cards in the stock. The paper development and appl. Of milk run distribution system in express industry based on saving algorithm [7] reviewed Millrun scheme can help to find shortest distance and lowest cost, but also there is problem, that they are solving just a shortest road and not a Millrun system as one unit. Paper [7] also provides information Millrun can improve the load factor, advance time efficiency, environment and time constrains are discussed too.

This paper uses Simplified steps of supply procurement planning process [8]. These steps are Volume calculation – Master road planning – Supplier negotiations – Master road revision – Dock scheduling – Daily route planning, but just in few steps and goes deeper into simplification of steps and more precisely defines each step to ensure applicability for other companies. Kopeček [9] made the conclusion that too sophisticated methods mostly fail due to the inaccuracy of the input data. This inaccuracy is more caused by dynamic changes in this process and its development in time than by insufficient quality of work in technical preparation of the production, inaccurate observations and measuring in the production process. Because of this, research and development activities have been focused more on simple and robust methods of optimization which could be both easier in implementation and resistant to changes of parameters of optimized systems.

Limitations of this paper are using of one type of supplying vehicle, multiple kinds of storage systems, unchanging train road (train is always using one road which does not change from the lap to lap) and stability of process (same process which does not change from lap to lap).

2. Methodology

The Milk run system setup methodology consists of the 4 steps that are depicted in Figure 2. Each of these steps is further discussed in the individual chapters of this article.

Fig. 2. Basic points of the methodology

2.1 Current state measurement

The methodology is different in the first step, depending on whether the Milk run method is already in place in the company, or whether this methodology is used to design a future state. If the Milk run loading system is already introduced in the company, it is advisable to measure the current supply system as a first step.
In essence, if the Milk run supply system is not implemented in the company, the first step of the methodology is skipped and the second step is carried out.

The basic idea behind this step of the methodology is to find out how the already implemented Milk run system works / does not work. It is also possible at this stage to identify potential waste in the supply system and it is also a step that makes it possible to design further improvements in addition to changes in the supply system.
The first step is to measure the current state by performing snapshots of an individual working day, more precisely, a Milkrunner. Ideally, at this step, it is advisable to measure all Milkrunner activities during work shift, so that measurements can be made on several shift focused on different Milkrunners and in a combination of shifts (morning, afternoon, night). Within the measurement it is necessary to pay attention not only to how long the activity takes, but also how many transport units are handled, what warehouse, what warehouse technology, based on what storage principle the material is handled and what all the activities are carried out during the loading. In this way, it is possible to record all processes and times for the current state. However, in this step it is also necessary to record the route that the trains run through the spaghetti diagram.

Subsequently, the measurement results should be analyzed to obtain information on all the trajectories traveled by the trains, how many transport units are loaded on average in that round, what is the difference between the loaded and unloaded packaging units (quantity) and how long did the loading cycle take.

2.2 Current state measurement

The second step of this methodology is the same if the Milk run system is already implemented and also if the Milk run system is not implemented yet. As can be seen from the title of this step, this is an analysis of loading materials. More specifically, it is necessary to determine which materials are or will be loaded by the Milk run method. An ABC-XYZ analysis, which clearly defines what material is to be loaded by the loading method, can be used to identify these materials. However, it always depends on the total quantity of materials in circulation and what other methods of loading are used in the company. After selecting materials that are or will be loaded by the Milk run method, it is necessary to get the basic data for further work with the data and these are

- Monthly material consumption (ideally for one or more years)
- Yearly material consumption
- Possibly line tact and composition of production (BOMs, decompositions, composition of annual production, etc.)
- Number of pieces in one storage unit
- Type of storage unit (KLT, pallet, freely, etc.)
- Place of material storage
- Type of technology used for storage (lift system, catch shelf, pallet rack, etc.)

From this data, the hourly consumption of the material is calculated based on monthly material consumption:

$$\text{Average hourly consumption} = \frac{\text{Monthly consumption (pcs)}}{\text{Shift time} * \text{Number of shifts per month}}$$

This calculates the values for all the available months and searches for fluctuations in the consumption of the material. If there are no fluctuations in consumption, there is no problem. If material consumption fluctuations are recorded, it is appropriate to continue to calculate the highest value for the consumption of the material, but only after checking the proportion to the average material consumption.
It is also appropriate to compare the average hourly consumption of material from the monthly consumption to the average hourly consumption of the annual material consumption.

$$\text{Average hourly consumption} = \frac{\text{Annual consumption (pcs)}}{\text{Shift time} * \text{Number of shifts per year}}$$

From the other data listed at the beginning of this chapter, data on the quantity of pieces in a single storage unit is used to calculate the processing time for a single storage unit. Basically, it's a simple calculation:

$$\text{The value of holding a package (hours)} = \frac{\text{Package size (pcs)}}{\text{Average consumption } \left(\frac{\text{pcs}}{\text{hour}}\right)}$$

Other data will primarily be used for further calculations. However, it is still appropriate to compare the consumption of monthly and annual data with data on consumption based on the line tact:

$$\text{Hourly consumption} = \left(\frac{3600}{\text{Time for 1 pc}}\right) * \text{Pcs per 1 product}$$

The main output of this step is to determine the average time consumption of a storage unit of monthly consumption and to associate information about the type of storage unit, the storage location of the material and the type of technology used for storage.

2.3 Calculation of accompanying activities

In this step, it is possible to use the data from step 1 to rationalize the current loading or the MOST method to determine the times of the partial and accompanying loading activities. The best way to define what this time is about is by using the following formula, which is further used as the simulation output:

$$t_k = t_{pick} + t_{unload} + t_{load} + t_{ride}$$

t_k – Total lap time for one load cycle in minutes
t_{pick} – The time required to pick up the material from the warehouse in minutes
t_{unload} – The time required for the removal of material in production at workplaces
t_{load} – The time required to store empty packaging in the production in minutes
t_{ride} – Train time in minutes

Now for the calculation of the accompanying activities - it is more precisely the picking time, the loading time, the unloading time that is related to one transport unit. The relationship to one transport unit is necessary for further work with these data in static simulation. Another accompanying time is the ride time of a train, which is logically not related to a single transport unit, but to a single round. As already explained in the introduction to this chapter, it is necessary to analyze the measurement results and possibly use the MOST method to determine the partial times for

- Time for pick – This value must be related to the type of packaging unit and the storage technology. If the material is picked from the vertical lift system, the time is different from when the material is picked from the drop shelf. It also takes a different time to remove one palette and pick up KLT, which is logical information. Time includes finding the material, picking up and putting the material into the train.
- Unloading time is the time that involves getting off from the train at the production line, picking up the material from the train, and placing the material in the storage position at the workplace.
- The loading time includes grabbing the empty storage unit and placing the unit on the train.
- Train travel time includes the total time of the train journey from the exit from the warehouse to the whole cycle and re-exit. This time can no longer be acquired through the MOST method, but it is possible to use the time form measurement (if the circuit is well set). If it is not well set up, it must be designed just like designing a new Milk run supply system by applying the route to the layout (the total transport distance is determined) and the conversion from the average speed of the train is used, however, it is necessary to not use the speed of the train from the train manufacturer's catalog, but reduce the speed to approximately half (turning, accelerating, braking, etc.) According to current information and measurement it is advisable to use a speed of about 3-5 m / s.

2.4 Calculation of accompanying activities

The static simulation must be done separately for each type of storage technology in combination with the type of storage unit. The materials are separated according to this combination and the following procedure is used:

- Enter the identification numbers and materials names horizontally into the first two lines.
- Insertion of the hourly hold value of one package of the material under the concrete material type.
- The times are also generated when the material is reprocessed, always taking into account the total simulated time (the minimum time to be generated for the fastest consumer must be greater than the time of the last simulated round).

In this way, a time axis that will show the total consumption of all materials for a given combination of storage technology and storage unit, is generated, see Figure 3, where X indicates the consumption time of one package for material one, and Y denotes the consumption time of one package for material 2. The visualization for all other materials would look the same way.

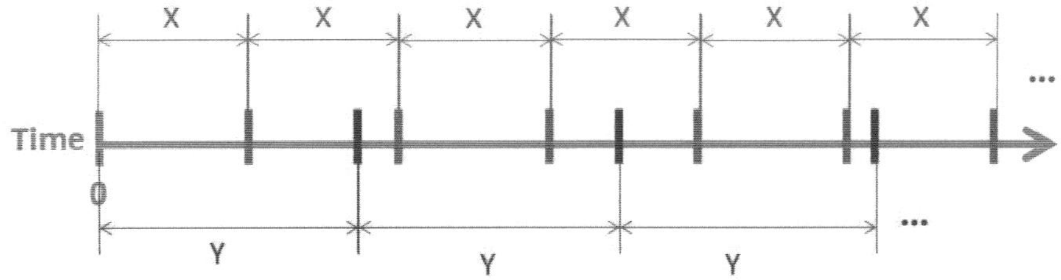

Fig. 3. Visualization of consumption

After generating these all occurrence times, it is necessary to sum up the number of packaging units in one round by creating a table of times indicating the beginning and end of one scheduled round (ideally referring to one cell that recalculates all values when the change occurs). See the example in Table 4.

Time				Round time	0,5
From	To				
0	0,5	0	0	0	0
0,5	1	0	0	0	0
1	1,5	0	0	0	0
1,5	2	0	0	0	0
2	2,5	0	0	0	0
2,5	3	0	0	0	0
3	3,5	0	0	0	0
3,5	4	0	0	0	0
4	4,5	0	0	0	0
4,5	5	0	0	0	0
5	5,5	0	0	0	0
5,5	6	0	0	0	0
6	6,5	0	0	0	0

Table 1. Simulation example

From these basic data, countif function will then determine how many materials will be processed in a given timeframe. More specifically, the function may look as follows:

(=COUNTIFS(C$4:C$841;">="&$A850;C$4:C$841; "<"&$B850)

If we then add up the number of simulated packaging units in the round, so we can proceed with the calculation of one round time, which is actually a combination of all of the data from step 3 and 4. The basic formula for calculating the round time is still:

$$t_k = t_{pick} + t_{unload} + t_{load} + t_{ride}$$

When:

$$t_{pick} = \sum_{i=1}^{n} time\ for\ pick_i * SKUs\ in\ round_i$$

In which is i = the type of storage units in storage technology, and n is the total number of combinations of storage unit and technology. However, the picking time must be taken from the values of the number of packing units in the round preceding the values in the other quantities.

$$t_{unload} = \sum_{k=1}^{n} unload\ time_k * packages\ in\ the\ round_k$$

In which k = given type of storage unit and n is the total number of types of storage units.

$$t_{load} = \sum_{k=1}^{n} load\ time_k * packages\ in\ the\ round_k$$

The ride time is the same for each round and the calculation method is in the previous chapter. In this way, it is appropriate to generate a sufficient number of occurrences so that at least twice the spin of the most slowly consumed material occurs. As a result, the end result is a histogram showing the frequency of occurrence of each round time - an output that uniquely determines whether it will be possible to prosecute the line loading in a combination of materials, processes and all possible aspects affecting the supply, is created.

3. Case study

The case study is based on the described methodology and for this reason only the outputs of the individual steps of the methodology will be discussed.

1st step Current state measurement

From the measurement of the current state it appeared that the different workers are traveling in a different way, the exact timetable is not specified, the ordering of the material in the electronic ordering system is not working correctly, the ordering frequency of the material is higher at the beginning and at the end of the shift and there is the wrong storage of some KLT.

						Time with ride			
Round	Number od picked SKUs empty	Picked SKUs	Supplied SKUs	Picking	Picking + kardex picking	Clean ride time	Round	Start time	End time
1	16	8	6	0:02:46	0:07:48	0:08:20	00:17:52	08:25:06	08:42:58
2	14	6	7	0:01:17	0:01:17	0:01:08	00:07:41	08:42:58	08:50:39
3	32	17	19	0:06:59	0:24:50	0:20:03	00:38:02	08:50:39	09:28:41
4	14	31	31	0:19:28	0:33:36	0:37:23	01:06:09	09:28:41	10:34:50
5	31	17	20	0:03:07	0:15:09	0:12:16	00:32:55	10:34:50	11:07:45
6	13	20	18	0:08:42	0:23:12	0:21:57	01:24:57	11:07:45	12:32:42
7	37	16	20	0:06:35	0:19:46	0:28:04	01:22:28	12:32:42	13:55:10
8	31	15	14	0:21:00	0:36:46	0:42:26	01:09:10	13:55:10	15:04:20
Suma	**188**	**130**	**135**	**1:09:54**	**2:42:24**	**2:51:37**	**6:39:14**		

Table 2. Demonstration of measurement outputs

2nd step Loading materials analysis

The materials for the Milk run have been identified in the company, so analyzes needed for further data processing have been performed. The company also used 3 warehouse technologies and one kind of handling unit.

Part number	Consumption / year	Consumption / mounth	Base unit	Avrg. Compsumtion / year	Avrg. Compsumtion / mounth	Compsumtion / tact	Pieces in SKU	Turn over time
0009145814	28000,000	2333,333	ST	8,889	7,778	1,579	7000	7,778
0009381980	5905,832	738,229	ST	1,875	2,461	3,158	6000	2,461
0009073317	2776,044	347,006	ST	0,881	1,157	3,158	5000	1,157
0009145614	8912,000	742,667	ST	2,829	2,476	1,579	4400	2,476
9292003012	21210,604	2651,326	ST	6,734	8,838	1,579	4000	8,838

7777700906	20154,224	2519,278	ST	6,398	8,398	3,158	4000	8,398
0009073315	7467,481	933,435	ST	2,371	3,111	1,579	3200	3,111
0009642111	332,000	110,667	ST	0,105	0,369	6,316	3000	0,369
9062316126	3180,918	397,615	ST	1,010	1,325	3,158	3000	1,325
1909146510	-2592,000	-216,000	ST	0,823	0,720	1,579	3000	0,720
0009643709	7089,804	886,226	ST	2,251	2,954	3,158	2400	2,954
0009552037	44000,000	3666,667	ST	13,968	12,222	1,579	2000	12,222
7914990928	2495,788	311,974	ST	0,792	1,040	6,316	2000	1,040

Table 3. Sample data analysis

3rd step Calculation of accompanying activities

According to the third step, it was determined on the basis of measurements that the times of the individual activities were, Time for pick - 8 seconds per the supermarket, 44 seconds per the cardex and 37 seconds per the virtual cardex. 4 second loading time, 8 second unloading time, Train travel time was determined when designing the route for the shortest route for handling all loaded workplaces

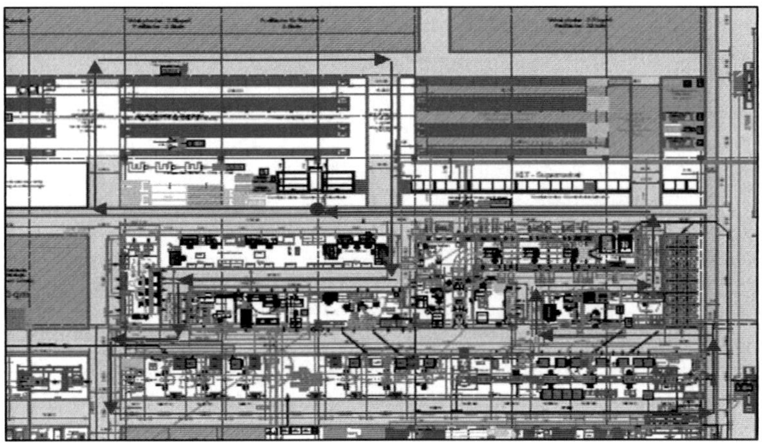

Fig. 4. Designed train route

4th step Static loading simulation

A static simulation showed that loading would be most convenient every 30 minutes, and the operator should handle all aspects of the cycle. The static simulation outputs have been compared with reality, so that the average amount of loading KLTs was complied with the long-term statistics that the company had processed.

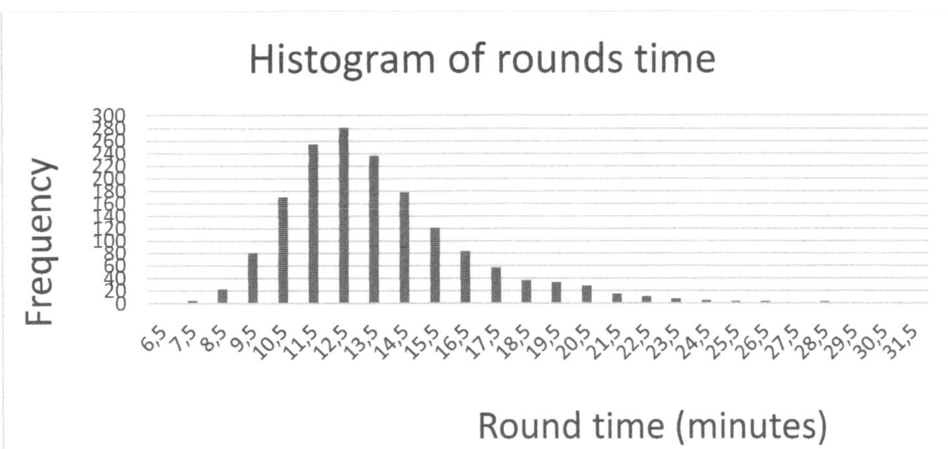

Fig. 5. Histogram of rounds times

4. Conclusion

The problem solved in this paper was uncertainty in setting up the Milkrun supplying system and lack of prediction in setting up the new Milkrun system or evaluation of already implemented Millrun system. Problem have been solved by setting up four step methodology which comes from Measuring current state of the art, Analysis of material supplying, Calculation of accompanying activities and Static Millrun simulation based on MS Excel. The Static simulation is the main part of the whole methodology because it brings prediction for amount of supplying materials.
The case study confirmed that Methodology and static simulation is sufficiently precise for evaluation and setting up. The confirmation is based on number of materials needed to be supplied from Static simulation and real data from company MES system. The Methodology has been used only in two companies so steps for the future are evaluation in larger group of companies and confirmation of generality for using this methodology.
This article describes the methodology that is used to implement and rationalize the Milk run loading system. The methodology consists of 4 basic steps:

1st step- Current state measurement
2nd step – Loading material analysis
3rd step – Calculation of accompanying activities
4th step – Static loading simulation

The conclusion of the article was a case study that used this methodology and at the same time confirmed that this methodology was applicable in practice. The outcome of the case study was the rationalization of the current Milk run loading system, and this study helped to identify waste and problems in the supply system on the Time distribution of the mulkrunner workload, Creating a model for continuous control of the Milk run supply functionality in the company, Setting a fixed train timetable, Determining a rationalized trail, Replacement of storage on pallets with a racking shelf, Changing the placement of material in the warehouse, Training train staff and assembly on ordering material and Improvement in an electronic ordering system.

5. Acknowledgments

This article was created with the support of the internal grant of the University of West Bohemia, SGS-2015-065 entitled Development of Sustainable Production System Parameters.

6. References

[1] Šimon, M., Troblová, P., (2005), Logistics of the products in the net businnes Proceeding of the 4th DAAAM international conference on advance technologies for developing countries, ISBN: 953-6048-29-9
[2] Edl, M., Kudrna, J., (2013), Methods of industrial engineering . 1. vyd. Pilsen : Smart Motion, s.r.o., ISBN: 978-80-87539-40-8
[3] Best industrial [Online], (2016) , Available from: http://bestindustrial.cz/clanky/system-milk-run-2/ Accessed: 3 July 2017
[4] citynetevents [Online], (2015) Available from: http://www.citynetevents.com/hanoi/courses/course-implementation-system-guidance-tpm-total-productive-maintenance-october-19th-2012. Accessed: 8 July 2017
[5] Gyulai, D., (2013) , Milk run Vehicle Routing Approach for Shop-floor Logistics, Forty Sixth CIRP Conference on Manufacturing Systems, ISBN: 978-1-62993-594-2
[6] Toshinori nemoto, katsuhiko hayashi, masataka hashimoto, (2010), Milk run logistics by Japanese automobile manufactures in Thailand: procedia social and behavioural sciences 2, ISBN:5980-5989
[7] Zhenlai you and yang jiao(yanshan university china), (2014), Development and appl. Of milk run distribution system in express industry based on saving algorithm:, hindawi publication corp. Mathematical problem appl. Vol.
[8] Theeratham, M., and Lohatepanont, M., (2010) "Vehicle Routing in Milk Run Operations: A Column Generation Based Approach"
[9] Kopeček, P., (2013) , Selected Heuristic Methods used in Industrial Engineering, 24th DAAAM International Symposium on Intelligent Manufacturing and Automation, ISSN 1877-7058

DOI: 10.2507/28th.daaam.proceedings.160

FRAMEWORK TO IMPLEMENT COLLABORATIVE ROBOTS IN MANUAL ASSEMBLY: A LEAN AUTOMATION APPROACH

Ali A. Malik & Arne Bilberg

This Publication has to be referred as: Malik, A[li] A[hmad] & Bilberg, A[rne] (2017). Framework to Implement Collaborative Robots In Manual Assembly: A Lean Automation Approach, Proceedings of the 28th DAAAM International Symposium, pp.1151-1160, B. Katalinic (Ed.), Published by DAAAM International, ISBN 978-3-902734-11-2, ISSN 1726-9679, Vienna, Austria
DOI: 10.2507/28th.daaam.proceedings.160

Abstract

The recent proliferation of smart manufacturing technologies has emerged the concept of hybrid automation for assembly systems utilizing the best of humans and robots in a combination. Based on the ability to work alongside human-workers the next generation of industrial robots (or robotics 2.0) are referred to as collaborative robots or cobots. This paper presents a systematic framework based on Roozenburg's engineering design cycle for the deployment of cobots in existing assembly cells for enhanced productivity. A model for defining evaluation-parameters for a cobot are presented. The effectiveness of virtual simulations is discussed for validation and optimization of human-robot work environment.

Keywords: Cobot; framework; collaborative; assembly system; cobot selection; HMLV; Lean automation; HRC

1. Introduction

The increasing demand of customization and shortened product life-cycles are pushing manufacturers to develop variant-oriented production systems [1]. The production paradigms are changing towards increasing flexibility and responsiveness of production processes, facilities, and networks. The same production system is expected to adapt to frequent changes in products, components and tasks [2]. Whereas the classical assembly methods for high-mix low-volume (HMLV) production are incompatible with the increasingly globalized industry. It is well agreed in the research literature that the flexibility and changeability are key elements to deal with the challenges of a global market [3] [4]. Besides high wage rates in Northern Europe, humans are responsible for 90% of the production tasks in final assembly [5]. Fasth [4] described the high degree of intelligence required in the assembly process as a challenge for automation of the final assembly systems and humans having a higher degree of intelligence are more suited for such intelligent tasks. Thus, the mainstream automation has always been far from combining the human and artificial resources. In conventional automated production systems, humans' tasks are designed to be as rigid as the rest of the automation [2] and it, thereby, keeps the humans and robots away from each other. In the last half century, industrial robots have assisted in huge productivity improvements but they remained in caged environments, away from any physical interaction with humans.

In the new industrial robotics paradigm (or robotics 2.0) robots are not fenced in cages but they are expected to become a fellow of humans in their work environment. Various researches have been made to realize the complementary strengths of humans and robots and combine them for collaborative work. Based on the paradigm," factories are long-life and complex products", this research uses product design theory to design and develop a human-robot collaborative production system. In this research paper a structured procedure is developed by analyzing the business needs, evaluating the requirements, fusing the infrastructural elements and translate this into a functional human-robot collaborative (HRC) solution.

2. What is Cobot?

The concept of a Collaborative Robot was first presented by Colgate [6] as an Intelligent Assist Device (IAD) which manipulates objects in direct collaboration with a human operator. The unique characteristic of cobots are their direct interaction with human operators, shared control of motion and provision of virtual surfaces to constrain and guide worker's motion (see Fig. 1). The resulted benefits are expected as enhanced productivity, better ergonomics and improved safety [7][8]. Kruger [28] described cobot as a mechanical device for human-machine cooperation in assembly lines through direct physical contact. The cobots being developed today are articulated with multiple rotational joints developing a high degree of flexibility and dexterity to reach every coordinate of their workspace in multiple configurations [9]. These cobots provide guidance while human operator offers motive power, flexibility and intelligence to complete a task. Cobots are designed to have direct interaction with fellow humans as well as with other robots and share the workload considering the abilities and strengths of each team member. The possible advantages are expected in productivity and improved ergonomics [5].

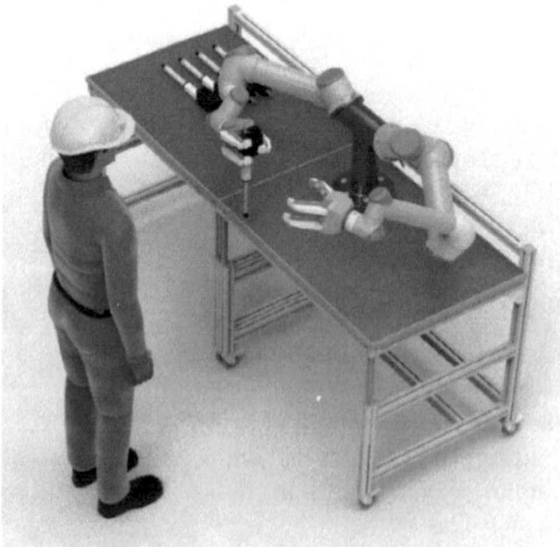

Fig. 1. A two-arm collaborative robot (Cobot)

3. Cobot as an enabler of lean automation in assembly cells

Production in high-wage economies demands a high degree of automation. A general limitation of mainstream automation is that it addresses low-mix high-volume production. The relationship between production volume, flexibility, automation and product variety is shown in (see Fig. 2). Heilala [10] had also presented a similar relationship between flexibility, volume, variants and batch size. Due to the need of variability in assembly-tasks, final assembly systems are designed to be manual systems. Hence the most flexible elements in an industrial assembly system are the humans. The natural intelligence of humans enables them to easily adapt to production changes and requirements [11]. To enhance the productivity of the production systems, in the last few decades, Lean Production principles gained popularity that promise a continuous improvement by avoiding the waste and focus on the value adding activities [12]. Already in 1990 approaches were presented to combine lean production principles into automation technologies resulting in non-complex and less over engineered automation solutions. Dulchlnos [13] defined lean automation as, *"Lean automation is a technique which applies the right amount of automation to a given task. It stresses robust, reliable components and minimizes overly complicated solutions"*. With the emerge of Industry 4.0 concepts an increasing trend is being observed to incorporate robotics and automation in increasingly more fields of human activity in industrial assembly processes [14]. The human-robot integrated production systems combine the creativity, intelligence, knowledge, flexibility and skill of humans with the electronics and physical power, speed and accuracy of a machine. The approach enables an assembly system to produce complex on-demand products at reduced costs [15].

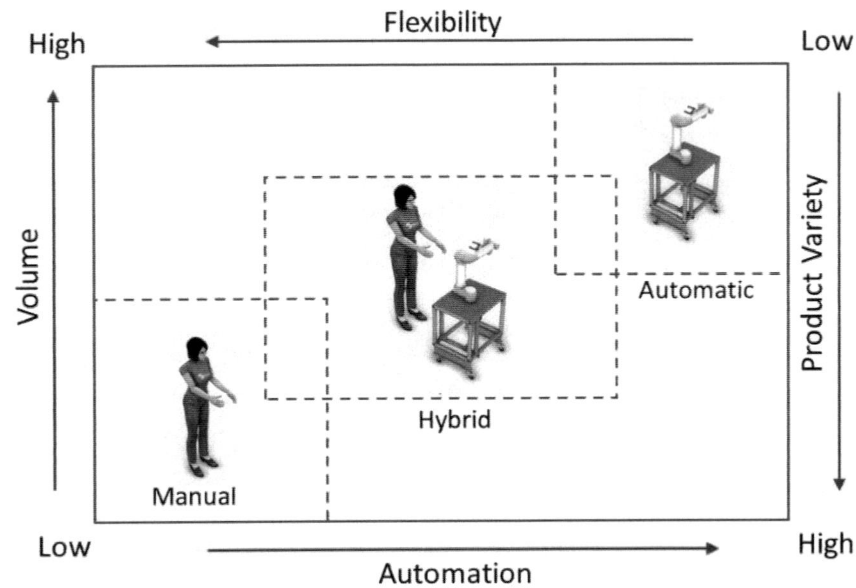

Fig. 2. Flexibility and automation (Modified from Heilala [10] and Rosati [31])

4. Factory as a product

Westkamper presented the paradigm, "factory is a long life and complex product". The basis of this paradigm is the argument that, from a system theory perspective, manufacturing systems can be modelled as complex technical systems [16]. Later, many researchers have used product-design-theories for manufacturing-system-design, thus strengthening the

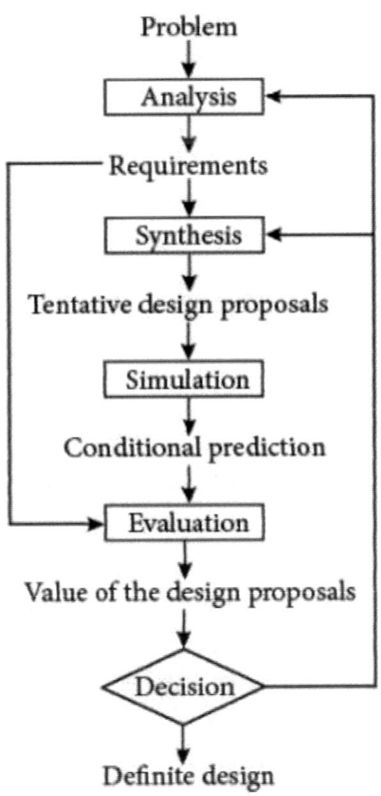

Fig. 3. Engineering design cycle by Roozenburg [18]

paradigm of "factories-are-products". Wiendahl [17] further described the process of planning a production system as a systematic, goal-oriented, and structured design process that supports activities from the first idea to the start of production. In this regard, Roozenburg's engineering design cycle [eekels1991methodological] is an iterative systematic approach to design technical systems as products (see Fig. 3).

The product design structure, as suggested by Roozenburg, is a sequence of empirical activities, in which a solution is achieved by cyclic progression in knowledge of the problem as well as of the solution. The process starts with identification of the problem and gradually reaches to the stage where a definite design is achieved. In a later research Francalanza [16] used the engineering product design cycle by Roozenburg and derived a systematic approach to design changeable manufacturing systems.

5. A systematic framework for cobot deployment in assembly cells

Based on above research works, this paper derives a systematic approach to implement cobots in an assembly system and enlists associated activities in developing a hybrid assembly work-cell (see Fig. 4). This process is divided into three phases of concept development phase, exploration phase and decision phase. The concept phase involves analyzing the business needs and production process, thus defining the needs and requirements for a hybrid assembly system.

Exploration phase highlights the activities of defining the requirements for the functional elements (e.g. cobot and worker) of the production system, layout design and designing the interaction structure of these functional elements. Decision phase involves concluding a provisional design, developing virtual prototypes of the design and achieving a final design. Each phase has an analysis component to analyze and optimize the design and make it aligned with the previous design stages. The design components within each of the three phases are described below:

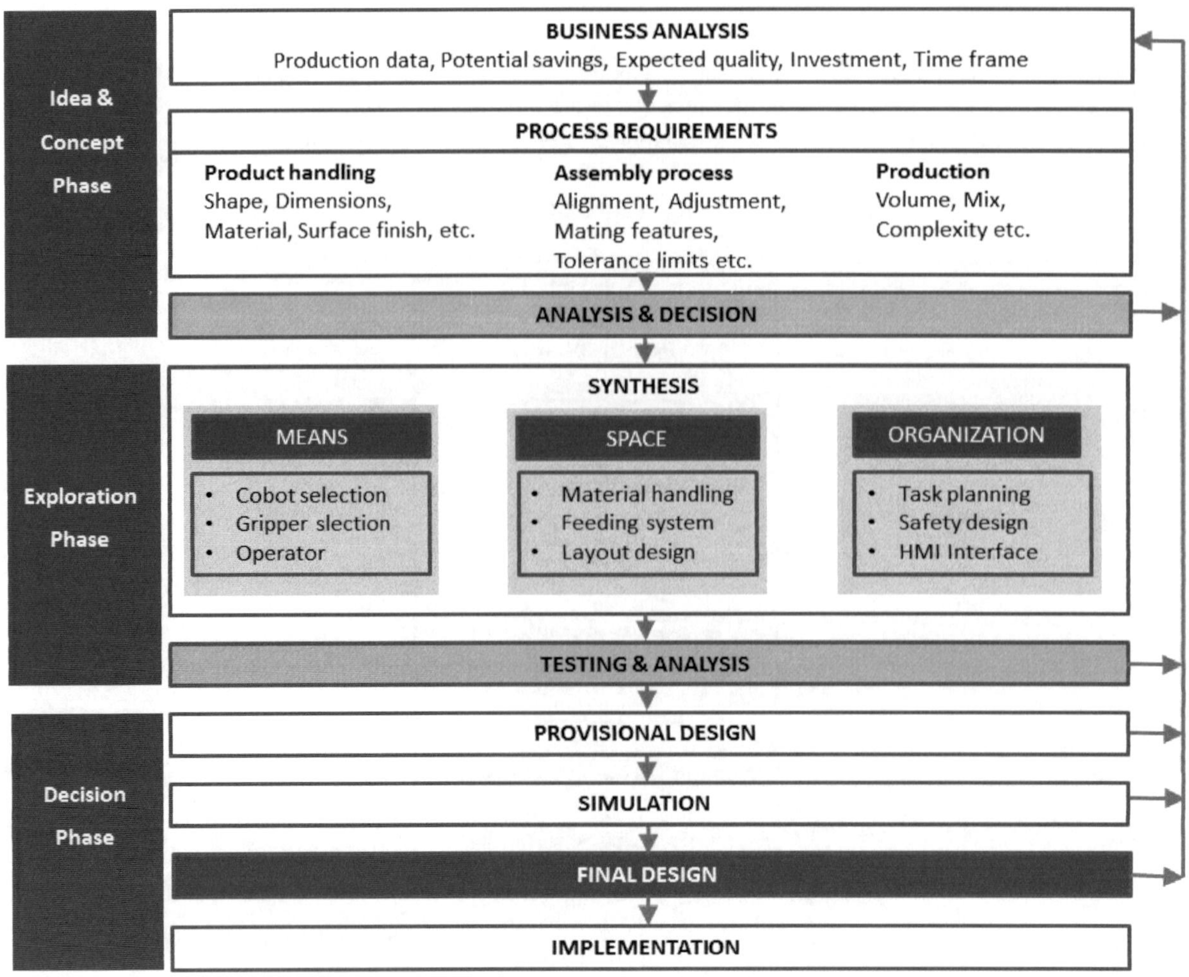

Fig. 4. Systematic framework for cobot implementation

5.1. Problem and requirements

Innovative factory planning processes emerge from the business strategy of an enterprise [19]. Hence investigating the associated potential business gains are the first step towards a new production strategy. This step is a product of the Investment and Performance Planning stage and defines the goals and targets to be achieved with hybrid flexible assembly automation. e.g. 'Cost per part', 'investment cost', etc. [16].

5.2. Requirements of the hybrid production system

The next stage in the process of implementing cobot to develop a human-robot collaborative assembly system is the analysis of the assembly-process. This includes analysis of the product design and assembly sequence for which cobot is being proposed for hybrid lean automation. Chryssolouris [20] stated the basic goal of a production system is to produce the required products at a required time, cost, quality and flexibility. Hence, the requirements of a human-robot automation may fall within the categories of product handling requirements, process requirements and production requirements:

- *Product handling requirements:* define the physical properties of the parts and components that make-up as product(s). It is required to analyze the properties (e.g. form, shape, dimensions, tolerances, material, weight, size and surface finish etc.) of the parts and components to be handled by the cobot during the assembly process.

- *Process and insertion requirements:* The sequences of the assembly process, alignment of parts, angle of rotation, mating features and fastening methods need to be investigated. This also includes need of fixturing and testing of sub-assemblies [21].

- *Production requirements*: These are the requirements for required production volume, cycle time, variety in the product (s) and associated complexity of the process.

5.3. Analysis of the concept phase

The above three types of assembly requirements within the categories of product, process and production, adds to the complexity of the assembly process for automation. A possible solution is to make structured complexity assessment based on above data for automation. Many researchers have attempted to define the manufacturing complexity and methods to quantify it.

The complexity assessment work by ElMaraghy is of significance to quantify manufacturing operational complexity [22], products assembly complexity [23] and complexity of an assembly system [24]. However, it is proposed that a more comprehensive complexity model for the objective of hybrid-automation be developed that comprehends all above important parameters of an assembly process. These analyses will also define the set of criteria for evaluation of the hybrid assembly system to be developed at the later stage.

5.4. Synthesis

Synthesis is described as an activity of mixing ideas, influences and/or things to form a connected whole. This is the stage where manufacturing designers form solutions for a collaborative flexible manufacturing system. A production system is "a place of adding value by production with the help of production factors" [17]. It is a complex socio-technical system that consists of elements and objects.

For the purpose of defining factory objects Nyhuis [25], classified the factory objects as means, space and organization and placed them at four different levels of station, system, segment and site levels. However, this research will focus only on the technical requirements at the station level. The means, organization and space in this classification refer to production equipment, operational structure and workspace design respectively [25].

5.4.1. Means

Means of production at the station-level define the production equipment and resources. In the case of a hybrid human-robot work system, the production equipment is cobot, cobot-grippers and worker.

a. Selection of cobot

There are several commercially available cobots with varying capabilities and strengths. It is critical to decide a cobot that best suits to the needs of the assembly system. Many researchers have presented the evaluation and selection of an industrial robot as a multiple criteria decision making (MCDM) problem. Chatterjee [26] classified the attributes or properties of an industrial robots as objective (numerically defined e.g. load, cost etc.), subjective (qualitative in nature e.g. programming flexibility, vendor's service quality etc.), beneficial (whose higher values are desirable e.g. load carrying capacity) and non-beneficial attributes (whose lower values are desirable e.g. cost, repeatability etc.) [26]. Multiple studies are made to achieve the objective of scoring and evaluating the selection parameters through various scientific analysis e.g. mathematical, statistical, simulations etc. Finally, the scores are normalized to same-units to help make a comparison. Mortensen [27] presented a literature review of 19 scientific studies for robot evaluation that enlists the set of parameters used and the evaluation method. However, the determination of selection-criteria in robot selection process is an overlooked issue in the research literature [27]. Additionally, the parameters discussed so far by researchers are associated with the basic functionality of a robotic manipulator and are not considering the fact that a cobot, as a hybrid-automation tool, needs additional parameters for its performance evaluation e.g. safety, social interaction, and ease of use. The "Domain Theory" by Andersen [28] suggests that multiple perspectives are associated with any product structure. Based on this theory, a multi-perspective view is presented for determining an evaluation criteria for a cobot (see Fig. 5). These parameters are:

Fig. 5. Different views to define cobot selection-parameters

- *Functional view:* describes the properties that help the cobot to perform its basic functionalities i.e. payload, degrees of freedom, accuracy, repeatability etc.
- *Human-interaction view:* The elements that define the ease and suitability for human interaction of a cobot e.g. ease of programming.
- *Flexibility view:* A cobot is supposed to ensure flexibility to realize transformation of the production scenarios. For this, it must carry specific characteristics referred to as change enablers. Wiendahl [17] described change enablers as specific characteristics that can be activated at a specific time to create a design change e.g. modular design of cobot to various reach and payload conditions.
- *Economic view:* These elements are defining the investment related aspects of a cobot.

b. Selection of gripper

Since the targeted objective of building a human-robot collaboration is to have a fine balance between flexibility and productivity, therefore the material feeding systems must exhibit high flexible [29]. However, the technology to develop flexible grippers that can adapt to several shapes, sizes and material conditions is still developing. Kruger [29] has presented a spectrum of gripping systems based on flexibility of a gripper and costs. The spectrum consists of two extremes positions. At one end, it has a cost effective two fingers gripper with lower flexibility and on the other end is a complex hand-like gripper with several DoF but the price can exceed the price of a robot system (see Fig. 6).

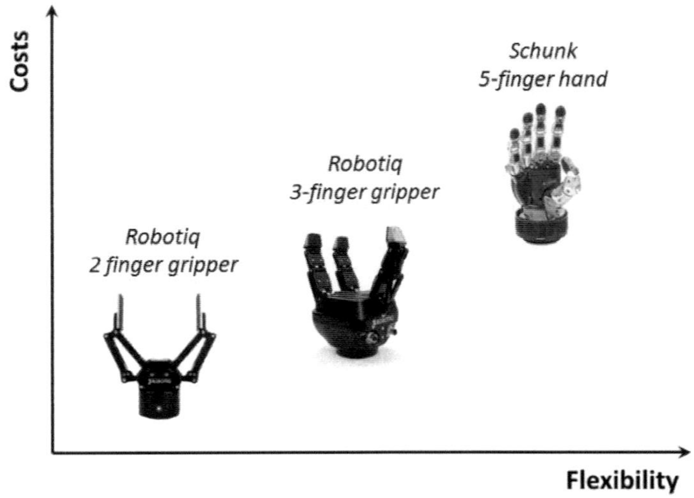

Fig. 6. Spectrum of robotic gripping systems [29]

c. Training need assessment for human-worker

Human-operator is at the heart of a lean automation system [30]. The production design elements for human operator are related to skill requirements analysis and training. In a robot-collaborative production system, the operator needs to have certain skills to interact with the robot. The operators may also face social challenges while working with robotics. A training assessment is developed to train the operator to work alongside robots.

5.4.2. The workspace

The elements of workspace design are material handling system, feeding system and layout design. For a flexible assembly system, a big challenge lies in the flexible feeding of parts and components into the assembly process. Mainstream automation in this area is quite rigid and deals with specialized feeder for each type of parts [29]. While presenting fully flexible assembly systems Rosati [31] presented flexible feeding of parts into an assembly operation with vision systems. An approach for flexible automated material handling can be automated guided vehicles (AGV). AGVs are being used in industries since many years but now they are getting smarter and flexible with low cost advanced electronics and autonomous navigation systems [29]. Modern assistive collaborative robots are another possible form of flexible material handling in smart workplaces [32] (see Fig. 7).

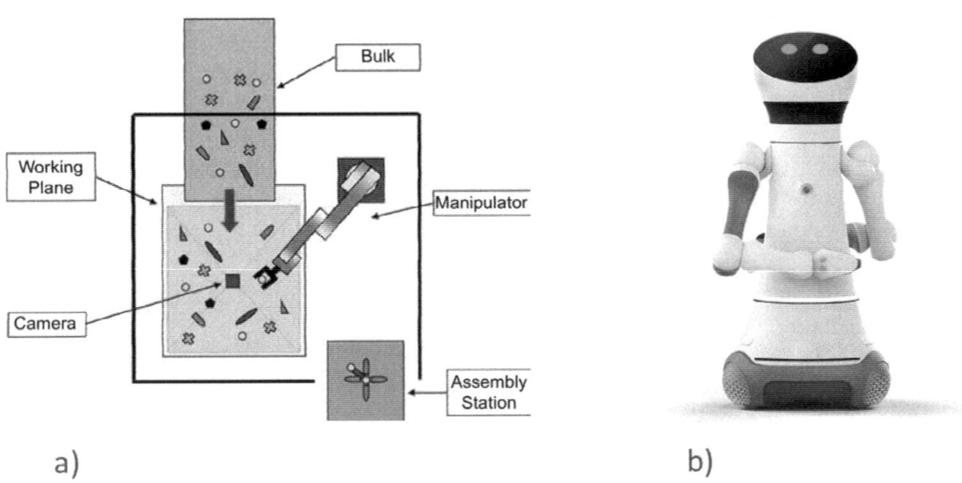

Fig. 7. (a) Vision based flexible material feeding [31] (b) Care-o-bot for flexible material handling [32]

5.4.3. Organization of the production elements in the workspace

The interface between operator and cobot consists of activities that make the manufacturing system collaborative. These include effective distribution of tasks, safety system design, and human-machine interface (HMI). HMI is important for instructing the robot and making changes to an existing program of the robot. Furthermore, in a factory environment, the question of task-allocation arises when both, human and robot, can perform the same task [33]. The major challenge is the lean-distribution of tasks relying on the best attributes of robots and humans in a combination. Careful planning is needed to design an optimal workstation. In a common observation, the designers automate everything possible and leave the rest to be performed by the human [4]. A modified form of task distribution presented in [29] is shown in Fig. 8. The research realizes that safety design is a comprehensive area of work when developing an HRC solution. There are various industrial standards that define the safety requirements of an industrial machine or robot. These standards are also being modified to the needs of HRC workplaces. However, considering the wide scope of safety design this research remains away from describing any details of the safety system design for HRC.

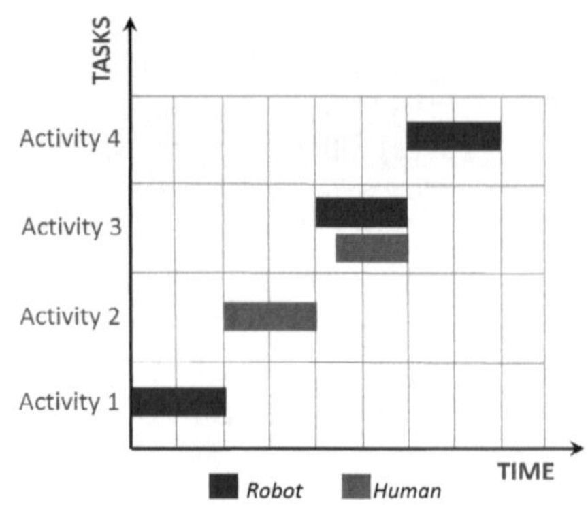

Fig. 8. The distribution of tasks between human and robot

5.5. Virtual simulation for testing and validation of the production system

Simulation phase involves construction of simulation models, test alternatives, optimize solutions, deducing hypothetical observations and predictions out of it before implementation in the real world. The accuracy of the simulation model is highly important to make observations as correct as possible. Though the results will not satisfy all but can help investigate most of the aspects of the proposed solution. A major advantage of having digital prototype of a production system offers the possibility to experiment a range of scenarios every time a parameter is changed. The results are concluded as a reduction in the overall production-system development time and cost [34]. Several virtual production simulation tools are available of which discrete event simulation (DES) are most popular in production environments.

A major challenge towards designing an HRC production system is the activity and task planning between human and robot. The times to complete each task are highly important to be calculated to design an optimal balance between the two means of production. Simulations can also prove to be a helpful tool in quantifying the processing times for both the operator and the robot. There is a limited availability of HRC simulations both in commercially available software and published research. Additionally, HRC simulations can play a vital role in safety assessment of human-robotic work cells. Simulators that can integrate physical phenomena of HRC can make great impact to safe HRC design [35]. The results will further translate into designing the optimal workstation layout. Backman [9] used simulation software, Industrial Path Solutions (IPS) from the Fraunhofer - Chalmers Center and Intelligently Moving Manikins (IMMA) developed by Dassault Systems to simulate robot and the human in an isolated simulation environment and the results were implemented for a cobot-human production system.

In this research DES software Tecnomatix Process Simulate by Siemens is used for design and validation of the human-robot automation. Tecnomatix provides an HRC environment for combined simulation of robot and operator.

The simulation helped to:

- Assess the cycle times for each activity of the assembly process
- Make reach tests for both cobot and the operator thus helping in the layout design of the assembly station
- Use the robot simulation for offline programming and transfer the code to a physical robot

The results from the simulation are then evaluated against the criteria defined at the analysis and requirements phase. It is then quantified to examine how well this satisfies the evaluations/ requirements criteria.

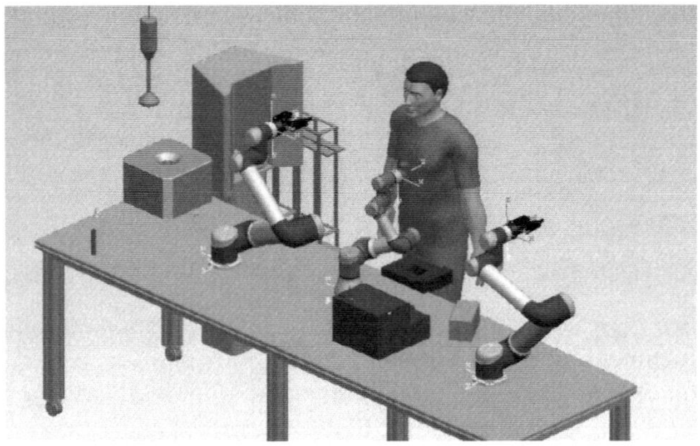

Fig. 9. Simulation of human and robot in Tecnomatix Process Simulate

5.6. Final design

The decision phase will conclude a design of the production system from all the previous analysis. It will also involve comparing the results with the requirements and criteria defined in the concept phase. This will help to determine whether the design is feasible enough to be moved to the implementation phase or if it needs modifications.

6. Conclusion

The research identifies cobots for lean-automation and strengths based distribution of tasks between robot and human in a common work environment. In the form of a systematic framework a mapping of activities is presented for implementation of cobots in assembly cells. It is proposed that a comprehensive generalized task-distribution methodology is developed considering the various attributes of an assembly process. The cobot evaluation model is presented that is not only helpful for manufacturing companies to evaluate market available cobots but also can help cobot-developers to define cobot specifications considering the future needs of manufacturing companies.

7. References

[1] Park, H. (2013). From automation to autonomy - a new trend for smart manufacturing, Chapter 03 in DAAAM International Scientific Book 2013, pp. 075-110, B. Katalinic & Z. Tekic (Eds.), Published by DAAAM International, ISBN 978-3-901509-94-0, ISSN 1726-9687, Vienna, Austria.

[2] Wang, X.V.; Kemény, Z.; Váncza, J. & Wang, L. (2017). Human-robot collaborative assembly in cyber-physical production: Classification framework and implementation, CIRP Annals-Manufacturing Technology.

[3] Mulc, T.; Udiljak, T. & Ciglar, D. (2015). Structure of reconfigurable manufacturing systems, Chapter 32 in DAAAM International Scientific Book 2015, pp.369-390, B. Katalinic (Ed.), Published by DAAAM International, ISBN 978-3-902734-05-1, ISSN 1726-9687, Vienna, Austria.

[4] Fasth-Berglund, Å. & Stahre, J. (2013). Cognitive automation strategy for reconfigurable and sustainable assembly systems, Assembly Automation, Vol. 33 (3), page numbers (294–303).

[5] Fasth-Berglund, Å.; Palmkvist, F.; Nyqvist, P.; Ekered, S. & Åkerman, M. (2016). Evaluating cobots for final assembly in Procedia CIRP, Vol. 44 page numbers (175–180).

[6] Peshkin M.; C.E.E.; Edward, J.; Peshkin, M.A. & Wannasuphoprasit, W. (1996). Cobots.

[7] Wannasuphoprasit, W.; Akella, P.; Peshkin, M. & Colgate, J.E. (1998). Cobots: A novel material handling technology, in Proceedings of International Mechanical Engineering Congress and Exposition, Anaheim, ASME 98-WA/MH-2.

[8] Grigorescu, S.; Vatau, S. & Dobra, A. (2010). Dedicated robot-robot cooperation, Chapter 55 in DAAAM International Scientific Book 2010, pp. 633-644, B. Katalinic (Ed.), Published by DAAAM International, ISBN 978-3-901509-74-2, ISSN 1726-9687, Vienna, Austria

[9] Djuric, A.M.; Urbanic, R. & Rickli, J. (2016). A framework for collaborative robot (CoBot) integration in advanced manufacturing systems, SAE International Journal of Materials and Manufacturing, Vol. 9 (2016-01-0337), page numbers (457–464).

[10] Heilala, J. & Voho, P. (2001). Modular reconfigurable flexible final assembly systems, Assembly Automation, Vol. 21 (1), page numbers (20–30).

[11] Peschl, M.; Roening, J. & Link, N. (2012). Human integration in task-driven flexible manufacturing systems, Proceedings of the 23rd International DAAAM Symposium, Volume 23, No.1, ISSN 2304-1382 ISBN 978-3-901509-91-9, CDROM version, Ed. B. Katalinic, Published by DAAAM International, Vienna, Austria.
[12] Kolberg, D. & Zühlke, D. (2015). Lean automation enabled by industry 4.0 technologies, IFAC-PapersOnLine, Vol. 48 (3), page numbers (1870–1875).
[13] Dulchlnos, J. & Massaro, P. (2005). The time is right for labs to embrace the principles of industrial automation, Drug World Discovery, Winter-Issue, Vol. 2006 page numbers (25–28).
[14] Sekoranja, B.; Basi'c, D.; Svaco, M.; Suligoj, F. & Jerbi'c, B. (2014). Human-robot interaction based on use of capacitive sensors, 24th DAAAM International Symposium on Intelligent Manufacturing and Automation, Vol. 69 page numbers (464–468).
[15] Bernhardt, R.; Surdilovic, D.; Katschinski, V. & Schröer, K. (2007). Flexible assembly systems through workplace sharing and time sharing human-machine cooperation - PISA, IFAC Proceedings Volumes, Vol. 40 (3), page numbers (247–251).
[16] Francalanza, E.; Borg, J. & Constantinescu, C. (2014). Deriving a systematic approach to changeable manufacturing system design in Procedia CIRP, Vol. 17 page numbers (166–171).
[17] Wiendahl, H.-P.; ElMaraghy, H.A.; Nyhuis, P.; Zäh, M.F.; Wiendahl, H.-H.; Duffie, N., et al. (2007). Changeable manufacturing - classification, design and operation, CIRP Annals - Manufacturing Technology, Vol. 56 (2), page numbers (783 – 809).
[18] Eekels, J. & Roozenburg, N.F. (1991). A methodological comparison of the structures of scientific research and engineering design: their similarities and differences, Design Studies, Vol. 12 (4), page numbers (197–203).
[19] Politze, D.; Bathelt, J.; Reinhard, M.; Jufer, N. & Kunz, A. (2010). Interfacing product development and factory planning-towards a unified life cycle management, in Proceedings of the 7th International DAAAM Baltic Conference Industrial Engineering, Tallinn, Estonia.
[20] Chryssolouris, G.; Mavrikios, D. & Mourtzis, D. (2013). Manufacturing systems: skills & competencies for the future, in Procedia CIRP, Vol. 7 page numbers (17–24).
[21] Hu, S.J.; Ko, J.; Weyand, L.; ElMaraghy, H.; Lien, T.; Koren, Y.; et al. (2011). Assembly system design and operations for product variety, in CIRP Annals-Manufacturing Technology, Vol. 60 (2), page numbers (715–733).
[22] ElMaraghy, W. & Urbanic, R. (2004). Assessment of manufacturing operational complexity, in CIRP Annals-Manufacturing Technology, Vol. 53 (1), page numbers (401–406).
[23] Samy, S. & ElMaraghy, H. (2010). A model for measuring products assembly complexity, International Journal of Computer Integrated Manufacturing, Vol. 23 (11), page numbers (1015–1027).
[24] Samy, S. & ElMaraghy, H. (2012). Complexity mapping of the product and assembly system, Assembly Automation, Vol. 32 (2), page numbers (135–151).
[25] Nyhuis, P.; Heinen, T. & Brieke, M. (2007). Adequate and economic factory transformability and the effects on logistical performance, International Journal of Flexible Manufacturing Systems, Vol. 19 (3), page numbers (286–307).
[26] Chatterjee, P.; Athawale, V.M. & Chakraborty, S. (2010). Selection of industrial robots using compromise ranking and outranking methods, Robotics and Computer-Integrated Manufacturing, Vol. 26 (5), page numbers (483–489).
[27] Ic, Y.T.; Yurdakul, M. & Dengiz, B. (2013). Development of a decision support system for robot selection, Robotics and Computer-Integrated Manufacturing, Vol. 29 (4), page numbers (142–157).
[28] Mortensen, N.H. & Hansen, C.T. (1999). Structuring as a basis for product modelling. Crit. Entusiasm, Trondheim: Tapir, page numbers (111–128).
[29] Krüger, J.; Lien, T.K. & Verl, A. (2009). Cooperation of human and machines in assembly lines, in CIRP Annals-Manufacturing Technology, Vol. 58 (2), page numbers (628–646).
[30] Zühlke, D. (2009). SmartFactory- a vision becomes reality, IFAC Proceedings Volumes, Vol. 42 (4), page numbers (31–39).
[31] Rosati, G.; Faccio, M.; Carli, A. & Rossi, A. (2013). Fully flexible assembly systems (F-FAS): a new concept in flexible automation, Assembly Automation, Vol. 33 (1), page numbers (8–21).
[32] Kittmann, R.; Fröhlich, T.; Schäfer, J.; Reiser, U.; Weißhardt, F. & Haug, A. (2015). Let me introduce myself: I am Care-o-bot 4, a gentleman robot, Mensch Und Computer 2015-Proceedings.
[33] Hancock, P.A.; Jagacinski, R.J.; Parasuraman, R.; Wickens, C.D.; Wilson, G.F. & Kaber, D.B. (2013). Human-automation interaction research past, present, and future, Ergonomics in Design: The Quarterly of Human Factors Applications, Vol. 21 (2), page numbers (9–14).
[34] Maurice, P.; Padois, V.; Measson, Y. & Bidaud, P. (2017). Human-oriented design of collaborative robots, International Journal of Industrial Ergonomics, Vol. 57 page numbers (88–102).
[35] Guiochet, J.; Machin, M. & Waeselynck, H. (2017). Safety-critical advanced robots: a survey, Robotics and Autonomous Systems.

DOI: 10.2507/28th.daaam.proceedings.161

GENERIC CHALLENGES AND AUTOMATION SOLUTIONS IN MANUFACTURING SMES

David Grube, Ali A. Malik & Arne Bilberg

This Publication has to be referred as: Grube Hansen, D[avid]; Malik, A[li] A[hmad] & Bilberg, A[rne] (2017). Generic Challenges and Automation Solutions in Manufacturing SMEs, Proceedings of the 28th DAAAM International Symposium, pp.1161-1169, B. Katalinic (Ed.), Published by DAAAM International, ISBN 978-3-902734-11-2, ISSN 1726-9679, Vienna, Austria
DOI: 10.2507/28th.daaam.proceedings.161

Abstract

Evermore research is conducted on smart manufacturing, digital manufacturing and other aspects of what is expected from the fourth industrial revolution known as Industry 4.0. Most of the research of Industry 4.0 is currently a better fit for large corporations than for SMEs, which in Europe however represent 98% of the manufacturing industry. In general, SMEs produce high-mix low-volume products, which require a high degree of flexibility. Historically flexibility and automation have been contradictory, but as automation becomes smarter, digitalized and less expensive, this may change which this paper is going to challenge in the context of Lean automation. Through a two-year automation research project where several automation solutions were built and implemented, it was verified that SMEs face very different challenges. In this paper the analysis of the companies, and automation integrators of the project, identifies a correlation between the challenges, age and size of the companies. The identified correlation lay ground for an Industry 4.0 light concept, targeting the identified generic challenges of companies employing 10-50 people. The solutions presented are based on cloud computing, Internet of Things and horizontal integration to simulate digitalized order handling and traceability in smaller companies.

Keywords: SME; Lean Automation; Industry 4.0 light; HMLV manufacturing; IoT

1. Introduction

SMEs generally operate in the paradigm of high mix low volume (HMLV) production and often act as a specialized supplier to larger companies. SMEs are characterized with production systems that are less automated and rely more on human aspect of production and hence exhibit higher flexibility. Consequently, they are in better position to adapt quickly to customers' demands and offer customized products, the challenge however is their productivity and level of automation especially in high salary countries in Europe. Globalization, market changes and not least digitalization technologies challenge manufacturing businesses to think in smart innovative solutions [1] [2]. A range of manufacturing development initiatives are available in industrialized economies such as Industrie 4.0 in Germany, Advanced Manufacturing Partnership in the United States and Manufacturing Innovation 3.0 in the Republic of Korea [3]. These initiatives are harvesting solutions for manufacturing companies around digitalization and internet connectivity.

The new production landscape is envisioned as getting full benefits of cyber physical system (CPS), Internet of Things (IoT) and digital twins. It is evident from the research and industrial innovation projects that these technologies are being cultivated around large multinational companies and there is not much research available for the development of SMEs in context of digitalization and smart manufacturing. Now scholars, politicians and industrial leaders are talking about a fourth industrial revolution (I4.0) that will elevate the productivity of manufacturing to unpreceded levels. The transition will utilize state of the art technologies within automation and digitalization of both products and entire value chains [4]. Industry 4.0 is gaining momentum and is widely known by both SMEs and larger corporation, as an example, 85% of the companies in Denmark are aware of I4.0 and are paying attention to it [5]. Despite the considerable attention, the share of research and development into Industry 4.0 for SMEs, compared to larger enterprises, is limited. This makes it hard for the smaller companies to grasp the knowledge, which however is important because SMEs are also in the middle of the fourth industrial revolution. SMEs may in the near future face technologies and solutions that they have to be aware of, to be able to take advantages of new functionalities and services in relationship to existing products and businesses. If not aware of these new opportunities in digitalization and automation, this may in worst case disrupt their businesses. It is well known that production systems and control hierarchy of SMEs are considerably different when comparing with larger enterprises [6] and recent studies conclude that company size is closely related to preparedness towards Industry 4.0, automation and digitalization [7].

Research Question: *How can we make use of digitalization and Industry 4.0 enablers to develop generic solutions that can remedy challenges in several SMEs?*

The scientific contribution of this paper is to:

- Investigate the challenges for flexible automation in SMEs in industrialized countries
- Propose flexible automation solution(s) to increase productivity of manufacturing SMEs
- Present the technical solution with the help of a learning-factory model

2. Literature review

The literature review is describing both what has been identified as challenges to manufacturing SMEs and how I4.0 is expected to impact manufacturing SMEs. In the literature complimented with case studies, it becomes clear that there are many variables that determines the challenges SMEs are facing, and also that the challenges may be different according to company size, history, business area, geographical location etc. This form of difference is also visible when SME related challenges are evaluated, and often the studies are formed in a local context [8].

2.1. SMEs for sustainable society

The annual report of European SMEs [9] exhibits the data on the role of SMEs in Europe. It states that SMEs have a share of 99.8% of total enterprises in Europe, are employing 66.9% of all enterprise' employees and add 57.8% value of all enterprises. In the past decades, much of the productivity gain is enabled by lean and similar productivity enhancing philosophies. Thus, the lean philosophy did not only gain attention in business-companies, but also in academia where the challenges, benefits and prospects of lean manufacturing were described. Considering the published research, the papers count for lean manufacturing is more than 70.000 articles and the share for SMEs related lean-manufacturing is almost 17.000 papers. But despite the great attention from both industry and academia, most of the SMEs remained struggling to implement lean methods in their production systems. Some of the main challenges in implementing lean were the lack of time and resources. Another challenge being faced by SMEs was the uniqueness of SMEs in their nature thus it was hard to form generalized lean tools to fit the company needs [6]. This gives rise to the hypothesis that SMEs will again face even bigger challenges in implementing automation and digitalization, considering the fact that the solutions are still not tailored to their needs and the fact that automation historically has remained away from SMEs.

2.2. The next industrial revolution

It is a human characteristic to struggle for improvements in their environment and quality of life. Factories, as a strong component of economic wellbeing of the society, have progressed rapidly in-line with these improvement related expectations. This is evident from the fact that humans have already experienced three big transformation or revolutions within industrial paradigms [10]. There have always been some technological enablers that gave rise to this notion of industrial revolutions i.e. steam engine, electricity, and PLCs for the 1st, 2nd, and 3rd industrial revolutions respectively. Now the recent developments in the field of information and communication technologies (ICT) foretell us to enter into a new industrial paradigm named Industry 4.0 (I4.0) or the fourth industrial revolution. The emergence of the term of Industry 4.0 happened in the last decade in Germany as a strategic initiative of the German government as part of the "High-Tech Strategy 2020 Action Plan" in 2011, describing it as the increasing ICT integration in production and manufacturing [11]. Industry 4.0 now refers to a network approach where components and machines ("things") are becoming intelligent and act as modular bricks (like LEGO bricks) to setup larger systems [12].

2.3. SMEs in era of lean automation and digitalized manufacturing

There has been a varied view towards the impact of Industry 4.0 on SMEs. The initial reports on Industry 4.0 stated that Industry 4.0 will benefit manufacturing SMEs to a great extent [4] whereas many other articles dealing with Industry 4.0 state that SMEs will become victims of this new digitalized manufacturing landscape [7]. In the relatively new literature stream of Industry 4.0 it has become evident that SMEs are under-represented in research field i.e. only 6,5% of the literature related flexible automation target SMEs, despite the fact that SMEs hold for 98% of the companies in the European Union.

According to Kolberg [12] Industry 4.0 phenomena has a close match with Lean philosophy meaning that increased integration of ICT will improve Lean production thus translating Industry 4.0 from science to reality. Some scholars state that Lean and automation contradict each other [13] and therefore earlier industrial automation did not try to couple Lean and automation. But in our research and experience in SME's, it is observed that Lean and automation can be closely related and ought to be combined in manufacturing strategies. Flexibility, as stated earlier is one of the crucial elements in manufacturing. There are many definitions of flexibility, but in the research context, it is about the ability to respond to changing demands from customers, which is different in SME's. Often automation is understood as robotics, but this is not the only and often not the most important solution, when talking SME's flexibility.

Lean Automation means in this context that automation has to focus at the customer, it has to create value and eliminate waste. The Toyota Production System operates with the expression Jidoka, known as "automation with a human touch". This is very much what Lean Automation is about, to use robust, reliable components build into non-complicated, not over engineered solutions. A challenge for Lean Automation is to determine the appropriate level and type of automation, where Bilberg and Hadar [14] presented lean semi-automated cells, where the automation is within the manufacturing process, while the planning, the material flow and part of the inspection is handled through the flexibility of the workers. According to Seifermann [15] traditional automation especially in western countries often become complex and expensive, and therefore the conclusion, especially when addressing SME's, is that low cost Lean Automation solutions may be the right concept.

According to Jackson [13] the challenges in Lean Automation are:

- The ability to choose the right level of automation as well as the right automation solution
- The ability to develop automation solutions which are flexible and reconfigurable, that enables the system being changed and adapted to new demands during its lifecycle
- The ability to handle complex equipment without being an expert. The requirements are intended to reduce the perceived level of complexity, possibly through intuitive user interfaces
- The ability to change and implement changes in a given automation system due to changes in the products

Besides challenges, as discussed above, Mrugalska [16] has described the strengths of SMEs towards innovation. These strengths distinguish them from large companies and may even place them in a stronger position for digital transformation.

The strengths are:

- Fast decision making
- Fewer layers of management
- Simple and direct communication
- SMEs are close to customers and get faster feedback and response quickly to customer's needs
- Higher flexibility at all levels
- Easy to introduce change and new methods

Jaione [17] has developed a three-stage model for SMEs to mature and become capable of developing the preconditions for an Industry 4.0 setup, they are well aware of some of the challenges regarding the concept of Industry 4.0, but at the same time, they believe that their three-step model can ease the road towards Industry 4.0. First step is to envision Industry 4.0 into a vision, second is to enable Industry 4.0 by means of a roadmap and third step is to enact Industry 4.0 by running simple bottom-up project towards Industry 4.0.

3. Empirical case study – the AutoSyd Project

In order to investigate the potential for SMEs based on automation and smart technologies, a research project was conducted targeting SMEs in the region of Southern Denmark. The aim was proliferation of knowledge and innovation-based-automation into the SMEs. For this, an action research approach was adopted to engage with industries, investigate the problems, create knowledge and contribute to scientific literature in a broader sense [18]. The project has been in contact with more than 100 SMEs in the southern part of Denmark. Forty of these SMEs were analyzed in-detail for mapping the current state and to point at future automation potentials.

The research carried out an extensive study to analyze the needs and thus develop automation solutions for 14 manufacturing companies. Consequently, 10 SMEs were selected to test the development plan. The solutions were developed following an iterative process with partner companies to develop possible frameworks for automation to address their manufacturing challenges.

The researchers designed a framework for how to get an insight of the company. The procedure adopted consisted of an introduction by the owner/director of the company, discussion on how they operated, what was important to them, and who their customers were. The introduction-session ended with an extensive and informative guided tour through the production facility where all significant processes were described, measured and captured in an internally developed scheme. The scheme was developed based on well-known operations management tools presented e.g. by Heizer and Render [18] and Lean literature [19]. Later, recommendations on what and how to automate was handed over to the company accompanied by a roadmap to proceed in the iterative process.

Observations from a two-year automation research project point towards the fact that SMEs in the span of 1 to 205 employees have indeed very different challenges, but also generic challenges were observed, and many of the SMEs seem to be equally ready to implement technology and automation to overcome challenges. Hence the question, how can SMEs be targeted with generic digitalized automation?

4. Findings in SMEs

4.1. Challenges with respect to evolution of an SME

Based on our observations, SME experiences different types of challenges at different phases of its maturity, where the number of employees has also close impact on this trend. In the figure (Fig. 1.) a summary of challenges is presented in relation to the number of employees as an indicator for the progressive evolution. For this work, SMEs are defined according to EU standard i.e. below 250 employees and up to € 50 million in turnover. In the AutoSyd project companies from one employee to 201 employees were observed, plus we have information from bigger companies i.e. above 250 employees. In brief, it is often observed that for micro companies the challenges are the products, markets and the financing. The smaller companies observe challenges in procedures, documentation and productivity. The medium sized get challenges with the size of organisation, communication and considerations according to automation. This paper research the opportunities for SMEs to start automation even earlier i.e. for small SMEs.

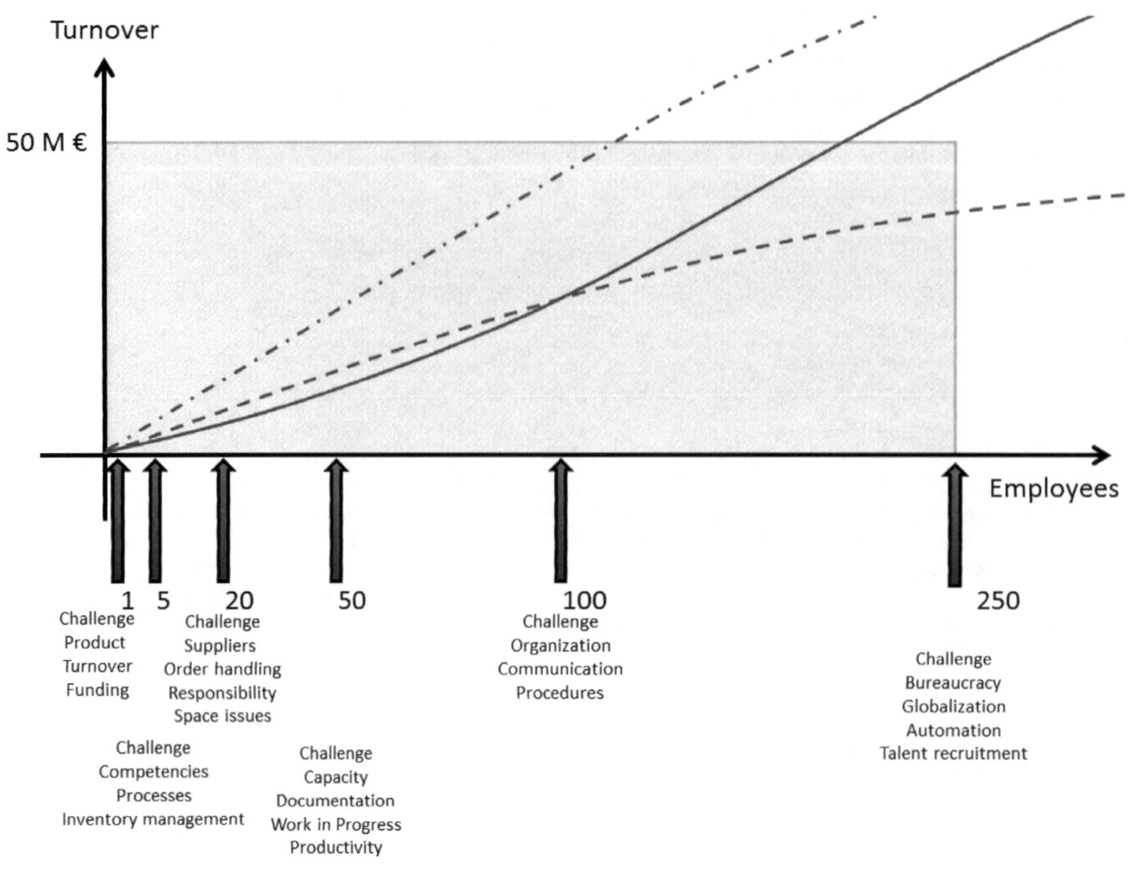

Fig. 1. Generic challenges of SMEs as identified from the AutoSyd research project

4.2. Generic challenges for SMEs – findings from the research project

Investigating the mentioned SMEs, general issues recur. SMEs are often within niche and specialized production, based on key technologies, which often results in High-Mix Low-Volume (HMLV) production. Most SMEs in this research did not have a clear strategy for their production and investments. Machines and equipment are fitted in the production, where space is available, without long term planning.

The inventories and flow of materials was unnecessary complex, with high work in progress -WIP, long lead times and many "dead" components in production. The planning and operations were often done manually without proper tools and procedures, which makes it hard to prioritize and keep the overview. Generally, there were high variety in volume (high/low volume), product/services and lead times. Quality, flexibility and adaptability has high attention in the SME's, but is challenged by many manual processes, complexity in structures and lack of overview.

The identified challenges are summarized to:

- Most SME's in this research did not have a strategy for their production and investments
- Machines and equipment is fitted in the production, where space is available, without long term planning
- The inventories and flow of materials was unnecessary complex, with high work in progress -WIP
- Long lead times and many "dead" components in production
- The planning and operations are often manual.
- High variety in volume (high/low volume) and mix of product/services
- Quality, flexibility and adaptability has high attention in the SME's, but is challenged by many manual processes
- Complexity in structures and lack of overview

Based on the experience throughout the project and findings above, the overall conclusion is that:

- SME's must work strategically with production that makes more long-term planning possible and consistent.
- The planning and control systems must be simple and efficient solutions to support personal at the shop floor, to make right decisions.
- Focus at Lean tools to reduce waste and create flow, to prepare for automation.
- SME's must think smart in simple Lean Automation solutions, that may be specific to the individual companies, but still there are some general learnings and benchmarking between companies.

5. Generic solutions for SMEs

The research and collaboration in the consortium between the university and the automation providers, has resulted in three generic concepts for automation solutions to SME's in line with the approach of simple Lean Automation. The solutions applied in this project, are all modular and can be multiplied to larger scale and adapted to new situations. The simple Lean Automation solutions are flexible and can be reconfigured.

5.1. The Production Feeder

Most SMEs have CNC machines (milling, turning etc.) with manual material handling. In this concept, a flexible feeder system has been developed to feed CNC machines automatically. The innovative element in this solution lies in the feeder, that is flexible and adaptable to any type of robot and flexible to new products. The feeder has a flexible magazine that by a human can be filled/emptied with products (while CNC-robot is working), the robot and CNC machine can then work for several hours without interference of a human.

The feeder is based on flexible modules, i.e. it can be rebuilt and reconfigured to new tasks. The feeder is mounted at wheels, so that it can move between machines. In this way, the CNC machine can also be operated manually if required. This concept increases flexibility and Overall Equipment Efficiency -OEE.

5.2. The Flexible Assembly cell

Most manufacturing SMEs have assembly tasks and demand more flexible assembly, that are easy to change and setup for new products. This concept is built of flexible modules that through different configurations can be operated by the operators themselves. The concept can be configured for different levels of automation, where assembly and handling may be by any combination of humans and/or robots in collaboration.

The robots will in the final setup be moveable and operate as helping devices for simpler work tasks, where the humans handle the more flexible and complicated tasks. The assembly cell concept provides a high degree of flexibility and scalability, combining the best characteristic from both automation and human operators.

5.3. Module based Pick and Place cube

This Pick and Place cube or cell, is designed to only take up the space of a regular production worker, it is highly modular, meaning that the capacity can be scaled up and down by simply adding more cubes. The hard wiring is simple, one power connection and one pneumatic connection and the system is up and running. The data is cloud based, and can be seamlessly up and downloaded, the data will then be presented on a standard of the shelf screen. On the same screen the Human Machine Interface (HMI) is made intuitive, for example the function for programming consists of dragging and dropping elements. It is possible continuously to get access to the helpdesk programmers for keeping the uptime at a high level. The cube itself has been designed for portability by means of a simple low lifter, this means that an SME with more production lines can move the cube very easily between lines and then plug in the connections.

The three examples of products/concepts described above are solutions/products coming out of the research and collaboration, where most likely none of these solutions would have been developed without the complementary competences between university and industrial system providers. There were also three other solutions that most likely will be turned into products in the near future. One of them is a simple planning system based on cloud technology and internet communication on the direction of internet of things.

6. Technologies for future research in SME context

The smart manufacturing paradigm is highlighting various technological enablers to transform manufacturing landscape. As already discussed, it is significantly hard for SMEs to research and develop the architecture for those technologies on their own. However, we are highlighting below technological enablers, that are part of the Industry 4.0 paradigm and carry a benefit potential for SMEs.

The technologies identified are in accordance with the challenges investigated in SMEs, development effort and maturity level of the technologies. The future research by the authors will investigate further into these technologies in cooperation with regional SMEs. In this regard an Innovation-Lab is setup at the Mads Clausen Institute of University of Southern Denmark. The Innovation-Lab is a state-of-the-art facility that offers collaboration opportunities between academic researchers and industrial partners to develop innovative solutions for manufacturing and production enterprises.

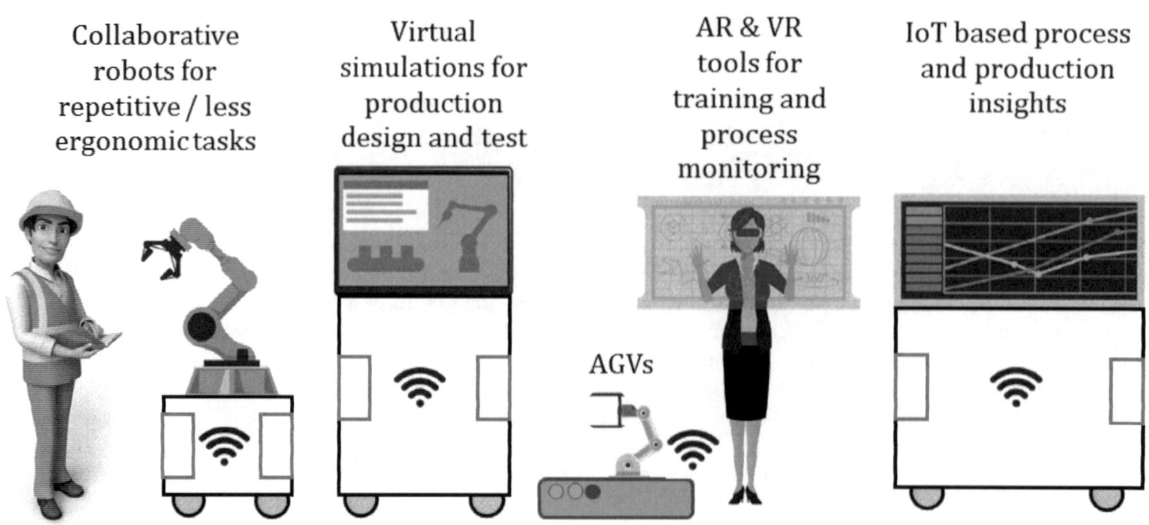

Fig. 2. Industry 4.0 technological enablers for SMEs

For the future research internet and connectivity of the “things” is prioritized. Schönsleben [20] described Internet of Things (IoT) as an internet connected network of goods and objects that can exchange data. IoT presents a new era for businesses to analyze, measure and conclude thus contributing towards strategical planning. This approach can greatly help SMEs to improve their relation with customers [21]. A major enabler for IoT, in case of SMEs, is RFID. RFID act as an electronic barcode to communicate and transfer data of anything it is attached to [22] thus transforming manufacturing objects as information careers.

Many advantages of RFID tags can be conceived in SMEs in material movement, resource planning and logistics. The data that is collected with the network of "things" is high in volume and because of its evaluation complexity, it is referred to as Big Data. The aim of doing so is to improve decision making process and reduce risks [20]. The investment to use this technology is huge however Cloud technology and Hadoop systems can be an effective solution both in terms of investment and number of employees required to manage and analyze the data.

At the site level of a production system, a big opportunity can be investigated in the movement of the material. Conventionally automated guided vehicles (AGVs) are used to be guided by cables mounted in the floor, and only big companies have had access to this type of internal transport. However now much more flexible solutions are becoming available, which are simple less expensive moveable robots and are easy programmable. So, the target customer also may be SMEs. We will see collaborative AGVs like the robots in near future. Another emerging trend in production automation is towards collaborative robots or cobots. In a collaborative environment, the production tasks can be distributed based on strengths of robots and humans.

The distinguishing features of collaborative robots are their safe and effective interaction with human workers and ease of deployment and operation. Productivity, ergonomics, and safety conditions are potential areas of getting benefit from collaborative robots [23][24]. Although SMEs are characterized with low volume of production with high mix and hence robotic automation has remained far from SMEs, but with the concept of lean automation the repetitive and less ergonomic tasks at the shop floor can well be assigned to collaborative robots.

The design and optimization of new as well as existing production system around these technologies will be a challenge, and in this regard, manufacturing simulations are an effective way to support the decision-making process. Creating a virtual model of a system that imitates the characteristics of the physical model is far cheaper to experiment before implementing in the real world. The modern simulation tools are getting much stronger in simulating complex scenarios with easier to use interfaces.

7. Smart Factory - the scaled model

A scaled production facility has been built-up to showcase and demonstrate the application of simple, easy to use smart technologies in SMEs. The inspiration to this scaled learning factory is SmartFactory [25] setup built up at DFKI, Germany, which is a hybrid production facility as a demonstration and development platform of Industry 4.0 related technologies[26]. The machinery and components used in the SmartFactory are similar to those used in industrial application and offer the same complexity as an industrial equipment [27].

A similar concept is used to build up learning factories in other research institutes e.g. Process Learning Factory at the Institute of Production Management, Technology and Machine Tools (TU Darmstadt), The IFA Production Training for teaching lean production methods at the Institute of Production Systems and Logistics (Leibniz University Hanover) and Experimental and Digital Factory (EDF) at the Department of Factory Planning and Factory Management (Chemnitz University of Technology) [28]. Wagner [28] is of the views that these experimental environments are an effective way to develop and demonstrate innovative concepts and educate industry practitioners as well as students about their potential.

In the AutoSyd project, contrary to learning factories in practice, a scaled model approach is adopted. The purpose is to build a low-cost factory setup with easy to use wireless "plug and produce" components to demonstrate innovative technologies to SMEs representatives. This model can be easily setup and rebuilt when required. The developed showcase is based on cheap and simple digitalization components such as RFID tags, embedded microcontrollers, Rasberry Pie LEGO bricks AND Little Bits, combined in a simple but illustrative setup that monitors and delivers the exact machine hours, man hours, material usage and process traceability. The scale model has caught great attention, because of the simplicity, it can be operated stand-alone or integrated to existing systems, and the model solely provides the information for handling orders, monitor and control operations until creating the invoices.

The learnings from working with the scale model with SMEs is that, instead of trying to explain the entire philosophy behind Industry 4.0 and the technologies, a simple setup that utilizes digitalization technologies, in this case, RFID, embedded microcontrollers and cloud solutions, elevated the willingness towards embracing the technologies which the politicians, industry leaders and consultancy companies appraises. In other words, a simple bottom-up solution to demonstrate part of Industry 4.0, where other of the enabling technologies described above are expected to be demonstrated in a similar setup.

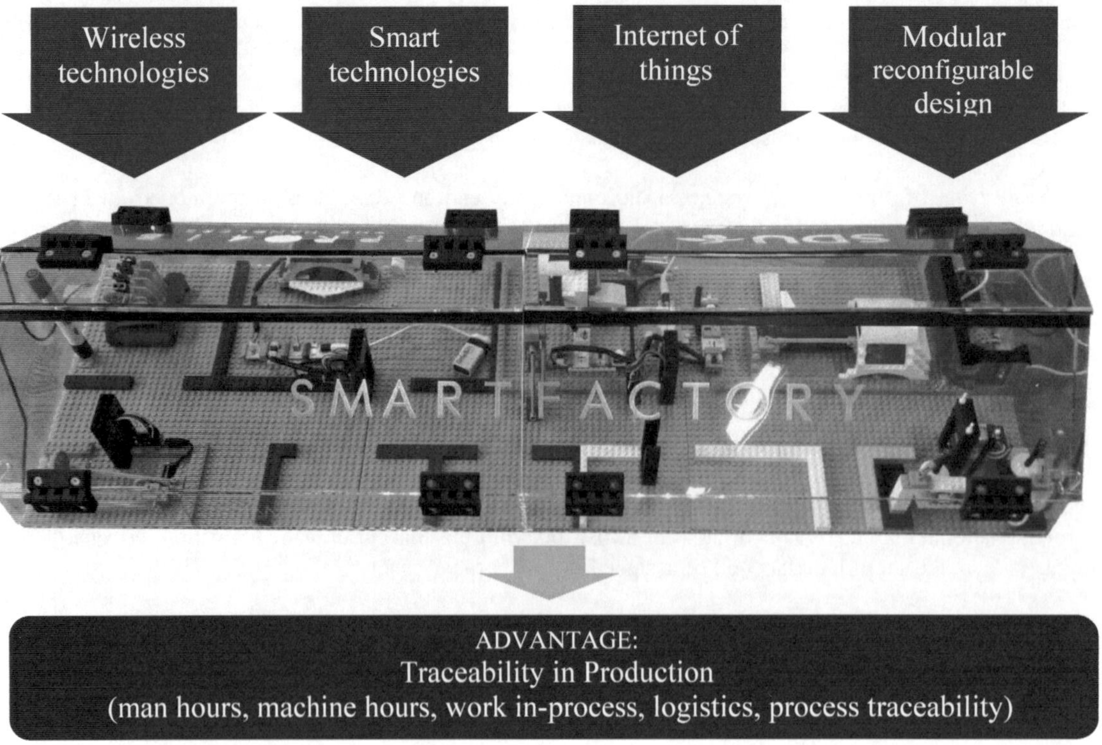

Fig. 3. A 1:50 scaled Smart Factory

8. Conclusion

Despite the fact that SMEs play a major role in production businesses of Europe, only limited research exists on how to increase productivity and automate SMEs. This is causing the SMEs to fall behind their contemporary large sized enterprises on the way to digitalization and smart automation. Based on literature and practical cases this paper is pointing at generic challenges and solutions to SMEs focusing at HMLV production, however the challenges and solutions are different depending on the evolution stage of an SME. Lean and automation are investigated as ingredients in automation solutions for SMEs, where automation with a human touch or "autonomation" is in focus. It is presented that strategic focus at production with "simple" smart technology concepts can enable SMEs to become more adaptable to the changing and dynamic environment e.g. collaborative robots and AGVs in a reconfigurable environment can adapt to changing environment. The future approach will be to investigate further into Industry 4.0 enablers and for Industry 4.0 -light approaches that foster Smart Lean automation solutions to SMEs. The research will further aim to ensure competitiveness by creating sustainable digitalization and automation solutions that are easy to use, cost competitive and flexible to HMLV production.

9. References

[1] Shah, R.; Park, H. & Lee, G. B. (2016). Design for assembly: an approach to increase design efficiency of electronics home appliance, Proceedings of the 26th DAAAM International Symposium, pp.0877-0882, B. Katalinic (Ed.), Published by DAAAM International, ISBN 978-3-902734- 07-5, ISSN 1726-9679, Vienna, Austria.

[2] Mulc, T.; Udiljak, T. & Ciglar, D. (2015). Structure of reconfigurable manufacturing systems, Chapter 32 in DAAAM International Scientific Book 2015, pp.369-390, B. Katalinic (Ed.), Published by DAAAM International, ISBN 978-3-902734-05-1, ISSN 1726-9687, Vienna, Austria.

[3] Jun, C.; Yoon, J.-S. & Kim, B.H. (2017). Applications' integration and operation platform to support smart manufacturing by small and medium-sized enterprises, in: Modena, Italy.

[4] Kagermann, H.; Helbig, J.; Hellinger, A. & Wahlster, W. (2013). Recommendations for implementing the strategic initiative INDUSTRIE 4.0: Securing the future of German manufacturing industry; final report of the Industrie 4.0 Working Group. Forschungsunion.

[5] Group, T.B.C. (2016). Winning the Industry 4.0 race; how ready are Danish manufacturers.

[6] Bejlegaard, M.; Brunoe, T.D.; Andersen, A.-L. & Nielsen, K. (2016). Reconfigurable manufacturing potential in small and medium enterprises with low volume and high variety, 3rd International Conference on Ramp-up Management (ICRM).

[7] Sommer, L. (2015). Industrial revolution-industry 4.0: Are German manufacturing SMEs the first victims of this revolution?, Journal of Industrial Engineering and Management, Vol. 8 (5), page numbers (1512).

[8] Kumalo, N. & van der Poll, J.A. (2015). The role of cloud computing in addressing SME challenges in South Africa. in: Cloud Computing Research and Innovation (ICCCRI), 2015 International Conference on, pp. 139–147.

[9] Association, E.F. of the F.R. & others (2013). Multi annual roadmap for the contractual PPP under Horizon 2020.

[10] Wang, S.; Wan, J.; Li, D. & Zhang, C. (2016). Implementing smart factory of Industrie 4.0: an outlook, International Journal of Distributed Sensor Networks, Vol. 2016 page numbers (7).

[11] Zezulka, F.; Marcon, P.; Vesely, I. & Sajdl, O. (2016). Industry 4.0- An introduction in the phenomenon, IFAC-PapersOnLine, Vol. 49 (25), page numbers (8–12).

[12] Kolberg, D. & Zühlke, D. (2015). Lean automation enabled by industry 4.0 technologies, IFAC- PapersOnLine, Vol. 48 (3), page numbers (1870–1875).

[13] Jackson, M.; Hedelind, M.; Hellström, E.; Granlund, A. & Friedler, N. (2011). Lean automation: requirements and solutions for efficient use of robot automation in the Swedish manufacturing industry, International Journal of Engineering Research & Innovation, Vol. 3 (2), page numbers (36–43).

[14] Bilberg, A. & Hadar, R. (2012). Adaptable and reconfigurable LEAN automation - a competitive solution in the Western industry, in: FAIM 2012 22nd International Conference on Flexible Automation and Intelligent Manufacturing, Helsinki.

[15] Seifermann, S.; Böllhoff, J.; Metternich, J. & Bellaghnach, A. (2014). Evaluation of work measurement concepts for a cellular manufacturing reference line to enable low cost automation for lean machining, Procedia CIRP, Vol. 17 page numbers (588–593).

[16] Mrugalska, B. & Wyrwicka, M.K. (2017). Towards Lean production in Industry 4.0, Procedia Engineering, Vol. 182 page numbers (466–473).

[17] Jaione, G. (2016). Three stage maturity model in SME's towards Industry 4.0, Journal of Industrial Engineering and Management, 2016 – 9(5): 1119-1128 – Online ISSN: 2013-0953 – Print ISSN: 2013-8423.

[18] Krajewski, L.J. & Ritzman, L.P. (2005). Operations management: processes and value chains, Prentice Hall.

[19] Bicheno, J. & Holweg, M. (2000). The Lean toolbox, PICSIE books Buckingham.

[20] Schönsleben, P.; Fontana, F. & Duchi, A. (2017). What benefits do initiatives such as Industry 4.0 offer for production locations in high-wage countries?, Procedia CIRP, Vol. 63 page numbers (179–183).

[21] Abazi, B. (2016). An approach to the impact of transformation from the traditional use of ICT to the Internet of Things: How smart solutions can transform SMEs, IFAC-PapersOnLine, Vol. 49 (29), page numbers (148–151).

[22] Gubbi, J.; Buyya, R.; Marusic, S. & Palaniswami, M. (2013). Internet of things (IoT): A vision, architectural elements, and future directions, Future Generation Computer Systems, Vol. 29 (7), page numbers (1645–1660).

[23] Malik, A. A.; Bilberg, A.; (2017). Framework to implement collaborative robots in manual assembly: A Lean automation approach, Proceedings of the 28th DAAAM International Symposium, B. Katalinic (Ed.), Published by DAAAM International, ISSN 1726-9679, Vienna, Austria, (in press).

[24] Grigorescu, S.; Vatau, S. & Dobra, A. (2010). Dedicated robot-robot cooperation, Chapter 55 in DAAAM International Scientific Book 2010, pp. 633-644, B. Katalinic (Ed.), Published by DAAAM International, ISBN 978-3-901509-74-2, ISSN 1726-9687, Vienna, Austria.

[25] Zuehlke, D. (2010). SmartFactory—Towards a factory-of-things, Annual Reviews in Control, Vol. 34 (1), page numbers (129 – 138).

[26] Radziwon, A.; Bilberg, A.; Bogers, M. & Madsen, E.S. (2014). The Smart Factory: Exploring adaptive and flexible manufacturing solutions, Procedia Engineering, Vol. 69 page numbers (1184–1190).

[27] Zuehlke, D. (2008). Smartfactory-from vision to reality in factory technologies. in: Proceedings of the 17th IFAC World Congress, pp. 82–89.

[28] Wagner, U.; AlGeddawy, T.; ElMaraghy, H. & Myller, E. (2012). The state-of-the-art and prospects of learning factories. Procedia CIRP, Vol. 3 page numbers (109–114).

DOI: 10.2507/28th.daaam.proceedings.162

ONTOLOGY ENRICHMENT FOR MULTI-DOMAIN KNOWLEDGE AND EXPERTISE REPRESENTATION

Harry T Yani Achsan, Heru Suhartanto & Wahyu Catur Wibowo

This Publication has to be referred as: Achsan, H[arry] T[. Yani]; Suhartanto, H[eru] S[eok] & Wibowo, W[ahyu] C[atur] (2017). Ontology Enrichment for Multi-Domain Knowledge and Expertise Representation, Proceedings of the 28th DAAAM International Symposium, pp.1170-1177, B. Katalinic (Ed.), Published by DAAAM International, ISBN 978-3-902734-11-2, ISSN 1726-9679, Vienna, Austria
DOI: 10.2507/28th.daaam.proceedings.162

Abstract

Expertise is specialized skills and knowledge. It can be shared or disseminated if stated explicitly using expertise representation. Since people have multi area/domain expertise, we need multi-domain expertise representation. Someone's expertise can be recognized by his/her publication. In this article we proposed a new expertise representation by utilizing bibliography (publication metadata) and annotating ontology (ontology enrichment). The expertise representation is multi-domain that can be integrated semantically as easy as single-domain ontology.

Keywords: expertise; knowledge representation; semantic heterogeneity; multi-domain ontology; bibliography; library catalogue

1. Introduction

Tacit knowledge cannot be transferred to other people unless it is stated explicitly. Since stone age many media have been developed to state knowledge explicitly e.g. slate, leather, paper, microfiche, magnetic tape, magnetic disk, optical disk, flash disk, solid state disk, etc. Some media can be used to transfer knowledge directly like slate, leather and paper, but some other media can transfer knowledge by using analogue or digital equipment only.

Recently, the trend of transferring knowledge uses digital environment [1]. Storing and transferring knowledge digitally has many advantages; one of them is that the knowledge stored digitally is not only can be presented to human but also can be processed by machine. It means that the machine has the ability to read and to interpret knowledge to be manipulated, synthesized, and presented.

In supporting and improving readability, the digitally stored knowledge should be understood by machine. Unfortunately, most of knowledge are represented in such a way that human can read easily and not intended for machine to understand it. Tim Berners-Lee, the founder of WWW, stated that the goal of WWW was not only for human-human communication, but also for machines to be able to participate and help. Thus, he proposed the Semantic Web two decades ago [2]. It is an extension of HTML that enables description of contents to be machine-readable and was developed based on Ontology [3]. Therefore, it can be inferred that ontology is a knowledge representation intended for both human and machine.

Ontology is now used widely in many domains like in health [4, 5, 6], engineering [7, 8], software [9, 10], commerce [11, 12, 13], law [14, 15], etc. Those Ontologies use different Upper Ontology. Consequently, reasoning with many different domains of Ontology cannot be done automatically since it is labour intensive and costly.

Reasoning is the process of thinking about something in a logical way in order to form a conclusion or judgment [16]. One important goal of constructing Ontology is to synthesize new knowledge by using reasoning. Reasoning of Ontology in a multi-domain environment (a.k.a. semantic heterogeneity) has to be done manually because it needs Ontology merging [17, 18, 19]. Ontology merging is an intentional process of creating new ontology by mapping and matching concepts from two Ontologies or more.

Ontology merging from multi-domain needs multi-domain Ontology, a multi-domain knowledge representation. By using multi-domain Ontology, merging of multi-domain Ontology can be treated as single-domain/domain-specific Ontology merging that can be done automatically/semi-automatically.

This paper proposed novelty knowledge representation in multi-domain environment by utilizing Ontology enrichment. Ontology enrichment is annotating Ontology with other metadata. The metadata used here are library classification codes. These codes are used to classify books, journals, proceedings, and other library items.

Library items are authored by experts. An expert is a person who has extensive skill or knowledge in a particular field [20]. Since library items contain expertise (a written opinion by an expert [21]), the enriched Ontology can also be used to represent expertise.

2. Related Research Work

2.1. Knowledge representation

Several knowledge representations have been developed since the emergence of computer. Early knowledge representation that uses modern notation is Classical logic, one of Formal Logical Systems. Some members of classical logic which are still being used up until now are Propositional Logic [22, 23, 24], First-Order Logic [25, 26, 27, 28], and Second-Order Logics [29, 30, 31] among others. These logical systems are still being used and developed because their capability in knowledge reasoning. Unfortunately, these knowledge representations are easily read by mathematician but not for others.

Some graphical knowledge representations have been developed for people who are not familiar with mathematics notations. Graphical knowledge representations have gained attention by researchers after Margareth Masterman introduced Semantic Network in 1961 [32]. Semantic Network are implemented in many research areas like Brain and Cognition [33, 34, 35], Health Care [36, 37], Education [38, 39], Communication [40], Computer Science [39, 41, 42], etc. Other graphical knowledge representations are: Association Diagram introduced by Rakesh Agrawal [43, 44], Conceptual Graph introduced by John F. Sowa [45], Tree Diagram, and Bayesian Networks.

Graphical knowledge representations have some advantage over classical logics since they can be understood by "common people" and can be interpreted faster. However, it also has a big drawback since it was not designed for computation. Modern knowledge representation should be designed for human and for machine. Therefore, it enables machine to process the knowledge, increasing its versatility in many applications like: coal mining [46], information security [47], medical [48], user profiling [49], and product designing [50].

2.2. Ontology

Ontology is a term in computer science adopted from philosophy. In philosophical study, ontology is about existence. In computer science, an ontology is an explicit specification of a conceptualization [51]. The term ontology used in computer science began when Amitabha Ghose wrote articles about semantic of programming languages [52, 53] in 1970s. Ontology for knowledge representation and knowledge sharing introduced about two decades later by Thomas R. Gruber [51], Nicola Guarino [54], O'Leary [55], etc.

Ontology is categorized by Nicola Guarino into Upper/Top Level Ontology, Domain Ontology, Task Ontology, and Application Ontology. Top Level Ontology describes general concepts which are independent of specific domain. Domain Ontology and Task Ontology specialize concepts found in the Upper Ontology into to a generic domain or task respectively. Application Ontology describe concepts introduced by both Domain and Task Ontology.

Ontology as a knowledge representation is intended to share common understanding among people or software agents [56, 57]. The most successful implementation of ontology as a knowledge representation is Semantic Web although the progress is very slow. At least 2.5 billion web pages have implemented Semantic Web in a sample of over 12 billion web pages, but it was only 21 percent [58]. Even though the progress is very slow, individuals and organizations insist to still implement Semantic Web, because it can decrease the page viewed per download by 50% and increase revenue by 20% [59].

Based on 627 surveys conducted by Jorge Cardoso in 2007 [59], he found that mostly people (68.2%) used Protégé as their ontology editor. Ontology languages mostly used were OWL & RDFS (75.9% & 64.9% respectively), while most used reasoning engine was Jena (53.6%). He also discovered that industries represented their knowledge with ontologies were mostly Education (31.0%), Computer and Software (28.5%), and Government (17%). Most of them (60.0%) did not employ any methodology or method to develop their ontologies.

The next finding was about the purpose of developing ontologies. Majority of participants (69.9%) used ontologies to share common understanding of the structure of information among people and software agents, and some 56.3% use ontologies to enable reuse of domain knowledge.

Two-third of surveyed subjects used ontology mapping to manipulate ontologies. Ontology mapping is a process of relating two terms from two ontologies that has the same meaning. Ontology mapping is crucial part in semantic integration, a process of interrelating information from diverse sources [60].

2.3. Expertise

Expertise is part of person profile, facts about someone. It can be developed by: a) web log analysis using Association Rule [61], fuzzy classification [62], and fuzzy clustering [63, 64], b) semantic enrichment on twitter messages and on emails, and c) tags analysis of social media. Since expertise is specialized skills and knowledge [64], it can be expressed as knowledge representation.

3. Bibliography

3.1. Metadata

Metadata is data to describe other data. Bibliography is metadata of library items. Bibliography in scholarly work is a list of books consisting metadata of books: the book title, year published, authors name, etc. Metadata of book can be found in libraries, specifically in the catalogue cards. Recently, most librarians do not use catalogue cards anymore since they have digitalized them. Searching library items is more convenient and more effective by using digitalized catalogue.

3.2. Sources

There are many sources of bibliography in the digital era. Online book stores provide bibliography of the items they sell. School and higher education libraries also serve online catalogue to search their items. Other sources of online catalogues are bibliographic databases like DOAJ (Directory of Open Access Journals), Scopus, and WOS (Web of Science). The amount of online catalogue or bibliography is very large. For example, a library can hold more than a million records. Scopus, a bibliographic database, hold more than 55 million records [61]. The largest bibliographic database is WorldCat that holds 2 billion items [62].

3.3. Mining

Collecting bibliography is very exhaustive and resource intensive if done manually because of its enormous size. It must be done automatically by using focused web-crawler that can collect data much faster than human can do manually [63].

3.4. Classification

Book classification has been employed in library for centuries. There are two major classification used by librarians all around the world which are Library of Congress Classification (LCC) and Dewey Classification Code (DCC). DDC is the most adopted classification code, but most of large libraries implement LCC because LCC uses alphanumeric and DDC uses numeric codes. Therefore, LCC has much more classes for the same length of codes. This is the reason why this paper focused on LCC.

Those book classifications are organized hierarchically such as tree diagram. There is no root node (level 0) because the root node is virtual. The first code is alphabetic, the second is alphanumeric, and the rest are numeric. There are 21 LCC nodes on level 1; some are mnemonics, like G (Geography), M (Music), P (Philology), and T (Technology). Level 2 has 213 nodes, where node M (Music) has 2 nodes on level 2: ML (Literature on music) and MT (Instruction and study), and node T has 16 children on level 2. Each level 1 node can be seen as representation of a domain, but a domain can be represented in more than one level 1 nodes. Nodes C to F are about history.

3.5. Semantics

A word in one domain can have different meaning in another domain. The word “conductor” have four meanings depends on the domain (the meanings were taken from WordNet, a lexical database), see Table 1. A word that has multiple meanings is called polysemy, and the science which study about meaning of words in a language is semantics.

Domain	Meaning	LCC Codes	
Music	The person who leads a musical group	M1-5000	Music
		MT1-960	Instruction and study
		MT85	Conducting. Score reading and playing
Physics	A substance that readily conducts e.g. electricity and heat	QC1-999	Physics
		QC251-338.5	Heat
		QC310.15-319	Thermodynamics
		QC501-766	Electricity and magnetism
Transportation	The person who collects fares on a public conveyance	G1-922	Geography (General)
		GT1-7070	Manners and customs (General)
		GT5220-5286	Customs relative to transportation and travel
Electronics	A device designed to transmit electricity, heat, etc.	TK1-9971	Electrical engineering. Electronics. Nuclear
		TK3001-3521	Distribution or transmission of electric power

Table 1. The meanings of "conductor" in some domains and its corresponding LCC codes.

Table 1 shows the corresponding LCC classes of the word related to its domain. Since every book in library should be classified and the class of a book can be related to its domain, therefore the domain of the book can be known. Thus, a book of an academic journal or a conference proceeding should consist of papers for the same domain. For example, the proceeding of DAAAM (Danube Adria Association for Automation & Manufacturing) contains papers in the domain of Automation & Manufacturing only, and Nature Medicine journal consist articles in the domain of medicine only.

Semantically, the meaning of words can be known if the class of the book that contains the words is stated. For instance, a journal with LCC code ML468 must be Music Literature in the domain of music and more specifically about instruments and instrumental music. The domain can be found by searching code ML in M-outline.pdf file that can be downloaded from https://goo.gl/NVPQ7h. The meaning of a keyword "conductor" in the book with classification code precedes by M (Music) is most likely the person who leads a musical group. However, if the class is more specific to MT85 (Music – Instruction and study – Conducting. Score reading and playing) then the meaning of word "conductor" must be the person who leads a musical group.

From ontological point of view, class MT85 is related to Domain, Task, and Application Ontology. First level of class MT85 is M that refers to domain Music, the second level is T (MT) which associates to task Instruction and Study, and the third level is 85 (MT85) that refers to Application Ontology of conducting, score reading and playing. Although the class MT85 matches to ontology categorization, not all classes in LCC fulfil all ontology categorization. Some classes even have no Task Ontology.

4. Expertise Representation Based on Bibliography

4.1. Multi-Domain Expertise Representation

Based on their publications, most researchers have multi-domain (multi-discipline) expertise. For an example is Wisnu Jatmiko, a researcher from Faculty of Computer Science – University of Indonesia. He has published his works in some domains (subject areas): Computer Science, Engineering, Social Sciences, Mathematics, Genetics and Molecular Biology, Biochemistry, Medicine, Decision Sciences, Physics and Astronomy, Business, Management and Accounting, and Health Professions, as stated by Scopus.

To represent such expertise, a multi domain knowledge representation is needed. The representation should be as simple as possible that can be understood easily not only by human but also by machine. It has to be adopted from a proven knowledge representation and has been used widely. Ontology meets those requirements; it is simple and widely used as semantic web. Unfortunately, ontology by default is used to represent knowledge from a domain. It is not designed for multi-domain knowledge representation. Ontology can be used as expertise representation, but it has to be annotated first to make it multi-domain expertise representation. This process is well-known as ontology enrichment.

We proposed ontology enriched by LCC codes for multi-domain expertise representation. The proposed enriched ontology is developed by annotating the concepts and/or the relations by using LCC codes.

Fig. 1 shows enriched ontology of our proposal. The diagrams represent these sentences: (a) Harry is a conductor, (b) conductor holds stick made of metal, (c) Harry is singer with Metal genre, (d) metal is a conductor, and (e) metal conducts electricity. There are two words that have more than 1 definition depends on its domain. The first word is metal with LCC codes T, TN and M. This word in domain of Technology (T) and Metallurgy (TN) means a solid, shiny and hard material that conducts heat and electricity. In domain music, metal is a type of music. The other word is Conductor whose definition has been discussed in the previous section.

Someone's expertise can be predicted by his/her publications. To determine the expertise based on documents published, we should analyse metadata of the publications. The metadata among others are: document title, author keywords, indexed keywords, main heading, and LCC code. Those metadata except LCC code can be retrieved from WorldCat web site (worldcat.org) freely, by using search engine provided. LCC code can be searched using the search engine in classify.oclc.org. We do not explain the prediction of someone's expertise here.

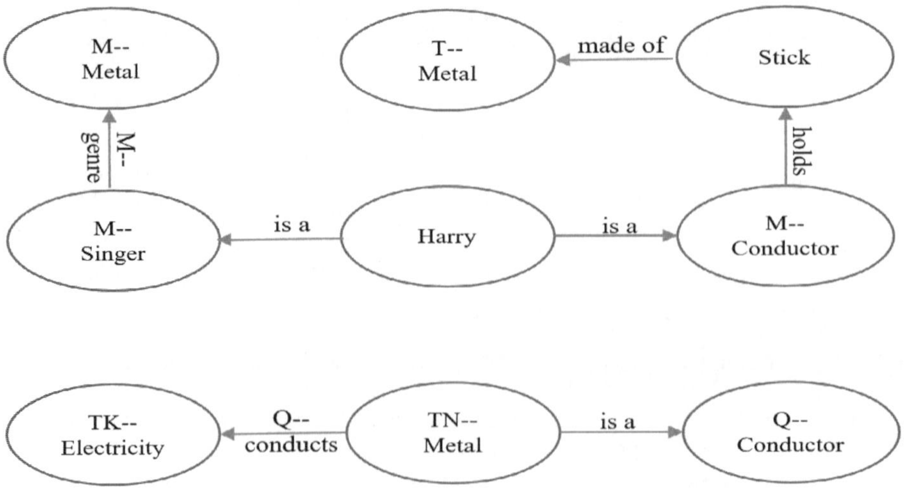

Fig. 1. Ontology enriched/annotated using LCC codes.

4.2. Semantic Integration

Semantic integration related to ontology is the process of interrelating concepts from different ontology. The process is very costly because it can only be done manually. Automatic semantic integration for multi domain ontology is not reliable. However, for single domain the precision and recall of automatic semantic integration can reach 92% and 96% respectively [65].

Semantic integration of multi-domain enriched ontology we proposed can be done automatically similar to single domain. The two ontologies in

Fig. 1 lied on two domains, technology domain and music domain, but can be integrated easily because the concepts has been annotated to distinguish each other. As can be seen in the figure, the two ontologies can be integrated because the two domains have interrelated concepts that is Metal (LCC code TN and T). If the ontologies have not been annotated, the two ontologies can make an invalid ontology if integrated semantically. This is caused by overlapping interpretation since metal in the music domain can be interrelated to metal in the metallurgy domain and conductor in the physics domain can be interrelated to conductor in the music domain.

Semantic integration is intended to generate new knowledge. The integration of the two ontologies in

Fig. 1 has interrelated TN--Metal and T--Metal to generate the following sentence: stick made of metal conducts electricity. Consequently, Harry should be careful while holding the stick because he may get electric shock. It seems that the new knowledge is obvious, because the sample ontologies are chosen as simple as possible.

4.3. Limitations

The proposed knowledge/expertise representation has some fundamental challenges to be implemented. The first is LCC codes of the concepts and the relationships of the ontology should be known. Currently, there is no reliable automatic document classification using LCC classes. Frank & Paynter got maximum accuracy of 38% from their experiments on automatic document classification [66] and Ávila-Argüelles et.al. got accuracy 36% [67]. The largest accuracy was achieved by Larson [68], 46.6%, but his experiments used only class "Z", 1 of 21 available classes in LCC.

Manually assigning LCC code to documents is very costly. To conduct document classification we have to know the main subject of the document to be classified. The classifier should be an experienced librarian who is used to classify books using LCC codes. Even for experienced librarian, it will take a while to look for an appropriate LCC code for a document. It will take much more time to classify concepts and relationships in an ontology, because a book only needs one class while an ontology needs many classes depending on the number of the sentences in the ontology.

The second challenge in implementing our proposed expertise representation is choosing the level of LCC code for concepts in the ontology. Fig. 1. shows concept "conductor" with different LCC code levels. One is in first level, LCC code T, and the other is in the second level, LCC code TN. LCC code T is for Technology and TN is for Metallurgy.

The latter is more specific than the first one, but for ontology the more general classes is the better. For the case of conductor we should choose T instead of TN. Terms for concepts and relations are more precise if using a more general LCC code, and the upper ontology which consists of very general terms need no LCC code.

4.4. Implementation

The most suitable implementation of the proposed knowledge/expertise representation is to represent knowledge or expertise retrieved from documents in which the LCC codes of the documents are already known. Although the LCC code of the document has already been known, the retrieved ontology does not necessarily use the code to all concepts from the document. For illustration, we present a sentence "A mouse cable has been bitten by a mouse.". The mouse words have different meaning and it is possible to have different LCC codes. We suggest to use the LCC code of the document only for concepts provided by librarian and the author. Concepts from librarian usually are controlled vocabularies. Concepts provided by the author, despite not controlled vocabularies, are key concepts of the document. The concepts provided by librarian and also the author among others are listed in book indices, author keywords, indexed keywords, and subject headings.

5. Conclusion and Future Works

Someone by nature has expertise in multiple subject areas or multiple domains. Engineers are not only knowledgeable about manufacturing only. They normally also have other expertise in other subject areas like information technology, material science, business, and/or any other else. This research has answered the research question of how to represent knowledge/expertise in multi-domain environment.

The expertise suited to the proposed representation is expertise which is retrieved from document which the LCC code has already been known. It is expected to increase the accuracy of multi-domain semantic integration that can be done automatically with high precision and recall. Reasoning also can be implemented to the expertise representation because the representation is using ontology.

Following this research, there are some advanced researches needed. First is research on how to retrieve expertise from bibliography. There are several ways to do it such as retrieving expertise from millions of online documents in the forms of books, proceeding papers and journal papers. Secondly, the result from the first suggestion can be used for a research in expert finding by utilizing learning to rank methods. Thirdly, since the prerequisite of the proposed expertise representation is LCC code for the documents, it is necessary to do research on automatic documents classification using LCC codes.

6. Acknowledgments

Part of this research was funded by PITTA Grants (hibah Publikasi Internasional Terindeks untuk Tugas Akhir) fiscal year 2017 from Universitas Indonesia.

7. References

[1] Islam, M. & Ikeda, M. (2014). Convergence issues of knowledge management in digital libraries: Steps towards state-of-the-art digital libraries. VINE, pp. 140-159.
[2] Berners-Lee, T. (1998). Semantic Web Road map. World Wide Web Consortium.
[3] Taye, M. M. (2010). Understanding Semantic Web and Ontologies: Theory and Applications. Journal of Computing, vol. 2, no. 6, pp. 182-192.
[4] Chen, R.-C., Huang, C.-Y. & Ting, Y.-H. (2017). A chronic disease diet recommendation system based on domain ontology and decision tree. Journal of Advanced Computational Intelligence and Intelligent Informatics, vol. 21, no. 3, pp. 474-482.
[5] Corral, M., Antonelli, L. & Sánchez, L. (2017). Health Ontology and Information Systems: A Systematic Review. IEEE Latin America Transactions, vol. 15, no. 1, pp. 103-120.
[6] Pozza, G., Borgo, S., Oltramari, A., Contalbrigo, L. & Marangon, S. (2017). Information and organization in public health institutes: An ontology-based modeling of the entities in the reception-analysis-report phases. Journal of Biomedical Semantics, vol. 7, no. 1.
[7] Ait-Ameur, Y., Baron, M., Bellatreche, L., Jean, S. & Sardet, E. (2017). Ontologies in engineering: the OntoDB/OntoQL platform. Soft Computing, vol. 21, no. 2, pp. 369-389.
[8] Zhou, J., Liu, J. & Yang, H. (2016). Dynamic ontology for engineering knowledge management of complex product research & development. Journal of Computer-Aided Design and Computer Graphics, vol. 28, no. 11, pp. 1957-1964.
[9] Gonzalez-Perez, C., Henderson-Sellers, B., McBride, T., Low, G. & Larrucea, X. (2016). An Ontology for ISO software engineering standards: 2) Proof of concept and application. Computer Standards and Interfaces, vol. 48, pp. 112-123.

[10] Dermeval, D., Vilela, J., Bittencourt, I., Castro, J., Isotani, S., Brito, P. & Silva, A. (2016). Applications of ontologies in requirements engineering: a systematic review of the literature. Requirements Engineering, vol. 21, no. 4, pp. 405-437.

[11] García, M., García-Nieto, J. & Aldana-Montes, J. (2016). An ontology-based data integration approach for web analytics in e-commerce. Expert Systems with Applications, vol. 63, pp. 20-34.

[12] Nederstigt, L., Aanen, S., Vandic, D. & Frasincar, F. (2014). FLOPPIES: A Framework for Large-Scale Ontology Population of Product Information from Tabular Data in E-commerce Stores. Decision Support Systems, vol. 59, no. 1, pp. 296-311.

[13] Rosaci, D. & Sarnè, G. (2014). Multi-agent technology and ontologies to support personalization in B2C E-Commerce. Electronic Commerce Research and Applications, vol. 13, no. 1, pp. 13-23.

[14] Gonźalez-Conejero, J., Figueroa, R., Mũnoz-Gomez, J. & Teodoro, E. (2014). Organized crime structure modelling for european law enforcement agencies interoperability through ontologies. Lecture Notes in Computer Science, pp. 217-231.

[15] Hage, J. & Verheij, B. (1999). The law as a dynamic interconnected system of states of affairs: A legal top ontology. International Journal of Human Computer Studies, vol. 56, no. 1, pp. 1043-1077.

[16] Merriam-Webster. (2017). Merriam-Webster Dictionary, Available from: https://www.merriam-webster.com/dictionary/reasoning Accessed 2017-07-24.

[17] Maree, M. & Belkhatir, M. (2015). Addressing semantic heterogeneity through multiple knowledge base assisted merging of domain-specific ontologies. Knowledge-Based Systems, vol. 73, pp. 199-211.

[18] Pai, F.-P., Yang, L.-J. & Chung, Y.-C. (2017). Multi-layer ontology based information fusion for situation awareness. Applied Intelligence, vol. 46, no. 2, pp. 285-307.

[19] Li, L., Wei, Y. & Tian, F. (2017). A Framework for Ontology-Based Top-K Global Schema Generation. Journal on Data Semantics, vol. 6, no. 1, pp. 31-53.

[20] Dictionary.com. Expert, Available from: http://www.dictionary.com/browse/expert Accessed 2017-07-24.

[21] Dictionary.com. Expertise, Available from: http://www.dictionary.com/browse/expertise?s=t Accessed 2017-07-24.

[22] Horn, K. Van. (2017). From propositional logic to plausible reasoning: A uniqueness theorem. International Journal of Approximate Reasoning, vol. 88, pp. 309-332.

[23] Chajda, I. & Kolařík, M. (2017). Reduced axioms for the propositional logics induced by basic algebras. Soft Computing, pp. 1-5.

[24] Caridroit, T. K. S. M. P. (2017). Contraction in propositional logic. International Journal of Approximate Reasoning, vol. 80, pp. 428-442.

[25] YiZhoua & YanZhang. (2017). A progression semantics for first-order logic programs. Artificial Intelligence, vol. 250, p. 58–79.

[26] Perikos, I., Grivokostopoulou, F., Kovas, K. & Hatzilygeroudis, I. (2016). Automatic estimation of exercises' difficulty levels in a tutoring system for teaching the conversion of natural language into first-order logic. Expert Systems, vol. 33, no. 6, pp. 569-580.

[27] Paes, A., Zaverucha, G. & Costa, V. (2017). On the use of stochastic local search techniques to revise first-order logic theories from examples. Machine Learning, vol. 106, no. 2, pp. 197-241.

[28] Bruynooghe, M., Denecker, M. & Truszczyński, M. (2016). First-order logic with inductive definitions for model-based problem solving. AI Magazine, vol. 37, no. 3, pp. 69-80.

[29] Figueira, S., Gorín, D. & Grimson, R. (2014). Independence friendly logic with classical negation via flattening is a second-order logic with weak dependencies. Journal of Computer and System Sciences, vol. 80, no. 6, pp. 1102-1118.

[30] Kelk, S., Ierse, L. van, Scornavacca, C. & Weller, M. (2016). Phylogenetic incongruence through the lens of monadic second order logic. Journal of Graph Algorithms and Applications, vol. 20, no. 2, pp. 189-215.

[31] Elberfeld, M. G. M. T. T. (2016). Where first-order and monadic second-order logic coincide. ACM Transactions on Computational Logic, vol. 17, no. 4, p. 25.

[32] Masterman, M. (1961). Semantic Message Detection for Machine Translation, using an Interlingua, International Conference on Machine Translation of Languages and Applied Language Analysis, Teddington, UK.

[33] Knaus, T., Burns, C., Kamps, J. & Foundas, A. (2017). Atypical activation of action-semantic network in adolescents with autism spectrum disorder. Brain and Cognition, vol. 117, pp. 57-64.

[34] Ching, T. & Williams, M. (2017). Association splitting of the sexual orientation-OCD-relevant semantic network. Cognitive Behaviour Therapy, pp. 1-17.

[35] Griffis, J. C., Nenert, R., Allendorfer, J. B., Vannest, J., Holland, S., Dietz, A. & Szaflarski, J. P. (2017). The canonical semantic network supports residual language function in chronic post-stroke aphasia. Human Brain Mapping, vol. 38, no. 3, pp. 1636-1658.

[36] Marván, M., Chrisler, J., Gorman, J. & Barney, A. (2017). The meaning of menarche: A cross-cultural semantic network analysis. Health Care for Women International, pp. 1-12.

[37] Deyne, S. De, Elvevåg, B., Hui, C., Poon, V. & Chen, E. (2016). Rich semantic networks applied to schizophrenia: A new framework. Schizophrenia Research, pp. 454-455.

[38] Gutiérrez-Villar, B., Alcaide-Pulido, P. & Carbonero-Ruz, M. (2017). How do pre-university students see private universities? An exploratory study by combination of natural semantic networks and multidimensional unfolding. Revista de Investigacion Educativa, vol. 35, no. 2, pp. 519-535.

[39] Mosharraf, M., Taghiyareh, F. & Alaee, S. (2017). Investigating elearning research trends in Iran via automatic semantic network generation. Journal of Global Information Technology Management, vol. 20, no. 2, pp. 91-109.

[40] Kwon, K., Bang, C., Egnoto, M. & Rao, H. Raghav. (2016). Social media rumors as improvised public opinion: semantic network analyses of twitter discourses during Korean saber rattling 2013. Asian Journal of Communication, vol. 26, no. 3, pp. 201-222.

[41] Ertek, G., Tokdemir, G., Sevinç, M. & Tunç, M. (2017). New knowledge in strategic management through visually mining semantic networks. Information Systems Frontiers, vol. 19, no. 1, pp. 165-185.

[42] Kim, J., Yoo, M., Lee, K. & Seo, H. (2017). The innovation of the internet: a semantic network analysis of the Internet of Things. Asian Journal of Technology Innovation, vol. 25, no. 1, pp. 129-139.

[43] Agrawal, R., Imielinski, T. & Swami, A. (1993). Mining Association Rules between Sets of Items in Large Databases, Proceedings of the 1993 ACM SIGMOD Conference, Washington.

[44] Agrawal, R. & Srikant, R. (1994). Fast Algorithms for Mining Association Rules, International Conference on Very Large Data Bases, San Francico.

[45] Sowa, J. F. (1976). Conceptual Graph for for a Data Base Interface. IBM Journal of Research and Development, vol. 20, no. 4, pp. 336-357.

[46] Achsan, H. T. Y. & Barcah, Q. K. (2015). Indonesia Coal Trade System: A knowledge-based application software. Energy Procedia, vol. 100, no. C, pp. 622-629.

[47] Ruiz, J., Arjona, M., Maña, A. & Rudolph, C. (2017). Security knowledge representation artifacts for creating secure IT systems. Computers and Security, vol. 64, pp. 69-91.

[48] Kobrinskii, B. (2016). Approaches to the construction of cognitive linguistic–image models of knowledge representation for medical intelligent systems. Scientific and Technical Information Processing, vol. 43, pp. 289-295.

[49] Achsan, H. T. Y. (2011). Web mining based on user profile and preferences, International DAAAM Symposium, Vienna.

[50] Fang, W., Guo, Y., Liao, W. & Wang, F. (2016). Knowledge representation and annotation method based on ontology for complex products' design. Jisuanji Jicheng Zhizao Xitong/Computer Integrated Manufacturing Systems, CIMS, pp. 2063-2071.

[51] Gruber, T. R. (1993). Toward Principles for the Design of Ontologies Used for Knowledge Sharing. International Journal Human-Computer Studies, vol. 43, pp. 907-928.

[52] Ghose, A. (1976). Semantics of lisp and the ontology of functions. Kybernetes, vol. 5, no. 2, pp. 97-99.

[53] Ghose, A. (1978). Remarks on theories of computation and semantics of programming languages. Kybernetes, vol. 7, no. 4, pp. 301-304.

[54] Guarino, N. (1995). Formal ontology, conceptual analysis and knowledge representation. International Journal of Human-Computer Studies, vol. 43, pp. 625-640.

[55] O'Leary, D. E. (1998). Using AI in knowledge management: Knowledge bases and ontologies. IEEE Intelligent Systems and Their Applications, vol. 13, no. 3, pp. 34-39.

[56] Gruber, T. R. (1993). A Translation Approach to Portable Ontology Specification. Knowledge Acquisition, vol. 5, pp. 199-220.

[57] Musen, M. A. (1992). Dimensions of knowledge sharing and reuse. Computers and Biomedical Research, vol. 25, pp. 435-467.

[58] Zaino, J. Schema.Org: The Fire's Been Lit. 15 September 2014, Available from: http://www.dataversity.net/schema-org-fires-lit/ Accessed 2017-August-27.

[59] Cardoso, J. (2007). The Semantic Web Vision: Where are We?. IEEE Intelligent Systems, pp. 22-26.

[60] Wikipedia. Scopus, Available from: https://en.wikipedia.org/wiki/Scopus Accessed 2017-08-28.

[61] OCLC. WorldCat. OCLC Accessed 2017-08-28.

[62] Achsan, H. T. Y. & Wibowo, W. C. (2014). A fast distributed focused-web crawling. Procedia Engineering, vol. 69, pp. 492-499.

[63] Faraj, S. & Sproull, L. (2000). Coordinating Expertise in Software Development Teams. Management Science, pp. 1554 - 1568.

[64] R.Puerta, H. E. A. & A.Musen, M. (1994). Generation of knowledge-acquisition tools from domain ontologies. International Journal of Human-Computer Studies, vol. 41, no. 3, pp. 425-453.

[65] Shi, L., Li, S., Yang, X., Qi, J., Pan, G. & Zhou, B. (2017). Semantic Health Knowledge Graph: Semantic Integration of Heterogeneous Medical Knowledge and Services. BioMed Research International, vol. 2017.

[66] Frank E. & Paynter, G. W. (2004). Predicting Library of Congress Classifications From Library of Congress Subject Headings. Journal of The American Society for Information Science and Technology.

[67] Ávila-Argüelles, R., Calvo, H., Gelbukh, A. & Godoy-Calderön, S. (2010). Assigning Library of Congress Classification codes to books based only on their titles. Informatica (Ljubljana), vol. 34, no. 1, pp. 77-84.

[68] Larson, R. R. (1992). Experiments in Automatic Library of Congress Classification. Journal of American Society for Information Science, pp. 130-148.

DOI: 10.2507/28th.daaam.proceedings.163

OBJECT RECOGNITION IN IMAGES

Adámek Milan, Neumann Petr, Pospíšilík Martin,
Španko Adrian, Vašek Vladimír

This Publication has to be referred as: Adamek, M[ilan]; Neumann, P[etr]; Pospisilik, M[artin]; Spanko, A[drian] & Vasek, V[ladimir] (2017). Object Recognition in Images, Proceedings of the 28th DAAAM International Symposium, pp.1178-1184, B. Katalinic (Ed.), Published by DAAAM International, ISBN 978-3-902734-11-2, ISSN 1726-9679, Vienna, Austria
DOI: 10.2507/28th.daaam.proceedings.163

Abstract

Perception of image perceptions is one of the most important features of every person. With the help of our eyesight, we can perceive the world around us, acquiring information about our surroundings and recognize different objects in them. Thanks to the recipes of our eye, we can capture it clear information and transferred information to the brain where this information is further processed and evaluated. Based on that, we can recognize what we see.The detection of various objects or shapes in an image is a very important element of all operations involving Image Processing. It is not only used in the processing of Medical Imagery data, (e.g. searching for structures in tomographic images); but also, in the Security field, (e.g. access control); as well as in the human sector – i.e. in cameras and cameras, (e.g. face detection, smiles, etc.). This article deals with possible detection methods of objects in images.

Keywords: image processing; neural network; Matlab; pre-processing; training data

1. Introduction

Computer Vision Systems are used in many areas of human activities. In the Security Technology field, it is possible to detect objects; or, eventually, even to identify individuals, as captured by a security camera. In most cases, this often involves the detection of faces, their physical build, or the identification of a person by means of studying and comparing their physio-motoric mannerisms and manifestations. In the Industrial Production field, one can – with the assistance of video-recording, to assure product quality control; video-recordings are very often components of Robot Control Systems.

Prior to detecting objects in a recorded image, the "captured record" must first be converted from an analogue to a digital form. Before the actual segmentation of the image into sub-segments - the recorded image is very often adjusted in such a way as to make it possible to better search for the object, (that is), being searched for. This adjustment very often includes geometric transformations, brightness transformations, filtration, and focusing.

2. Image Segmentation

Image Segmentation is a set of processes, the course of which leads to the subdivision of the image into its sub-components, which - at least to a certain extent, are related to reality. Mutual Indivisibility is a requisite for all areas of an area in a given set. The degree of matching is sought between the segments and the actual objects that are discovered. Depending on their degree of accord, the segmentation can be further sub-divided into Complete and Partial. It is logical that Partial Segmentation has a lower degree of consistency, unlike Complete Segmentation [1].

An image is obtained in the course of Partial Segmentation, which is divided into several independent parts. The greatest advantage of this type of segmentation is the ability to elaborate even complex scenes; and further, to reduce the volume of the processed data. The result of this type of segmentation is a set of homogeneous regions, which have a certain colour and brightness. The fundamental disadvantage of this type of segmentation is the need to use subsequent steps, which help to acquire the relevant results [1, 4].

Image Segmentation can be broken-down, according to principles used, into three basic methods. The first, is based on an overall general knowledge of images, or their parts. This method is represented with the aid of histograms. The second, exploits algorithms that specify the algorithms between objects. The third, is based on the algorithms that these areas create [3, 6].

2.1. Object Description

It is possible to describe an object in an image in two ways. The first, is based on the Quantitative approach - this approach uses a set of numerical characteristics. As a rule, under the term "set of numerical characteristics", it is possible to consider and evaluate the object size, colour scattering. or comparability. The second, is to describe the object using a Qualitative approach. This approach to description is based on the description of the relationships between selected objects; and describes their shape properties and characteristics. When comparing and recognising objects, these descriptions are considered as input information [2, 5].

2.2. Classification

The task of classification is to include an object found in the image into a group of known classes. The Classification methods are divided into two groups; Symptomatic and Structural. Symmetric Classification, uses groups of characteristic object numbers, which describe its properties; like, for instance, location or size. Quantitative Description of the object is used for the description of the object. Cluster Analysis can also be considered as a form of Symptom Analysis. This analysis classifies objects into cluster groups in such a way as that objects in a group have identical, or very similar properties/characteristics.

Structural Classification works on the basis of the assignment of certain properties that are characteristic for the given object. This type of classification works with the Qualitative Description of the given object. The properties of the object are, through the intermediary of algorithms subjected to by word-breakdown processing, which describe the object - and subsequent to control, the language, grammar, and alphabet can be defined [7, 8].

3. A Programme Application Designed to Search Objects in Images

The Matlab programme environment was used for the creation of a programme designed for searching for an object in an image. The following Image/Fig. (1), shows how to create a GUI in Matlab.

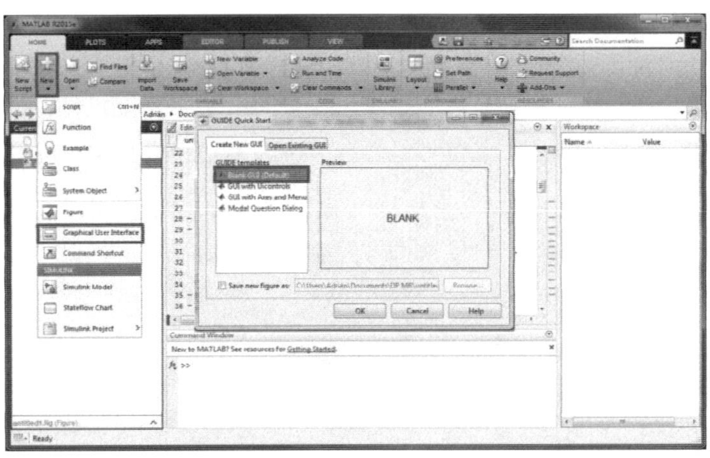

Fig. 1. Creating a graphic user environment in the Matlab programme

4. Programme Application Functions

4.1. Loading/Reading Images

The recorded image is loaded/read in this application by using the button, which uses the loadBotton_Callback function. It is necessary to define the place where the image is to be opened; on a standard basis, in a new window. The performance of the above steps is accomplished through the intermediary of the following function:

```
function loadButton_Callback ( hObject , eventdata , handles )
[ filename , pathname ] = uigetfile({'*.bmp' ; '*.jpg' ; '*.gif' ; ' * .*'},
'Vyber obrázok ');
S~= imread ( [ pathname , filename ] ) ;
uiwait (msgbox ( ' Trénovacie dáta načítané ! ' , 'Dáta načítané ! ' ) ) ;
axes ( handles . mainAxes ) ;
imshow(S ) ;
handles.S~= S ;
set ( handles.vyberButton , 'enable' , ' on ' ) ;
guidata ( hObject , handles ) ;
```

The following function can be used to record/capture images:

```
kamButton_Callback:
function kamButton_Callback ( hObject , eventdata , handles )
vid = videoinput ( ' winvideo ' , 1 , 'RGB24 640x480 ' ) ;
axes ( handles . mainAxes ) ;
vidRes = get ( vid, ' VideoResolution ' ) ;
nBands = get ( vid, 'NumberOfBands ' ) ;
hImage = image( zeros ( vidRes ( 2 ) , vidRes ( 1 ) , nBands ) ) ;
```

4.2. Selecting and Cropping Images

The Object Selection function initiates the process of selecting an object in an image, in such a way as one would know what object to search for. This function can also be used in the case where there are multiple objects in the image. The Display a Selection Rectangle function serves to indicate the object. The following functions can be used for selection purposes:

```
function vyberButton_Callback ( hObject , eventdata , handles )
S~= handles.S ;
axes ( handles.mainAxes ) ;
if isfield ( handles , 'api ' )
handles . api . delete ( ) ;
rmf ield ( handles , ' api ' ) ;
rmf ield ( handles , ' hRect ' ) ;
axes ( handles.mainAxes ) ;
imshow(S ) ;
end
axes ( handles . mainAxes ) ;
s z = size (S ) ;
handles.hRect = imrect (gca , [ round( sz ( 2 ) / 2 ) round( sz ( 1 ) / 2 ) 50 50 ] ) ;
handles.api = iptgetapi ( handles . hRect ) ;
set ( handles.orezButton , ' enable ' , ' on ' ) ;
guidata ( hObject , handles ) ;
```

4.3. Pre-processing

Another function is the Pre-processing feature, i.e. Image Pre-processing. After starting it, the Boot_Callback command is called. After recording and loading the cropped image, this is converted into shades of grey. Thereby, colour tones and saturation are removed; but image brightness is conserved/retained. This is followed by the "thresholding" of the image using the graythresh and im2bw commands. This approach encapsulates the following features:

```
function pripravaButton_Callback ( hObject , eventdata , handles )
img crop = handles.img crop ;
imgGray = rgb2gray ( img_crop ) ;
prah=graythresh ( imgGray ) ;
bw = im2bw( img_crop,prah ) ;
```

This is followed by an automatic cropping feature, which crops the object right up to its limits in such a way that there are no faint surfaces on it. This is accomplished by the following script:

```
bw2 = auto_orez (bw) ;
axes ( handles.mainAxes ) ;
imshow(bw2 ) ;
handles.bw2 = bw2 ;
```

This is followed by cycles for detecting white spots in the image. The cycles function on the principle of the sum of the values of the elements in the matrix columns. The white space search cycles are as follows: The following are the cycles for detecting the white spots in the image. Cycles work on the sum of values of the elements in the matrix columns. The white gap space search cycles are as follows:

```
pocB=1;
while (sum(bw( : , pocB))==y2pom)
x1=x1+1;
pocB=pocB+1;
end
pocB=1;
while (sum(bw(pocB, : ) )==x2pom)
y1=y1+1;
pocB=pocB+1;
end
pocB=x2pom;
while (sum(bw( : , pocB))==y2pom)
x2=x2-1;
pocB=pocB-1;
end
pocB=y2pom;
while (sum(bw(pocB, : ) )==x2pom)
y2=y2-1;
pocB=pocB-1;
end
```

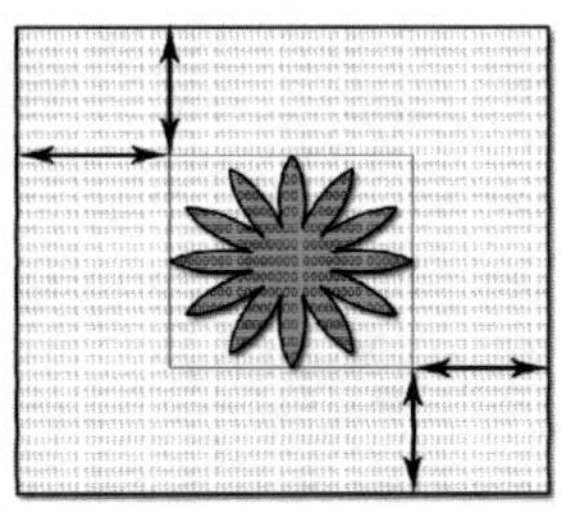

Fig. 2. Searching for White Spaces and the Creation of an Automatic Clipping of Sectors.

Subsequently, using the imcrop command, the final version of the crop is realised, and the image is the subsequent output of the given script. The cropping function is as follows:

```
bw2=imcrop (bw, [ x1 , y1 , ( x2-x1 ) , ( y2-y1 ) ] )
```

The NNdata variable creates a matrix of an image, which must be the same size for Neural Network Training purposes. When inserting the image into a database, the image is rotated by 360 °; and the rotations take place every 45°.

Each rotation is saved as a new image, the reason is for improving the learning of the neural network purposes. The whole function appears as follows:

```
if get ( handles.radiotren , ' value ' ) == 1
for i ~= 1 : 8
imshow(bw2)
pause ( 1 ) ;
handles.NNdata ( : , : , handles.pocobj )=imresize (bw2 , [ 80 , 80 ] ) ;
bw2 = (bw2 == 0 ) ;
bw2=imrotate (bw2 , 4 5 , ' bilinear ' ) ; bw2 = (bw2 == 0 ) ;
bw2 = auto_orez (bw2 ) ;
handles.pocobj=handles.pocobj+1;
end
else
handles.NNdata ( : , : , handles.cnt )=imresize (bw2 , [ 80 , 80 ] ) ;
end
handles.cnt=handles.cnt+1;
```

4.4. Training Neuron Networks

After processing and inserting the objects into a database, one can begin to train a neural network. The *treningButton_Callback* function can be used to train a neural network:

```
function treningButton_Callback ( hObject , eventdata , handles )
hasField = isfield ( handles , 'NNdata ' )
if hasField
NNdata = handles.NNdata ;
[ NNinput NNtarget ]=traindata (NNdata , handles.cnt , handles.pocobj ) ;
```

The "traindata" script contains two parameters at input; the first is the NNdata variable, (which contains the matrix of embedded images) - and the other is the total number of objects in the database. The script is composed of three cycles. The first distinguishes individual image matrices; the second determines the processing of the lines; and the third controls the processing of the columns in the given row. In essence, this functions by moving the whole matrix, sequentially, into blocks - and counting all of the values in each block. The dimensions of the matrix determine the total number of learning objects, (i.e. number of columns), and number of object types, i.e. (number of rows). Processing the matrix block by block, is shown in the following figure.

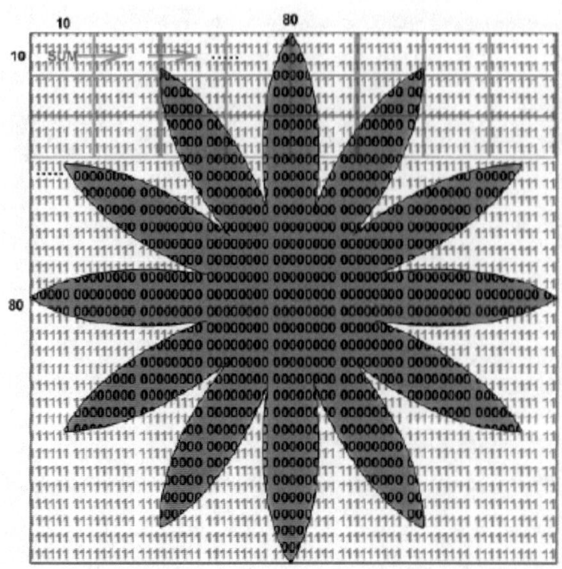

Fig. 3. Processing the matrix block by block [8]

The script entitled "*createnn*" creates the neural network and trains it in using the input data:

```
[ net , tr ] = creatnn (NNinput , NNtarget , handles.cnt )
handles . net=net ;
```

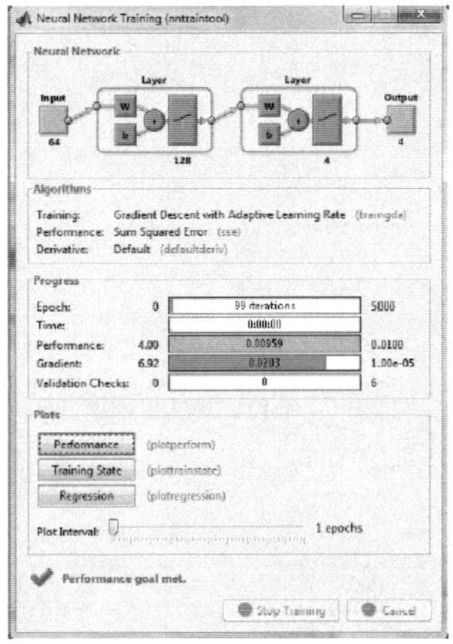

Fig. 4. The Neural Network Training Tool

5. Object Recognition

When using the designed GUI, one needs to first select the Recognition Mode. The "*radiotest_Callback*" function will then start. The function verifies whether there are NNdata and net variables, i.e. whether there are database objects and a neural network. Recognition Mode. Here is the function for switching over into the object recognition regime:

```
function radiotest_Callback ( hObject , eventdata , handles )
set ( handles.radiotest , ' value ' , 1 ) ;
set ( handles.radiotren , ' value ' , 0 ) ;
set ( handles.radiotren , ' enable ' , ' off ' ) ;
ha sField = isfield ( handles , 'NNdata ' )
hasField2 = isfield ( handles , ' net ' )
i f hasField && hasField2
handles = rmf ield ( handles , 'NNdata ' )
handles.cnt=1;
set ( handles.treningButton , ' enable ' , ' off ' ) ;
set ( handles.ulozbutton , ' enable ' , ' off ' ) ;
set ( handles.nacitajbutton , ' enable ' , ' off ' ) ;
else
warningMessage = sprintf ( ' Varovanie : Rozpoznávací režim je neprístupný,
najskôr uskutočni tréning siete. ' ) ;
uiwait ( warndlg ( warningMessage ) ) ;
set ( handles.radiotren , ' enable ' , ' on ' ) ;
```

6. Saving and Retrieving Training Data

Data from the training set can be saved using the Save function, which is executed by running the "*ulozButton_Callback*" function:

```
function ulozButton_Callback ( hObject , eventdata , handles )
state .NNdata= getfield ( handles , 'NNdata ' )
state. cnt= getfield ( handles , ' cnt ' )
state. pocobj= getfield ( handles , ' pocobj ' )
state. menaObj= getfield ( handles , ' menaObj ' )
[ filename , pathname ] = uiputfile ({' *.mat ' }, 'Vyber objekty ' ) ;
save ( [ pathname , filename ] , ' state ' )
uiwait (msgbox ( ' Trénovacie dáta sú uložené ! ' , 'Dáta uložené ! ' ) ) ;
guidata ( hObject , handles ) ;
```

In order to retrieve data from the training set, the Data Retrieval function was designed – which uses the following function:

```
function nacitajButton_Callback ( hObject , eventdata , handles )
[ filename , pathname ] = uigetfile ({ ' *.mat ' } , 'Vyber objekty ' ) ;
load ( [ pathname , filename ] , ' state ' )
uiwait (msgbox ( ' Trénovacie dáta sú načítané ! ' , 'Dáta načítané ! ' ) ) ;
handles .NNdata=state .NNdata ;
```

7. Creating a Standalone Executable Application

You can use the "mcc" command to create a standalone executable application. Parameter "m" in the "mcc" command creates a standalone executable application from the "main.m" file.

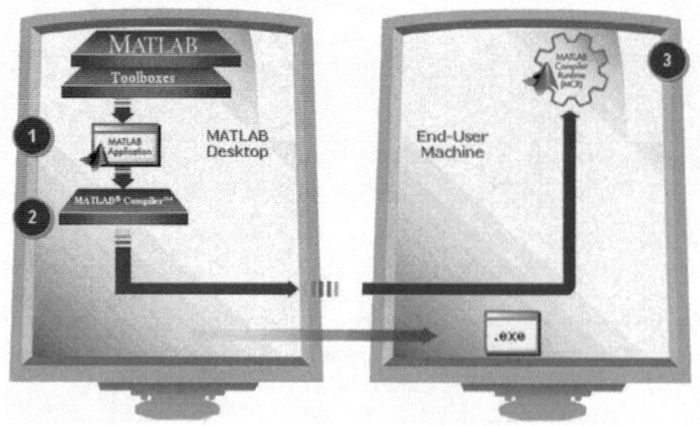

Fig. 5. Schema of the creation of a standalone executable application [8]

8. Conclusion

In the course of creating a programme, it is also necessary to take possible error events into account. These may be caused by incorrect handling of the programme application. The majority of states that could lead to the origin of error events are resolved by means of a GUI that does not allow some buttons to run in the programme. Sin view of the fact that it is not possible to resolve all the situations by means of the buttons in the GUI, the programme application also has error messages. An example of this is, for example, an error message that appears in the case where a user starts the recognition mode without prior training of the neural network.

9. Acknowledgments

This work was supported by the Ministry of Education, Youth and Sports of the Czech Republic within the National Sustainability Programme project No. LO1303 (MSMT-7778/2014) and also by the European Regional Development Fund under the project CEBIA-Tech No. CZ.1.05/2.1.00/03.0089.

10. References

[1] Gonzales, R.; Woods, R. (2002). Digital Image Processing. 2nd edition. Upper Saddle River, New Jersey 07458 : Prentice-Hall, 2002. 793 p. ISBN 0-20-118075-8

[2] Šonka, M,; Hlaváč, V. & Boyle, R. (2008). Image processing, analysis, and machine vision. 3rd ed.. Toronto: Thomson. 829 p. ISBN 978- 0-495-08252-1.

[3] Parker J. R.(2011) Algorithms for Image Processing and Computer Vision. Wiley Publishing, Indianapolis. ISBN 978–0–470–64385–3.

[4] Dobeš, M. (2008). Image processing and algorithms in C#. Prague. BEN. 144 p. ISBN 978-80-7300-233-6.

[5] Hlaváč, V. & Sedláček, M. (2005). Processing of signals and images. Prague. 255 p. ISBN 80-010-3110-1.

[6] Říha, K.(2007). Advanced image processing techniques. Brno. 109 p.

[7] Karban, P. (2006). Calculations and simulations in Matlab and Simulink. Brno, Computer Press. 220 p. ISBN 80-251-1301-9.

[8] Žára, J. (2004). Modern computer graphics. Praha. Computer Press. p. 542-546. ISBN 80-251-0454-0.

DOI: 10.2507/28th.daaam.proceedings.164

EMOTION RECOGNITION IN VIDEO WITH OPENCV AND COGNITIVE SERVICES API: A COMPARISON

Luis Antonio Beltrán Prieto & Zuzana Komínková Oplatková

This Publication has to be referred as: Beltran Prieto, L[uis] A[ntonio] & Kominkova Oplatkova, Z[uzana] (2017). Emotion Recognition in Video with OpenCV and Cognitive Services API: A Comparison, Proceedings of the 28th DAAAM International Symposium, pp.1185-1190, B. Katalinic (Ed.), Published by DAAAM International, ISBN 978-3-902734-11-2, ISSN 1726-9679, Vienna, Austria
DOI: 10.2507/28th.daaam.proceedings.164

Abstract

Emotions are people's reactions to certain stimuli. Most common way to detect an emotion is by facial expression analysis. Machine learning algorithms combined with other artificial intelligence techniques have been developed in order to identify expressions found in images and videos. Support Vector Machines, along with Haar Cascade classifiers can be used for efficient emotion recognition. OpenCV, an open-source library for machine learning, makes it possible to develop computer-vision applications. Cognitive Services is a free set of APIs which easily integrate artificial intelligence in applications. In this paper a comparison between two implementations of Emotion Recognition algorithms, namely SVM and Cognitive Services API, was carried out to compare their performance. For this research, 500 tests were performed per experiment. The SVM implementation in OpenCV obtained the best performance, with an 84% accuracy, which can be boosted by increasing the sample size per emotion.

Keywords: Support Vector Machine; OpenCV; Cognitive Services; Face Detection; Emotion Recognition; Haar Cascades

1. Introduction

Facial emotion detection can be defined as the process of recognizing the feeling that a person is expressing at a particular moment. Potential applications of emotion recognition include the improvement of student engagement [1], the built of smart health environments [2], the analysis of customers' feedback [3], and the evaluation of quality in children's games [4], among others. Face recognition within multimedia elements, such as images and videos, has been one of the challenges in the artificial intelligence field. Several powerful techniques have been examined exhaustively in the search of performance and accuracy improvement, for instance, Convolutional Neural Networks (CNN) [5], Deep Belief Networks (DBN) [6], and Support Vector Machines (SVM) [7], just to name a few. CNN and DBN are deep learning techniques. Deep Learning is a novel area of research in machine learning which focuses on learning high-level representations and abstractions of data, such as images, sound, and text by using hierarchical architectures, including neural networks, convolution networks, belief networks, and recurrent neural networks in several artificial intelligence areas, some of which are image classification [8], speech recognition [9], handwriting recognition [10], computer vision [11], and natural language processing [12].

Identifying the sentiment expressed by a person is one of the outcomes after achieving face detection. Recent research [13] has proven that emotion recognition can be accomplished by implementing machine learning and artificial intelligence algorithms. To assist in this task, several open-source libraries and packages, being OpenCV, TensorFlow, Theano, Caffe and the Microsoft Cognitive Toolkit (CNTK) some of the most notorious examples, cut down the process of building deep-learning-based algorithms and applications. Emotions such as anger, disgust, happiness, surprise, and neutrality can be detected.

The aim of this paper is to compare the performance of two emotion-recognition implementations from video sources. The first one is a Python-based application which uses OpenCV libraries with Support Vector Machine. The second implementation is a C# application which sends requests to the Emotion recognition API from Cognitive Services. 8000 facial expressions from the Radboud Faces Database were examined in different phases of the experiments for training and evaluation purposes.

This paper is organized as follows. Background information introducing emotion recognition, Support Vector Machines, OpenCV, Cognitive Services, and the Radboud Faces Database is presented firstly. Afterwards, the problem solution is described by explaining the methods and methodology that were used for this comparison. Evaluation results are shown subsequently. Finally, conclusions are discussed at the final section of the paper.

2. Background information

2.1. Emotion recognition

Emotions are strong feelings about people's situations and relationships with others. Most of the time, humans show how they feel by using facial expressions. Speech, gestures, and actions are also used to describe a person's current state. Emotion recognition can be defined as the process of detecting the feeling expressed by humans from their facial expressions, such as anger, happiness, sadness, deceitfulness, and others. Even though a person can automatically identify facial emotions, machine learning algorithms have been developed for this purpose. Emotions play a key role in decision-making and human-behaviour, as many actions are determined by how a person feels at some point.

Typically, these algorithms use either a picture or a video (which can be considered as a set of images) as input, then they proceed to detect and focus their attention on a face and finally, specific points and regions of the face are analysed in order to detect the affective state. Machine Learning algorithms, methods and techniques can be applied to detect emotions from a picture or video. For instance, a deep learning neural network can perform effective human activity recognition with the aid of smartphone sensors [14]. Moreover, a classification of facial expressions based on Support Vector Machines was developed for spontaneous behaviour analysis [15].

2.2. Support Vector Machines

Support Vector Machines were introduced [16] as a technique aimed to solve binary classification problems; due to their solid theoretical fundaments, SVMs have been used to answer regression, clustering and multi-classification tasks [17] along with practical applications in several fields, including computer vision [18], text classification [19], and natural language processing [20], among others. It can be defined as a discriminative classifier which works with labelled training data to output an optimal hyperplane used to categorize new examples.

While several learning techniques focus on minimizing the error rate generated by the model based on the training samples, SVMs attempt to minimize the so-called structural risk. The idea is to choose a separation hyperplane which is equidistant from the nearest examples of each class in order to obtain a maximum margin from each side of the hyperplane. Furthermore, when defining the hyperplane, only those training samples which are near the border of these margins are considered. These examples are known as support vectors. From a practical point of view, the maximum margin separating hyperplane has demonstrated to achieve a good generalization capacity, thus avoiding the overfitting of the training set.

Given a set of separable samples $S = \{(x_1, y_1), \ldots, (x_n, y_n)\}$, where $x_i \in \mathbb{R}^d$ and $y_i \in \{+1, -1\}$, a separation hyperplane, as shown in Fig. 1, can be defined as a linear function capable of split both sets without errors, according to (1), where w and b are real coefficients.

$$D(x) = (w_1x_1 + \cdots + w_dx_d) + b = < w, x > +b \quad (1)$$

The separation hyperplane meets the constraints expressed in (2) for all x_i from the examples set.

$$\begin{cases} < w, x_i > +b \geq 0 & if \quad y_i = +1 \\ < w, x_i > +b \leq 0 & if \quad y_i = -1, i = 1, \ldots, n \end{cases} \quad (2)$$

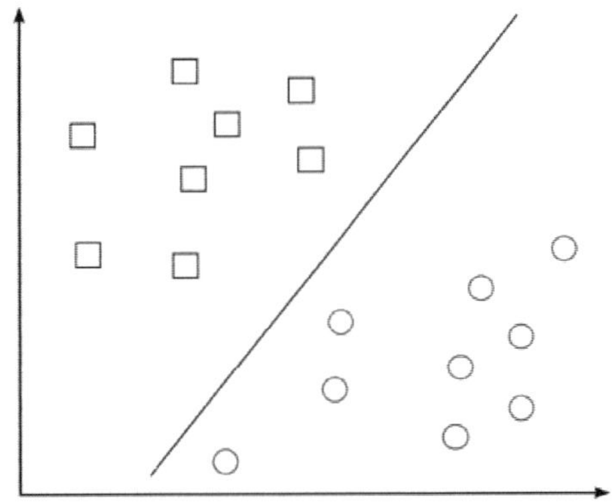

Fig. 1. Separation hyperplane in a bidimentional space from a set of examples separable in two classes

2.3. OpenCV

OpenCV [21] is a free, yet powerful, open-source library developed by Intel Corporation which has been widely used in computer vision and machine learning tasks, such as image processing, real-time image recognition, and face detection. With more than 2500 optimized algorithms included, this library has been extensively used for research and commercial applications from both global and small entrepreneurs. OpenCV contains an optimized set of libraries written in C language, with bindings to other languages and technologies, including Python, Android, iOS, and CUDA (for GPU fast processing), and wrappers in other languages, such as C#, Perl, Haskell, and others. Moreover, it works under Windows and Linux.

2.4. Cognitive Services

Cognitive Services [22] are a set of machine learning algorithms developed by Microsoft which are able to solve artificial intelligence problems in several fields, such as computer vision, speech recognition, natural language processing, machine learning search, and recommendation systems, among others. These algorithms can be consumed through Representational State Transfer (REST) calls over an Internet connection, allowing developers to use artificial intelligence research to solve problems. These services are open-source and can be consumed by many programming languages, including C#, PHP, Java, Python, and implemented in desktop, mobile, console, and web applications.

The Computer Vision API of Cognitive Services provides access to machine learning algorithms capable of performing image processing tasks. Either an image stream is uploaded or an image URL is specified to the service so the content can be analyzed for label tagging, image categorization, face detection, color extraction, text detection, and emotion recognition. Video is also supported as input. The Emotion API analyses the sentiment of a person in an image or video and returns the confidence level for eight emotions mostly understood by a facial expression, including anger, contempt, disgust, fear, happiness, neutrality, sadness and surprise.

2.5. The Radboud Faces Database

The Radboud Faces Database (RaFD) [23] is a high-quality set of pictures of 67 models (between male, female, Caucasian, Moroccan, children and adults) displaying 8 emotional expressions (anger, disgust, fear, happiness, sadness, surprise, contempt, and neutrality) in which each emotion was shown looking to three directions (left, front, and right), with five camera angles, as presented in Fig. 2. In total, the database contains 28709 faces. This initiative by the Behavioural Science Institute of the Radboud University Nijmegen is available for research purposes upon request.

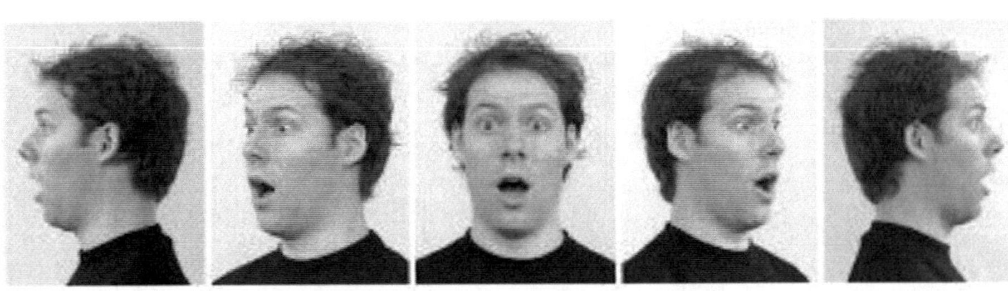

Fig. 2. Sample images from the RaFD

3. Methods and Methodology

The objective of this experiment is to compare the performance of a couple of emotion-recognition implementations in video. The first analysis (Experiment A) is a Python-based application which makes use of the OpenCV machine learning algorithms combined with Support Vector Machines for facial and emotion detection. The second study (Experiment B) is a C# mobile application which sends requests to a Cognitive Services API for emotion detection. In both cases, the Radboud Faces Database was used as input for the analysis.

For Experiment A, 8000 high resolution sequences which actually show a relevant expression, i.e. from a neutral feeling to the emotion itself, with 1000 images per emotion were taken into consideration. First step is then to obtain both the neutral and emotional faces. From this subset, OpenCV library is used to detect the face on each picture by using a custom Haar-filter. A successful object detection is possible because of the Haar feature-based cascade classifiers proposed in [24], which is an approach based on machine learning which involves a training from both positive and negative images, i.e. pictures with faces and without them, respectively. OpenCV already includes several Haar-filter libraries. Thereafter, all images were standardized by converting them to grayscale and resized to the same dimensions, an array of 48x48 grey-scale images. A Support Vector Machine was used in the training process of the classifier, consisting of the implementation of a non-linear support vector classification model with kernel radial basis function. The training process consists of getting the characteristics of each face along with the labelled emotion expressed by the person. Then, evaluation of the classifier proceeds by comparing the outcome of its predict function of each face with the actual labelled emotion. Fig. 2 presents the output of a video analysis of this experiment.

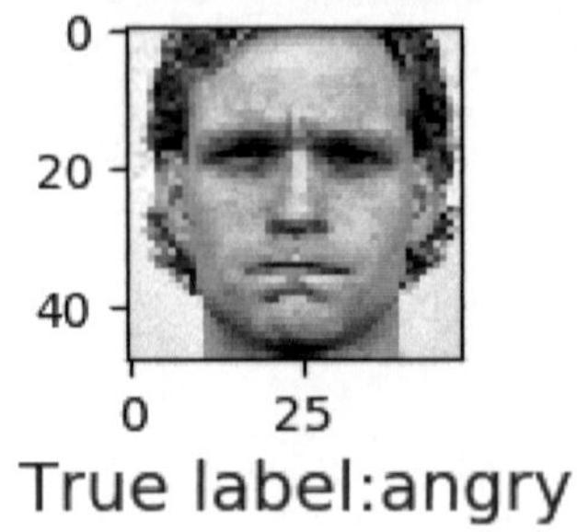

Fig. 3. Analysis of emotions detected on a video by the SVM implementation with OpenCV.

Experiment B starts with the 8000 images obtained from the dataset. 10 random groups consisting each of 500 sequences were generated for evaluation purposes. For each test, every video was submitted to the Cognitive Services API for its evaluation. A C# mobile application was developed with Xamarin cross-platform technology in order to execute this experiment. The service returns the information in a JSON-based format, containing the score for each emotion. The highest score means the facial expression detected by the service, and this result was compared with the actual labelled emotion for evaluation purposes. Fig. 4 shows the application developed in C# for this experiment. It extracts a picture from a video stream, then finds a face on it and finally sends a request to the service in order to detect the emotions expressed by the person. The analysis is performed by the Emotion Cognitive Service Video API.

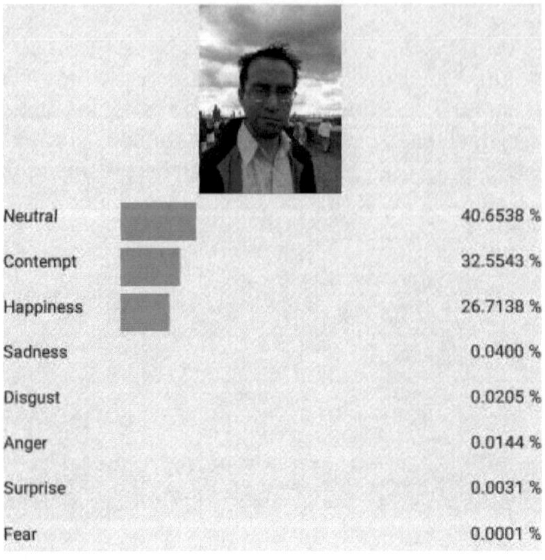

Fig. 4. Analysis of emotions detected on a video by the Emotion Cognitive Service.

4. Results

After running each of the 10 tests from Experiment A, the results which are presented in Table 1 were obtained. An average of 84.02 % correct predictions was calculated as an outcome. Likewise, Table 2 illustrates the outcome of each test in Experiment B. As a result, a 68.93 % average efficiency was accomplished after running 10 tests of this implementation.

Test Number	Correct predictions (%)	Incorrect predictions (%)
1	405 (81.00 %)	95 (19.00 %)
2	430 (86.00 %)	70 (14.00 %)
3	407 (81.40 %)	93 (18.60 %)
4	416 (83.20 %)	84 (16.80 %)
5	429 (85.80 %)	71 (14.20 %)
6	428 (85.60 %)	72 (14.40 %)
7	410 (82.00 %)	90 (18.00 %)
8	426 (85.20 %)	74 (14.80 %)
9	418 (83.60 %)	82 (16.40 %)
10	432 (86.40 %)	68 (13.60 %)

Table 1. Evaluation results from Experiment A

Test Number	Correct predictions (%)	Incorrect predictions (%)
1	339 (67.93%)	161 (32.06%)
2	362 (72.51%)	138 (27.48%)
3	328 (65.64%)	172 (34.35%)
4	347 (69.46%)	153 (30.53%)
5	336 (67.17%)	164 (32.82%)
6	343 (68.70%)	157 (31.29%)
7	362 (72.51%)	138 (27.48%)
8	355 (70.99%)	145 (29.00%)
9	321 (64.12%)	179 (35.87%)
10	351 (70.22%)	149 (29.77%)

Table 2. Evaluation results from Experiment B

The findings of the experiments show that the Python-based implementation in OpenCV with Support Vector Machines returned a higher accuracy than the Cognitive Services implementation in C# by approximately a 15% difference. Minor mistakes occurred in Experiment A when trying to predict emotions that are similar, particularly fear and contempt, which were wrongly classified as sadness and neutral, respectively. Likewise, a neutral face most of the time was identified as a sad face.

This behaviour was also found in Experiment B, in which neutral emotions were spotted as either contempt or sadness; taking a look at the scores obtained by the Cognitive Services API, a minimal difference between the incorrect predicted emotion and the real one was detected. Thus, in most cases, the second-best prediction was correct. However, for evaluation purposes of this experiment, this incorrect assumption was considered as a failure.

5. Conclusions

The objective of this experiment was to compare the performance of a couple of different implementations of emotion-recognition applications in faces from videos by using OpenCV and Support Vector Machines in the first case, while considering a C#-based solution which sends requests to a Cognitive Services API for Emotion detection in video for the second solution.

While the first implementation got the best results, the performance could be improved by increasing the sample size of those emotions with few faces, so the training phase gets benefited. Future research will include emotion recognition in faces performed by Deep Learning techniques, including Convolutional Neural Networks and Deep Belief Networks to name a few, which while more complex, are also more suitable for difficult tasks in terms of both performance and computational time.

6. Acknowledgments

This work was supported by the Ministry of Education, Youth and Sports of the Czech Republic within the National Sustainability Programme project No. LO1303 (MSMT-7778/2014) and also by the European Regional Development Fund under the project CEBIA-Tech No. CZ.1.05/2.1.00/03.0089, further it was supported by Grant Agency of the Czech Republic—GACR 588 P103/15/06700S and by Internal Grant Agency of Tomas Bata University in Zlin under the project No. IGA/CebiaTech/2017/004. L.A.B.P author also thanks the doctoral scholarship provided by the National Council for Science and Technology (CONACYT) and the Council for Science and Technology of the State of Guanajuato (CONCYTEG) in Mexico.

7. References

[1] Garn AC, Simonton K, Dasingert T, Simonton A. Predicting changes in student engagement in university physical education: Application of control-value theory of achievement emotions, Psychology of Sport and Exercise, Vol.29, pp. 93-102.

[2] Fernandez-Caballero A, Martinez-Rodrigo A, Pastor JM, Castillo JC, Lozano-Monasor E, Lopez MT, Zangroniz R, Latorre JM, Fernandez-Sotos A. Smart environment architecture for emotion detection and regulation, Journal of Biomedical Informatics, Vol.64 pp-57-73.

[3] Felbermayr A, Nanopoulos A. The Role of Emotions for the Perceived Usefulness in Online Customer Reviews, Jounal of Interactive Marketing, Vol.36, pp. 60-76.

[4] Gennari R, Melonio A, Raccanello D, Brondino M, Dodero G, Pasini M, Torello S. Children's emotions and quality of products in participatory game design, International Journal of Human-Computer Studies, Vol.101, pp. 45-61.

[5] LeCun Y, Bengio Y. Convolutional networks for images, speech, and time-series. The Handbook of Brain Theory and Neural Networks. MIT Press, 1995.

[6] Hinton GE, Osindero S, The Y. A Fast Learning Algorithm for Deep Belief Nets. Neural Computation. Vol. 18, 2016 No. 7 pp. 1527-1554

[7] Chen D, Tian Y, Liu X. Structural nonparallel support vector machine for pattern recognition. In Pattern Recognition, Vol. 60, 2016, pp. 296-305

[8] Zhang Y, Zhang E, Chen W. Deep neural network for halftone image classification based on sparse auto-encoder, In Engineering Applications of Artificial Intelligence, Vol 50, 2016. pp. 245-255

[9] Huang Z, Siniscalchi SM, Lee C. A unified approach to transfer learning of deep neural networks with applications to speaker adaptation in automatic speech recognition, In Neurocomputing, Vol. 218, 2016. pp. 448-459

[10] Elleuch M, Maalej R, Kherallah M. A New Design Based-SVM of the CNN Classifier Architecture with Dropout for Offline Arabic Handwritten Recognition, In Procedia Computer Science, Vol. 80, 2016. pp. 1712-1723

[11] Liu H, Lu J, Feng J, Zhou J. Group-aware deep feature learning for facial age estimation, Pattern Recognition, Vol.66, pp. 82-94.

[12] Bayer AB, Riccardi G. Semantic language models with deep neural networks, In Computer Speech & Language, Vol. 40, 2016, pp. 1-22

[13] Yogesh CK, Hariharan M, Ngadiran R, Adom AH, Yaacob S, Polat K, Hybrid BBO_PSO and higher order spectral features for emotion and stress recognition from natural speech, Applied Soft Computing, Vol.56, pp. 217-232.

[14] Ronao CA, Cho S, Human activity recognition with smartphone sensors using deep learning neural networks, Expert Systems with Applications, Vol.59, pp. 235-244.

[15] Bartlett MS, Littlewort G, Frank M, Lainscsek C, Fasel I, Movellan J, Recognizing facial expression: machine learning and application to spontaneous behavior. 2005 IEEE Computer Society Conference on Computer Vision and Pattern Recognition (CVPR'05). Vol. 2. pp. 568-573

[16] Boser BE, Guyon IM, Vapnik VN. A training algorithm for optimal margin classifiers. In Proceedings of the 5th Annual Workshop on Computational Learning Theory. pp. 144-152

[17] Ding S, Zhang X, An Y, Xue Y. Weighted linear loss multiple birth support vector machine based on information granulation for multi-class classification, In Pattern Recognition, Vol. 67, 2017. pp. 32-46

[18] Cha Y, You K, Choi W. Vision-based detection of loosened bolts using the Hough transform and support vector machines. In Automation in Construction. Vol. 71, Part 2, 2016, pp. 181-188

[19] Ramesh B, Sathiaseelan JGR. An Advanced Multi Class Instance Selection based Support Vector Machine for Text Classification. In Procedia Computer Science. Vol. 57, 2015, pp. 1124-1130

[20] Yang H, Lee S. Robust sign language recognition by combining manual and non-manual features based on conditional random field and support vector machine. In Pattern Recognition Letters. Vol. 34, Issue 16, 2013. pp. 2051-2056.

[21] OpenCV library. http://opencv.org [Online: accessed 01-Jun-2017]

[22] Cognitive Services – Intelligence Applications. http://microsoft.com/cognitive [Online: accessed 03-Jun-2017]

[23] Langner O, Dotsch R, Bijlstra G, Wigboldus DHJ, Hawk ST, van Knippenberg A. (2010). Presentation and validation of the Radboud Faces Database. Cognition & Emotion. Vol. 24 No. 8. pp. 1377-1388.

[24] Turk M, Pentland A. Eigenfaces for Recognition. Journal of Cognitive Neuroscience. Vol.3, No. 1, pp. 71-86.

DOI: 10.2507/28th.daaam.proceedings.165

INTELLIGENT ADVISER MODULE: PROPOSALS AND ADAPTIVE LEARNING CAPABILITIES

Damir Haskovic, Branko Katalinic, Ilija Zec,
Ilya Kukushkin & Alina Zavrazhina

This Publication has to be referred as: Haskovic, D[amir]; Katalinic, B[ranko]; Zec, I[lija]; Kukushkin, I[lya] & Zavrazhina, A[lina] (2017). Intelligent Adviser Module: Proposals and Adaptive Learning Capabilities, Proceedings of the 28th DAAAM International Symposium, pp.1191-1196, B. Katalinic (Ed.), Published by DAAAM International, ISBN 978-3-902734-11-2, ISSN 1726-9679, Vienna, Austria
DOI: 10.2507/28th.daaam.proceedings.165

Abstract

This paper presents the Intelligent Adviser Module (IAM) as a decision support tool for the system operator in Bionic Assembly System (BAS). Here, the system operator is the main human decision maker. To achieve high work efficiency of BAS, he needs to make high quality decisions in limited time and with incomplete information about the actual system state and its components. This time shortage and fragmented information brings stress to the system operator and forces him to make lower quality decisions. In such situations he uses the proposals from the IAM where he can accept, reject or inspect them. These proposals are based on many parameters and also on the digitally recorded data from all shop floor elements. The IAM collects this data and from it produces new system specific knowledge. The accuracy of IAM proposals heavily depends on the validity and usability of such acquired data. A new algorithm is developed and described which enables the IAM to collect, index and verify new data through adaptive learning. As a result, any new shop floor element behaviour can be compensated. This demonstrates the adaptive nature of the IAM as an integral part of the BAS control structure.

Keywords: Intelligent Adviser Module; Bionic Assembly System; hybrid control structure; proposals; adaptive learning

1. Introduction

Current product development has introduced new challenges for modern assembly systems. These include increased product complexity and variety as well as a shorter product lifetime [1]. As an answer, a Bionic Assembly System (BAS) concept based on task distribution, intelligence and self-organization was introduced [2].

BAS represents a next generation of hybrid assembly systems [3]. It is a part of a natural development within Industry 4.0 [4] [13]. BAS is a complex production system. Here, the human system operator is the main decision maker [5]. To achieve functionality and high work efficiency, the system operator must be able to make high quality decisions in limited time with fragmented and incomplete information about the actual system states and its components.

Quality of decisions and time needed to reach them by the system operator are variable, due to his human nature (concentration, stress, fatigue and other) [6] [7] [12]. Because of this, the quality of decisions can be lower, and the time needed to reach them can be longer than objectively possible. To improve that, the Intelligent Manufacturing Systems (IMS) group from Vienna University of Technology has introduced a decision support tool for the system operator. This tool is named as the Intelligent Adviser Module (IAM). As an integral part of BAS control system, the IAM should take into consideration actual system states, past system states, external data from manuals and other documentation, human experts and past system behaviour [8].

All BAS shop floor elements (assembly stations, mobile robots, shop floor operators) produce digital data which is recorded. One of the advantages of the IAM is its ability to use this recorded data to produce and accumulate new system specific knowledge [14]. This makes it possible for the quality of IAM proposals to constantly improve. When the IAM gives more precise proposals, the system operator can make better quality decisions in shorter time and avoid or minimise lost machine time [9].

However, it can oftentimes happen that the individual shop floor elements are replaced with new ones. When a new shop floor element is introduced, the main issue is, how does the IAM compensate for the potential new element behaviour? In other words, how to be sure that the used, already accumulated knowledge is up to date and useful? This is very important because if the IAM is using inadequate or obsolete knowledge, the proposals could be incorrect and cause even more serious problems.

Therefore, the main aim of this paper is to develop an algorithm which enables the IAM to collect, index and verify new data through adaptive learning. As a result, any new shop floor element behaviour can be compensated. The paper starts by describing the main digitally recorded BAS data stream as well as the position of the IAM in contrast to the BAS control structure. Middle part of the paper describes types of IAM proposals and their use by the system operator. The paper concludes with the presented IAM adaptive learning algorithm.

2. BAS data stream

During his shift in BAS, the system operator needs to constantly reach decisions. These decisions are based on fragmented and incomplete information about the actual system states and its components. On one side, the main source of information is based on his communication with other human workers and on the other side, from the audio / visual feedback as shown in Fig. 1.

The main digitally recorded data stream in BAS represents a vast collection of data. It is formed between the control and the controlled systems. This includes all the commands, responses and status messages.

It is impractical and moreover, impossible for the system operator to completely utilize all this data. For this reason, only a small part of this data stream is actively presented to the system operator in real time. The main purpose of this data selection is to give data about actual system states at the system operator's disposal. In order for the IAM to give relevant and useful information to the system operator, the target and the actual state of the system are constantly monitored.

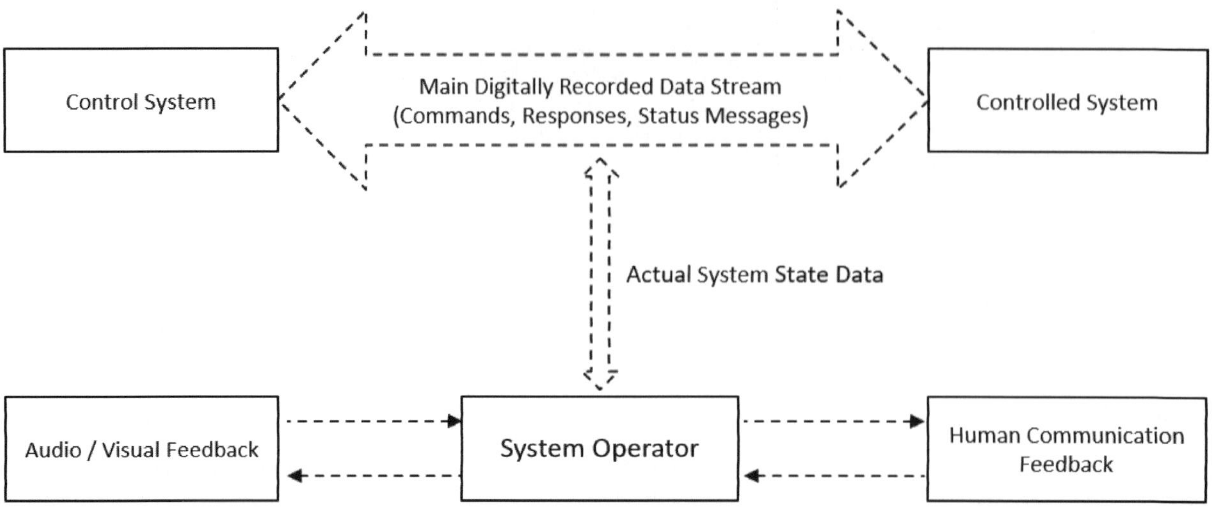

Fig. 1. Main sources of information for the system operator

As started the IAM is an integral part of the BAS control structure. The primary rule is that the IAM must not interfere with the operation of the BAS control system or have any direct influence on it. Therefore, the IAM needs to be implemented exclusively as a data collection system. That means that the data stream has a single direction (read-only) from the BAS control system to the IAM as shown in Fig. 2.

Fig. 2. IAM - Read only access to BAS control system

Otherwise, if the IAM would have the ability to send data directly to the control system, it would stop being a decision support system. and become a decision-making system [10]. This is not the goal of IAM in the context of BAS as a human centric system. On the other side, bidirectional exchange of information (data) takes place between the system operator and the IAM. Here, the system operator can receive feedback information in form of graphs, tables, data and of course, proposals.

3. Types of proposals and their use by the system operator

For the system operator, the most useful outputs from the IAM are proposals. He can use them in various scenarios. Based on problem complexity, information accuracy and knowledge amount, there are two principle types of proposals which can be defined:

- Implicit proposals – If a problem is very complex, or there is insufficient data or the IAM Knowledge Base does not contain any applicable knowledge, the IAM outputs implicit proposals. These can be in form of an overview, statistics, comparisons, etc. The main goal of such proposals is to give an inspiration or an idea to the system operator in which case they can be more generalized and open to interpretation (For example: the battery health of MR1 is at 93% which is higher when compared to the battery health of MR2 which is 15%). This information does not specify what the problem might be exactly, but it might be useful to the system operator to come closer to a solution.

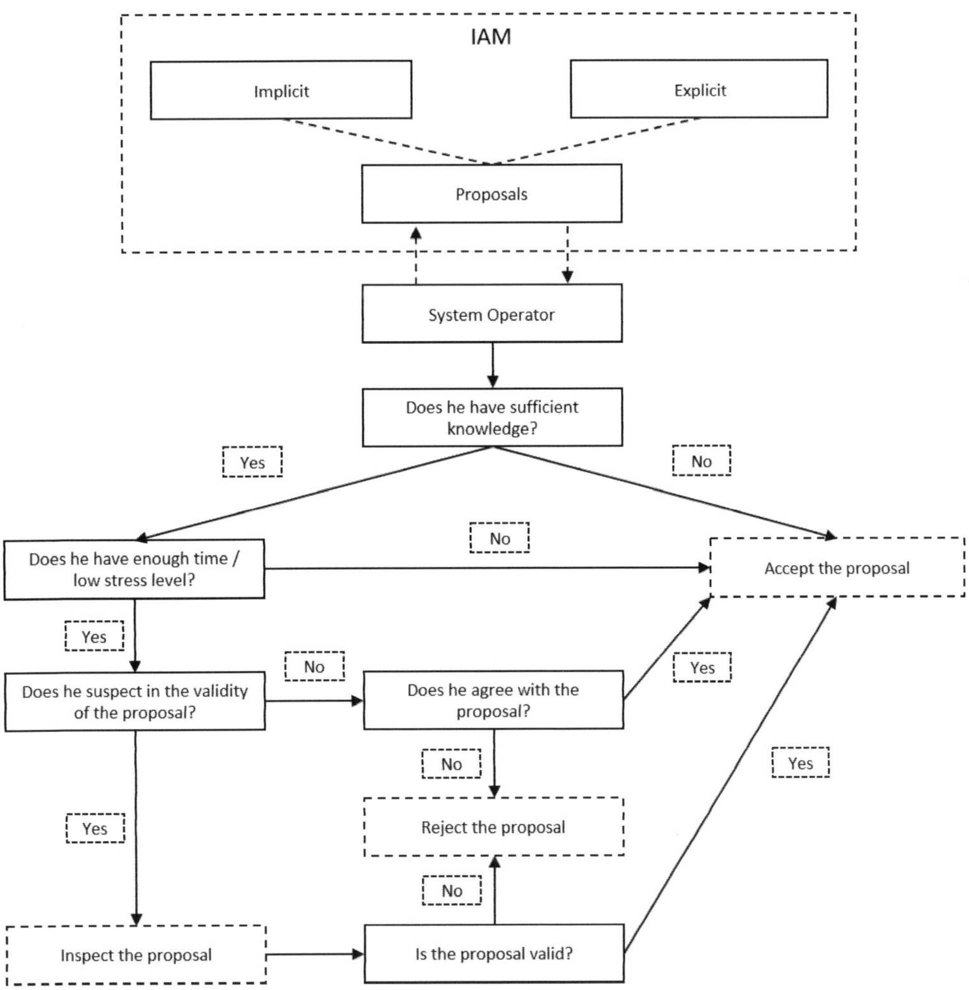

Fig. 3. Types of proposals and their use by the system operator

- Explicit proposals – these proposals are a result of the IAM's applicable Knowledge Base and sufficient information. In this case, the IAM is able to specify what the problem is and how to resolve it. As a result, the system operator is given explicit proposals. These can be in form of necessary steps, precise descriptions or problem definitions. The main goal of such proposals is to free the system operator from active contemplating in cases where he does not have the time, knowledge or will.

Fig. 3. shows the position of the system operator in contrast to proposals. In principle, he can accept, reject or inspect the validity of proposals. Which action he will choose, depends on the situation context and other certain factors. When the system operator is faced with a new problem, the first factor which comes in play is does he have sufficient level of knowledge to recognize and solve it. If no, he does not have any choice but to accept the proposal.

If yes, then the context of the situation plays a vital role. In other words, is he exposed to low stress levels or does he have enough time to contemplate about a decision. If no, he does not have any choice but to accept the proposal. If yes, does he suspect in the validity of the proposal. If no, does he agree with the proposal. If he has the knowledge, time and agrees with the proposal, then naturally he accepts it. On the other hand, if he does not agree, he will reject it.

If the system operator suspects the validity of the proposal, that means that he has sufficient knowledge and time to inspect the proposal. According to the inspection, if the proposal is valid it is accepted and if it is not valid, it is rejected.

4. IAM Adaptive learning

BAS shop floor elements represent a set of comparable hardware and personnel. This includes same or at the very least, similar mobile robots and assembly stations as well as similarly trained shop floor operators. During the execution of BAS working scenarios, these elements can be performing similar tasks respective to their function. Additionally, during their simultaneous operations they can have similar performance and statuses.

During their operational lifetime in BAS, these elements produce digital data which is recorded. This data is used to produce and accumulate new knowledge. However, hardware can be replaced with new type of machines, each shop floor operator can have individual work habits or methods etc.

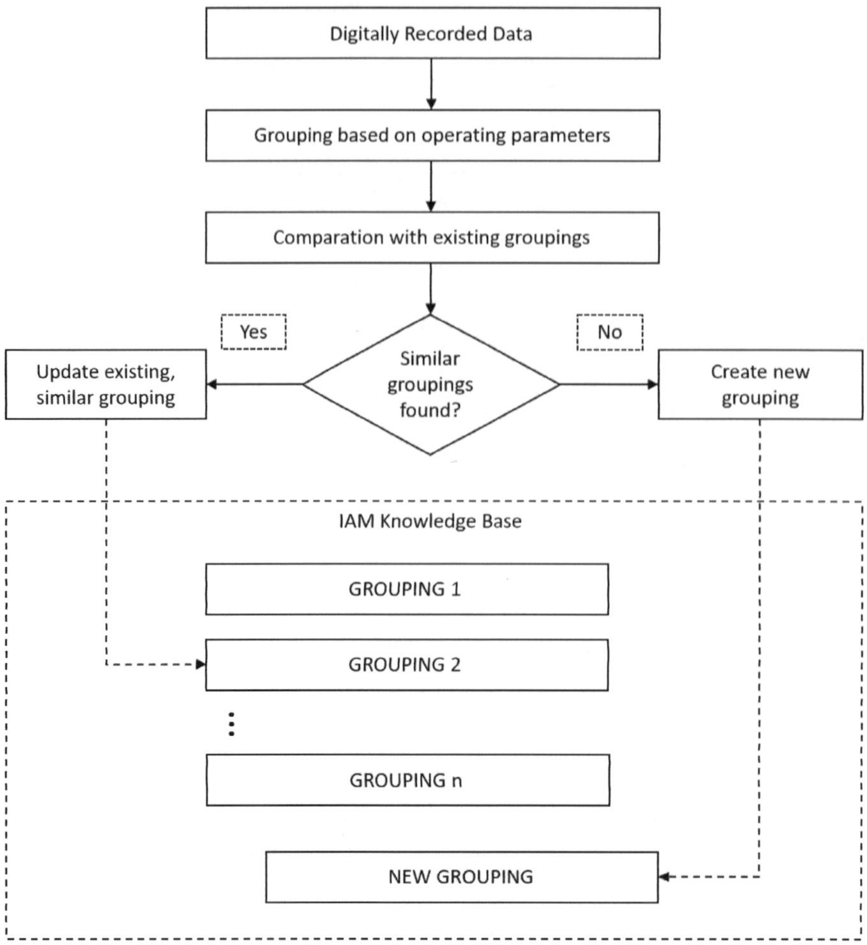

Fig. 4. Intelligent Adviser Module Adaptive Learning Algorithm

The issue here is – from where does the new knowledge accumulation begin? When a new shop floor element is introduced how does the IAM compensate for the potential new element behaviour? In other words, how to be sure that the used, already accumulated knowledge is up to date and useful? This is very important because if the IAM is using inadequate knowledge the proposals could be wrong and cause even more serious problems.

In order to avoid this, all new sources of knowledge have to be documented, indexed and verified through adaptive learning as shown in Fig. 4. IAM's adaptive learning algorithm automatically forms groupings from digitally recorded data. These groupings are based on similar operating parameters from individual shop floor elements and their working conditions. All groupings are contained and organized within the IAM Knowledge Base.

This algorithm continuously compares the newly created input data with existing groupings and tries to find the most similar grouping to that input. This search can have the following two results:

- Similar grouping found. The element which produced this input data will be associated with the performance of the identified grouping. That means its behaviour is within specified tolerances and previous knowledge can be applied. Any possible, minor differences between this input data and the grouping are used to update the grouping.
- Similar grouping not found. It means that the element which produced this data is operating in a new manner and previous knowledge should not be applied. When this happens, this input data becomes a base for a new grouping. When there are enough of such out-of-bound inputs it means a new behaviour is emerging from that specific shop floor element. This defines the adaptive nature of the IAM where its accumulated knowledge base can be used even when elements are changing [11].

However, one issue would still need to be resolved – determination of the Behaviour Tolerance Line (BHL). BHL specifies the tolerance or the amount of allowed differences between these compared shop floor parameters. The key is to find a balanced BHL.

If the BHL is set too low, even the slightest differences between the data inputs would classify as "out of bound". As a result, there could be a large number of newly formed groupings when in fact there would not be that many new behaviour characteristics for a specific shop floor element. This would unnecessarily increase data consumption as well as processing resources.

On the other hand, if the BHL is set too high, the crucial differences between data inputs would not be registered. That means there would be a small number of groupings and any protentional new shop floor element behaviour would not be detected. As a consequence, the IAM proposals could be less accurate.

5. Conclusion

A new algorithm for the IAM adaptive learning was proposed and developed. Initial results show that the inclusion of such a learning algorithm could further improve the IAM as a decision support tool for the system operator. Adaptive learning allows to collect, index and verify digitally recorded data which is generated by all shop floor elements in BAS. Such validated data is used to forms groupings based on similarly operating parameters. These groupings describe the different behaviours of the shop floor elements. This allows the IAM to compensate if a new element is added or the already existing ones are operating differently. As a result, new system specific knowledge is produced and accumulated. This knowledge is a basis for more accurate IAM proposals. Accurate proposals allow the system operator to make better quality decisions in shorter time and to avoid or minimise lost machine time in BAS. Without adaptive learning, proposals would be based on data which is not applicable and would even worsen an emergency situation during the execution of working scenarios. The limitation of this research is that the development and the implementation phases are separated. This means that without access to real world dynamic data, results cannot be verified. Future research will focus on methods for determining the Behaviour Tolerance Line (BTL). Additionally, next step includes experiments with laboratory generated data in order to confirm the applicability of the proposed adaptive learning algorithm for the IAM.

6. Acknowledgments

This paper would not be possible without the mentorship and encouragement from Univ. Prof. Dr.sc. Dr.mult.h.c. Prof.h.c. Branko Katalinic.

7. References

[1] Ming, X. G., Yan, J. Q., Lu, W. F., & Ma, D. Z. (2005). Technology solutions for collaborative product lifecycle management–status review and future trend. Concurrent Engineering, 13(4), 311-319

[2] Katalinic, B., Kukushkin, I., Pryanichnikov, V., & Haskovic, D. (2014). Cloud Communication Concept for Bionic Assembly System, Procedia Engineering, Volume 69, 2014, Pages 1562-1568, ISSN 1877-7058, doi:10.1016/j.proeng.2014.03.156

[3] Katalinic, B., Kukushkin, I. K., Cesarec, P., & Kettler, R. (2012). Hybrid control structure and scheduling of bionic assembly system. In Proc. 8th International Conference of DAAAM Baltic, Industrial Engineering (pp. 483-489). Tallinn, Estonia

[4] Post, P. (2014). "Smart Systems for Intelligent Manufacturing - Industry 4.0" Plenary Lecture, 25th DAAAM International Symposium, Vienna, Austria

[5] Haskovic, D., Katalinic, B. & Kukushkin, I. (2015). Role of the Adviser Module in the Hybrid Assembly Subordinating Control Structure, Procedia Engineering, Volume 100, 2015, Pages 1706-1713, ISSN 1877-7058, http://dx.doi.org/10.1016/j.proeng.2015.01.546

[6] Getty, David J., et al. (1995). System operator response to warnings of danger: A laboratory investigation of the effects of the predictive value of a warning on human response time, Journal of Experimental Psychology: Applied 1.1 (1995): 19

[7] Brake, D. J., & Bates, G. P. (2001). Fatigue in industrial workers under thermal stress on extended shift lengths. Occupational Medicine, 51(7), 456-463

[8] Haskovic, D., Katalinic, B., Kildibekov, A. & Kukushkin, I. (2016). Intelligent Adviser Module for Bionic Assembly Control System: Functions and Structure Concept, Proceedings of the 26th DAAAM International Symposium, pp.1158-1165, B. Katalinic (Ed.), Published by DAAAM International, ISBN 978-3-902734-07-5, ISSN 1726-9679, Vienna, Austria

[9] Haskovic, D., Katalinic, B., Zec, I., Kukushkin, I. & Zavrazhina, A. (2016). Structure and Working Modes of the Intelligent Adviser Module, Proceedings of the 27th DAAAM International Symposium, pp.0866-0875, B. Katalinic (Ed.), Published by DAAAM International, ISBN 978-3-902734-08-2, ISSN 1726-9679, Vienna, Austria, DOI: 10.2507/27th.daaam.proceedings.125

[10] Aronson, J. E., Liang, T. P., & Turban, E. (2005). Decision support systems and intelligent systems. Pearson Prentice-Hall

[11] Lee, J., Kao, H. A., & Yang, S. (2014). Service innovation and smart analytics for industry 4.0 and big data environment. Procedia Cirp, 16, 3-8

[12] Shumsky, A.; Zhirabok, A. & Kalinina, N. (2017). Human-Operator Behaviour Monitoring in IT-Systems, Proceedings of the 28th DAAAM International Symposium, pp.0160-0167, B. Katalinic (Ed.), Published by DAAAM International, ISBN 978-3-902734-11-2, ISSN 1726-9679, Vienna, Austria, DOI: 10.2507/28th.daaam.proceedings.021

[13] Nikolic, B.; Ignjatic, J.; Suzic, N.; Stevanov, B. & Rikalovic, A. (2017). Predictive Manufacturing Systems in Industry 4.0: Trends, Benefits and Challenges, Proceedings of the 28th DAAAM International Symposium, pp.0796-0802, B. Katalinic (Ed.), Published by DAAAM International, ISBN 978-3-902734-11-2, ISSN 1726-9679, Vienna, Austria, DOI: 10.2507/28th.daaam.proceedings.112

[14] Achsan, H. T.; Suhartanto, H. S. & Wibowo, W. C. (2017). Ontology Enrichment for Multi-Domain Knowledge and Expertise Representation, Proceedings of the 28th DAAAM International Symposium, pp.1170-1177, B. Katalinic (Ed.), Published by DAAAM International, ISBN 978-3-902734-11-2, ISSN 1726-9679, Vienna, Austria, DOI: 10.2507/28th.daaam.proceedings.162

DOI: 10.2507/28th.daaam.proceedings.166

DESIGN THE CONTACTLESS CHARGER AND CONTACTLESS DATA TRANSFER BETWEEN UNDERWATER ROBOT-SATELLITS AND UNDERWATER 6-LEGGED VEHICLE

Aleksander Ya. Ksenzenko, Evgeny A. Prysev,
Valentin .E. Pryanichnikov & Vadim V. Chernyshev

This Publication has to be referred as: Ksenzenko, A[lexander]; Prysev, E[vgeny]; Pryanichnikov, V[alentin] E[.] & Chernyshev, V[adim] (2017). Design the Contactless Charger and Contactless Data Transfer between Underwater Robot-Satellits and Underwater 6-Legged Vehicle, Proceedings of the 28th DAAAM International Symposium, pp.1197-1201, B. Katalinic (Ed.), Published by DAAAM International, ISBN 978-3-902734-11-2, ISSN 1726-9679, Vienna, Austria
DOI: 10.2507/28th.daaam.proceedings.166

Abstract

The work is devoted to prototyping of the 6-legged walking underwater vehicle as base station with contactless charger and contactless data transfer between itself and a number of autonomous underwater robot-sputniks. An architecture is proposed to build the robotic complex as interacting parts had shown the efficiency. The separation of slow-moving, but well passable on the bottom legged platform, connected by cable with a remote supervisory control station on a boat, and highly maneuverable, fully self-contained units – direct executors for underwater operations significantly extend the operational capabilities of the complex. This hardware build simplified the organization of the underwater operations due to of rapid charging of autonomous modules' batteries, directly in the area of the works, without their ascent to the surface, as well as the separation of the control channel into two parts: the cable and sonar lines. The work is executed at partial financial support of RFBR, projects 16-07-01264, 16-07-00811, 16-07-00935.

Keywords: underwater robotics; underwater contactless charger; underwater contactless data transfer; remote supervisory control

1. Introduction

Exploration of unmanned robotic devices in the study and practical use of spaces that are hidden under water becomes increasingly important because the problem of ensuring human security under water is avoided. The underwater robots constructed for this purpose can be conditionally divided into two groups: (1) fully autonomous underwater vehicles and (2) tethered devices connected by a cable to a control station that is located in a safe place (such as on the coast). Each of these groups has certain advantages and disadvantages. Despite significant advantages of autonomous underwater robots, their widespread use is hindered by the problem of uninterrupted power supply and the complexity of arranging a remote underwater control channel by the operator. Because of this, fully autonomous unmanned underwater vehicles turn out to be rather large, complex, and expensive in terms of their construction and operation.

To avoid this contradiction, we propose a configuration of the robotic system consisting of three separate yet interacting devices. The first device is a limited mobile base station connected by a cable to the control station that is located, for example, on the coast. This cable ensures power supply for all underwater components of the system. The cable serves for the transfer of control commands by both the base station and autonomous modules through a hydroacoustic communication channel.

The second device is a group of fully autonomous underwater modules that are relatively small in size, have no external connecting cable, and provide increased maneuverability; they are responsible for the direct execution of underwater operations several dozens of meters away from the base station. The modules are to be controlled by the operator station through a communication channel consisting of two parts: by a cable extending from the control panel to the base station and through a short hydroacoustic communication channel extending to each autonomous module. In addition, each module is equipped with a device for contactless operational recharging of its batteries from the base station just in the workplace (without ascending).

The third device is a control station (with operators) located on the coast or on an accompanying boat. The operator controls the movement of the base station along the bottom into the workplace with the help of on-board video sensors through a wire communication channel. The hydroacoustic channel and the base station are used by the operator to conduct all the manipulations with the autonomous modules required for relevant actions in the workplace.

This architecture of hardware for the underwater system makes it possible to combine the operational advantages of fully autonomous robotic unmanned robots with the convenience in organizing a channel for their remote control.

The underwater robotic system consists of several complex interconnected functional systems. This study considers the most important of these systems that go through the stages of prototyping and testing. Further, we consider a device of 6-legged walking construct of the base station, a contactless system for recharging on-board batteries of autonomous modules directly in the workplace without the need for their ascent, and a system for exchanging large data amounts between the base station and autonomous modules through a high-speed radio channel in salty water.

2. Composition and structure of the base station

The base station is functionally an intermediate link between the actuator operating in extreme conditions and the operator situated in rather comfortable conditions at the central control station. In terms of construction, the base station is designed as a self-moving 6-legged walking device, with the following equipment to be installed in sealed boxes:

- two identical stepped modules (right and left boards);
- movers of each board;
- front and rear view television cameras;
- powerful LED searchlights with visible and infrared light;
- devices for mooring of autonomous modules with the contactless recharging system;
- hydroacoustic antennas for channels of control over autonomous modules;
- an ultrasonic beacon for orienting the autonomous modules under reduced visibility;
- an electronic compass outputting current information about the location azimuth.

Each board has movers each consisting of three cyclic pacing mechanisms with an adjustable trajectory of the reference point. The pacing mechanisms in the mover are linked kinematically and act in antiphase: two of them act in phase and the third acts in antiphase. The step size and height can be operationally changed from 80 to 120 cm and from 20 to 60 cm, respectively. The electric drives of pacing mechanisms are made on asynchronous electric motors with a frequency regulation of speed and a total power of around 1500 W. The motion speed of the base station along the flat bottom reaches 5 km/h and depends basically on the bottom surface features and the visibility conditions of the route. Electricity and control signals arrive from the control station through a single-wire coaxial hoist cable; the length of this cable can reach 120 m. The base station can be rotated driven by different speeds of mover drives of each board. The weight of the experimental operating model of the base station is around 170 kg.

The movers of the left and right boards are made according to cyclic pacing mechanisms acting in antiphase. The cyclic movers make it possible to maintain the stability of pacing and do without a controlled adaptation system. It is known that the pacing machines with cyclic-type movers have simple kinematic drives and are simpler than their analogs with adaptive control of pacing. Unfortunately, the fact that the trajectory of reference points for cyclic pacing mechanisms is invariant does not allow the mover adaptation capabilities to be fully explored when the character of motion path profile varies. Therefore, the overcoming of obstacles makes allowance for controlling the step length and height as well as changing the position of reference points when turning to reduce the moment of resistance to rotation. The trajectory of reference points is adjusted by additionally introducing a special crank with a linear electric drive into the pacing mechanism; in this case, a single mechanism can adjust the trajectory of all three pacing mechanisms of its board. The control of the trajectory of reference points of the pacing mechanism is reduced to a discrete change in the angular position of the additional crank. This leads to a shift in the arm suspension point and a transformation of the base trajectory of marching motion into the trajectory of obstacle handling with increased step height and length [1–3].

The contactless recharging of the power of onboard batteries is conducted through a device based on the principle of electrodynamic induction. The radiating part of this device is installed inside the sealed casing of the base station and contains a generator of high-frequency oscillations with a power of around 95 W and a directional antenna. The generator is supplied with power from the base station network. High-frequency voltage is supplied to the radiating antenna that is located under the radio-transparent cover of the device. The hydro pump built into the mooring mechanism provides adequate positioning and holding of the mobile module in a required place during the whole period of contactless power transmission, which can last up to several hours.

Due to its own movers and video cameras, the base station can be moved by operator's commands to the workplace and fixed at this place. The base station has a hydroacoustic antenna to control the operation of autonomous modules. When the bottom is muddy, the visibility is reduced to less than a meter; in view of this, the base station should contain an ultrasonic beacon along with powerful searchlights.

3. Composition and structure of the autonomous modules

There should be at least two autonomous modules (highly maneuverable underwater robotic devices). The equipment of the module can be entirely placed in a sealed casing. An autonomous module must be equipped with:

- movers;
- on-board batteries;
- lamps and video cameras;
- a receiver of the system of contactless recharging of onboard batteries;
- a receiver/transmitter of the system of operational exchange of large data amount;
- a proper navigation system;
- a hydroacoustic system of communication through the base station with the central control station;
- an on-board computer;
- auxiliary actuators installed on the module for specific operations.

The actual efficiency of underwater operations by automated devices depends essentially on the maximum possible time of their continuous operation without ascending to the surface. Periodic interruptions are caused by the need to recharge the onboard battery of the power supply system as well as to promptly adjust and update the onboard computer software. To this end, one has to stop the operations, ascend the device to the water surface, unseal it, recharge or replace the batteries, reseal the device, execute control operations, submerge, search, move and position the device in the workplace. With an increase in the depth of operations, the importance of increased length of continuous operation grows substantially.

The method of power recharging of on-board batteries of autonomous modules makes it possible to operate directly in the workplace and avoid ascending to the surface. To solve this problem taking into account the large depth and salty water medium, the autonomous module is equipped with a receiver of the contactless energy transfer system. The autonomous module uses its navigation system, hydroacoustic beacons, and video cameras to approach the radiating antenna of the base station. This can be conducted both by the operator situated at the control post and in a fully automatic mode using a special-purpose program. Since the grouping of autonomous modules interacts with the base station, the operations are not interrupted during the charging of the moored autonomous module: another module with a previously charged battery is activated.

The receiver of the autonomous module is located behind the radio-transparent wall of the casing. The electromagnetic field radiated by the transmitting antenna of the base station is captured by the receiving antenna of the autonomous module and converted into electric current. A special-purpose device converts this current into a normalized current of battery charge. This device provides an optimal mode of battery charging and protects it from recharging. The degree of readiness of the battery for operation is controlled by the on-board computer and can be visualized on the control station display [4, 5].

The autonomous modules should be controlled with the help of a navigation system with a hydroacoustic duplex communication channel. Being relatively narrow-band, this channel cannot provide high quality television imagery; nevertheless, the quality is sufficient to meet the technical requirements of remote control. When using the range of ultrasonic frequencies of around several hundred kilohertz for the transmission of a good quality video signal and the reception of control commands, the acceptable distance of the mobile module to the base station is approximately 20–40 m. Thus, the control station operator can remotely perform underwater operations with the help of a manoeuvrable mobile module, using a high-quality cable communication channel to the base station and a short hydroacoustic section of the communication channel to the mobile module.

Autonomous modules can collect information from underwater objects of different origin and purpose. Normally, this is information is rather extensive: telemetry for a long period of observations, high-quality photo and video materials, etc. To this end, the autonomous module and the base station must be equipped with a system that can exchange large data amounts through a broadband radio channel capable of operating in a salty water medium. With a broadband radio channel at a frequency of 2.4 GHz, the attenuation in salty water is around 85 dB per meter.

Despite this fact, the signal at a distance of several centimetres was sufficiently powerful to transfer data efficiently from the module to the base, in the mode of base station docking during battery recharge (Fig. 1). This communication channel can also be used to remotely update the computing software at facilities of the autonomous module (Fig. 2). The exchange channel of large data arrays organized in this way is highly secure against unauthorized reading of transmitted data.

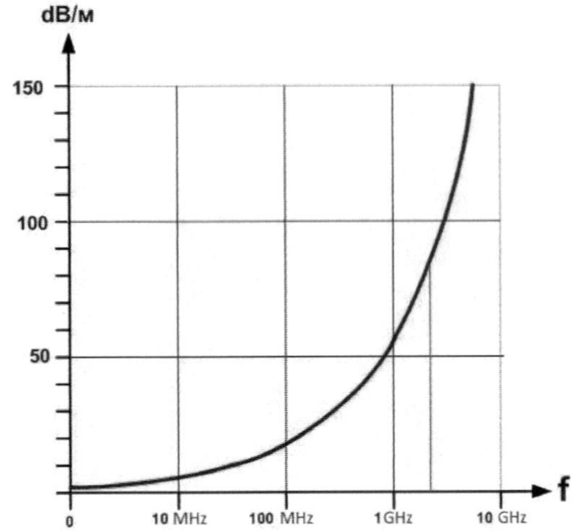

Fig. 1. Typical dependence of seawater radio transparency on frequency

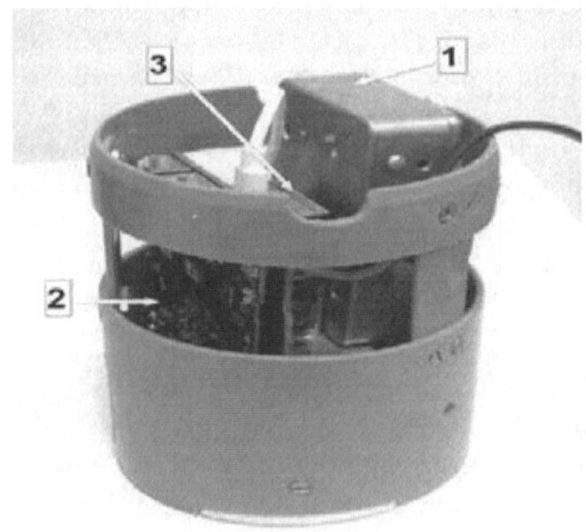

Fig. 2. Fragment of the radiating device for contactless charging and the receiver/transmitter system of the radio channel of exchange of large data amounts.
(1) Generator of high-frequency voltage for the system of contactless charging, (2) Board of the receiver/ transmitter system of the radio channel of exchange of large data amounts, (3) Auxiliary voltage inverter of the power receiver/ transmitter.

4. The principles of control between elements of robotic system.

The software of the robotic system should serve the mobile module control system and the base station control system. This is done by on-board computers that provide supervisory control, motion programming, and telemetry analysis. The possibility of motion programming without interrupting the current operations is ensured by using a technology that uses of an interpreted dynamic language for the upper level of control.

This allows one to promptly edit the on-board computer programs and organize exchange of messages between software components with the help of the actor model; this makes it possible to partially solve the problem of recovery after a network failure in data exchange and accumulate the results of telemetry in a non-relational database. A similar technology was successfully developed on Amur ground-based mobile robots.

5. Full-scale testing.

Tests of pacing characteristics of the current prototype of the pacing base station were conducted under different bottom conditions at a depth of up to 20 m in the Kandalaksha Gulf of the White Sea (Fig. 3). Different modes of straight-line motion and rotations were considered with various foot designs.

The design tests revealed that the pacing machine in terms of passing ability along the sea bottom is much superior to wheel and caterpillar movers. The advantages of the pacing mover manifest themselves most conclusively when moving along a complex rocky bottom.

The tests indicated that the system of contactless recharging of on-board batteries is efficient, with the coefficient of efficiency (taking into account the inaccuracy of positioning of docking nodes and water salinity) being in the range of 30 to 45% (Fig. 4).

Fig. 3. Motion of the pacing machine of the base system along the bottom at a small depth.

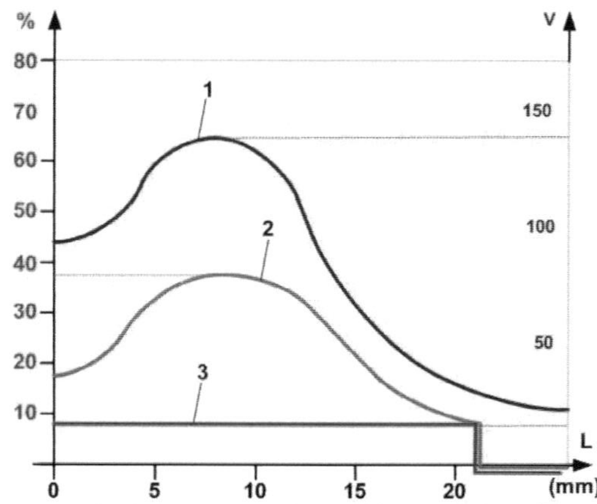

Fig. 4. Efficiency of contactless power transfer for underwater devices.
(1) Voltage at the output of receiving antenna,
(2) Coefficient of efficiency,
(3) Output voltage of charging device.

6. Conclusion.

The hardware construction principles for robotic systems developed here by combining the low-maneuverable (but highly passable along the bottom) tethered base station with several highly maneuverable autonomous mobile modules as well as by using a contactless method of battery charging make it possible to construct vehicle with qualitatively new operational capabilities (specifically, when operating on complex and almost inaccessible bottom reliefs and in submerged constructs).

7. References.

[1] Chernyshev V.V., Arykantsev A.E., Gavrilov, A., Control of the motion of underwater pacing devices moving along the bottom, Inv. YuFU, Technical Sciences, Proc. of All-Russian Conf. "Prospective Systems and Control Problems", 2016, no. 1, pp. 141–155.

[2] Chernyshev V.V., Simulation of the dynamics of pacing machine with cycling movers as a system of rigid bodies with elastic–dissipative coupling, Izv. VolgGTU: Ser. Important problems of control, computers, and informatics in technical systems, 2010, no. 11, pp. 32–35.

[3] Chernyshev V., Arykantsev V., Kalinin Y., Gavrilov A., Sharanov N., Development of the walking mover for underwater walking vehicle, Proccedings of the 26th DAAAM International Symposiumon, 2016, Katalinic B. (Ed.), Published by DAAAM International, Vienna, Austria, pp. 1143–1148.

[4] Bielich T., Kirsanov K.B., Ksenzenko A.Ya., Pryanichnikov V.E., Increasing the time of continuous operation of the underwater inspectoral robotic system, Information–measuring and control systems, Moscow: Radiotekhnika, 2015, vol. 13, no. 7, pp. 57–65.

[5] Pryanichnikov V.E., Ksenzenko A.Ya., Kuvshinov S.V., Poduraev Yu.V., Prysev E.A., Khelemendik R.V., Eprikov S.R. Intelligent rodotronics: hardware–software complexes of rodotariums. 25th DAAAM International Symposium on Intelligent Manufacturing and Automation 2014, Procedia Engineering, 2015, vol. 69, no. 1, pp. 35–40. DAAAM International, Vienna, Austria.

DOI: 10.2507/28th.daaam.proceedings.167

TECHNOLOGY OF MULTI-AGENT CONTROL FOR INDUSTRIAL AUTOMATION WITH LOGICAL PROCESSING OF CONTRADICTIONS

Valentin E. Pryanichnikov, Aleksander A. Aryskin, Stanislav R. Eprikov, Kirill B. Kirsanov, Roman V. Khelemendik, Aleksander Ya. Ksenzenko, Evgeny A. Prysev & Aleksander S. Travushkin

This Publication has to be referred as: V[alentin] E. Pryanichnikov, A[leksander] A. Aryskin, S[tanislav] R. Eprikov, K[irill] B. Kirsanov, R[oman] V. Khelemendik, A[leksander] Ya. Ksenzenko, E[vgeny] A. Prysev & A[leksander] S. Travushkin (2017). Technology of Multi-Agent Control for Industrial Automation with Logical Processing of Contradictions, Proceedings of the 28th DAAAM International Symposium, pp.1202-1207, B. Katalinic (Ed.), Published by DAAAM International, ISBN 978-3-902734-11-2, ISSN 1726-9679, Vienna, Austria
DOI: 10.2507/28th.daaam.proceedings.167

Abstract

The paper related to the problem of creation the new technology of multi-agent control for industrial automation – creating the automated system, realising multi-recipes production of a large number of types with logical processing of contradictions. This required to find a proper software decisions for distant access, reprogramming and monitoring the production line even during it's functioning. The created system of logical processing, basing on brunching time logics, and parallel software, controlling production line as well as the imitation modelling, are a new way for intelligent industry 4.0.

Keywords: industry 4.0; intelligent manufacturing; sensor systems; cyber-physical systems; Internet of Things; logical contradictions; industrial automation

1. Introduction

Technological advances had significantly, rapidly increase the productivity last time. To describe shot that global industrial revolutions I-IV, we can use the one-word characteristics: mechanization – electrification – automation - intellectualization. The basis for the concept of Industry 4.0 are: “Internet of Things”, “Big Data”, "Cyber-physical system". Under this concept, one of the main tasks is to ensure effective releasing of even minimal parties out-coming production with the rapidly changing structure of the production lines. Using built-in sensors/actuator network access leads to the necessity of creating new methods of optimisation for resource conservation in a production environment. It was necessary to create a mechanism for predicting the faults of the components or equipment, as well as the possibility of integrating the process of self-repair, finding alternative solutions to current production problems, the resolution of the relevant repair or maintenance.

If more recently, consumers of automation did not provide the manufacturer access to the purchased equipment. In recent years, it is appeared a large demand for life support equipment. One of the leaders in implementation of this concept is the concern FESTO AG, working in more, than hundred countries with 300,000 users and in practice, providing just-in-time production.

In the Russian Federation, not all businesses are ready to introduce the concept of "Smart production", as this is a difficult, long and expensive process. At the same time enterprise, created "from scratch", more receptive to such innovations. International laboratory "Sensorika", together with specialists and graduate students from MSTU "Stankin", INET RSUH, KIAM Russian Academy of sciences, JSC "TechInvest" has developed a technology for implementation and deployment of "Smart production" by building a new automatic line for manufacturing the foam-glass and products from it in the city Kimry. The system automation implemented in the framework of this project, is based on the previously patented methods for automatically resolve logical contradictions (refined to adapt to the needs and tasks). This technique allows to predict the probable failures of the system in certain situations and is one of the foundations of the reliability of the project. Automation implemented into the project is a symbiosis of decentralised systems integrated into a digital network with the possibility of switching to automated program or to manual control, if necessary. The developed concept allows to perform non-trivial tasks for the intelligent control of executive elements of technological process, for example, switching valves, dampers, motor control for input/output, for moving components in the production lines according with the flexible manufacturing process. In this process are involved specialised transport system, developed with the participation of the International laboratory "Sensorika". It is used for drying and moving the produced details and finished products in the production space about 1400 m2. This system is made on the basis of individual robotic trucks, which are driven by electro engines and a set of sensor devices. One such transport unit is able to move along a complicated paths the internal elements and finished products, weighing up to 2-3 tons in hard conditions with high humidity. Part of the overall management system of the production cycle is the interface, that provides access to it with the remote monitoring, testing and upgrading processes. Developed control system for transport and manipulation robotic devices, initially as a set of peripheral devices, which don't have a characteristics of flexibility in our project. At the same time it provides "intelligent data entry" through a system of sensors and feedback lines, controlling the operation of all actuators with logical analysis and the detection of incorrect manufacturing processes, which significantly reduces the probability of violation of the technological cycle and emergency. The listed requirements to the modern production, that we are implementing, to some extent fits into the concept of the next industrial revolution, called "Industry 4.0".

2. Experience in implementing the flexible production line.

The control system ensures fully automatic operation of the production, realised in two separate parts of the workshop. In one of them the preparation of a sodium glass substance with the appropriate additives is made, in the another, it is handled by shape, drying, final processing and packaging as the finished product, followed by transportation system for shipment to the consumer or to the warehouse of finished products.

The technological scheme for preparing the substance is shown in Fig. 1. It's work is formalised/described with around 200 logical variables. In reactor 1-3 is operating with large capacity, the main components of the substance are mixed in the required proportions with observance of the temperature regime. At the initial stage, this process requires the initial forced heating of the mixture to a temperature above 100 degrees. For this purpose, a special steam generator with a capacity of 250 kg / h with a steam temperature up to160 degrees is present in the circuit. The prepared mixture enters the storage tanks stock. During the mixing process, an exothermic chemical reaction develops, which requires forced cooling to a temperature of less degrees by means of a heat exchanger. For its' functioning, tanks with a reserve of cold and hot water are provided. The cooled mixture enters the vessels/tanks, after which it is envisaged to add a special fiberglass suspension to it from the containers, which provides the required consumer characteristics of the future product.

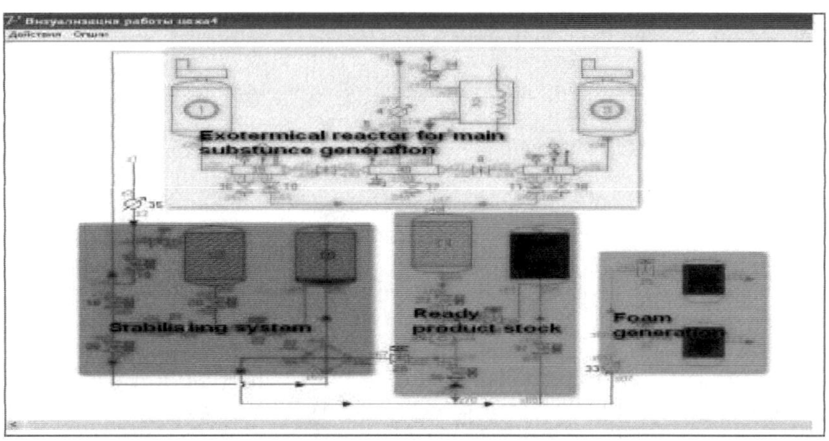

Fig. 1. The graphical interface of the simulation program - technological scheme with logical control of failures.

All valves have remote control mechanisms. The executive and regulating elements of the technological scheme are equipped with microprocessors for controlling the execution of control commands. The temperature regime, the level of filling of tanks and the pressure in them is controlled by means of suitable sensors. At the most critical nodes of the technological scheme, duplication of monitoring sensors is applied. The system ensures the execution of the required opening and closing time intervals, valves through which the components are fed. The developed control concept allows performing complex tasks of influence on the actuating elements in the process flow diagram. The preparation of each recipe for the preparation of a substance with the tracking time intervals for further execution of the control program is initially performed by the plant's technologist. After preparing the pilot batch of the product and passing the quality control, a decision is taken to write the recipe into the database. Then, this recipe can be called up by the operator on duty and executed without the help of a technologist.

Spillage of the finished substance with added components is made in the drying-filling part of the shop, Fig. 2. Its technological scheme is based on the use of transport trolleys, driven by gear racks, which, in turn, are moved by pairs of electric motors with a power of 1.5 kW through retarding reducers. The operational information support of the control system of this shop and the control of the execution of the specified trajectory of moving the carriages is provided by a set of induction and optical sensors, as well as a video monitoring system. A special additional system avoids the collision of loaded and empty trolleys.

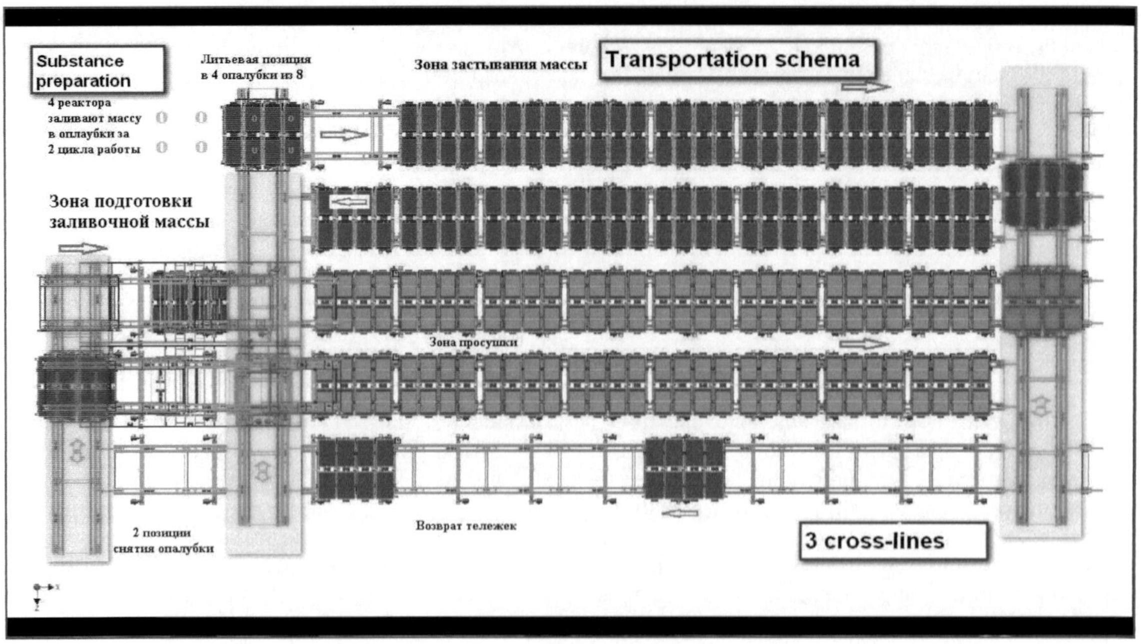

Fig. 2. Diagram of the reconfigurable transport system (fragment of drying-packing workshop).

The automatic control system of the actuating elements is located in the immediate vicinity of the workshop in a room protected by a glass partition. There is also a manual control panel and a duty operator's workplace.

The control system is a symbiosis of decentralised systems, combined into a common digital network with the possibility of switching to hard software or manual control, if necessary.

3. Development of methods of heuristic deduction for the representation of technological knowledge.

The development of the control system (basing on the experts-technologists knowledge) was implemented, using our IGEC technology, presented in [1-4]. IGEC technology is a combination of four main directions of mathematical approaches conventionally called Ancient Greek, Ancient Egyptian, Ancient Indian and Ancient Chinese. This allows to use its strengths for each direction, and to supplement and harmonise the results obtained, for example, to eliminate contradictions (conflicts) arising from different traditions (approaches). This technology has been successfully used, for example, in the development of the basic software and hardware elements of the network of associative laboratories and operational centres in the framework of the project "Intellectual Robotronics", initiated by the Central European Branch of the International Engineering Academy in cooperation with the universities of Russia, Croatia and Austria. With the use of IGEC, a search was made for the implementation of the intellectual level of software and adequate knowledge representation for mobile robots, the construction of heuristic deductive reasoning and decision making.

In accordance with [1-4], it was suggested that the ancient Indian (I) tradition presupposes reliance on the senses and represents contemplation, intuition as research methods and evidence.

Ancient Greek (G) tradition is based on logical conclusions, rational reasoning and uses the methods of the natural sciences. Ancient Egyptian (E) tradition presupposes reliance on authority: its affirmations are by all means the truth, and the exact execution of installations is the main subject of the method. Ancient Chinese (C) tradition relies on careful execution of a large number of small operations. Its method is numerous calculations-manipulations, which are typical for modern computer technology. The integrated application and integration of all approaches led to the creation of the so-called IGEC technology, which is a combination of conventionally designated "ancient Indo-Greco-Egyptian-Chinese" traditions and methods of constructing evidence. The specific original methods of each approach (each tradition) and examples of the applications, making informal knowledge computational (it's contained in [2-4]).

Using this analogy, a special four-dimensional cube was introduced in [4] to more accurately characterise the work with all four IGEC approaches, and Pentalogics was used to study its two-dimensional faces (see Fig.3).

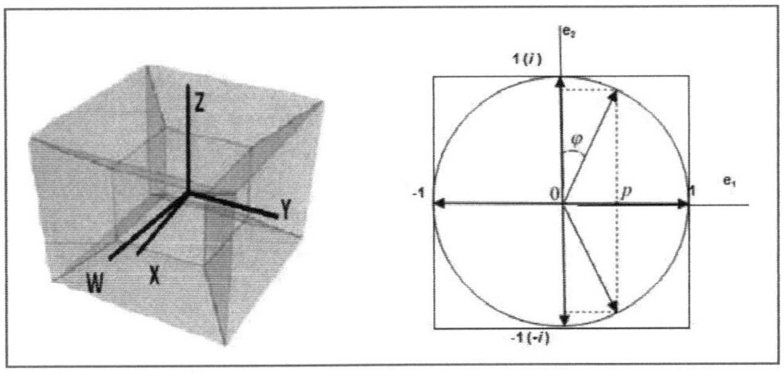

Fig. 3. The four-dimensional cube for IGEC technology and the plane of expert assessments in Pentalogics

The vertices of the cube are numbered by natural numbers: (1), (2), ..., (16) and have the following coordinates: (1) = <-1, -1, -1, -1>, (2) = <1, - 1, -1, -1>, ..., (16) = <1,1,1,1>. Let's interpret the axes of the cube's coordinates as follows: OX (length) - Ancient Greek measurement, OY (width) - Ancient Egyptian, OZ (height) - Ancient Chinese, OW (fourth axis) - Ancient Indian. Thus, each of the 4 coordinates changes in the range [-1..0] U [0..1], which shows the "success" of the corresponding approach for a task, that is characterised by a point or region of this cube. Pentalogics was proposed in [3] as a separate apparatus with the idea of formalising pairs of expert estimates in the ranges [-1..0] U [0..1], interpreted as the coordinate on (complex) plane as follows. The estimation along the abscissa axis corresponds to the level of knowledge: "I know-false, I do not know, I know-true", and the ordinate axis shows the so-called imaginary component of the assessment: "I do not believe, uncertainty, I believe."

Let us now consider the issues of the appearance of contradictions and their overcoming. As noted above, the results of each of the approaches for specific tasks may somehow contradict each other. Note that in this case we are not talking about logical contradictions and ways to overcome them, considered in [1-3], but "information", having different nature: according to one source of information (indicating the degree of reliability), the fact P holds, and according to another not-P . To clarify and overcome such information contradictions, we extend Pentalogics to the two-dimensional faces of the four-dimensional cube in the following way. Fix some two (for example, putting them equal to 0 from 4 coordinates and get a square for the estimation of the role of the two remaining approaches - Fig. 3 (right-hand side). In this way, analogous to the individual study of each of the 4 approaches, a method of investigating 6 pairs of their combinations appears, which can be done with the help of a Pentalogics apparatus. By interpreting the axes of coordinates with the values, of the corresponding approaches, we obtain specific Pentalogics for the formalisation of expert estimates. It is easy to verify that, up to a permutation of the coordinate axes, we have 6 pairs of Pentalogics, which we denote as follows: EG (Egyptian-Greek), EC (Egyptian-Chinese), IG (Indo-Greek), IC (Indo-Chinese), GI (Greek-Indian), GC (Greek-Chinese). So, if, for example, OX is an Ancient Egyptian dimension, and OY is Ancient Greek, then we have the so-called EG-Pentalogic, in which, in particular, the relations between the two approaches discussed above can be formalised. Here, the level of scientific knowledge can be compared on the abscissa axis: "science knows-lie, science does not know, science knows-true", and evaluation on the ordinate axis is an authoritative component of the assessment: "authority does not approve, uncertainty, authority approves").

4. Explanations to the industrial application of IGEC technology.

In the process of application of IGEC-technology to actual practical tasks there is a structuring, specification and clarification of each of its approaches, their coordination and development. At the same time, each of the four approaches of the IGEC using a logical device is "tuned" to specific areas of the subject area, and the use of Pentalogics makes it possible to detail their "inter-traditional junctions" and combine the results into a single whole. So, it seems to be promising application of IGEC in the process of designing, creating and maintaining a management system for the workshop, working on the technology "Industry 4.0". In this task, the most important problems are the synthesis, testing, designing a program control for this equipment, as well as formalisation and construction of a suitable imitation models. We indicate the ways of the solution with the use of IGEC-technology.

Logical formalisation and algorithms allow constructive, correct and complete solutions written in the language of production rules, Boolean functions, together with their deductive proofs. The visualisation of results, the construction of a strategy maps, allows you to structure and understand arrays of different types of information. Building Boolean functions leads to unification of measurement results, which is important for compatibility with other digital technologies. Logical scaling of the results, performed, for example, by means of propositional logic, and visualisation (including in terms of tasks terms), allows you to drill down into large amounts of data and bring it to a clear understanding: expert, logical, intuitive. The coordinated application of all IGEC approaches, together with the use of Pentalogics, helps diagnose and eliminate errors, smooth out contradictions ("see one" and "the program gives out another", the program conflicts with the expert, the results of the program are not supported by evidence, etc.) and iteratively refine the concepts in all planes on the axes of traditional approaches.

On the I-approach, visualisation of the main processes is carried out, implemented on a special layout. According to the G-approach, we distinguish the key "positions" in the state of this equipment - normal (effective), to which one should strive; dangerous, which must be avoided; and also describe illegal (impossible) positions. Applying the E-approach, we refine and supplement the concepts with the help of expert technologists. Starting the C-approach, we obtain estimates of other (not key) positions of interest to us (illegal, legal-normal, legal-dangerous). At the stage of verification and presentation of results, we again apply all approaches (we especially note the generation of product rules <situation, action>) by analogy with the application of IGEC to chess endings. Finally, with the help of Pentalogics, we identify and eliminate contradictions in all 6 planes of traditional approaches. So, for example, the discrepancies in the results of inference, computational experiments and the conclusions of a technologist are studied with the help of Pentalogics in the GC, EG and EC planes. And the addition of visualization and modelling on the layout allows you to eliminate the contradictions in the IE, IR and GI planes. The implementation of such a software-hardware production control system based on IGEC technology is in the prototyping stage.

When the work of the workshop is formalised, for each numbered object in the diagram in Fig. 1, sets of Boolean variables that characterise their state are introduced. Thus the additional numerical variables z_i are introduced, which allow diagnosing pipe states in the production scheme and object connections. The entered variables and their specific values allow us to describe the current state of the system, called configurations or positions, via the function f(x), where x is a set of values for variables. For the class of configurations in which the workshop works properly ("successful" configurations), we assume that $f(x) = 1$. For the class of configurations in which the workshop is off, we assume that $f(x) = 0$. If the shop fails during it's functioning, then we assume, that $f(x) = -1$. Thus, at this stage, the functioning of the workshop is represented by three-valued logic, which can be translated into special Boolean algebra or propositional logic by special translation. One of the principal problems is the large number of Boolean variables used in the model of this complex industrial production, which generates a significant amount of configurations to be studied and classified. To overcome this problem, special software is used and adapted to work with a big-data [12].

Problems of the second type include the elimination of contradictions, that have a non-logical, but "versatile" nature, which were described above, when describing the four-dimensional space of IGEC technology. The main tasks are: coordination of the data from different sensors, harmonisation of the recommendations of the technologist and automatic programs, harmonisation of various equipment standards, harmonisation of classical and modern service traditions, etc. All this is displayed on the corresponding IGEC-plane in order to understand the reason for such an "induced" contradiction and to eliminate it.

It is envisaged to enter data from a system of sensors and feedbacks. that control the operation of all actuators with logical analysis and identify the incorrect progress of production, which ensures the prediction of the state of assemblies and units in order to prevent disruption of the technological cycle.

The developed system for control the production cycle includes an interface that provides round-the-clock remote access to it, through cloud services and the Internet, for the purpose of remote monitoring, testing and operational modernisation of the technological processes. The interface allows you to create and merge several production processes into a single system.

The described production includes a set of fairly common technological operations, which make it possible to clearly illustrate the application of the principles of the concept Industry-4.0.

The work was carried out with the partial financial support of the RFBR, projects 16-07-00935, 16-07-01264, 16-07-00811.

5. References

[1] Kirilchenko AA, Pryanichnikov VE, Rogozin K.V. Limits of reliability and reliability of evidence. Skepticism in mathematics, functions, traditions // Information-measuring and control systems. - 2013. Vol. 11, No. 4. - P. 57-65.

[2] Pryanichnikov V.E., Katalinic B., Kirilchenko A.A., Khelemendik R.V., Kuvshinov S.V., Vician D., Uglesic A. New Creative Educational Technologies for Inter-university Network // 25th DAAAM International Symposium on Intelligent Manufacturing and Automation. 2014. / Procedia Engineering. 2015. Vol. 100, - P.259-268. - URL: http://www.sciencedirect.com/science/article/pii/S1877705815003938

[3] Pryanichnikov VE, Bieleich T., Vitsan D., Catalinich B., Kirsanov KB, Kuvshinov SV, Marzanov Yu.S., Poduraev Yu.V., Khelemendik RV, Prysev EA, Ugleshik A., Kharin K.V. Development of educational technologies and a network of associated laboratories-robotariums // Information-measuring and control systems. - 2015. - T. 13, No. 7. - C.7-25.

[4] Pryanichnikov VE, Khelemendik R.V. Information technology IGEC and Pentalogics / / Computer Science and Information Technology: Materials of International. Scientific Conf. - Saratov: The publication Centre "Science", 2016, pp. 331-333.

[5] Uspensky V.A. Seven Reflections on the Themes of the Philosophy of Mathematics // Regularities in the Development of Modern Mathematics, Moscow: Nauka, 1987, pp. 106-155.

[6] Frank S.L. Religion and Science // Religion, Philosophy and Science. 1953. - No. 1. - P. 1-26.

[7] Gorbunov-Posadov, M.M. Painless development of the program // Open systems. - 1996. - № 4. - P. 65-70. - URL: http://www.keldysh.ru/gorbunov/evolution.htm

[8] Frankl V. Man in search of meaning. Collection. / Translated into English. - Moscow: Progress, 1990. - 368p.

[9] Chess endings: Pawn. - 2 ed., Ext. / Ed. Yu.L. Averbakh. - Moscow: Physical training and sports, 1983. - 303s.

[10] Khelemendik R.V. On the solution of chess positions with the help of formulas for the logic of branching time, Izv. Sarat. un-ta. New Ser. Ser. Mathematics. Mechanics. Computer science. - 2011. T. 11., Iss. 1. P. 111-121.

[11] Endgame Nalimov Tablebases Online [Electronic resource]. URL: http://chessok.com/?page_id=361 (circulation date April 10, 2017).

[12] Khelemendik R.V. SFECalc: Program for constructing minimised circuits from functional elements, analysing them and synthesising them with specified parameters using large Boolean formulas / Certificate of official registration of the computer program No. 2013615516, registered in the computer program registry.

DOI: 10.2507/28th.daaam.proceedings.168

APPLICATION OF THE ARTIFICIAL NEURAL NETWORKS FOR DIESEL DIAGNOSTICS

Sergey Ovcharenko, Vitaliy Minakov

This Publication has to be referred as: Ovcharenko, S[ergey] & Minakov, V[italiy] (2017). Application of the Artificial Neural Networks for Diesel Diagnostics, Proceedings of the 28th DAAAM International Symposium, pp.1208-1212, B. Katalinic (Ed.), Published by DAAAM International, ISBN 978-3-902734-11-2, ISSN 1726-9679, Vienna, Austria
DOI: 10.2507/28th.daaam.proceedings.168

Abstract

This paper provides results of application of the artificial neural networks for diesel diagnostics based on the results of a spectral analysis of engine oil. Artificial neural network has been created and trained to identify pre-failure condition of the details of cylinder-piston group and crank-and-rod mechanism using the diesel locomotive Д49 as an example.

Keywords: diesel; technical diagnostics; spectral analysis of engine oil; artificial neural network

1. Introduction

A practice of diesel locomotive operation on Russian railways shows a diesel is the least reliable element of a locomotive which accounts for more than 40% of all unplanned repairs. Failure of the parts of crank-and-rod mechanism (CRM) and cylinder-piston group (CPG) of the diesel is more than 18% of the total number of locomotive failures. When organizing the repair of locomotives «on technical condition», it is important to have information about its technical condition while in operation. To do this, it is needed to apply different methods of diagnostics. One of the most effective methods of in-place parts diagnostics of CRM and CPG of the diesel is a method based on monitoring the current values of concentration of the wear products in a lubricant.

Engine oil is the most valuable information carrier, since during the diesel operation there is a process of wearing parts, and metal particles from the parts fall into the engine oil. New estimation technique of technical condition of CPG and CRM was developed and was proposed using the mathematical apparatus of artificial neural networks (ANN) as a result of the research. The proposed technique allows to identify the beginning of intensive wear of parts to modify the speed of increase in the concentration of wear products in engine oil by the control periods.

2. Research objectives

The research objective is to create and to train ANN, that allows to establish the fact of the beginning of intensive wear of the controlled groups of diesel parts, which characterizes its pre-failure state. Simulation results of accumulation of wear products in engine oil of diesel at various wear speed of certain groups of parts are used as initial information.

3. Creation and training of artificial neural network

Artificial neural networks are increasingly being used to solve a wide range of tasks including diagnosing technical objects. An important feature of ANN is the ability to learn and, as a result of learning, to increase its own productivity. To obtain a high-quality neural network, it is important to have sufficient amount of data for its training.

Training sample is determined by the results of modelling a process of accumulation of wear products in engine oil for groups of the chemical elements. According to the data [2], the sample was chosen for the training of ANN, consisting of the distribution of chemical elements for groups of controlled parts of diesel Д49 (CB – cylinder bushing, PP – piston pin, ARKP – articulated rod knuckle pin, PH – piston head, CR – compression rings, OSR – oil scrapper rings, CN – crankshaft neck, RD – rod journal, CBL – crankshaft bearing liner, CRL – connecting rod liner, PS – piston skirt, BBPI – bronze bearings of piston insert, BBACR – bronze bearings of articulated connecting rod). Amount of data is more than 32 thousand values. Figures 1 and 2 show the results of modeling a flow of Fe and Cu from the controlled parts at normal wear rate.

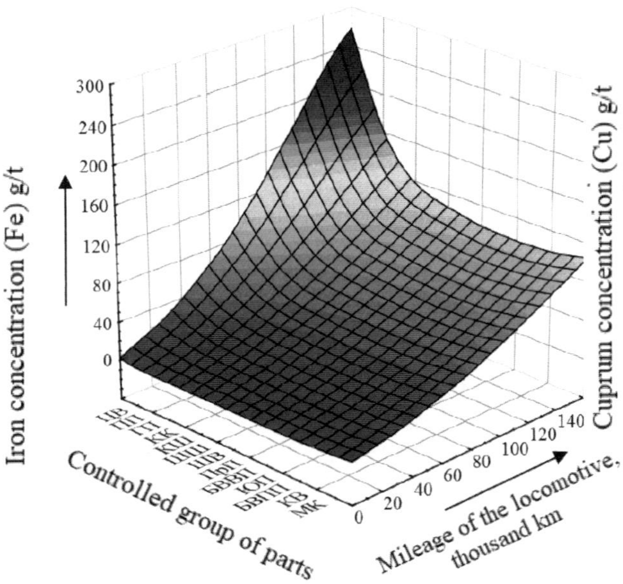

Fig. 1. Distribution of iron (Fe)

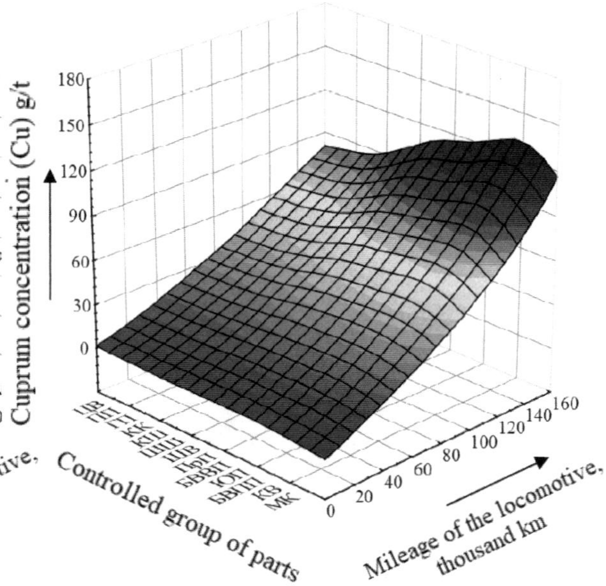

Fig. 2. Distribution of cuprum (Cu)

The most optimal network architecture for data analysis was determined as a result of the research and it is called a multilayer perceptron. The analysis of the obtained training results has showed that the highest performance indicators are provided by the ANN МЛП 9-20-14. The network has 9 neurons of input layer: the values of concentration of wear products in engine oil, a hidden layer, 14 output values corresponding to the groups of controlled parts (see Figure 3).

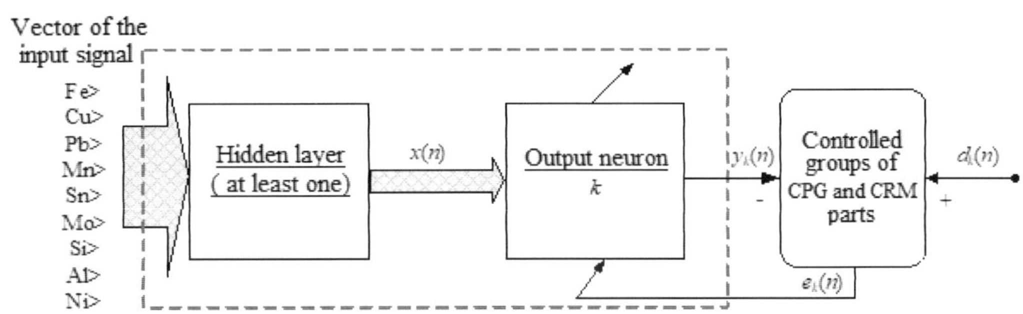

Fig. 3. Architecture of ANN

The goal of neural network training was minimization of difference between output y_i and desired d_k signal. The difference between output and desired signal is defined as an error signal [3,4]:

$$E(w)=\frac{1}{2}\sum_{k=1}^{M}\left[f\left(\sum_{i=0}^{K}w_{ki}^{(2)}v_i\right)-d_k\right]^2=\frac{1}{2}\sum_{k=1}^{M}\left[f\left(\sum_{i=0}^{K}w_{ki}^{(2)}\cdot f\left(\sum_{j=0}^{N}w_{ij}^{(1)}x_i\right)\right)-d_k\right]^2 \quad (1)$$

Where v, x – input signals of the layer

$f\left(\sum_{i=0}^{K}w_{ki}^{(2)}v_i\right)=y_k$ – output value of k signal.

To minimize the error signal of the network output values, we made an adjustment in the weight coefficients until the error signal doesn't adopt a steady state. Minimization of the error signal was carried out according to a delta rule, where the change in the weight coefficients of the neuron is given by:

$$\Delta w_i=\eta(d_i-y_i)\cdot x_i \quad (2)$$

where η – constant determining a speed of training

The output signals used in the paper as the wear rate of controlled parts are calculated as follows:

$$y=f((\sum_i w_i\cdot x_i)-\theta), \quad (3)$$

where f – nonlinear activation function;

θ – quieting level of the neuron.

The training sample was divided into teaching, test and control subsets. The training subset contained 75% of the total amount of data. The test subset required to evaluate a performance of the network is 15% of the sample. The control subset was used to verify a degree of network training and contained 15% of the sample. Block diagram of the ANN training is shown in Fig. 4.

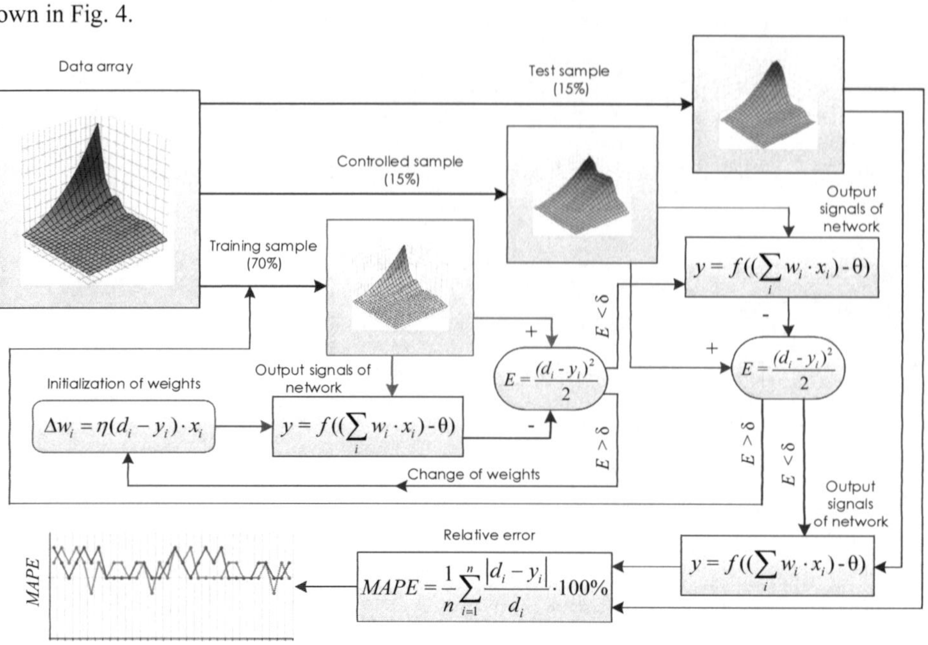

Fig. 4. Block diagram of the ANN training

MLP 9-20-14 neural network is obtained as a result of training. The network has a training efficiency of more than 94% and a training error of 5% for all parts groups (see Table 1).

Type of network	**Multilayer perceptron**
Architecture	9-20-14
Training efficiency	94,09
Control efficiency	92,099
Test efficiency	93,485
Error fuction	Sum of squares
Activation function of hidden neurons	Logistic
Activation function of output neurons	Logistic

Table 1. Results of ANN training

In the ANN the logistic activation function was used for hidden layer of neurons and output neurons. Logistic form of the sigmoidal nonlinearity in general form is defined as follows:

$$f_i(n) = \frac{1}{1+e^{(-\alpha \upsilon_j(n))}},$$

$$\alpha > 0, \qquad (4)$$

$$-\infty < \upsilon_i(n) < +\infty,$$

where $\upsilon_j(n)$ – local field of j neuron.

A level of trust to the results of diagnosis characterizes the fact that parameters of the trained neural network are within ist confidence interval. Confidence probability is usually denoted as $1 - \alpha$ and is selected from the values of 0,9; 0,95; 0,99. [5].

Fig. 5 provides a three-dimensional graph showing the confidence level of the ANN. The dark dots in the figure (control observations) characterize the belonging of the level of trust to the values of the objective and output functions received during the training of the ANN. Deviations of the levels do not exceed 7%, accepted for the solution of the problem.

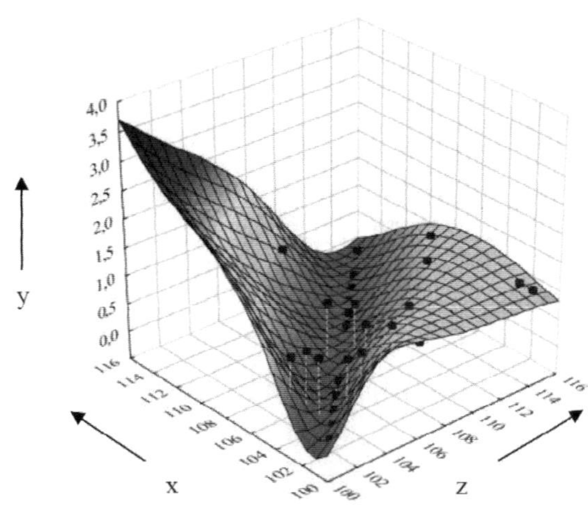

Fig. 5. The confidence level of the ANN:
y – rest; x – objective function; z – output function

Number of properly defined predicted values serves as an estimate of the performance of developed neural-network model. Fig. 6 presents the data obtained during the test of the ANN for groups of controlled parts.

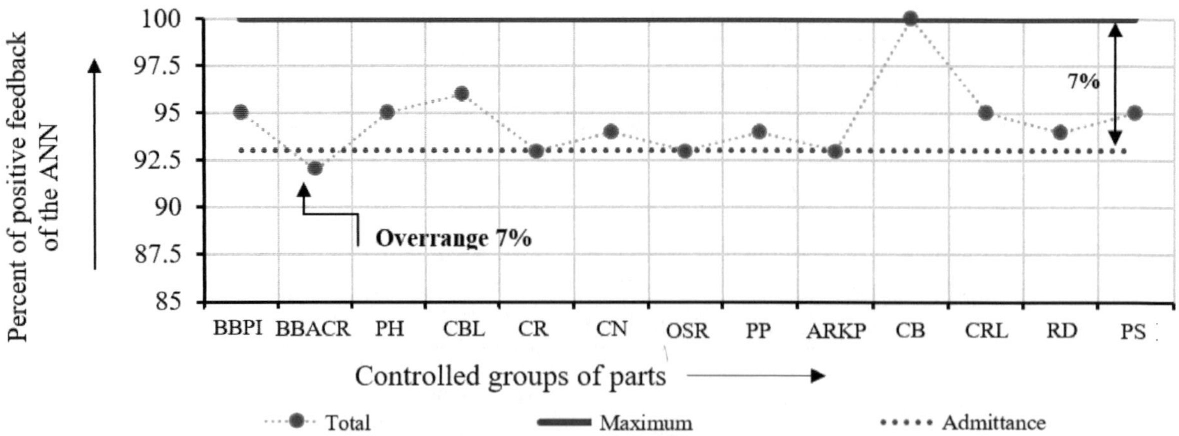

Fig. 6. Indicators of the trust level of the ANN МЛП 9-20-14

4. Conclusion

1. The ANN was developed for an analysis of dynamics of the wear products concentration in engine oil, which established the moment of intensive wear of a certain part or group of parts.
2. The diagnostics error does not exceed 10%.
3. The proposed method of evaluation of the technical state of diesel in the process of operation allows to solve the problem of transition to renovation according to technical condition.

5. References

[1] Ovcharenko S.M. (2007). Enhancing the effectiveness of diagnostics system of diesel locomotive. Dissertation in support of candidature for a technical degree. Omsk, 2007. 368 p.

[2] Minakov V.A. (2013). Simulation of the process of wear products accumulation in engine oil of diesel locomotive Д49 / Ovcharenko S.M., Minakov V.A. // Trans-siberian news: Scientific and Technical Journal / Omsk State Transport University. Omsk, 2013, №3 (15). 55-61 pp.

[3] Osovski S. (2004). Neural networks for information processing / Translation from Polish by I.D. Rudinsky. – Moscow: Finance and Statistics, 2004. 344 p.

[4] Galushkin A.I. (2010). Neural networks: basis of the theory. Galushkin. M.: 2010. 496 p.

[5] Ovcharenko S.M., Minakov V.A. Evaluation of the technical state of diesel parts using neural-network data processing / S.M. Ovcharenko, V.A. Minakov // Control. Diagnostics.: Scientific and technical journal of the Russian Society for nondestructive testing and technical diagnostics / Publishing House "Spektr", 2 (212) February 2016. 46-50 p.

DOI: 10.2507/28th.daaam.proceedings.169

CONNECTIVITY OF QUALITY MANAGEMENT SYSTEMS AND ENVIRONMENTAL MANAGEMENT SYSTEMS

Miroslav Rusko, Iveta Pavlova, Jaromira Vanova & Ruzena Kralikova

This Publication has to be referred as: Rusko, M[iroslav]; Paulova, I[veta]; Vanova, J[aromira] & Kralikova, R[uzena] (2017). Connectivity of Quality Management Systems and Environmental Management Systems, Proceedings of the 28th DAAAM International Symposium, pp.1213-1220, B. Katalinic (Ed.), Published by DAAAM International, ISBN 978-3-902734-11-2, ISSN 1726-9679, Vienna, Austria
DOI: 10.2507/28th.daaam.proceedings.169

Abstract

Environmental management is a set of voluntary environmental policies allowing the introduction of a systematic approach for organizations in solving environmental issues. Economic, social and environmental situation in the last century has resulted in the activities represented by the concept of sustainable development In order for development to have a long-term positive effect, it must be sustainable. Sustainable development is about ensuring that the resources of the Earth, which we use to promote economic growth, will be available also to future generations. The attention that is currently paid to the threats posed by global climate change is an example of how human freedom and human development can be undermined when the interests of economic growth and development are separated from the interests of the environment. The current environmental problems' solution is not just a question of technical solution, which until recently was caused by either a response to an environmental problem accrued or within the more progressive approach as a preventive approach. It should be a change in the value of the orientation of a human, as an individual, but also of society and all mankind. As today's consumers are beginning to become increasingly more aware of the problems of environmental pollution, so do producers and trades are beginning to modify their thinking and behavior towards pro-environmental direction and this way try to address their consumers.

Keywords: Environment; development; sustainable development; product

1. Introduction

Excessive consumption of natural resources and reckless environment pollution are among significant global problems of our civilisation. The idea of infinite natural resources that can be used without limits as we wish, the idea that we can sustainably manage them on both global as well as regional level, the idea that the economic growth means progress, that science and technologies will solve all our existing and future problems is dominant with a part of the population. Contrarily, other part of the population radically refuses such approach to solving the current environmental problems. A solution is somewhere in between these two extremes. To contribute to sustainable development as regards not only the humanity but also all life forms on the planet means to reassess the system of our priorities. [8]

It is necessary to limit the excessive polluting of the environment, reduce the population growth, in particular in developing countries, and to limit the excessive consumption in industrialized countries. Over-consumption in industrialized countries and under-consumption in developing countries (dialectically related to over-production and lagging-behind) is the main source of the global instability. [8] The current product policy sees a significant trend in focusing on a value-driven product strategy, i.e. a constant process of developing a product programme based on the value category for the customer. [7]

2. Environmental management systems and quality management systems

The requirements of the legislation in the area of environmental protection is continually tightening and their implementation requires ever greater demands on the security of the resources. To ensure compliance with individual legislation, however, does not guarantee a comprehensive and effective approach to environmental protection over the approach to the protection of the environment based on the system management approach. One of the ways to ensure a more effective and efficient approach to the protection of the environment in the company, to obtain a higher environmental credit and competitive advantage on the national and international market, is to introduce and certify an environmental management system according to the ISO 14001. EMS is the way of the implementation of environmental requirements, an indicator of competitiveness, a tool for fostering of environmental sustainability.

With the growing interest in maintaining and improving the quality of the environment, many organizations are turning with increased rate their attention to the potential environmental impacts of their activities. The interest in achieving and demonstrating of their good environmental performance through the management of the impacts of their activities, including the products and services, they consider the direction of their environmental policy and the objectives of development.

In the context of a tightening of the environmental laws for the development of economic policy, of the achievement of the objectives and priorities of the state environmental policy, of growing in the interests of the stakeholders in environmental issues, including sustainable development, as well as the increase of competitive performance, the organization is forced to reconsider its status and put in place an effective and efficient system the implementation of such, it reaches the meeting of these requirements.

EMS are applicable to the entire spectrum of organizations, from the productive of an industrial scale, in agriculture, in services, in the health sector and trade, in the region having regard to its local needs as well as conditions. An integral part of the building of the EMS is applying of such methods and procedures that allow, with their content and focus, on the basis of the latest findings from the fields of science, technology and engineering, to include a solution for alleviating and eliminating unwanted environmental impacts arising from the activities of the organization. The system affects the organisational structure, planning and control activities, resources, responsibilities, work practices, processes, and resources for the implementation of the environmental policy, objectives and ensuring the continuous improvement of care for the environment of the organization.

EMS is a voluntary instrument in the field of environmental management in an organization, introduced in order to achieve compliance with legal requirements. It includes organizational structure, planning, responsibilities, processes, proceedings and resources for preparing, implementing, reviewing and maintaining the environmental policy of the organization. It allows systematic management to achieve a level of performance that the organization has established by itself through the accepted policy of process and documentation within established environmental management system (EMS). EMS is a tool that brings together the approaches to the protection of the environment with the overall management of the organization in order to achieve environmental and business objectives of the management of the organization. It is applicable for any type of organization in industrial sector, in agriculture, in services, in the health sector and trade, in the financial sector or the public sector.

Slovakia's entry into the European Union has affected many changes in the business. For increasing competitiveness, there are created various tools for effective management of the organization and ensuring the quality of desired products outcomes in organizations. By increased demands and pressures from all stakeholders, we need to find a way how to react flexibly to their needs. Due to the trend of integration of management systems and to facilitate inerrability of individual management systems and for increasing the flexibility of management systems of enterprises, the Annex SL to Regulation ISO/EIC "Consolidated Supplement - Specific Procedures to ISO" was founded in 2012. According to this regulation, all standards for management systems that arise respectively will be revised after 2013 must have the same structure. All these aspects are reflected in the draft revision of ISO 9001:2015 and ISO 14001:2014. Organizations will have three years to be able to incorporate into their existing system the requirements and comprehensive approach defined in ISO 9001:2015 and ISO 14001:2014. From 2019, it will be necessary to fully comply with the requirements defined in these standards revisions. Availability of the draft ISO 9001:2015 and ISO 14001:2014 showed major shift in attitude and perception of quality and environment management. Some of the approaches are transferred from the requirements of the standard TS 16949 (automotive industry) or TQM models (EFQM excellence model).

The most important voluntary instrument of environmental care includes:

- environmental management systems,
- environmental assessment and labelling of products,
- green public procurement. [10]

In terms of the most elaborated tools with high efficiency it is currently recommended in building and implementing the EMS the use of two standardized tools: EMAS III and ISO 14001: 2015. A system approach to environmental management can be by:

- EMS (Environmental Management Systems. Requirements with guidance for use) according to ISO 14001: 2015
- EMAS (Eco-Management and Audit Scheme) - EU Regulation 761/2001 of 19.3.2001 on the voluntary participation by organizations in a eco-management and audit scheme (EMAS II) amended the EC Regulation 1836/1993 (EMAS I) on the voluntary participation of industry at Community program of environmental management and assessment approved by the Council of Ministers EC 29.6.1993 (entered into force on 13.7.1993 with effect for individual Member States of the EU from 13.4.1995). It was subsequently amended by Regulation of the European Parliament and Council Regulation (EC) No. 1221/2009 on the voluntary participation by organizations in a Community eco-management and audit scheme - EMAS III. The SR was to promote EMAS, the Act NR SR. 351/2012 on environmental verification and registration of organizations in the EU Eco-Management and Audit Scheme and on amendments to certain laws.
- Voluntary tools of an environmentally oriented product policy can include tools standardised by:
    - legal regulation (e.g. EMAS – legal regulation within the EU as well as SR)
    - legal regulation and standard (e.g. type I environmental labelling – legal regulation within the EU as well as SR and ISO 14024)
    - standard (e.g. lifecycle assessment, standards of ISO 14040 series)
    - procedure specifying the certification conditions (e.g. PEFC), or the respective calculation (e.g. ecological, carbon or water footprint)
- non-standardised tools (e.g. environmental marketing, environmental benchmarking).[11]

Review of the EMS certification according to the ISO 14001 development of annual increase in organizations with a certified EMS in Slovak Republic are given in Table 1.

Year	Number of certificates	Year	Number of certificates
1990	1	2006	103
1996	1	2007	109
1998	8	2008	170
1999	13	2009	150
2000	17	2010	118
2001	23	2011	91
2002	33	2012	59
2003	41	2013	69
2004	51	2014	94
2005	77	2015	62

Table 1. EMS certification in Slovak Republic

3. Sustainable production and consumption

The issue of sustainable consumption and Eco labelling was included into the agenda of a special meeting of the UN General Assembly – Earth Summit +5, the World Summit on Sustainable Development in Johannesburg, as well as into other important international activities. Poverty and environment degradation are closely interrelated. While poverty results in certain kinds of environmental stress, the major cause of the continued deterioration of the global environment is the unsustainable pattern of consumption and production, particularly in industrialised countries, which have serious consequences, aggravating poverty and imbalances. Measures to protect and improve the environment have to take into account the current imbalances in the global patterns of consumption and production. Development should include an effort to achieve sustainable consumption patterns.

At the same time, the society needs to develop effective ways of dealing with the problems of disposing of mounting levels of waste products and materials. Institutions should make a concerted effort necessary to reduce generation of wastes by:

- encouraging recycling in the industrial processes and at the consumed level;
- reducing wasteful packaging of products;
- encouraging the introduction of more environmentally sound products.[4]

Governments themselves play an important role as regards consumption, particularly in countries where the public sector plays a large role in the economy and can have a considerable influence on both corporate decisions and public perceptions. They should therefore review the purchasing policies of their agencies and departments so that they improve, where possible, the environmental content of government procurement policies, without prejudice to the international trade principles. An appeal for sustainable consumption was already made during the Earth Summit in Rio de Janeiro in 1992. Chapter 4 – "Changing Consumption Patterns" of the AGENDA 21 document concerns two programme areas: unsustainable patterns of production and consumption, and developing national policies and strategies to encourage changes in unsustainable consumption patterns. [4] Contrarily to the idea of sustainable development focusing in particular on the business area, sustainable consumption is a task especially for the consumers - individuals or the state. It is necessary for the consumers to change their purchase behaviour. However, they cannot do it just by themselves. Sometimes they lack information about how to live "more sustainably".

The European Parliament and the Council of the EU, having regard to the Treaty establishing the European Community, and in particular Article 175(3) thereof, stated that the Community's fifth environmental action programme (5EAP) "Towards Sustainability" ended on 31 December 2000 delivered a number of important improvements, and that continued effort is required to meet short-term and long-term environmental objectives as set out by EC.

The 6th EAP relates to a period of ten years (2002 – 2012). The respective initiatives to meet long-term objectives in various policy areas consist of a whole series of measures, including legal regulations and strategic approaches. The document, besides other, states that a prudent use of natural resources and the protection of the global eco-system together with economic prosperity and a balanced social development are a condition for sustainable development. The 6th EU Environment Action Programme (6EAP) states that the consumers need information to help them chose more environmentally-friendly products with the aim of positively affecting the market. [13]

The 7th Environment Action Programme (EAP) will be guiding European environment policy until 2020.[1]

In order to give more long-term direction it sets out a vision beyond that, of where it wants the Union to be by 2050: "In 2050, we live well, within the planet's ecological limits. Our prosperity and healthy environment stem from an innovative, circular economy where nothing is wasted and where natural resources are managed sustainably, and biodiversity is protected, valued and restored in ways that enhance our society's resilience. Our low-carbon growth has long been decoupled from resource use, setting the pace for a safe and sustainable global society."[2]

It identifies three key objectives:

- to protect, conserve and enhance the Union's natural capital
- to turn the Union into a resource-efficient, green, and competitive low-carbon economy
- to safeguard the Union's citizens from environment-related pressures and risks to health and wellbeing.

Four so called "enablers" will help Europe deliver on these goals:

- better implementation of legislation
- better information by improving the knowledge base
- more and wiser investment for environment and climate policy
- full integration of environmental requirements and considerations into other policies.

Two additional horizontal priority objectives complete the programme:

- to make the Union's cities more sustainable
- to help the Union address international environmental and climate challenges more effectively.

The programme entered into force in January 2014. It is now up to the EU institutions and the Member States to ensure it is implemented, and that priority objectives set out are met by 2020.

4. Environmentally business policy

Business and industry, including international companies, play a decisive role in the social and economic development of a country. A stable political regime creates suitable conditions for business and industry, allowing to work responsibly and efficiently and to implement long-term strategies. Business and industrial activities are the foremost contributors to the prosperity growth, which is the main goal of development. Business companies - small and big, formal and informal - create the majority of working and business opportunities, as well as opportunities to make a livelihood. Suitable business and industrial policy and its practical implementation which makes use of more efficient production procedures, applies preventive strategies, uses cleaner technologies and processes in the whole life-cycle of a product and thus reduces generation of wastes, can help reduce the negative impact on the resources and the environment. Within the environment policy we can distinguish - regional environmental policy, environmental policy concerning investments, environmental policy concerning waste, product policy, environmentally oriented consumption policy and substance flow policy (see Figure 1).[9]

Management of some enterprises understood the context and introduced a system of responsible care and programmes of product stewardship which lead to openness and dialogue with employees and the public; they carry out environmental audit and evaluate fulfilment of individual programmes. A growing number of such management members voluntarily and at their own initiative take over the responsibility and manage their activities so as to minimise their impact on human health and the environment.

The situation was helped by regulation measures adopted by individual countries, such as higher awareness of consumers and the public, and awareness of the management in business and industry. A positive approach of business and industry to sustainable development can be increased by economic tools such as free-market mechanisms which allow for the product and service prices to reflect the environmental costs of their inputs, production, use, recycling and final disposal in a growing degree and in compliance with the specific conditions in individual countries.[4]

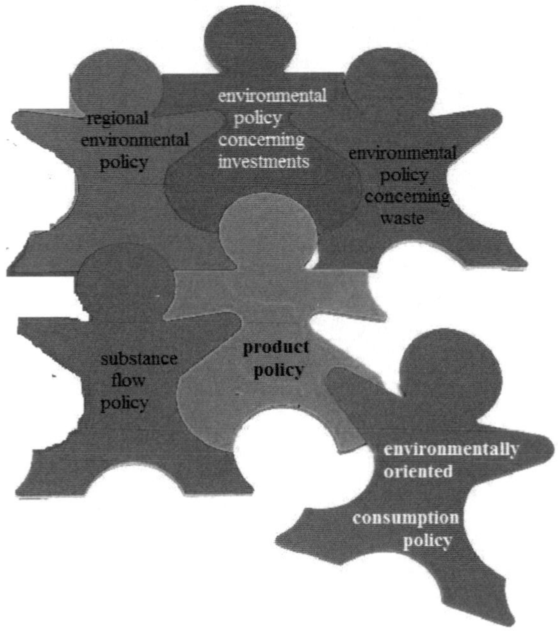

Fig. 1. Suitable business and industrial policy

Green purchasing is an example for the broad public and private sector, and it affects the market. Introduction of a policy of green public procurement increases the demand for environmentally sound products, which encourages the consumers of such products. Promotion of green public procurement gives incentives for development of industry, development of green technologies and products and promotes them on the market. Small and medium enterprise, in particular, can gain profit from green public procurement as it offers opportunities to find markets for their innovative solutions and products.

Green public procurement (GPP) is a way of integrating environmental requirements into the process of public procurement by public bodies through technical requirements or criteria for tender evaluation. The importance of green procurement is especially in the fact that it creates a market demand for the environmentally friendly products and thus promotes offer of these products. Introduction of an environmental point of view into public tenders can motivate producers to develop new products and technologies with less negative impacts on the environment as well as motivate suppliers to introduce environmental management systems. GPP is one of the set of efficient tools for promotion of production processes which are environmentally friendlier and stimulate supply of goods and services with the high value which is characteristic by its environmental benefit throughout the product´s lifecycle. Follow the steps of a simple management model for systematic implementation of GPP in the organisation, Figure 2.

In October 2004 the first international conference of ecological sellers took place in the town of Sendai (Japan). More than 1,000 representatives gathered to confirm the importance of an ecological sale and joint international activities in the area. In the Sendai Declaration adopted at the conference, all stakeholders agreed upon establishment of an international organisation (International Green Purchasing Network - IGPN) with the main task of promoting ecological purchasing at global level. IGPN was established next year by various shareholders from international organisations, associations, government organisations, regional authorities and NGO and it took over the initiative to implement ecological purchasing.

IGPN collects and delivers information on global green purchasing activities, case studies on development of new eco-products and eco-services and know-how, purchasing guidelines, information on products, research outcomes and the latest trends, as well as on international conferences and workshops focusing on promotion of ecological purchasing around the world. IGPN develops internationally harmonised principles of green purchasing, tools for product evaluation, constructs a database of eco-services and products, a database of environmental directives and regulations, training tools, and measurement assessment methods. The objective of IGPN is to spread the green purchasing activities globally to establish ecological supplier networks, which shall contribute to establishment of a sustainable society.

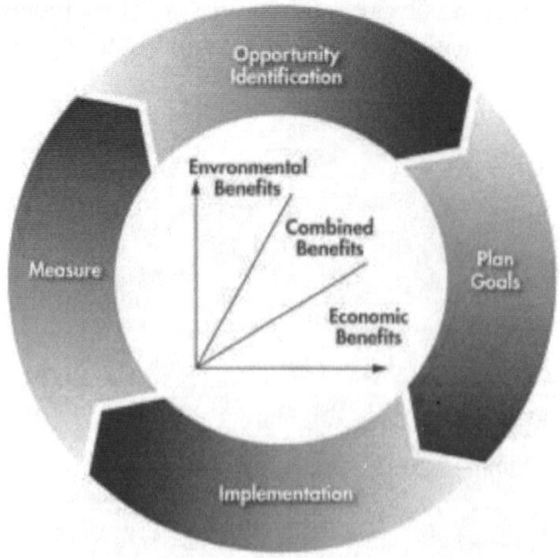

Fig. 2. Simple management model for implementation of GPP

5. Product policy and environmental tools of its implementation

Environmentally oriented product policy is in general determined by the relationships between its aims – subjects – objects - tools. It is based on the integrated approach to the product life cycle, which anticipates an enormous amount of information. It has to solve the questions of the international trade as well as the rules of the International Trade Organization. New forms of preventive environmental strategies and especially Green Marketing are being introduced helping to solve environmental problems and environmental motivation of producers. Many producers are faced with great attention of the public to their approach to the environment. Despite the fact that the customers buy products fairly prudently and their behaviour is markedly affected by prices, a particular part of the population prefers product that do not burden the environment. This brings about a situation, in which the producers within their mutual competition and in relation to customers are enforced to behave responsibly.

Product policy means analysis, planning, implementation and control of activities (measures, decisions) related to the product as a marketing tool, for instance the external design of a product, composition of a product series, brand policy etc. It is part of a marketing mix. It is closely related to the quality policy. Strategic considerations within the product policy of a company in a competitive environment are dominated by structural and company factors which determine the competition. Success of a company is nowadays based on the philosophy of preparation of products and their launch onto the market where they satisfy needs, wishes and requirements of consumers, or they themselves create these needs, and on the business philosophy based on searching of customers for the existing products offered on the market. Strategic decision-making of businesses in conditions of competition and market globalisation requires knowledge of markets, orientation on the quality, cost reduction and innovations. Understanding of these processes requires obtaining knowledge and being able to apply it in a creative way. The product and its three levels are the basis of the marketing mix and a means to achieve profit from a long-term perspective. The theory on the product and quality can be partially divided into two areas as regards the subject they examine:

- a general theory of product examining the product in all its various forms as regards the requirements and possibilities of its origin, its efficient value, usefulness, quality, processes, activities and their outcomes,
- a specially-focused theory of product examining individual groups of products, internally differentiated or interlinked to individual elements of processes and activities of market entities and the value chain of an entrepreneur company. [5]

The product policy means, in particular, the following 4 areas of interest:

- development and launch of new products on the market
- adaptation of the offered quantity of products to the real needs (based on monitoring the frequency of needs)
- the so-called "product production" (adaptation of technical features, design etc. according to the customer requirements and needs)
- monitoring of the life span of products. [11]

Instruments of the product policy can be applied at the level of the government product policy or the so-called product management which expresses the social responsibility of businesses, consumers and the state.[11]

A company has to define what sustainability means for its business (from the environmental, economic, social as well as ethical point of view) because its approach to the product issues will depend on its vision, obligations and type of business, i.e. the area of products or services.[3]

Instruments of the product policy can include taxes and subsidies supported by a legal regulation:

- price ones supported by a legal regulation
- economic and regulatory ones supported by a legal regulation
- public procurement supported by a legal regulation
- distribution ones (consumption, logistics)
- communication ones
- technical normalisation
- voluntary (standardised, non-standardised).

There are tendencies in the environmental product policy to extend indirect instruments based to a rather great degree on voluntary activities of producers and a growing environmental awareness of people. Systems of environmental assessment and labelling are implemented to declare that a product meets the parameters complying with the current knowledge, minimises or even eliminates adverse effects on the environment. There is also an effort to assess the product throughout its whole life cycle, i.e. from raw material, through its production, up to its disposal or recycling. Application of these systems results not only from the effort and requirements of state environmental bodies but also from the interest of producers themselves and the consumers.

Labelling of products that are less pollutant for the environment by a state-guaranteed environmental label is a respected and motivating indirect instrument of the product policy. Consumers are thus provided with guaranteed information on the products, the use of which contributes to the environment protection. [6] "Ecolabelling" is a voluntary method of environmental performance certification and labelling that is practised around the world. An "ecolabel" is a label which identifies overall, proven environmental preference of a product or service within a specific product/service category. There are different classifications of label. In contrast to "green" symbols, or claim statements developed by manufacturers and service providers, the most credible labels are based on life cycle considerations; they are awarded by an impartial third-party in relation to certain products or services that are independently determined to meet transparent environmental leadership criteria. Environmental labelling exists as standardised or without any voluntary or obligatory regulation. It is a natural manifestation or product of at least two significant interest groups – a part of producers or businessmen who want to increase the competitive features of their products through qualified labelling on one hand, and a part of consumers, involved environmental activists, researchers and state officers who want to provide or ensure an option to a common as well as industrial consumer to choose those products from the products on the market that have more reliable environmental characteristics.[12]

The LCA (Life-Cycle Assessment) method is one of the most important information tools for environmental policy. It is used for determining adverse environmental impacts of any system (product or service). Information that is provided by the LCA method enables both experts in eco-design and enterprise marketing to select new products and employees of the state administration and in the area of Eco labelling to stipulate regulations and criteria for the support of environmentally friendly products. Last but not least, this information is significant for further scientific and technical development as it realise the causes of harmful effects on the environment. The LCA method can then be summarily defined (according to ISO 14040) as: compilation and evaluation of the inputs, outputs and the potential environmental impacts of a product system throughout its life-cycle. It should be emphasized that the concept of a „product life-cycle“ is not identical with the concept of the „service life of a product“. The service life of a product is the period during which a product can be used for the purpose for which it was manufactured and thus constitutes only a small part of the life-cycle of a product.

The eco-design is a systematic process of design and development of a product, that apart from the classic properties as a functionality, economically, security, ergonomically, technical feasibility, aesthetically, a so one, accentuates on the achievement of minimal negative impact of a product on the environment namely in term of whole life cycle. The estimate of the environmental profile of a product has a significant role, it´s a specification of all significant factors, by which the product affects on the environment during its whole life cycle. The eco-design is one of the important prevention oriented voluntary regulative tools of the environmental policy.

The preventive bearing of the eco-design is given by that the request for the release of negative impact of a product on the environment is integrated already to the product reflections beginning, so it´s going on the most preventive and in term of the practise the most economic approach to the release of negative impacts of products on the environment.

6. Conclusions

From the analysis of different ways of the environmental protection in 80´s of the last century arose that the most effective and the most economic ways of the environmental protection are based on prevention, then a research and a liquidation of causes, which evocate the contamination of the environment. The application of this preventive strategy for the production processes conduced to the formation of the cleaner production. Over the past decades the European Union has put in place a broad range of environmental legislation. As a result, air, water and soil pollution has significantly been reduced. Chemicals legislation has been modernised and the use of many toxic or hazardous substances has been restricted. However, many challenges persist and these must be tackled together in a structured way.

Successive environment action programmes have provided the framework for Union action in the field of the environment since 1973. Suitable business and industrial policy and its practical implementation which makes use of more efficient production procedures, applies preventive strategies, uses cleaner technologies and processes in the whole life-cycle of a product and thus reduces generation of wastes, can help reduce the negative impact on the resources and the environment. A positive approach of business and industry to sustainable development can be increased by economic tools such as free-market mechanisms which allow for the product and service prices to reflect the environmental costs of their inputs, production, use, recycling and final disposal in a growing degree and in compliance with the specific conditions in individual countries.

Directions of further work in the field of research task will to Increase of environmental consciousness level among students to study the realities of the work of company (carrying out of audits within the confines of tutorials), starting the project, which could be a collaboration between university and industry in the field of environmental protection and also could help to solve ecological problems worrying Slovak companies. The main focus of next research will be oriented too on two factors determine the success of labelling systems: one is the awareness of consumers and their ability to understand their message, and the other one is the business community's willingness to accept the systems.

7. Acknowledgements

This paper was written in frame of the work on the project VEGA 1/0990/15 The readiness of industrial enterprises to implement the requirements of standards for quality management systems ISO 9001:2015 and environmental management systems ISO 14001: 2014 and project KEGA 048TUKE-4/2015Transfer of knowledge from scientific - research activities into multimedia education process in the subject "Environment and manufacturing".

8. References

[1] Decision No. 1386/2013/EU o General Union Environment Action Program to 2020 „Living well, within the limits of our planet". Available from: http://eur-lex.europa.eu/legal-content/EN/TXT/?uri=celex% 3A32013D1386. Accessed: 2017-02-04

[2] Environment Action Program to 2020. Available from: http://ec.europa.eu/environment/action-programme/ Accessed: 2015-03-07

[3] Charter, M. (2001) Ecodesign management. Environmental marking. 2, 3., Prague, pp. 9-11, ISSN 1212-476

[4] Klinda, J.(1996) Agenda 21 and sustainable indicators. Bratislava, 517 p., ISBN 80-88833-03-5

[5] Kollar, V. et al. (2003) Product policy. Bratislava: Sprint VFRA, 560 p. ISBN 80-88848-96-2

[6] Sustainable Purchasing Available from: http://oxmansafety.com/green_hotels/sustainable_purchasing/ Accessed: 2015-07-09

[7] Nuij R.(2001) Eco-innovation: Helped or hindered by Integrated Product Policy. The Journal of Sustainable Product Design, Vol. 1(1). pp 49-51. ISSN: 1367-6679

[8] Prochazkova, D. ; Wessely, E. ; Rusko, M. & Kralikova, R. (2011) Human System Safety Management and Environmental Management Relation. In DAAAM International Scientific Book. Chapter 09 in DAAAM . pp.103-118. ISSN 1726-9687.

[9] Rubik, F.; Petschow, U. & Scholl, G.(1995) Basic outline the environmental oriented product policy. Workshop: Instruments of the Product Oriented Environmental Policy of the Middle and Eastern European Countries 20.-22.11.1995, Warsaw. Heidelberg/Berlin: IŐW, p. 13

[10] Rusko, M. & Kralikova, R. (2012) Implementation of environmental management systems in the Slovak enterprises. In: DAAAM International Scientific Book 2012. Vol. 11, no. 31 (2012), p. 369-378. ISSN 1726-9687.

[11] Rusko, M.; Kralikova, R.; Mikulova, M. & Ilko, J. (2016) Labeling of Products from the Context of Environment, Quality and Safety, Chapter 37 in DAAAM International Scientific Book 2016, pp.419-434, B. Katalinic (Ed.), ISSN 1726-9687, Vienna, Austria. DOI: 10.2507/daaam.scibook.2016.37

[12] Suchanek, Z.(1998) ISO standards for environmental labeling - an effective environmental management tool or a conflicting topic ? In RUSKO,M.[Ed.] Environmental management (EMS, EMAS), Banska Bystrica.1998

[13] Tresnak, P.(2001) 6th EU Environment Action Program. In: Environmental aspects of business, Prague, pp. 4.

DOI: 10.2507/28th.daaam.proceedings.170

KNOWLEDGE MANAGEMENTS FOR IMPROVEMENT THE COMPETITIVENESS OF ORGANIZATION

Iveta Paulova, Jaromira Vanova, Miroslav Rusko,
Edita Hekelova & Ruzena Kralikova

This Publication has to be referred as: Paulova, I[veta]; Vanova, J[aromira]; Rusko, M[iroslav]; Hekelova, E[dita] & Kralikova, R[uzena] (2017). Knowledge Managements for Improvement the Competitiveness of Organization, Proceedings of the 28th DAAAM International Symposium, pp.1221-1226, B. Katalinic (Ed.), Published by DAAAM International, ISBN 978-3-902734-11-2, ISSN 1726-9679, Vienna, Austria
DOI: 10.2507/28th.daaam.proceedings.170

Abstract

The aim of this contribution is to present some results of the survey in a particular organization that has implemented and certified an integrated management system, and more recently was recertified in accordance with the requirements of new standards ISO 9001: 2015 Quality Management System, ISO 14001: 2015 Environmental Management System and OHSAS 18001 Occupational Health and Safety Assessment Specification. In current approaches to quality and environment management there is accentuated the need of knowledge of an organization and of its management. A novelty in the standards is the so called area of the context and the knowledge of an organization, a part of that is also the change of the organization to the knowledge one in order it to be competitive and unique. Based on these results, we have proposed some measures that would help to the effective fulfilment of the requirements and accession to the knowledge management that could lead to way to be competitive.

Keywords: knowledge; management; competitiveness; quality; system; environment

1. Introduction

Currently, we are exposed to large turbulence of changes, the stimulus of that are often unexpected changes, in particular, from the external environment. If the organizations want to win the competition and at the same time to handle the pressures, they must focus in particular on human potential that is leading to the so-called learning organization and this is only a step towards the knowledge management.

Today, there is a growing recognition in the business community about the importance of managing knowledge as a critical source for competitive advantage. Researchers in the field of sustainable competitive advantage have discovered that knowledge, that includes what the organization knows, how it uses what it knows, and how fast it can know something new is the only thing that offers an organization a competitive edge. Knowledge is the thermonuclear competitive weapon of our time; knowledge and its management are more valuable and more powerful than natural resources, big factories, or fat bankrolls [5].

For the restriction of negative influences on sustainable development assurance, some relevant measures are adopted, that have as a task to inform, mobilize, and regulate an activity to limit or eliminate its influences [7]. A positive approach of business and industry to sustainable development can be increased by economic tools [10]. It requires the environmental oriented monitoring in the manufacturing company and the optimization of management [14]. Information is a specific type of asset that represents value for the company (base for decision making processes) and has to be treated carefully [2]. The concept of knowledge management takes into account all of the above-mentioned areas [6].

The concept of the learning organization is connected with the name of Peter Senge. According to him, the learning organization is the place: "where the people still improve their ability to achieve the desired results, where a flexible and systemic thinking is supported, where people learn and work together" [11].

Learning organization is the one that creates for its people all the conditions, so they can learn anywhere and constantly [1]. However, it is not about an organization that has drawn up a plan of education, and in which all employees are involved in management training. It is the organization's ability to cope with rapid changes, its readiness to respond flexibly and successfully to survive in changing conditions [9].

2. Knowledge of an organization in the context of the standard ISO 9001: 2015 requirements

Our present era is characterized by the so-called information revolution, that motivated the governments to recognize its importance and its implications on the development of social, economic, cultural and political [12]. The organization, that is characterized by how readily it is adapting to the market, by transferring the responsibility of deciding on their staff and by readiness to change, can be described as a learning organisation that results in the knowledge management. An organization that learns with the correction of its behaviour and decisions, and in doing so it is based on the information obtained from both the internal and external environment. Internal environment can be, for example, the innovation of new products, the results of the analysis of corporate climate, suggestions or criticism on briefings, financial reports. Important are also some information from the external environment - customer complaints, opening of new markets, new technologies existence, competition proceedings, changes in the customer's interests [8]. The learning organization requires systemically managing knowledge that are specific to the organization. The big positive is the fact that these approaches are already enshrined in the new requirements of the standard ISO 9001: 2015.

2.1. Approach of managers in a learning organization

The philosophy of a learning organization requires major changes in the thinking of executives - of managers, at which the role of a manager is changing to the one of a leader. If the manager decides on everything, he does not have to deal with more time-demanding informing the people, with collecting of their views and ideas. This has the advantage for subordinates that, in turn, it is not required of them to be critical or creative thinking; they have just to carry out regulations. With that there is closely related an art to communicate. The keystone of success in good communication is the clarity and intelligibility, for to get things done in a real time [3]. We were based on these ideas when we were performing an analysis of the level of perception of the employees in these areas.

Organizations are increasingly realizing that the learning and development activities must be based on the strategy of the organization. This means that the management is increasingly going to have to put more emphasis on the organizational needs and to allow people to develop so as to promote the implementation of the principles of a learning organization, and not to aim on short-term training activities. Such a learning requires continued and meaningful interaction between people and such a working environment that supports and stimulates learning, and hence the growth of the organization and the individuals in it. It is, however, necessary to make changes, especially in organizations where there is prevalent a tradition of rules, in which the structure, systems and procedures were introduced to maintain stability by preserving the status quo, at any cost.

2.2. Standard requirements for knowledge management of organisation

The current approach to the quality management system can be described as an approach to the pursuit of the success and competitiveness of organizations. The chapter 7.1.6 The knowledge of the organization in ISO 9001: 2015 standard is devoted to this area. This is a new requirement of the standard from that it is clear that knowledge of the organization are an important source. According to the ISO 9001:2015: "Organizations shall determine the knowledge that is necessary for the functioning of its processes and for achievement conformity of products and services. In addressing some changing needs and trends, the organizations must consider their current knowledge and determine how to acquire all the necessary knowledge and required updates or how to ensure an access to them. Knowledge of organization can be based on internal resources (intellectual property, knowledge gained through experience, lessons learned from the mistakes and successful projects, collecting and sharing of non-documented knowledge and experience, etc.). It also can be based on external sources, such as participation in conferences, collaboration with academic communities, and the acquisition of knowledge from customers or from other interested parties."

Management of the organization includes, in particular, the setting of the entire management system – setting values and rules of the organization, setting up an organizational structure, resource management, processes and performance settings. In the interest of the organization is to determine all the necessary management skills of the organization, for knowledge of its employees to be continuously developing and ensured and to be available to all who need them. [15]

3. Implementation of the requirements of knowledge management and learning organization on the example of a particular production organization

In the framework of the research projects we carried out, inter alia, a survey of the level of the requirements fulfilment of the knowledge management in a particular production organization. There was determined the level from the perspective of the individual options of personal growth, some opportunities to develop the skills of staff as well as the possibility of using diverse modes of learning.

3.1. Methodology of the survey

Analysis (survey) of selected areas, that touches on elements of a learning organization making for knowledge management, was made in the form of a questionnaire in a manufacturing organization with a staff of 320, that means a large organization is concerned. The organization has implemented the system of quality management (STN EN ISO 9001: 2016) [13], of environment (ISO 14001:2015) and of security (OHSAS 18001) management. The survey was attended by 270 employees, i.e. by 75,7 % and was carried out following the recertification to new standards in the field of quality and environment management.

3.2. Results of the survey and a summary of the results

The results of the survey (Fig. 1. and Fig. 2.) show that the staff perceives the possibility of personal individual growth and the further professional growth as critical or very important, and demonstrate a level satisfactory of the satisfaction. Figures are tied up or very closely related to each other.

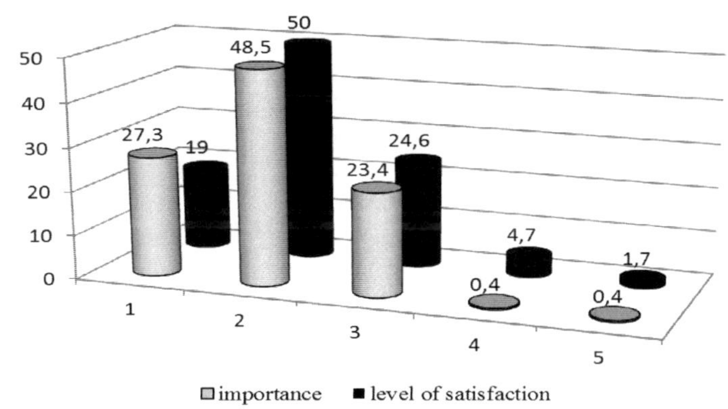

Fig. 1. The ability of an individual's personal growth (date are in %)

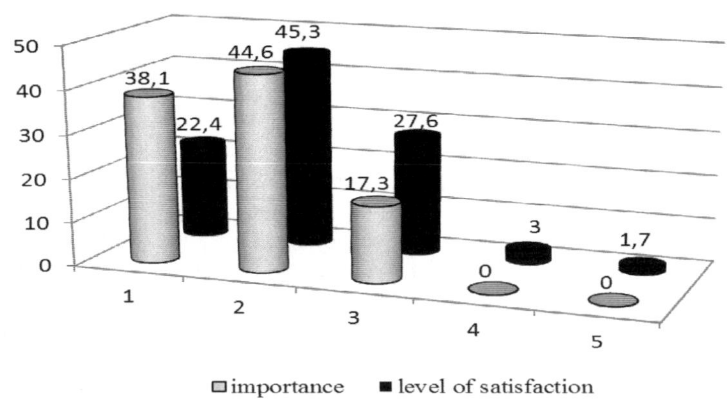

Fig. 2. The ability of professional growth (date are in %)

An excellent result - very satisfying – is said by about 20% of the employees. A third of employees, found the standard condition (they expressed neutrally). From that results that they are not interested in their personal and professional growth. At a more thorough analysis of the individual questionnaires (importance vs. satisfaction rate and education level) there was manifested a kind of a direct link between:

- the importance of personal and professional growth and the level of education,
- the satisfaction with personal development and further education and the level of education.

Direct dependence mostly appeared also between the importance and level of satisfaction (assessment of the importance and degree of satisfaction was almost the same in all cases), however, we believe that employees questioned misunderstood the difference between the importance and satisfaction. For this reason, we focus on the level of satisfaction.

A considerable portion of the interviewed marked a degree of satisfaction – rather satisfied (45– 50%)) and one third of the questioned has declared as neutral (neither satisfied nor unsatisfied). At a more detailed analysis, it was found that the university-educated inquired ones were less satisfied with a personal and further professional development, as confirmed in the responses to the other related issues. The sample given has an interest in constant education and to be competent to carry out their activities and to progress in career growth. The responses at the level of "very happy" have been identified in the sectors that are the support for the production management, that is, in the area of services and administration (mostly covered by education in the field of legislation and laws). Less satisfactory ratings of satisfaction were also identified in the director-general section, where there is included also the top and middle management of organization.

Next survey was focused on the results of the various target groups (depending on the function and level of educational attainment), where they showed direct proportion between greater satisfaction with the diversity of learning and its achievement. Some dependence can also be observed depending on the age of the asked. The survey showed that younger respondents with a university degree expect more possibilities of method or form of learning and they are interested in new lessons also from other areas than professional ones and that could help also to personal development. This is a very good potential of employees to approaching a learning organisation and knowledge management. Similar results we found in the other survey, where most of the organizations (72 %) planned to create the conditions for knowledge management, as a basis for the fulfilment of new approaches to the quality and environment management, leading to an increase in the competitiveness.

Positive outcome of the survey is the fact that the organizations created a good environment for the development of knowledge and experience. There exist some opportunities for further individual education, however, it is not about the systemic approach, but the ad hoc solution. If an organization has an interest in learning organization as a factor of competitiveness, it is necessary to create the systemic approach to the development of knowledge and experience on the basis of certain rules, which the organization determines by itself.

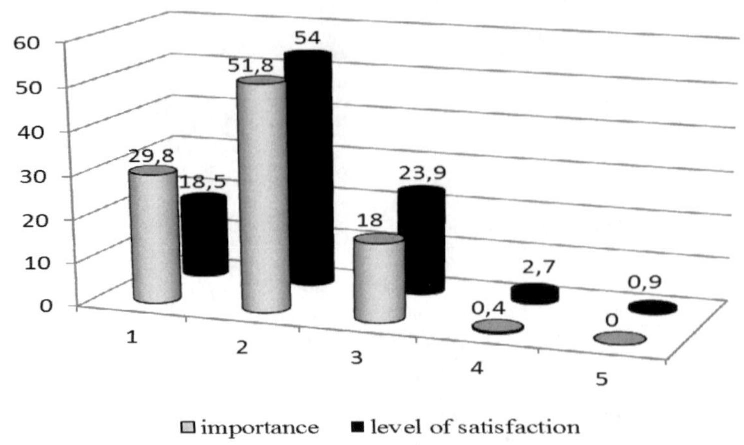

Fig. 3. The use of the knowledge gained by educating

The Figure 3 reflects the attitudes that have been analysed from previous pictures. This figure just confirms the areas and their direct dependence on each area leading to the knowledge management. By comparing the results of individual questions in the survey in a particular organization it has been found and confirmed that the organization that has a culture of a learning organization and has a plan to manage knowledge as a tool for competitiveness, must at first create an environment in organization, that will allow the key employees to meet the expectations in the field of personal development, of continuing professional training, the use of existing knowledge and provide the power to decide at the activities related to the improvement of processes and activities. Key staff should be an example to others, thus ensuring the gradual spread of knowledge to all levels of management and to all employees.

The key role is primarily of top managers, who have to change their role from managers to the leaders. The basis is, of course, made up of a strategy that is directed toward learning organization as one of the deciding factors of competitiveness.

By comparing the current trends that are processed in specialist publications of significant experts in the field given, as well as the requirements of the new standards on quality and environment management systems, it can be stated, that knowledge management is a way, which is important for improving the competitiveness of the organization.

4. Opened questions and direction for further research

The key question is a "will" to create and to record knowledge, because this path is a concept of several time-consuming steps. The real problems are high and exaggerated expectations of the system itself and also the fact that the concept in this area is still being formed, so it is difficult to define whether the offered system is a knowledge system. Just because someone claims that their information system includes document management features, artificial intelligence and knowledge management does not mean that it is really doing so. [4]

The further way of routing the research in this area should aim to address the following questions:

- Is the management aware of the need for knowledge of the organization as a basis for its growth and competitiveness?
- Are there, in general, identified knowledge in business subjects that is stored in a way so that they are available to all who need them?
- Which requirements and priorities are considered by management when deciding on the knowledge of the organization?
- What is important at the departure of the employees of the company, who are the holders of the key knowledge? What is necessary to ensure for the organization to protect this resource?

We do not consider these questions to be complete and final, but as some starting point of the solution of the issue in theory and practice, as well as a way of adapting to the requirements of the ISO 9001:2015 standard.

5. Conclusion

Within the resource-based perspective, information and knowledge have become increasingly recognized as competitive differentiators. In the knowledge-based view, organizational knowledge - such as operational routines, skills or know-how - is acknowledged as the most valuable asset. Currently, an industrial society is transformed into an information society.

The main task of knowledge management is to obtain the "correct" knowledge in the required time for specific employees. For this reason, it is an inevitable requirement of the learning organisations approaching the knowledge management to develop the concept or a system, resp., to ensure a feedback from employees and to listen to their "voice". The organisation must have a corporate culture and create an environment for the possibility of development of knowledge. This fact should be reflected in the values of the organization. If the organization wants to be competitive and manage knowledge sharing for the effective fulfilment of the objectives, it must be reflected in the strategies of the organization, in the form and management methods of its managers (leading to the change of managers to the leaders), in the organization culture, and in the processes in the organization as well. The knowledge management approaching the competitiveness is a matter for the people, of the employees of the particular organization, not the single technology or the use of a quantity of tools and methods that the staff is not able to use and to be uselessly absorbed in the chaos of non-systemic solutions.

Researchers in the field of sustainable competitive advantage have discovered that knowledges of organisation that offers an organization a competitive edge. Knowledge is the competitive force of our time, knowledge and its management are frequently more valuable than natural resources or fat bankrolls.

Knowledge management is based on the fact that the achievement of a sustained competitive advantage depends on firm's capacity to develop and deploy its knowledge-based resources. However, not all resources are equally important for the achievement of this competitive edge. Advance should be based an integrative framework of human capital combining from three different areas of research: knowledge management, intellectual capital, and strategic human resource management. Further research may totally change the focus of knowledge management, but only time will tell about this.

6. Acknowledgements

The paper is a part of VEGA 1/0990/15 research project The readiness of industrial enterprises to implement the requirements of ISO 9001: 2015 standards for quality management systems and of ISO 14001: 2014 standards for environmental management systems. in the frame of the project KEGA 048TUKE-4/2015 "Transfer of knowledge from scientific-research activities into multimedia education process" provided by the Ministry of Education, Youth and Sports of the Slovak Republic.

7. References

[1] Author's team. (2002) Knowledge management systems. Theory and practice. Thomson Learning. Barnes S. [Ed.] British Library Cataloguing-in-Publication Data. ISBN 1-86152-616-4.

[2] Cerna, M. (2013) Aspects of Information Management in Context with IS Selection by SME. – In: 24th DAAAM International Symposium on Intelligent Manufacturing and Automation, Katalinic, B. [Ed.] 1877-7058, Published by Elsevier Ltd. Selection and peer-review under responsibility of DAAAM International Vienna doi: 10.1016/j.proeng.2014.03.050

[3] Easterby-Smith, M., Lyle M.A. (2011) Handbook of organizational learning and Knowledge management.. - WILEY &Sons Ltd, Oxford, U.K. ISBN 9780470972809.

[4] Gregor, M., Grznar, P, Pedan, M., Cudrakova, M. (2016) Knowledge in Healthcare. – In: Proceedings of the 26th DAAAM International Symposium, pp.1115-1121, Katalinic, B. [Ed.] Published by DAAAM International, ISBN 978-3-902734-07-5, ISSN 1726-9679, Vienna, Austria. DOI: 10.2507/26th.daaam.proceedings.157

[5] Hurt, U., Tauno O., Kõrbe Kaare K., Koppel O. (2013) New Approach to Knowledge Transfer Environment Development. In: Procedia Engineering, Volume 69/2014, ISSN 1877-7058, Zadar, Croatia, Published by Elsevier, Ltd., p. 273-281.

[6] Mishra,J.K. (2009) Knowledge management: Complexity, Learning and Sustainable. - India. Global India Publications Pvt Ltd. ISBN 978-93-80228-15-0, p. 11.

[7] Moldovan, L. (2015) Sustainability Assessment Framework for VET Organizations. - Sustainability 2015, 7, 7156-7174. ISSN 2071-105,0 doi:10.3390/su7067156

[8] Petrikova, R. a kol. (2010). Modern knowledge management. - Professional Publishing. Praha. ISBN 978-80-7431-011-9. [in Czech]

[9] Rhem,A.J (2017). Knowledge management in Practice. CRC Press. Taylor and Francis Group, New York. ISBN 987-1-4665-6252-3.

[10] Rusko, M., Kralikova, R. (2012) Implementation of environmental management systems in the Slovak enterprises. - In: DAAAM International Scientific Book 2012. Vol. 11, no. 31 (2012), p. 369-378. ISSN 1726-9687.

[11] Senge, P.M. (2014) The Fifth Discipline Field book: Strategies and Tools for Building a Learning organizations. - Crown Publishing Group. New York. ISBN 9780804153164.

[12] Shudayfat, E; Moldoveanu, F., Moldoveanu, A. D. B. (2012) A 3D virtual learning environment for teaching chemistry in high school. Annals of DAAAM for 2012 & Proceedings of the 23rd International DAAAM Symposium, ISSN 1726-9687, ISBN 978-3-901509-86-5, Katalinic, B. [Ed.], Published by DAAAM International, Vienna.

[13] STN EN ISO 9001: 2016: Poziadavky. System manazerstva kvality. [Quality management systems - Requirements] – Bratislava: UNMS SR, 2016 [in Slovak]

[14] Varga, G. &, Kundrak, J. (2017) Effects of technological parameters on surface characteristics in face milling. In: Solid State Phenomena, 261 SSP, pp. 285-292.

[15] Zgodavova, K., Slimak, I. (2011) Focus on success. - Quality Innovation Prosperity, XV/1 - 2011, pp. 1–4, Accessed July 18th, 2013. [online] Available at: http://www.qip journal.eu/index.php/QIP/article/view/36

DOI: 10.2507/28th.daaam.proceedings.171

VERIFICATION OF COMBINED CUTTING TOOL WITH A LINEAR CUTTING EDGE TOOL

Jozef Nosaľ, Katarína Brezíková & Vladimír Simkulet

This Publication has to be referred as: Nosal, J[ozef]; Brezikova, K[atarina] & Simkulet, V[ladimir] (2017). Verification of Combined Cutting tool with a Linear Cutting Edge tool, Proceedings of the 28th DAAAM International Symposium, pp.1227-1233, B. Katalinic (Ed.), Published by DAAAM International, ISBN 978-3-902734-11-2, ISSN 1726-9679, Vienna, Austria
DOI: 10.2507/28th.daaam.proceedings.171

Abstract

This article discusses how to increase the productivity of machining-turning when changing tool geometry and using an integrated experimental tool that consist from both roughing and finishing in one operation. In the first part there is a brief overview of the problem., in this part is described the patent file from Escher and Lindl, which consisted in creating a second tip that aligns the roughness created by the first tip Another solution is Wiper technology, which is now widely used to ensure maximum surface quality. The second part of this article deals with experimental tests carried out on the universal lathe SU 50, the workpiece material is 100CrMn6. The first experiment compared the quality of surface, using an integrated tool for roughing (tip radius: r_ε = 0,8mm) and finishing (tip radius: r_ε = 6mm). Results of the experimental showed improvement in quality of machined surface, but in higher feed rate the quality gradually increased. In the second experiment, the integrated tool was compared with cutting tool with oblique cutting edge, with rake angle $\lambda_s = \pm 45^\circ$. The results showed that the quality of machined surface has improved throughout entire feed rate even at higher feed rates using tool with rake angle $\lambda_s = \pm 45^\circ$.

Keywords: turning; integrated cutting tool; feed rate; oblique cutting; rake angle

1. Introduction

Present engineering production is characterized by rising level of automation, the incoming application of new, generally light and high strength materials that are technologically demanding to process. This results in a significant impact on the productivity of the production and the quality of the machined surface. Automation has greatly reduced the lead times and cutting tools manufactures are trying to improve surface quality by varying tool geometry adjustments. Therefore, the contribution is an example of a proposal for the improvement of technological methods, specifically turning [8], [7].

The influence of technology on the manufacturing process was dealt by some authors [4], [5], [6], in this paper, questions are asked about technologies to improve the quality of machined surface at high feed turning, comparing special tools with differ tool geometry. This comparing will lead to find optimal cutting tool with specific tool geometry for high feed turning.

The following two examples are a selection of solutions for improving the quality of the machined surface during turning when changing tool geometry.

The first solution was submitted by Escher and Lindl [3] according to

Fig. *1*. The solution consists in creating another tip that aligns the machined surface created by the first tip. The position of the second point is defined by the distance (1), where *n* is an odd positive number. This causes the second point to reach the largest elevation of the surface formed by the first wedge. The disadvantage of such a solution is the complicated tip sharpening. There is also uneven wear of both spikes [3].

$$k = \frac{n.f}{2} \tag{1}$$

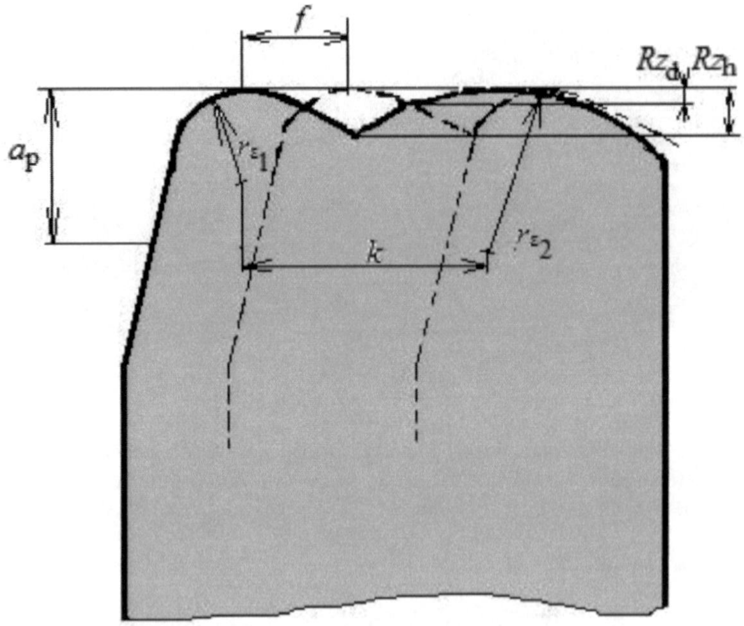

Fig. 1. Tool with two tip radius [3]

One of the other options is a technology which is focused on increasing productivity by adjusting the geometry of the cutting plate-

Fig. *2*. with this it is possible to increase the quality of machined surface by the same displacement [9].

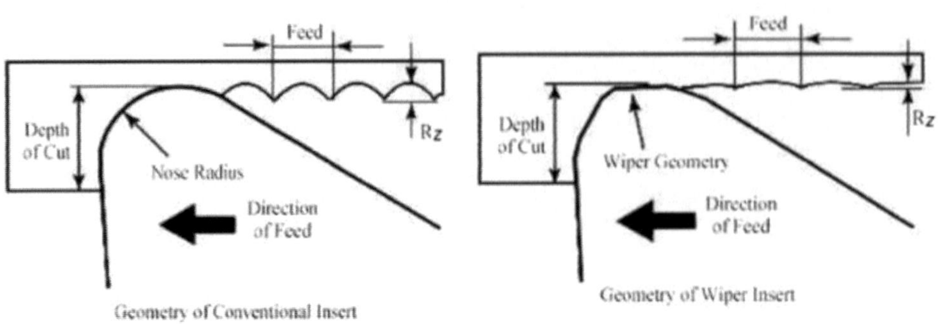

Fig. 2. Difference between conventional and wiper insert cutting tool nose radius [10], [11]

The principle of Wiper improving the quality of machined surface in the same displacement, its means, that with same displacement is obtained the twice quality of the surface as well as achieve better different cutting splinters. The service life of the cutting edges is often improvement due to the reduction in the time in the cut, the consequence leads to cost savings [1], [2].

Wiper technology has a wide range of applications to ensure maximum productivity. When changing parameters during cutting are also available alternative Wiper plates which improve the quality of finishing surface and eliminating expensive grinding operations [9].

2. Used machine and measuring instrument for roughness measurement

For this experiment, was selected workpiece material 100CrMn6. The machine for performing the experiments was the Universal lathe SU 50- Fig. 3. To measure the surface roughness, the Mitutoyo Surfest SJ-301 was used- Fig. 4.

Fig. 3. Universal Lathe machine SU 50

Fig. 4. Mitutoyo roughness meter

3. Experimental setup

For these experiments, two tools were selected. The first tool is integrated cutting tool with two nose radius. With this tool will test the quality of machined surface at higher feed rate. Experimenting conditions: Depth of cut $a_{ph} = 1$ mm, tip radius $r_{eh} = 0,8$ mm, the finishing tool worked at a depth of cut $a_{pd} = 0,2$ mm and a tip radius was $r_{ed} = 6$ mm.

Second tool is cutting tool with linear cutting edge with rake angle $\lambda_s = \pm 45^{\circ}$. This tool will be compared with integrated cutting tool for purposes of surface quality at higher feed rate.

3.1. *Verification of integrated tool for turning*

In Fig, 5. is shown an experimental cutting tool and in Fig. 6. the principle of operation of this tool.

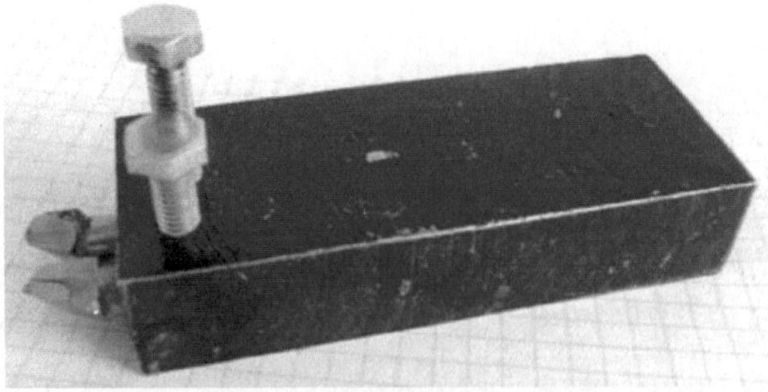

Fig. 5. Experimental integrated cutting tool

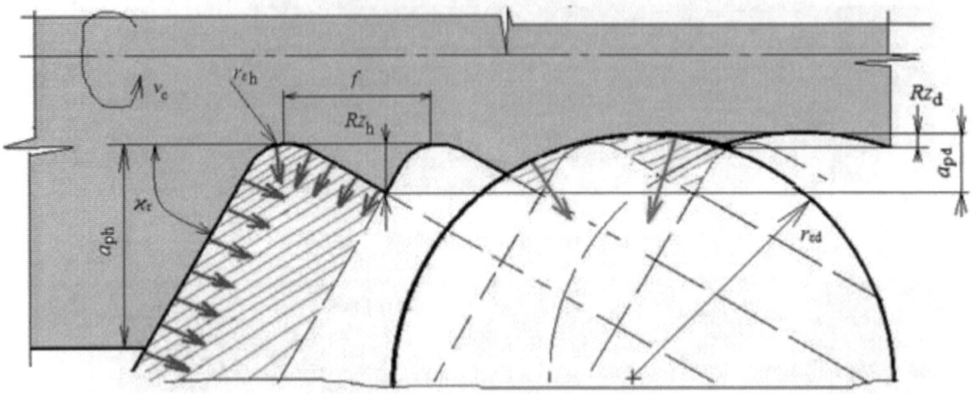

Fig. 6. Scheme of roughing and finishing using integrated cutting tool: v_c – cutting speed, f – feed rate, a_{ph} – depth of cut after roughing, a_{pd} – depth of cut after finishing, r_{eh} – tip radius for roughing, r_{ed} – tip radius for finishing, Rz_h – surface roughness

From the Fig. 6 is see, that in the first the tip radius for roughing r_{eh} we can see on the points of the cutting edge velocity vectors. The chip wants to get in perpendicular way to cutting edge because the cutting edge has a tip radius. By counting every single velocity vector we get the outcome of final velocity vector of movement of chip. First tool work is under the condition of bound rust. The depth of cut of second tip radius a_{pd} is equal or slightly larger than Rz_h ($a_{pd} \geq Rz_h$). In this case, the finishing tool r_{ed} removes all the roughness created by the first tip radius r_{eh}. With the finishing tool, the direction of the chip is identical with direction of movement of individual elements of chip. This case responds to the terms of the free cut, because the higher the tip radius is, and then is higher possibility for achieving free cutting conditions. The results of the experiment are shown in Fig. 7.

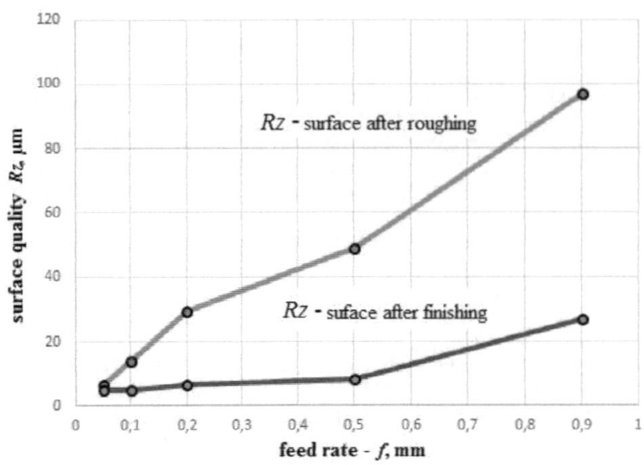

Fig. 7. Experimental results of Rz-f (f) by applying integrated turning

It can be seen from Fig. 7. that using the experimental tool improves the quality of the machined surface when finishing. At the lowest displacement, the *Rz* value do not differ significantly, changing from 0,1 mm. With the increase in feed rate, surface quality deteriorates using a tool with r_{eh} = 0,8 mm, it improves with finishing tip radius r_{ed} = 6 mm.

From this experiment, using a large tip radius, in this case r_{ed} = 6 mm, results in a better machined surface at higher feed rates compared to a smaller tip radius r_{eh} = 0,8 mm. It can be argued that the larger the tip radius is, than we are approaching conditions of free cutting. However, the problem of increasing the tip radius is that leads to reduce the thickness of the cut layer, which can lead to the oscillation of the technological system.

This problem can be solved by the inclination of the rake angle λ_s. The following chapter compares this tool with an integrated tool for turning.

3.2. *Verification of integrated tool for turning with a linear cutting edge tool with rake angle λ_s*

In this experiment, the same workpiece material and the same technological conditions were used. In Fig. 9. are results of the experiment. The tool is shown in Fig. 8.

Fig. 8. Tool with a linear cutting edge with rake angle λs = ± 45°

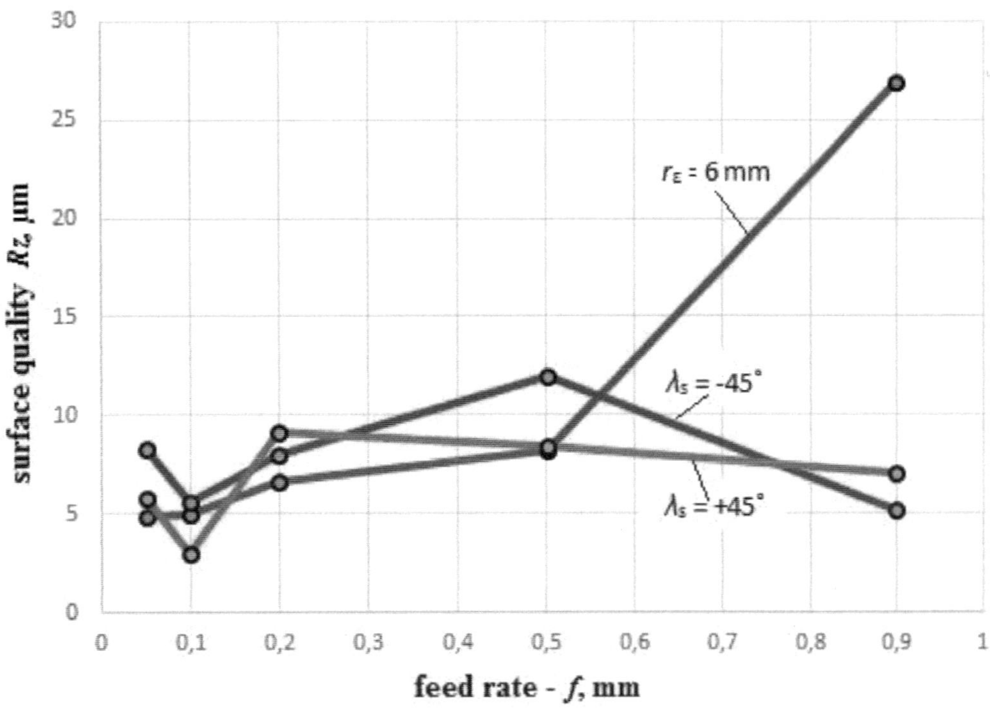

Fig. 9. Experimental results using a linear cutting edge with rake angle λ_s and a tool with large tip radius r_ε = 6 mm

It can be seen from Fig. 9. that when the tool is used $r_\varepsilon = 6$ mm, the quality of machined surface deteriorates at $f = 0,5$ mm. Vice versa, when using a linear cutting edge tool with rake angle $\lambda_s = \pm 45°$, quality improvement can be monitored, especially in the area of higher feed rate.

3.3. Experimental design of integrated turning tool with linear cutting edge

When applying a tool with an extremely large tip radius at a large feed rate, there is a problem that such a tool can not be used at high depth of cut. There is an increase in the length of the cutting edge in the engagement and the generation of vibration of the whole technological system- machine, tool and workpiece.

The possibility of solving this problem lies in the application of integrated tool for turning Fig. 10. This tool will work at a large feed rate and the finishing cutting edge with rake angle $\lambda_s = \pm 45°$will be working in slightly higher depth of cut than the first cutting edge with smaller tip radius.

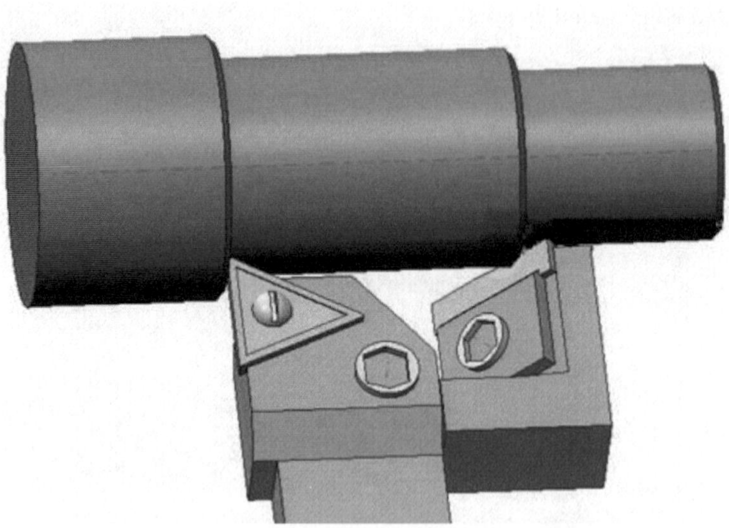

Fig. 10. Principal scheme of roughing and finishing in one operation, using secondary cutting edge with rake angle λs

4. Conclusion

This work analysed the use of an integrated cutting tool for turning, which consisted of two tips. The first tip radius was smaller and served as a roughing, and the second tip radius was larger. The experiments performed with this tool pointed to the improvement of the machined surface, but at higher feed rate the quality deteriorated quadratic, which this can be as negative aspect for high feed turning.

To improve the quality of the machined surface at higher feed rate, for high-shear machining, it is possible to use a tool with a linear cutting edge with rake angle $\lambda_s = \pm 45°$. The experiments performed with this tool have pointed to improving the quality of machined surface, especially in the area of higher feed rate.

These experimental tests were performed on a conventional lathe machine, so further experimental testing would be appropriate on CNC machines and use this kind of tool for other frequently used materials in manufacturing practice.

For future research on improving machined surface using these two tools, it will also be necessary to track the wear of cutting edges for machining with higher feed rate and to monitor the quality of machined surface at a certain degree of wear, this allow for precise design of the material for the cutting inserts under given machining conditions on different materials in manufacturing practice.

5. Acknowledgments

This research work was supported by the Cultural and Educational Grant Agency of the Ministry of Education, Science, Research and Sport of the Slovak Republic (KEGA No.036TUKE-4/2017).

6. References

[1]. Abbas, A. T. M. (2016). Comparative Assessment of Wiper and Conventional Carbide Inserts on Surface Roughness in the Turning of High Strength Steel, Journal of Materials Science Research, Vol.5, No.1, 2016, pp.32-45, ISSN 1927-0585, DOI: 10.5539/jmsr.v5nlp32

[2]. Astakhov, V. P. (2010). Geometry of single-point Turning Tools and Drills Fundamentals and Practical Applications, Springer. Vol.22. p.565. ISBN 978-2-84996-052-6

[3]. Esher, S.; Lindtl, K. (1976). Verfahren und Vorrichtung zur spananhebenden. Werkstatt und Betrieb, West German Patent No.2610097

[4]. Novak-Marcinčin, J.; Barna, J.; Nováková-Marcinčinová, Ľ. & Fečová, V.(2011). Principles of Product Assembly Visualization by Augmented Reality Usage, In: Annals of DAAAM for 2011 & Proceedings of the 22nd International DAAAM Symposium on „Intelligent Manufacturing & Automation: Power of Knowledge and Creativity": 23-26th November 2011. Vienna, Austria. Vienna: DAAAM, 2011 P. 1049-1050. ISBN 978-3-901509-83-4. ISSN 1726-9679

[5]. Orlovský, J.; Zajac, J.; Monka, P.P.; Michalik, P. & Simkulet,V.(2011). Mathematical modelling of spray drying technological process. In Annals of DAAAM for 2011 & Proceedings of the 22nd International DAAAM Symposium on „Intelligent Manufacturing & Automation: Power of Knowledge and Creativity": 23-26th November 2011. Vienna, Austria. Vienna: DAAAM International, 2011 P. 1067-1068. ISBN 978-3-901509-83-4. ISSN 1726-9679

[6]. Simkulet, V.; Hatala, M.; Orlovský, I.; Zajac, J.; Monka, P.P. & Michalik, P.(2011). Microhardness evaluation of steel EN S233J2G3 in heat affected zone after plasma ARC cutting. In Annals of DAAAM for 2011 & Proceedings of the 22nd International DAAAM Symposium on „Intelligent Manufacturing & Automation: Power of Knowledge and Creativity": 23-26th November 2011. Vienna, Austria. Vienna: DAAAM International, 2011 P.1103-1104. ISBN 978-3-901509-83-4. ISSN 1726-9679

[7]. Trent, E. M.; Wright, P.K. (2000). Metal cutting. Butterworth Heinemenn. Boston. p. 446. ISBN 0-7506-7069-X

[8]. Vasilko, K.; Vasilková, D. (2000). Metódy zlepšovania morfológie povrchov súčiastok. ISBN 80-7099-495-9. Prešov

[9]. Wiper. Sandvik Coromant. [online]. 2017. Available at: http://www.sandvik.coromant.com/pl-pl/knowledge/general_turning/getting_started/choice_of_insert/wiper

[10]. Kiyak, M.; Sahin. I. & Cakir. O. (2016). Application of wiper insert in cutting tool technology. Proceedings of 1st ICAS 2016. Istanbul. pp. 60-65, Available at: https://www.researchgate.net/publication/307964609_APPLICATION_OF_WIPER_INSERT_I_CUTTING_TOOL_TECHNOLOGY

[11]. http://artwi.blogspot.sk/2005/04/wiper-inserts-for-turning.html. Access in October 2017

DOI: 10.2507/28th.daaam.proceedings.172

SERVICE ROBOTS INTEGRATING SOFTWARE AND REMOTE REPROGRAMMING

Davydov D.V., Eprikov S.R., Kirsanov K.B., Pryanichnikov V.E.

This Publication has to be referred as: Davydov, D[enis]; Eprikov, S[tanislav]; Kirsanov, K[irll] & Pryanichnikov, V[alentin] (2017). Service Robots Integrating Software 4nd Remote Reprogramming, Proceedings of the 28th DAAAM International Symposium, pp.1234-1240, B. Katalinic (Ed.), Published by DAAAM International, ISBN 978-3-902734-11-2, ISSN 1726-9679, Vienna, Austria
DOI: 10.2507/28th.daaam.proceedings.172

Abstract

This paper presents a solution to the problems of redundancy and organisation of the software architecture inherent in robotic systems and its' simulators. The analysis and comparison of these solutions with the existing ones are given. It was designed a new software architecture for controlling mobile robotic complexes on the base of two models. Developed simulators and usage of the actor model were the base for creating the technological platforms for distributed software control, providing the simultaneous access of several users to a group of mobile robots and their virtual models.

Keywords: building a network of robotarium; distributed control of mobile robots and their simulation software for mobile robot; reservation methods in robotics; dynamic reprogramming

1. Main problems

When developing intelligent robots, one of the problems is the lack of effective software to create systems of group control. For example, in the well-known software packages (ROS, MRS, and others) does not included specific development tools for control of distributed mechatronic systems, especially in remote mode. They also have a common disadvantage, that is the unreasonably high costs associated with installing, upgrading and maintaining equipment and working in this software environment.

This paper presents the results of the development and testing of the new software-information platform to provide the user with unified access to the software environment to control a variety of robots, including operational and remote changes of executing programs during their functioning.

This enables the following: reducing the cost of setting up executing programs; implementation of operational adjustments during operation of the robots; improving the efficiency of the developed technical tools for teaching and learning processes of the students (see Fig. 1), specialised in robotics and IT- industry, by increasing the use of training equipment due to the paralleling access to it.

Fig. 1. Virtual model of Robocup Logistic competition for educational purposes.

Developed simulators and usage of the actor model were the base for creating the technological platforms for distributed software control, providing the simultaneous access of several users to a group of mobile robots and their virtual models. It's the first time, then under a single software platform are combined programming and control in parallel with a virtual models of mobile robots and their real prototypes “implemented in metal” (see Fig. 2).

Fig. 2. Real robotarium (right side) and their virtual model (left side) with mobile robots “Robotino”.

Platform allows you quickly and remotely reprogram systems and significantly reduce requirements to the level of training of specialists-software developers, to ensure the scalability of each component system on physically distributed computers. Thus, significantly decrease the costs of adaptation programs for specific tasks, and radically simplifies development of control systems and user interfaces.

This approach was tested on mobile robots with different kinematic schemes, different types of microprocessor control with the use of sensors, operating on different physical principles. Testing was conducted on robots: AMUR-105, AMUR-107, AMUR-205, three Robotino robots and two small mobile robots, placed in 4 laboratories-robotariums, as well as the simultaneous launch with 80 virtual simulators, operating in real-time. The experiments confirm all these advantages.

Future plans include the expansion of the use of these approaches for teaching and scientific experiments and development of a network of robotariums created by MINOT Russian State University Humanitas, Chair of Sensory and control systems MGTU "STANKIN" (basic department with KIAM Russian Academy of Sciences) jointly with universities in Russia, Croatia (Zadar University) and Austria (Vienna Technical University), including the association with the educational network "Synergy" (project of Concern "Festo"). It is also envisaged to improve further characteristics of embedded in the software platform physical engine, as well as the testing platform in the development of the system production processes with integrated logical control, for example, in automation workshop, implements the concept of "Industry 4.0" in Kimry, realising by International Laboratory “Sensorika”.

2. Redundancy processes

The development of mobile robots and robotic complexes is closely connected with the development and implementation of software (SW) specialised and common use SW. In this case SW, focused on mainframes for PC, finds its' applications in mobile robots and mobile robototic complexes (RTC).

Unfortunately, most of this modern software can't just be implemented in software without additional modification/processing, so the problem of migration from the world of mainframes in the RTC remains relevant. This paper presents a solution to the problem of redundancy of work processes in mobile robotic complex as well as a new software architecture for geographically-distributed management of a group of mobile robots.

To date, the capacity of modern electronic computing machines (computers) and micro-controllers (MC) allow us to solve a huge number of computational problems. In addition, the size of modern micro-processors and their architecture (ARM, MIPS) allows you to perform a complex calculations in such limited dimensions of the objects as any MRC. Unfortunately, computers and MC are not insured against loss computed results during operation. Reasons for the loss of information can serve as the lifting of external power supply, failure of some hardware, unauthorised system access, error, and other unforeseen situations. To prevent the loss of calculated results are the backup, and periodically saves the data in permanent memory, which can later play to continue after termination of the computation. Such mechanisms include hardware (*RAID arrays, redundant clusters*) and software (*rsync, rdiff-backup*). A brief look at the mechanisms of hardware and software redundancy. *RAID* arrays (redundant array of independent disks, [9]) is a redundant mass of independent disks.

These arrays serves to increase the reliability of data storage and/or to increase the speed of read/write information (or both simultaneously). Today, there are a few of the most frequent types of such arrays:

1. *RAID 1* - mirrored disk array
2. *RAID 2* - arrays using the Hamming code
3. *RAID 3 and 4* - disk arrays with striping and a dedicated parity disk
4. *RAID 5* - disk array with striping and no dedicated parity disks.
5. *RAID 6* - disk array with striping, using two checksums computed by two independent methods;

Failover clustering High-Availability cluster (HA-cluster) is a group of servers, that is designed in accordance with the methods of providing high availability and guaranteeing minimal downtime due to hardware redundancy. Without clustering server failure leads to the fact, that it supports applications or network services are not available to restore its status. Failover clustering corrects this situation, restarting the application on other nodes in the cluster without administrator intervention in case of detecting hardware or software failures. The restart process known as failover. In this process, software clustering may optionally configure the node before starting the application on it.

In addition to hardware redundancy, there are also tools to backup data by program tools. Such tools include many modern software like ***Acronics Disk Director, rsync, filezilla,*** etc. Such software is responsible for making copy of all processes to self-preserve information from permanent memory. Unfortunately, the processes that are running in memory of a computing device are not responsible for it.

To save the state of the computer, there are hardware solutions for full redundancy. The advantages of hardware redundancy before the software is fully clone all running processes on the remote machine. This requires the availability of accurate copies of the computer, where you stored the entire state of a redundant computer. If you are reserving the N devices, that will require another N exactly the same devices, which entails the problem of excessive use of electricity, large employment areas, which are the device and the cost of the computing complex. When considering computer complex, consisting of several of mobile robots (MR) this option cannot be applied due to its complexity (in front of a personal computer), as when you clone a built-in operating system (OS) the MR will not be taken into account many additional parameters such as sensors readings, the wheels positions because of the inability of the robots at the same time in the same environment and with identical coordinates. In addition, many embedded systems of MR does not have the possibility of hardware redundancy. A possible solution could be the introduction of virtual machines on MR, but this may be difficult due to the large size of the software/hardware and software requirements of MR, which can severely affect the performance of the MR during operation. Software redundancy was also considered in [1], but required a complete restart of the OS that require a long resume.

The new approach of the redundancy, in contrast to the above decisions is that the system redundancy (SR) MR (Master) periodically sends the process state to the neighbouring MR (Slave) directly over a local area network. MR (Slave) monitors the MR (Master) and will either run a copy of the process or will wait for the resumption of MR (Master) from which came the information to send the received information back to resume this process at the MR (Master).

However, the decision about who is the primary (master) object and who is the backup (slave) is able to get out SR or operator. For example, you can assign one of MR major, that if it was not found another similar MR. Thus, when the second MR will be on the network it will go into standby mode. With the simultaneous choice of roles MR (in case of simultaneous receiving the role, each MR) the advantage of choosing the main role will be given to MR, whose name will have less weight W. For example, MR with names Робот1 and Робот2 will have a main and standby role, respectively if the condition W(Robot1) < W(Робот2).

In this approach, the SR periodically retrieves full information about running process and sends it either to MR (Master), that can play the running programs, or saves on a remote server, that when you resume, MR (Master) will send the accumulated information on this MR (Master). That will allow MR (Master) to continue its work from where you left off. The relevance of this decision is to eliminate the need for hardware redundancy running processes for the OS as well as the possibility of software redundancy running process or multiple processes. This allows you to save the state of processes on the fly without involving a backup device. This approach is also relevant for the redundancy of individual modules in a heterogeneous modular mobile robots [2,3]. However, when transferring the backup information is not desirable the use wireless networks due to the low bandwidth, and also due to less secure wireless networks, which is described in detail in [4] [5].

To be able to reserve a work processes offered two solutions. The first is to add to the process possibility to retrieve the data and send this data to the backup device (MR, RTC server or the other computer), as well as the ability to read data from the backup device to resume operation. The second approach is to extract information on the process (*sections bss, heap, stack,* etc.) by means of the kernel and then sending of location information to MR (Slave). The following is a comparison of these backup options:

Option	Advantages	Disadvantages
Redundant process sends data itself	Not required to retrieve data using OS	The need to change source code of the redundant process. A lack of universality.
Separated the fallback mechanism	There is no need to work with the source code of the program. Versatility.	The complexity of the implementation.

Table 1. Comparing options software redundancy.

The first option does not provide mechanisms to work with the Linux kernel, but it needs an access to the source code of the reserved program. While inside of process needs to be as mechanisms of discharge data and algorithms for reading the data for the possibility of resumption of work. Option becomes impossible if the source code of the program is not available.

The second option: there is redundancy processes, without editing its source code, making this approach universal to any process without providing the source code, but increases the complexity of the implementation of this mechanism with the possibility of running a program in multiple threads, open files, read status registers, cache, etc. In this embodiment, due to the complexity of the implementation, we consider only the redundant single-threaded programs without interacting with the devices input/output and files.

Now fully worked out the first option because of it's relative easy implementation.

3. The architecture of the mobile robotic complex

To refer the software (designed to interface application software with the network and providing unification of interaction between programs in terms of the heterogeneity of the computing platforms) is used the term "middleware software". This type of software acts as subsystems in larger. If in modern database technology (DB) and Internet middleware is studied in detail, as applied to robotics, these studies have not reached the same level. Apparently, it is connected with the fact that until recently, the software for robotic systems was quite simple (from the point of view of software engineering) and does not go beyond organisation of the simplest interactions between the objects of control.

In the process of implementation for system of geographically distributed group control mobile robotic devices [6-8], we analysed the existing systems that provide similar functionality, namely, Microsoft Robotics Developer Studio (MRDS), Robotics Operating System (ROS), and their predecessors, such as Player Project, LAAS GenoM and URBI (see Wikipedia). These systems can be roughly classified into "large", i.e. includes many components for solving various problems: SLAM (Simultaneous Localisation and Mapping), the calibration of sensors, image processing, etc. However, there are "small" robotic frameworks, for example, the Concurrency and Coordination Runtime (included in MRDS), which carries out only the transfer of messages between components. These systems are designed to work within the local area network (LAN) with a stable connection and contain virtually no funds for proper recovery in a data channel. But this problem always arises when it is necessary to provide mobile robotic systems (RTC). Thus, in the process of creating the technology of territorially distributed group management mobile RTC was required to develop a brand new software architecture.

The proposed architecture is based on the principle of organisation of information-measuring and control systems (IMCR) of the robotic complex in the form of a local area network with mobile nodes [4] and the "concept of drivers". Under this concept, every node of such LAN is electronic (e.g., microcontroller) or mechatronic device (e.g., actuators with control system).

Thus a mobile robot (MR) is represented as a composition of several mechatronic devices and their drivers. On the onboard computer [2] mobile robot driver for each mechatronic device is a separate computing process, and collaboration of these processes it is necessary to organise and synchronise.

To solve these problems, we use the following approaches:

- Each command or data from the sensors are time-stamped when passing each node of the LAN;
- Each driver is equipped with its own "Service name", which stores information about the adjacent (from the point of view of the network topology) drivers;
- Each driver is dynamically constructed by the interaction graph, which stores information about the neighbours.
- Uses a text data exchange format.
- Data sharing enables transmission of not only individual teams, but entire programs.

The system of differentiation of access (LDAP — Lightweight Directory Access Protocol) allows you to control the access rights of user groups. Adding time stamps to messages, allows you to accurately synchronise the drivers and take into account the time delays and the difference in time between multiple geographically - distributed nodes in the network. "Name service" is a specialised software that every 20 seconds, scans your network and looks for similar services, and to discover and shares with them all available information in the algorithm is close to distributed hash tables – Distributed Hash Table (DHT). This makes the quality of decentralisation. For example, the robotic system is configured within a specific LAN, no changes can be disconnected from it, and thus will not be disturbed its normal operation. To ensure the integration MRTU in a geographically distributed group, it was necessary not only to implement communication between electromechanical part of the robot and the onboard computer, but also to integrate the IMCR directly into the General management hierarchy.

For a formal description of interaction in multi-agent networks was applied the actor model. As a mechanism of message passing library is used ZeroMQ (http://zeromq.org/), which serves as a high-level wrapper over standard network sockets operating system. This mechanism includes a number of functions such as auto-reconnection after breakage with no loss of context, the possibility of tunneling through the lower layer protocols: TCP, IPC, or through PGM (Pragmatic General Multicast). Furthermore, unlike more complex protocols (e.g., Concurrency and Coordination Runtime), ZeroMQ is quite simple, enabling you to deliver it directly into the Arduino microcontroller module c EthernetShield, but with some limitations related to its low productivity. So, when using this microcontroller maximum frequency command in the test implementation was 20 Hz.

In figure 3 is shown the general scheme of interaction: two mechatronic devices rc1 and rc2 (RobotComponents) asynchronously pass the procedure of registration in the name Service ns (NameServer). Then the control system (CS) is requesting a list of registered devices, and performs synchronous and asynchronous communication with the mechatronic devices. Filename in this case is the IP address and port, arbitrary text, e.g. "engine", or a path in the hierarchy, e.g., "University→Laboratory→Robotino→ Engines→3→PWM".

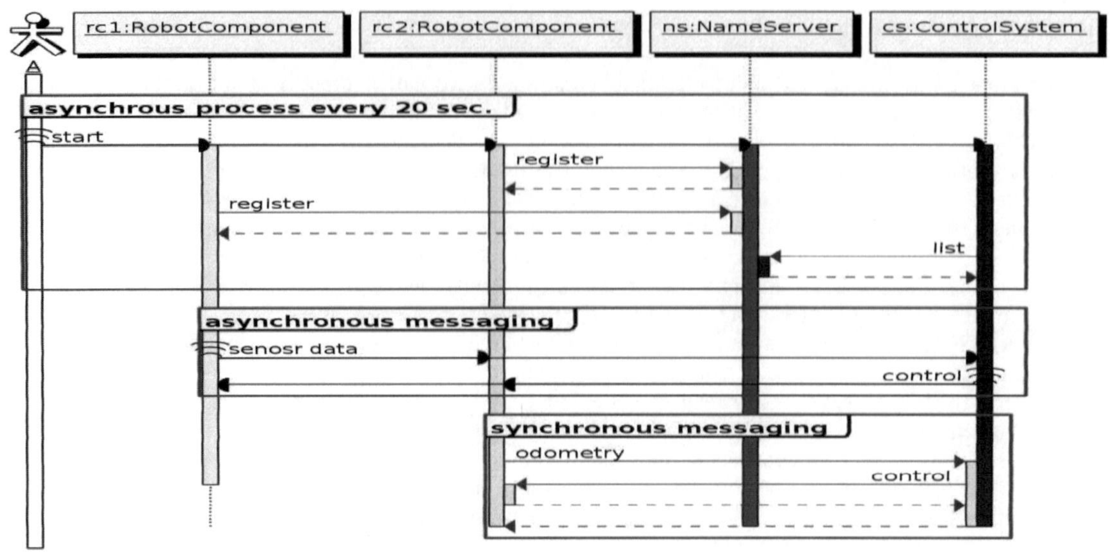

Fig. 3. The mechanism of network interaction in the framework of the proposed architecture.

Driver for each device provides:

- naming and identification (method of obtaining access to the device by name);
- transmit and receive commands, including the other drivers.

Commands and data are transmitted in JSON format (see Fig.4), which is quite simple and can be implemented on low-power processors, including AVR (Arduino). JSON is a text (ASCII) format for the presentation of structured data, which allows you to group a list (vector, array) and associative arrays, i.e. sets of pairs "key-value", which gives the possibility manually generate control commands, without resorting to specialised libraries, as required in other similar systems. Also, there is the opportunity to annotate transmitted data with additional information without modifying the host software, or to arbitrarily change the order of fields within a single message. This makes it possible to organise the transfer of not only individual teams, but entire programs, that immediately on receipt executed by the interpreter, operating on the onboard computer [2]. Thus arranged dynamic re-programming of mechatronic devices, i.e. without stopping or "reloading" on-Board computer of the MR device, you can update it or correct it's work. This greatly simplifies the simultaneous support of different versions of the same driver in a network of mechatronic devices in the task of geographically distributed control. The redundancy of information, that occurs due to its text representation, has no significant effect on throughput, since teams usually compact even in the textual representation and are in comparable size to the header of the IP packet (20 bytes).

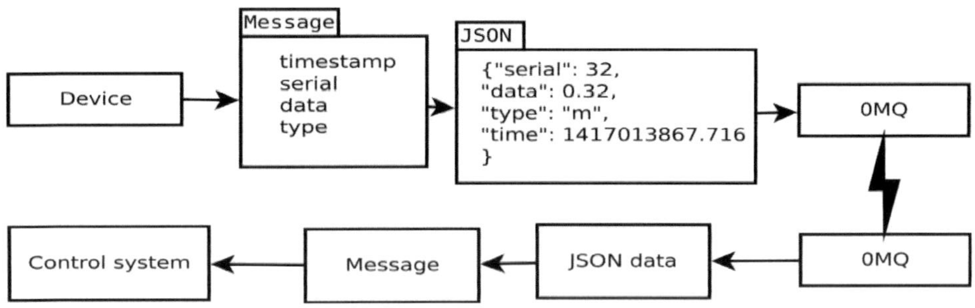

Fig. 4. Diagram - transmission of messages: device generates a messages, which are encoded in JSON and are passing to ZeroMQ.

In that case, when the amount of JSON-encoded data begins to have a significant impact on processing speed on the data transfer, we can use format is binary AVRO [8], however its implementation on the AVR is not yet. For visualisation of discovered network components, display operator interface control software was created. It represents all the detected components as a tree view, as shown in Fig.5.

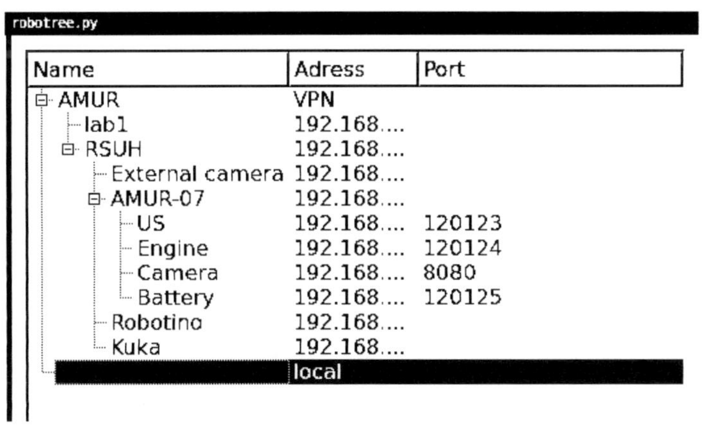

Fig. 5. One of the windows on-screen interface presents (in a tree view) detected robot AMUR-7 components (located in lab1 RSUH): ultrasonic sensors (US), electrical motors (Engine), TV-camera (Camera) and rechargeable battery (Battery).

4. Conclusion

In the course of work was drawn up two models of software redundancy process: as to modifying the source code of reserved process and without it. This technique allows to get rid of the need for hardware redundancy and to increase the ability of software to back up data on a computer and MP running on the Linux OS. At the moment a check process without editing its source code can be applied only to processes with a single flow and is not working with file descriptors or devices because of the complexity of the redundancy module, therefore, the support of multithreaded programs and programs, that work with other OS resources such as files, devices, I/O will be developed in the future.

The proposed software architecture has proven its effectiveness in the creation of geographically distributed control systems academic group of mobile robots, consisting of 2 robots of the type "AMUR" (the Autonomous Mobile Universal Robot, International Laboratory "Sensorika") and 3 robots "Robotino" (the German company Festo). The system operates between MSTU "STANKIN" (Moscow) and the International Institute of new educational technologies (IINET) Russian state University for Humanities (Moscow). The same channel also operates between Keldysh IAM RAS (Moscow) and the Far-Eastern Federal University (Vladivostok) [10-14]. The information and communication architecture can be used in the system of distance education [4,10] to organise the access of students to real mechatronic systems of different models and manufacturers.

5. Acknowledgments

The work was carried out with the partial financial support of the RFBR, projects №16-07-01264, 16-07-00935, 16-07-00811.

6. References

[1] V.Ya. Hecha, S.G. Sementsov, S.A. Kozlov, A.M. Konoplev. The security Concept for Linux-like operating systems in control systems and protection of nuclear reactors // Problems of electromechanics. Proceedings VNIIEM, 2004.

[2] (В.Я. Геча, С.Г. Семенцов, С.А. Козлов, А.М. Коноплев. Концепция обеспечения безопасности Linux-подобных операционных систем в системах управления и защиты реакторов АЭС // Вопросы электромеханики. Труды ВНИИЭМ, 2004).

[3] Kirill Kirsanov. The Architecture of Robotics Control Software for Heterogeneous Mobile Robots Network / Kirsanov Kirill // Procedia Engineering 01/2014; 69:216-221. DOI: 10.1016/j.proeng.2014.02.224.

[4] Andreev V. P. Group control of mobile robots by means of Ethernet technology / P. Andreev V., Pryanichnikov V. E., Prysev E. A. // Proceedings of XXI International scientific-technical conference Extreme robotics. – SPb.: Publishing house "Politekhnika-SERVIS", 2010. – P. 427 – 430.

[5] Education on the basis of virtual learning robotics laboratory and group-controlled robots / Victor Andreev, Sergey Kuvshinov, Valentin Pryanichnikov, Yury Poduraev // 24th DAAAM International Symposium, 2013. Procedia Engineering, 2014. – V. 69. – P. 35 – 40.

[6] Andreev, V.P., Kirsanov K.B Technology Supervisory control of mobile robots through Internet// Proceedings Int.conf. "Extreme robotics-2015". -SPb, "Politechnika-Servis", 2015.-P.270-272.

[7] Wolfgang Emmerich, The impact of research on the development of middleware technology / Wolfgang Emmerich, Mikio Aoyama, Joe Sventek // ACM Transactions on Software Engineering and Methodology. – N. Y.: ACM, 2008. – T. 17. – No 4. – P.19 – 48.

[8] Zareshin S.V., Shustova L.I., Shestakova N.Y. Comprehensive analysis of wireless networks security in moscow central district //Young Researchers in Electrical and Electronic Engineering (EIConRus), 2017 IEEE Conference of Russian. – IEEE, 2017. – P. 240-241.

[9] Anastasia A.V. et al. Analysis of security of public access to Wi-Fi networks on Moscow streets //Young Researchers in Electrical and Electronic Engineering (EIConRus), IEEE Conference, Russian. – IEEE, 2017. – C. 105-110.

[10] Description of RAID arrays - https://ru.wikipedia.org/wiki/RAID, 01.11.2017.

[11] V.E. Pryanichnikov, B. Katalinic, A.A. Kirilchenko, R.V. Khelemendik, S.V. Kuvshinov, D. Vician, A. Uglesic. New Creative Educational Technologies for Inter-university Network. Elsevier. // Procedia Engineering Volume 100, 2015, Pages 259–268. 25th DAAAM International Symposium on Intelligent Manufacturing and Automation, 2014. URL: http://www.sciencedirect.com/science/article/pii/S1877705815003938

[12] Kirillchenko A.A., Pryanichnikov V.E., Rogozin K.V. Limits of validity and reliability of proofs. Scepticism in mathematics, functions, and traditions // Information-Measuring and Control Systems. 2013. V. 11. № 4. P. 57–65.

[13] Frolov A.A., Biryukova E.V., Bobrov P.D., Mokienko O.A., Platonov A.K., Pryanichnikov V.E., Chernikova L.A. Principles of neurorehabilitation based on the brain–computer interface and biologically adequate control of the exoskeleton // Human Physiology. 2013. V. 39. № 2. P. 196–208.

[14] Katalinic B., Pryanichnikov V., Ueda K., Cesarec P., Kettler R., et al. Bionic Assembly System: hybrid control structure, working scenario and scheduling // Proceedings of 9th National Congress on Theoretical and Applied Mechanics. Brussels, 2012. P. 101–108.

[15] Vladimir F. Filaretov, Valentin E. Pryanichnikov. Autonomous mobile university robots AMUR: Technology and applications to extreme robotics. // Elsevier. Procedia Engineering Volume 100, 2015, Pages 269–277. 25th DAAAM International Symposium on Intelligent Manufacturing and Automation, 2014. URL: http://www.sciencedirect.com/science/article/pii/S187770581500394X